D1029189

KIRK-OTHMER

ENCYCLOPEDIA OF CHEMICAL TECHNOLOGY

FOURTH EDITION

VOLUME **15**

LASERS
TO
MASS SPECTROMETRY

EXECUTIVE EDITOR
Jacqueline I. Kroschwitz

EDITOR
Mary Howe-Grant

KIRK-OTHMER

ENCYCLOPEDIA OF CHEMICAL TECHNOLOGY

FOURTH EDITION

VOLUME **15**

LASERS
TO
MASS SPECTROMETRY

A Wiley-Interscience Publication
JOHN WILEY & SONS
New York • Chichester • Brisbane • Toronto • Singapore

r
TP9
.E685
1992
vol. 15

This text is printed on acid-free paper.

Copyright © 1995 by John Wiley & Sons, Inc.

All rights reserved. Published simultaneously in Canada.

Reproduction or translation of any part of this work beyond that permitted by Sections 107 or 108 of the 1976 United States Copyright Act without the permission of the copyright owner is unlawful. Requests for permission or further information should be addressed to the Permissions Department, John Wiley & Sons, Inc., 605 Third Avenue, New York, NY 10158-0012.

Library of Congress Cataloging-in-Publication Data

Encyclopedia of chemical technology/executive editor, Jacqueline I. Kroschwitz; editor, Mary Howe-Grant.—4th ed.
p. cm.
At head of title: Kirk-Othmer.
"A Wiley-Interscience publication."
Includes index.
Contents: v. 15, Lasers to mass spectrometry
ISBN 0-471-52684-3 (v. 15)
1. Chemistry, Technical—Encyclopedias. I. Kirk, Raymond E. (Raymond Eller), 1890–1957. II. Othmer, Donald F. (Donald Frederick), 1904– . III. Kroschwitz, Jacqueline I., 1942– . IV. Howe-Grant, Mary, 1943– . V. Title: Kirk-Othmer encyclopedia of chemical technology.
TP9.E685 1992 91-16789
660′.03—dc20

Printed in the United States of America

10 9 8 7 6 5 4 3 2 1

CONTENTS

EDITORIAL STAFF FOR VOLUME 15

Executive Editor: **Jacqueline I. Kroschwitz**
Editor: **Mary Howe-Grant**
Associate Managing Editor: **Lindy Humphreys**
Copy Editor: **Lawrence Altieri**

CONTRIBUTORS TO VOLUME 15

Dorothy Aiken, *Colin A. Houston & Associates, Inc., Mamaroneck, New York,* Market and marketing research

Eric M. Bergtraun, *National Semiconductor, Santa Clara, California,* Maintenance

Jerry Boisvert, *Magnetrol International, Inc., Downers Grove, Illinois,* Liquid level measurement

E. R. Booser, *Consultant, Scotia, New York,* Lubrication and lubricants

Joseph Breen, *United States Environmental Protection Agency, Washington, D.C.,* Industrial toxicology (under Lead compounds)

Irena Bronstein, *Tropix, Inc., Bedford, Massachusetts*, Chemiluminescence (under Luminescent materials)

Joseph C. Burnett, *Huntsman Specialty Chemical Corporation, St. Louis, Missouri*, Maleic anhydride, maleic acid, and fumaric acid

Dodd S. Carr, *Consultant, Pittsburgh, Pennsylvania*, Lead salts (under Lead compounds)

Boyce Carsella, Jr., *Magnetrol International, Inc., Downers Grove, Illinois*, Liquid level measurement

X. K. Chen, *University of Edinburgh, Scotland*, Machining methods, electrochemical

Kevin Chronley, *Hammond Lead Products, Inc., Pittsburgh, Pennsylvania*, Lead salts (under Lead compounds)

Ken G. Claus, *The Dow Chemical Company, Freeport, Texas*, Magnesium and magnesium alloys

Peter J. Collings, *Swarthmore College, Swarthmore, Pennsylvania*, Liquid crystalline materials

Stephen C. DeVito *United States Environmental Protection Agency, Washington, D.C.*, Industrial toxicology (under Lead compounds); Mercury

James A. Doncheck, *Bio-Technical Resources L.P., Manitowoc, Wisconsin*, Malts and malting

James Downing, *Consultant, Ellicottville, New York*, Manganese and manganese alloys

Matthew R. Earlam, *The Dow Chemical Company, Freeport, Texas*, Magnesium and magnesium alloys

Timothy R. Felthouse, *Huntsman Specialty Chemicals Corporation, St. Louis, Missouri*, Maleic anhydride, maleic acid, and fumaric acid

Chester H. Gelbert, *E. I. du Pont de Nemours & Company, Inc., Louisville, Kentucky*, Latex technology

Richard S. Givens, *University of Kansas, Lawrence*, Chemiluminescence (under Luminescent materials)

Michael C. Grady, *E. I. du Pont de Nemours & Company, Inc., Philadelphia, Pennsylvania*, Latex technology

Kenneth A. Gutschick, *National Lime Association, Kensington, Maryland*, Lime and limestone

James E. Hillis, *The Dow Chemical Company, Freeport, Texas*, Magnesium and magnesium alloys

Katsumi Hioki, *Kuraray Company, Ltd., Okayama, Japan*, Leather-like materials

L. C. Jackson, *Martin Marietta Magnesia Specialties, Inc., Baltimore, Maryland*, Magnesium compounds

Conrad W. Kamienski, *Consultant, Gastonia, North Carolina*, Lithium and lithium compounds

Robert Kessler, *Textron Defense Systems, Everett, Massachusetts*, Magnetohydrodynamics

David J. Kiemle, *SUNY-Syracuse, Syracuse, New York*, Magnetic spin resonance

Michael King, *ASARCO, Inc., Salt Lake City, Utah*, Lead

Fred Kish, *Hewlett-Packard Company, San Jose, California*, Light-emitting diodes (under Light generation)

Larry J. Kricka, *University of Pennsylvania, Philadelphia*, Chemiluminescence (under Luminescent materials)

Katharine Ku, *Stanford University, Palo Alto, California*, Licensing

Stuart E. Lebo, Jr., *LignoTech USA, Inc., Rothchild, Wisconsin*, Lignin

S. P. Levings, *Martin Marietta Magnesia Specialties, Inc., Baltimore, Maryland*, Magnesium compounds

Stephen Y. Lin, *LignoTech USA, Inc., Rothchild, Wisconsin*, Lignin

M. L. Maniocha, *Martin Marietta Magnesia Specialties, Inc., Baltimore, Maryland*, Magnesium compounds

Louis R. Matricardi, *Consultant, Tonawanda, New York*, Manganese and manganese alloys

Daniel P. McDonald, *Consultant, Belmont, North Carolina*, Lithium and lithium compounds

J. A. McGeough, *University of Edinburgh, Scotland*, Machining methods, electrochemical

C. A. Mintmier, *Martin Marietta Magnesia Specialties, Inc., Baltimore, Maryland*, Magnesium compounds

Scott F. Mitchell, *Huntsman Specialty Chemicals Corporation, St. Louis, Missouri*, Maleic anhydride, maleic acid, and fumaric acid

Colin Moore, *Consultant, Woodbury, Connecticut*, Mass spectrometry

Michael Mummey, *Huntsman Specialty Chemicals Corporation, St. Louis, Missouri*, Maleic anhydride, maleic acid, and fumaric acid

Kenneth Pisarczyk, *Carus Chemical Company, LaSalle, Illinois*, Manganese compounds

Peter Pollak, *Lonza Ltd., Basel, Switzerland*, Malonic acid and derivatives

R. David Prengaman, *RSR Corporation, Dallas, Texas*, Lead alloys

Venkoba Ramachandran, *ASARCO, Inc., Salt Lake City, Utah*, Lead

John F. Ready, *Honeywell Systems and Research Center, Edina, Minnesota*, Lasers

A. H. Reyes, *Martin Marietta Magnesia Specialties, Inc., Baltimore, Maryland*, Magnesium compounds

Gérard Romeder, *Lonza Ltd., Basel, Switzerland*, Malonic acid and derivatives

P. E. Scheerer, *Martin Marietta Magnesia Specialties, Inc., Baltimore, Maryland*, Magnesium compounds

Wayne Shannon, *Magnetrol International, Inc., Downers Grove, Illinois*, Liquid level measurement

David M. Smith, *Martin Marietta Magnesia Specialties, Inc., Baltimore, Maryland*, Magnesium compounds

Thomas F. Soules, *General Electric Company, Cleveland, Ohio*, Phosphors (under Luminescent materials)

William C. Spangenberg, *Hammond Lead Products, Inc., Pittsburgh, Pennsylvania*, Lead salts (under Lead compounds)

Alok M. Srivastava, *General Electric Company, Cleveland, Ohio*, Phosphors (under Luminescent materials)

Marshall W. Stark, *FMC Corporation, Bessemer City, North Carolina*, Lithium and lithium compounds

Steven G. Streitel, *Day-Glo Color Corporation, Cleveland, Ohio*, Fluorescent pigments (daylight) (under Luminescent materials)

Edward Tarnell, *Colin A. Houston & Associates, Inc., Mamaroneck, New York*, Market and marketing research

R. Ray Taylor, *Phillips Petroleum Company, Bartlesville, Oklahoma*, Liquefied petroleum gas

Henryk Temkin, *Colorado State University, Fort Collins*, Semiconductor lasers (under Light generation)

Thomas C. Thorstensen, *TSG, North Cholmsford, Massachusetts*, Leather

Karl S. Vorres, *Argonne National Laboratory, Argonne, Illinois*, Lignite and brown coal

M. T. Wajer, *Martin Marietta Magnesia Specialties, Inc., Baltimore, Maryland*, Magnesium compounds

M. D. Walter, *Martin Marietta Magnesia Specialties, Inc., Baltimore, Maryland*, Magnesium compounds

Armin Wendel, *Rhône-Poulenc Rorer, Cologne, Germany*, Lecithin

Jack Wernick, *Murray Hill, New Jersey*, Bulk; Thin films and particles (both under Magnetic materials)

Ted Williams, *Magnetrol International, Inc., Downers Grove, Illinois*, Liquid level measurement

Clifford B. Wilson, *The Dow Chemical Company, Freeport, Texas*, Magnesium and magnesium alloys

William T. Winter, *SUNY-Syracuse, Syracuse, New York*, Magnetic spin resonance

J. T. Witkawski, *Martin Marietta Magnesia Specialties, Inc., Baltimore, Maryland*, Magnesium compounds

NOTE ON CHEMICAL ABSTRACTS SERVICE REGISTRY NUMBERS AND NOMENCLATURE

Chemical Abstracts Service (CAS) Registry Numbers are unique numerical identifiers assigned to substances recorded in the CAS Registry System. They appear in brackets in the *Chemical Abstracts* (CA) substance and formula indexes following the names of compounds. A single compound may have synonyms in the chemical literature. A simple compound like phenethylamine can be named β-phenylethylamine or, as in *Chemical Abstracts*, benzeneethanamine. The usefulness of the *Encyclopedia* depends on accessibility through the most common correct name of a substance. Because of this diversity in nomenclature careful attention has been given to the problem in order to assist the reader as much as possible, especially in locating the systematic CA index name by means of the Registry Number. For this purpose, the reader may refer to the CAS Registry Handbook—Number Section which lists in numerical order the Registry Number with the *Chemical Abstracts* index name and the molecular formula; eg, **458-88-8**, Piperidine, 2-propyl-, (*S*)-, $C_8H_{17}N$; in the *Encyclopedia* this compound would be found under its common name, coniine [*458-88-8*]. Alternatively, this information can be retrieved electronically from CAS Online. In many cases molecular formulas have also been provided in the *Encyclopedia* text to facilitate electronic searching. The Registry Number is a valuable link for the reader in retrieving additional published information on substances and also as a point of access for on-line data bases.

In all cases, the CAS Registry Numbers have been given for title compounds in articles and for all compounds in the index. All specific substances indexed in *Chemical Abstracts* since 1965 are included in the CAS Registry System as are a large number of substances derived from a variety of reference works. The CAS Registry System identifies a substance on the basis of an unambiguous computer-language description of its molecular structure including stereochemical detail. The Registry Number is a machine-checkable number (like a Social Security number) assigned in sequential order to each substance as it enters the registry system. The value of the number lies in the fact that it is a concise and unique means of substance identification, which is independent of, and therefore bridges, many systems of chemical nomenclature. For polymers, one Registry Number may be used for the entire family; eg, polyoxyethylene (20) sorbitan monolaurate has the same number as all of its polyoxyethylene homologues.

Cross-references are inserted in the index for many common names and for some systematic names. Trademark names appear in the index. Names that are incorrect, misleading, or ambiguous are avoided. Formulas are given very frequently in the text to help in identifying compounds. The spelling and form used, even for industrial names, follow American chemical usage, but not always the usage of *Chemical Abstracts* (eg, *coniine* is used instead of *(S)-2-propylpiperidine*, *aniline* instead of *benzenamine*, and *acrylic acid* instead of *2-propenoic acid*).

There are variations in representation of rings in different disciplines. The dye industry does not designate aromaticity or double bonds in rings. All double bonds and aromaticity are shown in the *Encyclopedia* as a matter of course. For example, tetralin has an aromatic ring and a saturated ring and its structure

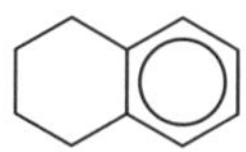

appears in the *Encyclopedia* with its common name, Registry Number enclosed in brackets, and parenthetical CA index name, ie, tetralin [*119-64-2*] (1,2,3,4-tetrahydronaphthalene). With names and structural formulas, and especially with CAS Registry Numbers, the aim is to help the reader have a concise means of substance identification.

CONVERSION FACTORS, ABBREVIATIONS, AND UNIT SYMBOLS

SI Units (Adopted 1960)

The International System of Units (abbreviated SI), is being implemented throughout the world. This measurement system is a modernized version of the MKSA (meter, kilogram, second, ampere) system, and its details are published and controlled by an international treaty organization (The International Bureau of Weights and Measures) (1).

SI units are divided into three classes:

BASE UNITS

length	meter[†] (m)
mass	kilogram (kg)
time	second (s)
electric current	ampere (A)
thermodynamic temperature[‡]	kelvin (K)
amount of substance	mole (mol)
luminous intensity	candela (cd)

SUPPLEMENTARY UNITS

plane angle	radian (rad)
solid angle	steradian (sr)

[†]The spellings "metre" and "litre" are preferred by ASTM; however, "-er" is used in the *Encyclopedia*.

[‡]Wide use is made of Celsius temperature (t) defined by

$$t = T - T_0$$

where T is the thermodynamic temperature, expressed in kelvin, and $T_0 = 273.15$ K by definition. A temperature interval may be expressed in degrees Celsius as well as in kelvin.

DERIVED UNITS AND OTHER ACCEPTABLE UNITS

These units are formed by combining base units, supplementary units, and other derived units (2–4). Those derived units having special names and symbols are marked with an asterisk in the list below.

Quantity	Unit	Symbol	Acceptable equivalent
*absorbed dose	gray	Gy	J/kg
acceleration	meter per second squared	m/s^2	
*activity (of a radionuclide)	becquerel	Bq	1/s
area	square kilometer	km^2	
	square hectometer	hm^2	ha (hectare)
	square meter	m^2	
concentration (of amount of substance)	mole per cubic meter	mol/m^3	
current density	ampere per square meter	$A//m^2$	
density, mass density	kilogram per cubic meter	kg/m^3	g/L; mg/cm^3
dipole moment (quantity)	coulomb meter	C·m	
*dose equivalent	sievert	Sv	J/kg
*electric capacitance	farad	F	C/V
*electric charge, quantity of electricity	coulomb	C	A·s
electric charge density	coulomb per cubic meter	C/m^3	
*electric conductance	siemens	S	A/V
electric field strength	volt per meter	V/m	
electric flux density	coulomb per square meter	C/m^2	
*electric potential, potential difference, electromotive force	volt	V	W/A
*electric resistance	ohm	Ω	V/A
*energy, work, quantity of heat	megajoule	MJ	
	kilojoule	kJ	
	joule	J	N·m
	electronvolt†	eV†	
	kilowatt-hour†	kW·h†	
energy density	joule per cubic meter	J/m^3	
*force	kilonewton	kN	
	newton	N	$kg{\cdot}m/s^2$

†This non-SI unit is recognized by the CIPM as having to be retained because of practical importance or use in specialized fields (1).

Quantity	Unit	Symbol	Acceptable equivalent
*frequency	megahertz	MHz	
	hertz	Hz	1/s
heat capacity, entropy	joule per kelvin	J/K	
heat capacity (specific), specific entropy	joule per kilogram kelvin	J/(kg·K)	
heat-transfer coefficient	watt per square meter kelvin	$W/(m^2·K)$	
*illuminance	lux	lx	lm/m^2
*inductance	henry	H	Wb/A
linear density	kilogram per meter	kg/m	
luminance	candela per square meter	cd/m^2	
*luminous flux	lumen	lm	cd·sr
magnetic field strength	ampere per meter	A/m	
*magnetic flux	weber	Wb	V·s
*magnetic flux density	tesla	T	Wb/m^2
molar energy	joule per mole	J/mol	
molar entropy, molar heat capacity	joule per mole kelvin	J/(mol·K)	
moment of force, torque	newton meter	N·m	
momentum	kilogram meter per second	kg·m/s	
permeability	henry per meter	H/m	
permittivity	farad per meter	F/m	
*power, heat flow rate,	kilowatt	kW	
radiant flux	watt	W	J/s
power density, heat flux density, irradiance	watt per square meter	W/m^2	
*pressure, stress	megapascal	MPa	
	kilopascal	kPa	
	pascal	Pa	N/m^2
sound level	decibel	dB	
specific energy	joule per kilogram	J/kg	
specific volume	cubic meter per kilogram	m^3/kg	
surface tension	newton per meter	N/m	
thermal conductivity	watt per meter kelvin	W/(m·K)	
velocity	meter per second	m/s	
	kilometer per hour	km/h	
viscosity, dynamic	pascal second	Pa·s	
	millipascal second	mPa·s	
viscosity, kinematic	square meter per second	m^2/s	
	square millimeter per second	mm^2/s	

Quantity	Unit	Symbol	Acceptable equivalent
volume	cubic meter	m^3	
	cubic diameter	dm^3	L (liter) (5)
	cubic centimeter	cm^3	mL
wave number	1 per meter	m^{-1}	
	1 per centimeter	cm^{-1}	

In addition, there are 16 prefixes used to indicate order of magnitude, as follows:

Multiplication factor	Prefix	Symbol	Note
10^{18}	exa	E	
10^{15}	peta	P	
10^{12}	tera	T	
10^{9}	giga	G	
10^{6}	mega	M	
10^{3}	kilo	k	
10^{2}	hecto	h[a]	[a]Although hecto, deka, deci, and centi are SI prefixes, their use should be avoided except for SI unit-multiples for area and volume and nontechnical use of centimeter, as for body and clothing measurement.
10	deka	da[a]	
10^{-1}	deci	d[a]	
10^{-2}	centi	c[a]	
10^{-3}	milli	m	
10^{-6}	micro	μ	
10^{-9}	nano	n	
10^{-12}	pico	p	
10^{-15}	femto	f	
10^{-18}	atto	a	

For a complete description of SI and its use the reader is referred to ASTM E380 (4) and the article UNITS AND CONVERSION FACTORS which appears in Vol. 24.

A representative list of conversion factors from non-SI to SI units is presented herewith. Factors are given to four significant figures. Exact relationships are followed by a dagger. A more complete list is given in the latest editions of ASTM E380 (4) and ANSI Z210.1 (6).

Conversion Factors to SI Units

To convert from	To	Multiply by
acre	square meter (m^2)	4.047×10^3
angstrom	meter (m)	1.0×10^{-10}†
are	square meter (m^2)	1.0×10^{2}†

†Exact.

To convert from	To	Multiply by
astronomical unit	meter (m)	1.496×10^{11}
atmosphere, standard	pascal (Pa)	1.013×10^{5}
bar	pascal (Pa)	$1.0 \times 10^{5\dagger}$
barn	square meter (m^2)	$1.0 \times 10^{-28\dagger}$
barrel (42 U.S. liquid gallons)	cubic meter (m^3)	0.1590
Bohr magneton (μ_B)	J/T	9.274×10^{-24}
Btu (International Table)	joule (J)	1.055×10^{3}
Btu (mean)	joule (J)	1.056×10^{3}
Btu (thermochemical)	joule (J)	1.054×10^{3}
bushel	cubic meter (m^3)	3.524×10^{-2}
calorie (International Table)	joule (J)	4.187
calorie (mean)	joule (J)	4.190
calorie (thermochemical)	joule (J)	$4.184^{\dagger}$
centipoise	pascal second (Pa·s)	$1.0 \times 10^{-3\dagger}$
centistokes	square millimeter per second (mm^2/s)	$1.0^{\dagger}$
cfm (cubic foot per minute)	cubic meter per second (m^3/s)	4.72×10^{-4}
cubic inch	cubic meter (m^3)	1.639×10^{-5}
cubic foot	cubic meter (m^3)	2.832×10^{-2}
cubic yard	cubic meter (m^3)	0.7646
curie	becquerel (Bq)	$3.70 \times 10^{10\dagger}$
debye	coulomb meter (C·m)	3.336×10^{-30}
degree (angle)	radian (rad)	1.745×10^{-2}
denier (international)	kilogram per meter (kg/m)	1.111×10^{-7}
	$tex^{\ddagger}$	0.1111
dram (apothecaries')	kilogram (kg)	3.888×10^{-3}
dram (avoirdupois)	kilogram (kg)	1.772×10^{-3}
dram (U.S. fluid)	cubic meter (m^3)	3.697×10^{-6}
dyne	newton (N)	$1.0 \times 10^{-5\dagger}$
dyne/cm	newton per meter (N/m)	$1.0 \times 10^{-3\dagger}$
electronvolt	joule (J)	1.602×10^{-19}
erg	joule (J)	$1.0 \times 10^{-7\dagger}$
fathom	meter (m)	1.829
fluid ounce (U.S.)	cubic meter (m^3)	2.957×10^{-5}
foot	meter (m)	$0.3048^{\dagger}$
footcandle	lux (lx)	10.76
furlong	meter (m)	2.012×10^{-2}
gal	meter per second squared (m/s^2)	$1.0 \times 10^{-2\dagger}$
gallon (U.S. dry)	cubic meter (m^3)	4.405×10^{-3}
gallon (U.S. liquid)	cubic meter (m^3)	3.785×10^{-3}
gallon per minute (gpm)	cubic meter per second (m^3/s)	6.309×10^{-5}
	cubic meter per hour (m^3/h)	0.2271

†Exact.

‡See footnote on p. xiv.

To convert from	To	Multiply by
gauss	tesla (T)	1.0×10^{-4}
gilbert	ampere (A)	0.7958
gill (U.S.)	cubic meter (m^3)	1.183×10^{-4}
grade	radian	1.571×10^{-2}
grain	kilogram (kg)	6.480×10^{-5}
gram force per denier	newton per tex (N/tex)	8.826×10^{-2}
hectare	square meter (m^2)	$1.0 \times 10^{4\dagger}$
horsepower (550 ft·lbf/s)	watt (W)	7.457×10^{2}
horsepower (boiler)	watt (W)	9.810×10^{3}
horsepower (electric)	watt (W)	$7.46 \times 10^{2\dagger}$
hundredweight (long)	kilogram (kg)	50.80
hundredweight (short)	kilogram (kg)	45.36
inch	meter (m)	$2.54 \times 10^{-2\dagger}$
inch of mercury (32°F)	pascal (Pa)	3.386×10^{3}
inch of water (39.2°F)	pascal (Pa)	2.491×10^{2}
kilogram-force	newton (N)	9.807
kilowatt hour	megajoule (MJ)	$3.6^{\dagger}$
kip	newton (N)	4.448×10^{3}
knot (international)	meter per second (m/S)	0.5144
lambert	candela per square meter (cd/m^3)	3.183×10^{3}
league (British nautical)	meter (m)	5.559×10^{3}
league (statute)	meter (m)	4.828×10^{3}
light year	meter (m)	9.461×10^{15}
liter (for fluids only)	cubic meter (m^3)	$1.0 \times 10^{-3\dagger}$
maxwell	weber (Wb)	$1.0 \times 10^{-8\dagger}$
micron	meter (m)	$1.0 \times 10^{-6\dagger}$
mil	meter (m)	$2.54 \times 10^{-5\dagger}$
mile (statute)	meter (m)	1.609×10^{3}
mile (U.S. nautical)	meter (m)	$1.852 \times 10^{3\dagger}$
mile per hour	meter per second (m/s)	0.4470
millibar	pascal (Pa)	1.0×10^{2}
millimeter of mercury (0°C)	pascal (Pa)	$1.333 \times 10^{2\dagger}$
minute (angular)	radian	2.909×10^{-4}
myriagram	kilogram (kg)	10
myriameter	kilometer (km)	10
oersted	ampere per meter (A/m)	79.58
ounce (avoirdupois)	kilogram (kg)	2.835×10^{-2}
ounce (troy)	kilogram (kg)	3.110×10^{-2}
ounce (U.S. fluid)	cubic meter (m^3)	2.957×10^{-5}
ounce-force	newton (N)	0.2780
peck (U.S.)	cubic meter (m^3)	8.810×10^{-3}
pennyweight	kilogram (kg)	1.555×10^{-3}
pint (U.S. dry)	cubic meter (m^3)	5.506×10^{-4}
pint (U.S. liquid)	cubic meter (m^3)	4.732×10^{-4}

†Exact.

To convert from	To	Multiply by
poise (absolute viscosity)	pascal second (Pa·s)	0.10†
pound (avoirdupois)	kilogram (kg)	0.4536
pound (troy)	kilogram (kg)	0.3732
poundal	newton (N)	0.1383
pound-force	newton (N)	4.448
pound force per square inch (psi)	pascal (Pa)	6.895×10^3
quart (U.S. dry)	cubic meter (m^3)	1.101×10^{-3}
quart (U.S. liquid)	cubic meter (m^3)	9.464×10^{-4}
quintal	kilogram (kg)	1.0×10^{2}†
rad	gray (Gy)	1.0×10^{-2}†
rod	meter (m)	5.029
roentgen	coulomb per kilogram (C/kg)	2.58×10^{-4}
second (angle)	radian (rad)	4.848×10^{-6}†
section	square meter (m^2)	2.590×10^6
slug	kilogram (kg)	14.59
spherical candle power	lumen (lm)	12.57
square inch	square meter (m^2)	6.452×10^{-4}
square foot	square meter (m^2)	9.290×10^{-2}
square mile	square meter (m^2)	2.590×10^6
square yard	square meter (m^2)	0.8361
stere	cubic meter (m^3)	1.0†
stokes (kinematic viscosity)	square meter per second (m^2/s)	1.0×10^{-4}†
tex	kilogram per meter (kg/m)	1.0×10^{-6}†
ton (long, 2240 pounds)	kilogram (kg)	1.016×10^3
ton (metric) (tonne)	kilogram (kg)	1.0×10^{3}†
ton (short, 2000 pounds)	kilogram (kg)	9.072×10^2
torr	pascal (Pa)	1.333×10^2
unit pole	weber (Wb)	1.257×10^{-7}
yard	meter (m)	0.9144†

†Exact.

Abbreviations and Unit Symbols

Following is a list of common abbreviations and unit symbols used in the *Encyclopedia*. In general they agree with those listed in *American National Standard Abbreviations for Use on Drawings and in Text* (*ANSI Y1.1*) (6) and *American National Standard Letter Symbols for Units in Science and Technology* (*ANSI Y10*) (6). Also included is a list of acronyms for a number of private and government organizations as well as common industrial solvents, polymers, and other chemicals.

Rules for Writing Unit Symbols (4):

1. Unit symbols are printed in upright letters (roman) regardless of the type style used in the surrounding text.
2. Unit symbols are unaltered in the plural.
3. Unit symbols are not followed by a period except when used at the end of a sentence.
4. Letter unit symbols are generally printed lower-case (for example, cd for candela) unless the unit name has been derived from a proper name, in which case the first letter of the symbol is capitalized (W, Pa). Prefixes and unit symbols retain their prescribed form regardless of the surrounding typography.
5. In the complete expression for a quantity, a space should be left between the numerical value and the unit symbol. For example, write 2.37 lm, *not* 2.37lm, and 35 mm, *not* 35mm. When the quantity is used in an adjectival sense, a hyphen is often used, for example, 35-mm film. *Exception:* No space is left between the numerical value and the symbols of degree, minute, and second of plane angle, degree Celsius, and the percent sign.
6. No space is used between the prefix and unit symbol (for example, kg).
7. Symbols, not abbreviations, should be used for units. For example, use "A," not "amp," for ampere.
8. When multiplying unit symbols, use a raised dot:

   $$\mathrm{N{\cdot}m} \quad \text{for} \quad \text{newton meter}$$

   In the case of W·h, the dot may be omitted, thus:

   $$\mathrm{Wh}$$

   An exception to this practice is made for computer printouts, automatic typewriter work, etc, where the raised dot is not possible, and a dot on the line may be used.
9. When dividing unit symbols, use one of the following forms:

   $$\mathrm{m/s} \quad \textit{or} \quad \mathrm{m{\cdot}s^{-1}} \quad \textit{or} \quad \frac{\mathrm{m}}{\mathrm{s}}$$

   In no case should more than one slash be used in the same expression unless parentheses are inserted to avoid ambiguity. For example, write:

   $$\mathrm{J/(mol{\cdot}K)} \quad \textit{or} \quad \mathrm{J{\cdot}mol^{-1}{\cdot}K^{-1}} \quad \textit{or} \quad \mathrm{(J/mol)/K}$$

   but *not*

   $$\mathrm{J/mol/K}$$

10. Do not mix symbols and unit names in the same expression. Write:

joules per kilogram *or* J/kg *or* $J{\cdot}kg^{-1}$

but *not*

joules/kilogram *nor* joules/kg *nor* $joules{\cdot}kg^{-1}$

ABBREVIATIONS AND UNITS

A	ampere
A	anion (eg, HA)
A	mass number
a	atto (prefix for 10^{-18})
AATCC	American Association of Textile Chemists and Colorists
ABS	acrylonitrile–butadiene–styrene
abs	absolute
ac	alternating current, *n.*
a-c	alternating current, *adj.*
ac-	alicyclic
acac	acetylacetonate
ACGIH	American Conference of Governmental Industrial Hygienists
ACS	American Chemical Society
AGA	American Gas Association
Ah	ampere hour
AIChE	American Institute of Chemical Engineers
AIME	American Institute of Mining, Metallurgical, and Petroleum Engineers
AIP	American Institute of Physics
AISI	American Iron and Steel Institute
alc	alcohol(ic)
Alk	alkyl
alk	alkaline (not alkali)
amt	amount
amu	atomic mass unit
ANSI	American National Standards Institute
AO	atomic orbital
AOAC	Association of Official Analytical Chemists
AOCS	American Oil Chemists' Society
APHA	American Public Health Association
API	American Petroleum Institute
aq	aqueous
Ar	aryl
ar-	aromatic
as-	asymmetric(al)
ASHRAE	American Society of Heating, Refrigerating, and Air Conditioning Engineers
ASM	American Society for Metals
ASME	American Society of Mechanical Engineers
ASTM	American Society for Testing and Materials
at no.	atomic number
at wt	atomic weight
av(g)	average
AWS	American Welding Society
b	bonding orbital
bbl	barrel
bcc	body-centered cubic
BCT	body-centered tetragonal
Bé	Baumé
BET	Brunauer-Emmett-Teller (adsorption equation)
bid	twice daily
Boc	*t*-butyloxycarbonyl
BOD	biochemical (biological) oxygen demand
bp	boiling point
Bq	becquerel

C	coulomb
°C	degree Celsius
C-	denoting attachment to carbon
c	centi (prefix for 10^{-2})
c	critical
ca	circa (approximately)
cd	candela; current density; circular dichroism
CFR	Code of Federal Regulations
cgs	centimeter-gram-second
CI	Color Index
cis-	isomer in which substituted groups are on same side of double bond between C atoms
cl	carload
cm	centimeter
cmil	circular mil
cmpd	compound
CNS	central nervous system
CoA	coenzyme A
COD	chemical oxygen demand
coml	commercial(ly)
cp	chemically pure
cph	close-packed hexagonal
CPSC	Consumer Product Safety Commission
cryst	crystalline
cub	cubic
D	debye
D-	denoting configurational relationship
d	differential operator
d	day; deci (prefix for 10^{-1})
d	density
d-	*dextro*-, dextrorotatory
da	deka (prefix for 10^{1})
dB	decibel
dc	direct current, *n.*
d-c	direct current, *adj.*
dec	decompose
detd	determined
detn	determination
Di	didymium, a mixture of all lanthanons
dia	diameter
dil	dilute
DIN	Deutsche Industrie Normen
dl-; DL-	racemic
DMA	dimethylacetamide
DMF	dimethylformamide
DMG	dimethyl glyoxime
DMSO	dimethyl sulfoxide
DOD	Department of Defense
DOE	Department of Energy
DOT	Department of Transportation
DP	degree of polymerization
dp	dew point
DPH	diamond pyramid hardness
dstl(d)	distill(ed)
dta	differential thermal analysis
(*E*)-	entgegen; opposed
ϵ	dielectric constant (unitless number)
e	electron
ECU	electrochemical unit
ed.	edited, edition, editor
ED	effective dose
EDTA	ethylenediaminetetra-acetic acid
emf	electromotive force
emu	electromagnetic unit
en	ethylene diamine
eng	engineering
EPA	Environmental Protection Agency
epr	electron paramagnetic resonance
eq.	equation
esca	electron spectroscopy for chemical analysis
esp	especially
esr	electron-spin resonance
est(d)	estimate(d)
estn	estimation
esu	electrostatic unit
exp	experiment, experimental
ext(d)	extract(ed)
F	farad (capacitance)
F	faraday (96,487 C)
f	femto (prefix for 10^{-15})

FAO	Food and Agriculture Organization (United Nations)
fcc	face-centered cubic
FDA	Food and Drug Administration
FEA	Federal Energy Administration
FHSA	Federal Hazardous Substances Act
fob	free on board
fp	freezing point
FPC	Federal Power Commission
FRB	Federal Reserve Board
frz	freezing
G	giga (prefix for 10^9)
G	gravitational constant = 6.67×10^{11} $N \cdot m^2/kg^2$
g	gram
(g)	gas, only as in $H_2O(g)$
g	gravitational acceleration
gc	gas chromatography
gem-	geminal
glc	gas–liquid chromatography
g-mol wt; gmw	gram-molecular weight
GNP	gross national product
gpc	gel-permeation chromatography
GRAS	Generally Recognized as Safe
grd	ground
Gy	gray
H	henry
h	hour; hecto (prefix for 10^2)
ha	hectare
HB	Brinell hardness number
Hb	hemoglobin
hcp	hexagonal close-packed
hex	hexagonal
HK	Knoop hardness number
hplc	high performance liquid chromatography
HRC	Rockwell hardness (C scale)
HV	Vickers hardness number
hyd	hydrated, hydrous
hyg	hygroscopic
Hz	hertz
^{i} (eg, Pr^i)	iso (eg, isopropyl)
i-	inactive (eg, *i*-methionine)
IACS	International Annealed Copper Standard
ibp	initial boiling point
IC	integrated circuit
ICC	Interstate Commerce Commission
ICT	International Critical Table
ID	inside diameter; infective dose
ip	intraperitoneal
IPS	iron pipe size
ir	infrared
IRLG	Interagency Regulatory Liaison Group
ISO	International Organization Standardization
ITS-90	International Temperature Scale (NIST)
IU	International Unit
IUPAC	International Union of Pure and Applied Chemistry
IV	iodine value
iv	intravenous
J	joule
K	kelvin
k	kilo (prefix for 10^3)
kg	kilogram
L	denoting configurational relationship
L	liter (for fluids only) (5)
l-	*levo*-, levorotatory
(l)	liquid, only as in $NH_3(l)$
LC_{50}	conc lethal to 50% of the animals tested
LCAO	linear combination of atomic orbitals
lc	liquid chromatography
LCD	liquid crystal display
lcl	less than carload lots
LD_{50}	dose lethal to 50% of the animals tested

LED	light-emitting diode
liq	liquid
lm	lumen
ln	logarithm (natural)
LNG	liquefied natural gas
log	logarithm (common)
LOI	limiting oxygen index
LPG	liquefied petroleum gas
ltl	less than truckload lots
lx	lux
M	mega (prefix for 10^6); metal (as in MA)
M	molar; actual mass
$\overline{M}_w$	weight-average mol wt
$\overline{M}_n$	number-average mol wt
m	meter; milli (prefix for 10^{-3})
m	molal
m-	meta
max	maximum
MCA	Chemical Manufacturers' Association (was Manufacturing Chemists Association)
MEK	methyl ethyl ketone
meq	milliequivalent
mfd	manufactured
mfg	manufacturing
mfr	manufacturer
MIBC	methyl isobutyl carbinol
MIBK	methyl isobutyl ketone
MIC	minimum inhibiting concentration
min	minute; minimum
mL	milliliter
MLD	minimum lethal dose
MO	molecular orbital
mo	month
mol	mole
mol wt	molecular weight
mp	melting point
MR	molar refraction
ms	mass spectrometry
MSDS	material safety data sheet
mxt	mixture
μ	micro (prefix for 10^{-6})
N	newton (force)
N	normal (concentration); neutron number
N-	denoting attachment to nitrogen
n (as n_D^{20})	index of refraction (for 20°C and sodium light)
^{n} (as Bu^n), *n*-	normal (straight-chain structure)
n	neutron
n	nano (prefix for 10^{9})
na	not available
NAS	National Academy of Sciences
NASA	National Aeronautics and Space Administration
nat	natural
ndt	nondestructive testing
neg	negative
NF	*National Formulary*
NIH	National Institutes of Health
NIOSH	National Institute of Occupational Safety and Health
NIST	National Institute of Standards and Technology (formerly National Bureau of Standards)
nmr	nuclear magnetic resonance
NND	New and Nonofficial Drugs (AMA)
no.	number
NOI-(BN)	not otherwise indexed (by name)
NOS	not otherwise specified
nqr	nuclear quadruple resonance
NRC	Nuclear Regulatory Commission; National Research Council
NRI	New Ring Index
NSF	National Science Foundation
NTA	nitrilotriacetic acid
NTP	normal temperature and pressure (25°C and 101.3 kPa or 1 atm)

NTSB	National Transportation Safety Board
O-	denoting attachment to oxygen
o-	ortho
OD	outside diameter
OPEC	Organization of Petroleum Exporting Countries
o-phen	*o*-phenanthridine
OSHA	Occupational Safety and Health Administration
owf	on weight of fiber
Ω	ohm
P	peta (prefix for 10^{15})
p	pico (prefix for 10^{-12})
p-	para
p	proton
p.	page
Pa	pascal (pressure)
PEL	personal exposure limit based on an 8-h exposure
pd	potential difference
pH	negative logarithm of the effective hydrogen ion concentration
phr	parts per hundred of resin (rubber)
p-i-n	positive-intrinsic-negative
pmr	proton magnetic resonance
p-n	positive-negative
po	per os (oral)
POP	polyoxypropylene
pos	positive
pp.	pages
ppb	parts per billion (10^9)
ppm	parts per million (10^6)
ppmv	parts per million by volume
ppmwt	parts per million by weight
PPO	poly(phenyl oxide)
ppt(d)	precipitate(d)
pptn	precipitation
Pr (no.)	foreign prototype (number)
pt	point; part
PVC	poly(vinyl chloride)
pwd	powder
py	pyridine
qv	quod vide (which see)
R	univalent hydrocarbon radical
(*R*)-	rectus (clockwise configuration)
r	precision of data
rad	radian; radius
RCRA	Resource Conservation and Recovery Act
rds	rate-determining step
ref.	reference
rf	radio frequency, *n.*
r-f	radio frequency, *adj.*
rh	relative humidity
RI	Ring Index
rms	root-mean square
rpm	rotations per minute
rps	revolutions per second
RT	room temperature
RTECS	Registry of Toxic Effects of Chemical Substances
^{s} (eg, Bu^s); *sec*-	secondary (eg, secondary butyl)
S	siemens
(*S*)-	sinister (counterclockwise configuration)
S-	denoting attachment to sulfur
s-	symmetric(al)
s	second
(s)	solid, only as in H_2O(s)
SAE	Society of Automotive Engineers
SAN	styrene-acrylonitrile
sat(d)	saturate(d)
satn	saturation
SBS	styrene–butadiene–styrene
sc	subcutaneous
SCF	self-consistent field; standard cubic feet
Sch	Schultz number
sem	scanning electron microscope(y)
SFs	Saybolt Furol seconds
sl sol	slightly soluble
sol	soluble

soln	solution
soly	solubility
sp	specific; species
sp gr	specific gravity
sr	steradian
std	standard
STP	standard temperature and pressure (0°C and 101.3 kPa)
sub	sublime(s)
SUs	Saybolt Universal seconds
syn	synthetic
^{t} (eg, Bu^t), *t*-, *tert*-	tertiary (eg, tertiary butyl)
T	tera (prefix for 10^{12}); tesla (magnetic flux density)
t	metric ton (tonne)
t	temperature
TAPPI	Technical Association of the Pulp and Paper Industry
TCC	Tagliabue closed cup
tex	tex (linear density)
T_g	glass-transition temperature
tga	thermogravimetric analysis
THF	tetrahydrofuran
tlc	thin layer chromatography
TLV	threshold limit value
trans-	isomer in which substituted groups are on opposite sides of double bond between C atoms
TSCA	Toxic Substances Control Act
TWA	time-weighted average
Twad	Twaddell
UL	Underwriters' Laboratory
USDA	United States Department of Agriculture
USP	*United States Pharmacopeia*
uv	ultraviolet
V	volt (emf)
var	variable
vic-	vicinal
vol	volume (not volatile)
vs	versus
v sol	very soluble
W	watt
Wb	weber
Wh	watt hour
WHO	World Health Organization (United Nations)
wk	week
yr	year
(*Z*)-	zusammen; together; atomic number

Non-SI (Unacceptable and Obsolete) Units		Use
Å	angstrom	nm
at	atmosphere, technical	Pa
atm	atmosphere, standard	Pa
b	barn	cm^2
bar†	bar	Pa
bbl	barrel	m^3
bhp	brake horsepower	W
Btu	British thermal unit	J
bu	bushel	m^3; L
cal	calorie	J
cfm	cubic foot per minute	m^3/s
Ci	curie	Bq
cSt	centistokes	mm^2/s
c/s	cycle per second	Hz

†Do not use bar (10^5 Pa) or millibar (10^2 Pa) because they are not SI units, and are accepted internationally only for a limited time in special fields because of existing usage.

Non-SI (Unacceptable and Obsolete) Units		Use
cu	cubic	exponential form
D	debye	C·m
den	denier	tex
dr	dram	kg
dyn	dyne	N
dyn/cm	dyne per centimeter	mN/m
erg	erg	J
eu	entropy unit	J/K
°F	degree Fahrenheit	°C; K
fc	footcandle	lx
fl	footlambert	lx
fl oz	fluid ounce	m^3; L
ft	foot	m
ft·lbf	foot pound-force	J
gf den	gram-force per denier	N/tex
G	gauss	T
Gal	gal	m/s^2
gal	gallon	m^3; L
Gb	gilbert	A
gpm	gallon per minute	(m^3/s); (m^3/h)
gr	grain	kg
hp	horsepower	W
ihp	indicated horsepower	W
in.	inch	m
in. Hg	inch of mercury	Pa
in. H_2O	inch of water	Pa
in.-lbf	inch pound-force	J
kcal	kilo-calorie	J
kgf	kilogram-force	N
kilo	for kilogram	kg
L	lambert	lx
lb	pound	kg
lbf	pound-force	N
mho	mho	S
mi	mile	m
MM	million	M
mm Hg	millimeter of mercury	Pa
mμ	millimicron	nm
mph	miles per hour	km/h
μ	micron	μm
Oe	oersted	A/m
oz	ounce	kg
ozf	ounce-force	N
η	poise	Pa·s
P	poise	Pa·s
ph	phot	lx
psi	pounds-force per square inch	Pa
psia	pounds-force per square inch absolute	Pa
psig	pounds-force per square inch gage	Pa
qt	quart	m^3; L
°R	degree Rankine	K
rd	rad	Gy
sb	stilb	lx
SCF	standard cubic foot	m^3
sq	square	exponential form
thm	therm	J
yd	yard	m

BIBLIOGRAPHY

1. The International Bureau of Weights and Measures, BIPM (Parc de Saint-Cloud, France) is described in Appendix X2 of Ref. 4. This bureau operates under the exclusive supervision of the International Committee for Weights and Measures (CIPM).
2. *Metric Editorial Guide* (*ANMC-78-1*), latest ed., American National Metric Council, 5410 Grosvenor Lane, Bethesda, Md. 20814, 1981.
3. *SI Units and Recommendations for the Use of Their Multiples and of Certain Other Units* (*ISO 1000-1981*), American National Standards Institute, 1430 Broadway, New York, 10018, 1981.
4. Based on *ASTM E380-89a* (*Standard Practice for Use of the International System of Units* (*SI*)), American Society for Testing and Materials, 1916 Race Street, Philadelphia, Pa. 19103, 1989.
5. *Fed. Reg.*, Dec. 10, 1976 (41 FR 36414).
6. For ANSI address, see Ref. 3.

R. P. LUKENS
ASTM Committee E-43 on SI Practice

Continued

LASERS

Lasers are sources of light, a form of electromagnetic radiation which propagates at a velocity of 3×10^{10} cm/s and is characterized by an oscillating electric field. For visible light, the frequency of oscillation, denoted ν, is on the order of 10^{15} Hz. The distance between peak values of the electric field is the wavelength. The wave-like nature of light is shown clearly by phenomena such as optical interference and diffraction. However, in some experiments, such as photoelectric emission, light has particle-like characteristics, behaving as if it consisted of discrete bundles of energy, called photons. The energy of a photon, E_p, is equal to $h\nu$, where h is Planck's constant, numerically equal to 6.63×10^{-34} J·s. Because light sometimes exhibits wave-like and sometimes particle-like properties, it is said to have a dual nature, referred to as the duality of light.

The many types of lasers produce light at different wavelengths in the visible, infrared, and ultraviolet regions of the spectrum. Light from lasers has many properties different from those of light from conventional light sources. Laser light can be highly monochromatic, well-collimated, coherent, and in some cases can have extremely high power. These unusual properties lead to a wide variety of applications for lasers in science, engineering, and industry.

The term laser is an acronym constructed from light amplification by stimulated emission of radiation. The first operating laser was produced in 1960 (1). This laser, which used a crystal of ruby [*12174-49-1*], chromium-doped alumina, Al_2O_3:Cr, and emitted a pulsed beam of collimated red light, immediately aroused scientific interest.

The first continuous laser, the helium–neon laser, was operated in 1961 (2). This laser, the first to employ a gaseous medium, operated in the near-infrared spectrum, but by 1962 its operation had been extended to the orange-red portion of the visible spectrum (3), making it also the first continuous visible laser. During the same period, the semiconductor laser, which used a tiny piece of gallium arsenide [*1303-00-0*], GaAs, as its active material, was invented (4–6) (see LIGHT GENERATION, SEMICONDUCTOR LASERS). Almost all the basic types of lasers available as of this writing (ca 1994) were invented in the 1960s. The development of lasers has been well documented (7).

The practical applications of lasers were slow to develop, largely because early lasers were experimental devices having limited reliability and durability. By the late 1960s, existing laser types were longer-lived and more reliable, and in the 1970s lasers began to be used economically in science and industry to perform practical tasks. Lasers as of this writing are familiar tools for applications such as alignment, measurement, as instrumental sources, and in industrial material processing. Use in conjunction with optical fibers has radically changed telecommunications (see FIBER OPTICS). Lasers also form the basis of many consumer products such as supermarket scanners, compact-disk players, and laser printers.

In the 1990s, significant developments in laser technology continue. These include semiconductor laser technology and semiconductor laser-pumped solid-state lasers. These developments should lead to smaller, more efficient laser devices. Future applications include laser-assisted thermonuclear fusion (see FUSION ENERGY) and laser-assisted separation of isotopes. Among the most important chemical applications are new spectroscopic techniques (see SPECTROSCOPY), the monitoring of transients in chemical reactions (see KINETIC MEASUREMENTS), and state-selective chemistry, in which the course of a chemical reaction is controlled by selectively exciting certain molecular states. These applications have all been demonstrated on a research scale.

Fundamentals of Lasers

Laser light is produced from transitions between atomic or molecular energy levels. Generation of light requires two energy levels, E_1 and E_2, separated by the photon energy E_p of the light that is to be produced.

$$E_p = h\nu = hc/\lambda = E_2 - E_1 = \Delta E \tag{1}$$

where h is Planck's constant, c is the velocity of light, and E_1 and E_2 are, respectively, the lower and higher energy levels. The relevant energy levels may be those of atoms or molecules in a gas, as in the case of the helium–neon laser, or of ions embedded in a solid host material, as in a ruby laser, or they may be energy levels that belong to a crystalline lattice as a whole, as in the aluminum gallium arsenide [*37382-15-3*], AlGaAs, semiconductor laser.

The interaction between the light and the energy levels to produce laser operation relies on the phenomenon of stimulated emission. Stimulated emission was first predicted by Albert Einstein on theoretical grounds in 1917 (8). For stimulated emission to occur an atom or molecule must be in one of its excited levels and have a vacant energy level of lower energy. Then incoming light of the proper frequency can trigger a transition from the upper level to the lower level. The photon energy of the incident light must equal the energy difference between the two levels. The light can then stimulate the atomic or molecular system to make the transition. At the same time, the energy stored in the atomic or molecular system is emitted as light, so that there is an increase in the light intensity. The emitted light has the same photon energy as the incident light, travels in the same direction as the incident light, and remains in phase with it. It is this last property which gives rise to many of the other important properties of lasers, including coherence and directionality.

The three requirements for a laser are a material that possesses an appropriate set of energy levels (the active medium), some means for excitation or pumping the atoms or molecules to excited upper energy levels while at the same time leaving lower lying energy levels empty, and some means of resonant feedback to allow the light to pass back and forth through the active medium. During these passes, the light is amplified by the stimulated emission process and increases in intensity.

Active Media. Potentially useful laser materials must have energy levels that are coupled radiatively, ie, there must be the possibility of fluorescent emission of light by atoms or molecules in excited energy levels. Materials having suitable energy levels include solids, such as ruby and certain glasses and crystals, atomic gases, ionized gases, molecular gases, semiconductors, and some dye materials in liquid solution.

Excitation. Pumping or excitation is necessary to drive the atomic or molecular system into an excited upper energy level from which stimulated emission can occur. Stimulated emission is exactly the inverse of absorption. Absorption, the process by which light energy is absorbed, drives the system from a lower to an upper energy level. The transition probability for absorption is numerically equal to the transition probability for stimulated emission. If there are more atoms or molecules in the lower energy level than in the upper level, there should be a net absorption of light energy and the light intensity subsequently decreases as the light travels through the material. Therefore, for a laser there must be more atoms or molecules in the upper energy than in the lower level. This situation, known as a population inversion, is not the usual condition. According to thermodynamic considerations, the lower energy levels should be more populated at thermal equilibrium, and the equilibrium must be disturbed in order to have the higher energy levels more populated than the lower ones.

There are a variety of methods for assuring population inversion. The most common are optical pumping and pumping by an electrical discharge. Additional methods include excitation through chemical reaction, through transfer of potential energy to kinetic energy during expansion of a gas, through radio-frequency reactions, and by nuclear radiation.

Figure 1 illustrates some common methods for atomic systems. The basic mechanism of optical pumping is shown in Figure 1**a**. The atoms ordinarily reside in the ground level, level 1. Illumination of the material with light at the proper frequency, ν_p, corresponding to the energy difference between levels 1 and 3, causes the atoms to absorb the light and make the transition to level 3. From level 3, the atoms make a rapid, radiationless transition to level 2. If the pumping is intense enough, most of the atoms can be raised from level 1 to level 2, producing a population inversion between level 2 and level 1. Laser operation can then occur between level 2 and level 1 at frequency ν_L. This is called the three-level system. The drawback is that the terminal level for laser operation is

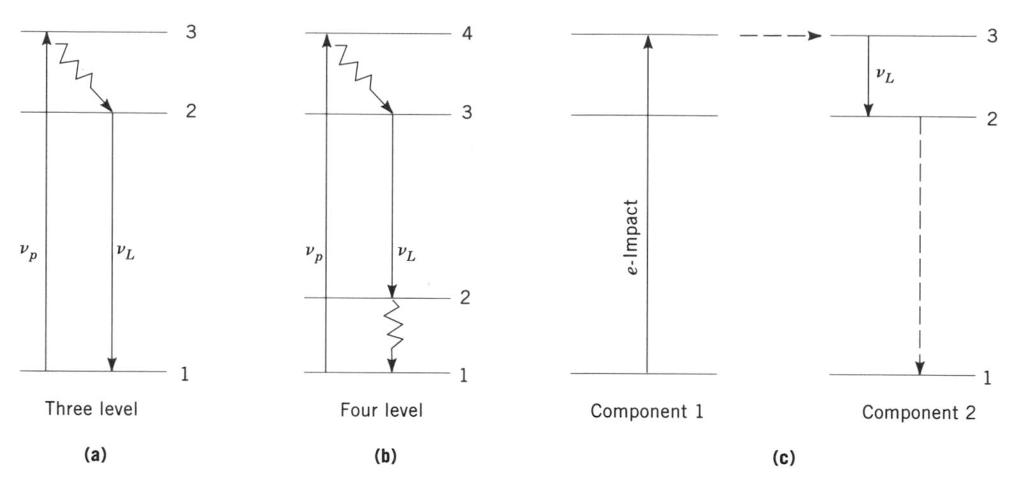

Fig. 1. Pumping methods for lasers where ν_p is the pump light frequency and ν_L is the laser frequency, wavy lines represent radiationless transitions, and the dashed line collisions: (**a**) optical pumping in three-level systems; (**b**) optical pumping in four-level systems; (**c**) pumping by electron impact and resonant transfer of energy.

the ground level, which is ordinarily fully populated. Thus a large percentage of the atoms must be moved to level 2 in order to produce the population inversion. Such lasers require intense pumping light.

Figure **1b** shows a four-level system. The terminal level, level 2, is ordinarily empty. Atoms are optically pumped to level 4. From level 4, the atoms make a rapid radiationless transition to level 3. The first few atoms to arrive begin to contribute to the population inversion. Therefore, laser operation can begin with much less intense pumping light. After the laser transition, the atoms return to the ground state (level 1) by a radiationless transition.

A third pumping method (Fig. **1c**) uses an electrical discharge in a mixture of gases. It relies on electronic excitation of the first component of the gas mixture, so that those atoms are raised to an upper energy level. The two components are chosen so that there can be a resonant transfer of energy by collisions from the upper level of the first component to level 3 of the second component. Because there are no atoms in level 2, this produces a population inversion between level 3 and level 2. After laser emission, the atoms in the second component return to the ground state by collisions.

Feedback. The third requirement for a laser is a structure that provides feedback of the laser light into the active material. This generally consists of two mirrors, one at each end of the active material, which form what is called a resonant cavity. These mirrors provide for multiple passes of the light through the material. Amplification of the laser light via stimulated emission can occur over a greater distance through the material. The cavity must satisfy the following condition:

$$N\lambda = 2D \tag{2}$$

where λ is the wavelength, D is the distance between the mirrors, and N is an integer. This same equation holds for resonant microwave cavities, in which the wavelength is similar in magnitude to the dimension of the cavity (see MICROWAVE TECHNOLOGY). In a microwave cavity, N is a small integer, typically one or two. In a laser, the wavelength is much smaller than the cavity dimension, which is the distance between the mirrors. If $D = 0.5$ m and $\lambda = 1\ \mu$m, then N need be around 1×10^6. There are many combinations of integer N and wavelength λ that satisfy equation 2. Each such combination is called a resonant longitudinal mode of the laser. A laser may operate simultaneously at several different wavelengths, each of which satisfies equation 2. This fact affects the monochromatic properties of the laser.

The earliest lasers used two plane mirrors which required exact alignment, or the laser light would wander from the edges of the mirrors after several passes through the material. Plane parallel mirrors were also very sensitive to vibration. Spherical mirrors were found to provide resonant paths for the light (9). Such mirrors offer less sensitivity to misalignment than plane parallel mirrors. Most modern lasers use two spherical mirrors, each having a radius of curvature much greater than the distance between them. This configuration offers good stability, plus reentrant paths that effectively fill the volume of active material.

The high power of the laser rapidly damaged the metallic mirrors used for early lasers. Most modern lasers use mirrors formed by multilayer dielectric coatings. These mirrors are fabricated by vacuum deposition of thin layers of dielectric materials onto a transparent substrate. Some typical materials include silicon dioxide [*7631-86-9*], SiO_2; titanium dioxide [*13463-67-7*], TiO_2; zirconium dioxide [*1314-23-4*], ZrO_2; and thorium fluoride [*13709-59-6*], ThF_4. The mirrors are made by evaporating alternate layers of a material having a high index of refraction and one having low index of refraction. Each layer has optical thickness (physical thickness multiplied by the index of refraction) equal to one quarter of the wavelength at which it is to be used. Proper choice of indexes of refraction and number of layers allows production of a mirror having the desired reflectivity. These mirrors are less susceptible to damage by high power laser light, although in high power lasers, the periodic replacement of mirrors is still required.

Usually the laser cavity consists of one mirror that is almost 100% reflecting and one mirror that is partially reflecting and partially transmitting, in order to allow emission of some of the light as the useful output of the laser.

Properties of Laser Light

The light emitted by a laser has a number of unusual properties that distinguish it from light emitted by conventional light sources. Unusual properties include a high degree of collimation, a narrow spectral linewidth, good coherence, and the ability to focus to an extremely small spot. Because of these properties, there are many possible applications for which lasers are better suited than conventional light sources. For example, alignment applications utilize the collimation of laser light. Spectroscopic and photochemical applications depend on the narrow spectral linewidth of tunable laser sources. Interferometric and holographic applications require a high degree of coherence (see HOLOGRAPHY).

Collimation. One of the most important characteristics of laser radiation is its highly directional collimated beam. Although the beam divergence angle is very small, it is not zero. The limitation on beam divergence angle is set by the fundamental physical phenomenon of diffraction, which provides a lower limit to the divergence of the beam. This lower limit is given by the equation

$$\theta = 2.44\ \lambda/d \tag{3}$$

where θ is the minimum beam divergence angle in radians (the full angle of the beam), d is the diameter of the aperture through which the light emerges, and λ is the wavelength of the light. A laser beam that equals or approaches this minimum possible value is said to be diffraction-limited. The limiting aperture may often be considered to be the diameter of the laser. As an example, according to the above equation, a helium–neon laser having a diameter of 1.5 mm operating at a wavelength of 0.6328 μm has a beam divergence angle of 1×10^{-3} rad.

The value of the beam divergence angle given by equation 3 corresponds to that emerging from the laser. The beam may be collimated further by a telescope. The improvement of the collimation is the inverse of the magnification of the telescope:

$$\theta_i d_i = \theta_f d_f \tag{4}$$

where the subscript i refers to the initial beam and the subscript f refers to the final beam emerging from the telescope, and where d and θ are, respectively, the beam diameter and divergence angle. Thus the beam divergence is decreased and the beam diameter is increased in the collimation process.

Spectral Linewidth. Laser light is highly monochromatic; that is, it has a very narrow spectral linewidth. This linewidth is not zero, but is typically much less than that of conventional light sources. The discussion of linewidth is complicated by the presence of the resonant longitudinal cavity modes. The spacing between the longitudinal modes is given by $c/2D$, where c is the velocity of light and D is the length of the laser. Several longitudinal modes may be present simultaneously in the laser output. The most intense ones are near the center of the fluorescent line, where the gain is highest, therefore the spectral output of the laser may have the rather complicated form shown in Figure 2.

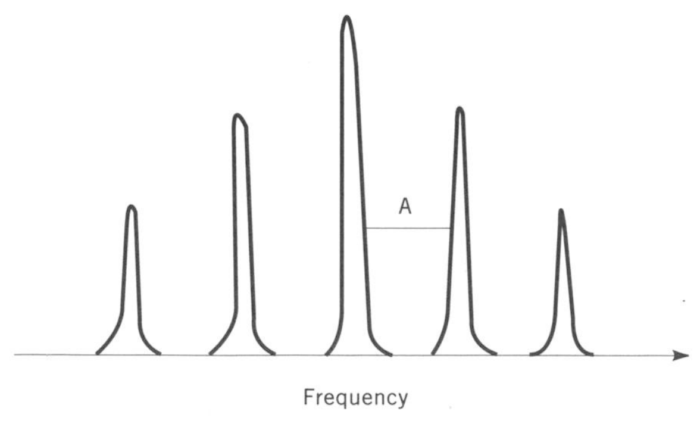

Fig. 2. Representation of longitudinal modes in a laser. Where line A $= c/2D$, c is the velocity of light, and D is the distance between the laser mirrors.

For example, in a typical commercial helium–neon laser, there may be four or five longitudinal modes in the output. For a laser tube 100-cm long, the spacing between the longitudinal modes would be 1.5×10^8 Hz. The total linewidth for five longitudinal modes would be 6×10^8 Hz.

However, the widths of the individual longitudinal modes can be much narrower. The frequency spectrum can be narrowed by a number of techniques to provide operation in a single longitudinal mode. Frequency-stabilized single-longitudinal mode lasers are commercially available for some types of laser such as the helium–neon laser. A commercial frequency-stabilized helium–neon laser operating on a single longitudinal mode may have a line width around 10^7 Hz (10). An experimental stable gas laser constructed in a wine cellar to provide good vibration isolation achieved short-term frequency stabilities as low as 20 Hz (11). This represented a frequency accuracy around one part in 10^{14}.

Single-mode operation is usually achieved at the expense of output power. Very high power lasers are usually not available as single-mode lasers, and an increase of power is usually accompanied by an increase in the spectral linewidth.

Coherence. Coherence implies a regular, orderly progression of the vibrations of the electromagnetic radiation. Coherence can be understood only by reference to a considerable body of mathematical development, a detailed discussion of which can be found in the literature (12). Conventional light sources do not share the property of coherence; rather these sources have a certain amount of randomness or irregularity in the oscillations of the light waves. Radio waves exhibit coherence, or orderliness, and because of coherence properties information can be impressed on radio waves for communications applications. The advent of lasers made available coherent light sources of reasonably high intensity in the visible portion of the spectrum. Coherence is important in any application in which the laser beam is to be split into parts, such as interferometry and holography (qv). Such applications rely on the high coherence of the laser beam.

The time period Δt in which the light wave undergoes random changes is called the coherence time. It is related to the linewidth $\Delta \nu$ of the laser by the equation

$$\Delta t \simeq \frac{1}{\Delta \nu} \tag{5}$$

Measurements made in a time that is short compared to the coherence time have high coherence. If multiple longitudinal modes are present in the output of the laser, the spectral width is broadened and the coherence time reduced. This has the effect of reducing fringe contrast in an interferometric or holographic experiment. Thus single-mode lasers are often used for applications such as interferometry and holography (see INFRARED TECHNOLOGY AND RAMAN SPECTROSCOPY).

The coherence of the beam in continuous visible lasers produces an interesting effect. When the beam reflects from a surface like a sheet of paper, it has a granular, speckled appearance. The speckle pattern results from interference of the light reflected from minute surface irregularities. When the observer's head moves, the speckles appear to be in motion. With incoherent light, randomness in the light waves causes the interference to average to a uniform level within a time that is short compared to the response of the eye. With coherent light, the interference pattern is stable, and the speckles are easily observed. A visible laser would thus make a poor reading lamp.

Focusing Laser Light. One of the most important properties of laser radiation is the ability to collect all of the radiation using a simple lens and to focus it to a spot. It is not possible to focus the laser beam down to a mathematical point; there is always a minimum spot size, set by the physical phenomenon of diffraction. A convenient equation is

$$D_m = F\theta \tag{6}$$

where D_m is the smallest focal diameter that can be obtained, F is the focal length of the lens, and θ is the beam divergence angle of the laser beam. According to equation 3 the diffraction-limited beam divergence is approximately given by λ/d where d can be taken as the diameter of the lens; thus the minimum focal spot size becomes

$$D_m = F\lambda/d \tag{7}$$

The ratio F/d is the F number of the lens. For F numbers much less than unity, spherical aberration precludes reaching the ultimate diffraction-limited spot size. Therefore a practical limit for the minimum spot size obtainable is approximately the wavelength of the light. Commonly this is expressed as the statement that laser light may be focused to a spot with dimensions equal to its wavelength.

The minimum spot size expressed by equation 6 cannot be achieved if lens aberrations are present or if there is poor optical engineering. The smallest spot sizes are obtained using beams of small beam divergence angle and lenses of short focal length. Conventional light sources, which have much larger beam divergence angles, cannot be focused to as small a spot. The focusing characteristics are important because of the high power per unit area obtainable for a focused laser beam, leading to possibilities for applications such as material processing.

Spatial Profiles. The cross sections of laser beams have certain well-defined spatial profiles called transverse modes. The word mode in this sense should not be confused with the same word as used to discuss the spectral linewidth of lasers. Transverse modes represent configurations of the electromagnetic field determined by the boundary conditions in the laser cavity. A full description of the transverse modes requires the use of orthogonal polynomials.

Only some of the configurations that commonly occur in low power gas lasers are described herein. The modes are denoted by the nomenclature TEM_{mn} where the term TEM stands for transverse electromagnetic, and where m and n are small integers. Figure 3 shows some TEM_{mn} modes that are commonly seen in laser outputs. The notation can be interpreted as meaning that m and n specify the number of low intensity points (nulls) in the spatial pattern in each of two perpendicular directions, transverse to the direction of beam propagation. Thus the TEM_{10} mode has one null in the horizontal direction and none in the vertical direction. Such patterns are commonly observed in the output of gas lasers. In some cases a superposition of a number of modes can be present at the

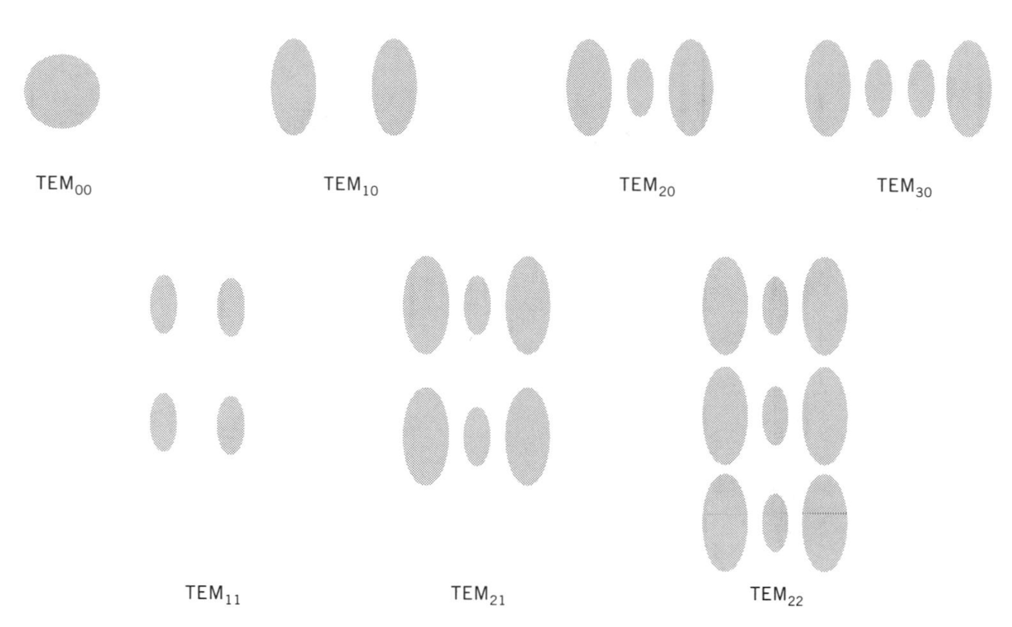

Fig. 3. Spatial profiles of TEM transverse modes.

same time, making the radiation profile complicated. This generally occurs as the laser power increases. The simple patterns shown in Figure 3 are sacrificed to obtain high power.

The TEM_{00} mode is a specially desirable case. This transverse mode, the Gaussian mode, is symmetric and has no nulls. It is preferred for many applications. Its intensity I as a function of radius r from the center of the beam is given by the equation

$$I(r) = I_0 \exp(-2r^2/r_0^2) \tag{8}$$

where I_0 is the intensity of the beam at the center and r_0 is the Gaussian beam radius, the radius at which the intensity is reduced from its peak value by a factor of e^2.

Temporal Characteristics. Laser operation may be characterized as either pulsed or continuous. There are a number of distinctive types of pulsed laser operation having widely different pulse durations.

The earliest solid-state lasers, such as ruby lasers, were simply pulsed by discharging a capacitor through a flashlamp, without any attempt to control the duration of the laser output. Such lasers typically had durations around 1 ms. The pulses were called normal or free-running laser pulses (Fig. 4**a**). The normal pulse frequently exhibits a considerable amount of substructure (13). Figure 4**b** shows a small portion of the pulse on an extended time scale. There are a number of subpulses called relaxation oscillations having microsecond duration. These pulses arise when a threshold value for population inversion is satisfied momentarily. The energy stored in the population inversion is extracted rapidly, the population inversion falls below the threshold level, and laser operation ceases. Continued repumping by the flashlamp restores the threshold condition,

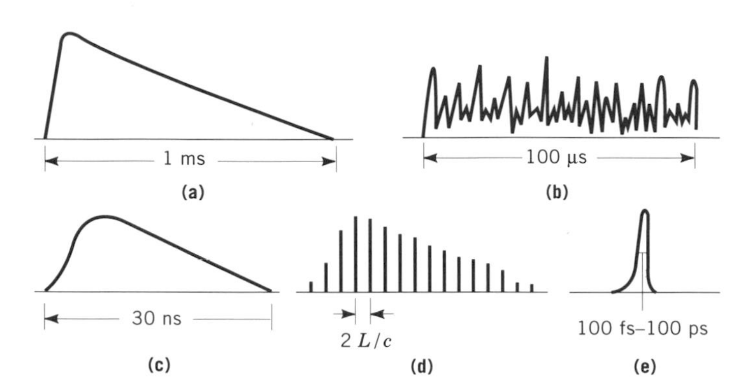

Fig. 4. Temporal pulse characteristics of lasers: (**a**) millisecond laser pulse; (**b**) relaxation oscillations; (**c**) Q-switched pulse; (**d**) mode-locked train of pulses, where L is the distance between mirrors and c is the velocity of light for $L = 37.5$ cm, $2\ L/c = 2.5$ ns; (**e**) ultrafast (femtosecond or picosecond) pulse.

and another pulse can occur. The oscillatory character is common for solid-state lasers.

In order to produce pulses having smoother temporal characteristics and also to increase the peak power, methods of pulse control were developed. The most common method is called Q-switching. In a Q-switched laser, a switchable shutter is inserted between the laser material and one of the mirrors (14,15). When the laser is first pumped, the shutter is closed. No light can reach the mirror and the process of amplification through stimulated emission cannot build up. The laser medium is excited to the point that the population inversion considerably exceeds the threshold. Then the shutter is rapidly switched to the open position. Because of the large population inversion, the gain can be very high, and the laser pulse develops rapidly. The energy stored in the population inversion can be swept out in a single pulse of high peak power. The pulse duration is much shortened, from the millisecond region to the region of perhaps 30 ns, as illustrated in Figure 4**c**. The total energy release in the laser pulse is reduced by the Q-switching operation. Because the pulse duration is decreased by many orders of magnitude, the peak power can be much higher than for a normal pulse.

The output of a Q-switched laser appeared temporally smooth when viewed by phototubes and oscilloscopes in the early 1960s. However by the late 1960s, it was realized that there was substructure in the Q-switched pulse as well (16,17). Under many conditions, the output of the Q-switched laser could consist of a train of very short pulses, as illustrated in Figure 4**d**. The individual pulses in the pulse train were separated by the round-trip transit time of the cavity, $2\ L/c$. The widths of the individual pulses were very short, on the order of tens of picoseconds. This train of pulses is commonly called a mode-locked train, and the individual pulses are called mode-locked or picosecond pulses. This behavior arises because of interference between the different longitudinal modes present in the laser cavity (18,19). The phases of the longitudinal modes lock together to produce the output shown. A temporally smooth Q-switched pulse can be obtained by constraining the laser to operate in a single longitudinal mode. In that case there can be no phase locking of different modes, and the output is temporally smooth.

It is also possible to switch a single picosecond pulse out of the train of mode-locked pulses using an electrooptic switch. It is possible to obtain a single pulse having duration in the picosecond regime or even less. Pulses with durations in the regime of a few hundred femtoseconds (10^{-15} s) are also available (Fig. 4**e**).

Laser Types

There are many types of lasers, having a wide variety of methods of construction and based on many different classes of materials. The properties of some commercially available lasers are summarized in Table 1. Typical available characteristics are given. More detailed compilations of the properties of commercially available lasers are available (20,21).

Gas Lasers. Many important lasers utilize a gaseous medium as the active material, for example, the first six entries in Table 1 represent lasers based on gaseous materials. A gas has many advantages over solid or liquid materials. First, the gas is homogeneous, in contrast to solid materials which often present problems with optical homogeneity. Second, the removal of heat, an important consideration in laser design, is relatively easy, because the heated gas can flow out of the region where laser action occurs. Also, the volume can be increased almost at will, in order to increase the laser output. In contrast, solid lasers depend on the production of crystals or glasses, which are limited in size.

Figure 5 shows a prototypical gas laser design. A wide variety of methods of construction may be employed, based on different classes of materials. An electrical discharge excites the gas between the electrodes, which may be in side arms attached to the main tube. The side arms may also serve as a reservoir of gas, to make up for small losses from leakage. In Figure 5, the electrical discharge is along the length of the tube. Whereas this is a common feature in many gas lasers, it is not necessary. Some gas lasers have transverse excitation, ie, the electrical discharge is perpendicular to the long axis of the laser. The mirrors shown are spherical mirrors, a situation common for gas lasers. The mirrors may be sealed directly onto the ends of the gas-filled tube, or mounted externally to the tube, where the ends of the tube are sealed with windows, as shown. The end windows are tilted at an angle, called Brewster's angle, θ_B (22). The effect of tilting the windows at this angle is to minimize the reflection loss for one component of polarization. The angle θ_B is given by the equation

$$\theta_B = \arc \tan n \tag{9}$$

where n is the index of refraction of the window material. As shown in the inset, the component of the polarization in the plane of the paper is transmitted without loss when the window is tilted at Brewster's angle. The component of polarization perpendicular to the paper suffers loss owing to reflection. The effect is to encourage the operation of the laser with a polarization parallel to the plane of the paper.

Table 1. Common Commercial Lasers

Laser	CAS Registry Number	Formula	Wavelength, μm	Operation	Output, W	Typical application
			Gas lasers			
helium–neon		He–Ne	0.6328	continuous (CW)	0.0005–0.035	alignment, holography, printers, bar-code readers, measurement
carbon dioxide	[*124-38-9*]	CO_2	10.6	continuous	to 25,000	welding, drilling, heat treating
				pulsed	500–1,000[a]	
argon	[*7440-37-1*]	Ar	0.4880, 0.5145 and other lines	continuous	to 25	measurement of speed, displays, entertainment
krypton	[*7439-90-9*]	Kr	0.6471 and other lines	continuous	to 16	displays, entertainment
helium–cadmium		He–Cd	0.442, 0.325	continuous	to 0.0150	photoresist exposure
krypton fluoride	[*34160-02-6*]	KrF	0.249	pulsed	to 4[b]	semiconductor processing
			Solid-state lasers			
neodymium yttrium-aluminum garnet (YAG)		$Y_3Al_5O_{12}$:Nd	1.06, 0.532	continuous	to 1800	welding, drilling, trimming, marking
				pulsed	10[a]	
ruby	[*12174-49-1*]	Al_2O_3:Cr	0.6943	pulsed	to 400[b,c]	spot welding
titanium sapphire		Al_2O_3:Ti	tunable 0.7–0.78	continuous	3.5	spectroscopy
				pulsed	3.5	
			Semiconductor laser			
aluminum gallium arsenide	[*37382-15-3*]	AlGaAs	0.8–0.95	continuous	to 20	communications, ranging, compact disk
				pulsed	0.050[a]	
indium gallium arsenide phosphide	[*112957-92-3*]	InGaAsP	1.3, 1.55	continuous	to 0.150	communications
				pulsed	to 0.0150	
aluminum indium gallium phosphide	[*115493-44-2*]	AlInGaP	0.63–0.69	continuous	to 250	replacement for He–Ne
				pulsed	to 250	
			Liquid laser			
dye			tunable, 0.25–1	continuous	to 2[d]	spectroscopy, photochemistry
				pulsed		

[a]Most common wattage. [b]Units are J/pulse. To convert J to cal, divide by 4.184. [c]Most common value is 10 J/pulse (2.4 cal/pulse). [d]At selected wavelengths.

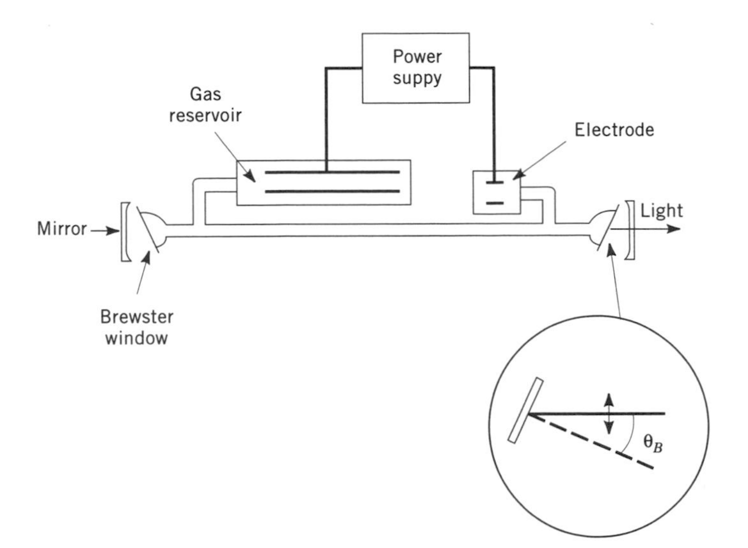

Fig. 5. Schematic of a gas laser construction.

The most familiar gas laser is the helium–neon laser (23,24). Sales of commercial helium–neon lasers exceed 400,000 units per year. The helium–neon laser is a compact package that produces a continuous beam of orange-red light. The inside diameter of the tube is commonly around 1.5 mm. The output of helium–neon lasers available commercially ranges from a fraction of a milliwatt to more than 35 mW. They have many applications in the areas of alignment, supermarket scanning, educational demonstrations, and holography.

The helium–neon laser employs an unionized mixture of helium and neon gases. Many other lasers use ionized gas as the active medium. The most common example is the argon laser (25,26). The argon laser consists of a plasma-filled tube having an electric discharge at high current density (see PLASMA TECHNOLOGY). The energy levels relevant for laser operation are those of the argon ion rather than the argon atom. Argon lasers require high electrical current density. Construction of the argon laser requires high temperature materials that can resist erosion by the electrical current. Materials such as tungsten [*7440-33-7*] and beryllium oxide [*1304-56-9*], BeO, have been used in the bore of argon lasers.

The complicated energy level structure of the argon ion means that a number of laser lines may be present simultaneously in the output of the laser. The most important lines, at 0.4880 and 0.5145 μm, contain most of the laser power. Single-line operation on any one of the lines may be obtained by inserting a wavelength-selecting element within the laser cavity, but this reduces the output power compared to the multiline operation. Many manufacturers specify the output of their argon lasers as the total multiline value. In single-line operation, the output may be one-third or less of the multiline value. The argon laser is also capable of operation at several wavelengths in the ultraviolet. These lines are produced only when special mirrors having high reflectivity in the ultraviolet are used. Commercial argon lasers are available having continuous multiline output up to 25 W.

Krypton lasers are also ionized gas lasers and are very similar in general characteristics to argon lasers (27). Krypton lasers having total multiline output up to 16 W are available commercially. The strongest line at 0.6471 μm is notable because it is in the red portion of the spectrum, and thus makes the krypton laser useful for applications such as display and entertainment.

The helium–cadmium laser, which has emission at 0.442 and 0.325 μm, is a somewhat different type of ionized gas laser (28). It operates using the ionized states of cadmium, produced by heating cadmium in a furnace. The output of continuous, commercially available helium–cadmium lasers ranges up to 150 mW.

Whereas the gas lasers described use energy levels characteristic of individual atoms or ions, laser operation can also employ molecular energy levels. Molecular levels may correspond to vibrations and rotations, in contrast to the electronic energy levels of atomic and ionic species. The energies associated with vibrations and rotations tend to be lower than those of electronic transitions; thus the output wavelengths of the molecular lasers tend to lie farther into the infrared.

The most important type of molecular laser is the carbon dioxide laser, which uses a mixture of carbon dioxide, nitrogen [*7727-37-9*] and helium [*7440-59-7*] (29–31). Operation of the carbon dioxide laser involves the excitation of vibrational levels of the nitrogen laser by collisions with electrons in the electrical discharge, followed by resonant energy transfer to a vibrational level of the carbon dioxide molecule. This is followed by laser transitions in two bands, centered at 9.6 and 10.6 μm, respectively. Because the 10.6-μm transition has higher gain, that wavelength is preferred unless other factors are present. Each of the two bands contains much substructure because of the presence of rotational sublevels. Because the higher gain is near the center of the 10.6-μm manifold, laser operation usually occurs near that wavelength. However, if a wavelength-selective element such as a grating is inserted into the laser cavity, operation on any of the individual lines may be obtained, at reduced power, over the wavelength range 9.3–11.0 μm.

In the earliest and most simple configuration, carbon dioxide lasers consisted of a long glass or quartz tube where the electric discharge and the gas flowed along the length of the tube, as shown in Figure 6**a**. The output of such lasers is scaled as the length of the tube. These lasers are capable of producing approximately 80 W/m of length. Thus it is possible to obtain continuous-output power of several hundred watts. Carbon dioxide lasers in the few hundred-watt range are still constructed in this fashion. But to obtain an output in the kilowatt range an unreasonably long tube is required, and in order to increase the power, a number of innovations developed. First, the gas flow can be perpendicular to the long axis of the laser, in other words, transverse to the optical path (see Fig. 6**b**). The power increases with gas-flow rate. Because the impedance to gas flow is reduced with transverse flow, the power output can be increased (32). Also, the electrical discharge may be perpendicular to the optical path. It becomes easier to supply the optimum electric field (voltage per unit length) when the length of the discharge is reduced. Thus, in many modern lasers, the electric field is also supplied perpendicular to the optical path in a configuration such as that shown in Figure 6**b**. Extraction of several kW of continuous power

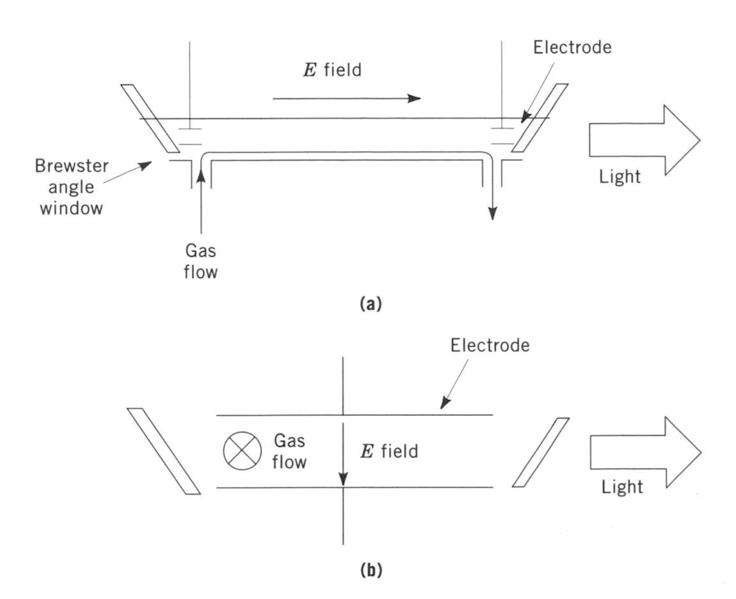

Fig. 6. Carbon dioxide laser construction: (**a**) original configuration having longitudinal electric (E) field and gas flow; (**b**) advanced configuration having transverse electric field and gas flow.

becomes possible in a reasonably sized device. Models of multikilowatt carbon dioxide lasers having continuous power up to 25 kW have become available.

The transversely excited atmospheric-pressure (TEA) laser, inherently a pulsed device rather than a continuous laser, is another common variety of carbon dioxide laser (33,34). Carbon dioxide–TEA lasers are an important class of high power pulsed lasers. Pulse durations are in the submicrosecond regime; peak powers exceed 10 MW.

A final type of gas laser is the excimer laser (35,36). Excimer lasers represent a significant thrust of laser technology into the ultraviolet portion of the spectrum, offering the capability for pulsed short-wavelength lasers having high peak power. An excimer is a compound that has no stable ground state but which may have excited states that are temporarily bound. Most excimer lasers involve molecules containing the noble gases, which do not form stable chemical compounds. A leading example of excimer lasers is the krypton fluoride laser. A gas mixture containing krypton and fluorine is excited in a pulsed electrical discharge. In a chain of complicated processes the metastable excited state KrF* is produced. The excited state is bound for a short time and then dissociates according to the following reaction:

$$KrF^* \longrightarrow Kr + F + h\nu \tag{10}$$

where $h\nu$ has energy corresponding to a wavelength of 0.249 μm. Because there is no stable ground state for KrF, the population inversion necessary for laser operation is easily obtained. Excimer lasers are inherently pulsed devices. Pulse durations are in the nanosecond regime. Commercial excimer lasers are available for a variety of wavelengths ranging from 0.157 to 0.351 μm, and having pulse energy typically in the range 0.1–2 J.

Solid-State Lasers. Solid-state lasers (37) use glassy or crystalline host materials containing some active species. The term solid-state as used in connection with lasers does not imply semiconductors; rather it applies to solid materials containing impurity ions. The impurity ions are typically ions of the transition metals, such as chromium, or ions of the rare-earth series, such as neodymium (see LANTHANIDES). Most often, the solid material is in the form of a cylindrical rod with the ends polished flat and parallel, but a variety of other forms have been used, including slabs and cylindrical rods with the ends cut at Brewster's angle.

The original solid-state laser was the ruby laser. Ruby, the material in which laser operation was first attained (38), is crystalline aluminum oxide [*1344-28-1*], Al_2O_3, in which a small (typically 0.05%) percentage of the aluminum is replaced by chromium ions. The chromium(III) ion [*16065-83-1*] resides in the aluminum oxide host. The ruby laser is a three-level system and requires rather intense pumping. Figure 7 shows the configuration of a typical ruby or other solid-state laser. The laser rod is excited by light from an adjacent flashlamp. The capacitor is charged by the power supply. Next the flashlamp is triggered by a high voltage pulse. The capacitor then discharges through the flashlamp with high current density accompanied by emission of light at wavelengths which lie in the pumping bands of the laser material. The flashlamp pulse may be shaped to be approximately rectangular by a pulse-forming network containing inductance and capacitance.

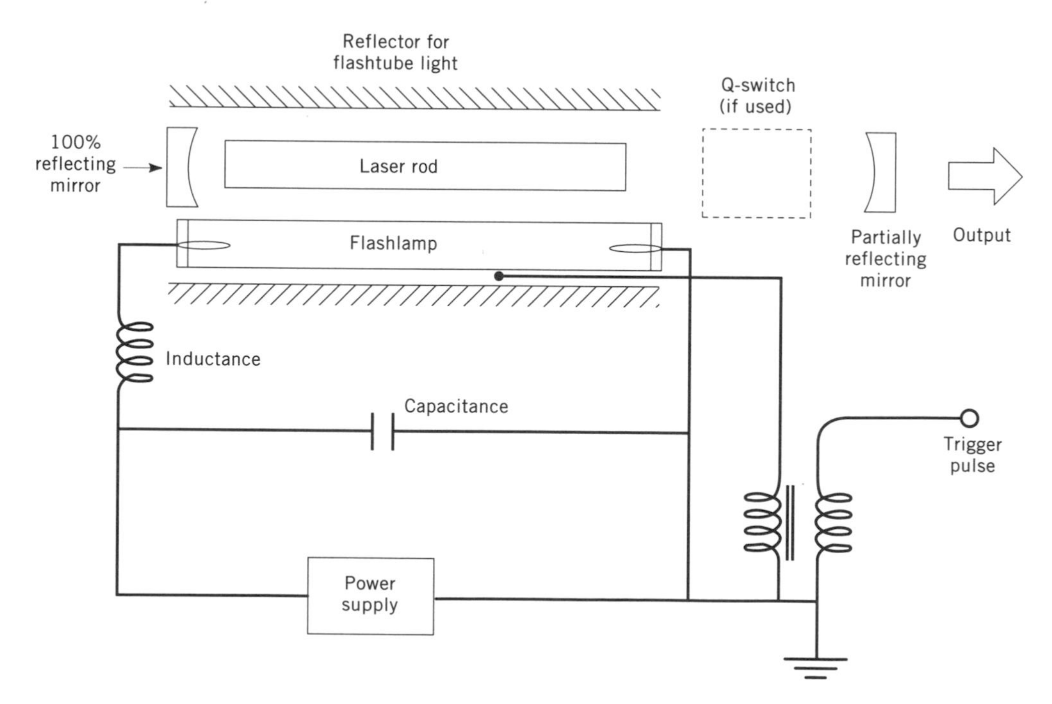

Fig. 7. Diagram of a solid-state laser.

Ruby lasers are frequently operated in the normal pulse mode, ie, pulse durations are around 1 ms and pulse energy up to tens of joules, or in the Q-switched mode, ie, pulse duration are on the order of a few tens of nanoseconds and peak power in excess of 10^9 W.

Other common solid-state lasers use neodymium ions as the active material. The neodymium ions are impurities in a crystalline or glassy host material. The laser is pumped by absorption of light in the wavelength region around 800 nm and emits laser light at a wavelength of 1.06 μm. The terminal level of the laser operation is an excited level. The neodymium laser is a four-level system (see Fig. 1) capable of operation with much lower threshold than the ruby laser. The most common laser material which incorporates neodymium [*7440-00-8*] is yttrium aluminum garnet [*12005-21-9*], $Y_3Al_5O_{12}$, commonly referred to as YAG. The neodymium-doped yttrium aluminum garnet laser is commonly designated Nd:YAG. The neodymium is present as a trivalent ion at concentrations around 2%. YAG has high thermal conductivity, and because it is a four-level system the relatively small amount of heat may readily be removed. Thus Nd:YAG lasers are capable of operation at high average power levels and may be operated continuously or at high (tens of kilohertz) pulse repetition rates.

This is in contrast to lasers based on ruby or neodymium in glass, which operate at much lower pulse-repetition rates. Nd:YAG lasers are often operated as frequency-doubled devices so that the output is at 532 nm. These lasers are the most common type of solid-state laser and have dominated solid-state laser technology since the early 1970s. Nd:YAG lasers having continuous output power up to 1800 W are available, but output powers of a few tens of watts are much more common.

The light source for excitation of Nd:YAG lasers may be a pulsed flash-lamp for pulsed operation, a continuous-arc lamp for continuous operation, or a semiconductor laser diode, for either pulsed or continuous operation. The use of semiconductor laser diodes as the pump source for solid-state lasers became common in the early 1990s. A variety of commercial diode-pumped lasers are available. One possible configuration is shown in Figure 8. The output of the diode is adjusted by composition and temperature to be near 810 nm, ie, near the peak of the neodymium absorption. The diode lasers are themselves relatively efficient and the output is absorbed better by the Nd:YAG than the light from flashlamps or arc lamps. Thus diode-pumped solid-state lasers have much

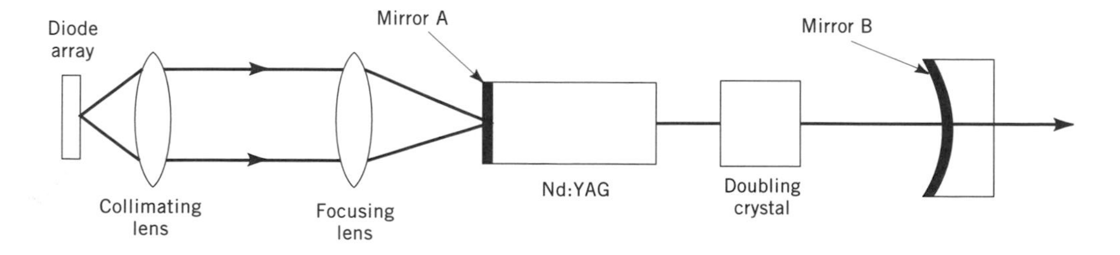

Fig. 8. Diagram of intracavity-doubled diode-pumped Nd:YAG laser. Both mirrors have high reflectance at 1064 nm. Mirror A has high transmission at 808 nm; Mirror B at 532 nm.

higher efficiency than conventionally pumped devices. Correspondingly, there is less heat to remove. Thus diode-pumped solid-state lasers represent a laser class that is much more compact and efficient than earlier devices.

Neodymium has been incorporated into many different host materials, but only a few have become available commercially. These include yttrium lithium fluoride, commonly called YLF, at a wavelength of 1.047 μm, and a variety of glasses. Silicate, phosphate, and fluorophosphate glass have been used. The exact wavelength when glass hosts are employed is near 1.06 μm, but is dependent on the glass composition. Glass can be produced in large sizes with good optical quality, compared to crystals, where sizes are limited. Therefore neodymium-doped glass lasers (Nd:glass) have been used when large amounts of energy are to be extracted in a pulse. Nd:glass lasers have also been used in the laser-assisted thermonuclear fusion program. In order to extract large amounts of energy, these are used in a master-oscillator power amplifier (MOPA) configuration. The laser pulse is first generated in a relatively small laser that has external mirrors and generates a beam of good quality. The pulse then passes through a number of stages of amplifiers, which are rods or disks of Nd:glass. The amplifiers are pumped optically using flashlamps but have no end mirrors and do not produce their own laser pulses. When the pulse from the oscillator passes through the amplifier stages, it sweeps out the energy stored in the population inversion in a single pass by stimulated emission. The diameters of the amplifier sections become progressively larger as the pulse passes through the system. This minimizes the possibility of damage by the high power optical pulse. In the MOPA configuration, the pulse energy can be made very large without increasing the pulse duration (see FUSION ENERGY).

Although neodymium-based lasers are the most common form of solid-state laser, a variety of other materials and wavelengths are available. These include alexandrite [*12252-02-7*] (a chromium-based crystalline system) having emission tunable over the range 0.72–0.78 μm, erbium [*7440-52-0*]-doped YAG (2.94 μm), holmium [*7440-60-0*]-doped YAG (2.1 μm), thulium [*7440-30-4*]-doped YAG (2.01 μm), and titanium [*7440-32-6*]-doped sapphire [*1317-82-4*], Ti:sapphire, where the output is tunable from 0.7 to 1.1 μm. Ti:sapphire lasers are notable for the wide tuning range and the generation of extremely short pulses having durations in the subpicosecond regime.

Organic Dye Lasers. Organic dye lasers represent the only well-developed laser type in which the active medium is a liquid (39,40). The laser materials are dyestuffs, of which a common example is rhodamine 6G [*989-38-8*]. The dye is dissolved in very low concentration in a solvent such as methyl alcohol [*67-56-1*], CH_3OH. Only small amounts of dye are needed to produce a considerable effect on the optical properties of the solution.

Most commercial dye lasers use flowing streams of liquid. Because dye molecules can be trapped in an excited state and take a relatively long time to return to the ground state, a steady flow of fresh dye is necessary to sustain continuous laser operation. The pump source is usually another laser, often argon, excimer, or frequency-doubled Nd:YAG. In the dye laser configuration of Figure 9, which represents a prototypical continuous dye laser configuration, the pump laser beam is focused on the high velocity stream of liquid and is incident on the stream at Brewster's angle so as to minimize reflection losses. Many types

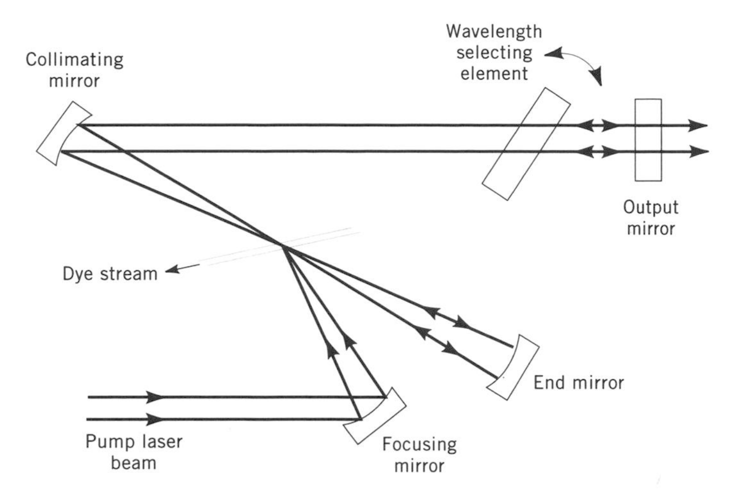

Fig. 9. Diagram of a continuous dye laser.

of tuning wedges, birefringent filters, and prisms have been used as wavelength selecting elements. Birefringent filters split light waves into two waves having mutually perpendicular vibration directions. In practical commercial devices, dye laser construction takes many forms.

The dye laser is of interest primarily because of its tunability, illustrated in Figure 10 for some common dye materials. Using a continuous operation argon laser pump, the tuning ranges for the individual materials overlap to cover the range from 375 to 920 nm. The pump wavelength must be shorter than the laser wavelength; thus the ultraviolet lines of the argon laser must be used for the dyes at short wavelengths. For longer wavelength operation, the visible argon laser lines are used. Using excimer laser pumping, an even wider range (ca 320 to 1100 nm) of tuning is possible. Choice of the proper dye material, pump laser, and wavelength selection make possible laser operation at any wavelength from the ultraviolet through the visible and into the near-infrared. Frequency doubling can extend the range of wavelengths into the ultraviolet to wavelengths as short as 190 nm.

The output power depends on the particular dye used and on the wavelength of the pump laser. The dye laser emission is at a wavelength longer than that of the pump laser. For dyes operating at wavelengths longer than 515 nm, the strong 488 and 514.5 nm lines of the argon laser may be used, and relatively large power may be obtained. Dyes having outputs at wavelengths shorter than 488 nm, pumped using the weaker lines from the argon laser, have reduced power. For any given dye, the output is greatest near the middle of its tuning range. Rhodamine 6G, one common dye material, can produce over 2 W of continuous power at the center of its tuning range.

One variation in dye laser construction is the ring dye laser. The laser cavity is a reentrant system, so that the laser light can circulate in a closed loop. The ring structure provides a high degree of stability and a narrow spectral width. The spectral width of a conventional dye laser on the order of 40 GHz is narrowed to a value as small as a few MHz. Such systems offer very high resolution in spectroscopic applications.

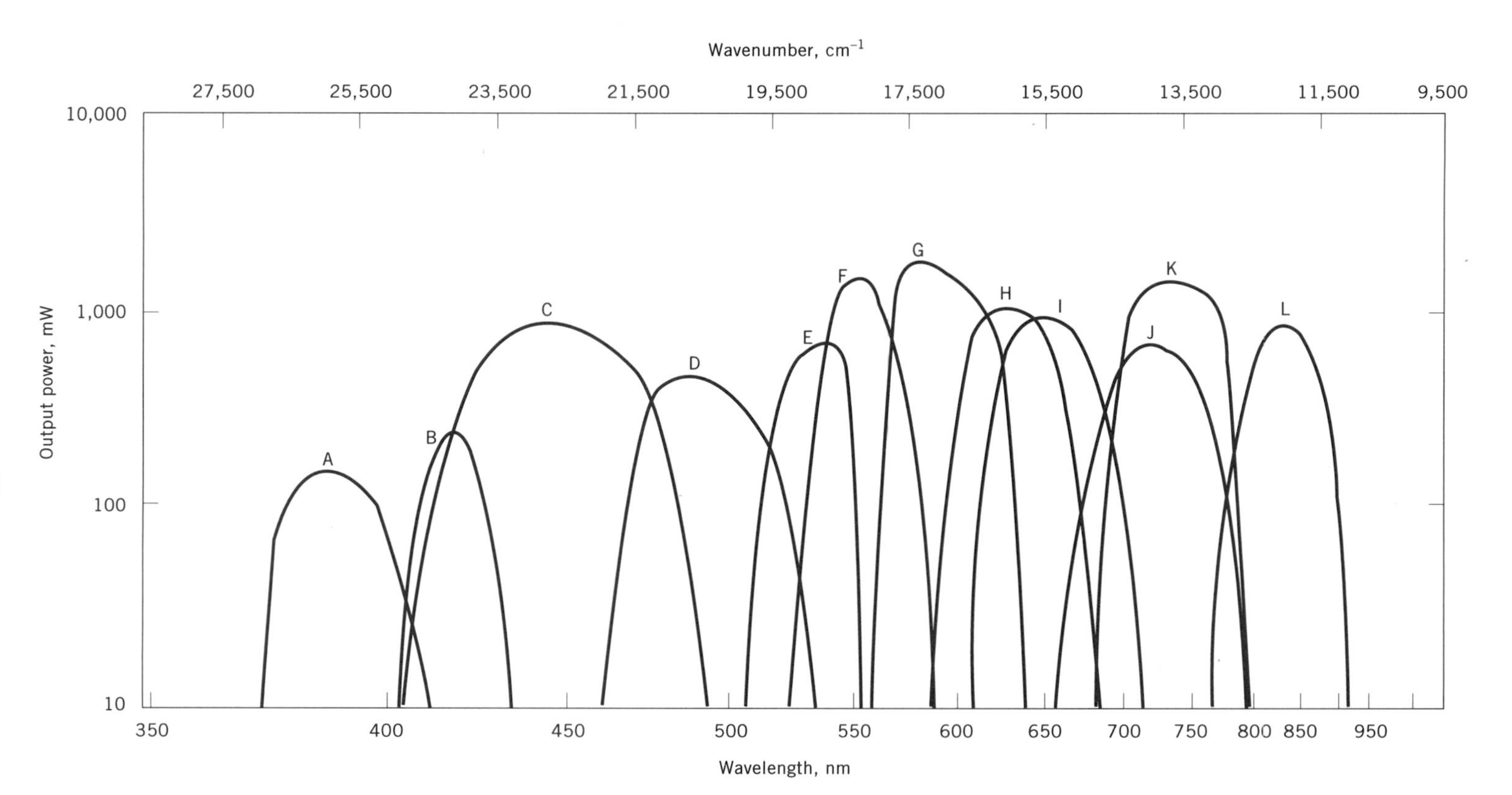

Fig. 10. Tuning curves (41) for a continuous argon laser pumped dye laser, where A represents Polyphenyl 2; B, Stilbene 1 [*74758-59-1*]; C, Stilbene 3 [*27344-41-8*]; D, Coumarin 102 [*41267-76-9*]; E, Coumarin 6 [*38215-36-0*]; F, Rhodamine 110 [*13558-31-1*]; G, Rhodamine 6G; H, Sulforhodamine B [*3520-42-1*]; I, DCM Special; J, Pyridine 2 [*89846-21-9*]; K, Rhodamine 700; and L, Styryl 9 [*82988-08-7*].

Another variation is the mode-locked dye laser, often referred to as an ultrafast laser. Such lasers offer pulses having durations as short as a few hundred femtoseconds (10^{-15} s). These have been used to study the dynamics of chemical reactions with very high temporal resolution (see KINETIC MEASUREMENTS).

Because of the tunability, dye lasers have been widely used in both chemical and biological applications. The wavelength of the dye laser can be tuned to the resonant wavelength of an atomic or molecular system and can be used to study molecular structure as well as the kinetics of a chemical reaction. If tunability is not required, a dye laser is not the preferred instrument, however, because a dye laser requires pumping with another laser and a loss of overall system efficiency results.

Semiconductor Lasers. Semiconductor lasers (42,43) use small crystals of a semiconductor, typically having submillimeter dimensions. These are grown with a junction region between *p*-type, where a majority of the carriers are holes, and *n*-type material in which a majority of carriers are electrons. Such a junction forms a diode. Electrical current passing through the junction region produces a population inversion (see SEMICONDUCTORS). Thus the *p–n* junction region in a semiconductor can serve as the active medium for a laser. Such semiconductor lasers are called diode lasers or injection lasers (see LIGHT GENERATION).

A simple diagram of a semiconductor laser is shown in Figure 11, which illustrates the orientation and size of the junction as well as the light emission. The junction region is extremely narrow. The structure is mounted on a heat sink, to which electrical leads are attached. The end mirrors are the polished or cleaved ends of the diode, which typically have reflectivity around 36%. Light emerges from the edge of the structure, through the narrow junction, and diffraction effects (see eq. 3) cause spreading of the light. Thus the emission is not well collimated. Because the limiting aperture defined by the height of the junction is smaller than the aperture along the width of the junction, the emission is in the form of an astigmatic, fan-shaped beam. Research directed

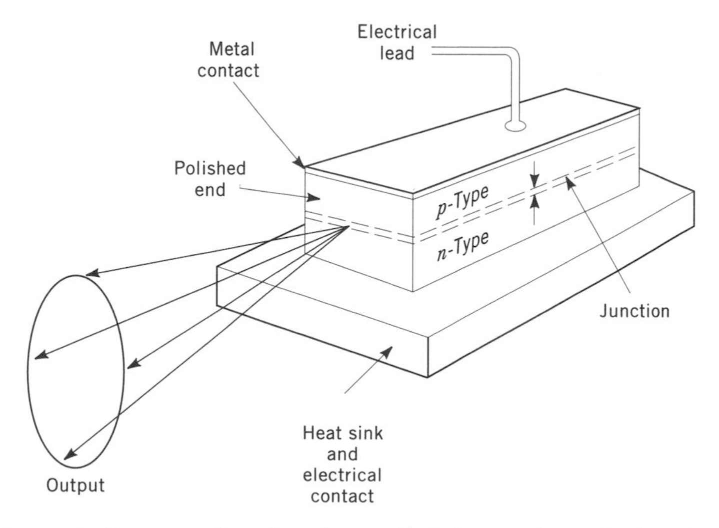

Fig. 11. Schematic diagram of semiconductor diode laser where the junction is ca 1 μm. Other dimensions are $\leq$ 1 mm.

toward the production of diode lasers having improved beam quality is under way.

The first semiconductor lasers, fabricated from gallium arsenide material, were formed from a simple junction (called a homojunction because the composition of the material was the same on each side of the junction) between the *p*-type and *n*-type materials. Those devices required high electrical current density, which produced damage in the region of the junction so that the lasers were short-lived. To reduce this problem, a heterojunction structure was developed. This junction is formed by growing a number of layers of different composition epitaxially. This is shown in Figure 12. There are a number of layers of material having different composition is this ternary alloy system, which may be denoted $Al_xGa_{1-x}As$. In this notation, x is a composition parameter which is controlled during growth and which may be varied continuously from zero to unity. The properties of aluminum gallium arsenide [*37382-15-3*], $Al_xGa_{1-x}As$, such as the index of refraction and the bandgap, are a function of x. The active region is surrounded by layers which have a higher index of refraction. This leads to a waveguide situation which tends to confine the light in the region of the junction. Because of this confinement, the power threshold for laser operation is reduced, the required current becomes less, and the lifetime becomes much longer. Aluminum gallium arsenide lasers may have lifetimes of many decades, if operated at low power levels.

Because there are two changes in material composition near the active region, this represents a double heterojunction. Also shown in Figure 12 is a stripe geometry that confines the current in the direction parallel to the length of the junction. This further reduces the power threshold and makes the diffraction-limited spreading of the beam more symmetric. The stripe is often defined by implantation of protons, which reduces the electrical conductivity in the implanted regions. Many different structures for semiconductor diode lasers have been developed.

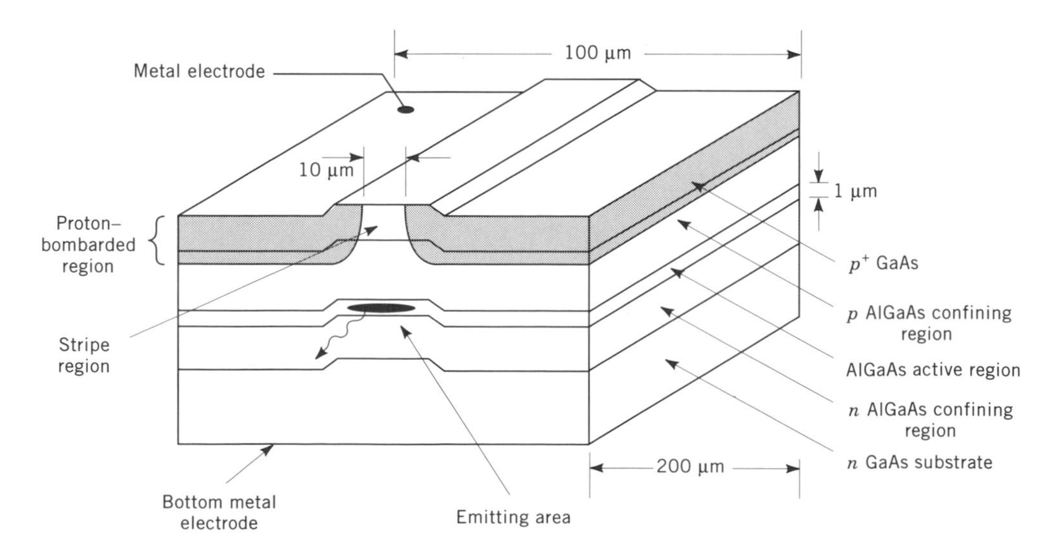

Fig. 12. Details of an aluminum gallium arsenide semiconductor diode laser.

The center wavelength for the aluminum gallium arsenide laser depends on the value of x in the active region. The range of laser parameters commercially available is shown in Figure 13. A given laser may be tuned about its center wavelength a small amount by varying the temperature and the operating current.

The $Al_xGa_{1-x}As$ and indium gallium arsenide phosphide [*12645-36-2*], $In_xGa_{1-x}As_yP_{1-y}$ lasers have undergone intensive development for optical fiber telecommunications applications. $Al_xGa_{1-x}As$ lasers operating near 0.8 μm and $In_xGa_{1-x}As_yP_{1-y}$ lasers operating at 1.3 and 1.55 μm are readily available. These three wavelengths represent minima in the loss of optical fibers. The lasers in Figure 13 operate at room temperature and somewhat above. These three classes of lasers represent highly mature technology, and in the lower ranges of output power are inexpensive.

The lasers in the 670-nm region, from the aluminum indium gallium phosphide [*107102-89-6*] system are available at center wavelengths from 635 to 690 nm. These wavelengths lie at the red end of the visible spectrum. Such lasers, which may compete for applications with the helium–neon laser, are under intensive development and represent less mature technology than the other lasers.

An important development in the 1980s was the multiple stripe laser, capable of emission of high output powers. A number of stripes are placed on a bar perhaps 1 cm wide; the output of the different stripes is coupled so that the device may be regarded as a single laser. Bars having continuous output up to 20 W are available in the aluminum gallium arsenide system. A number of

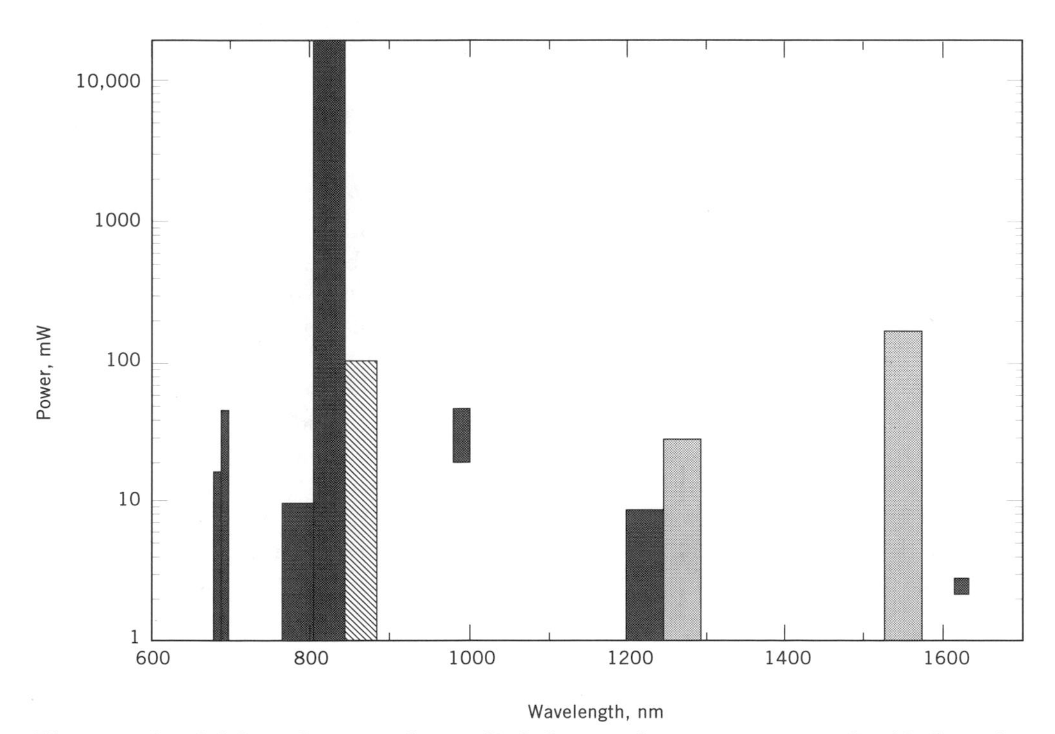

Fig. 13. Availability of semiconductor diode lasers where ▧ represents the $Al_xGa_{1-x}As$ system, ▩ $In_xGa_{1-x}As_yP_{1-y}$, and ■ $AlInGaP_y$.

bars may then be stacked to form two-dimensional arrays with high values of output power.

The diagram in Figure 12 shows an edge-emitting laser, where the light emerges through the end of the very thin junction. It would be desirable to have the laser emission emerge perpendicular to the junction region. Intensive research is being devoted to lasers of such a configuration, called vertical cavity surface-emitting lasers (VCSEL). These offer the advantage of having the light emerge through a broader aperture, improving beam quality and reducing the probability of optical damage. Commercial VCSELs are beginning to become available.

Other semiconductor lasers which operate in the longer infrared regions are based on lead compounds, such as lead selenide sulfide [*110987-37-6*], $PbSe_{0.6}S_{0.4}$. These represent much less developed technology and are much more expensive than the other semiconductor lasers described. They operate only at cryogenic temperatures and are available in versions having center wavelengths from 3.3 to 27 μm. They too may be tuned by varying the temperature and the current, and have been widely employed for spectroscopic applications in the far infrared.

Other Lasers. There are two other types of lasers which as of this writing are not at the same stage of maturity as those already discussed.

Chemical Lasers. Chemical lasers (44) produce a population inversion by a chemical reaction that leaves the product in an excited state. One example is the set of reactions leading to production of excited-state hydrogen fluoride [*7664-39-3*], HF, according to

$$\begin{aligned} F + H_2 &\longrightarrow HF^* + H \\ H + F_2 &\longrightarrow HF^* + F \\ HF^* &\longrightarrow HF + h\nu \end{aligned} \tag{11}$$

where HF^* denotes an excited state of the HF molecule. The free fluorine atom required to initiate the reactions may be provided by an electrical discharge or a pulse of light. Once the reaction is initiated, it can be maintained simply by continuing to supply fluorine and hydrogen gases to the reaction volume. In principle, after initiation the laser should be operatable purely by chemical reaction, without any external sources of electrical power. In practice, most chemical lasers do use a sustaining source of electrical power.

The product of the reactions in equation 11 is HF in a vibrationally excited state. Because there are no ground-state HF molecules present initially, the necessary population inversion is produced easily. Laser operation in the HF chemical laser occurs on a number of lines near 2.7 μm. A similar set of reactions involving deuterium and fluorine yields laser operation near 3.8 μm. The most common chemical lasers are the HF and deuterium fluoride [*14333-26-7*], DF, lasers. Because such lasers can be scaled to very large size and high output, there has been interest in these for military applications. A few models are available commercially, offering wavelengths in the mid-infrared where there are few other sources. But a variety of problems, including the corrosive nature of the reactants, have kept them from attaining wide application.

Free-Electron Lasers. The free-electron laser (FEL) directly converts the kinetic energy of a relativistic electron beam into light (45,46). Relativistic electron beams have velocities which approach the speed of light. The active medium is a beam of free electrons. The FEL, a specialized device having probably limited applications, is a novel type of laser with high tunability and potentially high power and efficiency.

A FEL is constructed using a series of magnets called wigglers, which provide an alternating magnetic field. The wavelength of the FEL is determined by a number of factors, including the wiggler spacing, the magnetic field, and the electron velocity. Typical spacings of the wiggler magnets may be a few centimeters; a total device length is a few meters. Variation of the electron velocity can tune the wavelength of the laser emission. Operation of FELs at wavelengths from the ultraviolet to the millimeter wave region has been demonstrated.

FELs offer some desirable properties, including an essentially unlimited tuning range and scalability to high power. There is no material medium to be damaged. Heat rejection is easy and the beam quality can be excellent. Because of these properties, there is high interest in FEL technology, and a number of laboratories and universities have constructed and operated FELs.

The limitation of FELs is that they require a high quality beam of relativistic electrons having low angular spreading and a small range of velocities. Such sources tend to be large and expensive, and the resource requirements for a FEL are large. As an example, in the early 1990s, the Advanced FEL, located at Los Alamos National Laboratory, began operation after three years of design and construction (47). In early operation, the laser operated at wavelengths in the 3–9-μm region, but it should also be capable of operating in the visible and ultraviolet. FEL development is a large-scale project; thus FELs are not likely to come into widespread use for many applications. The applications may include research at the national laboratories, military usage, and possibly medical applications, for which FELs could be maintained at large medical complexes.

Tunable Lasers. Tunability is an important feature for many spectroscopic and chemical applications. The availability of tunable lasers has been reviewed (48).

The leading tunable laser in the near ultraviolet, and visible and near infrared regions has been the dye laser, offering reasonably high power, narrow linewidth and a broad tuning range. It has been employed for many studies of molecular structure and chemical reactions. Tunable solid-state lasers, such as titanium-doped sapphire, Ti:sapphire, which offer a broad tuning range without the necessity to change dye materials, have begun to compete strongly in the visible and near infrared regions and have displaced dye lasers for some applications. Using frequency doubling, these cover most of the range from 0.35 to 1.1 μm.

In the near-infrared, $Al_xGa_{1-x}As$ and $In_xGa_{1-x}As_yP_{1-y}$ lasers, and in the far-infrared lead compound semiconductor lasers are tunable by varying temperature and operating current. Many excellent spectroscopic studies have been performed using them. However, they do have relatively limited tuning ranges for any one device.

Optical parametric oscillators (OPOs) represent another tunable solid-state source, based on nonlinear optical effects. These have been under development for many years and as of this writing (ca 1994) are beginning to become commercially available. These lasers may be tuned by temperature or by rotating a crystal. Models available cover a broad wavelength range in the visible and infrared portions of the spectrum. One commercial device may be tuned from 410 to 2000 nm.

The free-electron laser (FEL) offers the ultimate in tunability, in principle being unlimited in its tuning range. A FEL represents a very large investment, however.

Laser Safety

The beam from a laser can inflict damage on various parts of the human body. In addition, there are other hazards associated with the use of lasers. Therefore, a well-conceived and well-organized safety program is required for the use of lasers, particularly those of high power.

Effects of Laser Radiation. The structure of the body most easily damaged by laser light is the retina, the photosensitive surface at the back of the eyeball. Because the lens and cornea are transparent to light in the wavelength region 0.4–1.4 μm, laser light in this wavelength region can reach the retina, and if the light intensity is high enough, damage can occur. Because of its directionality, laser light can be collected efficiently by a lens and focused to a very small focal spot. Thus a high power per unit area can be produced, even if the total power in the beam is modest. The unusual properties of directionality and good focusability, advantages in many applications, increase the hazard to the retina. Because of these factors, a laser is a greater eye hazard than a conventional light source of the same power.

The lens of the eye can collect and focus laser light to a high power per unit area onto the retina. This can easily lead to burns in the photosensitive surface and cause permanent blind spots on the retina. The exact effect depends strongly on the wavelength of the light, the diameter of the focal spot, the exposure time, and the position of the retina that is illuminated. Investigations on experimental animals have delineated the effect of the various parameters (49–51). It is well established that exposure on the eye to a beam from a pulsed laser, or a transient passage of a continuous beam across the eye, can lead to serious eye damage, including blind spots on the retina and loss of visual acuity.

Lasers operating at wavelengths outside the region 0.4–1.4 μm do not deliver energy to the retina. Therefore, the increase in power per unit area because of focusing is not a factor. Such lasers are not as great hazards to the eye as lasers in the 0.4–1.4 μm region. However, light outside this range can be absorbed by the frontal structures of the eye. If the power per unit area in the incident beam is high enough, serious burns to the cornea or to the other frontal structures can result (52). The threshold for such damage occurs at values of power considerably higher than the threshold for damage to the retina.

High power lasers also can cause serious skin burns (53,54). The hazard is less than that of retinal burns, because the power per unit area is not increased by focusing, but the high power lasers in use for industrial applications could inflict extremely serious skin burns.

Associated Hazards. There are other possible hazards associated with the use of lasers outside of the hazards produced by the optical beam. The most serious is the possibility of electrical shock associated with the high voltage electrical power supplies. Such shocks are potentially lethal. In addition, the poisonous or corrosive substances used either in the laser itself, such as dye materials and solvents, or in equipment used in association with lasers, such as modulators, present serious hazards. Vapors, produced when the laser beam is used to vaporize certain types of target material, can be a safety concern. The cryogenic fluids used to cool lasers can also produce burns. A complete program for laser safety must include discussion of such associated hazards, as well as the hazards of the optical beam.

Safety Standards. Protection from laser beams involves not allowing laser radiation at a level higher than a maximum permissible exposure level to strike the human body. Maximum permissible exposure levels for both eyes and skin have been defined (55–57). One of the most common safety measures is the use of protective eyewear. Manufacturers of laser safety eyewear commonly specify the attenuation at various laser wavelengths. Under some conditions safety eyewear has been known to shatter or to be burned through (58), and it is not adequate to protect a wearer staring directly into the beam.

One of the most significant laser safety standards is that developed by the Z-136 committee of the American National Standards Institute (ANSI) (55). Although it is voluntary, many organizations use the ANSI standard. It contains a number of items including a recommendation for maximum permissible levels of exposure to laser radiation for various wavelengths, exposure durations, and different parts of the body; separation of lasers into four different classes according to the level of hazard they present; and recommendation of safety practices for lasers in each of the classes.

For evaluation of a particular laser installation, the standard should be consulted to determine the classification of the laser and appropriate safety measures. The maximum permissible exposure for the particular laser also should be determined in order to select the appropriate protective eyewear.

The U.S. Food and Drug Administration (FDA) adopted a legally binding standard, which took the form of a performance standard for laser products (56,57). The standard provides a classification scheme for lasers similar to the ANSI classification. All lasers sold after August 2, 1976 must comply with its provisions. The standard requires incorporation of safety-related labeling and protective equipment according to the class of the laser. The primary impact of the FDA standard is on laser manufacturers and scientific supply firms.

Several state governments have also passed laws regulating and controlling lasers. Provisions of the laws vary greatly from state to state.

Nonlinear Optics

The electromagnetic field of a light beam produces an electrical polarization vector in the material through which it passes. In ordinary optics, which may be termed linear optics, the polarization vector is proportional to the electric field vector $\vec{E}$. However, the polarization can be expanded in an infinite series:

$$P = \chi E(1 + a_2E + a_3E^2 + a_4E^3 + \ldots) \tag{12}$$

where P is the polarization, E is the electric field, χ is the linear polarizability, and the a_i are constants. Equation 12 is a simplified scalar representation of a tensor equation. The nonlinear coefficients a_i are very small compared to unity. For reasonably small values of electric field, only the first term in the equation is important and in the prelaser era polarization was approximated to be linearly proportional to electric field. When high power lasers became available, the electric fields became much higher and the product a_iE^{i-1} could become large enough to be observable. The second term in equation 12 is of the form χa_2E^2. If E has a sinusoidal variation of the form $E_0\cos\omega t$, the resulting components of polarization are of the form

$$a_2\chi E_0^2\cos^2\omega t = 0.5\ a_2\chi E_0^2(1 + \cos 2\omega t) \tag{13}$$

The second term on the right-hand side, a component oscillating at frequency 2ω, represents the second harmonic of the incident beam. This component of the polarization vector can radiate light at the frequency 2ω. Observation of the second harmonic generation was demonstrated in the early 1960s using ruby lasers (59).

As the incident radiation at frequency ω and the second harmonic radiation of frequency 2ω propagate through the material, the intensity of the second harmonic radiation builds. However, because of dispersion, the two waves eventually get out of phase and the intensity of the second harmonic radiation decreases. This limitation is avoided by use of a technique called phase matching (60). Some birefringent crystals offer combinations of directions of propagation and direction of polarization for the two beams so that the indexes of refraction of the incident light and the second harmonic are equal. Then the intensity of the second harmonic radiation can build up in phase with the incident light. Exact phase matching can be achieved by varying the angular orientation of the crystal or by varying its temperature.

Only certain types of crystalline materials can exhibit second harmonic generation (61). Because of symmetry considerations, the coefficient a_2 must be identically equal to zero in any material having a center of symmetry. Thus the only candidates for second harmonic generation are materials that lack a center of symmetry. Some common materials which are used in nonlinear optics include barium sodium niobate [*12323-03-4*], $Ba_2NaNb_5O_{15}$; lithium niobate [*12031-63-9*], $LiNbO_3$; potassium titanyl phosphate [*12690-20-9*], $KTiOPO_4$; beta-barium borate [*13701-59-2*], β-BaB_2O_4; and lithium triborate [*12007-41-9*], LiB_3O_5 (62,63) (see NONLINEAR OPTICAL MATERIALS).

Second harmonic generation works best in pulsed lasers having high peak power, because the polarization which produces it is proportional the square of the electric field. Thus the most common frequency-doubled lasers are Q-switched devices. Many commercial frequency-doubled lasers have become available. The most popular of these is the frequency-doubled Nd:YAG laser, having output in the green at 532-nm wavelength. Frequency-doubled tunable dye lasers, where the original output is in the visible, provide tunable output in the ultraviolet. Frequency doubling is the most common manifestation of nonlinear optics.

The third term in equation 12 corresponds to frequency tripling, which leads to an output at 355 nm for a Nd:YAG laser. The coefficient a_3 may be nonzero even in centrosymmetric crystals, so that effects that depend on mixing of three waves (a_3E^3) may be observed in such crystals.

Another nonlinear optical effect of particular significance is parametric amplification. Laser light is incident on the crystal. The incident light of angular frequency, ω_p, is called the pump light. Because of the nonlinear interaction in the crystal, two new frequencies build up, the so-called signal frequency, ω_s, and an idler frequency, $\omega_\iota = \omega_p - \omega_s$. Once again, to obtain reasonable efficiency, phase-matching methods must be used. However, there is an additional dimension of freedom in that only the sum of the signal and idler frequencies must equal the pump frequency. The value of the signal frequency can be varied by changing the temperature or the angular orientation of the crystal. Then the particular frequency that is phase-matched builds up with high efficiency. Efficiencies as high as 70% have been demonstrated in such parametric devices using lithium niobate. This technique offers another significant type of tunable source. Most of the demonstrations have been in the red portion of the visible spectrum and in the near-infrared.

A wide variety of other nonlinear optical effects also have been demonstrated. According to equation 12, if two light beams having frequency ω_1 and ω_2 are combined in a material with a nonzero value of a_2, light waves of frequency $\omega_1 + \omega_2$ and $\omega_1 - \omega_2$ are produced. A combination of such effects, used in conjunction with tunable lasers such as dye lasers and parametric oscillators, can produce laser light at any desired wavelength from the uv to about 25 μm in the infrared. These narrow-linewidth, widely tunable light sources are extremely useful for photochemical applications (see PHOTOCHEMICAL TECHNOLOGY). The wavelength of the light may be adjusted to match a resonant absorption of the molecules of interest.

Uses

The widespread interest in lasers is based on practical applications. Lasers are used to perform many useful functions in science, engineering, industry, and education. Additionally, many other applications are under development. An application such as welding (qv) depends on the collimation and focusing ability of the beam which leads to high power density. A conventional light source, even one having higher total radiant power than a welding laser, cannot be focused well and thus is not useful for welding.

Most applications for which lasers are used were originally demonstrated using conventional light sources. In many cases, the application was only marginally successful using conventional sources and required the development of laser light sources to be practical.

Material Processing. Laser radiation can be used for a variety of material processing functions such as welding, cutting, shaping, marking, and drilling

(64–66). Even a fairly modest pulsed laser can be focused to produce a power density greater than 10^9 W/cm^2; lasers can melt and vaporize materials. Spot welding using pulsed lasers was demonstrated in the 1960s (67). Seam welding is possible by overlapping pulses, or using continuous lasers. However, the penetration depth for such welding is limited. The energy is deposited at the surface of the workpiece. For complete penetration of the weld through the workpiece, thermal conduction has to carry heat energy through the entire thickness of the specimen. In practice, this factor limited maximum weld depths to perhaps 0.1 cm, depending on the thermal conductivity of the sample.

The advent of multikilowatt carbon dioxide lasers in the early 1970s eased the 0.1-cm restriction (68,69). The beam can drill a hole into the workpiece, then, as the beam is translated, the hole moves through the material. Molten material flows into the region behind the moving hole and resolidifies. Thus a continuous seam is produced. Because the energy is deposited at the bottom of the hole, limitations because of slowness of thermal conduction do not apply. Welds having much greater depth can be produced by this technique, called deep penetration laser welding. Welds as deep as 5 cm have been produced in stainless steel (70). Laser welding has become widely used in practical industrial processing (71–73).

Many types of steel and iron can be hardened by heating above a critical temperature, followed by rapid cooling. If a laser beam of relatively low (ca 10^4 W/cm^2) power density is scanned across such a surface, it heats the surface without melting the metal. The heated region, a thin layer at the surface, cools rapidly by thermal conduction after the beam moves on. The hardness of the surface layer can be increased considerably (74,75) and thus the wear resistance of the surface can be substantially increased. Typically, carbon dioxide lasers having output greater then 1 kW are used. Laser heat treating has been employed for applications such as hardening steering gear assemblies and diesel engine cylinder liners in the automotive industry.

High power laser beams can vaporize material, leading to applications such as hole drilling and cutting. Higher power densities are used for these applications than for hardening and welding. Hole drilling in ceramic materials has become common. There is a need for small (less than 0.5-mm) holes in the alumina substrates used in many electronic applications (see INTEGRATED CIRCUITS). Holes drilled before the ceramic is fired, tend to change dimension during firing. After firing, the material is very hard, brittle, and difficult to drill by conventional techniques. Laser drilling offers an improved technique for producing holes in the fired ceramic, and has become an important application for lasers in the electronics industry.

High power lasers have impressive capabilities for cutting both metallic and nonmetallic materials. The cutting rates can be increased by the presence of gases. Typically oxygen is used but other gases have sometimes been employed. A jet of oxygen or air is blown on the workpiece at the position where the beam strikes it. The exothermic chemical reaction with the oxygen can greatly increase the cutting rate for reactive materials. Laser cutting is widely used in many practical applications, such as cutting titanium in the aerospace industry, cutting steel [*12597-69-2*] plates in the automotive industry, and cutting fabric in the garment industry.

Another important application involving material removal is trimming of resistors. Thick-film resistors used in many electronic circuits must be trimmed to the final desired value. This can be done by fabricating the resistor using a low value of resistance and then cutting part way through it with a laser. The resistance can be monitored as cutting proceeds. Trimming is terminated when the desired value is reached. Repetitively Q-switched Nd:YAG lasers are commonly used for resistor trimming. Laser trimming has become standard procedure in the electronics industry.

Measurement Applications. Lasers have been used for measurement of many physical parameters. These include length and distance, velocity of fluid flow and of solid surfaces, dimensions of manufactured goods, and the quality of surfaces, including flaw detection and determination of surface finish.

Lasers are used in measurement of lengths in a variety of ways (76–77). A very short pulse of high power laser light may be directed toward a target the distance of which is desired. A photodetector viewing the target collects a signal resulting from light reflected by the target. The time t between the outgoing pulse and the detection of the reflected signal, ie, the time for the light to make a round trip from the laser to the target and back, is measured. Because the velocity of light is known, the range R to the target is

$$R = ct/2n \tag{14}$$

where n is the index of refraction of the air through which the pulse travels. Because the index of refraction depends on air temperature, pressure, altitude, and other factors, this method is subject to some uncertainties. It has been used for accurate measurement of the distance from the earth to the moon (78).

For higher accuracy, a method involving amplitude modulation of a continuous laser beam is used. Again, a detector receives light reflected from the object where the distance is to be measured. The phases of the modulation in the outgoing beam and in the reflected return are compared. For a total phase shift $\Delta\phi$ between the two signals, the range R is

$$R = c\Delta\phi/4\pi f_m n \tag{15}$$

where f_m is the frequency of the modulation. Again, the measurement is affected by variations in the index of refraction of the air. Because there exists an ambiguity in the measurement of Δf by an integral multiple of 2π, the measurement is usually performed at several different values of f_m. Measurements using this phase comparison method can be performed out of doors, at ranges up to several kilometers, and with accuracies around 1 ppm of the distance being measured.

For greater accuracy, an interferometric technique is employed. Although this measurement has high inherent sensitivity, it is restricted to indoor use, with a maximum practical range of several meters; otherwise, the interferometric fringes are degraded by turbulence and inhomogeneity in the air. In its simplest form, a mirror is attached to the object where the distance is to be measured. The return beam reflected from the mirror is combined with part of the original beam to form an interference pattern. When the mirror moves by half of the laser wavelength, the round-trip path to the mirror changes by one wavelength. At a particular point in the interference pattern, the light intensity fluctuates through a full cycle, eg, bright to dark to bright. A detector monitors the fluctuations, and logic circuitry converts the number of cycles to distance moved. Thus this measurement is not an absolute distance measurement, but rather a determination of distance from an arbitrary zero point. In more complicated versions, the laser interferometer measures the velocity of the moving mirror by means of a Doppler technique. This reduces problems with fluctuations caused by air turbulence. This version is used in modern commercial laser–interferometric measurement equipment.

Interferometric systems, usually helium–neon lasers, offer precise distance measurement over a scale of distances < 100 m and in an indoor environment. Such devices are suitable for dimensional control of machine tools (see MACHINING MATERIALS, ELECTROCHEMICAL).

A method which competes with interferometric distance measurement is laser Doppler displacement. In this approach the Doppler shift of the beam reflected from a target is measured and integrated to obtain displacement. This method also is best suited to use indoors at distances no more than a few hundred meters. Table 2 compares some of the characteristics of these laser-based methods of distance measurement.

Dimensional measurement using lasers is illustrated by measurement of wire diameter. When a fine wire is inserted in the highly collimated beam of a laser, the light is diffracted by the wire to form a distinct line of spots perpendicular to the length of the wire. The spots appear at angles θ_n with respect to the direction of the laser beam. θ_n is given by

$$\sin\,\theta_n = n\lambda/d \qquad (16)$$

where λ is the wavelength of the light, d the diameter of the wire, and $n = 0, 1, 2, 3\ldots.$ For a known wavelength, measuring the angular separation of the diffracted spots gives a noncontact method of determining the wire diameter. The angular separation of the spots is unaffected by vibration, as long as the wire remains within the laser beam. The method is useful as a continuous monitor for the diameter of moving or heated wires. Automated systems have been developed that use an array of photodiodes to measure the spot separation. The photodiode signals are processed by a microprocessor, which then, for example, could control the extrusion process in which the wire was being produced.

Table 2. Distance Measurement Using Lasers

Method	Laser type	Range	Typical application
round-trip time-of-flight	Q-switched solid state	many km	military ranging, satellite ranging
phase comparison of amplitude-modulated beam	He–Ne, AlGaAs	km	surveying
interferometry	frequency-stabilized He–Ne	few meters	machine-tool control
laser Doppler displacement	He–Ne	to 120 m	machine-tool control

The ring laser gyroscope is an interesting application of lasers to measurement of angular rotation rate. This gyroscope utilizes the Sagnac effect, which generates a phase difference for light traveling in a closed path in a rotating system. The ring laser gyroscope typically is constructed using three small helium–neon gas laser tubes arranged in an equilateral triangle. Mirrors at the points of the triangle direct the light around the triangular path.

Light may traverse this path in two opposite directions of travel. If the structure is still, these two counterpropagating light beams have the same frequency. But if the structure is rotating, the frequencies of the two beams are slightly different. If the angular rotation has a component of frequency ω perpendicular to the plane of the triangle, then the frequency shift Δf is given by

$$\Delta f = (3)^{1/2} \omega b / 6\lambda \qquad (17)$$

where b is the length of one side of the triangle and λ is the wavelength of the light. Because Δf is linearly proportional to ω, ω can be determined simply by measuring Δf. A portion of each of the two beams is combined at a detector. The output of the detector contains a fluctuating component of frequency Δf which can be directly measured. This fluctuating component is called the beat note. It is analogous to the acoustic beating observable when two tuning forks with slightly different frequencies are struck.

For a three-axis system, three triangles in mutually perpendicular planes may be used. For navigational purposes, the output of the laser gyroscope may be integrated to determine the heading of an aircraft. Laser-based navigation systems have been in use on commercial and military aircraft since the early 1980s.

A final example of laser measurement techniques is measurement of the velocity of fluid flow (79) (see FLOW MEASUREMENT). If a laser beam is directed into a transparent fluid, some of the light may be scattered with a frequency shift Δf given by

$$\Delta f = 2nv \sin\ (\theta/2)/\lambda \qquad (18)$$

where n is the index of refraction of the fluid, v the component of its velocity perpendicular to the laser light, λ the wavelength of the light, and θ the angle between the incident light beam and the scattered light. A detector receives the scattered light, along with a portion of the original light beam with no frequency shift. The detector output contains an oscillatory component at frequency Δf, the beat frequency. Measurement of Δf can yield the fluid velocity v according to equation 18. This application depends on the narrow spectral linewidth of the laser, which should be less than Δf.

This procedure offers the possibility of remote noncontact velocity measurement, where no probes disturb the flow. It is thus compatible for use with hot or corrosive gases. Commercial laser velocimeters have become well-developed measurement tools. Examples of laser velocimetry include remote measurement of wind velocity, measurement of vortex air flow near the wing tips of large aircraft, and *in vivo* measurement of the velocity of blood flow.

Holography. Holography (qv), a photographic process that yields three-dimensional images, was invented in the prelaser era in 1948 (80). Light sources available at that time were not adequate for making good holograms and holography became practical only after the invention of the laser (81). The properties of coherence and monochromaticity offered by lasers are important for holography.

Holography is a two-step process. First, a hologram is recorded. The recording material is usually photographic film, but for specialized applications other media such as thermoplastics, photopolymers, or dichromated gelatin may be used. A typical arrangement for holographic recording is illustrated in Figure 14. The laser light is split into two beams, which are recombined at the photographic film. One beam, called the object beam, strikes the object to be recorded. Light reflected from the object reaches the photographic film. The reference beam travels directly to the film, arriving at an angle to the object beam. In the simplest form, no lenses are required for this type of photography, although some variations may use lenses.

After the photographic film is exposed to the light, it is developed. Conventional photographic developing techniques can produce a perfectly acceptable hologram but there are also other techniques, such as a bleaching technique, that can lead to brighter images. The developed film itself looks like a piece of fogged photographic film. It contains no visual clues of the scene recorded in it.

After the film is developed, it is illuminated by a laser to release the image stored in the film. The production of the image is called reconstruction of the hologram. A laser beam strikes the hologram in the same direction as the original reference beam (Fig. 15). Two images of the original object are produced.

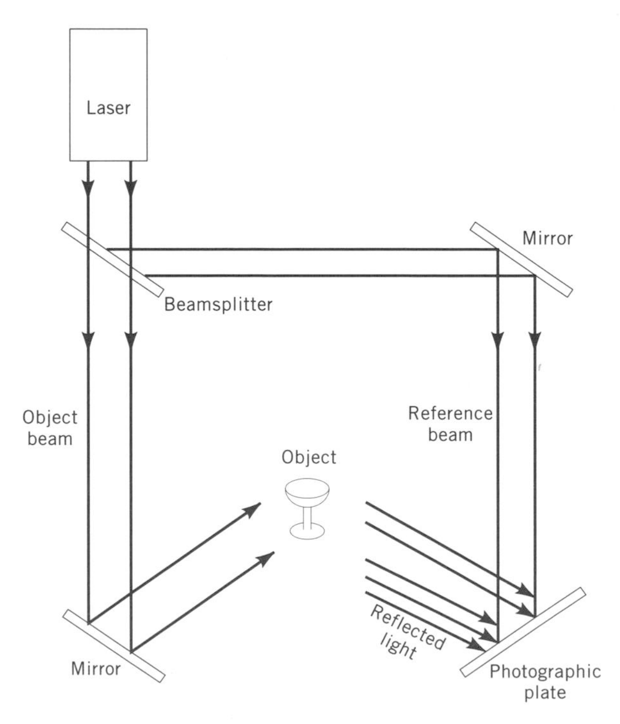

Fig. 14. Schematic arrangement for recording a hologram.

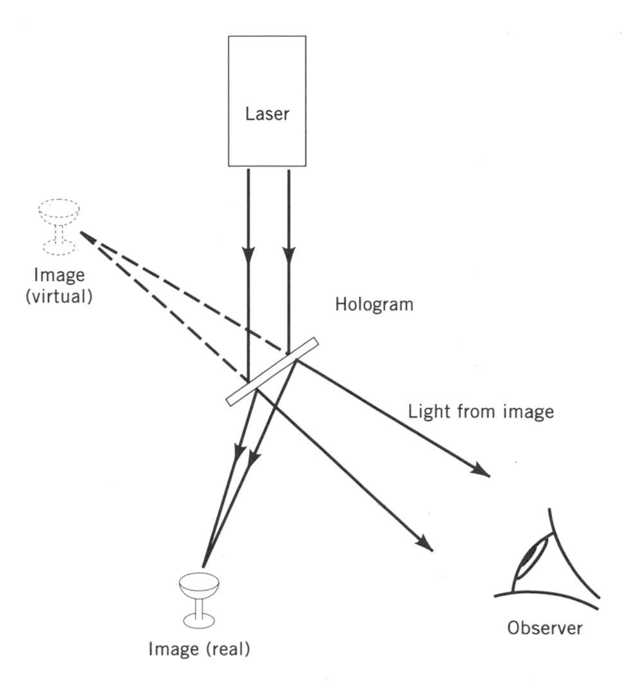

Fig. 15. Reconstruction of holographic images.

One, called the virtual image, appears behind the hologram, in the position of the original object. This image needs a lens, such as the lens of the human eye, to view it. A second image, called the real image, is formed in front of the hologram. This image can be projected onto a screen. Light from the two images has an angular separation, so that they may be observed separately.

The virtual image has all the properties of light coming from the original object. As the observer's eye moves, the sides of the object come into view, just as if one were viewing the original object. The striking three-dimensional images produced have led holography to become a form of artistic expression. Another important application is holographic interferometry. Two holograms are made of an object, before and after a slight change, which may be induced by heat, pressure, or other means. Both holograms are stored in the same storage medium. When the holograms are reconstructed, any change is easily detected by the presence of interference fringes. Holographic interferometry is useful in nondestructive evaluation (qv), defect detection, and the analysis of vibration patterns and stress analysis (82,83).

Medical Applications. Lasers have been used for a variety of medical applications (84–88). Perhaps the repair of tears and holes in the retina may be the best known example. The beam is focused on the retina by the lens of the eye. The high power density in the beam photocoagulates some of the tissue in the retina, thus sealing off or reattaching the defective area. Pulsed ruby lasers were used in early studies of retinal photocoagulation, but now argon lasers seem to be preferred because of the high absorption of green laser light by hemoglobin. Many patients with retinal defects have been clinically treated with laser radiation. Laser photocoagulation also has been widely used to treat diabetic retinopathy, one of the leading causes of blindness in the United States.

Laser removal of tattoos and of colored birthmarks has been widely studied. A high power pulsed laser at a wavelength absorbed by the pigment is used to vaporize the pigment and to bleach the colored area. Ruby, Nd:YAG, and dye lasers are favored for this purpose.

High power carbon dioxide lasers have been used for surgery. Absorption by tissue is very high at the 10-μm wavelength of the carbon dioxide laser. Laser surgery provides a possibility for simultaneous cutting and cauterization. Use of carbon dioxide lasers for bloodless surgery has been widely studied, especially for blood-rich organs such as the liver. Laser surgery has also been widely used for ear, nose, and throat procedures.

Laser angioplasty, a procedure in which laser energy is delivered through an optical fiber inserted into a blood vessel to a location of a blockage in the vessel and used to burn through the blockage, was studied extensively during the 1980s. Early enthusiasm for the procedure as a treatment for coronary disease seems to have waned because blockages can recur at the same locations.

The development of so-called photodynamic therapy uses lasers for treatment of cancer. The patient is injected with a substance called hematoporphyrin derivative [*68335-15-9*] which is preferentially localized in cancerous tissues. The patient is later irradiated with laser light, often with a dye laser at a wavelength around 630 nm. The light energy catalytically photooxidizes the hematoporphyrin derivative, releasing materials which kill the nearby cancerous tissue. Normal tissue which did not retain the chemical is not harmed. Photodynamic therapy offers promise as a new form of cancer treatment.

Communications. The advent of the laser improved prospects for optical communications enormously. The coherence of the laser meant that techniques developed in the radio portion of the electromagnetic spectrum could be extended to the optical portion of the spectrum. Because lasers operate at frequencies near 10^{15} Hz, they offer a potentially wide bandwidth, equal to about 10^7 television channels of width (ca 10^8 Hz). It has not proved possible to take advantage of this full bandwidth because devices such as modulators capable of operating at 10^{15} Hz are not available.

Laser communication systems based on free-space propagation through the atmosphere suffer drawbacks because of factors like atmospheric turbulence and attenuation by rain, snow, haze, or fog. Nevertheless, free-space laser communication systems were developed for many applications (89–91). They employ separate components, such as lasers, modulators, collimators, and detectors. Some of the most promising applications are for space communications, because the problems of turbulence and opacity in the atmosphere are absent.

The thrust of laser communications changed dramatically around 1970 with the development of optical fibers having low loss of signal (see FIBER OPTICS). Light can be guided along the length of a glass fiber of small diameter. Medical instruments using bundles of glass fibers were developed for viewing portions of the interior of the human body. However, because the fibers had a high signal loss, their applications were limited. Light propagating along the length of the fiber suffered attenuation greater than 1000 dB/km. Therefore practical applications could employ fibers at most a few meters long.

Beginning around 1970, fibers were produced having much lower signal loss. Optical fibers with losses as low as 0.2 dB/km are available. This made possible long-distance communication with optical fibers. Using a repeater station every few kilometers, light can be propagated for great distances in fiber-optic lightguides. System performance is not affected by atmospheric factors. Fibers with continuous lengths of many kilometers are produced, in continuous fiber-drawing processes. Semiconductor diode lasers are employed as the light source. The early systems used AlGaAs lasers operating near 0.85 μm, because the fiber loss was first reduced to low values near that wavelength. As the fiber loss was reduced at wavelengths near 1.3 and 1.55 μm, InGaAsP lasers operating at those wavelengths have become widely used. Most installed fiber-optic communications systems operate at 1.3 μm, but future installations are likely to be dominated by 1.55-μm operation.

The most important application of fiber-optic laser-based communication is in long-distance telecommunications (92,93). Fiber-optic systems offer very high capacity, low cost-per-channel, light weight, small size, and immunity to crosstalk and electrical interference.

Fiber-optic systems have data rates of hundreds to thousands of megabits per second over distances of tens of kilometers. This represents extremely high information capacity in links that are smaller, lighter, and less expensive than copper wire. Fiber optic communication links are commonly characterized by a distance–bandwidth product, in units of MHz·km. The distance–bandwidth product is the product of the data rate and the length of the communications link between repeater stations. In current long-distance systems, a repeater station is installed at intervals, typically around 40 km. The signal is detected, regenerated, amplified, and retransmitted using optoelectronic components. In the future, all-optical repeaters, based on fiber-optic amplifiers, will be used.

In the early 1990s, there were more than 9×10^6 km of fiber-optical telecommunication links in practical use in the United States. In addition, many other countries, notably Canada, Japan, and western Europe, have installed extensive fiber-optic communication systems. There are several transoceanic fiber-based telephone cables. Fibers are in use for intracity telephone links, where bulky copper [*7440-50-8*] wire is replaced by thin optical fibers. This allows crowded conduits in large cities to carry more messages than if copper wire were used. Fiber optics are used for intercity long-haul telephone links, for interoffice trunk lines, and have replaced many microwave communication links.

However, optical fiber communications are not useful only for long-distance communication links. Fiber-optic data links are also used in a variety of short-distance systems, for example in computer–computer links and for internal communications on ships and aircraft. Figure 16 shows some possible applications for fiber-optic communications, with respect to length and bit rate. The common carrier applications, like telephone links, lie to the upper right of the dashed line labeled 100 MHz·km. However, a wide variety of other lower performance applications, illustrated to the lower left of the dashed line, are in use or under development.

These applications include many of the shorter-distance communication links, such as the automated operation of all the environmental controls in a

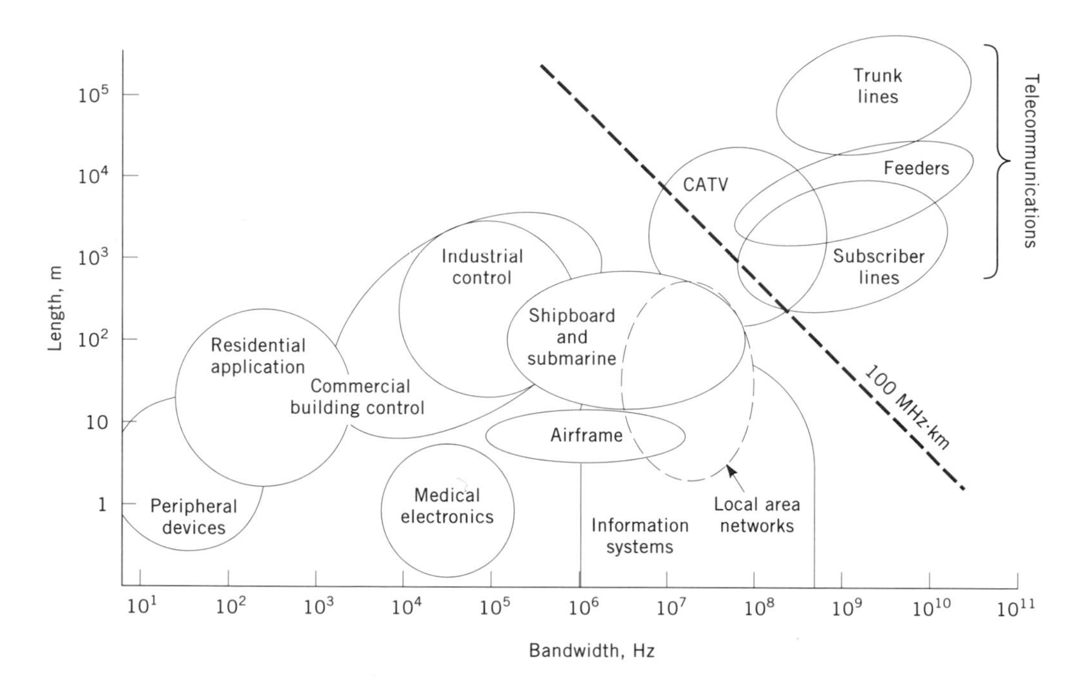

Fig. 16. Applications of fiber-optic communication systems as a function of length between repeater stations and system bandwidth. Applications to the upper right of the dashed line may be regarded as common carrier applications.

large commercial building. Fiber optics can replace copper wire at a savings in space and cost for many of these applications.

In summary, laser-based fiber-optic telecommunications has had a revolutionary impact on long-distance telephone communication and is now expanding into many new applications areas.

Consumer Products. Laser-based products have emerged from the laboratories and become familiar products used by many millions of people in everyday circumstances. Examples include the supermarket scanner, the laser printer, and the compact disk. The supermarket scanner has become a familiar fixture at the point of sale in stores. The beam from a laser is scanned across the bar-code marking that identifies a product, and the pattern of varying reflected light intensity is detected and interpreted by a computer to identify the product. Then the information is printed on the sales slip. The use of the scanner can speed checkout from places like supermarkets. The scanners have usually been helium–neon lasers, but visible semiconductor lasers may take an impact in this application.

Laser printers have been used to produce high quality documents in a process called electrophotography (qv) or xerography. Familiar xerographic machines, usually used for copying documents, form an image of the document on a drum coated with a photoconductive material. The image is transferred after several steps to a piece of paper. In the laser printer, the original document need not exist in the form of a hard copy. The information, stored in a computer, is used to modulate a laser beam which scans across the surface of the drum. In this way, a document created on a computer may be printed directly, with high resolution. This application was at first dominated by helium–neon lasers, but 780-nm aluminum gallium arsenide lasers are widely used for low end applications, like personal printers. Large, fast printers use helium–neon lasers, but visible semiconductor lasers may become the laser of choice for them.

The compact disk player has become a very widespread consumer product for audio reproduction. The information is stored along tracks on the disk in the form of spots of varying reflectivity. The laser beam is focused on a track on the surface of the disk, which is rotated under the beam. The information is recovered by detecting the variations in the reflected light. The compact disk offers very high fidelity because there is no physical contact with the disk. This application has usually employed a semiconductor laser source operating at a wavelength of around 780 nm. Tens of millions of such compact disk players are produced worldwide every year.

Spectroscopic Applications. Laser technology has led to a revolution in spectroscopic techniques. Absorption spectroscopy can be carried out at much higher resolution than is available from conventional spectrometers. Spectroscopy can be performed in a simple, straightforward manner using a tunable laser that covers the wavelength region of interest. A laser beam that passes through the sample is tuned over the selected wavelength interval and the spectrum recorded. Laser spectroscopy has revealed a great new wealth of detail in absorption spectra (94–97) (see SPECTROSCOPY).

For the visible and near-ultraviolet portions of the spectrum, tunable dye lasers have commonly been used as the light source, although they are being replaced in many application by tunable solid-state lasers, eg, titanium-doped sapphire. Optical parametric oscillators are also developing as useful spectroscopic sources. In the infrared, tunable laser semiconductor diodes have been employed. The tunable diode lasers which contain lead salts have been employed for remote monitoring of pollutant species. Needs for infrared spectroscopy provide an impetus for continued development of tunable infrared lasers (see INFRARED TECHNOLOGY AND RAMAN SPECTROSCOPY).

Beside conventional absorption spectroscopy, laser technology offers many other spectroscopic possibilities (98). Laser-induced fluorescence (LIF) has developed as a highly useful diagnostic tool for probing the dynamics of chemical reactions. The approach involves measuring the spontaneous emission from an energy level which has been populated by resonant absorption of laser radiation. If a volume of reactant material is probed by short (picosecond or subpicosecond) laser pulses at various times after initiation, the progress of the reaction, including the buildup and decay of intermediate reaction products, can be determined with very high temporal resolution. Most applications of LIF to date have employed tunable dye lasers. LIF has been employed for a variety of applications, such as measurement of species concentrations in flames and combustors, remote sensing of gaseous species, and determination of flow patterns in chemical reactions.

Raman spectroscopy, long used for qualitative analysis, has been revitalized by the availability of laser sources. Raman spectroscopy is based on scattering of light with an accompanying shift in frequency. The amount by which the frequency is shifted is characteristic of the molecules that cause the scattering. Hence, measurement of the frequency shift can lead to identification of the material.

Lasers offer narrow spectral line width and high brightness for Raman spectroscopy (see INFRARED TECHNOLOGY AND RAMAN SPECTROSCOPY). Raman spectroscopy with lasers provides higher resolution and lower threshold for detection compared to conventional light sources. Laser Raman spectroscopy is often carried out with argon lasers or with tunable dye lasers. It has been useful for many applications, including identification of drugs, metabolic by-products of drugs, vitamins, medicines, minerals, atmospheric pollutants, and petroleum products. Raman spectroscopy is especially useful for detecting small amounts of impurities in chemicals.

An interesting variation of Raman spectroscopy is coherent anti-Stokes Raman spectroscopy (CARS) (99). If two laser beams, with angular frequencies ω_1 and ω_2 are combined in a material, and if $\omega_1 - \omega_2$ is close to a Raman active frequency of the material, then radiation at a new frequency $\omega_3 = 2\omega_2 - \omega_1$ may be produced. Detection of this radiation can be used to characterize the material. Often one input frequency is fixed and the other frequency, from a tunable laser, varied until $\omega_1 - \omega_2$ matches the Raman frequency. CARS has the capability for measurements in flames, plasmas, and other hostile environments.

The availability of lasers having pulse durations in the picosecond or femtosecond range offers many possibilities for investigation of chemical kinetics. Spectroscopy can be performed on an extremely short time scale, and transient events can be monitored. For example, the growth and decay of intermediate products in a fast chemical reaction can be followed (see KINETIC MEASUREMENTS).

A typical arrangement could include a picosecond-duration pulse tuned to a frequency that is absorbed by the molecule being studied. This pulse drives the molecule to some transient state. A second picosecond pulse arrives after a chosen time delay. The time delay is provided by optical means, such as a variable path length. The second pulse is tuned to a frequency that is absorbed by the intermediate state or the intermediate molecular species. This probing pulse monitors the transient state. Absorption by the transient state causes the intensity of the transmitted probe pulse to vary as the delay time is varied. Many ultrafast chemical processes have been monitored on a picosecond time scale. Among the processes so studied are orientational relaxation in liquids, solvation of free electrons, transitions and decay processes of excited large molecules, photoexcitation of complex molecules such as visual pigments, intermolecular vibrational energy transfer, photodissociation, and isomerization.

An example (100) illustrating the capability of laser spectroscopy for diagnosis of chemical processes involves the probing of the formation and decay of a biophysical compound called prelumirhodopsin [*61489-81-4*]. This is considered the initial step in the photo-induced processes which constitute vision. Vision begins when rhodopsin [*9009-81-8*], a photosensitive pigment in the retina, is exposed to light. The compound prelumirhodopsin is formed very rapidly, faster than conventional methods can monitor. Rhodopsin was exposed to a 6-ps pulse of light at a wavelength of 530 nm. A train of pulses with 20-ps interpulse spacing at a wavelength of 560 nm, which corresponds to an absorption band of prelumirhodopsin, was used to monitor the strength of the absorption band. The results showed that the rise of the prelumirhodopsin concentration occurred less than 6 ps after the start of the 530-nm exciting pulse and indicated that production of prelumirhodopsin is the primary photochemical event. This example shows the power of laser spectroscopy for monitoring the dynamics of chemical reactions.

Laser Photochemistry. Photochemical applications of lasers generally employ tunable lasers which can be tuned to a specific absorption resonance of an atom or molecule (see PHOTOCHEMICAL TECHNOLOGY). Examples include the tunable dye laser in the ultraviolet, visible, and near-infrared portions of the spectrum; the titanium-doped sapphire, Ti:sapphire, laser in the visible and near infrared; optical parametric oscillators in the visible and infrared; and line-tunable carbon dioxide lasers, which can be tuned with a wavelength-selective element to any of a large number of closely spaced lines in the infrared near 10 μm.

Because of the narrow line width, absorption of laser energy can excite one specific state in an atom or molecule. The laser is tuned so that its wavelength matches an absorption corresponding to the desired state, which may be an electronic state or vibrational state. Absorption of laser energy can lead to excitation of specified states much more effectively than absorption of light from conventional light sources.

This leads to the possibility of state-selective chemistry (101). An excited molecule may undergo chemical reactions different from those if it were not excited. It may be possible to drive chemical reactions selectively by excitation of reaction channels that are not normally available. Thus one long-term goal of laser chemistry has been to influence the course of chemical reactions so as to yield new products unattainable by conventional methods, or to change the relative yields of the products.

Despite extensive research in this area, there are relatively few good examples of chemical reactions which have been driven selectively with lasers (102), and apparently none that have reached the status of commercial exploitation. Many of the studies have emphasized the excitation of vibrational energy levels corresponding to one or more quanta of vibrational energy, in the hope that selective excitation of vibrational states can drive the course of a reaction in a direction different from thermal energy. Some difficulties have kept state-selective chemistry from developing more rapidly. One difficulty is that the time scale for which the energy remains localized in a specific vibrational mode must be longer than the time between reactive collisions. A second difficulty is the apparently limited efficiency with which vibrational energy of a molecule influences the course of its reactions.

Perhaps the best example of bond-specific chemistry driven by absorption of laser light has been the set of reactions involving heavy water [*14940-63-7*], HOD:

$$H + HOD \longrightarrow OD + H_2 \tag{19}$$

$$H + HOD \longrightarrow OH + HD \tag{20}$$

When laser light at 720-nm wavelength from a dye laser was absorbed by the HOD molecule, it raised the molecule into a vibrationally excited state with four quanta of O—H stretching excitation. Then the formation of OD and H_2 (reaction 19) occurred at least two orders of magnitude faster than the formation of OH and HD (reaction 20) (103). In a similar type of experiment (104), excitation of a single quantum of O—H stretch vibration by an optical parametric oscillator operating near 2.63-μm wavelength resulted in the exclusive formation of OD and H_2 (reaction 19). Excitation of a single quantum of O—D vibration, with the optical parametric oscillator operating near 3.57 μm, resulted in the exclusive formation of OH and HD (reaction 20). These results indicate that through selective excitation, a specified bond in a polyatomic molecule can be broken.

The use of state-selective chemistry has been an experimental tool to elucidate the dynamics of chemical reactions, but its application to practical chemical process control to enhance yields of specific products is in the developmental stage.

Another area of research in laser photochemistry is the dissociation of molecular species by absorption of many photons (105). The dissociation energy of many molecules is around 4.8×10^{-19} J (3 eV). If one uses an infrared laser with a photon energy around 1.6×10^{-20} J (0.1 eV), about 30 photons would have to be absorbed to produce dissociation (Fig. 17). The curve shows the molecular binding energy for a polyatomic molecule as a function of interatomic distance. The horizontal lines indicate bound excited states of the molecule. These are the vibrational states of the molecule. For energies above the dissociation limit, the atoms can move apart to indefinitely large distances. In other words, the molecule can dissociate. The lowest energy levels are approximately evenly spaced. If an intense laser is tuned to the wavelength corresponding to this spacing, the molecule can make many transitions in rapid succession, rising part way up the ladder of excited states, as indicated by the vertical arrows. The first steps up the ladder are resonant, which means the laser wavelength must be tuned to the proper value to drive the molecule up the evenly spaced

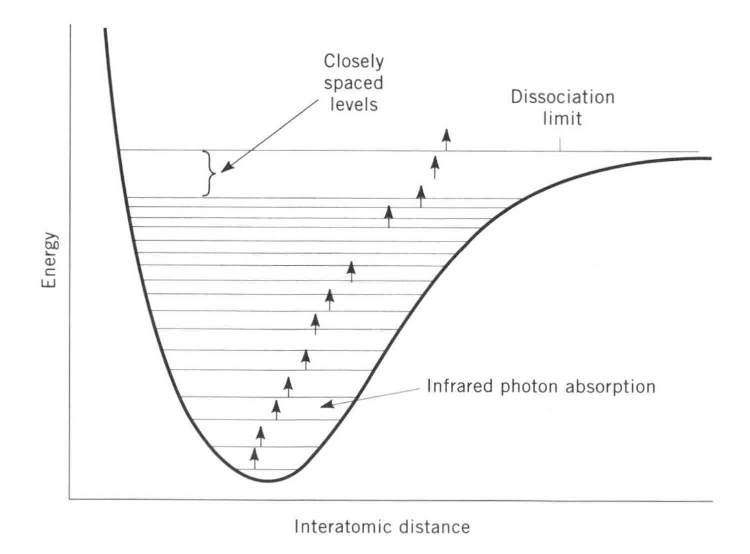

Fig. 17. Vibrational energy levels for a polyatomic molecule.

steps. Above the first steps, the spacing becomes unequal, because of anharmonicity of the potential well. A tuned infrared laser can still drive the system upward to dissociation, as indicated by the arrows. This is because of the presence of additional rotational and translational states, which tend to fill the gaps between the vibrational states forming a quasicontinuum. Absorption can continue even though the laser wavelength no longer matches the spacing of the vibrational states.

One example of unimolecular dissociation is the breakup of gaseous sulfur hexafluoride [*2551-62-4*], SF_6, according to reaction 21 (106):

$$SF_6 + n\,h\nu \longrightarrow SF_5 + F \tag{21}$$

The integer n represents the number of laser photons needed to supply the dissociation energy. In this case, 35 photons are required.

This reaction has been carried out with a carbon dioxide laser line tuned to the wavelength of 10.61 μm, which corresponds to the spacing of the lowest few states of the SF_6 ladder. The laser is a high power TEA laser with pulse duration around 100 ns, so that there is no time for energy transfer by collisions. This example shows the potential for breakup of individual molecules by a tuned laser. As with other laser chemistry, there is interest in driving the dissociation reaction in selected directions, to produce breakup in specific controllable reaction channels.

Another type of laser chemistry involves substitution of lasers for conventional light sources in photochemical applications. An example is the synthesis of vitamin D [*1406-16-2*]. The usual photochemical process involves photochemical conversion of 7-dehydrocholesterol [*434-16-2*] to previtamin D_3 [*1173-13-3*], a process which has low yield. A laser-driven process, involving a two-step irradiation with KrF laser light at 248 nm and nitrogen laser light at 337 nm suppressed competing side reactions and increased the yield to 90% (107). This substitution of light sources represented a significant improvement in the synthesis of vitamin D (see VITAMINS).

Purification of Materials. Purification of materials is another possibility for laser chemistry. One example is purification of silane [*7803-62-5*], SiH_4 (108,109). An argon fluoride excimer laser operating at a wavelength of 193 nm was used to photolyze impurities such as phosphine [*7803-51-2*], PH_3, arsine [*7784-42-1*], AsH_3, and boron hexahydride [*19287-45-7*], B_2H_6, in gaseous silane. These impurities were selectively removed from the silane, permitting production of silane of high purity. It was possible to obtain impurity levels < 1 ppm, whereas these gaseous impurities are difficult to remove by other means. This application is important because silane is used as the starting material for making the high purity silicon used in microelectronics (see ELECTRONIC MATERIALS).

Deposition of Thin Films. Laser photochemical deposition has been extensively studied, especially with respect to fabrication of microelectronic structures (see INTEGRATED CIRCUITS). This procedure could be used in integrated circuit fabrication for the direct generation of patterns. Laser-aided chemical vapor deposition, which can be used to deposit layers of semiconductors, metals, and insulators, could define the circuit features. The deposits can have dimensions in the micrometer regime and they can be produced in specific patterns. Laser chemical vapor deposition can use either of two approaches.

The first is a pyrolytic approach in which the heat delivered by the laser breaks chemical bonds in vapor-phase reactants above the surface, allowing deposition of the reaction products only in the small heated area. The second is a direct photolytic breakup of a vapor-phase reactant. This approach requires a laser with proper wavelength to initiate the photochemical reaction. Often ultraviolet excimer lasers have been used. One example is the breakup of trimethyl aluminum [*75-24-1*] gas using an ultraviolet laser to produce free aluminum [*7429-90-5*], which deposits on the surface. Again, the deposition is only on the localized area which the beam strikes.

A typical example might involve use of a krypton fluoride excimer laser operating at 249 nm with a pulse duration around 100 nanoseconds and a pulse repetition rate which can be varied up to 200 Hz. For metal deposition, energy densities in the range from 0.1 to 1 J/cm^2 per pulse are typical.

A variety of reactions pertinent to laser-assisted chemical vapor deposition have been investigated (110–112). All the operational steps required to fabricate an integrated circuit are available. Materials, including metals, insulators, and semiconductors can be deposited. Doping operations, eg, etching and stripping of insulating materials, can be performed. All these operations can be done in a highly controlled fashion, depositing a material only on the areas where it is desired, as the laser beam is scanned across the substrate surface and modulated on and off to form the desired patterns.

Despite extensive studies, laser photochemical deposition has not yet replaced the conventional processes for fabrication of integrated circuits, except in selected applications such as making and breaking of selected links for customization of circuits and in the repair of defects. The reason is that laser processing of an entire wafer has remained more expensive than the conventional processes.

The examples given above represent only a few of the many demonstrated photochemical applications of lasers. To summarize the situation regarding laser photochemistry as of the early 1990s, it is an extremely versatile tool for research and diagnosis, providing information about reaction kinetics and the dynamics of chemical reactions. It remains difficult, however, to identify specific processes of practical economic importance in which lasers have been applied in chemical processing. The widespread use of laser technology for chemical synthesis and the selective control of chemical reactions remains to be realized in the future.

Isotope Separation. The wavelengths of spectral absorption lines of atoms and molecules have slightly different values, depending on their isotopic constitution. This phenomenon is known as the isotope shift. In most cases isotope shifts are small. The narrow spectral linewidth available from tunable lasers allows energy to be deposited by absorption in a molecule with one isotopic composition, but not in the same molecule with a different isotopic composition. For example, if a line-tunable carbon dioxide laser is tuned to a wavelength of 10.61 μm, the light is absorbed in sulfur hexafluoride labeled with ^{32}S [*21110-12-3*], $^{32}SF_6$, causing it to dissociate (see PHOTOCHEMICAL TECHNOLOGY). If the laser is tuned to 10.82 μm, the light is absorbed by sulfur hexafluoride labeled with ^{34}S [*31719-72-9*], $^{34}SF_6$, but not by $^{32}SF_6$. Thus by choosing the laser wavelength, one may selectively dissociate either $^{32}SF_6$ or $^{34}SF_6$.

The existence of isotope shifts and of tunable lasers with narrow linewidth leads to the possibility of separating isotopes with laser radiation (113,114). This can be of importance, because isotopically selected materials are used for many purposes in research, medicine, and industry. In order to separate isotopes, one needs a molecule that contains the desired element and has an isotope shift in its absorption spectrum, plus a laser that can be tuned to the absorption of one of the isotopic constituents. Several means for separating isotopes are available. The selected species may be ionized by absorption of several photons and removed by application of an electric field, or photodissociated and removed by chemical means.

Laser isotope separation techniques have been demonstrated for many elements, including hydrogen, boron, carbon, nitrogen, oxygen, silicon, sulfur, chlorine, titanium, selenium, bromine, molybdenum, barium, osmium, mercury, and some of the rare-earth elements. The most significant separation involves uranium, separating uranium-235 [*15117-96-1*], ^{235}U, from uranium-238 [*7440-61-1*], ^{238}U (see URANIUM AND URANIUM COMPOUNDS). The separation is important because isotopically enriched ^{235}U is needed for light-water reactors. The isotope ^{235}U is the fissionable isotope that provides energy release. Needs for uranium enriched in ^{235}U are supplied by gaseous diffusion plants which are large and expensive to operate (see DIFFUSION SEPARATION METHODS). A successful laser-based separation process would increase the supply of ^{235}U available for use in nuclear reactors (qv) and reduce its cost.

The leading contender for this application is the atomic vapor laser isotope separation (AVLIS) process which has been under development for a number of years at Lawrence Livermore National Laboratory (115). Metallic uranium is vaporized by an electron beam to produce a vapor stream that is irradiated by dye lasers emitting orange-red light at wavelengths that will selectively ionize ^{235}U. The ^{235}U absorbs the light, whereas the ^{238}U does not. The resulting ^{235}U ions are removed by an electromagnetic field and collected. The ^{238}U isotopes pass through the collector because they are not ionized. The development of the 2.8-kW dye laser system and the copper [*7440-50-8*] vapor lasers which pump it has been the subject of intensive development work (116).

The exact status of the development of the AVLIS process is subject to security classification. It is believed that the process is ready for transition to an industrial operator for commercial development.

Laser-Assisted Thermonuclear Fusion. An application with great potential importance, but which will not reach complete fruition for many years, is laser-assisted thermonuclear fusion (117) (see FUSION ENERGY). The concept involves focusing a high power laser beam onto a mixture of deuterium [*7782-39-0*] and tritium [*10028-17-8*] gases. The mixture is heated to a temperature around 10^8 K (10 keV) (see DEUTERIUM AMD TRITIUM). At this temperature the thermonuclear fusion reaction

$$^{2}H + {}^{3}H \longrightarrow {}^{4}He + n \qquad (22)$$

can occur with a net release of energy. In practice, the gas mixture is contained in glass microspheres with a diameter around 100 μm. A number of laser beams are incident on the sphere from different directions. The pressure produced by absorption of the light drives an implosion. A high pressure shock front travels into the spherical volume. When the implosion reaches the center, the material is compressed to high density and high pressure, which are conditions suitable for the thermonuclear reaction. In experiments carried out at a number of laboratories worldwide, generation of neutrons from reaction 22 has been observed. A condition referred to as scientific breakeven, the point at which the release of thermonuclear energy equals the laser energy, has not been achieved as of the early 1990s. The generation of enough thermonuclear energy to overcome all the losses and inefficiencies of the process and to provide a net total energy gain is still further away.

Laser-assisted thermonuclear fusion experiments have been dominated by neodymium-doped glass lasers. The largest such laser is the Nova laser located at Lawrence Livermore National Laboratory (118). This is a 10-beam system based on neodymium-doped phosphate glass, capable of operation at 1.05 μm and at the second and third harmonics (0.53 and 0.35 μm, respectively). The shorter wavelengths are desirable because coupling of the laser energy to the thermonuclear plasma improves at the shorter wavelengths. The Nova laser has demonstrated output at 1.05 μm of 125 kJ of laser energy in a 2.5-ns pulse and 80 kJ in a 1-ns pulse. At 0.35 μm, 45 kJ in a 2-ns pulse has been obtained. This laser has made possible much greater understanding of the physics of laser interactions with the targets. A proposed next step is the development of a National Ignition Facility, also based on neodymium-doped glass and capable of delivering 1–2 MJ of energy in the ultraviolet, and leading toward the demonstration of scientific breakeven (119).

Although practical generation of energy by laser-assisted thermonuclear fusion remains well in the future, the program has provided some of the most exacting requirements for laser technology and has led to advances in laser equipment that have been adopted in other areas. Thus the research and development associated with thermonuclear fusion work has helped to spur advances in laser technology useful for many other applications.

BIBLIOGRAPHY

"Lasers" in *ECT* 3rd ed., Vol. 14, pp. 42–81, by J. F. Ready, Honeywell Technology Center.

1. T. H. Maiman, *Nature* **187**, 493 (1960).
2. A. Javan, W. R. Bennett and D. R. Herriott, *Phys. Rev. Lett.* **6**, 106 (1961).
3. A. D. White and J. D. Rigden, *Proc. IEEE* **50**, 1697 (1962).
4. R. N. Hall and co-workers, *Phys. Rev. Lett.* **9**, 366 (1962).
5. M. I. Nathan and co-workers, *Appl. Phys. Lett.* **1**, 62 (1962).
6. T. M. Quist and co-workers, *Appl. Phys. Lett.* **1**, 91 (1962).
7. J. L. Bromberg, *The Laser in America, 1950–1970*, The MIT Press, Cambridge, Mass., 1991.
8. A. Einstein, *Phys. Z.* **18**, 121 (1917).
9. G. D. Boyd and J. P. Gordon, *Bell Syst. Tech. J.* **40**, 489 (1961).
10. K. D. Mielenz and co-workers, *Appl. Opt.* **7**, 289 (1968).
11. T. S. Jaseja, A. Javan, and C. H. Townes, *Phys. Rev. Lett.* **10**, 165 (1963).

12. M. Born and E. Wolf, *Principles of Optics*, 6th ed., Pergamon Press, Tarrytown, N.Y., 1980.
13. R. J. Collins and co-workers, *Phys. Rev. Lett.* **5**, 303 (1960).
14. F. J. McClung and R. W. Hellwarth, *J. Appl. Phys.* **33**, 828 (1962).
15. R. W. Hellwarth, in A. K. Levine, ed., *Lasers, A Series of Advances*, Vol. 1, Marcel Dekker, Inc., New York, 1966, p. 253.
16. H. W. Mocker and R. J. Collins, *Appl. Phys. Lett.* **7**, 270 (1965).
17. A. J. DeMaria, C. M. Ferrar, and G. E. Danielson, *Appl. Phys. Lett.* **8**, 22 (1966).
18. M. A. Duguay, S. L. Shapiro, and P. M. Rentzepis, *Phys. Rev. Lett.* **19**, 1014 (1967).
19. A. J. DeMaria, W. H. Glenn, M. J. Brienza, and M. E. Mack, *Proc. IEEE* **57**, 2 (1969).
20. *Laser Focus World Buyers Guide*, PennWell Publishing Co., Tulsa, Okla., 1993.
21. *Lasers & Optronics Technology & Industry Reference*, Gordon Publications, Inc., Morris Plains, N.J., 1993.
22. F. A. Jenkins and H. E. White, *Fundamentals of Optics*, 4th ed., McGraw-Hill Book Co., Inc., New York, 1976.
23. A. L. Bloom, *Gas Lasers*, John Wiley & Sons, Inc., New York, 1968.
24. D. C. Sinclair and W. E. Bell, *Gas Laser Technology*, Holt, Rinehart and Winston, New York, 1969.
25. W. B. Bridges, *Appl. Phys. Lett.* **4**, 128 (1964).
26. W. B. Bridges, A. N. Chester, A. S. Halsted, and J. V. Parker, *Proc. IEEE* **59**, 724 (1971).
27. W. B. Bridges and A. N. Chester, *Appl. Opt.* **4**, 573 (1965).
28. J. P. Goldsborough, *Appl. Phys. Lett.* **15**, 159 (1969).
29. C. K. N. Patel, *Phys. Rev. A* **136**, 1187 (1964).
30. W. W. Duley, *CO_2 Lasers, Effects and Applications*, Academic Press, Inc., New York, 1976.
31. W. J. Witteman, *The CO_2 Laser*, Springer-Verlag, Berlin, 1987.
32. W. B. Tiffany, R. Targ, and J. D. Foster, *Appl. Phys. Lett.* **15**, 91 (1969).
33. A. J. Beaulieu, *Appl. Phys. Lett.* **16**, 504 (1970).
34. A. J. Beaulieu, *Proc. IEEE* **59**, 667 (1971).
35. J. J. Ewing, *Phys. Today*, 32 (May 1978).
36. C. K. Rhodes, ed., *Excimer Lasers*, Springer-Verlag, Berlin, 1984.
37. W. Koechner, *Solid-State Laser Engineering*, 3rd ed., Springer-Verlag, Berlin, 1992.
38. T. H. Maiman, R. H. Hoskins, I. J. D'Haenens, C. K. Asawa, and V. Evtuhov, *Phys. Rev.* **123**, 1151 (1961).
39. F. J. Duarte and L. W. Hillman, *Dye Laser Principles*, Academic Press, San Diego, Calif., 1990.
40. U. Brackmann, *Laser-Grade Dyes*, Lambda Physik, Göttingen, Germany, 1986.
41. Technical data, Lambda Physik Inc., Göttingen, Germany, 1991.
42. H. Kressel and J. K. Butler, *Semiconductor Lasers and Heterojunction LEDs*, Academic Press, Inc., New York, 1977.
43. G. A. Acket, ed., *Semiconductor Lasers, SPIE Proceedings*, Vol. 1025, SPIE, Bellingham, Wash., 1989.
44. A. N. Chester, *Proc. IEEE* **41**, 413 (1973).
45. C. A. Brau, *Free-Electron Lasers*, Academic Press, San Diego, Calif., 1990.
46. D. Prosnitz, ed., *Free-Electron Lasers and Applications, SPIE Proceedings*, Vol. 1227, SPIE, Bellingham, Wash., 1990.
47. D. C. Nguyen and co-workers, Paper CThN1, 1993 Conference on Lasers and Electro-Optics, Baltimore, Md., May 2–7, 1993.
48. A. Mooradian, T. Jaeger, and P. Stokseth, eds., *Tunable Lasers and Applications*, Springer-Verlag, Berlin, 1976.
49. W. T. Ham and co-workers, *Acta Ophthal.* **43**, 390 (1965).

50. W. T. Ham, H. A. Mueller, and D. H. Sliney, *Nature* **260**, 153 (1976).
51. W. T. Ham and H. A. Mueller, *J. Laser Appl.* **3**, 19 (1991).
52. N. A. Peppers and co-workers, *Appl. Opt.* **8**, 377 (1969).
53. A. S. Brownell, *Arch. Environ. Health* **18**, 437 (1969).
54. D. H. Sliney and B. C. Freasier, *Appl. Opt.* **12**, 1 (1973).
55. *Safe Use of Lasers*, ANSI Standard Z136.1-1993, Laser Institute of America, Orlando, Fla., 1993.
56. *Fed. Reg.* **40**, 148 (1975).
57. J. E. Dennis and D. M. Edmunds, *J. Laser Appl.* **3**, 7 (1991).
58. K. R. Ervall and R. Murray, *Evaluation of Commercially Available Laser Protective Eyewear*, HEW Publication (FDA) 79-8086, U.S. Dept. of Health, Education, and Welfare, Bureau of Radiological Health, Rockville, Md., 1979; R. L. Elder, *Science* **182**, 1080 (1973).
59. P. A. Franken, A. E. Hill, C. W. Peters, and G. Weinreich, *Phys. Rev. Lett.* **7**, 118 (1961).
60. J. A. Giordmaine, *Phys. Rev. Lett.* **8**, 19 (1962).
61. P. A. Franken and J. F. Ward, *Rev. Mod. Phys.* **35**, 23 (1963).
62. J. T. Lin and C. Chen, *Lasers Optronics*, 59, (Nov. 1987).
63. C. L. Tang and co-workers, *Laser Focus World*, 87 (Sept. 1990).
64. J. F. Ready, *Proc. IEEE* **70**, 533 (1982).
65. S. S. Charschan, ed., *LIA Guide to Laser Materials Processing*, Laser Institute of America, Orlando, Fla., 1993.
66. G. Chryssoulouris, *Laser Machining, Theory and Practice*, Springer-Verlag, Berlin, 1991.
67. F. P. Gagliano, R. M. Lumley, and L. S. Watkins, *Proc. IEEE* **57**, 114 (1969).
68. E. V. Locke and R. A. Hella, *IEEE J. Quantum Electron.* **QE-10**, 179 (1974).
69. A. S. Kaye, A. G. Delph, E. Hanley, and C. J. Nicholson, *Appl. Phys. Lett.* **43**, 412 (1983).
70. C. M. Banas, *Opt. Eng.* **17**, 210 (1978).
71. T. VanderWert, in D. Belforte and M. Levitt, eds., *The Industrial Laser Annual Handbook—1986 Edition*, PennWell Books, Tulsa, Okla., 1986, p. 58.
72. C. Banas, in Ref. 71, p. 69.
73. I. J. Spalding, in D. Belforte and M. Levitt, eds., *The Industrial Laser Annual Handbook—1990 Edition*, PennWell Books, Tulsa, Okla., 1990, p. 49.
74. R. A. Hella, *Opt. Eng.* **17**, 198 (1978).
75. M. J. Yessik, *Opt. Eng.* **17**, 202 (1978).
76. J. C. Owens in M. Ross, ed., *Laser Applications*, Vol. 1, Academic Press, Inc., New York, 1971, p. 62.
77. A. V. Jelalian, *Laser Radar Systems*, Artech House, Boston, Mass., 1992.
78. E. C. Silverberg, *Appl. Opt.* **13**, 565 (1974).
79. F. Durst, A. Melling, and J. H. Whitelaw, *Principles and Practices of Laser Doppler Anemometry*, Academic Press, Inc., New York, 1976.
80. D. Gabor, *Nature* **161**, 777 (1948).
81. E. N. Leith and J. Upatnieks, *J. Opt. Soc. Amer.* **54**, 1295 (1964).
82. P. Hariharan, *Optical Holography—Principles, Techniques and Applications*, Cambridge University Press, New York, 1986.
83. C. M. Vest, *Holographic Interferometry*, John Wiley & Sons, Inc., New York, 1979.
84. L. Goldman, *Biomedical Aspects of the Laser*, Springer, New York, 1967.
85. I. J. Constable and A. S. Lim, *Laser: Its Clinical Uses in Eye Diseases*, Churchill Livingstone, New York, 1990.
86. T. A. Fuller, ed., *Surgical Lasers: A Clinic Guide*, Pergamon Press, Inc., Tarrytown, N.Y., 1987.

87. J. Nelson and M. Berns, *J. Laser Appl.* **1**, 9 (Mar. 1989).
88. M. Barat and R. H. Ossoff, *J. Laser Appl.* **1**, 45 (Fall, 1988).
89. M. Ross, *Opt. Eng.* **13**, 374 (1974).
90. G. White, *Proc. IEEE* **58**, 1779 (1970).
91. F. E. Goodwin, *Proc. IEEE* **58**, 1746 (1970).
92. *AT&T Tech. J.* **66**, (Jan./Feb. 1987).
93. *AT&T Tech. J.* **71**, (Jan./Feb. 1992).
94. K. W. Nill, F. A. Blum, A. R. Calawa, and T. C. Harmon, *Appl. Phys. Lett.* **19**, 79 (1971).
95. E. D. Hinkley, *Appl. Phys. Lett.* **16**, 351 (1970).
96. E. D. Hinkley and P. L. Kelley, *Science* **171**, 635 (1971).
97. G. A. Antcliffe and J. F. Wrobel, *Appl. Opt.* **11**, 1548 (1972).
98. L. J. Radziemski, R. W. Solarz, and J. A. Paisner, eds., *Laser Spectroscopy and Its Applications*, Marcel Dekker, Inc., New York, 1986.
99. M. D. Levenson, *Phys. Today* **30**, 44 (May 1977).
100. G. E. Busch and co-workers, *Proc. Nat. Acad. Sci.* **69**, 2802 (1972).
101. S. Kimel and S. Speiser, *Chem. Rev.* **77**, 437 (1977).
102. H. R. Bachmann, H. Noth, R. Rinck, and K. L. Kompa, *Chem. Phys. Lett.* **29**, 627 (1974).
103. A. Sinha, M. C. Hsiao, and F. F. Crim, *J. Chem. Phys.* **92**, 6333 (1990).
104. M. J. Bronikowski, W. R. Simpson, B. Girard, and R. N. Zare, *J. Chem. Phys.* **95**, 8647 (1991).
105. N. Bloembergen and E. Yablonovitch, *Phys. Today* **31**, 23 (May 1978).
106. M. J. Coggiola, P. A. Schulz, Y. T. Lee, and Y. R. Shen, *Phys. Rev. Lett.* **38**, 17 (1977).
107. V. Malatesta, C. Willis, and P. A. Hackett, *J. Amer. Chem. Soc.* **103**, 6781 (1981).
108. J. H. Clark and R. G. Anderson, *Appl. Phys. Lett.* **32**, 46 (1978).
109. A. Hartford, E. J. Huber, J. L. Lyman, and J. H. Clark, *J. Appl. Phys.* **51**, 4471 (1980).
110. D. Baüerle, *Chemical Processing with Lasers*, Springer-Verlag, Berlin, 1986.
111. R. M. Osgood, S. R. J. Brueck, and H. R. Schlossberg, eds., *Laser Diagnostics and Photochemical Processing for Semiconductor Devices*, North-Holland, New York, 1983.
112. I. W. Boyd, *Laser Processing of Thin Films and Microstructures*, Springer-Verlag, Berlin 1987.
113. R. G. Denning, *Phys. Technol.* **9**, 242 (1978).
114. R. N. Zare, *Sci. Amer.* **236**, 86 (Feb. 1977).
115. J. A. Paisner, *Appl. Phys. B* **46**, 253 (1988).
116. I. L. Bass, R. E. Bonanno, R. P. Hackel, and P. R. Hammond, *Appl. Opt.* **31**, 6993 (1992).
117. C. Yamanaka, *Introduction to Laser Fusion*, Gordon & Breach, Reading, UK, 1991.
118. C. Bibeau and co-workers, *Appl. Opt.* **31**, 5799 (1992).
119. J. R. Murray, J. H. Campbell, and D. N. Frank, *Beamlet Project: Technology Demonstration for a National Inertial Confinement Fusion Ignition Facility*, Paper CTuC1, 1993 Conference on Lasers and Electro-Optics, Baltimore, Md., May 2–7, 1993.

General References

J. F. Ready, *Industrial Applications of Lasers*, Academic Press, New York, 1978.

J. F. Ready, *Effects of High Power Laser Radiation*, Academic Press, New York, 1971.

S. S. Charschan, ed., *Lasers in Industry*, Van Nostrand Reinhold Co., Inc., New York, 1972.

A. E. Siegman, *Lasers*, University Science Books, Mill Valley, Calif., 1986.

D. K. Evans, ed., *Laser Applications in Physical Chemistry*, Marcel Dekker, Inc., New York, 1989.

D. L. Andrews, *Lasers in Chemistry*, Springer-Verlag, Berlin, 1990.

A. N. Chester, V. S. Letokhov, and S. Martellucci, *Laser Science and Technology*, Plenum Publishing Corp., New York, 1988.
M. A. El-Sayed, ed., *Laser Applications to Chemical Dynamica, SPIE Proceedings*, Vol. 742, SPIE, Bellingham, Wash., 1987.
J. I. Steinfeld, *Laser-Induced Chemical Processes*, Plenum, New York, 1981.
J. P. Fouassier and J. F. Rabak, eds., *Lasers in Polymer Science and Technology: Applications*, CRC Press, Boca Raton, Fla., 1989.
T. Okoshi, *Optical Fibers*, Academic Press, Inc., New York, 1982.
J. T. Luxon and D. E. Parker, *Industrial Lasers and their Applications*, Prentice-Hall, Inc., Englewood Cliffs, N.J., 1985.
A. B. Marchant, *Optical Recording, A Technical Overview*, Addison-Wesley Publishing Co., Inc., Reading, Mass., 1990.
R. M. Wood, *Laser Damage in Optical Materials*, Adam Hilger, Bristol, U.K., 1986.
R. J. Collier, C. B. Burckhardt, and L. H. Lin, *The Applications of Holography*, Wiley-Interscience, New York, 1971.
A. Yariv, *Quantum Electronics*, John Wiley & Sons, Inc., New York, 1989.
C. Albright, ed., *Laser Welding, Machining and Materials Processing*, Springer-Verlag, Berlin, 1986.
D. Sliney and M. Wolbarsht, *Safety with Lasers and Other Optical Sources*, Plenum Publishing Corp., New York, 1980.
R. W. Boyd, *Nonlinear Optics*, Academic Press, San Diego, Calif., 1992.
M. C. Tobin, *Laser Raman Spectroscopy*, Krieger Pub. Co., Melbourne, Fla., 1982.
J. H. Eberly and P. W. Milonni, *Lasers*, John Wiley & Sons, Inc., New York, 1988.
W. F. Price and J. Uren, *Lasers in Construction*, Van Nostrand Reinhold Co., Inc., New York, 1988.
F. P. Schaefer, ed., *Dye Lasers*, Springer-Verlag, Berlin, 1990.
S. Lugomer, *Laser Technology, Laser Driven Processes*, Prentice-Hall, Inc., Englewood Cliffs, N.J., 1990.
P. K. Cheo, *Fiber Optics, Devices and Systems*, Prentice-Hall, Inc., Englewood Cliffs, N.J., 1985.
R. Grisar, H. Preier, G. Schmidtke, and G. Restelli, eds., *Monitoring of Gaseous Pollutants by Tunable Diode Lasers*, D. Reidel Publishing Co., Dordrecht, Germany, 1987.

JOHN F. READY
Honeywell Technology Center

LATEX. See LATEX TECHNOLOGY; ELASTOMERS, SYNTHETIC; RUBBER.

LATEX TECHNOLOGY

Latex technology encompasses colloidal and polymer chemistry in the preparation, processing, and conversion of natural and synthetic latices into useful products.

Historically latex was the milky liquid drawn from any of 200 plants, most notably the *Hevea brasiliensis* tree of South America (1). Early applications of natural rubber, 93–95% *cis*-1-4-isoprene, included waterproofing and strengthening of other materials (see ISOPRENE; RUBBER, NATURAL). Vulcanization and the automobile increased natural rubber demand through the end of the nineteenth century (2). Large rubber plantations in Malaya, Ceylon, Indonesia, and Indochina increased the world's natural rubber production to 200,000 metric tons by 1920. Enhanced supply led to rapid growth of natural rubber products and improvements in latex processing. The Allied blockade of Germany during World War I led to the first process for making synthetic latex (3). Soon thereafter United States companies began to produce commercial synthetic latices: Buna S (butadiene–styrene copolymer), also known as Government Rubber–Styrene or GR–S rubber; Neoprene (polychloroprene) (4); and Thiokol (polysulfides) (see POLYMERS CONTAINING SULFUR; STYRENE–BUTADIENE RUBBER). Japan's seizure of the Southeast Asia rubber plantations during World War II led to intensive research in synthetic rubber production (5). Synthetic latex production accounts for 65% of the 15×10^6-t total rubber market (6). Worldwide concern over the AIDS epidemic has sharply increased the demand for latex used in the preparation of rubber gloves and similar dipped goods. Estimates are for a 5–7% annual growth rate in this market.

Many synthetic latices exist (7,8) (see ELASTOMERS, SYNTHETIC). They contain butadiene and styrene copolymers (elastomeric), styrene–butadiene copolymers (resinous), butadiene and acrylonitrile copolymers, butadiene with styrene and acrylonitrile, chloroprene copolymers, methacrylate and acrylate ester copolymers, vinyl acetate copolymers, vinyl and vinylidene chloride copolymers, ethylene copolymers, fluorinated copolymers, acrylamide copolymers, styrene–acrolein copolymers, and pyrrole and pyrrole copolymers. Many of these latices also have carboxylated versions.

Traditional applications for latices are adhesives, binders for fibers and particulate matter, protective and decorative coatings (qv), dipped goods, foam, paper coatings, backings for carpet and upholstery, modifiers for bitumens and concrete, and thread and textile modifiers. More recent applications include biomedical applications as protein immobilizers, visual detectors in immunoassays (qv), as release agents, in electronic applications as photoresists for circuit boards, in batteries (qv), conductive paint, copy machines, and as key components in molecular electronic devices.

Synthetic Latex Manufacture

Early efforts to produce synthetic rubber coupled bulk polymerization with subsequent emulsification (9). Problems controlling the heat generated during bulk polymerization led to the first attempts at emulsion polymerization. In emulsion polymerization hydrophobic monomers are added to water, emulsified by a surfactant into small particles, and polymerized using a water-soluble initiator. The result is a colloidal suspension of fine particles, 50–1000-nm in diameter, usually comprising 30–50 wt % of the latex product. By 1935 emulsion polymerization became the method of choice in making synthetic rubber because of its many advantages (10): (*1*) the reaction mass viscosity remains low throughout the polymerization, providing for improved heat transfer, agitation, and product handling; (*2*) the sensible heat of the water in the emulsion balances the heat of reaction generated by free-radical polymerization; and (*3*) the rate of reaction is rapid, while producing very high molecular weight.

Kinetics and Mechanisms. Early researchers misunderstood the fast reaction rates and high molecular weights of emulsion polymerization (11). In 1945 the first recognized qualitative theory of emulsion polymerization was presented (12). This mechanism for classic emulsion preparation was quantified (13) and the polymerization separated into three stages.

Stage I: Particle Nucleation. At the start of a typical emulsion polymerization the reaction mass consists of an aqueous phase containing small amounts of soluble monomer, small spherical micelles, and much larger monomer droplets. The micelles are typically 5–30-nm in diameter and are saturated with monomer emulsified by the surfactant. The monomer droplets are larger, 1,000–10,000-nm in diameter, and are also stabilized by the surfactant.

Water-soluble initiator is added to the reaction mass, and radicals are generated which enter the micelles. Polymerization starts in the micelle, making it a growing polymer particle. As monomer within the particle converts to polymer, it is replenished by diffusion from the monomer droplets. The concentration of monomer in the particle remains as high as 5–7 molar. The growing polymer particles require more surfactant to remain stable, getting this from the uninitiated micelles. Stage I is complete once the micelles have disappeared, usually at or before 10% monomer conversion.

Radicals generated from water-soluble initiator might not enter a micelle (14) because of differences in surface-charge density. It is postulated that radical entry is preceded by some polymerization of the monomer in the aqueous phase. The very short oligomer chains are less soluble in the aqueous phase and readily enter the micelles. Other theories exist to explain how water-soluble radicals enter micelles (15). The micelles are presumed to be the principal locus of particle nucleation (16) because of the large surface area of micelles relative to the monomer droplets.

However, in the case of mini- and microemulsions, processing methods reduce the size of the monomer droplets close to the size of the micelle, leading to significant particle nucleation in the monomer droplets (17). Intense agitation, cosurfactant, and dilution are used to reduce monomer droplet size. Additives like cetyl alcohol are used to retard the diffusion of monomer from the droplets to the micelles, in order to further promote monomer droplet nucleation (18). The benefits of miniemulsions include faster reaction rates (19), improved shear stability, and the control of particle size distributions to produce high solids latices (20).

An expression for the number of particles formed during Stage I was developed, assuming micellar entry as the formation mechanism (13), where k is a constant varying from 0.37 to 0.53 depending on the relative rates of radical adsorption in micelles and polymer particles, r_i is the rate of radical generation, m is the rate of particle growth, a_s is the surface area covered by one surfactant molecule, and S is the total concentration of soap molecules.

$$N_p = k(r_i/m)^{0.4}(a_s S)^{0.6} \tag{1}$$

During Stage I the number of polymer particles range from 10^{13} to 10^{15} per mL. As the particles grow they adsorb more emulsifier and eventually reduce the soap concentration below its critical micelle concentration (CMC). Once below the CMC, the micelles disappear and emulsifier is distributed between the growing polymer particles, monomer droplets, and aqueous phase.

The Smith-Ewart expression (eq. 1) accurately predicts the particle number for hydrophobic monomers like styrene and butadiene (21), but fails to predict the particle number (22) for more hydrophilic monomers like methyl methacrylate and vinyl acetate. A new theory based on homogeneous particle nucleation, called the HUFT theory after Hansen, Ugelstad, Fitch, and Tsai, was developed (23) to explain the hydrophilic monomer data yielding more accurate particle number predictions. The HUFT theory has been extended to include precursor and mature latex particle formation (24,25). In homogeneous coagulation theory (24), very small-diameter precursors, containing only small amounts of monomer, transform into mature particles through coagulation. Mature particles form only near the end of the nucleation stage and positively skewed particle-size distributions are the result (25).

The debate as to which mechanism controls particle nucleation continues. There is strong evidence the HUFT and coagulation theories hold true for the more water-soluble monomers. What remains at issue are the relative rates of micellar entry, homogeneous particle nucleation, and coagulative nucleation when surfactant is present at concentrations above its CMC. It is reasonable to assume each mechanism plays a role, depending on the nature and conditions of the polymerization (26).

Whatever the nucleation mechanism the final particle size of the latex is determined during Stage I, provided no additional particle nucleation or coalescence occurs in the later stages. Monomer added during Stages II and III only serves to increase the size of the existing particles.

Stage II: Growth in Polymer Particles Saturated With Monomer. Stage II begins once most of the micelles have been converted into polymer particles. At constant particle number the rate of polymerization, R_p, as given by Smith-Ewart kinetics is as follows (27) where k_p is the propagation rate constant, $[M]$ is the concentration of monomer in the particle, N is the concentration of growing polymer particles, $\overline{n}$ is the average number of radicals per particle, and N_A is Avogadro's number.

$$R_p = \frac{10^3 N k_p [M] \overline{n}}{N_A} \tag{2}$$

During Stage II the growing particles maintain a nearly constant monomer concentration. The concentration of monomer is particle-size dependent, with smaller particles having lower concentrations (28).

During Stages II and III the average concentration of radicals within the particle determines the rate of polymerization. To solve for $\overline{n}$, the fate of a given radical was balanced across the possible adsorption, desorption, and termination events. Initially a solution was provided for three physically limiting cases. Subsequently, $\bar{n}$ was solved for explicitly without limitation using a generating function to solve the Smith-Ewart recursion formula (29). This analysis for the case of very slow rates of radical desorption was improved on (30), and later radical readsorption was accounted for and the Smith-Ewart recursion formula solved via the method of continuous fractions (31).

As the particles grow, they require more soap to remain stable. If soap is not available the particles can destabilize, causing product and process problems.

Stage III: Growth in Polymer Particles With a Decreasing Monomer Concentration. Stage III begins once the monomer droplets disappear. The rate of polymerization decreases with reduced particle monomer concentration. At high monomer conversion, diffusion control of termination can cause an apparent increase in the rate of polymerization (32). Still further conversion can lead to diffusion control of propagation and a marked reduction in rate of polymerization (33). High free monomer in the final latex can result (34) causing product odor and handling problems.

The Smith-Ewart kinetics described assume homogeneous conditions within the particle. An alternative view, where monomer polymerizes only on the surface of the particle, has been put forth (35) and supported (36). The nature of the intraparticle reaction environment remains an important question.

Basic Components. The principal components in emulsion polymerization are deionized water, monomer, initiator, emulsifier, buffer, and chain-transfer agent. A typical formula consists of 20–60% monomer, 2–10 wt % emulsifier on monomer, 0.1–1.0 wt % initiator on monomer, 0.1–1.0 wt % chain-transfer agent on monomer, various small amounts of buffers and bacteria control agents, and the balance deionized water.

Water. Latices should be made with deionized water or condensate water. The resistivity of the water should be at least $10^5\ \Omega$. Long-term storage of water should be avoided to prevent bacteria growth. If the ionic nature of the water is poor, problems of poor latex stability and failed redox systems can occur. Antifreeze additives are added to the water when polymerization below 0°C is required (37). Low temperature polymerization is used to limit polymer branching, thereby increasing crystallinity.

Monomers. A wide variety of monomers can be used, and they are chosen on the basis of cost and ability to impart specific properties to the final product. Water solubilities of industrially important monomers are shown in Table 1 (38). The solubility of the monomer in water affects the physical chemistry of the polymerization. Functional monomers like methacrylic and acrylic acid, infinitely soluble in water, are also used. These monomers impart long-term shelf stability to latices by acting as emulsifiers. The polymerization behavior of some monomers, such as methacrylic acid, as well as the final latex properties are influenced by pH. For optimum results with these acids, polymerization is best performed at a pH of ca 2. After polymerization, the latex is neutralized to give adequate shelf stability at tractable viscosities.

Table 1. Water Solubilities of Monomers Common to Latex Production

Monomer	CAS Registry Number	Solubility in water at 25°C, mM
n-octyl acrylate	[*2499-59-4*]	0.34
dimethylstyrene		0.45
vinyltoluene	[*25012-15-4*]	1.0
n-hexyl acrylate	[*2499-95-8*]	1.2
styrene	[*100-42-5*]	3.5
n-butyl acrylate	[*141-32-2*]	11
chloroprene	[*126-99-8*]	13
butadiene	[*106-99-0*]	15
vinylidene chloride	[*75-35-4*]	66
ethyl acrylate	[*140-88-5*]	150
methyl methacrylate	[*80-62-6*]	150
vinyl chloride	[*75-01-4*]	170
vinyl acetate	[*108-05-4*]	290
methyl acrylate	[*96-33-3*]	650
acrylonitrile	[*107-13-1*]	1600
acrolein	[*107-02-8*]	3100

When monomers of drastically different solubility (39) or hydrophobicity are used or when staged polymerizations (40,41) are carried out, core–shell morphologies are possible. A wide variety of core–shell latices have found application in paints, impact modifiers, and as carriers for biomolecules. In staged polymerizations, spherical core–shell particles are made when polymer made from the first monomer is more hydrophobic than polymer made from the second monomer (42). When the first polymer made is less hydrophobic then the second, complex morphologies are possible including voids and half-moons (43), although spherical particles still occur (44).

Surfactants. Surfactants perform many functions in emulsion polymerization, including solubilizing hydrophobic monomers, determining the number and size of the latex particles formed, providing latex stability as particles grow, and providing latex stability during post-polymerization processing.

Emulsification is the process by which a hydrophobic monomer, such as styrene, is dispersed into micelles and monomer droplets. A measure of a surfactant's ability to solubilize a monomer is its critical micelle concentration (CMC). Below the CMC the surfactant is dissolved in the aqueous phase and does not serve to solubilize monomer. At and above the CMC the surfactant forms spherical micelles, usually 50 to 200 soap molecules per micelle. Many properties, such as electrical conductivity, interfacial tension, surface tension, refractive index, and viscosity, show a sudden decrease at the CMC (45). The CMC is temperature- and chain-length dependent for a given class of surfactants (46). The CMCs of nonionic surfactants are higher than those of ionic surfactants (47).

Surfactants also stabilize the growing polymer particles by overcoming the attractive forces between particles. Anionic and cationic surfactants use electrostatic repulsion forces to negate the attraction. Nonionic surfactants use steric forces to repel the attraction. Figure 1 compares the two stabilizing mechanisms. The ability of a given surfactant to stabilize latex particles is dependent on many

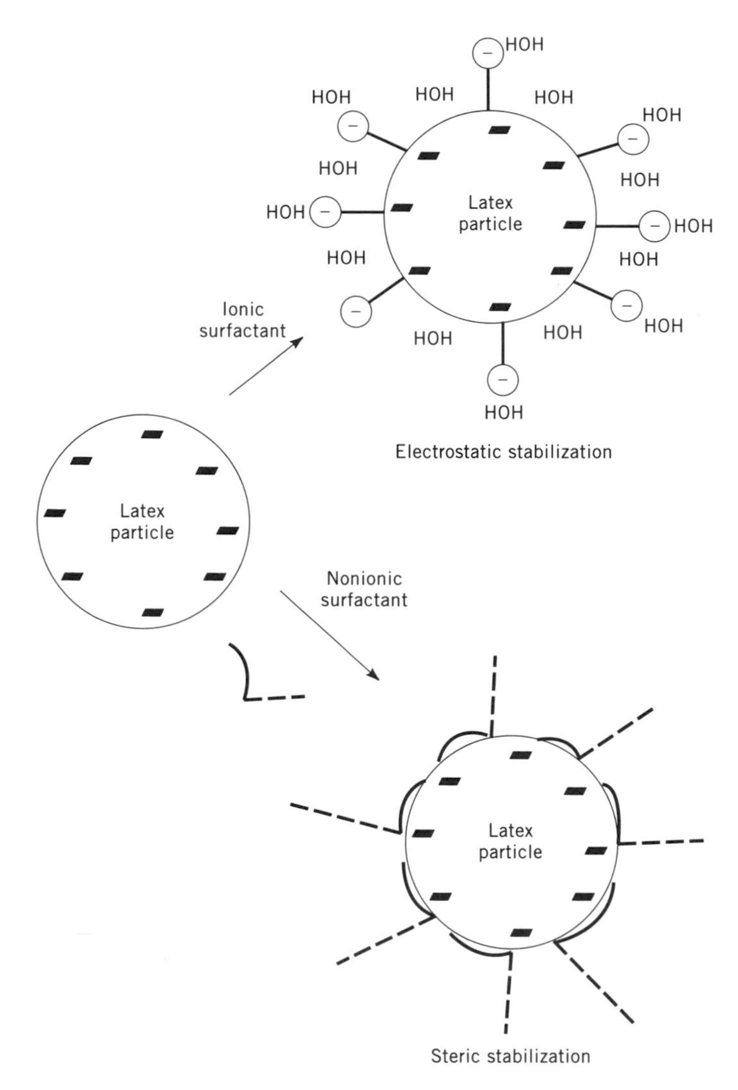

Fig. 1. Surfactant stabilization mechanisms.

factors (48), including surfactant type and concentration, aqueous solubility of the monomer and polymer, agitation and shear rate, temperature, surface tension, ionic strength, and concentration of the monomer and polymer (49).

An *a priori* method for choosing a surfactant was attempted by several researchers (50) using the hydrophile–lipophile balance or HLB system (51). In the HLB system a surfactant soluble in oil has a value of 1 and a surfactant soluble in water has a value of 20. Optimum HLB values have been reported for latices made from styrene, vinyl acetate, methyl methacrylate, ethyl acrylate, acrylonitrile, and their copolymers and range from 11 to 18. The HLB system has been criticized as being imprecise (52).

Three generations of latices as characterized by the type of surfactant used in manufacture have been defined (53). The first generation includes latices made with conventional (*1*) anionic surfactants like fatty acid soaps, alkyl carboxylates, alkyl sulfates, and alkyl sulfonates (54); (*2*) nonionic surfactants like poly(ethylene oxide) or poly(vinyl alcohol) used to improve freeze–thaw and shear stability; and (*3*) cationic surfactants like amines, nitriles, and other nitrogen bases, rarely used because of incompatibility problems. Portland cement latex modifiers are one example where cationic surfactants are used. Anionic surfactants yield smaller particles than nonionic surfactants (55). Often a combination of anionic surfactants or anionic and nonionic surfactants are used to provide improved stability. The stabilizing ability of anionic fatty acid soaps diminishes at lower pH as the soaps revert to their acids. First-generation latices also suffer from the presence of soap on the polymer particles at the end of the polymerization. Steam and vacuum stripping methods are often used to remove the soap and unreacted monomer from the final product (56).

The second generation includes latices made with functional monomers like methacrylic acid, 2-hydroxyethyl acrylate [*818-61-1*], acrylamide [*79-06-1*], 2-dimethylaminoethylmethacrylate [*2867-47-2*], and sodium *p*-vinyl-benzenesulfonate [*98-70-4*] that create *in situ* polymeric emulsifier. The initiator decomposition products, like the sulfate groups arising from persulfate decomposition, can also act as chemically bound surfactants. These surfactants are difficult to remove from the latex particle.

The third generation are latices made with independently prepared surfactant to mimic the *in situ* prepared functional monomer surfactant. These emulsifiers are often A–B block polymers where A is compatible with the polymer and B with the aqueous phase. In this way surface adsorption of the surfactant is more likely. These emulsions are known to exhibit excellent properties.

Initiators. The initiators most commonly used in emulsion polymerization are water soluble although partially soluble and oil-soluble initiators have also been used (57). Normally only one initiator type is used for a given polymerization. In some cases a finishing initiator is used (58). At high conversion the concentration of monomer in the aqueous phase is very low, leading to much radical–radical termination. An oil-soluble initiator makes its way more readily into the polymer particles, promoting conversion of monomer to polymer more effectively.

The most common water-soluble initiators are ammonium persulfate, potassium persulfate, and hydrogen peroxide. These can be made to decompose by high temperature or through redox reactions. The latter method offers versatility in choosing the temperature of polymerization with −50 to 70°C possible. A typical redox system combines a persulfate with ferrous ion:

$$S_2O_8^{2-} + Fe^{2+} \longrightarrow Fe^{3+} + SO_4^{2-} + SO_4^{-}$$

Reducing agents are employed to return the Fe^{3+} to Fe^{2+}. By starting at a lower temperature, the heat of reaction can be balanced by the sensible heat of the water in the emulsion. Temperature profiles from 20 to 70°C are typical for such systems. Care must be taken when working with redox systems to eliminate oxygen from the reactor before beginning the polymerization. The effectiveness of the redox system can be pH-dependent with the optimum pH range depending on the type of the redox system (59). For higher temperature polymerizations, eg, above 70°C, thermal decomposition of the initiator is used.

A third source of initiator for emulsion polymerization is hydroxyl radicals created by γ-radiation of water. A review of radiation-induced emulsion polymerization detailed efforts to use γ-radiation to produce styrene, acrylonitrile, methyl methacrylate, and other similar polymers (60). The economics of γ-radiation processes are claimed to compare favorably with conventional techniques although worldwide industrial application of γ-radiation processes has yet to occur. Use of γ-radiation has been made for laboratory study because radical generation can bc turned on and off quickly and at various rates (61).

The ionic nature of the radicals generated, by whatever technique, can contribute to the stabilization of latex particles. Soapless emulsion polymerizations can be carried out using potassium persulfate as initiator (62). It is often important to control pH with buffers during soapless emulsion polymerization.

Chain-Transfer Agents. The most commonly employed chain-transfer agents in emulsion polymerization are mercaptans, disulfides, carbon tetrabromide, and carbon tetrachloride. They are added to control the molecular weight of a polymer, P_n, by transferring a propagating radical to the chain transfer agent AX (63):

$$P_n\cdot\ + AX \longrightarrow P_nX + A\cdot$$

The newly formed short-chain radical A· then quickly reacts with a monomer molecule to create a primary radical. If subsequent initiation is not fast, AX is considered an inhibitor. Many have studied the influence of chain-transfer reactions on emulsion polymerization because of the interesting complexities arising from enhanced radical desorption rates from the growing polymer particles (64,65). Chain-transfer reactions are not limited to chain-transfer agents. Chain-transfer to monomer is in many cases the main chain termination event in emulsion polymerization. Chain transfer to polymer leads to branching which can greatly impact final product properties (66).

Other Ingredients. During polymerization and post-processing, the pH of the emulsion is important. Increasing pH to improve latex stability is achieved usually by adding sodium hydroxide, potassium hydroxide, or ammonia. To avoid causing any localized flocculation due to a rapid increase in electrolyte in a confined area, these ingredients must be added as dilute solutions of around 3% with mild agitation. In some cases some surfactant may be required to be added along with the dilute alkali. Antimicrobial agents are added for protection against bacteria attack (see INDUSTRIAL ANTIMICROBIAL AGENTS).

Process. Commercial processes manufacturing latex can be divided into batch, semibatch, and continuous methods. A schematic of typical equipment is shown in Figure 2. The reactor is usually glass-lined, including agitator and thermowell. The remaining tanks are constructed of stainless steel. The reactor is jacketed to allow for heating and cooling between 0 and 100°C. Reactor agitation is chosen to provide adequate mixing while avoiding shear-induced coagulation. The reactor is equipped with a small condenser; reflux is to be avoided to prevent coagulum from forming. A monomer–soap solution is emulsified by a centrifugal pump and fed to the reactor along with initiator, using suitable flow control. Premixing monomer and initiator is to be avoided to prevent premature

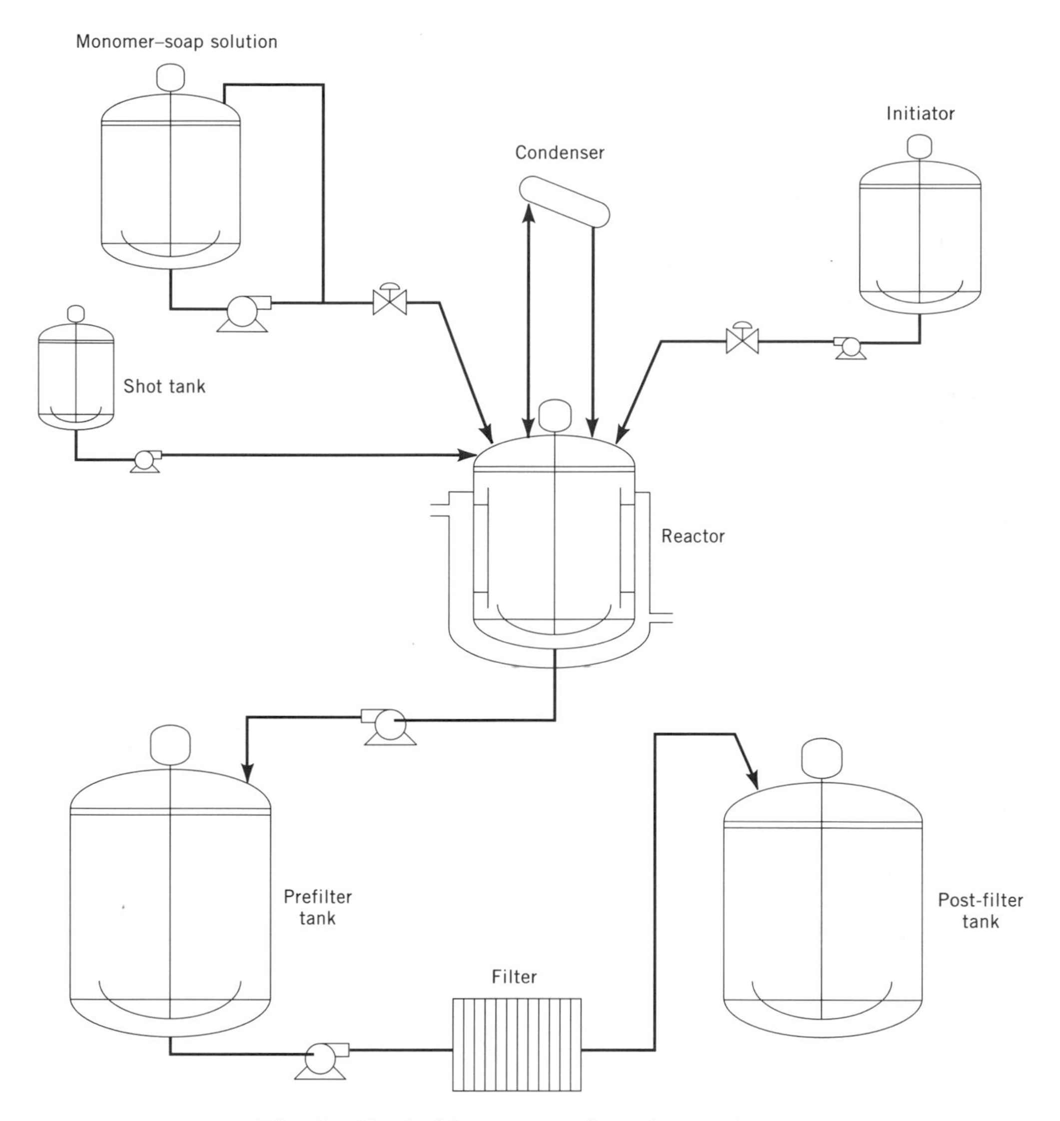

Fig. 2. Typical latex manufacturing equipment.

polymerization in the feed tank. A shot tank is usually required to allow for addition of ingredients in minor amounts.

In the most common production method, the semibatch process, about 10% of the preemulsified monomer is added to the deionized water in the reactor. A shot of initiator is added to the reactor to create the seed. Some manufacturers use master batches of seed to avoid variation in this step. Having set the number of particles in the pot, the remaining monomer and, in some cases, additional initiator are added over time. Typical feed times are 1–4 h. Lengthening the feeds tempers heat generation and provides for uniform comonomer sequence distributions (67). Sometimes skewed monomer feeds are used to offset differences in monomer reactivity ratios. In some cases a second monomer charge is made to produce core–shell latices. At the end of the process pH adjustments are often made. The product is then pumped to a prefilter tank, filtered, and pumped to a post-filter tank where additional processing can occur. When the feed rate of monomer during semibatch production is very low, the reactor is said to be monomer starved. Under these conditions monomer droplets are not present, and intraparticle polymerization takes place under high polymer–low monomer concentrations. These conditions can lead to branched polymer with bi- and trimodal molecular weight distributions (68).

The batch process is similar to the semibatch process except that most or all of the ingredients are added at the beginning of the reaction. Heat generation during a pure batch process makes reactor temperature control difficult, especially for high solids latices. Seed, usually at 5–10% solids, is routinely made via a batch process to produce a uniform particle-size distribution. Most kinetic studies and models are based on batch processes (69).

Continuous processes have been developed for many of the larger volume synthetic latices (70). Most of these processes involve the use of several continuously stirred tank reactors (CSTR) in series. The exponential residence time distribution of a CSTR is broad relative to a batch reactor, leading to broad particle size distributions. By placing many CSTRs in series the effective residence time and corresponding particle size distributions are narrowed. CSTR processes can also suffer from sustained oscillations and multiple steady states, leading to poor reactor and product performance (71). The cause of the oscillations is related to new particle formation. To avoid such oscillations many processes use a seed latex in the feed stream (72). If premanufacture of seed is not desirable a tubular reactor can be used to produce seed of uniform particle size (73). Tubular reactors have also been used as loop reactors, where feeds enter and leave a tubular loop in which the circulating flow is much greater than the throughput (74). Cooling water is sprayed directly on the tube to control the reaction temperature. Recently coagulum has been successfully controlled during tubular production of a latex using pulsatile flow (75).

Foaming represents a persistent problem in the processing and handling of latices. The most effective way to eliminate the presence of foam is to simply avoid its generation. Ways to reduce foam include eliminating the free-fall of latex by using dip-pipes, not stirring air into the latex by not agitating with exposed impellers, and not adding dry ingredients laden with adsorbed air directly to the latex. There are many proprietary additives for minimizing generation of foam (antifoaming agents) or eliminating foam (defoaming agents) but there is no one type that works for all latices (see DEFOAMERS). Only by trial and error can the most effective agent be found for a given compound or process. With any of them, the minimum amount required should be used since their addition tends to cause localized flocculation, poor wetting, lower water resistance, and "fisheyes" in films. Many antifoam agents get absorbed into the polymer or other ingredients in the compound and thus lose their effectiveness over time, particularly if the compound is being recirculated during processing. These conditions necessitate augmentation with additional amounts of antifoaming agent to counter these effects.

Process Modeling. The complexity of emulsion polymerization makes reliable computer models valuable. Many attempts have been made to simulate the emulsion polymerization process for different monomer systems (76–78).

Other Routes. *High Solids Emulsions.* Latices are made at the highest possible solids content consistent with acceptable viscosity. Latices solids can be increased by centrifuging, creaming, electrodecantation, or evaporation. The latter two techniques, however, are not of commercial importance. Natural rubber latex, 25–40 wt % solids to start, is concentrated to about 60 wt % solids by centrifuging with milk/cream separator equipment. Creaming is commonly used with synthetic polymer latices. Creaming is accelerated by adding solutions such as ammonium alginate and surfactant to the latex. Depending on the initial solids content and type of latex, solids contents of 58–65 wt % are possible.

Soap-starved recipes have been developed that yield 60 wt % solids low viscosity polymer emulsions without concentrating. It is possible to make latices for application as membranes and similar products via emulsion polymerization at even higher solids (79). Solids levels of 70–80 wt % are possible. The paste-like material is made in batch reactors and extruded as product.

Inversion of Nonaqueous Polymers. Many polymers such as polyurethanes, polyesters, polypropylene, epoxy resins (qv), and silicones that cannot be made via emulsion polymerization are converted into latices. Such polymers are dissolved in solvent and inverted via emulsification, followed by solvent stripping (80). Solid polymers are milled with long-chain fatty acids and diluted in weak alkali solutions until dispersion occurs (81). Such latices usually have lower polymer concentrations after the solvent has been removed. For commercial uses the latex solids are increased by techniques such as creaming.

Latex Properties

The observable properties of a latex, ie, stability, rheology, film properties, interfacial reactivity, and substrate adhesion, are determined by the colloidal and polymeric properties of the latex particles. Important colloidal properties include ionic charge, stability, particle size and morphology distribution, viscosity, solids, and pH. Important polymer properties include molecular weight distribution, monomer sequence distribution, glass-transition temperature, crystallinity, degree of cross-linking, and free monomer. Methods for analyzing each of these properties exist, depending on the end use of the product. An overview of the various polymer colloid characterization methods is available (82) (see also COLLOIDS).

Stability. For a latex to be a useful product, control of polymer isolation is crucial. The individual polymer particles must be stable enough to avoid coagulation resulting from perturbances like high temperature, freeze–thaw cycles, high shear in handling, electrolyte addition, and organic solvent addition during processing, but not so stabilized that polymer isolation is impossible. Stability is related to the surface properties of the latex particles, and these are usually determined during latex manufacture. Visual detection of coagulation is easy; more sophisticated optical techniques are possible (83). The types of initiator, emulsifier, and monomers used are the key determinants.

Electrostatic Stabilization. The electrical charges on the surface of a latex particle are balanced by an electrical double layer of oppositely and then similarly charged counterions. The outer layer is known as the diffuse electrical double layer, and its potential controls the colloidal stability of the latex particles

(84). The diffuse electrical double-layer potential is closely related to the zeta-potential. Electrophoresis measurements at various pH and ionic strengths are a means of calculating the zeta-potential (85). As two particles approach, their individual diffuse double layers begin to overlap and the particles repulse one another. As the particles get even closer, attractive forces build and the particles coagulate. The energy required to overcome the repulsive forces depends on the ionic strength of the bulk phase, the temperature, and the nature of the ions balancing the latex particle charge. One way to coagulate particles is to reduce the energy barrier between particles by adding electrolyte. The Derjaguin-Landau-Verwey-Overbeek (DLVO) theory (86) is a method for predicting how much electrolyte is needed to coagulate a latex by first predicting the critical coagulation concentration. DLVO theory does not account for the chemical nature of the electrolyte nor the interaction of the ions with water. Aluminum compounds, like aluminum alginate, coagulate latices with varying effectiveness depending on the pH of the latex system (87).

Steric Stabilization. Nonionic surfactants, usually containing ethylene oxide units, are able to stabilize particles by adsorbing their hydrocarbon chain ends on the hydrophobic polymer zones of the surface of the latex particle. The ethylene oxide groups extend into the water phase. These compounds suffer from temperature and dilution sensitivity. A–B block copolymers where A is compatible with the latex polymer and B with the dispersion medium offer a more robust method of steric stabilization. Reactive macromonomers with affinity toward the dispersion medium copolymerized with hydrophobic monomer offer a chemical way of grafting the stabilizer to the surface of the latex particle (88). Organic solvent addition is used to coagulate sterically stabilized latices (89). Addition of soluble homopolymer can lead to reversible and irreversible flocculation by depletion and bridging flocculations (90) (see FLOCCULATING AGENTS).

Other Stabilizers. In addition to anionic, cationic, and nonionic surfactants used in manufacture of the latex, other specialized surfactants can be added or compounded to the latex after manufacture to increase the stability of the latex. For example, amphoteric surfactants like *c*-cetyl betaine are used to improve the mechanical stability of low pH anionic compounds. Quaternary ammonium salts are used to improve the mechanical stability of cationic latices and their compounds. They are used either alone or in combination with a nonionic surfactant. Sequestrants (91) such as sodium silicate, sodium polyphosphate, and the sodium salt of ethylene diamine tetraacetic acid are added to anionic latices to retard the destabilizing action of cations that get leached slowly from compounding ingredients containing multivalent ions. The addition of these ingredients is also beneficial if the use of hard water is unavoidable in making and using the anionic latex compound. Care must be taken to avoid shocking the latex during compounding with these additional surfactants.

Rheology. Flow properties of latices are important during processing and in many latex applications such as dipped goods, paint, inks (qv), and fabric coatings. For dilute, nonionic latices, the relative latex viscosity is a power–law expansion of the particle volume fraction. The terms in the expansion account for flow around the particles and particle–particle interactions. For ionic latices, electrostatic contributions to the flow around the diffuse double layer and enhanced particle–particle interactions must be considered (92). A relative viscosity relationship for concentrated latices was first presented in 1972 (93). A review of empirical relative viscosity models is available (92). In practice, latex viscosity measurements are carried out with rotational viscometers (see RHEOLOGICAL MEASUREMENT).

It is possible to increase the viscosity of a latex after manufacture using thickeners. Thickening occurs through increases in medium viscosity or polymer particle aggregation. If considerable aggregation occurs without a corresponding increase in medium viscosity, undesirable separation or creaming occurs. Methylcellulose, caseinates, and polyacrylate salts are typical thickeners. Ease of adding the thickener, ability to maintain viscosity, and undesirable side effects must be considered when selecting a thickener. Some thickeners slowly hydrolyze in the latex and lose their effectiveness over time. The full range of the effects of adding thickener develops over time, some of them much faster than others. To avoid exceeding the desired viscosity, it is advisable to add thickener in small increments, waiting after each for the viscosity to reach equilibrium before adding the next one.

The viscosity of the latex can also be dependent on pH. In the case of some latices, lowering the pH with a weak acid such as glycine is an effective method for raising the viscosity without destabilizing the system. Latices made with poly(vinyl alcohol) as the primary emulsifier can be thickened by increasing the pH with a strong alkali.

Particle Size. The particle-size distribution of a latex is a determinant of its performance in application. Particles are almost always spherical, although nonspherical particles are possible (94). Particle-size distributions are now routinely measured using scattering methods, light scattering being the most common and most effective on particles >300 nm. Small-angle neutron scattering (sans) is useful for concentrated latices and smaller (<ca 300-nm) particle sizes (95). Microscopic techniques are also a reliable method. Two of the newer techniques are sedimentation field-flow fractionation (96) and capillary hydrodynamic fractionation (97). The particle-size distribution has an impact on stability, rheology, morphology, and film-forming properties. Stability is predicted for larger particles, with broadly distributed particle sizes showing a greater tendency to coagulate owing to interactions between the smaller particles (98). Viscosity effects are weighted to the larger particles. Morphology is affected by particle size, as in the case where the particle size of the polystyrene latex used to seed polystyrene core–poly(ethyl methacrylate) shell particles determined the morphology of the final latex. When the seed was less than 200 nm in diameter, the final particles were prolate spheroids of near-hemispherical polystyrene and poly(ethyl acrylate) domains. When the seed was 300 nm, spherical polystyrene particles embedded with circular patches of poly(ethyl methacrylate) were the result (99).

Film Properties. Dehydration (100) at temperatures above the polymer glass-transition temperature is the principal means for forming film from latex. Interdiffusion of the polymer chains between particles is thought to be the limiting step in film formation (101). The final properties of the film depend on the polymer in the latex particles. Significant differences can exist between films made from solution polymer and latex polymer, respectively, because of the colloidal debris remaining on the latex polymer. Important film properties include hardness, flexibility, clarity, conductance, impact strength, and toughness. Core–shell particles have seen extensive use as rubber toughening agents.

Improving Properties Through Compounding. The potential value of most polymers can be realized only after proper compounding. Materials used to enhance polymer properties or reduce polymer cost include antioxidants (qv), cross-linking agents, accelerators, fillers (qv), plasticizers (qv), adhesion promoters, pigments (qv), etc. Antioxidants are essential to retard degradation in unsaturated polymers. Cross-linking agents are used to build modulus, resistance to permanent deformation, and greater solvent resistance in many types of polymers. Accelerators are frequently used to reduce the time and temperature required to affect the cross-linking. Fillers, such as carbon black (qv) and clays (qv), do not reinforce latex polymers as they do their dry polymer counterparts. Rather, they are used in most latex applications to adjust processing rheology and to lower raw materials costs of the product, or to impart specific effects, eg, aluminum trihydrate to increase resistance to flame degradation, or carbon black to increase resistance to uv degradation. Plasticizers and oils are used to soften and increase flexibility at lower temperatures, improve resistance to crystallization, or depress the brittle point of the product. Hydrocarbon process oils, glycols, vegetable oils, ester plasticizers, and low melting point resins are some of the common materials used. Many types of resins are added to enhance the tackiness of polymers. Generally, within a class of tackifying resins, the lower the melting point, the greater the tack developed in the compounded polymer. The optimum amount of any resin for maximum tack depends on the type of polymer to which it is added. Resins added as solvent-cut emulsions rather than as solventless emulsion or dispersions develop more tack in the polymer, because the residual solvent in the polymer contributes to the tack of the polymer resin blend. Pigments and dyes are used to impart color. Some pigments with some polymers also impart other effects such as improved water resistance or reduced flammability.

Latex Applications

Adhesives. Latices are used as adhesives (qv) in the construction market, in tires and belt fabrication, in furniture manufacture, in packaging, and in tapes, labels, envelopes, and bookbinding. The adhesives are used in wet or dry laminations. In wet lamination, the adhesive is not dried before assembly; hence at least one of the substrates must be porous to allow for the water to evaporate. Wet lamination has the advantage that the surfaces to be adhered can be repositioned during assembly, provided the solids content remains below the level at which the adhesive begins to form a film. The disadvantage of wet lamination is slower development of cohesive strength in the adhesive film and the need for closer control of timing when the two substrates are to be brought together; if this occurs too soon, the adhesive film is too weak to hold the assembly together, and if it occurs too late, the adhesive is too dry to effect a satisfactory bond.

Binders. Latices are used as fiber binders by the paper and textile industries. The two principal methods of application are (*1*) wet-end addition, wherein the ionic latex is added to a fiber slurry and then coagulated in the slurry prior to sheet formation, and (*2*) saturation of the latex into a formed fiber web wherein the latex is coagulated by dehydration. Latices are also used as binders for particulate matter such as rubber scrap.

Coatings. Latices are used in residential and industrial paints (qv), coated paper and paperboard, fabric coatings, backing for carpet, upholstery, and drapery, as basecoats for wallpaper and flooring, and in insulation coatings (see COATED FABRICS; COATINGS). Application methods include brushing, squeegee, spraying, dipping, and frothing (see COATING PROCESSES).

Dipped Goods. Latices are used in various dipping processes to produce balloons, bladders, gloves, extruded thread, and tubing. Manufacturing techniques include multiple dip and dry, and coagulant dipping employing a colloidal destabilizer (102).

Foam Products. Latices are made into foams for use in cushioning applications. The latices are frothed with air and then chemically coagulated for thick applications, or heated to induce coagulation for thinner applications. The latter method allows for infinite pot life during production (see FOAMED PLASTICS).

Modifiers. Latices are added to bitumens, mortars, and concrete to improve impact resistance and reduce stress cracking. Key to the use of latices in these technologies is compatibility between the latex and the construction materials.

Newer Applications. Latices are used in electronic chip production as ceramic and metal powder dispersants, allowing for thin-film application. The organic material is burned away after the powder has been fused to the substrate. Latices are mixed with inorganic material to create conductive polymers for use in photo imaging, conductive coatings, batteries, and molecular electronic devices (103). In biomedical applications, latices act as a support for immobilizing protein, serving as a visual colloidal indicator. Tests for AIDS antibodies, hepatitis B surface antigen, and other materials have been developed using latices as the supports (104).

BIBLIOGRAPHY

"Latex Technology" in *ECT* 3rd ed., Vol. 14, pp. 82–97, by A. Klein, Lehigh University.

1. M. Morton, *Introduction to Rubber Technology,* Reinhold Publishing Corp., New York, 1959.
2. P. G. Cook, *Latex–Natural and Synthetic,* Reinhold Publishing Corp., New York, 1956.
3. U.S. Pat. 1,864,078 (1932), C. Hueck (to I.G. Farbenindustrie AG).
4. W. H. Carothers, G. J. Berechet, and A. M. Collins, *J. Am. Chem. Soc.* **54**, 4066 (1932). W. H. Carothers, I. Williams, and J. E. Kirby, *J. Am. Chem. Soc.* **53**, 4203 (1931).
5. G. G. Winspear, ed., *The Vanderbilt Latex Handbook,* R. T. Vanderbilt Co., New York, 1954.
6. *Rubber Statistical Bulletin,* International Rubber Study Group, London, 1992.
7. J. W. Vanderhoff, *Chem. Eng. Sci.* **48**(2), 203–217 (1993).
8. E. Daniels, E. D. Sudol, and M. S. El-Aasser, eds., *Polymer Latexes: Preparation, Characterization and Applications,* ACS Symposium Series, Vol. 492, American Chemical Society, Washington, D.C., 1992.

9. J. C. Carl, *Neoprene Latex,* E. I. du Pont de Nemours & Co., Inc., Wilmington, Del., 1962.
10. G. Odian, *Principles of Polymerization,* 2nd ed., John Wiley & Sons, Inc. New York, 1981, p. 319.
11. H. Fikentscher, *Angew. Chem.* **51**, 433 (1938); J. H. Baxendale and co-workers, *J. Polym. Sci.* **1**, 466 (1946); W. P. Hohenstein and H. Mark, *J. Polym. Sci.* **1**, 549 (1946).
12. W. D. Harkins, *J. Chem. Phys.* **13**(9), 381 (1945); *J. Amer. Chem. Soc.* **69**, 1428 (1947).
13. W. V. Smith and R. H. Ewart, *J. Chem. Phys.* **16**(6), 592 (1948).
14. R. G. Gilbert and G. T. Russell, Course Notes from a *Short Course on Emulsion Polymers* at Sydney University, Sydney, Australia, 1992, p. 18.
15. F. K. Hansen, *Chem. Eng. Sci.* **48**(2), 437–444, (Jan. 1993).
16. M. S. El-Aasser, in F. Candau and R. H. Ottewill, eds., *Scientific Methods for the Study of Polymer Colloids and Their Applications,* Kluwer Academic Publishers, the Netherlands, 1990, pp. 12–15.
17. J. Ugelstad and co-workers, *J. Polym. Sci. Polym. Lett.* **111**, 503 (1973).
18. P. L. Tang and co-workers, in Ref. 8, pp. 72–98.
19. K. Fontenot and F. J. Schork, *J. Appl. Polym. Sci.* **49**(4), 663–655 (1993); D. T. Barnett and F. J. Schort, *Chem. Eng. Commun.* **80**, 113–125 (1989); E. Elbing and co-workers, *Aust. J. Chem.* **42**(12), 2085–2094 (1989).
20. M. S. El-Aasser, in F. Candau and R. H. Ottewill, eds., *An Introduction to Polymer Colloids,* Kluwer Academic Publishers, the Netherlands, 1990, p. 15.
21. H. Gerrens, *Fortschr. Hochpolym. Forsch.* **1**, 234–238.
22. V. I. Yeliseyeva, in I. Piirma, ed., *Emulsion Polymerization,* Academic Press, New York, 1982, p. 248.
23. R. M. Fitch and C. H. Tsai, *Polym. Lett.* **8**, 703 (1970).
24. D. H. Napper and R. G. Gilbert, *Makromol. Chem., Macromol. Symp.* **10/11**, 503 (1987).
25. D. H. Nappera and R. G. Gilbert, in Ref. 16, p. 167.
26. A. S. Dunn, in Ref. 8, pp. 45–54.
27. W. V. Smith, *J. Am. Chem. Soc.* **70**, 3695 (1948).
28. M. Morton, S. Kaizerman, and M. W. Altier, *J. Colloid Sci.* **9**, 300 (1954).
29. W. H. Stockmayer, *J. Polym. Sci.* **24**(106), 314 (1957).
30. J. T. O'Toole, *J. Appl. Polym. Sci.* **9**, 1291 (1965); *J. Polym. Sci., Part C* **27**, 171 (1969).
31. J. Ugelstad, P. C. Mork, and J. O. Aasen, *J. Polym. Sci., Part A-1* **5**, 2281 (1967).
32. G. T. Russell and co-workers, *Macromolecules* **21**, 2133 (1988).
33. M. J. Ballard and co-workers, *Macromolecules* **19**, 1303 (1986).
34. I. A. Maxwell, E. M. F. J. Verdurmen, and A. L. German, *Makromol. Chem.* **193**, 2677–2695 (1992).
35. S. S. Medvedev, *Proc. Int. Symp. Makromol. Chem. Prague 1957,* Pergamon Press, New York, 1958, p. 174; S. S. Medvedev, *Kinet. Mech. Polyreactions,* IUPAC Int. Symp. Makromol. Chem., Plenary Main Lect., 1969, p. 39; S. S. Medvedev and co-workers, *J. Macromol. Sci. Chem.* **A7**, 715 (1973).
36. M. R. Grancio and D. J. Williams, *J. Polym. Sci., Part A-1* **8**, 2617 (1971).
37. D. C. Blackley, *Emulsion Polymerization,* Applied Science Publishers Ltd., London, 1975.
38. E. R. Blout and co-workers, *Monomers,* Interscience Publishers, New York, 1949.
39. Y. C. Chen, V. Dimonie, and M. S. El-Aasser, *Pure Appl. Chem.* **64**(11), 1691–1696 (1992); Y. C. Chen, V. L. Dimonie, O. L. Shaffer, and M. S. El-Aasser, *Polym. Int.* **30**, 185–194 (1993).
40. S. Lee and A. Rudin, in Ref. 8, pp. 234–254.
41. S. Laferty and I. Piirma, in Ref. 8, pp. 255–271.

42. M. Okubo, A. Yamada, and T. Matsumoto, *J. Polym. Sci. Polym. Chem. Ed.* **16**, 3219 (1980).
43. T. J. Min, A. Klein, M. S. El-Aasser, and J. W. Vanderhoff, *J. Polym. Sci. Polym. Lett. Ed.* **21**, 2845 (1981).
44. Y. C. Chen, V. L. Dimonie, O. L. Shaffer, and M. S. El-Aasser, *Polym. Int.* **30**, 185–194 (1993).
45. H. Gerrens and G. Hirsch, in J. Brandrup and E. H. Immergut, eds., *Polymer Handbook,* 2nd ed., John Wiley & Sons, Inc., New York, 1975, p. II-483.
46. B. D. Flockhart, *J. Colloid. Sci.* **16**, 484–492 (1961); J. K. Weil and co-workers, *J. Am. Oil Chem. Soc.* **40**, 538–540 (1963).
47. K. Shinoda, T. Nakagawa, B. I. Tamamushi, and T. Isemura, *Colloidal Surfactants,* Academic Press, Inc., New York, 1963.
48. R. H. Ottewill, in Ref. 22, pp. 1–6.
49. K. Fontenot and F. J. Schork, *Ind. Eng. Chem. Res.* **32**, 374 (1993).
50. M. P. Merkel, M.S. dissertation, Lehigh University, Bethlehem, Pa., 1982.
51. W. C. Griffin, *Off. Dig. Fed. Paint Varnish Prod. Clubs,* 28 (June 1956).
52. M. S. El-Aasser, in Ref. 16, p. 17.
53. J. W. Vanderhoff, *Chem. Eng. Sci.* **48**(2), 212 (1993).
54. Ref. 10, p. 333.
55. A. S. Dunn, in Ref. 8, p. 51.
56. Jpn. Pat. 4-239506 (1992), S. Takao, H. Nakano, and H. Kobayashi.
57. M. Nomura and K. Fujita, *Makromol. Chem., Rapid Commun.* **10**, 581–587 (1989); M. Nomura, J. Ikoma, and K. Fujita, in Ref. 8, pp. 55–71.
58. A. E. Hamielec and J. F. MacGregor, in Ref. 22, p. 330.
59. H. Warson, in I. Piirma and J. L. Gardon, eds., *Emulsion Polymerization,* ACS Symposium Series, Vol. 24, American Chemical Society, Washington, D.C., 1976, p. 228.
60. V. T. Stannett, in Ref. 22, pp. 415–450.
61. Ref. 14, p. 59.
62. A. R. Goodall, J. Hearn, and M. C. Wilkinson, *Brit. Polym. J.* **10**, 141 (1978); *J. Polym. Sci. Polym. Chem. Ed.* **17**, 1019 (1979).
63. Ref. 10, pp. 226–242.
64. H. Lee and G. W. Pohlein, *Polym. Process. Eng.* **5**(1), 37–74 (1987).
65. B. C. Wang and co-workers, *J. Polym. Sci. Polym. Lett. Ed.* **18**, 711 (1980).
66. P. A. Lovell, T. H. Shah, and F. Heatley, in Ref. 8, pp. 188.
67. K. Chujo and co-workers, *J. Polym. Sci., Part C* **27**, 321 (1969).
68. J. Guillot, A. Guyot, and C. Pichot, in Ref. 16, p. 103.
69. W. H. Ray, ACM Symposium Series No. 226, 1983.
70. G. W. Pohlein, in Ref. 22, pp. 357–382.
71. G. Ley and H. Gerrens, *Chem. Ing. Tech.* **43**, 693 (1971); A. W. DeGraff, Ph.D. dissertation, Lehigh University, Bethlehem, Pa., 1970; F. J. Schork, Ph.D. dissertation, University of Wisconsin–Madison, 1981; R. K. Greene, R. A. Gonzales, and G. W. Pohlein in Ref. 59, pp. 367–378; C. Kiparissides, Ph.D. dissertation, McMaster University, Hamilton, Ontario, Canada, 1978.
72. H. C. Lee and G. W. Pohlein, *Chem. Eng. Sci.* **41**(4), 1023–1030 (1986).
73. R. A. Gonzalez, M.S. dissertation, Lehigh University, Bethlehem, Pa., 1974.
74. U.S. Pat. 3,551,996, R. Lanthier (to Gulf Oil Canada, Ltd.).
75. D. A. Paquet, Ph.D. dissertation, University of Wisconsin–Madison, 1993.
76. K. W. Min and W. H. Ray, *J. Appl. Polym. Sci.* **22**, 89 (1978); J. B. Rawlings and W. H. Ray, *Polym. Eng. Sci.* **28**(5), 237 (1988).
77. M. Morbidelli, G. Storti, and S. Carra, *J. Appl. Polym. Sci.* **28**, 961 (1983).
78. J. R. Richards, J. P. Congalidis, and R. G. Gilbert, *J. Applied Polym. Sci.* **37**, 2727–2756 (1989).

79. E. Ruckenstein and F. Sun, *J. Applied Polym. Sci.* **46**, 1271–1277 (1992).
80. Ref. 59; J. Ugelstad and co-workers, in Ref. 22, pp. 383–413.
81. S. P. Suskind, *J. Appl. Polym. Sci.* **9**, 2451 (1965).
82. R. H. Ottewill and co-workers, in M. S. El-Aasser and R. M. Fitch, eds., *Future Directions in Polymer Colloids,* Martinus Nijhoff Publishers, Dordrecht, Germany, 1987, pp. 243–251.
83. M. S. El-Aasser and co-workers, in M. S. El-Aasser and J. W. Vanderhoff, eds., *Emulsion Polymerization of Vinyl Acetate,* Applied Science Publishers, London, 1981, p. 215.
84. R. H. Ottewill, in Ref. 16, p. 131.
85. R. L. Rowell, in Ref. 16, pp. 201–204.
86. E. J. W. Verwey and J. Th. G. Overbeek, *Theory of the Stability of Lyophobic Colloids,* Elsevier, Amsterdam, the Netherlands, 1948; B. V. Derjaguin and L. Landau, *Acta Physiochim. URSS.* **14**, 633 (1941).
87. C. G. Force and E. Matijevic, *Kolloid. Z. Z. Polym.* **224**, 51 (1968).
88. D. H. Napper, *Polymeric Stabilization of Colloidal Dispersions,* Academic Press, London, 1983.
89. D. H. Napper, *J. Colloid Interface Sci.* **58**, 390 (1977).
90. A. P. Gast, C. K. Hall, and W. B. Russell, *Faraday Discuss. Chem. Soc.* **76**, 189 (1983).
91. Ref. 37, p. 406.
92. J. W. Goodwin, in Ref. 16, pp. 212–215.
93. I. M. Krieger, *Adv. Colloid Interface Sci.* **3**, 111 (1972).
94. J. W. Vanderhoff, H. R. Sheu, and M. S. El-Aasser, in Ref. 16, pp. 529–565; B. S. Casey, *Mater. Forum* **16**, 117–122 (1992).
95. R. H. Ottewill, in Ref. 82, pp. 253–276; M. F. Mills and co-workers, *Macromolecules* **26**, 3553–3562 (1993).
96. S. Lee, M. N. Meyers, R. Beckett, and J. C. Gidding, *Anal. Chem.* **60**, 1129 (1988).
97. J. G. DosRamos and C. A. Silebi, *J. Colloid Interface Sci.* **135**, 165 (1990).
98. J. N. Shaw and R. H. Ottewill, *Discuss. Faraday Soc.* **42**, 154 (1966); A. Kithara and H. Ushiyama, *J. Colloid Soc.* **43**, 73 (1972); B. A. Mathews and C. T. Rhoades, *J. Colloid Soc.* **32**, 332 (1970).
99. J. W. Vanderhoff, in Ref. 82, pp. 43–44.
100. G. L. Brown, *J. Polym. Sci.* **22**, 423 (1956); D. P. Sheetz, *J. Appl. Polym. Sci.* **9**, 3759 (1965); J. W. Vanderhoff, E. B. Bradford, and W. K. Carrington, *J. Polym. Sci., Part C* **41**, 155 (1973).
101. K. Hahn, G. Ley, and R. Oberthur, in Ref. 16, pp. 463–479.
102. C. H. Gelbert and H. E. Berkheimer, *Paper C in Educational Symposium No. 18 on Latex Technology*, Rubber Division of the American Chemical Society, Montreal, Canada, 1987.
103. R. E. Partch and co-workers, in Ref. 8, pp. 368–386; P. Espiard and co-workers, in Ref. 8, pp. 387–404.
104. P. J. Tarcha and co-workers, in Ref. 8, pp. 347–367.

CHESTER H. GELBERT
MICHAEL C. GRADY
E. I. du Pont de Nemours & Co., Inc.

LAUNDERING. See DETERGENCY; SURFACTANTS.

LAURIC ACID. See CARBOXYLIC ACIDS.

LAURYL ALCOHOL. See ALCOHOLS, HIGHER ALIPHATIC.

LAWRENCIUM. See ACTINIDES AND TRANSACTINIDES.

LAXATIVES. See GASTROINTESTINAL AGENTS.

LEACHING. See EXTRACTION, LIQUID–SOLID; METALLURGY, EXTRACTIVE; MINERAL RECOVERY AND PROCESSING.

LEAD

Lead [*7439-92-1*], Pb, is an essential commodity in the modern industrial world, ranking fifth in tonnage consumed after iron (qv), copper (qv), aluminum (see ALUMINUM AND ALUMINUM ALLOYS), and zinc (see ZINC AND ZINC ALLOYS). In 1993, the United States accounted for 30% of the 4,450,000 metric tons of refined lead consumed by the Western world. Slightly over half of the lead produced in the world now comes from recycled sources (see RECYCLING, NONFERROUS METALS).

Lead has outstanding properties: low melting point, ease of casting, high density, softness, malleability, low strength, ease of fabrication, acid resistance, electrochemical reaction with sulfuric acid, and chemical stability in air, water, and earth. The principal uses of lead and its compounds in descending order are storage batteries (qv), pigments (qv), ammunition, solders, plumbing, cable covering, bearings, and caulking. In addition, lead is used to attenuate sound waves, atomic radiation, and mechanical vibration. In most of these applications lead is not used in its pure state, but rather as an alloy (see LEAD ALLOYS).

Lead, copper, silver, and gold were the metals first used by ancient humans. The Egyptians probably used lead as early as 5000 BC. Simplicity of reduction from ores, low melting point, and ease of fabrication presumably led to its use. Lead was also widely used by the Greeks and Romans. Segments of the fluted columns common to Greek architecture are pinned together by iron rods fitted into sockets which were filled with molten lead. Lead water pipes have been found in the ruins of Rome and Pompeii, confirming the use of lead in antiquity. Some pipes of this period still function in Britain as of this writing (ca 1994).

Use of lead in modern industrial society results from its unique physical and chemical properties. By the middle of the nineteenth century, world production of lead had risen to 1×10^5 metric tons per year, passed 1×10^6 t/yr early in the twentieth century, and reached 1.5×10^6 t/yr by midcentury. Lead production is expected to reach 5.6×10^6 t/yr by the year 2000.

Lead and its compounds are cumulative poisons and should be handled with recommended precautions. These materials should not be used in contact with food and other substances that may be ingested (see also LEAD COMPOUNDS).

Occurrence and Ores

In comparison to aluminum and iron, the most abundant metals in the earth's crust, lead is a rare metal. Even copper and zinc are more abundant by factors of five and eight, respectively. However, the occurrence of concentrated and easily accessible lead ore deposits is unexpectedly high, and these are widely distributed through the world. The most important ore mineral is galena [*12179-39-4*], PbS (87% Pb), followed by anglesite [*7446-14-2*], $PbSO_4$ (68% Pb), and cerussite [*14476-15-4*], $PbCO_3$ (77.5% Pb). The latter two minerals result from the natural weathering of galena.

Galena is found in fissure veins and replacement bodies, and may be associated with sphalerite (zinc sulfide), pyrite (iron sulfide), marcasite, chalcopyrite, tetrahedrite, cerussite, anglesite, dolomite, calcite, quartz, and barite, as well as the valuable metals gold, silver, bismuth, and antimony. The formation of lead ore deposits is thought to have occurred during the emplacement of igneous rock masses with the solidification of silicates from molten magma. Components of the molten magma, such as metal sulfides, were concentrated in the liquid remaining from the crystallization and forced by the pressure from the growing silicate rocks into available channels such as fault fissures.

Most (88%) lead mined in the United States comes from eight mines in Missouri. The rest comes from 11 mines in Colorado, Idaho, Montana, Alaska, Washington, and Nevada. Ores of the Southeast Missouri lead belt and extensive deposits such as in Silesia and Morocco are of the replacement type. These deposits formed when an aqueous solution of the minerals, under the influence of changing temperature and pressure, deposited the sulfides in susceptible sedimentary rock, usually limestone and dolomites. These ore bodies usually contain galena, sphalerite, and pyrite minerals, but seldom contain gold, silver, copper, antimony, or bismuth.

Lead and zinc minerals are so intimately mixed in many deposits that they are mined together and then separated. Silver minerals are frequently found in association with galena.

The concentration of lead in ore bodies of commercial interest generally ranges from 2 to 6%; the average is 2.5%. Improvements in ore-dressing techniques have made possible the exploitation of deposits having lead contents even less than 2%.

The world reserves of lead are estimated at 71×10^6 t and scattered around the world (1). Over one-third (25×10^6 t) of this total is located in North America where the United States has, in units of 10^6 t, 14; Canada, 7; Mexico, 3; and other sources, 1. South America has 2; Europe, 11; Africa, 4; and Australia, 14×10^6 t.

In Asia, the former Soviet Union has 9 and the People's Republic of China has 6×10^6 t. The recovery of lead from scrap is of prime importance in supplying U.S. demands so that the entire reserve base is estimated at 120×10^6 t. Total world resources are estimated at 1.4×10^9 t.

Physical Properties

Lead, atomic number 82, is a member of Group 14 (IVA) of the Periodic Table. Ordinary lead is bluish grey and is a mixture of isotopes of mass number 204 (15%), 206 (23.6%), 207 (22.6%), and 208 (52.3%). The average atomic weight of lead from different origins may vary as much as 0.04 units. The stable isotopes are products of decay of three naturally radioactive elements (see RADIOACTIVITY, NATURAL): ^{206}Pb comes from the uranium series (see URANIUM AND URANIUM COMPOUNDS), ^{208}Pb from the thorium series, and ^{207}Pb from the actinium series (see ACTINIDES AND TRANSACTINIDES). The crystal structure of lead is face-centered cubic; the length of the edge of the cell is 0.49389 nm; the number of atoms per unit cell is four. Other properties are listed in Table 1.

Chemical Properties

Lead forms two series of compounds corresponding to the oxidation states of +2 and +4. The +2 state is the more common. Compounds of lead(IV) are regarded as covalent, those of lead(II) as primarily ionic. Lead is amphoteric, forming plumbous (Pb(II)) and plumbic (Pb(IV)) salts as well as plumbites and plumbates, respectively.

Lead is one of the most stable of fabricated materials because of excellent corrosion resistance to air, water, and soil. An initial reaction with these elements results in the formation of protective coatings of insoluble lead compounds. For example, in the presence of oxygen, water attacks lead, but if the water contains carbonates and silicates, protective films or tarnishes form and the corrosion becomes exceedingly slow.

Because of its position relative to hydrogen in the electromotive series, theoretically lead should replace hydrogen in acids. However, the potential difference is small and the high hydrogen overvoltage prevents replacement. Reaction with oxidizing acids releases oxidants which combine with hydrogen to depress the overvoltage, resulting in replacement.

Acid Oxidation. Reactions of lead with acid and alkalies are varied. Nitric acid, the best solvent for lead, forms lead nitrate; acetic acid forms soluble lead acetate in the presence of oxygen; sulfuric acid forms insoluble lead sulfate. Sulfuric acid is stored in containers with chemical or acid-grade lead. Lead dissolves slowly in HCl, but in the presence of aqueous alkalies forms soluble plumbites and plumbates.

There are three common oxides of lead. Lead oxide [*1317-36-8*], PbO, also known as litharge, is formed by heating lead in air or blowing air into molten lead. It is used in batteries (qv), glass (qv), and ceramics (qv). Lead tetroxide [*1314-41-6*], Pb_3O_4, is formed by controlled oxidation of litharge at about 450°C, and lead dioxide [*1309-60-0*], PbO_2, is formed by electrolytic oxidation of lead salts or by strong oxidizing agents. Lead dioxide is used in the manufacture of

Table 1. Physical Properties of Lead

Property	Value
at. wt	207.2
melting point, °C	327.4
boiling point, °C	1770
specific gravity, °C	
20	11.35
327 (solid)	11.00
327 (liquid)	10.67
specific heat, J/(kg·K)[a]	130
latent heat of fusion, J/g[a]	25
latent heat of vaporization, J/g[a]	860
vapor pressure, kPa[b]	
at 980°C	0.133
1160°C	1.33
1420°C	13.33
1500°C	26.7
1600°C	53.3
thermal conductivity,[c] W/(m·K)	
at 28°C	34.7
100°C	33.0
327°C (solid)	30.5
327°C (liquid)	24.6
coefficient of linear expansion at 20°C, $°C^{-1}$	29.1×10^{-6}
electrical resistivity, $\mu\Omega$/cm	
at 20°C	20.65
100°C	27.02
330°C	96.73
specific conductance,[d] $(\Omega \cdot cm)^{-1}$	
at 0°C	5.05×10^4
18°C	4.83×10^4
327.4°C (liquid)	1.06
normal electrode potential, vs standard hydrogen electrode = 0, V	0.22
electrochemical equivalent of Pb^{2+}, g/(A·h)	3.8651
velocity of sound in lead, cm/s	122,700
surface tension at 360°C, mN/m(=dyn/cm)	442
viscosity, mPa·s(=cP)	
at 440°C	2.12
550°C	1.70
845°C	1.19
magnetic susceptibility at 20°C, m^3/kg[e]	-0.29×10^{-6}
hardness	
Mohs'	1.5
Brinell	
common lead	3.2–4.5
chemical lead	4.5–6
Young's modulus, GPa[f]	16.5
tensile strength, common lead, kPa[f]	
at –100°C	42,000
20°C	14,000
150°C	5,000
elongation in 5-cm gauge length, %	50–60

[a]To convert J to cal, divide by 4.184. [b]To convert kPa to mm Hg, multiply by 7.5. [c]Thermal conductivity relative to Ag = 100 is 8.2. [d]Electrical conductivity relative to Cu = 100 is 7.8. [e]To convert m^3/kg to emu/g, multiply by 79.3. [f]To convert GPa to psi, multiply by 145,000.

dyes (see DYES AND DYE INTERMEDIATES), rubber substitutes, and pyrotechnics (qv).

Processing

Lead is usually processed from ore to refined metal in four stages. These are ore dressing, smelting, drossing, and refining (see MINERAL RECOVERY AND PROCESSING).

Ore Dressing. The principal lead mineral, galena, in most crude ores, is separated from the valueless components, or gangue. Other valuable minerals that are present in the ore may be recovered either together with the lead, or in a separate step (2,3). Occasionally, the ores are sufficiently rich in lead and low in impurities to be smelted directly.

The principal steps in ore dressing are crushing, grinding, and concentration (beneficiation) (Fig. 1). Crushing and grinding, collectively called size reduction (qv) or comminution, of the mined ore are necessary to liberate the galena and other desired minerals from interlocking gangue, and also to bring the ore to the size appropriate for the concentration step. Size reduction is carried out in stages. Primary (gyratory or jaw) crushers receive the mined ore, and the product is fed to the secondary (cone or roll) crushers. Sometimes tertiary crushers are used. Vibratory screens separate the finer material in the crusher products and allow return of the oversized pieces for further crushing. Finer size reduction is accomplished by wet grinding in horizontal tumbling mills containing steel grinding media: rods for the first stage grind, and balls for the second. Autogenous grinding, wherein the large lumps of the ore itself are used as the grinding media, is employed occasionally. The finer (≤ 0.2 mm) fraction of the grinding mill discharge is separated using either mechanical (rake or spiral) classifiers or the more widely used hydrocyclone. The coarser fraction is fed back for further grinding.

Gravity concentration, ie, the separation of ore from gangue based on the differences in specific gravities, using jigs, heavy–medium separators, or spiral concentrators for example, is applicable for lead ores. However, the predominant beneficiation technique used in modern plants is the bubble or froth flotation (qv) process (4,5).

In froth flotation, the fine ore slurry is discharged to a conditioning tank in which it is mixed with frothers. It is then passed to the flotation cells where air is pumped up through the slurry, forming bubbles to which the selected mineral, eg, galena, adheres. The mineralized bubbles rise and accumulate as an ore-laden froth bed. A concentrate is obtained by skimming the froth from the cell. The adherence to the air bubbles, and therefore the recovery of selected minerals, depends on the action of the following types of reagents.

Collectors are the most important class of reagents in froth flotation. These selectively bind to the surface of a selected ore mineral particles to render the particles hydrophobic (water repellant). Collectors are heteropolar organic molecules having a reactive polar group and a nonpolar, hydrophobic hydrocarbon tail. The polar group reacts and adsorbs on the surface of the target mineral, forming a one molecule thick, water repellant film. The water repellency is conferred by the tail of the collector, and causes the ore particles to preferentially

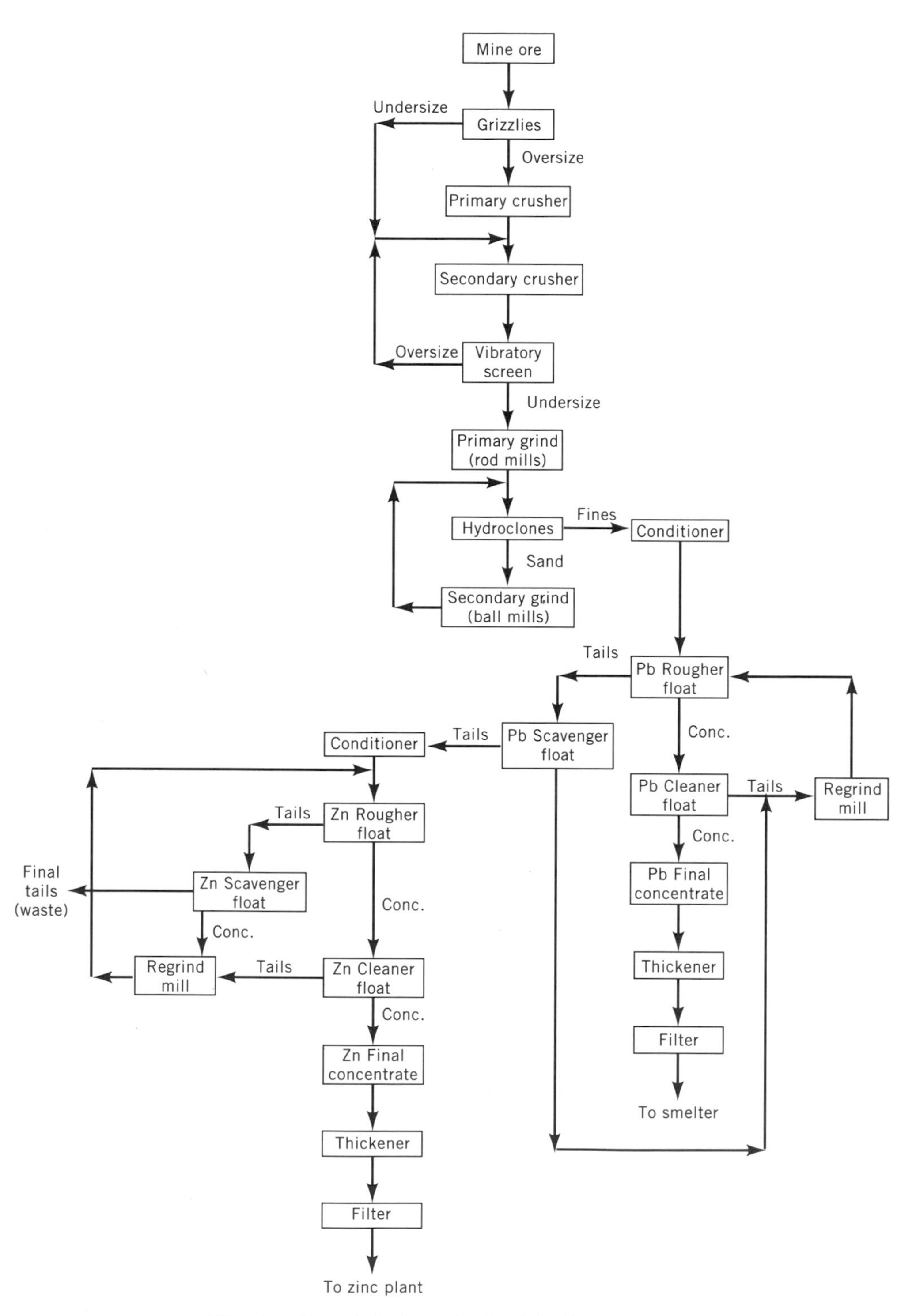

Fig. 1. Beneficiation of mixed lead–zinc ores.

adhere to air bubbles. Xanthates are the most common collector used in galena flotation.

Depressants are reagents that selectively prevent the reaction between a collector and a mineral, thus preventing its flotation. For example, sodium cyanide [*143-33-9*] depresses sphalerite [*12169-28-7*] (zinc sulfide) and pyrite [*1309-36-0*] (iron sulfide) but not galena. It thus enhances selective flotation of the galena.

Activators promote the reaction of the collector with some minerals. For example, ordinarily xanthates do not bind to sphalerite, but pretreatment of the sphalerite using copper sulfate enables it to adsorb the xanthate. Thus it is possible to float the sphalerite from lead–zinc ores after the galena has been recovered.

Frothers such as pine oil [*8006-88-0*], cresylic acid [*1319-77-3*], polyglycols, and long-chain alcohols stabilize the mineralized froth bed long enough for it to be gathered. Conditioners are used for pH control which is important in flotation. If the pH of the slurry is too high, the hydroxyl ion competes with the xanthate. If the pH is too low, xanthates become insoluble. A lead flotation slurry should be slightly alkaline. The pH is adjusted, when necessary, using sodium carbonate or calcium oxide.

The lead concentrate from rougher flotation cells is upgraded by additional flotation steps. The final concentrate is dewatered by settling in thickeners to a moisture content of 50%. Vacuum filtering further decreases the moisture level to 15%.

If sphalerite is present in the ore the tailings from the lead flotation cells are discharged to another conditioner where copper sulfate is added and then fed to the zinc flotation cells. The final tailings are discarded as waste.

The analysis of partially dried lead concentrate, ready to be treated in a series of processes to produce a commercial grade of lead, is presented in Table 2.

Smelting. Since the late 1800s the smelting of lead has been dominated by sinter-blast furnace technology adopted from the iron and steel (qv) industries, and in the early 1960s the Imperial smelting furnace (ISF) was developed to handle mixed lead–zinc concentrates. Several newer technologies, from work in the 1970s, 1980s, and 1990s, have been developed to directly smelt lead ores and bypass the sintering process. Some of these technologies, ie, the Kivcet, Isasmelt, and Queneau, Schumann, Lurgi (QSL), came into commercial operation in the 1980s, but as of this writing sinter-blast furnace and ISF processes are still dominant worldwide.

Table 2. Lead Concentrate Composition

Constituent	Wt %	Constituent	Wt %
Pb	45–75	Sb	0.1–2
Zn	0–15	Fe	1.0–8.0
Au	0–0.01	insolubles	0.5–4
Ag	0–0.15	CaO	trace–3.0
Cu	0–3	S	10–30
As	0.01–0.4	Bi	trace–0.1

Blast-Furnace Smelting. *Sintering.* The charge for sintering is prepared by blending selected concentrates, smelter by-products, returned sinter, flue dust, and when required, additional fuel such as coke breeze. The blend is then pelletized in preparation for sintering.

The blended charge is conditioned in a rotating pug mill where water is added to bring the moisture content between 6 and 8%. The feed is then pelletized (0.5 cm in size) in a disk or drum pelletizer.

A typical up-draft sinter machine (Fig. 2) has an endless belt of malleable iron pallets with grate bottoms upon which the charge is evenly spread. Beneath the pallets, wind boxes produce an up-draft of air through the charge. At the feed end, an ignition box starts the roasting. The combustion products, mostly SO_2 and SO_3, are collected, usually for sulfuric acid production (see SULFURIC ACID AND SULFUR TRIOXIDE).

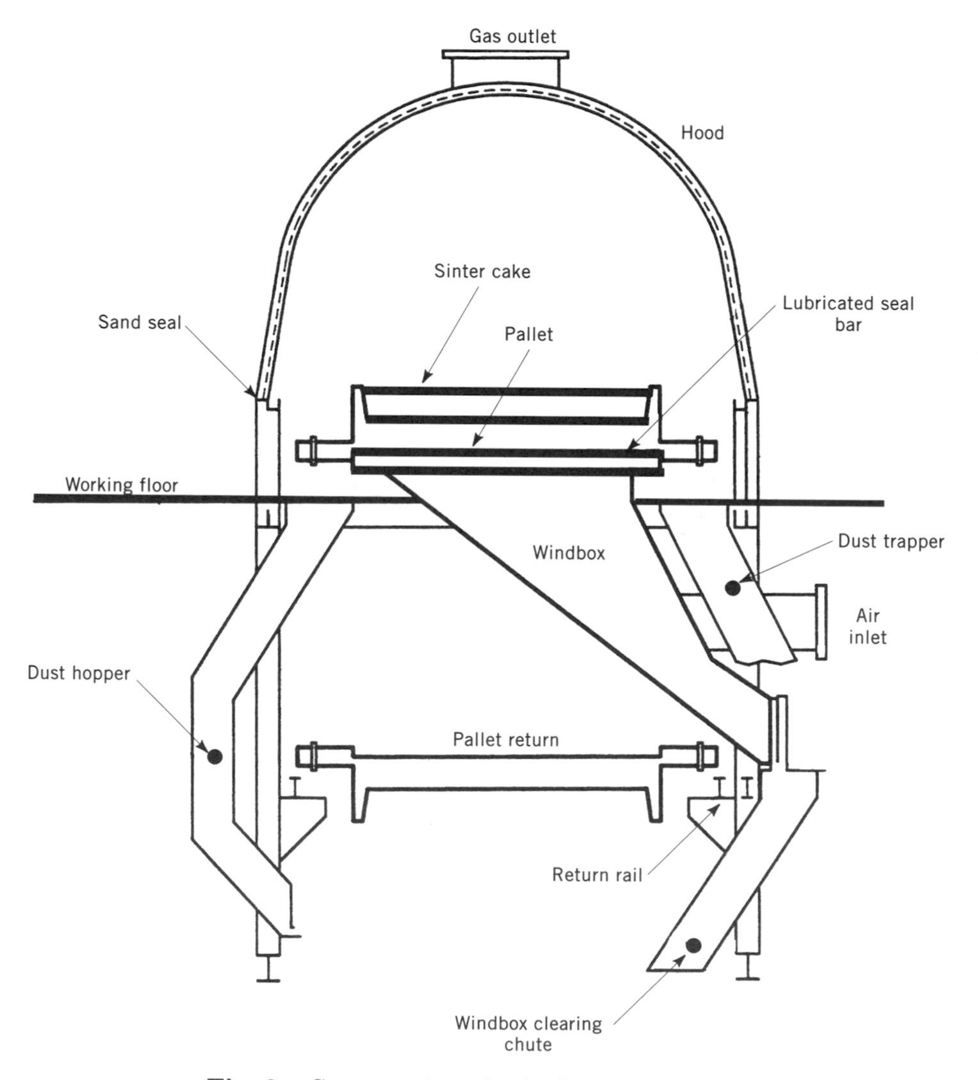

Fig. 2. Cross section of a lead sinter machine.

The sintering operation takes place at approximately 600°C and involves the following reactions:

$$2\ PbS + 3\ O_2 \longrightarrow 2\ PbO + 2\ SO_2$$

$$2\ FeS + 3\ O_2 \longrightarrow 2\ FeO + 2\ SO_2$$

$$2\ SO_2 + O_2 \longrightarrow 2\ SO_3$$

$$PbO + SO_3 \longrightarrow PbSO_4$$

$$PbS + 2\ PbO \longrightarrow 3\ Pb + SO_2$$

$$PbS + PbSO_4 \longrightarrow 2\ Pb + 2\ SO_2$$

The oxides combine with silica [*7631-86-9*] and form low melting complex silicates that tend to bind the concentrate particles together; for example

$$PbO + SiO_2 \longrightarrow PbO{\cdot}SiO_2$$

$$FeO + SiO_2 \longrightarrow FeO{\cdot}SiO_2$$

Some lead oxide is reduced by carbon or carbon monoxide, as shown.

$$2\ PbO + C \longrightarrow 2\ Pb + CO_2$$

$$PbO + CO \longrightarrow Pb + CO_2$$

The details of specific plant operations have been well documented (6).

Smelting. In the blast furnace (7), lead and other metal oxides, not reduced during sintering, are reduced to metals; the molten lead is coalesced in the hearth; and the gangue material is separated into a molten slag. The molten lead serves as a solvent for the valuable metallic impurities that are recovered during the refining process, separated, and ultimately purified for marketing. The contaminating oxides, silicates, and sulfides are removed by reagents that induce formation of stratified liquid phases, ie, slag, matte, and speiss in the crucible.

A typical slag (specific gravity 3.6) contains complex silicates of iron, calcium, zinc, magnesium, and aluminum oxides. The quantity of fluxes (compounds added to lower the melting point) added depends on the feed composition, and is calculated to ensure a fluid slag at the operating temperature.

Lead smelters often treat both ores and zinc-plant residues high in zinc oxide content. The zinc oxide is collected in the blast-furnace slag, and can be subsequently recovered in a slag-fuming furnace. Lead is unavoidably entrapped in the slag, therefore its volume should be kept to a minimum. Slags are classified according to the ratio of calcium oxide to iron oxide, and are identified as quarter, half, or one to one.

Copper is frequently a main impurity in blast-furnace charges, and its limited solubility in molten lead as copper sulfide requires that the excess be removed by chemical reaction with components of the charge. For this reason enough sulfur is left in the sinter to form a copper sulfide matte layer having a specific gravity of 5.2.

If antimony and arsenic are present in the feed, copper and iron react to form the respective antimonides and arsenides known as speiss (specific gravity 6.0). If it is preferred to remove copper in a speiss layer, the sulfur in the sinter must be reduced and the addition of scrap iron may be necessary to encourage speiss formation. Matte and speiss are usually sent to a copper smelter for recovery of the metals.

The contents of the crucible are tapped through a Roy tapper (8,9) into external settlers for layer separation. The tapper removes blast furnace products as they are made, giving a more uniform blast-furnace performance. A typical bullion analyzes in wt %, 1.0–2.5 Cu, 0.6–0.8 Fe, 0.7–1.1 As, 1.0–3.0 Sb, 0.01–0.03 Bi, 0.2 Ag, and 0.0003 Au. A typical blast furnace slag contains 25–33 wt % FeO, 10–17 wt % CaO, 20–22 wt % SiO_2, 1–2 wt % Pb, and 13–17 wt % Zn. A section across the width of the blast furnace is shown in Figure 3.

Imperial Smelting Process. The Imperial smelting process has been well documented (10,11). This smelting process is used particularly for mixed

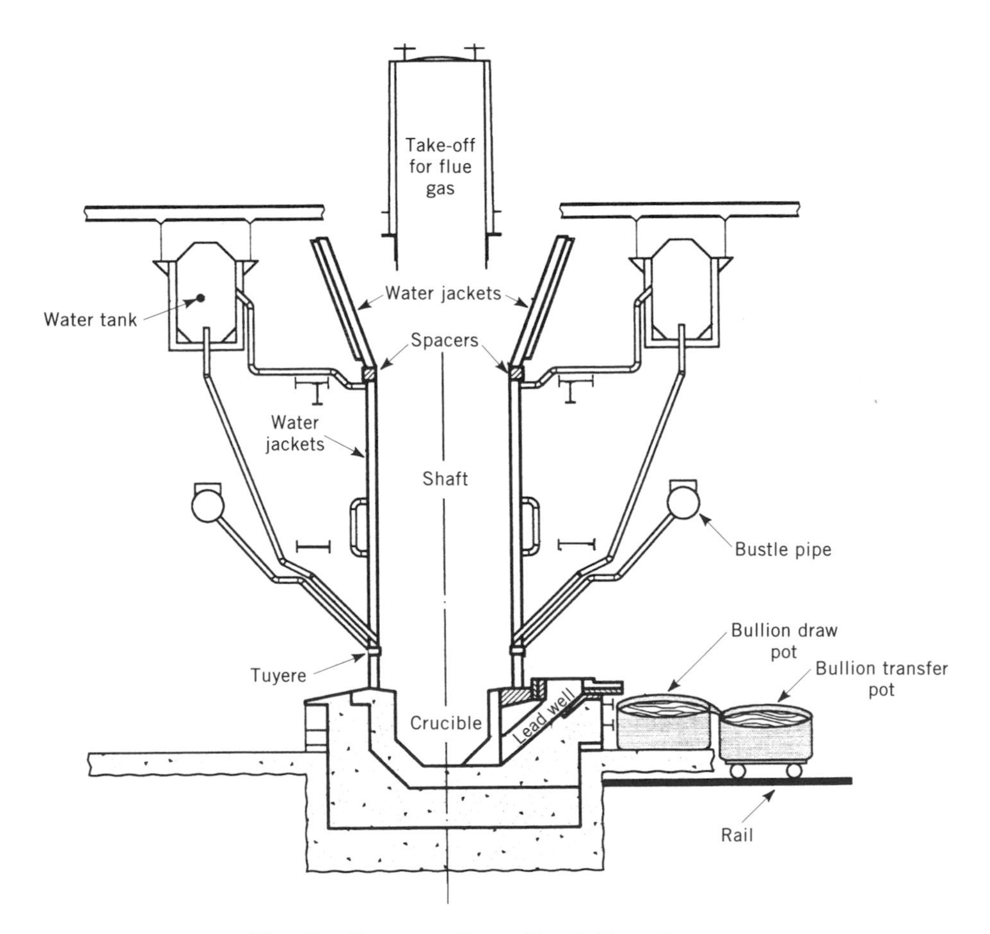

Fig. 3. Cross section of lead blast furnace.

zinc–lead ores, when the mineralization is such that separation of the lead and zinc portions is difficult. High grade galena concentrates and oxidized lead ores also can be added to the charge. Feed material is often a sinter of mixed zinc–lead concentrates, zinc blende [*12169-28-7*], and galena concentrates. Recoveries from concentrate to slab zinc and salable lead are 90–93% and 92–94%, respectively.

As of 1992, eight smelters used the Imperial smelting process (Fig. 4), but none within the United States. The essential reactions taking place for smelting are

$$2\ C + O_2 \longrightarrow 2\ CO$$

$$2\ CO + O_2 \longrightarrow 2\ CO_2$$

$$CO_2 + C \longrightarrow 2\ CO$$

$$ZnO + CO \longrightarrow Zn\ (g) + CO_2$$

$$PbO + CO \longrightarrow Pb\ (l) + CO_2$$

The furnace charge consists of zinc–lead sinter, metallurgical coke, and recirculating metallic drosses and flux. The charge cycle is fully automatic.

Blast air, preheated to 650°C, is delivered by centrifugal blowers through a refractory-lined bustle main to the furnace. Zinc vapor from the reduced sinter is carried out with the furnace gases to a condenser fitted with mechanical rotors that are partly immersed in a shallow pool of molten lead. The lead flows countercurrently to the gas and is vigorously agitated by the rotors to create an intense shower of lead droplets throughout the condenser.

The furnace gases are shock-cooled by the lead spray. The zinc is cooled rapidly through the critical temperature range in which zinc vapor reverts to zinc oxide. The liquid zinc is taken up by the lead. The remaining gases pass to a wet scrubber system for the removal of suspended oxides of lead and zinc called blue powder. The lead-containing zinc flows to an external separation system. The slag and molten lead are tapped intermittently into a forehearth where the lead bullion is separated from the slag by a syphon. The slag flows to a cast-iron launder and is granulated.

The lead and zinc are separated into two molten phases by progressive cooling. Following ammonium chloride treatment to remove dross, the lead is returned to the condenser. Zinc is cast into ingots after dissolved lead is removed by cooling.

Other Lead Smelting Processes. Stricter regulations concerning lead emissions and ambient lead in air levels (see AIR POLLUTION), and the necessity to reduce capital and operating costs have encouraged the development of alternative lead smelting processes to replace the sinter plant–blast furnace combination.

Four lead smelting processes have reached the stage of being promoted as commercially viable for lead concentrates (Table 3). In general these processes offer the following potential advantages: ability to meet proposed future in-plant hygiene requirements; reduction in energy costs through the utilization of the heat of combustion of the concentrates; reduction in capital and operating costs through the use of compact, high intensity vessels; and the production of low volumes of process gas of high SO_2 content through the use of large amounts of (tonnage) oxygen.

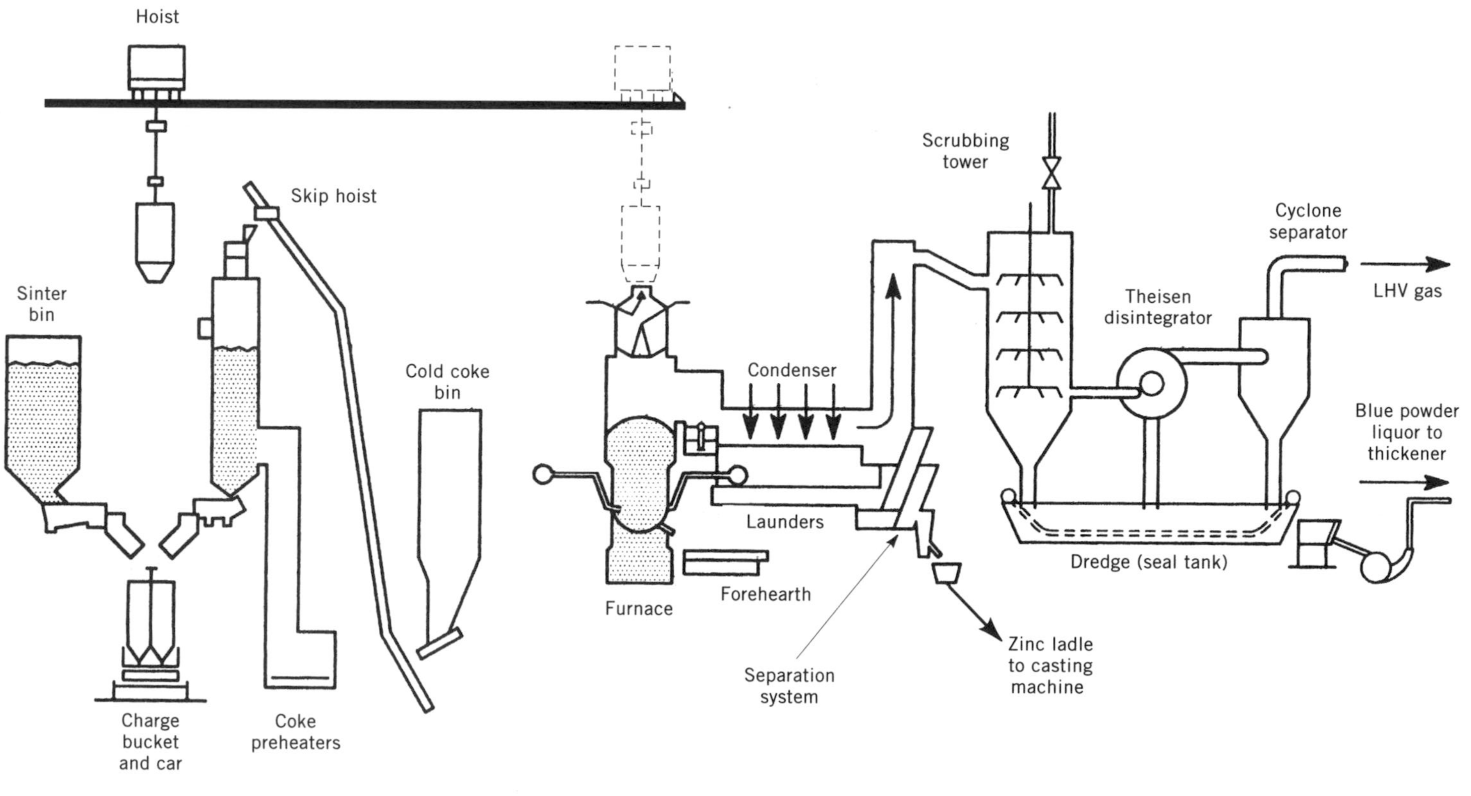

Fig. 4. Schematic illustration of an Imperial smelting furnace plant. LHV = low heating value.

Table 3. Alternative Lead Smelting Processes

Process name	Smelting type	Special features
	Continuous process	
Kivcet	reaction between gases and solids, ie, flash	integral electric furnace for slag reduction; water-cooled wall separates the smelting and reduction atmospheres
QSL	reactions between liquids, ie, bath	shielded tuyeres; vessel can be rotated to protect tuyeres; internal wall separates smelting and reduction zones
Isasmelt	bath	air-cooled lance from top; oxidation and reduction vessels connected by a launder
	Batch process	
Boliden Kaldo	flash and some bath reactions	top-blown vessel using a water-cooled lance; vessel rotates during smelting and reduction, and can be tilted for tapping

Kivcet Process. The Kivcet process (12) is based on the roasting–smelting of the lead-bearing charge in an oxygen atmosphere, and in the formation of sulfur dioxide gas and an oxide smelt. The oxide smelt is reduced by means of an agent such as coal (qv) or coke, to produce lead. Zinc can also be recovered as zinc oxide by fuming the slag.

These operations are all conducted in the single unit of the Kivcet furnace, which consists of a smelting shaft, gas removal shaft, and electrothermic part. A schematic of the Kivcet process is shown in Figure 5 (13). The electrothermal part is separated from the smelt shaft in the gas space by a partition wall which is partially immersed into the smelt. Raw materials are received from the stockpile, proportioned, mixed, and dried in the rotary drier before being fed into the furnace. The furnace charge is dispersed by oxygen.

Ignition of the charge takes place in this oxygen atmosphere. Melt material is reduced during its passage through the layer of reducing agent floating on the metal bath. The reducing agent is fed into the furnace with the lead-bearing charge. Metallic lead resulting from reduction settles at the bottom of the furnace.

Gases generated in the charge roasting are rich in sulfur dioxide, and contain metal oxide vapors and high metal dust concentrations. The gases are cooled in a waste heat boiler installed above the gas removal shaft. During the cooling, the sulfur dioxide reacts with the metal oxides to form the corresponding metal sulfates.

Cooled dust-laden gas is dedusted in an electrostatic precipitator and sent to the cleaning unit to remove impurities such as arsenic, fluorine, and chlorine before being sent on to the sulfuric acid production plant.

In the electrothermic part of the furnace, electrical energy introduced via three carbon electrodes, keeps the bath molten and completes the lead oxide

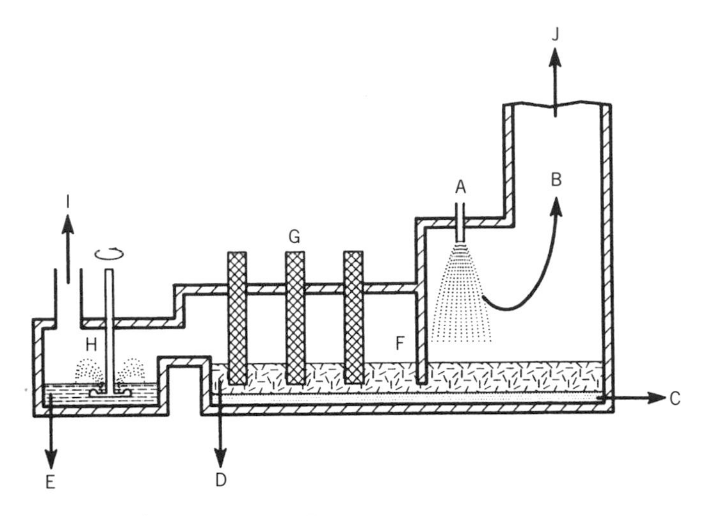

Fig. 5. Kivcet process: A, concentrate burner; B, fume-laden gas; C, lead tap; D, slag tap; E, zinc tap; F, water-cooled underflow wall; G, electrodes; H, lead-splash condenser; I, CO off-gas; and J, to electrostatic precipitator.

reduction. Fumes generated in the electrothermic section are oxidized in a post-combustion chamber by adding ambient air, before the vapor is cooled, dedusted, and released to the atmosphere.

The cooling of gases is performed in a vertical tube-type exchanger, and the hot air produced is used as a heating medium in the charge drying unit. Fumes having lead and zinc oxides are trapped in the heat exchanger and dedusting equipment. These dusts are normally treated in an electrolytic zinc plant.

Crude lead flows continuously from the Kivcet furnace through a syphon and is sent to two kettles where some copper, in dross form, is separated and subsequently processed; the lead bullion is cast into one ton ingots. Slag is tapped intermittently from the furnace through a sidewall tapping hole and sent to a slag granulation unit before being discharged for disposal.

For environmental reasons, the entire process is handled by enclosed equipment. Lead recoveries of 96% can be obtained from the raw materials, and sulfur dioxide gas released in the process is used to produce sulfuric acid. Four plants are in operation as of 1994. Three are in Russia and one is in Italy.

QSL Process. The QSL process (14) is a continuous single-step process having great flexibility in regard to the composition of the raw materials. In this process the highly exothermic complete oxidation, ie, the roasting reaction, can be avoided to some extent in favor of a weakly exothermic partial oxidation directly producing metallic lead. However, the yield of lead as metal is incomplete due to partial oxidation of lead to lead oxide.

The remaining lead must be oxidized and later can be reduced from the slag using carbon. The ratio of metallic lead to lead oxide which depends in part on the type of raw materials to be processed, can be adjusted within certain limits by varying the degree of oxidation. In treating lead-rich concentrates having a lead content of approximately 70%, more than 75% of the lead can be obtained directly as metallic lead.

The entire QSL process takes place in a single reactor as shown in Figure 6 (15). The reactor consists of an almost horizontal, refractory-lined cylinder, which can be tilted by 90° when operation is interrupted. Concentrates, fluxes, recirculated flue dust, and normally a small amount of coal, depending on the type of concentrate, are pelletized. The pelletizer ensures that the raw materials are mixed to the required degree of uniformity.

The moist green pellets, easier to handle than the nonagglomerated dusty material, are charged at the reactor top into the oxidation zone without further treatment. The pellets fall into a melt mainly consisting of primary slag. Tonnage oxygen is blown into the melt through submerged gas-cooled bottom nozzles called Savard-Lee injectors. The roasting reactions take place at 1000–1100°C, producing metallic lead, primary slag with a high (30%) content of lead oxide, and an off-gas rich in sulfur dioxide.

The lead bullion, ie, crude metallic lead, is discharged via a syphon, a gas-tight seal, whereas the primary slag passes into the reduction zone separated from the oxidation zone by a weir. Pulverized coal (qv) is injected into the reduction zone from the bottom by carrier air, and some is burnt by oxygen using the same injectors. The lead contained in the primary slag is gradually recovered by reduction. The low lead final slag is tapped and discarded or further

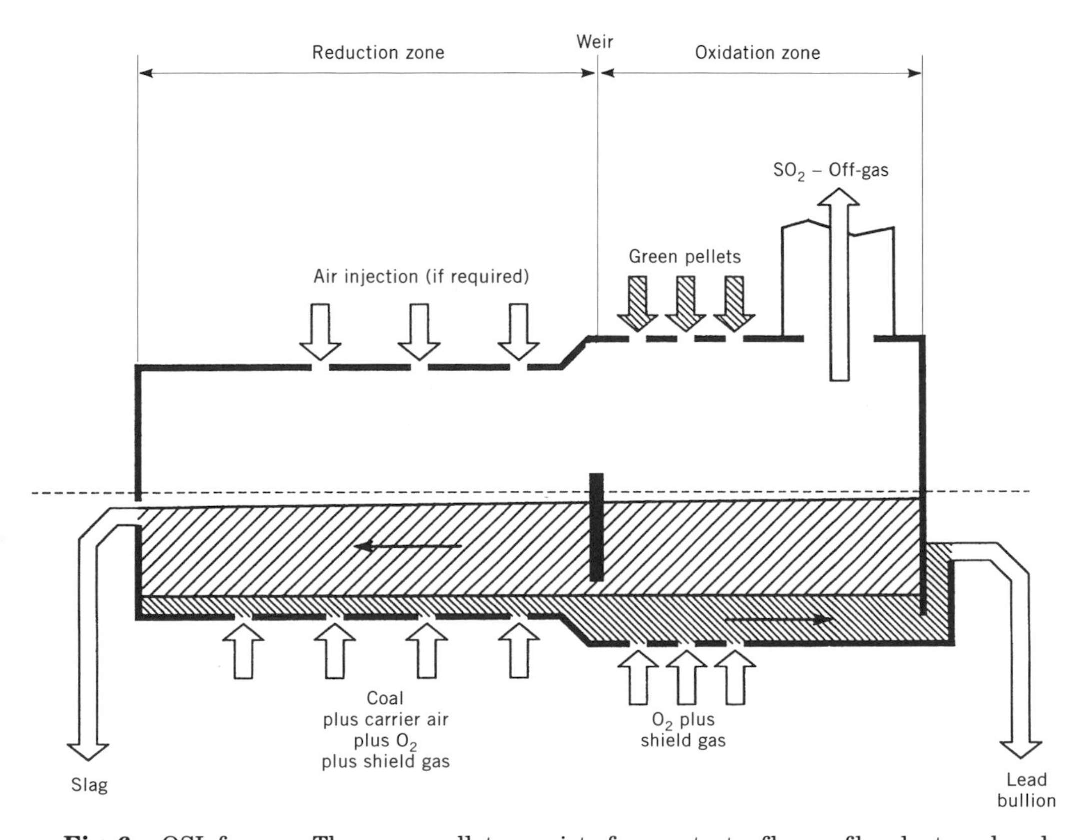

Fig. 6. QSL furnace. The green pellets consist of concentrate, fluxes, flue dust, and coal.

processed to recover zinc. The lead settles to the bottom of the reduction zone and flows back into the oxidation zone to recombine with the primary lead bullion.

The QSL process yields only a small off-gas volume having a high concentration of sulfur dioxide owing to the use of tonnage oxygen. The combined off-gas from reduction and oxidation leaves the reactor through a vertical uptake at 1100–1150°C. The uptake, part of the waste heat recovery system, cools the off-gas to approximately 750–800°C. The process gas passes through a waste heat boiler for further cooling and a hot gas electrostatic precipitator for dedusting before being transferred to the sulfuric acid plant.

The combined flue dust from waste heat boiler and electrostatic precipitator, including dust from the ventilation system, is collected in a bin and recirculated to the mixing and pelletizing step, where it is used as a binding reagent.

Boliden's Kaldo Process. The top blown rotary converter (TBRC) (16) is a highly flexible metallurgical unit which has been incorporated into processes producing steel, copper, and nickel. In the early 1970s Boliden Metall AB began work to adapt Kaldo technology to the treatment of lead-bearing dusts and later the smelting of lead concentrates (Fig. 7).

Dried lead concentrate, flux, and return dust are added to the converter through a lance during the smelting cycle. Oxygen enriched air is injected at the same time to carry out the smelting reactions and maintain the temperature. Once smelting is completed, the air is shut off and the reduction carried out using an oxygen–fuel mixture fed through the lance system. This is a batch process where both smelting and reduction are carried out in the same vessel. After reduction, slag having less than 3% Pb is skimmed for disposal. The slag

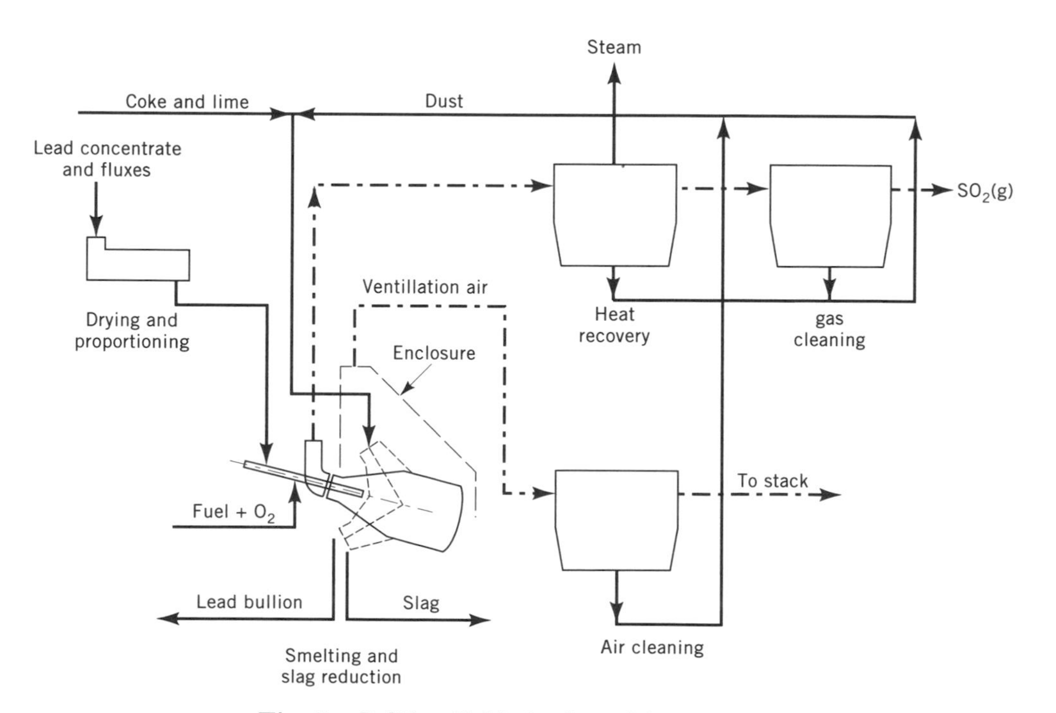

Fig. 7. Boliden Kaldo lead smelting process.

contains most of the zinc in the charge and could be transferred to a slag fuming plant for recovery of zinc and residual lead. Lead bullion is then cast for refining by conventional methods.

Owing to the cyclic nature of the TBRC operation, waste heat recovery from the off-gases is not practical and the SO_2 content of the gas varies with the converter cycle. In order to supply a relatively uniform flow and strength SO_2 gas to a sulfuric acid plant, a system has been installed at Ronnskär whereby the SO_2 from fluctuating smelter gases is partially absorbed in water. During smelter gas interruption, SO_2 is stripped with air and the concentrated gas delivered to the acid plant.

The Kaldo process offers some significant advantages from an emission viewpoint. The reactor can be completely enclosed in a vented enclosure throughout the operating cycle. Operator exposure is then reduced to a minimum because the operation, including charging, slag skimming, and bullion tapping, is controlled from a central control room.

The process is flexible and permits treatment of a wide variety of plant feed materials. Overall lead recovery is in the range of 96–98%. The operation is, however, cyclic which increases the cost of the sulfur fixation plant, and any zinc contained in the concentrate is lost in the slag unless slag fuming is added or already available at the site.

Isasmelt Process. The Isasmelt process (17) is a two-stage lead smelting process developed jointly by Mount Isa Mines Limited (Mt. Isa) and Commonwealth Scientific & Industrial Research Organization (Australia). In the first stage, lead sulfide concentrates are smelted to form a lead oxide-rich slag, which is then reduced by coal in the second stage to form lead bullion and a discard slag.

The Isasmelt process uses a simple stationary, cylindrical, refractory-lined reaction vessel. The typical blended feedstock to the vessel is of a moist lumpy consistency and this feature greatly simplifies environmental control equipment requirements.

The use of the Sirosmelt lance enables the process to be operated on ambient air or using oxygen enrichment. Coal, coke breeze, and fuel oil have all been demonstrated as viable supplementary sources of energy for smelting or reduction. The concept of the Isasmelt process is shown in Figure 8. A schematic diagram of a typical Isasmelt lead smelting process is shown in Figure 9 (18).

Lead concentrate slurry is gravity fed from a stock tank and dewatered on a drum vacuum filter. The resulting filter cake is mixed with silica and limestone fluxes, coke breeze, and recycle fume in a twin-shaft paddle mixer. This agglomerate, containing typically 10% moisture, is fed by conveyor to the smelting furnace where it is dropped directly into the molten slag bath. Coke breeze is used as fuel. Smelting air is injected into the bath through a submerged Sirosmelt lance to fully oxidize the feed and burn the coke. The smelting air is enriched to 30% oxygen or higher to reduce the off-gas volumes so that both furnaces can be operated simultaneously. Smelting temperatures are maintained in the range from 1170 to 1200°C to decrease the heat load on the reduction furnace.

The high lead slag from the smelting furnace is tapped continuously and transferred down a heated launder directly into the reduction furnace through a port in the side of the vessel. Lump coal for reduction is fed continuously to

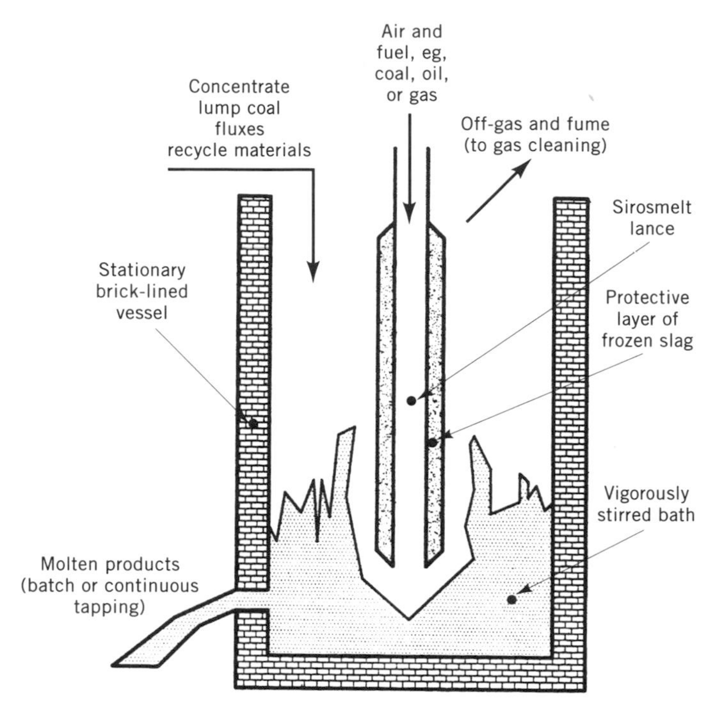

Fig. 8. The Isasmelt concept.

the furnace by conveyor and dropped directly into the bath. Heating for the endothermic reduction reactions is provided by oil injected down the lance. The combustion air stoichiometry is set at 95% of that required for complete oil combustion. Air is injected into the top of the furnace to afterburn the volatile materials from the coal and provide additional heat to the top of the furnace. Reduction temperatures range from 1170 to 1200°C to maintain slag fluidity.

The crude lead and discard slag from the reduction furnace are tapped continuously through a single taphole into molds. The discard slags have a 1 to 2% lead oxide content and 1 to 2% of lead metal prills (pellets).

The off-gas from each furnace is cooled in an evaporative gas cooler and cleaned in a reverse pulse baghouse before being either vented to atmosphere or used in manufacturing sulfuric acid. The baghouse dust from both the smelting and reduction furnaces is combined and recycled through the smelting furnace.

A commercial-scale lead Isasmelt reactor designed to treat 20 t/h concentrate and produce 60,000 t/yr lead has been in operation at Mt. Isa since 1991.

Outokumpu Lead Smelting Process. The Outokumpu flash smelting process (16), used to treat copper and nickel concentrates, has been adapted to flash smelting of lead concentrates. Extensive pilot-plant testing was carried out in the 1960s and 1970s, but was discontinued primarily because of high fume carryover and associated problems. Tests were run in the mid-1980s after redesign of the uptake. Typical pilot-plant design rate was 3–5 t/h.

The dried charge is pneumatically fed to a burner mounted on the roof of the reaction shaft. Normally oxygen is used to get an autogenous smelt. The off-

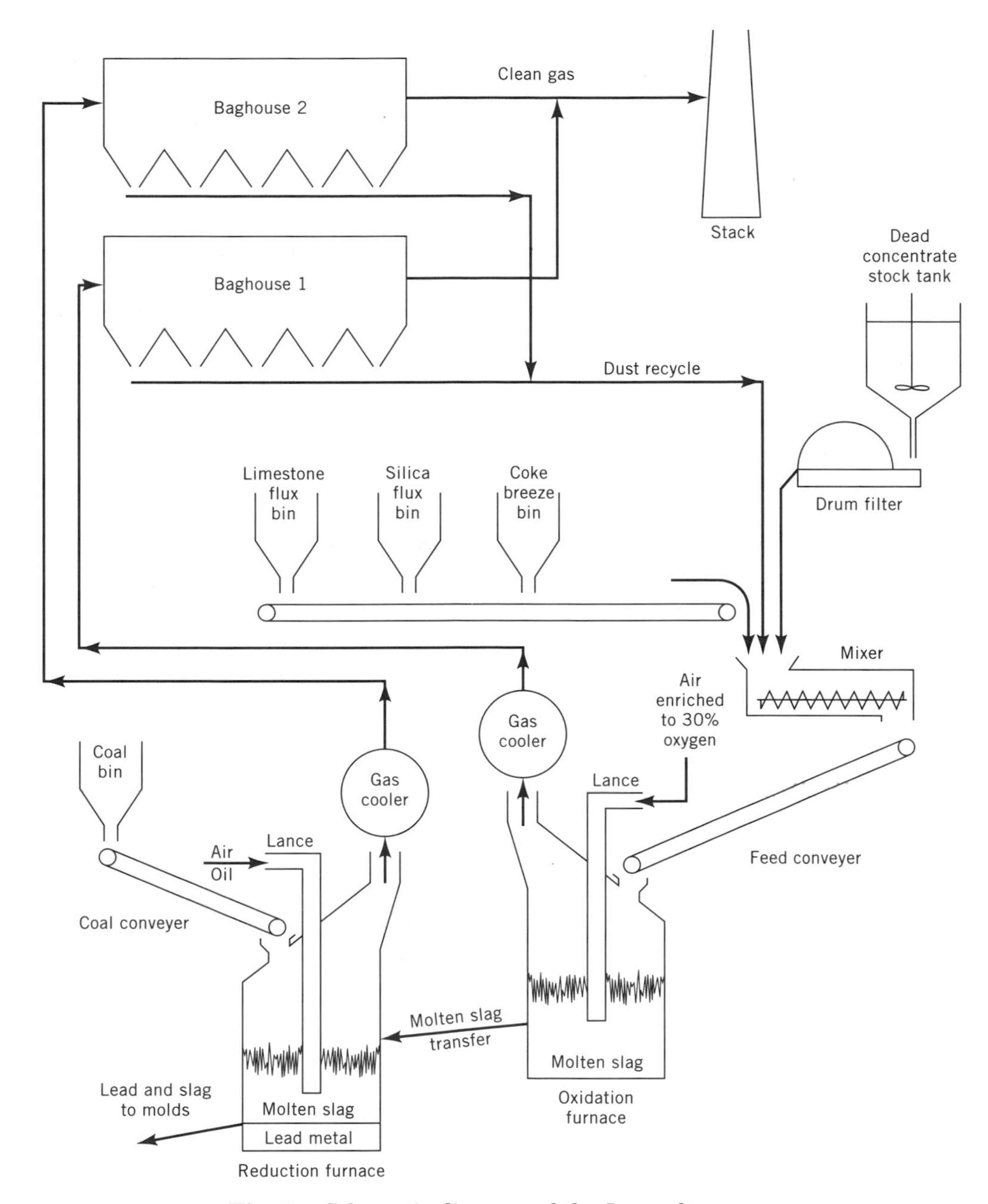

Fig. 9. Schematic diagram of the Isasmelt process.

gases analyze around 70% SO_2 and can be used for the manufacture of sulfuric acid after conventional gas cleaning.

Flash furnace slag containing 20–40% Pb is reduced in an electric furnace using coal injection. Discard slags have 1–3% Pb and CaO:FeO:SiO_2 in the ratio of 0.8:1.0:1.0.

Drossing. The impure lead bullion, produced from any of the smelting processes, is cooled to remove dissolved copper prior to the refining operation. The operation is referred to as copper drossing, and is performed in one or two 250 t cast-iron kettles. The process consists of skimming off the dross, stirring the lead, and reskimming.

The first drossing after the furnace tapping is a rough, high temperature skimming to remove the floating scum on the lead. After this step has removed most of the insoluble dross, the molten lead is transferred to a second kettle in which the temperature is lowered as close to the melting point of lead as practical to further reduce the solubilities of the contaminates, particularly copper. After skimming at this lower temperature, a vortex is created in the lead bath, and sulfur, at a concentration of approximately 10 kg/t lead, is added to the vortex. The sulfur further decreases the solubility of copper. The stirrer is then stopped, and the lead skimmed. The dross should contain from 0.004 to 0.04% copper.

The lead bullion, ready to be shipped to the refinery, contains in solution impurities such as silver, gold, copper, antimony, arsenic, bismuth, nickel, zinc, cadmium, tin, tellurium, and platinum metals.

The dross from this operation contains considerable quantities of copper and lead as well as other valuable metals. Separation and recovery is economically imperative. The dross is treated to produce readily separated stratified layers of slag, speiss, matte, and lead. Two processes are primarily used.

Continuous Drossing. A continuous copper-drossing process was developed and patented by The Broken Hill Associated Smelters Pty. Ltd. (Port Pirie, Australia) (19). A top view of the furnace is shown in Figure 10. The hearth is divided into four compartments by submerged walls of varying heights that do not, however, reach up into the matte layer. These control the movement of bullion and direct it to an external cooling system and discharge pot.

When a ladle of blast-furnace bullion is poured into the furnace, the bullion moves over a submerged wall toward the flue end, and is brought into contact with cooled bullion circulating through the cooling launder; the reduction in temperature leads to copper being rejected as mixed sulfide crystals which float to the top of the bath and are then melted to matte by the burners. As each ladle of blast furnace bullion is charged, an equivalent amount of drossed bullion flows through the discharge channel to the delivery pump pot and is pumped to a 100-t kettle for sulfur drossing.

The success of the process results from the fact that nowhere inside the furnace is heat extracted from the copper-saturated blast furnace bullion through a solid surface. The problem of accretion formation (metal build-up), which has plagued many other attempts to establish a copper drossing operation of this type, does not arise. In the cooling launder, lead-rich matte and slag accumulate on the water-cooled plates, but these are designed so that when they are lifted from the bullion stream, the dross cracks off and is swept into the furnace via the cooled lead pot.

Soda Process. Use of a soda smelting process for treating copper drosses in the reverberatory furnace increases the copper to lead ratios in the matte and speiss, and allows lower operating temperatures. A flow sheet describing this process is shown in Figure 11.

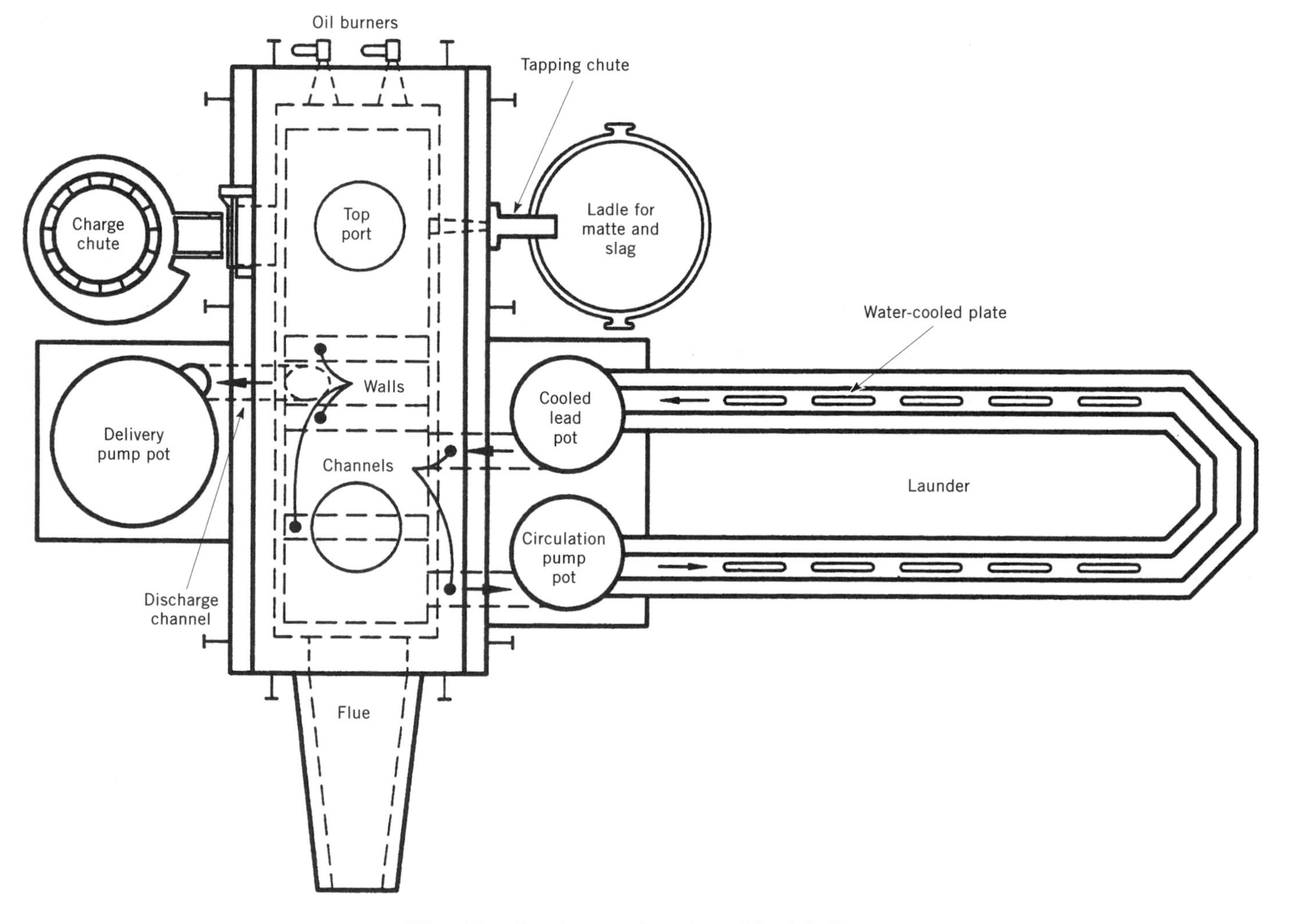

Fig. 10. Continuous drossing of lead bullion.

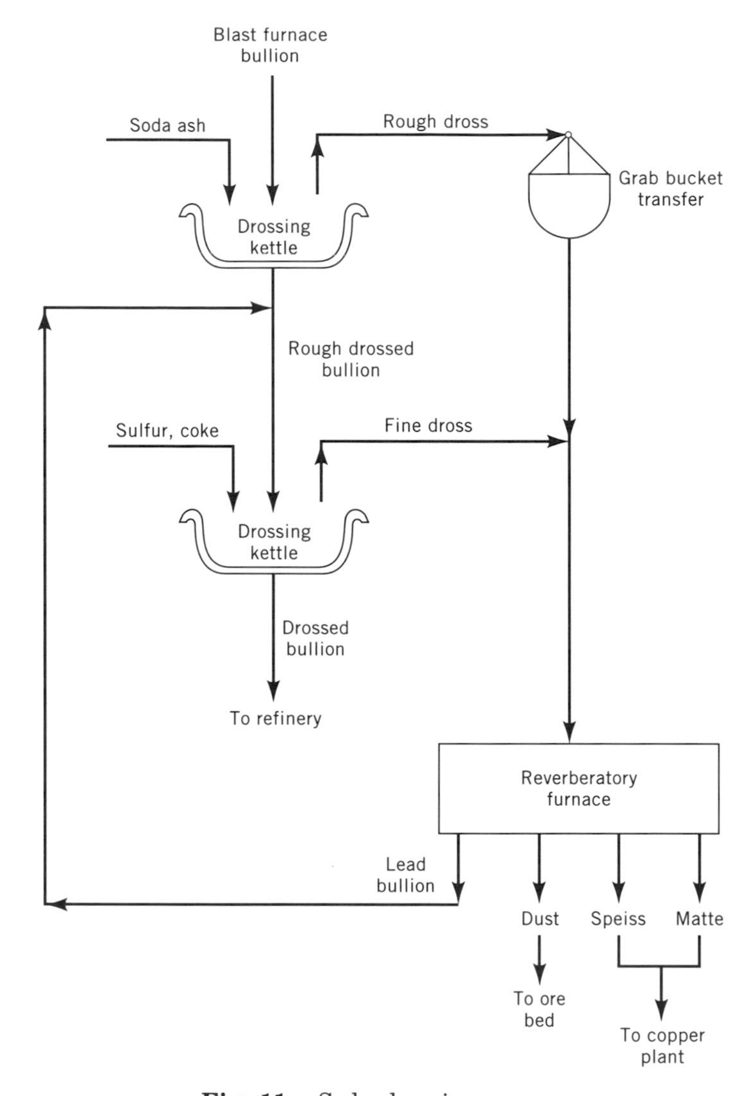

Fig. 11. Soda drossing process.

Soda ash, Na_2CO_3; baghouse fume (mostly PbO); coke; and sulfur react to form a low melting matte of sodium sulfide that serves as a collector of iron sulfide. Because no slag formation is required to take care of the iron, the furnace can operate at lower temperatures, just high enough to keep the speiss and matte molten. The reaction of iron and sulfur tends to keep the furnace in good condition and relatively free of magnetite accretions. A comparison of products after drossing using or not using the soda process is shown in Table 4.

Refining. *Pyrometallurgical Methods.* To prepare blast furnace bullion for commercial sale, certain standards must be met either by the purity of the ores and concentrates smelted or by a series of refining procedures (6–8,20,21). These separated impurities have market value and the refining operations serve not only to purify the lead, but also to recover valuable by-products.

Table 4. Composition of Products from Copper Dross Treatment

Constituent	Composition, wt %	
	Using soda process[a]	Without soda process[b]
Au	0.002	0.005
Ag	0.085	0.075
Pb	9.0	17.7
Cu	52.5	39.6
As	6.6	8.4
Sb	1.2	1.9
Ni	0.1	0.1

[a]The Cu:Pb ratio is 5.8:1.
[b]The Cu:Pb ratio is 2.2:1.

The pyrometallurgical processes, ie, furnace-kettle refining, are based on (*1*) the higher oxidation potentials of the impurities such as antimony, arsenic, and tin, in comparison to that of lead; and (*2*) the formation of insoluble intermetallic compounds by reaction of metallic reagents such as zinc with the impurities, gold, silver and copper, and calcium and magnesium with bismuth (Fig. 12).

Removal by Oxidation. The oxidizing process used to remove antimony, arsenic, and tin has been termed softening because lowering these impurities results in a readily detectable softening of the lead.

In the furnace/kettle batch process, a charge of drossed blast furnace bullion is treated in a reverberatory furnace or a kettle (see Fig. 12). Oxygen is supplied in the form of compressed air or as lead oxide blown into the bath through submerged pipes. The formation of lead oxide serves by mass action to assure the removal of the impurities to the desired low concentrations. The softening reactions are

$$2\ Pb + O_2 \longrightarrow 2\ PbO$$

$$4\ As + 3\ O_2 \longrightarrow 2\ As_2O_3$$

$$4\ Sb + 3\ O_2 \longrightarrow 2\ Sb_2O_3$$

$$2\ Sb + 3\ PbO \longrightarrow Sb_2O_3 + 3\ Pb$$

$$2\ PbO + Sb_2O_3 \longrightarrow 2\ Pb + Sb_2O_5$$

$$Sn + O_2 \longrightarrow SnO_2$$

In a typical run on a low hardness charge of 0.3 to 0.5%, after the softening slag from the previous charge has been removed, the bath is agitated by compressed air from pipes extending through the side of the furnace thus softening the bath to 0.03% antimony. This slag has a high oxygen content and is left in the furnace for softening the next batch of lead. The softened lead underneath it is either tapped or pumped to the desilverizing kettle.

The softening of a high hardness charge (3% or greater) follows the same procedure, except that after the first slag is removed, litharge, ie, PbO, is added to hasten the reduction of the hardness to 0.5%. The bath is then blown with air to a concentration of 0.03% antimony.

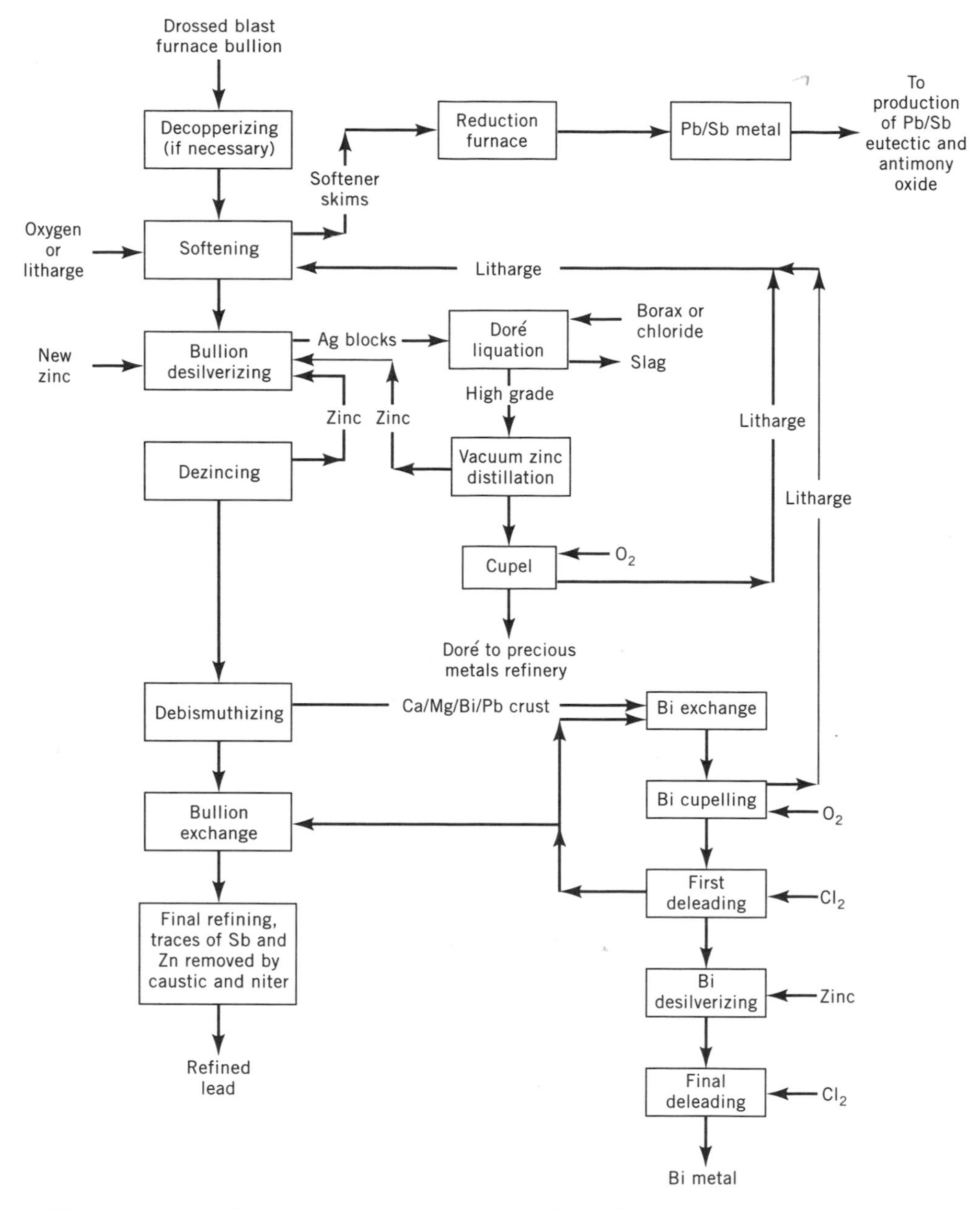

Fig. 12. Flow diagram of a furnace-kettle and/or all-kettle lead refining process.

The continuous softening process used by The Broken Hill Associated Smelters Pty., Ltd. is particularly suitable for lead bullion of fairly uniform impurity content. The copper-drossed blast furnace bullion continuously flows in the feed end of a reverberatory furnace at 420°C, and the softened lead leaves the opposite end at 750°C. Oxidation and agitation is provided by compressed air blown through pipes extending down through the arch of the furnace into the bath.

The continuous removal of the antimony and arsenic into a slag floating on the bath is possible because equilibrium between the lead and the antimony–arsenic-rich slag is rapidly reached at the operating temperature. The rich antimony–arsenic slag overflows through a notch at the flue end of the furnace. A portion of the slag is also discharged at the feed end into an adjacent reverberatory furnace where coke is added to reduce the lead content.

The Harris softening process (20) (Fig. 13) depends on the interaction of lead, niter [*7631-99-4*], $NaNO_3$, sodium hydroxide, and the impurities:

$$5\,Pb + 6\,NaOH + 4\,NaNO_3 \longrightarrow 5\,Na_2PbO_3 + 2\,N_2 + 3\,H_2O$$

$$5\,Na_2PbO_3 + 4\,As + 2\,NaOH \longrightarrow 4\,Na_3AsO_4 + 5\,Pb + H_2O$$

$$2\,Na_2PbO_3 + 2\,Sn \longrightarrow 2\,Na_2SnO_3 + 2\,Pb$$

$$5\,Na_2PbO_3 + 3\,H_2O + 4\,Sb \longrightarrow 4\,NaSbO_3 + 6\,NaOH + 5\,Pb$$

The formation of the metallic salts is a pyrometallurgical process, and is commonly referred to as the dry process. The separation of the salts from each other is accomplished by selective dissolution in water, and is named the wet process.

Dry Process. The softening equipment consists of a 100–225 t kettle in which a reagent cylinder is supported and submerged. The process is carried out in batches. Molten caustic is pumped into the reagent cylinder along with molten lead from the kettle. The latter prevents freezing of the caustic soda. The discharge from the pump is arranged to spray the lead to assure maximum exposure of lead surfaces.

The niter and fresh caustic soda, required to maintain the fluidity of the salt bath in the reactor chamber, are added gradually. When the color of the saturated salts turns from a dark gray to white, the impurity metals are at their highest state of oxidation, and the lead content of the spent salts is very low. In a modification, the arsenic and tin are selectively removed as sodium arsenate and sodium stannate, followed by the removal of antimony as sodium antimonate.

Wet Process. The sodium arsenate and stannate slag are treated by a leach and precipitation process to produce calcium arsenate, calcium stannate, and a sodium hydroxide solution for recycle. The sodium antimonate filtercake containing selenium, tellurium, and indium is treated in a special metals refinery to recover indium and tellurium.

Removal of Other Impurities. After softening, the impurities that may still remain in the lead are silver, gold, copper, tellurium, platinum metals, and bismuth. Whereas concentrations may be tolerable for some lead applications, the market values encourage separation and recovery. The Parkes process is used for removing noble metals and any residual copper, and the Kroll-Betterton process for debismuthizing.

In the Parkes process, a quantity (1–2%) of zinc is added to lead which is in excess of the saturation value. This creates insoluble intermetallic compounds consisting of zinc and the noble metals that precipitate from the lead on cooling (22).

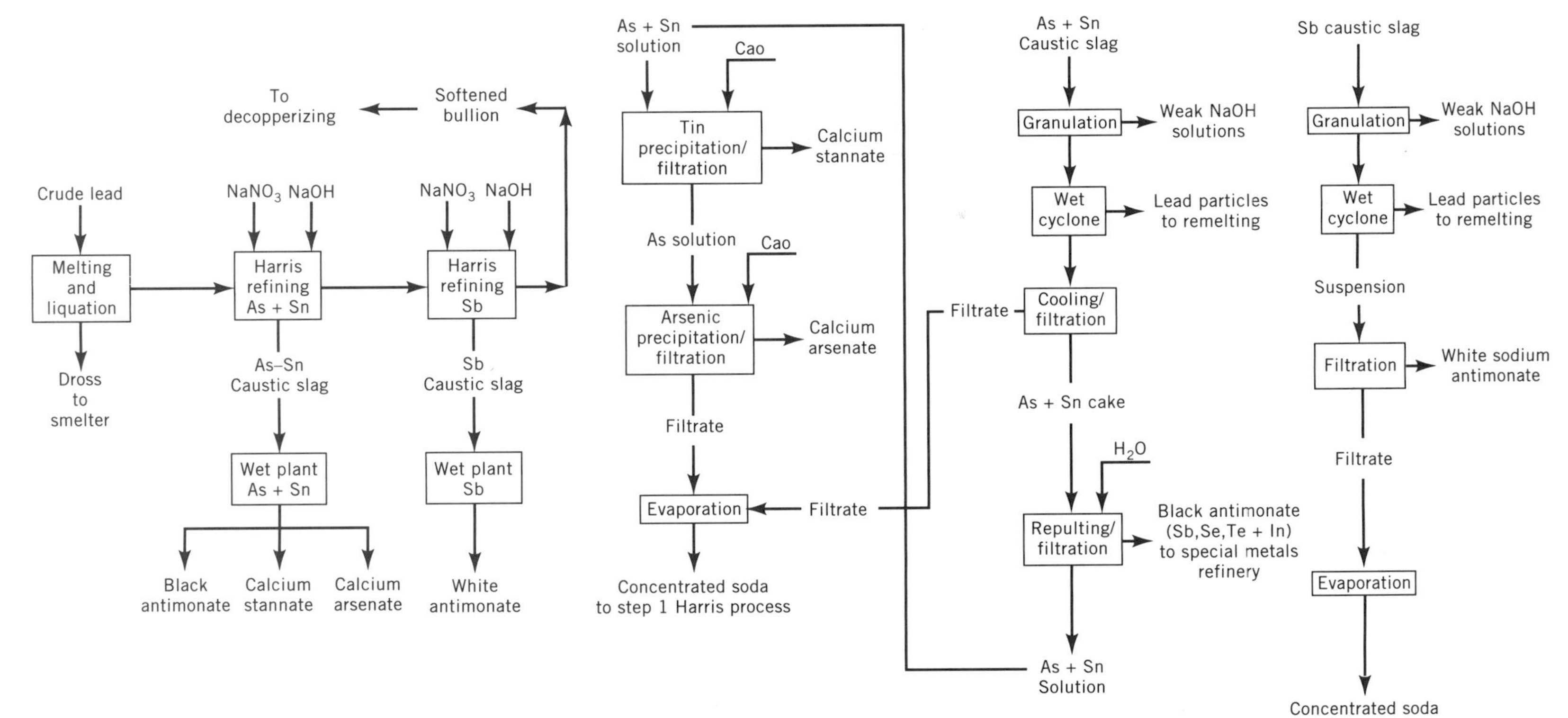

Fig. 13. Harris process for refining lead.

This process is a two-step procedure. In the first desilverizing kettle, a high silver–zinc crust is generated by adding to the kettle the low silver crust from the preceding second-kettle desilverizing. This enriching step reduces the amount of zinc per unit of silver to be subsequently recovered, as well as the amount of entrained lead that eventually must be removed in cupelling.

Batch desilverizing is preformed in open steel kettles. The low silver-content crust from the previous charge's second skimming is placed into the first kettles where new, soft lead is pumped. The temperature is raised to approximately 460°C where alloying takes place. A silver-enriched crust forms and rises to the surface. The silver–zinc crust discharged from treating bullion with 0.3% silver content contains about 12% silver. For the second desilverizing step, zinc recovered from the treatment of Parkes crust is added to the melt and stirred in at about 450°C. The quantity of zinc added is calculated to leave less than 3 ppm silver. The kettle is allowed to cool, and the precipitating silver–zinc crystals are skimmed as crust and held for addition to the next charge. The molten lead, which upon cooling to 320°C is given a final skimming leaving less than 3 ppm silver and copper, is then pumped to the next kettle for dezincing.

The crusts from the Parkes process are then treated to recover the zinc contained in the dross for reuse in desilverizing. This is done by distillation (qv) and more commonly by vacuum technology. A continuous adaptation of the Parkes process is carried out at The Broken Hill Associated Smelters Pty. (Port Pirie, Australia). The chemistry of this operation is the same as that for batch desilverizing (22,23).

Recovery of Metals from Parkes Crust. Parkes crust contains significant amounts of zinc and lead which must be recovered in addition to the silver and gold contained in the crust. Until the 1980s this was generally done by first using Faber du Faur-type retorts to distill zinc from the crust, followed by cupellation of the lead–silver metal to produce a litharge slag and doré metal.

In 1986 Britannia Refined Metals (Northfleet, U.K.) introduced technology for the treatment of Parkes crust, a triple alloy of Ag, Zn, Pb, which by 1992 had been adopted by seven lead refineries (22). The technology consists of a three-stage process in which the silver-rich crust is first liquated to reduce its lead content, then placed in a sealed furnace where the zinc is removed by vacuum distillation and, finally, the silver–lead metal is treated in a bottom blown oxygen cupel (BBOC) to produce a litharge slag and doré metal.

Liquation. The rich crust is transferred to a cast-iron kettle installed in a refractory-lined furnace. The kettle is first filled with lead, which is melted by burners firing at both the top and bottom of the furnace. Firing is then continued only in the top zone to maintain a temperature of 720°C. The bath is covered with an eutectic sodium chloride–calcium chloride flux to minimize oxidation of zinc. Borax-based fluxes may also be used as cover.

Blocks of rich crust are added periodically and allowed to melt. As melting takes place, the lead-rich phase sinks to the bottom and is withdrawn from the kettle by a syphon. The lighter silver–zinc phase rises and floats on the surface of the lead. After sufficient silver–zinc alloy has accumulated, it is tapped from the top section of the kettle. In this manner it is possible to achieve a 120:1 concentration of the silver in the crust which is passed on for retorting. The lead removed from the bottom of the kettle typically contains 0.5% silver and 2% zinc.

Slags are also removed from the liquation kettle periodically. When chloride slags are formed these are cast into chunks and transferred to a wet plant for leaching to remove chlorides and treatment of the metallic residue. If borax slags are employed, metal values in the slag are recovered by further pyrometallurgical treatment.

Typically, for every tonne of Parkes crust charged to a liquation kettle, about 0.25 t of triple alloy and 0.1 t of slag are generated. The triple alloy generally contains 30% silver, 10% lead, and 60% zinc.

Vacuum Retorting. Classical Faber du Faur retorting of the triple alloy, in which small charges of crust are heated in clay graphite crucibles, has inherent disadvantages: low zinc recovery, high labor requirements because of the need to use multiple furnaces, and hot and arduous working conditions. A process whereby the crust was heated under vacuum in a refractory-lined vessel was developed by Pennaroya. The bath of this process is quiescent, however, and the distillation rate of the zinc is low owing to only convectional agitation of the charge. During the 1970s Metallurgie Hoboken Overpelt in conjunction with Otto Junker developed a new process in which the charge is heated under vacuum using electric induction coils. The turbulence created in the bath by the induction currents results in a much higher distillation rate.

In practice, triple alloy is added to a clay graphite crucible in a refractory-lined vacuum-tight chamber (Fig. 14). Power input is controlled by adjusting the applied voltage until the charge is melted. A refractory cover is placed over the crucible and sealed with sand. The furnace cover contains an opening which mates with a port connecting to a condenser.

The process operates at 1 kPa (10 mbars) and 450 kW of power. When the condenser temperature reaches 580°C, the power is reduced to 350 kW. Cooling water is applied to the condenser, throughout distillation, by means of sprays. Normally distillation takes 10–12 hours and the end point is signified by an increase in furnace temperature and a decrease in vapor temperature to 500–520°C. At this point the power is turned off and the vacuum pump is shut down. Nitrogen is then bled into the system to prevent oxidation of zinc.

The taphole of the condenser is then opened and the molten zinc released is cast into molds. The zinc is returned to the refinery for reuse in the desilverizing process. After the vacuum furnace is disconnected from the condenser, it is tilted to pour the residual retort bullion in molds. The zinc content of the bullion is

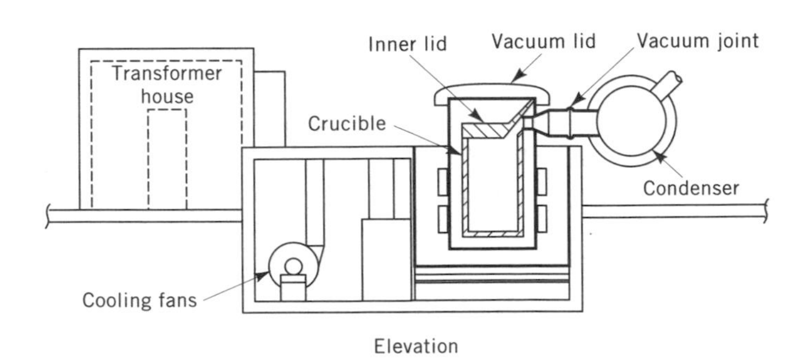

Fig. 14. Hoboken vacuum retort.

<3%. The residual bullion contains about 96% of the silver and 80% of the lead charged to the vacuum furnace. The balance of these elements is distributed between the distilled zinc metal and a retort slag. Vacuum retorting raises the silver content of the bullion from 30 to 75%, with lead essentially making up the balance.

Cupellation. The final stage in the recovery of silver (and gold) is the cupellation of the retort bullion. Until the 1980s this was carried out in small reverberatory furnaces (cupels) fired by oil and gas burners, in which air was blown across the surface of the bath through tuyeres. The furnace was periodically tilted to allow lead oxide (litharge) to flow out. This process suffers from several disadvantages: formation of zinc dross when the bullion is initially melted, high (8%) silver losses to the skimmed litharge, low fuel efficiency, and poor reaction rates between the oxygen in the tuyere air and lead in the bath.

These deficiencies were overcome with the development of the bottom blown oxygen cupel (BBOC), a refractory-lined cylindrical furnace having a hydraulically tilted mechanism (Fig. 15). The furnace is top fired using a natural gas–air burner and is also top charged. The key feature is the consumable lance through which oxygen is blown to convert lead to lead oxide. The lance is also shrouded with nitrogen, and as it is worn down it is automatically advanced to maintain a constant position in the furnace. The design of the furnace allows for tilting to the optimum positions for both blowing and slag removal. Oxygen is virtually totally absorbed and there is little turbulence on the charge surface.

In practice, blocks of vacuum retort bullion are added to the furnace while in the vertical position and melted by the natural gas–air burner. The furnace is

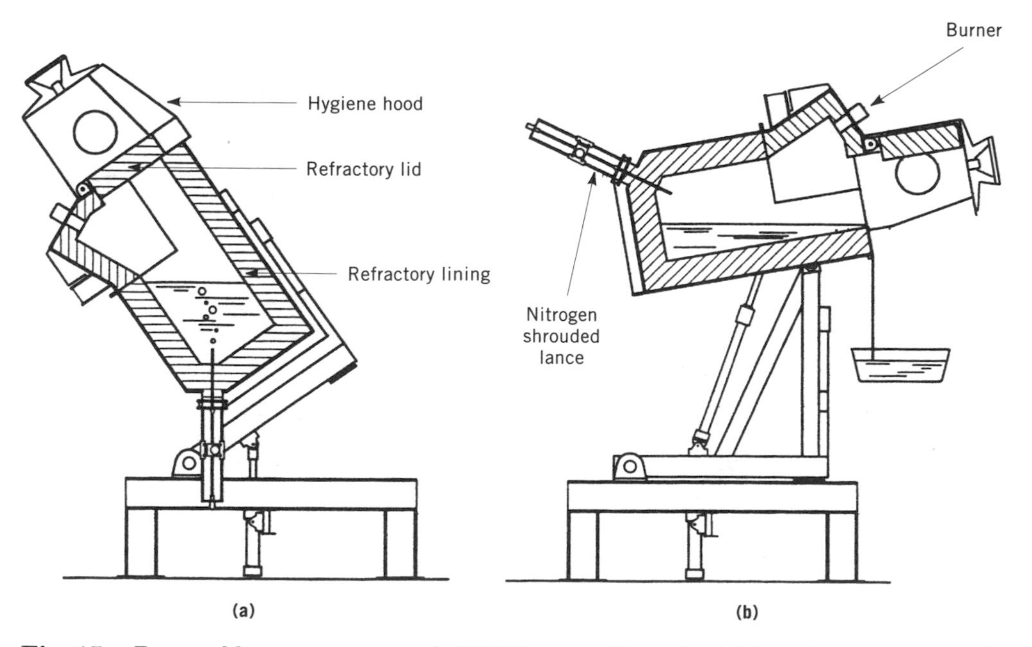

Fig. 15. Bottom blown oxygen cupel (BBOC) in positions for (**a**) blowing oxygen and (**b**) tapping slag.

tilted (Fig.15**a**) and blowing commenced. The burner is turned off, but blowing is continued. The zinc in the charge is oxidized first to form a viscous slag, but as blowing continues the litharge formed dilutes the zinc content and a fluid slag is formed. After one and a half hours the furnace temperature is stabilized at about 900°C and the first slag can be tapped. The furnace is tilted (Fig. 15**b**) and slag is poured into molds.

Oxygen blowing is continued to remove the remainder of the lead and the burner is used on low fire because the heat of reaction is not sufficient as the lead content of the bullion decreases. Litharge is again tapped and the process continued until the silver concentration exceeds 99.0%. At this point residual copper is removed by the addition of ingots of lead. The bath is maintained at a temperature of 960°C while sufficient lead is added and converted to lead oxide, which in turn carries copper oxide with it into the litharge slag. It is possible to reduce copper levels in the silver to <0.04% by this method. When the copper content is reduced to the desired level, the oxygen is turned off and the bath purged with nitrogen from the lance to reduce the oxygen content of the silver. The furnace is then tilted to the pouring position and the silver is cast into molds. Using the BBOC it is possible to recover 97% of the silver in the charge as doré metal. The balance of the silver is in the litharge, and is recovered by recycling the litharge to the softening operation.

Bismuth concentrates in the doré until the last stages of cupellation, when it is oxidized and removed with the litharge. After the last litharge has been removed, it is often necessary to add bars of refined lead to provide more litharge to carry off the last traces of bismuth. The doré metal is then cast into bars for marketing.

The good litharge is used as the oxidizer in batch reverberatory/kettle softening, and the other by-products, such as excess good litharge, coppery litharge, and niter skims and fume, are transferred to the smelting departments for recovery of the metal content.

Dezincing. The lead from the desilverizing contains 0.5–0.6% zinc which must be removed to meet standard specifications. This may be accomplished by vacuum distillation, or oxidation using caustic soda, ie, the Harris process.

The advantage of vacuum dezincing of desilverized bullion is that over 90% of the zinc is recovered as metallic zinc. The equipment consists of a steel kettle of suitable capacity and a water-cooled condenser unit which sits over the lead bath and mates with the kettle to make a vacuum-tight enclosure. A vacuum is maintained inside the dezincing chamber which is equipped with a water-cooled condensing pan extending over the top of the chamber. The molten lead is agitated by an impeller.

The dezincing chamber is set first in the drossed lead bath, then water connections are immediately made in order to prevent the formation of steam within the water jacket. While the temperature is being raised, the vacuum pump is placed in operation and the agitator started. The temperature is then raised to 600°C and held throughout the operation.

After several hours, depending on the size of the unit, the vacuum is broken and the dezincing condenser unit is lifted from the bath. The metallic zinc condenses during the operation into a coherent crystalline mass and adheres to the underside of the water-cooled condenser. It is removed from each charge by manually applying a steel rake to dislodge the zinc deposit. The zinc drops off into suitable containers and is used again in the desilverizing process.

Bismuth Removal. When the bismuth level is >0.05%, it is removed from the lead to meet commercial specifications and bismuth is recovered as a product (24,25). The procedure takes advantage of the extremely low solubility of an intermetallic compound, $CaMg_2Bi_2$, that is formed between the reagents and the impurity. The bismuth content can be reduced to <0.005% (see BISMUTH AND BISMUTH ALLOYS).

In the Betterton-Kroll process the dezinced lead is pumped to the debismuthizing kettle, in which special care is taken to remove drosses that wastefully consume the calcium and magnesium. The skimmed blocks from the previous debismuthizing kettle are added to the bath at 420°C and stirred for a short time to enrich the dross with the bismuth being extracted from the new charge. This enriched dross is skimmed to blocks and sent to the bismuth recovery plant.

Following the removal of the enriched dross, the required quantities of calcium, as a lead–calcium alloy and magnesium in the form of metal ingots, are added. The bath is stirred about 30 min to incorporate the reagents and hasten the reaction. The molten lead is cooled gradually to 380°C to permit the precipitate to grow and solidify. The dross is skimmed for use with the next lot of lead to be treated.

The lead contains residual calcium and magnesium that must be removed by chlorination or treatment with caustic and niter. The molten lead is pumped or laundered to the casting kettles in which it is again treated with caustic and niter prior to molding. After a final drossing, the refined lead is cast into 45-kg pigs or 1- and 2-t blocks.

Electrolytic Refining. Electrolytic refining (26,27), used by Cominco Ltd. (Trail, B.C., Canada) and Cerro de Pasco Corp. (La Oroya, Peru), as well as by several refineries in Europe and Japan, removes impurities in one step as slimes. The impurities must then be separated and purified. Before the development of the Betterton-Kroll process, electrolytic refining was the only practical method of reducing bismuth to the required concentrations.

Decopperized blast furnace bullion is softened to reduce impurities below 2% before casting as anodes. The electrolyte is a solution of lead fluosilicate [*25808-74-6*], $PbSiF_6$, and free fluosilicic acid [*16961-83-4*]. Cathode starting sheets are made from pure electrolytic lead. The concrete electrolytic cells are lined with asphalt or a plastic material such as polyethylene.

The electrolyte is prepared by dipping granulated lead, suspended in a basket, into and out of the fluosilicic acid. The lead oxidizes during the operation and lead fluosilicate is formed:

$$2\ H_2SiF_6 + 2\ Pb + O_2 \longrightarrow 2\ PbSiF_6 + 2\ H_2O$$

A typical electrolyte has a specific gravity of 1.21 and the following analysis: lead, 67 g/L; free H_2SiF_6, 95 g/L; total acid, 142 g/L. The addition reagents added to the electrolyte are a combination of glue with either Goulac or Binderine (1 kg/t of Pb).

The temperature in the cell is 40°C. Most electrolyte cells are equipped with 24 anodes spaced approximately 10 cm apart, center to center; 25 cathode starting sheets are used, one at each end and others evenly spaced between the anodes. Current density is typically 15 mA/cm^2 of cathode area; cell voltage ranges from 0.30 to 0.70 V, and a current efficiency of 90–95% is usually realized.

The deposition time should be limited to a four-day period to avoid increased weight and roughness of the deposit. Deposited cathodes, weighing approximately 70 kg each, are thoroughly washed by dipping or by a high pressure water spray. These are charged to large melting kettles, heated to ca 550°C, drossed, and cast into ingots.

Slimes Treatment. After the corroded anodes are washed, and the adhering slimes scraped off, filtered, and dried, approximately 8% moisture is left to prevent dusting. The general practice is to smelt the slimes in a small reverberatory furnace, which produces a slag 10–12% by weight of the slimes (Fig. 16). This slag is taken to a second small reverberatory furnace in which it is partially reduced to remove precious metals; the remainder is transferred to the smelting department for production of antimonial lead. The reduced portion, containing the precious metals, is returned to the original slimes-smelting furnace. By agitating the bath in the slimes-smelting furnace with air, the bulk of the arsenic and part of the antimony is fumed off to either a baghouse or a Cottrell unit.

The metal bath containing gold, silver, bismuth, copper, etc, together with the lead and antimony remaining after blowing, is transferred to a cupel furnace. The surface of the bath is blown by low pressure air, and the antimony in the bath is removed in a lead–antimony slag. As succeeding stages of oxidation take place, litharge containing increasing amounts of copper and bismuth is removed until the metal arrives at the doré stage and the bath has the appearance of metallic silver. If selenium and tellurium are present, niter is added to the bath and, by agitation and oxidation, the sodium salts of selenium and tellurium are formed and skimmed from the furnace. Upon completion of the cupellation stage the fineness of the doré is 996, ie, 996 parts gold plus silver per thousand parts metal. The doré metal is cast into suitable shapes for parting (separating gold from silver), followed by electrolytic refining of gold and silver.

Slag and litharge formed during cupellation are segregated and reduced to a metal containing 20–25% ore more bismuth, depending on the bismuth content of the original bullion, and transferred to a bismuth recovery plant.

Secondary Lead. The emphasis in technological development for the lead industry in the 1990s is on secondary or recycled lead. Recovery from scrap is an important source for the lead demands of the United States and the rest of the world. In the United States, over 70% of the lead requirements are satisfied by recycled lead products. The ratio of secondary to primary lead increases with increasing lead consumption for batteries. Well-organized collecting channels are required for a stable future for lead (see BATTERIES, SECONDARY CELLS; RECYCLING NONFERROUS METALS).

The principal types of scrap are battery plates and paste, drosses, skimmings, and industrial scrap such as solders, babbitts, cable sheathing, etc. Some of this material is reclaimed by kettle melting and refining. However, most scrap is a combination of metallic lead and its alloying constituents mixed with compounds of these metals, usually oxides and sulfates. Therefore, recovery as metals requires reduction and refining procedures.

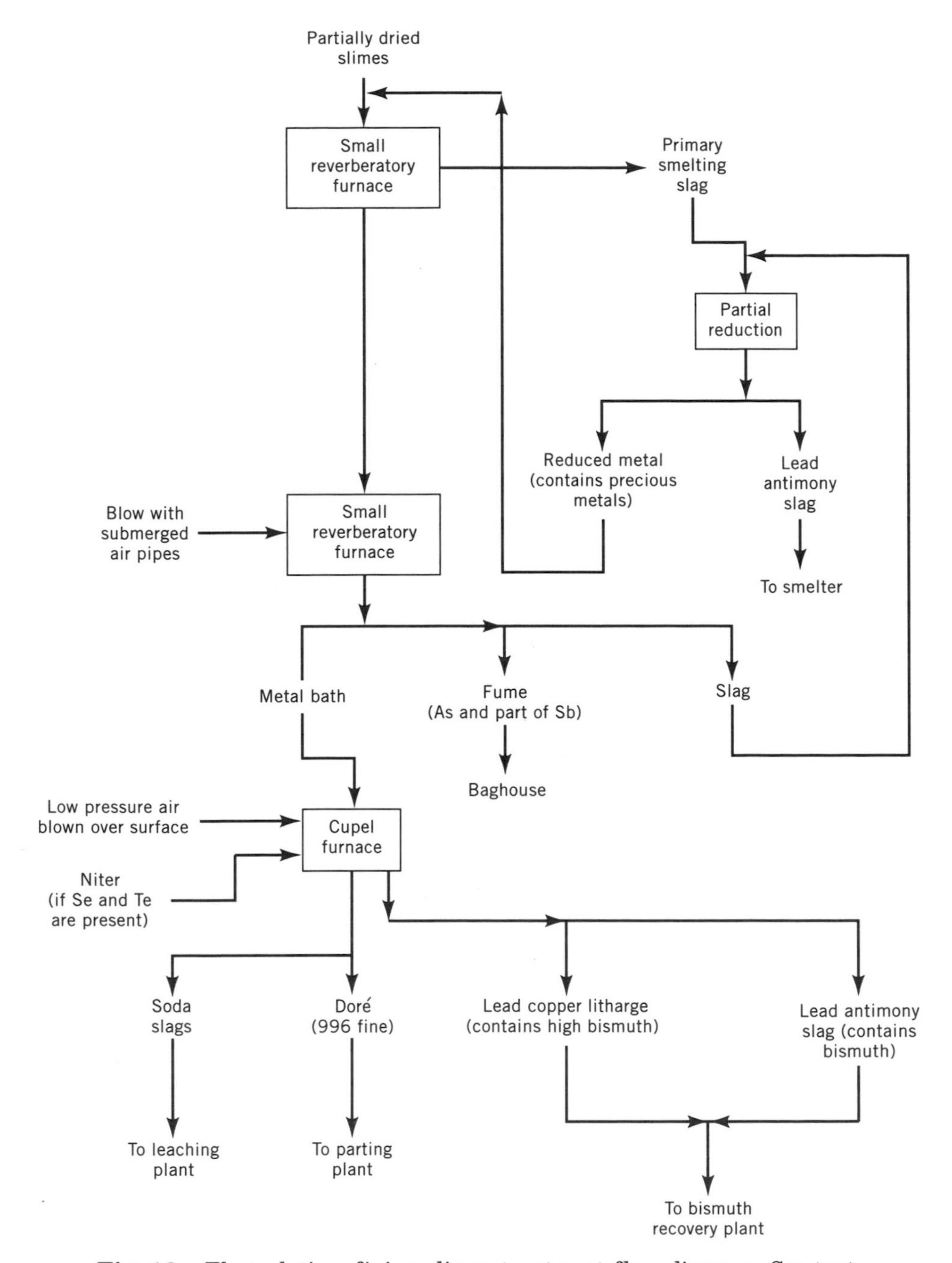

Fig. 16. Electrolytic refining slimes treatment flow diagram. See text.

Because about 80% of the lead consumed in the United States is for use in lead–acid batteries, most recycled lead derives from this source of scrap. More than 95% of the lead is reclaimed. Hence, the bulk of the recycling industry is centered on the processing of lead battery scrap.

The lead-bearing components are released from the case and other nonlead-containing parts, followed by the smelting of the battery plates, and refinement to pure lead or specification alloys. The trend toward battery grid alloys having little or no antimony, increases the ability of a recovery process to produce soft lead (refined). As required in the production of primary lead, each step in the secondary operations must meet the environmental standards for lead concentration in air (see AIR POLLUTION; LEAD COMPOUNDS, INDUSTRIAL TOXICOLOGY).

A typical automobile battery weighs 16.4 kg and consists of 3.5 kg metallic lead, 2.6 kg lead oxides, 4.0 kg lead sulfate, 1.3 kg polypropylene, 1.1 kg PVC, rubber and other separators, and 3.9 kg electrolyte. Including acid and water, the lead-bearing parts represent 61 wt %, ie, 21 wt % of lead alloy (2% Sb) and 40 wt % lead oxides and sulfate. Nonlead-bearing parts constitute the remaining 39%: the case (hard rubber or polypropylene) and separators (PVC) at 15 wt % and the electrolyte at 24 wt %.

Battery Breaking. There are well-defined, sophisticated technologies for the recovery of all materials of value in a battery (28). Technologies developed by Engitec (Milan, Italy) and M.A. Industries (Atlanta, Georgia) are generally employed. Figure 17 shows the flow sheet for a typical battery breaking system.

Battery breaking technologies use wet classification to separate the components of crushed batteries. Before crushing, the sulfuric acid is drained from

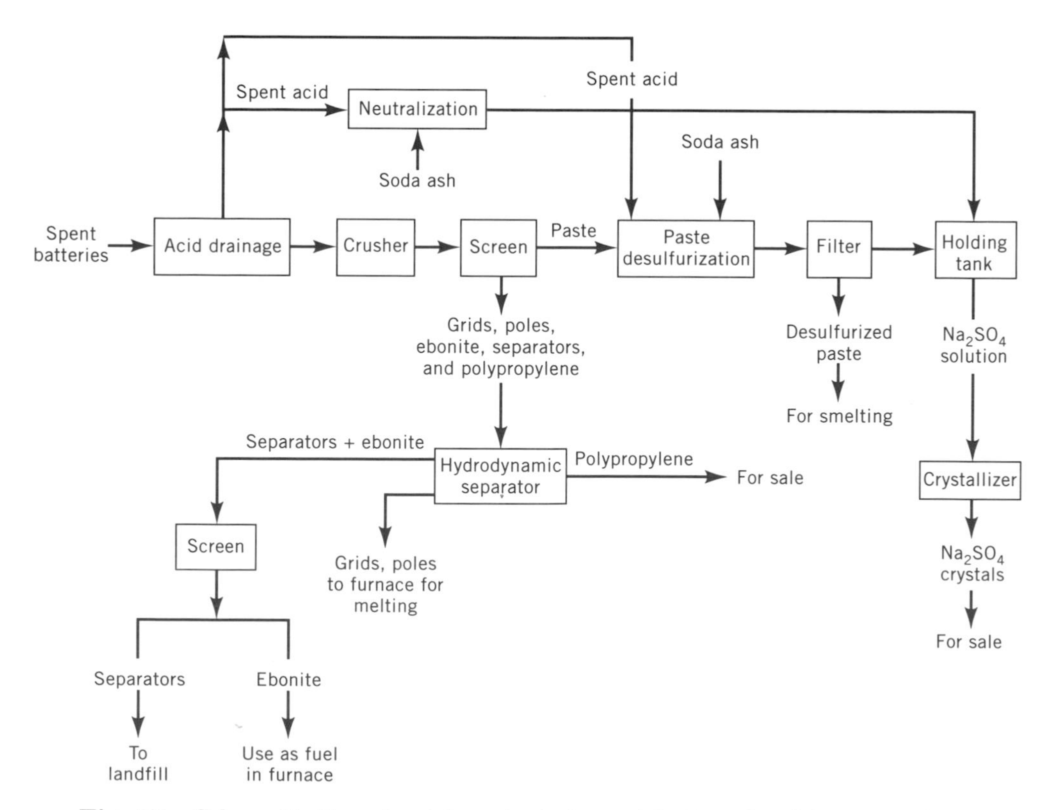

Fig. 17. Schematic flow sheet for a typical spent battery breaking operation.

the batteries. The sulfuric acid is collected and stored for use at a later stage in the process, or it may be upgraded by a solvent extraction process for reuse in battery acid.

After acid removal, scrap batteries are fed to a hammer mill in which they are ground to ≤ 5 cm particles. The ground components are fed to a conveyor and passed by a magnet to remove undesirable contamination. The lead scrap is then classified on a wet screen through which fine particles of lead sulfate and lead oxide pass, and the large oversize solid particles are passed on to a hydrodynamic separator. The fine particles are settled to a thick slurry and the clarified washwater recirculated to the wet screen.

The large particles sent to the hydrodynamic separator consist of metallic lead from the grids and poles in the battery together with ebonite [*12765-25-2*] and plastics from the cases and cell separators. In the hydrodynamic separator the feed is divided into three fractions: polypropylene chips in the upper section, metallics in the lower section, and ebonite and separators in the middle. The polypropylene chips are floated off, the metallics conveyed from the bottom, and the ebonite and separators are carried through in the water to a screen for further separation. The polypropylene is available for recycling, very often as battery cases, and the metallic lead is transferred to furnaces or kettles for refining (see RECYCLING, PLASTICS).

The paste recovered in the initial screening stage is not in a form suitable for lead smelting because of its high sulfate content. Thus the first step is desulfurization by a soda ash treatment at 60–70°C.

$$PbSO_4\ (s) + Na_2CO_3\ (aq) \longrightarrow PbCO_3\ (s) + Na_2SO_4\ (aq)$$

The products of reaction are pumped to a filter press for separation into a sodium sulfate solution and a filter cake having a low moisture content. The filter cake is then ready to be processed for the recovery of lead. The filtrate from the process contains an excess of sodium carbonate, and can be neutralized using the sulfuric acid drained from the batteries.

$$Na_2CO_3 + H_2SO_4 \longrightarrow Na_2SO_4 + H_2O + CO_2$$

The sodium sulfate solution may be discharged or further treated to recover sodium sulfate crystals or converted to other sodium salts for sale.

Almost all battery scrap and paste is converted to impure lead or lead alloys by pyrometallurgical processes employing blast, reverberatory, rotary, Isasmelt, or electric furnaces. In many plants, a furnace combination is used. Electrowinning technologies have also been developed but as of this writing none is yet in full commercial operation.

Smelting. Blast Furnace. If whole battery scrap is the feed, a blast furnace must be used (29). The furnace is constructed similarly to the cupola employed in the foundry industry (see COPPER ALLOYS). The melting zone is water jacketed and the shaft above is refractory lined. A normal charge consists of lead-bearing materials, ie, battery plates, drosses, slags, and lead scrap; and limestone, coke, silica, and scrap iron. The scrap iron is added to combine with the sulfur which is present as sulfate in the filler, forming an iron matte high (5–15%) in lead and usually discarded.

Use of a blast furnace is preferred if a regular supply of a charge of coarse and consistent quality is available. However, the blast furnace is not suitable for treating finely divided feed material.

Reverberatory Furnace. Using a reverberatory furnace, a fine particle feed can be used, the antimony content can be controlled, and batch operations can be carried out when the supply of scrap material is limited. However, the antimony-rich slags formed must be reduced in a blast furnace to recover the contained antimony and lead. For treating battery scrap, the reverberatory furnace serves as a large melting facility where the metallic components are liquefied and the oxides and sulfate in the filler material are concurrently reduced to lead metal and the antimony is oxidized. The furnace products are antimony-rich (5 to 9%) slag and low antimony (less than 1%) lead.

The overall recovery of the metallic components of scrap in plants having both reverberatory and blast furnaces is over 95%.

Rotary Furnace. The rotary furnace, which has more flexibility than either the blast or reverberatory furnace, can produce either a single metal product or a high and a low antimonial alloy. The rotary furnace, like the reverberatory furnace, allows for the option of producing low antimony lead for further refinement.

However, rotary furnaces tend to produce more exhaust gas and fumes, require more skillful manipulation, and are more labor intensive. Also, the slags produced in the rotary furnaces, soda or fayalite [*13918-37-1*] slags, normally do not pass the toxic characteristic leach procedure (TCLP) test and pose a disposal problem.

Electric Furnace for Secondary Slags. The slag from the reverberatory furnace is treated in a submerged arc electric furnace (30). Slag, reducing agents, and fluxes are blended. The mixed charge is fed to the furnace via a steel pan conveyor which discharges the blended slag charge to a rotary dryer fired by the hot off-gases from the electric furnace. The slag is dried and heated prior to charging into the furnace. In the electric furnace, Sb, As, and Sn are reduced from the slag. The CO from the reduction process is burned as it exits the furnace to provide heat for the dryer. The furnace is operated on a batch basis. When the furnace is fully charged (about 30 t of slag) the slag is held at constant temperature for a period of time to reduce the metal content to the desired level and produce a nonhazardous throwaway slag. The typical slag composition in wt % is CaO, 25–34; SiO_2, 22–35; FeO, 10–23; S, 1–3; Pb, <2; Sb, <1.5; Sn, <1.5; and As, <0.5.

The slag is batch tapped into a receiving room where it is cooled and broken up for disposal. The metal bullion is tapped from the furnace periodically via a siphon into 2-t cast-iron molds. Typical bullion content from an electric furnace in wt % is Sb, 13–18; Sn, 1–2; As, 0.5–1; Cu, 0.3–0.4; and Ni, 0.05–0.1. The balance is lead.

Gases from the furnace, metal tap, slag tap, and feed system are combined and fed to a six-cell pulse baghouse containing 864 high temperature Teflon bags. The dust from the electric furnace system is fed continuously back to the reverberatory furnace in a close screw conveyor.

Isasmelt Furnace. The Isasmelt furnace, originally developed for the production of primary lead, has also been adapted to treat desulfurized or non-desulfurized paste, to produce an impure bullion and a litharge slag. The slag is reduced in a rotary furnace to produce bullion and a discard slag. Also, the grid and plates produced during the battery breaking operation can be melted in an Isasmelt furnace to produce hard lead (31).

Newer Developments. The Engitec Impianti Co. (Milan, Italy), has developed a process (28) for the production of electrolytic lead from battery paste. The desulfurized paste is leached using a solution of fluoboric acid and lead fluoborate, ie, spent electrolyte from lead electrowinning. Hydrogen peroxide is used to convert Pb^{4+} to Pb^{2+}. The lead solution is treated by electrowinning in cascade cells using a specially designed proprietary composite anode.

The cathode material is stainless steel. The lead produced by this method analyzes 99.99+ %. The overall power consumption is less than 1 kWh/kg of lead, so that the electrolytic process for treating spent batteries has much less of an environmental impact than the conventional pyrometallurgical process.

In another development (32), the sodium sulfate solution produced during the desulfurization of paste with caustic soda is electrolyzed in a membrane cell to produce caustic soda and high purity sulfuric acid. The caustic soda is recycled to the desulfurization stage; the sulfuric acid, after concentration, can be reused in battery production.

In a similar development, RSR Corp. of the United States has developed an environmentally clean process (33) to recover high purity lead as electrowon cathode. The process consists of decomposing the desulfurized paste to PbO at 325°C and leaching the PbO with fluosilicic acid, ie, spent electrolyte from electrowinning, to produce the soluble $PbSiF_6$. Lead is electrowon from the leach solution using a PbO_2-coated graphite anode. The cathode material is sheet lead. The purity of lead exceeds 99.99%. The power consumption for electrowinning is expected to be 660 kWh/t.

The electrowinning process developed by Ginatta (34) has been purchased by M.A. Industries (Atlanta, Georgia), and the process is available for licensing (qv). MA Industries have also developed a process to upgrade the polypropylene chips from the battery breaking operation to pellets for use by the plastics industry. Additionally, East Penn (Lyons Station, Pennsylvania), has developed a solvent-extraction process to purify the spent acid from lead–acid batteries and use the purified acid in battery production (35).

Economic Aspects

A world production summary of primary and refined lead is given in Table 5. U.S. consumption for the same period is given in Table 6.

The principal U.S. lead producers, ASARCO Inc. and The Doe Run Co., account for 75% of domestic mine production and 100% of primary lead production. Both companies employ sintering/blast furnace operations at their smelters and pyrometallurgical methods in their refineries. Domestic mine production in 1992 accounted for over 90% of the U.S. primary lead production; the balance originated from the smelting of imported ores and concentrates.

Table 5. World Mine and Refined Lead Production, 10^3 t

Region	Production									
	Mine					Refined				
	1988	1989	1990	1991	1992	1988	1989	1990	1991	1992
Europe	373	348	340	324	290	1693	1652	1586	1633	1534
Africa	206	181	169	174	171	169	162	152	155	149
Asia	123	114	111	106	111	607	600	557	554	598
America										
Canada	367	276	241	276	342	268	243	184	212	255
United States	394	420	497	477	408	1091	1253	1291	1195	1158
Mexico	178	163	187	168	163	249	249	238	236	288
Peru	149	193	188	199	194	54	74	669	76	84
other	70	63	58	60	57	146	132	117	108	111
Total	*1158*	*1115*	*1171*	*1180*	*1164*	*1808*	*1951*	*1899*	*1827*	*1896*
Australia	457	498	556	579	575	209	215	229	244	232
other countries										
former USSR[a]	520	500	490	460		795	750	730	670	
China	312	341	315	320		241	302	287	296	
Korea	90	80	70	80		70	70	65	65	
others[a]	133	142	118	108		297	282	224	144	
Total	*1055*	*1063*	*993*	*968*		*1403*	*1404*	*1306*	*1175*	
World total	*3372*	*3319*	*3340*	*3331*		*5889*	*5984*	*5729*	*5588*	

[a]Estimated from Ref. 36 data.

Table 6. Lead Consumption of the United States, t[a]

Forms of lead	1988	1989	1990	1991	1992
metal products	1,112,941	1,182,559	1,178,381	1,124,676	1,116,524
ammunition	52,709	62,940	58,209	59,800	63,275
bearing metal	6,035	2,586	2,876	2,851	3,086
brass and bronze	9,993	9,610	9,941	10,599	10,300
cable covering	17,787	24,435	19,941	18,991	16,289
casting metals	15,829	16,175	14,844	13,927	14,201
pipes, traps, bends	11,192	9,817	9,281	8,646	8,592
sheet lead	17,457	20.987	21,013	20,818	20,262
solder	19,064	17,009	16,489	15,831	15,269
storage batteries[b]	955,259	1,012,150	1,019,634	967,210	959,177
terne metal[c]	7,617	6,851	6,153	6,003	6,070
other oxides	62,524	57,985	56,484	57,241	61,132
miscellaneous uses[d]	55,261	42,684	40,363	29,741	15,534
Total consumption	*1,230,727*	*1,283,228*	*1,275,228*	*1,211,658*	*1,193,190*

[a]Ref. 37.
[b]Includes lead oxides.
[c]Includes both type metal and other metal products.
[d]Includes use as gasoline additive.

As of this writing (ca 1994) none of the newer lead smelting technologies has been adopted in the United States. Also, no new lead mine has opened in the United States since the early 1980s. Most of the known U.S. reserves for lead are located in federally owned land in Missouri; future mine development depends on the outcome of the U.S. government's intent to reform the Mining Law of 1872.

Secondary lead production made up over 70% of the lead produced in the United States in 1992 vs 54% in 1980. The amount of secondary lead produced was 698×10^3 t in 1988, 888×10^3 t in 1990, and 878×10^3 t in 1992. Of the 1.2×10^6 t of lead consumed in the United States in 1992, approximately 880,000 t were produced from the recycling of lead–acid batteries and 350,000 t from primary sources. A similar trend exists worldwide. In 1992, for the first time, slightly over half (51%) of the lead produced in the world came from secondary sources.

World consumption of lead grew steadily at about 3% per annum through the mid-1980s until 1989, at which point it leveled off at about 4.5×10^6 t. U.S. consumption in that time period grew at about 5% per annum, but consumption actually declined in the early 1990s before rising again in 1993 to match the 1989 peak level of 1.3×10^6 t in 1993. A change in supply occurred in 1989 with the opening of the Communist Bloc production to Western markets. In 1980 the Communist Bloc imported 140,000 t of lead from the West; by 1993 Communist Bloc countries supplied 180,000 t to the West.

The dramatic change in the market/supply situation in the early 1990s impacted the price of lead as shown in Figure 18. After a period of rising lead prices in the late 1980s which matched rising consumption, price declined beginning in 1990 to under 40¢/kg in 1993. During the growth period of the late 1980s, the price of lead ranged between 65 and 90¢/kg in 1993 U.S. $, adjusted for inflation. Long-term trends in the price of lead are dependent on the overall

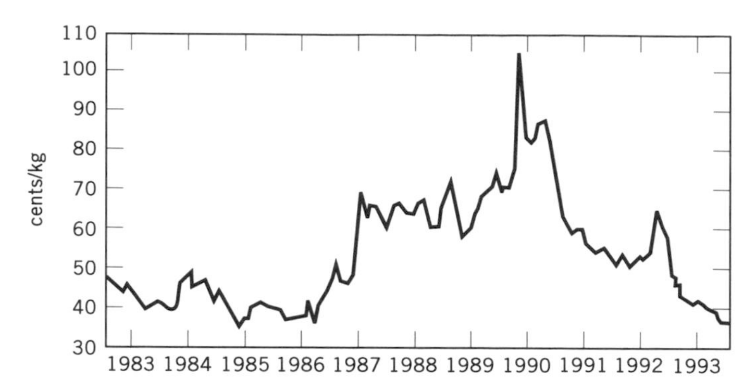

Fig. 18. London Metal Exchange lead prices.

world economy as well as an industrial infrastructure for the former Communist Bloc countries that consume the current surplus of lead.

The future use of lead may be decided by the resolution of an environmental paradox. Some markets for lead are being phased out because of environmental concerns, eg, the use of tetraethyllead as a gasoline additive. However, a 1990 State of California law and similar laws in nine eastern U.S. states require that 2% of new cars meet zero-emission standards in 1998. By 2003 this requirement rises to 10% of new vehicles. Zero emission vehicles are generally accepted to mean electric, ie, battery powered cars, and there is considerable research effort to bring suitable electric vehicles to market by 1998.

Whereas many battery systems are being investigated for powering electric vehicles, the lead–acid battery, despite environmental concerns, is by far the most mature and accepted. There is considerable potential to adapt and develop lead–acid batteries from the role in automobiles for lighting and ignition to that of the source of automotive power. The determination of the components of the zero-emission vehicle is expected to determine the use of lead in the twenty-first century. If lead–acid battery technology is adopted, the demand for lead is expected to increase strongly. The established world resources of 71×10^6 t can meet the demand for electric vehicles for a long time. Without usage for electric vehicles, the slowly increasing trend evidenced in the use of lead in the United States in the 1990s, as shown in Figure 19, is likely to go into decline because of environmental regulations limiting the applications of lead in uses other than batteries.

Analytical Methods

Lead is detected qualitatively in solution by the formation of inorganic precipitates between lead and the following reagents: sulfuric acid or a soluble sulfate, giving a white precipitate; hydrochloric acid or a soluble chloride, giving a white crystalline precipitate; a soluble iodide, chromate, or dichromate, giving a yellow precipitate. Lead may also be detected upon formation of precipitates with many organic reagents such an oxalate and tannic acid. Macro amounts of lead are determined quantitatively by gravimetric methods, by weighing the lead as one of the above mentioned precipitates, or by electrolytic deposition on the anode as

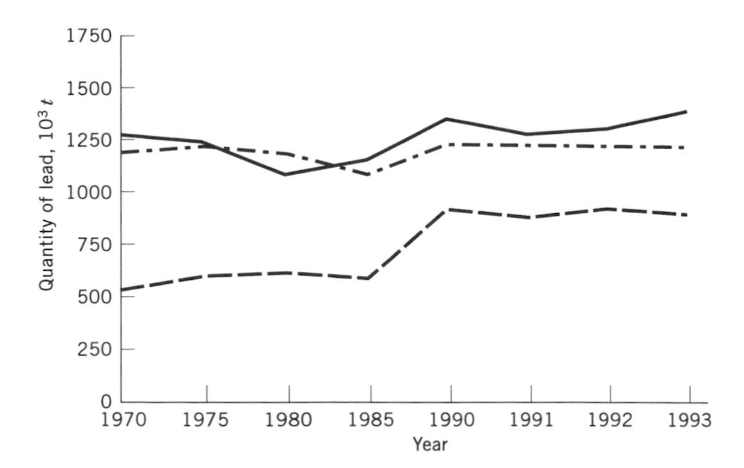

Fig. 19. Trends in the lead industry in the United States: (—), total consumption; (-·-·-), total production; and (— — —), secondary production. The difference between total and secondary production represents primary production.

lead dioxide. There are also numerous titrimetric methods for determining macro amounts of lead quantitatively, mainly titration with molybdate or EDTA.

In determining the purity or percentage of lead in lead and lead-base alloys, the impurities or minor components are determined and the lead content calculated by difference. Quality control in lead production requires that the concentration of impurities meet standard ASTM specifications B29 (see Table 7). Analyses of the individual impurities are performed using various wet chemical procedures and instrumental methods such as emission spectroscopy.

Health and Safety Factors

Exposure to excessive amounts of lead over a long period of time (chronic exposure) increases the risk of developing certain diseases. The parts of the body which may be affected include the blood, nervous system, digestive system, reproductive system, and kidneys. These effects include anemia, muscular weakness, kidney damage, and reproductive effects, such as reduced fertility in both men and women, and damage to the fetus of exposed pregnant women.

Lead enters the body through inhalation and ingestion, is absorbed into the circulatory system from the lungs and digestive tract, and excreted via the urine and feces. Normally, intake of lead approximately equals output. However, excessive exposure and intake can cause tissue concentrations to increase to the point where illness can result.

Workers who produce or use lead should be aware of possible hazards. Symptoms of chronic lead poisoning include fatigue, headache, constipation, uneasy stomach, irritability, poor appetite, metallic taste, weight loss, and loss of sleep. Most of these same symptoms also occur in many common illnesses, such as the flu, thus a physician must rely on tests, such as blood lead analysis, to determine chronic lead poisoning.

The particle size and chemical composition of lead and lead compounds affect the readiness with which lead is absorbed from the lungs and digestive tract

Table 7. Standard Specifications for Refined Lead[a,b]

	Grade			
Metal, wt %	Low bismuth, low silver, pure lead[c]	Refined pure lead[d]	Pure lead	Chemical copper–lead[e]
Sb	0.0005	0.0005	0.001	0.001
As	0.0005	0.0005	0.001	0.001
Sn	0.0005	0.0005	0.001	0.001
Sb, As, and Sn			0.002	0.002
Cu	0.0010	0.0010	0.0015	0.040–0.080
Ag	0.0010	0.0025	0.005	0.020
Bi	0.0015	0.025	0.05	0.025
Zn	0.0005	0.0005	0.001	0.001
Te	0.0001	0.0001		
Ni	0.0002	0.0002	0.001	0.002
Fe	0.0002	0.001	0.001	0.002
lead[f]	99.995	99.97	99.94	99.90
UNS number	L50006	L50021	L50049	L51121

[a]ASTM B29-92. Values given are maximum allowable unless range is given.
[b]By agreement between the purchaser and the supplier, analyses may also be required and limits established for elements or compounds not specified.
[c]This grade is intended for chemical applications requiring low silver and bismuth contents.
[d]This grade is intended for lead–acid battery applications.
[e]This grade is intended for applications requiring corrosion protection and formability.
[f]Value for lead is minimum value, determined by difference.

(see LEAD COMPOUNDS, INDUSTRIAL TOXICOLOGY). Larger particles and compounds having low aqueous solubility are less hazardous than finely divided particles and compounds of higher solubility.

Because lead may be ingested and inhaled, and because particle size and chemical composition affect its absorption, it is important that the concentration of lead in the blood be determined on a periodic basis for any person occupationally exposed to lead. Measures to control exposure to lead include the provision of proper ventilation, application of proper work practices, following rules of good hygiene, and wearing respirators and protective clothing (see AIR POLLUTION CONTROL METHODS; INDUSTRIAL HYGIENE). Detailed regulations by OSHA and other United States agencies governing occupational exposure to lead may be found in Part 1910 of Title 29 of the *Code of Federal Regulations.*

Environmental Standards. Lead in the environment is regulated in the United States because of its potential occupational impact, as well as concern about the impact lead may have on the cognitive and physical development of young children. Standards have been set for lead in air, water, and other environmental media.

The U.S. ambient air standard has been established as 1.5 $\mu g/m^3$ as a quarterly arithmetic mean. Some state standards are more restrictive, eg, Colorado has a monthly mean. The OSHA permissible exposure limit (PEL) for lead is 50 $\mu g/m^3$ at an action level (local authorities must investigate) of 30 $\mu g/m^3$. The American Conference of Government Industrial Hygienists' (ACGIH) threshold limit value (TLV) is set at 150 $\mu g/m^3$. This level, not enforceable in the United States, is adopted by such countries as Canada and Australia as a standard.

OSHA has set a standard to keep blood levels in the occupational work force below 40 $\mu g/dL$. ACGIH has set a goal relating to a biological exposure index of 50 $\mu g/dL$ for lead in blood and 150 $\mu g/g$ creatinine for lead in urine.

Lead in drinking water has been regulated in the United States to achieve a maximum contaminant level goal (MCLG) of zero and an action level of 15 $\mu g/L$ (see GROUNDWATER MONITORING). No standards have been established for lead in soil, but 500–1000 ppm is generally accepted as a level not leading to elevated blood lead levels in young children. The Center for Disease Control (CDC) has selected 10 $\mu g/dL$ of lead in blood as the upper area above which risk to young children is indicated. Follow-up samples are recommended when blood lead levels exceed 10 $\mu g/dL$.

Lead-based paint (qv) in old structures has been identified as an environmental risk because the chemical form of lead in paint is readily available biologically. Lead in paint is measured using a surface analysis of walls, windowsills, etc, using x-ray fluorescence spectrometry. The following criteria have been established by the United States CDC for lead in surficial materials: <0.07 mg/cm^3, negative; 0.7–2.9 mg/cm^3 low; 3.0–5.9 mg/cm^3 moderate; and ≥ 6.0 mg/cm^3 high.

Another level of regulatory significance is the toxic characteristic leach procedure (TCLP) limit of a characteristic waste. A material which is a waste because of the TCLP is hazardous if a liquor resulting from an 18-h leach in an acetic acid buffer exceeds 5 ppm (mg/L) lead in the leach liquor.

Preventive Measures. The intake uptake biokinetic model (IUBK) projects the impact of lead in the environment on blood lead. This model assumes conservatively high levels of intake and cannot account for chemical speciation, thus over-predictions of blood lead levels often occur. Nonetheless, because of the allegations of the impact of blood lead and neurobehavioral development, blood lead levels in children are being reduced administratively to below 10 $\mu g/dL$. In order to do so, soil leads are being reduced to a level of between 500–1000 ppm where remediation is required.

The different forms of lead have different bioavailability and this ultimately impacts cleanup levels. Mine tailings, slag, and other such residues have limited impact on blood lead levels because these materials contain lead in the form of lead sulfide, which has limited biological reactivity and uptake.

Uses

The uses of lead are many and varied. U.S. consumption is summarized in Table 6. Some of the main uses are in the manufacture of storage batteries, ammunition, nuclear and x-ray shielding devices, cable covering, ceramic glazes, and noise control materials (see NOISE POLLUTION AND ABATEMENT METHODS). Lead is also used in bearings (see BEARING MATERIALS), in brass and bronze (see COPPER ALLOYS), in casting metals, for pipes, traps, and bends, and in some solders (see SOLDERS AND BRAZING ALLOYS).

BIBLIOGRAPHY

"Lead" in *ECT* 1st ed., Vol. 8, pp. 217–253, by T. D. Jones, ASARCO Inc.; in *ECT* 2nd ed., Vol. 12, pp. 207–247, by J. E. McKay, The Bunker Hill Co.; in *ECT* 3rd ed., Vol. 14, pp. 98–139, by H. E. Howe, Consultant.

1. P. Crowson, *Minerals Handbook, 1992–93, Statistics and Analyses of the World's Minerals Industry,* Stockton Press, New York, p. 130.
2. F. M. Randall and R. A. Arterburn, in *AIME World Symposium on Mining and Metallurgy of Lead and Zinc,* Vol. 1, American Institute of Mining, Metallurgical, and Petroleum Engineers (AIME), New York, 1970, Chapt. 24, pp. 453–465.
3. A. W. Griffith, in Ref. 2, Chapt. 26, pp. 483–498.
4. H. H. Haman, in Ref. 2, Chapt. 41, pp. 852–869.
5. E. N. Doyle, in Ref. 2, Chapt. 40, pp. 814–851.
6. R. B. Paul, in *AIME World Symposium on Mining and Metallurgy of Lead and Zinc,* Vol. 2, AIME, New York, 1970, pp. 777–789.
7. D. H. Beilstein, in Ref. 6, Chapt. 24, pp. 702–737.
8. F. W. Gibson, in Ref. 6, Chapt. 25, pp. 738–776.
9. U.S. Pat. 2,890,951 (June 16, 1959), J. T. Roy (to ASARCO).
10. R. M. Sellwood, in Ref. 6, Chapt. 21, pp. 581–618.
11. C. F. Harris and co-workers, in *Lead–Zinc–Tin 1980, Symposium of the American Institute of Metallurgical Engineers*, AIME, Las Vegas, Nev., pp. 247–260.
12. *Met. Bull. Month.,* 54–55 (Sept. 1987).
13. T. R. A. Davey, in J. M. Cigan and co-workers, eds., *Advances in Lead, Zinc and Tin Technology: Projections for the 1980 Lead, Zinc–Tin 1980 TMS Symposium,* pp. 48–65.
14. *Lead Production by the QSL Process,* technical bulletin, Lurgi Corp.
15. *Eng. Min. J.,* 47–49 (July 1983).
16. J. H. Reimers and J. C. Taylor, in H. Y. Sohn and co-workers, eds., *Advances in Sulfide Smelting,* Vol. 2, AIME Symposium, San Francisco, Feb. 1983, pp. 529–551.
17. W. J. Errington and co-workers, in *Extractive Metallurgy '85, Institute of Mining and Metallurgy Symposium,* IMM, London, Sept. 1985, pp. 199–218.
18. S. P. Matthew, in T. S. Mackey and R. D. Prengaman, eds., *Lead–Zinc '90, AIME Symposium,* Anaheim, Calif., Feb. 1990, pp. 889–901.
19. W. H. Peck and J. McNiCol, *Metals* **18**, 1027 (Sept. 1966); Aus. Pat. 256,533 (May 11, 1961), T. R. H. Davey (to Broken Hill Smelters Pty. Ltd.).
20. P. Van Negen and co-workers, in Ref. 18, pp. 933–951.
21. P. J. Dugdale, "Oxygen Softening of Lead Bullion," paper presented at *16th Annual Conference of Metallurgists,* Canadian Institute of Mining & Metallurgy, Vancouver, B.C., 1977.
22. K. R. Barrett and R. P. Knight, in Ref. 17, pp. 683–708.
23. T. R. A. Davey, *J. Met.* **6**, 838 (1954).
24. J. O. Betterton and Y. E. Lebedeff, *Trans. AIME,* 205 (1936).
25. T. R. A. Davey, *J. Met.* **8**, 341 (1956).
26. C. A. Aranda and P. J. Taylor, in Ref. 2, Chapt. 31, pp. 891–915.
27. C. J. Krauss, *J. Met.* **28**, 4 (Nov. 1976).
28. R. M. Reynolds, E. K. Hudson, and M. Olper, in Ref. 18, pp. 1001–1022.
29. R. D. Prengaman, in Ref. 11, pp. 985–1002.
30. R. D. Prengaman, *International Conference of the International Lead Zinc Study Group*, Rome, Italy, June 1991, p. 437.
31. K. Ramus and P. Hawkins, "Lead Acid Battery Recycling and the New Isasmelt Process," *3rd European Lead Battery Conference*, Munich, Germany, Oct. 1992.
32. M. Olper, in Ref. 30, pp. 79–89.

33. R. D. Prengaman and H. McDonald, in Ref. 18, pp. 1045–1056.
34. M. V. Ginatta, "G.S. Electrolytic Process for the Recovery of Lead from Spent Electric Storage Batteries," paper presented at the *Annual AIME Meeting,* 1975, New York.
35. U.S. Pat. 4,971,780 (Nov. 20, 1990), R. A. Spitz (to East Penn Mfg.).
36. *Annual bulletin*, International Lead Zinc Study Group, London, 1993.
37. Technical data, American Bureau of Metal Statistics Inc., U.S. Bureau of Mines, Washington, D.C.

General References

"Analytical Chemistry of the Elements," in *Treatise on Analytical Chemistry,* Vol. 6, Part 2, Wiley-Interscience, New York, 1964.
J. L. Bray, *Nonferrous Production Metallurgy,* John Wiley & Sons, Inc., New York, 1947.
D. M. Liddell, "Recovery of the Metals," in *Handbook of Nonferrous Metallurgy,* Vol. 2, 2nd ed., McGraw-Hill Book Co., Inc., New York, 1945, Chapts. 7 and 13.
P. McIlroy, *Availability of U.S. Primary Lead Resources,* IC8646, U.S. Bureau of Mines, Washington, D.C., 1974.
Mineral Industry Survey, U.S. Bureau of Mines, Washington, D.C., Lead Reports 1991–1992.
Nonferrous Metal Data 1992, American Bureau of Metals Statistics Inc., New York.
W. Hofmann, *Lead and Lead Alloys, Properties and Technology,* 2nd ed., English trans., Springer-Verlag, New York, 1970.
D. R. Lide, ed., *Handbook of Chemistry and Physics,* 74th ed., CRC Press, Boca Raton, Fla., 1993.
H. M. Callaway, *Lead, A Material Survey,* IC8083, U.S. Bureau of Mines, Washington, D.C., 1962.
ASM Handbook, Vol. 2, 1991.
R. D. Pehlke, *Unit Process of Extractive Metallurgy,* Elsevier Publishing Co., New York, 1973.
R. L. Amistadi, "Whither Lead," paper presented at *Independent Battery Manufacturers Association,* Chicago, Ill., Oct. 20, 1993.

MICHAEL KING
VENKOBA RAMACHANDRAN
ASARCO Inc.

LEAD ALLOYS

Lead (qv) is a heavy, soft, bluish gray metal having a low melting point and a high boiling point. The density, coefficient of thermal expansion, malleability, lubricity, and flexibility are high. The tensile and compressive strength, hardness, elastic modulous, elastic limit, creep resistance, and yield strength are quite low. Lead has excellent resistance to corrosion in a wide variety of media, and is easily alloyed with many other metals. Lead alloys, which have low melting points, can be cast into many shapes by using a variety of molding materials and casting processes.

Lead is easily cast and formed. It is one of the oldest known metals, used before 3000 BC. Early civilizations used lead extensively for ornamental and structural uses, and lead pipes used for the transportation of water by the Romans have endured.

About 50% of lead is used as pure lead, lead oxides, or lead chemicals (see LEAD COMPOUNDS); the remainder is used in the form of lead alloys. The principal uses of lead alloys are in lead–acid batteries (qv); for ammunition; cable sheathing; building construction in sheets, pipes, and solders; bearings; gaskets; specialty castings; anodes; fusible alloys; shielding; and weights (see BUILDING MATERIALS, SURVEY; BEARING MATERIALS; METAL ANODES; SOLDERS AND BRAZING ALLOYS).

Lead and its alloys are generally melted, handled, and refined in cast-iron, cast-steel, welded-steel, or spun-steel melting kettles without fear of contamination by iron (qv). Normal melting procedures require no flux cover for lead. Special reactive metal alloys require special alloying elements, fluxes, or covers to prevent dross formation and loss of the alloying elements.

Lead is ductile and malleable, and can be fabricated into various shapes by rolling, extruding, forging, spinning, and hammering. The low tensile strength and very low creep strength of lead make it unsuitable for use without the addition of alloying elements. The principal alloying elements used to strengthen lead are antimony, calcium, tin, copper (qv), tellurium, arsenic, and silver. Minor alloying elements are selenium, sulfur, bismuth, cadmium, indium, aluminum, and strontium.

Lead–Antimony Alloys

Properties. Lead–antimony alloys are the most widely used lead alloys. The lead–antimony phase diagram is shown in Figure 1 (1). Antimony is relatively soluble in molten lead. The lead–antimony system contains a eutectic point at 11.1% antimony and 252°C. The solubility of antimony in solid lead decreases from 3.5 wt % at the eutectic temperature of 252°C to 0.25 wt % at 25°C. The reduced solubility with temperature makes the lead–antimony alloys age hardenable. The grain structure of a typical cast lead–antimony alloy is shown in Figure 2. It consists of a lead matrix (the dark particles in Fig. 2) surrounded by a network of antimony particles (white). These particles provide immediate strength upon casting and also provide high temperature strength. The addition of arsenic dramatically increases the rate of aging and final strength. Addition of tin increases the fluidity and, in combination with copper and arsenic, reduces the rate of oxidation of molten lead–antimony alloys (2).

The mechanical properties of lead–antimony alloys containing arsenic, tin, and copper are shown in Table 1. Most lead–antimony alloys of commercial importance contain 11 wt % or less antimony. High (> 5 wt %) antimony alloys are strengthened primarily by the eutectic antimony particles produced during solidification of the alloy. Lower antimony alloys are strengthened by a combination of the eutectic particles and precipitation hardening during aging.

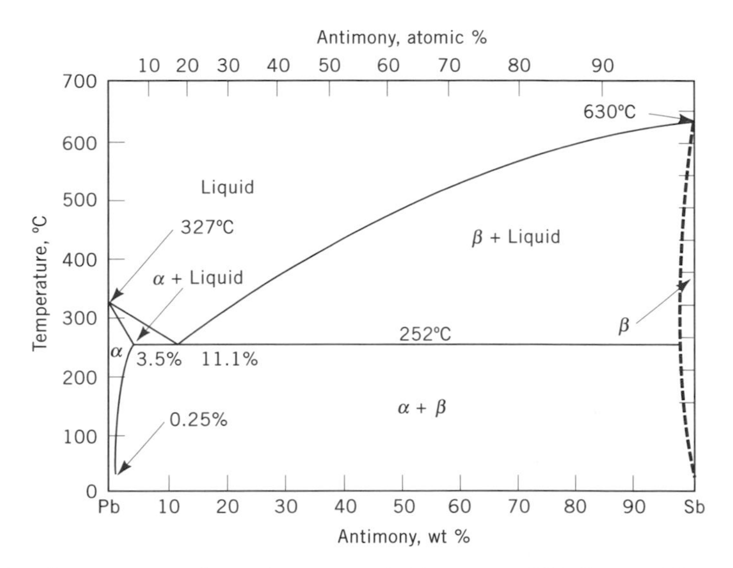

Fig. 1. Lead–antimony phase diagram (1). Courtesy of McGraw-Hill Book Co., Inc.

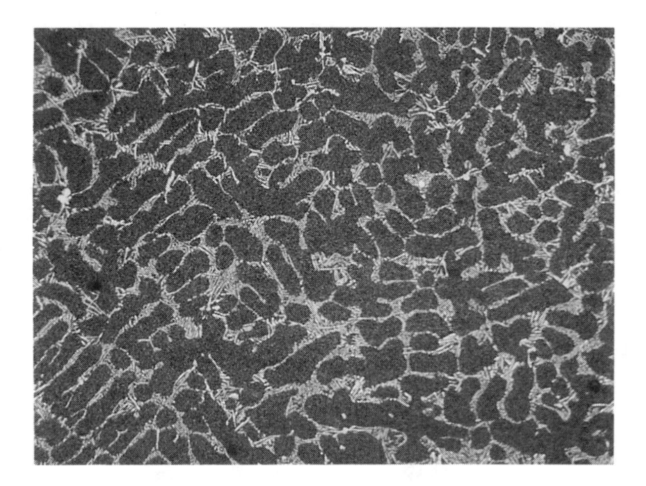

Fig. 2. Grain structure of cast lead–6 wt % antimony alloy showing the dark lead, white antimony eutectic particles at a magnification of 160×.

The decrease in solubility of antimony in lead with decreasing temperature allows the lead–antimony alloys to be precipitation-hardened by proper heat treatment. Differences in the rate of cooling of castings to room temperature can give wide variation in the mechanical properties owing to differences in the amount and type of antimony precipitates produced during cooling. The effects of quenching temperature on the properties of lead–3 wt % antimony alloy aged 1 day are as follows.

Quenching temperature, °C	Tensile strength, MPa (psi)
238	75 (10,875)
230	56 (8,120)
215	38 (5,510)
200	34 (4,930)

Table 1. Mechanical Properties of Lead–Antimony Alloys[a]

Antimony content, wt %	Yield strength after aging, MPa[b,c]		Tensile strength, MPa[b]	Elongation, %
11	68.9	74.4	75.9	5
6	55.2	71.0	73.8	8
3	34.0	55.2	65.5	10
2	24.1	37.9	46.9	15
1	13.8	19.3	37.9	20
0	3.5	3.5	11.7	55

[a]After 30 days, unless otherwise noted.
[b]To convert MPa to psi, multiply by 145.
[c]Column on left is after one day.

Uses. *Lead–Acid Batteries.* The primary use for lead–antimony alloys is as electrodes, connectors, and terminals of lead–acid batteries (see BATTERIES, SECONDARY CELLS, LEAD–ACID). The lead–antimony alloys, which are strong, creep resistant, and corrosion resistant in sulfuric acid, resist structural changes caused by charge–discharge cycles in the battery. Antimony as an alloying element in the positive electrode grid aids in retaining the active material in the battery and also aids in recharge of the battery when it is deeply discharged. Deep discharge industrial, load-leveling, and motive-power batteries utilize lead alloys containing 5–11 wt % antimony. Large tubular grids use 9–11 wt % antimony.

In lead–acid batteries antimony has one disadvantage. In use, the positive lead grid is oxidized during battery charging. The antimony, oxidized from the positive grid, enters the electrolyte and plates on the negative grid as metallic antimony. The antimony reduces the hydrogen overvoltage at the negative grid causing breakdown of the water in the electrolyte during charging into hydrogen and oxygen, and consumes water in the process. The amount of water loss and subsequent need to add water to the battery decreases as the antimony content of the positive grid decreases (3). Low gassing or maintenance-free batteries have antimony contents of 1.5–2.75 wt %. At this level, the migration of antimony is greatly decreased.

Automobile battery grids employ about 1–3 wt % antimony–lead alloys. Hybrid batteries use low (1.6–2.5 wt %) alloys for the positive grids and nonantimony alloys for the negative grids to give reduced or no water loss. The posts and straps of virtually all lead–acid batteries are made of alloys containing about 3 wt % antimony.

Low antimony alloys decrease water loss when used as positive grids. In alloys having less than 3.5 wt % antimony, however, the increased solidification temperature range and the reduced or negligible amount of eutectic liquid present at the final freezing temperature produces structures highly susceptible to solidification–shrinkage porosity and cracking. The grain size of lead–antimony alloys increases with decreasing antimony content as seen in Figure 3.

As the antimony content of the lead–antimony alloys decreases, the alloys tend to solidify into large, oriented, columnar grains. Figure 4**a** shows the large grains associated with a cast low antimony alloy battery grid. The large grains present areas for preferential mechanical cracking and penetrating corrosion. To prevent the formation of large grains, nucleants are added to lead–antimony alloys. Small amounts of sulfur, copper, or selenium serve as sites for growth of lead crystals during solidification and produce fine, rounded grain structures which are resistant to cracking (2,4–6).

By using nucleants, fine-grained structures, such as that shown in Figure 4**b**, can be produced in cast alloys independent of the antimony content. The molten metal must be kept at a temperature high enough to assure complete solubility of the nucleants prior to casting the alloy. In the United States primarily copper and sulfur are used as nucleants; in Europe and Asia selenium is used. At very low (1.0–1.6 wt %) antimony contents selenium is used exclusively.

Ammunition. Lead shot is produced by dropping molten lead–antimony alloys containing 0.5–8 wt % antimony through holes in pans into water. The shot alloys contain arsenic in an amount equal to about 20–30 wt % of the antimony content. Arsenic permits the molten drop of lead to become round during freefall. Tin produces elongated shot if present, therefore it is restricted to less than 0.0005 wt %. Shot up to 5.8 mm in diameter is dropped from towers; larger shot must be cast in molds. Lead–antimony (0.5–3.0 wt %) alloys are also used for cast or swaged bullets.

Cable Sheathing. Lead–antimony alloys containing 0.5–1.0 wt % antimony are used to form a barrier sheath in high voltage power cables. Lead–copper and lead–tin–arsenic alloys are also used. Antimony containing alloys are used when higher strength and resistance to vibration is required. British standard BS-801 alloy B contains 0.85 wt % antimony. Lead–antimony

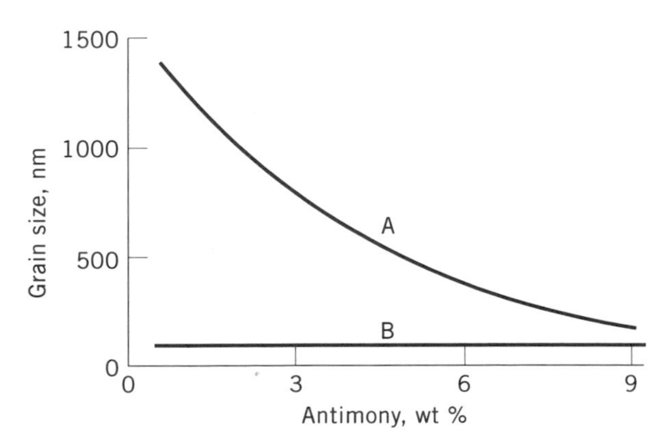

Fig. 3. Grain size as a function of antimony content: A, without grain refiners; B, with addition of selenium (4). Courtesy of the Electrochemical Society.

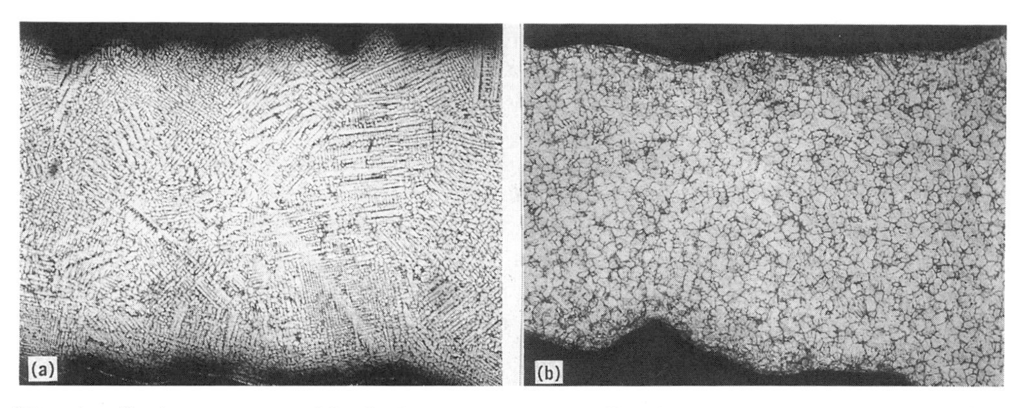

Fig. 4. Grain structure of lead–2 wt % antimony alloy battery grid at a magnification of 50×: (**a**) no nucleants; (**b**) containing 0.025 wt % selenium as a grain refiner.

alloys having 0.6 wt % or less antimony can be continuously extruded, are impervious to oil and moisture, and remain pliable indefinitely. Lead–antimony (1.0 wt %) is generally used as sleeves for cable splices.

Anodes. Lead–antimony (6–10 wt %) alloys containing 0.5–1.0 wt % arsenic have been used widely as anodes in copper, nickel, and chromium electrowinning and metal plating processes. Lead–antimony anodes have high strength and develop a corrosion-resistant protective layer of lead dioxide during use. Lead–antimony anodes are resistant to passivation when the current is frequently interrupted.

Wrought Lead–Antimony. Most lead–antimony alloys are used in the cast form. Rolling, extruding, or drawing breaks up the as-cast strengthening antimony eutectic particles, thus wrought lead–antimony alloys are weaker and have higher creep rates than cast lead–antimony alloys, but greater strength than lead–copper alloys. These can be easily fabricated into most lead products. Extruded and formed lead–4–6 wt % antimony alloys are used to produce a variety of products for radiation protection. Because of excellent resistance to corrosion by a variety of chemicals, lead–antimony (4–8 wt %) alloys are used for tank linings, pumps (qv), valves, and heating and cooling coils, particularly where sulfuric acid or sulfate solutions are handled at elevated temperature. Arsenic is generally added to increase resistance to creep at elevated temperatures.

Lead–Antimony–Tin Alloys. Lead–antimony–tin alloys are used for printing, bearings, solders, slush castings, and specialty castings. These alloys have low melting points, high hardness, and excellent high temperature strength and fluidity. Printing alloys generally contain more than 11 wt % antimony and 3–14 wt % tin. The alloys shown in Table 2 have low melting points and high hardness for wear resistance in printing. Replication of mold details is particularly important.

Excellent antifriction properties and good hardness (qv) make lead–antimony–tin alloys suitable for journal bearings. The alloys contain 9–15 wt % antimony and 1–20 wt % tin and may also contain copper and arsenic, which improve compression, fatigue, and creep strength important in bearings. Lead–antimony–tin bearing alloys are listed in ASTM B23-92 (7).

Table 2. Printing Alloys

Process	Composition, wt %			Temperature		Brinell hardness
	Sn	Sb	Pb	Liquidus, C°	Solidus, C°	
linotype	4	11.5	84.5	243	239	22
electrotype	3	3	94	299	245	14
stereotype	5	14	81	256	240	23
monotype	7	16.5	76.5	275	240	26
foundry type	14	24	62	318	240	32

Specialty castings include belt buckles, trophies, casket trim, miniature figures, and hollow-ware. Slush castings are produced by pouring an alloy into a mold, permitting a given thickness to solidify, and pouring out the remaining liquid to produce a hollow, light, high detail casting. The near-eutectic lead alloy, containing 11 wt % antimony, 1 wt % tin, and 0–0.5 wt % arsenic, has a low melting point which permits silicon rubber molds to be used in spin-casting processes. Lead–antimony–tin alloys for die casting are listed in ASTM B102-93 (7).

Low (2–5 wt %) antimony, low (2–5 wt %) tin lead alloys are used for automobile body solder. Special lead–antimony alloys containing 1–4 wt % antimony are used for wheel-balancing weights, battery cable clamps, collapsible tubes, and highly machined isotope pots.

Lead–Calcium Alloys

Lead–calcium alloys are replacing lead–antimony alloys for many applications. Most U.S. original equipment automotive batteries are constructed of lead–calcium grids, whereas most U.S. replacement batteries utilize lead–calcium alloys for the negative grid and lead–antimony alloys for the positive grid. Lead–calcium is used worldwide for standby power, submarine, and specialty sealed batteries. Lead–calcium alloy batteries do not require addition of water and can therefore be sealed. Lead–calcium alloys are used for electrowinning anodes, cable sheathing and sleeving, specialty boat keels, and lead alloy tapes.

Binary Lead–Calcium Alloys. The lead–calcium phase diagram is shown in Figure 5 (1). The phase diagram is a peritectic having a peritectic temperature of 328.3°C. The maximum solubility of calcium is 0.10 wt % at 328.3°C and rapidly decreases to 0.02 wt % at 200°C and 0.01 wt % at 25°C. The large decrease in solubility with decreasing temperature permits precipitation strengthening of binary lead–calcium alloys by the compound Pb_3Ca. This compound is also formed in molten alloys containing more than 0.07% calcium.

Below 0.10 wt % calcium the alloys solidify into a cellular, dendritic grain structure as seen in Figure 6 (8). Impurities segregate to the intercellular boundaries and are thus accented in the microstructure. Reverse segregation causes some areas of the casting to contain 0.10 wt % calcium regardless of the original calcium content of the melt. Precipitation of calcium occurs by a discontinuous precipitation reaction (8). In this reaction, grain boundaries move through the matrix, the original cast grain is destroyed, and the characteristic serrated grain boundaries of lead–calcium alloys seen in Figure 7 are produced. The calcium

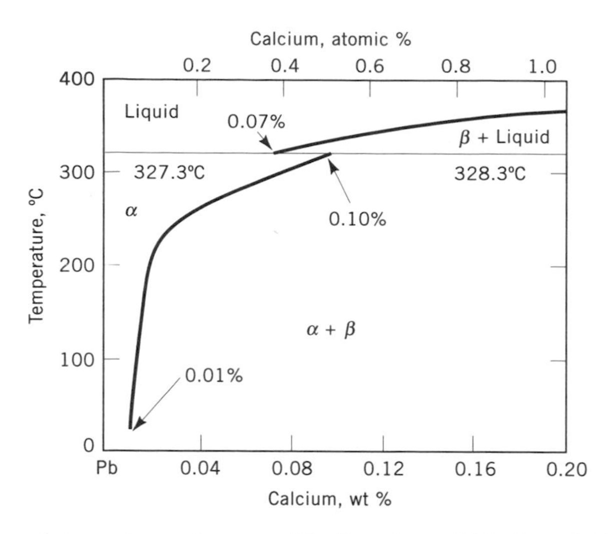

Fig. 5. Lead–calcium phase diagram (1). Courtesy of McGraw-Hill Book Co., Inc.

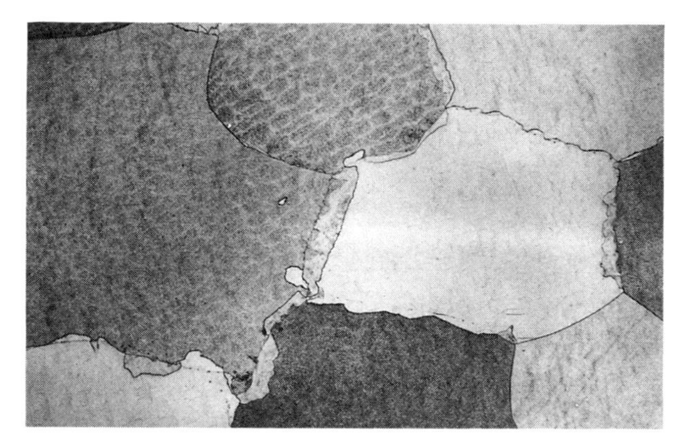

Fig. 6. Large columnar as-cast grain structure of lead–calcium alloys at a magnification of 80×.

content and the rate of grain boundary movement determine the final grain size and shape.

Lead–calcium (0.04%) alloy, used as cable sheaths and sleeves for cable splices, is significantly stronger and has greater creep resistance than lead–copper (0.06 wt %) or lead–antimony (1 wt %). Table 3 compares the mechanical properties of lead–calcium alloy to these alloys and to pure lead. The lead–calcium alloys have outstanding creep and fatigue resistance as well as relatively good ductility. Aging after production may cause a permanent set and difficulty in uncoiling; hence, lead–calcium is used primarily in straight lengths or as sleeves.

The main use of binary lead–calcium alloys is for the grids in large, stationary standby power batteries. These batteries use alloys containing 0.03–0.07 wt % calcium. The alloy has sufficient strength for the application, but it is used principally because of its resistance of self-discharge and because it reduces wa-

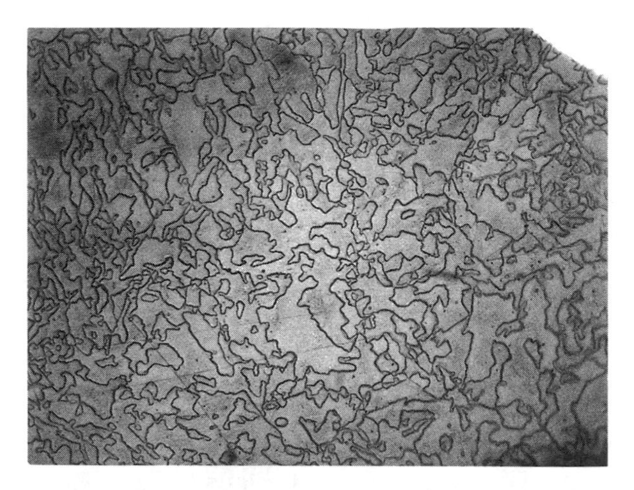

Fig. 7. Grain structure of lead–0.07 wt % calcium alloy aged for seven days showing serrated grain boundaries at a magnification of 320×.

Table 3. Mechanical Properties of Pure Lead and Lead Alloys

		Lead alloys		
Property	Pure lead	0.06 wt % Cu	1 wt % Sb	0.04 wt % Ca
tensile strength, MPa[a]	12.2	17.4	20.9	27.9
elongation, %	55	55	35	35
fatigue strength, MPa[a]	2.75	4.90	6.28	9.02

[a]To convert MPa to psi, multiply by 145.

ter loss. The grids are large grained and resistant to growth and creep during the long battery life.

Reactivity of Lead–Calcium Alloys. Precise control of the calcium content is required to control the grain structure, corrosion resistance, and mechanical properties of lead–calcium alloys. Calcium reacts readily with air and other elements such as antimony, arsenic, and sulfur to produce oxides or intermetallic compounds (see CALCIUM AND CALCIUM ALLOYS). In these reactions, calcium is lost and suspended solids reduce fluidity and castibility. The very thin grids that are required for automotive batteries are difficult to cast from lead–calcium alloys.

A rapid method to determine the calcium content of lead alloys is a liquid-metal titration using lead–antimony (1%) (9). The end point is indicated by a gray oxide film pattern on the surface of a solidified sample of the metal when observed at a 45° angle to a light source. The basis for the titration is the reaction between calcium and antimony. The percentage of calcium in the sample can be calculated from the amount of antimony used. If additional calcium is needed in the alloy, the melt is sweetened with a lead–calcium (1 wt %) master alloy.

Lead–Calcium–Aluminum Alloys. Lead–calcium alloys can be protected against loss of calcium by addition of aluminum. Aluminum provides a protective oxide skin on molten lead–calcium alloys. Even when scrap is remelted, calcium content is maintained by the presence of 0.02 wt % aluminum. Alloys without aluminum rapidly lose calcium, whereas those that contain 0.03 wt % aluminum exhibit negligible calcium losses, as shown in Figure 8 (10). Even with less than optimum aluminum levels, the rate of oxidation is lower than that of aluminum-free alloys.

Producing lead–calcium–aluminum alloys is difficult. Calcium and aluminum can be added simultaneously to lead using a calcium (73 wt %)–aluminum (27 wt %) master alloy (11) (see ALUMINUM AND ALUMINUM ALLOYS). Using this method, the calcium and aluminum contents can be precisely controlled. Pressed pellets of metallic aluminum and metallic calcium are also used.

Lead alloys containing 0.09–0.15 wt % calcium and 0.015–0.03 wt % aluminum are used for the negative battery grids of virtually all lead–acid batteries in the United States and are also used in Japan, Canada, and Europe. If the molten alloy is held at too low a temperature, the aluminum precipitates from solution, rises to the surface of the molten alloy as finely divided aluminum particles, and enters the dross layer atop the melt.

Lead–Calcium–Tin Alloys. Tin additions to lead–calcium and lead–calcium–aluminum alloys enhances the mechanical (8) and electrochemical properties (12). Tin additions reduce the rate of aging compared to lead–calcium binary alloys. The positive grid alloys for maintenance-free lead–calcium batteries contain 0.3–1.2 wt % tin and also aluminum.

Cast lead–calcium–tin alloys usually contain 0.06–0.11 wt % calcium and 0.3 wt % tin. These have excellent fluidity, harden rapidly, have a fine grain

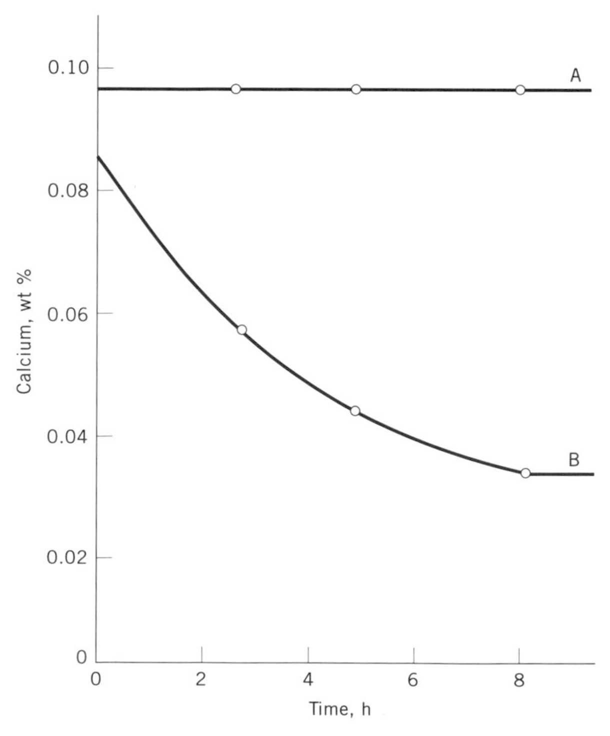

Fig. 8. Effect of aluminum on loss of calcium: A, with 0.03 wt % aluminum; B, without aluminum (10). Courtesy of the Electrochemical Society.

structure, and are resistant to corrosion. Table 4 lists the mechanical properties of cast lead–calcium–tin alloys and other alloys.

Table 4. Mechanical Properties of Cast Lead Alloys

	Lead alloy components, wt %						
Property	Cu 0.06	Ca 0.06	Ca 0.04, Sn 0.50	Ca 0.08, Sn 0.30	Ca 0.08, Sn 0.50	Ca 0.08, Sn 1.0	Sb 6.0
tensile strength, MPa[a]	17.4	34.8	41.8	46.0	48.8	59.7	59.2
yield strength, MPa[a]	9.0	24.3	27.9	34.8	38.2	46.0	55.8
elongation, %	55	30	15	20	15	15	15
time to failure,[b] h		10	30	50	100	450	> 10, 000

[a]To convert MPa to psi, multiply by 145.
[b]At 20.9 MPa.

Wrought lead–calcium–tin alloys contain more tin, have higher mechanical strength, exhibit greater stability, and are more creep resistant than the cast alloys. Rolled lead–calcium–tin alloy strip is used to produce automotive battery grids in a continuous process (13). Table 5 lists the mechanical properties of rolled lead–calcium–tin alloys, compared with lead–copper and rolled lead–antimony (6 wt %) alloys.

Lead Anodes. A principal use for lead–calcium–tin alloys is lead anodes for electrowinning. The lead–calcium anodes form a hard, adherent lead dioxide layer during use, resist corrosion, and greatly reduce lead contamination of the cathode. Anodes produced from cast lead–calcium (0.03–0.09 wt %) alloys have a tendency to warp owing to low mechanical strength and casting defects.

Wrought lead–calcium–tin anodes have replaced many cast lead–calcium anodes (14). Superior mechanical properties, uniform grain structure, low corrosion rates, and lack of casting defects result in increased life for wrought lead–calcium–tin anodes compared to other lead alloy anodes.

Table 5. Mechanical Properties of Rolled Lead Alloys

	Lead alloy components, wt %					
Property	Cu 0.06	Ca 0.06	Ca 0.04, Sn 0.50	Ca 0.065, Sn 0.70	Ca 0.06, Sn 1.30	Sb 6.0
tensile strength, MPa[a]	17.4	32.8	48.8	62.8	69.6	30.6
yield strength, MPa[a]	9.0	25.1	46.0	59.2	66.2	19.5
elongation, %	55	35	15	10	10	35
time to failure,[b] h		7	850	3,000	>10,000	1.5

[a]To convert MPa to psi, multiply by 145.
[b]At 20.9 MPa.

Lead–Copper Alloys

The lead–copper phase diagram (1) is shown in Figure 9. Copper is an alloying element as well as an impurity in lead. The lead–copper system has a eutectic

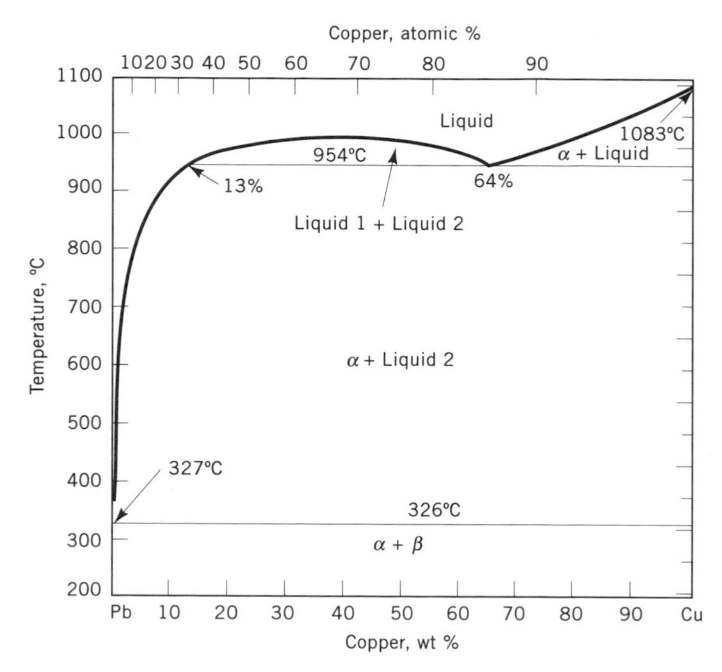

Fig. 9. Lead–copper system (1). Courtesy of McGraw-Hill Book Co., Inc.

point at 0.06% copper and 326°C. In lead refining, the copper content can thus be reduced to about 0.08% merely by cooling. Further refining requires chemical treatment. The solubility of copper in lead decreases to about 0.005% at 0°C.

High copper–lead alloys generally contain 60–70% copper. The mechanical properties of sand-cast copper–lead alloys are shown in Table 5. The high copper alloys are difficult to cast and are susceptible to extensive segregation. Cast lead–copper (60–70 wt %) alloys are used as bearing and bushings for high temperature service (see HIGH TEMPERATURE ALLOYS). More recently, the cast alloys have been replaced by sintered copper powder products infiltrated with lead to produce more uniform distribution of the lead.

Only lead alloys containing copper below 0.08% have practical applications. Lead sheet, pipe, cable sheathing, wire, and fabricated products are produced from lead–copper alloys having copper contents near the eutectic composition. Lead–copper alloys in the range 0.03–0.08 wt % copper are covered by many specifications: ASTM B29-92 (7), QQL 171 (United States), BS 334, HP2 Type 11 (Canada), DIN 1719 (Germany), and AS 1812 (Australia).

Lead–copper alloys are specified because of superior mechanical properties, creep resistance, corrosion resistance, and high temperature stability compared to pure lead. The mechanical properties of lead–copper alloys are compared to pure lead, and to lead–antimony and lead–calcium alloys in Tables 4 and 5.

Lead–copper alloys are the primary material used in the continuous extrusion of cable coverings for the electrical power cable industry in the United States. Other alloys, containing tin and arsenic as well as copper, have also been developed for cable sheathing in the United States to provide higher fatigue strength.

Extruded or rolled lead–copper alloys contain a uniform dispersion of copper particles in a lead matrix. Because the solid solubility of copper in lead is very low, copper particles in the matrix remain stable up to near the melting point of lead, maintaining uniform grain size even at elevated temperature.

Copper-containing lead alloys undergo less corrosion in sulfuric acid or sulfate solutions than pure lead or other lead alloys. The uniformly dispersed copper particles give rise to local cells in which lead forms the anode and copper forms the cathode. Through this anodic corrosion of the lead, an insoluble film of lead sulfate forms on the surface of the lead, passivating it and preventing further corrosion. The film, if damaged, rapidly reforms.

Lead–copper alloys are also used as tank linings, tubes for acid mist precipitators, steam heating pipes for sulfuric acid or chromate plating baths, and flashing and sheeting (see TANKS AND OTHER PRESSURE VESSELS).

Lead–Silver Alloys

Silver readily forms alloys with lead. Lead is often used as a base metal solvent for silver recovery processes. The lead–silver system is a simple eutectic having the eutectic point at 2.5 wt % silver and 304°C. The solid solubility of silver in lead is 0.10 wt % at 304°C, dropping to less than 0.02 wt % at 20°C.

Lead–silver alloys show significant age hardening when quenched from elevated temperature. Because of the pronounced hardening which occurs using small amounts of silver, the content of silver as an impurity in pure lead is restricted to less than 0.0025 wt % in most specifications. Small additions of silver to lead produces high resistance to recrystallization and grain growth.

The principal uses for lead–silver alloys are as anodes and high temperature solders. Only lead–silver alloys containing less than 6 wt % silver are used commercially. Lead alloys containing 0.75–1.25 wt % silver are used as insoluble anodes in the electrowinning of zinc and manganese. Some zinc refineries have reduced the silver content of the anodes to as low as 0.25 wt %. Lead–calcium–silver anodes are also used for zinc electrowinning. Silver promotes the formation of very hard, dense, electrically conducting layers of lead dioxide on the surface of the anode.

Silver reduces the oxygen evolution potential at the anode, which reduces the rate of corrosion and decreases lead contamination of the cathode. Lead–antimony–silver alloy anodes are used for the production of thin copper foil for use in electronics. Lead–silver (2 wt %), lead–silver (1 wt %)–tin (1 wt %), and lead–antimony (6 wt %)–silver (1–2 wt %) alloys are used as anodes in cathodic protection of steel pipes and structures in fresh, brackish, or seawater. The lead dioxide layer is not only conductive, but also resists decomposition in chloride environments. Silver-free alloys rapidly become passivated and scale badly in seawater. Silver is also added to the positive grids of lead–acid batteries in small amounts (0.005–0.05 wt %) to reduce the rate of corrosion.

Lead–silver alloys are used extensively as soft solders; these contain 1–6 wt % silver. Lead–silver solders have a narrower freezing range and higher melting point (304°C) than conventional solders. Solders containing 2.5 wt % silver or less are used either as binary alloys or combined with 0.5–2 wt %

tin. Lead–silver solders have excellent corrosion resistance. The composition of lead–silver solders is listed in ASTM B32-93 (solder alloys) (7).

Lead–Tellurium Alloys

Tellurium is often used in lead alloys when high mechanical strength at minimal alloy content is required. It is used for pipes and sheets, shielding for nuclear reactors (qv), and cable sheathing. Lead alloys containing 0.035–0.10 wt % tellurium generally also contain copper in amounts of 0.03–0.08 wt %. The U.S. Federal Specifications QQL-201F and ASTM B749-91 (7) cover the use of wrought lead–tellurium alloys. Lead–tellurium alloys for cable sheathing are specified by German DIN 17 640.

Cold-rolled alloys of lead with 0.06 wt % tellurium often attain ultimate tensile strengths of 25–30 MPa (3625–5350 psi). High mechanical strength, excellent creep resistance, and low levels of alloying elements have made lead–tellurium alloys the primary material for nuclear shielding for small reactors such as those aboard submarines. The alloy is self-supporting and does not generate secondary radiation.

Wrought or extruded lead–tellurium (0.035–0.10 wt %) alloys produce extremely fine grains. The binary alloy is, however, susceptible to recrystallization. The addition of copper or silver reduces grain growth and retains the fine grain size. Because tellurium is a poison for sealed lead–acid batteries, the tellurium content of lead and lead alloys used for such purposes is usually restricted to less than 1 ppm.

Lead–Tin Alloys

Lead alloys with tin in all proportions, providing a series of alloys that have wide application in many industries. The phase diagram in Figure 10 shows a eutectic composition of 61.9 wt % Sn–39.1 wt % Pb melting at 183°C (1). The solid solution of tin in lead decreases from 19.0 wt % at 183°C to 1.9 wt % at 20°C, and the solubility of lead in tin ranges from 2.5 wt % at 183°C to <0.3 wt % at 20°C.

Solders. The principal use of lead–tin alloys is as solders for sealing and joining metals. Solders range in composition from 20 to 98 wt % lead, the remainder being tin. Lead–tin (2 wt %) solder is used to seal the side seams of steel (tin) cans. Additions of 0.5 wt % silver to the alloy significantly improves the creep strength which is essential in pressurized cans. Pre-tinned cans allow the use of low tin solder. High speed manufacturing of cans requires an alloy having a very narrow freezing range.

Lead–tin alloys containing 40–50 wt % tin are used for general-purpose soldering. Low melting lead–tin solder having 63 wt % tin is used for electronic soldering, particularly for printed circuit boards. Lead alloys having 15–30 wt % tin are used for soldering automobile radiators and other types of heat exchangers. A lead alloy containing 2.5 wt % tin and 0.5 wt % silver is used when high temperature fatigue and creep strengths are required, for example, in soldered connections in heat exchangers. Solders having 35–40 wt % tin have a wide plastic range and are used for wiping lead joints. Sometimes 0.5–2.0 wt % antimony

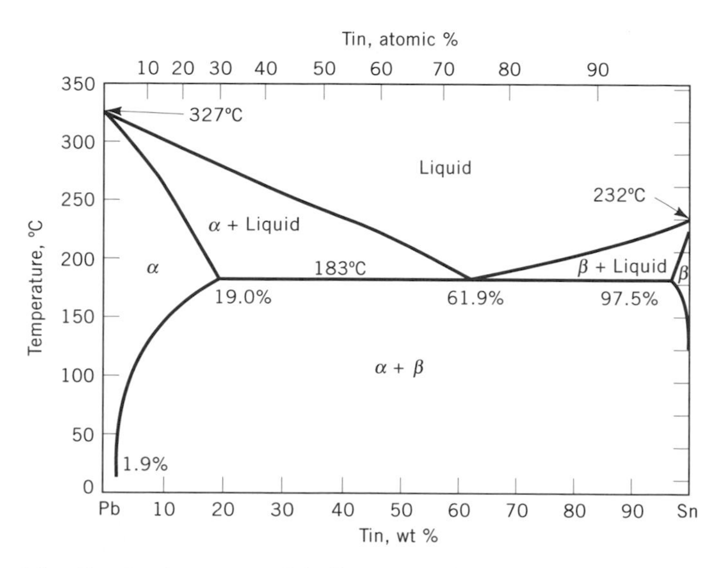

Fig. 10. Lead–tin system (1). Courtesy of McGraw-Hill Book Co., Inc.

is added to improve mechanical properties. Antimony contents of 0.2–0.5 wt % prevent brittle phase transformations at low temperature. Lead–tin solder alloys are listed in ASTM B32-93 (7).

Corrosion Protection. Lead–tin alloys are used for corrosion-resistant coatings on steel and copper (see METALLIC COATINGS). Terne steel is steel sheet coated with a lead alloy containing 15–20 wt % tin. This alloy contains sufficient tin to alloy with the surface of the steel. Terne sheets are produced flat or coiled, carrying different coating weights. These coated sheets are used for radio and television chassis, roofs, fuel tanks, air filters, oil filters, gaskets, metal furniture, gutters, and downspouts. Terne steel has good corrosion resistance, draws well, is relatively cheap, and offers a good base for painting.

A similar coating, containing 4 wt % tin, is applied to copper sheet and is used primarily for building flashings. Other lead–tin alloys, usually with 50 wt % tin, are applied as coatings to steel and copper electronic components for corrosion protection, appearance, and ease of soldering. Lead–tin alloys can be built up easily to any desired thickness by electroplating (qv) from a fluoborate solution. Electroplated coatings are not recommended for corrosion protection because these tend to be porous (see ELECTRONICS, COATINGS).

Tin is also used as an alloying element in lead–antimony alloys to improve fluidity and to prevent drossing, in lead–calcium alloys to improve mechanical properties and enhance electrochemical performance, in lead–arsenic alloys to maintain a stable composition, and as an additive to low melting alloys.

Lead–tin (1.8–2.5 wt %) is used both as a cable sheathing alloy (BS 801 alloy A and DIN 17640) and as a battery connector alloy in sealed lead–calcium–tin batteries (15). Tin is generally added to lead–arsenic cable alloys in small amounts. The arsenic alloys have excellent creep resistance and mechanical properties, but are unstable and lose arsenic readily by oxidation. The addition of small amounts of tin (0.10–0.20 wt %) eliminates arsenic loss. Lead alloys having 0.4 wt % tin and 0.15 % cadmium, which are used for cable sheathing, do not age harden, show excellent corrosion and creep resistance, and are very ductile.

Other Alloys

Low Melting Alloys. Lead alloys having large amounts of bismuth, tin, cadmium, and indium that melt at relatively low (10–183°C) temperatures are known as fusible or low melting alloys. The specifications of many of these alloys are listed in ASTM B774-87 (7).

These alloys are used as fuses, sprinkler system alloys, foundry pattern alloys, molds, dies, punches, cores, and mandrels where the low melting alloy is often melted out of a mold. The alloys are also used as solders, for the replication of human body parts (see PROSTHETIC DEVICES), and as filler for tube bending. Lead–indium alloys are often used to join metals to glass.

Reactive Lead Alloys. Strontium–lead alloys behave similarly to lead–calcium alloys in terms of mechanical properties and performance. Lead–strontium (0.08–0.20 wt %) alloys, also containing aluminum, have been used as battery grid alloys. These have excellent fluidity, harden rapidly, and have excellent resistance to corrosion. However, these alloys over-age rapidly, resulting in significant loss of mechanical strength within several days. The addition of 1 wt % or more of tin is required to resist over-aging. The high cost of lead–strontium alloys compared to lead–calcium alloys has restricted usage. Lead–strontium–tin alloys are used as anodes for copper electrowinning.

Lead–lithium and lead–lithium–tin alloys have been proposed as alloys for lead–acid battery grids because of very rapid aging and very high mechanical properties. These alloys are, however, susceptible to grain boundary corrosion. Lead–lithium alloys containing strontium, barium, and calcium have been used for bearings. The high melting point and the retention of mechanical properties and stability at high temperatures make these alloys particularly attractive for bearings. The addition of aluminum to prevent oxidation permits casting at high temperatures without oxidation.

Occupational Health and Safety

Because of the toxicity of lead, special care must be taken when working with lead alloys. Lead and its inorganic compounds are neurotoxins which may produce peripheral neuropathy. For an overview of the effects of lead exposure, see Occupational Exposure to Lead, Appendix A (29 CRF 1910.1025) (see LEAD COMPOUNDS, INDUSTRIAL TOXICOLOGY).

Most lead alloys are first melted during processing. The melting or fusing of lead alloys may generate lead oxides, drosses, or fumes which can present a health hazard. The principal routes for absorption of lead metal are inhalation and ingestion. Inhalation can be avoided by use of ventilators and respirators where the exposure is above the permissible exposure limit (PEL) or the threshold limit value (TLV) specified by OSHA (United States) or other local, municipal, or federal regulations. These limits vary considerably by industry and country but generally consist of a PEL of ca 50–200 $\mu g/m^3$ of lead in air, and lead levels of 50–80 $\mu g/dL$ in whole blood.

Coveralls or other full-body clothing should be worn when working with lead alloys and properly laundered after use. Hard hats, safety glasses, safety boots, and other safety equipment should be worn as appropriate for the industrial environment where the lead alloys are used. Hands, face, neck, and arms should be washed before eating or smoking.

Lead–antimony or lead–arsenic alloys must not be mixed with lead–calcium (aluminum) alloys in the molten state. Addition of lead–calcium–aluminum alloys to lead–antimony alloys results in reaction of calcium or aluminum with the antimony and arsenic to form arsenides and antimonides. The dross containing the arsenides and antimonides floats to the surface of the molten lead alloy and may generate poisonous arsine or stibine if it becomes wet. Care must be taken to prevent mixing of calcium and antimony alloys and to ensure proper handling of drosses.

If the temperature of a molten lead–calcium (tin)–aluminum alloy is not kept sufficiently high, finely divided aluminum particles may precipitate and float to the top of the melt. These may become mixed with oxides of lead in the dross. The finely divided aluminum particles can react violently with the oxides in the dross if ignited. Ignition can occur if attempts are made to melt or burn the dross away from areas of buildup with a torch. The oxides in the dross can supply oxygen for the combustion of aluminum once ignited.

Despite the benefits of lead and lead alloys, the use of these materials is declining rapidly, owing primarily to environmental health and safety factors. For many years, lead alloys were the materials of choice for many corrosive environments, but are now being replaced by stainless steel, plastics, and exotic metals. The toxic nature of lead requires special precautions and handling not necessary with other materials. These requirements have reduced the usage of lead alloys.

Other Uses for Lead Alloys

Antifriction. Lead has excellent antifriction properties which make it a good base element in bearings. When combined with other metals such as antimony, tin, copper, calcium, or aluminum, it provides excellent antifriction properties with the other metal providing strength. Powdered lead has extensive use as antiseize material for oilfield and mining industries because the lead does not adhere to the mating surfaces and prevents galling (see MINERAL RECOVERY AND PROCESSING; PETROLEUM).

Sound Attenuation. The high density and low strength of lead makes it an excellent barrier material for reducing the transmission of noise from adjacent areas (see NOISE POLLUTION AND ABATEMENT METHODS). The excellent sound attenuation properties of lead have been used by combining powdered lead with plastics and fibers to produce curtain walls or lead-bearing composites. In most of these applications, lead is alloyed with a small amount of copper.

Radiation Shielding. Because of the relatively high (11.35 g/cm^3) density of lead, it is used extensively for shielding of x-rays and gamma-rays in the form of sheets, shot, brick, or special castings. Lead sheet is most commonly used for lining x-ray rooms. The flat sheet lead is bonded to the supporting wall panels and the seams are covered with additional lead to prevent radiation leakage.

Lead bricks are generally used as temporary shields for radiation sources at nuclear power stations, research institutes, hospitals, and fuel reprocessing plants. Flat, rectangular bricks require a double layer with staggered seams whereas the interlocking bricks require only one course. Lead shot can be poured into inaccessible areas like a liquid.

Lead-loaded plastics containing up to 90 wt % lead are used in x-ray protection as aprons and temporary shields in medical and industrial applications. Leaded glass is used to attenuate radiation where viewing the ongoing process is required. Steel-jacketed containers filled with lead or special lead containers are used to transfer, ship, and store fuel rods, radioactive sources, and nuclear waste. Lead is generally used where space is limited.

Corrosion Resistance. Lead and many of its alloys exhibit excellent corrosion resistance owing to the rapid formation of a passive, impermeable, insoluble protective film when the lead is exposed to the corrosive solution (see CORROSION AND CORROSION CONTROL).

Lead shows excellent resistance to phosphoric and sulfuric acid in almost all concentrations and at elevated temperatures, as well as to sulfide, sulfite, and sulfate solutions. The corrosion film is insoluble lead sulfate which rapidly reforms if it is damaged. Lead is also resistant to chlorides, fluorides, and bromates at low concentrations and low temperatures. However, because lead is soluble in nitric and acetic acids, it is not resistant to these acids.

Because the corrosion resistance of lead and lead alloys is associated with the formation of the protective corrosion film, removal of the film in any way causes rapid attack. Thus the velocity of a solution passing over a surface can lead to significantly increased attack, particularly if the solution contains suspended particulate material. Lead is also attacked rapidly in the presence of high velocity deionized water. The lack of dissolved minerals in such water prevents the formation of an insoluble protective film. In most solutions, lead and lead alloys are resistant to galvanic corrosion because of the formation of a nonconductive corrosion film. In contact with more noble metals, however, lead can undergo galvanic attack which is accelerated by stray electrical currents.

The excellent corrosion-resistant lead dioxide, PbO_2, film formed on anodes and lead–acid battery positive grids in sulfuric acid has enabled lead insoluble anodes and lead–acid batteries to maintain the dominant positions in their respective fields.

Economic Aspects

Worldwide lead consumption was about 4.2 million in 1993. About 50% was produced from mined ore and 50% was produced from recycled lead-bearing products, primarily from used lead–acid batteries. In the United States, about 1.4 million t of lead was consumed in 1993; about 80% of the lead consumed was for the production of lead–acid batteries. Worldwide about 50% of the lead consumed is used for lead–acid batteries. In the United States, about 97% of the lead used in lead–acid batteries is recycled, making lead the most recycled of any metal in the world (see RECYCLING, NONFERROUS METALS). The lead usage throughout the world is about 50% pure lead, primarily for active materials of lead–acid batteries, gasoline additives, pigments, and glasses and glazes. The remaining material is consumed in the form of lead alloys.

The principal manufacturers of lead and lead alloys in North America are Tonolli, NOVA, Cominco, and Noranda in Canada; Doe Run, ASARCO, RSR Corp., Exide, Schuylkill, Sanders, GNB, and East Penn in the United States; and Penoles in Mexico. In Europe, the principal manufacturers are Union Meniere in Belgium; Metallgellschaft in Germany; Metalleurop in Germany–France; Brittania Refined Metals and H. J. Enthoven in the U.K.; Nuova SAMIM in Italy; and Boliden in Sweden. In Asia and Australia the principals are MIM and Pasminco in Australia; Mitsui, Mitsubishi, and Toho Zinc in Japan; and Korea Zinc in Korea.

The worldwide price for lead is determined daily by trading on the London Metal Exchange (LME). During the past several years, lead and lead alloy prices have been depressed by excess stocks in LME warehouses. These stocks of lead have come from the former Eastern Bloc countries, primarily from government stockpiles but also as exports to generate foreign exchange. LME average lead price for the year 1987 was $0.59/kg and for the years 1988 through 1992, $0.65, $0.67, $0.81, $0.56, and $0.54/kg, respectively. Prices for lead alloys generally reflect the additional cost of the metals used to produce the lead alloys above the cost of lead.

BIBLIOGRAPHY

"Lead Alloys" in *ECT* 1st ed., Vol. 8, pp. 253–266, by C. H. Hack, National Lead Co.; "Lead Alloys, Lead Utilization" in *ECT* 2nd ed., Vol. 12, pp. 247–266, by C. H. Hack, National Lead Co.; "Lead Alloys" in *ECT* 3rd ed., Vol. 14, pp. 140–160, by R. D. Prengaman, RSR Corp.

1. M. Hansen, *Constitution of Binary Alloys*, McGraw-Hill Book Co., Inc., New York, 1958.
2. V. Heubner and A. Ueberschaer, in *Proceedings 6th International Conference Lead*, Lead Development Association, London, 1974, p. 59.
3. R. D. Prengaman, in *Proceedings 8th International Conference Lead*, Lead Development Association, London, 1983, p. 69.
4. B. E. Kallup and D. Berndt, in K. R. Bullock and D. Pavlov, eds., *Advances in Lead Acid Batteries*, The Electrochemical Society, Pennington, N.J., 1984, p. 214.
5. R. D. Prengaman, *Battery Man*, 29 (Oct. 1983).
6. R. D. Prengaman, *Batteries Int.* **10**, 52 (Jan. 1992).
7. *Annual Book of ASTM Standards*, Sect. 2, Vol. 02.04, ASTM, Philadelphia, Pa., 1993.
8. R. D. Prengaman, *Proceedings 7th International Conference Lead*, Lead Development Association, London, 1980, p. 34.
9. G. M. Bouton and G. S. Phipps, *Trans. Electrochemical Society Prepr.*, 92-B (1947).
10. R. D. Prengaman, in Ref. 4, p. 201.
11. U.S. Pat. 4,439,398 (1984), R. D. Prengaman (to RSR Corp.).
12. H. K. Giess, in Ref. 4, p. 241.
13. U.S. Pat. 3,891,459 (1975), C. P. McCartney and N. C. Williams (to Delco Remy Div. of General Motors Corp.).
14. R. D. Prengaman, in *Proceedings 9th International Conference Lead*, Lead Development Association, London, 1986, p. 47.

15. R. D. Prengaman, in *Proceedings 3rd International Lead–Acid Battery Seminar*, ILZRO, Research Triangle Park, N.C., 1989, p. 1.
16. OSHA, *U.S. Fed. Reg.* (Nov. 17, 1978).

General References

W. Hofmann, *Lead and Lead Alloys*, Springer-Verlag, New York, 1970.
A. Worchester and J. O'Reilly, in *Metals Handbook*, Vol. 2, 10th ed., ASM International, Metals Park, Ohio, 1990, p. 543.

R. DAVID PRENGAMAN
RSR Corporation

LEAD COMPOUNDS

LEAD SALTS

Lead (qv) is a member of Group 14 (IVA) of the Periodic Table because it has four electrons in its outer, or valence, shell. However, the usual valence of lead is +2, rather than +4. The two *s* electrons have higher ionization energies. As a result, tetravalent lead exists as a free, positive ion only in minimal concentrations. Furthermore, the bivalent or plumbous ion differs from the other Group 14 bivalent ions, such as the stannous ion of tin, because Pb^{2+} does not have reducing properties.

In general, the chemistry of inorganic lead compounds is similar to that of the alkaline-earth elements. Thus the carbonate, nitrate, and sulfate of lead are isomorphous with the corresponding compounds of calcium, barium, and strontium. In addition, many inorganic lead compounds possess two or more crystalline forms having different properties. For example, the oxides and the sulfide of bivalent lead are frequently colored as a result of their state of crystallization. Pure, tetragonal α-PbO is red; pure, orthorhombic β-PbO is yellow; and crystals of lead sulfide, PbS, have a black, metallic luster.

The carbonates, sulfates, nitrates, and halides of lead (except the yellow iodide) are colorless. Bivalent lead forms a soluble nitrate, chlorate, and acetate; a slightly soluble chloride; and an insoluble sulfate, carbonate, chromate, phosphate, molybdate, and sulfide. Highly crystalline basic lead salts of both anhydrous and hydrated types are readily formed. Tetrabasic lead sulfate [*52732-72-6*], $4PbO{\cdot}PbSO_4$, and the hydrated tribasic salt [*12397-06-7*], $3PbO{\cdot}PbSO_4{\cdot}H_2O$, for example, may be formed by boiling suspensions of lead oxide and lead sulfate in water. In addition, complex mixed salts, such as white lead, $2PbCO_3 \cdot Pb(OH)_2$, are readily formed.

A clean lead surface is not attacked by dry air, but in moist air the surface quickly becomes coated with a film of lead monoxide, PbO, which may be hydrated and quickly combine with carbon dioxide from the atmosphere to produce a lead carbonate. Most of this combined carbon dioxide can be driven off by heating to 250°C, but traces remain even after heating to higher (650°C) temperatures for long periods. The ease with which lead monoxide combines with silicon dioxide to form a low melting silicate has been utilized in the ceramics industry in the preparation of glazes and in the manufacture of certain types of glasses (see GLASS-CERAMICS; SILICON COMPOUNDS).

Tetravalent lead is obtained when the metal is subjected to strong oxidizing action, such as in the electrolytic oxidation of lead anodes to lead dioxide, PbO_2; when bivalent lead compounds are subjected to powerful oxidizing conditions, as in the calcination of lead monoxide to lead tetroxide, Pb_3O_4; or by wet oxidation of bivalent lead ions to lead dioxide by chlorine water. The inorganic compounds of tetravalent lead are relatively unstable; eg, in the presence of water they hydrolyze to give lead dioxide.

The lead storage battery, the largest single user of lead and its compounds, is made possible by the high degree of reversibility, both chemical and physical, in the fundamental chemical reaction

$$Pb + PbO_2 + 2\ H_2SO_4 \rightleftharpoons 2\ PbSO_4 + 2\ H_2O$$

The reaction is especially useful because of the high emf (ca 2.2 V) of the Pb/PbO_2 couple in dilute sulfuric acid (see BATTERIES, SECONDARY).

All lead-containing compounds are produced from pig lead through a series of suitable steps, except for the small amount of lead in leaded zinc oxide for which high grade lead ore is used. Most lead compounds are prepared directly or indirectly from lead monoxide, PbO, commonly known as litharge. The physical nature of the oxide, as to particle size, and its exact method of preparation have profound effects on the suitability of any particular lead monoxide product for use in any specific process. In general, lead compounds may be formed by one or more of three methods: (*1*) reaction between a slurry of litharge, or a similar lead compound such as the hydroxide or carbonate, and the desired acid, or solution thereof in the case of an organic acid, or soluble salt of that acid; (*2*) reaction between the solution of a lead salt and the desired acid, or solution thereof in the case of an organic acid, or soluble salt of the acid. These reactions are facilitated by the fact that the desired lead compound usually is relatively insoluble, thus forming as a precipitate; and (*3*) fusion or calcination of litharge and the desired oxide, such as B_2O_3, SnO_2, ZrO_2, and TiO_2, resulting in lead borate, lead stannate [*1344-41-8*], lead zirconate, ZrO_2, and lead titanate. This method is particularly applicable with the oxides of the elements in Groups 14 (IVA), 15 (VA), and 16 (VIA) of the Periodic Table.

Most uses of lead in chemical compounds other than in storage batteries are dissipative. The greater part of the lead used in other forms is recoverable.

Halides

Lead Fluoride. Lead difluoride, PbF_2, is a white orthorhombic salt to about 220°C where it is transformed into the cubic form; some physical properties are given in Table 1. Lead fluoride is soluble in nitric acid and insoluble in acetone and ammonia. It is formed by the action of hydrofluoric acid on lead hydroxide or carbonate, or by the reaction between potassium fluoride and lead nitrate.

Lead fluoride has been used in low power fuses (1), as a catalyst for the manufacture of picoline (see PYRIDINES AND PYRIDINE DERIVATIVES) (2), in glass coatings for infrared reflection (3), in low melting glasses (4), in phosphors for television-tube screens (see LUMINESCENT MATERIALS) (5), in activators for electroless plating (qv) of nickel on glass (6), in electrooptical coatings (7), and in zinc oxide varistors (8).

Lead Chloride. Lead dichloride, $PbCl_2$, forms white, orthorhombic needles; some physical properties are given in Table 1. Lead chloride is slightly soluble in dilute hydrochloric acid and ammonia and insoluble in alcohol. It is prepared by the reaction of lead monoxide or basic lead carbonate with hydrochloric acid, or by treating a solution of lead acetate with hydrochloric acid and allowing the precipitate to settle. It easily forms basic chlorides, such as $PbCl_2 \cdot Pb(OH)_2$ [*15887-88-4*], which is known as Pattinson's lead white, an artist's pigment.

Lead dichloride is the starting material for a number of organolead compounds (9). It has been used in asbestos clutch or brake linings (see BRAKE LININGS AND CLUTCH FACINGS) (10), as a co-catalyst for acrylonitrile production (11), as a catalyst for polymerization of olefins to highly crystalline, stereoregular polymers (12), as a cathode for magnesium–lead dichloride seawater batteries (13), to make rectifying junctions on gallium arsenide (14), as a flame retardant in polycarbonates (15) and nylon-6,6 wire coatings (16), as a flux for the galvanizing of steel (17), as a solid-phase chemical scrubber for ozone and hydrogen sulfide removal from gas (18), as a photochemical-sensitizing agent for metal patterns on printed circuit boards (19), and as a sterilization indicator on tapes that darken with zinc sulfide at 121°C in moist air (20).

Lead Bromide. Lead dibromide, $PbBr_2$, forms white orthorhombic crystals; some physical properties are given in Table 1. Lead(II) bromide is slightly soluble in ammonia and highly soluble in potassium bromide solutions owing to complex formation, soluble in acetic acid, but insoluble in alcohol. On exposure to

Table 1. Physical Properties of Lead Halides

Property	$PbBr_2$	$PbCl_2$	PbF_2	PbI_2
CAS Registry Number	[*10031-22-8*]	[*7758-95-4*]	[*7783-46-2*]	[*10101-63-0*]
mol wt	376.04	278.1	245.21	461.05
mp, °C	373	501	855	402
bp, °C	916	950	1290	954
d, g/cm^3	6.66	5.85	8.24	6.16
soly, g in 100 mL H_2O				
at 0°C	0.455	0.673		0.044
20°C		0.99	0.064	0.063
100°C	4.71	3.34		0.41

light, lead dibromide decomposes slowly and darkens because of release of lead. It is prepared from lead monoxide or carbonate and hydrobromic acid, or lead diiodide and lead(IV) bromide, or by treating an aqueous solution of lead nitrate with hydrobromic acid or a soluble metal bromide and allowing the precipitate to settle.

Lead bromide is a photopolymerization catalyst for acrylamide monomer and is used in photoduplication at exposures of 365-nm radiation (21). Black-gray positive images are obtained on a white background by applying a methyl alcohol solution of lead bromide in a poly(vinyl butyral) binder to a suitable substrate and then exposing the coating to light through a negative film (22). In another photographic process, the latent image is developed by reduction of $PbBr_2$ with a sulfur-containing reducing agent, such as mercaptoacetic acid (23). Lead dibromide used as an inorganic filler in fire-retardant polypropylene, polystyrene, and acrylonitrile–butadiene–styrene (ABS) plastics reduces the requirements of chlorinated hydrocarbon flame-resistant additives (24) (see FLAME RETARDANTS). For welding (qv) aluminum or aluminum-base alloys to other metals, such as iron, nickel, copper, zinc, or their alloys, an aqueous paste containing $PbBr_2$ serves as an excellent general-purpose welding flux (25).

Lead Iodide. Lead diiodide, PbI_2, forms a powder of yellow hexagonal crystals; some physical properties are given in Table 1. Lead diiodide is soluble in alkalies and potassium iodide, and insoluble in alcohol. It is made by treating a water-soluble lead compound with hydroiodic acid or a soluble metal iodide. It is readily purified by recrystallization in water.

Lead iodide decomposes when exposed to green light at about 180°C, thereby making it possible to record optical images on thin lead iodide films (26). other applications of lead iodide include photographic emulsions with thiols (27), aerosols for cloud seeding to produce rain artificially (28), asbestos brake linings (29), primary thermal batteries with iodine (30), mercury-vapor arc lamps (31,32), thermoelectric materials (33,34), lubricating greases (35), electrosensitive recording papers (36), and filters for far-infrared astronomy (37).

Oxides

Lead forms two simple oxides, PbO and PbO_2, where it is divalent and tetravalent, respectively. Lead also forms a mixed oxide, Pb_3O_4, and a black oxide which normally comprises 55–85% lead monoxide, the remainder being finely divided metallic lead. The largest market for lead chemicals is the use of lead oxides in lead–acid storage battery electrodes (see BATTERIES, SECONDARY). The ceramics industry is the next largest consumer of lead oxides, for use in glasses, glazes, and vitreous enamels for metal coating and glass decoration (see CERAMICS), followed by the rubber industries. Lead also forms reacted mixed oxides with other metals, eg, the ferrites (qv), useful as ferrimagnetic materials, and the titanates or zirconates, which are ferroelectric materials (see FERROELECTRICS). Lead monoxide, PbO, is used widely as a component of heat stabilizers (qv) (basic lead salts) for poly(vinyl chloride) resins (38). Some physical properties of lead oxides are given in Table 2.

Total consumption of lead in the United States in 1993 reached 1,318,800 t. Of this, 766,000 t (58%) is allocated to battery use supplied as either a mixed

Table 2. Physical Properties of Lead Oxides

Property	PbO	PbO_2	Pb_2O_3	Pb_3O_4
CAS Registry Number	*[1317-36-8]*	*[1309-60-0]*	*[1314-27-8]*	*[1314-41-6]*
mol wt	223.21	239.21	462.42	685.63
mp, °C	897[a]			830[b]
dec, °C	1472[c]	290	370	500
d, g/cm^3				
α	9.53	9.375		9.1
β	9.6			
crystal structure				
α	tetragonal	orthorhombic (columbite)		spinel[d]
β	orthorhombic	rutile		

[a]Begins to sublime before melting.
[b]When decomposition is prevented by oxygen pressure.
[c]Boiling point.
[d]Unit cell contains four Pb_3O_4 groups.

oxide or as metal. Approximately 95% of batteries are recycled and the lead recovered. In 1993, 908,000 t of lead came from secondary smelters and refiners compared to 350,000 t originating in primary mines and smelters (39). Approximately 51,000 t of lead was consumed in U.S. production of all oxides and chemicals applicable to all industries other than batteries. Estimates include 8000 t for plastics, 6000 t for gasoline additives, 2000 t for rubber, and 30,000 t for ceramics, glass, and electronics. Lead is not used to any extent in dispersive applications such as coatings.

Lead Monoxide. Lead monoxide (litharge), PbO, occurs as a reddish alpha form, which is stable up to 489°C where it transforms to a yellow beta form (massicot). The latter is stable at high temperatures. The solubility of α-PbO in water is 0.0504 g/L at 25°C; the solubility of the β-PbO is 0.1065 g/L at 25°C (40). Lead monoxide is amphoteric and dissolves in both acids and alkalies. In alkalies, it forms the plumbite ion PbO_2^{2-}. The monoxide is produced commercially by the reaction of molten lead with air or oxygen in a furnace. Black or gray oxide is manufactured by the Barton process, by the oxidation of atomized molten lead in air, as well as by the ball mill process, in which metallic lead balls of high purity are tumbled in the mill to form partially oxidized lead particles.

Lead monoxide is used primarily as plates for electric storage batteries of the lead–sulfuric acid type. It is also widely used in optical, electrical, and electronic glasses, as well as in the glazing of fine tableware. For use in glazes and vitreous enamels, lead oxides are often converted to lead mono-, bi-, and trisilicate frits to render the lead compounds insoluble (see ENAMELS, PORCELAIN OR VITREOUS). Litharge is also used in rubber and plastics as a vulcanizing agent; in lead soaps employed in the past as driers in varnishes; as a high temperature lubricant; as a neutralizing agent in organic syntheses; and as an intermediate material in the manufacture of pigments (41).

Lead Dioxide. Lead dioxide (lead peroxide, plattnerite), PbO_2, is a brownish black crystalline powder consisting of fine crystalline flakes in either α-

or β- form. Lead dioxide decomposes rather easily to the lower oxide, releasing oxygen when heated to 290°C and above. It is practically insoluble in water or alkaline solutions, slowly soluble in acetic acid or ammonium acetate, and more rapidly soluble in hydrochloric acid and in a mixture of nitric acid and hydrogen peroxide. Lead dioxide can be produced by anodic oxidation of solutions of lead salts or, commercially, by the oxidation of red lead, Pb_3O_4, in alkaline slurry with chlorine.

Lead dioxide is electrically conductive and is formed in place as the active material of the positive plates of lead–acid storage batteries. Because it is a vigorous oxidizing agent when heated, it is used in the manufacture of dyes, chemicals, matches (qv), pyrotechnics (qv), and liquid polysulfide polymers (42) (see POLYMERS CONTAINING SULFUR).

Lead Sesquioxide. Lead sesquioxide (lead trioxide), Pb_2O_3, is an amorphous, orange-yellow powder soluble in cold water. It decomposes in hot water and in acids to lead salts plus PbO_2. Lead sesquioxide can be prepared from lead dioxide by hydrothermal dissociation (43).

Lead sesquioxide is used as an oxidation catalyst for carbon monoxide in exhaust gases (44,45) (see EXHAUST CONTROL), as a catalyst for the preparation of lactams (46) (see ANTIBIOTICS, β-LACTAMS), in the manufacture of high purity diamonds (47) (see CARBON, DIAMOND–NATURAL), in fireproofing compositions for poly(ethylene terephthalate) plastics (48), in radiation detectors for x-rays and nuclear particles (49), and in vulcanization accelerators for neoprene rubber (50).

Lead Tetroxide. Lead tetroxide (red lead; minium; lead orthoplumbite), Pb_3O_4, is a brilliant orange-red pigment which accounted for U.S. shipments of 17,780 t in 1977, mainly to the ceramics and storage battery industries (40). U.S. shipments in 1993 amounted to approximately 12,000 t. The decrease in usage since 1973 (19,000 t) is attributable to discontinued use in the paint and coatings (qv) industry, and alterations in rubber and ceramics (qv) markets. It is insoluble in water and alcohol, and dissolves in acetic acid or hot hydrochloric acid. Red lead is manufactured by heating lead monoxide in a reverberatory furnace in the presence of air at 450–500°C until the desired oxidative composition is obtained.

Red lead was used as a pigment in anticorrosion paints for steel surfaces (51,52), such as bridges (53) and reinforcements of concrete (54), and in adhesives for polyester tire cords (55). It continues to be used in positive battery plates (56), in colored glass for fiber optics (qv) (57), in electrically conductive polymer compositions (58), in catalysts for combustion of carbon monoxide in exhausts (59), in explosives for metal forming (60), in photochromic glass (see CHROMOGENIC MATERIALS) (61), in low melting glass-ceramics (62), in propellants for inflation of automotive safety bags (63,64), in radiation shields for gamma rays (65), in the vulcanization of polyether rubbers (66), in sealing glasses for color television picture tubes (67), in foaming agents for porcelain building materials (68), as an inhibitor of zinc dendrite growth in alkaline storage cells (69), in rubber sheets for x-ray protection (70), and in waterproofing putty for ship hulls (71).

Lead Hydroxide. Lead hydroxide [*19781-14-3*], $Pb(OH)_2$, mol wt 241.23, starts to dehydrate at about 130°C, and decomposes to lead monoxide at 145°C. It is only sparingly soluble in water (0.0155 g/100 mL at 20°C; slightly more in hot water), soluble in acids and alkalies, but insoluble in acetone. Lead hydroxide is prepared by adding alkali to a solution of lead nitrate or by electrolysis of an alkaline solution with a lead anode.

Lead hydroxide is used in sealed nickel–cadmium battery electrolytes (72), in oxidation catalysts for cyclododecanol (73), in electrical insulating paper (74), in gel stabilizers for petroleum well plugging (75), in the manufacture of porous glass (76), in wastewater filters for chromate removal (77), in building radiation shielding (78), in lubricating grease (79), with a thiourea derivative in photothermographic copy sheets (80), and in uranium recovery from seawater (81).

Sulfide and Telluride

Lead Sulfide. Lead sulfide [*1314-87-0*] (galena, lead glance), PbS, mol wt 239.25, mp 114°C, $d = 7.57–7.59$ g/cm^3, is metallic black and crystallizes in the cubic system. It has a hardness of 2.5–2.75 on the Mohs' scale. Lead sulfide is sparingly soluble in water (0.01244 g/100 mL at 20°C), but dissolves easily in dilute nitric acid where the sulfur is oxidized to the elemental state. Concentrated hydrochloric acid decomposes lead sulfide, liberating hydrogen sulfide. Lead sulfide is photoconductive. It is the chief ore of lead and is prepared by heating metallic lead in sulfur vapor. It is available in technical and high purity (99.999%) grades.

Lead sulfide is used in photoconductive cells, infrared detectors, transistors, humidity sensors in rockets, catalysts for removing mercaptans from petroleum distillates, mirror coatings to limit reflectivity, high temperature solid-film lubricants, and in blue lead pigments (82).

Lead Telluride. Lead telluride [*1314-91-6*], PbTe, forms white cubic crystals, mol wt 334.79, sp gr 8.16, and has a hardness of 3 on the Mohs' scale. It is very slightly soluble in water, melts at 917°C, and is prepared by melting lead and tellurium together. Lead telluride has semiconductive and photoconductive properties. It is used in pyrometry, in heat-sensing instruments such as bolometers and infrared spectroscopes (see INFRARED TECHNOLOGY AND RAMAN SPECTROSCOPY), and in thermoelectric elements to convert heat directly to electricity (33,34,83). Lead telluride is also used in catalysts for oxygen reduction in fuel cells (qv) (84), as cathodes in primary batteries with lithium anodes (85), in electrical contacts for vacuum switches (86), in lead-ion selective electrodes (87), in tunable lasers (qv) (88), and in thermistors (89).

Sulfates

Lead forms a normal and an acid sulfate and several basic sulfates. Basic and normal lead sulfates are fundamental components in the operation of lead–sulfuric acid storage batteries. Basic lead sulfates also are used as pigments and heat stabilizers (qv) in vinyl and certain other plastics.

Lead Sulfate. Lead sulfate, $PbSO_4$, is soluble in concentrated acids and alkalies, forming hydroxyplumbites; some physical properties are given in Table 3. It is prepared by treating lead oxide, hydroxide, or carbonate with warm sulfuric acid, or by treating a soluble lead salt with sulfuric acid. Lead sulfate forms in lead storage batteries during discharge cycles. It has been used in photography

Table 3. Physical Properties of Lead Sulfates

Property	$PbSO_4$	$PbO \cdot PbSO_4$
CAS Registry Number	[*7446-14-2*]	[*12765-51-4*]
mol wt	303.25	526.44
mp, °C	1170[a]	977
d, g/cm^3	6.2	6.92
soly, g/100 mL H_2O		
at 25°C	4.25×10^{-3}	4.4×10^{-3}[b]
40°C	5.6×10^{-3}	
crystal structure	orthorhombic, monoclinic	monoclinic

[a]Decomposes above 900°C.
[b]At 0°C.

in combination with silver bromide and is used in the stabilization of clay soil for adobe structures, earth-fill dams, and roads (90).

Monobasic Lead Sulfate. Monobasic lead sulfate, $PbO \cdot PbSO_4$, is very slightly soluble in hot water and slightly soluble in sulfuric acid; some physical properties are given in Table 3. Basic lead sulfate can be prepared by fusing PbO and $PbSO_4$ or by boiling aqueous suspensions of these two components. The resultant white solid is filtered and dried. It is available in 225-kg multiwall bags at ca $2.25/kg (1994 price). Basic lead sulfate is used in paints as a white pigment, in poly(vinyl chloride) (PVC) plastics as a heat stabilizer, in rubbers as an inert filler, and as additives in textile dyeing and printing (91).

Dibasic Lead Sulfate. Dibasic lead sulfate [*12036-76-9*], $2PbO \cdot PbSO_4$, is a white powder, mol wt 749.70, mp 961°C. The dibasic compound can be prepared by fusion of the two components. It has been sold as a PVC stabilizer in Japan and is sold in Europe in combination with dibasic lead phosphite.

Tribasic Lead Sulfate. Tribasic lead sulfate [*12202-17-4*], $3PbO \cdot PbSO_4 \cdot H_2O$, is a fine white powder, mol wt 890.93, sp gr 6.9, refractive index 2.1, lead oxide content 90.1%, sieve analysis, 99.8% through 44 μm (325 mesh) (wet), water solubility 0.0262 g/L at 18°C. Tribasic lead sulfate is by far the most widely used basic lead sulfate for the stabilization of PVC polymers. It may be prepared by boiling aqueous suspensions of lead oxide and lead sulfate. The anhydrous compound decomposes at 895°C. Tribasic lead sulfate provides efficient, long-term heat stability in both flexible and rigid PVC compounds, it is easily dispersible and has excellent electrical insulation properties, and it is an effective activator for azodicarbonamide-type blowing agents for vinyl foams. Applications for tribasic lead sulfate stabilizers include thermal stabilization of flexible PVC wire insulation compounds containing phthalate-type plasticizers, wire insulation designed to meet Underwriters' Laboratories specifications through 80°C, rigid and flexible PVC foams, rigid vinyl profiles, and PVC plastisols. The usual range of tribasic lead sulfate required in PVC is between two and seven parts per hundred of resin (2–7 phr), depending on the intended application of the vinyl product.

Tetrabasic Lead Sulfate. Tetrabasic lead sulfate [*12065-90-6*], $4PbO \cdot PbSO_4$, mol wt 1196.12, sp gr 8.15, is made by fusion of stoichiometric quantities of litharge (PbO) and lead sulfate ($PbSO_4$); heat of formation,

$\Delta H^\circ = -1814$ kJ/mol (-434.1 kcal/mol). Alternatively, tetrabasic lead sulfate may be prepared by boiling the components in aqueous suspensions. At about 70°C, tribasic hydrate reacts with lead oxide to form tetrabasic sulfate. At 80°C, this transformation is complete in $\sim$ 20 hours. Tetrabasic lead sulfate is used in limited quantities in Europe as a PVC stabilizer. However, in the United States, lead–acid batteries have been developed by Bell Telephone Laboratories, which contain tetrabasic lead sulfate. Such batteries are used for emergency power at telephone switchboard stations and have an anticipated service life of over 50 years.

Lead Nitrate

Lead nitrate [*10099-74-8*], $Pb(NO_3)_2$, mol wt 331.23, sp gr 4.53, forms cubic or monoclinic colorless crystals. Above 205°C, oxygen and nitrogen dioxide are driven off, and basic lead nitrates are formed. Above 470°C, lead nitrate is decomposed to lead monoxide and Pb_3O_4. Lead nitrate is highly soluble in water (56.5 g/100 mL at 20°C; 127 g/100 mL at 100°C), soluble in alkalies and ammonia, and fairly soluble in alcohol (8.77 g/100 mL of 43% aqueous ethanol at 22°C). Lead nitrate is readily obtained by dissolving metallic lead, lead monoxide, or lead carbonate in dilute nitric acid. Excess acid prevents the formation of basic nitrates, and the desired lead nitrate can be crystallized by evaporation.

Lead nitrate is used in many industrial processes, ranging from ore processing to pyrotechnics (qv) to photothermography. Thus lead nitrate is used as a flotation agent in titanium removal from clays (92); in electrolytic refining of lead (93); in rayon delustering (94); in red lead manufacture (95); in matches, pyrotechnics, and explosives (96); as a heat stabilizer in nylon (97); as a coating on paper for photothermography (98); as an esterification catalyst for polyesters (99); as a rodenticide (see PESTICIDES) (100); as an electroluminescent mixture with zinc sulfide (101); as a means of electrodepositing lead dioxide coatings on nickel anodes (102); and as a means of recovering precious metals from cyanide solutions (103).

Phosphite

In commercial applications of poly(vinyl chloride) polymers where weathering resistance, thermal stability, and electrical insulating properties are required, a stabilizer system base on dibasic lead phosphite provides a unique balance of properties. Its plasticizer reactivity is in the same range as dibasic lead phthalate, its electrical properties are superior, and it is the only stabilizer known that can provide the required electrical properties and weathering resistance in the absence of carbon black pigmentation. A properly formulated PVC electrical insulation compound containing a dibasic lead phosphite stabilizer, in combination with rutile-type titanium dioxide, remains in serviceable condition for up to 20 years. This superior performance results from the high absorption of the ultraviolet portion of sunlight, as well as the antioxidant activity of the phosphite anion (see ANTIOXIDANTS). The high PbO content of this dibasic lead salt makes it a very effective acid acceptor for HCl during PVC processing.

Dibasic Lead Phosphite. Dibasic lead phosphite [*12141-20-7*], $2PbO \cdot PbHPO_3 \cdot {}^1/_2H_2O$, is a white crystalline powder, mol wt 742.63, sp gr 6.9, refractive index 2.25, lead oxide content 90.2%, sieve analysis, 99.8% through 44 μm (325 mesh) (wet), water solubility, nil. Fields of application for dibasic lead phosphite stabilizers in PVC include garden hose, flexible and rigid vinyl foams (as high temperature activator for azodicarbonamide-type blowing agents), coated fabrics, plastisols, electrical insulation, and extruded profiles for outdoor use. In general, at five parts per hundred of resin (5 phr), dibasic lead phosphite provides good heat stability and superior outdoor weathering properties in PVC. This stabilizer should be stored in closed containers, away from open flames, and at temperatures below 200°C. Exposure to sparks or static electricity should be avoided by grounding all electrical equipment and using wooden scoops.

Lead Azide

Lead azide [*13424-46-9*], $Pb(N_3)_2$, mol wt 291.23, crystallizes as colorless needles. It is a sensitive detonating agent, exploding at 350°C (see EXPLOSIVES AND PROPELLANTS). Lead azide should always be handled and shipped submerged in water to reduce sensitivity. Its water solubility is very low (0.023 g/100 mL at 18°C and 0.09 g/100 mL at 70°C) and it is insoluble in ammonium hydroxide but very soluble in acetic acid. Lead azide is commonly prepared by the reaction between dilute solutions of lead nitrate and sodium azide. For safety, it is stirred vigorously to prevent formation of large crystals, which may detonate. Lead azide is usually precipitated with a protective material, such as gelatin, and then granulated (104). Lead azide is also used to prepare electrophotographic layers (105) and for information storage on styrene–butadiene resins (106).

Lead Antimonate

Lead antimonate [*13510-89-9*] (Naples yellow), $Pb_3(SbO_4)_2$, mol wt 993.07, d = 6.58 g/cm^3, is an orange-yellow powder that is insoluble in water and dilute acids, but very slightly soluble in hydrochloric acid. Lead antimonates are modifiers for ferroelectric lead titanates, pigments in oil-base paints, and colorants for glasses and glazes (see COLORANTS FOR CERAMICS). They are made by the reaction of lead nitrate and potassium antimonate solutions, followed by concentration and crystallization.

Acetates

Anhydrous Lead Acetate. Anhydrous lead acetate [*301-04-2*] (plumbous acetate), $Pb(C_2H_3O_2)_2$, is a white, crystalline solid that decomposes on heating above its melting point; some physical properties are given in Table 4. Because of its high solubility in water, lead acetate is often used for the preparation of other lead salts by the wet method. Lead acetate is made by dissolving lead monoxide (litharge) or lead carbonate in strong acetic acid. Several types of basic salts are formed when lead acetates are prepared from lead monoxide in dilute acetic acid or at high pH. The basic salts of lead acetate are white crystalline compounds, which are highly soluble in water and dissolve in ethyl alcohol.

Table 4. Physical Properties of Lead Acetates

Property	Anhydrous	Basic	Trihydrate	Tetraacetate
mol wt	325.28	807.69	379.33	443.77
mp, °C	280	75 (200 dec)	75 (200 dec)	175
d, g/cm^3	3.25		2.55	2.228
refractive index, n_D			1.567[a]	
soly, g/100 mL H_2O				
at 15°C	44.3[b]	6.25	45.61	
100°C	221[c]	25	200	

[a]Along the β-axis.
[b]At 20°C.
[c]At 50°C.

Basic Lead Acetate. Basic lead acetate [*1335-32-6*] (lead subacetate), $2Pb(OH)_2 \cdot Pb(C_2H_3O_2)_2$, is a heavy white powder which is used for sugar analyses. Some physical properties are given in Table 4. Reagent grade is available in 11.3-kg cartons and in 45- and 147-kg fiber drums.

Lead Acetate Trihydrate. Lead acetate trihydrate [*6080-56-4*] (plumbous acetate trihydrate), $Pb(C_2H_3O_2)_2 \cdot 3H_2O$, is a white, monoclinic crystalline solid; some physical properties are given in Table 4. Upon heating it loses some of its water of crystallization, and after melting, decomposes at 200°C. The trihydrate is highly soluble in water but insoluble in ethyl alcohol. It has an intensely sweet taste, hence it is sometimes called sugar of lead, but it is poisonous. The trihydrate is made by dissolving lead monoxide in hot dilute acetic acid solution; on cooling, large crystals separate, sometimes up to 60-cm long.

Lead acetate trihydrate, the usual commercial form, is used in the preparation of basic lead carbonate and lead chromate, as a mordant in cotton dyes, as a reagent for the manufacture of lead salts of higher fatty acids, as a water repellant, as a component in combined toning and fixing baths for daylight printing papers, and as a means of treating awnings and outdoor furniture to prevent removal of mildew- and rot-proofing agents by rain or laundering. Other uses include preparation of rubber antioxidants; processing agent in the cosmetic, perfume, and toiletry industries; component of coloring agents for adhesives; and preparation of organic lead soaps as driers of paints and inks. The trihydrate is available in technical and reagent grades in 11.3-kg cartons, and 45- and 181-kg fiber drums. The high purity salts demand a premium price of about $1.10/kg.

Lead Tetraacetate. Lead tetraacetate [*546-67-8*] (plumbic acetate), $Pb(C_2H_3O_2)_4$, is a colorless, monoclinic crystalline solid that is soluble in chloroform and in hot acetic acid, but decomposes in cold water and in ethyl alcohol. Some physical properties are given in Table 4. Lead tetraacetate can be prepared by adding warm, water-free, glacial acetate acid to red lead, Pb_3O_4, and subsequent cooling. The salt decomposes with the addition of water to give PbO_2, but the yield can be improved by passing in chlorine gas. Lead tetraacetate is available in laboratory quantities as colorless to faintly pink crystals stored in glacial acetic acid.

Oxidation with lead tetraacetate is often used in organic syntheses, because the lead salt is highly selective in the splitting of vicinal glycols. The rate of oxidation of cis glycols is more rapid than of the trans isomers, a property widely used in the structural determination of sugars and other polyols. Lead tetraacetate readily cleaves α-hydroxy acids as oxalic acid at room temperature. Another use is the introduction of acetoxy groups in organic molecules, as in the preparation of cyclohexyl acetate and the acetoxylation of cyclohexanol. At high temperature, methylation takes place. In these reactions, the organic molecule must contain double bonds or activating substituents (107).

Lead Benzoate

Lead benzoate monohydrate [*6080-57-5*], $Pb(C_6H_5CO_2)_2 \cdot H_2O$, mol wt 467.43, is a white crystalline powder that loses its water of hydration when heated to 100°C. It is slightly soluble in cold water (0.16 g/100 mL at 20°C) and somewhat more soluble in warm water (0.31 g/100 mL at 49.5°C). The salt may be prepared by adding benzoic acid to a slurry of litharge, PbO, or by the reaction between solutions of sodium benzoate and a soluble lead compound. Lead benzoate is used as an antioxidant in organolead engine lubricants (108), as a catalyst in a blowing agent for polyethylene foams (109), and in fluorescence quenching of organic phosphors (110).

Carbonates

Lead Carbonate. Lead carbonate [*598-63-0*], $PbCO_3$, mol wt 267.22, d = 6.6 g/cm^3, forms colorless orthorhombic crystals; it decomposes at about 315°C. It is nearly insoluble in cold water (0.00011 g/100 mL at 20°C), but is transformed in hot water to the basic carbonate, $2PbCO_3 \cdot Pb(OH)_2$. Lead carbonate is soluble in acids and alkalies, but insoluble in alcohol and ammonia. It is prepared by passing CO_2 into a cold dilute solution of lead acetate, or by shaking a suspension of a lead salt less soluble than the carbonate with ammonium carbonate at a low temperature to avoid formation of basic lead carbonate.

Lead carbonate has a wide range of applications. It catalyzes the polymerization of formaldehyde to high molecular weight crystalline poly(oxymethylene) products (111). It is used in poly(vinyl chloride) friction liners for pulleys on drive cables of hoisting engines (112). To improve the bond of polychloroprene to metals in wire-reinforced hoses, 10–25 parts of lead carbonate are used in the elastomer (113). Lead carbonate is used as a component of high pressure lubricating greases (114), as a catalyst in the curing of moldable thermosetting silicone resins (115), as a coating on vinyl chloride polymers to improve their dielectric properties (116), as a component of corrosion-resistant, dispersion-strengthened grids in lead–acid storage batteries (117), as a photoconductor for electrophotography (qv) (118), as a coating on heat-sensitive sheets for thermographic copying (119), as a component of a lubricant–stabilizer for poly(vinyl chloride) (120), as a component in the manufacture of thermistors (121), and as a component in slip-preventing waxes for steel cables to provide higher wear resistance (122) (see COMPOSITE MATERIALS, POLYMER-MATRIX).

Basic Lead Carbonate. Basic lead carbonate [*1319-46-6*] (white lead), $2PbCO_3 \cdot Pb(OH)_2$, mol wt 775.67, d = 6.14 g/cm^3, forms white hexagonal crystals; it decomposes when heated to 400°C. Basic lead carbonate is insoluble in water and alcohol, slightly soluble in carbonated water, and soluble in nitric acid. It is produced by several methods, in which soluble lead acetate is treated with carbon dioxide. For example, in the Thompson-Stewart process (123), an aqueous slurry of finely divided lead metal or monoxide, or a mixture of both, is treated with acetic acid in the presence of air and carbon dioxide. High quality, very fine particle-size basic lead carbonate is produced, ranging in carbonate content from 62 to 65% (vs 68.9% $PbCO_3$, theoretical).

Although white lead was the oldest white hiding pigment in paints, it has been totally replaced by titanium dioxide, which has better covering power and is nontoxic (see PIGMENTS). Nevertheless, basic lead carbonate has many other uses, including as a catalyst for the preparation of polyesters from terephthalic acid and diols (124), a ceramic glaze component, a curing agent with peroxides to form improved polyethylene wire insulation (125), a pearlescent pigment (126), a color-changing component of temperature-sensitive inks (127), a red-reflecting pigment in iridescent plastic sheets (128), a smudge-resistant film on electrically sensitive recording sheets (129), a lubricating grease component (130), a component of ultraviolet light reflective paints to increase solar reflectivity (131), an improved cool gun-propellant stabilizer which decomposes and forms a lubricating lead deposit (132), a heat stabilizer for poly(vinyl chloride) polymers (133,134), and as a component of weighted nylon-reinforced fish nets made of poly(vinyl chloride) fibers (135).

Phthalates

Two commercial forms of lead phthalates, both dibasic, are widely used as heat stabilizers (qv) in poly(vinyl chloride) (PVC) polymers and copolymers. During processing, usually extrusion, and in actual service, thermal degradation of PVC occurs principally by a dehydrochlorination mechanism. Thus one of the primary functions of dibasic lead phthalate stabilizers is to neutralize and inactivate the resulting hydrogen chloride. Such stabilizers are ideally suited for high temperature applications of PVC because of their low reactivity with plasticizers, particularly of the polyester type. Moreover, dibasic lead phthalates provide the long-term stability and retention of elongation required in 90 and 105°C Underwriters' Laboratories classes of wire insulation.

Dibasic Lead Phthalate. Dibasic lead phthalate [*17976-43-1*], $2PbO \cdot Pb(O_2C)_2C_6H_4 \cdot {}^1/_2H_2O$, is a white, crystalline powder, mol wt 826.87, sp gr 4.6, lead oxide content 79.8% PbO, moisture loss (2 h at 105°C), 0.3%; sieve analysis, 99.9% through 44 μm (325 mesh); water solubility, nil. In PVC, it provides excellent heat stability; excellent processibility, allowing high extrusion rates; excellent electrical properties over a wide range of temperatures and insulation classifications; good compatibility and low reactivity with a broad range of plasticizers; and for vinyl foams, it is an effective activator for azodicarbonamide-type blowing agents. Other applications include flexible extruded and molded PVC compounds, where it provides good resistance to early color development during processing. In vinyl plastisols, it provides low viscosity build-up on aging.

For vinyl compounds, the general range of dibasic lead phthalate stabilizer usage is between 4 and 4 phr resin. In 105°C electrical insulation PVC stocks, approximately 7 phr is required. For vinyl plastics and foams, between 3 and 5 phr of lead stabilizer is recommended.

Coated. Dibasic lead phthalate, coated, is a fluffy white powder, sp gr 3.5–3.9, lead oxide content 72–75% PbO, moisture loss 0.3–0.4%, fineness 99.5–99.9% through 44 μm (325 mesh) (wet), water solubility, nil. In PVC, it offers high resistance to early color development during processing, compatibility with organic ester and polyester-type plasticizers, higher heat stability and electrical properties, higher processing temperatures and production rates, improved retention of physical properties on aging, lower specific gravity, easier dispersion, and extra lubricity with reduced frictional heat development during processing.

Applications of dibasic lead phthalate, coated grade, include 90 and 105°C rated PVC electrical insulation, plastisols, profile extrusions, calendered sheet, and molded products. The recommended range of usage in vinyl electrical insulation is 5–7 parts per hundred resin, depending on the particular insulation classification to be met. In general-purpose extruded and molded PVC stocks, approximately 3–6 phr of coated dibasic lead phthalate is suggested.

Silicates

Lead forms acid, basic, and normal, or metasilicates. Commercial lead silicates (frits) are made to specific $PbO{:}SiO_2$ ratios for the glass and ceramics industries, the rubber industry as vulcanizing agent, and the plastics industry as a heat and light stabilizer (136). These are supplied as granular or pulverized fusion products. Some physical properties are given in Table 5.

Lead Monosilicate. Lead monosilicate [*10099-76-0*] (lead pyrosilicate), $1.5PbO{\cdot}SiO_2$, is a light yellow trigonal crystalline powder, insoluble in water. Its composition, by weight, is 85% PbO and 15% SiO_2. Lead monosilicate is commercially available as granular, <1.68 mm (10 mesh), and ground, 97% through 44 μm (325 mesh). It provides the most economical method of introducing lead into a ceramic glaze. It is also used as a source of PbO in the glass industry.

Lead Bisilicate. Lead bisilicate [*11120-22-2*], $PbO{\cdot}0.03Al_2O_3{\cdot}1.95SiO_2$, is a pale yellow powder, insoluble in water. Its composition, by weight, is 65% PbO, 1% Al_2O_3, and 34% SiO_2. Lead bisilicate is available as granular, <1.68 mm (10 mesh), and ground, 88% through 44 μm (325 mesh). It was developed as a low solubility source of lead in glazes, where its high viscosity and low volatility are equally important.

Tribasic Lead Silicate. Tribasic lead silicate [*12397-06-7*], $3PbO{\cdot}SiO_2$, is a reddish yellow powder, sparingly soluble in water. Its composition by weight is 92% PbO and 8% SiO_2. Tribasic lead silicate is available as granular, <1.68 mm (10 mesh), and ground, 99% through 44 μm (325 mesh). It is used primarily by glass and frit manufacturers and has the lowest viscosity of the three commercial

Table 5. Physical Properties of Lead Silicates

Property	Monosilicate	Bisilicate	Tribasic silicate
mol wt	294.85	343.37	729.63
mp, °C	700–784	788–816	705–733
d, g/cm^3	6.50–6.65	4.60–4.65	7.52
refractive index, n_D	2.00–2.02	1.72–1.74	2.20–2.24

lead silicates. Commercial lead silicates are generally prepared by melting lead monoxide and silica in the desired ratio.

Borate

Lead borate monohydrate [*14720-53-7*] (lead metaborate), $Pb(BO_2)_2 \cdot H_2O$, mol wt 310.82, $d = 5.6$ g/cm^3 (anhydrous) is a white crystalline powder. The metaborate loses water of crystallization at 160°C and melts at 500°C. It is insoluble in water and alkalies, but readily soluble in nitric and hot acetic acid. Lead metaborate may be produced by a fusion of boric acid with lead carbonate or litharge. It also may be formed as a precipitate when a concentrated solution of lead nitrate is mixed with an excess of borax. The oxides of lead and boron are miscible and form clear lead–borate glasses in the range of 21 to 73 mol % PbO.

The main use of lead metaborate is in glazes on pottery, porcelain, and chinaware, as well as in enamels for cast iron. Other applications include as radiation-shielding plastics, as a gelatinous thermal insulator containing asbestos fibers for neutron shielding, and as an additive to improve the properties of semiconducting materials used in thermistors (137).

Lead Titanate

Lead titanate [*12060-00-3*] (lead metatitanate), $PbTiO_3$, mol wt 302.09, $d =$ 7.52 g/cm^3, forms yellow tetragonal crystals below 490°C and cubic crystals above 490°C. It is insoluble in water. In hydrochloric acid, lead titanate decomposes into $PbCl_2$ and TiO_2. It can be formed by calcining an equimolecular mixture of lead monoxide and titanium dioxide and has been used in surface coatings as a pigment in outdoor paints (138), as a component of ceramic electrical insulators (139), in ceramic ferroelectric–piezoelectric compositions (140), in ceramic electrical capacitors (141) (see CERAMICS AS ELECTRICAL MATERIALS), in ceramic glazes (142), in transducers (143), in low melting glass sealants (144), and in oxidation catalysts for manufacturing acrylonitrile from propylene and nitrous oxide (145).

Lead Zirconate

Lead zirconate [*12060-01-4*], $PbZrO_3$, mol wt 346.41, has two colorless crystal structures: a cubic perovskite form above 230°C (Curie point) and a pseudotetragonal or orthorhombic form below 230°C. It is insoluble in water and aqueous alkalies, but soluble in strong mineral acids. Lead zirconate is usually prepared by heating together the oxides of lead and zirconium in the proper proportion. It readily forms solid solutions with other compounds with the ABO_3 structure, such as barium zirconate or lead titanate. Mixed lead titanate–zirconates have particularly high piezoelectric properties. They are used in high power acoustic-radiating transducers, hydrophones, and specialty instruments (146).

Other salts include lead arsenates and lead arsenites (see INSECT CONTROL TECHNOLOGY), lead chromates and lead silicochromates (see PIGMENTS), lead cyanide (see CYANIDES), lead 2-ethylhexanoate (see DRIERS AND METALLIC SOAPS), and lead fluoroborate (see FLUORINE COMPOUNDS, INORGANIC).

Health and Safety Factors

Lead is poisonous in all forms, but to different degrees, depending on the chemical nature and solubility of the lead compound. Exposure may be acute or chronic. Because the symptoms of lead poisoning may be similar to those of other ailments, they should be checked with blood and urine tests. Lead is one of the most hazardous toxic metals because the poison is cumulative, and its toxic effects are many and severe. Prolonged absorption of lead or its inorganic compounds can cause the onset of lead poisoning symptoms or plumbism, including weakness, lassitude, weight loss, insomnia, hypotension, and anemia. Associated with these may be gastrointestinal disturbances. Physical signs are usually facial pallor, malnutrition, abdominal tenderness, and pallor of the eye grounds. On gingival tissues, a line or band of blue-black pigmentation (lead line) may appear, but only in the presence of poor dental hygiene.

The alimentary symptoms may be overshadowed by neuromuscular dysfunction, accompanied by signs of motor weakness that may progress to paralysis of the exterior muscles or the wrist (wrist drop), and less often, of the ankles (foot drop). Encephalopathy, the most serious result of lead poisoning, frequently occurs in children as a result of pica, ie, ingestion of inorganic lead compounds in paint chips; this rarely occurs in adults. Nephropathy has also been associated with chronic lead poisoning (147). The toxic effects of lead may be most pronounced on the developing fetus. Consequently, women must be particularly cautious of lead exposure (148). The U.S. Center for Disease Control recommends a blood level of less than 10 μm per 100 mL for children.

Lead is absorbed into the human body after inhalation of the dust or ingestion of lead-containing products. Contamination of smoking materials in the work area leads to inhalation of lead fumes and constitutes a main factor in lead absorption.

OSHA regulations (149) limit exposure to inorganic lead compounds of an employee without a respirator to 50 $\mu g/m^3$ air as a time-weighted average (TWA) in an eight-hour shift. This standard went into effect on March 1, 1979.

Measurement of airborne lead concentrations may be made with a gravimetric dust-sampling kit. The apparatus draws in air from the breathing zone of a worker at a controlled rate. This air is passed, with contaminants, through a filter, where the particulate matter is trapped. After a specified length of time, the amount of lead on the filter is determined by analysis. If initial air sampling shows exposure for all employees to be below an action level of 30 $\mu m/m^3$ of air, averaged over an eight-hour period, only rechecking is needed when any change takes place which might affect lead exposure. On the other hand, if initial air sampling is between 30 and 50 $\mu g/m^3$ of air, averaged over an eight-hour period, then testing must be continued every six months.

However, if air sampling establishes that the lead exposure concentration is excessive, engineering controls (such as improved ventilation), administrative controls (such as job rotation), and work practices (such as improved personal hygiene of workers) have to be applied to comply with the permissible exposure limit (PEL) of the OSHA standard.

In addition to limits on airborne lead, an OSHA regulation provides for biological monitoring and places limits on blood lead levels in workers of 50 $\mu g/100$ g of whole blood.

Physical examinations, employee training and educational programs, medical protection, and record keeping, among others, are required. The regional OSHA office should be consulted for the latest rules and regulations.

In most cases, proper precautions provide worker safety. According to conditions in the workplace, all workers handling inorganic lead compounds should avoid creating dust, avoid inhaling or swallowing dust, wash thoroughly before eating or smoking, and keep inorganic lead compounds away from animal feed and food products. Adequate care and attention paid to safe handling practices can effectively minimize or eliminate any health risks associated with the handling, storage, use, and disposal of lead compounds.

The OSHA limits, regulations, and recommendations apply to in-plant air quality. Improperly filtered exhaust air may cause a plant to be in violation of the EPA standard, therefore these data should not be confused with the EPA limit for airborne lead, 1.5 μg lead/m^3, measured over a calendar quarter, which pertains to the exterior plant environment and emissions. The installation and proper maintenance of exhaust filtration systems enables most plants to comply with the EPA limits for airborne lead (see LEAD COMPOUNDS, INDUSTRIAL TOXICOLOGY).

BIBLIOGRAPHY

"Lead Compounds (Inorganic)" in *ECT* 1st ed., Vol. 8, pp. 267–274, by A. P. Thompson, The Eagle-Picher Co.; "Inorganic Compounds" under "Lead Compounds" in *ECT* 2nd ed., Vol. 12, pp. 266–282, by A. P. Thompson, Eagle-Picher Industries, Inc.; "Lead Salts" under "Lead Compounds" in *ECT* 3rd ed., Vol. 14, pp. 160–180, by D. S. Carr, International Lead Zinc Research Organization.

1. S. Afr. Pat. 68 04,061 (Dec. 18, 1968), J. Prior and A. Florin (to Dynamit Nobel A.-G.).
2. Ger. Offen. 1,903,879 (Oct. 30, 1969), S. Cane and L. E. Cooper (to BP Chemicals (U.K.) Ltd.).
3. Ger. Offen. 1,421,872 (Feb. 19, 1970), W. Reichelt and H. Eligehausen (to W. C. Heraeus GmbH).
4. Jpn. Pat. 69 18,745 (Aug. 15, 1969), M. Mikoda and T. Hikino (to Matsushita Electric Industrial Co., Ltd.).
5. Ger. Offen. 2,106,118 (Sept. 2, 1971), F. Auzel.
6. Jpn. Kokai 74 27,442 (Mar. 11, 1974), K. Morimoto and M. Kuroda (to Matsushita Electric Industrial Co., Ltd.).
7. U.S. Pat. 3,745,044 (July 10, 1973), E. C. Letter (to Bausch and Lomb, Inc.).
8. Jpn. Kokai 74 14,996 (Feb. 8, 1974), N. Ichinose and Y. Yokomizo (to Tokyo Shibaura Electric Co., Ltd.).
9. D. Greninger, V. Kollonitsch, and C. H. Kline, *Lead Chemicals,* International Lead Zinc Research Organization, Inc., New York, 1975, p. 173.
10. Brit. Pat. 1,235,100 (June 9, 1971), (to Toyota Central Research and Development Laboratories, Inc.).
11. Fr. Pat. 1,556,127 (Jan. 31, 1969), (to Imperial Chemical Industries, Inc.).
12. Brit. Pat. 1,078, 854 (Aug. 9, 1967), (to Mitsui Petrochemical Industries, Ltd.).
13. U.S. Pat. 3,468,710 (Sept. 23, 1969), D. Krasnov and J. Goodman (to Nuclear Research Associates, Inc.).
14. U.S. Pat. 3,484,312 (Dec. 16, 1969), F. Ermanis and B. Schwartz (to Bell Laboratories).
15. U.S. Pat. 3,475,372 (Oct. 28, 1969), C. L. Gable (to Mobay Chemical Co.).

16. U.S. Pat. 3,468,843 (Sept. 23, 1969), W. F. Busse (to E. I. du Pont de Nemours & Co., Inc.).
17. Ger. Offen. 2,051,925 (May 6, 1971), J. Tanaka and M. Watanabe (to Senju Metal Industry Co., Ltd.).
18. U.S. Pat. 3,495,944 (Feb. 17, 1970), J. J. McGee, T. J. Kelly, and J. N. Harman (to Beckman Instruments).
19. U.S. Pat. 3,562,944 (Feb. 9, 1971), M. A. DeAngelo and D. J. Sharp (to Western Electric Co.).
20. U.S. Pat. 3,360,337 (Dec. 26, 1967), M. I. Edenbaum and M. I. Hampton (to Johnson & Johnson).
21. U.S. Pat. 3,346,383 (Oct. 10, 1967), R. W. Baxendale (to Eastman Kodak Co.).
22. Ger. Offen. 1,956,513 (June 18, 1970), W. A. Van den Heuvel and co-workers (Agfa-Gevaert A.G.).
23. Ger. Offen. 1,956,713 (June 18, 1970), J. E. Vanhalst, E. M. Brinckman, and W. A. Van den Heuvel (to Agfa-Gavaert A.G.).
24. Fr. Pat. 2,039,700 (Jan. 15, 1971), J. A. Peterson and H. W. Marciniak (to Hooker Chemical Corp.).
25. U.S. Pat. 3,287,540 (Nov. 22, 1966), T. J. Connelly (to Allied Chemical Corp.).
26. Ref. 9, p. 183.
27. U.S. Pat. 3,377,169 (Apr. 9, 1968), R. K. Blake (to E. I. du Pont de Nemours & Co., Inc.).
28. Czech. Pat. 121,444 (Jan. 15, 1967), L. Kacetl, J. Pantoflicek, L. Sramek, and F. Anyz.
29. Brit. Pat. 1,235,100 (June 9, 1971), (to Toyota Central Research and Development Laboratories, Inc.).
30. U.S. Pat. 3,511,715 (May 12, 1970), J. C. Angus (to Valley Co., Inc.).
31. F. R. Pat. 1,467,694 (Jan. 27, 1967), (to N. V. Philips Gloeilampenfabrieken).
32. USSR Pat. 457,121 (Jan. 15, 1975), S. G. Ashurkov, G. S. Sarychev, and E. F. Fufaev.
33. U.S. Pat. 3,467,555 (Sept. 16, 1969), C. M. Henderson, E. R. Beaver, and L. J. Reitsma (to Montana Research Corp.).
34. U.S. Pat. 3,460,996 (Aug. 12, 1969), I. Kudman (to Radio Corporation of America).
35. U.S. Pat. 3,201,347 (Aug. 17, 1965), J. J. Chessick and J. B. Christian (to U.S. Dept. of the Air Force).
36. U.S. Pat. 3,713,996 (Jan. 30, 1973), E. C. Letter (to Bausch and Lomb, Inc.).
37. J. J. Wijnbergen, W. H. Moolenaar, and G. DeGroot, *Astrophys. Space Sci. Libr.* **30**, 243 (1972).
38. G. Smoluk, *Mod. Plast.* **56**(9), 74 (1979).
39. *Mineral Review Industry Survey,* U.S. Bureau of Mines, Washington, D.C., Apr. 1994.
40. Ref. 9, p. 52.
41. Ref. 9, pp. 59–60.
42. Ref. 9, p. 69.
43. Jpn. Kokai 73 28,396 (Apr. 13, 1973), E. Torikai, Y. Kawami, and Y. Maeda.
44. Ger. Offen. 2,156,414 (July 13, 1972), Y. Kuniyasu, T. Sakai, and T. Ogami (to Mitsui Mining and Smelting Co., Ltd.).
45. Ger. Offen. 2,142,001 (Apr. 6, 1972), T. Sakai, S. Kobayashi, K. Miyazaki, and M. Yamamoto (to Mitsui Mining and Smelting Co., Ltd.).
46. Fr. Pat. 1,511,984 (Feb. 2, 1968), (to Teijin Ltd.).
47. Ger. Offen. 2,124,145 (Dec. 30, 1971), A. A. Shul'zhenko and F. Get'man (to Ukrainian Scientific Research Institute of Instruments and Extrahard Materials).
48. U.S. Pat. 3,847,861 (Nov. 12, 1974), T. Largman and H. Stone (to Allied Chemical Corp.).
49. Ger. Offen. 2,053,706 (May 10, 1972), S. Roth and R. Willig (to Siemens A.G.).

50. Jpn. Kokai 76 20,248 (Feb. 18, 1976), H. Kato (to Dainichi-Nippon Cables, Ltd.).
51. B. Anderson and G. Eckwall, *FATIPEC Congr.* **8**, 45 (1966).
52. L. V. Nitsberg, L. A. Bobina, and O. Y. Khenven, *Zashch. Korroz. Gidrotekh. Sooruzhenii Rechn. Vodakh,* 182 (1968).
53. Jpn. Kokai 74 27,525 (Mar. 12, 1974), A. Minoshi, H. Tsugukuni, and Y. Chikazoe (to Dai Nippon Toryo Co., Ltd.).
54. Austrian Pat. 324,922 (Sept. 25, 1975), (to R. Avenarius Chemische Fabrik).
55. Jpn. Kokai 74 61,269 (June 13, 1974), I. Ogasawara and S. Kawashima (to Unitika Ltd.).
56. USSR Pat. 400,932 (Oct. 1, 1973), V. V. Novoderezhkin and co-workers.
57. Jpn. Kokai 73 56,712 (Aug. 9, 1973), T. Yamada and M. Tachibana (to Nippon Sheet Glass Co., Ltd.).
58. Jpn. Kokai 76 39,742 (Apr. 2, 1976), K. Ohtsuki and K. Eguchi (to Dainippon Printing Co., Ltd.).
59. Ger. Offen. 2,034,053 (July 15, 1971), S. Kobayashi, K. Miyazaki, and M. Yamamoto (to Mitsui Mining and Smelting Co., Ltd.).
60. Jpn. Pat. 69 10,115 (May 12, 1969), Y. Ishitani and S. Akimaru (to Asahi Chemical Industry Co., Ltd.).
61. USSR Pat. 224,022 (Aug. 6, 1968), V. V. Vargin and co-workers.
62. Jpn. Pat. 71 03,473 (Jan. 28, 1971), K. Minakawa (to Japan Electric Co., Ltd.).
63. Ger. Offen. 2,063,586 (July 22, 1971), M. Hamasaki, I. Kurokawa, and N. Izawa (to Asahi Chemical Industry Co., Ltd.).
64. Jpn. Kokai 74 46,586 (May 4, 1974), T. Shiki, K. Kozaki, and T. Harada (to Asahi Chemical Industry Co., Ltd.).
65. D. M. M. Ibrahim and G. M. Gad, *Sprechsaal Keram. Glas Email Silik.* **103**, 768 (1970).
66. Ger. Offen. 1,954,887 (May 6, 1970), J. T. Oetzel (to B. F. Goodrich Co.).
67. U.S. Pat. 3,907,585 (Sept. 23, 1975), J. Francel and J. E. King (to Owens-Illinois, Inc.).
68. Jpn. Kokai 72 15,414 (Aug. 22, 1972), O. Sato.
69. J. W. Diggle and A. Damjanovic, *J. Electrochem. Soc.* **119**, 1649 (1972).
70. Jpn. Kokai 74 01,719 (Jan. 9, 1974), K. Suszuki, A. Kanayama, and T. Shimomura (to Tanabe Seiyaku Co., Ltd.; Nitto Electric Industrial Co., Ltd.).
71. Brit. Pat. 1,281,527 (July 12, 1972), W. G. Wink.
72. Jpn. Kokai 75 16,045 (Feb. 20, 1975), Y. Morioka (to Sanyo Electric Co., Ltd.).
73. Jpn. Pat. 74 29,164 (Aug. 1, 1974), S. Ono, K. Nakamura, and Y. Mizoguchi (to Asahi Chemical Industry Co., Ltd.).
74. USSR Pat. 333,238 (Mar. 21, 1972), B. G. Milov, S. K. Kitaeva, and M. A. Chagina (to Central Scientific-Research Institute of Paper).
75. U.S. Pat. 3,766,984 (Oct. 23, 1973), K. H. Nimerick (to Dow Chemical Co.).
76. Ger. Offen. 2,128,845 (Feb. 17, 1972), H. E. Meissner and S. D. Stookey (to Corning Glass Works).
77. U.S. Pat. 3,791,520 (Feb. 12, 1974), G. J. Nieuwenhuls.
78. Ger. Offen. 1,913,099 (July 2, 1970), G. Tanaka and M. Shono (to Giken Kogyo K.K.).
79. Ger. (GDR) Appl. 21,328 (Aug. 4, 1958), E. Boeck and G. Keil.
80. U.S. Pat. 3,260,613 (July 12, 1966), E. C. Otto (to Interchemical Corp.).
81. N. Ogata and H. Kakihana, *Nippon Genshiryoku Gakkaishi* **11**(2), 82 (1969).
82. Ref. 9, pp. 140–141.
83. Ref. 9, p. 153.
84. D. Baresel, W. Sarholz, P. Scharner, and J. Schmitz, *Ber. Bunsenges. Phys. Chem.* **78**, 608 (1974).
85. U.S. Pat. 3,877,988 (Apr. 15, 1975), A. N. Dey and P. Bro (to P. R. Mallory and Co., Inc.).

86. S. Afr. Pat. 73 07,956 (Nov. 8, 1972), N. Habler and H. Schreiner (to Siemens A.-G.).
87. Ger. Offen. 2,210,525 (Nov. 2, 1972), K. Higashiyama and H. Hirata (to Matsushita Electric Industrial Co., Ltd.).
88. F. A. Blum and K. W. Nill, *Proceedings International Conference, Laser Spectroscopy* (AD-777 751), 1974, pp. 449–509.
89. U.S. Pat. 3,851,291 (Nov. 26, 1974), A. Sommer (to Ceramic Magnets, Inc.).
90. Ref. 9, p. 69.
91. Ref. 9, p. 256.
92. Fr. Pat. 1,425,881 (Jan. 21, 1966), (to English Clays Lovering Pochin & Co., Ltd.).
93. D. N. Gritsan, D. S. Shun, and L. N. Serpukhova, *Zh. Prikl. Khim.* **34**, 1528 (1961).
94. Brit. Pat. 827,646 (Feb. 10, 1960), P. H. Haycock, J. J. Ryan, and M. Tuson (to Tootal Broadhurst Lee Co., Ltd.).
95. Brit. Pat. 812,785 (Apr. 29, 1959), J. d'Ans, H. E. Freund, and H. J. Schuster.
96. Ref. 9, p. 235.
97. Brit. Pat. 1,004,309 (Sept. 15, 1965), (to Kurashiki Rayon Co., Ltd.).
98. U.S. Pat. 3,238,047 (Mar. 1, 1966), R. D. Murray and E. Berman (to Itek Corp.).
99. Jpn. Pat. 61 13,944 (Feb. 24, 1959), T. Isojima and H. Terada (to Mitsubishi Rayon Co., Ltd.).
100. I. Puhac, N. Hrgovic, M. Stankovic, and S. Popovic, *Acta Vet. (Belgrade)* **13**, 3 (1963).
101. Jpn. Pat. 63 16,627 (Sept. 2, 1963), Y. Nakahara (to New Japan Electric Co., Ltd.).
102. C. Drotschmann, *Batterien* **17**, 569 (1964).
103. U.S. Pat. 3,033,675 (May 8, 1962), N. Hedley (to American Cyanamid Co.).
104. Ref. 9, p. 201.
105. Neth. Appl. 69 02,804 (Aug. 29, 1969), (N. V. Phillips' Gloeilampenfabrieken).
106. U.S. Pat. 3,298,833 (Jan. 17, 1967), J. Gaynor (to General Electric Co.).
107. Ref. 9, p. 296.
108. Fr. Pat. 1,356,569 (Mar. 27, 1964), R. E. Hatton and L. R. Stark (to Monsanto Co.).
109. Neth. Appl. 64 08,827 (Feb. 2, 1965), (to Wallace and Tiernan, Inc.).
110. S. Kobayashi and S. Hayakawa, *Jpn. J. Appl. Phys.* **4**(3), 181 (1965).
111. Jpn. Pat. 18,963 (Sept. 4, 1964), S. Futami (to Teijin Ltd.).
112. Fr. Pat. 1,404,162 (June 25, 1965), (to Płekhanov Mining Institute).
113. U.S. Pat. 3,112,772 (Dec. 3, 1963), R. E. Connor and J. C. Kitching (to H. K. Porter Co., Inc.).
114. Neth. Appl. 65 11,631 (Mar. 9, 1966), (to Shell Internationale Research Maatschappij N.V.).
115. Neth. Appl. 65 04,192 (Oct. 4, 1965), (to Dow Corning Corp.).
116. Belg. Pat. 639,509 (May 4, 1964), (to Farbwerke Hoechst A.-G.).
117. U.S. Pat. 3,253,912 (May 31, 1966), J. L. Rooney and J. P. Badger (to Electric Auto-Lite Co.).
118. Brit. Pat. 982,564 (Feb. 10, 1965), J. G. Jarvis (to Kodak Ltd.).
119. U.S. Pat. 3,260.613 (July 12, 1966), E. C. Otto (to Interchemical Corp.).
120. Jpn. Pat. 63 03,910 (Apr. 20, 1963), F. Kato and Y. Machino.
121. Fr. Addn. (Nov. 3, 1961), to Fr. Pat. 1,165,582 (to Compagnie Generale de Telegraphic Sans Fil).
122. Neth. Appl. 64 08,291 (Jan. 21, 1966), (to Mining Institute, Leningrad).
123. Ref. 9, p. 216.
124. Jpn. Pat. 64 20,533 (Sept. 19, 1964), K. Nuruchina, Y. Takehisa, and J. Ichikawa (to Toyo Rayon Co., Ltd.).
125. U.S. Pat. 3,039,989 (June 19, 1962), W. O. Eastman (to General Electric Co.).
126. U.S. Pat. 2,950,981 (Aug. 30, 1960), H. A. Miller and L. M. Greenstein (to Francis Earle Labs, Inc.).
127. Fr. Pat. 1,335,076 (Aug. 16, 1963), (to Tempil Corp.).

128. U.S. Pat. 3,231,645 (Jan. 25, 1966), R. A. Balomey (to Mearl Corp.).
129. U.S. Pat. 3,138,547 (June 23, 1964), B. L. Clark (to 3M Corp.).
130. Neth. Appl. 65 09,144 (Jan. 18, 1966), (to Shell Internationale Research Maatschappij N.V.).
131. K. K. Chowdhry, *LABDEV* (*Kanpur, India*) **3**, 219 (1965).
132. U.S. Pat. 3,116,190 (Dec. 31, 1963), F. A. Zihlman, C. N. Bernstein, and F. C. Thames (to the U.S. Navy).
133. B. Levi and I. Mekjavic, *Tehnika* (*Belgrade*) **16**, 1461 (1961).
134. Brit. Pat. 917,082 (Jan. 30, 1963), (to National Lead Co.).
135. Norw. Pat. 105,779 (Jan. 9, 1965), T. Ringvold.
136. Ref. 9, pp. 230–231.
137. Ref. 9, p. 210.
138. Ref. 9, p. 94.
139. USSR Pat. 262,201 (Jan. 26, 1970), O. S. Didkovskaya, V. V. Klimov, and Y. N. Venevtsev.
140. U.S. Pat. 3,463,732 (Aug. 26, 1969), H. Banno and T. Tsunooka (to NGK Spark Plug Co., Ltd.).
141. Fr. Pat. 1,533,488 (July 19, 1968), S. A. Long and C. L. Fillmore (to Globe-Union Inc.).
142. U.S. Pat. 3,405,002 (Oct. 8, 1968), F. W. Martin (to Corning Glass Works).
143. U.S. Pat. 3,517,093 (June 23, 1970), J. J. Wentzel (to U.S. Dept. of the Navy).
144. Jpn. Kokai 76 13,820 (July 25, 1974), I. Matsuura and F. Yamaguchi (to Nippon Electric Glass Co., Ltd.).
145. V. M. Belousov and co-workers, *Katal. Katal.* (10), 37 (1973).
146. Ref. 9, p. 96.
147. M. H. Proctor and J. P. Hughes, *Chemical Hazards of the Workplace,* J. B. Lippincott Co., Philadelphia, Pa., 1978, p. 308.
148. P. B. Beeson and W. McDermott, *Textbook of Medicine,* W. B. Saunders Co., Philadelphia, Pa., 1975, Vol. 1, p. 58.
149. *Code of Federal Regulations,* Title 29, Part 1910.1025, OSHA, Washington, D.C., 1979.

General References

D. Greninger, V. Kollonitsch, and C. H. Kline, *Lead Chemicals,* International Lead Zinc Research Organization, Inc., New York, 1975.

J. W. Mellor, *Inorganic and Theoretical Chemistry,* Vol. III, Longmans, Green & Co., New York, 1930, pp. 636–888.

A. T. Wells, *Structural Inorganic Chemistry,* 3rd ed., Clarendon Press, Oxford, U.K., 1962, particularly pp. 475–479, 902–903.

DODD S. CARR
Consultant

WILLIAM C. SPANGENBERG
KEVIN CHRONLEY
Hammond Lead Products, Inc.

INDUSTRIAL TOXICOLOGY

Lead (qv), lead alloys (qv), and lead compounds have been used for thousands of years for a number of purposes. The toxic effects of these materials have been known or suspected for almost as long. Serious efforts intended to reduce occupational and population exposure to lead have been made only since the late 1960s. Although these efforts, largely in the form of regulatory actions, voluntary actions, and increased public awareness, have been highly successful, concern about lead as a significant public health problem has increased. Epidemiological and experimental evidence regarding adverse health effects at successively lower levels of lead exposure have led to downward revision of criteria for acceptable blood lead concentrations. The U.S. Environmental Protection Agency (EPA) has designated 10 μg/dL as a target level for regulatory development and enforcement/clean-up purposes (1). As of this writing, the EPA considers lead to be a significant and widespread health hazard in the United States (2).

Exposure

Exposure to lead can occur from a variety of occupational or nonoccupational sources. A comprehensive treatise on human lead exposure is available (3).

Nonoccupational. Lead enters the biosphere both through natural and human (anthropogenic) activities. Lead enters the atmosphere from natural occurrences such as erosion of the earth's crust, volcanic eruptions and emissions, and forest fires. The main source of atmospheric lead, however, is from anthropogenic lead emissions. Combustion of petroleum (qv) and its derivatives, especially leaded gasoline, accounts for the majority of atmospheric lead worldwide (4). The greatest increase in global emissions started in the 1920s when lead began to be used as an additive in gasoline (5) (see GASOLINE AND OTHER MOTOR FUELS). Because of human health concerns, the use of leaded gasoline has declined sharply especially in the United States. Production of lead, zinc, copper (qv), iron (qv), and steel (qv) also contributes significantly to atmospheric lead, and a wide range of lead concentrations in air have been reported (4). Most urban areas in the United States have air lead concentrations of < 0.5 $\mu g/m^3$ on average (5) (see also AIR POLLUTION).

Lead is a natural constituent of soils, ranging from 10 to 50 μg/g in uncontaminated areas. Urban soils, however, may contain lead from a variety of sources including automobile exhaust, industrial activity, and deteriorating or intentionally removed leaded paint. The lead content of roadside soils has been found to be as high as 2000 μg/g within two meters of the curb (4). These levels generally decline exponentially to background levels within 10–25 meters of the curb. Within three meters of houses coated with lead-based paints, soil lead concentrations may exceed 1000 μg/g. Exposure to soil lead may occur from vegetable consumption and by direct or indirect ingestion of soils or soil dust, particularly by children.

Lead is present in trace amounts in surface and ground waters. Because lead forms insoluble salts or is sorbed onto sediment particles, atmospheric lead deposition or surface runoff generally does not affect overall lead concentration. Most natural waters are low ($< 5\ \mu g/L$) in lead content (4). Drinking water (qv), however, can contain markedly higher lead concentrations owing to contact with lead-bearing pipes, solder, and plumbing fixtures. Waters that are relatively acidic or soft may contain elevated lead levels at the tap (4), and often represent a principal source of lead exposure. In the United States, the maximum contaminant level for lead in drinking water is 15 $\mu g/L$. The use of lead solder for materials contacting residential water supplies was banned in the United States in accordance to the 1986 Amendments to the Safe Drinking Water Act.

Ingestion of foods represents the majority of the daily intake of lead for most people, although this is not the route responsible for marked elevations of an individual's lead burden (4,5). The lead content of food can be attributed to such factors as foliar uptake of atmospheric lead by plants; contamination during harvesting and processing; use of water containing lead during food processing (qv) or preparation; lead solder in food cans; lead fallout from urban dusts in the kitchen; and the use of lead-containing glazes in pottery. The awareness of these causes for lead contamination of foods, and subsequent regulatory and voluntary actions, have resulted in a marked reduction in the lead in foods, from pre-1975 values of $> 250\ \mu g/d$ to $\leq 50\ \mu g/d$ in 1993 (4).

Occupational. The toxicity of lead following occupational exposure to the metal has been known since antiquity. In developed countries, regulations, engineering advances, and medical surveillance have greatly reduced the incidence of clinical lead poisoning in lead-based industries (6). As of this writing (ca 1994) occupational lead poisoning, defined as a whole blood lead concentration $> 50\ \mu g/dL$ (7), continues to be a problem, however (6,7), and severe exposures to lead continue to occur in workers in a variety of manufacturing processes (5–7). Lead smelting and refining have the greatest potential for occupational exposure to inorganic lead, because lead fumes are generated and dust containing lead oxide is deposited in the workplace. Other occupations associated with lead exposure include lead storage battery manufacture (see BATTERIES); autobody and auto radiator manufacture and repair; lead recycling; stained glass work; leaded paint removal; welding (qv); demolition and restoration; and plumbing installation or repair. More detailed listings of occupations and occupational settings with potential for lead exposure are available (6,7). The U.S. National Institute for Occupational Safety and Health (NIOSH) has published guidelines for minimizing occupational exposure to lead (7). At a minimum, airborne lead concentrations in the workplace should not exceed the Occupational Safety and Health Administration's (OSHA) permissible exposure level (PEL) of 50 $\mu g/m^3$.

Elevated blood lead levels have occurred in children of workers who are occupationally exposed to lead (5,6). Because children may be exposed to lead dust brought home on workers' clothing, in the United States the OSHA lead standard requires employers in traditional lead-related occupations to provide showers, changing facilities, and laundered work clothes. Moreover, workers must shower and change out of work clothes before leaving the workplace. However, these and other OSHA lead standards do not apply to workers in the less traditional lead-related occupations such as construction, plumbing, painting, demolition, and restoration.

Absorption, Distribution, and Excretion

Detailed discussions on the absorption, distribution, and excretion of lead in humans are available (8–10). The principal routes of lead absorption are from the gastrointestinal and respiratory tracts. Gastrointestinal absorption of lead varies with age. Adults absorb approximately 10% of ingested lead; children absorb up to 40%. Little is known about lead transport across the gastrointestinal mucosa, but it has been suggested that divalent lead and calcium compete for a common transport mechanism. There is an inverse relationship between the dietary content of calcium and lead absorption. Absorption of inhaled lead varies with form and concentration. Up to 90% of inhaled lead particles from ambient air may be absorbed. Once absorbed, lead is distributed among the blood, soft tissue, and skeleton. Approximately 90% of the lead in blood is bound to the red blood cells. The overall half-life of blood lead is approximately 36 days: lead is redistributed and deposited in bone, teeth, and hair. Once in bone, the half-life of lead can be as long as 30 years. In humans, the majority of the lead that is excreted is mainly through the urine, and the concentration of urine lead is directly proportional to that in the blood. However, because most lead in blood is bound to red blood cells, urinary excretion of lead generally does not cause a marked decrease in blood lead concentration. About 95% of the body lead burden is eventually found in bone.

Toxicity

Reviews on the occurrence, biochemical basis, and treatment of lead toxicity in children (11) and workers (3,12,13) have been published. Approximately 17% of all preschool children in the United States have blood lead levels $> 10\ \mu g/dL$. In inner city, low income minority children the prevalence of blood lead levels $> 10\ \mu g/dL$ is 68%. It has been estimated that over two million American workers are at risk of exposure to lead as a result of their work. Public health surveillance data document that each year thousands of American workers occupationally exposed to lead develop signs and symptoms indicative of lead poisoning (3).

Lead is toxic to the kidney, cardiovascular system, developing red blood cells, and the nervous system. The toxicity of lead to the kidney is manifested by chronic nephropathy and appears to result from long-term, relatively high dose exposure to lead. It appears that the toxicity of lead to the kidney results from effects on the cells lining the proximal tubules. Lead inhibits the metabolic activation of vitamin D in these cells, and induces the formation of dense lead–protein complexes, causing a progressive destruction of the proximal tubules (13). Lead has been implicated in causing hypertension as a result of a direct action on vascular smooth muscle as well as the toxic effects on the kidneys (12,13).

Lead-induced anemia results from impairment of heme biosynthesis and acceleration of red blood cell destruction (10,13). Lead-induced inhibition of heme biosynthesis is caused by inhibition of 5-aminolevulinic acid dehydratase and ferrochelatase which starts to occur at blood lead levels of 10 to 20 $\mu g/dL$ and 25 to 30 $\mu g/dL$, respectively (10,13). Anemia, however, is not manifested until higher levels are reached.

In the peripheral nervous system, motor axons are the principal target. Lead-induced pathological changes in these fibers include segmental demyelination and axonal degeneration (13). Extensor muscle palsy with wrist drop or ankle drop is the classic clinical manifestation of this toxicity. Lead-induced central nervous system toxicity, or lead encephalopathy, is the most serious manifestation of lead toxicity. It is more common in children than in adults. The toxicity is believed to arise from demyelination and nerve degeneration. Early signs of lead encephalopathy include clumsiness, vertigo, ataxia, falling, headaches, insomnia, restlessness, and irritability. If untreated, delerium, tonic–clonic convulsions, mental retardation, and even death may ensue (8,13).

Lead is known to cause reproductive and developmental toxicity. Decreased sperm counts and abnormal sperm development have been reported in male workers heavily exposed to lead. Increased incidences of spontaneous abortion have been reported in female lead workers as well as in the wives of male lead workers (13). Lead crosses the placenta and has been found to cause irreversible neurologic impairment to the fetus at maternal blood levels as low as 15 to 20 μg/dL (13).

Detection and Treatment of Lead Poisoning

The classic clinical manifestations of lead poisoning include gastrointestinal disturbances (eg, vomiting, diarrhea, constipation, or intestinal spasms), anemia, renal failure, hyperuricemia or gout, hypertension, reproductive impairment, peripheral nerve injury manifested by wrist or ankle drop, and neuropsychological impairment. In children the classic signs are delayed growth and deficits in psychometric intelligence, speech and language processing, attention, and classroom performance. Overt lead encephalopathy may occur in children having high lead exposure. The gastrointestinal syndrome is more prevalent in adults; the central nervous system syndrome is usually more common in children. Because all of these symptoms occur with other illnesses, misdiagnosis of lead poisoning may occur unless it is suspected and a detailed patient history is obtained.

More specific indications of lead poisoning are increased blood and urine levels of lead or 5-aminolevulinic acid [*106-60-5*]. Measurement of erythrocytic protoporphyrin or zinc protoporphyrin has been described as being a diagnostic test in the detection of lead poisoning (10). However, as epidemiological and experimental data continually indicate that the toxic effects of lead occur at blood levels lower than previously believed, these protoporphyrin assays are not sensitive enough for reliable quantification of the lower levels of lead believed to be clinically significant. Protoporphyrin assays are more useful as indicators of recent, high lead exposure. Measurement of bone lead content is a more definitive approach for assessing chronic lead exposure and total body lead burden. X-ray fluorescence, a relatively new analytical technique for determining bone lead content, is a noninvasive and relatively rapid approach for assessing chronic lead exposure and total body burden of lead (13). It is not useful for determining acute lead exposure. From a practical standpoint, blood lead is most useful in determining exposures.

The treatment of lead poisoning has been extensively reviewed (8,11,12). Lead in the blood is removed by administration of chelating agents (qv) such as *meso*-2,3-dimercaptosuccinic acid [*304-55-2*], $C_4H_6O_4S_2$, ethylenediaminetetraacetic acid [*60-00-4*] (EDTA), $C_{10}H_{16}N_2O_8$, dimercaprol [*59-51-9*], $C_3H_8OS_2$, or D-penicillamine [*52-67-5*], $C_5H_{11}NO_2S$. These agents form stable lead chelates, which are then excreted in the urine or feces. Chelation therapy, however, should not be used indiscriminately because there are potential hazards associated with the use of these agents. Other treatment measures include proper nutrition, eg, lead absorption is facilitated by iron deficiency, and removal of the patient from the source of lead to prevent continued exposure.

This article has been reviewed by the Office of Pollution Prevention and Toxics (U.S. EPA) and approved for publication. Approval does not signify that the contents necessarily reflect the views and policies of the Agency, nor does mention of commercial products constitute endorsement or recommendation for use.

BIBLIOGRAPHY

"Lead Poisoning" in *ECT* 1st ed., Vol. 8, pp. 281–288, by F. Princi, University of Cincinnati; "Health and Safety Factors" under "Lead Compounds," in *ECT* 2nd ed., Vol. 12, pp. 299–301 by A. P. Thompson, Eagle-Picher Industries, Inc.; "Control of Industrial Hazards," in *ECT* 2nd ed., Vol. 12, pp. 301–303, by D. G. Fowler, Lead Industries Association, Inc.; "Industrial Toxicology" under "Lead Compounds" in *ECT* 3rd ed., Vol. 14, pp. 196–200, by G. Ter Haar, Ethyl Corp.

1. J. M. Davis, R. W. Elias, and L. D. Grant, *NeuroToxicol.* **14**, 15–28 (1993).
2. J. J. Breen and C. R. Stroup, eds., *Lead Poisoning: Exposure, Abatement, Regulation*, Lewis Publishers/CRC Press, Boca Raton, Fla., 1995.
3. H. L. Needleman, ed., *Human Lead Exposure*, CRC Press, Inc., Boca Raton, Fla., 1992.
4. C. I. Davidson and M. Rabinowitz, in Ref. 3, pp. 65–86.
5. R. D. Schlag, in L. Fishbein, A. Furst, and M. A. Mehlman, eds., *Advances in Modern Environmental Toxicology*, Vol. XI, *Gentoxic and Carcinogenic Metals: Environmental and Occupational Occurrence and Exposure*, Princeton Scientific Publishing Co., Inc., Princeton, N.J., 1987, pp. 211–244.
6. T. D. Matte, P. J. Landrigan, and E. L. Baker, in Ref. 3, pp. 155–168.
7. "Preventing Lead Poisoning in Construction Workers," *NIOSH ALERT,* DHHS (NIOSH) publication no. 901-116, Cincinnati, Ohio, Aug. 1991.
8. C. D. Klaassen, in A. G. Gilman, T. W. Rall, A. S. Nies, and P. Taylor, eds., *Goodman and Gilman's The Pharmacological Basis of Therapeutics*, 8th ed., Pergamon Press, Inc., New York, 1990, pp. 1593–1598.
9. P. J. Landrigan, J. R. Froines, and K. R. Mahaffey, *Top. Environ. Health* **7**, 421–451 (1985).
10. R. L. Boeckx, *Anal. Chem.* **58**, 274A–287A (1986).
11. C. R. Angle, *Ann. Rev. Pharmacol. Toxicol.* **32**, 409–434 (1993).
12. R. A. Goyer, in M. O. Amdur, J. Doull, and C. D. Klaassen, eds., *Casarett and Doull's Toxicology: The Basic Science of Poisons*, 4th ed., Pergamon Press, Inc., New York, 1991, pp. 639–646.
13. P. J. Landrigan, *Environ. Health Perspect.* **91**, 81–86 (1991).

General References

For an excellent, comprehensive review of the chemistry, environmental, and anthropogenic release, environmental fate, and environmental and human health effects of lead, see *Air Quality Criteria for Lead*, Vol. I–IV, EPA-600/8-83/028a-dF, U.S. Environmental Protection Agency, Washington, D.C., June 1986, and *Air Quality Criteria for Lead: Supplement to the 1986 Addendum*, EPA-600/8-89/049F, U.S. Environmental Protection Agency, Washington, D.C., Aug. 1990.

Health Effects Assessment for Lead, EPA-540/1-86/055, U.S. Environmental Protection Agency, Washington, D.C., Sept. 1984.

An Exposure and Risk Assessment for Lead, EPA-440/4-85/010, U.S. Environmental Protection Agency, Washington, D.C., Aug. 1982.

For a bibliography of over 5000 publications pertaining to lead (mostly articles dealing with the health effects of lead published since 1977), see *A Bibliography for Lead*, EPA-600/9-85/035F, U.S. Environmental Protection Agency, Washington, D.C., Apr. 1986 (NTIS # PB86-197175).

An Exposure and Risk Assessment for Lead, EPA-440/4-85/010, U.S. Environmental Protection Agency, Washington, D.C., Aug. 1982 (NTIS # PB85-220606).

See publications of R. Lilis for additional studies related to the toxicity of lead following occupational exposure. These publications have been compiled into a volume entitled *Studies on Lead Exposed Occupational Groups 1967–1989*, and are available from the Dept. of Community Medicine, Mount Sinai School of Medicine, New York, 1991.

The Relative Contribution of Lead from Anthropogenic Sources to the Total Human Lead Exposure in the United States, EPA-600/D-86/184, U.S. Environmental Protection Agency, Washington, D.C., Aug. 1986 (NTIS # PB86-241015).

STEPHEN C. DEVITO
JOSEPH BREEN
U.S. Environmental Protection Agency

LEAD COMPOUNDS, CONTROLS OF INDUSTRIAL HAZARDS.

See LEAD COMPOUNDS, INDUSTRIAL TOXICOLOGY.

LEAD COMPOUNDS, HEALTH AND SAFETY FACTORS.

See LEAD COMPOUNDS, INDUSTRIAL TOXICOLOGY.

LEAD POISONING.

See LEAD COMPOUNDS.

LEAD STORAGE BATTERIES.

See BATTERIES, SECONDARY CELLS (LEAD–ACID).

LEATHER

Tanning and leather working are ancient skills dating from the earliest hunter/gatherer societies when the practical value of animal skins were adapted to human usage for clothing and foot protection allowing an existence outside of tropical climates. Because preservation of a hide is not practical during a hunt, it can be assumed that animals were either taken back to the camp without skinning or the skins were removed, then brought to the camp.

The temporary preservation of leather is an important practical problem. When an animal is killed and skinned, bacterial degradation starts immediately, but temporary preservation by drying or salting does not result in a usable leather. Treatment to prevent decay is not the same as tanning, which is a slow process and in modern leather production involves a series of chemically interdependent steps. Tanning not only preserves the hide or skin, but also makes the leather resistant to cracking from flexing. The desired properties of leather depend on use. From the same hide, leathers may be obtained for garments, shoes, or mechanical applications. There are, however, limitations to the possible use of a particular hide or skin. For example, sheep skin, and most other fur skins, has very little value for shoes. The heavy skin of alligator would not be satisfactory for gloves.

Tanning of hides by any of the primitive methods was a dirty job with little assurance of success. The tanner was a specialist generating little respect in the community. In some parts of the world, this ancient prejudice against tanners exists even as of this writing (ca 1994). The tanner in primitive societies was limited to making leather using the hides or skins available locally. Tanning materials were also limited to those in local supply. Techniques were dependent on the climate as well as availability of materials. The leather made was for local needs.

The Eskimos of Alaska have access to animals such as the seal or the polar bear. Seal skins make good furs that are essential for survival in the far northern climate. Because originally there were no tanning materials available, Eskimo tannage was done by scraping the skin to a minimum thickness and then exposing the skin to smoke. The smoke aids in the oxidation of some of the fats of the skin to aldehydes (qv) that help to preserve the skin. This system, which is satisfactory for tanning in frigid climates, would be of limited value in a tropical one.

In desert or tropical climates skin dries into a hard mass that does not soften even when soaked in water. By shaping a skin to a desired form and then drying, however, useful articles such as drum heads and shields can be made. Tanning for useful leather is not a simple skill.

In temperate climates extracts from some plants were found to be excellent preservatives for hides and skins. The hides, with or without hair, were placed in pits in the ground, then covered with alternating layers of bark or leaves and skins. Water was added and later, ie, days or months depending on the thickness of the hide, the hides could be removed, washed, and oiled. The resulting leather is flexible and lasts essentially forever. This procedure was used well into the seventeenth century as the most common method of tanning. In some isolated primitive societies, the method is used in the 1990s.

Hides and Skins

Structure. The structures of hides and skins are dependent on the needs of the animal and its environment. The functions of an animal's skin include protection from predators and infection, and maintenance of body temperature. The relative importance of these functions depends on the animal. Methods by which the skin accomplishes these functions is the same for most mammals.

Leather technologists have adopted the same histological techniques for the study of hide and skin structure as that used by the medical profession for the study of the structure and functions of human skin (1,2).

In Figure 1, the lower edge of the drawing is the flesh side of the hide. The hide, as it is removed from an animal, has body fat and a thin membrane separating the hide from the fat and flesh of the body of the animal. The area near the inside of the hide is made up of the heaviest fibers of the hide. These fibers, intertwined in directions somewhat parallel to the surface of the skin, vary in size and shape and may be ~ 0.01 mm in diameter. The length of these fibers vary but may be several centimeters long. The fibers are the ultimate in nonwoven structure that give leather its remarkable strength and flexibility.

Above the heavy fibers there are glands and hair follicles. Mammals are warm blooded and the temperature of the body must be maintained within a very narrow range. The function of the sweat glands is to release moisture to

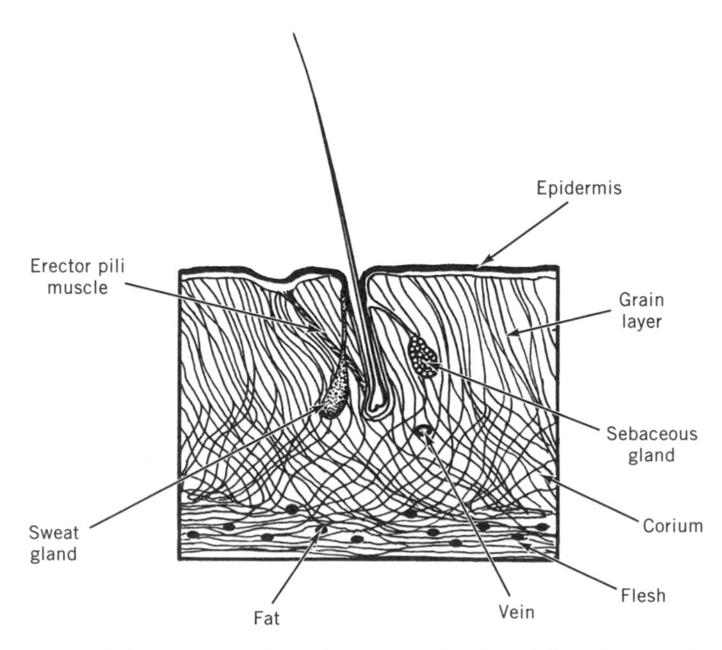

Fig. 1. Schematic of the cross section of a mammal's skin. The relative size and function of the parts depend on the species and breed of the animal. For goats, where the wool or hair is sparse because it is not needed for warmth, the skin is dense to provide protection; for sheep protected primarily by heavy wool, the skin contains more oil (sebaceous) glands to lubricate the wool; for cattle, both the hair and the heavy hide structure protect the animal (3). Courtesy of Krieger Publishing Co.

the surface of the skin to cool the body by evaporative cooling. The sebaceous glands are oil glands that have two functions, ie, to slow the evaporative cooling for temperature control, and to lubricate the hair of the animal. The hair grows from the hair follicle as soft, low molecular weight proteins and changes into hard inert hair protein.

The fiber structure is very fine near the surface of the skin and this fine structure imparts a silky feel to the leather. The smaller the animal of a given species the finer the surface fibers. The value of the skin or hide is then dependent in part on this smoothness (2). Calfskin leathers are smoother and have a silkier feel than cattle hide leathers. Calfskin leather is used in the shoe uppers of high quality, expensive men's and women's shoes, and specialty items.

Sheep and fur animals are protected primarily by their wool or hair. The fiber structure of the skin is very fine and has less strength than calfskin or other nonfur mammals. The sheepskin has a high concentration of hair follicles and sebaceous glands. When tanned the fur skin has an open structure, is soft, and lacks strength relative to many other leathers.

Goats, the animal of choice in areas of harsh climates and limited food supply, are particularly suited to warm and sometimes arid climates. Although goats differ greatly through breeding and environmental adaptation, they have some common characteristics. The skin of the goat is thin and made of tight strong fibers. The hair is coarse and functions more for protection from predators than for warmth. The structure of the skin determines the suitability of the leather for a given purpose. Tanning methods to accentuate the desirable characteristics of each of the leathers is required.

Chemical Composition. From the point of view of leathermaking, hides consist of four broad classes of proteins: collagen, elastin, albumen, and keratin (3). The fats are triglycerides and mixed esters. The hides as received in a tannery contain water and a curing agent. Salt-cured cattle hides contain 40–50% water and 10–20% ordinary salt, NaCl. Surface dirt is usually about 2–5 wt %. Cattle hides have 5–15% fats depending on the breed and source. The balance of the hide is protein (1).

Leathermaking occurs in three broad steps: (*1*) removal of all materials that are not a part of the final product, (*2*) rendering the remaining hide substance biorefractive, ie, tanning, and (*3*) treating the stabilized tanned material to impart the characteristics desired in the final product. In all steps the solubility and reactivity of the components form the basis of the treatments employed.

The hide proteins differ in amino acid composition and physical structure. The principal amino acids (qv) of the hide proteins are listed in Table 1. Of particular importance is the difference in the water solubility of the proteins. All of the proteins are soluble in water when heated, and upon the addition of either strong acids or bases. Proteins (qv) are amphoteric, possessing both acid and base binding capacity.

Proteins have reactive carboxyl groups and amino groups on the polypeptide chain (4,5). The more acid groups the greater the base-binding capacity of the protein. Conversely, the more basic groups the greater the acid-binding capacity; ie, the solubility characteristics of proteins can often be predicted from the acid–base amino acid composition.

Table 1. Amino Acid Content of Proteins, %[a]

Amino acid	Collagen	Elastin	Keratin	Albumin
		Nonpolar		
glycine	20	22	5	2
alanine	8	15	3	6
valine	3	12	5	6
leucine	5	10	7	12
other	4	15	7	9
Total	*40*	*74*	*27*	*35*
		Acid		
aspartic acid	6	0.5	7	11
glutamic acid	10	2.5	15	17
Total	*16*	*3*	*22*	*28*
		Basic		
arginine	8	1	10	6
lysine	4	0.5	3	13
other	2	0.5	1	4
Total	*14*	*2*	*14*	*23*
		Others		
serine	2	1	8	4
cystine			14	
proline–hydroxyproline	25	15	6	5
other	3	5	9	5
Total	*30*	*21*	*37*	*14*

[a]Ref. 3.

Albumen has the largest number of acid and basic groups. It is the most soluble of the proteins present in a hide. The albumen is not a fibrous material, however, and therefore has no value in the leather. Keratin is the protein of the hair and the outermost surface of the hide. Unless the hair is desired for the final product it is removed by chemical and/or physical means. The elastin has little acid- or base-binding capacity and is the least soluble of the proteins present. The lack of reactivity of the elastin is a detriment for most leather manufacture. The presence of elastin in the leather greatly limits the softness of the leather.

Collagen, the principal protein of the hide, is the material that is made into leather. The reactivity of the collagen toward tanning agents and the dyeability, as well as the strength, flexibility, and durability of the collagen when tanned, are all important to making leather a material of choice for utility and fashion.

Treatment. In each of the steps of leathermaking the chemistry of the system is designed to selectively react with the various proteins, particularly collagen (6). In the first step, the collagen is separated from the components that are not a part of the final product so that there is as little collagen degradation as possible. In the second step, tanning, collagen reacts with tanning materials to stabilize the protein to resist chemical, thermal, and biological degradation. The third and last step is to impart the desired mechanical and aesthetic properties to the product.

Leather Manufacture

The manufacture of leather follows the same general steps for a great variety of leathers (Fig. 2). The largest category of hides tanned is cattle hides. Of the cattle hides chrome tanning of unhaired hides is by far the dominant system used throughout the world. The tanning of other types of hides and skins requires variations in the systems used for cattle hides (3).

Curing. The temporary preservation of hides or skins is known as curing. Curing is not tanning in that the hides or skins are not stabilized to a biorefractive state (2). Cured hides immediately begin to decay if there is a change in conditions that removes the curing agent or permits bacterial action. The curing of hides or skins must meet several requirements: (*1*) the curing system must preserve the skin under the anticipated conditions of storage and transportation; (*2*) the curing agent must not damage the valuable leathermaking material; (*3*) the curing system must be reversible so that the hide or skin can be easily brought to a desired condition for tanning; (*4*) the curing system must be fast enough to be effective before the onset of biodegradation; and (*5*) the curing system must be economically practical.

Air Drying. Air drying, probably the oldest curing system, has the advantages of simplicity, speed, and low cost. The disadvantages are possible case hardening and excessive adhesion of the fibers. Drying (qv) involves the migration of water to the surface of the skin and subsequent evaporation. If the hide is thick and evaporation fast, a sealing of the surface of the hide may occur preventing continued migration of the water. As a result the water may be trapped in the hide and bacterial degradation can thus occur in the center of the hide.

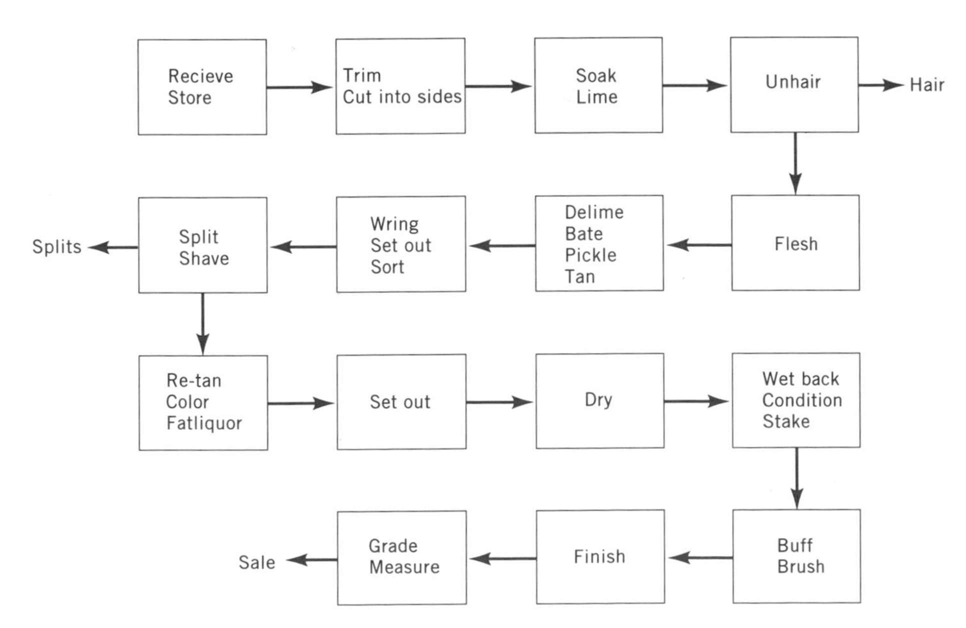

Fig. 2. Process flow diagram for the production of chrome-tanned cattle hide leather. Adapted from Ref. 3.

When drying cattle hides in tropical arid climates, the hides are scraped clean of flesh, then stretched on racks, and dried slowly in the shade. An application of an insecticide may be used. This method is slow, labor intensive, and unreliable for heavy hides. The system is only used in rural tropical areas and is not applicable to modern commercial cattle hide production (3).

Drying has a place in modern technology primarily for the curing of furs and goatskins. Fur skins are thin and the fur side is protected by the thick mass of hair. The flesh side of the skin is scraped clean of adhering flesh and the skin is stretched on a form of flat board, flesh out. The skin is slowly dried with or without the addition of salt. The advantage of this system is quick curing with a minimum chance of bacterial damage. The fur skins are valuable and therefore the labor required is acceptable.

Drying methods vary according to the type and source of skins (1). The significant commercial sources of goatskins are in Africa, Asia, and South America. The methods of cure include air drying, salt drying, and salting depending on the local conditions.

Salt Curing. Salt curing is the generally accepted curing method for cattle and other large animal hides. Curing can be done using solid salt or brine. The solid salt method is the more general and is used worldwide.

In salt curing the hides are piled on a concrete floor and salt is applied to them. To effect the cure the salt must be in contact with both the flesh and hair side of the hide. Thus salt is spread between the hides, and the hides piled. The amount of salt needed is about 50–60% based on the weight of the hides, although this amount of salt may vary in specific cases. The hides remain in the curing pile for a minimum of about 20 days to effect the cure. At the end of the cure time the hides are removed, the excess salt brushed off, and the hides folded individually into bundles for shipping. If water and excessive heat is avoided, the hides are effectively preserved for commercial purposes, including international shipping.

In modern mechanized slaughter houses, particularly in North America, the commercial practice has shifted to brine curing. In conventional salt curing, the curing time of 20 or more days presents a problem of space as well as an investment in inventory. The modern practice is to chill the hides in cold water while washing off dirt and manure. The hides are trimmed to remove the parts such as ears and tails not suitable for leather, fleshed by machine, then placed in a saturated agitated brine bath. The brine cure takes from 24 to 48 hours depending on the type of hides and the equipment used. At the end of the cure time the hides are removed, wrung by machine, and bundled for shipping.

The sale of hides is on a weight basis. The value of a hide depends on the type of animal, the seasonal characteristics, the location of the slaughter, the type of cure, and the market conditions. Cost of the hides is about 50% of the sale price of the leather, so an accurate knowledge of the hides and the leathermaking potential of the hides is critical to commercial success. In the case of furs and exotic skin leather production, the value of the pelt is by far the most important factor (7).

Alternative Methods of Curing. The presence of salt in hides and skins has been a problem for tanners. Moreover, the quantity of salt, usually 10–20%

by weight in cured hides, is an environmental problem. In the discharge of the water from the rewet and washing, the soluble solids are normally greater than 20,000 mg/L. This amounted to 2000–4000 mg/L in the combined waste stream of the tannery.

Several methods have received considerable research attention as alternatives to salt curing. These include use of sodium bisulfite as a disinfectant to allow preservation with or without decreased salt in a brine cure; use of disinfectants such as quatenary amines for temporary preservation in direct shipping to the tannery from the packing plant (see DISINFECTANTS AND ANTISEPTICS); preservation of hides by radiation sterilization (see STERILIZATION TECHNIQUES); and substitution of materials such as potassium chloride for sodium chloride. These methods have found only limited commercial success.

Pre-Production Handling. Salt-cured cattle hides, when received at the tannery, are individually bundled to prevent excessive moisture loss. The bundles are tied with ropes that are later cut and removed; the hides may be sorted for different weight or quality classification at this point. It is best to have hides of similar size and thickness in a given production batch to assure an even reactivity of the processing chemicals and to avoid frequent adjustments in the machinery to compensate for size and thickness variations. In the modern large tannery, the size/quality classification is not necessary because the hides arrive in carload quantities under specifications as to size, type, and month of slaughter.

Soaking. The hides are weighed and counted into production batches. For pre-fleshed and trimmed cattle hides the batches are about 3–5 t when the processing is in drums. If the tannery has hide processors the batches may be up to 10 t. Water is added to cover the hides and allow free movement of the load. The drum is turned intermittently during the normal 8 to 16-hour soaking period.

The quantity of water is two to three times the weight of the hides. The salt from the cure dissolves in the water and the reverse of the curing takes place. The water is drawn into the hides by osmotic forces. The concentration of the salt solution is about 3–5 g/100 mL. At this concentration some of the soluble proteins disperse. The soak water removes the salt, some proteins, some loose fat, blood, dirt, and manure.

If the hides were not fleshed before curing, the soaked hides are usually fleshed and trimmed at this time. If the hides are not to be trimmed or fleshed, they are drained and washed to decrease the salt concentration, drained, and the drum refilled with cold water.

Unhairing. Unhairing can be done either by a hair save or a hair pulp system. The hair pulp system is preferred by most tanners for its speed and labor efficiency. In the hair pulp system the hides are treated with sodium sulfide (sulfhydrate) and lime (calcium hydroxide). The hair is quickly destroyed by the strong alkaline reducing conditions.

The differences in the amino acid chemistry of the hide collagen and the hair keratin are the basis of the lime–sulfide unhairing system. Hair contains the amino acid cystine. This sulfur-containing amino acid cross-links the polypeptide chains of mature hair proteins. In modern production of bovine leathers the quantity of sulfide, as Na_2S or NaSH, is normally 2–4% based on the weight of the hides. The lime is essentially an unlimited supply of alkali buffered to pH 12–12.5. The sulfide breaks the polypeptide S–S cross-links by reduction. Unhairing without sulfide may take several days or weeks. The keratin can be easily hydrolyzed once there is a breakdown in the hair fiber structure and the hair can be removed mechanically. The collagen hydrolysis is not affected by the presence of the sulfides (1–4,7).

After about 4 to 12 hours the hides are drained, then floated with lime without additional sulfides. The drum or hide processor is run at slow speeds intermittently during the unhairing process. The continued action of the lime solution on the hides brings the pH to 12–13. At this pH the hides swell to about twice the original thickness. This swelling has a beneficial opening of the fibers to permit better tannage (Fig. 3**a**).

The mechanical and physical characteristics of the leather are also effected by the swelling (8). The charge on the protein is important for each step in leather production. When liming is completed, usually in about 12–20 hours, the hides are washed in running water. This washing removes the surface dirt, degraded hair lime, and sulfides. The unhairing and the wash water from the hair pulping system create the most serious environmental problems of the industry.

Trimming and Fleshing. The limed hides not fleshed before curing are usually trimmed and fleshed at this point in the production. The trimming is done by hand to remove any portions of the hide that could interfere with the subsequent machine processes, eg, the shanks, ears, and snout.

The fleshing is done on a multiroller machine that pulls the hide over a rotating blade, similar to a milling machine that cuts off the flesh from the inside of the hide. The machine includes a rubber roller that holds the hide near the rotating fleshing blades. Through the adjustment of the clearances or the thickness of the hides and the resilience of the rubber roller the flesh is effectively removed regardless of the differences in the thickness of the hide from back to belly and flanks.

The flesh and trimmings may be discarded as a waste in small tanneries. In the larger tanneries, the economies of size warrant the recovery of the fleshings for rendering or for glue or gelatin.

Splitting. In most modern large tanneries that make upholstery leather, and in some that make shoe uppers leather, the hides are split in the lime

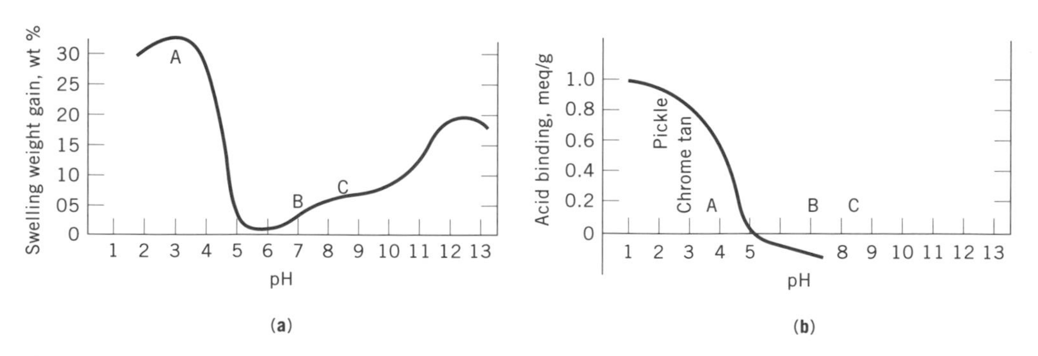

Fig. 3. The pH dependence, where A, B, and C represent regions corresponding to the pK_as of glutamic and aspartic acids, lysine, and argenine, respectively, of (**a**) protein swelling, and (**b**) protein acid-binding capacity. Adapted from Ref. 3.

condition. In splitting the hides are cut to the desired thickness with a horizontal belt knife. The hides are fed into the machine grain up. The clearance between the grain and the blade is maintained by a series of narrow rollers supported by a rubber roller underneath the spacing rollers. The grain layer is then cut to the thickness desired to an accuracy of about 0.1 mm.

The grain layer is the most valuable part of the hide and serves as the outside of the shoe. Splitting allows the grain layer to spread to the maximum area yield and also allows an efficient use of the valuable tanning chemicals.

The split is of variable thickness, and some of the split is too thin for leather production. The thin parts are trimmed off and discarded for collagen recovery. The trimmed split is tanned separately.

Deliming and Bating. The limed hides have a pH around 12. Because chrome tanning is done at pH 2–4, the lime must be removed for pH adjustment. In addition, the undesirable materials in the hide, ie, both natural and the degradation products from the unhairing, must be removed (7,9).

For deliming, ammonium salts and acids are used. The proportion of ammonium salts to acids and the type of acids employed is a matter of the tanner's choice. The acid neutralizes the lime, $Ca(OH)_2$, thereby adjusting the pH. The ammonium salts have two functions: to buffer the solution to a pH required for bating, and to form calcium ammonium complexes. The acidity and the complex formation solubilize the calcium and serve to bring the hide to the desired pH.

Bating is a part of the deliming step in cattle hide leather production. The hides contain some elastin proteins which are very inert to the action of acids and bases and react to tanning chemicals in a limited manner. If the elastin is not broken down sufficiently, the leather may be too firm and stiff for the desired use, and the grain may not be as smooth as desired. The bating materials are proteolytic enzymes having specific properties desired in the deliming. Commercial bating products are specialty dry chemicals supplied with enzyme carriers and sometimes ammonium salts. The bating enzymes and the pH adjustment disperse much of the degradation products from the unhairing. The resulting hides are clean and flacid.

Pickling. Bated hides or skins are at a near neutral pH and thus are immediately processed, because under these conditions the protein is subject to bacterial degradation. Pickling is the term used for acidification of the hides. For chrome tanning, the desired pH is about 2.0, thus the hides are placed in a solution of salt and acid (2,3,7,9). Proteins are amphoteric, and at neutral pH hides are flacid and have little acid or base bound. At pH $\sim$ 2.0 the acid absorbed is about 95 meq/100 g of hide protein (see Fig. 3**b**). A high ionic strength is necessary to prevent osmotic swelling which would cause damage to the delicate hide fibers and destroy the leathermaking properties of the hide. Swelling can be avoided by raising the ionic strength of the solution by the addition of salt. In practical application, the salt, NaCl, is kept above 3–4%. An excess of salt is less objectionable than a deficiency. The acid is usually sulfuric, and about 1–2% of the hide weight is used, just enough to reach the acid-binding capacity of the hide protein.

Once pickled, the hides can be drained and stored indefinitely. The possibility of bacterial or mold damage is always present, but under proper pickling and storage the hides or skins can be traded on the world market. There is, however, little trade in pickled hides except for sheepskins where commerce in pickled skins offers commercial advantages. Salt curing of sheepskins is risky because the salt may penetrate slowly owing to the heavy wool. Sheepskins also contain a large quantity of fat that may inhibit the penetration of the salt.

The usual procedure in the manufacture of chrome-tanned leathers is the use of a continuous bate, pickle, and tan method. The hides or skins remain in the drum from the lime washing through the chrome tanning stages. The entire process usually takes about 22–24 hours. The bate, pickle, and tan can be done faster, but most tanners find that a one-day cycle fits well into production scheduling and results in a quality leather.

The degree of pickle, ie, the amount and type of acid used, depends on the type of hides or skins and the tannage to be used. The pickle is in preparation for tanning and the chemistry of the subsequent tannage determines the pickle method.

Chrome Tanning. Chrome tanning is the most widely used tanning system worldwide. Chrome tanning has the advantages of light color, speed of processing, low costs, and great stability of the resulting leather. Chrome-tanned leather is so stable that exposure to boiling water for short ($< 2-3$ min) periods of time usually shows no adverse effect (1–4,6,7,9).

Tanning refers to a specific reaction of the tanning chemicals combining with the hide or skin to stabilize the protein and make it resistant to bacterial degradation. Tanned leather is so biorefractive that leathers even centuries old found under adverse natural storage conditions are often in almost usable condition. Prior to the 1900s essentially all leather was tanned by the vegetable tanning method.

Chrome tanning is done in a drum similar to that used for deliming. The salt solution from the pickle is present and the solution is at about pH 2.0. The chrome-tanning material is usually a basic chromium sulfate. The general formula for the most common commercial chrome-tanning product is $2Cr(OH)SO_4 \cdot Na_2SO_4$. After the addition of the chrome-tanning salts, the pH of the solution rises to about 2.5. At this pH the chromium salt is taken up by the hide and the tannage begins.

Control of chromium penetration, essential to permit tannage of the center of the hide, is accomplished by pH adjustment. At a pH ≥ 3.0 the reactivity of the hide to the chromium complex is greatly increased. The pH is therefore raised gradually to the desired point by addition of a mild alkali, usually sodium bicarbonate. The chemistry of chrome tanning involves competing reactions that must be controlled for satisfactory results.

The chrome-tanned leather is removed from the drum and wrung to remove the absorbed tanning solution. The leather is then inspected for quality of the grain and other characteristics of importance for the leather being made. In large tanneries where very uniform hides are worked, the leather may be trimmed and split to the desired thickness. Hides split before tanning need no splitting at this point.

Following splitting the leather is further brought to the desired thickness by shaving. Shaving is done on a machine having a multibladed rotating cylinder. This machine is similar to the fleshing machine, but is much more precise in the accuracy of the thickness of the leather.

Retanning, Coloring, and Fatliquoring. Chrome-tanned leather is a light blue in color. The fibers are only stabilized against microbial action and do not have the feel of leather. If the leather were dried at this point only a stiff unattractive product would result. The characteristics of desired leather result from the retanning, coloring, and fatliquoring.

The application of vegetable tanning materials has an additive effect on the leather. The more vegetable tannins applied the more the leather becomes like vegetable-tanned leather. The color is changed, the fullness of feel increases, and the leather can be worked and embossed like vegetable leather.

In retanning, vegetable tannins may be used in conjunction with or may be entirely replaced by synthetic tanning agents called syntans. The syntans and other specialty chemicals allow the creation of leathers not possible using vegetable tannins alone.

Modern leather can be made in any color desired. The dyeing of the chrome-tanned leather is done normally as part of the retan, color, and fatliquoring steps where synthetic organic acid dyes are applied as needed. Both acid and basic dyes are used. Traditional examples are Acid Blue 2B [*6408-78-2*] and Bismarck Brown G [*10114-58-6*], respectively. The penetration of the dye is important. For suede leather or any leather subjected to rough wear the color should be deep into the leather to assure uniform color during use. For most shoe upper leather, or any leather that is finished with an opaque surface coating, the dye need be only surface coloring. The cost of the dyes and the final use of the leather are the determining factors in the choice of the dye system used.

Fatliquoring is the term applied to the oiling of leather. Heavy leathers, such as work boot leather or harness leather, may be lubricated by the application of oils and greases directly in a drum. The greases are melted and applied to the leather without additional water. This system is not used for light chrome-tanned leather; most leather is lubricated by the application of emulsified oils. The oils comprise sulfated or sulfonated animal or fish oils, synthetic oils, or specialty products developed for the industry that can penetrate the leather as desired and then be strongly adsorbed on the leather fibers. The temperature and pH of the system during fatliquoring is important.

Drying. The retanned leather is stretched to increase the area for the best yield and to produce a flat leather surface. The leather is then dried. There are several drying (qv) systems used which depend on the type and thickness of the leather. Thin garment leathers, made soft by retanning and fatliquoring, may be dried by hanging in a dry loft. Soft leathers can be reworked by mechanical means to the softness desired when dry. For shoe uppers the leather should be held in a stretched condition. Two or three types of drying are commonly in use in the industry: toggle drying, paste drying, and vacuum drying. In toggle drying the leather is held in an extended condition with clamps on the edge of the leather stretched on a screen. The screens are placed in a temperature and humidity controlled dryer until the leather is dry.

In paste drying the leather is spread on glass or porcelain plates and held in place with a low strength water-soluble paste. The plates are on a conveyor and the drying is done in a drying tunnel. The dryer usually has several temperature- and humidity-controlled zones to control the rate of drying and to prevent overdrying.

Vacuum drying is done by spreading the leather grain on a smooth, heated, stainless steel surface. A cover is placed over the drying surface and the drying chamber evacuated. The advantage of vacuum drying is the speed at which each piece is dried using low temperatures. The vacuum drying system is beneficial for the production of some types of quality shoe leather. The equipment is expensive, however, and is only found in modern, well-financed tanneries.

Staking. Dried leather is somewhat stiff because of the slight adhesion of the fibers to one another. Staking is a mild flexing of the leather to bring it to the desired final softness. The ultimate softness of the leather is controlled by the retannage and fatliquoring and cannot be built into the leather by mechanical flexing. The methods of flexing include hand staking in cottage industry tanneries or machine staking by any of a number of commercial machines. Modern tanneries use a staking machine employing a number of vertically opposing oscillating rounded rods. The leather is fed through the machine between soft rubber belts. The machine can be adjusted for clearance and the desired degree of softness.

Buffing. Most leather is buffed before finishing. The buffing step consists of a light sandpapering of the grain. If the grain surface of the leather is free of blemishes it may be good enough for full grain where the full beauty of the natural hide surface is visible. Full-grain leathers are preferred and therefore are more expensive.

Buffing may be only to remove surface blemishes such as insect bites or minor healed scratches, or the buffing may be to make a suede surface. Newbuck is a light suedeing of the grain of cattle hide leather. Sheepskin suede is usually buffed on the flesh side of the skin. Splits may be buffed to make buck shoe leather or heavy garment leather, often called ranch hide. The variations depend on the desires of the customer.

Finishing. The finishing of leather is parallel in some ways to the finishing of wood (qv). In both cases the application of the finish results in a protective and/or decorative coat. The finishing of leather has changed greatly because of the development of resin systems in the coating industry (see COATINGS). Leather finishes must minimally be abrasion resistant and flexible, and must adhere to the leather. Formation of a tough water-resistant film is also desirable.

Leather finishes penetrate to a greater or lesser extent and have a profound effect on the grain or wrinkle characteristics of the leather. Penetration of the resin into the leather tightens the grain but may produce a surface stiffness and a tendency toward grain cracking upon flexing. The development of leather finish resins and the application of these finishes is done by specialty houses.

The application of the finishes can be by brush, roll coater, spray gun, or by flow coater. In small tanneries hand finishing without machinery is common. In larger tanneries the most common method is by spraying. The spray-finishing machines are usually multistage and have temperature-controlled drying stages in line. The spray guns may be actuated by photoelectric cells to spray onto the leather and not an empty conveyer belt. Roll coaters and flow coaters are not as common as spray lines but have great value in the production of some specialty leathers (see COATING PROCESSES).

The finishing usually requires several coats. The first coats are water-based latex finishes containing pigments or dyes. The final coats are liquids for protection and sheen as desired. Two or three pigment coats and one or two top coats are normal. For expensive aniline-type leathers the grain surface is visible through the finish; the color arises from the dye in the leather. For pigment-finished products the surface is colored with the pigment in the finish, which is similar to painting. Top dressing adds protection and the desired surface reflectance.

As a part of the finishing, the leather is almost always pressed with a warm press. The press plate may be an embossing plate to give a decorative surface, thus the grain may be made to simulate an exotic animal such as alligator or ostrich; even the familiar football grain can be made. Many poor-grade hides may be embossed. Heavy embossing covers hide defects and also upgrades the hide to novelty leather.

Hot pressing with a smooth plate has an advantage in smoothing the grain, and the heat can be used to cure the resin of the finish. The hot pressing is anticipated in the design of the finish system and in the choice of the resins by the finish manufacturer.

Grading, Measuring, and Shipping. Prior to shipment each hide, side, or skin is inspected and sorted. The grading is done on the basis of hide defects, shape of the skin, manufacturing defects, or any other factors of importance to the specifications of the sale.

The leather in the grade to be shipped is measured for area. In a cottage industry the area may be measured by placing a wire grid over the hide and estimating the area by counting the squares. In modern tanneries an electronic measuring machine is used. In the electronic system the leather passes under a bank of lights where measurement is carried out by photoelectric cells. The area of the leather is calculated and the area stamped on the back of each piece of leather. The degree of sophistication of the measurement and correlation with materials used and other manufacturing costs depend on the computer system of the company.

Chemistry of Tanning

Chrome Tanning. The original chrome tanning was a two-bath process. The unhaired hides, delimed and bated, were treated with a solution of sodium bichromate [*10588-01-9*]. The amount of bichromate used was about 3–5% based on the weight of the hides. The bichromate was absorbed or adsorbed into the hide, the solution drained, and the hides refloated. Sodium bisulfite was added and two important reactions resulted in the formation of a basic chromium and colloidal sulfur in the hide. This gave a chrome tannage and also helped to fill the hide with the solid sulfur. This crude system, which continued in the industry in some types of leather for over 50 years, is obsolete.

Modern chrome-tanning methods are well controlled and employ an extensive knowledge of the chemistry of the system. The most common chromium-tanning material used is basic chromium sulfate [*12336-95-7*], $Cr(OH)SO_4$, made by the reduction of sodium bichromate with sulfur dioxide or by sulfuric acid and a sugar.

$$Na_2Cr_2O_7 + 3\ SO_2 + H_2O \longrightarrow 2\ CrOHSO_4 + Na_2SO_4$$

Chromium sulfate is described as being 33% basic and in solution gives a pH of ca 2.5.

Ionization of the basic chromium salt results in the formation of complex ions such as

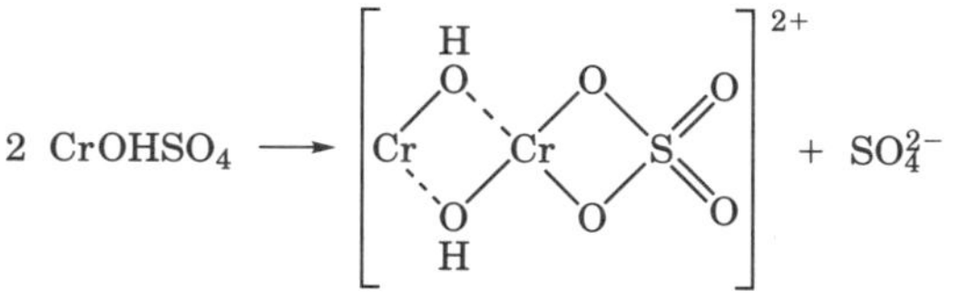

$$2\ CrOHSO_4 \longrightarrow [Cr(OH)_2Cr(O_2SO_2)]^{2+} + SO_4^{2-}$$

The basic complex structure penetrates the hide at pH <3.0. There is a fixation of the chromium to the hide protein primarily by reaction of the chromium and the carboxyl groups of the hide.

After penetration of the hide by the chromium the pH is raised to about 3.5–4.0. At this higher pH a change occurs in the chromium complexes as the basicity of the chromium increases and binding to the protein becomes possible. Chromium binds firmly to the protein forming a cross-link species, and as the pH increases the hydrogen is removed from the complex forming a stable structure.

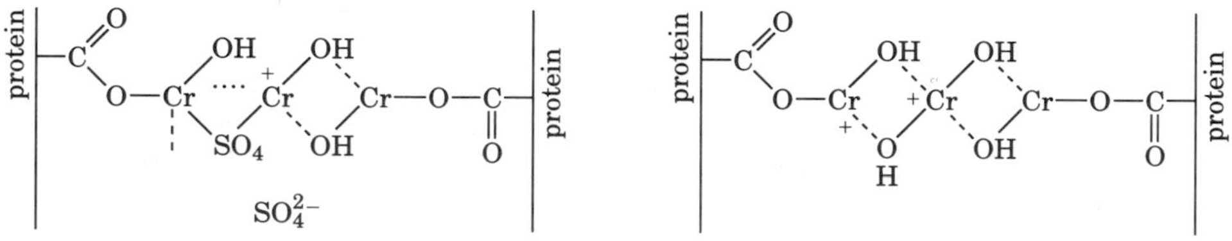

protein—C(=O)—O—Cr(OH)(SO₄)Cr(OH)Cr—O—C(=O)—protein
SO_4^{2-}

protein—C(=O)—O—Cr(OH)(O–H)Cr(OH)Cr—O—C(=O)—protein

The chromium can be stabilized in a limited way to prevent surface fixation by addition of formate ions. The formate displaces the sulfate from the complex and masks the hydroxyl ions from forming the larger higher basicity complexes. This stabilization can then be reversed in the neutralization to a pH of about 4.0 and tannage becomes complete. This simple formate addition has decreased the time of chrome tanning by about 50% and has greatly increased the consistent quality of the leather produced.

Vegetable Tanning. Vegetable tanning materials became a significant commodity in international trade from the 1700s through the early 1900s. The demand for vegetable tanning has decreased but remains a principal factor in the production of heavy leathers. Vegetable tannins have the general structure of polyphenolic compounds. There are two general classes of vegetable tannins: the hydrolyzable and the condensed. The hydrolyzable tannins are derivatives of pyrogallol [*87-66-1*] (1,2,3-trihydroxybenzene). The condensed tannins are derivatives of catechol [*120-80-9*], also known as pyrocatechol, *ortho*-dihydroxy-benzene, and 1,2-benzenediol (1,2,4,7,10).

Any vegetable tanning extract used commercially is a complex mixture of related substances. The individual tanning properties of the extracts have been extensively studied and are well known in the industry.

The mechanism of the tannage is accepted to be largely one of replacement of the bound water molecules by the phenolic groups of the tannin and subsequent formation of hydrogen bonds with the peptide bonds of the protein. The effect of this bonding is to make the leather almost completely biorefractive.

The size of the vegetable tanning molecules and the colloidal nature of the system result in the fixation in the hide of filling materials. The filling action is essentially an impregnation of the hide to form a dense firm leather. These properties are greatly desired in sole and mechanical leathers.

Vegetable tanning does not significantly cross-link the protein chain. Thus this leather can be mechanically shaped and easily embossed or tooled to a desired shape. Vegetable-tanned leather is the leather of choice for handicraft work. Shoe soles made of vegetable-tanned leather break in to the wearer's foot shape and become comfortable even though the sole may be thick and heavy. The use of vegetable extracts in the retanning of chrome leather adds these vegetable leather characteristics.

The vegetable-tanning materials are sold on a tannin analysis. The materials are dissolved to a specified solution strength, then portions are filtered (10). From the weight of the total solids, and that of the dissolved solids, the solids content and the soluble substances are calculated on a percentage basis. The tannins are measured by tanning a chrome-tanned hide powder, prepared as specified, then the solution is filtered and dried. The tannins are determined by difference to yield percent solids, insolubles, nontans, and tannin. This system can be used on both liquid and dried products.

Vegetable-tanning materials in commercial quantities come from many different countries. Quebracho is a principal tanning material from South America. Wattle or Mimosa is supplied from several African sources. India and other Asian countries supply a variety of materials including Myrabolans, Gall Nuts, and Tara Pods (3).

The vegetable-tanning materials are commercially extracted using hot water. The extraction is normally done in countercurrent extractors that permit the final removal of the extracts with fresh water. The dilute extracts are then evaporated to the desired concentration in multiple effect evaporators. Some extracts may be further dried by spray drying or any other means that proves effective without overheating the extract. Extract preparation depends on the type of extract, the size of the operation, and the desired concentration of the final product.

The tanning rates and the distribution of the tannins in the leather is dependent on the characteristics of the tanning material and the colloidal conditions in the solution and inside the hide (1,2,7). The tanning materials, extracted at high temperature, are applied at ambient or slightly raised temperatures, usually $<40°C$. Penetration is controlled by pH and temperature conditions. The lower the pH the more the weak-acid, negatively charged, colloidal particles coagulate and combine with the hide. In conventional vegetable tannages the pH was historically controlled by the choice of the extracts and the degree of fermentation of the tanning materials. The advent of accurate pH measurements has shortened the time of tannages.

The speed of the tannage is important. At the end of the nineteenth century tanning times of several months were normal for vegetable tanning; modern production using chemical controls and drum tannages has cut the time in process to a few days. Pre-tanning the hide by chrome tanning stabilizes the hide to eliminate the effects of swelling, and frees much of the bound water. The hides, pre-tanned with chrome, can be drum tanned using vegetable-tanning materials to make leather that was formerly tanned only by vegetable tanning. The leathers more commonly made with this system are upholstery, bag, case, and strap leathers. Many of the sole leathers are still made using only vegetable tanning. In these leathers the filling of the leather to make a firm yet moldable material is the governing factor.

Dyeing of Leather

Leather dyeing for any desired color can be done using the dyes developed for the textile industry (see DYES AND DYE INTERMEDIATES; TEXTILES). The penetration of the dye depends on the pH of the leather and the tannage. For leathers that are given a heavy finish, such as patent leather, only a surface dye is desired. The cost of the dyes is an important factor so if only surface dyeing is desired, placement of the dye is significant. For suede leathers or leathers that may be subject to scratching during use, a penetrating dye is desired (2,3,6).

Chrome-tanned leather has chromium bonded to the leather fibers. This chromium can act as a mordant for acid dyes resulting in fast colors and intense shading at the surface of the leather.

The dyeing of shearlings or wool sheepskins illustrates some of the technology. Shearlings, often made for the production of garments, have a sueded outside and a wool fur inside. The leather is tanned by chrome tanning leaving the wool on. After drying in the crust condition the wool is clipped with a precision machine to the desired wool length. Dyeing of the leather can be accomplished without any significant coloring of the wool. Because the skin has been tanned by chrome tanning, dyes are used that require mordanting and the chromium of the sueded leather aids in the development of the desired shade. The wool is thus unaffected.

Environmental Aspects

The processing of hides and skins into leather results in a large quantity of waste materials (9). The hide in the salt-cured condition contains salt in a crystalline form, water as salt solution, and as hide liquid components, flesh, blood, manure, and surface dirt from the animal.

Pollution control technology is a concern for the world tanning industry. Environmental considerations have been a primary factor in the expansion of the leather industry in developing countries. Whereas cottage industries continue to operate without significant pollution control, regulations are being enforced gradually in these countries.

The processing wastes come from the hides and processing chemicals. The hide wastes are the largest problem. For each metric ton of hides received at the tannery the following wastes are generated:

Organic waste	Quantity, kg/t of hides
fleshings	50–200
salt	100–200
surface dirt	10–20
manure	5–20
hair	5–20
oil and grease	30–50

Chemical waste	Quantity, kg/t of hides
salt	50–100
sulfides, as Na_2S	20–40
chromium(III)	10–20
suspended solids	50–100
biochemical oxygen demand	30–60

Additionally, the pH of the wastes can be from 2.0–12.5, depending on the place in the process, and the water usage is from 30–100 m^3/t of hides. There are pollution control requirements covering tannery wastes in all developed countries and almost all Third World ones as well. The cost of pollution control is very high and may require a capital investment of from 10 to 50% of the total value of the plant and production equipment. As a result many tanners, particularly in the developed countries, have stopped the operations of unhairing and tanning, limiting operations to the retanning and finishing of hides tanned elsewhere.

The environmental problems associated with the leather industry have spurred the development of some process changes to affect more efficient fixation of chemicals and the replacement of some toxic substances used in processing (Fig. 4). Pollution control within the industry has been largely concentrated in the areas of waste treatment techniques. Separate treatment of concentrated waste streams has been most successful and cost effective. These treatments include air oxidation of sulfide wastes from the unhairing process, high exhaustion processing of the trivalent chromium which is then recycled or precipitated, primary waste treatment for high suspended solids, coprecipitation and secondary waste treatment to lower the biological oxygen demand (BOD), and use of the solid wastes for fertilizers or other animal by-products. Moreover, a shift away from highly volatile solvents in the finish drying process to a low solvent system has been effected.

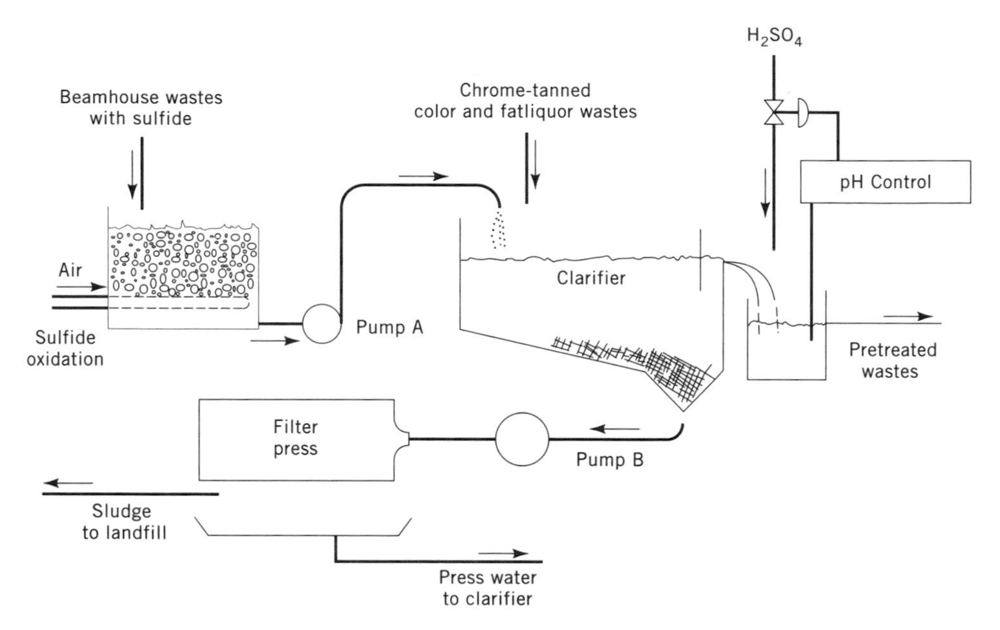

Fig. 4. Schematic of a leather tanning facility fitted with a wastewater treatment plant. Treatment of the combined wastes using sulfide oxidation and waste effluent pH adjustment greatly decreases the suspended solids and BOD loading (3). Courtesy of Krieger Publishing Co.

BIBLIOGRAPHY

"Leather and Tanning" in *ECT* 1st ed., Vol. 6, pp. 289–309, by F. O'Flaherty, University of Cincinnati; "Tanning Materials" in *ECT* 1st ed., Vol. 13, pp. 578–599, by F. L. Hilbert, U.S. Process Corp., and R. L. Stubbings, Lehigh University; "Leather" in *ECT* 2nd ed., Vol. 12, pp. 303–343, by F. O'Flaherty, University of Cincinnati, and R. L. Stubbings, Institute of Leather Technology; in *ECT* 3rd ed., Vol. 14, pp. 200–231, by D. G. Bailey, P. R. Buechler, A. L. Everett, and S. H. Feairheller, U.S. Dept. of Agriculture.

1. J. A. Wilson, *Modern Practice in Leather Manufacture,* Reinhold Publishing Co., New York, 1941.
2. F. O'Flaherty, W. T. Roddy, and R. M. Lollar, *The Chemistry and Technology of Leather,* ACS Monograph Series No. 134, Reinhold Publishing Co., New York, 1956–1965.
3. T. C. Thorstensen, *Practical Leather Technology,* 4th ed., Kreiger Publishing Co., Malabar, Fla., 1983.
4. K. H. Gustavson, *The Chemistry of Tanning Processes,* Academic Press, Inc., New York, 1956.
5. Cohn and Edsall, *Proteins, Amino Acids, and Peptides,* ACS Monograph Series No. 90, Reinhold Publishing Co., New York, 1943.
6. K. Bienkiewicz, *Physical Chemistry of Leather Making,* Kreiger Publishing Co., Malabar, Fla., 1983.
7. G. D. McLaughlin and E. R. Theis, *The Chemistry of Leather Manufacture,* Reinhold Publishing Co., New York, 1945.
8. J. Tancous, W. Roddy, and F. O'Flaherty, *Skin, Hide and Leather Defects,* Western Hills Publishing Co., Cincinnati, Ohio, 1959.

9. T. C. Thorstensen, *Fundamentals of Pollution Control for the Leather Industry,* Shoe Trades Publishing Co., Cambridge, Mass., 1993; K. H. Gustavson, *The Chemistry and Reactivity of Collagen,* Academic Press, Inc., New York, 1956.
10. H. R. Procter, *Leather Industries Laboratory Book,* London, 1893.

THOMAS C. THORSTENSEN
TSG

LEATHER-LIKE MATERIALS

Leather (qv) has been employed for many uses since ancient time on account of the convertibility of an easily decomposed substance into one which resists putrefaction. Leather is still an important material with its unique structure; it is so dense in texture that it resists wind and water while retaining breathability and flexibility, which makes the resulting goods comfortable. Early attempts to imitate leather included application of oil, rubber, or soluble cotton onto paper or fabrics. However, very little progress had been made until the era of synthetic resins began in the twentieth century.

In the second quarter of the twentieth century, with the development of poly(vinyl chloride), nylon, polyurethane, and other polymers, many new and improved leather-like materials, so-called coated fabrics (qv), were placed on the market. Shortages of leather after World War II led to the expansion of these leather-like materials ("man-made" leathers) to replace leather in shoes, clothing, bags, upholstery, and other items. Durability and waterproof qualities superior to leather made coated fabrics advantageous, in spite of imperfection in breathability and flexibility. Demands for shoes, clothing, and other items are still increasing due to growing world population and urbanization.

During the third quarter of the twentieth century, with improved nonwoven fabrics, man-made leathers finally succeeded in simulating leather to such an extent that they are nearly identical in appearance, physical properties, and structure. These leathers have enjoyed success in all leather-use areas. With the technology of microfibers, they continue to evolve both in quality and quantity.

Types of Leather-Like Materials

Leather-like materials now important in the market are of three main classes: (*1*) vinyl-coated fabrics, (*2*) urethane-coated (synthetic) fabrics, and (*3*) man-made leathers. To appreciate their leather-replacement capabilities it is necessary to know the structure of natural leather.

Leather. Natural leather is made from hides, which are salted and cured, then tanned. Through the preparing process, useless matter which cannot be tanned, such as outerskin (epidermis) and flesh, are removed, leaving the true skin (corium). In the tanning process, the fluid matter which maintains the skin in a flexible and moist condition is removed, and there remains nothing but the fibrous portion to be acted on by the tanning chemicals (1–3).

As a result, leather is made up of interlaced bundles of collagen fibers (Fig. 1). A schematic model of collagen bundles in leather is shown in Figure 2 (4). A collagen bundle (about 80 μm in diameter) is made up of collagen fibers (1–4 μm), composed of microfibrils (0.08–0.1 μm). Furthermore, a microfibril consists of many protofibrils (about 1.5 nm), which consist of several bundles of polypeptide chains.

Such a unique hierarchical structure gives leather several advantages: (*1*) transformability into any desired shape, (*2*) resistance to penetration of wind, water, and other materials, (*3*) breathability (water vapor and air permeability, and water absorption), (*4*) flexibility, and (*5*) processibility into finished forms having a grain or suede surface.

Vinyl-Coated Fabrics. Leather substitutes are designed to imitate the appearance of leather with its grain surface. This requirement has been accomplished by coating substances that are capable of forming a uniform film, and was first met by plasticized poly(vinyl chloride) (PVC). A leather-like material termed vinyl-coated fabric was developed in the 1930s in the United States and Germany. Shortages of leather after World War II spurred the expansion of this material.

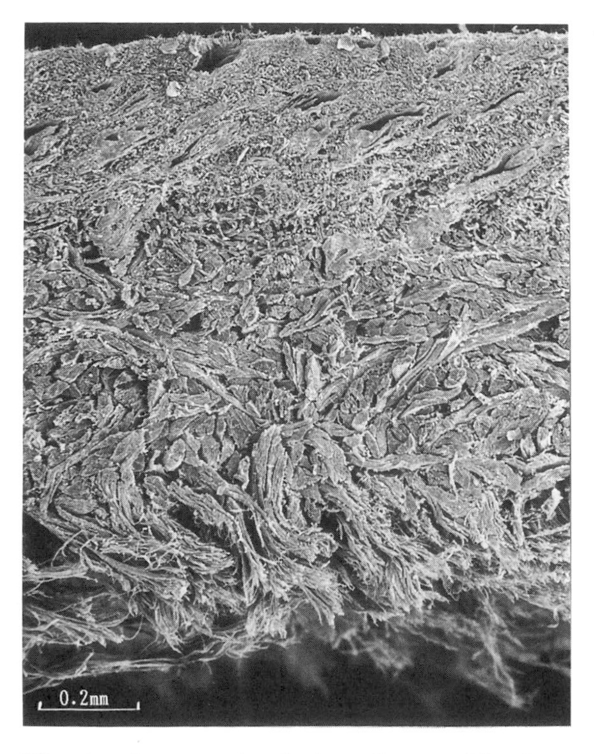

Fig. 1. Cross-sectional view of natural leather.

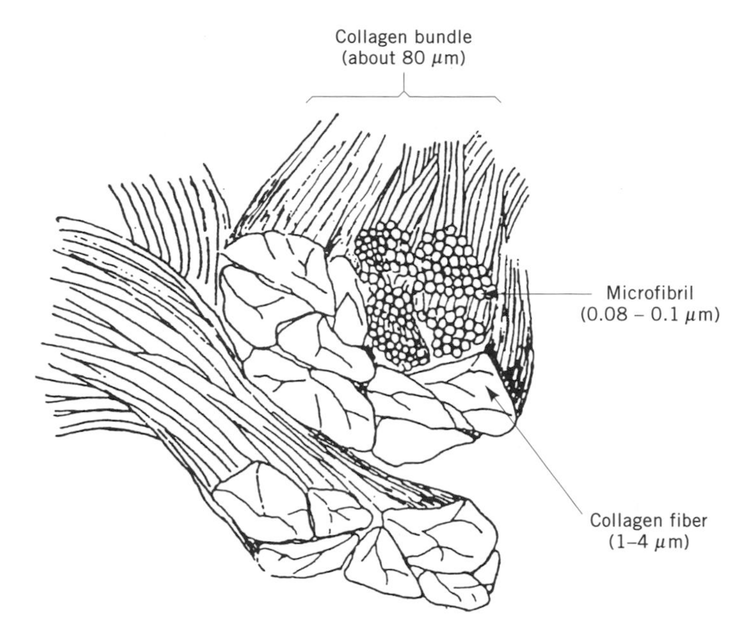

Fig. 2. Schematic model of collagen fiber bundle in natural leather.

The construction of vinyl-coated fabrics varies according to its application. A vinyl-coated fabric used for automobile seat covers is shown in Figure 3; a woven fabric is the substrate. The material is durable but stiff and heavy. Incorporating an expanded foam structure into the coating layer reduces the weight (Fig. 4), and replacing the woven substrate fabric with a soft knit fabric improves flexibility.

Vinyl-coated fabrics exhibit high density, extremely low water vapor and air permeability, cold touch, poor flex endurance, and plasticizer migration. However, they have good scratch resistance and colorability and are inexpensive.

Urethane-Coated Fabrics. Urethane-coated fabrics, termed synthetic leather, were developed in the 1960s by applying a coat of polyurethane (PU) onto a woven or knit fabric. Polyurethane is flexible at room temperature without a plasticizer due to its low glass-transition point ($<0°C$). Urethane-coated fabrics are manufactured by drying a cast polyurethane solution to form a film which is laminated onto a substrate. An improvement in appearance, feel, and resistance to grain break is achieved by using a brushed fabric as the substrate, which is laminated with a cast polyurethane film. Further improvement in flexibility is achieved by the introduction of poromerics, which provide polyurethane with a porous structure by using a solvent that permits coagulation with a nonsolvent. Thus poromeric urethane-coated fabrics are produced by applying an organic solvent solution of polyurethane to a brushed woven fabric and then immersing the fabric in a nonsolvent bath for coagulation, followed by coating with a cast polyurethane film (Fig. 5). With poromerics, urethane-coated fabrics can be employed for many uses. On the other hand, woven or knit fabric substrates still limit their application, due to low conformability. In making three-dimensional shaped goods, such as shoes,

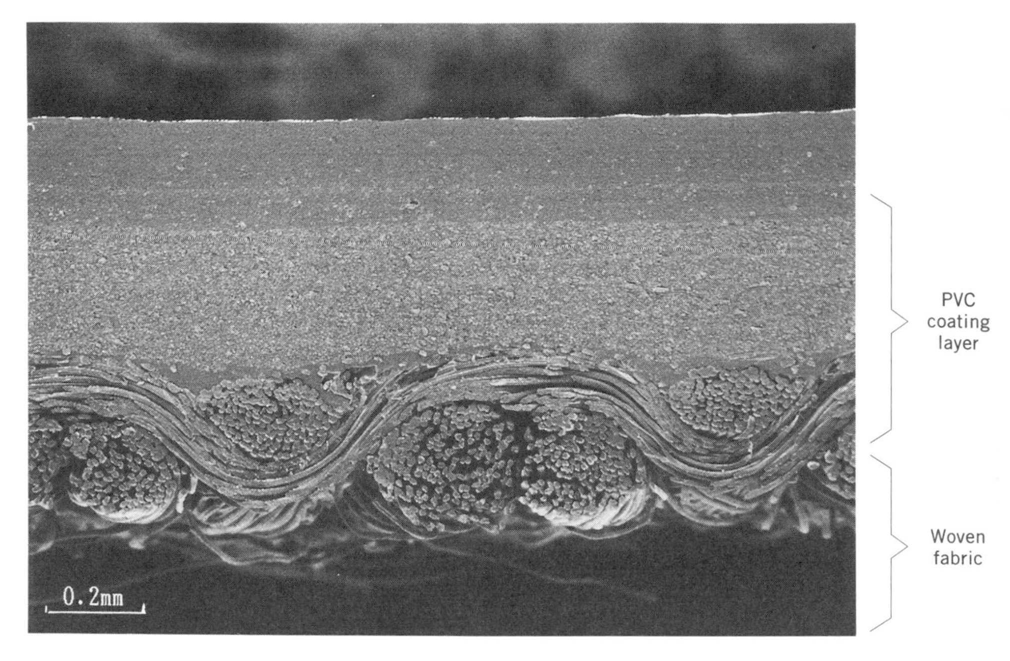

Fig. 3. Cross-sectional view of vinyl-coated fabric with PVC coating layer.

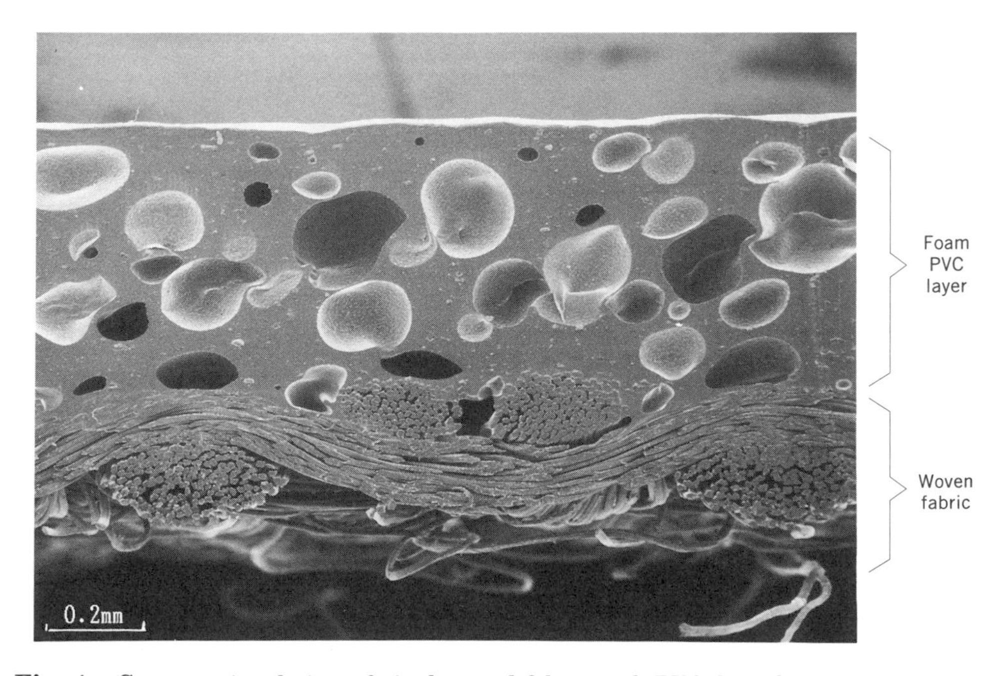

Fig. 4. Cross-sectional view of vinyl-coated fabric with PVC foam layer.

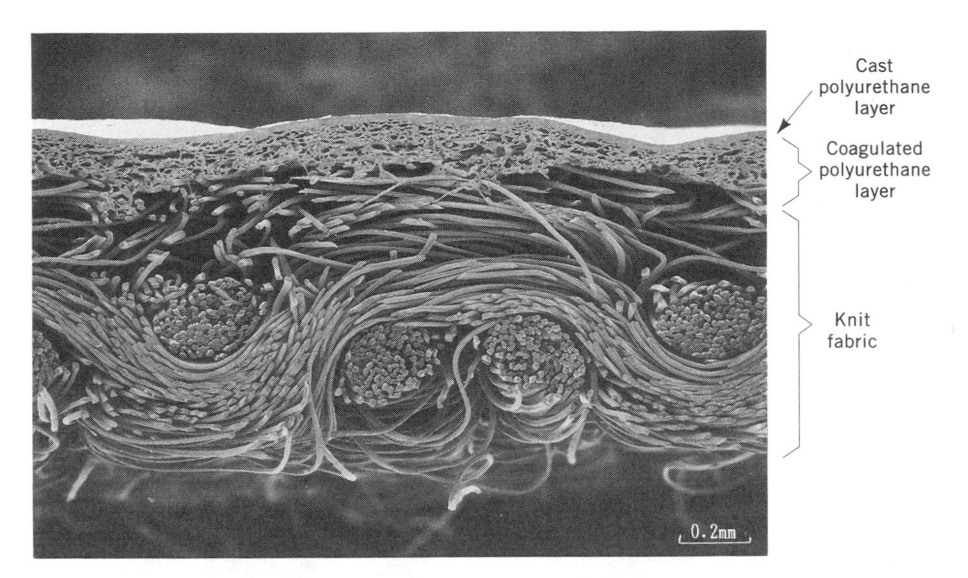

Fig. 5. Cross-sectional view of urethane-coated fabric with coagulated polyurethane layer.

the material must conform to the shape of a last (shoe form) and maintain it after the last has been removed.

Man-Made Leathers. These materials were developed in the 1960s by combining the technologies of poromerics and nonwoven fabrics. Substitution of nonwoven fabric for knit or woven fabric increased conformability, and eliminated this limitation in application (5,6). Most early man-made leathers, such as Du Pont's Corfam, had a three-layer structure containing a thin woven fabric inserted between a nonwoven fabric and a coating layer to compensate for the unevenness of the nonwoven fabric. However, this woven fabric stiffens the resulting three-layer products, which were hence not widely accepted. Clarino (Kuraray Co., Ltd.) has a two-layer structure utilizing a novel nonwoven fabric impregnated with polyurethane and coated with a layer of continuous micro-pores (Fig. 6). Grain-type man-made leathers are mostly of this two-layer structure.

Significant improvement in the fiber structure of leather is finally achieved by using microfibers as fine as 0.001–0.0001 tex (0.01–0.001 den). With this microfiber, a man-made grain leather Sofrina (Kuraray Co., Ltd.) with a thin surface layer (Fig. 7), and a man-made suede Suedemark (Kuraray Co., Ltd.) with a fine nap (Fig. 8) were first developed for clothing, and have expanded their uses. Ultrasuede (Toray Industries, Inc.) also uses microfibers with a rather thick fineness of 0.01 tex (0.1 den). Contemporary (1995) man-made leathers employ microfibers of not more than 0.03 tex (0.3 den) to obtain excellent properties and appearance resembling leather.

Physical and Chemical Properties

The properties of leather-like materials depend on the polymer used for substrate and coating layer. Feel, hand, and resistance to grain break are affected by the

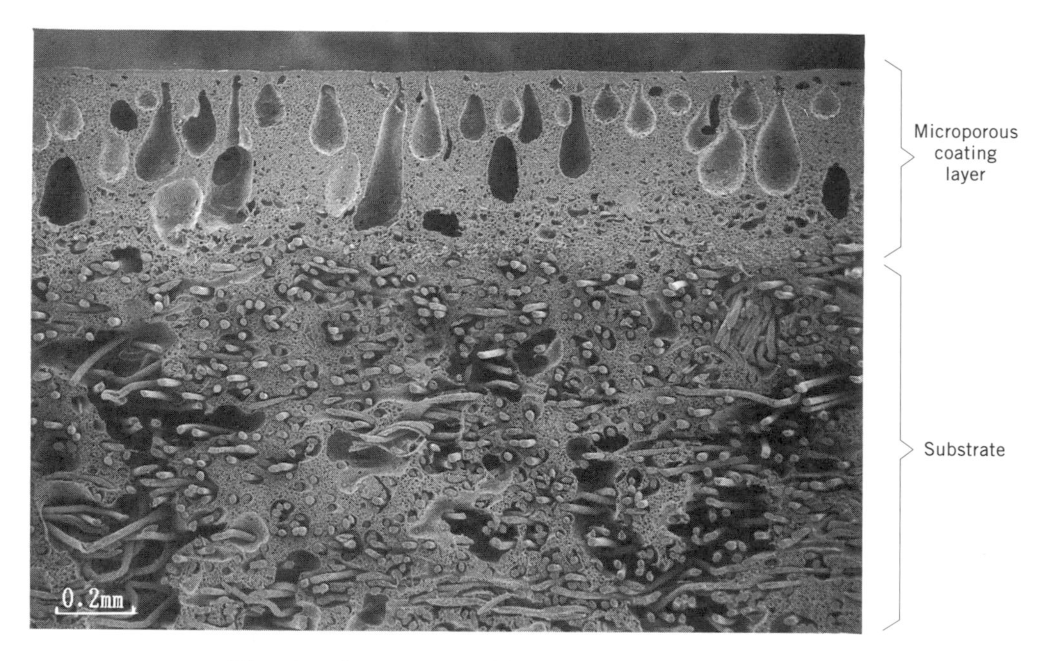

Fig. 6. Cross-sectional view of Clarino.

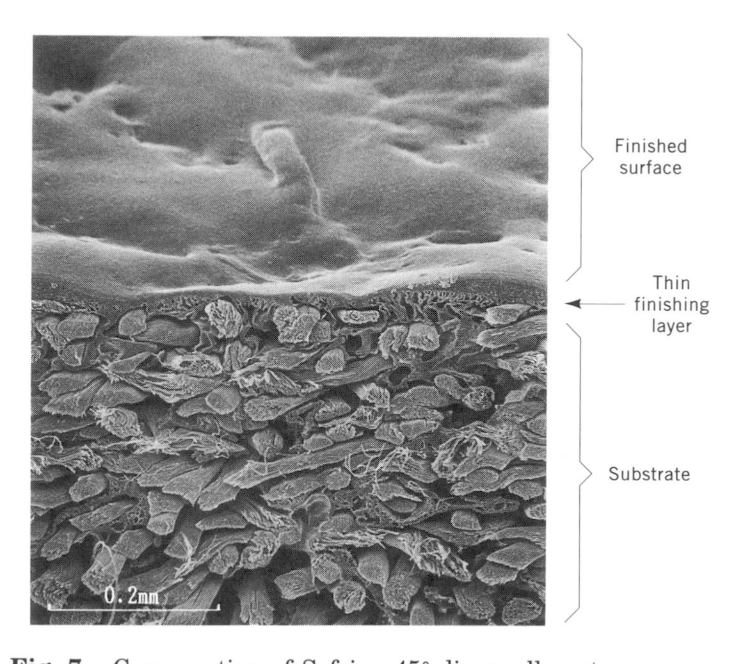

Fig. 7. Cross section of Sofrina 45° diagonally cut.

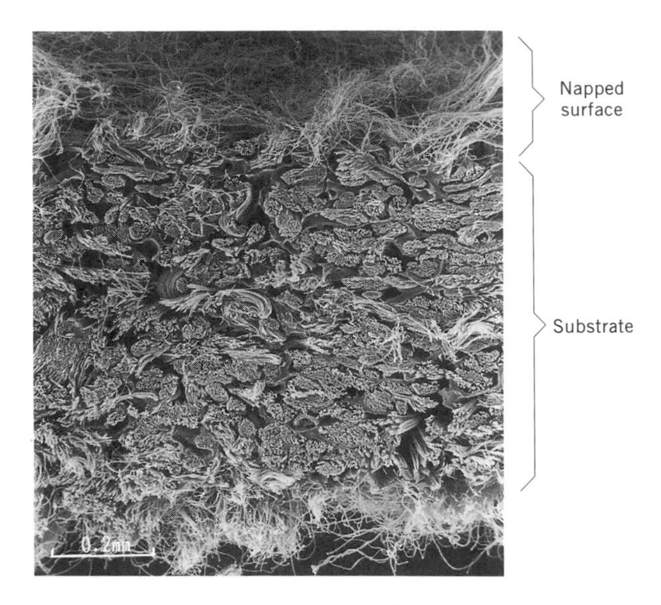

Fig. 8. Cross section of Suedemark 45° diagonally cut.

construction. The polymers and constructions of leather-like materials are shown in Table 1. Physical properties of leather and leather-like materials are shown in Table 2.

Weight, Thickness, and Density. The thickness differs according to application. Vinyl-coated fabrics are very heavy and dense. Urethane-coated fabrics and man-made leather are very light and two-thirds to one-half leather in density.

Strength and Elongation. The tensile strength and elongation depend mostly on the substrates. Woven fabric substrates give a steep slope in the stress–strain (S–S) curve with a small elongation at break, and knit and nonwoven fabric substrates give a gentle slope with large elongation. In making shoes which need lasting, a good balance of strength and stretch and a high conformability is desirable. Nonwoven fabrics, especially those comprising poromerics with microfibers, are the best for these shoes.

Table 1. Constructions of Leather-Like Materials

Material	Coating[a] layer polymer	Substrate	
		Structure	Polymer
man-made leather	polyurethane (segmented)	nonwoven	nylon, polyester, acrylics
urethane-coated	polyurethane (segmented)	knit or woven	cotton, rayon, nylon, acrylics, polyester, and others
vinyl-coated	PVC	knit or woven	cotton, rayon, acrylics, nylon, polyester, and others

[a]Structure is solid and/or porous.

Table 2. Physical Properties of Leather and Leather-Like Materials

Product	Substrate/fiber[a]	Coating layer	Weight, g/m^2	Thickness, mm	Density, g/mL	Tensile strength, N/mm^2 [b,c]	Elongation at break, %[c]	Tear strength, kN/m[c,d]	Water-vapor permeability,[e] mg/(cm·h)	Flex endurance,[e] 10^3 cycles
				Man-made leather						
Clarino[f]	nonwoven/porous	thick foam	580	1.50	0.39	8 (5)	45 (85)	30 (33)	2.6	>1000
	nonwoven/micro	thin solid	550	1.30	0.42	15 (13)	80 (115)	69 (68)	3.8	>1000
Sofrina[f]	nonwoven/micro	thin solid	220	0.50	0.44	12 (14)	115 (120)	100 (80)	5.0	>1000
Suedemark[f]	nonwoven/micro	suede	120	0.35	0.34	9 (7)	70 (106)	43 (43)	12.4	>1000
				Urethane-coated fabrics						
solid-type	woven/regular	thin solid	393	0.90	0.44	12 (7)	20 (28)	45 (66)	1.5	300
foamed-type	woven/regular	thick foam	404	1.13	0.36	11 (8)	7 (29)	30 (22)	1.5	400
				Vinyl-coated fabrics						
solid-type	knit/regular	thick solid	912	0.91	1.00	9 (7)	33 (229)	31 (35)	0.3	50
foamed-type	knit/regular	thick foam	839	1.03	0.81	6 (4)	31 (200)	24 (27)	0.4	150
				Leather						
carf		grain	450	0.80	0.56	11 (10)	55 (75)	25 (25)	10.1	>1000
side		grain	1000	1.50	0.67	23 (13)	70 (96)	73 (93)	7.5	500

[a] Product is nylon in man-made leathers; cotton in urethane- and vinyl-coated fabrics.
[b] To convert N/mm^2 to kgf/cm^2, multiply by 10.2; to psi, multiply by 145.
[c] Numbers in parentheses are crosswise.
[d] To convert kN/m to ppi, multiply by 5.71.
[e] At 20°C.
[f] Registered trademark of Kuraray Co., Ltd.

Water-Vapor Permeability. Water-vapor permeability depends on the polymer used for the coating layer and its structure. Vinyl-coated fabrics have little water-vapor permeability due to the coating layer. Although polyurethane polymer is water-vapor permeable, urethane-coated fabrics also have low permeability values due to their solid layer structure. On the other hand, man-made leathers have good permeability values as high as that of leather due to their porous layer structure. The permeability of grain-type is lower than that of suede-type, influenced by finishing method.

Durability. Flex endurance is correlated with water-vapor permeability, and man-made leathers have the best durability. Scratch resistance is inversely correlated with water-vapor permeability, and vinyl-coated fabric has the best performance. Durability for aging depends on the polymer used for the coating layer, because the polymer of the substrate fabric is generally more durable than that of the layer. Vinyl-coated fabrics are sufficiently durable. The durabilities of urethane-coated fabrics and man-made leathers vary to a large extent depending on the polyurethanes used because the physical and chemical properties of segmented polyurethanes markedly vary according to the segment type (see URETHANE POLYMERS). Polyurethanes degrade because of water, oxygen, NO_x, SO_x, and other chemical substances. Hydrolysis, degradation by H_2O, is the most important factor in the durability of polyurethanes, and it depends on the soft segments used. Soft segments are classified into three groups: polycarbonates, polyethers, and polyesters, and have better resistance to hydrolysis in this order. The components of a polyurethane can be selected according to the end use. Polyesters and polyethers are mainly used; polycarbonates are used for items especially requiring durability, such as in automobiles.

Other Properties. With respect to dry-cleanability, vinyl-coated fabrics are worse than the others due to dissolution of plasticizer in the cleaning solvent. Only man-made leathers with poromeric nonwoven can be skived, split, and cut in the same manner as leather.

Manufacture and Processing

Vinyl-Coated Fabrics. Manufacturing methods for vinyl-coated fabrics now available are calendering and extrusion for thick layer, and paste coating for thin layer. Both solid and foam vinyl-coating layers are used.

In the calendering method, a PVC compound which contains plasticizers (qv) (60–120 phr), pigments (qv) (0–10 phr), fillers (qv) (20–60 phr), stabilizers (10–30 phr), and other additives, is kneaded with calender rolls at 150–200°C, followed by extrusion between clearance-adjusted rolls for bonding onto the substrate. This method is employed for products with thick PVC layers, ie, of 0.05–0.75 mm thickness. The main plasticizer used is di-2-ethylhexyl phthalate (DOP). For filler to reduce cost, calcium carbonate is mainly used. A woven or knit fabric made of cotton, rayon, nylon, polyester, and their blend fiber is used as substrate. For foamed vinyl-coated fabrics, the bonded materials are heated in an oven to decompose the foam-blowing chemicals. Most foam-blown vinyl-coated fabrics are finished to have a solid coating layer to improve scratch resistance.

Another method is extrusion; the PVC compound is kneaded in an extruder, and then extruded through a T-die for bonding onto a substrate.

In the paste coating method, a PVC paste, which contains emulsion-polymerized PVC and additives, is applied onto a substrate and heated to gelation before fusion to produce a coating layer. This method is employed for products with a thin layer, ie, of 0.007–0.05 mm thickness. For foamed vinyl-coated fabrics, a substrate is laminated onto a transfer paper on which a PVC paste containing a foam-blowing agent has been applied and gelled. After removal of the transfer paper, the paste is blown.

These processes may be followed by heat treatment and pressing with engraved rolls to produce the desired grain surface.

Urethane-Coated Fabrics. Manufacturing methods for urethane-coated fabrics are the dry system and the wet system.

In the dry system, the coating layer consists of two or three layers, for which a solvent-soluble linear polyurethane and a two-component cross-linkable polyurethane are employed. The former is used for the top and/or middle layers, and the latter for the bottom layer, ie, adhesive layer. The solvent contains dimethylformamide (DMF), methyl ethyl ketone (MEK), 2-propanol, toluene, or other solvents, to accelerate drying. Manufacture proceeds by the following sequence: (*1*) 100 g/m^2 of a 10% polyurethane solution is applied onto a transfer paper which carries a grain pattern, and dried in an oven; (*2*) 100 g/m^2 of a 40% two-component polyurethane solution which is composed of a polymer diol and a polyisocyanate, is applied on the first layer and then slightly dried to a tacky state, to form an adhesive layer; (*3*) a substrate is laminated onto the adhesive layer thus formed and passed through an oven and rolled up, and the roll is cured at 40–60°C for 2–3 days; and (*4*) after completion of the cross-linking, the transfer paper is removed from the finished urethane-coated fabric. There are several modifications: (*1*) the first operation is repeated for a middle layer before the second operation; (*2*) drying is omitted in the second operation; (*3*) a linear polyurethane solution is used for second operation; and (*4*) the fourth operation is followed by a finishing process if required, such as color shading. For the substrate, a woven or knit fabric made of cotton (qv), rayon, nylon, polyester, or their blends is used.

In the wet system, manufacture proceeds as follows: (*1*) a 7–20% polyurethane solution of DMF is applied onto a fabric and immersed in water containing 0–10% of DMF for coagulation; (*2*) the coated fabric is washed and dried; (*3*) the surface is finished by the dry system. For the substrate, a woven or knit fabric which has been brushed on its surface is often used to improve appearance, resistance to grain break, and feel.

Man-Made Leathers. These materials contain a nonwoven fabric which is impregnated with a polyurethane to improve flexibility, processibility, and conformability (Fig. 9). Advanced man-made leathers contain microfibers as fine as 0.03 tex (0.3 den) or less to imitate collagen fiber bundles, thereby attaining the soft feel and appearance essential for soft leather use. Polyurethane in the substrate is usually provided with porous structure by poromeric technology. The coating layer is also porous in the two-layer type man-made leathers (5–10).

A special fiber has been developed for Clarino (5,7,8). Two polymers of different solubility are mixed for spinning. In the resulting fiber, the two polymers are separated by an islands-in-the-sea structure. The islands-in-the-sea structure is controlled by the polymers, their proportion, and spinning conditions

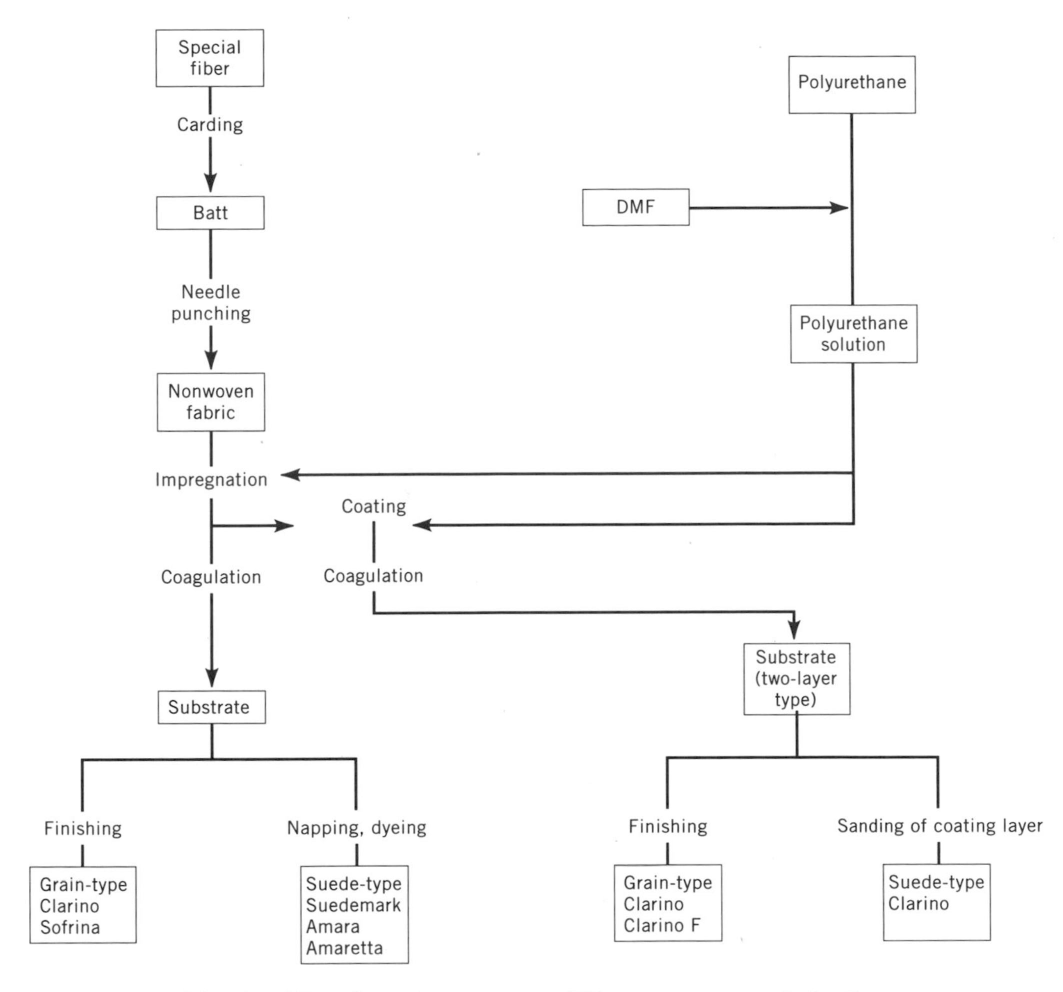

Fig. 9. Manufacturing process of Kuraray man-made leather.

(5,7,8,13). Solvent extraction of either component gives a porous fiber or a microfiber bundle (Fig. 10). The islands component ranges in fineness from about 0.01 tex (0.1 den) to about 0.0001 tex (0.001 den), or less, and in the number of islands from 100 to 1000 or more (5). Porous fibers reduce the weight of products (Fig. 11). As seen in Figure 11**b**, fine fibers obtained by this method show a thickness distribution most suited to simulate the collagen fiber of leather. Another method is used for Ultrasuede; a special nozzle is used to produce a multi-islands-conjugate fiber (5,14). The fiber fineness is limited to about 0.01 tex (0.1 den) and the number of islands to about 50, due to the structural limit of the nozzle.

In the method shown in Figure 9, manufacture proceeds by the following steps. (*1*) A special fiber is made by melt spinning and cut into 25–60 mm length. (*2*) The fibers are carded and cross-lapped (15) to form a batt. (*3*) The batt is needle-punched (16) with a barbed needle (17) and entangled to improve physical strength (18); the batt is then subjected to sizing and pressing (19) to become a nonwoven fabric with adequate thickness. (*4*) The nonwoven

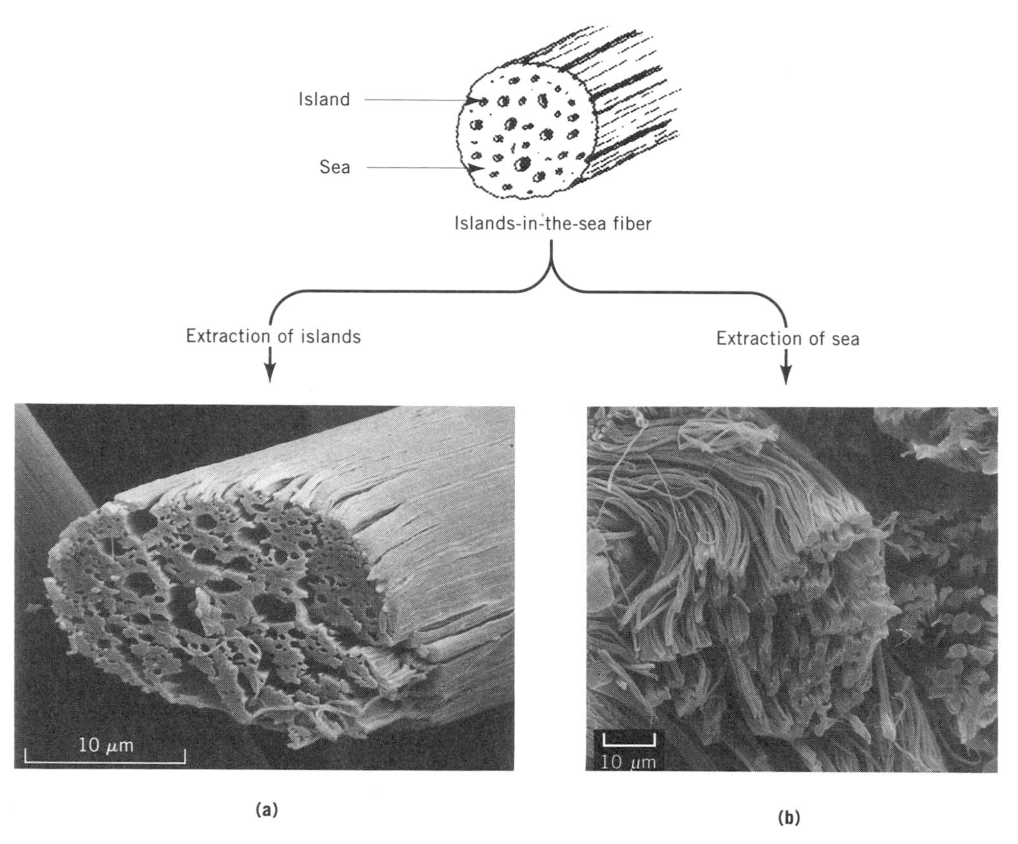

Fig. 10. Formation of fibers used in Kuraray man-made leather: (**a**) porous fiber, and (**b**) a bundle of microfibers.

fabric is impregnated with polyurethane and immersed in a DMF–water bath for coagulation to create a substrate with porous polyurethane structure. For the two-layer substrate, a polyurethane coating process must come between impregnation and coagulation. (*5*) The substrate is prefinished by coating or sanding for grain or suede surface, followed by, as required, embossing, dyeing, and other finishings.

Most manufacturing methods now available are similar to this but with the following modifications: in the first step, the polymers for fibers are mainly made of polyester, nylon, or their blends. Acrylics and polypropylene are also sometimes employed. A regular fiber as thick as 0.01–0.4 tex (0.1–4 den) may sometimes be used instead of the special fiber to imitate the hard leather.

In the second step, a papermaking method is also used for the fine fibers, less than 0.1 tex (1 den). This process is usually followed by a high pressure water jet process instead of the third step. In the fourth step, to obtain the required properties in specific applications, a polyurethane is selected out of the segmented polyurethanes, which comprises a polymer diol, a diisocyanate, and a chain extender (see URETHANE POLYMERS). A DMF–water bath for coagulation is also controlled to create the adequate pore structure in combination with fibers.

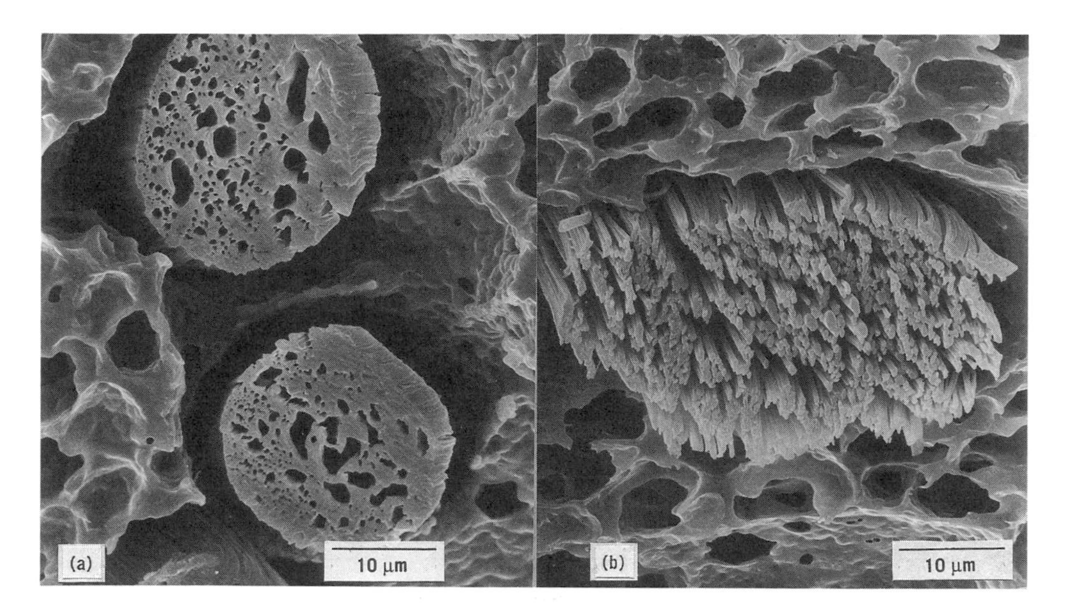

Fig. 11. (**a**) Cross-sectional view of substrate with porous fibers and polyurethane sponge. (**b**) Cross-sectional view of substrate with bundle of microfibers and polyurethane sponge.

In the fifth step for man-made leathers, a grain surface is given by embossing with heated rolls engraved with the desired leather pattern, color-printing with gravure rolls or color spraying, and dyeing. For the grain type with thin layer, such as Sofrina, the finishing includes application of a polyurethane thin layer before embossing. For suede type, finishing includes napping before dyeing.

Economic Aspects and Application

The demand for leather products is expected to increase steadily due to growing world population and urbanization. On the other hand, leather has a limit in its supply because it is a by-product of the meat industry.

The production of coated fabrics including vinyl- and urethane-coated in the early 1980s is shown in Table 3 (4). Only production in the Far East showed an increase. The increase of urethane-coated fabrics mainly supports this tendency (Table 4).

The production of man-made leather has increased rapidly due to its high quality (Table 5). Production was first started in Japan, and is expanding to the world. Up to 90% is produced in the Far East, and approximately 50% is exported to the United States and European countries.

The big three applications of vinyl-coated fabrics are (*1*) automotive (36%), (*2*) bags (17%), and (*3*) interiors (10%). Those of urethane-coated fabrics are (*1*) clothing (18%), (*2*) shoes (18%), and (*3*) accessories of shoes (11%). Those of synthetic leathers are (*1*) shoes (46%), (*2*) clothing (13%), and (*3*) bags (10%) (18).

Table 3. Worldwide Production of Coated Fabrics[a]

	10^6 m^2/year	
Location	1978	1983
Far East	480	700
Eastern Europe	330	330
Western Europe	700	700
North America	505	505
South America	115	115
Africa	35	35
Total	*2165*	*2385*

[a]Ref. 5.

Table 4. Urethane-Coated Fabrics Production in the Far East[a]

	10^6 m^2/year		
Country	1982	1987	1992
China (Taiwan)	71	113	166
Japan	46	49	46
Korea (South)	36	64	71
Thailand	8	7	15
China			25
others	10	8	12
Total	*171*	*241*	*335*

[a]Ref. 20.

Table 5. Worldwide Production of Man-Made Leather[a]

			10^6 m^2/yr	
Location	Company	Product	1982	1992
Far East			18	51
Japan	Kuraray	Clarino, Sofrina, Suedemark, Amara, Amaretta	18	33
	Toray	Ultrasuede		
	Teijin	Cordley		
	Kanebo	Bellace		
	Asahi	Lamous		
	Mitsubishi	Glore		
others				16
Europe	Iganto	Alcantara	4	6
	Porvair	Porvair		
Total			*22*	*55*

[a]Ref. 20.

Test Methods

Test methods vary from maker to maker. In the United States, test methods for the physical properties of coated fabrics are given in ASTM D751-79. Flex endurance can be measured by ASTM D813-52T. ASTM for fabrics, such as D1683-78, and ASTM for leather, such as D2346-68, can also be used. In Japan JIS for fabrics, such as L1096, and for leather, such as K6545, are available. In Europe BS and DIN, such as D75202 for automobile, are available.

BIBLIOGRAPHY

"Poromeric Materials" in *ECT* 2nd ed., Vol. 16, pp. 345–360, by J. L. Hollowell, E. I. du Pont de Nemours and Co., Inc.; "Leatherlike Materials" in *ECT* 3rd ed., Vol. 14, pp. 231–249, by F. P. Civardi and G. F. Hutter, Inmont Corp.

1. *The Encyclopedia Americana,* Vol. 17, Americana Corp., New York, 1964, pp. 155, 166, 167.
2. L. Davidow and co-workers, *Our Wonderful World,* Vol. 9, Spencer Press Inc., Chicago, 1962, p. 408.
3. J. H. Thornton, *Textbook of Footwear Materials,* The National Trade Press Ltd., London, 1955, pp. 1–7.
4. Y. Saito and A. Kubotsu, in Sen-I Gakkai ed., *Zusetsu Sen-I no Keitai* (*A Diagram of Fiber Structure*), Asakura-shoten Co., Ltd., Tokyo, 1983, p. 254.
5. K. Nagoshi, in J. I. Kroschwitz, ed., *Encyclopedia of Polymer Science and Engineering,* Vol. 8, 2nd ed., John Wiley & Sons, Inc., New York, 1987, pp. 677–697.
6. A. R. Payne, *Poromerics in the Shoe Industry,* Elsevier Publishing Co., Ltd., New York, 1970, pp. 16, 17.
7. T. Hayashi, *Chemtech* **5**, 28 (Jan. 1975).
8. T. Yasui and co-workers, *Polyester,* The Textile Institute, Manchester, U.K., 1993, pp. 210, 211.
9. O. Fukushima, *J. Coated Fabr.* **5**, 3 (1975).
10. K. Nagoshi and co-workers, *International Progress in Urethanes,* Vol. 3, Technomic Publishing Co., Inc., Lancaster, Pa., 1981, pp. 193–217.
11. Can. Pat. 881,219 (July 29, 1963), O. Fukushima and co-workers (to Kurashiki Rayon Co., Ltd.); Can. Pat. 897,249 (Apr. 1, 1964), O. Fukushima and T. Yasui (to Kurashiki Rayon Co., Ltd.); U.S. Pat. 3,330,899 (July 11, 1967), O. Fukushima and H. Hayanami (to Kurashiki Rayon Co., Ltd.); U.S. Pat. 3,334,153 (Aug. 1, 1967), O. Fukushima and H. Hayanami (to Kurashiki Rayon Co., Ltd.).
12. U.S. Pat. 3,424,604 (Jan. 28, 1969), O. Fukushima and H. Hayanami and K. Nagoshi (to Kurashiki Rayon Co., Ltd.).
13. O. Fukushima and co-workers, *Melliand Textilber.* **57**, 673 (1976).
14. U.S. Pat. 3,531,368 (Sept. 29, 1970), M. Okamoto and co-workers (to Toyo Rayon Kabushiki Kaisha); U.S. Pat. 3,692,423 (Sept. 19, 1972), M. Okamoto and co-workers (to Toray Industries, Inc.,); U.S. Pat. 4,008,344 (Feb. 15, 1977), M. Okamoto and co-workers (to Toray Industries, Inc.).
15. U.S. Pat. 3,183,557 (May 18, 1965), J. L. Hollowell (to E. I. du Pont de Nemours & Co., Inc.).
16. U.S. Pat. 2,974,393 (Mar. 14, 1965), J. L. Hollowell (to E. I. du Pont de Nemours & Co., Inc.).
17. U.S. Pat. 3,432,896 (Mar. 18, 1969), R. Meagher (to E. I. du Pont de Nemours & Co., Inc.).

18. U.S. Pat. 2,954,113 (Nov. 1, 1960), H. G. Lauterbach (to E. I. du Pont de Nemours & Co., Inc.).
19. U.S. Pat. 3,483,283 (Dec. 18, 1965), O. Fukushima and co-workers (to Kurashiki Rayon Co., Ltd.).
20. Y. Sonobe, private communication, Clarino Division, Kuraray Co., Ltd., Osaka, Japan, Sept. 30, 1993.

KATSUMI HIOKI
Kuraray Company, Ltd.

LECITHIN

The name lecithin was first used (1) to describe a sticky, orange-colored, phosphorus- and nitrogen-containing material, initially isolated from egg yolk in 1847 and in subsequent years from brain, blood, bile, and other organic materials. The word is derived from the Greek *lekithos* meaning in the feminine form "egg yolk" and in the masculine form "yellow pea soup." Subsequently, it was demonstrated in 1867–1868 (2) that the nitrogen-containing component of egg lecithin is choline, an organic base already known to be present in the bile. Lecithin and other phospholipids are of universal occurrence in living organisms. They are constituents of biological membranes and are involved in permeability, oxidative phosphorylation, phagocytosis, and chemical and electrical excitation.

Lecithin [*8002-43-5*] is not only used in the strict scientific sense to describe pure phosphatidylcholine (Fig. 1), but also to describe crude phospholipid mixtures containing phosphatidylcholine (PC), phosphatidylethanolamine (PE), phosphatidylinositol (PI), other phospholipids, and a variety of other compounds such as fatty acids, triglycerides, sterols, carbohydrates, and glycolipids. Structures having only one acyl group at the glycerol backbone, predominantly in position 1, are called lysolecithin [*9008-30-4*]. The International Lecithin and Phospholipid Society (ILPS) has published the following definition for lecithin and phospholipid (4): lecithin is a mixture of glycerophospholipids obtained from animal, vegetable, or microbial sources, containing a variety of substances, such as sphingosylphospholipids, triglycerides, fatty acids, and glycolipids. The pure phospholipids, which can be isolated ultimately from the mixture, are defined as lipids containing phosphoric acid. The term phospholipids is no longer recommended but is still used in the technical literature. Commercial lecithin is currently available in more than 40 different formulations varying from crude oily extracts from natural sources to purified and synthetic phospholipids. Many of these products are defined according to the stage of the purification process from which they are obtained and fall into three broad categories (Table 1) varying in their constituents both qualitatively and quantitatively.

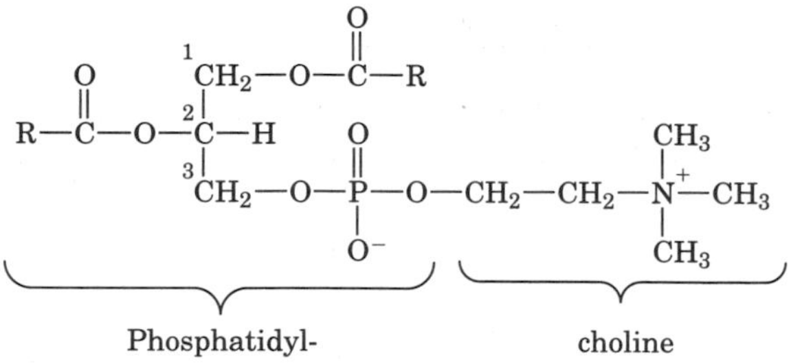

```
                        O
                        ||
        O      1CH2—O—C—R
        ||      2|
   R—C—O—C—H     O                 CH3
                3|     ||                 | +
                 CH2—O—P—O—CH2—CH2—N—CH3
                        |                  |
                        O−                 CH3
   \_______________________/  \_______________/
          Phosphatidyl-            choline
```

(1)

```
        +                     OH   OH                    +
—CH2—CH2—NH3     —[cyclohexane ring]—OH, OH     —CH2—CH—NH3     —CH2—CH—CH2OH
                          OH                          |               |
                                                     COOH             OH
```

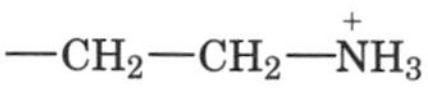

(2) 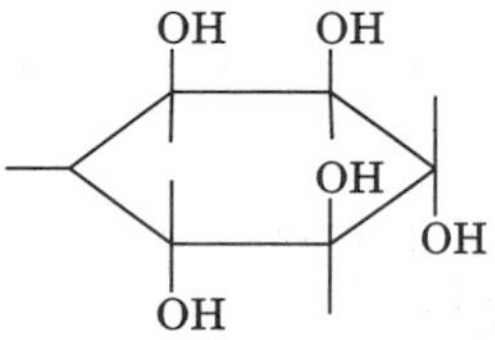 (3) 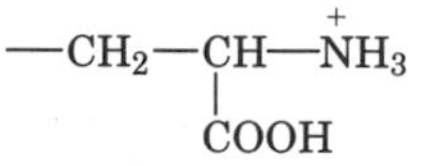 (4) 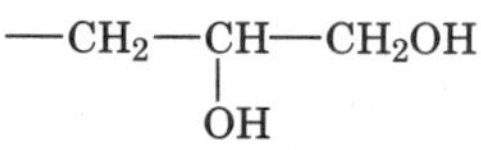 (5)

Fig. 1. Chemical structure of phosphatidylcholine (PC) (**1**) and other related phospholipids. R—C(=O)—O— represents fatty acid residues. The choline fragment may be replaced by other moieties such as ethanolamine (**2**) to give phosphatidylethanolamine (PE), inositol (**3**) to give phosphatidylinositol (PI), serine (**4**), or glycerol (**5**). If H replaces choline, the compound is phosphatidic acid (**6**). The corresponding IUPAC-IUB names are (**1**), 1,2-diacyl-*sn*-glycero(3)phosphocholine; (**2**), 1,2-diacyl-*sn*-glycero(3)phosphoethanolamine; (**3**), 1,2-diacyl-*sn*-glycero(3)phosphoinositol; (**4**), 1,2-diacyl-*sn*-glycero(3)phospho-L-serine; and (**5**), 1,2-diacyl-*sn*-glycero(3)phospho(3)-*sn*-glycerol.

Table 1. Categories of Commercial Lecithin

Natural	Refined	Modified
Plastic	*Deoiled*	*Physically*
unbleached		custom-blended
bleached		natural and refined
doubled-bleached		
Fluid	*Fractionated*	*Chemically*
unbleached	alcohol-soluble	
bleached	alcohol-insoluble	
double-bleached		*Enzymatically*

Industrial lecithins from a variety of sources are utilized (Tables 2 and 3). The main sources include vegetable oils (eg, soy bean, cottonseed, corn, sunflower, rapeseed) and animal tissues (egg and bovine brain). However, egg lecithin and in particular soy lecithin (Table 4) are by far the most important in terms of quantities produced. So much so that the term soy lecithin and commercial lecithin are often used synonymously.

Table 2. Composition of Lecithins, Oil-Free Basis, %

Phospholipid	Soybean lecithin	Corn lecithin	Sunflower-seed lecithin	Rapeseed lecithin	Peanut lecithin	Egg lecithin	Bovine brain lecithin
phosphatidylcholine	21	31	14	37	23	69	18
phosphatidyl-ethanolamine	22	3	24	29	8	24	36
phosphatidylinositol	19	16	13	14	17		2
phosphatidic acid	10	9	7		2		2
phosphatidylserine	1	1				3	18
sphingomyelin						1	15
glycolipids	12	30		20	12		

Table 3. Fatty Acid Composition of Oil-Free Lecithins, %[a]

Fatty acids[b]		Soybean lecithin	Rapeseed lecithin	Sunflower-seed lecithin	Egg lecithin
palmitic acid	C16:0	18.4	5.0	8.0	37.0
stearic acid	C18:0	4.0		2.0	9.0
oleic acid	C18:1	10.7	63.0	20.0	32.3
linoleic acid	C18:2	58.0	20.0	67.8	16.7
linolenic acid	C18:3	6.8	9.0	0.5	
arachidic acid	C20:0			0.5	
arachidonic acid	C20:4				5.0
others		2.1	3.0	1.2	0

[a]Percent of total fatty acid content.
[b]See FATS AND FATTY OILS.

Table 4. Composition of Commercial Soy Lecithin[a] and Egg Lecithin, wt %

Compound	Soy lecithin	Egg lecithin
phosphatidylcholine	10–15	65–70
phosphatidylethanolamine	9–12	9–13
phosphatidylinositol	8–10	
phosphatidylserine	1–2	
phosphatidic acid	2–3	
lysophosphatidylcholine	1–2	2–4
lysophosphatidylethanolamine	1–2	2–4
phytoglycolipids	4–7	
phytosterines	0.5–2.0	
other phosphorus-containing lipids	5–8	
sphingomyelin		2–3
saccharose	2–3	
free fatty acids	max 1	max 1
mono-, diglycerides	max 1	traces
water	max 1.5	max 1.5
triglycerides	ca 35–40	10–15

[a]Acetone-insoluble matter ca 60.

Physical Properties

Commercial crude lecithin is a brown to light yellow fatty substance with a liquid to plastic consistency. Its density is 0.97 g/mL (liquid) and 0.5 g/mL (granule). The color is dependent on its origin, process conditions, and whether it is unbleached, bleached, or filtered. Its consistency is determined chiefly by its oil, free fatty acid, and moisture content. Properly refined lecithin has practically no odor and has a bland taste. It is soluble in aliphatic and aromatic hydrocarbons, including the halogenated hydrocarbons; however, it is only partially soluble in aliphatic alcohols (Table 5). Pure phosphatidylcholine is soluble in ethanol.

In water, a particle of lecithin exhibits myelin growth, ie, cylindrical sheets that are formed by bilayers and are separated by water which may break up into liposomes (vesicles with a single bilayer of lipid enclosing an aqueous space). Phospholipids more generally form multilamellar vesicles (MLV) (5). These usually are converted to unilamellar vesicles (ULV) upon treatment, eg, sonication. Like other antipolar, surface-active agents, the phospholipids are insoluble in polar solvents, eg, ketones and particularly acetone. Acetone does, however, dissolve the triglyceride carrier, and this difference in solubility provides a convenient means of separating, purifying, and measuring the phospholipids.

Commercial lecithin is soluble in mineral oils and fatty acids but is practically insoluble in cold vegetable and animal oils. However, it melts and disperses readily in hot vegetable and animal oils, but on cooling it separates unless a considerable percentage of mineral oil, fatty acid, or another coupling agent is added. Commercial plastic lecithin is converted into fluid lecithin by increasing the free fatty acid content. This softening or liquification of the commercial product also may be accomplished by other acids, eg, glycerophosphoric acid, phosphoric acid, or other mineral acid, and by almost any other organic or inorganic acid that is soluble or dispersible in lecithin. It may be liquified by complexing with divalent salts, eg, calcium chloride and by adding lower aliphatic esters of fatty acids.

Commercial lecithin is insoluble but infinitely dispersible in water. Treatment with water dissolves small amounts of its decomposition products and adsorbed or coacervated substances, eg, carbohydrates and salts, especially in the presence of ethanol. However, a small percentage of water dissolves or disperses in melted lecithin to form an imbibition. Lecithin forms imbibitions or absorbates with other solvents, eg, alcohols, glycols, esters, ketones, ethers, solutions of almost any organic and inorganic substance, and acetone. It is remarkable that the classic precipitant for phospholipids, eg, acetone, dissolves in melted lecithin

Table 5. Solubility[a] of Lecithin and Various Phospholipids

Phospholipids[b]	Hexane	Benzene	Ethanol	Acetone
lecithin	+	+	±	−
phosphatidylcholine	+	+	+	−
phosphatidylethanolamine	+	+	±	−
phosphatidylinositol	+	+	−	−
phytoglycolipid	±	+	+	−
lysophospholipids	±	+	+	+

[a]Soluble (+), insoluble (−), and partially soluble (±).
[b]Lysophospholipids are soluble in water; the others are dispersible in water.

readily to form a thin, uniform imbibition. Imbibition often is used to bring a reactant in intimate contact with lecithin in the preparation of lecithin derivatives.

When commercial lecithin is mixed with water it readily hydrates to a thick yellow emulsion. Upon dilution and agitation with water the emulsion may be thinned to almost any desired dilution. Emulsions of commercial lecithin are subject to microbial attack and must be preserved if they are to be stored for extended periods of time. In special cases, the emulsifying power of lecithin may be improved by using an alcohol or glycol imbibition of lecithin instead of lecithin. Such an imbibition is particularly useful in emulsifying hydroxy fatty acid-containing glycerides, eg, castor oil and blown marine oils. Lecithin emulsions tend to be precipitated by the addition of acids or salts, but such precipitation may be hindered by using a suitable synthetic detergent.

Commercial lecithin is a wetting and emulsifying agent inasmuch as its constituents, eg, fatty acid-containing phospholipids, are amphiphatic in chemical structure, having strongly lipophilic, fat-forming acid nuclei at one end of the molecule and a strongly hydrophilic amino or phosphoric acid nucleus at the opposite end. Lecithin is one of the very few natural and edible surface-active agents of this type that is soluble or dispersible in oil. Phosphatidylcholine and phosphatidylethanolamine are cationic and anionic at the same time, ie, they are zwitterions or amphoteric compounds. However, phosphatidic acid and the phosphoinositides are quite strong acids and therefore are anionic. If present, the phosphosphingosides and galactosphingosides are anionic and nonionic, respectively. It is evident, therefore, that commercial lecithin has the structural aspects of an anionic interface modifier.

The phospholipids present in commercial soybean lecithin fall into two groups according to molecular size. The alcohol-soluble fraction has a lower molecular size and has a higher monomer content. It is more reactive than the alcohol-insoluble fraction which has a higher molecular size and more of the nature of a polymer phase. In the colloidal system represented by the mixed phospholipids, the micelles represent the higher molecular weight constituents as the dispersing phase, and the lower molecular weight constituents represent the continuous or intermicellar phase.

Because of the zwitterion formation, mutual buffering action, and the presence of strongly acid components, soybean phospholipids have an overall pH of about 6.6 and react as slightly acidic in dispersions-in-water or in solutions-in-solvents. Further acidification brings soybean phospholipids to an overall isoelectric point of about pH 3.5. The alcohol-soluble fraction tends to favor oil-in-water emulsions and the alcohol-insoluble phospholipids tend to promote water-in-oil emulsions.

Pure soybean phospholipids are hygroscopic and subject to oxidation. On the other hand, shelf-life properties are good where residual soybean oil and tocopherols are present, eg, with commercial lecithin containing 30–40% neutral oil as a carrier, or where the substantially oil-free phospholipids contain 1–2% residual oil and a fractional percentage of tocopherols. Unlike the glycerides, commercial lecithin is not resistant to high temperatures. When heated above 80°C under anhydrous conditions, it darkens in proportion to the time of heating and decomposes as the temperature increases above 120°C; however, it is somewhat less sensitive as a minor ingredient and especially in aqueous systems.

Chemical Properties

In general, the presence of fatty acid groups in the phospholipid molecule permits reactions such as saponification, hydrolysis, hydrogenation, halogenation, sulfonation, phosphorylation, elaidinization, and ozonization (6).

Hydrolysis. The first effect of either acid hydrolysis or alkaline hydrolysis (saponification) is the removal of the fatty acids. The saponification value of commercial lecithin is 196. Further decomposition into glycerol, phosphoric acid, and head groups (ie, choline, ethanolamine, etc) may follow prolonged heating. Lecithin may also be hydrolyzed by enzymes.

Acyl Side-Chain Reactions. Many reactions occur in the R group of the fatty acid residue (see CARBOXYLIC ACIDS; FATS AND FATTY OILS).

Hydrogenation. Lecithin can be hydrogenated. The resulting lecithins have only saturated fatty acid residues (palmitic or stearic acid) and are more or less colorless and crystalline.

Hydroxylation. Commercial lecithin can be hydroxylated at the unsaturated fatty acid chains by treatment with concentrated hydrogen peroxide and acids like lactic or acetic acid.

Autoxidation. The autoxidation (7) of unsaturated fatty acids in phospholipids is similar to that of free acids. Primary products are diene hydroperoxides formed in a free-radical process.

Browning Reactions. The fluorescent components formed in the browning reaction (8) of peroxidized phosphatidylethanolamine are produced mainly by interaction of the amine group of PE and saturated aldehydes produced through the decomposition of fatty acid hydroperoxides.

Other Reactions of Phospholipids. The unsaturated fatty acid groups in soybean lecithin can be halogenated. Acetic anhydride combined with the amino group of phosphatidylethanolamine forms acetylated compounds. Phospholipids form addition compounds with salts of heavy metals. Phosphatidylethanolamine and phosphatidylinositol have affinities for calcium and magnesium ions that are related to interaction with their polar groups.

Manufacture and Processing

Crude soy lecithin is obtained as a by-product during the degumming process of soy oil. The phosphorus-containing compounds are removed to improve the stability of the oil.

Only a minor proportion of the total lecithin that is potentially available in the vegetable processing industry is produced. If the phospholipids are not to be made into commercial lecithin, they may be left in the crude oil or, if they are to be separated from the crude oil as wet gum, they may be mixed into soybean meal for animal feed.

The lipids are initially extracted from the soy beans with hot hexane. The resulting miscella is then filtered to remove fines (minute particles of flaked seed, protein, metal impurities, etc). Hexane is removed by distillation with steam, and the crude soybean oil is treated with water, swells, and the oil-insoluble aggregates of phospholipids precipitate out. Typical processing is shown in Figure 2. After drying and before being filled into drums, the commercial lecithin is held

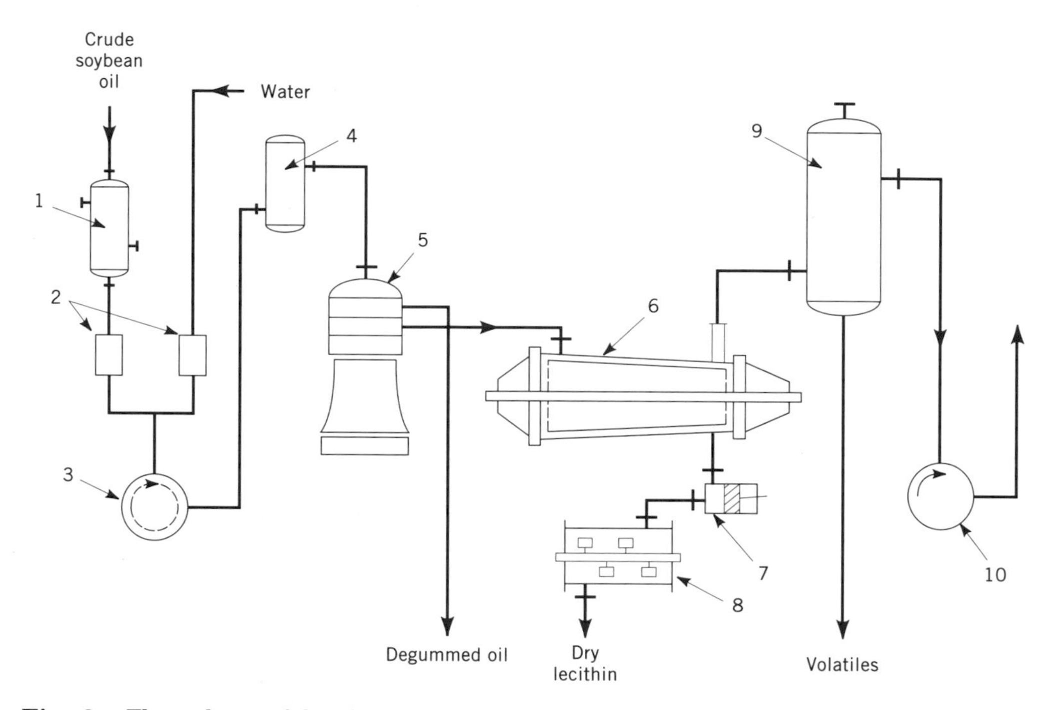

Fig. 2. Flow sheet of lecithin producing unit. Crude soybean oil is heated in the preheater, 1, to 80°C, mixed with 2% water in the proportion control unit, 2, and intensively agitated in 3. The mixture goes to a dwelling container, 4, and is then centrifuged after a residence time of 2–5 min. The degummed oil flows without further drying to the storage tanks. The lecithin sludge is dried in the thin-film evaporator, 6, at 100°C and 6 kPa (60 mbar) for 1–2 min and is discharged after cooling to 50–60°C in the cooler, 8. 9 and 10 are the condenser and vacuum pump, respectively.

in a work tank where addition of fluidizing agents, bleaching agents, or other material may be made.

Purification Processes. Separation of neutral and polar lipids, so-called deoiling, is the most important fractionation process in lecithin technology (Fig. 3). Lecithin is fluidized by adding 15–30% acetone under intensive agitation with acetone (fluidized lecithin:acetone, 1:5) at 5°C. The mixture goes to a separator where it is agitated for 30 minutes. The agitator is then stopped and the lecithin separates. The oil micella is removed and the acetone evaporated. After condensation the acetone is returned into the process.

Depending on the deoiling rate, the lecithin in the separation tank is again treated with acetone in the same way. This is repeated two to three times. The residual acetone is then removed in container 5 at < 80°C, dried, and separated to powder and granular lecithin in the classifier. Newer processes using less acetone and achieving deoiled lecithin having acetone insoluble matter as high as 99.9% have been described (9).

Due to possible environmental problems with acetone, new technologies are being developed for the production of deoiled lecithins involving treatment of lipid mixtures with supercritical gases or supercritical gas mixtures (10–12). In this process highly viscous crude lecithin is fed into a separation column at several

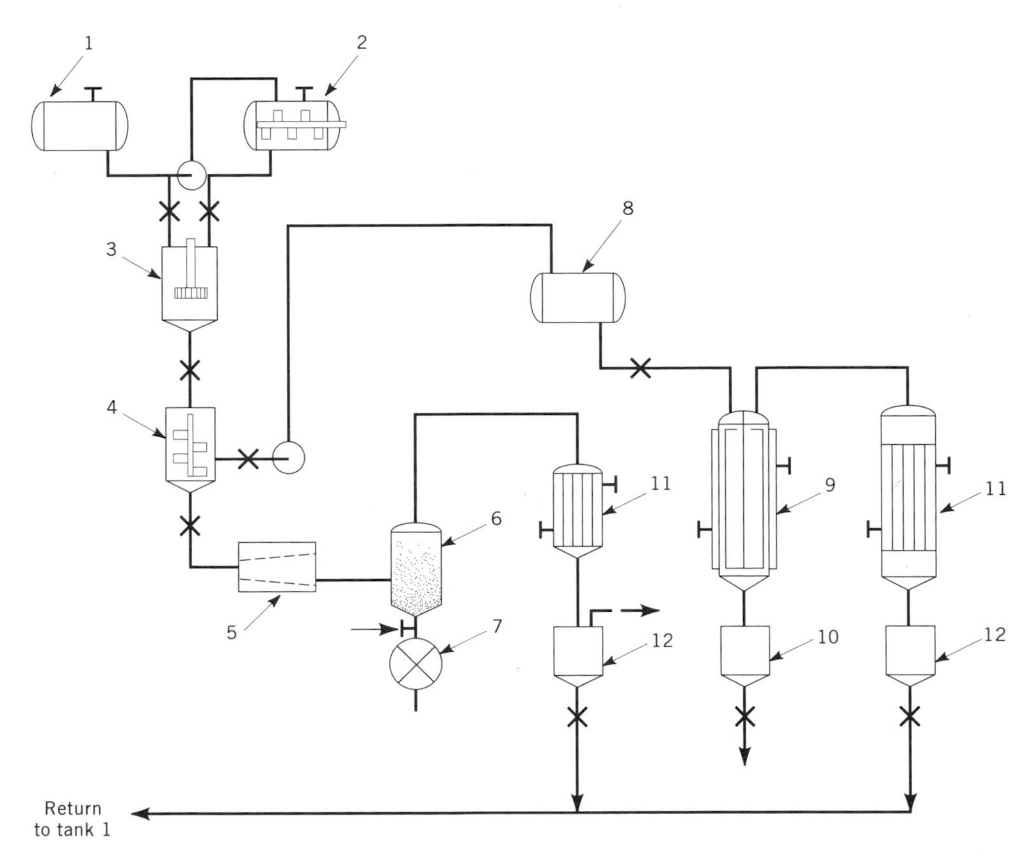

Fig. 3. Discontinuous deoiling of soy lecithin. 1, Acetone storage tank; 2, lecithin storage tank; 3, mixer; 4, separation tank; 5, filter/decanter; 6, dryer; 7, classifier; 8, oil miscella tank; 9, evaporator; 10, oil extract tank; 11, condenser; and 12, acetone storage tank.

levels. The supercritical extraction solvent flows through the column upward at a pressure of 8 MPa (80 bar) and temperature between 40 and 55°C. The soy oil dissolves together with a small amount of lecithin.

The mixture of propane, carbon dioxide, soy oil, and lecithin leaves the separation column and enters the first regeneration column. By increasing the temperature to 75°C, the lecithin is selectively precipitated. Due to its higher density, liquid lecithin flows down through the Sulzer packings of the regeneration column. It is drawn off and pumped back to the top of the separation column as reflux. The practically lecithin-free extraction agent leaves the first regeneration column at the top and is expanded in another regeneration column. The dissolved soy oil is precipitated at 6 MPa (60 bar) and 100°C, and continuously drawn off. The regenerated hot extractant leaves the second regeneration column at the top, is cooled, recompressed to 8 MPa (80 bar), and fed anew to the bottom of a separation column. In the separation column, lecithin flows down through the Sulzer packings, countercurrent to the extraction solvent. On its way down it comes into contact with increasingly pure solvent and is completely deoiled. The liquid mixture of propane, carbon dioxide, and lecithin is collected at the bottom of the separation column. It can be continuously drawn off through the product vessel or through an expansion valve.

Alcohol fractionation redistributes the phospholipids according to their respective hydrophilic and lipophilic properties (13). A process to produce fractionated phospholipids with a phosphatidylcholine (PC) content of more than 30% and a PC/PE (phosphatidylethanolamine) quotient of ca 4 has been developed. With this process it is possible to produce 1000 t per year.

The crude lecithin is thermostated to ca 30°C and mixed with a 5–10% sunflower monoglyceride (ca 50% mono content) and treated with 30% by weight ethanol, 90% by volume. This is then mixed with enough ethanol of the same or higher concentration that solvent and lecithin are in a ratio of 3:1 (vol/wt). The mixture is cooled to ca 20°C, stirring at such a rate that no emulsion is formed, and transferred into a settling vessel. The phases separate very rapidly. The slightly turbid upper phase is fed to a disk centrifuge by an intermediate vessel. After clarification it is passed by another intermediate vessel; after the addition of neutral oil, by a prewarmer into a downflow evaporator where the main part of the ethanol is removed. The remainder of the solvent is removed from the preconcentrated extract in a film evaporator. The lower phase, containing the ethanol-insoluble phospholipids, is collected; the fraction separated in the centrifuge is concentrated either using a Bollmann evaporator or a horizontal film evaporator. The ethanol in the condenser receivers is combined with

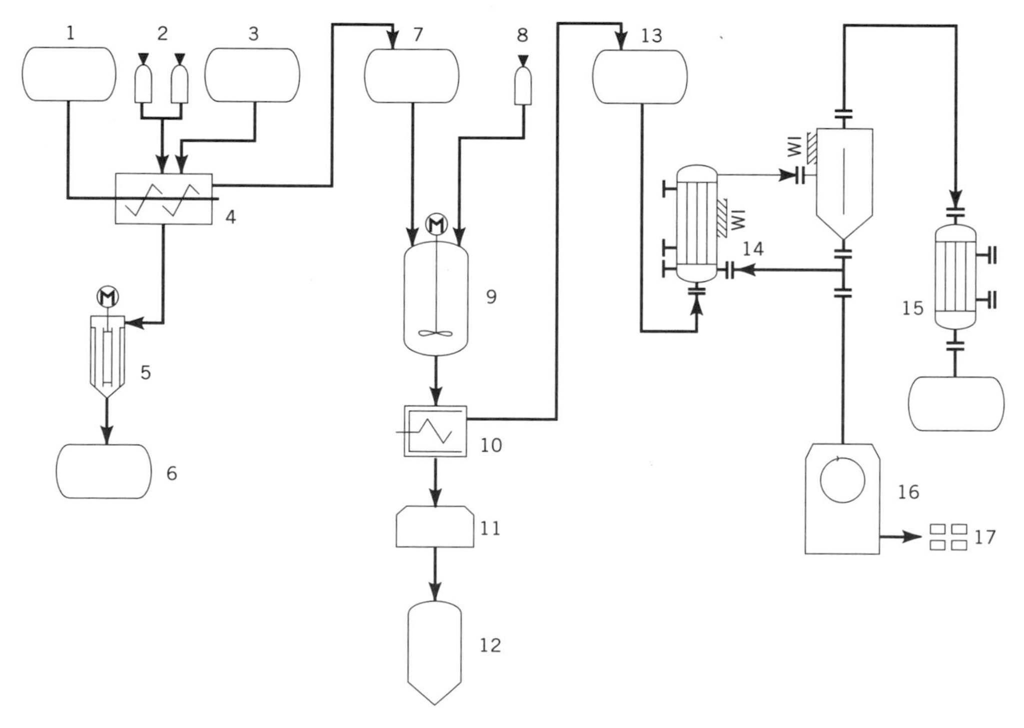

Fig. 4. Batch process for producing phosphatidylcholine fractions. 1, Ethanol storage tank; 2, deoiled lecithin; 3, solubilizer; 4, blender; 5, film-type evaporator; 6, ethanol-insoluble fraction; 7, ethanol-soluble fraction; 8, aluminum oxide; 9, mixer; 10, decanter; 11, dryer; 12, aluminum oxide removal; 13, phosphatidylcholine solution; 14, circulating evaporator; 15, cooler; 16, dryer; and 17, phosphatidylcholine.

that obtained from the recovery plant and adjusted to the required concentration before being used. The efficiency of the process is ca 35%. Increasing the ethanol–lecithin ratio to 5:1 allows this to be raised to 44% without the PC/PE ratio being reduced.

To produce highly purified phosphatidylcholine there are two industrial processes: batch and continuous. In the batch process for producing phosphatidylcholine fractions with 70–96% PC (Fig. 4) (14,15) deoiled lecithin is blended at 30°C with 30 wt % ethanol, 90 vol %, eventually in the presence of a solubilizer (for example, mono-, di-, or triglycerides). The ethanol-insoluble fraction is separated and dried. The ethanol-soluble fraction is mixed with aluminum oxide 1:1 and stirred for approximately one hour. After separation, the phosphatidylcholine fraction is concentrated, dried, and packed.

In the continuous process for producing phosphatidylcholine fractions with 70–96% PC at a capacity of 600 t/yr (Fig. 5) (16), lecithin is continuously extracted with ethanol at 80°C. After separation the ethanol-insoluble fraction is separated. The ethanol-soluble fraction runs into a chromatography column and is eluted with ethanol at 100°C. The phosphatidylcholine solution is concentrated and dried. The pure phosphatidylcholine is separated as dry sticky material. This material can be granulated (17).

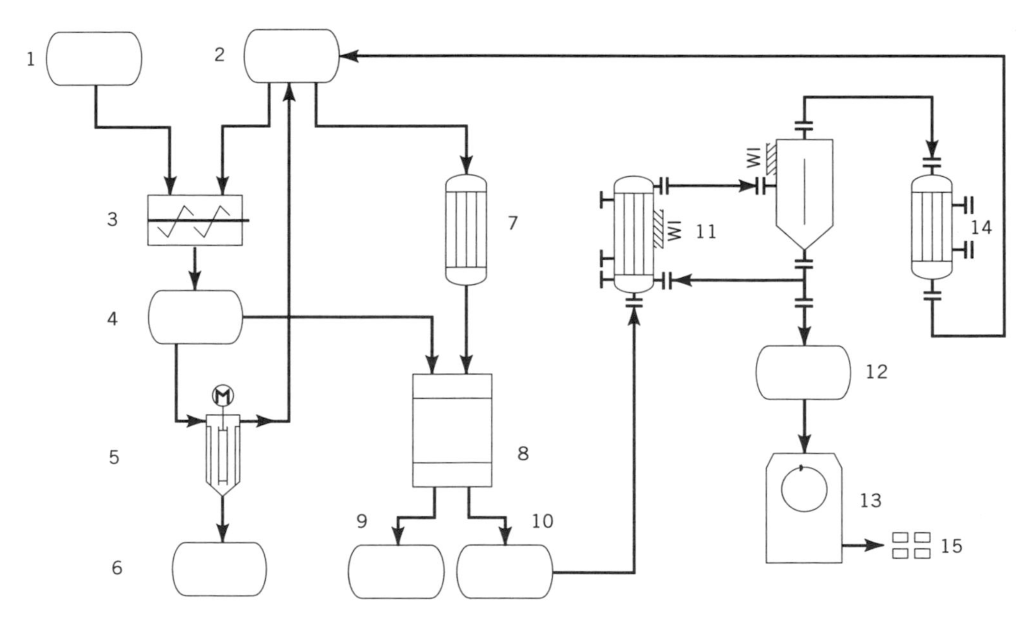

Fig. 5. Continuous process for producing phosphatidylcholine. 1, Lecithin; 2, ethanol; 3, blender; 4, diffuser; 5, thin-type evaporator; 6, ethanol-insoluble fraction; 7, heat exchanger; 8, chromatography column (SiO_2); 9, prestream; 10 and 12, phosphatidylcholine solution; 11, circulating evaporator; 13, dryer; 14, cooler; and 15, phosphatidylcholine.

Commercial Grades

There are six common commercial grades (18) of lecithin available (Table 6) including (*1*) clarified lecithins found either in the full miscella, crude oil, or directly as lecithin; (*2*) fluidized lecithins wherein the fluidization is done by adding calcium chloride, fatty acids, vegetable oil, or special diluents; (*3*) compounded lecithins which are special-purpose products made by direct addition of emulsifiers like sorbitan esters, polysorbates, or other surface-active agents; deoiled lecithin may be combined with selected additives to improve handling and performance; (*4*) hydroxylated lecithins which are highly water dispersible and made by the reaction of hydrogen peroxide and lecithin in the presence of a weak acid-like lactic or acetic acid; (*5*) deoiled lecithins for which the resulting dried product is available in different particle size with free flowing properties; it can be blended with other free flowing carriers; and (*6*) fractionated lecithins made from crude lecithin or deoiled lecithin by extraction with alcohol, resulting in alcohol-soluble and alcohol-insoluble fractions with different functionality. Fractions with different phosphatidylcholine content are commercially available. Besides these common commercial grades, more special products are available, eg, enzymatically modified lecithin and phospholipids, semisynthetic phospholipids, and acetylated lecithins.

Economic Aspects

The total commercial lecithin potential if all vegetable oils were degummed worldwide would be 552,000 t (Table 7). Although soybean, sunflower, and rape lecithins are available in the market, the principal commercial interest is only in soybean lecithin. The annual worldwide production is 130,000 t (Table 8).

Specification and Standards

The specifications of the *Food Chemical Codex* (FCCIII) and several other agencies are given in Table 9. The product to which they refer is defined as food-grade lecithin from soybeans and other plant sources consisting of a complex mixture of acetone-insoluble phospholipids (mainly phosphatidylcholine, phosphatidylethanolamine, and phosphatidylinositol) combined with various amounts of other substances such as triglycerides, fatty acids, and carbohydrates in varying proportions. The specified product contains 50% or more of phospholipids of differing grades, forms, and color.

The *U.S. Pharmacopeia* (USP XXII) or *National Formulary* (NFXVII) (20) also provide a similar description; however, the peroxide value is not defined (Table 9). These specifications are also given in the *Handbook of Pharmaceutical Excipients* (HPE), published jointly by the American Pharmaceutical Association and The Pharmaceutical Society of Great Britain (21), which defines lecithins both from plants and eggs. The *Merck Index* (22) specifies a slightly lower acid value. The *Japanese Monograph* (ISCI-II) (23) specifies a slightly lower acetone-insoluble matter and a lower heavy-metal content.

The European Community specifications (Guideline 78/664/EWG) for lecithins (E322) (24) are also included in the *Monographs for Emulsifiers for Foods* (second edition) of the European Food Emulsifier Manufacturer's Association (EFEMA) (25). These differ in some respects from the U.S. specifications. E322 distinguishes between lecithins, which may be from plant sources or from eggs, and hydrolyzed lecithins. The latter are permitted a higher acid value and somewhat lower content of acetone-insoluble matter than nonhydrolyzed

Table 6. Commercial Lecithins, % Composition

Component	Crude	Deoiled	Alcohol-soluble fraction	Alcohol-insoluble fraction	PC 70[a]	PC 90[b]
phosphatidylcholine	10–15	20–25	40–55	8–12	75–80	90–98
phosphatidylethanol-amine	9–12	22	21	24	10	
phosphatidylinositol	8–10	16	2	27		
polar lipids, others	8–12	15	9	16	7	3
soybean oil	40	3	3	3	3	3
carbohydrates	5	8	3	10		
sterols, glycosides	3	6	7	4		
moisture	1	1	1	1	1	1
miscellaneous	4	4	1	4	4	
trademarks[c]	Alcolec (ALC)	Alcolec (ALC)	Alcolec (ALC)	Alcolec (ALC)	Phospholipon 80 (ALC/NP)	Phospholipon 90 (ALC/NP)
	Nathin (NP)	Phospholipon 25 (NP)	Nathin (NP)	Nathin (NP)	Phosal (ALC/NP)	Phosal (ALC/NP)
	Centrocap (CS)	Centrolex (CS)	Phosal (ALC/NP)			
	Yelkin (ADM)	Yelkinol (ADM)				
	Leciprime (RI)	Lecigran (RI)				

[a]PC > 70%

[b]PC > 90%

[c]Names of companies (suppliers) are given in parentheses. ADM = Archer Daniels Midland; ALC = American Lecithin Co.; CS = Central Soya; NP = Nattermann Phospholipid GmbH; RI = Riceland.

Table 7. Commercial Lecithin Potential from Vegetable Oils

Type of oil	World production,[a] 10^6 t	Hydratable lecithin, %	Lecithin yield, t
soybean	16.44	2.2	361,680
sunflower-seed	7.37	0.5	36,850
rapeseed	9.37	1.0	93,700
cottonseed	4.32	0.8	33,840
peanut	3.45	0.4	13,800
corn	1.00	1.2	12,000
Total			*551,870*

[a]1991/1992.

Table 8. Lecithin Worldwide Production

Type	World capacity, t	Average sales price, $/kg
crude lecithin	132,000	0.62
deoiled lecithin	12,000	4.40
phospholipid fraction PC 35	1,000	7.48
phospholipid fraction PC 70	600	128.90
phosphatidylcholine PC > 90	50	253.00

Table 9. U.S., British, Japanese, and European Specifications for Purity[a]

Parameter[b]	FCC III	USP XXII/NF XVII + HPE	E322	E322 and EFEMA	ISCI-II
definition	lecithin vegetable	lecithin	lecithin	hydrolyzed lecithin	soybean phospholipid
acetone-insoluble matter, %	≥50	≥50	≥60	≥56	≥40
hexane-insoluble matter, %	≤0.3	≤0.3	≤0.3[c]	≤0.3[c]	≤0.3
volatile matter, %			≤2[d]	≤2[d]	≤2
H_2O, %	≤1.5	≤1.5	≤2	≤2	≤2
As, ppm	3	3	3	3	2
Pb, ppm	≤10	≤10	≤10	≤10	
heavy metals, ppm	40	40	50	50	20
peroxide value	≤100		≤10	≤10	≤10
acid value	≤36	≤36	≤35	≤45	≤40
iodine value					
liquid		95–100			
granule		82–88			

[a]Food-grade lecithins.
[b]Packing and storage should be in well-closed containers.
[c]Toluene-insoluble.
[d]At 105°C for 1 h.

lecithin, though both European specifications for acetone-insoluble matter are stricter than those of the FDA. This also applies to the specification for the peroxide value, though the U.S. specifications are marginally stricter with regard to water and heavy-metal content. E322 also specifies that lecithin should contain not more than 2% volatile matter by drying at 105°C for one hour.

Analytical and Test Methods

The standard methods (26) of analysis for commercial lecithin, as embodied in the Official and Tentative Methods of the American Oil Chemists' Society (AOCS), generally are used in the technical evaluation of lecithin (27). For example, the AOCS Ja 4-46 method determines the acetone-insoluble matter under the conditions of the test, free from sand, meal, and other petroleum ether-insoluble material. The phospholipids are included in the acetone-insoluble fraction. The substances insoluble in hexane are determined by method AOCS Ja 3-87.

Acid value (AOCS Ja 6-55) is the number of milligrams of potassium hydroxide necessary to neutralize the acids in one gram of sample. Peroxide value (AOCS Ja 8-87) is the number of milliequivalents of peroxide per 1000 grams of sample, which oxidize potassium iodide under the conditions of the test. The oxidizing substances are generally assumed to be peroxides or other similar products of fat oxidation. The Karl Fischer (AOCS Ja 26-87) method determines the actual water content of lecithin by titration with Fischer reagent which reacts quantitatively with water. Other methods for moisture determination are in AOCS Ja 2a-46. The Gardner Color (AOCS Ja 9-87) method determines lecithin color by comparison of an undiluted sample to standards of a specified color. Method AOCS Ja 10-87 determines Brookfield viscosity in mPa(=cP) of a fluid lecithin at a specified temperature. Another method is the viscosity of transparent liquids by bubble time (AOCS Ja 11-87). The iodine value (AOCS Cd 1-25) is a measure of the unsaturation of fats and oils and is expressed in terms of the number of centigrams of iodine absorbed per gram of sample (% iodine absorbed).

The total phosphorus content of the sample is determined by method AOCS Ja 5-55. Analysis of phospholipid in lecithin concentrates (AOCS Ja 7-86) is performed by fractionation with two-dimensional thin-layer chromatography (tlc) followed by acid digestion and reaction with molybdate to measure total phosphorous for each fraction at 310 nm. It is a semiquantitative method for PC, PE, PI, PA, LPC, and LPE. Method AOCS Ja 7b-91 is for the direct determination of single phospholipids PE, PA, PI, PC in lecithin by high performance liquid chromatography (hplc). The method is applicable to oil-containing lecithins, deoiled lecithins, lecithin fractions, but not applicable to lyso-PC and lyso-PE.

To determine the phospholipid and fatty acid compositions chromatographic methods (28) like gas chromatography (gc), thin-layer chromatography (tlc), and high performance liquid chromatography (hlpc) are used. Newer methods for quantitative determination of different phospholipid classes include ^{31}P-nmr (29).

Health and Safety Factors

Environmental considerations encourage degumming of crude vegetable oils. A large part of the soybean oil produced is degummed but only some of the wet gums (lecithin hydrate) are processed to finished lecithin; however, this proportion increases as the demand for commercial lecithin grows. The phospholipids are biodegradable, but their presence in streams and water resources, especially in the form of soap stock, is undesirable. Fatty acid recovery from phospholipids is less than with neutral oils because of the lower fatty acid content. There are no known health hazards involved in the production of commercial lecithin from crude vegetable oils because the phospholipids are nonvolatile and are a nonirritating food material. Care must be exercised in the use of small quantities of benzoyl peroxide which is required in the manufacture of highly bleached grades and in the acetone or alcohol fractionation of crude lecithin in the manufacture of purified grades.

The safety of lecithin is also confirmed by the World Health Organization (WHO). WHO has not set any acceptable daily intake (ADI) to lecithin as a foodstuff, but the FDA has awarded it GRAS status (generally recognized as safe; CFR No. 182.1400/184.1400).

Uses

The worldwide uses of lecithin break down as follows: margarine, 25–30%; baking/chocolate and ice cream, 25–30%; technical products, 10–20%; cosmetics, 3–5%; and pharmaceuticals, 3%.

Animal Feed. In animal feeds (1–3% lecithin) lecithin is an emulsifier; wetting and dispersing agent; energy source; antioxidant; surfactant; source of choline, organically combined phosphorus and inositol; and lipotropic agent. It is used in a milk replacer formula for calves (approximately 10,000 t of lecithin are used for this purpose) and for veal production, in mineral feeds, poultry feeds, fish foods, pet foods, and feeds for fur-bearing animals (30).

Baking Products. In baking products and mixes (0.1–1% lecithin) lecithin is an emulsifier, stabilizer, conditioning and release agent, and antioxidant. In yeast-raised doughs it improves moisture absorption, ease of handling, fermentation tolerance, shortening value of fat, volume and uniformity, and shelf life. In biscuits and crackers, pies, and cakes (1–3% based on shortening), it promotes fat distribution and shortening action, facilitates mixing, and acts as a release agent (31). Frozen doughs with liposomal encapsulated yeast show substantially more volume (32). Phospholipid fractions can replace chemical emulsifiers (33) and also function as fat reducers (34).

Candy. In confections (1% lecithin) made with oil or fat, lecithin emulsifies and distributes fat in caramels, nut brittles, nougats, etc; it also prevents fat separation and greasiness. It has a fixative action for flavors (35) (see FLAVORS AND SPICES). Also, lecithin is an emulsifier and conditioning agent for chewing gum base.

Chocolate (0.3–0.5% lecithin) lecithin is a wetting agent and emulsifier. It facilitates mixing, saves processing time and power, saves cocoa butter, stabilizes viscosity, increases shelf life, counteracts moisture thickening, and aids release of molded goods (see CHOCOLATE AND COCOA).

Cosmetics and Soaps, One to five percent lecithin moisturizes, emulsifies, stabilizes, conditions, and softens when used in products such as skin creams and lotions, shampoos and hair treatment, and liquid and bar soaps.

Since the introduction of Capture in 1986, liposomes produced from phospholipids are commercially available worldwide (36,37).

Food. Lecithin is a widely used nutritional supplement rich in polyunsaturated fatty acids, phosphatidylcholine, phosphatidylethanolamine, phosphatidylinositol, and organically combined phosphorus, with emulsifying and antioxidant properties (38).

In dehydrated foods (0.05–0.3% lecithin) lecithin is a release agent in drying, and it aids in rehydration. In instant foods (0.5–3% lecithin) lecithin is used for its wetting, dispersing, emulsifying, and stabilizing properties in beverage powders and mixes including milk powders, dessert powders, powdered soups, etc.

Ice Cream. Lecithin (0.15–0.5%) emulsifies, stabilizes, improves smoothness and melting properties, and counteracts sandiness in storage. Lecithin is also used as an emulsifier in whipped toppings.

Macaroni and Noodles. Lecithin (0.25–5%) is used as a conditioning agent and antioxidant, improves machining, counteracts disintegration and syneresis, and improves color retention.

Margarine. Lecithin (0.15–0.5%) is an emulsifier and an antispattering and browning agent; it improves frying properties and spreadability and shortening action in table margarine. It is also used in bakers' margarine.

Edible Oils and Fats. Lecithin (0.01–2%) is used as an emulsifier, wetting agent, and antioxidant; it extends shelf life, especially of animal fats; increases lubricity (shortening value); improves stability of compound shortenings; and lowers cloud point of vegetable oils.

Inks and Dyes. Lecithin (0.5–3%) is a wetting, dispersing, and suspending agent promoting uniformity, color intensity, and ease of remixing (especially printing inks). In dyes, lecithin (0.5–2%) is a coupling agent, especially for water-soluble colors in fatty media.

Liposomes. Lecithin, and more specifically purified phospholipids, are used to produce liposomes (39) for the food (40), cosmetics, pharmaceutical, agrochemical, and technical fields.

Paints. Lecithin (0.5–5% of pigment) is a wetting agent, dispersing agent, suspending agent, emulsifier, and stabilizer in both oil-base and water-base (latex- and resin-emulsion) paints. It facilitates rapid pigment wetting and dispersion, saves time in grinding and mixing, permits increased pigmentation, stabilizes viscosity, aids in brushing, and improves remixing after storage.

Petroleum Products. Lecithin (0.005–2%) is used as an antioxidant, detergent, emulsifier, and anticorrosive agent, and for lubricity and antiwear. It is added to gasoline to stabilize tetraethyllead and for its inhibition and anticorrosive effects. After reaction with aliphatic amines it is used as a detergent in motor oils for inhibition, detergent, and lubricity effects. Also, it is used in miscellaneous oils including household lubricants and cutting oils, in fuel oils for surfactant and inhibition effects, and in drilling muds as an emulsifier.

Pharmaceuticals. Lecithin is used especially as a dietetic source of phosphatidylcholine required in lipid metabolism including enzyme systems involved in cholesterol metabolism, for the metabolism of fats in the liver, and as a precursor of brain acetylcholine neurotransmitter. Dosage ranges from 2 g/d for substantially pure soybean phosphatidylcholine to as high as 60 g/d for whole, mixed, and substantially oil-free soybean phospholipids (containing about 25% phosphatidylcholine) in liver disorders, in cardiovascular disease, in neurologic disease associated with impaired acetylcholine function, in skin disorders involving deranged lipid metabolism (eg, psoriasis), in telangiectasia, in nervous tension derived from extended physical fatigue, and as an emulsifier and wetting agent, eg, in penicillin dispersions and in emulsions for intravenous alimentation. Highly purified soybean phosphatidylcholine also is given intravenously when solubilized in water and is effective in some conditions in smaller amounts than the polyenephosphatidylcholine given orally.

Lecithin and especially lecithin fractions with high phosphatidylcholine content are used as excipients and as an active drug. Phosphatidylcholine with a high content of unsaturated fatty acids (polyenephosphatidylcholine) are on the market in several countries in both oral and parenteral form as a lipid lowering agent (Lipostabil) and liver protector (Essentiale) (41,42). These phospholipids show membrane protective effects and have gastroprotective capability (43). Phospholipids in the form of liposomes are interesting tools for drug targeting and drug delivery (44).

Plant Protection. Lecithin (0.5–10%) and phospholipid fractions are used in fertilizers (qv), herbicides (qv), insecticides, and fungicides as emulsifiers or to increase the effectiveness of the active ingredient (45). In insecticides (0.5–5% lecithin), lecithin is used for improved emulsification, spreading, penetration, and adhesion (see INSECT CONTROL TECHNOLOGY).

Plastics. Lecithin (0.5–1.5%) is used for pigment dispersion and as a slip or release agent. It also may be sprayed on molds. It has surfactant effects in organosols and plastisols (see SURFACTANTS).

Release Agents. Lecithin (2–10%) is used as a surfactant and antisticking agent in sprays for cookware and in lubricants and release agents for general food application and industrial purposes.

Elastomers. Lecithin is a wetting and dispersing agent and mold-release agent in rubber. It increases plasticity and facilitates working. It emulsifies latex mixes and aids in preparing solvent dispersions and in vulcanizing. In sealing and caulking compounds it is used for wetting, dispersing, and plasticizing effects (see ELASTOMERS, SYNTHETIC).

Textiles. Lecithin (0.2–0.5%) is used for emulsifying, wetting, softening, and conditioning especially in sizing and finishing. It imparts soft, smooth handle and also is used as a spray to reduce cotton (qv) dust.

BIBLIOGRAPHY

"Lecithin" in *ECT* 1st ed., Vol. 8, pp. 309–326, by J. Stanley, Joseph Stanley Co.; in *ECT* 2nd ed., Vol. 12, pp. 343–361, by P. Sartoretto, W. A. Cleary Corp.; in *ECT* 3rd ed., Vol. 14, pp. 250–269, by J. Eichberg, American Lecithin Co.

1. M. Gobley, *J. Pharm. Chem. (Paris)* **17**, 401–417 (1850).
2. C. Diakanow, *Med. Chem. Untersuch.* **2**, 221–227 (1867); C. Diakanow, *Zbl. Med. Wiss.* **2**, 434–435 (1868).
3. *Chem. Physics Lipids* **21**, 141–158 (1978).
4. T. R. Watkins, in J. Hanin and G. Pepeu, eds., *Phospholipids*, Plenum Press, New York, 1990, p. 301.

5. R. R. C. New, ed., *Liposome. A Practical Approach*, IRL Press, Oxford, U.K., 1990.
6. C. R. Scholfield, in B. F. Szuhaj, ed., *Lecithins*, AOCS, Champaign, Ill., 1989, pp. 7–15.
7. N. A. Porter and C. R. Wagner, *Adv. Free Radical Biol. Med.* **2** 283–323 (1986).
8. F. J. Hidalgo and co-workers, *Fat Sci. Technol.* **5**, 185–188 (1990).
9. U.S. Pat. 4,803,016 (Feb. 7, 1989), M. D. Bindermann and J. N. Casey (to Central Soy Co.).
10. S. Peter and co-workers, *Chem. Eng. Technol.* **10**, 37–42 (1987).
11. E. Stahl and K. W. Quirin, *Fette Seifen Anstrichmittel* **87**, 219–224 (1985).
12. H. Wagner and R. Eggers, *Fat Sci. Technol.* **95**(2), 75–80 (1993).
13. H. Pardun, *Fette Seifen Anstrichmittel* **86**(2), 55–62 (1984).
14. U.S. Pat. 3,031,478 (Apr. 24, 1962), H. Eickermann (Nattermann & Cie).
15. U.S. Pat. 3,544,605 (Dec. 1, 1970), H. Betzing (Nattermann & Cie).
16. U.S. Pat. 4,425,276 (Jan. 10, 1984), B. R. Günther (Nattermann & Cie).
17. Eur. Pat. Appl. 0 521 398 (Jan. 7, 1993), R. Losch and co-workers (Natterman & Cie).
18. B. F. Szuhaj, *JAOCS* **60**(2), 258A–261A (1983).
19. G. Schuster, *Emulgatoren für Lebensmittel*, Springer-Verlag, Berlin, 1984.
20. Committee of Revision, *Lecithin, The United States Pharmacopeia*, U.S. Pharmacopeial Convention, Rockville, Md., 1990, p. 1572.
21. "Lecithin," *Handbook of Pharmaceutical Excipients*, American Pharmaceutical Association, Washington, D.C., 1986, p. 165.
22. S. Budavari and co-workers, "Lecithin," in *The Merck Index*, 11th ed., Merck & Co. Inc., Rahway, N.J. 1989, p. 854.
23. Yakuji Nippo, Ltd., *The Japanese Standards of Cosmetic Ingredients*, 2nd ed., 1983, pp. 418–419.
24. *Off. J. Eur. Commun.*, L 297/31 (Oct. 23, 1982).
25. *Lecithins, Monographs for Emulsifiers for Foods*, 2nd ed., European Food Emulsifier Association, Brussels, 1985, pp. 1–5.
26. K. H. Gober and co-workers, in G. Cevc, ed. *Phospholipid Handbook*, part 1, Marcel Dekker, Inc., New York, 1993, Chapt. 3.
27. R. A. Lantz, in Ref. 6.
28. W. W. Christie, *Z. Lebensm. Unters. Forsch.* **181**, 171–182 (1985).
29. N. Sotirhos, *Dev. Food Sci.***17**, 443–452 (1988).
30. F. W. Kullenberg, in Ref. 6, pp. 237–252.
31. W. H. Knightly, in Ref. 6, pp. 174–176.
32. R. Silva, *Cereal Food World* **35**(10) 1008–1012 (1990).
33. H. D. Jodlbauer, W. Freund, and J. Senneka, *Getreide. Mehl Brot.* **6**, 174–177 (1992).
34. U.S. Pat. 5,120,561 (June 9, 1992) R. F. Silva (American Lecithin Co.).
35. R. C. Appl, in Ref. 6, p. 207.
36. B. F. Haumann, *Inform* **3**(11), 1172–1178 (1992).
37. A. Wendel and M. Ghyczy, *SCCS* **6**, 32–37 (1990).
38. F. Orthoefer and S. U. Gurkin, *Food Marketing Technol.* **12**, 11–14 (1992).
39. G. Gregoriadis, ed., *Liposome Technology*, Vols. 1, 2, 3, 2nd ed., CRC Press, Inc., Boca Raton, Fla., 1993; D. D. Lasic, *Liposomes: From Physics to Applications*, Elsevier, New York, 1993.
40. J. C. Verillemard, *J. Microencapsulation* **8**(4), 547–562 (1991).
41. A. I. Archakov and K. J. Gundermann, eds., *Phosphatidylcholine on Cell Membranes and Transport of Cholesterol*, wbn-Verlag-Bingen, Rhein, Germany, 1988.
42. K. J. Gundermann and R. Schumacher, *50th Anniversary of Phospholipid Research (EPL)*, wbn-Verlag-Bingen, Rhein, Germany, 1989.
43. B. S. Dunjic and co-workers, *Scand. J. Gastrol.* **28**, 89–94 (1993).
44. G. Gregoriadis and A. T. Florence, *Drugs* **45**(1), 15–28 (1993).

45. U.S. Pat. 4,681,617 (July 21, 1987), U.S. Pat. 4,567,161 (Jan. 28, 1986), U.S. Pat. 4,506,831 (Mar. 26, 1985), U.S. Pat. 4,576,626 (Mar. 18, 1986), M. Ghyczy and co-workers (to A. Nattermann & Cie).

General References

I. Hanin and G. B. Ansell, eds., "Lecithin: Technological, Biological and Therapeutic Aspects," *Advances in Behavioral Biology*, Vol. 33, Plenum Press, New York, 1987.

I. Hanin and G. Pepeu, eds., *Phospholipids: Biochemical, Pharmaceutical, and Analytical Considerations*, Plenum Press, New York, 1990.

B. F. Szuhaj and G. R. List, eds., *Lecithins*, American Oil Chemist's Society, Champaign, Ill., 1985.

B. F. Szuhaj, ed., *Lecithins: Sources, Manufacture and Uses*, American Oil Chemist's Society, Champaign, Ill., 1989.

H. Pardun, *Die Pflanzenlecithine*, Verlag für Chem. Industrie H. Ziolkowsky KG, Augsburg, Germany, 1988.

G. Cevc, ed., *Phospholipids Handbook*, Marcel Dekker, Inc., New York, 1993.

ARMIN WENDEL
Rhône-Poulenc Rorer

LEUCO BASES. See DYES AND DYE INTERMEDIATES.

LEUKOTRIENES. See ANTIASTHMATIC AGENTS; PROSTAGLANDINS.

LEVULOSE (FRUCTOSE). See CARBOHYDRATES; SUGAR.

LICENSING

A license is an agreement between two or more parties which conveys certain intellectual property rights (eg, patent, copyright, trademark, or know-how) from the licensor (the holder or owner of the intellectual property rights) to the licensee (the recipient of the intellectual property rights). For the sake of simplicity, herein a licensor (the entity which is licensing-out) and licensee (the entity which is licensing-in) are assumed to be companies; however, an individual, the government, or a nonprofit entity such as a university may also be a licensor or licensee. Because a license is a legal contract or agreement between two or more parties, a discussion about licenses could emphasize the legal aspects of the terms and conditions; however, a license is more often entered into because of business reasons, hence this article focuses more on the business aspects of licensing.

Reasons for Licensing

In any contractual arrangement, each party has a particular set of reasons to enter into the agreement. In a licensing arrangement, the licensor holds rights to certain intellectual property that may be of value; the licensee would like access to those property rights in order to make, use, or sell products covered by such intellectual property rights.

Licensor's Perspective. Recognizing that its intellectual property has value, the owner may want to allow others to exploit the intellectual property through a licensing arrangement for a variety of reasons. (*1*) The licensor may not wish to take on the financial and resource risk of developing the product itself, knowing that it takes more money and time than the company is willing to invest. (*2*) Even if the licensor is interested in developing new products, the company may not have enough resources or expertise to develop the product itself. (*3*) The licensor may not want to incur the liability that would be associated with the product, eg, product liability for a vaccine. (*4*) The licensor does not know or understand the marketplace involved in the particular intellectual property and wishes to focus on those areas it knows best. (*5*) The licensor wishes to generate royalty income. (*6*) Although the patentee or copyright holder is often in a strong position with respect to the enforcement of intellectual property rights, the licensor wishes to avoid potential litigation by granting a license to a company which is using the licensor's intellectual property. (*7*) The licensor is selling a product protected by certain intellectual property rights, but would like to allow a second source of the product to come into being, and therefore grants a license to another company. Potential licensors may not want to grant licenses, because the company wants to keep all of the profits or wants to prevent competitors from entering the marketplace. Ultimately, the decision whether or not to license-out should be part of a strategic business decision.

Licensee's Perspective. A potential licensee may be interested in entering into a license agreement for a number of reasons. (*1*) The licensee would like the opportunity to augment its own research and development efforts by acquiring technology from other parties through licensing. (*2*) The licensee would like to acquire new ideas which fall into the licensee's research or business areas of expertise. (*3*) The licensee wants to avoid litigation. (*4*) There is often a window of opportunity via licensing to enter new markets, develop new products, etc, so as to be the first in the marketplace.

On the other hand, the technical and market risk involved in licensing technology, the difficulty in assessing the value of a particular intellectual property, and the fact that the technology is developed outside the license company often makes licensing less than desirable. Like licensing-out, licensing-in should be a strategic business decision.

Overall Perspective. For both parties, some of the factors which affect the decision to license-out or license-in within a strategic business plan might include (*1*) an assessment of competing products, (*2*) an assessment of alternative approaches, (*3*) an assessment of the real and perceived features, advantages, and benefits of the technology, (*4*) projections of market potential, and (*5*) the life span of the product. In addition, the question should be asked, "Is the technology evolutionary?" in which case a potential licensee would have to decide whether or not the technology is worth the investment, or is it "revolutionary," in which case the potential licensee might not want to "miss the revolution." In all cases, both the licensee and licensor must see a benefit from the license agreement in order for the business relationship to be successful.

Components of a License Agreement

Every license agreement is a negotiated contract between parties, and therefore there are no hard-and-fast rules as to what is required in a license agreement. Since every provision is negotiable, the license agreement should be viewed in its entirety, rather than as component parts. Nevertheless, for purposes of illustration, select component parts will be discussed.

Diligence. Although the financial terms of a license agreement generally receive the most attention, "diligence" provisions, ie, those provisions which specify the obligations of the licensee to develop the technology diligently, are often more important. The diligence provisions should set forth the expectations and understanding of both parties as to the developmental time frame and milestones to be met by the licensee. These provisions are important because they allow both parties to know if the intent of the license is being met. It gives the licensor a means of measuring the progress of the licensee in bringing the technology to the marketplace, and it can allow for termination if the milestones have not been met. In most licensing arrangements, the issue which should be of greatest concern to a licensor is that the licensee is diligently developing the technology. Otherwise, the licensor will not reap the potential benefits from having licensed the technology.

An example of a diligence provision might be as follows:

> As an inducement to Licensor to enter into this Agreement, Licensee agrees to use all reasonable efforts and diligence to proceed with the development, manufacture, and sale of the Licensed Product and to diligently develop markets for the Licensed Product. Unless Licensee has a Licensed Product available for commercial sale prior to ____, Licensee agrees that Licensor may terminate this Agreement. In addition, Licensee agrees: (*1*) to have a prototype developed by ____; (*2*) to have tested the prototype in ____; (*3*) to have developed a marketing plan; (*4*) to have sold *xxx* units by ____; and (*5*) to have generated $____ in sales by ____.

Financial Terms. *Royalty-Free Licenses.* Although most licenses involve a financial transaction, it should be noted that royalty-free licenses can be and often are granted, allowing a licensee to practice an invention or obtain rights under copyrights without making compensation. Royalty-free licenses allow the licensor to retain ownership or other control over the intellectual property but do not provide a financial return to the intellectual property owner.

Royalty-Bearing Licenses. A royalty-bearing license presumes a financial exchange between the parties, ie, the licensor grants certain intellectual property rights to the licensee for monetary consideration. Some royalty-bearing licenses are known as fully paid, in that once the payment is made no further payment is expected or due. Other royalty-bearing licenses specify payments throughout the term of the agreement. For licenses requiring payments throughout the term of the agreement, payments may include a license issue fee or upfront payment, annual payments, milestone payments, and earned royalties based on sales.

License issue fees, also known as upfront payments, are often but not always required from licensees as a gesture of good faith and serious intention by the licensee to develop the technology. It can be viewed as a down payment based on the perceived or actual value of the technology. Licensees generally prefer to pay as little as possible prior to sales; licensors prefer to generate income as soon as possible. The licensor and licensee can structure the financial terms such that upfront payment can be credited, ie, prepaid, against earned royalties, thereby meeting the desire of the licensor to obtain more cash upfront but also meeting the desire of the licensee to leverage the upfront payment against future royalty obligations.

Annual payments are often viewed as a financial diligence provision or as a fee for maintaining the license. If a licensee must make a yearly payment, then the licensee will make a considered decision as to whether or not it is still interested in developing the technology. Some licensors prefer to have a greater time period, eg, two to three years, between payments, with more significant payments due at the end of the period. The risk for the licensor, however, is that if during the no-payment period the licensee has lost interest in the technology, the licensor has then also lost a time opportunity to find another licensee. Once the licensed product is being sold, annual maintenance payments often become minimum annual payments to reflect a minimum level of sales, eg, one-quarter of expected sales, acceptable to the licensor.

Milestone payments are most often found in exclusive licenses and are generally linked to a licensee's meeting developmental milestones, such as a payment when the first prototype is made, when the first clinical trial is completed, when the product is approved for marketing, or even when the patent is issued. The milestone payment often reflects a sharing of the reduced developmental risk or increased value of the technology as the potential product moves through the developmental cycle. As the risk is reduced for the licensee, these milestone payments often increase monetarily. Milestone payments are also a means of enforcing diligence. A typical milestone/diligence clause might read, "Licensee agrees to have a product for sale by January 1, 1995 or pay Licensor $\$x$ in order to maintain the license."

Earned royalties are typically a percentage of the net selling price of the licensed product, although they can be a set amount of money per sale, regardless of the sales price. If a licensee has successfully developed a licensed product, the licensor will realize the greatest return from the license through earned royalties. The earned royalty rate is determined by negotiation between the parties and can range from 0.1 to 20–30% or more, depending on the technology, stage of development, risk involved, value to the licensee or licensor, value added to the technology, etc. The earned royalty rate is also dependent on industry norms: earned royalties of 10–15% are not unusual for software programs, whereas earned royalties of 0.1–0.5% are not unusual in the semiconductor industry. In addition, the earned royalty rate should reflect the other financial considerations of the license agreement, for example, the licensee may be willing to pay significant upfront and milestone payments in exchange for lower earned royalty payments. Earned royalties can also be calculated as a percentage of profits, although this is a less common method of determining royalties.

Exclusivity and Nonexclusivity. Intellectual property can be licensed exclusively or nonexclusively, with many variations in between. Typically, an exclusive license gives the licensee an incentive to develop the intellectual property. Exclusivity is most valuable for technology that requires significant expense and long development time because it encourages the licensee to invest the risk capital to develop and market the invention. An exclusive patent license is only valuable if the patent has meaningful broad claims and is not easy to invent around. An exclusive license to copyrighted work is of limited value, since the copyright protects only the actual "original work fixed in a tangible medium" and others can express the same ideas in different ways.

An exclusive license is generally more valuable, and therefore more expensive, than a nonexclusive license because exclusivity provides an incentive and protection for the licensee's investment in the technology. Licensors are limiting their future choices by granting an exclusive license, and therefore there is an opportunity cost for the licensor. Although most licensees prefer a worldwide life-of-the-patent exclusive license for all fields of use, exclusivity can be limited in time, geography (territory), field of use, application, etc. In granting an exclusive license, the licensor should be reasonably confident that the exclusive licensee can develop a marketable product; otherwise, the licensor may be losing an opportunity to exploit the invention. The licensor can grant the exclusive licensee the right to grant sublicenses to sublicensees, who in turn develop the technology.

From the perspective of the licensee, a nonexclusive license can be viewed as either a tax on the licensee's use of the patent or insurance against a patent infringement lawsuit. The licensor may want to make the technology broadly available to many parties in order to allow many parties to better exploit the technology.

Varying degrees of exclusivity and nonexclusivity can be negotiated, eg, co-exclusive licenses, limited number of total licensees, etc.

Field of use restrictions are common in license agreements. For example, a licensor can limit the licensed field of use for the same compound to research reagents, diagnostic products, or therapeutic products. In this instance, a licensor can grant at least three exclusive field of use licenses for one technology.

Other Terms and Conditions. The license agreement should always have a specified term and include provisions for termination. Other standard provisions can be included to address such issues as reporting, infringement, indemnity/warranties, governing law, assignment, or notices.

Option Agreement and Right of First Refusal

An *option agreement* is a contract between two parties in which one party acquires the right (option) to acquire a license to specified intellectual property, usually within a given time period. Typically, an option agreement provides for an evaluation period and consideration for the option agreement at a lesser price than in a license agreement. Occasionally, the prospective license agreement is negotiated along with the option but can be postponed until the option is exercised.

An agreement which grants one party, the potential licensee, the right of first refusal is often interpreted to mean that the party has the right to match any other offer for the technology. If the party matches the offer, then the licensor must grant the license to the first, matching party. Potential licensees are able to use other parties to help gauge the value of the technology under the right of first refusal. For the licensor, this type of grant makes negotiations with other parties more difficult since the first party can always license the technology.

Types of Intellectual Property Licenses

Patent License. A patent license usually grants to the licensee the right to make, use, and sell products which otherwise would infringe a patent or pending patent application. Because a patent allows the patentee to exclude others from making, using, or selling the patented invention, a licensee is ensured immunity from a lawsuit by the patent holder.

Copyright License. In very simple terms, a copyright owner has the exclusive right to (*1*) reproduce, (*2*) prepare derivative works of, (*3*) distribute copies of, (*4*) perform, and (*5*) display the copyrighted work. A copyright license, therefore, grants to the licensee any or all of the rights associated with a copyright.

Software licenses are particular versions of a copyright license in which the holder of copyrighted software grants rights to another party to use the software. Shrink-wrap licenses are those which, when a package is opened, bind the opener of the package to use the software in the manner specified by the shrink-wrap.

Trademark Licenses. Certain drugs or chemicals may have a trademark associated with the compound and, as such, have a certain value. A trademark licensee has the right to use the trademark for the licensed products. Because a trademark reflects the source and quality of goods associated with the trademark, trademarked products and their licensees need to be monitored to ensure that the trademarked products meet the specifications of the licensor. Trademark licensors typically monitor licensed products by requiring the licensee to send representative samples to the licensor or allow the licensor to inspect the manufacturing facilities of the licensee. Trademark licenses also require that licensed products be properly marked to indicate the existence of the license (see COPYRIGHTS AND TRADEMARKS).

Tangible Property Licenses and Material Transfer Agreements. Licenses which grant the licensee the right to use certain tangible property, such as cell lines and vectors, are very common in the biotechnology/pharmaceutical industry. These licenses are often used for nonpatented biological material and can be royalty-bearing and exclusive, as with other licenses. The termination clause in a tangible property license should include a provision for the return or destruction of the biological material if the license is terminated. In addition, there should be a provision that the licensee not distribute the material to others except with the written consent of the licensor.

Material transfer agreements are generally short letter agreements governing the exchange of biological materials between researchers. Typically, such material transfer agreements restrict the use of the biological material to a certain researcher and ask that the recipient of the material not distribute the material to others without prior consent. Material transfer agreements sometimes provide for the disposition of intellectual property rights which might arise from the use of the biological material.

Valuing Technology

The most commonly asked questions about licenses revolve around pricing and valuing the technology, eg, what is a reasonable royalty? There is no easy answer because many factors must be taken into consideration; some of these are discussed below. In the end, the value of the technology depends on what the licensee is willing to pay and what the licensor is willing to accept for the license. Licenses are usually a negotiated and unique agreement between two parties.

The value of the technology to be licensed depends on its potential technological and market competitive position; its features, advantages, and benefits; and whether the technology is evolutionary or revolutionary. Any evaluation of a technology should include an assessment of other competing products already being sold and who is selling a similar product. Another consideration is the cost or savings for each party. If there are viable alternative technologies or methods for obtaining the same or a similar end result, or if a company can easily invent around the technology, the license is less valuable to the licensee. Included in an evaluation should be market projections and an estimate of the life of the product; the larger the market and the longer the life, the more valuable the technology. A strong proprietary position also adds value to the technology. Lastly, there are often important intangible benefits which accrue to both parties and increase the perceived value of the technology.

Every industry has its own sense of what a reasonable royalty might be. Whereas earned royalties are usually based on net sales, a very rough rule of thumb is that an earned royalty reflects approximately 25% of profits earned on the particular technology. Minimum royalties are often based on one-fourth to one-third of a conservative projection of sales.

In pricing and packaging a license, the licensor must be very conscious of the perspective of the potential licensee. A successful licensor is prepared to make the license terms attractive to the licensee, emphasizing the features, advantages, and benefits of the technology and providing incentives for the licensee to take a license. The license is the basis for a long-term relationship between the licensor and licensee, and therefore both parties must feel that the agreement is fair and mutually beneficial.

BIBLIOGRAPHY

General References

1. E. L. Andrews, "Inventions, Patents, Markets: Orphans No More", *Venture* **10**(10), 42–45 (1988).
2. F. J. Contractor, "Technology Licensing in U.S. Companies: Corporate and Public Policy Implications", *Columbia J. World Bus.* **18**(3), 80–88 (1983).
3. B. Cronin, "Licensing Patents for Maximum Profits", *Int. J. Tech. Mgmt.* **4**(4,5), 411–420 (1989).
4. N. T. Gallini and B. D. Wright, *Rand J. Econ.* **21**(1), 147–160 (1990).
5. R. Goldscheider, "The Art of Licensing Out", *Technology Management: Law, Tactics, Forms,* Clark Boardman Callaghan, New York, 1988, Chapt. 6.
6. M. J. Hyland, "How to License Process Technology", *Chem. Eng. Suppl.* **100**(9) 7–11 Sept. 1993.
7. J. P. Killing, "Manufacturing Under License", *Bus. Q.* **42**(4), 22–29 (1977).
8. J. D. Major, "Some Practical Intellectual Property Aspects of Technology Transfer", *Int. J. Tech. Mgmt.* **3**(1,2), 43–49 (1988).
9. A. F. Millman, "Licensing Technology", *Mgmt. Decis.* **21**(3), 3–16 (1983).
10. R. M. Pandia, "Transfer of Technology: Techniques for Chemical and Pharmaceutical Projects", *Proj. Mgmt. J.* **20**(3), 39–45 (1989).
11. G. Stuart, "Technology Transfer—Patents and Licenses", *Productiv. Technol.* 4, 10–13 (1979).

KATHARINE KU
Stanford University

LIGHT-EMITTING DIODES AND SEMICONDUCTOR LASERS.

See LIGHT GENERATION.

LIGHT GENERATION

LIGHT-EMITTING DIODES

The discovery of the point contact transistor (1) marked the beginning of the modern electronics revolution. Developments in semiconductor transistors and Si-integrated circuits (ICs) (qv) have produced profound advances in electronic signal processing and led to the realization of semiconductor optoelectronic devices capable of interconverting electrical and optical energy. One of the most fundamental optoelectronic devices is the light-emitting diode (LED). The LED consists of a p–n junction which emits light (ultraviolet, visible, or infrared radiation) in response to a forward current passing through the diode. The wavelength (color) of light emitted is characteristic of the materials employed in the active (light-emitting) region of the device. Such materials typically consist of compound semiconductors (qv) from Groups 13 (III) and 15 (V) of the Periodic Table (III–V semiconductors), such as GaAs, GaP, GaAsP, AlGaAs, InGaAsP, AlGaInP, and InGaN. In addition, IV–IV (Groups 14–14) and II–VI (Groups 12–16) semiconductors may also be employed to form LEDs. The exact material choice is dictated by the desired wavelength of emission, performance, and cost of the device (see ELECTRONIC MATERIALS).

Although light emission in semiconductors has been known since the 1900s, the discovery (2) that Zn-diffused junction gallium arsenide [*1303-00-0*], GaAs, LEDs served as efficient emitters of ir radiation resulted in the commercial introduction of ir LEDs in the early 1960s. These LEDs typically exhibited external quantum efficiencies, ie, photons/electrons, of ~1% (into air). As of this writing (~1994), commercial LEDs typically exhibit external quantum efficiencies of 0.1 to 30%. Infrared LEDs are available that are capable of being modulated at speeds up to 1000 megabits per second (Mbits/s), making these devices useful for a variety of applications involving the optical transmission of data. In the visible regime, research (3) led to the commercial introduction by General Electric in 1962 of visible (red) LEDs based on GaAsP. The demonstration of efficient light emission by GaAsP was significant in establishing the feasibility of compound semiconductor alloys as viable candidates for LEDs. In fact, the highest performance visible LEDs as of 1994 were based on similar alloys, eg, AlGaInP and InGaN. Commercial visible LEDs exhibit performances ranging from 1 to 20 lm/W. Devices as high as 40 lm/W have been demonstrated in the research laboratory. Although the typical input power of LEDs is ~ 0.1 W, the increasing luminous efficiency of these devices has resulted in expanding usage from conventional indicators to various lighting applications such as large-area displays and vehicular lighting.

Materials and Device Physics

Semiconductors. The basic material employed in LEDs is the semiconductor, a solid which possesses a conductivity intermediate between that of a conductor and an insulator. Unlike conductors, semiconductors and insulators possess an energy gap, E_g, between two energy bands, the conduction band (CB) and the valence band (VB) (Fig. 1). In a semiconductor, this energy gap is typically on the order of 0.3–6.0 eV; in an insulator the energy gap is significantly higher. The magnitude of the energy gap is determined by the semiconductor material. At temperatures above absolute zero, ie, 0 K, electrons may be thermally excited across the energy gap in a semiconductor, occupying some of the states in the conduction band and leaving others empty in the valence band. The empty states in the valence band are referred to as holes. Holes behave similarly to electrons, but are positively charged and thus have potential energy of opposite sign. Thus holes are also capable of conducting electrical current. The

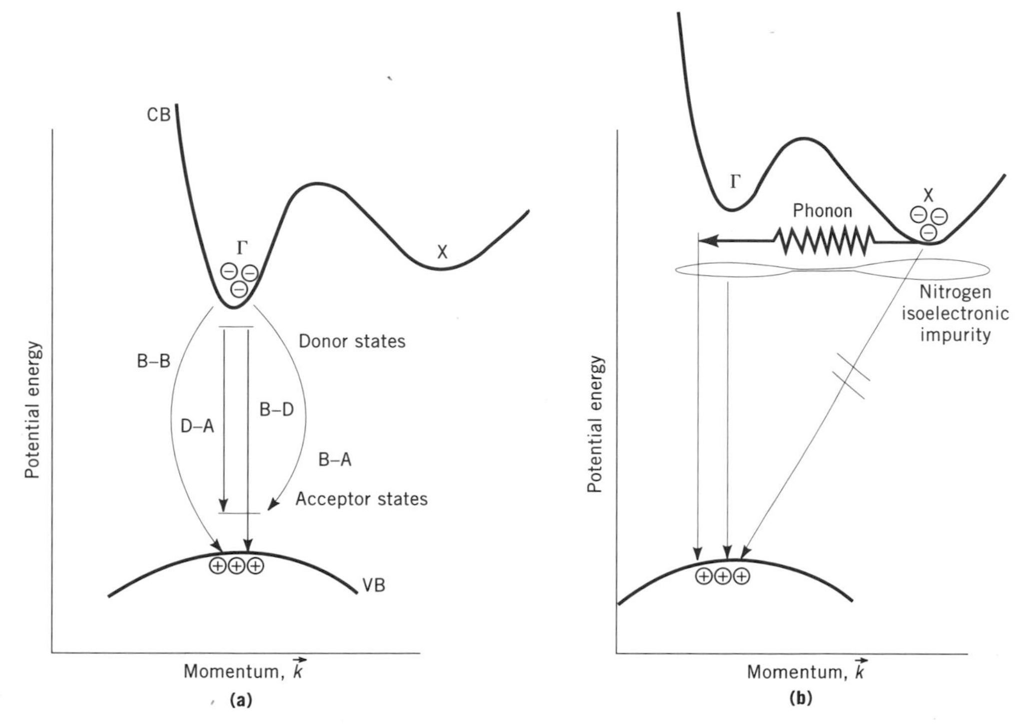

Fig. 1. Schematic diagram of semiconductor materials showing band gaps where CB and VB represent the conduction band and valence band, respectively and ⊕ and ⊖, mobile charge. The height of the curve represents the probability of finding an electron with a given momentum bound to an N-isoelectronic impurity. (**a**) Direct band gap: the conduction band minimum, Γ, is located where the electrons have zero momentum, ie, $\vec{k} = 0$. The couples B–B, D–A, B–D, and B–A represent the various routes for radiative recombination. See text. (**b**) Indirect band gap: the conduction band minimum, X, is located where electrons have nonzero momentum, $\vec{k} \neq 0$.

presence of partially occupied energy bands results in enhanced conductivity and is the distinguishing property between semiconductors and insulators.

Virtually all LEDs are formed from crystalline semiconductors wherein the atoms are arranged in a regular structure similar to that of diamond. This zinc blende crystal structure consists of two interpenetrating face-centered cubic (fcc) lattices, each consisting of different atoms. A few LED materials, eg, SiC and InGaN, are prepared in a different structure, the wurtzite crystal structure. This structure consists of two hexagonal closed-packed lattices composed of different atoms, each lattice interpenetrating the other.

In a semiconductor, the energy gap is generally much larger than thermal energy at room temperature (~25 meV), resulting in limited conductivity. However, the conductivity can be varied over several orders of magnitude by introducing impurities into the crystal structure. These impurities can either contain an excess or deficiency of valence electrons required for bonding to the host crystal. Such impurities are referred to as dopants and are classified as n-type donors (excess electrons) or p-type acceptors (deficient electrons or excess holes). The energy levels of these dopants generally lie within the energy gap of the semiconductor relatively close (<40 meV) to the energy band edges. Consequently, in n-type material sufficient thermal energy exists at room temperature to ionize electrons from the donor states to the conduction band creating mobile electrons. Similarly, in p-type material electrons from the valence band are excited to acceptor states within the gap resulting in the creation of mobile holes in the valence band. These mobile species (electrons in n-type material and holes in p-type material) are referred to as majority carriers.

***p–n* Junction Diode.** A p–n junction diode is formed when two adjacent areas of a single-crystal semiconductor are doped using opposite type impurities. At the junction interface, the carrier concentration gradient causes the majority carriers to diffuse across the junction where an equilibrium is reached between the drift and diffusion of the carriers. This space charge region results in a built-in voltage typically on the order of the band gap energy of the semiconductor. This built-in field opposes the motion of the mobile majority carriers and bends the conduction and valence bands. An applied forward bias, positive voltage on p-type material, diminishes this barrier, allowing electrons to be injected from the n-region and holes from the p-region. A forward bias nearly equivalent to the built-in potential is required to facilitate appreciable injection, and consequently current flow through the diode (Fig. 2**a**).

Radiative Recombination. After the carriers are injected across the junction, the carriers no longer constitute the majority of the mobile charge carriers, and thus are referred to as minority carriers. These minority carriers come into thermal equilibrium near the band edges on the opposing sides of the junction wherein they recombine with a majority carrier. The recombination process can either be radiative, ie, generating light, or nonradiative, ie, generating heat. Radiative transitions typically occur from/to the band edges or nearby impurity states. Some common radiative transitions are depicted in Figure 1**a**, and include recombination from band-to-band (B–B), band-to-acceptor (B–A), band-to-donor (B–D), and donor-to-acceptor (D–A). Generally, one of these recombination mechanisms is preferred relative to the others, resulting in spectral emission over a narrow bandwidth. The wavelength, λ, of the photon emitted in this process is given by equation 1:

$$\Delta E = \frac{hc}{\lambda} \tag{1}$$

where ΔE is the difference in the energy levels involved in the transition, h is Planck's constant, and c is the speed of light. The energy of the transition is most easily varied by varying the semiconductor composition (energy gap). Minority carriers may also combine nonradiatively, generating heat at crystal defects (dislocations, lattice vacancies, interstitials, impurities) or through quantized lattice vibrations (phonons).

The internal quantum efficiency of a LED is governed by the relative radiative and nonradiative recombination rates. The total recombination rate, R, for electrons is the sum of the radiative and nonradiative recombination rates, and is given by equation 2:

$$R = \frac{n}{\tau} = \frac{n}{\tau_r} + \frac{n}{\tau_{nr}} = n\left(\frac{1}{\tau_r} + \frac{1}{\tau_{nr}}\right) \tag{2}$$

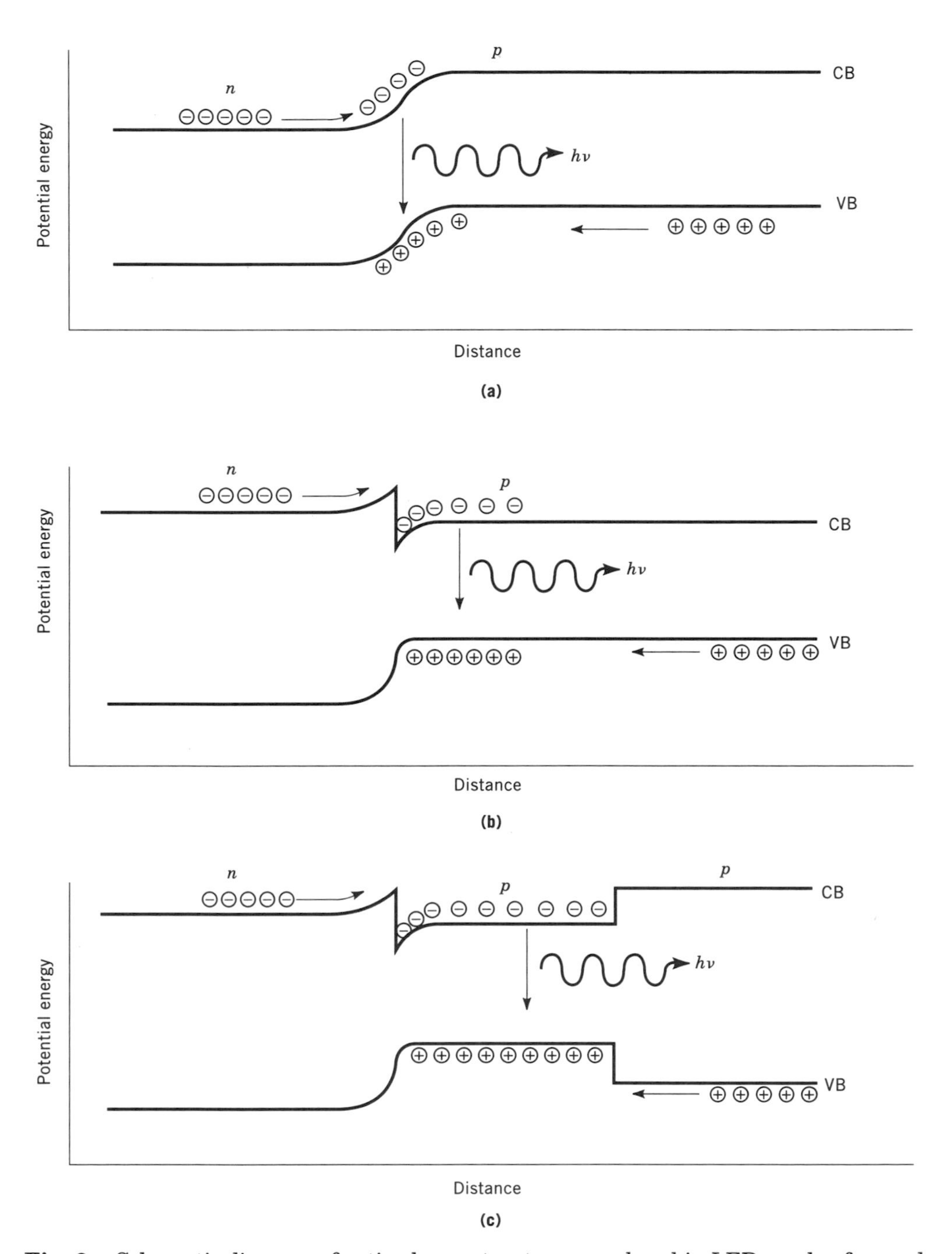

Fig. 2. Schematic diagram of active layer structures employed in LEDs under forward bias showing the conduction band (CB) and valence band (VB). The simplest devices employ (**a**) a homostructure active layer wherein the bandgap is constant throughout the device. More advanced structures consist of (**b**) single and (**c**) double heterostructures. Heterostructures facilitate the confinement and injection of carriers in the active region where the carriers may recombine to emit a photon.

where n is the electron minority carrier density and τ_r and τ_{nr} are the radiative and nonradiative lifetimes, respectively.

In LEDs, electrons are generally injected into a p-type active layer. The internal quantum efficiency, η_{int}, ie, the photons per injected electrons, is given by the radiative recombination rate divided by the total recombination rate:

$$\eta_{\text{int}} = \frac{n/\tau_r}{n/\tau} = 1 + \frac{1}{\tau_r/\tau_{nr}} \tag{3}$$

Thus high internal quantum efficiency requires short radiative and long nonradiative lifetimes. Nonradiative lifetimes are generally a function of the semiconductor material quality and are typically on the order of microseconds to tens of nanoseconds for high quality material. The radiative recombination rate, n/τ_r, is given by equation 4:

$$\frac{n}{\tau_r} = Bnp; \ \tau_r = 1/Bp \tag{4}$$

where B is a recombination rate constant dependent on the material system and $n(p)$ are the electron (hole) concentrations. For moderately doped ($10^{17}-10^{19}/\text{cm}^3$) materials, typical radiative lifetimes are 100 to 1 ns for direct materials and 1 ms to 1 μs for indirect materials. Equation 4 implies that short radiative lifetimes result from high hole concentrations. However, if the hole concentration is made excessively high ($>10^{19}$ cm^{-3}), the material quality is compromised and the nonradiative lifetime decreases, decreasing the internal quantum efficiency. This tradeoff is typically employed in high speed LEDs wherein heavy active layer dopings are utilized to reduce τ_r, increasing their speed. However, the corresponding decrease in τ_{nr} decreases the internal quantum efficiency. This approach results in a performance tradeoff between speed and power.

Direct and Indirect Energy Gap. The radiative recombination rate is dramatically affected by the nature of the energy gap, E_g, of the semiconductor. The energy gap is defined as the difference in energy between the minimum of the conduction band and the maximum of the valence band in momentum, $\vec{k}$, space. For almost all semiconductors, the maximum of the valence band occurs where holes have zero momentum, $\vec{k} = 0$. Direct semiconductors possess a conduction band minimum at the same location, $\vec{k} = 0$, Γ point, where electrons also have zero momentum as shown in Figure **1a**. Thus radiative transitions that occur in direct semiconductors satisfy the law of conservation of momentum.

In an indirect semiconductor (Fig. **1b**), the minimum of the conduction band occurs at a location, X or L point, where electrons have nonzero momentum, $\vec{k} \neq 0$. Consequently, a single recombination event cannot satisfy the law of conservation of momentum, and thus is forbidden. Radiative recombination can occur upon the additional interaction of a phonon (quantized lattice vibration). The emission or absorption of a phonon allows momentum and energy to be conserved when combined with a radiative recombination event. In this situation, the emitted photon possesses an energy equal to that of the energy gap minus that of the phonon energy. This two-step process has a much lower probability of occurring than a single recombination event. As a result, indirect semiconductors possess a much longer radiative lifetime (lower internal quantum efficiency) than direct semiconductors where internal quantum efficiencies may approach 100%.

In many semiconductors employed for LEDs, and especially in mixed alloys, the direct and indirect minima are separated by smaller energies than those of the purely direct and indirect semiconductors. As a result, finite electron concentrations exist within both minima. The total electron concentration, n, is given by equation 5:

$$n = n_I + n_\Gamma \tag{5}$$

where n_I and n_Γ are the electron concentrations in the indirect and direct minima, respectively (see Fig. 1). These concentrations are related to each other by Boltzman statistics:

$$\frac{n_I}{n_\Gamma} = Ne^{(E_\Gamma - E_I)/kT} \tag{6}$$

where N is the relative density of states in the valence and conduction bands, E_Γ and E_I are the respective energies of the direct and indirect minima, k is the Boltzmann constant, and T is temperature in Kelvin. Combining equations 5 and 6 with a modified version of equation 3, to account for radiative recombination in the direct minima and nonradiative recombination in the direct and indirect minima, gives an expression for the internal quantum efficiency:

$$\eta_{\text{int}} = \frac{1}{1 + \left(\frac{\tau_r}{\tau_{nr}}\right)\left(1 + Ne^{(E_\Gamma - E_I)/kT}\right)} \tag{7}$$

As a result, the internal quantum efficiency decreases exponentially as the separation between the direct and indirect minima decreases. This property strongly influences the useful compositional ranges of mixed alloys of direct and indirect materials.

Isoelectronic Centers. The internal quantum efficiency of some indirect materials may be drastically improved by the addition of isoelectronic impurities to the semiconductor lattice which serve as radiative centers, ie, N or (Zn, O) in GaP. These impurities are located substitutionally within the lattice and contain the precise number of valence electrons required for bonding to the host crystal. The presence of these impurities causes a high degree of local strain within the crystal which serves to tightly bind an electron in the vicinity of the isoelectronic impurity. The localization of the electron in real space results in a wide distribution in momentum space as a result of the Heisenberg uncertainty principle. Consequently, these trapped electrons have significant probability of possessing zero momentum and can thus recombine without the assistance of a phonon (Fig. **1b**). As a result, the radiative lifetime is reduced and the internal quantum efficiency is substantially improved. Although there is significant probability that the trapped electrons have zero momentum, there is also similar probability that the electrons have nonzero momentum. These electrons cannot recombine via a sole event. As a result, indirect semiconductors having isoelectronic traps are not as efficient as direct semiconductors.

Active Layer Structures. The structure of the LED active (light-emitting) layer may also strongly influence the efficiency of the device. The simplest LED structure is a homojunction wherein the epitaxial semiconductor layer consists of a single semiconductor material or alloy. Such a structure under forward bias is depicted in Figure 2**a**. The p–n junction is typically formed by introducing impurities during the epitaxial growth process or alternatively by a post-growth diffusion. Generally, the junction should be of sufficient depth to minimize the diffusion of carriers to the semiconductor surface where they combine nonradiatively, and to allow current spreading from the top contact. However, in order to minimize self-absorption, the junction should be placed as close to the surface as possible. In direct gap semiconductors, the large (10^4 cm^{-1}) absorption coefficient results in an optimal junction depth of <3 μm. In indirect semiconductors, the absorption coefficient is typically an order of magnitude smaller (10^3 cm^{-1}). In addition, indirect semiconductors typically employ isoelectronic impurities which emit radiation below the band gap of the overlying semiconductor. As a result, optimal junction depth is typically deeper for LEDs based on indirect materials, generally >10 μm.

Although homojunctions are relatively simple and cost-effective to produce, they possess the disadvantage of allowing the injected carriers to diffuse away from the junction where nonradiative recombination can occur. In addition, efficient carrier injection in these structures requires heavy doping which can compromise the material quality and decrease the nonradiative lifetimes of the device. A solution to these problems is the use of heterojunctions to confine and inject carriers in the active (light-emitting) region (4,5). In a heterojunction, two different semiconductor materials of different band gaps are placed adjacent to each other. Offsets in the valence and conduction bands at the heterojunction serve to confine carriers and make it energetically favorable to inject carriers to the lower band gap material. A standard LED single-heterojunction (SH) structure consists of a n-type wide gap injection region and a low gap p-type active layer. A band diagram of such a device under forward bias is depicted in Figure 2**b**. The wide gap n-type layer serves to inject electrons into the p-type active layer. Also, the band offset in the valence band serves to confine holes from diffusing away from the junction before they can recombine. An additional advantage facilitated by the SH is the implementation of a nonabsorbing wide gap window layer above the active region. This window layer is important for light extraction.

The highest efficiency LED structure employs a double-heterostructure (DH) active layer wherein a low gap active layer is surrounded by wide gap confining layers on each side. Typically, the thickness of the active layer ranges from 0.3–3 μm. A schematic structure of a standard DH active layer under forward bias is shown in Figure 2**c**. Accordingly, both electrons and holes are injected from the wide gap n-type and p-type layers, respectively. In addition, both electrons and holes are confined within the active region, preventing them from diffusing away from the junction and recombining nonradiatively. This confinement yields a higher carrier density of electrons and holes in the active layer and fast radiative recombination. Thus LEDs used in switching applications tend to possess thin DH active layers. The increased carrier density also may result in more efficient recombination because many nonradiative processes tend to saturate. The increased carrier confinement and injection efficiency facilitated by heterojunctions yields increasing internal quantum efficiencies for SH and DH active layers. Similar to a SH, the DH also facilitates the employment of a window layer to minimize absorption. In a structure grown on an absorbing substrate, the lower transparent window layer may be made thick (>100 μm), and the absorbing substrate subsequently removed to yield a transparent substrate device.

A further type of heterostructure active layer which has been employed in LEDs is a quantum well structure wherein an extremely thin (<0.02 μm) active layer or series of layers is surrounded by wide gap confining layers. This structure is potentially advantageous in facilitating even higher carrier densities than double heterostructures, and thus even more efficient recombination. In addition, the thin active layer serves to further minimize self-absorption. The quantization of energy states in the thin layer also reduces the spectral width of emitted light which can be advantageous for certain applications. However, the potential advantages of these structures have not generally produced significantly better results to justify the increased cost of controlling growth of a very thin active layer. An exception is the emerging resonant-cavity LED (RCLED) structure.

The design of heterostructures having abrupt interfaces generally necessitates that the active layer and surrounding confining layers be lattice matched, ie, possess essentially the same spacing between atoms. Lattice matching serves to minimize electrically and optically active defects and dislocations near the active layer which can severely impede device efficiency. Lattice matching of the epitaxial layers and substrates is feasible for a number of mixed alloys, eg, AlGaAs/GaAs, AlGaInP/GaAs, and InGaAsP/InP, making the use of heterostructures advantageous in these systems. In lattice mismatched alloys, abrupt changes in composition result in the generation of defects at the interface, making the use of heterostructures undesirable in such situations. Extremely thin (<0.2 μm) layers are an exception to this rule. High levels of strain can be tolerated in such pseudomorphic layers. However, similar to quantum wells, these devices are not widely used as a result of limited improvements in performance compared to the significantly increased complexity and cost of manufacturing these structures. Although not as critical as the lattice constant, the thermal expansion coefficients of the materials used in epitaxial growth should also be relatively well matched in order to minimize strain experienced during the high temperature epitaxial growth cycles.

Light Extraction. Light-emitting diode performance is limited by the ability to extract the photons generated within the high refractive index semiconductor chip (typically $n_1 \sim 3.5$) into the outside world ($n_2 = 1$ for free space or $n_2 \sim 1.6$ for epoxy typically utilized in encapsulation). The large difference in refractive indexes results in a critical angle, θ_c, for internal reflection given by Snell's law:

$$\theta_c = \sin^{-1}(n_2/n_1) \tag{8}$$

of 17° and 27° for LED chips in air and epoxy, respectively. The relatively small critical angle for internal reflection combined with internal absorption within the LED chip results in the external quantum efficiency being substantially lower than the internal one. The ratio of the external to internal quantum efficiency is defined as the extraction efficiency, C_{ex}. The extraction efficiency of a single surface (escape cone) is given by equation 9:

$$C_{ex} \cong \frac{1 - \cos\theta_c}{2} \times \frac{(n_2/n_1 - 1)^2}{(n_2/n_1 + 1)^2} \tag{9}$$

where the second term corrects for Fresnel reflection losses (assuming normal incidence). This equation gives a theoretical extraction efficiency of 0.04 per surface (escape cone) for emission into epoxy.

The extraction efficiency of a LED chip is most strongly influenced by the structure of the chip. A variety of LED chip structures and the corresponding escape cones of light defined by the critical angle are depicted schematically in Figure 3. For an absorbing substrate (AS) LED chip, the number of escape cones is strongly affected by the thickness of the transparent window layers. An AS LED having thin (<10 μm) transparent window layers possesses only a single-top escape cone (Fig. 3**a**). However, if the thickness of the window layers is increased to >40 μm, the number of cones increases to three as a result of contributions from the sides of the chip (Fig. 3**b**). A distributed Bragg

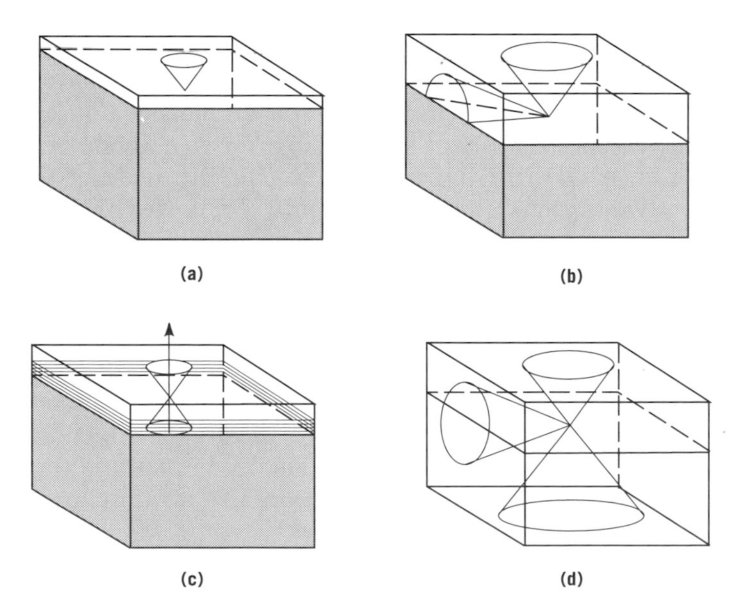

Fig. 3. Depiction of the light extraction, ie, escape cones of light emission, for various LED chip structures consisting of absorbing substrate devices having (**a**) thin window layers (top cone); (**b**) thick window layers (top cone and four one-half side cones); (**c**) thin window plus the implementation of a distributed Bragg reflector between the active layer and the substrate (top and bottom cone). Also shown is (**d**), the optimal structure for light extraction, a transparent substrate device (top and bottom cones as well as four side cones).

reflector (DBR), a multilayer stack of high and low index semiconductor layers, may be employed between the light-emitting active region and the substrate to reflect incident light, thus minimizing the effect of absorption from the substrate. However, DBRs only reflect light of near-normal incidence, ie, only light differing by $<25°$ from normal is typically reflected. In a LED, light is emitted isotropically from the active region. Consequently, light of all angles is incident on the mirror, and only a portion of it is reflected. The remainder passes into the absorbing substrate. Thus only a portion of the bottom escape cone is captured. The LED structures having a DBR employed as of this writing possess thin transparent windows, resulting in a maximum of two escape cones (Fig. 3**c**).

The ultimate chip structure for light extraction is a transparent substrate (TS) LED wherein six escape cones are present (Fig. 3**d**). In a conventional packaged LED, the bottom and side cones are captured by virtue of the fact that the chips are typically mounted using reflective Ag-loaded epoxy in a reflective mold cup (Fig. 4), reflecting and collecting light from these surfaces. Similar arguments are not necessarily applicable to LED chips that are packaged differently. For example, surface emission may be more important in some fiber-coupled devices (Fig. 5**a**) making the extraction of light from the edges of the chip unimportant.

The effects of randomization of the light (photon) path or absorption by layers other than the substrate must also be considered. In chips having thick transparent window regions (especially TS LEDs), photons may make multiple passes to the semiconductor surface without being absorbed. Randomization of the direction of internally reflected light may occur as a result of scattering at the chip surface or within the chip, allowing more light to escape than that predicted by the single-pass model of Figure 3. These effects can be significant. For example, the calculated extraction efficiency based on equation 9 for a TS LED (six escape cones) is 0.24, whereas experimentally estimated ratios as high as 0.3 have been measured for TS AlGaAs LEDs (6).

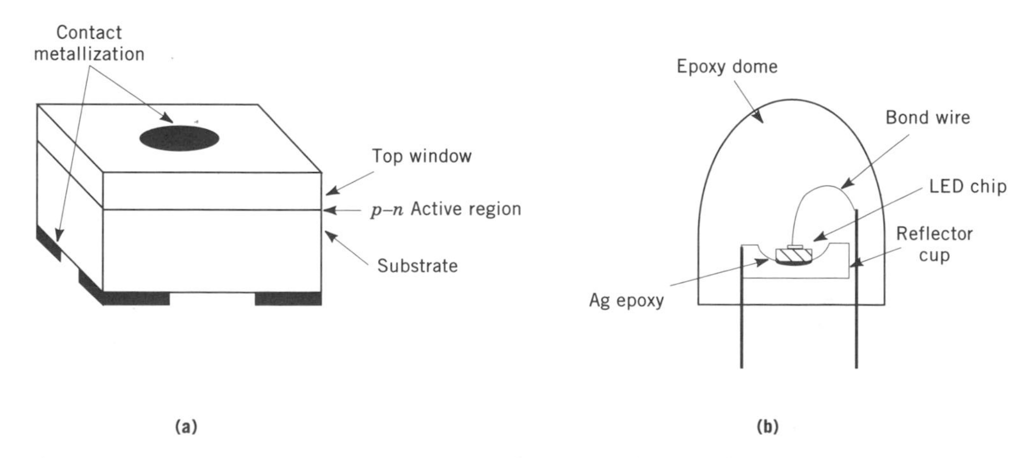

Fig. 4. Schematic diagrams of conventional LED (**a**) chips and (**b**) lamp packages where after fabrication, the p–n junction LED chip is mounted as shown, and encapsulated in epoxy to improve the light extraction of the device.

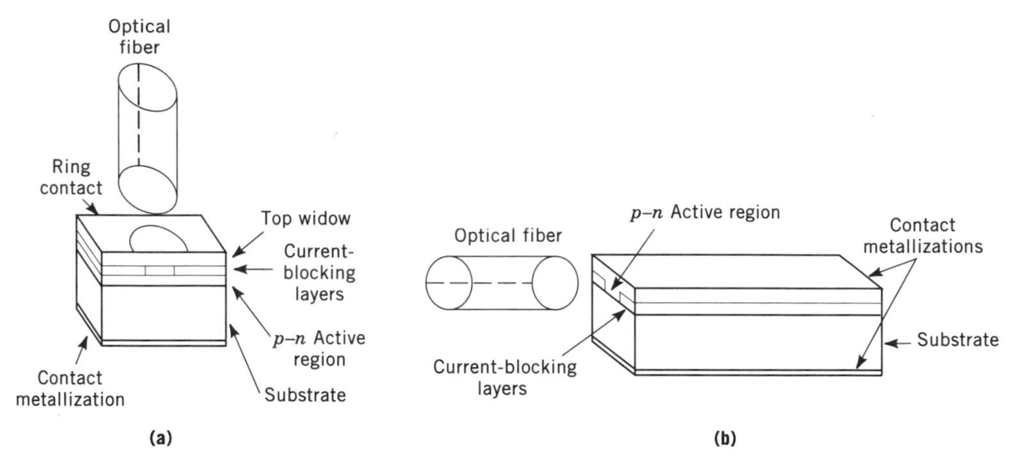

Fig. 5. Schematic diagrams of small emission area LED chip structures. (**a**) Cross section of a surface-emitting LED which typically possesses a 20–60 μm diameter emission area for coupling to optical fibers. (**b**) An edge-emitting device which generally possesses a smaller (3–5 μm) emission width and is useful for coupling to single-mode fibers.

Device Design and Fabrication

Although the LED is one of the most basic optoelectronic devices, there exists a variety of complex and interacting material and structural considerations in designing these devices. These include the choice of materials for emission wavelength of the LED as well as the geometry and fabrication methods of the device. The principal structural properties of commercially available LEDs are summarized in Table 1.

Growth Techniques. The epitaxial layers employed in LEDs are grown on semiconductor substrates that are chosen based on suitable lattice constant, defect density (material quality), optical transparency, and cost. Most commercial substrates employed in LEDs consist of GaAs; gallium phosphide [*12063-98-8*], GaP; or indium phosphide [*22398-80-7*], InP. These materials are formed by methods of liquid-encapsulated Czochralski, a crystal growth method similar to that employed in the growth of bulk Si crystals (see SILICON AND SILICON ALLOYS). Silicon carbide [*409-21-2*], SiC, and alumina [*1344-28-1*], Al_2O_3, substrates are being utilized to produce the commercially available blue emitters. These LEDs are based on SiC and indium gallium nitride, InGaN, respectively. Most other substrates used in LED fabrication yield inferior cost/performance in comparison.

The semiconductor epitaxial layers are grown on the bulk substrates by a variety of growth techniques at temperatures typically ranging from 500–1600°C, depending on the epitaxial materials and growth technique (see THIN FILMS, DEPOSITION TECHNIQUES). The highest volume growth techniques are vapor-phase epitaxy (VPE) and liquid-phase epitaxy (LPE), suitable for simultaneous growth on a large number of wafers. The LPE process is used to grow GaP; GaAs; aluminum gallium arsenide, AlGaAs; and indium gallium arsenic phosphide, InGaAsP, by passing a molten solution of Ga saturated with other

Table 1. Commercial LED Structures

Emission λ, nm	Active layer material[a]	Structure[b]	Window layer material	Substrate[c]	Lattice matched	Growth technique[d]	Other
450–510	$In_xGa_{1-x}N$ ($x \lesssim 0.2$)	DH	GaN	Al_2O_3 (TS)	no	MOCVD	
480	SiC, 6 H[e]	homo	SiC	SiC, 6 H (TS)	yes	MOCVD	
555	GaP[e]	homo	GaP	GaP (TS)	yes	LPE	
565	GaP:N[e]	homo	GaP	GaP (TS)	yes[f]	LPE	N-isoelectronic trap
570–630	$GaAs_xP_{1-x}$:N[e] ($x \lesssim 0.35$)	homo	GaAsP	GaP (TS)	no	VPE	diffused junction N-isoelectronic trap
560–640	$(Al_xGa_{1-x})_{0.5}In_{0.5}P$ ($x \lesssim 0.6$)	DH	AlGaAs or GaP	GaAs (AS)	yes	MOCVD[g]	
560–640	$(Al_xGa_{1-x})_{0.5}In_{0.5}P$ ($x \lesssim 0.6$)	DH	GaP	GaP (TS)	yes	MOCVD[g]	wafer-bonded GaP substrate (GaAs removed)
650	$GaAs_{0.6}P_{0.4}$	homo	GaAsP	GaAs (AS)	no	VPE	diffused junction
650–880	$Al_xGa_{1-x}As$ ($x \lesssim 0.45$)	SH	AlGaAs	GaAs (AS)	yes	LPE	
650–800	$Al_xGa_{1-x}As$ ($x \lesssim 0.45$)	DH	AlGaAs	GaAs (AS)	yes	LPE	
650–880	$Al_xGa_{1-x}As$ ($x \lesssim 0.45$)	DH	AlGaAs	AlGaAs (TS)	yes	LPE	thick epitaxial-grown substrate (GaAs removed)

Table 1. ***(Continued)***

Emission λ, nm	Active layer material[a]	Structure[b]	Window layer material	Substrate[c]	Lattice matched	Growth technique[d]	Other
700	GaP:(Zn,O)[e]	DH	GaP	GaP (TS)	no	LPE	(Zn,O) isoelectronic trap
880	AlGaAs:Si	homo	AlGaAs	AlGaAs (TS)	yes	LPE	thick epitaxial-grown substrate (GaAs removed)
820–880	$Al_xGa_{1-x}As$ ($x \lesssim 0.1$)	DH	AlGaAs	GaAs (AS)[h]	yes	LPE or MOCVD	small emission area geometry
900	GaAs:Zn	homo	GaAs	GaAs (AS)	yes	horizontal bridgeman	diffused junction
940	GaAs:Si	homo	GaAs	GaAs (TS)	yes	LPE	
1300–1550	InGaAsP	DH	InP	InP (TS)	yes	LPE or MOCVD	small emission area geometry

[a] All bandgaps are direct unless otherwise noted.
[b] DH = double heterostructure, homo = homostructure, and SH = single heterostructure.
[c] AS = absorbing substrate, TS = transparent substrate.
[d] LPE = liquid-phase epitaxy, MOCVD = metalorganic chemical vapor deposition, VPE = vapor-phase epitaxy.
[e] Band gap is indirect.
[f] Some small lattice mismatch occurs as a result of high nitrogen doping level.
[g] GaP window grown by VPE.
[h] GaAs substrate may be locally removed to form small transparent emission region (21).

source materials in direct contact with the growth substrates (7–9). Epitaxial films are grown as the melt is cooled. Additional layers can be grown by sliding the substrates across melts of various compositions.

Emitters based on GaAsP are typically grown by VPE (10). The VPE process consists of placing substrates in the cold zone of a heated reactor chamber, transporting the growth species to the substrates in the gaseous (vapor) phase, and depositing the epitaxial materials by chemical vapor deposition (CVD). The source materials for this technique generally consist of phosphine [*7803-51-2*], arsine [*7784-42-1*], and gallium chlorides formed by passing HCl over pure gallium [*7440-55-3*]. The VPE process is a hot wall growth process and thus is not suitable for using Al-bearing gases which tend to attack the quartz walls of the reactor vessel.

The Al-bearing compounds may be successfully grown by a modified form of vapor-phase epitaxy known as metalorganic chemical vapor deposition (MOCVD) (11,12). The MOCVD process is similar to VPE with the exceptions that metalorganic precursors are employed instead of metallic sources and the growth is pyrolitic facilitating cold wall reactor chambers. This technique has been employed to grow low volume communication emitters based on GaAs, AlGaAs, and InGaAsP, and more recently, high volume emitters based on aluminum gallium indium phosphide, AlGaInP (13,14). Generally, the growth rates used in MOCVD are substantially smaller than those of LPE or VPE technologies. It is thus difficult to grow thick layers for current spreading and light extraction using MOCVD.

Molecular beam epitaxy (MBE) is a radically different growth process which utilizes a very high vacuum growth chamber and sources which are evaporated from controlled ovens (15,16). This technique is well suited to growing thin multilayer structures as a result of very low growth rates and the ability to abruptly switch source materials in the reactor chamber. The former has impeded the use of MBE for the growth of high volume LEDs.

The materials used in the epitaxial layers for LEDs depend on the desired wavelength of emission, performance, and cost. The emission wavelength is selected by choosing the appropriate semiconductor compound or alloy with the corresponding energy gap. In addition, the highest performance devices are obtained by utilizing direct gap semiconductors that are lattice matched. A plot of common III–V semiconductor compounds and their alloys as a function of lattice parameter is shown in Figure 6**a**. A similar plot for the III–V nitride compounds, employed to fabricate bright blue and blue-green LEDs, is shown in Figure 6**b**. The nitride alloys exhibit band gap energies that correspond to emission in the ultraviolet to red portions of the spectrum.

During epitaxial growth, the semiconductor layers must be doped to form the p–n junction and conductive current spreading window layers. For III–V materials, zinc, Zn; beryllium, Be; carbon, C; magnesium, Mg; and silicon, Si are commonly employed as p-type dopants, whereas tellurium, Te; selenium, Se; sulfur, S; Si; and germanium, Ge, are employed as n-type dopants. Silicon is an amphoteric dopant, ie, Si can be utilized as either a p-type or n-type dopant depending on how it is incorporated into the semiconductor. Different dopants are employed for IV–IV and II–VI materials because the number of bonding valence electrons differs in these materials. For some materials, eg, GaAsP, the

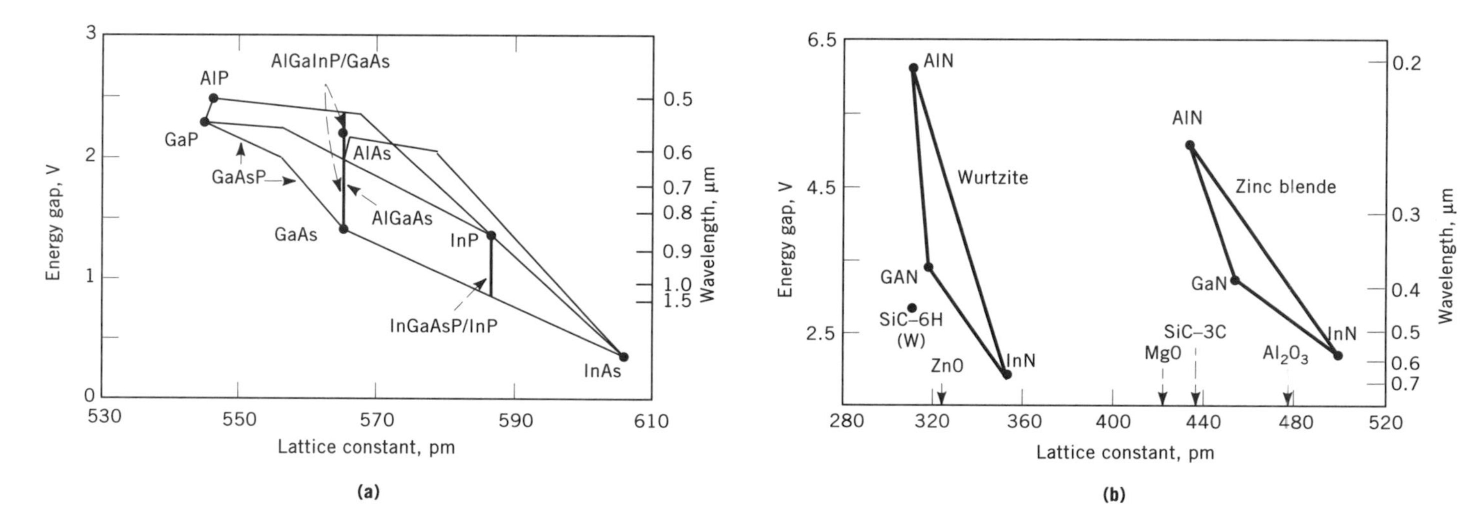

Fig. 6. Plot of band gap energy vs lattice parameter for (**a**) common III–V materials employed for LEDs where (—) corresponds to direct and (—) to indirect band gaps. Both $Al_xGa_{1-x}As$ and $(Al_xGa_{1-x})_{0.5}In_{0.5}P$ are lattice matched to GaAs, whereas $In_xGa_{1-x}As_yP_{1-y}$ can be matched to InP. (**b**) III–V nitride compounds suitable for fabricating blue/uv emitters. Also shown is the lattice parameter of various materials proposed as substrates.

p–n junction is formed after the epitaxial growth via the selective diffusion of Zn into the crystal. The surface of the wafer may be masked so that the junction (light-emitting region) is formed only in select regions.

The optimal structure for light extraction in a LED is a transparent substrate chip, usually obtained by direct growth on a substrate which is optically transparent to the light emitted by the active region of the material. However, lattice matched growth on a transparent substrate is not possible for some compounds, eg, AlGaAs or AlGaInP. In the case of AlGaAs, TS structures are formed by growing a thick ($>100\ \mu m$) transparent epitaxial layer between the active layer and the lattice matched GaAs absorbing growth substrate. After growth, the GaAs substrate is selectively removed by chemical etching, leaving a TS structure (17). For AlGaInP grown on GaAs, a newer approach has been developed wherein the GaAs substrate is selectively removed, and a GaP substrate is substituted in its place by compound semiconductor wafer bonding under applied uniaxial pressure and elevated temperature (18). The GaP substrate is applied within a few micrometers of the active layer and facilitates current spreading and light extraction beneath the device.

Wafer Fabrication and Assembly. After epitaxial growth and other processing, the LED wafers are metallized to form ohmic contacts to the device. Typically, alloyed gold–beryllium, AuBe, and gold–zinc, AuZn, are employed as p-type contacts whereas gold–germanium, AuGe, is used for n-type contacts. Additional metallization layers consisting of Au or Al are often employed to facilitate wirebonding to the device. The use of such multilayer metallizations frequently requires the use of a barrier metallization, usually tungsten, W, or related materials. The alloyed ohmic contacts are patterned in the transparent light-emitting regions to maximize light extraction because the contact area is generally absorbing as a result of the alloy process.

The wafers are then sawn (or cleaved) into individual dice ($\sim 250 \times 250\ \mu m^2$) and packaged (see PACKAGING; SEMICONDUCTORS AND ELECTRONIC MATERIALS). The conventional LED chip (Fig. 4**a**) is typically packaged in lamp form (Fig. 4**b**) wherein the chip is die-attached using reflective Ag-loaded epoxy into a leadframe with a reflector cup. A wire bond is made from the leadframe to the die and the entire package is encapsulated in low index ($n \sim 1.6$) epoxy to improve the light extraction from the device and create a robust package. The epoxy may be transparent to produce a tightly controlled radiation patterned or diffused, by the addition of glass particles, to spread the light into a wide field of view. Often the epoxy is tinted to match the emission color of the LED. Other types of packaging which are commonly employed include packages designed for surface mounting the LEDs directly into a printed circuit board and multichip display modules wherein LED chips are utilized as individual pixels in the display.

The LED chip geometry employed for high speed communication emitters (InGaAsP or AlGaAs) is substantially different. These devices are typically operated at high current densities to improve speed, and are fiber coupled. Both of these properties require that the light-emission area be made small (19–21). This is typically accomplished by additional epitaxial growth and/or wafer fabrication processing to restrict current flow within the device. These methods often include either dopant diffusion or the regrowth of reverse-biased junctions to define areas of current flow. Other methods, eg, mesa etching, proton implantation, or dielectric masking, may also be employed. Two conventional small area emission geometries are depicted in Figure 5. The emission area of the surface-emitting LED (Fig. **5a**) is typically 20–60 μm in diameter. These devices are often employed for coupling to multimode optical fibers. In some instances, the optical fiber is coupled to the substrate side of the device. This is possible if the substrate is transparent or locally removed to create an optical access port (22).

For coupling to single-mode optical fibers, edge-emitting geometries are often employed (Fig. **5b**). In this geometry, current is restricted to a single 3–5 μm stripe and the ends of the device are cleaved for coupling to the ca 7-μm single-mode fiber core. Double heterostructure active regions are generally used in these structures to increase the speed of the device and provide optical confinement of the emitted optical field in the plane of the p–n junction. Typically, current is not injected into the entire stripe region to prevent the devices from lasing (see LIGHT GENERATION, SEMICONDUCTOR LASERS). Because both the surface emitters and edge emitters are operated at high current densities, the p–n junction is usually mounted in close proximity to a conductive heat sink. The optical fiber is then coupled to the LED by standard techniques.

Device Performance

The emission wavelength of a LED is determined primarily by the energy gap of the semiconductor material employed in the active layer. In addition, radiative recombination from impurities, ie, donors, acceptors, or isoelectronic traps, increases the peak wavelength of LED emission because electronic states of impurities generally lie within the energy gap of the semiconductor. Radiative transitions in LEDs occur over a range of energies, and are governed by the properties of the semiconductor material, carrier density of the recombining species, and the density of states of the associated energy levels. As a result, the spectral emission of LEDs possesses a finite width. The full-width at half-maximum (FWHM) of the spectral intensity profile generally ranges from 15–100 nm. Typical emission spectra for various visible LEDs are shown in Figure 7. Ordinarily, LEDs which depend on band-to-band transitions have narrower spectral widths than those where emission depends on deep level impurities. The effect of the N-isoelectronic trap on GaP emission, GaP:N, is clearly evident. The addition of N shifts the peak to longer wavelengths and broadens the spectral emission. The curves for the AlGaInP LEDs represent devices of three different alloy compositions, all exhibiting recombination for the conduction band direct minimum. The emission spectrum of the blue InGaN LED exhibits uniquely broad emission, most likely as a result of recombination via deep Zn impurities levels (23).

Performance is described in a variety of ways depending on the desired application of the LED. The power efficiency, P_E, of a LED is the ratio of the output power to the power input to the device, and can be expressed as

$$P_E = C_{\text{ex}}\,\eta_{\text{int}}\left(\frac{hc}{\lambda}\right)\left(\frac{1}{V}\right) \tag{10}$$

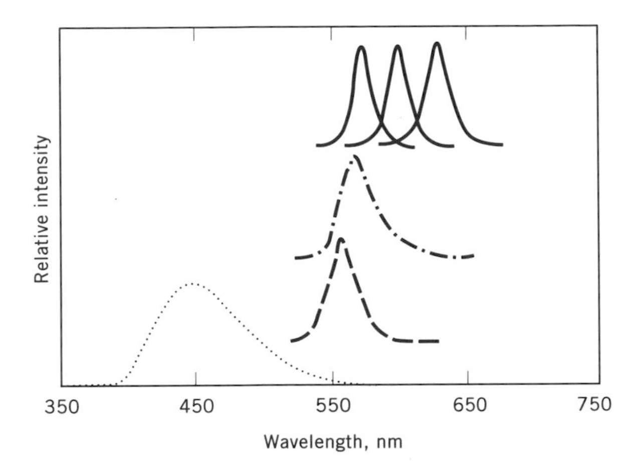

Fig. 7. Light-emission spectra of various LEDs: emission from (—) AlGaInP LEDs results from direct gap recombination, whereas that of (---) GaP is from indirect mechanisms; (– · –) represents GaP with an N-isoelectronic trap, GaP:N, and (····) represents InGaN:Zn where the broadening in the spectrum results from recombination originating from deep levels within the energy gap.

where C_{ex} is the extraction efficiency, η_{int} is the internal quantum efficiency, and V is the applied voltage to the device. The applied voltage is given by the sum of the junction voltage, approximately equivalent to the built-in potential, and the voltage drop given by the dynamic resistance of the diode. Typically, this resistance is dynamic and on the order of 1–3 Ω for communication LEDs and may be as high as 10 Ω for visible/display LEDs.

For LEDs utilized in visible/display applications, the human eye serves as the detector of radiation. Thus a key measure of performance is luminous efficiency which is weighted to the eye sensitivity (CIE) curve. The relative eye sensitivity, $V(\lambda)$, peaks in the green at $\lambda \approx 555$ nm where it possesses a value of 1.0. It drops sharply as the wavelength is shifted to the red or blue, reaching a value of 0.5 at 510 and 610 nm. The luminous efficiency, in units of lm/W, of an LED is given by equation 11:

$$\text{luminous efficiency} = 680\ V(\lambda)\ P_E \tag{11}$$

where 680 is a factor present as a result of the definition of a lumen (1 W = 680 lm at 555 nm). In some cases, the luminous efficiency is expressed in lumen/amp wherein the value in equation 11 is multiplied by the operating voltage.

Typical light output versus current (L–I) and efficiency curves for double heterostructure TS AlGaAs LEDs lamps are shown in Figure 8. The ir LED (Fig. 8**a**) is typically used for wireless communications applications. As a result, the light output is measured in radiometric units (mW) and the efficiencies of interest are the external quantum efficiency ($\eta_{ex} = C_{ex}\,\eta_{int}$, photons out/electrons in) and power efficiency. As a result of the direct band gap and high quality of the LED material, the L–I curve is essentially linear over the operating range depicted. In addition, although the lamp package possesses a thermal resistance

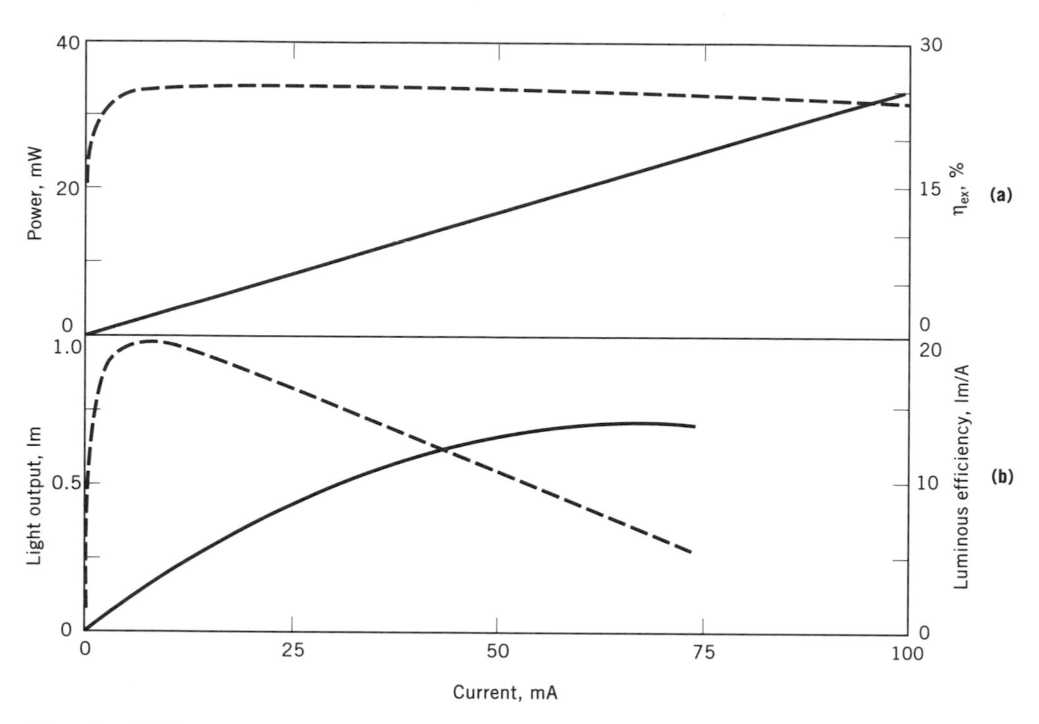

Fig. 8. Light output (—) vs current (L–I) and efficiency curves (---) of double heterostructure TS AlGaAs LED lamps where the composition of the active layer is adjusted for emission at (**a**) 875 nm and (**b**) 650 nm. See text.

of 220°C/W, little heating is observed in the L–I curve as a result of the low series resistance of the device.

The red TS AlGaAs LED (Fig. 8**b**) is typically used for high brightness display/lighting applications. The light output is thus measured in lumen and the efficiency in lm/A. The energy gap of this device is direct. However, it is very close to the direct–indirect transition for this materials system. As a result, nonradiative centers are present in the material as evidenced by the superlinear behavior of the L–I characteristic at low currents. Such behavior is characteristic of the saturation of nonradiative processes (traps). The L–I curve is linear at moderate currents, but begins to exhibit sublinear behavior as a result of heating (high dynamic resistance) in the device at higher currents.

One of the main advantages of LEDs over other lighting sources is reliability. Radiative recombination is a natural process which does not necessarily damage the crystal. Consequently, LEDs typically exhibit degradation rates in the range of 5–20% during the first 1000 hours of operation and do not reach half-brightness for more than 100,000 hours. In most cases, LEDs are used in applications wherein the output is not sensitive to changes in brightness by a factor of two. The degradation that occurs in LEDs often occurs from the relatively high energy radiation (1–2 eV) inducing or causing the migration of defects within the crystal. Such degradation is accelerated at high current densities (23). As a result, most conventional style LEDs (Fig. 4) are operated at current densities <50 A/cm^2; however, small emission area LEDs (Fig. 5) may be operated at current densities as high as 10 kA/cm^2. In some materials, eg, AlGaAs, the operation of LEDs at high current densities (>300 A/cm^2) may result in the formation of networks of dislocations, called dark line defects, around existing defects in the crystal. These defects act as nonradiative recombination sites, resulting in degradation of the internal quantum efficiency of the device (19). Migration of metal from the ohmic contacts to the active region of the LED can create nonradiative centers or electrical leakage, both of which reduce the internal efficiency of the device. The LED package may also induce degradation. At low temperatures, the encapsulation epoxy contracts and can exert extreme compressive stress on the chip, inducing defects and degradation therein. Also, in outdoor applications, the encapsulation epoxy may discolor as a result of exposure to ultraviolet radiation. This discoloration causes a decrease in the light output of the packaged device, yielding yet another form of degradation. Despite these problems, LEDs are some of the most reliable light sources available. Their robustness is a primary reason these devices are employed in many applications.

Visible/Display Emitters

The luminous efficiency of some of the more important commercial technologies of visible LEDs is shown in Figure 9 (see also Table 1). The material employed for an LED has the greatest effect on the cost and performance of the device. Light-emitting diodes based on GaAsP, the first visible LEDs introduced (2), are among the highest volume, lowest cost devices available. Red devices grown on GaAs (10)

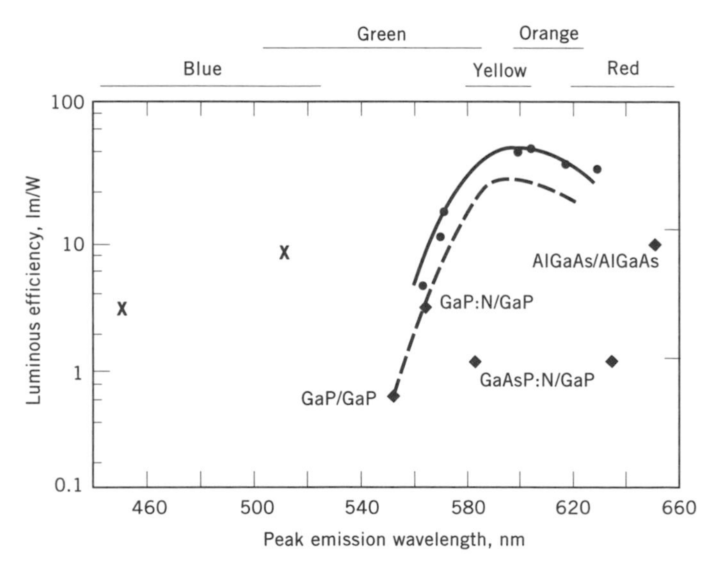

Fig. 9. Luminous efficiency vs peak emission wavelength for (◆) conventional commercial LED technologies. Also shown are data for the emerging technologies (---) AS AlGaInP/GaAs devices, (●) TS AlInGaP/GaP fitted to (——), a theoretical curve, and (**X**) $InGaN/Al_2O_3$. Values shown for established technologies are commercial averages, whereas those of the AlGaInP and InGaN devices are the best reported results as of this writing (18,24,29).

achieve performance levels of 0.15 lm/W. The junction is formed in these devices by Zn-diffusion, making these devices also useful for low performance small and multiple emission area applications. The low efficiency of these devices primarily results from the lattice mismatched structure and optically absorbing substrate. Growth of this material on a transparent GaP substrate and the addition of the N-isoelectronic trap (25,26) raises the performance to 1 lm/W in both the red and yellow spectral regime. The relatively low cost and high performance makes these devices useful for many indicator applications.

The first LEDs to utilize the N-isoelectronic trap were based on GaP:N (27). These devices, grown by LPE, emit in the yellow-green portion of the spectrum and have efficiencies of ~2.5 lm/W. The improved efficiency relative to GaAsP:N results from a much lower dislocation density facilitated by lattice matched growth. The higher efficiency makes these devices useful for a variety of applications, including indicators and moderate performance lighting/display applications. The emission of GaP may be extended further into the green (555 nm) by utilizing devices with band-edge emission, ie, without N-isoelectronic impurities. The resulting devices exhibit performance on the order of 0.6 lm/W. Another deep level, the (Zn,O) center, lies deeper in the energy gap of GaP than N, resulting in GaP:(Zn,O) LEDs that emit in the deep red (700 nm) (28). Although the external quantum efficiency of these devices is relatively high (2%), the luminous efficiency is relatively low (0.4 lm/W) because the emission wavelength is in a portion of the spectrum where the eye sensitivity is very low.

The AlGaAs materials system is used to produce high performance visible LEDs. This system possesses a direct energy gap in the red to near-ir spectral regime. The system is lattice matched to GaAs (8), and using LPE growth techniques can be employed to form heterostructure devices in high volumes (9). The direct band gap facilitates the fabrication of LEDs having typical luminous efficiencies ranging from 2–10 lm/W. The highest performance is obtained for a double heterostructure TS structure (17). These structures were the first to exceed red-filtered 60-W incandescent sources (3–4 lm/W) in luminous efficiency, opening new markets for LEDs. These high performance LEDs are relatively complex and costly to produce, and thus are only used in applications where high efficiencies are required.

The direct–indirect transition in AlGaAs limits the useful wavelength of these high brightness devices to the 650-nm band, ie, the red. Innovations in lattice matched $(Al_xGa_{1-x})_{0.5}In_{0.5}P/GaAs$ LEDs grown by MOCVD (13,14,29) have resulted in emitters which have extended the useful high brightness spectral regime into the orange, yellow, and green. More recently, TS $(Al_xGa_{1-x})_{0.5}In_{0.5}P/GaP$ structures have been realized using the technique of compound semiconductor wafer bonding, doubling the efficiency of these devices in the green to red spectral regime (18). Performance exceeds 40 lm/W at λ ~602 nm (see Fig. 9). Commercial averages are approximately half of the best reported results for the newer AS and TS devices for yellow and red-orange LEDs. These efficiencies are expected to increase substantially as material and device improvements are made. The TS AlGaInP devices were the first commercial LEDs to surpass the efficiency of unfiltered 60-W incandescent sources, making these LEDs useful for a wide variety of high performance lighting applications.

Another advance in LED technology has been the commercial introduction of blue emitters. In 1989, SiC emitters, IV–IV semiconductors having wurtzite crystal structure, were introduced commercially. These devices, based on indirect material, are very dim (0.04 lm/W), and thus are only useful for low performance indoor display and indicator applications (30). More significant was the introduction in 1994 of blue emitters based on the InGaN system (22). Unlike other III–V LEDs, relatively high internal quantum efficiencies can be achieved even in the presence of highly lattice mismatched growth, ie, high dislocation densities. Such mismatched growth is required because no lattice matched substrate is available for the nitride system (Fig. 6**a**). Bulk large area single crystals of the nitrides have not as of this writing been prepared, and all other alternatives result in a large lattice mismatch. However, high brightness blue and blue-green AlGaN/GaN/InGaN double heterostructure devices grown by MOCVD on sapphire, Al_2O_3, substrates are commercially available. The best devices exhibit luminous efficiencies of 2.5 and 6.5 lm/W in the blue (450 nm) and blue-green (510 nm) spectral regime, respectively (24). These devices use a deep level impurity (Zn) to increase the internal efficiency and increase the emission wavelength (Fig. 7). Furthermore, despite the high dislocation density, these devices possess reliability characteristics similar to other commercial LEDs. As of this writing, these complex emitters are very costly to produce and available in limited volumes. More development is required to improve the cost and availability of these devices.

Progress in the area of II–VI blue and blue-green light-emitter development led to the first injection blue laser diode. This device employed a CdZnSe/ZnSe/ZnSSe quantum well heterostructure having N as the *p*-type dopant (31). These advances have led to the demonstration of very high efficiency (8 lm/W) ZnSeTe/ZnSe blue-green (510 nm) double heterostructure TS LEDs grown on zinc selenide [*1315-09-9*], ZnSe (32). Although these devices exhibit excellent initial performance, they are plagued by reliability problems. In addition, these LEDs require more exotic fabrication techniques (MBE growth, ZnSe substrates) than those developed for the high volume manufacture of LEDs as of this writing. The advances in the InGaN system make the development of blue or blue-green II–VI LEDs unlikely unless fundamental limitations to the high volume production of InGaN emitters are encountered or the performance/manufacturability of II–VI light emitters drastically improves.

Communication Emitters

Figure 10 shows the typical commercial performance of LEDs used for optical data communication. Both free-space emission and fiber-coupled devices are shown, the latter exhibiting speeds of <10 ns. Typically there exists a tradeoff between speed and power in these devices, however performance has been plotted as a function of wavelength for purposes of clarity. In communication systems, photodetectors (qv) are employed as receivers rather than the human eye, making radiometric power emitted by the devices, or coupled into an optical fiber, an important figure of merit.

The first commercially available ir emitters consisted of Zn-diffused GaAs junctions (1). These devices, which exhibit relatively low power output by 1994

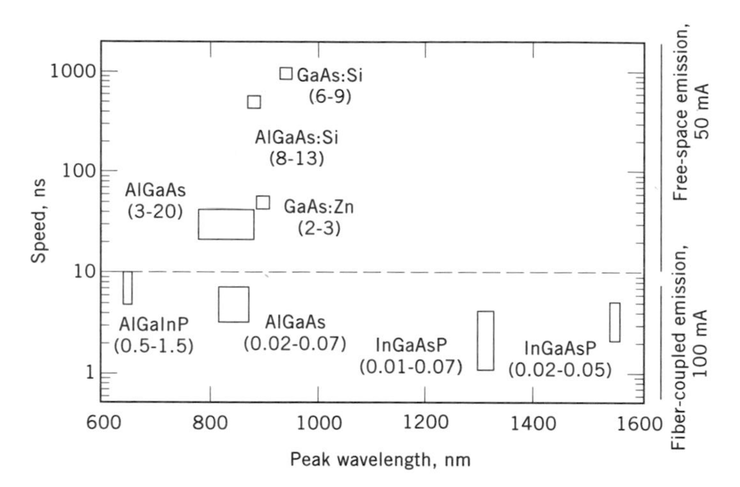

Fig. 10. Speed (90–10% fall time) vs peak wavelength for commercial communication LED emitters. Output power levels in mW are given in parentheses. Generally, these exists a tradeoff between speed and output power in these devices. The data for fiber-coupled devices are based on emission into a 50/125-μm fiber, except for AlGaInP which is for 1-mm plastic optical fiber coupling.

standards, are no longer widely utilized. They have been replaced by GaAs:Si (7) and AlGaAs:Si (33) emitters grown by LPE. In these latter materials, Si acts as an amphoteric dopant, incorporating an n-type impurity at growth temperatures above 820°C and p-type below this temperature. Thus a p–n junction can be formed by using Si doping throughout the entire structure and cooling through the 820°C transition temperature during the LPE process. Radiative recombination in these devices occurs between deep states within the energy gap, resulting in the surrounding GaAs or AlGaAs layers being largely transparent to the emitted light. In the AlGaAs:Si emitters, the absorbing GaAs substrate is removed after epitaxial growth to yield a TS structure. These devices produce high output powers and are of relatively low cost as a result of the simple growth process. However, the radiative recombination in these devices is rather slow, making them useful only for free-space communications at speeds <200 kbits/s.

The AlGaAs materials system possesses an energy gap that facilitates emission from the red to near-ir (650–880 nm). In the purely direct band gap regime, internal quantum efficiencies that approach 100% have been realized (34) resulting in relatively high power ir emitters. In addition, AlGaAs LEDs can be grown lattice matched to GaAs to form heterostructures to increase device speed. In the ir wavelength range, high quality active layers can be grown by LPE or MOCVD. However, LPE is generally employed for conventional chip geometry devices as a result of its significantly higher growth rates.

A variety of fiber-coupled small emission area LEDs are available for moderate data rate (<1000 Mbits/s) communications for distances ranging up to ~20 km. Generally these devices require thinner layers and thus are grown by either LPE or MOCVD. The wavelength of emission and materials used in these devices is chosen to correspond to absorption minima in the optical fiber. Such minima occur at ~850 nm (AlGaAs), ~1300 nm (InGaAsP), and ~1550 nm

(InGaAsP) for glass fibers and ~650 nm (AlGaInP (35)) for plastic fibers. These devices are operated at high (up to 10 kA/cm^2) current densities in order to reduce the radiative lifetime and increase the speed of the devices. In addition, special biasing techniques, eg, short pulse superimposed on a longer pulse or prebiasing, are often utilized to increase speed. Data rates over 1 Gbit/s have been attained in the research laboratory using LEDs.

The selection of the small emission chip geometries, whether surface- or edge-emitting, is dependent on the application. Generally, surface-emitting devices are employed to couple to larger, multimode fibers because of large emission spot size. Similarly, edge emitters, having smaller (3–5 μm) emission widths, are employed to couple to single-mode fibers. In addition edge emitters generally possess a narrower (by ~30%) spectral width as a result of increased self-absorption of the higher energy light by the active region. The narrower spectral width allows for less wavelength dispersion in an optical link. Furthermore, single-mode data transmission exhibits less modal dispersion. As a result, single-mode fiber-coupled edge-emitting devices are capable of transmitting data over longer link lengths than multimode fiber-coupled surface-emitting devices for coupling to fibers having the same attenuation.

A newer form of communications LED has been introduced which utilizes the fact that the radiative recombination process in a LED may be significantly altered when the light-emitting region is placed within an optical cavity that is on the dimensions of the wavelength of emitted light (36). These devices, referred to as resonant cavity LEDs (RCLEDs), exhibit unique and advantageous operating characteristics (37). This surface-emitting device structure typically employs a quantum well active region with mirrors on each side. Typically, the bottom and top mirrors consist of distributed Bragg reflectors (DBRs), which unlike vertical cavity surface-emitting lasers have lower reflectivity products to preclude stimulated emission. A metal (reflective) layer is, however, sometimes employed for one of the mirrors. In order to obtain the desired operating characteristics, the exact placement of the quantum well within the active region is critical. Consequently, these devices require high control of the device layer thicknesses, and thus are grown by MOCVD or MBE. The width of the emission spectrum can be substantially decreased in these structures, such that FWHM <1 nm in some devices. As a result, these devices can be employed as emitters in systems having greatly reduced chromatic dispersion and significantly enhanced communications bandwidth (38). These devices are being considered for implementation in high speed optical data links.

Markets and Applications

Light-emitting diodes are the most commercially important compound semiconductor devices in terms of both dollar and volume sales. The 1991 worldwide compound semiconductor device market totaled $2.8 billion (39). Light-emitting diodes accounted for ca $1.9 billion of this market. Visible and ir LEDs represented 37 and 30%, respectively. These markets are expected to grow as LEDs are increasingly employed in advanced applications.

Light-emitting diodes are utilized in a variety of markets (Table 2). Generally, visible/display LEDs can be grouped into two principal classifications: low

Table 2. LED Markets and Technologies

Market application	Primary technologies	Secondary technologies	Emerging technologies
	Visible/display emitters		
consumer	GaP:Zn,O, GaAsP:N, GaP:N	GaAsP, red AlGaAs	
industrial/instrumentation	GaAsP:N, GaP:N	GaAsP, GaP:Zn,O, red AlGaAs	
automotive			
interior	GaAsP:N, GaP:N		AlGaInP, InGaN
exterior	red AlGaAs		AlGaInP
signs			
indoor	GaAsP:N, GaP:N	red AlGaAs	AlGaInP, SiC, InGaN
outdoor	GaP:N, red AlGaAs	GaAsP:N	AlGaInP, InGaN
electrophotographic	GaAsP	red AlGaAs	
	Communication emitters		
optocouplers	GaAs:Si, AlGaAs:Si, ir TS AlGaAs[a], GaAsP		
sensors	ir AlGaAs[a], GaAsP		red AlGaInP[a]
free space–ir wireless, <115 kbits/s[c]	ir TS AlGaAs[b], AlGaAs:Si, GaAs:Si		ir TS AlGaAs[b,c]
plastic optical fiber-coupled, <5 Mbits/s	GaAsP, red AlGaAs[b]		red AlGaInP[a,c]
850-nm fiber-coupled, <200 Mbits/s, <2 km	ir AlGaAs[a]		ir AlGaAs RCLED
1300-nm fiber-coupled, 50–1000 Mbits/s, 1–20 km	InGaAsP[a]		InGaAsP RCLED
1550-nm fiber-coupled (50–1000 Mbits/s, 5–20 km)	InGaAsP[a]		InGaAsP RCLED

[a]Small emission area geometry.
[b]Conventional geometry.
[c]Emerging technology <5–10 Mbits/s.

end indicators and high brightness emitters. The low end indicators are typically utilized in such markets as consumer, industrial/instrumentation, automotive interiors, and indoor signs. The consumer and industrial/instrumentation markets employing individual LED lamps are multichip displays having relatively low performance. However, more complex and battery driven applications which require high brightness devices are emerging. In addition, LEDs are being employed to backlight liquid crystal displays (LCDs) further requiring the use of high performance devices. Indicators are also used in automotive interior applications on dashboards within cars. In another application, arrays of LEDs are used in electrophotography (qv) and employed to form write bars and erase bars for printing information (see PRINTING PROCESSES).

The developments in the red AlGaAs, red-orange-yellow-green AlGaInP, and blue and blue-green InGaN systems are, as of this writing, on the threshold of invoking a revolution in high performance lighting. Figure 11 shows the luminous performance of these devices compared to conventional high performance lighting sources. Red, orange, yellow, and yellow-green AlGaInP devices have been produced that exceed the efficiency of conventional 60-W tungsten sources. The brightest of these devices also exceeds 30-W halogen lamps. Also, single TS AlGaInP large area devices have been demonstrated in the laboratory that emit luminous fluxes of 84 lm, roughly within an order of magnitude of

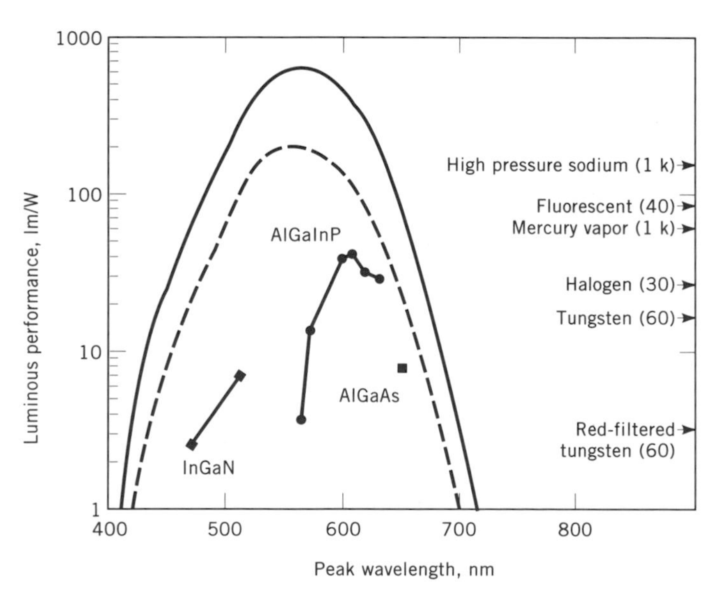

Fig. 11. Luminous performance vs peak emission wavelength for the best reported high brightness LEDs. Some high performance lighting sources are indicated. The numbers in parentheses correspond to the source wattage. Also shown is the (—) eye response curve (as defined by the Commission International de L'Eclairage, CIE) and a similar curve (- - -), 0.3 × CIE. These curves represent the performance of an ideal 100% efficient monochromatic source and indicate an upper limit of potential LED performance of devices exhibiting 100% internal quantum efficiency and having 100 and 30% extraction efficiency, respectively. The approximation of an LED as a monochromatic source is roughly valid for direct-gap emitters with a narrow spectral output (<20 nm FWHM).

conventional light bulbs (900 lm) (40). Accordingly, LEDs can be expected to ultimately replace high performance lighting sources, especially as the efficiency of the AlGaInP and InGaN LEDs continues to improve. Already, high brightness red AlGaAs and red-orange-yellow AlGaInP devices are being employed in markets previously reserved for filtered incandescents, including outdoor large area displays and vehicular lighting, eg, center high mount stop lights, turn-tail stop lights, and side markers.

Light-emitting diodes are used in communications to transmit data optically. These LEDs can generally be divided into two categories: free-space emitters and fiber-coupled devices. More conventional free-space emitters are employed for remote controls; optocouplers, wherein an optical emitter/detector pair provides electrical isolation; and sensors (qv), eg, bar-code wands, slot interrupters for measuring the presence of an object, etc. Typically, these devices are relatively slow. However, an emerging application is the ir wireless market wherein much higher data rates can be achieved using properly designed TS AlGaAs devices. Devices are commercially available that operate at 115 kbits/s. These are typically being used for wireless computer–computer, computer–peripheral, and consumer electronics communications. As of this writing, higher performance commercial devices are being introduced and are capable of transmitting data at rates of 5–10 Mbits/s. The projected use of these devices is for wireless communications in facsimile (FAX) machines, local-area networks (LANs), and image transfer. These emerging markets are expected to expand as the demand for short-distance wireless communications increases.

Much higher data rates can be obtained for small emission area fiber-coupled devices. The selection of the device/fiber depends on the desired application. Plastic optical fiber-coupled devices emitting at 650 nm are available for short haul low cost data links employed for industrial controls, medical instrumentation, and computer–peripheral communication (see FIBER OPTICS). Glass optical fiber-coupled emitters are employed to achieve higher data rates and longer link lengths. Devices emitting at 850 nm are typically employed for LANs, office switching links, and computer–peripheral communications at moderate (<200 Mbits/s) data speeds and link distances up to 2 km. High end performance is obtained by employing 1300 and 1550 nm fiber-coupled emitters wherein commercial devices can transmit data at rates up to 1000 Mbits/s. Typically, 1300-nm LED links are employed for telecommunication switches, video links, and high speed LANs. The 1550-nm devices are generally reserved for telecommunication applications wherein single-mode coupling is required for high data rates at distances up to 20 km. The applications for these devices are expected to grow as wires are replaced with optical fibers for short and moderate distance communications.

BIBLIOGRAPHY

"Light-Emitting Diodes and Semiconductor Lasers" in *ECT* 3rd ed., Vol. 14, pp. 269–294, by M. B. Parish, Bell Laboratories.

1. J. Bardeen and W. H. Brattain, *Phys. Rev.* **74**, 230 (1948).
2. R. J. Keyes and T. M. Quist, *Proc. IRE* **50**, 1822 (1962).

3. N. Holonyak, Jr. and S. F. Bevacqua, *Appl. Phys. Lett.* **1**, 82 (1962).
4. H. Kroemer, *Proc. IEEE* **51**, 1782 (1963).
5. Zh. I. Alferov and co-workers, *Sov. Phys. Semicond.* **9**, 305 (1973).
6. M. G. Craford and F. M. Steranka, in G. Trigg, ed., *Encyclopedia of Applied Physics*, VCH, New York, 1994, pp. 485–514.
7. H. Rupprecht, J. M. Woodall, K. Konnerth, and G. D. Pettit, *Appl. Phys. Lett.* **9**, 221 (1966).
8. H. Rupprecht, J. M. Woodall, and G. D. Pettit, *Appl. Phys. Lett.* **11**, 81 (1967).
9. J. Nishizawa and K. Suto, *J. Appl. Phys.* **48**, 3484 (1977).
10. A. H. Herzog, W. O. Groves, and M. G. Craford, *Appl. Phys. Lett.* **40**, 1830 (1969).
11. H. M. Manasevit and W. I. Simpson, *J. Electrochem. Soc.* **116**, 1725 (1969).
12. R. D. Dupuis and P. D. Dapkus, *Appl. Phys. Lett.* **31**, 466 (1977).
13. C. P. Kuo, R. M. Fletcher, T. D. Osentowski, M. C. Lardizabel, and M. G. Craford, *Appl. Phys. Lett.* **57**, 2937 (1990).
14. H. Sugawara, M. Ishikawa, and G. Hatakoshi, *Appl. Phys. Lett.* **58**, 1010 (1991).
15. J. R. Arthur, *J. Appl. Phys.* **39**, 4032 (1968).
16. A. Y. Cho, *J. Vac. Sci. Technol.* **8**, S31 (1971).
17. L. W. Cook, M. D. Camras, S. L. Rudaz, and F. M. Steranka, *Proceedings of the 14th International Symposium on GaAs and Related Compounds, 1987*, Institute of Physics, Bristol, U.K., 1988, pp. 777–780.
18. F. A. Kish and co-workers, *Appl. Phys. Lett.* **64**, 2839 (1994).
19. M. Fukuda, *Reliability and Degradation of Semiconductor Lasers and LEDs*, Artech House, Boston, Mass., 1991.
20. T. P. Pearsall, ed., *GaInAsP Alloy Semiconductors*, John Wiley & Sons, Inc., New York, 1982.
21. C. A. Burrus and R. W. Dawson, *Appl. Phys. Lett.* **17**, 97 (1970).
22. S. Nakamura, T. Mukai, and M. Senoh, *Appl. Phys. Lett.* **64**, 1687 (1994).
23. L. C. Kimerling and D. V. Lang, in F. A. Huntley, ed., *Lattice Defects in Semiconductors*, Institute of Physics Conference, Proceedings No. 23, Bristol, U.K., 1975.
24. Technical data, Nichia Chemical Industries, Ltd., Aran, Tokushima, Japan, 1994
25. W. O. Groves, A. H. Herzog, and M. G. Craford, *Appl. Phys. Lett.* **19**, 184 (1971).
26. A. H. Herzog, D. L. Keune, and M. G. Craford, *J. Appl. Phys.* **43**, 600 (1972).
27. R. A. Logan, H. G. White, and W. Wiegmann, *Appl. Phys. Lett.* **13**, 139 (1968).
28. R. H. Saul, J. Armstrong, and W. H. Hackett, Jr., *Appl. Phys. Lett.* **15**, 229 (1969).
29. K. H. Huang and co-workers, *Appl. Phys. Lett.* **61**, 1045 (1992).
30. J. A. Edmond, H.-S. Kong, and C. H. Carter, Jr., in C. Y. Yang, M. M. Rahman, and G. L. Harris, eds., *Amorphous and Crystalline Silicon Carbide IV*, Springer Verlag, Berlin, 1992, pp. 344–351.
31. M. A. Haase, J. Qiu, J. M. DePuydt, and H. Cheng, *Appl. Phys. Lett.* **58**, 1272 (1991).
32. D. B. Eason and co-workers, *Electron. Lett.* **30**, 1178 (1994).
33. L. R. Dawson, *J. Appl. Phys.* **48**, 2485 (1977).
34. Zh. I. Alferov, V. M. Andreev, D. Z. Garbuzov, and V. D. Rumyantsev, *Sov. Phys. Semicond.* **6**, 1930 (1975).
35. B. V. Dutt, J. H. Racette, S. J. Anderson, F. W. Scholl, and J. R. Shealy, *Appl. Phys. Lett.* **53**, 2091 (1988).
36. E. Yablonovitch, *Phys. Rev. Lett.* **58**, 2059 (1987).
37. E. F. Schubert, Y.-H. Wang, A. Y. Cho, L.-W. Tu, and G. J. Zydzik, *Appl. Phys. Lett.* **60**, 921 (1992).
38. N. E. J. Hunt and co-workers, *Appl. Phys. Lett.* **63**, 2600 (1993).
39. *1992 Report*, Strategies Unlimited, Mountain View, Calif.
40. F. A. Kish and co-workers, *Electron. Lett.* **30**, 1790 (1994).

General References

A. A. Bergh and P. J. Dean, *Light-Emitting Diodes*, Clarendon, Oxford, U.K., 1976.
H. C. Casey and M. B. Panish, *Heterostructure Lasers*, Academic Press, Inc., Orlando, Fla., 1978.
M. G. Craford, in L. E. Tannas, Jr., ed., *Flat Panel Displays and CRTs*, Van Nostrand Reinhold, New York, 1985, pp. 289–331.
K. Gillessen and W. Schairer, *Light-Emitting Diodes—An Introduction*, Prentice Hall, Englewood Cliffs, N.J., 1987.
R. H. Saul, T. P. Lee, and C. A. Burrus, in W. T. Tsang, ed., *Semiconductors and Semimetals*, Vol. 22, Academic Press, Inc., Orlando, Fla., 1985, pp. 193–237.
B. G. Streetman, *Solid-State Electronic Devices*, Prentice Hall, Englewood Cliffs, N.J., 1980.

FRED KISH
Hewlett-Packard Company

SEMICONDUCTOR LASERS

Diode lasers have evolved from a laboratory curiosity to devices used in a variety of everyday applications, from consumer products such as the compact disc (CD) audio players to high speed telecommunication networks based on optical fibers (see FIBER OPTICS). This evolution was made possible by a series of rapid advances in materials preparation, laser design, and fabrication. Progress in materials growth and device design has been particularly impressive (1–4). Most of the diode lasers emitting in the red and near-infrared (nir) parts of the electromagnetic spectrum are based on compound semiconductors (qv) composed of elements of Group 13 (III) and 15 (V) of the Periodic Table. Some of the newer green and blue emitting lasers (qv) use compound semiconductors composed of Group 12 (IIB) and 16 (VI) of the Periodic Table. Compounds containing as many as four different elements are often used in order to control the laser wavelength, electrical properties, and other desired characteristics.

Semiconductors

Many of the optical and electrical characteristics of semiconductors are conveniently described by energy level diagrams. Electrons in atoms are restricted to sets of discrete energy states and these states are separated by gaps in which electrons are not allowed. In solids formed by bringing isolated atoms together, the allowed energy levels of discrete atoms spread into essentially continuous energy bands. Two such bands are of particular importance in semiconductors: the low lying valence band (VB) and the upper band, known as the conduction band (CB). The conduction and valence bands are separated from each other by an energy gap, E_g. In an ideal undoped semiconductor there are no energy states within the energy gap. At a temperature of absolute zero all of the states in the valence band are occupied by electrons, which thus cannot contribute to conductivity, and all of the conduction band states are empty. Thermal excitation of electrons across the energy gap becomes possible at higher temperatures establishing a net electron concentration in the conduction band as illustrated in Figure 1**a**. The excited electrons leave behind empty states in the valence band called holes which behave like positively charged electrons. Both the electrons in the conduction band and holes in the valence band are free to move in space and participate in electrical conductivity.

In semiconductors of interest for light-emitting sources the band gap energy is considerably larger than room temperature thermal energy ($kT = 0.025$ eV) and the electrical conductivity resulting from the thermally excited electrons is negligibly small. In order to obtain appreciable conductivities these materials must be intentionally doped. Doping is accomplished by adding small amounts of impurity atoms having either more or fewer valence electrons than required for bonding into the host crystal lattice. The impurity atoms deficient in electrons tend to attract an electron and become ionized. The impurity atoms having an excess electron tend to donate it to the lattice and become ionized. These two different kinds of impurity atoms are known, respectively, as acceptors and donors. The regions of semiconductor doped with the donor impurities contain more ionized donors than acceptors and are called n-type. The p-type regions have more acceptor than donor impurities. The boundary region between the p-type and n-type regions is called a p–n junction. Such a region behaves electrically like a rectifying diode in which current flow with applied voltage is not symmetric with respect to voltage polarity. All of the so-called injection lasers are based on p–n diodes.

The impurity atoms used to form the p–n junction form well-defined energy levels within the band gap. These levels are shallow in the sense that the donor

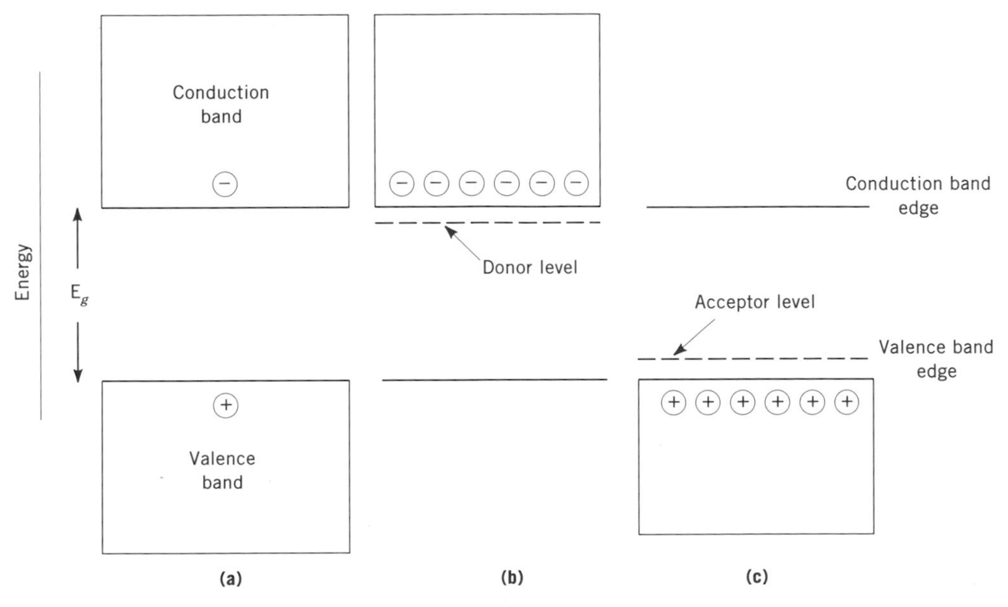

Fig. 1. Band-edge energy diagram where the energy of electrons is higher in the conduction band than in the valence band: (**a**) an undoped semiconductor having a thermally excited carrier; (**b**) n-type doped semiconductor having shallow donors; and (**c**) a p-type doped semiconductor having shallow acceptors.

levels lie close to the conduction band (Fig. **1b**) and the acceptor levels are close to the valence band (Fig. **1c**). The thermal energy at room temperature is large enough for most of the dopant atoms contributing to the impurity levels to become ionized. Thus, in the p-type region, some electrons in the valence band have sufficient thermal energy to be excited into the acceptor level and leave mobile holes in the valence band. Similar excitation occurs for electrons from the donor to conduction bands of the n-type material. The electrons in the conduction band of the n-type semiconductor and the holes in the valence band of the p-type semiconductor are called majority carriers. Likewise, holes in the n-type, and electrons in the p-type semiconductor are called minority carriers.

At thermal equilibrium characterized by temperature, T, the distribution of electrons over the allowed band of energies is given by a Fermi-Dirac distribution function:

$$f(E) = [1 + e^{(E-E_f)/kT}]^{-1}$$

where $k = 8.62 \times 10^{-5}$ eV/K is the Boltzmann's constant. The function, $f(E)$, describes the probability of a state with an energy, E, being occupied at a temperature, T. The quantity, E_f (the Fermi level), denotes the energy level with the occupation probability of 1/2. At $T = 0$ all the available states below E_f are filled with electrons, and all the states above E_f are empty. The function, $f(E)$, is symmetrical about E_f for all temperatures. Thus $f(E)$ is an expression for an average number of fermions in a single orbital of energy, E. The Fermi level is then the chemical potential μ.

Instead of plotting the electron distribution function in a band energy level diagram, it is convenient to indicate the Fermi level. For instance, it is easy to see that in p-type semiconductors the Fermi level lies near the valence band.

The spatial distribution of carriers in the immediate vicinity of the p–n junction is very important. Some of the majority carriers at the p–n junction neutralize each other resulting in a thin region depleted of free carriers, known as the space–charge region. In this region the fixed negative charges on the p-side repel the mobile electrons from the n-side of the junction. Similarly, the mobile holes from the p-side are repelled by the fixed positive charges at the p-side. The result is a built-in electric field which inhibits carrier diffusion across the p–n junction. The potential V_D associated with this field bends the conduction and valence bands in the space–charge region by an amount called the barrier height, as illustrated in Figure 2**a**. The potential difference, V_D, is also called the contact potential.

The current flow across the p–n junction can be accomplished only by the application of an external voltage bias which opposes the V_D and reduces the barrier height. This bias, called a forward bias, supplies electrons at the n-contact and holes at the p-contact. These carriers flow, as majority carriers, toward the p–n junction. Because the barrier height is similar in magnitude to the band gap energy, the external voltage bias needs to be fairly small, also on the order of the band gap energy divided by the electron charge. That is, the applied voltage needs to be less than the numerical value of the band gap. As electrons move up on the energy band diagram their energy increases. Holes increase their energy when moving down.

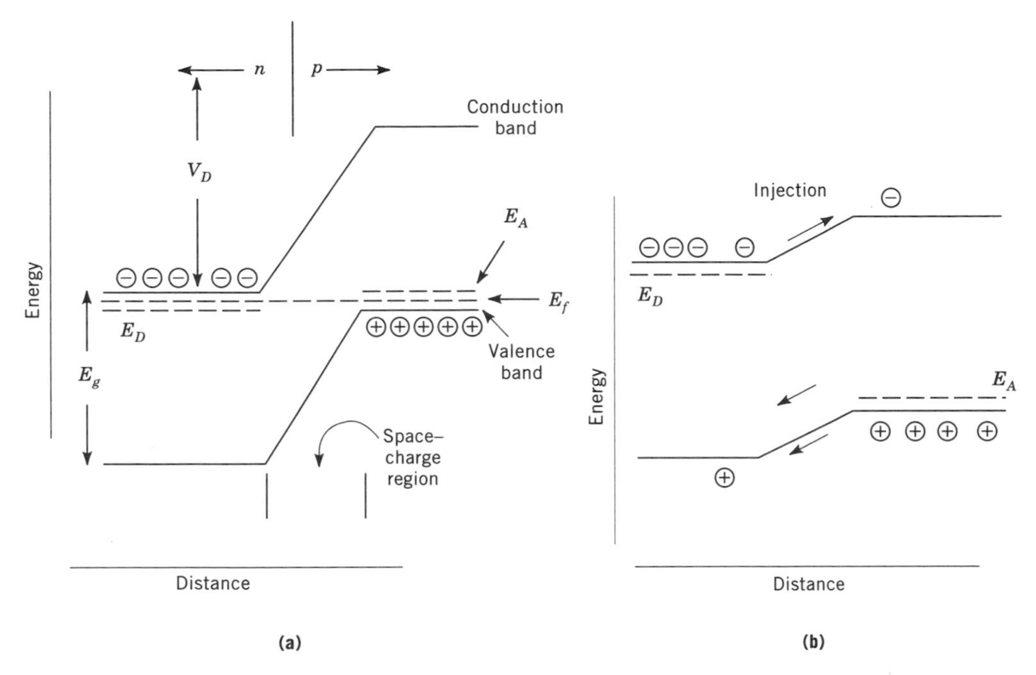

Fig. 2. Representation of the band edges in a semiconductor p–n junction where shallow donor, acceptor energies, and the Fermi level are labeled E_D, E_A, and E_f, respectively. (**a**) Without external bias; V_D is the built-in potential of the p–n junction; (**b**) under an applied forward voltage V_B.

Under a forward bias the majority carriers cross the p–n junction and become minority carries, eg, holes on the n-side of the junction. These holes rapidly come into thermal equilibrium with the lattice of the semiconductor and reach the energy approximately equal to the energy of the valence band edge on the n-side of the junction. The minority electrons reach an equilibrium energy at the p-side of the junction. The minority carriers are not in thermodynamic equilibrium with the majority carriers and must give up their excess energy. The hole on the n-side is said to recombine with a majority electron which then loses energy about equal to E_g. The minority electron loses a similar energy and recombines with a majority hole. In semiconductors of interest herein the energy given up by the minority carriers results in emission of light. This process is known as the radiative recombination or spontaneous emission. Recombination can also occur through a range of competing nonradiative processes, which occur without the emission of light. These depend on defects and imperfections in the host lattice and must be carefully controlled in order to assure efficient emission of light in forward biased p–n junctions.

The distributions of excess, or injected, carriers are indicated in band diagrams by so-called quasi-Fermi levels for electrons, F_n; or holes, F_p. These functions describe steady-state concentrations of excess carriers in the same form as the equilibrium concentration. In equilibrium $F_n = F_p = F_f$.

Light, ie, a photon, is emitted in the transition from a higher to a lower energy state. The energy difference between the two states, ΔE, is converted to a photon having wavelength, λ, of

$$\lambda = hc/\Delta E$$

where h is Planck's constant and c is the velocity of light. A practical conversion when the wavelength of light, λ, is in units of μm and ΔE is in eV is $\lambda = 1.24/\Delta E$.

Direct and Indirect Band Gap Semiconductors

Semiconductors can be divided into two groups: direct and indirect band gap materials. In direct semiconductors the minimum energy in the conduction band and the maximum in the valence band occur for the same value of the electron momentum. This is not the case in indirect materials. The difference has profound consequences for the transitions of electrons across the band gap in which light is emitted, the radiative transitions, of interest here.

In direct band gap materials, recombination of an electron and a hole occurs without the need for momentum transfer making it possible to achieve high rates of radiative recombination across the band gap. In addition, in band-to-band transitions the energy of the photon needed to stimulate the transition is the same as that emitted in the transition. In indirect semiconductors the conduction–valence band transition can occur only if momentum is conserved. This can be achieved by the simultaneous emission and absorption of phonons, ie, quanta of lattice vibration. However, the rates for indirect radiative recombination processes are low, and competing nonradiative transitions can be much more important. The radiative rates can be somewhat adjusted by the intentional incorporation of deep levels within the band gap, a process useful in some types of light-emitting diodes. The energy of the photon emitted in this process is considerably lower than that of the band gap. Indirect band gap materials cannot be used in lasers owing to their low efficiency of converting injected carriers into light (low quantum efficiency).

Semiconductor Laser

The operation of a laser is based on stimulated emission. A photon generated in the forward biased junction, in the process of spontaneous emission, can be absorbed in the semiconductor or cause an additional transition by stimulating an electron to make a transition to a lower energy state. In this latter process another photon is emitted. The rate of stimulated emission is governed by a detailed balance between absorption and spontaneous and stimulated emission rates. That is, stimulated emission occurs when the probability of a photon causing a transition of an electron from the conduction to the valence band, with the probability emission of another photon, is greater than for the upward transition of an electron from the valence to conduction band upon absorption of the photon. These rates are commonly described in terms of Einstein's theory of A and B coefficients (5,6). For semiconductors, a simple condition describing the carrier density necessary for stimulated emission has been derived (7). Lasing can start when the density of electrons injected into the conduction band exceeds the hole density in the valence band. This is a condition of population inversion which occurs when the separation of the quasi-Fermi levels for the holes, F_p, and the electrons, F_e, is greater than the energy of the emitted photon,

$$F_e - F_p > h\nu$$

and the photon energy $h\nu$ must be at least equal to the band gap energy. Thus in semiconductor lasers stimulated emission occurs between distributions of states in the conduction and valence bands. In most other lasers (qv), such as gas or glass lasers, this transition occurs between discrete energy levels.

The distributions of states in conduction and valence bands are commonly described by the effective density of states. The concentration of electrons, n, in the conduction band can be calculated as

$$n = \int_{E_c}^{\infty} f(E)N(E)dE$$

where $N(E)dE$ is the density of states (in cm^{-3}) in the energy range dE. A simpler way of calculating n is to represent all of the electron states in the conduction band by an effective density of states, N_c, at the energy, E_c (band edge). The electron density is then simply

$$n = N_c f(E_c)$$

Population inversion is easier to achieve when the effective density of states in the conduction band is low.

The processes contributing to laser action are illustrated in Figure 3 which shows absorption, emission, and the gain spectra of a semiconductor (8). Significant absorption occurs only for the photon energies greater than that of the band gap. The spectrum of spontaneous emission is determined by the absorption spectrum, at the low energy side, and the electron energy distribution at the high energy side. The gain spectrum represents the difference between the spontaneous emission and absorption. Lasing occurs at the energy of the gain peak.

A semiconductor laser requires a means of generating spatially localized high concentrations of minority carriers, a medium to provide the gain, and a means of providing some feedback to the stimulated emission. The medium is the semiconductor structure arranged in a way which helps to confine the carriers and light. Light is generated in this structure by means of a p–n junction which injects electrons from the valence to the conduction band and thus provides the population inversion. This is followed by recombination of electrons with holes and emission of light. Further recombination can be stimulated (stimulated emission) by light already present in the medium. This optical feedback is carefully arranged by forming a cavity which has two mirrors parallel to each other. Light generated within the cavity is then partially reflected back into the

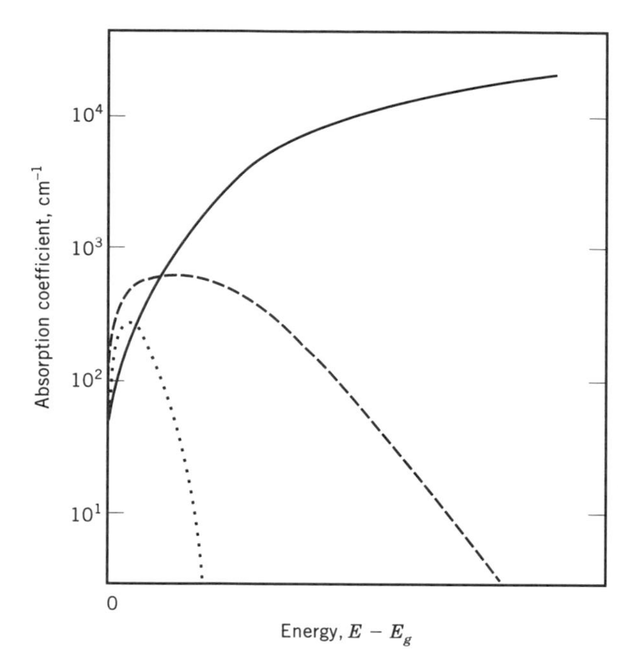

Fig. 3. Spectra showing absorption coefficient as a function of the photon energy in a direct band gap semiconductor where (—) represents absorption, (– – –) the spontaneous emission, and (····) gain.

crystal. Such mirrors can be formed in most compound semiconductors lasers by cleaving two ends of a waveguide.

Heterostructures

In the p–n junction illustrated in Figure 2 both sides are made of the same type of semiconductor and have the same energy gap. Such a junction is called a homojunction. The disadvantage of this type of junction is that the minority carriers are free to move away from the junction. It is thus very difficult to achieve the high carrier density needed in a laser. Higher carrier densities can be obtained by introducing an energy barrier at, or very near to, the p–n junction. The energy barrier arises when two semiconductors, having different band gaps, are joined together (9,10). Barriers in the conduction band (CB) and valence band (VB), the band-edge discontinuities ΔE_c and ΔE_v, respectively, can be formed by changing the composition of the semiconductor. A junction of a small and large band gap semiconductor is called a single heterojunction. It confines one type of minority carrier (electrons or holes) to the p–n junction region.

A more effective carrier confinement is offered by a double heterostructure in which a thin layer of a low band gap material (the active layer) is sandwiched between larger band gap layers. The physical junction between two materials of different band gaps, and chemical compositions, is called a heterointerface. A schematic representation of the band diagram of such a structure is shown in Figure 4. Electrons injected under forward bias across the p–N junction into the lower band gap material encounter a potential barrier, ΔE_c, at the p–P junction

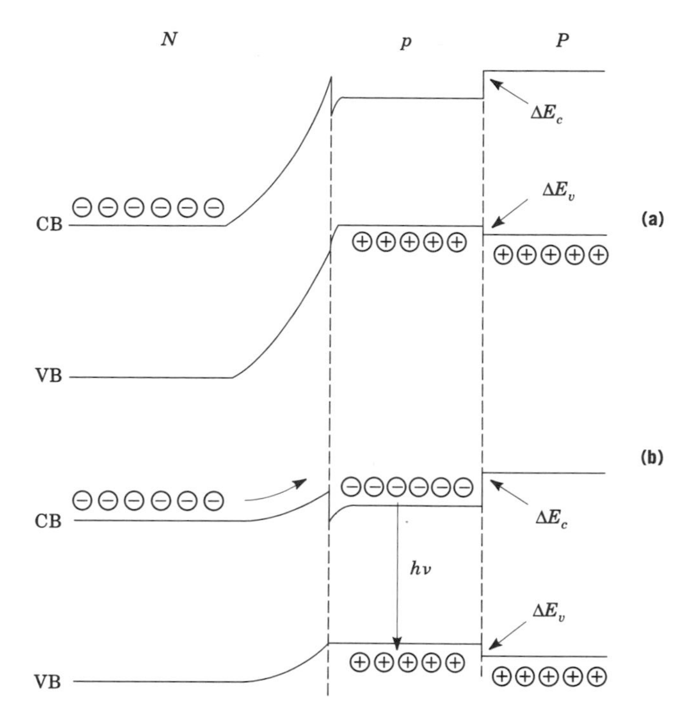

Fig. 4. Schematic cross section and the band diagram of a double heterostructure showing the band-edge discontinuities, ΔE_c and ΔE_v, used to confine carriers to the smaller band gap active layer. (**a**) Without and (**b**) with forward bias. See text.

which inhibits their motion away from the junction. The holes see a potential barrier of ΔE_v at the N–p heterointerface which prevents their injection into the N region. The result is that the injected minority carriers are confined to the thin narrow band gap region. If this region is thinner than the average distance the carrier can move before recombination (called diffusion length), on the order of a few micrometers in most compound semiconductors, a very high density of injected carriers can be obtained in a forward biased diode. In laser diodes this thin low band gap layer is called the active layer. In most of the conventional double heterostructure lasers the thickness of this layer is about 0.1 μm. The thinner the active layer, the easier it is to achieve the carrier density needed for stimulated emission. However, as a matter of practice, the thickness of the active layer may be limited by the precision of the epitaxial growth technique.

In addition to the carrier confinement, the active region is crucial in providing light confinement and waveguiding because the index of refraction of this low band gap layer is larger than the indexes of the surrounding layers. Such light confinement is absent in homostructure or single heterostructure lasers. In most modern laser structures special waveguide layers having a band gap, and thus the index of refraction, intermediate to that of the active and confining layers, are included in order to assure adequate light confinement. In such structures the carriers are confined to the narrow section of the active layer having the lowest band gap and the highest index of refraction, and light is confined to a wider region defined by the waveguides on each side of the active layer. This type of structure is known as a separate confinement heterostructure.

Lasing occurs whenever the gain arising from stimulated emission exceeds the cavity losses. Internal losses, α_i, result from absorption and scattering of light. The reflectivity, R, of the mirror facet must be <1 and this contributes a loss term of $(1/L)\ln(1/R)$, where L is the cavity length. At threshold, the gain, g, is equal to losses and

$$g\Gamma = \alpha_i + \frac{1}{L}\ln\frac{1}{R}$$

where Γ is the confinement factor, a fraction of the light intensity within the active region. It is typically <1 because the active layer (a region of carrier confinement) is thinner than the waveguide layers (regions of light confinement). The threshold current density, J_{th}, in A/cm^2 can be written (6) as

$$J_{th} = 4.5 \times 10^3(d/\eta) + (20d/\eta\Gamma)\left\{\alpha_i + \frac{1}{L}\ln\left(\frac{1}{R}\right)\right\}$$

where d, the thickness of the active region written, is in μm, and η is the efficiency with which current is converted to photons.

Large threshold current density of homostructure lasers results mainly from poor carrier confinement and the resulting large active region thickness d. This is because the diffusion length of electrons is fairly long in most semiconductors discussed herein, on the order of many micrometers. For $d = 3\ \mu$m and $L =$ 500 μm the threshold current density is as large as 15 kA/cm^2. In practice threshold current densities are much larger because the homostructure also suffers from small Γ and high α_i. Very high threshold current density limits operation to short pulses and cryogenic temperatures.

In a double heterostructure laser having a gallium arsenide, GaAs, active region, $d \sim 0.1\ \mu$m, and $Al_xGa_{1-x}As$ waveguide layers, threshold current density drops to less than 0.5 kA/cm^2. Good light confinement results in Γ close to 1 and the internal losses are less than 30 cm^{-1}. Such lasers readily achieve continuous operation at room temperature and are capable of high power output.

Quantum Wells

Epitaxial crystal growth methods such as molecular beam epitaxy (MBE) and metalorganic chemical vapor deposition (MOCVD) have advanced to the point that active regions of essentially arbitrary thicknesses can be prepared (see THIN FILMS, FILM DEPOSITION TECHNIQUES). Most semiconductors used for lasers are cubic crystals where the lattice constant, the dimension of the cube, is equal to two atomic plane distances. When the thickness of this layer is reduced to dimensions on the order of 0.01 μm, between 20 and 30 atomic plane distances, quantum mechanics is needed for an accurate description of the confined carrier energies (11). Such layers are called quantum wells and the lasers containing such layers in their active regions are known as quantum well lasers (12).

The uncertainty principle, according to which either the position of a confined microscopic particle or its momentum, but not both, can be precisely measured, requires an increase in the carrier energy. In quantum wells having abrupt barriers (square wells) the carrier energy increases in inverse proportion to its effective mass (the mass of a carrier in a semiconductor is not the same as that of the free carrier) and the square of the well width. The confined carriers are allowed only a few discrete energy levels (confined states), each described by a quantum number, as is illustrated in Figure 5. Stimulated emission is allowed to occur only as transitions between the confined electron and hole states described by the same quantum number.

The two-dimensional carrier confinement in the wells formed by the conduction and valence band discontinuities changes many basic semiconductor parameters. The parameter important in the laser is the density of states in the conduction and valence bands. The density of states is greatly reduced in quantum well lasers (11,12). This makes it easier to achieve population inversion and thus results in a corresponding reduction in the threshold carrier density. Indeed, quantum well lasers are characterized by threshold current densities as low as 100–150 A/cm^2, dramatically lower than for conventional lasers. In the quantum well lasers, carriers are confined to the wells which occupy only a small fraction of the active layer volume. The internal loss owing to absorption induced by the high carrier density is very low, as little as $\alpha_i \sim 2$ cm^{-1}. The output efficiency of such lasers shows almost no dependence on the cavity length, a feature useful in the preparation of high power lasers.

The incorporation of quantum wells into the material has other subtle consequences. The valence band in direct bulk semiconductors is degenerate at the maximum energy point (valence band maximum). At this point there are two types of holes of the same energy, the heavy and light ones. The effective mass of the light hole is similar to that of the electron, whereas that of the heavy hole is about 10 times larger than that of the electron. In quantum wells the two hole bands become separated in energy, and it is said that the degeneracy is lifted (12). The energy shift owing to quantum well confinement is larger for the light holes. The lasing transition is more likely to occur between the confined electron and heavy hole states, as is illustrated in Figure 5.

All the layers of conventional heterostructure lasers, that is, lasers not based on quantum wells, must precisely replicate the lattice structure of the substrate. These layers are fairly thick on the order of 0.1 μm, and any lattice mismatch large enough to alter electronic properties invariably results in generation of defects. Strain can be introduced in quantum wells, with the thickness on the order of 0.01 μm or less, without generating defects (13). In such thin layers strain is accommodated without any change in the in-plane lattice constant, as long as the thickness of the well is lower than a certain critical thickness (14). Depending on the material system, strain in the quantum well can be either compressive, with the well's lattice constant trying to be larger than that of the substrate, or tensile. Compressive strain increases the separation of the light and heavy hole bands already occurring because of quantum confinement. The tensile strain initially reduces the light and heavy hole band separation and thus counteracts the quantum confinement. However, as the tensile strain is increased further the two hole bands can cross and the light hole band becomes the

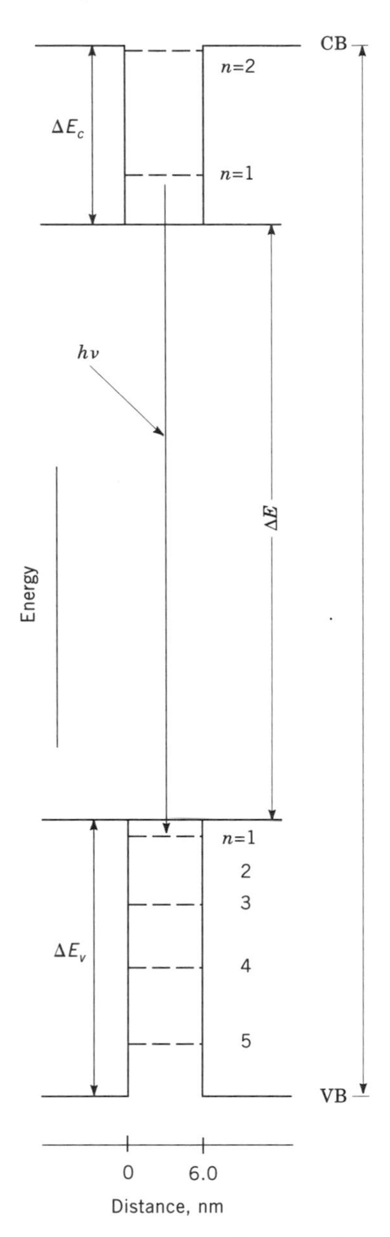

Fig. 5. Energy levels of electrons and heavy holes confined to a 6-nm wide quantum well, $In_{0.53}Ga_{0.47}As$, with InP valence band, ΔE_v, and conduction band, ΔE_c, barriers. In this material system approximately 60% of the band gap discontinuity lies in the valence band. Lasing occurs between the confined electron from the upper level and heavy hole states in the lower level.

lowest energy band in the valence band. Laser transition then occurs between the confined electrons and light holes. Both of these strain directions can result in lower density of states and the threshold current density even lower than that of lattice matched quantum well lasers (15). The detailed manipulation of the electronic states of the band gap is now a subject of research.

Materials

Several compound semiconductor systems permit the growth of high quality, thin-layered crystal structures that have large and abrupt changes in the band gap and the index of refraction required in heterostructure lasers. The layer-to-layer changes result from changes in composition. Several of the Group 13–15 (III–V) material systems used for the preparation of laser structures are shown in the band gap–lattice constant plots of Figure 6. All of these materials have cubic structures known as zinc blende. The special nature of the binary semiconductor compounds such as gallium arsenide [*1303-00-0*], GaAs, or indium phosphide [*22398-80-7*], InP, arises from their availability as the substrate materials needed for the growth of lasers.

Two of the materials systems shown in Figure 6 are of particular importance. These are the ternary compounds formed from the Group 13 (III) elements such as Al and Ga in combination with As (6) and quaternary compounds formed from Ga and In in combination with As and P (16–18). The former, aluminum gallium arsenide, $Al_xGa_{1-x}As$, grown on GaAs, is the best known of the general class of compounds $A_x^{III}B_{1-x}^{III}C^V$. The latter, gallium indium arsenide phosphide, $Ga_xIn_{1-x}As_{1-y}P_y$, grown on InP, is of the general class $A_x^{III}B_{1-x}^{III}C_y^VD_{1-y}^V$. In general, the lattice constants, band gaps, indexes of refraction, and a number of other parameters of these materials depend on the values of x and y. These properties are intermediate between those of the parent binary compounds.

$Al_xGa_{1-x}As$ grown on GaAs was the first material system used for the preparation of injection lasers and as of this writing, a great majority of the lasers in use are based on it. These include the popular compact disc (CD) lasers and most of the high power semiconductor lasers. The lattice constants, a, of

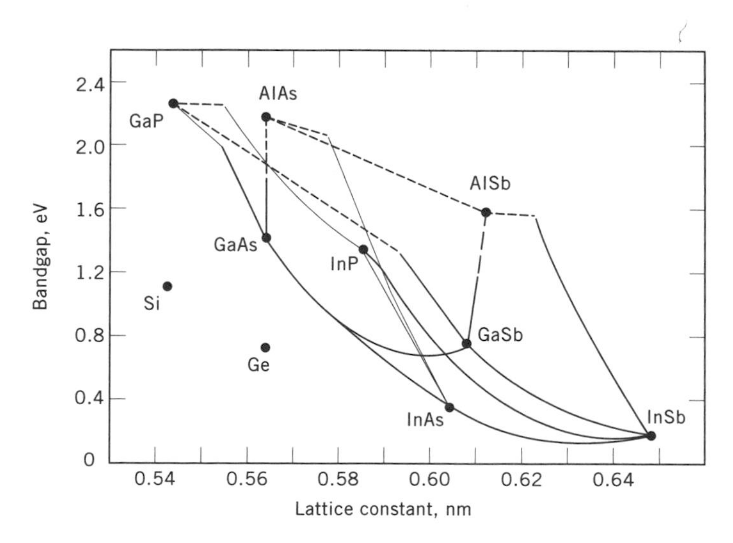

Fig. 6. Band gap versus lattice constant for Group 13–15 (III–V) semiconductors where (—) denotes direct gap and (- - -) indirect. Lines joining the binary semiconductors indicate possible compositions of ternary compounds. The quaternary compound $Ga_xIn_{1-x}As_{1-y}P_y$ may have any band gap and lattice constant that lie in the region of the plane bound by InP, GaP, GaAs, and InAs. Only the compositions where $x = 2.2\,y$ are lattice matched to InP.

GaAs and AlAs are almost identical, being 0.565 and 0.566 nm, respectively. Thus aluminum atoms can be substituted for Ga atoms in the GaAs lattice to form $Al_xGa_{1-x}As$ without any significant change in the lattice constant. It is possible to vary the bandgap from 1.43 eV, the E_g of GaAs to 2.16 eV, and the E_g of AlAs, simply by adjusting the Al fraction x in the epitaxial layer. This feature of $Al_xGa_{1-x}As$ is quite unique among the compound semiconductors. The band gap of the ternary alloy changes from direct to indirect for Al concentrations where $x > 0.43$. Compounds having lower Al concentrations can be used in active layers. Lasers having a large range of emission wavelengths, from nearly visible (red) to infrared, can be prepared in this way.

An even wider range of wavelength, toward the infrared, can be covered with quantum well lasers. In the $Al_xGa_{1-x}As$ system, compressively strained wells of $Ga_xIn_{1-x}As$ are used. This ternary system is indicated in Figure 6 by the line joining GaAs and InAs. In most cases the Al fraction is quite small, $x < 0.2$. Such wells are under compressive strain and their thickness must be carefully controlled in order not to exceed the critical layer thickness. Lasers prepared in this way are characterized by unusually low threshold current density, as low as ca 50 A/cm^2 (19).

The usual acceptor and donor dopants for $Al_xGa_{1-x}As$ compounds are from Groups 2 and 12 (II), 14 (IV), and 16 (VI) of the Periodic Table. Group 2 and 12 elements are acceptors and Group 16 elements are donors. Depending on the growth conditions and the growth method, Si and Ge (Group 14) can be either donors or acceptor, ie, be amphoteric, which is of special interest in light-emitting diodes. Doping levels in lasers range from mid-10^{17} cm^{-3} in the buffer and cladding layers to mid-10^{19} cm^{-3} in the contact layers. In quantum well-based lasers it is also desirable to vary the doping level very abruptly, ie, by two to three orders of magnitude, over dimensions as small as 0.01 μm. This requires dopants which do not move significantly (diffuse) at the growth temperature.

Another principal material system is $Ga_xIn_{1-x}As_{1-y}P_y$ grown on InP. A summary of compositions and doping levels of a laser heterostructure using such layers is presented in Table 1. Quaternary compounds are not naturally lattice matched to that substrate. In order to avoid generation of defects care must be taken in adjusting the ratio of x and y to maintain the lattice matched composition of $x = 2.2\,y$. The available band gaps range from that of InP to that of $In_{0.53}Ga_{0.47}As$, the ternary compound shown in Figure 6 by a line joining GaAs and InAs. The particular advantage of this system lies in the excellent match of available band gaps with the optical properties of fused silica optical fibers (17).

Table 1. Composition of a Double Heterostructure Laser[a,b]

Layer	Width, μm	Composition	Comment
contact	0.2	InGaAs	$p = 2 \times 10^{19}$
cladding layer	2	InP	$p = 2 \times 10^{18}$
waveguide	0.05	InGaAsP	$E_g = 1.15\ \mu m$
active	0.075	InGaAsP	$E_g = 1.30\ \mu m$
buffer	1	InP	$n = 2 \times 10^{18}$

[a]Based on InGaAsP/InP system.
[b]n-type InP substrate.

Quantum well lasers in this system typically use ternary $In_{0.53}Ga_{0.47}As$ wells and binary InP barriers. All quaternary lasers, ie, lasers in which both the wells and barriers are formed by quaternary compounds, are also being developed. These structures can be lattice matched or strained.

The materials discussed yield lasers operating in the infrared and near visible spectral ranges. Many applications of lasers, such as printing or high density memories, require as short a wavelength as possible. The III–V system most suitable for short wavelength visible operation is the $(Al_xGa_{1-x})_yIn_{1-y}P$ system (Fig. 7) (18). Direct band gap compositions of $(Al_xGa_{1-x})_{0.5}In_{0.5}P$ are lattice matched to GaAs and span the range from 2.25 eV (for $x = 0.3$), to 1.92 eV for $Ga_{0.5}In_{0.5}P$. Room temperature lasing up to 580 nm is possible, the shortest emission wavelength of any lattice matched III–V heterostructure. Because in a double heterostructure some of the bandgap must be used for carrier confinement, the shortest room temperature emission wavelength achieved as of this writing in continuous wave (CW) operation is 633 nm which matches that of a HeNe laser.

More exotic Group 13–15 (III–V) materials such as gallium nitride [*25617-97-4*] are being investigated for shorter lasing wavelengths. One of the prime candidates is the InGaN/GaN system in which blue (~410 nm) light-emitting

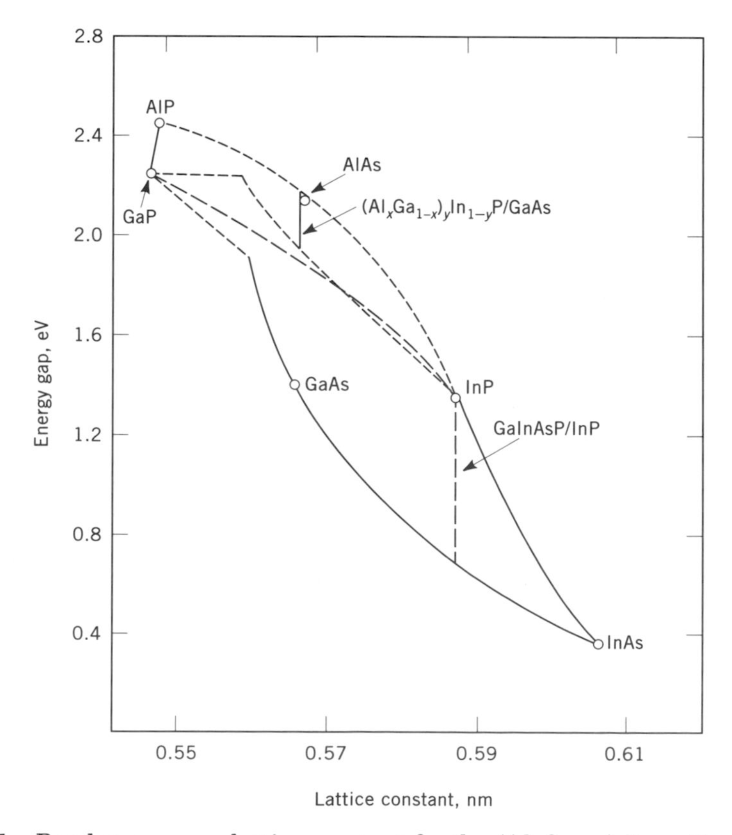

Fig. 7. Band gap versus lattice constant for the $(Al_xGa_{1-x})_yIn_{1-y}P$ system.

diodes have been prepared (20). The search is on for other large band gap III–V-like semiconductor systems.

Injection lasers based on Group 2 or 12–16 (II–VI) compounds have also been demonstrated (21). These large band gap compounds make green lasers a reality and promise an extension of the lasing wavelength into the blue. The first green lasers were based on a cadmium zinc selenide, $Cd_{0.2}Zn_{0.8}Se$ quantum well sandwiched between cladding layers consisting of zinc sulfur selenide, $ZnS_{0.07}Se_{0.93}$, and zinc selenide [*1315-09-9*], ZnSe, waveguide layers. The band gap–lattice constant plots for these materials are shown in Figure 8. The lattice constant of the ternary $ZnS_{0.07}Se_{0.93}$, but not that of ZnSe, matches that of (100) GaAs which serves as a substrate for this structure. The first lasers operated at cryogenic temperatures (77 K) and had an emission wavelength of 490 nm. Pulsed room temperature operation, and CW operation for very brief times (seconds) was also reported at 509 nm (22).

The absence of lattice match between the ZnSe waveguide and the substrate is a serious problem. In order to be effective as a buffer or waveguide this layer must be fairly thick, on the order of 0.2–1 μm. The strain is then too large to be accommodated elastically and the layer relaxes. The relaxation occurs through the generation of defects, ie, misfit dislocations, at the $ZnSe/ZnS_{0.07}Se_{0.93}$ interface. The presence of such defects significantly increases the threshold current density. This problem can be solved by using a quaternary compound, zinc magnesium sulfur selenide, $Zn_{1-x}Mg_xS_ySe_{1-y}$, also shown in Figure 8 (23). Cladding and waveguide layers can be grown lattice matched to GaAs, and their composition adjusted to yield a desired sequence of the band gap steps. Larger band gap range available for the cladding layer allows for increased band gap of the active layer and a shorter emission wavelength. Room temperature laser operation having a threshold current density as low as 500 A/cm^2 has been reported at 516 nm (24). This quaternary II–VI compound appears to have the flexibility analogous to that already exploited in InGaAsP.

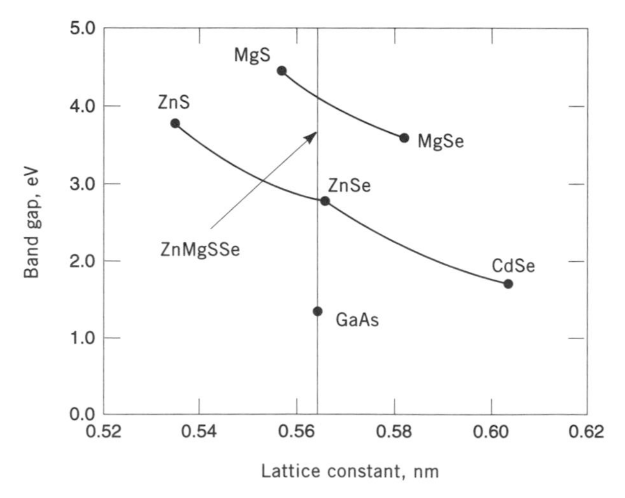

Fig. 8. A plot of band gap versus lattice constant for Groups 2–16 and 12–16 (II–VI) semiconductors used for the preparation of green and blue lasers.

The potential importance of II–VI compounds as visible laser materials has been recognized for some time. Injection lasers are difficult to prepare because of the inability to introduce dopants, and particularly *p*-type dopants. A convenient *n*-type dopant is chlorine, which can be introduced from $ZnCl_2$ sources. Fairly high doping levels, ca $10^{18}/cm^3$, can be readily achieved. Historically, the most common *p*-type dopants were lithium, Li, and nitrogen, N, the latter most often from NH_3. These dopant elements can be introduced into the semiconductor but only a small fraction becomes electrically active, ie, contributes to electrical conductivity. The *p*-type concentrations in large band gap materials had not even reached ca $10^{17}/cm^3$, a level too low to form *p–n* junctions or contact layers with low electrical resistance, until the impasse was broken by the discovery that nitrogen free-radicals produced by a high frequency plasma source could be incorporated up to a level of $p \sim 10^{18}/cm^3$ (25) (see PLASMA TECHNOLOGY). This allowed for successful preparation of diode lasers.

Laser Structures

Early injection lasers were small rectangular parallelepipeds made by cutting a wafer of GaAs. Feedback was provided by mirrors polished on two edges or by cleaving. The wafer had a *p–n* junction incorporated into it and broad area or stripe contacts were provided. Laser structures have since evolved to satisfy a wide range of application specific requirements.

$Al_xGa_{1-x}As$ lasers show outstanding performance in two areas: high power lasers and vertical cavity surface emitting lasers. High power lasers are based on phase coupled arrays (26) of the type illustrated in Figure 9. The individual lasers are formed simply by providing a narrow stripe contact, 5–10 μm in width. Proton implantation, which turns GaAs highly resistive, is used to confine the current to a narrow region close to the stripe (see ION IMPLANTATION). These lasers have no lateral index step to guide the light and are said to be gain guided.

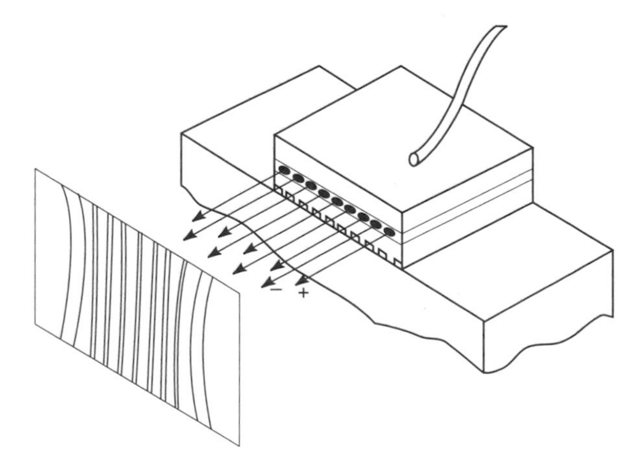

Fig. 9. High power array of phase coupled GaAs/AlGaAs lasers mounted *p*-side down on a thermal heat sink. The $\pi/2$ shift of the neighboring lasers is indicated by the + and − signs. The output pattern consists of two dominant peaks, each associated with the lasers of the same phase, and much weaker peaks in between.

Lasing occurs in the areas under and very near the stripe in which the carrier density necessary for threshold is reached. The array is formed by placing a number of individual lasers close enough, 5–10 μm from each other, so that their radiation fields influence each other, hence the name phase coupled. The advantage is a very narrow angle of emission in the plane of the array and high power output. The narrow stripe of individual lasers results in a fairly wide angle of in-plane emission, approximately 14° for a 5-μm wide stripe (6). An array of interacting lasers may operate as a spatially coherent unit. The wide aperture of such a unit results in a very low divergence beam. Because the electric field must go through zero between lasers, it changes its phase from laser to laser by $\pi/2$; the most likely emission pattern consists of two narrow beams each from all the lasers operating in phase. These two beams are emitted at a small angle from each other. The typical values are ca 1–1.5° for the width of each beam and about 4° for the separation between beams, respectively. The exact values depend strongly on the dimensions of the array and the current drive.

The large emission aperture and tight current confinement in laser arrays are important for high power operation. In GaAs lasers the output power is limited by catastrophic facet damage, the damage caused by a high light flux at the facet. It must be realized that the submicrometer active layer thickness of double heterostructure and the narrow stripe width result in radiative power fluxes on the order of MW/cm^2. Facet damage is known to occur at a flux of approximately 10 MW/cm^2 (27). The large aperture of laser arrays is very beneficial in this respect. Power output exceeding 100 mW/facet was achieved in a coupled array consisting of five lasers in 1978. The output powers have been vastly increased. For instance, a large array of 1716 lasers, arranged in a two-dimensional pattern, reached a peak power output of 1 kW at a current drive of 60 A (28). The power output of such large arrays is a matter of thermal management and proper heat sinking more than the individual laser properties. Finally, in these arrays the current is confined by a narrow stripe and the lasing filament is well defined. It stays reasonably well defined at current values many times greater than the threshold current. The result is a light output pattern stable as a function of power level.

GaAs-based lasers also excel in new applications directed toward optical computing and optical data processing (see also INFORMATION STORAGE MATERIALS). Such applications require two-dimensional arrays of lasers having a very low threshold current, high beam quality, and high speed of operation. These requirements are well served by vertical cavity surface emitting lasers (VCSELs). In the lasers discussed previously light propagates in the plane of the wafer, in a carefully constructed waveguide, and is emitted through a cleaved facet. The edge-emitting lasers are quite long, the typical length exceeds 250 μm, and the facet reflectivity is about 30%, the natural reflectivity of the semiconductor–air interface arising from the differences in respective indexes of refraction. It is also possible to rearrange the geometry of the laser to produce light emission in the direction perpendicular to the plane of the p–n junction, surface emission (29,30). In such lasers the cavity length is quite small (< 1 μm) and the lower gain must be compensated for by much higher facet reflectivity.

VCSELs are formed by placing the active layer between two highly reflecting mirrors, as illustrated in Figure 10. The Bragg mirrors, as they are called,

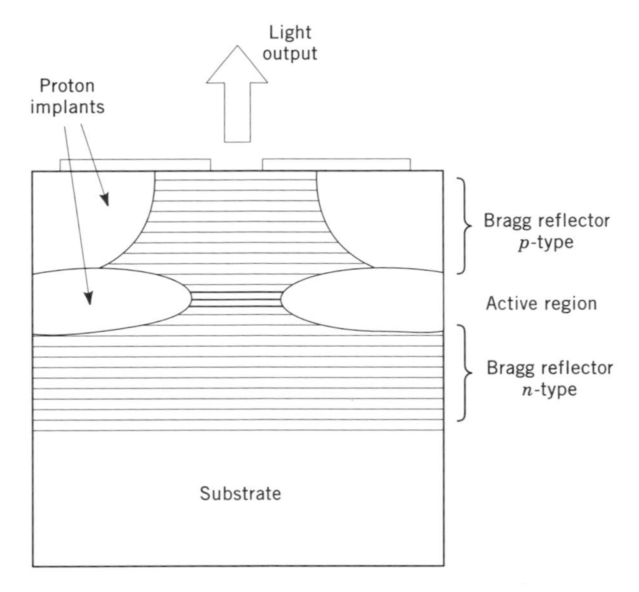

Fig. 10. Cross-sectional drawing of a vertical cavity surface emitting laser (VCSEL). Proton implantation is used to channel the current through a small active region. Light is emitted in the direction perpendicular to the plane of the wafer. This makes preparation of two-dimensional arrays quite easy.

are formed from alternating layers of semiconductor having high and low indexes of refraction, such as GaAs and AlAs. The thickness of each layer is chosen to be equal to one-fourth of the wavelength of light in the semiconductor, ie, $\lambda/(4n)$ where n is the index of refraction at the lasing wavelength. Because the index changes with the wavelength this condition can be strictly satisfied only at a single wavelength. The choice of thicknesses of individual layers results in a maximum reflectivity for each pair of layers at the lasing wavelength. Because the composition-induced index variation is not large, as many as 30 pairs must be used in order to produce a mirror having the needed reflectivity greater than 99%. The total active layer thickness is chosen to be λ/n, where n is the effective index of refraction. This condition allows for a standing wave to be formed between the lower and the upper mirrors. VCSELs emit light in a narrow mode at the wavelength at which a single-λ cavity is formed. Many of their properties are thus determined by the interplay between the complicated reflectivity spectra of the mirrors and the gain spectrum. This may result in very low temperature dependence of the threshold current and other desirable characteristics. These lasers show very low (< 1 mA) threshold currents, and modulation bandwidths, ie, the speed with which they can be turned on and off, can possibly reach 70 GHz (31). High reflectivity of the facet limits the light output to a few mW. The exact value depends strongly on the laser diameter.

Lasers fabricated from alloys of InGaAsP operate in the 1.3–1.55 μm wavelength range and are used in fiber optic communications (17,18) (see FIBER OPTICS). These wavelengths correspond, respectively, to the minimum dispersion and loss spectral windows of fused silica fibers. The lasers are designed to be coupled to optical fibers and must have excellent beam properties, narrow spectral width, and very high speed. All of these characteristics can be achieved at the same time but with a considerable increase in the structural complexity and difficulty of preparation. An example of a device designed for such demanding application is a buried heterostructure laser shown in Figure 11. This laser structure is known as capped mesa buried heterostructure (CMBH). It is prepared in three (or more) epitaxial growth cycles, all typically carried out by metalorganic chemical vapor deposition (MOCVD). The planar base structure containing the active, waveguide, and confining layers is grown first. It may be based on a conventional bulk active layer design or quantum wells, or contain a distributed feedback structure. The active layer mesa is formed by masking a narrow stripe using a dielectric layer, such as SiO_2, and etching off the remainder of the first growth. The width of the mesa is kept at approximately 0.8 μm in order to assure low threshold current and a narrow angle of emission. The narrow mesa width can be achieved using conventional wet etching only if the thickness of the InP cladding layer is < 1 μm. The current blocking layers are grown in the second epitaxial growth cycle in the exposed areas of the wafer but not on the active mesas which are protected by a dielectric mask. The blocking structure typically consists of a n–i–n–p doped sequence of layers of InP. Here the i-layer is semi-insulating, prepared by doping with iron which acts as a deep acceptor. The p-type InP layers on each side of the i-layer are intended to prevent injection of holes and electrons into the semi-insulating layer even at high injection levels or high operating temperatures. In the third cycle of growth the

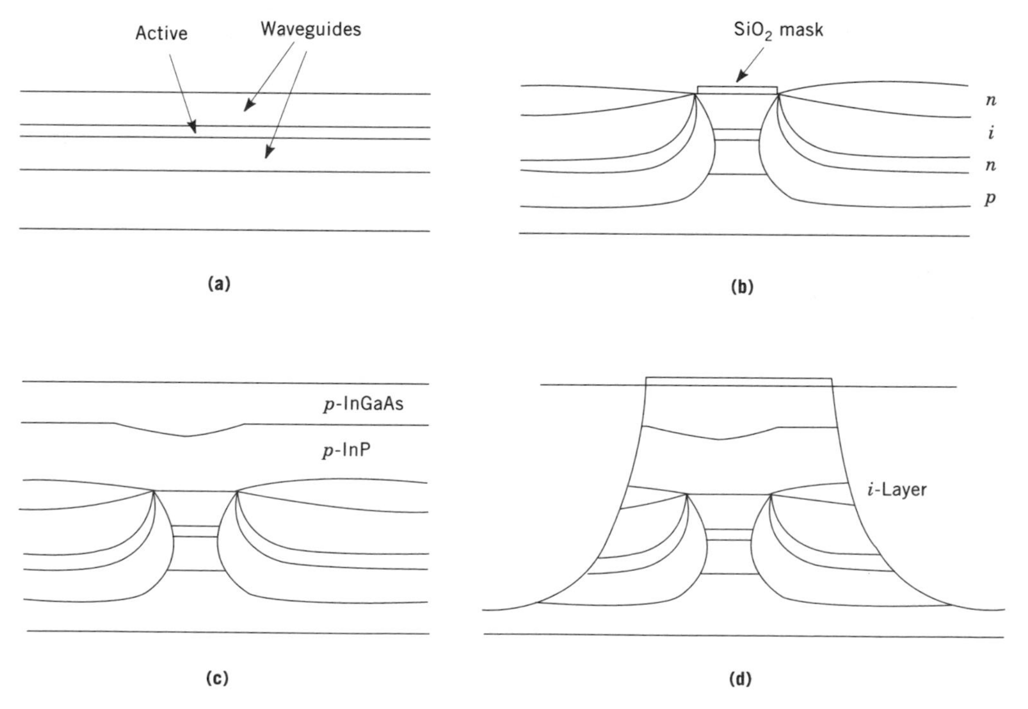

Fig. 11. Schematic diagram illustrating the preparation sequence of an InGaAsP/InP-based buried heterostructure laser. (**a**) The active and waveguide layers; (**b**) the SiO_2 mask; (**c**) the p-type regions; and (**d**) the injection layer. See text.

stripe mask is removed and *p*-type InP and GaInAs layers are grown over the entire wafer. The *p*-type layer of InP increases the cladding layer thickness to the final desired value and the *p*-type GaInAs contact layer assures low contact resistance.

The complex current blocking structure is effective in confining the current to the narrow active layer but it adds a parasitic capacitance, a feature undesirable in high speed operation. A fourth growth step may be added at this point. Most of the current blocking structure is removed by etching, leaving only a 20–50 μm wide ridge. The removed regions are then backfilled with Fe-doped InP in order to maintain planarity of the device.

The resulting laser offers excellent performance. The threshold currents are usually much lower than 10 mA and the light output is linear at least up to 20 mW. Specialized high power devices are capable of producing power levels in excess of 250 mW and having excellent beam characteristics. The modulation bandwidth of these lasers exceeds 25 GHz. Another feature of the buried heterostructure lasers fabricated from InGaAsP is their outstanding long-term stability. These lasers are reliably used in fiber optic communication cables spanning the oceans.

Most of the lasers discussed operate in a small number of discrete longitudinal modes, the Fabry-Perot modes. The individual modes are very narrow, much less than 0.01 nm, but are separated by spectral distances of ca 1.0 nm. Thus the overall width of the laser spectrum may exceed 4–5 nm. This is unacceptable in communication systems operating at 1.55 μm, the minimum loss region of optical fibers, over distances on the order of 100 km and data rates as high as 10 gigabites per second (Gb/s). This is because light of different wavelengths travels in optical fibers at slightly different speeds. This property of optical fibers results in time spreading of optical pulses traveling down the fiber and carrying information in digital form. Eventually it is not possible to distinguish between the individual pulses. The only way to avoid this problem is to work with lasers emitting light in a single and very narrow mode. The most effective of such devices is the distributed feedback (DFB) laser shown in a cross section of Figure 12.

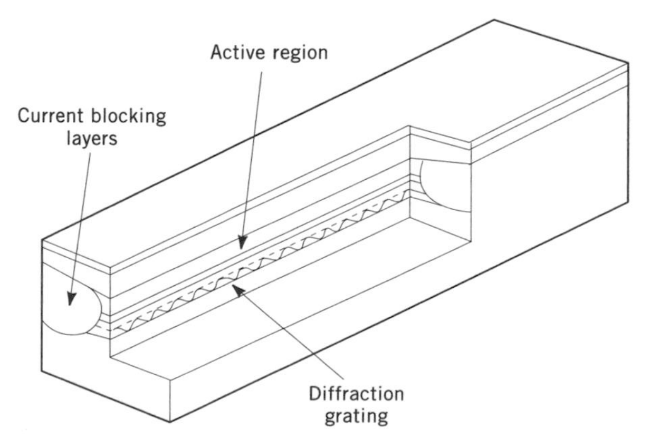

Fig. 12. Cut-out drawing of a distributed feedback (DFB) laser showing the active region and a diffraction grating, under the active layer, which produces the feedback.

Selection of a single longitudinal mode is accomplished by the introduction of a diffraction grating in one of the waveguide layers of a separate confinement cavity (17,32). The grating extends throughout the length of the cavity and produces feedback, hence the name of this laser, at a single wavelength. The wavelength depends on the period of the grating and the effective index of the cavity. Lasing at other wavelengths would require much higher gain and thus cannot be reached. The grating, which requires a fairly fine pitch and a period of ca 230 nm, is usually prepared by holographic methods (see HOLOGRAPHY). A successful addition of such a grating requires a great deal of sophistication in the crystal growth. The grating must be carefully preserved in the overgrowth. At the same time the overgrown layer must become planar after less than 0.1–0.2 μm of growth. If this is not accomplished the resulting undulations in the active layer introduce high loss at all wavelengths. What is desired is a periodic perturbation in the index of refraction of the active layer and not the width of the layer itself. After the DFB active layer structure is prepared it is then incorporated into a CMBH laser of the type discussed above.

When properly done the incorporation of a grating changes only the laser spectrum. The light-current characteristics of a DFB laser are illustrated in Figure 13**a**. The laser has low threshold current and the output power is highly linear to at least 50 mW. The spectrum, shown in Figure 13**b**, is characterized by a single narrow line at the feedback frequency of the grating. All other modes are suppressed; their intensities are at least a thousand times weaker than the main line. In DFB lasers based on strained quantum wells, the width of the laser

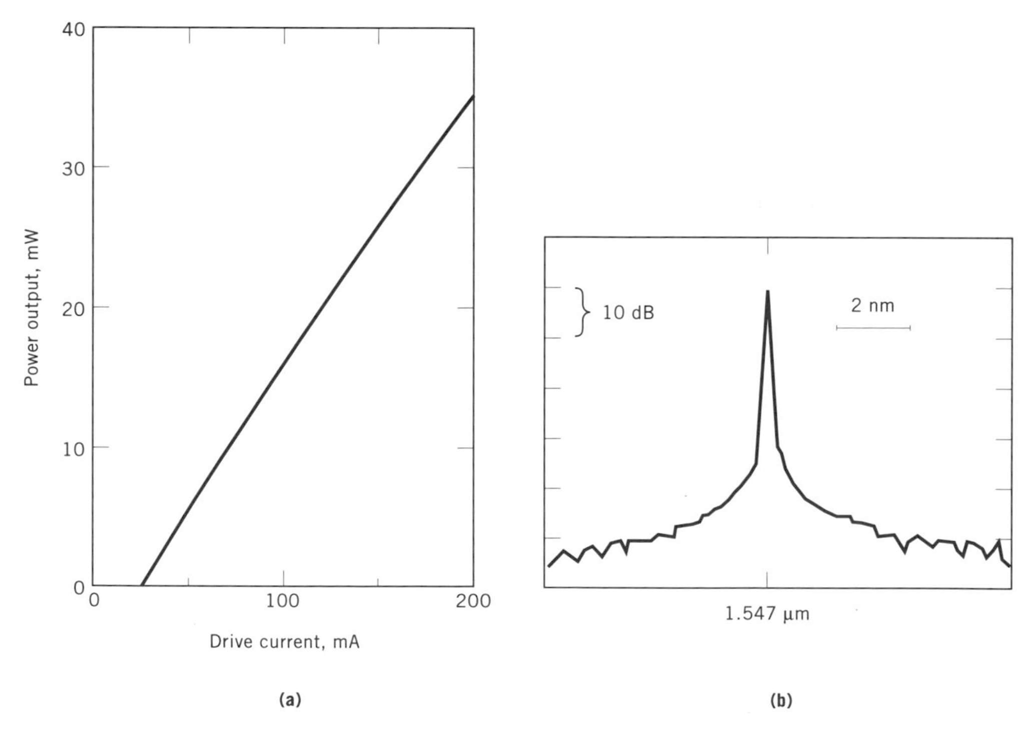

Fig. 13. Characteristics of a 50-μm long DFB laser. (**a**) Light-current properties, (**b**) spectral intensity plotted on a logarithmic scale to better illustrate suppression of undesired modes.

line is so narrow that it must be measured in units of frequency. In the best lasers of this type the linewidth is lower than 250 kHz or approximately 2×10^{-6} nm.

BIBLIOGRAPHY

"Light-Emitting Diodes and Semiconductor Lasers" in *ECT* 3rd ed., Vol. 14, pp. 269–294, by M. B. Parish, Bell Laboratories.

1. A. S. Grove, *Physics and Technology of Semiconductor Devices,* John Wiley & Sons, Inc., New York, 1967.
2. S. M. Sze, *Physics of Semiconductor Devices,* John Wiley & Sons, Inc., New York, 1981.
3. B. G. Streetman, *Solid State Electronic Devices,* Prentice Hall, Englewood Cliffs, N.J., 1990.
4. M. Shur, *Physics of Semiconductor Devices,* Prentice Hall, Englewood Cliffs, N.J., 1990.
5. A. Einstein, *Phys. Z.* **18**, 121 (1917).
6. H. C. Casey, Jr. and M. B. Panish, *Heterostructure Lasers,* Academic Press, Inc., New York, 1978.
7. M. G. A. Bernard and G. Duraffourg, *Phys. Stat. Solidi.* **1**, 699 (1961).
8. C. H. Henry, in Y. Yamamoto, ed., *Coherence, Amplification, and Quantum Effects in Semiconductor Lasers,* John Wiley & Sons, Inc., New York, 1991.
9. H. Kroemer, *Proc. IEEE* **51**, 1782 (1963).
10. USSR Pat. 181,737 (1963), Zh. I. Alferov and R. F. Kazarinov.
11. G. Bastard, *Wave Mechanics Applied to Semiconductor Heterostructures,* Halsted Press, New York, 1988.
12. P. S. Zory, Jr., ed., *Quantum Well Lasers,* Academic Press, Inc., Boston, 1993.
13. G. C. Osbourn, *Phys. Rev.* **B27**, 5126 (1983).
14. J. W. Matthews and A. E. Blakeslee, *J. Cryst. Growth* **27**, 118 (1974).
15. E. Yablonovitch and E. O. Kane, *J. Lightwave. Technol.* **LT-4**, 504 (1986).
16. T. P. Pearsall, ed., *GaInAsP Alloy Semiconductors,* John Wiley & Sons, Inc., New York, 1982.
17. G. P. Agrawal and N. K. Dutta, *Semiconductor Lasers,* van Nostrand Reinhold, New York, 1993.
18. M. B. Panish and H. Temkin, *Gas Source Molecular Beam Epitaxy,* Springer Verlag, Berlin, 1994.
19. N. Chand and co-workers, *Appl. Phys. Lett.* **58**, 1704 (1991).
20. S. Nakamura, M. Senoh, and T. Mukai, *Appl. Phys. Lett.* **62**, 2390 (1993).
21. M. A. Haase and co-workers, *Appl. Phys. Lett.* **59**, 1272 (1991).
22. A. Salokatve and co-workers, *Electron. Lett.,* (Nov. 1993).
23. H. Okuyama and co-workers, *J. Cryst. Growth* **117**, 139 (1992).
24. J. M. Gaines and co-workers, *Appl. Phys. Lett.* **62**, 2462 (1993).
25. R. M. Park and co-workers, *Appl. Phys. Lett.* **57**, 2127 (1990).
26. D. Scifres, R. D. Burnham, and W. Streifer, *Appl. Phys. Lett.* **33**, 1015 (1978).
27. C. H. Henry and co-workers, *J. Appl. Phys.* **50**, 3721 (1979).
28. A. Rosen and co-workers, *IEEE Photonics Techn. Lett.* **1**, 43 (1989).
29. K. Iga, F. Koyama, and S. Kinoshita, *IEEE J. Quant. Electron.* **24**, 1845 (1988).
30. J. Jewell and co-workers, *IEEE J. Quant. Electron.* **27**, 1332 (1991).
31. D. Tauber and co-workers, *Appl. Phys. Lett.* **62**, 325 (1993).
32. H. Temkin and co-workers, *Appl. Phys. Lett.* **57**, 1295 (1990).

HENRYK TEMKIN
Colorado State University

LIGNIN

The word lignin is derived from the Latin word *lignum* meaning wood. It is a main component of vascular plants. Indeed, lignin is second only to polysaccharides in natural abundance, contributing 24–33% and 19–28%, respectively, to dry wood weights of normal softwoods and temperate-zone hardwoods.

According to a widely accepted concept, lignin [*8068-00-6*] may be defined as an amorphous, polyphenolic material arising from enzymatic dehydrogenative polymerization of three phenylpropanoid monomers, namely, coniferyl alcohol [*485-35-5*] (**2**), sinapyl alcohol [*537-35-7*] (**3**), and *p*-coumaryl alcohol (**1**).

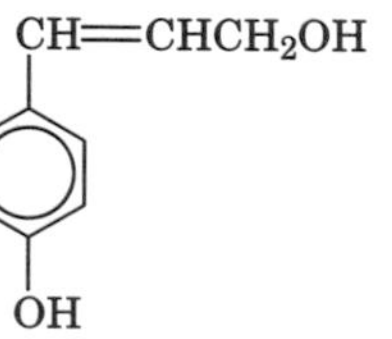

$CH{=}CHCH_2OH$ / OH (**1**)

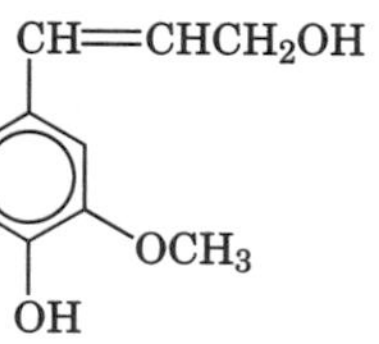

$CH{=}CHCH_2OH$ / OCH_3 / OH (**2**)

$CH{=}CHCH_2OH$ / CH_3O / OCH_3 / OH (**3**)

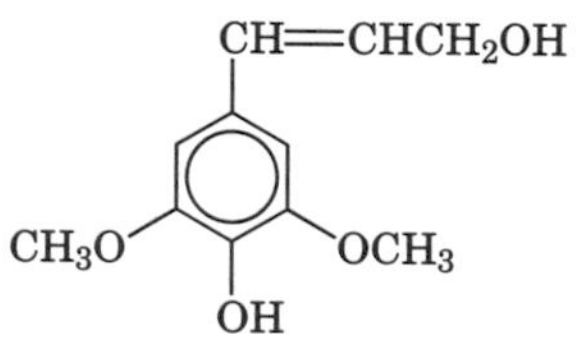

The biosynthesis process, which consists essentially of radical coupling reactions, sometimes followed by the addition of water, of primary, secondary, and phenolic hydroxyl groups to quinonemethide intermediates, leads to the formation of a three-dimensional polymer which lacks the regular and ordered repeating units found in other natural polymers such as cellulose and proteins.

Normal softwood lignins are usually referred to as guaiacyl lignins because the structural elements are derived principally from coniferyl alcohol (more than 90%), with the remainder consisting mainly of *p*-coumaryl alcohol-type units. Normal hardwood lignins, termed guaiacyl–syringyl lignins, are composed of coniferyl alcohol and sinapyl alcohol-type units in varying ratios. In hardwood lignins, the methoxyl content per phenylpropanoid unit is typically in the range of 1.2–1.5 (1). Grass lignins are also classified as guaiacyl–syringyl lignins. However, unlike hardwood lignins, grass lignins additionally contain small but significant amounts of structural elements derived from *p*-coumaryl alcohol. Grass lignins also contain *p*-coumaric, hydroxycinnamic, and ferulic acid residues attached to the lignin through ester and ether linkages (2).

The distribution of lignin in individual cells of lignified wood has been well examined. The lignin concentration is rather uniform across the secondary wall, but there is a significant increase in lignin concentration at the boundary of the middle lamella and primary wall region (3). This pattern of lignin distribution, with the highest concentration in the interfiber region and a lower, uniform concentration in the bulk of the cell walls, is typical for most wood cells. Thus lignin serves the dual purpose of binding and stiffening wood fibers through its distribution between and in the cell walls.

Lignin performs multiple functions that are essential to the life of the plant. By decreasing the permeation of water across the cell wall in the conducting xylem tissues, lignin plays an important role in the internal transport of water, nutrients, and metabolites. It imparts rigidity to the cell walls and acts as a binder between wood cells, creating a composite material that is outstandingly resistant to compression, impact, and bending. It also imparts resistance to biological degradation.

In commercial chemical pulping of wood, the reverse process in nature is performed to isolate fibers for papermaking. In the process, wood is delignified by chemically degrading and/or sulfonating the lignin to water-soluble fragments. The industrial lignins thus obtained are used in many applications.

Structure and Reactions

The structural building blocks of lignin are linked by carbon–carbon and ether bonds (4,5). Units that are trifunctionally linked to adjacent units represent branching sites which give rise to the network structure characteristic of lignin (see Figs. 1 and 2). The types and frequencies of several prominent interunit lignin linkages are summarized in Table 1.

Because the interunit carbon–carbon linkages are difficult to rupture without extensively fragmenting the carbon skeleton of the lignin, solvolysis of the ether linkages is often utilized as the best approach for degrading lignin. Of the functional groups attached to the basic phenylpropanoid skeleton, those having the greatest impact on reactivity of the lignin include phenolic hydroxyl, benzylic hydroxyl, and carbonyl groups. The frequency of these groups may vary according to the morphological location of lignin, wood species, and method of isolation.

Electrophilic Substitution. The processes by which the aromatic ring in lignin is modified by electrophilic substitution reactions are chlorination, nitration, and ozonation. Chlorination, widely used in multistage bleaching sequences for delignifying chemical pulps, proceeds by a rapid reaction of elemental chlorine with lignin in consequence of which the aromatic ring is nonuniformly substituted with chlorine. In nitration, nitro groups are introduced into the aromatic moiety of lignin with nitrogen dioxide (9). As one of several competing processes, electrophilic attack of ozone on lignin ultimately leads to ring hydroxylation (10).

Conversion of Aromatic Rings to Nonaromatic Cyclic Structures. On treatment with oxidants such as chlorine, hypochlorite anion, chlorine dioxide, oxygen, hydrogen peroxide, and peroxy acids, the aromatic nuclei in lignin typically are converted to *o*- and *p*-quinoid structures and oxirane derivatives of quinols. Because of their relatively high reactivity, these structures often appear as transient intermediates rather than as end products. Further reactions of the intermediates lead to the formation of catechol, hydroquinone, and mono- and dicarboxylic acids.

Aromatic rings in lignin may be converted to cyclohexanol derivatives by catalytic hydrogenation at high temperatures (250°C) and pressures (20–35 MPa (200–350 atm)) using copper–chromium oxide as the catalyst (11). Similar reduction of aromatic to saturated rings has been achieved using sodium in liquid ammonia as reductants (12).

Conversion of Cyclic to Acyclic Structures. Upon oxidation, the aromatic rings of lignin may be converted directly to acyclic structures, eg, muconic acid derivatives, or indirectly by oxidative splitting of *o*-quinoid rings. Further oxidation creates carboxylic acid fragments attached to the lignin network.

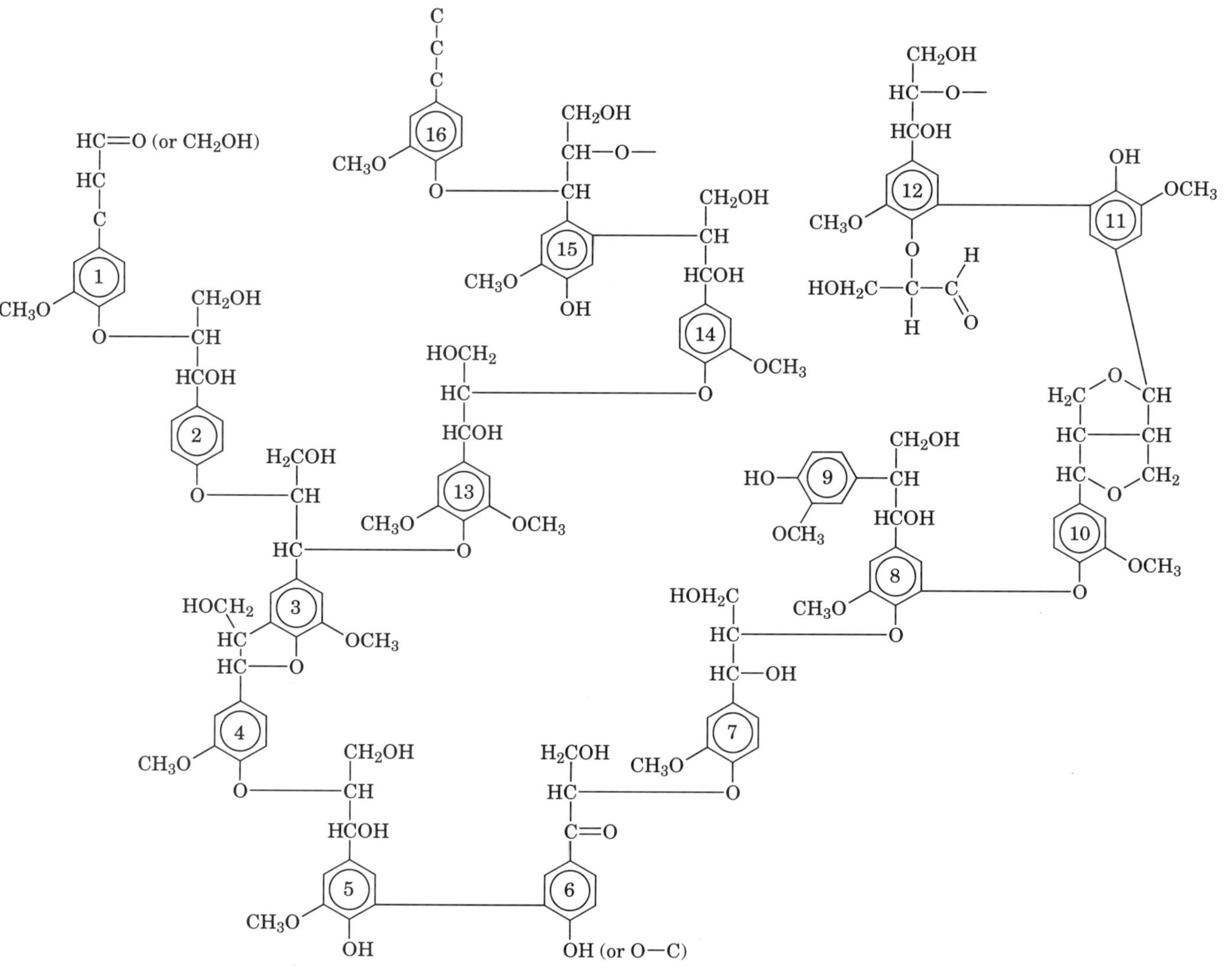

Fig. 1. Structural model of spruce lignin (6).

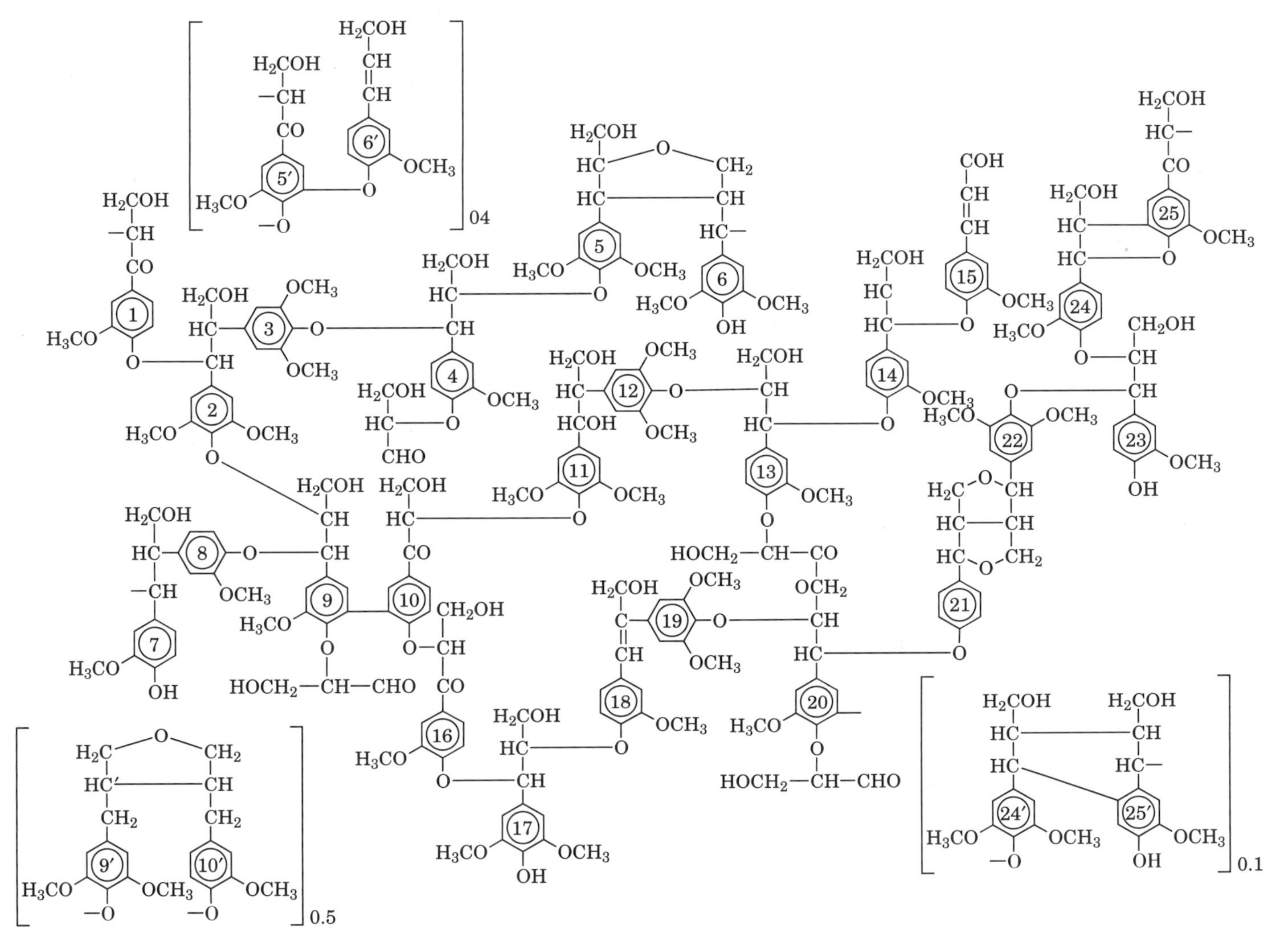

Fig. 2. Structural model of beech lignin (7).

Table 1. Types and Frequencies of Interunitary Linkages in Softwood and Hardwood Lignins (Number of Linkages per 100 C_9 Units)

Linkage	Softwood lignin[a]	Hardwood lignin[b]
β-O-4	49–51	65
α-O-4	6–8	
β-5	9–15	6
β-1	2	15
5-5	9.5	2.3
4-O-5	3.5	1.5
β-β	2	5.5

[a] Ref. 8.
[b] Ref. 7.

Ring Coupling and Condensation Reactions. Many oxidants, eg, ClO_2, O_2, generate free radicals in lignin. Coupling of such reactive radicals ultimately leads to diphenyl structures. In alkaline media, phenolic units may react with formaldehyde forming methylol derivatives that condense with themselves or with other phenols. This formaldehyde condensation reaction is the basis for using technical lignins in the preparation of adhesives.

Cleavage of Ether Bonds. Ether linkages at the α- and β-positions are the most abundant functional groups on the propanoid side chain of lignin. Under acid conditions these linkages undergo solvolytic cleavage initially forming secondary alcohols which are converted to carbonyl, ethylene, and carboxyl structures through a combination of dehydrations and allylic rearrangements, leading eventually to fragmentation of the side chain (13).

The alkali-promoted cleavage of α- and β-ether linkages, an important step in alkaline pulping processes, is mainly responsible for the fragmentation and dissolution of lignin in the pulping liquor. Addition of bisulfide ion to the aqueous alkaline media, as in the case of kraft pulping, enhances the rate and extent of β-aryl ether cleavage in phenolic units (14).

Cleavage of Carbon–Carbon Bonds. Under appropriate conditions, the propanoid side chain in lignin may be ruptured to form three-, two-, or one-carbon fragments. This carbon–carbon fragmentation occurs in a variety of laboratory treatments and technical processes such as in bleaching of chemical pulps with Cl_2, ClO_2, and O_2, in microbial degradation (15), and in photooxidation (16).

Substitution Reactions on Side Chains. Because the benzyl carbon is the most reactive site on the propanoid side chain, many substitution reactions occur at this position. Typically, substitution reactions occur by attack of a nucleophilic reagent on a benzyl carbon present in the form of a carbonium ion or a methine group in a quinonemethide structure. In a reversal of the ether cleavage reactions described, benzyl alcohols and ethers may be transformed to alkyl or aryl ethers by acid-catalyzed etherifications or transetherifications with alcohol or phenol. The conversion of a benzyl alcohol or ether to a sulfonic acid group is among the most important side chain modification reactions because it is essential to the solubilization of lignin in the sulfite pulping process (17).

Formation and Elimination of Multiple Bond Functionalities. Reactions that involve the formation and elimination of multiple bond functional groups may significantly effect the color of residual lignin in bleached and unbleached pulps. The ethylenic and carbonyl groups conjugated with phenolic or quinoid structures are possible components of chromophore or leucochromophore systems that contribute to the color of lignin.

Reduction of ring-conjugated carbonyl groups to the corresponding primary and secondary alcohols is generally achieved by reaction with sodium borohydride. Ring-conjugated olefinic groups may be converted to their saturated components by hydrogenation.

Analytical Methods

Detection of Lignin. The characteristic color-forming response of lignified tissue and some lignin preparations on treatment with certain organic and inorganic reagents was recognized in the early nineteenth century. More than 150 color reactions have now been proposed for the detection of lignin (18). Reagents used in these reactions may be classified into aliphatic, phenolic, and heterocyclic compounds, aromatic amines, and inorganic chemicals. Among the important reactions are the Wiesner and Mäule color reactions.

The Wiesner Reaction. The reaction of lignified tissue and phloroglucinol–hydrochloric acid gives a visible absorption spectrum with a maximum at 550 nm. This has been attributed to coniferaldehyde units in lignin as the groups responsible for the color formation.

RO–C$_6$H$_3$(OCH$_3$)–CH═CHCHO $\xrightarrow[\text{HCl}]{\text{phloroglucinol}}$ RO–C$_6$H$_3$(OCH$_3$)–CH═CH$\overset{+}{\text{C}}$H–C$_6$H$_2$(OH)$_3$

λ_{max} 550 nm

The Mäule Color Reaction. The procedure for this test consists basically of three sequential treatments of lignified material with 1% potassium permanganate, 3% hydrochloric acid, and concentrated ammonium hydroxide. A red-purple color develops for hardwoods and a brown color for softwoods. The steps comprising the Mäule reaction may be portrayed as follows (19,20):

[syringyl unit: –C– ring with CH$_3$O, OCH$_3$, OH] $\xrightarrow{\text{KMnO}_4\text{–HCl}}$ [–C– ring with Cl, CH$_3$O, OH, OH] $\xrightarrow{\text{NH}_4\text{OH}}$ [–C– chloro-o-quinone with Cl, CH$_3$O, O, O]

λ_{max} 515–517 nm

Determination of Lignin Content. Lignin content in plants (wood) is determined by direct or indirect methods (21). The direct method includes measurement of acid-insoluble (ie, Klason) lignin after digesting wood with 72% sulfuric acid to solubilize carbohydrates (22). The Klason lignin contents of representative lignified materials are shown in Table 2.

In contrast to the direct determination of lignin content, indirect methods do not involve the isolation of a lignin residue. These include spectrophotometric methods and procedures that are based on oxidant consumption. A uv microspectrophotometric method has been used to determine the distribution of lignin in the various cell wall regions of softwoods (23). Supplementing the uv-microscopic technique is a method in which lignin is brominated and the bromine uptake, which is proportional to the lignin content, is determined by a combination of scanning or transmission electron microscopy (sem or tem) and energy dispersive x-ray analysis (edxa) (24). A number of spectral methods for determining lignin content are based on totally dissolving the sample in a suitable solvent and measuring the uv absorbance of the solution. Among the solvents used to dissolve lignocellulosic material are sulfuric acid, phosphoric acid, nitric acid, cadoxene, and acetyl bromide in acetic acid. The acetyl bromide method appears to have gained the most widespread acceptance (25).

The methods of oxidant consumption are used exclusively in the analysis of residual lignin in unbleached pulps. These procedures are all based on the common principle that lignin consumes the applied oxidants at a much faster rate than the carbohydrates, and oxidant consumption under carefully specified conditions can be regarded as a measure of lignin concentration in the pulp.

Two oxidants commonly used are chlorine and potassium permanganate. The Roe chlorine number, the uptake of gaseous chlorine by a known weight of unbleached pulp (ie, Technical Association of the Pulp and Paper Industry (TAPPI) Standard Method T202 ts-66) has been superseded by the simpler hypo number (ie, TAPPI Official Test Method T253 om-86), eg, chlorine consumption in treatment of the pulp with acidified sodium or calcium hypochlorite.

Table 2. Klason Lignin Contents of Lignified Materials[a]

Material	Klason lignin, %
softwoods	26–28.8
hardwoods	22
nonwood fibers	
bagasse	19.6
bamboo	22.2
wheat straw	17.0
kenaf	10.9
sorghum	7.9
pulp	
pine kraft	4.8
birch kraft	5.0
spruce kraft	2.8
birch acid sulfite	3.2
birch bisulfite	4.0

[a]Ref. 21.

By far the most commonly used oxidation method is the corrected permanganate number test (26) in which the number of mL of 0.1 N $KMnO_4$ consumed by 1 g of oven-dried pulp under specified conditions (kappa number) is determined (TAPPI Historical Method T236 hm-85). Typical kappa numbers for representative pulps are shown in Table 3.

Characterization of Lignin. Lignin is characterized in the solid state by Fourier transform infrared spectroscopy (ftir), uv microscopy, interference microscopy, cross polarization/magic angle spinning nuclear magnetic resonance spectroscopy (cp/mas nmr), photoacoustic spectroscopy, Raman spectroscopy, pyrolysis-gas chromatography–mass spectroscopy, and thermal analysis. In solution, lignins are characterized by spectral methods such as uv spectroscopy, ftir, proton nmr, carbon-13 nmr, electron spin resonance spectroscopy (esr), and by several chemical degradation methods such as acidolysis, nitrobenzene and cupric oxidations, permanganate oxidation, thioacidolysis, hydrogenolysis, nuclear exchange reaction, and ozonation. The details of these characterization methods have been discussed (27).

Ftir spectroscopy is a versatile, rapid, and reliable technique for lignin characterization. Using this technique, the *p*-hydroxyphenyl, guaiacyl, and syringyl units, methoxyl groups, carbonyl groups, and the ratio of phenolic hydroxyl to aliphatic hydroxyl groups can be determined. The uv microscopy method is best suited for investigating the topochemistry of lignin in wood, namely, for determining the concentration and chemical structure of lignin in different layers of the cell wall. Cp/mas nmr spectroscopy provides for another spectral technique whereby lignin can be characterized in the solid state. Results obtained by cp/mas nmr are in good agreement with Klason lignin contents for softwoods.

In solution, lignin is most conveniently analyzed qualitatively and quantitatively by uv spectroscopy. Typical absorptivity values, D, at 280 nm for milled wood (MW) lignins and other types of lignins are listed in Table 4. These values are used for quantitative determination of the lignins in suitable solvents.

Proton and carbon-13 nmr spectroscopy provides detailed information on all types of hydrogen and carbon atoms, thus enabling identification of functional groups and types of linkages in the lignin structure. Detailed assignments of signals in proton and carbon-13 nmr spectra have been published (29,30).

Table 3. Kappa Numbers for Typical Pulps[a]

Pulp	Kappa number range
kraft (bleached grade)	
softwood	25–35
hardwood	14–18
neutral sulfite semichemical (softwood)	80–100
bisulfite (softwood)	30–50
acid sulfite	
softwood	16–22
hardwood	14–20
kraft (chlorinated and alkali extracted)	
softwood	5–8
hardwood	3–6

[a]Ref. 21.

Table 4. Absorptivity Values, *D*, of Lignin at 280 nm[a]

Lignin	D, $L(g \cdot cm)^{-1}$	Solvent
spruce MW	16.7	2-methoxyethanol
spruce MW	20.7	formamide
spruce MW	19.5	dioxane
pine MW	18.8	2-methoxyethanol/ethanol
beach MW	13.3	formamide
maple MW	12.9	2-methoxyethanol/ethanol
poplar dioxane	12.6	dioxane
spruce lignosulfonate	11.9	water
beech lignosulfonate	10.4	water
pine kraft	24.6	water
	26.4	2-methoxyethanol/water

[a]Ref. 28.

Electron spin resonance (esr) or electron paramagnetic resonance (epr) spectroscopy is an essential tool for the study of structure and dynamics of molecular systems containing one or more unpaired electrons. These methods have found application as a highly sensitive tool for the detection and identification of free-radical species in lignin and lignin model compounds (31,32). Milled wood lignin generally exhibits a singlet esr signal with a g-value of 2.0023 and a line width of 1.6 mT (16 gauss), typical of a phenoxy radical.

Among the chemical degradation methods, acidolysis, nitrobenzene and cupric oxide oxidations, permanganate oxidation, thioacidolysis, and hydrogenolysis are all based on a common principle of chemically degrading lignin polymers to identifiable low molecular weight products through side-chain cleavages and maintaining the aromatic nature of the lignin units. By these methods, the make-up of monomeric units in the lignin (eg, guaiacyl–syringyl–*p*-hydroxyphenyl ratio) is determined. In addition, the identification of dimeric and trimeric degradation products reveals the types of linkages existing in the lignin.

A technique based on ozonation, in contrast, provides information on the structure of the lignin side chain by degrading the aromatic rings (33). Thus the side chain of the dominant structure in all native lignins, the arylglycerol–β-aryl ether moiety, can be obtained in the form of erythronic and threonic acids. Ozonation proves to be an elegant method for determination of the stereospecificity in lignin.

The quantities of noncondensed and condensed phenyl nuclei in various lignins and in the morphological regions of cell walls are determined by a nucleus exchange method (34). The data obtained from this method indicate that lignin in the middle lamella is more condensed than lignin in the secondary wall and that hardwood lignin is less condensed than softwood lignin. By combining nucleus exchange with nitrobenzene oxidation, the methylol groups formed in the condensation of lignin with formaldehyde can be directly measured without isolation of the lignin.

Functional Group Analysis. The total hydroxyl content of lignin is determined by acetylation with an acetic anhydride–pyridine reagent followed by saponification of the acetate, and followed by titration of the resulting acetic acid with a standard 0.05 N sodium hydroxide solution. Either the Kuhn-Roth (35) or the modified Bethge-Lindstrom (36) procedure may be used to determine the total hydroxyl content. The aliphatic hydroxyl content is determined by the difference between the total and phenolic hydroxyl contents.

The phenolic hydroxyl group is one of the most important functionalities affecting the chemical and physical properties of lignin. It promotes the base-catalyzed cleavage of interunitary ether linkages and oxidative degradation. It has a pronounced influence in the reactivity of lignin polymers in various modification reactions such as methylolation with formaldehyde. A method called the periodate method is based on the oxidation of a phenolic guaiacyl group with sodium periodate to orthoquinone structures, wherein nearly one mole of methanol per mole of phenolic hydroxyl group is released (37). Measurement of the methanol formed is approximately equivalent to the phenolic hydroxyl content. In an aminolysis procedure (36) consisting of acetylation of lignin and aminolysis with pyrrolidine to remove acetyl groups such as 1-acetylpyrrolidine, the amount of removed acetyl is a measure of the phenolic hydroxyl content of lignin. These and other procedures for determining phenolic hydroxyl groups have been compared (38). Table 5 lists the total phenolic and aliphatic hydroxyl contents of some representative milled wood, bamboo, and technical lignins.

As early as 1922, the presence of carbonyl groups in spruce lignin was postulated (40). Coniferaldehyde [*458-36-6*] has definitely been identified as a building block in lignin, and the α-carbonyl content has been found to increase in the milling of wood and during pulping processes. The total carbonyl content of lignin is determined by a borohydride or hydroxylamine hydrochloride method (41), and the α-carbonyl content from analysis of uv alkaline difference spectra.

The method of choice for determining carboxyl groups in lignin is based on potentiometric titration in the presence of an internal standard, *p*-hydroxybenzoic acid, using tetra-*n*-butylammonium hydroxide as a titrant (42). The carboxyl contents of different lignins are shown in Table 6. In general, the carboxyl content of lignin increases upon oxidation.

Methoxyl groups are determined by the Viebock and Schwappach procedure (44). In treatment of lignin with hydroiodic acid, the methoxyl group is cleaved forming methyl iodide which is quantitatively stripped from the reaction mixture and collected in a solution of sodium acetate and glacial acetic acid containing bromine. The bromine reacts with methyl iodide to form alkyl bromide and iodine bromide. The iodine thus produced is titrated with a dilute standard sodium thiosulfate solution using 1% starch solution as an indicator. The

Table 5. Phenolic and Aliphatic Hydroxyl Contents of Milled Wood and Technical Lignins[a]

	Hydroxyl content, mol/C_9 unit		
Lignin	Total	Phenolic	Aliphatic
spruce MWL	1.46	0.28	1.18
bamboo MWL	1.49	0.36	1.13
pine kraft lignin	1.35	0.58	0.77
bamboo kraft lignin	1.00	0.44	0.56

[a]Ref. 39.

Table 6. Carboxyl Contents of Various Lignins[a]

Lignin	COOH, meq/g
hardwood kraft	1.44
hardwood native	0.92
lignosulfonates	0.31–2.08
wheat straw MWL	0.81
spruce MWL	0.12
decayed spruce	0.55
softwood kraft	0.80

[a]Ref. 44.

methoxyl content can be quantitatively determined with high accuracy based on the quantity of iodine recovered.

Finally, the sulfonate content of lignin is determined by two main methods: one typified by conductometric titration in which sulfonate groups are measured directly, and the other which measures the sulfur content and assumes that all of the sulfur is present as sulfonate groups. The method of choice for determining the sulfonate content of lignin samples that contain inorganic or nonsulfonate sulfur, however, is conductometric titration (45).

Properties

Molecular Weight and Polydispersity. Because it is not possible to isolate lignin from wood without degradation, the true molecular weight of lignin in wood is not known. Different methods for measuring the molecular weight of isolated lignins give various results, and aggregation of lignin molecules may prevent determination of real molecular weight. The weight-average molecular weight, $\overline{M}_w$, of softwood milled wood lignin is estimated to be 20,000; lower values have been reported for hardwoods (46).

Table 7 shows $\overline{M}_w$ and $\overline{M}_n$ (number-average molecular weight) values of kraft lignins and lignosulfates determined by light scattering and vapor pressure osmometry, respectively. Kraft lignins invariably have lower molecular weights than lignosulfonates, indicative of a more extensive degradation of the lignin during the kraft pulping process.

The polydispersity of softwood milled wood lignin, as measured by $\overline{M}_w/\overline{M}_n = 2.5$, is high compared with that of cellulose and its derivatives. Other lignins show different polydispersity as demonstrated by high pressure size exclusion chromatograms (47). The polydispersity of lignosulfates is much greater, with $\overline{M}_w/\overline{M}_n$ ratios in the range of 6–8 (48).

Table 7. $\overline{M}_w$ and $\overline{M}_n$ for Kraft Lignin and Lignosulfonate

Lignin	$\overline{M}_w$	$\overline{M}_n$
kraft		
pine	5,600–25,000	1,400–2,500
slash pine	13,700–48,300	2,300–6,100
lignosulfonate	7,900–126,000	

Solution Properties. Lignin in wood behaves as an insoluble, three-dimensional network. Isolated lignins (milled wood, kraft, or organosolv lignins) exhibit maximum solubility in solvents having a Hildebrand's solubility parameter, δ, of $20.5 - 22.5\ (\mathrm{J/cm^3})^{1/2}$ $(10 - 11\ (\mathrm{cal/cm^3})^{1/2}$, and $\Delta\mu$ in excess of 0.14 micrometer where $\Delta\mu$ is the infrared shift in the O–D bond when the solvents are mixed with CH_3OD. Solvents meeting these requirements include dioxane, acetone, methyl cellosolve, pyridine, and dimethyl sulfoxide.

Thermal Properties. As an amorphous polymer, lignin behaves as a thermoplastic material undergoing a glass transition at temperatures which vary widely depending on the method of isolation, sorbed water, and heat treatment (49). Lignin stores more energy than cellulose (qv) in wood. For example, the glass-transition temperature, T_g, and heat capacity at 350 K for dioxane lignin are 440 K and 1.342 J/(g·K), respectively (50). Thermal softening of lignin at elevated temperatures accelerates the rate of delignification in chemical pulping and enhances the bond strength of fibers in paper- and boardmaking processes. In commercial thermomechanical pulping, a pretreatment of wood chips with sulfite lowers the glass-transition temperature of lignin to 70–90°C (50), thus decreasing the power consumption in defibration. Other physical properties of lignin have been comprehensively reviewed (51).

Chemical Properties. Lignin is subject to oxidation, reduction, discoloration, hydrolysis, and other chemical and enzymatic reactions. Many are briefly described elsewhere (51). Key to these reactions is the ability of the phenolic hydroxyl groups of lignin to participate in the formation of reactive intermediates, eg, phenoxy radical (**4**), quinonemethide (**5**), and phenoxy anion (**6**):

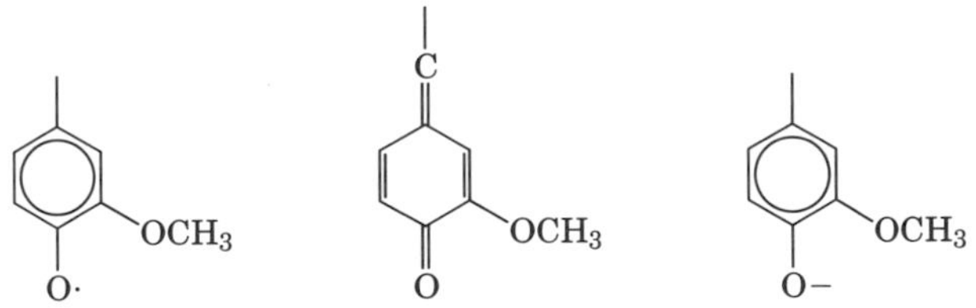

OCH_3 OCH_3 OCH_3

(**4**) (**5**) (**6**)

The free-radical intermediate initiates light-induced discoloration (yellowing) and enzymatic degradation of lignin (32,52). Nucleophilic addition occurs at the quinonemethide center, of which the most important reactions are the addition of sulfonate groups to the α-carbon during sulfite pulping and the sulfide assisted depolymerization in kraft pulping (Fig. 3).

The significance of phenoxy anions is well recognized in the isolation of kraft and other water-insoluble technical lignins by acid precipitation. The ionization of phenolic hydroxyl groups coupled with the reduction of molecular size renders native lignin soluble in the aqueous pulping solution, thus enabling its separation from the polysaccharide components of wood.

The aromatic ring of a phenoxy anion is the site of electrophilic addition, eg, in methylolation with formaldehyde (qv). The phenoxy anion is highly reactive

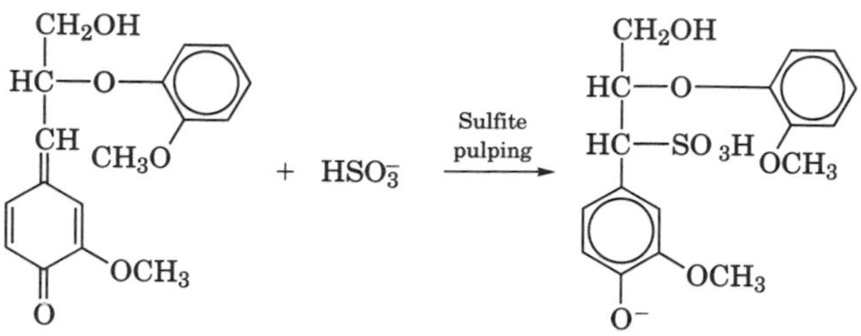

(a)

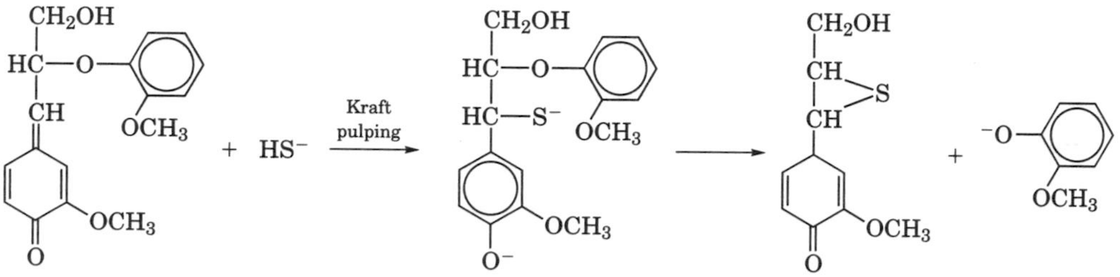

(b)

Fig. 3. Reactions at the quinonemethide center during pulping: (**a**) sulfite pulping, and (**b**) kraft pulping.

to many oxidants such as oxygen, hydrogen peroxide, ozone, and peroxyacetic acid. Many of the chemical modification reactions of lignin utilizing its aromatic and phenolic nature have been reviewed elsewhere (53).

Industrial Lignins

Industrial lignins are by-products of the pulp and paper industry. Lignosulfonate [*8062-15-5*], derived from sulfite pulping of wood, and kraft lignin [*8068-05-1*], derived from kraft pulping, are the principal commercially available lignin types. Organosolv lignins derived from the alcohol pulping of wood are also reported to be available commercially, but their quantities are limited (54).

The production capacity of lignin in the Western world is estimated to be ca 1.1×10^6 t/yr (Table 8). Although the production of lignosulfonates has been declining, kraft lignin production has increased. Of the companies listed in Table 8, LignoTech Sweden and Westvaco produce kraft lignins. The rest produce lignosulfonates.

Advances in technology have increased the importance of lignin products in various industrial applications. They are derived from an abundant, renewable resource, and they are nontoxic and versatile in performance.

Lignosulfonates. Lignosulfonates, also called lignin sulfonates and sulfite lignins, are derived from the sulfite pulping of wood. In the sulfite pulping process, lignin within the wood is rendered soluble by sulfonation, primarily at benzyl alcohol, benzyl aryl ether, and benzyl alkyl ether linkages on the side chain of phenyl propane units (55). Some demethylation also occurs during

Table 8. European and American Lignin Manufacturers

Producer	Country	Annual capacity, t/yr
Borregaard LignoTech	Norway	160,000
LignoTech Sweden	Sweden	50,000
LignoTech Deutschland	Germany	50,000
LignoTech Iberica	Spain	30,000
LignoTech Finland	Finland	50,000
Avebene	France	80,000
Attisholz	Switzerland	100,000
Georgia Pacific	United States	190,000
LignoTech USA	United States	65,000
Westvaco	United States	45,000
others		290,000
Total		*1,110,000*

sulfite pulping, and this leads to the formation of catechols and methane sulfonic acid (see Fig. 3).

Depending on the type of pulping process, lignosulfonates of various bases, including calcium [*904-76-3*], sodium [*8061-51-6*], magnesium [*8061-54-9*], and ammonium lignosulfonates [*8061-53-8*], can be obtained. Typical compositions for hardwood and softwood spent sulfite liquors are given in Table 9. In addition to whole liquor products, commercial forms of lignosulfonates include chemically modified whole liquors, purified lignosulfonates, and chemically modified forms thereof.

Isolation of Lignosulfonates. Various methods have been developed for isolating and purifying lignosulfonates from spent pulping liquors. One of the earliest and most widely used industrial processes is the Howard process, where calcium lignosulfonates are precipitated from spent pulping liquor by addition of excess lime. Lignin recoveries of 90–95% are obtainable through this process. Other methods used industrially include ultrafiltration and ion-exclusion (55), which uses ion-exchange resins to separate lignin from sugars.

Table 9. Compositions of Spent Sulfite Liquors[a]

	Percentage of total solids	
Component	Softwood	Hardwood
lignosulfonate	55	42
hexose sugars	14	5
pentose sugars	6	20
noncellulosic carbohydrates	8	11
acetic and formic acids	4	9
resin and extractives[b]	2	1
ash	10	10

[a]Ref. 48.
[b]For example, polyphenolic oils and tall oils.

Laboratory methods for isolating lignosulfonates include dialysis (56,57), electrodialysis (58), ion exclusion (58,59), precipitation in alcohol (60,61), and extraction with amines (62–64). They can also be isolated by precipitation with long-chain substituted quaternary ammonium salts (65–67).

Physical and Chemical Properties of Lignosulfonates. Even unmodified lignosulfonates have complex chemical and physical properties. Their molecular polydispersities and structures are heterogeneous. They are soluble in water at any pH but insoluble in most common organic solvents.

Typical C_9 formulas reported for isolated softwood and hardwood lignosulfonates are $C_9H_{8.5}O_{2.5}(OCH_3)_{0.85}(SO_3H)_{0.4}$ and $C_9H_{7.5}O_{2.5}(OCH_3)_{1.39}(SO_3H)_{0.8}$, respectively. These correspond to monomer unit molecular weights of 215 for softwood lignosulfonates and 256 for hardwoods. Polymer molecular weights are polydisperse and difficult to determine precisely. However, a range of from 1,000–140,000 has been reported for softwood lignosulfonates (68) with lower values reported for hardwoods (69).

A number of different functional groups are present in lignosulfonates. ^{13}C-nmr analysis of a purified sulfonated lignin from Western hemlock revealed 2.0% phenolic hydroxyl, 17.5% sulfonate, 12.5% methoxyl, and 0.6% carboxyl groups per unit weight of lignosulfonates (70). Additional studies indicate that lignosulfonates also contain limited numbers of olefinic, carbonyl, and catechol groups (71).

Lignosulfonates exhibit surface activity but have only a slight tendency to reduce interfacial tension between liquids. When compared to true surface-active agents, they are not effective in reducing the surface tension of water or for forming micelles (72). Their surface activity can be improved, however, by introducing long-chain alkyl amines into the lignin structure (73), by ethoxylation of lignin phenolic structures (74), or by conversion to oil-soluble lignin phenols (75).

Lignosulfonate Uses. Large-volume uses include production of vanillin (qv) and DMSO (76). Commercially, softwood spent sulfite liquors or lignosulfonates can be oxidized in alkaline media by oxygen or air to produce vanillin [*121-33-5*]. Other oxidizing agents, such as copper(II) hydroxide, nitrobenzene, and ozone, can also be used.

$HCSO_3^-$ / OCH_3 / OH $\xrightarrow{[O]}$ CHO / OCH_3 / OH

vanillin

Through reaction with sulfide or elemental sulfur at 215°C, lignosulfonates can also be used in the commercial production of dimethyl sulfide and methyl mercaptan (77). Dimethyl sulfide produced in the reaction is further oxidized to dimethyl sulfoxide (DMSO), a useful industrial solvent (see SULFOXIDES).

$$\text{(HCSO}_3\text{Na, OCH}_3\text{, OH aromatic unit)} + HS^- \longrightarrow \text{(HCSO}_3\text{Na, OH, OH aromatic unit)} + CH_3S^- \xrightarrow{\text{(OCH}_3\text{, OH aromatic unit)}} (CH_3)_2S \xrightarrow{NO_2} (CH_3)_2SO$$

dimethyl sulfoxide

Other relatively large-volume uses of lignosulfonates include animal feed pellet binders, water reducing agents in concrete admixtures, dispersants (qv) for gypsum board manufacture, thinners/fluid loss control agents for drilling muds, dispersants/grinding aids for cement (qv) manufacture, and in dust control applications, particularly road dust abatement.

Lignin technology has advanced significantly, and increased research and development efforts have resulted in specialty uses in several key market areas.

Dye Dispersants. One such area is dye manufacture, where lignosulfonates act as primary dispersants, extenders, protective colloids, and grinding aids. Products produced by the reaction of lignosulfonates with benzyl alcohols have low azo dye reduction properties, low fiber-staining properties, high dispersion efficiency, good grinding aid qualities, and increased heat stability (53). Their superior performance together with their low cost allow them to dominate the dye market.

Pesticide Dispersants. Modified lignosulfates are used in the formulation of pesticides. In wettable powders, suspension concentrates, and water dispersible granules, they act as dispersants and prevent sedimentation. They also act as binders in the production of granular pesticides. Typical usage levels in these types of products range from 2–10%.

Carbon Black Dispersant. Specially modified lignosulfonates are used in a wide range of pigment applications to inhibit settling and decrease solution viscosity. Applications include dispersants for dark pigment systems used to color textile fibers, coatings, inks (qv), and carbon black. Lignosulfonates are also used as grinding aids and binders in the pelletizing of carbon black (qv).

Water Treatment/Industrial Cleaning Applications. Boiler and cooling tower waters are treated with lignosulfonates to prevent scale deposition (78). In such systems, lignosulfonates sequester hard water salts and thus prevent their deposition on metal surfaces. They can also prevent the precipitation of certain insoluble heat-coagulable particles (79). Typical use levels for such applications range from 1–1000 ppm.

In industrial cleaning formulations, lignosulfonates function as dirt dispersants and suspending agents (80). Rinsing properties are improved, corrosivity is reduced and the amount of wetting agent needed is lowered when lignosulfonates are added to acid and alkaline industrial cleaning formulations. Typical use levels in such formulations range from 0.05–2.0%.

Complexing Agent for Micronutrients. Complexes of lignosulfonates and iron, copper, zinc, manganese, magnesium, boron, or combinations of such are used to provide essential micronutrients to plants growing in metal-deficient soils. In most instances application of such complexes is by foliar spray. When applied in this manner the micronutrients can be readily absorbed by the plant without undesired leaf burn (81). Lignosulfonate complexes can also be used in soil treatment where they maintain availability longer than if metals are applied alone (82).

Lignosulfonate–metal complexes are weaker complexes than those formed from amine-based complexing agents such as ethylenediaminetetracetic acid (EDTA). They are compatible with most pesticides/herbicides, but their use in phosphate fertilizers is not recommended.

Other Uses. Other uses of specially modified lignosulfonates include leather (qv) tanning (83), retarders for oil-well drilling cements (84), as expanders for lead–acid batteries (qv) (85), as flotation and wetting aids in ore processing, as sacrificial agents in enhanced oil recovery (86), as precipitating agents in protein recovery (87,88), in deicing formulations (89), and as wood preservatives (90). Medicinally, lignosulfonates have been purported to have value as antithrombotic (91) and antiviral (92,93) agents.

Interest in acrylic-graft copolymers of lignosulfonates is also growing. Commercially such products have found use as dispersants/fluid loss control agents for oil-well drilling muds and cements (94,95), as scale control agents in water treatment (96), as water reducing agents in the manufacture of bricks and ceramic materials (97), and as low inclusion animal feed binders (98).

Toxicology of Lignosulfonates. Rather extensive testing has shown that lignosulfonates are nontoxic. In most cases, LD_{50} values are greater than 5 g/kg. The safe use of lignosulfonates in the manufacturing and processing of a wide variety of food and food packaging applications is covered under the following United States Food and Drug Administration regulations: (*1*) as adjuvants in pesticide formulations exempt from the requirements of tolerance when applied pre- or post-harvest (21 CFR 182.99, 40 CFR 180.1001); (*2*) as dispersant or stabilizers in pesticides applied pre- or post-harvest to bananas (21 CFR 172.715); (*3*) as a boiler water additive used in the preparation of steam that will contact food (21 CFR 173.310); (*4*) as components of paperboard or paper in direct contact with moist, fatty, or dry food (21 CFR 176.170, 178.3120, 176.120, 176.180); (*5*) as component in food packaging (qv) adhesives (21 CFR 175.105); (*6*) as component of defoamers (qv) used in manufacturing food packaging-grade paper or paperboard (21 CFR 176.210); (*7*) in animal feed as pelleting or binder aid (limit of 4%), surfactant in molasses (limit of 11%), source of metabolizable energy (limit 4%) (21 CFR 573.600).

Kraft Lignins. Kraft lignins, also called sulfate or alkali lignins, are obtained from black liquor by precipitation with acid. Generally, acidification is conducted in two steps. In the first step, carbon dioxide from the waste gases of boiler fires or from lime kilns is used to reduce the pH of the liquor from 12 to 9–10. About three quarters of the lignin is precipitated in this step as a sodium salt. After isolation, the material thus obtained can be used as is or further refined by washing. By suspending the salt in water and lowering the pH to 3 or less with sulfuric acid, refined lignin is obtained.

Typical compositions for softwood and hardwood kraft black liquors are shown in Table 10. Most commercial kraft lignins are sulfonated kraft lignins or lignin amines. A few nonsulfonated products are, however, available.

Physical and Chemical Properties of Kraft Lignins. Kraft lignins are soluble in alkali (pH > 10.5), dioxane, acetone, dimethyl formamide, and methyl cellosolve. They are insoluble in water at neutral and acidic pH, have number average molecular weights in the 2000–3000 range, and are less polydisperse than lignosulfonates. A C_9 formula of $C_9H_{8.5}O_{2.1}S_{0.1}(OCH_3)_{0.8}(CO_2H)_{0.2}$ has been reported for softwood kraft lignin corresponding to a monomer molecular weight of about 180 (99). Due to the high degree of degradation during pulping, they also have a large number of free phenolic hydroxyl groups (4.0%).

The aromatic rings of kraft lignins can be sulfonated to varying degrees with sodium sulfite at high temperatures (150–200°C) or sulfomethylated with formaldehyde and sulfite at low temperatures (< 100°C). Oxidative sulfonation with oxygen and sulfite is also possible.

Many of the chemical reactions used to modify lignosulfonates are also used to modify kraft lignins. These include ozonation, alkaline–air oxidation, condensation with formaldehyde and carboxylation with chloroacetic acid (100), and epoxysuccinate (101). In addition, cationic kraft lignins can be prepared by reaction with glycidylamine (102).

The physical and chemical properties of kraft lignin differ greatly from those of lignosulfonates. A summary of these differences is presented in Table 11.

Applications of Kraft Lignins. Because of the high fuel value of black liquor, kraft lignin products are generally used in high value applications. In many applications, the base lignin must be modified (ie, through sulfonation or oxidation) prior to use. Once modified, kraft lignins can be used in most of the same applications in which lignosulfonates are used. These include usage as emulsifying agents/emulsion stabilizers (103), as sequestering agents, as pesticide dispersants (104), as dye dispersants (53,105,106), as additives in alkaline cleaning formulations (107), as complexing agents in micronutrient formulations, as flocculants (108), and as extenders for phenolic adhesives (109). In addition, kraft lignins can also be used as an extender/modifier, and as a reinforcement pigment in rubber compounding (110,111).

Table 10. Compositions of Kraft Black Liquors[a]

	Total solids, %	
Component	Softwood	Hardwood
kraft lignin	45	38
xyloisosaccharinic acid	1	5
glucoisosaccharinic acid	14	4
hydroxy acids	7	15
acetic acid	4	14
formic acid	6	6
resin and fatty acids	7	6
turpentine	1	
others	15	12

[a]Ref. 48.

Table 11. Properties of Kraft Lignins and Lignosulfonates[a]

Property	Kraft lignins	Lignosulfonates
molecular weight	2,000–3,000	20,000–50,000
polydispersity	2–3	6–8
sulfonate groups, meq/g	0	1.25–2.5
organic sulfur, %	1–1.5	4–8
solubility	soluble in alkali (pH > 10.5), acetone, dimethylformamide, methyl cellosolve	soluble in water at all pHs; insoluble in organic solvents
color	dark brown	light brown
functional groups	many phenolic hydroxyl, carboxyl, and catechol groups; some side-chain saturation	fewer phenolic hydroxyl, carboxyl, and catechol groups; little side-chain saturation

[a]Ref. 48.

Organosolv Pulping Lignins. In organosolv pulping processes, hardwood chips are batch cooked for set times at appropriate temperatures and pH in an aqueous ethanol or methanol liquor. In the process lignin, hemicelluloses, and other miscellaneous components of the wood are extracted into the alcoholic pulping liquor forming a black liquor. Organosolv lignin is recovered from the black liquor by precipitation, settled, centrifuged or filtered, and dried (112,113). The resulting lignin is a fine, brown, free-flowing powder.

Physical and Chemical Properties. Organosolv lignins are soluble in some organic solvents and in dilute alkali. They are insoluble in water at neutral or acidic pH. They have number average molecular weights lower than 1000 and polydispersities between 2.4 and 6.3 (54). A C_9 formula of $C_9H_{8.53}O_{2.45}(OCH_3)_{1.04}$ has been reported for one organosolv lignin (54). This corresponds to a monomer molecular weight of 188.

Applications. These materials are still in developmental infancy. Current production is limited to one commercial process in Europe and a demonstration-scale process in North America. The lignins produced in these processes have potential application in wood adhesives, as flame retardants (qv), as slow-release agents for agricultural and pharmaceutical products, as surfactants (qv), as antioxidants (qv), as asphalt extenders, and as a raw material source for lignin-derived chemicals.

Other Lignins. In addition to main commercial lignins, there are a number of other lignins of no or limited commercial value. One of these is produced almost exclusively in the former Soviet Union where wood is used to produce glucose by acid hydrolysis. The lignin isolated as a by-product of this process is called acid hydrolysis lignin. It is claimed that in modified form such lignins can be used as rubber filling agents, as binders in wood adhesives, as additives in fabric treating compounds where they impart decay resistance, and as flotation aids in ore processing.

Numerous lignins have also been isolated in the laboratory including milled wood lignins, dioxane lignins, and enzymatically liberated lignins. These laboratory prepared lignins have different chemical and physical properties depending on the chemical modifications undergone during their isolation. Soda lignins derived from the soda pulping process are not available commercially.

BIBLIOGRAPHY

"Lignin" in *ECT* 1st ed., Vol. 8, pp. 327–338, by A. Pollak, Consultant; in *ECT* 2nd ed., Vol. 12, pp. 361–381, by D. W. Goheen, D. W. Glennie, and C. H. Hoyt, Crown Zellerbach Corp.; in *ECT* 3rd ed., Vol. 14, pp. 294–312, by D. W. Goheen and C. H. Hoyt, Crown Zellerbach Corp.

1. K. V. Sarkanen and H. L. Hergert, in K. Sarkanen and C. Ludwig, eds., *Lignins: Occurrence, Formation, Structure, and Reactions,* Wiley-Interscience, New York, p. 43.
2. P. Lewis and M. Paice, eds., *Plant Cell Wall Polymers: Biogenesis and Degradation,* ACS Symposium Series, Washington, D.C., 1989, p. 299.
3. B. J. Fergus, "The Distribution of Lignin in Wood as Determined by Ultraviolet Microscopy," Ph.D. thesis, McGill University, Montreal, Canada, 1968.
4. J. Gierer, *Wood Sci. Technol.* **19**, 289 (1985).
5. J. Gierer, *Wood Sci. Technol.* **20**, 1 (1986).
6. E. Adler, *Wood Sci. Technol.* **11**, 169 (1977).
7. H. H. Nimz, *Angew. Chem. Int. Ed.* **13**, 313 (1974).
8. M. Erickson, S. Larsson, and G. E. Miksche, *Acta Chem. Scand.* **27**, 903 (1973).
9. S. I. Andersson and O. Samuelson, *Sven. Papperstidn.* **88**, R102 (1985).
10. J. Gierer, *Holzforschung* **36**, 43 (1982).
11. E. E. Harris and H. Adkins, *Pap. Trade J.* **107**(20), 38 (1938).
12. N. N. Shorygina, T. Y. Kafeli, and A. F. Samechkina, *J. Gen. Chem.* **19**, 1558 (1949).
13. A. F. A. Wallis, in Ref. 1, p. 345.
14. J. Gierer, *Wood Sci. Technol.* **14**, 241 (1980).
15. T. K. Kirk and R. L. Farrell, *Ann. Rev. Microbiol.* **41**, 465 (1987).
16. G. Gellerstedt and E. L. Pettersson, *Acta Chem. Scand.* **B29**, 1005 (1975).
17. G. Gellerstedt, *Wood Sci. Technol.* **14**, 241 (1976).
18. J. Nakano and G. Meshitsuka, in S. Y. Lin and C. W. Dence, eds., *Methods in Lignin Chemistry,* Springer-Verlag, Berlin, 1992, p. 23.
19. G. Meshitsuka and J. Nakano, *Mokuzai Gakkaishi* **24**, 563 (1988).
20. K. Iirama and R. Pant, *Wood Sci. Technol.* **22**, 167 (1988).
21. C. W. Dence, in Ref. 18, p. 37.
22. *TAPPI Test Method T222 om-83*, Atlanta, Ga., 1988.
23. B. J. Fergus and co-workers, *Wood Sci. Technol.* **3**, 117 (1969).
24. S. Saka, R. J. Thomas, and J. S. Gratzl, *Tappi* **61**(1), 73 (1978).
25. K. Iiyama and A. F. A. Wallis, *Wood Sci. Technol.* **22**, 271 (1988).
26. J. E. Tasman and V. Berzins, *Tappi* **40**, 691 (1957).
27. S. Y. Lin and C. W. Dence, eds., in Ref. 18.
28. S. Y. Lin, in Ref. 18, p. 217.
29. K. Lundquist, in Ref. 18, p. 242.
30. D. Robert, in Ref. 18, p. 250.
31. C. Steelink, *Adv. Chem. Ser.* **59**, 51 (1966).
32. K. P. Kringstad and S. Y. Lin, *Tappi* **53**, 2296 (1970).
33. K. V. Sarkanen, A. Islam, and C. D. Anderson, in Ref. 18, p. 387.
34. M. Fumaoka, I. Abe, and V. L. Chiang, in Ref. 18, p. 369.
35. R. Kuhn and H. Roth, *Ber. Dtsch. Chem. Ges.* **66**, 1274 (1933).
36. P. Mansson, *Holforschung* **37**, 143 (1983).
37. E. Adler, S. Hernestam, and I. Wallden, *Sven. Papperstidn.* **61**, 640 (1958).

38. O. Faix, C. Gruenwald, and O. Beinhoff, *Holforschung* **46**, 425 (1992).
39. C.-L. Chen, in Ref. 18, p. 409.
40. P. Klason, *Ber. Dtsch. Chem. Ges.* **55**, 448 (1922).
41. C.-L. Chen, in Ref. 18, p. 446.
42. H. Pobiner, *Anal. Chim. Acta.* **155**, 57 (1983).
43. C. W. Dence, in Ref. 18, p. 458.
44. F. Viebock and A. Schwappach, *Ber. Dtsch. Chem. Ges.* **63**, 2818 (1930).
45. S. Katz, R. P. Beatson, and A. M. Scallan, *Sven. Papperstidn.* **87**, R48 (1984).
46. T. I. Obiaga, "Lignin Molecular Weight and Molecular Weight Distribution during Alkaline Pulping of Wood," Ph.D. thesis, University of Toronto, Canada, 1972.
47. T. Yamasaki and co-workers, *The Ekman Days 1981 Proc. Int. Symp. Wood Pulp Chem. SPCI,* Vol. 2, Stockholm, 1981, p. 34.
48. S. Y. Lin and I. S. Lin, in *Ullmann's Encyclopedia of Industrial Chemistry*, 5th ed., Vol. 15, VCH, Weinheim, Germany, 1990, p. 305.
49. H. Hatakeyama, K. Kubota, and J. Nakano, *Cellulose Chem. Technol.* **6**, 521 (1972).
50. D. Atack, C. Heitner, and M. I. Stationwala, *Sven. Papperstidn.* **81**, 164 (1978).
51. D. A. I. Goring, in Ref. 18, p. 695.
52. T. K. Kirk, H. H. Yang, and K. Keyser, *Dev. Ind. Microbiol.* **19**, 51 (1978).
53. S. Y. Lin, *Progress in Biomass Conversion,* Vol. 4, Academic Press, Inc., Orlando, Fla., 1983, p. 31.
54. J. H. Lora and co-workers in W. G. Glasser and S. Sarkanen, eds., *Lignin - Properties and Materials,* ACS Symposium Series 397, Washington, D.C., 1989, p. 312.
55. H. Schneider and co-workers, *Int. Sugar J.* **77**, 259 (1975).
56. W. Q. Dean and D. A. I. Goring, *Tappi* **47**, 16 (1964).
57. A. E. Markham, Q. P. Peniston, and J. L. McCarthey, *J. Am. Chem. Soc.* **71**, 3599 (1949).
58. G. A. DuBey, T. R. McElhinney, and A. J. Wiley, *Tappi* **48**, 95 (1965).
59. V. F. Felicetta and J. L. McCarthey, *Tappi* **40**, 851 (1957).
60. J. Benko, *Tappi* **44**, 771 (1961).
61. J. L. Gardon and S. G. Mason, *Can. J. Chem.* **33**, 1477 (1955).
62. E. E. Harris and D. Hogan, *Ind. Eng. Chem.* **49**, 1393 (1957).
63. Y. Kojima and co-workers, *Jpn. Tappi* **15**, 607 (1961).
64. S. Y. Lin, in Ref. 18, p. 76.
65. L. Sato, *Science (Japan)* **13**, 403 (1943).
66. I. Croon and B. Swan, *Svensk. Papperstidn.* **66**, 812 (1963).
67. G. R. Quimby and O. Goldschmid, *Tappi* **49**, 562 (1966).
68. W. Q. Yean and D. A. I. Goring, *Svensk Papperstidn.* **55**, 563 (1952).
69. E. Sjostrom and co-workers, *Svensk Papperstidn.* **65**, 855 (1962).
70. C. H. Ludwig and W. T. Zdybak, paper presented at the *185th National Meeting of the American Chemical Society, Cellulose, Paper and Textile Division,* Seattle, Wash., Mar. 23, 1983.
71. D. W. Glennie, in Ref. 1, p. 614.
72. K. F. Keirstead, *Colloid Interface Sci.* **III**, 431 (1976).
73. U.S. Pat. 4,562,236 (1985), S. Y. Lin (to Reed Lignin, Inc.).
74. U.S. Pat. 5,094,296 (1990), M. G. DaGue (to Texaco Inc.).
75. U.S. Pat. 5,095,986 (1990), D. G. Naae and C. A. Davis (to Texaco Inc.).
76. R. A. Northey, *Emerging Technology of Materials and Chemicals from Biomass,* ACS Symposium Series 476, Washington, D.C., 1992.
77. U.S. Pat. 2,840,614 (1958), D. W. Gohen (to Crown Zellerbach Corp.).
78. U.S. Pat. 2,826,552 (1958), P. W. Bonewitz, E. H. Fults and S. W. Hockett (to Bonewitz Chem., Inc.).
79. U.S. Pat. 3,317,431 (1967), S. Kaye (to Wright Chemicals).

80. U.S. Pat. 3,247,120 (1966), J. A. Von Pless (to Cowles Chem. Co.).
81. U.S. Pat. 3,244,505 (1966), C. Adolphson and R. W. Simmons (to Georgia Pacific).
82. A. Wallace and R. T. Ashcroft, *Soil Sci.* **82**(3), 233 (1956).
83. U.S. Pat. 3,447,889 (1969), R. W. Simmons (to Georgia Pacific).
84. U.S. Pat. 32,895 (1989), W. J. Detroit and M. E. Sanford (to Reed Lignin, Inc.).
85. G. J. Szava, *J. Power Sources* **28**(1–2), 149 (1989).
86. S. A. Hong and J. H. Bae, *SPE Reservoir Eng.* **11**, 467 (1990).
87. R. J. Sherman, *J. Food Tech.* **33**(6), 50 (1979).
88. E. I. Tonseth and H. B. Berridge, *Effluent Water Treat. J.* **3**, 124 (1968).
89. U.S. Pat. 4,824,588 (1989), S. Y. Lin (to Reed Lignin, Inc.).
90. U.S. Pat. 4,988,576 (1991), S. Y. Lin and L. L. Bushar (to Daishowa Chemicals, Inc.).
91. Eur. Pat. 0 303 236 A2 (1988), R. H. Samson and J. W. Hollis (to Reed Lignin, Inc.).
92. Jpn. Pat. 02,262,524 (1989), S. Toda and co-workers (to Noda Shokukin Kogyo).
93. Jpn. Pat. 03,206,043 (1991), H. Sakagami, Y. Kawazoe, and K. Konno (to Kawakami).
94. U.S. Pat. 4,676,317 (1987), S. E. Fry and co-workers (to Halliburton Co.).
95. Brit. Pat. 2,210,888 A (1988), C. D. Williamson (to Nalco, Inc.).
96. U.S. Pat. 4,891,415 (1990), S. Y. Lin and L. L. Bushar (to Daishowa Chemicals, Inc.).
97. U.S. Pat. 4,871,825 (1989), S. Y. Lin (to Reed Lignin, Inc.).
98. U.S. Pat. 4,952,415 (1989), T. S. Winowiski and S. Y. Lin (to Daishowa Chemicals, Inc.).
99. S. Y. Lin and W. J. Detroit, *Ekman-Days 1981 Int. Symp. Wood Pulp. Chem.* **4**, 44 (1981).
100. U.S. Pat. 3,841,887 (1974), S. I. Falkenhag and C. W. Bailey (to Westvaco).
101. U.S. Pat. 3,956,261 (1976), S. Y. Lin (to Westvaco).
102. U.S. Pat. 4,728,728 (1988), S. Y. Lin and L. H. Hoo (to Reed Lignin, Inc.).
103. U.S. Pat. 3,123,569 (1964), M. J. Borgfeldt (to Chevron Research and Technology).
104. U.S. Pat. 3,986,979 (1976), H. H. Moorer and C. W. Sandefur (to Westvaco).
105. U.S. Pat. 4,670,482 (1987), P. Dilling (to Westvaco).
106. U.S. Pat. 4,740,590 (1988), P. Dilling (to Westvaco).
107. U.S. Pat. 3,803,041 (1974), M. S. Dimitri (to Westvaco).
108. U.S. Pat. 4,781,840 (1988), P. Schilling (to Westvaco).
109. K. Kratzl and co-workers, *Tappi* **45**, 113 (1962).
110. G. V. Rao and co-workers, *Ind. Pulp Paper,* 11 (June–July, 1978).
111. S. I. Falkehag and co-workers, *ACS Symp. Renewable Resources for Plastics,* Philadelphia, Pa., 1975, p. 68.
112. J. H. Lora and S. Aziz, *Tappi* **68**(8), 94 (1985).
113. P. N. Williamson, *Pulp Paper Can.* **88**(12), 47 (1987).

STEPHEN Y. LIN
STUART E. LEBO, JR.
LignoTech USA, Inc.

LIGNITE AND BROWN COAL

Lignite and brown coal are common names for coals having properties intermediate between peat and bituminous coal as a result of limited coalification (see COAL). In general, brown coal designates a geologically younger, ie, less coalified, material than the firmer, fibrous lignite. In the ASTM classification (1), both kinds of coal are classified as lignite. In many English-speaking countries, the consolidated coals are termed lignite, and unconsolidated coals are termed brown coal. In Australia, and in Germany and a number of other European countries, the generic term brown coal is used for the whole class, including some coals that are included in the ASTM classification as subbituminous. Lignite signifies the firmer, fibrous, woody variety. Herein lignite is used as the comprehensive term.

Selection of coal for a particular use requires a knowledge of composition greater than that supplied from the ASTM classification. Progress is being made toward classifying all kinds of coal, including lignite, by correlating properties with composition and other qualities (2).

The primary use of lignite is combustion in steam (qv) generation of electric power (see POWER GENERATION). Lesser amounts generally in the form of briquettes are used for industrial and domestic heating outside of the United States. Briquettes are pressed and often carbonized at low temperatures to provide a smokeless fuel. The by-product tars obtained from briquette production have been used for liquid fuels and chemical manufacture (see TAR AND PITCH). Lignite is also converted by gasification to synthesis gas for motor fuels, chemicals, and ammonia-based fertilizers (qv) in large integrated plants (see COAL CONVERSION PROCESSES; FEEDSTOCKS, COAL CHEMICALS AND FEEDSTOCKS; FUELS, SYNTHETIC).

Worldwide production of lignite was over 1.13×10^9 tons in 1990 (U.S. production was 8×10^7 t). U.S. production has risen at ca 2%/yr in the early 1990s. The world's proved lignite reserves can be recovered under local economic conditions with existing available technology, and were over 3.8×10^{11} t in 1990, representing almost 300 years of production in 1990 terms (3,4).

Lignite is less valuable than coals of higher rank, primarily because its much higher (30–70% as mined) water content (5) and high chemically combined oxygen content result in a relatively low heating value (LHV). In the past, the expense of shipping limited the market largely to the vicinity of the mine. However, in the United States the low sulfur content of lignite has made long distance shipments economically feasible, in order to limit sulfur oxide emissions at electric power generation plants. The increasing worldwide demand for energy together with desire for national self-sufficiency has increased the importance of low heating value coals.

Geology

Lignite was deposited relatively recently (ca $2.5–60 \times 10^6$ yr ago), mainly during the Tertiary era. U.S. deposits include those in the Dakotas, Alaska, Montana, and Wyoming. Other deposits exist in Saskatchewan and northwestern Canada, Germany, Asiatic Russia, Pakistan, northern India, Borneo, Sumatra, and Manchuria. The Miocene period provided the brown coal deposits that are up to 300 m thick in the Latrobe Valley of Victory in Australia. In addition, deposits in Venezuela, Mexico, southern Germany, the Volga region, and northern China were laid down during this period (6). The oldest deposits, which occur in the Moscow basin, were formed in the lower Carboniferous period, ca 200×10^6 years ago.

The Pliocene lignites in Alaska, southeastern Europe, and southern Nigeria are the youngest coals. A number of peat accumulations in different parts of the world representing a range of climates indicate that the process of coal formation continues to take place.

Classification

Several classification schemes have been used in different places and for different purposes. In the United States the ASTM method is used for all kinds of coal, from lignite through anthracite (1). The criterion for classification of lignite through high volatile B bituminous is moist, mineral–matter-free energy content. The term moist refers to bed moisture only, and the bed samples must be collected as described in ASTM Standard D388. In this method of classification, lignite and brown coal have moist energy < 19.3 MJ/kg (8300 Btu/lb). Before 1960 consolidated coals were called lignite and unconsolidated coals, brown coal. The 1994 ASTM Standard D388 distinguishes lignite A from lignite B. The heating values of dry lignite A range from 14.6 to 19.3 MJ/kg (6300–8300 Btu/lb); the values of dry lignite B range below 14.6 MJ/kg (6300 Btu/lb). In the United States, the terms soft coal and hard coal refer to bituminous and anthracite, respectively. However, in Europe soft coal refers to lignite and brown coal, whereas hard coal refers to bituminous. In 1958 the Coal Committee of the Economic Commission for Europe (ECE) officially adopted the classification of lignite and brown coal given in Table 1 (7). Total moisture, which can be correlated with heat value, provides a guide to the use of coal as a fuel. Tar yield provides a

Table 1. International Classification of Coals having a Gross Calorific Value below 23.8 MJ/kg[a–c]

Group number	Group parameter tar yield, %[d]	Code number					
40	25	1040	1140	1240	1340	1440	1540
30	20–25	1030	1130	1230	1330	1430	1530
20	15–20	1020	1120	1220	1320	1420	1520
10	10–15	1010	1110	1210	1310	1410	1510
00	≤10	1000	1100	1200	1300	1400	1500
class number		10	11	12	13	14	15
class parameter, %[e]		≤20	>20–30	>30–40	>40–50	>50–60	>60–70

[a]Moist, ash-free basis (30°C and 96% RH).
[b]Ref. 7.
[c]To convert MJ/kg to Btu/lb, multiply by 430.2.
[d]Dry, ash-free basis.
[e]Numbers represent % of total moisture of ash-free freshly mined coal.

Table 2. Classification of Coals by Rank, International Committee for Coal Petrology[a]

Rank stages[b]	Vitrinine reflectance, %	Microscopic characteristics	C in vitrinite, %
peat		large pores details of initial plant material still recognizable free cellulose	≤ 50
brown coal soft	0.3[d]	no free cellulose plant structures still recognizable (cell cavities frequently empty)	≤ 60
dull		marked gelification and compaction takes place	
bright	0.5[d]	plant structures still partly recognizable (cell cavities filled with collinite)	≤ 70
bituminous hard coal		exinite becomes markedly lighter in color[g]	≤ 80
	2.5[d]	exinite no longer distinguishable from vitrinite in reflected light	≤ 90
anthracite		reflectance anisotropy	
graphite	11.0		< 100

[a]Ref. 9.
[b]In order of increasing rank.
[c]Dry ash-free.
[d]Values are estimated.
[e]Ash-free.
[f]To convert MJ/kg to Btu/lb, multiply by 430.2.
[g]This represents a coalification jump.

measure of value as a raw material for the chemical industry. This nonscientific classification provides a guide for use. Another system based on a two-figure classification index derived from moisture content, tar yield, and petrographic values was adopted in 1961 as the International Classification for Brown Coals for the Peoples' Democracies (8).

In 1963 a classification of coals by rank (differing from the ECE scheme) was published by the International Committee for Coal Petrology (Table 2) (9). This includes a classification of brown coal that correlates a number of important properties including the percent reflectance of vitrinite in the coal. This is a simpler version of that used in German practice, which further subdivides soft

Table 2. (*Continued*)

Vitrite volatile matter, %[c]	H_2O *in situ*, %[d]	Vitrite heating value, MJ/kg[e,f]	Applicability of the different parameters for the determination of rank
	75		calorific value[e] or moisture *in situ* (moisture-holding capacity)
53[c]	35	16.7	reflectance of the vitrinites
49[d]	25	23.0	
45[d]	8–10	29.3	volatile matter[c]; carbon[c]
			x-ray diffraction (graphite lattice)
30			
		36.0	hardness[c]
10			
0			

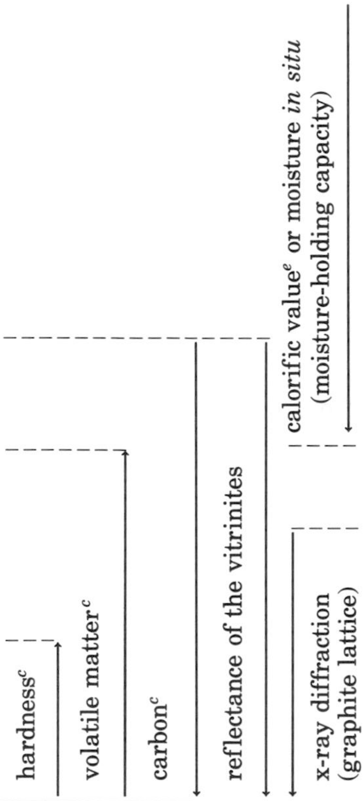

brown coals into foliaceous and earthy. Most brown coals belong to the latter group.

Other terms that have been used to differentiate types of lignitic coal include humic brown coals, referring to those having substantial amounts of extractable humic acids, and sapropelic coals, referring to more homogeneous coals often having a high concentration of individual plant components. Some additional methods of classification are under development that center on the use of lignite for combustion in utility boilers or electric power generation. Correlations based on the sodium concentration in the lignitic ash (10), or soluble Al concentration (11) are used. The classifications are often given in terms of the severity of boiler fouling.

Composition, Properties, and Analysis

Macroscopic Appearance. Lignitic coals vary from brown to dull black when moist, although the color may appear considerably lighter when the coal is dried. The freshly broken surface of the most common type, the unconsolidated humic variety, may be light reddish brown but darkens rapidly during oxidation. Breakage is easiest for the unconsolidated coals. Strength and toughness increase as coalification increases. Because of its weak structure and tendency to shrink and crack on drying, brown coal disintegrates more easily than more mature coals (12). Remains of plants can be seen in some of these coals.

Physicochemical Structure. Water-filled pores and capillaries of differing diameters permeate the organic gel material that makes up as-mined lignite. There is some retention of moisture on air drying (5). The void volume or porosity ranges up to about 44% for lignitic coals (13) and decreases as the rank increases. The pore diameters vary and include a significant amount of very small pores that limit the size of molecules that can enter or leave. Lignitic coals have properties of molecular sieves (qv) and the large pore volume is believed to be partially responsible for the high observed reactivity. The relative accessible internal surface area ranges from 100–200 m^2/g for lignitic coal and is about half this value for bituminous coals. In general, the internal surface area of coal is associated with capillary systems having 4-nm pores linked by 0.5–0.8-nm passages. About 75% of this free volume in lignite is associated with the larger pores (14). Mineral matter including salts of the humic acids is nonuniformly distributed.

Properties. The apparent density of lignite is 0.8–1.35 g/cm^3 (13), which is lower than values given for higher ranking coals. Therefore, greater volume is required for storage, transportation, and lignite reactors than is needed for an equivalent weight of more mature coals. Lignite generally has lower elasticity and greater plasticity than more mature coals. The plasticity index, ie, ratio of elastic energy to plastic energy, involved in compressing coals has been used to indicate the ease of briquette formation (14). Briquetting without a binder is possible only for softer, less mature coals.

Humic acids are alkali-extractable materials and total humic acid content is a term that refers to the humic acid content of coal that has had its carboxylate cations removed with sodium pyrophosphate. Values for some typical Australian brown coals range from 24–92% (13). Treatment of lignitic coals with mineral acid to release the alkali and alkaline cations may dissolve up to 20% of the coal. The naturally moist coals are slightly acidic and have a pH of 3.5–6.5.

Solvent extraction using nonreactive liquids, such as C_3- or C_4-alcohols, benzene, or benzene–alcohol mixtures, yields generally 5–20% wax or bitumen (15). The yield and composition of the product are determined primarily by the petrologic character of the coal, not its degree of coalification. Montan wax is extracted from suitable coals for a variety of purposes.

The tar yield is usually higher for lignite than for more mature coals. Tar yields are important in determining selection for carbonization and for liquid fuel production by pyrolysis.

Oxidation. The high reactivity of lignites with oxygen requires special care during mining, transportation, and storage to avoid spontaneous combustion from heat generation. Contact with basic ($pH > 8$) solutions or oxidizing agents results in slow oxidation and the subsequent formation of humic acids. Many lignitic coals are almost entirely soluble in alkali solution, providing a technique for distinguishing between lignitic and bituminous coal. Spontaneous ignition of briquettes and coal is observed after wetting and at freshly broken surfaces (16).

Analysis. Analyses of a number of lignitic coals are given in Table 3. Figure 1, a distribution plot of 300 U.S. coals according to ASTM classification by rank, indicates the broad range of fixed carbon values (18). According to the ASTM classification, fixed carbon for both lignite and subbituminous coals has an upper limit of 69%, but in practice this value rarely exceeds 61%.

The moisture content of freshly mined lignitic coals can be as high as 73%, but it is usually 30–65%. The more mature, consolidated coals have lower moisture contents and thus a higher heating value. Figure 2 shows moisture and ash contents, as well as net heating values for lignitic coals from the world's principal deposits (19).

Moisture content affects a number of applications. The grindability index, ASTM D409, measures the relative ease of pulverizing coals and theoretically helps determine the capacity of pulverizers. In practice, low values of grindability occur at moisture extremes and maximal grindability occurs at intermediate moisture content. A small pulverizer to test grinding conditions for design purposes has been developed (20).

Mineral matter content or ash yield varies widely, from $< 6\%$ in thick deposits as found in Australia and Germany, to $> 40\%$ in deposits in Turkey. The composition of these minerals varies with location and depth in the seam (21). In the United States, lignitic ashes tend to have higher CaO, MgO, and Na_2O contents than do the ashes of bituminous coals. Significant amounts of metals are also organically bound. For example, the uranium contents of some ashes from the northern Great Plains exceed 1000 ppm (22). The volatile matter of lignite and brown coals ranges from ca 40 to 55%, and hydrogen from ca 4.3 to 6.1%, both on a dry mineral–matter-free basis. The hydrogen-to-carbon atomic ratio for subbituminous coal in the ASTM classification is ca 0.8–1.0. For lignite, the range is from 0.8 to about 1.1. The oxygen-to-carbon ratio for subbituminous coals is ca 0.1–0.2, and for lignite it is 0.2–0.3 (23).

Resources and Production

The importance of a coal deposit depends on the amount that is economically recoverable by conventional mining techniques. The world total recoverable reserves of lignitic coals were 3.28×10^{11} metric tons at the end of 1990 (3), of which ca 47% was economically recoverable as of 1994 (Table 4). These estimates of reserves change as geological survey data improve and as the resources are developed.

The production of lignite and brown coal for nations having large published reserves is also indicated in Table 4. Total world production, which was 6.6×10^8 t in 1961, increased to 8.1×10^8 t in 1972, and to 1.13×10^9 t in 1990 (24).

A comparison of available resources and production shows that Germany, ranked second in resources, was ranked first in production in 1990. Indeed a number of central and eastern European countries are producing proved recoverable reserves at a rate that should lead to exhaustion of local deposits before the end of the twenty-first century. On the other hand, the massive Russian reserves could allow production for a much longer time.

Table 3. Analyses of Lignitic Coals[a]

Coal	Proximate analysis, %					Ultimate analysis, %					Heat value, gross dry, MJ/kg[c]
	Moisture	Ash	Volatile matter	Fixed carbon	Volatile matter[b]	C	H	S	N	O	
					United States						
North Dakota[d]	33.9–41.2	3.5–8.5	25.4–27.6	26.9–31.7	45.2–48.8	71.1–74.4	4.8–5.3	0.3–2.3	1.0–1.1	16.9–22.7	27.8–29.6
South Dakota	38.5	5.8	26.9	28.8	48.3	72.0	4.8	0.7	1.4	21.1	28.4
					Australia						
Victoria											
Yallourn	66.3	0.7	17.7	15.3	53.4	67.4	4.7	0.3	0.5	27.1	25.9
Yallourn north	50.0	2.0	26.0	22.0	54.2	68.3	4.9	0.3	0.6	25.9	25.9
					Germany						
Lower Rhine[e]	60.0	2.3	20.6	17.1	54.6	68.9	5.3	0.3	25.5		25.9
Geiseltal	50.4	6.3	27.3	20.0	57.7						26.8
Riebeck-Montan	49.2	6.7	26.5	17.6	60.1						28.9

[a]Ref. 17.
[b]Dry ash-free.
[c]To convert MJ/kg to Btu/lb, multiply by 430.2.
[d]Range for coals from five areas.
[e]Average for four similar coals.

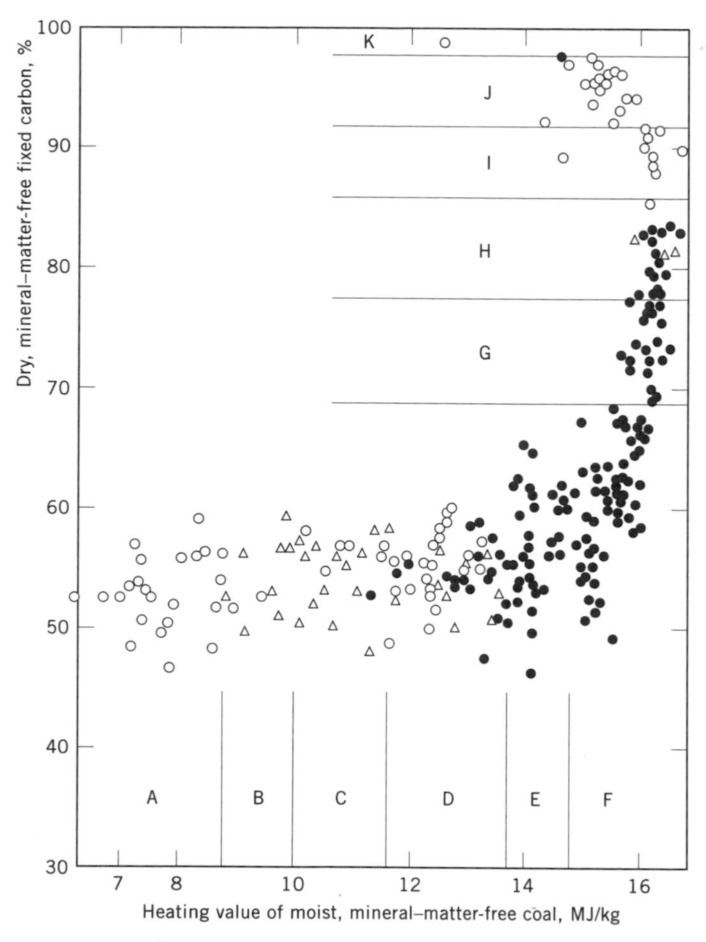

Fig. 1. Distribution plot for over 300 coals of the United States, illustrating ASTM classification by rank where (•) represents caking or agglutinating; (∘) noncaking; and (△) no information concerning caking properties. Region A corresponds to lignite; B, C, and D to subbituminous C, B, and A, respectively; D also corresponds to high volatile C bituminous; E corresponds to high volatile B bituminous; F, to high volatile A bituminous; G, to medium volatile bituminous; H, low volatile bituminous; I, semianthracite; J, anthracite; and K, *meta*-anthracite. To convert MJ/kg to Btu/lb, multiply by 430.2.

Rates of production of lignite have continued to increase since 1960. In 1980 374×10^6 tons of coal equivalent (tce) were produced. One tce is the amount of energy available from combustion of a metric ton of coal having a heat content of 29.3 GJ, ie, 29.3 MJ/kg (12,600 Btu/lb) (3). In 1989 this figure had risen to 460×10^6 tce. This 23% increase is somewhat less than the 28% increase in hard coal production during this period (see COAL). In 1990 the 1130×10^6 metric tons of lignite produced worldwide represented 24% of the total coal production.

The extent of lignite production is generally not proportional either to total resources or to known economic reserves. Lack of energy alternatives is a strong motive to developments in lignite production. For example, after World War I, Germany's concern over secure fuel supplies and development of synthetic

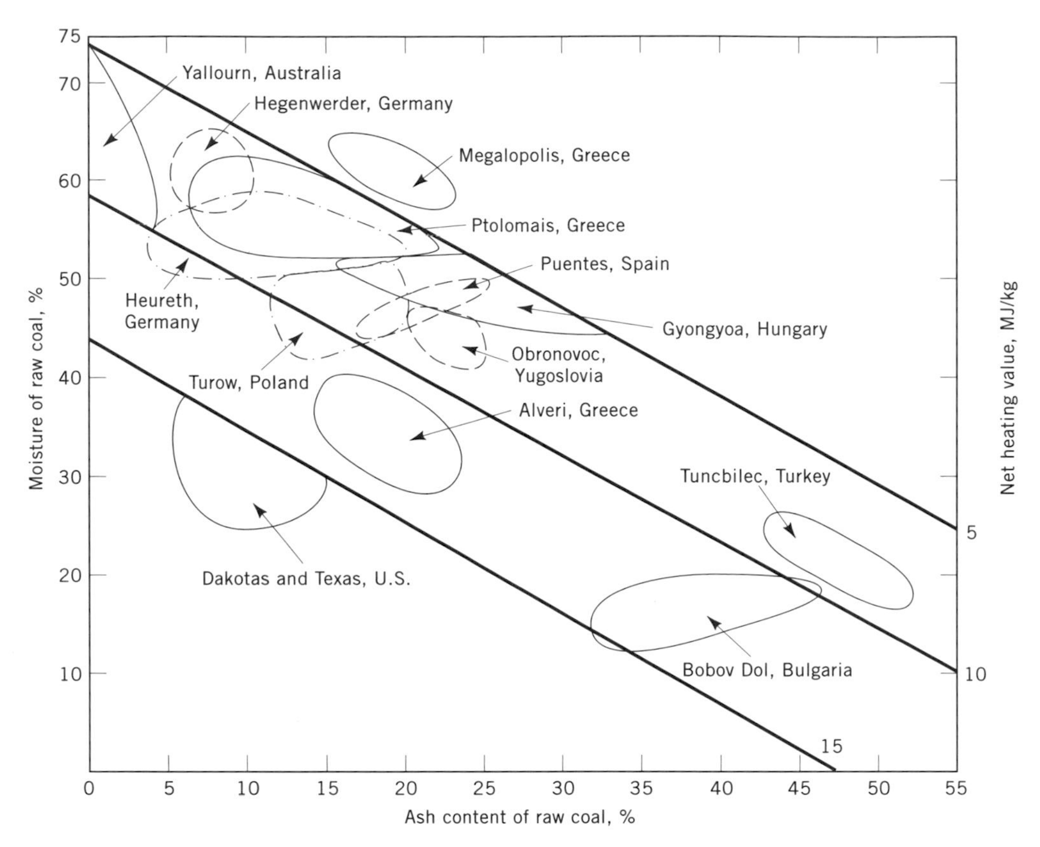

Fig. 2. Quality of lignitic coal. To convert MJ/kg to Btu/lb, multiply by 430.2.

fuels accelerated studies with brown coal, and led to production of fuels based on synthetic fuel (synfuel) technology. South Africa, which has essentially no oil, has plentiful supplies of subbituminous coal, and synthetic fuel plants in Sasolburg and Secunda are based on coal. These synfuel plants are the largest in the world (see FUELS, SYNTHETIC). Other areas with limited alternatives are Victoria, Australia; southern India; and some eastern European countries.

When energy alternatives are available, a compromise between cost and quality is often realized. Blending of coals can be used to achieve more desirable qualities. For example, lignite from the former Yugoslavia has been blended with, and even substituted for, the highly caking Rasa coal used for coke production in the iron (qv) and steel (qv) industries.

In the United States and increasingly in other parts of the world, environmental regulations prohibit the combustion of all but very low sulfur-content coals without sulfur oxide emission controls. The cost of installing sulfur oxide control equipment together with concern about equipment reliability have led to the shipment of the lower rank low sulfur coals from up to 1600 km away from the mining site.

Especially in countries having planned economies, higher quality fuels are reversed for domestic heating, industrial fuels, and chemical feedstocks, and the

Table 4. Worldwide Reserves and Production of Lignite[a]

	Resources, 10^6 tce[b,c]			Production, 10^3 tce[c]			
Country	Proved in place	Proved recoverable reserves	Added recoverable reserves[b]	1980	1987	1989	1990[d]
Russia	110,000	100,000		79,970	82,500	82,000	188,000
Germany	102,000	56,150		116,141	123,017	121,356	357,000
Australia	46,000	41,900	183,000	10,629	13,509	15,604	48,000
United States	40,828	31,963		19,884	32,382	48,186	80,000
Indonesia		24,047					
China	37,200	18,600	34,700	9,359			
Yugoslavia	16,000	15,000		16,075	22,409	23,470	76,000
Poland	12,900	11,600	10,200	9,901	19,867	19,893	68,000
Mongolia	12,000			1,315	2,346	2,425	7,200
Turkey	7,705	6,986	235	6,197	18,648	16,228	43,000
Czech and Slovak FR	6,100	3,500	1,000	40,382	41,904	38,549	86,000
Hungary	5,465	2,883	1,124	8,152	7,368	6,440	5,400
Greece	5,312	3,000		4,341	8,527	9,640	52,000
Bulgaria	4,418	3,700		14,973	18,311	17,053	32,000
Romania	2,463	2,307	1,325	8,944	13,727	17,229	34,000
Iran	2,295						
India	2,100	1,900	3,932	1,501	2,762	3,267	9,500
Canada	1,615	2,827		10,239	15,967	17,666	9,400
New Zealand	1,556	9	28	137	43	80	200
Thailand	1,179	829	117	572	2,573	3,319	11,000
United Kingdom	1,000	500					
World total		*328,284*					*1,130,000*

[a]Refs. 3,4,24.
[b]Estimated values.
[c]1 tce = 1 metric ton of coal equivalent = 29.3 GJ (2.78×10^7 Btu).
[d]Values are in units of 10^3 t.

lower quality lignitic coals for electric power generation. The high reactivity of lignites, coupled with relatively low cost at the mine, and the relatively lower cost of transporting product electricity or synfuels, favor the construction of facilities for lignite use that are adjacent to the mine site. In the United States, the Great Plains facility is an example (25).

Main Deposits and Production Areas. *Europe and Russia.* The eastern European reserves of lignitic coals provide the primary solid fuel for the eastern part of Germany, the former Czechoslovakia, Hungary, the former Yugoslavia, and Bulgaria. The importance of lignite as an energy source is great enough in Germany to permit long-range planning that includes removal and relocation of towns or villages situated on deposits in order to permit more complete recovery of the lignite resource. The mining is not considered complete until reclamation practices have demonstrated satisfactory crop production on the area that was mined (26).

Hard coal is more important in most of the western European countries with the exception of Austria and Italy. No lignitic coal production was indicated in 1989 for the Netherlands, Denmark, Belgium, Sweden, Norway, and the United Kingdom (24).

Much of the lignitic coal in Russia was laid down in the Lower Carboniferous age but was not covered deeply enough for the conversion to bituminous coal. Deposits in the Moscow basin, an area of ca 28,000 km^2, occur in discontinuous beds consisting of lenticular pockets of dull, laminated lignite. Some seams at the bottom of the deposit consist of a more consolidated bog-head coal. A wide range of coal ranks is found in the Kusnetz basin's largest field. Additional deposits are located in the eastern Urals (Sverdlov basin), in the central Urals (Kiselov basin), in the southern Ukraine, the Caucausus (Ahalzich district), the Bashkir Republic, the Kansk Atshinsky basin, the southern part of the Tungus basin, and the Far East (Artimov basin).

The German deposits are usually found in the southern part of the northern lowland and are mostly from the Eocene–Oligocene period. The two principal areas are the lower Rhineland and the central German fields. The deposits in central Germany are thick (ca 12 m) seams that are interconnected over large areas centered in the Leipzig–Halle area in the middle Elbe basin. The central German brown coal usually gives high yields of tar and coal extracts and is desirable for chemical processing. The coal in the Cottbus district has a lower bitumen and sulfur content and is used to make high temperature coke.

Coals of the western part of Germany usually occur in thinner seams and in more local areas. The main producing areas are west of Cologne with open-cast mines at Ville, Frechen, Garsdorf, and Frimmersdorf. This brown coal is important to electric power generation (27).

Reserves of high quality occur in Slovenia, Middle Bosnia, and Serbia. Larger quantities of lower quality coals are in West Slovenia (Velenj basin), northern Croatia (Zagorje basin), and eastern (Kolubara basin) and southern Serbia (Kosov basin).

Poland's deposits tend to be located primarily in the west as isolated, lens-shaped deposits. There are other deposits in the center of the country at Turow, Konin/Goslawice/Patnow, Turek/Adomow, and Rogozno. The production is used in power plants.

The highest quality lignite in central Europe is found in the former Czechoslovakia, where both open-cast and deep mining are used. The principal reserves are in the Eger Valley in northwest Bohemia, especially around Sokolov and Most. Other deposits are located at Grothau, Budweis, and Handlow. These deposits are being depleted at the greatest rate.

Hungary's coal production, which is mainly lignitic, is used for power generation. Coal is mined near Dorog in the western Bakony hills and in the northeastern hills. Bulgarian lignitic coal comes mostly from the Dimitrovo field (Pernik). Another field is near Dimitrovgrad (Maritsa basin). Greece's reserves, found mainly in the northwestern part of the country, are also used primarily for power generation.

North America. In the United States, lignite deposits are located in the northern Great Plains and in the Gulf states. Subbituminous coal is found along the Rocky Mountains. The western half of North Dakota has about 74% of the nation's resources, Montana 23%, Texas 2%, and Alabama and South Dakota about 0.5% each. The lignite resources to 914 m represent 28% of the total tonnage of all coal deposits in the United States. The lower cost and low sulfur content have contributed to rapid growth in production.

Overburden thicknesses in North Dakota range from 4–23 m, and seam thicknesses are 1–8 m. Clay partings split thick seams into several subseams. Some seams have been depleted by spontaneous combustion after exposure of the seam by erosion. Most of the U.S. lignite coal is woody, dark gray-black, splits readily along the bedding plane, and has bed moisture up to 40%, which dries to an equilibrium air-dried moisture of 15–20%. Sulfur content varies with location. Much of North Dakota lignite is low in sulfur and is used almost exclusively for electric power generation. Some, however, is used to fuel the Great Plains Gasification facility.

The lignite deposits of North Dakota and Montana extend into Canada as far as Saskatchewan. Canadian deposits are also located in Alberta, Yukon, the Northwest Territories, Ontario, and Manitoba. Production by open-cast mining, ca 3.5×10^6 t in 1975, was 10.8×10^6 t in 1989.

Other Regions. In Australia, Victoria has the largest reserves, although smaller ones occur in southern Australia, western Australia, Tasmania, and Queensland (28). The main deposit consists of many thick seams in about 500 km^2 of the Latrobe Valley. The Yallourn coal field provides most of the fuel. The top coal seam is 65 m thick and is covered with 13 m of overburden. Most of this coal is used for electric power generation. It is very moist (55–72%), but has less than 5% ash. The lignitic coal found in South Australia is, for the most part, too deep for economic recovery. However, some of the better deposits are mined for power generation.

New Zealand's reserves are situated in the South Island. These are poorer and more dispersed than those in Australia. Production was about 80,000 t in 1989.

South America's reserves are small. In Chile, which has most of the South American reserves and production, only 22,000 t was produced in 1989.

In the Far East, total lignite production is a small percentage of the world total, but there are several significant areas. India's resources rank sixteenth in world size. The largest deposits are at Neyveli about 233 km from Madras. Poorer quality lignite is found in Kashmir. Pakistan's only reserve in production is at Jhimpir-Metig, about 97 km from Karachi. The largest producer in the Far East is the Democratic People's Republic of Korea (North Korea). The estimated North Korean production was 7.8×10^6 tce in 1989. Japan produced only 8000 tce in 1989. China's lignite resources are about 18.6×10^9 t, found mostly in the Northwest, not counting Mongolia. There are lesser amounts in the Southwest, and smaller amounts in the central southern region (3,29).

No lignite production was reported in Africa. The only significant resources are in the Central African Republic and Nigeria. The latter deposit is unusual, having high hydrogen content, and coking provides unusually high yields of hydrocarbon-rich waxy tars.

Production. *Mining.* The mining or winning of lignitic coal typically involves deposits near the surface. The open-cast, open-cut, or strip-mining techniques employed involve mobile equipment built to provide a range of capacities to over 200,000 m^3/d. The rate of production can be increased rapidly, and the amount of labor per ton of coal mined is less than for underground mining. The quality of the coal, ratio of overburden thickness to seam thickness, stratigraphy, and distance to location of consumption are important in determining the cost to the consumer.

In modern practice, topsoil is first stockpiled for later application. The overburden, ie, sand, gravel, clay, etc, is then removed and the exposed coal is removed by bucket-wheel excavators, bucket-chain dredges, or draglines and shovels. Excavators having daily capacities of 2×10^5 m^3 of overburden or coal have been built in Germany. These machines are 83 m high, 220 m long, and weigh 13,000 t. Plans for the Hambach mine near Cologne call for reaching a depth of about 500 m with an overburden:coal ratio of more than 6:1. The bucket-wheel excavator works with a stacker of similar capacity to move overburden. The coal usually is moved by a conveyor belt and later by electric locomotives in 2000 t lots (27).

Preparing for mining in the United States involves studying cores drilled initially on 1.6 km centers and later at 1.5–3 cores/km^2 (4–8 cores/mi^2), depending on the occurrence of discontinuities. The actual mining takes into account the variation of deposit properties in terms of percent moisture, ash composition, ash content, and grindability as a function of moisture (9).

Seam thicknesses and depths vary tremendously. The most favorable deposits have shallow overburdens and thick seams that cover large areas. Acceptable stripping ratios, ie, overburden thickness to coal thickness, depend on the quality of the fuel. Ratios up to 10:1 have been used for bituminous coals, but lower ones are required for lignitic coals because of the lower heating value per unit weight.

A variety of measures must be taken to assure safe and continued operation. Because the natural water table is higher than the coal seams, or the seams are natural aquifers, it is necessary to pump water out of the pit or to drill wells around the mine and pump to reduce the water table. The Rheinische Braunkohlenwerke (Rheinbraun) pumps water at a rate of $1{-}1.2 \times 10^9$ m^3/yr. Part of this water is processed to provide drinking water for Neuss and Dusseldorf. The tendency of lignite to ignite spontaneously requires care in the amount of face that is exposed, especially in naturally dry, hot, windy climates.

Storage. Concern about spontaneous ignition has led some operators to try to match the mining and consumption rates, so that there is little if any reserve, as in minemouth power generation stations. When the coal must be stockpiled, careful stacking minimizes oxygen reaction and overheating. Uniform stacking in layers no more than 0.3 m thick avoids segregation of particle sizes, then compacting using earth-moving equipment, and covering the pile with finer material limits oxygen penetration, overheating, and ignition. By sloping (14°) the sides gradually, segregation is prevented and compaction is improved (30).

A smooth coal pile surface, coupled with the gradual slope, minimizes the differential wind pressures and consequent oxygen penetration. A $4{-}6 \times 10^6$ t lignite stockpile from the excavation for the Garrison Dam in North Dakota has been stable for many years as a result of this storage method.

To limit drying, spraying with cold water is useful. The spraying can be coupled with a straw covering. Four Hungarian stacks covered with 10 cm of straw decreased in heat value only 0.4–6% after spraying periodically for 10 months, but lost 6–20% if unsprayed. Underwater storage is sometimes used in drier climates such as Australia. Processed fuels, eg, briquettes, can be stored without difficulty since they are less permeable to air, and, depending on process conditions, are less oxidizable.

Transportation. For short distances from the mine, transportation (qv) is by truck or conveyor belt. Rail transportation is generally used for greater distances. Slurry pipelines (qv) are being considered as an alternative. Rail transport over hundreds of kilometers results in loss of surface material in uncovered cars and a tendency to overheat in bottom-dumping rail cars owing to air infiltration around the cracks (31). Proper sealing and covers permit shipping over hundreds of kilometers.

Drying. In many cases, the high moisture content of young coals dictates significant drying (qv) before use. In some cases, partial removal of mineral matter, especially water-soluble species, is desirable.

Drying can be accomplished by evaporative, hydrothermal, or other thermal processes. Evaporative drying reduces the moisture content substantially, but increases the tendency to spontaneous combustion and decrepitation, ie, breaking up and cracking. Power plants use combinations of heated shafts and hot flue gas passed through the size reduction (qv) equipment to remove moisture (10). Hydrothermal, ie, nonevaporative, drying or dewatering (qv) was originally developed in the 1920s and has been continually refined. The Fleissner process is used commercially at a 250,000 t/yr plant in Kosovo, in the former Yugoslavia. This approach removes a large part of the water in a low rank coal in the liquid form, without using the energy necessary to evaporate this water. Also, a notable part of the water-soluble inorganic species, such as sodium salts, can be removed with the expressed water, reducing the tendency for fouling the boiler during combustion. The process water is acidic (pH 3–4), and rich (up to 4500 ppm) in phenols. Hydrothermal treatment is carried out at as low a temperature as possible to remove the water without increasing the need for water treatment.

Thermal drying has been studied in conjunction with a rail shipment of ca 1200 km from North Dakota to Illinois. Oil was applied at 6–8 L/t to suppress dust loss, and cracks around the doors in the base of the car were sealed to prevent ignition. Stable shipment and stockpiling were then possible (31). Thermal drying may be carried out to further reduce the moisture content as required for briquetting or for more efficient pulverizing and combustion.

Health and Safety Factors

Because lignite mining is carried out by surface methods, the hazards associated with underground mining typically do not exist (see COAL). The principal hazards involve the tendency of the coal toward spontaneous combustion as the coal dries, especially at the exposed seam. Thus careful planning and continued reclamation efforts are required to cover faces. Adequate water for revegetation has been of some concern in the arid areas of the northwestern United States where vegetative growth is slow and reclamation is expected to take many years.

The lignitic coals of the northern United States tend to have low sulfur contents, making them attractive for boiler fuels to meet sulfur-emission standards. However, low sulfur content coals have impaired the performance of electrostatic precipitators. The ash of these coals tends to be high in alkaline earths (Ca, Mg) and alkalies (Na, K). As a result, the ash can trap sulfur as sulfites and sulfates (see AIR POLLUTION CONTROL METHODS).

Some North Dakota lignite ashes have also been observed to have above-average concentrations of uranium (21,22), leading to interest in processing the ash for uranium recovery. However, this ash may be classified as hazardous.

Economic Aspects

The price of lignite per mined ton or per heat unit is lower than that for higher rank coals. The market for all coals is primarily as boiler fuel for electric power production. Prices are generally established by contracts between utility and supplier before mining begins.

Because of its low sulfur content, lignite is becoming more important. The U.S. Clean Air Act Amendments of 1990 have resulted in economic premiums for low sulfur coal corresponding to \$10/t for emission allowances at \$500/t of SO_2 (32).

Uses

Most of the world's coal supply is used for combustion to generate steam for electric power production. This is especially true of lignitic coals. Other uses for lignite, such as briquetting, for domestic and industrial fuels; carbonization, to provide coke and liquid by-products; gasification, to provide gaseous fuels; chemical feedstocks, for making fertilizers and other liquid fuels; and direct liquefaction are being developed.

Briquetting. Lignite briquettes have long been preferred over unprocessed brown coal for residential and industrial heating (10,33,34). The extrusion press has been used for the production of most of the briquettes made from lignite coals since 1858. Design improvements have provided for multiple pressing, and for different channels, feeding methods, and press drives, all of which have significantly increased output. The demand for briquettes has been dropping as other fuels are used for residential heating. About 75% of the carbon originally present in the lignite appears in the product briquettes. A typical briquetting plant produces surplus electric power equivalent to a few percent of the energy content of the original lignite.

Plants for briquette production exist in the eastern part of Germany, Australia, and India. German transport costs per unit of heating value are about 40% less for briquettes than for lignite.

Briquetting of coal may be carried out with or without the addition of a binder. Binderless briquetting, the predominant method, is restricted to the relatively soft, unconsolidated lignites. The coal size distribution and moisture content must be carefully regulated for successful agglomeration and pelletting. To obtain close packing of the material during compression, the coal is generally crushed to a size below 4 mm, where 60–65% is < 1 mm. For maximum briquette strength, moisture contents are adjusted to optimum (between 11 and 18%) values which vary with the coal seam (5). At higher moisture, briquettes shrink and crack on equilibrating with the atmosphere, and at lower values the briquettes may swell and weaken.

After crushing, the lignite is dried thermally in a rotary dryer and cooled slowly to achieve uniform moisture distribution in the particles. When the temperature reaches the optimum range of 38–65°C, the lignite is pressed at ca 138 MPa (2000 psi). A reciprocating piston rams the coal through an increasingly restricted channel during the forward stroke and loads more lignite during the return strokes. A final gradual cooling is required. Pressing increases the briquette temperatures as much as 30°C internally and more at the surfaces. The commonly used Exter press forms $20 \times 6 \times 4$ cm briquettes, each weighing ca 550 g. The moisture in the coal is the binder and forms hydrogen bonds between the polar functional groups (5).

Harder, more mature coals generally have lower plasticity and greater elasticity. These coals are briquetted at higher pressures using a binder in ring-roll presses to give a harder carbonized product. Because briquettes have been used primarily for domestic heating and metallurgical and chemical processes, the binderless coking process has been predominant.

Briquettes must be transported carefully to avoid breakage. They are usually dumped into piles in sheds and frequently are screened to remove smaller broken particles. Hand-stacking significantly improves storage quality, and permits more material to be stored in a limited volume.

Briquettes have a heating value of 16.7–23.4 MJ/kg (7,200–10,000 Btu/lb) or from 2–3 times the value of a typical brown coal primarily because of the moisture loss.

Combustion. The combustion of lignite is used to generate energy steam for turbine generators to provide electric power. Most modern plants have a capacity of at least 100 MW, and frequently as much as 600 MW (35–38). A brown coal boiler needs to be 1.5 times larger than a bituminous coal boiler to produce the same amount of power because of the larger amount of inert materials therein, eg, water vapor and recycled flue gas. The fuel feed rates must be up to three times as great as for hard coal because of the high moisture content in the brown. The resulting higher capital costs are, however, offset by lower fuel costs.

The formerly common grate-supported combustion is used in some smaller units, but the dominant method of combustion involves suspension or pulverized coal firing. The high reactivity of lignite permits burning of coarser (65–70% mm (200 mesh)) particles than those of bituminous coal for a given furnace residence time. The average residence time of a coal particle in a large boiler may be 3–4 s, of which 0.25 s is required for combustion (see FURNACES, FUEL-FIRED).

The relative reactivities of lignite and other coals have been the subject of many studies. The burning profile or rate of weight loss as a function of temperature or time for a coal sample heated at a constant rate in air is shown in Figure 3 for a number of coals (39). The initial weight loss owing to moisture release is followed by a more significant weight loss resulting from oxidation of the bulk of the organic substance. The onset of oxidation occurs at a lower temperature for lignite than for other coals. Thus, the oxidation reaction is complete at a lower temperature because of the greater ease of lignite ignition. The completion temperature relates to residence time or size of the furnace cavity required for complete combustion.

Because of the wide variation in composition and properties of brown coal (see Table 3), efficient combustion of these fuels cannot be accomplished by a single system. The moisture content limits combustion efficiency because some chemical energy is required to convert liquid water to steam in the flue gases. The steam then increases the dew point of the gases, requiring higher temperatures to avoid condensation in the stack. For fuels up to 25% moisture content, 80% efficiency can be achieved. As the moisture content increases to 60%, the efficiency decreases to 70% and efficiency continues to decline about another 1% for each additional 1% moisture to 70%.

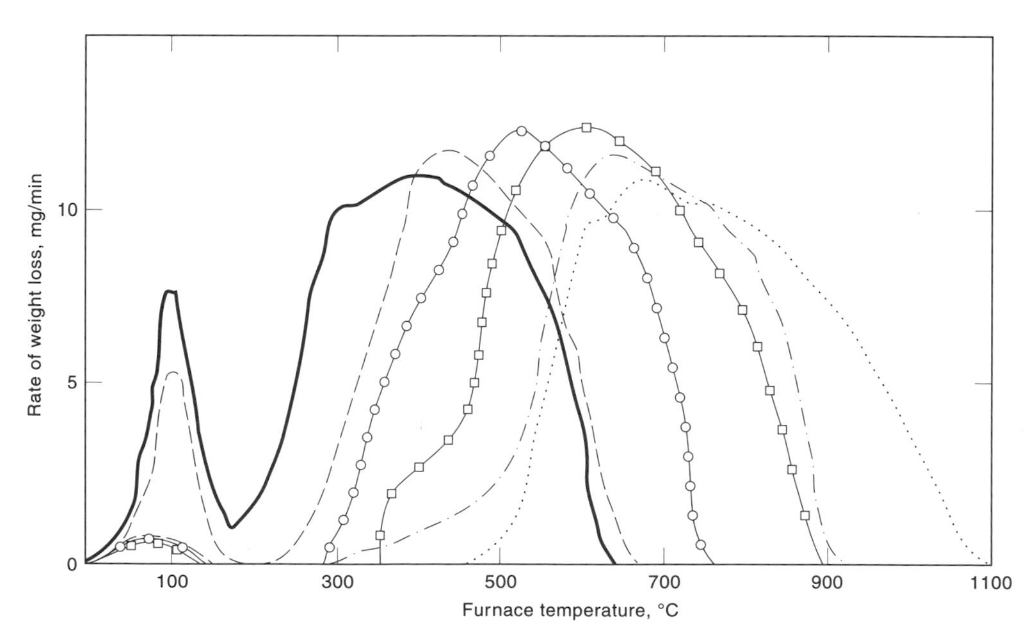

Fig. 3. Comparison of burning profiles for coals of different ranks where (· · ·) represents anthracite; (– · –) semianthracite; (□) LV bituminous; (○) HV bituminous; (–––) subbituminous; and (——) lignite. LV = low volatile; HV = high volatile.

Most utility boilers are of the dry-bottom (solid ash) design, although a few cyclone-burner, slag-tap design stations are in operation in the United States. A variety of firing systems have been developed to partially dry the lignite before combustion (Fig. 4) (19). The Niederaussem plant (Fig. 4**a**) uses a stream of hot flue gas to dry the raw coal before pulverizing, then cold air is added through the pulverizer and all of the products are fed into the boiler. The Megalopolis station (Fig. 4**d**) uses hot flue gas to dry the lignite. A cyclone separator and electrostatic precipitator permit rejection of some of the water vapor to the atmosphere rather than to the boiler. Another drying method uses a vertical shaft, heated by combustion gases, for partial drying prior to grinding.

A high moisture content necessitates finer grinding of coal for rapid water release. For such coals, the classifier in the pulverizer is set to return more of the oversize material for further grinding.

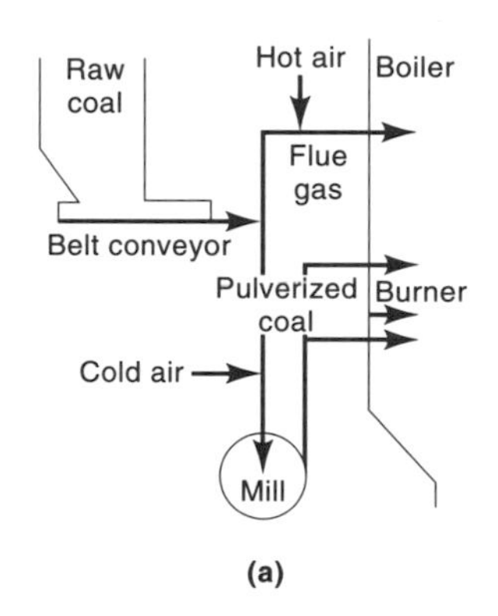

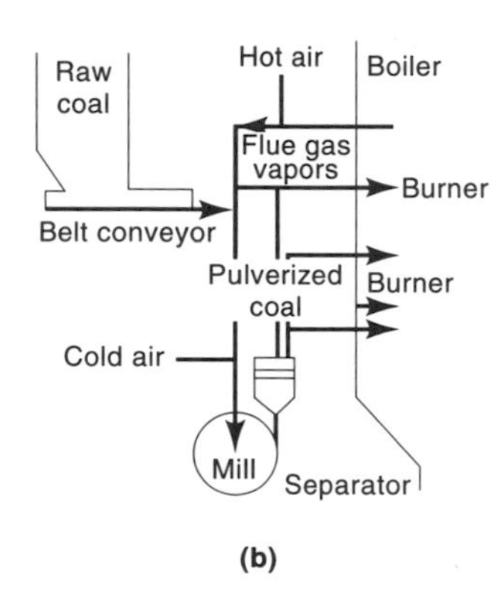

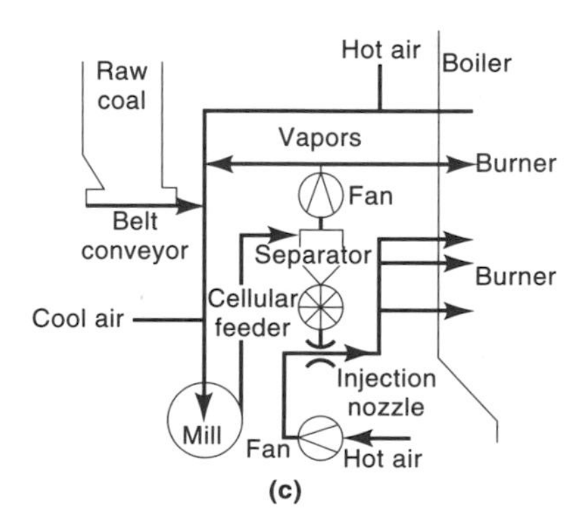

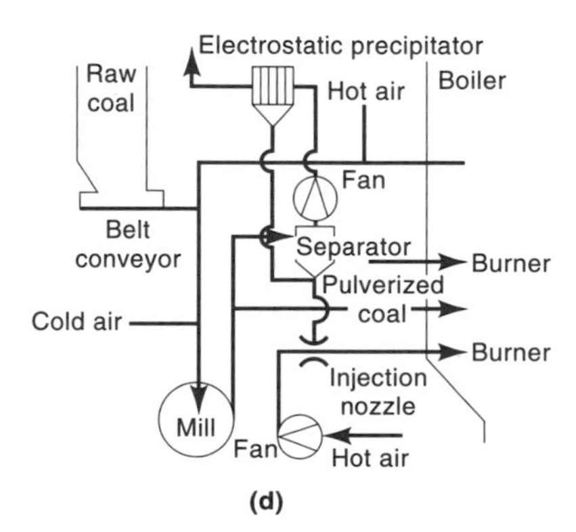

Fig. 4. Firing systems for lignitic coal. (**a**) Niederaussem, Germany: 600 MW; heating value (HV) = 7.7–10.7 MJ/kg; moisture: 52–57%; ash content: 2–13%. (**b**) Yallourn, Australia: 350 MW; HV: 4.2–8.8 MJ/kg; moisture: 58–73%; ash content: max 3%. (**c**) Tuncbilec, Turkey: 150 MW; HV: 8.4 MJ/kg; moisture 22–24%; ash content: 42%. (**d**) Megalopolis, Greece: 125 MW; HV: 3.8 MJ/kg; moisture: 64%; ash content: 13%. To convert MJ/kg to Btu/lb, multiply by 430.2.

Particulate removal following combustion is accomplished by electrostatic precipitators, which depend on the particles accumulating sufficient electric charge. Charge has been associated with the sulfur oxide content of the flue gas. The higher the SO_x, the more charge present. Larger precipitators must be used for low sulfur coals. Alternatively, an additive that can be adsorbed and readily ionized, eg, SO_3, can be injected upstream of the precipitator to improve the performance of existing units.

The ignition of moist lignitic coal requires up to five times the energy required for ignition of bituminous coal. This necessitates preheating, firing techniques, control of air supply, and preignition grates on chain-grate stokers. For grate firing, the lump coal must be evenly sized and distributed to allow uniform air flow and combustion. The disintegration of soft coal on heating produces a mixture of fine and lump coal, leading to flow resistant masses and some losses of incompletely burned fuel. The ash of some (German) coals tends to maintain the shape of the coal pieces; the small amount of ash from Australian (Victoria) coal does not.

Briquettes burn similarly to bituminous coal, although some tend to disintegrate on combustion. A low (<6–7%) ash content increases the possibility of disintegration. Normal combustion depletes the combined oxygen and volatile matter in the coal quickly, effectively changing its composition and combustion behavior, making control of combustion difficult.

During combustion, the alkali content (CaO, MgO, K_2O, Na_2O) of lignitic coals is released. Alkalies are normally sufficient to keep the sulfur in the coal during wet scrubbing (40). Sulfur capture is dependent primarily on the sodium, calcium, and the alkali:sulfur ratio (10,41). Loss of retention occurs if the silica and alumina content is high, because an unreactive alkali aluminosilicate is formed, complexing the alkali. As sulfur emission standards become more stringent, supplemental limestone may be required in some cases.

Pyrolysis. Heating in the absence of oxygen releases moisture at low temperatures, carbon dioxide at temperatures $> 200°C$, and a variety of gaseous products at very high temperatures. Acid washing of the raw coal is used to remove extractable cations, followed by treatment with selected cations. Yields of CO_2, CO, CH_4, H_2, and H_2O depend on the amounts of inorganic species in the coal (42).

Carbonization. Low temperature carbonization of brown coal, usually as briquettes, was originally used to produce a tar and oil mixture that could be further processed to motor fuels, and later to produce a low heat–energy heating gas and a solid smokeless fuel. High temperature carbonization of lignite briquettes produces a coke that could be used to supplement coke from metallurgical-grade bituminous coal. Typically, the gas by-product was burned for steam generation and for some electric power generation (10,43–46).

Low Temperature Carbonization. The Lurgi Spülgas process was developed to carbonize brown coal at relatively low temperatures to produce tars and oils (Fig. 5). A shaft furnace internally heated by process-derived fuel gas (Spülgas) is used. The product can range from a friable coke breeze to hard lump coal depending on the quality of the briquettes used in the feed. The briquettes, made in normal extrusion presses, break down into smaller sizes during carbonization.

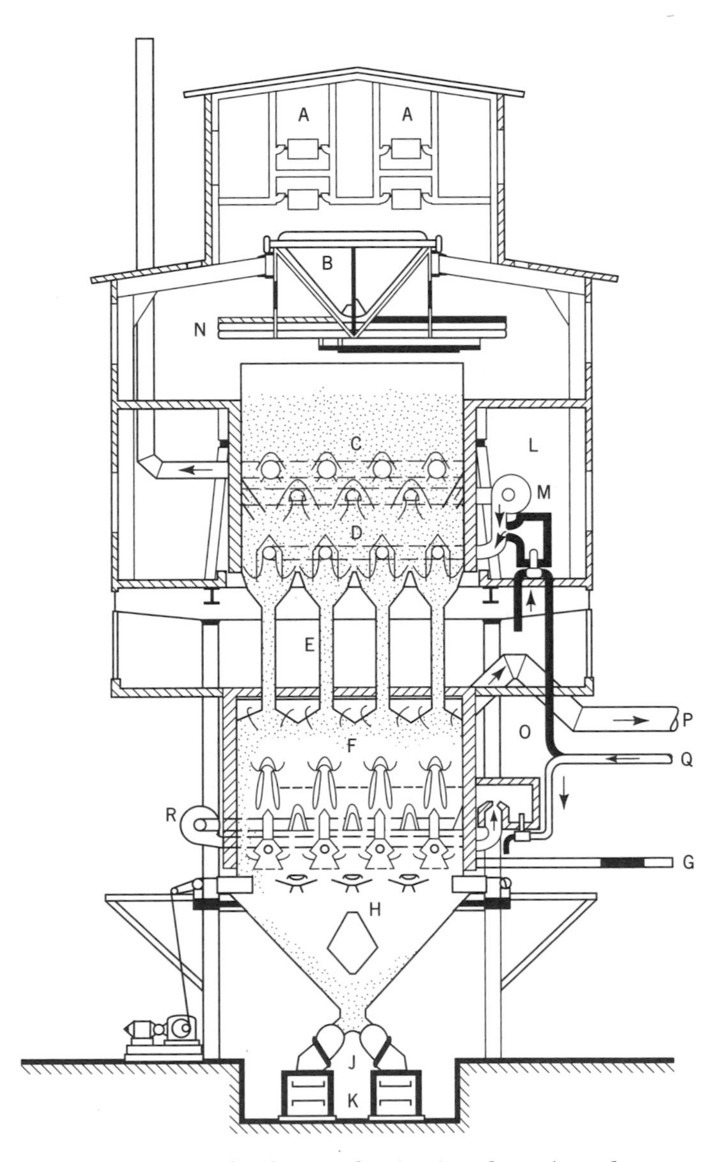

Fig. 5. Diagram of large Lurgi Spülgas carbonization plant: A, coal conveyor; B, movable distributor; C, coal bunker; D, drying zone; E, connecting shafts; F, carbonization zone; G, coke-cooling zone; H, coke extractors; J, coke hopper doors; K, coke conveyor; L, circulation fan for drying zone; M, combustion chamber for drying zone; N, stack for waste gases from drying zone; O, combustion chamber for carbonizing zone; P, offtake for mixture of carbonization gas and Spülgas; Q, intake for tar- and oil-free cooled Spülgas; R, fan for gas for coke quenching.

The briquettes are distributed over bunkers and dried for about 5 h at 150°C (0.5% moisture), then descend to the lower carbonization zone for another 5 h at 500–800°C. Heating is provided by burning the purified product gas with air using from 35–48% of the gas. The rest usually is used for steam generation. The char (product) is cooled by circulation of product gas. The char reactivity increases with time and must be stored carefully.

When denser briquettes are used for carbonizing, the product has 10–12% volatile matter, up to 20% ash, and a heat value ca 25 MJ/kg (11,000 Btu/lb). Overall yields for normal briquette carbonization are char 45%, tar 12.5%, and gas 130 m^3 (4600 ft^3) having a gross heating value of 8.4 MJ/m^3 (225 Btu/ft^3). The char was ca 30% above 20 mm, used for domestic and central heating; 50% from 6–20 mm, used for gasification; 30% less than 6 mm, used for steam generation and gasification. About 270 Lurgi Spülgas plants exist, mainly in Germany, Australia, India, and the Eastern European countries, and having a coal input of 10^5 t/d. Fluidized-bed and entrained carbonization of lignite have been studied but have not been commercialized (see FLUIDIZATION).

High Temperature Carbonization. Some lignitic coals can be blended with caking coals in coke production, primarily to produce a coke or coke breeze for iron ore reduction, or for carbide or phosphorus (qv) production (see CARBIDES). In Germany, small briquetted material under 1 mm is carbonized to 950°C for about 12 h. After dry cooling, the product coke is screened from 45 mm to <3 mm. A 24-retort battery has an output of 125–130 t/d (42% yield), more than half of the coke is large and strong enough to be used in low shaft furnaces. The rest is used in gasifiers, lime producers, and carbide furnaces. The large coke is mixed with conventional coke for iron ore smelting. The CaO/SiO_2 ratio is high enough to reduce typical lime additions.

Gasification. Gasification converts solid fuel, tars, and oils to gaseous products such as CO, H_2, and CH_4 that can be burned directly or used in synthesis gas (syngas) mixtures, ie, CO and H_2 mixtures for production of liquid fuels and other chemicals (47,48) (see COAL CONVERSION PROCESSES, GASIFICATION; FUELS, SYNTHETIC–GASEOUS FUEL; HYDROGEN).

The Lurgi process has been the most commercially accepted gasification method since its commercialization in 1936, and is used in the large plants in South Africa, in modified designs in Germany, and in the United States for the Great Plains facility (25,49,50).

The gasification process includes coal preparation (crushing or pulverizing), charging into a gasifier (using special equipment if the gasifier is pressurized), quench of the hot product gas, shift reaction or adjustment of the $CO:H_2$ ratio (depending on the product requirement), gas purification (removal of sulfur-bearing species, typically H_2S, and often CO_2), and catalytic conversion if a special product such as substitute natural gas (SNG) or a liquid fuel is desired. For fuel-gas production, the shift, CO_2 removal, and catalytic conversion can be eliminated. The Lurgi gasifier shown in Figure 6 is a fixed-bed reactor into which coal is charged through lockhoppers at the top. Steam and oxygen mixtures enter at the bottom, react with the coal at ca 3.0 MPa (30 atm) pressure in a countercurrent mode. Product gas is removed at the top and quenched (25).

One of the largest plants began operation in 1984, near the largest (12.2×10^6 t/yr) coal mine in North Dakota by the Dakota Gasification Co. In 1988 average production was 4.2×10^6 m^3/d (158×10^6 ft^3/d) of pipeline-quality gas (CH_4) (25). About 27,500 t/d of lignite are mined and crushed to pass a 5 cm screen. Slightly more than half (56%) is fed to the gasifiers, and the remaining fines are sold for boiler fuel. The plant consumes 2,800 t/d of oxygen and about 13,000 t/d of steam. The accompanying oxygen (qv) plant is the largest in the country. In 1990 a plant to recover krypton and xenon was added.

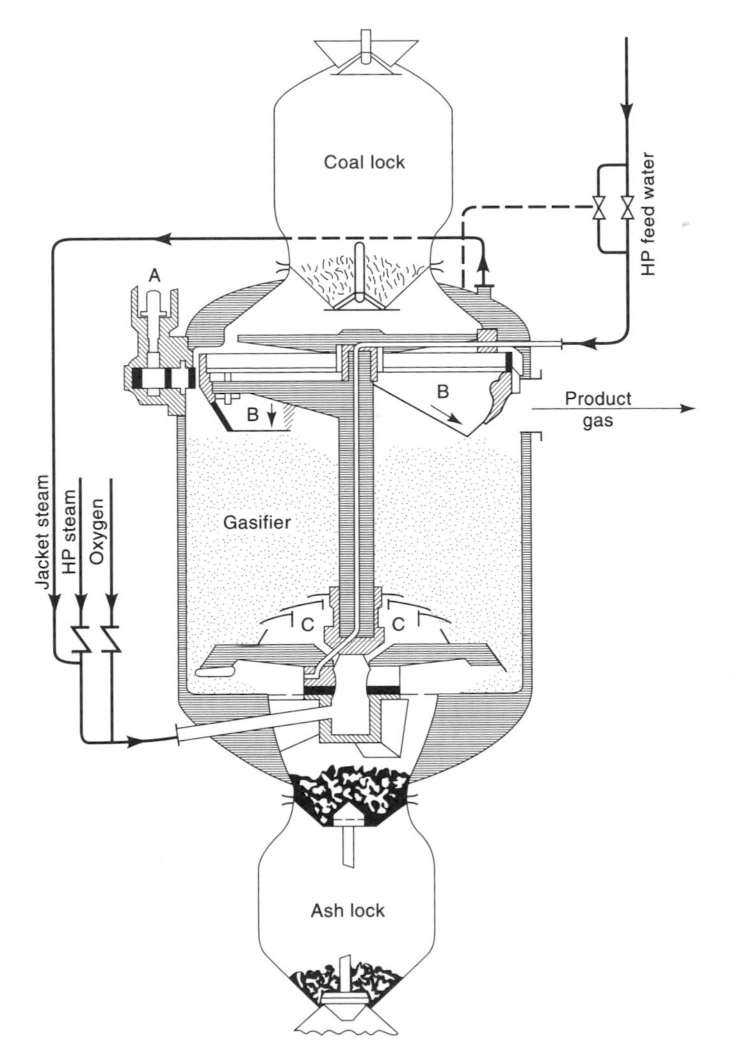

Fig. 6. Lurgi pressure gasifier: A = variable speed drive; B = coal distributor; C = rotating grate. HP = high pressure.

Fourteen Lurgi Mark IV (4 m) gasifiers are in operation at 3.07 MPa (430 psig). About 30% of the gas stream is shifted to provide the 3:1 overall H_2:CO ratio needed for catalytic conversion to methane, CH_4. The acid gases, CO_2 and H_2S, are removed using a Rectisol unit. The gas stream is catalytically converted to methane, dried, and compressed to about 9.9 MPa (98 atm) with a gross heating value of 36.4 MJ/m^3 (977 Btu/ft^3). The Phosam process recovers 113 t/d of anhydrous ammonia (qv) which is sold, and the tars, oils, phenols, and naphtha are used for boiler fuel. The Sulfolin process (Linde AG) recovers 106 t/d of sulfur from the gaseous products (90% from the gases fed to the process).

The British Gas Corp./Lurgi slagging gasifier has been tested in Scotland (300 t/d unit). This method is designed to achieve higher throughputs. Lower costs are expected from operating at temperatures high enough to melt the coal ash, after which it flows out the bottom of the gasifier as a molten slag.

The first commercial fluidized-bed systems used Winkler units in which lignite and its char are gasified at atmospheric pressure with air or oxygen and steam mixtures. Improvements led to high temperature Winkler (HTW) units, being tested in a 1 t/h pilot plant in Germany. A number of fertilizer plants use lignite in the Koppers-Totzek gasifier, an entrained or suspension gasifier in which pulverized coal reacts with a steam–oxygen mixture to produce synthesis gas and ammonia. The ash is removed as a molten slag.

Methods in Development. The U.S. Department of Energy (DOE) and the American Gas Association (AGA) have sponsored the development of more efficient gasification processes, but as of this writing the plentiful supply of inexpensive natural gas has precluded commercialization.

A medium heat (11.2 MJ/m^3 (300 Btu/ft^3)) gas for industrial application has been produced at Westinghouse Electric's pilot plant in Waltz Mill, Pennsylvania (50), where lignite is gasified in a pressurized fluidized bed which uses oxygen and steam.

The HYGAS process has been developed by the Institute of Gas Technology under AGA/DOE sponsorship for SNG production in a 3 t/h pilot plant in Chicago, Illinois. The coal (lignite or higher rank coals), prepared as a recycle-oil slurry, is pumped into the top of the gasifier. The slurry is then dried, and the coal contacts a rising stream of H_2 and other gases to go through rapid-rate methanation. Operation at pressures to 7 MPa (1000 psi) permits over 50% of the methane to be made in the gasifier, reducing the downstream processing requirements. The residual char is further gasified in a lower fluidized bed. The resultant char is gasified using a steam–oxygen mixture in the lower fluidized bed. Coal-to-methane conversion efficiencies are about 75%, but this method has not been commercialized.

The CO_2 Acceptor process, also developed under AGA/DOE sponsorship, by the Consolidation Coal Co., uses steam to gasify lignitic coal. Heat is supplied by the exothermic reaction between CO_2 and calcined dolomite [*17069-72-6*]. The dolomite is calcined in a separate fluidized bed. This process operates in a 40 t/d pilot plant, but there are no plans for commercialization as of this writing.

Underground gasification of lower rank coals has potentially lower gas costs and reduced mining hazards. High pressure air is used to increase permeability. Tests of this technology have been carried out in Texas under sponsorship that included the Texas Utility Commission. Moscow-region coals have been gasified underground since 1947. Coal beds are up to 5 m thick, but these cannot be gasified when only 1 m thick.

Liquefaction. The synthesis gas or hydrogen required for liquid fuels is produced by a process known as indirect liquefaction; ie, gasification to produce synthesis gas followed by catalytic conversion to liquid products, or by direct hydrogenation of lignite (see COAL CONVERSION PROCESSES, LIQUEFACTION). The Texaco partial oxidation process uses a slurry of coal in a high pressure gasifier (7 MPa (70 atm)) in the presence of oxygen. The slurry enters at the top of the reactor and is gasified as the coal descends. Ash is quenched in a water bath at the bottom. This gasifier has been successfully used in a number of plants.

The U-GAS process has been developed by the Institute of Gas Technology under DOE sponsorship with a 1 t/h pilot plant in Chicago. The gasifier is a single-stage fluidized bed able to accept lignitic or raw bituminous coal to produce a synthesis as at 1000°C. A unique feature is a venturi (short tube with tapering construction in the middle) at the base of the bed, which permits the discharge of ash agglomerates with high carbon utilization. The technology is being commercialized by Enviropower Inc. (Finland).

Liquid Fuels. Liquid fuels can be obtained as by-products of low temperature carbonization; by pyrolysis, solvent refining, or extraction; and gasification followed by catalytic conversion of either the coal or the products from the coal. A continuing interest in liquid fuels has produced activity in each of these areas (44–46). However, because crude oil prices have historically remained below the price at which synthetic fuels can be produced, commercialization awaits an economic reversal.

The conditions of pyrolysis either as low or high temperature carbonization, and the type of coal, determine the composition of liquids produced, known as tars. Humic coals give greater yields of phenol (qv) [*108-95-2*] (up to 50%), whereas hydrogen-rich coals give more hydrocarbons (qv). The whole tar and distillation fractions are used as fuels and as sources of phenols, or as an additive in carbonized briquettes. Pitch can be used as a binder for briquettes, for electrode carbon after coking, or for blending with road asphalt (qv).

Tars can be hydrogenated to produce liquid fuels. High hydrogen and low asphaltene, ie, benzene-soluble and pentane-insoluble, contents are desirable. The central German brown coals are attractive for this reason. The tars from the eastern part of Germany require much lower pressures and less hydrogen per unit of product than do brown coals near Cologne, which can require pressures up to 71 MPa (700 atm) (see PETROLEUM).

Integrated Projects. Large integrated complexes are employed to process lignitic coal into a variety of fuels and chemicals. Examples of these exist in South Africa, Germany, and India (43,49,51,52) (see FUELS, SYNTHETIC). The South African complexes at Sasolburg and Secunda, in operation since 1982, were planned to provide liquid transportation fuels from local coals. These plants draw coal from a 300 km^2 coal field, which is expected to last until at least 2060. The subbituminous coal has a gross heating value (dry basis) of 23.9 MJ/kg (10,300 Btu/lb). The composition is ash 21.5%, sulfur 1.3%, carbon 79.67%, and hydrogen 4.3% on a dry ash-free basis. The seam is almost horizontal and has a range of thickness of 2–7 m at 100–200 m depth. It is mined using essentially continuous and longwall mining techniques. Belt conveyors take the coal to the adjacent complex. Total production from the six mines in the complex is more than 32×10^6 t/yr (49). Annual coal consumption for Sasol II is about 13×10^6 t, two-thirds of this for gasification, and the remainder for steam and electric power generation.

Fine coal is not acceptable as Lurgi-type gasifier (Fig. 6) input. Power generation capacity of 240 MW was selected to utilize the fines produced during crushing and handling of the coal. Oxygen (8600 t/d) is needed for the 36 Lurgi 4.0 m diameter gasifiers. High pressure steam requirements are 1230 t/h. Raw gas production is about 1.65×10^6 m^3/h (1.4×10^9 ft^3/d). After quenching, this gas is fed to a Rectisol (cold methanol) purification plant to provide ca 1.2×10^6 m^3/h (1×10^9 ft^3/d) of pure gas (0.07 ppm S). The pure gas composition is about 1.5% CO_2, 84.1% H_2 + CO, 13.5% CH_4, 0.5% N_2, and 0.4% C_nH_m. An adjacent oxygen plant consists of six units of 2300 t/d capacity, each at 3.45 MPa

(500 psi). Steam generation involves six boilers producing 540 t/h of 430°C, 4 MPa (580 psi) steam each. The Rectisol plant discharges H_2S to a Claus unit which produces 99.97% pure sulfur (49) (see SULFUR REMOVAL AND RECOVERY).

The purified raw gas goes to a Synthol (Fischer-Tropsch) unit for catalytic conversion of CO and H_2 to liquid fuels. The tars and oils obtained from quenching the raw gas from the gasifiers go to a Phenosolvan plant to provide tar products for the refinery and ammonia for fertilizer. The Synthol plant has seven reactors, each with 1.9×10^6 m^3/h (1.6×10^9 ft^3/d) gas feed. Annual plant production is 1.5×10^6 t motor fuels, 185×10^3 t ethylene, 85×10^3 t chemicals, 180×10^3 t tar products, 1×10^5 t ammonia (as N), and 9×10^4 t sulfur, for a total of 2.14×10^6 t/yr of valuable products.

Another integrated project is the VEB Gaskombinat Schwarze Pumpe plant in eastern Germany. This complex consists of three briquetting plants, three power stations, one brown coal high temperature coking plant, and one pressure gasification plant. Open-cast mines near Welzaw-Sud and Nochten produce soft brown coal. The overburden ratio is about 5.2:1. The equipment can remove up to 60 m of overburden and has a capacity of 20,000 m^3/h (1.7×10^8 ft^3/d). The raw brown coal goes to both the briquetting plants and power stations. Some briquettes are produced directly for fuel, others go to the coke-oven plant to produce coke and liquid products or to the pressure gasification plant to produce gas and liquid products. The power station generates both electric energy and steam which is used on-site. The fuel gases from the coke-oven plant and the oil and coal gasification plants are blended, after purification, with product gases from the natural gas reformer to provide town gas. The capacity is estimated at 13×10^6 m^3/d (450×10^6 ft^3/d) town gas which has a heat content of 15.9 MJ/m^3 (ca 430 Btu/ft^3). The composition of the town gas is 1.3% CO_2, 12.3% CO, 0.84% O_2, 22.9% CH_4, 34.4% H_2, 26.2% N_2, 0.94% C_2H_6, 0.53% C_3H_8, 0.56% C_4H_{10}, and 0.03% C_2H_4. This gas is obtained by blending natural gas from eastern Germany and Russia, nitrogen (qv) from the adjacent air-separation plant, and reformed gas with the purified fuel gas stream from the plant.

Two air-separation plants provide oxygen to the gasifiers, 24 units with a 3.6 m internal diameter. These gasifiers, of a design similar to the Lurgi fixed-bed type, are located in an open-air structure for safety. After gasification of the briquettes, the raw gas is quenched to remove tars, oil, and unreacted fine particles. The product gas from the gasifiers and coke-oven plants is purified with refrigerated methanol. A Claus unit converts H_2S to elemental sulfur.

Operating parameters of this German plant, on the basis of one cubic meter of raw gas, include 0.139 m^3 O_2, 0.9 kg briquettes, 1.15 kg steam, 1.10 kg feed water, 0.016 kWh, and 1.30 kg gas liquor produced. Gasifier output is 1850 m^3/h and gas yield is 1465 m^3/t dry, ash-free coal. The coal briquettes have a 19% moisture content, 7.8% ash content (dry basis), and ash melting point of 1270°C. Thermal efficiency of the gas production process is about 60%, limited by the quality and ash melting characteristics of the coal. Overall efficiency from raw coal to finished products is less than 50%.

In the plant at Neyveli, India, the clay and sand overburden is used to make china clay, and the lignite is used for power generation, for gasification to provide feedstocks for fertilizer production, and for briquette production with by-product light oils. The mine area is 14 km^2 (5.41 mi^2), the recoverable lignite is ca 180 × 10^6 t, the overburden average thickness is 62 m, and the average lignite thickness is 13 m.

Power generation using pulverized coal produces up to 250 MW for export and 400 MW for internal consumption. Crushed lignite, dried to 8% H_2O, is gasified in Winkler generators with steam and oxygen. Ammonia is made from H_2 and N_2 from the air-separation plant. Further reaction of the NH_3 with CO_2 produces urea [*57-13-6*] for fertilizer. Extruded briquettes are formed by low temperature carbonization of crushed 12% H_2O coal. Lurgi Spülgas low temperature carbonization makes carbonized briquettes, light oils, and fuel for power generation.

Products from Synthesis Gas. Steam–oxygen gasification of coal produces syngas, a mixture of carbon monoxide [*630-08-0*] and hydrogen [*1333-74-0*]. Increasing pressure increases the amounts of methane [*74-82-8*] formed. This mixture can be used for a variety of products by Fischer-Tropsch synthesis. A 2:1::H_2:CO mixture over iron-based catalysts produces hydrocarbons and some alcohols, suitable for motor fuels. A similar gas mixture over zinc or copper-based catalysts yields methanol [*67-56-1*]. By the water gas shift reaction, a hydrogen stream can be obtained from the synthesis gas. The hydrogen can be used as a chemical feedstock in refineries or synthesis plants, such as those for ammonia. The Lurgi plants in South Africa are based on this technology (51,52).

In 1974 a 1000 t/d ammonia plant went into operation near Johannesburg, South Africa. The lignitic (subbituminous) coal used there contains about 14% ash, 36% volatile matter, and 1% sulfur. The plant has six Koppers-Totzek low pressure, high temperature gasifiers. Refrigerated methanol (−38°C, 3.0 MPa (30 atm)) is used to remove H_2S. A 58% CO mixture reacts with steam over an iron catalyst to produce H_2. The carbon dioxide is removed with methanol (at −58°C and 5.2 MPa (51 atm)). Ammonia synthesis is carried out at ca 22 MPa (220 atm) (53) (see AMMONIA).

Direct Hydrogenation. Direct hydrogenation of lignitic and other coals has been studied by many investigators. Lignite can be slurried with an anthracene-oil solvent, heated to a temperature of 460–500°C with 1:1 CO:H_2 synthesis gas at pressures to 28 MPa (280 atm) in a 2 kg/h reactor. The product liquids are separated, and in a commercial process, a suitable hydrogen-donor solvent would be recycled (54).

In a 1 t/d pilot plant utilizing the Exxon Donor Solvent process, coals from lignitic through bituminous are crushed, dried, and fed to the liquefaction reactor along with hydrogenated recycle solvent and hydrogen. The products are separated by distillation (qv). The recycle solvent is hydrogenated in a conventional reactor. The heavy bottoms are fed to a partial coker and gasifier to maximize liquid products and obtain a low heating value fuel gas. Hydrogen is generated by either steam reforming the gas or by gasifying unconverted coal (55).

British Coal Corp. is developing a gasoline-from-coal process at a facility at Point of Ayr (Scotland). This process involves treatment with liquid recycle solvents, digestion at 450–500°C, filtration to separate unconverted residues, and separation into two fractions. The lighter fraction is mildly hydrotreated, and the heavier one is hydrocracked (56).

Other methods have been tested by various companies and governmental agencies (57), but none have proven economically feasible.

Nonfuel Uses. *Montan Wax.* An important product of the direct extractive treatment of lignitic coals is montan wax [*8002-53-7*]. The term montan wax or Bergwachs refers strictly to the material obtained by solvent extraction of suitable German brown coals. The generic term for similar materials is montana wax. The small quantities made outside Germany are usually referred to as montan, prefixed by the country of origin (see WAXES).

Brown coals yield, on solvent extraction, 10–15% of a material that contains 60–90% light yellow or brown waxy substances. The remainder is a mixture of deep brown resinous and asphaltic substances. The yield may be increased by increasing the pressure during extraction, but this also adds dark colored dispersion products, and the resultant brown coal cannot be briquetted.

The crude wax is refined by extracting at 90–100°C with an azeotropic mixture of benzene and a mixture of alcohols, typically 85% benzene and 15% methanol (see DISTILLATION, AZEOTROPIC AND EXTRACTIVE). Distilling the solvent leaves a wax too darkly colored to be used without added refining.

Acid mixtures are used to oxidize and remove the dark materials. Proper control gives a series of bleached waxes. A white wax requires double refining and reduces the yield to about 30% of the crude wax input. A series of synthetic waxes is prepared by separating the acids and alcohols produced during saponification of the wax and reesterifying them with acids or alcohols selected to give desired properties of hardness, solubility, emulsification, and gloss.

In the United States, the most desirable lignites for wax production are those from California and Arkansas. A yield of ca 7% is obtained from these lignites; those from other states give only about 2%.

Miscellaneous. Activated carbon is made from low temperature char and superheated steam or by digesting brown coal with KOH, producing a hard product having a high surface area (57). Cation-exchange resins can be made by sulfonation or by nitration and reduction. Treatment with ethylene dichloride and ammonia gives an anion-exchange resin (see ION EXCHANGE).

The German and Australian brown coals are also excellent humus sources (58). Reclaimed loess soils initially have only 0.4–0.5% humus, as compared to 1.5–2.0% for normal agricultural soils. The addition of sewage sludge or peat to brown coal improves its soil additive properties, both soil moisture retention and pH. The reduction in normally alkaline pH releases phosphates and nitrogen compounds for plant use. Brown coal soil conditioners are being applied in reclamation and have become commercially available for gardens and vineyards (58). During World War II, and electrode carbon was produced in Germany from material extracted from lignite. The mineral matter was almost completely removed, and the extract was carbonized in coke ovens (59). Lignite ash used for cellular concrete blocks, as a concrete additive, soil stabilizer, cement raw material, and filler for construction and bricks was 2×10^6 t in 1967 (ca 4% of total production) (60).

Combinations of lignite flyash from North Dakota and hydrated lime can increase the strength and durability of soils. The lime content varies from 2–7% and lime:flyash ratio from 1:1 to 1:7 (61). Lignite flyash can also be used as a partial replacement for Portland cement to produce strong, durable concrete (62).

Low temperature (lt) tars of Fischer-Tropsch (FT) fractions provide reasonable substrates for growth of yeast for human or animal food supplements. Yeast growth yields were 99.8% (FT fraction), 95.2 and 84.2% (lt tar) of those from a petroleum-derived paraffin fraction (63) (see FOODS, NONCONVENTIONAL).

BIBLIOGRAPHY

"Lignite" in *ECT* 1st ed., Vol. 8, pp. 339–346, by L. H. Reyerson and R. E. Montonna, University of Minnesota; "Lignite and Brown Coal" in *ECT* 2nd ed., Vol. 12, pp. 381–414, by M. Vahrman, The City University, London; in *ECT* 3rd ed., Vol. 14, pp. 313–343, by K. S. Vorres, The Institute of Gas Technology.

1. "Gaseous Fuels. Coal and Coke," in *Annual Book of ASTM Standards*, Vol. 5.05, American Society for Testing and Materials, Philadelphia, Pa., published annually.
2. P. R. Solomon, T. H. Fletcher, and R. J. Pugmire, *Fuel* **72**, 587 (1993).
3. *1992 Survey of Energy Resources*, 16th ed., World Energy Council, London.
4. National Coal Association, *International Coal*, 1991 edition, with data from the World Energy Conference, Washington, D.C., 1989.
5. D. J. Allardice, in R. A. Durie, ed., *The Science of Victorian Brown Coal*, Butterworth, Heinemann, Oxford, 1991, Chapt. 3.
6. E. D. J. Stewart and C. S. Gloe, *Proceedings of the 6th World Power Conference*, Vol. 2, Melbourne, 1962, pp. 602–619.
7. United Nations, *Publ. No. 195711. E., Mln. 20*, New York, 1957.
8. B. Roga and K. Tomkow, *Przegl. Gorn.*, (7–8), 355 (1961).
9. *International Handbook of Coal Petrography*, 2nd ed., Centre National de la Recherche Scientifique, Paris, 1963.
10. A. F. Duzy and co-workers, "Western Coal Deposits, Pertinent Qualitative Evaluations Prior to Mining and Utilization," paper presented at *Ninth Annual Lignite Symposium*, Grand Forks, N.D., May 1977.
11. D. J. Allardice and B. S. Newell, in Ref. 5, Chapt. 12.
12. *Brown Coal in Victoria, the Resource and Its Development*, Ministry of Fuel and Power, Victoria, Australia, 1977.
13. F. Woskobenko, W. O. Stacy, and D. Raisbeck, in Ref. 5, Chapt. 4.
14. H. Gan, S. P. Nandi, and P. L. Walker, Jr., *Fuel* **51**, 272 (1972).
15. T. V. Verheyen and G. J. Perry, in Ref. 5, Chapt. 6.
16. M. F. R. Mulcahy, W. J. Morley, and I. W. Smith, in Ref. 5, Chapt. 8.
17. A. B. Edwards, in P. L. Henderson, ed., *Brown Coal*, Cambridge University Press, London, 1953.
18. *Steam: Its Generation and Use*, The Babcock and Wilcox Co., New York, 1978, Chapt. 5, p. 12.
19. *Brown Coal Utilization, Research and Development*, Rheinische Braunkohlenwerke Aktiengesellschaft, Cologne, Germany, 1976, p. 62.
20. *Steam, Its Generation and Use*, 40th ed., The Babcock and Wilcox Co., New York, 1992, Chapt. 12, p. 9.
21. D. J. Brockway, A. L. Ottrey, and R. S. Higgins, in Ref. 5, Chapt. 11.
22. E. A. Noble, *N. Dakota Geol. Survey Bull.* **63**, 80–85 (1973); D. G. Wyant and E. P. Beroni, *Reconnaissance for Trace Elements in North Dakota and Eastern Montana*, U.S. Geol. Survey TEI-61, 29, U.S. Technical Information Service, Oak Ridge, Tenn., 1950.
23. D. W. van Krevelen, *Coal*, Elsevier, Amsterdam, the Netherlands, 1961, pp. 113–120.
24. *United Nations Statistical Yearbook*, 37th issue, New York, 1992.
25. R. D. Doctor and K. E. Wilzbach, *J. Energy Resources Technol. (Trans. ASME)* **111**, 160 (1989); *Coal*, 23 (Nov. 1992).

26. E. A. Nephew, *Surface Mining and Land Reclamation in Germany*, Oak Ridge National Laboratory Rpt. ORNL-NSF-EP-16, Oak Ridge, Tenn., 1972.
27. E. Gaertner, *Trans. SME* **258**, 353 (1975).
28. C. S. Gloe, in Ref. 5, Chapt. 13.
29. *The Chinese Coal Industry*, Joseph Crosfield and Sons, Ltd., Warrington, U.K., 1961, part 1, section II.
30. W. S. Landers and D. J. Donaven, in H. H. Lowry, ed., *Chemistry of Coal Utilization*, suppl. vol., John Wiley & Sons, Inc., New York, 1963, Chapt. 7.
31. R. C. Ellman, L. E. Paulson, and S. A. Cooley, *Proceedings, 1975 Symposium on Technology and Use of Lignite*, Grand Forks, N.D., GFERC/IC-75/2, p. 312; R. Kurtz, in Ref. 19, pp. 92–102.
32. *PETC Review*, Pittsburgh Energy Technology Center, Pittsburgh, Pa., 1992, p. 8.
33. D. C. Rhys Jones, in Ref. 30, Chapt. 16; R. Kurtz, in Ref. 19, pp. 92–102.
34. P. Speich, in Ref. 19, pp. 18–26.
35. Ref. 18, Chapt. 6.
36. N. Berkowitz, *An Introduction to Coal Technology*, Academic Press, Inc., New York, 1979, Chapt. 10.
37. R. A. Sherman and B. A. Landry, in Ref. 30, Chapt. 18, pp. 773–819.
38. D. Schwirten, in Ref. 19, pp. 47–65.
39. C. L. Wagoner and A. F. Duzy, *Burning Profiles for Solid Fuels*, ASME Paper 67-WA/FU-4, Chicago, 1967; Ref. 20, Chapt. 8, p. 8.
40. F. Y. Murad, L. V. Hillier, and E. R. Kilpatrick, *Boiler Flue Gas Desulfurization by Flyash Alkali*, Morgantown Energy Research Center Report, Morgantown, W. Va., 1976, pp. 450–460.
41. G. M. Goblirsch, R. W. Fehr, and E. A. Sondreal, *Proceedings of the Fifth International Conference on Fluidized Bed Combustion*, Vol. 2, Washington, D.C., 1978, pp. 729–743.
42. Y. Otake and P. L. Walker, Jr., *Fuel* **72**, 139 (1993).
43. G. Seifert and G. Hubrig, *Sasol*, brochure by Sasol Limited, Johannesburg, South Africa, 1990, p. 291.
44. H. C. Howard, in Ref. 30, Chapt. 9, pp. 340–394.
45. P. J. Wilson, Jr., and J. D. Clendenin, in Ref. 30, Chapt. 10, pp. 395–460.
46. Ref. 18, Chapt. 11.
47. H. R. Linden and co-workers, *Ann. Rev. Energy* **1**, 65 (1976).
48. F. H. Franke, in Ref. 16, pp. 134–146.
49. *Sasol*, brochure by Sasol Ltd., Johannesburg, Rep. of S. Africa, 1990; see also *Research and Development Leads Sasol to the Future*, 1983.
50. *Chem. Week* **127**, 39 (Aug. 13, 1980).
51. J. C. Hoogendoorn, *Proceedings of Ninth Synthetic Pipeline Gas Symposium*, Oct. 31–Nov. 2, 1977, American Gas Association, Alexandria, Va., 1977, p. 301.
52. M. Heylin, *Chem. Eng. News* **57**(38), 13 (1979).
53. L. J. Partridge, *Coal Process. Technol.* **3**, 133 (1977).
54. W. G. Willson and co-workers, "Application of Liquefaction Process to Low-Rank Coals," paper presented at *10th Biennial Lignite Symposium*, Grand Forks, N.D., May 1979.
55. W. N. Mitchell, K. L. Trachte, and S. Zaczepinski, "Performance of Low-Rank Coals in the Exxon Donor Solvent Process," paper presented at *10th Biennial Lignite Symposium*, Grand Forks, N.D., May 1979.
56. *Gasoline from Coal*, British Coal Liquefaction Project, Point of Ayr, Scotland, 1992.
57. T. R. Verheyen and co-workers, *Energeia* **2**(3), 1 (1991).
58. E. Petzold and F. H. Kortmann, in Ref. 19, p. 73.
59. R. A. Glenn, in Ref. 30, Chapt. 23.

60. O. E. Manz, *Proceedings of Second Ash Utilization Symposium*, Bureau of Mines, Morgantown, W. Va., 1970, pp. 282–299.
61. T. R. Dobie, S. Y. Ng, and N. W. Henning, *Laboratory Evaluation of Lignite Flyash as a Stabilization Additive for Soils and Aggregates*, final report PB 242741, Twin City Testing and Engineering Lab., Inc., St. Paul, Minn., 1975.
62. T. R. Dobie and N. E. Henning, *Lignite Flyash as a Partial Replacement for Portland Cement in Concrete*, final report PB 247414, Twin City Testing and Engineering Lab., Inc., St. Paul, Minn., 1975.
63. M. P. Silverman, J. M. Novak, and I. Wender, *Prepr. Fuel Chem. Div. Am. Chem. Soc.* **9**(4), 55 (1965).

General References

R. A. Durie, ed., *The Science of Victorian Brown Coal: Structure, Properties and Consequences for Utilisation*, Butterworth Heinemann, Oxford, 1991. An excellent reference not only for Victorian Brown Coal, but for lignitic coals of the world.
The World Energy Council issues Conference reports on reserves, resources and production at six-year intervals. More limited reports are issued at two-year intervals. The next report is expected in the fall of 1997.
Symposia on the Technology and Use of Lignite have been held in conjunction with the University of North Dakota Energy and Environmental Research Center and the preceding organizations.
M. A. Elliott, ed., *The Chemistry of Coal Utilization*, 2nd suppl. vol., Wiley-Interscience, New York, 1981.
N. Berkowetz, *An Introduction to Coal Technology*, Academic Press, Inc., New York, 1979.

KARL S. VORRES
Argonne National Laboratory

LIGNOSULFONIC ACIDS. See LIGNIN.

LIME AND LIMESTONE

The elements calcium and magnesium, which are distributed very widely in the earth's crust, most commonly occur in carbonate forms of rock, generally classified as limestone [*1317-65-3*]. Although vast strata of this ubiquitous rock are buried so deeply as to be inaccessible, great tonnages of this stone are extracted for commercial use. Annual world production was estimated at 2×10^9 metric tons in 1990. Limestone, literally one of the most basic raw materials of industry and construction, occurs in varying degrees in nearly every country (1).

Limestone may be classified as to origin, chemical composition, texture of stone, and geological formation. Chemically it is composed primarily of calcium carbonate [*471-34-1*], $CaCO_3$, and secondarily of magnesium carbonate [*546-93-0*], $MgCO_3$, with varying percentages of impurities (see also CALCIUM

COMPOUNDS, CALCIUM CARBONATE). Although these carbonates occur in many other rocks, ores, and soils, in its broadest definition limestone is distinguished by a content of more than 50% total carbonate. More restrictive interpretations demand at least 75% or even 90%, depending on point of view. In a cursory manner, limestone can be distinguished from most other rock by applying a dilute hydrochloric acid solution to it. If the stone effervesces, it is a basic carbonate rock with a definite alkaline reaction. Limestone's most important chemical characteristic is that when subjected to high temperature it decomposes chemically into lime, calcium oxide [*1305-78-8*], CaO, with decarbonation occurring through the expulsion of carbon dioxide gas. This primary product, know as quicklime, can then be hydrated, or slaked, into hydrated lime, calcium hydroxide [*1305-62-0*], $Ca(OH)_2$, ie, the water is chemically combined with the calcium oxide in an equimolecular ratio.

Limestone was fashioned into many useful tools and implements in prehistoric time. The use of lime as a cementing and plastering material is probably almost as old as the history of fire. Lime is one of the oldest chemicals known, and the process of lime burning is one of the oldest chemical industries. Primitive kilns discovered by archeologists are believed to have been used during the Stone Age for burning lime. Lime plaster, still in good condition, has been found in Egyptian pyramids built over 4500 years ago. The pyramids themselves were built largely of nummulitic limestone and mortar. Lime plaster and mortar were used by the early Greeks, Romans, Etruscans, Arabians, and Moors. Lime is mentioned several times in the Old and New Testaments of the Bible. Vitruvius, a Roman architect under Augustus, wrote the first detailed lime specifications. Many public works were built by the Romans using hydraulic lime and limestone aggregate. Both were extensively used in building the road base and pavement of the Appian Way.

In North America, quicklime was produced locally as early as 1635 in Rhode Island. It was not until 1733, when lime was shipped by boatload from Rockland, Maine, to Boston, that lime manufacture was established as a significant industry in commerce. The commercial hydration of lime is a relatively recent development initiated in 1904. Technical progress has allowed the industry to advance rapidly during the latter part of the twentieth century.

Definitions. In addition to showing varying degrees of chemical purity, limestone assumes a number of widely divergent physical forms, including marble, travertine, chalk, calcareous marl, coral, shell, oolites, stalagmites, and stalactites. All these materials are essentially carbonate rocks of the same approximate chemical composition as conventional limestone (2–4).

Limestone is generally classified into the following types: (*1*) high calcium, in which the carbonate content is essentially calcium carbonate having no more and usually less than 5% magnesium carbonate; (*2*) magnesian, which contains both calcium and magnesium carbonates, and has a magnesium carbonate content of 5–20%; and (*3*) dolomitic, which contains > 20% but not more than 45.6% $MgCO_3$, the exact amount contained in a true, pure, equimolecular dolomite. The balance is $CaCO_3$. Similarly, limes calcined from these stone types are identified as high calcium, magnesian, and dolomitic limes. The magnesian lime and limestone are more prevalent in Europe and other countries than in the United States.

The carbonate minerals that comprise limestone are calcite [*13397-26-7*] (calcium carbonate), which is easily the most abundant mineral type; aragonite [*14791-73-2*] (calcium carbonate); dolomite [*17069-72-6*] (double carbonate of calcium and magnesium); and magnesite [*13717-31-5*] (magnesium carbonate). Individual limstone types are further described by many common names (1). Some of this nomenclature is repetitious and overlapping. The following terms are in common use in Europe and the United States.

Argillaceous limestone is an impure type containing considerable clay or shale, and as a result has a relatively high silica and alumina content.

Calcitic limestone is generally used by agronomists to denote a high calcium stone. This term can be misleading, however, because its use could suggest pure calcite, which calcitic limestone usually is not.

Carbonaceous limestone contains various types of organic material, such as peat, natural asphalt, and even oil shale (qv), as impurities. Such stone is often black and may exude a fetid odor.

Cementstone is an impure (usually argillaceous) limestone, possessing the ideal balance of silica, alumina, and calcium carbonate for Portland cement (qv) manufacture. When calcined it produces a hydraulic cementing material.

Chalk is a soft, fine-grained, fossiliferous form of calcium carbonate that varies widely in color, hardness, and purity. Its grain size is so minute that it appears amorphous, but actually it is cryptocrystalline with a very high surface area.

Chemical-grade limestone is a pure type of high calcium or dolomitic limestone used by the chemical-process industry or where exacting chemical requirements are necessary. It contains a minimum of 95% total carbonate. In a few areas of the United States this minimum may be extended to 97 or 98%.

Compact limestone is a general term depicting a dense, fine-grained, homogeneous, usually hard type of stone.

Dolomitic limestone contains considerable $MgCO_3$. A true dolomitic stone contains a ratio of 40–44% $MgCO_3$ to 54–58% $CaCO_3$. However, the term is more loosely used to denote any carbonate rock that contains more than 20% $MgCO_3$. It varies in color, hardness, and purity.

Ferruginous limestone contains considerable iron as an impurity and is yellow or red in color.

Fluxstone is a pure form of limestone used as flux or purifier in metallurgical furnaces. It can be high calcium, magnesian, or dolomitic, providing it contains at least 95% carbonate.

Fossiliferous limestone is a general term for any carbonate stone in which the fossil structure is visually evident.

High calcium limestone is a general term for stone that contains largely $CaCO_3$ and not much (2–5% max) $MgCO_3$. It occurs in varying degrees of purity.

Hydraulic limestone is an impure argillaceous carbonate somewhat akin to cementstone, except that it may contain more $MgCO_3$, and usually it produces cement-like materials of lower hydraulicity.

Iceland spar is the purest limestone, virtually pure calcite of about 99.9% $CaCO_3$. It is also known as optical calcite; its occurrence is rare.

Magnesian limestone is intermediate between high calcium and dolomitic, and contains 5–20% $MgCO_3$. It occurs in varying purity.

Marble is a metamorphic, highly crystalline rock that may be high calcium or dolomitic limestone of varying purity. It occurs in virtually every color in diverse mottled effects and is the most beautiful form of limestone. It is usually very hard and can be cut and polished to a very smooth surface.

Marl, an impure, soft, earthy, carbonate rock, contains varying amounts of clay and sand intermixed in a loosely knit crystalline structure.

Oolitic limestone is composed of small rounded grains of $CaCO_3$, precipitated in concentric laminates around a nucleus of $CaCO_3$ or silica. It is frequently very pure but may be impure.

Oyster shell, another of the many forms of fossiliferous limestones, is a relatively pure source of $CaCO_3$.

Phosphatic limestone is usually a high calcium type that contains appreciable percentages (up to 5%) phosphorus. It originates from invertebrate marine organisms.

Stalactites and *stalagmites* are conical, icicle-like shapes of pure $CaCO_3$ that form on roofs and floors, respectively, of caverns. These are precipitated from cold groundwater that drips from limestone crevices.

Travertine is a calcium carbonate formed by chemical precipitation from natural hot-water mineral springs. In appearance and use it is closely akin to marble.

Whiting at one time connoted only a very fine form of chalk of micrometer sizes but the term is now used more broadly to include all finely divided, meticulously milled carbonates derived from high calcium or dolomitic limestone, marble, shell, or chemically precipitated calcium carbonate. Unlike all of the above natural forms of limestone, it is strictly a manufactured product.

The term lime also has a broad connotation and frequently is used in referring to limestone. According to precise definition, lime can only be a burned form: quicklime, hydrated lime, or hydraulic lime. These products are oxides or hydroxides of calcium and magnesium, except hydraulic types in which the CaO and MgO are chemically combined with impurities. The oxide is converted to a hydroxide by slaking, an exothermic reaction in which the water combines chemically with the lime. These reversible reactions for both high calcium and dolomitic types are

Quicklime

$$\underset{\text{high calcium limestone}}{CaCO_3} + \text{heat} \rightleftharpoons \underset{\text{high calcium quicklime}}{CaO} + CO_2$$

$$\underset{\text{dolomitic limestone}}{CaCO_3 \cdot MgCO_3} + \text{heat} \rightleftharpoons \underset{\text{dolomitic quicklime}}{CaO \cdot MgO} + 2\ CO_2$$

Hydrated lime

$$\underset{\text{high calcium quicklime}}{CaO} + H_2O \rightleftharpoons \underset{\text{high calcium hydrate}}{Ca(OH)_2} + \text{heat}$$

$$CaO{\cdot}MgO + H_2O \rightleftharpoons Ca(OH)_2{\cdot}MgO + \text{heat}$$
dolomitic quicklime → dolomitic hydrate

$$CaO{\cdot}MgO + 2\,H_2O \rightleftharpoons Ca(OH)_2{\cdot}Mg(OH)_2 + \text{heat}$$
dolomitic quicklime → dolomitic hydrate

In most types of dolomitic quicklimes, when hydrated under atmospheric conditions, all the CaO component readily hydrates, but very little of the MgO slakes. The result is a dolomitic monohydrate or a combination of hydroxide and oxide. However, when dolomitic quicklime is hydrated under pressure or is subject to long retention periods, most of the MgO hydrates to form a so-called highly hydrated dolomitic lime. In the relatively pure commercial limes, the available lime contents of quicklime are 88–94%, and the total oxide content (CaO + MgO) is 92–98% in North America. The average purity of the limes in Europe, South America, and elsewhere, except for Japan, is generally not as high as in the United States and Canada, because the countries outside of North America possess fewer high grade limestone deposits. For reasons of necessity and economy, some submarginal (in quality) limestone deposits are exploited.

The common nomenclature for specific types and forms of lime, some of which is repetitious and overlapping, is as follows (1).

Air-slaked lime contains various proportions of the oxides, hydroxides, and carbonates of calcium and magnesium which result from excessive exposure of quicklime to air that vitiates its quality. It is partially or largely decomposed quicklime that has become hydrated and carbonated.

Autoclaved lime is a special form of highly hydrated dolomitic lime, used largely for structural purposes, that has been hydrated under pressure in an autoclave.

Available lime, the total free lime (ie, CaO) content in a quicklime or hydrate, is the active constituent of a lime. It provides a means of evaluating the concentration of lime.

Building lime may be quick or hydrated lime, but usually connotes the latter, where the physical characteristics make it suitable for ordinary or special structural purposes (see BUILDING MATERIALS, SURVEY).

Carbide lime is a waste lime hydrate by-product from the generation of acetylene from calcium carbide and may occur as a wet sludge or dry powder of widely varying purity and particle size. It is gray and has the pungent odor associated with acetylene (see HYDROCARBONS, ACETYLENE).

Chemical lime is a quick or hydrated lime used for one or more chemical or industrial applications. Usually chemical lime has a relatively high chemical purity.

Dead-burned dolomite is a specially sintered or double-burned form of dolomitic quicklime which is further stabilized by the addition of iron oxides. Historically, it was used as a refractory for lining steel furnaces, particularly open hearths, but as of this writing is used primarily in making dolomite refractory brick (see REFRACTORIES).

Fat lime is a pure lime (quick or hydrated), as distinct from an impure or hydraulic lime; it is also used to denote a lime hydrate that yields a plastic putty for structural purposes.

Finishing lime is a refined hydrated lime, milled to make it suitable for plastering, particularly the finish coat. Putty derived from this hydrate possesses unusually high plasticity.

Fluxing lime is lump or pebble quicklime used as flux in steel (qv) manufacture; the term may also be applied more broadly to include fluxing of nonferrous metals and glass (qv). It is a type of chemical lime.

Ground burnt lime refers to ground quicklime used for agricultural liming.

Hard-burned lime is a quicklime that is calcined at high temperature and is generally characterized by relatively high density and moderate-to-low chemical reactivity.

Hydraulic hydrated lime is a chemically impure form of lime with hydraulic properties of varying extent. It contains appreciable amounts of silica, alumina, and usually some iron, chemically combined with much of the lime. Hydraulic hydrated lime is employed solely for structural purposes.

Lime putty is a form of lime hydrate in a wet, plastic paste form, containing free water.

Lime slurry is a form of lime hydrate in aqueous suspension that contains considerable free water.

Lump lime is a physical shape of quicklime, derived from vertical kilns.

Mason's lime is a hydrated lime used in mortar for masonry purposes.

Milk-of-lime is a dilute lime hydrate in aqueous suspension which has the consistency of milk.

Pebble lime is a physical shape of quicklime.

Refractory lime is synonymous with dead-burned dolomite, an unreactive dolomitic quicklime, stabilized with iron oxides, that is used primarily for lining refractories of steel furnaces, particularly open hearths.

Slaked lime is a hydrated form of lime, available as a dry powder, putty, or aqueous suspension.

Soft-burned lime is a quicklime that is calcined at a relatively low temperature. It is characterized by high porosity and chemical reactivity.

Type S hydrated lime, also called special hydrated lime, is an ASTM designation that distinguishes a structural hydrate from a normal hydrated lime, designated type N. Type S lime meets specified plasticity and gradation requirements. It may be dolomitic or high calcium and is more precisely milled than type N hydrates. Type SA hydrated lime is a Type S hydrate to which a small amount of air entraining agent has been added, having up to 14% air content; it is used in mortar.

Unslaked lime is any form of quicklime.

Geology

Limestone, as a constituent of the earth's crust, is a rock of sedimentary origin (4) from material precipitated by chemical and organic action on drainage waters. Calcium, a common element, is estimated to comprise 3–4% of the earth's crust, and the calcium constituent of limestone must have come originally from igneous rocks. By the action of various eroding and corroding forces, including the solution of carbonic and other mineral acids, the rocks are disintegrated and the calcium is dissolved and removed in the drainage waters emptying into the sea. The amount of material removed in this manner is astonishing. It is estimated that the Thames River in England annually carries more than 550,000 metric tons of dissolved material, of which approximately two-thirds is calcium carbonate. This represents the removal of about 62 t/km^2 of limestone from the drainage area involved.

Upon reaching the ocean, some of the dissolved calcium carbonate may be reprecipitated because of its lower solubility in seawater. Surface evaporation and temperature changes may reduce the carbon dioxide content of the water, causing precipitation of calcium carbonate from saturated conditions. The calcium carbonate sedimented in this manner may give rise to limestone of purely chemical (sometimes termed physical) origin. Also of chemical origin, by a similar evaporation–deposition process around springs and streams, are limestones known as travertine and calcareous tufa and stalactities and stalagmites in caves. By far the largest part of the limestone in existence as of the twentieth and twenty-first centuries is of organic origin formed through skeleton-building processes of marine life. The lime remaining in solution following chemical precipitation is utilized by many different varieties of sea life, such as corals, foraminifera, mollusks, arthropods, and echinoderms, to form shells and skeletons, which ultimately accumulate on the sea bottom. The skeletal structures are almost pure calcium carbonate and are frequently found intact in such limestones as chalk and marl.

The calcareous sediment produced in either manner may become contaminated during deposition with argillaceous, siliceous, or ferruginous silts, which affect the chemical composition and nature of the resulting limestone. The size and shape of the calcareous particles, together with conditions of pressure, temperature, and solvent action to which a deposition is subsequently exposed, are factors that influence the physical characteristics of the stone. The degree of consolidation of the calcareous sediment ranges from little change, as exhibited by the soft marls and chalks in which the skeletal particles are loosely cemented, to the metamorphosed, dense, crystalline rock, marble, which shows no indication of its origin. Between these two extremes many types and kinds of limestone are known. Examples of limestone in the process of formation are the globigerina ooze, which covers vast areas of bottom at ocean depths of 1800–5500 m, and coral reefs in tropical seas.

The origin of magnesian and dolomitic limestone is uncertain; it is generally believed, and considerable evidence supports the theory, that it is formed by direct chemical replacement of the calcium in the limestone by magnesium from waters high in magnesium salts. Several small deposits of dolomite appear to have originated through the coprecipitation of both carbonates.

Geographical Occurrence. Limestone, present in the majority of geological formations, is widely distributed throughout the world in deposits of varying sizes and degrees of purity. Significant deposits in the U.K. are chiefly confined to the Devonian, lower Carboniferous, Jurassic, and Cretaceous systems. In the United States, the geological distribution is broader. Quantities of stone occur in the older Cambrian, Ordovician, and Silurian systems, as well as in the previously mentioned groups. Although some outcrops are known in the pre-Cambrian system, limestones as a whole are not as prevalent as in the older Paleozoic rocks.

The deposits of limestone in the United States, if no limitation as to quality is placed, are widespread, occurring in nearly every state, usually in tremendous amounts. It is estimated that 15–20% of the areas of the United States is underlain by limestone. Even though the deposits are extensive, they are frequently so overburdened that quarrying or mining is not economical. Only a small proportion of the total limestone is of a grade suitable to meet the high requirements demanded for either industrial lime or stone for metallurgical and chemical processes. The lower grade stones are generally suitable for agricultural and construction uses where the chemical composition is not a limiting factor.

Detailed information concerning the location and analysis of limestone deposits in the United States can be obtained from the various state geological surveys, the U.S. Bureau of Mines, and the U.S. Geological Survey. Descriptive summaries of the limestone deposits in the various states have been published (5,6).

Impurities. The chemical composition and properties of lime and limestone depend on the nature of the impurities and the degree of contamination of the original stone. The contaminating materials either were deposited simultaneously with the $CaCO_3$ or entered during some later stage (6).

Alumina in combination with silica is present in limestone chiefly as clay, though other aluminum silicates in the form of feldspar and mica may be found. When present in appreciable quantities, clay converts a high calcium limestone into a marl or argillaceous stone, which when calcined yields limes with hydraulic properties. Limestones containing 5–10% clayey matter yield feebly hydraulic limes; those containing 15–30% produce highly hydraulic limes.

Siliceous matter other than clay may occur in the free state as sand, quartz fragments, and chert, and in the combined state as feldspar, mica, talc (qv), and serpentine. Metallurgical and chemical limestones should contain less than 1% alumina and 2% silica.

Iron compounds (qv) in limestone are seldom injurious to a lime product unless a very pure lime is required. Normally, the iron compounds are in the form of limonite [*1317-63-1*] (ferric hydroxide) and pyrite [*1309-36-0*], FeS_2. Occasionally, hematite, magnetite, marcasite, and other forms of iron are found in limestone.

Sodium and potassium compounds are rarely present to any extent in limestone and are not objectionable unless a pure lime is required. When present in small proportions, these impurities are usually volatilized during burning. Carbonaceous matter is sometimes present in limestone. It is of little importance to the resulting lime because it burns and is lost during calcination. Sulfur and phosphorus compounds (generally sulfates and phosphates) are objectionable impurities in chemical lime and limestone. In metallurgical processes requiring relatively pure fluxing lime and limestone, these acidic components should not be present in quantities greater than 0.03% sulfur and 0.02% phosphorus.

Many lime plants are able to reduce the impurities in their lime product by careful screening and selecting of stone for burning. Because 9 kg of limestone produce only 5 kg of quicklime, the percentage of impurities in a quicklime is nearly double that in the original stone. Analyses of typical samples of high calcium, magnesian, and dolomitic limestones found in the United States are listed in Table 1.

Table 1. Composition of U.S. Limestones,[a] wt %

	Limestone[b]							
Component	1	2	3	4	5	6	7	8
CaO	54.54	38.90	41.84	31.20	29.45	45.65	55.28	52.48
MgO	0.59	2.72	1.94	20.45	21.12	7.07	0.46	0.59
CO_2	42.90	33.10	32.94	47.87	46.15	43.60	43.73	41.85
SiO_2	0.70	19.82	13.44	0.11	0.14	2.55	0.42	2.38
Al_2O_3	0.68	5.40	4.55	0.30	0.04	0.23	0.13	1.57
Fe_2O_3[c]	0.08	1.60	0.56	0.19	0.10	0.20	0.05	0.56
SO_3[d]	0.31		0.33			0.33	0.01	
P_2O_5			0.22		0.05	0.04		
Na_2O	0.16		0.31	0.06	0.01	0.01		
K_2O			0.72		0.01	0.03		
H_2O			1.55		0.16	0.23		
other			0.29		0.01	0.06	0.08	0.20

[a]Ref. 1.
[b]1 = Indiana high calcium stone; 2 = Lehigh Valley, Pa., cement rock; 3 = Pennsylvania cement rock; 4 = Illinois Niagaran dolomitic stone; 5 = Northwestern Ohio Niagaran dolomitic stone; 6 = New York magnesian stone; 7 = Virginia high calcium stone; and 8 = Kansas Cretaceous high calcium stone (chalk).
[c]Includes some Fe as FeO.
[d]Includes some elemental S.

Properties of Limes and Limestones

The chemical and physical properties of limestone vary tremendously, owing to the nature and quantity of impurities present and the texture, ie, crystallinity and density. These same factors also exert a marked effect on the properties of the limes derived from the diverse stone types. In addition, calcination and hydration practices can profoundly influence the properties of lime.

Color. The purest forms of calcite and magnesite are white, often with an opaque cast, but most conventional limestone, even relatively pure types, are gray or tan. The presence of carbonaceous impurities can render the grays dark, even approaching black. The presence of iron gives the tan, brown, pink, and buff colors of some limestones. Impurities, such as pyrite, marcasite, and siderite may alter the surface color through weathering. Marble and travertine have many brilliant colors of diverse mottled, variegated effects ranging to milky white.

Quicklime is usually white of varying intensity, depending on chemical purity; some species possess a slight ash-gray, buff, or yellowish cast. Invariably quicklime is lighter in color than the limestone from which it is derived. Hydrated limes, except for hydraulic and impure hydrates, are extremely white in color, invariably whiter than their quicklimes.

Odor. Except for highly carbonaceous species, most limestones are odorless. Quick and hydrated limes possess a mild odor that is characteristic but difficult to describe except that it is faintly musty or earthy, not offensive.

Texture. All limestones are crystalline, but there is tremendous variance in the size, uniformity, and arrangement of their crystal lattices. The crystals of the minerals calcite, magnesite, and dolomite are rhombohedral; those of aragonite are orthorhombic. The crystals of chalk and of most quick and hydrated limes are so minute that these products appear amorphous, but high powered microscopy proves them to be cryptocrystalline. Hydrated lime is invariably a white, fluffy powder of micrometer and submicrometer particle size. Commercial quicklime is used in lump, pebble, ground, and pulverized forms.

There is considerable variance in the porosity of limestones, thus the bulk densities generally are 2000–2800 kg/m^3 (125–175 lb/ft^3). The density of some chalk limestones are even less. Dolomitic stones average 2–3% higher densities than high calcium ones. Depending on the physical size of the quicklime particles and their divergent porosities, bulk densities are 770–1120 kg/m^3 (48–70 lb/ft^3), and densities of their hydrates 400–640 kg/m^3 (25–40 lb/ft^3). Again, dolomitic limes average about 4% denser than their high calcium counterparts. The severity of the calcination process largely determines the porosity of a quicklime; the higher the temperature of calcination and the longer its duration, the more the porosity declines.

Specific gravities of pure and commercial limes and limestones are shown:

Material	Specific gravity
calcite	2.7112
aragonite	2.929
high calcium limestones	2.65–2.75
dolomitic limestones	2.75–2.90
chalk	1.4–2.0
pure calcium oxide	3.34
high calcium quicklimes	3.2–3.4
dolomitic quicklimes	3.4–3.6
high calcium hydrated lime	2.3–2.4
dolomitic hydrated lime	2.4–2.9

Hardness. Most limestone is soft enough to be readily scratched with a knife. Pure calcite is standardized on Mohs' scale at 3; aragonite is harder, 3.5–4. Dolomitic limestone is generally harder than high calcium. Dead-burned or sintered limes are 3–4 on this scale, whereas most commercial soft-burned quicklimes are 2–3 (see HARDNESS).

Strength. The compressive strength of limestone varies tremendously, having values from 8.3 to 196 MPa (1,200–28,400 psi). Marble generally has the highest value and chalk and calcareous marl the lowest.

Luminescence. Limestone possesses only limited luminescent qualities, ranging from very faint or none with the impure types. However, quicklime is very luminescent at calcining temperatures, hence the term limelight.

Thermal Properties. Because all limestone is converted to an oxide before fusion or melting occurs, the only melting point applicable is that of quicklime. These values are 2570°C for CaO and 2800°C for MgO. Boiling point values for CaO are 2850°C and for MgO 3600°C. The mean specific heats for limestones and limes gradually ascend as temperatures increase from 0 to 1000°C. The ranges are as follows: high calcium limestone, 0.19–0.26; dolomitic quicklime, 0.19–0.294; dolomitic limestone, 0.206–0.264; magnesium oxide, 0.199–0.303; and calcium oxide, 0.175–0.286.

In the hydration of quicklime, considerable heat is generated: for $Ca(OH)_2$, 63.6 kJ/mol (488 Btu/lb) of quicklime; and for $Mg(OH)_2$, 32.2–41.8 kJ/mol (247–368 Btu/lb). The heat of solution for $Ca(OH)_2$ is 11.7 kJ/mol (67.8 Btu/lb). The value for MgO is immeasurable because MgO is virtually insoluble.

Solubility. High calcium limestone is only very faintly soluble in water. In cold CO_2-free water it is often regarded as insoluble. Between 17 and 25°C its solubility is only 14–15 mg/L. As the temperature increases, so does the solubility until at 100°C solubility reaches 30–40 mg/L. This faint solubility at elevated temperatures accounts for the accretion of a primarily $CaCO_3$ scale in steam boilers. Carbon dioxide exerts a mild solvent action on $CaCO_3$ increasing the solubility in direct proportion to the increase of CO_2 pressure as shown in Figure 1. The maximum solubility of $CaCO_3$ is 3.93 g/L at 5.7 MPa (56 atm) CO_2 pressure and 18°C. At a CO_2 pressure of 101.3 kPa (1 atm), the solubility ranges between 1.3 and 0.765 g/L between temperatures of 9–35°C, the converse of CO_2-free water, because the solubility declines as the temperature increases. Argonite solubility values are slightly greater than those of calcite. However, $MgCO_3$ is much more water soluble than $CaCO_3$, with or without CO_2. On an equivalence basis, the $MgCO_3$ is about 15–20 times more soluble.

There are no solubility values for quicklime because the oxide is hydrated to its hydroxide before dissolving. The magnitude of solubility of a high calcium hydrate on a CaO basis is 1.330 g/L (or 0.13%) of saturated solution at 10°C in distilled water. Thus lime is about 75 times as soluble as high calcium limestone on a comparable basis and can be regarded as slightly soluble at low temperatures (5). Contrary to limestone in CO_2-free water, the solubility of hydrate is in inverse proportion to temperature, decreasing with rising temperatures. Figure 2 displays how temperature changes influence lime's solubility. Fractional percents up to 5–10% of many inorganic salt solutions, such as $CaCl_2$, NH_4Cl, NaCl, etc, increase lime's solubility in varying degrees up to threefold. Alkalies, notably Na_2CO_3 and NaOH, exert an adverse effect, rendering lime almost totally insoluble at elevated temperatures. However, the greatest stimulants to solubility are certain organic compounds, such as glycerol, phenol, and sugar. In a 35% sugar solution at 25°C, 10.1 g of CaO/100 cm^3 can be dissolved, nearly 100 times the solubility of lime in distilled water.

Data on the solubility of magnesium hydroxide in water are not all in agreement, but the solubility is extremely low. The extent of $Mg(OH)_2$ solubility is 10 mg/L, which is about 1/100 the solubility of $Ca(OH)_2$. In concentrated solutions of NH_4Cl and NH_4CO_3, the solubility of $Mg(OH)_2$ is markedly increased, but

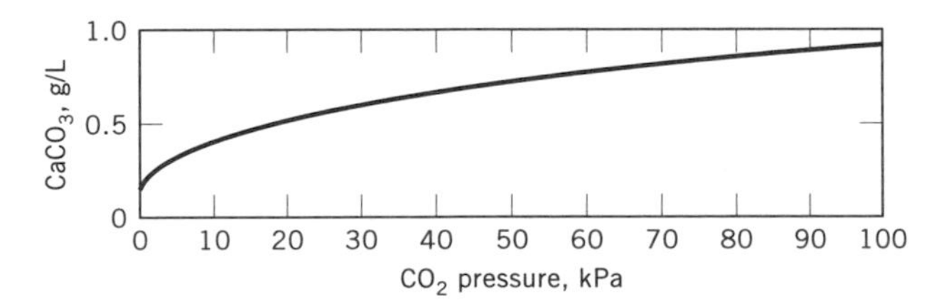

Fig. 1. Influence on aqueous $CaCO_3$ solubility of increasing fractional CO_2 pressure at constant temperature of 25°C. To convert kPa to mm Hg, multiply by 7.5.

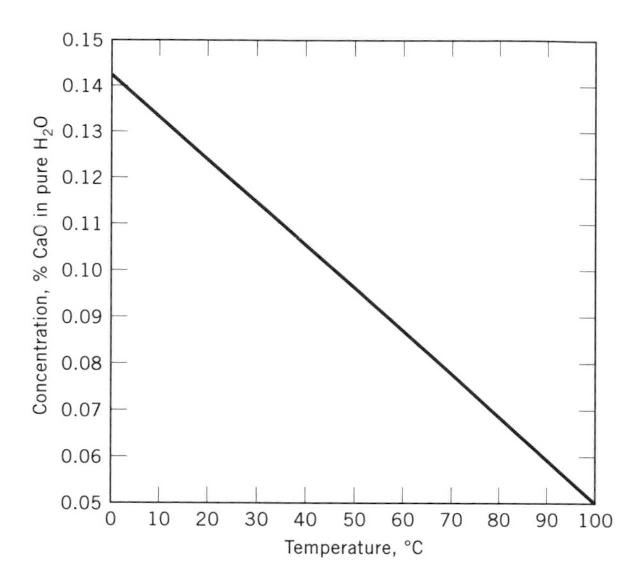

Fig. 2. The water solubility of lime which decreases with rising temperatures.

in no instance does its solubility equal that of $MgCO_3$ in water heavily permeated with CO_2. Dolomitic hydrates are slightly less soluble than high calcium hydrates, but much nearer the latter in value than $Mg(OH)_2$, because the presence of MgO and $Mg(OH)_2$ does not impede the dissolution of its $Ca(OH)_2$ constituent.

Plasticity. An innate characteristic of a lime putty of paste-like consistency is it plasticity or its ability to be molded under pressure and to retain its altered shape without deformation. This rheological property is important for structural uses of lime in masonry mortar and plaster (7). Although no completely satisfactory method for measuring plasticity has been developed, North America has a standardized Emley Plasticimeter test. This test simulates the action of a plasterer applying a lime finish coat to an absorbent base using a trowel. A special machine that measures the resistance of a lime putty of standard consistency against a rotating disk is used. If a lime has an Emley plasticity value of 200 or more, it is judged satisfactory for all structural purposes. Limes vary widely in the degree of plasticity they impart. Generally, dolomitic limes develop greater plasticity than high calcium types in the United States, but this marked superiority does not occur in other countries. The ASTM specification on physical tests of limes (C110) describes this test and other physical tests pertaining to building lime, such as water retentivity, pitting and popping, and soundness.

Stability. All calcitic and dolomitic limestones are extremely stable compounds, decomposing only in fairly concentrated strong acids or at calcining temperatures of 898°C for high calcium and about 725°C for dolomitic stones at 101.3 kPa (1 atm). A very mild destabilizing effect is caused by CO_2-saturated water, as described in the preceding section on solubility. Aragonite, however, is not as stable as calcite. In sustained contact with moisture, it tends to revert to calcite through recrystallization. At a temperature of 400°C it is transformed irreversibly to the more stable form of calcite. Dolomite and magnesite are the equal of calcite in stability.

Quicklime and hydrated lime are reasonably stable compounds but not nearly as stable as their limestone antecedents. Chemically, quicklime is stable at any temperature, but it is extremely vulnerable to moisture. Even moisture in the air produces a destabilizing effect by air-slaking it into a hydrate. As a result, an active high calcium quicklime is a strong desiccant (qv). Probably hydrate is more stable than quicklime. Certainly hydrated lime is less perishable chemically because water does not alter its chemical composition. However, its strong affinity for carbon dioxide causes recarbonation. Dolomitic quicklime is less sensitive to slaking than high calcium quicklime, and dead-burned forms are completely stable under moisture-saturated conditions. Except for dead-burned dolomite, all limes are much more reactive with acids than limestone. The high calcium types are the most reactive.

Chemical Reactions. *Neutralization.* In water, lime ionizes readily to Ca^{2+}, Mg^{2+}, and OH^-, forming a strong base or alkali. Both $Ca(OH)_2$ and $Mg(OH)_2$ are strong diacid bases neutralizing such strong monobasic acids as HCl and HNO_3, yielding neutral salts and heat.

$$Ca(OH)_2 + 2\ HCl \longrightarrow CaCl_2 + 2\ H_2O + 115\ kJ\ (27{,}400\ cal)$$

$$Mg(OH)_2 + 2\ HNO_3 \longrightarrow Mg(NO_3)_2 + 2\ H_2O + 115\ kJ\ (27{,}400\ cal)$$

For H_2SO_4, a dibasic acid, the molar ratio is one-to-one:

$$Ca(OH)_2 + H_2SO_4 \longrightarrow CaSO_4 + 2\ H_2O + \text{heat}$$

The neutralizing power of lime and limestone and other alkalies is compared in Table 2 (8). Of all these alkalies, MgO is the strongest base, followed by CaO. Thus neutralization of a given acid requires less dolomitic limestone or lime than high calcium limestone or lime.

pH. As Figure 3 indicates, lime solutions develop a high pH of slightly under 12.5 at 25°C and approach 13 at maximum solubility at 0°C. Using even the barest trace of lime a pH of 11 is easily achieved, causing a precipitous rise

Table 2. Basicity Factors of Common Alkaline Reagents[a]

Alkali	Descriptive formula	Basicity factor[b]
dolomitic quicklime	$CaO{\cdot}MgO$	1.110
high calcium quicklime	CaO	0.941
dolomitic normal hydrate	$Ca(OH)_2{\cdot}MgO$	0.912
dolomitic pressure hydrate	$Ca(OH)_2{\cdot}Mg(OH)_2$	0.820
high calcium hydrate	$Ca(OH)_2$	0.710
sodium hydroxide	NaOH	0.687
dolomitic limestone	$CaCO_3{\cdot}MgCO_3$	0.564
sodium carbonate	Na_2CO_3	0.507
high calcium limestone	$CaCO_3$	0.489

[a]These factors were determined on representative commercial samples, except those of sodium hydroxide and sodium carbonate, which were calculated.
[b]On a basis of pure CaO = 1.000.

on the pH scale from 7. The pH of limestone is much lower. Calcium carbonate attains a pH of 8–9 and dolomitic stone a pH of about 8.5–9.2.

Causticization. Lime, particularly the high calcium type, reacts with carbonates such as Na_2CO_3 and Li_2CO_3 to form other hydroxides and carbonates through double decomposition or metathesis reactions as follow:

$$Na_2CO_3 + Ca(OH)_2 \longrightarrow 2\,NaOH + CaCO_3$$

The $CaCO_3$ precipitate is easily separated from the other, soluble reactant.

Silica and Alumina. The manufacture of Portland cement is predicated on the reaction of lime with silica and alumina to form tricalcium silicate [*12168-85-3*] and aluminate. However, under certain ambient conditions of compaction with sustained optimum moisture content, lime reacts very slowly to form complex mono- and dicalcium silicates, ie, cementitious compounds (9,10). If such a moist, compact mixture of lime and silica is subjected to steam and pressure in an autoclave, the lime–silica reaction is greatly accelerated, and when sand and aggregate is added, materials of concrete-like hardness are produced. Limestone does not react with silica and alumina under any circumstances, unless it is first calcined to lime, as in the case of hydraulic lime or cement manufacture.

Other Reactions. Dry hydrated lime adsorbs halogen gases, eg, Cl_2 and F_2, to form hypochlorites and fluorides. It reacts with hydrogen peroxide to form calcium peroxide, a rather unstable compound. At sintering temperatures, quicklime combines with iron to form dicalcium ferrite.

Limestone Production. Because more than 99% of U.S. limestone is sold or used as crushed and broken stone, rather than dimension-stone, most of the description of limestone's extraction and processing herein focuses on the former (Fig. 4). Most stone is obtained by open-pit quarrying methods. Underground

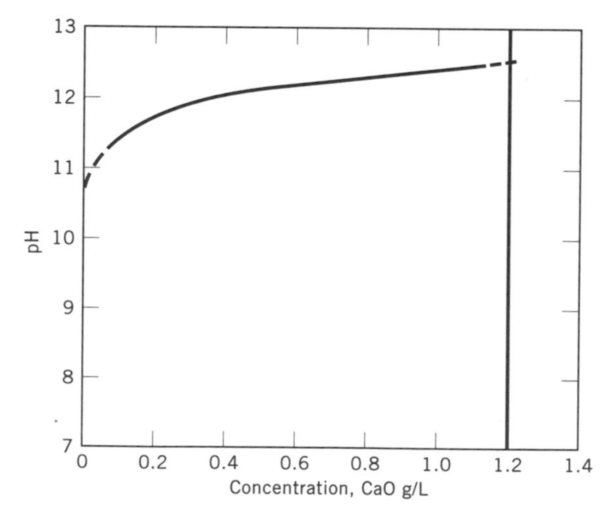

Fig. 3. The pH values of $Ca(OH)_2$ solutions of varying concentrations of CaO in water at 25°C. The solid vertical line represents the maximum solubility of $Ca(OH)_2$ solutions at 25°C.

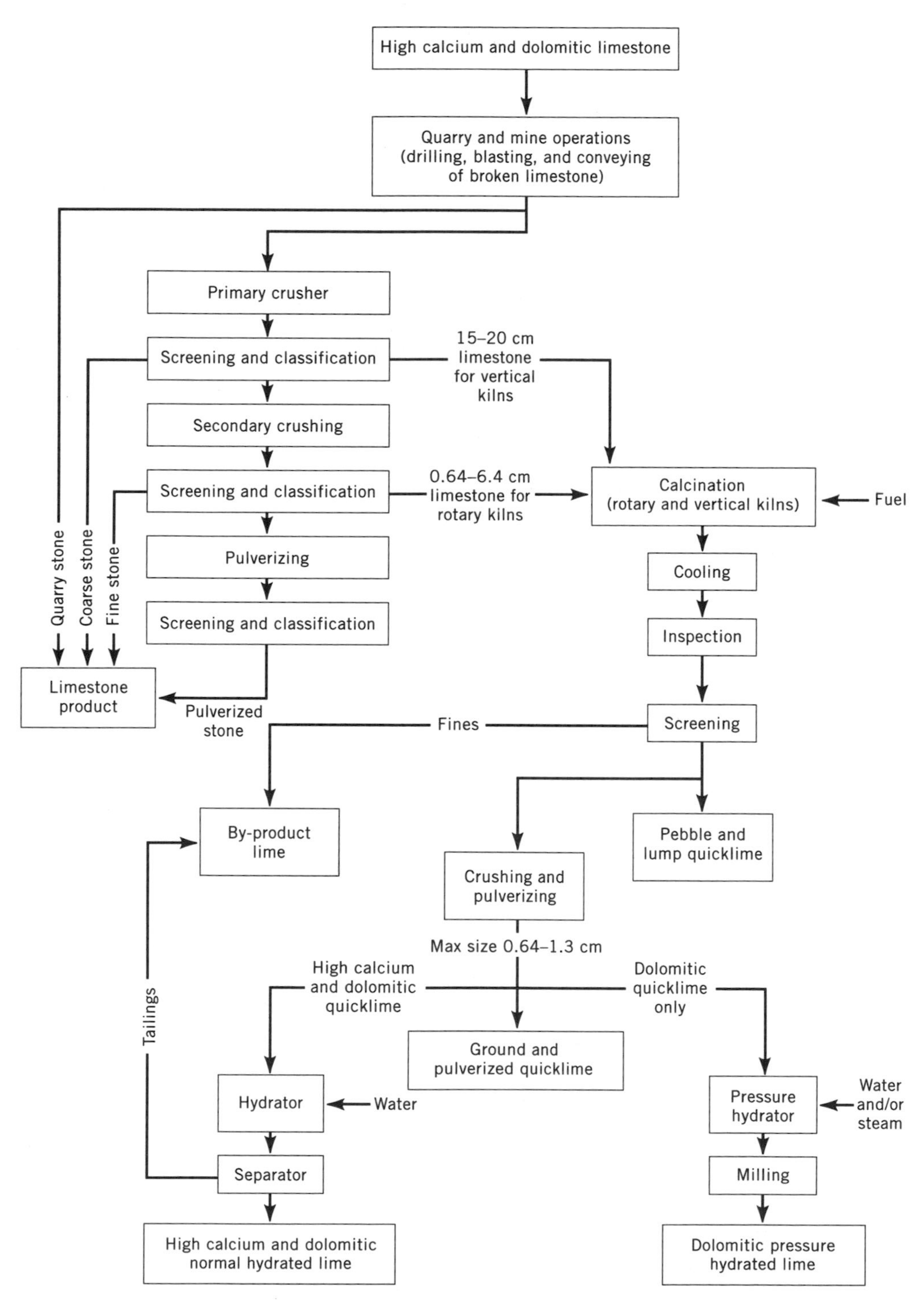

Fig. 4. Simplified flow sheet for lime and limestone products.

mining is pursued by some important operations, but the tonnage quarried exceeds that mined by nearly 20-fold. There is, however, a slight trend toward increased mining which should continue.

Stone Extraction. Quarries vary greatly in depth, from 4.5–7.5 m to deep quarries of up to 100 m. The first step in quarrying is stripping of the soil and loose rock that overlays the limestone deposit. Very rarely is limestone discovered in bare outcrop. If the overburden is thick, more than 5 m, irregular in thickness, and containing rock pinnacles and cavities, stripping can be so arduous and costly that mining may be more feasible. A variety of earth-moving equipment and methods are used in stripping. Washing soil away hydraulically was practiced historically, but because of cost and environmental constraints is seldom used. Much of the overburden is used for building roads, quarry ramps, and protective berms.

After the bedrock is exposed and clean, the next procedure is drilling. The productivity of large rotary and percussion drilling rigs is such that these rigs have largely replaced lower capacity well drills.

Diameters of the holes vary from 5–25.4 cm. Drilling perpendicularly to the deposit is preferable but in folding or tilted beds inclined drilling is often practiced. Spacing of the holes and borehole diameters depend on the hardness and fracturing characteristics of the stone, and desired top size for the primary crusher.

Most limestone quarries use either 100% ammonium nitrate [*6484-52-2*] (fertilizer grade) and fuel oil (ANFO), or a combination of ANFO and ammonium or gelatin dynamite, for blasting (see EXPLOSIVES AND PROPELLANTS, EXPLOSIVES). After blasting, oversized boulders usually are reduced to manageable sizes by drop ball cranes.

Quarries that excavate soft stone, notably marl or chalk, do not drill or blast, but extract the stone using heavy-duty rippers and scrapers. In the Middle West and Florida, lake marls and soft coralline limestone are dredged in a process much like stripmining.

Environmental regulations compel operations to abate dust, control wastewater discharge, and abide by noise-control regulations (11) (see AIR POLLUTION CONTROL METHODS; NOISE POLLUTION AND ABATEMENT METHODS). Drilling dust is eliminated by fabric filters that are a part of modern drilling rigs. Soil banks are planted with vegetation to reduce dust and erosion; trees, snow fences, and berms serve as wind barriers at vulnerable locations.

Most underground limestone mines (6) in the United States are relatively shallow room- and pillar-mines of 3–30 m depth. Thick 2 m supporting pillars are preserved at 7.5–10.6 m centers with 10.5–25 m high roofs or ceilings. However, some deep mines, reminiscent of coal (qv) operations, reach depths of 300 m with stope-mining techniques from a central vertical shaft. Because stone extracted from deep mines is often too costly to compete at the marketplace, it is generally mined for captive consumption, ie, for making higher priced products, eg, lime, chemicals.

Stone Processing. Next to blasting, primary crushing is the most effective method of reducing stone size (see SIZE REDUCTION). For maximum productivity it is essential that the primary crusher size be correlated with that of the bucket of the loader or shovel employed in the quarry or mine. Otherwise production is lost through large stone obstructions causing jamming delays in the crusher.

Primary crushers are of two basic types: compression or impact. Compression crushers typified by the jaw, cone, and gyratory primary crushers, operate by the slow application of pressure which causes the rock to crack and rupture. In contrast, impact crushers, on which impact breakers and hammer mills are predicated, subject the stone to sharp, rapid, repeated blows. Selection of the crusher type is contingent on the hardness and fracturing characteristics of the limestone, plant capacity, and the desired size or gradation. In general, compression types are favored for the hardest, most abrasive, and largest stone sizes, the impact type for the smaller sizes, and the gyratory type for plants of highest capacity. Depending on type, primary crushers yield product sizes of 3.8–25 cm. Crusher capacities range broadly from 90 to 450 t/h.

Impact crushers that rotate at speeds of 250–1000 rpm fracture the stone by collision of the rock against breaker bars or other rocks. Cubical shapes of stone predominate. The jaw and particularly the gyratory crushers yield a higher proportion of flat, elongated shapes. Often secondary crushing is unnecessary with impact types, but it is necessary with the larger top sizes used in jaw crushers and gyratory crushers. Special hammer mills, a modified gyratory and cone crusher, are often used for secondary crushing.

If a plant crushes to obtain stone of 0.6 and 7.6 cm for lime-kiln feed, coarse aggregate, or fluxstone, much undersized material is also produced. Oversized material can be reduced by recycling through the crusher system, but the subsized stone, called spall, is wasted in a spall pile. Such spalls have potential value as by-products for use as, eg, asphalt (qv) filler.

For fine pulverization, both dry and wet processes are utilized, but increasingly the dry process is more popular because wet grinding ultimately requires drying and is much more energy intensive. A sensitive fan swirls the dust sizes into the air separator and permits coarse particles to recycle to the grinding mill or be rejected as tailings; the fines are drawn into cyclones where the dust is collected.

Many different pulverizers are used by the industry. For sand-sized particles, hammer, cage, and rod mills are utilized; for soft stone, roller mills are preferred; for dust and very fine sizes, ball, tube, pebble, rod, and compartmented mills are used.

Crushed stone is conveyed by a rubber-belt conveyor and bucket elevator. Fine stone and dust are conveyed by enclosed screw conveyors, air slides, or pneumatic air systems into storage bins and tank trucks for shipment. For screening, changeable vibratory screens predominate for all sizes from 23 cm to 0.074 mm (200 mesh). Most stone is stored uncovered on the ground in conical stockpiles, supplied by radial belt conveyors. Such a conveyor can maintain four stockpiles of different sized stone. Large commercial plants typically stockpile stone in 10 sizes: 12.7 × 7.6 cm, 7.6 × 5.0 cm, 5.0 × 2.5 cm, 2.5 × 1.9 cm, 1.9 × 1.3 cm, 1.3 × 1.0 cm, 1.0 × 0.6 cm, 0.6 × 0.3 cm, limestone sand, 0.25–1 mm (10–60 mesh), and agstone or 90%–0.14 mm (100 mesh).

To comply with stringent specifications, some plants beneficiate the semiprocessed stone by removing clay and soil clinging to limestone. Several wet methods are used: washing, scrubbing, flotation, and heavy-media separation are used for removing silica, ie, chert and quartz. Optical mineral sorters use compressed air to deflect stone particles deviating from a preset color standard; and hand-picking from conveyor belts is still practiced by some plants.

Since 1960, portable stone-processing plants have grown steadily. Although such units are relatively small, companies utilize them to supplement the permanent facilities at times of peak demand or to provide stand-by capacity.

The main environmental problem is dust control, which requires collection of particulate emissions from point sources and suppression of fugitive dust from a multitude of areas (12). Rotary dryers for drying stone, now operated by few plants, produce by far the most visible and concentrated emission, requiring use of multiple cyclones plus a baghouse or high energy wet scrubber to meet standards. Other point sources of particulates are at each stone-transfer junction, ie, crushers, grinders, conveyors, screens, and loading. Such dust can be quelled by use of multiple-jet water sprays to keep the stone moist. Special wetting agents are applied to the water to enhance dustproofing while conserving water. Plants desiring a dry product install protective hoods, baffles, or enclosures at each transfer point. A few plants collect emissions from each transfer point and convey the dust to a single large wet scrubber for treatment. This is the most costly method.

Most fugitive dust is derived from spillage of stone fines and overburden soil from conveyors, bucket elevators, loading spouts, trucks, etc; from stockpiled-processed stone and spall piles that become air-dried and then wind-blown; and from truck traffic and wind on plant roads. Spillage can be minimized by a variety of practices including not overloading conveyor belts, elevators, and trucks; better coordination of the stone-feed flow; reducing conveyor gravity drops; use of enclosures at vulnerable transfer points; use of retractable loading spouts that fit tightly in circular ports of tank trucks; liberal application of rainbird-type jet-spray systems on stockpiles; watering unpaved areas; removing dust from paved surfaces with vacuum-cleaning equipment; reducing truck speed; or using pneumatic pumplines in place of trucks. Successful dust abatement requires a well-supervised, unremitting campaign, supported by all levels of management.

Dimension-Stone Production. The production of dimension stone for facing buildings, tombstones, and varied ornamental effects is totally different from that of crushed stone. No blasting is done. Stone is cut from the quarry floor in huge blocks of approximately $1.2 \times 15 - 30$ m having a depth of 3 m, using either a channeling machine or a wire saw. Skilled, experienced workers break the stone free by wedging and subdivide the block into smaller mill blocks of $3 \times 1.2 \times 0.9$ m by cutting and wedging. The smaller blocks are hoisted from the quarry with a crane and transported to the mill shop for further cutting and finishing or shipment to other stone finishers.

All products are specially made to comply with exacting specifications on dimensions, shape, finish, and appearance. Finishing stone requires precision sawing and such skills as planing, joining, milling, turning, fluting, and carving. Even packing and shipping the finished product requires a skilled artisan.

The limestone industry has achieved a high degree of mechanization since the 1960s, counteracting to some extent the general cost of inflation. The principal factors behind the productivity gains and cost saving have been use of larger earth-moving equipment in stripping operations; improvement in speed, diameter, and depth of drilling with large rotary- and percussion-type drilling rigs equipped for inclined drilling; expanded use of delayed action blasting; widespread substitution of ANFO for dynamite in blasting; use of mechanical or hydraulic breakage in place of secondary blasting; use of large shovels or buckets for quarry loading; greater use of front-end loaders for quarry loading instead of shovels; large capacity primary crushers; increasing use and efficiency of conveying systems, reducing truck hauling in quarries; vibratory screens replacing revolving screens; increased diesel conversion of the electric power generation for the whole plant; individual hoisting, earth-moving, and truck equipment; use of portable stone plants, providing greater production flexibility; and increased use of computerized control systems in stone processing.

Lime Manufacture

Most lime plants worldwide produce their own kiln feed from a contiguous quarry or mine, and are thus integrated lime producers. However, several unintegrated lime plants located on the Great Lakes obtain kiln feed by boat from large commercial quarries in northern Michigan. Most of these plants, among the largest in U.S. lime production, are situated in the Chicago and Detroit areas and in northern Ohio.

Theory of Calcination. The reversible reaction involved in the calcination and recarbonation of lime–limestone is one of the simplest and most fundamental of all chemical reactions. In practice, lime burning can be quite complex, however, and many empirical modifications are often necessary for efficient performance.

There are three essential factors in the thermal decomposition of limestone: (*1*) the stone must be heated to the dissociation temperature of the carbonates; (*2*) this minimum temperature (but in practice a higher temperature) must be maintained for a certain duration; and (*3*) the carbon dioxide evolved must be removed rapidly.

At calcination temperatures of 925–1340°C, dissociation of the limestone proceeds gradually from the outer surface of the stone particle inward, like a growing veneer or shell. Actually initial surface dissociation can occur under certain conditions, such as fractional atmospheric pressures below the dissociation point at 101.3 kPa (1 atm). However, for dissociation to penetrate the interior of the stone particle, attainment of temperatures often considerably higher than the dissociation point is necessary. Generally, the larger the diameter of the stone particle, the higher is the temperature required to calcine its center. The CO_2 expelled has a longer distance to travel and often considerable internal pressure is exerted as the gas forces its escape. If dissociation of the particle is incomplete, there remains in the center of the particle a core of uncalcined carbonate stone that may range in size from that of a grain of rice to that of an acorn, depending on the linear dimensions of the fraction and the extent to which calcination is complete. Such a core is usually not deleterious, but it does dissipate the concentration of the available lime. Its presence is inevitable with under-burned lime.

On the other hand, if the stone is calcined under severe calcining conditions, ie, high temperature and long retention, the lime may become hard-burned or even dead-burned at sintering temperatures. Under these conditions the stone shrinks by 25–50% of its original size. This shrinkage densifies the resultant lime, narrowing and occluding its micropores and fissures, so that the reactivity of the lime is reduced in varying degrees and extinguished with dead-burned dolomite. Chemical reactivity is usually measured by the rapidity with which lime hydrates in water. Densification of the quicklime particles is caused by the accretion of large crystallites from the stone's original minute crystals. As a result, in hard- and dead-burned limes the crystal lattice is much more compact, and the oxide molecules are in very intimate proximity (7,13).

Both of the above extremes are undesirable for most lime uses. Usually, the objective is to produce a completely calcined but soft-burned lime having no core or no more than 1–2% core. Such limes are more porous and chemically reactive.

Certain stifled calcination conditions can cause recarbonation in which CO_2 is readsorbed on the lime's surface. This can seriously diminish the quality and concentration of the lime. The possibility of recarbonation underscores the importance of rapid expulsion of the CO_2 gas during calcination.

Some limestones, more often the coarse crystalline types, can never be calcined successfully. Such stone tends to decrepitate during preheating or calcination into fine particles that interfere with this pyrochemical reaction. The adaptability of a stone for calcination can only be ascertained with surety by empirical methods. Possibly the greatest influence on lime quality is the size gradation of limestone. Narrow gradations, such as 10×20 cm, 2.5×5 cm, 1.3×0.6 cm, etc, or even narrower ranges, are more conducive to uniform calcination. As an example, if the size ranges between 1.25–15 cm, the small size would tend to be severely over-burned, or the large size would be incompletely calcined if the small size was properly burned. At constant temperature, the rate of calcination varies inversely with the size of the stone, increasing with smaller fractions.

Raising the temperature completes calcination more effectively than lengthening its duration. Generally, impurities complicate the process and impair lime quality more in a quantitative than qualitative manner. Impurities, such as silica, alumina, and iron, tend to combine chemically with lime, except at low calcining temperature, forming silicates, aluminates, and ferrites, and vitiating further the concentration of free lime. The total loss of available CaO, however, approximates 11–12% as a result of lime-fluxing these impurities.

Thermal Requirements. To produce a ton of lime theoretically requires 1.79 and 1.90 t of high calcium and dolomitic stone, respectively. Practically, at least 2 t of stone is needed to produce 1 t of lime, because some stone is lost as dust during the process. To heat the stone to the dissociation point, ca 1.70 GJ/t (1.46×10^6 Btu/short ton) for high calcium and 1.45 GJ/t (1.25×10^6 Btu/short ton) are required for dolomitic quicklimes. However, temperature, usually a higher than the theoretical one, must be maintained until dissociation is terminated. This additional thermal requirement is estimated at 3.22 and 3.02 GJ/t (2.77 and 2.60×10^6 Btu/short ton) for high calcium and dolomitic stones, respectively. In addition, some heat loss is inevitable in lime manufacture, such as heat of evaporation of moist limestone and/or coal, radiation and convection through the kiln structure, retention in the discharged lime, exhaust gases, and incombustible dusts. The thermal efficiency is given as

$$\frac{\text{theoretical heat requirement} \times \text{\% available oxide}}{\text{total thermal requirement}} = \text{\% thermal efficiency}$$

Often, maximum thermal operating efficiency is incompatible with optimum lime quality. Usually this problem is resolved by operating under compromise conditions between these two extremes.

Kilns. *Rotary Kilns.* As of this writing (ca 1994) in the United States, about 90% of commercial lime capacity and 60% of captive lime is calcined in rotary kilns. The rotary is not nearly as preeminent in most other countries although a company in Germany was the first to operate a 1000 t/d rotary kiln. That company currently operates four such kilns in addition to many vertical kilns. South Africa has a high percentage of rotary kiln production, including three of 1000 t/d, and one at 1500 t/d capacity. Belgium, France, Finland, and Canada also have large rotary kiln production. In the United States there is one producer operating a 1300 t/d rotary kiln, two other companies operate a total of four 1000 t/d rotary kilns, and several have 800 t/d rotaries. Paradoxically, as capacities have increased, the length of the new kilns have been markedly shortened from 90–127.5 to 45–60 m, but kiln diameters have increased from 2.1–3 to 4.8–5.1 m. The diameter–length ratio has decreased from 1:30–40 to 1:12 in modern rotary kilns.

Helping to propel capacities upward has been the advent of greatly improved preheaters, which partially calcine the stone and significantly improve thermal efficiency. Modern preheaters improve capacity by 15–20% and decrease fuel consumption a similar percentage. Other kiln appurtenances and accessories that enhance efficiency and lime quality are the contact coolers, and such kiln internals as metal refractory trefoil systems that act as heat exchangers, dams, and lifters.

The exterior of the kiln is heavy steel boiler plate, welded into sections; the interior is lined with 15–24 cm refractory brick. A preheater rotary kiln system is shown in Figure 5. Kilns are installed at 3–5° inclination on foundation piers and revolve on trunnions at 1–2 rpm. Limestone is fed into the elevated end of a kiln from the preheater or silos and is discharged as quicklime into the cooler at the lower end. Cooling air is induced into the cooler and from there to the calcining zone of the kiln next to the discharge end as secondary combustion air, providing heat recuperation. The hot air and gases are sucked countercurrent to the flow of kiln feed to the charging end where they provide recuperative heat for the preheater. Most U.S. rotaries are fired with pulverized coal, but are also adaptable to gas and oil firing. Only 10% of the kiln is filled with limestone–lime as the kiln feed tumbles gently through the kiln. Kiln feed ranges in size from 0.625–6.25 cm, but multiple rotary kiln plants use more restricted gradations of 0.625–1.88 cm, 1.88–3.75 cm, etc.

Rotary kilns produce the greatest output per hour and the highest, most uniform quality. There are, however, compensating disadvantages, such as the highest capital investment among kiln types, a higher average energy consumption, lack of flexibility in single-kiln plants, and the most complicated and

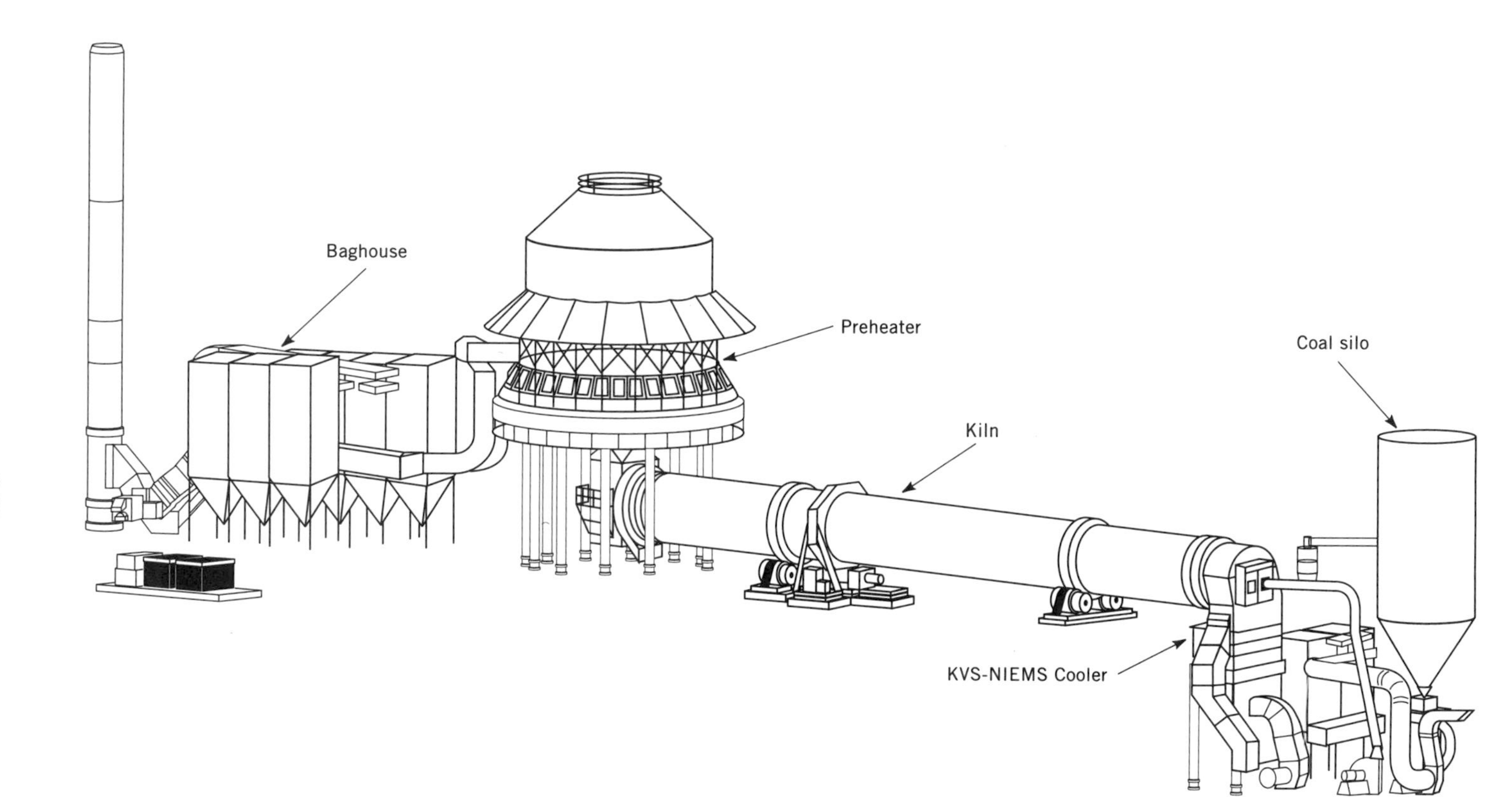

Fig. 5. Schematic of the Kennedy van Saun (KVS) patented low pressure drop (LPD) preheater/rotary kiln lime calcining system available in sites ≥1100 t/d. Courtesy of KVS.

expensive dust collection system. Energy consumption ranges widely from ca 6.4 to 9.3 GJ/t (5.5 to 8 $\times$ 10^6 Btu/short ton). The minimum energy value applies to high capacity rotary kilns with the most advanced preheaters, internal accessories, and baghouses. The highest value applies to old, low capacity kilns with no preheater and a minimum of internal accessories. Of these energy values, 90% is for fuel and 10% is electricity, mainly for dust collection.

The average dust loading in rotary kilns is 10% of the kiln feed or approximately 20 kg/t of lime, allowing for a ratio of limestone to lime of 2 on a weight basis. Primary collection of particulates is accomplished with multiple cyclones that entrap about 85% of the dust loading. A secondary system is necessary to abate most of the remaining dust. Of the secondary systems in use, the baghouse is the predominant type, followed by the wet scrubber, electrostatic precipitator, and gravel bed filter (see AIR POLLUTION CONTROL METHODS).

Vertical Kilns. Outside the United States, particularly in developing countries, the vertical kiln is the most commonly used. One reason for the decline in use in the United States is the energy crisis of the 1970s, when supplies of natural gas and fuel oil, the principal fuel for vertical kilns, became stringent and prices escalated rapidly. The vertical kiln has made a slight comeback in the United States, however, through the introduction of the Maerz parallel flow regenerative kiln.

There are many vertical kiln types and designs having widely varying efficiency. All of these kilns have four imaginary zones or sections, as depicted in Figure 6. In the preheating zone, recirculated exhaust gases preheat the stone, preparatory to calcination in the adjacent zone. The calcining zone is the calcination chamber where 95% of lime burning occurs. The lower portion of this zone is called the finishing zone where calcination is completed and where fuel ports for firing are situated. Cool air enters the cooling zone from the base of the kiln or discharge point and by natural, induced, or forced draft flows upward countercurrent to the lime descending through the kiln. The air cools the lime for discharge onto conveyors below, and the air recoups much heat from the red-hot lime as secondary combustion air for the calcining above.

Significant improvements in vertical kiln performance (14) during the 1970s and 1980s have increased capacities to 600–800 t/d, reduced energy consumption to 4.2–4.6 GJ/t (3.6–4.0 Btu/short ton) of lime, and improved product quality. Energy values include an estimated 5% for electricity. The most efficient vertical kilns are of Austrian and German origin: the Maerz or parallel-flow regenerative kiln (Austrian) and the double-inclined, Ring or Beckenbach annular shaft kilns (German). These kilns are widely used in Europe and Japan, but only two, the Maerz parallel-flow gas-fired kiln and the Beckenbach, are in operation in North America. These kilns develop tremendous heat recuperation.

The parallel-flow kiln operates with two or three independent shafts within one large refractory-lined shell. As one shaft is calcining, the waste hot gases are preheating the kiln feed in an adjoining shaft. Thus calcining, discharging, charging, and preheating are performed cyclically, programmed at preset intervals of 10–15 min. At each cycle, firing lances automatically are switched to an adjacent shaft with its preheated stone charge; ducts conveying the hot exhaust gases are similarly reversed. During each cycle's transition period, precise increments of kiln feed are provided by a mobile overhead weight hopper that

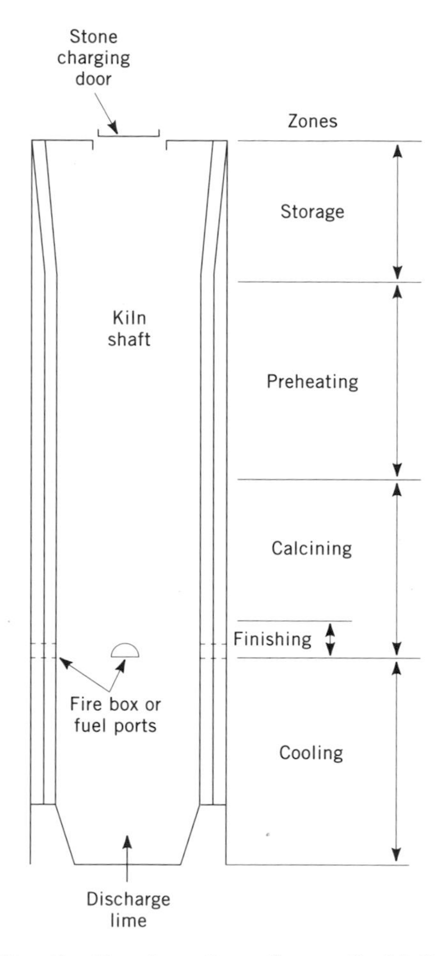

Fig. 6. Zonal section of a vertical kiln.

maintains constant levels. This kiln operates at relatively low temperatures of 950–1050°C with kiln feed of 2.5–15 cm, but usually a size ratio of only 1:3.

The double-inclined kiln calcines even smaller sized stone of 1.88–3.75 cm and at reduced capacity with stone of only 0.63 cm minimum size. Most of these kilns operate using gaseous or oil fuels, including propane. An exception is the double-inclined kiln, which appears to operate at optimum efficiency with a mixture of fuel, ie, 60–75% natural gas or oil and 40–25% coke, although it can operate on 100% gas or oil.

Another thermally efficient kiln is the modern mixed-feed vertical kiln in which coke is admixed with 8.5–20 cm lump limestone and charged into the top of the vertical kiln by a mobile, overhead charging system. However, use of this kiln is waning since the quality of mixed-feed kiln lime does not equal that of the other three kilns described above, owing to ash contamination from the coke and poorer reactivity and to the higher cost of coke in most areas.

Low capacity shaft kilns, direct-fired manually with bituminous coal, which were extant during 1900–1940, largely disappeared following World War II because of their high thermal and labor costs, poor lime quality, and offensive emission of black smoke. Another kiln that was phased out in the United States after World War II is the producer-gas (indirect-fired) kiln. The bituminous coal fuel was burned in a special firebox exterior to the kiln with low heat value gases charged into the calcining zone. These gases did not match natural gas in lime-burning performance, because they possess only 20–30% as much heat value as natural gas. After the energy crisis of 1973, this fuel was carefully reconsidered as a substitute for scarce natural gas and oil, but was judged uneconomical and unfeasible for lime production in the United States.

Many kilns that formerly were direct coal-fired or producer-gas verticals were retrofitted to natural gas firing with center-burners and after World War II, dramatically improving lime quality, kiln capacity, and fuel efficiency. By the 1960s, this improved vertical kiln had lost favor to rotary and other special kilns because of the supply and cost problems of oil and gas in the United States and the spectacular improvement in rotary kiln performance. Many natural gas-fired center burners were permanently closed and dismantled because they could not be converted to coal. However, the reverse occurred in Europe where the extensive oil and gas discoveries heightened interest in the new, advanced vertical kilns.

Miscellaneous Kilns. A U.S. kiln, the Fluo-Solids, appears to be another vertical kiln type, but this is its only similarity. It operates on a different principle. It utilizes as kiln feed only a discrete granulation of 0.225–2.4 mm (65–8 mesh) sizes. Delicately controlled by air and exhaust gas pressure, the kiln feed of granules is fluidized as a dense suspension. Because it is instrumented, this kiln can produce a very reactive lime at better than average thermal efficiency. The kiln, however, has limited utility because the cost of obtaining the kiln feed with many hard, compact limestones is prohibitive.

The Calcimatic is a patented kiln of Canadian origin that is radically different from other kiln types. It consists of a circular traveling hearth of variable speed, supported on two concentric tiers of rollers. Kiln feed of 12.7 mm is fed onto the hearth in a 2.5–10 cm bed from a preheater chamber. The kiln is usually fired with natural gas or fuel oil, although the option of using pulverized coal has also been developed. After great interest, resulting in sales of many units throughout the world, the popularity of the Calcimatic has ebbed because of disappointment in the unit's mediocre thermal efficiency.

Calcination Products. Tables 3 and 4 summarize the analyses and forms of commercial quicklime in the United States (1). In addition to conventionally calcined dolomitic quicklime, a special refractory lime is made by sintering or dead-burning granules of high quality dolomitic limestone in a rotary kiln at 1650°C. Iron oxide (5–8%) is added to the feed to stabilize the product against hydration. The grayish brown dustless granules produced are used only for lining open-hearth and electric steel furnaces or as raw materials for refractory brick and other products.

Rotary kilns and, to a lesser extent, Fluo-Solids kilns are used to calcine a wet precipitated calcium carbonate filter cake in the kraft or sulfate paper-pulp process (15). Lime is regenerated for use as a causticization reagent in recovering caustic soda for pulp digestion. Losses in lime recovery are replaced by purchased lime (see PAPER; PULP).

Table 3. Analyses of Commercial Quicklimes

Component	High calcium,[a] %	Dolomitic,[a] %	Component	High calcium,[a] %	Dolomitic,[a] %
CaO	93.25–98.00	55.50–57.50	Al_2O_3	0.10–0.50	0.05–0.50
MgO	0.30–2.50	37.60–40.80	H_2O	0.10–0.90	0.10–0.90
SiO_2	0.20–1.50	0.10–1.50	CO_2	0.40–1.50	0.40–1.50
Fe_2O_3	0.10–0.40	0.05–0.40			

[a]Values given are typical values and do not necessarily represent minimum and maximum percentages.

Table 4. Physical Forms and Sizes of Quicklime

Physical forms	Physical size, mm	(mesh size)	Derivation
lump	63–255		vertical kiln
pebble	6.3–63		rotary, speciality kilns; crushed lump
ground			
100%	< 2.38	(−8)	Fluo-Solids; screening, grinding
40–60%	< 1.49	(−10)	
pulverized			
100%	< 0.84	(−20)	pulverizing; screened dust
80–90%	< 1490	(−10)	
pellets (briquettes)	16.4–55.7[a]		screened fines and dust
dead-burned dolomite	0.84–9.5	(20–2)	iron-stabilized, dead-burned dolomite granules (screened)

[a]Size is given in cm^3.

Hydrated Lime Manufacture

Although most lime is sold as quicklime, production of hydrated lime is also substantial. This product is made by the lime manufacturer in the form of a fluffy, dry, white powder, and its use obviates the necessity of slaking. Small lime consumers cannot economically justify the additional processing step that hydration entails.

The manufacture of hydrated lime proceeds by the slow addition of water to crushed or ground quicklime in a premixing chamber or a vessel known as a hydrator, both of which mix and agitate the lime and water. The amount of water added is critical. Too much water makes it impossible, or too costly, to produce the desired dry form; too little water leaves hydration incomplete, causing degraded quality, namely, chemical instability and structural unsoundness.

More than the theoretical amount (24.5%) of moisture is necessary to counteract the loss of moisture as steam generated by the appreciable heat of hydration. In practice, about 50–65% water is added, depending on the degree of reactivity of the quicklime and its physical size. The finest particles hydrate most rapidly. After hydration, the slightly moist slaked lime is propelled by an enclosed screw-conveyor to an air separator, where the coarse fractions are largely removed as tailings. This step enhances the fineness of the powder, upgrading its chemical purity, and dries the powder further. The hydrate may be further refined, or conveyed to a silo for bagging in 22.7 kg paper sacks, or for bulk truck or rail-car shipment.

The above hydration procedure applies to high calcium and normal dolomitic hydrates (ASTM type N), but the latter is usually incompletely hydrated. Only 10–20% of its MgO component is actually hydrated. All of the CaO hydrates readily. To hydrate the MgO substantially, other hydration measures are employed to produce a highly hydrated dolomitic lime (ASTM type S (Special)). For some dolomitic hydrates, retention in silos in a moist condition for 2–3 d produces the highly hydrated condition. But for most such hydrates it is necessary to employ autoclaves in lieu of normal hydrator machines that operate at atmospheric pressure. At 276–690 kPa (40–100 psi), hydration is completed in this vaporous, hotter atmosphere in less than an hour. The steam and pressure catalyze the hydration of the MgO. After autoclaving, the hydrates are usually subjected to other milling practices in addition to air separation, such as tube milling to enhance the plasticity of the resulting hydrate.

In the United States most commercial hydrates easily conform to the minimum basic requirement of no more than 0.5% retained on 590 μm (30 mesh) screen and no more than 15% retained on a 74 μm (200 mesh) screen. Typical analyses of hydrates are contained in Table 5.

Table 5. Chemical Analyses of Commercial Hydrates

Component	High calcium hydrate, wt %	Highly hydrated dolomitic, wt %
CaO	71–74	45–41
MgO	0.5–2	25–30
H_2O	24–25	27–28
CO_2	0.3–0.7	0.3–0.7
SiO_2	0.2–0.5	0.2–0.5
R_2O_3	0.1–0.3	0.1–0.3

Economic Aspects

Limestone. In the crushed, broken, and pulverized stone industry, limestone (including dolomite) is predominant, accounting for 71% of the tonnage which includes granite, basalt, trap rock, sandstone, and other miscellaneous rock. Its total of 710×10^6 t in 1991 makes limestone the third greatest commodity in tonnage in the United States after coal (qv) and sand and gravel. In addition, three other sources of calcium carbonate are produced, including marl, shell, and marble, for which production was 5.7, 5.0, and 3.4×10^6 t in 1991, respectively. The United States is clearly the largest limestone producing country, followed by the former USSR, Japan, and Germany. Most countries produce their own limestone requirements.

The average value of limestone in 1991 was $4.73/t, an increase in value of about 50% from 1979. In 1995, the average value was $5.1/t. Table 6 gives the average value of the limestone according to usage. Use of crushed and broken U.S. limestone has grown rapidly since World War II. The all-time record year for U.S. limestone was 1987 when production reached 763×10^6 t. Construction aggregate, although declining in volume, still dominates all uses. Transportation

Table 6. U.S. Limestone Average Value, 1991

Material	Value, $/t	Material	Value, $/t
cement manufacture	3.45	limestone sand	4.90
riprap	4.52	asphalt filler	8.17
concrete and roadstone	5.03	glass manufacture	11.14
flux stone	5.17	coal-mine dusting	17.05
agricultural liming	5.21	mineral food	11.93
lime manufacture	6.09	SO_2 removal	6.14
limestone whiting	17.78		

plays a principal role in limestone cost and availability. In 1991, modes of transportation for all crushed stone (typical for limestone), based on 67% return, were truck, 72.8%; railroad, 5.5%; water, 6.3%; other modes, 2.5%; and used on-site, 12.9%.

Table 7 presents 1991 statistics on limestone and dolomite uses, and includes production from 2338 U.S. plants (16). Generally the growth markets for uses have been in construction aggregate, Portland cement and lime manufacture, industrial fillers, mineral feed, and stack-gas desulfurization that utilizes limestone wet scrubbing. Declining markets have been iron and steel fluxing, fertilizer filler, pulp and paper, rock-wool manufacture, refractories, and agricultural liming.

Lime. The total tonnage of quicklime and hydrated lime sold and used in 1992 was 16.2×10^6 t. Captive lime accounted for 11% of this total, nearly the same percentage as captive limestone. Captive lime production decreased drastically since the 1970s, falling from 6.7×10^6 t in 1973 to 1.82×10^6 in 1992. Whereas commercial lime production was at an all time high in 1992, total lime (commercial lime plus captive lime) was 18% below that of the all time record of 19.8×10^6 t, set in 1973.

More than 90% of all lime uses are in the chemical and metallurgical areas, and lime is the fifth largest chemical in the United States, behind sulfuric acid and certain industrial gases such as oxygen. Although steel manufacture is still the greatest use of lime, many established and developing environmental applications are growing rapidly.

There were 113 commercial and captive lime plants in the United States as of 1993. Of these plants, the 10 largest produce about 35% of total lime output. Ohio is the largest lime producing state, followed by Pennsylvania, Missouri, Kentucky, Alabama, Texas, Illinois, Indiana, Virginia, and Arizona.

The former USSR was the leading lime-producing country in 1991, producing 26×10^6 t; followed by China, 18.5×10^6 t; and the United States, 15×10^6 t; Germany, 9.2×10^6 t; Japan, 9.0×10^6 t; Mexico, 6.5×10^6 t; and Brazil, 5.5×10^6 t.

Except for two countries, there is very little world trade in lime. The largest importer is the Netherlands, which is nearly devoid of limestone and thus imports about 10^6 t annually from Belgium and Germany. The other net importer of consequence is the United States, which imports ca 150,000 t/yr or about 1% of U.S. production. About 85% of the U.S. imports are from Canada; the balance is from Mexico.

Categories for total lime usage in 1991 are presented in Table 8. Steel and environmental uses predominate, accounting for 30 and 24% of the total, respectively.

In 1991, the average value per metric ton of lime in bulk was reported to be $78.44 for agricultural lime, $68.12 for construction lime, $54.90 for chemical lime, and $82.62 for refractory lime. The average value for total lime was $57.02/t.

Table 7. Crushed Limestone and Dolomite Sold or Used by Producers in the United States, 1991

	Limestone		Dolomite	
Material	Quantity, t × 10^3	Value, $ × 10^3	Quantity, t × 10^3	Value, $ × 10^3
coarse (+1.25 cm) aggregate				
macadam	2,983	13,392	111	531
riprap and jetty stone	13,531	61,269	1,163	7,379
filter stone	4,590	21,259	122	653
other coarse aggregate	179	774	[a]	[a]
coarse aggregate, graded				
concrete aggregate, coarse	61,419	300,273	7,604	42,104
bituminous aggregate, coarse	40,923	208,579	7,137	43,572
bituminous surface-treatment aggregate	18,539	93,027	2,349	12,094
railroad ballast	4,133	18,216	1,009	4,392
fine (–0.95 cm) aggregate				
stone sand				
concrete	12,076	62,338	1,178	6,988
bituminous mix or seal	10,377	47,634	4,034	22,532
screening, undesignated	13,514	56,166	1,294	10,220
other fine aggregate	108	398		
coarse and fine aggregates				
graded road base or subbase	104,802	428,122	9,809	47,279
unpaved road surfacing	19,466	86,798	688	2,825
terrazzo and exposed aggregate	1,425	9,388	4	102
crusher run or fill or waste	21,010	88,367	2,038	8,624
other construction materials	9,673[b]	44,020[b]	3,497[c]	20,002[c]
roofing granules	755	4,657	[a]	[a]
agricultural				
agricultural limestone	14,163	74,041	2,287	20,371
poultry grit and mineral food	2,257	26,999	[d]	[d]
other agricultural uses	690	3,676	88	1,143
chemical and metallurgical				
cement manufacture	71,832	248,780		
lime manufacture	16,305	99,632	611	3,507
dead-burned dolomite manufacture			1,371	5,733
flux stone	3,635	18,803	752	2,751
chemical stone	294	1,614	[d]	[d]
glass manufacture	269	3,009	[d]	[d]
sulfur oxide removal	2,040	12,551		

Table 7. (*Continued*)

Material	Limestone Quantity, t × 10³	Limestone Value, \$ × 10³	Dolomite Quantity, t × 10³	Dolomite Value, \$ × 10³
special				
mine dusting or acid water treatment	659	11,250	200	1,542
asphalt fillers or extenders	1,448	11,861	482	4,159
whiting or whiting substitute	533	9,489	*d*	*d*
other fillers or extenders	2,753	41,851	297	3,593
other miscellaneous uses				
chemicals	505	3,120	*d*	*d*
magnesia (dolomite)	16	30		
other uses not listed	235[e]	3,672[e]	230	3,157
unspecified[f]				
actual	148,595	748,891	6,807	32,457
estimated	49,880	242,494	197	846
Total[g]	*655,622*	*3,106,400*	*55,338*	*308,600*

[a]Included with other construction materials.
[b]Includes other graded coarse aggregate, dam construction, drain fields, building products, pipe bedding, and waste material.
[c]Includes building products.
[d]Included in other uses not listed.
[e]Includes abrasives, paper manufacture, and refractory stone (including ganister).
[f]Includes production reported without a breakdown by end use and estimates for nonrespondents.
[g]Data may not add to totals shown because of independent rounding.

The price of lime doubled during the 1970s, owing to the rapidly escalating fuel prices brought on by the Arab oil embargo. This was followed by a leveling of prices during the 1980s, to around \$56.00/t (the highest rate was \$57.87/t in 1986). However, when comparing prices based on constant 1991 dollars, prices actually decreased steadily from 1979 to 1991, finishing the period down about 30%.

Uses

Limestone. *Construction.* Generally, for highway and building construction, limestone that is clean, strong, durable, sound (free from incipient cracks), and dense is preferred. This is particularly true with respect to use as a coarse aggregate. Most engineers purchase limestone on specifications based largely on the physical properties of the stone. Many of these specifications differ widely, particularly with respect to the exact application of the stone and to a lesser extent by geographical location. Many tests and specifications are published by ASTM, AASHTO, the National Institute of Standards and Technology, and others. Generally, the presence of alkalies, soluble sulfides, gypsum, and organic matter in the stone is objectionable to most users. Another important consideration is the physical size and range of particle gradation of the limestone. As a result, many types of limestone are rejected for one or more construction uses when unable to meet these requirements.

Table 8. Lime Sold or Used by Producers in the United States, t × 10³[a]

Industry	1990 Sold	1990 Used	1990 Total[b]	1991 Sold	1991 Used	1991 Total[b]
		Chemical and industrial				
alkalies	[c]	[c]	96	[c]	[c]	80
aluminum and bauxite	141		141	145		145
copper ore concentration	[c]	[c]	338	[c]	[c]	371
food products, animal or human	19		19	20		20
glass	90		90	152		152
magnesia from seawater or brine	[c]	[c]	592	[c]	[c]	523
oil well drilling	11		11	11		11
oil and grease	[c]	[c]	17	[c]	[c]	[c]
ore concentration, other	300		300	343		343
paper and pulp	[c]	[c]	1,148	[c]	[c]	1,125
precipitated calcium carbonate	[c]	[c]	230	[c]	[c]	270
steel						
BOF	[c]	[c]	3,660	[c]	[c]	3,770
electric	[c]	[c]	884	835		835
open-hearth	111		111	44		44
sugar refining	27	569	596	26	669	694
tanning	24		24	19		19
other[d]	8,091	1,069	2,194	7,116	982	1,958
Total[b]	*8,814*	*1,637*	*10,452*	*8,711*	*1,651*	*10,360*
		Environmental				
acid water, mine or plant	259		259	314		314
sewage treatment	424		424	473		473
flue gas sulfur removal	1,518		1,518	1,565		1,565
water purification	972		973	1,032		1,032
other[e]	587	9	596	408	5	413
Total[b]	*3,758*	*9*	*3,768*	*3,792*	*5*	*3,797*
		Construction and other				
soil stabilization	610		610	648		648
finishing lime	145		145	96		96
mason's lime	[c]	[c]	230	[c]	[c]	168
other[f]	268		268	269		269
Total[b]	[c]	[c]	*1,253*	*1,013*	[c]	*1,181*
agriculture	43		43	48		48
refractory lime (dead-burned dolomite)	[c]	[c]	342	[c]	[c]	308
Grand total[b]	*14,040*	*1,818*	*15,858*	*13,875*	*1,819*	*15,694*

[a]Excludes regenerated lime; includes Puerto Rico.
[b]Data may not add to totals shown because of independent rounding.
[c]Included in "other" category.
[d]Includes briquetting, brokers, calcium carbide, chrome, citric acid, commercial hydrators, desiccants, ferroalloys, fiber glass, glue, insecticides, ladle desulfurizing, magnesium metal, metallurgy, pelletizing, pharmaceuticals, petrochemicals, rubber, silica brick, soap wire drawing, and uses[c] in chemical and industrial lime only.
[e]Includes industrial solid waste treatment, industrial wastewater treatment, scrubber sludge solidification, and other environmental uses.
[f]Includes asphalt antistripping.

Table 9 shows the dominance of limestone over other types of crushed stone as an aggregate.

Concrete Aggregate. A versatile application for coarse limestone aggregate is in Portland cement concrete, either job-mixed or ready-mixed for a wide variety of concrete applications, such as footings, poured foundations, paving, curbs, structural products, etc. Limestone sand also provides a satisfactory fine aggregate, but usually it is more costly than conventional sand from local pits.

Roadstone. The greatest tonnage of aggregate is consumed in highway construction for the subgrade or subbase and base course, surface aggregate for unpaved macadam roads, coarse and fine aggregate for asphalt paving mixes of varying thicknesses for the wearing surface and/or black base, and bituminous macadam and other types of asphalt surfaces, such as sheet asphalt and cold-mix. Screenings of < 6 mm are favored in water-bound macadam roads. Most of this limestone aggregate is subject to the Los Angeles abrasive resistance test for durability.

Railroads also require coarse and fine aggregate of physical strength and durability similar to concrete and roadstone for road beds. Railroads prefer gradation of 19–63.5 mm as coarse aggregate.

Riprap. Heavy, irregular limestone fragments, ranging in size from 25–30 cm to large boulders are used for dam spillways; construction of docks, piers, and breakwaters; and rustic massive rock barriers and dry rock retaining walls. When the size of riprap is specified, its price rises sharply.

Asphalt Filler. One of the preferred mineral-dust fillers for asphalt-paving mixtures is pulverized limestone meeting a fineness requirement of at least 60–75% minus 0.074 mm (200 mesh). The limestone fines greatly reduce the size and extent of voids in the mixture, providing a more stable, water-resistant, denser-graded aggregate for mixing with the asphalt cement. Usually, it enhances toughness and resistance to softening in hot weather (see ASPHALT).

Limestone Sand. A discrete gradation of substantially 2.38–0.225 mm (8–65 mesh) size provides a versatile fine aggregate or sand for road mixtures, concrete, plaster, or any construction use supplied by silica sand. The only disadvantage is that in many areas conventional sand is less costly.

Table 9. Crushed Stone Sold or Used in the United States, 1991

Stone	Number of quarries	Quantity, $t \times 10^3$	Value, $\$ \times 10^3$	Unit value, $/t
limestone[a]	2,201	655,578	3,106,444	4.74
dolomite	137	55,336	308,556	5.58
calcareous marl	15	5,175	13,769	2.66
shell	14	4,959	44,376	8.95
marble	46	3,400	83,335	24.51
granite	492	149,050	864,374	5.80
traprock	561	75,593	485,223	6.42
sandstone and quartzite	235	25,855	137,683	5.32
miscellaneous stone	312	25,616	143,061	5.58
Total	*4,013*	*1,000,534*	*5,186,800*	*5.18*

[a]Includes reporting for limestone–dolomite when no distinction is made between the two.

Miscellaneous. Granules of limestone and oyster or clam shell are generally a second-choice mineral aggregate for roofing granules. Siliceous rocks and slag, however, provide most of this requirement. Larger aggregate is used in built-up roofs. Limestone wastes or spalls are used as fill for swamps and low lying areas.

Building Materials. Portland Cement Manufacture. The second greatest use of limestone is as raw material in the manufacture of Portland cement (10). The average limestone factor per ton of Portland cement is 1.0–1.1 t of pulverized limestone. The limestone, calcined to lime in the cement rotary kiln, combines with silica and alumina to form tricalcium silicate and tricalcium aluminate and other lesser cementing compounds (see CEMENT). Most cement companies operate captive limestone quarries.

A second form of limestone is finely pulverized limestone for masonry cement. This form is achieved by intergrinding roughly equal parts of limestone and cement clinker in a ball mill to which an air-entraining agent is added.

Concrete Products. Limestone aggregate is used competitively with other aggregate in the manufacture of molded, reinforced, and prestressed concrete products in the form of block, brick, pipe, panels, beams, etc.

Insulation. Impure siliceous limestone and blast-furnace slag are the main raw materials for making rock-wool insulation bats and pellets (see INSULATION, THERMAL).

Fillers. Micronized carbonate whiting is the preferred mineral fill for putty and caulking compounds based on linseed oil or plastic, and vinyl-based floor coverings. It comprises 20–60% of the raw material mix (see FILLERS).

Calcium Silicate Products. Quicklime is used as the cementitious binder in making autoclaved building materials based on calcium silicate, ie, sand–lime brick, cellular (or foam) concrete block and insulation material, and the former USSR extremely dense and reinforced silicate–concrete building materials. Lime reacts with fine silica in the autoclave under steam and pressure to form strong dicalcium silicate cement compounds, representing a type of cementless concrete because no Portland cement is used in the mix. Use of these building materials is growing steadily in Germany, Scandinavia, the Netherlands, the eastern European countries, Australia, Canada, and Japan, but little use is made of them in the United States. They are economically useful, can be mass produced, and are of stable quality. By increasing the foaming agent in cellular concrete, an extremely lightweight insulation, microporite, can be produced. Only lime (no limestone) is used in these products.

Mineral Feed. Mineral feed supplements for domestic animals and fowl usually contain a pure form of pulverized limestone. In fact, some state laws require the supplement to be at least 35% available calcium. Other sources of calcium are bone meal and dicalcium phosphate. Use as mineral feed has been a steadily growing market for limestone. The material is ground to 90% minus 0.15 mm (100 mesh) or 80% minus 0.9074 mm (200 mesh), is low in silica, and has strict tolerances on arsenic and fluorine (see FEEDS AND FEED ADDITIVES).

Lime. *Building.* Lime, in hydrated form, is used in lime–cement–sand mortar of various proportions to provide plasticity, high water retentivity, and bond strength (17). In the United States, nearly all of the mason's lime used is

ASTM Type S hydrated lime. It is applied to mortar in several ways: (*1*) one bag each of lime and Portland cement with six parts sand by volume is added to the mortar mixer at the job site; (*2*) 5–10% of hydrated lime is added to some masonry cements and packaged for later mixing with sand (in a 1:3 proportion) and water; (*3*) at a central mixing plant, lime putty made from quicklime is mixed with a measured amount of sand and water, and delivered to the project as ready-mixed mortar or ready to use (mixed) in drums; (*4*) lime, cement, and sand are dry-mixed and delivered to the job in bulk, optionally with all mortar handling and mixing equipment on a rental basis; (*5*) pulverized quicklime is slaked at the job site, and the resulting putty is then mixed with cement and sand.

Any of these mortars can be used for unit masonry or for stucco (exterior plaster). The finish coat in conventional interior plastering is composed of either neat lime putty or a sanded putty, gauged with Keene's cement or gypsum-gauging plaster. The former is called a whitecoat finish; the latter a sand-float finish.

Lime–Soil Stabilization. Lime–soil stabilization (18) has been successfully applied to all types of roads from farm to market to interstate freeways; to parking lots, airport runways, building foundations; in embankments and earth dams, railroad beds, and irrigation canal linings; and to river levees. Lime's reaction is with silica, derived from clay in the soil, drastically lowering the soil's plasticity index and increasing the compacted soil's stability and strength. Lime also expedites construction by drying up saturated subgrades. It is applied in the subbase (subgrade) and/or base course. Where soils are deficient in clay, lime is applied with flyash which generates the desired lime–silica reaction (19). Lime, flyash, and aggregate are premixed in a central mixing plant and then spread on roadways like base material. Compaction and moist-curing follow (see SOIL STABILIZATION).

Lime stabilization originated in Texas after World War II, and now it is used throughout the world. Lime is most commonly applied at a 4 wt % application or ca 11 kg/m^2 (20 lb/yd^2) for 15 cm of compacted depth. It can be applied dry as hydrated lime or granular quicklime or as a wet slurry. Distribution of the latter form is dustless. Copious amounts (as much as 5–10%) of water are always needed in excess of the optimum moisture content of the soil. Then, a requisite for success is intimate mixing with a rotary mixer, followed by compaction to a minimum of 95% Proctor density.

Technology has developed for lime to stabilize soil at a considerable depth (20). The U.S. method is called lime slurry pressure injection, with lime slurry being injected *in situ* to 3–6.6 m depth through lances on a grid-pattern but there is no mixing of lime and soil. Both Japan and Sweden have developed techniques using specially constructed machines to inject dry quicklime and then mixing the lime, soil, and water to depths of 11 m or more under critical hydraulic conditions.

Lime–Asphalt Treatment. Hydrated lime has become an important additive for use in hot mix asphalt pavements. The lime is added at a 1–1.5% rate to the asphalt mix, primarily as an antistripping agent to reduce moisture damage. In this application, lime helps provide a more permanent bond between the asphalt and aggregate. In addition, lime also reduces long-term age hardening and low temperature cracking, thereby adding to the durability of the pavement.

Agricultural Uses. *Liming Soils.* Over 95% of limestone's consumption in agriculture is direct application of limestone to the soil (21). Acid soils are neutralized (sweetened) with lime, raising the pH values of the soil to a 6–7 range, the optimum level for most crops (see FERTILIZERS). Liming also provides two important plant nutrients, calcium and magnesium, for the soil as well as beneficial trace elements contained in limestone. Liming counteracts acidity generated by some nitrogen fertilizers such as ammonium nitrate and sulfate, improves soil tilth, and elevates the organic content of the soil.

Rates of applications of lime vary greatly depending on many factors, but range from 0.1 to 0.9 kg/m^2 (0.5 to 4 short ton/acre). Liming is beneficially practiced during all seasons of the year. In evaluation of a liming material, two critical factors are percentage CaO and particle size. Purity signifies a higher concentration of lime and fine particles, ie, 0.14–0.25 mm (60–100 mesh) react with soil acids much faster than coarse particles.

More than 99% of liming in the United States occurs in the eastern half of the country; in the west, soils are alkaline. In 1991, 16.4×10^6 tons of liming material was used in the United States, about half the amount used during the record year of 1976, when federal subsidies for liming were in effect.

Fertilizer Filler. Fertilizer-mixing plants use ground limestone as a filler. Unlike inert fillers, limestone and dolomite provide calcium and magnesium as plant nutrients and acid-neutralizing values to the mix. Dolomitic stone is preferred because it is less reactive than high calcium limestone.

Chemical and Industrial Uses. *Iron and Steel Metallurgy.* As a flux in the refining of metals, limestone is essential in producing pig iron. In the blast furnace, limestone reacts with impurities in the iron ore and fuel, mainly silica and other acid oxide components, creating a blast-furnace slag that is separated from the molten iron. Pulverized fluxstone is contained in a self-fluxing sinter, a concentrated iron ore agglomerate. Owing to improved iron ore beneficiation methods, the limestone to pig iron factor is only about 120 kg/t (240 lb/short ton) (see IRON; STEEL).

In producing steel from pig iron (hot metal) and steel scrap, quicklime is the main flux employed. In the basic oxygen furnace (BOF) which comprises about two-thirds of the steel in the United States, the average factor of lime in a ton of BOF steel is about 60 kg. These furnaces employ a mixture of high calcium and dolomitic pebble quicklime, averaging about 20% dolomitic, the latter being used primarily to extend lining life. In the Q-BOP furnace, pulverized quicklime is charged into the bottom of this oxygen-blown furnace. The other principal lime consuming furnace is the electric furnace which is charged with steel scrap. The basic open-hearth furnaces have been largely phased out; these use more limestone than lime as flux. In the United States, about 30% of total lime is used as steel flux. The percentage is somewhat higher in Japan and most European steel-producing nations.

In Germany and Japan, pulverized quicklime is used in making self-fluxing sinters, partially replacing limestone. Granular dead-burned dolomite is still used to protect the refractory lining of open-hearth and electric furnaces, but not the basic oxygen furnace. Refractory lime has declined with the obsolescence of the open hearth, and is primarily used in making tar bonded refractory brick for the BOFs. Other minor uses of lime by the steel industry include coating pig and slag-casting molds with lime whitewash to prevent sticking; as a lubricant for coating wire in wire drawing; in pickle-liquor and plating-waste treatment, as a neutralizer and precipitant; and in ammonia recovery from coke-oven gases.

Nonferrous Metallurgy. Lime and limestone are required in many strategic nonferrous metallurgical processes (22). All seawater, brine, or bittern processes for magnesium metal and magnesia manufacture require either high calcium or dolomitic quicklime (see MAGNESIUM AND MAGNESIUM ALLOYS). In the Bayer process, lime is used for causticization and desilification in the manufacture of alumina for reduction to aluminum metal (see ALUMINUM AND ALUMINUM ALLOYS). Limestone is also used instead of lime in an alumina process adaptation called the sinter process. The second largest metallurgical use for lime is in the beneficiation of copper ore by flotation (qv), where it is used for neutralization and to maintain proper pH control (see COPPER). Limestone serves as a flux in smelting copper, lead, zinc, ferrosilicon, and antimony from their ores. Lime is the key reagent for recovering uranium from gold slimes in South Africa, and in Canada and the United States lime neutralizes acid wastewater in acid extraction of uranium from its ore. Lime also aids in the recovery of nickel and tungsten by chemical processes after smelting, in the flotation of gold and silver, and in the sintering of low carbon ferrochrome.

In gold and silver production, lime is used to control the pH in both heap and vat leaching processes which utilize sodium cyanide solutions. Lime helps maintain the pH of the cyanide solution between 10 and 11, thereby maximizing gold recovery and preventing the formation of dangerous hydrogen cyanide gas. Because the United States is now one of the world's leading gold producers, this development has spurred lime usage in the western states and also in South Carolina.

Environmental Uses. Next to steel fluxing, environmental uses of lime are the biggest market, accounting for 24% of total lime consumption. These uses include air pollution control, water, sewage, and industrial wastewater treatment, hazardous waste treatment, etc.

Air Pollution Control. Both lime and limestone are increasingly and competitively used for desulfurizing stack gases from utility and industrial plants that operate coal-burning boilers (23). This developing technology has resulted in stack gas scrubbing becoming the second largest market for lime, comprising about 1.75×10^6 t in 1992. The use of high purity limestone for scrubbing has also increased considerably, largely owing to its much lower material cost than lime. However, lime has nearly twice the SO_2 neutralizing capacity and produces less waste because the stoichiometrics of lime use are 100–110% compared to 140–150% for limestone. Lime is also more chemically reactive. Thus capital investment costs for limestone scrubbers are considerably higher than for lime scrubbers. More and larger treatment equipment would be required for the former, as well as the need for larger waste disposal areas.

Most of the stack gas scrubbing is handled by wet scrubbers utilizing slurries containing 10–15% lime solids, and attaining SO_2 efficiencies as high as 99%. Dry scrubbers are also being used in which lime slurry is pumped into a spray dryer. These are used with low sulfur coals, produce a dry waste product, and operate at 70–90% sulfur removal efficiency. Recently, two other techniques have been tried, namely, dry injection of hydrated lime into the top of the boiler, and also into the downstream exhaust ducts. The passage of the Clean Air Act Amendments should lead to a large increase in use of lime for scrubbing by the mid-1990s (see also COAL CONVERSION PROCESSES, CLEANING AND DESULFURIZATION).

Other developing or potential applications for lime are neutralization of tail gas from sulfuric acid plants, neutralization of waste hydrochloric and hydrofluoric acids and of nitrogen oxide (NO_x) gases, scrubbing of stack gases from incinerators (qv), and of course, from small industrial coal-fired boilers.

Water Treatment. Potable water treatment (24) is the third largest tonnage use of lime. Approximately 1.1×10^6 t/yr are used for water softening, purification through coagulation, and high pH retention. In water softening, lime removes the temporary (bicarbonate) hardness, and when soda ash is added, permanent hardness is also removed. In clarification using alum and iron salts as coagulants, lime maintains the optimum pH for removing turbidity from river and lake water sources. Similar water treatment methods are applicable to industrial boiler and process waters and to the recovery of waste-process water for recycling.

In recent years, lime treatment has been advocated for corrosion control by removing lead and copper from distribution systems, mainly by raising the pH to around 7.5, which prevents these heavy metals from solubilizing. This type of treatment is applicable to all water supplies, and especially for small systems. It involves the use of hydrated lime, generally delivered in bags (see WATER).

Industrial Wastewater Treatment. Industrial wastewaters require different treatments depending on their sources. Plating waste contains toxic metals that are precipitated and insolubilized with lime (see ELECTROPLATING). Iron and other heavy metals are also precipitated from waste-pickle liquor, which requires acid neutralization. Akin to pickle liquor is the concentrated sulfuric acid waste, high in iron, that accumulates in smokeless powder ordinance and chemical plants. Lime is also useful in clarifying wastes from textile dyeworks and paper pulp mills and a wide variety of other wastes. Effluents from active and abandoned coal mines also have a high sulfuric acid and iron oxide content because of the presence of pyrite in coal.

For neutralization, both lime and limestone are used, but limestone is only effective in under-neutralization to pH 6–6.5. In complete- or over-neutralization to pH 9–10, necessary for precipitating ferrous iron and other heavy metals, only lime is effective (25).

Sewage Treatment. In the abatement of stream pollution, wastewater from sewage plants must meet stringent standards that increasingly require chemical treatment, usually including lime. Lime treatment precipitates phosphates and most heavy metals. It also aids clarification by coagulating a high percentage of solid and dissolved organic compounds, thereby reducing biological oxygen demand (BOD). When lime raises the pH to 11–12, most bacteria and viruses are destroyed, as well as odor. The high pH also helps to volatilize ammonia, a nitrogen-plant nutrient (26).

Solid-Waste Disposal. Heavy lime treatment of sewage sludge quells obnoxious odors, sterilizes or destroys pathogens, and precipitates toxic heavy metals. Thus sludge is stabilized for safe disposal as landfill or for beneficial disposal on agricultural land. This use of lime in sewage sludge treatment is being spurred on by passage in 1992 of the Federal 503 Sludge Regulation. This law recognizes lime pasteurization (in Class A treatment) in which the sludge is treated to pH $>$ 12 for two hours, at a temperature of 70°C for 30 minutes, this can be accomplished by using quicklime. The sludge can then be used as a safe soil amendment.

Hydrated lime is also used to stabilize the calcium sulfite–sulfate sludge derived from thickeners at SO_2 scrubbing plants that use limestone–lime. Hydrated lime (2–3%) is added to react with the gypsum sludge and flyash or other added siliceous material. Under ambient conditions the lime and silica serve as a binder by reacting as calcium silicates so that the material hardens into a safe, nonleaching, stable, sanitary landfill or embankment fill.

Filter Beds. The sprinkling filter beds of sewage plants are composed of closely screened mineral aggregate of 38–64 mm or 51–76 mm that meet a severe soundness test. Much of this requirement is supplied by dense, fine-grained, hard limestone and dolomite.

Chemicals Manufacture. Lime Manufacture. Limestone is consumed at the rate of 32–34 $\times$ 10^6 t/yr in the manufacture of lime. About 75% of the tonnage is captively produced in the United States. The balance is shipped from quarries in northern Michigan and British Columbia in large ore boats or barges. The limestone:lime ratio is ca 2:1.

Alkalies. In the 1960s, 3.2–34 $\times$ 10^6 t/yr of lime was captively produced by the U.S. alkali industry for manufacturing soda ash and sodium bicarbonate via the Solvay process. Electrolytic process caustic soda and natural soda ash (trona) from Wyoming have largely replaced the Solvay process. Three of the trona producers in Wyoming now purchase quicklime for producing caustic soda.

Calcium Carbide. Until the 1940s, calcium carbide, which is made by interacting quicklime and coke in an electric furnace, was the only source of acetylene. Although much more acetylene is now derived from natural gas, calcium carbide is still being produced, using 0.9–1.0 t of quicklime to make 1 t of carbide (see CARBIDES).

Plastics. The fastest-growing use of whiting (microcarbonate fillers) is in the plastics industry where dry, pulverized limestone is used intensively for most types of plastics. Other carbonate fillers, precipitated calcium carbonate, oyster shell, marble, and wet-ground limestone, are also used.

Miscellaneous. Both whiting and hydrated lime are used as diluents and carriers of pesticides, such as lime–sulfur sprays, Bordeaux, calcium arsenate, etc. The most widely used bleach and sterilizer, high test calcium hypochlorite, is made by interacting lime and chlorine (see BLEACHING AGENTS). Calcium and magnesium salts, such as dicalcium phosphate, magnesium chloride, lithium salts, etc, are made directly from calcitic and dolomitic lime and limestone. Two types of magnesia, caustic-calcined and periclase (a refractory material), are derived from dolomitic lime. Lime is required in refining food-grade salt, citric acid, propylene and ethylene oxides, and ethylene glycol, precipitated calcium carbonate, and organic salts, such as calcium stearate, lactate, caseinate, etc.

Other Industrial Uses. Pulp and Paper. Limestone is the traditional alkaline medium for the sulfite pulp process in preparing the calcium bisulfite pulp-cooking liquor. The main application of quicklime is in the sulfate pulp process for causticizing the waste black liquor to regenerate caustic soda for reuse in digesting the pulp. The sulfate pulp industry captively produces several million tons of lime per year from the waste calcium carbonate sludge resulting from the causticizing reaction, with the commercial lime industry supplying about one million tons per year as make-up lime.

The latest development in paper making has been the switch from the acid to alkaline pulping process, which reduces production costs and results in stronger, longer lasting paper. Although ground limestone is used in the alkaline process, the growth is primarily the result of the installation of economical satellite precipitated calcium carbonate (PCC) plants at the paper mills (27). The PCC serves as both a filler and coating material for making fine white grades of paper. There are about 40 satellite PCC plants in the United States and Canada built since 1986 and producing more than 0.8 t/yr of PCC. This use is now being extended overseas.

Glass Manufacture. Both high calcium limestone and dolomite are high volume batch ingredients for many types of glass (qv), television picture tubes, flat glass (for windows, automotive glass, and mirrors), light bulbs, food and beverage containers, glass tableware, and glass fiber (for reinforcement and insulation). The glass fiber industry generally uses quicklime, whereas dried, double-screened limestone or dolomite is generally used in other glass products. Because magnesium oxide enhances durability and weatherability, dolomite (or dolomite along with limestone) is specified for nearly all glass types except one-use nonreturnable beverage containers. Glass companies require raw materials that are consistent in sizing and chemistry after shipment. They have established very strict limits for both the iron content of the ore itself (typically $< 0.1\%$, but as low as 0.06%), and for contaminating metallic material like stainless steel or aluminum.

Industrial Fillers. Whiting is widely used in paints as a filler and pigment extender. A high reflective white color is a requisite for paint. Both ultrafine and relatively coarse carbonate fillers are incorporated in rubber products. The latter type of a nominal 0.074 mm (200 mesh) size is for inexpensive rubber products.

Coal-Mine Dusting. A steady market for pulverized limestone is in dusting coal mines with a noncombustible mineral dust as a federally mandated mine safety requirement. Of all mineral dusts, limestone is by far the most commonly used (60–75% 0.074 mm (200 mesh)). Consumption in 1991 was 0.7×10^6 t.

Sugar. Lime is an essential reagent in refining sugar (qv). There is a vast difference in the use of lime in beet and cane sugar. In beet sugar, the lime factor is 0.25 t/t of sugar; in cane sugar, it is only 4.5–5.4 kg/t. Because all beet sugar refineries require an abundance of carbon dioxide in the process, invariably every plant has on-site lime kilns for captive lime and CO_2 even though the kilns are operated only in the autumn after the beet harvest for ca 1–2 mo. Most beet-sugar plants purchase their limestone kiln feed.

Petroleum. Apart from its use in petrochemicals manufacture, there are a number of small, scattered uses of lime in petroleum (qv) production. These are in making red lime (drilling) muds, calcium-based lubricating grease, neutralization of organic sulfur compounds and waste acid effluents, water treatment in water flooding (secondary oil recovery), and use of lime and pozzolans for cementing very deep oil wells.

Other. Lime is also used in leather (qv) tanning for dehairing hides, in the manufacture of some paint and ceramic pigments, and for glue and gelatin from packinghouse wastes, in controlled atmospheric storage of fresh fruit as a CO_2 absorbent, in making monocalcium phosphate for baking powder, and as a binder for making refuse-derived fuel pellets.

BIBLIOGRAPHY

"Lime and Limestone" in *ECT* 1st ed., Vol. 8, pp. 346–382, by R. S. Boynton and F. K. Jander, National Lime Association; in *ECT* 2nd ed., Vol. 12, pp. 414–460, by R. S. Boynton, National Lime Association; in *ECT* 3rd ed., Vol. 14, pp. 343–382, by R. S. Boynton, National Lime Association.

1. R. S. Boynton, *Chemistry and Technology of Lime and Limestone,* John Wiley & Sons, Inc., New York, 1966; *Ibid.,* 2nd rev. ed., 1979.
2. A. B. Searle, *Limestone and Its Products,* E. Benn Ltd., London, 1935.
3. N. V. S. Knibbs, *Lime and Magnesia,* E. Benn Ltd., London, 1924.
4. F. Pettijohn, *Sedimentary Rocks,* Harper & Row, New York, 1957.
5. R. T. Haslam and E. C. Hermann, *Ind. Eng. Chem.* **18**, 960 (1926).
6. D. D. Carr and L. F. Rooney, "Limestone and Dolomite," in *Ind. Min. Rocks,* 5th ed., AIME, New York, 1987, pp. 833–868.
7. R. Hedin, *Plasticity of Lime Mortars, Bull. No. 3,* National Lime Association, Washington, D.C., 1963.
8. *Chemical Lime Facts,* 6th ed., National Lime Association, Washington, D.C., 1992.
9. F. Lea, *Chemistry of Cement and Concrete,* 3rd ed., E. Arnold, London, 1971.
10. J. Ames and W. Cutcliffe, "Cement and Cement Raw Materials," in *Ind. Min. Rocks,* 5th ed., AIME, New York, 1987, pp. 133–159.
11. L. J. Minnick, *APCA J.* **21**, 195 (Apr. 1971).
12. P. Rivers-Moore, "Environmental Control of Lime Plants and Quarries in S. Africa," *4th International Lime Congress Proceedings,* Hershey, Pa., 1978.
13. J. A. Murray and co-workers, *J. Am. Ceram. Soc.* **37**, 323 (1954).
14. M. M. Miller, *Lime, 1991*, U.S. Bureau of Mines, Washington, D.C., Nov. 1992.
15. L. Bingham, *TAPPI Proc.,* 265 (1977).
16. V. V. Tepordai, *Crushed Stone, 1991*, U.S. Bureau of Mines, Washington, D.C., Mar. 1993.
17. *Mortar Technical Notes*, bulletins 1–5, National Lime Association, Washington, D.C., 1967.
18. "State of the Art Report 5: Lime Stabilization", *Transp. Res. Bd.*, 1987.
19. E. Barenberg, "Lime-Fly Ash Aggregate Mixtures in Pavement Construction," *National Ash Association Bulletin*, 1972.
20. K. A. Gutschick, in Ref. 12.
21. *100 Questions and Answers on Liming Land,* 6th ed., National Lime Association, Washington, D.C., 1976.
22. C. M. von Staden, in Ref. 12.
23. National Lime Association, *Acid Rain Retrofit Seminar Proceedings,* Philadelphia, Pa., Jan. 1991.
24. M. Riehl, *Water Supply and Treatment,* 11th ed., bull. 211, National Lime Association, Washington, D.C., 1976.
25. C. J. Lewis, "Acid Neutralization with Lime," *National Lime Association Bulletin 216*, Washington, D.C., 1976.
26. R. Bernhoff, *3rd International Lime Congress Proceedings*, Berlin, 1974.
27. L. Gorbaty and co-workers, "Fine Ground and Precipitated Calcium Carbonate," *CEH Product Review*, 1993.

General References

Pit and Quarry Handbook, Pit & Quarry publications, Chicago, Ill., 1982, compendium of quarry and lime plant equipment.

Pit and Quarry Magazine, Cleveland, Ohio, numerous articles on production of limestone and lime.

Rock Products Magazine, Chicago, Ill., numerous articles on production of limestone and lime.

National Stone Association, Washington, D.C., proceedings of conventions on operating problems and discussions, 1946–1993.

National Lime Association, Arlington, Va., proceedings of conventions and operating meetings.

K. A. GUTSCHICK
National Lime Association

LINCOSAMINIDES. See ANTIBIOTICS, LINCOSAMINIDES.

LINEN. See FIBERS, VEGETABLE.

LINOLEIC ACID, LINOLENIC ACID. See CARBOXYLIC ACIDS.

LINSEED. See FIBERS, VEGETABLE.

LINSEED METAL, LINSEED CAKE. See FEEDS AND FEED ADDITIVES.

LINT. See COTTON.

LINTERS. See CELLULOSE; VEGETABLE OILS.

LIPASES. See ENZYME APPLICATIONS, INDUSTRIAL.

LIPIDS. See FATS AND FATTY OILS; VEGETABLE OILS.

LIPOSOMES. See DRUG DELIVERY SYSTEMS.

LIPSTICK. See COSMETICS.

LIQUEFIED PETROLEUM GAS

Liquefied petroleum gas (LPG) is a subcategory of a versatile class of petroleum products known as natural gas liquids (NGLs) that are produced along with and extracted from natural gas (see GAS, NATURAL). LPG is also produced from the refining of crude oil (see PETROLEUM). Although LPG is commercially defined as propane [*74-98-6*], butane [*106-97-8*], and butane–propane mixtures, commercial availability is primarily limited to propane (see HYDROCARBONS). There are two grades of specification propane, propane HD-5 and special-duty propane. The primary difference in the two grades is that the propylene [*115-07-1*] content of propane HD-5 is restricted to a maximum of 5 vol %. Propylene (qv) is found only in refinery-produced propane. The principal uses of LPGs are as fuels and feedstocks (qv) for the production of motor gasoline and a wide variety of chemicals (see FUELS, SYNTHETIC; GASOLINE AND OTHER MOTOR FUELS).

Other natural gas liquids include natural gasoline [*8006-61-9*], which is composed of the pentanes and heavier components of the natural gas stream, and ethane [*74-84-0*]. Most recently ethane has become the principal product of natural gas processing plants.

Properties

In general, LPG specifications involve limits for physical properties. Consequently, the composition of the commercial-grade products varies between wide limits. Physical properties of the principal components of LPG are summarized in Table 1 (1).

Table 1. Physical Properties of LPG Components[a]

Component	CAS Registry Number	Molecular formula	Boiling point, 101.3 kPa,[b] °C	Vapor pressure, 37.8°C, kPa[b]	Liquid density, g/L[c]
ethane	[*74-84-0*]	C_2H_6	−88.6		354.9
propane	[*74-98-6*]	C_3H_8	−42.1	1310	506.0
isobutane	[*75-28-5*]	C_4H_{10}	−11.8	498	561.5
n-butane	[*106-97-8*]	C_4H_{10}	−0.5	356	583.0
1-butene	[*106-98-9*]	C_4H_8	−6.3	435	599.6
cis-2-butene	[*590-18-1*]	C_4H_8	3.7	314	625.4
trans-2-butene	[*624-64-1*]	C_4H_8	0.9	343	608.2
n-pentane	[*109-66-0*]	C_5H_{12}	36.0	107	629.2

[a]Ref. 1.
[b]To convert kPa to psi, multiply by 0.145.
[c]At saturation pressure.

Manufacture and Processing

LPG recovered from natural gas is essentially free of unsaturated hydrocarbons, such as propylene and butylenes (qv). Varying quantities of these olefins may be found in refinery production, and the concentrations are a function of the refinery's process design and operation. Much of the propylene and butylene are removed in the refinery to provide raw materials for plastic and rubber production and to produce high octane gasoline components.

LPG is recovered from natural gas principally by one of four extraction methods: turboexpander, absorption (qv), compression, and adsorption (qv). Selection of the process is dependent on the gas composition and the degree of recovery of ethane and LPG, particularly from large volumes of lean natural gas.

Turboexpander Process. Ethane has become increasingly desirable as a petrochemical feedstock resulting in the construction of many plants that recover the ethane from natural gas at −73 to −93°C. Combinations of external refrigeration and liquid flash-expansion refrigeration with gas turboexpansion cycles are employed to attain the low temperatures desired for high ethane recovery. Figure 1 is a flow diagram of a one-expander cycle having external refrigeration (see REFRIGERATION AND REFRIGERANTS).

Dry inlet gas that has been dehydrated by molecular sieves (qv) or alumina beds to less than 0.1 ppm water is split into two streams by a three-way control valve. Approximately 60% of the inlet gas is cooled by heat exchange with the low pressure residue gas from the demethanizer and by external refrigeration. The remainder of the inlet gas is cooled by heat exchange with the demethanized bottoms product, the reboiler, and the side heater. A significant amount of low level refrigeration from the demethanizer liquids and the cold residue gas stream is recovered in the inlet gas stream.

The two portions of the feed stream recombine and flow into the high pressure separator where the liquid is separated from the vapor and is fed into an intermediate section of the demethanizer with liquid level control. The

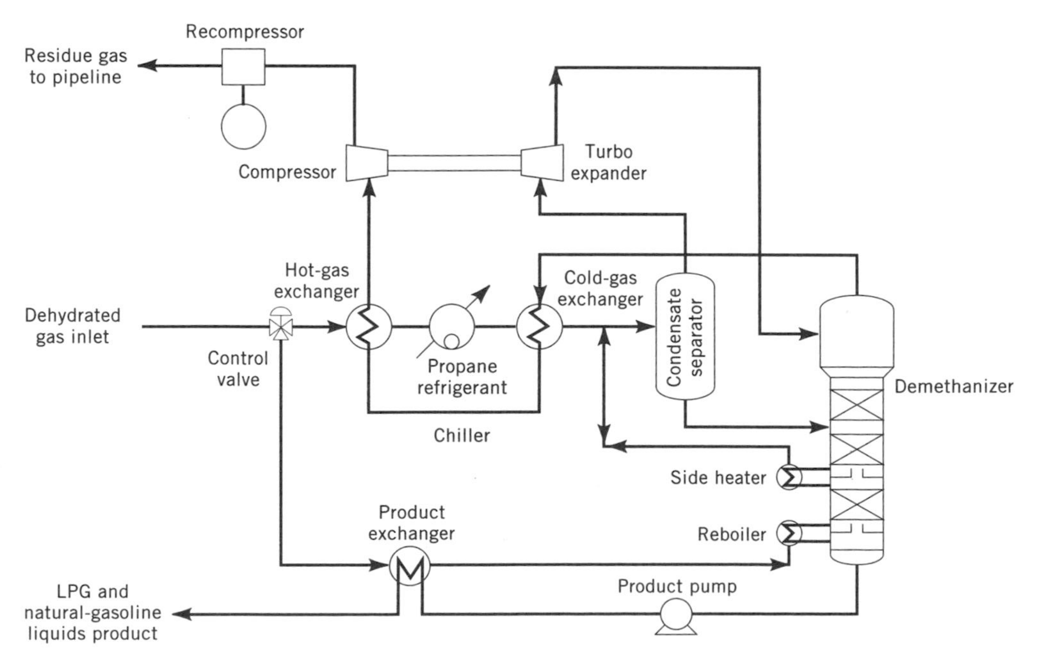

Fig. 1. One-expand cycle with external refrigeration for high ethane recovery in the hydrocarbon liquid product.

decrease in pressure across the level-control valve causes some of the liquid to flash which results in a decrease in the stream temperature. The pressure of the vapor stream is decreased by the way of a turboexpander to recover power, thus to achieve more cooling than would be possible by Joule-Thompson expansion. The outlet of the turboexpander then is fed into the top of the demethanizer where the separation of liquid and vapor occurs. The vapor is passed as cold residue to the heat exchanger and the liquid is distributed to the demethanizer top tray as reflux.

Essentially all of the methane [*74-82-8*] is removed in the demethanizer overhead gas product. High recovery of ethane and heavier components as demethanizer bottoms products is commonplace. The work that is generated by expanding the gas in the turboexpander is utilized to compress the residue gas from the demethanizer after it is warmed by heat exchange with the inlet gas. Recompression and delivery to a natural gas pipeline is performed downstream of the plant. A propane recovery of 99% can be expected when ethane recoveries are in excess of 65%.

Recoveries of 90–95% ethane have been achieved using the expander processes. The liquid product from the demethanizer may contain 50 liquid vol % ethane and usually is delivered by a pipeline to a central fractionation facility for separation into LPG products, chemical feedstocks, and gasoline-blending stocks.

Absorption. Oil absorption is another process used for recovery of LPG and natural gas liquids from natural gas. Recovery is enhanced by lowering the absorption temperature to −45°C and by keeping the molecular weight of the absorption oil down to 100. Heat used to separate the product from the absorption oil contributes to the cost of recovery. Therefore, this process has become less competitive as the cost of energy has increased. A simplified flow diagram of a typical oil-absorption process is shown in Figure 2.

The natural gas feed exchanges heat with the residue gas from the absorber overhead. Ethylene glycol [*107-21-1*] is injected as an antifreeze and the stream is cooled further by refrigeration to −37°C. The gas from the cooler enters a glycol separator where the glycol that contains water is separated from the natural gas as a liquid phase. A dry glycol is recovered for recycling to the injection point by distilling the water. The gas and any higher boiling hydrocarbons that are present pass to the base of the absorber where the gas comes into contact with absorption oil which enters at the top of the absorber at −37°C. Approximately 85% of the propane and essentially all of the higher boiling hydrocarbons are absorbed in the oil. The overhead residue gas from the absorber (−34°C) is heat exchanged with the inlet gas; at 4°C and 4.24 MPa (600 psig), the gas flows to the booster where the pressure is increased to that of the natural gas line.

The rich oil from the absorber is expanded through a hydraulic turbine for power recovery. The fluid from the turbine is flashed in the rich-oil flash tank to 2.1 MPa (300 psi) and −32°C. The flash vapor is compressed until it equals the inlet pressure before it is recycled to the inlet. The oil phase from the flash passes through another heat exchanger and to the rich-oil deethanizer. The ethane-rich overhead gas produced from the deethanizer is compressed and used for producing petrochemicals or is added to the residue-gas stream.

The bottoms, consisting of absorption oil, absorbed propane, and higher boiling hydrocarbons, are fed to the lean-oil fractionator. The LPG and the

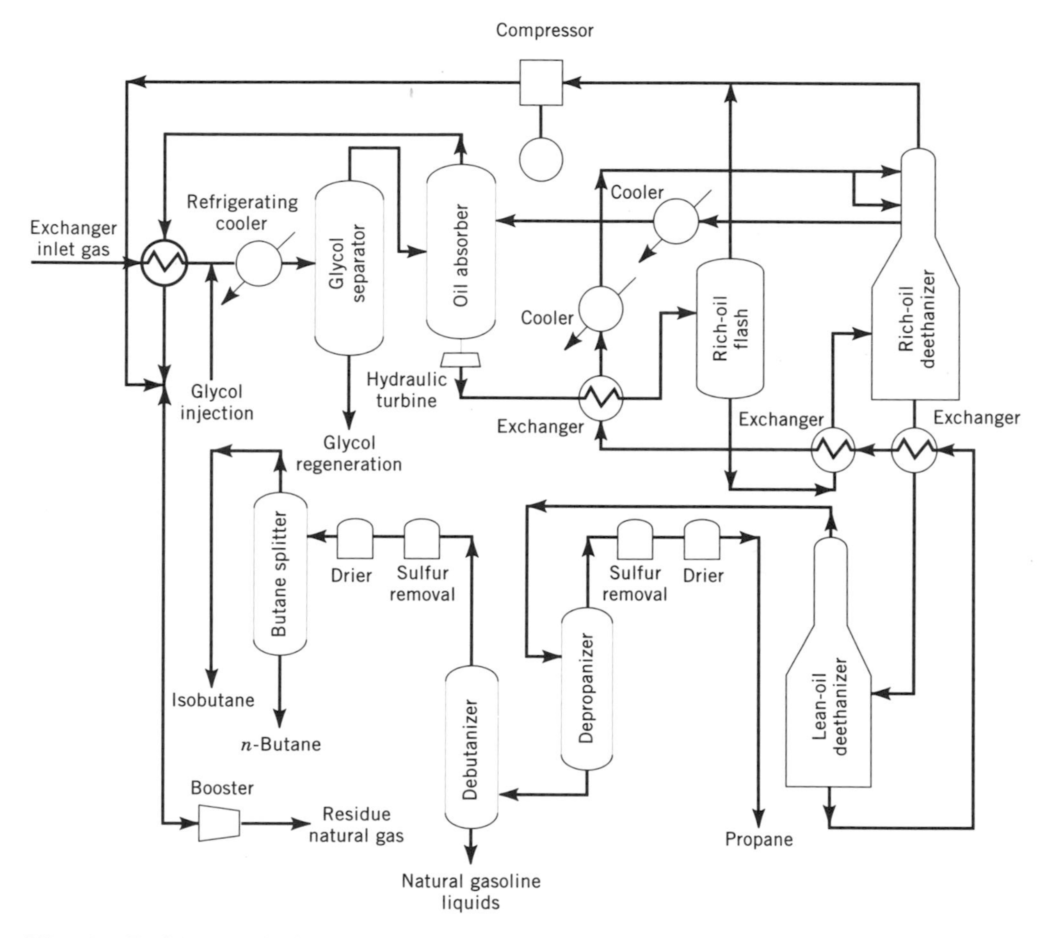

Fig. 2. Refrigerated absorption process for the production of LPG and natural gasoline liquids.

natural gas liquids are removed as the overhead product from the absorption oil which is removed as a kettle-bottom product.

The lean oil from the lean-oil fractionator passes through several heat exchangers and then through a refrigerator where the temperature is lowered to −37°C. Part of the lean oil is used as a reflux to the lower section of the rich-oil deethanizer. Most of the lean oil is presaturated in the top section of the deethanizer, is cooled again to −37°C, and is returned to the top of the absorber, thus completing the oil cycle.

The overhead product from the lean-oil fractionator, consisting of propane and heavier hydrocarbons, enters the depropanizer. The depropanizer overhead product is treated to remove sulfur and water to provide specification propane. The depropanizer bottoms, containing butane and higher boiling hydrocarbons, enters the debutanizer. Natural gasoline is produced as a bottom product from the debutanizer. The debutanizer overhead product is mixed butanes, which are treated for removal of sulfur and water, then fed into the butane splitter.

Isobutane is produced as an overhead product from the splitter and n-butane is produced as a bottoms product.

Compression. Compression is the simplest and the least effective of the four recovery methods. It was the first process used for the recovery of hydrocarbon liquids from natural gas but is used only in isolated cases. The most significant application of the compression process is for gas-cycling plants where the natural gas liquids are removed and the remaining gas is returned to the production formation. Figure 3 is a schematic of a typical gas-cycle plant.

The pressure used in producing gas wells often ranges from 690–10,300 kPa (100–1500 psi). The temperature of the inlet gas is reduced by heat-exchange cooling with the gas after the expansion. As a result of the cooling, a liquid phase of natural gas liquids that contains some of the LPG components is formed. The liquid is passed to a set of simple distillation columns in which the most volatile components are removed overhead and the residue is natural gasoline. The gas phase from the condensate flash tank is compressed and recycled to the gas producing formation.

Condensable liquids also are recovered from high pressure gas reservoirs by retrograde condensation. In this process, the high pressure fluid from the reservoir produces a liquid phase on isothermal expansion. As the pressure decreases isothermally the quantity of the liquid phase increases to a maximum and then decreases to disappearance. In the production of natural gas liquids from these high pressure wells, the well fluids are expanded to produce the optimum amount of liquid. The liquid phase then is separated from the gas for

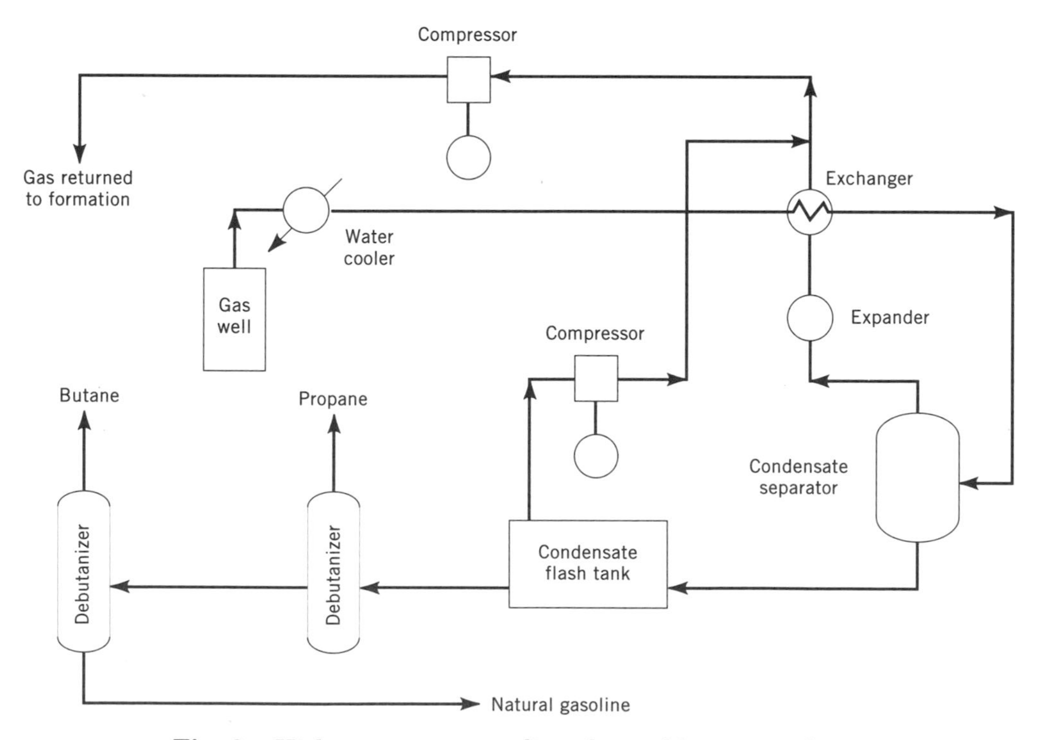

Fig. 3. High pressure gas-cycling plant with compression.

further processing. The gas phase is used as a raw material for one of the other recovery processes, as fuel, or is recompressed and returned to the formation.

Adsorption. Adsorption processes have been used to recover hydrocarbons that are heavier than ethane from natural gas. Although the adsorption process has applications for the recovery of pentane and heavier hydrocarbons from lean gas, the percentage recovery of LPG components in these plants usually is low compared to the normal recovery of LPG in modern turboexpander or oil-absorption plants.

A simplified flow diagram for the fast-cycle adsorption plant is shown in Figure 4. Activated carbon, alumina gel, and silica gel are used as adsorbents. Use of internal insulation in the adsorption towers affords less cycle time. The complete process cycle consists of three phases: regeneration, cooling, and adsorption. In Figure 4, the inlet gas is divided. The larger portion flows directly to the bed in the adsorption cycle; the smaller portion flows first through the bed in the cooling phase of the cycle, then through the regeneration cycle. The effluent gas from the cooling cycle is heated further to 260–315°C in a separate fired furnace before it is injected into the bed in the regeneration cycle. The effluent from the regeneration cycle contains the condensable hydrocarbons that have been stripped from the adsorbent and that are removed in the high pressure separator after the gas that is leaving the regeneration cycle is cooled. The liquid from the high pressure separator is flashed through to remove light hydrocarbons. The flash vapor is compressed and is mixed with the vapor from the high pressure separator. The pressure on the compressed gas stream is boosted to the inlet gas pressure and the gas is recycled to the feed to the adsorption cycle. The liquid from the low pressure flash, which contains LPG and natural gas liquid components, is processed further in a series of distillation towers to produce propane, butanes, and natural gasoline.

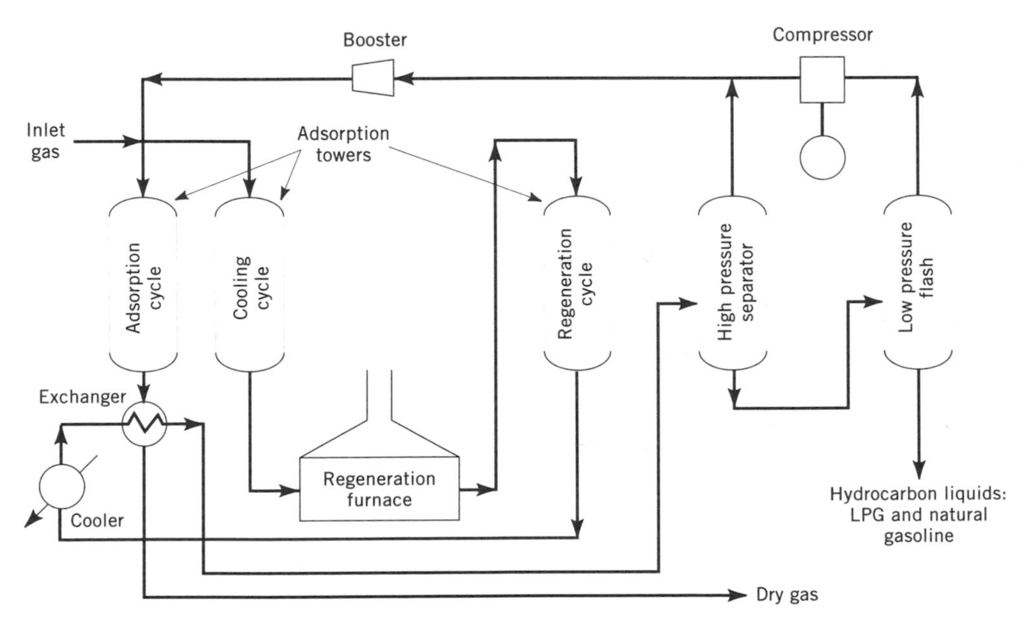

Fig. 4. A fast-cycle absorption unit for recovery of hydrocarbon liquids from natural gas.

Less propane and butanes are produced compared to natural gas liquids by the adsorption process than are obtained normally for the same gas by the oil-absorption process. Because adsorption efficiency increases with a decrease in temperature, the adsorption cycle should operate at the lowest temperature that is economically feasible.

Purification. The LPG generally requires treatment for removal of hydrogen sulfide [*7783-06-4*], H_2S, organic sulfur compounds, and water in order to meet specifications. Several methods are used.

Amine Treatment. The LPG is brought into contact with a 15–20 wt % solution of ethanolamine [*141-43-5*] in water, which removes H_2S to specification levels. The foul amine solution that contains the H_2S is regenerated in a stripper at low pressure using indirect stream stripping, and the stripped amine solution is returned to the LPG contactor. The amine solution generally has little affect on organic sulfur compounds, although diethanolamine [*111-42-2*] (DEA) and diglycolamine [*929-06-6*] (DGA) have been used to obtain acceptable carbonyl sulfide [*463-58-1*], COS, levels in propane.

Caustic Treatment. Amine treatment may be followed by a caustic treatment step in which the LPG is brought into contact with 10–20 wt % caustic solution to remove any residual H_2S and to remove mercaptans. The mercaptans may be stream-stripped from the caustic solution, after which the stripped caustic is recycled to the LPG caustic contactor. Caustics that contain H_2S must be discarded. Various promoters may be added to the caustic to improve the efficiency of the mercaptan removal.

Coalescing. Sand towers or cartridge-type coalescers may be used to separate any undissolved water from the LPG. Removal of the undissolved water meets the specification moisture limit for butanes. However, this step does not produce specification propane.

Solid-Bed Dehydration. Silica gel, bauxite, activated alumina, or molecular sieves can be used for removing dissolved water to meet propane specifications. The solid-bed dehydrators are used in a cyclic adsorption process. After an adsorption cycle has completed, the bed is heated with a purge gas or a vaporized liquid-product stream for regeneration. If the latter is used, the liquid product is condensed, separated from the free water, and returned to the process. After the beds are regenerated, they are cooled and returned to the adsorption cycle.

Molecular Sieve Treatment. Molecular sieve treaters can be designed to remove H_2S, organic sulfur compounds (including carbonyl sulfide), and water in one step. Solid-bed units are utilized and regeneration occurs in the same manner as simple, solid-bed dehydrators.

Solid-Bed Caustic Treatment. Solid-bed caustic units utilizing methanol [*67-56-1*] injection into the LPG feed stream can be used for carbonyl sulfide removal. The methanol–caustic solution must be drained periodically from the beds and discarded. When the solid bed is exhausted, the spent caustic must be discarded and replaced. The LPG from the treater has a low enough water content to meet the propane specification.

Fractionation. Direct fractionation also can be used to remove dissolved water from LPG. The water-rich overhead vapor from the dryer fractionator is returned to the fractionator as reflux and the water phase is discarded. A dry LPG product that meets either propane or butane water specifications is produced as a kettle product from the fractionator.

Production and Shipment

Historically, about two-thirds of the LPG produced in the United States came from natural gas processing and one-third was produced from refinery operations (2). In 1991, this ratio was 61% from natural gas processing and 39% from refinery operations. Total production of LPG in 1991 was 76.85×10^6 m^3 (294.19×10^6 bbl) from natural gas processing and 30.08×10^6 m^3 (189.23×10^6 bbl) produced from refinery operations.

Ethane production for 1991 was 30.74×10^6 m^3 (193.32×10^6 bbl) from natural gas operations and 1.49×10^6 m^3 (9.34×10^6 bbl) from refinery operations for a total production of 32.21×10^6 m^3 (202.66×10^6 bbl). A summary of total natural gas liquids (NGL) including LPG supply and demand in the United States is shown in Table 2.

The progress of LPG utilization has been closely related to progress in transportation (qv) and storage of this fuel. Large volumes of LPG must be transported from the producing plants to centers of consumption, and transportation costs are a principal factor in the cost of LPG to the consumer. Large volumes of LPG usually are transported by high pressure pipelines (qv). As of this writing, this use of pipelines is increasing rapidly. Large quantities of LPG are transported in railroad tank cars which have an average capacity of 113.5 m^3/car (3×10^4 gal/car), although this use is decreasing. Tank-truck transports having

Table 2. 1991 U.S. NGL Supply and Demand, 10^6 m^3[a]

Places	Ethane	Propane	*n*-Butane	*i*-Butane	Pentanes[b]	Total
			Supply			
gas plants	30.74	28.28	8.72	9.78	18.79	96.30
refineries	1.49	25.27	4.56	0.25		31.57
imports	0.48	5.29	2.04	0.73	1.31	9.85
from (to) stocks[c]	0.35	−0.27	0.06	0.28	−0.18	0.23
Total	*33.05*	*58.57*	*15.37*	*11.04*	*19.92*	*137.95*
			Demand			
chemicals	32.09	24.36	5.23		4.81	66.49
RES and COM[d]		20.16	0.00			20.16
engine fuel		2.05	0.00			2.06
industrial	0.05	4.61	1.70			6.36
utility	0.61	0.26	0.02			0.89
gasoline			5.53	11.04	12.93	29.50
farm		4.29	0.00			4.29
export	0.01	1.61	0.80		0.03	2.45
other	0.30	1.23	2.08		2.15	5.76
Total	*33.06*	*58.57*	*15.37*	*11.04*	*19.91*	*137.95*

[a]To convert m^3 to bbl, divide by 0.159.
[b]Includes C_x compounds where $x \geq 5$.
[c]Negative sign indicates supply to stocks (inventory) in order to balance supply/demand.
[d]RES and COM = research and commercial.

capacities of about 17.8 m^3 (10^4 gal) also move large quantities of product from producing plants to distribution centers and from the pipeline terminals to points of distribution or consumption. Delivery trucks having capacities from 3.8–11.4 m^3 (1000–3000 gal) generally are used for the final delivery from the distribution bulk plant to the storage plants of the larger consumers. Smaller quantities are shipped in metal cylinders having capacities from 9–45 kg LPG. These cylinders are filled at the distribution bulk plant and are delivered by truck to the consumer or to the cylinder dealer or cylinders may be filled for the customer at small distribution stations.

Tankers and barges are also used for transporting LPG. Tankers and barges are designed for both high pressure ambient temperature and for low pressure refrigerated transportation. For larger volumes and long distances, low pressure refrigerated tankers almost always are used (2).

Ethane usually is transported in high pressure pipelines from the point of production to the point of consumption. However, for small quantities of ethane, multitube trunk trailers are used.

Economic Aspects

The production and consumption of LPG in the United States increased dramatically from its early beginnings in the 1930s until the international energy crises of the 1970s when rising prices and regulatory restraints resulted in reduced domestic production. However, total consumption, including imports, resumed a modest growth characteristic after that time. In 1984, total LPG consumption in the United States was 76.58×10^6 m^3 (481.71×10^6 bbl); by 1991, total consumption was 84.98×10^6 m^3 (534.50×10^6 bbl). The principal growth segment has been the increasing use of LPG for petrochemical feedstocks as can be seen from Table 3.

Specifications and Standards

Specifications for the principal LPG products are summarized in Table 4. Detailed specifications and test methods for LPG are published by the Gas Processor's Association (GPA) (3) and ASTM (4). The ASTM specification for special-duty propane and GPA specification for propane HD-5 apply to propane that is intended primarily for engine fuel. Because most domestic U.S. LPG is handled through copper tubing, which could fail if corroded, all products must pass the copper strip corrosion test. A test value of No. 1 represents a LPG noncorrosive to the copper.

Storage

Large volumes of LPG are stored to meet peak demand during cold seasons. LPGs are both volatile and flammable and must be stored and handled in special equipment. Standards for storing and handling LPG are published by the National Fire Protection Association (5) and API (6).

Four main types of storage are used: high pressure storage above ground, low pressure refrigerated storage above ground, frozen earth storage, and

Table 3. Sales of LPG 1984–1991, 10^6 m^3 [a,b]

Year	Chemical	Residential and commercial heating	Engine fuel	Industrial	Utility gas	Motor fuel blending	Farm	Export	Other	Total
1984	47.49	15.05	2.80	10.06	1.45	35.71	3.77	2.93	11.82	2,115.08
1985	51.57	16.97	2.78	7.66	0.91	37.09	4.38	3.72	8.90	2,118.98
1986	48.72	16.62	2.48	8.23	2.73	36.36	4.29	2.59	2.90	2,110.92
1987	54.45	18.39	2.39	8.35	0.54	37.69	4.08	2.33	2.96	2,118.17
1988	60.24	18.25	2.21	8.19	1.23	35.15	4.03	2.96	3.97	2,124.23
1989	55.52	20.41	2.20	9.90	1.14	33.26	4.44	2.37	6.74	2,123.99
1990	57.57	18.84	2.01	5.80	1.06	32.71	4.30	2.16	3.78	2,118.23
1991	66.49	20.16	2.06	6.36	0.89	29.50	4.29	2.45	5.76	2,128.95

[a] Ref. 2.

[b] To convert m^3 to bbl, divide by 0.159.

Table 4. Commercial Liquefied Petroleum Gas Specifications[a,b]

Parameter	Propane	Butane	Butane–propane mixture	Propane HD-5
composition	predominantly propane and/or propylene	predominantly butane and/or butylenes	butanes and/or butylenes plus propane and/or propylene	≥ 90 liquid vol % propane, ≤ 5 liquid vol % propylene
vapor pressure at 37.78°C, kPa[c]	1434	482.6	1434	1434
temperature at 95% evaporation,[d] °C	−38.3	2.2	2.2	−38.3
total sulfur, ppmw	185	140	140	123
moisture	pass			pass
free water		none	none	

[a]Refs. 3 and 4.
[b]Corrosion copper strip test, maximum = No. 1.
[c]To convert kPa to psi, multiply by 0.145.
[d]Values are maximum.

underground cavern storage. The capacities of the storage unit vary from 500 mL pressure cylinders to 1.9×10^6 m^3 (500×10^6 gal) underground storage caverns. Economic factors determine the proper storage for any given requirement.

Above ground pressure-storage tanks usually are designed for a 1720 kPa (250 psi) working pressure for propane and 860 kPa (125 psi) for butane. Refrigerated, aboveground storage tanks usually are designed for a few kilopascals of pressure. These tanks must be coupled with refrigeration systems to cool the product that is to be stored to a temperature equal to the product's boiling point at the operating pressure of the tanks. Vapors generally are recondensed by refrigeration and returned to the tanks.

In frozen earth storage of propane, the walls and bottom of a pit in the ground are frozen and a dome is constructed over the pit. The pressure in the storage cavern is maintained at nearly atmospheric pressure by refrigeration systems that cool the product to its boiling point at storage pressure. Heat leaks into the cavity and vaporizes some of the propane. The vapor that is formed is compressed, cooled, and returned to the pit as a liquid by the refrigeration system. Because this storage must operate at temperatures considerably below the freezing point of wet earth and at atmospheric pressure, it cannot be used for butane storage.

Underground storage caverns, which operate at approximately formation temperatures and at the corresponding LPG vapor pressure, may be either mined underground storage caverns or cavities that have been produced in a salt formation by solution mining. The underground caverns must be of sufficient depth to develop an overburden pressure greater than the vapor pressure of the stored liquid. Mined storage caverns are 60–152 m deep, whereas salt formation caverns may be from 106–1524 m deep. Underground as compared to aboveground storage is much more economical for storage of large volumes, ie, more than 2785 m^3 (10^7 gal) of LPG. A washed-out salt cavern costs only 10–50% as much to develop as typical mined cavern storage. In 1987, underground storage capacity for LPG in the United States was 79×10^6 m^3 (493×10^6 bbl) (2) and 90% of this was in salt formations.

Uses

About 35% of total U.S. LPG consumption is as chemical feedstock for petrochemicals and polymer intermediates. The manufacture of polyethylene, polypropylene, and poly(vinyl chloride) requires huge volumes of ethylene (qv) and propylene which, in the United States, are produced by thermal cracking/dehydrogenation of propane, butane, and ethane (see OLEFIN POLYMERS; VINYL POLYMERS).

Residential and commercial fuel demands represent about 24% of total U.S. LPG consumption. Although this market demand is weather dependent, it has assumed the characteristics of a mature market. Growth is related to the general economic trends.

Nearly two-thirds of total butane supply, about 20% of total LPG, is consumed in the manufacture of motor gasoline. However, the environmental mandates of the early 1990s have had a negative impact on this market segment. These mandates have reduced gasoline volatility requirements, effectively reducing the value of the butanes as blending stocks. However, normal butane can be used as feedstock for production of isobutylene, a key ingredient of ether blendstocks, such as methyl *tert*-butyl ether [*1634-04-4*] (MTBE) for motor gasoline. Shifts in U.S. use patterns can be seen in Table 3.

BIBLIOGRAPHY

"Liquefied Petroleum Gas" in *ECT* 2nd ed., Vol. 12, pp. 470–480, by L. Pollack, Phillips Petroleum Co.; in *ECT* 3rd ed., Vol. 14, pp. 383–394, by F. E. Selim, Phillips Petroleum Co.

1. *Engineering Data Book,* 10th ed., Gas Processors Supplier's Association, Tulsa, Okla., 1987.
2. *LP-Gas Market Facts*, National LP-Gas Association, Oak Brook, Ill., 1977.
3. *Liquefied Petroleum Gas Specifications and Test Methods*, Gas Processors Association, GPA Publication 2140-92, Tulsa, Okla.
4. *ASTM Standard D1835-91*, American Society for Testing and Materials, Philadelphia, Pa., 1992.
5. *Storage and Handling of Liquefied Petroleum Gases*, National Fire Protection Association, NFPA 58, Boston, Mass., 1989.
6. *Design and Construction of LP-Gas Installations at Marine Terminals, Natural Gas Processing Plants, Refineries, Petrochemical Plants, and Tank Farms*, API Standard 2510, 4th ed., American Petroleum Institute, Washington, D.C., Dec. 1978.

General References

Petroleum Products Handbook, McGraw-Hill Book Co., Inc., New York, 1960.

C. C. McKee, "The Supply/Demand Outlook for LP-Gas," *Proceedings of the 58th Annual Convention 1979,* Gas Processors Association, Tulsa, Okla.

Magic Formula, LP-Gas, Duluth, Minn., Jan. 1980.

Stayin' Alive, LP-Gas, Duluth, Minn., Jan. 1980.

R. RAY TAYLOR
Phillips Petroleum Company

LIQUID CRYSTALLINE MATERIALS

Liquid crystals represent a state of matter with physical properties normally associated with both solids and liquids. Liquid crystals are fluid in that the molecules are free to diffuse about, endowing the substance with the flow properties of a fluid. As the molecules diffuse, however, a small degree of long-range orientational and sometimes positional order is maintained, causing the substance to be anisotropic as is typical of solids. Therefore, liquid crystals are anisotropic fluids and thus a fourth phase of matter. There are many liquid crystal phases, each exhibiting different forms of orientational and positional order, but in most cases these phases are thermodynamically stable for temperature ranges between the solid and isotropic liquid phases. Liquid crystallinity is also referred to as mesomorphism.

Many thousands of organic substances, some rigid-rod polymers, and other macromolecules exhibit liquid crystallinity. The general common molecular feature is either an elongated or flattened, somewhat inflexible molecular framework, which is usually depicted as either a cigar- or disk-shaped entity. Certain macromolecules and some amphiphilic molecules, containing both hydrophilic and oliophilic moieties, adopt liquid crystalline structures in solution. The orientational and positional order in a liquid crystal phase is only partial, with the intermolecular forces striking a very delicate balance involving both attractive and repulsive interactions. As a result, liquid crystals are extraordinarily sensitive to external perturbations, eg, temperature, pressure, electric and magnetic fields, shearing stress, or foreign vapors. For this reason, liquid crystals are used to design practical devices to either monitor ambient changes of various kinds or to transduce an environmental fluctuation into a useful electrical or optical output.

Besides being used in the scientific study of cooperative phenomena and complex fluid phases, liquid crystalline phenomena have received a good deal of attention due to the possibility of practical applications. Liquid crystals are widely used in electrooptic displays, eg, digital watches, calculators, portable computers, televisions, and electronic instrumentation. Other applications include radiation and pressure sensors, optical switches and shutters, and thermography. The liquid crystalline structures formed by amphiphilic molecules form the basis for emulsions and are studied thoroughly by researchers in the food, drug, and oil industries. Polymers that form an anisotropic fluid phase are important in the fabrication of lightweight, ultrahigh strength, and temperature-resistant fibers, and are beginning to be used in electrooptic displays. Liquid crystals also appear to play an important role in the structure and biochemical function of living tissue, where the characteristic combination of order and flow mobility is particularly suited to life processes. Certain disease states, eg, atherosclerosis, sickle cell anemia, or cancer, may be associated with physical changes in the liquid crystalline order within biological structures.

Orientational and Positional Order in Fluids

Conventionally, matter exists in one of three distinct states of aggregation: the solid state, where constituent molecules or atoms execute small vibrations about firmly fixed lattice positions but cannot rotate or translate; the liquid state, characterized by hindered rotation and translation but no long-range order; and the gaseous state, where particles move freely through the entire volume of the container, with almost no constraint to rotation or translation. The melting of normal solids involves the abrupt collapse of the overall positional and orientational order of the lattice and marks the onset of hindered rotation and translation of the molecules. Short-range correlations of the position and orientation of molecules in the liquid phase are all that remain of the long-range order of the solid phase.

Solids of mesogenic (liquid crystal forming) molecules melt to form fluids in which some of the long-range molecular order is retained. At the simplest level, the elongation or flattening of the mesogenic molecules prevents the immediate dissolution of the parent, solid-state order. Certain intermolecular attractions, enhanced in these elongated or flattened molecules, are operative, but geometric effects in the packing of these nonspherical molecules play just as important, if not more important, a role in this behavior. The loss of positional order of the centers of mass of the molecules in the parent solid may be either partial or complete upon melting, but some degree of orientational order is always retained. Thus long-range correlations of the positions of the centers of mass of the molecules may or may not be present in the melted solid, but long-range correlations of the orientations of one or more of the molecular axes always exist. The fluid retains many solid-like properties, which are finally eliminated when the substance passes into the normal, isotropic liquid phase at a higher temperature (a second melting point). Solid-like features return if the substance is cooled from the isotropic state; this intermediate state is usually thermodynamically reversible, but in some cases it only forms upon cooling. Partial dissolution of solid-state order also may occur in certain substances by the use of solvents. In this case the molecules either orientationally order in the solvent (some macromolecules), or form aggregates, in which the molecules exhibit long-range positional and/or orientational order. Liquid crystals that are established solely by the adjustment of temperature are referred to as thermotropics, whereas those that form through the addition of a solvent are called lyotropics. The residual order in lyotropics is also dependent on temperature, and usually can be broken down completely at high enough temperature. One basic feature of lyotropic liquid crystals is that they are always multicomponent systems.

Orientational Distribution Function and Order Parameter. In a liquid crystal a snapshot of the molecules at any one time reveals that they are not randomly oriented. As shown in Figure 1, there is a preferred direction for alignment of the long molecular axes. This preferred direction is called the director, and it can be used to define an orientational distribution function, $f(\theta)$, where $f(\theta)\sin\theta d\theta$ is proportional to the fraction of molecules with their long axes within the solid angle $\sin\theta d\theta$. Figure 2 illustrates how such an orientational distribution function differs from one phase to another. In a crystalline solid where the molecules are constrained to point in a certain crystallographic direction, $f(\theta)$ is a highly peaked function about $\theta = 0$. In an isotropic liquid, the molecules point in random direction, so $f(\theta)$ is a constant function. In a liquid crystal, $f(\theta)$ is peaked around $\theta = 0$, but the peak is quite broad.

It is useful to describe the amount of orientational order with a single quantity. Since orientational order is related to the projection of the long molecular axes on the director, averaging $\cos\theta$ over all the molecules might seem to be a solution. However, the average of $\cos\theta$ for all molecules in the sample is zero, because there are as many molecules with long axes pointing in the up direction as in the down direction. In fact, the director can be defined as pointing either up or down without changing the situation. On the other hand, the average of

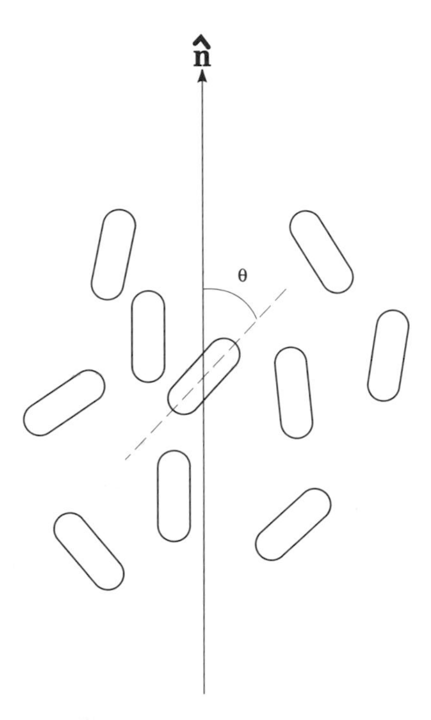

Fig. 1. Orientational order of the molecules in a liquid crystal. θ is the angle between the long axis of a molecule and the direction of preferred orientation (director), $\hat{\mathbf{n}}$.

$\cos^2\theta$ is not zero in a liquid crystal, so it can be used to describe the amount of orientational order. For convenience, the average of the second Legendre polynomial is used to define an order parameter S, where the brackets indicate the average of the enclosed quantity.

$$S = \langle(3\cos^2\theta - 1)/2\rangle \tag{1}$$

$$\langle(\ldots)\rangle = \int (\ldots)f(\theta)\sin\theta d\theta \Big/ \int f(\theta)\sin\theta d\theta \tag{2}$$

If all the molecules are perfectly parallel, S would equal 1. In an isotropic liquid, $f(\theta)$ is constant so that $\langle\cos^2\theta\rangle$ equals 1/3 and S is therefore zero. The order parameter for liquid crystals falls somewhere between these limits and decreases somewhat with increasing temperature.

X-ray, uv, optical, ir, and magnetic resonance techniques are used to measure the order parameter in liquid crystals. Values of S for a typical liquid crystal are shown in Figure 3. The compound, *p*-methoxybenzylidene-*p*′-*n*-butylaniline (MBBA) is mesomorphic around room temperature. The order parameter ranges from 0.7 to 0.3 and discontinuously falls to zero at T_c, which is sometimes called the clearing temperature (1).

Positional Distribution Function and Order Parameter. In addition to orientational order, some liquid crystals possess positional order in that a snapshot at any time reveals that there are parallel planes which possess a higher density of molecular centers than the spaces between these planes. If the normal to

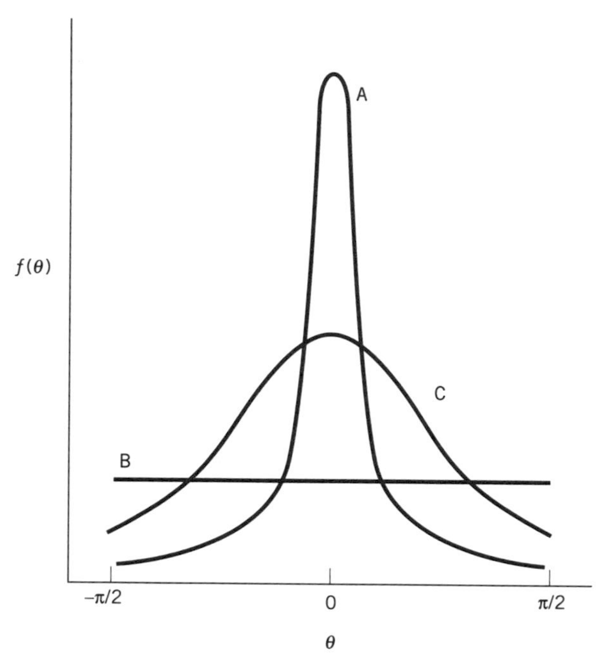

Fig. 2. Schematic representation of the orientational distribution function $f(\theta)$ for three classes of condensed media that are composed of elongated molecules: A, solid phase, where $f(\theta)$ is highly peaked about an angle (here, $\theta = 0°$) which is restricted by the lattice; B, isotropic fluid, where all orientations are equally probable; and C, liquid crystal, where orientational order of the solid has not melted completely.

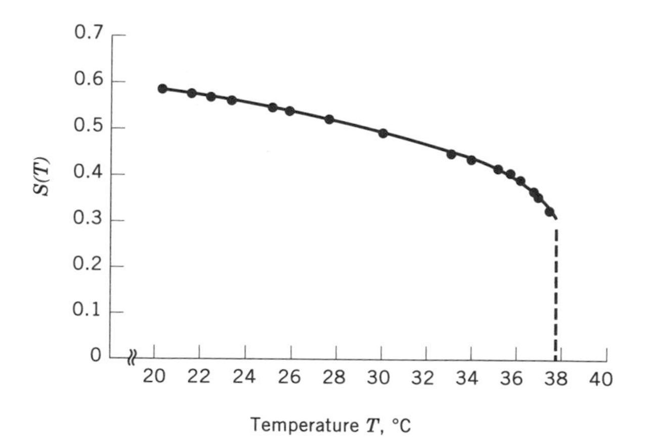

Fig. 3. Order parameter as a function of temperature for *p*-methoxybenzylidene-*p′*-*n*-butylaniline (MBBA), a room temperature nematic liquid crystal. $S(T)$ is determined from the polarization of the absorption (dichroism) of small quantities of a dye molecule of similar structure (*trans*-dimethylaminonitrostilbene) which has been dissolved in the liquid crystal host (1).

these planes is defined as the z-axis, then a positional distribution function, $g(z)$, can be defined, where $g(z)dz$ is proportional to the fraction of molecular centers between z and $z + dz$. Since $g(z)$ is periodic, it can be represented as a Fourier series (a sum of a sinusoidal function with a periodicity equal to the distance between the planes and its harmonics). To represent the amount of positional order, the coefficient in front of the fundamental term is used as the order parameter. The more the molecules tend to form layers, the greater the coefficient in front of the fundamental sinusoidal term and the greater the order parameter for positional order.

There are transition temperatures in some liquid crystals where the positional order disappears but the orientational order remains (with increasing temperature). The positional order parameter becomes zero at this temperature, but unlike S, this can either be a discontinuous drop to zero at this temperature or a continuous decrease of the order parameter which reaches zero at this temperature.

In some liquid crystal phases with the positional order just described, there is additional positional order in the two directions parallel to the planes. A snapshot of the molecules at any one time reveals that the molecular centers have a higher density around points which form a two-dimensional lattice, and that these positions are the same from layer to layer. The symmetry of this lattice can be either triangular or rectangular, and again a positional distribution function, $h(x,y)$, can be defined. This function can be expanded in a two-dimensional Fourier series, with the coefficients in front of the two fundamental sinusoidal terms used as order parameters. Since these materials possess positional order in three dimensions, they should properly be called disordered crystals, but it is not unusual for them to be referred to as liquid crystalline phases.

Bond Orientational Order. In some cases, although the lattice of points of high density of molecular centers parallel to the planes are not correlated from layer to layer, the two principal directions of the lattice are the same for all layers. In these materials, the interactions between the planes do not prevent the planes from translating relative to each other, but do prevent them from rotating relative to each other. Lines drawn between the molecules and parallel to the planes preferentially lie along directions showing hexagonal symmetry. An orientational distribution function, $p(\phi)$, can be defined, where ϕ is an azimuthal angle lying in the plane. This also can be expanded in a Fourier series, with the coefficient in front of the $\cos(6\phi)$ term used as an order parameter for bond orientational order. A wide variety of liquid crystal phases occur due to the many combinations of orientational and positional order possible.

Thermotropic Liquid Crystals

Liquid crystals may be divided into two broad categories, thermotropic and lyotropic, according to the principal means of breaking down the complete order of the solid state. Thermotropic liquid crystals result from the melting of mesogenic solids due to an increase in temperature. Both pure substances and mixtures form thermotropic liquid crystals. In order for a mixture to be a thermotropic liquid crystal, the different components must be completely miscible. Table 1 contains a few examples of the many liquid crystal forming compounds (2). Much

Table 1. Thermotropic Liquid Crystalline Compounds

Formula	Name	CAS Registry Number	Liquid crystalline range, °C
	Nematic liquid crystals		
CH_3O–[ring]–CH=N–[ring]–$(CH_2)_3CH_3$	*p*-methoxybenzylidene-*p*′-*n*-butylaniline (MBBA)	[*26227-73-6*]	21–47
CH_3O–[ring]–N=N(→O)–[ring]–OCH_3	*p*-azoxyanisole (PAA)	[*1562-94-3*]	117–137
n-C_6H_{13}–[ring]–[ring]–CN	*p*-*n*-hexyl-*p*′-cyanobiphenyl	[*4122-70-7*]	14–28
CH_3O–[ring]–OC(=O)–[cyclohexane ring]–C(=O)O–[ring]–OCH_3	di-4-methoxyphenyl-*trans*-1,4-cyclohexane-dicarboxylate	[*24707-00-4*]	143–242
[ring]–[ring]–[ring]–[ring]–[ring]	*p*-quinquephenyl	[*3073-05-0*]	401–445
	Cholesteric[a] liquid crystals		
cholesteric esters[b]			78–90
CH_3O–[ring]–CH=N–[ring]–CH=CHC(=O)$OCH_2CH(CH_3)C_2H_5$	(–)-2-methylbutyl 4-(4′-methoxybenzylideneamino)cinnamate	[*24140-30-5*]	53–97
	Smectic liquid crystals		
smectic *A*			
[ring]–[ring]–CH=N–[ring]–$COOC_2H_5$	ethyl 4-(4′-phenylbenzylideneamino)benzoate	[*3782-80-7*]	121–131

Structure	Name	CAS Registry Number	Temperature range, °C
smectic *B*			
$C_2H_5O-C_6H_4-CH{=}N-C_6H_4-CH{=}CHCOOC_2H_5$	ethyl 4-(4′-ethoxybenzylideneamino)cinnamate	*[2863-94-7]*	77–116
smectic *C*			
$n\text{-}C_8H_{17}O-C_6H_4-COOH$	*p*-*n*-octyloxybenzoic acid	*[2493-84-7]*	108–147
smectic *D*			
$n\text{-}C_{18}H_{37}O-C_6H_3(NO_2)-C_6H_4-COOH$	4-(4′-*n*-octadecyloxy-3′-nitrophenyl) benzoic acid	*[21351-71-3]*	159–195
smectic *E*			
$C_2H_5OOC-C_6H_4-C_6H_4-C_6H_4-COOC_2H_5$	diethyl *p*-terphenyl-*p,p″*-carboxylate	*[37527-56-3]*	173–189
smectic *F* and *G*			
$n\text{-}C_5H_{11}O-C_6H_4-C_4H_2N_2-C_6H_4-(CH_2)_4CH_3$	2-(*p*-pentylphenyl-5-(*p*-pentyloxyphenyl)-pyrimidine	*[34913-07-0]*	[c]
smectic *H*			
$C_4H_9O-C_6H_4-CH{=}N-C_6H_4-C_2H_5$	4-ethyl-4′-butyloxybenzylideneaniline	*[29743-15-5]*	40.5–51

[a]Spontaneously twisted nematic. [b]Eg, cholesteryl nonanoate [*1182-66-7*]. [c]Smectic *F*, 103–114°C; Smectic *G*, 79–103°C.

more is known about calamitic (rod-like) liquid crystals then discotic (disk-like) liquid crystals, since the latter were discovered only recently. Therefore, most of this section deals exclusively with calamitics, with brief coverage of discotics at the end.

Nematic. In a nematic liquid crystal, the long axes of the molecules remain substantially parallel, but the positions of the centers of mass are randomly distributed. Therefore, there is orientational order and a nonzero orientational order parameter, but there is no positional order. A nematic phase is depicted in Figure 4, where the preferred direction of orientation (director) is labeled by the unit vector **n**. In the absence of external orienting influences such as an electric field or boundary, the direction of **n** varies continuously throughout the sample. In fact, thermal energy causes fluctuations in the orientation of the molecules as they diffuse throughout the sample, which causes the director to fluctuate also. These director fluctuations modulate the refractive index of the material on a microscopic level and lead to strong light scattering, which gives the nematic phase a turbid appearance. Because physical properties of the material are not the same in all directions, the nematic phase is anisotropic. For example, the refractive indexes for light polarized parallel and perpendicular to n are different; the nematic phase is birefringent. This means that a liquid crystal appears bright when placed between crossed polarizers, except where the director is parallel to the axis of one of the polarizers. When viewed under a microscope between crossed polarizers, these dark regions appear as lines ending at point-like singularities where n is undefined (Fig. 5).

Nematic liquid crystal molecules can be substantially oriented by a nearby surface. For example, a glass slide with microscopic grooves running in one direction or a stretched polymer film applied to the glass causes the molecules to align in the plane of the surface pointing along the grooves or stretching direction. In thin samples, this order near the surface can cause the director to point in this direction all the way across the sample. So persistent is the influence of the surface that the material can be made to adopt a helicoidal or screw structure by twisting the top piece of glass relative to the bottom one. This twisted

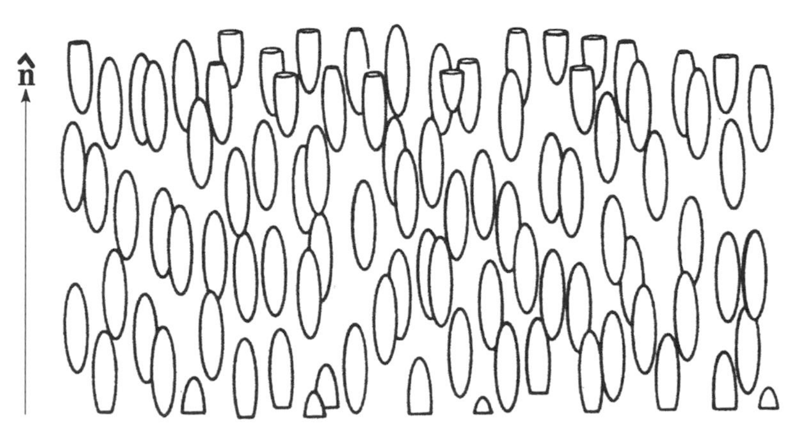

Fig. 4. The nematic liquid crystal structure.

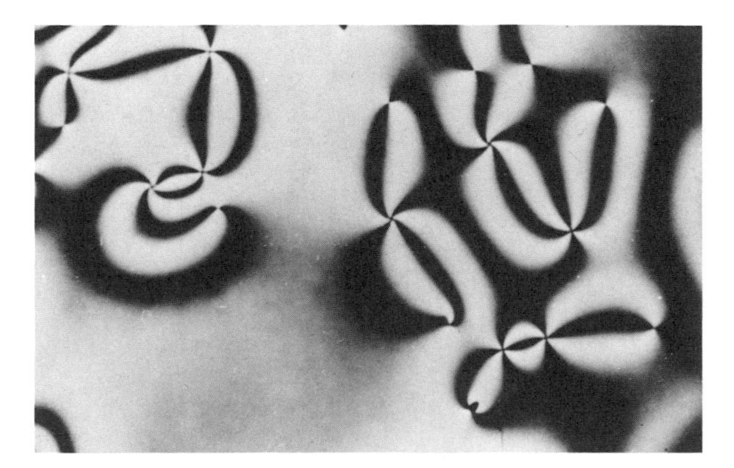

Fig. 5. Nematic schlieren texture observed between crossed polarizers. Courtesy of G. H. Brown, Liquid Crystal Institute, Kent State University.

nematic structure is extremely useful in liquid crystal devices. Alternatively, coating the glass with other surfactants can cause the molecules to orient perpendicular to the glass surface and thus establish a director perpendicular to the glass surface throughout the sample.

Orientation of nematic liquid crystals may be achieved easily in electric or magnetic fields. Depending on the sign of the dielectric anisotropy $\Delta\epsilon = \epsilon_{\|} - \epsilon_{\perp}$ of the material, nematics orient parallel ($\Delta\epsilon > 0$) or perpendicular ($\Delta\epsilon < 0$) to the applied field direction. For liquid crystals with permanent dipole moments along the long axis of the molecule, only a few volts are required for a distortion to occur in samples tens of micrometers thick. The diamagnetic anisotropy $\Delta\chi = \chi_{\|} - \chi_{\perp}$ is usually positive, thereby permitting parallel alignment due to a magnetic field.

Molecules of nematic liquid crystals also are aligned in flow fields which results in a viscosity that is lower than that of the isotropic liquid; the rod-shaped molecules easily stream past one another when oriented. Flow may be impeded if an electric or magnetic field is applied to counter the flow orientation; the viscosity then becomes an anisotropic property.

All distortions of the nematic phase may be decomposed into three basic curvatures of the director, as depicted in Figure 6. Liquid crystals are unusual fluids in that such elastic curvatures may be sustained. Molecules of a true liquid would immediately reorient to flow out of an imposed mechanical shear. The force constants characterizing these distortions are very weak, making the material exceedingly sensitive and easy to perturb.

Chiral Nematic. If the molecules of a liquid crystal are optically active (chiral), then the nematic phase is not formed. Instead of the director being locally constant as is the case for nematics, the director rotates in helical fashion throughout the sample. This chiral nematic phase is shown in Figure 7, where it can be seen that within any plane perpendicular to the helical axis the order is nematic-like. In other words, as in a nematic there is only orientational order in chiral nematic liquid crystals, and no positional order. Keep in mind, however, that there are no planes of any sort in a chiral nematic liquid crystal, since the

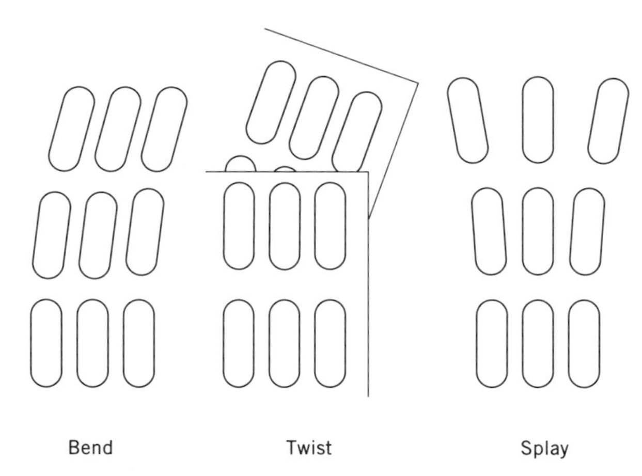

Fig. 6. The three basic curvature deformations of a nematic liquid crystal: bend, twist, and splay. The force constants opposing each of these strains are different. The figure does not mean to imply smectic layering of the molecules.

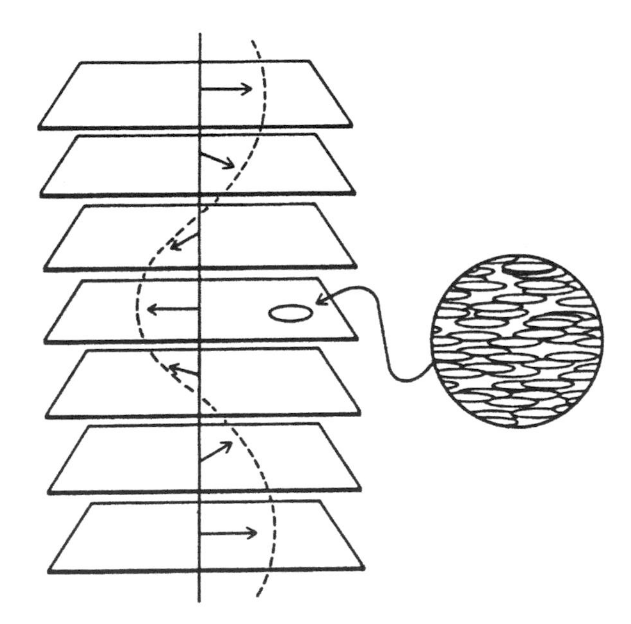

Fig. 7. The chiral nematic liquid crystal structure. The director (arrow) traces out a helical path within the medium. Since the rotation of the director is continuous, the figure does not mean to imply the existence of layers perpendicular to the helical axis.

director rotates continuously about the helical axis. The pitch of the helix formed by the director, ie, the distance it takes for the director to rotate through 360°, can range from 100 nm to as large a distance as can be measured. Chiral nematic liquid crystals are often called cholesteric liquid crystals, since this phase was first discovered in derivatives of cholesterol.

Chiral nematic liquid crystals show strong optical activity, much more than could ever be accounted for on the basis of the rotatory power of the constituent molecules. Optical rotations greater than 300°/mm are standard, and can reach as high as several thousand°/mm (3). More important, perhaps, is what occurs when the pitch is equal to the wavelength of light in the chiral nematic liquid crystal. In this case the periodicity of the director acts like a diffraction grating, reflecting most and sometimes all of the light incident on it. In fact, only one circular polarization is reflected, depending on whether the material forms a right- or left-handed helix. If white light is incident on such a material, only one color of light is reflected and it is circularly polarized. This phenomenon is called selective reflection and is responsible for the iridescent colors produced by chiral nematic liquid crystals. The precise color observed depends on the actual compound, the angle of observation, and the temperature. The temperature affects the pitch and hence the wavelength of light reflected at any specific angle. This temperature dependence can be very strong, showing a noticeable color change for a temperature change as small as 0.001°C (4). The color is also affected by mechanical disturbances, eg, pressure, jarring or shear, and by traces of foreign vapors (5).

Chiral nematic liquid crystals are sometimes referred to as spontaneously twisted nematics, and hence a special case of the nematic phase. The essential requirement for the chiral nematic structure is a chiral center that acts to bias the director of the liquid crystal with a spontaneous cumulative twist. An ordinary nematic liquid crystal can be converted into a chiral nematic by adding an optically active compound (4). In many cases the inverse of the pitch is directly proportional to the molar concentration of the optically active compound. Racemic mixtures (1:1 mixtures of both isomers) of optically active mesogens form nematic rather than chiral nematic phases. Because of their twist encumbrance, chiral nematic liquid crystals generally are more viscous than nematics (6).

Smectic. Smectic liquid crystals are distinguished from nematics by the presence of some positional order in addition to orientational order. When the solid melts, the lateral interactions are strong enough so the molecules spend more time in planes or layers than they do between these planes or layers. Usually the orientational order parameter is quite high throughout a smectic phase, reaching values greater than 0.9 (7). There are deviations from planarity in that the smectic layers can be splayed. Since the layers cannot twist or bend, these distortions do not occur in smectic liquid crystals. A smectic phase in which the director is perpendicular to the layers, called the smectic *A* phase, is shown in Figure 8. Keep in mind that the positional order of Figure 8 is much larger than found in just about all smectic liquid crystals. A snapshot of the molecules would show only a small tendency for more molecules to lie in the layers as opposed to between the layers. Fluidity is maintained by the gliding of the layers past each other; since large-scale movement in other directions is difficult, smectic phases are typically quite viscous (6). Under the microscope smectic phases usually adopt the focal conic texture, a complicated texture of fan-shaped areas and polygonal lines and curves. Such a texture is shown in Figure 9.

The smectic *A* phase depicted in Figure 8 is the simplest one in that the director is normal to the layers and there is no positional order within the layers. Many other smectic phases have been identified, varying from one another in

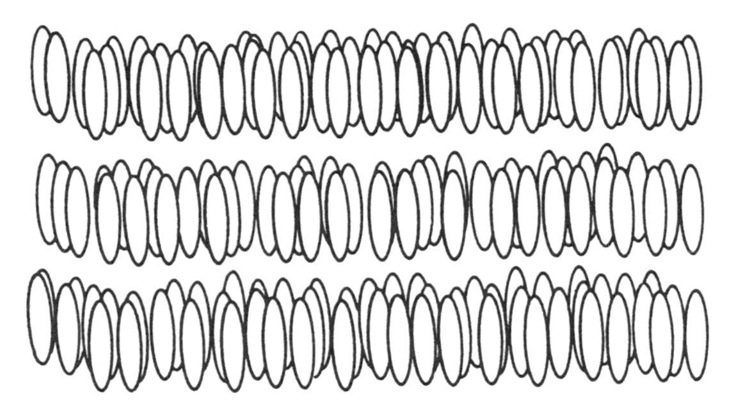

Fig. 8. The smectic liquid crystal structure.

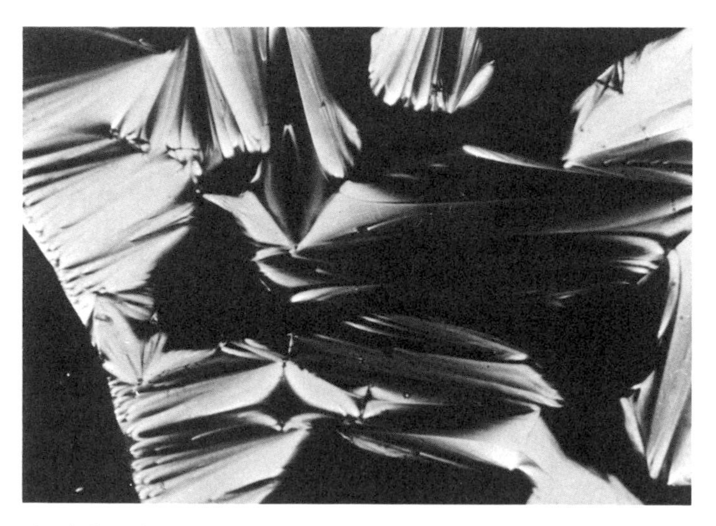

Fig. 9. Smectic *A* fan-shaped texture observed between crossed polarizers. Courtesy of G. H. Brown, Liquid Crystal Institute, Kent State University.

three ways. First, the director may be normal to the layers or tilted relative to the normal to the layers. Second, the positional order within the layers, if present, may take on a number of forms. Third, the layers may be formed by single molecules or by pairs of molecules. These various smectic phases have been assigned letters of the alphabet according to the order in which they were discovered. The smectic *C* phase is similar to the smectic *A* phase except that the director is not normal to the layers (ie, the molecules are tilted within the layers). There is no positional order within the layers, so both of these smectic phases can be thought of as positionally ordered in only one dimension. The hexatic smectic *B*, the smectic *F*, and the smectic *I* phases possess long-range bond orientational order. The packing within the layers is hexagonal, with the positional order being short-range for the hexatic smectic *B* phase and longer range (but not truly long-range) in the smectic *F* and smectic *I* phases. The director in the hexatic smectic *B* phase is normal to the layers; the director in the smectic F and smectic I phases is tilted relative to the normal to the layers with the tilt of the director being toward nearest neighbors in the lattice (smectic I) or toward the midpoint between two nearest neighbors (smectic F). The smectic B, E, G, H, J, and K phases are really disordered crystal or soft solid phases in that positional correlations occur not just within the layers but also from layer to layer. They differ from crystalline solids by the presence of a large amount of disorder and the constant reorientation of the molecules. The smectic B and E phases have a director which is normal to the layers, while in the smectic G, H, J, and K phases the director is tilted relative to the layer normal. The smectic D phase is probably not a layered structure at all, but some sort of cubic arrangement of the molecules (8).

Polar Smectic. Molecules with permanent dipole moments along the length of the molecule often form smectic phases which are different from those just described. In the smectic A_2 phase, for example, successive layers show antiferroelectric order, with the direction of the permanent dipole alternating from layer to layer. In the smectic A_d phase, each layer is identical but is composed of molecular dimers in which the two molecules partially overlap lengthwise. Unlike the smectic phases of nonpolar molecules, the distance over which the structure repeats is roughly two molecular lengths in the smectic A_2 phase and about one and a half molecular lengths in the smectic A_d phase. Tilted versions of these two phases exist, which are called the smectic C_2 and C_d phases. Other smectic phases in polar compounds exist which show a polarization lattice extending within and across the layers (9).

Chiral Smectic. In much the same way as a chiral compound forms the chiral nematic phase instead of the nematic phase, a compound with a chiral center forms a chiral smectic C phase rather than a smectic C phase. In a chiral smectic C liquid crystal, the angle the director is tilted away from the normal to the layers is constant, but the direction of the tilt rotates around the layer normal in going from one layer to the next. This is shown in Figure 10. The distance over which the director rotates completely around the layer normal is called the pitch, and can be as small as 250 nm and as large as desired. If the molecule contains a permanent dipole moment transverse to the long molecular axis, then the chiral smectic phase is ferroelectric. Therefore a device utilizing this phase can be intrinsically bistable, paving the way for important applications.

Frustrated Phases. Chiral molecules normally form chiral phases, but in some cases this is done in an interesting way. For example, it is not unusual for a chiral molecule to form a smectic A phase, which is not chiral. If the molecule is highly chiral, however, twist is sometimes introduced into the smectic A phase by an array of grain boundaries which are perpendicular to the smectic A layers and parallel to the director. The normal to the layers is rotated by roughly 17° on either side of a grain boundary and the grain boundaries are separated by about 24 nm, giving this twist grain boundary (TGB) phase a pitch of a little more than 500 nm (10). In a sense the frustration of an achiral phase of chiral molecules has been relieved by the introduction of these twist grain boundaries.

A similar effect occurs in highly chiral nematic liquid crystals. In a narrow temperature range (seldom wider than 1°C) between the chiral nematic phase and the isotropic liquid phase, up to three phases are stable in which a cubic lattice of defects (where the director is not defined) exist in a

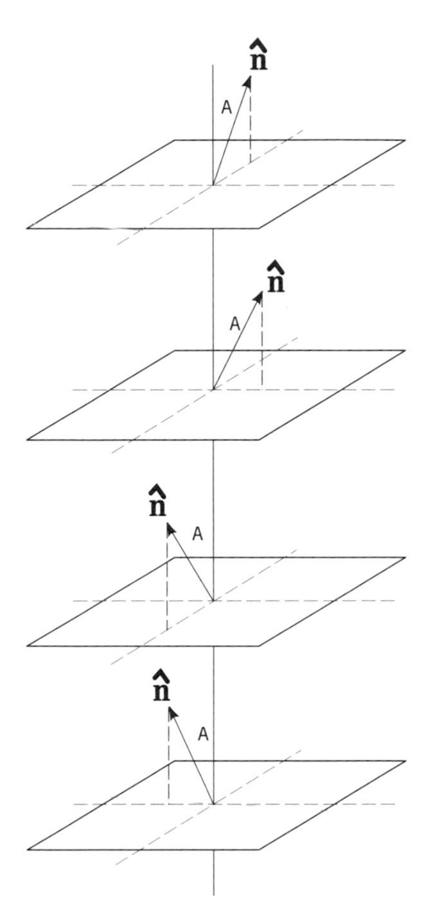

Fig. 10. The chiral smectic *C* liquid crystal structure. The director, **n̂**, rotates about the normal to the smectic layers, keeping the tilt angle, A, constant.

complicated, orientationally ordered twisted structure (11). Again, the introduction of these defects allows the bulk of the liquid crystal to adopt a chiral structure which is energetically more favorable than both the chiral nematic and isotropic phases. The distance between defects is hundreds of nanometers, so these phases reflect light just as crystals reflect x-rays. They are called the blue phases because the first phases of this type observed reflected light in the blue part of the spectrum. The arrangement of defects possesses body-centered cubic symmetry for one blue phase, simple cubic symmetry for another blue phase, and seems to be amorphous for a third blue phase.

Discotic Phases. Molecules which are disk-shaped rather than elongated also form thermotropic liquid crystal phases. Usually these molecules have aromatic cores and six lateral substituents, although the predominance of six lateral substituents is solely historical; molecules with four lateral substituents also can form liquid crystal phases. Although the flatness of these molecules creates a steric effect promoting alignment of the normal to the disks, the fact that disordered side chains are also necessary for the formation of these phases (as is often the case for liquid crystallinity in elongated molecules) should not be ignored.

The most simple discotic phase is the nematic phase, in which the normal to the disks are preferentially aligned along a single direction (director). Such a phase is shown in Figure 11**a**. A discotic phase which sometimes exists below the nematic is the columnar phase, which is depicted in Figure 11**b**. In this phase the molecules are preferentially arranged in columns which show either hexagonal or rectangular positional order. There may or may not be long-range positional order within the columns, but in either case the molecules in one column are not positionally ordered along the column relative to a molecule in another column. This means that the columnar phase does not show three-dimensional order. An example of a typical discotic liquid crystal, a derivative of triphenylene, is shown in Figure 12, along with a description of its phases.

If the molecules are chiral or if a chiral dopant is added to a discotic liquid crystal, a chiral nematic discotic phase can form. The director configuration in this phase is just like the director configuration in the chiral nematic phase formed by elongated molecules (12). Recently, discotic blue phases have been observed.

Metallomesogens. It is also possible to synthesize compounds based on metal atoms which possess liquid crystal phases. The series based on dithiolene complexes (**1**), where M = Ni, Pd, or Pt, contains a number of compounds which show the liquid crystal phases typical of rod-like molecules (13,14).

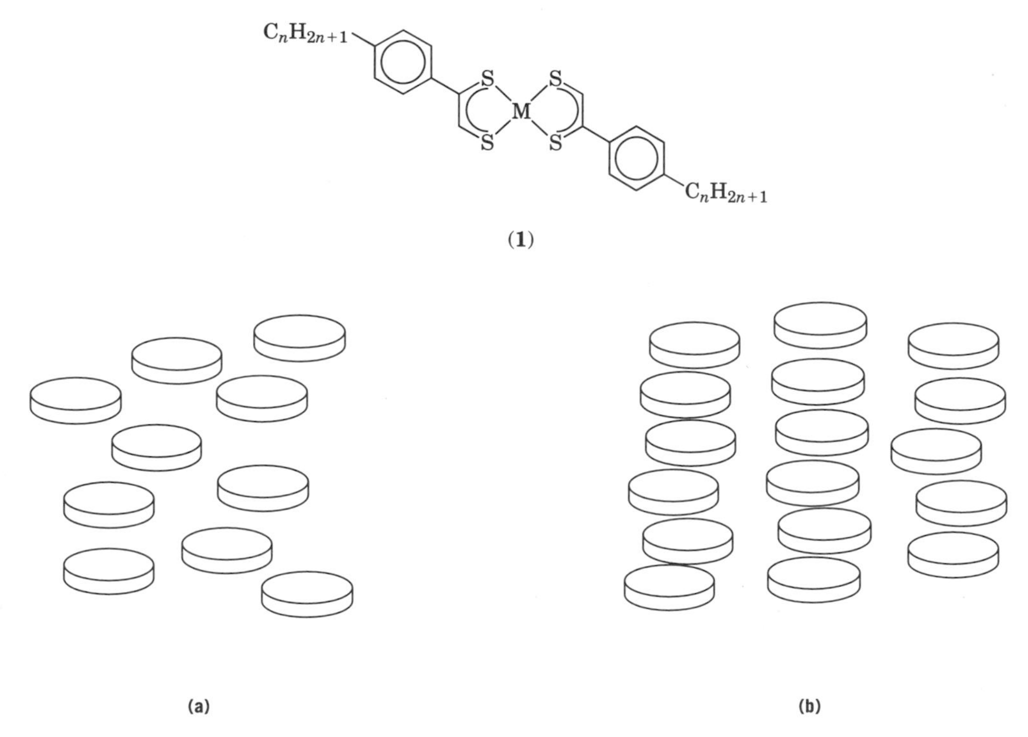

Fig. 11. Orientational order in discotic liquid crystal phases: (**a**) nematic phase; (**b**) columnar phase.

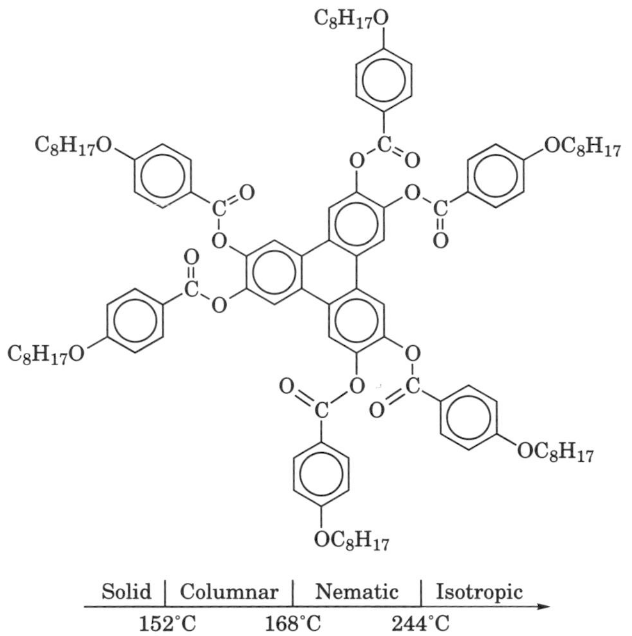

Fig. 12. Molecular structure and phases of a typical discotic liquid crystal.

Disk-shaped molecules based on a metal atom possess discotic liquid crystal phases. An example is octasubstituted metallophthalocyanine. Finally, metallomesogens which combine both rod-like and disk-like features into a single molecule adopt the biaxial nematic phase. In addition to there being a preferred direction for orientation of the longest molecular axis as is true for the nematic phase, perpendicular to this direction is another preferred direction for orientation of the shortest molecular axis (12). Nonmetallomesogens which combine both rod- and disk-like features into a single molecule also adopt a biaxial nematic phase, but at least in one case the amount of biaxiality is very small (15).

Lyotropic Liquid Crystals

Some molecules in a solvent form phases with orientational and/or positional order. In these systems, the transition from one phase to another can occur due to a change of concentration, so they are given the name lyotropic liquid crystals. Of course temperature can also cause phase transitions in these systems, so this aspect of thermotropic liquid crystals is shared by lyotropics. The real distinctiveness of lyotropic liquid crystals is the fact that at least two very different species of molecules must be present for these structures to form.

Amphiphilic Molecules. In just about all cases of lyotropic liquid crystals, the important component of the system is a molecule with two very different parts, one that is hydrophobic and one that is hydrophilic. These molecules are called amphiphilic because when possible they migrate to the interface between a polar and nonpolar liquid. Soaps such as sodium laurate and phospholipids such as α-cephalin [*5681-36-7*] (phosphatidylethanolamine) (**2**) are important examples of amphiphilic molecules which form liquid crystal phases (see LECITHIN; SOAP).

$$
\begin{array}{l}
\qquad\qquad\qquad\qquad\qquad\qquad\qquad\qquad\quad O \qquad\quad O^- \\
\qquad\qquad\qquad\qquad\qquad\qquad\qquad\qquad\quad \| \qquad\quad\; | \qquad\quad + \\
\quad CH_2\ CH_2\ CH_2\ CH_2\ CH_2\ CH_2\ CH_2\ C \quad CH_2OPO(CH_2)_2NH_3 \\
H_3C \quad CH_2\ CH_2\ CH_2\ CH_2\ CH_2\ CH_2\ CH_2\ OCH \quad O \\
\qquad CH_2\ CH_2\ CH_2\ CH_2\ CH_2\ CH_2\ CH_2\ OCH_2 \\
H_3C \quad CH_2\ CH_2\ CH_2\ CH_2\ CH_2\ CH_2\ C \\
\qquad\qquad\qquad\qquad\qquad\qquad\qquad\qquad\quad \| \\
\qquad\qquad\qquad\qquad\qquad\qquad\qquad\qquad\quad O
\end{array}
$$

(2)

When a typical soap is mixed with a polar solvent such as water, the hydrophobic parts of the molecules, which are usually hydrocarbon chains, collect together so that the polar part of the molecule is in contact with the water. Such a structure is shown in Figure 13**a** and is called a micelle. If phospholipids are mixed with water, a slightly different structure forms which is shown in Figure 13**b** and is known as a vesicle. Again, the polar heads of the molecules are in contact with the water, but now there is water both inside and outside of the structure. The reason this structure is preferred is due to the strong tendency of the phospholipid molecules to form bilayers. If either of these molecules are mixed with a nonpolar solvent, reversed micelles and reversed vesicles form. In both of these structures, it is the hydrocarbon chains which are in contact with the solvent, with the polar heads either in the center of the micelle or in the middle of the bilayer. The amphiphilic molecules are both orientationally and

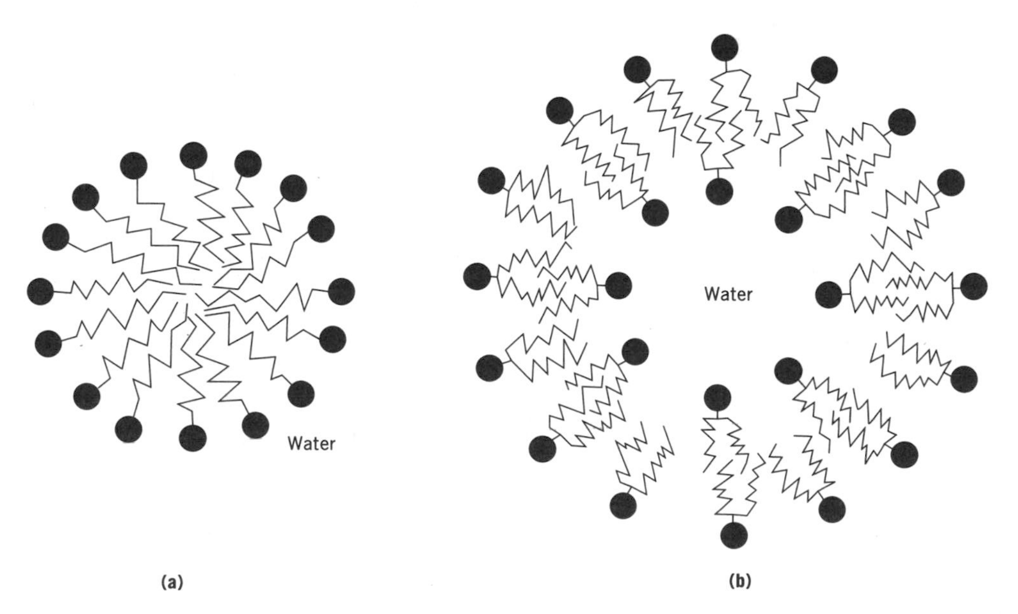

Fig. 13. Lyotropic liquid crystal structures: (**a**) micelle formed by a typical soap; (**b**) vesicle formed by a typical phospholipid.

positionally ordered when these structures form, but the molecules also diffuse throughout the structure. For this reason they are referred to as liquid crystal phases. The positional order can simply be the one-dimensional order of the layer, or can result from additional positional packing within the layers.

At even higher concentrations of amphiphilic compounds, the micelles combine and form larger structures. For example, the micelles of soap molecules can deform into very long cylinders which themselves pack in a hexagonal array. If the concentration is increased even more, sometimes a lamellar phase forms, where the soap molecules form large bilayers separated from each other by water. Like thermotropic liquid crystals, the amphiphilic molecules are free to diffuse throughout the sample, maintaining both orientational and positional order during the diffusion process. Unlike thermotropic liquid crystals, however, the concentration of amphiphilic molecules changes drastically in going from the cylinder or bilayers to the water-rich volumes in between.

Emulsions. Even more interesting phenomena occur when amphiphilic compounds are put into water–oil mixtures. If the oil concentration is low, the amphiphilic molecules form micelles and the oil collects inside the micelles. As the oil concentration is increased, the micelles continue to swell with oil until it is safe to say that the system is really composed of volumes of water and volumes of oil separated by a single amphiphilic layer. This type of system is called an emulsion, and thus amphiphiles can serve as emulsifiers. Amphiphilic compounds have a long history of being used as emulsifiers, with common examples being detergents and food emulsifiers. More recent applications include the flooding of oil-bearing strata with a water–detergent mixture in order to collect oil trapped in the rock (16), and the containment of drugs within vesicles so they may be delivered to the proper point within the body before activation (17) (see EMULSIONS).

Phases Formed By Micelles. When a highly polar liquid, a slightly polar liquid, and an amphiphile are mixed together at the right temperature and in the right concentrations, the micelles which form are not spherical (18). Within this vary narrow concentration range, the micelles are rod-shaped for one part of this range and disk-shaped for another part. In either case the micelles themselves orient their symmetry axes (the long axis for the rod-shaped micelles and the short axis for the disk-shaped micelles) just like a thermotropic liquid crystal. The rod-shaped micelles therefore form a phase with the same structure as a nematic liquid crystal; the same is true for the disk-shaped micelles and a discotic nematic phase. A biaxial nematic phase, similar to the one formed by metallomesogens, is also found in these systems. Although it is easy to imagine the micelles moving about maintaining a preferred orientation, it is more difficult to keep in mind the orientational and positional order of the amphiphilic molecules as they diffuse about the surface of the micelles and less frequently move into the space inside or outside of the micelle.

Polymorphism

A liquid crystal compound in more cases than not takes on more than one type of mesomorphic structure as the conditions of temperature or solvent are changed. In thermotropic liquid crystals, transitions between various phases occur at definite temperatures and are usually accompanied by a latent heat. For example, bis-(4′-*n*-heptyloxybenzylidene)-1,4-phenylenediamine [*24679-01-4*] exhibits seven liquid crystal phases: the nematic and six smectic phases (19). The order of appearance of the phases as the temperature is increased is usually consistent with a decrease in the long-range positional and orientational order of the molecules. For example, ethyl 4-(4′-ethoxybenzylideneamino)cinnamate [*2863-94-7*] (**3**) has three liquid crystal phases (20).

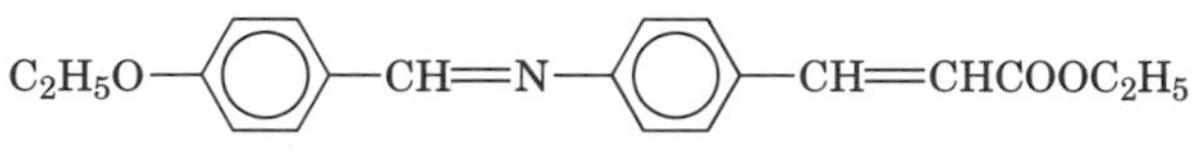

$C_2H_5O-C_6H_4-CH{=}N-C_6H_4-CH{=}CHCOOC_2H_5$

(**3**)

solid $\xrightleftharpoons{81°C}$ smectic *B* $\xrightleftharpoons{119°C}$ smectic *A* $\xrightleftharpoons{157°C}$ nematic $\xrightleftharpoons{159°C}$ isotropic

The positional order of the molecules within the smectic layers disappears when the smectic *B* phase is heated to the smectic *A* phase. Likewise, the one-dimensional positional order of the smectic *A* phase is lost in the transition to the nematic phase. All of the transitions given in this example are reversible upon heating and cooling; they are therefore enantiotropic. When a given liquid crystal phase can only be obtained by changing the temperature in one direction (ie, the mesophase occurs below the solid to isotropic liquid transition due to supercooling), then it is monotropic. An example of this is the smectic *A* phase of cholesteryl nonanoate [*1182-66-7*] (**4**), which occurs only if the chiral nematic phase is cooled (21). The transitions are all reversible as long as crystals of the solid phase do not form.

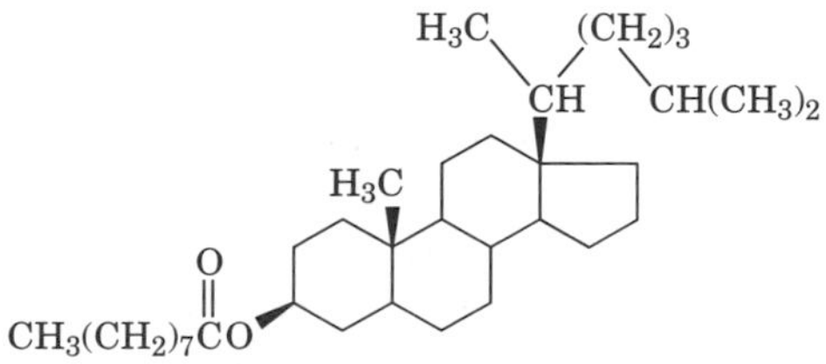

$CH_3(CH_2)_7\overset{O}{\overset{\|}{C}}O$ (cholesteryl skeleton with H_3C, H_3C, CH, $(CH_2)_3$, $CH(CH_3)_2$)

(**4**)

solid $\xrightarrow{78.6°C}$ chiral nematic $\xrightleftharpoons{91.2°C}$ isotropic

chiral nematic $\rightleftarrows$ (75.0°C) smectic *A*

smectic *A* $\rightarrow$ solid

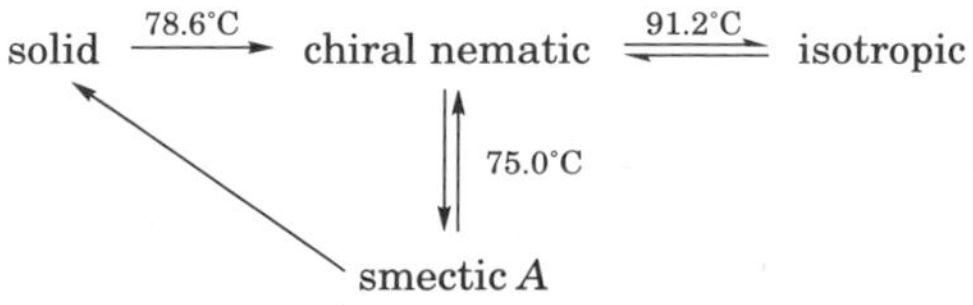

An exception to the rule that lowering the temperature causes transitions to phases with increased order sometimes occurs for polar compounds which form the smectic A_d phase. Decreasing the temperature causes a transition from nematic to smectic A_d, but a further lowering of the temperature produces a transition back to the nematic phase (called the reentrant nematic phase) (22). The reason for this is the unfavorable packing of the molecules in the smectic A_d phase due to overlap of the molecules in the center of the layers. As the temperature is lowered, the steric interactions overpower the attractive forces, causing the molecules to pack much more favorably in the nematic phase. The reentrant nematic phase can also be produced from the smectic A_d phase by increasing the pressure (23).

Electric or magnetic fields also may induce mesomorphic phase transitions. If a chiral nematic liquid crystal is composed of molecules with positive dielectric or diamagnetic susceptibility, an applied field tends to align the director along the field direction. At sufficiently high field strengths, a transition to the nematic phase can occur as the helical structure is unwound. This chiral nematic to nematic transition is continuous and occurs at a critical field strength inversely proportional to the pitch of the chiral nematic in zero field. For a thermotropic chiral nematic with a pitch in the micrometer range, the critical magnetic field is several teslas and the critical electric field can be anywhere between 1 and 5 million volts per meter. For certain lyotropic polymer solutions with a longer pitch, the critical field is one to two orders of magnitude lower (24).

Synthesis

Just because a molecule is long, narrow, and meets the requirement of geometric anisotropy does not ensure that it will have a liquid crystal phase. For example, the *n*-paraffins and homologues of acetic acid are not liquid crystalline. The forces of attraction between these molecules are not sufficiently strong for an ordered, parallel arrangement to be retained after the melting of the solid. In addition, steric effects which promote parallel alignment are not strong in molecules which are extremely flexible and thus deviate significantly from an elongated shape. The particular phase structure that occurs in a compound, ie, smectic, nematic, or chiral nematic, not only depends on the molecular shape but is intimately connected with the strength and position of the polar or polarizable groups within the molecule, the overall polarizability of the molecule, and the presence of chiral centers.

Molecular interactions that lead to attraction include dipole–dipole interactions, dipole-induced dipole interactions, dispersion forces, and hydrogen bonding. Dispersion forces alone, at least in simple aliphatic flexible compounds, apparently are inadequate to achieve the degree of molecular order necessary for liquid crystallinity, eg, the straight-chain paraffins which melt to form normal liquids.

In order for dipole–dipole and dipole-induced dipole interactions to be effective, the molecule must contain polar groups and/or be highly polarizable. Ease of electronic distortion is favored by the presence of aromatic groups and double or triple bonds. These groups frequently are found in the molecular structure of liquid crystal compounds. The most common nematogenic and smectogenic molecules are of the type shown in Table 2.

A central core of benzene rings is linked by a functional group X. The most common end groups at the para sites, R_1 and R_2, are alkyl ($—C_nH_{2n+1}$) or alkoxy ($—OC_nH_{2n+1}$), or acyl chains; $C{\equiv}N$; NO_2; cinnamate ($—CH{=}CHCOOC_nH_{2n+1}$); or halogens. Cyclohexane rings can sometimes replace one or more of the benzene rings without loss of liquid crystallinity.

In general, if the X link is rigid, a liquid crystal phase is favored; branching on the end chains usually is unfavorable (25). The influence of the R_1 and R_2 chain lengths is more subtle. If the rigid part of the molecule gives rise to strong anisotropic interactions, increasing the chain length destabilizes the liquid crystal phase. If the rigid part does not give rise to strong angular correlations increasing the chain length favors anisotropic interactions.

The importance of unsaturation is illustrated by the fact that 2,4-nonadienoic acid [*21643-39-0*] forms a liquid crystal phase, whereas the *n*-aliphatic carboxylic acids do not. The two double bonds enhance the polarizability of the molecule and bring intermolecular attractions to a level that is suitable for mesophase formation. The overall linearity of the molecule must not be sacrificed in potential liquid crystal candidates. For example, whereas *trans-p-n*-alkoxycinnamic acids (**5**) are mesomorphic, the cis isomers (**6**) are not, a reflection of the greater anisotropy of the trans isomer.

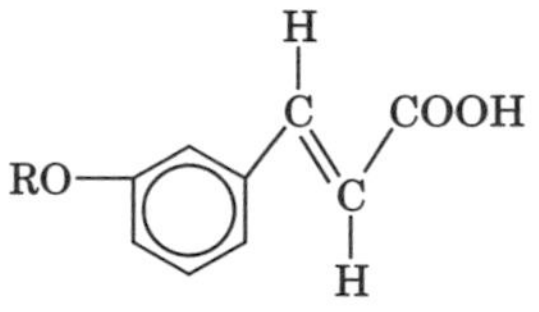

RO—C₆H₄—CH=CH—COOH (trans)

(**5**)

RO—C₆H₄—CH=CH—COOH (cis)

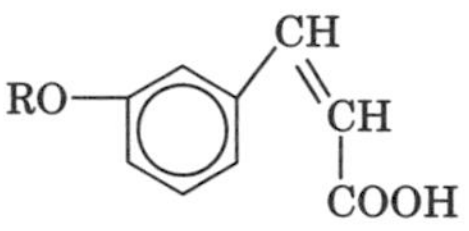

(**6**)

Table 2. Some Central Linkages Found in Liquid Crystalline Compounds

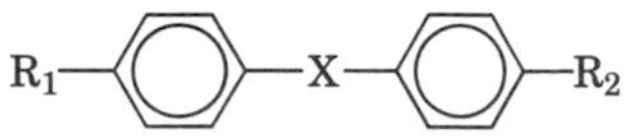

$R_1—C_6H_4—X—C_6H_4—R_2$

X	Series name
$—CH{=}N—$	Schiff bases
$—N{=}N—$	diazo compounds
$—N{=}N(\rightarrow O)—$	azoxy compounds
$—CH{=}N(\rightarrow O)—$	nitrones
$—CH{=}CH—$	stilbenes
$—C{\equiv}C—$	tolans
$—OC({=}O)—$	esters
— (nothing)	biphenyls

Freedom of rotation about the double methylene bridge in the compound (**7**) (dimethyl 4,4′-(1,2-ethanediyl)bisbenzoate [*797-21-7*]) destroys the rod shape of the molecule and prevents liquid crystal formation. The stilbene derivative (**8**) (dimethyl 4,4′-(1,2-ethenediyl)bisbenzoate [*10374-80-8*]), however, is essentially linear and more favorable for liquid crystal formation.

$CH_3OOC-C_6H_4-CH_2CH_2-C_6H_4-COOCH_3$

(**7**)

$CH_3OOC-C_6H_4-CH{=}CH-C_6H_4-COOCH_3$

(**8**)

Bulky, even if highly polarizable, functional groups or atoms that are attached anywhere but on the end of a rod-shaped molecule usually are less favorable for liquid crystal formation. Enhanced intermolecular attractions are more than countered as the molecule deviates from the required linearity. For example, the inclusion of the bromine atom at position three of 4-decyloxy-3-bromobenzoic acid [*5519-23-3*] (**9**) prevents mesomorphic behavior. In other cases the liquid crystal phases do not disappear, but their ranges are narrower.

$C_{10}H_{21}O-C_6H_3(Br)-COOH$

(**9**)

In the case of carboxylic acids, hydrogen bonding can induce liquid crystal phases by lengthening the molecular unit through dimerization:

$$R-C(=O\cdots H-O)(-O-H\cdots O=)C-R$$

On the other hand, hydrogen bonding may lead to nonlinear molecular associations that disrupt the parallelism, eg, phenolic compounds generally are not mesomorphic. Hydrogen bonding associations may also be so strong that by the time the solid reaches its melting point the thermal energy is too intense to permit substantial order to remain within the fluid; in this case, the solid passes directly into the isotropic liquid. Such reasoning could explain the absence of liquid crystallinity in cholesterol and its presence in the esters of cholesterol. 4-Amino-4″-nitro-*p*-terphenyl [*38190-45-3*] (**10**) melts at 300°C and is not mesomorphic. Both of the amine hydrogens in this compound participate in intermolecular associations. The substituted nitro compound [*75802-59-4*] (**11**), however, melts at only 218°C and is a liquid crystal. In this case, one of the hydrogens is internally associated with the NO_2 group.

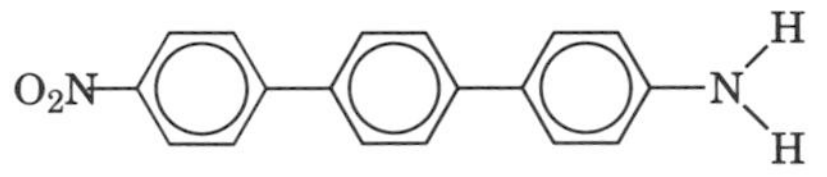

(10)

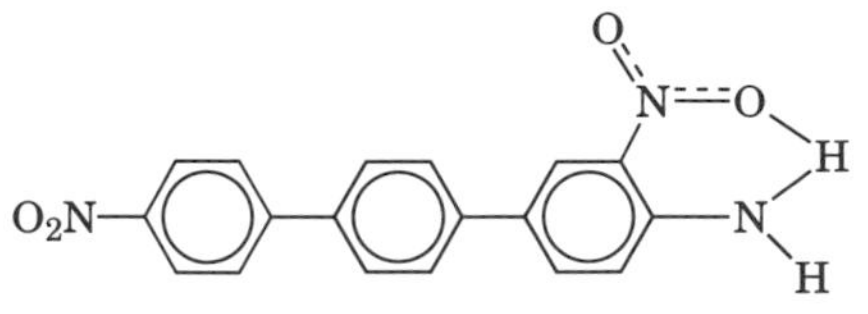

(11)

Although it is difficult to predict exactly which type of liquid crystal phase will be formed by a molecule meeting the general requirements, rough trends can be recognized. The presence of functional groups that lead to strong lateral interactions, eg, dipoles operating across the long molecular axis, favor the layered smectic structure. When these structural elements are not present but the molecule is otherwise suitable for mesomorphism, ie, is long and narrow, the nematic phase is likely. Longer terminal groups favor the smectic phase over the nematic phase. An asymmetric center on the molecule causes the chiral nematic and chiral smectic *C* phases in place of the nematic and smectic *C* phases.

Goals in liquid crystal synthesis include the design of room temperature thermotropics which are stable, colorless liquid crystalline over a wide range of temperature, and operate at low voltage and power levels. The number of compounds of commercial importance is actually not very large; representative ones are shown in Table 3 (26). Extended mesomorphic temperature ranges are obtained by using eutectic mixtures, since obtaining a wide room temperature liquid crystal phase in a single compound is extremely difficult. Large positive dielectric anisotropy is achieved by attaching strongly dipolar terminal groups, although the effect is often reduced due to antiparallel association between pairs of molecules. Large negative dielectric anisotropy is much more difficult to obtain, due to difficulties in reducing the longitudinal dipole moment, the fact that the electric field points along all directions perpendicular to the long axis of the molecules, and strong dipole moments from lateral substituents frequently affect the stability and viscosity adversely. The birefringence can be controlled by adjusting the number of aromatic rings and π-bonded terminal or linking groups. As a rule, molecules with low polarity and polarizability, with short terminal groups and no lateral substituents, have the lowest viscosity and are best suited for nematic displays.

A good deal of synthesis effort has been devoted to chiral liquid crystals, especially those with chiral smectic *C* phases. The chiral smectic *C* phase is ferroelectric, which gives it properties quite useful for applications. Perhaps the most important property of these phases is that a lateral dipole can produce a spontaneous polarization. Since the usefulness of these materials depends on creating a large spontaneous polarization, a great amount of work has gone into synthesizing chiral smectic *C* liquid crystals with high spontaneous polarizations. The most important factors influencing the size of the spontaneous polarization

Table 3. Technologically Important Nematic Liquid Crystals

Compound type	N–I,[a] °C	Δn^b	$\Delta\epsilon^c$	$k_{33}/k_{11}{}^d$
R–C_6H_4–C_6H_4–CN	35	0.18	11.5	1.3
R–C_6H_{10}–C_6H_4–CN	50	0.1	9.7	1.6
R–(pyrimidine, N,N)–C_6H_4–CN	50	0.18	19.7	1.2
R–C_6H_4–CO_2–C_6H_4–CN	55	0.15	19.7	1.7
R–C_6H_4–CO_2–C_6H_4–R′	20	0.13	0.5	1.2
R–C_6H_{10}–CO_2–C_6H_4–OR′	75	0.07	−1.0	1.3
R–C_6H_{10}–C_6H_{10}–CN	85	0.06	4.4	1.5
R–C_6H_4–CO_2–C_6H_3(F)–CN	30	0.14	48.9	1.7
R–(1,3-dioxane, O,O)–C_6H_4–CN	50	0.09	13.3	1.4

[a]Nematic to isotropic liquid transition temperature.
[b]Anisotropy in the index of refraction.
[c]Anisotropy in the dielectric constant.
[d]Ratio of bend to splay force constant.

are the angle between the layer normal and the director, the magnitude of the lateral dipole, and the amount of rotation of the chiral center about the long molecular axis.

Examples of chiral smectic *C* liquid crystals range from 2-methylbutyl (*S*)-4-*n*-decyloxybenzylideneaminocinnamate (**12**), the first ferroelectric liquid crystal discovered,

$$C_{10}H_{21}O\text{–}C_6H_4\text{–}CH{=}N\text{–}C_6H_4\text{–}CH{=}CHCOOCH_2\overset{*}{C}H(CH_3)C_2H_5$$

(**12**)

to (*R*)-4-(methoxycarbonyl-1-ethylcarboxyl)-3-hydroxyphenyl 4-(octyloxyphenyl)-benzoate (**13**), a newly synthesized material with a high spontaneous polarization (27).

$$C_8H_{17}O-C_6H_4-C_6H_4-\overset{O}{\overset{\|}{C}}O-C_6H_3(OH)-\overset{O}{\overset{\|}{C}}O\overset{*}{C}H(CH_3)\overset{O}{\overset{\|}{C}}OCH_3$$

(13)

The purpose of the hydroxyl group is to achieve some hydrogen bonding with the nearby carbonyl group and therefore hinder the motion of the chiral center. Another way to achieve the chiral smectic *C* phase is to add a chiral dopant to a smectic *C* liquid crystal. In order to achieve a material with fast switching times, a chiral compound with high spontaneous polarization is sometimes added to a mixture of low viscosity achiral smectic *C* compounds. These dopants sometimes possess liquid crystal phases in pure form and sometimes do not.

Polymer Liquid Crystals

Both polymer melts and polymer solutions sometimes form phases with orientational and positional order. Thermotropic polymer liquid crystals possess at least one liquid crystal phase between the glass-transition temperature and the transition temperature to the isotropic liquid. Lyotropic polymer liquid crystals possess at least one liquid crystal phase for certain ranges of concentration and temperature. Thermotropic polymer liquid crystals fall into two classes depending on whether the rigid section, which usually resembles a low molecular weight liquid crystal molecule, is incorporated into the main chain of the polymer or appended as side chains. Lyotropic polymer liquid crystals usually result when a polymer arranges itself in elongated assemblies when in solution. Large macromolecules which adopt a single- or double-stranded helix are typical examples. The solid phase of many polymers, which is usually amorphous, can possess liquid crystalline order if placed under stress or when under the influence of a nearby interface.

Polymers would seem predisposed to liquid crystallinity since they may adopt an extremely elongated shape. In many cases, however, the incorporation of rigid polarizable segments into the polymer chain results in increased thermal stability and decreased solubility. The polymers decompose at the high temperatures that are needed to produce a fluid state and/or prove intractable in conventional solvents. It is possible, however, by design of the monomer, to optimize polymer rigidity and retain useful thermal or solubility properties. Temperatures for liquid crystal phase formation are generally lower in side-chain polymers than in main-chain polymers.

Polymer Melts. When a rigid, polarizable monomer forms either a main-chain polymer with flexible segments in between or a side-chain polymer with flexible segments between the rigid segments and the flexible main chain, liquid crystal phases are usually stable. Figure 14 illustrates the order present in the nematic and smectic phases of a main-chain polymer; Figure 15 shows a typical side-chain polymer along with the temperature ranges of its liquid crystal phases. Other liquid crystal phases such as chiral nematic, chiral smectic *C*, and

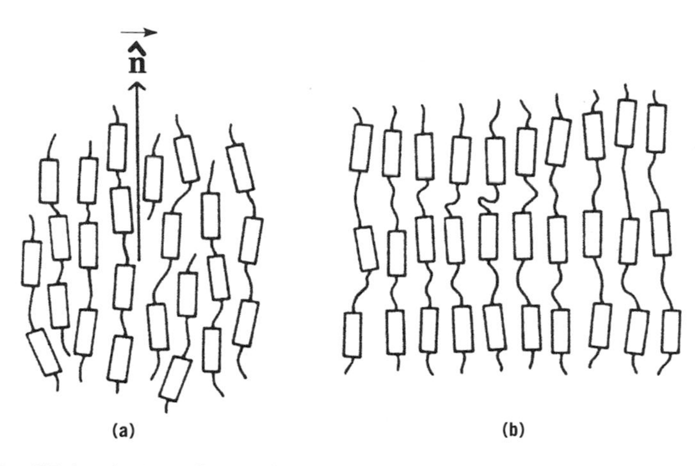

Fig. 14. Main-chain polymer liquid crystal phases: (**a**) nematic, (**b**) smectic.

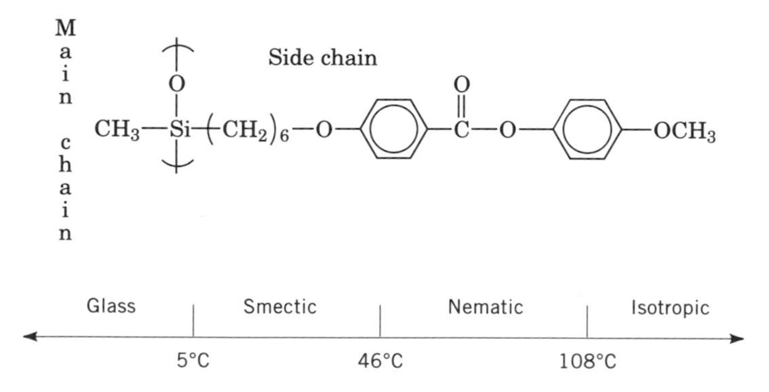

Fig. 15. Molecular structure and phases of a typical polysiloxane side-chain polymer liquid crystal.

even the blue phases are also formed by polymers. In both main- and side-chain polymer liquid crystals, it is the rigid, polarizable segments which are ordered; the flexible segments of the main-chain polymer and the entire flexible main chain of the side-chain polymer remain unordered. Order parameters can be defined for the rigid segments in much the same way as for thermotropic liquid crystals, and in fact have temperature dependences which are extremely similar to Figure 3.

Examples of polymers which form anisotropic polymer melts include petroleum pitches, polyesters, polyethers, polyphosphazines, α-poly-p-xylylene, and polysiloxanes. Synthesis goals include the incorporation of a liquid crystal-like entity into the main chain of the polymer to increase the strength and thermal stability of the materials that are formed from the liquid crystal precursor, the locking in of liquid crystalline properties of the fluid into the solid phase, and the production of extended chain polymers that are soluble in organic solvents rather than sulfuric acid.

Polymer Solutions. Perhaps the most extensively studied macromolecular liquid crystals are the synthetic polypeptides, such as poly(γ-benzyl L-glutamate) [*25513-40-0*] (PBLG). PBLG is a homopolymer of the L-enantiomorph of a single amino acid with the following repeat unit.

$$\left[\mathrm{NH\underset{\underset{\underset{\displaystyle COOCH_2-C_6H_5}{|}}{\displaystyle (CH_2)_2}}{\underset{|}{C}}H\overset{\overset{\displaystyle O}{\|}}{C}}\right]_n$$

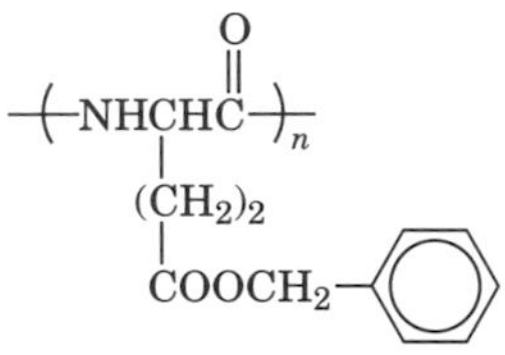

The polymer may be prepared in high degrees of polymerization ($n > 1000$) and has good solubility characteristics. It is an excellent model system because many variables, eg, molecular weight, supporting solvent character, concentration, and temperature, may be easily controlled for study over wide ranges.

PBLG adopts the α-helical conformation in a number of solvents as a result of intramolecular hydrogen bonding and favorable stacking of the pendent side chains. Thus the polymer assumes an extended, relatively rigid geometry and may become ordered spontaneously at sufficiently high concentrations. The formation of this lyotropic liquid crystal phase occurs at a critical volume fraction of polymer $\phi*$ which is inversely proportional to the length-to-diameter ratio of the macromolecule (28). The longer the α-helical rod, the lower the concentration that is necessary for spontaneous ordering. Solutions of PBLG and similar polypeptides usually possess nematic and chiral nematic phases (29).

A variety of aromatic and extended-chain polyamides that spontaneously form a mesophase in concentrated solutions also have been synthesized (30). Polybenzamide [*24991-08-0*], with the following repeat unit, is an example.

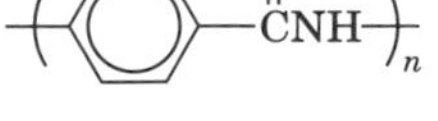

$$\left[\mathrm{C_6H_4-\overset{\overset{\displaystyle O}{\|}}{C}NH}\right]_n$$

The necessary molecular rigidity of polybenzamide undoubtedly results from the hindered bond rotation within the planar amide group.

The polyamides are soluble in high strength sulfuric acid or in mixtures of hexamethylphosphoramide, *N,N*-dimethylacetamide, and LiCl. In the latter, complicated relationships exist between solvent composition and the temperature at which the liquid crystal phase forms. The polyamide solutions show an abrupt decrease in viscosity which is characteristic of mesophase formation when a critical volume fraction of polymer $\phi*$ is exceeded. The viscosity may decrease, however, in the liquid crystal phase if the molecular ordering allows the rod-shaped entities to glide past one another more easily despite the higher concentration. The liquid crystal phase is optically anisotropic and the texture is nematic. The nematic texture can be transformed to a chiral nematic texture by adding chiral species as a dopant or incorporating a chiral unit in the main chain as a copolymer (30).

Applications. The polyamides have important applications. The very high degree of polymer orientation that is achieved when liquid crystalline solutions are extruded imparts exceptionally high strengths and moduli to polyamide fibers and films. Du Pont markets such polymers, eg, Kevlar, and Monsanto has a similar product, eg, X-500, which consists of polyamide and hydrazide-type polymers (31) (see HIGH PERFORMANCE FIBERS; POLYAMIDES, FIBERS).

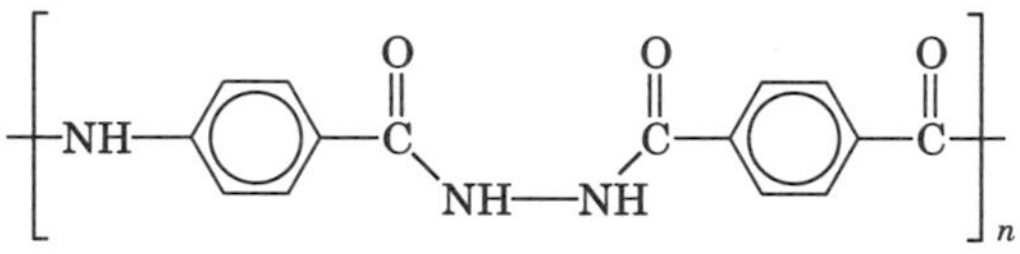

$$\left[-NH-C_6H_4-\overset{O}{\overset{\|}{C}}-NH-NH-\overset{O}{\overset{\|}{C}}-C_6H_4-\overset{O}{\overset{\|}{C}}- \right]_n$$

Because of the rotation of the N–N bond, X-500 is considerably more flexible than the polyamides discussed above. A higher polymer volume fraction is required for an anisotropic phase to appear. In solution, the X-500 polymer is not anisotropic at rest but becomes so when sheared. The characteristic viscosity anomaly which occurs at the onset of liquid crystal formation appears only at higher shear rates for X-500. The critical volume fraction $\phi*$ shifts to lower polymer concentrations under conditions of greater shear (32). The mechanical orientation that is necessary for liquid crystal formation must occur during the spinning process which enhances the alignment of the macromolecules.

Liquid crystal polymers are also used in electrooptic displays. Side-chain polymers are quite suitable for this purpose, but usually involve much larger elastic and viscous constants, which slow the response of the device (33). The chiral smectic *C* phase is perhaps best suited for a polymer field effect device. The ability to attach dichroic or fluorescent dyes as a proportion of the side groups opens the door to applications not easily achieved with low molecular weight liquid crystals. Polymers with smectic phases have also been used to create laser writable devices (30). The laser can address areas a few micrometers wide, changing a clear state to a strong scattering state or vice versa. Future uses of liquid crystal polymers may include data storage devices. Polymers with nonlinear optical properties may also become important for device applications.

If a modest number of cross-links between the polymer backbone are introduced, the polymer liquid crystal takes on elastomeric properties. The usefulness of these materials probably lies in the coupling of mechanical and optical effects.

Liquid Crystals in Biological Systems

Many biological systems exhibit the properties of liquid crystals. Considerable concentrations of liquid crystalline compounds have been found in many parts of the body, often as sterol or lipid derivatives. A liquid crystal phase has been implicated in at least two degenerative diseases, atherosclerosis and sickle cell anemia. Living tissue, such as muscle, tendon, ovary, adrenal cortex, and nerve, show the optical birefringence properties that are characteristic of liquid crystals. The liquid crystal state has been identified in many pathological tissues, particularly in areas of large lipid deposits. Massive deposits of liquid crystalline cholesterol derivatives have been found in the kidneys, liver, brain, spleen, marrow, and aorta walls. Certain living sperms possess a liquid crystalline state. Solutions of tobacco mosaic virus (TMV), collagen, hemoglobin of sickle cell anemia, native protein, nucleic acid genetic material, and fibrinogen also show resemblances to the liquid crystal state.

Cell Membrane. The fluid mosaic model of the cell membrane is one in which the phospholipids provide the basic order and integrity of the cell through amphiphilic interaction with the aqueous environment. The assumed structural feature is one where a lipid bilayer, which eventually forms a closed vesicle, separates the internal from the external, thereby defining the cell. Globular and glycoproteins and cholesterol moieties are solubilized or attached to the surface of the bilayer in the model. Some may penetrate and act as pores for ion transport across the membrane. One particularly striking observation is the high degree of mobility in the hydrocarbon chains, which is similar to that of a liquid state at room temperature in spite of the high degree of order of the phospholipid polar group. The lipid bilayer is fluid also in the sense that the integral proteins can move freely; this has been shown by incorporating fluorescent labels on the membrane surface and analyzing their migration spectroscopically (34).

Microfilaments and Microtubules. There are two important classes of fibers found in the cytoplasm of many plant and animal cells that are characterized by nematic-like organization. These are the microfilaments and microtubules which play a central role in the determination of cell shape, either as the dynamic element in the contractile mechanism or as the basic cytoskeleton. Microfilaments are proteinaceous bundles having diameters of 6–10 nm that are chemically similar to actin and myosin muscle cells. Microtubules also are formed from globular elements, but consist of hollow tubes that are about 30 nm in diameter, uniform, and highly rigid. Both of these assemblages are found beneath the cell membrane in a linear organization that is similar to the nematic liquid crystal structure.

There is also a correlation between the type of subsurface organization of these fibers and gross cell shape in tissue other than muscle. Flattened but elongated cells have sheets of striated, parallel fibers below the plasma membrane. Elongated but branched cells have fibrous bundles that separate at branching junctions. An extreme expression of the development of cytoplasmic microtubules is found in Heliozoa, an organism exhibiting numerous axial spikes that are supported internally by microtubules. The effect of microfilament–microtubule subsurface organizations on the change of shape of malignant cells has been noted; highly malignant cells of irregular external shape do not contain the normal internal oriented structures (35).

Liquid Crystalline Structures. In certain cellular organelles, deoxyribonucleic acid (DNA) occurs in a concentrated form. Striking similarities between the optical properties derived from the underlying supramolecular organization of the concentrated DNA phases and those observed in chiral nematic textures have been described (36). Concentrated aqueous solutions of nucleic acids exhibit a chiral nematic texture *in vitro* (29,37).

Liquid crystalline behavior occurs in the exocuticle of certain classes of beetles. The bright iridescent colors that are reflected from the surface of Scarabaeid beetles originates from a petrified chiral nematic structural arrangement of chitin crystallites in the exocuticle (38). It is suggested that this chiral nematic texture forms spontaneously in a mobile, liquid crystal phase that is present during the initial stages of the exocuticle growth cycle.

Viruses such as TMV form liquid crystal phases in solution. As an aqueous solution of TMV is concentrated, birefringent, liquid crystalline tactoids appear suspended in the isotropic phase. In this thixotropic gel, the tactoids coalesce into a homogeneous, liquid crystalline phase with an increase in the TMV concentration and/or a change in the solution pH. Also, fd-type filamentous bacterial viruses form a birefringent phase in concentrated solution (39). The gel-like solution adopts a chiral nematic structure in the liquid crystal phase. Iridescent colors are reflected from highly concentrated gels of the fd virus; in dry gels, the chiral nematic pitch is comparable to the wavelength of light. The chiral nematic structure is caused by the asymmetry in the intermolecular forces between solute particles. The fd virus tertiary structure is chiral; the viruses are rod-like with a chiral arrangement of essentially α-helical proteins on the exterior. The chiral nematic structure also is the most commonly seen structure in lyotropic liquid crystal solutions of synthetic polypeptides.

Diseases. Liquid crystals have been implicated in a number of disease conditions in the human body. A complex cholesterol–phospholipid–lipoprotein liquid crystal phase has been identified in the initiation and maintenance of atheromatous deposits on the aortic intima in dissected human and rabbit arteries (40). The paracrystalline nature of this precursor to plaque buildup with the resultant loss of arterial elasticity and atherosclerotic stenosis is evident when viewed through a polarization microscope. The familiar liquid crystal sperulitic structures exhibit the birefringent and optical rotation properties that are characteristic of the chiral nematic phase. Even in the healthy individual, the arterial wall has a liquid crystalline component composed of cholesterol and phospholipids. These molecular types occur in a 1:1 ratio in the adult (41). The resultant medium may enable the artery to expand during the systolic phase of the heartbeat, and to contract during the diastolic phase.

Liquid crystal accumulations have been noted in pathological lipid and cholesterol deposits in some rare metabolic diseases, eg, cholesterol ester storage disease, Tangiers', Farbers', Neimann-Pick, Gauchers', Krabbes', Fabrys', and Tay Sachs' diseases, and in gallstone formation (42).

The presence of hemoglobin-S (Hb-S) in red blood cells leads to the formation of liquid crystalline aggregates inside the cell under conditions of low oxygen tension (43,44). The morbid aggregates ultimately arrange themselves into a gel-like material composed of long fibers that extend the entire length of the cell and distort its usual shape.

Liquid crystalline modifications are singularly well suited to provide the delicate balance of organization and lability that is characteristic of life processes. It seems certain that liquid crystallinity has important biological consequences, both because of its presence in living material and its unusual dependence on slight changes in composition and in the physical and chemical environment for its formation, continuation, or cessation. The unusual combination of lability and lateral cohesion makes the liquid crystal phase biologically useful.

Applications

Since the early 1970s, the potential applications of liquid crystals have been a strong motivation for both pure and applied research. Most important of these applications is the use of liquid crystals to make displays, especially ones which require much less power to operate than conventional displays. The first use for the liquid crystals display (LCD) was in small, battery-operated equipment, but LCD technology has made so many advances that flat panel LCD displays will soon control more of the television and computer markets than the cathode ray tube. Liquid crystal devices are also becoming more important as optical components in communications systems and optical computing research. Another application for liquid crystals which has been profitable since at least the 1970s has been as temperature sensors. These range from toy thermometers to medical thermal imaging systems. Finally, liquid crystalline materials have been used as pressure and chemical vapor sensors, as electric and magnetic field sensors, as anisotropic solvents, and as separation media.

Liquid Crystal Displays. The workhorse of the LCD field is an electrooptic device called the twisted nematic display. A typical diagram of this device is shown in Figure 16. A small amount of chiral dopant is added to a room temperature nematic liquid crystal mixture with high dielectric anisotropy and low viscosity. This mixture is then introduced between two flat pieces of glass, the inside surfaces of which have a transparent metallic coating of indium–tin oxide covered by a surfactant which promotes alignment of the liquid crystal director parallel to the surface. The direction of parallel alignment on one piece of glass is perpendicular to the direction of parallel alignment on the other piece

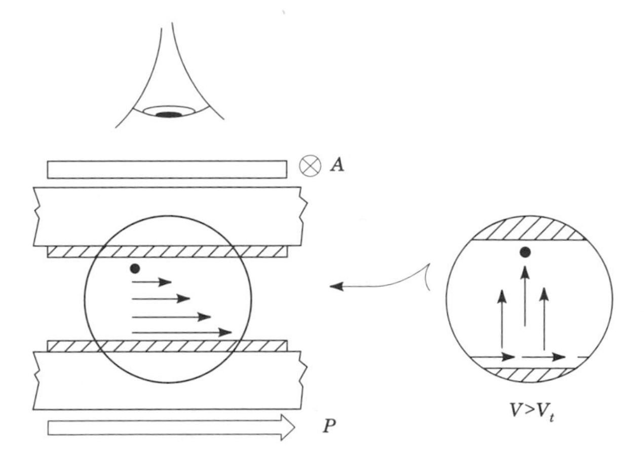

Fig. 16. Twisted nematic LCD showing the $V < V_t$ bright state (left) and the $V > V_t$ dark state (right), where V_t is the threshold voltage of the cell. P = polarizer (axis lies in the page): A = analyzer (axis lies perpendicular to the page).

of glass. The space between the pieces of glass is roughly 10 μm. Under these conditions, the liquid crystal mixture spontaneously adopts a twisted structure, in which the director rotates by 90° in going from one glass surface to the other. The two pieces of glass have polarizing films deposited on their outside surfaces, with the polarization direction identical to the alignment direction on the inside surface. When fabricated in this way, light incident on the cell is polarized by the first polarizer, rotated by 90° by the twisted structure of the liquid crystal, and passes through the analyzer. The cell therefore appears bright or transparent. When a voltage is applied, the anisotropy of the liquid crystal molecules causes the director to align with the field, except for a very thin layer next to the glass surfaces. This untwisted structure no longer rotates the polarization direction of the light as it passes through the device, so the analyzer extinguishes it. The cell appears dark.

The characteristics of twisted nematic LCDs are excellent. The contrast between the dark and light states is large; the device can be used with a backlight in a transmissive mode or with a mirror in a reflective mode; the voltage necessary to switch the display is under 5 volts; the switching times are in the millisecond range; colored filters allow for pleasing color displays; and the liquid crystal mixtures are both chemically and photochemically stable. In addition, the viewing angle can be made quite wide and a reflective display does not wash out under bright ambient light conditions. The display itself consumes extremely little power; reflective displays using ambient light are therefore perfect for battery operated instruments. Backlighted displays obviously consume significantly more power, but are still preferable in many circumstances.

A slight variant of the classic twisted nematic display has been introduced due to its wider viewing angle and improved switching characteristics. This supertwisted nematic display utilizes a twist of the liquid crystal director of 270° within the cell rather than 90° (45). The basic operation of the cell is unchanged in that the effect of the analyzer on light which has been rotated by 270° is the same as for 90° rotation.

Many other types of LCDs have been invented. In the dynamic scattering device, an otherwise clear liquid crystal mixture is made highly scattering due to hydrodynamic turbulence caused by the application of an electric field. If a chiral nematic mixture is used in a dynamic scattering device, the display can have storage capability. This means that the "on" state remains even after the voltage has been removed. The clear "off" state returns if a higher voltage is applied to the cell. In guest–host LCDs using dichroic dyes, the reorientation of the liquid crystal by the applied voltage causes the absorption characteristics of the dye molecules to switch from low to high. Certain birefringent LCDs operate like twisted nematic displays, but create the 90° shift of the polarization direction not by rotation but by the birefringence of the liquid crystal layer. Smectic liquid crystals can also be used in displays. The most interesting example of these is the surface stabilized ferroelectric liquid crystal device. In this display, a chiral smectic *C* liquid crystal is switched between two alignments by an electric field in one direction or the other. Proper use of the polarizers causes one state to be bright and the other dark. This display has one significant advantage, namely, shorter switching times.

Simple alphanumeric LCDs are made by creating patterns in one of the transparent electrodes and addressing the segments independently to create both numbers and letters. This works well as long as the characters are fairly large and the number of characters is small. The creation of a flat panel display for either a computer screen or television is a much more difficult task, as millions of individual pixels must be addressed in order to create the small alphanumeric characters or the highly resolved video image. Individual connections to each pixel are out of the question, so a multiplexing scheme is used. Here individual pixels are not addressed at all times, but once per cycle. During the part of the cycle in which an individual pixel is not addressed, it receives varying small voltages. This means the electrical characteristics of a pixel must be engineered so that it responds correctly when addressed, but holds that state during the rest of the cycle when it is not being addressed. The superior electrooptic features needed for multiplexing are usually achieved by mounting an active electrical device such as a thin-film transistor on one of the substrates at each pixel. The liquid crystal cell is then driven by control voltages to the active device. Such a scheme is known as active matrix addressing. By using colored filters to create red, green, and blue pixels at each matrix location, which are addressed separately, a color display can be formed.

Two new forms of liquid crystal displays utilize polymers along with a thermotropic liquid crystal. In the first kind, called a polymer dispersed liquid crystal (PDLC) display, droplets of liquid crystals are randomly distributed through an isotropic but solid polymer matrix (46). The liquid crystal mixture is one in which the index of refraction for light polarized perpendicular to the director is equal to the index of refraction of the polymer matrix. When no electric field is applied, the orientation of the liquid crystal inside the droplets is random as shown in Figure 17**a**. The mismatch between the polymer refractive index and refractive index of the liquid crystal for light polarized in any direction but perpendicular to the director causes a large amount of scattering. The display appears opaque. When an electric field is applied, the director inside the droplets orients parallel to the field as shown in Figure 17**b**, so there is no index mismatch between the polymer matrix and the liquid crystal, since the electric field of the light is always perpendicular to its direction of propagation. The display is

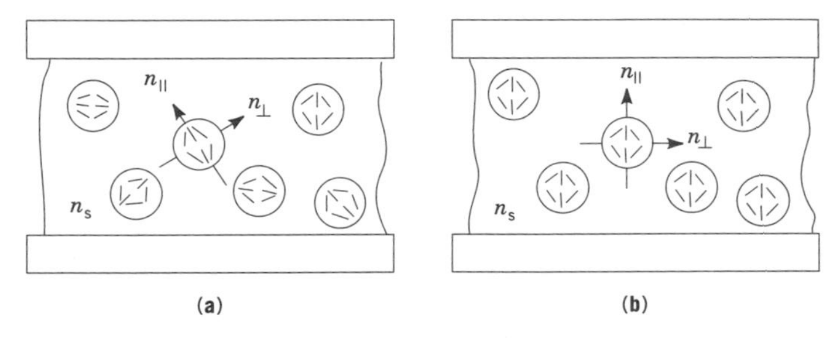

Fig. 17. Polymer dispersed liquid crystal display (PDLC). (**a**) $V < V_t$ clear state, where V_t is the threshold voltage of the cell. $n_{\|}$ and $n_{\perp}$ represent the indexes of refraction for light polarized parallel and perpendicular to the director of the liquid crystal; n_s represents the index of refraction of the isotropic polymer matrix.

now transparent. PDLC displays are currently used for switchable windows in housing and automobiles; alphanumeric PDLC displays are being developed. The second kind of display is a polymer stabilized cholesteric texture (PSCT) display. In this even newer display, a small amount of cross-linking polymer is added to a chiral nematic display (47). The cross-linking is photoinitiated once the display is in its proper zero field texture. The advantages of this is that the time for the liquid crystal to relax to its zero field texture is much shorter, and the selective reflection of the chiral nematic texture produces excellent color. Use of a polymer to stabilize a liquid crystal texture is a new idea, which may find use in a wide variety of LCDs, both conventional and new.

Optical Elements. Devices which control the amount of light which passes through individual pixels within a small area are important for a number of reasons. Such a device is crucial for the operation of projection-type displays, a possibility for the switching components in communications systems, and the basis for optical computing. A device which performs this function is called a spatial light modulator, and liquid crystals are being investigated along with other technologies for this purpose. Not only have devices been fabricated which control the amount of light transmitted through the device, but image transfer devices have been designed in which an image is "written" on the device by one light beam and subsequently "read" at a later time by a second light beam. Image transfer devices combine a liquid crystal display and some type of electrical device, eg, a photoconductor or charge-coupled device (CCD). One possible use for this type of device is in large-screen projection television systems, where an image is written by an electron beam on one side of the device, while light is projected from a source to the screen through the other side of the device.

Liquid Crystal Sensors. Because chiral nematic liquid crystals reflect light of a single wavelength, they are easily fabricated into colorful displays. The wavelength reflected is determined by the pitch of the chiral nematic liquid crystal, which can be made very sensitive to temperature. In this way a display can vividly reflect colors ranging from violet to red due to temperature changes of only a few tenths of a degree. By proper use of different liquid crystal mixtures, thermometers of all sorts have been made, including thermometers for measuring body temperature, fish tank temperature, and room temperature. This same principle has been used to manufacture "mood" rings and sophisticated medical thermal imaging systems.

The selective reflection of chiral nematic liquid crystals has also been used to develop sensors for pressure, radiation (especially infrared), wind shear over surfaces, structural fatigue, and foreign chemical vapor (48). Other types of liquid crystals have been used to make sensors to measure both electric and magnetic fields.

Other Applications. Liquid crystals have found a number of applications in various research fields. Molecules that are dissolved in nematic liquid crystals solvents (an anisotropic environment) give a very highly resolved nmr spectrum exhibiting intermolecular dipole–dipole fine structure (49). Analysis of the spectra of molecules in liquid crystal solvents yields information regarding the anisotropy of chemical shifts, direct magnetic dipole–dipole interactions, indirect spin–spin couplings, bond angles and bond lengths, molecular order, and relaxation processes. High resolution nmr spectroscopy on such partially oriented molecules has provided a new method of determining molecular structure, particularly for small organic molecules. Some liquid crystals have been used in chromatographic separations (50), as solvents to direct the course of chemical reactions or as a medium to study molecular rearrangements and kinetics (51), and also as anisotropic host fluids for visible, uv, and ir spectroscopy of organic molecules (49). Chiral nematic liquid crystals have also found their way into the visual arts (52) and the cosmetic and clothing industries (48).

Availability and Safety

The amount of liquid crystals produced each year for applications is several tens of tons, with the vast majority designed specifically for display applications. Several of the largest producers of commercial liquid crystals are E. Merck, Hoffmann-LaRoche Inc., and Chisso. E. M. Chemicals (Hawthorne, New York) is the distributor for E. Merck in the United States and Chisso America Inc. has an office in New York. Hoffmann-LaRoche and Di Nippon Inc. have joined forces to form a new company, Rodic.

Liquid crystalline compounds are not very dangerous and only basic precautions should be used in handling them. They are not poisonous or carcinogenic, and do not cause problems when in contact with skin (see also BIPHENYLS AND TERPHENYLS; CINNAMIC ACID; STILBENE DYES).

BIBLIOGRAPHY

"Liquid Crystals" in *ECT* 3rd ed., Vol. 14, pp. 395–427, by D. B. Du Pré, University of Louisville.

1. D. B. DuPré and L. L. Chapoy, *J. Chem. Ed.* **56**, 759 (1979).
2. G. H. Brown, *J. Colloid Interface Sci.* **58**, 534 (1977).
3. J. W. Goodby, *J. Mater. Chem.* **1**, 307 (1991).
4. D. G. McDonnell, in G. W. Gray, ed., *Thermotropic Liquid Crystals*, John Wiley & Sons, Inc., New York, 1987, Chapt. 5.
5. W. Elser and R. D. Ennulat, in G. H. Brown, ed., *Advances in Liquid Crystals*, Vol. 2, Academic Press, Inc., New York, 1976.
6. R. S. Porter and J. F. Johnson, in F. Eirich, ed., *Rheology*, Vol. 4, Academic Press, Inc., New York, 1967.
7. J. W. Doane and co-workers, *Phys. Rev. Lett.* **28**, 1694 (1972).
8. G. Etherington and co-workers, *Liq. Cryst.* **1**, 209 (1986).
9. A. J. Leadbetter, in Ref. 4.
10. K. J. Ihn and co-workers, *Science* **275** (1992).
11. T. Seideman, *Rep. Prog. Phys.* **53**, 659 (1990).
12. S. Chandrasekhar, *Liquid Crystals*, 2nd ed., Cambridge University Press, Cambridge, U.K., 1992.
13. A. M. Giroud-Godquin and P. M. Maitlis, *Angew. Chem. Int. Ed. Engl.* **30**, 375 (1991).
14. S. A. Hudson and P. M. Maitlis, *Chem. Rev.* **93**, 861 (1993).
15. S. M. Fan and co-workers, *Chem. Phys. Lett.* **204**, 517 (1993).
16. M. Kahlweit, *Science* **240**, 617 (1988).
17. M. Ostro and P. R. Cullis, *Am. J. Hosp. Pharm.* **46**, 1576 (1989).
18. L. J. Yu and A. Saupe, *J. Am. Chem. Soc.* **102**, 4879 (1980).
19. W. Helfrich and C. S. Oh, *Mol. Cryst. Liq. Cryst.* **14**, 289 (1971).

20. M. Lambert and A. M. Levelut, in T. Riste, ed., *Anharmonic Lattices, Structural Transitions and Melting*, Noordhoff, Leiden, the Netherlands, 1974.
21. W. L. McMillan, *Phys. Rev. A* **6**, 936 (1972).
22. P. E. Cladis, *Phys. Rev. Lett.* **35**, 48 (1975).
23. P. E. Cladis and co-workers, *Phys. Rev. Lett.* **39**, 720 (1977).
24. R. W. Duke and D. B. DuPre, *J. Chem. Phys.* **60**, 2759 (1974).
25. K. J. Toyne, in Ref. 4, Chapt. 2.
26. I. Sage, in Ref. 4, Chapt. 3.
27. H. Taniguchi and co-workers, *Jpn. J. Appl. Phys.* **27**, 452 (1988).
28. P. J. Flory, *Proc. R. Soc. Sect. A* **234**, 73 (1956).
29. C. Robinson, *Tetrahedron* **13**, 219 (1961).
30. A. M. Donald and A. H. Windle, *Liquid Crystalline Polymers*, Cambridge University Press, Cambridge, 1992.
31. W. B. Black and J. Preston, *High Modulus Aromatic Polymers*, Marcel Dekker, Inc., New York, 1979.
32. G. Alfonso and co-workers, *Polym. Prepr.* **18**, 179 (1977).
33. H. Finkelmann, in Ref. 4, Chapt. 6.
34. S. J. Singer and G. L. Nicolson, *Science* **175**, 720 (1972).
35. E. J. Ambrose, in S. Friberg, ed., *Lyotropic Liquid Crystals*, American Chemical Society, Washington, D.C., 1976, Chapt. 10.
36. Y. Bouligand, M.-O. Soyer, and S. Puiseux-Dao, *Chromosoma* **24**, 251 (1961).
37. M. Spencer and co-workers, *Nature* **194**, 1014 (1962).
38. A. C. Neville and S. Cabeney, *Biol. Rev.* **44**, 531 (1969).
39. J. LaPoint and D. A. Marvin, *Mol. Cryst. Liq. Cryst.* **19**, 269 (1973).
40. G. T. Stewart, in N. H. G. W. Gray and P. A. Winsor, eds., *Liquid Crystals and Plastic Crystals*, Vol. 1, Ellis Horwood Ltd., Chichester, U.K., 1974.
41. D. M. Small and G. G. Shipley, *Science* **185**, 222 (1974).
42. R. T. Holzback, M. Marsh, and P. Tang, in S. Matern and co-eds., *Advances in Bile Acid Research*, F. K. Schattauer Verlag, Berlin, 1974.
43. J. T. Finch and co-workers, *Proc. Nat. Acad. Sci. USA* **70**, 718 (1973).
44. K. Moffat, *Science* **185**, 274 (1974).
45. T. Scheffer and J. Nehring, in B. Bahadur, ed., *Liquid Crystals: Applications and Uses*, Vol. 1, World Scientific, Singapore, 1990.
46. J. W. Doane, in Ref. 45.
47. D. K. Yang and J. W. Doane, *SID Digest of Technical Papers* **23**, 759 (1992).
48. I. Sage, in B. Bahadur, ed., *Liquid Crystals: Applications and Uses*, Vol. 3, World Scientific, Singapore, 1992, Chapt. 20.
49. C. L. Khetrapal, R. G. Weiss, and A. C. Kunwar, in B. Bahadur, ed., *Liquid Crystals: Applications and Uses*, Vol. 2, World Scientific, Singapore, 1991, Chapt. 25.
50. Z. Witkiewicz, in Ref. 49, Chapt. 26.
51. W. J. Leigh, in Ref. 49, Chapt. 27.
52. D. Makow, in Ref. 49, Chapt. 21.

General References

B. Bahadur, *Liquid Crystals: Applications and Uses*, Vols. 1–3, World Scientific, Singapore, 1990–1992.

S. Chandrasekhar, *Liquid Crystals*, 2nd ed., Cambridge University Press, Cambridge, U.K., 1992.

P. J. Collings, *Liquid Crystals: Nature's Delicate Phase of Matter*, Princeton University Press, Princeton, N.J., 1990.

P. G. deGennes, *The Physics of Liquid Crystals*, Clarendon Press, Oxford, U.K., 1974.

W. H. deJeu, *Physical Properties of Liquid Crystalline Materials*, Gordon and Breach Science Publishers, New York, 1980.
D. Demus and L. Richter, *Textures of Liquid Crystals*, Verlag-Chemie, Berlin, 1978.
S. Friberg, ed., *Lyotropic Liquid Crystals*, American Chemical Society, Washington, D.C., 1976.
J. W. Goodby and co-workers, *Ferroelectric Liquid Crystals: Principles, Properties, and Applications*, Gordon and Breach Publishers, Philadelphia, Pa., 1991.
G. W. Gray, ed., *Thermotropic Liquid Crystals*, John Wiley & Sons, Inc., New York, 1987.
G. W. Gray and J. W. G. Goodby, *Smectic Liquid Crystals*, Leonard Hill, London, 1984.
H. Kelker and R. Hatz, *Handbook of Liquid Crystals*, Verlag-Chemie, Berlin, 1980.
G. R. Luckhurst and G. W. Gray, eds., *The Molecular Physics of Liquid Crystals*, Academic Press, Inc., New York, 1979.
S. Martellucci and A. N. Chester, eds., *Phase Transitions in Liquid Crystals*, Plenum Press, New York, 1992.
G. Vertogen and W. H. deJeu, *Thermotropic Liquid Crystals, Fundamentals*, Springer-Verlag, Berlin, 1988.
A. M. White and A. H. Windle, *Liquid Crystalline Polymers*, Cambridge University Press, Cambridge, U.K., 1992.

PETER J. COLLINGS
Swarthmore College

LIQUID LEVEL MEASUREMENT

Level gauging has existed at least as long as written history; markings on the walls of Egyptian temples show that as early as 3000 years ago humans tracked the level of water for hydrologic data. These nilometers are some of the first recorded uses of systems to measure level (1).

The four process control parameters are temperature, pressure, flow, and level. Modern process level detection systems are varied and ubiquitous; in modern chemical plants there are thousands of processes requiring liquid level indication and liquid level control. From accumulators to wet wells, the need for level devices is based on the need for plant efficiency, safety, quality control, and data logging. Unfortunately, no single level measurement technology works reliably on all chemical plant applications. This fact has spawned a broad selection of level indication and control device technologies, each of which operates successfully on specific applications.

Measurement vs Control

Level devices can be divided into two broad groups: those that indicate level and those that provide means to control level. Indication devices report where the level is at any given point in the process. Control devices provide supervision of the process and can be used to initiate other devices to control process levels.

Sight Glass Gauges. These indication devices provide visual indication of the process level by means of a clear, nonmetallic, vertical tube piped to the vessel at top and bottom. As the level rises in the vessel, it maintains the same level in the sight glass (Fig. **1a**). For elevated temperatures and pressures, reflex and refractive-type sight glasses are available (Fig. **1b**). Increased solids and dense liquids can restrict movement of the liquid in and out of the glass making these sight devices unreliable. Clean liquids with no color can cause a problem determining if the glass is full or empty if no intermediate level is apparent.

Dip Stick Indicators. Visual level indication can be obtained by dropping a weighted cable or rigid dip stick into the media until it reaches the bottom of the vessel. Graduations are marked on the cable or stick. Upon retrieval the operator looks for the point of dry vs wet indicating the depth of the media. This method of level indication is useful in ambient/atmospheric applications in nonhazardous environments. It is not recommended for other applications. Measurements are accurate only to the extent of the skill of the operator.

Magnetic Liquid Level Indicators. Where it is necessary to isolate the process due to pressure or toxic, lethal, volatile, or corrosive liquid, a magnetic level indicator is available. These devices utilize a float mounted inside a pipe which is connected to the process. The float has strong magnets internally mounted around its diameter. A sealed isolated sight glass with a magnetic indicator is mounted on the float tube. As level changes the float follows the level change and the magnetic coupling between float and level indicator keeps the

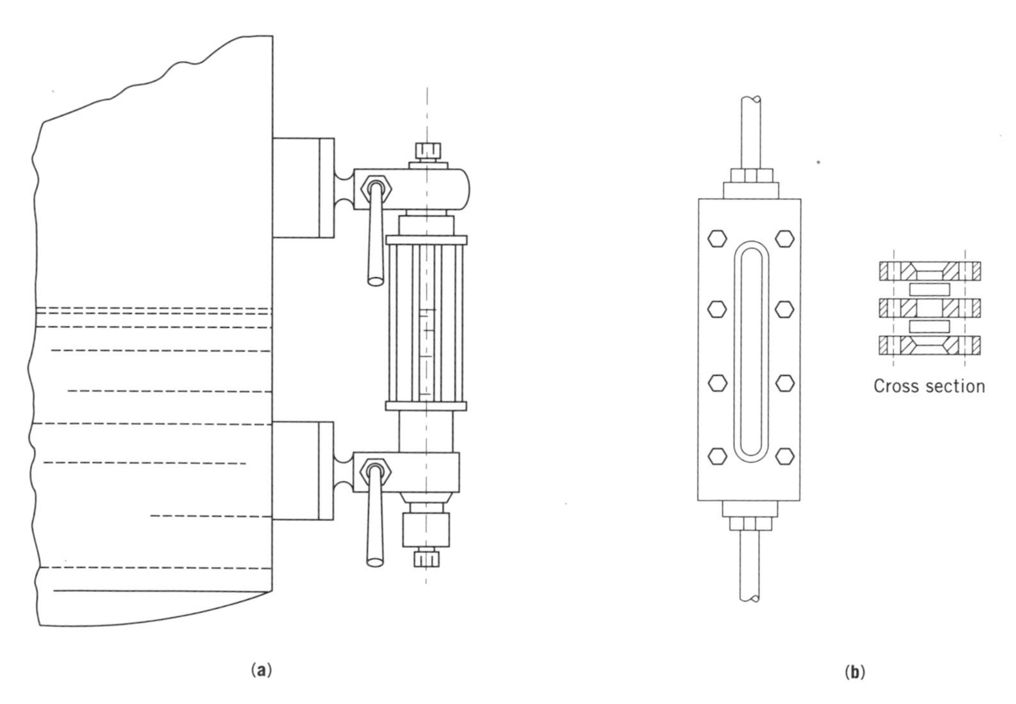

Fig. 1. Sight glass gauges: (**a**) tubular, (**b**) reflex. Courtesy of John C. Ernst Co.

level indicator equal with the process liquid. Level indicators are either shutters that flip with level change or fluorescent indicators (Fig. 2). In addition to process isolation, these devices also allow visual level detection at much greater distances than sight glasses. Solids and heavier liquids can cause float hang-up, making the units unreliable. They are available in a variety of metals, plastics, and connection arrangements. Options for magnetically coupled level switches and transmitters are also available.

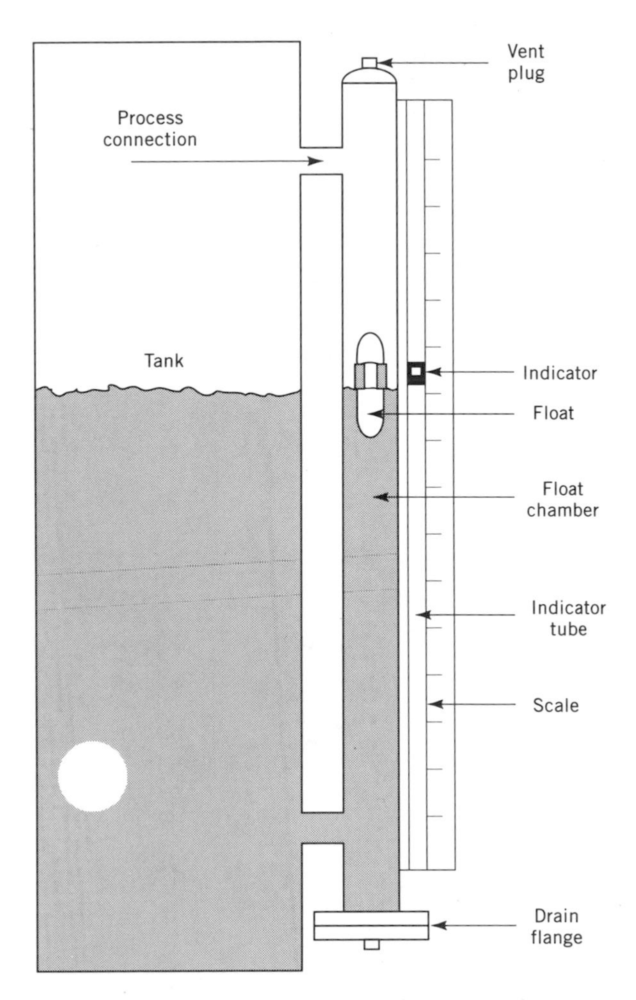

Fig. 2. Magnetic level gauge. Courtesy of ProMag.

Level Control Devices

There are three basic requirements that liquid level control devices are designed to satisfy: alarm functions, pump/valve control, and transmitted output signal to track level continuously. Alarm devices provide warning or shutdown functions when process levels pass a predetermined point in the vessel; pump/valve control devices turn on/off pumps or open/close valves at predetermined levels in the vessel; and transmitters provide a proportional output signal over a predetermined span to send to a local meter or signal back to a control room.

In a discussion of the various level technologies, it is important to know the differences between the various level requirements. The implementation of a particular technology, such as ultrasonic, is different for a single alarm device than it is for a transmitter.

Floats. Float level switches are suitable for clean liquid applications, primarily for alarm function (Fig. 3). A float follows level change moving a stem and magnetic attraction sleeve within a nonmagnetic enclosing tube. When the attraction sleeve enters the field of the magnet, the magnet pulls in actuating a switch mechanism. The magnetic coupling allows complete isolation of switch mechanism from process. Float level switches must be mounted at the desired set point. No field calibration is required. Units can be mounted outside of the vessel in a separate cage or can be mounted directly to the vessel at top or side. A broad selection of mounting arrangements, switch mechanisms and housings, and materials of construction are available to provide compatibility with clean liquid processes. Float switches provide total isolation of switch mechanism from process, require no power to operate, are intrinsically safe, and are accurate to 0.5 cm. Float level switches can also monitor interface between two immiscible

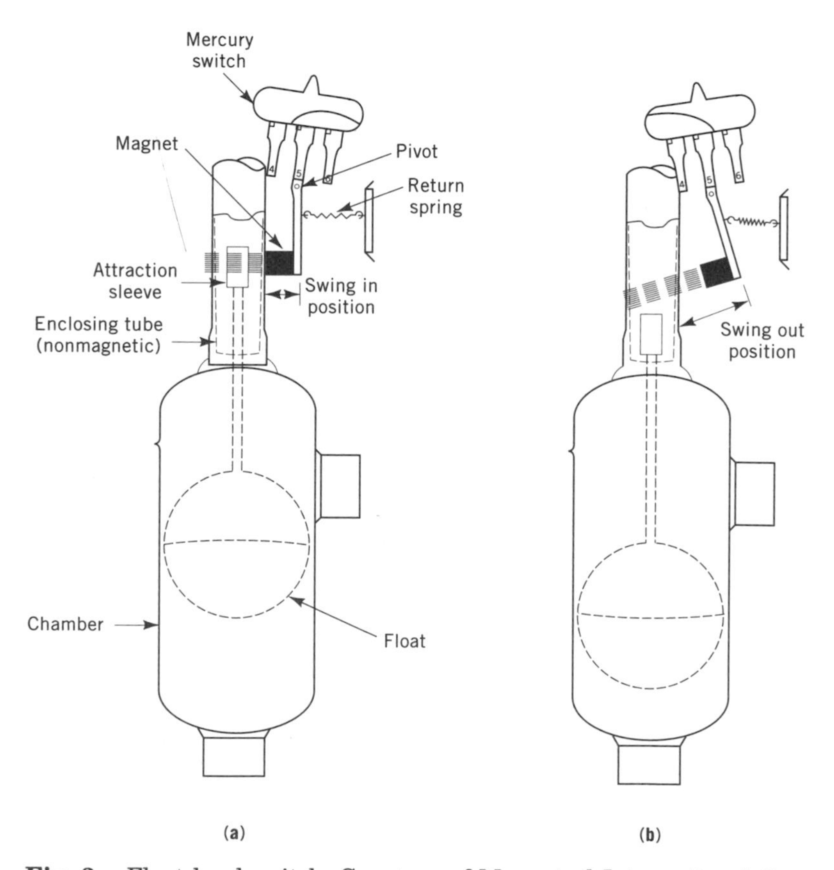

Fig. 3. Float level switch. Courtesy of Magnetrol International, Inc.

fluids using specially weighted floats that sink in the upper fluid and float in the lower fluid. The float must be covered with liquid at all times to maintain calibration.

Displacers. Displacer level switches are suitable for clean and dirty fluids and are used principally to control sump pumps where shifting specific gravity, turbulent surface, and foam are common problems. Displacer(s) are suspended from a range spring connected to a stem and attraction sleeve. With change in level the spring senses the change in buoyancy causing the stem and attraction sleeve to move within a nonmagnetic enclosing tube. When the attraction sleeve enters the field of the magnet the switch mechanism is actuated. The enclosing tube totally isolates the switch mechanism from process (Fig. 4). A magnetic coupling exists between the attraction sleeve and switch mechanism. Units can be mounted in the vessel top or externally mounted in a separate cage piped to the vessel.

Displacer units are tolerant of specific gravity shifts within their range, require no power to operate, are intrinsically safe, and can be easily field calibrated. A single displacer unit can control up to three alarm set points or up to three pumps with different spans. They can be calibrated for set points or pump spans from 0.15–30 m from the top of the vessel mounting.

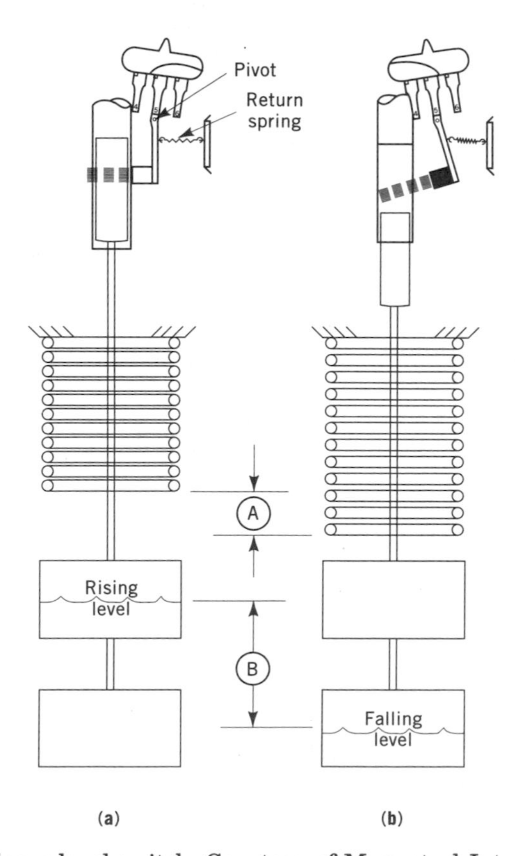

Fig. 4. Displacer level switch. Courtesy of Magnetrol International, Inc.

Buoyancy. Buoyancy level controllers are used to control a process having continuous flow through the vessel. The primary application is in reactors, feedwater heaters, deaerators, and similar processes having boiling and turbulent conditions. A hollow cylinder (displacer) is suspended from a range spring or torque tube. Change in level causes a change in buoyancy which creates a force–balance shift which is transmitted through a metallic seal into the control housing (Fig. 5). Level change causes the controller to increase or decrease the output signal to a proportional control valve. The valve is opened or closed to control flow through the vessel, thus maintaining level within a predetermined span. Units are available with a pneumatic 21–103 kPa (3–15 psi) or electronic (4–20 mA d-c) output. Spans are available from 36 to 305 cm. Mounting arrangements include top, side, and external cage with selections of process connection size, style, and materials.

Buoyancy level controllers/transmitters are tolerant of turbulent level and can effectively control high temperature and high pressure applications. Specific gravity shifts within the unit range do not reduce the unit's ability to maintain control of the process. Pressure changes are ignored and temperature shifts within the unit range produce minimal shifts of span. Range spring models

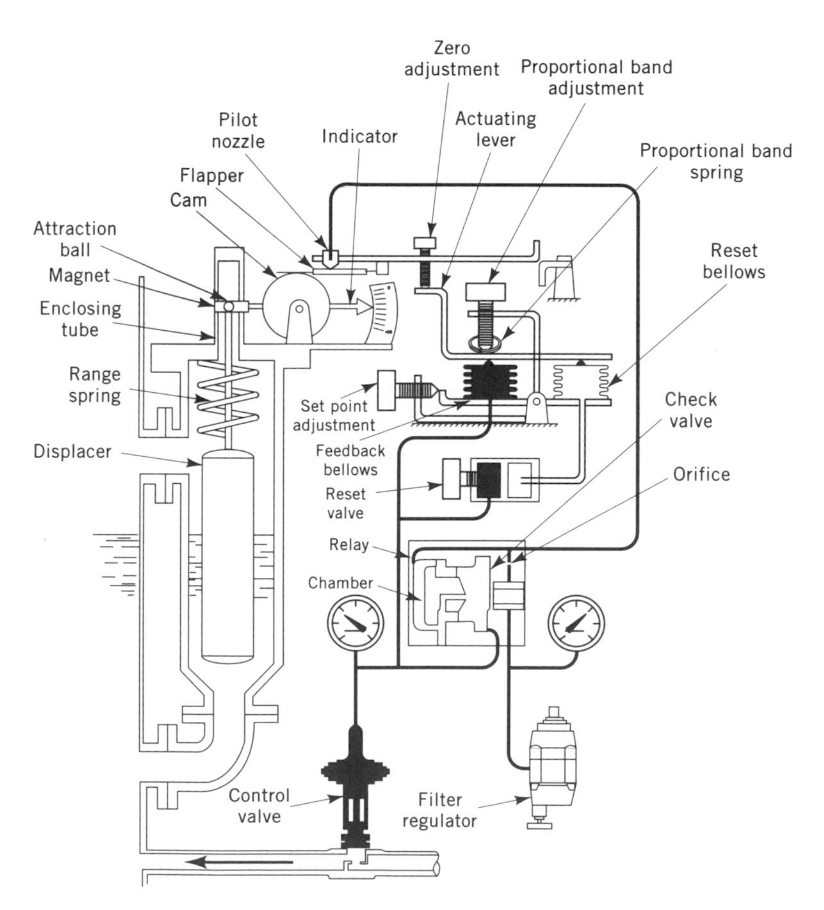

Fig. 5. Buoyancy level controller. Courtesy of Magnetrol International, Inc.

have a more stable output signal on turbulent level, and can be field calibrated without moving process level.

Conductivity. Conductivity level switches are generally limited to applications with low pressure, conductive fluids (high pressure models are available). They are alarm or pump control devices. Metal rods are inserted into the vessel with low level a-c voltage applied (Fig. 6). The conductive liquid serves to complete the electrical path from probe to ground. Multiple set points may be obtained by insertion of multiple rods. Set point is determined by rod length. Conductance level switches are not recommended on applications with volatile fluids or explosive atmospheres. Contamination of the rods can cause the unit to malfunction. Materials forming a conductive coating between the two rods cause the unit to continually show a high level. Nonconductive materials coating the rods insulate the rods from the process level causing the units to fail in a low level condition.

Capacitance. Capacitance-based measurement devices, like conductivity devices, utilize the electrical properties of the medium to derive its measurement. Unlike conductivity devices, capacitance can be used to measure either conductive or nonconductive media (dielectric value: nonconductive $< 10 >$ conductive). Capacitance is developed when an oscillator impresses a high frequency a-c signal across two conductive plates separated by an insulating material, or dielectric (Fig. 7). The amount of capacitance generated is dependent on the frequency of the a-c signal, the size of the conductive plates, the distance between the conductive plates, and the dielectric value of the insulating material. In industrial process measurement, a metal probe is one plate of the capacitor while the metal tank wall is the other plate, or ground reference (Fig. 8). When a tank is empty it is actually full of air which has a dielectric value of 1. All materials on earth have a dielectric value greater than 1; therefore, as the tank is filled with material having a dielectric greater than air, the amount of capacitance developed by the system increases. This increase or decrease in capacitance, proportional to the level change, can then be used with on/off or analogue devices.

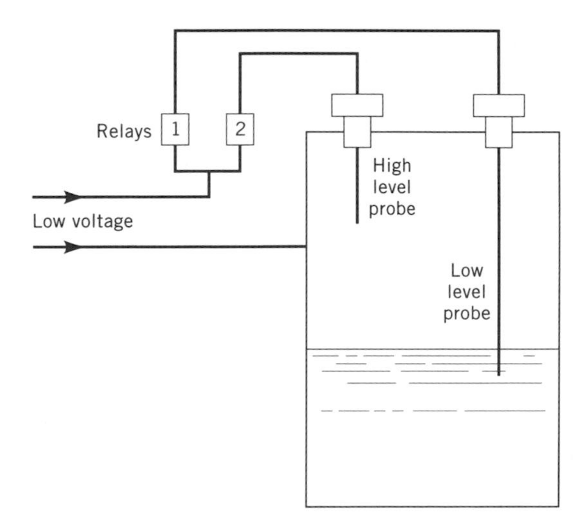

Fig. 6. Electrode system for detecting high and low levels in a conductive field.

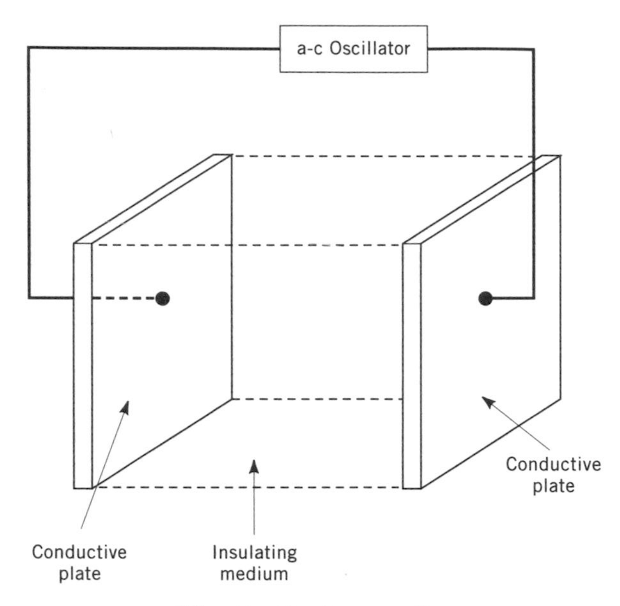

Fig. 7. Basic capacitor.

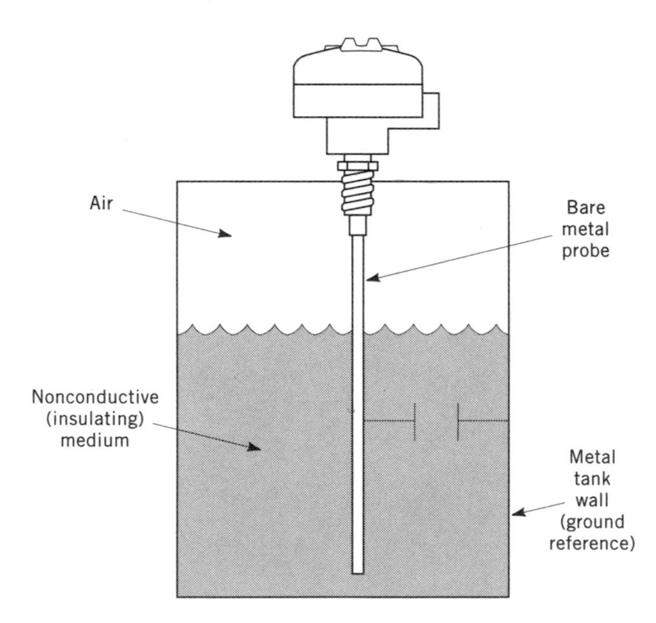

Fig. 8. Capacitance-based level measurement, nonconductive media.

When the process medium is electrically conductive (dielectric values > 10), the capacitor developed above does not work; the insulating material needed between the two conductive plates is lost. The conductive liquid surrounding the probe acts as a short circuit to the tank wall (second plate of the capacitor). To reestablish the dielectric (insulating material), the probe can be insulated with a nonconductive material such as tetrafluoroethylene (TFE), poly(vinylidene fluoride) (PVDF), poly(vinyl chloride) (PVC), etc. The capacitor exists between the probe rod, through the thickness of the insulation (dielectric), to the conductive liquid which is now acting as the second plate of the capacitor, or ground reference (Fig. 9).

In nonmetallic vessels, the second plate of the capacitor is missing and must be supplied. A stillwell probe, one with a concentric metal tube, is utilized. The concentric tube supplies the second plate. Stillwell probes have numerous other uses. In applications of nonconductive media, a stillwell probe is more sensitive and supplies a greater amount of capacitance because the ground reference is so close to the probe. Further, if a tank wall offers a ground reference that is a varying distance to the probe, eg, a horizontal cylinder, the stillwell offers a much more consistent (linear) ground reference.

Capacitance is an extremely flexible level measurement technology. Although the approach is electronic, the sensing probe is a mechanical object serving an electronic function in the circuit. Because it has no electronic parts, it can be constructed to withstand extremes of temperature, pressure, corrosive media, etc. Temperatures of 538°C (1000°F) and pressures of 35 MPa (5000 psi) are not uncommon. Probes can be constructed as rigid rods to lengths of 3–6 m, and as flexible cables to lengths of many hundreds of feet.

Capacitance-based systems can be utilized in hazardous environments in two ways, explosion-proof housings and intrinsically safe circuitry. Explosion-proof housings can be used to contain an explosion that might ignite inside the enclosure. These electronics are usually qualified with certain probes so any explosion does not escape the housing into the atmosphere or back into the vessel. Intrinsically safe circuitry precludes an explosion from occurring by limiting the energy impressed on the probe. At these low energy levels, there is not enough energy to cause ignition. An approved intrinsically safe barrier is installed in the loop in a nonhazardous location to preclude dangerous energy levels from entering the hazardous area even during component failure.

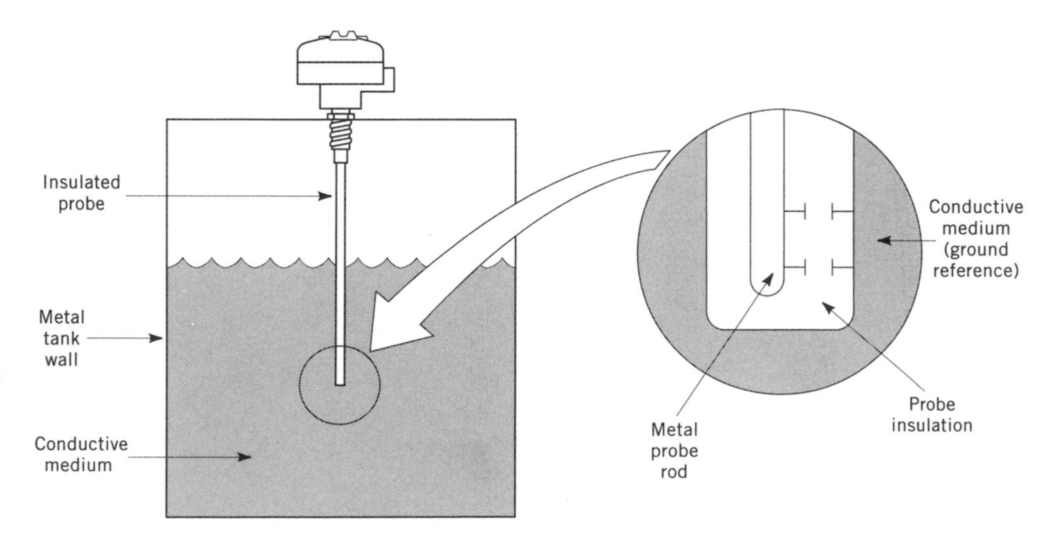

Fig. 9. Capacitance-based level measurement, conductive media.

There are two general weaknesses associated with capacitance systems. First, because it is dependent on a process medium with a stable dielectric, variations in the dielectric can cause instability in the system. Simple alarm applications can be calibrated to negate this effect by calibrating for the lowest possible dielectric. Multipoint and continuous output applications, however, can be drastically affected. This is particularly true if the dielectric value is less than 10. Secondly, buildup of conductive media on the probe can cause the system to read a higher level than is present. Various circuits have been devised to minimize this problem, but the error cannot be totally eliminated.

Static Pressure. The static pressure system offers an inferential method of measuring liquid level that is flexible and convenient, especially when there is considerable change in level. It is based on the fact that the static pressure exerted by a liquid is directly proportional to the height of the liquid above the point of measurement regardless of the volume in the tank, provided that the specific gravity remains constant. For a given type of liquid, the specific gravity is a function of the temperature; therefore, in using the static pressure method a correction must be made for the temperature of the liquid before the exact height can be determined. However, if the mass of the liquid above the point of measurement is desired, it is necessary only to multiply the pressure measurement by the average cross-sectional area of the vessel. Any measurement, then, that measures pressure can be calibrated in terms of the height or mass of a given liquid and can be used to measure either of these variables in vessels under atmospheric pressure (see PRESSURE MEASUREMENT). Differential pressure measuring instruments should be used when the liquids are in closed vessels and under nonatmospheric pressure.

Open Vessels. When a pressure gauge is used to measure liquid level in a vessel under atmospheric pressure, the pressure tap is located at the approximate minimum level line of the vessel. If the gauge is not at the same elevation as the pressure tap, it can usually be recalibrated to compensate for the head effect on the gauge line. If the liquid being measured contains entrained solids or would have a corrosive effect on the gauge, a seal pot can be used, as shown in Figure 10. The sealing liquid should have a higher specific gravity than the liquid that is being measured. When an instrument measuring the level in an elevated, open tank is located at ground level, the total hydrostatic head at

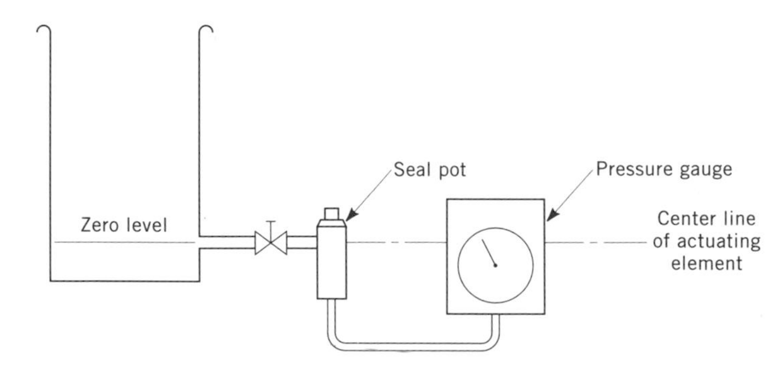

Fig. 10. Pressure gauge, calibrated in terms of liquid height, for measurement of level in open tanks.

the instrument is the sum of the head of the liquid in the elevated tank plus the elevation head of the tank above the ground. For such applications, instruments are available with pressure elements designed to compensate for the elevation head and to measure only the liquid head in the tank. For example, a measuring element can give full-scale indication for a level change of 6 m of water at an elevation of 30 m.

When the pressure gauge cannot be located at the minimum tank level, a diaphragm box is used. It contains a relatively large amount of air compared with that in the pressure measuring element. The pressure exerted against the underside of the diaphragm by the liquid head compresses the air within the box until the air pressure is equal to the liquid pressure. The gauge measures the air pressure, but is calibrated in terms of liquid level. Figure 11 shows the open and closed diaphragm box. The open type is immersed in the liquid in the vessel, and the closed type is mounted externally and connected to the vessel by a short length of piping. The former is used with liquids containing some suspended material, whereas the latter is for clear liquid only. Neither type should be located more than 15 m from the gauge.

An air purge or bubbler system used for corrosive liquids or slurries where the gauge can be located up to 30 m from the point of measurement is shown in Figure 12. Corrosive liquids and those with relatively large amounts of suspended material can be handled. As shown in the diagram, an air line is immersed in the liquid to the minimum level, whereas the pressure and volume of air supply are controlled by a regulator to give a small bubbling of air when the vessel is full. The pressure in the air line is then equal to the back pressure exerted by the hydrostatic head of the liquid. The measurement of this air pressure is, therefore, equal to the measurement of the static pressure of the liquid, and thus of the liquid level. The air–purge supply should always be connected at or near the point of measurement, not at the gauge, in order to minimize error caused by friction loss in the air flow through the connecting tubing. The use of a differential regulator and a rotometer is recommended to obtain the same rate of flow regardless of level. A high flow rate can cause errors through friction losses.

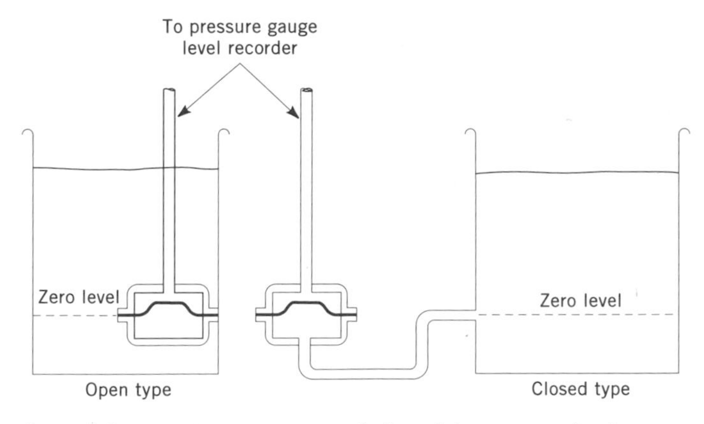

Fig. 11. Diaphragm-box systems, open and closed types, used where pressure gauges cannot be located at the minimum tank level.

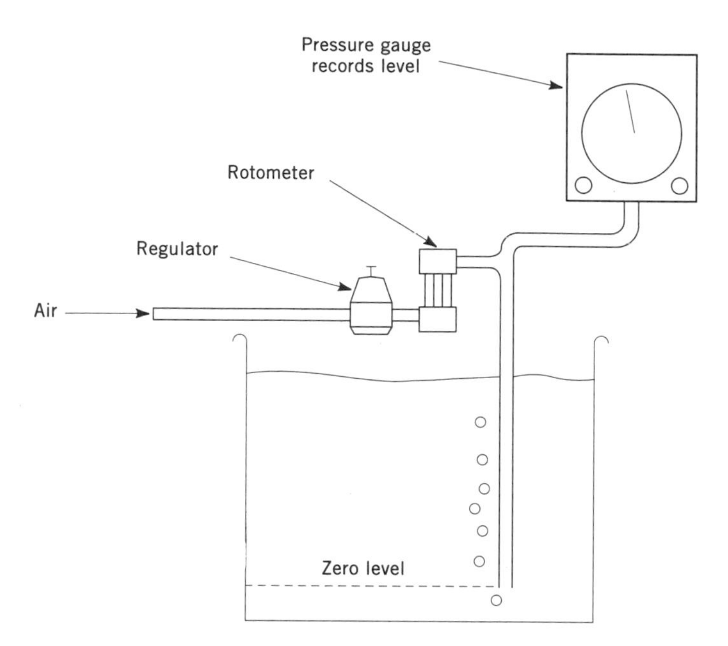

Fig. 12. Air–purge or bubbler system.

Closed Vessels. Liquid level can be measured by the static pressure method also at nonatmospheric pressures. However, in such cases the pressure above the liquid must be subtracted from the total head measurement. Differential pressure measuring instruments that measure only the difference in pressure between the pressure tap at the bottom of the tank and the pressure in the vapor space are used for this purpose. At each tap, the pressure detected equals the liquid head pressure plus the vapor pressure above the liquid. Since the pressure above the liquid is identical in both cases, it cancels out. Therefore, the change in differential pressure measured by the instrument is due only to the change in head of liquid in the vessel. It is independent of the pressure within the tank and is an accurate measure of the level.

Figure 13 shows a typical installation of a differential pressure instrument for closed tanks. Connections from the instruments are made to taps in the vessel at minimum and maximum levels. Between the instrument and the maximum level tap is a constant reference leg. This leg is filled with liquid until its head is equivalent to the head of the liquid in the vessel at maximum level. The reference leg must remain constant, with no formation of vapor under varying ambient conditions. On some applications it may be necessary to fill the reference leg with a liquid, such as water or a light oil, that remains stable. If the liquid used in the reference leg has a higher specific gravity than the liquid in the tank, the resulting difference in head must be corrected for in the instrument. Most differential pressure measuring instruments are equipped mechanically to suppress this difference.

For applications where a second liquid cannot be used in the constant reference leg, a self-purging system can be installed, as shown in Figure 14. Here the piping is filled with the vapor of the measured liquid instead of the

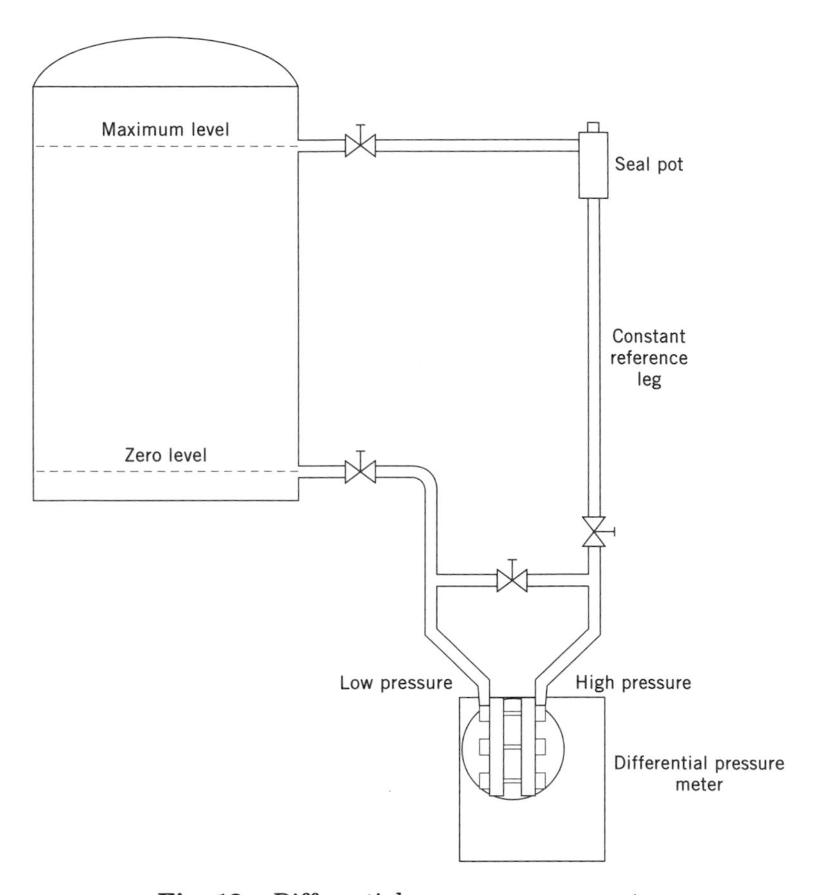

Fig. 13. Differential pressure gauge system.

liquid itself. The differential pressure transmitter is located at the upper vessel connection. The pressure at the high pressure side is now the vapor pressure in the vessel plus the head of the liquid; that at the low pressure is the vapor pressure alone. The difference between the two pressures is equal to the liquid level. The horizontal pipe, L, must be long enough to prevent a liquid head from forming in the vertical connection to the transmitter, as this would result in a low reading. If ambient temperature causes a head of liquid to accumulate, in spite of the length L, the connecting piping requires steam or electric tracing.

A differential pressure transmitter with remote diaphragm seals offers another convenient method of measuring liquid level in closed tanks (Fig. 15). A sensing diaphragm at the upper flange connection transmits the vapor pressure at the tank top through a liquid-filled stainless steel capillary to one side of the differential meter. Another sensing diaphragm at the flange near the bottom of the tank transmits the total vapor pressure plus the variable liquid head $H1$ through another oil-filled capillary to the other side of the differential meter. The meter measures the difference between these two pressures, which is the variable head $H1$. The high pressure side of the meter is connected to the upper flange and, when the tank is empty, an elevated zero adjustment on the meter is used to balance out the fixed-liquid reference leg $H2$ in such a way that the

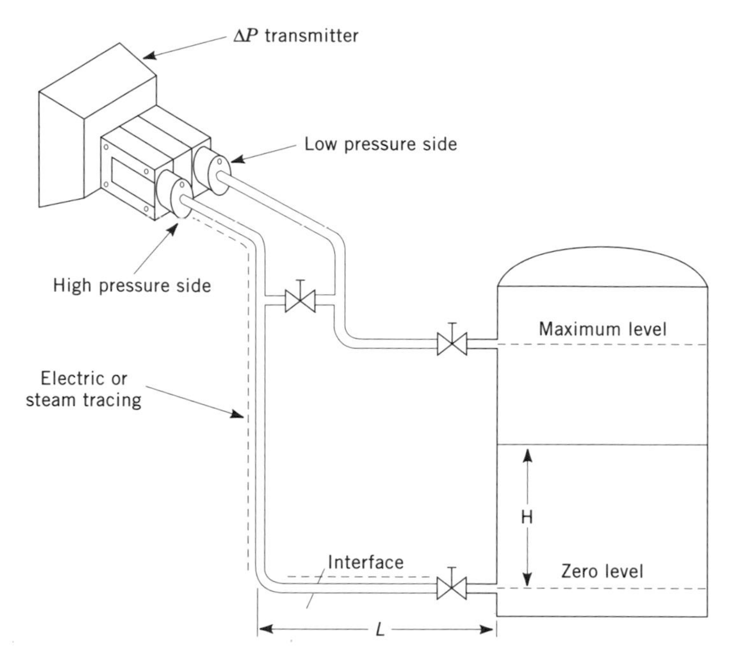

Fig. 14. Self-purging system with the differential pressure (ΔP) transmitter located at the upper level connection.

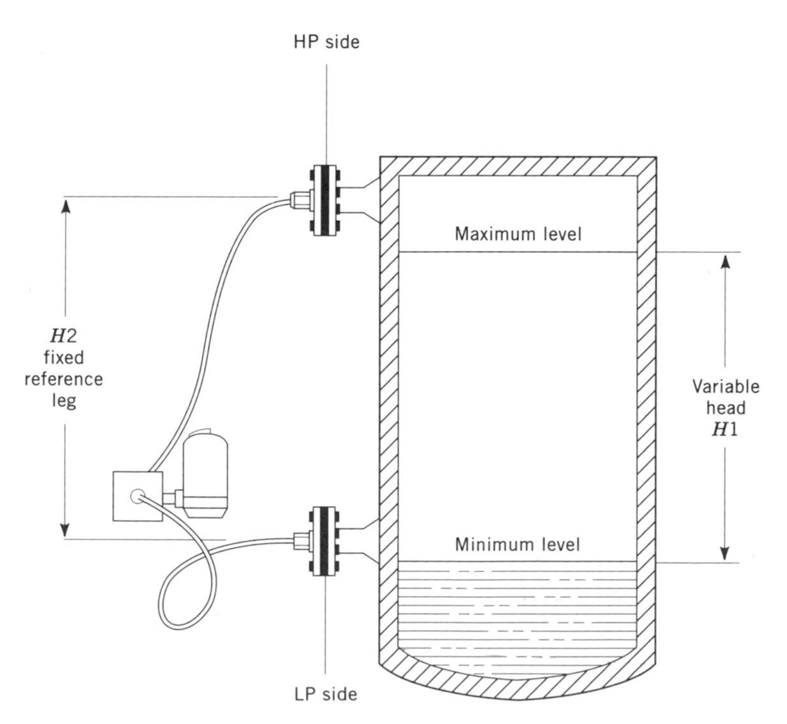

Fig. 15. Differential pressure transmitter with remote seals.

output of the meter reads zero. When the liquid rises in the tank, the actual differential pressure across the meter body decreases. However, the electronic circuit of the meter is arranged for reverse action and, therefore, the output increases as the level rises. The advantage of this system is that no seal pots or heat-traced lines are required and there is no chance for solid materials, such as in a slurry, to plug up the connecting lines to the meter.

Differential Pressure. Differential pressure transmitters designed for liquid level measurements use solid-state electronics and have a two-wire 4–20 mA d-c output.

The Series 1151 differential pressure transmitter manufactured by Rosemount (Minneapolis, Minnesota) uses a capacitance sensor in which capacitor plates are located on both sides of a stretched metal-sensing diaphragm. This diaphragm is displaced by an amount proportional to the differential process pressure, and the differential capacitance between the sensing diaphragm and the capacitor plates is converted electronically to a 4–20 mA d-c output.

Foxboro's Model 823 transmitter uses a taut wire stretched between a measuring diaphragm and a restraining element. The differential process pressure across the measuring diaphragm increases the tension on the wire, thus changing the wire's natural frequency when it is excited by an electromagnet. This vibration (1800–3000 Hz) is picked up inductively in an oscillator circuit which feeds a frequency-to-current converter to get a 4–20 mA d-c output.

The Honeywell ST 3000 transmitter contains a solid-state sensing element. A Wheatstone bridge resistance circuit is diffused into a single-crystal silicon chip, creating an integrated piezoresistive sensor. An area on the back side of the chip is etched out to a precise thickness, in such a way that a given differential process pressure across the chip gives the desired bridge output change. The bridge output is amplified and converted to a 4–20 mA d-c output signal.

All these devices are filled with silicone oil and have low gradient, corrosion-resistant barrier diaphragms on both the high and low pressure sides of the sensor.

Ultrasonic. Ultrasonic level devices are based on measuring the propagation of inaudible sound waves through air, liquids, or metals at a frequency range of 20 kHz to 4 mHz. This is a mechanical process of compression and expansion initiated by a vibrating material. This vibration is induced by a piezoelectric crystal with an alternating current of a frequency equal to the frequency at which the material vibrates most easily. The piezoelectric crystals are typically made from a lead zirconate or barium titanate compound which converts electrical energy to mechanical energy and vice versa.

The use of ultrasonic energy is different in on/off switches and in transmitters. Switches act on the attenuation of the acoustic signal in the gap between two crystals, while transmitters measure the time of flight of the ultrasonic pulse.

Switches. An ultrasonic point sensor is constructed from two piezoelectric crystals mounted opposite each other in a plastic or metallic body, and separated by a gap (Fig. 16). One crystal is connected to the input of an amplifier and transmits acoustical energy across the transducer gap, while another identical crystal is connected to the output of the same amplifier and becomes the receiver. This technique transmits the acoustical energy at frequencies from 1 to 4 mHz. At this high frequency, air becomes a deterrent to the transmission of the signal,

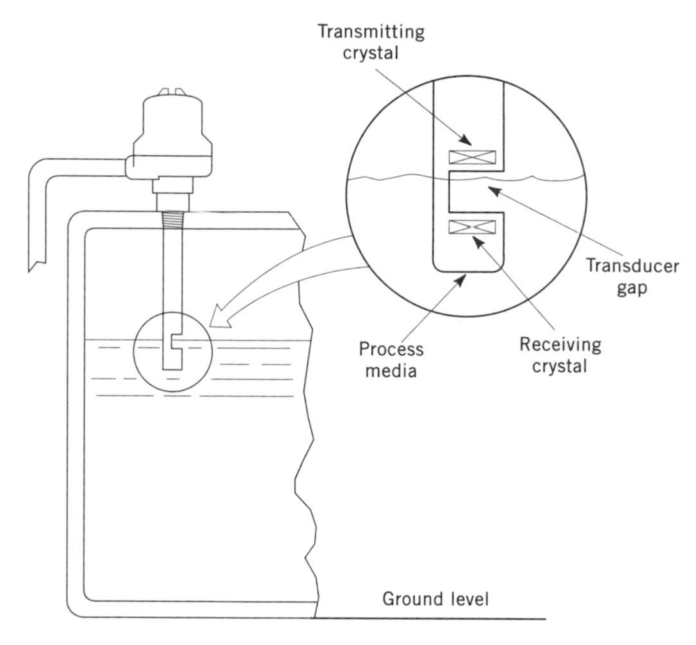

Fig. 16. Typical gap sensor application.

attenuating the acoustical energy. When the gap is filled with air (or gas), the acoustical signal is not allowed to transmit, and the amplifier remains idle. Conversely, when the gap is filled with liquid, a coupling path is provided to propagate the signal. Once the receiver signal is detected, the amplifier becomes an oscillator which causes a relay circuit or current shift output to indicate a wet condition. As soon as liquid is removed from the gap, the amplifier returns to the idle state. The ultrasonic sensors are positioned in a tank at a point where level is to be controlled. The sensor design incorporates single or multiple gaps to provide optimum level control, eg, high alarm, pump control, etc. This type of measurement is unaffected by changes in temperature viscosity, specific gravity, dielectric constants, or conductivity. Installation is extremely easy and needs little calibration.

The sensors can be constructed in a wide variety of materials and lengths to suit most application conditions. These sensors can be used in almost any liquid that does not have the ability to form a crystallized coating on the sensor face. The coating can, in some instances, cause false attenuation of the acoustical signal. In addition, severe aeration, liquid temperatures above 260°C, and a high percentage of solids may also attenuate the signal to a degree that causes a false level indication. Self-test circuits can be employed to indicate the integrity of the unit.

Transmitters. The use of sonic or ultrasonic sound pulses to measure level on a continuous basis is known as air sonar. In its most elementary form, an electronic circuit applies multiple bursts of high voltage energy to a transducer crystal. The burst of electrical pulses causes the transducer crystal to generate an acoustical pulse at a specific oscillating frequency, typically 20 to 55 kHz. The pulse propagates through the air (or vapor) and is reflected back to the transducer from the liquid surface (Fig. 17). At the transducer, the acoustical pulse is converted back into electrical signals by the transducer and receiver circuits. Based on the microprocessor's counter/timer, the instrument knows the precise time when the crystal was charged and the elapsed time between transmission and reception of the acoustical signal. This time function is represented by the relationship $t = 2d/V_a$, where t represents time, d is the distance between the transducer and the liquid surface, and V_a is the velocity of sound in air (or gas). Solving for d, $d = V_a\ t/2$. The total transit time down and up through the air space is proportional to the distance from the transmitter to the liquid surface.

This method of measuring level is highly desirable because it is a noncontacting technique. There are no mechanical parts and the acoustical signal is typically not affected by the physical properties of the liquid. An adjustment must

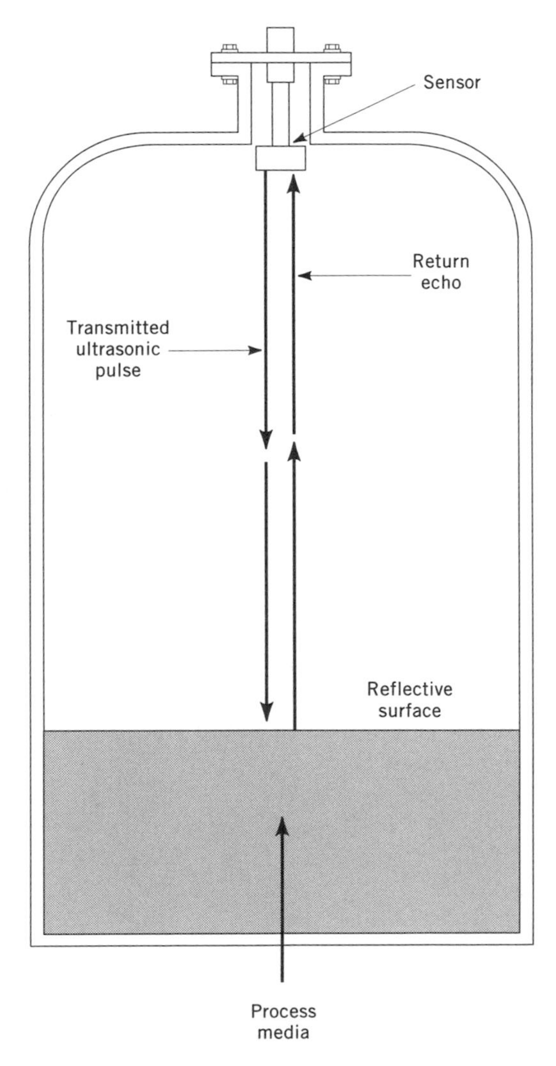

Fig. 17. Ultrasonic air sonar.

be provided in the electronic circuit to correct for the changes in temperature of the vapor space through which the signal passes. Temperature variances affect the speed (V_a) of the acoustical signal so a temperature compensation circuit is used. The circuit may be internally potted in the transducer head near the crystal or a separate temperature probe can be used which provides temperature information to the electronics. Acoustical signals transmit much faster and more efficiently at high temperatures, slower at low temperatures. This acoustical rate change caused by the varying air temperatures necessitates the need for a temperature compensation circuit. A few other considerations should be taken into account when applying this technology. Dust particles and vapors that affect the speed of sound, high temperatures, and operating pressures exceeding ~700 kPa (100 psig) affect the measurement. Since the calculation in the microprocessor is based on the speed of sound in air, vapors with higher or lower densities impede the transmission of the sound wave to a certain degree. The ultrasonic signal slows down or speeds up and therefore induces some error into the measurement. If the liquid surface has foam or excessive turbulence, the acoustical signal may not have a good reflective target (this type of level measurement should not be used in mechanically operated tanks). There are certain foams that absorb the acoustical signal and others that have a good reflective surface.

Microwave. Microwave devices utilize high frequency energy to make their measurement. Microwave is defined as being electromagnetic energy in the high frequency spectrum between 1 GHz and 1 THz. One significant advantage is it can be made nonintrusive. Microwave energy has the ability to be transmitted through a nonconductive window (process seal) thereby maintaining the integrity of the vessel. The implementation of microwave energy, as with ultrasonic, is different between on/off (presence or absence) switches and transmitters (continuous measurement).

Switches. Microwave switches are low energy devices that send a high frequency (usually 5.8 to 24 GHz) signal at a target (process level) and measure its return. It utilizes the strength (amplitude) of the return signal as an indicator of the presence or absence of a process level. The dielectric value of the medium determines the amount of energy reflected back to the unit. Air, having a dielectric value of 1, returns very little energy. When the process is above the unit, the strength of the return signal is greater, signaling a high level condition.

Transmitters. Microwave transmitters for process level measurement are radar-type devices. Radar (radio detection and ranging) devices, like ultrasonic (air sonar) units, bounce a high frequency signal off the process level and measure its time of flight. Radar devices, however, use high frequency electromagnetic energy (in the 5.8–24 GHz range). Aviation radar simply sends a pulse of energy out at the speed of light, and times its return signal. When measuring over long distances (miles) this is a valid technique. For the relatively short distances used in process level measurement, the timing of such a signal is impractical. Because it is extremely difficult to measure such short time sequences, the frequency modulated continuous wave (FMCW) technique is utilized. FMCW transmits a continuous stream of energy swept across a certain bandwidth, eg, 1 GHz. The return signal cannot be measured by simple timing circuits since it will never be the exact frequency as the transmitted signal. However, if the rate of the sweep is known, distance can be derived by measuring the difference in frequency between the transmitted and received signals.

Radar transmitters have a number of advantages. Among them are the ability to sense through nonconductive process seals, operate in heavy vapors and dust, and excellent accuracy (±1 mm). The greatest disadvantage to date (1995) is high cost. Process accuracy (±0.25%) devices range from $6500 to $10,000, while inventory accuracy (±1 mm) is greater than $10,000.

Fiber Optic. Fiber optic level switches are normally limited to free flowing, noncoating fluids at low temperatures and pressures. They are alarm devices utilizing nonmetallic sensors. Pulsed light signals from a light-emitting diode (LED) source are transmitted via fiber optics (qv) to a prism where they are reflected back to a photodiode receiver through another fiber optic cable (Fig. 18). When a liquid with a refractive index higher than 1.4 starts to cover the prism, the pulsed light is refracted rather than reflected by the prism, preventing the light pulses from reaching the photodiode (see LIGHT GENERATION, LIGHT-EMITTING DIODES). This change is detected by a control monitor actuating a switch.

Thermal Dispersion. Thermal dispersion level switches are used on applications where multiple shifts in liquid characteristics are present. The unit is responsive only to a change in the thermal conductivity of the liquid and ignores shifts in specific gravity, dielectric, density, temperature, and pressure. Units are used for alarm signal; however, pump control may be obtained using two units with a latching relay.

The thermal switch consists of a dual element-sensing assembly wired to an electronics package. Each element of the sensor assembly contains a miniature resistance temperature detector (RTD) tightly encased in a tube. One element provides a reference to the process conditions thus providing temperature compensation over the operating range. The second element is internally heated to establish a temperature differential above the process temperature. When the sensors are dry, the temperature differential is greatest. A cooling effect on the heated RTD, caused by the presence of level, decreases the differential temperature between the two elements. This temperature is converted to actuation of the switch by the electronics package (Fig. 19). Units must be mounted at the set point location. Calibration of the unit is required after installation. Thermal switches also provide for self-checking to verify functionality and also include a

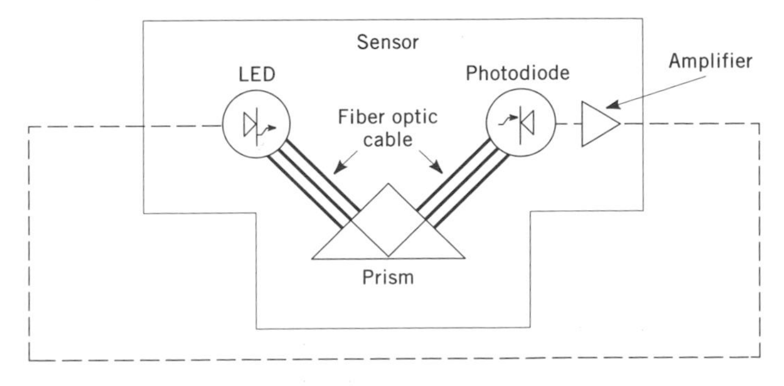

Fig. 18. Point-level device using fiber optics.

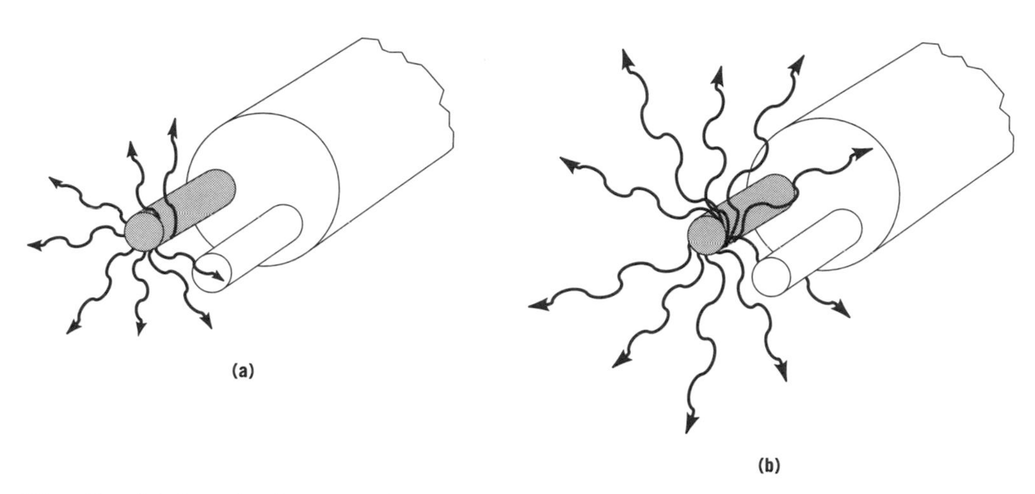

Fig. 19. (**a**) Low level sensor. In the absence of media, the heated sensor tip causes a temperature differential between the two sensors. (**b**) High level sensor. As media contacts the sensing assembly, heat is dissipated and temperature differential decreases.

time delay relay to prevent switch chatter on a turbulent level. Thermal switches may also be used to monitor the interface of two immiscible fluids. Units must be calibrated to the lower thermal conductivity fluid.

Continuous Level Monitoring. The thermal dispersion technique can also be utilized as a continuous level monitor providing an analogue output of the level in the vessel. This is accomplished utilizing an insertion probe as indicated in Figure 20 along with a separate electronics section. The probe consists of two separate sections: the level measuring section and the dry compensator section. This probe contains a reference sensor and an active sensor. The reference sensor is a continuous RTD which detects the temperature in the vessel providing self-compensation for changes in process temperature. The active sensor consists of a continuous RTD and a heater which is energized with a constant current. Both sensors extend the entire length of the probe. When the active sensor is dry, the heater increases the temperature and the resistance of the active RTD creating a high temperature difference relative to the reference sensor. As the probe is immersed in fluid, heat is dissipated into the fluid media reducing the temperature and resistance of the active RTD. This reduction in resistance is proportional to the insertion depth of the sensor. The electronics measure the difference in resistance between the active and reference sensors and convert this to an analogue output signal. Cooling of the active RTD can also occur due to changes in the thermal conductivity of the air above the liquid level. The thermal conductivity of the air is dependent on various factors including temperature, pressure, and humidity. The dry compensator section of the probe compensates for changes in the thermal conductivity of air ensuring that the measured change in resistance is due only to immersion in the fluid.

Magnetostrictive. When a ferromagnetic material is subjected to a magnetic field, it expands or contracts in a predictable fashion. This phenomenon is the basis for magnetostrictive measurement. The liquid level gauge consists of three primary parts: a ferromagnetic waveguide protected by a solid outer rod,

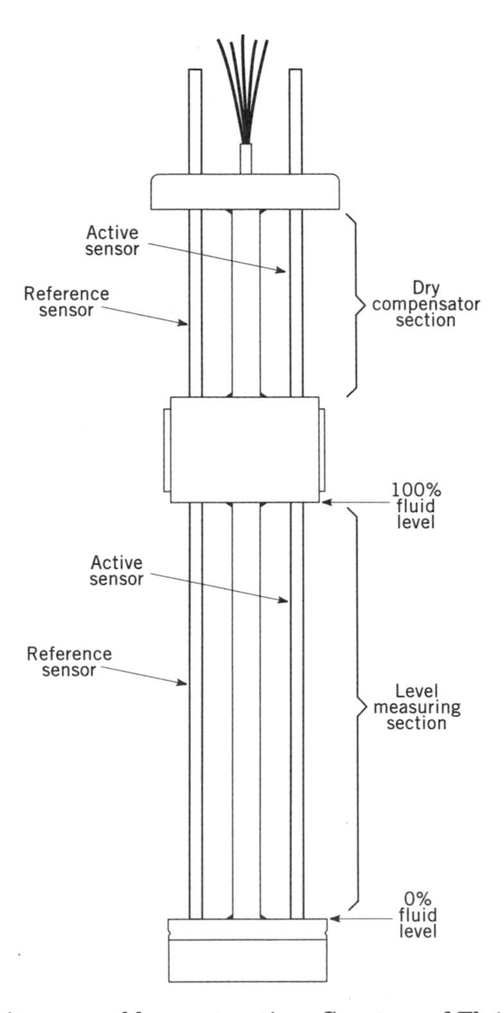

Fig. 20. Monitor assembly construction. Courtesy of Fluid Components.

an electronics assembly that determines the product level based on the waveguide behavior, and a float containing a set of magnets that ride the outside of the gauge's outer rod.

Magnetostrictive gauges typically operate in the following manner. (*1*) The electronics assembly initiates a short, low current pulse onto a wire that runs through the center of the waveguide material. A timer starts simultaneously. (*2*) The pulse, along with the magnetic field it generates, travels the length of the gauge. (*3*) When the pulse reaches the float, the magnetic field from the pulse interacts with the magnetic field generated from the float (Fig. 21) and initiates a torsional twist in the waveguide material. (*4*) The physical twist creates a sonic wave that travels along the waveguide in both directions and is detected by the strain gauge in the electronics assembly. The timer is stopped as soon as this return signal is detected. The distance from the float (magnet) can be determined accurately based on the time and on the signal transmission properties of the

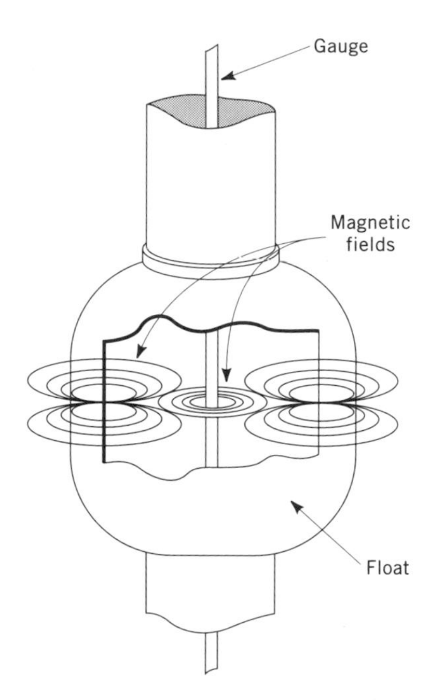

Fig. 21. Magnetostrictive level sensors measure the intersection of two magnetic fields: one in the float, the other in the gauge.

individual waveguide material. (*5*) If a second float is present on the gauge, a second twist can be detected and recorded as the interface level.

Phase Tracking. The principle of phase tracking uses a high frequency transmission line as a sensor. The sensor is comprised of two parallel conductors and hangs vertically in the tank. The electronics transmit a high frequency electrical sine wave down the sensor. This wave creates an electromagnetic field which simultaneously travels around both conductors. The signal travels at a constant velocity to the surface of the stored product. At the surface, the signal is reflected and travels back to the sensor at the same constant velocity.

The signal is reflected from the product surface because there is an abrupt impedance change in the sensor at the air–product interface. Because the electromagnetic field extends outside the two sensor conductors, the sensor impedance depends on the dielectric constant of the surrounding medium. In air (or vacuum) the dielectric constant is unity (1). In all other materials the dielectric constant is always greater than unity. Hence, there is always an echo at the air–product interface because of the difference in dielectric constant. The minimum difference is about 0.5, ie, a minimum dielectric of about 1.5 in the product.

Servo Gauge. Servo gauges are high accuracy, electromechanical devices that are used on inventory control applications where accountability is mandated for custody transfer of liquids. The large, million barrel, bulk terminal vessels are where these devices originally found a niche.

Servo gauges use a displacer as a primary element. The displacer is critically sized and weighted for optimum detection of primary and interface liquids. The displacer is suspended on a cable that is wound on a precision drum located in a housing at the top of the vessel (Fig. 22). The drum is magnetically coupled

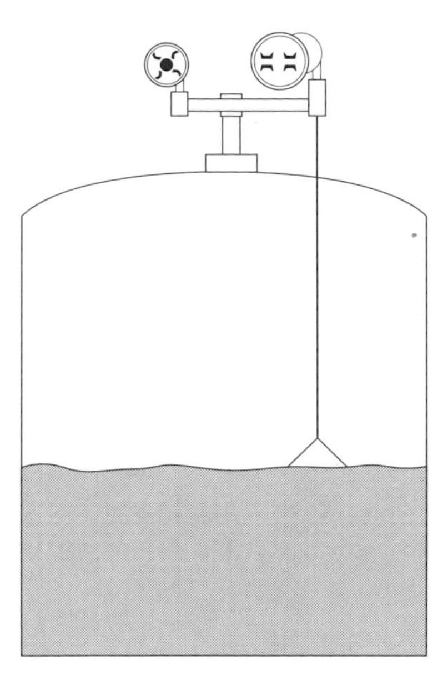

Fig. 22. Servo gauge.

to the drive shaft. An isolation barrier separates the process from the electronics housing protecting the components from the tank vapors. Within the servo housing, a precision stepper motor is contained in a beam assembly. The beam counterbalances the apparent weight of the displacer in air and in the various products being gauged. As the level changes, the corresponding change in the weight of the displacer causes the beam to rotate and reestablish equilibrium. An optical encoder senses the rotational location of the beam and transmits this information to the microprocessor. The level is calculated from the length of the wire in the vessel which is determined from measuring the drum circumference.

In many applications temperature compensation is added to calculate level (or volume) to an industry standard value, usually the American Petroleum Institute (API).

Some servo gauges also have the ability to measure interface. This can be very important when water accumulates in the bottom of the vessel over time (water bottoms). In this way, the user receives information on the accumulation of water (which will eventually need to be pumped out), and also gets a more accurate reading of the real level of the product being stored.

Radiation. Nuclear radiation level switches and level transmitters are primarily used where process contamination is not allowed, process media prohibits use of other technologies, or where high temperatures prohibit use of other devices. The chief advantage of the nuclear unit is that all elements are completely external to the vessel. Radiation cell(s) are positioned outside of the vessel at the set point. A detecting cell is positioned outside the vessel opposite the radiating cell (Fig. 23). The gamma rays emitted by the radiation cell are partially absorbed by rising liquid. The radiation received by the detecting cell decreases proportionally to the change in level and the unit electronics convert the change to a switch action or a proportional output signal. Calibration of the unit is

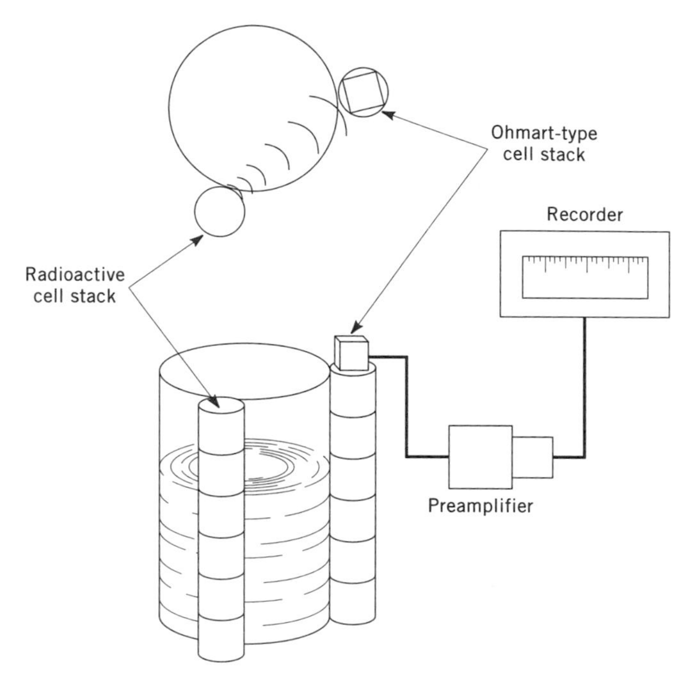

Fig. 23. Nuclear radiation level gauge, using an Ohmart-type cell stack.

required after installation. Cobalt-60, cesium-137, and radium-226 are the source materials normally used. Source decay can cause signal shift unless automatic compensation is provided. There are many considerations required prior to selecting radiation level devices. The device manufacturer should be consulted for all applications. A license from appropriate federal and state agencies is normally required.

Economic Aspects

Following is a list of suppliers for level sensing technologies.

Technology	Supplier	Technology	Supplier
sight glasses	John C. Ernst Co.	ultrasonic	Bestobell-Mobray
	Penberthy Inc.		Endress & Hauser
			Magnetrol International, Inc.
dip sticks	B&K, Inc.		Milltronics
	Bagby Gage Pole Co.		Sensall
magnetic liquid level indicators	Champ Tech	microwave	Endress & Hauser
	K-Tek		Krohne
			Magnetrol International, Inc.

Technique	Manufacturers
floats	Magnetrol International, Inc. SOR, Inc.
displacers	Magnetrol International, Inc. SOR, Inc.
buoyancy	Fisher Controls. International, Inc. Magnetrol International, Inc.
conductivity	B&W Controls Warrick Controls Yarway
capacitance	Bindicator Drexelbrook Endress & Hauser Magnetrol International, Inc. Princo Robertshaw Controls
pressure/differential pressure	Foxboro Rosemount Smar Honeywell
microwave	Saab TN-Canonbear Vega
fiber optic	Besta Genelco (Bindicator) Honeywell Microswitch Moore Technologies
thermal dispersion	FCI Kurz Magnetrol International, Inc. Sierra
magnetostrictive	MTS-Temposonics Magnetek Petrovend
phase tracking	CTI-Celtek
servo gauge	Enraf-Nonius L&J Whessoe-Varec
radiation	KayRay-Sensall Ohmart Ronan Texas Nuclear

BIBLIOGRAPHY

"Liquid-Level Measurement" in *ECT* 2nd ed., Vol. 12, pp. 481–499, by R. C. Whitehead, Jr., Honeywell, Inc.; in *ECT* 3rd ed., Vol. 14, pp. 427–448, by R. C. Whitehead, Jr., Honeywell, Inc.

1. *Stevens Water Resource Data Book*, 5th ed., 1991, pp. 13–15.

General References

R. C. Whitehead, *Liquid Level Measurement*, Vol. 12, 2nd ed., pp. 481–499.

V. N. Lawford, *Instrumentation Technol.* **21**(12), 30 (1974).

W. W. Schoop, *Instrumentation Control Systems* **46**(5), 73 (1973).

G. D. Anderson, *Power* **116**(9), 35 (1972).

P. S. Buckley, *ISA Transcript* **12**(1), 45 (1973).

C. F. Cusick, *Combustion* **41**(11), 23 (1969).

T. S. Imsland, *Instrumentation Control Systems* **42**(5), 120 (1969).

I. H. Cohn and W. E. Dunn, *Control Engineering* **15**(1), 51 (1968).

JERRY BOISVERT
BOYCE CARSELLA, JR.
WAYNE SHANNON
TED WILLIAMS
Magnetrol International, Inc.

LITERATURE ON CHEMICAL TECHNOLOGY.
See INFORMATION RETRIEVAL; PATENT LITERATURE.

LITHIUM AND LITHIUM COMPOUNDS

Lithium [*7439-93-2*], Li, an element with unique physical and chemical properties, is useful in a wide range of applications. The estimated increase in future demand has led to the development of lower cost resources as well as additional plant openings. Capacity as of this writing (ca 1994) is in excess of demand.

Lithium was first identified in 1817 by Johan August Arfvedson working in the laboratory of Jöns Jakob Berzelius (1). The new element was named *lithos* from the Greek meaning stone. Minute amounts of metallic lithium were first prepared in 1818. Lithium was prepared in larger quantities simultaneously in 1855 by Robert Bunsen (Germany) and Augustus Matthieson (England).

Many of the properties of lithium are similar to those of magnesium and other of the alkaline-earth metals (see MAGNESIUM AND MAGNESIUM ALLOYS). This is in accord with the diagonal relationship principle of the Periodic Table. Resemblance to magnesium includes the high solubility of the halides (except the fluoride) in both water and polar organic solvents and the high solubility of the alkyls in hydrocarbons; the low aqueous solubility of the carbonate, phosphate, fluoride, and oxalate; the thermal instability of the carbonate and nitrate; the formation of the carbide and nitride by direct combination; and the reaction with oxygen to form the normal oxide.

The first commercial use of lithium occurred toward the end of World War I when small amounts were used in an aluminum–zinc alloy, Scleron. After that war lithium was used as a hardener in a lead alloy-bearing material, Bahnmetall. Between World Wars I and II there was little production of lithium materials. Most lithium trade was as ores sold as additives to frit and glass (qv) formulators. In 1942 lithium chemicals were needed for wartime efforts and commercial production of lithium salts, the metal, and metal derivatives increased. Lithium hydride [*7580-67-8*], LiH, was used in military sea rescue equipment to provide a source of hydrogen upon reaction with water. The hydrogen inflated rescue balloons to carry the radio antenna needed for the SOS signal broadcast. During this period, all-purpose temperature-resistant greases using lithium stearate [*4485-12-5*] were produced for military application.

Whereas new applications of lithium compounds were developed, commercial growth was slow. In 1953 worldwide sales of lithium products, expressed as lithium carbonate, were only ca 1000 metric tons (2). In 1954 the U.S. lithium industry underwent a sudden, very large expansion when the U.S. Atomic Energy Commission required large amounts of lithium hydroxide [*1310-65-2*] for its nuclear weapons program (see NUCLEAR REACTORS). Three domestic producers built 4500-t/yr plants to meet contract commitments with the U.S. government. When these government contracts ended in 1960, capacity exceeded demand and several operations were discontinued.

There has been significant growth in lithium and lithium compounds usage since 1960. Applications of lithium metal are mainly in batteries (qv) and in the manufacture of lithium derivatives such as the hydride (see HYDRIDES), amide, nitride, and organolithium compounds (see ORGANOMETALLICS, σ-BONDED ALKYLS AND ARYLS). By 1992 the lithium industry supplied an estimated 39,000 t carbonate equivalent (CE) market, 10,000 t CE as ores, and 29,000 t CE as chemicals (lithium and derivatives). From 1970 through 1993 the demand for lithium chemicals grew from 15,000 to 29,000 t CE, a compound annual growth rate of 3.0%. General reviews are available (3–5).

Geochemistry

Lithium is widely distributed in nature. Trace amounts are present in many minerals, in most rocks and soils, and in many natural waters. The lithium content of the earth's crust is estimated to be from 20 to 70 ppm by weight. This compares with the more familiar lead constituent, estimated at only 16 ppm of the earth's crust, and zinc at 1 ppm. Ocean water contains about 0.18 ppm Li, whereas many natural brines have several hundred ppm and a few brines contain more than 1000 ppm Li. Typical values for igneous rocks range from 35 ppm for granites to less than 1 ppm for ultramafic rocks. Lithium does not readily substitute for calcium or magnesium in carbonate rocks; thus lithium content in limestones and dolomites is low, typically about 8 ppm. Lithium is contained mostly in the accessory minerals, especially in clays (qv). The typical lithium content of shales and clays is about 70 ppm, although some clays and clay minerals contain from several hundred to 5000 ppm Li. In geochemical behavior, lithium differs from the other alkali metals. It forms no feldspar structure and shows very limited ability to substitute for sodium and potassium in feldspars. Rather, lithium shows a marked tendency to substitute for Mg^{2+} and Fe^{2+}. Thus, in igneous rocks, the lithium that is present generally is concentrated in the dark ferromagnesian minerals such as biotite, amphiboles, and pyroxenes (see ASBESTOS).

Lithium-bearing minerals occur mainly in granitic pegmatites, which are coarse-grained igneous rocks composed largely of quartz, feldspar, and mica. Accessory minerals often are present in these pegmatites that contain less abundant elements, such as lithium, tin, and beryllium. Only spodumene [*1302-37-0*], $LiAlSi_2O_6$, and petalite [*1302-66-5*], $LiAlSi_4O_{10}$, are important lithium sources from minerals. Lithium also is present in some sedimentary deposits or in brines associated with granitic pegmatites. It is possible that the concentrations resulted from the extraction of lithium from igneous rocks or volcanic ash by hydrothermal action, with the eventual formation of lithium minerals in the sediments or lithium concentrations in the associated brines. The latter hypothesis is in agreement with the cation-exchange properties of lithium and the high solubility of lithium salts.

Ore Minerals and Deposits. Spodumene, a lithium aluminum silicate, $LiAlSi_2O_6$, has a theoretical lithia [*12057-24-8*], Li_2O, content of 8.03%. It is the most important lithium ore mineral and is characterized by high lithium content, extensive deposits, and processing advantages. It is the principal mineral source for the production of lithium chemicals and, as beneficiated mined ore, it is an important source of lithium in ores used directly in the glass and ceramics (qv) industry. Run of mine ore generally contains 1–2% Li_2O with some deposits giving concentration as high as 4.0%. Concentration to 5–6% is necessary for lithium extraction. Ores for sale are beneficiated to 4.8% for bulk container glass usage and as high as 7.6% for use in tableware, ceramics, frits, and glazes (6). Concentration of run of mine ore is achieved by froth flotation (qv) of the spodumene away from feldspathic sands and other tailings. Natural spodumene is a monoclinic pyroxene having a density of 3.16 g/cm^3. On heating to about 1000–1100°C, natural α-spodumene undergoes an irreversible phase change to the tetragonal β-spodumene, density 2.400 g/cm^3. The open lattice of β-spodumene is much more amenable to chemical attack than is α-spodumene. In North America, principal deposits are in North Carolina, Quebec, Manitoba, and the Northwest Territories. The most extensive deposit worldwide is in Australia. Other deposits are in Brazil, the CIS, People's Republic of China, and Zaire (see MINERAL RECOVERY AND PROCESSING).

Petalite, also a monoclinic lithium aluminum silicate, $LiAlSi_4O_{10}$, has a theoretical Li_2O content of 4.88%. Commercial ores usually contain 3.5–4.5% Li_2O without concentration and are a preferred source of lithia for use in ceramics and specialty glazes. Petalite is monoclinic and has a density of 2.4–2.5 g/cm^3. Heating to high temperature results in an irreversible phase change to a β-spodumene–SiO_2 solid solution that could provide an extractable source for the production of lithium chemicals. Large deposits of petalite occur at Bikita in southern Zimbabwe, Namibia, Brazil, Australia, and the CIS.

Lepidolite [*1317-64-2*] and amblygonite [*1302-58-5*] have selective use in supplying lithia to special glasses, ceramics, and glazes by direct addition of the ore. These are not, however, used for the manufacture of lithium chemicals. Lepidolite is a complex lithium mica of variable composition. Li_2O content is from 3–4%. One formula is $K(Li,Al)_3(Si,Al)_4O_{10}(F,OH)_2$. Rubidium and cesium are usually present in lepidolite, which is an ore source for rubidium (see RUBIDIUM AND RUBIDIUM COMPOUNDS). Commercial grades of amblygonite, $Li(F,OH)AlPO_4$, contain 8.0% Li_2O and 20% P_2O_5 and act as a source of fluorine as well as lithium. A commercial grade of 7.0% Li_2 and 8% P_2O_5 is also available (7).

Brine Sources. Lithium occurs naturally in brines from salars, saline lakes and seawater, oil-field waters, and geothermal brines. Of these sources, lithium is produced only from brines of two salars.

Salars and Lakes. Brines having high lithium concentration are found in salars of northern Chile, southwestern Bolivia, and northwestern Argentina.

Brines of lower lithium concentration are found in salars in the western United States and the Tibetan Plateau. Brines pumped from beneath the surface of the Salar de Atacama (Chile) and Silver Peak (Clayton Valley, Nevada) are used for commercial production of lithium utilizing solar evaporation (see CHEMICALS FROM BRINES). The concentration of selected ions in brines from salars and lakes of potential commercial interest worldwide are shown in Table 1.

The salars, or playas, in South America are dried lake beds within a closed or restricted drainage basin. These are normally composed of a salt crust that is interspersed with varying amounts of sands, clays, and other detritus. The largest constituent of the evaporite in these deposits is sodium chloride (see SODIUM COMPOUNDS, SODIUM HALIDES, SODIUM CHLORIDE). Whereas these salars may have been sea-filled at one time, the mineral contents are not marine in origin, but come from other primary source material, such as erosion of surrounding hills, or from volcanic or other underground activity. The salt crust is usually porous, more so near the surface, and the interstices contain the salt brines. South American salars are at high elevations, from 2300–3600 m. The largest, Uyuni, covers approximately 9000 km^2.

Oil Fields. Oil field waters in the United States containing lithium have been identified in 10 states. The greatest concentrations are in waters from the Smackover formation of southern Arkansas and eastern Texas. Concentrations from this formation have been measured from 300–600 ppm in waters originating at a 2500–3300 m depth. Recovery of lithium from this resource would only be commercially feasible if a selective extraction technique could be developed. Lithium as a by-product of the recovery of petroleum (qv), bromine (qv), or other chemicals remains to be exploited (12).

Geothermal Sources. Geothermal brines in the United States containing lithium from 55–238 ppm occur near the Salton Sea in southern California in fluids obtained at a depth of about 1500 m. Lower concentrations of lithium occur in other geothermal brines in the Pacific rim countries. Recovery of lithium from geothermal brines would be secondary to the intent of the development of geothermal resources, namely, power generation (qv) and production of fresh water (GEOTHERMAL ENERGY) (13).

Table 1. Salar Ion Concentrations

Country	Salar or lake	Ion composition, wt %				Reference
		Li	Mg	K	Na	
Chile	Salar de Atacama	0.15	0.96	1.80	7.6	8
Bolivia	Salar de Uyuni	0.096	2.0	1.67	9.1	9
Argentina	Salar del Hombre Muerto	0.062	0.089	0.61	10.4	10
United States	Great Salt Lake, Utah	0.006	0.8	0.4	7.0	7
	Salton Sea, Calif.	0.022	0.028	1.42	5.71	7
	Searles Lake, Calif.	0.0083	0.034	2.30	15.20	7
	Silver Peak, Nev.	0.03	0.04	0.8	6.20	7
Israel–Jordan	Dead Sea	0.002	3.40	0.6	3.00	7
People's Republic of China	Lake Zabuye	0.097	0.001	2.64	10.80	11

Clays and Other Sources. Sedimentary deposits, especially lithium-bearing clays found in the western United States, offer an additional source of lithium. These clays contain lithium-bearing trioctahedral smectites, of which hectorite [*12173-47-6*], $Na_{0.33}(Mg,Li)_3Si_4O_{10}(F,OH)_2$, is one mineral. Hectorite usually contains 0.3–0.6% Li or 0.7–1.3% Li_2O. Deposits are found in Nevada, California, Utah, Oregon, Wyoming, Arizona, and New Mexico. The most significant deposits are in southern Nevada, in the Lake Mead area, and in the McDermitt caldera complex on the Nevada–Oregon border. In the McDermitt caldera, lithium probably originated from volcanic sedimentary rocks deposited in the caldera moat. There is evidence that areas of the caldera were hydrothermically active contributing to enrichment of lithium (14). This and other similar deposits are not economically viable as of this writing. These deposits do represent a significant lithium reserve, however, whenever large expansion in demand occurs.

Flint clays and other related rocks are another potential lithium source. These are high alumina clays that are composed largely of well-crystallized kaolinite [*1318-74-1*] and are used for the manufacture of refractories (qv). The lithium content ranges from < 100 to 5000 ppm. Deposits occur in many states, including Missouri, Pennsylvania, and Ohio. Lithium (at ca 1.3%) is present in a chlorite mineral that is similar to cookeite [*1302-92-7*]. High lithium contents may be the reason why some deposits are unsatisfactory for refractory use.

World Reserves and Resources. A summary of demonstrated and inferred worldwide reserves of lithium is given in Table 2. The reserve base and the lithium equivalent represent resources in the ground. An overall lithium recoverability of 65% for pegmatites and 33% for brines must be applied to these values to establish the amount of lithium that could eventually reach the marketplace (15).

Demonstrated reserve quantities are established by measurements including drilling, surface sampling, etc. Inferred reserves are those derived from geological survey information, not by measurement of the extent of the particular reserve. Not included herein are identified marginal and speculative resources, such as the oil-field and geothermal brines and lithium-bearing clays. These latter reserves are speculative as to extent, not existence. Total undiscovered clays in the western United States are speculatively estimated at 15×10^6 t lithium (16). More detailed lists of reserves are also available (15,17).

Recovery

Recovery from Ores and Clays. The preferred method of extraction of lithium from spodumene ore is the sulfuric acid process (18), used on ore concentrates of 5–6% Li_2O, representing 62–74% pure spodumene. Methods suitable for extraction from spodumene also can be used for petalite, because the latter mineral converts to β-spodumene–SiO_2 solid solution on heating to a high temperature.

Sulfuric Acid Process. Natural α-spodumene is virtually unattacked by hot sulfuric acid; thus the first step in the extraction process is the conversion of α-spodumene to the much more reactive β-spodumene by heating to 1075–1100°C in a brick-lined rotary kiln. Although the melting point of β-spodumene is

Table 2. Lithium Reserve and Resource Summary, 1993[a]

	Reserves, t				
	Demonstrated		Inferred		
Country	Total, 10^6	As lithium, 10^3	Total, 10^6	As lithium, 10^3	Reference
	Pegmatites				
Australia					
Greenbushes	33.5	456	35.3	480	2,7
Canada					
Bernic Lake	6.6	92			7
Quebec	14.5	84	10	58	7
Yellow Knife	49.1	319			7
others	12.5	71	11	60	7
United States					
N.C.	55.1	369	62.5	419	2
Zaire	30	180	30	180	15
Zimbabwe	4.1	57			7
	Brines				
Argentina					
Hombre Muerto	221	137	640	400	
Bolivia					
Uyuni	15,600	5,500			9
Chile					
Atacama	4,230	4,400	4,134	4,300	2,9
People's Republic of China					
Zabuye			80[b]	55.1[b]	11
United States	130	31	103	31	2

[a]The sizes of the reserves in the CIS and People's Republic of China are not known.
[b]Derived from information in Ref. 11.

greater than 1400°C, the maximum ore temperature that is reached must be well below 1400°C, otherwise, the gangue minerals present in the ore concentrate could form low melting mixtures with spodumene. The kiln discharge is cooled, ground by ball milling to $< 149\ \mu m$ (-100 mesh), and mixed with 93% (66°Bé) sulfuric acid (density 1.84) in an amount equivalent to the lithium present plus about 35% excess. The mixture is roasted in a small rotary kiln to about 250°C. An exothermic reaction results, starting at about 175°C, in which hydrogen ions from the acid replace lithium ions in the β-spodumene structure to form soluble lithium sulfate [*10377-48-7*], leaving an insoluble ore residue.

The kiln discharge is leached with water, resulting in an impure lithium sulfate solution that contains the excess sulfuric acid and small amounts of aluminum, iron, and other alkali sulfates. The excess sulfuric acid is neutralized with ground limestone. The slurry is then filtered to separate the ore residue, giving a mixed alkali sulfate solution that is free of iron and aluminum but that is saturated with calcium sulfate. The solution also contains magnesium ions derived mainly from the limestone. Magnesium is precipitated using hydrated lime, followed by precipitation of calcium using soda ash or mother liquor containing sodium carbonate generated in subsequent precipitation of by-product sodium sulfate decahydrate. After filtration, the solution is adjusted using sulfuric acid to pH 7–8, followed by concentration in a multiple-effect evaporator to alkali sulfate concentration of 350 g/L, 200–250 g/L of this being lithium sulfate. After a clarifying filtration, lithium carbonate is precipitated at 90–100°C with a 28 wt % soda ash solution. The precipitated lithium carbonate is centrifuged, washed, and dried. Approximately 15% of the lithium remains in the mother liquor, along with residual sodium carbonate and large amounts of sodium sulfate. Cooling to about 0°C separates the greater part of the sodium sulfate as the decahydrate, which is centrifuged and converted to the anhydrous salt for by-product sale. The mother liquor from the sodium sulfate decahydrate precipitation is recycled for lithium and soda ash values.

Other Recovery Processes. Most other processes that have been described for the extraction of lithium can be classified as being either alkaline or ion-exchange (qv). The alkaline methods (19,20) are based on the reaction of the ore with limestone at high temperature in a rotary kiln, sometimes with the addition of salt such as a calcium sulfate or calcium chloride. Sintering occurs in these reactions and it is necessary to grind resulting clinkers prior to leaching. Water leaching following the reaction of ore and limestone alone yields an impure lithium hydroxide solution, whereas the use of additives results in lithium sulfate or chloride solutions. The required high weight ratio of limestone to ore, often nearly 4:1, is a disadvantage as is the necessity of grinding the clinker prior to leaching. As of this writing, the alkaline process is used only in the CIS and in the People's Republic of China.

A great number and variety of methods comprise the ion-exchange group although none is used commercially as of this writing (ca 1994). These methods involve heating the ore, usually β-spodumene and generally at relatively moderate temperatures, in contact with the desired reagent, which is either an acid or an alkali metal salt. The latter may be either fused or in aqueous solution. Ion exchange occurs. The lithium ions in the ore are replaced by hydrogen, sodium, or potassium ions, thus forming the soluble lithium salt of the anion that was used.

One ion-exchange process, which was used for several years by Quebec Lithium Corp., is based on the reaction of β-spodumene with an aqueous sodium carbonate solution in an autoclave at 190–250°C (21). A slurry of lithium carbonate and ore residue results, and is cooled and treated with carbon dioxide to solubilize the lithium carbonate as the bicarbonate. The ore residue is separated by filtration. The filtrate is heated to drive off carbon dioxide resulting in the precipitation of the normal carbonate.

Another ion-exchange method involves contacting β-spodumene with a strongly acidic cationic-exchange resin in the presence of water at 90–150°C and subsequently separating the resin and eluting the lithium values (22). In another process, β-spodumene reacts with a sodium chloride or sodium sulfate solution in an autoclave at 100–300°C to form the corresponding lithium salt in the solution (23). In another ion-exchange process, β-spodumene reacts with fused sodium or potassium acetate; double the amount of salt equivalent to the amount of lithium in ore is used (24). The mixture is heated slowly to 324°C, cooled, the reaction mass ground, and lithium acetate leached countercurrently with hot water. Lithium carbonate is precipitated from the acetate solution by hot concentrated soda ash. Sodium acetate is recovered from the mother liquor and recycled.

Recovery from Clays. Limestone–gypsum roasting and selective chlorination have been demonstrated to be applicable to extracting lithium from clays containing hectorite. However, chlorination techniques do not give lithium extraction recovery above 20% (25). With these processes, lithium silicate in the clay is converted to either water-soluble lithium sulfate or chloride. The limestone–gypsum roast-water leach process (26) is similar to the alkaline roast process for ores. In this process a mixture of clay, limestone, and gypsum is pelletized, dried, and fed to a roasting furnace where lithium in the clay converts at 900°C to water-soluble lithium sulfate. Water leaching of the calcine produces a mixed alkali sulfate and a small amount of calcium sulfate. This solution, together with recycle streams, is concentrated by evaporation. Calcium precipitates as the carbonate and is filtered. Lithium carbonate is precipitated by soda ash addition. The mother liquor from this precipitation is cooled to crystallize glazerite and glaubers salt that is separated. This liquor is recycled for recovery of the lithium values. Lithium recovery in excess of 80% has been demonstrated.

Recovery from Brines. Natural lithium brines are predominately chloride brines varying widely in composition. The economical recovery of lithium from such sources depends not only on the lithium content but on the concentration of interfering ions, especially calcium and magnesium. If the magnesium content is low, its removal by lime precipitation is feasible. Location and availability of solar evaporation (qv) are also important factors.

The Salar de Atacama in the Chilean Andes is the largest commercial source for lithium from brines. Brines of 0.15% lithium, 28% total dissolved solids, are pumped from 30-m deep wells into a series of solar evaporation ponds where concentration is increased and composition altered (27). High solar evaporation rates are achieved owing to strong solar radiation, low humidity, moderately intense winds, and low (25-mm/yr) average rainfall. Freshwater evaporation rates are about 3000 mm/yr (28). In processing, a calcium chloride brine is added, causing precipitation of the sulfate ion as gypsum. Magnesium, sodium, and potassium precipitate as halite, sylvanite, carnallite, and bischofite (29). To maintain brine volume in the ponds, precipitated salts are harvested regularly. More concentrated lithium-containing brines are recovered from precipitated salts by use of thickeners and centrifugation. Lithium concentration of 4–6% is attained in the produced chloride brine, which contains some residual borate and magnesium. Prior to precipitating lithium carbonate, borate is removed by extraction. Magnesium is then removed by two-stage precipitation, using sodium carbonate and lime, respectively. The purified brine is reacted with saturated soda ash solution to precipitate lithium carbonate. Even though the magnesium content of the brines from the Atacama is 6.4 times greater than the lithium content, the lithium concentration is the highest known. This is a very economical operation, partly because of the favorable conditions for solar evaporation and despite the remote location of the salar.

Lithium is also produced from brines at Clayton Valley, Nevada, where lithium is only present at 0.03% in brines pumped from wells of 90–210 m. Magnesium content of this brine is low (0.04%) allowing removal by precipitation with lime. Lime is added in the fourth of nine ponds, raising the pH to 11 and causing hydrated magnesia and gypsum to precipitate. Most of the remaining calcium precipitates in subsequent ponds owing to carbon dioxide absorption. Solar evaporation increases concentrations to saturation and precipitation of sodium chloride and other salts follows. The lithium concentration increases to 0.5%. The resultant brines from the evaporation ponds are then treated with lime and soda ash to remove remaining magnesium and calcium prior to reaction with saturated soda ash solution to precipitate lithium carbonate (30).

The Clayton Valley brine contrasts with other lithium-bearing brines which often contain higher magnesium content. Brine of the Great Salt Lake, for example, contains 0.006% lithium and 0.8% magnesium. Magnesium removal with lime is not feasible because of high reagent cost and the sheer volume of a magnesium precipitate. Instead, lithium chloride would need to be selectively removed from the brine. Various ion-exchange and liquid–liquid extraction methods have been suggested but these methods are not used commercially. In one proposed procedure, lithium chloride is converted to lithium tetrachloroferrate [*15274-95-0*] and is extracted using a water-insoluble organic solvent (31). Another approach involves precipitation of a lithium aluminate from the brine by adding freshly precipitated aluminum hydroxide or by adding aluminum chloride followed by pH adjustment (32). This method is also suggested for use in recovery of lithium from geothermal brines (33).

The lithium-bearing oil-field waters of southern Arkansas and eastern Texas contain high concentrations of calcium chloride. The application of ion-exchange technology has been proposed where lithium is selectively absorbed from brines high in calcium and sodium, and then recovered through regeneration using an aqueous solution of low lithium concentration. Ion adsorption resins for this application are prepared by forming crystalline $LiOH{\cdot}2Al(OH)_3{\cdot}nH_2O$ by reaction of $Al(OH)_3$ suspended in the resin with aqueous LiOH at elevated temperatures. The crystalline $LiOH{\cdot}2Al(OH)_3$ thus formed can be beneficially converted to $LiX{\cdot}2Al(OH)_3$, where X is a halide. These lithium aluminates are useful in selectively recovering lithium ions from solution if the amount of lithium in the aluminate structure is first reduced to a lower concentration, but not completely removed, leaving space in the crystal for taking up a lithium salt until the crystal is again loaded with that salt (34).

Lithium Metal

Properties. Lithium [*7439-93-2*], an alkali metal, has a silvery luster, an atomic number of 3, an atomic weight of 6.941, and a $1s^2\ 2s^1$ electronic configuration. It is the first metallic member of Group 1 (IA) in the Periodic Table and is preceded only by hydrogen and helium. Two stable isotopes are present in natural lithium: ^7Li having an abundance of 92.4 at. % and ^6Li, 7.6 at. %. Lithium, density = 0.531 g/cm^3 at 20°C, is the lightest of all solid elements. In general, the properties of lithium are similar to those of the other alkali metals, eg, ease of oxidation to form a univalent ion, the strongly basic property of the hydroxide, etc. In the alkali metal group, lithium has the highest melting point, boiling point, and heat capacity, and the smallest ionic radius, ie, 60 pm. The ionic radius and the resulting high ionic charge density largely account for the unusual properties and effects of lithium such as the powerful fluxing action of lithia [*12057-24-8*], Li_2O, in ceramic compositions. It remains untarnished in dry air but in moist air, its surface becomes coated with a mixture of LiOH [*1310-65-2*], $LiOH \cdot H_2O$ [*1310-66-3*], Li_2CO_3 [*554-13-2*], and Li_3N [*26134-62-3*] (35).

Thin films (qv) of lithium metal are opaque to visible light but are transparent to uv radiation. Lithium is the hardest of all the alkali metals and has a Mohs' scale hardness of 0.6. Its ductility is about the same as that of lead. Lithium has a bcc crystalline structure which is stable from about −195 to −180°C. Two allotropic transformations exist at low temperatures: bcc to fcc at −133°C and bcc to hexagonal close-packed at −199°C (36). Physical properties of lithium are listed in Table 3.

The reaction of hydrogen and lithium readily gives lithium hydride [*7580-67-8*], LiH, which is stable at temperatures from the melting point up to 800°C. Lithium reacts with nitrogen, even at ordinary temperatures, to form the reddish brown nitride, Li_3N. Lithium burns when heated in oxygen to form the white oxide, Li_2O.

Table 3. Physical Properties of Lithium

Property	Value	Reference
at. vol, cm^3	13.0	
mp, °C	180.5	37
bp, °C	1336	38
first ionization potential, kJ/mol[a]	519	39
electron affinity, kJ/mol[a]	52.3	39
crystal structure	bcc	
lattice constant, pm	350	39
radius, pm		
metallic	122.5	39
ionic	60	
d_{20}, g/cm^3	0.531	40
specific heat, J/g[a]		
at 25 °C	3.55	37
liquid at mp	4.39	37
heat of fusion, J/g[a]	431.8	37
heat of vaporization, kJ/g[a]	ca 21.3	41
electrical resistivity at 20°C, $\mu\Omega \cdot cm$	9.446	42
characteristic spectrum lines, nm		
red	670.8	
orange	610.4	
vapor pressure, kPa[b]		37
at 702°C	0.065	
802°C	0.376	
902°C	1.61	
1002°C	5.47	
1052°C	9.4	
1077°C	12.13	

[a]To convert J to cal, divide by 4.184.
[b]More extensive vapor pressure data for lithium and other metals are given in Ref. 43. To convert kPa to mm Hg, multiply by 7.5.

Lithium reacts with water, evolving hydrogen, but does not ignite unless the metal is finely divided. In this respect, it is less reactive than other alkali metals. Lithium amide [*7782-89-01*], $LiNH_2$, is formed by reaction with ammonia. This direct synthesis is not used for manufacture of commercial quantities, however, because of the difficulties of temperature control and competing alternative reactions. Carbon dioxide reacts with lithium only at high temperatures. Halogens react readily with lithium with consequent emission of light (44). In general, lithium reacts violently with inorganic acids, although it reacts slowly with cold sulfuric acid.

Like the other alkali metals (45), lithium has appreciable solubility in liquid ammonia. A saturated solution at −33.2°C contains 15.7 mol lithium in 1000 g of ammonia, and at 19°C has a density of 0.477, lower than that of any other known liquid. Lithium reacts readily in liquid ammonia to form compounds, eg, lithium sulfide [*12136-58-2*], Li_2S; lithium selenide [*12136-60-6*], Li_2Se; and the mono- [*1111-64-4*] and diacetylide [*1070-75-3*]. Lithium reacts with many organic halides to form organolithium compounds (46), and with carbon to form lithium carbide [*1070-75-3*], Li_2C_2, at temperatures above 800°C (47).

Manufacture. An electrolytic process devised in 1893 (48) resembles the one generally used for lithium production. Molten salt electrolysis from a lithium chloride–potassium chloride mixture is performed using graphitized carbon rod as the anode and a carbon steel cell body as the cathode. Modern U.S. installations employ a 55 wt % LiCl–45 wt % KCl electrolyte at about 460°C. Two grades are produced by electrolysis. The essential difference is in sodium concentration. Metal having about 0.6% sodium is used as a reactant in synthesizing inorganic derivatives, organolithiums, and other organic intermediates. Metal having 100 ppm sodium is used for manufacture of high energy batteries and for alloying with other metals. Sodium concentration is determined by sodium in the LiCl feed, which reports stoichiometrically to the metal. Metallic impurities are less than 100 ppm. Contamination by oxygen and nitrogen is controlled by minimizing the exposure of the liquid metal to the atmosphere and by keeping the cell temperature as low as possible. Cells are fabricated of steels of 0.25–0.3% carbon or of low carbon steels. Low carbon steels are more resistant to decarbonization from the produced lithium, although they are less forgiving of excessive temperatures. Typical current efficiency is 80–85% and lithium recovery is > 98% based on charged chloride.

Lithium metal can also be produced by electrolysis in nonprotic solvents. Other methods based on reduction by metals, eg, aluminum or silicon from the chloride or by magnesium from the oxide, have not had practical commercial application.

Health and Safety. Lithium metal, UN No. 1415, is classified by the United States Department of Transportation as "Dangerous When Wet." The required shipping label which shows this classification identifies the key hazards: emission of flammable gases on reaction with water, corrositivity to eyes and skin, and solid flammability. Lithium does not need the same degree of stringent safety precautions as do the other alkali metals because it does not burn spontaneously upon reaction with water. However, lithium can ignite when its surface films dissolve or if heated, with resulting vigorous fires; hence, contact with moisture should be avoided. Lithium can be handled easily at temperatures up to 225°C in argon, helium, or carbon dioxide.

Liquid lithium is easily ignited in air and, once it has begun to burn, requires special techniques to extinguish. Dry-powdered lithium chloride is an effective extinguisher. Copper powder (Navy 125F sold by The Ansul Co. and others) has UL approval as an extinguishing agent for lithium fires. Another material, LITH-X (The Ansul Co.), is a compound of a special graphite base that extinguishes lithium fires by excluding air and conducting heat away from the burning mass. Water, sand, carbon tetrachloride, and carbon dioxide cannot be used. Lithium is usually stored under mineral oil, or with an oil coating to passivate the surface, in metal drums. If the drum is not airtight, a slow deterioration of the surface occurs and lithium nitride [*26134-62-3*] forms. Under airtight conditions, lithium can be stored indefinitely.

Analysis. Lithium can be detected by the strong orange-red emission of light in a flame. Emission spectroscopy allows very accurate determination of lithium and is the most commonly used analytical procedure. The red emission line at 670.8 nm is usually used for analytical determinations although the orange emission line at 610.3 nm is also strong. Numerous other methods for lithium determinations have been reviewed (49,50).

Economic Aspects. Lithium metal is available commercially in ingots, special shapes, shot, and dispersions. Ingots are sold in 0.11-, 0.23-, 0.45-, and 0.91-kg sizes. Special shapes include foil, wire, and rod. Lithium is available in hermetically sealed copper cartridges and in sealed copper tubes for use in treating molten copper and copper-base alloys. Shot is sold in 1.19–4.76 mm (16–4 mesh) sizes. Lithium dispersions (30% in mineral oil) of 10–50-μm particle size are used primarily in organic chemical reactions. Dispersions in other solvents and of other size fractions can be supplied.

The price of lithium metal decreased drastically between 1925 ($143/kg) and 1965 ($16.50/kg), but because of increased energy and chemicals costs, and other inflationary pressures, the price has actually increased. Prices for lithium ingots ($/kg) were $72.05 in 1993.

Uses. The largest use of lithium metal is in the production of organometallic alkyl and aryl lithium compounds by reactions of lithium dispersions with the corresponding organohalides. Lithium metal is also used in organic syntheses for preparations of alkoxides and organosilanes, as well as for reductions. Other uses for the metal include fabricated lithium battery components and manufacture of lithium alloys. It is also used for production of lithium hydride and lithium nitride.

Metallurgy. Lithium forms alloys with numerous metals. Early uses of lithium alloys were made in Germany with the production of the lead alloy, Bahnmetall (0.04% Li), which was used for bearings for railroad cars, and the aluminum alloy, Scleron. In the United States, the aluminum alloy X-2020 (4.5% Cu, 1.1% Li, 0.5% Mn, 0.2% Cd, balance Al) was introduced in 1957 for structural components of naval aircraft. The lower density and structural strength enhancement of aluminum lithium alloys compared to normal aluminum alloys make it attractive for uses in airframes. A distinct lithium–aluminum phase (Al_3Li) forms in the alloy which bonds tightly to the host aluminum matrix to yield about a 10% increase in the modules of elasticity of the aluminum lithium alloys produced by the main aluminum producers. The density of the alloys is about 10% less than that of other structural aluminum alloys.

Lithium magnesium alloys, developed during World War II, have found uses in aerospace applications. Lithium alters the crystallization of the host magnesium from the normal hexagonal structure to the body-centered cubic structure, with resultant significant decreases in density and increases in ductibility.

Lithium is used in metallurgical operations for degassing and impurity removal (see METALLURGY). In copper (qv) refining, lithium metal reacts with hydrogen to form lithium hydride which subsequently reacts, along with further lithium metal, with cuprous oxide to form copper and lithium hydroxide and lithium oxide. The lithium salts are then removed from the surface of the molten copper.

Electrochemical Applications. Lithium batteries (qv) are used in numerous consumer, medical, industrial, and military applications. The advantages of lithium batteries include not only the light weight and high energy density, but also short voltage delays and low self-discharge rates (51). Consumer uses include uses in electronic devices such as watches, cameras, and calculators as well as CMOS-RAM memory backups. Lithium batteries are especially adaptable to unusual configurations and miniaturization, which make them especially useful in small electronics. Medical uses include cardiac pacemakers. Industrial uses include power for remote sensing devices such as oil-well logging devices. Lithium batteries have uses in military applications ranging from communication devices to stand-by power for missile systems.

Most primary lithium batteries contain a lithium anode, although use of lithium alloys, especially Li–Al alloys, is preferred in many applications requiring higher temperatures. Most cathode materials are lithium intercalation compounds such as metal oxides or sulfides, although carbon monofluoride, CF_x, and liquid cathode materials such as thionyl chloride [*7719-09-7*], $SOCl_2$, and sulfur dioxide [*7446-09-5*], SO_2, are also extensively used. The electrolytes are usually a lithium salt dissolved in an organic electrolyte, although lithium ion-conductive solids and glasses are used in many applications. Common electrolyte salts are lithium bromide [*7550-35-8*], LiBr; lithium hexafluoroarsenate [*29935-35-1*], $LiAsF_6$; lithium hexafluorophosphate [*21324-40-3*], $LiPF_6$; lithium iodide [*10377-51-2*], LiI; lithium perchlorate [*7791-03-9*], $LiClO_4$; lithium tetrachloroaluminate [*14024-11-4*], $LiAlCl_4$; lithium tetrafluoroborate [*14283-07-9*], $LiBF_4$; and lithium trifluoromethane sulfonate [*33454-82-9*] (triflate), $LiCF_3SO_2$, which are soluble in a variety of organic solvents and are not easily reduced by lithium. Lithium reaction with the solvent or the electrolyte salt is prevented by formation of a thin lithium ion-conductive film at the lithium–electrolyte interface, formed by reaction of lithium with either the solvent or the salt.

The attractiveness of lithium rechargeable (secondary) batteries has long been recognized. The high energy density, light weight, and other performance and environmental factors make secondary lithium batteries desirable for applications ranging from portable electronic devices to power supply for traction vehicles. The first attempts to commercialize secondary lithium batteries employed lithium metal anodes, which have largely been discontinued owing to safety hazards resulting in fires, especially under abuse conditions. Polymeric lithium ion-conductive electrolytes are one method of preventing the occurrence of this problem. The low conductivity of current polymeric electrolytes such as $LiPF_6$ in poly(ethylene oxide) require use of high surface area, thin-film lithium in order to achieve satisfactory performance. Another method of preventing this problem has been to use lithium ion intercalation compounds for both the anode and the cathode. Such batteries are known as lithium ion or "rocking chair" batteries (52) because the lithium ion swings between intercalation compounds at the anode and cathode during the charge and discharge cycles. A large number of intercalation compounds are suitable for such applications, especially for the cathode, but as of this writing commercially available designs use lithium–carbon anodes. Another significant secondary battery technology is the high temperature molten salt electrolyte battery developed by Argonne National Laboratory. This battery is based on the high melting lithium aluminum alloy anode, a molten salt electrolyte consisting of lithium halides, and other alkali metal halides, and an iron sulfide cathode. This battery is especially attractive for traction vehicles and for load leveling storage in electrical power generation.

An emerging electrochemical application of lithium compounds is in molten carbonate fuel cells (qv) for high efficiency, low polluting electrical power generation. The electrolyte for these fuel cells is a potassium carbonate–lithium carbonate eutectic contained within a lithium aluminate matrix. The cathode is a lithiated metal oxide such as lithium nickel oxide.

Inorganic Lithium Compounds

The unique properties of the lithium ion result in part from its small size and correspondingly high charge density. The ^6Li isotope which has a large neutron capture radius finds uses in thermonuclear devices, neutron shielding, and tritium production. All of these are especially important for the fusion reactors proposed for future power generation (see FUSION ENERGY). Lithium is also used for treating affective mood disorders and has been proposed for other medical applications (see PSYCHOPHARMACOLOGICAL AGENTS) (53). The size and charge density of this lightest metal ion are expected to lead to additional uses in emerging technologies.

Lithium Acetate. Lithium acetate [*546-89-4*] is obtained from reaction of lithium carbonate or lithium hydroxide and acetic acid. Crystalline lithium acetate dihydrate [*6108-17-4*], $CH_3CO_2Li{\cdot}2H_2O$, melts congruently in its own water of crystallization at 57.8°C. The anhydrous salt [*546-89-4*] melts without decomposition at 291°C. Aqueous solubility, expressed as the anhydrous salt, is 31% at 25°C and 66% at 100°C. Uses of lithium acetate are in polyester fiber (see FIBERS, POLYESTER) manufacture and in certain catalytic processes.

Lithium Amide. Lithium amide [*7782-89-0*], $LiNH_2$, is produced from the reaction of anhydrous ammonia and lithium hydride. The compound can also be prepared by the removal of ammonia from solutions of lithium metal in the presence of catalysts (54). Lithium amide starts to decompose at 320°C and melts at 375°C. Decomposition of the amide above 400°C results first in lithium imide, Li_2NH, and eventually in lithium nitride, Li_3N. Lithium amide is used in the production of antioxidants (qv) and antihistamines (see HISTAMINE AND HISTAMINE ANTAGONISTS).

Lithium Benzoate. Lithium benzoate [*553-54-8*], $LiC_7H_5O_2$, is formed by reaction of benzoic acid and lithium hydroxide in aqueous solution with subsequent drying. The salt is very soluble (29.9 g/100 g soln at 25°C) in water. It is used as an alkaline catalyst for curing epoxides and as a crystal nucleation modifier in polypropylene.

Lithium Borates. Lithium metaborate [*13453-69-5*], $LiBO_2 \cdot 2H_2O$, is prepared from reaction of lithium hydroxide and boric acid. It is used as the fluxing agent for the matrix for x-ray fluorescence analytical techniques and in specialty glasses and enamels. The anhydrous salt melts at 847°C.

Lithium tetraborate [*1303-94-2*], $Li_2B_4O_7$, is used as a flux in ceramics and in x-ray fluorescence spectroscopy. The salt has also been proposed for nonlinear optic applications. The salt melts at 917°C (see NONLINEAR OPTICAL MATERIALS).

Lithium Carbonate. Lithium carbonate [*554-13-2*], Li_2CO_3, is produced in industrial processes from the reaction of sodium carbonate and lithium sulfate or lithium chloride solutions. The reaction is usually performed at higher temperatures because aqueous lithium carbonate solubility decreases with increasing temperatures. The solubility (wt %) is 1.52% at 0°C, 1.31% at 20°C, 1.16% at 40°C, 1.00% at 60°C, 0.84% at 80°C, and 0.71% at 100°C. Lithium carbonate is the starting material for reactions to produce many other lithium salts, including the hydroxide. Decomposition of the carbonate occurs above the 726°C melting point.

Lithium carbonate addition to Hall-Heroult aluminum cell electrolyte lowers the melting point of the eutectic electrolyte. The lower operating temperatures decrease the solubility of elemental metals in the melt, allowing higher current efficiencies and lower energy consumption (55). The presence of lithium also decreases the vapor pressure of fluoride salts.

Lithium carbonate is used to prepare lithium aluminosilicate glass ceramics which have low thermal coefficients of expansion, allowing use over a wide temperature range. It also finds uses in specialty glasses and enamels.

The molten carbonate fuel cell uses eutectic blends of lithium and potassium carbonates as the electrolyte. A special grade of lithium carbonate is used in treatment of affective mental (mood) disorders, including clinical depression and bipolar disorders. Lithium has also been evaluated in treatment of schizophrenia, schizoaffective disorders, alcoholism, and periodic aggressive behavior (56).

Lithium Halides. Lithium halide stability decreases with increasing atomic weight of the halogen atom. Hence, the solubility increases from the sparingly soluble lithium fluoride to the very soluble bromide and iodide salts. The low melting points of lithium halides are advantageous for fluxes in many applications.

Lithium Fluoride. Lithium fluoride [*7789-24-4*], LiF, is produced from the reaction of lithium carbonate or hydroxide and hydrofluoric acid (see FLUORINE COMPOUNDS, INORGANIC–LITHIUM). The salt melts at 848°C and boils at 1681°C. It is a strong flux for glasses, enamels, and glazes and imparts a lower coefficient of expansion to glasses. It is also an important component of fluoride glasses in infrared optical applications. Lithium fluoride is used in brazing or soldering fluxes and in welding (qv) rod coatings. Lithium fluoride, prepared *in situ* from reaction of lithium carbonate and aluminum fluoride, is the active lithium component in improving electrolysis efficiency in aluminum reduction cells. The salt is only sparingly soluble in water (0.133 g/100 g water at 25°C), similarly to magnesium or calcium fluoride.

Lithium Chloride. Lithium chloride [*7447-41-8*], LiCl, is produced from the reaction of lithium carbonate or hydroxide with hydrochloric acid. The salt melts at 608°C and boils at 1382°C. The 41-mol % LiCl–59-mol % KCl eutectic (melting point, 352°C) is employed as the electrolyte in the molten salt electrolysis production of lithium metal. It is also used, often with other alkali halides, in brazing flux eutectics and other molten salt applications such as electrolytes for high temperature lithium batteries.

The salt is extremely hydroscopic and is used in dehumidification applications. It is very soluble in water (Table 4). The hydrates $LiCl \cdot 2H_2O$ [*16712-19-9*] and $LiCl \cdot H_2O$ [*16712-20-2*] precipitate at temperatures below 100°C. The anhydrous salt precipitates at 100°C. The salt has appreciable solubility in alcohols and amines.

Lithium Bromide. Lithium bromide [*7550-35-8*], LiBr, is prepared from hydrobromic acid and lithium carbonate or lithium hydroxide. The anhydrous salt melts at 550°C and boils at 1310°C. Lithium bromide is a component of the low melting eutectic electrolytes in high temperature lithium batteries.

The salt is extremely soluble in water (Table 4), crystallizing from aqueous solution as the hydrates $LiBr \cdot H_2O$ [*23303-71-1*], $LiBr \cdot 2H_2O$ [*13453-70-8*], and $LiBr \cdot 3H_2O$ [*76082-04-7*]. The anhydrous salt is obtained by drying under vacuum at elevated temperatures.

The high solubility of the salt and resultant low water vapor pressure (58) of its aqueous solutions are useful in absorption air conditioning (qv) systems. Lithium bromide absorption air conditioning technology efficiencies can surpass that of reciprocal technology using fluorochlorocarbon refrigerants.

Table 4. Lithium Halide Solubility in Water[a]

Temperature, °C	Lithium salt, wt %	Solid-phase formula
	LiCl	
0	40.2	$LiCl \cdot 2H_2O$
19.4	45.2	$LiCl \cdot 2H_2O + LiCl \cdot H_2O$
20	45.2	$LiCl \cdot H_2O$
50	48.3	$LiCl \cdot H_2O$
80	53.0	$LiCl \cdot H_2O$
95	56.0	$LiCl \cdot H_2O + LiCl$
100	56.3	LiCl
	LiBr	
0	56.7	$LiBr \cdot 3H_2O$
3	58.3	$LiBr \cdot 3H_2O + LiBr \cdot 2H_2O$
20	60.2	$LiBr \cdot 2H_2O$
30	62.0	$LiBr \cdot 2H_2O$
42.7	65.0	$LiBr \cdot 2H_2O + LiBr \cdot H_2O$
50	65.5	$LiBr \cdot H_2O$
80	67.8	$LiBr \cdot H_2O$
100	69.5	$LiBr \cdot H_2O$

[a]Ref. 57.

Basic solutions of lithium bromide can react with oxygen to form hypobromites and bromates under certain conditions.

Lithium Iodide. Lithium iodide [*10377-51-2*], LiI, is the most difficult lithium halide to prepare and has few applications. Aqueous solutions of the salt can be prepared by careful neutralization of hydroiodic acid with lithium carbonate or lithium hydroxide. Concentration of the aqueous solution leads successively to the trihydrate [*7790-22-9*], dihydrate [*17023-25-5*], and monohydrate [*17023-24-4*], which melt congruently at 75, 79, and 130°C, respectively. The anhydrous salt can be obtained by careful removal of water under vacuum, but because of the strong tendency to oxidize and eliminate iodine which occurs on heating the salt in air, it is often prepared from reactions of lithium metal or lithium hydride with iodine in organic solvents. The salt is extremely soluble in water (62.6 wt % at 25°C) (59) and the solutions have extremely low vapor pressures (60). Lithium iodide is used as an electrolyte in selected lithium battery applications, where it is formed *in situ* from reaction of lithium metal with iodine. It can also be a component of low melting molten salts and as a catalyst in aldol condensations.

Lithium Hydride. Lithium hydride [*7580-67-8*], LiH, is prepared by the reaction of lithium metal and hydrogen (qv) in an extremely exothermic reaction. The reaction initiates above the melting temperature of lithium and is then controlled by the hydrogen addition rate to a temperature of about 700°C, slightly above the 686.4°C melting point of lithium hydride (61). Lithium hydride crystallizes in a cubic lattice and has a density of 0.78 g/cm^3. It is insoluble in organic solvents but reacts vigorously with protic solvents. The reaction with water produces lithium hydroxide and large volumes of hydrogen, and thus has found applications as a hydrogen source in inflatable devices such as balloons carrying emergency beacons. The lack of solubility of lithium hydride makes it unattractive in most organic reductions. Lithium tetrahydridoaluminate [*16853-85-3*], $LiAlH_4$, lithium tetrahydridoborate [*16949-15-8*], $LiBH_4$, or other complex hydrides are used instead (61) (see HYDRIDES). The largest industrial use of lithium hydride has been the production of high purity silane from the reaction of silicon halides in a lithium chloride–potassium chloride eutectic (see SILICON COMPOUNDS, SILANES).

Lithium Hydroxide. Lithium hydroxide monohydrate [*1310-66-3*], $LiOH\cdot H_2O$, is prepared industrially from the reaction of lithium carbonate and calcium hydroxide in aqueous slurries. The calcium carbonate is subsequently separated to yield a lithium hydroxide solution from which lithium hydroxide monohydrate can be crystallized. Lithium hydroxide is the least soluble alkali hydroxide, and solubility varies little with temperature.

Temperature, °C	Solubility of $LiOH\cdot H_2O$ in water, wt %
0	18.8
10	18.9
20	19.1
40	19.8
60	21.0
80	23.0
100	25.9

Lithium hydroxide can be used for preparation of numerous lithium salts. The dominant use is the preparation of lithium stearate [*4485-12-5*], which is added to lubricating greases in amounts up to about 10% by weight. This salt has very low water solubility and extends the acceptable viscosity for the grease to both low and high temperatures (see LUBRICATION AND LUBRICANTS). Lithium hydroxide is also used in production of dyes (62) and has been proposed as a source of lithium ion for inhibition of alkali-aggregate expansive reactivity in concrete (63).

Anhydrous lithium hydroxide [*1310-65-2*], LiOH, is obtained by heating the monohydrate above 100°C. The salt melts at 462°C. Anhydrous lithium hydroxide is an extremely efficient absorbent for carbon dioxide (qv). The porous structure of the salt allows complete conversion to the carbonate with no efficiency loss in the absorption process. Thus LiOH has an important role in the removal of carbon dioxide from enclosed breathing areas such as on submarines or space vehicles. About 750 g of lithium hydroxide is required to absorb the carbon dioxide produced by an individual in a day.

Lithium Hypochlorite. Lithium hypochlorite [*13840-33-0*], LiOCl, is obtained from reaction of chlorine and an aqueous solution of lithium hydroxide. The solid is usually obtained as a dry stable product containing other alkali halides and sulfates (64). A product containing 35% available chlorine is used for sanitizing applications in swimming pools and in food preparation areas where its rapid and complete dissolution is important. The salt can also be obtained in higher purity by reaction of lithium hydroxide and hypochlorous acid (65).

Lithium Niobate. Lithium niobate [*12031-64-9*], $LiNbO_3$, is normally formed by reaction of lithium hydroxide and niobium oxide. The salt has important uses in switches for optical fiber communication systems and is the material of choice in many electrooptic applications including waveguide modulators and sound acoustic wave devices. Crystals of lithium niobate are usually grown by the Czochralski method followed by infiltration of wafers by metal vapor to adjust the index of refraction.

Lithium Nitrate. Lithium nitrate [*7790-69-4*] is prepared by neutralization of nitric acid using a lithium base. The nitrate is extremely soluble, 43 wt % at 20°C, in water and is hydroscopic. This salt melts at 251°C and forms very low melting eutectics with a large number of other salts.

Lithium Nitride. Lithium nitride [*26134-62-3*], Li_3N, is prepared from the strongly exothermic direct reaction of lithium and nitrogen. The reaction proceeds to completion even when the temperature is kept below the melting point of lithium metal. The lithium ion is extremely mobile in the hexagonal lattice resulting in one of the highest known solid ionic conductivities. Lithium nitride in combination with other compounds is used as a catalyst for the conversion of hexagonal boron nitride to the cubic form. The properties of lithium nitride have been extensively reviewed (66).

Lithium Oxide. Lithium oxide [*12057-24-8*], Li_2O, can be prepared by heating very pure lithium hydroxide to about 800°C under vacuum or by thermal decomposition of the peroxide (67). Lithium oxide is very reactive with carbon dioxide or water. It has been considered as a potential high temperature neutron target for tritium production (68).

Lithium Perchlorate. Lithium perchlorate [*7791-03-9*], $LiClO_4$, is prepared from neutralization of perchloric acid using lithium hydroxide or carbonate.

The salt crystallizes from aqueous solution first as the trihydrate. The anhydrous salt, which can be obtained by vacuum drying of the trihydrate at 130°C, melts at 236°C. The salt is kinetically stable up to 400°C (69) decomposing to lithium chloride and oxygen above that temperature. If mixed with catalysts, decomposition can occur at lower temperatures. Lithium perchlorate has been proposed as a solid rocket fuel oxidant but the most common use is as a lithium battery electrolyte salt. Lithium perchlorate catalyzes Diels-Alder reactions (70), but must be used with precautions (71).

Lithium Peroxide. Lithium peroxide [*12031-80-0*], Li_2O_2, is obtained by reaction of hydrogen peroxide and lithium hydroxide in ethanol (72) or water (73). Lithium peroxide, which is very stable as long as it is not exposed to heat or air, reacts rapidly with atmospheric carbon dioxide releasing oxygen. The peroxide decomposes to the oxide at temperatures above 300°C at atmospheric pressure, and below 300°C under vacuum.

Lithium Phosphate. Lithium phosphate [*10377-52-3*], Li_3PO_4, is prepared from the neutralization of lithium hydroxide using phosphoric acid. Partial neutralization can lead to precipitation of lithium dihydrogen phosphate [*13453-80-0*], LiH_2PO_4, but the dilithium monohydrogen phosphate salt cannot be obtained pure from aqueous solution. Lithium phosphate, sparingly soluble (0.039 wt %) in aqueous solution, is used in specialty glasses, including ion-conductive glasses. Lithium phosphate is also a catalyst for the isomerization of epoxides to unsaturated alcohols, most notably the conversion of propylene oxide to allyl alcohol (74).

Lithium Sulfate. Lithium sulfate [*10377-48-7*], Li_2SO_4, is produced from neutralization of sulfuric acid using lithium hydroxide or lithium carbonate. Subsequent evaporation of water from the solution yields first the monohydrate [*10102-25-7*], $Li_2SO_4 \cdot H_2O$, and then the anhydrous salt. The dehydration of $Li_2SO_4 \cdot H_2O$ initiates at temperatures as low as 72°C but complete drying can extend over a wide temperature range and is described by a contracting volume model (75). The salt is a component of many photographic developer solutions (see PHOTOGRAPHY).

Lithium Silicate. Lithium silicate [*10102-24-6*], Li_2SiO_3, is formed from calcination of lithium carbonate or hydroxide using finely ground silica. The product melts at 1201°C and is insoluble in water. It is used in production of enamels and glazes (see ENAMELS, PORCELAIN OR VITREOUS). Lithium silicate solutions [*12627-14-4*] containing excess silica can be prepared from the reaction of lithium hydroxide and silica gel. These solutions typically contain a Si:Li mole ratio between four and six and a solids content of around 20%. The solids can be precipitated from solution by adding other salts which, with drying, form insoluble lithium silicate films.

Other Lithium Salts. A wide range of other lithium salts are emerging in specialized industrial applications. Lithium aluminate [*12003-67-7*], $LiAlO_2$, has important applications in molten carbonate fuel cells and may be useful in fusion reactors. A series of lithium metal oxides are useful in lithium ion batteries, including lithium cobaltite [*12190-79-3*], $LiCoO_2$; lithium nickelate [*12031-65-1*], $LiNiO_2$; and various stoichiometries of lithium manganese oxides. Lithium chromate [*14307-35-8*], Li_2CrO_4, and lithium molybdate [*13568-40-6*], Li_2MoO_4, are both important for corrosion inhibition in lithium halide absorption air conditioning and dehumidification devices. Lithium titanate [*12031-82-2*], $LiTiO_2$, and lithium zirconate [*12031-83-3*], Li_2ZrO_3, have important ceramic applications.

Organolithium Compounds

Organolithium compounds are organometallic compounds in which the lithium is bonded directly to carbon. Because of the substantial covalent character in these bonds, many of the compounds exist as liquids or as low melting solids and are soluble in organic solvents, such as ethers and liquid hydrocarbons. These compounds are reactive to oxygen and moisture and may ignite spontaneously in the pure state or in concentrated solutions on exposure to air. Organolithium compounds are useful in many Grignard-type reactions employed in the synthesis of pharmaceutical and agricultural products (see GRIGNARD REACTIONS). These compounds are also used as initiators of the stereospecific polymerization of conjugated dienes and vinylaromatic compounds to produce rubbery polymers and plastics.

A simple and economical route involving the reaction of lithium metal and organic halides was developed in 1930 to prepare organolithium compounds such as *n*-butyllithium (76):

$$RX + 2\ Li \longrightarrow RLi + LiX$$

This direct method of preparing organolithium compounds is commonly used in commercial processes.

Organolithium compounds generally are less reactive than the other members of the organoalkali series but are more reactive than the Group 2 (IIA) organometallics. However, many differences in type and degree of reaction distinguish organolithiums from other organometallics and make the lithium compounds intermediates in syntheses. Several excellent reviews concerning the preparation, uses, and applications of organolithium compounds are available (77–89), as are yearly reviews (90,91).

***n*-Butyllithium.** The most important organolithium compound is *n*-butyllithium [*109-72-8*], $CH_3CH_2CH_2CH_2Li$. *n*-Butyllithium is a clear, colorless-to-pale yellow, slightly viscous liquid that exists in solution in an associated form. Physical properties are given in Table 5.

n-Butyllithium decomposes thermally to 1-butene and lithium hydride at elevated temperatures. Dilute solutions of *n*-butyllithium in hydrocarbon solvents possess a negligible rate of decomposition at ordinary handling temperatures and under an inert atmosphere, eg, argon or nitrogen. For example, the decomposition rate is $< 0.01\%$ active material per day at 25°C for a 15 wt % solution in hexane (96). However, at elevated temperatures and high concentrations of *n*-butyllithium, the losses can be significant. For example, the same 15 wt % solution loses about 0.05% to its active material per day at 45°C, and an 85 wt % solution loses about 0.2 wt %/d at the same temperature. The pyrolysis of alkyllithiums was studied both in neat form (97) and in decane solution (98). A 15 wt % solution of *n*-butyllithium in hexane is 50% decomposed after 5 h at 130°C and after 50 min at 150°C (98). The decomposition rate constants double as the

Table 5. Physical Constants of *n*-Butyllithium

Property	Value	Reference
distillation range at 13.3 mPa,[a] °C	80–90	92
d_{25}, g/cm^3	0.765	93
viscosity, mPa·s(= cP)		
at 25°C	34.6	
0.1°C	119.6	93
heat of formation at 25°C, kJ/mol[b]	−133.9 + 7.1	94
	−131.4 ± 2.9	95
vapor pressure, mPa[a]		
at 60°C	58.3	96
80°C	592	96
95°C	2590	96

[a]To convert mPa to mm Hg, multiply by 7.5×10^{-6}.
[b]To convert J to cal, divide by 4.184.

percentage of carbon-bound lithium decreases from 96 to 76% by the addition of lithium *n*-butoxide, the oxidation product of *n*-butyllithium.

Although *n*-butyllithium is quite soluble in ether solvents, its solutions in these solvents are not stable and thus cannot be stored or shipped. A solution of *n*-butyllithium in diethylether decomposes to give

$$C_4H_9Li + C_2H_5OC_2H_5 \longrightarrow C_2H_5OLi + C_4H_{10} + C_2H_4$$

Ethylene also is a by-product of the cleavage reaction with tetrahydrofuran (99). The rate of loss for *n*-butyllithium in a variety of ether solvents is given in Table 6.

Butyllithium is available in hydrocarbon media, eg, hexane, heptane, cyclohexane, and toluene in several concentrations, eg, 15, 25, and 90 wt %. It is shipped commercially in 10-, 19-, 106-, 216-, and 454-L cylinders and 1900-L (500-gal) tanks and in bulk in 19,000-L (5,000-gal) tank trucks and 30,300-L (8,000-gal) rail tank cars. For shipment purposes, *n*-butyllithium is described as a pyrophoric fuel and is classified as a flammable liquid. As such, full precautions should be taken according to handling instructions (100).

n-Butyllithium solutions are routinely assayed by the following methods.

Table 6. Decomposition of *n*-Butyllithium in Ethers

Ether	$t_{1/2}$[a]	Temperature, °C
ethyl	6	25
	31[b]	35
isopropyl	18	25
glycol dimethyl	10[c]	25
tetrahydrofuran	23.5[b]	0
	5	−30

[a]Units are days unless otherwise noted.
[b]Units are hours.
[c]Units are minutes.

(*1*) The double titration method, which involves the use of benzylchloride, 1,2-dibromoethane, or allyl bromide, determines carbon-bound lithium indirectly (101,102). One sample of the *n*-butyllithium is hydrolyzed directly, and the resulting alkalinity is determined. A second sample is treated with benzylchloride and is then hydrolyzed and titrated with acid. The second value (free base) is subtracted from the first (total base) to give a measure of the actual carbon-bound lithium present (active base).

(*2*) The oxidimetric method, which involves the use of solid vanadium pentoxide as oxidant (103). The vanadium is reduced quantitatively by butyllithium and is determined potentiometrically by titration with standard sulfatoceric acid [*17106-39-7*]. This method gives a direct measure of the actual carbon-bound lithium present when compared to the total titrated alkalinity.

(*3*) Other analytical methods which include, among many, a thermometric method (104), a high frequency titration (105), and a colored indicator method (106).

Uses for *n*-butyllithium are mainly as an anionic initiator in solution polymerization to produce elastomeric products and plastics, predominantly of the styrene–butadiene type (107–109) (see ELASTOMERS, SYNTHETIC; INITIATORS, ANIONIC; STYRENE). Butyllithium-initiated polymers of this type can be varied widely in composition and molecular weight in a highly controlled manner, leading to tailor-made products such as footwear, molded and extruded goods, hose and tubing, mechanical rubber goods, adhesives (qv), sealants (qv), and packaging film wrap. A number of reviews are available on organolithium-initiated polymerizations (110–115).

Another large-volume use for organolithium compounds is in the synthesis of pharmaceutical and agricultural chemicals, eg, antibiotics (qv), antihistamines, antidepressants, anticoagulants, vasodilators, tranquilizers, analgesics, fungicides, and pesticides (116–119).

***sec*-Butyllithium.** *sec*-Butyllithium [*598-30-1*], $CH_3CH_2CH(Li)CH_3$, is a clear, colorless-to-pale yellow, pyrophoric liquid, $d_{25} = 0.783$ g/cm^3, viscosity$_{25}$ = 20.1 mPa·s(=cP) that can be obtained by vacuum distillation of solvent from its hydrocarbon solutions. It can be distilled at about 90°C and 6.7 Pa (0.050 mm Hg) (119). The association state of *sec*-butyllithium is tetrametric (119). *sec*-Butyllithium is available as a 12% solution in *n*-hexane or cyclohexane. These solutions are less stable thermally than those of *n*-butyllithium. Elimination of lithium hydride can occur in a number of ways.

$$CH_3CH_2CH(Li)CH_3 \xrightarrow{\text{in octane solution}} \underset{40\%}{CH_3CH_2CH{=}CH_2} + \underset{34\%}{CH_3CH{=}CHCH_3} + \underset{26\%}{CH_3CH{=}CHCH_3} + LiH$$

A cis-elimination mechanism has been postulated for this decomposition which follows first-order kinetics (120). The rate is accelerated by addition of lithium *sec*-butoxide [*4111-46-0*], and other bases, and by an increase in temperature

(120). Pyrolysis of *sec*-butyllithium in the presence of added alkoxide is one-half order in alkyllithium and first order in alkoxide (120). Thermal decomposition of *sec*-butyllithium at 0.18% alkoxide at 25, 40, 50, and 60°C is 0.1%, 0.6%, 2.0%, and 6.8%/d, respectively (121).

Like *n*-butyllithium, *sec*-butyllithium is infinitely soluble in most hydrocarbons, such as pentane and hexane. Its solutions in hexane are flammable and pyrophoric and therefore should be handled like *n*-butyllithium (96,100). *sec*-Butyllithium also is very soluble in ethers, but the ether solutions must be kept cold because ether cleavage is more rapid than in the presence of *n*–butyllithium (122). *sec*-Butyllithium has a $t_{1/2}$ of 2 d at 25°C in di-*n*-butyl ether and of 1 d at 25°C in di-*n*-hexyl ether.

Assay methods for *sec*-butyllithium essentially are identical to those listed for *n*-butyllithium, except that the titrations are performed at lower temperatures because of the lower thermal stability of *sec*-butyllithium.

sec-Butyllithium is available in commercial quantities and is regarded as the second most important organolithium compound. Its uses generally are the same as those for *n*-butyllithium. *sec*-Butyllithium often is the initiator of choice in styrene–butadiene block copolymerizations to produce thermoplastic elastomers. It initiates the polymerization of styrene much faster than *n*-butyllithium, thus permitting the formation of low molecular weight, narrow molecular weight distribution styrene end blocks, with concomitant complete consumption of initiator (123).

***tert*-Butyllithium.** *tert*-Butyllithium [*594-19-4*], $(CH_3)_3CLi$, is a colorless, crystalline solid that can be sublimed at 70–80°C and 13.3 Pa (0.1 mm Hg). Ebullioscopic measurements of the molecular weight in benzene and hexane solution show the average degree of association of *t*-butyllithium to be about four (124).

t-Butyllithium is available as a 15–20 wt % solution in *n*-pentane or heptane. Noticeable decomposition occurs after a 1 h reflux in heptane (bp 98.4°C) but not after a 15 min reflux in benzene (bp 80.1°C) or hexane (bp 68°C). *t*-Butyllithium in pentane or heptane is more stable than *n*-butyllithium in hexane (125). Solutions of *t*-butyllithium in pentane and heptane are flammable liquids and are considered pyrophoric. The *t*-butyl compound is more reactive than either the *n*- and *sec*-butyl. Di-*n*-butylether is cleaved by *t*-butyllithium in 4–5 h at 25°C, compared to the 2 d for *sec*-butyllithium and 32 d for *n*-butyllithium (126). *t*-Butyllithium can be assayed by all of the techniques used for *n*-butyllithium. *t*-Butyllithium is a useful reagent in syntheses where the high reactivity of the carbon–lithium bond and small size of the lithium atom promote the synthesis of sterically hindered compounds, eg, *t*-butyldimethylchlorosilane (117), di-*t*-butyldichlorosilane, and *t*-butyldiphenylchlorosilane.

Hexyllithium. Hexyllithium [*21369-64-2*], $CH_3CH_2CH_2CH_2CH_2CH_2Li$, is soluble in hexane and other hydrocarbons in high concentrations. Hexyllithium has advantages over butyllithium in metalations where volatility of the by-product hydrocarbon is a factor (127).

Methyllithium. Methyllithium [*917-54-4*], CH_3Li, crystallizes from benzene or hexane solution giving cubic crystals that have a salt-like constitution (128). Crystalline methyllithium molecules exist as tetrahedral tetramers (129). Solutions of methyllithium are less reactive than those of its higher homologues.

Methyllithium is stable for at least six months in diethyl ether at room temperature. A one-molar solution of methyllithium in tetrahydrofuran (14 wt %) and cumene (83 wt %) containing 0.08 *M* dimethylmagnesium as stabilizer loses only 0.008% of its activity per day at 15°C and is nonpyrophoric (117).

Methyllithium solutions can be assayed by the methods used for *n*-butyllithium, by gas chromatography (qv), or by nmr. Reaction with dimethyl(phenyl)chlorosilane gives the silane $(CH_3)_3SiC_6H_5$, which is determined quantitatively using a silicone fluid 710 X Chromosorb P column with cumene as an internal standard (130). Methyllithium is available in commercial quantities. It is a useful reagent in the synthesis of vitamins A and D and in the synthesis of various analgesics and steroids (qv) (see ANALGESICS, ANTIPYRETICS, AND ANTIINFLAMMATORY AGENTS; VITAMINS).

Lithium Acetylide. Lithium acetylide–ethylenediamine complex [*50475-76-8*], $LiC{\equiv}CH{\cdot}H_2NCH_2CH_2NH_2$, is obtained as colorless-to-light-tan, free-flowing crystals from the reaction of *N*-lithoethylenediamine and acetylene in an appropriate solvent (131). The complex decomposes slowly above 40°C to lithium carbide and ethylenediamine. Lithium acetylide–ethylenediamine is very soluble in primary amines, ethylenediamine, and dimethyl sulfoxide. It is slightly soluble in ether, THF, and secondary and tertiary amines, and is insoluble in hydrocarbons.

Stabilized lithium acetylide is not pyrophoric or shock-sensitive as are the transition-metal acetylides. Among its uses are ethynylation of halogenated hydrocarbons to give long-chain acetylenes (132) and ethynylation of ketosteroids and other ketones in the pharmaceutical field to yield the respective ethynyl alcohols (133) (see ACETYLENE-DERIVED CHEMICALS).

Lithium acetylide also can be prepared directly in liquid ammonia from lithium metal or lithium amide and acetylene (134). In this form, the compound has been used in the preparation of β-carotene and vitamin A (135), ethchlorvynol (136), and *cis*-3-hexen-1-ol (leaf alcohol) (137). More recent synthetic processes involve preparing the lithium acetylide *in situ*. Thus lithium diisopropylamide, prepared from *n*-butyllithium and the amine in THF at 0°C, is added to an acetylene-saturated solution of a ketosteroid to directly produce an ethynylated steroid (138).

Phenyllithium. Phenyllithium [*591-51-5*], C_6H_5Li, forms colorless, monoclinic, pyrophoric crystals that do not melt before decomposition at 150°C. It can be obtained from its halide-free solutions in cyclohexane and ethylether by vacuum distillation to remove the ether. The usual preparative method is by reaction of chloro- or bromobenzene and lithium metal in ethyl ether or in a mixture of ethyl ether and cyclohexane.

Phenyllithium can be used as a solution in ethyl ether, but because of its limited stability ($t_{1/2}$ = 12 d at 35°C) it is commercially available in solution in mixtures, usually 70:30 wt % cyclohexane:ethyl ether (117). In this particular mixture of solvents, a 20 wt % solution, free of chlorobenzene, is stable for at least four months under an inert atmosphere (argon or nitrogen) in sealed containers at room temperature. Phenyllithium is also available in dibutyl ether solution (117). It is classified as a flammable liquid.

Phenyllithium can be used in Grignard-type reactions involving attachment of phenyl group, eg, in the preparation of analgesics and other chemotherapeutic agents (qv). It also may be used in metal–metal interconversion reactions leading, eg, to phenyl-substituted silicon and tin organics.

Other Organolithium Compounds. Organodilithium compounds have utility in anionic polymerization of butadiene and styrene. The lithium chain ends can then be converted to useful functional groups, eg, carboxyl, hydroxyl, etc (139). Lewis bases are required for solubility in hydrocarbon solvents.

Many organic syntheses require the use of sterically hindered and less nucleophilic bases than *n*-butyllithium. Lithium diisopropylamide (LDA) and lithium hexamethyldisilazide (LHS) are often used (140–142). Both compounds are soluble in a wide variety of aprotic solvents. Presence of a Lewis base, most commonly tetrahydrofuran, is required for LDA solubility in hydrocarbons. A 30% solution of LHS can be prepared in hexane. Although these compounds may be prepared by reaction of the amine with *n*-butyllithium in the approprite medium just prior to use, they are also available commercially in hydrocarbon or mixed hydrocarbon–THF solvents as 1.0–2.0 *M* solutions.

Other, even milder bases than LDA and LHS, such as lithium methoxide and lithium *t*-butoxide, may be used in organic syntheses (143,144). Lithium methoxide is available commercially as a 10% solution in methanol and lithium *t*-butoxide as an 18% solution in tetrahydrofuran (145). Lithium *t*-butoxide is also soluble in hydrocarbon solvents (146). Both lithium alkoxides are also available as solids (147) (see ALKOXIDES, METAL).

Handling and Toxicity of Lithium Compounds

Lithium ion is commonly ingested at dosages of 0.5 g/d of lithium carbonate for treatment of bipolar disorders. However, ingestion of higher concentrations (5 g/d of LiCl) can be fatal. As of this writing, lithium ion has not been related to industrial disease. However, lithium hydroxide, either directly or formed by hydrolysis of other salts, can cause caustic burns, and skin contact with lithium halides can result in skin dehydration. Organolithium compounds are often pyrophoric and require special handling (53).

BIBLIOGRAPHY

"Lithium" under "Alkali Metals and Alkali Metal Alloys" in *ECT* 1st ed., Vol. 1, pp. 431–435, by E. H. Burkey, J. A. Morrow, and M. S. Andrew, E. I. du Pont de Nemours & Co., Inc.; "Lithium Compounds" in *ECT* 1st ed., Vol. 8, pp. 467–477, by J. J. Kennedy, Maywood Chemical Works; "Lithium and Lithium Compounds" in *ECT* 2nd ed., Vol. 12, pp. 529–556, by R. O. Bach, C. W. Kamienski, and R. B. Ellestad, Lithium Corp. of America, Inc.; in *ECT* 3rd ed., Vol. 14, pp. 448–476, by R. Bach and J. R. Wasson, Lithium Corp. of America.

1. M. E. Weeks, *J. Chem. Ed.* **6**, 484 (1956).
2. Unpublished data, FMC Corp., Lithium Division, Bessemer City, N.C.; H. J. Andrews, "The Lithium Industry," presented June 28, 1991 to Club de Mineria, La Paz, Bolivia.
3. "Lithium," Syst.-Nr.20 in *Gmelin's Handbuch der Anorganischen Chemie*, 8th ed., Suppl. Vol., Verlag Chemie, Berlin, 1920, Syst.-Nr.20, Weinheim, Germany, 1960.
4. P. Pascal, in *Nouveau Traité de Chimie Minerale*, Vol. II, Mason & Cie, Paris, 1966, pp. 21–83.

5. W. A. Hart and O. F. Beumel, Jr., in J. C. Bailer, Jr., and co-eds., *Comprehensive Inorganic Chemistry*, Vol. I, Pergamon Press, Oxford, U.K., 1973, pp. 331–367.
6. D. J. McCraken and M. Haig, *Ind. Minerals*, 56 (June 1992).
7. *Ind. Minerals*, 23 (June 1987).
8. M. Coad, *Ind. Minerals*, 28 (Oct. 1984).
9. R. D. Crozier, *8th Industrial Minerals International Congress*, Boston, 1988, Metal Bulletin, PLC, London, 1988, p. 59.
10. R. K. Nesheim, unpublished data, FMC Corp., Lithium Division, Bessemer City, N.C., 1993.
11. H. D. Holland, G. I. Smith, H. W. Jannasch, A. G. Dickson, Z. Mianping, and B. Tiping, *Gewiss. Schaft.* **9**(2), 38,43 (1991).
12. A. G. Collins, in J. D. Vine, ed., *Lithium Resources and Requirements by the Year 2000*, Professional Paper 1005, U.S. Geological Survey, Washington, D.C., 1976, pp. 116–123.
13. C. E. Berthold, D. H. Baker, Jr., in Ref. 12, pp. 61–66.
14. L. Crocker and co-workers, *Lithium and It's Recovery from Low-Grade Nevada Clays*, U.S. Dept. of the Interior Bulletin 691, Washington, D.C., 1988, Chapt. 3.
15. J. D. Vine, Open-File Rpt 80-1234, U.S. Geological Survey, Washington, D.C., 1980, pp. 85–92.
16. Ref. 7, p. 35.
17. R. K. Evans, in S. S. Penner, ed., *Lithium Needs and Resources*, Pergamon Press, Oxford, U.K., 1978, pp. 379–385.
18. U.S. Pat. 2,516,109 (July 25, 1950), R. B. Ellestad and K. M. Leute (to Metalloy Corp.).
19. *Chem. Eng.* **62**, 113 (1955).
20. U.S. Pat. 2,940,820 (June 14, 1960), H. Mazza, S. L. Cohen, and G. H. Schaefer (to American Lithium Chemicals, Inc.).
21. U.S. Pat. 3,112,171 (Nov. 26, 1963) M. Archambault (to Department of Natural Resources, Province of Quebec, Canada).
22. U.S. Pat. 3,295,920 (Jan. 3, 1967), R. D. Goodenough, G. D. Jones, and R. E. Anderson (to The Dow Chemical Co.).
23. U.S. Pat. 2,331,838 (Oct. 12, 1943), A. R. Lindblad, S. J. Wallden, and K. A. Sivander (to Bolidens Gruvaktiebolag).
24. U.S. Pat. 2,816,007 (Dec. 10, 1957), A. V. Kroll (to Geomines).
25. L. Crocker, R. H. Lien, C. F. Davidson and J. T. May, in Ref. 14, Chapt. 4.
26. L. Crocker, R. H. Lien, and V. E. Edlund, in Ref. 14, Chapt. 5.
27. Ref. 8, p. 29.
28. Ref. 7, p. 30.
29. G. E. Foltz, in J. J. McKelta and W. A. Cunningham, eds., *Encyclopedia of Chemical Processing and Design*, Vol. 28, Marcel Dekker Inc., New York, 1988, pp. 324–344.
30. W. T. Barrett and B. J. O'Neill, in J. L. Rau and L. E. Dellwig, eds., *Third Symposium on Salt*, Vol. 2, Northern Ohio Geological Society, Inc., Cleveland, Ohio, 1970, pp. 47–50.
31. U.S. Pat. 3,537,813 (Nov. 3, 1970), J. R. Nelli and T. E. Arthur (to Lithium Corp. of America).
32. U.S. Pat. 2,964,381 (Dec. 13, 1960), R. D. Goodenough (to The Dow Chemical Co.).
33. C. E. Berthold and D. H. Baker, Jr., in Ref. 12, p. 64.
34. U.S. Pat. 4,812,245 (Mar. 14, 1983), J. L. Burba, III and W. C. Bauman (to The Dow Chemical Co.).
35. M. M. Markowitz and D. A. Boryta, *J. Chem. Eng. Data* **7**, 586 (1965); P. F. Adams, P. Hubberstey and R. J. Pulham, *J. Less-Common Metals* **42**, 1 (1975); R. M. Yonco, E. Veleckis and V. A. Maroni, *J. Nucl. Matl.* **57**, 317 (1975).

36. I. A. Gindin and co-workers, *Fiz. Melal. Metalloved.* **10**, 472 (1960).
37. T. B. Douglas, L. F. Epstein, J. L. Dever and W. H. Howland, *J. Am. Chem. Soc.* **77**, 2144 (1955).
38. H. Hartman and R. Schneider, *Z. Anorg. Chem.* **180**, 275 (1929).
39. W. A. Hart and O. F. Beumel, Jr., in J. C. Bailer, Jr., and co-eds., *Comprehensive Inorganic Chemistry*, Vol. I, Pergamon Press, Oxford, U.K., 1973, pp. 331–367.
40. D. D. Snyder and D. J. Montgomery, *J. Chem. Phys.* **27**, 1033 (1957).
41. *Janaf Thermochemical Tables*, 2nd ed., Nat. Stand. Ref. Data Ser., National Bureau of Standards, Washington, D.C., 1971.
42. F. Roelich and F. Tebber, *J. Electrochem. Soc.* **3**, 234 (1965).
43. J. Bohdansky and H. E. J. Schins, *J. Phys. Chem.* **71**, 215 (1967).
44. P. Davidovits and D. L. McFadden, eds., *Alkali Halide Vapors*, Academic Press, Inc., New York, 1979.
45. W. L. Jolly, *Prog. Inorg. Chem.* **1**, 235 (1959).
46. B. J. Wakefield, *The Chemistry of Organolithium*, Pergamon Press, Oxford, U.K., 1974.
47. R. Juza and V. Wehle, *Naturwissenchaften* **52**(19), 537 (1965).
48. A. Guntz, *Compt. Rend.* **177**, 732 (1893).
49. S. Kallman, in I. M. Koltroff, P. J. Elving, and E. B. Sandell, ed., *Treatis on Analytical Chemistry*, Vol. 1, Wiley-Interscience, New York, 1961, pp. 301–447.
50. F. A. Lowenstein, in F. D. Snell and L. S. Ettre, eds., *Encyclopedia of Industrial Chemical Analysis*, Vol. 15, John Wiley & Sons, Inc., New York, 1973, pp. 260–289.
51. G. Pistola, *Bull. Electrochem.* **7**(11), 524–528 (Nov. 1991).
52. B. Scorsati, *Proc. Electrochem. Soc.* **92**(12), 70–79 (1992).
53. R. O. Bach, *Lithium: Current Applications in Science, Medicine and Technology*, John Wiley & Sons, New York, 1985, pp. 337–407.
54. U.S. Pat. 4,206,191 (June 3, 1980), R. C. Morrison and R. O. Bach (to Lithium Corporation of America).
55. X. Wang, R. D. Peterson, and N. E. Richards, *Light Metals*, 323 (1991).
56. NIH Publication No. 93-3476, U.S. Department of Health and Human Services, Washington, D.C., January 1993.
57. Unpublished data, FMC Corp., Lithium Division, Bessemer City, N.C.; J. J. Kessis, *Bull. Soc. Chim. France* **1**, 48 (1965). D. A. Boryta, *J. Chem. Eng. Data* **15**, 142 (1970).
58. D. A. Boryta, A. J. Mass, and C. B. Grant, *J. Chem. Eng. Data* **20**, 316 (1975).
59. R. T. Ellington, G. Kunst, R. E. Peck, and J. F. Reed, *The Absorption Cooling Process*, Institute of Gas Technology, Chicago, Ill., 1957.
60. R. O. Bach and W. W. Boardman, Jr., *ASHRAE J.* **9**(11), 33 (1967).
61. R. O. Bach in F. Korte, H. Zimmer, and K. Niedenszu, eds., *Methodicum Chimicum*, Vol. 7, Part A, Academic Press, Inc., New York, 1977, p. 9; Technical Bulletin, "Lithium Aluminum Hydride Industrial Use," Chemetall GmbH, 1993.
62. Ger. Pat. 3,443,305 (May 28, 1986), F. Meiminger, I. Schdafer (to Hoechst AG).
63. D. Sharp, and S. Diamond, *SHPP-C-343: Eliminating or Minimizing Alkali-Silica Reactivity*, National Research Council, Washington, D.C., 1993.
64. U.S. Pat. 3,171,814 (Mar. 2, 1965) G. J. Orazem, R. B. Ellestad and J. R. Nelli (to Lithium Corp. of America).
65. U.S. Pat. 5,028,408 (July 2, 1991), B. L. Duncan, L. D. Carpenter, and L. R. Osborne (to Olin Corp.).
66. A. Rabenau, *Festkorperprobleme* **18**, 77 (1978).
67. U.S. Pat. 4,732,751 (Mar. 22, 1988), Dennis J. Salmon (to Lithium Corp. of America).
68. G. W. Hollenberg, R. C. Knight, P. J. Densley, L. A. Pember, C. E. Johnson, R. B. Poeppel, and L. Yang, *J. Nucl. Mater.*, 141–143, 271–274 (1986).

69. H. F. Cordes and S. R. Smith, *J. Phys. Chem.* **78**(8), 773 (1973).
70. P. A. Grieco, J. J. Nunes, and M. D. Gaul, *J. Am. Chem. Soc.* **112**, 4595 (1990).
71. R. A. Silva, *Chem. Eng. News*, 2, (Dec. 21, 1992).
72. P. Pierron, *Bull. Soc. Chim. Fr.* **6**(51), 235 (1939).
73. U.S. Pat. 3,185,546 (May 25, 1965), R. O. Bach and W. W. Boardman (to Lithium Corp. of America).
74. U.S. Pat. 2,986,585 (Apr. 2, 1961), W. I. Denton (to Olin Mathieson Chemical Corp.).
75. N. Koga, H. Tanaka, and A. K. Galwey, *J. Chem. Soc. Faraday Trans.* **86**(3), 531–537 (1990).
76. K. Ziegler and H. Colonius, *Liebigs Annalen der Chemie* **479**, 135 (1930).
77. G. E. Coates, *Organometallic Compounds*, 3rd ed., Methuen & Co., Ltd., London, 1960.
78. U. Schollkopf, in *Houben/Weyl Methoden der Organischen Chemie*, Series IV, Vol. 13/1, Thieme-Verlag, Stuttgart, Germany, 1970, p. 93.
79. C. W. Kamienski, in F. Korte, H. Zimmer, and K. Niedenszu, eds., *Methodicum Chimicum*, Vol. 7, part A, Academic Press, Inc., New York, 1977, p. 23.
80. B. J. Wakefield, *The Chemistry of Organolithium Compounds*, Pergamon Press, New York, Pt. I.
81. B. J. Wakefield, in G. Wilkinson, F. G. A. Stone, and E. W. Abel, eds., *Comprehensive Organometallic Chemistry*, Pergamon Press, New York, 1982, Vol. 7, Chapt. 44.
82. B. J. Wakefield, in *Best Synthetic Methods*, Academic Press, New York, 1988, p. 1.
83. J. L. Wardell, in F. R. Hartley, ed., *The Chemistry of the Metal-Carbon Bond*, Vol. 4, John Wiley & Sons, Inc., New York, 1987, Pt. 1, Sect. 1.
84. D. Seebach and K. H. Geiss, in D. Seyferth, ed., *New Applications of Organometallic Reagents in Organic Synthesis*, Elsevier Science, New York, 1976, Chapt. 1.
85. E. Negishi, *Organometallics in Organic Synthesis*, John Wiley & Sons, Inc., New York, 1980.
86. J. C. Stowell, *Carbanions in Organic Synthesis*, John Wiley & Sons, Inc., New York, 1979.
87. T. A. Hase, *Umpoled Synthons, A Survey of Sources and Uses in Synthesis*, John Wiley & Sons, Inc., New York, 1987.
88. H. Fieser, in *Fieser and Fieser's Reagents for Organic Synthesis*, Vol. 1–14, John Wiley & Sons, Inc., New York, 1967–1989.
89. D. J. Cram, *Fundamentals of Carbanion Chemistry*, Academic Press, Inc., New York, 1965.
90. D. Seyferth and R. King, *Annual Surveys of Organometallic Chemistry*, Elsevier, New York; F. G. A. Stone and R. West, *Advances in Organometallic Chemistry*, Academic Press, Inc., New York, 1964–1980.
91. *Organomet. Chem. Rev. Sect. B*, annual surveys.
92. K. Ziegler and H. G. Gellert, *Liebigs Annalen der Chemie* **567**, 179 (1950).
93. D. H. Lewis, W. S. Leonhardt, and C. W. Kamienski, *Chimia (Aarau)* **18**, 134 (1964).
94. Yu A. Lebedev, E. A. Miroshnichenko, and A. M. Chaikin, *Dokl. Akad. Nauk SSSR* **145**, 1288 (1962); *Proc. Acada. Sci. USSR Chem. Sect. (Engl. Transl.)* **145**, 1288 (1962).
95. P. A. Fowell and C. T. Mortimer, *J. Chem. Soc.*, 3793 (1961).
96. Technical Data Bulletin 103, Foote Mineral Co., Kings Mountain, N.C., May 1968.
97. W. H. Glaze, J. Lin, and E. G. Felton, *J. Org. Chem.* **31**, 2643 (1966).
98. R. A. Finnegan and H. W. Kutta, *J. Org. Chem.* **30**, 4138 (1965).
99. R. Bates, L. Kropolski, and D. Potter, *J. Org. Chem.* **37**, 560 (1972).
100. Product Bulletin, P-492, FMC Lithium Division, Gastonia, N.C.; Technical Data Bulletin, 107A, Cyprus-Foote Mineral Co., Kings Mountain, N.C.; Chemical Safety Data Sheet SD-91, MCA, Washington, D.C., 1966.

101. Technical Data Bulletin 109, Cyprus-Foote Mineral Co., Kings Mountain, N.C.; ASTM E-233-73, American Society for Testing and Materials, Philadelphia, Pa., 1973.
102. H. Gilman and F. Cartledge, *J. Organometal. Chem. (Amsterdam)* **2**, 477 (1964).
103. P. F. Collins, C. W. Kamienski, D. L. Esmay, and R. B. Ellestad, *Anal. Chem.* **33**, 468 (1961).
104. W. L. Everson, *Anal. Chem.* **36**, 854 (1964).
105. S. C. Watson and J. F. Eastham, *Anal. Chem.* **39**, 171 (1967).
106. S. C. Watson and J. F. Eastham, *J. Organometal. Chem.* **9**, 1965 (1967).
107. *Solprene Elastomers*, bulletins, Phillips Chemical Co., Bartlesville, Okla.
108. *Specialized Elastomers*, bulletins, Firestone Synthetic Rubber & Latex Co., Akron, Ohio.
109. *KRATON Thermoplastic Elastomers*, bulletins, Shell Chemical Co., Houston, Tex.
110. M. Morton and L. J. Fetters, *Rubber Chem. Technol.* **48**(3), 359 (1975).
111. H. L. Hsieh and W. H. Glaze, *Rubber Chem. Technol.* **43**(1), 22 (1970).
112. A. F. Halasa, Rubber Rev., *Rubber Chem. Technol.* **54**, 627 (1981).
113. G. Allen and J. C. Bevington, eds., *Comprehensive Polymer Science*, Pergamon Press, New York, 1989.
114. T. Hogen-Esch and J. Smid, eds., *Recent Advances in Anionic Polymerization*, Elsevier, New York, 1987.
115. S. Bywater, in J. I. Kroschwitz, ed., *Encyclopedia of Polymer Science and Engineering*, 2nd ed., Vol. 2, Wiley-Interscience, New York, 1985.
116. *n-Butyllithium in Organic Synthesis*, bulletin, FMC Lithium Division, Gastonia, N.C.
117. *Organometallics in Organic Synthesis*, bulletin, FMC Lithium Division, Gastonia, N.C.
118. U.S. Pat. 4,127,580 (Nov. 28, 1978), E. Braye (to PARCOR).
119. C. A. Hendrix, *Tetrahedron Rpt.* **34**, 1 (1978).
120. W. H. Glaze, J. Lin, and E. G. Felton, *J. Org. Chem.* **30**, 1258 (1965).
121. *Technical Data Bulletin No. 114*, Foote Mineral Co., Kings Mountain, N.C.
122. H. Gilman, A. H. Haubein, and H. Hartzfed, *J. Org. Chem.* **19**, 1034 (1954).
123. I. Kuntz and A. Gerber, *J. Polym. Sci.* **42**, 299 (1960).
124. M. Weiner, G. Vogel, and R. West, *Inorg. Chem.* **I**, 654 (1962).
125. *Organometallics Commercial Product Data*, FMC Lithium Division, Gastonia, N.C.; *Butyllithium—Properties and Uses*, Chemetall GmbH Lithium Division, Frankfurt, Germany; *t-Butyllithium in Heptane*, FMC Lithium Division, Gastonia, N.C.
126. *tert-Butyllithium Data D-501*, Lithium Corp. of America, Gastonia, N.C.
127. *n-Hexyllithium*, Chemetall GmbH, Frankfurt, Germany; *1-Hexyllithium in Hexane*, FMC Lithium Division, Gastonia, N.C.
128. T. L. Brown and J. T. Rogers, *Acta Cryst.* **10**, 465 (1957).
129. E. Weiss and E. A. C. Lucken, *J. Organometal. Chem. (Amsterdam)* **2**, 197 (1964).
130. H. O. House and W. L. Respess, *J. Organometal. Chem. (Amsterdam)* **4**, 95 (1964).
131. O. F. Beumel, Jr., and R. F. Harris, *J. Org. Chem.* **28**, 2775 (1963).
132. R. E. A. Dear and F. L. M. Pattison, *J. Am. Chem. Soc.* **85**, 622 (1963).
133. O. F. Beumel, Jr., and R. F. Harris, *J. Org. Chem.* **29**, 1872 (1964).
134. K. R. Martin, C. W. Kamienski, M. H. Dellinger, and R. O. Bach, *J. Org. Chem.* **33**, 778 (1968); W. Ried, *Angew. Chem.* **76**, 933 (1964).
135. O. Isler and co-workers, *Helv. Chim. Acta.* **39**, 249 (1965); U.S. Pat. 2,917,539 (Dec. 15, 1959), O. Isler and co-workers (to Hoffmann-La Roche).
136. W. M. McLamore and co-workers, *J. Org. Chem.* **20**, 109 (1955); U.S. Pat. 2,746,900 (May 22, 1956), A. Bavley and W. M. McLamore (to Chas. Pfizer & Co.).
137. *cis-e-Hexanol*, technical data, Hoffmann-La Roche, Nutley, N.J.
138. U.S. Pat. 4,526,720 (July 2, 1985), W. H. Rheenen and D. Y. Cha (to the UpJohn Co.).

139. U.S. Pat. 4,039,593 (Aug. 2, 1977), C. W. Kamienski and R. C. Morrison (to Lithium Corp.).
140. U.S. Pat. 3,674,836 (1972), P. L. Creger (to Parke-Davis & Co.).
141. A. S. Kende and P. Fludzinski, *Org. Synth.* **64**, 68 (1985).
142. D. J. Hart and co-workers, *J. Org. Chem.* **48**, 289 (1983).
143. U.S. Pat. 5,075,439 (1991), F. R. Busch and co-workers (to Pfizer, Inc.).
144. U.S. Pat. 4,320,061 (1982), H. W. Ohlendorf and co-workers (to Kali-Chemie Pharma).
145. *Lithium Methoxide Product Bulletin*, FMC Lithium Division, Gastonia, N.C.; *Lithium Methoxide Product Bulletin*, Chemetall GmbH, Frankfurt, Germany.
146. C. W. Kamienski and D. H. Lewis, *J. Org. Chem.* **30**, 3498 (1965).
147. *Lithium t-Butoxide Product Bulletin* and *Lithium Methoxide Product Bulletin*, Chemetall GmbH, Frankfurt, Germany.

CONRAD W. KAMIENSKI
Consultant
DANIEL P. MCDONALD
Catawba Valley Community College
MARSHALL W. STARK
FMC Corporation

LITHOGRAPHY. See RESIST MATERIALS.

LUBRICATION AND LUBRICANTS

The primary purpose of lubrication is separation of moving surfaces to minimize friction and wear. Although the fundamental principles were discovered by da Vinci, general understanding of the science of lubrication developed only in the latter part of the nineteenth century (1). Oil film lubrication was discovered in 1885 during studies of railroad car journal bearings in England, and this led almost immediately to the still current theoretical understanding by Reynolds.

Tallow was used to lubricate chariot wheels before 1400 BC. Although vegetable and animal oils were used in following years, significant production of petroleum oils and greases only followed the founding of the modern petroleum industry with the Drake well in Titusville, Pennsylvania in 1859 (2). Production reached 9500 m^3/yr (2,500,000 gal/yr) in the following 20 years. Worldwide production is now nearly 1000 times that volume and petroleum lubricants constitute about 98% of total oil and grease production volume.

Lubrication Principles

Several distinct regimes are commonly employed to describe the fundamental principles of lubrication. These range from dry sliding to complete separation of two moving surfaces by a fluid lubricant, with an intermediate range involving partial separation in boundary or mixed lubrication. When elastic surface deflections exert a strong influence on the nature of lubrication of a concentrated contact, as in a ball or roller bearing, a regime of elastohydrodynamic lubrication is encountered with its distinctive characteristics.

Dry Sliding. When two surfaces rub, the real area of contact involves only sufficient asperities of the softer material so that their yield pressure balances the total load (3). As the initial load W increases, the real contact area A illustrated in Figure 1 increases proportionately according to the relation

$$A = W/p \tag{1}$$

Yield pressure p of the asperities is about three times the tensile yield strength for many materials. The real area of contact is frequently a minute fraction of the total area. With a typical bearing contact stress of 3 MPa and a bronze bearing asperity yield pressure of 500 MPa, for instance, less than 1.0% of the nominal area would involve asperity contact.

Friction during dry sliding primarily involves a force F required to displace interlocking asperities of the softer material with shear strength s.

$$F = As \tag{2}$$

Although this shearing of asperity junctions often accounts for 90% or more of the total friction force, other factors may contribute. A lifting force may be needed to raise asperities over the roughness of the mating surface. Scratching by dirt and wear particles, or by sharp asperities, may introduce ploughing resistance. Internal damping, surface charges, and chemical films also play a role.

Combining the two previous relations for contact area and friction force gives Amonton's law:

$$F = Ws/p \tag{3}$$

Coefficient of friction f, the ratio of friction force to applied load, is

$$f = F/W = s/p \tag{4}$$

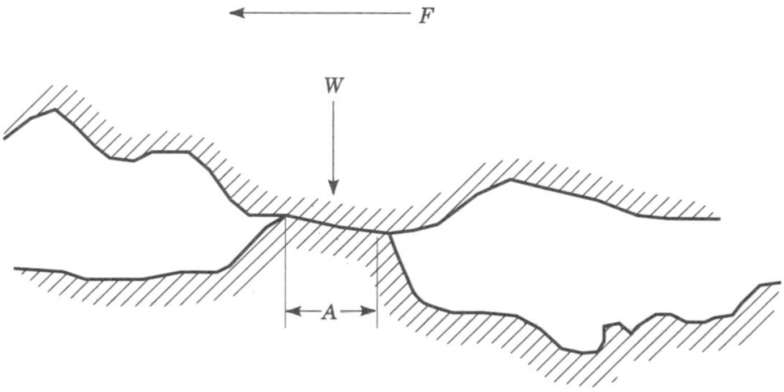

Fig. 1. An asperity contact between two rubbing surfaces.

Because shear and compressive strengths s and p depend in a similar way on material properties such as lattice structure and bond strength, f is often in a rather narrow range of about 0.20–0.35 for a wide variety of materials. The following are typical data for sliding on steel with bearing materials varying several hundredfold in yield pressure:

Material	f
carbon–graphite	0.19
lead babbitt	0.24
bronze	0.30
aluminum alloy	0.33
polyethylene	0.33

With some low surface energy materials, such as polytetrafluoroethylene, f may drop to 0.04–0.10 at low sliding velocities. With bulk welding and material transfer at the other extreme, as with lead sliding on lead, f exceeds 1.0. Coefficient of friction usually drops somewhat with increasing load and speed. Surface roughness variations usually introduce surprisingly small changes for dry sliding.

A thin surface layer of soft solid or adsorbed lubricant controls the coefficient of friction for a structural metal backing according to the following relation:

$$f = s_f / p_m \tag{5}$$

where s_f is the shear strength of the surface film and p_m the yield pressure of the backing metal. Minimum friction is provided by a low shear strength film on a hard substrate used to maintain a small contact area. A soft film of indium [*7440-74-6*] applied as a 10-μm thick solid lubricant on steel, for instance, gives f as low as 0.04 (4) (see INDIUM AND INDIUM COMPOUNDS).

The volume V of wear fragments can be related for adhesive contacts to sliding distance x as follows:

$$V = kWx/3p \tag{6}$$

Some dimensionless wear coefficients k are given in Table 1 (3).

Although equations 1–6 serve as useful guides, they are applicable only in very general terms. Local temperature rise in contacts influences the complex processes at asperities. High surface temperatures at high loads and speeds may lead to failure of adsorbed lubricant films or bonded solid-film lubricants. Events may be further complicated by work hardening, surface fatigue, welding, recrystallization, oxidation, and hydrolysis.

The goal of lubrication is elimination of this wear and minimizing friction otherwise encountered in dry sliding. This is accomplished ideally with complete separation of the rubbing surfaces with a full film of lubricant. When complete full-film separation is impossible, surface chemical effects of a lubricating oil and its additives, or solid-film lubricants such as graphite and molybdenum sulfide, can assist.

Table 1. Wear Coefficients for Various Sliding Combinations[a]

Combination	Wear coefficient, $k = 3p \cdot V/(W \cdot x)$
zinc on zinc	0.16
low carbon steel on low carbon steel	0.045
copper on copper	0.032
stainless steel on stainless steel	0.021
copper on low carbon steel	0.0015
low carbon steel on copper	0.0005
phenolic resin on phenolic resin	0.00002

[a]Ref. 3.

Fluid-Film Lubrication. In this regime, the moving surfaces are completely separated by a film of liquid or gaseous lubricant. A load-supporting pressure is commonly generated hydrodynamically in the film by pumping action in a converging, wedge-shaped zone, as in Figure 2**a**. Both the pressure and the frictional power loss in this film are functions of the lubricant viscosity in combination with the geometry and shear rate imposed by the bearing operating conditions.

The squeeze-film action illustrated in Figure 2 is encountered in dynamically loaded bearings in reciprocating engines and under shock loads. Because time is required to squeeze the lubricant film out of a bearing, much higher loads can be supported than with steady, unidirectional loads such as are common in electric motors and generators (Table 2). The much lower load capacity of bearings lubricated with low viscosity fluids, eg, water and gases, is also indicated in Table 2.

When normal hydrodynamic and squeeze-film action gives inadequate load support, the fluid may be pressurized externally before being introduced into the bearing film in the manner of Figure 2**b** and **c**. Such a procedure is common for starting and slow speeds with heavy machines, or with low viscosity fluids.

Detailed performance analyses for a wide variety of fluid-film bearings provide formal viscous flow determinations of fluid-film thickness, power loss, flow rate, temperature rise, and the influence of changes in operating parameters (5–8). In computer codes for carrying out these analyses, methods have also been developed for estimating the higher power loss and thicker film resulting from turbulent fluid-film behavior in many large and high speed bearings (9). Much attention has been given to the dynamic response of these fluid-film bearings and their effect on machinery vibration; in many rotating machines, about half of the rotor system elasticity and most of the damping may be found in the fluid film.

Boundary Lubrication. As the severity of operating conditions increases, it becomes impossible for the load to be carried completely by the oil film. High spots, or asperities, of the mating surfaces then contact to share in load support and the lubrication shifts, as indicated in Figure 3, from full film with a coefficient of friction of about 0.001, to mixed film and boundary lubrication where the coefficient of friction rises to 0.03–0.1, and finally to complete loss of film support where friction may rise to the range of 0.2–0.4 typical for dry sliding. The shift from full film to boundary lubrication may result from any

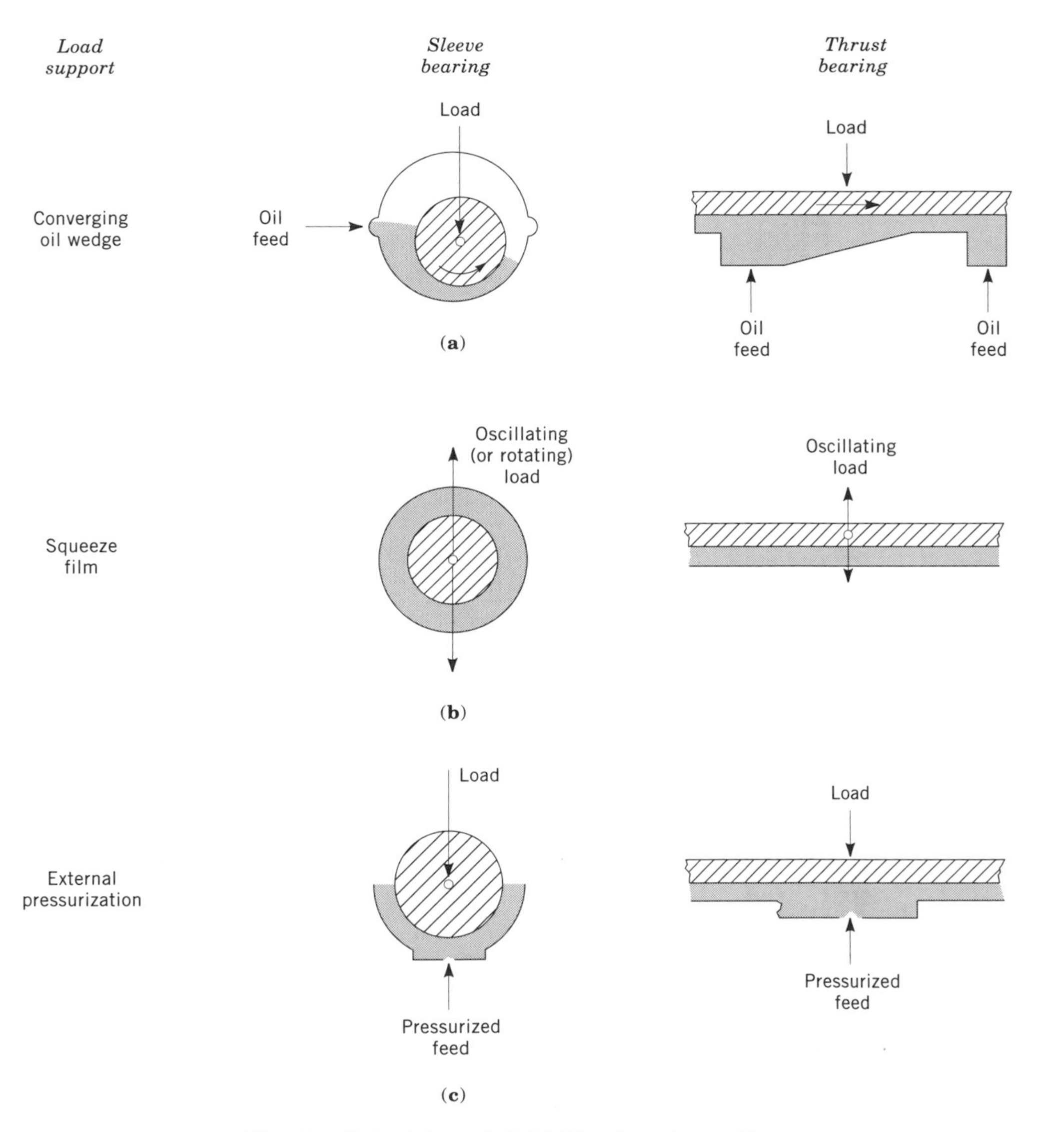

Fig. 2. Principles of fluid-film bearing action.

one or a combination of the following conditions: high load, high temperature, low speed, low lubricant viscosity, rough surfaces, misalignment, or inadequate supply of lubricant. With boundary lubrication, chemical additives in the oil and chemical, metallurgical, and mechanical factors involving the two rubbing surfaces determine the extent of wear and the degree of friction (see BEARING MATERIALS).

In boundary lubrication, some asperity contacts begin to penetrate through the fluid film and adherent surface films. With increasing loads, more asperity contacts occur with more plastic deformation of the contacting surfaces, higher temperatures, and welding. Surface tearing and seizure finally occur on a gross scale.

Table 2. Typical Design Limits for Fluid-Film Hydrodynamic Bearings

Item	Load on projected area, MPa[a]
oil lubrication	
steady load	
electric motors	1.4
steam turbines	2.1
railroad car axles	2.4
dynamic load	
automobile engine main bearings	24
automobile connecting-rod bearings	34
steel-mill roll necks	34
water lubrication	0.2
gas bearings	0.02

[a]To convert MPa to psi, multiply by 145.

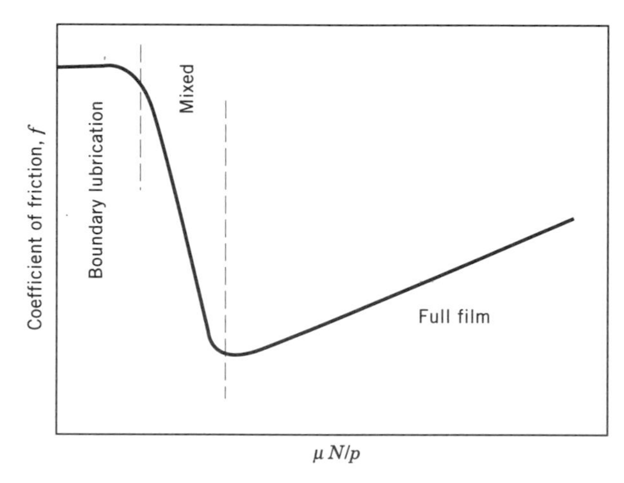

Fig. 3. Stribeck curve relating friction coefficient to absolute viscosity μ, speed N in rpm, and unit load p.

Hypoid gears in automobile differentials are particularly susceptible to this damage since they impose severe sliding conditions in combination with high contact stress. Intense heat then leads to ineffectiveness of the organic lubricant film normally present. Antiwear and extreme pressure (EP) lubricants prevent welding under these conditions by reacting at the high contact temperatures to form protective low shear strength surface films on the metal surfaces. These antiwear and EP additives generally consist of organic sulfur, phosphorus, and chlorine compounds dissolved in the oil, or less frequently a dispersion of fine particles of graphite, molybdenum disulfide, or polytetrafluoroethylene (PTFE).

Effectiveness of these EP oils can be evaluated by a number of laboratory test units such as those shown in Figure 4. While the American Society for Testing and Materials (ASTM) procedures describe a number of standard test procedures (10), the operating conditions and test specimen materials should be chosen to simulate as nearly as possible those in an application.

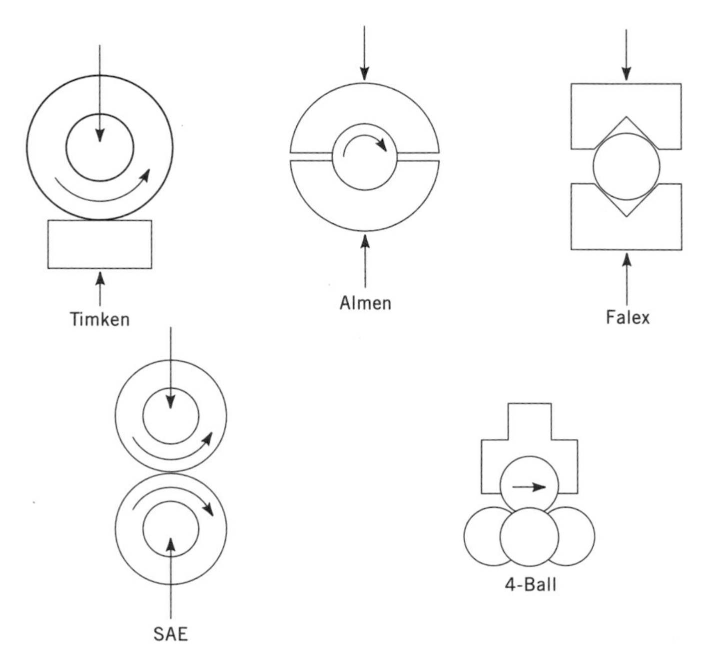

Fig. 4. Operating principles of various laboratory lubricant-test machines.

Because EP additives are effective only by chemical action, their general use should be avoided to minimize possible corrosion difficulties and shortened lubricant life in any application where they are not necessary. For long-time operation of machines, conversion from boundary to full-film operation is desirable through changes such as higher oil viscosity, lowered loading, or improved surface finish.

Elastohydrodynamic Lubrication (EHL). Lubrication needs in many machines are minimized by carrying the load on concentrated contacts in ball and roller bearings, gear teeth, cams, and some friction drives. With the load concentrated on a small elastically deformed area, these EHL contacts are commonly characterized by a very thin separating hydrodynamic oil film which supports local stresses that tax the fatigue strength of the strongest steels.

Pressure distribution in an EHL rolling contact takes on the elliptical pattern of Figure 5 (11). Overall oil-film thickness (often ca 0.1–0.5 μm) is primarily set by oil viscosity, film shape, and velocity at the entry to the contact zone. Film thickness then remains nearly uniform over most of the length along the contact. The high contact pressure leads to very high oil viscosity, and the pressure distribution approximates that of the Hertz pattern for simple static elastic contact with no oil film. Increasing load causes increased elastic deformation and larger contact area, but only gives a slight reduction in EHL film thickness. A sharp pressure spike at the end of the contact zone and an associated local constriction of about 25% in oil-film thickness result from a combination of accelerating flow caused by the exiting pressure gradient together with elastic expansion of the bearing surface as the contact pressure drops.

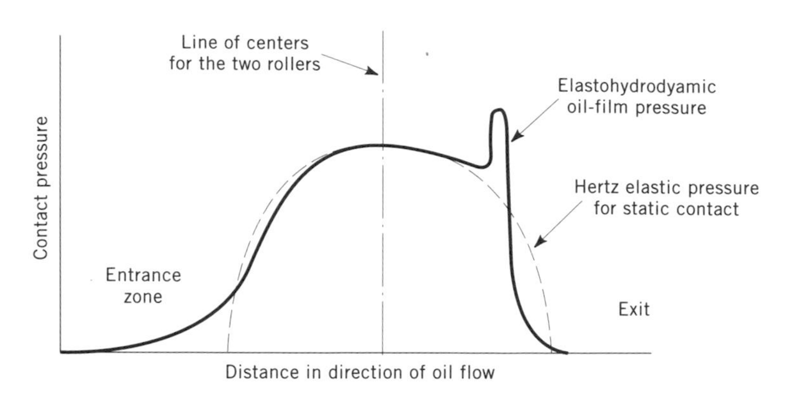

Fig. 5. Pressure distribution between two rollers under load.

In rolling contacts with full separation by an EHL oil film, load capacity is limited primarily by fatigue strength of the metal. Fatigue cracks and spalling under too heavy load are eventually generated by repeated working of grain boundaries about 20–50 μm beneath the contact surface where shear stress is maximum. Surface flaking then occurs with the thickness of loosened particles representing the depth to this zone of maximum shear stress. If the full-film lubrication in the rolling contact is lost under some combination of low speed, high load, low viscosity, or lubricant starvation, increased tangential traction transfers the maximum shear stress out to the metal surface. Surface wear and flaking then occurs (12). In this region of boundary lubrication, lubricant composition and additives may have a pronounced influence, either positive or negative, on fatigue life.

Only very small amounts of oil, less than one drop with most small and medium-sized ball and roller bearings, are sufficient to provide a full EHL film (5). In such cases, a small amount of grease or oil mist balances lubricant loss by vaporization, creepage, and throw-off. With high surface speeds and heavy loads, however, much larger lubricant feed is needed for cooling and makeup.

Petroleum Lubricants

Petroleum (qv) products dominate lubricant production with a 98% share of the market for lubricating oils and greases. While lower cost leads to first consideration of these petroleum lubricants, production of various synthetic lubricants covered later has been expanding to take advantage of special properties such as stability at extreme temperatures, chemical inertness, fire resistance, low toxicity, and environmental compatibility.

Petroleum oils generally range from low viscosity with molecular weights as low as 250 to very viscous lubricants with molecular weights up to about 1000. Typical molecular structures of the complex mixtures of hydrocarbon molecules involved are indicated in Figure 6 (13). Physical properties and performance characteristics depend heavily on the relative distribution of paraffinic, aromatic, and alicyclic (naphthenic) components. For a given molecular size, paraffins have relatively low viscosity, low density, and higher freezing temperatures. Aromatics have higher viscosity, rapid change in viscosity with temperature,

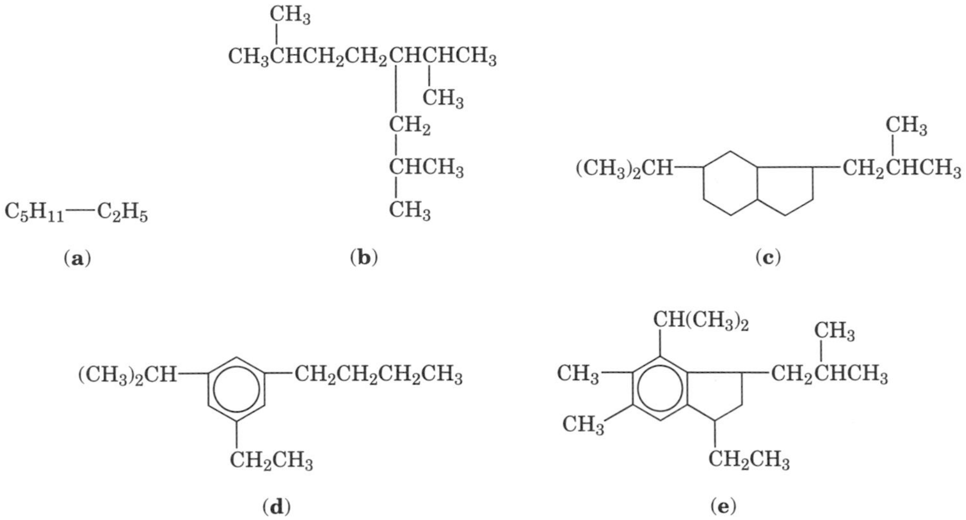

$C_5H_{11}—C_2H_5$

(**a**)

$CH_3CHCH_2CH_2CHCHCH_3$ (with CH_3, CH_3, CH_2CHCH_3, CH_3 branches)

(**b**)

$(CH_3)_2CH$—[ring]—CH_2CHCH_3 (CH_3)

(**c**)

$(CH_3)_2CH$—[ring]—$CH_2CH_2CH_2CH_3$, CH_2CH_3

(**d**)

$CH(CH_3)_2$, CH_3, CH_3—[ring]—CH_2CHCH_3, CH_3, CH_2CH_3

(**e**)

Fig. 6. Typical structures in lube oil: (**a**) *n*-paraffin, (**b**) isoparaffin, (**c**) cycloparaffin, (**d**) aromatic hydrocarbon, and (**e**) mixed aliphatic and aromatic ring (13).

higher density, and darker color. Although aromatics have a high degree of oxidation stability, they oxidize to form insoluble black sludge at high temperature. Alicyclic oils are characterized by low pour point, low oxidation stability, and other properties intermediate to those of the paraffins and aromatics.

Almost all premium lubricants are so-called paraffinic oils composed primarily of both paraffinic and alicyclic structures, with only a minor portion of aromatics. When stabilized with an oxidation inhibitor and fortified with other appropriate additives, these paraffinic–alicyclic compositions provide nonsludging oils that are satisfactory for almost any type of service.

The first step in producing a lubricating oil involves distillation (qv) of the crude petroleum (14). The lower boiling gasoline, kerosene, and fuel oils are removed first, and the lubricating oil fractions are then divided by boiling point into several grades of neutral distillates and a final more viscous residuum. Subsequent refining steps remove undesirable aromatics and the minor portion of sulfur, nitrogen, and oxygen compounds. Although solvent extraction or sulfuric acid treatment, followed by activated clay to absorb dark-colored and unstable molecules, had been used for this purification step, hydrogen treatment at high pressure and in the presence of a catalyst was introduced in 1955. Mild hydrofining involves primarily only the removal of color and some nitrogen, oxygen, and sulfur compounds. More severe hydrofining or hydrocracking at temperatures in the 500–575°C range further alters the chemical structures to convert aromatics to paraffins and alicyclics in oils of very high viscosity index (VHVI).

Low temperature filtration (qv) is a common final refining step to remove paraffin wax in order to lower the pour point of the oil (14). As an alternative to traditional filtration aided by a propane or methyl ethyl ketone solvent, catalytic hydrodewaxing cracks the wax molecules which are then removed as lower boiling products. Finished lubricating oils are then made by blending these refined stocks to the desired viscosity, followed by introducing additives needed to provide the required performance. Table 3 lists properties of typical commercial petroleum oils. Methods for measuring these properties are available from the ASTM (10).

Viscosity. The viscosity of an oil is its stiffness or internal friction, as illustrated in Figure 7. With a surface of area A moving at velocity V at a distance ΔX from an equal parallel area moving at velocity $V + \Delta V$, force F is required to maintain the velocity difference according to the equation 7:

$$F/A = \mu \Delta V / \Delta X \tag{7}$$

Constant μ is the viscosity of the liquid separating the two surfaces. Viscosity may also be defined as the ratio of shear stress F/A to rate of shear $\Delta V/\Delta X$. For example, a bearing surface moving 100 m/s, 0.0001 m from a stationary surface produces a shear rate of 10^6 s^{-1} in the lubricating oil. A liquid has a viscosity of 0.1 Pa·s(=dyn·s/cm^2 or 1 Poise) when a force of 0.1 N (1 dyn) is required to move 1 cm^2 of area past a parallel area 1 cm away at a velocity of 1 cm/s. The common engineering unit in the British system is the reyn (lbf·s/$in.^2$) with dimensions of 6.8947×10^6 mPa·s(=cP). With gravity providing the driving force in most laboratory capillary viscometers, flow time of the oil is proportional to kinematic viscosity, the absolute viscosity divided by oil density with units of m^2/s. The unit mm^2/s(=cSt), equal to mPa·s(= cP) divided by oil density in g/cm^3, is commonly used to avoid decimal values.

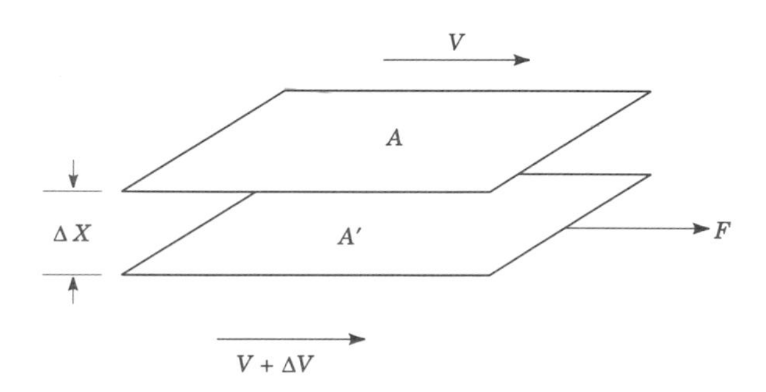

Fig. 7. Diagram to illustrate the definition of viscosity.

A number of arbitrary viscosity units have also been used. The most common has been the Saybolt Universal second (SUs) which is simply the time in seconds required for 60 mL of oil to empty out of the cup in a Saybolt viscometer through a carefully specified opening. Detailed conversion tables appear in ASTM D2161, approximation of kinematic viscosity ν in mm^2/s(=cSt) can be made from the relation shown in equation 8:

$$\nu = 0.22\,(\text{SUs}) - 180/(\text{SUs}) \tag{8}$$

Wide range of viscosity in commercial petroleum oils is illustrated by the representative types listed in Table 3. Despite this range, the largest proportion of oils are in the 25–75 mm^2/s at 40°C viscosity range. Oils in this range combine generally adequate hydrodynamic load capacity with low power loss, low volatility, and satisfactory low temperature properties.

Table 3. Representative Petroleum Lubricating Oils

Type	Viscosity, mm^2/s(=cSt)		Flash point, °C	Pour point, °C	Sp gr, at 15°C	Viscosity index	Common additives[a]	Uses
	at 40°C	at 100°C						
automobile (SAE)								
10W	28	4.9	204	−28	0.878	106	R,O,D,VI,P, W,F,M	automobile, truck, and marine reciprocating engines
20W	48	7.0	218	−24	0.884	103		
30	93	10.8	228	−20	0.890	100		
40	134	13.7	238	−16	0.895	97		
50	204	17.8	250	−10	0.901	94		
10W-30	62	10.3	208	−36	0.880	155		
20W-40	138	15.3	246	−21	0.897	114		railroad diesels
15W-40	108	15.0	218	−27	0.885	145		diesels
gear (SAE)								
80W-90	144	14.0	192	−22	0.900	93	EP,O,R,P,F	automotive and industrial gear units
85W-140	416	27.5	210	−14	0.907	91		
automatic transmission	38	7.0	188	−40	0.867	140	R,O,W,F,VI, P,M	automotive hydraulic systems

Table 3. (*Continued*)

Type	Viscosity, mm^2/s(=cSt)		Flash point, °C	Pour point, °C	Sp gr, at 15°C	Viscosity index	Common additives[a]	Uses
	at 40°C	at 100°C						
turbine								
light	31	5.4	206	−10	0.863	107	R,O	steam turbines, electric motors, industrial circulating systems
medium	64	8.7	220	−6	0.876	105		
heavy	79	9.9	230	−6	0.879	103		
hydraulic fluids								
light	30	5.3	206	−24	0.868	99	R,O,W	machine tool hydraulic systems
medium	43	6.5	210	−23	0.871	98		
heavy	64	8.4	216	−22	0.875	97		
extra low temp	14	5.1	96	−62	0.859	370	R,O,W,VI,P	aircraft hydraulic systems
aviation								
grade 65	98	11.2	218	−23	0.876	100	D,P,F	reciprocating aircraft engines
grade 80	139	14.7	232	−23	0.887	105		
grade 100	216	19.6	244	−18	0.893	100		
grade 120	304	23.2	244	−18	0.893	95		

[a]R, rust inhibitor; O, oxidation inhibitor; D, detergent–dispersant; VI, viscosity-index improver; P, pour-point depressant; W, antiwear; EP, extreme pressure; F, antifoam; and M, friction modifier.

Viscosity Classifications. The general ISO international viscosity classification system for industrial oils is given in Table 4 from ASTM D2422 (American National Standard Z11.232). For high speed machines, ISO viscosity-grade 32 turbine and hydraulic oils are a common choice. ISO grades 68 and 100 are applied for more load capacity in slower speed machines where power loss and temperature rise are less of a question.

The Society of Automotive Engineers (SAE) viscosity grades for automotive engine oils are also given in Tables 4 and 5 (15). With the addition of viscosity-index improvers, oils are available that meet requirements of more than one SAE grade. The common 10W-30 oils, for example, combine the low temperature viscosity of the SAE 10W classification for easy low temperature starting, with SAE 30 high temperature viscosity for better load capacity in bearings at the normal engine running temperature. SAE 30, 40, and 50 grades containing additives for severe service are used in industrial, railroad, and marine diesel engines. Although automotive oils are widely distributed, they should be used only with caution in industrial applications. Their detergent additives may cause problems with foam and water emulsions, and the viscosity-index improving additives slowly lose their thickening power under high shear rates.

Turbine oils are the premium products commonly used in circulating systems for steam turbines, steel mills, paper mills, and electric motors and generators. These oils contain rust and foam inhibitors, plus an oxidation inhibitor for long life. Hydraulic oils intended for circulating systems of factory machine tools also contain a zinc dithiophosphate additive to minimize wear in high pressure hydraulic pumps. General-purpose oils with no additives are used to minimize expense for once-through lubrication with mist, drip feed, etc, in factory machines.

Gear oils are generally formulated for industrial applications in AGMA grades. These oils are supplied in the viscosity grades shown in Table 4 either with simply a rust and oxidation (R&O) inhibitor for lightly loaded spur and helical gears, compounded with about 3 to 10% fatty additive for worm gears, or with extreme pressure (EP) additives for hypoid gears and heavily loaded and low speed spur and helical gears. SAE automotive gear oils are also sometimes used for industrial gearing and often have higher EP performance than those formulated for the AGMA specifications.

Oil viscosity grades have also been developed with suitable additives for use in a variety of specific applications in two-cycle engines, refrigeration and air conditioning, oil mist lubricators, low outdoor temperatures, instruments, and office machines as partially reflected in Table 3. Equipment manufacturers and lubricant suppliers provide recommendations for individual cases.

Viscosity–Temperature. Oil viscosity decreases with increasing temperature in the general pattern shown in Figure 8, an example of ASTM charts which are available in pad form (ASTM D341). A straight line drawn through viscosities of an oil at any two temperatures permits estimation of viscosity at any other temperature, down to just above the cloud point. Such a straight line relates kinematic viscosity ν in $mm^2/s(=cSt)$ to absolute temperature T (K) by the Walther equation,

$$\log\log(\nu + 0.7) = A + B\log T \tag{9}$$

Table 4. ASTM D2422 ISO Viscosity System for Industrial Oils

ISO-VG Grade	Viscosity, $mm^2/s(=cSt)$			Former ASTM SUs grades	SAE crankcase oil grades[a]	SAE Aircraft oil grades[a]	SAE Gear lube grades[a]	AGMA[b] gear lube grades		Typical fuels and base oils
	Minimum	Typical	Maximum					Regular	EP	
2	1.98		2.42							kerosine
3	2.88		3.52							#2 fuel
5	4.14		5.06							
7	6.12		7.48							
10	9.00		11.0							
15	13.5	14.2	16.5	75						#4 fuel
22	19.8	20.9	24.2	105	5W					100 neutral
32	28.8	30.4	35.2	150	10W					150 neutral
46	41.4	43.6	50.6	215	20W		75W	1		200 neutral
68	61.2	64.5	74.8	315	20			2	2 EP	300 neutral
100	90.0	94.8	110	465	30	65		3	3 EP	450 neutral
150	135	143	165	700	40	80	80W-90	4	4 EP	600 neutral
220	198	209	242	1000	50	100	90	5	5 EP	
320	288	304	352	1500		120		6	6 EP	
460	414	436	506	2150			85W-140	7 comp	7 EP	150 bright stock
680	612	644	748	3150				8 comp	8 EP	175 bright stock
1000	900	948	1100	4650				8A comp	8A EP	190 bright stock
1500	1350	1421	1650	7000			250			

[a]Comparisons are nominal since SAE grades are not specified at 40°C viscosity; VI of lubes could change some of the comparisons.
[b]American Gear Manufacturers' Association.

Table 5. SAE Viscosity Grades for Engine Oils

SAE Viscosity grade	Viscosity, mPa·s(=cP) at °C	Borderline pumping temperature, °C	100°C Viscosity, mm^2/s(=cSt) Minimum	Maximum
0W	3250 at −30	−35	3.8	
5W	3500 at −25	−30	3.8	
10W	3500 at −20	−25	4.1	
15W	3500 at −15	−20	5.6	
20W	4500 at −10	−15	5.6	
25W	6000 at −5	−10	9.3	
20			5.6	<9.3
30			9.3	<12.5
40			12.5	<16.3
50			16.3	<21.9
60			21.9	<26.1

where A and B are constants for any given oil. The constant 0.7 increases gradually for viscosities below 2.0 mm^2/s encountered for very low viscosity fluids (ASTM D341). For individual lubricants and extrapolation to low temperatures, the 0.7 value can be further modified for better correlations (16).

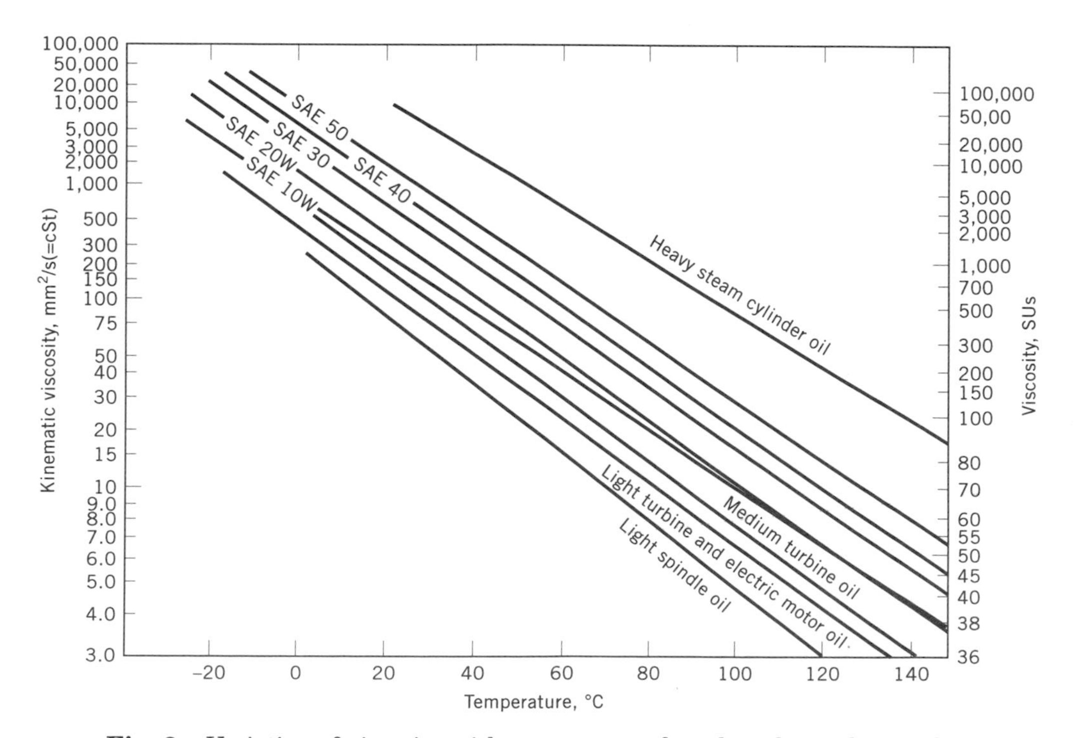

Fig. 8. Variation of viscosity with temperature for selected petroleum oils.

The viscosity index (VI), although empirical, is the most common measure of the relative decrease in oil viscosity with increasing temperature. A series of Pennsylvania petroleum oils exhibiting a relatively small decrease in viscosity with increasing temperature is arbitrarily assigned a VI of 100, whereas a series of Gulf Coast oils having viscosities that change relatively rapidly is assigned a VI of 0. From viscosity measurements at 40 and 100°C, the VI of any oil sample can be obtained from detailed tables published by ASTM (ASTM D2270). Figure 9 indicates the relation between 40 and 100°C viscosities for oils of varying VI.

Oils having a VI above 80 to 90 are generally desirable. These oils are composed primarily of saturated hydrocarbons of the paraffinic and alicyclic types which give long life, freedom from sludge and varnish, and generally satisfactory performance when they are compounded with proper additives for a given application. Lower VI oils sometimes are useful in providing low pour point for outdoor applications in cold climates and for some refrigeration and compressor applications.

Although the viscosity index is useful for characterizing petroleum oils, other viscosity–temperature parameters are employed periodically. Viscosity temperature coefficients (VTCs) give the fractional drop in viscosity as temperature increases from 40 to 100°C and is useful in characterizing behavior of silicones and some other synthetics. With petroleum base stocks, VTC tends to remain constant as increasing amounts of VI improvers are added. Constant B in equation 9, the slope of the line on the ASTM viscosity–temperature chart, also describes viscosity variation with temperature.

Viscosity–Pressure. The great increase in viscosity with pressure in Figure 10 (17) indicates the dramatic effects to be expected in elastohydrody-

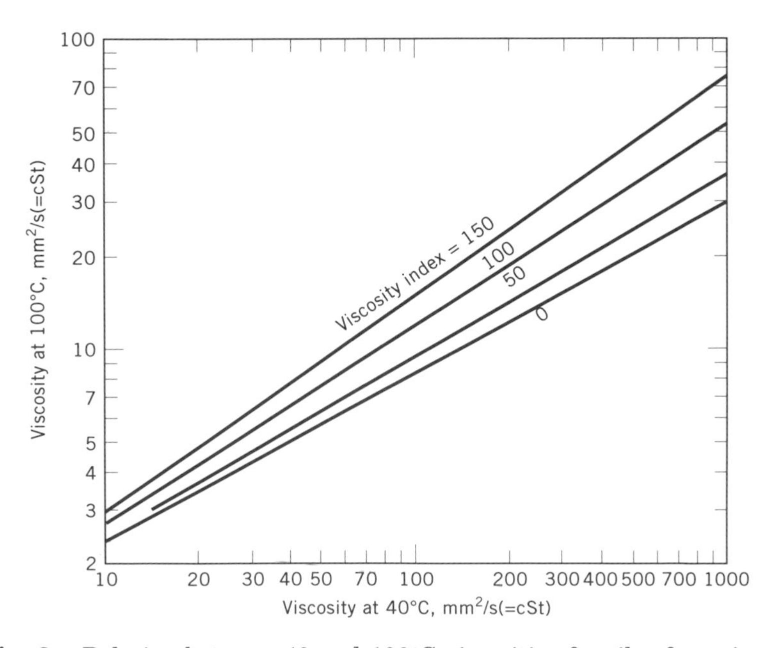

Fig. 9. Relation between 40 and 100°C viscosities for oils of varying VI.

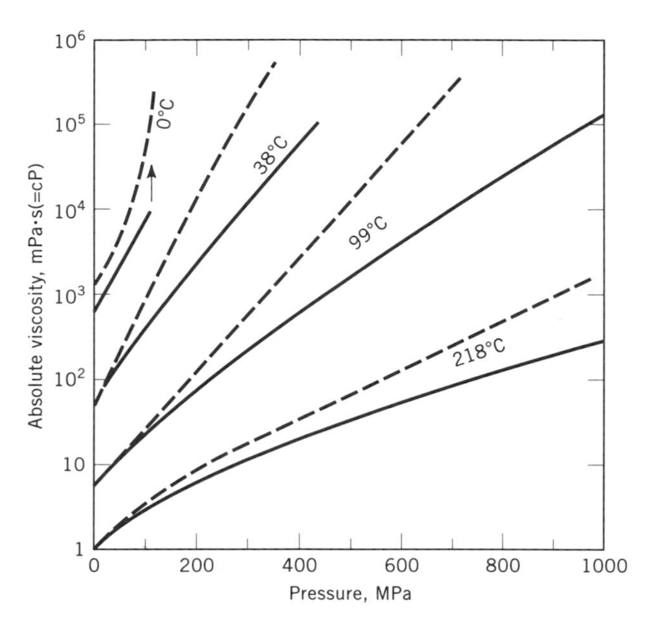

Fig. 10. Viscosity–pressure curve for typical petroleum oils: (—) paraffinic; (- - -) alicyclic; and (↑) solid. To convert MPa to atm, divide by 0.101.

namic contacts in rolling bearings, gears, and cams at pressures ranging up to 2000 to 3000 MPa (300,000 to 450,000 psi). In the lower pressure range of Figure 10, the following relationship can be applied for many oils:

$$\mu_p = \mu_0 e^{\alpha p} \tag{10}$$

where μ_0 is the viscosity at atmospheric (essentially zero) pressure and μ_p is the viscosity at pressure p. The pressure coefficient α at the low entry pressure in elastohydrodynamic contacts is then used in calculating oil film thickness (11,12).

Generalized pressure–temperature–viscosity relations have been developed from the extensive data for petroleum and synthetic oils (18,19). Interestingly, lubricating oils may drop slightly in viscosity as they are exposed to high pressures in equilibrium with nitrogen and some other gases. The thinning effect of the dissolved gas tends to balance the increase in viscosity which normally occurs with increased pressure (20).

Additives

With chemical additives being used in almost all lubricants, their worldwide production has grown to be a $5 billion segment of the chemical industry. Typical volume percentages applied in commercial petroleum lubricants, with lubricants for internal combustion engines accounting for about 72% of the market volume, include automotive and diesel engine oils, ie, straight, single SAE grade, 12%, or multigrade, 20%; automotive gear and transmission oils, 12%; hydraulic and turbine oils, 0.75%; and greases 4%.

Comprehensive reviews of additive practices are available in the literature (18,21–23) and in extensive patent coverage. The common types of additives are discussed in approximate order of the frequency of their use.

Oxidation Inhibitors. Oxidation of petroleum oils is the most common form of degradation. Three stages are normally involved: generation of free radicals under the accelerating influence of heat and metal catalysts, a propagation stage in which these free radicals react with oxygen and the lubricant to form hydroperoxides and other free radicals in a chain reaction, and termination when radicals combine or react with oxidation inhibitors (22). Some hydroperoxides decompose to give alcohols, aldehydes, ketones (qv), and organic acids which may then polymerize or break down further to viscous soluble polymers, insoluble sludge, and eventually darkened varnish-like deposits.

Oxidation inhibitors function by interrupting the hydroperoxide chain reaction. At temperatures up to ca 120°C, di-*tert*-butyl-*p*-cresol, 2-naphthol, 1-naphthyl(phenyl)amine, and related hindered phenols and amines effectively act as free-radical scavengers. Kinetic studies have both raised questions and elucidated some details of this mechanism (24). These inhibitors are commonly used at 0.5–1.0% concentration in highly refined paraffinic oils for lubrication of steam and gas turbines, electric motors, and hydraulic equipment and instruments. Additives of this type, and sulfur and phosphorus compounds, can also function as hydroperoxide decomposers to break the propogation process. Selective polar additives are effective in inactivating ions of iron, copper, and other metals which would otherwise catalyze the oxidation reaction (22).

Zinc dialkyl dithiophosphates are the primary oxidation inhibitors in combining these functions with antiwear properties in automotive oils and high pressure hydraulic fluids. Their production volume is followed by aromatic amines, sulfurized olefins, and phenols (22).

Rust Inhibitors. These are surface-active additives which preferentially adsorb as a film on iron or steel surfaces to prevent their corrosion by moisture, as suggested in Figure 11. For mild conditions with a small amount of water present in a large quantity of circulating oil, long-chain amines, alkyl succinic acids, and other mildly polar organic acids find use. For more severe conditions in shipping and storage of machinery, and in outdoor weather, more strongly

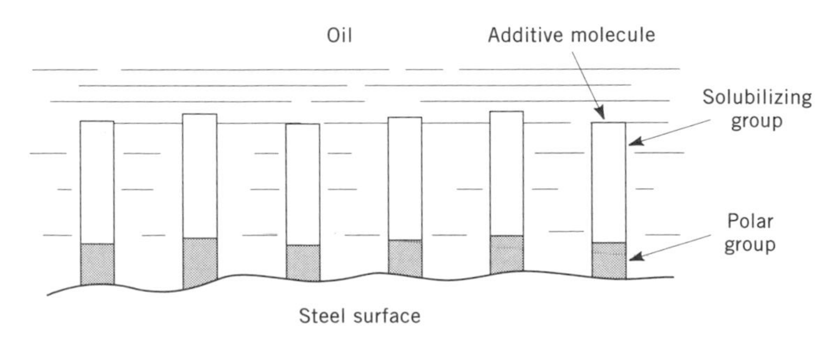

Fig. 11. Use of rust inhibitor to protect steel surface from attack by moisture.

adherent sodium and calcium sulfonates, organic phosphates, and polyhydric alcohols are used. When incorporated in vapor-space inhibited oils, dicyclohexylamine and related amines with modest vapor pressure provide rust protection above the oil level during extended shutdown periods for machinery.

For protection against nonferrous and copper alloy corrosion, thiadiazole and triazole derivatives have been found especially useful (22).

Antiwear and Extreme Pressure Agents. Zinc dialkyl dithiophosphates are the most widely used antiwear agents. These are commonly produced by reaction of an alcohol with phosphorus pentasulfide and neutralization of the resulting dithiophosphoric acid with zinc oxide. Although these additives give remarkable results in reducing wear in cams, gears, and high pressure hydraulic components, they lead to corrosion and deposits under some conditions which promote their hydrolysis. Because thermal breakdown above about 150–200°C generates hydrogen sulfide and other degradation products which may soften electrical insulation, these antiwear oils are generally avoided in electric motors and generators.

In steel-on-steel lubrication with a zinc dialkyl dithiophosphate additive, a complex surface paste appears to form first of zinc particles and iron dithiophosphate. The iron dithiophosphate then thermally degrades to a brown surface film of ZnS, ZnO, FeO, plus some iron and zinc organophosphates (25). Tricresyl phosphate is also an effective antiwear and extreme pressure agent which reacts at high temperature rubbing contacts to form protective metal phosphite or phosphate protective films (22).

For extreme rubbing conditions involving severe metal-to-metal contact, active sulfur compounds are used to generate low shear strength protective surface layers. The iron sulfide coating then prevents destructive welding, excessive metal transfer, and severe surface breakdown in hypoid gears, machine tool slideways, and various metal-cutting operations. Alkyl and aryl disulfides and polysulfides (synthesized from olefins), dithiocarbamates, and sulfurized fats are common additives. Chlorine compounds, such as chlorinated paraffins with 40 to 70% chlorine, were popular to generated protective metal chloride films, but environmental concerns now minimize their use (22).

Since surface reactions involved with antiwear and EP additives depend not only on the type of rubbing materials but also on operating temperature, surface speed, and corrosion questions, selection should be carefully integrated with the oil type, machine design, and operating conditions.

Friction Modifiers. These additives have found increasing use, especially in automotive applications, as mild EP agents in boundary lubrication conditions. They have been especially helpful during start-up and shutdown of heavily loaded sliding metal surfaces as suggested in Figure 12 (21). With their aid in lubricant film formation at these low speeds, the friction modifiers prevent stick-slip oscillations and noises (squawking) in automatic transmissions. They also conserve energy in their widespread use in automotive engine and drivetrain lubricants (21), and are applied in metalworking fluids.

The primary products used are fatty acids with 12–18 carbon atoms and fatty alcohols, or esters of fatty acids such as the glycerides of rapeseed and lard oil (18). Fatty acid amines and amides are used in metal working, particularly in emulsions (18).

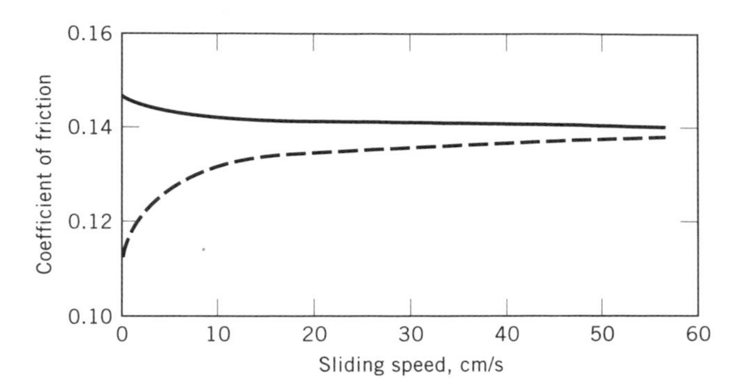

Fig. 12. Effect of friction modifier in automatic transmission fluid (21).

Detergents and Dispersants. Widely used at 2–20% concentration, detergent additives reduce high temperature deposits in internal combustion engines of oil-insoluble sludge, varnish, and carbon from fuel combustion. The detergent both exerts a surface cleaning action and also adsorbs on any insoluble particles to maintain them as a suspension in the bulk oil to minimize deposits on rings, valves, and cylinder walls. Dispersants serve much the same function in suspending oil-insoluble resinous oxidation products and particulate contaminants in the bulk oil to minimize deposits and wear (22).

Detergents are metal salts of organic acids used primarily in crankcase lubricants. Alkylbenzenesulfonic acids, alkylphenols, sulfur- and methylene-coupled alkyl phenols, carboxylic acids, and alkylphosphonic acids are commonly used as their calcium, sodium, and magnesium salts. Calcium sulfonates, overbased with excess calcium hydroxide or calcium carbonate to neutralize acidic combustion and oxidation products, constitute 65% of the total detergent market. These are followed by calcium phenates at 31% (22).

A dispersant molecule usually contains a nitrogen- or oxygen-based polar group attached to an oil-solubilizing aliphatic hydrocarbon chain containing from 70 to 200 or more carbon atoms. Polybutenylsuccinic acid derivatives are commercially the most commonly used. In their manufacture maleic anhydride condenses with olefin polymers, eg, polybutene of 500–2000 mol wt. The resulting alkenyl succinic anhydrides and acids then react with polyamines (21). Succinate esters, high molecular weight amines, alkyl hydroxyl benzene polyamines, and phosphonic acid derivatives also find use. In addition to their primary use in internal combustion engine oils, dispersants also are employed in automatic transmission fluids and gear oils.

Detergents generally are avoided in oils other than for internal-combustion engines since they may introduce foaming and emulsion problems.

Pour-Point Depressants. The pour point of a low viscosity paraffinic oil may be lowered by as much as 30–40°C by adding 1.0% or less of polymethacrylates, polymers formed by Friedel-Crafts condensation of wax with alkylnaphthalene or phenols, or styrene esters (22). As wax crystallizes out of solution from the liquid oil as it cools below its normal pour point, the additive molecules appear to adsorb on crystal faces so as to prevent growth of an interlocking wax network which would otherwise immobilize the oil. Pour-point depressants become less effective with nonparaffinic and higher viscosity petroleum oils where high viscosity plays a dominant role in immobilizing the oil in a pour-point test.

Viscosity (Viscosity-Index) Improvers. Oils of high viscosity index (VI) can be attained by adding a few percent of a linear polymer similar to those used for pour-point depressants. The most common are polyisobutylenes, polymethacrylates, and polyalkylstyrenes; they are used in the molecular weight range of about 10,000 to 100,000 (18). A convenient measure for the viscosity-increasing efficiency of various polymers is the intrinsic viscosity η, as given by the function

$$(\eta) = (\ln \eta/\eta_o)/\phi \tag{11}$$

where η is the viscosity of the polymer-thickened oil, η_o is the viscosity of the oil without the additive, and ϕ is the volume fraction of additive in the oil (26). Intrinsic viscosity usually is sufficiently independent of the oil base, polymer concentration, and temperature to serve as a useful measure of the viscosity-increasing efficiency of a polymer.

These polymer viscosity improvers seem to function primarily by thickening a light oil to a higher viscosity while retaining the original viscosity–temperature coefficient. This is of particular advantage with petroleum oils where the lower viscosity fractions from a crude have by far the lowest viscosity–temperature coefficients. This effect can provide a VI for an oil of 50 units or more above the value obtained with a higher molecular weight fraction from the same crude. Figure 13 illustrates the effect of several VI improvers on a SAE 10W base stock which is thickened to give a 10W-50 product (27).

Viscosity improvers are primarily used in multigrade automotive engine oils, automatic transmission oils, power steering fluids, and gear oils. They also find use in aircraft and some industrial hydraulic fluids for low temperature use.

Caution should be observed in relying on the higher viscosity obtained with a VI-improving additive. Shear rates of the order of 1,000,000 s^{-1} or higher slowly break down mechanically the polymer thickener, and the viscosity then gradually approaches that of the base oil. The degradation is minimized by using polymers of relatively low molecular weight. Oxidation and cavitation damage to the polymer additive can also result in loss of the added viscosity initially given to the oil. Nevertheless, outstanding automotive and aviation oils with VIs of 150 and higher have been formulated by using VI additives.

Foam Inhibitors. Methyl silicone polymers of 300–1000 mm^2/s(=cSt) at 40°C are effective additives at only 3–150 ppm for defoaming oils in internal combustion engines, turbines, gears, and aircraft applications. Without these additives, severe churning and mixing of oil with air may sometimes cause foam to overflow from the lubrication system or interfere with normal oil circulation. Because silicone oil is not completely soluble in oil, it forms a dispersion of minute droplets of low surface tension that aid in breaking foam bubbles.

Synthetic Oils

In 1929, polymerized olefins were the first synthetic oils to be produced commercially in an effort to improve on the properties of petroleum oils. Interest in

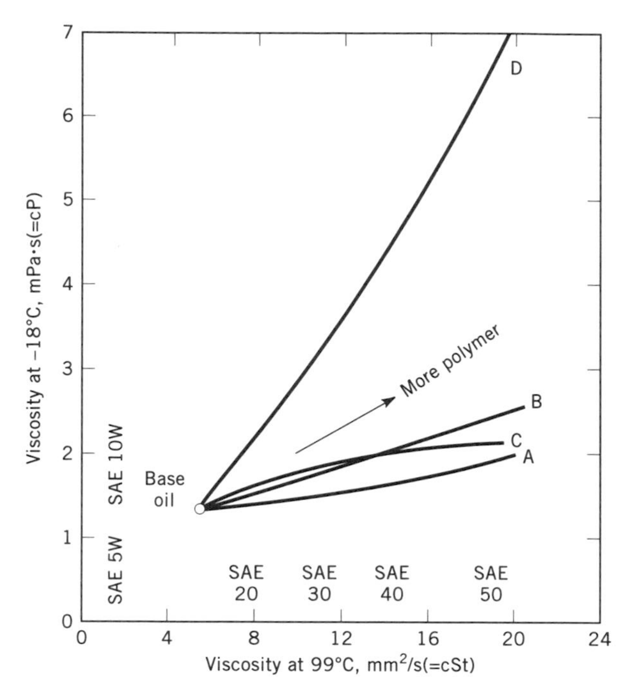

Fig. 13. Thickening of 10W base stock to multigraded oil with polymer additives. A, high mol wt poly(alkyl methacrylate); B, low mol wt poly(alkyl methacrylate); C, ethylene–propylene copolymer; and D, polyisobutylene (27).

esters as lubricants appears to date back to 1937 in Germany, and their production and use expanded rapidly during and following World War II to meet the needs of the military and the newly developed jet engines (2).

Alkylated aromatic lubricants, phosphate esters, polyglycols, chlorotrifluoroethylene, silicones, and silicates are among other synthetics that came into production during much that same period (28,29). Polyphenyl ethers and perfluoroalkyl polyethers have followed as fluids with distinctive high temperature stability. Although a range of these synthetic fluids find applications which employ their unique individual characteristics, total production of synthetics represent only on the order of 2% of the lubricant market. Poly(α-olefin)s, esters, polyglycols, and polybutenes represent the types of primary commercial interest.

Typical chemical structures and representative sources of different classes of synthetics are given in Table 6. Properties and uses of representative synthetics follow in Table 7. In addition to considering their physical properties, selection is needed of appropriate paints, seals, hoses, plastics, and electrical insulation to avoid problems with the pronounced solvency and plasticizing action of many of these synthetic oils.

Synthetic Hydrocarbons. Primary production of synthetic hydrocarbon lubricants now focuses on poly(α-olefin)s (PAO). These are manufactured in a two-step process in a variety of molecular weights starting with ethylene available from petroleum cracking (28). The first step is polymerization to a mixture of low molecular weight oligomers. Further oligomerization with a boron trifluoride catalyst to the final PAO is dominated by decene-derived materials,

Table 6. Typical Structures and Temperature Ranges for Synthetic Oils[a]

Chemical class	Typical structure	Manufacturer	Approximate continuous temperature range, °C	
			Minimum	Maximum
petroleum	[b]		−15	120
poly(α-olefin)s	$-\!\!\left(CH_2-CH(CH_3)-CH_2\right)\!\!-_n$	Mobil, Ethyl, Chevron, Uniroyal	−50	140
esters	$C_8H_{17}-O-C(=O)-C_8H_{16}-C(=O)-O-C_8H_{17}$	ICI, Henkel, Exxon, Mobil, Quaker Chemical	−50	180
poly(alkylene glycol)s	$-\!\!\left(O-CH_2-CH(CH_3)\right)\!\!-_n$	Union Carbide, ICI, Nippon Oil, Mobay, BASF, BP Chemical	−30	170
phosphate esters	$O=P\left(O-C_6H_4-C_3H_7\right)_3$	FMC, Bayer AG, Monsanto, Akzo	−20	180
polybutenes	$-\!\!\left(C(CH_3)_2-CH_2\right)\!\!-_n$	Amoco, BP Chemicals, Exxon, Lubrizol	−10	120

Table 6. (*Continued*)

Chemical class	Typical structure	Manufacturer	Approximate continuous temperature range, °C	
			Minimum	Maximum
alkylated benzene	$C_6H_5-(CH(R')-R)_2$	Vista Chemical, Shrieve Chemical	−25	160
chlorotrifluoroethylene	$-(CF(Cl)-CF_2)_n-$	Halocarbon Products, Occidental Chemical, Autochem SA	−50	140
silicones	$-(Si(CH_3)_2-O)_n-$	General Electric, Dow Corning	−70	230
perfluoroalkyl polyether	$-(CF(CF_3)-CF_2-O)_n-$	Du Pont, Montedison, Daiken, NOK	−30	280
polyphenyl ethers	$-(O-C_6H_4)_n-$	Monsanto	30	310

[a] Refs. 28 and 29.
[b] See Fig. 6.

Table 7. Properties of Representative Synthetic Oils

Type	Viscosity, $mm^2/s(=cSt)$ at 100 °C	at 40 °C	at −54 °C	Pour point, °C	Flash point, °C	Typical uses
synthetic hydrocarbons						
Mobil 1, 5W-30[a]	11	58		−54	221	auto engines
SHC 824[a]	6.0	32		−54	249	gas turbines
SHC 629[a]	19	141		−54	238	gears
organic esters						
MIL-L-7808	3.2	13	12,700	−62	232	jet engines
MIL-L-23699	5.0	24	65,000	−56	260	jet engines
MIL-L-6085	3.2	12	10,000	−68	232	aircraft hydraulics and instruments
Synesstic 68[b]	7.5	65		−34	266	air compressors, hydraulics
polyglycols						
LB-300-X[c]	11	60		−40	254	rubber seals
50-HB-2000[c]	70	398		−32	226	water solubility
phosphates						
tricresyl phosphate	4.3	31		−26	240	fire-resistant fluids for die casting, air compressors, and hydraulic systems
Fyrquel 150[d]	4.3	29		−24	236	
Fyrquel 220[d]	5.0	44		−18	236	
Skydrol 500B-4[e]	3.8	11	3,100	−65	182	aircraft hydraulic fluid
silicones						
SF-96 (50)[f]	16	37	460	−54	316	hydraulic and damping fluids
SF-95 (1000)[f]	270	650	7,000	−48	316	
F-50	16	49	2,500	−74	288	aircraft and missiles
polyphenyl ether						
OS-124[e]	13	373		4	288	radiation resistance and high temperatures
silicate						
Coolanol 45[e]	3.9	12	2,400	−68	188	aircraft hydraulics and cooling
fluorochemical						
Halocarbon 27[g]	3.7	30		−18	none	oxygen compressors, liquid-oxygen systems
Krytox 103[h]	5.2	30		−45	none	

[a]Mobil Oil Corp.
[b]Exxon Corp.
[c]Union Carbide Chemicals Co.
[d]Akzo Chemicals.
[e]Monsanto Co.
[f]General Electric Co.
[g]Halocarbon Products Corp.
[h]Du Pont Co.

commonly involving combination of about three to five decene units. Higher viscosity PAO in the 40–100 cSt range at 100°C is manufactured using alkyl aluminum catalysts in conjunction with an organic halide, or by use of $AlCl_3$.

Properties provided by the branched hydrocarbon chain structure of these PAO fluids include high viscosity index in the 130–150 range, pour points of −50 to −60°C for ISO 32 to 68 viscosity range (SAE 10W and SAE 20W, respectively), and high temperature stability superior to commercial petroleum products. In their use in automotive oils such as Mobil 1, some ester synthetic fluid is normally included in the formulation to provide sufficient solubility for the approximately 20% additives now employed in many automotive oils.

In addition to their automotive use, PAO oils also find application in industrial and aircraft hydraulic fluids, gear oils, compressors, and environmentally sensitive applications. They are also used in multipurpose greases for army, navy, nuclear, and industrial applications. Expanding use has led to a growth rate of PAO production from 1985 to 1990 of 19% per year (28).

Other synthetic hydrocarbon lubricants have generally been employed in rather specific uses. Polybutene oils are produced from isobutylene in petroleum catalytic cracking gases from gasoline production and are available commercially with viscosities from ~ 1 to 45,000 $mm^2/s(=cSt)$ at 100°C. Since these oils generally have lower viscosity index, higher pour point, and lower flash point than PAO and ester oils, polybutenes find use where other properties such as very low deposit formation, low toxicity, and shear stability are of concern. Such applications are as oils in two-stroke engines, electrical transformers and cables, refrigerator compressors, and metal working. Higher molecular weight polymers are also used as viscosity index improvers for gear and hydraulic oils (30).

Alkylated aromatics also find some use, primarily based either on shortages of petroleum products or special requirements for high or low temperatures (31). During the shortage of petroleum in Germany during World War II, for instance, alkylnaphthalene lubricants were produced in Germany at a 3600 t/yr scale. These were made by alkylation of excess naphthalene with chlorinated aliphatic hydrocarbons. The advantageous properties of dialkylbenzene oils at low temperatures led to their use for the military and oil prospecting in Alaska and Canada during the 1960s and during the construction of the Alaska pipeline during the 1970s. Formulated lubricants made from dialkylbenzene base stocks included year-round engine crankcase oil, torque converter fluid, hydraulic oil, and greases. Whereas expanded use of PAO lubricants has displaced dialkylbenzenes from many applications, current worldwide production of about 15,000 tons finds significant use in refrigeration and air conditioning. Their advantages include good low temperature miscibility with refrigerants, low wax separation temperature, and good thermal stability. Use is also found as electrical insulating oils, as a heat-transfer agent, and in water emulsions for metal working (31).

Esters. Search during and following World War II for wide temperature range lubricants for military equipment led to extensive application of diesters in MIL-L-6085 instrument oils, and multipurpose greases. Large-scale production of diester lubricants under the MIL-L-7808 specification then followed during the 1950s to match the continuing development of jet engines in Britain and the United States. Polyol ester lubricants meeting the MIL-L-23699 specification then followed in the 1960s with higher viscosity for larger and more demanding engines. While ester production volume has been eclipsed by PAO synthetic hydrocarbons, market for esters is expanding about 8% per year with their use extended to automotive and marine engines, two-cycle engines, compressors, hydraulic fluids, metal rolling, and gear oils (32,33).

Diesters have been produced primarily by esterification of a C_6-C_9 branched-chain alcohol with adipic (C_6), azelaic (C_9), or sebacic (C_{10}) diacid. Di(2-ethylhexyl)sebacate [*122-62-3*] was quite generally used in military greases and MIL-L-7808 jet engine oil, but more recent demands and price competition have led to use of a variety of diesters.

Polyol ester turbine oils currently achieve greater than 10,000 hours of no-drain service in commercial jet aircraft with sump temperatures ranging to over 185°C. Polyol esters are made by reacting a polyhydric alcohol such as neopentyl glycol, trimethylol propane, or pentaerythritol with a monobasic acid. The prominent esters for automotive applications are diesters of adipic and azelaic acids, and polyol esters of trimethylolpropane and pentaerythritol (34).

The esterification reaction in making ester oils is commonly carried out with a catalyst at about 210°C while removing excess water as it forms (32). Excess acid or alcohol is then stripped off, and unreacted acid is neutralized with calcium carbonate or calcium hydroxide before final vacuum drying (qv) and filtration (qv).

Ester fluids are modified with additives in much the same manner as petroleum oils. They are stabilized with an oxidation inhibitor, eg, 0.5 wt % phenothiazine. Improved load capacity for gears and rolling bearings in aircraft engines is provided by 1–5% tricresyl phosphate. Zinc dialkyldithiophosphate additives are used for automotive engine oils (34). The relatively low viscosity of diester fluids at high temperatures is increased and higher viscosity index is obtained with about 5% of added polymethacrylate.

Although poly(α-olefin)s (PAO) and esters are the prominent synthetic base stocks for automotive applications, combinations of the two are becoming the choice in offering a balance of properties such as additive solubility, sludge control, and elastomer compatibility (34).

Esters generally tend to be readily biodegradable. This is advantageous for two-cycle engine oils which are discharged to the surroundings from power-driven recreational boats and various portable power units around the home.

Poly(alkylene glycol)s. While these can be made from polymerization of any alkylene oxide, they are usually prepared either from propylene oxide as the water-insoluble type, or as water-soluble copolymers of propylene oxide and up to 50% ethylene oxide (35,36) (see POLYETHERS, PROPYLENE OXIDE POLYMERS). Current worldwide production is estimated to be about 45,000 t.

The polyalkylene glycol polymer employs a starter that consists of a relatively reactive alcohol and a smaller amount of its potassium or sodium salt. With propylene oxide, for instance, initiation of the polymerization then involves the starter in the following steps:

$$ROH + ROM + \overset{O}{\overbrace{CH_2-CH_2}} \longrightarrow ROH + ROCH_2CH_2OM \rightleftharpoons ROM + ROCH_2CH_2OH$$

The epoxide monomers react with the metal salts of the alcohol much faster than with the alcohol. Once each starter alcohol has reacted with an epoxide, all molecules in the system have approximately the same reactivity. The fast exchange of metal salt between the growing polymer chains then results in a relatively narrow distribution of molecular weights which can range up to 20,000 and with no significant fraction of volatile components. Propylene oxide polymers that are commonly used in lubrication are started with butanol; water-soluble polymers are started either from butanol or a diol such as ethylene glycol. With a diol, one polyether chain grows out from each of the two hydroxyl groups of the starter.

Preparation of the polymer can be carried out in glass equipment at atmospheric pressure at temperatures typically above 100°C, but the higher pressures in an autoclave result in much faster reaction rates. Each polymer molecule which used butanol as a starter contains one hydroxyl end group as it comes from the reactor; diol-started polymers contain two terminal hydroxyls. Whereas a variety of reactions can be carried out at this remaining hydroxyl to form esters, ethers, or urethanes, this is normally not done and therefore lubricant fluids contain at least one terminal hydroxyl group (36).

Poly(alkylene glycol)s have a number of characteristics that make them desirable as lubricants. Compared to petroleum lubricants, they have lower pour points, a higher viscosity index, and a wider range of solubilities including water, compatibility with elastomers, less tendency to form tar and sludge, and lower vapor pressure (35).

First use of poly(alkylene glycol)s was in combination with 35–60% water for fire-resistant hydraulic fluids, and this use continues in foundries, steel mills, and mines. Other applications are as brake fluids for automobiles; textile fiber and textile machine lubricants because they are nonstaining and easily washable (see TEXTILES); compressor lubricants for ethylene, natural gas, helium, nitrogen, and the new automotive air-conditioning refrigerants; lubricants in food processing equipment; and nonsludging lubricants for bearings and gears in mills and calenders used by the rubber, textile, paper, and plastics industry up to temperatures of 175°C (35).

Poly(alkylene glycol)s are also used as lubricity additives in water-based synthetic cutting and grinding fluids (36), and in aqueous metalworking fluids. Under the high frictional heating at the tool or die contact with the workpiece, the polyalkylene glycol comes out of solution in fine droplets which coat the hot metal surfaces.

Phosphate Esters. A variety of phosphate esters are used as synthetic lubricants, particularly because of their good fire resistance. They have the general formula $OP(OR)_3$, where R may represent a variety of aryl or alkyl hydrocarbon groups containing four or more carbon atoms to give three broad classes: triaryl, trialkyl, and aryl alkyl phosphates (37,38).

Triaryl phosphates are produced by reaction of phosphorus oxychloride with phenolic compounds at 100–200°C with magnesium or aluminum chloride catalyst. Past use of cresols and xylenols from coal tar or petroleum is replaced for lower toxicity and cost by synthetic phenolics, primarily isopropyl phenol, *t*-butyl phenol, and phenol itself. A range of viscosities is achieved by selection and proportioning of the phenols and their isomers used for the starting material.

Inefficiencies in the reaction with $POCl_3$ leads to alternative production of trialkyl phosphates by employing the sodium alkoxide rather than the alkyl alcohol itself. Dialkyl aryl phosphates are produced in two steps. The low molecular weight alcohol involved (eg, butyl) first reacts with excess $POCl_3$. The neutral phosphate ester is then completed by the intermediate chloridate reacting with excess sodium arylate in water.

Phosphate ester fluids are the most fire resistant of moderately priced lubricants, are generally excellent lubricants, and are thermally and oxidatively stable up to 135°C (38). Fire-resistant industrial hydraulic fluids represent the largest volume commercial use. Applications are made in air compressors and continue to grow for aircraft use (tributyl and/or an alkyl diaryl ester) and in hydraulic control of steam turbines in power generation (ISO 46 esters).

Triaryl phosphates of ISO 32 viscosity show promise for the main bearing lubricants of steam and gas turbines (39,40). An interesting possibility involves unique delivery of phosphate ester vapor to lubricate the piston ring zone of low heat rejection (adiabatic) diesel engines (41).

Hydrolysis is a significant threat to phosphate ester stability as moisture tends to cause reversion first to a monoacid of the phosphate ester in an autocatalytic reaction. In turn, the fluid acidity can lead to corrosion, fluid gelation, and clogged filters. Moisture control and filtration with Fuller's earth, activated alumina, and ion-exchange resins are commonly used to minimize hydrolysis. Toxicity questions have been minimized in current fluids by avoiding triorthocresyl phosphate which was present in earlier natural fluids (38).

Perfluoroalkylpolyethers. While high cost has limited general use, these fluids are remarkably stable chemically, have good viscosity–temperature characteristics, low pour points, and quite low vapor pressures (42,43).

The perfluoroalkylpolyethers (PFPE) are commonly produced by fluoride–ion catalyzed polymerization of hexafluoropropylene epoxide at around −40°C. A reactive acid fluoride end group is stabilized by reaction with elemental fluorine. Polymer with a molecular weight range of 435 to 13,500 is then fractionated into viscosity grades by vacuum distillation. Synthesis of the PFPE can also be carried out by photochemical-catalyzed polymerization of tetrafluoroethylene or hexafluoropropylene in the presence of oxygen at low temperature. Lewis acid-catalyzed ring-opening polymerization of 2,2,3,3-tetrafluorooxetane is also used (43).

Depending on their structural type, PFPE oils are stable up to 300–400°C in air. Pure oxygen in a test bomb at 13 MPa (1886 psi) at temperatures up to 400°C was tolerated with no ignition (43). Densities at 20°C vary from 1.82 to 1.89 g/mL, and viscosities from 10 to 1600 mm^2/s. The pour point for low temperature operation usually ranges from −30 to −70°C, and the viscosity index varies from about 50 for low viscosity grades up to 150 for more viscous oils and considerably higher for fully linear polymers (43).

High cost has generally limited use of PFPE lubricants to severe applications for which they have unique capabilities. These have included a variety of aircraft and aerospace instrument and accessory bearings, industrial ovens, plasma etching equipment, and pump bearings with oxygen, chlorine, and missile fuels. Efforts are underway to develop soluble rust and lubricity additives needed for suitable performance in further aerospace applications (44). A unique and important use has been for the lubrication of magnetic data disks in computers (45).

Silicones. Silicone fluids consist of an alternating silicon–oxygen backbone (siloxane), with two organic side groups branching off from each of the silicon atoms. Although there are many possibilities, methyl and phenyl side chains have been the most common (46,47).

Commercial silicone production starts with the reaction of methyl chloride or phenyl chloride vapor with finely ground silicon metal in a fluid-bed reactor (47). The silicon is converted to a crude mixture of chlorosilanes separated by fractional distillation to provide $(CH_3)_3SiCl$, $(CH_3)_2SiCl_2$, $(CH_3)SiCl_3$, corresponding phenyl compounds, and $(CH_3)HSiCl_2$. The last intermediate, methylhydrogendichlorosilane, is a versatile starting point for lubricants since the hydrogen can add across the double bond of various organic molecules and of silanes and siloxanes containing vinyl groups (46).

After polymerization is carried out by blending mono- and difunctional chlorosilanes in excess water, the siloxanes are separated from the water and neutralized. Ratio of the mono-chain stopper to di-chain extender controls the length of the polymer. Once an equilibrium mixture of chain lengths is catalytically formed, volatile light ends are removed and the desired product results.

Most common of the silicones are the various grades of dimethylpolysiloxane which are available in a wide range of viscosities. Figure 14 indicates their uniquely low change in viscosity with temperature. They also have superior low and high temperature behavior in providing a temperature operating range of about −70 to 230°C (−100 to 450°F). Low toxicity, high compressibility, and low surface tension are other unique characteristics. The methyl phenyl type of fluids give somewhat increased thermal stability and are common base fluids for ball bearing greases.

Whereas the traditional dimethyl siloxane fluids provide very poor lubrication for steel on steel and other common metals, thin films on glass reduce

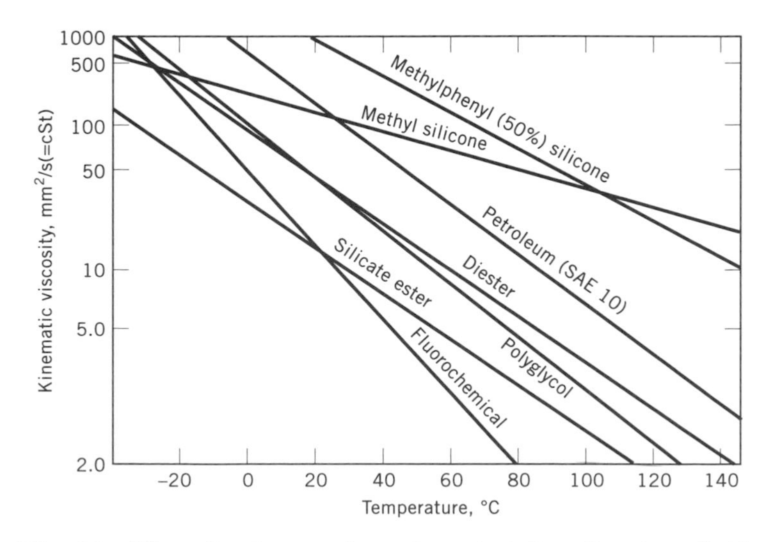

Fig. 14. Viscosity–temperature characteristics of various fluids.

handling damage, small amounts in plastic composites bleed to the surface for self-lubrication, and they provide a superior lubricant for rubber surfaces.

Improved lubrication is provided by a number of modifications. Replacement of the methyl groups by a longer chain alkyl gives a much thicker boundary lubricating layer for effective lubrication of steel, bronze, aluminum, glass, plastics, and Monel. Including halogens in fluoropropyl siloxanes and in chlorophenyl silicones gives a variation of extreme pressure lubrication (46). With the compounding of methyl phenyl siloxane fluids into lubricating greases, ball and roller bearing lubrication has been satisfactory under extreme temperature conditions.

Chlorotrifluoroethylene. The CTFE oils are polymers of chlorotrifluoroethylene varying from 2 to about 10 monomer units. Nonflammability and oxygen compatibility are primary characteristics. Available in viscosity grades ranging from 0.8 to 1000 mm^2/s(=cSt) at 38°C, these oils have a useful temperature range from 204°C down to their pour point which is in the −30 to −70°C range for 95 to 27 mm^2/s(=cSt) viscosity grades (48).

CFTE oil inertness leads to their use in vacuum pump oils and with a wide variety of chemicals including oxygen, chlorine, hydrogen peroxide, sodium chlorate, ammonium perchlorate, and mineral acids. They are used as lubricants in the oxidizer section of U.S. space and missile engines, and in the oxygen loading systems for space shuttles. Contact is to be avoided, however, with metallic sodium and potassium and with aluminum and magnesium under rubbing contact conditions which might induce ignition with the halogens of the oil.

Polyphenyl Ethers. These very stable organic structures have been synthesized in a search for lubricants to meet the needs of future jet engines, nuclear power plants, high temperature hydraulic components, and high temperature greases (49). A typical formula is $C_6H_5\text{-}(OC_6H_4)_n\text{-}OC_6H_5$ [*25718-67-6*], all connections being para.

One liquid in this class intended for aircraft engine use is described in military specification MIL-L-87100 for operation from +15 to 300°C. Limitations of this class of synthetics are pour points of +5°C and higher, relatively poor lubricity, and high cost of $265/L ($1000+/gal) (44). Polyphenyl ether greases are available with good radiation resistance for applications in the temperature range of +5 to 288°C.

Greases

A grease is a lubricating oil that is thickened with a gelling agent, eg, a soap (qv). For design simplicity, decreased sealing requirements, and less need for maintenance, greases are almost universally given first consideration as lubricants for ball and roller bearings in electric motors, household appliances, automotive wheel bearings, machine tools, aircraft accessories, and railroad apparatus. Greases are also used for lubrication of small gear drives and for many slow speed sliding applications.

Oils in Greases. Essentially the same type of oil is used in compounding a grease as would normally be selected for oil lubrication. Petroleum oils are used in about 99% of the grease produced and commonly are in the SAE 20–30 viscosity range with about 100–130 mm^2/s viscosity at 40°C. Such oils provide low volatility for long life at elevated temperatures (50) together with low torque down to subzero temperatures.

Some quite viscous oils in the 450–650 mm^2/s are employed for high temperatures. Less viscous oils, down to 25 mm^2/s and lower at 40°C, are used in special greases for low temperatures. The maximum oil viscosity in a grease for starting medium torque equipment is about 100,000 $mm^2/s(=cSt)$ (4). Extrapolations for various oils can be made on viscosity–temperature charts, as shown in Figure 8, to estimate this approximate low temperature limit.

Thickeners. Common gelling agents are the fatty acid soaps of lithium, calcium, sodium, and aluminum in concentrations of 6–25 wt %. Use of lithium soaps has expanded from their introduction in 1942 to comprise about 65% of the total market (51). Fatty acids used are usually oleic, palmitic, stearic, and other carboxylic acids derived from tallow, hydrogenated fish oil, castor oil, and, less often, wool grease and rosin. The relatively low upper temperature limit with calcium and aluminum greases has been significantly raised through new complex soap formulations. Calcium-complex greases commonly include a minor portion of calcium acetate [*62-54-4*] to provide multipurpose greases with dropping points above 260°C. Aluminum-complex grease can be made from reaction of a combination of stearic and benzoic acids with a reactive aluminum compound such as aluminum isopropoxide (18,52).

Finely divided clay particles of the bentonite and hectorite types are also commonly used as grease thickeners after being coated with an organic material such as quaternary ammonium compounds (qv) (see CLAYS, USES). Many of these clay-thickened greases are manufactured by simple mixing to provide high melting points, excellent water resistance, and long life for multipurpose use in industrial, automotive, and agricultural equipment. Carbon black and amorphous silica are used as thickeners in some high temperature petroleum and synthetic greases. Arylurea compounds are used in petroleum greases for ball bearings at temperatures up to about 150–170°C. Polytetrafluoroethylene, indanthrene [*81-77-6*], phthalocyanines, and ureides are among other organic powders which have also been used at elevated temperatures.

Gelling action of these thickening agents varies. Oil is believed to be held in the grease structure by a combination of capillary forces, adsorption on the gel-forming molecules, and physical entrapment within fibrous interlacing crystallites in the case of fatty acid soaps. Relative importance of each of these mechanisms depends on the type and degree of dispersion of the thickener, type and solvency of the oil, and the influence of any stabilizing agents and additives. The wide variation in characteristics of petroleum greases using various thickener types is indicated in Table 8.

Additives. Chemical additives similar to those used in lubricating oils also are added to grease to improve oxidation resistance, rust protection, and extreme pressure properties (18,53). Although 1-naphthylamine is the common choice as an oxidation inhibitor at about 0.1–1.0% concentration, other amine, phenolic, phosphite, and sulfur inhibitors are also used. A common procedure involves testing a number of commercial additives in varying concentration to determine the least expensive means for obtaining satisfactory oxidation inhibition for the Norma-Hoffmann bomb test (ASTM D942) at 99°C and 0.76 MPa (110 psi)

Table 8. Typical Characteristics of Petroleum Greases

Base	Texture	Dropping point, °C	Max temp for continuous use, °C	Water resistant	Mechanical stability
		Soap			
aluminum	smooth and stringy	90	65	yes	poor
barium	buttery or fibrous	200+	120	yes	good
calcium	smooth and buttery	100	80	yes	fair
lithium	buttery to stringy	200	120	yes	good to poor
sodium	buttery or fibrous	200	120	no	good to poor
strontium	buttery or fibrous	200	120	yes	good
complex soaps	smooth and buttery	200+	120	yes	good
		Nonsoap			
modified clay	smooth	260+	140	yes	fair
silica gel	smooth	260+	140	some	poor
carbon black	smooth	260+	140	yes	good
polyurea	smooth	260+	140	yes	good

oxygen pressure. Often, 0.2–0.3% of an amine metal deactivator is also added to minimize any catalytic effect of copper on oxidation of the grease.

Although most greases offer some inherent protection against rusting, additives, eg, amine salts, sodium sulfonate, cycloparaffin (naphthenate) salts, esters, and nonionic surfactants (qv), are often used to provide added protection against water and salt-spray corrosion. A dispersion of sodium nitrite has been particularly effective in some multipurpose greases.

EP additives are not needed in greases for most ball bearing applications nor for general-purpose industrial use, but they are necessary to minimize bearing wear under shock loads in steel rolling mills, for many gear applications, and for sliding conditions that involve boundary lubrication. Various sulfur and phosphorus additives are employed for this purpose. Solid powders added as fillers for extreme conditions of boundary lubrication include molybdenum disulfide, graphite, zinc oxide, and talc.

Glycerol is also present in many greases. Frequently the glycerol remains after the formation of the metallic soap thickener when natural fats are used as a raw material. Even with some soaps that are produced from fatty acids, glycerol may be added in combination with a small amount of water for its stabilizing effect on the soap structure. A few parts per million dimethyl silicone oil is frequently added to minimize foaming during grease manufacture; this appears to have no effect on subsequent performance characteristics of the finished grease.

Synthetic Grease. Although all of the synthetic oils mentioned previously have been used in formulating lubricating greases, synthetic production appears to be only about 1% of the total grease market; this reflects the ability of petroleum greases to meet most operating requirements of ball and roller bearings.

Synthetics are commonly employed only when their higher cost is justified by extreme temperatures or by need for special properties which cannot be achieved with petroleum greases. Severe temperature and operating requirements have led to a broad range of synthetic greases for military use (54). Comparison of typical temperature limits are given in Table 9.

Volume production of synthetic hydrocarbon and diesters is greatest among the synthetics. MIL-G-81322 grease incorporating poly(α-olefin) synthetic hydrocarbon oil with a bentonite clay thickener is used in the −55 to 150°C range for a variety of navy and other military applications. Other SHC greases are finding broadening use in steel mills, paper machines, ovens, and nuclear plant accessories.

With good performance in the range of about −55 to 125°C, lithium and nonsoap ester greases conforming to MIL-G-23827 are ideally suited for a wide range of aircraft and military equipment and for refrigeration and outdoor industrial equipment. While the upper temperature limit with these diester oils, eg, di(2-ethylhexyl) sebacate, is largely governed by evaporation rate, ester greases comprised partially of polyol ester fluids for the MIL-G-25760 specification provide a minimum life of 400 hours at 177°C.

Silicone greases can be used over an even broader temperature range than synthetic hydrocarbon or ester types. Methyl phenyl silicone oil–lithium soap grease covers the range from about −35 to 170°C; arylurea and indanthrene thickeners in MIL-G-25013 greases extend to the upper end of the range to 230°C. MIL-G-83261 is a polytetrafluoroethylene-thickened fluorosilicone grease used from −70 to 230°C and is characterized by low wear in gears and heavily loaded ball and roller bearings (55).

Perfluoroalkyl ether greases thickened with polytetrafluoroethylene (MIL-G-38220 and MIL-G-27617) are used from −40 to 200°C in missiles, aircraft, and applications where fuel, oil, and liquid oxygen resistance is needed (55). Polyphenyl ether greases find special use from 10 to 315°C in high vacuum diffusion pumps and for radiation resistance.

Table 9. Characteristics of Synthetic Greases

Grease type	Maximum temp for 1000-h life, °C	Lowest temp for 1000-g/cm torque in 204 bearing, °C
petroleum	145	−28
diester	125	−56
polyester	160	−46
synthetic hydrocarbon	145	−40
conventional silicone	170	−35
special silicone	230	−73
perfluoroalkyl polyether	260	−35
polyphenyl ether	280	10

Mechanical Properties. Greases vary in consistency from soap-thickened oils that are fluid at room temperature to hard brick-type greases that are cut with a knife.

The most common measurement of consistency employs a standard penetrometer cone; its depth of penetration into the grease in 5 s at 25°C is measured in tenths of a millimeter (ASTM D217). This penetration depth is usually measured both on the original grease and after working 60 strokes with a perforated disk plunger. Worked penetration is the basis for the consistency classification in Table 10 developed by the National Lubricating Grease Institute (NLGI). Also tabulated are the approximate yield values for a grease in each penetration range. A yield value of 9.81 mN/cm^2 (981 dyn/cm^2) indicates that a grease layer (with a density of 1 g/cm^3) 1 cm high is just able to support its own weight without slumping. Thus yield values can serve as a guide in selecting greases of suitable mechanical strength for various sizes of ball bearings and their housings.

Grade 2 greases are the most commonly used. They generally are sufficiently stiff to avoid mechanical churning which would break down their gel structures, and are adequately soft and oily to provide the lubrication needs of most bearings. Softer greases (down to Grade 000) are used where greater feeding is necessary, as with multiple row roller bearings and various gear mechanisms. Stiffer greases of Grade 3 consistency are used for prepacked ball bearings where the grease is held by the bearing seals in close proximity with the ball complement. Hard brick greases are applied as blocks that are inserted directly in the sleeve-bearing box, eg, in a paper mill.

Apparent viscosity of a grease at low shear rates, eg, below about 10 s^{-1}, is approximately equal to the yield stress divided by the shear rate. This apparent viscosity drops rapidly as the shear rate is increased to about 1000 s^{-1}. Apparent viscosity vs rate of shear for greases of three soap types is given in Figure 15. The statically stiff grease is seen to provide nearly the same viscosity as the oil in the grease under the high shear rates in most rolling bearings.

It is important that any grease selected for a given application maintain its desired properties during operation. For lack of adequate bench tests, trial operation in the actual installation usually is desirable with the bearing overpacked with grease, at high temperature, for water washout, or under any other

Table 10. Consistency Classification of Greases

NLGI number	ASTM worked penetration, mm/10 at 25°C	Approximate yield value, Pa[a]
000	445–475	
00	400–430	90
0	355–385	130
1	310–340	180
2	265–295	300
3	220–250	560
4	175–205	1300
5	130–160	
6	85–115	

[a]To convert Pa to dyn/cm^2, multiply by 10.

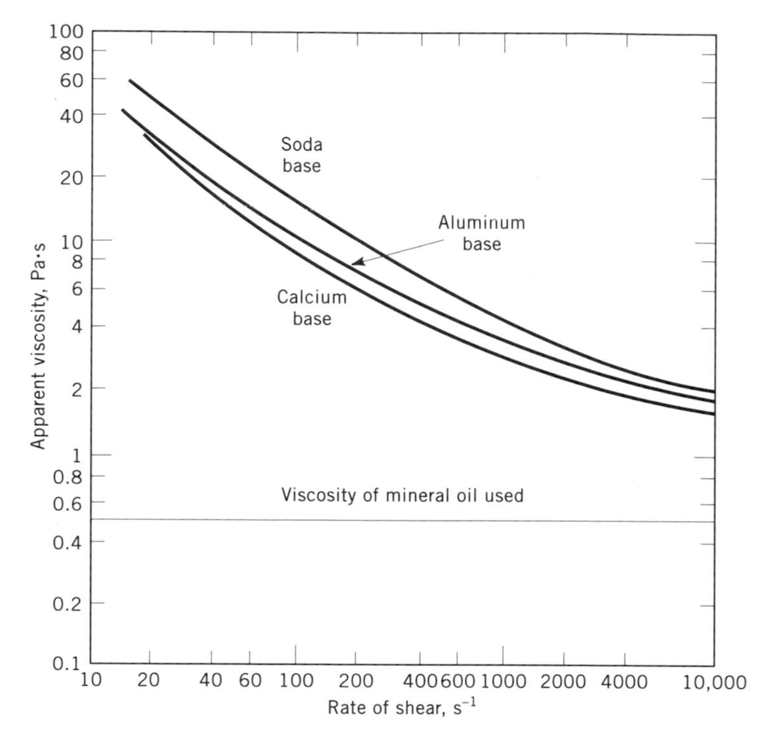

Fig. 15. Apparent viscosity vs rate of shear of three greases (53). To convert Pa·s to P, multiply by 10.

potentially severe conditions. Abnormally high power loss, high temperature rise, leakage, or noisy operation might indicate performance inadequacy. During use, periodic checks for changes in color, oil content, and acidity should help in further evaluations.

Solid-Film Lubricants

These provide thin films of a solid, or a combination of solids, interposed between two moving surfaces to reduce friction and wear. They are coming into more general use for high temperatures, vacuum, nuclear radiation, aerospace, and other environments that prohibit use of oils and greases.

The wide range of solid lubricants can generally be classified as either inorganic compounds or organic polymers, both commonly used in a bonded coating on a matching substrate, plus chemical conversion coatings and metal films. Since solid-film lubricants often suffer from poor wear resistance and inability to self-heal any breaks in the film, search continues for improved compositions.

Inorganic Compounds. The most important inorganic materials are layer–lattice solids in which the bonding between atoms in an individual layer is by strong covalent or ionic forces and those between layers are relatively weak van der Waal's forces. Because of their high melting points, high thermal stabilities, low evaporation rates, good radiation resistance, and effective friction lowering ability, molybdenum disulfide [*1317-33-5*], MoS_2, and graphite [*7782-42-5*] are the preferred choices in this group. Among other layer–lattice solids that find occasional use are tungsten disulfide [*12138-09-9*], WS_2; tungsten diselenide [*12067-46-8*], WeS_2; niobium diselenide [*12034-77-4*], $NbSe_2$; calcium chloride [*10108-64-2*]; cadmium iodide [*7790-80-9*], CdI; and graphite fluoride [*11113-63-6*] (Table 11).

Graphite is widely used as a dry powder or as a colloidal dispersion in water, petroleum oil, castor oil, mineral spirits, or other solvents. The water dispersions are used for lubricating dies, tools, metalworking molds, oxygen equipment, and wire drawing. Graphite dispersed in solvents is used for drawing, extruding, and forming aluminum and magnesium, as a high temperature lubricant for conveyors, and for a variety of industrial applications. Graphite alone is ineffective in vacuum since adsorbed water normally plays a decisive role in lubrication by graphite. Its film-forming ability can be restored, however, by mixing with cadmium oxide or MoS_2 and most organic materials, so that graphite may offer effective lubricating action when bonded to the surface with organics. Oxidation by air commonly sets a limit of about 550°C, and high friction may occur in air with water desorption above 100°C (57).

Molybdenum disulfide has increasingly supplanted graphite for three reasons: consistent properties in rigid specifications, independence from need for adsorbed vapors in providing lubrication, and superior load capacity (57). Like graphite, MoS_2 has a layer–lattice structure in which weak sulfur–sulfur bonds allow easy sliding between each sulfur–molybdenum–sulfur layer. MoS_2 covered by MIL-M-7866 is the most common lubricant grade: it is purified from molybdenite ore and is essentially free of abrasive constituents (56).

Petroleum oil and grease dispersions of MoS_2 are used extensively in automotive and truck chassis lubrication and in general industrial use. Dispersions are also made with 2-propanol, poly(alkylene glycol)s, other synthetic oils, and water for airframe lubrication, in wire drawing, and for splines, fastenings, gears, and fittings. Above 400°C, the MoS_2 is oxidized to molybdenum trioxide, which may be abrasive. As rubbed films, both MoS_2 and graphite may accelerate corrosion: MoS_2 by hydrolysis to form corrosive acids and graphite by galvanic action.

Various other soft materials without the layer–lattice structure are used as solid lubricants (58), eg, basic white lead or lead carbonate [*598-63-0*] used in thread compounds, lime [*1305-78-8*] as a carrier in wire drawing, talc [*14807-96-6*] and bentonite [*1302-78-9*] as fillers for grease for cable pulling, and zinc oxide [*1314-13-2*] in high load capacity greases. Graphite fluoride is effective as a thin-film lubricant up to 400°C and is especially useful with a suitable binder such as polyimide varnish (59). Boric acid has been shown to have promise as a self-replenishing solid composite (60).

Organic Polymers. These self-lubricating polymers are used primarily in three ways: as thin films, as self-lubricating materials (see BEARING MATERIALS), or as binders for lamellar solids (57,61). Coatings are typically applied in powder or dispersion form at coating thickness ranging upward from 25 μm. The polymer is then fused to the surface as a coating which provides lubricity, abrasion and chemical resistance, or release properties.

PTFE is outstanding in this group. In thin films it provides the lowest coefficient of friction (0.03–0.1) of any polymer, is effective from –200 to 250°C, and

Table 11. Common Solid Lubricants[a]

Material	Acceptable usage temperature, °C: Min, In air	Min, In N_2 or vacuum	Max, In air	Max, In N_2 or vacuum	Av friction coefficient, f: In air	In N_2 or vacuum	Remarks
molybdenum disulfide, MoS_2	−240	−240	370	820	0.10–0.25	0.05–0.10	low f, carries high load, good overall lubricant, can promote metal corrosion
polytetrafluoroethylene (PTFE)	−70	−70	290	290	0.02–0.15	0.02–0.15	lowest f of solid lubricants, load capacity moderate and decreases at elevated temp
fluoroethylene–propylene copolymer (FEP)	−70	−70	200	200	0.02–0.15	0.02	low f, lower load capacity than PTFE
graphite	−240		540	unstable in vacuum	0.10–0.30	0.02–0.45	low f and high load capacity in air, high f and wear in vacuum, conducts electricity
niobium diselenide			370	1320	0.12–0.40	0.07	low f, high load capacity, conducts electricity in air or vacuum
tungsten disulfide, WS_2	−240	−240	430	820	0.10–0.20		f not as low as MoS_2, temp capability in air a little higher
tungsten diselenide			370	1320			same as for WS_2
lead sulfide, PbS			480		0.10–0.30		very high load capacity, used primarily as additive with other solid lubricants
lead oxide			650		0.10–0.30		same as for PbS
calcium fluoride–barium fluoride eutectic	430	430	820	820	0.10–0.25 above 540°C 0.25–0.40 below 540°C	same as in air	can be used at higher temp than other solid lubricants, high f below 540°C
antimony trioxide							high load capacity, used as corrosion inhibitor in MoS_2 lubricants

[a] Ref. 56.

is generally unreactive chemically. The low friction is attributed to the smooth molecular profile of PTFE chains which allows easy sliding (57). Typical applications include chemical and food processing equipment, electrical components, and as a component to provide improved friction and wear in other resin systems.

Other polymers finding self-lubricating use are fluorinated ethylene–propylene copolymer (FEP), perfluoroalkoxy resin (PFA), ethylene–chlorotrifluoroethylene alternating copolymer (ECTFE), and poly(vinylidene fluoride) (PVDF) (61). With a useful temperature range up to 200°C, outstanding weatherability, and low friction, FEP finds use in chemical process equipment, roll covers, wire and cable, and as powder in resin bonded products. PFA provides somewhat better mechanical properties than PTFE and FEP at temperatures up to 250°C. ECTFE provides superior strength, wear resistance, and creep resistance from cryogenic temperatures to about 165°C. Although fairly expensive, it is effective in its common use as a corrosion-resistant coating. Also having superior mechanical properties, PVDF is more commonly used for lining chemical piping and reactor vessels than as a lubricant.

Bonded Solid-Film Lubricants. Although a thin film of solid lubricant that is burnished onto a wearing surface often is useful for break-in operations, over 95% are resin bonded for improved life and performance (62). Use of adhesive binders permits applications of coatings 5–20 μm thick by spraying, dipping, or brushing as dispersions in a volatile solvent. Some commonly used bonded lubricant films are listed in Table 12 (62) with a more extensive listing in Reference 61.

For many moderate-duty films for operating temperatures below 80 to 120°C, MoS_2 is used in combination with acrylics, alkyds, vinyls, and acetate room temperature curing resins. For improved wear life and temperatures up to 150–300°C, baked coatings are commonly used with thermosetting resins, eg, phenolics, epoxies, alkyds, silicones, polyimides, and urethanes. Of these, the MIL-L-8937 phenolic type is being applied most extensively.

Inorganic binders are used, usually with graphite or MoS_2, for extreme conditions such as high vacuum, liquid oxygen, radiation resistance, and high temperatures (61). The most common binder systems are silicates, phosphates, and aluminates. Some silicon and titanate metallo-organics used for high temperature binders become inorganic on curing. An emerging class of ceramic bonded materials for aerospace applications use either graphite, a CaF_2–BaF_2 eutectic, or proprietary systems, often with a glass frit binder which is fused into a continuous film (61). Plasma spray coating avoids overheating damage to the substrate metal while achieving the melting point of at least one component in high temperature film compositions (57,59).

The solid lubricant-to-binder ratio is a principal performance factor. High lubricant content usually gives minimum friction, while high binder content tends to give better corrosion resistance, hardness, durability, and a glossy finish (62). With commonly used MoS_2-graphite and organic resin binders, the optimum lubricant:binder ratio usually is from 1:1 to 4:1. With inorganic binding agents, the ratio is from 4:1 to as high as 20:1 and increases with high temperatures.

Substrate Properties. It is clear from equation 5 that higher hardness of the substrate lowers friction. Wear rate of the film also is generally lower. Phosphate undercoats on steel considerably improve wear life of bonded coatings

Table 12. Performance Properties of Typical Solid-Film Lubricants[a]

Property	Organic								
	Thermoset				Air dry		Inorganic		
specification	MIL-L-8937	MIL-L-46010			MIL-L-23398	MIL-L-46009	MIL-L-81329	AMS2525A	AMS2526A
composition									
lubricant	MoS_2	MoS_2/metallic oxide	MoS_2/graphite	PTFE	MoS_2	MoS_2/graphite	MoS_2/graphite	graphite	MoS_2
binder	phenolic	epoxy	silicone	phenolic			silicate		
application	spray	spray	spray	spray	spray	aerosol	spray		impingement
cure, °C	149	204	260	204	ambient	ambient	204	149	149
operating temperature									
air (high), °C	260	260	371	260	176	204	649	1093	400
air (low), °C	−220	−220	−157	−220	−220	−185	−240	−240	−220
vacuum, Pa[b]	10^{-4}	10^{-7}	10^{-7}	na	na	na	10^{-5}	10^{-7}	10^{-7}
load capacity									
force test	1130 kg Falex	1130 kg Falex	1130 kg Falex	68 kg LFW-1	1130 kg Falex	1130 kg Falex			
wear life									
load	454 kg	454 kg	454 kg	68 kg	454 kg		454 kg	50	50
test	Falex	Falex	Falex	LFW-1	Falex		Falex	Falex	Falex
time, min	60	450	60	120	120		70	2	5
coefficient of friction	<0.1	<0.1	<0.1	<0.1	<0.1	<0.1	<0.1	<0.1	<0.1
corrosion resistance	good	very good	fair	excellent	good		fair		

[a] Ref. 62.
[b] To convert Pa to mm Hg, multiply by 0.0075.

by providing a porous surface which holds reserve lubricant. The same is true for surfaces that are vapor- or sandblasted prior to application of the solid-film lubricant. A number of typical surface pretreatments are given in Table 13 to prepare a surface for solid-film bonding (61).

Optimum surface roughness usually is 0.05–0.5 μm; a very smooth surface contains very little lubricant within its depressions, whereas rough peaks penetrate the lubricant to promote wear. Improved corrosion resistance may be obtained with a suitable subcoating surface conversion treatment or by inclusion of inhibitors in the coating.

Chemical Conversion Coatings. These involve inorganic surface compounds developed by chemical or electrochemical action (see METAL SURFACE TREATMENTS). One of the best known treatments for steel is phosphating to coat the surface with a layer of mixed zinc, iron, and manganese phosphates. Other films are anodized oxide coatings on aluminum, oxalate on copper alloys, and various sulfides, chlorides, and fluorides. Although many of these films are not strictly solid lubricants, they are often effective for short-term wear resistance. For long-term effectiveness, they often provide a porous reservoir for liquid lubricants and increased life of organically bonded coatings.

Diffusion provides an alternative procedure for generating a chemically modified surface, eg, sulfide surface films can be formed by immersing steel in molten mixtures of sulfur-containing salts such as sodium thiosulfate or sodium sulfide. Similar processes are employed for carburizing, nitriding, boriding, or siliconizing. Metalliding can introduce a new element into many metal surfaces from a molten fluoride bath. A number of hardening treatments, as well as flame sprayed tungsten and titanium carbides, provide excellent wear resistance. Some of these also provide good bases for low shear strength films.

Metal Films. In many respects, soft metals such as those listed in Table 14 are ideal solid lubricants (58). They have low shear strength, can be bonded strongly to substrate metal as continuous films, have good lubricity, and have

Table 13. Typical Pretreatments for Various Substrates[a]

Substrate	Pretreatment
aluminum	vapor degrease plus anodize
	light abrasive blast plus chromate conversion
copper and its alloys	vapor degrease
	abrasive blast (light)
	chromate conversion
iron and steel	vapor degrease
	abrasive blast
	phosphate
stainless steel	vapor degrease
	sandblast
	passivate
titanium	alkaline cleaning
	abrasive blast
	fluoride phosphate or alkaline anodize

[a]Ref. 61.

Table 14. Properties of Soft Metals

Metal	CAS Registry Number	Mohs' hardness	Melting point, °C
gallium	[7440-55-3]		30
indium	[7440-74-6]	1	155
thallium	[7440-28-0]	1.2	304
lead	[7439-92-1]	1.5	328
tin	[7440-31-5]	1.8	232
gold	[7440-57-5]	2.5	1063
silver	[7440-22-4]	2.5–3	961

high thermal conductivity. Metal films can be applied by electroplating or by vacuum processes, eg, evaporation, sputtering, and ion plating (see FILM DEPOSITION TECHNIQUES; METALLIC COATINGS).

Melting points of gallium, indium, and tin are too low, and those of thallium and lead are borderline when high surface temperatures are generated by high speeds and loads. Gallium is a special case, ie, it is above its melting point under most conditions and is too reactive with many metals. It is effective, however, when applied in a vacuum with AISI 440C stainless steel and with ceramics (qv) such as boron carbide or aluminum oxide which can be applied as undercoats (63).

A number of metal films are used industrially. Copper and silver are electroplated on the threads for bolt lubrication. Slurries of powders of nickel, copper, lead, and silver are also used in commercial bolt lubricants. Tin, zinc, copper, and silver coatings are used as lubricants in metalworking where toxicity has virtually eliminated lead as a lubricating coating (64). Gold and silver find limited use on more expensive workpiece materials such as titanium. Silver films are useful in a variety of other sliding and rolling contacts in vacuum and at high temperatures since silver forms no alloys with steels and is soft at high temperatures.

Under severe conditions and at high temperatures, noble metal films may fail by oxidation of the substrate base metal through pores in the film. Improved life may be achieved by first imposing a harder noble metal film, eg, rhodium or platinum–iridium, on the substrate metal. For maximum adhesion, the metal of the intermediate film should alloy both with the substrate metal and the soft noble-metal lubricating film. This sometimes requires more than one intermediate layer. For example, silver does not alloy to steel and tends to lack adhesion. A flash of hard nickel bonds well to the steel but the nickel tends to oxidize and should be coated with rhodium before applying silver of 1–5 μm thickness. This triplex film then provides better adhesion and greatly increased corrosion protection.

Metalworking Lubrication

Metalworking commonly involves one of two processes: cutting or machining, which includes drilling, turning, grinding, honing, lapping, milling, and broaching; or deformation to change shape without melting or cutting. Deformation includes rolling, drawing, extrusion, forging, stamping, and spinning (see METAL TREATMENTS) (64,65). In both cutting and forming operations, metal-to-metal contact causes wear and frictional heating. The purpose of metalworking fluids is both to remove heat from the tool and workpiece and to minimize friction and wear by providing good lubricity.

In cutting at high speeds where cooling is a principle requirement, water-base fluids are commonly used containing 5% or less of additives for improved lubrication, rust protection, and better wetting. Straight oils containing sulfur compounds and other film strength and EP additives are more common for slower cutting speeds and relatively deep cuts, or where surface finish and tolerances are important. A generalized severity matrix for metal cutting fluid selection has been given (65). Commercial cutting lubricant suppliers commonly fill in recommended products for a particular operation and workpiece material, and give the water:oil dilution ratio for water-base fluids. In general, straight oils would be used for more severe operations, water-base for less severe. In high speed machining, as for cast iron and superalloys with boron nitride compacts, no cutting fluid is quite adequate (66).

Most metal forming employs petroleum or synthetic oil fortified with additives to provide as much lubricating film support as possible for the high stresses involved at the workpiece contact (65). Polyol esters are finding broadening use in rolling steel, and poly(alkylene glycol)s and polybutenes are used as dispersions in solvents for cold rolling aluminum foil (67). Water emulsions are used where cooling is required. Semifluid pastes are applied in cold pressing sheet metal, and water or oil dispersions of graphite and other solid lubricants are used as spray lubricants in hot forging (65).

Extreme Ambient Conditions

Gas Lubrication. Despite severe limitations, gas lubrication of bearings has received intensive consideration for its resistance to radiation, for high speeds, temperature extremes, and use of the working fluid (gas) in a machine as its lubricant. A primary limitation is, however, the very low viscosity of gases (Fig. 16 (68)) which leads to a limiting load of only 15–30 kPa (2.2–4.4 psi) for most self-acting (hydrodynamic) gas bearings and up to 70 kPa (10 psi) for operation with external gas-lifting pressure in hydrostatic operation.

Gases that have been used for bearing lubrication include air, hydrogen, helium, nitrogen, oxygen, uranium hexafluoride [*7783-81-5*], carbon dioxide, and argon [*7440-37-1*]. A useful property of gases is that their viscosity, and hence their capacity to generate hydrodynamic pressure P, increases with temperature as indicated in Figure 16, whereas the opposite is true for liquids. Gas viscosity is usually independent of pressure up to about 1 MPa (10 atm).

Hydrodynamic principles for gas bearings are similar to those involved with liquid lubricants except that gas compressibility usually is a significant factor (8,69). With gas employed as a lubricant at high speeds, start–stop wear is minimized by selection of wear-resistant materials for the journal and bearing. This may involve hard coatings such as tungsten carbide or chromium oxide flame plate, or solid lubricants, eg, PTFE and MoS_2.

Because of the very small bearing clearances in gas bearings, dust particles, moisture, and wear debris (from starting and stopping) should be kept to a minimum. Gas bearings have been used in precision spindles, gyroscopes, motor

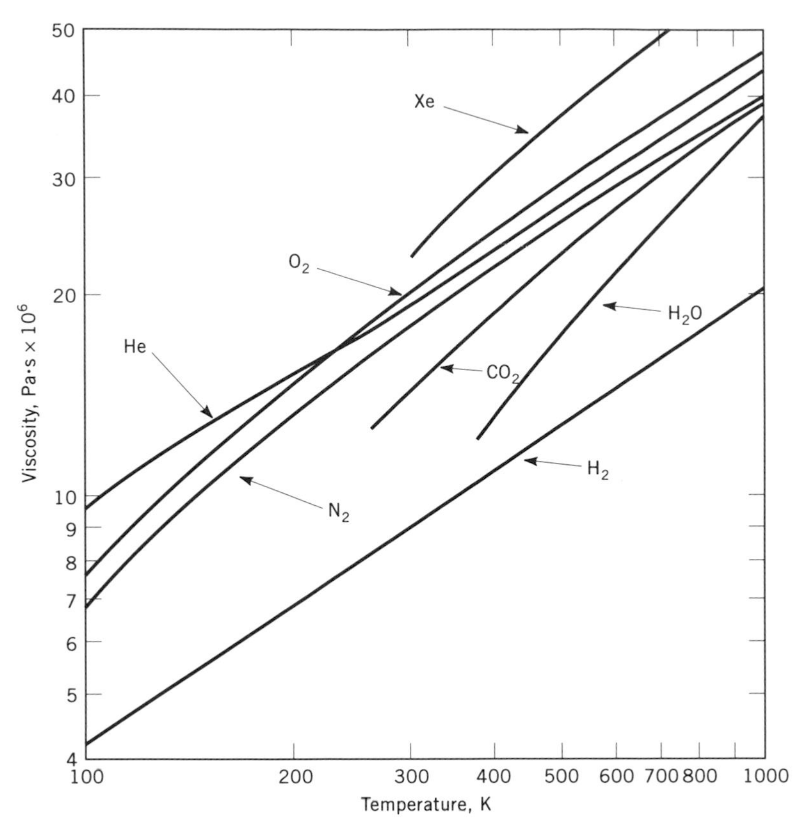

Fig. 16. Viscosity of several gases at P = 0.1 MPa (1 bar) (68). To convert Pa·s to P, multiply by 10.

and turbine-driven circulators, compressors, fans, Brayton cycle turbomachinery, environmental simulation tables, and memory drums.

Liquid Metals. If operating temperatures rise above 250–300°C, where many organic fluids decompose and water exerts high vapor pressure, liquid metals have found some use, eg, mercury for limited application in turbines; sodium, especially its low melting eutectic with 23 wt % potassium, as a hydraulic fluid and coolant in nuclear reactors; and potassium, rubidium, cesium, and gallium in some special uses.

Liquid metal selection is usually limited to the lower melting point metals in Table 15. Figure 17 shows that liquid metal viscosity generally is similar to water at room temperature and approaches the viscosities of gases at high temperature. Hydrodynamic load capacity with both liquid metals and water in a bearing is about 1/10 of that with oil, as indicated in Table 2.

The sodium–potassium eutectic is commercially available for use as a liquid over a wide temperature range. Because of its excessive oxidizing tendency in air, however, its handling and disposal is hazardous; it can be used only in closed vacuum or in an inert gas atmosphere of helium, argon, or nitrogen. In addition to the oxidation problem, bearing material selection is critical for liquid metal bearings. Tungsten carbide cermet with 10–20 wt % cobalt binder gave

Table 15. Liquid Metal Lubricants

Liquid metal	CAS Registry Number	Liquid temperature range, °C	
		Mp	Bp[a]
cesium	[*7440-46-2*]	28	670
gallium	[*7440-55-3*]	30	1980
mercury	[*7439-97-6*]	−39	360
potassium	[*7440-09-7*]	62	760
rubidium	[*7440-17-7*]	39	700
sodium	[*7440-23-5*]	98	880
sodium–potassium		−11	780

[a]Values rounded to nearest 10°C.

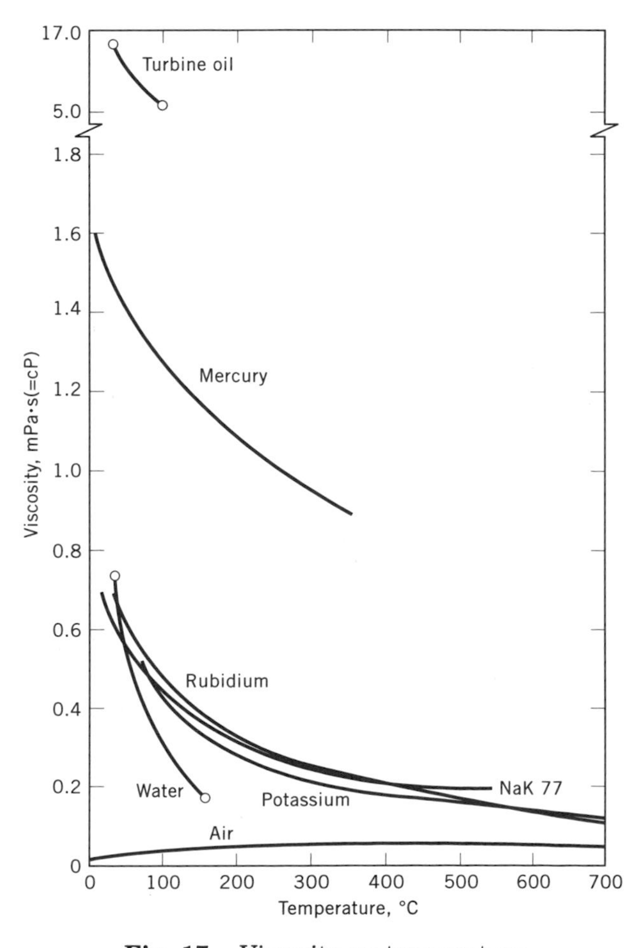

Fig. 17. Viscosity vs temperature.

superior performance when running against molybdenum under heavy loads at low speeds at temperatures up to 815°C (70).

A low melting (5°C) gallium–indium–tin alloy has been the choice for small spiral-groove bearings in vacuum for x-ray tubes at speeds up to 7000 rpm (71). Surface tension 30 times that of oil avoids leakage of the gallium alloy from the ends of the bearings.

Cryogenic Bearing Lubrication. Cryogenic fluids, such as liquid oxygen, hydrogen, or nitrogen, are used as lubricants in liquid rocket propulsion systems, turbine expanders in liquefaction and refrigeration, and pumps to transfer large quantities of liquefied gases. Properties of typical cryogenic fluids are given in Table 16 (see CRYOGENICS). Bearings operating in cryogenic fluids are amply cooled from the standpoint of dissipating the heat generated from friction, but unfortunately the low viscosity of the fluids leads to marginal lubrication.

For wear resistance and low friction, coatings of PTFE or MoS_2 generally have been satisfactory. Use of low thermal expansion filler in PTFE helps minimize cracking and loss of adhesion from metal substrates with their lower coefficients of expansion.

Because of the low viscosities of cryogenic liquids, rolling element bearings seem better suited than hydrodynamic bearings for turbo pumps. AISI 440C stainless balls and rings generally are preferred for their corrosion resistance over the more commonly used AISI 52100 steel.

Nuclear Radiation Effects. Components of a nuclear reactor system that require lubrication include control-rod drives, coolant circulating pumps or compressors, motor-operated valves, and fuel handling devices, and, of course, are exposed to varying amounts of ionizing (14).

Degree of damage suffered by a lubricant depends primarily on the total radioactive energy absorbed, whether it is from neutron bombardment or from gamma radiation. The common energy unit for absorbed dosage, the gray (Gy), is equal to 1×10^{-5} J (100 ergs) absorbed per gram of material, or 0.01 Gy = 1 rad. The first changes observed with petroleum oils (at about 10^4 gray dosage) is evolution of hydrogen and light hydrocarbon gas as fragments from the original molecule. Unsaturation results in decreased oxidation stability, cross-linking, polymerization, or scission (72).

The trend for increasing viscosity with increased dose is shown in Figure 18 for several petroleum oils (72). For many lubricant applications, a dose that gives a 25% increase in 40°C viscosity can be taken as a tolerance limit. Lower radiation absorption seldom changes the lubricant sufficiently to interfere with

Table 16. Cryogenic Liquids

Liquid	Freezing point, °C	Bp, °C	Viscosity at bp, mPa·s(=cP)
helium	<−272	−269	0.005
hydrogen	−259	−253	0.013
nitrogen	−210	−196	0.016
fluorine	−219	−188	0.26
argon	−189	−186	
oxygen	−219	−183	0.19
methane	−182	−161	

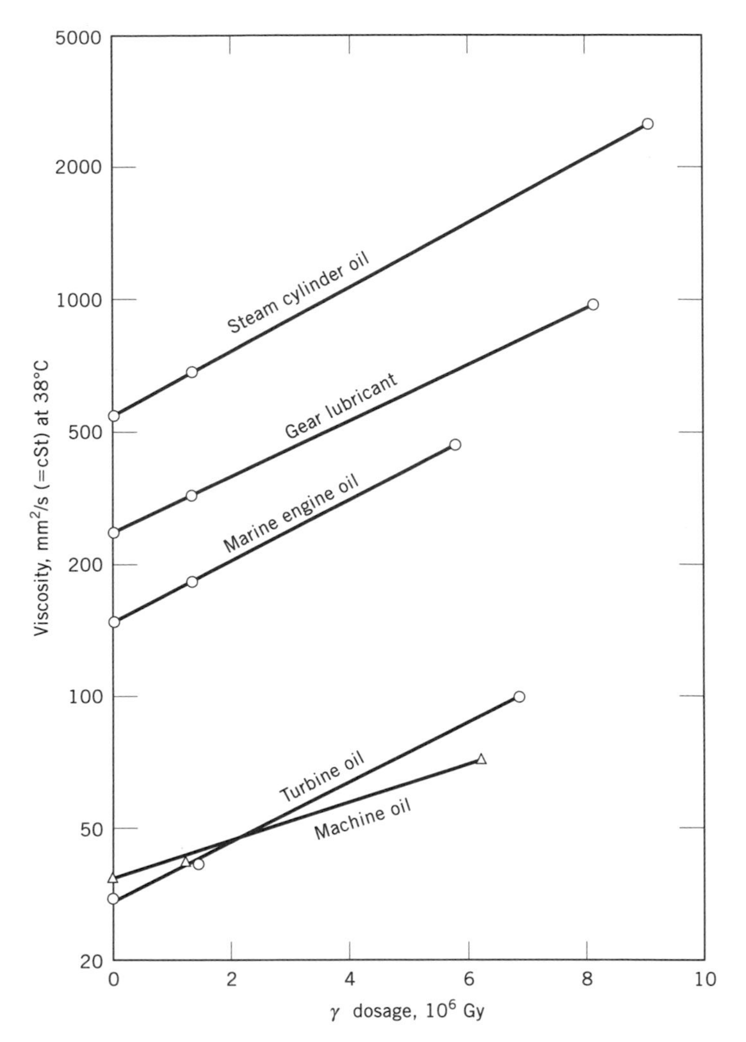

Fig. 18. Viscosity change of industrial petroleum oils with irradiation (72). To convert Gy to rad, multiply by 100.

its performance. Greater dosage results in more rapid thickening, sludging, and operating trouble (73).

The general range of tolerance limit of 1 to 4 $\times$ 10^6 Gy (1–4 $\times$ 10^8 rads) for petroleum oils in Table 17 tends to be somewhat higher than for synthetic oils (74). This is surprising in view of the excellent thermal and oxidative stability of methyl silicones, diesters, silicates, and some other synthetics. An exception is the high order of stability with synthetic oils consisting of aromatic hydrocarbons in which much of the absorbed energy appears to be transferred into harmless resonance in the aromatic ring structure. This reduces the degree of damaging ionization and free-radical formation which occurs on a more general basis with the chain-like structures in paraffinic oils or in the saturated ring structure of alicyclic oils. Oil life is reduced by both the radiation dose and by oxidation if oxygen (air) is present at high temperature.

Table 17. Radiation Tolerance Limits of Several Oil Types

Oil	Tolerance limit, 10^6 Gy[a] for 25% increase in 40°C viscosity
petroleum	1–4
synthetic	
diester MIL-L-6085	1.1
synthetic hydrocarbon	2.5–4.5
phosphate ester	0.4–0.6
poly(propylene oxide)	1.0
alkylbenzene	5
dimethyl silicone	<1
methyl phenyl silicone	1
tetraaryl silicate	0.6

[a] To convert Gy to rad, multiply by 100.

Conventional greases consisting of petroleum oils thickened with lithium, sodium, calcium, or other soaps suffer significant breakdown of the soap gel structure at doses above about 10^5–10^6 Gy (10^7–10^8 rad). Initial breakdown commonly involves increased softening of the grease to the point where it may become fluid. At even higher doses, polymerization of the oil phase eventually leads to overall grease hardening. Some greases with radiation-resistant components, eg, polyphenyl ether oil and nonsoap thickeners, maintain satisfactory consistency for lubrication purposes up to 10^7 Gy (10^9 rad).

Lubrication with Glass. Softening glass (qv) is used as a lubricant for extrusion, forming, and other hot working processes with steel and nickel-base alloys up to about 1000°C, for extrusion and forming titanium and zirconium alloys, and less frequently for extruding copper alloys (64). Principal types of glasses used are pure fused silica [*7631-86-9*], 96% silica–soda–lime, borosilicates, and aluminosilicates [*1327-36-2*]. The glass composition is selected for proper viscosity, typically 10–100 Pa·s (1,000–10,000 P) at the mean temperature of the die and workpiece, to serve as a true hydrodynamic lubricant. Glass may be applied as fibers or powder to the die or hot workpiece, or as a slurry with a polymeric bonding agent to the workpiece before heating. Another method involves rolling heated steel billets across glass sheets, where the glass then wraps itself around the billet before passing to a die extrusion chamber.

The U.S. Bureau of Mines has employed glass for forming ceramic materials at high temperatures (75). The viscosity curve for a soda–lime–silica glass in Figure 19 indicates the high viscosity available at hot forming temperatures.

Production

Total yearly production of lubricants in the United States has been fairly stable since the 1960s. The production peak of 11.2×10^6 m^3 (70.7×10^6 bbl) in 1974 gradually declined to 8.9×10^6 m^3 (55.9×10^6 bbl) in 1991, which is about 30% of worldwide production. Automotive lubricants make up about 56% of U.S. production, industrial lubricants 38%, and greases 2%. Future growth rate of the market is expected typically to be 1–3% per year.

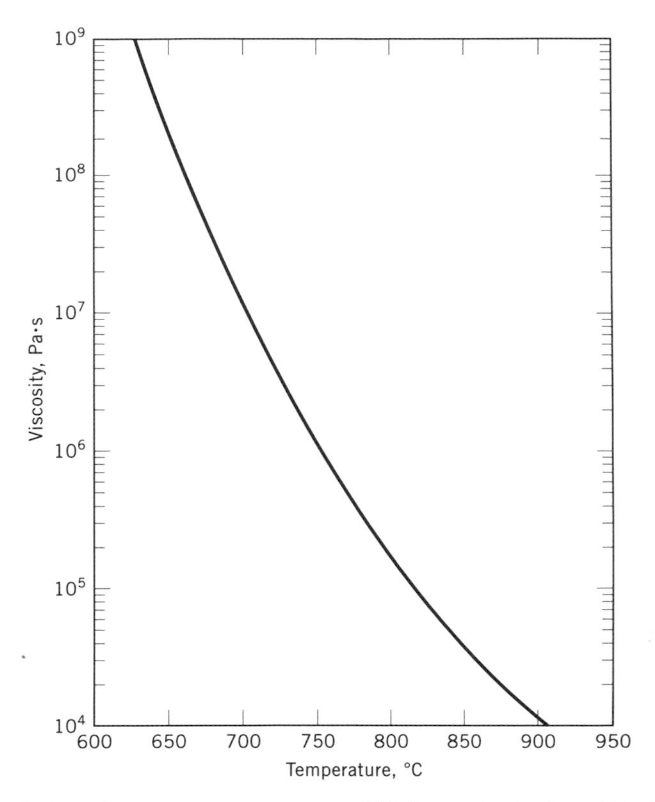

Fig. 19. Plot of viscosity vs temperature for NBS standard glass No. 710. To convert Pa·s to P, multiply by 10.

Oil additives account on average for 7–8% of lubricant production volume (automotive 13%, industrial 3%) (67). Additive production volumes have largely mirrored overall lubricant production. New standards for automotive and diesel engine oils are requiring higher additive levels and more expensive chemistry.

Although synthetic lubrication oil production amounts to only about 2% of the total market, volume has been increasing rapidly (67). Growth rates of the order of 20% per year for poly(α-olefin)s, 10% for polybutenes, and 8% for esters (28) reflect increasing automotive use and these increases would accelerate if synthetics were adopted for factory fill of engines by automotive manufacturers. The estimated production of poly(α-olefin)s for lubricants appears to be approximately 100,000 m^3/yr, esters 75,000, poly(alkylene glycol)s 42,000, polybutenes 38,000, phosphates 20,000, and dialkyl benzene 18,000 (28,67). The higher costs reflected in Table 18 (18,28) have restricted the volume of silicones, chlorotrifluoroethylene, perfluoroalkylpolyethers, and polyphenyl ethers.

Environmental and Health Factors (Toxicology)

Conservation, health, safety, and environmental pollution concerns have led to the creation of wide-reaching legislation such as the U.S. Congress Energy Policy

Table 18. Relative Cost of Synthetic Oils, 1993

Oil type	Approximate relative cost
petroleum	1
poly(α-olefin)	3
polybutene	3
diester	5
polyglycol	5
dialkylbenzene	5
polyol esters	7
phosphates	8
silicone	25
polyphenyl ether	250
fluorocarbon	300

and Conservation Act, Toxic Substances Control Act of 1976 (76), Resources Conservation and Recovery Act of 1976, the Oil Recycling Act of 1980, and subsequent implementation of many rules and regulations such as the OSHA Hazard Communication Standard. Continuing publications of new and proposed rules and regulations are available from the U.S. Environmental Protection Agency (Washington, D.C.) and from the National Technical Information Service of the U.S. Department of Commerce (Springfield, Virginia).

Regulations generally prohibit disposal of lubricants in streams, chemical dumps, or other environmental channels. Over half of disposed lubricants are burned as fuel, usually mixed with virgin residual and distillate fuels (77).

Waste aqueous metalworking fluids may be successfully treated by conventional means for removal of tramp oil, surfactants, and other chemical agents to provide suitable effluent water quality (78).

Lubricant Recycling. Considerable effort is underway to improve and expand recycling (qv) of lubricating oils. Although typical processes result in 80–90% yield, questions remain regarding initial collection and the separation of used oil from water and other contaminants. Recycling treatment varies from simple cleaning to essentially the complete refining process used with virgin oil. The following are typical steps involved in purifying used petroleum lubricating oil and are indicated schematically in Figure 20 (79).

Reclamation. This involves simple separation of contaminants by gravity settling of water and dirt, centrifuging, filtering, and membrane techniques. With water-soluble cutting oils containing only a few percent oil, chemical emulsion breakers are first added which consist of sulfuric acid and then aluminum sulfate as a coagulant. Polymers are sometimes added to speed the process. The separated oil then is decanted, skimmed, or centrifuged and commonly is burned. Generally, 1–5% reprocessed waste oil may be added to fuel and still meet U.S. EPA industrial furnace limits, ie, <5 ppm arsenic by mass, <2 ppm cadmium, <10 ppm chromium, <100 ppm lead, and <4000 ppm halogens (19).

Reprocessing. The simplest operation involves flash distillation in an evaporator at about 100–200°C in partial vacuum to remove water and low boiling contaminants, eg, gasoline and solvents. This is followed by treatment with fuller's earth or other activated clay for removing oxidation products and most additives to produce a purified, light-colored oil which, with suitable additives,

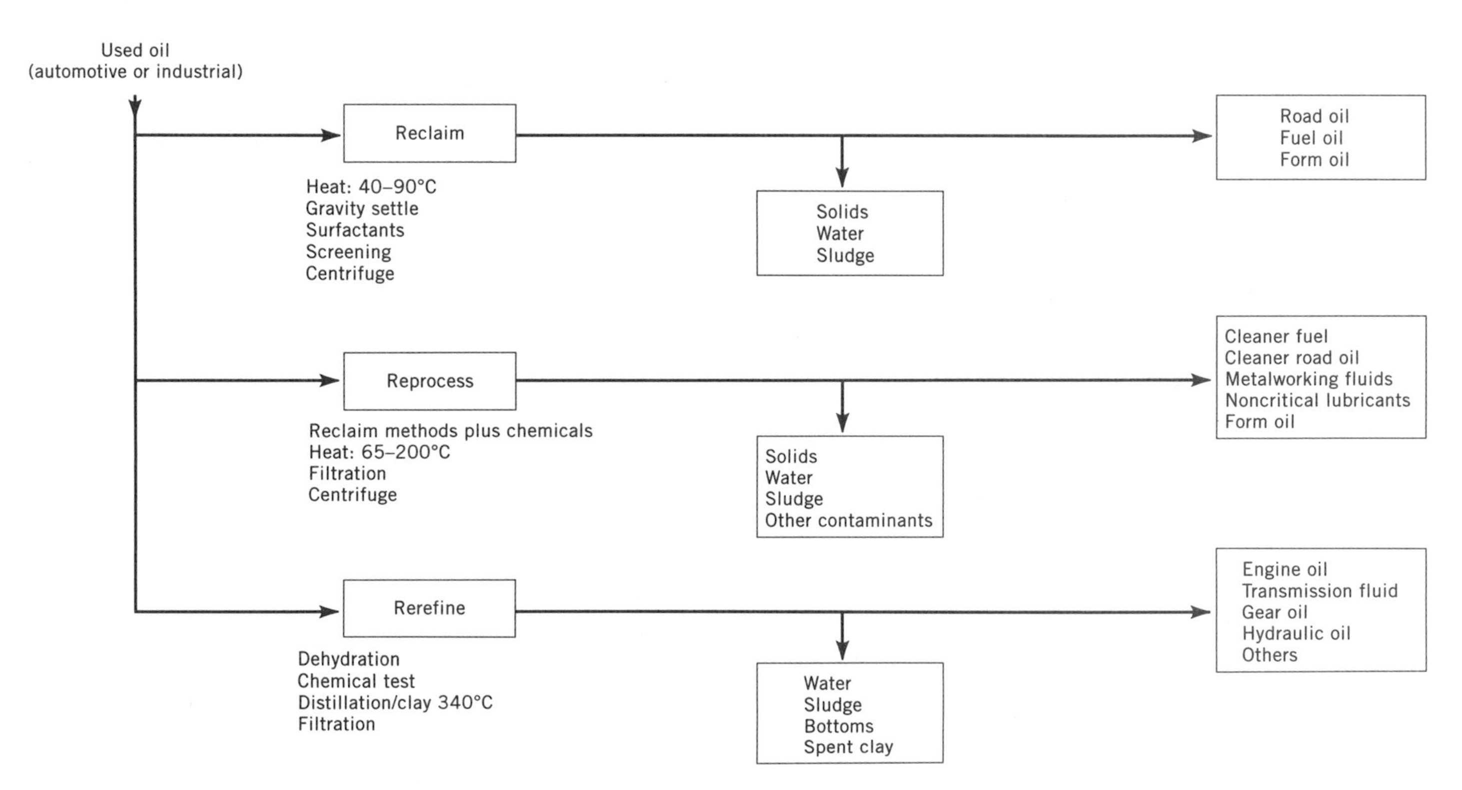

Fig. 20. Oil recycling flow diagram (79).

is satisfactory for use as fuel, metalworking base stocks, noncritical lubricants, and concrete form oil. Some used oils, eg, hydraulic and transformer oils, can be reprocessed with a portable unit of 200–400 L/h capacity to their original oil quality directly at the equipment in which they are being used.

Rerefining. The technology currently attracting most attention for producing original quality lubricating oil depends on distillation in thin-film evaporators (TFE) (80). TFE processes usually involve a scheme similar to that shown at the bottom of Figure 20. The preliminary removal of water, solvents, and fuel is done in the same fashion as in most other recycling. Coking and fouling during distillation is avoided because the maximum temperature is maintained for only 2–5 s as the oil flows down the TFE wall under the influence of moving wiper blades; this is a small fraction of residence time in the packing or plates of a more traditional distillation tower. TFE variations involve batch operation with a single unit, or sequential distillation in multiple units to produce several lube fractions.

Older rerefining units used 2–5 kg/L of activated clay at 40–70°C and higher temperatures in place of TFE to clean the oil (80). More elaborate chemical and hydrotreating of used engine oils without a distillation step has been developed by Phillips Petroleum for processing 40,000 m^3/yr (10×10^6 gal/yr). Establishment of a reliable feedstock supply is a critical consideration for larger rerefining plants.

Toxic and Hazardous Constituents. Questionable constituents of lubricating oils are polycyclic aromatics in the base oil plus various additives (81). Of refining steps used in preparing lubricating oil base stocks from toxic distillates, only effective solvent extraction, severe hydrogenation, or exhaustive fuming sulfuric acid treatment appear adequate to eliminate carcinogenicity. Conventional hydrofinishing or light solvent extraction reduces carcinogenic potential but does not necessarily eliminate it. Synthetic hydrocarbon poly(α-olefin)s are not expected to be carcinogenic since no polycyclic aromatics are present.

Most additives for lubricants present little risk, but the following involve significant hazards: lead compounds (qv), phenyl 2-naphthylamine, sodium nitrite plus amines, tricresylphosphate high in the *ortho*-cresol isomer, and chlorinated naphthalenes. A number of sulfur compounds used as additives cause skin irritation; however, properly refined base oils containing usual concentrations of these additives have a low degree of toxicity. Used motor oil has displayed increased carcinogenic activity over its new counterpart (81). Users should also avoid contact with lubricants, metalworking oils, and quench oils which are highly degraded, were in service at extremely high temperatures, or are contaminated with toxic metals or bacteria.

The latest government regulations set forth under the Toxic Substances Control Act and in Public Health Service publications should be checked before formulating new lubricants. Users of lubricants should request Material Safety Data Sheets for each substance involved plus certification of compliance from vendors. Lubricant compounders should insist on similar information from their suppliers for any additive packages. Manufacturers of both additives and lubricants commonly make toxicity checks on commercial products.

Food Processing. To ensure safe processing of edible products in the food and beverage industries, two federal agencies control use of food-grade lubricants: the U.S. Department of Agriculture (USDA) regulates meat and poultry plants, whereas the U.S. Food and Drug Administration (FDA) monitors other food as well as drug manufacturers (see FOOD PROCESSING).

Upon satisfactory determination of nontoxicity of a lubricant, the USDA issues one of two ratings: H1 for use where there is incidental or possible food contact as by splashing or dripping from machinery above an edible product, or H2 for no food contact as in sealed gear boxes or machinery below a product line (82–84). These classes include a number of petroleum and synthetic oils and greases.

For severe requirements where lubricants contact food on a regular basis, the FDA publishes a list of authorized ingredients in the *Codes of Federal Regulations* (85). These are included in three classes: (*1*) white mineral oils (21 CFR 172.878) used, for example, as release agents in bakery products, confections, dehydrated fruits and vegetables, and egg whites; (*2*) petrolatums (21 CFR 172.880) used in applications similar to white mineral oils; and (*3*) technical white oils (21 CFR 178.3620) used in processing aluminum foil for food packaging (qv), in manufacture of animal feed and fiber bags, and on food machinery.

These are highly purified petroleum products which have been fully refined by either acid treatment or hydrogenation to remove all unsaturates, aromatics, and coloring materials to meet *United States Pharmacopoeia* (USP) requirements.

BIBLIOGRAPHY

"Lubrication and Lubricants" in *ECT* 1st ed., Vol. 8, pp. 495–540, by R. G. Larsen and A. Bondi, Shell Development Center; in *ECT* 2nd ed., Vol. 12, pp. 557–616, by R. E. Lee, Jr., and E. R. Booser, General Electric Co.; in *ECT* 3rd ed., Vol. 14, pp. 477–526, by E. R. Booser, General Electric Co.

1. D. Dowson, *History of Tribology*, Longman Group Limited, London, 1979.
2. W. A. Zisman, in R. C. Gunderson and A. W. Hart, eds., *Synthetic Lubricants*, Reinhold Publishing Co., New York, 1962, pp. 6–60.
3. E. Rabinowicz, *Friction and Wear of Materials*, John Wiley & Sons, Inc., New York, 1965.
4. F. P. Bowden and D. Tabor, *The Friction and Lubrication of Solids*, Oxford University Press, London, 1950.
5. D. F. Wilcock and E. R. Booser, *Bearing Design and Application*, McGraw-Hill Book Co., Inc., New York, 1957.
6. D. D. Fuller, *Theory and Practice of Lubrication for Engineers*, 2nd ed., John Wiley & Sons, Inc., New York, 1984.
7. A. Raimondi and A. Z. Szeri, in E. R. Booser, ed., *Handbook of Lubrication*, Vol. 2, CRC Press Inc., Boca Raton, Fla., 1983, pp. 413–462.
8. W. A. Gross, ed., *Fluid Film Lubrication*, John Wiley & Sons, Inc., New York, 1980.
9. H. J. Sneck and J. H. Vohr, in Ref. 7, pp. 69–91.
10. *Annual Book of ASTM Standards: Petroleum Products and Lubricants*, American Society of Testing Materials, Philadelphia, Pa., issued annually.
11. B. J. Hamrock, *Fundamentals of Fluid Film Lubrication*, NASA Reference Publication 1255, U.S. Government Printing Office, Washington, D.C., 1991.
12. E. V. Zaretsky, ed., *Life Factors for Rolling Bearings*, Society of Tribologists and Lubrication Engineers, Park Ridge, Ill., 1992.

13. N. W. Furby, in P. M. Ku, ed., *Interdisciplinary Approach to Liquid Lubricant Technology*, NASA SP-318, NTIS N74-12219-12230, 1973, pp. 57–100.
14. J. G. Wills, *Lubrication Fundamentals*, Marcel Dekker, Inc., New York, 1980.
15. *SAE Handbook*, Society of Automotive Engineers, Warrendale, Pa., issued annually.
16. M. Sanchez-Rubio and co-workers, *Lubr. Eng.* **48**, 821–826 (1992).
17. *Viscosity and Density of over 40 Lubricating Fluids of Known Composition at Pressures to 150,000 psi and Temperatures to 425°F*, American Society of Mechanical Engineers, New York, 1952.
18. D. Klamann, *Lubricants and Related Products*, Verlag Chemie, Weinheim, Germany, 1984.
19. R. S. Fein, *ASM Handbook*, Vol. 18, ASM International, Metals Park, Ohio, 1992, pp. 81–88.
20. E. E. Klaus and E. J. Tewksbury, in Ref. 7, pp. 229–254.
21. J. A. O'Brien, in Ref. 7, pp. 301–315.
22. S. Q. A. Rizvi, in Ref. 19, pp. 98–112.
23. T. V. Liston, *Lubrication Engineering* **48**, 389–397 (1992).
24. M. Hunter, E. E. Klaus, and J. L. Duda, *Lubrication Engineering* **49**, 492–498 (1993).
25. J. M. Georges and co-workers, *Wear* **53**, 9–34 (1979).
26. A. Bondi, *Physical Chemistry of Lubricating Oils*, Reinhold Publishing Co., New York, 1951.
27. W. J. Bartz and N. Nemes, *Lubrication Engineering* **33**, 20–32 (1977).
28. R. L. Shubkin, *Synthetic Lubricants and High-Performance Functional Fluids*, Marcel Dekker, Inc., New York, 1993.
29. E. R. Booser, ed., *Handbook of Tribology and Lubrication*, Vol. 3, CRC Press, Inc., Boca Raton, Fla., 1994.
30. J. D. Fotheringham, in Ref. 28, pp. 271–318.
31. H. Dressler, in Ref. 28, pp. 125–144.
32. S. J. Randles, in Ref. 28, pp. 41–65.
33. J. M. Perez and E. E. Klaus, in Ref. 29, Chapt. 15.
34. B. J. Beimesch, in Ref. 29, Chapt. 13.
35. P. L. Matlock and N. A. Clinton, in Ref. 28, pp. 101–123.
36. W. L. Brown, in Ref. 29, Chapt. 16.
37. M. P. Marino, in Ref. 28, pp. 67–100.
38. M. P. Marino and D. G. Placek, in Ref. 29, Chapt. 17.
39. Electric Power Research Institute, *Evaluation of Fire-Retardant Fluids for Turbine Bearing Lubricants*, report NP-6542, Palo Alto, Calif., 1989.
40. W. D. Phillips, *Lubrication Engineering* **42**, 228–235 (1986).
41. M. Groeneweg and co-workers, *Lubrication Engineering* **47**, 1035–1039 (1991).
42. T. W. DelPesco, in Ref. 28, pp. 145–172.
43. T. W. DelPesco, in Ref. 29, Chapt. 18.
44. C. E. Snyder and L. J. Gschwender, in Ref. 29, Chapt. 11.
45. B. Bushan, in Ref. 29, Chapt. 20.
46. E. D. Brown, in Ref. 29, Chapt. 19.
47. D. H. Denby, S. J. Stoklosa, and A. Gross, in Ref. 28, pp. 183–203.
48. D. A. Ruesch, in Ref. 28, pp. 173–181.
49. C. L. Mahoney and E. R. Barnum, in Ref. 2, pp. 402–463.
50. E. R. Booser and A. E. Baker, *NLGI Spokesman* **50**, 60–65 (1976).
51. R. H. Boehringer, in Ref. 19, pp. 123–131.
52. P. R. McCarthy, in Ref. 13, pp. 137–185.
53. C. J. Boner, *Manufacture and Application of Lubricating Greases*, Reinhold Publishing Corp., New York, 1954.
54. I. W. Ruge, in Ref. 7, pp. 255–267.
55. H. Schwenker, in Ref. 13, pp. 180–184.

56. I. C. Lipp, *Lubrication Engineering* **32**, 574–584 (1976).
57. J. K. Lancaster, in Ref. 7, pp. 269–290.
58. W. E. Campbell, in F. F. Ling, E. E. Klaus, and R. S. Fein, eds., *Boundary Lubrication: An Appraisal of World Literature*, American Society of Mechanical Engineers, New York, 1969.
59. H. F. Sliney, in Ref. 19, pp. 113–122.
60. A. Erdemir and co-workers, *Lubrication Engineering* **47**, 179–184 (1991).
61. R. M. Gresham, in Ref. 29, Chapt. 10.
62. R. M. Gresham, *Lubrication Engineering* **44**, pp. 143–145 (1988).
63. D. H. Buckley and R. L. Johnson, *ASLE Transactions* **6**, 1 (1963).
64. J. A. Schey, *Tribology in Metalworking: Friction, Lubrication and Wear*, ASM International, Metals Park, Ohio, 1983.
65. J. T. Laemmle, in Ref. 19, pp. 139–149.
66. R. Komanduri and D. G. Flom, in Ref. 29, Chapt. 24.
67. E. I. Williamson, in Ref. 28, pp. 245–582.
68. D. F. Wilcock, in Ref. 7, pp. 291–300.
69. M. Khonsari, L. A. Matsch, and W. Shapiro, in Ref. 29, Chapt. 30.
70. A. J. Baumgartner, in R. A. Burton, ed., *Bearing and Seal Design in Nuclear Power Machinery*, American Society of Mechanical Engineers, New York, 1967.
71. J. Gerkema, *ASLE Trans.* **28**, 47–53 (1985).
72. J. G. Carroll and S. R. Calish, *Lubrication Engineering* **13**, pp. 388–392 (1957).
73. E. R. Booser, in J. J. O'Connor and J. Boyd, eds., *Standard Handbook of Lubrication Engineering*, McGraw-Hill Book Co., Inc., 1968, Chapt. 44.
74. R. O. Bolt, in E. R. Booser, ed., *Handbook of Lubrication*, Vol. 1, CRC Press, Inc., Boca Raton, Fla., 1983, pp. 209–223.
75. J. E. Kelley, T. D. Roberts, and H. M. Harris, *A Penetrometer for Measuring the Absolute Viscosity of Glass*, report 6358, U.S. Bureau of Mines, Washington, D.C., 1964.
76. *Toxic Substances Control Act*, Public Law No. 469, 94th U.S. Congress Chemical Substance Inventory, U.S. Environmental Protection Agency, Office of Toxic Substances, Washington, D.C., 1975.
77. J. W. Swain, Jr., in Ref. 74, pp. 533–549.
78. S. Napier, *Lubr. Eng.* **41**, 361–365 (1985).
79. J. W. Swain, Jr., *Lubr. Eng.* **39**, 551–554 (1983).
80. D. W. Brinkman, *Lubr. Eng.* **43**, 324–328 (1987).
81. T. M. Warne and C. A. Halder, *Lubr. Eng.* **42**, 97–103 (1986).
82. G. Arbocus, in Ref. 74, pp. 359–371.
83. J. Brown, *Power Trans. Des.*, 39–42 (Oct. 1991).
84. *List of Chemical Compounds Authorized for Use under USDA Inspection and Grading Programs*, Supt. of Documents, U.S. Government Printing Office, Washington, D.C.
85. *21 Code of Federal Regulations*, U.S. Government Printing Office, Washington, D.C.

General References

E. R. Booser, ed., *Handbook of Lubrication*, Vols. 1, 2, 3, CRC Press, Boca Raton, Fla., 1983, 1984, 1994.

D. Klamann, *Lubricants and Related Products*, Verlag Chemie, Deerfield Beach, Fla., 1984.

Friction, Lubrication, and Wear Technology, ASM Handbook, Vol. 18, ASM International, Metals Park, Ohio, 1992.

E. R. BOOSER
Consultant

LUMINESCENT MATERIALS

CHEMILUMINESCENCE

Chemiluminescence is the emission of light from chemical reactions at ordinary temperatures (1). Chemiluminescent reactions produce a reaction intermediate or product in an electronically excited state, and radiative decay of the excited state is the source of the light (2). When the excited state is a singlet, the radiative process is identical to fluorescence; when the excited state is a triplet, phosphorescent emission results (3). Electronically excited states can emit ultraviolet (uv) or infrared (ir) radiation as well as visible light, and the definition of chemiluminescence is no longer restricted to visible light emission. Moreover, the formation of electronically excited reaction products can be detected by their photochemical reactions, even when radiation is negligible. Thus chemiluminescence is a special case of the more general process of chemiexcitation. It is observed in liquid-, gas-, and solid-phase reactions.

Chemiexcitation is uncommon because most chemical reactions follow a ground-state potential energy surface and release chemical energy as vibrational excitation of ground-state products, which is observed as heat (4). Chemiluminescence is even less common because most electronically excited states decay to ground states by nonradiative processes (3). Nevertheless, a significant number of chemiluminescent reactions are known and a few, such as firefly, *Cypridina*, or bacterial bioluminescence, and peroxyoxalate chemiluminescence combine high chemiexcitation efficiency with high fluorescence yield to provide substantial light production.

Chemiluminescence has been studied extensively (2) for several reasons: (*1*) chemiexcitation relates to fundamental molecular interactions and transformations and its study provides access to basic elements of reaction mechanisms and molecular properties; (*2*) efficient chemiluminescence can provide an emergency or portable light source; (*3*) chemiluminescence provides means to detect and measure trace elements and pollutants for environmental control, or clinically important substances (eg, metabolites, specific proteins, cancer markers, hormones, DNA); and (*4*) classification of the bioluminescent relationship between different organisms defines their biological relationship and pattern of evolution.

Mechanism

The mechanism of chemiluminescence is still being studied and most mechanistic interpretations should be regarded as tentative. Nevertheless, most chemiluminescent reactions can be classified into (*1*) peroxide decomposition, including bioluminescence and peroxyoxalate chemiluminescence; (*2*) singlet oxygen chemiluminescence; and (*3*) ion radical or electron-transfer chemiluminescence, which includes electrochemiluminescence.

In principle, one molecule of a chemiluminescent reactant can react to form one electronically excited molecule, which in turn can emit one photon of light. Thus one mole of reactant can generate Avogadro's number of photons defined as one einstein (ein). Light yields can therefore be defined in the same terms as chemical product yields, in units of einsteins of light emitted per mole of chemiluminescent reactant. This is the chemiluminescence quantum yield Qc which can be as high as 1 ein/mol or 100%.

The theoretical yield is approached by the firefly reaction, discussed later, which is reported to have a Qc of 88% (5). In practice, however, most chemiluminescent reactions are inefficient, with Qc values on the order of 1% or much less. The factors influencing yields can be discussed in terms of a generalized three-step chemiluminescent mechanism (6):

$$\begin{array}{c} \text{chemiluminescent} \\ \text{reactant +} \\ \text{other reactants} \end{array} \xrightarrow{\textit{step 1}} \text{key intermediate} \xrightarrow{\textit{step 2}} \text{excited-state product} \xrightarrow{\textit{step 3}} \begin{array}{c} \text{ground-state} \\ \text{product} + h\nu \end{array}$$

As in any multistep process the overall yield Qc is the product of the yields of the separate steps. In most reactions, step 1 is subject to competitive side reactions which not only reduce the yield but can also obscure the nature of the chemiluminescent reaction itself. Step 2 requires a special chemistry which is discussed below. Step 3 is reasonably well understood from fluorescence and phosphorescence studies (7). Most molecules have low or negligible fluorescence quantum yields, Q_f, and phosphorescence is always inefficient in liquids. Thus efficient chemiluminescence requires a selective reaction producing a key intermediate and efficient conversion of the key intermediate to the singlet excited state of a highly fluorescent product. The yield of the second step is called the excitation yield and the product of the yields of the first two steps is the yield of excited state. Interest focuses on the chemiexcitation, step 2. In general, efficient excitation requires a large energy release in a single-reaction step (8) and a reaction pathway that promotes crossing of the ground-state potential energy surface to an electronically excited potential energy surface (4).

Energy Requirement. Visible light has an energy content of 167 kJ/ein (40 kcal/ein) (red) to 293 kJ/ein (70 kcal/ein) (blue), and an excited state radiating visible light must have that same energy with respect to its ground state. The excitation energy requirement is met by the sum of reaction enthalpy and activation energy. Few chemical reactions are sufficiently energetic. Moreover, release of the entire chemical energy requirement must occur in a single reaction step because excitation must be essentially instantaneous (8). Energetic two-step reactions where a part of the energy is released in each step cannot be chemiluminescent in solution because the energy released in the first step will be lost as vibrational energy to the solvent before the second step can raise the energy level to the excitation requirement.

The Chemiluminescent Pathway. Theory regarding the crossing of ground- to excited-state potential energy surfaces is incomplete; several potential criteria related to efficient chemiexcitation have been considered. First, since the energy released by a reaction can evolve as either vibrational or electronic excitation energy, small or rigid product molecules, which have relatively few vibrational degrees of freedom, should favor electronic excitation (6,9). Most likely, the conversion of substantial chemical energy to low energy vibrational excited states is a "forbidden" process analogous to the low probability of the transfer of excitation energy to vibrational energy when the energy gap between available electronic and vibrational quantum states is large. Thus a large energy release combined with a paucity of vibrational modes should favor electronic excitation and chemiluminescence.

Second, excited-state molecular geometry is often different from ground-state geometry. A reaction producing, eg, a bent carbonyl group, may favor chemiluminescence because the carbonyl excited state configuration is unfavorable compared to the planar ground state. Electronic excitation would then be preferred because it requires less molecular motion in the transition state (10,11). Orbital symmetry conservation and spin-orbit coupling may also be factors (12,13).

Third, singlet excitation, which is required for efficient chemiluminescence, may be favored over triplet excitation when the developing excited state is $\pi \rightarrow \pi^*$ rather than $n \rightarrow \pi^*$ (14). Finally, electron transfer between an anion radical–cation radical pair can produce a neutral excited-state–ground-state product pair, and it has been suggested that reactions of certain peroxides with electron-rich fluorescers can produce an ion-radical pair comprising the fluorescer cation radical and a carbonyl anion radical derived from the peroxide. Electron transfer within the solvent cage then provides the electronically excited fluorescer (15). Alternatively, it has been suggested that electron-rich fluorescers form charge-transfer complexes with such peroxides, and that reversal of charge during peroxide decomposition is related to fluorescer excitation (6,9).

Liquid-Phase Chemiluminescence

PEROXIDE DECOMPOSITION

In many chemiluminescent reactions of peroxides, two carbonyl groups are formed simultaneously by decomposition of an intermediate such as compound (**1**):

$$\text{(anhydride, } C_6H_5, C_6H_5) + H_2O_2 \longrightarrow \text{(1)} \longrightarrow [\text{(2)}]^* + 2\,CO_2 + H_2O$$

(**1**) (**2**)

In such reactions the substantial heat of the simultaneous (concerted) formation of the carbonyl groups produced meets the energy requirement (8,16). In the reaction shown (8), the product is the highly fluorescent excited state of 9,10-diphenylanthracene [*1499-10-1*] (**2**). It is not necessary for the new carbonyl groups to be a part of the structure of the excited product, only that the excited state be formed synchronously with two carbonyl groups.

1,2-Dioxetanes. Simple dioxetanes (**3**) decompose thermally near or below room temperature to generate excited states of carbonyl products (17).

$$\underset{(\mathbf{3})}{\begin{array}{c}\mathrm{O{-}O}\\ |\quad\;|\\ \mathrm{R_2C{-}CR_2}\end{array}} \longrightarrow \mathrm{R_2C{=}O} + [\mathrm{R_2C{=}O}]^*$$

Excitation appears to be general for this reaction but yields of excited products vary substantially with the substituent R. The highest yield reported is from tetramethyl-1,2-dioxetane [*35856-82-7*] (TMD) where the yield of triplet acetone is 50% of total acetone formed (18,19). Probably only one carbonyl of the two produced can be excited by the thermal decomposition, and TMD provides 100% of the possible yield of triplet acetone. Singlet excited acetone is also formed, but at the low yield of 0.1–0.3% (17–21). Other tetraalkyldioxetanes behave similarly to TMD (22).

Because the fluorescence and phosphorescence radiative yields from acetone are very low, chemiluminescence is weak, with a quantum yield of 1×10^{-6} ein/mol (21). Light emission increases significantly when a fluorescent acceptor, such as 9,10-diphenylanthracene, is added to trap the singlet excited state by energy transfer and provide an efficient singlet emitter, or when a triplet acceptor, such as 9,10-dibromoanthracene, is added to trap some of the triplet product (17,19,20,23,24). Neither fluorescer changes the decomposition rate. In the latter case the heavy bromine substituent permits moderately efficient energy transfer from triplet acetone to moderately fluorescent singlet dibromoanthracene by weakening the spin conservation rule through spin-orbit coupling. Even with the addition of fluorescent acceptors, however, chemiluminescence efficiencies remain low because of the inefficient transfer processes at attainable fluorescer concentrations. The highest chemiluminescence quantum yield from TMD in the presence of diphenylanthracene is about 0.1% (20).

Most other dioxetanes investigated provide lower triplet yields, but some provide higher yields of excited singlets. Tetramethoxy-1,2-dioxetane [*28793-21-7*] gives only one-third the triplet yield of TMD but gives a 1% yield of excited singlet dimethyl carbonate (20). The higher singlet yield of the methoxy derivative may relate to its higher heat of decomposition (−502 kJ/mol (−120 kcal/mol) vs −372 kJ/mol (−89 kcal/mol)) for TMD. Despite its higher heat of decomposition, the methoxy derivative is more stable than TMD with a half-life of 14 h at 80°C compared to 14 min for TMD (20). Because the activation energies for both are nearly the same (~ 117 kJ/mol (28 kcal/mol)), the methoxy derivative clearly has a low activation entropy which may also relate to the higher singlet yield.

The following reaction was reported to give an excited singlet yield of 0.9%. The singlet yield is increased substantially by silica gel, which also increases the reaction rate (25). Dioxetanes that can decompose to *N*-methylacridone have been reported to give excited singlet yields as high as 25% (25).

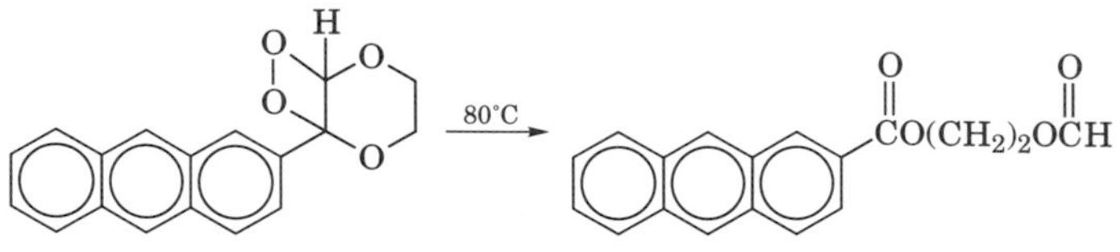

Formation of the carbonyl group does not appear to be concerted with O–O bond cleavage in 1,2-dioxetane decomposition, since replacement of methyl in the 3-position with phenyl, which would conjugate with the forming carbonyl group and stabilize it, does not change the activation energy (26). Singlet excitation, however, may involve concerted decomposition.

Yields of excited states from 1,2-dioxetane decomposition have been determined by two methods. Using a photochemical method (17,18) excited acetone from TMD is trapped with *trans*-1,2-dicyanoethylene (DCE). Triplet acetone gives *cis*-1,2-dicyanoethylene with DCE, whereas singlet acetone gives 2,2-dimethyl-3,4-dicyanooxetane. By measuring the yields of these two products the yields of the two acetone excited states could be determined. The yields of triplet ketone (**6**) from dioxetanes are determined with a similar technique.

In principle, the excitation energy would be expected to be distributed between ketones (**5**) and (**6**) in a ratio dependent on the substituent R, and the distribution would be expected to favor the ketone having the lowest triplet excitation energy.

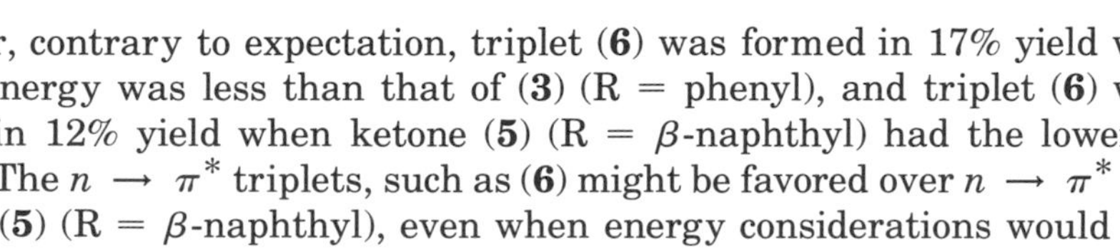

However, contrary to expectation, triplet (**6**) was formed in 17% yield when its triplet energy was less than that of (**3**) (R = phenyl), and triplet (**6**) was still formed in 12% yield when ketone (**5**) (R = β-naphthyl) had the lower triplet energy. The $n \rightarrow \pi^*$ triplets, such as (**6**) might be favored over $n \rightarrow \pi^*$ triplets, such as (**5**) (R = β-naphthyl), even when energy considerations would indicate the opposite (26).

Excited-state yields from several dioxetanes have been determined (19,20,24) by the chemiluminescent method (6). Singlet yields are determined by measuring the increasing chemiluminescence yields obtained on adding increasing concentrations of a singlet acceptor, such as 9,10-diphenylanthracene, and extrapolating the chemiluminescence yield to infinite fluorescer concentration where all of the singlet product is trapped. Triplet yields are determined similarly by adding the triplet acceptor 9,10-dibromoanthracene [*523-27-3*].

1,2-Dioxetanes were isolated in 1969 (27). Previously, they had been expected to be thermally unstable in view of their high decomposition energies of 270–500 kJ/mol (65–120 kcal/mol). Many are kinetically stable at room temperature with activation energies near 96 kJ (23 kcal), but most decompose below 80°C, and careful synthesis and storage are necessary. An exception is adamantylidineadamantane-1,2-dioxetane [*35544-39-9*] (**7**) which decomposes at about 140°C to give a 2% yield of singlet and 15% yield of triplet adamantanone (28).

1,2-Dioxetane, synthesized by photooxygenation of adamantylideneadamantane, demonstrates that two adamantyl groups do not sterically inhibit peroxide formation. The unsymmetrically substituted adamantyl 1,2-dioxetane (9-(2-adamantylidene)-*N*-methylacridan-1,2-dioxetane [*66762-85-4*]) (**8**) when heated, generates chemiluminescence attributed exclusively to the excited singlet state of *N*-methylacridone (29). It was proposed that dioxetanes that contain electron-donating substituents decompose by charged intermediates in an electron-transfer process (30). Subsequently, the syntheses and stabilities of several unsymmetrically substituted adamantyl 1,2-dioxetanes were described; the stabilization mechanism for these compounds is quite complex, depending on conformation isomerism and other factors (31). The application of chemically functionalized 1,2-dioxetanes as potentially useful chemotherapeutic agents has been proposed (32).

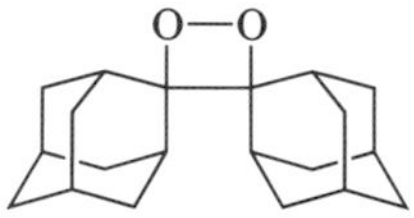

O—O

(**7**)

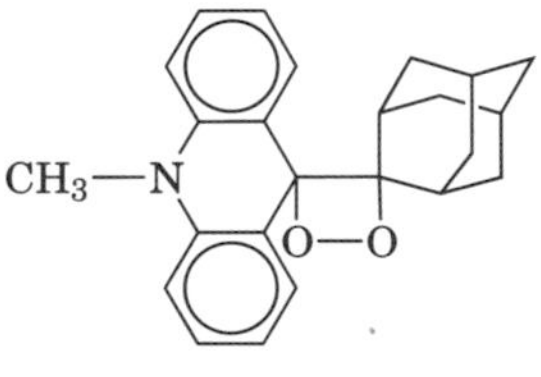

CH_3—N

O—O

(**8**)

Chemical off–on switching of the chemiluminescence of a 1,2-dioxetane (9-benzylidene-10-methylacridan-1,2-dioxetane [*66762-83-2*] (**9**)) was first described in 1980 (33). No chemiluminescence was observed when excess acetic acid was added to (**9**) but chemiluminescence was recovered when triethylamine was added. The off–on switching was attributed to reversible protonation of the nitrogen lone pair and modulation of chemically induced electron-exchange luminescence (CIEEL). Base-induced decomposition of a 1,2-dioxetane of 2-phenyl-3-(4′-hydroxyphenyl)-1,4-dioxetane (**10**) by deprotonation of the phenolic hydroxy group has also been described (34).

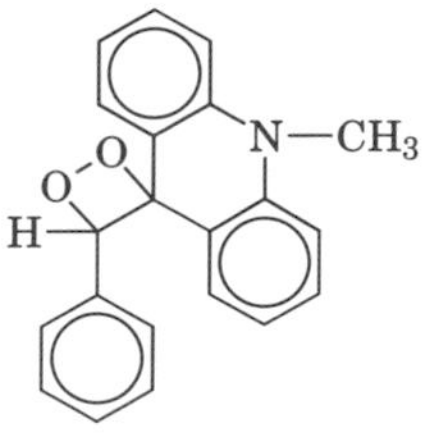

O—O N—CH_3 H

(**9**)

O O O–O OH

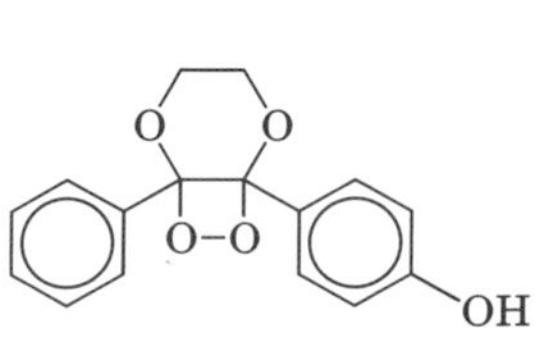

(**10**)

A range of adamantyl 1,2-dioxetanes have been synthesized that have enzyme (EH) cleavable groups (**11**, **14**, where X = 7-$OOCCH_3$, 6-$OOCCH_3$, or 7-OPO_3Na_2, and **15**) (35–37). Enzyme-catalyzed decomposition produces a metastable phenoxide intermediate (eg, **12**) which decomposes to produce light; the emitting moiety is the methyl 3-oxybenzoate anion (**13**). 5-Substituents (eg, Cl, Br, HO) (**15**) on the adamantyl ring increase the rate of light emission from the phenoxide intermediate most likely, in part, due to a hyperconjugation effect (38).

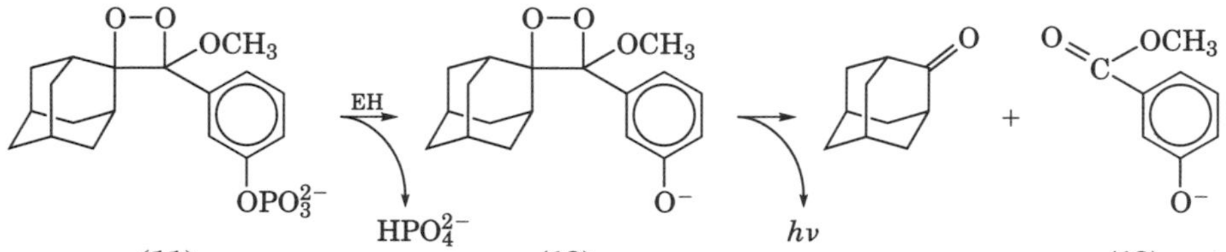

(**11**) (**12**) (**13**)

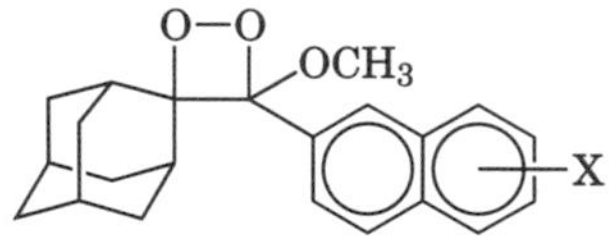

(**14**) 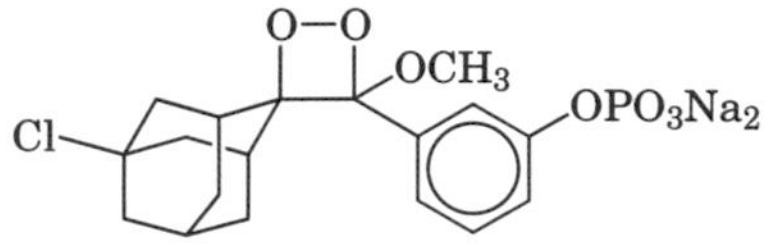 (**15**)

In addition to ready thermal decomposition, 1,2-dioxetanes are also rapidly decomposed by transition metals (39), amines, and electron-donor olefins (10). However, these catalytic reactions are not chemiluminescent as determined by the temperature drop kinetic method.

Ultraviolet light also initiates decomposition. Photolysis tends to give a higher yield of singlet products than thermolysis, with the singlet yield increasing with the energy of the exciting light. On the other hand, triplet sensitized photolysis gives exclusively triplet excited products (17). The triplet ketone from thermal decomposition also sensitizes dioxetane decomposition, except in the presence of a triplet quencher such as oxygen (19,24,40).

1,2-Dioxetanes are obtained from an α-halohydroperoxide by treatment with base (41), or reaction of singlet oxygen with an electron-rich olefin such as tetraethoxyethylene or 10,10′-dimethyl-9,9′-biacridan [*23663-77-6*] (**16**) (25,42).

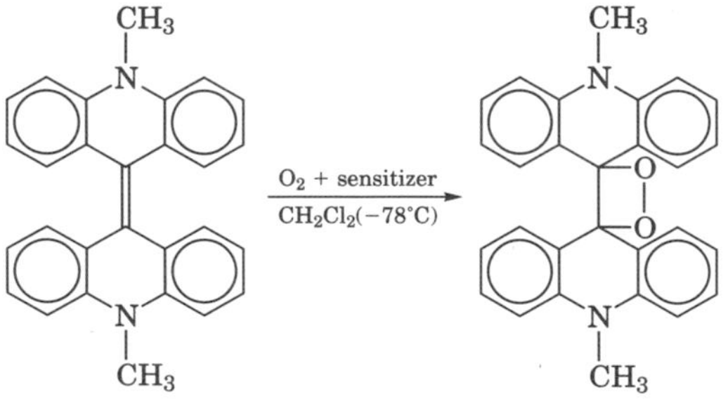

(**16**) (**17**)

A number of chemiluminescent reactions may proceed through unstable dioxetane intermediates (12,43). For example, the classical chemiluminescent reactions of lophine [*484-47-9*] (**18**), lucigenin [*2315-97-7*] (**20**), and transannular peroxide decomposition. Classical chemiluminescence from lophine (**18**), where $R = C_6H_5$, is derived from its reaction with oxygen in aqueous alkaline dimethyl sulfoxide or by reaction with hydrogen peroxide and a cooxidant such as sodium hypochlorite or potassium ferricyanide (44). The hydroperoxide (**19**) has been isolated and independently emits light in basic ethanol (45).

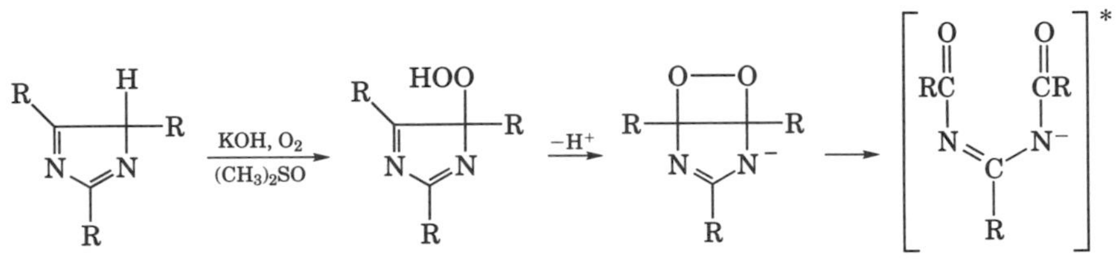

KOH, O_2 / $(CH_3)_2SO$; $-H^+$

(**18**) (**19**)

Classical chemiluminescence from lucigenin (**20**) is obtained from its reaction with hydrogen peroxide in water at a pH of about 10; *Qc* is reported to be about 0.5% based on lucigenin, but 1.6% based on the product *N*-methylacridone which is formed in low yield (46). Lucigenin dioxetane (**17**) has been prepared by singlet oxygen addition to an electron-rich olefin (**16**) at low temperature (47). Thermal decomposition of (**17**) gives a *Qc* of 1.6% (47).

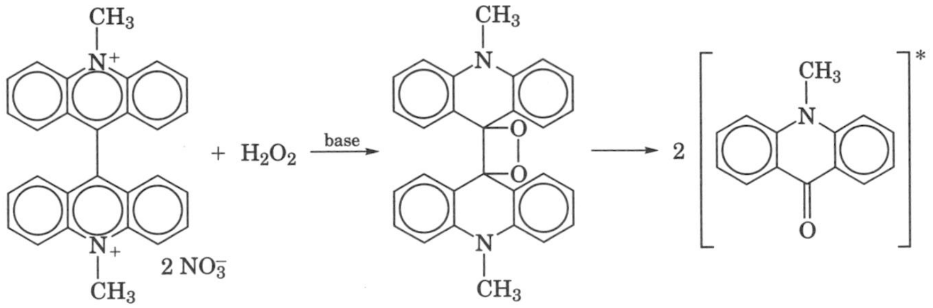

$+ H_2O_2$ base ; $2\ NO_3^-$; 2

(**20**) (**17**)

Tetrakis(dimethylamino)ethylene [*26640-54-0*] (TMAE) reacts spontaneously with oxygen to generate light (48–52) with TMAE itself being the emitting excited state. This is in agreement with an energy-transfer process from an excited state produced by dioxetane decomposition to a second molecule of TMAE (50). Although the reaction rate is first order in TMAE and oxygen, the chemiluminescence intensity and *Qc* are second order in TMAE, indicating that the second TMAE molecule is involved in a fast, nonrate-determining step. An alternative mechanism involving a chemical dimerization and regeneration of excited TMAE through decomposition of the dimer has also been suggested to account for these results (48). Chemiluminescence quantum yields ranging from

10^{-5} to 10^{-3} ein/mol have been reported under varying conditions (48–50). The reaction products are primarily tetramethylurea and tetramethyloxamide (51), which increasingly quench chemiluminescence as the reaction proceeds (49).

Although *Qc* for the reaction is low, a significant yield can be maintained at high TMAE concentrations, and moderate brightness and lifetime can be achieved. Several syntheses of TMAE and analogues have been described (52).

Dioxetane decomposition has also been proposed to account for chemiluminescence from other reactions (43), including gas-phase reactions of singlet oxygen with ethylene and vinyl ethers (53).

α-Peroxylactones (1,2-Dioxetanones). Alkyl-substituted 1,2-dioxetanones (**21**) are prepared using low temperature techniques (54,55). The α-hydroperoxy acids (**22**) can be prepared in high yield and cyclized to the dioxetanone (**21**) with dicyclohexylcarbodiimide in carbon tetrachloride at low temperatures.

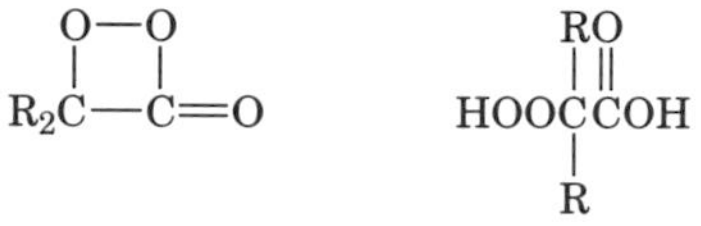

(**21**) (**22**)

Dioxetanones decompose near or below room temperature to aldehydes or ketones (56). The decomposition reactions are weakly chemiluminescent (*Qc* ca 10^{-7} ein/mol) because the products are poorly fluorescent. However, addition of 10^{-3} *M* rubrene provides a *Qc* ca 10^{-3} ein/mol, and a *Qc* on the order of 3–7% was calculated at rubrene concentrations above 10^{-2} *M* after correcting for yield loss factors (57). The decomposition rates are first order in (**21**) and are independent of added fluorescer at concentrations below 10^{-3} *M*, where *Qc* is about 10^{-3} ein/mol. At higher fluorescer concentrations, where *Qc* increases strongly, rubrene substantially increases the decomposition rate (57).

Long before 1,2-dioxetanones were isolated, they were proposed as key intermediates in bioluminescence (58–60). This idea led to the discovery of a number of new chemiluminescent reactions. For example, (**23**) reacts with H_2O_2 to give (**25**). The hydroperoxide (**24**) has been isolated and is independently chemiluminescent under basic conditions (43). The reaction is efficient with a *Qc* of 3% reported for (**23**) (X = Cl) (6). Discovery of (**23**) (X = Cl) was based independently on the dioxetanone (61) and concerted peroxide decomposition (6,8,62) theories. Possible examples of dioxetanones in bioluminescence are discussed later.

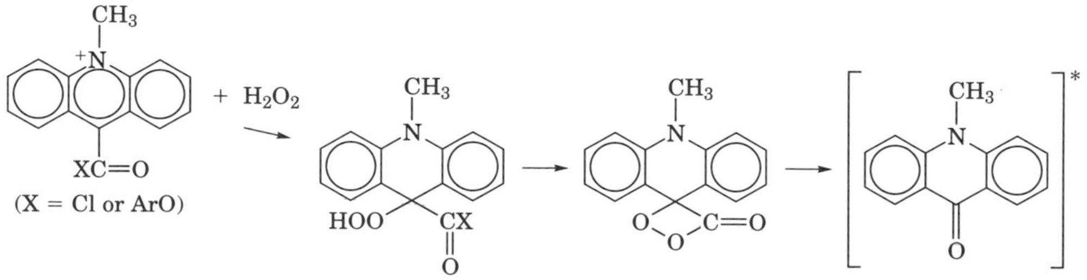

(X = Cl or ArO)

(**23**) (**24**) (**25**)

Peroxyoxalate. The chemical activation of a fluorescer by the reactions of hydrogen peroxide, a catalyst, and an oxalate ester has been the object of several mechanism studies. It was first proposed in 1967 that peroxyoxalate (**26**) was converted to dioxetanedione (**27**), a highly unstable intermediate which served as the chemical activator of the fluorescer (flr) (6,9).

$$\underset{(\mathbf{26})}{\mathrm{HOO{-}\overset{O}{\overset{\|}{C}}{-}\overset{O}{\overset{\|}{C}}{-}OAr}} \longrightarrow \underset{(\mathbf{27})}{\left[\text{dioxetanedione}\right]} \xrightarrow{\text{flr}} [\text{complex}] \longrightarrow 2\,CO_2 + \text{flr}^*$$

$$\text{flr}^* \longrightarrow \text{flr} + h\nu$$

Subsequent studies (63,64) suggested that the nature of the chemical activation process was a one-electron oxidation of the fluorescer by (**27**) followed by decomposition of the dioxetanedione radical anion to a carbon dioxide radical anion. Back electron transfer to the radical cation of the fluorescer produced the excited state which emitted the luminescence characteristic of the fluorescent state of the emitter. The chemical activation mechanism was patterned after the CIEEL mechanism proposed for dioxetanones and dioxetanes discussed earlier (65). Additional support for the CIEEL mechanism, was furnished by demonstration (66) that a linear correlation existed between the singlet excitation energy of the fluorescer and the chemiluminescence intensity which had been shown earlier with dimethyldioxetanone (67).

The first detailed investigation of the reaction kinetics was reported in 1984 (68). The reaction of bis(pentachlorophenyl) oxalate [*1173-75-7*] (PCPO) and hydrogen peroxide catalyzed by sodium salicylate in chlorobenzene produced chemiluminescence from diphenylamine (DPA) as a simple time–intensity profile from which a chemiluminescence decay rate constant could be determined. These studies demonstrated a first-order dependence for both PCPO and hydrogen peroxide and a zero-order dependence on the fluorescer in accord with an earlier study (9). Furthermore, the chemiluminescence quantum efficiencies (Qc) are dependent on the ease of oxidation of the fluorescer, an unstable, short-lived intermediate ($\tau = 0.5\ \mu s$) serves as the chemical activator, and such a short-lived species "is not consistent with attempts to identify a relatively stable dioxetane as the intermediate" (68).

The lack of independent evidence for dioxetanedione (**27**) (69) and later results (66,68) have diminished the likelihood that (**27**) plays any significant role in the chemical excitation process and attention has been redirected to peroxyoxalate (**26**) and its isomers. More recent studies suggest that more than one intermediate may be required (70); ie, a pool of intermediates has been suggested.

HO, O
ArO—C—C
(26) ⇌ | |
O—O
⇅ ⇅ (28) ——→ 2 CO_2 + HOAr
flr flr*
HO OH $h\nu$
ArO—C—C—OAr
| |
O—O

"pooled intermediates"

Time–intensity emission profiles have provided quantitative mechanistic information on the effects of the oxalate structure, catalyst, oxidant, and solvent on the rate of reaction. For the bis(2,4,6-trichlorophenyl) oxalate [*1165-91-9*] (TCPO), the profile changes from a complex bimodal time–intensity emission curve to a simple rise and fall of the chemiluminescence emission as the concentration of the catalyst is increased (70) (Fig. 1).

The bimodal profile observed at low catalyst concentration has been explained by a combination of two light generating reactive intermediates in equilibrium with a third dark reaction intermediate which serves as a way station or delay in the chemiexcitation processes. Possible candidates for the three intermediates include those shown as "pooled intermediates". At high catalyst concentration or in imidazole-buffered aqueous-based solvent, the series of intermediates rapidly attain equilibrium and behave kinetically as a single kinetic entity, ie, as pooled intermediates (71). Under these latter conditions, the time–intensity profile (Fig. 2) displays the single maximum as a biexponential rise and fall

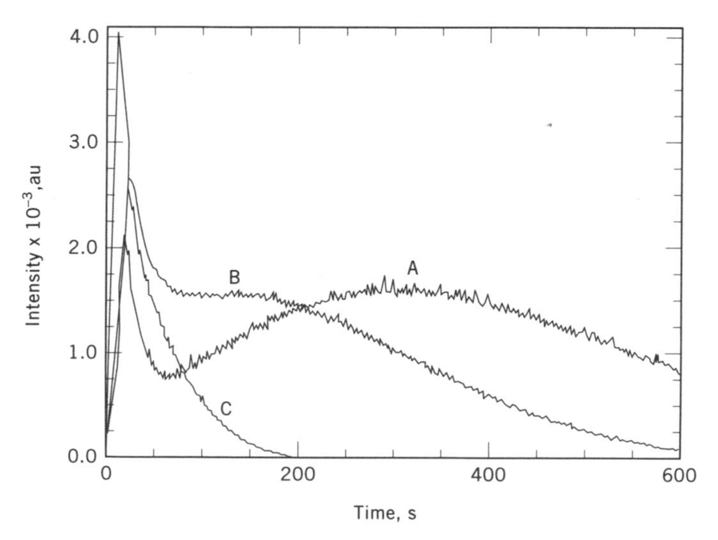

Fig. 1. Time courses of the chemiluminescence intensity from oxalate–hydrogen peroxide systems in ethyl acetate as solvent, 0.7 mM TCPO. The curves correspond to the following concentrations of triethylamine (TEA) catalyst: A, 0.05 mM; B, 0.10 mM; and C, 0.20 mM (70).

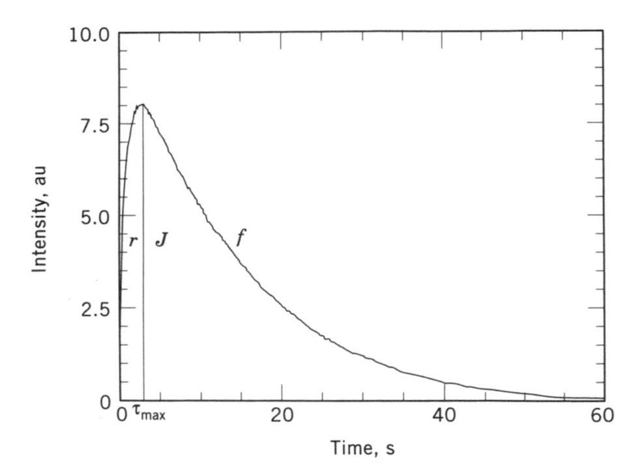

Fig. 2. Schematic of the fraction of chemiluminescence time profile observed in a flowing stream detector (71). See text.

of the intensity which is readily modeled as a typical irreversible, consecutive, unimolecular process:

$$A \xrightarrow{r} B \xrightarrow{f} zC$$

The kinetic expression for the time–intensity profile (I_t vs t) for this model is given by the following

$$I_t = d(h\nu)/dt = [Mr/(f - r)][e^{-rt} - e^{-ft}]$$

where I_t is the intensity at time t, M is the theoretical maximum intensity, and f and r are the first-order rise and fall rate constants. This model has been used to determine the effects of structure, temperature, solvent, and catalyst on the rates of reaction (72). The kinetic model has subsequently been useful in determining the quantum efficiencies $Qc = \int I_t dt = M/f$ and the optimum conditions and substrates necessary to maximize light yields for analytical applications, eg, $\tau_{max} = [\ln(f/r)]/(f - r)$ and $J = M(r/f)$ (71–75).

Peroxyoxalate chemiluminescence is the most efficient nonenzymatic chemiluminescent reaction known. Quantum efficiencies as high as 22–27% have been reported for oxalate esters prepared from 2,4,6-trichlorophenol, 2,4-dinitrophenol, and 3-trifluoromethyl-4-nitrophenol (6,76,77) with the fluorescers rubrene [*517-51-1*] (78,79) or 5,12-bis(phenylethynyl)naphthacene [*18826-29-4*] (79). For most reactions, however, a quantum efficiency of 4% or less is more common with many in the range of 10^{-4} to 10^{-8} ein/mol (80). The inefficiency in the chemiexcitation process undoubtedly arises from the transfer of energy of the activated peroxyoxalate to the fluorescer. The inefficiency in the CIEEL sequence derives from multiple side reactions available to the reactive intermediates in competition with the excited state producing back-electron transfer process.

Highly efficient peroxyoxalate chemiluminescence requires oxalic acid derivatives with good leaving groups (6,9,76,81). This feature may facilitate the closure of the peroxyoxalate to the putative dioxetanone intermediate (**28**). Efficient chemiluminescent oxalates include electronegatively substituted aromatic and aliphatic esters (9,76–82), amides (83–85), sulfonamides (83,86), *O*-oxalyl hydroxylamine derivatives (87), mixed oxalcarboxylic anhydrides (88), and oxalyl chloride (89). Among the amides, bis(1-(1-*H*)-2-pyridonyl)glyoxal is particularly efficient with a quantum yield of 17% reported (84).

Because the fluorescer is independent of the key intermediate, a variety of fluorescers can be used to provide emission spectra encompassing the visible (78,79,90,91) and near infrared regions (92). Excitation yields generally decrease as the excitation energy of the fluorescer increases and *Qc* for higher energy (blue) fluorescers tends to be relatively low (6). The chemiluminescence efficiency for the peroxyoxalate reaction generally increases with decreasing singlet energy or oxidation potential of the fluorescer between E_{ox} values of 1.2 eV (DPA) and 0.87 eV (rubrene) vs standard calomel electrode (93).

Most peroxyoxalate chemiluminescent reactions are catalyzed by bases and the reaction rate, chemiluminescent intensity, and chemiluminescent lifetime can be varied by selection of the base and its concentration. Weak bases such as sodium salicylate or imidazole are generally preferred (94). Alternatively, weak acids and certain salts have been found to extend the lifetimes of inherently rapid reactions which occur with highly reactive esters, such as bis(2,4-dinitrophenyl) oxalate (95). A chemiluminescent demonstration based on the oxalic ester reaction has been described (96) and the reaction has been developed into a practical lighting system.

Several new oxalates have been developed for use in analytical applications. Bis(2,6-difluorophenyl) oxalate (72) and bis(4-nitro-2-(3,6,9-trioxadecylcarbonyl)phenyl) oxalate (97) have been used in flow injection and high performance liquid chromatography (hplc) as activators for chemiluminescence detectors. These oxalates are generally more stable and show better water solubility in mixed aqueous solvents yet retain the higher efficiencies (*Qc*) of the traditional oxalates employed for chemiluminescence.

A number of other chemiluminescent reactions appear to be related to peroxyoxalate chemiluminescence although their mechanistic details may vary. For example, various chlorinated esters and ethers react with H_2O_2 and a fluorescer to emit light (98–101). Other examples have been given (8,102,103).

Peroxyoxalate chemistry has been used to carry out photochemical reactions but does not appear to produce triplet excited states (91).

Luminol (Phthalhydrazide). Chemiluminescence from luminol [*521-31-3*] (3-aminophthalhydrazide) (**29**), isoluminol [*3682-14-2*] (4-aminophthalhydrazide), and analogues has been studied extensively (104–106).

(**29**) —oxidant, base→ (**30**) —H_2O_2, base→ [(**31**)] → [(**32**)]* + N_2

Reaction takes place in aqueous solution with hydrogen peroxide and catalysts such as Cu(II), Cr(III), Co(II), ferricyanide, hemin, or peroxidase. Chemiluminescent reaction also takes place with oxygen and a strong base in a dipolar aprotic solvent such as dimethyl sulfoxide. Under both conditions *Qc* is about 1% (light emission, 375–500 nm) (105,107).

The mechanism appears to follow the equation above (105,108). Dianion (**32**) has been shown to be the emitting fluorescer (109), and reaction of luminol (**29**) with oxygen-18 in KOH–dimethylsulfoxide produced one labeled oxygen in each carboxylate group (110). A kinetic study of the reaction of (**29**) with aqueous alkaline persulfate indicated a one-step, two-electron oxidation of the mono anion of structure (**29**) to the azoquinone (**30**) (99), and the presence of structure (**30**) has been demonstrated during the chemiluminescent reaction (111). Compound (**30**) and several analogues have been synthesized (112–113) and have been shown to be chemiluminescent under luminol conditions. A charge-transfer mechanism has been proposed (15).

A substantial effort has been applied to increasing *Qc* by structural modification (114), eg, the phthalazine-1,4-diones (**33**) and (**34**) which have chemiluminescence quantum yields substantially higher than luminol (115,116). The fluorescence quantum yield of the dicarboxylate product from (**34**) is 14%, and the yield of singlet excited state is calculated to be 50% (116). Substitution of the 3-amino group of luminol reduces the CL efficiency >10-fold, whereas the opposite effect occurs with the 4-amino isomer (117). A series of pyridopyridazine derivatives (**35**) have been synthesized and shown to be more efficient than luminol (118).

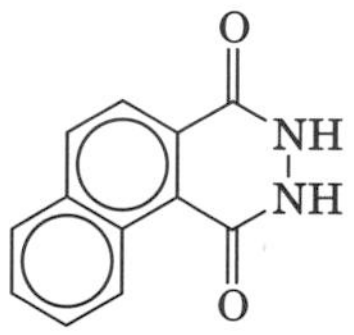

(**33**) 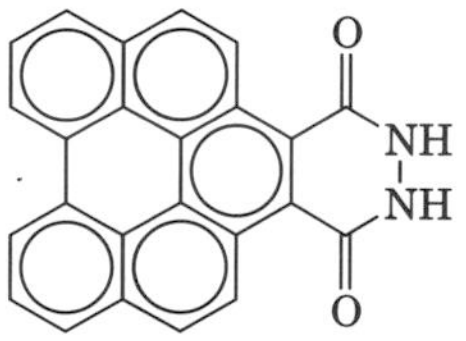 (**34**) 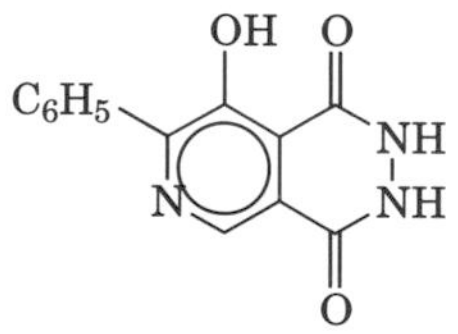 (**35**)

The emission yield from the horseradish peroxidase (HRP)-catalyzed luminol oxidations can be increased as much as a thousandfold upon addition of substituted phenols, eg, *p*-iodophenol, *p*-phenylphenol, or 6-hydroxybenzothiazole (119). Enhanced chemiluminescence, as this phenomenon is termed, has been the basis for several very sensitive immunometric assays that surpass the sensitivity of radioassay (120) techniques and has also been developed for detection of nucleic acid probes in dot-slot, Southern, and Northern blot formats (121).

It is suggested that the phenol serves as a one-electron source in the enhanced chemiluminescence process, replacing one of the functions of HRP and thereby protecting the HRP from degradation. The quantum efficiencies for light production are not increased by the presence of the *p*-iodophenol; however, the length of effective emission is greatly extended from a few seconds to an hour or more, and the background emission that occurs in the absence of HRP is decreased (119,120).

Organometallics. Arylmagnesium halides, especially bromides, react with oxygen in ether to generate light in the range 330–360 nm (reaction enthalpy 378 kJ/mol) (122,123). *p*-Chlorophenylmagnesium bromide [*873-77-8*] is the most efficient with Qc ca 10^{-4} ein/mol (122). ArOOMgX is probably an intermediate and electron spin resonance (esr) studies indicate the presence of free-radical species (122,123). The emitting species are brominated biphenyls (123), and since these are only weakly fluorescent, the excitation yield must be high. Alkyl Grignard reagents are weakly chemiluminescent in reaction with oxygen (124). Weak chemiluminescence is also seen in reactions of aryl Grignard reagents with benzoyl peroxide (125) and nitro compounds (126).

Lithium diphenylphosphide [*4541-02-2*] and related organophosphides are chemiluminescent in reaction with oxygen (127). Chemiluminescence is observed from the solid phosphides.

Autooxidation. Liquid-phase oxidation of hydrocarbons, alcohols, and aldehydes by oxygen produces chemiluminescence in quantum yields of 10^{-8} to 10^{-10} ein/mol (128–130). Although the efficiency is low, the chemiluminescent reaction is important because it provides an easy tool for study of the kinetics and properties of autooxidation reactions including industrially important processes (128,131). The light is derived from combination of peroxyl radicals (132), which are primarily responsible for the propagation and termination of the autooxidation chain reaction. The chemiluminescent termination step for secondary peroxy radicals is as follows:

$$2\,R_2CHOO\cdot \;\longrightarrow\; R_2C\overset{O-O\cdots O}{\underset{H\cdots O-CHR_2}{<}} \;\longrightarrow\; [R_2CO]^* + O_2 + HOCHR_2$$

(**36**) (**37**)

Because ground-state oxygen is a triplet, spin conservation during the decomposition of the transition state (37) must lead to an excited triplet-state ketone or (excited) singlet oxygen. The emitting species has been found to be the triplet excited state of the carbonyl product (132), although singlet oxygen has also been detected.

Because the chemiluminescence intensity can be used to monitor the concentration of peroxyl radicals, factors that influence the rate of autooxidation can easily be measured. Included are the rate and activation energy of initiation, rates of chain transfer in cooxidations, the activities of catalysts such as cobalt salts, and the activities of inhibitors (128).

Tertiary peroxyl radicals also produce chemiluminescence although with lower efficiencies. For example, the intensity from cumene autooxidation, where the peroxyl radical is tertiary, is a factor of 10 less than that from ethylbenzene (132). The chemiluminescent mechanism for cumene may be the same as for secondary hydrocarbons because methylperoxy radical combination is involved in the termination step. The primary methylperoxyl radical terminates according to the chemiluminescent reaction just shown for (**36**), ie, R = H.

Addition of fluorescent energy acceptors such as 9,10-dibromoanthracene substantially increases chemiluminescence intensity by transferring excitation energy (132,133), as is the case with dioxetanes.

Other Oxidation Reactions. Dihydrophthaloyl cyclic peroxides such as (**38**) appear to be efficient chemiluminescent key intermediates (134). Relatively weak chemiluminescence obtained from the excited singlet state of *p*-terphenyl (**39**) is strongly intensified by addition of an energy accepting fluorescer, such as 9,10-diphenylanthracene or perylene, which changes the emission spectrum to that of the added fluorescer. By determining *Qc* as a function of perylene concentration, the yield of excited singlet (**39**) was estimated to be about 1.5%. The triplet yield was estimated by a similar technique to be 35%. After correction for side reactions, the yield of excited (**39**) from (**38**) was estimated to be about 60%. It was pointed out that concerted decomposition of (**38**) by a sterically favored but electronically forbidden $2s + 2a + 2s$ path is predicted by orbital symmetry theory to produce excited (**39**) rather than the normally expected ground state.

$+ H_2O_2 \xrightarrow[THF]{R_3N}$ (**38**) $\longrightarrow$ (**39**)* $+ 2\,CO_2$

Decomposition of diphenoylperoxide [*6109-04-2*] (**40**) in the presence of a fluorescer such as perylene in methylene chloride at 24°C produces chemiluminescence matching the fluorescence spectrum of the fluorescer; *Qc* with perylene was reported to be 10 ± 5% (135). The reaction follows pseudo-first-order kinetics with the observed rate constant increasing with fluorescer concentration according to $k_{obs} = k_1 + k_2[\text{flr}]$. Thus the fluorescer acts as a catalyst for peroxide decomposition, with catalytic decomposition competing with spontaneous thermal decomposition. An electron-transfer mechanism has been proposed (135).

(**40**) + flr $\longrightarrow$ $+ CO_2 + [\text{flr}]^*$

Weak to moderate chemiluminescence has been reported from a large number of other liquid-phase oxidation reactions (1,128,136). The list includes reactions of carbenes with oxygen (137), phenanthrene quinone with oxygen in alkaline ethanol (138), coumarin derivatives with hydrogen peroxide in acetic acid (139), nitriles with alkaline hydrogen peroxide (140), and reactions that produce electron-accepting radicals such as HO· in the presence of carbonate ions (141). In the latter, exemplified by the reaction of iron(II) with H_2O_2 and $KHCO_3$, the carbonate radical anion is probably a key intermediate and may account for many observations of weak chemiluminescence in oxidation reactions.

SINGLET OXYGEN

The electronically excited singlet state of oxygen (142) can be produced by passing ground-state (triplet) oxygen through a microwave discharge (143,144), by reaction of hydrogen peroxide with hypochlorite ion (145), by energy transfer from triplet excited states formed by irradiation to ground-state oxygen (146), and by low temperature thermal decomposition of the triphenyl phosphite–ozone complex (147). Two singlet states are produced: $^1\Delta g$, a relatively long-lived, low-energy (92 kJ, 22 kcal) state with all its electrons paired, and $^1\Sigma g^+$, a short-lived, higher energy (159 kJ, 38 kcal) state with two singly occupied orbitals (148). Both singlet states are chemiluminescent in the gas phase (148). Emission from $^1\Delta g$ occurs principally at 1269, 634, and 703 nm with the latter two bands derived from simultaneous decay of two $^1\Delta g$ molecules. Infrared emission from $^1\Sigma g^+$ is observed at 762 nm. Green emission is observed at 478 nm from simultaneous decay of a $^1\Delta^- g^1\Sigma g^+$ pair. The red band at 634 nm is prominently visible in the aqueous hydrogen peroxide–hypochlorite reaction where the $^1\Delta g$ oxygen pair emits from the bubbles of product oxygen (148). The intensity of the red band is proportional to the square of the Δg oxygen concentration as would be expected for a double-molecule decomposition (148,149).

Chemiluminescence from $^1\Delta g$ oxygen can be strong at high concentrations, but addition of 5×10^{-4} M violanthrone increases the intensity 100-fold (150). The emission spectrum, centered at 630 nm, is identical to violanthrone fluorescence and the intensity remains proportional to the square of the $^1\Delta g$ concentration. Energy pooling of two $^1\Delta g$ molecules is clearly also required for violanthrone excitation, but it appears that a different process is involved than in singlet oxygen emission itself. Neither singlet oxygen emission nor violanthrone fluorescence are quenched by oxygen, but the $^1\Delta g$-violanthrone chemiluminescence is strongly oxygen quenched. It was suggested that one $^1\Delta g$ molecule excites violanthrone to its triplet state and that the triplet is excited further to the singlet by a second $^1\Delta g$ molecule in a subsequent step (143,150). Chemiluminescence from singlet oxygen and rubrene (151) probably involves the same mechanism.

Most likely singlet oxygen is also responsible for the red chemiluminescence observed in the reaction of pyrogallol with formaldehyde and hydrogen peroxide in aqueous alkali (152). It is also involved in chemiluminescence from the decomposition of secondary dialkyl peroxides and hydroperoxides (153), although triplet carbonyl products appear to be the emitting species (132).

ELECTRON-TRANSFER CHEMILUMINESCENCE

Electron-transfer reactions appear to be inherently capable of producing excited products when sufficient energy is released (154–157). This ability may be related to the speed of electron transfer, which is fast relative to atomic motion, so that vibrational excitation is inhibited (158).

Examples include luminescence from anthracene crystals subjected to alternating electric current (159), luminescence from electron recombination with the carbazole free radical produced by photolysis of potassium carbazole in a frozen glass matrix (160), reactions of free radicals with solvated electrons (155), and reduction of ruthenium(III)tris(bipyridyl) with the hydrated electron (161). Other examples include the oxidation of aromatic radical anions with such oxidants as chlorine or benzoyl peroxide (162,163), and the reduction of 9,10-dichloro-9,10-diphenyl-9,10-dihydroanthracene with the 9,10-diphenylanthracene radical anion (162,164). Many other examples of electron-transfer chemiluminescence have been reported (156,165).

Stable anion radicals are easily prepared from aromatic hydrocarbons, eg, 9,10-diphenylanthracene, by electrochemical reduction in acetonitrile or dimethylformamide-containing electrolytes such as tetrabutylammonium perchlorate. Reversal of electrode polarity generates cation radicals from the hydrocarbon, and their reaction with the anion radical reservoir generates chemiluminescence (154,155,166). More simply, an alternating current may be used so that cation and anion radicals are continuously formed and annihilated to produce light and regenerate the original hydrocarbon (167). When hydrocarbons with stable ion radicals are used and impurities reactive with ion radicals are rigorously excluded, long-lasting electrochemiluminescence can be achieved. The oxidation–reduction potentials of the hydrocarbon can be determined by cyclic voltammetry and the stabilities of the ion radicals assessed (166).

Electrochemiluminescence is somewhat complicated in that three processes can produce light, depending on the energy released by the electron-transfer process and the excitation energy of the aromatic hydrocarbon (154,155,166). In each case a charge-transfer complex between the oppositely charged radicals is probably formed. If sufficient energy is available, the complex can dissociate to one ground-state molecule and one excited singlet molecule, and luminescence is relatively efficient. If only enough energy is available for triplet excitation, a triplet excited state results that can produce excited singlets by triplet–triplet annihilation. If insufficient energy is released even for triplet excitation, luminescence can still be produced by excimer (excited dimer) emission from the complex itself (168). In the first two cases, the luminescence spectrum matches the normal fluorescence spectrum of the hydrocarbon, whereas in the latter case typical, red-shifted, broad-band excimer emission results. Excitation energy transfer from an excimer produced by electrochemiluminescence to a europium chelate has been reported to produce narrow band europium emission (169).

Electron-transfer reactions producing triplet excited states can be diagnosed by a substantial increase in luminescence intensity produced by a magnetic field (170). The intensity increases because the magnetic field reduces quenching of the triplet by radical ions (157).

Under optimum conditions electron transfer can produce excited states efficiently. Triplet fluoranthrene was reported to be formed in nearly quantitative yield from reaction of fluoranthrene radical anion with the 10-phenylphenothiazine radical cation (171), and an 80% triplet yield was indicated for electrochemiluminescence of fluoranthrene by measuring triplet sensitized isomerization of *trans*- to *cis*-stilbene (172).

Electrochemiluminescence quantum yields of 8–10% from 9,10-diphenylanthracene and 14–20% from the 9,10-diphenylanthracene anion–thianthrene cation combination have been reported using the rotating ring disk electrode technique (157,173).

Gas-Phase Chemiluminescence

Gas-phase chemiluminescence is illustrated by the classic sodium–chlorine cool flame (174):

$$Na + Cl_2 \longrightarrow NaCl + Cl\cdot$$

$$Cl\cdot + Na_2 \longrightarrow NaCl + [Na]^*$$

Intense sodium D-line emission results from excited sodium atoms produced in a highly exothermic step (175). Many gas-phase reactions of the alkali metals are chemiluminescent, in part because their low ionization potentials favor electron transfer to produce intermediate charge-transfer complexes such as $[Cl^- \cdot Na_2^+]$ (176). There appears to be an analogy with solution-phase electron-transfer chemiluminescence in such reactions.

Excitation energy can be provided by kinetic (translational) and vibrational energies as well as from reaction enthalpy as demonstrated by molecular beam experiments. An alkali metal vapor can be accelerated to energies above 20 eV and passed through a chamber containing a substrate gas at low temperature. The cross section for reaction can then be determined, as a measure of the probability for an excitation interaction, as a function of the beam energy and the substrate structure (176). Such reactions between an alkali metal M and substrate S involve no net chemical change and simply convert kinetic energy to light via excited alkali metal atoms.

$$M + S \rightleftharpoons [M^+ \cdot S^-] \longrightarrow [M]^* + S$$

The importance of electron transfer was indicated by the threshold translational energy required to excite substrates with high electron affinities which was lower than, eg, that required for noble gas substrates (177).

Investigation of excitation probability as a function of substrate structure revealed that the reaction was efficient for such gases as N_2, NO, CO, O_2, SO_2, and olefinic and aromatic hydrocarbons (176). Such molecules contain a vacant, weakly antibonding orbital which can accept electron transfer. The bond length in the charge-transfer complex intermediate is short and a substantially higher energy (ca 3 eV) is required to produce the free ions $M^+ + S^-$ than to excite the alkali metal (1.6 eV for potassium and 2.1 eV for sodium). Thus strong chemiluminescence is produced at kinetic energies between 1.6 and 3 eV in the reaction between potassium and SO_2; however, the chemiluminescence weakens as ions are formed above 3 eV. A second set of substrates, including H_2, HCl,

Cl_2, and saturated hydrocarbons is inefficient in exciting alkali metals because the lowest lying vacant orbitals are strongly antibonding, σ bond lengths in the intermediates are long, and the energy required for free-ion formation is only slightly above the energy required for excitation (176,178).

A study of interaction of sodium atoms with vibrationally excited nitrogen molecules at the intersection of two gas beams shows that conversion of vibrational energy to electronic excitation is substantially more efficient than conversion of translational energy (179). This has also been indicated in other reactions (180).

The use of molecular and atomic beams is especially useful in studying chemiluminescence because the results of single molecular interactions can be observed without the complications that arise from preceding or subsequent energy-transfer collisions. Such techniques permit determination of active vibrational states in reactants, the population distributions of electronic, vibrational, and rotational excited products, energy thresholds, reaction probabilities, and scattering angles of the products (181).

A number of chemiluminescent reactions have been studied by producing key reactants through pulsed electric discharge, by microwave dissociation, or by observing the reactions of atoms and free radicals produced in the inner cone of a laminar flame as they diffuse into the flame's cool outer cone (182,183). These are either combination reactions or atom-transfer reactions involving transfer of chlorine (184) or oxygen atoms (181,185–187), the latter giving excited oxides.

The rates and chemiluminescent intensities of atom-transfer reactions are proportional to the concentrations of the reactants, but the intensity is inversely proportional to the concentration of inert gas present. The latter quenches the excited state through collision with an efficiency dependent on the structure of the inert gas. Chemiluminescence Qc increases with temperature, indicating that excitation has a higher activation energy than the ground state reaction (183). Such reactions generally provide banded, but very broad, emission spectra.

Electronic excitation from atom-transfer reactions appears to be relatively uncommon, with most such reactions producing chemiluminescence from vibrationally excited ground states (188–191). Examples include reactions of oxygen atoms with carbon disulfide (190), acetylene (191), or methylene (190), all of which produce emission from vibrationally excited carbon monoxide. When such reactions are carried out at very low pressure (13 mPa (10^{-4} torr)), energy transfer is diminished, as with molecular beam experiments, so that the distribution of vibrational and rotational energies in the products can be discerned (189). Laser emission at 5 μm has been obtained from the reaction of methylene and oxygen initiated by flash photolysis of a mixture of SO_2, C_2H_2, and SF_6 (186).

Recombination reactions are highly exothermic and are inefficient at low pressures because the molecule, as initially formed, contains all of the vibrational energy required for redissociation. Addition of an inert gas increases chemiluminescence by removing excess vibrational energy by collision (192,193). Thus in the nitrogen afterglow chemiluminescence efficiency increases proportionally with nitrogen pressure at low pressures up to about 33 Pa (0.25 torr) (194). However, inert gas also quenches the excited product and above about 66 Pa (0.5 torr) the two effects offset each other, so that chemiluminescence intensity becomes independent of pressure (192,195).

White Phosphorus Oxidation. Emission of green light from the oxidation of elemental white phosphorus in moist air is one of the oldest recorded examples of chemiluminescence. Although the chemiluminescence is normally observed from solid phosphorus, the reaction actually occurs primarily just above the surface with gas-phase phosphorus vapor. The reaction mechanism is not known, but careful spectral analyses of the reaction with water and deuterium oxide vapors indicate that the primary emitting species in the visible spectrum are excited states of $(PO)_2$ and HPO or DPO. Ultraviolet emission from excited PO is also detected (196).

Solid-Phase Chemiluminescence

Siloxene. Siloxene, obtained from reaction of calcium silicide with hydrochloric acid (197), is a yellow polymer with the basic formula $(Si_6H_6O_3)_n$. The silicon atoms are arranged in hexagonal rings joined in a laminar polymeric structure by oxygen atoms (198). The basic structure appears to be substituted randomly by hydroxyl and chlorine groups that affect the chemiluminescence spectrum and efficiency.

Siloxene is fluorescent and red chemiluminescence results from oxidation with ceric sulfate, chromic acid, potassium permanganate, nitric acid, and several other strong oxidants. The chemiluminescence spectrum peaks at 600 nm and has been reported (199) to give a maximum brightness of $3.43\ cd/m^2$ (1 footlambert).

Bioluminescence

Bioluminescence is characteristic of numerous marine and a few land organisms ($\sim$ 666 genera from 13 phyla) (2,10,14,56,200,201), extending from single-cell microorganisms such as bacteria and dinoflagellates, to marine vertebrates, such as the hatchet fish. Certain fish, such as the flashlight fish which has a light organ under its eyes, use photobacteria symbiotically to generate light (202). Marine bioluminescence includes sponges, worms, crustaceans, corals, snails, squids, clams, shrimp, and jellyfish. Bioluminescent land species include fungi, centipedes, millipedes, worms, beetles, and fireflies (Table 1).

Bioluminescence functions in mating (fireflies, the Bahama fireworm), in the search for prey (angler fish, *Photinus* fireflies), camouflage (hatchet fish, squid), schooling (euphausiid shrimp), and to aid deep water fish (flashlight fish, *Photoblepharon*) to see in the dark ocean depths.

The intensity of bioluminescence emission is $>2 \times 10^9$ photon/s $\cdot$ cm^2 in the dinoflagellate *Gonyaulax*, and the spectrum of light emission ranges from 450–490 nm (blue) in deep sea species, 490–520 nm (green) in coastal water species, and 510–580 nm (yellow-green) in terrestrial and freshwater species.

The chemistry of bioluminescence is complex and in general the reactions involve oxygen, a luciferin and a luciferase enzyme (eg, firefly luciferase) or a photoprotein (eg, apoaequorin) (2). The reactants can be isolated from the organisms, and bioluminescence can be demonstrated and studied *in vitro*. Structures and composition of luciferins (compare structures (**41**) and (**44**)) and luciferases

Table 1. Bioluminescent Organisms

Organism	Common name	Wavelength[a] of light emission, nm
	Marine	
Aequorea aequorea	jelly fish	500–523
Agyropelecus affinis	hatchet fish	ca 480
Gonyaulax polyedra	dinoflagellate	479
Malacosteus niger	stomiatoid fish	469–702
Mneniopsis leidyi	sea comb	485
Obelia geniculata	sea fir	475
Pholas dactylus	piddock	490
Photoblepheron palpebratus	flashlight fish	490
Renilla reniformis	sea pansy	480
Vargula hilgendorfi	sea firefly	460
Vibrio fischeri	marine bacterium	489
	Terrestrial	
Arachnocampa luminosa	glow worm	
Diplocardia longa	earthworm	500
Photinus pyralis	firefly	530–590
Pleurotus japonicus	moon night mushroom	524
Quantula striata	land snail	

[a]Maximum range.

(Table 2) from different species exhibit considerable diversity. The genes for firefly (*luc*) and marine bacterial luciferase (*lux*) were cloned in the 1980s (203,204) and subsequently there has been an explosive growth in knowledge of the molecular biology of bioluminescence. Photoprotein genes (*phot*) for apoaequorin have also been cloned. In some cases the genes are being used to produce recombinant proteins (firefly luciferase, apoaequorin) and fusion proteins (apoaequorin-IgG) for analytical purposes (205).

Most studies have been carried out with the American firefly (*Photinus pyralis*), the click beetle (*Pyrophorus plagiophthalamus*), the crustacean *Vargula*

Table 2. Luciferases and Photoproteins

Organism	Protein (gene)	M_r subunits, kDa
	Marine	
Aequorea victoria	apoaequorin (*phot*)	ca 20
Gonyaulax polyedra	luciferase (*luc*)	420[a]
Obelia geniculata	apoobelin (*phot*)	ca 20
Renilla reniformis	luciferase (*phot*)	35
Vargula hilgendorfi	luciferase (*luc*)	ca 60 (6,10)
Vibrio fisheri	luciferase (*lux*)	77 (2,40 + 37)
	Terrestrial	
Photinus pyralis	luciferase (*luc*)	100 (2,62)[b]
Pyrophorus plagiophthalamus	luciferase (*luc*)	62

[a]140-kD monomer, also an active 35-kD proteolytic fragment.
[b]Sodium dodecyl sulfate electrophoresis, gives one band, 62 kDa.

(formerly *Cypridina hilgendorfi*), the coelenterates *Renilla reformis* (sea pansy) and *Aequorea*, the dinoflagellate *Gonyaulax polyedra*, and the marine bacteria *Vibrio fisheri* and *Photobacterium phosphoreum*.

Firefly. Firefly luciferase (EC 1.13.12.7) is a homodimeric enzyme (62 kDa subunit) that has binding sites for firefly luciferin and Mg ATP^{2-}. Amino acid sequence analysis has indicated that beetle luciferases evolved from coenzyme A synthetase (206). Firefly bioluminescence is the most efficient bioluminescent reaction known, with *Qc* reported to be 88% (5), and λ_{max} at 562 nm (56). At low pH and in the presence of certain metal ions (eg, Pb^{2+}, Hg^{2+}, Zn^{2+}, Cd^{2+}) light emission is shifted to the red (610–615 nm). *In vitro* a flash of light is produced (<1 s) that decays rapidly. Glow-type emission is obtained in the presence of detergents (Triton X-100), polymers (PEG 6000), coenzyme A, inorganic pyrophosphate, and cytidine nucleotides (206,207).

Luciferin (**41**) (R = H) reacts with adenosine triphosphate (ATP) and the enzyme to give a complex (**42**) of the adenylate ester (**41**) R = AMP·luciferase (AMP = adenosine monophosphate). Reaction of the complex with oxygen produces the excited state of the highly fluorescent and inhibitory oxyluciferin (**43**) and carbon dioxide. In common with the specificity of other enzyme reactions, only the D(-)-enantiomorph of luciferin produces light (208,209). Luciferin [*2591-17-5*] (**41**) has been synthesized (210) and its structure established. A number of firefly species appear to use the same luciferin (211), but the color of the emitted light can differ because of variations in enzyme structure as demonstrated with click beetle luciferases (212,213).

(**41**) + ATP + luciferase $\xrightarrow{Mg^{2+}}$ (**42**) $\xrightarrow{O_2}$ (**43**) + CO_2 + AMP

The carbonyl compound (**43**) has also been synthesized, and its fluorescence spectrum has been shown to match the bioluminescence spectrum under equivalent conditions (214). The details of the excitation step are unclear and a dioxetanone mechanism (59,215) may apply to the reaction.

Chemiluminescence is also obtained by anionic autooxidation of (**41**) with oxygen in alkaline dimethyl sulfoxide (DMSO) (216). *Qc* has been reported to be 10% and ketone (**43**) and CO_2 are obtained. Several analogues of luciferin have been prepared that are also chemiluminescent when they react with oxygen in alkaline DMSO (62).

Vargula. In the ostracod *Vargula* (formerly *Cypridina*) *hilgendorfi* the luciferin (**44**) reacts with oxygen and its luciferase enzyme to generate the excited state of amide (**46**) and by-products (**47**) and (**48**). The amide is formed in 86% yield (217), Qc is 30% (218) and λ_{max} is at 465 nm (48). Luciferin (**44**) also produces chemiluminescence by reaction with oxygen in alkaline diglyme; however, Qc is only about 3% (219).

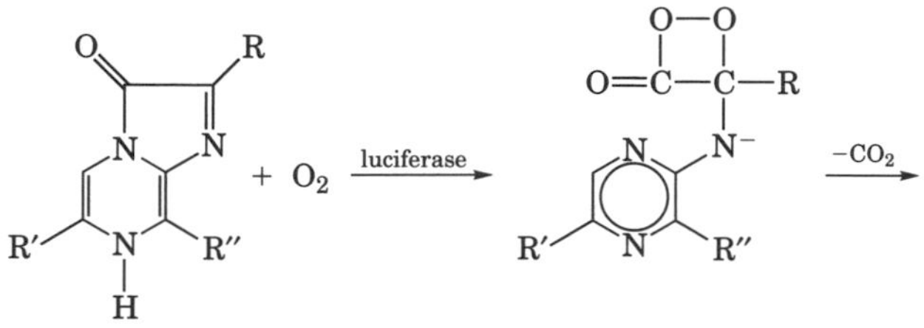

(**44**) + O_2 —luciferase→ (**45**) —$-CO_2$→

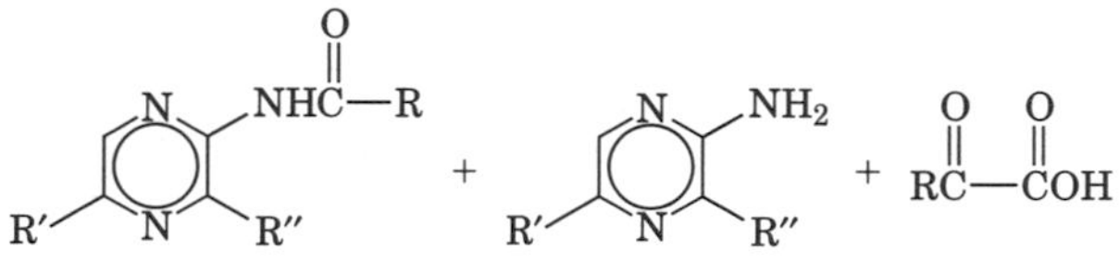

(**46**) + (**47**) + (**48**)

where R′ = [indol-3-yl], R″ = $CH_2CH_2CH_2NHC(=NH)NH_2$, R = $CH(CH_3)CH_2CH_3$

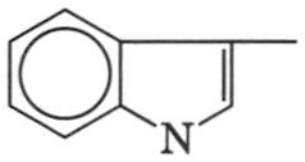

When oxygen-18 is used, 80% of the label is incorporated into carbon dioxide in the aqueous, enzyme-catalyzed reaction, and 62% of the label is incorporated during chemiluminescence in DMSO with potassium *tert*-butoxide (219). This result is in accord with the dioxetanone intermediate (**45**). Both luciferin (**44**) (220), and the amide (**46**) (221) have been synthesized. Analogues of structure (**44**) have been synthesized and shown to be chemiluminescent in organic solvents with base and oxygen (219,222).

Coelenterate. Coelenterates *Renilla reformis* (sea pansy) and *Aequorea forskalea* (jelly fish) produce bioluminescence by similar processes (223). The basic luciferin structure is (**49**) (224) and excited amide (**50**) is the emitter. The structural relationship to *Vargula* is evident. A structural analogue where R = CH_3 is active in bioluminescence. The quantum yield is about 4% (223), with λ_{max} at 509 nm (56). This reaction involves a charge transfer between green fluorescent protein and the excited-state coelenterate oxyluciferin.

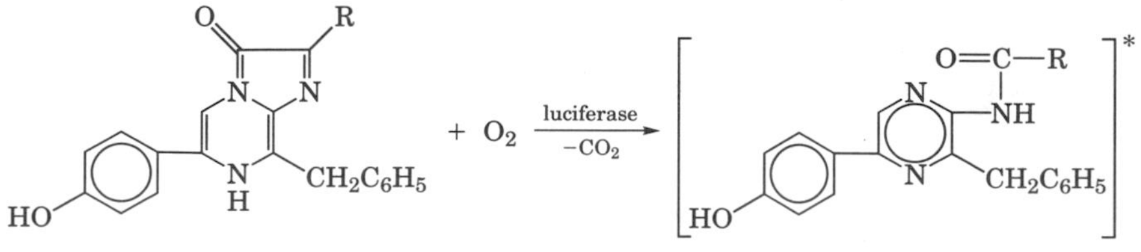

(**49**) + O_2 —luciferase, $-CO_2$→ [(**50**)]*

where (R = $CH_2C_6H_4OH$)

The jellyfish *Aequorea* can be extracted to yield a photoprotein, a complex of luciferin and apoaequorin that generates light (rapid flash, 460 nm) on treatment with calcium ions or other metal ions (eg, Sr, Ba, L, Yb) (225). The cloned genes for *Renilla* luciferase and apoaequorin provide a source of the pure recombinant proteins (226).

Bacteria. The luminous bacteria (*Vibrio, Photobacterium,* and *Xenorhabdus*) are found principally in a marine environment, but freshwater (*Vibrio albenis*) and terrestrial species (*Xenorhabdus luminescens*) are also known. Marine luminous bacteria are free-living, or as symbionts in fish (eg, Teleost fish and *Photobacterium leiognathi*). Luminous bacteria enzymatically oxidize reduced flavin mononucleotide and a long-chain aldehyde to flavin mononucleotide and the carboxylic acid corresponding to the aldehyde (227–229). The quantum yield has been reported as 10–20% (230–232). The quantum yield increases with aldehyde chain length up to about eight carbon atoms. The emitting fluorescer is the protonated flavin mononucleotide reaction product (233) formed by decomposition of a 4-α-peroxy flavin intermediate. The reactions leading to light emission are as follows:

$$\text{RCOOH} + \text{ATP} + \text{NAD(P)H} \xrightarrow{\text{fatty acid reductase complex}} \text{NADP} + \text{AMP} + \text{PP}_i + \text{RCHO}$$

$$\text{NAD(P)H} + \text{FMN} \xrightarrow{\text{FMN:NAD(P)H oxidoreductase}} \text{NAD} + \text{FMNH}_2$$

$$\text{FMNH}_2 + \text{RCHO} \xrightarrow{\text{bacterial luciferase}} \text{FMN} + \text{H}_2\text{O} + \text{RCOOH} + \text{light}$$

Genes (*lux*) encoding luciferase and the other proteins involved in the bioluminescent reaction have been cloned. *LuxA* and *luxB* genes code for the luciferase subunits, and the fatty acid reductase complex is coded by the *luxCDE* genes.

During the growth of luminous marine bacteria light emission lags behind cell growth, because of the requirement for an autoinducer to accumulate in the medium. The autoinducer in *Vibrio fisheri* is β-ketocaproyl homoserine lactone, and in *Vibrio harveyi* is β-hydroxybutyryl homoserine lactone, respectively (227).

Gonyaulax. *Gonyaulax polyedra* produces flashes of light (0.1 s) from numerous small (0.5 μm) scintillons (234). These contain an open tetrapyrrole-type luciferin bound to a luciferin-binding protein (dimer, 72 kDa subunits). A change in the pH (8 to 5.7) releases luciferin that reacts with luciferase (420 kDa) to produce a flash of light (479 nm). The genes encoding the luciferase and the luciferin-binding protein have been cloned (235). Light emission exhibits a circadian rhythm in which cells emit light at night but not during the daytime, and this is achieved by translational control of luciferin and luciferin-binding protein synthesis.

Latia. The freshwater snail *Latia* has been reported to provide bioluminescence by the following reaction (236).

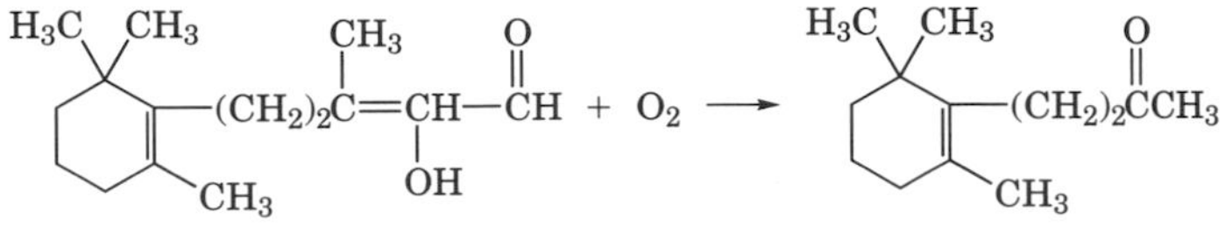

$$\text{(H}_3\text{C)(CH}_3\text{)(CH}_3\text{)C}_6\text{-}(CH_2)_2C(CH_3){=}C(OH){-}CH{=}O + O_2 \longrightarrow \text{(H}_3\text{C)(CH}_3\text{)(CH}_3\text{)C}_6\text{-}(CH_2)_2C(=O)CH_3$$

(51)

Applications of Chemical Light

Marking and Illumination. Chemical light is well suited for lighting applications where distributed electric power is unavailable or hazardous (237). A substantial amount of light can be generated from a small, lightweight package easily transported and stored, and the cold light of chemiluminescence can be used safely in situations where a thermal light could cause fire or explosion. The many uses of chemical light include as an emergency light for power failures in homes, office buildings, theaters, and factories; for disabled vehicles and aircraft; for lifeboats and lifejackets; as a marker light for pedestrians and bicycles on dark streets; and as a portable light for hikers, campers, and military units. The product requirements for such applications include adequate brightness and lifetime, convenient utilization, low toxicity, high flash point, long shelf-life, and low cost (237).

The brightness and lifetime limits of a formulation are determined by its light capacity, L_{cap} in (lm·h)/L, which is defined as the integral of luminous intensity I, in lumens, with respect to reaction time t in hours for one liter of formulation (237):

$$L_{cap} = \int_0^{\infty} I\ dt/\text{vol} \qquad L_{cap} = 4.07 \times 10^4 C \cdot Qc \cdot P$$

C is the concentration of limiting reactant in mol/L, Qc is the chemiluminescence quantum yield in ein/mol, and P is a photopic factor that is determined by the sensitivity of the human eye to the spectral distribution of the light. Because the human eye is most responsive to yellow light, where the photopic factor for a yellow fluorescer such as fluorescein can be as high as 0.85, blue or red formulations have inherently lower light capacities.

High quantum yields are uncommon and, moreover, the quantum yield almost always decreases as the concentration of chemiluminescent reactant approaches practical levels. Thus even reactions with high inherent quantum yields at low concentrations, such as the firefly reaction, do not necessarily provide high light capacities (237).

The theoretical limit of light capacity has been estimated for an ideal reaction that provides yellow light with a photopic factor of 0.85 in a quantum yield of one at 5 M concentration as 173,000 (lm·h)/L, equivalent to the light output of a 40-W bulb burning continuously for two weeks (237). The most efficient formulation available, based on oxalic ester chemiluminescence, produces about 0.5% of that limit, with a light capacity of 880 (lm·h)/L (237).

Because chemiluminescent brightness decreases as the reactants are consumed, not all of the light capacity is actually useful. Some light at the onset of reaction is emitted at intensities brighter than required by an application, and some light near the end of reaction is emitted below a threshold brightness requirement. A term called decay curve efficiency is used to quantify the fraction of total light capacity that is effective in practical use. The decay curve efficiency is the ratio of the area of the largest rectangle that can be drawn under the intensity–time graph of an emission to the total area under the curve. Useful brightness and lifetime increase with this ratio for any given light capacity. Maximizing decay curve efficiency is an important goal in applied chemiluminescence. Efficiencies as high as 70% have been reported (238).

Several chemiluminescent reactions have been considered for illumination and marking. Luminol formulations have been proposed for air–sea rescue signaling (239) and a one-package, solid-state formulation comprising luminol, potassium persulfate, potassium perborate, potassium carbonate, and sodium fluorescein has been patented as a chemical light tablet which generates yellow light when added to water (240). The maximum light capacity of current luminol formulations, however, is only about 1.3 (lm·h)/L, because of low inherent Q_c and the decrease in Q_c at practical luminol concentrations (241).

Chemical light formulations based on air oxidation of tetrakis(dimethylamino)ethylene (TMAE) have been used on aircraft emergency escape slides (242). The reaction is convenient to use, since the components only have to be exposed to air for activation. Maintenance of an oxygen-free environment during storage, however, can be difficult unless the application permits the use of sealed metal or glass containers (242).

The quantum yield for TMAE chemiluminescence is only about 0.1% (48–50), but unlike other chemiluminescent reactions light emission can be obtained at high TMAE concentrations (242). Light capacities of about 60 (lm·h)/L for practical high concentration formulations have been estimated. Chemiluminescence of TMAE is quenched by its reaction products and is affected by moisture. A number of formulations have been devised to minimize these problems. In particular, solvents have been recommended (243) that dissolve TMAE but not the quenching tetramethylurea and tetramethyloxamide reaction products. Formulations that include alkali or alkaline-earth group metal salts, such as lithium chloride, give improved performance in humid environments (244). The use of dehydrating agents (eg, calcium chloride) in combination with thickeners, such as finely divided silica, also improves high humidity performance (245). A formulation containing sulfolane is described as providing superior subzero temperature performance (246). A wax formulation (247), a polyethylene sponge impregnated with TMAE and lithium chloride (248), and an aerosol TMAE device have been patented (249).

The only chemical light reaction with significant use at present for illumination is the Cyalume lightstick (237). This product was originally developed by American Cyanamid Co. The technology has been sold to Omniglow. The Cyalume chemical light product is based on oxalic ester chemiluminescence (76):

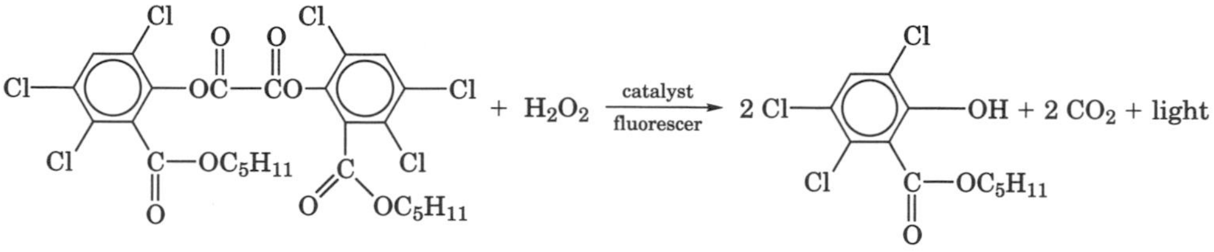

$$(\mathbf{52}) + H_2O_2 \xrightarrow[\text{fluorescer}]{\text{catalyst}} 2\ \text{Cl}_3C_6H(COOC_5H_{11})\text{—OH} + 2\ CO_2 + \text{light}$$

(**52**)

This reaction has four special features which promote its utility: (*1*) the quantum yield is high, about 14% in practical formulations, and this high yield can be maintained at ester concentrations as high as 0.2 *M* with certain fluorescers such as 1-chloro-9,10-bis(phenylethynyl)anthracene and certain esters, such as (**52**). Light capacities as high as 880 (lm·h)/L have been reported (90). A number of high light capacity esters and fluorescers have been described (81,82,90). (*2*) The fluorescer is independent of the oxalic ester and can be selected separately to provide a desired color and to accommodate other practical requirements such as solubility, stability under the oxidizing conditions, excitation efficiency, fluorescence quantum yield, and shelf-life. Moreover, since the fluorescer acts as a catalyst and is recycled, it can be formulated in low concentrations (below 10^{-2} *M*) to avoid efficiency loss through fluorescence self-quenching, while permitting use of the high ester concentrations required for high light capacity. A number of fluorescers have been disclosed, including 9,10-diphenylanthracene (blue) (79), 9,10-bis(phenylethynyl)anthracene (green) (79,81,94), 1-chloro-9,10-bis(phenylethynyl)anthracene (yellow) (90), rubrene (orange) (78), and 5,12-bis(phenylethynyl)tetracene (red) (79). (*3*) Since the reaction is accelerated by weak bases, such as sodium salicylate, the brightness–lifetime performance of the reaction can be varied to meet the needs of an application by varying the catalyst or its concentration (94). (*4*) The reaction proceeds efficiently in a number of solvents facilitating selection in regard to safety requirements. Formulations using dimethyl or dibutyl phthalates are preferred because these solvents are essentially nontoxic and have high flash points (81,90,237).

The Cyalume lightstick is a 15-cm long, translucent, flexible plastic tube (250) containing a thin-walled glass ampoule (251), which floats in a solution of oxalate ester and fluorescer. The ampoule contains a dilute solution of hydrogen peroxide and catalyst. The plastic tube is easily bent to break the ampoule and mix the reactants. Light emission is immediate and depending on the formulation can last up to 12 hours or longer. The Cyalume green lightstick provides an intensity above 1 lumen up to one hour after activation and an intensity above 0.1 lumen up to three hours. The shelf-life is well in excess of two years. Water causes the formulation to deteriorate and the lightstick must be packaged in a hermetic aluminum foil plastic laminate wrapper. As in any chemical reaction, the rate of oxalic ester chemiluminescence is temperature dependent, so that the lightstick is brighter warm than cold. An activated lightstick can be deactivated and preserved for many days by placing it in a freezer.

A number of patents disclose practical refinements to the oxalic ester system related to increasing curve-shape efficiency and light output (94,95,252), brightness-lifetime control (94,95) and lengthening of shelf-life (253). Variations in lightstick design (254), use of lightsticks in emergency lighting devices (255), and other means of packaging (256), displaying (257), or dispensing (258) chemical light formulations have been described.

A modified oxalic ester reaction that is activated by air rather than hydrogen peroxide has been provided by combining a 9,10-dihydroxyanthracene or benzoin with the ester and fluorescer (259). Oxygen from air is converted to hydrogen peroxide by the dihydroxyanthracene.

Analytical Applications. Chemiluminescence and bioluminescence are useful in analysis for several reasons. (*1*) Modern low noise phototubes when properly instrumented can detect light fluxes as weak as 100 photons/s (1.7×10^{-22} eins/s). Thus luminescent reactions in which intensity depends on the concentration of a reactant of analytical interest can be used to determine attomole–zeptomole amounts (10^{-18} to 10^{-21} mol). This is especially useful for biochemical, trace metal, and pollution control analyses (93,260–266) (see TRACE AND RESIDUE ANALYSIS). (*2*) Light measurement is easily automated for routine measurements as, for example, in clinical analysis.

Flow Injection Analysis and High Performance Liquid Chromatography. Analytical applications of chemiluminescence for analyte detection for flow injection analysis and hplc-based methods were first reported in 1976 (260) for the assay of H_2O_2 employing the peroxyoxalate chemiluminescent reaction. Addition of TCPO, triethylamine, and a fluorescer (perylene) to the column effluent just prior to the detector produced chemiluminescence in proportion to the concentration of H_2O_2 (71). A limit of detection of 10 nM of H_2O_2 and a dynamic range of 10 nM to 1 mM were realized. The method has been extended to include a variety of additional analytes, including labeled amino acids, oligopeptides, catecholamines, and polyaromatic hydrocarbons (267–269). In order to attain maximum sensitivity, these methods require that the maximum of the analyte chemiluminescence intensity (τ_{max}) coincide with the passage of the analyte through the detector (71). When this is realized, sensitivity of detection can approach very low levels, eg, 10^{-15} M of the analyte. Analytical applications are limited, however, to systems that are chemiluminescent.

Direct Metal Analyses. Calcium ion can be detected to a lower limit of 10^{-7} M by *Aequorea* bioluminescence. Strontium interferes to a minor extent (270,271).

Divalent copper, cobalt, nickel, and vanadyl ions promote chemiluminescence from the luminol–hydrogen peroxide reaction, which can be used to determine these metals to concentrations of 1–10 ppb (272,273). The light intensity is generally linear with metal concentration of 10^{-5} to 10^{-9} M range (272). Manganese(II) can also be determined when an amine is added to increase its reduction potential by stabilizing Mn(III) (272). Since all of these ions are active, ion exchange must be used for determination of a particular metal in mixtures (274).

Chromium(III) can be analyzed to a lower limit of 5×10^{-10} M by luminol–hydrogen peroxide without separating from other metals. Ethylenediaminetetraacetic acid (EDTA) is added to deactivate most interferences. Chromium(III) itself is deactivated slowly by complexation with EDTA; measurement of the sample after Cr(III) deactivation is complete provides a blank which can be subtracted to eliminate interference from such ions as iron(II), iron(III), and cobalt(II), which are not sufficiently deactivated by EDTA (275).

Iron(II) can be analyzed by a luminol–air reaction in the absence of hydrogen peroxide (276). Iron in the aqueous sample is reduced to iron(II) by sulfite; other metals which might interfere are also reduced to valence states that are inactive in the absence of hydrogen peroxide. The detection limit is 10^{-10} M.

Titration Indicators. Concentrations of arsenic(III) as low as 2×10^{-7} M can be measured (272) by titration with iodine, using the chemiluminescent iodine oxidation of luminol to indicate the end point. Oxidation reactions have been titrated using siloxene; the appearance of chemiluminescence indicates excess oxidant. Examples include titration of thallium (277) and lead (278) with dichromate and analysis of iron(II) by titration with cerium(IV) (279).

Hydrogen Peroxide Analysis. Luminol has been used for hydrogen peroxide analysis at concentrations as low as 10^{-8} M using the cobalt(III) triethanolamine complex (280) or ferricyanide (281) as promoter. With the latter, chemiluminescence is linear with peroxide concentration from 10^{-8} to 10^{-5} M.

Luminol-based chemiluminescence methods have also been employed for detection of analytes in flowing stream analytical techniques such capillary electrophoresis (282), flow injection analyses, and hplc (267). Applications of the enhanced luminol methodology to replace radioassay methods have been developed for a number of immunological labeling techniques (121,283).

Hydrogen peroxide has also been analyzed by its chemiluminescent reaction with bis(2,4,6-trichlorophenyl) oxalate and perylene in a buffered (pH 4–10) aqueous ethyl acetate–methanol solution (284). Using a flow system, intensity was linear from the detection limit of 7×10^{-8} M to at least 10^{-3} M.

Clinical Analysis. A wide range of clinically important substances can be detected and quantitated using chemiluminescence or bioluminescence methods. Coupled enzyme assay protocols permit the measurement of kinase, dehydrogenase, and oxidases or the substrates of these enzymes as exemplified by reactions of glucose, creatine phosphate, and bile acid in the following:

$$\text{glucose} + O_2 \xrightarrow{\text{glucose oxidase}} \text{gluconolactone} + H_2O_2$$

$$H_2O_2 + \text{luminol} \xrightarrow{\text{microperoxidase}} \text{light}$$

$$\text{creatine phosphate} + \text{ADP} \xrightarrow{\text{creatine kinase}} \text{creatine} + \text{ATP}$$

$$\text{ATP} + \text{firefly D-luciferin} \xrightarrow{\text{firefly luciferase/Mg}^{2+}} \text{light}$$

Light production is proportional to the limiting component in the initial reaction (eg, glucose or glucose oxidase). The enzymes involved in this type of coupled enzyme assay can also be co-immobilized onto the same solid support (eg, Sepharose beads, nylon reactor tube) (285). The close proximity of the enzymes increases the efficiency of substrate transfer between successive reactions and hence improves the overall reaction efficiency. Table 3 lists examples of substrates and enzymes assayed by the coupled enzyme techniques. Generally, the assays have good precision and are highly sensitive, eg, detection limit for serum

Table 3. Clinically Important Substances Detected Using Coupled Enzyme Reactions

Enzymes	Substrates
Chemiluminescent assay	
glucose oxidase	acetylcholine, cholesterol, choline, glucose, oxalate, urate
Bioluminescent firefly luciferase assay	
acetate kinase, adenylate kinase, apyrase, ATP-ase, creatine kinase, pyruvate kinase	1,3-diphosphoglycerate, glycerol, phosphocreatine, phosphoenolpyruvate, pyrophosphate
Bioluminescent bacterial luciferase assay	
alcohol dehydrogenase, lactate dehydrogenase, glucose, 6-phosphate dehydrogenase, lipase, phospholipase	ethanol, galactose, glucose 6-phosphate, glycerol, lactate, oxaloacetate, pyruvate, sorbitol

bile acids is 0.5 pmol; NADH reacts further to produce light.

$$\text{bile acid} + \text{NAD} \xrightarrow{\text{hydroxysteroid dehydrogenase}} \text{7-oxo bile acid} + \text{NADH}$$

Cellular chemiluminescence (enhanced by luminol or lucigenin) is a key analytical method in the study of phagocytosis by polymorphonuclear neutrophils (286). Chemiluminescence and bioluminescence methods have also been developed for measuring antioxidants (qv) in biological fluids (287), detecting blood stains for forensic applications (288), quantitating reporter gene expression (*luc, lux, GAL, GUS*) (289), and for post-column detection in hplc and flow injection analysis of complex biological mixtures (290).

Immunoassay. Chemiluminescence compounds (eg, acridinium esters and sulfonamides, isoluminol), luciferases (eg, firefly, marine bacterial, *Renilla* and *Vargula* luciferase), photoproteins (eg, aequorin, *Renilla*), and components of bioluminescence reactions have been tested as replacements for radioactive labels in both competitive and sandwich-type immunoassays. Acridinium ester labels are used extensively in routine clinical immunoassay analysis designed to detect a wide range of hormones, cancer markers, specific antibodies, specific proteins, and therapeutic drugs. An acridinium ester label produces a flash of light when it reacts with an alkaline solution of hydrogen peroxide. The detection limit for the label is 0.5 amol.

Chemiluminescence and bioluminescence are also used in immunoassays to detect conventional enzyme labels (eg, alkaline phosphatase, β-galactosidase, glucose oxidase, glucose 6-phosphate dehydrogenase, horseradish peroxidase, microperoxidase, xanthine oxidase). The enhanced chemiluminescence assay for horseradish peroxidase (luminol-peroxide-4-iodophenol detection reagent) and various chemiluminescence adamantyl 1,2-dioxetane aryl phosphate substrates, eg, (**11**) and (**15**) for alkaline phosphatase labels are in routine use in immunoassay analyzers and in Western blotting kits (261–266).

Nucleic Acid Assays. A series of routine clinical DNA probe assays for infectious agents (eg, *Chlamydia*) have been developed using a chemiluminescent acridinium ester label in a nonseparation hybridization protection assay format

(291). The acridinium ester-labeled DNA probe bound to target DNA hydrolyzes slowly whereas any unbound labeled probe hydrolyzes rapidly (million-fold faster) to form nonchemiluminescent products. In this way bound and free fractions are readily differentiated and upon addition of an alkaline peroxide solution the light emission is due entirely to the specifically bound labeled probe. Numerous research DNA probe assays, and Southern (DNA detection) and Northern (RNA detection) blotting assays, based predominantly on alkaline phosphatase (chemiluminescence detection), horseradish peroxidase (chemiluminescence detection), and glucose 6-phosphate dehydrogenase (bioluminescence detection), have also been developed and some are used routinely in biomedical research (292).

Bacteria and Biomass Determination. Firefly bioluminescence specifically requires adenosine triphosphate (ATP) and has been used to determine ATP by its reaction with firefly luciferin, luciferase, and magnesium(II) in buffered aqueous oxygenated solutions (293). Since ATP is specific to living organisms, its analysis provides a rapid method for determining biomass concentrations. The emitted light is proportional to ATP concentration which in turn is dependent on biomass. The method has been suggested as a means for detecting life on other planets.

As little as 10^{-14} g of ATP can be detected with carefully purified luciferase. Commercial luciferase contains enough residual ATP to cause background emission and increase the detection limit to 10^{-12} g (294). The method has been used to determine bacterial concentrations in water. As few as 10^4 cells/mL of *Escherichia coli*, which contains as little as 10^{-16} g of ATP per cell, can be detected (294). Numerous species of bacteria have been studied using this technique (293–295).

Other applications of firefly bioluminescence include measurement of the activity of bacteria in secondary sewage treatment activated sludge (296,297), detection of bacteria in clean rooms and operating rooms, measurement of bacteria in bottled foods, beverages (298), and pharmaceuticals (299), determination of the antimicrobial activity of potential drugs (300), determination of the viability of seeds (301), and measuring marine biomass concentrations as a function of ocean depth or geographical location (302).

Bacterial concentrations have also been determined by using the enzyme-catalyzed chemiluminescent reaction of reduced flavin mononucleotide (FMN) with oxygen and aldehydes. The detection limit was reported to be 10^5 cells of *E. coli*, which contains 7×10^{-17} g of FMN per cell (303).

Luminol chemiluminescence has also been recommended for measuring bacteria populations (304,305). The luminol–hydrogen peroxide reaction is catalyzed by the iron porphyrins contained in bacteria, and the light intensity is proportional to the bacterial concentration. The method is rapid, especially compared to the two-day period required by the microbiological plate-count method, and it correlates well with the latter when used to determine bacteria in cooling tower water, paper pulp, activated sludge in secondary water treatment plants, drinking water, and air. The limit of detectability depends somewhat on the water source, varying from 5×10^4 cells/mL for cooling water to 5×10^5 cells/mL in paper pulp. Using a concentration technique, bacterial counts as low as 6×10^3 cells/mL could be determined in distilled water (304).

Bioluminescence *in vitro* chemosensitivity assays are now used to assess the sensitivity of tumor cells (obtained by surgical or needle biopsy) to different drugs and combinations of drugs. Cells are grown in microwell plates in the presence of the drugs at various concentrations. If the tumor cells are sensitive to the drug then they do not grow, hence total extracted cellular ATP, measured using the bioluminescence firefly luciferase reaction, is low. This method has been used to optimize therapy for different solid tumors and for leukemias (306).

Oxidation Analyses. Polymers, petroleum fuels, lubricating oils, foods, and other materials are degraded by air oxidation. Lifetimes depend on the conditions of use and on the inherent oxidizability of the material. The latter is conventionally determined by accelerated, high temperature techniques, since failure under use conditions may require many years of use. However, the autooxidation mechanism can change with temperature, so that the results of accelerated high temperature tests do not always correlate well with practical performance. A method based on a computer controlled, photon-counting apparatus measures the oxidizability of liquids and solids under use conditions by determining the minute chemiluminescence which accompanies autooxidation (307). Chemiluminescence from oxidation reactions as slow as 10^{-9} mol/yr in solids and 10^{-14} mol/yr in liquids can be detected. Since the chemiluminescence intensity is a function of oxidation rate, the latter can be determined to estimate the time to failure of the material. The method is also useful in determining the activities of antioxidants (qv).

Air Pollution Analyses. Ozone. Air pollution (qv) levels are commonly estimated by determining ozone through its chemiluminescent reaction with ethylene. A relatively simple photoelectric device is used for rapid routine measurements. The device is calibrated with ozone from an ozone generator, which in turn is calibrated by the reaction of ozone with potassium iodide (308). Detection limits are 6–9 ppb with commercially available instrumentation (309).

Nitrogen oxides are significant pollutants produced by cars, trucks, and fossil fuel power plants. Nitric oxide (NO) can be determined by its chemiluminescent reaction with ozone using a photomultiplier detector. Total nitrogen oxides can be determined by thermally dissociating nitrogen dioxide (NO_2) to NO before measurement. Ammonia, which interferes with the latter determination, can be removed from the gas stream by oxidation with dichromate. The method thus permits continuous monitoring of NO, NO_2, and NH_3. Nitrogen oxide concentrations as low as 0.02 ppm are easily detected, and response is linear with concentration up to several percent (310); detectable limits are 22 $\mu g\ m^{-3}$ (311).

Sulfur dioxide concentrations as low as 40 mg/m^3 in air have been determined by passing air samples through an aqueous solution of tetrachloromercurate, which converts SO_2 to the dichlorosulfitomercurate complex. Oxidation of the complex by potassium permanganate is chemiluminescent and the intensity, as measured by a photomultiplier, is proportional to sulfur dioxide concentration (312).

Total sulfur in air, most of which is sulfur dioxide, can be measured by burning the sample in a hydrogen-rich flame and measuring the blue chemiluminescent emission from sulfur atom combination to excited S_2 (313). Concentrations of about 0.01 ppm can be detected.

Hydrocarbons. The gas-phase chemiluminescent reaction of oxygen atoms with hydrocarbons has been proposed as a method for measuring hydrocarbon concentrations in automobile exhaust (314). Olefinic, saturated, and aromatic hydrocarbons are detected, although the method is most sensitive to olefins. The detection limit is 0.2 ppm ethylene equivalent and response is linear with concentration to >1000 ppm. Contaminants such as CO, CO_2, and NO_2 do not interfere nor do methane or acetylene, which are thought not to be important contributers to photochemical smog. The method involves combining atomic oxygen, generated by passing an oxygen–helium mixture through a microwave discharge, with the hydrocarbon sample in a reaction zone monitored by a photomultiplier tube through 308.9- and 312.3-nm interference filters. Emission from olefins is detected at 308.9 nm, emission from acetylene is detected at both wavelengths. The 312.2-nm signal is subtracted from the 308.9-nm signal to eliminate the contribution at 308.9 nm from acetylene. Emission at both wavelengths arises from electronically excited hydroxyl radicals, but the vibrational excitation differs for the two reactions, causing a spectral difference. The concentration of atomic oxygen for calibration was also measured by chemiluminescence, using the reaction of atomic oxygen with excess NO, where intensity is proportional to oxygen concentration.

Nickel Carbonyl. The extremely toxic gas nickel carbonyl can be detected at 0.01 ppb by measuring its chemiluminescent reaction with ozone in the presence of carbon monoxide. The reaction produces excited nickel(II) oxide by a chain process which generates many photons from each pollutant molecule to permit high sensitivity (315).

BIBLIOGRAPHY

"Chemiluminescence" in *ECT* 3rd ed., Vol. 5, pp. 416–450, by M. M. Rauhut, American Cyanamid Co.

1. E. N. Harvey, *A History of Luminescence*, American Philosophical Society, Philadelphia, Pa., 1957.
2. K. D. Gunderman, *Top. Current Chem.* **46**, 61 (1974); E. H. White and co-workers, *Angew. Chem. Intern. Ed. Engl.* **13**, 229 (1974); A. K. Campbell, *Chemiluminescence*, Ellis Horwood, Chichester, U.K., 1988; P. J. Herring, *J. Biolumin. Chemilumin.* **3**, 147 (1987); K.-D. Gundermann and F. McCapra, *Chemiluminescence in Organic Chemistry*, Springer-Verlag, Weinheim, Germany, 1987; K. Van Dyke, *Luminescence Immunoassay and Molecular Applications*, CRC Press, Boca Raton, Fla., 1990.
3. R. M. Hochstrasser and G. B. Porter, *Quart. Rev. (London)* **14**, 146 (1960); C. A. Parker, *Photoluminescence of Solutions*, Elsevier, New York, 1968; G. G. Guilbault, *Practical Fluorescence; Theory, Method and Techniques*, Marcel Dekker, New York, 1973.
4. M. G. Evans, H. Eyring, and J. F. Kincaid, *J. Chem. Phys.* **6**, 349 (1938).
5. H. H. Seliger and W. D. McElroy, *Arch. Biochem. Biophys.* **88**, 136 (1960).
6. M. M. Rauhut, *Acc. Chem. Res.* **2**, 80 (1969).
7. C. A. Parker, *Advan. Photochem.* **2**, 305 (1964).
8. M. M. Rauhut and co-workers, *Photochem. Photobiol.* **4**, 1097 (1965); U.S. Pats. 3,637,784 (Jan. 25, 1972); 3,914,255 (Oct. 21, 1975), D. Sheehan (to American Cyanamid).
9. M. M. Rauhut and co-workers, *J. Am. Chem. Soc.* **89**, 6515 (1967).

10. D. C. Lee and T. Wilson in M. J. Cormier, D. M. Hercules, and J. Lee, eds., *Chemiluminescence and Bioluminescence*, Plenum Press, New York, 1973, p. 265.
11. R. W. Dixon, *Diss. Faraday Soc.* **35**, 105 (1963).
12. F. McCapra, *Chem. Commun.*, 155 (1968).
13. W. H. Richardson and H. E. O'Neal, *J. Am. Chem. Soc.* **94**, 8665 (1972); D. R. Kearns, *Chem. Rev.* **71**, 395 (1971); M. J. S. Dewar and S. Kirshner, *J. Am. Chem. Soc.* **96**, 7578 (1974); C. Eaker and J. Hinze, *Theor. Chem. Acta* **40**, 113 (1975); D. R. Roberts, *Chem. Commun.*, 683 (1974); N. J. Turro and A. Devaquet, *J. Am. Chem. Soc.* **97**, 3859 (1973).
14. J. W. Hastings and T. Wilson, *Photochem. Photobiol.* **23**, 461 (1976).
15. J. Koo and G. B. Schuster; *J. Am. Chem. Soc.* **99**, 6107 (1977); **100**, 4496 (1978); K. A. Zalika, Schaap, *J. Am. Chem. Soc.* **100**, 4916 (1978); F. McCapra, *Chem. Commun.*, 946 (1977); J. Michl, *Photochem. Photobiol.* **25**, 141 (1977).
16. F. McCapra and D. G. Richardson, *Tetrahedron Lett.* 3167 (1964).
17. N. J. Turro and P. Lechtken, *Pure Appl. Chem.* **33**, 363 (1973); N. J. Turro and co-workers, *Acc. Chem. Res.* **7**, 97 (1974); W. Adams, *Adv. Heterocyclic Chem.* **21**, 438 (1977).
18. N. J. Turro and P. Lechtken, *J. Am. Chem. Soc.* **94**, 2886 (1972); P. Lechtken, A. Yekta, and N. J. Turro, *J. Am. Chem. Soc.* **95**, 3027 (1973).
19. N. J. Turro and co-workers, *J. Am. Chem. Soc.* **96**, 1623 (1974).
20. T. Wilson and co-workers, *J. Am. Chem. Soc.* **98**, 1086 (1976).
21. W. Adam, *J. Am. Chem. Soc.* **97**, 5464 (1975).
22. E. J. Becharct, A. L. Baumstark, and T. Wilson, *J. Am. Chem. Soc.* **98**, 4648 (1976).
23. N. J. Turro and H. Steinmetzer, *J. Am. Chem. Soc.* **96**, 4679 (1974).
24. T. Wilson and A. P. Schaap, *J. Am. Chem. Soc.* **93**, 4126 (1971).
25. A. P. Schaap, P. A. Burns, and K. A. Zaklika, *J. Am. Chem. Soc.* **99**, 1270 (1977); *Ibid.*, **100**, 318 (1978).
26. W. H. Richardson, M. B. Yelvington, and H. E. O'Neal, *J. Am. Chem. Soc.* **94**, 1619 (1972); H. E. Zimmerman and G. E. Keck, *J. Am. Chem. Soc.* **97**, 3527 (1975).
27. K. R. Kopecky and C. Mumford, *Can. J. Chem.* **47**, 709 (1969).
28. G. B. Schuster and co-workers, *J. Am. Chem. Soc.* **97**, 7110 (1975); J. H. Wieringa and co-workers, *Tetrahedron Lett.* 169 (1972).
29. F. McCapra, I. Beheshti, A. Burford, R. A. Hann, and K. A. Zaklika, *J. Chem. Soc. Chem. Commun.*, 944 (1977).
30. F. McCapra, *J. Chem. Soc. Chem. Commun.*, 946 (1977).
31. W. Adam, L. Encarnacionand, and K. Zinner, *Chem. Ber.* **116**, 839 (1983).
32. W. Adam, V. Bhushan, T. Dirnberger, and R. Fuchs, *Synthesis*, 330 (1986).
33. C. Lee and L. A. Singer, *J. Am. Chem. Soc.* **102**, 3823 (1980).
34. A. P. Schaap and S. D. Gagnon, *J. Am. Chem. Soc.* **104**, 3504 (1982).
35. I. Bronstein, B. Edwards, and J. C. Voyta, *J. Biolumin. Chemilumin.* **4**, 99 (1989).
36. A. P. Schaap, R. S. Handley, and B. P. Giri, *Tetrahedron Lett.* **28**, 935 (1987); A. P. Schaap, M. D. Sandison, and R. S. Handley, *Tetrahedron Lett.* **28**, 1159 (1987).
37. I. Bronstein and J. C. Voyta, *Clin. Chem.* **35**, 1856 (1989); G. Thorpe, I. Bronstein, L. J. Kricka, B. Edwards, and J. C. Voyta, *Clin. Chem.* **35**, 2319 (1989); I. Bronstein, J. C. Voyta, K. G. Lazzari, O. J. Murphy, B. Edwards, and L. J. Kricka, *BioTechniques* **8**, 310 (1990); I. Bronstein, J. C. Voyta, K. G. Lazzari, O. J. Murphy, B. Edwards, and L. J. Kricka, *BioTechniques* **9**, 160 (1990).
38. V. R. Bodepudi and W. J. LeNoble, *J. Org. Chem.* **56**, 2001 (1991).
39. T. Wilson and co-workers, *J. Am. Chem. Soc.* **95**, 4765 (1973); P. D. Bartlett, A. L. Baumstark, and M. E. Landis, *J. Am. Chem. Soc.* **96**, 5557 (1974).
40. N. J. Turro and P. Lechtken, *J. Am. Chem. Soc.* **94**, 2888 (1972); P. Lechtken, A. Yekta, and N. J. Turro, *J. Am. Chem. Soc.* **95**, 3027 (1973).

41. K. R. Kopecky and co-workers, *Can. J. Chem.* **53**, 1103 (1975); *Ibid*, **51**, 468 (1973).
42. P. D. Bartlett and A. P. Schaap, *J. Am. Chem. Soc.* **92**, 3223, 6055 (1970); S. Mazur and C. S. Foote, *J. Am. Chem. Soc.* **92**, 3225 (1970); W. H. Richardson, M. B. Yelvington, and H. E. O'Neal, *J. Am. Chem. Soc.* **94**, 1619 (1972).
43. F. McCapra in W. Carruthers and J. K. Sutherland, eds., *Progress in Organic Chemistry*, Vol. 8, John Wiley & Sons, Inc., New York, 1973, p. 231; *Pure Appl. Chem.* **24**, 611 (1970).
44. E. W. Evans, *J. Chem. Ed.* **14**, 236 (1937); G. E. Philbrook and M. A. Maxwell, *Tetrahedron Lett.*, 1111 (1964); I. Nicholson and P. Poretz, *J. Chem. Soc.* 3067 (1965); T. Hayashi and K. Maeda, *Bull. Chem. Soc. Jpn.* **36**, 1052 (1963); *Bull. Chem. Soc. Jpn.* **35**, 2057 (1962); *Bull. Chem. Soc. Jpn.* **33**, 565 (1960).
45. E. H. White and M. J. Harding, *J. Am. Chem. Soc.* **86**, 5686 (1964); J. Sonnenberg and D. M. White, *J. Am. Chem. Soc.* **86**, 5685 (1964).
46. J. R. Totter, *Photochem. Photobiol.* **3**, 231 (1964).
47. K. Lee, L. A. Singer, and K. D. Legg, *J. Org. Chem.* **41**, 2685 (1976).
48. A. N. Fletcher and C. A. Heller, *J. Phys. Chem.* **71**, 1507 (1967).
49. H. E. Winberg, J. R. Downing, and D. D. Coffman, *J. Am. Chem. Soc.* **87**, 2054 (1965); J. P. Paris, *Photochem. Photobiol.* **4**, 1059 (1965).
50. W. H. Urry and J. Sheeto, *Photochem. Photobiol.* **4**, 1067 (1965).
51. N. Wilberg and J. W. Buckler, *Z. Naturforsch.* **19b**, 5 (1964).
52. R. L. Pruett and co-workers, *J. Am. Chem. Soc.* **72**, 3646 (1950); N. Wiberg and J. W. Bucher, *Naturwissenschaften* **19b**, 953 (1964); H. E. Winberg and co-workers, *J. Am. Chem. Soc.* **87**, 2055 (1965); U.S. Pat. 3,239,519 (Mar. 8, 1966), H. E. Winberg (to E. I. du Pont de Nemours & Co., Inc.).
53. D. J. Bogan, R. S. Shienson, and F. W. Williams, *J. Am. Chem. Soc.* **98**, 1034 (1976); D. J. Bogan and co-workers, *J. Am. Chem. Soc.* **97**, 2560 (1975).
54. W. Adam and J. C. Liu, *J. Am. Chem. Soc.* **94**, 2894 (1972); W. Adam and H. C. Steinmetzer, *Angew. Chem. Intern. Ed. Engl.* **11**, 540 (1972); W. Adam and co-workers, *J. Am. Chem. Soc.* **99**, 5768 (1977).
55. N. J. Turro and co-workers, *J. Am. Chem. Soc.* **99**, 5836 (1977); W. Adam and co-workers, *J. Am. Chem. Soc.* **99**, 5768 (1977).
56. W. Adam, *J. Chem. Ed.* **52**, 138 (1975).
57. W. Adam, C. A. Simpson, and F. Yang, *J. Phys. Chem.* **78**, 2559 (1974); G. B. Schuster and S. P. Schmidt, *J. Am. Chem. Soc.* **100**, 1966, 5559 (1978); W. Adam, O. Cueto, and F. Yang, *J. Am. Chem. Soc.* **100**, 2587 (1978).
58. T. A. Hopkins and co-workers, *J. Am. Chem. Soc.* **89**, 7148 (1967).
59. F. McCapra, T. C. Chang, and V. P. Francois, *Chem. Commun.*, 22 (1968).
60. E. H. White and co-workers, *J. Am. Chem. Soc.* **91**, 2178 (1969).
61. F. McCapra, D. G. Richardson, and Y. C. Chang, *Photochem. Photobiol.* **4**, 1111 (1965).
62. M. M. Rauhut and co-workers, *J. Org. Chem.* **30**, 3587 (1965); U.S. Pat. 3,539,574 (Nov. 11, 1970) (to American Cyanamid).
63. F. McCapra, *Prog. Org. Chem.* **8**, 231 (1973).
64. P. Lechtken and N. J. Turro, *Mol. Photochem.* **6**, 95 (1974).
65. G. B. Schuster, *Acc. Chem. Res.* **12**, 366 (1979).
66. F. McCapra, K. Pirring, R. J. Hart, and R. A. Hahn, *Tetrahedron Lett.*, 5087 (1981).
67. G. B. Schuster and S. P. Schmidt, *Adv. Phys. Org. Chem.* **18**, 187 (1982); S. P. Schmidt and G. B. Schuster, *J. Am. Chem. Soc.* **102**, 306 (1980).
68. C. L. R. Catherall, T. F. Palmer, and R. B. Cundall, *J. Chem. Soc. Faraday Trans.* **80**, 823 and 827 (1984).
69. M.-M. Chang, T. Saji, and A. J. Bard, *J. Am. Chem. Soc.* **99**, 5399 (1977).
70. F. J. Alvarez and co-workers, *J. Am. Chem. Soc.* **108**, 6435 (1986).

71. R. S. Givens and R. L. Schowen, in J. W. Birks, ed., *Chemiluminescence and Photochemical Reaction Detection in Chromatography*, VCH, New York, 1989, Chapt. 5, p. 125.
72. M. Orlović and co-workers, *J. Org. Chem.* **54**, 3606 (1989).
73. R. S. Givens and co-workers, *J. Pharm. Biomed. Anal.* **8**, 477 (1990).
74. K. Imai, in H. Lingeman and W. J. M. Underberg, eds., *Detection Oriented Derivatization Techniques in Liquid Chromatography*, Vol. 40, Marcel Dekker, New York, 1990, Chapt. 10, p. 359.
75. G. J. de Jong and P. J. M. Kwakman, *J. Chromatogr.* **492**, 319 (1989).
76. U.S. Pat. 3,597,362 (Aug. 3, 1971), L. J. Bollyky and M. M. Rauhut (to American Cyanamid).
77. U.S. Pat. 3,704,231 (Nov. 28, 1972), L. J. Bollyky (to American Cyanamid).
78. U.S. Pat. 3,701,738 (Oct. 11, 1972), B. G. Roberts and M. M. Rauhut (to American Cyanamid).
79. U.S. Pats. 3,729,426 (Apr. 24, 1973); 3,557,233 (Jan. 19, 1971), A. Zweig and D. R. Maulding (to American Cyanamid).
80. I. Kamiya and T. Sugimoto, *Bull. Chem. Soc. (Japan)*, 2442 (1977); I. Kamiya, *Bull. Chem. Soc. (Japan)*, 2447 (1977); R. S. Givens and G. Orösz, unpublished results, 1994.
81. U.S. Pat. 3,749,679 (July 31, 1973), M. M. Rauhut (to American Cyanamid).
82. U.S. Pat. 3,781,329 (Dec. 25, 1973), L. J. Bollyky (to American Cyanamid); 3,816,326 (June 11, 1974), L. J. Bollyky (to American Cyanamid).
83. D. R. Maulding and co-workers, *J. Org. Chem.* **33**, 250 (1968).
84. L. J. Bollyky and co-workers, *J. Org. Chem.* **34**, 836 (1969); U.S. Pat. 3,843,549 (Oct. 22, 1974), L. J. Bollyky (to American Cyanamid).
85. U.S. Pat. 3,442,815 (May 6, 1969), M. M. Rauhut and L. J. Bollyky (to American Cyanamid).
86. U.S. Pat. 3,400,080 (Sept. 3, 1968), D. R. Maulding (to American Cyanamid).
87. L. J. Bollyky, R. H. Whitman, and B. G. Roberts, *J. Org. Chem.* **33**, 4266 (1968); U.S. Pat. 3,909,440 (Sept. 30, 1975), L. J. Bollyky and R. H. Whitman (to American Cyanamid); U.S. Pat. 3,978,079 (Aug. 31, 1976), L. J. Bollyky and R. H. Whitman (to American Cyanamid).
88. L. J. Bollyky and co-workers, *J. Am. Chem. Soc.* **89**, 5623 (1967); U.S. Pats. 3,399,137 (Aug. 27, 1968), 3,804,891 (Apr. 16, 1974), M. M. Rauhut and L. J. Bollyky (to American Cyanamid).
89. E. A. Chandross, *Tetrahedron Lett.* 761 (1963); M. M. Rauhut, B. G. Roberts, and A. M. Semsel, *J. Am. Chem. Soc.* **88**, 3604 (1966); U.S. Pat. 3,325,417 (June 13, 1967), M. M. Rauhut (to American Cyanamid); U.S. Pats. 3,442,813 (May 6, 1969), 3,644,517 (Feb. 22, 1972), L. J. Bollyky (to American Cyanamid).
90. U.S. Pat. 3,888,786 (June 10, 1975), D. R. Maulding (to American Cyanamid).
91. P. Lechtken and N. J. Turro, *Mol. Photochem.* **6**, 95 (1974); H. Guston and E. E. Ullman, *Chem. Commun.*, 28 (1977).
92. M. M. Rauhut and co-workers, *J. Org. Chem.* **40**, 330 (1975); U.S. Pat. 3,630,941 (Dec. 28, 1971), W. R. Bergmark (to American Cyanamid).
93. K. Imai, K. Miyaguchi, and K. Honda, in K. van Dyke, *Bioluminescence and Chemiluminescence: Instruments and Applications*, Vol. II, CRC Press, Boca Raton, Fla., 1985, p. 65. See also S. P. Schmidt and G. B. Schuster, *J. Am. Chem. Soc.* **100**, 1966 (1978); *Ibid.*, **102**, 306 (1980).
94. U.S. Pats. 3,775,336 (Nov. 27, 1973); 3,704,231 (Nov. 18, 1972), L. J. Bollyky (to American Cyanamid).
95. U.S. Pat. 3,691,085 (Sept. 12, 1972), B. G. Roberts and M. M. Rauhut (to American Cyanamid).

96. A. G. Mohan and N. J. Turro, *J. Chem. Ed.* **51**, 528 (1974).
97. M. Sugiura, S. Kanda, and K. Imai, *Biomed. Chromatogr.* **7**, 149 (1993).
98. D. R. Maulding and B. G. Roberts, *J. Org. Chem.* **37**, 1458 (1972).
99. U.S. Pat. 3,677,957 (July 18, 1972), D. R. Maulding (to American Cyanamid).
100. U.S. Pats. 3,697,432 (Oct. 10, 1972), 3,894,050 (July 8, 1975), D. R. Maulding (to American Cyanamid).
101. U.S. Pat. 3,749,677 (July 31, 1973), D. R. Maulding (to American Cyanamid).
102. U.S. Pat. 3,425,949 (Feb. 4, 1969), M. M. Rauhut and G. W. Kennerly (to American Cyanamid).
103. U.S. Pat. 3,329,621 (July 4, 1967), M. M. Rauhut and A. M. Semsel (to American Cyanamid).
104. W. R. Vaughan, *Chem. Rev.* **43**, 496 (1948); A. Berannose, T. Bremer, and P. Goldfinger, *Bull. Soc. Chem. Belges.* **56**, 269 (1947); *Dis. Faraday. Soc.* **15**(2), 221 (1947); *Bull. Soc. Chim. Fr.* **15**, 946 (1948).
105. M. M. Rauhut, A. M. Semsel, and B. G. Roberts, *J. Org. Chem.* **31**, 2431 (1966).
106. D. F. Roswell and E. H. White, *Methods Enzymol.* **57**, 409 (1978).
107. J. Lee and H. H. Seliger, *Photochem. Photobiol.* **4**, 1015 (1965).
108. H. D. Drew, *Trans. Faraday Soc.* **35**, 207 (1939).
109. E. H. White and M. M. Bursey, *J. Am. Chem. Soc.* **86**, 941 (1964).
110. E. H. White and co-workers, *J. Am. Chem. Soc.* **86**, 940 (1964).
111. Y. Omote, T. Miyake, and N. Sugiyama, *Bull. Chem. Soc. Jpn.* **40**, 2446 (1967).
112. E. H. White and D. F. Roswell, *Acc. Chem. Res.* **3**, 54 (1970); E. H. White and R. B. Brundrett in Ref. 10, p. 231.
113. K. D. Gundermann, H. Fiege, and G. Klockenbring, *Ann.* **738**, 140 (1970); **734**, 200 (1971).
114. K. D. Gundermann and M. Drawert, *Chem. Ber.* **95**, 2018 (1962); H. D. K. Drew and F. H. Pearman, *J. Chem. Soc.*, 586 (1937); H. D. K. Drew and R. E. Garwood, *J. Chem. Soc.*, 836 (1939); B. E. Cross and H. D. Drew, *J. Chem. Soc.*, 1532 (1949).
115. K. D. Gundermann, W. Horstman, and G. Bergman, *Ann.* **684**, 127 (1965).
116. C. C. Wei and E. H. White, *Tetrahedron Lett.*, 3559 (1971).
117. H. R. Schroeder and F. M. Yeager, *Anal. Chem.* **50**, 1114 (1978).
118. Eur. Pat. Appl. 491,477 (1992), H. Masuya, K. Kondo, Y. Aramaki, and Y. Ichimori (to Takeda Chemical Industries, Ltd.).
119. G. H. G. Thorpe and L. J. Kricka, *Methods in Enzym.* **133**, 331 (1986); T. P. Whitehead, G. H. G. Thorpe, T. J. N. Carter, C. Groucutt, and L. J. Kricka, *Nature* **305**, 158 (1983); G. H. G. Thorpe and L. J. Kricka, in J. Schölmerich, R. Andreesen, A. Kapp, M. Ernst, and W. G. Woods, eds., *Bioluminescence and Chemiluminescence, New Perspectives*, John Wiley and Sons, Inc., New York, 1987, p. 199.
120. G. H. G. Thorpe, L. J. Kricka, S. B. Moseley, and T. P. Whitehead, *Clin. Chem.* **31**, 1335 (1985).
121. I. Durrant, in L. J. Kricka, ed., *Nonisotope DNA Probe Techniques*, Academic Press, Inc., San Diego, Calif., 1992, p. 167.
122. R. T. Dufford, S. Calvert, and D. Nightingale, *J. Am. Chem. Soc.* **45**, 2058 (1923); *J. Opt. Soc. Am.* **9**, 405 (1924); W. V. Evans and E. M. Diepenhorst, *J. Am. Chem. Soc.* **48**, 715 (1926); R. L. Bardsley and D. M. Hercules, *J. Am. Chem. Soc.* **90**, 4545 (1968).
123. P. H. Bolton and D. R. Kearns, *J. Am. Chem. Soc.* **96**, 4651 (1974).
124. R. T. Dufford, *J. Am. Chem. Soc.* **50**, 1822 (1928).
125. H. Gilman and C. E. Adams, *J. Am. Chem. Soc.* **47**, 2816 (1925).
126. H. Gilman, J. McGlumphy, and R. E. Fothergill, *Rec. Trav. Chem.* **49**, 526, 726 (1930).
127. K. Issleib and A. Tzschach, *Chem. Ber.* **92**, 1118 (1959); R. A. Strecker, J. L. Snead, and G. P. Sallot, *J. Am. Chem. Soc.* **95**, 210 (1973).
128. V. Ya. Shlyapintokh and co-workers, *Chemiluminescence Techniques in Chemical Reactions*, Consultants Bureau, New York, 1968.

129. V. Ya. Shlyapintokh and co-workers, *J. Chim. Phys. (Russ.)* **57**, 1113 (1960).
130. R. E. Kellogg, *J. Am. Chem. Soc.* **91**, 5433 (1969).
131. A. A. Vichutinskii, *J. Phys. Chim. (Russ)* **38**, 1242 (1964); O. N. Karpukhin, V. Ya. Shlyapintokh, and I. V. Mikhailov, *J. Phys. Chim. (Russ)* **38**, 81 (1964).
132. V. A. Belyakov and R. F. Vassil'ev, *Photochem. Photobiol.* **11**, 179 (1970); R. F. Vassil'ev, *Optics Spect.* **18**, 131,254 (1965); R. F. Vassil'ev and A. A. Vichutinskii, *Nature* **194**, 1276 (1962).
133. R. F. Vassilev, *Nature* **196**, 668 (1962); **200**, 773 (1963). G. Lundeen and R. Livingston, *Photochem. Photobiol.* **4**, 1085 (1965); V. A. Belyakov and R. F. Vassil'ev, *Photochem. Photobiol.* **6**, 35 (1967); S. R. Abbott, S. Ness, and D. M. Hercules, *J. Am. Chem. Soc.* **92**, 1128 (1970); J. H. Helberger and D. B. Hever, *Chem. Ber.* **72B**, 11 (1939); P. Rothemund, *J. Am. Chem. Soc.* **60**, 2005 (1938); G. D. Dorough, J. R. Muller, and F. F. Huennekens, *J. Am. Chem. Soc.* **73**, 4315 (1951); H. Linschitz and E. W. Abrahamson, *Nature* **72**, 909 (1953).
134. G. B. Schuster, *J. Am. Chem. Soc.* **99**, 651 (1977).
135. J. Koo and G. B. Schuster, *J. Am. Chem. Soc.* **99**, 6107 (1977); J. P. Smith and G. B. Schuster, *J. Am. Chem. Soc.* **100**, 2564, 4496 (1978).
136. K. D. Gundermann, *Top. Current Chem.* **46**, 61 (1974); *Naturwissenschaften* **56**, 62 (1969); *Chemilumineszenz Organisher Verbindungen*, Springer-Verlag, New York, 1968; *Angew. Chem. Int. Ed. Engl.* **11**, 566 (1965).
137. E. Wasserman, L. Barash, and W. A. Yager, *J. Am. Chem. Soc.* **87**, 4974 (1965).
138. B. Lachowicze, *Chem. Ber.* **16**, 332 (1883).
139. W. Dilthey and W. Hoschen, *J. Prakt. Chem.* **38**(27), 42 (1933).
140. E. McKeown and W. A. Waters, *Nature* **203**, 1063 (1964).
141. J. Stauff, V. Sander, and W. Jaeschke in Ref. 10, p. 131.
142. C. S. Foote, *Acc. Chem. Res.* **1**, 104 (1968); D. R. Kearns, *Chem. Rev.*, **71**, 395 (1971).
143. E. A. Ogryzlo and A. E. Pearson, *J. Phys. Chem.* **72**, 2913 (1968).
144. E. J. Corey and W. C. Taylor, *J. Am. Chem. Soc.* **86**, 3881 (1964); R. P. Wayne, *Adv. Photochem.* **7**, 311 (1969).
145. R. J. Browne and E. A. Ogryzlo, *Can. J. Chem.* **43**, 2915 (1965); C. S. Foote and S. Wexler, *J. Am. Chem. Soc.* **86**, 3879 (1964).
146. C. S. Foote, S. Wexler, and W. Ando, *Tetrahedron Lett.* **4111** (1965).
147. E. Wasserman and co-workers, *J. Am. Chem. Soc.* **90**, 4160 (1968); R. W. Murray and M. L. Kaplan, *J. Am. Chem. Soc.* **90**, 4161 (1968); **91**, 5385 (1969).
148. S. J. Arnold, E. A. Ogryzlo, and H. Witzke, *J. Chem. Phys.* **40**, 1769 (1964); R. J. Browne and E. A. Ogryzlo, *Proc. Chem. Soc. (London)* **117** (1964); A. V. Khan and M. Kasha, *J. Am. Chem. Soc.* **92**, 3293 (1970).
149. S. J. Arnold and E. A. Ogryzlo, *Can. J. Phys.* **45**, 2053 (1967).
150. R. J. Browne and E. A. Ogryzlo, *Can. J. Chem.* **43**, 2915 (1965).
151. T. Wilson, *J. Am. Chem. Soc.* **91**, 2387 (1969).
152. E. J. Bowen and R. A. Lloyd, *Proc. Chem. Soc. (London)* **305** (1963).
153. I. Stauff, H. Schmidkunz, and G. Hartman, *Nature* **198**, 281 (1963); E. McKeown and W. A. Waters, *Nature* **203**, 1063 (1964); E. J. Bowen, *Nature* **201**, 180 (1964); R. A. Lloyd, *Trans. Faraday Soc.* **61**, 2173 (1965); R. F. Vasilev, *Prog. Reaction Kinetics* **4**, 304 (1967); J. A. Howard and K. U. Ingold, *J. Am. Chem. Soc.* **90** 1057 (1968); S. R. Abbott, S. Ness, and D. M. Hercules, *J. Am. Chem. Soc.* **92**, 1128 (1970).
154. D. M. Hercules, *Acc. Chem. Res.* **2**, 301 (1969); A. Zweig, *Adv. Photochem.* **6**, 425 (1968).
155. E. A. Chandross, *Trans. New York Acad. Sci. Ser. II* **31**, 571 (1969).
156. A. Weller and K. Zachariasse in Ref. 10, p. 181, p. 193.
157. A. J. Bard and co-workers in Ref. 10, p. 193.

158. E. A. Chandross and F. I. Sonntag, *J. Am. Chem. Soc.* **88**, 1089 (1966); D. M. Hercules, R. C. Lansbury, and D. K. Roe, *J. Am. Chem. Soc.* **88**, 4578 (1966); R. A. Marcus, *J. Chem. Phys.* **43**, 2654 (1965).
159. H. P. Kallmann and M. Pope, *J. Chem. Phys.* **36**, 2482 (1962); W. Helfrich and W. G. Schneider, *J. Chem. Phys.* **44**, 2902 (1966).
160. G. N. Lewis and J. Bigeleisen, *J. Am. Chem. Soc.* **65**, 2424 (1943); H. Linschitz, M. G. Berry, and D. Schweitzer, *J. Am. Chem. Soc.* **76**, 5833 (1954).
161. J. E. Martin and co-workers, *J. Am. Chem. Soc.* **94**, 9238 (1972).
162. E. A. Chandross and F. F. Sonntag, *J. Am. Chem. Soc.* **86**, 3179 (1964); **88**, 1089 (1966).
163. U.S. Pat. 3,391,069 (July 2, 1968), M. M. Rauhut and G. W. Kennerly (to American Cyanamid); T. D. Santa Cruz, D. L. Akins, and R. L. Birke, *J. Am. Chem. Soc.* **98**, 1677 (1976).
164. U.S. Pat. 3,391,068 (July 21, 1968), M. M. Rauhut (to American Cyanamid); C. P. Keszthelyi and A. J. Bard, *J. Org. Chem.* **39**, 2936 (1974).
165. A. Weller and K. Zachariasse, *J. Chem. Phys.* **46**, 4984 (1967); *Chem. Phys. Lett.* **10**, 424 (1971).
166. D. M. Hercules, *Science* **145**, 808 (1964); D. M. Hercules and F. F. Lytle, *J. Am. Chem. Soc.* **88**, 4745 (1966); J. Chang, D. M. Hercules, and D. K. Roe, *Electrochem. Acta* **13**, 1197 (1968); R. E. Visco and E. A. Chandross, *J. Am. Chem. Soc.* **86**, 5350 (1964); *Electrochem. Acta* **13**, 1187 (1968). S. V. Santhanum and A. J. Bard, *J. Am. Chem. Soc.* **87**, 3259 (1965); A. Zweig and co-workers, *J. Am. Chem. Soc.* **88**, 2864 (1966); **89**, 4091 (1967); D. L. Maricle and A. H. Maurer, *J. Am. Chem. Soc.* **89**, 188 (1967). A. Zweig, A. H. Maurer, and B. G. Roberts, *J. Org. Chem.* **32**, 1322 (1967); A. Zweig and co-workers, *J. Am. Chem. Soc.* **89**, 473 (L967); A. Zweig and co-workers, *Chem. Commun.* 106 (1967).
167. U.S. Pat. 3,319,132 (May 9, 1967), E. A. Chandross and R. E. Visco (to Bell Telephone Laboratories); U.S. Pat. 3,654,525 (Apr. 4, 1972), D. L. Maricle and M. M. Rauhut (to American Cyanamid); U.S. Pat. 3,816,795 (June 11, 1974); D. L. Maricle and M. M. Rauhut (to American Cyanamid); U.S. Pat. 3,900,418 (Apr. 19, 1975), A. J. Bard and N. E. Takvoryan (to Bell-Northern Research, Ltd.).
168. E. A. Chandross, J. W. Longworth, and R. E. Visco, *J. Am. Chem. Soc.* **87**, 3259 (1965); A. Weller and K. Zachariasse, *J. Chem. Phys.* **46**, 4984 (1967); S. M. Park and A. J. Bard, *J. Am. Chem. Soc.* **97**, 2978 (1975).
169. R. E. Hemingway, S. M. Park, and A. J. Bard, *J. Am. Chem. Soc.* **97**, 200 (1975).
170. L. R. Faulkner and A. J. Bard, *J. Am. Chem. Soc.* **91**, 209, 6495, 6497 (1969). L. R. Faulkner, H. Tachikawa, and A. J. Bard, *J. Am. Chem. Soc.* **94**, 691 (1972); *Chem. Phys. Lett.* **26**, 246, 568 (1974); P. W. Atkins and G. T. Evans, *Mol. Phys.* **29**, 921 (1975).
171. D. J. Freed and L. R. Faulkner, *J. Am. Chem. Soc.* **93**, 2097, 3565 (1971); **94**, 6324 (1972).
172. D. J. Freed and L. R. Faulkner, *J. Am. Chem. Soc.* **94**, 4790 (1972).
173. C. P. Keszthelyi, N. E. Tokel-Takvoryan, and A. J. Bard, *Anal. Chem.* **47**, 249 (1975).
174. M. G. Evans and M. Polanyi, *Trans. Faraday Soc.* **35**, 178, 192, 195 (1939).
175. W. S. Strive, T. Kitagawa, and D. R. Herschback, *J. Chem. Phys.* **54**, 2959 (1971).
176. D. R. Herschback in Ref. 10, p. 29.
177. R. W. Anderson, V. Aquilante, and D. R. Herschback, *Chem. Phys. Lett.* **4**, 5 (1969); V. Kempter and co-workers, *Chem. Phys. Lett.* **6**, 97 (1970); **11**, 353 (1971).
178. K. Lacmann and D. R. Herschback, *Chem. Phys. Lett.* **6**, 106 (1970).
179. J. E. Mentall and co-workers, *Diss. Faraday Soc.* **44**, 157 (1967); *J. Chem. Phys.* **56**, 4593 (1972).
180. W. Braun and M. J. Kurylo, *J. Chem. Phys.* **61**, 461 (1974).

181. C. Ottinger and R. N. Zare, *Chem. Phys. Lett.* **5**, 243 (1970); D. M. Manos and J. M. Parsons, *J. Chem. Phys.* **63**, 3575 (1975); F. Engelke, R. K. Sandar, and R. N. Zare, *J. Chem. Phys.* **65**, 1146 (1976); L. C. Loh and R. R. Herm, *Chem. Phys. Lett.* **38**, 263 (1976).
182. M. F. Golde and B. A. Thrush, *Adv. At. Mol. Phys.* **11**, 361 (1975).
183. B. A. Thrush, *Chem. Br.* **2**, 287 (1966).
184. D. O. Ham, *Diss. Faraday Soc.* **55**, 313 (1973); *J. Chem. Phys.* **60**, 1802 (1974).
185. J. C. Greaves and D. Garvin, *J. Chem. Phys.* **30**, 348 (1959); M. A. A. Clyne, B. A. Thrush, and R. P. Wayne, *Trans. Faraday Soc.* **60**, 359 (1964); P. N. Clough and B. A. Thrush, *Trans. Faraday Soc.* **63**, 915 (1967).
186. H. L. Welsh, C. Cumming, and E. J. Stansburg, *J. Opt. Soc. Am.* **41**, 712 (1951).
187. C. J. Halstead and B. A. Thrush, *Nature* **204**, 992 (1964); *Photochem. Photobiol.* **4**, 1007 (1965); B. A. Thrush, *Annual Rev. Phys. Chem.* **19**, 371 (1968).
188. I. W. M. Smith in Ref. 10, p. 43.
189. A. G. Anlauf and co-workers, *J. Chem. Phys.* **53**, 4091 (1970); P. E. Charters and J. C. Polanyi, *Diss. Faraday Soc.* **33**, 107 (1962); I. W. M. Smith, *Acc. Chem. Res.* **9**, 161 (1976).
190. G. Hancock and I. W. M. Smith, *Chem. Phys. Lett.* **3**, 469 (1969); *Trans. Faraday Soc.* **67**, 2856 (1971); *Chem. Phys. Lett.* **8**, 41 (1971); *Appl. Optics* **10**, 1827 (1971).
191. M. C. Lin in Ref. 10, p. 61.
192. F. Kaufman in Ref. 10, p. 83.
193. B. A. Thrush and M. F. Golde in Ref. 10, p. 73.
194. M. Jeunehomme and A. B. F. Duncan, *J. Chem. Phys.* **41**, 1692 (1964); K. H. Becker, W. Groth, and D. Thran, *Chem. Phys. Lett.* **6**, 583 (1970); **15**, 215 (1972).
195. R. A. Young and R. L. Sharpless, *Chem. Phys. Lett.* **39**, 1071 (1963); I. M. Campbell and B. A. Thrush, *Chem. Commun.*, 250 (1965).
196. R. J. VanZee and A. V. Khan, *J. Am. Chem. Soc.* **96**, 6805 (1974).
197. F. Kenny and R. B. Kurtz, *Anal. Chem.* **22**, 693 (1950).
198. H. Kautsky and E. Gaubatz, *J. Anorg. Chem.* **191**, 384 (1930).
199. J. L. Dyer and W. Lusk, *U.S. National Technical Information Service, AD 631458*, Washington, D.C.
200. T. Goto, *Pure Appl. Chem.* **17**, 421 (1968).
201. W. D. McElroy and H. H. Seliger, *Sci. Am.*, 76 (Dec. 1962); F. H. Johnson, ed., *The Luminescence of Biological Systems*, AAAS Press, Washington, D.C., 1961; F. H. Johnson and Y. Haneda, eds., *Bioluminescence in Progress*, Princeton University Press, Princeton, N.J., 1966; H. H. Seliger and W. D. McElroy, *Light, Physical and Biological Action*, Academic Press, Inc., New York, 1965; W. D. McElroy and B. Glass, eds., *Light and Life*, Johns Hopkins Press, Baltimore, Md., 1961.
202. J. G. Morin and co-workers, *Science* **190**, 74 (1975); J. E. McCosker, *Sci. Am.* **236**, 106 (Mar. 1977).
203. J. R. deWet, K. V. Wood, D. R. Helinski, and M. A. DeLuca, *Proc. Natl. Acad. Sci. USA* **82**, 7870 (1985); J. R. deWet, K. V. Wood, D. R. Helinski, M. A. DeLuca, and S. Subramani, *Mol. Cell Biol.* **7**, 725 (1987).
204. R. Belas and co-workers, *Science*, 791 (1982).
205. J. Casadei, M. J. Powell, and J. H. Kenten, *J. Biolumin. Chemilumin.* **4**, 346 (1989).
206. K. V. Wood, in P. E. Stanley and L. J. Kricka, eds., *Bioluminescence and Chemiluminescence: Current Status*, John Wiley & Sons, Ltd., Chichester, U.K., 1991, p. 11.
207. A. Lundin, in A. Szalay, L. J. Kricka, and P. Stanley, eds., *Bioluminescence and Chemiluminescence: Status Report*, John Wiley, & Sons, Ltd., Chichester, U.K., 1993, p. 291; F. R. Leach, S. R. Ford, M. S. Hall, and K. D. Hooper, *J. Cell. Biol.* **107**, 189a (1988); L. J. Kricka and M. A. DeLuca, *Arch. Biochem. Biophys.* **217**, 674 (1982).
208. T. Goto and Y. Kishi, *Angew. Chem. Int. Ed. Engl.* **7**, 407 (1968).

209. W. D. McElroy and M. DeLuca, in Ref. 10, p. 285.
210. E. H. White, F. McCapra, and G. F. Field, *J. Am. Chem. Soc.* **83**, 2402 (1961); **85**, 337 (1963).
211. H. H. Seliger and W. D. McElroy, *Proc. Nat. Acad. Sci. U.S.A.* **52**, 75 (1964).
212. O. Shimomura, F. H. Johnson, and T. Masugi, *Science* **164**, 1299 (1969).
213. K. V. Wood, Y. A. Lam, H. H. Seliger, and W. D. McElroy, *Science* **244**, 700 (1989).
214. E. H. White and co-workers, *Bioorg. Chem.* **1**, 92 (1971); N. Suzuki and T. Goto, *Tetrahedron* **28**, 4075 (1972); *Tetrahedron Lett.* **2021** (1971).
215. F. McCapra and Y. C. Chang, *Chem. Commun.*, 1011 (1967).
216. F. McCapra, *Acc. Chem. Res.* **9**, 201 (1976).
217. O. Shimomura and F. H. Johnson, in Ref. 10, p. 337.
218. O. Shimomura and F. H. Johnson, *Photochem. Photobiol.* **12**, 291 (1970).
219. T. Goto, S. Inoue and S. Sugiura, *Tetrahedron Lett.*, 3873 (1968); O. Shimomura and F. H. Johnson, *Biochem. Biphys. Res. Commun.* **44**, 340 (1971); **51**, 558 (1973).
220. Y. Kishi and co-workers, *Tetrahedron Lett.*, 3427 (1966).
221. T. Goto and co-workers, *Tetrahedron Lett.*, 4035 (1968); T. Goto, S. Inoue, and S. Sugiura, *Chem. Commun.*, 3873 (1968); T. Goto and co-workers, *Chem. Commun.*, 4035 (1968); J. G. Morin and J. W. Hastings, *J. Cell. Physiol.* **77**, 305 (1971).
222. F. McCapra and co-workers, in Ref. 10, p. 313.
223. M. J. Cormier, J. E. Wampler, and K. Hori, *Fortschr. Chem. Org. Naturst.* **30**, 1 (1973); M. J. Cormier, K. Hori, and J. M. Anderson, *Biochem. Biophys. Acta* **346**, 137 (1974); O. Shimomura and F. H. Johnson, *Nature* **256**, 236 (1975).
224. K. Hori and M. J. Cormier, *Proc. Nat. Acad. Sci. U.S.A.* **70**, 120 (1973).
225. O. Shimomura, F. H. Johnson, and Y. Saiga, *J. Cell. Comp. Physiol.* **59**, 223 (1962); **62**, 19 (1963); O. Shimomura and F. H. Johnson, *Biochemistry* **8**, 3991 (1969).
226. W. W. Lorenz, R. O. McCann, and M. J. Cormier, *Proc. Natl. Acad. Sci. USA* **88**, 4438 (1991).
227. E. A. Meighen, *Microbiol. Rev.* **55**, 123 (1991).
228. E. A. Meighen, in Ref. 206, p. 3.
229. J. W. Hastings and co-workers, *Proc. Nat. Acad. Sci. U.S.A.* **70**, 3468 (1973); J. W. Hastings and C. J. Balny, *J. Biol. Chem.* **250**, 7288 (1975).
230. J. Lee, *Biochemistry* **11**, 3350 (1972).
231. O. Shimomura, F. H. Johnson, and Y. Kohama, *Proc. Nat. Acad. Sci. U.S.A.* **69**, 2086 (1972); J. E. Becvar and J. W. Hastings, *Proc. Nat. Acad. Sci. U.S.A.* **72**, 3374 (1975).
232. J. Lee and C. L. Murphy, *Biochemistry* **14**, 2259 (1975).
233. M. Eley and co-workers, *Biochemistry* **9**, 2902 (1970).
234. J. W. Hastings, *J. Biolumin. Chemilumin.* **4**, 12 (1989).
235. E. Roux, P. M. Milos, and J. W. Hastings, *Plant Physiol.* **80**, 16 (1986).
236. O. Shimomura and F. H. Johnson, *Biochemistry* **7**, 1734 (1968).
237. M. M. Rauhut, in Ref. 10, p. 451.
238. U.S. Pat. 3,775,336 (Nov. 27, 1973), L. J. Bollyky (to American Cyanamid).
239. U.S. Pat. 2,420,286 (May 6, 1947), H. T. Lacey, H. E. Millson, and F. H. Heiss (to American Cyanamid); U.S. Pat. 2,453,578 (Nov. 9, 1948), H. T. Lacey and R. E. Brouillard (to American Cyanamid).
240. U.S. Pat. 3,366,572 (Jan. 30, 1968), J. M. W. Scott and R. F. Phillips (to American Cyanamid).
241. M. M. Rauhut and A. M. Semsel, unpublished work.
242. U.S. Pat. 3,239,406 (Mar. 8, 1960), D. D. Coffman and H. E. Winberg (to E. I. du Pont de Nemours & Co., Inc.).
243. U.S. Pat. 3,264,221 (Aug. 2, 1966), H. E. Winberg (to E. I. du Pont de Nemours & Co., Inc.).
244. U.S. Pat. 3,728,271 (Apr. 17, 1973), W. S. McEwan, H. B. Jonassen, and C. H. Morley (to U.S. Government); U.S. Pat. 3,728,270 (1973), E. M. Bens and C. H. Morley (to

U.S. Government); U.S. Pat. 3,558,502 (Jan. 26, 1971), A. F. Tatyrek and B. Werbel (to U.S. Government).

245. U.S. Pat. 3,311,564 (Mar. 28, 1967), E. T. Cline (to E. I. du Pont de Nemours & Co., Inc.).
246. U.S. Pat. 3,726,802 (Apr. 10, 1973), C. H. Morley and E. M. Bens (to U.S. Government).
247. U.S. Pat. 3,392,123 (July 9, 1968), H. E. Winberg (to E. I. du Pont de Nemours & Co., Inc.).
248. U.S. Pat. 3,729,425 (Apr. 24, 1973), C. A. Heller, H. P. Richter, and W. S. McEwan (to U.S. Government).
249. U.S. Pat. 3,697,434 (Oct. 10, 1972), S. Shafler (to U.S. Government).
250. U.S. Pat. 3,539,794 (Nov. 10, 1970), M. M. Rauhut and G. W. Kennerly (to American Cyanamid); U.S. Pat. 3,576,987 (May 1, 1971), H. K. Voight and R. L. Myers (to American Cyanamid).
251. U.S. Pat. 3,752,406 (Aug. 14, 1973), P. A. McDermott and A. M. Semsel (to American Cyanamid).
252. U.S. Pat. 3,969,263 (July 13, 1976), H. P. Richter, C. A. Heller, and R. E. Tedrick (to U.S. Government); U.S. Pat. 3,994,820 (Nov. 30, 1976), D. R. Maulding and M. M. Rauhut (to American Cyanamid).
253. U.S. Pat. 3,718,599 (Feb. 27, 1973), M. M. Rauhut (to American Cyanamid); U.S. Pat. 3,974,086 (Aug. 10, 1976), M. M. Rauhut (to American Cyanamid); U.S. Pat. 3,948,797 (Apr. 6, 1976), M. L. Vega (to American Cyanamid).
254. U.S. Pat. 3,813,534 (Sept. 2, 1974), C. W. Gilliam (to U.S. Government); U.S. Pat. 3,819,925 (June 25, 1974), H. P. Richter and R. E. Tedrick (to U.S. Government); U.S. Pat. 3,764,796 (Oct. 9, 1973), C. W. Gilliam and T. N. Hall (to U.S. Government).
255. U.S. Pat. 3,829,678 (Aug. 13, 1974), G. B. Holcombe; U.S. Pat. 3,940,604 (Feb. 24, 1976), M. M. Rauhut (to American Cyanamid).
256. U.S. Pat. 3,816,325 (June 11, 1974), M. M. Rauhut and A. M. Semsel (to American Cyanamid); U.S. Pat. 3,800,132 (June 28, 1974), R. H. Postal (to American Cyanamid); U.S. Pat. 3,500,033 (Mar. 10, 1970), W. T. Cole and B. K. Daubenspek (to Remington Arms); U.S. Pat. 3,578,962 (May 18, 1971), R. L. Gerber (to Remington Arms); U.S. Pat. 3,671,450 (June 20, 1972), M. M. Rauhut and A. M. Semsel (to American Cyanamid).
257. U.S. Pat. 3,808,414 (May 30, 1974), B. G. Roberts (to American Cyanamid); U.S. Pat. 3,893,938 (July 8, 1975), M. M. Rauhut (to American Cyanamid); U.S. Pat. 3,875,602 (Apr. 8, 1975), R. Miron (to American Cyanamid); U.S. Pat. 3,934,539 (Jan. 27, 1976), S. M. Little and co-workers (to U.S. Government); U.S. Pat. 3,933,118 (Jan. 27, 1976), J. H. Lyons, S. M. Little, and V. J. Esposito (to U.S. Government); U.S. Pat. 3,895,455 (July 22, 1975), C. J. Johnson.
258. U.S. Pat. 3,511,612 (May 12, 1970), G. W. Kennerly and M. M. Rauhut (to American Cyanamid); U.S. Pat. 3,584,211 (June 8, 1971), M. M. Rauhut (to American Cyanamid).
259. U.S. Pat. 3,850,836 (Nov. 26, 1974), H. P. Richter and co-workers (to U.S. Government).
260. D. C. Williams, G. F. Huff, and W. R. Seitz, *Anal. Chem.* **49**, 432 (1976); G. Scott, W. R. Seitz, and G. Ambrose, *Anal. Chim. Acta* **115**, 221 (1980); M. L. Grayeski and W. R. Seitz, *Anal. Biochem.* **136**, 155 (1984).
261. L. J. Kricka, *Clin. Chem.* **37**, 1472 (1991).
262. P. E. Stanley and L. J. Kricka eds., *Bioluminescence and Chemiluminescence: Current Status*, John Wiley & Sons, Chichester, U.K., 1991.

263. M. Pazzagli, E. Cadenas, L. J. Kricka, A. Roda, and P. E. Stanley, eds., *Bioluminescence and Chemiluminescence: Studies and Applications in Biology and Medicine*, John Wiley & Sons, Chichester, U.K., 1989.
264. L. J. Kricka, *Anal. Biochem.* **175**, 14 (1988).
265. O. Nozaki, L. J. Kricka, and P. E. Stanley, *J. Biolumin. Chemilumin.* **7**, 223 (1992); L. J. Kricka and G. H. G. Thorpe, *Analyst* **108**, 1274 (1983).
266. A. Szalay, L. J. Kricka, and P. Stanley, eds., *Bioluminescence and Chemiluminescence: Status Report*, John Wiley & Sons, Chichester, U.K., 1993.
267. K. Imai, *Methods in Enzymol.* **133**, 435 (1986).
268. S. Kobayashi and K. Imai, *Anal. Chem.* **52**, 424 (1980); S.-I. Kobayashi, H. Seking, K. Honda, and K. Imai, *Anal. Biochem.* **112**, 99 (1981); Y. Watanabe and K. Imai, *Anal. Chem.* **55**, 1786 (1983); T. Toyo'oka and K. Imai, *J. Chromatogr.* **282**, 495 (1983).
269. G. L. de Jong, N. Lammers, F. J. Spruit, U. A. Th. Brinkman, and R. W. Frei, *Chromatographia* **18**, 129 (1984); G. L. de Jong, N. Lammers, F. J. Spruit, R. W. Frei, and U. A. Th. Brinkman, *J. Chromatogr.* **353**, 249 (1986).
270. O. Shimomura, F. H. Johnson, and Y. Saiga, *Science* **140**, 1339 (1963); E. B. Ridgway and C. C. Ashley, *Biochem. Biophys. Res. Commun.* **29**, 229 (1967).
271. O. Shimomura and F. H. Johnson in E. W. Chappelle and G. L. Picciolo, eds., *Analytical Applications of Biochemiluminescence and Chemiluminescence, NASA-SP-388*, NASA, Washington, D.C., 1975, p. 89.
272. W. R. Seitz and D. M. Hercules in Ref. 10, p. 427.
273. A. K. Babko and N. M. Lukovskaya, *Zh. Anal. Khim.* **17**, 50 (1962); *Zavod. Lab.* **29**, 404 (1963); A. K. Babko and L. I. Dubovenko, *Z. Anal. Chem.* **200**, 428 (1964); A. K. Babko and I. E. Kalinichenko, *Ukr. Khim. Zh.* **31**, 1316 (1965).
274. K. A. Krause in J. H. Yoe and H. J. Koch, eds., *Trace Analysis*, John Wiley & Sons, New York, 1957, pp. 34–101.
275. W. R. Seitz, W. W. Suydam, and D. M. Hercules, *Anal. Chem.* **44**, 957 (1972).
276. W. R. Seitz and D. M. Hercules, *Anal. Chem.* **44**, 2143 (1972).
277. I. Buyas and L. Erdey, *Talanta* **10**, 467 (1963).
278. F. Kenny and R. B. Kurtz, *Anal. Chem.* **25**, 1550 (1953).
279. F. Kenny and R. B. Kurtz, *Anal. Chem.* **23**, 382 (1951).
280. V. Patrovsky, *Talanta* **23**, 553 (1976).
281. D. T. Bostick and D. M. Hercules, *Anal. Chem.* **47**, 447 (1975); J. P. Auses, S. L. Cooks and J. T. Maloy, *Anal. Chem.* **47**, 244 (1975); D. C. Williams, G. F. Huff, and W. R. Seitz, *Clin. Chem.* **22**, 372 (1976).
282. W. G. Baeyens, B. L. Ling, K. Imai, A. C. Calokerinos, and S. G. Schulman, *J. Microcol.*, 195 (1994).
283. D. V. Pollard-Knight in Ref. 206, p. 83.
284. D. C. Williams, C. F. Huff, and W. R. Seitz, *Anal. Chem.* **48**, 1003 (1976).
285. G. Wienhausen and M. A. DeLuca, *Methods Enzymol.* **133**, 198 (1986).
286. R. C. Allen, *Methods Enzymol.* **133**, 449 (1986).
287. T. P. Whitehead, G. H. G. Thorpe, and S. R. J. Maxwell, *Anal. Chim. Acta* **266**, 265 (1992).
288. T. E. Yeshion, in Ref. 206, p. 379.
289. I. Bronstein, J. Fortin, P. E. Stanley, G. S. A. B. Stewart, and L. J. Kricka, *Anal. Biochem.* **219**, 169 (1994).
290. S. W. Lewis, D. Price, and P. J. Worsfold, *J. Biolumin. Chemilumin.* **8**, 183 (1993).
291. L. J. Arnold, Jr., P. W. Hammond, W. A. Wiese, and N. C. Nelson, *Clin. Chem.* **35**, 1588 (1989).
292. L. J. Kricka, ed., *Nonisotopic DNA Probe Techniques*, Academic Press, Inc., San Diego, Calif., 1992.
293. P. E. Stanley, *J. Biolumin. Chemilumin.* **4**, 375 (1989).

294. E. W. Chapelle and G. V. Levin, *Biochem. Med.* **2**, 41 (1968).
295. R. D. Hamilton and O. Holm-Hansen, *Limnol. Oceanog.* **12**, 319 (1967).
296. J. W. Patterson, P. L. Brezonik, and H. D. Putnam, *Environ. Sci. Technol.* **4**, 569 (1970).
297. G. V. Levin, J. R. Schrot, and W. C. Hess, *Environ. Sci. Technol.* **9**, 961 (1975).
298. M. W. Griffiths, in Oxford, 1989, p. 167.
299. G. L. Picciolo and co-workers, in Ref. 271, p. 1.
300. H. Vellend and co-workers, in Ref. 271, p. 43.
301. T. M. Cheng, in Ref. 271, p. 49.
302. O. Holm-Hansen, *Limnol. Oceanogr.* **14**, 740 (1969); *Plant Cell Physiol.* **11**, 689 (1970).
303. E. W. Chapelle, G. L. Picciolo, and R. H. Altland, *Biochem. Med.* **1**, 252 (1967).
304. N. D. Searle, in Ref. 271, p. 95.
305. U.S. Pat. 3,959,081 (1976), S. Witz and W. H. Hartung (to Apzonia Corp.).
306. P. E. Andreotti and co-workers, in Ref. 266, p. 271.
307. W. Worthy, *Chem. Eng. News*, 30 (Nov. 24, 1975); R. A. Nathan and co-workers, *Ind. Res.*, 62 (Dec. 1975); G. D. Mendenhall, *Angew. Chem. Int. Ed. Engl.* **16**, 225 (1977).
308. *Fed. Reg.* **36**, Appendix D, 84 (Apr. 30, 1971).
309. H. C. McKee, *J. Air Pollut. Control Assoc.* **26**, 124 (1976).
310. J. E. Sigsby and co-workers, *Environ. Sci. Technol.* **7**, 51 (1973).
311. P. A. Constant, M. C. Sharp, and G. M. Scheil, *Report EPA-650/4-75-013* U.S., National Technical Information Service, Springfield, Va., PB-246843, Feb. 1975.
312. J. Stauff and W. Jalschke, *Atmos. Enviro.* **9**, 1038 (1975).
313. R. K. Stevens, A. E. O'Keeffe, and G. C. Ortman, *Environ. Sci. Technol.* **3**, 652 (1969).
314. A. Fontijn and R. Ellison, *Environ. Sci. Technol.* **9**, 1157 (1975).
315. D. H. Stedman and D. A. Tommuro, *Anal. Lett.* **9**, 81 (1976).

IRENA BRONSTEIN
Tropix, Inc.

LARRY J. KRICKA
University of Pennsylvania

RICHARD S. GIVENS
University of Kansas

PHOSPHORS

Luminescence is the process of producing light in excess of thermal radiation following an excitation. A solid material exhibiting luminescence is called a phosphor. Phosphors are usually fine inorganic compound powders of a high degree of purity and a median particle size of 3–15 micrometers but may be large single crystals, used as scintillators, or glasses or thin films. Phosphors may be excited by high energy invisible uv radiation (photoluminescence), x-rays (radioluminescence), high energy electrons (cathodoluminescence), a strong electric field (electroluminescence), or in some cases infrared radiation (up-conversion), chemical reactions (chemiluminescence), or even stress (triboluminescence). Figure 1

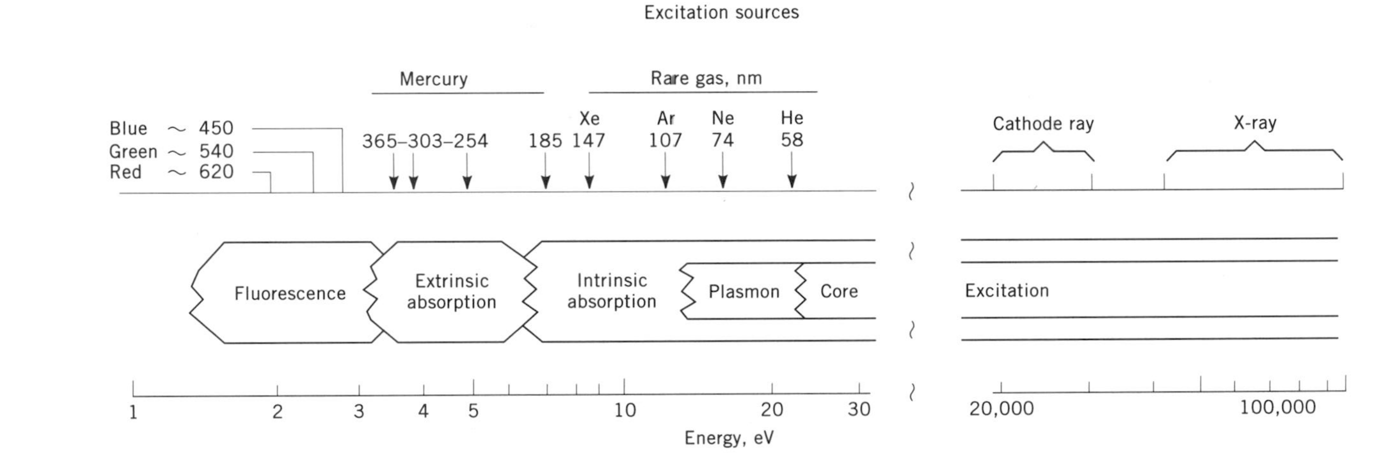

Fig. 1. Common phosphor excitation sources, energies, and the type of solid-state excitation caused by these sources. Following the initial excitation, high energy excitons may create many hole-electron pairs.

shows the electromagnetic energy spectrum indicating some of the common energy forms that excite phosphors. Because phosphors convert the exciting energy to visible radiation, they have many everyday applications; phosphors are responsible for the light generated by fluorescent lamps, televisions, computer terminals, etc.

Phosphors usually contain activator ions in addition to the host material. These ions are deliberately added in the proper proportion during the synthesis. The activators and their surrounding ions form the active optical centers. Table 1 lists some commonly used activator ions. Some solids, made up of complexes such as calcium tungstate [*7790-75-2*], $CaWO_4$, are self-activated. Also in many photoluminescence phosphors, the primary activator does not efficiently absorb the exciting radiation and a second impurity ion is introduced known as the sensitizer. The sensitizer, which is an activator ion itself, absorbs the exciting radiation and transfers this energy to the primary activator.

The optical properties of a phosphor are measured on relatively thick plaques of the phosphor powder. An important optical property for the application of the phosphor is its emission spectrum, the variation in the intensity of the emitted light versus wavelength. Fluorescent lamps must have phosphors which produce white light of high luminous efficiency and with good color rendering properties. Because individual activator centers generally emit in a relatively narrow region of the spectrum producing a colored light, more than one activator or phosphor must be used. Similarly colored televisions employ three phosphors in separate closely spaced dots; one dot contains a phosphor which emits in the blue, one in the green, and one in the red region of the spectrum. Saturated colors are needed in order for the screen to be able to reproduce nearly all colors using these emissions in different relative proportions. In other applications, such as x-ray screens, it is desirable to have an emission spectrum concentrated near the peak in the sensitivity of the receptor, such as the x-ray photographic film. The reflectance spectrum is a graph of the percentage of radiation reflected and absorbed by the powder plaque versus wavelength. The excitation spectrum gives the variation of the light output from the phosphor as a function of the changes

Table 1. Common Activator Ions

Type	Important examples	Color range	Others
$s^2 \longrightarrow sp$	Sb^{3+}	blue-green	Tl^+, Ga^+
broad band	Sn^{2+}	visible	Bi^{3+}, In^+
$d \longrightarrow f$	Eu^{2+}	blue-green	
broad (50 nm)	Ce^{3+}	uv-green	
$O \longrightarrow M$	WO_4^{2-}	460–520 nm	MoO_4^{2-}
very broad (100 nm)	VO_4^{3-}	480–580 nm	NbO_4^{3-}
$d_t \longrightarrow d_e$	Mn^{2+}	510–580 nm	Mn^{4+}, Fe^{3+}
broad and narrow		green-orange	Cr^{3+}, Ni^{2+}
$f \longrightarrow f$	Eu^{3+}	red	Pr^{3+}, Nd^{3+}
narrow	Tb^{3+}	green	Tm^{3+}, Dy^{3+}
			Er^{3+}, Ho^{3+}

in the wavelength of the exciting radiation. When normalized by the reflectance spectrum, the excitation spectrum measures the relative quantum efficiency of a photoluminescent phosphor as a function of the exciting radiation wavelength. The quantum efficiency of the phosphor is the number of quanta emitted by the phosphor divided by the number of quanta absorbed. The methods and precautions needed in measuring these spectra have been discussed (1).

Theory of Luminescence

The Configuration Coordinate Model. To illustrate how the luminescent center in a phosphor works, a configurational coordinate diagram is used (2) in which the potential energy of the luminescent or activator center is plotted on the vertical axis and the value of a single parameter describing an effective displacement of the ions surrounding the activator, Q, is plotted on the horizontal axis (Fig. 2). At low temperatures, near room temperature and below, the activator is in the lowest vibrational level of the ground electronic state. Absorption of energy results in transition to an upper electronic state and since electronic transitions occur rapidly the lattice ions cannot rearrange during the transition according to the Frank-Condon principle, and the transition is seen as a vertical line on the configuration coordinate diagram. Following excitation the activator ion is in an excited nonequilibrium vibrational level of the excited state. The center then undergoes severe anharmonic vibrational motion and relaxes to a new equilibrium shape or size. The weak electromagnetic radiation that occurs during this relaxation is called hot luminescence. Relaxation generally occurs within less than one hundred vibrations or tens of picoseconds with the result that hot luminescence can only be detected by measuring radiation emitted within picoseconds after excitation. Ordinary luminescence takes place after the luminescent center has relaxed to the lowest vibrational level of the excited state.

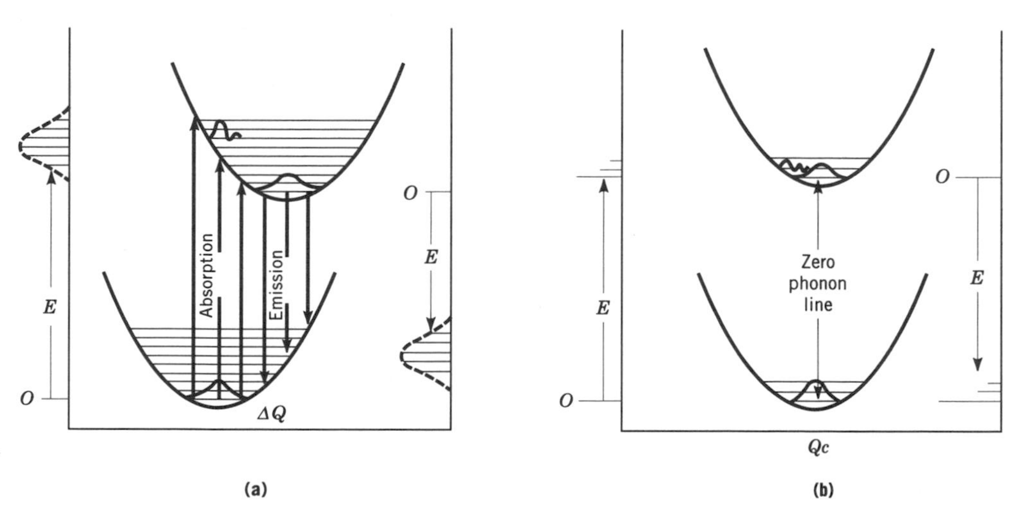

Fig. 2. General configurational–coordinate diagrams for (**a**) broad-band absorbers and emitters, and (**b**) narrow-band or line emitters. The ordinate represents the total energy of the activator center and the abscissa is a generalized coordinate representing the configuration of ions surrounding the activator.

Activator ions that are very strongly coupled to the lattice exhibit a large Stokes' shift. The Stoke's shift is the decrease in energy and increase in wavelength of the emitted radiation compared to the exciting radiation. This is shown on the y-axis of Figure 2**a**. The energy difference goes into heating the lattice. In such cases both the absorption and emission bands are broad and approximately Gaussian in shape when intensity is plotted against radiation energy. On the other hand, sharp spectral emission lines are indicative of activator ions which do not change shape or size significantly between the ground and excited states (see Fig. 2**b**, Table 1). In addition to the Stoke's shift and the general shape of absorption and emission bands, configurational coordinate models explain a number of other experimental observations associated with phosphor spectra. The diagrams themselves can be constructed from careful measurements of optical properties of the phosphor. Approximate configurational coordinate diagrams can be calculated using a molecular-orbital theory, assuming a molecular cluster consisting of the activator and its surroundings.

Nonradiative Decay. To have technical importance, a luminescent material should have a high efficiency for conversion of the excitation to visible light. Photoluminescent phosphors for use in fluorescent lamps usually have a quantum efficiency of greater than 0.75. All the exciting quanta would be reemitted as visible light if there were no nonradiative losses.

The occurrence of nonradiative losses is classically illustrated in Figure 3. At sufficiently high temperature the emitting state relaxes to the ground state by the crossover at B of the two curves. In fact, for many broad-band emitting phosphors the temperature dependence of the nonradiative decay rate P_{nr} is given by equation 1:

$$P_{nr} = A\exp(-E^*/\kappa T) \quad (1)$$

where E^* is the activation energy as shown in Figure 3. However, the occurrence of nonradiative processes is better explained by including quantum mechanical tunneling to the ground state as shown by the dashed arrows C (3). The nonradiative decay rate by C can be orders of magnitude greater than the radiative decay rate and occurs even at moderate temperatures, when only the lowest vibrational levels of the excited state are significantly populated. This rate has been shown to depend exponentially on the energy separation between two electronic levels (4).

$$P_{nr} = A\exp(-\Delta E) \quad (2)$$

If the magnitude of ΔE is less than a few vibrational frequencies, radiationless relaxation always occurs. Hence exciting energy cascades down from one electronic level to another, heating the lattice, until it reaches a state for which E^* is large and ΔE is more than a few vibrational frequencies; then the excitation is trapped. If trapped for times long enough, radiation decay dominates, however, although the temperature dependence of the nonradiative decay rate and its dependence on the energy separation of the two parabolas is in good agreement with theory, the nature of the nonadiabatic interaction between an emitting state

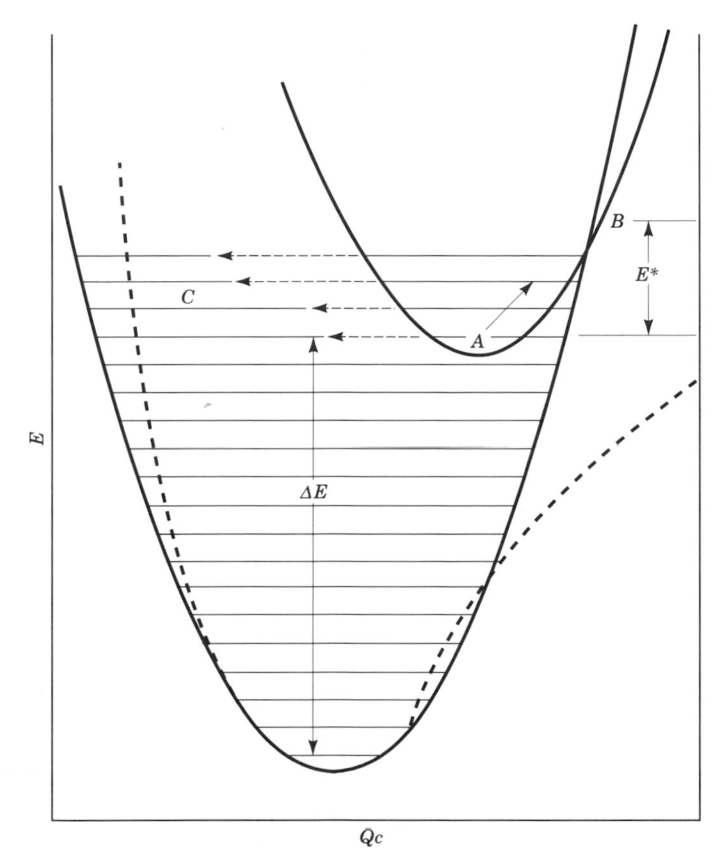

Fig. 3. A configurational–coordinate diagram showing mechanisms of radiationless decay to the ground state. Nonradiative decay to the ground-state vibrational manifold can occur by the semiclassical path $A \longrightarrow B$ or by quantum mechanical tunneling $A \longrightarrow C$ (3). (– – –) represents a more realistic Morse potential for the ground state.

and the ground state is often unknown. Because energy transfer to the ground state curve may occur at a very high vibrational level where anharmonicity in the ground state has a large influence, a quantitative understanding of the important phenomenon of radiationless decay, and the ability to predict which host plus activator systems result in efficient phosphors, appears hopeless (5).

Energy Transfer. In addition to either emitting a photon or decaying nonradiatively to the ground state, an excited sensitizer ion may also transfer energy to another center either radiatively or nonradiatively, as illustrated in Figure 4.

Nonradiative energy transfer is induced by an interaction between the state of the system, in which the sensitizer is in the excited state and the activator in the ground state, and the state in which the activator is in the excited and the sensitizer in the ground state. In the presence of radiative decay, nonradiative decay, and energy transfer the emission of radiation from a single sensitizer ion decays exponentially with time, t.

$$I(t) = I(0)\exp(-P_{\text{rad}}t - P_{nr}t - P_{et}t) \quad (3)$$

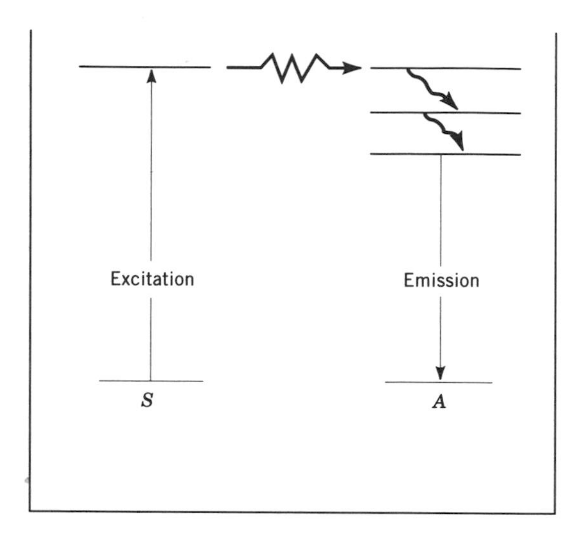

Fig. 4. A schematic diagram showing energy transfer from sensitizer S to activator A followed by relaxation from one electronic level to another and then emission.

P_{et} can be written as a series of terms which depend on the distance between the sensitizer and activator ions. Contributions include a leading term, called the dipole–dipole interaction, which is proportional to R^{-6}; a term called electric dipole–electric quadrupole interaction, proportional to R^{-8}; the electric quadrupole–quadrupole, proportional to R^{-10}; an exchange interaction caused by the overlap of the wave functions of the sensitizer and activator ions that varies as $\exp(-cR)$; and an electric dipole–magnetic dipole interaction due to the presence of a magnetic field at the activator ion caused by the motion of the sensitizer electrons. Because different sensitizer ions have different distributions of activators around them, the statistical average of equation 3 gives an observed radiation decay which is nonexponential and when carefully measured can be analyzed to determine the mechanism of energy transfer. Because of the importance of nonradiative energy transfer in many commonly used phosphors, these mechanisms have been extensively studied. In addition to its dependence on the distance between the sensitizer and activator ions, P_{et} is proportional to the overlap of the emission band of the sensitizer with absorption bands of the activator so that energy is conserved in the transfer (6).

Lamp Phosphors

In fluorescent lamps, phosphors are coated on the inside of the lamp tube using a slurry containing the powder and a liquid which is either poured down through the tube, up-flushed, or in some cases the tubes are filled and then drained. Because of concerns over having volatile organic solvents in the air, the liquid medium containing the powder is usually water with an added agent, a thickener, to increase the viscosity of the suspension, such as poly(methacrylic acid) or poly(propylene oxide). Other additives are included, such as dispersants, in order to improve the dispersion of the powder, defoamers (qv), and sometimes powder adherence additives, such as fumed alumina, Alon, or boric oxide.

It is important to dry the coating quickly; usually hot air flowing in and around the bulb is used in order to prevent the powder from flowing off the bulb. This creates a layer of powder of relatively uniform thickness on the tube, although generally the coating is thinner at the top and thicker at the bottom. Phosphor coatings for fluorescent lamps in the center region of the lamp are about four particles thick. Such a thickness is sufficient to cover the glass tube so that statistically there is little area, not more than 1 or 2% of which is uncovered. The visible reflectance of such a coating is around 55–65%. A still thicker coating would waste phosphor material, and by further increasing the visible reflectance of the coating would make it more difficult for visible light generated near the inner surface of the phosphor layer to get out of the lamp.

During lamp operation mercury atoms are ionized and excited, and after being excited emit their characteristic resonance uv radiation predominately at 254 and 185 nm. The conversion of electrical energy in this way into invisible uv radiation is quite efficient, about 70%. A phosphor for fluorescent lamps must be capable of absorbing the uv and converting it through relaxation between energy levels and the Stoke's shift to visible light of a suitable white color with high luminous efficiency and good color rendering ability. It must also be capable of withstanding the environment of mercury discharge.

The Calcium Halophosphate Phosphors. Early fluorescent lamps used various combinations of naturally occurring fluorescent minerals. The development of the calcium halophosphate phosphor, $Ca_5(PO_4)_3(Cl, F){:}Sb^{3+}, Mn^{2+}$, in the 1940s was a significant breakthrough in fluorescent lighting (7). As is often the case in new phosphor discoveries, this phosphor was found accidentally while searching for phosphors for radar screens.

In the halophosphate phosphor Sb^{3+} sensitizer ions absorb the uv radiation from the discharge thereby undergoing a transition from the ^1S ground state to the excited singlet ^1P and triplet ^3P state of the $5s5p$ configuration. Relaxation occurs to the lower lying triplet ^3P state. Antimony ions emit part of this energy in a band peaking near 480 nm. Energy is also transferred to Mn^{2+} ions which emit near 580 nm (Fig. 5). By increasing the concentration of the Mn^{2+} activator, more and more of the emission is in the orange Mn^{2+} band allowing attainment of a range of whitish colors from near blue to orange. The energy transfer from Sb^{3+} to Mn^{2+} is key to application of this phosphor. By careful measurement of the decay of the sensitizer Sb^{3+} ion, and a comparison with theory, a comprehensive study of this energy transfer was made which indicates that the energy transfer occurs by an exchange mechanism rather than longer range dipole–dipole or dipole–quadrupole mechanisms (8). A further variation in color can be achieved by changing the F:Cl ratio. As the concentration of F increases, the Mn^{2+} band shifts toward the green. The total range of colors possible with halophosphate phosphors in typical fluorescent lamps, which include visible mercury line radiation, is shown on the CIE (Commission de L'Eclairage) color diagram in Figure 6.

The crystal structure of the calcium fluoroapatite has two different crystallographic sites for the Ca^{2+} ion. The Ca(I) site has a threefold axis of symmetry and is coordinated to six oxygen ions at the vertices of a distorted trigonal prism. The Ca(II) ions are located at the corners of equilateral triangles centered around a fluoride ion. The site symmetry for the Ca(II) site is C_{1h}. Epr data suggest that

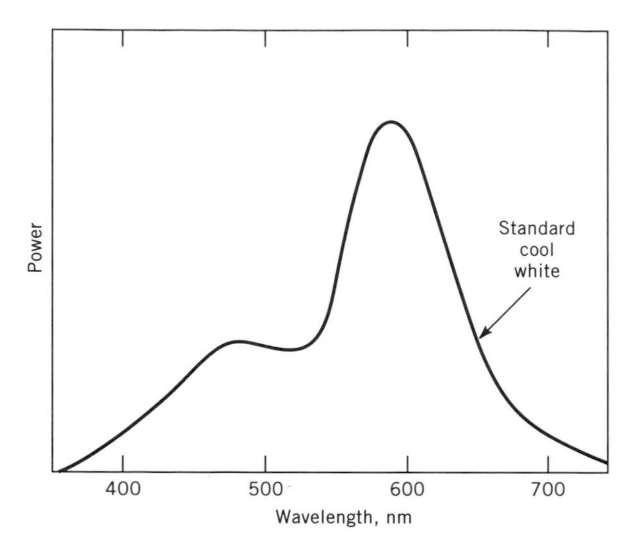

Fig. 5. The emission spectrum of a typical cool white halophosphate phosphor showing the Sb^{3+} emission band around 480 nm and the Mn^{2+} emission band around 580 nm.

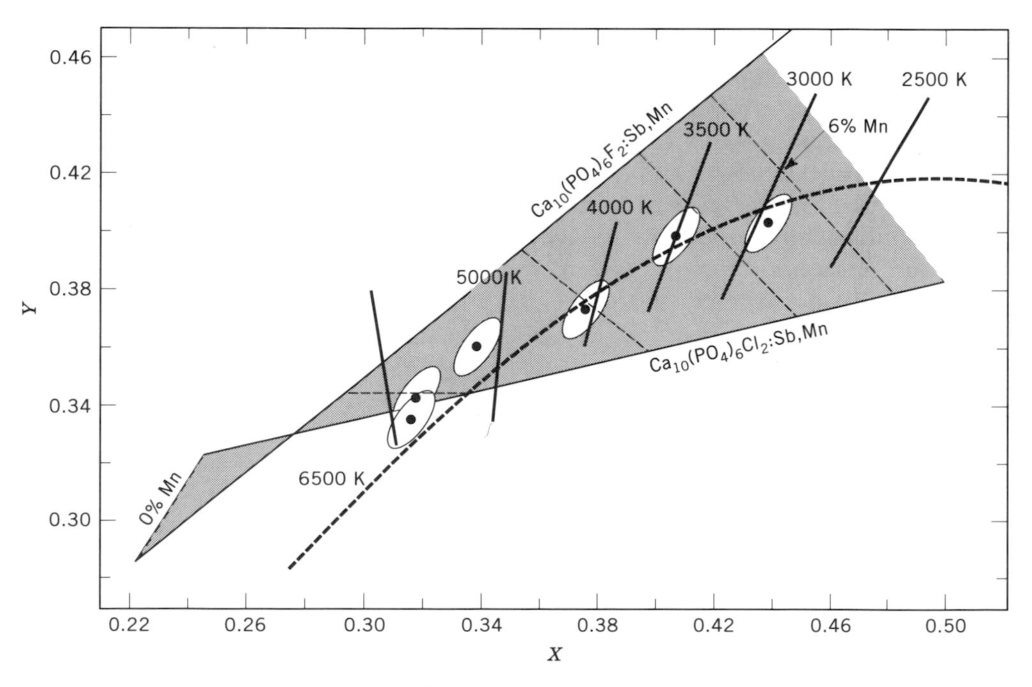

Fig. 6. A portion of the CIE color diagram where X and Y define the color. The shaded area shows the range of color of fluorescent lamps using calcium halophosphate phosphors. Also shown are standard white lamp colors. The locus of colors of Planckian radiators (– – –) and the correlated color temperatures in Kelvin (—) are indicated.

the Mn^{2+} ions occupy both the Ca(I) and Ca(II) sites in the lattice with a preference for the Ca(I) site at low Mn^{2+} concentrations (9). Optical measurements support the epr data. The Mn^{2+} ions on the Ca(I) sites have been shown to be responsible for most of the observed luminescence, whereas emission from the Mn^{2+} ions on the Ca(II) sites occur only for higher Mn^{2+} concentrations. Neutron diffraction studies have also confirmed these results (10).

Because Sb^{3+} does not have any unpaired electrons, its site preference is ambiguous and much more difficult to determine. It has been speculated that the Sb^{3+} ions would occupy the Ca(II) sites with charge compensation achieved by an oxygen replacing an adjacent halogen ion. This model was further supported by comparing predictions of molecular orbital calculations with the excitation spectra for the Sb^{3+} ion in the fluoroapatite (11). A marked change in the excitation spectrum, which occurs when oxygen is excluded from providing charge compensation, is also in agreement with the calculations. X-ray data (12) support the location of the Sb^{3+} ions on the Ca(II) sites and the mentioned nature of charge compensation. Nevertheless an opposing view and some experimental measurements suggest that Sb^{3+} ions may occupy *P* sites in the structure (13).

The halophosphate phosphors are synthesized by first thoroughly blending $CaHPO_4$, $CaCO_3$, CaF_2, NH_4Cl, $MnCO_3$, and Sb_2O_3 powders. The calcium-containing intermediates are generally prepared and purified by the phosphor manufacturer. In particular, the $CaHPO_4$ is often recrystallized after dissolving the initially precipitated $CaHPO_4{\cdot}2(H_2O)$ in an acid solution heated to around 80–90°C. The recrystallized $CaHPO_4$ is of a high degree of purity and consists of clear well-formed plates the size of which determines the size of the final phosphor.

The intermediates are blended in large blenders. Surprisingly, it is important to avoid formulations of the blend which have the stoichiometric cation:phosphate ratio and halide:phosphate ratio for two reasons. First, the NH_4Cl and Sb_2O_3 powders react to form volatile antimony oxychloride compounds, hence an excess of these species is required. If the chloride drops below the stoichiometry due to this volatilization then some calcium orthophosphate is formed and the Mn^{2+} activator ions enter this undesired phase giving the cake a pink color. On the other hand a large excess of the chloride results in greater losses due to volatilization and some additional sintering. The losses due to volatilization depend on the amount of excess chloride, the final temperature of synthesis, cake depth, etc. A loosely fitting cover for the crucible in which the blend is fired must be used to prevent variation in the stoichiometry due to volatilization. The second important consideration in the blend formulation is the use of slightly less $CaCO_3$ than is required to meet the stoichiometry of the material. The phase diagram indicates that for such a blend composition a small amount of the calcium pyrophosphate phase is formed. The activator Mn^{2+} ions do not enter this relatively inert second phase; however, a large excess of this phase generally results in excessive sintering and shrinkage resulting in a hard cake which is difficult to process further. If on the other hand the amount of the pyrophosphate phase formed is too little or does not exist at all then the cake is very soft and there is a risk of forming CaF_2 and CaO as undesired second phases due to normal weighing errors in manufacturing. Typically the blend is formulated to give 1–3 wt % of the calcium pyrophosphate phase.

After thorough blending the powder is placed in a suitable crucible and heated in a furnace at temperatures between 1075 and 1150°C. During the heating, H_2O and CO_2 are given off at 450 and 850°C, respectively. The halophosphate phosphor starts to form around 1000°C. At this stage some manufacturers lightly mill the material and second fire it in a nitrogen atmosphere to ensure the full incorporation of the manganese ions in the divalent state.

After firing, the powder is washed in water typically with a small amount of complexing agent such as ethylenediaminetetraacetic acid (EDTA), sodium EDTA, or a weak acid such as citric acid to remove the excess chloride, volatile antimony oxychlorides which have recondensed on the phosphor during cooling, and manganese compounds which are not incorporated in the halophosphate lattice. The powder is then ready for suspension.

The morphology of the resulting halophosphate phosphor closely resembles the plate-like morphology of the starting $CaHPO_4$ but with an important difference: very small submicrometer grains of the calcium halophosphate can be seen. These grains are fused together and oriented but not epitaxial. There are also holes in the particles apparently where water and carbon dioxide have escaped. At least one manufacturer (Nichia, Japan) appears to have been able to make single-crystal particles.

Because it is still by far the most commonly used phosphor in fluorescent lamps, calcium halophosphate total production far exceeds that of all other phosphors put together, in excess of 1000 metric tons per year.

Deluxe Phosphors. Because the two complementary emission bands from the calcium halophosphate phosphor do not fill the visible region of the spectrum and in particular are deficient in the red region of the spectrum, colors are distorted under these lamps compared to their appearance under blackbody radiator sources or sunlight. This distortion is measured by color rendition indexes for different colors and by the average color rendition index *Ra*. To improve color rendition, blends of other broad-band emitting phosphors are used in some fluorescent lamps. In particular, strontium or strontium:calcium:barium orthophosphate activated with tin is commonly used to provide a broad-band red emission peaking near 620–630 nm. This phosphor is blended with strontium halophosphates. The strontium halophosphates are completely analogous to the calcium halophosphates but with the Mn^{2+} band shifted toward 560 nm and the Sb^{3+} band shifted toward the green. Using a blend of the broad-band green strontium halophosphate phosphors together with the broad red emission of the orthophosphate there is continuous emission of radiation across the visible region of the spectrum and good color rendering Delux-type lamps can be made. However, because the broad-band emissions extend outside of the range of the eye sensitivity function and because of rather low quantum efficiencies of these phosphors, the lamps only have about two-thirds of the light output of standard halophosphate lamps.

Triphosphors. The lighting industry underwent a revolution in the 1970s following theoretical work (14,15) which demonstrated that improved efficiency and a much improved color rendering ability was possible with a spectrum having three emission bands: red at 610 nm, green at 545 nm, and blue at 450 nm. These wavelengths are near peaks in the CIE tristimulus functions which are used to define colors. Further work has shown that it is particularly important to have a narrow emission in the red near 610 nm. If the red emission is moved to longer wavelengths or broadened, color rendition improves but luminous efficiency decreases. If it is moved to shorter wavelengths the color rendering ability of the lamp drops sharply. On the other hand there is more flexibility in choice of green and blue emission bands.

The optimum spectra were made a reality by the discovery in the mid-1970s of a class of rare-earth phosphors which provided these emissions. Activator ions chosen from the rare-earth elements are ideal for providing narrow emission bands in selected wavelength regions due to their many $4f$–$4f$ optical transitions. These levels are weakly coupled to host material and so the ground and excited state curves illustrated in Figure 2 are almost directly above each other. This also explains the good thermal characteristics of these phosphors because thermally induced nonradiative relaxation occurs at a much higher energy. Typically the rare-earth ions relax from one electronic state to another and then emit from a level well separated in energy from any level directly below. In suitable hosts Eu^{3+} emits red-orange radiation from the $^5D_0 \rightarrow F_2$ transition, which is used for the red emission; the Tb^{3+} $^5D_4 \rightarrow F_4$ transition is used for the green; and divalent europium, Eu^{2+}, $d \rightarrow f$, for the blue band (Fig. 7). The emission spectrum of a plaque of a typical triphosphor blend is shown in Figure 8.

The use of rare-earth phosphors in fluorescent lamps has also resulted in improved maintenance as compared to the halophosphate phosphor throughout

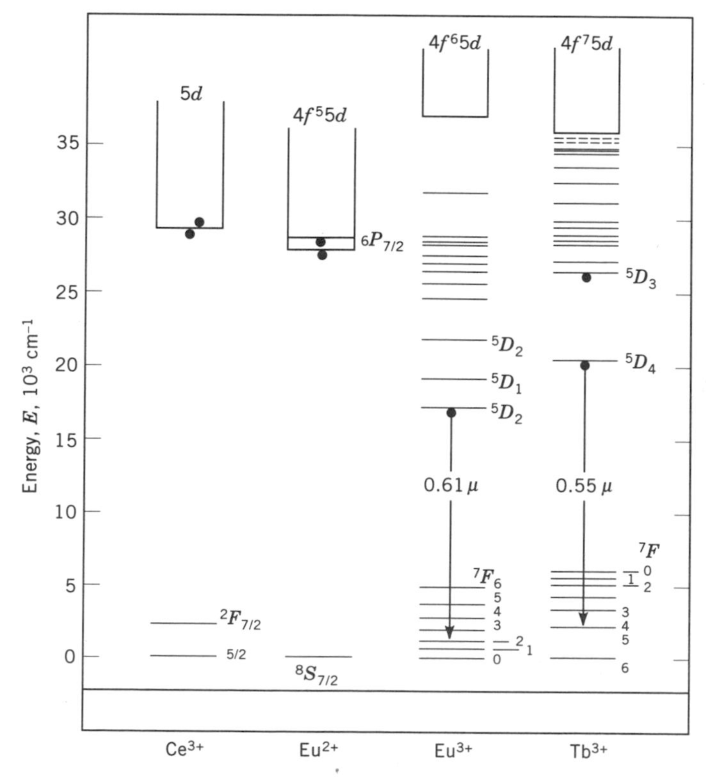

Fig. 7. Energy levels of the most commonly used rare-earth activators.

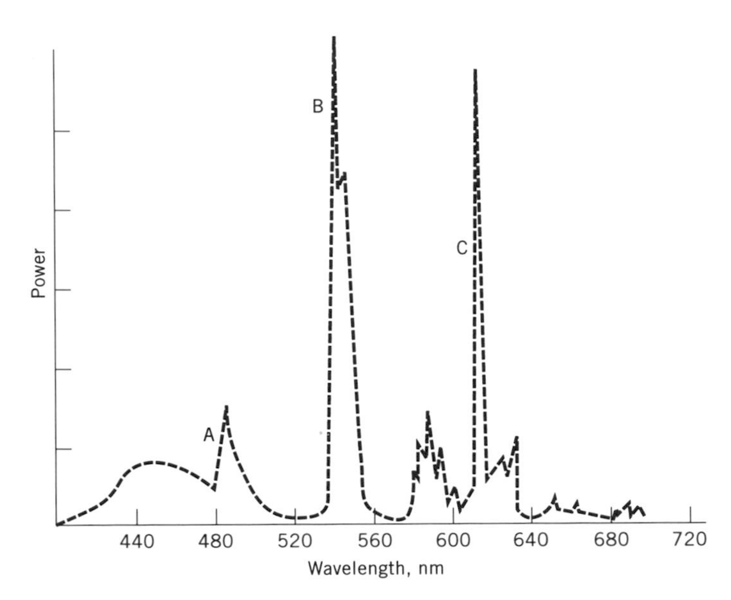

Fig. 8. The emission spectrum of a triphosphor blend where A is $BaMg_2Al_{16}O_{27}:Eu^{2+}$; B, $CeMgAl_{11}O_{19}:Tb^{3+}$; and C, $Y_2O_3:Eu^{3+}$.

the life of a lamp. Phosphor lumen maintenance is determined by the resistance of the material to higher energy 185-nm radiation from the mercury discharge and resistance to mercury ion bombardment. The compact fluorescent lamps which have gained popularity in the 1990s owe their existence to the ability of these phosphors to resist degradation even under high loading over the life of the lamp.

The principal disadvantage of the rare-earth activated phosphors is their high cost. Ores containing the rare-earth elements are found principally in China, where they are particularly plentiful, Australia, and the United States. However, the high cost of rare-earth elements for phosphors is primarily due to the extensive number of liquid separation steps required to isolate individual rare-earth elements from the ore which generally contains many or most of the lanthanides (qv). As a result of the demand for these phosphors for television, x-ray screens, and fluorescent lamps, this technology has been greatly improved and automated thereby providing a ready source of the individual rare-earth elements with only a few to around 10 parts per million or less of the other individual rare-earth elements. Prices have dropped dramatically from, for example, over $500/kg for $Y_2O_3:Eu^{3+}$ precursor rare-earth oxide precipitate in the mid 1970s to well under $100/kg in 1994. The principal companies involved in separating rare-earth elements for phosphors include several plants in China, Shin-Etsu in Japan, Rhône-Poulenc in France, and Molycorp in the United States.

The cost of rare-earth phosphors in fluorescent lamps is often reduced by double coating the lamps. The rare-earth phosphor blend is coated over a base layer of the inexpensive halophosphate phosphor (Fig. 9). In this configuration it absorbs a disproportionate amount of the uv discharge. For example, about 70% of the uv is absorbed in the inner coating with only one layer of triphosphor particles on the inside.

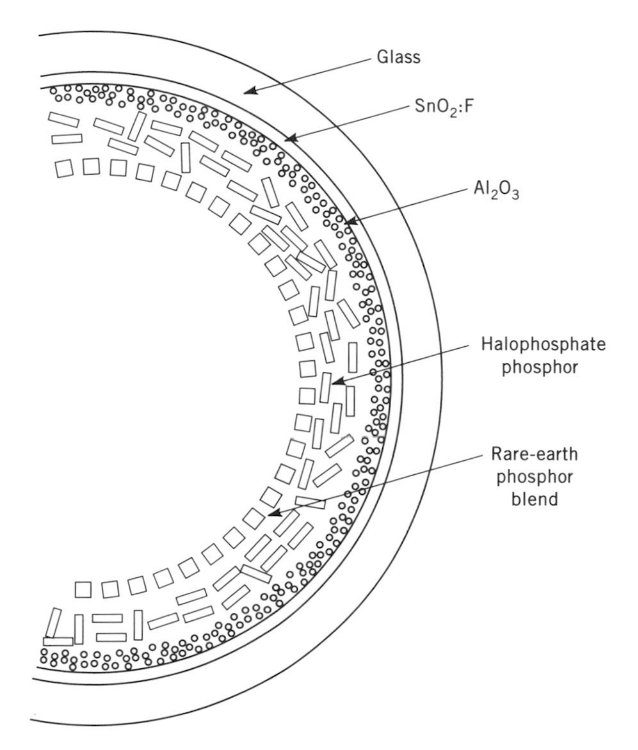

Fig. 9. A modern fluorescent lamp coating including a conductive layer of SnO_2:F, then a protective coating of finely divided alumina, followed by the inexpensive halophosphate phosphor, and finally a thin layer of the triphosphor rare-earth blend.

Further cost reduction can be achieved by decreasing the particle size of the phosphors. A covering of four layers of phosphor powder requires proportionately less material with decreasing particle size. A decrease in particle size can continue as long as the individual particles remain good absorbers of the ultraviolet radiation. Typically halophosphate phosphors have a particle size distribution which is log normal with a median around 11 micrometers, although at least one manufacturer (General Electric Lighting) has reduced its standard product median size to 8–9 μm.

The Red-Emitting Triphosphor. Eu^{3+}-activated Y_2O_3 phosphor is the universally used red-emitting triphosphor for lamps. The emission spectrum of this phosphor is almost ideal being dominated by one strong line at 611 nm. This $^5D_0 \rightarrow F_2$ transition is called hypersensitive because if, for example, europium occupies a site with a strict center of symmetry then only the magnetic dipole transition $^5D_0 \rightarrow F_1$ is expected, which is in the orange and is not useful. On the other hand, if the site symmetry deviates greatly from inversion symmetry, a significant amount of radiation will come from the $^5D_0 \rightarrow F_{4,6}$ transitions which are in the far red and infrared.

The crystal structure of Y_2O_3 is of the rare-earth sequisoxide *C*-type in which each of the cations are coordinated to six oxygen ions at the corners of the cube. Two of the corners of the cube have anion vacancies, and these vacancies can either be located along the body diagonal or along the face diagonal of the cube, S_6 and C_2, respectively. The S_6 site has inversion symmetry, and since the desired $^5D_0 \rightarrow F_2$ transition dominates in the optical spectra of Eu^{3+} ion only if the crystallographic site for the rare-earth ion lacks a center of symmetry, the bulk of Eu^{3+} emission in Y_2O_3 originates from the C_2 site. It is not easy to distinguish the emission from the two sites. The concentration of the S_6 site is only one-third that of the C_2 site and the rare-earth transitions on this site exhibit low oscillator strength due to the presence of the inversion symmetry. Complex energy-transfer pathways have been determined between the two centers at liquid helium temperatures when exciting within the $4f^6$ levels of the Eu^{3+} ion in Y_2O_3 (16).

In the yttrium oxide europium phosphor, a uv photon is absorbed at the Eu^{3+} activator center and emission comes directly from this same center after relaxation. This fact contributes to the high quantum efficiency, in excess of 90%, of this phosphor. The excitation band is due to a ligand (O^{2-}) to metal (Eu^{3+}) charge-transfer transition and has a maximum at roughly 230 nm. As a result, the absorption of 254 nm uv radiation is not high with plaques reflecting 25–35% of this radiation, depending on the europium concentration. Because the green and blue phosphor components of the blend are generally good absorbers of 254-nm radiation and because a lot of red emission, the amount depending on the desired color of the lamp, is required in triphosphor blends to achieve white light, the yttrium oxide phosphor is the most expensive component of the blend.

In order to make an efficient $Y_2O_3{:}Eu^{3+}$, it is necessary to start with well-purified yttrium and europium oxides or a well-purified coprecipitated oxide. Very small amounts of impurity ions, particularly other rare-earth ions, decrease the efficiency of this phosphor. Ce^{3+} is one of the most troublesome ions because it competes for the uv absorption and should be present at no more than about one part per million. Once purified, if not already coprecipitated, the oxides are dissolved in hydrochloric or nitric acid and then precipitated with oxalic acid. This precipitate is then calcined, and fired at around 800°C to decompose the oxalate and form the oxide. Finally the oxide is fired usually in air at temperatures of 1500–1550°C in order to produce a good crystal structure and an efficient phosphor. This phosphor does not need to be further processed but may be milled for particle size control and/or screened to remove agglomerates which later show up as dark specks in the coating.

The Green-Emitting Phosphor. The usual green-emitting phosphors for triphosphor blends contain Ce^{3+} as a sensitizer to absorb the uv from the discharge, and the green-emitting Tb^{3+} ion for the activator. The Ce^{3+} ion both absorbs the uv and emits some broad-band radiation in this phosphor at around 350 nm in the uv, and the Ce^{3+}-only phosphor is sometimes used for suntanning lamps.

$CeMgAl_{11}O_{19}{:}Tb^{3+}$. The first triphosphor lamps were made possible with the discovery of this phosphor and the class of aluminate-based hosts, the crystal structures of which are the magnetoplumbite ($PbFe_{12}O_{19}$) and β-alumina types. These closely resemble alumina and consist of spinel blocks separated, in the case of magnetoplumbite, by layers of cations including the Ce^{3+} and Tb^{3+} ions, and in the case of β-alumina by large cation–oxygen pairs. A variety of

ionic substitutions are possible in the layers between the spinel blocks, and the thickness of the spinel layers also can be varied (17,18).

The absence of concentration quenching of the Ce^{3+} luminescence up to the stoichiometric composition suggests that there is no energy migration over Ce^{3+} ions in $CeMgAl_{11}O_{19}$. With increasing terbium the emission of Ce^{3+} decreases. However, the energy transfer is not particularly efficient so that the Ce^{3+} emission is effectively quenched and the Tb^{3+} reaches its maximum at about $Ce_{0.65}Tb_{0.35}MgAl_{11}O_{19}$.

The synthesis of this phosphor requires both a strongly reducing atmosphere and a high temperature. The starting materials are the rare-earth oxides CeO_2 and Tb_4O_7, and $Al(OH)_3$ and MgO or basic magnesium carbonate. The firing temperature must be higher than 1400°C and for the stabilization of the trivalent state of the rare-earth ions a strongly reducing atmosphere containing a high concentration of hydrogen is required. Any residual tetravalent oxidation state of the rare-earth ions is detrimental to phosphor performance. A remarkable property of the luminescence in this structure is the maintenance of high quantum efficiency of luminescence at temperatures as high as 500°C.

$LaPO_4$:Ce^{3+}, Tb^{3+}. The green luminescence of a $LaPO_4$ lattice activated with Ce^{3+} and Tb^{3+} has been known since the 1970s, but only in the 1990s has the material gained importance as a green-emitting fluorescent lamp phosphor (19). The crystal structure of mineral monazite $LaPO_4$ is monoclinic where the site symmetry of the rare-earth ions is C_1.

The energy-transfer process between the Ce^{3+} and Tb^{3+} ions in a typical commercial composition, $La_{0.60}Ce_{0.27}Tb_{0.13}PO_4$, is different from that in the aluminate material previously described. At the 27 mol % Ce^{3+} doping in $LaPO_4$ there is considerable migration of the absorbed energy within the Ce^{3+} ions. The activator Tb^{3+} ions capture the migrating excitation energy and then give their characteristic green luminescence.

When properly manufactured the efficiency of this phosphor is 5–6% higher than the aluminate phosphor in most lamp applications. However, perhaps because of the energy transfer between cerium ions and the ease with which cerium can be oxidized, the preparation of this phosphor with good efficiency in a lamp has been elusive. One problem is that after being coated on a lamp and dried, the lamp passes through a lehring oven which heats it to ~600°C in air. This is necessary to decompose all traces of the organic agents used in the suspension. This lehring, perhaps by oxidizing some of the Ce ions, often has an adverse effect on the efficiency of this phosphor.

The lanthanum phosphate phosphor is usually prepared by starting with a highly purified coprecipitated oxide of lanthanum, cerium, and terbium blended with a slight excess of the stoichiometric amount of diammonium acid phosphate. Unlike the case of the aluminate phosphor, firing is carried out in an only slightly reducing or a neutral atmosphere of nitrogen at a temperature ~1000°C. Also this phosphor is typically made with the addition of a flux, which provides a molten salt medium that dissolves the rare-earth oxides and allows small well-formed crystals of the phosphate phosphor to grow from the melt as it cools. After firing, the flux must be removed by thorough washing in hot water.

$GdMgB_5O_{10}$:Ce^{3+},Tb^{3+}. A new class of ternary monoclinic pentaborate compounds with the general composition $LnMgB_5O_{10}$ has been synthesized and

characterized (20). The crystal structure of the compounds contains the rare-earth ions coordinated by 10 oxygen ions. Asymmetric rare-earth–oxygen polyhedra share edges to form isolated zig-zag chains. The shortest intra-Ln–Ln distances are ~0.4 nm, whereas the shortest interchain distance is ~0.64 nm. The structure thus displays one dimensionality of the Ln–Ln chains which lead to interesting energy migration studies in this material (21). The visible quantum efficiency of the borate phosphor is high and the phosphor displays excellent stability in fluorescent lamps.

Energy transfer first occurs from Ce^{3+} to Gd^{3+} ions. The energy then migrates along the one-dimensional Gd chains until it reaches a Tb^{3+} ion. This intermediate role of the Gd^{3+} ions in transporting energy from the sensitizer to the activator was first demonstrated in a number of stoichiometric Gd^{3+} compounds (22).

The $GdMgB_5O_{10}$:Ce^{3+}, Tb^{3+} is synthesized by a solid-state firing of the rare-earth coprecipitated oxide plus boric acid and $MgCO_3$ at ~ 900°C in a slightly reducing atmosphere. As in the case of the lanthanum phosphate phosphor, a flux is usually used. The synthesis of this phosphor is further complicated, however, by the fact that it is a ternary system and secondary phases such as gadolinium borate form and must then react to give the final phosphor.

Blue-Emitting Triphosphor Components. Two blue-emitting phosphors are commonly used in triphosphor systems. One is the phosphor $BaMg_2Al_{16}O_{27}$: Eu^{2+} which has the β-alumina structure previously discussed. This phosphor composition can be written $(Ba_xEu_yMg_{1-x-y}O)(Al_2O_3)_z$, which shows the alumina spinel blocks and the layers containing the divalent cations and oxygen atoms and illustrates the variability in composition which is possible. Activation with Eu^{2+}, which occupies the Ba^{2+} sites of the host lattice, yields a highly efficient phosphor with emission maximum near 450 nm. The full width at half maximum is ~50 nm so that the emission is useful in supplying narrow-band blue emission in the phosphor blend. This host can also be activated with Mn^{2+} and the incorporation of manganese at the magnesium sites results in efficient energy transfer from europium to manganese with the emission from manganese providing a very saturated green peaking around 515 nm.

The synthesis of this phosphor is similar to that described for the green-emitting aluminate phosphor. $Al(OH)_3$, $BaCO_3$, $MgCO_3$, and Eu_2O_3 are blended and typically fired at ~1400°C in a mildly reducing atmosphere to maintain the divalent state of the europium ion. Somewhat lower temperatures are possible if BaF_2 and MgF_2 are used as sources of some of the cations to provide fluxing action. Some manufacturers also use zinc oxide and fluoride. The zinc is then incorporated in the phosphor in magnesium locations.

$Sr_{5-x-y}Ba_xCa_y(PO_4)_3Cl$:$Eu^{2+}$ is the second commercially important blue-emitting phosphor. This phosphor is a halophosphate activated with europium which goes into the alkaline-earth sites. The pure strontium halophosphate phosphor gives a narrow blue emission peaking near 450 nm. However, the color rendition of the lamp can be slightly improved by adding barium which results in a second band peaking at somewhat longer wavelengths. Some calcium appears to help the stability of this phosphor but the amount of calcium used should not exceed $y = 1$.

Since there are no volatile components this halophosphate phosphor is prepared with close to the stoichiometric amounts of $SrHPO_4$, $SrCO_3$, $CaCO_3$, $BaCO_3$, $SrCl_2$, or NH_4Cl and Eu_2O_3. The blend is fired under an atmosphere containing 1–2% hydrogen at 1100°C. A small excess of chloride provides some fluxing action and gives well-formed crystals of apatite. The chlorapatites are dimorphous: one modification is hexagonal and the other monoclinic.

Other Lamp Phosphors. There are a number of other phosphors used in lighting for special applications. For example, there are several uv-emitting phosphors employed in industrial photochemical applications such as polymer curing, skin-tanning lamps used in tanning salons, so-called black lights used to cause dyes in fabrics to fluoresce, and in insect traps; insects can see the ultraviolet emission and are attracted to it. $BaSi_2O_5:Pb^{2+}$ has been used as an all purpose uv emitter in the past but because of poor maintenance it is largely supplanted. $Sr_4B_4O_7:Eu^{2+}$ emits in a narrow band centered at 368 nm. This phosphor is made by mixing $SrCO_3$ and Eu_2O_3 with an excess of H_3BO_3. The material is fired at around 400°C to dehydrate the H_3BO_3 to B_2O_3 and then fired again at 900°C to form the phosphor. The excess B_2O_3 serves as a fluxing agent and must be removed from the final product by washing. Another uv emitter as mentioned above is the green aluminate phosphor without Tb^{3+}.

Fluorescent lamps used in photocopiers are configured so that part of the lamp acts as a reflector. The lamp is first coated with a high visibly reflecting material, such as finely divided titania used in paints or alumina. It is then coated with phosphor and a window is scraped off to allow the light to escape through the aperture giving the lamp a highly directional output. The phosphor emission is chosen to match the action spectrum of the photosensitive element on the copier drum. Typically a blue-green emission around 490 nm is desired. A commonly used but expensive phosphor for this purpose is magnesium gallate activated with manganese.

Fluorescent lamps for showing plants use a blue-white phosphor blended with a deep red-emitting phosphor. This more closely corresponds to the action spectrum for plant growth; because there is little green in the spectrum, African violets, for example, have leaves which appear more purple in color. The deep red emitter which is commonly used is magnesium fluorogermanate activated by Mn^{4+}.

There are still other fluorescent lamps having improved color rendering ability compared to the triphosphor lamps. These are sometimes called four- and five-band lamps since they use four and five emission bands or phosphors compared to the three of triphosphor lamps. To improve color rendering slightly without losing efficiency, some Japanese lamps include a blue-green phosphor added to their triphosphor blends. The phosphors $SrAl_{14}O_{25}:Eu^{2+}$, $Sr_6P_5BO_{20}:Eu^{2+}$, and $BaAl_8O_{13}:Eu^{2+}$ emit near 490 nm and have high quantum efficiencies making them good candidates for the blue-green emission (23). To get very high color rendering close to an incandescent or blackbody radiator, the color rendition index (Ra) ~95, which is useful for color critical applications such as display lighting, it is necessary to use a deeper red-emitting phosphor than $Y_2O_3:Eu^{3+}$. European versions of these very high Ra lamps use the pentaborate phosphor discussed above activated with Mn^{2+} giving a broad red emission peaking at ~ 620 nm. This deeper red emitter is combined with a blue-green phosphor, halophosphate, and triphosphor green and blue phosphors. For lower color temperatures, it is necessary to remove some of the visible blue mercury line emission which can be done either with the magnesium germanate phosphor or with a Ce^{3+} activated aluminate phosphor, yttrium aluminum garnet (YAG). In YAG, Ce^{3+} absorbs both blue and near-uv radiation and emits in the green region of the spectrum.

Finally, phosphors are also used for applications in high pressure mercury discharge lamps. The most commonly used phosphor for this application is yttrium vanadate and yttrium vanadate phosphate activated with Eu^{3+}. In this phosphor the near-uv from the lamp is absorbed by vanadate groups near the surface of the phosphor particle and then the excitation undergoes a random walk migrating from one vanadate group to an adjacent group until transfer occurs to a nearby Eu^{3+} ion. The addition of phosphate groups to the lattice impedes this migration and the vanadate emission increases. The Eu^{3+} ion emits in the red region of the spectrum and provides color correction for the mercury lamp. These lamps sometimes also employ other phosphors including the near-uv absorbing blue and green phosphors, $BaMg_2Al_{16}O_{27}{:}Eu^{2+}$ and $BaMg_2Al_{16}O_{27}{:}Eu^{2+}, Mn^{2+}$, respectively, and may use the filtering action of the Ce^{3+}-activated garnet.

X-Ray Excited Phosphors

X-ray intensifying screens make use of phosphors that convert the high energy x-ray photons to visible radiation which sensitizes a photographic film. In order to be useful as an x-ray phosphor the material must have high x-ray absorption, high density, and the activator must emit efficiently in the blue or green spectral region to match the sensitivity of the film. Conventional screens have used $CaWO_4$ as a broad-band emitter in the uv–blue region of the spectrum.

Divalent europium-activated BaFCl was the first rare-earth-activated x-ray phosphor (24). The advantage of $BaFCl{:}Eu^{2+}$ over the conventional $CaWO_4$ material is in the higher x-ray absorption and better x-ray-to-visible light conversion. The problem with BaFCl for x-ray application is in the lower density (4.56 g/cm^3 vs $\sim$ 6 g/cm^3 for $CaWO_4$) and plate-like morphology.

Another x-ray phosphor is LaOBr activated with Tm^{3+}. The density of the host lattice is high (6.13 g/cm^3) and the emission of the Tm^{3+} is in the blue spectral region which matches the sensitivity of the blue photographic film. This phosphor is widely used but is being replaced in some applications by yttrium tantalate–niobate which emits in a very broad band in the blue region of the spectrum. The green-emitting $Gd_2O_2S{:}Tb^{3+}$ phosphor with a physical density of 7.34 g/cm^3 has been described (25). This phosphor exhibits excellent x-ray-to-visible light conversion and is able to efficiently sensitize a green-sensitive film.

Scintillators are phosphor materials made in the form of single crystals or optically transparent polycrystalline ceramic or glass rods. These serve as detectors in computer-aided tomography (CAT) and other applications. The ceramic rod is excited close to the surface on one side or end of the rod. The light generated is detected on the other side by means of a photoconductor, such as a silicon detector. An optically clear dense ceramic of $(Y, Gd)_2O_3{:}Eu^{3+}$ is used for this application. Because the detector must be repeatedly excited, an important consideration is the elimination of afterglow caused by the recombination of electrons and holes which are created by the x-ray beam. The afterglow is minimized by adding other ions to the structure which remove shallow traps. Other rare-earth phosphors, such as $Gd_2O_2S:Pr^{3+}$, have also been used in CAT applications.

Fuji Corp. commercialized an x-ray photostimulable storage phosphor screen around 1985. In this device the bombardment of the phosphor screen by high energy x-rays generates free electrons and holes which are subsequently trapped. The stored energy can later be released by either thermal or optical stimulation. The stimulation releases the trapped charge carriers which then combine, transferring the recombination energy to a luminescent center, typically Eu^{2+}, which decays radiatively. The intensity of luminescence is proportional to the x-ray dosage. The luminescence can be measured by a photomultiplier tube and the information can be stored in a computer. At present the x-ray storage phosphor used in nearly all commercial systems is $BaFBr:Eu^{2+}$. Other Eu^{2+}-activated materials have been proposed as x-ray storage phosphors including $Ba_2B_5O_9Br:Eu^{2+}$ (26) and $Ba_4OBr_6:Eu^{2+}$ (27).

Phosphors for Cathode Ray Tubes

In colored cathode ray tubes (CRTs), such as those used in televisions and computer terminals, three electron gun beams are focused on three different sets of phosphor dots on the front face of the tube. The dots are produced by using a complicated photolithography process. The phosphor dots are produced by settling the three different phosphors, each of which emits one of the primary saturated colors, red, green, or blue. Each phosphor is deposited separately and the three dots in each set are closely spaced so that the three primary colors are not resolved at normal viewing distances. Instead the viewer has the impression that there is only one color, the color achieved when the three primary colors are added together.

The first red-emitting phosphor used in color televisions was Mn^{2+}-activated $Zn_3(PO_4)_2$. The Mn^{2+} emission is a broad band peaking at ~630 nm. In 1960 RCA introduced the all sulfide screen which utilized $(Zn, Cd)S:Ag^+$ as the red-emitting component. Because they have a low band gap energy and accompanying high cathode ray excitation efficiency, sulfides have enjoyed predominance as cathode ray phosphors. Some of these phosphors have energy conversion efficiencies of 20% or more which implies that one electron with thousands of volts of energy must excite hundreds of thousands of elementary excitations in the solids which in turn excite activator ions. In the case of the red phosphor, the sulfide is roughly three times as efficient as the earlier phosphate. The main problem, however, with both of these materials is that the broad Mn^{2+} emission does not provide a saturated red and is of low luminous efficiency. In 1964 Sylvania introduced Eu^{3+}-activated YVO_4 as the red-emitting component of the color television CRT. The emission spectrum of this phosphor is dominated by strong lines at 620 nm and its brightness exceeds that of the sulfide. However, the vanadate phosphor has been largely superseded in the United States for color television application by two other Eu^{3+}-activated phosphors: $Y_2O_3:Eu^{3+}$ and more recently $Y_2O_2S:Eu^{3+}$ which has a high efficiency and a nearly ideal spectrum. Most of its emission is concentrated in lines around 620–630 nm which is saturated enough to look red and not orange and yet is of relatively high luminous efficiency.

For the green-emitting component, the U.S. green phosphor (Zn,Cd)S:Cu,Al is used. One drawback of this sulfide-based green phosphor is that it saturates under a high electron current which prevents achievement of a very high brightness screen. This is particularly important, for example, in projection television screens. There are several rare-earth activated phosphors that could be used instead of the sulfide including $Y_4(SiO_4)_3$:Tb^{3+}, La_2O_2S:Tb^{3+} (27), CaS:Ce^{3+} (28), and $SrGa_2S_4$:Eu^{2+} (29). Terbium-activated cathodoluminescent phosphors, such as La_2O_2S:Tb^{3+} and yttrium silicate, do not provide a saturated enough green color. Two approaches have been taken to improve this. One is to use a filter or interference coating to absorb or reflect some of the orange emission from Tb^{3+}. Alternatively the Tb^{3+}-activated phosphor may be blended with a more saturated green, such as the Mn^{2+}-coactivated barium–magnesium–aluminate.

The blue-emitting component of most television screens and computer terminals is another sulfide, ZnS:Ag,Al. Although rare-earth activated blue-emitting phosphors ZnS:Tm^{3+} and $Sr_5(PO_4)_3Cl$:Eu^{2+} (30) have also been evaluated for this application, the search for a good blue phosphor that does not saturate at high current densities and maintains well continues.

Light-Emitting Diodes and Electroluminescence

A phosphor which generates light directly when an applied electric field is impressed across it is most desirable for flat panel displays. There are two ways this can be done with present materials. The first is to use a light-emitting diode (LED). These are single crystals usually of GaP doped with trace amounts of nitrogen. GaP is a wide band gap semiconductor. It is coated on alternate sides with metallic and transparent electrodes. When an applied d-c voltage of only a couple of volts is placed across the semiconductor, electrons tunnel into it and radiatively recombine with holes in the valence band near the surface. Energy efficiencies are currently on the order of a few percent but significant improvements have been reported. The emission is in the red. Unfortunately efficient green and blue LEDs remain elusive (see LIGHT GENERATION, LIGHT-EMITTING DIODES).

The second way to directly convert electric energy into light is with an electroluminescent phosphor. By far the best electroluminescent phosphor is ZnS:Mn^{2+}. Electroluminescent (EL) devices using this phosphor can have very high surface brightness and generate about 6 lumens per watt of input power. By comparison fluorescent lamps with triphosphors have efficiencies about 90 lumens per watt. Modern EL panels were developed originally by Sharp Corp. in the mid-1970s. These employ thin films. The ZnS:Mn^{2+} phosphor is deposited between two layers of dielectrics, such as Y_2O_3. The dielectric layers in turn are in contact with the top and bottom electrodes, one of which is transparent.

In contrast to the LEDs, electroluminescent phosphors operate under high voltage and very high electric fields. The Sharp devices are ac and purely capacitively coupled to the external circuit. As the applied voltage is increased, carriers tunnel out of interface states and are accelerated into the phosphor until some reach energies sufficient to excite the luminescent centers and generate light. The field polarizes the device and light output ceases until the field polarity reverses and carriers are reaccelerated back across the cell in the opposite direction. The Mn^{2+} emission in ZnS is in the yellow region of the spectrum so that the first displays using this technology were always yellow. A concerted effort has been going on to develop other colors. Planar has displayed a colored electroluminescent screen using a filtered $ZnS{:}Mn^{2+}$ phosphor for the red, $ZnS{:}Tb^{3+}$ for the green, and $CaGa_2S_4{:}Ce$ for the blue. As electroluminescent phosphors improve they will become more and more important in the search for a true flat panel display that directly converts electricity to light.

BIBLIOGRAPHY

"Luminescent Materials" in *ECT* 1st ed., Vol. 8, pp. 540–553, by G. R. Fonda, Consultant, General Electric Co.; in *ECT* 2nd ed., Vol. 12, pp. 616–631, by E. F. Apple, General Electric Co.; "Phosphors" under "Luminescent Materials" in *ECT* 3rd ed., Vol. 14, 527–545, by T. F. Soules and M. V. Hoffman, General Electric Co.

1. W. A. Thornton, *J. Electrochem. Soc.* **116**, 286 (1969).
2. K. H. Butler, *Fluorescent Lamp Phosphors Technology and Theory*, The Pennsylvania State University Press, University Park, 1980, Chapt. 12, pp. 135–151; B. DiBartolo, *Optical Interactions in Solids*, John Wiley & Sons, Inc., New York, 1968, for a quantum mechanical description.
3. C. W. Struck and W. H. Fonger, *J. Lumin.* **10**, 1 (1975).
4. R. Reisfeld, in B. Jezowska-Trzebiatowska, J. Legendziewicz, and W. Strk, eds., the *1st International Symposium on Rare Earth Technology*, World Scientific, Singapore, 1985.
5. Orbach, *Optical Properties of Ions in Solids*, Plenum Press, New York, 1975, p. 370.
6. D. L. Dexter, *J. Chem. Phys.* **21**, 836 (1953).
7. U.S. Pat. 2,448,733 (1949), A. H. McKeag and P. W. Ranby; H. G. Jenkins, A. H. McKeag, and P. W. Ranby, *J. Electrochem. Soc.* **96**, 1 (1949).
8. T. F. Soules, R. L. Bateman, R. A. Hewes, and E. R. Kriedler, *Phys. Rev.* **B7**, 1657 (1973).
9. F. W. Ryan and co-workers, *Phys. Rev.* **B2**, 2341 (1971).
10. P. R. Switch, J. L. LaCout, A. Hewat, and R. A. Young, *Acta Cryst.* **B41**, 173 (1985).
11. T. F. Soules, T. S. Davis, and E. R. Kridler, *J. Chem. Phys.* **55**, 1056 (1971).
12. B. G. DeBoer, A. Sakthivel, J. R. Cagel, and R. A. Young, *Acta Cryst.* **B47**, 683 (1991).
13. K. C. Mishra, R. J. Patton, E. A. Dale, and T. P. Das, *Phys. Rev. B* **35**, 1512 (1987); E. A. Dale and J. K. Berkowitz, *170th Electrochemical Society Meeting*, Oct. 1986, Abs. 708.
14. W. A. Thorton, *J. Opt. Soc. Am.* **61**, 1155 (1971).
15. M. Koedam and J. J. Opstelten, *Lighting Res. Tech.* **3**, 205 (1971).
16. R. G. Pappalardo and R. B. Hunt, Jr., *J. Electrochem. Soc.* **132**, 721 (1985).
17. N. Iyi, S. Takekawa, and S. Kimura, *J. Solid St. Chem.* **83**, 8 (1989).
18. J. L. Sommerhijik and A. L. N. Stevels, *Philips Tech. Rev.* **37**, 221 (1977).
19. R. C. Ropp, *J. Electrochem. Soc.* **115**, 531 (1968); J. C. Bourcet and F. K. Fong, *J. Chem. Phys.* **60**, 34 (1974).
20. B. Saubat, M. Vlasse, and C. Foussier, *J. Solid St. Chem.* **34**, 271 (1980).

21. C. Foussier, B. Saubet, and P. Hagenmuller, *J. Lumin.* **23**, 405 (1981); M. Buijs and G. Blasse, *J. Lumin.* **34**, 263 (1981); M. Bujis, J. P. M. Van Vliet, and G. Blasse, *J. Lumin.* **35**, 213 (1986).
22. J. Th. W. deHair, *J. Lumin.* **18/19**, 797 (1979); J. Th. W. deHair and W. L. Konijnendijk, *J. Electrochem. Soc.* **127**, 161 (1980).
23. B. M. J. Smets, *Mat. Chem. Phys.* **16**, 283 (1987).
24. C. Foussier, B. Laourette, J. Portier and P. Hagenmuller, *Mat. Res. Bull.* **11**, 933 (1976).
25. M. Tecotzky, *Electrochemical Society Meeting*, Boston, May 1968.
26. A. Meijernik and G. Blasse, *J. Phys.* **D24**, 626 (1991).
27. S. P. Wang, O. Landi, H. Lucks, K. A. Wickersheim, and R. A. Buchanan, *IEEE Trans. Nucl. Sci.* **NS17**, 49 (1979).
28. W. Lehman and F. M. Ryan, *J. Electrochem. Soc.* **119**, 275 (1972).
29. T. E. Peters and J. A. Baglio, *J. Electrochem. Soc.* **119**, 230 (1972).
30. F. C. Palilla and B. E. O'Reilly, *J. Electrochem. Soc.* **115**, 1076 (1968).

General References

Advances in Solid State Phosphors, Solid State Luminescence, Academic Press, Inc., New York, 1993.

Reference 2.

H. M. Crosswhite and H. W. Moos, eds., *Optical Properties of Ions in Crystals*, Wiley-Interscience, New York, 1967.

D. Curie, *Luminescence in Crystals*, Methuen, London, 1963.

ALOK M. SRIVASTAVA
THOMAS F. SOULES
General Electric Company

FLUORESCENT PIGMENTS (DAYLIGHT)

There are many types of luminescent materials, some of which require a special source of excitation such as an electric discharge or ultraviolet radiation. Daylight-fluorescent pigments, in contrast, require no artificially generated energy. Daylight, or an equivalent white light, can excite these unique materials not only to reflect colored light selectively, but to give off an extra glow of fluorescent light, often with high efficiency and surprising brilliance. These pigments can also be excited with both short- and long-wave ultraviolet light. The use of a black light markedly increases the brilliance of the pigments, which makes them useful as tracers in many different applications.

Fluorescent pigments are comprised of dyed organic polymers. These polymers are clear and colorless and are formulated to be a solvent for the fluorescent dyestuff. There are many different chemical types of polymers and dyestuffs. In this article, the term dye applies to any organic substance that exhibits strong absorption of light in the visible or even ultraviolet region of the spectrum without regard to any affinity for textile fibers, paper, or other substrates (see DYES, APPLICATION AND EVALUATION).

A fluorescent substance is one that absorbs radiant energy of certain wavelengths and, after a fleeting instant, gives off part of the absorbed energy as quanta of longer wavelengths. In contrast to ordinary colors in which the absorbed energy degrades entirely to heat, light emitted from a fluorescent color adds to the light returned by simple reflection to give the extra glow characteristic of a daylight fluorescent material. This fluorescence phenomenon can lead to reflectance values greater than 100% in a specific part of the spectrum.

History

Fluorescent Dyestuffs. Very few dyes are of use in making daylight-fluorescent products. Of the dyes discovered up to 1920, only the brilliant red and salmon dyes of the rhodamine and rosamine classes are used in fluorescent materials in the 1990s. The first of these, Rhodamine B, was discovered in 1877. Fluorescence excited by both uv and visible light components in daylight was formally recognized as a notable property of certain dyed fabrics by the 1920s (1).

The early yellow dyes, including Auramine O and Thioflavine T, were found to be extremely fugitive to light. However, in 1927 the first of the naphthalimide yellows, Brilliant Sulfoflavine FF [*2391-30-2*] (CI Acid Yellow 7), was discovered at I.G. Farben (2), followed by Azosol Brilliant Yellow 6GF [*2478-20-8*] (CI Solvent Yellow 44) (3). These two greenish yellow fluorescing dyes produced for the first time workable, bright, clean, fluorescent colors in the yellow range. They were by no means lightfast but at least were a great improvement over the earlier yellows. In addition, in mixtures with rhodamines, they formed bright orange and orange-red lacquers.

In the 1930s, fluorescent-colored lacquers were used in theaters for special effects under black light (long-wave uv light), and the outstanding brilliance of the colors in simple daylight as well as their potential for outdoor advertising was recognized. This was followed by the preparation of dyed resinous pigments.

An important advance with regard to light stability was made with a group of yellow coumarin dyes with heterocyclic systems attached to the coumarin nucleus (4), eg, a greenish yellow cationic dye that is sold under the name Maxilon Brilliant Flavine 10 GFF [*12221-86-2*] (Blue Wool #4), designated CI Basic Yellow 40, available from several manufacturers.

Benzothioxanthene and benzoxanthene dyes, discovered by Farbwerke Hoechst AG (5–8) in the mid-1960s, have been employed to some extent in fluorescent pigments. Some members of this group have good color strength and very good lightfastness. Table 1 lists some of the important dyestuffs still used for daylight-fluorescent pigments.

Fluorescent Pigments. The first patents for daylight fluorescent products were issued in 1947 (9,10), describing fluorescent dyed cellulose acetate fabrics with several barrier coats to improve long-term stability. These fabrics were brilliantly fluorescent and were widely used during World War II as signal panels.

Later, the manufacture of a thermoset pigment was patented (11). The polymer was dyed with fluorescent dyes and pulverized into a fine powder that could be used as a fluorescent pigment. The following procedure was employed: Solvent Yellow 44 was dissolved in a solution of butanol-modified urea–formaldehyde,

Table 1. Important Dyestuffs for Daylight-Fluorescent Pigments

Colour Index name	CAS Registry Number	Selected manufacturers
Basic Violet 11	[*2390-63-8*]	BASF, others
Basic Violet 10	[*81-88-9*]	BASF
Basic Red 1	[*989-38-8*]	L. B. Holliday, BASF, others
Acid Red 52	[*3520-42-1*]	Sandoz Chemical
Solvent Yellow 44	[*2478-20-8*]	L. B. Holliday
Basic Yellow 40	[*12221-86-2*]	Ciba-Geigy, L. B. Holliday
Solvent Yellow 135		Day-Glo Color Corp., L. B. Holliday
Solvent Yellow 160:1	[*61902-43-0*]	Day-Glo Color Corp. Bayer AG

the resin slowly heated to 90°C and gelled by the addition of acid, followed by post-curing at 140–145°C. The thermoset product was ball- or hammer-milled to the required fineness. This type of fluorescent thermoset resinous material had poor lightfastness, was difficult to grind, and was highly sensitive to environmental moisture.

Early fluorescent pigments were promoted and adopted for use in screen inks for poster boards and paints for safety applications. These thermoset pigments were not well-suited because of their poor lightfastness. Also, because of their relatively coarse particle size, their use in thinner film applications, such as gravure or flexo, was limited.

The performance of daylight-fluorescent pigments has steadily improved since their first large-scale manufacture. Modification with toluenesulfonamide yielded thermoplastic resins (12). These pigments were fused and condensed completely in the reaction kettle and, after cooling, were finely ground into pigments. This procedure greatly improved both water resistance and lightfastness.

These thermoplastic pigments found application in a much wider range of finished products. New, large-volume applications included coated paper for labels and point of purchase signage, gravure for soap box cartons and bright textiles for fashion, and safety applications.

Recent pigment technology has yielded a wide range of products which are much more specialized for individual end use applications. New polymers have been combined with improved dyestuffs to yield fluorescent pigments with better performance properties and economics, and more desirable environmental characteristics.

Availability

The primary manufacturers of daylight-fluorescent pigment at the present time are Dane and Co. (London); Day-Glo Color Corp. (Cleveland, Ohio); Nippon Keiko Kagaku Co. Ltd. (Tokyo); Nippon Shokubai (Osaka); Lawter Chemical Corp. (Skokie, Illinois); Radiant Color, Division of Magruder (Elizabeth, New Jersey); Sinloihi Co., Ltd. (Kamakura, Japan); and U.K. Seung (Busan, Korea). Smaller regional manufacturers are located in China, India, Russia, and Brazil.

Table 2 lists fluorescent product lines marketed by various manufacturers, where similarities are known. New products are introduced and old products obsoleted on a regular basis.

Table 2. Commercial Fluorescent Products

Pigment characteristics	Manufacturer's code				
	Day-Glo	Lawter	Radiant	Sinloihi	Dane
single strength[a]	A	B-3500	R-105, R-103	FZ-2000	A
high strength[a]	AX	D-3000	R-106, R-104	FZ-4000	E
single strength[a] thermoset	T		P-1600	FZ-3000	T
high strength thermoset[a]	GT	TS	P-1700	FZ-3040	GT
high heat resistance, high lightfastness for plastics	ZQ, NX	TC	K-600		ZN
toners soluble in polar solvents	HM, HMS	HVT	GF	FM-10, FM-70	ST-EBT
paste concentrate bases for litho and letterpress	SFB	Optichrome 8600	Visiprint, VF	BO	OLC
high strength for vinyl	VC		K-500		
alkali water-soluble toners	WST		Aquabest		
fine particle size pigments	EP/EPX	SG-2500	JST	SW	

[a]Standard fineness.

Theory of Fluorescence

Structure. Virtually all important dyes contain aromatic rings in their structures. According to the theory, groups called chromophores have to be present on benzenoid rings in order for compounds to have appreciable light absorption or color. Certain basic groups, so-called auxochromes, are also necessary to bring out or intensify the color. The electronic theory of atoms and molecules, wave mechanics, and the theories of valence-bond resonance and molecular orbitals have increased the understanding of colored and fluorescent substances and their interaction with light. The benzene ring with its six π-electrons can act in conjunction with electron-donating groups (auxochromes) and electron-accepting (unsaturated) groups to produce strong absorption in the uv or visible regions which may give rise to fluorescence. Such a system of atoms, responsible for significant absorption of photons in the uv or visible regions, is referred to as a chromogen. A chromogen that absorbs in the uv as a rule can be modified chemically to absorb visible light, thus becoming colored. This is often accomplished by adding benzene rings to the molecule or introducing a long unsaturated chain of atoms (see CHROMOGENIC MATERIALS, PHOTOCHROMIC).

Chromogens. Organic dyes can be divided into four classes, depending on the type of chromogen or unsaturated system present: (*1*) $n \rightarrow \pi^*$ chromogens, (*2*) donor–acceptor chromogens, (*3*) cyanine-type chromogens, and (*4*) acyclic and cyclic polyene chromogens (13). Almost all strongly fluorescent dyes fall into classes (*2*) and (*3*), whereas only a few have cyclic polyene chromogens of groups (*4*). The chromogens of class (*1*) are detrimental to fluorescence.

Donors and Acceptors. Table 3 lists common electron–donor groups, and electron–acceptor groups selected from a large number capable of evoking fluorescence.

In dyes in which a particular benzene ring carries a donor and an acceptor group, these groups are introduced in positions ortho or para to each other. In condensed ring systems such as naphthalene, conjugated-bond paths between donors and acceptors are necessary for interaction. These conditions are the rule rather than the exception also for nonfluorescent dyes. For example, the donor amino group in 7-amino-4-methylcoumarin [*26093-31-2*] (**1**) shares its lone pair of electrons with the benzene ring; the α,β-unsaturated ester group accepts electrons from the benzene ring. If the donor and the acceptor are in a para position to each other, as they are here, a partial electron displacement toward the carbonyl oxygen atom takes place. Absorption of a photon of uv light carries this displacement even further toward the high energy polarized form (**2**). Fluorescence occurs if the molecule emits a photon of light in a transition from its first excited singlet state. The singlet state refers to a condition of paired spins for an even number of electrons in a molecule.

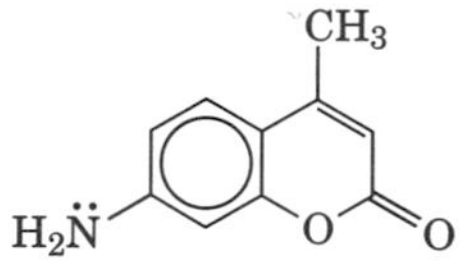

(**1**) 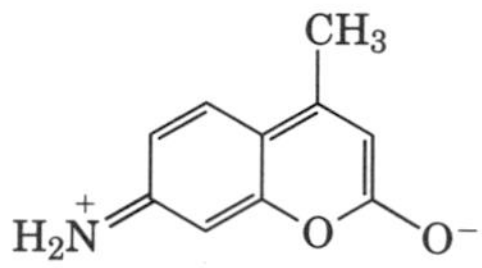 (**2**)

Table 3. Donor and Acceptor Groups in Fluorescent Dye Molecules

Electron–donor groups		Electron–acceptor groups	
strong		cyano	—C≡N
amino	H_2N—	carbonyl	>C=O
alkylamino	RHN—	vinylene	—CH=CH—
dialkylamino	R_1R_2N—	styryl	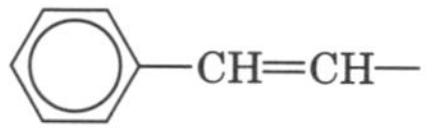 —CH=CH—
oxido	—O—	acrylic ester	—CH=CH—COOR
weak		β-methacrylic ester	—$C(CH_3)$=CH—COOR
hydroxy	HO—	benzoxazolyl	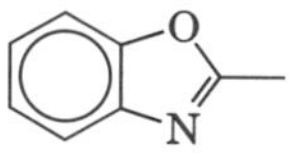
alkoxy	RO—	benzothiazolyl	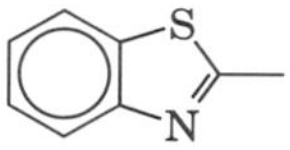
		benzimidazolyl	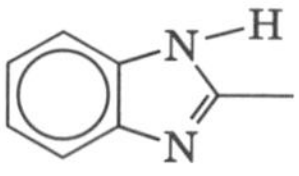

Donor–acceptor molecules are often sensitive to pH changes. For example, if acid is added to a solution of (**1**), the lone pair of electrons on the nitrogen takes up a hydrogen ion forming a substituted ammonium salt. In the salt form, the nitrogen cannot conjugate with the benzene ring, and hence the fluorescence capability is strongly reduced. Addition of an alkali to the salt causes removal of the hydrogen ion and immediately restores the fluorescence under ultraviolet light.

Alkylation of the amino group to a mono- or dialkyl form strengthens the uv absorption and also increases the wavelength of the fluorescent light; two alkyl groups are more effective than one and ethyl groups are more powerful than methyl groups.

Hydroxy groups are very weak donors to the benzene ring. However, on treatment with alkali they form strongly donating ionized oxido groups (or phenolate ions when attached to the benzene ring). The strongly fluorescent alkaline solutions of 7-hydroxycoumarin [*93-35-6*] and fluorescein [*2321-07-5*] are examples. Neutralizing with acid quickly destroys the fluorescence almost completely. Even dilution with water causes gradual weakening of the fluorescence by hydrolysis.

If the hydrogen atom of a hydroxy group becomes hydrogen bonded to an ortho-substituted group, appreciable or even strong fluorescence can be evoked. For example, the crystalline yellow fluorescent pigment, the azine of 2-hydroxynaphthaldehyde-1 (14), is sold as Lumogen Light Yellow L [*2387-03-3*] by BASF. It performs well in applications under ultraviolet light, but is not stable enough for daylight exposure.

Alkylation of donor amino groups strengthens them as donors but usually reduces lightfastness. CI Solvent Yellow 44 [*2478-20-8*] (**3**) has already been mentioned as a dye for daylight-fluorescent pigments, but its lightfastness is only fair (Blue Wool Scale 2–3). It has been a component of green, yellow, orange, and red pigments for many years. The monoalkylamino compound, CI Solvent Yellow 43 [*12226-96-9*] (**4**), is also available. It is appreciably redder and stronger than unsubstituted (**3**) but poorer in light stability (Blue Wool Scale 2.0).

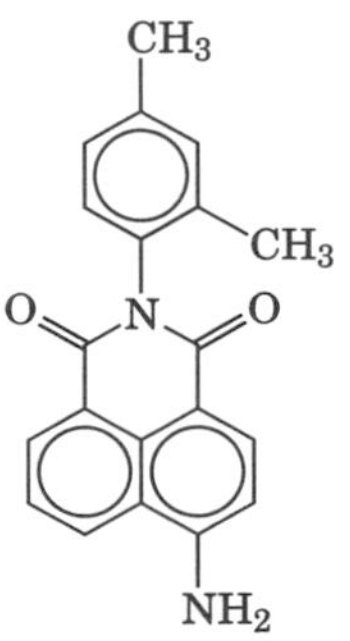

CH_3 CH_3 O N O NH_2

(**3**)

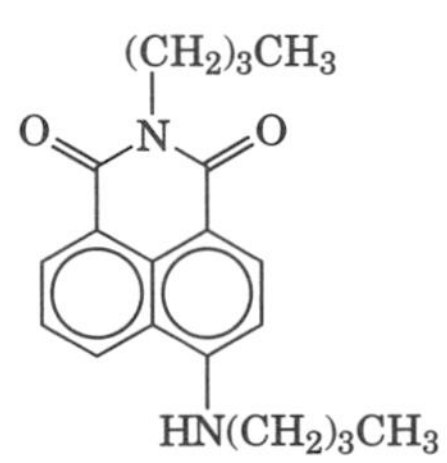

$(CH_2)_3CH_3$ O N O $HN(CH_2)_3CH_3$

(**4**)

In these dyestuffs, the two carbonyl groups are acceptors; the aryl group is that of naphthalene. The one obvious para conjugation between the amino group and the carbonyl group is hardly sufficient to explain the moderately strong absorption in the blue and violet region that occurs in these dyes. There is probably some conjugation between the groups on the 4 and 8 positions of the naphthalene structure.

Greater depth of color in a chromogen can be obtained by providing a more complex acceptor group or linear addition of acceptor groups. Thus the 7-diethylaminocoumarin [*20571-42-0*] (**5**) is colorless with a blue fluorescence, whereas Coumarin 7 [*27425-55-4*] (**6**) is yellow with a green fluorescence. The conjugation in structure (**6**) includes the benzene ring fused to imidazole. By methylating both nitrogen atoms of the imidazole ring, a yellow of better light stability for nonalkaline substrates is obtained, Basic Yellow 40 [*12227-86-2*] (**7**).

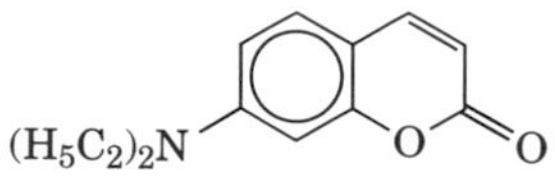

$(H_5C_2)_2N$ ... O ... O (**5**) $(H_5C_2)_2N$ ... N, NH ... O ... O (**6**) $(H_5C_2)_2N$ ... CH_3, ^+N, N, CH_3 ... O ... O X^- (**7**)

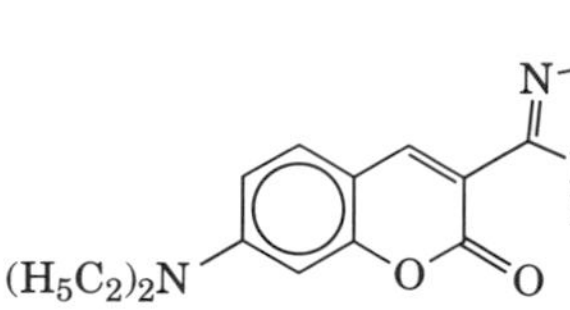

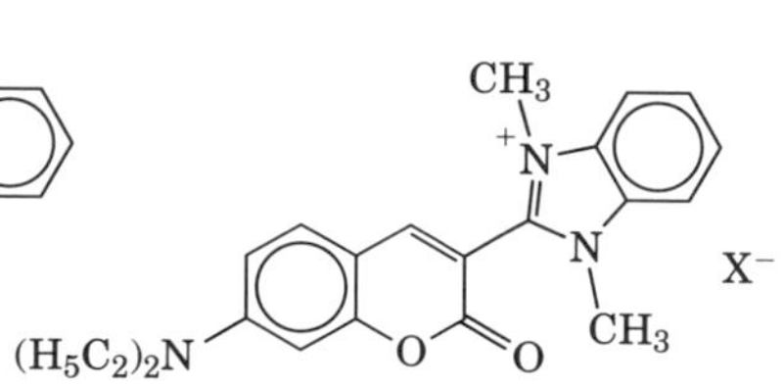

Donor–acceptor chromogens in solution are often strongly affected by the nature of the solvent or the resinous substrate in which they are dissolved. The more polar the solvent or resin, the longer the wavelength of the fluorescent light emitted. Progressing from less polar to more polar solvents, the bathochromic, or reddening, effect of the solvents on the dye increases in the order of aliphatics < aromatics < esters < alcohols < amides.

Cyanine Types. Cyanine-type chromogens are odd-alternate systems that have at least two equivalent or nearly equivalent resonance forms (13). Open-chain cationic chromogens of this class are represented by (**8**) and can be viewed as true cyanines (see CYANINE DYES). If the two terminal atoms were oxygen instead of nitrogen, the system would be a cyanine-type chromogen. In each of these forms, one oxygen atom would carry a negative charge.

$$R_2\ddot{N}{-}(CH{=}CH)_n{-}CH{=}\overset{+}{N}R_2\ X^- \longleftrightarrow R_2\overset{+}{N}{=}CH{-}(CH{=}CH)_n{-}\ddot{N}R_2\ X^-$$

(**8**)

In most true cyanines the nitrogen atoms are part of heterocyclic rings. Astra-phloxine FF [*6320-14-5*] (**9**), CI Basic Red 12, is a true cyanine in the modern sense and is strongly fluorescent in the red region. Its light stability is not as good as that of Rhodamine B. Chromogens of the odd-alternate types such as (**8**) and (**9**) have nonbonding molecular orbitals extending along the conjugated chain. The cyanines thus have completely delocalized electrons, a situation quite different from that in the donor–acceptor chromogens.

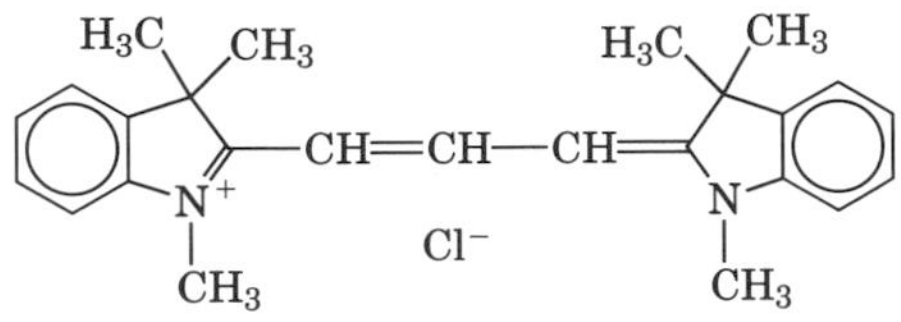

(9)

There are good reasons to consider extending the class of cyanine-type chromogens to the older and well-known group of basic dyes (15). In a diphenylmethane dye such as Michler's Hydrol Blue (**10**), one ring contributes two double bonds in order to give an odd-alternate system. Aromatic-ring systems can be made part of the conjugated chain of a cyanine without altering the characteristic properties of the chromogen (13,16).

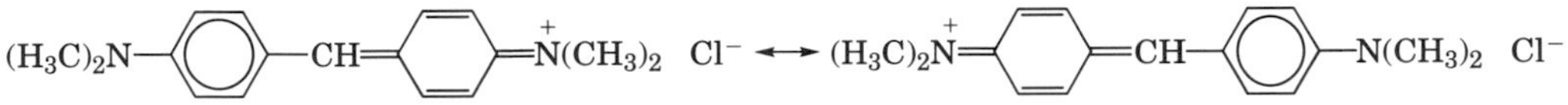

(10)

None of the simple di- and triarylmethane dyes, with the exception of the auramines, have strong fluorescence (see TRIPHENYLMETHANE DYES). For example, the well-known indicator phenolphthalein [*518-57-4*] has a nonfluorescent strongly red alkaline form (**11**). However, if an oxygen bridge is introduced between the upper benzene rings, the intensely green fluorescing alkaline form of fluorescein disodium salt [*578-47-8*] (**12**) results.

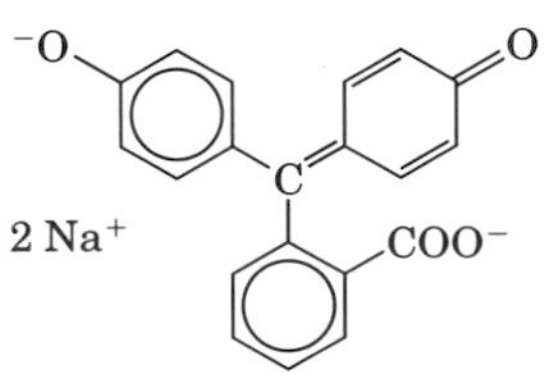

(11) 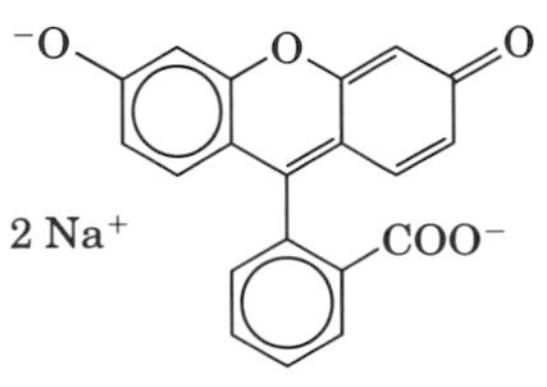 **(12)**

The alkyl and dialkylamino compounds are much more important. If the structure of Brilliant Green [*633-03-4*] (**13**) (CI Basic Green 1) is altered by inserting an oxygen bridge and a carboxyl group on the lower ring, the brilliant red-fluorescing Rhodamine B [*81-88-9*] (**14**) is formed.

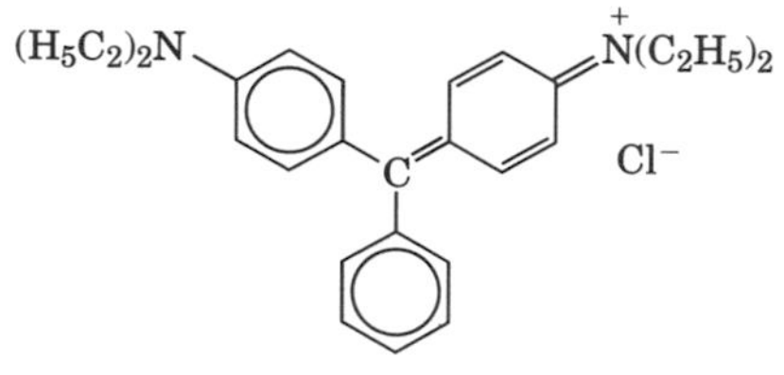

$(H_5C_2)_2N$ $\overset{+}{N}(C_2H_5)_2$ Cl^-

(13)

$(H_5C_2)_2N$ O $\overset{+}{N}(C_2H_5)_2$ Cl^- COOH

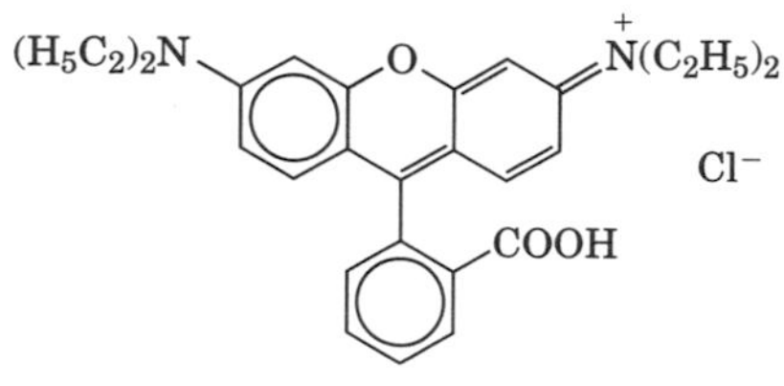

(14)

In Rhodamine 6G, also sold as Rhodamine F5G [*989-38-8*] (**15**), the carboxy ester group prevents free rotation of the lower phenyl group. Its position is roughly perpendicular to the plane of the other three rings. Retention of color strength is good because there is less electronic interaction between the lower ring and the rest of the molecule.

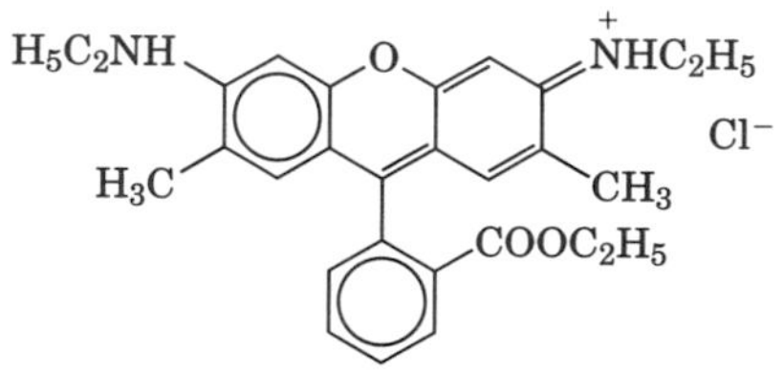

H_5C_2NH O $\overset{+}{N}HC_2H_5$ Cl^- H_3C CH_3 $COOC_2H_5$

(15)

Cyclic Polyenes. Polycyclic aromatic hydrocarbons, including cyclic polyene chromogens, are not represented among current daylight-fluorescent dyes. Compounds with more than two or three rings are often carcinogenic. However, derivatives with electron-withdrawing groups offer some definite possibilities, as, for example, the commercial dye Thermoplast Brilliant Yellow 10G [*2744-50-5*] (**16**) (CI Solvent Green 5) which gives brilliant green fluorescent solutions.

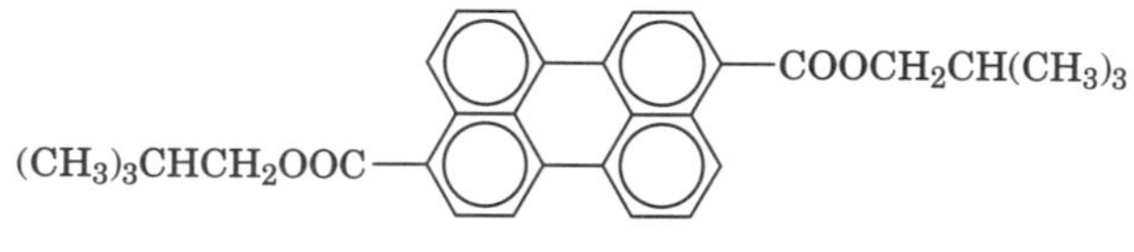

$COOCH_2CH(CH_3)_3$ $(CH_3)_3CHCH_2OOC$

(16)

Groups Detrimental to Fluorescence. Strong fluorescence is prevented by nitro and azo groups, the latter being very common in dyes (see AZO DYES). Because of its high electronegativity, the fluorine atom has a strong electron-withdrawing effect. If the fluorine is present on a benzene ring carrying a donor group, the inductive action is sufficient to practically eliminate the possibility of fluorescence. A trifluoromethyl group substituted for methyl attached to the acceptor group, however, increases the light stability of several laser dyes (17). Bromine and iodine atoms in almost any position weaken fluorescence by accepting electronic energy of excitation and converting it into heat.

Sulfonation increases the water solubility of dyes. A sulfonic group is electron withdrawing but has a weakening effect on the fluorescence only if it is ortho to the donor group. For example, in Brilliant Sulfoflavine FF [*2391-30-2*] (**17**), where R = *p*-tolyl, —$C_6H_4CH_3$, it has a weakening effect, although the cleanness of the fluorescent color is somewhat improved compared to that of the unsulfonated compound.

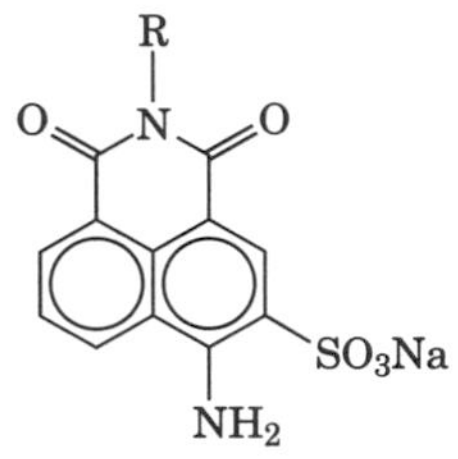

R, O, N, O, SO_3Na, NH_2

(**17**)

Dyes containing a dominant benzoquinone group are generally nonfluorescent. The anthraquinones, for example, are almost completely without fluorescence (see DYES, ANTHRAQUINONE).

Rigidity and Fluorescence. The more rigid the molecule, the less likely that low energy vibrations are initiated by transfer of energy from the first excited singlet state before fluorescence can take place. A small class of dyes, the pyrimidanthrones, in which one of the quinone oxygen atoms is replaced by nitrogen and the latter made part of a rigid ring, for example, are strongly fluorescent. One of them, Lumogen L Red Orange [*6871-91-6*] (**18**) is a crystalline substance with a deep red-orange fluorescence.

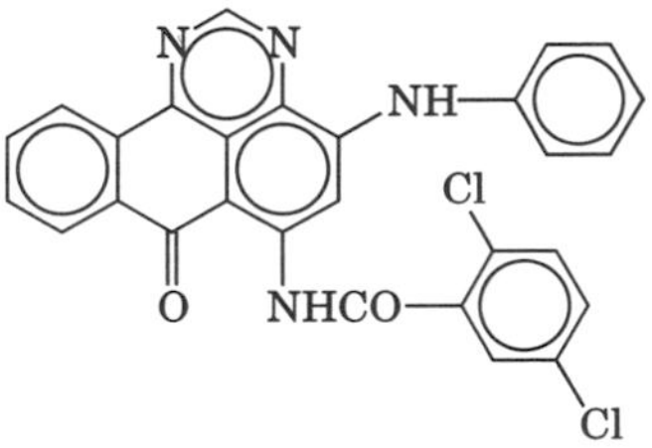

N, N, NH, Cl, O, NHCO, Cl

(**18**)

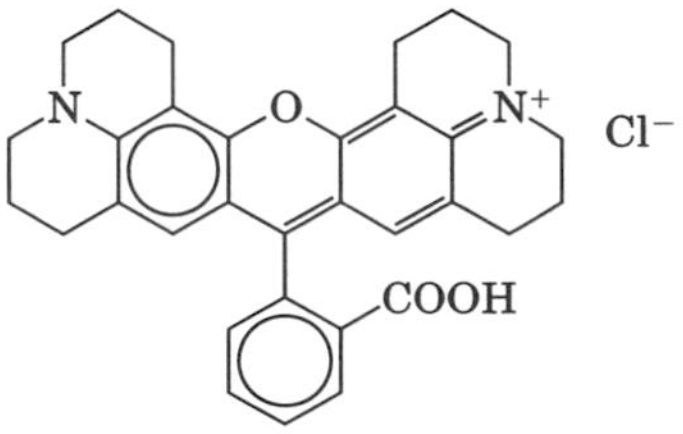

N, O, N^+, Cl^-, COOH

(**19**)

The effect of forming a more rigid structure in fluorescent dyes of the rhodamine series has been clearly demonstrated (18) with the remarkable dye designated Rhodamine 101 [*41175-43-3*] (**19**). This dye has its terminal nitrogen atoms each held in two rings and has a fluorescence quantum yield of virtually 100% independent of the temperature.

Energy Levels and Light Absorption. A dye molecule of about 50 atoms would have ca 150 normal vibrations of the molecular skeleton (15). Figure 1 shows the typical transition between various energy states that the π-electrons of a dye molecule can undergo. The singlet ground state of the π-electrons in the molecule is designated S_0 and represents the lowest electronic energy level possible for the molecule. The molecule can be excited to higher electronic states such as S_1 or S_2 with an associated set of vibrational energy levels represented by a series of lines above the particular electronic level.

In the process of excitation, the dye molecule absorbs a quantum of uv or visible radiation. The quantum has an energy $E = hv$, where h is Planck's constant and v is the frequency of the radiation. The higher the frequency of the quantum, the shorter the wavelength λ, with $v \cdot \lambda = c$, where c is the velocity of light in a vacuum.

If, as is usually the case, it is a π-electron that is raised in energy by the quantum or photon from the S_0 state to the S_1 state, the electron is said to undergo a $\pi \rightarrow \pi^*$ transition. Because the quantum is totally absorbed, the π-electron is raised in energy by exactly the same amount as the energy of the quantum.

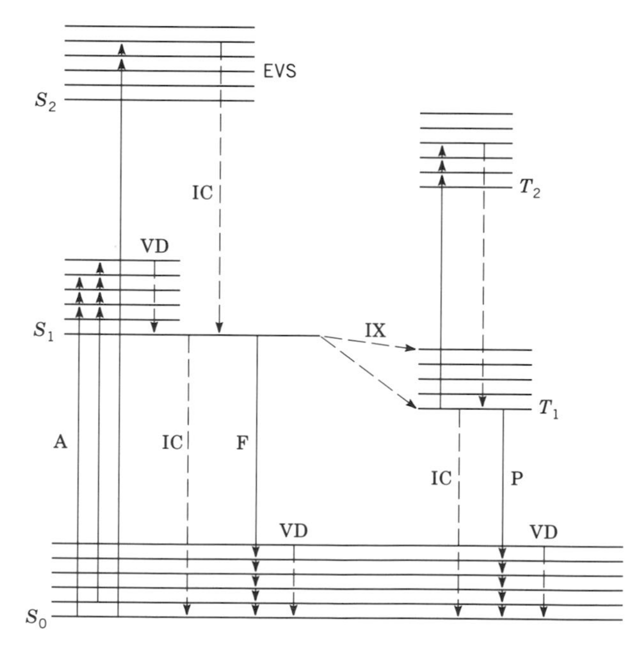

Fig. 1. Schematic energy-level diagram for a dye molecule. Electronic states: S_0 = ground singlet state; S_1 = first excited singlet state; S_2 = second excited singlet state; T_1 = first excited triplet state; T_2 = second excited triplet state; EVS = excited vibrational states. Transitions: A = absorption to excited states; VD = vibrational deactivation; IC = internal conversion; F = fluorescence; IX = intersystem crossing; and P = phosphorescence.

The new orbital of the π-electron in the π^*-state is larger than it was in the unexcited state. The two adjacent atoms with which the electron was associated in the ground state may be partially held by the electron in its expanded π^*-orbital. The atoms, in adjusting to the new binding condition, must move farther apart. They absorb the energy necessary to do this from the electron in its π^*-orbital. An additional vibrational amplitude is attained by the two atoms as a system. Some added energy is transmitted to other atoms of the conjugated molecule. These atomic vibrational adjustments take place very quickly, in 10^{-13} to 10^{-11} s.

In addition to interatomic vibrations, groups of atoms in the molecule can have a number of rotational energy levels which add to the number of possible levels of total energy of the chromogenic system. Association of various atoms of the molecule with polar solvents, for example, can also alter the degree of conjugation of the interrelated system. These added influences, which can also change from instant to instant, result in an enormous number of different energy levels in the billions of molecules of a particular irradiated sample. The excited states are also affected by the same influences. As a result, the molecule can be in any one of a very large number of total energy levels.

Because, in principle, transitions can occur on light absorption to any of the many possible energy levels of the excited state from any one of the many possible energy levels of the ground state, the absorption spectrum of a chromogen at room temperature or above is virtually continuous.

Fluorescence from the Excited S_1 State. In Figure 1, after absorption (A) and vibrational deactivation (VD) occur, the lowest or nearly lowest level of the singlet excited state S_1 is reached. If the molecule is fluorescent with a high quantum efficiency, fluorescent emission of a quantum of light generally occurs, indicated by fluorescence (F). Readjustment of some bond lengths of the chromogenic skeleton very likely would have to occur before reaching the lowest possible vibrational state consistent with the temperature of the surrounding medium. Thus the photon of light given off would not in general be equal to the quantity $S_1 - S_0$ but would be less, because the emission of the photon would complete the process.

If the incident photon has sufficient energy to excite the electron to the second state S_2, another process comes into play. The S_2 state has a very short lifetime of less than 10^{-11} s. Part of the energy of the S_2 state is converted rapidly to interatomic vibrations which quickly lose energy to other parts of the molecule or to surrounding molecules. This process is called internal conversion (IC) and is nonradiative. However, the electron energy loss is not total. The system generally attains the S_1 state from which fluorescence or other transitions can occur. The S_3 and higher singlet states are also subject to internal conversion, all such states having the possibility of losing energy to give the S_1 state from which a fluorescence quantum can be given off. The S_1 state can last as long as 10^{-8} s, a period long enough for fluorescence to occur.

Because fluorescence takes place only after the system has reached the S_1 state, regardless of the energy of the absorbed radiation, the fluorescence emission spectrum is independent of the wavelength of the absorbed light. The two requirements for the incident radiation is that it be high enough in energy to excite the dye molecule to the S_1 state but not so high that it causes decomposition of the dye.

Because the drop in energy of the molecule that occurs on emitting a fluorescent photon is generally less than the energy of the absorbed photon, the wavelength of the fluorescent light is almost invariably longer than the wavelength of the exciting light. This is Stoke's law, and applies in almost all cases of fluorescence.

Fluorescent photons can vary widely in energy, even if emission occurs from the same type of dye molecule. As a result, the emission spectrum of a typical dye is quite broad, frequently extending for 150 nm.

Competing Processes from the S_1 State. It can be seen from Figure 1 that transitions other than fluorescence can take place from the S_1 state. The molecule can lose electronic energy by internal conversion, passing through a higher vibrational level of the S_0 state, before undergoing vibrational deactivation by surrounding molecules. The electronic excitation energy can also pass by intersystem crossing (IX) to one of the levels of the first excited triplet state, characterized by two unpaired electrons with parallel spins.

The triplet state has the relatively long lifetime of 10^{-4} s or more. When emission occurs from the triplet state, it is called phosphorescence (P). Inorganic materials can exhibit phosphorescence after a delay as long as several hours but with dyestuffs in resins any delay is quite short. That the excited triplet state persists as long as it does in a phosphorescent organic compound is due to the fact that one of the electrons must reverse its spin before or during its transition to the ground state. The process $T_1 \rightarrow S_0$ is not an allowed transition according to quantum mechanics. As a result, the probability of occurrence is small and the lifetime of the triplet state is relatively long. Because of the relatively long lifetime of the first excited triplet state, the possibility of chemical reaction or photochemical breakdown of the molecule takes on added weight for those dyes that are prone to triplet-state formation. Such dyes could have poor lightfastness.

In addition to the processes that can compete with fluorescence within the molecule itself, external actions can rob the molecule of excitation energy. Such an action or process is referred to as quenching. Quenching of fluorescence can occur because the dye system is too warm, which is a very common phenomenon. Solvents, particularly those that contain heavy atoms such as bromine or groups that are detrimental to fluorescence in a dye molecule, eg, the nitro group, are often capable of quenching fluorescence as are nonfluorescent dye molecules.

A high concentration of the fluorescent dye itself in a solvent or matrix causes concentration quenching. Rhodamine dyes exhibit appreciable concentration quenching above 1.0%. Yellow dyes, on the other hand, can be carried to 5 or even 10% in a suitable matrix before an excessive dulling effect, characteristic of this type of quenching, occurs. Dimerization of some dyes, particularly those with ionic charges on the molecules, can produce nonfluorescent species.

A nearby molecule with a conjugated system may rob the dye molecule of its electronic energy. On the other hand, a fluorescent dye can pick up electronic energy from such a substance, called a sensitizer, with increased fluorescence.

Radiation, both in the uv and in the visible region, can have a highly destructive effect by decomposing the dye molecule. Other substances, particularly water, can reinforce the photochemical effect of light. Once the dyed material fades, its original condition usually cannot be restored. Sometimes a darker material forms that bleaches on further exposure to the incident light. In any case, reasonable stability to light is one of the most important requisites of a useful fluorescent pigment, but the most difficult to achieve. An interesting discussion of the products of photolysis of 4-methyl-7-diethylaminocoumarin [*91-44-1*] is given in Reference 19.

The effect of the molecular matrix in which the dye is dissolved is of great significance. Matrices vary greatly in their effect in making a dye light resistant. Rigid polymeric media can have a beneficial effect on a fluorescent dye by reducing intermolecular motion and energy transfer.

Whereas the earliest fluorescent-dye pigments would last only 20 days outdoors in a screen-ink film, fade resistance has been improved to such an extent that some modern daylight-fluorescent coated panels still have useful color after nine months or more in Florida sunlight in a 45° exposure rack facing south. The fluorescent layer is usually coated with an acrylic film containing a uv absorber. Indoor-accelerated exposure equipment is, of course, invaluable in the development of such systems. Better dyes and resins very likely will make possible far more stable coatings in the future.

When measuring the lightfastness of fluorescent materials it is important to keep in mind several criteria. Lightfastness is dependent upon the type of pigment used, the concentration of the pigment in the test specimen, the thickness of the specimen, the composition of the specimen in which the pigment is dispersed or dissolved, any additives to the specimen, the substrate the specimen is coated on if any, and any overcoats used to protect the specimen. In general, the greater the concentration of the pigment the greater the lightfastness, the thicker the specimen the greater the lightfastness, and in some cases additives such as light stabilizers and antioxidants (qv) can have beneficial effects. Such additives are most beneficial when used in an overcoat to protect the pigment. Additives which have a detrimental effect are metals such as iron and zinc, and oxidizing chemicals.

Color Formation

Spectral-Energy-Ratio Curves. Figure 2 shows the spectral-energy-ratio curves of three daylight-fluorescent dyes in pigment drawdowns, ie, paint samples drawn down with a bar on special panels, and a curve for a nonfluorescent ink. The lower left part of each of the first three curves is essentially the same as the transmittance or reflectance spectrum of the dye. If these were nonfluorescent dyes, the absorbed energy of the incident daylight would all be lost as heat and at no point could the curve rise above 100%. With a strongly fluorescent substance, however, most of the absorbed energy is stored in the S_1 excited state and is largely given off as fluorescent light of longer wavelengths covering a considerable range.

For example, the greenish yellow coumarin dye, Alberta Yellow (curve 1) has a nonfluorescent reflection of 75 to 80%, but emitted fluorescent light adds to reflected light to give a peak radiance factor of 177%. This means that there is 77% more green light of 525 nm returned from the sample than there is green light of that wavelength in the white light falling on the sample. The result is

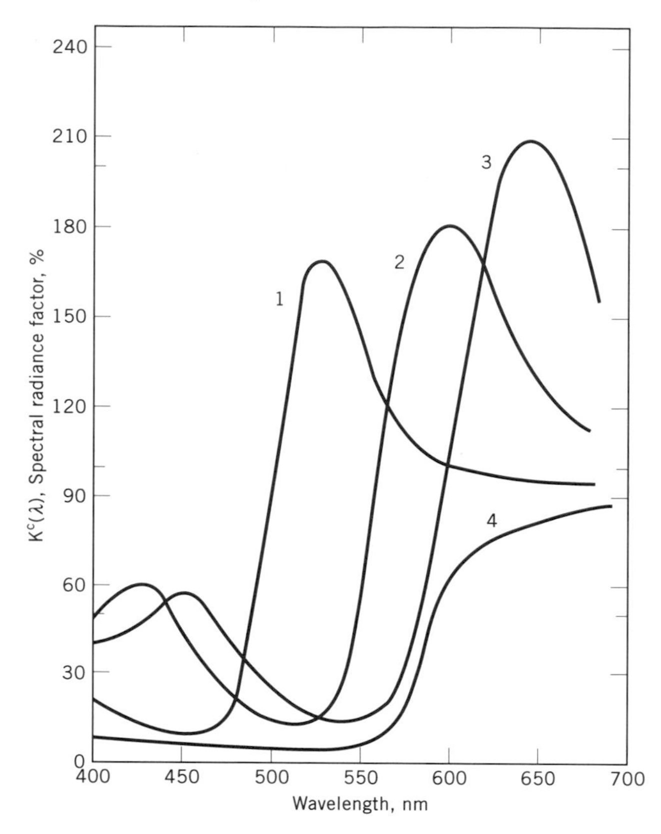

Fig. 2. Curves 1, 2, and 3 show the spectral radiance factor for equivalent coatings of separate toluenesulfonamide–melamine–formaldehyde Day-Glo pigments containing 0.5% of a dye, either Alberta Yellow, Rhodamine F5G, or Rhodamine B Extra. Curve 4 is for a bright nonfluorescent red-orange printing ink. The illuminant was Source C. A magnesium oxide-coated block was used as a comparison white.

that the human eye and brain which adds the effects of light of all wavelengths in the visible region sees a very bright greenish yellow. Because the eye is especially sensitive in the yellow-green region, the luminance factor, which takes into account the sensitivity of the eye to all wavelengths, is 123%, or 23% brighter than the purest white. The same yellow dye is the starting point for a bright fluorescent green pigment. Simple addition of a nonfluorescent phthalocyanine green toner in aqueous dispersion in the pigment-forming process cuts out important orange and red components of the yellow dye resulting in a bright daylight fluorescent green.

Effect of Two or More Dyes. A most remarkable effect in daylight fluorescence, the transfer of energy from one fluorescent dye to another, can be used to produce colors more brilliant for their particular spectral regions than one dye alone could produce. If Rhodamine F5G (Fig. 2, curve 2) is added to a pigment system based on Alberta Yellow and the finished pigment drawdown exposed to white light, the energy stored momentarily by the yellow dye passes to the F5G dye either by emission and reabsorption or by direct energetic excitation. If the latter occurs, the yellow dye would be acting as a sensitizer for the F5G. A much greater peak height of the radiance factor curve results. At the same time, the yellow dye absorbs so powerfully in the violet region that the blue-violet component of the salmon curve 2 is eliminated. A very bright orange is formed. Addition of a small amount of Rhodamine F3B can displace the peak further toward the red, giving a color with the highest peak of all of the colors, for example, Day-Glo Blaze Orange. This phenomenon of energy transfer from one molecule to another gives the fluorescent pigments their brilliance but also leads to problems in using conventional color matching equipment to give accurate values to fluorescent pigments.

Rhodamine F3B (Fig. 2, curve 3) can also participate strongly in the energy-transfer effect, receiving energy from both the greenish yellow and the salmon dyes. The blue and violet components of the Rhodamine F3B are also reduced, resulting in strong bright red-oranges and reds. Pinks and deep bluish reds are produced essentially with Rhodamine F3B. For comparison, curve 4 (Fig. 2) shows the reflectance of a bright nonfluorescent printing ink. It is possible for multiple-dye-pigmented coatings applied over a white surface to show peak radiance factors of nearly 300%.

Methods of Manufacturing

Bulk Pigment Polymerization. Most fluorescent pigments are of the toluenesulfonamide–melamine–formaldehyde resin matrix type. The following general technique is used (12). A mixture of *o*- and *p*-toluenesulfonamide, paraformaldehyde, and a B-stage unmodified melamine–formaldehyde resin is heated at ca 170°C for about 15 minutes (see AMINO RESINS). Rhodamine F3B (CI Basic Violet 11), Rhodamine 6GDN Extra (Basic Red 1), and Brilliant Yellow 6G Base (Solvent Yellow 44) are added and heating continued. Upon cooling, the completely condensed resin solidifies at about 115°C. The finished colored resin is clear, brittle, and friable. It has a fiery red-orange color and is highly daylight fluorescent. The resin mass can be pulverized by impact milling to a fine powder, which is the final fluorescent pigment and is used in a wide variety of applications.

Post-curing and chemical modification improves chemical and solvent resistance (20). Paraformaldehyde and acetylene diurea are added to a hot borax solution. Toluenesulfonamide (*p* and *o*), a few drops of phosphorous acid, Brilliant Yellow 6G [*2429-76-7*], Rhodamine F3B, and Rhodamine 6GDN [*989-38-8*] are added. After heating, the mass is cured in an oven at 150°C. The resulting cured resin is thermoset but can be ground to fine particle sizes.

A polyamide-type resin matrix is claimed (21) to have superior heat resistance and be especially suitable for use in coloring plastics. The matrix is made of isophoronediamine and isophthalic acid; Rhodamine F3B is added above 200°C. Upon cooling, the mass solidifies to a friable resinous solid. This product, after grinding, is highly suitable for coloring plastics processed by injection molding. Products of this type are reported to be resistant to temperatures as high as 315°C for 10 minutes.

A polyester-type fluorescent resin matrix (22) is made by heating trimellitic anhydride, propylene glycol, and phthalic anhydride with catalytic amounts of sulfuric acid. Addition of Rhodamine BDC gives a bright bluish red fluorescent pigment soluble in DMF and methanol. It has a softening point of 118°C. Exceptional heat resistance and color brilliance are claimed for products of this type, which are useful for coloring plastics.

Another fluorescent pigment class (23) is based on a urethane-type resin; the primary raw materials are isocyanates, amines, and hydroxy compounds.

Several new types of pigments have been introduced commercially which are based on polymers that do not contain formaldehyde. These pigments have some different characteristics but have the advantage of not giving off formaldehyde. Most of the primary manufacturers provide these types of pigments and more are being developed.

Emulsion–Suspension Polymerized Pigment Ink. Polymerization of a polar prepolymer as the internal phase in an oil-based external phase (24) gives a fluorescent ink base in which spherical fluorescent particles are dispersed. This base is suitable for litho and letterpress inks (qv). An emulsion is formed from paraformaldehyde, *o*- and *p*-toluenesulfonamide, Solvent Yellow 44, and melamine with heating in an oil vehicle which forms the continuous external phase of the finished product. Rhodamine F3B and Rhodamine 6GDN are added under high speed agitation and heating. The oil vehicle is formed by blending 60 parts of a 50% styrenated-type alkyd solution at 50% solids in a high boiling ink oil (Magie 535) with 40 parts of bodied linseed oil.

Soluble Fluorescent Polymers. Several pigment manufacturers have developed fluorescent polymers intended to be used as a solution for application to various substrates. These toners come in both solvent soluble and alkaline water-soluble forms.

For example, Day-Glo HM Series toners come in a range of colors suitable for flexographic and gravure inks of the solvent-base type. The Radiant GF Series and Lawter HVT Series are also suitable for this type of application (see Table 2). Generally, these toners are of the formaldehyde–sulfonamide type and require oxygenated solvents, primarily alcohol–ester blends, for proper solution. For applications such as flexographic printing on film, these materials are modified with other resins such as nitrocellulose or polyamides in the finished ink.

In addition to the solvent soluble toners, alkali water-soluble toners have been produced. These types include WST produced by Day-Glo and Aquabest produced by Radiant Color. These toners are dissolved in water which contains a portion of ammonia and, if necessary, some isopropyl alcohol. These toners can be used as binders or additional binders and other additives can be added to give the ink the desired properties. These toners are condensation-type polymers other than the formaldehyde types.

Nippon Shokubai and U.K. Seung are producing a fluorescent polymer claimed to be made from a co-condensation of benzoguanamine and formaldehyde. Fine highly thermoset particles are manufactured in solution and later dried. It is useful in a wide range of applications, specifically plastics, and markets where bleed is a problem.

Economic Aspects

The market price of fluorescent pigments varies from ca $9/kg for certain grades of material that might find application in the textile and paper coating industries, to over $18/kg for special, high technology products with applications ranging from flexo and felt tip markers to plastics. Growth in the primary market segments such as packaging, safety, signage, toys, etc, approximate GNP growth in most of the world's regions, and new markets that are opening have seen substantially better growth.

The total world marketplace can vary widely from year to year, however, due to the cyclic nature of the textile and collateral applications which can be sizable but short-lived. Perhaps this is the reason that there are no published estimates for the world market for fluorescent color. Also, competition has forced most producers to develop higher strength fluorescent materials that offer greater color yield and a better money value. This reduces unit volume and, in many cases, dollar sales.

Application Properties and Uses

Fluorescent colors are remarkable for their extremely high visibility and their ability to attract attention, and applications utilizing these properties have gained the greatest acceptance. Advertising offers one of the main uses of fluorescent colors. Fluorescent billboards along highways, fluorescent packages at supermarkets, and fluorescent advertisements in magazines have all proven to be extremely effective ways of promoting consumer goods. The presence of the fluorescent color attracts the attention of the viewer earlier, holds their attention longer, and brings the consumer back for a second look more frequently than conventional color.

Another large field for fluorescent color is for safety uses. Fluorescent color has been particularly effective on vehicles, such as fire trucks, ambulances, and rescue equipment. Fluorescent color is far more visible than the equivalent hue of conventional color under all lighting conditions and is particularly valuable at dawn, dusk, and in overcast conditions. Fluorescent colors are used to mark off the boundaries of danger areas at construction sites and around heavy equipment, highlight safety equipment in industrial operations, as well as for highway traffic cones, flagging, and on worker safety vests. A large safety application for fluorescent colors is for hunters' and woodsmen's clothing during hunting season; in most states this is mandatory.

Fluorescent color also is used in the optical-sensing field and in the coding and tracing of documents and other items. This is not strictly a daylight-fluorescent use since there are many ultraviolet-responsive products that have virtually no daylight fluorescence. Daylight-fluorescent color has, however, become important in this field. Fluorescent inks are extremely useful in the high speed handling of documents, such as automatic sorting where a fluorescent mark on each document allows proper orientation and provides much improved efficiency compared with manual sorting.

The brilliance of daylight-fluorescent color finds use in most of the color consuming markets because of advances in use technologies. Markets served are injection molded toys, blow molded bottles, high speed sheet-fed and web-offset printing, gravure and flexo printing (water and solvent inks), industrial paint to tempra colors, felt tip pens, paper coating, textile printing and dyeing, plus a variety of specialty applications.

Table 4 shows daylight fluorescent pigments with approximately equivalent colors manufactured by U.S. manufacturers. In addition to the colorants listed, other colors are available such as purples and shades which are stronger and between the shades listed.

Table 4. Commercial Pigment Colors, Standard Strength Pigments[a]

Day-Glo A-Series[b,c]	Lawter B-3500 Series[d]	Radiant R-105 Series[e]
A-17-N, Saturn Yellow	B-3539, lemon yellow	R-105-810, chartreuse
A-18-N, Signal Green	B-3545, green	R-105-811, green
A-16-N, Arc Yellow	B-3515, gold-yellow	R-105-812, orange-yellow
A-15-N, Blaze Orange	B-3514, yellow-orange	R-105-81, orange
A-14-N, Fire Orange	B-3513, red-orange	R-105-814, orange-red
A-13-N, Rocket Red	B-3534, red	R-105-815, red
A-12, Neon Red	B-3530, cerise red	R-105-816, cerise
A-11, Aurora Pink	B-3522, pink	R-105-817, pink
A-21, Corona Magenta	B-3554, magenta	R-103-G-188, magenta
A-19, Horizon Blue	B-3556, vivid blue	R-103-G-119, blue

[a]Similar colors are listed horizontally but are not exact color matches.
[b]These names are trademarks of Day-Glo Color Corp.
[c]Thermoplastic pigments for use in paint, screen ink, plastisol, gravure ink, paper coatings, and many other applications.
[d]Multipurpose pigments for paint, gravure ink, screen ink, paper coatings, plastisol, candles, plastics, and many other applications.
[e]Multipurpose pigments for paint, screen ink, paper coatings, plastisol, gravure ink, plastics, and many other applications.

Commercial Properties of Fluorescent Pigments and Colorants

The largest percentage of commercial fluorescent pigments are made by bulk polymerization and are mechanically ground. The sulfonamide melamine–formaldehyde pigments generally have a density of 1.3–1.4 g/mL and average particle sizes from 2.5 to 6 μm. Melting points of the thermoplastic types range from 110–140°C; the thermosets do not melt but soften in the range of 150–170°C. These pigments decompose at about 200°C.

The other mechanically ground pigments are of the ester, amide, and other condensation-type chemistries. The average particle sizes are similar to the formaldehyde types except for toners and some plastic pigments which are coarsely ground. Melting points range from 70 to 170°C depending on the chemistry; however, the density tends to be less than the formaldehyde, in the 1.15 to 1.25 g/mL range. Decomposition of these resins tends to occur between 250 and 300°C but color degradation starts at about 200°C. Special pigments for high temperature applications such as Radiant K and Day-Glo ZQ have been developed which have better color retention properties at elevated temperature. Various other pigments are available with many different properties, eg, particle sizes from the submicrometer range to coarse sand-like materials.

Plastics. Most manufacturers of fluorescent pigments offer special products for coloring thermoplastic molding resins. Low and high density polyethylene, high impact and general-purpose polystyrene, ABS, and various acrylic polymers are best suited for these pigments. The pigment, 1–2% of the total weight of the plastic, is added either as a dry-blended material or first formulated into a color-concentration pellet which is blended into the uncolored resin before molding into a finished article.

Products suitable for this type of use are Day-Glo ZQ-Series, vinyl colorant, and NX-Series pigments; Dane's ZN Series, Radiant's K-600 and K-7000 pigments; and Lawter's TC Series (see Table 2). These products are dramatic improvements over the toluenesulfonamide-modified melamine–formaldehyde-type resins, and are heat resistant up to 300°C when used in injection molding equipment; however, there is significant color shift with increasing temperature. They also exhibit improved light resistance and do not have the troublesome formaldehyde odor long associated with the use of fluorescent pigments in plastics. The NX-Series and ZQ-Series have dramatically reduced plateout over other fluorescent pigments, which makes them more like conventional pigments in certain plastic applications.

Paint. Fluorescent pigments in various types of paint offer an effective way to impart fluorescence. Because of the inferior lightfastness of fluorescent products in thin layers, the paint is generally applied in a 75–150-μm thick layer to optimize the resistance to exterior fading. For maximum color effect and durability, fluorescent paints should be applied over a high grade white substrate and overcoated with a clear uv-absorbing coating that virtually doubles the life of the color effect. The most commonly used paint systems are alkyd enamels or acrylic lacquers. For these paint systems, Day-Glo A or D, and AX Series, Radiant R-103-G, R-105, and R-203-G Series, and Lawter B-3500 Series are recommended. A formulation for an acrylic lacquer follows; it is thinned to spraying viscosity with xylene. The acrylic vehicle is 35% Acryloid B-66 (Rohm & Haas) and 65% SC-100.

High speed mix component	Wt %
acrylic vehicle	54.0
Cab-O-Sil M-5 (Cabot Corp.)	0.5
fluorescent pigment	38.0
SC-100	7.5
Total	*100.0*

Aerosol Paint. For aerosol paints, fluorescent pigments are available, including Day-Glo A and AX Series, Radiant R-105 and R-106 Series, and Lawter G-3000 Series. With new regulations from the U.S. Environmental Protection Agency (EPA) tightening the allowed VOCs, more recent formulas contain oxygenated solvents and can therefore require the use of a thermoset pigment such as the Day-Glo T or GT, Radiant P-1600 or P-1700, or Lawter TS Series. A typical starting formulation for a fluorescent aerosol paint concentration is given below. The finished canned material should consist of 55% thinned paint and 45% propellant.

High speed mix component	Wt %
Acryloid F-10, 40% in mineral thinner (Rohm & Haas)	45.5
fluorescent pigment	35.0
lactol spirits	17.7
toluene	1.8
Total	*100.0*

Water-Based Paint. Fluorescent pigments also have some use in water-based paint systems, with many being sensitive to aqueous media, especially at pH above 7.5. For extended shelf stability, Radiant P-1600 or P-1700 Series, Day-Glo T and GT pigments, and Sinloihi FZ-3000 Series are recommended.

Gravure Ink. Fluorescent pigments are used in liquid gravure inks. A fine particle sized pigment is incorporated into a vehicle system composed of a soluble nonoxidizing binder in volatile solvent. Pigments suitable for this type of application, depending on the nature of the solvent used, are Day-Glo A and AX Series pigments, the Radiant R-105 and R-106 Series, and Dane A and E Series pigments. These products are useful for A-type gravure where aliphatic and small amounts of aromatic solvents are used.

In C- and T-type gravure systems where oxygenated and aromatic solvents are used, the Radiant P-1700 Series and Day-Glo GT and STX pigments are recommended. A typical formulation for an A-type gravure ink is 30% Acryloid NAD-10 (Rohm & Haas), 50% fluorescent pigment, 5% toluene, and 15% heptane (as thinner).

Flexographic Inks. Fluorescent toners such as the Radiant GF, Lawter HVT, and Day-Glo HM and HMS Series toners are used in flexographic ink formulations. These products are soluble in blends of alcohol (80%) and ester solvents (20%) and are compatible with modifying materials such as nitrocellulose resins and acrylic solution polymers. Flexographic inks of this type are used most commonly to print products such as cellophane and polyethylene film for packaging, and also to print paper products such as gift wrap and price labels.

Water-based flexo inks can be formulated with either a soluble toner or with the Day-Glo EPX Series which is a true pigment and can be formulated like a conventional pigment dispersion. The Radiant Aquabest or the Day-Glo WST can be formulated in an alkaline water-soluble system to yield strong inks. They have limited shelf life and inferior fade, but do not necessarily require additional binder. Day-Glo EPX must be formulated with a binder such as a hard resin or can be used with one of the soluble toners such as WST. The EPX Series has several advantages over soluble toners such as much superior fade, excellent ink stability, and some hiding power over kraft-type papers. A disadvantage of the EPX is its lower tinctorial strength than other fluorescent toners.

Screen Inks. Among the earliest uses for products containing daylight-fluorescent pigment was screen printing inks. The process involves pushing ink through a coarse screen leaving a heavy deposit. This process is used to make outdoor and point of purchase advertising. By using daylight-fluorescent pigments in the ink, previously unattainable brilliance was achieved which provided longer viewing hours, particularly under low light conditions, such as at dawn, dusk, and on cloudy days.

This remains a good market for fluorescent pigments for the same reasons, and many boards have been fitted with ultraviolet lights or filtered white light to further increase the impact and extend the viewing time through the nighttime hours. Lawter B-3500 Series, Day-Glo A and AX Series pigments, and Radiant R-105 and R-106 Series are recommended. Formulations for screen ink employ complex vehicles and about 35% pigment.

Lithographic and Letterpress Printing Inks. Fluorescent pigments of fine enough particle size for litho and letterpress printing inks could not be obtained by standard techniques when starting with dry pigment. However, manufacturing spherical fluorescent particles *in situ* in a paste-ink vehicle made acceptable printing properties possible. Typical products include Day-Glo Starfire Series printing-ink bases, Radiant VF and Visiprint Series printing-ink bases, and Dane OLC Series bases. These bases are similar in nature to one another, although modifications of the ink formulation are needed depending on the specific product. The following is a typical finished formulation for a fluorescent lithographic ink starting with one of the above-mentioned printing-ink bases. Certain types of driers accelerate drying but may cause darkening of the ink, such as cobalt driers.

Component	Wt %
fluorescent-ink base	84.0
polyethylene wax compound	6.0
gloss varnish (tung oil-based)	3.6
tridecyl alcohol	4.2
driers	2.0
antioxidant	0.2
Total	*100.0*

Vinyl Products. The use of fluorescent colorants in vinyl products, especially calendered and cast films, has opened up another important market. Because of the thickness of vinyl sheeting and cast articles, excellent exterior durability can be achieved. In Europe there is a very large market for fluorescent vinyl film for safety marking and high visibility applications such as on emergency vehicles and construction equipment, and there is a growing market in the United States and Asia.

For vinyl plastisol, organosol products and calendering, Day-Glo Color Corp. offers T, D, VC, and AX-Series pigments, Lawter Chemical offers the B-3500 and G-3000 Series, and Radiant Color offers P-1600 and R-203-G Series. In addition, Day-Glo offers VC Series for vinyl calendering where nonformaldehyde products are needed.

Health and Safety Factors

Good safety practices are recommended when handling fluorescent pigments, including a respirator and dust collecting equipment. The pigments present no unusual fire or explosion hazards. They clean up easily with detergent and water or solvents appropriate for the coating vehicle system.

Daylight fluorescent pigments (qv) are considered to be nontoxic. Since they are combinations of polymers and dyestuffs, the combined effect of the ingredients must be taken into account when considering the net toxic effect of these materials. Table 5 gives results of laboratory animal toxicity tests of standard modified melamine–formaldehyde-type pigments, the Day-Glo A Series, and the products recommended for plastic molding, Day-Glo Z-series.

In heavy-metal analysis of the same pigments, metals found were present in only trace amounts. The data listed place the products tested in the category of nontoxic materials. The Radiant Color Co. has conducted toxicity tests on its own products similar to the A-Series and has found them to be nontoxic. Heavy metals were found only in trace amounts in these tests.

Table 5. Results of Laboratory Animal Toxicity Tests

Test	Day-Glo A-Series	Day-Glo Z-Series
acute oral toxicity LD_{50}, g/kg	16.0	15.38
acute dermal toxicity LD_{50}, g/kg	23.0	3.0
acute dust inhalation LC_{50}, mg/L air	4.4[a]	2.88
eye irritation	no significant	mildly irritating

[a]Four hours.

BIBLIOGRAPHY

"Fluorescent Pigments (Daylight)" in *ECT* 2nd ed., Vol. 9, pp. 483–506, by R. W. Voedisch, Lawter Chemical, Inc., and D. W. Ellis, University of New Hampshire; "Luminescent Materials (Fluorescent Daylight)" in *ECT* 3rd ed., Vol. 14, pp. 546–569, by E. L. Kimmel and R. A. Ward, Day-Glo Color Corp.

1. P. Krais, *Melliand Textilber.*, 10 (1929).
2. U.S. Pat. 1,796,011 (Mar. 10, 1931), W. Eckert (to General Aniline Works).
3. U.S. Pat. 1,836,529 (Dec. 15, 1931), W. Eckert and C. E. Muller (to General Aniline Works).
4. U.S. Pat. 3,014,041 (Dec. 19, 1961), H. Hausermann and J. Voltz (to J. R. Geigy).
5. Brit. Pat. 1,095,784 (Dec. 20, 1967), (to Farbwerke Hoechst, AG).
6. Brit. Pat. 1,112,726 (May 8, 1968), (to Farbwerke Hoechst, AG).
7. Brit. Pat. 1,345,176 (Jan. 30, 1974), (to Farbwerke Hoechst, AG).
8. U.S. Pat. 3,853,884 (Dec. 10, 1974), (to Farbwerke Hoechst, AG).
9. U.S. Pat. 2,417,383 (Mar. 11, 1947), J. L. Switzer.
10. U.S. Pat. 2,606,809 (Aug. 12, 1952), J. L. Switzer and R. A. Ward.
11. U.S. Pat. 2,498,592 (Feb. 21, 1950), J. L. Switzer and R. C. Switzer.
12. U.S. Pat. 2,938,873 (May 31, 1960), Z. Kazenas (to Switzer Brothers, Inc.).
13. J. Griffiths, *Colour and Constitution of Organic Molecules*, Academic Press, Inc., London, 1976.
14. Gattermann and von Horlacher, *Ber.* **32**, 286 (1819).
15. F. P. Schafer, ed., *Dye Lasers*, 2nd ed., Vol. 1, *Topics in Applied Physics*, Springer-Verlag, New York, 1977, p. 16.
16. G. Hallas, *J. Soc. Dyers Colour.* **86**, 237 (1970).
17. E. J. Schimitscheck and co-workers, *Opt. Commun.* **11**(4), (1974).

18. K. H. Drexhage, in Ref. 15, p. 148.
19. B. H. Winters, H. I. Mandelbert, and W. B. Mohr, *Appl. Phys. Lett.* **25** (Dec. 15, 1974).
20. U.S. Pat. 3,412,036 (Nov. 19, 1968), M. D. McIntosh (to Switzer Brothers, Inc.).
21. U.S. Pat. 3,915,884 (Oct. 28, 1975), Z. Kazenas (to Day-Glo Color Corp.).
22. U.S. Pat. 3,922,232 (Nov. 25, 1975), A. K. Schein (to Hercules, Inc.).
23. U.S. Pat. 3,741,907 (June 26, 1973), H. P. Beyerlin (to Siegle GmbH).
24. U.S. Pat. 3,412,035 (Nov. 19, 1968), M. D. McIntosh, Z. Kazenas, and J. L. Switzer (to Switzer Brothers, Inc.).

General References

R. Donaldson, "Spectrophotometry of Fluorescent Pigments," *Brit. J. Appl. Phys.* **5**(6), 120 (1954).
K. Venkataraman, *The Chemistry of Synthetic Dyes*, Vol. 3, Academic Press, Inc., New York, 1970, pp. 169–221.
F. Forster, *Fluoreszenz Organischer Verbindungen*, Vandenhoeck and Ruprecht, Gottingen, 1951.
Colour Index, 3rd ed., American Association of Textile Chemists and Colorists, Triangle Park, N.C., 1971–1976.
D. B. Judd and G. Wyszecki, *Color in Business, Science and Industry*, 2nd ed., John Wiley & Sons, Inc., New York, 1967.
G. Wyszecki and W. S. Stiles, *Color Science*, Sect. 1, John Wiley & Sons, Inc., New York, 1967, pp. 1–63.

STEVEN G. STREITEL
Day-Glo Color Corporation

LUTETIUM. See LANTHANIDES.

LYSOXYGENASES. See ANTIBACTERIAL AGENTS, SYNTHETIC.

MACHINING METHODS, ELECTROCHEMICAL

Electrochemical machining (ECM) is an electrolytic process. Metal removal is achieved by electrochemical dissolution of an anodically polarized workpiece which is one part of an electrolytic cell. The first significant developments occurred in the 1950s, when ECM was investigated as a method for shaping high strength, heat-resistant alloys which were difficult to cut by established methods (see HIGH TEMPERATURE ALLOYS). As of the 1990s, ECM is employed in many ways, eg, by automotive, offshore petroleum, and medical engineering industries, as well as by aerospace firms, which are its principal user (1–3).

Theoretical Background

Electrolysis. An example of an anodic dissolution operation is electropolishing, in which the item to be polished is made the anode in an electrolytic cell. Irregularities on the surface of this anode are dissolved preferentially so that on removal the surface becomes flat and polished. A typical current density in this operation is 10^{-1} A/cm^2, and polishing is usually achieved on the removal of irregularities as small as 10^{-2} μm. In both electroplating (qv) and electropolishing, the electrolyte is either in motion at low velocities or unstirred. ECM is similar to electropolishing in that ECM also is an anodic dissolution process. The rates of metal removal offered by the polishing process are, however, considerably less than those needed in metal machining practice. In the electrolysis of iron, the anode, in aqueous sodium chloride, several possible reactions can occur when a potential difference is applied across the electrodes. The probable anodic reaction is dissolution of iron, eg,

$$Fe \longrightarrow Fe^{2+}(aq) + 2\,e^-$$

At the cathode, the reaction is likely to be the generation of hydrogen gas and the production of hydroxyl ions:

$$2\,H_2O + 2\,e^- \longrightarrow H_2 + 2\,OH^-$$

The net reaction is thus:

$$Fe + 2\,H_2O \longrightarrow Fe(OH)_2\,(s) + H_2$$

The ferrous hydroxide may react to form ferric hydroxide:

$$4\,Fe(OH)_2 + 2\,H_2O + O_2 \longrightarrow 4\,Fe(OH)_3$$

although this last reaction, albeit an oxidation–reduction reaction, does not form part of the electrolysis. For this metal–electrolyte combination, the electrolysis involves dissolution of iron from the anode, and the generation of hydrogen at the cathode. No other action takes place at the electrodes.

Two observations relevant to ECM can be made. (*1*) Because the anode metal dissolves electrochemically, the rate of dissolution (or machining) depends, by Faraday's laws of electrolysis, only on the atomic weight A and valency z of the anode material, the current I which is passed, and the time t for which the current passes. The dissolution rate is not influenced by hardness (qv) or any other characteristics of the metal. (*2*) Because only hydrogen gas is evolved at the cathode, the shape of that electrode remains unaltered during the electrolysis. This feature is perhaps the most relevant in the use of ECM as a metal-shaping process (4).

Characteristics of ECM. By use of Faraday's laws if m is the mass of metal dissolved, and because $m = \eta\rho_a$, where η is the corresponding volume and ρ_a the density of the anode metal, the volumetric removal rate of anodic metal $\dot{\eta}$ is given by

$$\dot{\eta} = \frac{AI}{zF\rho_a} \tag{1}$$

where F, the Faraday constant, equals 96,487 C. If a machining operation has to be carried out on an iron workpiece at a typical rate of 2.6×10^{-8} m^3/s, for this removal rate to be achieved by ECM, the current in the cell must be about 700 A because $A/zF = 29 \times 10^{-8}$ kg/C and $\rho_a = 7860$ kg/m^3 for iron. Currents used in ECM are minimally of this magnitude, and often higher by as much as an order of magnitude. The corresponding average current densities are typically 50 to 150 A/cm^2.

The means by which high current densities are obtained can be understood from an examination of the electrolyte conductivity and the interelectrode gap width. These parameters are related to the current through Ohm's law, which states that the current I flowing in a conductor of resistance R is directly proportional to the applied voltage V:

$$V = IR \tag{2}$$

In ECM, electrolytes serve as conductors of electricity and Ohm's law also applies to this type of conductor. The resistance of electrolytes may amount to hundreds of ohms.

The resistance R of a uniform conductor is directly proportional to its length h and inversely proportional to its cross-sectional area A. Thus

$$R = \frac{h\rho}{A} \tag{3}$$

where ρ is the constant of proportionality. If the conductor is a cube of side 10 mm, then $R = \rho$ and ρ is termed the specific resistance or resistivity of the conductor. The reciprocal of the specific resistance is the specific conductivity, often denoted by the symbol κ.

If equations 2 and 3 are combined, relationships between the average current density J, current I, surface area to be machined A, applied potential difference, gap width h, and electrolyte conductivity κ_e, are

$$J = \frac{I}{A} = \frac{\kappa_e V}{h} \tag{4}$$

In practice J is often about 50 A/cm^2. To obtain a current density of this magnitude, a cell could be devised having high values for κ_e and V, and low values for h. Even for strong electrolytes, however, κ_e is small. If the current is high, power requirements, amongst other considerations, restrict the use of high voltages, and in practice the voltage is usually about 10 to 20 V. If values of 0.2 $(\Omega\cdot\text{cm})^{-1}$ and 10 V are taken for κ_e and V, respectively, then for J to be 50 A/cm^2, the gap h must be 0.4 mm. A gap of this size is also necessary for accurate shaping of the anode. As dissolution of the anode proceeds, this gap is maintained by mechanical movement of one electrode, say the cathode, toward the other. To maintain the gap of 0.4 mm, a cathode feed rate of about 0.02 mm/s is needed if the values given for the other process variables are retained.

Accumulation within the small machining gap of the metallic and gaseous products of the electrolysis is undesirable. If growth were left uncontrolled, eventually a short circuit would occur between the two electrodes. To avoid this crisis, the electrolyte is pumped through the interelectrode gap so that the products of the electrolysis are carried away. The forced movement of the electrolyte is also essential in diminishing the effects both of electrical heating of the electrolyte, resulting from the passage of current and hydrogen gas, which respectively increase and decrease the effective conductivity. The Joule heating effect provides a simple, convenient way of estimating a typical electrolyte velocity. Without forced agitation to control the increase in the electrolyte temperature, boiling eventually occurs in the gap. If all heat caused by the passage of current remains in the electrolyte, the temperature increase δT in a length $\delta\chi$ of gap is, from Joule's and Ohm's laws,

$$U\delta T = \frac{J^2 \delta\chi}{\kappa_e \rho_e C_e \Delta T} \tag{5}$$

where U is the electrolyte velocity, ρ_e the electrolyte density, and C_e its specific heat.

If the increase with temperature of the electrolyte conductivity is neglected, integration of equation 5 yields

$$U = \frac{J^2 L}{\kappa_e \rho_e C_e \Delta T} \tag{6}$$

where L is the electrode length and ΔT is the temperature difference of the electrolyte between points at inlet and outlet to the gap.

Using the typical values of $J = 50$ A/cm^2, $L = 10^2$ mm, $\kappa_e = 0.2$ $(\Omega \cdot \text{cm})^{-1}$, $\rho_e = 1.1$ g/cm^3, $C_e = 4.18$ J/(g·C), and keeping ΔT to 75°C, to avoid boiling at the exit point, and letting the inlet temperature be 25°C, then from equation 6, the velocity U to maintain this condition is calculated to be about 3.6 m/s. Velocities of the electrolyte solution through the gap in ECM usually range from 3 to 30 m/s. The pressures required to achieve these velocities can be calculated to be ca 600–700 kN/m^2, and the flow is usually found to be turbulent, on the basis of the usual criterion that the Reynolds number > 2300.

Working Principles. ECM has been founded on the principles outlined. As shown in Figure 1, the workpiece and tool are the anode and cathode, respectively, of an electrolytic cell, and a potential difference, usually fixed at about 10 V, is applied across them. A suitable electrolyte, eg, aqueous NaCl, is chosen so that the cathode shape remains unchanged during electrolysis. The electrolyte, where the conductivity is about 0.2 $(\Omega \cdot \text{cm})^{-1}$, is also pumped at a rate $\sim$ 3–30 m/s through the gap between the electrodes to remove the products of machining and to diminish the unwanted effects, such as those that arise from cathodic gas generation and electrical heating. The rate at which metal is then removed from the anode is approximately in inverse proportion to the distance between the electrodes. As machining proceeds and with the simultaneous movement of the cathode at a typical rate, eg, 0.02 mm/s, toward the anode, the gap width along the electrode length gradually tends to a steady-state value. Under these conditions, a shape complementary to that of the cathode is reproduced on the anode. A typical gap width should be about 0.4 mm and the average current density on the order of 50–150 A/cm^2. Moreover, if a complicated shape

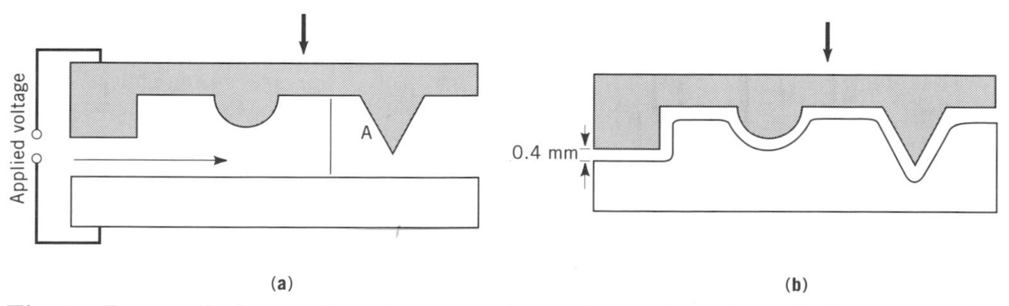

Fig. 1. Brass cathode tool (▩) and anode workpiece (□) configurations for ECM where the heavy arrows indicate the direction of cathode feed at a feed rate of 0.02 m/s. (**a**) Initial positions where A is the initial interelectrode gap of height h, and (→) is the direction of electrolyte flow (10 m/s). The applied voltage is 10 V. (**b**) Final positions showing the 0.4-mm steady-state gap between the electrodes.

is to be formed on a workpiece of hard material, the complementary shape can first be produced on a cathode of softer metal. The latter electrode is then used electrochemically to machine the workpiece.

The main advantages of ECM are that the rate of metal machining does not depend on the hardness of the material, complicated shapes can be machined on hard metals, and there is no tool wear.

Electrochemical Machining

Machine Components. Industrial electrochemical machines work on the principles outlined. Particular attention has to be paid to the stability of the electrochemical machine tool frame, and to the machining table which should also be stable and firm. The electrolyte has to be filtered carefully to remove the products of machining and often has to be heated in its reservoir to a fixed temperature, for instance 30°C, before entering the machining apparatus. This procedure is used to provide constant operating conditions. During machining the electrolyte heats up from the passage of current. Precautions must be taken to avoid a high electrolyte temperature which can cause changes in the electrolyte specific conductivity, and subsequent undesirable effects on machining accuracy (5).

Rates of Machining. Faraday's laws, embodied in equation 7, can be employed to calculate the rates at which metals can be electrochemically machined:

$$m = \frac{AIt}{zF} \tag{7}$$

where m is the mass of metal electrochemically machined by a current I, in amperes, passed for a time t in seconds. The quantity A/zF, called the electrochemical equivalent of the anode–metal, corresponds to the atomic weight of the dissolving ions over the valency times the Faraday's constant.

Table 1 shows the metal machining rates theoretically obtained when a current of 1000 A is used in ECM. The values in Table 1 assume that all the current is used to remove metal. That is not always the case, however, as some metals are more likely to machine at the Faraday rates of dissolution than others.

Many factors other than current influence the rate of machining. These involve electrolyte type, rate of electrolyte flow, and other process conditions. For example, nickel machines at 100% current efficiency, defined as the percentage ratio of the experimental to theoretical rates of metal removal, at low current densities, eg, 25 A/cm^2. If the current density is increased to 250 A/cm^2 the efficiency is reduced typically to 85–90%, by the onset of other reactions at the anode. Oxygen gas evolution becomes increasingly preferred as the current density is increased.

If the ECM of titanium is attempted in sodium chloride electrolyte, very low (10–20%) current efficiency is usually obtained. When this solution is replaced by some mixture of fluoride-based electrolytes, to achieve greater efficiencies (>60%), a higher voltage (ca 60 V) is used. These conditions are needed to break down the tenacious oxide film that forms on the surface of titanium. It is this

Table 1. Theoretical ECM Removal Rates for a 1000-A Current

				Removal rate	
Metal	Atomic weight	Valency	Density, kg/m^3 × 10^3	10^{-6} kg/s	10^{-6} m^3/s
aluminum	26.97	3	2,670	95	0.035
beryllium	9.0	2	1,850	50	0.025
chromium	51.99	2	7,190	250	0.038
		3		200	0.025
		6		90	0.013
cobalt	58.93	2	8,850	305	0.035
		3		205	0.023
niobium (columbium)	92.91	3	9,570	320	0.034
		4		240	0.025
		5		195	0.020
copper	63.57	1	8,960	660	0.074
		2		330	0.037
iron	55.85	2	7,860	290	0.037
		3		195	0.025
magnesium	24.31	2	1,740	125	0.072
manganese	54.94	2	7,430	285	0.038
		4		140	0.019
		6		95	0.013
		7		80	0.011
molybdenum	95.94	3	10,220	330	0.032
		4		250	0.024
		6		165	0.016
nickel	58.71	2	8,900	305	0.034
		3		205	0.023
silicon	28.09	4	2,330	75	0.031
tin	118.69	2	7,300	615	0.084
		4		305	0.042
titanium	47.9	3	4,510	165	0.037
		4		125	0.028
tungsten	183.85	6	1,930	315	0.016
		8		240	0.012
uranium	238.03	4	1,910	620	0.032
		6		410	0.022
zinc	65.37	2	7,130	340	0.048

film which accounts for the corrosion resistance of titanium, and together with its toughness and lightness, make this metal so useful in the aircraft engine industry.

If the rates of electrolyte flow are kept too low, the current efficiency of even the most easily electrochemically machined metal is reduced. Insufficient flow does not allow the products of machining to be so readily flushed from the machining gap. The accumulation of debris within the gap impedes further dissolution of metal; the build-up of cathodically generated gas can lead to short circuiting between the tool and workpiece, causing termination of machining and damage to both electrodes if the feed mechanism is not stopped. The inclusion of proper flow channels to provide sufficient flow of electrolyte so that machining can be efficiently maintained remains a primary exercise in ECM practice. When complex shapes have to be produced the design of tooling incorporating the right kind of flow ports becomes a considerable problem.

Various expressions have been derived from which corresponding rates for alloys can be calculated. All these procedures are based on calculating an effective value for the chemical equivalent of the alloy. Thus for Nimonic 75, a typical nickel alloy used in the aircraft industry, a chemical equivalent of 25.1 may be derived (4). The Nimonic alloy is given to have, on a basis of wt %, 72.5 Ni, 19.5 Cr, 5.0 Fe, 0.4 Ti, 1.0 Si, 1.0 Mn, and 0.5 Cu (see NICKEL AND NICKEL ALLOYS).

Surface Finish. As well as influencing the rate of metal removal, electrolytes also affect the quality of surface finish obtained in ECM. Depending on the metal being machined, some electrolytes leave an etched finish. This finish results from the nonspecular reflection of light from crystal faces electrochemically dissolved at different rates. Sodium chloride electrolyte tends to produce a kind of etched, matte finish when used for steels and nickel alloys. A typical surface roughness average, Ra is about 1 μm.

In many applications, a polish is desirable on machined components. The production of an electrochemically polished surface is usually associated with the random removal of atoms from the anode workpiece, the surface of which has become covered with an oxide film. These conditions are determined by the particular metal–electrolyte combination being used. The mechanisms controlling high current density electropolishing in ECM are not completely understood. For example, using nickel-based alloys, the formation of a nickel oxide film seems to be necessary. A polished surface, having roughness of 0.2-μm Ra is claimed for a Nimonic (nickel only) machined in saturated sodium chloride solution. Surface finishes as fine as 0.1 μm Ra have been reported when nickel–chromium steels are machined in sodium chlorate solution. The formation of an oxide film on the metal surface is considered the key to these conditions of polishing.

Sometimes the formation of oxide films on the metal surface hinders efficient ECM, and leads to poor surface finish. For example, the ECM of titanium is rendered difficult in chloride and nitrate electrolytes because the oxide film formed is so passive. Even when higher (eg, ca 50 V) voltage is applied, to break the oxide film, its disruption is so nonuniform that deep grain boundary attack of the metal surface occurs.

Occasionally, metals that have undergone ECM have a pitted surface, the remaining area being polished or matte. Pitting normally stems from gas evolution at the anode electrode; the gas bubbles rupture the oxide film causing localized pitting.

Peripheral pitting and etching associated with the low current densities arising outside the main machining zone occur when higher current densities of 45–75 A/cm^2 are applied. This is a recurrent difficulty when high alloy, particularly those containing about 6% molybdenum, titanium alloys are electrochemically machined.

The occurrence of pitting seems to stem from the differential stability of the passive film that forms on the titanium alloy. This film does not break down uniformly even when the electrolytes are fluoride and bromide based. The pitting can be so severe that special measures are needed to counteract it.

Process variables also play a significant part in determination of surface finish. For example, the higher the current density, generally the smoother the finish on the workpiece surface. Tests using nickel machined in HCl solution show that the surface finish improves from an etched to a polished appearance when the current density is increased from ca 8 to 19 A/cm^2 and the flow velocity is held constant. A similar effect is achieved when the electrolyte velocity is increased. Bright smooth finishes are obtained over the main machining zone using both NaCl and $NaNO_3$ electrolyte solutions and current densities of 45–75 A/cm^2.

The distribution of the electric current lines leads to rounding of edges, thus very sharp corners cannot be produced by ECM. Tolerances of ca 0.127 mm are typical, although accuracies to 0.013 mm have been claimed under special circumstances. Reports on micro-ECM reveal that accuracy of ECM can be improved by special shielding and masking in order to direct the current flow only to required areas (6).

Pulsed ECM (PECM) may be a promising way to improve dimensional accuracy control and also to simplify tool design. Accuracies as fine as 0.002 mm have been quoted using current pulse lengths of ca 0.2 to 2.0 ms, at current densities of 55 A/cm^2. Pulse offtimes are from 1 to 2 ms (7).

Accuracy and Dimensional Control. Electrolyte selection plays an important role. Sodium chloride, for example, yields much less accurate components than sodium nitrate. The latter electrolyte has far better dimensional control owing to its current efficiency/current density characteristics. Using sodium nitrate electrolyte, the current efficiency is greatest at the highest current densities. In hole drilling these high current densities occur between the leading edge of the drilling tool and the workpiece. In the side gap there is no direct movement between the tool and workpiece surface, so the gap widens and the current densities are lower. The current efficiencies are consequently lower in the side gap and much less metal than predicted from Faraday's law is removed. Thus the overcut in the side gap is reduced with this type of electrolyte. If another electrolyte such as sodium chloride solution were used instead, then the overcut could be much greater. Using soldium chloride solutions, its current efficiency remains steady at almost 100% for a wide range of current densities. Thus, even in the side gap, metal removal proceeds at a rate which is mainly determined only by current density, in accordance with Faraday's law. A wider overcut then ensues.

Shaping. Most metal-shaping operations in ECM utilize the same inherent feature of the process whereby one electrode, generally the cathode tool, is driven toward the other at a constant rate when a fixed voltage is applied between them. Under these conditions, the gap width between the tool and the workpiece becomes constant. The rate of forward movement between the tool and the workpiece becomes constant. The rate of forward movement of the tool is matched by the rate of recession of the workpiece surface resulting from electrochemical dissolution.

Some useful expressions can be derived for the variation of the interelectrode gap width, h (Fig. **1a**). If the electrodes constitute a set of plane–parallel electrodes having a constant voltage V applied across them, and the cathode tool is driven mechanically toward the anode workpiece at a constant rate f, then from Faraday's law the rate of change of gap width h relative to the tool surface is

$$\frac{dh}{dt} = \frac{AJ}{ZF\rho_a} - f \tag{8}$$

where A and Z are the atomic weight and valency, respectively, of the dissolving ions, F is Faraday's constant, ρ_a is the density of the anode workpiece metal, and J is the current density. The electrolyte flow is not expected to have any significant effect on the specific conductivity of the electrolyte κ_e, which is assumed to stay constant throughout the ECM operation. Also, all the current passed is taken to be used in the removal of metal from the anode, ie, no other reactions occur there.

From Ohm's law, the current density J is given by equation 4 and substitution of equation 4 into 8 gives

$$\frac{dh}{dt} = \frac{A\kappa_e V}{ZF\rho_a h} - f \tag{9}$$

Three practical cases are of interest in considering solutions to equation 9. (*1*) When the feed rate, $f = 0$, ie, no tool movement occurs, equation 9 has the solution for gap $h(t)$ at time (t):

$$h^2(t) = h^2(0) + \frac{2\,A\kappa_e Vt}{ZF\rho_a} \tag{10}$$

where $h(0)$ is the initial machining gap. That is, the gap width increases indefinitely with the square root of machining time t. This condition is often used in deburring by ECM when surface irregularities are removed from components in a few seconds, without the need for mechanical movement of the electrode. (*2*) When the feed rate is constant, ie, the tool is moved mechanically at a fixed rate toward the workpiece, equation 9 has the following solution:

$$t = \frac{1}{f}\left\{h(0) - h(t) + h_e \ln_e \frac{h(0) - h_e}{h(t) - h_e}\right\} \tag{11}$$

The gap width tends to a steady value given by

$$h_e = \left(\frac{A\kappa_e V}{ZF\rho_a f}\right) \tag{12}$$

This inherent feature of ECM, whereby an equilibrium gap width is obtained, is used widely in ECM for reproducing the shape of the cathode tool on the workpiece. (*3*) Under short-circuit conditions the gap width goes to zero. If process conditions such as too high a feed rate arise the equilibrium gap may be so small that contact between the two electrodes ensues. This condition causes a short circuit between the electrodes and hence premature termination of machining.

The equilibrium gap is applied widely in the shaping process. Studies of ECM shaping are usually concerned with three distinct problems: (*1*) the design of a cathode tool shape needed to produce a required profile geometry of the anode workpiece; (*2*) for a given cathode tool shape, prediction of the resultant anode workpiece geometry, eg, hole drilling by ECM; and (*3*) specification of the shape of the anode workpiece, as machining proceeds. This is most readily predicted for the smoothing of surfaces (Fig. 2), although for actual shaping of components by ECM, estimates of the machining times as the shape develops provide useful information about the process.

Applications

Smoothing of Rough Surfaces. The simplest and a very common application of ECM is deburring. An example is given in Figure 2**a**, where a plane cathode tool is placed opposite a workpiece. The current densities at the peaks of the surface irregularities are higher than those elsewhere. The former are therefore removed preferentially, and the workpiece becomes smoothed (8).

Electrochemical deburring is a fast process. Typical process times are 5 to 30 s for smoothing the surfaces of manufactured components. Owing to its speed and simplicity of operation, electrochemical deburring can often be performed using a fixed, stationary cathode tool. The process is used in many applications, and is particularly attractive for the deburring of the intersectional region of cross-drilled holes.

ECM can also be used to smooth the surface of dies and molds, the shapes produced by electrodischarge machining. The latter process uses a liquid dielectric, such as light oil, as its working medium. Sparks generated between the die to be formed and a tool electrode reproduce the die shape accurately; however, the surface is rough and has a heat-affected zone. By use of the same electrode, and replacement of the dielectric fluid with an aqueous electrolyte solution, ECM can be employed to smooth the die surface, and the heat-affected zone removed.

Hole Drilling. Hole drilling is another popular way of using ECM. As indicated in Figure 3 a tubular electrode is used as the cathode tool. Electrolyte is pumped down the central bore of the tool, and out through the side gap

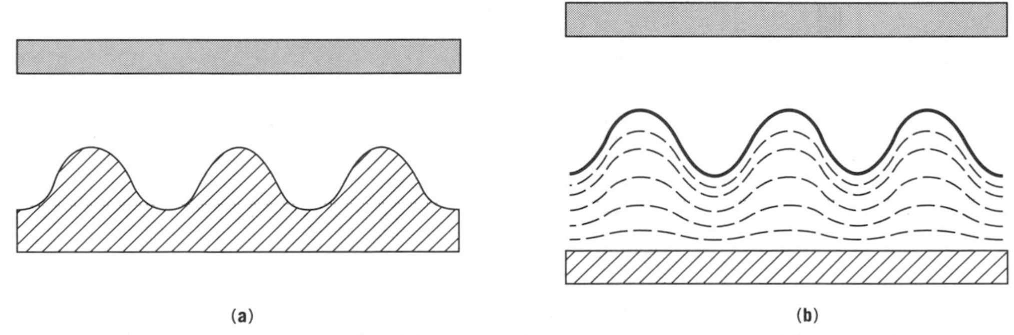

Fig. 2. Smoothing by ECM. Specification of a machined anode profile over time where (▒) is the cathode tool and (—) is (**a**) the initial irregular anode surface, and (**b**) the final anode surface; (– – –) is successive anode profiles over time.

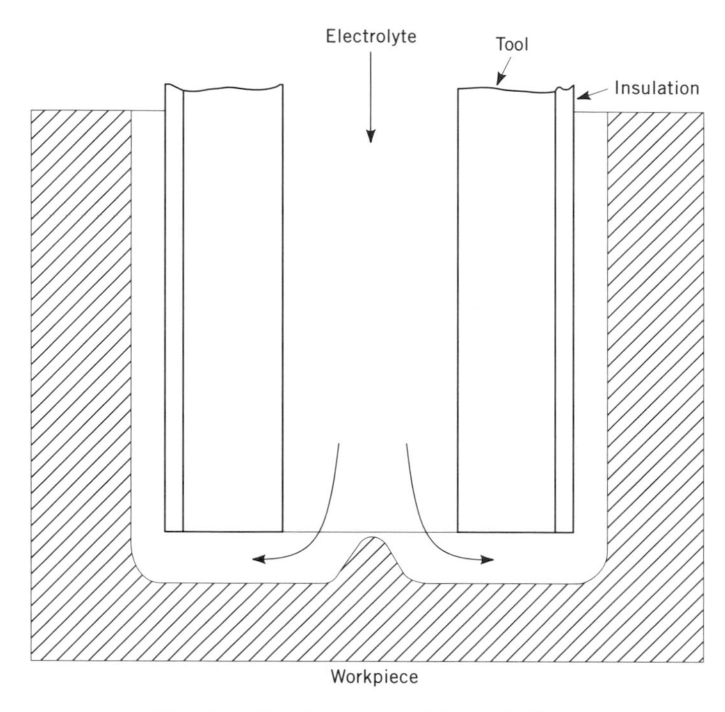

Fig. 3. Electrochemical hole drilling.

formed between the wall of the tool and the hole electrolytically dissolved in the workpiece (2,9,10).

The main machining action is carried out in the interelectrode gap formed between the leading edge of the drill tool and the base of the hole in the workpiece. ECM also proceeds laterally between the side walls of the tool and component. Because the lateral gap width is larger than that at the leading edge of the tool, the side ECM rate is lower. The overall effect of the side ECM is to increase the diameter of the hole produced. The local difference between the radial length between the side wall of the workpiece and the central axis of the cathode tool, and the external radius of the cathode, is known as the overcut. The amount of overcut can be reduced by several means. A common method is reversal of the direction of flow of the electrolyte, which can be pumped down between the outer wall of the cathode tool, across the main machining gap, and upward through the central bore of the drilling tool. This procedure removes the gaseous products of electrolysis from the machining zone without having them reach the side gap. The overcut can also be reduced by electrical insulation of the external tool walls to stop side current flow. A third method of reducing overcut is by the choice of electrolyte such as sodium nitrate, which permits high, efficient rates of metal removal in the leading edge where current density is high, and lower rates in the side gap where current density is lower (11).

A wide range in hole sizes can be drilled. Diameters as small as 0.05 mm to ones as large as 20 mm have been reported (5). Drilling by ECM is not restricted to round holes. The shape of the workpiece is determined by that of the tool electrode, thus a cathode drill having any cross section produces a corresponding shape on the workpiece.

There are significant differences between drilling and smoothing. In the former, forward mechanical movement of one of the electrodes, eg, the tool, toward the other is usually necessary in order to maintain a constant equilibrium gap width in the main machining zone between the leading face of the drill and the workpiece. In smoothing, mechanical drive can often be avoided. A typical feed rate in ECM drilling is 1–5 mm/min.

Full-Form Shaping. The third application of ECM, full-form shaping, utilizes a constant gap across the entire workpiece, and a constant feed rate in order to produce the type of shape used for the production of compressor and turbine blades. In this procedure, current densities as high as 100 A/cm^2 are used, and the current density remains high across the entire face of the workpiece.

Electrolyte flow plays an even more influential role in full-form shaping than in drilling and smoothing. The entire large cross-sectional area of the workpiece has to be supplied by the electrolyte as it flows between the electrodes. The larger areas of electrodes involved mean that comparatively higher pumping pressures and volumetric flow rates are needed.

Electrochemical Grinding. The main feature of electrochemical grinding (ECG) is the use of a grinding wheel in which an insulating abrasive, such as diamond particles, is set in a conducting material (see ABRASIVES). This wheel becomes the cathode tool. the nonconducting particles act as a spacer between the wheel and workpiece, providing a constant interelectrode gap, through which electrolyte is flushed.

When a voltage of ca 4–8 V is applied between the wheel and the workpiece, current densities of ca 120–240 A/cm^2 are created, removing metal mainly by ECM, although mechanical action of the nonconducting particles accounts for an additional 5–10% of total metal removal. The rate of machining is typically 1600 mm^3/min. The surface finish produced by ECG is about 0.2 μm, depending on the metal being ground.

Accuracies achieved by ECG are usually about 0.125 mm, although some claims have been made for accuracies an order of magnitude better. A drawback of ECG is the loss of accuracy when inside corners are ground. Because of the electric field effects, radii better than 0.25–0.375 mm can seldom be achieved.

A wide application of electrochemical grinding is the production of tungsten carbide [*12070-12-1*] cutting tools (see CARBIDES; TOOL MATERIALS). ECG is also useful in the grinding of fragile parts such as hypodermic needles and thin-wall tubes.

A more recent application of the technique has arisen in the offshore drilling industry, for the removal of fatigue crack from underwater steel structures. Seawater itself is an electrolyte composed mainly of sodium chloride solution of approximate 3.5% salinity. Although its specific conductivity is about one-fifth that of electrolytes normally used in ECM, it is a suitable vehicle for ECG processes used in the North Sea. The diamond particles embedded in the grinding tool are used to remove nonconducting materials, such as organic sea growth on the surface of the steel, before the ECG action properly starts. Holes about 25 mm in diameter, in steel 12–25-mm thick, have been produced by ECG at the end of fatigue cracks to stop further development of the crack and to enable the removal of specimens for metallurgical inspection. Further offshore application of ECM (12) in which wire or hollow tube configurations of cathode electrode are employed (13) have been the subject of investigations of this aspect of the process.

Electrochemical Arc Machining. A process that relies on electrical discharges in electrolytes (14–16), thereby permitting metal erosion as well as ECM in that medium, has been developed. Because this process relies on the onset of arcs rather than sparks, it has been named electrochemical arc machining (ECAM). A spark has been defined as a sudden transient and noisy discharge between two electrodes; an arc is a stable thermionic phenomenon. Duration discharges of approximately 1 μs to 1 ms are described as sparks, whereas for durations of about 0.1 s said discharges can be considered arcs. Because in the ECAM process the duration, energy, and time of ignition of sparks are under control, it is valid to regard them as arcs.

An attraction of the ECAM technique is the very fast rates of metal removal attainable by the combined effects of sparking and ECM. For example, in comparison to hole drilling rates for EDM and ECM, respectively 0.1 and 5.0 mm/min^1, rates of 15–40 mm/min may be achieved by ECAM. The ECAM technique can be applied in all the ways discussed for ECM, thus surfaces can be smoothed and drilled. Turning is also possible, as is wire machining (17).

One form of this process relies on a pulsed direct current, ie, full-wave rectified a-c, power supply that is locked in phase with a vibrating tool head. The oscillation of the tool (100 Hz) in phase with the pulsed d-c power supply, a slight phase difference is permissible, gives rise to a set of conditions whereby ECM occurs over each wave cycle. As the tool vibrates over one cycle, the interelectrode gap narrows. During the same period the current rises until, for conditions of comparatively smaller gap and higher current, sparking takes place by breakdown of the electrolyte and/or generation of electrolytic gas or steam bubbles in the gap, the production of which aids the discharge process (18).

For drilling, the discharge action occurs at the leading edge of the tool, whereas ECM takes place on the side walls between the tool and the workpiece. The combined spark erosion and ECM action yields fast rates of metal removal. Because ECM is still possible, any metallurgical damage to the components caused by the sparking action can be removed by a short period (eg, 15 s) of ECM after the main ECAM action. Currents of 250 A at 30 V are typically used in the process.

Economic Aspects

The industrial sectors utilizing ECM technology fall into five main categories: tool and die, automotive, aerospace, power generation, and oil and gas industries. Leading the world's principal machine tool manufacturing nations in production and export of tools in the 1980s were Japan followed by the former West Germany. The United States led in imports and consumption; consumption was high for both Japan and Germany, as well. Unconventional machine tools including ECM are generally considered to account for only ca 1% of total tool production. Electrodischarge machining (EDM) holds the largest share, possibly as much as

50%, and ECM about 15%, lagging behind laser processes which are ca 20% (see LASERS).

Manufacturing engineers wishing to use ECM processes in industry need to address the challenge of proper tool design. The cost of design can be as much as 20% of the cost of an electrochemical machine for complex components. Predictability of overcuts obtained for specific applications and the particular electrolytes to be used for the alloy metals that have to be machined must also be considered along with specific controls and limits on the ECM equipment needed.

Computer-controlled equipment and sensors (qv) are available for electrochemical machining systems. However, in the 1990s practical ECM systems are often favored because the amount of control and/or monitoring of the process is far less than that which was required in the 1960s and 1970s. Thus machines are used successfully in which electrical spark detection is eliminated and machining products control, eg, pH monitoring, is nonexistent. The measures in most industrial countries to protect the environment, however, is expected to lead to increased control of electrochemical machining products (normally called sludge), gas generation, and disposal of spent electrolyte solutions.

BIBLIOGRAPHY

"Electrolytic Machining" in *ECT* 2nd ed., Vol. 7, pp. 866–873, by J. Crawford, Ingersoll Milling Machine Co.; "Electrolytic Machining Methods" in *ECT* 3rd ed., Vol. 8, pp. 751–763, by J. P. Hoare and M. A. LaBoda, General Motors Research Laboratories.

1. G. Bellows, *Non-Traditional Machining Guide 26 Newcomers for Production*, Metcut Research Associates Inc., Cincinnati, Ohio, 1976, pp. 28–29.
2. G. Bellows and J. D. Kohls, *Am. Machin.*, 178–183 (1982).
3. J. Kaczmarek, *Principles of Machining by Cutting, Abrasion and Erosion*, Peter Peregrinus, Stevenage, U.K., 1976, pp. 487–513.
4. J. A. McGeough, *Principles of Electrochemical Machining*, Chapman and Hall, London, 1974.
5. J. A. McGeough, *Advanced Methods of Machining*, Chapman and Hall, London, 1988.
6. M. Datta, R. V. Shenoy, and L. T. Romankiw, *Am. Soc. Mech. Eng. PED.* **64**, 675–692 (1993).
7. K. P. Rajurkar, J. Kozak, and B. Wei, *Annals CRIP*, **42**(1), 231–234 (1993).
8. D. Graham, *Prod. Eng.* **61**(6), 27–30 (1982).
9. A. DeSilva and J. A. McGeough, *Proc. Inst. Mech. Eng.* **200**(B4), 237–246 (1986).
10. T. Drake and J. A. McGeough, *Proceedings of the Machine and Tool Design and Research Conference*, Macmillan, New York, 1981, pp. 362–369.
11. V. K. Jain and V. N. Nanda, *Prec. Eng.* **8**(1), 27–33 (1986).
12. D. Clifton, M. B. Barker, R. W. Gusthart, and J. A. McGeough, *Underwater Technol.* **17**(4), 9–17 (1991).
13. S. R. Ghabrail and C. F. Noble, *Proceedings of the 24th International Machine Tool Design and Research Conference*, 1982, pp. 323–328.
14. M. Kubota, Y. Tamura, J. Omori, and Y. Hirano, *J. Assoc. Electro-Mach.* **12**(23), 24–33 (1978).
15. M. Kubota, Y. Tamura, H. Takahahi, and T. Sugaya, *J. Assoc. Electro-Mach.* **13**(26), 42–57 (1980).

16. I. M. Crichton, J. A. McGeough, W. Munro, and C. White, *Prec. Eng.* **3**(3), 155–160 (1981).
17. A. B. M. Khayry and J. A. McGeough, *Proc. Roy. Soc. A* **412**, 403–429 (1987).
18. X. Ni, J. A. McGeough, and C. A. Greated, *J. Electrochem. Soc.* **140**(12), 3505–3512 (1993).

J. A. McGeough
X. K. Chen
University of Edinburgh

MACROLIDE ANTIBIOTICS. See ANTIBIOTICS, MACROLIDES.

MAGNESIUM AND MAGNESIUM ALLOYS

Magnesium [*7439-95-4*], atomic number 12, is in Group 2 (IIA) of the Periodic Table between beryllium and calcium. It has an electronic configuration of $1s^22s^22p^63s^2$ and a valence of two. The element occurs as three isotopes with mass numbers 24, 25, and 26 existing in the relative frequencies of 77, 11.5, and 11.1%, respectively.

Magnesium occurs widely in nature in the minerals dolomite [*17069-72-6*], magnesite [*13717-00-5*], olivine [*1317-71-1*], brucite [*1317-43-7*], and carnallite [*1318-27-0*], and in the form of magnesium chloride [*7786-30-3*] in seawater, underground natural brines, and salt deposits (see also CHEMICALS FROM BRINE; MAGNESIUM COMPOUNDS; OCEAN RAW MATERIALS). Metallic magnesium is produced by electrolysis of molten magnesium chloride or thermal reduction of magnesium oxide [*1309-48-4*] (1).

Elemental magnesium is silvery white. Having a specific gravity of 1.74, it is the lightest structural metal. For engineering applications, it is alloyed with one or more elements, ie, aluminum [*7429-90-5*], manganese [*7439-96-5*], rare-earth metals, lithium [*7439-93-2*], silver [*7440-22-4*], thorium [*7440-29-1*], zinc [*7440-66-6*], and zirconium [*7440-67-7*] to produce alloys having very high strength-to-weight ratios. The outstanding characteristics of magnesium and the advantages its use brings to both the fabricator and the user are described in Reference 2.

Magnesium alloys are available in a variety of metal forms, including cast ingots, slabs, and billets; sand, permanent-mold, die, and investment castings; forgings; extruded bars, rods, tubes, structural and special hollow and solid shapes; and rolled sheet and plate. Magnesium alloys are used widely in a great variety of applications.

In contrast to predictions of eventual exhaustion of high grade domestic ores of many common metals, seawater is a virtually unlimited source of magnesium. It has been estimated that 1.306×10^6 metric tons of magnesium are present in each cubic kilometer (2.6×10^{11} gal) of seawater and there is an estimated 1.3×10^9 km^3 (3.4×10^{20} gal) of seawater on earth (3).

In 1808, Sir Humphry Davy reported the production of magnesium in the form of an amalgam by electrolytic reduction of magnesium oxide using a mercury [*7439-97-6*] cathode. Some years later, the French scientist Bussy fused magnesium chloride with metallic potassium [*7440-09-7*] and produced free metallic magnesium for the first time, whereas Faraday in 1833 first produced magnesium by electrolytic reduction from the chloride (4). For many years, however, the metal remained a laboratory curiosity. In the late nineteenth century, magnesium was produced on a commercial scale in Germany by electrolysis of fused magnesium chloride. Although The Dow Chemical Company initiated the first domestic production in 1916, the magnesium industry did not develop in the United States until World War II. Until the early 1940s, the source of magnesium chloride was brine pumped from deep wells.

In 1941, The Dow Chemical Company began operation of a plant for the manufacture of magnesium by electrolysis using seawater as a source for magnesium ions. At the same time, Permanente Metals Corp. started production of magnesium using a carbothermic process. In addition, the U.S. Government constructed plants. By 1943, 13 other facilities were in operation under the management of 11 different companies. Of these plants, six were electrolytic (two used seawater, three dolomite, and one calcined magnesite as raw materials to form magnesium chloride), and the remainder used thermal reduction technology. The magnesite-based plant had a rated capacity of 45,390 metric tons and was the largest capacity magnesium facility constructed at that time.

Properties

Table 1 gives some of the physical properties of 99.9% pure magnesium. Magnesium is high in the electrochemical series, having a standard potential of −2.4 V. Like most metals, it is resistant to atmospheric and chemical attack because of a stable protective film, ie, oxide, carbonate, sulfate, fluoride, and others. The oxide film can be more protective if it is produced by an anodizing process. It is relatively easy to ignite fine powers or thin films of magnesium resulting in a strongly dazzling light. This property is exploited in the production of some fireworks and flares (see PYROTECHNICS). With pieces over 3-mm thick, combustion is difficult to sustain due to rapid heat transfer of the metal and the refractory nature of the magnesium oxide produced. Metal produced in casting processes must be protected from air at metal temperatures above 400°C, because the metal can sustain combustion above this temperature.

Metallic magnesium and water [*7732-18-5*] react. Under normal atmospheric conditions or in pure or chloride-free water of high pH, the reaction is suppressed by the formation of an insoluble magnesium hydroxide [*1309-42-8*] film.

$$Mg + 2\ H_2O \longrightarrow Mg(OH)_2 + H_2 \qquad (1)$$

Table 1. Properties of Magnesium

Properties	Value	Refs.
atomic weight	24.31	
melting point, °C	650	
boiling point, °C	1103	
crystal structure	close-packed hexagonal, no phase transformations	
lattice parameters at 20°C, nm		
a_0	0.3203	
c_0	0.5199	
axial ratio	1.624	5
density, g/cm^3		
at 20°C	1.738	6
400°C	1.682	
650°C (s)	1.61	
650°C (l)	1.58	
volume contraction, %		
at solidification	3.97–4.2	7,8
from 650 to 20°C as solid	1.8–2	7,8
electrical resistivity, $\Omega \cdot m \times 10^{18}$		
at 0°C	4.10	8
20°C	4.45	
50°C	4.92	
300°C	9.02	
600°C	14.12	
heat of fusion, 20°C, kJ/kg[a]	386	8
heat of sublimation, 20°C, kJ/kg[a]	6109	
heat of vaporization, 20°C, kJ/kg[a]	5272	
specific heat, J/(kg·K)[a]		
at 20°C	1025	
100°C	1034	
surface tension, 20°C, mN/m(=dyn/cm)	563	7
viscosity at melting, mPa·s(=cP)	1.25	8

[a]To convert J to cal, divide by 4.184.

If conditions are such that the film does not form, such as in the case of acids, then the reaction proceeds until all the metal is consumed. The reaction of magnesium with hydrofluoric acid [*7664-39-3*] is an exception to this rule, because a stable fluoride film forms.

The ability of magnesium metal to reduce oxides of other metals can be exploited to produce metals such as zirconium, titanium [*7440-32-6*], and uranium [*7440-61-1*] (see ZIRCONIUM AND ZIRCONIUM COMPOUNDS; TITANIUM AND TITANIUM ALLOYS; URANIUM AND URANIUM COMPOUNDS). These reactions are

$$2\ Mg + TiCl_4 \longrightarrow 2\ MgCl_2 + Ti \quad (2)$$

$$2\ Mg + UF_4 \longrightarrow U + 2\ MgF_2 \quad (3)$$

$$2\ Mg + ZrCl_4 \longrightarrow Zr + 2\ MgCl_2 \quad (4)$$

The most significant reactions of magnesium with air are

$$2\ Mg + O_2 \longrightarrow 2\ MgO \tag{5}$$

$$3\ Mg + N_2 \longrightarrow Mg_3N_2 \tag{6}$$

The magnesium nitride [*12057-71-5*] produced does not form a stable film. If sufficient nitrogen is present this reaction can be self-sustaining. The nitride produced can react with water to form ammonia [*7664-41-7*].

Manufacturing

Magnesium metal can be manufactured by electrolytic and metallothermic reduction. The method of choice depends on several variables including raw material availability, location, and integration into other chemical facilities. Producers and corresponding capacities are shown in Table 2 (see also ELECTROCHEMICAL PROCESSING, INORGANIC).

Electrolytic Reduction. The largest manufacturers of magnesium use processes based on the electrolytic reduction of magnesium chloride [*7786-30-3*] to form magnesium and chlorine [*7782-50-5*]. Several variations exist in the raw materials, the method of preparing magnesium chloride, the design of the cells used for the electrolysis, and the utilization of the by-product chlorine.

Sources for magnesium ions include seawater, natural brines, magnesium-rich brines from potash production, magnesite [*13717-00-5*], $MgCO_3$, carnallite,

Table 2. Magnesium Producers and Capacity[a]

Producer	Process	Nominal capacity, t $\times 10^2$
Dow Magnesium	electrolytic–seawater	108
Magcorp	electrolytic–brine	33
Northwest Alloys	Magnetherm	33
Chromasco	Pidgeon	12
Magcan[b]	electrolytic–magnesite	10
Norsk Hydro, Canada	electrolytic–magnesite	44
Total North America		*240*
Ube, Japan	Pidgeon	8
Japan Chem. and Met.[b]	Magnetherm	5
Total Japan		*13*
Norsk Hydro, Norway	electrolytic–seawater, brine, magnesite	55
Pechiney, France	Magnetherm	17
Siam, Italy[b]	resistance	12
Magnrohm, Yugoslavia[b]	Magnetherm	5
Total Europe		*89*
Brasmag, Brazil	resistance	9
Total South America		*9*
Total		*351*

[a]Ref. 14.
[b]Temporarily inactive in 1993.

$MgCl_2 \cdot KCl \cdot 6H_2O$, and dolomite [*17069-72-6*], $MgCO_3 \cdot CaCO_3$. Attempts have been made to recover magnesium ion from sulfate minerals (9), but none of this technology has been commercialized. Patents describe the recovery of magnesium ions from magnesium silicates (10,11). Magnesium chloride is formed from either an aqueous liquor that is then dehydrated, or by direct chlorination of magnesium oxide formed by calcination of magnesium hydroxide, $Mg(OH)_2$, or oxide-based ores. The dehydration of magnesium chloride hexahydrate [*7791-18-6*], $MgCl_2 \cdot 6H_2O$, proceeds stepwise, first to the tetrahydrate, and then the dihydrate (12,13). Further dehydration produces some hydrolysis with concurrent formation of hydrogen chloride [*7647-01-0*].

$$MgCl_2 \cdot 2H_2O \longrightarrow MgCl_2 \cdot H_2O + H_2O \quad (7)$$

$$MgCl_2 \cdot 2H_2O \longrightarrow MgOHCl + H_2O + HCl \quad (8)$$

$$MgCl_2 \cdot H_2O \longrightarrow MgOHCl + HCl \quad (9)$$

$$MgCl_2 \cdot H_2O \longrightarrow MgCl_2 + H_2O \quad (10)$$

$$MgOHCl \longrightarrow MgO + HCl \quad (11)$$

These reactions have been studied extensively, and many variations exist in the use of HCl atmospheres to maximize the desired reactions. The dehydration of carnallite occurs with minimum hydrolysis, but results in a mixed $MgCl_2$–KCl cell feed. The use of magnesium chloride hexammoniate [*24349-22-2*], $MgCl_2 \cdot 6NH_3$, has been proposed, but is not practiced commercially. This salt readily deammoniates to anhydrous $MgCl_2$.

Molten magnesium chloride can be formed by the direct carbochlorination of magnesium oxide obtained from the calcination of magnesium carbonate ores or magnesium hydroxide [*1309-42-8*].

$$MgO + C + Cl_2 \longrightarrow MgCl_2 + CO \quad (12)$$

This reaction, carried out at high (700–800°C) temperatures, also converts several impurities in the ores to volatile chlorides, thus purifying the $MgCl_2$. A patent describes the carbothermal chlorination of magnesite directly (15).

The magnesium chloride is then fed into electrolytic cells that use a molten chloride electrolyte. The standard reduction potential of magnesium ion (−2.4 V) precludes its being extracted by electrochemical processes (electrowon) from aqueous systems. Commercial cells operate at about 700°C. Chlorine [*7782-50-5*] is formed as a by-product. Manufacturers have developed their own cell designs optimized to their specific process. The magnesium product is recovered molten, and is forwarded to a casting plant where it is cast into ingots or alloyed and then cast. The by-product chlorine is either recycled back to the process, or forwarded to a chlorine user. The theoretical decomposition potential is 2.5 V, which requires a minimum electrical input of 5.5 kWh/kg metal. Actual decomposition potentials in the electrolytes used commercially are 2.6 to 2.8 V. Actual energy consumption is also considerably higher because of cell current efficiencies of 75–90% and the Joule heating of the electrolyte, electrodes, bus bar, and connectors. This energy must be added to the energy needed to produce the magnesium chloride cell feed to obtain the total energy needed for the process.

Dow Chemical Company Seawater Process. Seawater is the primary source for magnesium in The Dow Chemical Company process begun in 1941 in Freeport, Texas. A schematic is shown in Figure 1. Magnesium is present in seawater at a level of 1272 ppm. This is precipitated as magnesium hydroxide using either caustic soda or the calcined form of dolomite, dolime [*50933-69-2*], $MgO{\cdot}CaO$, as an alkaline source in large agitated flocculators. The magnesium hydroxide is then settled in Dorr thickeners, ie, tanks equipped with a slowly rotating rake at the bottom, which move settled solids to the center. The spent overflow enters the wastewater system where it is treated and returned to the ocean. The thickened underflow is pumped to rotary drum filters, where it is dewatered, washed, and reslurried with water. This slurry is then pumped to neutralizers, where it reacts with HCl to form a $MgCl_2$ liquor. Sulfuric acid

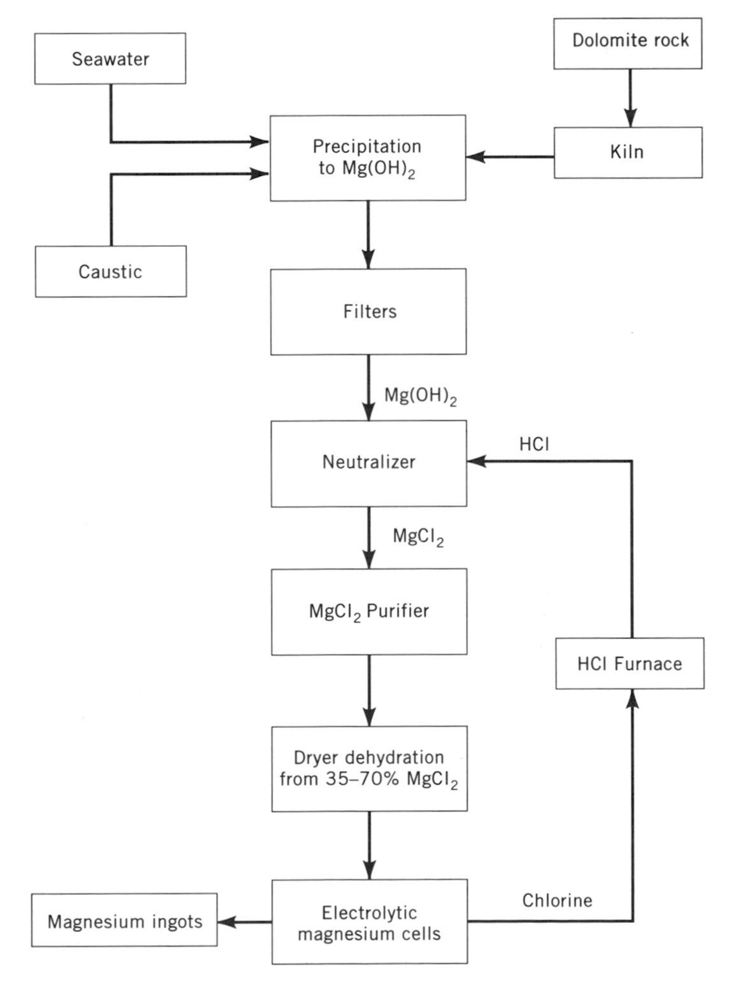

Fig. 1. The Dow seawater process.

[*7664-93-9*] is added to remove the excess calcium [*7440-70-2*], and the brine is filtered to remove calcium sulfate [*7778-18-9*] along with other solids such as clays and silica [*7631-86-9*]. This brine is purified to remove sulfate [*14808-79-8*] and boron [*7440-42-8*] and forwarded to the dryer. The purified brine is dried by direct contact with combustion gases in a fluid-bed dryer to produce granules of approximately 70% $MgCl_2$. Dow is the only magnesium manufacturer using water-based feed in electrolytic cells.

The Dow electrolytic cells (Fig. 2) are constructed of steel with cathodes welded to the tub-like container that holds the electrolyte. These are fitted with refractory covers through which cylindrical anodes pass. These anodes are independently suspended to allow them to be adjusted to maintain proper spacing and centering with respect to the cathode. Because the feed is aqueous, the graphite [*7782-42-5*] anodes and water or water decomposition products react to form carbon dioxide [*124-38-9*], and frequent adjustments of the anodes must be made while maintaining a tight seal on the cell system. The electrolyte is a mixture of molten chlorides at approximately 700°C. The bottom of the cell pot is surrounded by a gas-fired refractory chamber. This allows a great deal of flexibility in adjusting to various current loads, and the cells may even be restarted after a complete shutdown.

The cells are fed semicontinuously and produce both magnesium and chlorine (see ALKALI AND CHLORINE PRODUCTS). The magnesium collects in a chamber at the front of the cell, and is periodically pumped into a crucible car. The crucible is conveyed to the cast house, where the molten metal is transferred to

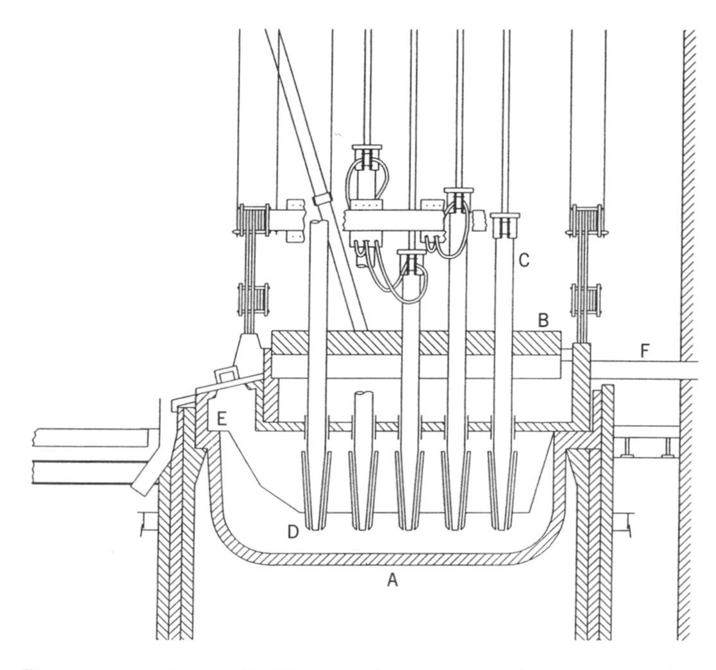

Fig. 2. The Dow magnesium cell. The steel container, A, is equipped with a ceramic cover, B, through which graphite anodes, C, pass. The magnesium is deposited on the cathode, D, and is diverted as it rises into the collection sump, E. The chlorine is withdrawn through a vent, F.

holding furnaces from which it is cast into ingots, or sent to alloying pots and then cast. The ingot molds are on continuous conveyors.

Norsk Hydro Process. Norsk Hydro operates two magnesium plants, one at Porsgrunn, Norway, and one at Becancour, Province of Quebec, Canada, with capacities of 55,000 and 45,000 t/yr, respectively. The plant at Porsgrunn went on line in 1951 using the I.G. Farben process (16,17). Later, Norsk Hydro developed its own process for the preparation of $MgCl_2$ cell feed and electrolysis (18–24). The basic steps of the two processes are shown in Figure 3.

In the chlorination process, caustic magnesia is extracted from dolomite and seawater, then mixed with carbon and aqueous $MgCl_2$ on a rotating disk to

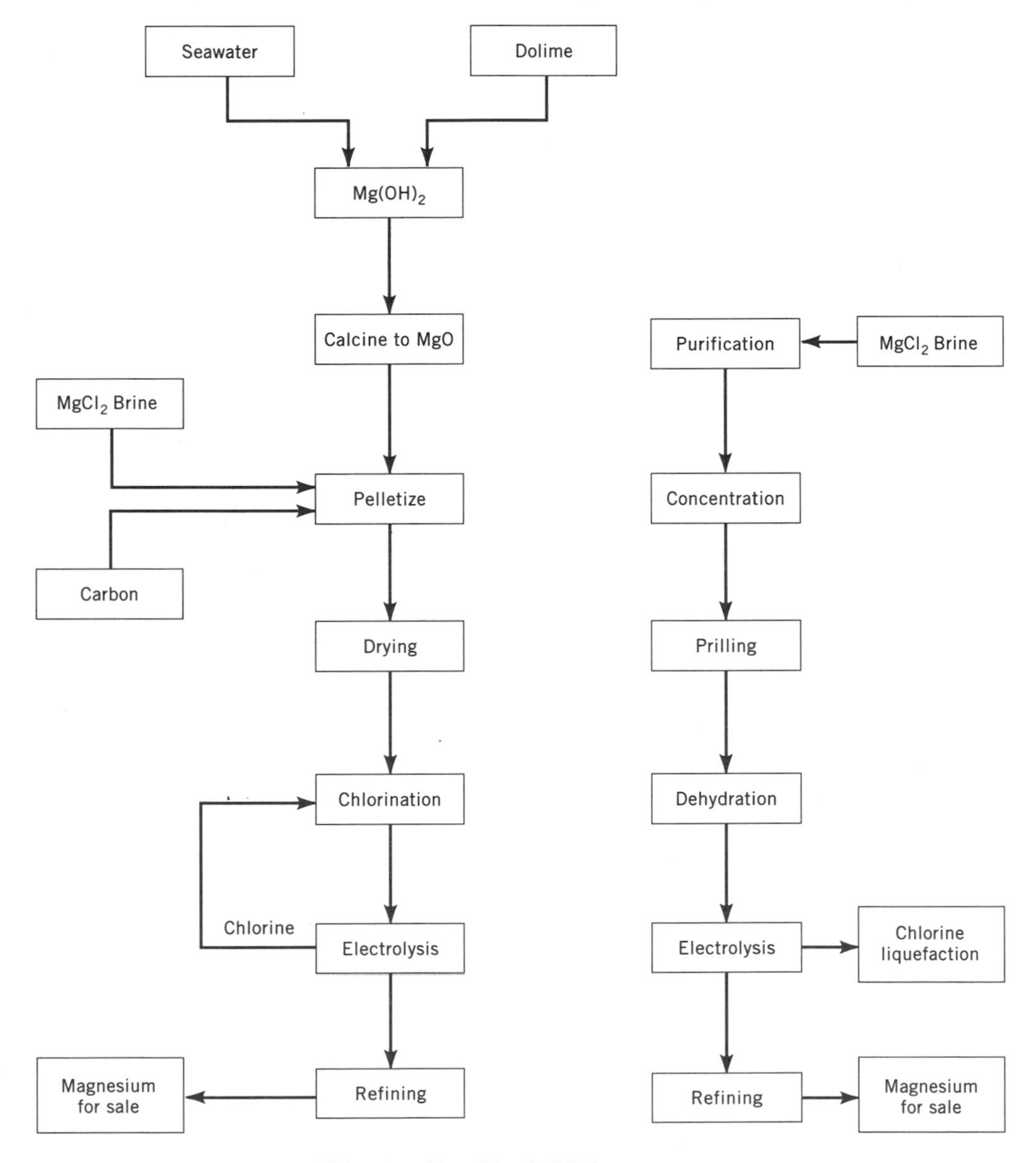

Fig. 3. The Norsk Hydro process.

form 5–10 mm diameter pellets. Hydrated oxide and oxychlorides act as binders. After a light drying, pellets containing approximately 50% MgO, 15–20% $MgCl_2$, 15–20% H_2O, 10% C, and a balance of alkali metal chlorides are conveyed to the chlorinators. The lower one-third of the brick-lined cylindrical shaft furnaces are filled with carbon blocks acting as resistors heated by carbon electrodes. The charge resting on the resistor bed reacts at 1000–1200°C with recycled chlorine gas introduced into the resistor-filled zone. Molten magnesium chloride collects in the bottom of the furnace. The molten material is tapped and transported to the electrolytic cells in closed containers. By-product gases at 100–200°C, containing air, carbon monoxide [*630-08-0*], carbon dioxide, traces of HCl, Cl_2, sulfur dioxide [*7446-09-5*], hydrogen sulfide [*7783-06-4*], and chlorinated hydrocarbons, are scrubbed in several stages before release to a stack. The washwater is filtered and chlorinated hydrocarbons removed before release. At intervals, magnesium silicate-rich slag is removed from the chlorinators. The magnesium yield is approximately 90% and the carbon consumption is 0.45 t/t magnesium produced. The addition of magnesium chloride solution to the reactants compensates for chlorine loss. The magnesium chloride produced contains typically less than 0.1% MgO, 0.1% SiO_2, and 20 ppm boron. Each chlorinator has an equivalent annual production capacity of the order of 2000 t Mg.

The dehydration process in Norway has as its raw material basis brine from the potash industry of the following average composition: 33% $MgCl_2$; 1–2% magnesium sulfate [*7487-88-9*], $MgSO_4$; 0.5% sodium chloride [*7647-14-5*], and 0.2% potassium chloride [*7447-40-7*].

Feedstock for the plant in Canada is obtained by dissolving magnesite in hydrochloric acid. The brine is chemically treated with sodium sulfide [*1313-82-2*], calcium chloride [*10043-52-4*], and barium chloride [*10361-37-2*] to remove heavy metals and sulfates by precipitation and filtration. Purified brine is preheated by waste heat from the process and concentrated to 45–50% $MgCl_2$ in steam heated evaporators, before prilling in a prilling tower. Prilling is a process by which pellet-sized crystals or agglomerates of material are formed by the action of upward blowing air on a falling hot solution. The size of the prills is kept within close tolerances and controlled physical shape to optimize the subsequent dehydration. The prills are converted to anhydrous $MgCl_2$ in a two-stage fluidized-bed dehydration process. In the first stage, the prills are dried to approximately magnesium chloride dihydrate [*19098-17-0*], $MgCl_2{\cdot}2H_2O$, with hot air. In the second stage, the dihydrate prills are contacted with a hot, anhydrous HCl gas stream to form anhydrous $MgCl_2$. Water and $MgCl_2$ dust in the off-gases from the dehydration are absorbed in a concentrated $MgCl_2$ solution. The HCl gas is provided by an extractive distillation process, dried, preheated, and returned to the HCl dehydration step. $MgCl_2$ prills containing less than 0.1% MgO are transported pneumatically to the electrolytic cells. Magnesium and chlorine recoveries in this continuous closed process are 97% or better.

Originally, Norsk Hydro operated I.G. Farben electrolytic cells with current loads of 32–62 kA. Since 1988, Norsk Hydro has been operating its own internally developed electrolytic cell. This cell, in full-scale operation since 1978, is a sealed brick-lined unit with two separate chambers, one for electrolysis and one for metal collection (Fig. 4). Densely packed and cooled graphite anode plates

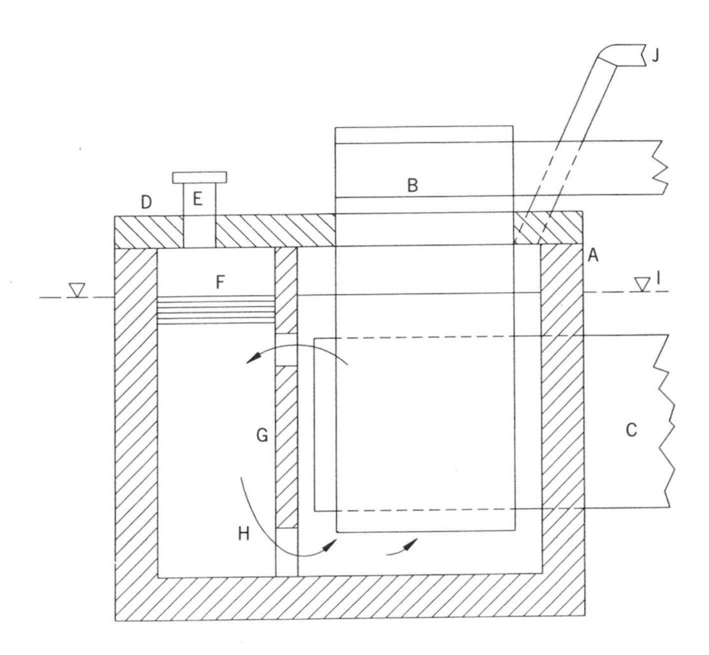

Fig. 4. The Norsk Hydro cell. Refractory material, A; graphic anode, B; steel cathode, C; refractory cover, D; metal outlet, E; metal, F; partition wall, G; electrolyte flow, H; electrolyte level, I; and chlorine outlet, J.

enter through the top, and double acting steel [*12597-69-2*] plate cathodes enter through the back wall of the cell. Chlorine of 98 wt % concentration is collected from one central pipe in the anode compartment. The circulation of the electrolyte is parallel to the electrodes bringing the metal to the collection chamber, from where it is extracted by vacuum and transported to the foundry. The cell is operated at 700–720°C with a current load of 350–400 kA. The energy consumption is 12–14 kWh/kg Mg. The cell life is on the order of five years.

Most of the magnesium is cast into ingots or billets. The refining of the molten metal extracted from the electrolysis is performed continuously in large, stationary brick-lined furnaces of proprietary design (25). Such installations have a metal yield better than 99.5% and negligible flux consumption.

Magnesium Corporation of America (Magcorp) Process. The magnesium facility in Rowley, Utah, originally built by NL Industries in 1972, was purchased by AMAX, Inc., in 1980, sold in 1989 to a private investment group, and was named Magcorp.

This magnesium plant (Fig. 5) utilizes brine from the Great Salt Lake as feed for solar evaporation ponds. The Great Salt Lake magnesium concentration is about 0.4%, about four times that of the world's oceans. The evaporated brine contains 7.5% magnesium, 4% sulfate, 0.5% sodium [*7440-23-5*], 0.7% potassium [*7440-09-7*], 0.1% lithium [*7439-93-2*], and 20% chloride [*16887-00-6*] (26). The brine-holding ponds contain enough brine to supply two years of raw material ready for processing. This brine is further concentrated and treated with $CaCl_2$ to remove sulfate. Solids such as calcium sulfate, potassium chloride, and sodium chloride are removed in thickeners and settling ponds. Boron is removed by

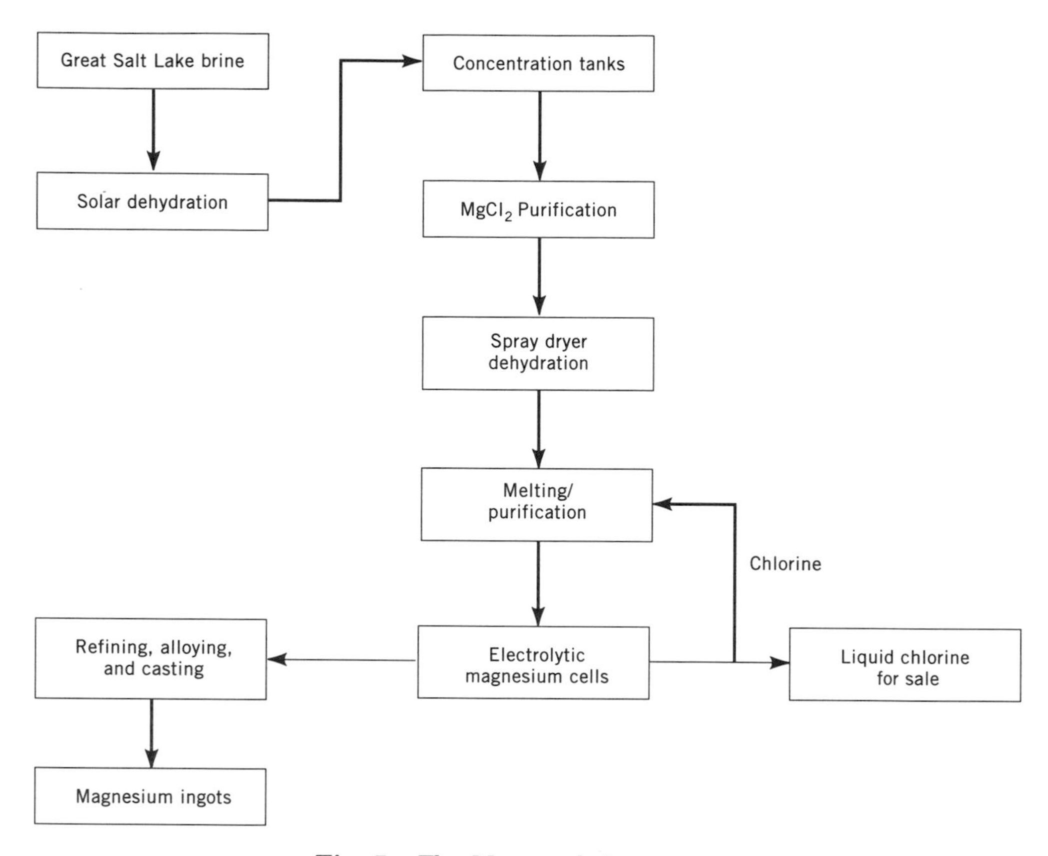

Fig. 5. The Magcorp brine process.

solvent extraction. Cogeneration is used to provide heat for further concentration of the brines and for drying the brine into powder. Spray dryers flash dry the solution into powder. The dry magnesium chloride powder contains mainly $MgCl_2$ along with about 4% MgO, 4% H_2O, and other salts which comprise the cell electrolyte. The dryers are heated with exhaust gases from gas-fired turbines that generate some of the power used to operate the electrolytic cells.

The spray dried $MgCl_2$ powder is melted in large reactors and further purified with chlorine and other reactants to remove magnesium oxide, water, bromine [*7726-95-6*], residual sulfate, and heavy metals (27,28). The molten $MgCl_2$ is then fed to the electrolytic cells which are essentially a modification of the I.G. Farben cell. Only a part of the chlorine produced is required for chlorination, leaving up to 1 kg of chlorine per kg of magnesium produced. This by-product chlorine is available for sale.

Liquid magnesium is removed from the electrolytic cells under vacuum and transferred to the cast house where it is refined, purified, and cast into a wide variety of shapes, sizes, and alloys.

Russian Process Technology. Magnesium production in the former Soviet Union is apparently done via molten chloride electrolysis (29,30). The basic process uses carnallite [*1318-27-0*], $MgCl_2{\cdot}KCl{\cdot}6H_2O$, either from natural deposits or as a by-product of processing natural salt deposits, as its raw material. Recrystallized carnallite has an average value of 32% $MgCl_2$. This material is dried in a two-step process (31–33). The first stage consists of a fluid-bed/moving-bed furnace with three distinct temperature zones. The temperature increases in each zone, going from 130 to 200°C, which results in a product containing 1–2% MgO and 3–6 H_2O. This product is then sent to a chlorinator, which operates at 700–750°C. The chlorinator is designed to remove most of the remaining impurities and settle out any solids. Considerable research effort is being done to improve the efficiency of the chlorination step. The molten $MgCl_2$/KCl is tapped and either sent directly to the cells while still molten, or cooled to a solid and then fed to the cells.

Up until the late 1960s, the cells used in Russia were basically I.G. Farben diaphragm cells. In the period of 1972–1977, production was converted to diaphragmless cells (34). These cells operate in the temperature range of 680–720°C and at currents of 150–200 kA. The electrolyte used depends on the feed. Using carnallite feed the electrolyte is 5–15% $MgCl_2$, 60–80% KCl, 8–20% NaCl, and <1% CaCl; using anhydrous $MgCl_2$ feed from titanium production, the electrolyte contains 8–18% $MgCl_2$, 30–55% KCl, 30–60% NaCl, and 0–10% $CaCl_2$. A small amount of calcium fluoride [*7789-75-5*], CaF_2, is maintained in the electrolyte in either case. Cell temperatures are somewhat higher and current efficiencies are somewhat lower in carnallite fed cells.

Russian production may be going to a flow line cell concept (35). In this process, dehydrated carnallite is fed to a chamber where it is mixed with spent electrolyte coming from the electrolytic cells. The spent electrolyte first enters a metal collection chamber, where the molten magnesium is separated. The electrolyte is then enriched with carnallite and any insoluble impurities are allowed to settle. The enriched electrolyte is then returned to the electrolytic cells. The result is that most of the remaining impurities are removed in the first electrolytic cell.

Titanium–Magnesium Chloride Recycle Processes. Titanium [*7440-32-6*] is manufactured by the Kroll process, where magnesium is used as a reducing agent for titanium tetrachloride [*7550-45-0*] (see TITANIUM AND TITANIUM ALLOYS) (see eq. 2). This reaction produces a pure anhydrous magnesium chloride that can then be used in an electrolytic cell to convert it back to magnesium metal. Electrolytic cells have been developed to take advantage of this type of feed. Alcan has developed both monopolar and multipolar cells for the production of magnesium (36,37). Alcan monopolar cells are used by Oremet Titanium (Albany, Oregon) and by Timet (Henderson, Nevada). The first Alcan multipolar cell was perfected at Osaka Titanium Co. (now Sumitomo Sitex) in Osaka, Japan. This cell operates between 80 and 140 kA and has an energy consumption of 9.5–10 kWh/kg of magnesium. The cell consists of a chamber for electrolysis and a chamber for metal collection. The electrolytic chamber has a central anode and a terminal cathode with several intermediate bipolar electrode assemblies. These bipolar electrodes are disposed in series in the path of the electrolytic current between the anode and the cathode. A bipolar electrode is an electrically conductive plate inserted between the anode and the cathode of a conventional electrolytic cell. The side of the plate facing the anode becomes cathodic and the opposite side becomes anodic. The bipoles are tapered to almost surround the anode, including the edges and bottom. A curtain wall partially submerged in the electrolyte separates the two collection chambers. The metal produced is swept by the anodically evolved chlorine to metal collection ducts that carry it to the metal collection chamber (38). All the Alcan cells show excellent operation characteristics with $MgCl_2$ from titanium manufacture, but none is in use with brine or ore-based feed processes.

The Ishizuka cell (39–41), another multipolar cell that has been in use by Showa Titanium (Toyama, Japan), is a cylindrical cell divided in half by a refractory wall. Each half is further divided into an electrolysis chamber and a metal collection chamber. The electrolysis chamber contains terminal and center cathodes, with an anode placed between each cathode pair. Several bipolar electrodes are placed between each anode–cathode pair. The cell operates at 670°C and a current of 50 kA, which is equivalent to a 300 kA monopolar cell.

Magcan Process. The Magcan process is based on a technology developed by MPLC of Great Britain and Alberta Natural Gas for the preparation of anhydrous $MgCl_2$ directly from magnesite ore (13). The Magcan plant (Aldersyde, Alberta, Canada) underwent start-up in 1990 but was shut down in 1991. In this process, magnesite ore is crushed, screened, and placed in the top of a vertical reactor. Carbon monoxide [*630-08-0*] and chlorine gas are injected into the bottom of the chlorinator. The magnesite passes down the reactor where it is chlorinated by the following reaction:

$$MgCO_3 + Cl_2 + CO \longrightarrow MgCl_2 + 2\ CO_2 \tag{13}$$

Molten anhydrous magnesium chloride is tapped from the bottom of the reactor. Iron, aluminum, and silicon-based impurities are also converted to their chlorides, which volatilize out of the reactor. Carbon monoxide is generated from coke, carbon dioxide, and oxygen. The magnesium chloride is sent to electrolytic cells. Russian diaphragmless cells purchased from the defunct American Magnesium Co. are used.

Dead Sea Works Process. The Dead Sea Works, a subsidiary of Israel Chemicals Ltd., announced plans in 1992 to construct a 25,000 t/yr magnesium plant at Beer-Sheva, Israel. The plant, to be based on Russian carnallite technology, is designed to use an existing potash plant as the source of carnallite. The chlorine by-product can be either liquefied and sold, or used in an existing bromine plant. Waste streams from the carnallite process, as well as spent electrolyte from the electrolytic cells, can be returned to the potash plant.

Queensland Metals Process. The Queensland Metals Corp. is developing a large magnesite deposit in Kunwarara, Queensland, Australia (42,43). The first phase of development of this ore is a plant to produce magnesia, MgO. Feasibility studies relating to producing magnesium metal from this deposit have also begun. Research efforts aimed at producing a high quality magnesium chloride are underway. An option to license the Alcan electrolytic cell has also been taken.

Noranda Process. Noranda, Inc. (Toronto, Canada) has developed technology to recover magnesium chloride from asbestos [*1332-21-4*] tailings (44,45). Several patents have been issued covering this process technology, which involves using HCl to leach magnesium from silicate materials at elevated temperatures followed by purification steps. This $MgCl_2$ solution can then be fed to a dryer. The chlorine from the electrolytic cell is used to make HCl and recycled to the leaching step.

Thermal Reduction. Magnesium metal can also be formed by the thermal reduction of magnesium oxide with a reactive metal, such as silicon [*7440-21-3*], which forms a stable oxide.

$$2\ MgO + Si \longrightarrow 2\ Mg + SiO_2 \tag{14}$$

Silicon is normally used in the form of a high grade ferrosilicon alloy, and the MgO is normally supplied in the form of dolime. Ferrosilicon [*8049-17-0*] is prepared by the carbothermal reduction of silica in the presence of iron:

$$SiO_2 + 2\ C + Fe \longrightarrow Si(Fe) + 2\ CO \tag{15}$$

Additional components such as alumina [*1344-28-1*] are also added to obtain more favorable thermodynamics, and to obtain a slag having favorable properties. Many different feed and slag compositions exist, as do alternative reductants for ferrosilicon. It is also theoretically possible to manufacture magnesium metal by the reduction of MgO with carbon.

$$MgO + C \longrightarrow Mg + CO \tag{16}$$

Three basic processes exist for the thermal reduction of magnesium oxide: the Pidgeon process, the Magnetherm process, and the Bolzano process.

Pidgeon Process. The Pidgeon (46–49) process (Fig. 6) was the first commercial thermal reduction process using silicon, and was developed in the 1940s. This process is used by Timminco (Haley, Ontario, Canada) and Ube Industries (Japan). The overall reaction for this process is

$$2\ MgO + 2\ CaO + Si \longrightarrow 2\ Mg + Ca_2SiO_4 \tag{17}$$

The raw materials for the process are dolomitic limestone that has been calcined to form the oxide, and ferrosilicon (65–85% Si). The dolime and ferrosilicon are finely ground, mixed, and compacted into briquettes. These briquettes are inserted into seamless stainless steel tubes 3 m long having a bore diameter of 28 cm and a capacity of 350 kg. The charge is distributed through the length of the reaction section of the horizontal retort and placed into a gas-fired or electrically heated furnace. The operating cycle consists of three phases. Initially, the retort is heated to drive off any CO_2 and water remaining in the charge. The retort is then sealed and operated at low vacuum to complete this burnoff. The retorts are then evacuated to 13 Pa (0.1 mm Hg) and brought to a temperature of 1200°C. The magnesium vapors condense to form a solid metal crown at the water-cooled collection end of the tube. The size of the retort tubes is limited by the heat transfer to the reactants from the furnace.

The retorts must be opened, the reaction products removed, and the retorts filled with raw materials and resealed. The typical cycle is 8–10 hours. Capacity is controlled by the number of retorts used and the number of furnaces available. The metal crowns are removed, remelted, and cast into ingots, or alloyed and then cast.

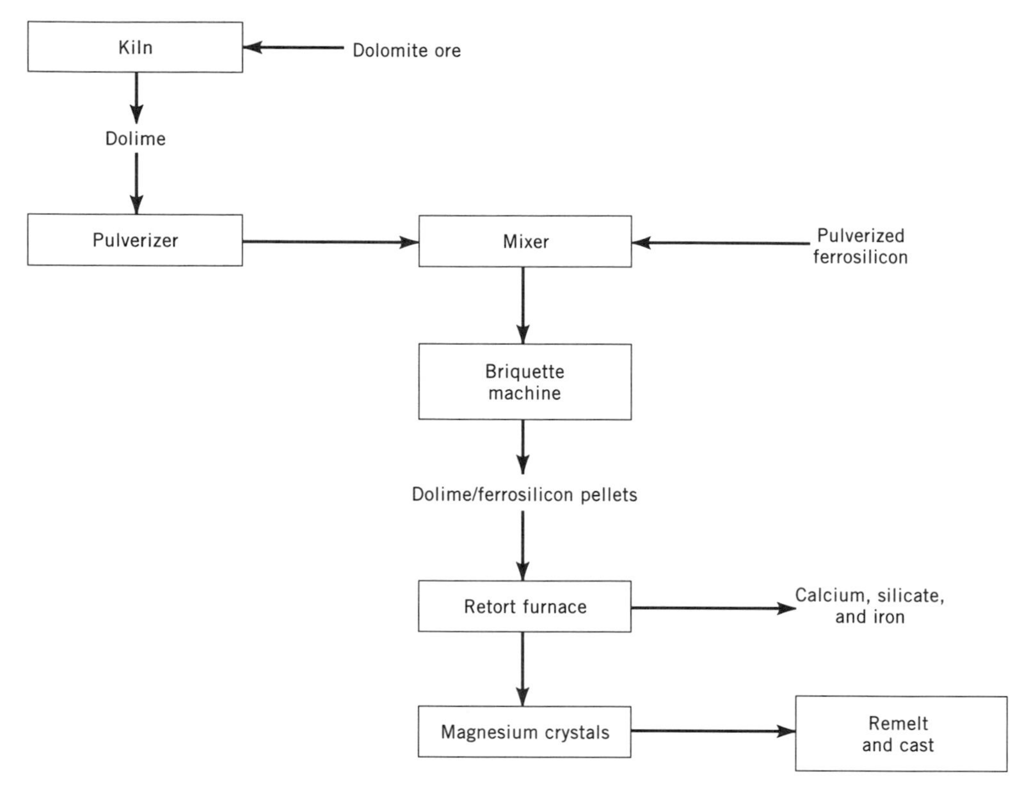

Fig. 6. The Pidgeon ferrosilicon magnesium process.

Magnetherm Process. The Magnetherm process (Fig. 7) was developed over a period from 1950 to 1963 by Group Pechiney Ugine Kuhlmann of France (17,50–53). This technology is practiced by Societe Francaise d'Electrometallurgie (SOFREM) (Marignac, France), Northwest Alloys (Addy, Washington), Magnrohm Co. (Bela Sterna, Yugoslavia), and Japan Metals (Takaoka, Japan). In this process, alumina is added to the reactor to maintain a liquid slag, allowing the reactor to be heated by the electrical resistance of the slag.

$$2\ CaO{\cdot}MgO + Al_2O_3 + (Fe)Si \longrightarrow Ca_2SiO_4{\cdot}Al_2O_3 + Fe + 2\ Mg \tag{18}$$

The magnesium ion source is normally a calcined dolomite or a mixture of calcined dolomite and calcined magnesite. Alumina, usually in the form of bauxite [*1318-16-7*], is added to control the properties of the slag. The reducing agent is mainly ferrosilicon, but some aluminum is also used, particularly by Northwest Alloys. The resulting calcium aluminosilicate [*1327-39-5*] slag remains liquid and the magnesium vapor is condensed.

The Magnetherm reactor used by Northwest Alloys is shown in Figure 8. The reactants are fed into a carbon-lined vessel, where the reaction is carried out in a range of 1300–1700°C and at a pressure of at least 2.7 kPa (20 mm Hg). Under these conditions, the metallic reducing agent reacts with the magnesium

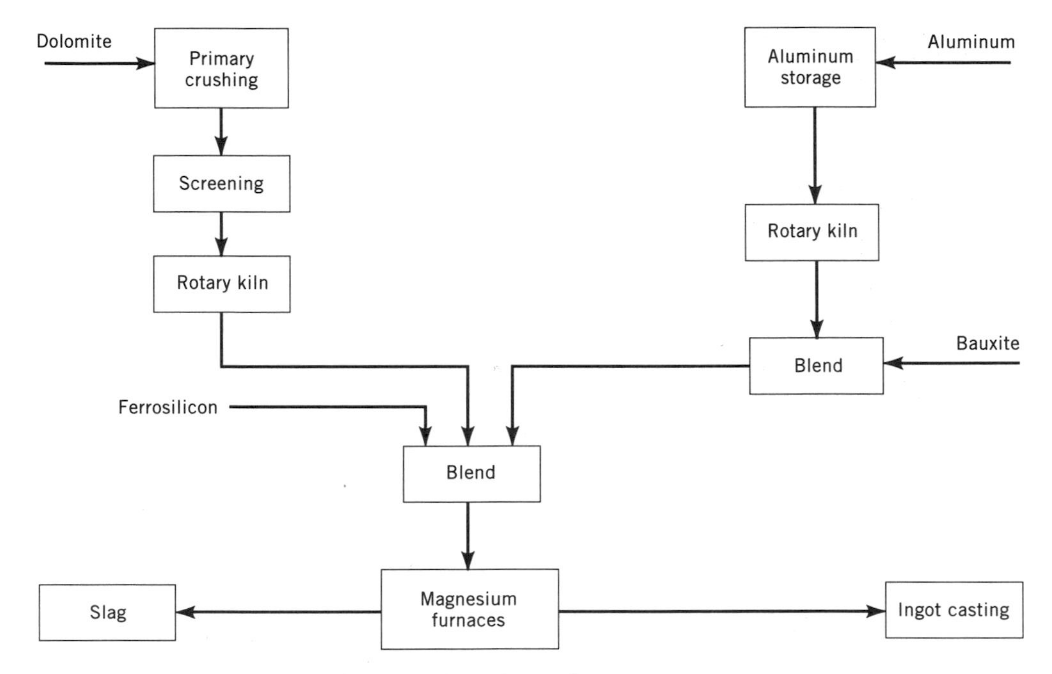

Fig. 7. The Magnetherm process.

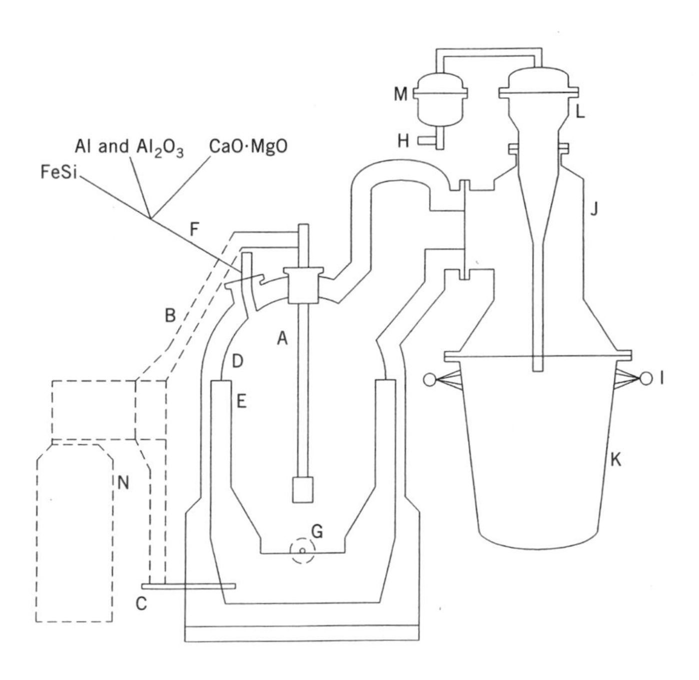

Fig. 8. Magnetherm reactor: central electrode, A; secondary circuit, B; grounding electrode, C; refractory lining, D; carbon lining, E; primary material feed, F; slag taphole to FeSi recovery, G; vacuum line, H; water spray ring, I; condenser, J; crucible, K; trap, L; filter, M; and transformer, N.

oxide in the calcium–silicon–aluminum–magnesium oxide slag to produce magnesium vapor. The composition of the slag is controlled to maintain a minimum calcium oxide [*1305-78-8*], CaO, to SiO_2 weight ratio of 1.68 and minimum Al_2O_3 to SiO_2 ratio of 0.44. The vapor is condensed as either a liquid or a solid.

The driving force for moving the magnesium vapor to the condenser is the volume change of the magnesium vapor going from a vapor to a liquid or solid state. A small amount of argon [*7440-37-1*] is purged to the reaction vessel through the feed system to minimize the condensation of magnesium metal on the colder parts of the feed system. The process operates on a 16–24 hour cycle, with the cycle split in two halves. The end of the first half is used to tap slag and refill the feed system with dolime. The end of the second half is used to replenish all feed materials, tap slag, and remove the filled magnesium crucible. Thus in every cycle, the condenser assembly shown in Figure 8 is removed and a clean, empty unit is attached to the reduction vessel for continued operation on the next cycle. The slag is removed from the furnace using an oxygen [*7782-44-7*] lance to penetrate a clay plug or a carbonaceous plug. The calcium aluminosilicate slag is a cementaceous product and can be used as a cement or liming agent.

The quantity of feed materials required are 1–1.05 kg of metallic reductant, 5.4 kg of dolime, and 0.35 kg of calcined bauxite or alumina to produce 1 kg of magnesium. The quantity of slag produced depends on the feed material composition and may vary from 5.2 to 5.9 kg/kg of magnesium.

Bolzano Process. The Bolzano process was developed by Societa Italiana per Magnesio e Lehge do Magnesio of Bolzano, Italy (54,55). In this process, the reactants are calcined dolomite and ferrosilicon that are compressed into solid blocks and stacked with electrical heating conductors between the blocks. The block assembly is placed in a refractory-lined furnace so that heat applied to the system goes only to the compressed reactants and not to the chamber. The furnace is the lower portion of the assembly, which also has an upper bell-shaped section joined to the lower portion by a flange. The top portion is cooled and acts as a condenser for the magnesium vapors formed in the furnace.

The process operates at 1200°C and <400 Pa (3 torr) and has a cycle time of 20–24 hours. The reactor is opened at the flange and the metal removed. Energy usage in the furnace is 7–7.3 kWh/kg magnesium. A similar process is used by Brasmag (Minas Givras, Brazil) (56).

Carbothermal Reduction. At temperatures above 1850°C, carbon monoxide is more stable than magnesium oxide. This allows the formation of magnesium vapor from the reaction of magnesium oxide and carbon [*7440-44-0*] at these temperatures. However, the reaction is reversible, and the product gas must be rapidly quenched to prevent the reforming of MgO. Shock cooling has been used, which produces a finely divided magnesium dust that is pyrophoric. A plant was operated by Kaiser (Permanente, California), during World War II, but it had difficulty separating this dust from the quenching medium and was shut down (57).

Refining and Casting of Magnesium. Most magnesium extraction processing is followed by a refining operation to remove impurities and to manufacture clean alloy compositions. The metal is then converted to ingots, slabs, and billets which are short, thick, cylindrical bars. Magnesium is also converted directly into granules for subsequent use in steelmaking. Some magnesium process slags are beneficiated to recover magnesium granules without a remelting operation. Typical impurities are nonmetallic inclusions, metallic impurities, and hydrogen.

Nonmetallic Inclusions. Inclusions in the form of magnesium oxide, magnesium nitride, and magnesium chloride impair the appearance and corrosion resistance of magnesium alloys. Impurities can have a detrimental effect on the performance of aluminum alloys that are made with magnesium. Alkali metal chlorides wet the surface of magnesium and the inclusion. Because of the higher density of the alkali metal chlorides relative to magnesium, the chlorides sink to the bottom of melts as a slag. Intermetallic compounds formed in the production of magnesium alloys as a rule have higher densities than the melt and also report to a slag layer. Fluxes added to magnesium melts have densities similar to the melt and are used to coat the surface of the melt to prevent oxidation (58).

Metallic Impurities. Iron [*7439-89-6*], nickel [*7440-02-0*], and copper [*7440-50-8*] reduce the corrosion resistance of magnesium and magnesium alloys significantly. These metals form very small galvanic islands which act as cathodes. These impurities are controlled by the choice of raw materials used to make up the feed to the electrolytic or metallothermic processes. Other common impurities found in magnesium are aluminum, manganese, zinc, silicon, and sodium. Aluminum is sometimes introduced by the reduction of alumina brick or tile used in the construction of magnesium cells and furnaces as well as alumina in the raw materials (59). Iron is soluble in molten magnesium in the range of 250–400 ppm at normal processing temperatures. Most electrolytically derived magnesium contains iron in this range. Intermetallic compounds of these impurities have also been found.

The iron concentration can be kept low by operating at low temperatures. Iron can be removed by addition of manganese [*7439-96-5*] or manganese chloride [*7773-01-5*] because it forms an intermetallic complex which is denser than the metal and becomes part of the slag (60). Addition of titanium tetrachloride gas to the metal reduces iron levels as well, but has a disadvantage of the possible discharge of titanium tetrachloride to the atmosphere above the furnace and subsequent conversion to titania and HCl (58). An alternative method is the addition of zirconium [*7440-67-7*] metal, which forms an intermetallic compound and settles. Excess zirconium is removed by additions of silicon (61). Sodium [*7440-23-5*], calcium, and strontium [*7440-24-6*] can be removed by the addition of magnesium chloride which is reduced producing the respective metal chloride.

Metallic impurities are also detrimental in applications where magnesium is used as a reductant such as in the Kroll process. The produced metal can be contaminated with boron rendering it useless in some nuclear applications.

Hydrogen. Hydrogen [*1333-74-0*] is formed when magnesium reacts with moist air (62–64). The normal level of hydrogen in magnesium is 1–7 ppm, which is acceptable for most purposes. Higher levels (~24 ppm) cause macroporosity in metal castings and there is some evidence that the intermediate levels can produce some microporosity in castings. Hydrogen can be completely removed by forcing an inert gas such as argon or chlorine into the melt. Nitrogen [*7727-37-9*] has also been used in this application although the formation of magnesium nitride may be of concern.

Equipment. The standard equipment in magnesium foundries consists of large stationary brick-lined reverbatory furnaces which can hold up to 10–15 t of molten magnesium. Reverbatory furnaces are furnaces in which heat is supplied by burning a fuel in a space between the reactants and the low roof. The stationary furnaces can be heated electrically or with gas or oil burners. Some furnaces heat the melt by submerged electrical heating systems. Magnesium is usually alloyed in steel pots which are used to formulate an alloy composition, remove hydrogen, and remove iron. The magnesium can then be transferred to a reverbatory furnace for settling of intermetallics (59).

Metallothermic magnesium is recovered in the solid state. Magnesium produced in this manner is then remelted and refined for subsequent casting.

Casting. The casting of magnesium can be highly automated. It normally involves the transfer of molten magnesium, introduction into a mold, protection of the cast surface, and handling of the solidified product. Molten metal transfer is accomplished by mechanical or electromagnetic pumping or by static displacement. Magnesium contracts about 4.2% in the liquid/solid-phase transformation and can lose up to 5% of its volume in cooling to room temperature (7). The surface can reveal shrink holes and surface cracks. These can be the result of change of volume in the process and the size and shape of the ingot mold used. Turbulent casting of the ingot can produce a sponge-like dross that detracts from the surface appearance of the ingot. This dross may be removed by skimming before solidification is complete. It is necessary to prevent oxidation of the metal during casting because magnesium spontaneously ignites in air in the molten state. To suppress oxidation, a number of techniques are employed. Casting of magnesium under a sulfur dioxide [*7446-09-5*] air mixture is one method used to suppress oxidation of the metal (65). Sulfur [*7704-34-9*] powder can also be used to provide the sulfur dioxide through application to the metal surface (66). A newer method is to use a mixture of sulfur hexafluoride [*2551-62-4*], carbon dioxide [*124-38-9*], and air to suppress oxidation (67–70). This method can be used to replace molten fluxes in casting pots. The necessity of a protective atmosphere increases the cost of magnesium casting practice over that of aluminum.

Billets, which are cylindrical bars, are cast in multistrand continuous or semicontinuous direct-chill casting (71) machines. These machines use water to cool the bar as it is produced. The water is then removed from the metal with rubber wipers. This process has the advantage of water removal from the surface without the use of a large pit to contain the water. Large pits containing water can be dangerous in continuous casting since metal run-outs from partially solidified metal bars can cause explosions due to the reaction of the molten magnesium or aluminum with water. Sulfur hexafluoride gas protection systems are used in this application since they eliminate the need for fluxes to prevent oxidation. The size of continuous casters for magnesium is limited compared to aluminum because of the low heat content of magnesium and its tendency to oxidize. Very large machines with 50 or more strands are used for aluminum casting. To avoid discoloration and surface corrosion during transport and handling, the metal is protected by plastic or treated paper wrappings.

Recycling. Substantial quantities of magnesium are recycled annually in the United States, as well as in many other countries. The largest single recycling (qv) effort in magnesium is in the area of aluminum beverage cans. Because these cans contain around 2% magnesium by weight, this represents approximately 25,000 metric tons of magnesium per year (72,73). Most of the remaining recycled magnesium comes from die castings and from scrap generated in the die casting process. This is estimated to be 9,000–11,000 metric tons annually worldwide and the quantity is expected to grow as the volume of die castings expands. Metal coming out of the recycling industry ends up mainly in magnesium alloys or metal going to steel desulfurization.

The energy required to melt and cast recycled magnesium is approximately 6.7 MJ/kg (2865 Btu/lb) (74) compared to the 267 MJ/kg (115,000 Btu/lb) to produce primary magnesium by the most efficient technology available.

Economic Aspects

There are five magnesium producers in the United States (see Table 2). Timet and the Oregon Metallurgical Corp. convert magnesium chloride from their titanium manufacturing process into metallic magnesium by electrolysis and return it to their titanium production operation. Northwest Alloys, a subsidiary of Alcoa, operates a Magnetherm reduction facility, mainly for their captive magnesium needs. The Dow Chemical Company and Magcorp produce magnesium via electrolytic processes for sale to magnesium customers for a wide variety of uses.

The largest growth segment of the magnesium market is in automotive die castings where the lighter weight of magnesium offers significant weight and fuel savings advantages. The ability to cast intricate shapes in magnesium also allows the metal to serve as replacement for some steel fabricated parts on a cost competitive basis.

In 1915 the price of magnesium was $11.00/kg of the year. On a constant dollar basis, the price of magnesium decreased steadily until the 1950s and has fluctuated mostly in the $3.00 to $4.00/kg range since that time. The price history of magnesium on a constant dollar basis from 1930 to 1990 is shown in Figure 9. The 1994 price for magnesium was $3.37/kg.

The magnesium industry serves a wide variety of structural and nonstructural uses. Consumption of the 257,300 t producer shipments in 1992 was 52% for aluminum alloying; 14% for steel desulfurization; 13% for die casting; 5% for manufacture of nodular iron; 4% for electrochemical processing; 3% each for metallic reductions, chemical usage, and wrought products; 1% for gravity casting; and 2% for other uses (72).

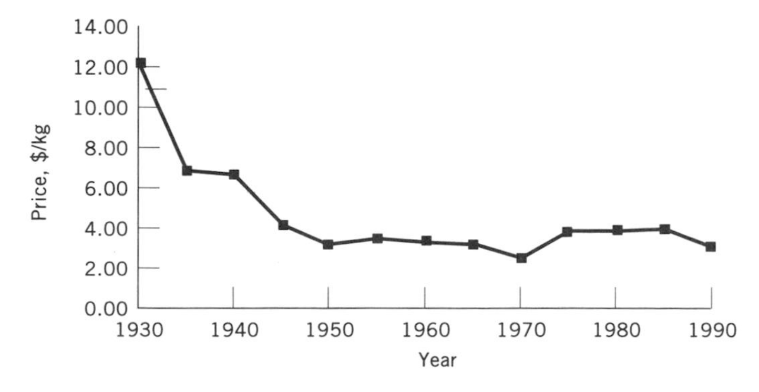

Fig. 9. Magnesium price in constant dollars.

Specifications

Commercial primary magnesium has a typical purity of 99.8%, which is sufficient for most chemical and metallurgical uses. A typical analysis might be expected to show about 0.003% each of aluminum and copper, 0.04% iron, 0.08% manganese, 0.001% nickel, and 0.005% silicon. Primary magnesium is available in five grades (Table 3). Considerably higher purity can be obtained by distillation.

Table 3. Standard Specifications for Magnesium Ingot[a]

	Grade				
Impurity, %[b]	9980A	9980B	9990A	9995A	9998A
aluminum			0.003	0.01	0.004
copper	0.02	0.02			0.0005
iron			0.04	0.003	0.002
lead	0.01	0.01			0.001
manganese	0.10	0.10	0.004	0.004	0.002
nickel	0.001	0.005	0.001	0.001	0.0005
silicon			0.005	0.005	0.003
sodium	0.006				
tin	0.01	0.01			
titanium				0.01	0.001
other impurities[c,d]	0.05	0.05	0.01[e]	0.005[e]	0.005[e]
magnesium[f]	99.80	99.80	99.90	99.95	99.98

[a]Ref. 75.
[b]Maximum value unless otherwise indicated.
[c]For specific applications, other minor impurities may be required to be controlled to limiting maxima by agreement between the purchaser and the seller.
[d]Includes elements for which no specific limit is shown.
[e]For nuclear applications, the cadmium and boron (high capture cross-section elements) shall be defined as cadmium, max % 0.0001 or 0.00005; boron, max % 0.00007 or 0.00003.
[f]Minimum value determined by difference.

Analytical Methods

Because the production of magnesium is a large-scale industrial process, fast and reliable methods for magnesium analysis have been developed for the quick turnaround times necessary in a production foundry (76,77). Referee methods which are more time consuming but have larger ranges and greater accuracy compared to the production methods have also been developed (78).

Production Methods. Analyses of magnesium or magnesium alloy batches are typically performed by atomic absorption (AA) spectroscopy, by emission spectroscopy employing an inductively coupled plasma (ICP) source, or direct spark emission spectroscopy. The atomic absorption method requires the dissolution of the sample in a standard solution followed by introduction into the flame of the AA unit. The absorption of light by the flame at specific wavelengths defines the concentration of the species. These methods require careful dilution of the sample and standard solutions since the response of the instruments are only linear over a small concentration range. The ICP emission instruments also require careful dilution but are linear over a wider range. Direct spark emission spectroscopy is a quick method of analysis, but is also subject to many sources of error. Sample preparation is important in this analytical method. The most severe limitation of this method is the absorption of light energy by atoms not completely excited. This is especially true in the case of aluminum where the response is linear over a small range and the response curve is not very steep.

Referee Methods. The American Society for Testing Materials (ASTM) has collected a series of standard referee methods for the analysis of magnesium and its alloys (78). These methods are accurate over a larger range of concentration than the production methods, but are time consuming in their application. The methods are based on potentiometric titration, photometric methods, or gravimetric methods. The photometric methods are most common and are relatively straightforward.

Health and Safety Factors

Magnesium articles or parts are difficult to ignite because of good thermal conductivity and high (>450°C) ignition temperatures. However, magnesium can be a fire hazard in the form of dust, flakes, or ribbon when exposed to flame or oxidizing agents. A magnesium powder or dust ignites readily, if suspended in air in concentrations above the lower explosive limit (0.04 g/L). Such ignition can result in a violent explosion (79,80).

Magnesium fires are readily extinguished with the appropriate metal-extinguishing powder. Magnesium fires do not flare up violently unless there is moisture present, therefore water is not recommended for extinguishing magnesium fires and must be avoided with molten magnesium or magnesium powders. Proper storage of magnesium products greatly reduces the risk of accidental ignition (81–83).

Magnesium is essential to most plant and animal life (see MINERAL NUTRIENTS). Dietary deficiency, rather than toxicity, is the more significant problem.

Uses

Magnesium is employed in a wide variety of applications, based on its chemical, electrochemical, physical, and mechanical properties. The International Magnesium Association (IMA) divides the markets for magnesium into 10 categories and tracks the volume of primary magnesium shipments to each market area on an annual basis.

Nonstructural Applications. *Aluminum Alloying.* Aluminum [*7429-90-5*] alloying is the single largest market for magnesium, accounting for 52% of reported shipments in 1992 (see ALUMINUM AND ALUMINUM ALLOYS). When magnesium is added to aluminum at levels of just a few percent or less, a significantly stronger alloy having good corrosion resistance results. The 5000 and 7000 series alloys of aluminum, sometimes referred to as the marine and aerospace alloys, contain up to 5.5 and 3.5% magnesium, respectively. The single largest application for magnesium-containing alloys of aluminum is the aluminum beverage can, which possesses a magnesium content of about 4.5% in the lid (alloy 5181 or 5182, UNS A95181 and A95182, respectively) and about 1.1% in the can body (alloy 3004, UNS A93004). Since the early 1980s, magnesium consumption in this market has grown at an average compound annual rate of 3.2%. Except for the significant increases in aluminum recycling, this rate might have been greater. More than 60% of aluminum beverage cans are recycled annually conserving both the aluminum and magnesium content of the alloys as well as the energy required to produce them (73,76,84,85).

Hot Metal Desulfurization. Magnesium plays a critical role in helping North American and European steel producers remain competitive in the world marketplace (see STEEL). Magnesium's unique affinity for sulfur [*7704-34-9*] allows it to be injected into molten iron (qv), where it vaporizes and reacts with high efficiency to form magnesium sulfide [*12032-36-9*], which floats to the surface as a readily separated phase (see SULFUR REMOVAL AND RECOVERY). This allows the steel producer the flexibility to use lower cost raw materials, while maintaining the ability to produce the high quality, low sulfur product required for high strength low alloy steels. The magnesium employed is often derived from low quality streams or alloy scrap which is then ground to a coarse powder and combined with lime prior to injection in the hot metal. Lime [*1305-78-8*] blends have been found to provide significantly improved efficiencies based on the magnesium required. Since the early 1980s, the desulfurization market has grown at an average annual compound rate of 16% in North America and Europe. Whereas magnesium's share of the North American desulfurization market is reported to be approaching 100%, there remain significant growth opportunities in Europe, Japan, and elsewhere (84,86).

Ductile Iron. Magnesium, in combination with ferrosilicon, is used in the production of ductile (nodular) iron [*7439-89-6*] because of the ability of magnesium to promote the formation of spheriodized (globular) graphite particles in place of the normal flake structure. This results in an iron (qv) product having improved toughness and ductility. Two principal applications for ductile iron are in the production of pipe and of automotive engine and drive train components. These markets are predicted to grow significantly. The projected overall growth for ductile iron is just 1.5% annually into the twenty-first century (84).

Chemical, Electrochemical, and Metal Reduction. In chemical applications, unalloyed magnesium is employed in the form of Grignard intermediates for the production of thousands of complex organic and organometallic chemicals employed in products such as sulfonated lubricating oils, silicones, pharmaceuticals (qv), and vitamins (qv) (see GRIGNARD REAGENTS; LUBRICATION AND LUBRICANTS; SILICON COMPOUNDS). In metal reduction, pure magnesium is used via thermal processes to produce metals such as titanium, zirconium, hafnium [*7440-58-6*], uranium, and beryllium [*7440-41-7*] from their respective halides (see BERYLLIUM AND BERYLLIUM ALLOYS; HAFNIUM AND HAFNIUM COMPOUNDS; TITANIUM AND TITANIUM ALLOYS; URANIUM AND URANIUM COMPOUNDS; ZIRCONIUM AND ZIRCONIUM COMPOUNDS). Magnesium is also employed in the production of elemental boron from boron oxide (see BORON, ELEMENTAL). The largest segment of the metal reduction market is in the production of titanium, followed at a distant second by zirconium production (87,88). Titanium producers normally recycle 90% of their by-product magnesium chloride back to magnesium metal by electrolytic reduction. In electrochemical applications, magnesium is used in alloys with manganese, or in aluminum–zinc–manganese alloys for the cathodic corrosion protection of steel structures (see CORROSION AND CORROSION CONTROL). Magnesium sacrificial anodes provide effective corrosion protection for household water heaters, underground pipelines, ship hulls, and storage tanks (see METAL ANODES). Magnesium alloy dry cell batteries (qv) are also used extensively for military applications. Unlike the more common zinc dry cell, the magnesium batteries retain high current capacity in storage for periods as long as 10 years even in tropical climates. All three of these markets have been flat since the middle 1980s. These combined markets accounted for 10% of the magnesium shipments reported in 1992 (76).

Other Applications. In finely divided form, magnesium is used in pyrotechnics (qv), either as the pure elemental metal or as an alloy with aluminum at contents up to 50%. Magnesium, alloyed with 3% aluminum and other minor elements, is used in the printing industry for photoengraving plates (see PRINTING PROCESSES). The magnesium alloy employed etches rapidly and uniformly in an acid medium to produce a sharp image. The combined shipments to these and other limited markets, plus shipments by the primary producers to secondary foundries, accounted for just 2% of the reported shipments in 1992 (72,76,89).

Structural Applications. Primary magnesium, like most metals, lacks sufficient strength in its elemental state to be used as a structural metal. Therefore, it must be alloyed with various other metals, such as aluminum, manganese, rare-earth metals, lithium, tin [*7440-31-5*], zinc, zirconium, silver, and yttrium [*7440-65-5*] (90–94). The combined market for structural applications accounted for 17% of reported shipments in 1992 and includes die cast, gravity, and wrought products. The primary reason for selecting magnesium for structural components is its light weight. Having a specific gravity of 1.74, it is the world's lightest structural metal. Aluminum weighs 1.5 times more; zinc weighs 4 times more; and iron and steel weigh more than 4.5 times more on an equivalent volume basis.

Die Castings. The single largest structural market for magnesium is die castings. Die casting is a metal casting process in which molten metal is forced under pressure into a permanent mold. Since the introduction of new high purity alloys having significantly improved corrosion resistance in 1982, the North American market has experienced an average annual compound growth rate of nearly 18% for the 10-year period ending in 1993. Automotive applications such as valve covers, engine brackets, clutch housings, steering columns, and instrument panel support members have accounted for most of this growth due to the need for weight reduction. Nonautomotive applications of magnesium die castings include chainsaws, lawnmower decks, archery bow handles, and power hand tool housings, as well as video camera, cellular phone, and computer components (89,95).

Gravity Castings. The low density of magnesium is especially important for gravity cast military and aerospace applications. Gravity castings are essentially all produced as sand castings with permanent mold and plaster casting representing a small segment of the alloy market. Typical applications include helicopter gear housings, aircraft canopy frames, air intakes, engine frames, speed brakes, and auxiliary component housings.

Wrought Products. Magnesium is also used in wrought form in products such as extrusions, forgings, sheet, and plate. Wrought magnesium alloys have been mechanically worked after casting. Applications for these products range from bakery racks, loading ramps, tennis rackets, and hand trucks to concrete finishing tools, computer printer platens, and nuclear fuel element containers and aerospace assemblies. Magnesium tooling plate is widely used in the production of fabricated products because of its combination of light weight, high dimensional stability, and ease of machining (76,89,96). The extrusion process yields an almost limitless variety of shapes. The dies are usually inexpensive enough that special shapes can be designed for specific uses.

A large number of magnesium alloys are in use for structural applications. The most common contain up to 9% aluminum, up to 2% zinc, and have minor amounts of manganese. Alloys of higher aluminum content are used for cast applications owing to their good strength. The lower aluminum content alloys are preferred for wrought applications owing to their reduced tendency to improper solidification and shrinkage. For gravity cast aerospace applications, special alloys containing zinc, zirconium, silver, yttrium, and rare-earth metals are used in order to meet operating temperature requirements as high as 300°C for extended periods of time.

Magnesium Alloys. *Alloy Designations.* Magnesium alloys are most commonly designated by a system established by ASTM which covers both chemical compositions and tempers (97,98). Tempers are treatments which usually improve toughness. The designations are based on the chemical composition, and consist of two letters representing the two alloying elements specified in the greatest amount, arranged in decreasing percentages, or alphabetically if of equal percentage. The letters are followed by the respective percentages rounded off to whole numbers, with a serial letter at the end. The serial letter indicates some variation in composition. Experimental alloys have the letter X between the alloy and a serial number. The following letters designate various alloying elements: A, aluminum; C, copper; D, cadmium; E, rare earths; H, thorium; K, zirconium; L, lithium; M, manganese; Q, silver; S, silicon; T, tin; W, yttrium; and Z, zinc.

Primary magnesium metal and alloys have also been assigned unified numbering system (UNS) designations according to the Standard Recommended Practice for Numbering Metals and Alloys (99). The UNS designation for a metal or alloy consists of a letter followed by five numbers. The UNS system is intended to provide a nationally accepted means of correlating the many alloy designation numbers used by various organizations and an improved system for indexing, record keeping, data storage and retrieval, and cross referencing. The numbers M10001 through M19999 have been reserved for magnesium and magnesium alloys. The letter M denotes a class of miscellaneous nonferrous metals and alloys. The magnesium primary grades and alloys registered with ASTM have been assigned UNS numbers and are listed in ASTM B275.

ASTM B296 defines the temper designations used and ASTM B661 defines the heat treatment schedules required to achieve the desired tempers for magnesium alloys. The temper designation is separated from alloy designation by a dash. The following describes the ASTM tempers commonly used for magnesium cast and wrought products (98):

-F	as fabricated
-O	annealed recrystallized (wrought products only)
-H	strain hardened
-H1	strain hardened only
-H2	strain hardened and then partially annealed
-H3	strain hardened and then stabilized
-T	thermally treated to produce stable tempers other than -F, -O, or -H
-T2	annealed (cast products only)
-T4	solution heat treated and naturally aged to a substantially stable condition
-T5	cooled and artificially aged only
-T6	solution heat treated and then artificially aged
-T7	solution heat treated and then stabilized
-T8	solution heat treated, cold worked, and then artificially aged

Composition and Properties of Selected Alloys. Table 4 shows the chemical compositions and physical properties of the magnesium alloys used most commonly in cast and wrought form. Typical mechanical properties at 20–25°C of selected magnesium alloys in various cast forms are given in Table 5, in extruded forms in Table 6, and in sheet and plate forms in Table 7. Alloys containing rare-earth metals, or rare earths and yttrium, have good strength retention at temperatures up to 315°C and higher. The strength of the other alloys diminishes rapidly above 150°C.

Properties of castings are determined using test bars cut from castings or on separately cast test bars. The properties of sections cut from actual castings may be only about 75% of the separately cast test bar values, due to slower solidification rates which result in large grain size and a consequent reduction of tensile properties. Properties of wrought alloy are determined on sections cut from extrusions, sheet, forgings, etc, in accordance with ASTM E55 sampling procedures. Magnesium alloys do not have the sharp yield point characteristic of carbon steels. Instead, the alloys yield gradually when stressed and the term yield strength is used. It has been defined as the stress at which the stress–strain curve intersects a line parallel to the modulus line offset 0.2% on the strain axis. The slope of this line is known as the Young's modulus. In cast form, the tensile and compressive yield strengths of magnesium alloys are substantially equal. In most wrought alloys, however, the compressive yield strength is less than that of the tensile yield.

Magnesium alloys have a Young's modulus of elasticity of approximately 45 GPa (6.5×10^6 psi). The modulus of rigidity or modulus of shear is 17 GPa (2.4×10^6 psi) and Poisson's ratio is 0.35. Poisson's ratio is the ratio of transverse contracting strain to the elongation strain when a rod is stretched by forces at its ends parallel to the rod's axis.

Sand and permanent-mold castings in magnesium alloys are produced in a large variety of sizes and shapes for many uses. Typical alloys with good casting qualities containing aluminum and zinc [*7440-66-6*] are AZ63A, AZ91C, and AZ92A. The last provides the optimum combination of high yield strength

Table 4. Chemical Compositions and Physical Properties of Magnesium Cast and Wrought Alloys[a]

Alloy			Nominal composition, %[b]					Physical properties			
ASTM	UNS	Temper	Al	Mn	RE[c]	Zn	Other	Density at 20°C, g/cm³	Melting point,[d] °C	W/(m·K)[e]	μΩ·cm[f]
Sand and permanent-mold castings											
AM100A	M10100	-T6	10.0	0.2				1.81	465	58.3	12.4
AZ63A	M11630	-F	6.0	0.2		3.0		1.82	455	59.2	12.2
		-T4								52.2	14.0
		-T5								65.1	11.0
		-T6								61.0	11.8
AZ81A	M11810	-T4	7.6	0.2		0.7		1.80	510	50.3	15.0
AZ91C,E	M11914,-18	-F	8.7	0.2		0.7		1.80	470	53.6	13.6
		-T4								44.3	16.2
		-T6								56.2	12.9
AZ92A	M11920	-F	9.0	0.2		2.0		1.83	445	52.2	14.0
		-T4								44.3	16.8
		-T5								58.3	12.4
		-T6								58.3	12.4
EZ33A	M12330	-T5			3.0	2.7	0.7 Zr	1.80	545	99.0	7.0
QE22A	M18220	-T6			2.2		2.5 Ag	1.82	550	102.2	6.8
WE43A	M18430	-T6			3.0		4.0 Y	1.84	543	51	14.8
WE54A	M18410	-T6			3.5		5.2 Y	1.85	549	52	17.3
ZE41A	M16410	-T5			1.2	4.2		1.84	510	123.1	5.6
ZE63A	M16630	-T6			2.6	5.7	0.7 Zr	1.87	515	123.1	5.6
ZK51A	M16510	-T5				4.6	0.7 Zr	1.81	550	108.3	6.4
ZK61A	M16610	-T6				6.0	0.8 Zr	1.83	520		

					Die castings						
AM50A	M10500	-F	5.0	0.4				1.78	543	62	12.5
AE42X1		-F	4.0	0.3	2.0			1.79	565	68	
AM60A,B	M10600,-02	-F	6.0	0.2				1.79	541	62	12.5
AS41A,B	M10410,-12	-F	4.2	0.3			1.0 Si	1.77	566	68	
AZ91B,D	M11912,-16	-F	9.0	0.2		0.6		1.80	470	51.2	14.1
					Sheet and plate						
AZ31B	M11311	-F	3.0			1.0		1.77	565	76.9	9.2
AZ31B,C	M11311,-12	-H24	3.0	0.3		1.0		1.77	565	76.9	9.2
		-H26	3.0	0.3		1.0		1.77	565	76.9	9.2
		-O	3.0	0.3		1.0		1.77	565	76.9	9.2
AZ61A	M11610	-F	6.5	0.2		1.0		1.80	510	57.9	12.5
		-O	6.5	0.2		1.0		1.80	510		
					Extruded bars, rods, solid and hollow shapes, and tubes						
AZ80A	M11800	-F	8.5			0.5		1.80	490	47.3	15.6
		-T5	8.5			0.5		1.80	490	59.2	12.2
ZK60A	M16600	-F				5.7	0.55 Zr	1.83	520	117.6	6.0
		-T5				5.7	0.55 Zr	1.83	520	121.0	5.7

[a]Refs. 93 and 94.
[b]Balance Mg.
[c]Rare earths.
[d]The solidus temperature (lower limit of alloy melting range).
[e]Thermal conductivity, at 20°C.
[f]Electrical resistivity at 20°C.

Table 5. Mechanical Properties of Magnesium Casting Alloys[a]

Alloy				Tensile yield	
ASTM	UNS	Temper	Tensile strength, MPa[b]	strength, MPa[b]	Elongation in 51 mm, %
		Sand and permanent-mold castings			
AM100A	M10100	-F	152	83	2
		-T4	276	90	10
		-T5	152	110	2
		-T61	276	152	1
AZ63A	M11630	-F	200	97	6
		-T4	276	97	12
		-T5	200	103	4
		-T6	276	131	5
AZ81A	M11810	-T4	276	83	15
AZ91C,E	M11914,-18	-F	165	97	2.5
		-T4	276	90	15
		-T6	276	131	5
AZ92A	M11920	-F	172	97	2
		-T4	276	97	10
		-T5	172	117	1
		-T6	276	152	3
EZ33A	M12330	-T5	159	110	3
QE22A	M18220	-T6	276	207	4
WE43A	M18430	-T6	252	190	7
WE54A	M18410	-T6	275	171	4
ZE41A	M16410	-T5	207	138	3.5
ZE63A	M16630	-T6	276	186	5
ZK51A	M16510	-T5	276	165	8
ZK61A	M16610	-T6	276	179	5
		Die castings			
AM60A,B	M10600,-02	-F	220	130	8
AS41A,B	M10410,-12	-F	210	140	6
AZ91B,D	M11912,-16	-F	230	158	3

[a]Properties determined on separately cast test bars using 0.2% offset method; Ref. 93.
[b]To convert MPa to psi, multiply by 145.
[c]To convert J to ft·lbf, divide by 1.356; Ref. 94.
[d]See HARDNESS.
[e]500-kg load, 10-mm ball.

and moderate elongation. Casting alloys containing zirconium and zinc, such as ZK51A and ZK61A, have been developed for their improved properties, including a reduced tendency for microporosity. Alloys containing rare earths, such as EZ33A and QE22, respectively, are specified for elevated temperature service. Yttrium–rare-earth alloys such as WE54A and WE43A, introduced in recent years, offer still further improved elevated temperature properties with the added advantage of enhanced corrosion performance in salt water exposures.

Heat Treatment. Heat treatment improves the properties (101) of magnesium castings. In solution heat treatment (ASTM T4), the casting is heated to the proper temperature and held long enough for the precipitated compound

Table 5. (*Continued*)

Compressive yield strength, MPa[b]	Bearing strength, MPa[b]	Bearing yield strength, MPa[b]	Shear strength, MPa[b]	Impact strength Charpy, J[c]	Hardness[d]	
					Brinell[e]	Rockwell E
			Sand and permanent-mold castings			
83			124	0.8	53	64
90	476	310	140	2.7	52	62
110					58	70
131	560	470	145	0.9	69	80
97	415	275	125	1.4	50	59
97	410	270	124	3.4	55	66
97	455	275	130	3.5	55	66
131	475	355	138	1.5	73	83
83	400	241	165	6.1	55	66
97	415	275		0.8	60	66
90	415	255	150	4.1	55	62
131	460	360	165	1.4	70	77
97	345	315	125	0.7	65	76
97	470	315	140	2.7	63	75
117	345	317	140		69	80
152	540	460	180	1.1	81	88
110	310	275	135		50	59
207					78	
187			162		85	
171			150		85	
138	485	355	150	1.4	62	72
165	485	350	150		65	77
					70	
			Die castings			
160			140	3	63	75

to dissolve. Air quenching upon completion of the solution heat-treatment cycle prevents precipitation or reforming of the constituents. Controlled precipitation by artificial aging following solution heat treatment is designated as T6 temper. This treatment increases tensile yield strength and hardness. An artificial aging treatment consists of simply heating the as-cast product for a few hours at a suitable intermediate temperature. This treatment, designated as T5, relieves internal stresses that are likely to result during cooling after casting. Such a treatment improves mechanical properties and decreases the possibility of warping after machining or permanent distortion when used at elevated temperatures.

Casting alloy M11630 (AZ63A) is heat treated in a furnace at 260°C until it is the same temperature as the furnace. The temperature is then raised gradually over a two hour period to 390°C and the casting held at this temperature for about 10 hours and then air cooled. This provides solution heat treatment (T4

Table 6. Mechanical Properties of Magnesium Extrusions[a]

Alloy ASTM	Alloy UNS	Temper	Least dimension, mm	Area, cm^2	Tensile strength, MPa[b]	Tensile yield strength, MPa[b]
			Bars, rods, and shapes			
AZ31B	M11311	-F	under 6.3		260	195
			6.3–38.1		260	200
			38.1–63.5		260	195
			63.5–127		260	195
AZ61A	M11610	-F	under 6.3		315	230
			6.3–63.5		310	230
			63.5–127		310	215
AZ80A	M11800	-F	under 6.3		340	250
			6.3–38.1		340	250
			38.1–63.5		340	240
			63.5–127		330	250
AZ80A	M11800	-T5	under 6.3		380	260
			6.3–38.1		380	275
			38.1–63.5		365	270
			63.5–127		345	260
ZK60A	M16600	-F		under 12.9	340	260
				12.9–19.3	340	255
				19.3–32.3	340	248
				32.3–259	330	255
ZK60A	M16600	-T5		under 12.9	365	305
				12.0–19.3	360	295
				19.3–32.3	350	290
			Hollow and semihollow shapes			
AZ31B	M11311	-F			250	165
AZ61A	M11610	-F			285	165
ZK60A	M16600	-F			315	235
		-T5			345	275
			Tube			
			wall thickness,[d] mm	OD,[e] cm[f]		
AZ31B	M11311	-F	0.7–6.3	15.2	250	165
			6.3–18.9	38.7	250	165
AZ61A	M11610	-F	0.7–18.9	38.7	285	165
ZK60A	M16600	-F	0.7–6.3	19.3	325	240
		-T5	0.7–6.3	19.3	345	275
		-T5	2.4–48	19.3–54.8	340	270

[a]Refs. 93 and 100.
[b]To convert MPa to psi, multiply by 145.
[c]500-kg load, 10-mm ball.
[d]Values are inclusive.
[e]OD = outer diameter.
[f]Values are maximum unless range is given.

Table 6. (*Continued*)

Elongation, in 5.08 mm, %	Compressive yield strength, MPa[b]	Shear strength, MPa[b]	Bearing strength, MPa[b]	Bearing yield strength, MPa[b]	Brinell hardness[c]
		Bars, rods, and shapes			
14	103	130	385	235	
15	95	130	385	230	49
14	95	130	385	230	
15	95	130	385	230	
17		160	450	260	
16	130	150	470	275	60
15	145	150	470	290	
12		150	470	330	60
11		150	470	330	60
11		150	470	330	60
9		150	470	330	60
8	235	165	415	395	82
7	240	165	415	400	82
6	220	165	455	370	82
6	215	165	470	365	82
14	230	165	525	385	75
14	195	165	525	345	75
14	185	165	525	340	75
9	160	165	515	305	75
11	250	180	545	405	82
12	215	180	540	365	82
14	205	170	530	360	82
		Hollow and semihollow shapes			
16	85				46
14	110				50
12	170				75
11	200				82
		Tube			
16	85				46
12	85				
14	110				50
13	170				75
11	205				82
12	180				

Table 7. Mechanical Properties of Magnesium Sheet and Plate[a]

Alloy										
ASTM	UNS	Temper	Thickness, mm	Tensile strength, MPa[b]	Tensile yield strength, MPa[b]	Elongation, in 5.08 mm, %	Compressive yield strength, MPa[b]	Shear strength, MPa[b]	Bearing strength, MPa[b]	Bearing yield strength, hardness[c]
					Sheet and plate					
AZ31B	M11311	-H24	0.4–6.3	290	220	15	180	200	530	47
			6.3–9.5	275	200	17	160	195	495	45
			9.5–12.7	270	185	19	130	185	485	40
			12.7–25.4	260	165	17	110	180	470	37
			25.4–50.8	255	160	14	95	180	455	35
			50.8–76.2	255	145	16	85	180	455	33
					Sheet and plate					
AZ31B	M11311	-O	0.4–1.5	255	150	21	110	180	455	37
			1.5–6.3	255	150	21	110	180	455	37
			6.3–12.7	250	150	21	90	170	448	34
			12.7–50.8	250	150	17	85	170	448	33
			50.8–76.2	250	145	17	75	170	448	32
					Plate					
AZ31B	M11311	-H26	6.3–9.5	275	205	16	165	195	495	46
			9.5–11.1	275	195	13	150	195	495	44
			11.1–12.7	275	195	13	150	195	495	44
			12.7–18.9	275	195	10	130	195	495	40
			18.9–25.4	270	180	10	125	195	485	39
			25.4–38.1	260	170	10	110	185	470	37
			38.1–50.8	260	170	10	100	180	470	36

[a] Refs. 93 and 100.
[b] To convert MPa to psi, multiply by 145.
[c] 500-kg load, 10-mm ball.

temper). For artificial aging (T6 temper), the solution-heat treated casting is heated further for 16 hours at 175°C. Heat treatment of AZ92A alloy castings requires different temperatures. For the solution-heat treatment, the alloy is soaked at 410°C for 18 hours, whereas for artificial aging the casting is heated at 230–290°C for 2–6 hours. Because of the higher temperature for M11920 (AZ92A) treatments, the furnaces should be equipped to circulate the heated air and should contain about 0.7% sulfur dioxide or 0.5% SF_6 to avoid excessive oxidation during the solution heat-treating cycle.

Alloy M16630 (ZE63A) which contains rare-earth metals and zinc, is designed to take advantage of a newer heat-treatment technique involving inward diffusion of hydrogen and formation of zirconium hydride [*7704-99-6*]. The alloy is heated in hydrogen at 480°C for 10, 24, or 72 hours for 6.3, 12.7, or 19 mm sections, respectively, to complete the formation of zirconium hydride. It is then quenched and aged for 48 hours at 140°C.

For the manufacture of die castings, the alloy M11916 (AZ91D), which contains aluminum and zinc, is most commonly used because of its excellent stiffness and strength at normal ambient temperatures. Alloys M10602 (AM60B) or AM50X1, which contain aluminum and manganese, are used where ductility is desired, and M10410 (AS41A), which contains silicon, or developmental alloy AE42 are used where good long-time yield strength (or creep resistance) is needed at elevated temperatures. Heat treating is not used on magnesium die castings.

For common wrought applications, alloys AZ31B, AZ61A, AZ80A, and ZK60A are employed. When wrought magnesium alloys are in tempers other than as fabricated, these are treated at the mill rather than by the customer. ZK60A and AZ80A are produced in the T5, artificially aged temper for maximum strength, whereas AZ31B sheet and plate is produced in the O (fully annealed) or the H24 or H26 (strain hardened and partially annealed) tempers. Magnesium is usually press-forged, although hammer forging is possible. Forged commercial alloys include AZ31B, AZ61A, AZ80A, and ZK60A.

In Europe, the same types of alloy systems as in the United States are employed.

Metallography. Most commercial magnesium alloys are either of the solid solution or hypoeutectic type, where intermediary phases are second constituents. The phase diagrams in Figures 10–12 show binary systems of some common alloying elements. Figure 13 is a phase diagram of the widely used ternary system comprised of magnesium, aluminum, and zinc (102).

Under equilibrium conditions, magnesium can contain as much as 12.7% aluminum in solid solution at the eutectic temperature. However, the slow diffusion of aluminum to the grain boundary leads to a coring effect in primary crystals and a hard-phase magnesium–aluminum compound(17:12) [*12254-22-7*], $Mg_{17}Al_{12}$, or β-(MgAl). Thus aluminum occurs in magnesium alloys both in solid solution and as the intermediary intermetallic phase. The latter is clear white and in slight relief in polished and etched samples. In as-cast alloys, the hard phase occurs in massive form, but when precipitated from solid solution a lamellar structure is formed similar to pearlite in steel. When produced by aging at low temperatures, it appears as fine particles.

In commercial alloys, zinc is usually dissolved in the magnesium matrix and in the hard magnesium–aluminum phase when aluminum is present. Zinc

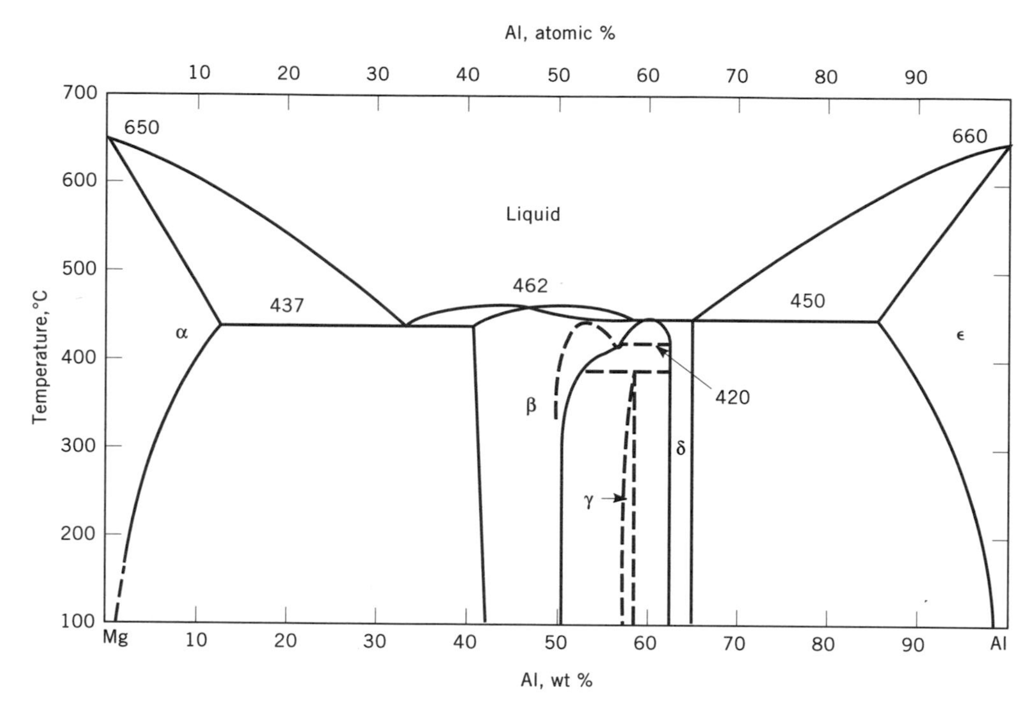

Fig. 10. Aluminum–magnesium phase diagram (102).

additions to magnesium–aluminum alloys change the eutectic structure to a so-called divorced eutectic, characterized by the presence of massive compound particles surrounded by a magnesium-rich solid solution.

Grain size and other structural characteristics of magnesium alloys are most easily determined by polishing and etching. These characteristics are classified by reference to a series of charts that have been devised to express numerically the various structural characteristics. Magnesium alloys deform plastically by slip and twinning mechanisms, and polished and etched samples indicate prior working by the presence of lenticular twins, ie, having the shape of a double convex lens, within grains of deformed metal. The success of heat-treating operations can be followed by examining samples for decrease of the β-(MgAl) constituent on solution heat treatment, and by noting the character of precipitated β-(MgAl) after aging treatments.

Manganese appears as bluish gray angular crystals in the binary alloy if the crystals are formed during solidification of the alloy. In magnesium–aluminum alloys, the manganese appears as a manganese–aluminum binary phase, or as a manganese–aluminum–iron ternary phase formed with any iron contamination present in the alloy. Silicon is an impurity usually present as an intermetallic, light blue compound, magnesium silicide [*22831-39-6*], Mg_2Si. In magnesium–aluminum alloys containing more than about 0.2% silicon, the Mg_2Si develops an appearance of Chinese script. This is the form commonly found in die cast alloy AS41B. Tin [*7440-31-5*] is soluble in magnesium, and may be found in alloys in solid solution or as a dull grayish blue constituent that becomes tan to brown when etched. Calcium, if present in amounts greater

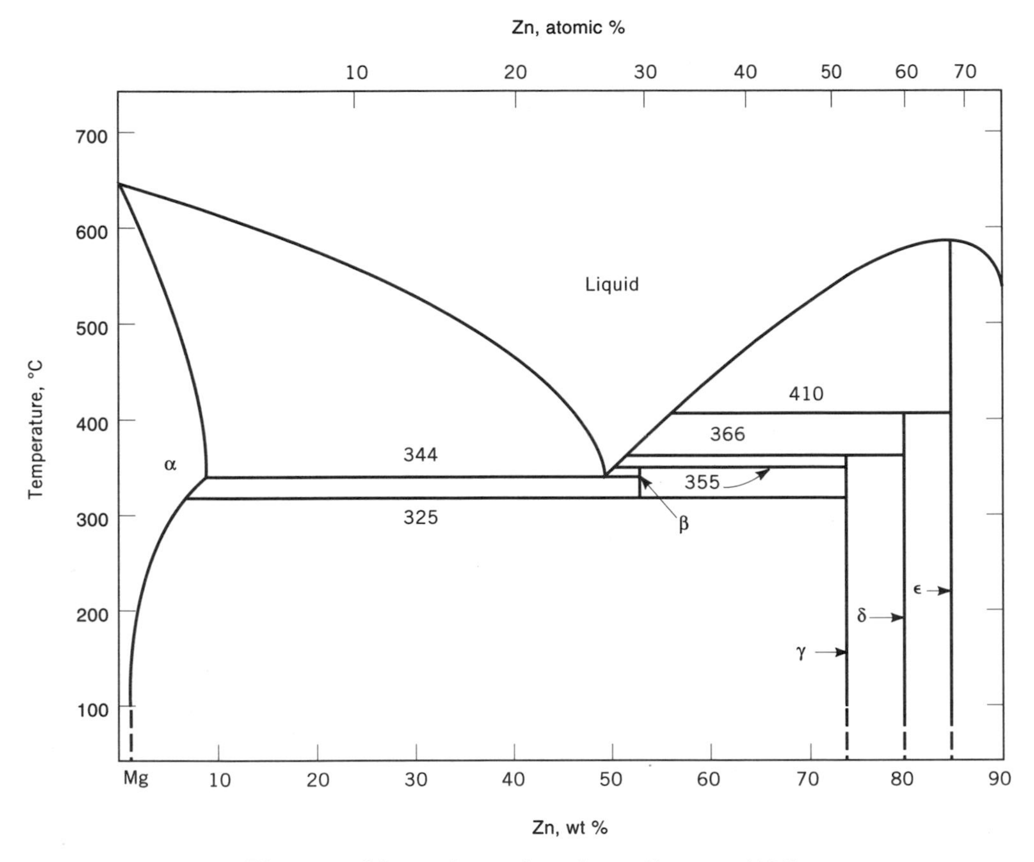

Fig. 11. Magnesium–zinc phase diagram (102).

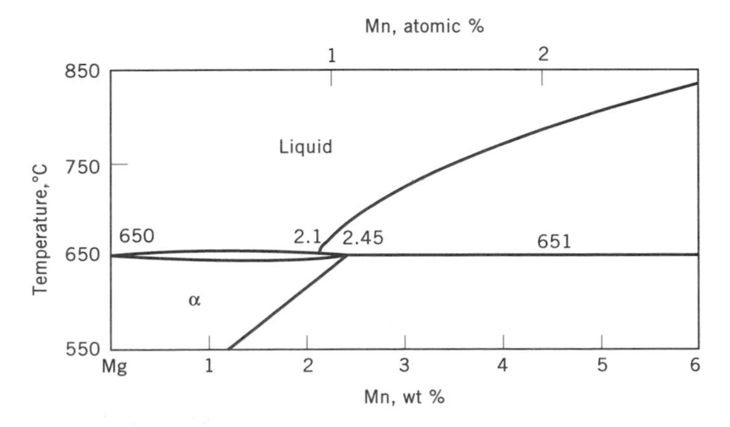

Fig. 12. Magnesium–manganese phase diagram (102).

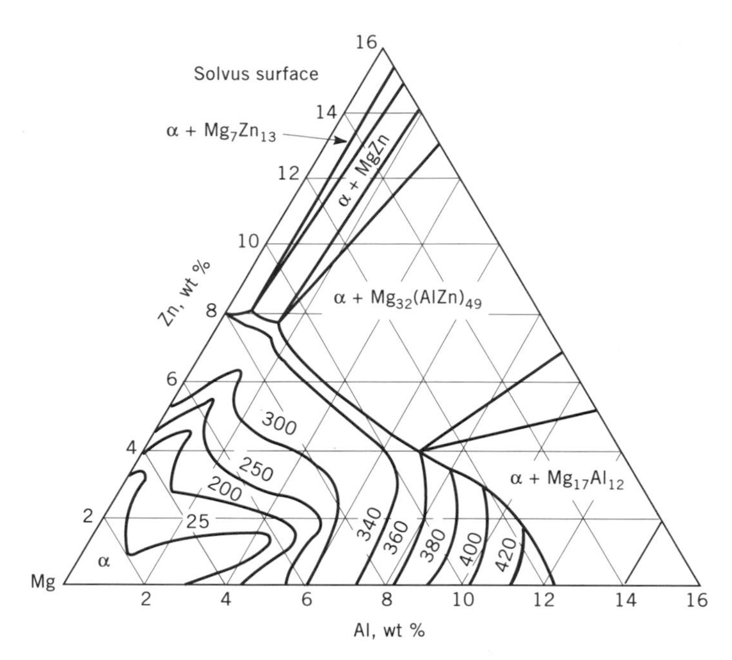

Fig. 13. Aluminum–magnesium–zinc phase diagram (102).

than 0.1%, occurs as magnesium calcium compound(2:1) [*12133-32-3*], Mg_2Ca, unless aluminum is present, in which case calcium aluminide(1:2) [*12042-65-8*], Al_2Ca, forms instead.

Fabrication. Magnesium alloys are fabricated by common methods, including melting followed by casting, rolling, extrusion, and forging. Total energy required for the manufacture of magnesium sheet or extrusions has been estimated at 75 MJ/kg (32,090 Btu/lb) and 83 MJ/kg (35,980 Btu/lb), respectively (103). Further fabrication includes forming, joining, and machining after which standard assembly methods are used.

Magnesium alloys are produced from molten magnesium directly from magnesium cells or by remelting magnesium pigs or ingots in oil- or gas-fired steel pots or electric-induction furnaces (104,105). Alloying metals, especially aluminum and zinc, may be placed in a perforated steel basket and suspended in the molten magnesium which is held at about 700°C. Movement of the basket facilitates alloying by flushing, which minimizes the possibility of segregation of heavy constituents. Manganese is added either in the metallic or chloride form for the reduction of iron content through the formation of an insoluble intermetallic. After the alloying ingredients are added, the melt is allowed to stand to settle out impurities, including the iron–manganese phase. The metal can then be cast in ingot molds to produce the alloy ingot used for the various casting processes. In high production alloying operations, the magnesium ingots are melted in multipot systems or reverberatory furnaces and the liquid metal is then transferred by pumping. A multipot system may consist of a complex of melting, alloying, settling, holding, and casting pots from which the metal is ladled or pumped into ingot molds.

Magnesium alloys containing zirconium and rare earths are often made up in the casting pot in the foundry because remelting can result in substantial loss of these alloying metals. Rare-earth metals generally are added as mischmetal (mixed rare-earth metals). Zirconium can be added as a magnesium–zirconium alloy called a zirconium hardener.

Salt-based fluxes are employed to prevent excessive oxidation of the melt and improve its purity. They contain a mixture of chlorides, including potassium chloride [*7447-40-7*] and magnesium, as well as calcium fluoride and magnesium oxide. Since the early 1980s, many foundries have adopted the use of sulfur hexafluoride, SF_6, at approximately 0.3 to 0.5% in air, or air/CO_2 mixtures, to inhibit excessive oxidation of magnesium melts. This practice, combined with developing fluxless refining methods, has dramatically improved the corrosion performance of cast products by eliminating the source of salt contamination which often leads to serious corrosion of machined castings (69,106,107).

The extrusion process starts with a cast extrusion billet that is made by casting in thick-walled iron or steel molds, or preferably by means of the direct-chill casting process which gives a fine grain and prevents compound segregation. Billets that have been machined to remove the casting skin are preheated at 315–455°C, depending on the alloy, placed in the container of the extrusion press, and then by means of a hydraulic ram are forced through a steel die to provide the desired shape. Certain magnesium alloys containing only small amounts of alloying constituent may be extruded at a rate as high as 30.5 m/min. Other alloys and more difficult configurations may have a limiting extrusion speed of only one or two meters per minute. The reduction in cross-sectional area from the original ingot to the final extruded section must be relatively high, 20:1 or greater, to obtain desired properties in the extrusion. Additional improvement in properties of certain alloys can be obtained after extrusion by a heat treatment, such as artificial aging, which causes precipitation of intermetallic compounds and thus improves yield strength but with some sacrifice in ductility.

Magnesium sheet and plate are fabricated by rolling slab which is made by the direct-chill casting process. Several separate operations are needed. First, in breakdown rolling, the slabs are heated in a two-step preheating oven to a temperature of 425–480°C. The slab is then reduced in thickness by repeated passes through the rolls of a hot mill. The original slab is reduced in thickness to about 6.4 mm without reheating. Following breakdown and reheating, the sheet is transferred to a finishing mill and rolled with intermittent steps to the final desired thickness and temper. Additional operations, such as shearing at various stages in production, acid cleaning, wire brushing, chrome pickling, or oil finishing, may be required depending on the final product.

Magnesium forgings usually are made by the press-forging process in closed dies or, less often, by hammer forging (105). The size of forgings is limited only by the size of available equipment. A magnesium canister forging made for the Echo satellite program weighed 104 kg.

The earliest method of fabrication used for magnesium alloys, still commonly used as of 1994, is sand casting (105,108). Inhibitor agents are added to green sand or resin-bonded sand to prevent excessive reaction with atmospheric nitrogen, oxygen, or moisture in the sand mold. Inhibitor agents usually include sulfur, boric acid [*10043-35-3*], diethylene glycol [*111-46-6*], and ammonium fluorosilicate [*16919-19-0*]. A typical molding sand might contain 0.7% sulfur, 1% boric acid, and 5% ammonium fluorosilicate. Sand cores may contain inhibitors mixed with the sand or the cores may be sprayed with inhibitors after baking.

A wide variety of parts are made by die casting (109,110). Magnesium's low heat content per unit volume and its limited tendency to wet the die permits high casting speeds with extended die life relative to aluminum die casting. The cold-chamber process has been used widely for many years but the hot chamber process has assumed a significant share of the magnesium die casting market during the 1990s. However, recent developments in metering devices allow the delivery of the exact amount of metal required to the shot well of cold chamber die casting machines. This development may support their competitive use for larger castings for some years to come. Typically, the energy required to produce a finished die casting, including normal processing efficiencies, is reported to be 21–45 MJ/kg (9,000–19,160 Btu/lb) (75,103).

Thixotropic injection molding offers a competitive technology for the magnesium die casting industry and has some unique advantages such as no molten metal handling or associated losses, rapid start-up and shutdown, improved die life, reduced porosity relative to die casting, and tighter part tolerances, and reduced scrap rates. The process, which is modeled after the plastics injection molding process, employs a granular alloy feed. The feed is heated under argon in the machine's screw injector to a temperature between the alloys' solidus and liquidus temperatures. Under the continuous mixing imparted by the screw, the material is converted to a thixotropic state at which point the material is injected into a mold of similar design to a standard high pressure die casting mold (111,112).

Magnesium sheet or extrusions can be formed by many methods (96, 100,113). Bending around large radii is usually involved. Relatively minor deformations can be performed at room temperature. However, the formability is greatly improved at 200–315°C, and most forming operations are carried out in this temperature range. This increased formability at elevated temperature is due to magnesium's hexagonal close-packed structure. When the metal is heated, additional slip planes become available within the metal lattice allowing improved plastic deformation characteristics. In high speed mechanical presses, draws are made in one step at up to 59% reduction and at speeds up to 24 m/min (114). Magnesium alloys are also formed by bending, stretch forming, spinning, impact extruding, drop-hammer forming, and other common methods.

Parts are assembled by joining methods including arc welding, electric-resistance welding, brazing, soldering, riveting, bolting, and adhesive bonding (115–119). Because of rapid heat transfer, magnesium cannot be satisfactorily cut with an oxy-fuel gas flame in the same manner as steel. Welding (qv) is best accomplished by the gas–metal arc processes which prevent the formation of magnesium oxide or nitrides. Helium [*7440-59-7*] or argon are the preferred protective gases. In gas–tungsten arc welding, a tungsten [*7440-33-7*] electrode maintains the arc which melts the magnesium filler rod. Gas–metal arc welding uses a coil of magnesium wire that functions as both an electrode and filler rod. This method is sometimes described as consumable electrode-arc welding. The gas–metal arc welding processes are preferred over gas welding because they do not use a corrosive flux. Magnesium can be welded with oxy-fuel gases but a chloride-base flux is necessary to prevent oxidation. This flux is corrosive and difficult to remove completely after welding. Therefore, gas welding of magnesium alloys is limited to emergency repairs. Spot welding is the most popular method of electric-resistance welding. Forge welding and stud welding can also be used. Magnesium alloys are also joined by furnace, flux-dip, and torch brazing. Soldering is avoided because of the brittle joint that is formed (see SOLDERS AND BRAZING ALLOYS).

Riveting is another method of making mechanical joints in magnesium. Aluminum alloy 5056 (A95056) rivets in the H32 temper are commonly used, although rivets of 6053-T61 (A96053) or 6061-T6 (A96061) can be substituted. These aluminum alloys minimize the possibility of galvanic corrosion. Rivets of steel, brass, copper, and certain aluminum alloys should not be brought in contact with magnesium because of serious galvanic corrosion. Rivet holes in magnesium should be drilled rather than punched and squeeze rivets are preferred over pneumatic riveting hammers. The latter are more likely to damage the magnesium sheet by overdriving. Many small rivets are preferred over a few of large diameter. Optimum rivet-joint design calls for a minimum spacing in any direction of three times the rivet diameter. Similarly, a minimum edge distance of 2.5 times the rivet diameter is suggested. Riveted joints should be protected with a sealing compound to prevent water entrapment. A coat of chromate-pigmented primer is often used for this purpose.

In some applications, adhesive bonding has become a popular method of joining magnesium sheet. For example adhesive bonding is used when joining material too thin to be effectively riveted or welded. A large variety of adhesives (qv) are available, most of which require elevated (≥90°C) temperature curing. Curing times range from a few minutes to an hour or more. Certain adhesives require the application of pressure but usually contact is sufficient. The shear strength of adhesive bonded joints ranges from about 7 to 28 MPa (1000–4000 psi). Most bonded joints hold up to 65°C. Some epoxy phenolic adhesives have good joint strength retention up to 260°C.

Machining. Magnesium is the easiest of all structural metals to machine (120). Because of this machinability, it is sometimes used in applications where a large number of machining operations are required. The machinability of magnesium alloys relative to that of other metals based on the lowest power required to remove 16 cm^3 of metal and when magnesium is assigned a value of unity, is (120):

Metal	Relative power required
magnesium alloys	1.0
aluminum alloys	1.8
brass	2.3
cast iron	3.5
mild steel	6.3
nickel alloys	10.0

Some of the advantages of excellent machinability include reduced machining time, resulting in higher productivity for the machine tools and thus lower capital investment; greatly increased tool life; an excellent surface finish with a single large cut; well-broken chips which minimize handling costs; and less tool buildup.

Dry machining is strongly encouraged owing to the fact that the value of turnings or chips produced are significantly higher as a result of ease with which these can be recycled through remelting or through use as a desulfurization reagent. Where chip removal is required, a combination of air- and screw-driven conveyors can be employed. If a coolant or cutting fluid must be used for a particular operation, mineral oil is usually preferred over water-based coolants. This is due to the fact that magnesium reacts with water to some degree, over time, producing hydrogen gas and magnesium hydroxide. As a result, wet chips should be treated with caution in order to prevent ignition of the evolved hydrogen, and to prevent partial drying of the wet mass, which may result in spontaneous ignition due to the heat evolved in the reaction in combination with poor heat-transfer characteristics of the partially dried mass. If wet chips are generated, the chips should be kept fully submerged in excess water and stored in a well-ventilated, remote location until they can be properly recovered or safely destroyed (79,81,83,120). Despite the hazards associated with water-based coolants, well-inhibited coolants have been employed in some applications for a number of years, with few problems.

Coolants or cutting fluids containing animal or vegetable oil must be avoided. The carboxylic acid functions present can undergo reaction with the magnesium on standing.

Corrosion and Finishing. With few exceptions, magnesium exhibits good resistance to corrosion at normal ambient temperatures unless there is significant water content in the environment in combination with certain contaminants. The reaction which typically occurs is described by the equation

$$Mg + 2\,H_2O \longrightarrow Mg(OH)_2\,(s) + H_2\,(g) \qquad (19)$$

In neutral and alkaline environments, the magnesium hydroxide product can form a surface film which offers considerable protection to the pure metal or its common alloys. Electron diffraction studies of the film formed in humid air indicate that it is amorphous, with the oxidation rate reported to be less than 0.01 μm/yr. If the humidity level is sufficiently high, so that condensation occurs on the surface of the sample, the amorphous film is found to contain at least some crystalline magnesium hydroxide (brucite). The crystalline magnesium hydroxide is also protective in deionized water at room temperature. The aeration of the water has little or no measurable effect on the corrosion resistance. However, as the water temperature is increased to 100°C, the protective capacity of the film begins to erode, particularly in the presence of certain cathodic contaminants in either the metal or the water (121,122).

In extended atmospheric exposures of magnesium and magnesium alloys, the reaction of the magnesium hydroxide with acid gases, such as CO_2 and SO_2, has been reported to play an important role in the stability and composition of the film present. X-ray diffraction analysis of the oxidation product present on unalloyed magnesium ingot reveals it consists of a mixture of crystalline hydroxycarbonates of magnesium such as hydromagnesite [*12275-04-6*], $MgCO_3 \cdot Mg(OH)_2 \cdot 9H_2O$, nesquehonite [*5145-46-0*], $MgCO_3 \cdot 3H_2O$, and lansfordite [*5145-47-1*], $MgCO_3 \cdot 5H_2O$. In the case of the common commercial wrought alloy, AZ31B (magnesium, 3% aluminum, 1% zinc, 0.2% manganese) analysis of the adherent corrosion product by x-ray diffraction revealed only two crystalline phases: hydromagnesite and hydrotalcite [*12304-65-3*], $Mg_6Al_2(OH)_{16}CO_3 \cdot 4H_2O$. In an industrial atmosphere with high SO_2 content, traces of magnesium sulfate hexahydrate [*10034-99-8*], $MgSO_4 \cdot 6H_2O$, and magnesium sulfite hexahydrate [*13446-29-2*], $MgSO_3 \cdot 6H_2O$, were detected in addition to the hydroxycarbonate products for unalloyed ingot. It was suggested that SO_2 exposures accelerate the corrosion of magnesium through the conversion of the protective hydroxide and carbonate compounds to the highly soluble sulfate and sulfite, which are then eroded (121,123).

Corrosion by Various Chemicals and Environments. In general, the rate of corrosion of magnesium in aqueous solutions is strongly influenced by the hydrogen ion [*12408-02-5*] concentration or pH. In this respect, magnesium is considered to be opposite in character to aluminum. Aluminum is resistant to weak acids but attacked by strong alkalies, while magnesium is resistant to alkalies but is attacked by acids that do not promote the formation of insoluble films.

With regard to salts, neutral or alkaline fluorides form insoluble magnesium fluoride, and consequently magnesium alloys are resistant to them. Chlorides are usually corrosive even in solutions having pH values above that required to form magnesium hydroxide. Acid salts are generally destructive but chromates, vanadates, and phosphates form films that usually retard corrosion except at elevated temperatures. Most mineral acids attack magnesium rapidly. Hydrofluoric acid, except at low concentration and elevated temperature, is an exception to this rule. Chromic acid [*11115-74-5*] has a low rate of attack except when chlorides or sulfates are present. Chromic acid solutions of about 20% are used to clean corroded samples, since this solution dissolves the magnesium oxidation products without significantly attacking the base metal (124). Most organic acids attack magnesium alloys readily.

Contaminant Effects. Magnesium alloys have long had a reputation for poor corrosion performance, particularly in salt water exposures. The importance of metal composition in determining the susceptibility to salt water attack has been recognized since the 1930s and 1940s. However, the first complete definition of the compositional limits required for some common commercial alloys was only established within the 1980s (125–129). A summary of the effects of 14 separate elements on the salt water corrosion performance of magnesium was published in 1942 (130). Of the four elements having the most degrading effects, iron, nickel, copper, and cobalt [*7440-48-4*], the first three are common contaminants of standard commercial alloys. It is these elements that in large part have led to magnesium's poor reputation for salt water corrosion durability (125–129). When these elements are controlled at levels lower than established limits, referred to as tolerance limits, for a specific alloy composition, the salt water corrosion rates may be reduced by more than 100-fold in accelerated ASTM B117 salt spray testing. Figure 14 illustrates the effect of reduced levels of iron, nickel, and copper on the corrosion rates of high pressure die cast AZ91 alloy, versus

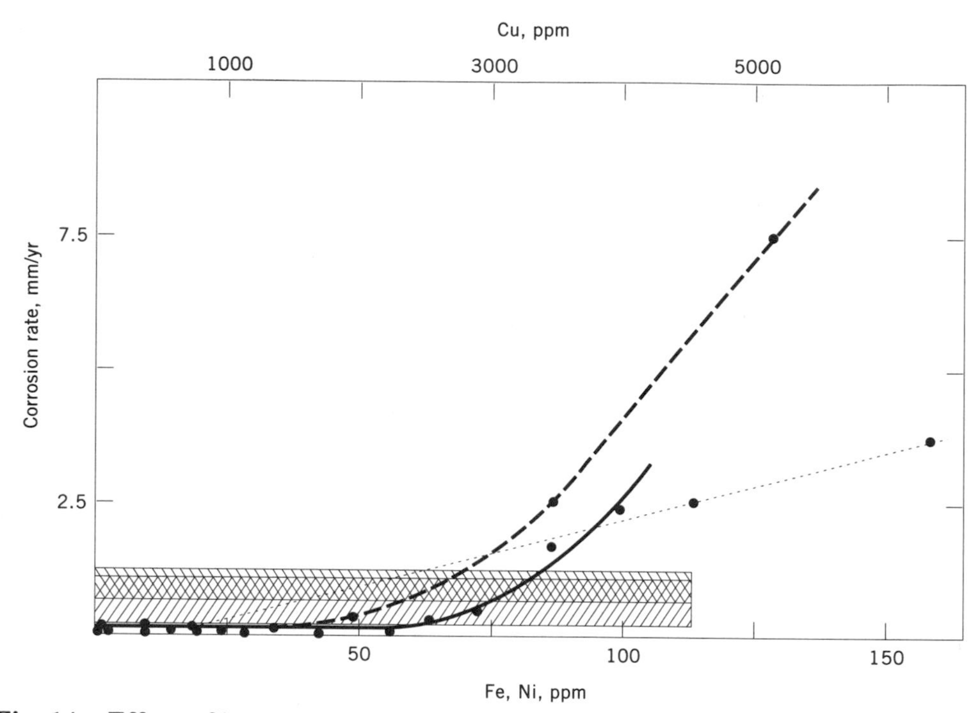

Fig. 14. Effects of iron (—), nickel (- - -), and copper (...) contaminant levels on the salt-water corrosion performance of magnesium AZ91 alloy containing 0.23% Mn. Corrosion of carbon steel (▧) and 380 die cast Al (▨) is also shown (102).

cold-rolled carbon steel, and 380 die cast aluminum alloy (125,126). As may be noted, the corrosion performance of the alloy having each of the three critical contaminants at low levels is better than that of both the carbon steel and die cast aluminum alloy which were simultaneously tested.

Since 1982, five high pressure die casting and three sand casting alloys have been introduced commercially having high purity corrosion performance. The die casting alloys are AZ91D, AM60B, AS41B, AM50A, and AE42X1 (see Table 4). The sand casting alloys are AZ91E, WE54A, and WE43A (see Table 5). The introduction of these alloys has played a significant part in the growth of the magnesium alloy business. In the North American die casting industry, alloys grew at an average compound rate of almost 18%/yr in the period 1983–1993.

Magnesium is resistant to pure alkalies in solutions of pH 10.2 or greater, even when metals cathodic to magnesium are present as impurities. Magnesium is resistant to dilute alkalies even in boiling solutions, and to 50% caustic solutions up to about 60°C. At higher temperatures there may be serious attack. Fruit juices and sour milk attack magnesium due to their acid content. Ethylene glycol [*107-21-1*] solutions are considered essentially noncorrosive at room temperature; however, the rate of attack increases with temperature and inhibitors are recommended above 100°C. Magnesium alloys have been used for oil and gasoline tanks, but pitting can result from the presence of tetraethyllead [*78-00-2*] and ethylene dibromide [*106-93-4*] when water is present, unless inhibitors are used (see CORROSION AND CORROSION CONTROL).

Table 8 indicates the compatibility of magnesium with a variety of chemicals and common substances. Because the presence of even small amounts of impurities in a chemical substance may result in significantly altered performance, a positive response in the table only means that tests under the actual service conditions are warranted (132). Other factors which may significantly alter magnesium compatibility include the presence of galvanic couples, variations in operating temperatures, alloy composition, or humidity levels.

Table 8. Behavior of Magnesium in Contact with Chemicals[a]

Chemical	Concentration, %[b]	Service test warranted[c]
acids, most	any	no
acid salts	any	no
alcohols, except methyl	100	yes
ammonia, gas or liquid	100	yes
arsenates, most	any	yes
benzene [*71-43-2*]	100	yes
brake fluids, most	100	yes
bromides, most	any	no
butter	100	no
camphor [*76-22-2*]	100	yes
carbon tetrachloride [*56-23-5*]	100	yes
carbonated water	any	no
cement	100	yes
chlorides, most	any	no
chromates, most	any	yes
chromic acid	any	yes
cyanides, most	any	yes
dry-cleaning fluids	100	yes
ethers	100	yes
fats, cooking, acid-free	100	yes
fluorides, most	any	yes
fruit juices and acids	any	no
gasoline	100	yes, if inhibited
gelatin	any	yes
glycerol, [*56-81-5*]	100	yes
hydrocarbons	100	yes
hydrofluoric acid	5–60	yes
hydrogen peroxide [*7722-84-1*]	any	no
iodides	any	no
kerosene	100	yes
lard	100	yes
lime	100	yes
methyl chloride [*74-87-3*]	100	yes
milk, fresh and sour	100	no
naphtha [*8030-30-6*]	100	yes
nitrates, all	any	no
oils, acid- and chloride-free	any	yes
oxygen	100	yes
permanganates, most	any	yes
phenol	100	yes
phosphates, most	any	yes

Table 8. (*Continued*)

Chemical	Concentration, %[b]	Service test warranted[c]
refrigerants, fluorinated	100	yes
rubber and rubber cements	100	yes
seawater	100	no
steam	100	no
sugar solutions, acid-free	any	yes
sulfur	100	yes
toluene [*108-88-3*]	100	yes
turpentine [*8006-64-2*]	100	yes
urea [*57-13-6*]	100	yes
vinegar	any	no
waxes, acid-free	100	yes

[a]Ref. 131.
[b]100% refers to pure substance in dry or liquid form; concentrations less than 100% refer to water solutions of the chemical.
[c]Resistance is indicated by yes, ie, laboratory tests have shown enough promise to warrant test under actual service conditions.

Organic compounds normally cause little or no corrosion of magnesium. Tanks or other containers of magnesium alloys are used for phenol [*108-95-2*], methyl bromide [*74-96-4*], and phenylethyl alcohol [*60-12-8*]. Most alcohols cause no more than mild attack, but anhydrous methanol attacks magnesium vigorously with the formation of magnesium methoxide [*109-88-6*]. This attack is inhibited by the addition of 1% ammonium sulfide [*12135-76-1*], or the presence of water.

Magnesium is not attacked seriously by dry chlorine [*7782-50-5*], iodine [*7553-56-2*], bromine, or fluorine [*7782-41-4*] gas. However, the presence of water promotes attack. Similarly, sulfur dioxide, ammonia, and fluorine-containing refrigerants do not attack in the absence of water.

Some tests indicate that magnesium alloys are resistant to loam soil. However, in the presence of chlorides, corrosive attack may be serious particularly if galvanic couples are present as a result of coupling to iron structures.

Galvanic Corrosion. Galvanic corrosion is an electrochemical process with four fundamental requirements: (*1*) an anode (magnesium), (*2*) a cathode (steel, brass, or graphite component), (*3*) direct anode to cathode electrical contact, and (*4*) an electrolyte bridge at the anode and cathode interface, eg, salt water bridging the adjacent surfaces of steel and magnesium components. If any one of these is lacking, the process does not occur (133,134).

The most common means of controlling galvanic attack on magnesium is through minimizing the electrochemical potential difference between the magnesium and the other metal. For steel fasteners, tin, cadmium [*7440-43-9*], and zinc electroplates have long been recognized for their ability to reduce the galvanic attack induced by the fastener when compared to bare steel. The relative effectiveness of these electroplates has generally been accepted to be in descending order as listed. Not all methods of deposition are equivalent, however. Certain proprietary zinc- and aluminum-filled polymers, as well as some ion vapor deposited aluminum coatings, applied to steel fasteners may actually produce more damage than the untreated steel fastener itself. This has been attributed to two possible causes: either the high surface area involved with each of the particulate coatings, or the contamination of the particulate surface with active cathodic contaminants such as iron, nickel, or graphite. Another point of interest is that a simple inorganic chromate treatment applied to a cadmium or zinc electroplate, used to preserve its brightness, has been found as effective in further retarding the galvanic attack on magnesium as was a coating of epoxy resin. This is consistent with the known inhibitive effects of chromate on the cathodic reduction process and has been observed with other fastener coatings as well (133–135).

In addition to selecting compatible metals or electroplates for use with magnesium assemblies, there are several other methods of controlling galvanic corrosion of magnesium parts. One method is to electrically isolate the magnesium component from the cathodic material. In many applications, however, this is not a practical alternative due to the mechanical, electrical, or cost requirements. Another alternative is to paint the cathodic material. This provides a barrier against electrolyte contact with the cathode. However, this is not easily accomplished in many cases because of the permeability of coatings by moisture and the generation of hydrogen and hydroxyl ion [*14280-30-9*] on the cathode surface. The combined effect is to strip all but the most alkali-resistant coatings. A third possibility is a design which prevents pooling of the electrolyte in critical areas of fasteners and interfaces with other cathodic components. This is an effective method of control since it both limits the electrolyte resistance and the duration of its presence because thin films of an aqueous electrolyte have both lower conductivity and higher rates of drying, so that the time of conductivity is minimized in applications where exposure to electrolytes is an intermittent event. A final alternative is to use compatible shims or washers in association with joints and fasteners. This control technique works well where exposure to the electrolyte is again intermittent and surfaces are well drained. By using a shim or washer that extends 3–6 mm beyond the interface of the cathodic material with the magnesium surface, the electrolyte path is extended to a range where the resistance of typical aqueous salt water electrolytes is sufficiently high to significantly retard the electrochemical activity. This compatible material may be a nonconductive ceramic, a polymeric material, or a compatible metal, such as an aluminum 5052 or 6061 alloy. The use of an aluminum washer beneath the head of a steel fastener that extends 3–6 mm beyond the radius of the bolt head is a simple effective method of control of the galvanic attack associated with fasteners in magnesium (132–135).

High Temperature Corrosion. The rate of oxidation of magnesium alloys increases with time and temperature. Additions of beryllium, cerium [*7440-45-1*], lanthanum [*7439-91-0*], or yttrium as alloying elements reduce the oxidation rate at elevated temperatures. Sulfur dioxide, ammonium fluoroborate [*13826-83-0*], as well as sulfur hexafluoride inhibit oxidation at elevated temperatures.

Finishing Requirements. Magnesium parts are finished by the usual steps of cleaning, chemical treatment, anodizing, electroplating, and painting (131,136) (see METAL SURFACE TREATMENTS). Cleaning may be either mechanical or chemical. Sand blasting, grit blasting, and hydroblasting are used on sand and permanent-mold castings. Sand and grit blasting must be followed by acid pickling in order to remove embedded blast media if optimum corrosion performance is desired. Grinding, sanding, wire brushing, and barrel or bowl abrading are other common mechanical cleaning methods. When dust is generated, an approved dust collector is required because mixtures of fine magnesium dust and air can burn or explode. Magnesium surfaces are cleaned by acid pickling, and numerous pickling bath compositions are available. Grease and oil are removed with solvent cleaners by dipping, vapor degreasing, or with emulsion-type cleaners. Another cleaning method is soaking in hot alkaline cleaner. Yet another is making the magnesium part to be cleaned the cathode in an alkaline bath and applying direct current of 105–403 A/m^2 at 6 V. Most magnesium alloys, unlike aluminum, are unaffected by strong alkaline cleaners, with the exception of alloy ZK60A, which is attacked in baths containing 2% or more sodium hydroxide.

Many chemical treatments are available. Some provide decorative finishes, others provide a limited degree of protection, or serve as a base for subsequent paint coats. The choice of a chemical treatment is influenced by cost, ease of application, metal loss during treatment, and durability. Traditional chemical treatments for magnesium are described in Military Specification MIL-M-3171 and ASTM D1732 (137,138). These treatments, however, are generally based on chromate solutions, which are under increasing environmental regulation. While chromate-based formulations remain the best choice overall for magnesium alloys, particularly in critical applications, they offer little advantage over selected phosphate formulations when employed with the new high purity alloys (139,140). Some effective anodic treatments are available for magnesium alloys (141): the conventional treatments are covered in Military Specification MIL-M-45202 (142), while some promising new anodizing treatments have been described (143–145). The popular chemical dip treatment, known as the ferric nitrate [*10421-48-4*] bright pickle, provides a thin decorative bright finish which has good conductivity. It consists of a simple dip, after cleaning, in a solution containing chromic acid, ferric nitrate, and potassium fluoride [*7789-23-3*]. Parts are immersed for 15 s to 3 min, depending on the amount of tarnish on the magnesium surface (131,138).

Electroplating of magnesium is used successfully in several commercial applications. Caution should be used when using these coatings on components exposed to corrosive environments or electrolytes. The process requires a preliminary dip in an aqueous solution containing zinc sulfate monohydrate [*7446-19-7*] and tetrasodium pyrophosphate [*7722-88-5*] to form a thin coating of metallic zinc on the magnesium surface. This is followed by a copper strike (a preliminary, very thin coating) after which parts may be electroplated in standard plating baths. Any metal that can be electrodeposited can be applied successfully to magnesium alloys. An electroless nickel process permits the deposition of a coating of nickel directly on magnesium by immersion in a special electroless nickel bath. No current is required. This process can be used as a final electroplated coating or as a strike coating over which standard electroplates such as bright nickel, chromium [*7440-47-3*], tin, cadmium, and zinc can be applied (see ELECTROLESS PLATING) (131,146).

Painting, like chemical treating, may be used to apply a decorative finish or as a means of protection against corrosion (see PAINT). Proper preparation of the metal surface and careful choice of priming materials are important factors for the performance of the paint coating. Primers for magnesium are based on such alkali-resistant vehicles as poly(vinyl butyral) [*9003-62-7*], polyvinyl, epoxy, polyurethane [*26778-67-6*], acrylic resins, and baked phenolic resins. Zinc chromate [*13530-65-9*] or titanium dioxide [*1317-70-0*] are often chosen as commercial pigments for the corrosion-inhibiting action they provide. Finishes are selected for their compatibility, ease of application, and performance in service. Baking rather than air drying is preferred since it improves adhesion of the primer and durability of the finish coats. Powder coatings are employed effectively for a growing number of magnesium applications. The adhesion and alkali resistance of many powder epoxy coatings are so good that in some applications parts are simply mechanically cleaned and powder coated without a chemical pretreatment (147). Porcelain enamels can be applied at relatively low temperatures to magnesium alloys (see ENAMELS, PORCELAIN OR VITREOUS). Such enamels are available in attractive colors and textures. They provide excellent adhesion and resist chemical attack and abrasion. The popular AZ31B alloy lends itself readily to the process. Porcelain enamels are not suggested for alloys that have low melting point eutectics, such as AZ61A, AZ80A, and ZK60A.

BIBLIOGRAPHY

"Magnesium and Magnesium Alloys" in *ECT* 1st ed., Vol. 8, pp. 554–593, by A. L. Tarr, Consultant; in *ECT* 2nd ed., Vol. 12, pp. 661–708, by W. H. Gross, Dow Chemical Co.; in *ECT* 3rd ed., Vol. 14, pp. 570–615, by L. F. Lockwood, G. Ansel, and P. O. Haddod, Dow Chemical USA.

1. *Magnesium: Light, Strong, Versatile*, International Magnesium Association, Dayton, Ohio, June 1978.
2. *Magnesium Characteristics and Advantages*, Bull. No. 141-310, The Dow Chemical Company, Midland, Mich., 1965.
3. H. U. Sverdrup, M. W. Johnson, and R. H. Fleming, *The Oceans*, Prentice-Hall, Inc., Englewood Cliffs, N.J., 1942, p. 176.
4. W. H. Gross, *The Story of Magnesium*, Techbook Series, 1st ed., American Society for Metals, Metals Park, Ohio, 1959.
5. R. S. Busk, *Trans. AIME* **188**, 1460 (1952).
6. R. S. Busk, *Trans. AIME* **194**, 207 (1952).
7. Kh. L. Strelets, *Electrolytic Production of Magnesium*, TT76-50003, U.S. Dept. of Commerce, Technical Information Service, Springfield, Va., translated by J. Schmorak, Keter Publishing House Jerusalem Ltd., 1977, p. 1.
8. R. S. Busk, *Magnesium Products Design*, Marcel Dekker, Inc., New York, 1987, pp. 150–151.
9. U.S. Pat. 4,298,379 (Nov. 3, 1981), A. R. Zambrano (to Hanna Mining Co.).
10. U.S. Pat. 4,800,003 (Jan. 24, 1989), J. G. Peacy and G. B. Harris (to Noranda, Inc.).
11. U.S. Pat. 5,091,161 (Feb. 25, 1992), G. B. Harris and J. G. Peacy (to Noranda, Inc.).
12. K. K. Kelly, *Energy Requirements and Equilibria in the Dehydration, Hydrolysis, and Decomposition of Magnesium Chloride*, technical paper 676, U.S. Dept. of Interior, Bureau of Mines, Washington, D.C., 1945.
13. G. J. Kipouros and D. R. Sadoway, *Advances in Molten Salt Chemistry*, Vol. 6, 1987, Elsevier, Amsterdam.
14. H. I. Kaplan, *Proceedings of the International Magnesium Association*, IMA, Washington, D.C., 1993, pp. 74–79.
15. U.S. Pat. 4,269,816 (May 26, 1981), C. E. Shackleton, A. J. Wickens, and J. H. W. Turner (to Mineral Process Licensing Corp.).

16. N. Hoy-Petersen, *J. Metals*, 43 (Apr. 1969).
17. N. Jarrett, *Metallurgical Treatises*, Metallurgical Society of AIME, Warrendale, Pa., 1981, pp. 159–169.
18. G. Mezzetta, *Light Metal Age*, **49**(5,6), 12 (June 1991).
19. U.S. Pat. 3,742,100 (June 26, 1973), O. Boyum and co-workers (to Norsk-Hydro AS).
20. U.S. Pat. 3,760,050 (Sept. 18, 1973), I. Blaker and co-workers (to Norsk-Hydro AS).
21. U.S. Pat. 3,907,651 (Sept. 23, 1975), K. A. Anderassen and K. B. Stiansen (to Norsk-Hydro AS).
22. U.S. Pat. 4,308,116 (Dec. 29, 1981), K. A. Andreassen and co-workers (to Norsk-Hydro AS).
23. U.S. Pat. 5,112,584 (May 12, 1992), G. T. Mejdell, H. M. Baumann, and K. W. Tveten (to Norsk-Hydro AS).
24. U.S Pat. 5,120,514 (June 9, 1992), K. W. Tveten, G. T. Mejdell, and J. B. Marcussen (to Norsk-Hydro AS).
25. U.S. Pat. 4,385,931 (May 31, 1983), O. Wallevik and J. B. Ronhaug (to Norsk Hydro AS).
26. E. W. Barlow, S. C. Johnson, and A. Sadan, in C. J. McMinn, ed., *Light Metals 1980*, The Metallurgical Society of AIME, Warrendale, Pa., 1979, p. 913.
27. U.S. Pat. 3,980,536 (Sept. 14, 1976), D. G. Braithwaite and W. P. Hettinger, Jr. (to NL Industries, Inc.).
28. U.S. Pat. 4,248,839 (Feb. 3, 1981), R. D. Toomey and co-workers (to NL Industries, Inc.).
29. Kh. L. Strelets, *Electrolytic Production of Magnesium*, TT76-50003, U.S. Dept. of Commerce, Technical Information Service, Springfield, Va., translated by J. Schmorak, Keter Publishing House Jerusalem Ltd., 1977.
30. A. N. Petrunko and V. S. Lobanov, *Light Metal Age*, **35**, 16, 18, 20, 21 (Oct. 1977).
31. P. A. Donskikh, Y. A. Korotkov, and E. F. Mikhailov, *Tsvetnye Metally* (Engl. trans.) **26**(6), 68–70 (1985).
32. L. N. Saburov, V. V. Teterin, E. F. Mikhailov, and A. V. Penskii, *Tsvetnye Metally* (Engl. trans.) **26**(8), 83–84 (1985).
33. I. L. Reznikov, G. Y. Sandler, V. P. Svidlo, and A. B. Krayukhin, *Tsvetnye Metally* (Engl. trans.) **26**(9), 51–53 (1985).
34. U.S. Pat. 4,058,448 (Nov. 15, 1977), K. D. Muzhzhavlev and co-workers.
35. U.S. Pat. 4,483,753 (Nov. 20, 1984), I. V. Zabelin and co-workers (to Vsesojuzny Nauchno-Issledovatelsky I Proektny Institut Titana; Ust-Kamenogorsky Titano-Magnievy).
36. O. G. Sivilotti, *Light Metal Age* **42**, 16, 18, 84 (Aug. 1984).
37. O. G. Sivilotti, in L. G. Boxall, ed., *Light Metals 1988*, The Metallurgical Society of AIME, Warrendale, Pa., 1987, p. 817.
38. U.S. Pat. 4,604,177 (Aug. 5, 1986), O. G. Sivilotti; U.S. Pat. 4,960,501 (Oct. 2, 1990), O. G. Sivilotti (to Alcan).
39. U.S. Pat. 4,647,355 (Mar. 3, 1987), H. Ishizuka; U.S. Pat. 4,495,037 (Jan. 22, 1985), H. Ishizuka.
40. U.S. Pat. 4,481,085 (Nov. 6, 1984), H. Ishizuka.
41. U.S. Pat. 4,334,975 (June 15, 1982), H. Ishizuka.
42. R. Brown, *Light Metal Age* **50**(9,10), 20 (Oct. 1992).
43. M. Burban and M. T. Frost, *Mater. Aust.* **32**(10), 18–21 (1991).
44. U.S. Pat. 4,800,003 (Jan. 24, 1989), J. G. Peacy and G. B. Harris (to Noranda, Inc.).
45. U.S. Pat. 5,091,161 (Feb. 25, 1992), G. B. Harris and J. G. Peacy (to Noranda, Inc.).
46. L. M. Pidgeon and co-workers, *Trans. Metall. Soc., AIME* **159**, 315 (1944).
47. A. Froats, in Ref. 26, pp. 969–979.
48. J. M. Toguri and L. M. Pidgeon, *Canadian J. Chem.* **40**, 1769 (1962).

49. J. M. Toguri and L. M. Pidgeon, *Canadian J. Chem.* **39**, 540 (1961).
50. C. Faure and J. Marchal, *J. Metals* **16**, 721 (1964).
51. F. Trocme, in T. G. Edgeworth, ed., *Light Metals 1971*, The Metallurgical Society of AIME, Warrendale, Pa., 1971, p. 669.
52. M. P. Lugane, *Erzmetall* **31**, 310–313 (1978).
53. R. A. Christini, in Ref. 26, pp. 981–995.
54. U.S. Pat. 4,238,223 (Dec. 9, 1980), C. Bettanini, S. Zanier, and M. Enrici (to Societa Italiana per il Magnesio e Leghe de Magnesio SPA, Bolzano, Italy).
55. U.S. Pat. 4,264,778 (Apr. 28, 1981), S. Ravelli and co-workers (to Societa Italiana per il Magnesio e Leghe de Magnesio SPA, Bolzano, Italy).
56. S. Afr. Pat. 8704237 (Dec. 17, 1987), (to Brasmag Cia Brasil).
57. U.S. Pat. 2,437,815 (Mar. 16, 1948), F. J. Hansrig (to The Permanent Metals Corp.).
58. E. F. Emley, *Principles of Magnesium Technology*, Pergamon Press, New York, 1966, p. 120.
59. Ref. 58, p. 78.
60. U.S. Pat. 2,267,862 (Dec. 30, 1941), J. D. Hanawalt, C. E. Nelson, and G. E. Holdeman (to The Dow Chemical Co.).
61. U.S. Pat. 5,147,450 (Sept. 15, 1992), B. A. Mikucki and J. E. Hillis (to The Dow Chemical Co.).
62. B. A. Mikucki and J. D. Shearouse, *Interdependence of Hydrogen and Microporosity in Magnesium Alloy AZ91*, SAE, SP-962, Detroit, Mich., Mar. 1993.
63. R. S. Busk and E. G. Boblek, *Trans. AIME* **171**, 261–276 (1947).
64. J. Keonenan and A. G. Metcalfe, *Trans. ASM* **51**, 1072–1082 (1959).
65. Ref. 58, p. 77.
66. Ref. 58, p. 81.
67. S. L. Couling, *Proceedings of the 36th Annual World Conference on Magnesium, Oslo, Norway*, June 24–28, 1979, IMA, Washington, D.C., 1979, pp. 54–57.
68. *Foundry Trade J.* **162**(3381), 944–945 (1989).
69. S. E. Housh and V. Petrovich, "Magnesium Refining: A Fluxless Alternative," *Society of Automotive Engineers International Congress and Exposition* Paper 920071, Detroit, Mich., 1992.
70. S. L. Couling, F. C. Bennett, and T. E. Leontis, *Light Metal Age* **35**(9,10), 12–21 (Oct. 1977).
71. U.S. Pat. 4,651,804 (May 24, 1987), R. Grimes and D. C. Martin (to Alcan International, Ltd.).
72. *Magnesium*, International Magnesium Association, McLean, Va., Mar. 1993.
73. J. Sirdeshpande, *Proceedings of the International Magnesium Association, Cannes, France*, IMA, Washington, D.C., 1990, pp. 24–28.
74. A. Bauer, *Proceedings of the International Magnesium Association, 31st Annual Meeting*, Paris, IMA, Washington, D.C., June 1974, pp. 37–44.
75. "Standard Specification for Magnesium Ingot and Stick for Remelting," specification B 92/B 92M-89, *Annual Book of ASTM Standards*, Vol. 2.02, American Society for Testing and Materials, Philadelphia, Pa., 1992.
76. H. I. Kaplan, *Proceedings of the International Magnesium Association*, Washington, D.C., 1993, pp. 74–79.
77. A. D. Apsher, *Appl. Spectrosc.* **32**(2), 212–215 (1978).
78. *Standard Test Methods For Chemical Analysis of Magnesium and Magnesium Alloys*, ASTM E-35 to 88, American Society for Testing and Materials, Philadelphia, Pa., 1992.
79. *Manufacture of Aluminum and Magnesium Powder*, 1987 ed., NFPA No. 651, National Fire Protection Association, Batterymarch Park, Quincy, Mass., 1987.
80. M. Jacobson, A. R. Cooper, and J. Nagy, *Explosivity of Metal Powders*, U.S. Bureau of Mines, Report of Investigations 6516, Washington, D.C., 1964.

81. *Storage, Handling and Processing of Magnesium 1987 Edition*, NFPA No. 480, National Fire Protection Association, Batterymarch Park, Quincy, Mass., 1987.
82. R. N. Lewis, Sr., *Sax's Dangerous Properties of Industrial Materials*, eighth ed., Van Nostrand Reinhold, New York, 1992.
83. S. E. Housh and J. S. Waltrip, *Safe Handling of Magnesium Alloys*, paper no. 900786, Society of Automotive Engineers, Detroit, Mich., 1990.
84. H. I. Kaplan, *Light Metal Age* **51**(3,4), 22–25 (Apr. 1993).
85. G. Chevalier, *Proceedings of the International Magnesium Association, Quebec, Canada*, IMA, Washington, D.C., 1991, pp. 32–37.
86. A. Rhomberg, *Proceedings of the International Magnesium Association*, Washington, D.C., 1993, pp. 59–61.
87. E. R. Poulsen and D. Sprayberry, *Proceedings of the International Magnesium Association, Chicago, Ill.*, IMA, Washington, D.C., 1992, pp. 79–86.
88. R. C. Bartcher, *Proceedings of the International Magnesium Association*, IMA, Washington, D.C., 1988, pp. 64–69.
89. *Magnesium, the Metal for Today*, publication of the International Magnesium Association, McLean, Va., 1991.
90. *Metals Handbook—Properties and Selection of Nonferrous Alloys and Special Purpose Materials*, 10th ed., Vol. 2, American Society for Metals, Metals Park, Ohio, 1990, pp. 455–516.
91. E. F. Emley, *Principles of Magnesium Technology*, Pergamon Press, New York, 1966.
92. *Magnesium In Design*, bull. no. 141-213, The Dow Chemical Co., Midland, Mich., 1967.
93. R. S. Busk, *Magnesium Product Design*, Marcel Dekker, Inc., New York, 1987; *Annual Book of ASTM Standards*, ASTM B80, ASTM B107, ASTM B90, American Society of Testing and Materials, Philadelphia, Pa., 1992.
94. *Annual Book of ASTM Standards*, ASTM E23-94a, Vol. 3.01, American Society of Testing and Materials, Philadelphia, Pa., 1994, pp. 140–160; ASTM E1236-91, pp. 795–800.
95. G. S. Cole, *Proceedings of the International Magnesium Association, Chicago, Ill.*, IMA, Washington, D.C., 1992, pp. 1–6.
96. S. O. Shook, *Proceedings of the International Magnesium Association, Chicago, Ill.*, IMA, Washington, D.C., 1992, pp. 44–49.
97. "Codification of Certain Nonferrous Metals and Alloys, Cast and Wrought," specification no. ANSI/ASTM B275-90, *Annual Book of ASTM Standards*, in Ref. 76.
98. "Temper Designations of Magnesium Alloys, Cast and Wrought," specification no. ANSI/ASTM B296-67, in Ref. 76.
99. "Standard Practice for Numbering Metals and Alloys," specification no. E527-83, in Ref. 76.
100. *Metals Handbook—Forming*, 8th ed., Vol. 4, American Society for Metals, Metals Park, Ohio, 1969, pp. 424–431.
101. *Heat Treating Sand and Permanent Mold Magnesium Castings*, bull. no. 141–552, The Dow Chemical Co., Midland, Mich., 1987.
102. C. S. Roberts, *Magnesium and Its Alloys*, John Wiley & Sons, Inc., New York, 1960.
103. S. L. Couling, *International Conference on Energy Conservation in Production and Utilization of Magnesium*, Massachusetts Institute of Technology, Cambridge, May 1977, pp. 29–43.
104. Ref. 92, pp. 342–418.
105. *Metals Handbook—Forging and Casting*, 8th ed., Vol. 5, American Society for Metals, Metals Park, Ohio, 1970, pp. 138–140, 433–444.
106. S. L. Couling, F. C. Bennett, and T. E. Leontis, *Proceedings of the International Magnesium Association, Columbus, Ohio*, IMA, Washington, D.C., 1977, pp. 16–20.

107. R. S. Busk, *Proceedings of the International Magnesium Association, Salt Lake City, Utah*, IMA, Washington, D.C., 1980, pp. 1–4.
108. *Molding and Core Practice for Magnesium Foundries*, bull. no. 141-29, The Dow Chemical Co., Midland, Mich., June 1957.
109. *Magnesium Properties and Applications for Automobiles*, SP962, Society of Automotive Engineers, Warrendale, Pa., 1993.
110. S. O. Shook, *Proceedings of the International Magnesium Association, Detroit, Mich.*, IMA, Washington, D.C., 1989, pp. 6–12.
111. S. C. Erickson, *Proceedings of the International Magnesium Association, Tokyo, Japan*, IMA, Washington, D.C., 1987, pp. 39–45.
112. N. L. Bradley and P. S. Frederick, *Proceedings of the International Magnesium Association, Detroit, Mich.*, IMA, Washington, D.C., 1989, pp. 63–65.
113. *Fabricating Magnesium*, Form No. 141-477, The Dow Chemical Co., Midland, Mich., 1984.
114. M. A. Molnar, *Met. Prod. News* **12**(2), 4 (1961).
115. "Magnesium and Magnesium Alloy," in *Welding Handbook*, 6th ed., Section 4, American Welding Society, Miami, Fla., 1972, Chapt. 70.
116. *Recommended Practices for Resistance Welding*, pub. AWS C1.1-66, American Welding Society, Miami, Fla., 1966.
117. "Magnesium and Magnesium Alloys," in *Brazing Manual*, 3rd rev. ed., American Welding Society, Miami, Fla., 1976, Chapt. 13.
118. *Joining Magnesium*, form 141-524, The Dow Chemical Co., Midland, Mich., 1990.
119. "Magnesium and Magnesium Alloys," in *Soldering Manual*, 2nd rev. ed., American Welding Society, Miami, Fla., 1978, Chapt. 16.
120. *Machining Magnesium*, form No. 141-480, The Dow Chemical Co., Midland, Mich., 1990.
121. M. R. Bothwell, in H. Godard, W. B. Jepson, M. R. Bothwell, and R. L. Kane, eds., *The Corrosion of Light Metals*, John Wiley & Sons, Inc., New York, 1967, pp. 259–311.
122. W. S. Loose, in H. H. Uhlig, ed., *The Corrosion Handbook*, John Wiley & Sons, Inc., New York, 1948, pp. 218–252.
123. G. D. Bengough and L. Whitby, *Trans. Inst. Chem. Eng.* **11**, 176–189 (1933).
124. "Recommended Practice for Preparing, Cleaning, and Evaluating Corrosion Test Specimens," specification G1, *Annual Book of ASTM Standards*, American Society for Testing and Materials, Philadelphia, Pa., 1992.
125. J. E. Hillis, *The Effects of Heavy Metal Contamination on Magnesium Corrosion Performance*, paper 830523, Society of Automotive Engineers, Detroit, Mich., 1983.
126. K. N. Reichek, K. J. Clark, and J. E. Hillis, *Controlling the Salt Water Corrosion Performance of Magnesium AZ91 Alloy*, paper 850417, Society of Automotive Engineers, Detroit, Mich., 1985.
127. J. E. Hillis and R. N. Reichek, *High Purity Magnesium AM60 Alloy: The Critical Contaminant Limits and the Salt Water Corrosion Performance*, paper 860288, International Congress and Exposition of the Society of Automotive Engineers, Detroit, Mich., 1986.
128. S. O. Shook and J. E. Hillis, *Composition and Performance of an Improved Magnesium AS41 Alloy*, paper 890205, International Congress and Exposition of the Society of Automotive Engineers, Detroit, Mich., 1989.
129. W. E. Mercer and J. E. Hillis, *The Critical Contaminant Limits and Salt Water Corrosion Performance of Magnesium AE42 Alloy*, paper 920073, Society of Automotive Engineers, Detroit, Mich., 1992.
130. J. D. Hanawalt, C. E. Nelson, and J. A. Peloubet, *Trans. AIME*, **147**, 273–299 (1942).
131. *Dow Magnesium Operations in Magnesium Finishing*, Form No. 141-479-90HYC, The Dow Chemical Co., Midland, Mich., 1990.

132. "Magnesium: Designing Around Corrosion," Form No. 141-396-R87, The Dow Chemical Co., Midland, Mich., 1987.
133. D. L. Hawke, "Galvanic Corrosion of Magnesium," paper G-T87-004, *Society of Die Casting Engineers 14th International Die Casting Congress and Exposition*, Toronto, Canada, 1987.
134. D. L. Hawke, J. E. Hillis, and W. Unsworth, *Preventive Practice for Controlling the Galvanic Corrosion of Magnesium Alloys*, International Magnesium Association, McLean, Va., 1988.
135. D. Hawke, J. Davis, and R. Fekete, *Field Corrosion Performance of Magnesium Powertrain Components in Light Truck*, Technical Paper 890206, Society of Automotive Engineers, Detroit, Mich., 1989.
136. R. W. Murray and J. E. Hillis, "Magnesium Finishing: Chemical Treatment and Coating Practices," *Society of Automotive Engineers International Congress and Exposition*, Paper 900791, Detroit, Mich., 1990.
137. *Magnesium Alloy, Processes for Pretreatment and Prevention of Corrosion On*, Military Specification MIL-M-3171, Dept. of Navy, Naval Air Engineering Center, Philadelphia, Pa., July 11, 1966.
138. "Standard Practices for Preparation of Magnesium Alloy Surfaces for Painting," standard D1732, in Ref. 76.
139. J. E. Hillis and R. W. Murray, "Finishing Alternatives for High Purity Magnesium Alloys," paper G-T87-003, in Ref. 134.
140. D. Hawke and K. Gaw, "Effects of Chemical Surface Treatments on the Performance of an Automotive Paint System on Die Cast Magnesium," Paper 920074, *Society of Automotive Engineers International Congress and Exposition*, Detroit, Mich., 1992.
141. G. R. Kotler, D. L. Hawke, and E. N. Aqua, *Proceedings of the International Magnesium Association, 33rd Annual Meeting*, Montreal, Quebec, Canada, May 1976, pp. 45–48.
142. *Magnesium Alloys, Anodic Treatment Of*, Military Specification MIL-M-45202, Dept. of the Army, Army Materials and Mechanics Research Center, Watertown, Mass., Oct. 3, 1968.
143. D. E. Bartak, T. D. Schleisman, and E. R. Woolsey, "Electrodeposition and Characteristics of a Silicon–Oxide Coating for Magnesium Alloys," Paper T-91-041, *North American Die Casting Association 16th International Die Casting Congress and Exposition*, Detroit, Mich., 1991.
144. J. H. Hawkins, *Proceedings of the International Magnesium Association*, Washington, D.C., 1993, pp. 46–58.
145. R. E. Brown, *Light Metal Age* **51**(3,4), 14–17 (Apr. 1993).
146. R. Ellmers and D. Maguire, *Proceedings of the International Magnesium Association*, Washington, D.C., 1993, pp. 28–33.
147. R. W. Murray and J. E. Hillis, "Powder Coatings on High Purity Die Casting Magnesium for Appearance and Performance," Paper T91-012, in Ref. 145.

CLIFFORD B. WILSON
KEN G. CLAUS
MATTHEW R. EARLAM
JAMES E. HILLIS
The Dow Chemical Company

MAGNESIUM COMPOUNDS

Magnesium is the eighth most abundant element in the earth's crust and the third most abundant element in seawater. More than sixty magnesium-containing minerals are known. The most important can be divided into three classes according to their commercial importance: the carbonates magnesite [*1317-61-9*], $MgCO_3$, and dolomite [*17069-72-6*], $CaCO_3 \cdot MgCO_3$; the salts and double salts found in oceanic deposits, including carnallite [*1318-27-0*], $KCl \cdot MgCl_2 \cdot 6H_2O$; kieserite, $MgSO_4 \cdot H_2O$; langbeinite, $K_2SO_4 \cdot 2MgSO_4$; and epsomite, $MgSO_4 \cdot 7H_2O$; and the silicates such as olivine [*1317-71-1*], (Mg,Fe)2SiO_4; serpentine [*12168-92-2*] (chrysotile), $Mg_6(OH)_6Si_4O_{11} \cdot H_2O$; antigorite [*12135-86-3*], $Mg_6(OH)_8Si_4O_{10}$; talc [*14807-96-6*], $Mg_3(OH)_2Si_4O_{10}$; and sepiolite [*15501-74-3*], $Mg_3Si_4O_{11} \cdot 4H_2O$. Magnesium also occurs as the hydroxide brucite [*1317-43-7*], $Mg(OH)_2$, and in combination with aluminum as spinel [*1302-67-6*], $MgAl_2O_4$. There are hundreds of magnesium compounds known, varying from chlorophyll [*1406-65-1*] to asbestos [*1332-21-4*]. Only those compounds produced commercially are discussed herein.

Magnesium Acetate

Anhydrous magnesium acetate [*142-72-3*], a white, crystalline, deliquescent solid, occurs in two forms: α-$Mg(C_2H_3O_2)_2$, formed by the reaction of MgO and concentrated acetic acid (13–33%) in boiling ethyl acetate, and β-$Mg(C_2H_3O_2)_2$ which is formed using 5–6% acetic acid. Of commercial interest is magnesium acetate tetrahydrate [*16674-78-5*], $Mg(C_2H_3O_2)_2 \cdot 4H_2O$, a colorless to white crystalline solid obtained from aqueous solution. The tetrahydrate is the only stable phase below 68°C, the transition point of the anhydrous salt. A monohydrate [*60582-92-5*], $Mg(C_2H_3O_2)_2 \cdot H_2O$, can be prepared from the reaction of MgO and acetic acid in slightly hydrated isobutyl alcohol. Physical properties of magnesium acetate and its hydrates are given in Table 1.

The solubility of magnesium acetate is shown in Figure 1. Aqueous solutions of magnesium acetate are characterized by very high viscosities that are generally attributed to acetate association. As the concentration of anhydrous magnesium acetate rises from 20.1 to 43.6%, the relative viscosity increases 45-fold at 25°C (4,5). Anhydrous magnesium acetate may be prepared by heating the tetrahydrate to 134°C. The anhydrous salt melts at about 320°C, though decomposition starts at about 300°C. Upon decomposition, the residue is magnesium oxide and the volatile products include acetone, acetic acid, carbon dioxide, and steam.

Uses. The largest use for magnesium acetate is in the production of rayon fiber, which is used for cigarette filter tow (see FIBERS, REGENERATED CELLULOSICS). During the acetylation of cellulose (qv), sulfuric acid is present as a catalyst and some cellulose is sulfated. In a subsequent hydrolysis step, the sulfate cellulose hydrolyzes first. Sulfuric acid is evolved and neutralized by the magnesium acetate, presumably through the formation of magnesium sulfate (see CELLULOSE ESTERS).

Magnesium acetate also has uses as a dye fixative in textile printing, as a deodorant, disinfectant, an antiseptic in medicine, and as a reagent chemical

Table 1. Physical Properties of Magnesium Acetates[a]

Property	α-Mg-$(C_2H_3O_2)_2$	β-Mg-$(C_2H_3O_2)_2$	$Mg(C_2H_3O_2)_2\cdot H_2O$	$Mg(C_2H_3O_2)_2\cdot 4H_2O$	β-Mg-$(C_2H_3O_2)_2\cdot 4H_2O$
mol wt	142.40	142.40	160.38	214.46	214.46
crystal system	orthorhombic	triclinic	orthorhombic	monoclinic	monoclinic
space group	$P2_12_12_1$		$P2_1cn$ or Pmcn	$P2_1/A$	
lattice constants, nm					
a	1.127	1.034	1.175	0.8550	1.296
b	1.501	1.295	1.753	1.1995	0.7647
c	1.100	0.7726	0.6662	0.4807	1.017
angle, degree					
α		112.02			
β		94.53		95.37	113.84
γ		95.80			
Z^b	12	6	8	2	4
density, g/cm^3					
calculated	1.524		1.553	1.453	1.545
observed	1.507	1.502		1.454	
mp, °C	323 dec			80	
color	white	white			

[a]Refs. 1–3.
[b]Number of formulas per unit cell.

(see DISINFECTANTS AND ANTISEPTICS; TEXTILES) (6). In the United States, Hoescht-Celanese and Tennessee Eastman are the principal producers of magnesium acetate. These companies make about 36,000 t/yr, which is largely used in-house for the production of cellulose acetate.

Handling and Safety. Magnesium acetate is hygroscopic and should be stored in a cool, dry place. Personal protective equipment to be used when handling magnesium acetate includes chemical safety goggles, chemical resistant gloves, and a NIOSH/MSHA approved respirator. To keep exposure to respirable dust to a minimum, mechanical exhaust is required. Although magnesium acetate is a relatively low hazard chemical, intravenous poisoning can occur if this material is not handled properly. Magnesium acetate is incompatible with strong oxidizers. When heated to decomposition, acrid smoke and irritating fumes may evolve (7,8).

Magnesium Alkyls

Magnesium alkyl compounds RMg, RMgR, or RMgR′, along with other compounds are useful as polymerization catalysts (see ORGANOMETALLICS) (9). These compounds should not be confused with alkyl magnesium halides or the much discussed ether solvated Grignard reagents. Magnesium alkyls may, however, be prepared from Grignard reagents (see GRIGNARD REACTIONS) (10).

Properties. Magnesium alkyls are white, crystalline, pyrophoric solids that react vigorously with water, alcohols, and other compounds containing an

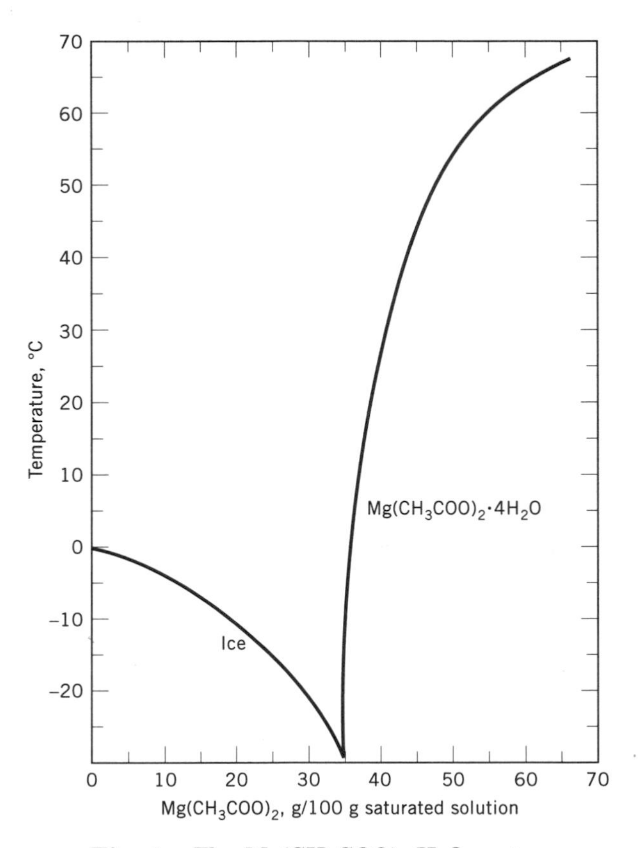

Fig. 1. The $Mg(CH_3COO)_2 \cdot H_2O$ system.

active hydrogen (11). Magnesium alkyls, soluble in ether solutions but insoluble in benzene and some alkane solutions, decompose at 170–200°C (12). The molecular weights of unsolvated compounds fall in the range of 100–200, but the molecular weights in solution, as determined by cryoscopic methods, are in the range of 1,000–10,000. The low solubility and high molecular weights in solution are attributed to extensive association resulting from the electron-deficiency of the magnesium (13).

Preparations. Magnesium alkyls may be prepared from a Grignard reagent according to the following disproportionation reaction:

$$2\ RMgX + \text{dioxane} \longrightarrow R_2Mg + MgX_2 \qquad (1)$$

Magnesium halide and alkyl magnesium halide precipitate and the alkyl magnesium compound remains in solution. Filtration (qv) followed by drying the filtrate yields solid magnesium alkyl (11). Another preparation method is that of metal exchange using mercury alkyl in ether.

$$RHgR + Mg \longrightarrow RMgR + Hg \qquad (2)$$

Magnesium alkyls can also be prepared by reaction of alkyl iodide and a calcium–magnesium alloy in ether.

Reactions. The most noted magnesium alkyl reactions involve the solvated Grignard reagents. The more common reactions involving unsolvated magnesium alkyl are as follows (14,15).

Oxidation	$RMg + O_2 \rightleftharpoons RMg{\cdot}O_2$	(3)
Hydrolysis	$R{=}RMg + 2\ H_2O \longrightarrow Mg(OH)_2 + \text{alkane}$	(4)
Peroxide addition	$RMg + ROOMg \longrightarrow 2\ ROMg$	(5)
Polymer formation	$2\ RMgR + \text{heat} \longrightarrow RMgRC{=}CH_2Mg + \text{alkane}$	(6)
Alkane formation	$RMgR \longrightarrow \text{alkane} + RC{=}CMg$	(7)
Magnesium hydride formation	$RMgR + \text{heat} \longrightarrow MgH_2 + \text{alkene}$	(8)

Uses. Magnesium alkyls are used as polymerization catalysts for alpha-alkenes and dienes, such as the polymerization of ethylene (qv), and in combination with aluminum alkyls and the transition-metal halides (16–18). Magnesium alkyls have been used in conjunction with other compounds in the polymerization of alkene oxides, alkene sulfides, acrylonitrile (qv), and polar vinyl monomers (19–22). Magnesium alkyls can be used as a liquid detergents (23). Also, magnesium alkyls have been used as fuel additives and for the suppression of soot in combustion of residual furnace oil (24).

Economic Aspects. Annual worldwide production of magnesium alkyls is <300 t. *n*-Butylethylmagnesium [*62202-86-2*], di-*n*-butylmagnesium [*1194-47-5*], di-*n*-hexylmagnesium [*37509-99-2*], *n*-butyl-*n*-octylmagnesium [*69929-18-6*], and *n*-butyl-*sec*-butylmagnesium [*39881-32-8*] are all commercially available (25). It is possible to purchase magnesium alkyls as a solution in heptane for approximately \$8–11/mL. The price is dependent on the cost of magnesium turnings (see MAGNESIUM AND MAGNESIUM ALLOYS) (26).

Magnesium Bromide

Occurrence. Magnesium bromide [*7789-48-2*], $MgBr_2$, is found in seawater, some mineral springs, natural brines, inland seas and lakes such as the Dead Sea and the Great Salt Lake, and salt deposits such as the Stassfurt deposits. In seawater, it is the primary source of bromine (qv). By the action of chlorine gas upon seawater or seawater bitterns, bromine is formed (see CHEMICALS FROM BRINE).

$$MgBr_2 + Cl_2 \longrightarrow MgCl_2 + Br_2 \tag{9}$$

Properties. Magnesium bromide hexahydrate [*13446-53-2*], $MgBr_2{\cdot}6H_2O$, which crystallizes from an aqueous solution at temperatures above 0°C, is highly hygroscopic and isomorphous with $MgCl_2{\cdot}6H_2O$. It is also formed by the reaction of magnesium carbonate and hydrobromic acid (27). Physical properties of anhydrous magnesium bromide and the hexahydrate are shown in Table 2; solubilities are shown in Figure 2 (31). The solubility of magnesium bromide is

Table 2. Physical Properties of Magnesium Bromide and Magnesium Bromide Hexahydrate[a]

Property	$MgBr_2$	$MgBr_2 \cdot 6H_2O$
mol wt	184.13	292.22
crystal system	hexagonal	monoclinic
space group	$P\bar{3}m1$	C_2/m
lattice constants, nm		
a	0.3822	1.0286
b		0.7331
c	0.6269	0.6211
angle, β degree		93.34
Z[b]	1	2
density, calculated, g/cm^3	3.855	2.076
mp, °C	711	174.2
color	white	colorless
heat of formation, ΔH_{298}, kJ/mol[c]	−524.3	−2410.0
free energy of formation, ΔG_{298}, kJ/mol[c]	−503.8	2056.0

[a]Refs. 28–30.
[b]Number of formulas per unit cell.
[c]To convert J to cal, divide by 4.184.

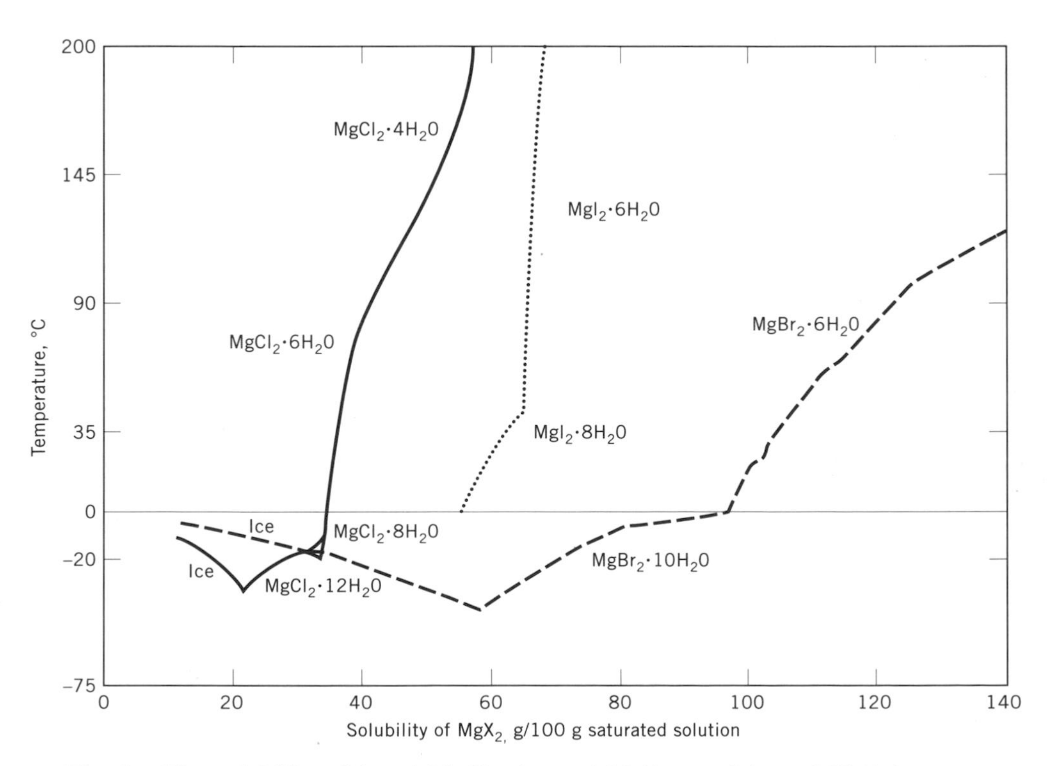

Fig. 2. The solubility of (——) $MgCl_2$, (– – –) $MgBr_2$, and (······) MgI_2 in water.

101 g/100 mL of water at 20°C; the solubility of the hexahydrate is 160 g/100 mL of 95% ethanol at 20°C (29).

Magnesium bromide is soluble in alcohols and forms addition compounds with numerous organic substances such as alcohols. The hexamethanolate, $MgBr_2 \cdot 6CH_3OH$, and the ethanolate, $MgBr_2 \cdot C_2H_5OH$, both exist. By gradually adding bromine to a cold mixture of magnesium powder and dry ether, the dietherate of magnesium bromide diethylether [*17950-53-7*], $MgBr_2 \cdot 2C_4H_{10}O$, is formed. Other compounds form with ammonia. For example, the compound magnesium bromide hexammoniate [*75198-46-8*], $MgBr_2 \cdot 6NH_3$, is easily prepared from anhydrous $MgBr_2$ and ammonia gas. Thermal decomposition of the hexammoniate yields the diammoniate [*75198-47-9*], $MgBr_2 \cdot 2NH_3$, and monoammoniate [*75198-48-0*], $MgBr_2 \cdot NH_3$.

Magnesium bromide forms double salts with potassium bromide, KBr, and ammonium bromide, NH_4Br, which are isomorphous with carnallite, $KCl \cdot MgCl_2 \cdot 6H_2O$. Basic bromides such as magnesium bromide oxide dodecahydrate [*75300-54-8*], $MgBr_2 \cdot 3MgO \cdot 12H_2O$, and magnesium bromide oxide hexahydrate [*75300-55-9*], $MgBr_2 \cdot 3MgO \cdot 6H_2O$, have been reported (32).

Uses and Economic Aspects. Magnesium bromide is used in medicine as a sedative in treatment of nervous disorders, in electrolyte paste for magnesium dry cells, and as a reagent in organic synthesis reactions. The price of magnesium bromide hexahydrate in January 1995 was $5.51/kg (33).

Magnesium Carbonate

Occurrence. Chemical reactions in the system $MgO-CO_2-H_2O$ result in a series of normal carbonates which include the following mineral species: magnesite [*13717-00-5*], $MgCO_3$; barringtonite [*5145-48-2*], $MgCO_3 \cdot 2H_2O$; nesquehonite [*14457-83-1*], $MgCO_3 \cdot 3H_2O$; and lansfordite [*61042-72-6*], $MgCO_3 \cdot 5H_2O$. These reactions also produce a series of basic, ie, hydroxyl-containing, magnesium carbonates having the general formula $x MgCO_3 \cdot y Mg(OH)_2 \cdot z H_2O$. This basic carbonate series includes artinite [*12143-96-3*], $MgCO_3 \cdot Mg(OH)_2 \cdot 3H_2O$; hydromagnesite [*12072-90-1*], $4MgCO_3 \cdot Mg(OH)_2 \cdot 4H_2O$; dypingite [*12544-02-4*], $4MgCO_3 \cdot Mg(OH)_2 \cdot 5H_2O$ (34); and an unnamed octahydrate [*75300-49-1*], $4MgCO_3 \cdot Mg(OH)_2 \cdot 8H_2O$ (35,36). Hydromagnesite is the most stable of the basic carbonates at ambient temperatures, humidities, and partial pressures of CO_2.

In its natural form magnesite is a member of the calcite group of rhombohedral carbonates. It is the most common species of the naturally occurring magnesium carbonate minerals. The other magnesium carbonate species are rare in occurrence compared to magnesite (37). Because of the similar sizes of the ionic radii of magnesium and iron, magnesite forms a complete solid solution series with the mineral siderite [*14476-16-5*], $FeCO_3$. The mineral which forms the species intermediate between magnesite and siderite is known as breunnerite, $(Mg,Fe)CO_3$.

Properties. The physical properties of the normal magnesium carbonates are given in Table 3, those of the basic magnesium carbonates in Table 4. Magnesium carbonate is insoluble in CO_2-free water. The solubility products, K_{sp}, for magnesium carbonate and some hydrates follow (38).

Table 3. Physical Properties of Magnesium Carbonates[a]

Parameter	Magnesite	Barringtonite	Nesquehonite	Lansfordite
mol wt	84.32	120.35	138.37	174.4
crystal system	hexagonal	triclinic	monoclinic	monoclinic
space group	R3c		$P2_1/n$	$P2_1/n$
lattice constants, nm				
a	0.46332	0.9115	1.2112	
b		0.6202	0.539	
c	1.5015	0.6092	0.77697	
angle, degree				
α		94.00		
β		95.53	90.42	
γ		108.87		
Z[b]	6	4	4	
density, calculated, g/cm^3	3.009	2.825	1.837	1.730
hardness, Mohs'	3.5–5.0		2.5	2.5
color	white	colorless	colorless to white	white
melting point, °C	402–480[c]			
index of refraction	1.510, 1.700	1.458, 1.473, 1.501	1.412, 1.501, 1.526	1.456, 1.476, 1.502
heat of formation, ΔH_{298}, kJ/mol[d]	−1095.8			
free energy of formation, ΔG_{298}, kJ/mol[d]	−1012.1		−1726.3	−2199.5

[a]Refs. 3,30, and 37–41.
[b]Number of formulas per unit cell.
[c]Material decrepitates.
[d]To convert J to cal, divide by 4.184.

Table 4. Physical Properties of Basic Magnesium Carbonates[a]

Parameter	Artinite	Hydromagnesite	Dypingite	Octahydrate
mol wt	196.70	467.67	485.69	539.74
crystal system	monoclinic	monoclinic	monoclinic	
space group	C_2 or C_2/M	$P2_1/C$		
lattice constants, nm				
a	1.656	1.011		
b	0.315	0.894		
c	0.622	0.838		
angle, β degree	99.15	114.58		
Z[b]	2	2		
density, calculated, g/cm^3	2.039	2.254		
hardness, Mohs'	2.5	3.5		
color	white	white	white	white to gray
index of refraction	1.488, 1.534, 1.556	1.458, 1.473, 1.501	1.412, 1.501, 1.526	1.456, 1.476, 1.502

[a]Refs. 3,30 and 37–41.
[b]Number of formulas per unit cell.

Compound	Solubility product, K_{sp}
magnesite	1.0×10^{-5}
$MgCO_3 \cdot H_2O$	2.7×10^{-5}
barringtonite	2.3×10^{-5}
nesquehonite	8.9×10^{-6}

The relatively high solubility of magnesium bicarbonate, $Mg(HCO_3)_2$, at low temperature and high partial pressures of CO_2 permits the separation of magnesium compounds from impurities. The high degree of solubility of magnesium bicarbonate was first utilized in the mid-nineteenth century in the Pattison process to prepare pure magnesium compounds from calcined dolomite [*17069-72-6*] (42).

Normal, hydrated, and the basic magnesium carbonates react with acids to yield salts which can be recovered by crystallization (43). Magnesium carbonate forms many double salts, some of which are listed in Table 5.

Because of this solid solution, natural magnesite contains varying amounts of iron which can affect the ultimate use of the material. Small amounts of calcium and manganese may also be present.

Production. Naturally occurring magnesite is widely distributed throughout the earth's crust and is used as a starting raw material for the production of magnesia, MgO, and other magnesium compounds (45,46). It occurs as crystalline magnesite, and in the cryptocrystalline forms. Economically exploitable deposits of natural magnesium carbonate occur in many countries. The world's leading producers of natural magnesite in 1991 included Australia, 231,000 metric tons; Austria, 667,000 t; Brazil, 245,000 t; Canada, 174,000 t; China, 1,050,000 t; Greece, 240,000 t; India, 257,000 t; Russia, 2,205,000 t; Spain, 205,000 t; Slovakia, 726,000 t; Turkey, 309,000 t; Ukraine, 1,345,000 t; United States, 100,000 t; the Yugoslav republics, 240,000 t; and the various republics of the Commonwealth of Independent States of the former Soviet Union (44). North Korea is reported to have a production capacity of 2,000,000 t/yr. In the Americas, natural magnesite deposits are currently being mined at Gabbs, Nevada; Mount Brussilof, British Columbia; and Brumado, Bahia, Brazil (47). Deposits of magnesite are also known to exist in Washington state and in the province of Quebec.

Benefication. The purity of naturally occurring magnesite ores are quite variable. The ores can range from nearly pure magnesite to magnesite intermixed with other carbonate mineral species, quartz, clay minerals, and other

Table 5. Double Salts of Magnesium Carbonate[a]

Salt	CAS Registry Number	Salt	CAS Registry Number
$MgCO_3 \cdot MgCl_2 \cdot 7H_2O$	[*11140-13-9*]	$MgCO_3 \cdot K_2CO_3 \cdot 8H_2O$	[*75198-51-5*]
$2MgCO_3 \cdot MgBr_2 \cdot 8H_2O$	[*75198-49-1*]	$MgCO_3 \cdot KHCO_3 \cdot 4H_2O$	[*19154-48-4*]
$MgCO_3 \cdot MgBr_2 \cdot 7H_2O$	[*75198-50-4*]	$MgCO_3 \cdot Rb_2CO_3 \cdot 4H_2O$	[*75198-52-6*]
$MgCO_3 \cdot NH_4CO_3 \cdot 4H_2O$	[*22450-55-1*]	$MgCO_3 \cdot Na_2CO_3$	[*19086-68-1*]

[a]Ref. 44.

aluminosilicates. The purity of the mined ore can be increased by several processes of mineral benefication (see MINERAL RECOVERY AND PROCESSING). Selective mining practices can produce a relatively good quality of industrial-grade magnesium carbonate. Ores can also be upgraded by means of heavy media separation. High quality magnesium carbonate is produced by means of froth flotation (qv) (48).

Manufacture. Synthetic forms of magnesium carbonate and basic magnesium carbonates can be produced by the carbonation of magnesium hydroxide slurries. Anhydrous magnesium carbonate, $MgCO_3$, can be prepared in aqueous systems only under very high partial pressures of carbon dioxide (qv), CO_2 (49). Magnesium carbonate trihydrate, synthetic nesquehonite, is formed under normal conditions of temperature and pressure regardless of the methods of preparation (50). If the carbonation of magnesium hydroxide slurry is carried out at 345–517 kPa (50–75 psi) and at temperatures below 50°C, the more soluble bicarbonate is formed and is then separated from the other reaction products by means of filtration with subsequent crystallization. The trihydrate form is recovered from the filtrate by decarbonation under vacuum or by filtration. Boiling the bicarbonate solution or drying the trihydrate precipitate at about 100°C produces the basic magnesium carbonate $4MgCO_3 \cdot Mg(OH)_2 \cdot 4H_2O$, a synthetic version of the naturally occurring mineral hydromagnesite.

Hydrated and basic magnesium carbonates can also be produced from calcined dolomite, $CaO \cdot MgO$, or half calcined dolomite, $MgO \cdot CaCO_3$, by means of slaking, carbonation, filtration, and decarbonation. Magnesium carbonate trihydrate can be produced from calcined dolomite dust by means of the Judd process (51). In this process the calcined dolomite kiln dust is first slaked to hydrate the CaO. Deliming of the solution is achieved through precarbonation. The magnesium-rich solution is then carbonated, pressure carbonated, decarbonated, filtered, and finally dried.

The principal producers of synthetic normal magnesium carbonates and basic magnesium carbonates are J.T. Baker Inc., Philipsburg, New Jersey; GTE Corp., Sylvannia Chemicals Division, Towanda, Pennsylvania; Mallinkrodt Specialty Chemicals, St. Louis, Missouri; Marine Magnesia, San Francisco, California; Morton Specialty Chemicals, Manistee, Michigan; and Ube Chemical Industries, Tokyo.

Economic Aspects. The 1991 world production of natural magnesites was 9.97 million metric tons; in 1992 world production of natural magnesite and other minor magnesium carbonates was estimated to be 12.5 million metric tons (52). In 1990 the United States consumed 2821 metric tons of synthetic, precipitated magnesium carbonate with a value of $703,000 (52). The prices quoted in July 1993, for light technical-grade magnesium carbonate ranged from $0.73 to $0.78 per pound and for USP grade the price range was $0.74 to $0.80 per pound (53).

Uses. Thermal decomposition of any of the species of normal and hydrated magnesium carbonates produces magnesia, MgO, the physical and chemical properties of which are determined by the purity of the original natural magnesite, the method of magnesite ore benefication, the temperature, and degree of calcination. Many of the Austrian magnesite deposits contain breunnerite, $(Mg,Fe)CO_3$. The presence of up to 6% iron as Fe_2O_3 limits the use of this form of natural magnesite to the production of low quality magnesia refractory brick. The physical properties of magnesia prepared by calcination of basic magnesium carbonate are available (54). Natural magnesites are calcined at temperatures approaching 2100°C (dead burned) for the production of synthetic periclase, MgO, a refractory material used in the steel industry. Magnesites that are calcined at 500–550°C (light burned) find use as magnesium supplements for animal feed (see FEEDS AND FEED ADDITIVES), as components of oxysulfate cement (qv), in sugar (qv) processing, and in the flue-gas scrubbing of gaseous SO_2 emissions (see SULFUR REMOVAL AND RECOVERY).

Natural magnesites are used in the pollution control industry as acid neutralizing agents and in gaseous SO_2 scrubbers (see AIR POLLUTION CONTROL METHODS). The only processing that the natural magnesite receives for these applications is crushing and screening to a uniform size.

Precipitated magnesium carbonate and basic magnesium carbonates are calcined to produce magnesia having surface areas of ca 200 m^2/g. These high surface area magnesias are used as thickening agents in the sheet molding of rubber and as scorch retarders in chloroprene rubbers (see RUBBER, COMPOUNDING). Precipitated carbonates are useful as extenders in paint (qv) and pigments (qv). Specially prepared spherical magnesium carbonate finds use as an extender in spray paints (55). Crystalline magnesium carbonate has found use in various optical applications. Precipitated carbonates are used in smoke suppressants, fire-extinguishing compounds, flooring products, ceramics (qv), specialty inks (qv) for high speed printers, and as granular filtering media (see FLAME RETARDANTS).

USP-grade anhydrous magnesium carbonate is used as a flavor impression intensification vehicle in the processed food industry (see FLAVORS AND SPICES). Basic magnesium carbonates are used as free flowing agents in the manufacture of table salt, as a bulking agent in powder and tablet pharmaceutical formulations, as an antacid, and in a variety of personal care products (see PHARMACEUTICALS).

Health and Safety Factors. Magnesium carbonates and its basic hydrated forms have minimal toxicological effects when encountered at normal exposure levels. However, in response to the possible adverse effects of long-term exposure to magnesium carbonate dust, the ACGIH has established a TLV–TWA of 10 mg/m^3 for magnesite dust that contains no asbestos (qv) fibers and less than 1% free silica (56). These compounds are best utilized within a conscientiously applied program of industrial hygiene (qv).

Magnesium Chloride

Properties. Magnesium chloride [*7786-30-3*], $MgCl_2$, is one of the primary constituents of seawater and occurs in most natural brines and salt deposits formed from the evaporation of seawater. It occurs infrequently in nature as the mineral bischofite [*13778-96-6*], $MgCl_2{\cdot}6H_2O$. Large deposits of oceanic origin contain the mineral carnallite [*1318-27-0*], $KCl{\cdot}MgCl_2{\cdot}6H_2O$. Magnesium chloride, one of the most commercially important magnesium compounds, is available in the anhydrous and hexahydrate forms. Both are deliquescent and form saturated solutions on standing in a moist atmosphere.

The physical properties of these compounds are given in Table 6. The step by step thermal dehydration of the hexahydrate goes through various hydrates (57–59): from 95–115°C $MgCl_2 \cdot 6H_2O \rightarrow MgCl_2 \cdot 4H_2O + 2\ H_2O$, $MgCl_2 \cdot 4H_2O \rightarrow MgCl_2 \cdot 2H_2O + 2\ H_2O$; from 135–180°C $MgCl_2 \cdot 4H_2O \rightarrow Mg(OH)Cl + HCl + 3\ H_2O$, $MgCl_2 \cdot 2H_2O \rightarrow MgCl_2 \cdot H_2O + H_2O$; from 185–230°C $MgCl_2 \cdot 2H_2O \rightarrow Mg(OH)Cl + HCl + H_2O$; and >230°C $MgCl_2 \cdot H_2O \rightarrow MgCl_2 + H_2O$, $MgCl_2 \cdot H_2O \rightarrow Mg(OH)Cl + HCl$, $Mg(OH)Cl \rightarrow MgO + HCl$.

Magnesium chloride also forms hydrates containing 8 and 12 molecules of water of hydration. The solubility for $MgCl_2$ in water is shown in Figure 2 (31) from which it can be seen that the hexahydrate is the only stable hydrate in the range of temperatures from 0 to 100°C.

Anhydrous magnesium chloride is soluble in lower alcohols. In 100 g of methanol, its solubility is 15.5 g at 0°C and 20.4 g at 60°C. In ethanol, the solubility is 3.61 g at 0°C and 15.89 g at 60°C. Upon cooling, anhydrous $MgCl_2$ forms addition compounds with alcohols of crystallization such as magnesium chloride hexamethanolate [*57467-93-0*], $MgCl_2 \cdot 6CH_3OH$, and magnesium chloride hexaethanolate [*16693-00-8*], $MgCl_2 \cdot 6C_2H_5OH$. Both of these alcoholates are deliquescent.

Magnesium chloride forms double salts with potassium and ammonium chlorides. Carnallite is an important source of $MgCl_2 \cdot 6H_2O$. Magnesium chloride hexammoniate [*68374-23-2*], $MgCl_2 \cdot 6NH_3$, can be formed by the reaction of anhydrous magnesium chloride and ammonia gas in a closed system, or by the action of ammonia on an aqueous solution of $MgCl_2$ and ammonium chloride, NH_4Cl, upon subsequent cooling to −30°C. Thermal decomposition of $MgCl_2 \cdot 6NH_3$ yields the diammoniate [*68374-24-3*], $MgCl_2 \cdot 2NH_3$, and the monoammoniate [*68374-25-4*], $MgCl_2 \cdot NH_4$.

Table 6. Physical Properties of Magnesium Chloride[a]

Property	$MgCl_2$	$MgCl_2 \cdot 6H_2O$
mol wt	95.22	203.31
crystal system	hexagonal	monoclinic
space group	$R\bar{3}m$	C_2/m
lattice constants, nm		
a	0.3632	0.9871
b		0.7113
c	1.7795	0.6079
angle, β degree,		93.74
Z^b	3	2
density, calculated, g/cm^3	2.333	1.585
mp, °C	708	116–118 dec
index of refraction	1.675, 1.59	1.498, 1.505, 1.525
color	white lustrous	colorless
heat of formation, ΔH_{298}, kJ/mol[c]	−641.3	−2499.0
free energy of formation, ΔG_{298}, kJ/mol[c]	−591.8	−2115.0

[a]Refs. 28–30
[b]Number of formulas per unit cell.
[c]To convert J to cal, divide by 4.184.

Preparation and Manufacture. Magnesium chloride can be produced in large quantities from (*1*) carnallite or the end brines of the potash industry (see POTASSIUM COMPOUNDS); (*2*) magnesium hydroxide precipitated from seawater; (*3*) by chlorination of magnesium oxide from various sources in the presence of carbon or carbonaceous materials; and (*4*) as a by-product in the manufacture of titanium (see TITANIUM AND TITANIUM ALLOYS).

Magnesium chloride is obtained from mother liquors resulting from the recovery of potassium chloride from carnallite. These liquors contain up to 28% magnesium chloride and have been regarded as a waste product because it is costly to obtain pure magnesium chloride from them. The liquor is purified by raising the concentration of the magnesium chloride through evaporation (qv) until potassium chloride, sodium chloride, sodium sulfate, and magnesium sulfate crystallize out and are removed. Ferrous iron is removed by oxidation using potassium chlorate at 158°C and precipitated with lime. Evaporation of the purified liquor to a density of 1.435 yields, upon cooling the solution, impure magnesium chloride as a glassy mass. The solution can be solidified on rotating cooling drums from which the salt layer can be removed by scrapers (60). Repeat recrystallization yields pure $MgCl_2 \cdot 6H_2O$.

In preparation of $MgCl_2$ from seawater, magnesium hydroxide, $Mg(OH)_2$, is first precipitated from seawater by the addition of dolime or lime. This is then treated with hydrochloric acid to produce a neutralized magnesium chloride solution. The solution obtained is evaporated and converted into solid magnesium chloride hexahydrate (60,61).

When magnesium oxide is chlorinated in the presence of powdered coke or coal (qv), anhydrous magnesium chloride is formed. In the production of magnesium metal, briquettes containing $CaCl_2$, KCl, NaCl, MgO, and carbon are chlorinated at a temperature such that the electrolyte or cell melt collects at the bottom of the chlorinator, enabling the liquid to be transferred directly to the electrolytic cells.

Another way of preparing anhydrous magnesium chloride is by a two-stage dehydration of magnesium chloride hexahydrate. In the first stage, concentrated magnesium chloride brine is spray-dried to a point where the residual water content is between 1 and 2 molecules. The partially dehydrated product can sometimes be chlorinated in the presence of carbon because the intermediate hydrate often contains some oxychloride. In the final stage of dehydration, the intermediate hydrate is treated with hydrogen chloride or chlorine gas. The anhydrous salt is then melted and cast into blocks which are packed so as to exclude all air (60).

A one-step, low energy process was developed by Magnesium International Corp. for producing anhydrous magnesium chloride. This process involved the reaction of magnesite and chlorine gas in the presence of carbon monoxide in a packed-bed reactor at 900°C to produce magnesium chloride and carbon dioxide. Carbon dioxide is withdrawn from above the packed bed and liquid magnesium chloride is collected at the bottom of the reactor and tapped periodically for transfer to electrolytic cells (61,62). The process can be represented by

$$MgCO_3 + CO + Cl_2 \longrightarrow MgCl_2 + 2\ CO_2 \qquad (10)$$

Economic Aspects. U.S. imports and exports for consumption of magnesium chloride are shown in Table 7 (61). In 1989, The Dow Chemical Co. ceased production of their $MgCl_2$ prill leaving only a 32–34% $MgCl_2$ solution in their product line. As a result, Great Salt Lake Minerals & Chemical Corp. (Utah) became the sole producer of dry magnesium chloride in the United States by 1992. At that time, Dead Sea Works Ltd. (Israel) completed an expansion of the production facility in Israel to produce 100,000 t/yr of $MgCl_2$ flakes for export. In April 1992, U.S. demand for $MgCl_2$ flakes was estimated at 10,000 to 15,000 t/yr (63). In January 1995, anhydrous magnesium chloride (92%) in flake or pebble drums (carload) was listed at \$0.281–\$0.33/kg. Hydrous magnesium chloride (99%) sold in flake bags (carload) was listed as \$0.32/kg (33). The Dow Chemical Co. sold their technical-grade $MgCl_2$ solution for \$0.485/dry kg fob. Freeport, Texas.

Uses. Anhydrous magnesium chloride is used mainly as a raw material in the production of magnesium metal. Most electrolytic recovery processes of magnesium metal use anhydrous magnesium chloride in the feedstock (see MAGNESIUM AND MAGNESIUM ALLOYS) with the exception of Dow Chemical Co. which uses hydrous magnesium chloride prepared from magnesium hydroxide (61). Electrolytic recovery processes differ from company to company. Essentially, magnesium chloride is fed to an electrolytic cell and is broken down into magnesium metal and chlorine gas by direct current at 700°C. Magnesium is removed from the cell and cast into ingots, and the chlorine gas is recycled or sold (61).

Another important use of magnesium chloride is in the preparation of oxychloride cements, $5Mg(OH)_2 \cdot MgCl_2 \cdot 8H_2O$, for flooring (nonsparking), wall plaster compositions, fire-resistant panels, fireproofing of steel beams, and grinding wheels. The cements are produced on-site by adding a 20% solution of $MgCl_2$ in water and a dry mix consisting of magnesium oxide, fillers (eg, wood fiber), and fine aggregates. After a few hours of setting, the cements form a dense but smooth-textured stonelike product (64). For example,

$$5\,MgO + MgCl_2 + 13\,H_2O \longrightarrow 5Mg(OH)_2 \cdot MgCl_2 \cdot 8H_2O \qquad (11)$$

The phase diagram in Figure 3 represents equilibria in the $MgO \cdot MgCl_2 \cdot H_2O$ system at room temperature (65).

Magnesium chloride is also used in the processing of sugar beets (see SUGAR, BEET SUGAR) and textiles (qv), in water treatment, as a fireproofing agent for wood, as a dust control agent in mines and on haul roads, as an ingredient of floor-sweeping compounds, refrigeration brines, and fire-extinguishing agents. Magnesium chloride flakes are also sold as a deicer (63).

Table 7. U.S. Imports and Exports of $MgCl_2$

	Imports		Exports	
Year	Quantity, t	Value, \$ $\times 10^3$	Quantity, t	Value, \$ $\times 10^3$
1989	5994	1202	2201	1812
1990	6914	1477	4763	6468

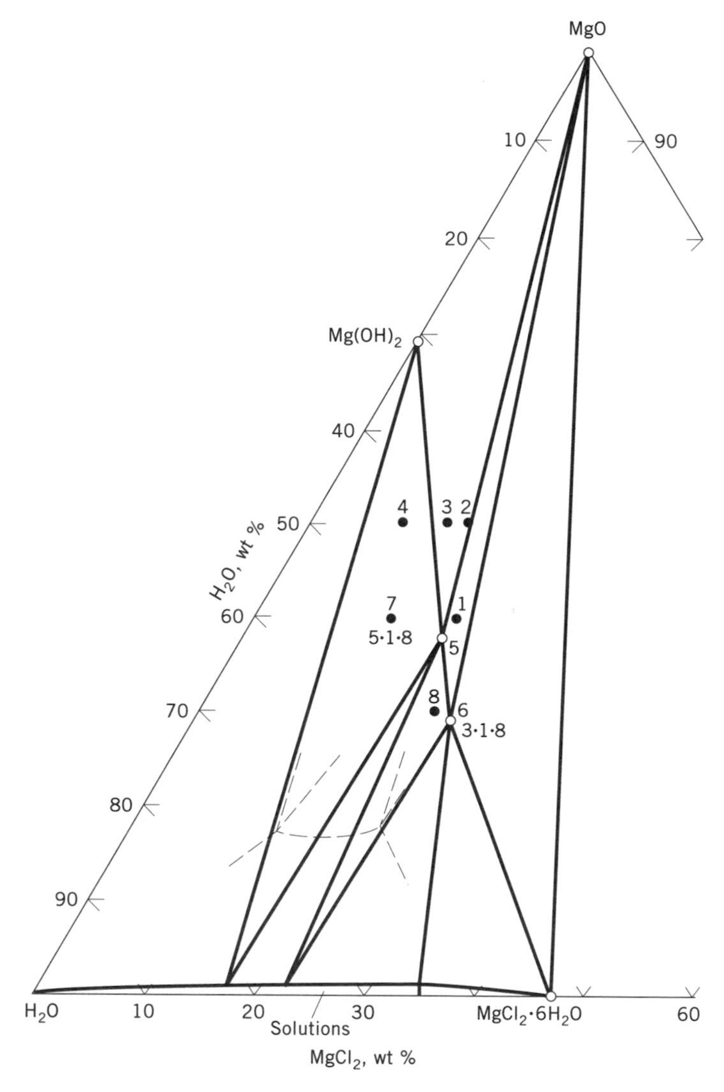

Fig. 3. Isothermal section at 23±3°C of the system $MgO{\cdot}MgCl_2{\cdot}H_2O$ showing phases in equilibrium with vapor phases in sealed containers (65).

Anhydrous magnesium chloride, along with magnesium bromide and magnesium iodide, is used in a process for producing organometallic compositions such as alkyllithium compounds used as reagents in the preparation of pharmaceuticals (qv) and special chemicals (66). Molten magnesium chloride has been used in the preparation of pure crystalline ceramic powders such as crystalline cordierite, forsterite, enstatite, and spinel (67). The introduction of $MgCl_2$-supported $TiCl_3$ (Ziegler-Natta) catalysts has changed the manufacturing technology of polypropylene because of greatly enhanced catalytic productivity (68) (see OLEFIN POLYMERS, POLYPROPYLENE). Also as a catalyst, $MgCl_2{\cdot}6H_2O$ is used in finishing to increase strength properties of cotton fabric (69).

Magnesium Hydroxide

Occurrence. Magnesium hydroxide [*1309-42-8*], $Mg(OH)_2$, occurs naturally as the mineral brucite [*1317-43-7*]. Brucite, usually found as a low temperature, hydrothermal vein mineral associated with calcite, aragonite, talc, or magnesite, appears as a decomposition product of magnesium silicates associated with serpentine, dolomite, magnesite, and chromite. Brucite also occurs as a hydrated form of periclase, and is found in serpentine, marble, chlorite schists, and in crystalline limestone. At one time brucite was recovered commercially from deposits at Wakefield, Quebec and Nye County, Nevada; both operations have since ceased.

Properties. The physical properties of magnesium hydroxide are listed in Table 8. The crystalline form of magnesium hydroxide is uniaxial hexagonal platelets (Fig. 4). Magnesium hydroxide begins to decompose thermally above 350°C, and the last traces of water are driven off at higher temperatures to yield magnesia.

Table 8. Physical Properties of Magnesium Hydroxide[a]

Property	Value
mol wt	58.32
crystal system	hexagonal
space group	P3m1
lattice constants, nm	
a	0.3147
c	0.4769
Z^b	1
density, g/cm^3	
$Mg(OH)_2$	2.36
brucite	2.38–3.40
index of refraction	1.559, 1.580
color	colorless to white
hardness, Mohs'	2.5
melting point, °C	
brucite	dec 268[c]
$Mg(OH)_2$	dec 350[c]
solubility,[d] mg/L	
25°C	11.7
100°C	4.08
solubility product, K_{sp}, at 25°C	5.61×10^{-12}
heat of formation,	
ΔH_{298}, kJ/mol[e]	–924.54
free energy of formation,	
ΔG_{298}, kJ/mol[e]	–833.58
$C_{p,298}$, J/(mol·K)[e]	77.03

[a]Refs. 3,39.
[b]Number of formulas per unit cell.
[c]Begins to lose H_2O.
[d]There is only fair agreement between data of various authors. See Ref. 70.
[e]To convert J to cal, divide by 4.184.

Fig. 4. Magnesium hydroxide. (**a**) ×20,000; (**b**) ×10,000. Courtesy of Martin Marietta Magnesia Specialties, Inc.

Upon exposure to the atmosphere, magnesium hydroxide absorbs moisture and carbon dioxide. Reactive grades are converted to the basic carbonate $5MgO \cdot 4CO_2 \cdot xH_2O$ over a period of several years. Grades that resist carbonization at high temperature and humidity have been reported (71).

Chemical Properties. The reactivity of magnesium hydroxide is measured primarily by specific surface area in units of m^2/g and median particle size in μm. Reactivity ranges from low, 1–2 m^2/g, 5 μm, eg, Kyowa's product; to high, 60–80 m^2/g, 5–25 μm, eg, Barcroft's CPS and CPS-UF powders. Higher reactivity tends to lower capital and operating costs for users as a result of shorter reaction times in their processes.

Manufacture and Processing. Most commercial-grade magnesium hydroxide is obtained from seawater or brine using lime or dolomitic lime (see LIME AND LIMESTONE). Calcination of this magnesium hydroxide yields synthetic magnesia. Producers of refractory-grade MgO products have placed increased emphasis on high purity magnesium hydroxide products. Demand for refractory magnesia has decreased and the production capacity formerly allotted to these materials has shifted to magnesium hydroxide products (52). Environmental applications of magnesium hydroxide, primarily wastewater treatment and SO_x scrubbing of flue gases, are increasing.

Production from Magnesium Salts. Magnesium hydroxide is produced from aqueous solutions of magnesium salts. To precipitate and recover magnesium hydroxide from solutions of magnesium salts, a strong base is added. The more commonly used base is calcium hydroxide [*1305-62-0*], derived from lime [*1305-78-8*], CaO, or dolime [*50933-69-2*], $CaO \cdot MgO$. Lime and dolime are calcination products of limestone and dolomite, respectively (see LIME AND LIMESTONE). Calcination of crushed and sized dolomite or limestone is done in rotary or shaft kilns controlled to drive off all CO_2 without deactivating the material. Theoretical precipitant requirements per kilogram of $Mg(OH)_2$ are approximately 0.97 kg lime or 0.83 kg dolime. Sodium hydroxide is used as a precipitant if a product having low CaO content is desired.

In seawater–dolime and brine–dolime processes, calcined dolomite or dolime, $CaO \cdot MgO$, is used as a raw material (Table 9). Dolime typically contains

Table 9. Production Processes for Magnesium Hydroxide

Process	Raw materials	Manufacturer	Facility location[a]
brine, deep well	dolime, $CaCl_2/MgCl_2$ brines	Dow Chemical	Ludington, Mich.
		Martin Marietta Magnesia Specialties	Manistee, Mich.
		Morton International	Manistee, Mich.
		Ned Mag BV	Veendam, Netherlands
		Penoles SA de CV	Quimica del Rey, Mexico
seawater	seawater, lime	Dow Chemical	Freeport, Tex.
		Redland Magnesia	Hartlepool, U.K.
		Sardamag SpA	Priolo, Sicily, Italy
		Ube Chemical Ind.	Japan
seawater dolomite	seawater, dolime	National Magnesia	Moss Landing, Calif.
		American Premier	Port St. Joe, Fla.
hydration of MgO[b]	MgO, water	Martin Marietta Magnesia Specialties	Pittsburgh, Pa.
		American Premier	

[a]Ref. 72 except as noted.
[b]Ref. 73.

58% CaO, 41% MgO, and less than 1% combined SiO_2, R_2O_3, and CO_2 where R is a trivalent metal ion, eg, Al^{3+} or Fe^{3+} (74). Roughly one-half of the magnesia is provided by the magnesium salts in the seawater or brine and the other half is from dolime (75). Plant size is thus reduced using dolime and production cost is probably lower.

Raw $MgCl_2$ brines at Ludington and Manistee, Michigan typically contain 51.6 g/L of equivalent MgO (74). This magnesium oxide equivalent represents approximately 4% $Mg(OH)_2$ by weight. Brine at Midland, Michigan typically contains one-third the equivalent MgO of Ludington and Manistee brines. The equivalent magnesia content of delivered brine may be 3 to 8% lower because of dilution at the well head to prevent salting out in the well tubulars (74). Seawater typically contains the equivalent of 2.2 g/L of MgO, present as $MgCl_2$ and $MgSO_4$ (75). Typical compositions of deep-well brine and seawater are compared in Table 10.

The reactions (76) for the brine process are (*1*) calcination of dolomite; (*2*) slaking of dolime; and (*3*) precipitation of $Mg(OH)_2$ (eqs. 12–14):

$$CaMg(CO_3)_2 \longrightarrow CaO + MgO + 2\ CO_2\ (g) \qquad \Delta H = +305.5\ \text{kJ}\ (73.02\ \text{kcal}) \qquad (12)$$

$$CaO + MgO + 2\ H_2O\ (l) \longrightarrow Ca(OH)_2 + Mg(OH)_2 \qquad \Delta H = -84\ \text{kJ}\ (-20\ \text{kcal}) \qquad (13)$$

$$Ca(OH)_2 + Mg(OH)_2 + MgCl_2\ (aq) \longrightarrow 2\ Mg(OH)_2 + CaCl_2\ (aq) \qquad \Delta H_{298} = -14.4\ \text{kJ}\ (3.44\ \text{kcal}) \qquad (14)$$

A flow diagram for the brine process is given in Figure 5. Dolime and brine are metered in fixed proportion to a slaking reactor where the temperature and

Table 10. Comparison of Deep-Well Raw Brine and Seawater

	Concentration, g/L	
Constituent	West central Michigan deep-well brine[a]	Seawater[b]
$MgCl_2$	122	4.176
$MgSO_4$		1.668
NaCl	47	27.319
$CaCl_2$	230	
$CaSO_4$		1.034
$Ca(HCO_3)_2$		0.122
Br	3.35	0.066[c]
B	50[d]	9[d]
others	14–15[e]	0.869 K_2SO_4, 0.008 SiO_2, and 0.022 R_2O_3[f]
specific gravity	1.28	1.024

[a]Ref. 74.
[b]Ref. 75.
[c]As $MgBr_2$.
[d]Concentration given in ppm.
[e]Constituents are KCl, LiCl, and SrCl.
[f]R_2O_3 = Al_2O_3 and Fe_2O_3.

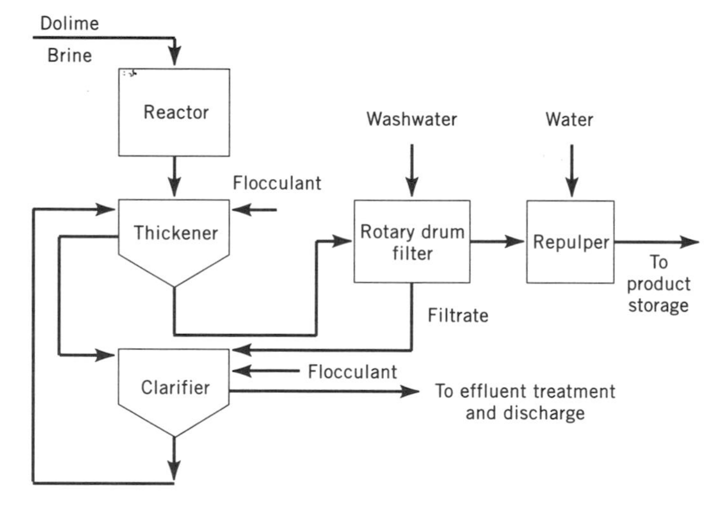

Fig. 5. Brine–dolime production of $Mg(OH)_2$.

chemistry are controlled to meet product purity and trace element specifications. A faster settling and more filterable precipitate results from a reduction in the reactivity of the precipitant, ie, lime or calcined dolomite, by control of the slaking conditions and degree of burn. Recirculation of a fraction of the precipitated slurry into the thickener or the continuous addition of flocculating agents (qv) boosts the settling rates of magnesium hydroxide. Improved settling and filtering characteristics reduce capital expenditures and operating costs.

Improvements in settling characteristics must be balanced against final product purity. This latter can be reduced by addition of settling aids (flocculants) or occlusion of residual unreactive lime. Clarification following thickening maximizes recovery of solids. Underflow slurry from the thickeners is dewatered, washed, and repulped to produce magnesium hydroxide of high bulk density and high purity.

Approximately 65 to 80% of the total $Mg(OH)_2$ yield is formed in the slaking reactor via the parallel reactions given in equations 13 and 14. The remaining 20 to 35% is formed by hydration of remaining dolime-borne MgO in equation 13 (74). The theoretical yield from 1 kg of dolime is approximately 1.2 kg of magnesium hydroxide, excluding impurities. Trace impurities include CaO, SiO_2, Fe_2O_3, Al_2O_3, SO_3^{2-}, Cl^-, and B. Sources of CaO are $CaCl_2$, $Ca(OH)_2$, $CaCO_3$, and $CaSO_4$, the last three being functions of the dolime. The brine is the source of the boron.

Calcium chloride in the spent brine includes the residual quantity in the raw brine plus $CaCl_2$ produced in equation 14. Approximately 0.84 kg of by-product calcium chloride are produced per kg of magnesium hydroxide via the brine–dolime process.

The key difference between the brine process and seawater process is the precipitation step. In the latter process (Fig. 6) the seawater is first softened by adding small amounts of lime to remove bicarbonate and sulfates, present as $MgSO_4$. Bicarbonate must be removed prior to the precipitation step to prevent formation of insoluble calcium carbonate. Removal of sulfates prevents formation of gypsum, $CaSO_4{\cdot}2H_2O$. Once formed, calcium carbonate and gypsum cannot be separated from the product.

An alternative pretreatment for seawater is acidification of the bicarbonate followed by degasification to remove the carbon dioxide generated. The precipitation step for the seawater process is given by (76):

$$2\,Ca(OH)_2 + 2\,Mg(OH)_2 + MgCl_2\,(aq) + MgSO_4 + 2\,H_2O\,(l) \longrightarrow 4\,Mg(OH)_2\,(s) + CaCl_2\,(aq) + CaSO_4{\cdot}2H_2O\,(s)$$

$$\Delta H = -22.6\ \text{kJ}\ (-5.40\ \text{kcal}) \qquad (15)$$

The softened seawater is fed with dry or slaked lime (dolime) to a reactor. After precipitation in the reactor, a flocculating agent is added and the slurry is pumped to a thickener where the precipitate settles. The spent seawater overflows the thickener and is returned to the sea. A portion of the thickener underflow is recirculated to the reactor to seed crystal growth and improve settling and filtering characteristics of the precipitate. The remainder of the thickener underflow is pumped to a countercurrent washing system. In this system the slurry is washed with freshwater to remove the soluble salts. The washed slurry is vacuum-filtered to produce a filter cake that contains about 50% $Mg(OH)_2$. Typical dimensions for equipment used in the seawater process may be found in the literature (75).

Other Processes. Dead Sea Periclase (DSP, Mishor Rotem, Israel) converts magnesium chloride into MgO by spray-roasting, then hydrates the MgO to $Mg(OH)_2$. The $Mg(OH)_2$ is washed and drum filtered. DSP purchases the brine from Dead Sea Works, which collects and stores enriched brine from the southern margins of the Dead Sea (77).

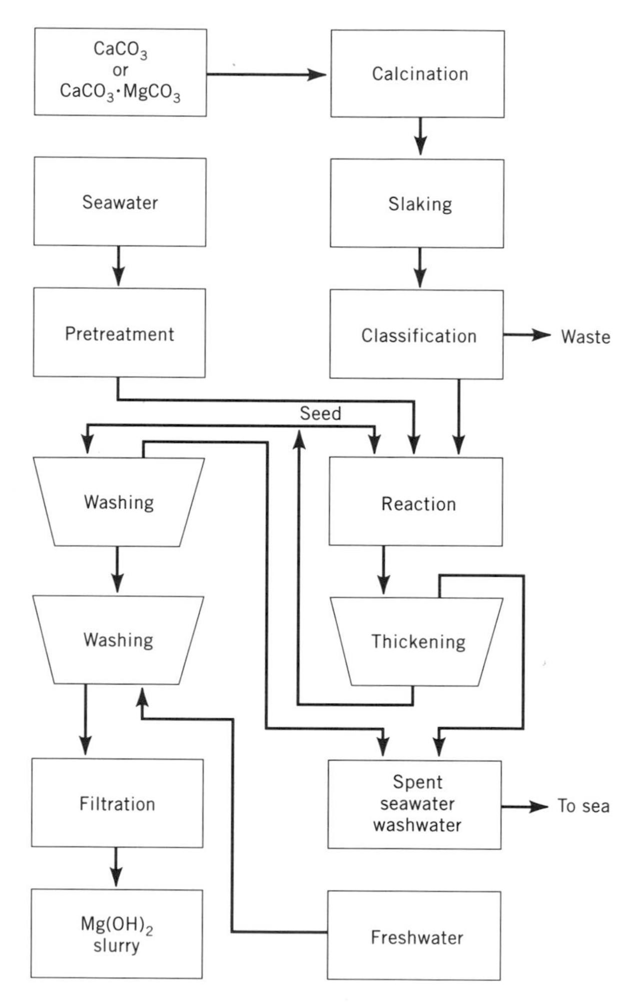

Fig. 6. Magnesia recovery from seawater.

The Austrian manufacturer Magnifin Magnesiaprodukte GmbH uses a magnesite acidification process to produce a high purity magnesium hydroxide for flame retardant/smoke suppressant applications (78). Magnesite is acidified to produce $MgCl_2$ and subsequent MgO of 99% purity, which is converted to magnesium hydroxide by hydration.

Magnesium hydroxide can also be produced by slaking or pressure hydrating various reactive grades of magnesium oxide. The reaction is highly exothermic ($\Delta H_{298} = -40.86$ kJ/mol (-9.77 kcal/mol)) to produce crystalline form at stoichiometric water addition; $\Delta H_{298} = -43.03$ kJ/mol (-10.28 kcal/mol) including heat of solution, at standard state $m = 1$) and may require a heat sink to prevent boiling of the reaction mixture. A 30% by weight suspension of MgO in 20°C water boils in the absence of any heat sink. The time to reach boiling is dependent on the reactivity of the MgO raw material, and this time can be only several hours for the more reactive grades of MgO. Investigations of the kinetics of formation of magnesium hydroxide by hydration of MgO have been reported (79).

Martin Marietta Magnesia Specialties, Inc. (Pittsburgh, Pennsylvania, formerly Clearwater, Inc.) hydrates MgO to produce magnesium hydroxide slurry containing 58% solids (73).

Production and Shipment. Magnesium hydroxide is produced and shipped in aqueous slurry or as dry powder.

Slurry. Dow Chemical was the first producer of a stabilized slurry product. National Magnesia, American Premier, and Martin Marietta Magnesia Specialties, Inc. have followed with slurry formulations. The slurry is shipped in totes, bulk trucks, or bulk railcars. Railcars equipped with air spargers for agitation are available from some manufacturers for improving unloading of slurry. Approximate container loadings are given in Table 11.

Bulk density is 1.49 kg/L for 57% solids slurry with a $Mg(OH)_2$ loading of 0.85 kg/L. The properties of some magnesium hydroxide slurries are given in Table 12.

Slurry Storage and Handling. Bulk quantities of slurry are stored in agitated or recirculated storage tanks. Agitation is necessary to resuspend settled solids. Slurry stability can be characterized by agitation requirements ranging from several hours per week for very stable products to 2 h/d for medium stability products to constant agitation for low stability products. Guidelines for storing

Table 11. $Mg(OH)_2$ Container Loadings

Container	Tote	Bulk truck	Bulk railcar
capacity, L	1,040	14,080	60,600
capacity, kg	1,560	21,100	90,700
$Mg(OH)_2$, kg	885	11,980	51,500

Table 12. Properties of Magnesium Hydroxide Slurries

Property	Normal grade[a]	Ultrastable grade[b]
	Slurry properties	
delivery/application	bulk	packaged
$Mg(OH)_2$, wt %	55–58	55–58
$Mg(OH)_2$, kg/L	0.799–0.844	0.799–0.844
bulk density, g/cm^3	1.451–1.499	1.451–1.499
viscosity, mPa(=cP)	100–300	200–400
heat capacity, J/g[c]	2.7	2.7
	Dry solids basis	
$Mg(OH)_2$, wt %	98.7	98.7
surface area, m^2/g	10–14	10–14
mean particle size, μm	1.5–3	1.0–3
	Particle size, cumulative wt %	
passing 325 mesh (−44 μm)	99.5	99.8

[a]Martin Marietta Magnesia Specialties, Inc. FloMag H.
[b]Martin Marietta Magnesia Specialties, Inc. FloMag HUS.
[c]To convert J to cal, divide by 4.184.

and agitating slurries such as magnesium hydroxide and for estimating agitation requirements have been presented in detail in the literature (80,81).

Instrumentation and control guidelines for processes utilizing magnesium hydroxide and other slurries have been outlined (82). An experimental determination of the accuracy of magnetic flow meters for magnesium hydroxide slurries flowing in pipelines (qv) has been reported (83).

Powder. Magnesium hydroxide powder can be produced by drying the $Mg(OH)_2$ filter cake from the various production processes. Powders having particle sizes comparable to the $Mg(OH)_2$ solids in the slurry can be obtained by spray drying the slurry. Slaking reactive MgO followed by spray drying also produces a $Mg(OH)_2$ powder. The minimum purity specification is typically 95–98% by weight $Mg(OH)_2$. The loose bulk density typically ranges from 0.4 to 0.6 g/cm^3. The properties of some commercial grades of magnesium hydroxide powders are given in Table 13.

Packaging of magnesium hydroxide powders is usually in paper bags of 10 to 25 kg size. Bulk shipments may be made in 1 t supersacks. Magnesium hydroxide powder is hygroscopic and should be protected against exposure to moisture by use of lined bags. Magnesium hydroxide powder can be dusty, and appropriate precautions against dust exposure should be taken during handling (see POWDERS, HANDLING).

Analytical and Test Methods. Many of the procedures for technical analyses of magnesium hydroxide are readily available from the principal producers. These procedures should be carefully reviewed. Site-specific variations in procedure steps and mechanics, especially for chemical activity, can bias results and inadvertantly disqualify an otherwise acceptable product.

Particle Size. Wet sieve analyses are commonly used in the 20 μm (using microsieves) to 150 μm size range. Sizes in the 1–10 μm range are analyzed by

Table 13. Properties of Commercial-Grade Magnesium Hydroxide Powders

Property	Technical grade[a]	Controlled particle size (CPS)[b]	CPS-ultrafine (UF)[c]	USP[d]
	Powder properties			
$Mg(OH)_2$, wt %[e]	95.0	95.0	95.0	95.0–100[f]
Ca, wt %[g]	1.0	1.0	1.0	0.7
moisture at 105°C, wt %[g]	2.0	1.6	1.6	2.0
mean particle size, μm	35–50	15–25	10 ± 5	
surface area, m^2/g	50–80	60–80	60–80	
	Particle size, cumulative wt %			
passing 100 mesh (–149 μm)	100	100	100	
passing 325 mesh (–44 μm)			99	

[a]Barcroft Technical Grade.
[b]Barcroft CPS Grade.
[c]Barcroft CPS-UF Grade.
[d]Barcroft USP Grade.
[e]Minimum value.
[f]On dry basis.
[g]Maximum value.

light-transmission liquid-phase sedimentation, laser beam diffraction, or potentiometric variation methods. Electron microscopy is the only reliable procedure for characterizing submicrometer particles. Scanning electron microscopy is useful for characterizing particle shape, and the relation of particle shape to slurry stability.

Surface Area. Surface area is measured by determining the quantity of nitrogen gas that adsorbs on the particle/crystal surfaces of a dry sample. Determination of surface area by measuring adsorption at gas–solid interfaces is covered extensively in the literature (84). Instruments such as the FlowSorb 2300 are used to control the adsorption/desorption within specific conditions of temperature and pressure.

Mineral and Chemical Composition. X-ray diffraction is used to determine the mineral composition of an $Mg(OH)_2$ sample. Induced coupled plasma (icp) spectrophotometry is used to measure the atomic concentrations present in a sample. X-ray fluorescence analysis is another comparative instrumental method of determining chemical composition.

Chemical Activity. The activity of magnesium hydroxide is measured using a citric acid activity (CAA) test. The CAA is determined by measuring the time required for a given weight of a particular magnesium hydroxide to provide hydroxyl ions sufficient to neutralize a given weight of citric acid.

Loss on Ignition. The loss on ignition (LOI) test is used to approximate the percent of magnesium hydroxide present in a sample and to determine extent of hydration. When the magnesium hydroxide sample is heated to 1000°C, all of the magnesium hydroxide and trace impurities of magnesium carbonate are converted to magnesium oxide. Some chlorides and sulfates are also decomposed. The relative difference in sample weight before and after heating approximately represents the quantity of water given up by the $Mg(OH)_2$. Assuming that complete dehydration of the $Mg(OH)_2$ occurs and the starting sample is dry, the theoretical LOI for a 100% sample of $Mg(OH)_2$ is 30.85%.

Slurry Viscosity. Viscosities of magnesium hydroxide slurries are determined by the Brookfield Viscometer in which viscosity is measured using various combinations of spindles and spindle speeds, or other common methods of viscometry. Viscosity decreases with increasing rate of shear. Fluids, such as magnesium hydroxide slurry, that exhibit this type of rheological behavior are termed pseudoplastic. The viscosities obtained can be correlated with product or process parameters. Details of viscosity determination for slurries are well covered in the literature (85,86).

Uses. The principal use of magnesium hydroxide is in the pulp (qv) and paper (qv) industries (52). The main captive use is in the production of magnesium oxide, chloride, and sulfate. Other uses include ceramics, chemicals, pharmaceuticals, plastics, flame retardants/smoke suppressants, and the expanding environmental markets for wastewater treatment and SO_x removal from waste gases (87).

Wastewater Treatment. Most magnesium hydroxide on the merchant market is used for the high growth area of water treatment. Substitution of $Mg(OH)_2$ for caustic soda or lime for wastewater water treatment occurred owing to increased emphasis on product handling safety and on reducing total treatment costs (88,89). The magnesium hydroxide serves as a neutralizing agent and as a heavy metals precipitant. Buffering ability, reduced sludge volume, and improved filterability of the sludge are potential advantages.

The lower equivalent weight of magnesium hydroxide compared to caustic soda, hydrated lime, and soda ash reduces the stoichiometric amounts necessary to neutralize a given amount of acid. Depending on relative alkali costs, magnesium hydroxide can offer the advantage of lower chemical costs.

Because of limited solubility and common ion effects, magnesium hydroxide treatment buffers most industrial wastewaters around pH 9, reducing the likelihood of pH overshoot and resolubilization of heavy metals being removed. Magnesium slats, the product of acid neutralization with magnesium hydroxide, are generally more soluble than calcium salts, eg, in sulfuric acid neutralization $MgSO_4 \cdot 7H_2O$ solubility is 71 parts per 100 parts, compared to $CaSO_4 \cdot 2H_2O$ solubility of 0.26 parts per 100 parts, both at 20°C. This results in lower sludge quantities and disposal costs for magnesium hydroxide relative to lime.

Magnesium hydroxide reacts more slowly in hydroxide precipitation of heavy metals compared to stronger alkalies such as caustic soda and lime, but the slower reaction produces larger crystallites, less entrained water in sludge, and lower sludge weights and disposal costs. The superior filterability of sludges formed in magnesium hydroxide treatment can reduce operating costs and maintenance costs of filtration and sludge drying equipment.

Flame-Retardant Filler. Demand has increased for $Mg(OH)_2$ as a nonhalogenated, flame-retardant filler for thermoplastics used in the aerospace, microelectronics, and cable and wire manufacturing industries (90). Producers of nonhalogenated, flame retardant fillers include Kyowa, Aluisuisse-Lonza (Magnifin product line), Morton, and a Dead Sea Periclase/Dead Sea Bromine joint venture (91).

Magnesium fillers combust endothermically and are free of toxic or corrosive decomposition products, mitigating the concerns of both personnel and equipment exposed to fire conditions. In some cases $Mg(OH)_2$ has replaced alumina trihydrate as a filler in insulation or sheathing for high temperature wire and cable. Compared to other halogen-free flame retardants, $Mg(OH)_2$ provides higher (up to 340°C) thermal stability. Higher thermal stability allows $Mg(OH)_2$ to be used in isopropylene and some nylon grades, whereas alumina trihydrate use is limited to poly(vinyl chloride) and polyethylene (91). The required $Mg(OH)_2$ loading may be up to 50 wt % in nonhalogenated thermoplastics. The loading must be optimized and effectively dispersed because high loading adversely affects physical properties, particularly for thermoplastics. Coating the wire or cable may provide an alternative to high loadings.

Numerous variations in composition, application method and surface treatment, and properties of $Mg(OH)_2$-containing flame-retardant fillers are disclosed in the patent literature (92–95). Smoke-suppressant foams incorporating $Mg(OH)_2$ are useful as roof insulation slabs (96). $Mg(OH)_2$ is contained in coated cigarette papers that produce less side-stream smoke (97–99).

Pharmaceutical Grade. Pharmaceutical-grade $Mg(OH)_2$ in a 30% paste is used to produce milk of magnesia (100). The $Mg(OH)_2$ concentration in milk of magnesia is typically 7.75 wt %. Unlike other magnesium hydroxide products, the $Mg(OH)_2$ in milk of magnesia is produced from basic magnesium carbonate. The fine particle structure formed in this reaction makes milk of magnesia a stable slurry. A pharmaceutical-grade powder is described in Table 13. The powder is used in antacid tablet formulations (100).

Other Applications. $Mg(OH)_2$ fibers can be added to polyolefin resins used in automobile bumpers and trims and electrical housings (101). $Mg(OH)_2$ is used in the manufacture of magnetic recording media to coat CrO_2 magnetic powders to suppress magnetization deterioration (102) (see INFORMATION STORAGE MATERIALS, MAGNETIC; MAGNETIC MATERIALS). $Mg(OH)_2$ is also used as a fuel oil additive to reduce SO_x and metals in the stack gas.

Economic Aspects. The quantity of magnesium hydroxide shipped and used in the United States decreased by 17% from 1990 to 1991, then decreased further from 1991 to 1993. These shipments, excluding material produced as an intermediate step in the manufacture of other magnesium compounds, are given in Table 14 (52,103).

Domestic shipments totaled 252,500 t in 1993. Martin Marietta Magnesia Specialties, Inc. expanded its production capacity at Manistee, Michigan from 10,000 t/yr to between 50,000 and 70,000 t/yr in 1994 (91,104). Dow Chemical increased capacity at Ludington, Michigan by 30,000 t/yr to 125,000 t/yr (91,105,106). Martin Marietta Magnesia Specialties also had a capacity of 30,000 t/yr at Pittsburgh, acquired from Clearwater, Inc., in late 1994 (73). Total domestic production capability of magnesium hydroxide excluding powders and pharmaceutical grade reached the low 200,000 t/yr range by 1994.

U.S. imports and exports of $Mg(OH)_2$ in 1989 and 1990, combined with those for magnesium peroxide, are listed in Table 15 (52). Individual import/export figures for $Mg(OH)_2$ are not reported.

Prices and Markets. In late 1993 the magnesium hydroxide market was reported to be slightly oversupplied but continued to grow. Magnesium hydroxide accounted for 1% of the 40×10^6 t/yr total U.S. market for alkalinity (100), and held a 1% share, or 36,000 t, of the growing wastewater treatment chemicals market in mid-1993 (106). Low caustic soda prices slowed growth of magnesium hydroxide in this sector (91) until the third quarter of 1994.

The price of magnesium hydroxide powder (National Formulary, freight-equalized) was $1.72/kg in the third quarter of 1993 (107), the same as at year

Table 14. Magnesium Hydroxide Shipments

Parameter	1990	1991	1992	1993
quantity,[a] t	366,016	302,244	267,087	252,471
value, $\$ \times 10^3$	69,280	82,768	64,445	61,585

[a] 100% $Mg(OH)_2$ basis.

Table 15. U.S. Imports and Exports of $Mg(OH)_2$ and MgO

	Imports		Exports	
Year	Quantity, t	Value, $\$ \times 10^3$	Quantity, t	Value, $\$ \times 10^3$
1989	1,792	3,091	12,072	3,673
1990	3,548	5,456	6,342	6,245

end for 1988 and 1987 (52). The price then rose to \$2.10–\$2.32/kg at year-end 1993. The price of technical-grade magnesium hydroxide in slurry form ranged from \$200 to \$220/t (fob works, 100% solids basis) in 1993 (100). The demand for a flame-retardant grade is estimated to range from 4500 to 9000 t/yr. Prices range from \$1.54/kg to \$2.65/kg. The price of pharmaceutical-grade powder ranged from as low as \$0.99/kg up to \$2.09/kg to \$2.32/kg. Thirty-percent paste was priced in the \$1.10/kg range (fob works, truckloads) (100).

Health and Safety Factors. Magnesium hydroxide is not absorbed by the skin. Dry magnesium hydroxide may irritate the eyes, skin, nasal passages, and respiratory tract. Routes of body entry are skin contact, eye contact, inhalation, and ingestion. No LD_{50} values for $Mg(OH)_2$ are available.

If $Mg(OH)_2$ is heated above 1700°C, magnesium oxide fumes may volatilize. Inhalation of freshly generated MgO fumes may result in metal fume fever. Ingestion of $Mg(OH)_2$ generally causes purging of the bowels, although swallowing a large amount of $Mg(OH)_2$ may lead to bowel obstruction. No data is available regarding chronic exposure to $Mg(OH)_2$.

$Mg(OH)_2$ powder is classified by OSHA as a nuisance dust. ACGIH categorizes the powder form as particulates not otherwise classified. Exposure limits are as follows (108): ACGIH 10 mg/m^3, OSHA 5 mg/m^3 (respirable), and 15 mg/m^3 (total). Magnesium hydroxide is reported in the EPA TSCA inventory (109).

Magnesium Iodide

Properties. Magnesium iodide [*10377-58-9*] can exist as two deliquescent and heat sensitive compounds: the octahydrate [*7790-31-0*], $MgI_2 \cdot 8H_2O$, and the hexahydrate [*75535-11-4*], $MgI_2 \cdot 6H_2O$. The octahydrate is crystallized as a white powder from solutions between 8°C and 43°C and the hexahydrate from

Table 16. Physical Properties of Magnesium Iodide and Hydrates[a]

Property	MgI_2	$MgI_2 \cdot 6H_2O$	$MgI_2 \cdot 8H_2O$
mol wt	278.12	386.21	422.24
crystal system	hexagonal	monoclinic	orthorhombic
space group	$P\bar{3}m$	C	Aca
lattice constants, nm			
a	0.4148	1.1159[b]	0.9948
b		0.7740	1.5652
c	0.6894	0.6323	0.8585
Z[c]	1	2	4
density, calculated, g/cm^3	4.496	2.353	2.098
mp, °C	637 dec		43.5
color	white	white	white
heat of formation, ΔH_{298}, kJ/mol[d]	–364.0		
free energy of formation, ΔG_{298}, kJ/mol[d]	–358.2		

[a] Refs. 28–30.
[b] The angle β is 93.12°.
[c] Number of formulas per unit cell.
[d] To convert J to cal, divide by 4.184.

solutions above 43°C. The physical properties of both compounds are shown in Table 16, solubilities in water are shown in Figure 2 (31).

The octahydrate can be obtained from a solution of magnesium hydroxide in hydriodic acid and evaporation at room temperature, followed by cooling to 0°C to increase the yield. By heating $MgI_2 \cdot 8H_2O$ in a current of dry hydrogen iodide, HI, followed by dry nitrogen, N_2, the anhydrous salt MgI_2 is formed. This salt is extremely hydroscopic and decomposes in air with the formation of free iodine.

Magnesium iodide is soluble in alcohols and many other organic solvents, and forms numerous addition compounds with alcohols, ethers, aldehydes, esters, and amines. One example is magnesium iodide dietherate [*29964-67-8*], $MgI_2 \cdot 2C_4H_{10}O$, prepared by gradual addition of iodine to a mixture of magnesium and dry ether. Magnesium iodide dietherate, which occurs as white, needle-like crystals, is very hygroscopic and becomes yellowish after several hours, and then brown after a day because of separation of iodine. The action of water upon magnesium iodide dietherate leads to the formation of the octahydrate salt, $MgI_2 \cdot 8H_2O$.

Uses. Magnesium iodide is used in the deoxygenation of oxiranes into olefins and iodine. This step is important to organic chemistry because it helps in the structure elucidation of complex organic molecules (110). For example,

$$R_1R_2C(-O-)CR_3R_4 + MgI_2 \longrightarrow \left[R_1R_2C(OH)-C(I)R_3R_4 \right] \longrightarrow R_1R_2C{=}CR_3R_4 + {}^{-}OMgI + 1/2\ I_2$$

Magnesium iodide is also an effective electrophilic reagent for the cleavage of C—O bonds in aromatic and aliphatic carboxylic esters (111). In organic catalysis, the addition of MgI_2 to a catalyst compound containing CuI_2 and CuI increases the activity of the catalyst for removing SO_2 from gases to reduce air pollution (112).

Anhydrous MgI_2 is used in a process for producing organometallic and organobimetallic compositions, which are important in the preparation of pharmaceutical and special chemicals. An organic halide, an alkali metal, and magnesium halide react in a liquid hydrocarbon solvent (66).

Economic Aspects. The bulk price of 98% MgI_2 in February 1995 was approximately \$2000/kg (113).

Magnesium Nitrate

Anhydrous magnesium nitrate [*10377-60-3*], $Mg(NO_3)_2$, is very difficult to isolate. The commercial product is the deliquescent hexahydrate [*13446-18-9*], $Mg(NO_3)_2 \cdot 6H_2O$. As illustrated in the solubility curve in Figure 7, the hexahydrate is the stable solid phase between −18 and 55–56°C. Properties are given in Table 17 (1–4). The unit cell contains two formula units and the calculated density is 1.643 g/cm^3.

Magnesium nitrate is prepared by dissolving magnesium oxide, hydroxide, or carbonate in nitric acid, followed by evaporation and crystallization at room

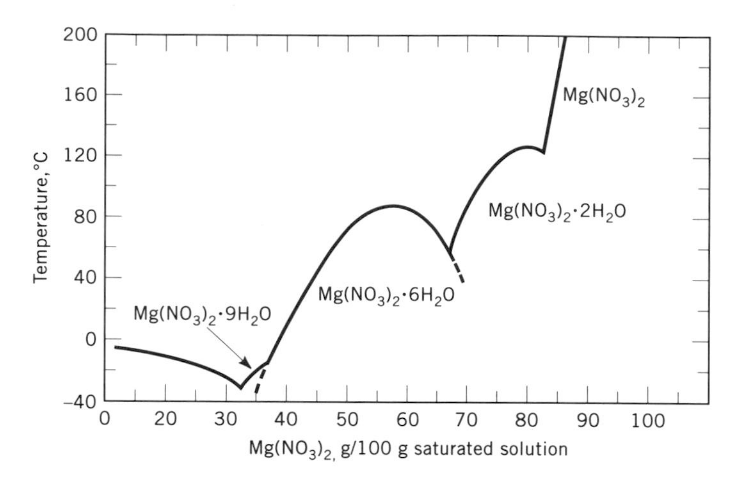

Fig. 7. The system $Mg(NO_3)_2 \cdot H_2O$.

Table 17. Physical Properties of Magnesium Nitrates[a]

Property	$Mg(NO_3)_2$	$Mg(NO_3)_2 \cdot 6H_2O$
mol wt	148.32	256.38
crystal system		monoclinic
space group		$P2_1/c$
mp, °C		89
heat of formation, ΔH_{298}, kJ/mol[b]	−790.7	−2613.3
free energy of formation, ΔG_{298}, kJ/mol[b]	−589.5	2080.7

[a]Refs. 1–3 and 38.
[b]To convert J to cal, divide by 4.184.

temperature. Impurities such as calcium, iron, and aluminum are precipitated by pretreatment of the solution with slight excess of magnesium oxide, followed by filtration. Most magnesium nitrate is manufactured and used on site in other processes.

Dehydration of the hexahydrate above its melting point is generally accompanied by hydrolysis and the formation of basic nitrates such as $Mg(NO_3)_2 \cdot 4Mg(OH)_2$ [*76190-42-6*]. At about 400°C magnesium nitrate is converted completely to magnesium oxide and oxides of nitrogen. All magnesium nitrates are soluble in methanol and ethanol and form addition compounds with urea, aniline, and pyridine. The basic nitrates $MgO \cdot 2Mg(NO_3)_2 \cdot 5H_2O$ [*75300-53-7*], $2MgO \cdot Mg(NO_3)_2 \cdot 5H_2O$ [*75300-52-6*], and $3MgO \cdot Mg(NO_3)_2 \cdot 11H_2O$ [*75300-51-5*] also exist.

Uses. A soluble form of magnesium nitrate is used as a fertilizer in states such as Florida where drainage through the porous, sandy soil depletes the magnesium (see FERTILIZERS). Magnesium nitrate is also used as a prilling aid in the manufacture of ammonium nitrate. A 0.25–0.50% addition of magnesium nitrate to the process improves the stability of the prills and also improves durability and abrasion resistance.

Another use for magnesium nitrate is as an alternative to sulfuric acid in the purification of nitric acid. A process developed by Hercules Powder (114) combines molten magnesium nitrate containing ca 70% anhydrous $Mg(NO_3)_2$ and ca 30% water, and 60% nitric acid. These are fed to the middle of a dehydration tower. A vapor containing 90–95% nitric acid is taken from the top of the tower and distilled. The solution of 50–55% magnesium nitrate is removed from the bottom of the tower, reconstituted to 72%, and recirculated through the dehydration tower (115,116). The tower and circulation equipment must be kept above 100°C, the freezing point of liquid magnesium nitrate (115).

Handling and Safety. Magnesium nitrate should be stored in a cool, dry place because it is hygroscopic. Magnesium nitrate is an acute skin, eye, and respiratory irritant which can be absorbed into the body via inhalation and ingestion. Ingestion of magnesium nitrate may lead to the formation of methemoglobin. Personal protection to be used when handling magnesium nitrate includes chemical safety goggles, chemical resistant gloves, and a NIOSH/MSHA approved respirator. To keep exposure to respirable dust to a minimum, mechanical exhaust is required.

Magnesium nitrate is a strong oxidizer and is incompatible with strong reducing agents and strong acids. A mixture of aluminum powder, water, and metal nitrates may explode owing to a self-accelerating reaction. Magnesium nitrate mixed with alkyl esters may explode owing to the formation of alkyl nitrates. Magnesium nitrate reacts violently with dimethylformamide, causing fire and explosion hazards. When heated to decomposition (above 330°C), magnesium nitrate may give off toxic and corrosive nitrous vapors (117,118).

Magnesium Oxide

Magnesium oxide, MgO, also known as magnesia [*1309-48-4*], occurs in nature only infrequently as the mineral periclase, most commonly as groups of crystals in marble. The principal commercial forms of magnesia are dead-burned magnesia (periclase), caustic-calcined (light-burned magnesia), hard-burned magnesia, and calcined dolomite. These materials are usually formed by the thermal decomposition or chemical reaction of various magnesium compounds including magnesite ore, magnesium hydroxide, magnesium chloride, and synthetic magnesium carbonate. Physical properties of periclase are given in Table 18 (30,38,119–122).

Properties. The properties of magnesia produced by thermal decomposition are determined by the calcination time, temperature, the nature of the magnesium-containing precursor, and other chemical compounds in the process. Increasing calcination time and temperature increases the crystallite size of the magnesia, simultaneously decreasing the surface area and reactivity of the product. Typical production conditions and resulting properties of magnesias produced from magnesium hydroxide are shown in Table 19.

Light-burned, ie, caustic-calcined magnesia is characterized by small crystallite size, relatively large surface area, and moderate to high chemical reactivity. It readily dissolves in dilute acids, and hydrates upon exposure to moisture or water. The most reactive grades combine with moisture and carbon dioxide eventually to form basic magnesium carbonates. Reactivity of light-burned magnesia is often quantified by specific surface area, iodine number, rate of reaction in acetic acid, and rate of reaction with citric acid.

Table 18. Physical Properties of Periclase

Property	Value
mol wt	40.304
crystal form	fcc
lattice constant, nm	0.42
density,[a] g/cm^3	3.581
index of refraction	1.732
hardness, Mohs'	5.5–6.0
melting point, °C	2827 ± 30
thermal conductivity at 100°C, J/(s·cm·°C)[b]	0.360
electrical resistivity, Ω·cm	
at 27°C	1.3×10^{15}
727°C	2×10^7
1727°C	4×10^2
specific heat, kJ/(kg·K)[b]	
at 27°C	0.92885
227°C	1.1255
727°C	1.2719
1727°C	1.3389
2727°C	1.3598
heat of fusion at 2642°C kJ/mol[b]	77.4
heat of formation, ΔH_{298}, kJ/mol	–601.7
free energy of formation, ΔG_{298}, kJ/mol[b]	–569.44
aqueous solubility, g/100 mL	
at 20°C	0.00062
30°C	0.0086

[a]Determined by x-ray.
[b]To convert J to cal, divide by 4.184.

Table 19. Production Conditions and Resultant Magnesia Properties

Magnesia	Calcination temperature, °C	Surface area,[a] m^2/g	Crystallite size, μm	Porosity, %
light-burned	<950	1–200+	<0.5	70–80
hard-burned	1090–1650	0.1–1	1–20	40–50
dead-burned	>1800	<0.1	>40	0–5

[a]Brunauer, Emmett, and Teller (BET) method.

Hard-burned magnesia is characterized by moderate crystallite size and moderately low chemical reactivity. Hard-burned magnesia is readily soluble only in concentrated acids.

Dead-burned magnesia, characterized by large crystallite size and very low chemical reactivity, is resistant to the basic slags employed in the metals refining industry. It reacts very slowly with strong acids, and does not readily hydrate or react with carbon dioxide unless finely pulverized.

Pure-fused magnesia is produced at extremely high (>ca 2750°C) temperatures using graphite electrodes in an electric arc furnace. Fused magnesia has extremely large crystal size and may have single crystals weighing 200 g or more. The chemical stability, strength, and abrasion resistance of fused magnesia surpass those of either light-burned or dead-burned magnesia.

Manufacture. There are many processes for producing magnesium oxide. Martin Marietta Magnesia Specialties, Inc. mines dolomitic limestone in Woodville, Ohio (see Fig. 8). The limestone is calcined at a high temperature under controlled conditions to produce calcined dolomite or dolime [*50933-69-2*] which upon reaction with magnesium chloride-rich brine produces magnesium hydroxide and calcium chloride. The insoluble magnesium hydroxide is then separated from the liquid calcium chloride carrier and calcined under controlled conditions. The various grades of magnesia range from very reactive light-burned to nonreactive dead-burned.

Another process, in use globally, involves the mining, crushing, sizing, and subsequent calcination of natural magnesite. The chemical purity of the magnesia produced is dependent on the mineralogical composition of the natural magnesite. This magnesia is often less pure than magnesia produced by other processes.

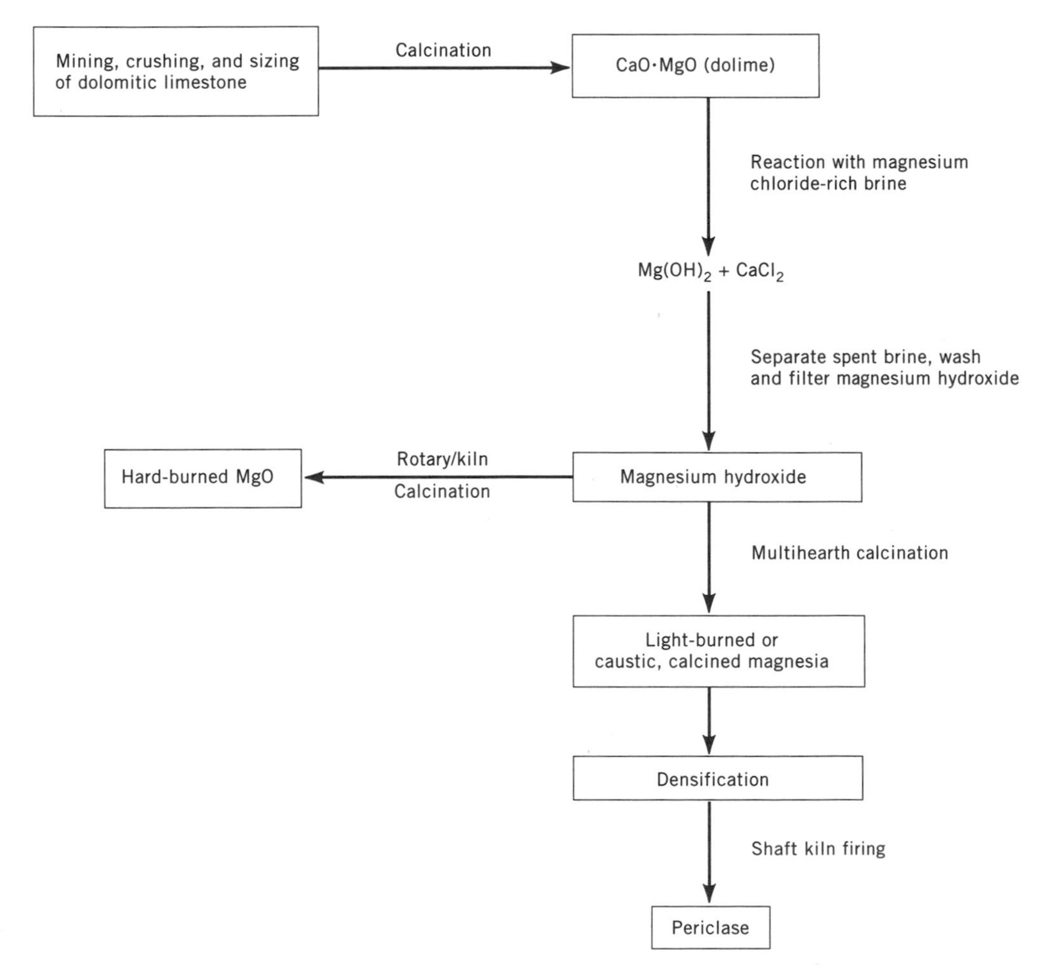

Fig. 8. Magnesia manufacturing process, Martin Marietta Magnesia Specialties, Inc.

The seawater process (45,123–126) used by American Premier, National Magnesia Chemicals, and others, involves decarbonating limestone or dolomite to the point where all CO_2 is removed without converting the resulting magnesia to a chemically inactive form. Reaction of filtered seawater, treated to remove bicarbonate and/or sulfate, and dolime is followed by seeding with magnesium hydroxide to promote crystal growth. Upon formation of magnesium hydroxide, flocculants are added and the magnesium hydroxide precipitate is allowed to settle while the spent seawater is disposed to the sea. The precipitate is washed, filtered, and dried to obtain magnesium hydroxide which is calcined to produce light-burned, hard-burned, or dead-burned magnesium oxide.

Dead Seas Periclase Ltd., on the Dead Sea in Israel, uses yet another process to produce magnesium oxide. A concentrated magnesium chloride brine processed from the Dead Sea is sprayed into a reactor at about 1700°C (127,128). The brine is thermally decomposed into magnesium oxide and hydrochloric acid. To further process the magnesia, the product is slaked to form magnesium hydroxide which is then washed, filtered, and calcined under controlled conditions to produce a variety of MgO reactivity grades. A summary of MgO purities, for the various processes is given in Table 20.

Economic Aspects. Economic data for magnesia are given in Table 21. The 1993 price for bulk synthetic dead-burned magnesia in carload lots was $364/t. The 1993 prices per metric ton for the various grades of caustic calcined (light-burned) magnesia were $1874 for technical, light, neoprene grade excluding that used in the production of refractory magnesia; $403 for synthetic, technical, chemical grade, including that used by producers; and $273.90 for natural, technical, heavy 85% passing 150 mesh (ca 100 μm) MgO (131).

Uses. Dead-burned magnesia is used extensively for refractory applications in the form of basic granular refractories and brick. Brick refractories may be used in cement kilns, furnaces, ladles, glass-tank checkers, and secondary refining vessels in the metals refining industry. Basic granular refractories are used primarily in the metals refining industry in furnace hearths, ladles, and as components of gunning mixtures. These applications involve the majority of magnesium compounds produced in the United States.

Table 20. Purity of Dead-Burned Magnesia from Various Processes[a]

Process	Composition, wt %					
	MgO	CaO	SiO_2	Fe_2O_3	Al_2O_3	B_2O_3
dolomitic limestone/brine	97.4–98.8	0.69–1.70	0.13–0.30	0.06–0.18	0.06–0.18	0.025
natural magnesite calcination[b]	90–98	0.35–3.5	0.35–2.5	0.15–0.8	0.1–0.4	0.007
seawater	98	0.85–1.8	0.25–0.40	0.07–0.30	0.07–0.30	0.1–0.25[c]
magnesium chloride pyrolysis	99	0.70	0.02	0.03	0.01	<0.01

[a] Ref. 129, except where noted.
[b] Data are for magnesia having at least 90% MgO content.
[c] As reported by analyses from Martin Marietta Magnesia Specialties, Inc.

Table 21. Worldwide Production and United States Economic Data for Magnesium Oxide[a]

Area	1989	1990	1991	1992	1993
United States					
caustic-calcined and specified magnesias[b]					
shipments[c]					
quantity, t × 10^3	135	135	154	130	131
value, \$ × 10^3	39,529	37,850	48,074	36,781	39,476
exports					
value,[d] \$ × 10^3	2,263	1,406	2,289	2,404	2,459
imports for consumption					
value,[d] \$ × 10^3	13,657	13,957	15,891	12,309	15,709
refractory magnesia					
shipped and used by producers					
quantity, t × 10^3	348	335	290	291	268
value, \$ × 10^3	97,673	94,962	85,292	80,736	77,716
exports					
value, \$ × 10^3	10,685	19,709	23,038	22,257	21,807
imports for consumption, \$	38,555	32,858	30,209	37,928	48,673
Worldwide					
crude magnesite production, t × 10^3	11,952	10,481	9,813	10,501	10,136

[a]Ref. 130.
[b]Excludes caustic-calcined magnesia used in the production of refractory magnesia.
[c]Includes magnesia used by producers.
[d]Caustic-calcined magnesia only.

Hard-burned magnesias may be used in a variety of applications such as ceramics (qv), animal feed supplements, acid neutralization, wastewater treatment, leather (qv) tanning, magnesium phosphate cements, magnesium compound manufacturing, fertilizer, or as a raw material for fused magnesia. A patented process has introduced this material as a cation adsorbent for metals removal in wastewater treatment (132).

The 1990 U.S. domestic shipments of caustic-calcined (light-burned) and specified magnesias were 36% for animal feeds and fertilizers; 19% for chemical processing; 18% for metallurgical uses such as refractories, electrical, water treatment, stack gas scrubbing, and foundry; 17% for manufacturing of rayon, fuel additives, rubber, pulp and paper, and uranium processing; 3% for construction including oxychloride and oxysulfate cement, general construction, and insulation; 3% for pharmaceuticals and nutrition, ie, medicinal and pharmaceutical usage, and use in sugar and candy; and 4% was unspecified (133).

Safety. Magnesium oxide (fume) has a permissible exposure limit (PEL) (134) (8 hours, TWA), of 10 mg/m^3 total dust and 5 mg/m^3 respirable fraction. Tumorigenic data (intravenous in hamsters) show a TD_{LO} of 480 mg/kg after 30 weeks of intermittent dosing (135), and toxicity effects data show a TC_{LO} of 400 mg/m^3 for inhalation in humans (136). Magnesium oxide is compatible with most chemicals; exceptions are strong acids, bromine pentafluoride, chlorine trifluoride, interhalogens, strong oxidizers, and phosphorous pentachloride.

Magnesium Peroxide

Industrial production of magnesium peroxide [*14452-57-4*], MgO_2, involves the reaction of magnesium oxide and hydrogen peroxide (qv). A product containing not more than 50% MgO_2 is obtained (137). Another process using magnesium hydroxide yields a product containing up to 60% magnesium peroxide (137). Most users prefer magnesium peroxide having residual $Mg(OH)_2$. Magnesium peroxide is stable in a range of oxygen partial pressures wider than those where MgO is stable. Magnesia is calculated to be stable at oxygen partial pressures $>3.6 \times 10^{-188}$ Pa (3.6×10^{-200} atm) at 25°C (119). Decomposition of MgO_2 has been observed to occur at temperatures from 245 to 375°C (137,138).

Uses. Magnesium peroxide is used mainly in medicine for treating hyperacidity in the gastric intestinal tract, and in the treatment of metabolic diseases such as diabetes and ketonuria. It is also used in the preparation of toothpaste and antiseptic ointments. All of these uses involve a mixture of magnesium peroxide, magnesium oxide, magnesium hydroxide, and an admixture of magnesium carbonate. Magnesium peroxide is also used in bleaching and agricultural applications (see BLEACHING AGENTS).

Magnesium Phosphate

An aqueous solution of monoammonium phosphate [*10361-65-6*] reacts with MgO to form ammonium magnesium phosphate hexahydrate [*15490-91-2*], $NH_4MgPO_4{\cdot}6H_2O$. Several other species of hydrated phosphates are created during this reaction which takes place quickly and produces compounds that have desirable properties as cementing agents. The hexahydrate is the most prevalent. Properties are given in Table 22.

Investment Castings. Magnesium phosphate compounds are used as cementing agents for the refractory material used in high temperature dental

Table 22. Physical Properties of Magnesium Phosphates[a]

Property	Farringtonite	Dittmarite	Struvite
CAS Registry Number	[*10043-83-1*]	[*16674-60-5*]	[*1309-48-4*]
molecular formula	$Mg_3(PO_4)_2$	$NH_4MgPO_4{\cdot}H_2O$	$NH_4MgPO_4{\cdot}6H_2O$
mol wt	262.85	155.33	245.40
crystal system	monoclinic	orthorhombic	orthorhombic
space group	$P2_1/n$	$Pmn2_1$	$Pm2_1n$
lattice constants, nm			
a	7.60	5.606	6.945
b	8.23	8.758	11.208
c	5.08	4.788	6.1355
Z[b]	2	2	2
density, calculated, g/cm^3	2.76	2.19	1.706
hardness, Mohs'		2	2
color	white to yellow	colorless	colorless to white
melting point, °C	1184		decrepitates
index of refraction	1.540, 1.544, 1.559	1.549, 1.569, 1.571	1.495, 1.496, 1.504

[a]Refs. 30,39 and 41.
[b]Number of formulas per unit cell.

investment castings (see DENTAL MATERIALS) (139,140). The initial ammonium magnesium phosphate compound develops strength within several hours of the onset of the cementing reaction. The casting is then dried in stages in order to develop further strength and to minimize shrinkage. Dehydration takes place initially at 50°C then proceeds up to 160°C, producing ammonium magnesium phosphate monohydrate [*16674-60-5*], $NH_4MgPO_4 \cdot H_2O$. The crystal structure changes during this drying step. Magnesium acts as the bonding atom between the corners of PO_4^{3-} tetrahedra. Further heating from 300 to 650°C drives off the ammonia to produce $Mg_2P_2O_7$ [*13446-24-7*] which results in a further reorientation of the crystal structure. Final heating of the casting to 1040°C develops the greatest strength as the compound reorients to anhydrous magnesium phosphate [*10043-83-1*], $Mg_3(PO_4)_2$ where three magnesium atoms act as the bonding atoms between two PO_4^{3-} tetrahedra. At this stage the processing of the dental investment is complete. These phosphate-bonded investment casting are used for the formation of noble metal alloy dental appliances. A discussion of magnesium phosphate investment cements is available (141).

Magnesium Phosphate Cements. The reaction of magnesia and various forms of ammonium phosphate has great utility in the production of fast setting concrete (142). Acceptable levels of concrete strength are developed within several hours of the initial reaction. Onset times can be varied by controlling the ratio of phosphate to polyphosphate in the starting aqueous mixture (143). The ultimate strength of the concrete is attained over a 7–28 day period through dehydration and ammonia loss.

Rapid development of concrete strength finds utility in the highway construction industry, where repairs on high volume thoroughfares may be accomplished quickly using this material. Magnesium phosphate cements are also amenable to application by pneumatic gunning equipment. Pneumatic gunning is useful for application of the cements to vertical, rounded, or irregular surfaces such as sewer pipes and exposed steel.

Additional Uses. Magnesium calcium phosphates find use as nutrient supplements in animal feeds. Magnesium pyrophosphate [*13446-24-7*], $Mg_2P_2O_7 \cdot 3H_2O$, spray dried at 100°C, takes on a fibrous form during the drying and dehydration process (144). This fibrous material is used as a strengthening agent in plastics. Granulated ammonium magnesium phosphate, $NH_4MgPO_4 \cdot H_2O$, is used as a slow-release fertilizer providing magnesium, nitrogen, and phosphorus.

Magnesium Sulfate

Magnesium sulfate [*7487-88-9*], $MgSO_4$, is found widely in nature as either a double salt or as a hydrate. The more important mineral forms are listed in Table 23.

Properties. Physical properties of anhydrous magnesium sulfate, kieserite, and epsomite, as well as physical properties of four less prominent hydrates, are listed in Table 24. The complexity of the $MgSO_4 \cdot H_2O$ system is apparent from Figure 9. Several metastable phases exist in very close proximity to stable phases, making equilibrium difficult to achieve. The magnesium sulfate hydrates tend to form supercooled solution and metastable solid phases, hence several hydrates may coexist in aqueous solution at a given

Table 23. Minerals Containing Magnesium Sulfate

Mineral name	CAS Registry Number	Formula
kieserite	[*14168-73-1*]	$MgSO_4 \cdot H_2O$
starkeyite	[*24378-31-2*]	$MgSO_4 \cdot 4H_2O$
pentahydrite	[*15553-21-6*]	$MgSO_4 \cdot 5H_2O$
hexahydrite	[*13778-97-7*]	$MgSO_4 \cdot 6H_2O$
epsomite	[*10034-99-8*]	$MgSO_4 \cdot 7H_2O$
vanthoffite	[*15557-33-2*]	$3Na_2SO_4 \cdot MgSO_4$
bloedite	[*15083-77-9*]	$Na_2SO_4 \cdot MgSO_4 \cdot 4H_2O$
langbeinite	[*13826-56-7*]	$K_2SO_4 \cdot 2MgSO_4$
leonite	[*15226-80-9*]	$K_2SO_4 \cdot MgSO_4 \cdot 4H_2O$
schoenite	[*15491-86-8*]	$K_2SO_4 \cdot MgSO_4 \cdot 6H_2O$
kainite	[*67145-93-1*]	$4KCl \cdot 4MgSO_4 \cdot 11H_2O$
polyhalite	[*15278-29-2*]	$K_2SO_4 \cdot MgSO_4 \cdot 2CaSO_4 \cdot 2H_2O$

temperature. Only the mono-, hexa-, and heptahydrates are stable. Ranges of stability are −5 to 48.2°C, 48.2 to 67.5°C, and >67.5°C, respectively. As can be seen from Figure 9, the monohydrate displays inverse aqueous solubility: solubility, g/100 g saturated solution, 37.1, 8, and 0.5, at temperatures of 67.5, 170, and 240°C, respectively. The hydrates begin thermal decomposition at approximately 150°C and are often hydrolyzed to oxysulfates which are analogous to the oxychlorides, concomitantly. Hydrates decomposing in the presence of concentrated sulfuric acid yield a stable anhydrous salt. This salt starts to decompose at about 900°C and decomposes completely at 1100°C, producing MgO, O_2, SO_2, and SO_3. $MgSO_4$ decomposes in the presence of carbon at about 750°C:

$$MgSO_4 + C \longrightarrow MgO + SO_2 + CO$$

Addition of alkali such as NaOH or $Ca(OH)_2$ to $MgSO_4$ solutions precipitates magnesium hydroxide. Addition of a soluble carbonate such as Na_2CO_3 precipitates nesquehonite [*14457-83-1*], $MgCO_3 \cdot 3H_2O$. Insoluble magnesium salts, eg, sulfite, phosphate, or stearate, are prepared by adding the appropriate soluble, (often alkali metal) reagent to a $MgSO_4$ solution.

The reaction of MgO and $MgSO_4$ solutions produces magnesium oxysulfate cements. Oxysulfate cements, defined as Mg salts precipitated in alkaline conditions, contain both hydroxide and sulfate salt components. The crystalline forms $MgSO_4 \cdot 3MgO \cdot 11H_2O$ [*65496-31-3*] and $MgSO_4 \cdot 5MgO \cdot 8H_2O$ have been isolated.

Magnesium sulfate forms many double salts, including naturally occurring minerals (Table 23). The sulfuric acid double salts $MgSO_4 \cdot H_2SO_4$ [*10028-26-9*], $MgSO_4 \cdot H_2SO_4 \cdot 3H_2O$ [*75198-53-7*], and $MgSO_4 \cdot 3H_2SO_4$ [*39994-66-6*] are crystallized from solutions of $MgSO_4$ in H_2SO_4. The amine double salts $MgSO_4 \cdot NH_3 \cdot 3H_2O$ [*75198-54-8*], $MgSO_4 \cdot 2NH_3 \cdot 4H_2O$ [*75198-56-0*], and $MgSO_4 \cdot 2NH_3 \cdot 2H_2O$ [*75198-55-9*] are products of $MgSO_4 \cdot 7H_2O$ and gaseous ammonia.

Manufacture and Processing. Anhydrous $MgSO_4$ can be prepared only by dehydration of a hydrate. Crystallization from aqueous solution is not possible. Aqueous solutions of $MgSO_4$ can be prepared by dissolving MgO, $Mg(OH)_2$, or

Table 24. Physical Properties of Magnesium Sulfates and Magnesium Sulfate Hydrates[a]

Property	$MgSO_4$	Kieserite	Epsomite	$MgSO_4 \cdot 3H_2O$	Starkeyite	Pentahydrate	Hexahydrate
mol wt	120.37	138.38	246.48	174.42	192.44	210.45	228.47
crystal system	orthorhombic	monoclinic	orthorhombic	orthorhombic	monoclinic	triclinic	monoclinic
space group	Cmcm	C_2/n	$P2_12_12_1$	PbCa	$P2_1/n$	$P\bar{1}$	A_2/a
lattice constants, nm							
a	0.5182	0.690	1.186	0.820	0.7902	0.6335	2.2442
b	0.7893	0.771	1.199	1.093	1.3594	1.055	0.7216
c	0.6506	0.754	0.6858	1.242	0.5920	0.6075	1.0119
angle, β degree		116.09			90.89	109.88	98.28
Z^b	4	4	4	8	4	2	8
density, g/cm^3							
calculated	2.908	2.571	1.678	2.082	2.009	1.904	1.718
observed	2.93		1.677			1.896	1.757
index of refraction	1.557, 1.582	1.520, 1.533, 1.584	1.4325, 1.4554, 1.4609	1.495, 1.497, 1.498	1.490, 1.491, 1.497	1.482, 1.492, 1.493	1.426, 1.453, 1.456
color	white	colorless	colorless		colorless	colorless	colorless
heat of formation, ΔH_{298}, kJ/mol[c]	−1284.9	−1602.1	−3388.6		−2496.6		−3087.0
free energy of formation, ΔG_{298}, kJ/mol[c]	−1170.7	−1428.8	−2871.9				−2632.2

[a]Refs. 3 and 39.
[b]Number of formulas per unit cell.
[c]To convert J to cal, divide by 4.184.

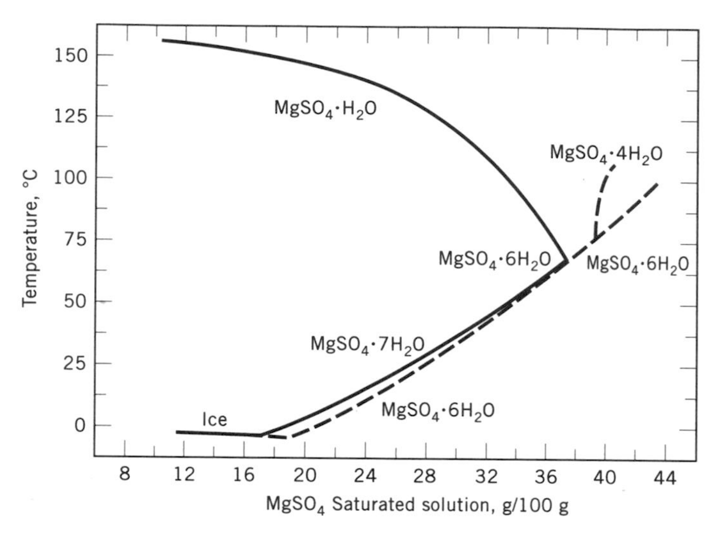

Fig. 9. The $MgSO_4{\cdot}H_2O$ system where the dashed lines represent metastable phases.

$MgCO_3$ in sulfuric acid; or absorbing SO_2 using a $Mg(OH)_2$ slurry to form the soluble bisulfite, $Mg(HSO_3)_2$, followed by air oxidation to SO_4^{-2}.

Technical-grade Epsom salt is prepared by dissolving MgO, $Mg(OH)_2$, or $MgCO_3$ in sulfuric acid. The reaction mixture is crystallized to separate the product. In one process $MgSO_4$ solution is recycled from crystallizers to a reaction vessel containing sulfuric acid and low reactivity MgO. After pH adjustment to slightly acidic conditions and a 4–5 h reaction time, a 34% $MgSO_4$ mother liquor at 82°C is produced. Iron is precipitated and insolubles are filtered from the mother liquor. Epsom salt is crystallized at 15°C and screened; the 24% $MgSO_4$ filtrate is recycled. The Epsom salt crystals are dried at low temperature in a rotary oven. Following filtration, the 34% mother liquor can be diluted to 24% and sold as a solution. The theoretical yield is 1 t of Epsom salt per ton MgO. The actual yield depends on particle size, reactivity, and purity of the MgO. The heat of reaction is often the determining factor for using MgO or $Mg(OH)_2$ as the reagent. The $Mg(OH)_2$ reaction generates only 65–75% of the heat that the MgO reaction does.

To prepare a USP-grade Epsom salt, higher purity MgO or $Mg(OH)_2$ is used. USP and food grades require low chloride levels, limiting allowable chloride content of the MgO to 0.08 wt %. Trace impurities including iron and aluminum are precipitated using excess MgO. Following crystallization, the Epsom salt is washed free of mother liquor.

Economic Aspects. Epsom salt is usually shipped in bulk or in 45-kg bags. Magnesium sulfate solution can be shipped in bulk, in either totes or drums. In January 1995 prices for a truckload in dollars per 100 kg were $MgSO_4{\cdot}7H_2O$ technical, 10% Mg, in bags \$37.49–\$40.79, works in bulk \$35.28; and USP, crystalline, works in 22.7 kg bags, \$39.69, in 45-kg bags, \$38.59–\$40.79 and, in bulk \$36.38. Year-end prices of 100 kg technical-grade Epsom salt for 1993, 1991, and 1990 were \$35.28, \$33.08, and \$33.08, respectively (52,103,130,145).

The quantities of magnesium sulfate (anhydrous and hydrous) shipped and used in the United States during the period 1989–1993 were (52,103,130):

Year	Quantity, t	Value, $ $\times 10^3$
1991	34,872	12,229
1992	37,889	13,919
1993	42,687	13,974

U.S. imports of magnesium sulfate, both natural kieserite and Epsom salts, from 1986 through 1988 were (52,103,146):

Year	Quantity, t	Value, $ $\times 10^3$
1986	24,648	1,711
1987	22,139	1,581
1988	30,405	1,865

U.S. imports and exports of natural kieserite and Epsom salts from 1989 through 1992 are given in Table 25 (52,103,146).

Uses. Magnesium sulfate is used primarily in the chemical and pharmaceutical industries (52). Smaller quantities are used for animal feed and for peroxide bleaching in pulp and paper manufacturing. In the peroxide bleaching process, magnesium sulfate protects the cellulose from being destroyed. Lower purity $MgSO_4$ is used as sizing and as a fireproofing agent (75). A process for manufacturing fire-proof construction material out of magnesium oxysulfate cements is depicted schematically in Figure 10.

Other uses include fertilizers, in medicine as a cathartic and analgesic, as a conditioning agent for wool and cotton in the textile industry, and as a mordant in dyeing. Magnesium sulfate is used as a coagulant in rubber and plastics, in plating baths, as a drying agent for organic solvents, in the manufacture of high fructose corn syrups, magnesium silicate adsorbents, citric acid, magnesium stearate [*557-04-0*], monosodium glutonate, and in photographic solutions.

Langbeinite is mined in the Carlsbad region of New Mexico. The International Minerals and Chemical Corp. (Germany) uses langbeinite to produce

Table 25. U.S. Imports and Exports of Natural Kieserite and Epsom Salts

	Imports		Exports	
Year	Quantity, t	Value, $ $\times 10^3$	Quantity, t	Value, $ $\times 10^3$
1989	166[a]	33[a]	95	96
1990	9,992	1,656	241	393
1991	6,524	1,265	809	497
1992	5,193	1,119	436	333

[a]Reported as kieserite only, does not include Epsom salts.

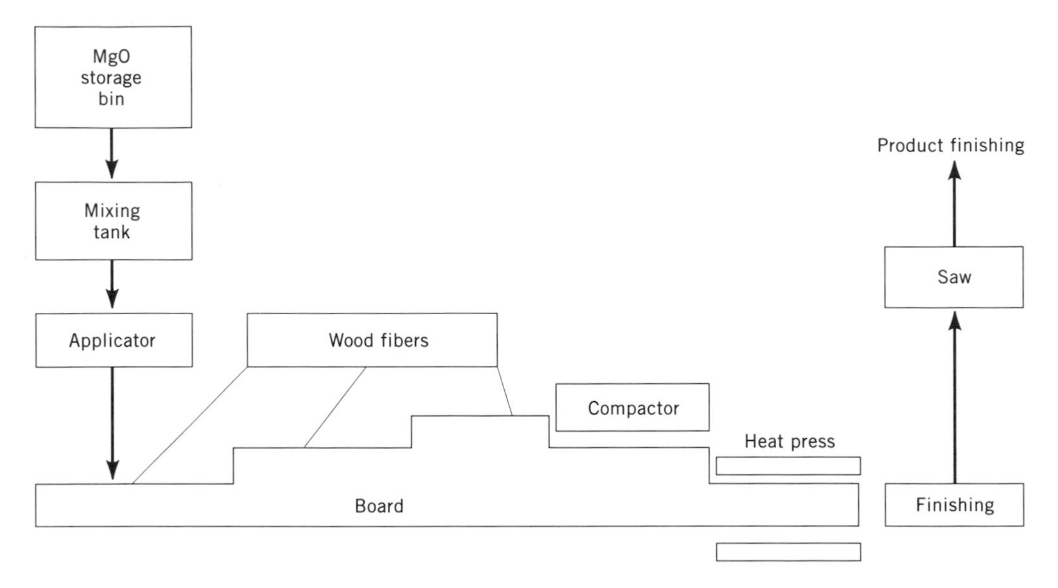

Fig. 10. Manufacturing of oxysulfate cement boards.

$MgCl_2$ by crystallizing out and decomposing carnallite, $KCl{\cdot}MgCl_2{\cdot}6H_2O$ (75). Because it also contains potassium sulfate, langbeinite is used widely as a fertilizer ingredient.

Magnesium Sulfite

Preparation and Properties. The white hexahydrate [*13446-29-2*], $MgSO_3{\cdot}6H_2O$, is prepared by adding an excess of sulfur dioxide, SO_2, to a suspension of magnesium hydroxide, $Mg(OH)_2$, or basic magnesium carbonate, [*12306-51-3*], $5MgO{\cdot}4CO_2{\cdot}5H_2O$. The formation of magnesium bisulfite (magnesium hydrogen sulfite), $MgHSO_3$, unisolable in solid form, in the presence of excess SO_2 increases the solubility of magnesium sulfite in the liquid phase. In dilute solutions of both magnesium sulfite and magnesium bisulfite, the solubility of magnesium sulfite increases with increasing temperature independent of $MgHSO_3$ concentration. The basic salt $11MgSO_3{\cdot}2Mg(OH)_2{\cdot}22H_2O$ forms as dilute solutions of magnesium sulfite are heated.

The $MgSO_3{\cdot}H_2O$ system is shown in Figure 11 and the properties of the tri- and hexahydrate are listed in Table 26.

Uses. *Flue Gas Desulfurization.* The system $Mg(OH)_2{\cdot}SO_2{\cdot}H_2O$ is employed in the scrubbing process for removing SO_x from flue gases (see SULFUR REMOVAL AND RECOVERY) (87). The equilibria involved in scrubbing has been studied in detail (147).

Typically flue gas containing 50–200 ppm SO_x is scrubbed in a venturi absorber. The absorbing medium is an aqueous slurry of $Mg(OH)_2$ also containing magnesium sulfite [*7757-88-2*], $MgSO_3$, and $MgSO_4$. Centrifuging can be used to remove magnesium sulfite (usually a mixture of the tri- and hexahydrate), some unreacted $Mg(OH)_2$, and some adhering $MgSO_4$ from the slurry. The solids

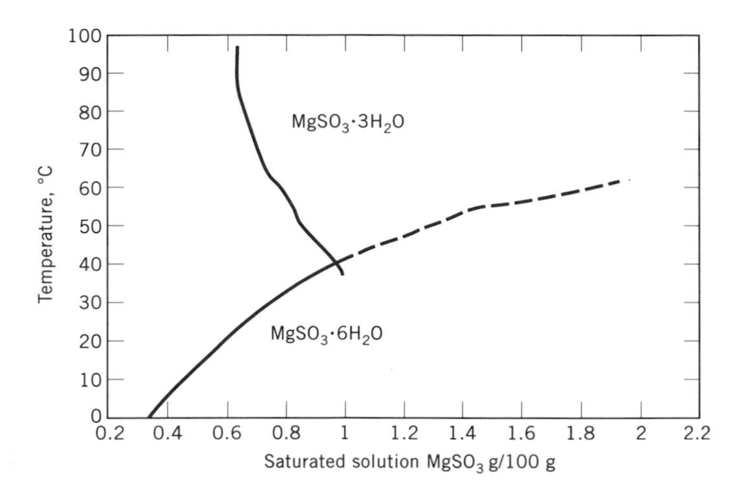

Fig. 11. The $MgSO_3 \cdot H_2O$ system (70).

Table 26. Properties of Magnesium Sulfite Tri- and Hexahydrates[a]

Property	$MgSO_3 \cdot 3H_2O$	$MgSO_3 \cdot 6H_2O$
CAS Registry Number	[*19086-20-5*]	[*13446-29-2*]
mol wt	158.42	212.47
crystal system	orthorhombic	hexagonal
space group	$Pbn2_1$	R3
lattice constants, nm		
a	0.939	0.88385
b	0.9584	
c	0.5523	0.9080
Z[b]	4	3
calculated density, g/cm^3	2.117	1.723
mp, °C		200 dec
index of refraction	1.552, 1.555, 1.595	1.464, 1.511
color	colorless	white
heat of formation,		
ΔH_{298}, kJ/mol[c]	–1931.8	–2817.5

[a]Refs. 3 and 39.
[b]Number of formulas per unit cell.
[c]To convert J to cal, divide by 4.184.

can be dried and calcined to recover MgO and SO_2 for reuse. MgO is reused in the scrubber, and SO_2 can be used to produce sulfuric acid (148).

Wood Pulping. The system $Mg(OH)_2 \cdot SO_2 \cdot H_2O$ is also used in acid bisulfite pulping. Compared to a calcium-based system which is not as amenable to regeneration of the pulping bisulfite (87), fewer technical problems are encountered in the digesters, evaporators, or recovery boiler of the Mg-based process. In the presence of excess SO_2, bisulfite forms in a 43% $MgSO_3$ solution, at 25°C and 101.3 kPa (1 atm) SO_2 pressure, to increase $MgSO_3$ solubility.

An acid bisulfite solution containing excess SO_2 is used as the pulping liquor in a few sulfite mills often devoted to dissolving pulp. By using excess sulfites, the lignin (qv) can be dissolved completely and removed leaving the cellulose behind. The black liquor that contains the inorganic salts and lignin after pulping can be concentrated to about 55% solids in multiple-effect evaporators and burned in a heat recovery boiler, at the same time recovering the pulping chemicals. Gases exiting the recovery boiler contain MgO that is recovered in cyclones and SO_2 that is absorbed in $Mg(OH)_2$ slurry.

Magnesium Sulfonates

Magnesium sulfonates are detergents containing magnesium carbonate or magnesium complexes as the metallic portion, and an oil-soluble magnesium-based substrate, dispersed as a colloid in petroleum oil. By definition a soap is commonly the sodium or potassium salt of a high molecular weight fatty acid. The term metallic soap refers to substitution of another metal for the sodium, in this case, magnesium.

Classification of detergents reflects their alkalinity. Magnesium sulfonates may be either neutral or overbased. Overbasing is the process of preparing oil-soluble detergents containing up to 30 times as much metal as the normal neutral detergents (145). These methods, discovered in the early 1950s, have led to the development of extensive patents. (More than 80 related U.S. patents have been issued since 1963.) The degree of overbasing is defined in part by total base number (TBN) which refers to the equivalent amount of potassium hydroxide, KOH, contained in the material. A base number of 100 corresponds to 2.14% Mg; for magnesium sulfonates a TBN of 300–400 is typical.

The overbasing process generally involves the addition of magnesium oxide to oil, solvent, a promoter, and water in a reflux vessel. Sulfonic acid is added to neutralize about 10% of the MgO. The resulting magnesium sulfonate acts as a crystal modifier to produce colloidal magnesium carbonate when carbon dioxide is bubbled through the mixture. The resulting product is equivalent to 10–16% MgO by weight. Approximately 90% of the magnesia is present as submicrometer, colloidal particles of $MgCO_3$, fine enough to preserve optical clarity. The magnesium sulfonate detergent accounts for the remaining 10% of magnesia. Filtration or centrifugation are used to remove any unreacted material or sludge followed by distillation to remove the solvent. In another process (146), noncarbonated magnesium sulfonates are prepared by heating a mixture of magnesium compounds, an oleophilic reagent containing a sulfonic acid, water, and an organic solubilizing agent above 30°C. The resulting product may be in liquid or in gel form.

Principal uses of magnesium sulfonates are as additives to engine oils, automatic transmission fluids, gear oils and industrial oils (see HYDRAULIC FLUIDS). In engine lubricating oils, in concentrations of 1–2%, the primary function is as a sludge dispersant and neutralizer of acidic contaminants from partially oxidized fuels, oil degradation products, and NO_x. The noncarbonated forms may be used also in corrosion-resistant coatings for metals, and as liquid fuel additives, in smoke suppression and in vanadium scavenging (146). North American producers of magnesium sulfonates include Lubrizol, Witco, and Amoco. In Europe magnesium salicyclates are an alternative detergent manufactured by Shell Chemical.

Magnesium Vanadates

Several forms of magnesium vanadates have been characterized. Some physical properties are summarized in Table 27 (28–30) (see also VANADIUM AND VANADIUM ALLOYS).

Fuels from areas having natural deposits of vanadium, such as Venezuela (147), may contain significant amounts of this metal which results in deposition of vanadium compounds in the boiler as the fuels are burned (see PETROLEUM). In 1963 the Long Island Lighting Co. (LILCO), Mineola, New York, began injecting magnesium oxide fuel additives to control corrosion and slagging problems in the power utility boilers and acid stack emissions. These additives produce magnesium vanadates and affect the type of slag produced, reducing slag adherence to heat-exchange surfaces, and facilitating removal (147).

The recovery of vanadium from these slags is of commercial interest because of the depletion of easily accessible ores and the comparatively low concentrations (ranging from less than 100 ppm to 500 ppm) of vanadium in natural deposits (147,148). In the LILCO applications the total ash contained up to 36% V_2O_5 (147). Vanadium is of value in the manufacture of high strength steels and specialized titanium alloys used in the aerospace industry (148,149). Magnesium vanadates allow the recovery of vanadium as a significant by-product of fuel use by electric utilities (see RECYCLING, NONFERROUS METALS).

Magnesium vanadates, as vanadium compounds in general, are known irritants of the respiratory tract and conjunctiva. The threshold limit value

Table 27. Physical Properties of Magnesium Vanadates[a]

Property	$Mg_{1.9}V_3O_8$	MgV_3O_8	$Mg_2V_2O_7$	$Mg_3V_2O_8$
CAS Registry Number		*[12181-49-6]*	*[13568-63-3]*	*[13568-68-8]*
mol wt	327.01	305.12	262.50	302.79
crystal system	monoclinic	orthorhombic	triclinic	orthorhombic
space group	C2/Cm/C2/m	IC2a/Icma	$P\bar{1}$	Aba2
lattice constants, nm				
a	1.0293	1.428	1.3767	0.831
b	0.853	0.840	0.5414	1.142
c	0.7744	0.986	0.4912	0.606
angle, degree				
α			81.42	
β	119.5		106.82	
γ			130.33	
Z[b]	2	4	2	4
density, g/cm3				
calculated	3.41	3.42	3.26	3.473
observed	3.37	3.39	3.1	
heat of formation				
ΔH_{298}, kJ/mol[c]			–2835.9	
free energy of formation				
ΔG_{298}, kJ/mol[c]			–2645.3	

[a] Refs. 28–30.
[b] Number of formulas per unit cell.
[c] To convert J to cal divide by 4.184.

(TLV) for vanadium compounds in air recommended by the National Institute of Occupational Safety and Health is 0.05 mg/m^3 based on a typical 8-h workday and 40-h workweek (7,147). Chronic inhalation can lead to lung diseases such as bronchitis, bronchopneumonia, and lobar pneumonia. These dust-related effects can be avoided by use of individual respirators in areas where exposure is likely.

BIBLIOGRAPHY

"Magnesium Compounds" in *ECT* 1st ed., Vol. 8, pp. 593–617, by G. H. Gloss, J. T. Baker Chemical Co.; in *ECT* 2nd ed., Vol. 12, pp. 708–736, by A. F. Boeglin and T. P. Whaley, International Minerals & Chemical Corp.; in *ECT* 3rd ed., Vol. 14, pp. 615–646, by S. N. Copp and R. Wardle, CE Basic.

1. *Powder Diffraction File*, Sets 1–29, JCPDS International Center for Diffraction Data, Swarthmore, Pa., 1985.
2. R. C. Weast, ed., *Handbook of Chemistry and Physics*, 70th ed., CRC Press Inc., Boca Raton, Fla., 1989.
3. J. A. Dean, ed., *Lange's Handbook of Chemistry*, 13th ed., McGraw-Hill Book Co., Inc., New York, 1985.
4. A. C. D. Rivett, *J. Chem. Soc.*, 1063 (1926).
5. E. A. Goode, N. S. Bayliss, and A. C. D. Rivett, *J. Chem. Soc.*, 1950 (1928).
6. N. I. Sax and R. J. Lewis, Sr., *Hawley's Condensed Chemical Dictionary*, 11th ed., Von Nostrand Reinhold Co., Inc., New York, 1987.
7. R. J. Lewis, Sr., *Sax's Dangerous Properties of Industrial Materials*, 8th ed., Vol. 3, Von Nostrand Reinhold Co., Inc., New York, 1992.
8. R. E. Lenga and K. L. Votoupal, *The Sigma-Aldrich Library of Regulatory and Safety Data*, Vol. 1, Sigma Chemical Co., St. Louis, Mo., and Aldrich Chemical Co., Inc., Milwaukee, Wis., 1993.
9. C. W. Kamienski and J. F. Eastham, *J. Organomet. Chem.* **34**, 116 (1969).
10. G. E. Coates and K. Wade, *Organometallic Compounds*, Vol. 1, Methuen & Co., Ltd. London, 1969, p. 97.
11. E. C. Rochow, D. T. Hurd, and R. N. Lewis, *The Chemistry of Organometallic Compounds*, John Wiley & Sons, Inc., New York, 1957, pp. 77–94.
12. G. E. Coates, *Organometallic Compounds*, Methuen & Co., Ltd. London, 1960, p. 54.
13. E. Wiess, *J. Organomet. Chem.* **2**, 314 (1964).
14. T. G. Brilinka and V. A. Shushunov, *Reactions of Organometallic Compounds with Oxygen and Peroxide*, CRC Press, Inc., Cleveland, Ohio, 1969, pp. 184–185.
15. G. E. Coates, *Organometallic Compounds*, Methuel & Co., Ltd. London, 1964.
16. Jpn. Pat. 90,173,105 A2, 02,173,105 (July 4, 1990), L. M. Chuang and S. J. Hu (to Formosa Plastics Corp.).
17. Eur. Pat. 177,689 A2 (Apr. 16, 1986), H. Moringa, S. Yamaoto, and T. S. Iwabuchi (to Nissan Chemical Industry, Ltd.).
18. C. Blombert, *J. Organomet. Chem.* **1**, 138 (1977).
19. U.S. Pat. 3,766,091 (Oct. 16, 1973), E. J. Vondenbert (to Hercules).
20. Brit. Pat. 1,401,920 (Aug. 5, 1975), A. Roggero and co-workers (to Snamprogetti).
21. K. Kamide, H. Ono, and K. Histani, *Polym. J.* **24**, 917 (1992).
22. V. Dimonie, *Makromol. Chem.* **93**, 171 (1973).
23. Eur. Pat. 487,169 A1 (May 27, 1992) G. J. Jakubicki and D. Warsheweski (to Colgate-Palmolive Co.).
24. G. Somasundaram and P. D. Sunavala, *Fuel* **68**, 921 (1989).
25. As quoted by Lithium Corp. of America, FMC, Gastonia, N.C., Aug. 1993.

26. As quoted by SAF Bulk Chemicals, Ronkonkoma, N.Y., Aug. 1993.
27. D. M. Considine, ed., *Scientific Encyclopedia*, Van Nostrand Reinhold, New York, 1989.
28. R. C. Weast, ed., *Handbook of Chemistry and Physics*, 60th ed., CRC Press Inc., Boca Raton, Fla., 1979.
29. J. A. Dean, ed., *Lange's Handbook of Chemistry*, 12th ed., McGraw-Hill Book Co., Inc., New York, 1979.
30. *Powder Diffraction File*, Sets 1–29, JCPDS International Center for Diffraction Data, Swarthmore, Pa., 1979.
31. A. Seidell, *Solubilities of Inorganic and Metal Organic Compounds*, 3rd ed., D. Van Nostrand Co., Inc., New York, 1940, and 4th ed., American Chemical Society, Washington, D.C., 1965.
32. American Chemical Society, Washington, D.C., 1993.
33. *Chem. Mark. Rep.*, (Jan. 1995).
34. G. Reade, *American Mineralogist* **55**, 1457 (1970).
35. J. Suzuki and J. Ito, *J. Assoc. Japanese Mineralogists* **68**, 639 (1973).
36. M. Fleisher, *Glossary of Mineral Species*, Mineralogical Record Publishing Co., Tucson, Ariz., 1991.
37. C. Hurlbut and C. Klein, *Dana's Manual of Mineralogy*, 20th ed., John Wiley & Sons, Inc., New York, 1988.
38. V. Parker, D. Wagman, and W. Evans, *NBS Technical Note 270-6*, U.S. Government Printing Office, Washington, D.C., 1971.
39. D. R. Lide, ed., *Handbook of Chemistry and Physics*, 74th ed., CRC Press, Inc., Boca Raton, Fla., 1993.
40. R. Boynton, *Chemistry and Technology of Lime and Limestone*, John Wiley & Sons, Inc., New York, 1980.
41. W. Roberts, ed., *Encyclopedia of Minerals*, Van Nostrand Reinhold Co., New York, 1989.
42. Brit. Pat. 9,102 (Sept. 24, 1841), H. L. Pattison.
43. C. Jacobson, ed., *Encyclopedia of Chemical Reactions*, Vol. 4, Reinhold Publishing Co., New York, 1951.
44. G. Clarke, *Industrial Minerals*, 45 (Apr. 1992).
45. D. Graf and J. Lamor, *Economic Geology* **50**, 639 (1955).
46. J. Wicken and L. Duncan, "Magnesite and Related Minerals," in *Industrial Minerals and Rocks*, 5th ed., American Institute of Mining, Metallurgical and Petroleum Engineers, New York, 1983.
47. B. Coope, *Mining Eng.* **45**, 575 (1993).
48. H. Williard and R. Gates, *Mining Eng.* **15**, 44 (1963).
49. Eur. Pat. 302,514 (Feb. 8, 1989). V. Bumbalek and co-workers (to Ustav pro Vyzkum Rvd).
50. Jpn. Pat. 89,232,830 (Apr. 23, 1989), T. Morifuji and co-workers (to Tokuyama Soda Co.).
51. U.S. Pat. 4,179,490 (Dec. 18, 1979), G. Judd (to Woodville Lime and Chemical Co.).
52. D. Kramer, "Magnesium and Magnesium Compounds," in *Minerals Yearbook*, U.S. Department of Interior, U.S. Government Printing Office, Washington, D.C., 1990.
53. *Chem. Mark. Rep.*, 40 (June 28, 1993).
54. N. Mansour, *Cemurgi Int.* **4**, 24 (1978).
55. Jpn. Pat. 01,224,218 (Sept. 7, 1989), Y. Takenaya, S. Sana, and K. Miharo (to Ube Chemical Industries, Ltd.).
56. *1992–1993 Threshold Limit Value for Chemical Substances and Physical Agents*, American Conference of Government Industrial Hygienists, Cincinnati, Ohio, 1992.
57. *Proceedings of the 101st AIME Annual Meeting*, San Francisco, Calif., Feb. 20–24, 1972.

58. J. L Reuss and J. T. May, *U.S. Bur. Mines Rep. Invest.*, RI-7922 (1974).
59. K. K. Kelley, *Energy Requirements and Equilibria in the Dehydration, Hydrolysis, and Decomposition of Magnesium Chloride*, Tech Paper 676, U.S. Dept. of Interior, Bureau of Mines, Washington, D.C., 1945.
60. *Chemical Technology: An Encyclopedic Treatment*, Vol. 1, Barnes & Noble, Inc., New York, 1968.
61. M. Lujan, Jr., *Minerals Yearbook*, U.S. Dept of Interior, Bureau of Mines, U.S. Government Printing Office, Washington, D.C., 1993.
62. U.S. Pat. 4,269,816 (May 26, 1981), E. E. Shackleton, A. J. Wickens, and J. H. W. Turner (to Mineral Process Licensing Corp.).
63. *Chem. Mark. Rep.* (Apr. 1992); as quoted by Dow Chemical Co., Ludington, Mich., Feb. 1995.
64. *Chemical Technology: An Encyclopedic Treatment*, Vol. 2, Barnes & Noble, Inc., New York, 1971.
65. L. Urwongse and C. A. Sorrell, *J. Am. Ceram. Soc.* **63**(9–10), 503 (1980).
66. U.S. Pat. 5,171,467 (Dec. 15, 1992), V. C. Mehta, R. C. Morrison, and C. W. Kamienski (to FMC Corp.).
67. E. I. Cooper and D. H. Kohn, *Ceramics Int.* **9**(2), 68–72 (1983).
68. J. C. W. Chien, *ACS Symp. Ser. 496, Catal. Polym. Synth.*, 27–55 (1992).
69. N. A. Ibrahim, R. Refai, and A. Hebeish, *Am. Dyestuff Rep.* **75**(7), 25–28 (1986).
70. W. F. Linke, ed., *Solubilities of Inorganic and Metal Organic Compounds*, 4th ed. Vol. 2 (K-Z), Amer. Chem. Society, Washington, D.C., 1965.
71. Jpn. Pat. 63,230,518 (Sept. 27, 1988), N. Keiichi and T. Kunio (to Asahi Glass KK).
72. G. Clarke, *Ind. Minerals*, 49 (Apr. 1992).
73. *Chem. Mark. Rep.* 7 (Jan. 24, 1994); *Chem. Mark. Rep.* (Oct. 3, 1994).
74. A. Richmond, *Manistee Plant Process and Products Seminar Study and Reference Manual*, Martin Marietta Magnesia Specialties, Inc., Baltimore, Md., 1992.
75. G. T. Austin, *Shreve's Chemical Process Industries*, 5th ed., McGraw-Hill Book Co., Inc., New York, 1984, pp. 189–190.
76. Ref. 75, p. 189.
77. Ref. 72, p. 63.
78. Ref. 72, p. 53.
79. G. L. Smithson and N. N. Bakhshi, *Canadian J. Chem. Eng.* **47**, 508–513 (Oct. 1969).
80. R. B. Corpstein, J. B. Fasano, and K. J. Myers, *Chem. Eng.* **101**(10), 138–144 (1994).
81. D. S. Dickey and R. R. Hemrajani, *Chem. Eng.* **99**(3), 82–94 (1992).
82. N. Brown and N. Heywood, *Chem. Eng.* **99**(9), 106–113 (1992).
83. N. I. Heywood and K. A. Mehta, *Proc. of Hydrotransport 11"*, Paper C2, Stratford, UK, 1988, pp. 131–156.
84. P. C. Hiemenz, *Principles of Colloid and Surface Chemistry*, Marcel Dekker, Inc., New York, 1977, pp. 306–351.
85. R. Darby, in N. P. Cheremisinoff, ed., *Encyclopedia of Fluid Mechanics*, Vol. 5, Gulf Publishing, Houston, Tex., 1986.
86. R. Darby, R. Mun, and D. Boger, *Chem. Eng.* **99**(9), 117–119 (1992).
87. A. Slack and G. Hollinden, *Sulfur Dioxide Removal from Waste Gases*, 2nd ed., Noyes Data Corp., Park Ridge, N.J., 1975, pp. 227–238.
88. J. Teringo III, *Products Finishing*, 48–55 (Aug. 1987).
89. S. Palmer and co-workers, *Metal/Cyanide Containing Wastes Treatment Technologies*, Pollution Technology Review No. 158, Noyes Data Corp., Park Ridge, N.J., 1988, pp. 373–377.
90. P. Hornsby and C. Watson, *Plas. Rubber Proc. Appl.* **6**(2), 169–175 (1986).
91. *Chemical & Engineering News*, 15–16 (Oct. 25, 1993).
92. U.S. Pat. 4,695,445 (Sept. 22, 1987), K. Nakaya and K. Tanaka (to Asahi Glass KK).

93. Eur. Pat. 537,013 (Apr. 14, 1993), K. A. Schryer (to General Electric Co.).
94. Jpn. Pat. 5,032,835 (Feb. 9, 1993), A. Nobuhiko, I. Hitoshi, and O. Yoshiji (to Tokuyama Soda KK).
95. Intl. Pat. WO 92/12097 (July 23, 1992), S. F. Mertz (to The Dow Chemical Co.).
96. U.S. Pat. 4,931,481 (June 5, 1990), N. Adam and R. Widermann (to Bayer AG).
97. U.S. Pat. 4,998,541 (Mar. 12, 1991), P. F. Perfetti and W. R. Cook (to R. J. Reynolds Tobacco Co.).
98. U.S. Pat. 5,060,675 (Oct. 29, 1991), E. Milford and co-workers (to R. J. Reynolds Tobacco Co.).
99. U.S. Pat. 4,915,118 (Apr. 10, 1990), C. Kaufman and R. Martin (to P. H. Glatfelter Co.).
100. *Chem. Mark. Rep.*, 35 (Aug. 30, 1993).
101. U.S. Pat. 4,987,173 (Jan. 22, 1991), K. Kazuaki, N. Manabu, and T. Ryuzo (to Idemitsu Petrochem Co.).
102. Jpn. Pat. 63,183,616 (July 29, 1988), H. Shigao (to Hitachi Maxell).
103. D. Kramer, *Annual Report—Magnesium and Magnesium Compounds–1991*, U.S. Dept. of Interior, Bureau of Mines, U.S. Government Printing Office, Washington, D.C., 1991, p. 18.
104. Ref. 100, p. 9.
105. *Chem. Mark. Rep.*, 5 (May 24, 1993).
106. Ref. 105, pp. 19–20.
107. *Chem. Mark. Rep.*, 34 (Aug. 16, 1993).
108. *1992–1993 Threshold Limit Values for Chemical Substances and Physical Agents*, American Conference of Governmental Industrial Hygienists, Cincinnati, Ohio, 1992.
109. Ref. 7, p. 2150.
110. P. K. Chowdhury, *J. Chem. Res.* **6**, 192 (1990).
111. A. G. Martinez and co-workers, *Tetrahedron Lett.* **32**(42), 5931–5934 (1991).
112. U.S.S.R. Pat. 822,883 (Apr. 23, 1981), Y. A. Dorfman and co-workers.
113. As quoted by SAF Bulk Chemicals, Ronkonkoma, N.Y., Feb. 1995.
114. *Chem. Eng. News*, **36**, 40 (1958).
115. T. J. W. Van Thoor, ed., Ref. 66.
116. J. A. Kent, *Riegel's Handbook of Industrial Chemistry*, 8th ed., Van Nostrand Reinhold Co., New York, 1983.
117. *Handling Chemicals Safely*, 2nd ed., Dutch Assoc. of Safety Experts, Dutch Chemical Industry Assoc., and Dutch Safety Inst., 1980.
118. L. Bretherick, *Handbook of Reactive Chemical Hazards*, 3rd ed., Butterworths, London, 1985.
119. H. A. Wriedt, *Bulletin of Alloy Phase Diagrams* **8**(3), 227–233 (June 1987).
120. C. Palache, H. Berman, and C. Frondel, *The System of Mineralogy of James Dwight and Edward Salisbury Dana*, 7th ed., John Wiley & Sons, Inc., New York, 1944, pp. 498, 636.
121. M. Neuberger and D. B. Carter, *Magnesium Oxide*, Electronic Properties Information Center, Culver City, Calif, 1969, p. 3.
122. J. R. Hague and co-workers, *Refractory Ceramics for Aerospace*, The American Ceramic Society, Inc., Columbus, Ohio, 1964, p. 227.
123. R. J. Hall and D. R. F. Spencer, *Interceram* **22**(3), 212 (1973).
124. W. C. Gilpin and N. Heasman, *Chem. Ind.* **14**, 567 (1977).
125. T. Glasscock, *Chem. Process.* **43**(1), 40 (1980).
126. *United Nations Publication No. E.17.II.B.25*, United Nations, New York, 1972.
127. Austrian Pat. 1,560,568 (Feb. 6, 1980), H. Grohmann and M. Grill (to Veitscher Magnesitwerke).
128. Brit. Pat. 2,023,563 (Jan. 3, 1980), M. Grill and H. Grohmann (to Veitscher Magnesitwerke).

129. G. Clark, *Ind. Minerals*, 74 (Apr. 1992).
130. D. Kramer, *Annual Report—Magnesium and Magnesium Compounds—1993*, U.S. Department of the Interior, Bureau of Mines, U.S. Government Printing Office, Washington, D.C., 1993, p. 19.
131. *Chem. Mark. Rep.*, 30 (Jan. 16, 1995).
132. U.S. Pat. 5,211,852 (May 18, 1993), R. Van de Walle and co-workers (to Martin Marietta Magnesia Specialties, Inc.).
133. Ref. 130, p. 718.
134. *Federal Register*, Vol. 54, U.S. Government Printing Office, Superintendent of Documents, Washington, D.C., 1989, p. 2923.
135. *Cancer Res.* **33**, 2209 (1973).
136. G. D. Clayton and F. E. Clayton, eds., *Patty's Industrial Hygiene and Toxicology*, 3rd rev. ed., Vol. 2A, John Wiley & Sons, Inc., 1981, p. 1745.
137. I. I. Vol'nov, *Peroxides, Superoxides, and Ozonides of Alkali and Alkaline Earth Metals*, Nauka, Moscow, 1964, in Russian, Trans. Plenum Press, New York, 1966.
138. B. Lorant, *Siefen-Ole-Fette-Wachse* **92**(20), 644–647 (1966).
139. *Guide to Dental Materials and Devices*, 8th ed., Amer. Dental Assoc., Chicago, 1978.
140. *Dental Phosphate-Bonded Casting Investments*, International Standard ISO 9694, American National Standards Institute, New York, 1988.
141. R. Neiman and A. Sarma, *J. Dental Res.* **59**, 1478 (1980).
142. *L. Cartz in Cements Research Progress 1975*, American Ceramics Society, Columbus, Ohio, 1976.
143. U.S. Pat. 4,059,455 (June 17, 1976), R. Russell and R. Limes, (to Republic Steel Corp.).
144. Jpn. Pat. 1,215,709 (Aug. 29, 1989), K. Otsuka (to Agency of Industrial Scientific Technology).
145. *Chem. Mark. Rep.* (Jan. 16, 1995).
146. National Trade Data Bank, *U.S. Merchandise Export and Import Trades, Titles 2,833,210,000* (*Magnesium Sulfate*) and *2,530,200,000* (*Kieserite, epsom salts [natural magnesium sulfate]*), U.S. Department of Commerce, Bureau of the Census, Washington, D.C, 1993.
147. C. H. Rowland and A. H. Abdulsattar, *Environ. Sci. Technol.* **12**, 1158 (1978).
148. *Flue Gas Desulfurization and Sulfuric Acid Production Via Magnesia Scrubbing*, U.S. Environmental Protection Agency, Technology Transfer, EPA 625/2-75/077, Washington, D.C., 1975.
149. E. R. Booser, ed., *CRC Handbook of Lubrication, Vol. 1, Application and Maintenance*, CRC Press Inc., Boca Raton, Fla., 1983, p. 18.

L. C. Jackson
S. P. Levings
M. L. Maniocha
C. A. Mintmier
A. H. Reyes
P. E. Scheerer
D. M. Smith
M. T. Wajer
M. D. Walter
J. T. Witkowski
Martin Marietta Magnesia Specialties, Inc.

MAGNETIC MATERIALS

BULK

All materials that are magnetized by, ie, exhibit a response in, a magnetic field are magnetic materials. Magnetism is classified according to the nature of the magnetic response, ie, diamagnetism, paramagnetism, ferromagnetism, antiferromagnetism, ferrimagnetism, metamagnetism, parasitic ferromagnetism, and mictomagnetism (spin glass). Although the observed response to an applied field is the same for ferromagnetism and ferrimagnetism, there is a distinct difference in the microscopic–atomistic source of the observed magnetic behavior. In a fully magnetized ferromagnet, the atomic–ion moments are all aligned in the same direction. This is not so for a ferrimagnet and hence the magnitude of the observed bulk magnetization is smaller. The extreme case of zero magnetization is called antiferromagnetic behavior. Most commercially important magnetic materials are ferromagnets and ferrimagnets (see also FERRITES).

Theory

Types of Magnetism. The two atomic origins of magnetism are the spin and orbital motions of electrons. Diamagnetism occurs when the orbital rotation of the electrons is induced electromagnetically by an applied field. This weak magnetism is characterized by magnetization that is directed opposite to the applied field. The susceptibility $\kappa = M/H$, where M is the magnetization and H is the magnetic-field strength, is ca 10^{-5} and, with few exceptions, is independent of temperature. Many metals and most nonmetals are diamagnetic. These substances usually contain electrons that constitute a closed shell such that the atom as a whole has no permanent magnetic moment.

In paramagnetism, the magnetization is aligned parallel to the applied field and the susceptibility is 10^{-3} to 10^{-5}. Most paramagnetic materials contain atoms or ions that have a permanent magnetic moment but which are unaligned except in the presence of an applied field. In the latter case, the susceptibility is independent of the applied field and is inversely proportional to the temperature; examples include many salts of the iron and rare-earth families, platinum and palladium metals, and ferromagnetic materials when these are above their Curie point. Conduction electrons, which form an energy band in metals, exhibit a Pauli paramagnetism the susceptibility of which is independent of temperature. Thus, in some substances, eg, the alkali metals and metals such as copper [*7440-50-8*] (qv), silver [*7440-22-4*], and gold [*7440-57-5*], both diamagnetism and Pauli paramagnetism exist and the net value of susceptibility depends on the difference of two comparable quantities. Susceptibility is positive for the alkali metals and negative for Cu, Ag, and Au.

Where the permanent magnetic moments are aligned as a result of a strong positive interaction among neighboring atoms or ions, the material exhibits a spontaneous magnetization and ferromagnetism results. Examples include iron [*7439-89-6*] (qv), nickel [*7440-02-0*], cobalt [*7440-48-4*], and their alloys as well as many of the rare-earth elements (see LANTHANIDES). The principal source of magnetism in Fe, Ni, and Co is the spin motion of the electrons, whereas orbital motion contributes substantially to the magnetism of the rare earths.

In the case where the permanent magnetic moments are aligned antiparallel as a result of a strong negative interaction, the complete cancellation of the neighboring atomic moments results in zero net magnetization called antiferromagnetism. Examples include chromium [*7440-47-3*], manganese [*7439-96-9*], manganese(II) oxide [*1344-43-0*], MnO, and nickel oxide [*1313-99-1*], NiO. In compounds containing two or more kinds of atoms or ions that exhibit different values of magnetic moment, the cancellation is incomplete. The result is a net value of magnetization, as in the case of ferromagnetism. Such incomplete antiferromagnetism is called ferrimagnetism and is exemplified by the ferrites (qv), metal oxides that have a spinel structure.

Metamagnetism refers to the appearance of a net magnetization resulting from a transition from antiferromagnetism to ferromagnetism by the application of a strong field or by a change of temperature. Manganese diauride [*12006-65-4*], $MnAu_2$ and iron(II) chloride [*7758-94-3*], $FeCl_2$, undergo the transition by field application, and heavy rare-earth metals, eg, terbium [*7440-27-9*], dysprosium [*7429-91-6*], and holmium [*7440-60-0*], undergo the transition by temperature change.

Parasitic ferromagnetism is a weak ferromagnetism that accompanies antiferromagnetism, eg, in α-ferric oxide [*1309-37-1*], α-Fe_2O_3. Possible causes include the presence of a small amount of ferromagnetic impurities, defects in the crystal, and slight deviations in the directions of the plus and minus spins from the original common axis.

Mictomagnetism, or spin glass, refers to the onset of short-range magnetic order in alloys that have spin orientations which are frozen at a critical low temperature. In contrast to ferromagnetism and antiferromagnetism, however, this magnetic order is not long range. Spin glass is associated with a cusp in the susceptibility at a critical temperature and generally occurs in alloys containing a small fraction of magnetic atoms embedded in a nonmagnetic matrix, eg, copper manganese (1:1) [*12272-98-9*], CuMn, or gold ferride (1:1) [*12399-92-2*], AuFe. Mictomagnetic behavior, however, includes other features, eg, field cooling effects, such as displacements of the magnetization curve from zero field, and occurs in more concentrated alloys where incidences of structural clusters of near-neighbor magnetic atoms are high (1,2).

Magnetic Domains. Magnetic domains are associated with ferromagnetic, antiferromagnetic, or ferrimagnetic solids. In the demagnetized condition, these materials do not possess a net magnetization in the bulk because there are domains, ie, local regions within which the magnetic moments of all atoms are aligned. The direction of these moments, however, changes from one domain to another such that the net magnetization is zero for the solid. The transition region between domains is the domain wall or boundary and is a region of high energy. Typical domain sizes are 10^{-1} to 10^{-5} cm and domain-wall thicknesses are ca 100 nm for Fe. Domains arise as a result of minimizing a total energy composed of three principal contributions. The exchange energy tends to align all magnetic moments in one direction which maximizes the magnetostatic energy. This energy can be reduced by subdividing the solid into domains. The direction of alignment of the magnetic moments inside each domain, however, is not arbitrary. Instead, alignment is directed along an easy axis, a direction of minimum anisotropy energy.

Magnetic Anisotropy Energy. There are several kinds of magnetic anisotropy energy and perhaps the most well known is the magnetocrystalline anisotropy. Only a crystalline solid has this property because the energy is dictated by the symmetry of the crystal lattice. For example, in bcc Fe, the easy axis is in a ⟨100⟩ direction and in fcc Ni, it is in a ⟨111⟩ direction.

Another kind of magnetic anisotropy is magnetostrictive anisotropy, through which the magnetic moments tend to respond to the application of a stress. Conversely, a change in dimension of the material accompanies the application of an applied field. If the alignment is parallel to a tensile-stress axis, the material has positive magnetostriction; if the alignment is perpendicular, the material has negative magnetostriction. Nickel is an example of the latter.

Shape anisotropy is related to the magnetostatic energy of a magnet. A needle-shaped sample tends to line the atomic moments along the needle axis and a disk-shaped sample tends to line these moments parallel to the disk surface.

Thermomagnetic anisotropy, also referred to as induced anisotropy or magnetic annealing anisotropy, is a uniaxial anisotropy that is developed in some materials, primarily alloys, by heat treating below the Curie temperature in the presence of a magnetic field. The origin is thought to be a short-range directional ordering of the nearest neighboring atom pairs (3).

Some materials that are atomically ordered also develop a slip-induced anisotropy as a result of plastic deformation. The origin is thought to be identical to that of thermomagnetic anisotropy, ie, short-range directional order, except that the order is brought on by deformation rather than by heat treatment in a field (3,4).

Magnetostrictive, shape, and thermomagnetic anisotropies are not properties solely of the crystalline lattice and, therefore, also can be manifested in amorphous materials. The mechanism of slip-induced anisotropy requires the existence of atomic ordering (long- or short-ranged) and, hence, a crystalline structure. In alloys such as silicon steel for transformer applications, magnetocrystalline energy predominates. In alloys such as the permalloys (Ni–Fe-based), several kinds of anisotropy energy can dominate depending on the work history and thermal treatment of the material.

Technical Magnetic Behavior. When a magnetic-field strength H is applied to a ferromagnetic or ferrimagnetic material, the latter develops a flux density or induction B as a result of orientation of the magnetic domains. The relation between B and H is

$$B = \mu_0(H + M) = \mu_0 H + J$$

where B, the number of lines of magnetic flux per unit of cross-sectional area, is in T ($= 1.0 \times 10^{-4}$ G) or Wb/m^2; H is in A/m (= Oe/79.58); M, the magnetization, is in A/m; and μ_0, the permeability of free space, is a constant equal to

$4\pi \times 10^{-7}$ (T·m)/A (1.0 G/Oe). The product $\mu_0 M$ is the magnetic polarization and is given in T, and generally is denoted by J. Discussions and recommendations regarding these systems are available (5–8).

A plot of induction vs field strength for an initially demagnetized material is shown in Figure 1. The curve is interpreted in terms of rearrangement of the domain structure. In the demagnetized state, the magnetizations of various domains are oriented randomly so that there is no net magnetization for the sample. When a small field is applied, domains that are favorably oriented, with respect to the field direction, grow at the expense of the unfavorable domains by the reversible motion of domain walls away from pinning points, eg, inclusions and grain boundaries. This motion results in a small rise in induction. At higher fields, the induction rises rapidly as the motion becomes irreversible when the walls break loose from their pinning points. Each grain then consists of a single domain with its magnetization directed along the local easy axis, which is dictated by the anisotropy energy, nearly parallel to H. At higher fields, the local magnetization rotates reversibly into the direction of H. The induction increases very slowly as the polarization reaches its saturation value J_s, also referred to as M_s, the saturation magnetization, which is an intrinsic characteristic of the material. For soft magnetic materials, J_s is reached at low fields, in which case the value of saturation induction B_s is practically equal to J_s.

As the applied field is reduced, the induction does not retrace curve 1 but follows curve 2 as the domains at first merely rotate back to the nearest local easy axis direction. The value of B at $H = 0$ is B_r, the remanent induction. At $H = -H_c$, which is the coercive-field strength of coercive force or coercivity, the induction is zero. Upon increase of field, the induction follows curve 3, completing the hysteresis loop. Commercial magnetic materials generally are classified into soft magnets, where $H_c \leq 10$ A/cm (12.6 Oe), and hard or permanent magnets,

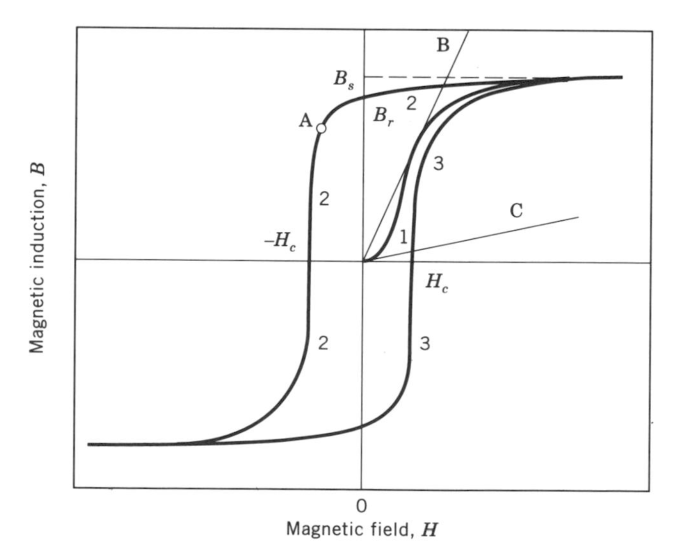

Fig. 1. Magnetic hysteresis loop of an initially demagnetized material (curve 1) where point A corresponds to $(BH)_{\max}$; and the slopes of lines B and C, with tangents to curves 1 and 2, represent $\mu_m = B/H$ and μ_i, respectively. Terms are defined in the text.

where $H_c \geq 100$ A/cm (126 Oe). A number of applications also make use of semihard magnets in which 10 A/cm $< H_c <$ 100 A/cm.

For permanent-magnet materials where the coercivity is large, the demagnetization curve, which corresponds to the second quadrant of the hysteresis loop, sometimes is plotted as the polarization $J(= B - \mu_0 H)$ vs $H(B - H$ *vs* $H)$ to show the intrinsic characteristics of the material. The value of H, for which $J = 0$, is the intrinsic coercivity H_{cJ}, whereas the usual coercivity, for which $B = 0$, is denoted by H_{cB} or H_c. For permanent magnets, the value $(BH)_{\max}$, the maximum energy product, is an important measure of quality. The value $(BH)_{\max}$, indicated in the second quadrant of Figure 1, represents the point of maximum efficiency where a given amount of magnetic flux is produced by the smallest amount of material.

The permeability $\mu = B/H$ is important information for soft magnetic materials. The most often quoted values are the initial permeability μ_i and the maximum permeability μ_m. These correspond to the initial and maximum slopes of the virgin magnetization curve, respectively (Fig. 1). Because the value of μ_i at $B/H = 0$ needs to be extrapolated from measurements at finite H, the value often is quoted in commercial catalogues at specific B or H. The quoted values usually are relative to the free-space value μ_0.

Other forms of permeability often are quoted as related to specific applications (9). A term closely related to the permeability is the volume susceptibility $\kappa = M/H$ which particularly characterizes diamagnetic and paramagnetic substances. A variety of definitions of susceptibilities is given in Reference 5.

When a ferromagnetic or ferrimagnetic substance is heated, the value of saturation polarization J_s decreases with temperature and reaches a zero value at the Curie temperature T_C or Curie point. Above this temperature, the material becomes paramagnetic because the magnetic moments of the atoms are prevented by thermal agitation from spontaneous alignment. The equivalent temperature for the transition from antiferromagnetism to paramagnetism is the Néel temperature.

Soft Magnetic Materials

Soft magnetic materials are characterized by high permeability and low coercivity. There are six principal groups of commercially important soft magnetic materials: iron and low carbon steels, iron–silicon alloys, iron–aluminum and iron–aluminum–silicon alloys, nickel–iron alloys, iron–cobalt alloys, and ferrites. In addition, iron–boron-based amorphous soft magnetic alloys are commercially available. Some have properties similar to the best grades of the permalloys whereas others exhibit core losses substantially below those of the oriented silicon steels. Table 1 summarizes the properties of some of these materials.

Iron and Low Carbon Steel. Research-grade (99.99+% pure) iron (qv) generally is obtained by zone refining (qv) or electrodeposition from high purity salts; total metallic and nonmetallic impurity levels are as low as 10–20 ppm. Such purity generally is too expensive for commercial use. Commercially pure iron, often referred to as Armco iron or magnetic ingot iron, is 99.9% pure.

Table 1. Magnetic Properties of Fully Annealed Iron and Iron Alloys[a]

Iron and alloys	B_s, T[b]	Density, g/cm^3	Resistivity, $\mu\Omega\cdot$cm	H_c (B_m = 1 T),[b] A/cm[c]	Permeability, A/cm[c]		Core loss (1.5 T,[b] 60 Hz), W/kg[d]		
					H = 0.8	H = 8	0.35 mm	0.46 mm	0.64 mm
magnetic ingot iron									
cast	2.15	7.85	10.7	0.68	3,500	1,500			
0.2-cm sheet	2.15	7.85	10.7	0.88	1,800	1,575			13.20
electromagnet iron, 0.2-cm sheet	2.15	7.85	12.0	0.81	2,750	1,575			
hydrogen-annealed iron	2.15	7.85	10.1	0.04	14,000	1,580			
low carbon steel, decarburized	2.14	7.85	12.5	0.70	2,000	1,530	8.10	9.2	11.44
cold-rolled									
M36 Si–Fe	2.04	7.75	41.0	0.36	7,400	1,485		3.85	4.73
M22 Si–Fe	1.98	7.65	49.0	0.31	8,100	1,450		3.63	4.29
M6 (110)[001] 3.2% Si–Fe	2.03	7.65	48.0	0.06	16,000	1,820	1.45		

[a] Ref. 10.
[b] To convert T to G, multiply by 10^4.
[c] To convert A/cm to Oe, divide by 0.7958.
[d] At thickness shown.

The chief impurity is oxygen (0.15 wt %) which is tied up as oxide inclusions and which does not greatly affect the magnetic properties. Typical magnetic properties of annealed sheet which has been prepared from magnetic ingot iron are shown in Table 1 (10) and Figure 2. The annealing is carried out at about 800°C. The carbon and nitrogen that are dissolved in the lattice at this temperature can precipitate at low temperature, causing magnetic aging which is manifested by a rise in coercivity. Nitrogen, in particular, precipitates slowly at room temperature, but the aging can be hastened by a heat treatment at ca 100°C which stabilizes the material.

Another way of stabilizing the iron is to add elements, eg, titanium and aluminum, to tie up the carbon and nitrogen in stable compounds. This type of iron is known as electromagnet iron and its magnetic properties are shown in Table 1. The magnetic properties of both types of commercially pure iron are nearly equivalent. The electromagnet iron has a somewhat larger resistivity and smaller coercivity. Iron having exceptionally low coercivity and high permeability can be prepared by careful annealing in pure H_2 at temperatures above 1300°C.

Low carbon steel for use as soft magnetic material contains <0.10 wt % C. It is considerably less expensive than the Si–Fe alloys and has reasonably good magnetic properties after decarburization. Therefore, it is widely used in low or intermittent duty motors where cost is of primary importance and loss is secondary. Typical magnetic properties of decarburized, low carbon steel sheets are shown in Table 1. Decarburization generally is carried out at ca 800°C in a

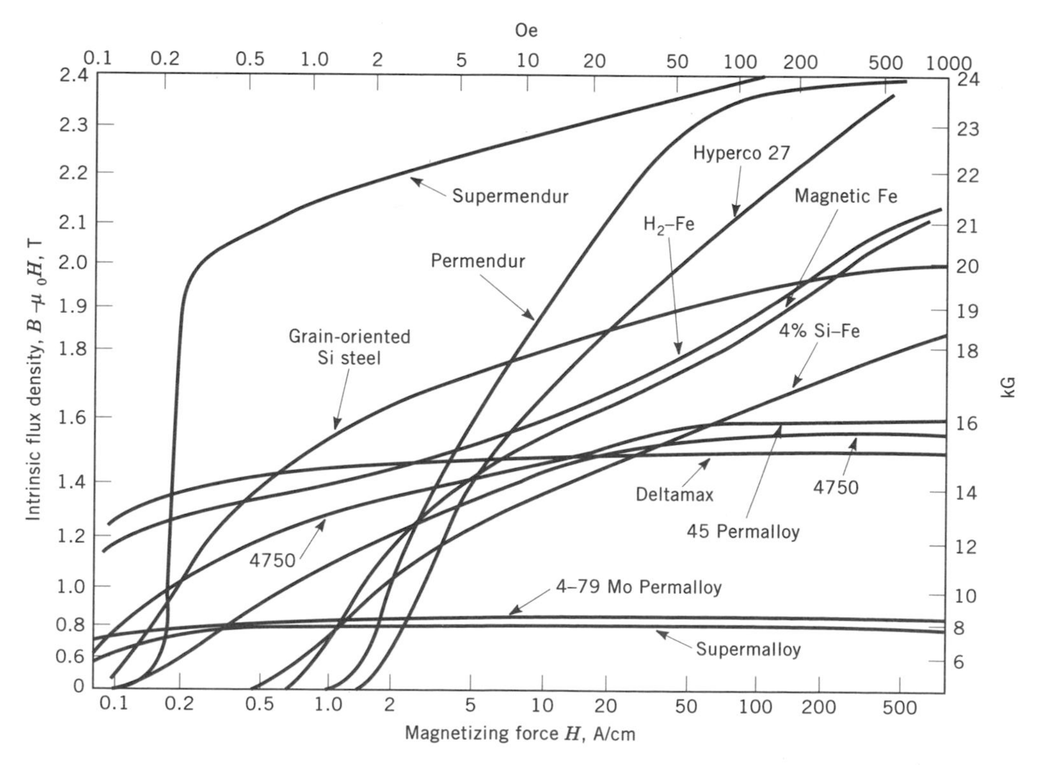

Fig. 2. Magnetization curves of commercial soft magnetic materials.

suitable decarburizing atmosphere, eg, wet hydrogen. The carbon level can be lowered to 0.002–0.005% to decrease coercivity. The presence and distribution of the insoluble carbon influences coercivity. For example, the coercivity of an SAE 1010 steel having a fine lamellar cementite, ie, Fe_3C, structure can be reduced by one half by a spherodization annealing at 680°C (11).

Commercial low carbon sheet steels for electrical applications are furnished in one of four grades according to the American Iron and Steel Institute (AISI) (12): cold-rolled lamination steel, type 1, ie, low grade material furnished to a controlled chemical composition in C, Mn, and P, which may be full-hard or annealed; cold-rolled lamination steel, type 2, ie, intermediate-grade material having magnetic properties improved by special mill processing; cold-rolled lamination steel, type 2S, ie, high grade material having guaranteed core-loss limits as referred to material decarburized to ca 0.005% C at 788°C for 1 h; and hot-rolled sheet steel which is used primarily for d-c applications and used as a low cost substitute for Armco iron.

Iron–Silicon Alloys. Iron–silicon alloys, commercially known as silicon steel, contain silicon up to about 4%. The addition of silicon to iron results in several beneficial effects: (*1*) silicon increases the electrical resistivity, thereby reducing the eddy-current loss and enhancing the a-c use of the alloy. This increase, together with the changes in some magnetic properties, is illustrated in Figure 3 (10). A fairly good approximation (13) relating the silicon content and resistivity ρ in $\mu\Omega\cdot$cm in commercial alloys is $\rho = 13.25 + 11.3(\%\ \text{Si})$. (*2*) Because Si reduces the size of the γ loop (fcc-phase field), high temperature heat treatment for purification and orientation control is possible without the deleterious effect of the α-γ-phase transformation. (*3*) Si decreases the value of the magnetocrystalline anisotropy energy K_1 (Fig. 3) and thus tends to enhance the permeability and to decrease the core loss. The decrease of the magnetostriction γ toward zero near 6% Si is particularly attractive in terms of lower core loss and transformer noise. However, the brittleness of alloys beyond about 4% Si has precluded utilization of such high silicon content.

The detrimental effects of Si addition are (*1*) Si increases the yield strength and decreases the ductility of iron such that commercial-grade materials are limited to ca 4% Si, and (*2*) as shown in Figure 3, the saturation induction and Curie temperature are decreased with increasing silicon content.

Silicon steel is used primarily at 60 or 50 Hz; thus the most important quality factor is core loss at a given induction level. Core loss has been improving. From 1900–1930, the improvement primarily resulted from advances in the art of steelmaking and processing. In 1934, the {110} ⟨001⟩ or cube-on-edge (COE) grain-oriented material was developed through a combination of cold work and recrystallization (14). In COE-oriented material a majority of grains have a {110} plane parallel to the sheet surface and a ⟨001⟩ direction parallel to the rolling direction. Because ⟨001⟩ is an easy axis for Si–Fe alloys, the domains are favorably oriented for high permeability and low loss when a field is applied in the rolling direction of the grain-oriented material. Guaranteed core-loss limits were critically specified at 1.5 T (1.5×10^4 G). Developments in manufacturing technology, eg, basic oxygen and electric furnaces for melting and refining and continuous hot and cold rolling of sheet stock using improved gauge and flatness control, have contributed to continuing improvements. Typical magnetic

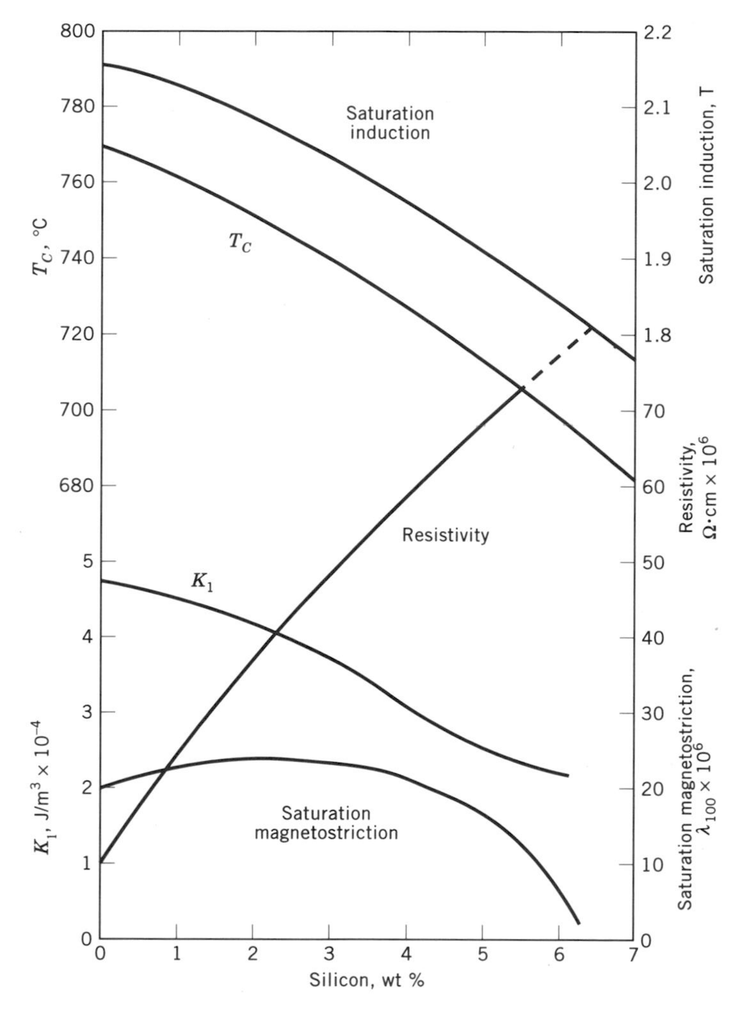

Fig. 3. Effect of silicon on properties of iron (10). T_C = Curie temperature; K_1 = magnetocrystalline anisotropy constant. To convert T to G, multiply by 10^4; to convert J/m^3 to erg/cm^3, multiply by 10.

properties of nonoriented (M36 and M22) and 3.2% Si-oriented material (M6) are given in Table 1 and in Figure 2.

Grades of high induction, low loss material of exceptionally sharp COE orientation were introduced commercially in 1968 (15). As a result, core-loss guarantees are quoted at 1.7 T (1.7 × 10^4 G). Values of core loss of these new high induction grades along with conventional oriented grades are given in Table 2.

The development of a sharp COE texture in the finished strip requires complex control of numerous variables. The conventional commercial process (18) involves hot-rolling a cast ingot at ca 1370°C to a thickness of about 2 mm, annealing at 800–1000°C, and then cold-rolling to a final thickness of 0.27–0.35 mm in two steps of 70 and 50%, respectively, with a recrystallization

Table 2. Core Loss of Grain-Oriented Silicon Steels, W/kg[a]

Grade	0.27 mm		0.30 mm		0.35 mm	
	50 Hz	60 Hz[b]	50 Hz[c]	60 Hz	50 Hz[c]	60 Hz
			High induction			
M0H	(0.99)	1.32	1.05	(1.41)		
M1H	(1.04)	1.40	1.11	(1.49)	1.16	(1.55)
M2H	(1.11)	1.49	1.17	(1.56)	1.22	(1.63)
M3H	(1.17)	1.57	1.23	(1.65)	1.28	(1.72)
M4H					1.37	(1.84)
			Conventional			
M4(27H076)[d]	1.27	1.67				
	0.89[e]	1.17[e]				
M5(30H083)[d]			1.39	1.83		
			0.97[e]	1.28[e]		
M6(35H094)[d]					1.57	2.07
					1.11[e]	1.45[e]

[a]At 1.7 T (1.7×10^4 G) unless otherwise noted. Data in parentheses calculated on basis of loss (60 Hz)/loss (50 Hz) = 1.34 typical of high induction material.
[b]Data for 0.28-mm thick samples; Ref. 16.
[c]Ref. 17.
[d]ASTM A725-75.
[e]Core loss at 1.5 T (1.5×10^4 G).

(800–1000°C) anneal in between. The cold-rolled strip is decarburized (800°C) to ca 0.003% C in mixtures of wet H_2 and N_2, a step which also results in a primary recrystallized structure containing grains of the COE orientation. It is finally box-annealed (1200°C) in dry hydrogen to form the COE texture by secondary recrystallization, ie, the growth of the COE-oriented grains at the expense of their neighbors. A very important concept in the development of the secondary recrystallized COE structure involves grain-growth inhibitors. In the conventional process, manganese and sulfur, which occur naturally in steelmaking, form manganese(II) sulfide, MnS, inclusions. These inclusions retard the motion of grain boundaries during primary recrystallization. At the secondary recrystallization step, which takes place at a higher temperature, the inclusions are dissolved, allowing the preferential growth of the COE-oriented grains.

The Nippon Steel HI-B process (18) differs from the conventional process in the use of aluminum nitride [*24304-00-5*], AlN, in addition to MnS as a grain-growth inhibitor and a one-stage cold reduction of large deformation. A large cold reduction results in a sharper COE texture in the final strip, although it also enhances the undesirable growth of the primary grains. However, AlN is a more potent grain-growth inhibitor than MnS, permitting greater cold reduction without undesirable primary grain growth.

In the Kawasaki Steel RG-H process (18), antimony is added with manganese selenide, MnSe, or MnS as a grain-growth inhibitor. It is thought that Sb inhibits primary grain growth by segregating to the grain boundaries and acts in addition to the role of inclusions from MnSe or MnS. The RG-H process also introduced a two-step box-annealing, ie, a low temperature, long-time annealing (820–900°C, 5–50 h), followed by the usual high temperature (1200°C) purification. The COE temperature becomes sharper if the secondary recrystallization is conducted first at the low temperature. Finally, a low thermal expansion inorganic phosphate coating has been developed to impart a large tensile stress to the finished strip. Losses are reduced in the presence of a tensile stress.

In the General Electric–Allegheny Ludlum (GE–AL) process (18), boron and nitrogen with sulfur or selenium are used as grain-growth inhibitors. It is thought that controlled amounts of B, N, and S or Se in solute form segregate to the grain boundaries; hence, inclusions are not involved.

Core-Loss Limits. In the United States, flat-rolled, electrical steel is available in the following classes (12): nonoriented, fully processed; nonoriented, semiprocessed; nonoriented, full-hard; and grain-oriented, fully processed. Loss limits are quoted at 1.5 T (1.5×10^4 G). The loss limits at 1.7 T (1.7×10^4 G) of the fourth class and of the high induction grades are shown in Table 2. Typical applications include use for transformers, generators, stators, motors, ballasts, and relays.

Iron–Aluminum and Iron–Aluminum–Silicon Alloys. The influence of aluminum on the physical and magnetic properties of iron is similar to that of silicon, ie, stabilization of the bcc phase, increased resistivity, decreased ductility, and decreased saturation magnetization, magnetocrystalline anisotropy, and magnetostriction. Whereas Si–Fe alloys are well established for electrical applications, the aluminum–iron alloys have not been studied commercially. However, small (up to ca 0.3%) amounts of Al have been added to the nonoriented grades of silicon steel, because the decrease in ductility is less with Al than with Si.

In the Fe–Al–Si ternary system, alloys close to the 9.5 Si, 5.6 Al composition exhibit very low magnetostriction and anisotropy. As a result, these show very high values of initial and maximum permeability. However, the ternary alloys are very brittle, a factor which limits their general usefulness.

A 16 wt % Al alloy (Alfernol, Vacodur 16) and the 9.5 wt % Si–5.6 wt % Al alloy (Sendust [*12606-95-0*]) are produced commercially. Their primary use is as recording-head material because of high hardness and high resistivity together with reasonably good permeability (see INFORMATION STORAGE MATERIALS). Typical property values of various recording-head materials are listed in Table 3. Both Sendust and ferrite generally are prepared by powder technology, although precision casting of Sendust and the use of ferrite single crystal also are being carried out commercially (see METALLURGY, POWDER).

Nickel–Iron Alloys. The Ni–Fe alloys in the permalloy range, from ca 35–90% Ni, probably are the most versatile soft magnetic alloys in use (see NICKEL AND NICKEL ALLOYS). Using suitable alloying additions and proper processing, the magnetic properties can be controlled within wide limits. Some exhibit high up to 100,000 μ_o (9.274×10^{-19} J/T) initial permeability and are ideal for high quality electronic transformers and for magnetic shielding. Others display a square hysteresis loop shape, desirable in inverters, converters, and other saturable reactors. Still others show low remanence combined with the constant permeability needed for unbiased unipolar pulse transformers.

For alloys that usually are cooled from high temperature to room temperature, the fcc γ-phase exists from ca 30–100% Ni. In the low nickel regime, the γ-phase undergoes a martensitic transformation to the bcc α-phase with consid-

Table 3. Properties of Materials Intended for Recording-Head Applications[a]

Property	4-79 Permalloy[b]	Tufperm[c] YEP-H	YEP-S	16% Al–Fe	Sendust	Ferrite
hardness, Vickers', H_V	120	230	290	290	480	580
μ, at 1 kHz, 0.2 mm	11,000	11,000	7,000	4,000	8,000	4,500
ρ, $\mu\Omega\cdot$cm	100	60	100	150	85	10^6-10^7
B_s, T[d]	0.8	0.5	0.5	0.8	1.0	0.4
H_c, A/cm[e]	0.02	0.01	0.02	0.03	0.02	0.06
T_C, °C	460	280	280	350	500	150

[a]Ref. 19.
[b]CAS Registry Number [*39323-53-0*].
[c]Ti–Nb–Mo Permalloy, Hitachi Magnetic Metals Designation.
[d]To convert T to G, multiply by 10^4.
[e]To convert A/cm to Oe, divide by 0.7958.

erable hysteresis. Of prime importance to the magnetic behavior, however, is the appearance of the long-range-ordered Cu_3Au-type $L1_2$ structure near the Ni_3Fe composition below ca 500°C. The principal magnetic parameters in the Ni–Fe system are illustrated in Figure 4. The most pronounced effect of the $L1_2$ ordering is a large change in the value of K_1, making it less positive or more negative as indicated by the dashed slowly cooled curve in Figure 4. The highest initial permeability in the Ni–Fe system occurs in the 78.5% Ni alloy, but rapid cooling below 600°C is necessary. Magnetic theory indicates that high permeability is achieved by minimizing K_1 and λ_s. This occurs for the 78.5% Ni alloy in the quenched condition. If the alloy is cooled slowly, ordering sets in, K_1 becomes highly negative, and the permeability is degraded severely. With the addition of molybdenum, the kinetics of ordering is slowed and simultaneous attainment of zero K_1 and λ_s is possible with moderate cooling. In this way, alloys of ca 4 wt % Mo and 79 wt % Ni reach very high values of permeability. The addition of molybdenum also has the beneficial effect of increased resistivity (lower eddy-current loss) but at the expense of lower saturation induction and Curie temperature.

In addition to magnetocrystalline anisotropy and magnetostriction, two other sources of magnetic anisotropy are important for Ni–Fe alloys. One is thermomagnetic anisotropy, which occurs when the material is annealed below its Curie temperature. Alloys near 60 wt % Ni where the Curie temperature is high (Fig. 4) are particularly responsive to magnetic annealing. This squares up the hysteresis loop in the field direction and conversely produces a skewed or flat loop of low remanence and constant permeability perpendicular to the field direction. The other source of anisotropy is slip-induced anisotropy, which is obtained by plastic deformation, eg, rolling or wire drawing. The effect is most pronounced when the material is atomically ordered prior to the deformation and also is manifested in a change in the shape of the hysteresis loop, although in a more complicated manner than by magnetic annealing. For Ni–Fe alloys, the thermomagnetic anisotropy energy is ca 0.1 kJ/m^3 (ca 10^3 erg/cm^3) and the slip-induced anisotropy energy is ca 10 kJ/m^3 (ca 10^5 erg/cm^3) both overlapping

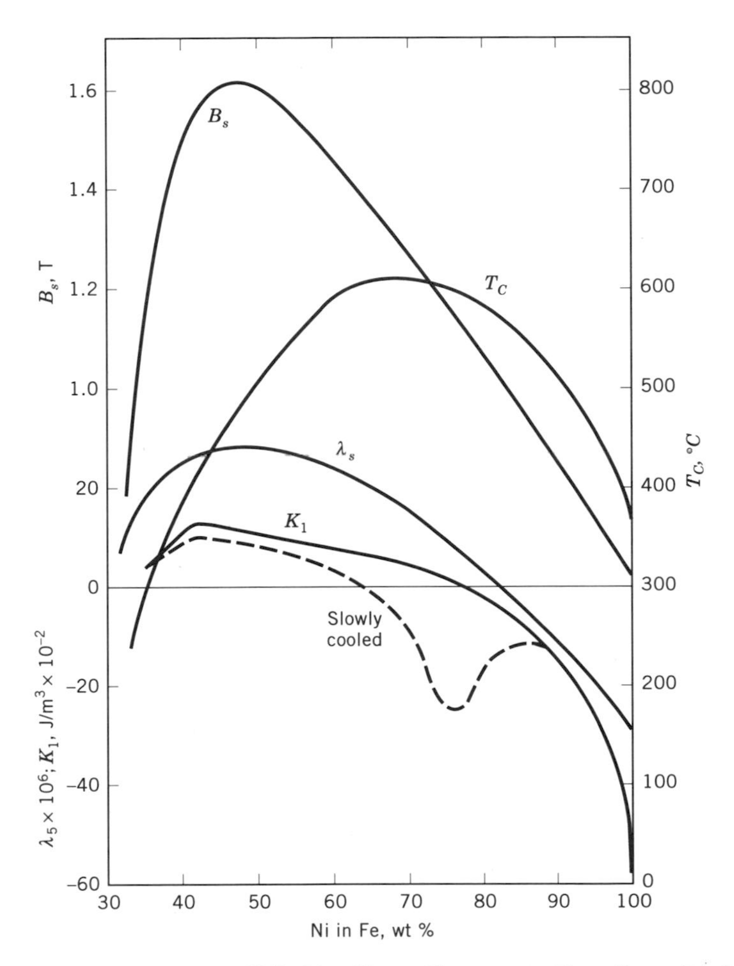

Fig. 4. Magnetic parameters of Ni–Fe alloys. To convert T to G, multiply by 10^4; to convert J/m^3 to erg/cm^3, multiply by 10.

the values of K_1 and λ_s. The richness of anisotropy energies of the Ni–Fe alloys permits custom tailoring of engineering magnetic properties through control of chemistry and processing.

The magnetic and physical properties of some commercial Ni–Fe alloys are compared to those of other iron alloys in Table 4 and Figure 2. The alloys are grouped into three types: high initial permeability, square-loop, and skewed or flat-loop alloys. The alloys generally are used in sheet form as laminations, cut-cores, and tape-wound (toroidal) cores. Laminations usually are 0.35 and 0.15 mm thick for operations of 60–400 Hz, whereas tape cores generally are wound from thin-gauge material ranging from 0.025 to 0.10 mm thick to enable operating frequencies up to ca 25 kHz for the 0.025-mm material of the 4-79 Mo–Permalloy type. Tape-wound cores generally are made from strips 0.025, 0.050, and 0.10 mm thick, although miniature bobbin cores made from material 0.003–0.025 mm thick are available for operations to 500 kHz.

Table 4. Materials for Laminations, Cut-Cores, and Tape-Wound Cores[a]

Alloy type	Trade names[b]	μ_i, $\mu_0 \times 10^{-3}$	μ_m, $\mu_0 \times 10^{-3}$	H_c, A/cm[c]	B_s, T[d]	B_r, T[d]	$\frac{B_r}{B_m}$	T_C, °C	Resistivity, $\mu\Omega \cdot$cm	Density, g/cm^3
			High initial μ							
36Ni	Permenorm 3601 K2, Hyperm 36M, Radiometal 36, 0.3 mm/50 Hz	3	20	0.16	1.3			250	75	8.15
48Ni	4750 Alloy, High Permeability 49, Hyperm 52, Alloy 48, Superperm 49, Super Radiometal, Permenorm 5000 H2, 0.15 mm/60 Hz	11	80	0.024	1.55			480	48	8.25
56Ni	Permax M, 0.15 mm/60 Hz	30	125	0.016	1.5			500	45	8.25
4Mo–80Ni	4-79Mo–Permalloy, HyMu 80, Round Permalloy 80, Superperm 80, 0.1 mm/60 Hz	40	200	0.012	0.8			460	58	8.74
4Mo–5Cu–77Ni	Mumetal Plus, Hyperm 900, Vacoperm 100, 0.1 mm/60 Hz	40	200	0.012	0.8			400	58	8.74

			High initial μ							
5Mo–80Ni	Supermalloy, HyMu 800, Hyperm Maximum, 0.1 mm/60 Hz	70	300	0.004	0.78			400	65	8.77
4Mo–5Cu–77Ni	Supermumetal, Ultraperm 10, 0.1 mm/60 Hz	70	300	0.004	0.8			400	60	8.74
			Square loop							
4Mo–80Ni	Square Permalloy, HyRa 80, Square Permalloy 80, Square 80, 0.05 mm/(0.4 A/cm) dc			0.024	0.8	0.66	0.80	460	58	8.74
4Mo–5Cu–77Ni	Ultraperm Z, Orthomumetal, 0.015 mm/(0.4 A/cm) dc			0.024	0.8	0.66	0.80	400	58	8.74
3Mo–65Ni	Permax Z, 0.05 mm/(0.4 A/cm) dc			0.02	1.25	1.05	0.94	520	60	8.50
50Ni	Deltamax, HyRa 49, Hyperm 50T, Orthonol, Square 50, HCR Alloy, Permenorm Z, 0.05 mm/(0.4 A/cm) dc			0.08	1.60	1.50	0.95	500	45	8.25

Table 4. (*Continued*)

Alloy type	Trade names[b]	μ_i, $\mu_0 \times 10^{-3}$	μ_m, $\mu_0 \times 10^{-3}$	H_c, A/cm[c]	B_s, T[d]	B_r, T[d]	$\frac{B_r}{B_m}$	T_C, °C	Resistivity, $\mu\Omega\cdot$cm	Density, g/cm^3
Square loop										
3Si	Silectron, Microsil, Magnesil, Hyperm 5T/7T, 0.1 mm/(2.4 A/cm) dc			0.32	2.03	1.63	0.85	730	50	7.65
2V–49Co	Supermendur, Vacoflux Z, Hyperm Co50, 0.1 mm/(2.4 A/cm) dc			0.16	2.30	2.00	0.90	940	26	8.15
Skewed (flat) loop										
4Mo–5Cu–77Ni	Ultraperm F			0.012	0.8	0.12		400	58	8.74
3Mo–65Ni	Permax F			0.10	1.25	0.15		520	60	8.50

[a]Ref. 20.

[b]Manufacturers for alloys: Arnold Engineering: 4-79 Mo Permalloy, Supermalloy, Square Permalloy, Supermendur, Deltamax, Silectron, 4750 alloy. Carpenter Technology: HyMu80, HyMu800, High Permeability 49, HyRa49, HyRa80. F. Krupp Widiafabrik: Hyperm 36M, Hyperm 52, Hyperm 50T, Hyperm 5T/7T, Hyperm 900, Hyperm Maximum, Hyperm Co50. Magnetics, Division of Spang, Inc.: Alloy 48, Round Permalloy 80, Square Permalloy 80, Orthonol, Supermendur, Magnesil. Magnetic Metals: Superperm 49, Superperm 80, Square 80, Square 50, Microsil. Telcon Metals: Radiometal 36, Superradiometal, Mumetal Plus, Supermumetal, HCR alloy, Orthomumetal. Vacuumschmelze: Permenorm 3601 K2, Permenorm 500 H2, Permenorm Z, Permax F, Permax M, Permax Z, Vacoperm 100, Ultraperm 10, Ultraperm F, Ultraperm Z, Vacoflex Z.

[c]To convert A/cm to Oe, divide by 0.7958.

[d]To convert T to G, multiply by 10^4.

High Permeability Alloys. In the United States, initial permeability μ_i values generally are given as μ_{40}, measured at $B = 4$ mT (40 G) and 60 Hz for 0.35-mm thick material. In Europe, μ_4, measured at $H = 4$ mA/cm (5 mOe) and 50 Hz for 0.1-mm thick material generally is used. All values of μ_i are in units of μ_0.

Two broad classes of alloys have been developed. One is ca 50% Ni and is characterized by moderate permeability (μ_{40} = ca 10,000) and high saturation (B_s = ca 1.5 T (1.5×10^4 G)). The other is near 79% Ni and is characterized by high permeability (μ_{40} = ca 50,000) but lower saturation (B_s = ca 0.8 T (0.8×10^4 G)). A few suppliers also offer alloys of ca 36 wt % Ni, which have lower permeability than 50 wt % Ni but higher resistivity and lower cost. However, a 2% Mo, 34.5% Ni alloy exhibits a remarkable permeability μ_4 of 55,000 and a large resistivity of 90 $\mu\Omega\cdot$cm (21).

The 50% Ni alloys generally range from 45–50 wt % Ni and may contain up to 0.5 wt % Mn and 0.35 wt % Si. The carbon level generally is kept below 0.03 wt %. Most alloys contain 48 wt % Ni, and some are available in two grades: rotor grade, μ_{40} = ca 6000, useful for rotors and stators in which the magnetic properties must be nondirectional; and transformer grade, μ_{40} = ca 10,000, where high permeability is achieved in directions parallel and perpendicular to the rolling direction. These latter generally are used in audio and instrument transformers, instrument relays, for rotor and stator laminations, and for magnetic shielding. Typical permeability–flux density curves for a 0.35-mm thick, transformer-grade lamination at various frequencies are given in Figure 5, and the magnetic properties of thin-gauge (0.1 mm) material suitable for cut-cores and tape-wound cores are illustrated in Figure 6. A three- to fourfold increase in initial and maximum permeability can be achieved by annealing alloys of ca 56–58 wt % Ni in the presence of a magnetic field after the usual high temperature anneal. The 56 wt % alloy listed in Table 4 is so treated.

For 79 wt % Ni, most modern alloys contain Mo (4–5 wt %) or Cu plus Mo (4 wt % Mo–4.5 wt % Cu–77 wt % Ni). Two grades are available: a standard grade, μ_{40} = ca 35,000, and a very high permeability grade, μ_{40} = ca 60,000. The latter is based on the development of Supermalloy (22) with $\mu_{40} > 100,000$ and where impurities, eg, C (ca 0.01 wt %) and Si (ca 0.15 wt %), are minimized and careful attention is paid to melting and fabrication practice. Practically all alloys of both grades are melted in vacuum. The general applications of the 79 wt % Ni alloys are the same as for the 50 wt % Ni alloys especially where superior qualities are sought and, in particular, where compactness and weight factors are important. Magnetic properties are compared with those of other alloys in Figures 2, 5, and 6.

Other developments include the attainment of high permeability at cryogenic temperatures (23) for shielding applications inside liquid nitrogen or helium chambers, and temperature stability of μ_i near room temperature (24) which is required in earth-leak transformers.

Square-Loop Alloys. Three classes of square-loop Ni–Fe alloys have been developed. In one class (50 wt % Ni), squareness is obtained through the development of cube texture, {100} ⟨001⟩, in the sheet. The texture results from extensive rolling (>95%) followed by primary recrystallization at ca 1100–1200°C. The use of cube texture to square the hysteresis loop also is possible in the 79 wt % Ni

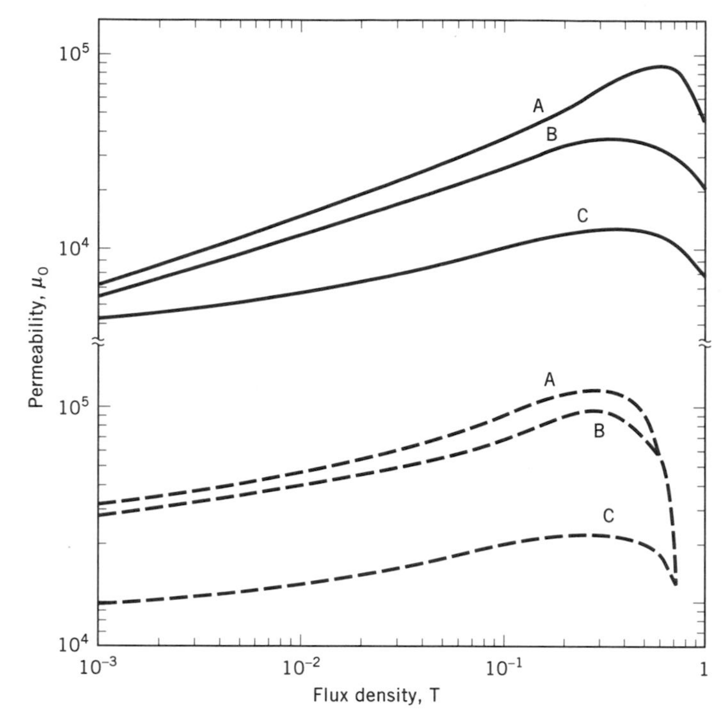

Fig. 5. Permeability-flux density curves for (—) Alloy 48, 0.35 mm (48% Ni) and (---) Permalloy 80, 0.35 mm (4 Mo, 80 Ni) where A represents dc; B, 60 Hz; and C, 400 Hz. To convert T to G, multiply by 10^4.

alloys, but the composition and heat treatment must be chosen such that $K_1 > 0$, ie, ⟨001⟩ is an easy axis of magnetization. The most recently developed class of square-loop alloys, which are centered around 65 wt % Ni, are characterized by a squareness that is like that of the 50 wt % Ni alloy and superior to that of the 79 wt % Ni alloys and by a saturation induction that is intermediate between the two. Representative hysteresis loops are illustrated in Figure 7; the magnetic properties are given in Table 4. The 3 wt % Mo–65 wt % Ni alloy (Mo is added for increased resistivity) is unoriented, and squareness is obtained by annealing in a magnetic field.

It also is possible to develop square hysteresis loops via the slip-induced anisotropy through plastic deformation. This technique had been employed in the commercial processing of Twistor memories (25) no longer used in telephone electronic systems.

Skewed-Loop Alloys. Skewed loops exhibiting Isoperm characteristics, ie, low remanence and constant permeability, are obtained by the development of a magnetic anisotropy where the easy axis is perpendicular to the direction of an applied field. Isoperm, a 50 wt % Ni alloy, is obtained by rolling a cube-textured material; the source of anisotropy is slip-induced. Ultraperm F and Permax F are obtained by magnetic annealing in a transverse field. The advantages of magnetic annealing vs rolling are lower coercivity and higher permeability.

Other. Molybdenum Permalloy powder of ca 2 wt % Mo–81 wt % Ni is used to prepare cores for applications, eg, high Q (quality factor) filters, loading

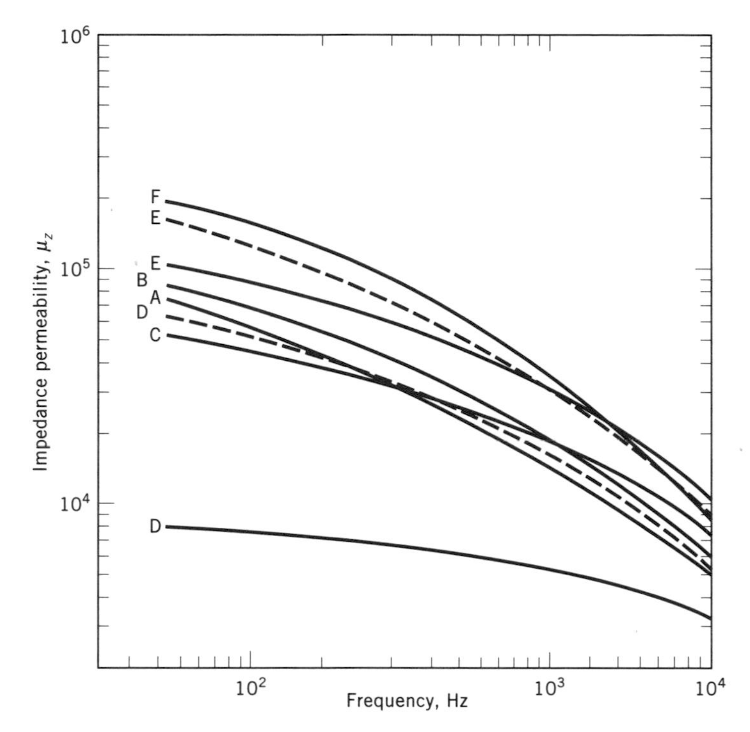

Fig. 6. Permeability-frequency curves of commercial thin-gauge tapes (0.1-mm thick) where A = Supermendur, 20 T; B = Magnesil, 1.0 T; C = Orthonol, 1.0 T; D = Alloy 48, (—) 0.004 T and (---) 1.0 T; E = Supermalloy, (—) 0.002 T and (---) 0.15 T; F = Sq permalloy 80, 0.60 T. To convert T to G, multiply by 10^4.

coils, resonant circuits, and radio frequency inductors (RFI) filters. Generally, a range of permeabilities from ca 15–550 is available commercially having operating frequencies as high as 300 kHz for the low permeability cores. As compared to ferrites, Permalloy cores have higher saturation and hence superior inductance stability after high d-c magnetization, but the Permalloy cores cannot compete at high frequencies because of their lower resistivity.

Nickel is being used in magnetostrictive transducers in some ultrasonic devices, eg, soldering irons and ultrasonic cleaners, because of its moderate magnetostriction and availability. This market, however, is dominated by piezoelectric transducers of lead zirconate–titanate (PZT) (see ULTRASONICS).

Ni alloys of 30–32 wt % are used as temperature-compensator alloys and are characterized by a steep decrease in magnetic permeability with temperature. These alloys are ideally suited in electrical circuits as shunts which maintain constant magnetic strength in devices such as electric meters, voltage regulators, and speedometers.

The thermal expansivity of Ni–Fe alloys vary from ca 0 at ca 36 wt % Ni (Invar [*12683-18-0*]) to ca 13×10^{-6}/°C for Ni. Hence, a number of compositions, which are available commercially, match the thermal expansivities of glasses and ceramics for sealing electron tubes, lamps, and bushings. In addition, the thermal expansion characteristic is utilized in temperature controls, thermostats, measuring instruments, and condensers.

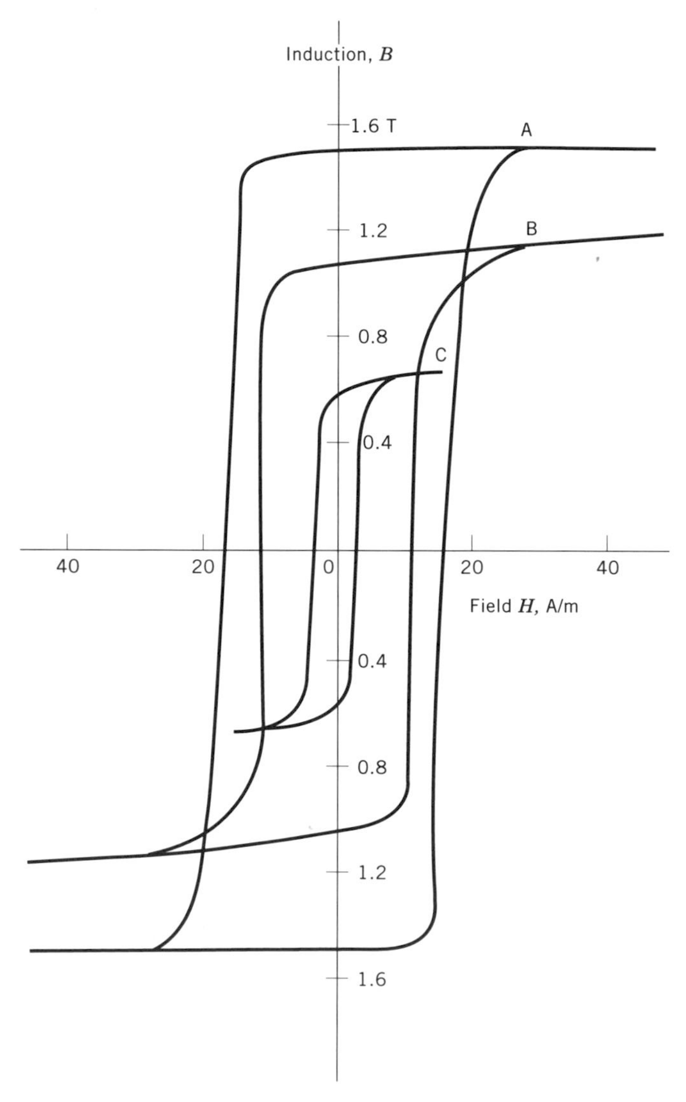

Fig. 7. Hysteresis loops at 60 Hz of three commercial Ni–Fe alloys: A, Orthonol; B, Alloy 48; and C, Permalloy 80. To convert A/m to Oe, divide by 79.58.

Iron–Cobalt Alloys. Iron–cobalt alloys have the highest values of saturation induction at room temperature, at 2.45 T (2.45×10^4 G) for the 35% Co alloy. Commercial alloys of ca 35% Co, with 0.5% Cr added to improve ductility and to increase the resistivity, are known as Hipercos. A 27 wt % Co Hiperco is being produced as of this writing. Somewhat lower values of saturation induction but substantially higher permeability are obtained in equiatomic alloys because of low magnetocrystalline anisotropy. Commercially important equiatomic alloys are 2V Permendur [*37188-80-0*] and Supermendur which contain 49 wt % Fe, 49 wt % Co, and 2 wt % V. Iron–cobalt alloys are costly because of the high price of cobalt and because fabrication is difficult. These alloys are used for specialized applications, eg, aircraft generators, where the high saturation induction permits weight and size savings. These alloys also are used for receiver coils, switching and storage cores, high temperature components, as diaphragms in telephone receivers because of the high incremental permeability over a wide induction range, and to a limited extent as magnetostrictive transducers because of the large magnetostriction at ca 50 wt % Co.

The Fe–Co alloys exist in the fcc structure above 912–986°C to ca 70 wt % Co. Below this temperature range, the structure changes to bcc. At ca 50 wt % Co, the material further orders to the CsCl-type B2 structure below about 730°C and becomes very brittle. The addition of V in Permendur retards the rate of ordering and imparts substantial ductility to the alloy, although quenching is necessary. Vanadium addition also increases the resistivity, eg, from 7–26 $\mu\Omega\cdot$cm using a 2% addition.

Supermendur is prepared by using the highest purity commercial ingredients, by melting in a controlled atmosphere furnace and by using wet and dry hydrogen treatments so as to remove interstitial impurities, and by annealing in a magnetic field (26). The result is large maximum permeability and low coercivity. A comparison of the properties of various commercial Co–Fe alloys is given in Table 5 and in Figure 2.

Although 2V Permendur is sufficiently ductile to permit extensive rolling to thin gauge, controlled heat treatments are required for the development of superior strength and ductility combinations which permit application in high speed rotors (28,29). The addition of 4.5 wt % Ni to the 2V–49Co alloy results in enhanced ductility and strength over a wide heat-treating temperature range, thus eliminating the need for relatively close control (30).

Soft Ferrites. The name ferrite without qualifiers implies the spinel oxide having the general formula $MO\cdot Fe_2O_3$. Commercially important ferrites contain two or more metallic elements M, and the iron content usually deviates from stoichiometry with consequent improvement in magnetic properties. Soft ferrites, ie, those having low coercivity, also include garnets typified by the formula $R_3Fe_5O_{12}$, where R is yttrium or a rare earth. The electrical resistivity of ferrites usually is 10^6 times that of metals. Ferrite components, therefore, have much lower eddy-current losses and, hence, are used at frequencies above about 10 kHz.

Table 5. Properties of Co–Fe Alloys[a]

Property	Hiperco 27	Hiperco 35	2V Permendur	Supermendur
composition	27Co–0.5Cr–Fe	35Co–0.5Cr–Fe	2V–49Co–49Fe	2V–49Co–49Fe
electrical resistivity, $\mu\Omega\cdot$cm	19	40	25	25
induction, T[b]				
saturation	2.36	2.4	2.4	2.4
remanent	1.0		1.5	2.2
μ_m, μ_0	2,800		8,000	92,500
H_c, A/cm[c]	2.0	2.0	4	0.16

[a] Ref. 27.
[b] To convert T to G, multiply by 10^4.
[c] To convert A/cm to Oe, divide by 0.7958.

The magnetic properties of ferrites are intricately related to composition, microstructure, and processing much more so than in the case of metals primarily because of the complex chemistry of the oxides and because of the ceramic processing required to produce the finished parts.

Inductors and Transformers. For communication transformers involving low flux densities, the most important magnetic requirement is high initial permeability. Ferrites operating in this linear range of the B/H curve are linear ferrites. The most widely used linear ferrite is MnZn ferrite having values of μ_i, typically measured at 10 kHz and $H = 0.4$ A/m (5 mOe), up to 18,000 available commercially. A value of $\mu_i = 40,000$ has been achieved in the laboratory (31). The relevant properties of some typical materials are listed in Table 6. Generally, the higher the permeability, the lower the useful frequency range is before the permeability drops off significantly.

Optimum permeability is achieved by choosing the composition where the anisotropy constants K_1 and λ are near zero, by using high purity raw materials, and by controlling the sintering process to achieve highly dense material with uniform large grain sizes without significant loss of ZnO. The composition region where high μ ferrites have been produced (33) showing regions of low temperature coefficient (α/μ_i), low loss (tan δ/μ_i), and low disaccommodation (D/μ_i) is illustrated in Figure 8. For transformers, eg, those used in switching power supplies for computers and other electronic products, a combination of high saturation induction, high permeability, and low loss is desired. Deflection yoke cores, flyback transformers, convergence coils, etc, for television receivers constitute the largest usage of ferrites in terms of material weight. The property requirements, however, are far less severe than those for telecommunications components.

For inductors, the important magnetic requirements are low loss in combination with moderate permeability, small temperature coefficient of permeability, and small disaccommodation. MnZn ferrites generally are used to ca 1 MHz, beyond which NiZn ferrites having electrical resistivities about 10^4-10^5 times higher than MnZn ferrites are used. In MnZn ferrites, the bulk resistivity is increased through additions, eg, Ca and Si, which segregate to the grain boundaries, thereby providing insulating films. Through carefully controlled additions and by starting with homogeneous, high purity powders prepared by chemical coprecipitation techniques, ferrites have been developed with losses as low as 0.8×10^{-6} for tan δ/μ_i at 100 kHz (34). The use of MnZn ferrite cores in power transformers has increased rapidly as operating frequencies of power supplies rises. Power ferrites exhibit power losses at 200 kHz of less than 500 mW/g at 100°C. The loss tangent tan $\delta = \mu''/\mu'$ is the ratio of the imaginary component μ'' of the complex permeability to the real component μ'. Losses as low as 0.3×10^{-6} tan δ/μ_i have been achieved using Co^{2+} additions to stabilize the domain walls (33). The low loss properties of both materials are easily degraded by magnetic and mechanical shock and, hence, are very difficult to achieve in manufactured components. The properties of some typical low loss commercial MnZn and NiZn ferrites are listed in Table 6. Domain-wall stabilization using Co^{2+} has also been successful in reducing loss in NiZn ferrites (35,36).

Disaccommodation denotes a change (generally a decrease) of permeability with time and is undesirable for filter inductors in terms of frequency instability.

Table 6. Characteristics of Ferrites[a]

Property	MnZn ferrites					
code[a]	H5A	H5B	H5C2	H5E	H6F	H6H3
practical frequency, MHz	<0.2	<0.1	<0.1	<0.01	0.2–2.0	0.01–0.8
initial permeability, μ_o	3,300	5,000	10,000	18,000	800	1,300
relative loss factor, $\tan \delta/\mu_i \times 10^6$, at (kHz)	<2.5 (10)	<6.5 (10)	<7.0 (10)		<17 (1,000)	<1.2 (100)
temperature coefficient of $\mu_i \times 10^6$ from −30 to 20°C, $(\mu_2 - \mu_1)/\mu_1^2(T_2 - T_1)$	−0.5 to 2.0	−0.5 to 2.0	−0.5 to 1.5	−0.5 to 2.0		0.3 to 2.0
Curie temperature, °C	>130	>130	>120	>115	>200	>200
saturation flux density, T[b]	0.41	0.42	0.40	0.44	0.40	0.47
disaccommodation factor, $D \times 10^6$ (from 1–10 min), $(\mu_1 - \mu_2)/\mu_1^2 \log(t_2/t_1)$ where t = time	<3	<3	<1	<1	<12	<5
resistivity, $\Omega \cdot m$	1	1	0.15	0.05	4	25
applications			transformers			inductors

Table 6. (*Continued*)

Property	MnZn ferrites			NiZn ferrites		
code[a]	H6K	H7C1	H7C2	K5	K6A	K8
practical frequency, MHz	0.01–0.3	<0.3	<0.2	<8	1–50	<200
initial permeability, μ_o	2,200	2,500	3,900	290	25	16
relative loss factor, $\tan \delta/\mu_i \times 10^6$, at (kHz)	<3.5 (100)			<28 (1,000)	<150 (10,000)	<250 (100,000)
temperature coefficient of $\mu_i \times 10^6$ from −30 to 20°C, $(\mu_2 - \mu_1)/\mu_1^2(T_2 - T_1)$	0.4 to 1.2			−4.0 to 2.0		
Curie temperature, °C	>130	>230	>200	>280	>450	>500
saturation flux density, T[b]	0.39	0.51	0.48	0.33	0.30	0.27
disaccommodation factor, $D \times 10^6$ (from 1–10 min), $(\mu_1 - \mu_2)/\mu_1^2 \log(t_2/t_1)$ where t = time	<2			<30	<20	
resistivity, $\Omega \cdot m$	8	10	2	20×10^5	2.5×10^5	1.0×10^5
applications			power supplies		inductors	

[a]Ref. 32.

[b]To convert T to G, multiply by 10^4.

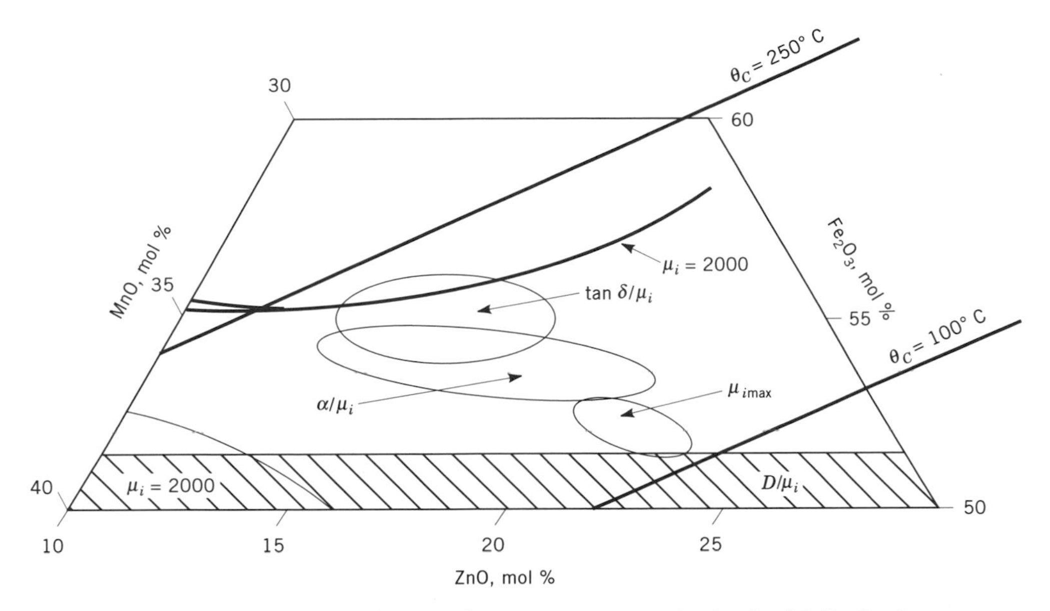

Fig. 8. Composition regions of optimal magnetic properties in the MnZn ferrite system (33). Values of μ_i in units of μ_0.

This phenomenon is attributed to the diffusion of Fe^{2+} ions; hence, both Fe^{2+} and vacancy contents are detrimental. In technical ferrites, a certain amount of Fe^{2+} is needed and, consequently, the disaccommodation is minimized by decreasing the vacancy content through controlled-atmosphere sintering and cooling and through the use of additives (37).

Microwave Ferrites. Microwave devices employing ferrites make use of the nonreciprocal propagation characteristics that are close to or at a gyromagnetic-resonance frequency at ca 1–100 GHz. The most important devices are isolators and circulators (see MICROWAVE TECHNOLOGY).

Properties that are important for microwave devices include saturation polarization J_s and its temperature dependence T_c, resonance line width ΔH, and dielectric constant. Materials having a range of J_s are needed, depending on the operating frequency ω of the device, since for resonance, $J_s < \omega/\lambda$ where λ is the gyromagnetic ratio. Thus, in the low frequency range of 1–5 GHz, garnets, eg, yttrium iron garnet (YIG), with various substitutions and having a range of $J_s = 0.02 - 0.18\ T$ (200–1800 G) are used. In the intermediate frequency range of 2–30 GHz, MgMn, MgMnZn, and MgMnAl ferrites with $J_s = 0.06$–0.25 T (600–2500 G) are used along with the garnets. At high frequencies in the millimeter wave region (30–100 GHz), NiZn ferrites having J_s up to 0.50 T (5000 G) are used. The garnets have particularly small linewidths. Single crystals have values of ΔH as low as 0.3 A/cm (0.4 Oe) as compared to ca 40 A/cm (50 Oe) for the MgMn ferrites. In polycrystalline material, grain boundaries and porosity also increase the linewidth. Hence, sintering to high density with large grain size is important, and values of ΔH = 0.6mA/cm (0.75 mOe) have been reported (38). Examples of commercial microwave ferrites are listed in Table 7.

Table 7. Selected Microwave Materials[a]

Material	J_s, T[b]	ΔH, A/cm[c]	ϵ'	Loss tangent ($\tan \delta$)	T_C, °C	Remarks
			Garnets			
Y	0.180	36	15.0	2×10^{-4}	280	
YAl	0.018	36	13.8	2×10^{-4}	105	decreasing
⋮	⋮	⋮	⋮	⋮	⋮	aluminum
YAl	0.120	36	14.8	2×10^{-4}	220	content
YGd	0.073	160	15.4	2×10^{-4}	280	decreasing Gd,
⋮	⋮	⋮	⋮	⋮	⋮	low $\Delta M_s/\Delta T$,
YGd	0.160	40	15.1	2×10^{-4}	280	constant T_c
YGdAl	0.040	52	14.2	2×10^{-4}	150	similar to
⋮	⋮	⋮	⋮	⋮	⋮	YAl, but
YGdAl	0.140	40	15.1	2×10^{-4}	265	lower
			Spinels			$\Delta M_s/\Delta T$,
MgMnAl	0.075	96	11.3	2.5×10^{-4}	90	decreasing
⋮	⋮	⋮	⋮	⋮	⋮	aluminum
MgMnAl	0.175	180	12.2	2.5×10^{-4}	225	content
MgMn	0.215	432	12.7	2.5×10^{-4}	320	
MgMnZn	0.250	416	12.9	2.5×10^{-4}	275	increasing
⋮	⋮	⋮	⋮	⋮	⋮	zinc
MgMnZn	0.280	432	13.1	2.5×10^{-4}	225	content
NiZn	0.400	272	12.3	2.5×10^{-3}	470	high J_s
NiZn	0.500	128	12.5	1.0×10^{-3}	375	high J_s
Li	0.375	520	15.0	2.5×10^{-3}	640	high T_c
LiZn	0.480	192	14.5	2.5×10^{-3}	400	high J_s
LiTi	0.100	240	18.0	2.5×10^{-3}	300	decreasing
⋮	⋮	⋮	⋮	⋮	⋮	titanium
LiTi	0.290	440	15.2	2.5×10^{-3}	600	content

[a]Ref. 39.
[b]To convert T to G, multiply by 10^4.
[c]To convert A/cm to Oe, divide by 0.7958.

Amorphous Soft Magnetic Alloys. Although most alloys solidify to a crystalline structure, several transition-metal (Fe, Co, and Ni)$_{80}$-metalloid (B, C, Si)$_{20}$ alloys become amorphous when quenched from the liquid state at a sufficiently rapid rate. These can be quenched into the form of thin continuous ribbons and sheets and are commercially available as Metglass (AlliedSignal Corp.) for use in power distribution transformers (up to 500 W) which are being installed in the United States at the rate of one million per year. The amorphous alloys are atomically disordered and thus exhibit a lower density, higher resistivity (about a factor of three, $\sim 150 \mu\Omega \cdot$cm), small macroscopic anisotropy, and the absence of grain boundaries than their crystalline counterparts such as grain-oriented silicon transformer steels. This results in a lower eddy-current core loss (41,42) leading to an annual saving of power exceeding several hundred million dollars per year. The material price, since 1978, has dropped by a factor of 100. For low power electronic applications, several amorphous alloys are comparable to the Ni–Fe crystalline alloys. There are at least nine commercially available Metglass (amorphous) alloys (see GLASSY METALS).

The useful amorphous magnetic alloys can be conveniently grouped into the three classes given in Table 8. The iron-based alloys having saturation inductions in the 1.6–1.8 T range are substituted for grain-oriented silicon steels in power distribution applications, and with appropriate modifications in composition and heat treatment, possess good high (up to 100 kHz) frequency properties comparable to or better than those of crystalline nickel–iron alloys. The amorphous iron–nickel-based alloys exhibit lower magnetostriction and better corrosion resistance than the iron-based alloys, but are more expensive. The amorphous cobalt-based alloys exhibit nearly zero stress sensitivity and the highest permeabilities, but are the most expensive.

Uses of Soft Magnetic Materials. Because of low resistivity, iron and low carbon steels tend to be used in static applications, eg, pole pieces for electromagnets and cores for d-c magnets or relays. Low carbon steels and the lower grade Fe–Si alloys are used in small motors and generators. The higher grade Fe–Si alloys have traditionally been used in power distribution transformers and large rotating machinery, but certain amorphous iron–metalloid alloys are increasingly being used in the manufacture of distribution transformers by General Electric, Westinghouse, and Osaka, for example. Fe–Al and Fe–Al–Si alloys are used primarily as recording head materials because of their high hardness, resistivity, and saturation magnetization. Ni–Fe alloys used widely in high quality relays, transformers, converters, and inverters in the electronics industry, have much higher permeability and lower loss than Fe–Si alloys. The Co–Fe alloys are used because of their higher saturation polarization (flux density) and Curie temperature, but have the disadvantage of poorer workability and higher cost. Because of their exceptionally higher resistivities, ferrites are particularly suitable for high frequency applications.

Hard Magnetic Materials

Hard or permanent magnetic materials are characterized by high coercivity and high energy product. The important commercial hard magnetic materials as of 1994 are ferrites, rare-earth (R)-cobalt alloys, and the ternary alloys based on $Nd_2Fe_{14}B$. The last exhibit the highest coercivities and energy products. The use of Alnico and the binary R–Co alloys has continually decreased because of the high cost of cobalt. These are being replaced by the ternary NdFeB materials. The total market for permanent magnet materials in 1987 was estimated to be about 1.46×10^9 U.S. dollars: ferrites had 65% of the market, R–Co alloys 18%, NdFeB 4%, Alnicos 11%, and others 2%. Because of the superior magnetic properties and cost advantage, the market share of NdFeB magnets had increased to 16% by 1990.

Rare-earth magnets exhibit coercivities in excess of 600 kA/m (7.5 kOe) and energy products up to 290 kJ/m^3 (36×10^6 G·Oe). The newer Cr–Co–Fe alloys have magnetic properties similar to the Alnicos but have the advantage of being cold formable. Thus these alloys are classed with Cunife and Vicalloy in the family of ductile hard magnets. Physical and magnetic properties of selected hard magnetic materials are given in Tables 9 and 10 (44–48). The progress in energy

Table 8. Properties of Amorphous Magnetic Alloys[a]

Alloy	Composition	Saturation induction, B_s, T[b]	Coercive force H_c, A/m	Magnetostriction, $\lambda_s \times 10^{-6}$	Resistivity, ρ, $\mu\Omega\cdot$cm	T_C, °C	Core loss 60 Hz, 1.4 T,[b] W/kg	Core loss 20 kHz, 0.2 T,[b] kW/m³
			Iron-based					
Metglas 2605SC	$Fe_{81}B_{13.5}Si_{3.5}C_2$	1.61	3.2	30	1.30	370	0.3	300
Metglas 2605S-2	$Fe_{78}B_{13}Si_9$	1.56	2.4	27	1.30	415	0.23	
Metglas 2605CO	$Fe_{67}Co_{18}B_{14}Si_1$	1.80	4.0	35	1.30	415	0.55	
Metglas 2605S-3	$Fe_{79}B_{16}Si_5$	1.58	8.0	27	1.25	405	1.2	58
			Iron–nickel-based					
Metglas 2826MB	$Fe_{40}Ni_{38}Mo_4B_{18}$	0.88	1.2	12	1.60	353		200
			Cobalt-based					
Metglas 2705M	$Co_{67}Ni_3Fe_4Mo_2B_{12}Si_{12}$	0.72	0.4	0.5	1.35	340		43

[a] Refs. 33 and 40.

[b] To convert T to G, multiply by 10^4.

Table 9. Magnetic Properties of Commercial Permanent Magnet Materials

Material	T_C, °C	$(BH)_{max}$, kJ/m^3 [a]	B_r, T [b]	H_c, kA/m [c]
Ferroxdure ($SrFe_{12}O_{19}$)	450	36	0.42	250
Alnico 9	850	72	1.05	120
$SmCo_5$	724	144	0.87	600
$Sm(Co_{0.68}Cu_{0.10}Fe_{0.21}Zr_{0.01})_{7.4}$	800	240	1.10	510
$Nd_2Fe_{14}B$	312	290	1.23	880

[a] To convert kJ/m^3 to G·Oe, multiply by 12.57×10^4.
[b] To convert T to G, multiply by 1×10^4.
[c] To convert kA/m to Oe, divide by 7.958×10^{-2}.

products for hard magnetic materials is shown in Figure 9 and representative demagnetization curves of some of these materials are illustrated in Figure 10.

Alnicos. Alnico is a family of iron-based alloys containing Al, Ni, and Co plus about 3% Cu. A few percent of Ti and/or Nb is added in the higher coercivity alloys (Alnicos 6–9). Alnicos 1–4 are isotropic, whereas Alnicos 5 and up are anisotropic as a result of being heat-treated in a magnetic field. The anisotropic

Table 10. Properties of Permanent (Hard) Magnet Materials[a]

Magnet material	Chemical composition	B_r, T [b]	H_c, kA/m [c]	$(BH)_{max}$, kJ/m^3 [d]
3½% Cr steel	3.5Cr,1C,bal Fe	1.03	5	2.4
3% Co steel	3.25Co,4Cr,1C,bal Fe	0.97	6	3.0
17% Co steel	18.5Co,3.75Cr,5W,0.75C,bal Fe	1.07	13	5.5
36% Co steel	38Co,3.8Cr,5W,0.75C,bal Fe	1.04	18	7.8
Alnico 1	12Al,21Ni,5Co,3Cu,bal Fe	0.72	37	11.0
Alnico 2	10Al,19Ni,13Co,3Cu,bal Fe	0.75	45	13.5
Alnico 4	12Al,27Ni,5Co,bal Fe	0.56	57	10.7
Alnico 5 DG[e]	8Al,14Ni,24Co,3Cu,bal Fe	1.33	53	52.0
Alnico 5 Col.[e]	8Al,14Ni,24Co,3Cu,bal Fe	1.35	59	60.0
Alnico 8[e]	7Al,15Ni,35Co,4Cu,5Ti,bal Fe	0.82	130	42.0
Alnico 9[e]	7Al,15Ni,35Co,4Cu,5Ti,bal Fe	1.05	120	72.0
Col. Alnico HC[e,f]	7Al,14Ni,40Co,3Cu,7.5Ti,bal Fe	0.97	150/155[g]	91.5
Col. Alnico HC[e,f]	7Al,14Ni,39Co,3Cu,8Ti,bal Fe	0.88	170/180[g]	77.0
sintered Alnico 6[e]	8Al,16Ni,24Co,3Cu,1Ti,bal Fe	0.94	63	23.0
sintered Alnico 8 HC[e]	7Al,14Ni,38Co,3Cu,8Ti,bal Fe	0.67	140	36.0
Ceramic 7[e,h]	$MO{\cdot}6Fe_2O_3$	0.34	260/320[g]	22.0
bonded ceramic[e,i]	flexible anisotropic ferrite	0.24	170/215[g]	11.0
Cunife 1[e]	60Cu,20Ni,20Fe	0.55	42	11.0
Vicalloy 1	10V,52Co,bal Fe	0.75	20	6.4
Remalloy	12Co,15Mo,bal Fe	0.97	20	8.0
rare-earth cobalt[e,j]	25.5Sm,8Cu,15Fe,1.5Zr,50Co	1.10	510/520[g]	240.0
Cr–Co–Fe[e,k]	23Co,31Cr,1Si,bal Fe	1.25	52	40.0
Cr–Co–Fe[e,l]	11.5Co,33Cr,bal Fe	1.20	60	42.0
Cr–Co–Fe[e,m]	5Co,30Cr,bal Fe	1.34	42	42.0

[a] All data adapted from Ref. 43 except where noted. [b] To convert T to G, multiply by 10^4. [c] To convert kA/m to Oe, divide by 7.958×10^{-2}. [d] To convert kJ/m^3 to G·Oe, multiply by 12.57×10^4. [e] Anisotropic. [f] Ref. 44. [g] Intrinsic coercive force, H_{cJ}. [h] M represents Ba or Sr. [i] TDK BQ A14 Rubber Magnet. [j] TDK REC-30. [k] Sumitomo CKS500. [l] Chromindur III; Ref. 46. [m] Ref. 47.

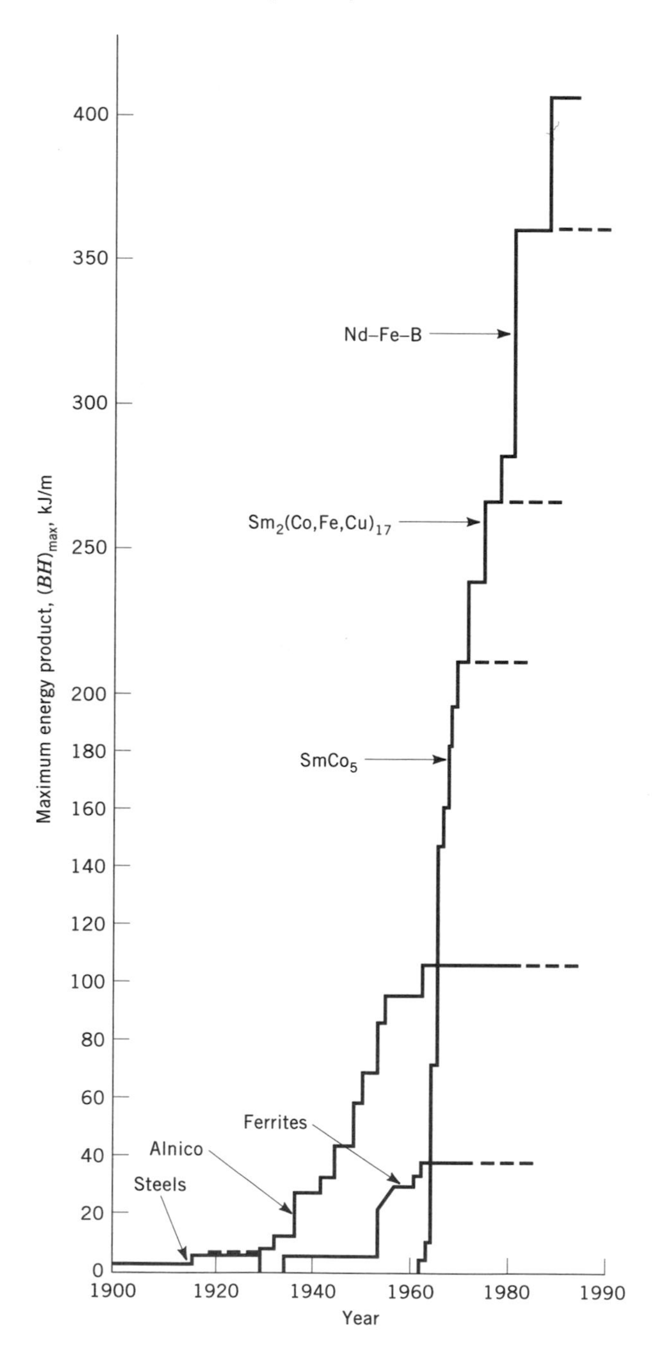

Fig. 9. Progress in energy product for hard magnetic materials. To convert J to cal, divide by 4.184.

magnets are known as Alcomax and Hycomax in the U.K. and as Ticonal in Holland. Alnico 5 is the most widely used of the family. Because of their brittle nature, the Alnicos are produced by casting or by power metallurgy. The sintered alloys generally are mechanically stronger but magnetically weaker as a result

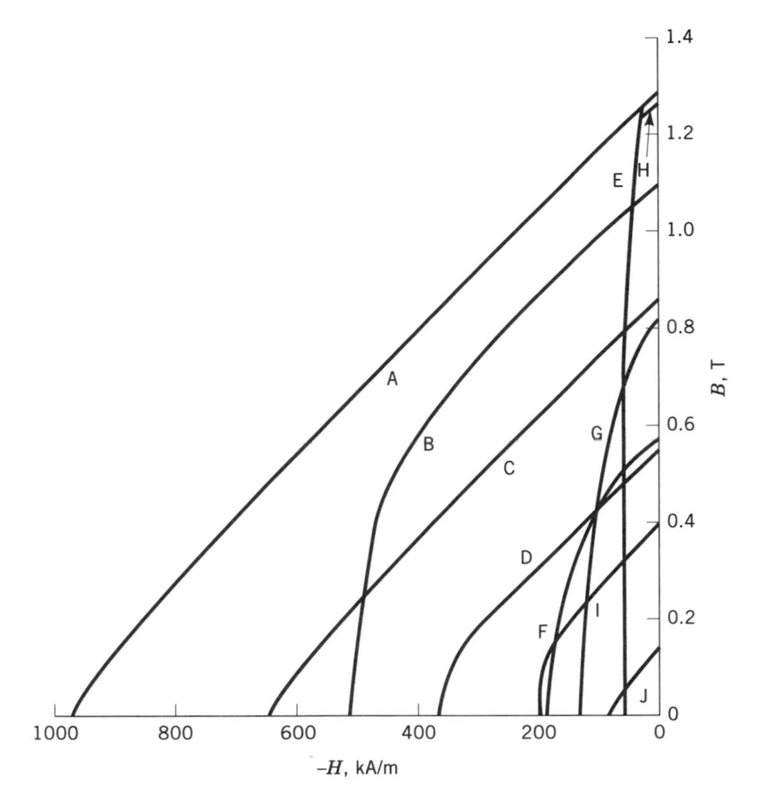

Fig. 10. Demagnetization curves of hard magnetic materials: A, $Nd_2Fe_{14}B$; B, Sm(Co, Cu,Fe,Zr$)_{7.4}$; C, $SmCo_5$; D, bonded $SmCo_5$; E, Alnico 5; F, Mn–Al–C; G, Alnico 8; H, Cr–Co–Fe; I, ferrite; J, bonded ferrite. To convert T to G, multiply by 10^4.

of, among other factors, incomplete densification. They are used primarily in large quantity production of small articles of complex shape because surface grinding is the only possible finishing method for the Alnicos.

Above the solution treatment temperature (ca 1250°C), the alloy is single phase with a bcc crystal structure. During cooling to ca 750–850°C, the solid solution decomposes spinodally into two other bcc phases α and α' which differ little in lattice parameter composition. The matrix α-phase is rich in Ni and Al and weakly magnetic as compared with α', which is rich in Fe and Co. The α'-phase tends to be rod-like in the $\langle 100 \rangle$ direction and ca 10 nm in diameter and ca 100 nm long. As the temperature is decreased, segregation of the elements becomes more pronounced and the difference between the saturation polarizations of the two phases increases.

It generally is accepted that the mechanism of coercivity in the Alnicos is incoherent rotation of single-domain particles of the α'-phase based on shape anisotropy. As coercivity increases, the larger the aspect ratio of the rods and the smoother their surface becomes; the difference between the saturation polarizations of the two phases also increases. It is thought that Ti increases the coercivity of Alnico because of an increased aspect ratio of the rods and a smoother surface.

If the decomposition reaction takes place below the Curie temperature, which is the case for anisotropic Alnicos, and if a magnetic field is applied during the decomposition, the α' rods tend to form along the ⟨100⟩ direction closest to the field direction so as to minimize the magnetostatic energy. This effect was first demonstrated in an Alnico 5 single crystal using electron microscopy (49). The result is an increase in H_c, B_r, and $(BH)_{max}$ in the field direction at the expense of the direction normal to it. This is the origin of improved magnetic properties in the anisotropic magnets shown in Table 10.

Some of the composition adjustments in the Alnicos result in a high Curie temperature so that the decomposition reaction can take place relatively rapidly below T_C. This is particularly true for Co, which is 24 wt % or greater for the anisotropic magnets. Another important consideration is the suppression of nonmagnetic fcc γ-phase which may appear at 1000–1100°C; in this regard, the amount of Al, which is a γ-suppressor, is critical. The formation of γ is pronounced if the Al content falls much below 7–8 wt %.

Isotropic Alnicos. Alnicos 1–4 in Table 10 are isotropic because the Curie temperatures are too low, as a result of the low Co content, for magnetic-field treatment to be effective. A typical heat treatment consists of solution treatment at 1250°C, controlled cooling to room temperature, and aging at ca 600°C with controlled slow cooling or step-aging at successively lower temperatures. The maximum energy is ca 12 kJ/m^3 (1.5×10^6 G·Oe).

Anisotropic Alnicos. Alloys composed of about 24% Co, ie, Alnicos 5 and 6, develop anisotropic properties by heat treatment in a magnetic field, generally ca 80 kA/m (1.0 kOe). This is accomplished by cooling from the solution treatment temperature at an average speed of about 1°C/s, by taking the magnets out of the solution treatment furnace and putting them in the field of an electromagnet, solenoid, or permanent magnet. The field needs to be only from 750–850°C while the spinodal decomposition occurs. The primary difference between Alnicos 6 and 5 is a 1% Ti addition in Alnico 6 which results in an increase in H_c at the expense of B_r and energy product.

High H_c Alnicos. The coercivities for Alnicos 8 and 8 HC of 120–160 kA/m (1.5–2.0 kOe) (vs ca 50 kA/m (0.6 kOe) for Alnico 5) are achieved by increasing the Co content to ≥35% and the Ti content to 5–8%. The coercivity of Alnico 8 HC is higher than that of Alnico 8 by virtue of the higher Ti content. Some of the Ti may be replaced by Nb; this is particularly true of alloys produced in the U.K. In contrast to Alnicos 5 and 6, the magnetic-field heat treatment of Alnicos 8 and 8 HC is conducted isothermally at ca 800°C instead of by continuous cooling. The isothermal heat-treatment temperature is critical ($\pm 2 - 3$°C) and is optimized for a given alloy.

Columnar Alnicos. Because the α' rods precipitate along ⟨100⟩ axes, the directional magnetic properties can be enhanced if the cast structure possesses a preferred ⟨100⟩ orientation among the grains. For the Alnicos, as for most materials with a cubic crystal structure, ⟨100⟩ is the preferred growth direction in the direction of heat flow during solidification. Hence, ⟨100⟩ textured material with columnar grain structure is obtained commercially by the use of heated molds and chill plates to achieve unidirectional heat flow. The result is a squaring of the hysteresis loop with concomitant improvements in B_r, H_c, and $(BH)_{max}$. Values of $(BH)_{max} = 64$ kJ/m^3 (8×10^6 G·Oe) are offered commercially for Alnico 5 Col. A less effective but less expensive technique for achieving partial orientation is by the use of chill plates without heated molds. These are designated Alnico 5 DG in Table 10.

It is very difficult to achieve columnar structure in the Ti-bearing, high H_c alloys. Raw materials should be of high purity (50) and the addition of 0.2% S helps promote columnar grain growth (51). The latter technique is most likely used in the commercial production of Alnico 9, which is the columnar version of Alnico 8 and contains 5% Ti. Columnar grain growth for the 8% Ti alloy is more difficult because of grain nucleation from oxides, nitrides, and carbides in the melt. The Magnicol process (52) has been successful for commercial production (44). The technique involves deoxidation of the melt using Si and C followed by the addition of sulfur. The properties of two grades (Col. Alnico HC) having peak values for $(BH)_{max}$ ca 92 kJ/m^3 (11.6 × 10^6 G·Oe) and H_{cB} ca 175 kA/m (2.2 kOe) (53) are listed in Table 10.

Hard Ferrites. Hard ferrites, general formula $MO{\cdot}6Fe_2O_3$, where M is Ba or Sr, are hexagonal in crystal structure. Most commercial ferrites are barium ferrites. These have dominated the market since their introduction in 1952. More recently, strontium ferrite is being produced in large quantities as it has superior properties compared with barium ferrite. Ferrites are produced by ceramic techniques and the magnets are often called ceramic magnets (Table 10). Ferrite powders are often bonded in plastic or rubber for low cost, large volume production, especially when complex shapes are required. These are known as bonded magnets and because they are flexible, they find wide use. The 3M Co. manufactures flexible magnet strip under the name Plastiform.

The origin of large coercivity in the hard ferrites lies in the large magnetocrystalline anisotropy (K_1 = ca 0.3 MJ/m^3 (3 × 10^6 erg/cm^3)) combined with low saturation polarization (J_s = ca 0.47T (4700 G)). Strontium ferrite has a somewhat larger K_1 and smaller J_s as compared with barium ferrite. Because the Curie temperature is rather low (ca 450°C), the magnetic properties are much more temperature sensitive than for the Alnicos. Remanent induction decreases with increasing temperature at a rate of 0.19%/°C and the intrinsic coercivity decreases with decreasing temperature at a rate of 0.2–0.5%/°C. Exposure to low temperatures can lead to serious demagnetization problems as a result of reduced H_{cJ}. Thus some of the commercial hard ferrites were developed to have high H_{cJ} values, eg, Ceramics 4 and 7 in Table 10.

A typical ferrite magnet is prepared by first mixing iron oxide and $BaCO_3$ or $SrCO_3$. The mixed powder is calcined at ca 1100–1200°C to form the ferrite. The aggregate is ball-milled to ca 1 μm, pressed in a die, and sintered to high density at ca 1200°C. In general, the grain size is finer and a lower sintering temperature is used resulting in a higher coercivity and lower remanence. Anisotropic magnets are prepared by aligning the powder in a magnetic field during the pressing operation. The exact composition of typical commercial ferrites (Table 10) usually is not given. Lead may be present in a small amount in Sr or Ba ferrite to facilitate sintering.

One of the fastest-growing areas in hard magnetic materials is the plastic-bonded ferrite where the milled calcined ferrite powder is embedded in a plastic (see EMBEDDING). Both isotropic and anisotropic grades are prepared. Although the magnetic properties are inferior (see Table 10), the plastic-bonded magnets are inexpensive, can be cut and drilled, and can be made flexible. Large quantities are used in door catches, wall magnets, refrigerator-door gaskets, board games, toys, and small motors.

Because one advantage of ferrites over the other hard magnetic materials is low price, innovations in the industry tend toward lowering production costs. Much attention has centered on the substitution of low cost, impure raw materials for the more expensive high quality ones. Celestite may be used as a cheap source of Sr (54); it is primarily $SrSO_4$ and contains varying small amounts of $CaSO_4$, $BaSO_4$, SiO_2, and Al_2O_3. Natural hematite is much cheaper than synthetic oxide, but contains small amounts of SiO_2 and Al_2O_3. Not only do the impurities affect the calcining and sintering behavior of the ferrite, but the size, shape, size distribution, and the chemical perfection of the starting materials also have a profound and complicated effect. Unlike other hard magnetic materials that contain Co and/or rare earths, the raw materials for the ferrites generally are inexpensive, plentiful, and nonstrategic. Thus the conversion of Alnico to ferrite in loudspeakers and d-c motors has accelerated.

Uses. Hard ferrites are used widely in electromechanical devices, eg, generators, relays, motors, and magnetos; electronic applications, eg, loudspeakers, traveling-wave tubes, and telephone ringers and receivers; antitheft tags, holding devices such as door closers, seals, and latches; and are perennial favorites in various toy designs. Loudspeakers are the largest use of permanent magnets (ca 50%). Strontum ferrites exhibit higher coercivities and are increasingly being produced.

Cobalt–Rare-Earth Magnets. The origin of magnetism in the rare-earth (R) transition-metal intermetallic compounds is the interatomic exchange between the spins of the two sublattices plus the spin-orbit coupling between the rare-earth atoms. The magnetocrystalline anisotropy also comes from two sources, one originating in the itinerant electrons of the Co sublattice in the R–Co phases and one caused by the crystalline electric field of the rare earths. Reviews of magnetism in the rare-earth transition-metal compounds are given in References 55–61. Of the large number of rare-earth transition-metal compounds only a few, ie, RCo_5, $R_2Co_{17}(RCo_{8.5})$, particularly the Sm series outside of the newest rare-earth ternary intermetallics based on $Nd_2Fe_{14}B$, are the most prominent high performance hard magnets.

As shown in Table 11, several of the cobalt–rare-earth alloys exhibit moderate values of saturation polarization J_s (ca 1 T (1×10^4 G)) and extremely high values of magnetocrystalline anisotropy ($K_1 > 1$ MJ/m^3 (10^7 erg/cm^3)). The latter is primarily responsible for the large coercivity. The R_2Co_{17} compounds have a hexagonal crystal structure with either the Th_2Ni_{17} (hexagonal) or the Th_2Zn_{17} (rhombohedral) modification. Both have RCo_5 as subcells and differ only in stacking sequence. These compounds generally exhibit a higher value of J_s than the corresponding RCo_5 series and result in a greater energy product. However, with the exception of Sm, Er, and Tm, all the 2–17 binary compounds have $K_1 < 0$, ie, an easy (0001) plane. Low coercivity is expected for the $K_1 < 0$ compounds.

The first rare-earth magnets, developed in the 1960s and commercially produced in the 1970s, were based on the intermetallic compounds RCo_5 and R_2Co_{17}. $SmCo_5$ has the hexagonal $CaCu_5$ structure, a moderate saturation induction (Table 11), and extremely high positive values of magnetocrystalline

Table 11. Magnetic Properties of RCo_5 and R_2Co_{17} Compounds

R	CAS Registry Number	J_s, T[a]	T_C, °C[b]	K_1, MJ/m^3[c–e]
		RCo_5		
Ce	[*12214-13-0*]	0.85[f]	374	5.3
Pr	[*12017-67-3*]	1.12[f]	612	8.1
Nd	[*12017-65-1*]	1.20[f]	630	0.7
Sm	[*12017-68-4*]	0.97	724	17.2
		R_2Co_{17}		
Ce	[*12014-88-9*]	1.15[g]	800	−0.6
Pr	[*12052-77-6*]	1.38[g]	890	−0.6
Nd	[*12052-76-5*]	1.39[g]	900	−1.1
Sm	[*12052-78-7*]	1.20[g]	920	3.3

[a]To convert T to G, multiply by 10^4.
[b]Ref. 56.
[c]Values at 25°C.
[d]To convert MJ/m^3 to G·Oe, multiply by 12.57×10^7.
[e]Ref. 57.
[f]Ref. 55.
[g]Ref. 58.

anisotropy. Magnets based on $SmCo_5$ are single phase. Laboratory magnets have attained values B_r = 1T (1×10^4 G) H_{cJ} = 3200 kA/m (40.2 kOe) and $(BH)_{max}$ = 200 kJ/m^3 (25 MG·Oe). Commercial magnets are characterized by $(BH)_{max}$ from 130–160 kJ/m^3 (16–20 MG·Oe). The coercivity of these magnets is based on nucleation of domains and wall pinning at grain boundaries. Plastic-bonded magnets are also in production.

Sm_2Co_{17}, also hexagonal, has a higher saturation induction but a lower value of K_1. Commercial magnets based on $SmCo_{8.5}$ contain, in addition to Co, the metals Cu, Fe, and Zr substituted for some of the Co. The good hard magnetic properties arise from microstructural precipitates which inhibit (pin) domain walls. A commercial alloy of $Sm(Co_{0.68}Cu_{0.10}Fe_{0.21}Zr_{0.01})_{7.4}$ exhibits values of B_r = 1.10T (1.10×10^4 G), H_{cJ} = 520 kA/m (6.5 kOe) and $(BH)_{max}$ = 240 kJ/m^3 (30 MG·Oe). Values of $(BH)_{max}$ = 264 kJ/m^3 (33.2 MG·Oe) have been achieved in the laboratory. Representative magnetic properties and typical demagnetization curves are given in Tables 9, 10 and 11 and Figure 10, respectively.

The mechanism of coercivity in single-phase $SmCo_5$ magnets is based on nucleation and/or pinning of domain walls at surfaces or grain boundaries. Large values of coercivity can be achieved only in fine particle form (1–10 μm) or in fine-grained sintered magnets prepared from such powders. Thus fabrication techniques are based on powder metallurgical procedures, including alignment in a magnetic field during the pressing operation.

Other RCo_5 single-phase hard magnets in addition to Sm have been prepared both commercially and in the laboratory. These include Sm in combination with Pr, Ce, and Ce-mischmetal (CeMM), this last a very low cost mixture containing about 55% Ce, 25% La, 13% Nd, and 5% Pr. $(Pr,Sm)Co_5$ magnets have higher values of saturation induction than $SmCo_5$, but are produced only in a limited way. The Sm–Ce and Sm–CeMM combinations lower the magnetic properties, but the price of raw materials is lower.

Partial substitution of Co by Cu in $SmCo_5$ and $CeCo_5$ results in fine-scale (ca 10 nm) precipitation at low temperatures (400–500°C) (55,60). The precipitate is a phase that is coherent with the RCo_5 structure that develops into R_2Co_{17} particles upon coarsening beyond the maximum coercive force point. Unlike the single-phase magnets, the coercivity mechanism here is based on homogeneous pinning of domain walls at the precipitate particle; thus magnetic hardening is possible with bulk material, and high energy product magnets can be produced in directionally solidified samples similar to the columnar Alnicos. However, commercial practice also is based on powder metallurgy because magnets that are prepared by powder metallurgical techniques have better mechanical properties and improved magnetic alignment.

The large values of maximum energy product and coercivity of the rare-earth magnets permit use in devices where small size and superior performance are desired (62). Magnets for electronic wristwatches and for traveling-wave tubes are largely made of rare-earth alloys. There are medical device applications which make use of the RCo_5 and R_2Co_{17}-base alloys. The temperature coefficient of polarization, which is typically −0.04%/°C, is too high for precision applications, such as gyros and accelerometers. However, polarization of the heavy RCo_5 compounds, eg, Co_5Dy, increases with temperature at ca 25°C. When these rare earths are mixed with Sm, the temperature coefficient can be reduced by partially replacing the Sm with a heavier rare earth such as Dy.

$Nd_2Fe_{14}B$ Family of Magnets and Related Materials. The high cost of cobalt and samarium stimulated investigation and development in the early 1980s of low cost permanent magnets to replace the Co–Sm-based materials. Ternary metal alloy systems containing Fe and the cheaper rare-earth metals Ce, Nd, and Pr were investigated. This led to the discovery of the ternary intermetallic boride $Nd_2Fe_{14}B$ (63,64). Developed in 1983 by Sumitomo special metals and General Motors (65,66), magnets based on the ternary $Nd_2Fe_{14}A$, where A = B, C, N, exhibit the highest magnetic performance (Table 8 and Figs. 9 and 10). These materials are largely replacing the commercial R–Co ones.

The theoretical maximum energy product for $Nd_2Fe_{14}B$ is approximately 64 MG·Oe (512 kJ/m^3) (67), 2.5 times greater than for $SmCo_5$. Energy products for the commercially available magnets (200–300 kJ/m^3) are less than theoretical but well in excess of the 130–160 kJ/m^3 exhibited by $SmCo_5$. $Nd_2Fe_{14}B$, like $SmCo_5$, is brittle and several manufacturing techniques have been developed for forming this material into usable shapes. As of this writing, permanent magnets based on this material are the best available.

The crystal structure of $Nd_2Fe_{14}B$ is tetragonal, the unit cell contains four formula units, and the space group is $P4_2mmm$ (68,69). There are crystal structure features similar to the hexagonal RCo_5 phases. The Nd–Fe atomic moments are coupled parallel, ie, ferromagnetically, similar to the Co-light R phases, but antiparallel to the heavy R-atoms leading to lower net magnetizations (ferrimagnetic). Thus the net magnetization of $Nd_2Fe_{14}B$ is approximately 1.6 T (16,000 G).

The energy product $(BH)_{max}$ is a measure of a magnet's resistance to demagnetization in reverse fields, thus a high coercivity along with the saturation magnetization is required for a large energy product. The easy axis of magnetization in $Nd_2Fe_{14}B$ is the crystallographic c-axis and it is the origin of the high coercivity exhibited by this material. To rotate the magnetization from the c-axis to an axis normal to it (a-axis) or hard axis, a high applied field must be used. The field required to do this is called the anisotropy field, H_a. This is the upper limit of the coercivity; the practical coercivity is sensitive to details of the microstructure, ie, effected by impurity phases and grain boundaries, and seldom exceeds 10–20% of the anisotropy field. The contribution of the Nd and Fe ions to H_a (65) are sufficiently high for development of high energy product magnets. The isostructural $Pr_2Fe_{14}B$ phase is the only one of the other ternary boride materials that exhibits high magnetization and H_a for development of practical hard magnetic materials.

The Curie temperature for $Nd_2Fe_{14}B$ (T_C = 312°C), as well as those for all of the other $R_2Fe_{14}B$ type-phases, are lower than those for the RCo_5 and R_2Co_{17} phases but higher than the R_2Fe_{17} phases (Table 12) (66). For many applications, high values of T_c are desirable, because this leads to a smaller temperature dependence of magnetic properties and higher temperature applications. Substitution of small amounts of Dy and Co for a part of the Nd and Fe, respectively, improves Curie temperature and temperature dependence but at the expense of lower magnetization and higher material price.

Research has been directed toward discovery of lower cost ternary permanent magnet materials having even better properties, ie, higher saturation coercive force and Curie temperatures, than $Nd_2Fe_{14}B$. A summary is shown in Table 12. $Nd_2Fe_{14}C$ exhibits a lower Curie temperature and smaller values of intrinsic magnetic properties and suffers from difficulties in preparing suitable castings by normal procedures prior to grinding. A promising nitride is $Sm_2Fe_{17}N_{2.7}$. As of this writing (ca 1994) development is under way for commer-

Table 12. Intrinsic Properties of Ternary R-Based Magnetic Materials[a]

Material	Curie temperature, T_C, °C	M_s, T[b]	Anisotropy field, H_a, MA/m[c]	$(BH)^{th}_{max}$, kJ/m^3 [d]
$Nd_2Fe_{14}B$	312	1.60	5.4	512
$Nd_2Fe_{14}C$	262	1.50	7.6	450
Fe_3B:Nd	512	1.60		512
$SmFe_{11}Ti$	312	1.16	7.4	268
$SmFe_{10}V_2$	337	1.10	4.8	240
$SmFe_{10}Mo_2$	187	0.97	>4.0	188
$Sm_2Fe_{17}C$	267	1.42	4.2	403
$Sm_2Fe_{17}N_{2.7}$	477	1.50	11.2	450
$SmCo_5$	747	1.14	24	258
$Sm(Co,Fe,Cu)_7$	827	1.3	4.0	336
$Nd_2Co_{14}B$	722	1.15	8.0	266

[a] Ref. 67.
[b] To convert T to G, multiply by 1×10^4.
[c] To convert MA/m to kOe, multiply by 12.407.
[d] To convert kJ/m^3 to MG·Oe, multiply by 0.125.

cialization of this material (70,71) (see NITRIDES). A summary of developments of other materials appears in Reference 72.

Sintered Magnets. $Nd_2Fe_{14}B$ is brittle, similar to the ferrites and RCo_5 materials, and the manufacture of useful shapes can be made by a conventional powder forming process. In general, melting and casting of the material are done in an inert atmosphere, usually argon, and the castings are crushed and milled to powder below 10 microns in size, then aligned in a magnetic field, pressed into desired shapes and sintered at about 1100°C and cooled rapidly. An enhancement in coercivity can be obtained by a post-sintering heat treatment. The particle alignment in a magnetic field results in a product referred to as anisotropic magnets. If liquid-phase sintering is done to enhance density and coercivity, the starting composition is richer in Nd and B and the final microstructure will have a layer of Nd-rich and Fe–B-rich second phases at the grain boundaries. The Nd-rich second phase appears to enhance coercivity.

Melt Spun Magnets. Melt spun magnets refer to those manufactured from rapidly quenched melts of $Nd_2Fe_{14}B$. The quenched material is in the form of ribbon, 15–50-μm thick and approximately 1.5-μm wide, because the molten alloy liquid stream is directed at a rotating cold cylinder in an inert atmosphere so as to prevent oxidation. Depending on the rate of cooling, the microstructure can be amorphous, finely crystalline, or coarsely crystalline. Amorphous ribbon (over quenched) exhibits weak intrinsic coercivity, thus the grain or crystallite size can be finer than that obtained by the above conventional process. The optimally quenched ribbon is then crushed to powder which is then blended with a polymer, such as epoxy, and pressed at 600–700 MPa (87,000–100,000 psi) into bonded magnets by various molding techniques. The volume fraction of the powder is approximately 80% and such magnets are isotropic, ie, the crystallite's easy axis of magnetization (hexagonal *c*-axis) are randomly oriented, similar to the grains in the ribbon, and the magnets are referred to as being isotropic. Such magnets exhibit remanent magnetizations in zero applied field of 0.7T (ca 7.0 kG) and $(BH)_{max}$ of ca 70kJ/m^3.

Hot pressing of melt spun powder at ca 100 MPa (14,500psi) and ca 670°C leads to nearly 100% dense magnets. Because some grain growth can occur during hot pressing, amorphous or overquenched powder is used as starting material to obtain the optimum grain size. Little grain texture results from this operation; only about 10% crystallographic alignment occurs and the magnets are essentially isotropic. Hot pressed isotropic magnets have been produced with energy products in the 80–160 kJ/m^3 (10–20 MG·Oe) range.

Hot pressing to produce substantial texture and magnetic anisotropy via plastic deformation is accomplished by a process referred to as die-upsetting (73–75). Initial hot pressed isotropic samples are placed in a larger die cavity and pressed at temperatures about 700°C. Substantial *c*-axis alignment along the compression direction occurs. The microstructure consists of individual plate-like grains lying normal to the compression axis. By this process, the remanence is increased to 0.8–1.2 T and maximum energy products to over 250 kJ/m^3 (31 MG·Oe). The effect of Nd content on induced anisotropy in hot deformed Fe–Nd–B magnets has been studied and shown that crystallographic alignment is only possible for magnets containing 12% Nd (76); a liquid phase is present at the grain boundaries during deformation. The die-upsetting route can be used to produce anisotropic bonded magnets. Die-upset magnets are reground to produce magnetically aligned powder subsequently encapsulated into plastic (74,77).

Uses. The commercial development of magnets based on this boride proceeded rapidly, and they are now being used in many diverse applications, for example, servo devices for machine tools, for over 30 d-c motors for fully equipped automobiles (windshield wipers, cooling fans, window and antenna lift motors, etc), magnetic resonance imaging (mri), computer disk drives, and medical device applications. Their largest use is in positioning motors for hard disk drives. New designs of electrical machines are taking place (78).

Chromium–Cobalt–Iron Alloys. In 1971, a family of ductile Cr–Co–Fe permanent-magnet alloys was developed (79). The Cr–Co–Fe alloys are analogous to the Alnicos in metallurgical structure and in permanent magnetic properties, but are cold formable at room temperature. Equivalent magnetic properties also can be attained with substantially less Co, thereby offering savings in materials cost.

Typical property values for Cr–Co–Fe magnets are B_r = 1.0–1.3 T ((1.0–1.3 G) $\times$ 10^4, H_c = 150–600 A/cm (190–753 Oe), and $(BH)_{max}$ = 10–45 kJ/m^3 (1.3 – 5.7 $\times$ 10^6 G·Oe). Some representative properties of commercially available materials and laboratory specimens are listed in Table 10. Very favorable permanent magnetic properties, equivalent to those of Alnico 5, were obtained in low cobalt (ca 10%) ternary alloys. As in the Alnicos, isotropic and anisotropic grades are available. The latter exhibits superior properties achieved by annealing in a magnetic field and/or uniaxial deformation between annealing and aging; uniaxial deformation is not possible with the Alnicos.

Developments up to 1978 are summarized in References 80 and 81. Metallurgically, the Cr–Co–Fe alloys possess the bcc (α) structure at elevated temperature (>1200°C). From ca 700–1200°C, the α-phase tends to coexist with the nonmagnetic γ-phase, ie, fcc. At higher Cr content, a brittle σ-phase also appears in this temperature range. The high temperature α-phase can be retained by quenching, in which case it can undergo spinodal decomposition to an Fe-rich α_1-phase and a Cr-rich α_2-phase with particles of ca 30 nm. This occurs at about 650°C, a temperature which falls off with decreasing Co content to ca 550°C for the Fe–Cr binary. Additions, eg, Zr, Nb, Al, V, and Ti, tend to enlarge the α-phase field at the expense of γ.

The mechanism for coercivity in the Cr–Co–Fe alloys appears to be pinning of domain walls. The magnetic domains extend through particles of both phases. The evidence from transmission electron microscopy studies and measurement of J_s, H_c, and anisotropy vs T is that the walls are trapped locally by fluctuations in saturation magnetization.

Particle shape is very important as evidenced by the elongation and alignment of particles resulting from magnetic annealing and/or uniaxial deformation and the consequent improvement of the magnetic properties, primarily B_r and squareness. Ternary alloys containing ca 12% Co and 33% Cr have achieved values of B_r = 1.2 T (1.2 $\times$ 10^4 G), H_c = 60 kA/m (750 Oe), and $(BH)_{max}$ = 42 kJ/m^3 (5.3 $\times$ 10^6 G·Oe) through deformation aging, whereby the alloy is first aged and then uniaxially deformed by drawing, followed by final aging (82). Such treatment also produced the highest energy product in the Cr–Co–Fe system to date, at 78 kJ/m^3 (9.8 $\times$ 10^6 G·Oe) for a 23Co–2Cu–33Cr alloy (83).

The record for magnetic annealing treatment is 76 kJ/m^3 (9.5 × 10^6 G·Oe), for a 15Co–3Mo–24Cr alloy with ⟨100⟩ columnar grain structure. Good permanent magnet properties could be developed even in the 2–5% Co range, with values of $(BH)_{max}$ = 42 kJ/m^3 (5.3 × 10^6 G·Oe) for a 5Co–30Cr alloy (84). These low Co alloys exhibit the highest ratio of $(BH)_{max}$ to Co content of all Co-based alloys. Thus Alnico 5 properties are achieved in Cr–Co–Fe alloys with a Co content less than half of that in Alnico 5, with added ductility. The very low Co alloys, however, require extremely long heat-treatment times because of the decreased kinetics of the spinodal decomposition. Deformation aged 23%Cr–23%Co–2%Cu exhibits $(BH)_{max}$ of 78 kJ/m^3 (9.75 MG·Oe) (85).

It seems likely that the Cr–Co–Fe alloys could replace the ductile Cunife and Vicalloys and some of the Alnicos. A member of the Cr–Co–Fe alloy family called Chromindur (12Co–20Mo–68Fe) is used as the bias-magnet material for a telephone receiver. The Co composition was reduced from 15 wt % (46) to 10.5% (85,86). The magnet must be formed into cup shape. The cold ductility of the Cr–Co–Fe alloy permits high speed room-temperature forming which is economically advantageous over the slow hot-forming (1250°C) operation required for Remalloy.

Copper–Nickel–Iron and Copper–Nickel–Cobalt Alloys. Cu–Ni–Fe permanent magnet alloys are ductile. They can be stamped even in the hard magnet state, ie, after aging. The nominal composition for a commercial alloy, Cunife, is 60 wt % Cu, 20 wt % Ni, 20 wt % Fe. The hard-magnet state is developed by quenching from 1040°C, tempering at 650°C, cold drawing or rolling, and aging at 650°C. Anisotropic properties are developed by extensive deformation, with $(BH)_{max}$ values up to 12 kJ/m^3 (1.5 × 10^6 G·Oe). Typical commercial values are given in Table 10.

Because of their cold ductility, Cunife magnets are used in speedometer and timing motors where parts are precision stamped at high speed. However, the ingot cannot be hot worked and ingot sizes are limited by segregation to ca 3 cm in diameter. Cunico, similar to Cunife, has composition 29 wt % Co, 21 wt % Ni, and 50 wt % Cu. However, it is no longer produced commercially.

Vanadium–Cobalt–Iron Alloys. V–Co–Fe permanent-magnet alloys also are ductile. A common commercial alloy, Vicalloy I, has a nominal composition: 10 wt % V, 52 wt % Co, and 38 wt % Fe (Table 10). Hard magnetic properties are developed by quenching from 1200°C for conversion to bcc α-phase followed by aging at 600°C (precipitation of fcc γ-phase). The resulting properties are isotropic, with $(BH)_{max}$ ca kJ/m^3 (0.75 MG·Oe). Because it can be rolled to very thin sheets, Vicalloy is used widely in antitheft labels in department store articles and library books.

The V–Co–Fe family with ca 50 wt % Co is rather versatile. The 2 wt % V alloy, 2V-Permendur and Supermendur, is a soft magnet, and alloys containing 3–5 wt % V (Remendur) develop semihard properties, ie, H_c = ca 14–50 A/cm (18–63 Oe). Some of the V may be replaced by Cr, as is done commercially in Germany.

Platinum–Cobalt. Until the advent of the rare-earth magnets, platinum–cobalt magnets represented the best combination of coercivity and energy product (H_{cJ} = 400 kA/m (5 kOe)) and $(BH)_{max}$ = 72 kJ/m^3 (9 MG·Oe)). Optimum properties are achieved near the equiatomic composition, ie, ca 77 wt % Pt.

The alloy is ductile, with the hard magnetic properties developed by controlled cooling from 1000°C followed by aging at ca 600°C. The principal applications have been in hearing aids, watches, and traveling-wave tubes. The rare-earth magnets have replaced, for the most part, PtCo in the latter applications because of the expense of platinum.

Molybdenum–Cobalt–Iron Alloys. Commercial magnets in the Mo–Co–Fe family often are called Remalloy or Comol or Comalloy. There are two important compositions: 17Mo–12Co–71Fe and 20Mo–12Co–68Fe. Magnetic hardening is developed through precipitation aging by oil quenching from 1250°C, followed by aging at 675°C. Values of $(BH)_{max}$ are about 10 kJ/m^3 (1.3×10^6 G·Oe) (Table 10).

The 20 wt % Mo Remalloy was used primarily in a single product, as the bias magnet in an armature-type telephone receiver which was produced in more than 10^7 units annually. Because hot forging is necessary in its manufacture, Remalloy receiver magnets have been replaced by the cold-formable Cr–Co–Fe magnets.

Magnet Steels. Magnet steels are carbon steels containing ca 1% C and various percentages of Co, W, and Cr. They are among the first steels made specifically for permanent magnets. The hard magnetic properties are developed by a quench from ca 1000°C to develop a martensitic structure. Hence, they also are known as martensitic steels. Properties of some representative magnet steels are listed in Table 10. These are used in considerable amounts for applications such as hysteresis motors.

Manganese–Aluminum–Carbon Alloys. Anisotropic Mn–Al–C permanent magnet alloys have been developed using warm working (87). Properties as high as B_r = 0.61 T (6100 G), H_{cB} = 220 kA/m (2.8 kOe) and $(BH)_{max}$ = 56 kJ/m^3 (7×10^6 G·Oe) have been obtained. A typical alloy composition is 70Mn–29.5Al–0.5C. After casting, a cylindrical ingot is solution-treated at 1100°C, quenched to 500°C, tempered at 600°C, and extruded at 700°C. Additional improvements were obtained by aging at 700°C following the extrusion. Typical values of magnetic properties are given in Table 10.

The Mn–Al–C magnets have good mechanical properties and can be machined readily. Their use could expand because manufacture does not require expensive raw materials. However, manufacture is restricted to warm extrusion, a relatively expensive process. Production, as of this writing, is limited to a single plant in Japan for internal consumption.

Semihard Alloys. Coercivities of semihard magnets are from 10–100 A/cm (12–126 Oe). A good number of them are used in hysteresis motors. The magnet steels Cunife and Vicalloy are commonly used. More recent development involves the use of semihard magnets in self-latching remanent-reed electrical contacts in the telecommunications industry (88). These consist of a pair of paddles of semihard magnetic material sealed in a glass tube. The contact between the pair can be opened or closed by current passing through an external coil which surrounds the contacts. The reed material must be sufficiently ductile to be drawn into wire of ca 0.50 mm in diameter, followed by flattening to ca a 0.2-mm thick paddle. Magnetic requirements call for high remanent induction ($B_r > 1.2$ T (12,000 G)) and controlled H_c (ca 25 A/cm (31 Oe)). Physical requirements call for good solderability and plateability (with

Table 13. Alloys for Application in Reed Contacts[a]

Alloy	Coercive force, A/cm[b]	Remanent induction, T[c]	Squareness ratio	Magnetostrictive coefficient $\times 10^6$
49Co–48Fe+3V (Remendur)	24	1.8	0.90	50
38Co–39Fe–20Ni–3Nb	24	1.6	0.95	~45
Co–Fe–Ni–Al–Ti (Vacozet 655)	32	1.4	0.90	48
85Co–12Fe–3Nb (Nibcolloy)	16	1.5	0.95	2
82Co–13Fe–5Mo	24	1.3	0.92	$\gtrsim 0$
84Co–12Fe–4Ti	18	1.4	>0.90	$\gtrsim 0$
82Co–12Fe–6Au	11	1.6	0.85	small
89Co–10Fe–1Be	24		0.90	small
54Fe–28Ni–18Co (Kovar)	<8	1.8	<0.70	
80Fe–16Ni–3Al–1Ti	24	1.3	0.90	0.4
76Fe–16Ni–5Cu–1W	28	1.4	0.95	
80Fe–18Cu–2Mn	24	1.5	0.90	

[a]Adapted from Ref. 85.
[b]To convert A/cm to Oe, divide by 0.7958.
[c]To convert T to G, multiply by 10^4.
[d]G = good; P = poor.

noble-metal contact material), sealability (with glass of appropriate wetting behavior and thermal expansivity), and, preferably, low magnetostriction. The alloys listed in Table 13 can be classified into equiatomic Co–Fe alloys, Co-based alloys, and Fe-based alloys. Of these, Remendur and Nibcolloy are perhaps the most widely used.

Experimental Methods

Field Production. Although magnetic fields from the gap of a permanent magnet often are used for fast qualitative indication of magnetic effects, controlled fields in laboratory instrumentation usually are provided by a solenoid or an electromagnet. Most solenoids are made with insulated copper wire which is wound around a tube; an axial field is generated with the passage of current. Because of resistance (I^2R) heating, field strengths from practical solenoids are limited to ca 80 kA/m (1.0 kOe). The Bitter magnet is used to produce very high fields. The winding is composed of thin perforated disks of copper, and water under high pressure is forced through the holes for cooling. A field of ca 10 MA/m (126 kOe) can be produced and fields twice that have been reached. Such installations, however, require an enormous amount of power and expense. Since the discovery in 1961 of high field Nb_3Sn superconductors, high fields up to ca 10 MA/m (126 kOe) are obtained routinely using superconducting solenoids that

Table 13. (*Continued*)

Thermal expansion coefficient at 20–400°C $\times 10^6$, C^{-1}	Formability[d]	Plateability[d]	Solderability[d]	Sealability[d]
10.2	G	G	G	G
9.8	G			
11	G	G	P	G
12.1	G	G	G	G
ca 12	G			
ca 12	G			
12.5	G			
ca 12	G			
5	G	G	G	G
12.5	G	G	P	
13	G	G	G	
11.5	G	G	G	

are compact and far less expensive. For fields up to ca 2 MA/m (25 kOe), electromagnets are very popular. An electromagnet consists of a pair of cylindrical iron cores, called pole pieces, around which is wrapped a coil of wire carrying a direct current which magnetizes the cores to saturation. The maximum field achieved in the air gap between the pole pieces, except for a small contribution from the coil, is J_s/μ_o or 1.7 MA/m (21 kOe) for iron pole pieces. Higher fields can be obtained by using tapered pole caps to concentrate the magnetic flux.

Field-Strength Measurement. The two most used methods of measuring magnetic fields are the ballistic method and the Hall-effect method. In the ballistic method, a search coil connected to a ballistic galvanometer or a fluxmeter is moved from the region of the field to a region outside the field. The induced emf produces a current through the galvanometer and causes a deflection which is proportional to the field strength. In the Hall-effect method, a small plate-shaped semiconductor sensing element that is connected to a current source develops an emf across the plate at right angles to the current i when placed in a field H normal to the plate. The magnitude of this Hall emf is equal to $R_H iH/tV$, where t is the thickness of the plate in meters and R_H is the Hall constant, eg, ca 10^{-9} $\Omega\cdot m^2/A$ in a semiconductor, eg, InSb.

Magnetization Measurement. For weakly magnetic specimens, the magnetic susceptibility usually is measured in a magnetic balance. Most modern systems make use of commercial automatic analytical balances, with the sample in a magnetic field. By way of cooling and heating, the susceptibility can be measured over wide temperature ranges.

A Foner vibrating sample magnetometer is a technique for the measurement of magnetization of samples that range from weakly magnetic to strongly magnetic. In this technique, current passing through a loudspeaker-type device forces the specimen to vibrate vertically. The a-c signal induced by the specimen in a pair of detection coils is amplified and compared with that induced by a known magnet. The output signal is proportional to the magnetic moment of the specimen.

For the measurement of magnetization in hysteresis-loop measurements, a search coil consisting of several-turn windings around a specimen is connected to a fluxmeter whose output can be recoded on the y-axis of an x,y plotter. The x-axis records the field strength of the primary coil around the specimen. The whole procedure can be done conveniently in a hysteresigraph, whereby the induction is recorded automatically as the field is varied continuously.

Economic Aspects

The value of the U.S. magnetics market in 1980 exceeded $\$2 \times 10^9$. The annual world magnetic materials market in 1988 was $>\$5 \times 10^9$. The value of the components made from these materials is perhaps 20 or more times greater than $5 billion.

The manufacturers of permanent magnets include Hitachi which produces Alnico Grades 5–9 (Table 10) in cast form in various shapes of Grades 2 and 5 in sintered form as well as the rare-earth–cobalt (Hicorex); IG Technologies Inc., which produces Alnico, cast grades (Hyflux) 5, 8, 9 in various shapes, sintered, grades 2, 5, 8, cunife magnets, ceramic magnets (Indox) grades 1 and 5, and the rare-earth–cobalt Incor; Crucible Magnetics, which produces Alnico, cast grades 5, 7, and 8, ceramic magnets (Ferrimag), and the rare-earth–cobalt Crucore; Arnold Engineering, which produces Alnico, cast grades 5 and 8, sintered grades 2, 5, 8; GM, Delco Remy Division which produces Nd–Fe–B, limited to several shapes only (Magnaquench), isotropic and anisotropic, many grades, under the name of Permag, in the form of disks, rectangles, and squares, supplied by the Magnetic Materials Division of the Dexter Corp.; Sumitomo,which produces Nd–Fe–B and sintered magnets; 3M Co., which produces magnetic oxides, ferrites rubber bonded to form flexible permanent magnet tape with or without adhesive (Plastiform); and AlliedSignal, which produces the amphorous magnetic alloys known as Metglas. Other manufacturers are listed in Tables 4 and 10.

BIBLIOGRAPHY

"Magnetic Substances" in *ECT* 1st ed., Vol. 8, pp. 639–659 by R. A. Chegwidden, Bell Telephone Laboratories, Inc.; "Magnetic Materials" in *ECT* 2nd ed., Vol. 12, pp. 737–772, by E. A. Nesbitt, Bell Telephone Laboratories; "Magnetic Materials, Bulk" in *ECT* 3rd ed., Vol. 14, pp. 646–686, by G. Y. Chin and J. H. Wernick, Bell Laboratories.

1. *J. Appl. Phys.* **49**, 1599 (1978).
2. P. A. Beck, *Prog. Mater. Sci.* **23**, 1 (1978).
3. S. Chikazumi and C. D. Graham, Jr., in A. Berkowitz and E. Kneller, eds., *Magnetism and Metallurgy*, Vol. 2, Academic Press, Inc., New York, 1969, Chapt. 12.
4. G. Y. Chin, *Adv. Mater. Res.* **5**, 217 (1971).
5. L. H. Bennett, C. H. Page, and L. J. Swartzendruber, *AIP Conf. Proc.*, (29), xix (1976).

6. C. D. Graham, Jr., *IEEE Trans. Mag.* **12**, 822 (1977).
7. M. McCaig, *Permanent Magnets in Theory and Practice*, John Wiley & Sons, Inc., New York, 1977.
8. *IEC Publ.* **50**, 901 (1973).
9. C. Heck, *Magnetic Materials and Their Applications*, Crane, Russak & Co., Inc., New York, 1974, pp. 23–35.
10. M. F. Littmann, *IEEE Trans. Mag.* **7**, 48 (1971).
11. J. W. Swisher, A. T. English, and R. C. Stoffers, *Trans. ASM* **62**, 257 (1969); J. H. Swisher and E. O. Fuchs, *J. Iron Steel Inst.*, 777 (Aug. 1970).
12. *Steels Products Manual—Flat Rolled Electrical Steel*, American Iron and Steel Institute (AISI), Washington, D.C., 1978.
13. Ref. 9, p. 317.
14. U.S. Pat. 1,965,559 (1934), N. P. Goss (to Cold Metal Process. Co.).
15. S. Taguchi, T. Yamamoto, and A. Sakakura, *IEEE Trans. Mag.* **10**, 123 (1974).
16. F. A. Malagari, *IEEE Trans. Mag.* **13**, 1437 (1977).
17. S. Taguchi, *Trans. ISIJ* **17**, 604 (1977).
18. G. Y. Chin and J. H. Wernick, in E. P. Wohlfarth, ed., *Ferromagnetic Materials*, Vol. 2, North-Holland Publishing Co., New York, 1980, Chapt. 2.
19. *Report E-172* (*Tufperm*), Hitachi Metals Ltd., Tokyo.
20. Ref. 18, p. 143.
21. F. Pfeifer and R. Cremer, *Z. Metall.* **64**, 362 (1973).
22. O. L. Boothby and R. M. Bozorth, *J. Appl. Phys.* **18**, 173 (1947).
23. F. Pfeifer, *Z. Angew. Phys.* **28**, 20 (1969).
24. F. Pfeifer and R. Boll, *IEEE Trans. Mag.* **5**, 365 (1969).
25. G. Y. Chin, T. C. Tisone, and W. B. Grupen, *J. Appl. Phys.* **42**, 1502 (1971).
26. H. L. B. Gould and D. H. Wenny, *AIEE Spec. Publ. T-97*, 675 (1957).
27. Ref. 18, pp. 175 and 179.
28. D. R. Thornburg, *J. Appl. Phys.* **40**, 1579 (1969).
29. B. Thomas, *Proceedings Conference Soft Magnetic Materials*, Cardiff, Wales, 1975, p. 109.
30. R. V. Major, M. C. Martin, and M. W. Branson, *Proceedings Conference Soft Magnetic Materials*, Cardiff, Wales, 1975, p. 103.
31. A. Beer and T. Schwartz, *IEEE Trans. Mag.* **2**, 470 (1966).
32. "Ferrite Cores—for Telecommunication and Industrial Fields," *TDK Data Book*, TDK Electronics Co., Ltd., Skokie, Ill., Feb. 1981.
33. C. H. Smith, *IEEE Trans. Mag.* **18**, 1376 (1982).
34. T. Akashi and co-workers, *Ferrites—Proceedings 1970 International Conference*, University of Tokyo Press, Japan, 1971, p. 183.
35. U.S. Pat. 3,609,083 (Sept. 28, 1971), P. I. Slick (to Bell Telephone Laboratories).
36. J. G. M. DeLau, *Philips Res. Rep. Suppl.*, (6) (1975).
37. P. I. Slick, in Ref. 18, Chapt. 3.
38. T. Inui and N. Ogasawara, *IEEE Trans. Mag.* **13**, 1729 (1977).
39. *Microwave Ferrite Specification Bulletin 1972–1979*, Trans-Tech, Inc., Gaithersberg, Md.
40. F. E. Luborsky and co-workers, *J. Appl. Phys.* **49**, 1769 (1978).
41. R. C. O'Handley, C.-P. Chou, and N. DeCristofaro, *Conference of Magnetics & Magnetic Materials*, Cleveland, Ohio, Nov. 1978; R. Hasegawa and R. C. O'Handley, *J. Appl. Phys.* **50** 1551 (1979).
42. S. Hatta, T. Egami, and C. D. Graham, Jr., *Appl. Phys. Lett.* **34**(1), 113 (1979).
43. *MMPA Standard 0100-78*, Magnetic Materials Producers Association, Evanston, Ill., 1978.
44. A. Hoffmann and P. Pant, *Tech. Mitt. Krupp. Forsch. Ber.* **28**, 117 (1970); *Tech. Mitt. Krupp. Fursch. Ber.* **33**, 25 (1975).

45. G. Y. Chin, J. T. Plewes, and B. C. Wonsiewicz, *J. Appl. Phys.* **49**, 2046 (1978); B. C. Wonsiewicz, J. T. Plewes, and G. Y. Chin, *IEEE Trans. Mag.* **15**, 950 (1979).
46. S. Jin, *IEEE Trans. Mag.* **15**, 1748 (1979).
47. M. L. Green and co-workers, *IEEE Trans. Mag.* **16**, 1053 (1980).
48. T. Ohtani and co-workers, *IEEE Trans. Mag.* **13**, 1328 (1977).
49. R. D. Heidenreich and E. A. Nesbitt, *J. Appl. Phys.* **23**, 352 (1952).
50. A. J. Luteijn and K. J. deVos, *Philips Res. Rep.* **11**, 489 (1956).
51. J. Harrison, *Z. Angew. Phys.* **21**, 101 (1966).
52. D. J. Palmer and S. W. K. Shaw, *Cobalt* **43**, 55 (1969).
53. P. Pant and H. Stablein, "Manufacture and Properties of Columnar Alnico Permanent Magnets," presented at the *World Electrotechnical Congress*, Moscow, Russia, June 21–25, 1977.
54. A. Cochardt, *J. Appl. Phys.* **37**, 1112 (1966).
55. E. A. Nesbitt, J. H. Wernick, and E. Corenzwit, *J. Appl. Phys.* **30**, 365 (1959); E. A. Nesbitt and co-workers, *J. Appl. Phys.* **32**, 342 (1961).
56. K. N. R. Taylor, *Adv. Phys.* **20**, 551 (1971).
57. A. Menth, H. Nagel, and R. S. Perkins, *Ann. Rev. Mater. Sci.* **8**, 21 (1978).
58. K. Strnat and co-workers, *J. Appl. Phys.* **37**, 1252 (1966).
59. E. A. Nesbitt and J. H. Wernick, *Rare Earth Permanent Magnets*, Academic Press, Inc., New York, 1973.
60. W. E. Wallace, *Rare Earth Intermetallics*, Academic Press, Inc., New York, 1973.
61. T. Ojima and co-workers, *Jpn. J. Appl. Phys.* **16**, 671 (1977); *IEEE Trans. Mag.* **13**, 1317 (1977); K. J. Strnat, ed., *Proceedings 3rd International Workshop on Rare Earth-Cobalt Permanent Magnets*, University of Dayton, Ohio, 1978, p. 406.
62. K. J. Strnat, in *Proceedings 4th International Workshop on Rare Earth-Cobalt Permanent Magnets*, Society Non-Traditional Technology, Tokyo, 1979, p. 8.
63. J. J. Croat and co-workers, *Appl. Phys. Lett.* **44**, 148 (1984).
64. D. J. Sellmyer and co-workers, *J. Appl. Physics* **55**, 2088 (1984).
65. M. Sagawa and co-workers, *Jpn. J. Appl. Phys.* **26**, 785 (1987).
66. K. J. Strnat, in E. P. Wohlfarth and K. H.-J. Buschow, eds., *Ferromagnetic Materials*, Vol. 4, Elsevier, Amsterdam, 1988.
67. K. H.-J. Buschow, *Rep. Prog. Phys.* **54**, 1123 (1991).
68. J. F. Herbst and co-workers, *Phys. Rev.* **B29**, 4176 (1984).
69. D. Givord, H. S. Li, and J. M. Moreau, *Solid State Commun.* **50**, 497 (1984).
70. J. M. D. Coey and H. Sun, *J. Magn. and Magn. Mat.* **87**, L251 (1990).
71. Y. Otani and co-workers, *J. Appl. Phys.* **69**(8), 5584 (1991).
72. H. H. Stadelmaier and E. T. Henig, *JOM* **43**(2), 32 (1991); *J. Mat. Eng. Perform.* **1**(2), 167 (1992).
73. J. J. Croat and J. F. Herbst, *MRS Bulletin*, 37 (June 1988).
74. J. D. Livingston, in Jan Evetts, ed., *Concise Encyclopedia of Magnetic and Superconducting Materials*, Pergamon Press Inc., Elmsford, N.Y., 1992, p. 344.
75. K. Raja and co-workers, *J. Appl. Phys.* **73**, 967 (1993).
76. M. Leonowicz and H. A. Davies, *Mat. Lett.* **19**, 275 (1994).
77. K. H.-J. Buschow, in Ref. 74.
78. D. Howe and T. S. Birch, in G. J. Long and F. Grandjean, eds., *Supermagnets, Hard Magnetic Materials*, Kluwer Dordrecht, the Netherlands, 1990, p. 679.
79. H. Kaneko, M. Homma, and K. Nakamura, *AIP Conf. Proc.* **5**, 1088 (1971).
80. G. Y. Chin, *J. Magn. Mag. Mat.* **9**, 283 (1978).
81. H. Zijlstra, *IEEE Trans. Magn.* **14**, 661 (1978).
82. S. Jin, *IEEE Trans. Magn.* **15**, 1748 (1979).
83. S. Jin, N. V. Gayle, and J. E. Bernardini, *IEEE Trans. Magn.* **16**, 1050 (1980).
84. M. L. Green and co-workers, *IEEE Trans. Magn.* **16**, 1053 (1980).

85. S. Jin and G. Y. Chin, *IEEE Trans. Magn.* **23**, 3187 (1987).
86. S. Jin, G. Y. Chin, and B. C. Wonsiewicz, *IEEE Trans. Magn.* **16**, 139 (1980).
87. T. Ohtani and co-workers, *IEEE Trans. Magn.* **13**, 1328 (1977).
88. M. R. Pinnel, *IEEE Trans. Magn.* **12**, 789 (1976).

General References

R. Ball, *Soft Magnetic Materials*, Heyden and Sun Ltd., London, 1979, handbook of soft magnetic materials.

R. M. Bozorth, *Ferromagnetism*, 5th printing, D. Van Nostrand Co., Princeton, N.J., 1951; reprinted by IEEE.

M. McCaig, *Permanent Magnets in Theory and Practice*, John Wiley & Sons, Inc., New York, 1977.

E. P. Wolfarth, ed., *Ferromagnetic Materials—A Handbook on the Properties of Magnetically Ordered Substances*, Vols. 1 and 2, Elsevier, New York 1980; E. P. Wolfarth and K. H.-J. Buschow, eds., Vols. 3 and 4, Elsevier, 1988.

C. W. Chen, *Magnetism and Metallurgy of Soft Magnetic Materials*, North-Holland, New York, 1977.

C. Heck, *Magnetic Materials and Their Applications*, Crane, Russak and Co., Inc., New York, 1974.

F. E. Luborsky, ed., *Amorphous Metallic Alloys*, Butterworths, London, 1983.

T. R. Anantharaman, *Metallic Glasses: Production, Properties and Applications*, Trans. Tech. Publications, Switzerland, 1983.

J. F. Herbst, "Permanent Magnets," *Am. Sci.*, 251 (May–June 1993).

JACK WERNICK

THIN FILMS AND PARTICLES

The largest use of magnetic films and particles, in the form of tapes and disks for recording and retention of audio, visual, and digital information, is in memory and storage technologies (see INFORMATION STORAGE MATERIALS, MAGNETIC). Price per bit of information, including the cost of the peripheral electronics, and performance, as denoted by access time, generally are used to characterize the various memory technologies. Power, modular capacity, reliability, nonvolatility, etc, are also factors describing the efficacy of memories.

Magnetic Properties and Structure

The static or low frequency magnetic properties pertinent to thin-film materials generally are utilized to characterize magnetic materials. As a first approximation, these properties serve to suggest utility for device applications. Saturation magnetization M_s and Curie temperature T_C are intrinsic (structure insensitive) properties and are equal to the bulk values when thick films are made properly. For very thin highly paramagnetic films, such as those of platinum, sandwiched between ferromagnetic, eg, Co, or antiferromagnetic, eg, Cr, thin films, in the form of multilayered structures being developed for recording heads, a magnetization can be induced in the normally paramagnetic material (see MAGNETIC

MATERIALS, BULK). The surface area-to-volume ratio of the individual layers is so large that the atomic moments at the interfaces play an important role.

The extrinsic or structure-sensitive properties depend on size, shape, and surface topography of the films; the size, shape, and orientation of crystallites in polycrystalline films; concentration and distribution of imperfections, impurities, and alloying elements; and state of residual stress. The extrinsic properties can be classified further as static or dynamic depending on whether or not the property displays a frequency dependence. Remanent magnetic induction, B_r; coercive force, H_c; and permeability, μ, are examples of static extrinsic properties. Eddy-current loss and resonance of spins and domain walls typify structure-sensitive dynamic properties. However, high frequency dynamic measurements generally are required for final evaluation of magnetic films. Thin-film preparation techniques, such as the deposition process employed, also highly influence magnetic properties (see THIN FILMS).

Shape anisotropy generally causes the magnetization M_s in thin films to be in the plane of the film. Otherwise a huge demagnetizing field H_d $(= 4\pi M_s)$ would act normal to the plane of the film if M_s were turned in that direction. Similarly, the magnetization in rod-like acicular particles of large aspect ratios may prefer to lie in the long direction. Domains in the films extend completely through the film thickness, and the walls between them are largely 180° and roughly parallel to the easy axis of the film. If the magnetization vector rotates about the wall normals, the walls are called Bloch walls and result in free poles existing on the surface. Bloch walls can exist in bulk materials. For very thin films, eg, ca 50 nm and less, the magnetostatic energy can be reduced if the magnetization vector forming the 180° wall rotates about the film normal. This wall is a Néel wall and free poles do not exist on the surface. Cross-tie walls also can exist in very thin films. A cross-tie wall is crossed at regular intervals by Néel-wall segments and its energy is less than both a Bloch wall and Néel wall in a certain range of film thicknesses. In general, films of technological interest are of such thickness that only Bloch walls are stable. A detailed discussion of domain walls in films and magnetization dynamics is available (1).

In addition to shape anisotropy, an induced anisotropy represented by the constant K_u can be present in films as a result of deposition in a magnetic field. These anisotropies result from short-range directional order or an anisotropic distribution of atom pairs. In the liquid-phase epitaxy (LPE) of single-crystal, mixed-garnet films, the uniaxial anisotropy also appears to be a result of short-range order resulting from the growth process. The induced anisotropy constant is related to the anisotropy field H_k, where $H_k = 2K_u/M_s$. H_k is the field required to rotate the magnetization from the easy axis into the hard directions.

In a polycrystalline film, the easy axis can vary from point to point about an average direction. This dispersion is represented by a dispersion angle, α, which is the dispersion of easy axes about the average direction. A film quality factor q, where $q = H_c/\alpha H_k$, is used as a measure of the usefulness of a material for device applications. It is desirable to maximize this ratio: higher H_c helps make the material more resistant to stray fields, and αH_k determines the drive currents for maximum output (2,3). The effects of all of the deposition variables on the structure and magnetic properties of films are discussed in the literature (3–9).

Permalloys, eg, 81.5% Ni–18.5% Fe, exhibit very low magnetocrystalline anisotropy and magnetostriction. Very low or zero magnetostriction is necessary for storage elements because dimensional changes, which can lead to stresses, are absent when the magnetization is switched. Substrate temperature and deposition rate influence the kinetics of film growth, the degree of impurity incorporation, and the residual stress distribution. For evaporated Permalloy films, H_c remains essentially constant up to a substrate temperature of ca 425°C; then H_c rises rapidly (6). Smaller deposition rates result in lower H_c and H_k for a given substrate temperature, although the deposition rate least affects H_c. Grain size increases with substrate temperature. H_k is affected markedly by the angle at which the metal atoms strike the substrate at low (<300°C) deposition temperatures, because the grains or crystallites are no longer equiaxed but elongated. Easy axis dispersion is relatively independent of rate or temperature up to ca 400°C and then increases sharply.

Sputtering offers advantages over vacuum evaporation. Using sputtering techniques, angle of incidence effects are absent, film composition is generally the same as target composition, and melt composition need not be periodically altered. Additionally deposition rate and film thickness are easily controlled. The effect of argon pressure and sputtering-power density on coercivity and magnetoresistance of 500-nm 80.6 wt % Ni–19.4 wt % Fe films has been studied (7), and the results are shown in Figure 1. Magnetoresistance behavior of Permalloy films is important because these are used for detection in bubble memory devices. As shown in Figure **1a**, resistivity and coercivity decrease with increasing sputtering-power density, whereas the magnetoresistance coefficient increases. The magnetoresistance coefficient is defined as $\Delta R(100)/R_o$, where R_o is the electrical resistance when current flows parallel to the magnetic easy axis and ΔR is the change of R_o in a field of 15.92 A/cm (20 Oe) applied perpendicular to the current flow. At constant power density, maximum magnetoresistance and minimum coercivity are obtained at lower argon pressure (Fig. **1b**). Films that are sputtered at lower power density (low deposition rate) contain more trapped impurities, eg, O_2 and N_2, from the Ar and from the chamber walls which account for the higher coercivity and zero field resistances. At lower Ar pressures, there are decreased deposition rates and fewer impurity atoms striking the film surface (see THIN FILMS, FORMATION TECHNIQUES).

The effect of applying a 100 V negative bias voltage relative to the anode is shown in Table 1 (8). H_c and H_k are independent of deposition rate if a negative bias voltage is applied. Also, large amounts of oxygen or nitrogen impurities in the film are tolerated before magnetic properties deteriorate.

Fabrication

Thermal Evaporation. Thermal evaporation in vacuum, the oldest and most economical method of thin-film preparation, is illustrated in Figure **2a** and consists of heating the material that is to be deposited to a temperature at which appreciable vapor pressure is developed. The vapor condenses onto an appropriately placed substrate which may be maintained at any temperature. Heating of the evaporant is accomplished either by use of a resistance heater or by an electron beam. Electron-beam heating is preferred because higher

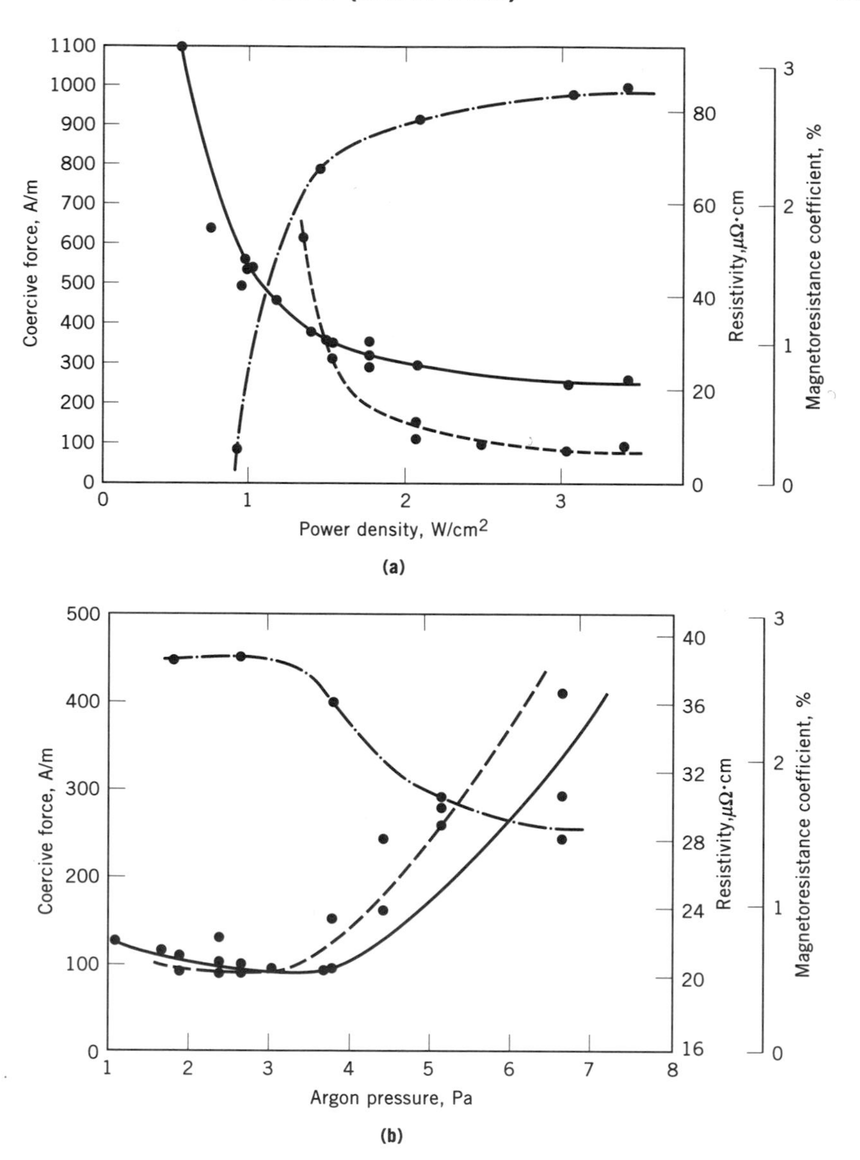

Fig. 1. The effect of (– – –) coercive force, (——) resistivity, and (—·—) magnetoresistance coefficient for Permalloy films (**a**) at a constant argon pressure of 2.9 Pa (2.2×10^{-2} mm Hg) and (**b**) argon pressure at a constant power density of 3 W/cm^2 (7). To convert A/m to Oe, divide by 79.58; to convert Pa to mm Hg, divide by 133.3.

melting metals can be evaporated without contamination, as occurs in crucibles. Water-cooled copper crucibles can be used, and multiple heaters can be used for deposition of alloys. Flash evaporation, a technique for evaporating an alloy where the constituents differ widely in vapor pressure, can be performed by placing a small amount of the evaporant on a very hot source where all of the evaporant flash evaporates. Ion plating is a variant of evaporation in which a high d-c or r-f voltage is used to ionize the evaporant and accelerate it onto the substrate. The substrate, which is the cathode, is in a high voltage gas

Table 1. Magnetic Properties of Sputtered Films[a]

	No bias			Negative bias[b]		
Rate, nm/s	H_c, A/cm[c]	H_k, A/cm[c]	$\Delta\alpha$	H_c, A/cm[c]	H_k, A/cm[c]	$\Delta\alpha$
2.70	1.51	2.47	2.0°	1.83	1.99	6.2°
1.65	1.99	3.10	2.5°	1.19	2.63	2.8°
1.20	2.63	3.42	3.6°	1.83	3.10	2.2°
0.75	5.41	4.78	3.9°	1.59	2.87	3.3°
0.40	12.74			1.19	3.42	1.3°

[a]Ref. 8.
[b]Bias of −100 V.
[c]To convert A/cm to Oe, divide by 0.7958.

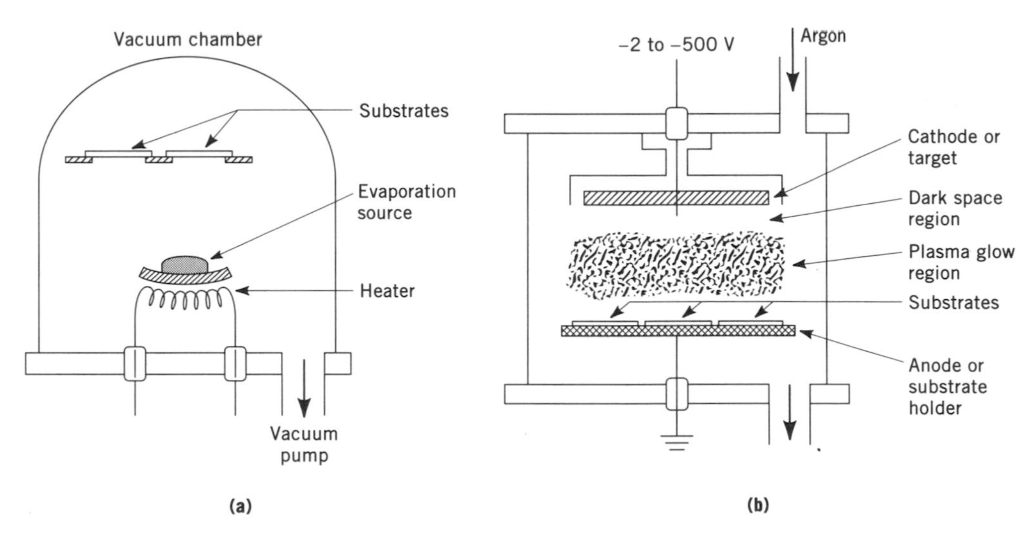

Fig. 2. Simple schematic representation of (**a**) vacuum evaporation and (**b**) cathodic sputtering.

discharge. Discussions of crucible materials that are used for supporting various evaporants, heaters, the cosine law of emission, film thickness measurement, and thickness distribution of evaporated films are included in Reference 9.

Sputtering. Cathodic sputtering processes have come into widespread production use. As schematically illustrated in Figure 2**b**, material is sputtered from a source target (cathode) by inert (argon) energetic ions and deposits on a substrate (anode). The glow discharge may be produced either by a d-c voltage (d-c sputtering) or under r-f conditions. The r-f diode is the most widely used. It permits sputtering of nonconducting or dielectric materials, is characterized by high deposition rates, can be scaled up to handle large substrate loads, the substrates can be sputter-etched clean prior to deposition, and the diode can be operated in a bias-sputtering mode. Another advantage of the sputtering process is that materials of low or zero vapor pressure can be deposited in thin-film form.

A primary source of substrate heating during sputtering is electron bombardment of secondary electrons produced by ion bombardment of the cathode. The secondary electrons can depress the deposition rates by biasing the substrate (anode) negatively, resulting in resputtering of the deposit by Ar ions. However, this problem is not encountered with ion-beam sputtering. The ion beam, which is generated from a separate source, strikes the target in the deposition chamber and the sputtered target atoms are deposited on a substrate in front of the target (10).

Magnetron Sputtering. In magnetron sputtering the discharge is contained near the anode and cathode. This type of sputtering results in high deposition rates, and low substrate heating. It is particularly important where sputtering is carried out at low Ar pressures, eg, ca 2.7 Pa (2.0×10^{-2} mm Hg), so as to minimize the effects of higher Ar pressures, such as gas entrapment in the film. The higher deposition rate results from increased ionizing efficiency caused by the magnetic field. The magnetic field increases the path length of the ionizing electrons. Another method of increasing sputtering rates at lower Ar pressures is by triode sputtering by which an increased quantity of ionizing electrons is supplied thermionically from a filament.

In reactive sputtering, the evaporant in the glow discharge is exposed to reactive gases, eg, oxygen. Reactive sputtering is used for depositing hard-to-form deposits, for example, oxide dielectrics, when starting with a metallic alloy target.

Minimization of contamination of the sputtered film by reactive residual gases, eg, O_2, N_2, H_2O, hydrocarbons, etc, in the system can be accomplished by bias sputtering (11) and getter sputtering (12). In the former process, a small negative bias is applied to the substrate or film being deposited by d-c sputtering. The reduced contamination appears to be a result of positive-ion bombardment of the film during deposition, and adsorbed gases, which could be trapped as impurities, are sputtered off. In a-c sputtering, the cathode and anode are bombarded alternately by ions. In the latter process, sputtering is confined within an anode can surrounding the cathode and anode and a portion of the sputtered material is used to purify or getter the argon before it reaches that part of the system where deposition occurs. Detailed discussions of cathode and r-f sputtering are given in References 13–15.

Off-Axis Magnetron Sputtering. When the substrate faces the target, the technique is referred to as on-axis sputtering. This technique gives the fastest film growth rate. However, this arrangement and low pressures generally yield poor quality films, particularly in terms of stoichiometry and defects for multicomponent materials. The off-axis geometry, schematically shown on Figure 3**a**, overcomes these problems and has been used to make high quality high temperature ternary superconducting oxide films (see SUPERCONDUCTING MATERIALS). The substrates are also mounted on a heater–holder which is outside the region of direct on-axis ion flux but still within the outer edge of the plasma region. The atoms that deposit on the substrate are generally low energy thermalized neutral atoms and the film stoichiometry matches the target. The off-axis arrangement convenient for the preparation of multilayer magnetic structures is shown in Figure 3**b**.

Pulsed Laser Evaporation. Laser evaporation or ablation consists of using a laser emitting at an appropriate wavelength, generally a KrF excimer laser, in a pulsed mode in a controlled atmosphere to deposit a thin film of a material the composition of which is that of the target (16–18) (see LASERS). The process can be modified to carry out plasma-assisted laser deposition by placing a bias

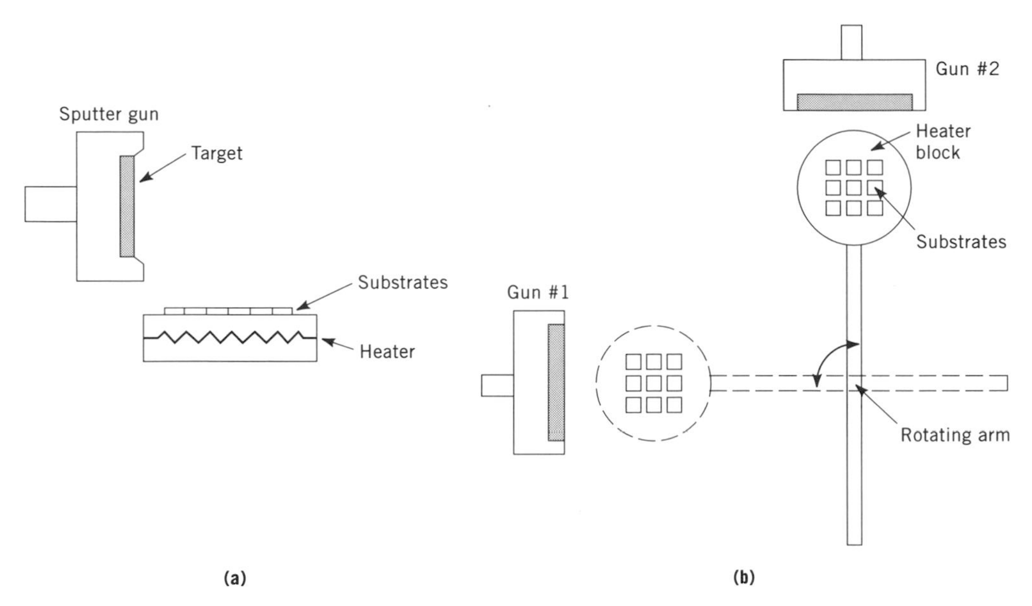

Fig. 3. Schematic configuration of (**a**) the 90° off-axis sputtering technique and (**b**) the system used to prepare multilayers.

voltage between a grid and the source. The role of the plasma is to enhance the chemical reactions leading to the formation of the film and to modify the growth and hence the properties of the film (see PLASMA TECHNOLOGY). Films of the high temperature oxide superconductors and dielectric and optical materials, for example $LiNbO_3$ and $Ba_{0.5}Sr_{0.5}TiO_3$, have been deposited on sapphire (α-Al_2O_3) and silicon.

Molecular Beam Epitaxy. Molecular beam epitaxy (MBE), a high vacuum process for producing high quality ultrathin single-crystal layers, was initially developed for the production of multilayer thin film semiconductor devices based on gallium arsenide, GaAs (19). The sources for the atomic beams of Group 3 (III) to Group 15 (V) elements were condensed phases of the elements located in effusion ovens (Fig. 4). Subsequently, the process was successfully extended to the use of gas-phase sources (20) which gave higher quality GaAs and indium phosphide, InP, films. Metal organic chemical vapor deposition (MOCVD) is a cold-wall process which utilizes volatile metal organic compounds, as sources for the Group 3 (III) to Group 15 (V) elements (21–23). These processes are under investigation for the preparation of Group (III) to (V) semiconductor films having magnetic materials.

Chemical Vapor Deposition. In chemical vapor deposition (CVD), often referred to as vapor transport, the desired constituent(s) to be deposited are in the form of a compound existing as a vapor at an appropriate temperature. This vapor decomposes with or without a reducing or oxidizing agent at the substrate–vapor interface for film growth. CVD has been used successfully for preparing garnet and orthoferrite films (24,25). Laser-assisted CVD is also practiced.

Electrolytic and Electroless Deposition. In electrodeposition, the substrate to be coated is the cathode of an electrolytic cell and the element(s) to be

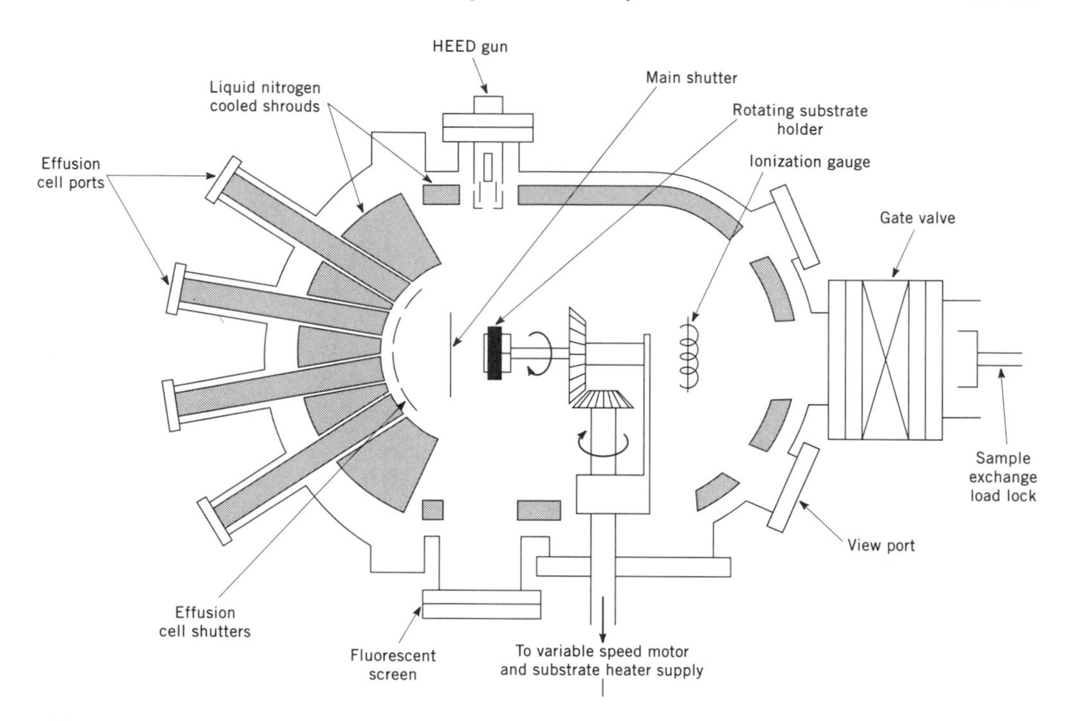

Fig. 4. Schematic of a high vacuum molecular beam epitaxy (MBE) chamber containing four effusion (Knudsen) cells. Also shown is a high energy electron diffraction (HEED) unit for monitoring the deposition as it occurs.

formed into a film are present as ions in an appropriate electrolyte (plating bath) which is in contact with the substrate. By application of an external voltage, the ions are reduced at the substrate surface, thereby forming the film. The quality of the film surface and composition, important in determining the intrinsic and extrinsic magnetic properties, is influenced greatly, among other things, by the external voltage, bath composition, nature of the anode, surface topography of the film, metallurgical structure, ie, grain size, etc, and cleanliness of substrate surface. Historically the largest application of electrodeposition in producing magnetic films was for plated-wire memories using Permalloy. However, the advent of larger capacity memories such as magnetic tape and disk, compact disk-read only, and semiconductor, has resulted in diminished importance of plated-wire memory, as well as planar memory, technology. Comprehensive discussions of electrodeposition of Permalloy films, influence of film structure on magnetic behavior, and plated-wire memories may be found in the literature (26–28).

In electroless deposition, the substrate, prepared in the same manner as in electroplating (qv), is immersed in a solution containing the desired film components (see ELECTROLESS PLATING). The solutions generally used contain soluble nickel salts, hypophosphite, and organic compounds, and plating occurs by a spontaneous reduction of the metal ions by the hypophosphite at the substrate surface, which is presumed to catalyze the oxidation–reduction reaction.

Growth from Solution. In solution growth, film growth is accomplished by immersing the substrate in a saturated or supersaturated solution containing the film constituents. If the solution is saturated initially, the temperature is lowered after immersion to a temperature at which it becomes supersaturated. Film growth then occurs by precipitation onto the substrate surface. If the solution is supersaturated initially, spontaneous growth occurs very soon after the substrate is immersed and film growth occurs at constant temperature until the solution becomes saturated at that temperature. If the substrate is a single-crystal slice and the film is to be a single crystal of identical crystal orientation as the substrate, the process is liquid-phase epitaxy (LPE).

Liquid-Phase Epitaxy. The LPE process (29), which was applied first to the preparation of semiconductor devices, is the growth process that was used for producing bubble memories based on magnetic garnets. The oxide components for the garnet film were dissolved in a molten oxide flux, eg, $PbO-B_2O_3$, contained in a Pt crucible, and the single-crystal, nonmagnetic garnet slice, generally gadolinium gallium garnet (GGG) immersed into the solution for film growth. Bubble memories are no longer produced.

Materials

Magnetic storage materials for storage of audio and video information as well as of digital data are in the form of tape and disks. Optical disks for data storage came into use in 1985; magnetooptic recording systems were being produced as of 1994. Recording materials, which generally exhibit coercivities in the range of 25–100 kA/m (300–1250 Oe), can be classified as semihard.

There are two states of remanent magnetization for recording: longitudinal, in which the magnetization is in the plane of the recording medium; and perpendicular, in which the magnetization is normal to the plane. For particulate media in which the acicular submicronic particles are single domain and embedded in plastic, the magnetization is confined in the direction of the long dimension. The length-to-width ratios are of the order of 5 to 1, with the long dimension approximately 1 μm. Subsequent recording applications led to the development of thin-film metallic coatings so as to achieve higher saturation magnetization and recording density and the ability to tailor, by alloying, the desired magnetic properties.

Multilayer materials exhibiting high magnetization and permeability are undergoing considerable research and development for advanced recording heads. The discovery of giant magnetoresistance in multilayered nano-thick magnetic materials is expected to become important for advanced read heads. Newer magnetooptical materials have the potential for increased storage density in magnetooptical recording. Magnetic and magnetooptic storage technology has been identified as one of the 22 critical technologies for the United States.

Particulate Materials. There are three principal classes of particulate magnetic materials: γ-ferric oxide, γ-Fe_2O_3, and its modifications; chromium dioxide [*12018-01-8*], CrO_2; and iron [*7439-89-6*]. A comparison of the remanent magnetization, B_r, and coercivity, H_c, for several γ-Fe_2O_3 material systems in commercial use or for those being commercialized, ie, barium ferrite [*11138-11-7*], $BaFe_{12}O_{19}$, is shown in Table 2. γ-Fe_2O_3, the most popular magnetic tape material, is used for general-purpose audio recording. Fe_3O_4 has been used instead of γ-Fe_2O_3 for tape. As can be seen from the table, solid solutions of

Table 2. Magnetic Properties of Common Magnetic Recording Media

Material	B_r, T[a]	H_c, kA/m[b]
γ-Fe_2O_3	0.11	26
Fe_2O_3–Fe_3O_4	0.15	37
Co–γ-Fe_2O_3	0.15	52
CrO_2	0.15	45
$BaFe_{12}O_{19}$	0.12	64
Fe	0.30	120

[a]To convert T to G, multiply by 10^4.
[b]To convert kA/m to Oe, divide by 7.958×10^{-2}.

γ-Fe_2O_3 and Fe_3O_4 result in increased H_c. γ-Fe_2O_3 containing Co exhibits even higher coercivities. CrO_2 containing a few percent of Co also results in increased coercivity. Although acicular particles of metals can be produced, problems of chemical stability such as resistance to oxidation, and production costs have prevented use in large amounts in recording media.

A larger B_r leads to heightened readout amplitude, whereas the larger coercive force and thinner film lead to greater recording density. Considerable progress has been made in increasing packing density, uniaxial behavior (acicularity), and homogeneity in size distribution of the γ-Fe_2O_3 particles, as well as attaining higher coercivities. Higher packing densities result in larger output signals at all frequencies. Homogeneity in shape and size results in higher squareness ratios (B_r/B_s) and in homogeneity in particle-switching fields. Higher coercivities are required for higher frequency operation and are a result of increased uniaxial anisotropy through shape and magnetocrystalline anisotropies.

One of the early problems associated with the development of CrO_2 for tape was its chemical instability in organic binders. The stability of CrO_2 particles has been enhanced by the use of a surface-reduction treatment, and high coercivity CrO_2 is available for high quality video tape. The beneficial effect of Co in increasing the coercivity of γ-Fe_2O_3 has been exploited in Japan where the deleterious effects of Co resulting from increased magnetoresistive and anisotropy changes have been minimized through preparation of surface-modified γ-Fe_2O_3, ie, formation of a thin surface layer of cobalt ferrite [*12052-28-7*], $CoFe_2O_4$, on γ-Fe_2O_3 particles (30).

Recording Heads. Materials that are suitable for read/write recording heads for tapes and disks are characterized by high saturation flux density, low remanent induction to avoid erasure of information when the writing current ceases, and low hysteresis and low eddy-current loss, particularly for high data rates or high frequency operation. In addition, because of the small air gap between the head and recording medium, the head material should be abrasion resistant. Dust particles and the magnetic attraction between head and tape can lead to abrasion.

The early recording heads were based on bulk Ni–Fe permalloy and manganese–zinc ferrite, which are soft magnetic materials (see FERRITES). For general-purpose audio recording, laminated Ni–Fe alloys exhibit the required high saturation and low remanence and eddy-current losses; moreover, abrasion is low. Head wear is improved by use of precipitation hardened material. The spinel structure oxides, manganese–zinc and nickel–zinc ferrites, exhibit good abrasion resistance and high frequency characteristics and in some cases are the preferred material despite their relatively low saturation.

For high quality audio and video recording where the recording medium is CrO_2 or Co impregnated γ-Fe_2O_3, sputtered Sendust alloy films (9.6 wt % Si, 5.4 wt % Al, balance Fe; see also Table 3) and ferrites are used as head materials. Sendust exhibits nearly the same resistivity as the ferrites, but the films of this material exhibit higher (1.1 T (11 kG)) saturation flux densities and effective permeabilities of 240 at 20 MHz. Wear resistance is also very good. Ni–Zn and Mn–Zn ferrites are generally used in video and high frequency recording because eddy currents are minimized by their high resistivity relative to metals.

Thin-Film Magnetic Metallic Media. Advanced magnetic recording media are in the form of thin films. The metallic media are typically sputtered films having carbon overcoats for protection. Cobalt-based alloys have been developed for use as longitudinal, ie, *c*-axis of the crystalline Co-alloy parallel to the plane of the substrate, magnetic recording media (51) (see COBALT AND COBALT ALLOYS). Magnetic disks are presently fabricated on nickel phosphide [*12035-46-0*], NiP, coated aluminum alloy disk substrates. An aluminum-base hard disk is illustrated in Figure 5. The NiP is used to provide a polishable surface for the subsequent sputtered Cr-underlayer. For most of the cobalt-based alloys used for recording, the underlayer of chromium is used so as to help in controlling the crystallite orientation (texture) of the film as well as controlling the isolation of the crystallites of the longitudinal recording Co-alloy subsequently deposited on it. Co–Cr–Ta/Cr and Co–Pt–Cr/Cr alloys (42) are being developed for high density longitudinal recording media. Table 3 lists magnetic metals investigated for thin-film devices.

Alloy composition, compositional inhomogeneities in the grains, and deposition parameters including substrate temperature in relation to the isolation of the individual crystallites, ie, the magnetic regions, and properties are all important (Fig. 6). It is generally believed that some isolation of the grains, ie, wider grain boundary regions, is necessary to weaken the exchange coupling among magnetic regions to reduce switching of adjacent crystallites and noise (42). The introduction of nonmagnetic phases at the grain boundaries, by segregation or precipitation for example, should reduce exchange coupling between grains and therefore reduce noise (52) by the formation of a paramagnetic region rich in Cr.

Cobalt–chromium films (20 at. % Cr) exhibiting strong perpendicular anisotropy, ie, hexagonal *c*-axis normal to the substrate surface, have been studied (53). Fifty nanometer films are composed of columnar crystallites and the domain size was found to be a few structural columns in diameter. Magnetization reversal was shown to occur by domain rotation in thick films. Thinner (ca 10-nm thick) films do not show the columnar crystallite morphology. These exhibit well-defined 180° domain walls and magnetization reversal occurs by domain wall motion. Material of intermediate thickness exhibits a microstructure which shows in-plane (longitudinal) and out-of-plane (perpendicular) components. Magnetization reversal by domain rotation in the columnar grains leads to high recording density.

Table 3. Magnetic Metals Investigated For Thin-Film Devices

Material	Composition, wt %	Application[a]	Reference
iron		MR	31
iron–nickel alloys	Permalloys	MR	31
		MD, T, RH	32
cobalt–nickel	82 Co, 18 Ni		
cobalt–phosphorus	98 Co, 2 P	R	31
cobalt–nickel–phosphorus	75 Co, 23 Ni, 2 P	R	31
iron–nickel–chromium	76 Fe, 12 Ni, 12 Cr 74 Fe, 8 Ni, 18 Cr		
Vicalloy II	13 V, 35 Fe, 52 Co	R	31
Cunife I	60 Cu, 20 Ni, 20 Fe	R	31
Cunife II	50 Cu, 20 Ni, 27.5 Fe, 2.5 Co	R	31
Cunico I	50 Cu, 21 Ni, 29 Co	R	31
Cunico II	35 Cu, 24 Ni, 41 Co	R	31
manganese bismuth (1:1)[b]		TH	33
manganese aluminum germanide[c]		TH	34
manganese gallium germanide[d]		TH	35
Sendust alloy	85 Fe, 9.6 Si, 5.4 Al	TS, 12-μm films, RH,	35
RCo(Fe) amorphous alloys[e]	variable	MRM	36–38
Co–Fe–Cr–P–C–B amorphous alloys	variable	SMB, $H_c < 8$ A/m[f]	39
Co–Cr	18–22 Cr	PR, LR	40,41
Co–Cr–Ta/Cr, Co–Pt–Cr/Cr		LR	42
Pt/Co or Pd/Co multilayers	ultrathin alternating layers	MRM	37,43,44
Pt/Fe epitaxial multilayers	3 nm Pt/2.3 nm Fe	MRM, PR	45
CoTa–Zr amorphous multilayers on Al_2O_3 separators	variable	HFRH	46
FeTaN multilayers on Al_2O_3	0.5 μm alloy/0.1 μm Al_2O_3	RH/HDTV	47
Co/Ni multilayers	variable	MRM, PR	48
Co/Au multilayers	variable	PR	49
$Co_{1-x}Pt_x$ multilayers	$x = 0.45 - 0.9$	PA, MRM	50

[a]MR = magnetic recording; MD = magnetoresistive detectors; T = transducers; RH = recording heads; R = recording; TH = thermomagnetic or Curie-point writing; TS = tetrode sputtering; MRM = magnetooptic recording media or magnetooptical recording; SMB = soft magnetic behavior; PR = perpendicular recording; LR = longitudinal recording; HFRH = high frequency recording heads; RH/HDTV = recording heads for high definition television; and PA = perpendicular anisotropy.
[b]The CAS Registry Number is [*12010-50-3*].
[c]The CAS Registry Number is [*12042-22-7*].
[d]The CAS Registry Number is [*37195-97-4*].
[e]Where R = Gd or Tb.
[f]8 A/m = 0.10 Oe.

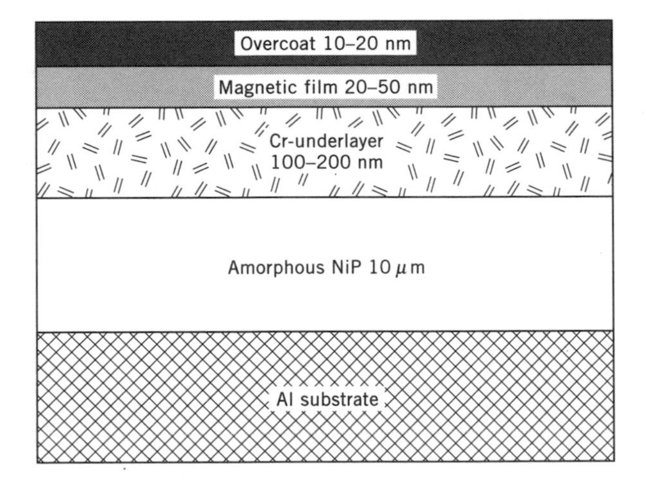

Fig. 5. Schematic drawing of the various layers in a hard disk (51).

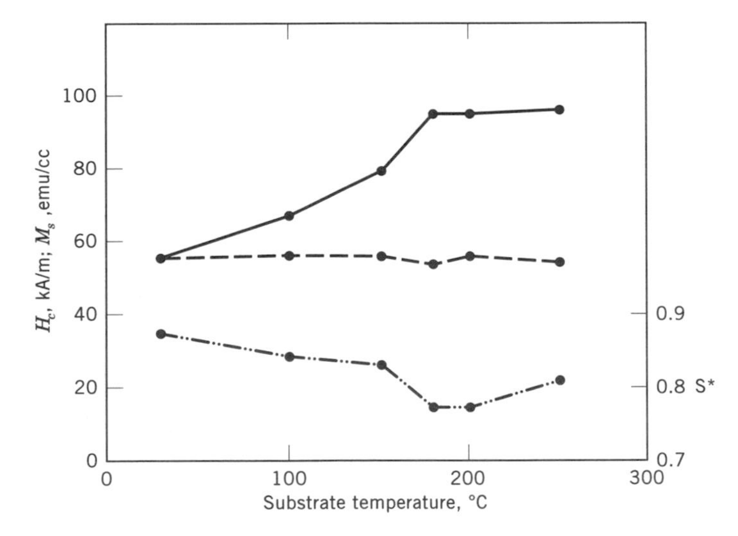

Fig. 6. Changes in (—) coercive force, H_c, (– – –) saturation magnetization M_s, and (–··–) coercive squareness ratio, S^*, of Co–Cr–Ta/Cr films deposited on Cr-underlayers on polymide substrates for longitudinal recording as a function of substrate temperature; d-c magnetron sputtering was used (42). To convert kA/m to Oe, divide by 7.958×10^{-2}.

Reference 37 provides excellent overviews of metallic films, materials science of thin magnetic recording materials, and the potential technological significance.

Magnetooptic Materials. The application of magnetooptic effects to optical memory systems, such as for laser beam writing and magnetooptic read, has been the subject of much research. Magnetooptic storage media offer the potential of storing over 120 Mbit/cm^2 of information without contact of the read/write head which would thus be very competitive to floppy disks and tape.

Memory systems based on laser writing and reading through the interaction of electromagnetic radiation, either through reflection utilizing the Kerr effect or by transmission utilizing the Faraday effect, have begun to appear in the marketplace. As of this writing, recording systems utilizing disks of 90 and 130 mm in diameter are appearing. Recording depends on creating regions of reverse magnetization in films which exhibit uniaxial magnetic anisotropy normal to the film and a relatively low (100–200°C) Curie temperature which results in a strong temperature dependence of coercivity and magnetization. Coercivities of the order of 80 kA/m (1000 Oe) are required at room temperature and about 8 kA/m (100 Oe) at 200°C. Thus the local heating of a submicrometer laser beam impinging on the film can easily heat a submicrometer region near or above the Curie temperature. The demagnetizing field of adjacent material causes a region of reverse magnetization which persists to room temperature. The recorded information can be read using the magnetooptic Kerr effect. Erasure of the recorded information is accomplished by irradiation of the film (heating) with the bias magnetic field in the opposite direction.

The magnetic storage media being employed are ternary amorphous alloys (Table 3) composed of the rare-earth elements gadolinium, Gd, and terbium, Tb, with Fe and Co for use in the near infrared. These materials are compatible with GaAs-based lasers (37). These alloys are ferrimagnetic, ie, the atomic moments of the rare-earth atoms, owing to the $4f$ electrons, couple antiferromagnetically to the atomic moments of Fe and Co resulting from their $3d$ electrons. This negative exchange interaction gives rise to ferrimagnetic behavior resulting in a small net moment or magnetization which is temperature dependent. There is a compensation temperature where the $4f$ and $3d$ magnetizations are equal but opposite in sign and cancel, resulting in a nonmagnetic material. This compensation temperature is below room temperature, but above this temperature the $3d$-electron magnetization is smaller than that which would be exhibited if no rare-earth atoms were in the alloy. Thus the magnetization of interest to magnetooptic recording can be tailored based on the above negative exchange interaction that occurs between some of the rare-earth and Fe-group atoms.

The rare-earth (R) garnets, $R_3Fe_5O_{12}$, which are ferrimagnetic, are being investigated for magnetooptic recording. Atomic substitutions for the trivalent R and Fe atoms result in garnets having the required high room temperature coercivity and Curie temperature range (150–260°C). An example is $Bi_2DyFe_{3.6}Al_{1.2}O_{12}$ referred to as a Bi-substituted garnet (54). Films of several garnets in the system $(GdPrTmBi)_3(Fe,Ga)_5O_{12}$ have been proposed for use in magnetooptic read heads (55). Further work on garnets and ultrathin Pt/Co or Pd/Co multilayers containing Pt or Pd of about 1-nm thick and Co layers about 0.35-nm thick is required to establish their possible use for magnetooptic recording (37,43,44).

Amorphous single-domain CoTaZr cores having Al_2O_3 interlayers where the CoTaZr thickness is from 0.23–0.9 μm, depending on the number of layers, and Al_2O_3 is 0.01-μm thick, were evaluated for use as thin-film heads (46). This material combination is attractive for low noise heads operating at frequencies up to 40 MHz (46). Similarly, the read/write characteristics of laminated cores consisting of four layers made of Fe–Ta–N films 0.5-μm thick separated by 0.1-μm Al_2O_3 layers are characterized by high magnetic saturation and permeability and are attractive for high definition television (HDTV) (47).

Magnetooptical recording in sputtered Co/Pt multilayers (0.4 nm Co–1.4 nm Pt) and $Gd_{17}Tb_8Fe_{75}$ disks were evaluated at wavelengths of 820, 647, and 458 nm, corresponding to AlGaAs, krypton, and argon lasers, respectively. At 458 nm, Co/Pt performed 3 dB better than GdTbFe. This was attributed to the higher Kerr signal exhibited by Co/Pt, low noise level, and quality of the written domain structure (56).

Synthetic magnetic superlattices can be considered important new materials. The perpendicular anisotropy developed by ultrathin layered superlattices, such as Co/Pt and Co/Pd, in contrast to the more usual tendency for the preferred direction of the magnetization to be in the film plane (shape anisotropy) because of lower magnetostatic energy, appears to arise from the interfacial areas of the composite. The magnetic anisotropy energy was shown to be proportional to the reciprocal of the thickness of the Co-layers and perhaps is a consequence of the atomic disorder at the interfaces. The surface area-to-volume ratio is large and thus should dominate the magnetic anisotropy below a certain thickness (57–59).

Magnetic Superlattices. The discovery in the late 1980s of giant magnetoresistance (GMR) in antiferromagnetically coupled Fe/Cr superlattices (60,61) stimulated great interest. Properties of metallic superlattices consisting of thin alternate single-crystal layers of different magnetic materials as well as alternate layers of magnetic and nonmagnetic materials were examined.

Epitaxically oriented single-crystal layers of Fe and Cr have been prepared by MBE on exceptionally clean (001) GaAs single-crystal substrates (60,61). The notation for one individual bilayer is ((001)Fe/Cr(001)). The individual layer thickness ranges from 0.9 to 9 nm and the total number of bilayers in a single film is around 30. Notation for a complete superlattice of 30 bilayers would be $(\text{Fe 3 nm/Cr 1.2 nm})_{30}$. Chromium, in bulk form, is antiferromagnetic. At low temperatures and in zero-applied field, the magnetization behavior of each successive layer is antiparallel and the electrical resistance of the film is high (Fig. 7**a**). When a magnetic field is applied parallel to the layers and is sufficient to overcome the antiparallel arrangement of the layer magnetizations (Fig. 7**b**), the electrical resistance of the film begins to decrease. When this occurs, the film is said to exhibit a negative magnetoresistance. The complete switching fields (H_S), for a constant negative magnetoresistance, for several Fe/Cr superlattices at 4.2 K, are shown in Figures 8 and 9. H_s for the saturation of magnetoresistance is a function both of composition of the individual layers and the number of bilayers. There is also an anisotropy in the magnetoresistance as illustrated in Figure 9 (60).

Magnetoresistive recording heads offer much more sensitivity than inductive heads (63) and there is strong evidence as of this writing (ca 1994) that such heads will be used exclusively by the year 2000. The trend in the development of head materials is toward thin-film media. Although Permalloy films ($Ni_{18}Fe_{19}$) are used for magnetoresistive sensors, the change in resistance is only about 2.5%. Higher magnetoresistive materials are needed. The Fe/Cr GMR system requires large (ca 1 T) switching fields to align the magnetization in the layers (60,61). Weaker interlayer interactions are needed to obtain the desired resistivity change at low fields and at room temperature. Table 4 summarizes some of the work along these lines. Theoretical discussions of the microscopic origin of GMR can be found in References 70–74. Materials-related aspects of multilayered materials are summarized in References 75 and 76.

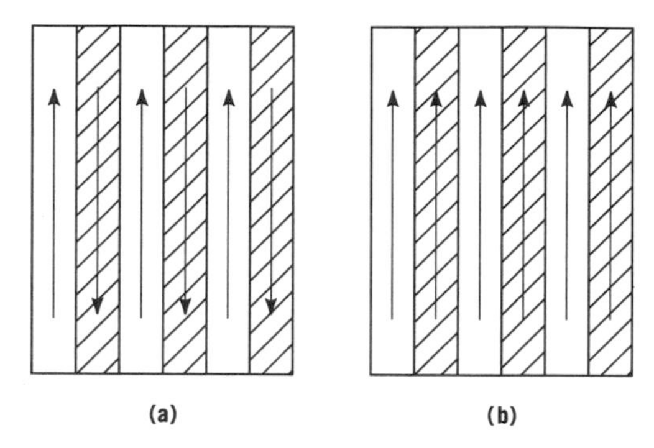

Fig. 7. Schematic representation of an antiferromagnetic magnetic multilayer superlattice in (**a**) zero-applied field ($H = 0$) at 4.2 K and (**b**) in an applied field H_s sufficient to establish saturated negative magnetoresistance (62).

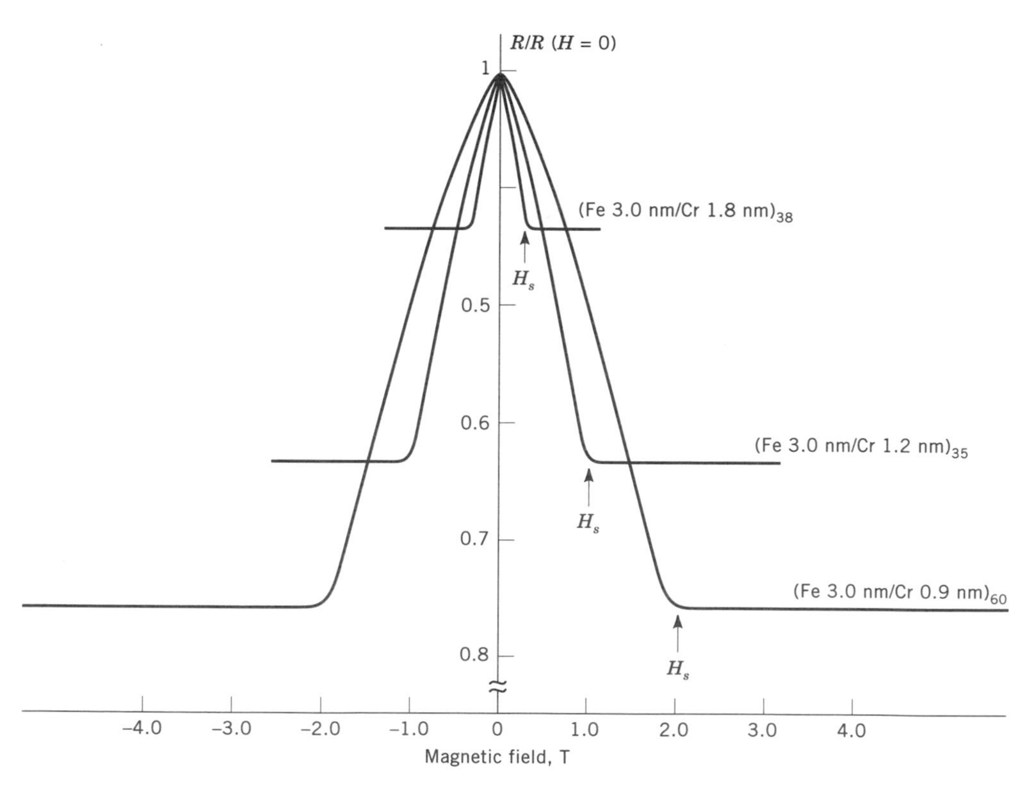

Fig. 8. Giant negative magnetoresistance $R/R(H = 0)$ behavior of three Fe/Cr superlattices as a function of applied field at 4.2 K showing the value of the complete switching field, H_s, for each lattice. The current and the applied field are along the (110) direction in the plane of the layers (60). To convert T to G, multiply by 10^4.

Table 4. Magnetic Superlattices Exhibiting GMR

Material system	Multilayer information	Result[a–c]	Reference
Fe/Cr	variable ((001) Fe/Cr(001))$_n$, 0.9–9.0-nm single-crystal layer thickness on single-crystal GaAs	resistivity lowered by as much as a factor of two at 4.2 K and switching fields of the order of 1 T are required	60,61
Co/Cu	variable	strong antiparallel coupling through the nonferromagnetic copper; high (>800 kA/m) fields required to change antiferromagnetic spin structure into a ferromagnetic one	61,64
Co–Fe/Cu	1.0 nm Co_9Fe/1.0 nm Cu ion-beam sputtering on MgO(110) substrates	Co–Fe/Cu grew having inplane uniaxial anisotropy and easy axis parallel to the cube direction in the MgO(110) plane; saturation field (240 kA/m) at room temperature for GMR = 45%	65
Ni,Fe/Cu	$Ni_{81}Fe_{19}$/Cu/Co	GMR dramatically enhanced by presence of the thin Cu layer; magnetoresistance of more than 17% for field changes of ±8 kA/m at room temperature	66
Ni,Fe/Cu/Co	[$Ni_{80}Fe_{20}$/Cu/Co/Cu] r-f diode sputtering on Si(100) single-crystal wafers at room temperature	resistance changes as large as 70% within a few tens of amperes per meter	67
NiFe/Cu/NiFe/FeMn		resistance changes of 3–4% in fields of 40–800 A/m	68
Co/Ag	Co(15 nm)/Ag(6.0 nm) electon-beam evaporation on top of a 5.0-nm Cr buffer layer on Si(111) substrates	interface roughness appears to be important in understanding the connection between GMR and antiferromagnetic coupling	69

[a] To convert T to G, multiply by 10^4.
[b] To convert kA/m to Oe, divide by 7.958×10^{-2}.
[c] GMR = giant magnetoresistance.

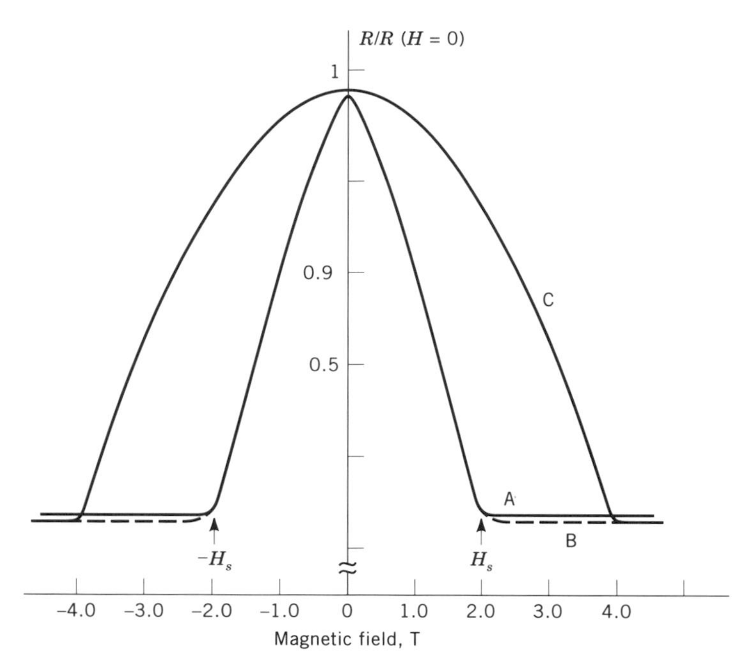

Fig. 9. Effect on (Fe 3.0 nm/Cr 0.9 nm)$_{40}$ magnetoresistance, $R/R(H = 0)$, of changing the direction of the applied field relative to the current direction of [110] in the plane, where A represents an applied field in the layer plane along the current direction; B, (– – –) the applied field in the layer plane, but directed perpendicular to the current direction; and C, the applied field perpendicular to the layer plane (60).

Magnetic Fluids. Magnetic fluids are stable colloidal suspensions of ferromagnetic particles, such as Fe_3O_4 and of subdomain size (ca 10 nm) in aqueous or organic bases (77,78). The fluid behaves as a homogeneous Newtonian liquid and reacts to a magnetic field. These materials are used in bearings, rotary-shaft seals, and feedthroughs (see BEARING MATERIALS). Other possible applications, eg, as jet inks and for float separation, have been studied (79) (see SEPARATION, MAGNETIC SEPARATION). Fluid properties, such as magnetic moment, viscosity, and density, can be tailored for various applications if the proper choice of particle concentration, composition and size, particle coatings, and liquid base is made.

BIBLIOGRAPHY

"Magnetic Materials, Thin Film" in *ECT* 3rd ed., Vol. 14, pp. 686–707, by J. H. Wernick and G. Y. Chin, Bell Laboratories.

1. B. D. Cullity, *Introduction to Magnetic Materials*, Addison-Wesley, Reading, Mass., 1972.
2. M. Prutton, *Ferromagnetic Films*, Butterworths, London, 1964.
3. J. F. Freedman, *IEEE Trans. Magn.* **MAG-5**, 752 (1969).
4. E. W. Pugh, in G. Haas, ed., *Physics of Thin Films*, Vol. I, Academic Press, Inc., New York, 1963.
5. R. E. Thun, *Rev. Elec. Comm. Lab.* **25**, 209 (Mar./Apr. 1977).

6. E. Pugh and T. O. Mohr, *Thin Films: Properties of Ferromagnetic Films*, ASM, Metals Park, Ohio, 1963, Chapt. 7.
7. T. Serikawa, in Ref. 5.
8. B. L. Flur, in *Physics of Thin Films*, Vol. 3, Academic Press, Inc., New York, 1966.
9. R. Glang, in L. I. Maissel and R. Glang, eds., *Handbook of Thin Film Technology*, McGraw-Hill Book Co., Inc., New York, 1970.
10. D. Bouchier and co-workers, *J. Appl. Phys.* **49**, 5896 (1978).
11. L. I. Maissel and P. M. Schaible, *J. Appl. Phys.* **36**, 237 (1965).
12. H. C. Theurer and J. J. Hauser, *Trans. AIME* **233**, 588 (1965).
13. L. I. Maissel, in Ref. 8.
14. J. L. Vossen and J. J. O'Neill, Jr., *RCA Rev.*, 149 (June 1968).
15. G. N. Jackson, *Thin Solid Films* **5**, 209 (1970).
16. A. M. Marsh and co-workers, *Appl. Phys. Lett.* **62**(9), 952 (Mar. 1, 1993).
17. R. C. Baumann, R. A. Rost, and T. A. Robson, *Mater. Res. Soc. Symp. Proc.* **200**, 25 (1990).
18. D. Roy and S. B. Kruopanidhi, *Appl. Phys. Lett.* **62**(10), 1056 (Mar. 8, 1993).
19. A. Y. Cho, *J. Appl. Physics* **41**, 2780 (1970).
20. M. B. Panish, *J. Electrochem. Soc.* **127**(12), 2729 (1980).
21. H. M. Manasvet and W. I. Simpson, *J. Electrochem.* **116**, 1725 (1969).
22. H. M. Manasvet, *J. Crystal Growth* **55**, 1 (1981).
23. G. B. Stringfellow, *J. Crystal Growth* **68**, 111–122 (1984).
24. J. E. Mee and co-workers, *IEEE Trans. Magn.* **MAG-5**, 717 (1969).
25. J. W. Nielsen, *Met. Trans.* **2**, 625 (1971).
26. I. W. Wolf, *J. Appl. Phys.* **33**, 1152 (1962).
27. R. Girard, *J. Appl. Phys.* **38**, 1423 (1967).
28. D. J. Wollons, *Introduction to Digital Computer Design*, McGraw-Hill Book Co., Inc., London, 1972.
29. E. A. Giess and R. Ghez, in *Epitaxial Growth*, Part A, Academic Press, Inc., New York, 1974, p. 183.
30. A. R. Corradi, *IEEE Trans. Magn.* **MAG. 14**, 655 (1978).
31. C. D. Mee, *The Physics of Magnetic Recording*, North Holland, Amsterdam, 1963; *Inst. Elec. Electron. Engrs. Trans. Commun. Electron.*, 399 (1964).
32. R. P. Hunt, *IEEE Trans. Magn.* **MAG-7**, 150 (1975).
33. R. W. Keyes, *Comments Solid State Phys.* **5**, 97 (1973).
34. R. C. Sherwood and co-workers, *J. Appl. Phys.* **42**, 1704 (1971).
35. E. Sawatzky and G. B. Street, *J. Appl. Phys.* **44**, 1789 (1973).
36. C. H. Bajorek and R. J. Kobliska, *IBM J. Res. Dev.*, 271 (May 1976).
37. M. H. Kryder, *Thin Solid Films* **216**, 174 (1992); S. D. Bader, *Proc. IEEE* **78**(6), 909 (1990).
38. D. Raasch, *IEEE Trans. Magn.* **29**(1), 34 (1993).
39. N. Heiman, R. D. Hempstead, and N. Kazama, *J. Appl. Phys.* **49**, 5663 (1978).
40. Y. Maeda and M. Takahashi, *J. Appl. Phys.* **68**, 4751 (1990).
41. Y. Maeda and M. Asahi, *J. Appl. Phys.* **61**, 1972 (1987).
42. Y. Maeda and K. Takei, *IEEE Trans. Magn.* **27**(6), 472 (1991).
43. H. Yamane, Y. Maeno, and M. Kobayashi, *Appl. Phys. Lett.* **62**, 1562 (1993).
44. F. J. A. Den Broeder and co-workers, *Appl. Physics* **A**, 507 (1989).
45. B. M. Lairson and co-workers, *Appl. Phys. Lett.* **62**, 639 (1993).
46. R. Arai and co-workers, *IEEE Trans. Magn.* **28**(5), 2115 (1992).
47. T. Okumura and co-workers, *IEEE Trans. Magn.* **28**(5), 2121 (1992).
48. F. J. A. Den Broeder and co-workers, *Appl. Phys. Lett.* **61**(12), 1468 (1992).
49. F. J. A. Den Broeder and co-workers, *Phys. Rev. Lett.* **60**(26), 2769 (1988).
50. D. Weller and co-workers, *Appl. Phys. Lett.* **61**(22), 2726 (1992); P. F. Garcia and Z. G. Li, *Appl. Phys. Comm.* **11**, 531 (1992).

51. D. E. Laughlin and B. W. Wong, *IEEE Trans. Magn.* **27**, 4713 (1991).
52. T. Chen and T. Yamashita, *IEEE Trans. Magn.* **24**, 2700 (1988).
53. B. G. Demczyk, *IEEE Trans. Magn.* **28**(2), 998 (1992).
54. K. Shono and co-workers, *Mater. Res. Symp. Proc.* **150**, 131 (1989).
55. B. Ferrand and co-workers, *IEEE Trans. Magn.* **24**, 2563 (1988).
56. W. B. Zeper and co-workers, *IEEE Trans. Magn.* **28**(5), 2503 (1992).
57. H. J. G. Draaisma, W. J. M. de Jonge, and F. J. A. de Broeder, *J. Magn. and Magn. Mat.* **66**, 351 (1987).
58. L. Neel, *J. Phys. Rad.* **15**, 225 (1954).
59. F. Hakkens and co-workers, *J. Mater. Res.* **8**(5), 1019 (1993).
60. M. N. Baibich and co-workers, *Phys. Rev. Lett.* **61**, 2472 (1988).
61. G. Binasch and co-workers, *Phys. Rev.* **B39**, 4828 (1989).
62. L. M. Falicov, *Physics Today*, 46 (Oct. 1992).
63. C. Tsang, *J. Appl. Phys.* **55**, 2226 (1984).
64. D. H. Mosch and co-workers, *J. Magn. Magn. Mat.* **94**, L1 (1991).
65. K. Inomata and Y. Saito, *Appl. Phys. Lett.* **61**(6), 726 (1992).
66. S. S. P. Parkin, *Appl. Phys. Lett.* **61**(11), 1358 (1992); *Phys. Rev. Lett.* **71**, 1641 (1993); *Mat. Lett.* **20**, 1 (1994).
67. T. Valet and co-workers, *Appl. Phys. Lett.* **61**(26), 3187 (1992).
68. B. Dieny and co-workers, *Phys. Rev.* **B43**, 1297 (1991).
69. L. F. Schelp and co-workers, *Appl. Phys. Lett.* **61**(15), 1858 (1992).
70. R. L. White, *IEEE Trans. Magn.* **28**(5), 2482 (1992).
71. D. M. Edwards, J. Mathon, and R. B. Muniz, *IEEE Trans. Magn.* **27**, 3548 (1991).
72. R. E. Camley and J. Barnas, *Phys. Rev. Lett.*, 63 (1989).
73. J. Barnas and co-workers, *Phys. Rev.* **B42**, 8110 (1990).
74. P. Grünburg and co-workers, *J. Magn. Magn. Mater.* **93**, 58 (1991).
75. *Mat. Res. Soc. Bull.* **15**(2), (Feb. 1990).
76. *Mat. Res. Soc. Bull.* **15**(3), (Mar. 1990).
77. R. E. Rosensweig, *J. Sci. Technol.* **48** (July 1966).
78. S. W. Charles and J. Poppewell, in E. P. Wohlfarth, ed., *Ferromagnetic Materials*, North Holland, Amsterdam, 1980.
79. *IEEE Trans. Magn.* **16** (Mar. 1980).

General References

M. H. Kryder, "Data Storage in 2000: Trends in Data Storage Technologies," *IEEE Trans. Magn.* **25**(6), 4358 (1989).

M. P. Sharrock, "Particulate Magnetic Recording Media: A Review," *IEEE Trans. Magn.* **25**(6), 4374 (1989).

P. Hansen and H. Heitmann, "Media for Erasable Magnetooptic Recording. *IEEE Trans. Mag.* **25**(6O

J. M. E. Harper, "Ion Beam Techniques in Thin Film Deposition," *Solid State Technol.*, 129 (Apr. 1987).

O. Kohmoto, "Recent Developments of Thin Film Materials for Magnetic Heads," *IEEE Trans. Magn.* **27**(4), 3640 (1991).

T. Kunieda, K. Shinohara, and A. Tomago, "Metal Evaporated Video Tape," *Proc. IERE* **59**, 37 (1984).

B. Heirich and J. A. C. Bland, eds., *Ultrathin Magnetic Structures*, Springer, Berlin, 1994.

JACK WERNICK

MAGNETIC PROPERTIES. See MAGNETIC MATERIALS.

MAGNETIC SPIN RESONANCE

Magnetic spin resonance techniques are among the most powerful methods available for determining primary structure, conformation, and local dynamic properties of molecules in liquid, solid, and even gas phases of organic and inorganic molecules. These methods measure the interactions between matter and external radio frequency (rf) fields in the presence of a static external magnetic field. Under suitable conditions the measurements also may be sufficiently quantitative for analytical applications. Although much of the theory is similar, spin resonance experiments are subdivided into nuclear magnetic (nmr) and electron spin (esr) resonance methods. The first of these techniques has led to a newer field of study, magnetic resonance imaging, covered elsewhere (see MEDICAL IMAGING TECHNOLOGY). Nuclear magnetic resonance originally was carried out in continuous wave (CW) mode where the external magnetic field was systematically varied to match the resonance conditions for individual nuclei. As of this writing almost all experiments are done using the pulse or Fourier transform (ft) mode of operation and that mode is discussed herein. The reader is encouraged to consult the references for more detailed information in specific areas. References, selected for both utility and clarity, may not be the original citation of primary work. Several excellent textbooks for both nmr (1,2) and esr (3) exist.

Theoretical Background

An unpaired electron or an atomic nucleus, unless it has even integral values for both atomic mass and atomic number, has a nonzero angular momentum and measurable magnetic moment. Both the electron and the hydrogen nucleus or proton, ^1H, have a spin, I, of 1/2. Other nuclei may have larger values of I (see Table 1). A given nucleus can exist in any of $2nI + 1$ states described by a nuclear spin quantum number, m_i, ranging from $+I$ (lowest energy) to $-I$ (highest energy) in unit steps. When placed in an external static magnetic field, B_0, the difference in energy, ΔE, between adjacent states is given by:

$$\Delta E = h\omega_0/2\pi = h\gamma B_0/2\pi = h\nu$$

where h is Planck's constant (6.626×10^{-34} J·s), γ is the magnetogyric ratio, and $\omega_0/2\pi = \nu$ is the Larmor frequency in Hertz. The value of γ for an electron is about 660 times larger than for a hydrogen atom. As shown in Figure 1 for the spin 1/2 case, the precise value of both the energy and Larmor frequency is proportional to the strength of the static external field. The relative population, N_β/N_α, of the two energy levels is described by the Boltzmann relation,

$$N_\beta/N_\alpha = \exp(-\Delta E/kT)$$

leading to population differences on the order of 5 parts in 10^5 for ^1H at 300 K and 7.05 T (70.5 kG) $= B_0$.

Table 1. Nmr Parameters for Less Commonly Used Nuclei

Isotope	Spin, I	Receptivity ratio[a]	Reference compound	Detection range, ppm	ν at 7.05 T,[b] MHz[c]	Q,[d] 10^{-28} m^2
^{2}D	1	0.008	$Si(CD_3)_4$	10 to 0	46.5	0.00027
^{6}Li	1	3.58	$LiCl_2/D_2O$	2 to −10	44.15	−0.0006
^{11}B	3/2	754.00	$BF_3/(C_2H_5)O$	65 to −130	96.21	0.0355
^{14}N	1	5.69	NH_3 (l)[e]	900 to 0	21.68	0.016
^{15}N	1/2	0.022	NH_3 (l)[e]	900 to 0	30.41	
^{17}O	5/2	0.061	H_2O	1700 to −50	40.67	−0.026
^{19}F	1/2	4730.00	$CFCl_3$	−276 to −280	282.23	
^{23}Na	3/2	525.00	1M $NaCl/H_2O$	10 to −65	79.35	0.12
^{25}Mg	5/2	1.54	$MgCl_2/H_2O$	50 to −25	18.36	0.22
^{27}Al	5/2	117.00	$Al(NO_3)_3$	240 to −240	79.70	0.149
^{29}Si	1/2	2.09	$Si(CH_3)_4$	80 to −380	59.60	
^{31}P	1/2	377.00	85% H_3PO_4	270 to −480	181.04	
^{35}Cl	3/2	20.2	NaCl in H_2O	1200 to −100	29.39	−0.08
^{39}K	3/2	3.69	KCl in H_2O	10 to −25	14.00	0.055
^{43}Ca	7/2	0.53	$CaCl_2$ in H_2O	40 to −40	20.19	−0.05
^{47}Ti	5/2	0.864	$(TiF_6)^{2-}$	1700 to 0	16.92	0.29
^{71}Ga	3/2	319.00	$Ga(H_2O)_6^{3+}$	650 to −800	72.00	0.178
^{75}As	3/2	143.00	$KAsF_6$	370 to −300	51.37	0.3
^{77}Se	1/2	2.98	$Se(CH_3)_2$	2000 to −800	57.21	
^{81}Br	3/2	277.00	$NaBr/H_2O$	100 to −500	81.02	0.28
^{87}Rb	3/2	277.00	$RbCl/H_2O$	50 to −230	98.16	0.12
^{89}Y	1/2	0.668	$Y(ClO_4)_3$	20 to −140	14.70	
^{99}Ru	5/2	0.83	1M RuO_4/CCl_4	3000 to −3300	13.84	0.076
^{113}Cd	1/2	7.6	$Cd(ClO_4)_2/H_2O$	850 to −100	66.53	
^{117}Sn	1/2	19.5	$Sn(CH_3)_4$	500 to −2300	58.62	
^{127}I	5/2	530.00	NaI/H_2O	4000 to −500	60.03	−0.79
^{129}Xe	1/2	31.80	$XeOF_4$ neat	2000 to −5300	82.97	
^{139}La	7/2	336.00	10^{-2} M $LaCl_3/H_2O$	310 to −20	42.38	0.21
^{195}Pt	1/2	19.1	Na_2PtCl_6	9000 to −4000	64.39	
^{199}Hg	1/2	11.8	$Pb(CH_3)_4$	11200 to −6000	62.57	

[a]The receptivity ratio is calculated by comparing the value of $a_x \gamma_x^3 l_x(l_x + 1)$ to the corresponding value for ^{13}C, where a_x = abundance of x, γ_x = magnetogyric ratio, and l_x = spin x. See text.
[b]To convert T to G, multiply by 1.00×10^4.
[c]Corresponding to 300 MHz for ^{1}H.
[d]Q = electric quadrupolar moment.
[e]At 25°C.

Nuclear Magnetic Resonance. The interaction of a nucleus with B_0 is usually described using vector notation and models as in Figure 2 where the bulk magnetization, M, and the static field B_0 are initially parallel to z. A radio frequency pulse is applied in the xy plane for a duration of t μs, tipping the magnetization by θ radians where $\theta = \gamma B_1 t$. The duration of the 90° pulse required to completely shift magnetization into the xy plane is thus inversely proportional to B_1. The change in orientation of individual spins with respect to the external field is referred to as resonance. Because the net magnetization is now inclined with respect to B_0 in response to the rf source, it precesses about

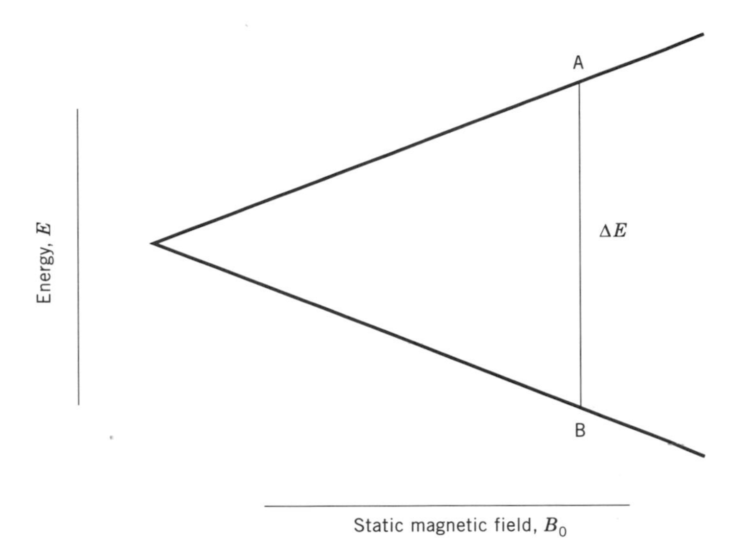

Fig. 1. Schematic representation of energy, E, vs external field strength, B_0, for a nucleus of spin, $I = 1/2$. A, spin $= -1/2$; B, spin $= 1/2$.

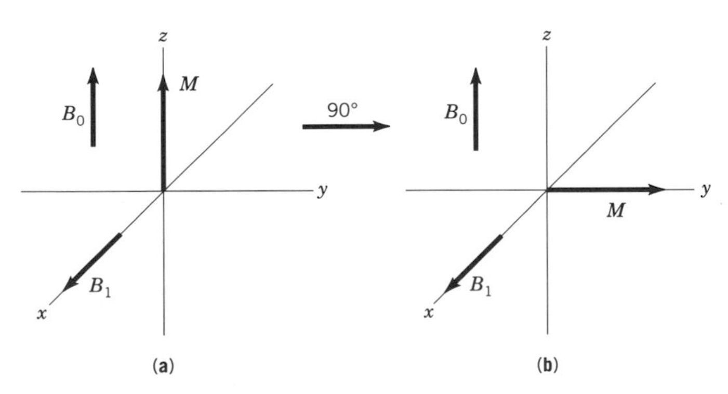

Fig. 2. Interaction of nucleus (electron) with static magnetic field, B_0, where the bulk magnetization, M, is (**a**) parallel to B_0 and to the z-axis, and (**b**), upon application of a 90° radio frequency pulse along x, M perpendicular to B_0 and to the z-axis. See text.

the z axis at the Larmor frequency. The concept of a rotating frame, in which a new set of axes, x' and y', rotate about z at the same frequency as the precessing nucleus, ie, γB_0, is conceptually convenient. Because the rotating frame is the standard reference system for describing magnetic resonance experiments, the prime designations for that system usually are omitted in most theoretical discussions. The application of an rf field, B_1, is considered to be along x' and this leads to a tipping of the magnetization vector in the $y'z'$ plane, as shown in Figure 2. The magnitude of the tipping is determined by the pulse width or duration, ie, the length of time, typically in μs, that B_1 is applied. For a $\pi/2$ or 90° pulse the magnetization is rotated completely into the $x'y'$ plane with no remaining component along z. This is equivalent to equalizing the population of the spin states. Continuation of the irradiation results in a continuing rotation of the magnetization about the x'-axis. It is possible to create π, $3\pi/2$, or any other pulse width by varying the duration of the irradiation. It is also possible to saturate a spin simply by continuing to irradiate with rf of the correct frequency. While a spin is saturated, it provides no resonance signal. When the rf is switched off, the system attempts to reestablish its equilibrium distribution of spins. Whereas 90° or $\pi/2$ pulses provide the maximum possible signal from a single pulse, shorter pulses are used in many cases when the rate of restoring magnetic equilibrium is slow.

The precessing magnetization induces a current in the detector coil which surrounds the sample and is centered on the laboratory y-axis. The time dependent fluctuation in this current decays in intensity as the spin distribution returns to its equilibrium distribution. The response to a single pulse is referred to as a free induction decay (FID) or transient and is converted to a digital form by an analogue-to-digital converter (ADC) for storage in a computer file. A principal advantage of the pulse nmr spectrometer over CW methods lies in the fact that the pulse experiment can be repeated after a suitable delay with the FID from each pulse train added to the accumulating time domain spectrum. The observed signal is a composite of a true signal from the experiment and a noise component. However, the increase in magnitude of the true signal is proportional to the number of transients which are acquired, whereas the noise increases with the square root of the number of transients. This additivity in the time domain permits experiments utilizing nuclei such as ^{13}C which have low natural abundance, ie, isotopically dilute spin systems.

In ft/nmr the detected signal is the sum of the individual decays for all of the nuclei. To convert this information back into a frequency-dependent form in which spectra are generally viewed, it is necessary to apply a Fourier transformation. Mathematically this is expressed as

$$f(\omega) = \int f(t)\exp(i\omega t)dt$$

where $f(\omega)$ is the frequency domain spectrum and $f(t)$ is the time-dependent fluctuation of the amplitude in the composite nmr signal. Because $f(t)$ is real, $f(\omega)$ has both real, absorption spectrum, and imaginary, dispersion mode components. When computing this transformation, the absolute value of the phase angle is initially unknown but can be determined from the boundary conditions. The best value for this angle makes each peak in the real component symmetric about its center and always positive. The Cooley-Tukey fast Fourier algorithm permits computation of the transformation for several thousand data points in about 1 s, which is essential to this and other ft methods (4) (see also CHEMOMETRICS).

The time constants associated with the restoration of magnetic equilibrium after application of a pulse are referred to as relaxation times and their reciprocals as relaxation rates. Relaxation may occur by any of several different mechanisms involving either longitudinal, T_1, or transverse, T_2, processes. Historically, the terms spin–lattice and spin–spin were used to refer to T_1 and T_2 relaxation times, respectively. The first of these involves the restoration of magnetization to the z direction and controls the rate at which the experiment can be repeated. The value of T_2 is determined by all processes contributing to loss of magnetization in the *xy* plane. For small, freely tumbling molecules in solution, the values of these two constants are often nearly equal and range from a few hundred ms to s in duration. For large molecules, eg, synthetic polymers, proteins (qv), and nucleic acids (qv), local magnetic field fluctuations establish very short T_2 values in both liquid and, especially, solid phases. Molecular scale rigidity, especially in crystalline systems, also results in significantly longer T_1 values which may exceed 100 s.

Because the energy differences between adjacent levels are very small in magnetic resonance experiments, the transition probabilities are also very small in the absence of local field fluctuations. Dipole–dipole interactions are the dominant mechanism for creating local field fluctuations and these decrease in intensity with the sixth power of the internuclear distance, so that only the closest nuclei have an appreciable effect on the relaxation rate for a given atom. The strength of the interaction also depends on the square of the product of the gyromagnetic ratios of the interacting nuclei. Other mechanisms, such as quadrupolar relaxation, may become important if either of the interacting nuclei is paramagnetic or the quadrupolar moment (see Table 1) is significant.

At a minimum, knowledge of relaxation times is important in determining the rate at which the pulse experiment can be repeated and successive transients added to the FID. The delay between acquisitions may be tuned for either minimum acquisition time, ie, short delays, or for uniform sensitivity by setting the delay to at least five times the longest T_1 in the system. Relaxation behavior can also be related to molecular structure and organization, although the process is less straightforward than for integrations, spin–spin couplings, or chemical shifts. The values of both T_1 and T_2 are dependent on the magnitude of the external field B_0. The relative values of T_1 and T_2 are also strongly influenced by the rate of molecular tumbling, which is usually measured as $(\tau_c)^{-1}$ where τ_c is the rotational correlation time. In the short τ_c regime ($< 10^{-10}$ s at 400 MHz for ^{1}H) T_1 and T_2 are nearly equal; in the long, ie, slowly tumbling nuclei ($\tau_c > 10^{-9}$ s), regime T_1 increases and T_2 decreases exponentially with increasing τ_c.

One additional phenomenon which plays an increasingly important role in nmr-based structure determination is the nuclear Overhauser effect (NOE) (5). The NOE originates in relaxations arising from dipole–dipole interactions through space and decreases in intensity with r_{IS}^6, where r_{IS} is the internuclear separation of two mutually relaxing spins. In organic molecules, this means that the effect is dominated by individual protons and their interaction with nearby nuclei in the same molecule. The effect is not evident in coupled ^{13}C or ^{1}H spectra, but can alter intensities in ^{1}H decoupled ^{13}C spectra causing enhancements of some peaks by up to a 200%. Inverse gated decoupling experiments where the ^{1}H decoupling is turned on for the acquisition period and off during a preparation period, which is long compared to any of the relaxation times, provides a method for eliminating NOE influences on the decoupled spectrum. The observation of an NOE between two nuclei implies that the nuclei are close together in space, typically separated by less than 0.45 nm. The failure to observe an NOE does not, however, mean that the nuclei are far apart, because the relaxation rate also depends on τ_c.

Resolution is an important consideration in any experiment involving a number of closely spaced maxima. The digital resolution (DR) achievable in an nmr spectrum is defined by the relation

$$\mathrm{DR} = 2\cdot(\text{sweep width})/(\text{number of data points})$$

The acquisition time during which the spectrum is acquired is the reciprocal of DR. In practice this resolution is only achievable if the signal from the sample persists throughout the acquisition time. When molecular motion is slow as in large molecules or viscous solutions, T_2 relaxation is very fast, leading to broad lines in the spectrum, and resolution is limited by the natural line width $1/T_2$. Because the mean instrumental noise amplitude is constant during acquisition, data points acquired at the beginning of the FID have a higher signal-to-noise ratio (SNR) than those at the end of the FID. Window (apodization) functions define the weights assigned to individual data points in the FID and may lead to SNR or resolution enhancement in the frequency spectrum. Multiplication of points in the FID by an exponentially decaying function is referred to as line broadening, whereas a negative exponential sharpens individual peaks. The sine–bell function, which places greatest importance on the middle region of the FID and less on either extreme, and its variations are widely used because of the ability to sharpen the spectrum while minimizing noise effects. It is common practice in nmr data processing to expand, or zero-fill, the FID from 2^N to 2^{N+1} data points before transformation.

If the actual signal has not vanished by the end of the acquisition period then the FID is truncated, producing artifacts in the spectrum. This can be corrected either by increasing the number of data points in the FID or by decreasing the sweep width and concomitantly increasing the acquisition time. The former method may be restricted by computing power and processing time. Both of these approaches result in increased digital resolution in the processed spectrum.

Electron Spin Resonance. Electron spin resonance (esr) also known as electron paramagnetic resonance (epr) is a second magnetic resonance technique which finds particular applications in the study of free radicals, paramagnetic species, and other molecules containing an unpaired electron. The electron has a spin of 1/2 and, because of its low mass, a gyromagnetic ratio of 1760 rad/(T·s). Field strengths required for esr are much lower, typically 0.34 T (3.4 kG) and the frequencies higher (9–35 GHz) than for nmr. Data are usually reported in terms of intensity, signal strength, versus energy where the energy for a transition is usually expressed as

$$\Delta E = g\beta H$$

where g is the dimensionless splitting constant equal to 2.0023 for a free electron, β is the Bohr magnetron, μ_B ($1\ \mu_B = 9.2732 \times 10^{-24}$ J/T), and H is the external field strength. Experimental data are usually reported as first derivative spectra rather than in the absorption mode typical of nmr. Peak positions are defined by the position where the spectrum changes sign. Couplings between the unpaired electron and nearby nuclei create hyperfine interactions with characteristic hyperfine splitting constants, ΔH, expressed in Gauss ($1\ \text{G} = 10^{-4}$ T). Line shape, as in nmr, is dependent on correlation times. Three τ_c regimes are recognized. The fast, slow, and very slow regions are recognized as $10^{-11}-10^{-9}$ s, $10^{-9}-10^{-7}$ s, and $10^{-7}-10^{-3}$ s, respectively. Because of the very short lifetimes of many organic free radicals and radical ions, measurements are often made at low, ie, liquid nitrogen or liquid helium, temperatures (see CRYOGENICS).

Equipment

Nuclear Magnetic Resonance. In 1994 there were three principal vendors of nmr instrumentation in the U.S., Bruker Instruments (Billerica, Mass.), JEOL USA, Inc. (Peabody, Mass.), and Varian Associates (Palo Alto, Calif.). Details of instrumentation are best obtained directly from manufacturers. A schematic illustrating the principal components of a ft/nmr spectrometer is shown in Figure 3.

Magnets in high field nmr spectrometers are cryostats having niobium alloy wound solenoids, which are superconducting at liquid helium temperatures (4 K) (see MAGNETIC MATERIALS; NIOBIUM AND NIOBIUM COMPOUNDS; SUPERCONDUCTING MATERIALS). Because the wire has no resistance at this temperature, the magnet does not require additional energy to maintain constant current and hence constant field strength. In modern nmr spectrometers the magnetic field strength, B, ranges from 1.4–17.6 T (14–176 kG), corresponding to observation frequencies for ^{1}H of 60–750 MHz. Three critical considerations for these magnets are field strength, field stability (drift), and field homogeneity. The advantages of higher field strength are better sensitivity and resolving power. However, in selecting a new instrument these properties must be balanced against cost. As of this writing (1994) a good rule of thumb for the cost of an nmr spectrometer up to 500 MHz is approximately $1000/MHz for a basic

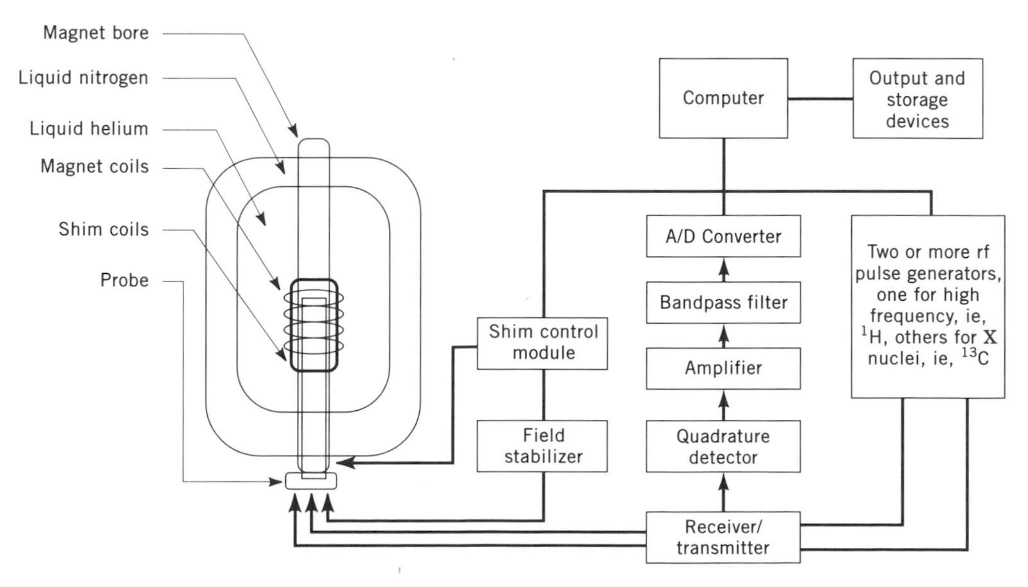

Fig. 3. A block diagram schematic representation of a Fourier transform nmr spectrometer, ie, a superconducting magnetic resonance system.

spectrometer. The 600-MHz and 750-MHz spectrometers typically cost about \$1.0 and \$2.5 $\times 10^6$, respectively.

Long-term field stability is maintained by the use of a deuterium lock signal. Because deuterated solvents have a constant Larmor frequency at a given field strength, these provide a convenient reference point. The deuterium lock system uses a feedback control loop to monitor the deuterium frequency and make small adjustments in the static field maintaining a constant position for the deuterium signal. To illustrate the importance of minimizing drift in the static field, carbon-13 resonances often have line widths at half height of 0.1 Hz. A static field drift during data acquisition of only 1/3 part in 10^9 or 0.1 Hz doubles the observed line width to 0.2 Hz.

Achieving and maintaining field homogeneity around the sample is perhaps the most important and, at times, most frustrating experimental aspect of high field nmr. Small electromagnets called shimming coils are used to bend and shape the static magnetic field in the vicinity of the sample. When correctly tuned to optimize the local field homogeneity, the sharpest possible Lorentzian lineshapes in the frequency domain spectrum are obtained. One method of optimizing the shim settings is to maximize the deuterium lock signal intensity by adjusting the current in the individual shims. Alternatively, the FID or the lineshape for a strong resonance from the sample can be monitored as the shims are adjusted. There are two sets of adjustable shims. The vertical field gradients, which are aligned along the static field, B_0, are referred to as spinning or Z shims because these are set while the sample tube is spinning at 0–20 Hz around the z axis. These strongly affect the lineshape and linewidth of the observed signals. Typically, there are five orders of Z shims (Z, Z^2, Z^3, Z^4, Z^5) and all are interactive during the adjustment process. The horizontal field gradients, or nonspinning shims, are generally corrected to the third or fourth order with combinations of X, Y, and Z gradients. Typical gradients are X, Y, XZ, YZ, XY, X^2-Y^2, XZ^2, and YZ^2. Correct setting of these shims is most important in minimizing the satellite images of each signal called spinning sidebands; however, the higher order nonspinning shims also affect the lineshape and linewidth. These shim settings also are interactive. The initial shimming of a new instrument is a difficult process. Once reasonable settings have been made for each probe, however, the values are stored in a file on the acquisition computer. Usually, only small adjustments of Z, Z^2, and Z^3 need thereafter to be made for each sample (1,6).

The nmr probe is used to transmit and receive rf energy to and from the sample. Each manufacturer supplies specifications defining the achievable performance for each of their specific probes. Probe performance may be characterized in terms of the sensitivity and lineshape. Sensitivity is measured in terms of the signal-to-noise ratio for a single-pulse experiment on a specified sample. Linewidth is measured in several ways but the simplest involves a comparison of the full linewidth at 50% of maximum intensity. Probe sizes are defined in terms of the diameter of the sample tube that they accommodate and range, from 2.5 to 20 mm or larger. Probes are limited by the bore of the magnet in which they will be installed. Large probe diameters permit large sample volumes and are used for such insensitive nuclei as natural abundance ^{15}N or for samples that are poorly soluble. Small diameter probes are preferred when sample quantity is the limiting factor. Typically, 5-mm probes are used for routine work as these have better lineshape specifications and shorter 90° pulse widths than the larger diameter probes. Variations in probe design and capabilities are numerous and are specific to particular types of analyses. Most probes contain one coil tuned to the deuterium lock frequency and two or more tunable coils. In broadband probes one of these coils can be tuned to one of the high frequency-range nuclei, such as ^{1}H, ^{3}H, or ^{19}F (see Table 1), whereas the other coil or coils each can be tuned to one of the other nuclei. Specifically tuned probes working at a single frequency or dual probes such as the $^{1}H/^{13}C$ probe are also in common use. In general, the broadband or tunable probes are less sensitive but more versatile than the specifically tuned probe. Specialized probes are also required for solid-state measurements (7) and for gas-phase (8) studies.

Precisely controllable rf pulse generation is another essential component of the spectrometer. A short, high power radio frequency pulse, referred to as the B_1 field, is used to simultaneously excite all nuclei at the Larmor frequencies. The B_1 field should ideally be uniform throughout the sample region and be on the order of 10 μs or less for the 90° pulse. The width, in Hertz, of the irradiated spectral window is equal to the reciprocal of the 360° pulse duration. This can be used to determine the limitations of the sweep width (SW) irradiated. For example, with a 90° hard pulse of 5 μs, one can observe a 50-kHz window; a soft pulse of 50 ms irradiates a 5-Hz window. The primary requirements for rf transmitters are high power, fast switching, sharp pulses, variable power output, and accurate control of the phase.

The analogue-to-digital converter (ADC) samples the fluctuating voltage produced in the coils of the probe at regular time intervals, storing each value as a binary encoded integer. The rate at which the ADC must sample the voltage is defined by the Nyquist theorem, which states that the sampling rate must be greater than or equal to twice the signal frequency. The maximum speed of the digitizer determines the maximum observable spectral width; the number of bits used in storing each data point and the number of bits in each computer word determine the dynamic range of intensities that can be observed.

Quadrature detection is employed in nmr spectrometers to distinguish between positive and negative frequencies. This capability allows placement of the reference frequency in the center of the spectrum, thus reducing the required operating frequency of the ADC by a factor of two while increasing the sensitivity. In quad detection, two phase-sensitive detectors, 90° out of phase with one another, collect the time-domain spectrum as two independently digitized spectra. Fourier transformation of the resulting set of complex numbers are used to determine the relative positions and signs of the signals with respect to the reference frequency.

A computer-controlled bandpass filter system controls the size of the acquired spectral window. Typically, this is set to about 120% of the desired sweep width. Only frequencies within these limits are allowed to reach the ADC. Those frequencies outside the limits would only contribute to the noise in the final spectrum. The need for this system is dictated by the nonselective nature of the excitation rf pulse.

Modern nmr spectrometers are almost completely controlled by one or more computers (see COMPUTER TECHNOLOGY). Criteria for a computer system should include speed, compatibility with other systems, and the availability of large amounts of rapidly accessible storage. The size of data arrays used in routine one-dimensional (1-D) nmr experiments is relatively small; however, when using 2-D or higher nmr experiments, the size is determined by the product of the number of points in each dimension and can become quite large, resulting in lengthy processing times. Most systems built up through the late 1980s contained integral computer systems that often contained vendor-specific memory structures. The trend as of the mid-1990s, however, is to employ standard personal computers or work stations made by computer companies. All of the principal systems now have separate software for acquisition and processing of spectra. By carrying out these functions on physically separate computers, ie, off-line processing, spectrometer use is not interrupted by lengthy processing steps. Software for off-line processing which can accept data from several different spectrometers is also available from third-party sources.

Electron Spin Resonance. Instrumentation for esr differs from nmr instrumentation principally as a consequence of the larger gyromagnetic ratio in this experiment. As a result, the external field is typically 0.35 T (3.5 kG). At this field strength the frequency associated with an unpaired electron is about 9.5 GHz (X-band). Systems operating at frequencies up to 34 GHz (Q-band) are also available. Fourier transform is becoming commonplace, but CW instruments also are widely used. Microwave radiation at a fixed frequency is produced by a Klystron valve at power levels ranging from less than 0.1 μW to several hundred mW and transmitted to the sample via rigid waveguides rather than the coaxial wire. The waveguide terminates in a reflection resonance cavity controlled by a microwave bridge. The function of this cavity is to focus the energy on the sample (see MICROWAVE TECHNOLOGY). In comparing systems, reference is made to the Q factor, ie, the ratio of the energy stored in the cavity to the energy lost in a cycle. Q factors of 5000–7000 are common. Esr signals typically become stronger at lower temperatures, and variable temperature systems capable of approaching liquid helium temperatures are available. Data are plotted in most applications as the first derivative spectrum. This presentation is more sensitive to changes in hyperfine structure than the absorption mode. Many esr spectrometers are also equipped to connect the electron nuclear double resonance (endor) experiment (9), which requires that the instruments supply rf at both the electron and nuclear frequencies.

Pulsed ft mode esr instruments have appeared beginning in the mid-1980s. These collect digitized time-domain spectra which may be processed into the frequency domain as are nmr data. Pulse durations are much shorter than in nmr with typical 90° times of 8–20 ns.

Health and Safety

Safety considerations for magnetic resonance (mr) experiments have received little attention except for the problems associated with the use of electronic devices such as pacemakers in the magnetic field. However, in a 1990 study of reproductive health involving more than 1900 women working in clinical mr facilities in the United States no substantial differences were reported between the group of women directly involved with mr equipment (280 individuals) and other working women (894 individuals) (10). Conclusions are restricted to exposure to the static external field.

The Nuclei of Nuclear Magnetic Resonance

Although nmr can be applied to many different nuclei, the overwhelming interest, especially for the organic chemist, continues to lie with ^1H and ^{13}C nmr. The hydrogen experiment is simpler because ^1H has a natural abundance of more than 99%, a large magnetogyric ratio, and a spin of 1/2. A three-stage model is commonly used to describe nmr experiments. Stages involve preparation, evolution, and detection periods.

Proton nmr. In the simplest experiment, the sample and a small amount of a reference compound such as tetramethylsilane [*75-76-3*] (TMS), $C_4H_{12}Si$, are placed in a tube, usually of 5-mm diameter. Typical samples may be a neat liquid or a solution containing as little solute as 0.01 mg/cm^3. The tube is positioned so that the sample lies in the coil. The spins are prepared by the application of a 90° rf pulse moving magnetization into the xy plane. In this experiment there is no evolution period, and detection commences as soon as the spins have been prepared. During detection the current in the detector coil is measured and digitized at regular time intervals and the results stored in a computer file. This cycle is one transient. After waiting for a suitable delay so that magnetic equilibration can occur, the entire cycle can be repeated. The results of the second experiment can be added to the first. After a suitable number of pulses have been acquired, the time-domain spectrum is transformed into a frequency-domain spectrum and phased so that the base line of the spectrum is flat and the peaks all fall on the same (positive) side of the base line. Information obtainable from this spectrum includes chemical shifts, coupling constants, and integrated peak areas.

Each hydrogen atom in a unique chemical environment is shielded differently from the external field and has a slightly different resonance frequency. The chemical shift in ppm, δ, is defined as

$$\delta = 10^6(\omega - \omega_{\text{ref}})/\omega_0$$

where ω is the resonance frequency of the signal, ω_{ref} the frequency for a standard reference material, eg, the hydrogen in TMS in the case of ^1H. See Table 1 for reference materials for other nuclei. The Larmor frequency of the nucleus at the spectrometer's field is ω_0. The advantage of expressing shifts in ppm is that the scale is independent of the value of B_0. The value of the chemical shift can be correlated to the electronic environment of the nucleus, and extensive tables of chemical shift correlations are available in many sources (11). More precise databases specific to particular classes of compounds are also available in many instances.

The relative number of equivalent nuclei associated with each chemical shift is obtained from the integrated spectrum by normalizing the areas so that the area corresponding to the smallest peak in the spectrum is defined as 1. This relation may not be exactly correct in ft experiments where signals may be affected by significant differences in relaxation times for nuclei in different environments.

When inequivalent nuclei are closely connected through covalent bonds, typically three or fewer, then the spin state of the proximate nuclei split the observed signal into multiple peaks through a process known as spin–spin coupling. In general, coupling of n equivalent nuclei of spin I gives rise to $2nI + 1$ lines where the intensity ratios follow the coefficients of a binomial expansion. The coupling pattern can be extremely useful in determining the chemical structure of an unknown compound. These patterns may become highly complex, however, when a nuclei couples to two or more inequivalent nuclei. In ethanol, for example, the methyl proton signal is a triplet having intensities in the ratio 1:2:1, the methylene proton signal is a quartet having intensity ratios 1:3:3:1, and, because of rapid exchange, the hydroxyl proton is a singlet. The triplet, for example, arises from the fact that the methylene protons, which are three bonds away, may have spins of either $+1/2$ or $-1/2$. This leads to four different combinations of spins. Two of these are degenerate because each has one proton at $+1/2$ and one proton at $-1/2$. The result is that the methyl protons are in three slightly different environments depending on the spins of the adjacent methylene group, and one of these states is twice as probable as either of the other two.

A second aspect of coupling is the separation between the lines in the multiplet. This spacing is equal between all lines in the multiplet if only a single coupling is involved. The strength of the coupling, ie, the coupling constant, J, is measured in Hz and is independent of B_0. An important consequence is that with increasing field strength, coupling patterns become more isolated from one another and thus easier to identify and interpret. Increasing the magnitude of B_0 simplifies spectral interpretation by isolating couplings from one another as shown in the similated spectra in Figure 4.

The magnitude of a coupling constant decreases as the number of intervening bonds between the coupled nuclei increases and typically three bonds, or ^3J, is the detectable limit. The ^3J case is of particular interest because the magnitude of these constants is dependent on the conformation angle defined by the three bonds (12). The use of equations and theoretical approaches to the calculation of coupling constants has been reviewed (13). In general, the value of ^3J is maximum when coupled species are trans to one another. Within particular classes of compounds trigonometric functions of the torsion angle such as the Karplus equation may be used to predict the angle from the magnitude of the ^3J.

The values of the time constants T_1 and T_2 are important in understanding both internal and overall motional behavior of the sample molecule. T_1 values are measured by the inversion recovery pulse sequence:

$$180° - \tau - 90° - \text{FID}$$

where 180° and 90° are rf pulses, and the preparation and detection steps, separated by a variable delay, τ, the evolution time. For each chemical shift the peak amplitude is measured as a function of τ and the value of T_1 is calculated from the following relation:

$$A_\tau = A[1 - 2\exp(-\tau/T_1)]$$

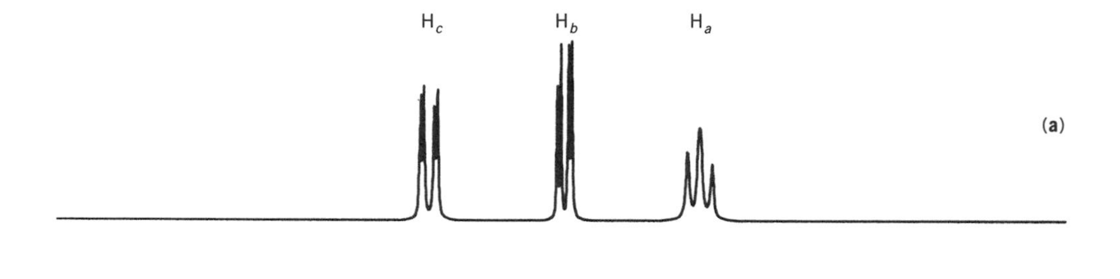

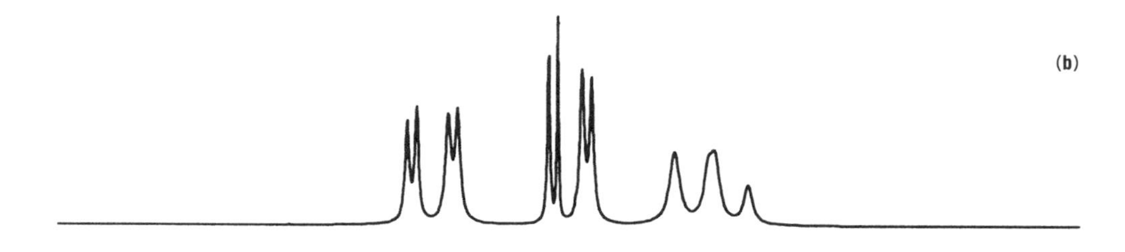

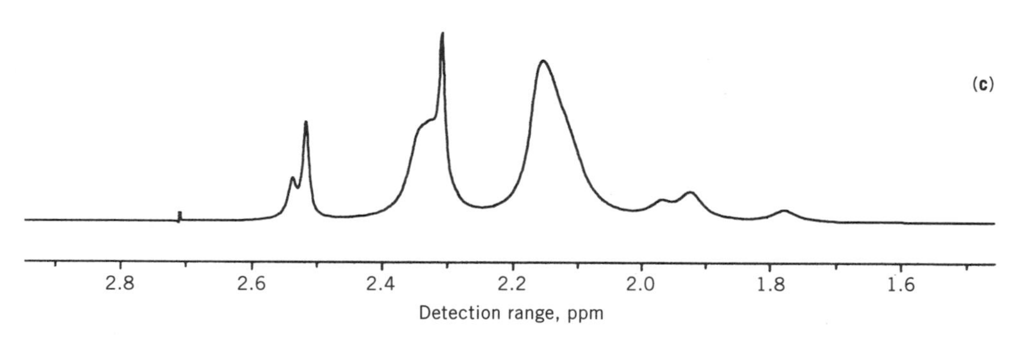

Fig. 4. Simulated ^{1}H nmr spectrum for a molecule having three interacting spin systems H_a, H_b, and H_c where $J_{ab} = 10$ Hz, $J_{bc} = -3$ Hz, and $J_{ac} = 12$ Hz at (**a**) 600 MHz, (**b**) 200 MHz, and (**c**) 60 MHz.

This process can be repeated thousands of times until adequate signal-to-noise is obtained providing that an initial recycling delay is inserted prior to the first pulse allowing restoration of magnetic equilibrium at the temperature of the experiment. This last delay should be 5–10 times longer than the longest T_1 which can be detected. The value of T_2 can be estimated from shapes of peaks in the absorption spectrum or measured exactly by the Carr-Purcell-Meiboom-Gill (CPMG) method (2). If these peaks are narrow, approaching the instrumentally limited linewidth, then $T_2 > T_1$, which is the case when the sample molecules are engaged in rapid tumbling such that $\omega_0\tau_c < 1$. As lines become broader, ie,

T_2 becomes shorter, its value can be estimated from the following relation:

$$\Delta\overline{\omega}_{1/2} = (1/\pi)T_2^*$$

where the left-hand side represents the actual measured line width and the asterisk T_2^* indicates that this is an apparent T_2 which includes effects arising from inhomogeneity of the field. Very short T_2 values map into large, rigid molecular structures. Details of these measurements may be found in the general references (1,2) and in more specialized works (14).

The addition of paramagnetic species, such as the metal ions Cu^{2+}, Mn^{2+}, or Cr^{3+}, can have dramatic effects on both the observed spectrum and the relaxation behavior of a molecule. The added ion reduces nuclear relaxation times, T_1 and T_2, permitting more rapid data collection. In addition, faster relaxation rates minimize NOE effects in the spectra, which can be useful in obtaining quantitative ^{13}C intensity data. The most widely used reagent for this purpose is chromium acetylacetonate [*13681-82-8*], $Cr(C_5H_7O_2)_3$, known as $Cr(acac)_3$. Practically speaking, the use of such reagents requires care, because at high concentrations they may increase relaxation times and absorb strongly on the walls of the nmr tube. The use of these agents is particularly beneficial in achieving separations of peaks when working at lower field strengths. The line-broadening effect of these ions increases with the square of B_0; separation in chemical shift increases linearly with B_0 (15).

Lanthanide ions are used as contact shift reagents (2). The binding of these to specific sites in a molecule causes large changes in the ^1H chemical shifts that become progressively smaller as the distance between the ion and the proton increase. Although originally thought to proceed by the same mechanism as the paramagnetic chemical shift reagents, some contact shift reagents do not show the same external field dependence behavior. Some of these ions do not inhibit the measurement of NOE effects (16).

NOE difference spectroscopy is used to identify short, throughspace $^1H-^1H$ interactions and measure these interactions to an upper limit of about 0.45 nm. The experiment uses a selective pulse to irradiate a single frequency during the preparation delay. This transfers magnetization through space to nearby nuclei. A 90° pulse is then applied and the data collected. The experiment requires the acquisition of two or more separate spectra for the same sample. One is a control spectrum where the selective pulse is applied to the open part of the spectrum, and the others, in turn, select individual protons. After transformation, a difference spectrum is computed by subtracting the control from each of the subspectra. The resulting spectra show NOE enhancements, either positive or negative, which measure the interaction between the irradiated signal and nearby hydrogen atoms. The magnitude of these effects is inversely proportional to the sixth power of the internuclear separation and can be used to estimate those distances.

Homonuclear decoupling, or spin decoupling, is a technique that simplifies the J coupling patterns of a spectrum (17). The pulse sequence uses two rf signals, ie, the normal hard 90° pulse irradiating the entire spectrum and a selective rf signal which saturates a particular spin. The selective pulse rapidly flips the energy states of the irradiated nucleus so that other coupled nuclei can only see the average orientation, and are therefore no longer coupled. This simplification of the J-coupling permits the identification of coupled spin systems and is particularly useful for determining coupling constants in a complex spin network. By sufficiently irradiating each multiplet in the spectrum sequentially the complete connectivity of the molecule can be determined. The two-dimensional (2-D) $^1H-^1H$ correlated spectroscopy (COSY) experiment yields the same information at a glance and is the preferred technique for direct determination of coupled spins.

The ID homonuclear Hartmann-Hahn (HOHAHA) experiment is an excellent way to determine complete coupled spin networks (18). The following pulse sequence is used:

$$\text{selective } 90^\circ - 1/(2J) - \text{spin lock}$$

The selective 90° pulse is applied to an isolated spin, thus aligning it along the y-axis; the $1/(2J)$ delay, where J is the coupling constant, allows the signal to rotate to the $-y$-axis. Then a spin-lock sequence is applied which maintains this orientation so that the spins no longer rotate about the z-axis, ie, no chemical shift differences exist. This allows the mixing or propagation of spins through the ^1H coupling network where directly and indirectly coupled spins can be observed. The spin-lock or mixing time can be varied to give subspectra with varying amounts of propagation. The longer the mixing time, the farther out from the source the spins propagate. This experiment can be used to make assignments of complex overlapped regions of a spectrum. A ^1H-edited spectrum is observed where the only observed spin network is the one which was selectively irradiated. The 1-D version has some advantages over the 2-D version of this experiment. By varying the spin-lock time from short to long periods, those spins which are 2, 3, 4, or even 5 positions removed from the source signal can be identified. However, in the 2-D experiment all signals are present, complicating the identification of propagation pathways.

The most common method of acquiring high resolution spectra of samples which are solids at the experiment temperature is solution nmr. If the nucleus under investigation is also found in the solvent, then the concentration of that nucleus in the solvent is usually far higher than in the sample, eg, pure water is 110 M in hydrogen. Such high concentrations can create severe dynamic range problems for detection of the sample. The historical method of solving this problem is to change the isotopic form of the solvent, for example by using deuterated solvents such as D_2O or $CDCl_3$. Alternatively, several techniques for solvent suppression are possible. If the solvent signal is a single sharp line, eg, for pure water, the simplest method is selective saturation of the solvent resonance by continuous irradiation. Differences in relaxation times may also be employed because ^1H and ^{13}C often have longer relaxation times in small solvent molecules than in the sample species. In the water eliminated Fourier transform (WEFT) (2) water suppression sequence, for example, an initial 180° pulse is followed by a delay, a 90° pulse, and acquisition of the FID. The value of the delay is set so as to minimize the height of the water peak. During this period, the magnetization of the sample is restored to $+z$ so that the net effect on the sample is the same as a 90° pulse. Other techniques for water suppression include spin echo techniques (19), Dante (2) sequences, and most recently gradient field methods such as WATERGATE (20).

Carbon-13 nmr. Carbon-13 [*14762-74-4*], ^{13}C, nmr (1,2,11) has been available routinely since the invention of the pulsed ft/nmr spectrometer in the early 1970s. The difficulties of studying carbon by nmr methods is that the most abundant isotope, ^{12}C, has a spin, I, of 0, and thus cannot be observed by nmr. However, ^{13}C has $I = 1/2$ and spin properties similar to ^{1}H. The natural abundance of ^{13}C is only 1.1% of the total carbon; the magnetogyric ratio of ^{13}C is 0.25 that of ^{1}H. Together, these effects make the ^{13}C nucleus ca 1/5700 times as sensitive as ^{1}H. The interpretation of ^{13}C experiments involves measurements of chemical shifts, integrations, and J-coupling information; however, these last two are harder to determine accurately and are less important to identification of connectivity than in ^{1}H nmr.

Observation of homonuclear $^{13}C-^{13}C$ coupling is very unlikely because of the low probability for bonding between two ^{13}C atoms. In addition, ^{13}C coupling is not easily observable in the ^{1}H nmr spectrum because of ^{13}C's low natural abundance. The J_{CC} couplings, which may range from 35–175 Hz, can be used to determine $^{13}C-^{13}C$ connectivity using experiments such as the 1-D INADEQUATE technique, which filters out the single-quantum signals, ie, those signals arising from uncoupled nuclei. Heteronuclear coupling from ^{1}H to ^{13}C is easily observed: $^{1}J_{CH}$ values are large, ranging from 110–330 Hz, typical $^{2}J_{CH}$ and $^{3}J_{CH}$ values are between 0 and 12 Hz. The observation of a completely coupled ^{13}C spectrum would have so many overlapping multiplets that it would be difficult to interpret. To simplify this problem, the energy levels of the ^{1}H nuclei are saturated by irradiation using a broadband decoupling frequency eliminating coupling to the ^{13}C atoms. This process, which yields a proton-decoupled ^{13}C spectrum, is the standard method for ^{13}C nmr. Decoupling has two effects: all multiplets are collapsed into singlets, one for each nonequivalent ^{13}C atom; and the transfer of NOE from the excited ^{1}H to the ^{13}C signals enhances some peaks by up to 200%. The relative intensities of individual peaks in a ^{13}C spectrum are also affected by differences in T_1 values of the atoms. Normally, the relative intensities of ^{13}C nmr signals diminish in the order of CH_3, CH_2, CH, C.

Quantitation of ^{13}C nmr intensities employs inverse gated decoupling experiments, where the ^{1}H decoupling rf is turned off during a preparation period and then turned on during the acquisition period, thereby eliminating NOE influences on the decoupled spectrum. Ideally, elimination of relaxation-time effects requires that the preparation period should be at least five times longer than the largest T_1 expected. Thus acquisition times for these experiments are quite long. When only qualitative information is sought, buildup of the NOE during the preparation time is actually desirable, because NOE increases the SNR in each transient. In addition the preparation time period can be decreased to 1–3 times that of T_1. Care must be employed to avoid saturation of quaternary ^{13}atoms.

Spectral editing refers to any method which alters the appearance of the spectrum so that specific structural features become readily identifiable. This may include elimination of signals from solvent molecules, and changes in disposition of peaks in a ^{13}C spectrum with respect to the baseline as a function of the number of directly coupled ^{1}H nuclei. Polarization transfer, or population transfer, as it is sometimes called, involves shifting characteristics, eg, energy and spin state, of the abundant spin magnetization to ^{13}C or some other nucleus and therefore is an important experimental approach to spectral editing. Two examples of spectral editing techniques are the INEPT and DEPT experiments. Simultaneous transfer of polarization from all ^{1}H nuclei to all of X nuclei can be accomplished using the insensitive nuclei enhanced by polarization transfer (INEPT) method (2). A serious drawback of INEPT is the fact that triplets and other multiplets are not immediately recognizable, because intensities are distorted. The distortionless enhancement by polarization transfer (DEPT) method eliminates this drawback and can directly distinguish among methyl, methylene, methine, and quaternary carbons (2). The technique uses polarization transfer from sensitive nuclei such as ^{1}H, ^{19}F, or ^{31}P to insensitive nuclei, ie, ^{13}C, ^{29}Si, or ^{15}N (see Table 1). One final advantage of polarization transfer-based methods is an increase in sensitivity of γ_S/γ_I where the subscripts refer to the sensitive and insensitive nuclei, respectively.

Other Nuclei. Although most nmr experiments continue to involve ^{1}H, ^{13}C, or both, many other nuclei may also be utilized. Several factors, including the value of I for the nucleus, the magnitude of the quadrupolar moment, the natural abundance and magnetogyric ratio of the isotope, or the possibility of preparing enriched samples, need to be considered. The product of the isotopic parameters can be compared to the corresponding value for ^{13}C, providing a measure of relative sensitivity or receptivity. Table 1 summarizes these factors for a number of isotopes. More complete information may be found in the literature (21,22).

For nuclei where $I > 1/2$, such as deuterium, ^{2}H, there are two important differences from spin $I = 1/2$ systems. First, different coupling patterns occur when $I \geq 1$, and secondly, significant line broadening is a consequence of medium-to-large quadrupolar moments. The simplest and most commonly occurring example of the coupling effects is evident when a proton-decoupled ^{13}C spectrum is obtained for a compound dissolved in $CDCl_3$. Because ^{2}D has $I = 1$, and there are $2nI + 1$ spin states, the carbon spectrum shows a 1:1:1 triplet at 77.0 ppm arising from deuterium–carbon coupling in the solvent. The nuclear electric quadrupolar moment, Q, which may be positive or negative, is usually reported in units of barn, where 1 barn $= 10^{-28}$ m^2. In general, values of $|Q|$ range between 6.4×10^{-32} m^2 (^{6}Li) and 3×10^{-28} (^{181}Ta). When present, the quadrupolar process is the most efficient of the relaxation processes available to a nucleus. Hence its rate controls the line width of the signal.

The fluorine-19 [*14762-94-8*], ^{19}F, nucleus has a sensitivity second only to that of ^{1}H. Moreover, the frequency is similar enough to ^{1}H's that often the signal can be detected by tuning the proton channel of a probe to the ^{19}F frequency. In addition to the usual chemical shift and coupling experiments, ^{19}F nmr of fluorinated ligands is widely applicable in kinetic studies of enzyme–substrate interactions (23) (see KINETIC MEASUREMENTS). The success of this approach depends on the larger enzyme-induced changes in ^{19}F chemical shifts, as compared to ^{1}H chemical shifts in the parent compound. Such protein-induced changes are typically 9–15 ppm in a downfield direction. The danger in interpreting such studies is that the replacement of hydrogen by fluorine may also have significant effects on the structure and stability of a complex. A second problem for fluorine is the large chemical shift anisotropy (CSA). The effect of this property on the magnetic-field dependence of both T_1 and T_2 increases the natural linewidth with increasing B_0.

Phosphorus-31 [*7723-14-0*], ^{31}P, also is readily accessible for nmr because of its spin of 1/2, 100% natural abundance, magnetogyric ratio of 10.8, 600-ppm chemical shift window, and moderate relaxation times. Early ^{31}P studies were done in CW mode (24,25). The availability of ft instrumentation led to improved sensitivity such that analysis of millimolar solutions in standard 5-mm tubes has become routine. Technically, the primary difficulty lies in achieving satisfactory digital resolution using this large chemical shift window. Both the large chemical shift literature and theoretical approaches to the ^{31}P chemical shift are discussed elsewhere (26). The occurrence of phosphorus in cell membrane phospholipids and in nucleotides has led to its use as a reporter on structure and dynamics in enzyme complexes, cell membranes, and nucleic acids as well as organophosphorous compounds (27). In nucleic acid studies the ^{31}P chemical shift is considered to report on both local conformation and sequence. The assignments of specific resonances to individual nucleotides usually involves HETCOR-type studies of $^{31}P-^{1}H$ correlations similar to $^{13}C-^{1}H$ studies of multidimensional nmr (28). Among the more unique applications of ^{31}P is its applicability to the study of *de novo* metabolic pathways for the synthesis of phosphatidylcholine and other membrane phospholipids (29,30). In these studies the samples are living perfused cells or organs, or intact organisms, and the probes are specifically designed for the application.

Although the natural abundance of nitrogen-15 [*14390-96-6*], ^{15}N, leads to lower sensitivity than for carbon-13, this nucleus has attracted considerable interest in the area of polypeptide and protein structure determination. Uniform enrichment of ^{15}N is achieved by growing protein synthesizing cells in media where $^{15}NH_4OH$ is the only nitrogen source. $^{15}N-^{1}H$ reverse shift correlation via double quantum coherence permits the complete assignment of all ^{15}N resonances. Because there is one such nitrogen in the main chain for each amino acid, complete assignments, ^{1}H scalar coupling, hydrogen exchange experiments, and NOE studies taken together can provide a detailed picture of the protein conformation and the kinetics of protein folding. An example of this approach is available in the literature (31). Details of the enrichment experiments and of the application of two-dimensional techniques such as COSY, TOCSY, and NOE experiments to the products have been reviewed (32). Because of the large number of resolvable distance-sensitive data which may be obtained by these experiments, the reliability of conformational studies by nmr for small and intermediate-sized proteins are likely to be dramatically improved.

The nmr relaxation behavior and chemical shift effects of metal ions in solution may provide considerable insight into metal ion-containing systems. The presence of such ions may induce significant changes in the ^{1}H spectrum and, if the ion is sitebound, the magnitude of such effects decreases sharply with increasing distance between the ion and the proton. Secondly the chemical shifts and line profiles of the ion itself are both sensitive to the binding state and to the distribution of ions between bound and free states providing a probe for the kinetics of binding. Finally, when the ion is paramagnetic, quadrupolar relaxation mechanisms become the predominant mechanism for loss of magnetization (33).

Multidimensional nmr

The first two-dimensional (2-D) nmr experiment was proposed in 1971 at an nmr conference (1). Multidimensional nmr underwent explosive growth in the 1980s and early 1990s and impinges on all aspects of modern nmr. In a typical 2-D experiment, a set of related FIDs are acquired using the same basic pulse sequence while systematically varying the evolution time. The experiment is illustrated schematically in Figure 5 which shows the steps in obtaining a correlated spectroscopy (COSY) spectrum. In the acquisition process, several FIDs are acquired sequentially under identical preparation and acquisition conditions, but at varying evolution times, t_1, resulting in a family of FIDs. By convention, the time axis is t_2 as shown in Figure **5b**. Fourier transformation of any single row leads to a frequency-domain spectrum along the F_2 axis. If all rows are processed using identical phasing constants and a vertical line is drawn through the spectra at any point f_2 on the F_2 axis, then the intensity of the signal at that point oscillates as a function of the evolution time (Fig. **5c**). A Fourier transform of this time-dependent signal leads to a second frequency-domain, F_1, spectrum. The resulting two-dimensional matrix contains maxima and minima corresponding to coupled interactions between the two frequency domains (Fig. **5d**). In a homonuclear experiment, the F_1 and F_2 domains are identical whereas this is not so in a heteronuclear experiment. The nature of the interaction creating these peaks and valleys depends on the particular pulse sequence employed. In general, the interactions may arise from J-coupling, through space interactions, or from polarization transfer. Finally, each pulse sequence (Fig. **5a**) could be extended using one or more evolution times and additional pulses leading to third or higher dimensions in the spectrum.

The first example of a multinuclear experiment is the homonuclear proton COSY experiment (1,2,34). A typical ^{1}H COSY spectrum is that of Figure **5d** where the conventional one-dimensional ^{1}H spectrum is evident along both the F_2 and F_1 axes, as well as along the diagonal. Starting from any cross-peak at position (f_1, f_2), lines parallel to F_1 and F_2 can be extended to the diagonal. Ideally, the pattern is symmetric across the diagonal and this leads to the formation of a box where the two symmetry-related cross-peaks are corners, and two J-coupled peaks on the diagonal are the other corners. Although the COSY spectrum can be generated in several ways, the simplest pulse sequence is $90° - t_1 - 90° - t_2 -$ FID where t_1 is the variable evolution time and t_2 is the acquisition time. Interpretation of COSY spectra, in simple cases, is straightforward and unambiguous in the identification of coupled proton spin systems. For strongly coupled systems, ie, where $[f_2 - f_1 \text{ (Hz)}/J\text{(Hz)}] < 10$, cross-peaks may occur indistinguishably close to the diagonal. Suppression of the true diagonal peaks by double-quantum filtering (DQF-COSY) may resolve such problems. Finally, quantitative measurements of the magnitude of the coupling constants is possible using the Z-COSY modification. These experiments are restricted to systems of abundant spins such as ^{1}H, ^{11}B, and ^{31}P, which have reasonably narrow linewidths.

A 2-D total correlated spectroscopy (TOCSY), or HOHAHA experiment (2) resembles a homonuclear COSY. The pulse sequence differs from the COSY in that the 90° pulse after the evolution time is replaced by a mixing time. This

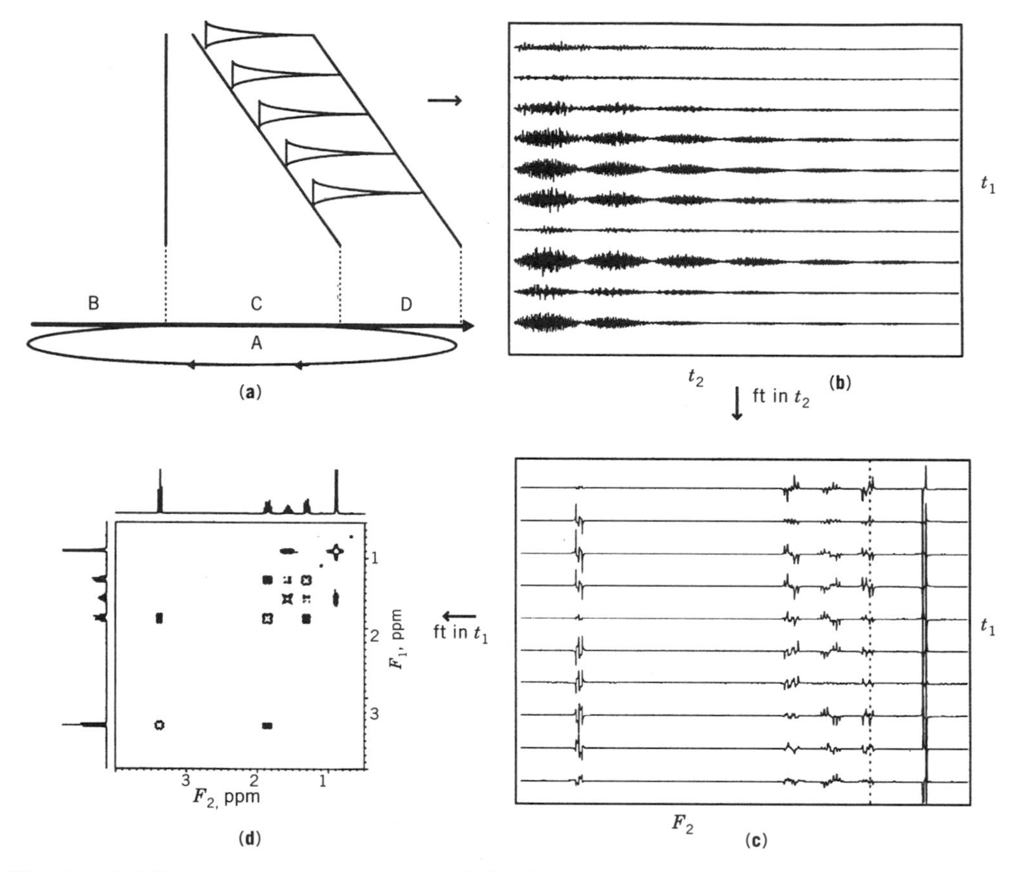

Fig. 5. A 2-D nmr experiment of 2-methyl-5-bromopentane [*626-88-0*], $C_6H_{13}Br$, where t_1 and t_2 correspond to evolution and acquisition time, respectively. (**a**) A set of pulsing experiments differing in evolution times leading to (**b**) a set of related FIDs to which a Fourier transform in t_2 is applied. Each FID leads to (**c**) a set of spectra; and a second Fourier transform along t_1 results in (**d**) the COSY spectrum. In (**a**), A indicates the progression of time through B, preparation, C, evolution and mixing, and D, acquisition. See text.

mixing time is used to lock the spins onto the *y*-axis which allows isotropic mixing of the spin systems. As in the 1-D HOHAHA experiment, choosing a short mixing time yields data very similar to a COSY, ie, directly coupled spins produce cross-peaks, whereas the use of long mixing times includes the complete spin system. An example of a 2-D-TOCSY having a mixing time of 120 ms is given in Figure 6**a**. As for the 1-D HOHAHA several 2-D TOCSYs can be run at different mixing times to determine the number of bonds over which the spins have propagated. The advantage of the 2-D experiment over the 1-D version is that all the information is acquired in one experiment. However, if only a small number of spin networks need to be determined, then the experiment of choice is the 1-D HOHAHA.

TOCSY data are acquired in the phase-sensitive mode using quadrature detection, and all the data phases are positive. This increases the SNR for the matrix, and the time required for the experiment is short because very little, if

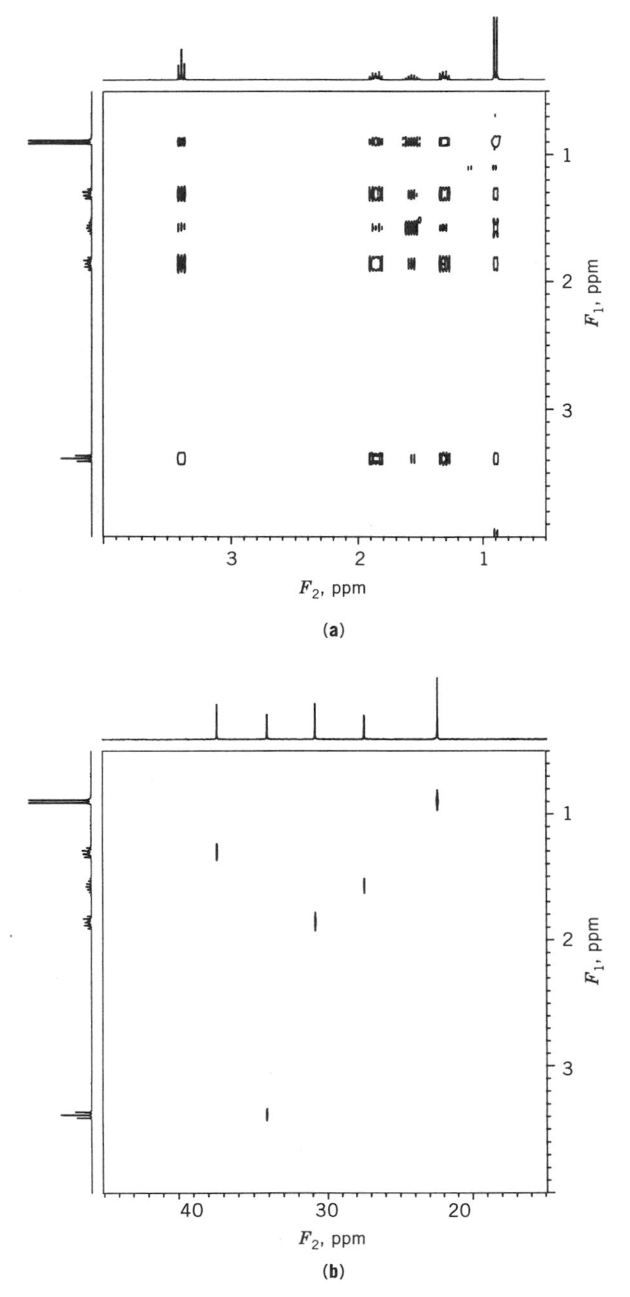

Fig. 6. Spectra of 2-methyl-5-bromopentane acquired using a Bruker 300AMX spectrometer: (**a**) TOCSY using a 5-mm $^{13}C-^{1}H$ dual probe and (**b**) HMQC using a 5-mm inverse detection broadband probe.

any, phase cycling is necessary. In some cases a single scan per FID suffices, and the data can be acquired in approximately 10 min.

Heteronuclear chemical shift-correlated spectroscopy, commonly called H-X COSY or HETCOR has, as the name implies, different F_1 and F_2 frequencies.

The experiment uses polarization transfer from the ^{1}H nuclei to the ^{13}C or X nuclei which increases the SNR. Additionally, the repetition rate can be set to 1–3 T_1 of the ^{1}H rather than the longer ^{13}C. Using the standard ^{1}H–^{13}C COSY, the amplitude of the ^{13}C signals are modulated by the J-coupled frequencies of the ^{1}H nuclei in the evolution time. The projection of the F_2 axis is an edited ^{13}C spectrum where only those carbons having attached protons are observed. The projection of the F_1 axis is an edited ^{1}H spectrum where only protons attached to ^{13}C are observed. Figure **6b** shows a standard HETCOR. The existence of a cross-peak indicates a direct correlation of the ^{1}H peak to the ^{13}C peaks. To identify these peaks, simply draw a line from the ^{13}C signal on the F_2 axis to the cross-peak and then a second line parallel to the F_1 axis identifies the directly coupled proton. The sample requirements are similar to those of a standard ^{13}C experiment. Interpretation of the ^{13}C spectrum is straightforward if the ^{1}H spectrum has been assigned and this is an excellent way to determine diastereotopic protons because these produce two ^{1}H signals for one ^{13}C signal.

An alternative way of acquiring the data is to observe the ^{1}H signal. These experiments are referred to as reverse- or inverse-detected experiments, in particular the inverse HETCOR experiment is referred to as a heteronuclear multiple quantum coherence (HMQC) spectrum. The amplitude of the ^{1}H nuclei is modulated by the coupled frequencies of the ^{13}C nuclei in the evolution time. The principal difficulty with this experiment is that the ^{13}C nuclei must be decoupled from the ^{1}H spectrum. Techniques used to do this are called GARP and WALTZ sequences. The information is the same as that of the standard HETCOR except that the F_1 and F_2 axes have been switched. The obvious advantage to this experiment is the significant increase in sensitivity that occurs by observing ^{1}H rather than ^{13}C.

Another important inverse experiment is the heteronuclear multiple bond correlation (HMBC) spectrum; its ^{13}C observed counterpart experiment is the COLOC (35), which suffers from lack of sensitivity. The HMBC detects long-range coupling from protons to carbon atoms separated by two to three bonds, while suppressing the directly attached proton–carbon correlations. This is useful when making assignments of isolated spin groups such as methoxy, carbonyl, and quaternary carbon groups, because there is typically at least one proton set that is within the two to three-bonds range for any given carbon in an organic molecule. The projection of either an HMBC or an HMQC spectrum contains ^{13}C chemical shift information, which can be useful when working with samples that are too dilute to permit the acquisition of a standard ^{13}C spectrum.

The 2-D incredible natural abundance double-quantum transfer experiment (INADEQUATE) was developed in the late 1980s (35,36). This experiment, as the name implies, is not applicable to most real samples because of the large quantity of sample required. The signals that are being observed are the ^{13}C satellites arising from the ^{13}C–^{13}C coupling. The natural abundance of ^{13}C is 1.1% of the total carbon, and therefore the likelihood of a covalent bond between two ^{13}C atoms is 1:10,000. Also, the normal single-quantum ^{13}C signals must be suppressed by phase cycling. Hence the experiment has low sensitivity and experiment times are long. The 2-D INADEQUATE yields a great deal of information, however. The F_2 axis contains the standard ^{13}C spectrum and F_1 is the double-quantum frequency of $\nu_1 + \nu_2$, where ν_1 and ν_2 are the frequencies of the coupled spins, measured in Hz from the center band frequency. When the experiment is set up correctly, its interpretation is straightforward, and the molecular carbon skeleton can be deduced. Highly concentrated solutions, as well as the addition of a relaxation reagent such as $Cr(acac)_3$ to shorten the preparation delay, are helpful.

The 2-D nuclear Overhauser effect spectroscopy (2-D-NOESY) experiment resembles the $^1H-^1H$ COSY; however, the cross-peaks arise from chemical exchange or from the NOE effects, not J-coupling. The experiment yields data similar to a series of transient NOE difference experiments for all chemical shifts. The pulse sequence is $90° - t_1 - 90° - t_m - 90°$, where t_1 is the evolution time, t_m is the mixing time, and t_2 is the detection time. The second 90° pulse restores the frequency-labeled transverse magnetization back to the z-axis. During the mixing time, t_m, the magnetization on the z-axis is allowed to mix and a 90° pulse is applied to read the information. The data are collected in the phase-sensitive mode. Positive NOEs are phased up, whereas the diagonal negative NOEs and chemical exchange cross-peaks are phased down. COSY-type peaks may also appear. Typically, a series of 2-D-NOESY spectra, each having a different mixing time, are collected and compared to the corresponding COSY spectrum to determine which signals arise from J-coupling, which from chemical exchange, or which from NOEs. NOE buildup curves are used to determine internuclear distance constraints (2). The intensity of the effect varies with r_{ij}^{-6}, and only interactions with $r_{ij} < 0.5$ nm are detectable. The NOESY experiment is particularly applicable to large molecules such as proteins, nucleic acid, polysaccharides, or synthetic polymers because on the nmr time scale, macromolecules tumble relatively slowly, producing large ($\leq$50%) negative NOEs suitable for quantification. Because of rapid tumbling, small molecules can have positive, negative, or no NOE even if the distances are within the proper magnitude. The rotating frame nuclear Overhauser effect spectroscopy (ROESY) (2,5) experiment is best suited for small- to intermediate-sized molecules. The ROESY replaces the evolution time with a spin-lock field so that the effective precession frequency is very small. This makes even the largest molecules appear to be tumbling rapidly and makes all NOE signals positive.

Given the possibility of making complete assignments of all chemical shifts in molecules as large as small proteins and the availability of both distance- and angle-dependent nmr parameters, attempts are being made to determine the three-dimensional structure of molecules from nmr, and nmr-restrained molecular modeling (qv) methods are being developed as an alternative to x-ray diffraction (36). One such approach is distance geometry (37), which uses the short, throughspace contacts identified in NOESY experiments to formulate a set of mathematical restraint equations which tend to force the selected interatomic distances to assume suitably short values. Using a sufficient number of such interactions for each residue in a polymer chain, it should be possible to correctly identify probable structures from the nmr data alone. Other computer programs elastically restrain torsion angles to values estimated from ^3J-coupling constants. A second and somewhat different approach involves using the nmr restraints to augment the molecular mechanics forcefield in a molecular dynamics calculation (38). In this approach the basic forcefield contains terms describing contributions to the potential energy arising from bond stretching, bond-angle deformation, torsional interactions, van der Waals interactions, and electrostatic interactions. The additional nmr term contains equations designed to satisfy coupling and NOE data for the molecule. Finally, a kinetic energy term is also included where the total kinetic energy at a temperature, T, is $3N(kT/2)$ where $3N$ is the number of degrees of freedom for a system of N atoms, and kT is the usual Boltzmann energy term. The entire system is then allowed to fluctuate in terms of atomic positions and velocities by solving the Newtonian equations of motion at very closely spaced intervals in time. All of the molecular dynamics algorithms including CHARMm, Discover, GROMOS, and AMBER permit the calculation of such restrained dynamics. Both the distance geometry approach and, to a lesser degree, the dynamics approach can have problems with structures becoming locked in local minima. A second problem is that whereas a crystal structure typically involves a unique set of atomic coordinates describing the position of each atom, the structure of a molecule in solution may involve several coexisting conformers contributing to a time- and space-averaged structure.

There are many instrumental considerations in multidimensional spectroscopy. These are important in defining the appropriate digital resolution in F_1 and F_2. Unlike 1-D nmr, the data size in 2-D experiments can become quite large very rapidly if the wrong parameters are used. For instance, using a 32-bit integer word size, a $2\ \text{k} \times 2\ \text{k}$ data set occupies at least 16 Mb and even more if phase-sensitive data is being acquired. The best way to determine the minimum number of data points required in F_2 is to run a routine 1-D spectrum and determine the minimum acceptable Hz per point from it. Because F_1 resolution is a function of the number of evolution time increments, ie, the number of separate FIDs which must be stored, the resolution in the F_1 dimension is much lower than in the F_2 dimension. The minimum necessary F_1 digitization can also be determined from the number of Hz per points needed to resolve signals in the 1-D spectrum. As for 1-D nmr, zero-filling and optimal window functions can be used to improve the apparent resolution in the 2-D matrix.

Time constraints are an important factor in selecting nmr experiments. There are four parameters that affect the amount of instrument time required for an experiment. A preparation delay of 1–3 times T_1 should be used. Too short a delay results in artifacts showing up in the 2-D spectrum whereas too long a delay wastes instrument time. The number of evolution times can be adjusted. This affects the F_1 resolution. The acquisition time or number of data points in t_2 can be adjusted. This affects resolution in F_2. Finally, the number of scans per FID can be altered. This determines the SNR for the 2-D matrix. In general, a lower SNR is acceptable for 2-D than for 1-D studies.

Window functions for 2-D nmr are necessary because of the poor digital resolution in both dimensions. The short acquisition time required by the small number of data points in the F_2 domain and the limited number of evolution times for F_1 domain means that the data is being truncated. The software provided by any of the manufacturers provides several different window functions including exponential decays, shifted sinebell, shifted sinebell squared, Gaussian, and trapezoidal functions. The best function to use depends on the type of information sought. Often it is best to process the data using several different functions to determine the best choice for that particular data set.

High Resolution Solid-State nmr

The ability to acquire high resolution nmr data from solid phases is critical to many applications where, for example, test samples are insoluble, exploration of a nascent structure is desired, or samples are cross-linked (39,40). However, as seen by comparing Figures 7**a** and 7**d**, the use of liquid-phase nmr experiments using solid samples leads to broad lines having little discernible structural detail. Three principal problems in obtaining nmrs of solids are chemical shift anisotropy (CSA), a high degree of dipolar interactions, and very long (10^2-10^3 s) T_1 values typical of rigid molecules. The first two characteristics have origins in the comparative absence of motional averaging in solid samples. The Hamiltonians describing both the chemical shift and dipolar interactions are actually tensors, 3×3 matrices, where the interaction is a function of the orientation of the molecular axis system with respect to the laboratory axis system. Motional averaging in solution reduces both to scalar quantities which are actually the trace, ie, the average of the diagonal components of the tensor, and all off-diagonal components average to 0 (2).

As originally proposed in 1962, magic-angle spinning (MAS), where the sample is rapidly rotated about an axis inclined to B_0 by 54.74°, the magic angle, produces considerable line sharpening. This reduction in linewidth is a

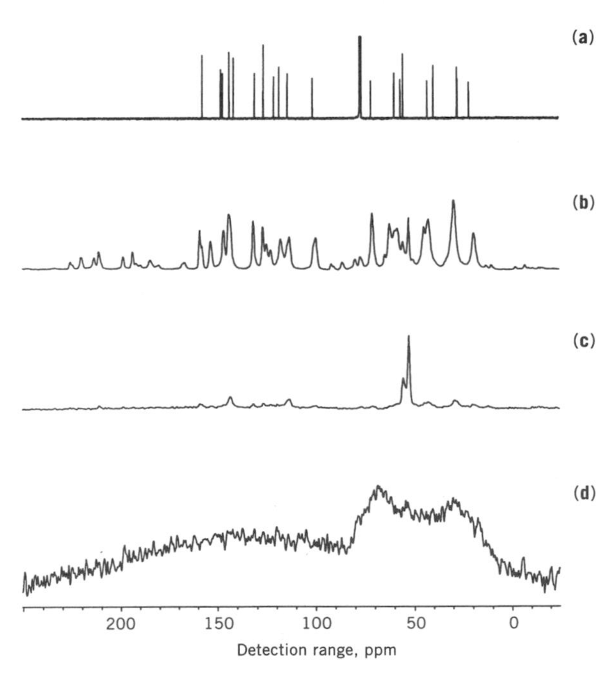

Fig. 7. Nmr spectra of quinine [*103-95-0*], $C_{20}H_{24}N_2O_2$, acquired on a Bruker 300AMX spectrometer using a Bruker broadband CP/MAS probe. (**a**) Proton-decoupled ^{13}C spectrum of quinine in $CDCl_3$; (**b**) the corresponding spectrum of solid quinine under CP/MAS conditions using high power dipolar decoupling; (**c**) solid-state spectrum using only MAS and dipolar decoupling, but without cross-polarization; and (**d**) solid quinine run using the conditions of (**a**).

consequence of the off-diagonal elements in both the CSA and dipolar interaction tensors leading to terms of the form $3\cos^2\theta - 1$ where θ is the angle between B_0 average orientation of dipoles in the sample. At the magic angle this term is equal to 0. The reduction in signal away from the central line is incomplete because it is rarely possible to spin the sample fast enough. Typical spinning speeds in MAS probes are 4–7 kHz although the newest probes can achieve rates of 27 kHz.

A further reduction in the dipolar interaction term is achieved by dipolar decoupling. In the simplest case this is accomplished by replacing the usual liquid-phase decoupler with a much higher power decoupler, ≥ 100 W. Using both of these techniques together results in a considerable sharpening of the lines, as shown in Figure 7**c**. The intensity, however, is quite weak and, owing to the long T_1 values, the preparation delay is extremely long. Hartman-Hahn matching for cross-polarization (CP), ie, polarization transfer, minimizes the time required for repolarization of the carbon atoms. In this technique a 90° pulse and spin-locking of the ^{1}H magnetization into the transverse, xy, plane is followed by simultaneous rf irradiation such that the Hartman-Hahn condition is met for a time period referred to as a mixing time or contact time, t_c, which is usually about 1 ms.

$$\gamma_H B_{1H} = \gamma_C B_{1C}$$

Because both spins are in the transverse plane and transition energy levels are matched, energy can be transferred from the protons to the ^{13}C nuclei. In this manner the rate of ^{13}C repolarization is controlled by T_{1H} rather than by T_{1C}. Because the protons can interchange energy by spin-diffusion only a single-proton T_1 exists and its value is usually on the order of 1 s. As a result the preparation delay can be reduced from 10^3 s to about 5 s increasing the number of transients, which can be acquired by two or more orders of magnitude.

One difficulty with this method is that the optimal mixing time for different carbon environments is not the same; as a result, the integrated intensities are even less quantitative than in liquid-phase ^{13}C nmr. A compromise value is selected by varying the mixing time systematically and looking for a value that provides an adequate signal for all resonances. A second difficulty is that the benefit derived from CP decreases with increasing proton–X nucleus distance so that it really only affects atoms having covalently bound hydrogen. Thus, quaternary carbon atoms and other X nuclei such as ^{31}P are insensitive to CP. The increase in intensities seen in Figure 7**b** as compared to 7**c** results from the use of CP. One additional feature apparent in Figure 7**b** is the existence of a number of additional lines not found in the liquid spectrum (Fig. 7**a**). These lines may originate from any of several sources. Because elimination of CSA is incomplete, the CSA is sampled by sidebands separated from each central resonance by a multiple of the spinning rate. These sidebands may be useful, as they sample the complete CSA, or they may be eliminated using sideband suppression modifications to the pulse sequence. A second source of additional lines results from different phases within the sample, ie, crystalline and amorphous phases, having somewhat different chemical shifts and line profiles for each resonance. Even in crystalline samples, if two different crystalline forms co-exist or if the crystal structure contains chemically identical molecules which are not related by symmetry, multiple peaks may occur.

One final technical improvement in solid-state nmr is the use of combined rotational and multiple pulse spectroscopy (CRAMPS) (2), a technique which also requires a special probe and permits the acquisition of high resolution ^1H and X nucleus nmr from solids. The combination of these methods permits adapting most of the 1-D and 2-D experiments previously described for liquids to the solid phase.

Applications of solid-state nmr include measuring degrees of crystallinity, estimates of domain sizes and compatibility in mixed systems from relaxation time studies in the rotating frame, preferred orientation in liquid crystalline domains, as well as the opportunity to characterize samples for which suitable solvents are not available. This method is a primary tool in the study of high polymers, zeolites (see MOLECULAR SIEVES), and other insoluble materials.

Electron Spin Resonance

Esr is also a powerful technique both alone or in combination with nmr in the endor method (9). Precise measurements of the hyperfine couplings and lineshape lead to detailed information concerning the structure and conformation of molecules containing unpaired electrons. It is perhaps the most powerful tool for studying structure in free-radical containing species and as such plays an important role in understanding reaction mechanisms. A second area of current importance is the study of radiation-induced changes in chemicals. The attachment of spin labels to different parts of a molecule also provides a method of probing local structure.

BIBLIOGRAPHY

"Analytical Methods" in *ECT* 3rd ed., Vol. 2, pp. 586–683, by E. Lifshin and E. A. Williams, General Electric Co.

1. A. E. Derome, *Modern NMR Techniques for Chemical Research*, Pergamon Press, Oxford, U.K., 1987.
2. J. K. M. Sanders and B. K. Hunter, *Modern NMR Spectroscopy: A Guide for Chemists*, 2nd ed., Oxford University Press, Oxford, U.K., 1993.
3. L. Kevan and M. K. Bowman, eds., *Modern Pulsed and Continuous-Wave Electron Spin Resonance*, John Wiley & Sons, Inc., New York, 1990.
4. R. N. Bracewell, *The Fourier Transform and Its Applications*, 2nd rev. ed., McGraw-Hill Book Co., Inc., New York, 1986.
5. D. Neuhaus and M. P. Williamson, *The Nuclear Overhauser Effect in Structural and Conformational Analysis*, Verlag Chemie, New York, 1989.
6. W. W. Conover, in G. C. Levy, ed., *Topics in Carbon-13 NMR Spectroscopy*, Vol. 4, John Wiley & Sons, Inc., New York, 1984, pp. 38–51.
7. E. Fukushima and S. B. W. Roeder, *Experimental Pulse NMR: A Nuts and Bolts Approach*, Addison-Wesley Publishing Co., Inc., Reading, Mass., 1981.
8. C. J. Jameson, *Chem. Rev.* **91**, 1375–1395 (1991).
9. C. F. Poole, Jr. and H. A. Farach, eds., *Handbook of Electron Spin Resonance: Datasources, Computer Technology, Relaxation and ENDOR*, American Institute of Physics, New York, 1994.

10. E. Kanal, J. Gillen, J. A. Evans, D. A. Savitz, and F. G. Shellock, *Radiology* **187**, 395–399 (1993).
11. R. M. Silverstein, C. Bassler, and T. C. Morrill, *Spectrometric Identification of Organic Compounds*, John Wiley & Sons, Inc., New York, 1991.
12. M. Karplus, *J. Chem. Phys.* **30**, 11 (1959).
13. R. H. Contreras and J. C. Facelli, *Ann. Rep. NMR Spectrosc.* **27**, 256–356 (1993).
14. D. J. Craig and G. C. Levy, in G. C. Levy, ed., *Topics in* ^{13}C *NMR Spectroscopy*, Vol. 4, John Wiley & Sons, Inc., New York, 1984, pp. 239–275.
15. A. Carrington and A. D. McLachlan, *Introduction to Magnetic Resonance*, Harper and Row, New York, 1967, pp. 221–229.
16. J. M. Bulsing, J. K. M. Sanders, and L. D. Hall, *J. Chem. Soc. Chem. Commun.*, 1201–1203 (1981).
17. R. A. Hoffman and S. Forsen, *Prog. N.M.R. Spectrosc.* **1**, 15–204 (1966).
18. D. G. Davis and A. Bax, *J. Am. Chem. Soc.* **107**, 7197–7198 (1985).
19. V. Sklenár and A. Bax, *J. Magn. Reson.* **75**, 378 (1987).
20. V. Saudek, M. Piotto, and V. Sklenár, *Bruker Report* **140**, 6–9 (1994).
21. P. Laszlo, *NMR of Newly Accessible Nuclei*, Vols. 1 and 2, Academic Press, New York, 1979.
22. C. Brevard and P. Granger, *Handbook of High Resolution Multinuclear NMR*, John Wiley & Sons, Inc., New York, 1981.
23. J. T. Gerig, *Methods Enzymol.* **177B**, 3–23 (1989).
24. W. C. Dickenson, *Phys. Rev.* **81**, 717 (1951).
25. H. S. Gutowsky and D. W. McCall, *Phys. Rev.* **82**, 748 (1951).
26. D. G. Gorenstein, ed., *Phosphorus-31 NMR: Principals and Applications*, Academic Press, San Diego, Calif., 1994.
27. D. G. Gorenstein, *Methods Enzymol.* **177B**, 295–316 (1989).
28. C. R. Calladine, *J. Mol. Biol.* **161**, 343 (1982).
29. S. M. Cohen, *Methods Enzymol.* **177B**, 417–434 (1989).
30. P. F. Daly, R. C. Lyon, P. J. Faustino, and J. S. Cohen, *J. Biol. Chem.* **262**, 14875 (1987); J. S. Cohen, R. C. Lyon, and P. F. Daly, *Methods Enzymol.* **177**, 435–452 (1989).
31. P. Varley, A. M. Gronenborn, H. Christensen, P. T. Wingfeld, R. H. Pain, and G. M. Clore, *Science* **260**, 1110–1113 (1993) and Ref. 6 therein.
32. D. C. Muchmore, L. P. McIntosh, C. B. Russell, D. E. Anderson, and F. W. Dahlquist, *Methods Enzymol.* **2177B**, 43–73 (1989).
33. C. R. Sanders II and M.-D. Tsai, *Methods Enzymol.* **2177B**, 317–333 (1989).
34. A. Bax and R. Freeman, *J. Am. Chem. Soc.* **104**, 1099 (1981).
35. W. E. Hull, in W. R. Croasmun and R. M. K. Carlson, eds., *Two-Dimensional NMR Spectroscopy: Applications for Chemists and Biochemists*, VCH Publishers, New York, 1987.
36. A. Bax, *Ann. Rev. Biochem.* **58**, 223–256 (1989).
37. G. M. Crippen and T. F. Havel, *Distance Geometry and Molecular Conformation*, John Wiley & Sons, Inc., New York, 1988.
38. J. DeVlieg and W. F. van Gunsteren, *Methods Enzymol.* **202**, 268–300 (1991).
39. J. L. Koenig, *Spectroscopy of Polymers*, American Chemical Society, Washington, D.C., 1992, pp. 197–256.
40. R. Komoroski, ed., *High Resolution NMR Spectroscopy of Synthetic Polymers in Bulk*, VCH Publishers, New York, 1986.

DAVID J. KIEMLE
WILLIAM T. WINTER
State University of New York, Syracuse

MAGNETIC TAPE. See INFORMATION STORAGE MATERIALS.

MAGNETOHYDRODYNAMICS

Magnetohydrodynamic (MHD) power generation is a method of generating electric power by passing an electrically conducting fluid through a magnetic field (see also POWER GENERATION). By means of the interaction of the conducting fluid with the magnetic field, the MHD generator transforms the internal energy of the conducting fluid into electric power in much the same way as a conventional turbogenerator does by means of the interaction of a solid conductor with a magnetic field. In principle, the working fluid can be any electrically conducting fluid, such as salt water, liquid metal, or hot ionized gas. For central station power generation applications, the most suitable working fluid is a hot ionized gas. This can be a relatively clean gas, eg, a noble gas (see HELIUM GROUP, GASES) heated in an externally fired heat exchanger, or it can be composed of combustion products from any fossil fuel. There are two basic types of MHD energy conversion systems. Closed cycle typically operates using a clean gas which is recycled; open cycle generally operates using combustion products which are then discarded.

In its simplest form the MHD generator consists of a duct through which the gas flows, driven by an applied pressure gradient, and a magnet, in which the duct is located. The generator operates in a Brayton cycle similar to that of a turbine. Because the MHD process requires no rotating machinery or moving mechanical parts, the MHD generator can operate at much higher temperature and hence, higher efficiency, than is possible for other power generation technologies. The system of most interest for central station power generation is the open cycle system using electrically conducting coal (qv) combustion products as the working fluid. Coal-burning central station MHD power plants promise to generate power at up to 50% greater efficiency and at a lower cost of electricity than can be achieved using conventional coal-burning power plants of the early to mid-1990s (see COAL CONVERSION PROCESSES).

In addition to potential advantages in efficiency, MHD power generation offers significant potential for reduced environmental intrusion. Effective control of pollutants is inherent in the basic design and operation of MHD power plants. Emissions of SO_x, NO_x, and particulates can be reduced to levels far below the New Source Performance Standards (NSPS) of 1979 without requiring expensive gas clean-up equipment. Furthermore, because of the higher efficiency and consequent lower fuel usage of MHD power plants, emissions of carbon dioxide (qv) are lower, heat rejection is reduced, and less solid waste is produced (see AIR POLLUTION; WASTES, INDUSTRIAL). The basic principle of MHD power generation was discovered in 1831 by Michael Faraday (1), who investigated electromagnetic interactions induced by the flow of the Thames River in the magnetic field of the earth. The first serious attempts at MHD power generation were made between 1936 and 1945 (2). The first successful MHD generator was built in 1959 (3–5). This device operated on argon heated to high temperatures by a plasma jet, and produced 11.5 kilowatts of power. A combustion-driven generator of approximately the same size was built in 1960 (6), and achieved continuous operation for about one hour. A much larger combustion-driven generator which produced about 1.5 megawatts of electrical power was built by Avco in 1962 (7). Measured performance was in close agreement with theoretically predicted results.

System Components

The implementation of the Faraday effect to produce MHD power is shown in Figure 1. An electrically conducting fluid flows with velocity $\vec{u}$, through a magnetic field $\vec{B}$, to induce an electric field $\vec{E}$ which is orthogonal to both the flow direction and the magnetic field direction. If the flow is contained in a duct, as shown, the two walls perpendicular to the electric field are at different potentials. If these two walls are electrically conducting and connected through an external resistance or load, the field causes a current to flow through the load, thus generating power. The conducting walls, through which current is extracted from the duct, are the electrode walls. The wall which emits electrons is designated the cathode; the other wall is the anode. The two walls separating the electrode walls are the insulator walls.

For central station power generation the open cycle system using electrically conducting coal combustion products as the working fluid is employed. The fuel typically is pulverized coal burned directly in the MHD combustor, although in some plant designs cleaner fuels made from coal by gasification or by beneficiation have been considered (8–10) (see FUELS, SYNTHETIC).

In application to electric utility power generation, MHD is combined with steam (qv) power generation, as shown in Figure 2. The MHD generator is used as a topping unit to the steam bottoming plant. From a thermodynamic point of view, the system is a combined cycle. The MHD generator operates in a Brayton cycle, similar to a gas turbine; the steam plant operates in a conventional Rankine cycle (11).

Starting with combustion products at a pressure of 0.5–1 MPa (5–10 atm), and a temperature sufficiently high (ca 3000 K) to produce a working fluid of adequate electric conductivity when seeded with an easily ionizable salt such as

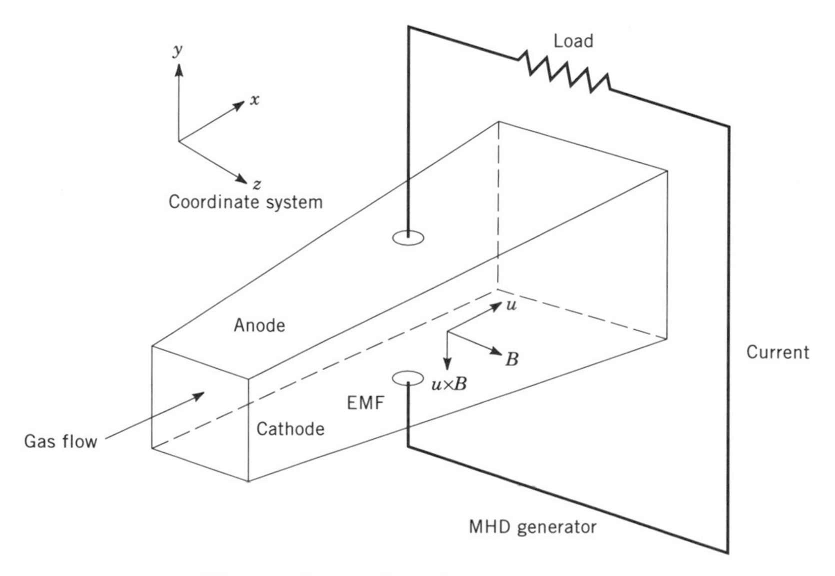

Fig. 1. Principles of MHD. See text.

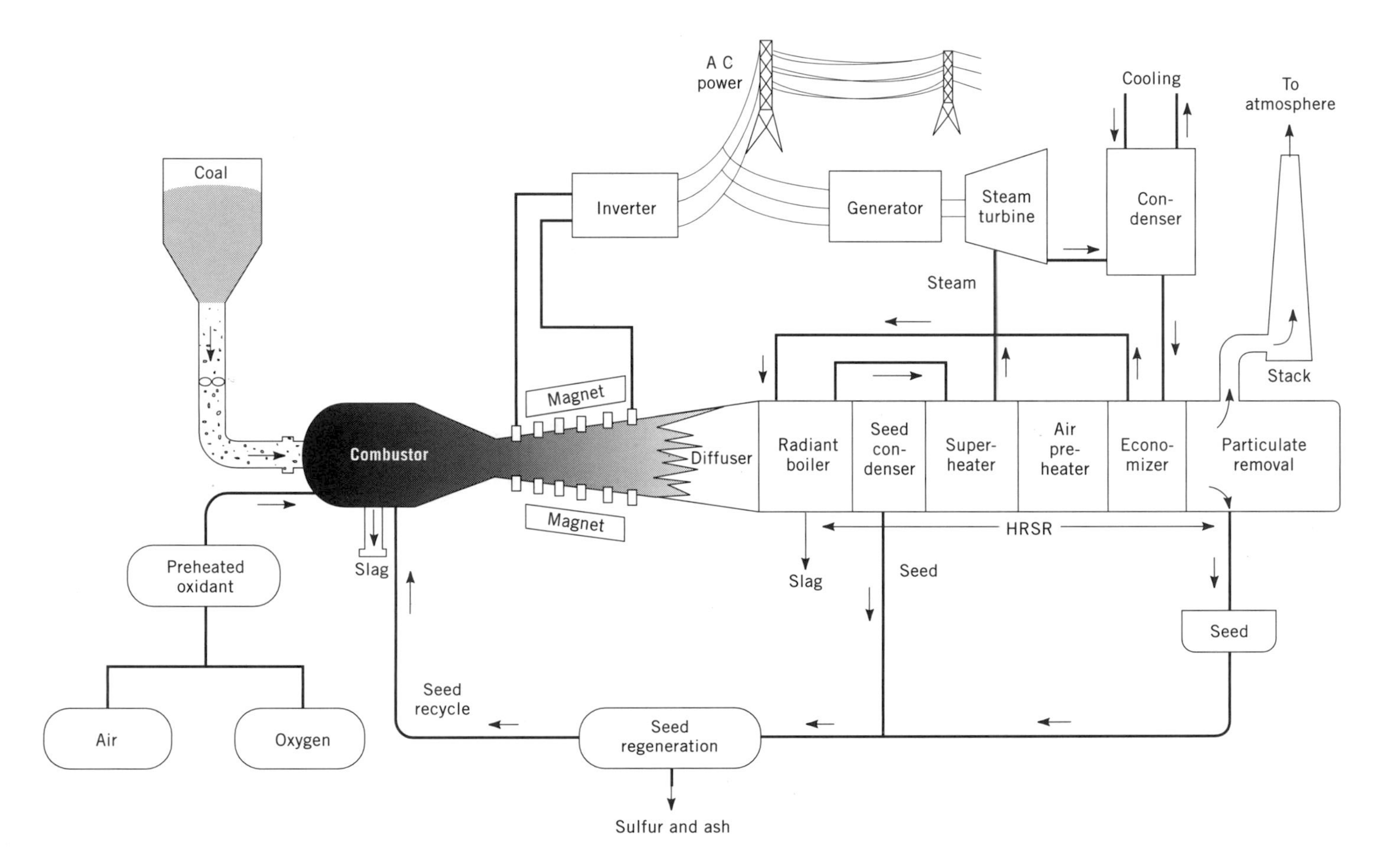

Fig. 2. MHD-steam power plant where HRSR is heat recovery seed recovery and the seed is an easily ionizable potassium salt. See text.

potassium carbonate or formate, the hot ionized gases flow through the MHD generator at approximately sonic velocity. The MHD generator duct, or channel, extracts energy from the gas, and the flow is expanded so that it can maintain its velocity against the decelerating forces resulting from its interaction with the magnetic field. The combination of energy extraction and flow expansion causes the gas temperature to drop. Energy is extracted until the gas temperature becomes too low (ca 2300 K) to have a useful electric conductivity.

The gases exhausting from the generator still contain significant useful heat energy. This energy is used in the bottoming plant to raise steam to drive a turbine and generate additional electricity in the conventional manner of a steam plant. The generator exhaust gases also contain potassium and sulfur from the coal. These combine to form primarily potassium sulfate, which condenses in the bottoming plant and is then delivered to a seed regeneration unit where it is converted back to potassium formate or carbonate and recycled to the MHD combustor.

The required high combustion temperatures can be achieved by preheating the combustion air to temperatures of 1650–1950 K. The highest efficiencies are obtained by direct high temperature preheat of the combustion air with the MHD exhaust gas, but this requires the use of high temperature refractory heat exchangers, which are not commercially available. The requisite combustion temperatures can also be attained by enriching the combustion air with oxygen and preheating the oxygen-enriched air to more moderate temperatures, which can be reached in conventional metal tubular heat exchangers (see HEAT-EXCHANGE TECHNOLOGY). The resultant plant efficiencies are high enough to make this latter method attractive for use in first generation commercial plants based on available technology. High temperature refractory heat exchangers would allow realization of the full efficiency potential of MHD as well as improved fuel utilization, and even lower energy costs (see REFRACTORIES).

Other MHD power cycles have been proposed in which the heat energy of the MHD generator exhaust gas is utilized in a bottoming gas turbine plant (12,13). The gas turbine working fluid in such a plant is clean air heated by the MHD generator exhaust gas. Efficiency advantages offered by the use of a high temperature gas (air) turbine instead of a steam turbine in the bottoming plant should improve the overall MHD power cycle accordingly. MHD power cycles of this type do not need cooling water for steam condensation and heat rejection.

The components of a combined MHD–steam power plant which are most directly associated with the MHD process are referred to collectively as the MHD power train, or as the topping cycle components. The rest of the plant consists of the steam bottoming plant, a cycle compressor, the seed regeneration plant, and the oxygen plant, if necessary. The MHD topping cycle components are the magnet, the coal combustor, nozzle, MHD channel, associated power conditioning equipment, and the diffuser. The magnetic field required in a commercial plant is typically 4.5–6 tesla ($4.5–6 \times 10^4$ G). Hence, the magnet is superconducting, as a conventional magnet would require impractical amounts of electric power (see MAGNETIC MATERIALS; SUPERCONDUCTING MATERIALS).

Power conditioning is necessary between the channel and the transmission grids for two reasons. First, the channel produces d-c electric power, because both the magnetic field and the channel flow are nominally time-invariant. Therefore, an inverter is required to convert the d-c MHD output to a-c for transmission. Voltage step-up is also required, as the generator output is typically at lower voltage (20–40 kV) than that required for transmission. Second, a large channel may have a large number of two terminal outputs, and circuitry is needed to consolidate the outputs from all the terminal pairs for delivery into the main load inverter.

The steam bottoming plant of the combined MHD–steam power plant consists basically of a heat recovery and seed recovery system (HRSR) and a turbine–generator for additional power production. The HRSR is essentially a heat recovery boiler and oxidant preheater which is fired by the exhaust gases from the MHD channel. In addition to generating steam, the HRSR system must also perform the functions of NO_x and SO_x control, slag tapping, seed recovery, and particulate removal (see AIR POLLUTION CONTROL METHODS; EXHAUST CONTROL, INDUSTRIAL).

Chemical Regeneration. In most MHD system designs the gas exiting the topping cycle exhausts either into a radiant boiler and is used to raise steam, or it exhausts into a direct-fired air heater and is used to preheat the primary combustion air. An alternative use of the exhaust gas is for chemical regeneration, in which the exhaust gases are used to process the fuel from its as-received form into a more beneficial one. Chemical regeneration has been proposed for use with natural gas and oil as well as with coal (14) (see GAS, NATURAL; PETROLEUM).

A coal-based system (Fig. 3) is described in References 15 and 16. The generator exhaust gas is used in a multistage process in which the incoming coal undergoes devolatilization, gasification, and partial combustion at atmospheric pressure to produce a low heating value (LHV) fuel gas for the primary combustor. The thermal energy recovered from the exhaust gas and stored in the fuel gas is roughly 40% of the combustion energy of the fuel into the primary combustor (15), so that a substantial increase in fuel energy is achieved by use of chemical regeneration. An alternative scheme would be to locate the regenerator upstream of the MHD generator so that gasification is done at peak cycle pressure. The advantage of this system is that a smaller volume of gas is processed. Variations of this scheme are described elsewhere (17–19). Comparisons of cycles which use the generator exhaust gas to preheat combustion air to cycles which use chemical regeneration can be found in Reference 15. Somewhat higher efficiencies can be obtained with the use of chemical regeneration than without, for a given cycle pressure ratio.

MHD Fundamentals

The basic principles of MHD power generation are shown in Figure 1. The working fluid is typically a slightly ionized gas, ie, mostly neutral atoms or molecules where a small (<0.1%) fraction are ions and electrons. The working fluid flows with velocity u, in the axial, x, direction, through the MHD channel, which is located in a magnetic field of strength B, in the z direction. An electric field, uB, is induced in the transverse ($-y$) direction, as shown. Because there exists also an internal electric field, E, owing to the load current, the electric field, E',

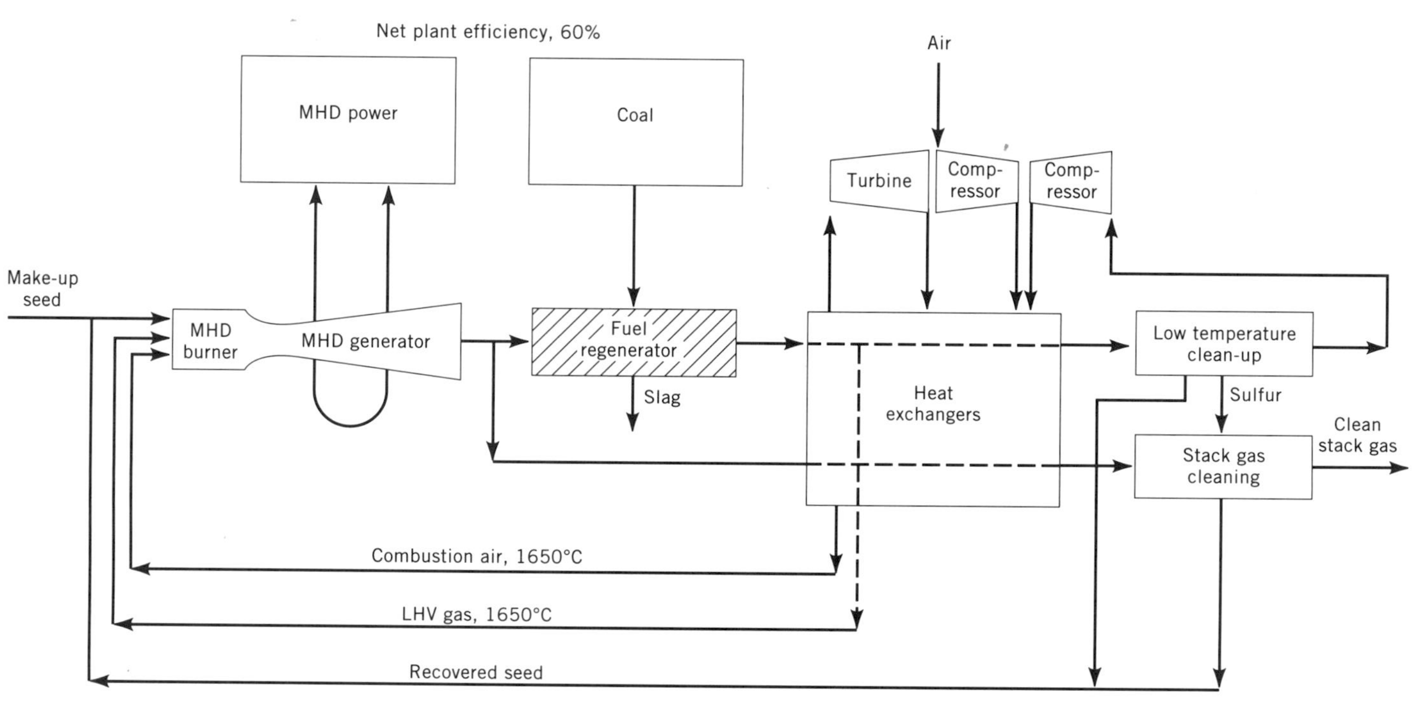

Fig. 3. MHD power plant design burning low heating value (LHV) gas produced from coal by chemical regeneration.

seen by the moving gas is

$$E' = E - uB \tag{1}$$

and the induced current per unit area, j_y, is

$$j_y = \sigma E' = \sigma(E - uB) \tag{2}$$

where σ is the electrical conductivity of the gas. If a load coefficient K is defined, such that

$$K \equiv E/uB \tag{3}$$

the magnitude, j, of the current density can be expressed as

$$j = \sigma uB(1 - K) \tag{4}$$

and the generator power per unit volume, P, is then

$$P = jE = \sigma u^2B^2K(1 - K) \tag{5}$$

Assuming that the current in the gas is carried mostly by electrons, the induced electric field uB causes transverse electron motion (electron drift), which, being itself orthogonal to the magnetic field, induces an axial electric field, known as the Hall field, and an axial body force, F, given by

$$F = jB = \sigma uB^2(1 - K) \tag{6}$$

This force acts in the $-x$ direction and retards the flow. The rate at which the gas does work in pushing itself against this force is

$$Fu = jBu = \sigma u^2B^2(1 - K) \tag{7}$$

The ratio of electrical power output (eq. 5) to the power required to push the gas (eq. 7) can be defined as the efficiency of the process, η_e:

$$\eta_e = jE/juB = K \tag{8}$$

The difference between the push power and the electric power output is

$$juB - jE = j^2/\sigma \tag{9}$$

The term j^2/σ is the rate of dissipation of energy per unit volume by joule heating. This occurs within the working fluid, and so represents a departure from thermodynamic irreversibility rather than an energy loss.

This simplified discussion has neglected the effects of axial current flow, ie, Hall current, induced by the axial field. At a local region in the channel, equation 1 can be written in more general form as

$$\vec{E}' = \vec{E} + \vec{u} \times \vec{B} \tag{10}$$

The electrons move with a drift velocity $\vec{v}_e$ relative to the gas. Then the electric field $\vec{E}''$ felt by the electrons is

$$\vec{E}'' = \vec{E} + \vec{u} \times \vec{B} + \vec{v}_e \times \vec{B} \tag{11}$$

The electron current density $\vec{j}_e$ is

$$\vec{j}_e = \sigma\vec{E}'' = \sigma(\vec{E} + \vec{u} \times \vec{B} + \vec{v}_e \times \vec{B}) \tag{12}$$

Electron current density can also be expressed as

$$\vec{j}_e = -n_e\vec{v}_e e \tag{13}$$

where n_e is the electron number density and e the electron charge. The equation has a minus sign because the electron charge is negative and current flow is conventionally defined as the flux of positive charges.

The electrical conductivity σ can be expressed in terms of the mobility μ_e:

$$\sigma = n_e\mu_e e \tag{14}$$

Using equations 13 and 14 in equation 12, the current density can be expressed as

$$\vec{j}_e = \sigma(\vec{E} + \vec{u} \times \vec{B}) - \mu_e(\vec{j}_e \times \vec{B}) \tag{15}$$

The first term on the right represents scalar conduction and the second term the Hall effect. This is generally expressed in terms of the Hall parameter $\beta = \mu_e B$, so that

$$\vec{j}_e = \sigma(\vec{E} + \vec{u} \times \vec{B}) - \frac{\beta}{B}(\vec{j}_e \times \vec{B}) \tag{16}$$

Equation 16 is a form of the generalized Ohm's law. Ion mobility, electron pressure gradients, and inertial effects have been neglected. The latter two phenomena are generally of no concern in MHD generator flows, but ion mobility is sometimes significant. Ohm's law including ion mobility, stated here without proof, is

$$\vec{j} = \sigma(\vec{E} + \vec{u} \times \vec{B}) - \frac{\beta}{B}(\vec{j} \times \vec{B}) + \frac{\beta\beta_i}{B^2}(\vec{j} \times \vec{B}) \times \vec{B} \tag{17}$$

where β_i, the Hall parameter, $= \mu_i B$; μ_i is the ion mobility. A derivation of equation 17 which is valid for weakly ionized gases, ie, the fraction of ionized atoms is small, can be found in Reference 20.

Equation 17 can be solved for the components of $\vec{j}$. In the coordinate system of Figure 1, with $\vec{B}$ in the z direction and $\vec{u}$ in the x direction, $u_x = u$, $u_y = u_z = 0$; $B_z = B$, $B_x = B_y = 0$; $E_z = 0$. Then,

$$j_x = \frac{\sigma}{(1 + \beta\beta_i)} E_x - \frac{\beta}{(1 + \beta\beta_i)} j_y \tag{18}$$

$$j_y = \frac{\sigma}{(1 + \beta\beta_i)} (E_y - uB) + \frac{\beta}{(1 + \beta\beta_i)} j_x \tag{19}$$

$$j_z = 0 \tag{20}$$

The ratio β_i/β is the ratio of ion mobility to electron mobility, μ_i/μ_e, which is approximately the square root of the ratio of electron mass to ion mass. For potassium seed material in combustion gases, μ_i/μ_e is about 0.003. Therefore $\beta_i/\beta \approx 0.003$, and, for typical values of β, $1 \lesssim \beta \lesssim 3$, $(1 + \beta\beta_i) \approx 1$. Equations 18 and 19 then reduce to the following:

$$j_x = \sigma E_x - \beta j_y \tag{21}$$

$$j_y = \sigma(E_y - uB) + \beta j_x \tag{22}$$

As shown in equation 6, an axial force jB, or, more generally, $\vec{j} \times \vec{B}$, is induced by the transverse current flow. Most of the current is carried by the electrons owing to their much greater mobility compared to the ions. The axial force is therefore felt primarily by the electrons, but is communicated to the ions via Coulomb interactions. As a result, the electrons and ions together begin to move with a slip velocity relative to the neutral particles in the flow, and a steady state is reached in which the slip momentum is transferred to the neutrals via collisions. This is the process by which the $\vec{j} \times \vec{B}$ force is transferred from the current carriers to the gas as a whole and is the mechanism through which the kinetic and thermal energy of the gas is converted to the electrical energy extracted from the flow. The $\vec{j} \times \vec{B}$ force is known as the Lorentz force and acts to oppose the gas flow.

MHD Generator Geometries. The basic requirement for any generator geometry is that the gas velocity have a component that is not parallel to the magnetic field. For efficiency, the gas velocity should be orthogonal to the magnetic field direction. Some of the possible geometries for achieving this are shown in Figure 4. Development work has focused mainly on the linear generator

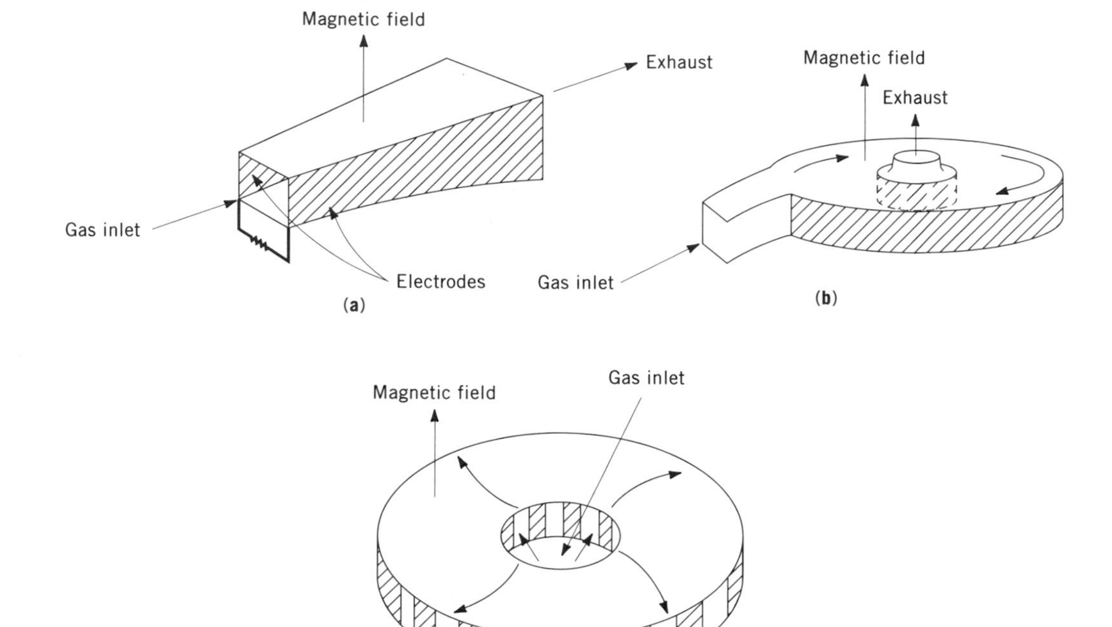

Fig. 4. MHD generator geometries: (**a**), linear; (**b**), vortex; and (**c**), disk having radial outflow.

(Fig. 4**a**). Detailed descriptions of the disk and vortex configurations can be found in References 21 and 22, respectively.

There are four basic variations of the linear MHD channel (Fig. 5) which differ primarily in their method of electrical loading. The simplest is the two-terminal Faraday or continuous electrode generator, Figure 5**a**, where a single pair of current-collecting electrodes spans the channel in the axial direction, short-circuiting the channel from end to end. Hence, for this configuration, $E_x = 0$, and j_y can be obtained from equations 21 and 22:

$$j_y = \frac{\sigma}{1 + \beta^2}(E_y - uB) \tag{23}$$

Additionally, there exists an axial or Hall current j_x given by

$$j_x = -\frac{\sigma\beta}{1 + \beta^2}(E_y - uB) \tag{24}$$

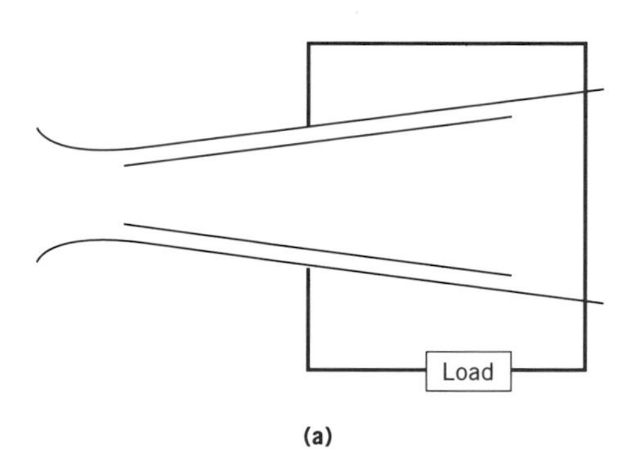

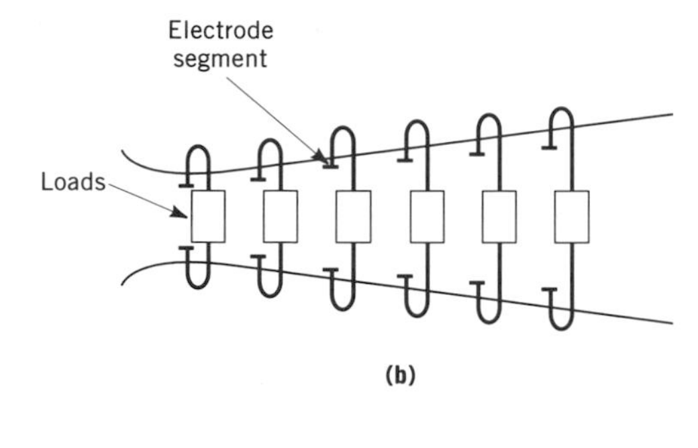

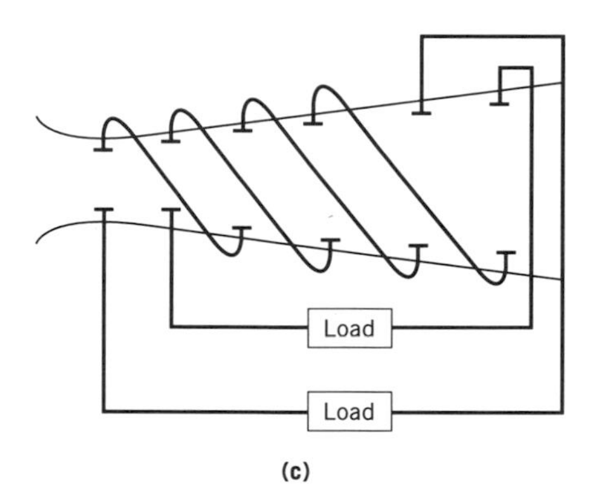

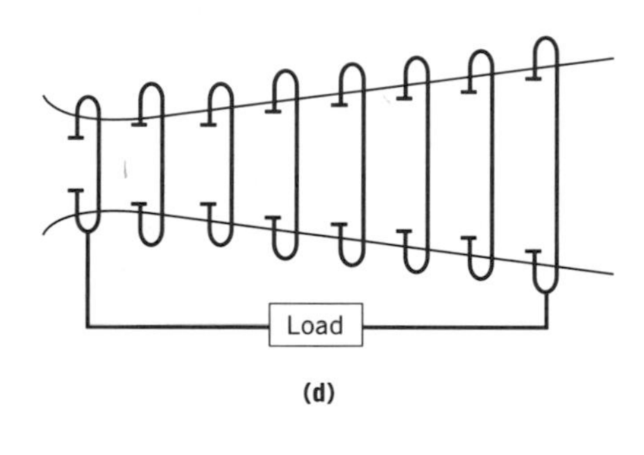

Fig. 5. Linear MHD generator configurations: (**a**), two-terminal Faraday or continuous electrode; (**b**), segmented Faraday; (**c**), diagonal connections; and (**d**), Hall geometry.

The generator power output per unit volume P is

$$P = -\vec{E} \cdot \vec{j} = -E_y j_y = -\frac{1}{1 + \beta^2}\sigma(E_y - uB)E_y \tag{25}$$

In terms of the load parameter, $K = E/uB$,

$$P = \frac{\sigma u^2 B^2}{1 + \beta^2} K(1 - K) \tag{26}$$

Compared to the expression of equation 5, having no axial current flow, power output is reduced by the factor $1/(1 + \beta^2)$. This is because part of the kinetic and thermal energy of the gas generates the axial current j_x which flows upstream in the gas and returns through the electrode walls. This current does not flow through the external load and so represents a loss.

The axial current flow can be eliminated, and so the power output increased, by axial segmentation of the electrodes, as shown in Figure 5**b**, the segmented Faraday generator. Each opposing pair of electrodes is connected to a single load. The axial electrode segments are separated by insulators, thus preventing axial current flow, ie, $j_x = 0$. Then the transverse current j_y becomes

$$j_y = \sigma(E_y - uB) \tag{27}$$

The power density is

$$P = \sigma u^2 B^2 K(1 - K) \tag{28}$$

and an axial field, E_x, is developed, given by

$$E_x = \beta(E_y - uB) \tag{29}$$

In practice, elimination of axial current flow requires relatively fine segmentation, eg, 1–2 cm, between electrodes, which means that a utility-sized generator contains several hundred electrode pairs. Thus, one of the costs paid for the increased performance is the larger number of components and increased mechanical complexity compared to the two-terminal Faraday generator. Another cost is incurred by the increased complexity of power collection, in that outputs from several hundred terminals at different potentials must be consolidated into one set of terminals, either at an inverter or at the power grid.

To alleviate these problems, the diagonally connected generator of Figure **5c** was devised (23). Those electrodes which would lie at the same potential in a normally operating segmented Faraday generator are connected, ie, the diagonal connection angle $\theta = \tan^{-1} E_y/E_x$. This generator has the advantage of requiring only a few output terminals while in principle operating at conditions which are the same as for the Faraday generator. If the connection angle θ is fixed, the power output and efficiency of the diagonally connected generator equal that of the segmented Faraday generator only at a single operating point, at which $E_y/E_x = \tan\theta$. In fact, E_y and E_x vary as the external load is changed, or as gas conditions change along the channel, and so the ratio E_y/E_x does not remain constant and equal to $\tan\theta$. In principle, performance is degraded as E_y/E_x varies from $\tan\theta$. In practice, generator performance is not extremely sensitive to moderate mismatches between the connection angle and the plasma conditions. The diagonally connected channel design is viewed as an acceptable compromise between optimum performance and mechanical/electrical simplicity. Methods can be incorporated to change the connection angle to match the changing fields along the channel and to match external load changes, but at the expense of mechanical or electrical complications.

Another configuration having power collection from only two terminals is the Hall generator shown in Figure **5d**. In this configuration, opposing electrode pairs are connected, and the current flow for power extraction is axial. $E_y = 0$, and

$$j_x = \frac{\sigma}{1 + \beta^2}(E_x + \beta uB) \tag{30}$$

at open circuit $E_x = -\beta uB$ and so the load parameter in this case is defined as $K' \equiv -E_x/\beta uB$. The power output is

$$P = -j_x E_x = \frac{\beta^2}{1+\beta^2} K'(1-K')\sigma u^2 B^2 \tag{31}$$

and the local conversion efficiency is

$$\eta = \frac{\beta^2 K'}{1+\beta^2 K'}(1-K') \tag{32}$$

Efficiency and power output approach those of a segmented Faraday generator only for large values of β, $\beta^2 \geq 1$. Figure 6 compares the relative efficiencies of the Hall and segmented Faraday generators. Although the maximum attainable efficiencies are comparable, the highest value in a Hall generator is obtained when the generator is heavily loaded, just the opposite of the situation in a Faraday generator.

In a channel operating under power plant conditions, the Hall parameter β varies from about 1 at the channel inlet to about 4 at the channel exit, typically, with an average value of 2–3. This is too high to make the two terminal Faraday configuration attractive, and too low to make the Hall configuration attractive. Thus the segmented Faraday (Fig. **5b**) and diagonal geometries (Fig. **5c**) are the configurations of choice, and the diagonal configuration is favored because of its simpler electrical loading.

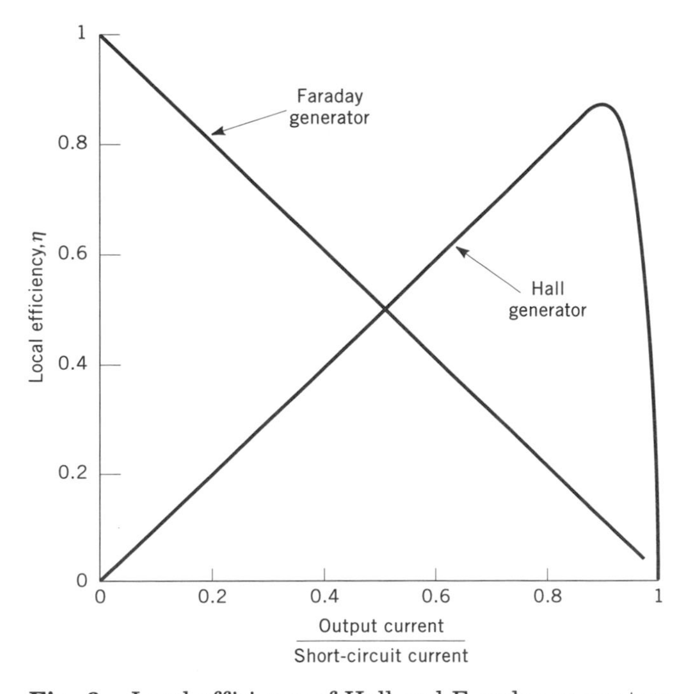

Fig. 6. Local efficiency of Hall and Faraday generators.

The disk generator, which has certain advantages for particular applications, has also received considerable attention. The disk generator offers a simpler structure, an electrode system having in principle only two electrodes, and a simpler magnet, which has a solenoidal configuration that is easier and less expensive to design and build than the more complex saddle coil configurations required for linear generators. Offsetting these advantages is the disadvantage that a disk generator has a much larger surface-to-volume ratio than a linear generator, and correspondingly larger frictional and heat losses. Thus the disk generator is in principle less efficient than the linear generator and so is less attractive for central station power generation applications. Consequently, the disk generator lags in development status. Nonetheless, it continues to receive attention, primarily for closed cycle power generation (24,25), but also for open cycle power generation (26,27).

Flow and Performance Calculations. Electrodynamic equations are useful when local gas conditions (u, σ, B) are known. In order to describe the behavior of the flow as a whole, however, it is necessary to combine these equations with the appropriate flow conservation and state equations. These last are the mass, momentum, and energy conservation equations, an equation of state for the working fluid, an expression for the electrical conductivity, and the generalized Ohm's law.

The conservation equations are as follows:

Mass
$$\rho u A = \text{constant} \quad (33)$$

Momentum
$$\rho \vec{u} \frac{du}{dx} + \nabla p = \vec{j} \times \vec{B} \quad (34)$$

Energy
$$\rho u \frac{d}{dx}\left(\frac{u^2}{2} + h\right) = \vec{j} \cdot \vec{E} \quad (35)$$

where ρ = density, p = pressure, A = flow cross-section area, and h = specific flow enthalpy. The right-hand term of the momentum equation is a body force, representing the electromagnetic drag force imposed on the flow by the induced currents. The right-hand term of the energy equation is the energy lost from the flow by means of electrical power extraction. The equations of state are

$$\begin{aligned} p &= p(\rho, T) \\ h &= h(\rho, T) \end{aligned} \quad (36)$$

where T = temperature. In addition, a relation for the conductivity σ is necessary:

$$\sigma = \sigma(\rho, T) \quad (37)$$

Relations for transport properties such as viscosity and thermal conductivity are also required if wall friction and heat-transfer effects are considered.

The most widely used approach to channel flow calculations assumes a steady quasi-one-dimensional flow in the channel core, modified to account for boundary layers on the channel walls. Electrode wall and sidewall boundary layers may be treated differently, and the core flow may contain nonuniformities.

In real channels the walls are rough and colder (ca 1700 K), even when covered with slag, than the gas in the bulk of the channel (ca 2500 K). Hence, velocity and thermal boundary layers exist at the channel walls. Their most important effect is to reduce the electrical conductivity at the electrode walls. Resultant voltage losses are caused by the higher electrical resistance of the cold boundary layers through which the induced current must pass. These cold boundary layers also force arc-mode current transport in the coldest regions, as current can no longer traverse in a diffuse mode. Wall friction and heat-transfer effects also exist. In addition, the slag layer on the walls changes the channel geometry somewhat and also allows some electrical leakage.

In the flow models these effects are confined to the boundary layers, maintaining the validity of the quasi-one-dimensional flow model. The flow is generally assumed to be developing, rather than fully developed, and is divided into an inviscid core region occupying most of the channel volume, and boundary layer and slag flow regions confined to the immediate vicinity of the channel walls. The behavior of the boundary layers determines the magnitude of the electrode voltage drops, the heat loss to the walls, the frictional stagnation pressure drop, the potential of axial shorting, the potential for flow separation, etc.

A technique for modeling boundary layers is described in detail in Reference 28. The boundary layer equations are self-consistent and use core flow and wall parameters as boundary conditions. Semiempirical factors are typically introduced to account for wall roughness, axial current leakage along slag layers, core flow nonuniformities, and electrode voltage drops in the presence of arc current transport. MHD effects substantially modify the boundary layer profiles, especially the electrode wall enthalpy profile and the insulating wall velocity profile. Because of the effects of MHD interactions and Prandtl number $\neq 1$, a more general expression between the boundary layer enthalpy and velocity profiles than that given by Crocco's relationship is used. Consequently, the thermal boundary layer and the velocity boundary layer are allowed to develop separately and are not forced to be of equal thickness. The enthalpy distribution affects the electrical conductivity distribution, and, hence, the boundary layer impedance. The conductivity distribution in turn responds to the excessive Joule dissipation near the electrode wall. In the sidewall boundary layers, where the favorable axial pressure gradients are not balanced by the Lorentz force near the wall, the velocity distribution may exhibit overshoot, which may result in negative momentum and displacement thicknesses.

Descriptions of various MHD generator flow models can be found in the literature (28–30). A typical procedure for performing actual channel calculations (29) is to start by specifying the composition of the reactants, from which thermochemical, thermodynamic, and electrical properties of the working fluid are generated (31). The principal input data required to proceed with the calculations are the total mass flow rate, the combustor stagnation pressure and enthalpy, and specified design conditions of magnetic field, electrical load parameter, and Mach number along the channel. It is implicitly assumed that the magnetic field can in fact be treated as a prescribed quantity, ie, it is not significantly influenced by the induced currents in the gas. More sophisticated, two- and three-dimensional computer codes have been developed to treat aspects of channel flow (32,33). Codes which can treat unsteady flows have also been developed for the analysis of end effects, transient flows, and flows with shock waves, and to determine conditions under which secondary flows or instabilities may occur.

These multidimensional analyses do not necessarily predict overall generator performance or operating characteristics significantly more accurately than do the quasi-one-dimensional analyses, which are more economical to run. Thus the latter are used for general channel design calculations, and the more sophisticated codes mainly to deal with more detailed aspects of channel operation. For example, current concentrations at electrode edges can be predicted by use of the more sophisticated codes. This allows appropriate electrode design for the condition.

Electrical Conductivity. In order to conduct electricity, the working fluid must contain charged particles, ie, it must be partially ionized. Some of the gas atoms or molecules must be stripped of one or more of their electrons. The energy required to accomplish this, called the ionization potential, is measured in electron volts. In MHD flows of interest, the required energy is supplied by heating the gas. Thus the ionization process is referred to as thermal ionization.

Most common gases, eg, air, combustion products such as CO, CO_2, H_2O, and the noble gases, have high (>10 eV) ionization potentials and ionize thermally only at very high (>4000 K) temperatures. Alkali metals, which have much lower (<5 eV) ionization potentials are ionized at much lower temperatures. Thus small amounts (~1%) of alkali metal salts are added to a gas at typical combustion temperatures (2500–3000 K) and enough ionization is obtained to achieve usable levels of electrical conductivity. This process is called seeding, and is a practical way of obtaining useful conductivities at temperatures that can be withstood in practical devices. The seed material used in MHD generators is most commonly a potassium salt, eg, potassium carbonate or potassium formate, both of which are readily available and economical. Cesium salts, which have an even lower ionization potential, are sometimes used when maximum conductivity is necessary, but are not economical for power plant use. Electrical conductivities in potassium-seeded combustion gases are in the range 1–10 mho/m.

The electrical conductivity σ of a gas is defined as the ratio of the current to the field, ie, from the most general form of Ohm's law. Neglecting ion mobility, this becomes equation 16, which can be written in terms of the current density components:

$$\begin{aligned} \vec{j}_x &= \sigma \vec{E}''_x - \frac{\beta}{B}(j_y B_z - j_z B_y) \\ \vec{j}_y &= \sigma \vec{E}''_y - \frac{\beta}{B}(j_z B_x - j_x B_z) \\ \vec{j}_z &= \sigma \vec{E}''_z - \frac{\beta}{B}(j_x B_y - j_y B_x) \end{aligned} \tag{38}$$

If the magnetic field is in the z direction, $B_z = B$ and $B_x = B_y = 0$, $E_z'' = 0$. Solving for the current density components:

$$j_x = \sigma E_x'' - \frac{\sigma\beta}{1 + \beta^2} E_y''$$
$$j_y = \frac{\sigma}{1 + \beta^2} E_y'' + \frac{\sigma\beta}{1 + \beta^2} E_x'' \qquad (39)$$
$$j_z = 0$$

Equations 39 have the form

$$j_i = \sum_{k=1}^{3} \sigma_{ik} E_k'' \qquad (40)$$

where the conductivity σ_{ik} is a tensor quantity the value of which depends on the magnetic field and on the orientation of the electric field to the magnetic field. If an electric field is aligned with the magnetic field, ie, in the z-direction, then $E'' = E_z''$ and $E_x'' = E_y'' = 0$. Then $j_x = j_y = 0$; $j_z = \sigma E_z''$ and the current density has the same direction as, and is proportional to, the electric field. The conductivity is independent of the magnetic field strength and is called the scalar conductivity. This is the conductivity of the plasma in the absence of a magnetic field (see PLASMA TECHNOLOGY).

When the electric field is not aligned with the magnetic field, the conductivity is not independent of the magnetic field. For example, if the total electric field E'' is perpendicular to the magnetic field, in the y-direction, then $E'' = E_y''$ and $E_x'' = E_z'' = 0$. Then equations 39 become

$$j_x = -\frac{\sigma\beta}{1 + \beta^2} E_y''$$
$$j_y = \frac{\sigma}{1 + \beta^2} E_y'' \qquad (41)$$
$$j_z = 0$$

Now the effective conductivity in the direction of the electric field is $\sigma/(1 + \beta^2)$, ie, the scalar conductivity reduced by a factor of $(1 + \beta^2)$ by the magnetic field. Also, the electric current no longer flows in the direction of the electric field; a component j_x exists which is perpendicular to both the electric and magnetic fields. This is the Hall current. The conductivity in the direction of the Hall current is greater by a factor of β than the conductivity in the direction of the electric field. The calculation of the scalar conductivity starts from its definition:

$$\sigma = \frac{j}{E} \qquad (42)$$

where j is the total conduction current

$$j = \sum_k n_k v_k Z_k e \tag{43}$$

The subscript k identifies the different species present, n_k is the number density of the kth species, v_k its drift velocity, and Z_k denotes the number of charges on a particle of species k,

$$n_k v_k = \int c f_k \, dc \tag{44}$$

where c is the random thermal speed of the particle and f_k is the distribution function of the kth species.

In order to calculate $n_k v_k$, the distribution function f_k must be obtained in terms of local gas properties, electric and magnetic fields, etc, by direct solution of the Boltzmann equation. One such Boltzmann equation exists for each species in the gas, resulting in the need to solve many Boltzmann equations with as many unknowns. This is not possible in practice. Instead, a number of expressions are derived, using different simplifying assumptions and with varying degrees of validity. A more complete discussion can be found in Reference 34.

A relatively simple derivation assumes that the current is carried primarily by electrons. Then equation 43 becomes equation 13, $j_e = n_e v_e e$, where v_e is the electron drift velocity. An electron with charge e in an electric field E experiences a force eE in the direction of the field, and an acceleration eE/m_e, where m_e is the electron mass. The electron also has a random thermal motion which causes it to collide with the other particles in the gas. Assuming that the electron drifts with its field-induced drift velocity, v_e, only between collisions, then

$$v_e = \left(\frac{Ee}{m_e}\right)\tau \tag{45}$$

where τ is the mean free time between collisions. This assumes that on each collision the electron loses all the directed motion acquired from the field. The mean free time between collisions is given by

$$\tau = 1/nQc_e \tag{46}$$

where n is the number density of atoms or molecules in the gas, Q their momentum transfer cross section, and c_e the mean random thermal speed of the electrons. Combining equations 43–46,

$$\sigma = \frac{n_e e^2}{m_e n Q c_e} \tag{47}$$

Only the quantities n_e, the electron number density, and Q, the collision cross section, present any difficulty in their calculation.

To calculate n_e, electron production must be balanced against electron depletion. Free electrons in the gas can become attached to any of a number of species in a combustion gas which have reasonably large electron affinities and which can readily capture electrons to form negative ions. In a combustion gas, such species include OH (1.83 eV), O (1.46 eV), NO_2 (3.68 eV), NO (0.09 eV), and others. Because of its relatively high concentration, its ability to capture electrons, and thus its ability to reduce the electrical conductivity of the gas, the most important negative ion is usually OH^-.

Assuming that the gas is electrically neutral over regions having dimensions larger than the Debye length, typically of the order 10^{-6} m in an MHD generator, the electron and ion densities in the bulk of the gas are equal.

The mass action law or Saha equation for thermal ionization of seed atoms is

$$\frac{n_e n_i}{n} = \frac{(2\pi m_e kT)^{3/2}}{h^3} \frac{2g_i}{g_0} \exp\left(-\frac{e\epsilon_i}{kT}\right) \tag{48}$$

where n_i = ion concentration, n = neutral seed atom concentration, h = Planck's constant, ϵ_i = ionization potential of seed atom, g_i = the statistical weight of the ground state of the ion, and g_0 = the statistical weight of the ground state of the neutral atom (35).

The quantities g_i and g_0 that appear in equation 48 are approximations for the complete partition function. For highest accuracy, above about 9000 K, the partition functions should be used.

A form of equation 48 which can be conveniently expressed in terms of equilibrium constants is

$$\log \frac{n_e n_i}{n_0 - n_i} = \frac{-5,040}{T}\epsilon_i - \frac{3}{2}\log\frac{5,040}{T} + 26.9366 + \log\frac{2g_i}{g_0} \tag{49}$$

where n_0 is the original concentration of seed atoms before ionization.

A derivation of equation 49 is given in Reference 36. In flows of interest in MHD power generation the total pressure is about 101 kPa (1 atm) and the partial pressure of seed is 1 kPa (0.01 atm). Also, it is usually possible to assume that $n_e = n_i$ and that only one species (seed atoms) ionize. In rare situations, more than one species may ionize, if so, equation 49 must be solved simultaneously for each species with the constraint that $n_e = n_{i1} + n_{i2} + \cdots n_{in}$.

Referring back to equation 47, the other quantity necessary in calculating the gas conductivity is the collision cross section, Q. Gases contain at least four types of particles: electrons, ionized seed atoms, neutral seed atoms, and neutral atoms of the carrier gas. Combustion gases, of course, have many more species. Each species has a different momentum transfer cross section for collisions with electrons. To account for this, the product nQ in equation 47 is replaced by the summation $\sum_k n_k Q_k$ where k denotes the different species present. This generalization also allows the conductivity calculation to include the total current in the gas, obtained from the summation of the currents carried by each type of charged particle. A separate term accounting for electron-ion collisions is added, as ions have a very large cross section for collisions with electrons owing to the

Coulombic field which surrounds them. Including the term for electron-ion cross sections, the expression for conductivity becomes

$$\sigma = \frac{n_e e^2}{m_e c_e}\left[\frac{1}{\sum_k n_k Q_k + 3.9 n_i\left(\frac{e^2}{8\pi\epsilon_0 kT}\right)^2 \ln\Lambda}\right] \tag{50}$$

where the summation over k includes all species except for ions, which are accounted for in the second term in the denominator (35).

Other quantities in equation 50 are ϵ_0 = permittivity of free space, and

$$\Lambda = \frac{12\pi}{n_e}\left(\frac{\epsilon_0 kT}{e^2}\right)^{3/2} \tag{51}$$

Equation 50 is adequate in most cases, but becomes inaccurate if the collision cross sections are strongly temperature dependent. In this case, integral expressions for cross section are necessary, and the expression for conductivity becomes more complicated, eg,

$$\sigma = \frac{4\pi}{3}\frac{n_e}{n_k}\frac{e^2}{kT_k}\left(\frac{m_e}{2\pi kT_k}\right)^{\frac{3}{2}}\int \frac{c_e^3}{Q_{ek}} e^{-m_e c_e^2/2kT_k}\, dc_e \tag{52}$$

The collision cross sections are functions of the electron energy and of the relationship which describes the forces between the heavy particles and the electrons (34).

A widely used method for calculating conductivity (37), uses the relationship between conductivity and electron mobility given by equation 14, $\sigma = n_e \mu_e e$. The electron mobility (38) is given by

$$\mu_e = -\frac{4\pi e}{3m_e}\int_0^\infty \frac{u^3}{v}\cdot\frac{d}{dv}f(v)\,dv \tag{53}$$

where $f(v)$ is the electron velocity distribution function,

$$f(v) = \left(\frac{m_e}{2\pi kT}\right)^{\frac{3}{2}}\exp\left(\frac{-m_e v^2}{2kT}\right) \tag{54}$$

The collision frequency, ν_k, for momentum transfer with a particular molecular or atomic species, k, is given by

$$\nu_k = v_e n_k Q_k \tag{55}$$

and the total collision frequency with all neutral species is $\nu_0 = \sum \nu_k$. If electron-ion collisions are also accounted for, with ν_{ei} being the electron-ion collision frequency, the total collision frequency ν is

$$\nu = \sum \nu_k + \nu_{ei} \tag{56}$$

The electron-ion collision frequency can be approximated by the Spitzer-Harm (39) formulation for a completely ionized gas:

$$\nu_{ei} = \frac{3.64 \times 10^6}{T^{1.5}} n_e \ln[(1.27 \times 10^7)T^{1.5}/n_e^{0.5}] \tag{57}$$

Electron-ion collisions become important when the degree of ionization exceeds about 0.1% and introduce additional complexity in calculating their effects. Methods for calculating conductivity in strongly ionized gases are given in Reference 40.

To evaluate the mobility, the reciprocal of the collision frequency is expressed as a power series in electron energy, allowing the integration in equation 53 to be performed analytically.

Conductivity is calculated according to equation 14 by combining the mobility results with the electron density obtained as described earlier. Results for the combustion products typical of a coal-fired MHD generator are shown in Figure 7. A very strong dependence on temperature can be seen. At a seed concentration of 1.2% by weight of potassium, the conductivity is about 8 mho/m at typical generator temperatures of 2500 K, whereas at 2300 K it has dropped to about 2 mho/m, a value too low to be useful for power generation. This reflects the temperature dependence of electron number density. The conductivity is not so sensitive to changes in seed concentration, varying only about 10% over a factor of three range in potassium concentration. This is because of the increase in the cross section and number density of KOH molecules as the potassium concentration is increased.

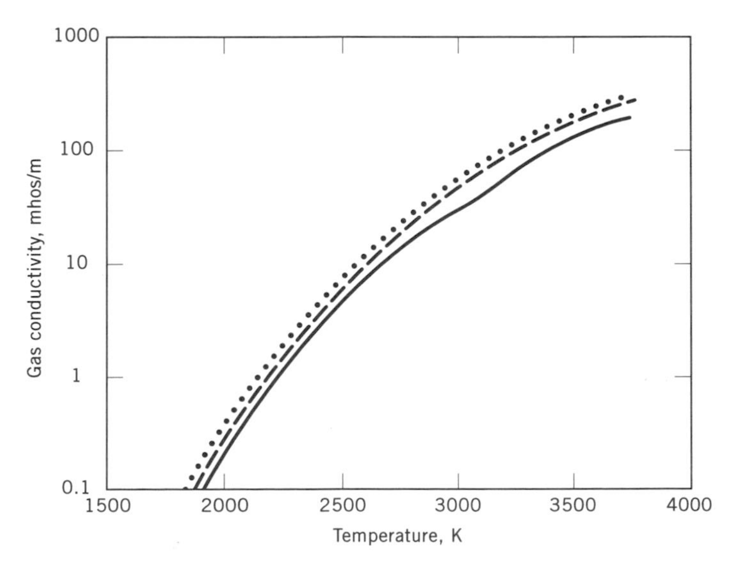

Fig. 7. Electrical conductivity of combustion products where the fuel is Montana subbituminous coal, the oxidizer, air + O_2, having a fuel:oxidizer stoichiometric ratio of N:O 0.65; and the pressure, 101 kPa (1 atm). (—), (- - -), and (· · ·) represent 0.5, 1.2, and 1.6 wt % potassium, respectively.

Efficiency and Economic Factors

Because the MHD generator has no moving mechanical parts, it can operate at a much higher combustion temperature than other power generating systems, allowing the combined MHD–steam cycle to achieve higher thermal efficiency than other systems. The high efficiencies together with competitive capital costs yield very attractive cost of electricity (COE) estimates for MHD. A comparison of about 20 advanced technology processes with a conventional steam plant (41,42) concluded that the coal-fired open-cycle MHD system has potentially one of the highest coal pile-to-busbar efficiencies as well as one of the lowest COEs among the systems studied. Figure 8 summarizes COE comparisons for a number of advanced power cycles (43). The cost of electricity was found to be lower for MHD than for any other advanced coal-fueled power system studied. The COE is about 25% lower for MHD than for a modern steam plant at a coal cost of $20/t ($0.97/MBtu) in 1978 dollars.

Higher thermal efficiency and the corresponding smaller influence of fuel cost leads to the COE advantage of MHD increasing as the cost of fuel increases. Studies summarized in Table 1 (44–46), predict that first generation MHD–steam power plants should have thermal efficiencies in the 42% range and as the technology matures, plant efficiency should increase to 55–60%. This may be compared with 33–38% for modern coal-fired steam plants hav-

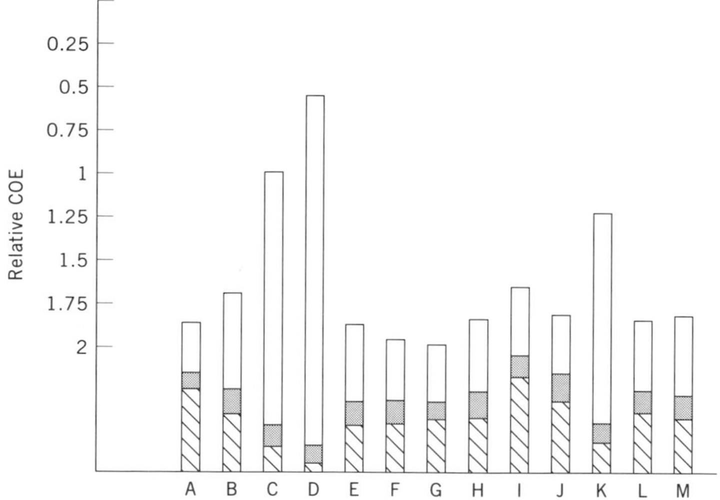

Fig. 8. Cost of electricity (COE) comparison where ▧ represents capital charges, ▩ operation and maintenance charges, and □ fuel charges for the reference cycles, A, steam, light water reactor (LWR), uranium; B, steam, conventional furnace, scrubber coal; C, gas turbine combined cycle, semiclean liquid; D, gas turbine, semiclean liquid, and advanced cycles; E, steam atmospheric fluidized bed, coal; F, gas turbine (water-cooled) combined low heating value (LHV) gas; G, open cycle MHD coal; H, steam, pressurized fluidized bed, coal; I, closed cycle helium gas turbine, atmospheric fluidized bed (AFB), coal; J, metal vapor topping cycle, pressurized fluidized bed (PFB), coal; K, gas turbine (water-cooled) combined, semiclean liquid; L, gas turbine (air-cooled) combined, LHV gas; M, steam (650°C reheat), AFB coal.

Table 1. Comparison of MHD and Conventional Steam Plants[a]

Type of cycle	Conventional pulverized coal steam	MHD steam	
		First commercial	Advanced direct-fired
net plant output, MW_e	954	212–492	953
net plant efficiency,[b] %	37.4[c]	40.2–41.7	58.1
combustor oxidant	air	air + oxygen[d]	air
air heater type	Lungstrom-type regenerative	recuperative metal	direct-fired regenerative refractory
combustion air preheat, K	589	922	1978
combustion pressure, kPa[e]	101	555	1460
MHD stress level	none	POC[f]	advanced
steam cycle:	current	current	current
throttle pressure, MPa[e]	24.0	16.5	24.0
outlet temperatures, K			
superheater	811	811	811
first reheat	839	811	839
second reheat	none	none	none

[a]Primary fuel for each plant is pulverized Illinois No. 6 coal.
[b]On a higher heating value basis.
[c]Obtainable only in modern plants. Average for U.S. plants is 32.8%.
[d]Total oxygen concentration is 32 mol %.
[e]To convert kPa to psi, multiply by 0.145.
[f]Similar to levels planned for proof of concept (POC) tests.

ing scrubbers. The first commercial MHD plants shown in Table 1 are to use oxygen-enriched air at a moderate (922 K) preheat temperature as the oxidant, and are 40–42% efficient (44). Conceptual design studies using economic analysis have shown that such first generation MHD power plants are economically attractive (44,45). The advanced MHD plant of the future (46) should achieve an efficiency of 58% mainly because it uses high temperature (1978 K) air heaters fired directly by the channel exhaust gas instead of oxygen enrichment, and has a more advanced topping cycle, a higher magnetic field, and a more highly stressed channel. This last plant operates at higher pressure and has improvements in other parts of the cycle.

Environmental Factors

Environmental intrusion from MHD plants is projected to be not only well below the mid-1990s acceptable limits, but also low enough to satisfy the more stringent requirements expected in the future. Emissions of SO_x, NO_x, and particulates can be reduced to levels well below the 1979 NSPS without requiring expensive exhaust gas cleanup systems. Pollutant control is inherent in the basic design of MHD power plants. Furthermore, because of the higher efficiency and consequent lower fuel usage of MHD plants, emissions of CO_2 are lower, heat rejection is reduced, and less solid waste is produced than from other less efficient plants. Table 2 (46) compares a mature MHD–steam power plant to a conventional plant. Emissions are lower from the MHD plants because of MHD's lower fuel usage.

Table 2. Environmental Intrusion Comparison[a]

Parameter	Conventional steam power plant	MHD–steam power plant	
		Early	Advanced
carbon dioxide, kg	844	703	522
solid wastes, kg	116	98	748
cooling tower heat rejection, GJ[b]	4.511	3.43	1.67
cooling water consumption, kg	2176	1436	905

[a]All data are per megawatt-hour.
[b]To convert GJ to millions of Btu, multiply by 0.9485.

At the high temperatures found in MHD combustors, nitrogen oxides, NO_x, are formed primarily by gas-phase reactions, rather than from fuel-bound nitrogen. The principal constituent is nitric oxide [*10102-43-9*], NO, and the amount formed is generally limited by kinetics. Equilibrium values are reached only at very high temperatures. NO decomposes as the gas cools, at a rate which decreases with temperature. If the combustion gas cools too rapidly after the MHD channel the NO has insufficient time to decompose and excessive amounts can be released to the atmosphere. Below about 1800 K there is essentially no thermal decomposition of NO.

Reactions of primary interest during the cooling process are as follows:

$$NO + N \rightleftharpoons O + N_2 \tag{58}$$

$$NO + O \rightleftharpoons N + O_2 \tag{59}$$

$$NO + H \rightleftharpoons N + OH \tag{60}$$

where equations 58 and 59 are the well-known Zeldovich reactions (47) and equation 60 is of particular interest in fuel-rich mixtures typical of MHD flows. Although these three are the dominant reactions, there are a large number of other reactions and species which can influence NO decomposition (48). Detailed studies of NO decomposition, both analytical and experimental, are described in the literature (49–52).

NO_x control is achieved by means of a two-stage combustion process. Primary combustion occurs in the coal combustor under fuel-rich conditions. Secondary combustion takes place in the heat recovery boiler after cooling the fuel-rich MHD generator exhaust gases in a radiant furnace which provides a residence time of at least two seconds at a temperature above 1800 K. Conditions in the radiant furnace allow the NO to decompose into N_2 and O_2. The cooling rate of the exhaust gas is the key element in the decomposition. The required residence time is provided by appropriate design of the radiant furnace. Using proper choices of primary stoichiometry, gas cooling rates, and secondary combustion temperatures, NO_x emissions are kept below the proposed European standard of 0.04 kg/GJ (0.1 lb/(Btu × 10^6)), as shown in Figure 9 (53).

Control of SO_x is intrinsic to the MHD process because of the strong chemical affinity of the potassium seed in the flow for the sulfur in the gas. Although the system is operated fuel-rich from the primary combustor to the

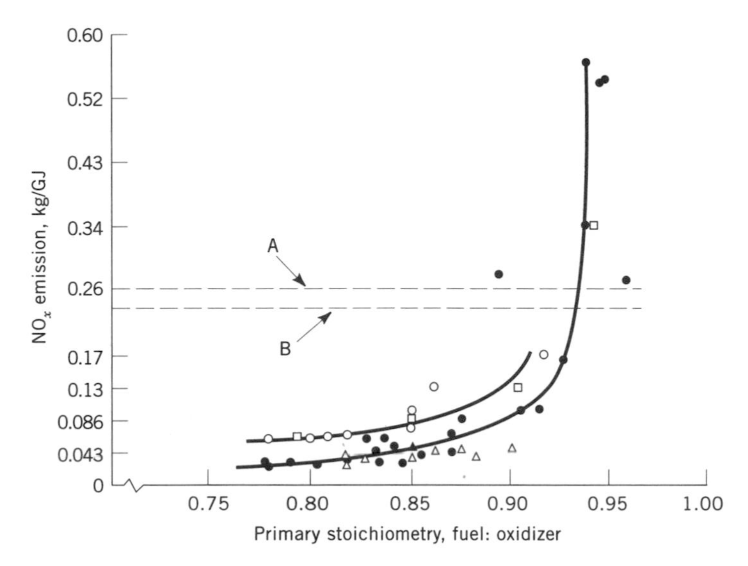

Fig. 9. Measured coal-fired flow facility (CFFF) NO_x emissions where (•) represents high sulfur coal, (∘) low sulfur coal, (△) low sulfur coal having $K_2/S = 1.15$, and (□) LMF5-G. A, Illinois No. 6 coal (3% S); B, Montana Rosebud coal (1% S), and the NSPS range is between the dotted lines. To convert kg/GJ to lb/(MBtu), multiply by 2.326.

secondary combustor, the predominant sulfur compound in the gas is sulfur dioxide (54,55). Hydrogen sulfide begins to form at gas temperatures below about 2000 K and about 10 mol % of the sulfur is present as H_2S at 1800 K. At lower temperatures SO_2 converts rapidly to H_2S. The primary factor affecting SO_2 removal is the potassium to sulfur molar ratio. At K_2/S ratios >1.4, SO_2 emissions are reduced to <0.04 kg/GJ (0.1 lb/MBtu) (see Fig. 10) (53).

The potassium combines with the sulfur to form potassium sulfate, which condenses as a solid primarily in the electrostatic precipitator (ESP) or baghouse.

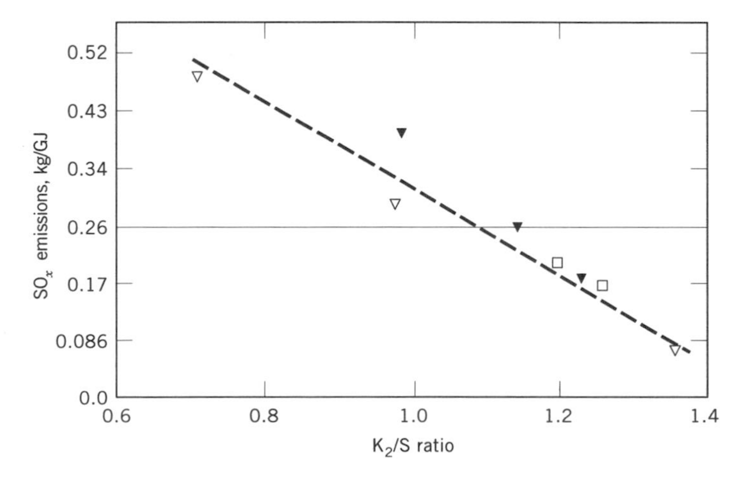

Fig. 10. Effect of potassium on SO_2 removal where (—) represents the NSPS limit, and (▽) represents LMF4-T, (▼) LMF4-U, and (□) LMF4-V.

The recovered potassium sulfate is then delivered to a seed regeneration unit where the ash and sulfur are removed, and the potassium, in a sulfur-free form such as formate or carbonate, is recycled to the MHD combustor. It is necessary also to remove anions such as Cl^- and F^- which reduce the electrical conductivity of the generator gas flow. These are present in the coal ash in very small and therefore relatively harmless concentrations. As the seed is recycled, however, the concentrations, particularly of Cl^-, tend to build up and to become a serious contaminant unless removed.

Several methods for reprocessing the potassium seed have been considered, including the double alkali process, the Engle-Precht process, the aqueous carbonate process, the Pittsburgh Energy Technology Center (PETC, U.S. Dept. of Energy) process, and the formate process, among others. A review (56) and more detailed descriptions of some of the processes are available (57–60).

Considerable attention has been given to the formate process where the primary reaction is

$$K_2SO_4 + Ca(OH)_2 + 2\ CO \xrightarrow[1\ \text{atm}]{70^\circ\text{C}} 2\ KHCOO + CaSO_4\ (s) \qquad (61)$$

A diagram for one implementation of this process (61,62) is shown in Figure 11. Recovered potassium sulfate is converted to potassium formate [*590-29-4*] by reaction with calcium formate [*544-17-2*], which is made by reacting hydrated lime, $Ca(OH)_2$, and carbon monoxide. The potassium formate (mp 167°C), in liquid form, is recycled to the combustor at about 170°C. Sulfur is removed as solid calcium sulfate by filtration and then disposed of (see SULFUR REMOVAL AND RECOVERY).

The cost penalty of this process on a commercial MHD plant operating on a high sulfur eastern coal is estimated to be in the range \$0.0154–0.0181/kWh, depending on seed loading. For a plant operating on a low sulfur western coal the cost penalty is estimated to be in the range \$0.0068–0.0112/kWh (62). A resin-based anion-exchange seed regeneration process has been suggested (60), which promises considerable process simplicity, less Cl^- contamination, and lower costs.

Particulates in the MHD exhaust gas stream are primarily (80–90%) K_2SO_4, the remainder being coal ash constituents. Because of the very high temperatures in the MHD combustor, most of the particles have undergone vaporization and condensation steps. Most of the slag is rejected into a slag tap upon entering the radiant furnace. The remaining slag forms particles, primarily, it is believed, by homogeneous nucleation. The potassium compounds form particles by both homogeneous and heterogeneous nucleation, with the condensed ash particles serving as nucleation sites for the heterogeneous nucleation. Because of the high combustion temperatures and the presence of a lower boiling species (potassium), MHD systems produce very small particles. The average mass mean diameter varies between 0.2 μm at K_2/S ratios near 1 and 0.7 μm at K_2/S ratios near 4. This is almost two orders of magnitude smaller than those found in typical utility operation. The particle size distribution appears to be unaffected by the type of coal used. Particulate mass loadings are in the range 11–18 g/m^3 (5–8 grains per dry standard cubic foot). The most commonly used particulate collection device on coal-fired power plants is the dry electrostatic precipitator

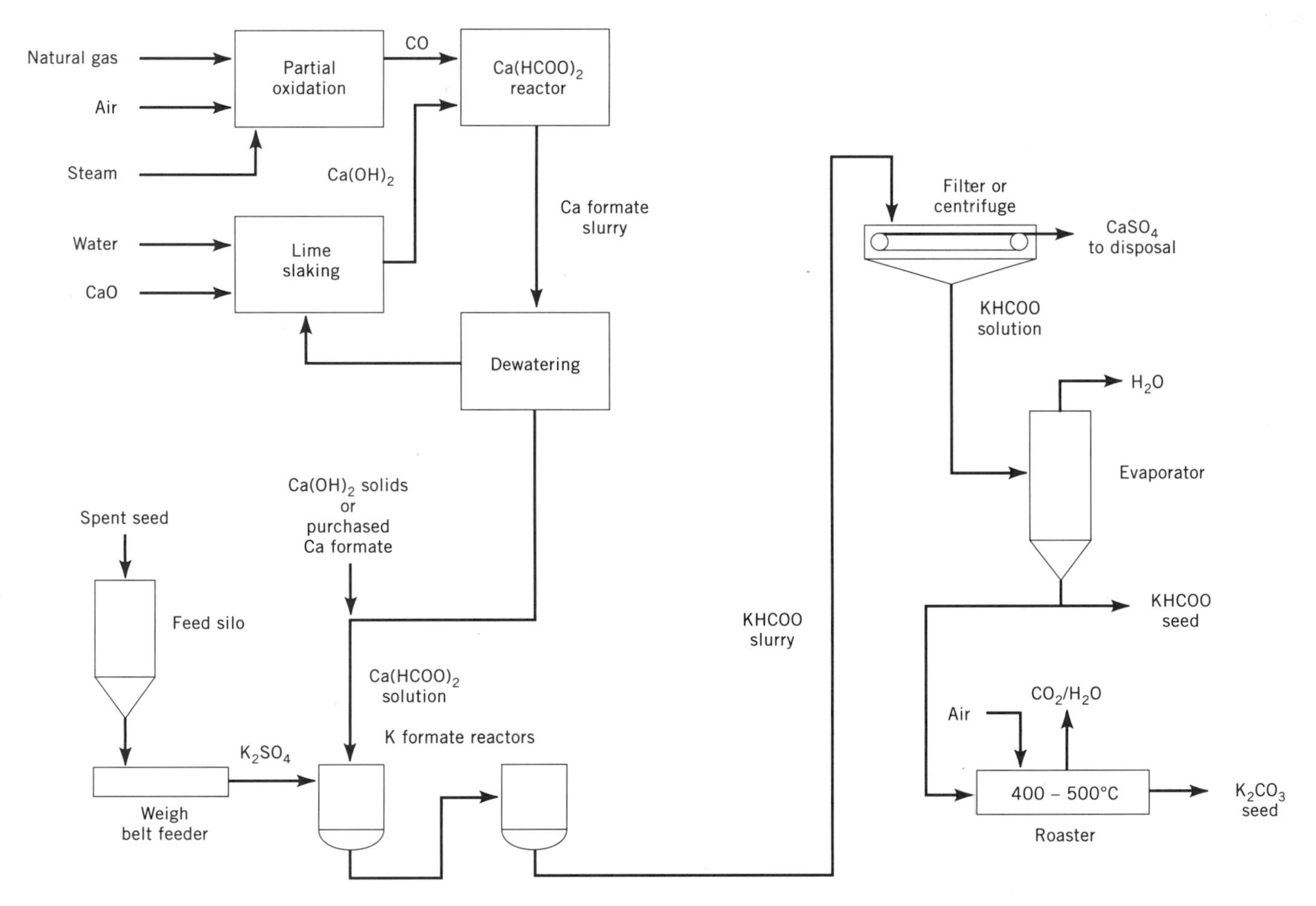

Fig. 11. TRW, Inc. econoseed process where the reactor in the first step is at a pressure from 8.3–8.9 MPa (1200–1290 psi) and 150–200°C.

(ESP). Because of the presence of potassium salts in the particulate, the resistivity is typically 10^9 Ω·cm, regardless of coal type. Thus ESP problems associated with high resistivity do not occur in MHD systems. Requirements set by NSPS are met or exceeded (53), although the specific collection area required for MHD may be somewhat higher than for conventional systems. Particulate collection performance using a baghouse has been measured at 0.0013–0.026 kg/GJ (0.003–0.006 lb/MBtu (53), well below NSPS limits (0.13 kg/GJ). Because of the submicrometer particulates, Gore-tex bag material may be necessary to eliminate the effects of fabric blinding.

The high temperatures in the MHD combustion system mean that no complex organic compounds should be present in the combustion products. Gas chromatograph/mass spectrometer analysis of radiant furnace slag and ESP/baghouse composite, down to the part per billion level, confirms this belief (53). With respect to inorganic priority pollutants, except for mercury, concentrations in MHD-derived fly-ash are expected to be lower than from conventional coal-fired plants. More complete discussion of this topic can be found in References 53 and 63.

Power Plants

A complete MHD power plant must integrate the MHD generator and the bottoming plant to maximize plant efficiency and minimize cost of electricity. Net plant efficiency is maximized by simultaneous optimization of net MHD power generation, ie, MHD generator power minus the cycle compressor power and oxygen plant compressor power, and of waste heat utilization in the steam bottoming plant. Compromises between performance and cost, particularly of the magnet but also of the oxygen plant and other ancillary equipment, must be made.

Power system designs and analyses have been performed on both standalone MHD-steam plants (44,46,64), and on retrofit plants (65). These latter are MHD topping cycles retrofitted to existing steam plants. In a first generation commercial MHD power plant producing 500 MW of electricity, and operating on Montana subbituminous coal, approximately half (261 MW) of the total plant output would be produced by the MHD channel (44). Overall design parameters are as follows: for MHD combustion, oxidizer O_2 content, 34 vol %; fuel moisture as fired, 5%; ash slag removal, 80%; oxidizer/fuel equivalence ratio, 0.90; and combustor coolant, high pressure boiling feed water. The MHD generator is a diagonal channel type having subsonic channel gas velocity and a peak magnetic field of ca −6 T (6×10^4 G); the gas seed concentration is 1.0% potassium and the seed regeneration process a formate one; the diffuser recovery factor, 0.6; the diffuser exit pressure 101 kPa (1.0 atm); and the channel coolant, low pressure boiler feed water. For the bottoming plant: the main steam is at 16.5 MPa (163 atm) at 807 K; the reheat steam at 807 K. The final MHD combustor gas O_2/fuel equivalence ratio is 1.05. The operating pressure ratio of the plant is 7.5. The net plant efficiency is 42.9% from the coal pile to busbar. The plant utilizes oxygen-enriched (34 vol % oxygen) combustion air preheated to 922 K in a metallic, recuperative-type, tubular heat exchanger which is part of the bottoming plant heat recovery system. The oxygen is produced at a purity of 80%, in an oxygen plant which is integrated with the power plant.

The Rankine cycle efficiency of the bottoming plant is 41.6%. To utilize the waste heat from the topping cycle, low pressure and low temperature feedwater is used for channel cooling, and high pressure and high temperature feedwater for cooling of the MHD combustor. The total feedwater temperature rise in the feedwater heater train is 178°C for the plant, which employs a total of six heaters. Cooling of the diffuser is incorporated as part of the evaporative circuit of the steam cycle. Heat recovered from the hot MHD generator exhaust gas is used for steam generation, oxidizer preheating to 922 K, feedwater heating in a split high pressure (HP) and low pressure (LP) economizer, coal drying, and preheating of secondary combustion air to 589 K. The secondary combustion air is introduced into the bottoming plant steam generator for afterburning, in order to achieve final oxidation of the fuel-rich MHD combustion gases. Flue gas at stack gas temperature is also utilized for spray drying in the seed regeneration system for effective utilization of waste heat.

The oxygen plant delivers 3996 t/d of contained oxygen at full load. The specific compressor power required for manufacturing of oxygen is 190 kWh/t of contained oxygen (or 203.5 kWh/t of equivalent pure oxygen), corresponding to 31.6 MW_e at nominal plant load conditions. The required compressor power for oxygen manufacturing is provided by steam turbines which are part of the bottoming plant steam cycle. High pressure steam is used for the turbine which drives the cycle compressor and oxygen plant compressor.

The resulting overall energy balance for the plant at nominal load conditions is shown in Table 3. The primary combustor operates at 760 kPa (7.5 atm) pressure; the equivalence ratio is 0.9; the heat loss is about 3.5%. The channel operates in the subsonic mode, in a peak magnetic field of 6 T. All critical

Table 3. Overall Energy Balance at 500 MW_e Nominal Load

Parameter	Value, MW
Fuel input	
MHD combustor	1162.0
gasifier for seed regeneration	16.0
Total	*1178.0*
Gross power outputs	
MHD power	261.0
steam power[a]	359.6
Total	*620.6*
Auxiliary and losses	
cycle compressor	57.7
O_2 plant compressor	31.6
auxiliaries	19.5
inverter and transformer	7.7
Total	*116.5*
net plant output	504.1
net plant efficiency, %	42.9[b]

[a]Includes power from recovery of available heat in seed system.
[b]This number is a percentage.

electrical and gas dynamic operating parameters of the channel are within prescribed constraints; the magnetic field and electrical loading are tailored to limit the maximum axial electrical field to 2 kV/m, the transverse current density to 0.9 A/cm^2, and the Hall parameter to 4. The diffuser pressure recovery factor is 0.6.

The channel length is 18 m, having a cross section of about 0.7×0.7 m^2 at the inlet and 1.8×1.8 m^2 at the exit. Channel performance (net MHD power, enthalpy extraction) could be improved by increasing the oxygen enrichment and the channel length. However, the overall system would then be penalized by the increased oxygen plant costs, from the increased oxygen enrichment; by higher magnet costs, from the increased channel length; and by the adverse impact on waste heat recovery and steam plant efficiency from increased channel heat loss. The channel cooling is limited to low temperature and low pressure feedwater in accordance with state-of-the-art channel technology.

An alternative configuration which could be considered is a supersonic channel in a lower (4–4.5 T) magnetic field. This configuration would suffer a relatively small penalty in MHD generator performance and net power output, but the magnet would be considerably smaller, having stored magnetic energy of half or less of that at subsonic operation. Thus, there is a significant reduction in magnet cost and risk to be weighed against the relatively small reduction in plant performance. The oxygen plant required for supersonic operation is somewhat larger than that required for subsonic operation, which increases its cost.

The plant is designed to satisfy NSPS requirements. NO_x emission control is obtained by fuel-rich combustion in the MHD burner and final oxidation of the gas by secondary combustion in the bottoming heat recovery plant. Sulfur removal from MHD combustion gases is combined with seed recovery and necessary processing of recovered seed before recycling.

The steam generator is a balanced draft, controlled circulation, multichamber unit which incorporates NO_x control and final burnout of the fuel-rich MHD combustion gases. The MHD generator exhaust is cooled in a primary radiant chamber from about 2310 to 1860 K in two seconds, and secondary air for afterburning and final oxidation of the gas is introduced in the secondary chamber where seed also condenses. Subsequent to afterburning and after the gas has been cooled down sufficiently to solidify condensed seed in the gas, the gas passes through the remaining convective sections of the heat recovery system.

The oxidant preheater, positioned in the convective section and designed to preheat the oxygen-enriched air for the MHD combustor to 922 K, is located after the finishing superheat and reheat sections. Seed is removed from the stack gas by electrostatic precipitation before the gas is emitted to the atmosphere. The recovered seed is recycled by use of the formate process. Alkali carbonates are separated from potassium sulfate before conversion of potassium sulfate to potassium formate. Sodium carbonate and potassium carbonate are further separated to avoid buildup of sodium in the system by recycling of seed. The slag and fly-ash removed from the HRSR system is assumed to contain 15–17% of potassium as K_2O, dissolved in ash and not recoverable.

The basic seed processing plant design is based on 70% removal of the sulfur contained in the coal used (Montana Rosebud), which satisfies NSPS requirements. Virtually complete sulfur removal appears to be feasible and can be considered as a design alternative to minimize potential corrosion problems related to sulfur in the gas. The estimated reduction in plant performance for complete removal is on the order of 1/4 percentage point. The size of the seed processing plant would have to be increased by roughly 40% but the corresponding additional cost appears tolerable. The construction time for the 500 MW_e plant is estimated to be ca five years.

In a more advanced MHD-steam plant (46), the main difference is the use of 1978 K air preheat, achieved by means of a regenerative air preheater fired directly by the exhaust gases from the MHD diffuser. No oxygen enrichment is used. Other differences are the use of a supercritical steam cycle, a higher peak magnetic field (10 T) and a channel operating at higher electrical fields and currents. For Illinois No. 6 coal as fuel, the net plant output is projected to be 953 MW for a net efficiency of 58.1%. Operating parameters are shown in Table 4.

Plant Economics. A power plant is evaluated economically in terms of capital costs ($/kW) and levelized costs of electricity (mills/kWh = $\$10^{-3}$/kWh). Important factors are the escalation of costs with time and the cost over time of the capital to build the plant. Table 5 gives the distribution of plant capital costs in percent of plant direct costs, and shows the relative cost of components and

Table 4. Advanced MHD Topping Cycle Parameters[a]

Parameter	Value
coal thermal (HHV) input,[b] MW_t	1641
combustion air preheat, K	1978
inlet total pressure, MPa[c]	1.47
combustor stoichiometry	0.90
seed, % potassium	1.0
combustor exit flow, kg/s	545.9
exit total temperature, K	2941
inlet	
Mach number	0.95
static temperature, K	2768
static pressure, MPa[c]	0.885
conductivity, $(\Omega\cdot\mathrm{m})^{-1}$	7.63
channel[d] length, m	26.7
inlet area, m^2	0.533
exit area, m^2	7.457
maximum magnetic field, T	10
maximum E_x, V/m	4000
maximum J_y, A/m^2	13,375
maximum Hall parameter	7.3
diffuser pressure recovery	0.85
d-c power output, MW_e	848
enthalpy extraction ratio, %	35.9

[a]Segmented Faraday (Fig. 5) MHD generator configuration. The fuel is Illinois No.6 coal.
[b]HHV = higher heating value.
[c]To convert MPa to psi, multiply by 145.
[d]The loading parameter varies along the channel.

Table 5. Cost Distribution for MHD–Steam Base Load Power Plants[a,b]

Parameter	Cost,[c] %	
	500 MW Plant	200 MW Plant
land, %	0.25	0.37
structures and improvements	9.54	11.77
boiler plant		
coal and ash handling	5.36	5.05
steam generator with oxidant heater	21.61	20.41
effluent control and other	4.50	4.42
steam turbine generator	9.24	10.35
accessory electric equipment	6.17	6.00
miscellaneous power plant equipment	0.47	0.63
MHD topping cycle		
combustion equipment	6.0	6.14
MHD generator	1.59	1.23
magnet subsystem	9.84	10.89
inverters	8.09	6.94
oxidizer supply subsystem	1.98	1.74
oxygen plant	9.20	7.83
seed subsystem	4.87	4.88
transmission plant	1.29	1.34
subtotal direct costs	100.00	100.00
engineering services and other costs	10.00	10.00
overnight construction costs	110.00	110.00
interest and escalation	9.00	8.00
total construction cost with IDC and EDC[d]	119.00	118.00
specific plant cost[e] with IDC and EDC,[d] $/kW	838	1090

[a]Ref. 44.

[b]Fuel costs are taken to be $1.00/GJ ($1.05/MBtu); the escalation and interest rates are 6.5% and 10%, respectively; and the factor used for calculating levelized fuel and operating and maintenance cost is 2.004.

[c]Construction time period is assumed to be 4.83 years and 4.33 years for the 500 and 200 MW plants, respectively.

[d]IDC = interest during construction; EDC = escalation during construction.

[e]In mid-1978 dollars.

subsystems specific to the MHD topping cycle for two first generation commercial MHD power plants of 200 and 500 MW output. The most costly item is the steam generator with oxidant preheater. The second most costly item is the superconducting magnet, which costs slightly more than the oxygen plant. A more recent estimate of capital costs gives a range of $1300–1320/kW for a 600 MW plant (66). Figure 12 shows the variation in plant capital costs with plant size (44). Costs are normalized with respect to the modified reference steam plant of 800 MW nominal capacity used in the Energy Conversion Alternatives Study (ECAS) (41,42) and are based on the first generation MHD plant costs (44). Comparative estimated costs for conventional steam power plants over the same size range are also shown. The estimated capital costs (Fig. 12**a**) of first generation MHD power plants are somewhat higher than the capital costs of conventional coal-fired steam plants of comparable output, for the range of plant sizes studied. Calculated levelized costs of electricity shown in Figure 12**b** are

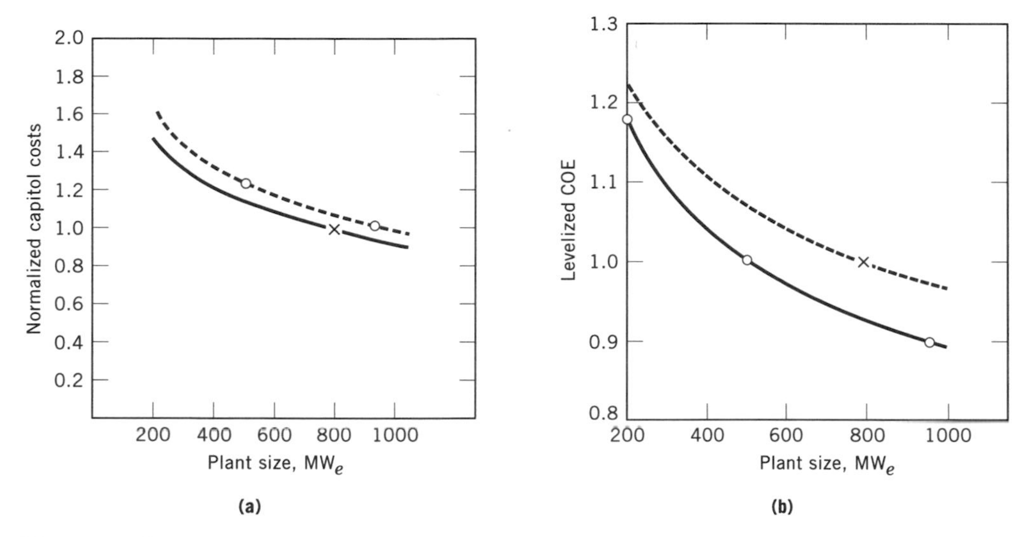

Fig. 12. Costs as a function of plant size for (—) coal-fired steam plants and (---) early MHD power plants where × corresponds to the ECAS reference plant costs. (**a**) Normalized capital (IDC and EDC are not included) and (**b**) levelized cost of electricity (normalized) where IDC and EDC are included and the LEV factor is 2.004 (44) (see Table 5).

lower for the MHD power plants than for comparably sized conventional steam power plants, by about 7.4% for the 950 MW plant, 7.1% for the 500 MW plant and 3.3% for the 200 MW plant. Not surprisingly, the savings become less as the plant becomes smaller. The costs in Figure 12**b** are based on the capital cost curve (Fig. 12**a**) and fuel costs based on the modified ECAS reference steam plant efficiency of 34.3%. The cost comparisons are based on a coal price of $1.00/GJ ($1.05/MBtu). Higher fuel costs would increase the attractiveness of MHD because of its more efficient use of the fuel.

Components and Subsystems

High Temperature Air Preheaters. Combustion air–oxidant preheating for open cycle generators is accomplished in one of two ways. One way is to use the heat energy of the MHD generator exhaust gas directly; in this case, the preheater is classified as directly fired and is located in the MHD generator exhaust as part of the bottoming plant (Fig. 3). The other way of preheating combustion air is to use a separate heat source using clean fuel. This type of preheater is classified as separately fired. Directly fired preheat offers the potential of higher cycle efficiencies than can be achieved with separately fired preheat, at the same oxidizer temperature. However, because of the severe difficulties associated with designing a directly fired preheater capable of operating with the seed and ash-laden gases flowing from the generator, first-generation commercial plants are to use separately fired preheat.

Air preheat temperature requirements of 2250–2300 K are anticipated for natural gas-fired systems, and about 2000 K for oil or coal-fired systems (11). Use of 32–40% oxygen enrichment lowers the preheat temperature requirement to a moderate 900–1000 K, which can be attained with conventional metal-type tubular heat exchangers. Depending on the cost of oxygen, this is a viable alternative to the use of separately fired high temperature preheaters.

More advanced MHD power plants of the future are expected to use preheat temperatures of up to 2000 K. These temperatures, to be achieved by direct firing, require the use of high temperature regenerative heat exchangers, where heat is transferred for a time from a hot fluid, eg, the MHD generator exhaust gas, to a medium which subsequently transfers the heat to a cool fluid, eg, the incoming combustion air. While heat is transferred to the incoming air, the MHD exhaust is directed to a second regenerative preheater operating identically. The two preheaters operate cyclically to provide continuous heating of the combustion air. The other type of heat exchanger is the recuperator, in which heat is transferred continuously from one fluid to another through a solid wall which separates the two fluids. Metallic recuperators are used widely in industry, but are limited for MHD use to about 1250 K, because of corrosive problems caused by seed and ash, as well as mechanical strength problems caused by the pressure requirements. Whereas ceramic recuperators can operate at higher temperatures, development of ceramic recuperators for MHD has not been pursued because of severe problems related to fabrication, fluid leakage, and mechanical strength.

Regenerative heat exchangers of both the fixed-bed and moving-bed types (67) have been considered for MHD use. The more recent efforts have focused on the fixed-bed type (68), which operates intermittently through recycling. A complete preheater subsystem for a plant requires several regenerators with switch-over valves to deliver a continuous supply of preheated air. The outlet temperature of the air then varies between a maximum and a minimum value during the preheat cycle.

Fixed-bed regenerators have been used in the glass (qv) and steel (qv) industries to preheat air to 1350–1650 K, but these operate with relatively clean gases compared to the MHD combustion gases. One design, fabrication and testing of a regenerator for MHD use, involved a regenerator matrix, or bed packing, of 8.5 m in height and 7.9 m in dia (68). The matrix geometry consisted of cored bricks, or checkers, made of fusion cast magnesia–spinel. Cycle times were 1280 seconds on MHD gas flow, 760 seconds on oxidant flow, with 360 seconds for switching. Operating times up to 1470 hours were achieved. System heat, leakage (oxidant mass loss to the MHD heating gas and MHD gas loss to the oxidant), and pressure losses were all within acceptable limits. High temperature refractory-lined water-cooled gate valves were also developed and tested for up to 1390 hours and 2100 cycles at temperatures of 1860–1925 K and inlet pressures to 800 kPa (8 atm).

For gas-fired systems the state-of-the-art is represented by the preheater described in Reference 69. A pebble bed instead of a cored brick matrix is used. The pebbles are made of alumina spheres, 20 mm in diameter. Heat-transfer coefficients 3–4 times greater than for checkerwork matrices are achieved. A prototype device 400 m^3 in volume has been operated for three years at an industrial blast furnace, achieving preheat temperatures of 1670 to 1770 K.

Combustor. In the majority of MHD plant designs the MHD combustor burns coal directly. Because MHD power generation is able to utilize pulverized coal in an environmentally acceptable fashion, there is usually no need to make cleaner fuels from coal, eg, by gasification or by beneficiation. A discussion of combustion techniques for MHD plants is available (70).

The function of the MHD combustor is to process fuel, ie, coal; oxidizer, ie, preheated air, possibly enriched with oxygen; and seed to generate the high temperature electrically conducting working fluid required for the MHD channel. There are several design requirements: (*1*) highly efficient combustion, ie, high carbon conversion and low heat losses, in order to achieve the temperature (2800–3000 K) required for adequate electrical conductivity; (*2*) innovative wall designs capable of extended life, to contain 500–1000 kPa (5–10 atm) of pressure in the presence of molten slag, seed, and heat fluxes up to 50 W/cm^2; (*3*) spatially and temporally homogeneous output flow, requiring sophisticated aerothermodynamic design; (*4*) low pressure drop through the combustor, because this directly affects the net power output of the MHD topping cycle; (*5*) effective seed utilization, which means minimizing slag–seed interactions which remove seed from the gas, and attaining uniform seed dispersion; (*6*) electrical isolation of the combustor and its ancillary systems at voltages of 20–40 kV below ground potential, because of the electrical contact of the combustor with the MHD channel (this is particularly challenging for the slag-rejection system); and (*7*) efficient slag rejection, up to ~50–70% of the ash content of the coal burned, as low slag rejection (high ash carry-over) increases seed recovery costs. These design requirements differ sufficiently from those of conventional coal combustors so as to require essentially new technology for the development of MHD coal combustors.

A process receiving considerable attention as a way of burning coal and rejecting ash as slag is that used in the cyclone furnace (71). Vortex flow in the chamber promotes efficient combustion, by maintaining continuous ignition. Strong radial accelerations promote separation of particles and slag droplets and tend to bring the unreacted oxidant (cool air) into contact with coal particles on the chamber wall. The wall is protected primarily by a layer of molten slag, which also reduces the heat flux. There would normally be considerable ash vaporization in a high temperature vortex coal combustor but this can be minimized by use of a two-stage configuration. The idea of the two-stage cyclone coal combustor for MHD systems was first introduced in 1963–1965 (72).

Combustors based on this concept have been built and operated by TRW, Inc. (73,74). A diagram of a 250 MW_t combustor design is shown in Figure 13. A precombustor stage is used to supply 1867 K to the first stage, which is a confined vortex flow chamber. The first stage is connected to the second stage via a deswirl section (not specifically identified in the figure), designed to provide uniform flow to the second stage. The second stage connects the combustor to the MHD channel.

The first stage is the slagging stage, in which slag separation and tap-off occur. Coal, having particle size distribution of 70% through 200 mesh (74 μm), is injected at the head end via multiple injection ports. Combustion of the coal particles is designed to occur in flight. The first stage operates essentially as a gasifier, and has an oxidizer/fuel ratio of about 0.6. The balance of the oxidizer is admitted to the deswirl section, immediately before the second stage, to bring the final stoichiometric ratio up to 0.85. Because the slag is tapped off in the first

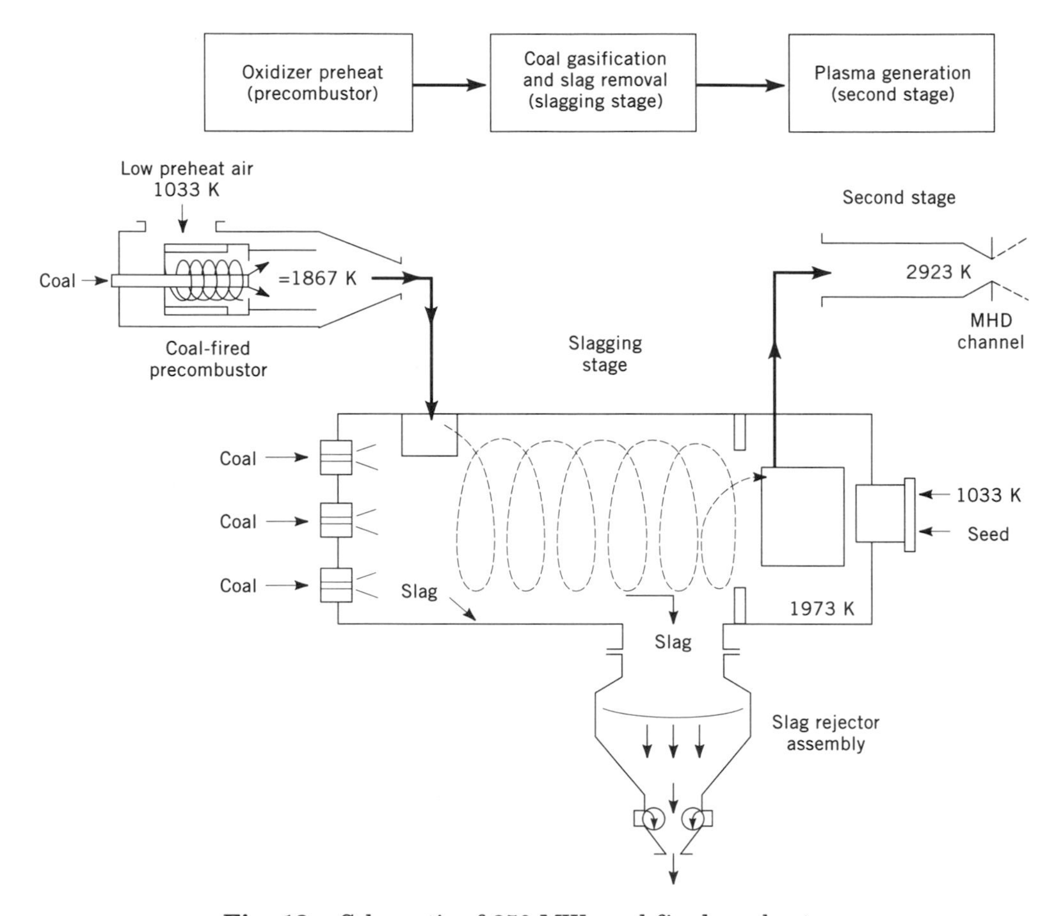

Fig. 13. Schematic of 250 MW_t coal-fired combustor.

stage, the fuel supplied to the second stage is largely ash-free. Hence, seed is also injected in the deswirl section, as the relative absence of slag here means that the removal of seed from the gas by means of slag–seed reactions is minimized.

The combustor is designed to operate at a pressure of 600 kPa (6 atm) with 1867 K preheated air. First stage heat loss of the 250 MW_t combustor is about 4.3% and the total heat loss is about 6%. The relative pressure drop is 3%. More complete discussions of the design and scale-up of the combustor are available (75).

The combustor is assembled of flanged, spool-shaped water-cooled metal components, each with its own water-cooling circuit and pressure shell. No ceramic linings are used. Gas pressure is contained by stainless steel outer shells and the internal surfaces subject to high heat fluxes are lined with low alloy water-cooled panels.

Other approaches to slag-rejecting coal combustors have been taken by Rocketdyne (76) and Avco (77), which have built and operated units at the 20 MW_t scale. The Rockwell design is a two-stage device where two first stage combustors fire tangentially into a ceramic-lined cyclone slag separator, and are followed by a water-cooled second stage. The Avco approach utilizes a single-stage cylindrical configuration with downward combustion flow, horizontal exit flow, and slag separation by means of toroidal vortices at the dome at the top of the cylinder. Descriptions of other slag-rejecting combustors are available (78–81) as is that of a system employing a combustor from which 100% of the slag passes downstream to the MHD generator (82).

The principal combustor ancillary systems are the systems for coal feed, slag rejection, water cooling, and high temperature oxidant supply. All are required to be electrically isolated from ground. Of particular interest are the first two systems. The coal feed systems required to feed the coal into the pressurized combustion chambers utilize dense phase coal transport having solids to gas mass ratios up to 200:1. These are in contrast to the more common dilute phase or slurry transports commonly used in industrial systems. Electrical and pressure isolation are achieved by use of batch mode material transfer from consecutive hoppers, separated from each other by air gaps. The slag rejection system operates in the same fashion and has the additional requirement that the slag must be kept from freezing solid and blocking flow passages. A detailed modeling of dense phase coal transport may be found in Reference 83. Systems in use are described elsewhere (84–87) as is a slag rejection system (88).

MHD Channel. The MHD channel, the heart of the MHD power generation system, is the component which produces the MHD power. Channel requirements determine the principal specifications for other components and subsystems of the MHD power plant. The basic requirements for channel development are governed by overall plant requirements of high plant reliability and availability, high coal-pile to bus-bar efficiency, and low cost of electricity. To satisfy these plant requirements, three primary MHD channel design criteria can be identified (11,44): (*1*) duration or operating time between maintenance periods; (*2*) fraction of thermal energy input extracted from the gas as electric power output (enthalpy extraction ratio); and (*3*) isentropic efficiency, the ratio of the actual enthalpy change of the gas flowing through the channel to the enthalpy change of an isentropic flow at the same pressure ratio. The isentropic efficiency is closely related to the local electrical efficiency defined in equation 8, but is somewhat lower for the typical Mach numbers of interest, in the range 1–2. For early commercial plants, the channel goals are operational for several thousand hours between scheduled maintenance, and enthalpy extraction of at least 15% at isentropic efficiency of 60% or greater.

Construction. From a construction and fabrication point of view, the channel must provide a secure means of containing the working fluid from the combustor and a means of conducting current from the working fluid to the external load, and have adequate durability to satisfy overall power system requirements. Issues related to durability have dominated the development of channel construction methods, particularly of those surfaces which face the hot conducting gas. These surfaces consist of electrodes, which are the current-carrying elements, and insulators, which separate the electrode walls. Durability issues and the resulting designs of gas-side surfaces for coal-fired channels differ from those for clean fuel-fired channels. Channel operating on clean fuels, eg, natural gas, can use a variety of high temperature ceramic materials for both electrode and insulator surfaces which cannot be used in coal-fired channels, because of the incompatibility with molten slag. This allows operation with hotter walls and reduced electrical and thermal losses compared to coal-fired channels. The latter are typically built using cooled metal walls better able to survive the environment. Natural gas-fired channels have been studied extensively (89). The emphasis, however, has shifted to coal-fired MHD systems, both in the United States and elsewhere.

Durability. Two lifetime-limiting mechanisms have been identified from long duration channel tests using slagging flows (90): (*1*) electrochemical erosion of channel gas-side surfaces, which occurs over relatively long durations at nominal channel operating conditions; and (*2*) localized electrical or thermal faults, which can cause serious damage to the channel walls. The mechanisms affecting channel gas-side erosion differ for anode, cathode, and insulator walls. Anodes are subject to electrochemically induced oxidation and/or attack by sulfur (91). The erosion is caused either by oxygen or sulfur which is driven to the anode surface by the electric field, or which is chemically bound in the slag and released by arc current transport through the slag layer.

Cathode and insulator walls are less subject to severe electrochemical attack. In the case of the cathode wall, this is because of the reducing conditions which prevail, and in the case of the insulator wall, because the wall nominally carries no current. However, certain surfaces of cathode and insulator walls are anodic with respect to other surfaces, because of the axial electric field present in the generator, and these surfaces do require protection against electrochemical attack.

Besides gas-side surface erosion, the other important life-limiting mechanism is damage caused by interelectrode faults. These manifest themselves in two ways. The first, axial leakage current between adjacent electrodes, from electrically positive to electrically negative surfaces, results in gradual electrolytic corrosion of the anodic (positive) surface. The second, and perhaps more serious, fault mechanism is arcing between adjacent electrodes, which results from complete breakdown of the interelectrode gap (92,93). This sharply increases the erosion of the corners of the affected electrodes and results in rapid destruction of the interelectrode insulator. Interelectrode arcs are particularly dangerous on anode walls, where they are driven by the Lorentz force into the wall structure and can cause severe damage.

The effects of interelectrode arcs are minimized, first, by limiting the power which can couple into such faults and second, by designing wall structures which can withstand the effects of the arcs.

For a given channel power density the fault power in the channel is proportional to the square of the electrode pitch, ie, the distance between the centers of adjacent electrodes, times the electrode length in the magnetic field direction (94). Hence, the most effective way to limit fault power in the channel is to minimize the electrode pitch. About 2 cm is the practical minimum value in large channels, limited by manufacturing constraints. Once the minimum electrode pitch has been established, fault power can be limited only by limiting the length of the electrode parallel to the magnetic field, ie, by transverse segmentation of the electrodes. In all cases the electrode current must be controlled to avoid large current overloads which can greatly increase the available fault power. Acceptable values of fault power are of the order of a few hundred watts for existing channel designs.

On a slag-covered cathode wall, leakage currents flow through the slag layer. This has the effect of electrically short-circuiting individual cathodes, typically in groups of 3–5, so that each shorted group acts like a single cathode (95). The accumulated Hall voltage, which would normally be divided approximately equally over each interelectrode gap, then appears across only one gap, that is, between one shorted group of cathodes and its downstream neighboring group. Such a pattern can be seen in Figure 14 for a Faraday-loaded generator, which shows clearly the effects of the slag layer on the cathode wall, and the contrast between anode and cathode walls. The large nonuniformities on the cathode walls are caused by the presence of electrically conducting slag constituents, such as potassium compounds or iron and iron oxide compounds, which are driven to the cathode wall by the electrical field in the channel (96). Cathode wall nonuniformities are not as harmful as anode wall nonuniformities. Locally, however, the insulator walls experience high electrical stresses. Also, in uncontrolled diagonal operation, cathode wall nonuniformities are reflected on to the anode wall through the diagonal cross-connection, accompanied by possible increase in anode interelectrode current leakage and the associated harmful effects. Current controls on the cross-connections are used to prevent this from occurring.

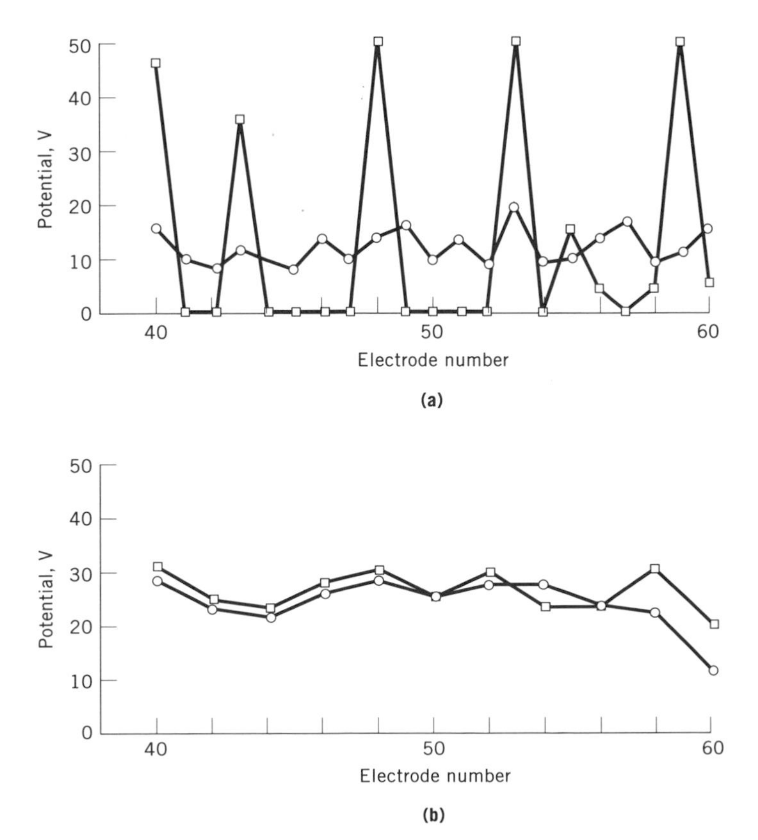

Fig. 14. Effects of slag on interelectrode walls for (□) slagged and (○) unslagged systems for (**a**) cathode and (**b**) anode interelectrode voltage.

Gas-Side Surface Design. Electrode Walls. Development of durable electrode walls, one of the most critical issues for MHD generators, has proceeded in two basic directions: ceramic electrodes operating at very high surface temperatures ($\geq$2000 K) for use in channels operating with clean fuels such as natural gas, and cooled metal electrodes with surface temperatures in the range 500–800 K for channels operating with slag or ash-laden flows.

The hot ceramic electrodes tend to operate with diffuse transport of current from the plasma to the electrode surface, reduced tendency for interelectrode breakdown, and reduced heat losses. The most common designs, developed for the U.S.S.R. High Temperature Institute gas-fired U-25 channel, use zirconia electrodes either brazed to metal substrates made of special stainless steel or chromium alloys, or else rammed on to metal substrates reinforced with wire mesh (97,98). The zirconia is doped with rare-earth oxides such as yttria or ceria; other oxides such as calcia have also been used, particularly in the formulations designed for ramming. Typical compositions are 0.88 ZrO_2 + 0.12 Y_2O_3, or, for the calcia-stabilized ceramics, $Zr_{0.85}Ca_{0.15}O_{0.15}$ (99). Electrical current in these ceramics is transported primarily by oxygen anions. Another class of ceramic electrodes is based on materials such as lanthanum chromite (100) or silicon carbide; in these materials current transport is electronic rather than ionic, and electrical conductivity is higher. Also, thermal conductivity is higher than in zirconia-based ceramics. A disadvantage is that their maximum operating temperatures are lower, in the range 1400–1600 K, compared to the 2000–2200 K capability of zirconia.

Although ceramic electrodes have received much attention (101), they have not been successful in channels operating with slag-laden flows, because of excessive electrochemical corrosion caused by the slag. Only well-cooled metallic elements have been used successfully in slagging environments.

An important feature of slag-covered metal electrodes is that current transport to both electrode walls, anode and cathode, is via arcs (95). Hence, a well-cooled structure having good thermal diffusivity is required. Water-cooled copper has been used successfully for many years in developmental channels. Metal electrode walls are designed to retain a slag coating so that a higher gas-side surface temperature (ca 1700 K) can be maintained than is possible for bare metal walls. This reduces electrode voltage drops and heat losses.

The main cause of anode wear is electrochemical oxidation or sulfur attack of anodic surfaces. As copper is not sufficiently resistant to this type of attack, thin caps of oxidation and sulfur-resistant material, such as platinum, are brazed to the surface, as shown in Figure 15**a**. The thick platinum reinforcement at the upstream corner protects against excessive erosion where Hall effect-induced current concentrations occur, and the interelectrode cap protects the upstream edge from anodic corrosion caused by interelectrode current leakage. The tungsten underlayment protects the copper substrate in case the platinum cladding fails.

Other cap materials have been tested, but in regions of the channel where the electrical and thermal stresses are the highest, the most successful, ie, longest lifetime, electrode design has platinum caps operated at low (ca 500 K) temperatures. Anodes of like design, but without the tungsten back-up layer, have been operated successfully for more than 1000 hours in a slag and sulfur-laden flow (102), at electrical and thermal stresses similar to those expected in commercial-sized generators.

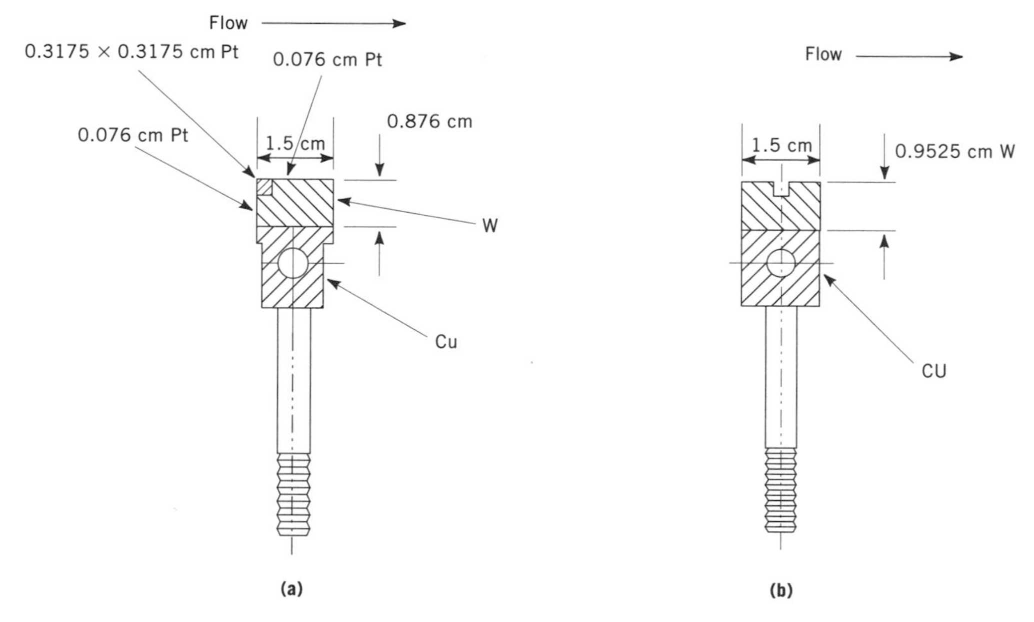

Fig. 15. Designs of an MHD (**a**) anode and (**b**) cathode.

Cathodes are less subject to erosion by electrochemical attack than are anodes because reacting ionic species are not released in the cathode slag layer. A viable cathode design (Fig. 15**b**) is a cooled copper substrate capped with tungsten–copper to resist microarcs and mechanical erosion by slag. An earlier design incorporated a cap of TD nickel on the upstream side, which is an anodic surface with respect to the neighboring upstream cathode. This design was operated successfully for 500 hours (103) and the erosion rates were low enough to indicate that much longer lifetimes can be expected. A modified version was operated for 250 hours, with even lower erosion rates.

The interelectrode insulators, an integral part of the electrode wall structure, are required to stand off interelectrode voltages and resist attack by slag. Well cooled, by contact with neighboring copper electrodes, thin insulators have proven to be very effective, particularly those made of alumina or boron nitride. Alumina is cheaper and also provides good anchoring points for the slag layer. Boron nitride has superior thermal conductivity and thermal shock resistance.

Insulator Walls. Because of the unavailability of electrically insulating materials which can withstand the harsh environment inside coal-fired channels, the insulator walls of the channel are typically made of metal elements which are insulated from each other to prevent any net flow of current. Like electrode walls, insulator walls are designed to operate with a slag coating.

Figure 16**a** shows a so-called peg wall design, in which thin insulators separate rectangular or square metallic elements, ie, pegs, typically 2–3 cm on a side. The advantage of the design is its superior electrical insulating properties under all operating conditions, and its electrical flexibility. The disadvantage is the mechanical complexity arising from the large number of small elements. However, using proper engineering and assembly procedures such walls can be made to operate reliably and have, in fact, been tested at 20 MW_{th} scale for hundreds of hours (103).

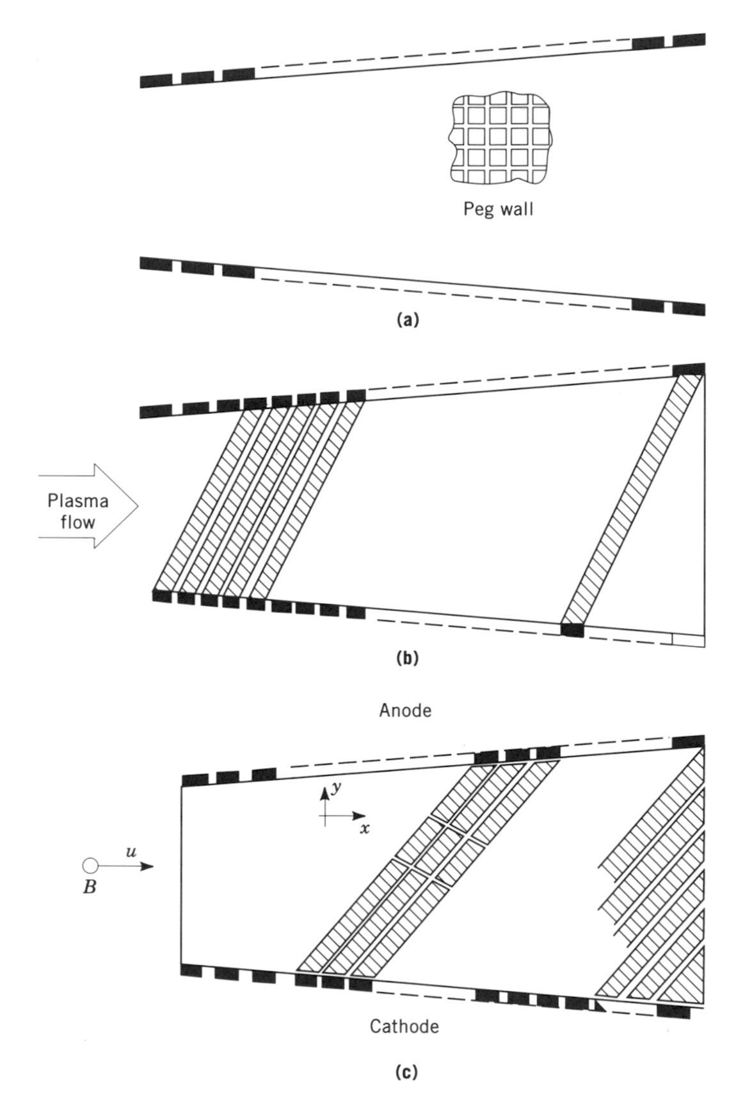

Fig. 16. Insulator wall designs: (**a**) peg wall; (**b**) conducting bar wall; and (**c**) segmented bar wall. The gas-side materials are tungsten and tungsten–copper composite, the base material, copper, and the insulators, boron nitride. Slagging grooves are shown.

To simplify insulator wall design, the continuously conducting sidewall, shown in Figure 16**b**, is used. The elements lie nominally along equipotential lines in the generator. This type of construction does not allow external current control or fault power control. Also, although the sidewall elements lie nominally along equipotential lines, exact alignment is not possible, as the equipotential lines are not straight, owing to boundary layer effects and load variation. The sidewalls in fact do collect some current which tends to concentrate near the channel corners, causing erosion at these locations (104).

To alleviate these problems, the continuous side bar is split, as shown in Figure 16**c**. Each sidewall segment is large enough to be individually water-cooled. In comparison to the single side bar design, the segmented bar requires a larger number of water hoses and penetrations of the pressure vessel, but far fewer than the peg wall design.

Mechanical and Thermal Design. The main objectives of channel mechanical and thermal design are to maintain structural and sealing integrity, to provide adequate cooling of gas-side surface elements, and to use efficiently the magnet bore volume, ie, to maximize the ratio of channel flow cross-section area to the magnet bore cross-section area. This last requirement affects not only the channel mechanical design but also the packaging of channel electrical wires, cooling hoses, and manifolds. In broad terms, MHD channels built to date have fallen into one of three types of construction categories: plastic box construction; window frame construction; or reinforced window frame construction.

In plastic box construction an example of which is shown in Figure 17, the channel is a four-wall assembly. Each wall consists of the individual gas-side surface elements mounted on an electrically insulating board, which is made typically of a fiber glass-reinforced material such as NEMA Grade G-11. The box formed by assembly of the four walls serves as the main structural member and the pressure vessel of the channel. Final gas-side contouring is done by varying the height of the electrodes and insulating wall elements. Gas sealing is done on the edges of the plastic wall, along the corners of the box. The Textron 1A4 channel (Fig. 17) (104) has operated for hundreds of hours. Other channels built in this manner include the Mark VI and Mark VII channels built by Avco (105) and the high performance demonstration experiment (HPDE) channel (106). The most extensive data base has been accumulated for this type of construction.

Plastic box construction has several advantages: it is readily scalable to large commercial sizes; readily separable walls make assembly, disassembly, and refurbishing of the walls relatively simple, fast, and inexpensive; noncurrent carrying sidewalls can be used, which permits the use of local current controls; gas sealing and interelectrode insulator functions are separated, thus minimizing the risk of plasma leakage in the event of interelectrode breakdown and arcing; and there are only four main gas seals, along the corners of the box, further minimizing the risk of plasma leakage. The main disadvantage is the large number of cooled elements that either carry current or must be electrically insulated from each other, and the associated large number of water hoses and electrical wires that are required.

The window frame channel design (Fig. 18) is made by stacking together metallic frames inclined at the same angle as the generator equipotentials. The frames serve as both the current-carrying elements and as the pressure vessel of the channel. Gas sealing is done around the perimeter of each frame, at some distance from the gas. Window frame construction was used for the Lorho generator (107), a large Hall generator built by Avco, and for another large channel built in the United States for use in the Russian U-25 facility (108).

Window frame channels offer several advantages: the continuous metal frames have good strength and can be assembled to form a rugged structure; electrical simplicity is achieved by using the frames as the current carrying elements, thus minimizing the amount of external wiring for this purpose; and hydraulic reliability is maximized by reducing the number of hydraulic circuits.

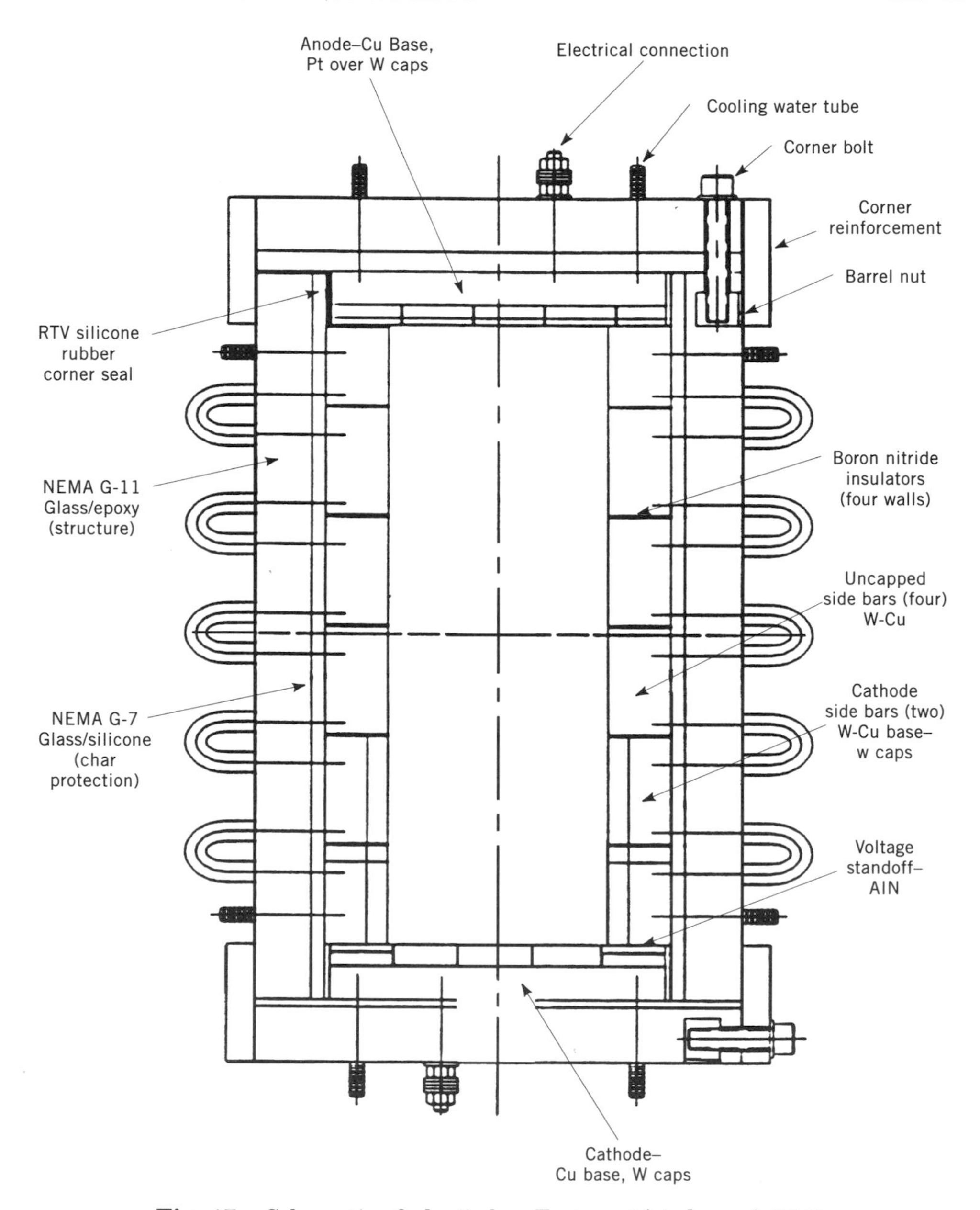

Fig. 17. Schematic of plastic box Textron 1A4 channel (104).

Offsetting these advantages are some serious disadvantages. First, the great length of sealing surface (equal to the frame perimeter times the number of frames), together with the fact that gas sealing and structural functions are combined, make this type of construction vulnerable to hot gas leaks. Second, scaling to commercial sizes is difficult because of the problems associated with fabrication of large window frame channels. A pilot-plant size channel may have about 500 frames, each about 2-cm thick and about 1-m long on a side, requiring

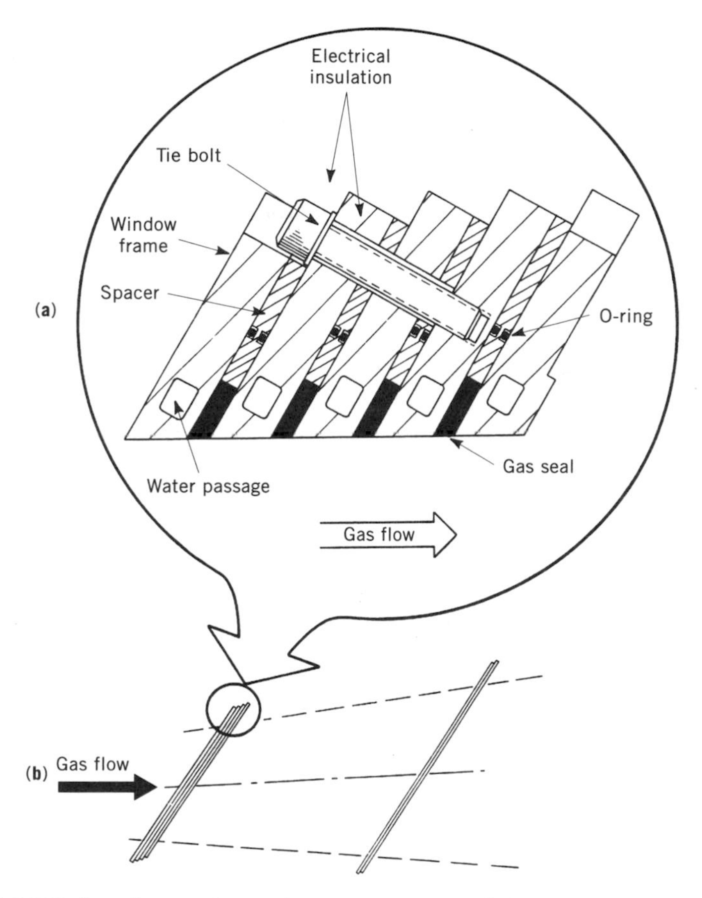

Fig. 18. (**a**) Window frame channel construction, and (**b**) extended view of channel.

great care to avoid bending and distortion during handling. Also, water passages are difficult to incorporate into such frames. Third, a commercial-scale window frame channel of minimum practical pitch, at conditions typical of full-scale operation, has a very high fault power because of the large continuous length of frame, and offers no possibility of fault power control either by segmentation or by frame current control.

The reinforced window frame channel, eg, the large Russian RM channel (109) and the smaller Russian U25B channel (110), is essentially a window frame channel inside a plastic box which serves as the pressure vessel and main structural member. It combines some features of both window frame and plastic box construction. Frames can be segmented, although with some difficulty, for fault power control. This construction is difficult to disassemble for inspection and/or refurbishment.

Channel thermal design, although requiring care, poses no significant problems. Heat fluxes from the gas to the walls of the channel can range from 50 W/cm^2 at the exit of a well slagged channel to about 500 W/cm^2 at the inlet of an unslagged channel. Coolant flow velocities in gas-side surface elements are typically in the range 2–5 m/s. Coolant hoses, manifolds, etc, must have adequate mechanical and thermal properties, and also be electrically insulating, in order to avoid electrical shorting of channel elements. These requirements limit the types of hoses and manifolds which can be used, and therefore the allowable cooling water pressure and temperature.

The operating durations achieved by various combustion-driven experimental channels is shown in Figure 19. The electrochemical and thermal stress levels of an Avco Mk VI channel, operated for 500 h, (103) were similar to those expected in commercial coal-burning plants. Results indicate that durability of properly designed and operated channels can be extrapolated to several thousand hours. This test conducted in two over 300 hour segments was performed prior to the availability of adequate MHD coal combustors. Coal-burning operating conditions were simulated by injecting ash and sulfur into an oil-fired MHD combustor. Another significant demonstration used actual coal-fired operation at the CDIF for over 300 hours. The generator was operated at power outputs up to 1.5 MW_e.

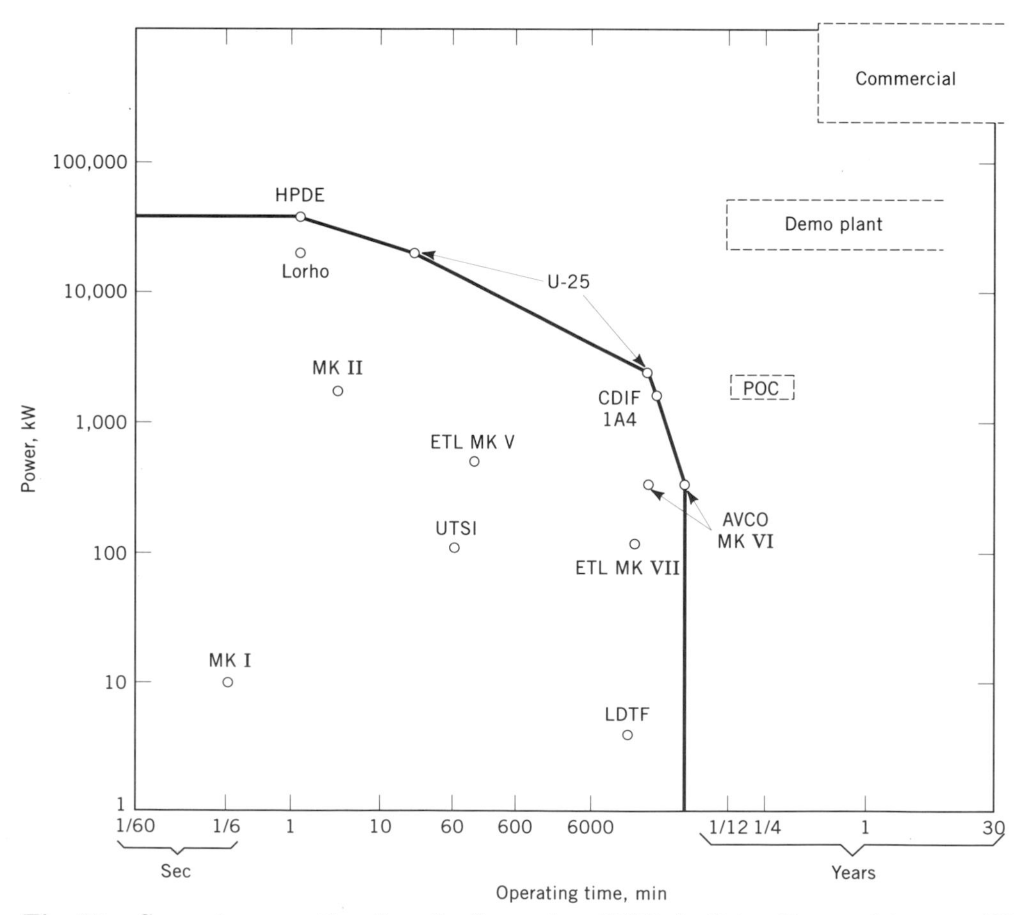

Fig. 19. Generator operating time for the various MHD facilities discussed in text: (○) achieved performance and (⬚) those planned. POC = proof of concept.

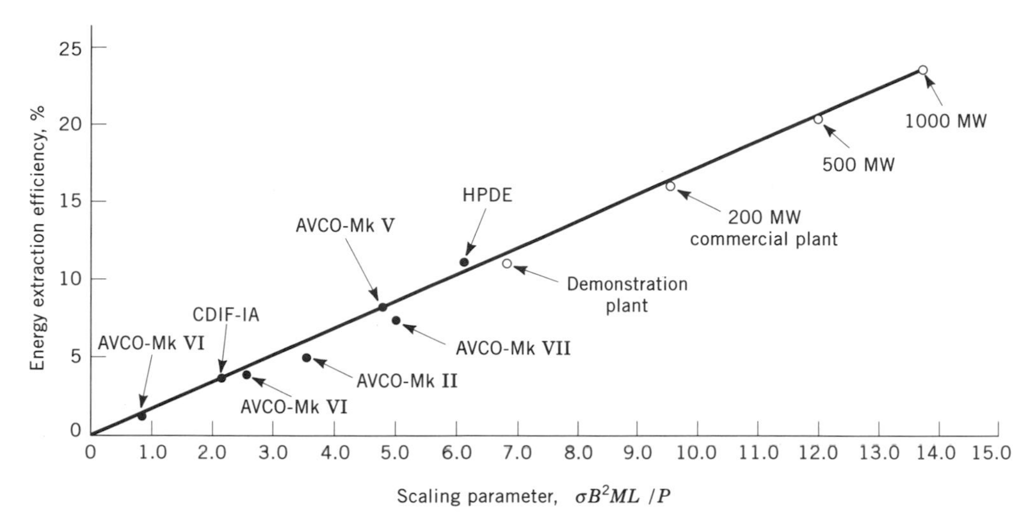

Fig. 20. Scaling of channel enthalpy extraction for the various MHD facilities discussed in text: (•) achieved, and (∘) predicted where σ is the average gas conductivity in mho/m; B, the average magnetic field in T; M, the channel Mach number; L, the average channel active length, m; and P, the average channel static pressure, kPa. To convert kPa to atm, divide by 101.3.

Figure 20 shows values of the channel enthalpy extraction ratio for a number of channels. Enthalpy extraction (111) equal to that required by a proposed demonstration plant (112) has been achieved. Channels have performed generally in accordance with predictions.

Whereas considerable progress has been made towards achieving acceptable channel performance (power and enthalpy extraction) and durability, as of this writing performance and durability have not been demonstrated simultaneously. A larger scale demonstration plant has been proposed in the United States by the MHD Development Corp. (112).

Further work is also required in the area of mechanical design and construction of large MHD channels, especially with respect to construction features which are scalable to commercial size channels. The mechanical design of channels is aimed first, at achieving electrical and structural integrity of the channel, and second, at achieving the most efficient use of the magnet bore in order to minimize the required volume of magnetic field, and hence, the magnet cost. This is done by compact packaging of channel structure, electrical wiring, water manifolds, hoses, etc. Additional important considerations are channel installation, maintenance, and repair.

Electrical Loading and Control. The function of the channel loading system is to extract from the channel the power generated in each plasma element with minimal losses (113,114). This means that the load circuit impedance must match as closely as possible the channel impedance at all axial locations along the channel, which is achieved by use of multiple power take-off points. Ultimately, power from the separate take-offs must be consolidated into a single terminal pair at the transmission grid, by means of appropriate circuitry. An inverter is necessary between the channel and the transmission grid in order to convert the relatively low voltage (20–40 kV) dc output of an MHD generator to ac at transmission line voltages (200–400 kV). In principle, the power consolidation function can be combined with the inversion function by use of common circuitry; in practice, it is simpler and less costly to separate these functions.

Segmented Faraday generators and multiloaded diagonal generators require that outputs from multiple terminals at different potentials be consolidated into one set of load terminals, at the inverter. The consolidation circuitry must be nondissipative and should not change the axial voltage gradient along the channel. For the segmented Faraday generator, consolidation circuitry is necessary at each electrode pair, whereas for the diagonal generator this circuitry is necessary only at the power take-off regions (115). This circuitry can also be used to perform control and safety functions, such as maintaining a prescribed electrode current distribution to prevent destructive current overload of the electrodes, and to prevent nonuniformities occurring in part of the channel, eg, cathode wall, from propagating to other parts of the channel, eg, anode wall. Hence, the circuits are used at each electrode pair even in diagonal generators. The combined functions of current control and consolidation are generally referred to as power conditioning or power management. Current consolidation and control was first proposed in the 1970s (116). A number of methods for its accomplishment have been proposed (117–119). The most well-developed of these is described in detail elsewhere (119,120). Extensive operational experience with this method, which maintains individual electrode currents at the average value of a group of electrodes, has been obtained on various channels (121).

For current consolidation, the basic circuits, used at each of the multiple power take-off points, are stacked into a Christmas tree topology to form a single power take-off terminal pair. Scale-up of these devices to commercial sizes is not expected to be a problem, as standard electrical components are available for all sizes considered. A different type of consolidation scheme developed (117), uses dc to ac converters to connect the individual electrodes to the consolidation point. The current from each electrode can be individually controlled by the converter, which can either absorb energy from or deliver energy to the path between the electrode and the consolidation point. This scheme offers the potential capability of controlling the current level of each electrode pair.

A comprehensive discussion of the design of inverters for MHD applications can be found in Reference 122. MHD inverters using both line commutation (122,123) and forced commutation (124,125) have been used. Line commutated systems require power factor correction and harmonic filtering, and are susceptible to loss of commutation caused by anomalous ac line disturbances. Forced commutated circuitry requires control of both real and reactive power between the inverter and the ac grid; also, its costs and losses increase with dc voltage ratio, restricting the practical voltage range. Line commutation is preferred, primarily because the technology is considered to be better developed.

Magnet. The magnetic field for utility scale MHD generators is provided by a superconducting magnet system (SCMS), for economic reasons, as the cost of electricity for a conventional magnet of the required size is prohibitive (see MAGNETIC MATERIALS; SUPERCONDUCTING MATERIALS). The SCMS consists of three principal subsystems: the main magnet and cryostat subsystem, the cryogenic refrigeration system, and the power supply and protection subsystem. Of these, the magnet subsystem is the most critical, having the majority of the design choices and requiring the bulk of the engineering and manufacturing effort.

The magnet is required to provide a field of the required magnitude and, in the case of linear channels, axial profile. Linear MHD systems require a sharp magnetic field reduction at the channel ends (Fig. 21). This is to reduce the induced electric fields at the channel power take-off regions and to minimize the magnetic field seen by nonpower generating components of the flow train, such as the combustor, the nozzle, and the diffuser, so as to minimize induced circulating currents which could cause erosion. Field uniformity requirements for MHD magnets are relatively modest; uniformity within ± 1% is adequate.

The magnet is wound from a composite of niobium–titanium and copper or aluminum. The Nb–Ti is the superconductor and the copper or aluminum serves to stabilize the conductor, by providing capacity for heat absorption from the joule heating which occurs in the event that the conductor undergoes a transition from its superconducting state into a normally conducting state. Liquid helium cooling to about 4 K is necessary to maintain the superconducting state.

The composite conductor is typically wound in the form of a cable, which can be cooled either internally by a forced helium flow or externally by immersion in a pool of helium. Large electromagnetic body forces, up to 500 t/m^2, are experienced by the conductor during operation. These are contained by a massive external structure, although designs have been proposed in which the conductor itself serves as its own force containment structure (126).

The entire magnet structural assembly must be cooled, requiring containment in a dewar with vacuum layer thermal barriers. With careful design the refrigeration power requirements can be kept to a few hundred kilowatts, even for an MHD plant producing hundreds of megawatts. Magnets of various coil geometries can be used. The solenoidal configuration, Figure 22**a**, is the simplest and least costly to fabricate. It is a suitable configuration for disk generators, but not for linear generators, which leave too large a fraction of the available magnetic field volume unused. In the racetrack geometry (Fig. 22**b**) two flat oval coils are placed opposite each other, one on either side of the channel, to provide the transverse field. This configuration is more efficient with linear channels than is the solenoidal configuration, but still wastes magnetic field volume at

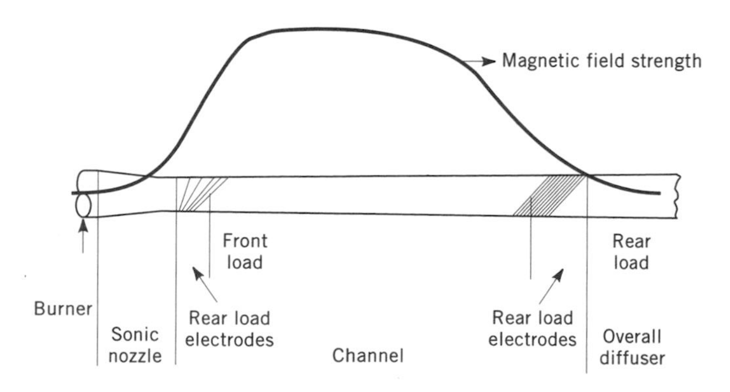

Fig. 21. Magnetic field distribution.

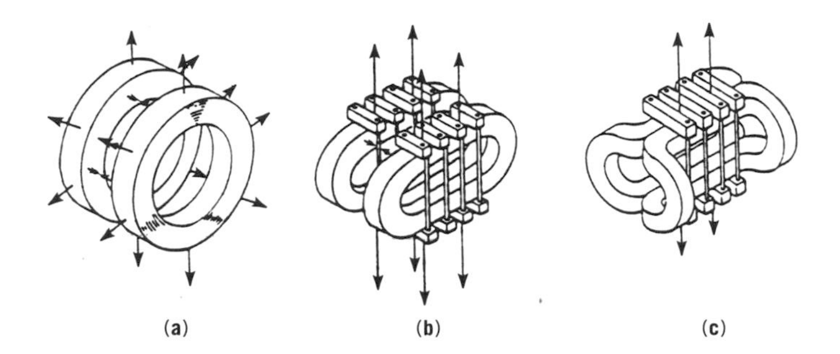

Fig. 22. Superconducting magnet configurations for MHD generators where the arrows represent the magnetic lines of force. (**a**) Solenoid; (**b**) racetrack; and (**c**) saddle.

the ends. Another drawback of the racetrack is that it is difficult to achieve with it the required axial field profile.

The saddle-shaped configuration (Fig. 22**c**) is the most efficient configuration for linear channels. The two saddle shaped coils are located parallel to each other; the MHD channel is located in the gap between the two coils. The longitudinal part of the conductors lies parallel to the direction of flow in the MHD channel, the direction of the field being transverse to the flow in the horizontal plane. This configuration is the most efficient in its use of magnetic field volume and can be readily designed to provide the required axial field profile. Its principal disadvantage is that it is more complicated to fabricate.

For power plants, magnetic fields of 4.5 to 6 T (4.5–6 $\times$ 10^4 G) are required, over warm bore volumes having typical dimensions of 3–4 m dia and 15–20 m long. Stored energies in such magnets are 2000 MJ (480,000 kcal) or greater. The external dimensions are of the order of 15 m dia by 25 m in length. Such large structures must withstand the high mechanical loads imposed by the magnetic fields while retaining cooling integrity so that internal temperatures of 4 K can be maintained without excessive heat loss. Descriptions of a number of different design approaches can be found elsewhere (127–129).

Some technology for large superconducting magnets has been developed, mainly for bubble chamber and fusion reactor applications, and magnets having stored energy up to 500 MJ (100,000 kcal) have been built. Winding and fabrication techniques for very large saddle-shaped coils needs further development. Magnets of commercial size are too large to transport. These require field assembly with the attendant need to develop suitable fabrication techniques.

MHD superconducting magnets have been built in Japan (130,131) and in the United States (132,133); the largest was built in 1981 (133). This magnet, designed to be cost-effective and scalable to commercial size, was successfully operated at its design field of 6 T. Characteristics of the superconducting magnet having a horizontal peak on-axis field of 6 T and stored energy of 210 MJ (50,000 kcal) are: inlet bore diameter, 0.8 m; outlet bore diameter, 1.0 m; active field length, 3.0 m; overall height and width, 4.9 m $\times$ 4.1 m; overall length, 6.4 m; and total weight, 173 T.

Studies of MHD superconducting magnets are in progress at the Plasma Fusion Center at MIT (126,127). These include studies of materials concerned with the properties of highly stressed structural members operating at temperatures near absolute zero, studies of superconductor configurations and winding techniques, studies of shipping and on-site assembly methods to establish the degree of modularity required, and studies of scaling factors and costs. Similar work is in progress as part of the Italian national superconducting magnet program (129) and also in India (134).

Heat Recovery and Seed Recovery System. Although much technology developed for conventional steam plants is applicable to heat recovery and seed recovery (HRSR) design, the HRSR has several differences arising from MHD-specific requirements (135,136). First, the MHD diffuser, which has no counterpart in a conventional steam plant, is included as part of the steam generation system. The diffuser experiences high 30–50 W/cm^2 heat transfer rates. Thus, it is necessary to allow for thermal expansion of the order of 10 cm (137) in both the horizontal and vertical directions at the connection between the diffuser and the radiant furnace section of the HRSR.

Secondly, inlet conditions are more severe in the MHD HRSR because the hot gas entering the HRSR is at considerably higher temperature and enthalpy than that entering a conventional boiler. Typical radiant furnace inlet gas temperatures for the MHD plant are between 2300–2400 K, and have about 3250 kJ/kg (777 kcal/kg) enthalpy. These values are about 300 K and 500 kJ/kg (120 kcal/kg) higher than the corresponding quantities for a conventional boiler. Hence, heat transfer rates are higher in the HRSR. In addition, the gas enters the MHD HRSR with a velocity of 100–200 m/s, even after deceleration in the diffuser. This is higher by a factor of 5–10 than the value encountered in conventional plants. A transition section is used between the diffuser and the furnace entrance to decelerate the gas further to prevent impingement of the particle-laden gas on the rear walls of the furnace, with the consequent risk of wall burn-out (138).

Third, design constraints are imposed by the requirement for controlled cooling rates for NO_x reduction. The 1.5–2 s residence time required increases furnace volume and surface area. The physical processes involved in NO_x control, including the kinetics of NO_x chemistry, radiative heat transfer and gas cooling rates, fluid dynamics and boundary layer effects in the boiler, and final combustion of fuel-rich MHD generator exhaust gases, must be considered.

Finally, the MHD HRSR conditions are more hazardous to the furnace materials because of the more corrosive environment (139,140) created by the hotter combustion gases, bearing potassium, which is not found in conventional plants. Of particular concern is the effect of condensed seed compounds, mainly K_2SO_4, on the boiler tubes and recuperative air preheater. Sulfur and ash are of course also present and detailed understanding of seed–sulfur chemistry is necessary for proper design. Research is in progress (141) to develop a detailed understanding of seed condensation, particle deposition, and fouling of heat transfer surfaces in order to develop techniques for removing slag and seed deposits from heat transfer surfaces.

Testing aimed at resolving issues of seed/ash removal and chemistry is in progress at the Coal Fired Flow Facility in Tullahoma, Tennessee, where a 28 MW_t HRSR, downstream of a coal-fired flow train, is being operated. More than 2500 hours of coal-fired testing has been performed and the measured NO_x and SO_x emissions (Figs. 9 and 10) are well below NSPS standards (53). Particulate emissions were also below NSPS standards using either a baghouse or an ESP for particulate removal. The concentration of priority pollutant organics in ash/seed samples was about the same or less as from conventional coal-burning plants. No difficulty is expected in disposing of ash or slag from commercial MHD plants.

Plant Control, Part Load Performance, and Availability. Conventional power plant control practices, ie, conventional boiler-following, turbine-following, or coordinated control strategies, are all applicable, with some modifications, to the operation of MHD–steam power plants. These can provide for attractive load following characteristics, plant stability, and safe operation (142). Special control actions to safeguard plant equipment during emergencies and abnormal operating conditions such as sudden loss of MHD generator load are part of the overall plant control strategy.

The MHD generator itself has practically instantaneous response characteristics compared to the steam bottoming plant. Thus the load following capabilities and dynamic response of the overall plant are dominated by the bottoming steam plant and oxidant supply system. However, stringent requirements are imposed on reactant mass flow rates because these affect power generation. Also, outputs from external electrical circuits are affected by inputs from gasdynamic variables internal to the MHD channel. These inputs cannot in all cases be measured directly, and so must be inferred from external measurements by use of computer algorithms (143).

Assessments of control, operability and part load performance of MHD–steam plants are discussed elsewhere (144 and 145). Analyses have shown that relatively high plant efficiency can be maintained at part load, by reduction of fuel input, mass flow, and MHD combustor pressure. In order to achieve efficient part load operation the steam temperature to the turbine must be maintained. This is accomplished by the use of flue gas recirculation in the heat recovery furnace at load conditions less than about 75% of full load.

Reliability, availability, and maintainability are critical factors in the use of MHD for electric utility power generation. The duration and lifetime requirements of the principal components and subsystems of an MHD–steam power plant have been assessed in relation to the requirements of acceptable overall plant reliability and availability (146). Duration and reliability of the MHD generator channel are critical factors. The concept of using a stand-by spare channel for scheduled replacement of the operating channel after a certain operating time is both practical and cost effective. Planned maintenance is expected to be a significant factor in avoiding forced outages and in achieving high availability of the MHD generator channel as well as other plant equipment and the overall power plant.

Development Programs

The U.S. national MHD development program (147) was funded by the U.S. Department of Energy (DOE), through the Pittsburgh Energy Technology Center. The program objectives were to establish, through proof of concept (POC) testing, an engineering database so that the risks and benefits of the technology can be evaluated by the private sector before proceeding with commercial demonstration. The POC program was aimed at performance and lifetime of principal components and subsystems. The national program also included conceptual designs of possible MHD pilot-scale plants.

The principal test programs were conducted at two DOE test facilities, the integrated topping cycle at the 50 MW_t Component Development and Integration Facility (CDIF), in Butte, Montana (148), and the integrated bottoming cycle activities at the 28 MW_t Coal Fired Flow Facility (CFFF), in Tullahoma, Tennessee (149). The CDIF has a complete coal-fired MHD power train, a 3 T iron core magnet, and an inverter which interfaces with the local utility grid. The CFFF has a complete HRSR system fired by a coal-fired flow train, but does not operate with a magnet.

Conceptual Design of MHD Repowered Plant. The first step toward MHD plant commercialization is a pilot-scale demonstration plant. Repowering of existing plants, actively under study both in the United States and elsewhere (65), allows use of existing systems at considerable cost savings compared to building a new plant from the ground up. It also ensures that the pilot scale demonstration occurs in a realistic utility environment and with utility participation. Existing power plant systems which can be used include the steam turbine and generator, principal parts of the cooling water system, the electrical transmission system, waste handling, auxiliary support systems such as fire protection, heating, ventilation and plant utilities, and, of course, the site itself.

Conceptual designs of two repowered existing coal-fired plants in the United States, the Montana Power Company's Corette plant in Billings, Montana, and Gulf Power's Scholz plant, in Sneads, Florida are described in References 150 and 151.

Integrated Topping Cycle. The objective of the integrated topping cycle program was to build and test, for a total of 1000 hours, an integrated coal-fired MHD flow train consisting of a combustor, nozzle, channel, associated power conditioning equipment, and diffuser. The flow train and the operating conditions are intended to be prototypical of hardware for commercial plants, so that design and operating data can be used to project component performance, lifetime, and reliability in commercial plants. The flow train operated at 50 MW_t. The program was conducted by a team led by TRW, Inc., and included Textron Defense Systems and Westinghouse Electric Corp. The integrated topping cycle operation followed an earlier program of component testing at the CDIF, and concluded in 1993, when program funding was terminated, with over 500 hours of accumulated operating time, of which over 300 hours were at design power-generating conditions (152,153).

In order to support the design and fabrication of the prototypical flow train, extensive component testing at the 50 MW_t level was performed at the CDIF, in addition to testing of electrodes and sidewalls and of coal-fired channels of 20 MW_t size at Textron Defense Systems.

Integrated Bottoming Cycle. The bottoming cycle program was conducted with a heat recovery and seed recovery system, located at the CFFF in Tullahoma, Tennessee. The system includes a radiant furnace, secondary combustor, air heaters, superheater modules, cyclone scrubbers, a baghouse, and an electrostatic precipitator. It was fired by a coal-fired flow train rated at 28 MW_t. Long-term testing was conducted, with the goal of obtaining 2000 hours of operation on each of two types of coal: Montana Rosebud, considered to be a representative low sulfur Western coal; and Illinois No. 6, a representative high sulfur Eastern coal. The goal of the integrated bottoming cycle program was to obtain an engineering database for the heat recovery and seed recovery (HRSR) components of the bottoming cycle, so that operational characteristics, reliability, maintainability, and materials performance applicable to commercial systems can be established. The program was conducted by the University of Tennessee Space Institute, with assistance from Babcock and Wilcox Corp. A small program to provide supplemental materials test data was in place at Argonne National Laboratory.

The bottoming cycle components used existing boiler technology and materials wherever possible (see COMBUSTION TECHNOLOGY). Appropriate gas-side conditions were provided by a coal-fired flow train which simulated as closely as possible the composition, temperature, and residence time, among other conditions, in a commercial MHD–steam plant. Specific aspects of system operation investigated included slagging, fouling, erosion, and corrosion of heat transfer surfaces; identification, control, and measurements of gaseous and particulate pollutants; waste management; seed recovery; heat transfer; and system integration and scaling characteristics.

Seed Regeneration. The objectives of the seed regeneration program were to provide experimental verification of the feasibility of one of the seed regeneration processes which have been selected for first commercial use. The specific process under evaluation was the Econoseed process (Fig. 11) (61). A 5 MW_t POC facility was in operation in Capistrano, California, and produced over 6 t of potassium formate seed, which was regenerated from spent seed, ie, K_2SO_4, obtained from CFFF tests.

International MHD Programs. A number of countries are conducting programs in coal-fired MHD power generation. Detailed descriptions of these programs can be found in Reference 65. A summary is given in Table 6.

Coal-fired power generation for hundreds of hours, at conditions representative of those projected for commercial plants, has been demonstrated at 50 MW_t scale at the CDIF test site in Butte, Montana. Satisfactory operation of the primary flow train components, the slag-rejecting coal combustor, and the channel has been demonstrated, as has operation of the primary ancillary systems: the coal feed system and the slag rejection system of the coal combustor, and the power conditioning, inversion, and transmission systems of the channel. Elsewhere, channel enthalpy extraction meeting pilot-scale requirements has been demonstrated, and large superconducting MHD magnets have been built and tested.

Thousands of hours of bottoming cycle operations at the CFFF (Tullahoma, Tennessee), have generated a substantial engineering database for the design of future commercial MHD bottoming plants. Operations with both high and low sulfur coals have verified projections that SO_x and NO_x emissions are well below current and expected future standards. A pilot-scale seed regeneration plant has been built and has demonstrated the technical viability of the formate seed regeneration process.

Table 6. National MHD Programs

Country	Organization	Location	Facility[a]	Thermal input, MW	Magnetic field, T[b]	Test duration, h	Comments
Australia	University of Sydney, School of Electrical Engineering	White Bay Power Station	disk generator	2	2.7	36[c,d]	basic generator research studies of coal-fired MHD in progress
China	Shanghai Power Equipment Research Institute	Shanghai		25	4.5	1000[e]	MHD–steam combined cycle generator proposed
		Shanghai	SMS	5		60	coal-fired bottoming cycle test facility
	Institute of Electrical Engineering	Beijing		25	1.9		coal-fired flow train in operation 4 T superconducting magnet under constrution
India	Bharat Heavy Electricals	Tiruchirapalli	pilot plant	5	2	up to 150[f]	5 MW pilot plant in operation; 3 MW coal combustor test facility also in operation
Italy	Industrial MHD Consortium/Ansaldo	Geneva	MDA	1	1.4		gas-fired test facility
	CNR/Ansaldo	Geneva					5 T, 62 M MHD superconducting magnet under construction
	Industrial MHD Consortium	University of Bologna					design of 230 MW_t MHD pilot plant in progress
Japan	Tokyo Institute of Tech	Yokohama	FUJI-1	2–3			closed cycle disk facility; 30% enthalpy extraction achieved
	Hokkaido University	Hokkaido	MDX-1	5	2.5		oil-fired open cycle test facility
	Kyoto University						analytical studies, computer simulations
	Toyohashi University of Technology	Toyohashi			2		experimental studies of electrode phenomena

Table 6. (*Continued*)

Country	Organization	Location	Facility[a]	Thermal input, MW	Magnetic field, T[b]	Test duration, h	Comments
Romania	ICPET	Bucharest	G-MHD-03	1–2.5			coal-fired MHD combustor studies
Russia	Institute of High Temp	Moscow	U-25G	25	3.5		integrated facility originally gas-fired, now coal-fired
		Moscow	U-02	5	1.5–2	100s	material test facility
	Krzhyzhanovski Power Institute	Moscow	M-25	25			
	Institute of Energy Saving Problems, Ukranian SSR Academy of Sciences	Kiev	K-1	15	1.8		channel development; electrophysical investigations of electrical discharge and interelectrode breakdown

[a] See text.
[b] To convert T to G, multiply by 10^4.
[c] Cumulative.
[d] Clean fuel was used.
[e] Continuous generation.
[f] LPG was used as fuel.

NOMENCLATURE

Symbol	Definition
A	flow area
B	flux density, magnetic field, magnetic induction
c	random thermal speed of particle
e	electron charge
$\vec{E}$	electric field in laboratory coordinates
$\vec{E}'$	sum of electrical and motional emf ($\vec{E}' = \vec{E} + \vec{u} \times \vec{B}$)
$\vec{E}''$	electric field felt by particles with drift velocity relative to the fluid
E_x	Hall field
F	body force
$f(v)$	velocity distribution function
g	statistical weight
h	specific flow enthalpy, Planck's constant
j	current density
k	Boltzmann constant, 1.38×10^{-23} J/K
K	generator load coefficient
K'	Hall generator load coefficient
m	mass
$\dot{m}$	mass flow rate
n	number density
p	pressure
P	power density
Q	collision cross-section
T	temperature
u	flow speed
$\bar{u}$	velocity
$\bar{v}$	mean drift velocity
MW_t	thermal megawatt
MW_e	electric megawatt
Z_k	number of charges on particle of species k
α	mole ratio at equilibrium
β	Hall parameter
ϵ	ionization potential
ϵ_o	permittivity of free space
η_e	efficiency
θ	diagonal connection angle
μ	mobility
ν	collision frequency
ρ	density
σ	electrical conductivity

Subscripts

e	electron
i	ion
k	species identification
x	axial direction
y	transverse direction
z	direction parallel to magnetic field

BIBLIOGRAPHY

"Power from Coal by Gasification and Magnetohydrodynamics" in *ECT* 2nd ed., Suppl. Vol., pp. 217–249, by S. Way, Westinghouse Electric Corp.; "Coal Conversion Processes (MHD)" in *ECT* 3rd ed., Vol. 6, pp. 324–377, by S. Way, Consultant.

1. M. Faraday, *Philos. Trans. R. Soc.*, 125–162 (1832).
2. B. Karlovitz and D. Halasz, in N. W. Mather and G. W. Sutton, eds., *Proc. of 3rd Symposium on Engineering Aspects of MHD*, Gordon and Breach Science Publishers, Inc., New York, 1964, pp. 187–204.
3. R. J. Rosa, *J. Appl. Phys.* **31**, 735–736 (Apr. 1960); also *Avco Everett Res. Lab. Rept. AMP 42* (Jan. 1960).
4. R. J. Rosa, *Phys. Fluids* **4**, 182–194 (Feb. 1961); also *Avco Everett Res. Lab. Rept. RR 69* (Jan. 1960).
5. S. C. Lin, E. L. Resler, and A. R. Kantrowitz, *J. Appl. Phys.* **26**(1), 83–95 (Jan. 1955); H. E. Petschek, *Approach to Equilibrium behind Strong Shock Waves in Argon*, Ph.D. dissertation, Cornell University, Ithaca, N.Y., 1955; R. M. Patrick, *Magnetohydrodynamics of a Compressible Fluid*, Ph.D. dissertation, Cornell University, Ithaca, N.Y., 1956; R. J. Rosa, *Engineering Magnetohydrodynamics*, Ph.D. dissertation, Cornell University, Ithaca, N.Y., 1956; J. Jukes, *Ionic Heat Transfer to the Walls of a Shock Tube*, Ph.D. dissertation, Cornell University, Ithaca, N.Y., (1956).
6. S. Way and co-workers, *J. Eng. Power*, **83A**, 397 (1961).
7. J. F. Louis, J. Lothrop, and T. R. Brogan, *Phys. Fluids* **7**(3), 362–374 (Mar. 1964); also *Avco Everett Res. Lab. Rept. RR 145* (Mar. 1963).
8. R. B. Boulay and co-workers, in *Proc. of 10th International Conference on MHD Electrical Power Generation*, Tiruchirapalli, India, Dec. 1989, pp. IX.139–IX.145.
9. B. Zaporowski and co-workers, in *Proc. of 11th International Conference on MHD Electrical Power Generation*, Beijing, Oct. 1992, pp. 106–113.
10. V. K. Rohatgi and co-workers, in *Proc. of 6th International Conference on MHD Electrical Power Generation*, Washington, D.C., June 1975, pp. 45–59.
11. M. Petrick and B. Ya. Shumyatsky, eds., *Open Cycle MHD Electrical Power Generation*, Argonne National Laboratory, Argonne, Ill., 1978, pp. 16–48.
12. S. Hamilton, in *Proc. of 2nd Symposium on Engineering Aspects of MHD*, Columbia University Press, New York, 1962, pp. 211–227.
13. K. Yoshikawa and co-workers, in *Proc. of 31st Symposium on Engineering Aspects of MHD*, Whitefish, Mont., June 1993, pp. IVb 3.1–IVb 3.7.
14. J. Carrasse, in *Electricity from MHD, Proceedings of the Salzburg Symposium*, Vol. 3, Salzburg, July 1966, p. 883.
15. F. Hals and R. Gannon, *Proc. of 13th Symposium on Engineering Aspects of MHD*, Stanford, Calif., Mar. 1973, pp. V.3.1–V.3.10.
16. R. E. Gannon and co-workers, in *Proc. of 14th Symposium on Engineering Aspects of MHD*, Tullahoma, Tenn., Apr. 1974, pp. II.2.1–II.2.8.
17. W. S. Brzozowski and co-workers, in Ref. 10, pp. 137–154.
18. V. I. Kovbasiuk and co-workers, in *Proc. of 30th Symposium on Engineering Aspects of MHD*, Baltimore, Md., June 1992, pp. IV.3.1–IV.3.5.
19. O. V. Bystrova and co-workers, in Ref. 9, pp. 114–121.
20. T. G. Cowling, *Magnetohydrodynamics*, Interscience Publishers Inc., New York, 1957, Chapt. 6.
21. J. Louis, "Studies on an Inert Gas Disk Hall Generator Driven in A Shock Tunnel," *Proc. of 8th Symposium on Engineering Aspects of MHD*, Stanford, Calif., Mar. 1967.
22. C. du P. Donaldson, in Ref. 12, pp. 228–254.
23. A. de Montardy, *Proc. of First International Symposium on MHD Electric Power Generation*, Newcastle on Tyne, U.K., 1962.

24. N. Harada and co-workers, in Ref. 18, pp. IX.5.1–IX.5.9.
25. N. Harada and co-workers, in Ref. 13, pp IV.2.1–IV.2.9.
26. V. G. Kirillov and co-workers, in Ref. 9, pp. 658–665.
27. S. W. Simpson and co-workers, in Ref. 18, pp. IV.6.1–IV.6.5.
28. J. Gertz and co-workers, in *Proc. of 18th Symposium on Engineering Aspects of MHD*, Butte, Mont., June 1979, pp. B.4.1–B.4.9.
29. C. C. P. Pian and A. W. McClaine, *Computers and Fluids* **12**(4), 319–338 (1984).
30. S. T. Demetriades and co-workers, in *Proc. of 12th Symposium on Engineering Aspects of MHD*, Argonne, Ill., Mar. 1972, pp. I.5.1–I.5.13.
31. S. Gordon and B. McBride, *Computer Program for Calculation of Complex Chemical Equilibrium Compositions*, NASA SP-273, National Aeronautics and Space Administration, Washington, D.C., 1971.
32. J. C. Cutting and co-workers, in *Proc. of 16th Symposium on Engineering Aspects of MHD*, Pittsburgh, Pa., May 1977, pp. VII.4.21–VII.4.26.
33. C. D. Maxwell and co-workers, in Ref. 32, pp. VII.3.13–VII.3.20.
34. G. W. Sutton and A. Sherman, *Engineering Magnetohydrodynamics*, McGraw-Hill Book Co., Inc., New York, 1965, Chapt. 5.
35. R. J. Rosa, *Magnetohydrodynamic Energy Conversion*, McGraw-Hill Book Co., Inc., New York, 1968, Chapt. 2.
36. M. W. Zemansky, *Heat and Thermodynamics*, 3rd ed., McGraw-Hill Book Co., Inc., New York, 1951, pp. 412–414.
37. L. S. Frost, *J. Appl. Phys.* **32**(10), 2029–2036 (Oct. 1961).
38. W. P. Allis, *Handbuch der Physik*, Vol. 21, Springer-Verlag, Berlin, 1956, pp. 413.
39. L. Spitzer and R. Harm, *Phys. Rev.* **89**, 977 (1953).
40. L. Spitzer, Jr., *Physics of Fully Ionized Gases*, 2nd ed., Interscience Publishers, Inc., New York, 1962, Chapt. 5.
41. G. Seikel and co-workers, in *Proc. of 15th Symposium on Engineering Aspects of MHD*, Philadelphia, Pa., May 1976, pp. III.4.1–III.4.22.
42. *Evaluation of Phase 2 Conceptual Designs and Implementation Assessment Resulting from the Energy Conversion Alternatives Study (ECAS)*, NASA TM X-73515, NASA Lewis Research Center, Cleveland, Ohio, Apr. 1977.
43. B. D. Pomeroy and co-workers, *Comparative Study and Evaluation of Advanced Cycle Systems*, Electric Power Research Institute, Palo Alto, Calif., Report AF-664, 1, Feb. 1978.
44. F. Hals and co-workers, in *Proc. of 20th Symposium on Engineering Aspects of MHD*, Irving, Calif., June 1982, pp. 1.1.1–1.1.9.
45. General Electric Co., *Definition of the Development Program for an MHD Advanced Power Train*, DOE Report No. DE-AC22-83PC60574, U.S. Department of Energy, Washington, D.C., Dec. 1984.
46. R. E. Weinstein and R. B. Boulay, in *Proc. of 26th Symposium on Engineering Aspects of MHD*, Nashville, Tenn., June 1988, pp. 8.5.1–8.5.9.
47. J. Zeldovich, *Acta. Physicochim.* **21**, 577 (1946).
48. F. A. Hals and P. F. Lewis, in Ref. 30, pp. VI.5.1–VI.5.10.
49. J. Klinger and co-workers, in *Proc. of 22nd Symposium on Engineering Aspects of MHD*, Starkville, Miss., June 1984, pp. 10.2.1–10.2.20.
50. J. W. Pepper and C. H. Kruger, in Ref. 15, pp. VII.2.1–VII.2.3.
51. J. W. Pepper, *Effect of Nitric Oxide Control on MHD-Steam Power Plant Economics and Performance*, SU-IPR Report No. 614, Institute for Plasma Research, Stanford University, Calif., Dec. 1974.
52. L. W. Crawford and co-workers, "Nitrogen-Oxide Control in a Coal-Fired MHD System," *Proc. of 21st Symposium on Engineering Aspects of MHD*, suppl. vol., Argonne, Ill., June 1983.

53. A. C. Sheth and co-workers, "MHD Can Clean Up The Environment," *Proceedings of the American Power Conference*, Chicago, Apr. 1993.
54. J. Lanier and co-workers, "Sulfur Dioxide and Nitrogen Oxide Emissions Control in a Coal-Fired MHD System," *ASME Winter Annual Meeting*, Atlanta, Ga., Dec. 1979.
55. D. G. Rasnake and co-workers, in *Proc. of 29th Symposium on Engineering Aspects of MHD*, New Orleans, La., June 1991, pp. V.1.1–V.1.8.
56. E. J. Lahoda and T. E. Lippert, in Ref. 28, pp. C.3.1–C.3.6.
57. P. Bergman and co-workers, in Ref. 32, pp. X.3.11–X.3.23.
58. J. K. Holt and co-workers, in Ref. 8, pp. XI.80–XI.85.
59. A. C. Sheth and co-workers, in Ref. 8, pp. IX.86–IX.93.
60. A. C. Sheth and co-workers, in Ref. 13, pp. IX.9.1–IX.9.12.
61. R. A. Meyers and co-workers, in Ref. 46, pp. 12.3.1–12.3.10.
62. J. L. Anastasi and co-workers, in Ref. 18, pp. II.4.1–II.4.12.
63. R. C. Attig and co-workers, in *Proc. of 27th Symposium on Engineering Aspects of MHD*, Reno, Nev., June 1989, pp. 1.2-1–1.2-9.
64. Ref. 8, pp. III.1–III.78.
65. Ref. 9, pp. 3–74.
66. J. N. Chapman and N. R. Johanson, "Design Considerations for a Class of 600 MWe MHD Steam Combined Cycle Plants," *28th Intersociety Energy Conversion Engineering Conference*, Atlanta, Ga., Aug. 1993.
67. J. B. Heywood and G. J. Womack, eds., *Open Cycle MHD Power Generation*, Pergamon Press, London, 1969, pp. 18–158.
68. D. P. Saari, in Ref. 21, pp. I.8.1–I.8.8.
69. Yu. A. Gorshkov and co-workers, in Ref. 9, pp. 1033–1039.
70. S. Way, *Combustion Technology, Some Modern Developments*, Academic Press, Inc., New York, 1974.
71. H. Seidl, *Proceedings Joint Symposium on Combustion*, Institute of Mechanical Engineers, London, (1955); H. Seidl, *Eleventh Coal Science Lecture*, Inst. of Civil Engineers, Publ. Gazette, 46, British Coal Utilization Research Assoc., Leatherhead, U.K., Oct. 1962.
72. D. Q. Hoover and co-workers, *Feasibility Study of Coal Burning MHD Generation*, Contract 14-01-001-476, Westinghouse Research Laboratories, Pittsburgh, Pa., (1966); S. Way and co-workers, internal company report, Westinghouse Research Labs, Pittsburgh, Pa., Nov. 5, 1963; U.S. Pat. 3,358,624 (Dec. 19, 1967), S. Way (to Westinghouse).
73. M. Bauer and co-workers, in Ref. 44, pp. 3.3.1–3.3.8.
74. G. Listvinsky and co-workers, in Ref. 8, pp. IX.42–IX.47.
75. G. Roy and A. Solbes, in Ref. 49, pp. 6:4:1–6:4:21; A. Solbes and G. Listvinsky, in Ref. 13, pp. XII.2.1–XII.2.17.
76. C. A. Hauenstein and co-workers, in *Proc. of 19th Symposium on Engineering Aspects of MHD*, Tullahoma, Tenn., June 1981, pp. 16.2.1–16.2.4.
77. J. O. A. Stankevics and co-workers, in Ref. 44, pp. 3.1.1–3.1.11.
78. Sha Ciwen and co-workers, in *Proc. of 23rd Symposium on Engineering Aspects of MHD*, Somerset, Pa., June 1985, pp. 109–113.
79. S. A. Arunachalam and co-workers, in Ref. 63, pp. 7.4-1–7.4-4.
80. W. Zheng and co-workers, in Ref. 8, pp. IX.7–IX.11.
81. M. Akai and co-workers, in Ref. 8, pp. IX-16–IX-23.
82. J. B. Dicks and co-workers, in Ref. 16, pp. II.1–II.1.10.
83. B. L. Liu and H. J. Schmidt, in Ref. 78, pp. 479–490.
84. T. V. Velikaya and G. E. Goryainov, in Ref. 8, pp. IX.52–IX.57.
85. J. E. Cox, in Ref. 49, pp. 6.2.1–6.2.16.
86. L. C. Farrar and co-workers, in Ref. 78, pp. 879–887.

87. R. T. Burkhart, in Ref. 18, pp. V.3.1–V.3.3.
88. G. Roy, in *Proc. of 24th Symposium on Engineering Aspects of MHD*, Butte, Mont., June 1986, pp. 131–136.
89. A. E. Sheindlin and co-workers, in *4th US–USSR Colloquium on MHD Electrical Power Generation*, Washington, D.C., Oct. 1978, pp. 87–104.
90. R. Kessler, *J. Energy*, 178–184, (1981).
91. S. W. Petty and co-workers, in Ref. 32, pp. VIII.1.1–VIII.1.12.
92. V. J. Hruby and P. Weiss, in Ref. 76, pp. 2.2.1–2.2.20.
93. I. Sadovnik and co-workers, in Ref. 44, pp. 4.2.1–4.2.9.
94. A. Solbes and A. Lowenstein, in Ref. 41, pp. IX.3.1–IX.3.7.
95. A. Demirjian and co-workers, in *Proc. of 17th Symposium on Engineering Aspects of MHD*, Stanford, Calif., Mar. 1977, pp. D1.1–D1.6.
96. J. K. Koester and R. M. Nelson, in Ref. 32, pp. VI.2.5–VI.2.12.
97. E. K. Keler, in *3rd US–USSR Colloquium on MHD Electrical Power Generation*, Moscow, Oct. 1976, pp. 405–412.
98. G. P. Telegin and co-workers, in Ref. 97, pp. 413–431.
99. Ya. P. Gokhstein and co-workers, in Ref. 89, pp. 637–685.
100. A. M. George and co-workers, in Ref. 41, pp. II.1.1–II.1.5.
101. Session XI, in Ref. 9, pp. 853–950.
102. V. J. Hruby and co-workers, in Ref. 44, pp. 4.3.1–4.3.6.
103. A. M. Demirjian and co-workers, in Ref. 28, pp. A.3.1–A.3.11.
104. S. W. Petty and co-workers, *Proceedings of the 29th Symposium on Engineering Aspects of MHD*, New Orleans, La., June 1991, pp. 11.1.1–11.1.10.
105. S. Petty and co-workers, in Ref. 41, pp. IV.5.1–IV.5.10.
106. R. F. Starr and co-workers, in *Proc. of 7th International Conference on MHD Electrical Power Generation*, Vol. 3, Cambridge, Mass., June 1980, pp. 203–217.
107. J. Teno and co-workers, *Proc. of 10th Symposium on Engineering Aspects of MHD*, Cambridge, Mass., Mar. 1969, pp. 194–200.
108. K. D. Kuczen and co-workers, in Ref. 106, Vol. 1, pp. 195–201.
109. A. E. Barshak in Ref. 95, pp. F.2.1–F.2.9.
110. V. A. Kirillin and co-workers, in Ref. 95, pp. F.1.1–F.1.12.
111. L. Whitehead and co-workers, paper presented at *MHD Contractors' Review Meeting*, Pittsburgh, Pa., Nov. 1983.
112. J. Sherick and co-workers, in Ref. 9, pp. 62–65; W. R. Owens and co-workers, in Ref. 9, pp. 65–74.
113. B. M. Antonov and co-workers, in Ref. 89, pp. 543–570.
114. A. Lowenstein in Ref. 76, pp. 17.1.1–17.1.6.
115. A. Lowenstein, in Ref. 95, pp. I.1.1–I.1.6.
116. R. J. Rosa, in Ref. 41, pp. VII.5.1–VII.5.4.
117. R. Putkovich, in Ref. 63, pp. 8.11.1–8.11.9.
118. Y. Inui and co-workers, in Ref. 8, pp. XIII.2–XIII.9.
119. V. J. Hruby and co-workers, in Ref. 78, pp. 804–824.
120. I. Sadovnik and V. J. Hruby, in Ref. 78, pp. 790–803.
121. J. Reich and co-workers, in Ref. 55, pp. IX.3.1–IX.3.8.
122. Ref. 11, pp. 275–318.
123. E. Ray and R. Schainker, in Ref. 106, Vol. 1, pp. 421–425.
124. B. M. Antonov and co-workers, in Ref. 106, Vol. 1, pp. 410–420.
125. A. Chaffee and co-workers, in Ref. 28, pp. F.1.1–F.1.10.
126. P. G. Marston and co-workers, in *Proc. 11th International Conference on Magnet Technology*, Vol. 2, MT-11, Elsevier Applied Science, London, 1989, pp. 920–925.
127. P. G. Marston and co-workers, *IEEE Trans. Mag.* **28**, 271–274 (1992).
128. R. W. Baldi, in Ref. 106, pp. 433–439.

129. F. Negrini and co-workers, in Ref. 18, pp. I.1.1–I.1.10.
130. T. Okamura and co-workers, in *International Workshop on MHD Superconducting Magnets*, Nov. 1991, pp. 13–15.
131. Y. Aiyama and co-workers, in *Proceedings of the Fourth International Cryogenics Engineering Conference*, Eindhoven, 1972, pp. 227–229.
132. V. A. Kirillin and co-workers, in Ref. 95, pp. F.1.1–F.1.12.
133. S.-T. Wang and co-workers, *Cryogenics* **22**, 335 (1982); Argonne National Laboratory, in Ref. 53, suppl. vol.
134. R. Rajaram, in Ref. 18, pp. 7.1–7.12.
135. Ref. 11, pp. 465–487.
136. K. V. S. Sundaram, in Ref. 55, pp. V.2.1–V.2.6.
137. F. Hals, "Conceptual Design Study of Potential Early Commercial MHD Power Plant," *NASA CR-165235*, NASA Lewis Research Center, Cleveland, Ohio, Mar. 1981, pp. 3–60.
138. G. F. Berry in Ref. 88, pp. 167–174.
139. M. K. White and M. Li, in *Proc. of 28th Symposium on Engineering Aspects of MHD*, Chicago, 1990, pp. VII.5.1–VII.5.12.
140. K. Natesan and W. M. Swift, in Ref. 63, pp. 3.2.1–3.2.10.
141. M. K. White, in Ref. 63, pp. 3.1.1–3.1.8.
142. D. A. Rudberg and co-workers, in Ref. 76, pp. 14.3.1–14.3.9.
143. D. Lofftus and co-workers, in Ref. 13, pp. X.6.1–X.6.14.
144. M. Ishikawa and co-workers, in Ref. 8, pp. III.36–III.43.
145. M. L. R. Murthy and co-workers, in Ref. 76, pp. 6.2.1–6.2.5.
146. F. D. Retallick and co-workers, in Ref. 28, pp. H.2.1–H.2.5.
147. R. J. Wright, in Ref. 9, pp. 1365–1371.
148. A. T. Hart and co-workers, in Ref. 63.
149. N. R. Johanson and J. W. Muehlhauser, "MHD Bottoming Cycle Operations and Test Results at the CFFF," paper presented at *Second International Workshop on Fossil Fuel Fired MHD*, Bologna, Italy, 1989.
150. R. Labrie and co-workers, in Ref. 8, pp. II.63–II.69.
151. L. Van Bibber and co-workers, in Ref. 8, pp. II.58–II.62.
152. C. C. P. Pian and co-workers, in Ref. 13, pp. II.2.1–II.2.11.
153. C. C. P. Pian and E. W. Schmitt, in *Proceedings of the 32nd Symposium on Engineering Aspects of MHD*, Pittsburgh, Pa., June 1994, Session 2.

General References

R. J. Rosa, *Magnetohydrodynamic Energy Conversion*, McGraw-Hill Book Co., Inc., New York, 1968.

G. W. Sutton and A. Sherman, *Engineering Magnetohydrodynamics*, McGraw-Hill Book Co., Inc., New York, 1965.

M. Petrick and B. Ya. Shumyatsky, eds., *Open Cycle MHD Electrical Power Generation*, Argonne National Laboratory, Argonne, Ill., 1978.

J. B. Heywood and G. J. Womack, eds., *Open Cycle MHD Power Generation*, Pergamon Press, London, 1969.

Published proceedings of: *Symposium on Engineering Aspects of Magnetohydrodynamics*, annually since 1961; *International Conference on Magnetohydrodynamic Electrical Power Generation*, every 3–4 years since 1962.

ROBERT KESSLER
Textron Defense Systems

MAINTENANCE

In this competitive world it is expected of managers to operate facilities without interruptions, shutdowns, utility outages, and environmental problems. Especially in manufacturing plants the cost of production equipment and production labor is so high that even a short business interruption can impact the profit and loss of business. The problem of operation increases with complicated automation, safety requirements, environmental protection, and energy conservation. In facility management all this can only be accomplished with increased and better maintenance efforts. One of the goals of good management is to operate with a predictive and preventive maintenance program that prevents unscheduled utility service interruptions and machine breakdowns. A reasonable cost of such a program contributes to the profitability of manufacturing plants and results in reduction of operating costs and increased productivity. This all can be accomplished by organizing, managing, and controlling maintenance.

Designing, Laying Out, and Constructing Facilities

Besides planning a facility around its products and occupancy, it is also important to design a facility with future sustaining maintenance in mind, eg, adequate space must be allowed around machinery, and heavy and large equipment should be placed next to doors and hallways for future moves. Heavy and large equipment on roofs should be placed as close to the edge of the building as possible. A freight elevator should be provided to handle moving of equipment and material to all levels including the roof.

Grouping like operations together in order for them to make use of central utility services, and placement near outside windows to take advantage of natural light, is important. Noisy equipment should be situated where it affects the least number of workers. Necessary heating, ventilating, and air conditioning (HVAC) requirements must be met. Utility distribution should be located so that they do not interfere with production, lighting, and general access. Adequate electric power outlets (properly sized for floor machines, welding machines, and general maintenance tools) for inside and outdoor maintenance, house vacuum outlets for janitorial services, water taps, and outlets for landscape maintenance are necessary.

Storage areas for maintenance, janitorial, and other service organizations must be provided. Safety items such as fire extinguishers, firehose cabinets, safety hoops on permanent ladders, guard rails, shielding for acid pumps, clearance for electric panel boards, etc, are needed. Manholes and cleanouts for sewer pipes within the facility as well as in the landscape and parking areas should be provided.

The layout should ensure that exhausts are not placed close to fresh air intakes and that fire sprinkler protection for present and future requirements, eg, under stairs, storage racks, overhangs, covered walks, etc, is available. All facility equipment must be structurally secured as well as freeze protected.

In a mature facility the design team for a new building or an addition to an old building should seek input from the maintenance department. Maintenance workers know what equipment and material is difficult to maintain, and the standard in the existing facility employed in order to prevent stocking of many different replacement parts.

Providing for Reliable Utilities and Their Uses

Most facilities have a variety of utilities connected to their buildings. Some of these utilities are piped into the facility from the outside, others are manufactured and distributed on site, and others are piped from storage tanks to various buildings: electric power, natural gas, burning fuels, compressed air, city water, deionized and distilled water, process cooling water, sanitary sewer, storm drain, fire-protection water supply, process gases, piped chemicals, etc. Interruption or failure of these utilities can easily shut down important parts or a complete facility. Because the prevention of work stoppage is the most important challenge for maintenance, reliability must be designed into the facility. In a large facility a master plan should include consideration for some important special buildings.

Central Utilities Building. This can be accomplished by separate smaller buildings or one building with separated sections for housing. Sections include the main electric incoming service and distribution panels, emergency generator with all its distribution gear; the main incoming water service, fire pumps, booster pumps, special water purification, neutralization equipment and instrumentation; boilers, hot water heaters, secondary fuels, and A/C chilled water plant. Providing a maintenance shop as part of this utility complex would facilitate supervision of equipment, meters, and facility alarms.

A master plan that provides for a centralized utility building or a cluster of various buildings of similar purpose should also have provision for a separate or dedicated facility maintenance building as part of the complex. By having a centralized facility building, maintenance departments can reduce traveling and reaction time. Such a building should house the centralized computer needed to operate the facility, and include maintenance and historical files, drawings, instruction books, technical facility library, and records as well as security and operational alarms, etc.

Central Chemical Storage and Supply Building. Such a building could also be combined with a water treatment plant and acid waste treatment.

Central Industrial Gases Building. The building could be adjacent to liquid gas storage tanks or an on-site process-gas generating plant, and house the instrumentation and valves for pipelines entering the manufacturing buildings, storage of industrial gas cylinders, parts (regulators, valves, etc) and a maintenance shop for work on regulators, valves, and controllers. In many cases such a shop needs to be a clean room.

Central Shipping and Receiving Building. Such a building could also include many other warehousing and distribution operations such as a mailroom, printing/copy center, stores, surplus warehouse, etc.

Stores Building and Surplus Warehouse. This could be a separate building or it could be combined with the central shipping and receiving building. In some companies such a building would require special security installations.

Back-Up Systems. When designing for reliable utilities another important item is to provide for dual and alternative services of important utilities. They should include dual electric power services, double-ended electrical substations that are fed from independent electric service feeders; back-up or multiple pumps for sumps, chilled-water, boiler water, cooling water, etc. Good building designs include back-up for any important utility or chemical supply to enable maintenance people to take machinery out of service for maintenance and accomplish repairs without disabling the facility. The back-up supplies should include chemical storage tanks, air scrubbers, air compressors, vacuum pumps, electric service equipment, emergency generators, electrical distribution equipment, water meters, backflow preventers, cooling towers, boilers, air handler equipment, chillers, etc.

Monitoring Systems. Well-designed facilities also have monitoring systems that provide facility operation people with reliable information about the status of operation equipment and early warning of any malfunctions and unacceptable conditions; this can include security as well as fire protection alarms. As an example, when a pump fails it triggers an alarm indicating that a storage tank is filling too high or has too low a level; when chemicals have temperature requirements, a high or low temperature alarm can be monitored; when chemicals are mixing wrong an alarm can indicate this incorrect mix; and effluent going to the sewer can be monitored and should have a pH alarm. Other important alarms are activation of the automatic fire sprinkler system or smoke alarms for computer floors. Especially in the chemical-process industry, the monitoring of gas concentrations, lack of exhaust, and activation of fire-protection systems should be monitored in the maintenance shops and backed up by the plant security with possible direct warning calls to the fire department.

Maintenance Considerations during Facility Construction or Renovation

Maintenance managers should be on board during the design state to ensure the maintainability of a facility. A maintenance manager's case is simple: the original cost of new construction is soon forgotten, but facility maintenance and operational costs continue for the life of a facility.

The maintenance manager oversees important issues during the design and construction phase. Some areas requiring special attention include checking the design from an operational point of view; checking machinery and equipment for operation and maintenance; checking that all equipment is safely installed and can be maintained; checking the marking of supply lines (air, water, gases, chemicals, etc); identifying the electrical circuitry and checking panel labels; reviewing manufacturers' suggestions for spare parts and tools; reviewing test results, ie, strength of materials, pipeline integrity, etc; reviewing balance reports for HVAC work, ie, air, water, electric load, etc; reviewing fire sprinkler alarm and water test; participating in housekeeping and sanitary support during construction; and checking utilities, including electric power, water, drainage, air supply, that will supply new areas without overloading the individual systems.

Also, maintenance workers should inspect roof penetrations and the installation of equipment on roofs during construction. Because the maintenance department is responsible when the roof leaks, their inspection of roof penetrations during construction can prevent future problems. Maintenance must also make sure that the roof was not damaged during construction.

Standardization of equipment and material is another area that lends itself for partnership between the design team and maintenance. When an existing facility is renovated or a new section/building is added, maintenance can help ensure that materials, equipment, and parts are used that match those that were used in the existing facility. Setting such standards helps to cut down inventory and reduce operation and maintenance costs as well. Items for standardization include mechanical equipment, such as HVAC equipment, compressors, and pumps; piping and valves; electric panel boards, motors, controls, and switches; instrumentation; restroom equipment, plumbing fixtures, and partitions; lighting fixtures and air conditioning outlets; ceiling and floor tiles; movable partitions and their electric distribution systems; door hardware and locks; and office furniture.

In selecting a supplier or manufacturer, the maintenance manager must be sure that the supplier can provide the support needed to operate and maintain the facility effectively. The supplier or manufacturer must be willing and able to service the facility in the future, to instruct and oversee installation, and to provide adequate maintenance instructions and training. Also, parts must be readily available and locally stocked.

It is essential that as-built architectural/construction drawings are given to maintenance at the completion of every job, together with all other job documents, such as equipment information, operating instructions, test data, supply information, etc. These documents are needed to operate and maintain the facility effectively. At the completion of any project, there should be an official acceptance of documentation by the maintenance/operation department. No job is complete without this formal acceptance and sign off.

Maintenance costs influence the bottom line of every balance sheet. As a result, maintenance managers can improve the entire facility by helping make sure that the planning and design of new construction and renovation projects takes place with maintenance in mind (1).

Developing a Simple Predictive and Preventive Maintenance Program

A PPM program is needed to avoid equipment failures, utility outages, and production interruptions. From a cost savings angle it is extremely important to do preventive maintenance in order to avoid breakdowns. Periodic inspections and a good lubrication program uncover conditions that could lead to breakdowns. When problems are found early, they can be taken care of without work interruption and costly repairs. Sometimes facility managers are so afraid of downtime that preventive maintenance is done too often. In other cases production does not allow adequate time to provide proper maintenance.

A more sophisticated program is a predictive maintenance system. Maintenance gets done when it is really needed. Special tools are required to pick the right intervals. One simple way is to perform preventive maintenance as suggested by the equipment manufacturer. When machinery is operated intermittently, a very cost-effective way is to install elapsed operating time indicators and to do maintenance only after the equipment has operated for a predetermined number of hours. Vibration analysis is another predictive maintenance tool. The instrument measures the machine vibration at different frequencies as measured by the RPM of the shaft for bearings, fans, or other equipment. Vibration spikes indicate an imminent machine problem. Instruments are available as portable equipment for maintenance workers to manually take readings and input into a computer, or data can be downloaded into a PC or database for analysis. This equipment is also available for stationary installation to continually monitor the vibration of important machinery to alert operation people when a problem is developing. For electric motors this is valuable since motor bearings can be monitored and can easily be fixed without causing shaft damage and failure of motor windings. There is really no excuse for neglecting motor bearings which can cause costly damage and production interruptions. Another way is to check the temperature of machines and equipment. Electric power distribution systems should be checked periodically with infrared heat-scanning, resistivity checks, and voltage insulation testing.

PM programs vary considerably and depend very much on how much a facility can spend and how skilled the technicians are that operate the systems. More expensive and sophisticated are computerized maintenance management systems (CMMS). These systems often include programs that are needed by a facility as well as manufacturing management. Technological developments have made it possible for CMMS to integrate facility, manufacturing, safety, security, and many other functions of a facility. Facility managers can order from an extended menu that includes machinery maintenance, work orders, projects, labor and material cost tracking, scheduling control, stores issues, inventory control, purchasing activity, drawings, reports, analysis, etc.

It is important for managers to know that such systems are available. The cost of workstations are dropping as PCs are becoming more popular. PCs are using software programs that are incorporating the latest client/server technology, using graphical interface and access to other department systems database, bar coding, and conditioning monitoring.

Nearly all facilities inventory their capital equipment; PM systems may involve compiling data on equipment records or in computer files. Equipment cards are filed by areas, equipment names, or manufacturers. Some facilities use visual files, but most use computerized equipment inventory. In conjunction with this inventory it is important to collect manufacturers' manuals, which usually include installation and preventive maintenance suggestions as well as spare parts inventory that maintenance should keep on hand. With this information a simple inspection and PM schedule, complete with check off procedures, can be the basis of starting an effective PM or PPM program. As facilities grow such a simple program can become more complex; financial and historical information is required to make future decisions of adding on and enlarging a facility. Management uses the historical and cost information to determine the cost relationship of PM and PPM programs vs the cost of unscheduled breakdowns and production interruptions, and can determine when old equipment should be replaced.

The keeping of good records is a significant tool in providing good facility maintenance. Especially in chemical-process plants, the quality of record keeping supports efficient operation. Records facilitate pinpointing equipment that causes excessive downtime, identify safety problems, establish manufacturing based cost accounting, and are often used for tax information. Many facilities use the latest computerized maintenance systems to attain faster cost reporting, fast reaction to work request, and a more energy efficient operation.

Supplementing In-House Organizations with Contract Maintenance

If a facility is large enough to support an in-house maintenance department, management of such a department has the responsibility to assure that its maintenance and operation's program is at all times cost effective. The greatest benefit of an in-house maintenance department must be its spirit of ownership. Workers should be trained to have pride of ownership, and must be given the tools to maintain this pride.

In-house departments must not be allowed to become too large. Departments that try to tool up to do everything in-house often lose cost effectiveness. It is difficult to accommodate a large department that handles all maintenance in addition to small repairs, a variety of small construction jobs, and minor breakdowns that occur no matter how good a maintenance program is performed. Having an in-house maintenance department should never rule out the use of contracted maintenance or repair services. For example, even if the in-house air conditioning operating and repair crew also does the required monthly PMs, the semiannual and/or annual A/C chiller plant service could be done by a factory authorized service shop. This would also provide a good quality check of the in-house performance.

Contracted services should be used for specialized work like elevator service, emergency generator testing, uninterruptible power supply (UPS) service, and maintenance. Many facilities have good mechanics who maintain pumps, air compressors, electric panel boards, but they contract out for large repairs as well as routine maintenance if there is danger of missing maintenance schedules. It is also more cost effective to contract out maintenance and repairs that require expensive tooling, instrumentation, and high technology services. Also, contract services should be used if there is a need for long periods of uninterrupted work. There is a need for contracted services to provide relief for peak maintenance loads, seasonal workloads, crash programs, vacation, and sick relief. When hiring a contracted maintenance or repair service, a specification for the work to be performed and expectations of the contractors should be written. The technical capability and financial strength of each contractor should be investigated. Maintenance consultants can clarify the legal and working relationship of owner and contractor. Prime consideration must also be given to safety and security as well as mutual insurance responsibility.

Complete Maintenance Contracts. Such complete service contracts can help especially smaller facilities that can only provide an on-site service administrator. In such a case the maintenance organization must be able to provide design of the program, complete scheduling and reporting services, adequate checks of the required material and its control, and furnish tools, instrumentation, and simple computerized reports showing what has been done. Facilities often start out with such a program; some never switch over to an in-house service and others continue to use parts of such a service even after they start their own.

Supplemental Maintenance Contracts. The contractor supplements an existing in-house maintenance program. This is used to relieve peak loading of the in-house maintenance department which may happen during vacation periods, seasonal work, when breakdowns occur, or for large rearrangement jobs.

This also gives management a chance to study how a contractor does the work and how it compares with the in-house work performance.

Manpower Labor Contracts. Some facilities prefer a strict manpower contract that supplies skilled or unskilled labor as required to fill in and supplement an existing workforce. Arrangements can be made at the outset for temporary workers to be hired by the facility, if the need for their service continues and they fit well into the in-house organization.

Specialty Skills Contracts. These contracts might handle work such as janitorial, window washing, painting, gardening, security guards, parking lot sweeping, weed and pest control, instrumentation maintenance and calibration, air conditioning filter service, water treatment, fire prevention equipment maintenance, lighting replacements and group relamping, public address and music service, reproduction equipment maintenance, vehicle fleet maintenance, waste disposal, laundry service, cafeteria operation, cafeteria equipment, and vending machines maintenance. This list includes only the more common contracted services. These contracts should be negotiated while a new building is being constructed, and the services must be operational when the building is ready for occupancy. A contractor of such services, if consulted during the building design and construction phase, can often make valuable suggestions that will save on maintenance and operating costs.

Motivating the Maintenance Workers

The motivation to work productively is an often neglected factor in U.S. industry. Management must motivate employees by creating a professional environment that includes enriching the work itself. Maintenance workers enjoy working in a professional atmosphere, with good shops and tools, readily available job information and material, and proper training so that jobs can be done right the first time. Enriching jobs means challenging employees, encouraging their initiative and creativity, and empowering them to improve productivity and quality performance. On the job training and allowing greater responsibilities provide maintenance workers with an atmosphere of growth in their profession.

Enriching the Job. The maintenance manager must organize the department to provide all the tools and equipment to do the work, a clear description of the work, an understanding of what is expected of the workers, and knowledge of who makes the decisions and who is available for further instructions. Management must delegate authority and establish priorities. It is also of prime importance that the engineers design and provide a facility that is easily maintainable and has adequate maintenance-shop space. The engineers also must provide blueprints and instruction manuals that enable the maintenance department to do a professional job.

Providing Growth. Management and employees grow with the company. Growth provides the employees with employment security, advancement, good facilities, high status, financial rewards, increased fringe benefits, etc. The maintenance manager provides the staff with good technical training, information about the company, opportunities for promotion within the department, and advancement in the company.

Giving Responsibility and Establishing Goals. For enrichment and growth it is desirable to make individuals responsible for specific areas. Many companies assign machines, specific areas, or buildings to individuals. The individuals are then responsible for the maintenance, cleanliness, safety, security, and productivity in their assigned areas. A good manager can define the objectives, set goals, and judge the results fairly. Such programs encourage worker identification with company goals and greater worker participation in the achievement of these goals. A good manager realizes that an important motivator is recognition of achievement.

Advancement. When the job itself is enriched, when provision is made for growth, when responsibilities are given and goals are established, when achievement is encouraged and provision is made for recognition, advancement occurs. When a maintenance manager delegates responsibilities to the workers and arranges meaningful training for them, workers can be advanced both within the department and also by transfers to other growing departments in the company, such as plant and product engineering, quality assurance, and safety. As trained maintenance workers move into advanced positions, other advancements open within the ranks of the maintenance group, giving helpers and mechanics a chance to improve their craft ratings. A good manager easily can develop a job-enrichment program (2).

Organizing Safety

It is the maintenance manager's responsibility to safeguard the facility and to avoid accidents that cause injuries and losses. The best way to accomplish this is to make every employee aware that safety first must become part of their daily work habits. According to the National Safety Council, the first step in setting up a safety program is to make policy crystal clear to every maintenance worker. Once this policy has been established and is enforced by upper management, maintenance managers can organize a good safety program that should include use of safety factors; forming a safety committee and scheduling safety meetings; using safety suggestion boxes; participation in a company's emergency response team (ERT); organizing safety training, either as an individual department or as part of a company-wide safety program and keeping training records; organizing fire protection training; enforcing the wearing of personal protective clothing and use of safety equipment; enforcing lock-out procedures when working on machinery or systems; establishing a hazard communication program to study and understand the impact of Federal regulations; providing Red Cross first-aid training and CPR training; establishing and keeping up records of any injuries and incidents of the facility that involve the maintenance department; establishing good housekeeping programs and habits; holding periodic safety and housekeeping inspections, complete with follow-up of corrective actions taken; initiating safety instructions for new and transferred employees as part of new employee indoctrinations; and establishing a safety library. Insurance carriers and local fire departments can help with some of these items. The greatest safety problems are apathy, indifference, and attitude toward safety.

Supervisors are responsible to see that every worker is adequate on the job. Physically, mentally, and emotionally inadequate workers are accident prone.

Personal hazards are lack of knowledge, conflict of motives, physical, and mental factors.

Mock OSHA Inspection. Maintenance can learn a lot about how the Occupational Safety and Health Administration (OSHA) trains their inspectors and what is emphasized in an OSHA inspection. Some of the training of OSHA inspectors follows a program involving the recognition of potential hazards, avoidance of these hazards, and prevention of accidents (RAP).

One of the first steps a maintenance department should take when putting a safety program in place is to give the entire facility a thorough mock OSHA inspection. Of course the same holds true for a Fire Marshall inspection or any inspection by a code enforcing agency. Potential hazards that lead to corrective action citations can be grouped in four categories. (*1*) *Safe access and movements* include adequate and unrestricted work areas; aisles and passageways clear of rubbish and obstructions; adequate emergency exits and lights; and protection for openings in floors and roofs. (*2*) *Work procedures* comprise safe handling of chemicals and tools; use of available safety equipment when required; electric lockouts in place where required; and drive guards in place where required. (*3*) *Tools*: all tools must be UL approved; proper electric cords used; and no extension cords or temporary wiring in use. Adequate tools and equipment must be supplied for each job; all hand tools should be checked for electric ground before issued; ladders and scaffolding must be in safe working condition; and adequate exhaust provided where needed. (*4*) *Material handling*: There must be adequate space available for safe work; proper use of cranes, hoists, elevators, trucks; safe methods for loading, unloading, and storing material; material securely stacked and stored; and good housekeeping in place to avoid tripping on wet or slippery floors, refuse control.

This list of basic items covered in a mock OSHA inspection appear to be obvious items that any facility should observe. However, nearly every safety or fire inspection uncovers these items during inspections, causing the facility to receive fines or warnings because maintenance people do not perform their own regular and thorough inspections of their facilities to avoid corrective actions by code enforcing agencies.

Use of Outside Contractors. When maintenance uses an outside contractor or a subcontractor, it is important that maintenance managers understand the legal and working relationship between the owner and the contractor. In all negotiations the maintenance manager represents the owner. Safety and fire prevention in form of a complete loss control is the combined responsibility of the maintenance management and the job management of each individual contractor. Their prime consideration must always be the safety of their workers, prevention of even minor accidents, and prevention of loss of material and property.

Tools and Equipment. Adequate tools and equipment must be provided for each part of the job and arrangements made for proper care of them. Tools should be checked before issue and periodic inspections of tools should be made on the jobsite while they are in use. Only OSHA-approved electrical extension-cords, plug-caps, and receptacles should be used. Adequate guards must be provided for exposed gears, sprockets, pulleys and flywheels, belt and chain drives, set screws, etc, of any power transmission equipment and adequate lighting and exhaust provided where needed. Protective clothing, and eye and ear protection, must be available where needed, and special care must be taken in the use of ladders and scaffolds. Special fire-safety precautions must be used for fluids in hydraulic-powered hand tools. All leased equipment must be checked to see that it complies with OSHA standards, and arrangements must be made to ensure that all hazardous shipments meet Department of Transportation (DOT) regulations. Finally, workers must get confined space entry procedure training before working in areas that are difficult to enter and leave, due to tight openings and restricted space availability, and that present serious hazards to the workers.

Employers who provide a nonstop quality program as part of their service to both customers and employees have found that total quality management (TQM) must include total quality safety. Such a program must include safety training for all employees as the most important item on the agenda. The benefits of a good safety program are fewer work interruptions, less down-time, and fewer injuries. Employees appreciate a safe and healthy work environment. A good safety program does not increase operating expenses; rather it drives costs down (3).

Including Environmental Care and Appreciation

The responsibility of maintenance managers must include strict compliance to environmental regulations. Pollution prevention is a continuous responsibility of the maintenance management and can only be achieved by educating every worker in the maintenance department and watching over outside contractors. Keeping up with the state-of-the-art of environmental regulation compliance is as important as keeping up with the knowledge of operating a facility. Making sure that every company is a good neighbor by preventing pollutants from the plants entering the environment is as important as the safeguarding of employees from hazardous exposure while at work.

An environmental requirement of the 1990s insists that formal facility plans must address the elimination of pollution from the plants by modifying manufacturing processes. Maintenance is usually in charge of scrubbers, air-exhaust systems, operation of the wastewater treatment before it gets discharged to the city sewage system, elimination of contamination from storm drains, hazardous waste removal, and other environmental chores. Making sure while doing all this that all regulations are strictly enforced is necessary to avoid citations resulting in heavy fines and unfavorable publicity.

The identification of pollution prevention options has become a maintenance requirement. In addition to these requirements, the National Institute of Occupational Safety and Health (NIOSH) performed its first investigation of indoor air quality. The U.S. Department of Energy (DOE) has also begun to research air quality.

Avoiding sick building syndrome requires special consideration of building construction and maintenance material used. Asbestos, organic solvents, paint sprays, dirty filters, moist environment caused by poorly maintained humidifiers, and dirty HVAC machinery have contributed to the deterioration of indoor air quality. Nonstop quality maintenance is required at all times.

The American Society of Heating, Refrigeration and Air Conditioning Engineers (ASHRAE) and the American National Standards Institute (ANSI) have provided standards for architects and engineers who specify HVAC systems. Sufficient and clean make up air must be provided at all times for a healthy building occupancy. It is the maintenance department that must make sure that the HVAC systems are clean, maintained properly, and operate within design specifications. Every maintenance worker must be empowered to care for the air handling and exhaust system of the facility as part of his or her ownership responsibility.

Developing an Energy Conservation Program

Although there are different opinions of how to combat the U.S. national energy shortage, it is agreed that in order to survive as much energy must be conserved as possible. Industrial energy consumption accounts for about half of the total energy used in the United States. Because the cost of fuel, electric power, and water has increased steadily and dramatically since the 1980s, it is of even greater importance to develop an ongoing conservation program. Plant engineers and maintenance managers must realize that such a program saves money and reduces manufacturing cost. For an existing plant, any energy program must start with a careful survey of energy uses and losses. For example, an inventory of electric power, water, steam, compressed air, and vacuum usually shows some obvious wastes resulting from equipment faults and equipment being on line when it is not needed. The maintenance management must have a program that provides for continuous monitoring of energy usage, survey systems, and equipment for energy losses, and must make sure that lighting, heating, ventilating, air-conditioning, and production machinery is not energized when it is not needed.

An effective energy-conservation program should include efficient use of lighting, including low energy-consuming lights; efficient use of heating, ventilating, and air conditioning; use of heat recovery; use of solar heating; reducing peak electric loads and shifting some energy uses to evening and weekends when power companies have electric power available at a lower cost; improving the power factor of the plant, using only energy efficient electric motors; making use of variable speed and frequency drives; turning off lighting, water, and machinery when it is not in use; improving the efficiency of heat transfer to reduce energy use; reuse water, ie, avoid waterwaste; and improve and change production processes to save energy. Industrial plants should have an energy-conservation committee with a person in charge who makes sure that conservation is carried out; has a plan, communicates it, and involves other employees; has goals and communicates achievements; continuously monitors usage, savings, and progress; convinces management that, in order to optimize achievement in energy conservation, a company must spend money for the work force and equipment.

In the past, energy was so plentiful and inexpensive that facilities were not designed with energy conservation in mind. In the 1990s, however, architects and engineers must design any new building with the following points in mind: careful orientation of the building, adequate insulation and choice of exterior glass, lighting level and use of low energy-consuming lighting, low energy-using HVAC systems that make use of heat recovery, and reuse of water.

When a new facility has been designed for energy conservation and when an older facility has been remodeled to make it more energy efficient, it becomes the responsibility of the maintenance management to provide a preventive and ongoing maintenance program that assures a continuous energy-conservation effort. When this maintenance effort is allowed to relax, energy expenditures tend to increase. Without a good maintenance program, even a well-designed, energy-conserving facility soon loses all its effectiveness. It is important to have a good ongoing preventive-maintenance program; it is even of greater importance to apply this preventive-maintenance program to energy conservation (4).

Creating a Partnership with Suppliers and Equipment Manufacturers

In order to be successful any maintenance department needs strong internal customer–supplier partnerships. Such partnerships require opportunity for mutual benefits, predictable performance by each partner, and communication across links.

This concept not only includes the many service contracts that maintenance departments negotiate, but also the daily supplies as well as capital equipment and installation contracts. In the maintenance department a close partnership must be in place between suppliers who help maintenance by providing nonstop quality service to their internal customers, which enables them to provide nonstop quality service to the external customers of the company.

In order to establish this type of service, the maintenance management should meet often with their outside supplier as well as with their inside customers to benefit the end user: the external customers. During these meetings the following steps should be taken to assure nonstop quality: listen to the customers, work with customers to clarify expectations, identify measurable indicators, exceed expectations, deliver products and services when customers need them, keep promises, design for ease of use, constantly improve, focus improvements on areas related to customer expectations, and respond quickly (5).

Cooperation with Other Service Groups of the Facility

Maintenance departments provide support to other functions of manufacturing management. Some of the corporate staff activities that are of concern to the plant maintenance organization are industrial relations, finance, material control, services, and manufacturing engineering.

Industrial Relations. Employment; compensation; motivation; training; employee eating facilities, eg, cafeteria and vending machines; recreational activities; health and safety; plant security; parking; and visitors reception are included here.

Finance. Cost accounting, budget control, payroll, records, accounts payable, accounts receivable, capital equipment control, taxes, customs, insurance, loss control, and management information are all part of the finance function.

Material Control. Production forecasting, scheduling, direct and indirect material procurement, purchasing, shipping and receiving, storerooms, storage, surplus equipment, and reclamation of materials are included.

Services. These include telephone system; FAX machines; mail; moves; deliveries; trucks and vehicle fleet management; janitorial and housekeeping services; landscaping and gardening; water treatment and waste disposal; reproduction, eg, copying machines, blue-line machines, and printing; furniture control; tool rental and maintenance; and instrument rental and calibration.

Manufacturing Engineering. Industrial engineering, equipment engineering, production-machine maintenance, instrumentation engineering and maintenance, tooling and operation of machine shop, chemical and environmental engineering, and the technical library are within this area.

Some of these functions are part of a well-organized, multiplant operation. Often, when a company or plant is newly started or a satellite plant is begun, some of these corporate staff functions have to be handled by the plant engineer and maintenance manager. Usually it is accepted that plant engineering and maintenance is the initial caretaker on a catch-all basis; however, this usually does not continue as a facility grows. Then other staff functions begin and it is always the duty and responsibility of the plant engineer and maintenance manager and their departments to support and cooperate with other service groups of the facility; often the growth and success of a company depends on this type of cooperation.

Controlling Cost and Assuring Budget Responsibility

A well-managed maintenance program results in a reduction of operating costs and a controlled budget that makes use of goals that were established to ensure the maximum return on the capital invested. This must be done despite rising costs, OSHA regulations, environmental protection, and the energy shortage. To be effective, every maintenance manager always must have the profits of the company in mind. Good managers must be able to account for every hour spent by their employees and contractors. As labor and material costs increase, it is management's duty to devise systems, procedures, and controls that cause a reduction in unproductive time, waste of material, and unnecessary expenses.

Planning is defined as the process of analyzing each job so as to determine the nature of the job and the results desired; specify the logical sequence of the job and apply humanpower and estimates for each sequential step; list predeterminable material, tools, and special equipment; and estimate the total cost to meet the required results.

Scheduling is the allocation of manpower and special equipment to jobs based on operational requirements; it includes the delivery of materials, tools, and equipment to the jobsite. The most cost-effective and productive approach to plant organization is outlined in planning, and the timing of jobs is determined by scheduling. Planning and scheduling should provide working environments that are not punctuated with interruptions, ie, all equipment and instruction should be available to the worker before he or she begins a job. The managers must have enough time at the jobsite to supervise and train the craftspeople. Planning and scheduling also should provide for adequate equipment examination and/or overhaul to ensure maximum reliability and, in essence, reduced cost (6).

In modern plant facilities, management controls costs and achieves budget responsibility by providing the above-mentioned programs and tools. The overriding management responsibility is to make everyone in the department aware of costs and budget responsibility and to provide a suitable working environment for all employees; this can be achieved by following the principles described in this article.

BIBLIOGRAPHY

"Maintenance" in *ECT* 3rd ed., Vol. 14, pp. 753–769, by E. M. Bergtraun, National Semiconductor Corp.

1. E. M. Bergtraun, "Beyond Renovation: The Role of Maintenance in Ensuring Facility 'Maintainability'," *Maintenance Solutions*, Nov./Dec. 1993.
2. E. M. Bergtraun, "The Importance of Job Satisfaction," *Plant Services*, Apr. 1990.
3. E. M. Bergtraun, "Organizing An Effective Facility Safety Program," *AIPE Facilities*, July/Aug. 1990.
4. A. Thumann, *Plant Engineers and Managers Guide to Energy Conservation*, Fairmont Press, Lilburn, Ga., 1989.
5. E. M. Bergtraun, "Nonstop Quality in Facilities Management," *AIPE Facilities*, July/Aug. 1993.
6. R. P. McFarland, in B. T. Lewis, *Management Handbook for Plant Engineers*, McGraw-Hill Book Co., Inc., New York, 1977, Sect. 4, Chapt. 1.

General References

The American Institute of Plant Engineers, Cincinnati, Ohio, publishes an information services catalog which features *Facility Management Library Reprint Services*.

G. H. Magee, *Facilities Maintenance Management*, R. S. Means, Kingston, Mass., 1988.

R. K. Mobley, *An Introduction to Predictive Maintenance*, Van Nostrand Reinhold, New York, 1990.

J. W. Criswell, *Planned Maintenance for Productivity and Energy Conservation*, Fairmont Press, Lilburn, Ga., 1989.

B. C. Langley, *Plant Maintenance*, Prentice Hall, New York, 1986.

W. Wrennall and Q. Lee, *Handbook of Commercial and Industrial Facilities Management*, McGraw-Hill Book Co., Inc., New York, 1993.

A. Thumann, *Plant Engineers and Managers Guide to Energy Conservation*, PE, CEM, Fairmont Press, Lilburn, Ga., 1989.

E. Teicholz, *Computer-Aided Facility Management*, McGraw-Hill Book Co., Inc., New York, 1992.

R. A. Carlson and R. DiG., *Understanding Building Automation Systems*, R. S. Means, Kingston, Mass., 1991.

R. C. Rosaler, *Standard Handbook of Plant Engineering*, 2nd ed., McGraw-Hill Book Co., Inc., New York, 1994.

ERIC M. BERGTRAUN
National Semiconductor Corporation

MALEIC ANHYDRIDE, MALEIC ACID, AND FUMARIC ACID

Maleic anhydride [*108-31-6*] (**1**), maleic acid [*110-16-7*] (**2**), and fumaric acid [*110-17-8*] (**3**) are multifunctional chemical intermediates that find applications in nearly every field of industrial chemistry. Each molecule contains two acid carbonyl groups and a double bond in the α,β position. Maleic anhydride and

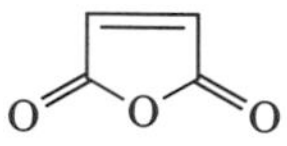

(**1**) 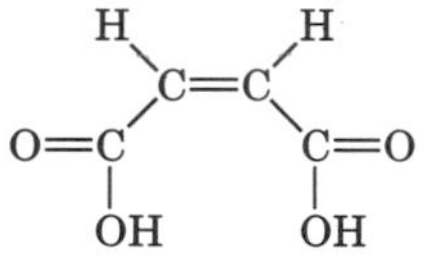 (**2**) 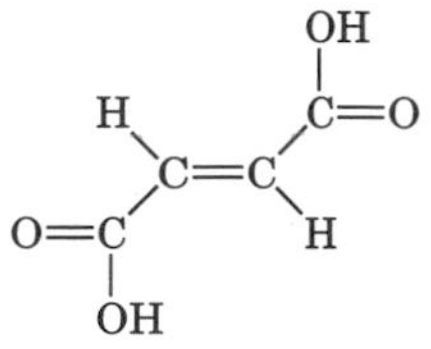 (**3**)

maleic acid are important raw materials used in the manufacture of phthalic-type alkyd and polyester resins, surface coatings, lubricant additives, plasticizers (qv), copolymers (qv), and agricultural chemicals (see ALKYD RESINS; POLYESTERS, UNSATURATED; LUBRICATION AND LUBRICANTS). Both chemicals derive their common names from naturally occurring malic acid [*6915-15-7*]. Other names for maleic anhydride are 2,5-furandione, toxilic anhydride, or *cis*-butenedioic anhydride. Maleic acid is also called (*Z*)-2-butenedioic acid, toxilic acid, malenic acid, or *cis*-1,2-ethylenedicarboxylic acid.

Fumaric acid occurs naturally in many plants and is named after *Fumaria officinalis*, a climbing annual plant, from which it was first isolated. It is also known as (*E*)-2-butenedioic acid, allomaleic acid, boletic acid, lichenic acid, or *trans*-1,2-ethylenedicarboxylic acid. It is used as a food acidulant and as a raw material in the manufacture of unsaturated polyester resins, quick-setting inks, furniture lacquers, paper sizing chemicals, and aspartic acid [*56-84-8*].

Maleic anhydride and the two diacid isomers were first prepared in the 1830s (1) but commercial manufacture did not begin until a century later. In 1933 the National Aniline and Chemical Co., Inc., installed a process for maleic anhydride based on benzene oxidation using a vanadium oxide catalyst (2). Maleic acid was available commercially in 1928 and fumaric acid production began in 1932 by acid-catalyzed isomerization of maleic acid.

Physical Properties

Physical constants (3–20) for maleic anhydride, maleic acid, and fumaric acid including solid and solution properties are given in Tables 1–3. From single crystal x-ray diffraction data (14,15), maleic anhydride is a nearly planar molecule with the ring oxygen atom lying 0.003 nm out of the molecular plane. A twofold rotation axis bisects the double bond and passes through the ring oxygen atom.

Table 1. Physical Properties of Maleic Anhydride, Maleic Acid, and Fumaric Acid

Property	Maleic anhydride	Maleic acid	Fumaric acid	Ref.
formula	$C_4H_2O_3$	$C_4H_4O_4$	$C_4H_4O_4$	
formula weight	98.06	116.07	116.07	
mp, °C	52.85	138–139[a]	287	3,4
		130–130.5[b]		
		144 (air)	282 (air)	5
bp, °C	202	ca 138 (dec)	290	4
sp gr, at 20/20°C, solid	1.48[c]	1.590	1.635	4
molar volume		81	79	7
heat of formation, kJ/mol[d]	−470.41	−790.57	−811.03	8
free energy of formation, kJ/mol[d]		−625.09	−655.63	
heat of combustion, kJ/mol[d]	−1390	−1358	−1335	9
heat of hydrogenation, kJ/mol[d]		−153.2	−130.3	10
heat capacity, kJ/(K·mol)[d]				
solid	0.119	0.137	0.142	6,11
liquid	0.164			
heat of vaporization, kJ/mol[d]	54.8			12
heat of fusion, kJ/mol[d]	13.65			3
heat of hydrolysis, kJ/mol[d]	−34.9			6
dipole moment, 10^{-30} C·m[e]	13.2	10.6	8.17	13
crystalline form	orthorhombic	monoclinic	monoclinic, prismatic, needles, or leaflets	14–16
space group	$P2_12_12_1$	$P2_1/c$		14–16
a, nm	0.7180	0.7473		
b, nm	1.1231	1.0098		
c, nm	0.539	0.7627		
β, deg		123.59		
dissociation constant, at 25°C				17
K_1		1.14×10^{-2}	9.57×10^{-4}	
K_2		5.95×10^{-7}	4.13×10^{-5}	
heat of neutralization, kJ/mol[d]	126.9			6

[a] Crystallized from water.
[b] Crystallized from alcohol; sublimes at 165°C at 0.23 kPa.
[c] Specific gravity at 70/70°C, molten = 1.3 (6).
[d] To convert kJ to kcal, divide by 4.184.
[e] In dioxane at 25°C. To convert C·m to debye, divide by 3.336×10^{-30}.

Table 2. Solubility[a] of Maleic Anhydride, Maleic Acid, and Fumaric Acid, g/100 g Solution

Solvent	Maleic anhydride[b]	Maleic acid[c]	Fumaric acid
Water[d]			
at 25°C		44.1	0.70
40°C		52.9	1.05
60°C		58.9	2.34
97.5°C		79.7	
100°C			8.93
acetone			
at 20°C		38.6	
25°C	227		
29.7°C		26.3	1.69
benzene, at 25°C	50	0.024	0.003
toluene, at 25°C	23.4		
o-xylene, at 25°C	19.4		
kerosene (bp 190–210°C), at 25°C	0.25		
methanol, at 22.5°C		41.0	
ethanol			
at 0°C		30.2	
22.5°C		34.4	
95% ethanol, at 29.7°C		41.1	5.44
1-propanol			
at 0°C		20.0	
22.5°C		24.3	
chloroform, at 25°C	52.5	0.11	0.02
carbon tetrachloride, at 25°C	0.6	0.002	0.027
diethyl ether, at 25°C		7.57	0.71
2-butenenitrile, at 50°C		4.38	0.034
ethyl acetate, at 25°C	112		

[a]Refs. 13 and 17; g/100 g solution unless otherwise noted.
[b]In g/100 g solvent.
[c]Heat of solution = 18.6 kJ/mol (4.45 kcal/mol) (6).
[d]At all temperatures given maleic anhydride hydrolyzes slowly.

Similar bond distances and angles for maleic anhydride were obtained using electron diffraction (21) and double resonance modulation microwave spectroscopic (22) techniques. The polarized Raman spectrum of single-crystal maleic anhydride has been reported (23).

Maleic and fumaric acids have physical properties that differ due to the cis and trans configurations about the double bond. Aqueous dissociation constants and solubilities of the two acids show variations attributable to geometric isomer effects. X-ray diffraction results for maleic acid (16) reveal an intramolecular hydrogen bond that accounts for both the ease of removal of the first carboxyl proton and the smaller dissociation constant for maleic acid compared to fumaric acid. Maleic acid isomerizes to fumaric acid with a derived heat of isomerization of −22.7 kJ/mol (−5.43 kcal/mol) (10). The activation energy for the conversion of maleic to fumaric acid is 66.1 kJ/mol (15.8 kcal/mol) (24).

Table 3. Other Properties of Maleic Anhydride

Property	Value	Property	Value
flash point,[a] °C		vapor pressure,[e] kPa[f]	
open cup	110	at 44.0°C	0.13
closed cup	102	63.4°C	0.67
flammable limits,[a,b] vol%		78.7°C	1.3
lower	1.4–3.4	95.0°C	2.7
upper	7.1	111.8°C	5.3
autoignition temperature,[a] °C	477	122.0°C	8.0
vapor density[a] (air = 1)	3.38	135.8°C	13.3
viscosity,[c] mPa·s(= cP)		155.9°C	26.7
at 60°C	0.61	179.5°C	53.3
90°C	1.07	202.0°C	100
150°C	0.6		
pH of water solutions[d]			
$1 \times 10^{-2} M$	2.42		
$5 \times 10^{-3} M$	2.62		
$1 \times 10^{-4} M$	3.10		
polarographic half-wave potential,[d] V			

[a]Ref. 18. [b]Ref. 19. [c]Ref. 17. [d]Ref. 20. [e]Ref. 13. [f]To convert kPa to mm Hg, multiply by 7.5.

Chemical Properties

The *General References* and two other reviews (17,25) provide extensive descriptions of the chemistry of maleic anhydride and its derivatives. The broad industrial applications for this chemistry derive from the reactivity of the double bond in conjugation with the two carbonyl oxygens.

Acid Chloride Formation. Monoacid chlorides of maleic and fumaric acid are not known. Treatment of maleic anhydride or maleic acid with various reagents such as phosgene [*75-44-5*] (qv), phthaloyl chloride [*88-95-9*], phosphorus pentachloride [*10026-13-8*], or thionyl chloride [*7719-09-7*] gives 5,5-dichloro-2(5*H*)furanone [*133565-92-1*] (**4**) (26). Similar conditions convert fumaric acid to fumaryl chloride [*627-63-4*] (**5**) (26,27). Noncyclic maleyl chloride [*22542-53-6*] (**6**) forms in 11% yield at 220°C in the reaction of one mole of maleic anhydride with six moles of carbon tetrachloride [*56-23-5*] over an activated carbon [*7440-44-4*] catalyst (28).

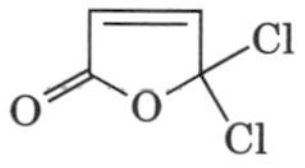

(**4**) 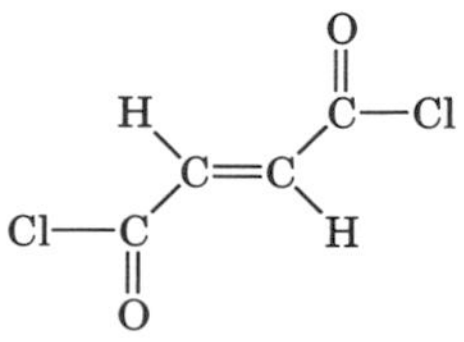 (**5**) 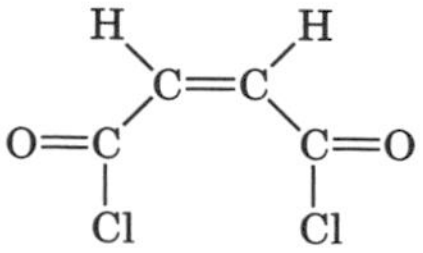 (**6**)

Acylation. In chlorinated solvents, maleic anhydride reacts with aromatic hydrocarbons (ArH) in the presence of aluminum chloride [*7446-70-0*], $AlCl_3$, to form β-aroylacrylic acids (29).

$$(\mathbf{1}) + ArH \xrightarrow[C_2H_2Cl_4]{AlCl_3} Ar\overset{O}{\overset{\|}{C}}CH{=}CH\overset{O}{\overset{\|}{C}}OH$$

Under Friedel-Crafts conditions, *trans*-1,2-dibenzoylethylene [*959-28-4*] is synthesized by the reaction of one mole of fumaryl chloride with two moles of benzene (30) (see FRIEDEL-CRAFTS REACTIONS).

Alkylation. Maleic anhydride reacts with alkene and aromatic substrates having a C–H bond activated by α,β-unsaturation or an adjacent aromatic resonance (31,32) to produce the following succinic anhydride derivatives.

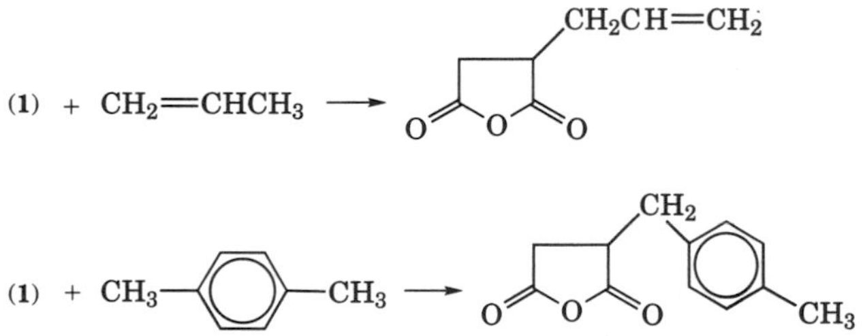

(1) + $CH_2{=}CHCH_3$ ⟶ [succinic anhydride substituted with $CH_2CH{=}CH_2$]

(1) + CH_3–C₆H₄–CH_3 ⟶ [succinic anhydride substituted with CH_2–C₆H₄–CH_3]

Typical reaction conditions are 150 to 300°C and up to 2 MPa pressure. Polyalkenyl succinic anhydrides are prepared under these conditions by the reaction of polyalkenes in a nonaqueous dispersion of maleic anhydride, mineral oil, and surfactant (33).

N-Alkylpyrroles react with maleic anhydride to give the electrophilic substitution product (**7**) and not the Diels-Alder addition product found for furan and thiophene compounds (34). However, the course of this reaction can be altered by coordination of the pyrrole compound to a metal center.

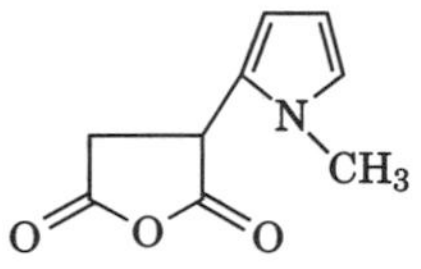

[succinic anhydride substituted with N-methylpyrrol-2-yl; N, CH_3]

(**7**)

Amidation. Reaction of maleic anhydride or its isomeric acids with ammonia [*7664-41-7*] (qv), primary amines (qv), and secondary amines produces mono- or diamides. The monoamide derivative from the reaction of ammonia and maleic anhydride is called maleamic acid [*557-24-4*] (**8**). Another monoamide derivative formed from the reaction of aniline [*62-53-3*] and maleic anhydride is maleanilic acid [*555-59-9*] (**9**).

$HOOC{-}CH{=}CH{-}CONH_2$ (**8**)

$HOOC{-}CH{=}CH{-}CONH{-}C_6H_5$ (**9**)

The reactions of primary amines and maleic anhydride yield amic acids that can be dehydrated to imides, polyimides (qv), or isoimides depending on the reaction conditions (35–37). However, these products require multistep processes. Pathways with favorable economics are difficult to achieve. Amines and pyridines decompose maleic anhydride, often in a violent reaction. Carbon dioxide [*124-38-9*] is a typical end product for this exothermic reaction (38).

Maleic hydrazide [*123-33-1*] (**10**) is one of a number of commercial agricultural chemicals derived from maleic anhydride. Maleic hydrazide was first prepared in 1895 (39) but about 60 years elapsed before the intermediate products were elucidated (40).

maleic anhydride + $N_2H_4H_2SO_4$ → (**10**)

Concerted Nonpolar Reactions. Maleic anhydride exemplifies the model dienophile for cycloaddition with dienes such as 1,3-butadiene [*106-99-0*] (**11**) (41). Tetrahydrophthalic anhydride [*85-43-8*] (**12**) or its derivatives are produced.

(**1**) + $HC_2{=}CH{-}CH{=}CH_2$ (**11**) → [transition state] → (**12**)

The success of the cycloaddition reaction of maleic anhydride varies greatly depending on which heterocyclic diene is used. The cycloaddition of maleic anhydride to furan [*110-00-9*] occurs in a few seconds under ambient conditions

(42,43). Although the endo adduct (**14**) is favored kinetically, the exo adduct (**13**) is isolated.

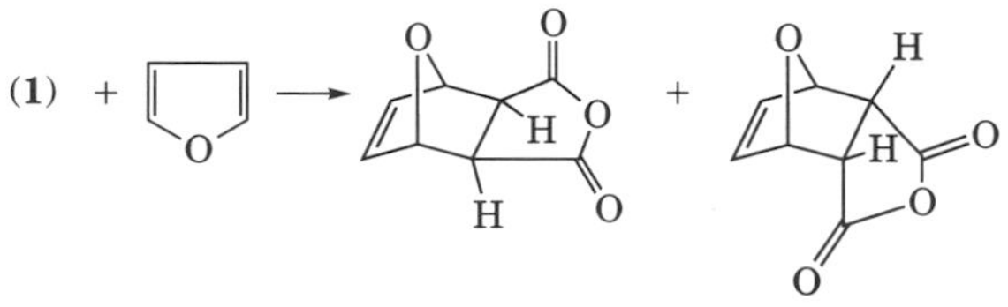

(**13**) (**14**)

Endo adducts are usually favored by interactions between the double bonds of the diene and the carbonyl groups of the dienophile. As was mentioned in the section on alkylation, the reaction of pyrrole compounds and maleic anhydride results in a substitution at the 2-position of the pyrrole ring (34,44). Thiophene [*110-02-1*] forms a cycloaddition adduct with maleic anhydride but only under severe pressures and around 100°C (45). Addition of electron-withdrawing substituents about the double bond of maleic anhydride increases rates of cycloaddition. Both α-(carbomethoxy)maleic anhydride [*69327-00-0*] and α-(phenylsulfonyl) maleic anhydride [*120789-76-6*] react with 1,3-dienes, styrenes, and vinyl ethers much faster than tetracyanoethylene [*670-54-2*] (46).

Metal-Induced Cycloadditions. The effect of coordination on the metal-induced cycloadditions of maleic anhydride and the isostructural heterocycles furan, pyrrole, and thiophene has been investigated (47). Each heterocycle is bound to an Os(II) center in the complex $[(NH_3)_5Os(2,3\text{-}\eta^2\text{-}L)]^{2+}$, where L = furan, pyrrole, and thiophene. Although neither the furan nor thiophene complexes react with maleic anhydride over a period of 10 days, the pyrrole complex (**15**) reacts rapidly at room temperature and 101.3 kPa to form a mixture of endo (**17**) and exo (**16**) complexes. An azomethine ylide intermediate was

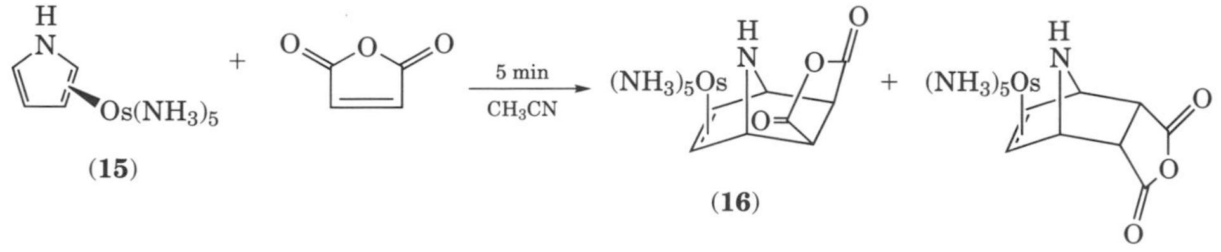

(**15**) (**16**) (**17**)

postulated as the key intermediate through which maleic anhydride added to the 2- and 5-positions of the coordinated pyrrole ring.

2 + 2 Cycloadditions. Cyclobutene adducts are formed from the reaction of acetylenic derivatives and maleic anhydride through a 2 + 2 cycloaddition (48). The reaction is photochemically catalyzed (see PHOTOCHEMICAL TECHNOLOGY).

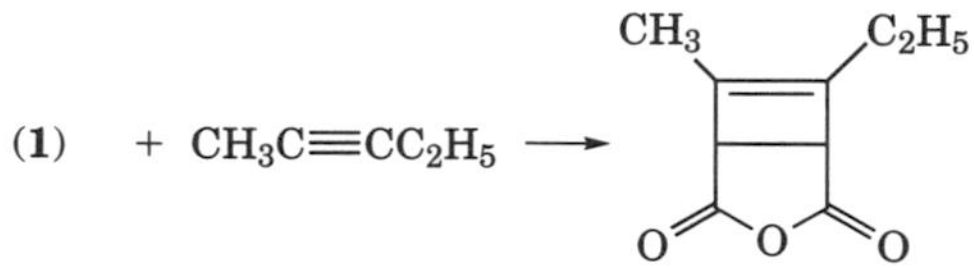

(1) + $CH_3C{\equiv}CC_2H_5$ ⟶

Cyclobutane derivatives are formed after exposing a mixture of alkenes and maleic anhydride to light. Photoadducts are formed by reaction of maleic anhydride with ethylene [*74-85-1*] and benzene (50).

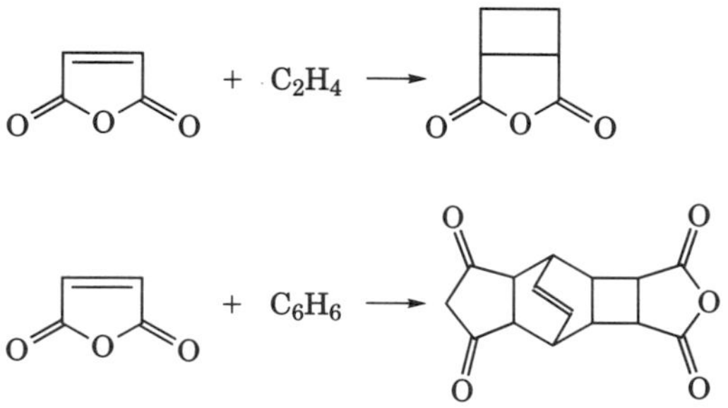

+ C_2H_4 ⟶

+ C_6H_6 ⟶

Ene Reaction. Maleic anhydride and maleate and fumarate esters participate in thermal ene reactions (51) with alkenes having an allylic hydrogen. An ene reaction is the thermal reaction of an alkene having an allylic hydrogen (an ene) with a compound containing a double or triple bond to form a new bond with migration of the ene double bond and 1,5-hydrogen shift. Alkenylsuccinic anhydrides are produced. ^{13}C-nmr spectroscopy has been used to determine regioselectivity, selectivity for endo and exo, and selectivity for cis and trans in the reaction (200°C, 16 h) of maleic anhydride with the nine linear decene isomers (52). The results show a slight preference for maleic anhydride addition to the least hindered end of the decene isomer. Similar reaction conditions were used to form ene adducts with mono- and disubstituted oligoisobutylenes (53,54).

Decomposition and Decarboxylation. Maleic anhydride undergoes anaerobic thermal decomposition in the gas phase in a homogeneous unimolecular reaction to give carbon monoxide, carbon dioxide, and acetylene [*74-86-2*] in equimolar amounts. The endothermic (ΔH = +142 kJ/mol (33.9 kcal/mol)) decomposition was studied in a quartz tube with and without quartz packing in the temperature range of about 370 to 490°C (55). The same linear Arrhenius plot was obtained for packed and unpacked reaction vessels. The same decomposition products were found during photolysis between 220 and 350 nm (55). Catalysts alter the decomposition profile for maleic anhydride. The decomposition of maleic acid in an aqueous solution over a bed of Y zeolite which contains copper and sodium at around 200°C occurs with 95% conversion. The selectivity of the zeolite catalyst to acrylic acid [*79-10-7*] is 91%.

Maleic anhydride is decomposed in the liquid phase by various nitrogen bases. Treatment of maleic anhydride in refluxing acetic acid with

2-aminopyridine [*504-29-0*] gives, after work-up in 4 *N* H_2SO_4 at 100°C, the decarboxylative dimerization product, 2,3-dimethylmaleic anhydride [*766-39-2*] (75% yield), and CO_2 (56). Homopolymers of maleic anhydride form in the liquid phase upon addition of pyridine [*110-86-1*] (57,58). At a maleic anhydride/pyridine ratio of 1.64 in acetone solution at 25°C, reproducible oligomers having molecular weights of 400 to 700 are formed (58). Exothermic decomposition of maleic anhydride can occur with amines and alkali (38,57). Explosions can result from this reaction (38).

Electrophilic Addition. Electrophilic reagents attack the electron-deficient bond of maleic anhydride (25). Typical addition reagents include halogens, hydrohalic acids, and water.

Esterification. Both mono- and dialkyl maleates and fumarates are obtained on treatment of maleic anhydride or its isomeric acids with alcohols or alkoxides (25). An extensive review is available (59). Alkyl fumarates (**18**) often are made from isomerization of the corresponding maleate (**19**) (60).

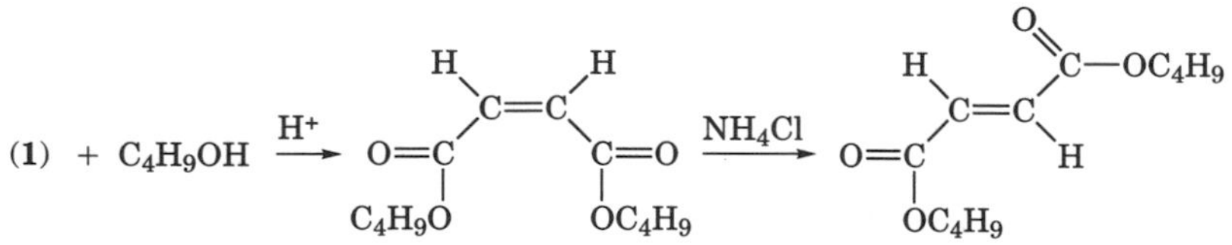

$$(\mathbf{1}) + C_4H_9OH \xrightarrow{H^+} \underset{(\mathbf{19})}{\text{cis-}C_4H_9OC(=O)CH{=}CHC(=O)OC_4H_9} \xrightarrow{NH_4Cl} \underset{(\mathbf{18})}{\text{trans-}C_4H_9OC(=O)CH{=}CHC(=O)OC_4H_9}$$

Glycols and epoxides react with maleic anhydride to give linear unsaturated polyesters (61,62). Ethylene glycol and maleic anhydride combine to form the following repeating unit. This reaction is the first step in industrially important polyester resin production (see POLYESTERS, UNSATURATED).

$$\text{+}CH_2CH_2OC(=O)CH{=}CHC(=O)O\text{+}_n$$

Free-Radical Reactions. Free-radical reactions of maleic anhydride are important in polymerizations and monomer synthesis. Nucleophilic radicals such as the one from cyclohexane [*110-82-7*] serve as hydrogen donors that add to maleic anhydride at the double bond to form cyclohexylsuccinic anhydride [*5962-96-9*] (**20**) (63).

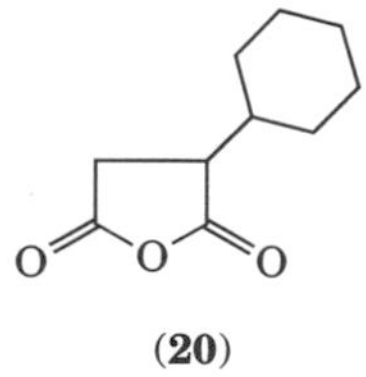

(**20**)

Free-radical reaction rates of maleic anhydride and its derivatives depend on polar and steric factors. Substituents added to maleic anhydride that decrease planarity of the transition state decrease the reaction rate. The reactivity decreases in the order maleic anhydride > fumarate ester > maleate ester.

Grignard-Type Reactions. Grignard reagents provide nucleophilic addition to the maleyl carbonyl groups but yields are often poor (25). Phenyl addition to dimethyl maleate has been demonstrated with a palladium-based catalyst system (63). A solution of dimethyl maleate, iodobenzene [*591-50-4*], triethylamine [*121-44-8*], and palladium diacetate [*19807-27-3*] (1 mol %) in acetonitrile [*75-05-8*] after reflux for 5 h gives 39% (*Z*)-dimethyl phenylmaleate [*29576-99-6*] and 54% (*E*)-dimethyl phenylmaleate [*29394-47-6*] without loss of unsaturation.

Halogenation. Halogens add directly to the double bond of maleic anhydride to give dihalosuccinic acids. However, different procedures are used for dihalomaleic anhydride derivatives. Fluorinated C_4 substrates offer access to difluoromaleic anhydride [*669-78-3*] (64). Hexafluoro-2,5-dihydrofuran [*24849-02-3*] is distilled into sulfur trioxide [*7446-11-9*] at 25°C. Addition of trimethyl borate [*121-43-7*] initiates a reaction which upon heating and distillation leads to a 53% yield of difluoromaleic anhydride. Dichloromaleic anhydride [*1122-17-4*] can be prepared with 92% selectivity by oxidation of hexachloro-1,3-butadiene [*87-68-3*] with SO_3 in the presence of iodine-containing molecules (65). Passing vaporized hexachlorobutadiene over a vanadium–phosphorus oxide catalyst also gives dichloromaleic anhydride (66). A benzene solution of tetrabromofuran [*32460-09-6*] can be photooxidized to dibromomaleic anhydride [*1122-12-9*] in 85% yield with uv light (67). Radical chain mechanisms are suggested.

Hydration and Dehydration. Maleic anhydride is hydrolyzed to maleic acid with water at room temperature (68). Fumaric acid is obtained if the hydrolysis is performed at higher temperatures. Catalysts enhance formation of fumaric acid from maleic anhydride hydrolysis through maleic acid isomerization.

Hydration of fumaric acid proceeds at high temperatures and pressures to give DL-malic acid [*6915-15-7*], $HOOCCH_2CHOHCOOH$ (25).

Maleic acid can be thermally dehydrated to maleic anhydride (69) or dehydrated through azeotropic distillation. Solvents such as xylenes (70) or dibutyl phthalate [*84-74-2*] (71) are preferred but conditions must be carefully adjusted to avoid isomerization to fumaric acid.

Hydroformylation. Esters of maleate and fumarate are treated with carbon monoxide and hydrogen in the presence of appropriate catalysts to give formyl derivatives. Dimethyl fumarate [*624-49-7*] is hydroformylated in 1:1 CO/H_2 at 100°C and 11.6 MPa pressure with a cobalt [*7440-48-4*] catalyst to give an 83% yield of dimethyl formylsuccinate [*58026-12-3*] product (72).

Isomerization. Maleic acid is isomerized to fumaric acid by thermal treatment and a variety of catalytic species. Isomerization occurs above the 130 to 140°C melting point range for maleic acid but below 230°C, at which point fumaric acid is dehydrated to maleic anhydride. Derivatives of maleic acid can also be isomerized. Kinetic data are available for both the uncatalyzed (73) and thiourea catalyzed (74) isomerizations of the cis to trans diacids. These data suggest that neither carbonium ion nor succinate intermediates are involved in the isomerization. Rather, conjugate addition imparts sufficient single bond character to afford rotation about the central C—C bond of the diacid (75).

Ligation to Metal Atoms. Maleic anhydride and its diacid isomers coordinate to metal atoms through either the double bond or the carboxylate oxygen atoms when the diacid is deprotonated. Generally, low valent (soft) metals prefer coordination to the double bond, while high valent (hard) metals coordinate through the carboxylate oxygen atoms of the deprotonated diacids (76). Examples of double-bond coordination in maleic anhydride, maleic acid, and fumaric acid include molybdenum bis(maleic anhydride) carbonyl complex, $Mo(CO)_2(2,3\text{-}\eta^2\text{-}C_4H_2O_3)_2(CH_3CN)_2$ (77), *cis*-tetracarbonyl (2-3-η^2-maleic acid) iron complex, $Fe(2,3\text{-}\eta^2\text{-}C_4H_4O_4)(CO)_4$ (78), and *trans*-tetracarbonyl (2-3-η^2-fumaric acid) iron complex, $Fe(2,3\text{-}\eta^2\text{-}C_4H_4O_4)(CO)_4$ (78), where these structures were determined by single-crystal x-ray diffraction data. In the case of the Mo(0) complex, the trans coordinated maleic anhydride molecules are almost mutually orthogonal (85.2°) to one another. 2,3-η^2-Maleic anhydride coordinated to Mo(0) has a central C—C bond length of 0.1420 nm (77) compared to a C—C double bond length of 0.1303 nm for maleic anhydride (15,21,22). This increase in the central C—C bond length suggests significant π-backbonding interaction (76). Singly deprotonated maleic acid coordinates through symmetrical carboxylate oxygens to Zn(II) in the structure zinc maleic acid complex, $Zn(O_2CCH{=}CHCO_2H)_2{\cdot}4H_2O$ (79).

Nucleophilic Addition. Nucleophilic reagents attack the β-carbon position in the conjugated maleic and fumaric frameworks. Basic reaction conditions favor these condensations for the addition of glycolate (**21**) to maleate [*142-44-9*] (**22**):

$$^{-}O_2CCH_2OH \; (\mathbf{21}) + HC(CO_2^-){=}CH(CO_2^-) \; (\mathbf{22}) \xrightarrow[\text{aqueous, 90°C}]{La^{3+},\ pH>6} {}^{-}O_2CCH_2OCH(CO_2^-)CH_2CO_2^-$$

This Michael-type addition is catalyzed by lanthanum(3+) [*16096-89-2*] (80). Ethylene glycol [*107-21-1*] reacts with maleate under similar conditions (81). A wide range of nucleophilic reagents add to the maleate and fumarate frameworks including alcohols, ammonia, amines, sulfinic acids, thioureas, Grignard reagents, Michael reagents, and alkali cyanides (25).

Thiols and phosphines add to maleic anhydride to give α-thiosuccinic anhydrides (82) and phosphoranylidene–maleic anhydride adducts (83). Triethyl phosphite [*122-52-1*] reacts with maleic anhydride to give the ylide structure (**23**) (84). Hydrolysis of this adduct (**23**) leads to succinic acid [*110-15-6*], maleic acid, triethyl phosphate [*78-40-0*], and diethyl phosphite [*762-04-9*].

$$(CH_3CH_2O)_3P^{+}\text{—}C\!\left(CH_2\text{—}C(=O)\text{—}O\text{—}\right){=}C\text{—}O^-$$

(**23**)

Oxidation. Maleic and fumaric acids are oxidized in aqueous solution by ozone [*10028-15-6*] (qv) (85). Products of the reaction include glyoxylic acid [*298-12-4*], oxalic acid [*144-62-7*], and formic acid [*64-18-6*]. Catalytic oxidation of aqueous maleic acid occurs with hydrogen peroxide [*7722-84-1*] in the presence of sodium tungstate(VI) [*13472-45-2*] (86) and sodium molybdate(VI) [*7631-95-0*] (87). Both catalyst systems avoid formation of tartaric acid [*133-37-9*] and produce *cis*-epoxysuccinic acid [*16533-72-5*] at pH values above 5. The reaction of maleic anhydride and hydrogen peroxide in an inert solvent (methylene chloride [*75-09-2*]) gives permaleic acid [*4565-24-6*], HOOC—CH=CH—CO_3H (88) which is useful in Baeyer-Villiger reactions. Both maleate and fumarate [*142-42-7*] are hydroxylated to tartaric acid using an osmium tetroxide [*20816-12-0*]/iodate [*15454-31-6*] catalyst system (89).

Polymerization. Maleic anhydride which contains a double bond and an anhydride group is used in both addition and condensation polymerization schemes. Research since the early 1960s has shown that homopolymerization occurs using γ and uv radiation, free-radical initiators, pyridine bases, and electrochemical initiation. Copolymers of maleic anhydride and its isomeric acids (or ester derivatives) are formed with a wide variety of monomers. Suitable monomers for copolymerization with maleic anhydride include styrene [*100-42-5*], vinyl chloride [*75-01-4*], vinyl esters, acrylonitrile [*107-13-1*], acrylic acid [*79-10-7*], acrylic and methacrylic esters, acrylamide [*79-06-1*], acrolein [*107-02-8*], vinylsulfonic acid [*1184-84-5*], allyl acetate [*591-87-7*], and alkenes such as ethylene, vinyl ketones, and carbon monoxide. Copolymers may be assembled, in random or alternating additions, by grafting maleic anhydride onto existing polymers or by condensations. An enormous amount of literature on these polymers exists, eg, see *General References*.

Aqueous ring-opening metathesis polymerization (ROMP) was first described in 1989 (90) and it has been applied to maleic anhydride (91). Furan [*110-00-9*] reacts in a Diels-Alder reaction with maleic anhydride to give exo-7-oxabicyclo[2.2.1]hept-5-ene-2,3–dicarboxylate anhydride [*6118-51-0*] (**24**). The condensed product is treated with a soluble ruthenium(III) [*7440-18-8*] catalyst in water to give upon acidification the polymer (**25**). Several applications for this new copolymer have been suggested (91).

aqueous ROMP

(**24**) (**25**)

Unsaturated polyester resins prepared by condensation polymerization constitute the largest industrial use for maleic anhydride. Typically, maleic anhydride is esterified with ethylene glycol [*107-21-1*] and a vinyl monomer or styrene is added along with an initiator such as a peroxide to produce a three-dimensional macromolecule with rigidity, insolubility, and mechanical strength.

Reduction. Heterogeneous catalytic reduction processes provide effective routes for the production of maleic anhydride derivatives such as succinic anhydride [*108-30-5*] (**26**), succinates, γ-butyrolactone [*96-48-0*] (**27**), tetrahydrofuran [*109-99-9*] (**29**), and 1,4-butanediol [*110-63-4*] (**28**). The technology for production of 1,4-butanediol from maleic anhydride has been reviewed (92,93).

(1) $\xrightarrow{H_2}$ (**26**) $\xrightarrow[-H_2O]{H_2}$ (**27**) $\xrightarrow{H_2}$ $CH_2OH-CH_2-CH_2-CH_2OH$ (**28**) $\xrightarrow{-H_2O}$ (**29**)

Survey of the patent literature reveals companies with processes for 1,4-butanediol from maleic anhydride include BASF (94), British Petroleum (95,96), Davy McKee (93,97), Hoechst (98), Huels (99), and Tonen (100,101). Processes for the production of γ-butyrolactone have been described for operation in both the gas (102–104) and liquid (105–108) phases. In the gas phase, direct hydrogenation of maleic anhydride in hydrogen at 245°C and 1.03 MPa gives an 88% yield of γ-butyrolactone (104). Du Pont has developed a process for the production of tetrahydrofuran back-integrated to a butane feedstock (109). Slurry reactor catalysts containing palladium and rhenium are used to hydrogenate aqueous maleic acid to tetrahydrofuran (110,111).

Sulfonation. Maleic anhydride is sulfonated to α-sulfomaleic anhydride [*40336-85-4*] (**30**) with sulfur trioxide [*7446-11-9*] (112,113). Uses for this monomer have not been published.

maleic anhydride + SO_3 $\longrightarrow$ (SO_3H-substituted maleic anhydride) (**30**)

Manufacture

Process Technology Evolution. Maleic anhydride was first commercially produced in the early 1930s by the vapor-phase oxidation of benzene [*71-43-2*]. The use of benzene as a feedstock for the production of maleic anhydride was dominant in the world market well into the 1980s. Several processes have been used for the production of maleic anhydride from benzene with the most common one from Scientific Design. Small amounts of maleic acid are produced as a by-product in production of phthalic anhydride [*85-44-9*]. This can be converted to either maleic anhydride or fumaric acid. Benzene, although easily oxidized to maleic anhydride with high selectivity, is an inherently inefficient feedstock since two excess carbon atoms are present in the raw material. Various C_4 compounds have been evaluated as raw material substitutes for benzene in production of maleic anhydride. Fixed- and fluid-bed processes for production of maleic anhydride from the butenes present in mixed C_4 streams have been practiced commercially. None of these processes is currently in operation.

Rapid increases in the price of benzene and the recognition of benzene as a hazardous material intensified the search for alternative process technology in the United States. These factors led to the first commercial production of maleic anhydride from butane [*106-97-8*] at Monsanto's J. F. Queeny plant in 1974. By the early 1980s the conversion of the U.S. maleic anhydride manufacturing capacity from benzene to butane feedstocks was well under way using catalysts developed by Monsanto, Denka, and Halcon. One factor that inhibited the conversion of the installed benzene-based capacity was that early butane-based catalysts were not active and selective enough to allow the conversion of benzene-based plant without significant loss of nameplate capacity. In 1983 Monsanto started up the world's first butane-to-maleic anhydride plant, incorporating an energy efficient solvent-based product collection and refining system. This plant was the world's largest maleic anhydride production facility in 1983 at 59,000 t/yr capacity, and through rapid advances in catalyst technology remains the world's largest facility with a capacity of 100,000 t/yr (1994). Advances in catalyst technology, increased regulatory pressures, and continuing cost advantages of butane over benzene have led to a rapid conversion of benzene- to butane-based plants. By the mid-1980s in the United States 100% of maleic anhydride production used butane as the feedstock.

Coincident with the rapid development of the butane-based fixed-bed process, several companies have developed fluidized-bed processes. Two companies, Badger and Denka, collaborated on the development of an early fluid-bed reaction system which was developed through the pilot-plant stage but was never commercialized. Three fluid-bed butane-based technologies were commercialized during the latter half of the 1980s by Mitsubishi Kasei, Sohio (British Petroleum), and Alusuisse. A second fluidized-bed technology for the oxidation of butane to maleic anhydride, known as transport bed, has been piloted by Du Pont. A world-scale plant in Spain for the production of THF by the hydrogenation of maleic acid is planned for start-up in 1996 (114).

Europe is converting from benzene-based to butane-based maleic anhydride technology. Growth in the worldwide maleic anhydride industry is exclusively in the butane-to-maleic anhydride route, often at the expense of benzene-based production. Table 4 shows 1993 and estimated 1995 worldwide maleic production capacity broken down in categories of fixed-bed benzene, fixed-bed butane, fluidized-bed butane, and phthalic anhydride coproduct. As can be seen from this table, both fixed- and fluidized-bed butane routes are expected to grow at the expense of benzene-based processes with the fixed-bed route adding 105,000 t/yr capacity compared to 65,000 t/yr expected to be added for the fluid-bed process. Only a few newer benzene-based fixed-bed processes have been built since the early 1980s and these were built where the availability of butane was limited. The fluidized-bed butane-based process is experiencing growth, but based on growth rates from Table 4 (115,116), it does not appear destined to surpass fixed-bed technology. The announcement from Huntsman Specialty Chemicals Corp., formerly Monsanto, and DWE (117) that they intend to cooperate in the development of catalyst and reactor technology to permit operation at 50% higher productivity than the standard nonflammable fixed-bed butane process

Table 4. World Maleic Anhydride Capacity by Reactor Type

	1993 Actual[a]		1995 Forecast[b]	
Reactor (feed)	10^3 t/yr	%	10^3 t/yr	%
fixed-bed				
butane	369	43.0	474	50.0
benzene	325	37.9	244	25.8
fluidized-bed				
butane	127	14.8	192	20.3
phthalic anhydride coproduct	37	4.3	37	3.9
Total	*858*	*100.0*	*947*	*100.0*

[a]Ref. 115.
[b]Ref. 116.

indicates that the largest companies in fixed-bed technology are confident that further advances are possible. Three fixed-bed processes, from Huntsman (118), Pantochim (119), and Scientific Design (120) and two fluidized-bed processes, from Alusiusse-Lummus (ALMA) (121) and BP Chemicals (122), are currently offered for license.

Butane-Based Catalyst Technology. The increased importance of the butane-to-maleic anhydride conversion route has resulted in efforts being made to understand and improve this process. Since 1980, over 180 U.S. patents have been issued relating to maleic anhydride technology. The predominant area of research concerns the catalyst because it is at the heart of this process. The reasons for this statement are twofold. First, there is the complexity of this reaction: for maleic anhydride to be produced from butane, eight hydrogen atoms must be abstracted, three oxygen atoms inserted, and a ring closure performed. This is a 14-electron oxidation which occurs exclusively on the surface of the catalyst. The second reason for the emphasis placed on the catalyst is that all of the commercial processes use the same catalyst. This catalyst is the only commercially viable system which selectively produces maleic anhydride from butane.

The catalyst used in the production of maleic anhydride from butane is vanadium–phosphorus–oxide (VPO). Several routes may be used to prepare the catalyst (123), but the route favored by industry involves the reaction of vanadium(V) oxide [*1314-62-1*] and phosphoric acid [*7664-38-2*] to form vanadyl hydrogen phosphate, $VOHPO_4 \cdot 0.5H_2O$. This material is then heated to eliminate water from the structure and irreversibly form vanadyl pyrophosphate, $(VO)_2P_2O_7$ (123,124). Vanadyl pyrophosphate is believed to be the catalytically active phase required for the conversion of butane to maleic anhydride (125,126).

The reaction of V_2O_5 with H_3PO_4 to form $VOHPO_4 \cdot 0.5H_2O$ can be carried out in either an aqueous or organic medium such as isobutyl alcohol [*78-83-1*] (70,123). Two possible routes are as follows.

$$V_2O_5 + H_3PO_4 \xrightarrow[\text{water}]{\text{reducing agent, HCl}} VOHPO_4 \cdot 0.5H_2O \xrightarrow{\Delta} (VO)_2P_2O_7$$

$$V_2O_5 + H_3PO_4 \xrightarrow[\text{isobutyl alcohol}]{\text{reducing agent}} VOHPO_4 \cdot 0.5H_2O$$

The use of an organic medium yields an increase in the surface area of the $VOHPO_4 \cdot 0.5H_2O$ (70,126). This increase in surface area is carried over to the resulting vanadyl pyrophosphate phase (123) and is desirable because a concurrent increase in activity toward butane oxidation is observed (70).

An additional effect of the use of an organic medium in the catalyst preparation is creation of more defects in the crystalline lattice when compared to a catalyst made by the aqueous route (123). These defects persist in the active phase and are thought to result in creation of strong Lewis acid sites on the surface of the catalysts (123,127). These sites are viewed as being responsible for the activation of butane on the catalyst surface by means of abstraction of a hydrogen atom.

There are two types of patented technologies for the transformation of $VOHPO_4 \cdot 0.5H_2O$ into $(VO)_2P_2O_7$ (70). In the first procedure, the vanadyl hydrogen phosphate is transformed through heating to 415°C in a controlled environment with combinations of nitrogen, air, and steam (128). The second procedure is an *in situ* process that transforms the vanadyl hydrogen phosphate into the active phase while the catalyst is placed in the maleic reactor (129). This procedure entails slow heating of the catalyst in air with gradual introduction of butane to the gas stream. The use of the *in situ* activation method requires little capital investment. However, the possibility of inhomogenities in the activated catalyst exists due to nonuniformities in reactor flow distribution. On the other hand, the controlled environment procedure reduces plant downtime and increases the homogeneity of the catalyst in the reactor.

Promoters are sometimes added to the vanadium phosphorus oxide (VPO) catalyst during synthesis (129,130) to increase its overall activity and/or selectivity. Promoters may be added during formation of the catalyst precursor ($VOHPO_4 \cdot 0.5H_2O$), or impregnated onto the surface of the precursor before transformation into its activated phase. They are thought to play a twofold structural role in the catalyst (130). First, promoters facilitate transformation of the catalyst precursor into the desired vanadium phosphorus oxide active phase, while decreasing the amount of nonselective VPO phases in the catalyst. The second role of promoters is to participate in formation of a solid solution which controls the activity of the catalyst.

The bulk structure of the catalytically active phase is not completely known and is under debate in the literature (125,131–133). The central point of controversy is whether $(VO)_2P_2O_7$ alone or in combination with other phases is the most catalytically active for the conversion of butane to maleic anhydride. The heart of this issue concerns the role of structural disorder in the bulk and how it arises in the catalyst (125,134,135). Most researchers agree that the catalysts with the highest activity and selectivity are composed mainly of $(VO)_2P_2O_7$ that exhibits a clustered or distorted platelet morphology (125). It is also generally acknowledged that during operation of the catalyst, the bulk oxidation state of the vanadium in the catalyst remains very close to the +4 valence state (125).

Only the surface layers of the catalyst solid are generally thought to participate in the reaction (125,133). This implies that while the bulk of the catalyst may have an oxidation state of 4+ under reactor conditions, the oxidation state of the surface vanadium may be very different. It has been postulated that both V^{4+} and V^{5+} oxidation states exist on the surface of the catalyst, the latter arising from oxygen chemisorption (133). Phosphorus enrichment is also observed at the surface of the catalyst (125,126). The exact role of this excess surface phosphorus is not well understood, but it may play a role in active site isolation and consequently, the oxidation state of the surface vanadium.

Vanadium phosphorus oxide-based catalysts are unstable in that they tend to lose phosphorus over time at reaction temperatures. Hot spots in fixed-bed reactors tend to accelerate this loss of phosphorus. This loss of phosphorus also produces a decrease in selectivity (70,136). Many steps have been taken, however, to alleviate these problems and create an environment where the catalyst can operate at lower temperatures. For example, volatile organophosphorus compounds are fed to the reactor to mitigate the problem of phosphorus loss by the catalyst (137). The phosphorus feed also has the effect of controlling catalyst activity and thus improving catalyst selectivity in the reactor. The catalyst pack in the reactor may be stratified with an inert material (138,139). Stratification has the effect of reducing the extent of reaction per unit volume and thus reducing the observed catalyst temperature (hot spot). These measures have minimized concerns of heat removal in the reactor and improved catalyst performance.

Fluidized-bed reactor systems put other unique stresses on the VPO catalyst system. The mixing action inside the reactor creates an environment that is too harsh for the mechanical strength of a vanadium phosphorus oxide catalyst, and thus requires that the catalyst be attrition resistant (121,140,141). To achieve this goal, vanadium phosphorus oxide is usually spray dried with colloidal silica [*7631-86-9*] or polysilicic acid [*1343-98-2*]. Vanadium phosphorus oxide catalysts made with colloidal silica are reported to have a loss of selectivity, while no loss in selectivity is reported for catalysts spray dried with polysilicic acid (140).

Benzene-Based Catalyst Technology. The catalyst used for the conversion of benzene to maleic anhydride consists of supported vanadium oxide [*11099-11-9*]. The support is an inert oxide such as kieselguhr, alumina [*1344-28-1*], or silica, and is of low surface area (142). Supports with higher surface area adversely affect conversion of benzene to maleic anhydride. The conversion of benzene to maleic anhydride is a less complex oxidation than the conversion of butane, so higher catalyst selectivities are obtained. The vanadium oxide on the surface of the support is often modified with molybdenum oxides. There is approximately 70% vanadium oxide and 30% molybdenum oxide [*11098-99-0*] in the active phase for these fixed-bed catalysts (143). The molybdenum oxide is thought to form either a solid solution or compound oxide with the vanadium oxide and result in a more active catalyst (142).

Butane-Based Fixed-Bed Process Technology. Maleic anhydride is produced by reaction of butane with oxygen using the vanadium phosphorus oxide heterogeneous catalyst discussed earlier. The butane oxidation reaction to produce maleic anhydride is very exothermic. The main reaction by-products are carbon monoxide and carbon dioxide. Stoichiometries and heats of reaction for the three principal reactions are as follows:

$$C_4H_{10} + 3.5\ O_2 \longrightarrow C_4H_2O_3 + 4\ H_2O \qquad \Delta H = -1236\ \text{kJ/mol}\ (-295.4\ \text{kcal/mol})$$

$$C_4H_{10} + 6.5\ O_2 \longrightarrow 4\ CO_2 + 5\ H_2O \qquad \Delta H = -2656\ \text{kJ/mol}\ (-634.8\ \text{kcal/mol})$$

$$C_4H_{10} + 4.5\ O_2 \longrightarrow 4\ CO + 5\ H_2O \qquad \Delta H = -1521\ \text{kJ/mol}\ (-363.5\ \text{kcal/mol})$$

Air is compressed to modest pressures, typically 100 to 200 kPa (~15–30 psig) with either a centrifugal or radial compressor, and mixed with superheated vaporized butane. Static mixers are normally employed to ensure good mixing. Butane concentrations are often limited to less than 1.7 mol % to stay below the lower flammable limit of butane (144). Operation of the reactor at butane concentrations below the flammable limit does not eliminate the requirement for combustion venting, and consequently most processes use rupture disks on both the inlet and exit reactor heads. A flow diagram of the Huntsman fixed-bed maleic anhydride process is shown in Figure 1.

The highly exothermic nature of the butane-to-maleic anhydride reaction and the principal by-product reactions require substantial heat removal from the reactor. Thus the reaction is carried out in what is effectively a large multitubular heat exchanger which circulates a mixture of 53% potassium nitrate [*7757-79-1*], KNO_3; 40% sodium nitrite [*7632-00-0*], $NaNO_2$; and 7% sodium nitrate [*7631-99-4*], $NaNO_3$. Reaction tube diameters are kept at a minimum 25–30 mm in outside diameter to facilitate heat removal. Reactor tube lengths are between 3 and 6 meters. The exothermic heat of reaction is removed from the salt mixture by the production of steam in an external salt cooler. Reactor temperatures are in the range of 390 to 430°C. Despite the rapid circulation of salt on the shell side of the reactor, catalyst temperatures can be 40 to 60°C higher than the salt temperature. The butane to maleic anhydride reaction typically reaches its maximum efficiency (maximum yield) at about 85% butane conversion. Reported molar yields are typically 50 to 60%.

Efficient utilization of waste heat from a maleic anhydride plant is critical to the economic viability of the plant. Often site selection is dictated by the presence of an economic use for by-product steam. The steam can also be used to

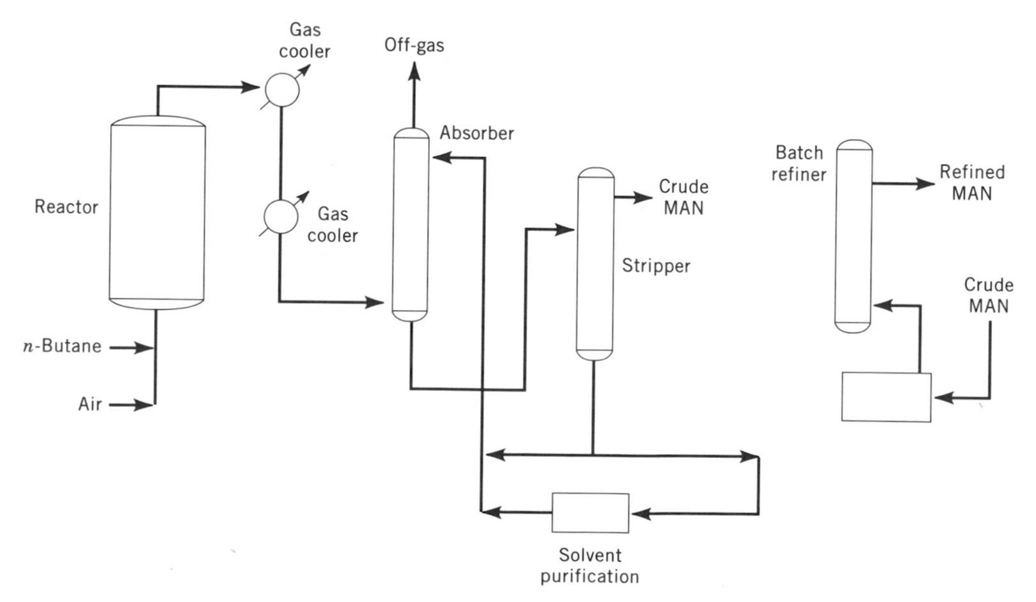

Fig. 1. Schematic flow diagram of the Huntsman fixed-bed maleic anhydride process. MAN = maleic anhydride.

drive an air compressor, generate electricity, or both. Alternatively, an energy consuming process, such as a butanediol plant, can be closely coupled with the maleic anhydride plant. Several such plants have been announced (145). Design and integration of the heat recovery systems for a maleic anhydride plant are very site specific. Heat is removed from the reaction gas through primary and sometimes secondary heat exchangers. In addition to the heat recovered from the reactor and process gas heat exchangers, additional heat can be recovered from the destruction of unreacted butane, the carbon monoxide by-product, and other by-products which cannot be vented directly to the atmosphere. This destruction is done typically in a specially designed thermal oxidizer or a modified boiler.

Reactor operation at 80 to 85% butane conversion to produce maximum yields provides an opportunity for recycle processes to recover the unreacted butane in the stream that is sent to the oxidation reactor. Patents have been issued on recycle processes (146,147) both with and without added oxygen. Pantochim has announced the commercialization of a partial recycle process (119). Operation of the butane to maleic anhydride process in a total recycle configuration can produce molar yields that approach the reaction selectivity which is typically 65 to 75%, significantly higher than the 50 to 60% molar yields from a single pass, high conversion process. The Du Pont transport bed process achieves its high reported yields at least partially through implementation of recycle technology. Recovery of the fuel value of the butane in the off-gas from a single pass configuration plant reduces the economic attractiveness of recycle operation.

Butane-Based Fluidized-Bed Process Technology. Fluidized-bed processes offer the advantage of excellent control of hot spots by rapid catalyst mixing, simplification of safety issues when operating above the flammable limit, and a simplified reactor heat-transfer system. Some disadvantages include the effect of back mixing on the kinetics in the reactor, product destruction and by-product reactions in the space above the fluidized bed, and vulnerability to large-scale catalyst releases from explosion venting.

A schematic flow diagram for the ALMA fluidized-bed process is shown in Figure 2 (121). Compressed air and butane are typically introduced separately into the bottom of the fluidized-bed reactor. Heat from the exothermic reaction is removed from the fluidized bed through steam coils in direct contact with the bed of fluidized solids. Fluidized-bed reactors exploit the extremely high heat-transfer coefficient between the bed of fluidized solids and the steam coils. This high heat-transfer coefficient allows a relatively small heat-transfer area in the fluid-bed process for the removal of the heat of reaction compared to the fixed-bed process. Gas flow patterns in a commercial scale fluid-bed reactor are generally backmixed, which can lead to maleic anhydride destruction. Patents have been issued for mechanical modifications to the reactor internals which claim to control backmixing (148). Other methods to reduce backmixing include introduction of catalyst fines (small particles of catalyst) to decrease bubble size and operation of the reactor in the turbulent, fast fluidization regime in an attempt to minimize bubbling (140). Fluidized-bed reactors require a significant amount of space above the catalyst level to allow the solids to separate from the gases. This exposure of the product to high temperatures at relatively long residence times can lead to side reactions and product destruction.

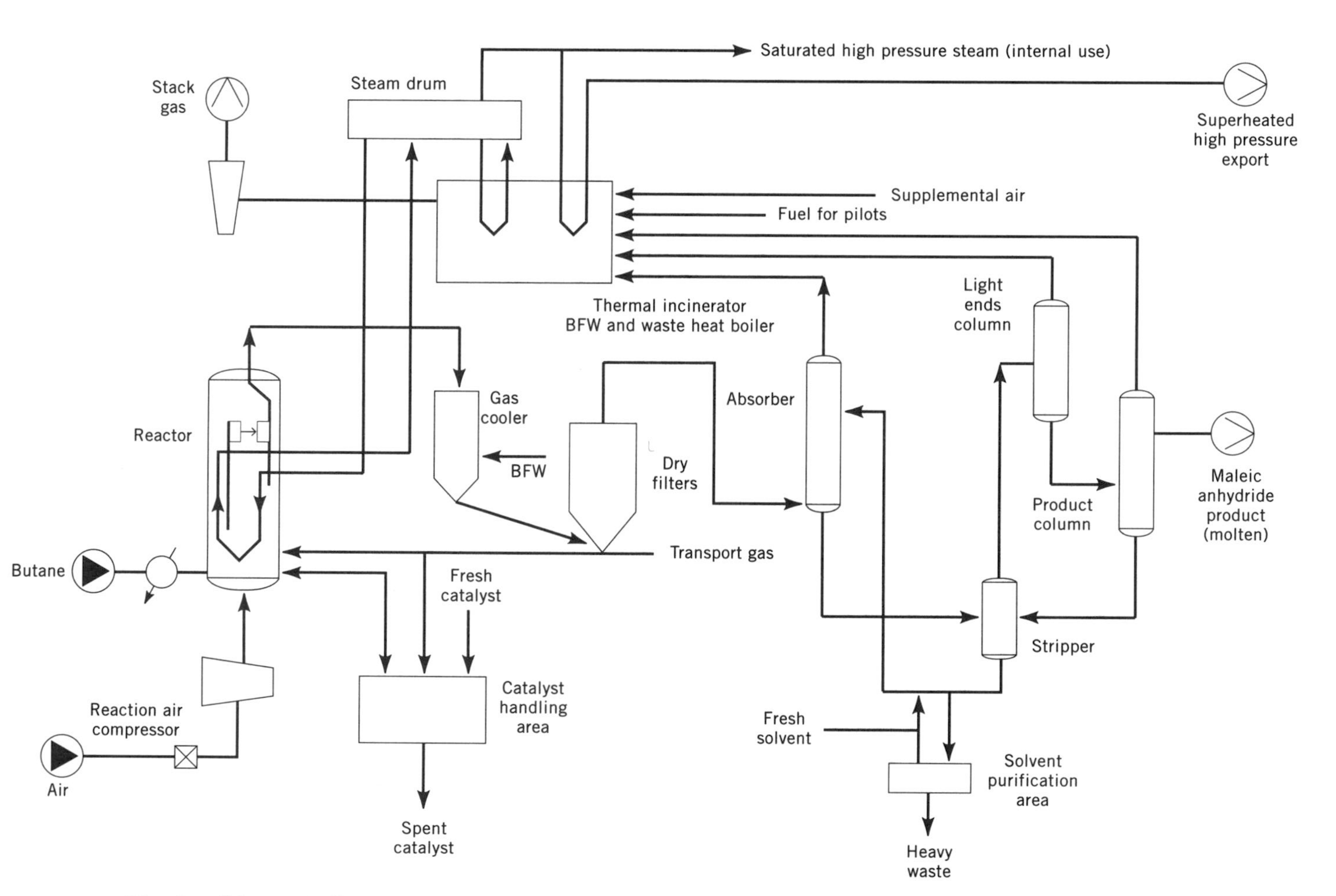

Fig. 2. Schematic flow diagram of the ALMA fluidized-bed process (121). BFW = boiler feed water.

Fluidized-bed processes are operated at high butane concentrations (121) but at longer gas residence times than fixed-bed processes.

The product stream contains gases and solids. The solids are removed by using either cyclones, filters, or both in combination. Cyclones are devices used to separate solids from fluids using vortex flow. The product gas stream must be cooled before being sent to the collection and refining system. The ALMA process uses cyclones as a primary separation technique with filters employed as a final separation step after the off-gas has been cooled and before it is sent to the collection and refining system (148). As in the fixed-bed process, the reactor off-gas must be incinerated to destroy unreacted butane and by-products before being vented to the atmosphere.

Fluidized-bed reaction systems are not normally shut down for changing catalyst. Fresh catalyst is periodically added to manage catalyst activity and particle size distribution. The ALMA process includes facilities for adding back both catalyst fines and fresh catalyst to the reactor.

Butane-Based Transport-Bed Process Technology. Du Pont announced the commercialization of a moving-bed recycle-based technology for the oxidation of butane to maleic anhydride (109,149). Athough maleic anhydride is produced in the reaction section of the process and could be recovered, it is not a direct product of the process. Maleic anhydride is recovered as aqueous maleic acid for hydrogenation to tetrahydrofuran [*109-99-9*] (THF).

The reaction technology known as transport bed is a circulating solids technology where the oxygen required in the oxidation of butane to maleic anhydride is provided by the VPO catalyst and the catalyst is reoxidized in a separate step. The exclusion of gas-phase oxygen from the reaction step is claimed to enhance selectivity (150,151). Separation of butane oxidation from catalyst reoxidation allows both steps to be independently optimized. The circulating solids system is similar to catalytic cracking as can be seen from an examination of Figure 3.

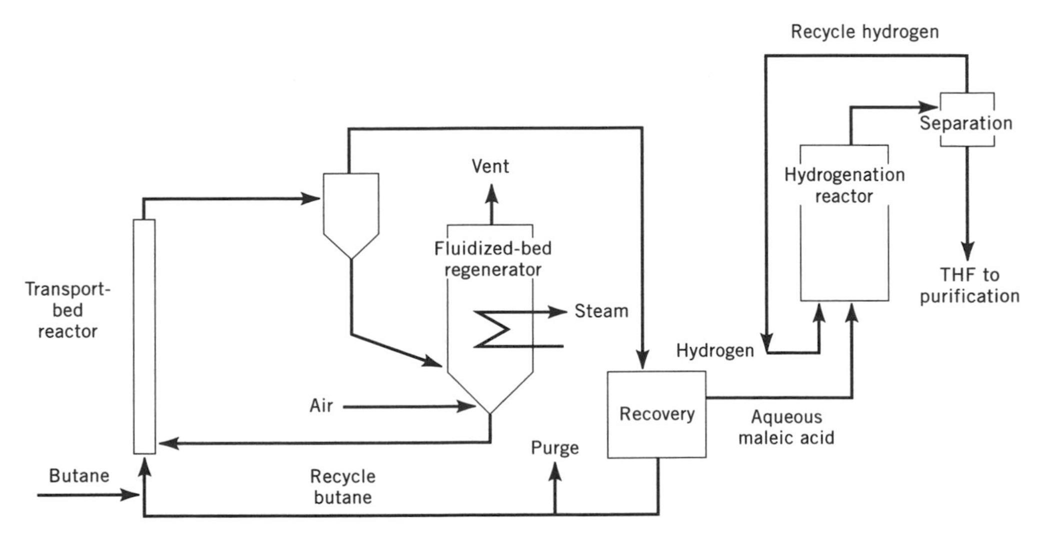

Fig. 3. Du Pont transport-bed process for making THF from butane (109).

Fresh butane mixed with recycled gas encounters freshly oxidized catalyst at the bottom of the transport-bed reactor and is oxidized to maleic anhydride and CO_x during its passage up the reactor. Catalyst densities (80–160 kg/m^3) in the transport-bed reactor are substantially lower than the catalyst density in a typical fluidized-bed reactor (480–640 kg/m^3) (109). The gas flow pattern in the riser is nearly plug flow which avoids the negative effect of backmixing on reaction selectivity. Reduced catalyst is separated from the reaction products by cyclones and is further stripped of products and reactants in a separate stripping vessel. The reduced catalyst is reoxidized in a separate fluidized-bed oxidizer where the exothermic heat of reaction is removed by steam coils. The rate of reoxidation of the VPO catalyst is slower than the rate of oxidation of butane, and consequently residence times are longer in the oxidizer than in the transport-bed reactor.

Maleic anhydride in the product stream is removed and converted to a maleic acid solution in a water scrubbing system. The maleic acid is sent to the hydrogenation to produce THF while the reactor off-gas after scrubbing is sent to the recycle compressor. A small purge stream is sent to incineration.

Benzene-Based Fixed-Bed Process Technology. The benzene fixed-bed process is very similar to the butane fixed-bed process and, in fact, the Scientific Design butane process has evolved directly from their benzene process. Benzene-based processes are easily converted to butane-based processes. Typically, only a catalyst change, installation of butane handling equipment, and minor modifications to the recovery process are required. The benzene reaction is a vapor-phase partial oxidation reaction using a fixed-bed catalyst of mixed vanadium and molybdenum oxides. The reactors used are the same multitubular reactors cooled by circulating a molten mixture of KNO_3–$NaNO_2$–$NaNO_3$ salts described in the previous section on the butane process. The benzene concentrations used are about 1.5 mol % or just below the lower flammable limit of benzene in air. Unlike the butane reaction, the reactor normally operates at conversions greater than 95% and molar yields greater than 70%. The benzene oxidation reaction runs a little cooler than the butane oxidation reaction with typical reactor temperatures being in the 350 to 400°C range.

The reactor off-gas is cooled by one or more heat exchangers and sent to the collection and refining section of the plant. Unreacted benzene and by-products are incinerated.

Recovery and Purification. All processes for the recovery and refining of maleic anhydride must deal with the efficient separation of maleic anhydride from the large amount of water produced in the reaction process. Recovery systems can be separated into two general categories: aqueous- and nonaqueous-based absorption systems. Solvent-based systems have a higher recovery of maleic anhydride and are more energy efficient than water-based systems.

The Huntsman solvent-based collection and refining system will be used as a generic model for solvent-based recovery systems (see Fig. 1). The reactor exit gas is cooled in two heat exchangers for energy recovery. The cooled gas product stream is passed to a solvent absorber where a proprietary solvent is used to absorb, almost completely, the maleic anhydride contained in the product stream. The solvent stream, coming from the bottom of the absorber with a high concentration of maleic anhydride, known as rich oil, is sent to a stripper where the rich oil is heated and maleic anhydride is vacuum stripped from the solvent. The vacuum-stripped maleic anhydride is typically greater than 99.8% purity, and is sent to the purification section of the plant where it is batch distilled to produce extremely pure maleic anhydride. A small slip stream of the solvent which has had the maleic anhydride removed by stripping is sent to the solvent purification section of the plant where impurities are removed.

The Scientific Design water-based collection and refining system is in broad use throughout the world in butane-based and benzene-based plants (120,152). The reactor off-gas is cooled from reaction temperatures in a gas cooler with generation of steam. The off-gas is then sent to a tempered water-fed aftercooler where it is cooled below the dew point of maleic anhydride. The liquid droplets of maleic anhydride are separated from the off-gas by a separator. The condensed crude is pumped to a crude tank for storage. The maleic anhydride remaining in the gas stream after partial condensation is removed in a water scrubber by conversion to maleic acid which accumulates in the acid storage section at the bottom of the scrubber. The acid solution is converted to crude maleic anhydride in a dual purpose dehydrator/refiner. Xylene [*1330-20-1*] is used as an azeotropic agent for the conversion of maleic acid to maleic anhydride. Water from the dehydration step is recycled to the scrubber. When the conversion of the acid solution to crude maleic anhydride is complete, condensed crude maleic anhydride is added to the still pot and a batch distillation refining step is conducted.

The UCB collection and refining technology (owned by BP Chemicals (122,153–155)) also depends on partial condensation of maleic anhydride and scrubbing with water to recover the maleic anhydride present in the reaction off-gas. The UCB process departs significantly from the Scientific Design process when the maleic acid is dehydrated to maleic anhydride. In the UCB process the water in the maleic acid solution is evaporated to concentrate the acid solution. The concentrated acid solution and condensed crude maleic anhydride is converted to maleic anhydride by a thermal process in a specially designed reactor. The resulting crude maleic anhydride is then purified by distillation.

Fumaric Acid. Fumaric acid for commerce is derived from maleic acid through catalytic isomerization. Purified maleic anhydride is the main source of maleic acid. Also, crude maleic acid may be used as recovered in benzene oxidation or phthalic anhydride manufacture (18). Numerous catalysts exist for maleic acid isomerization (73,74) but three types are widely used: mineral acids (156), peroxycompounds with bromides (157) and bromates (158), and sulfur-containing compounds such as thiourea [*62-56-6*] (74). Little change in process technology has occurred since the 1970s. Processing is simplest with purified maleic anhydride as the raw material. High purity fumaric acid is produced through crystallization of the aqueous mixture, washing, and drying (159). Decolorizing and crystallization techniques are used to treat impure maleic solutions (160).

Shipment

Molten maleic anhydride is shipped in tank rail cars, tank trucks, and isotanks (for overseas shipments). Tank rail cars are typically constructed of lined carbon steel and are insulated and equipped with steam coils. Tank rail cars containing as much as 80 m^3 (20,000 gal) are used. Tank trucks are typically constructed of stainless steel [*12597-68-1*], insulated, and equipped with steam coils. Tank trucks containing up to 17 m^3 (4500 gal) are used. Isotanks are typically constructed of stainless steel and are insulated and equipped with steam coils. Isotanks containing up to 17 m^3 are used.

Solid form maleic anhydride is produced from molten maleic anhydride as briquettes or pastilles weighing 0.5 to 20 g. Flaked maleic anhydride is used in some areas of the world but is not generally accepted in the United States because of the high dust content. Briquettes or pastilles are packed in 50-lb (25-kg) bags and supersacks of up to 2500 lbs (1136 kg). Outside the United States, the standard bag weight is 25 kg. Typical bag construction is either polyethylene or multi-ply paper with at least one polyethylene layer. Solid form maleic anhydride can be stored in bags for several months in a cool, dry location.

Fumaric acid is shipped in solid form, the particle size varying based upon the specification. The standard shipping containers are 50-lb bags (25 kg bags outside the United States), supersacks containing up to 2500 lbs, and occasionally bulk hopper truck.

Economic Aspects

Data for the production and sales of maleic anhydride and fumaric acid in the United States between 1979 and 1992 are shown in Table 5. Production of maleic anhydride during this time grew ~2% on average per year. Production of fumaric acid has declined during the same period as customers have switched to the less costly maleic anhydride when possible. All production of maleic anhydride in the United States in 1992 was from butane-based plants which used fixed-bed reactor technology as shown in Table 6. The number of fumaric acid producers has been reduced considerably since the early 1980s with only two producers left in the United States in 1992 as shown in Table 6. Pfizer shut down its fumaric acid plant at the end of 1993. However, Bartek of Canada will start up an expanded fumaric acid facility to supply the North American market for both their own and Huntsman's requirements.

Capacities of maleic anhydride facilities worldwide are presented in Table 7. The switch of feedstock from benzene to butane was completed in the United States in 1985, being driven by the lower unit cost and lower usage of butane in addition to the environmental pressures on the use of benzene. Worldwide, the switch to butane is continuing with 58% of the total world maleic anhydride capacity based on butane feedstock in 1992. This capacity percentage for butane has increased from only 6% in 1978. In 1992, 38% of the total world maleic anhydride capacity was based on benzene feedstock and 4% was derived from other sources, primarily phthalic anhydride by-product streams.

Another characteristic of the maleic anhydride supply picture has been the emergence of newer capacity, predominantly in Asia, which has resulted in a significant oversupply to that area of the world. The historically largest and second largest maleic anhydride markets are North America and Western Europe, respectively. Although the names of the producers in both of these areas of the

Table 5. Production and Sales in the United States[a,b]

	Maleic anhydride			Fumaric acid		
Year	Production, 10^3 t	Sales, 10^3 t	Unit value, \$/kg	Production, 10^3 t	Sales, 10^3 t	Unit value, \$/kg
1979	146.6	100.0	0.79	23.1	18.3	1.04
1980	137.7	99.1	0.84	18.1	13.0	1.10
1981	133.0	98.5	0.93	16.0	11.5	1.21
1982	117.7	93.9	0.84	10.9	9.8	1.23
1983	135.6	137.9	0.75	13.6	13.0	1.12
1984	162.9	142.3	0.90	13.8	13.7	1.19
1985	178.5	150.1	0.97	14.3	13.2	1.34
1986	163.0	137.1	0.97	14.7	12.8	1.32
1987	173.4	141.7	0.94	15.2	na	na
1988	194.4	162.3	0.95	16.1	14.4	1.25
1989	192.8	169.9	0.98	15.6	12.9	1.34
1990	176.9	160.2	0.97	15.2	12.3	1.33
1991	172.8	142.9	0.87	12.2	na	na
1992	189.1	na	na	na	na	na

[a]Refs. 161 and 162.
[b]na = not available.

Table 6. Producers of Maleic Anhydride and Fumaric Acid in the United States, 1992[a]

Producer	Location	Feedstock	Capacity, 10^3 t/yr
	Maleic		
Amoco	Joliet, Ill.	butane	40.8
Aristech	Neville Island, Pa.	butane	28.2
Ashland	Neal, W.Va.	butane	28.6
Huntsman[b]	Pensacola, Fla.	butane	100.0
Miles	Houston, Tex.	butane	29.5
Total			*227.1*
	Fumaric		
Miles[c]	Duluth, Minn.		2.7
Pfizer	Terre Haute, Ind.		11.4
Total			*14.1*

[a]Refs. 116, 163, and 164.
[b]Formerly Monsanto.
[c]Haarmann and Reimer.

world have changed rather dramatically between the late 1970s and the early 1990s, the total capacity in these world areas has only declined a small amount. Beginning in the late 1980s, small commercial plants based on butane fluidized-bed technologies were built in Asia. As of the end of 1992, fixed-bed butane-based technology accounted for 74% of the butane-based maleic anhydride facilities worldwide with butane-based fluidized-bed technology accounting for the remainder. The butane-based transport-bed process announced by Du Pont is scheduled to start up in Spain in 1996 (114).

Table 7. World Maleic Anhydride Capacities[a]

World area	1978 Capacity, 10^3 t/yr	1992 Capacity, 10^3 t/yr
North America	255	235
South and Central America	12	44
Western Europe	196	168
Eastern Europe	48	54
Middle East	na	10
Asia	91	315
Africa	na	10
Total	*602*	*836*

[a]Ref. 161.

Specifications

The general sales specification under which maleic anhydride is sold in the United States specifies a white fused mass or briquettes of 99.5% minimum assay and 52.5°C minimum crystallization point. The melt color specification is 20 APHA maximum with a maximum APHA color of 40 after two hours of heating at 140°C. Four grams of maleic anhydride in 10 milliliters of water are to be completely soluble. The resulting solution is to be colorless. The acidity resulting from maleic acid is allowed to be a maximum of 0.2%.

Fumaric acid is sold as resin-grade and food-grade. The general sales specification under which resin-grade fumaric acid is sold in the United States specifies white, crystalline granules with a minimum assay of 99.6% and maximum ash content of 0.05%. The moisture specification is 0.3% maximum with <10 ppm heavy metals. The color of a 5% solution in methanol is to be less than 10 APHA. Food-grade fumaric acid calls for somewhat lower impurity levels. Particle size and particle size distribution are important in many applications.

Analytical and Test Methods

The test methods used by industry to determine if a sample of maleic anhydride is within specifications (165) are ASTM methods D2930, D1493, and D3366. These methods describe procedures for the determination of maleic acid content, the crystallization point, and the color properties of the maleic anhydride sample, respectively. By quantitative determination of these properties, a calculation of the overall purity of the maleic anhydride sample can be made.

The free maleic acid content in maleic anhydride is determined by direct potentiometric titration (166). The procedure involves the use of a tertiary amine, *N*-ethylpiperidine [*766-09-6*], as a titrant. A tertiary amine is chosen as a titrant since it is nonreactive with anhydrides (166,167). The titration is conducted in an anhydrous solvent system. Only one of the carboxylic acid groups is titrated by this procedure. The second hydrogen's dissociation constant is too weak to titrate (166). This test method is not only used to determine the latent acid content in refined maleic acid, but also as a measure of the sample exposure to moisture during shipping.

The other two methods used by industry to examine the purity of maleic anhydride are the crystallization point (168) and color determination of the sample (169). These tests determine the temperature at the point of solidification of the molten sample and the initial color properties of the melt. Furthermore, the color test also determines the color of the sample after a two-hour heat treatment at 140°C. The purpose of these tests is to determine the deviation in properties of the sample from those of pure maleic anhydride. This deviation is taken as an indication of the amount of contaminants in the maleic anhydride sample.

Other analytical techniques are also available for the determination of maleic anhydride sample purity. For example, maleic anhydride content can be determined by reacting it with a known excess of aniline [*62-53-3*] in an alcohol mixture (170). The solution is then titrated with an acid to determine the amount of unconsumed aniline. This number is then used to calculate the amount of maleic anhydride reacted and thus its concentration. Another method of a similar type has also been reported (171).

Finally, the techniques of nmr, infrared spectroscopy, and thin-layer chromatography also can be used to assay maleic anhydride (172). The individual anhydrides may be analyzed by gas chromatography (173,174). The isomeric acids can be determined by polarography (175), thermal analysis (176), paper and thin-layer chromatographies (177), and nonaqueous titrations with an alkali (178). Maleic and fumaric acids may be separated by both gel filtration (179) and ion-exchange techniques (180).

Health and Safety Factors (Toxicology)

Maleic Anhydride. The ACGIH threshold limit value in air for maleic anhydride is 0.25 ppm and the OSHA permissible exposure level (PEL) is also 0.25 ppm (181). Maleic anhydride is a corrosive irritant to eyes, skin, and mucous membranes. Pulmonary edema (collection of fluid in the lungs) can result from airborne exposure. Skin contact should be avoided by the use of rubber gloves. Dust respirators should be used when maleic anhydride dust is present. Maleic anhydride is combustible when exposed to heat or flame and can react vigorously on contact with oxidizers. The material reacts exothermically with water or steam. Violent decompositions of maleic anhydride can be catalyzed at high temperature by strong bases (sodium hydroxide, potassium hydroxide, calcium hydroxide, alkali metals, and amines). Precaution should be taken during the manufacture and use of maleic anhydride to minimize the presence of basic materials.

Maleic Acid. Maleic acid is produced by the hydration of maleic anhydride. The hazards of its use are analogous to those of maleic anhydride. It is a skin and severe eye irritant. It is combustible when exposed to heat or flame. As discussed above, contamination with bases should be avoided where the material is to be handled at high temperature and high concentration because of the potential decomposition hazard.

Fumaric Acid. Fumaric acid is used to acidify beverages for human consumption and has many industrial uses. Its acidic properties can cause skin and eye irritation. Fumaric acid combusts when exposed to heat or flame and can react vigorously with oxidizing agents.

Table 8 lists toxicological data.

Table 8. Toxicological Data for Maleic Anhydride, Maleic Acid, and Fumaric Acid[a]

Test	Maleic anhydride	Maleic acid	Fumaric acid
oral, rat, LD_{50}, mg/kg	400	708	10,700
oral, mouse, LD_{50}, mg/kg	465	2400	
skin, rabbit, LD_{50}, mg/kg	2620	1560	20,000
intraperitoneal			
rat, LD_{50}, mg/kg	97		
LD_{LO},[b] mg/kg			587
subcutaneous, rat, TD_{LO},[c] mg/kg	1220		

[a]Ref. 181.
[b]The lowest dose reported to have caused death (lethal dose low).
[c]The lowest dose reported to produce any toxic effect (toxic dose low).

Uses

Maleic anhydride itself has few, if any, consumer uses but its derivatives are of significant commercial interest (161). The distribution of end uses for maleic anhydride is presented in Table 9 for the year 1992 (182). The majority of the maleic anhydride produced is used in unsaturated polyester resin (see POLYESTERS, UNSATURATED). Unsaturated polyester resin is then used in both glass-reinforced applications and in unreinforced applications as shown in Table 10 (183).

Fumaric acid and malic acid [*6915-15-7*] are produced from maleic anhydride. The primary use for fumaric acid is in the manufacture of paper sizing products (see PAPERMAKING ADDITIVES). Fumaric acid is also used to acidify food as is malic acid. Malic acid is a particularly desirable acidulant in certain beverage selections, specifically those sweetened with the artificial sweetener aspartame [*22839-47-0*].

Lubrication oil additives represent another important market segment for maleic anhydride derivatives. The molecular structures of importance are adducts of polyalkenyl succinic anhydrides (see LUBRICATION AND LUBRICANTS). These materials act as dispersants and corrosion inhibitors (see DISPERSANTS; CORROSION AND CORROSION CONTROL). One particularly important polyalkenyl succinic anhydride molecule in this market is polyisobutylene succinic anhydride (PIBSA) where the polyisobutylene group has a molecular weight of 900 to 1500. Other polyalkenes are also used. Polyalkenyl succinic anhydride is further derivatized with various amines to produce both dispersants and corrosion

Table 9. Uses for Maleic Anhydride in the United States, 1992[a]

Product	Demand, %
unsaturated polyester resins	57
fumaric and malic acid	10
lube oil additives	10
maleic copolymers	8
agricultural chemicals	5
miscellaneous	10
Total	*100*

[a]Ref. 182.

Table 10. Uses for Unsaturated Polyester in the United States, 1992[a]

Product	Demand, %
Glass-reinforced uses	
marine and marine accessories	21.4
general construction	14.2
bathroom components and fixtures	11.3
corrosion-resistant tanks, pipes, and ducts	10.0
transportation	7.8
gel coatings	5.0
consumer goods (appliances and recreational uses)	4.8
electrical components	1.9
other	2.2
Nonreinforced uses	
synthetic marble, onyx	8.7
other casting resins	3.6
auto repair putty	3.8
other	4.4
Total	*100.0*

[a]Ref. 183.

inhibitors. Another type of dispersant is a polyester produced from a polyalkenyl succinic anhydride and pentaerythritol [*115-77-5*].

Maleic anhydride is used in a multitude of applications in which a vinyl copolymer is produced by the copolymerization of maleic anhydride with other molecules having a vinyl functionality. Typical copolymers and their end uses are listed in Table 11.

The use of maleic anhydride in the manufacture of agricultural chemicals has declined in the United States since the early 1980s. Malathion [*121-75-2*]

Table 11. Maleic Anhydride Copolymers with Vinyl Functionality

Copolymer	CAS Registry Number	End uses
styrene–maleic anhydride copolymer	[*9011-13-6*]	engineering thermoplastics, paper treatment chemicals, floor polishes, emulsifiers, protective colloids, antisoil agents, dispersants
methyl vinyl ether–maleic anhydride copolymer	[*9011-16-9*]	thickeners, dispersants, stabilizing agents, adhesives, detergents, cosmetics, toiletries
diisobutylene–maleic anhydride copolymer		dispersing agent
acrylic acid–maleic anhydride copolymer		detergent ingredient
1,3-butadiene–maleic anhydride copolymer	[*25655-35-0*]	sizing agent
C_{18}-α-olefin–maleic anhydride copolymer		emulsification agent, paper coating

and Difolatan [*2425-06-1*] are no longer produced in the United States and Alar [*1596-84-5*] volumes have been significantly reduced by intense environmental scrutiny. Maleic hydrazide, Captan [*133-06-2*], Endothall [*145-73-3*], and several other maleic derivatives are still used in a number of agricultural functions such as plant growth regulation, fungicides, insecticides, and herbicides (see FUNGICIDES, AGRICULTURAL; INSECT CONTROL TECHNOLOGY; HERBICIDES).

There are numerous further applications for which maleic anhydride serves as a raw material. These applications prove the versatility of this molecule. The popular artificial sweetener aspartame [*22839-47-0*] is a dipeptide with one amino acid (L-aspartic acid [*56-84-8*]) which is produced from maleic anhydride as the starting material. Processes have been reported for production of poly(aspartic acid) [*26063-13-8*] (184–186) with applications for this biodegradable polymer aimed at detergent builders, water treatment, and poly(acrylic acid) [*9003-01-4*] replacement (184,187,188) (see DETERGENCY). Alkenylsuccinic anhydrides made from several linear alpha olefins are used in paper sizing, detergents, and other uses. Sulfosuccinic acid esters serve as surface active agents. Alkyd resins (qv) are used as surface coatings. Chlorendric anhydride [*115-27-5*] is used as a flame resistant component (see FLAME RETARDANTS). Tetrahydrophthalic acid [*88-98-2*] and hexahydrophthalic anhydride [*85-42-7*] have specialty resin applications. Gas barrier films made by grafting maleic anhydride to polypropylene [*25085-53-4*] film are used in food packaging (qv). Poly(maleic anhydride) [*24937-72-2*] is used as a scale preventer and corrosion inhibitor (see CORROSION AND CORROSION CONTROL). Maleic anhydride forms copolymers with ethylene glycol methyl vinyl ethers which are partially esterified for biomedical and pharmaceutical uses (189) (see PHARMACEUTICALS).

An important future use for maleic anhydride is believed to be the production of products in the 1,4-butanediol–γ-butyrolactone–tetrahydrofuran family. Davy Process Technology has commercialized a process (93) for producing 1,4-butanediol from maleic anhydride. This technology can be used to produce the product mix of the three molecules as needed by the producer. Another significant effort in this area is the tetrahydrofuran plant under construction in Spain by Du Pont in which butane is oxidized and recovered as maleic acid and the maleic acid is then reduced to tetrahydrofuran (109).

BIBLIOGRAPHY

"Maleic Acid, Fumaric Acid, and Maleic Anhydride" in *ECT* 1st ed., Vol. 8, pp. 680–696, by W. H. Gardner and L. H. Flett, National Aniline Division, Allied Chemical Dye Corp.; "Maleic Anhydride, Maleic Acid, and Fumaric Acid" in *ECT* 2nd ed., Vol. 12, pp. 819–837, by B. Dmuchovsky and J. E. Franz, Monsanto Co.; in *ECT* 3rd ed., Vol. 14, pp. 770–793, by W. D. Robinson and R. A. Mount, Monsanto Co.

1. J. Pelouze, *Ann.* **11**, 263 (1834); R. Winckler, *Ann.* **4**, 230 (1832).
2. J. R. Skeen, *Chem. Eng. News* **26**, 3684 (1948).
3. S. V. R. Mastrangelo, *Anal. Chem.* **29**, 841 (1957); H. G. M. De Wit and co-workers, *Thermochim. Acta* **65**, 43 (1983).
4. J. A. Dean, ed., *Lange's Handbook of Chemistry*, 14th ed., McGraw-Hill Book Co., Inc., New York, 1992.
5. Y. Suzuki, K. Muraishi, and K. Matsuki, *Thermochim. Acta* **211**, 171 (1992).

6. *Maleic Anhydride, Monsanto Technical Bulletin Pub. No. 9094*, Monsanto Chemical Co., St. Louis, Mo., 1988.
7. C. W. Davies and V. E. Malpass, *Trans. Faraday Soc.* **60**, 2075 (1964).
8. R. C. Wilhoit and D. Shiao, *J. Chem. Eng. Data* **9**, 595 (1964).
9. G. S. Parks, J. R. Mosley, and P. V. Peterson, Jr., *J. Chem. Phys.* **18**, 152 (1950).
10. T. Flitcroft, H. A. Skinner, and M. C. Whiting, *Trans. Faraday Soc.* **53**, 784 (1957).
11. R. C. Wilhoit, J. Chao, and K. R. Hall, *J. Phys. Chem. Ref. Data* **14**, 1 (1985).
12. L. O. Winstrom and L. Kulp, *Ind. Eng. Chem.* **41**, 2584 (1949).
13. H. Stephen and T. Stephen, eds., *Solubilities of Inorganic and Organic Compounds*, The MacMillan Co., New York, 1963.
14. M. Shahat, *Acta Cryst.* **5**, 763 (1952).
15. R. E. Marsh, E. Ubell, and H. E. Wilcox, *Acta Cryst.* **15**, 35 (1962).
16. M. N. G. James and G. J. B. Williams, *Acta Cryst.* **B30**, 1249 (1974).
17. L. S. Luskin, in R. H. Yokum and E. B. Nyquist, eds., *Functional Monomers*, Vol. 2, Marcel Dekker, Inc., New York, 1974, pp. 357–554.
18. *Maleic Anhydride, Chemical Safety Data Sheet SD-88*, Manufacturing Chemists Association, Washington, D.C., 1974.
19. F. W. Makison, R. S. Stricoff, and L. J. Partridge, Jr., eds., *NIOSH/OSHA Pocket Guide to Chemical Hazards*, DHEW (NIOSH) Publication No. 78-210, National Institute for Occupational Safety and Health, Cincinnati, Ohio, 1978.
20. *Monsanto Technical Bulletin No. RE-12*, Monsanto Co., St. Louis, Mo., 1962.
21. R. L. Hilderbrandt and E. M. A. Peixoto, *J. Mol. Struct.* **12**, 31 (1972).
22. O. L. Stiefvater, *Z. Naturforsch.* **33a**, 1480 (1978).
23. Y. Ishibashi, R. Shimada, and H. Shimada, *Bull. Chem. Soc. Jpn.* **55**, 2765 (1982).
24. K. Höjendahl, *J. Phys. Chem.* **28**, 758 (1924).
25. L. H. Flett and W. H. Gardner, *Maleic Anhydride Derivatives*, John Wiley & Sons, Inc., New York, 1952.
26. U.S. Pat. 2,653,168 (Sept. 22, 1953), S. M. Spatz (to Allied Chemical); U.S. Pat. 3,222,397 (Dec. 7, 1965), E. R. Talaty (to Pittsburgh Plate Glass Co.).
27. E. C. Horning, ed., *Organic Synthesis*, Collective Vol. 3, John Wiley & Sons, Inc., New York, 1955, p. 422.
28. V. N. Dubchenko and V. I. Kovalenko, *Zh. Prikl. Khim. (Leningrad)* **41**, 2568 (1968).
29. W. N. Marmer, D. E. Van Horn, and W. M. Linfield, *J. Am. Oil Chem. Soc.* **51**(4), 174 (1974).
30. R. E. Lutz, *Org. Synth.* **20**, 29 (1940).
31. K. Alder, F. Pasher, and A. Schmitz, *Ber.* **76B**, 27 (1943).
32. U.S. Pat. 3,409,638 (Nov. 5, 1968), C. M. Selwitz (to Gulf R & D Co.).
33. U.S. Pat. 4,496,746 (Jan. 29, 1985), J. C. Powell (to Texaco Inc.).
34. O. Diels, K. Alder, and D. Winter, *Ann.* **486**, 211 (1931).
35. U.S. Pat. 4,154,737 (May 15, 1979), G. G. Orphanides (to E. I. du Pont de Nemours & Co., Inc.).
36. Eur. Pat. Appl. EP 461,096 (June 5, 1990), A. Van Gysel, I. Vanden Eynde, and J. C. Vanovervelt (to UCB SA).
37. Jpn. Kokai Tokkyo Koho JP 04,202,178 (July 22, 1992), Y. Watabe and co-workers, (to Mitsui Toatsu Chemicals, Inc.).
38. W. R. Davie, *Chem. Eng. News* **42**(8), 41 (1964); L. Bretherick, ed., *Handbook of Reactive Chemical Hazards*, 3rd ed., Butterworths, London, 1985, p. 405.
39. T. T. Curtius and M. A. Feuterling, *J. Prakt. Chem.* **51**, 371 (1895).
40. H. Feuer, E. H. White, and J. E. Wyman, *J. Am. Chem. Soc.* **80**, 3790 (1958).
41. O. Diels and K. Alder, *Ann.* **460**, 98 (1928).
42. M. W. Lee and W. C. Herndon, *J. Org. Chem.* **43**, 518 (1978).
43. V. M. Zhudin and co-workers, *Bull. Acad. Sci. USSR, Div. Chem. Sci.* **38**, 2303 (1989).

44. K. Matsumoto and co-workers, *Heterocycles* **24**, 1835 (1986).
45. H. Kotsuki and co-workers, *Bull. Chem. Soc. Jpn.* **52**, 544 (1979).
46. M. Ramezanian and co-workers, *J. Org. Chem.* **54**, 2852 (1989).
47. R. Cordone, W. D. Harman, and H. Taube, *J. Am. Chem. Soc.* **111**, 5969 (1989).
48. R. M. Scarborough, Jr., and A. B. Smith, III, *J. Am. Chem. Soc.* **99**, 7085 (1977).
49. D. C. Owsley and J. J. Bloomfield, *J. Org. Chem.* **36**, 3768 (1971).
50. E. Grovenstein, Jr., D. V. Rao, and J. W. Taylor, *J. Am. Chem. Soc.* **83**, 1705 (1961).
51. B. B. Snider, in B. M. Trost, I. Fleming, and L. A. Paquette, eds., *Comprehensive Organic Synthesis*, Vol. 5, Pergamon Press, Oxford, U.K., 1991, pp. 1–26.
52. S. H. Nahm and H. N. Cheng, *J. Org. Chem.* **51**, 5093 (1986).
53. M. Tessier and E. Marechal, *Eur. Polym. J.* **20**, 269 (1984).
54. *Ibid.*, p. 281.
55. R. A. Back and J. M. Parsons, *Can. J. Chem.* **59**, 1342 (1981).
56. M. E. Baumann and co-workers, *Helv. Chim. Acta* **67**, 1897 (1984).
57. Brit. Pat. 933,102 (Aug. 8, 1963), V. C. E. Burnop (to Esso Research and Engineering Co.).
58. F. Severini and co-workers, *Br. Polym. J.* **23**, 23 (1990).
59. A. G. Gonorskaya and S. A. Marina, *Lakokras. Mater. Ikl Primen.*, (1) 18 (1984).
60. U.S. Pat. 3,953,616 (Apr. 27, 1976), P. D. Thomas (to Pfizer Inc.).
61. J. Selley, in J. I. Kroschwitz, ed., *Encyclopedia of Polymer Science and Engineering*, 2nd ed., Vol. 12, Wiley-Interscience, New York, 1988, pp. 256–290.
62. A. Fradet and P. Arlaud, in G. Allen and J. C. Bevington, eds., *Comprehensive Polymer Science*, Vol. 5, Pergamon Press plc, Oxford, U.K., 1989, pp. 331–344.
63. N. A. Cortese and co-workers, *J. Org. Chem.* **43**, 2952 (1978).
64. U.S. Pat. 5,112,993 (May 12, 1992), C. G. Krespan (to E. I. du Pont de Nemours & Co.).
65. Jpn. Tokkyo Koho JP 57 47,193 (Oct. 7, 1982), (to All-Union Research Institute of Chemicals for Plant Protection, Ufa, Japan).
66. USSR Pat. 1,616,893 (Dec. 30, 1990), A. D. Efendiev and co-workers (to Institute of Theoretical Problems of Chemical Technology, Azerbaidzhan, USSR).
67. C. W. Shoppee and W.-Y. Wu, *Aust. J. Chem.* **40**, 1137 (1987).
68. J. M. Rosenfeld and C. B. Murphy, *Talenta* **14**(1), 91 (1967).
69. Brit. Pat. 1,242,320 (Aug. 11, 1971), E. Weyens (to UCB, SA); E. Weyens, *Hydrocarbon Process.* **53**(11), 132 (1974).
70. J. C. Burnett, R. A. Keppel, and W. D. Robinson, *Catal. Today* **1**(5), 537 (1987).
71. U.S. Pat. 4,118, 403 (Oct. 3, 1978), J. E. White (to Monsanto Co.).
72. Jpn. Kokai 75,101,320 (Aug. 11, 1975), S. Umemura and Y. Ikeda (to Ube Industries, Ltd.).
73. M. F. Hughes and R. T. Adams, *Amer. Chem. Soc. Div. Pet. Chem. Prepr.* **17**(1), B109 (1972).
74. A. Aurrecoechea, L. Pallas, and M. C. Roces, *Afinidad* **31**, 827 (1974).
75. J. S. Meek, *J. Chem. Educ.* **52**, 541 (1975).
76. F. A. Cotton and G. Wilkinson, *Advanced Inorganic Chemistry*, 5th ed., John Wiley & Sons, Inc., New York, 1988, pp. 71–74, 1152–1156.
77. C.-H. Lai and co-workers, *Inorg. Chem.* **32**, 5658 (1993).
78. Y. Hsiou, Y. Wang, and L. K. Liu, *Acta Crystallogr., Sect. C: Cryst. Struct. Commun.* **C45**, 721 (1989).
79. A. S. Antsyshkina and co-workers, *Koord. Khim.* **8**, 1256 (1982).
80. J. van Westrenen and co-workers, *Tetrahedron* **46**, 5741 (1990).
81. C. Zhi and co-workers, *Inorg. Chem.* **29**, 5025 (1990).
82. F. B. Zienty, B. D. Vineyard, and A. A. Schleppnik, *J. Org. Chem.* **27**, 3140 (1962).
83. U.S. Pat. 4,048,141 (Sept. 13, 1977), G. A. Doorakian, D. L. Schmidt, and M. C. Cornell, III (to Dow Chemical Co.).

84. D. B. Denney and D. Z. Denney, *Phosphorus Sulfur* **13**, 315 (1982).
85. E. Gilbert, *Z. Naturforsch.* **32b**, 1308 (1977).
86. J. O. Oludipe and co-workers, *J. Chem. Tech. Biotechnol.* **55**, 103 (1992).
87. I. Ahmad and co-workers, *Hydrocarbon Process., Int. Ed.* **68**(2), 51 (1989).
88. R. W. White and W. D. Emmons, *Tetrahedron* **17**, 31 (1962).
89. G. P. Panigrahi and P. K. Misro, *Indian J. Chem. A* **16**, 201 (1978).
90. R. H. Grubbs and W. Tumas, *Science* **243**, 907 (1989).
91. T. Viswanathan and J. Jethmalani, *J. Chem. Educ.* **70**, 165 (1993).
92. S. Uihlein and G. Emig, *Chem. Ind. (Duesseldorf)* **116**, 44 (1993).
93. N. Harris and M. W. Tuck, *Hydrocarbon Process., Int. Ed.* **69**(5), 79 (1990).
94. Eur. Pat. Appl. EP 447963 (Sept. 25, 1991), U. Stabel and co-workers (to BASF AG).
95. U.S. Pat. 5,196,602 (Mar. 23, 1993), J. R. Budge, T. G. Attig, and A. M. Graham (to Standard Oil Co.).
96. Eur. Pat. Appl. EP 322140 (June 28, 1989), T. G. Attig and J. R. Budge (to Standard Oil Co.).
97. PCT Int. Appl. WO 8800937 (Feb. 11, 1988), K. Turner and co-workers (to Davy McKee Ltd.).
98. Ger. Offen. DE 77-2715667 (Apr. 7, 1977), D. Freudenberger and F. Wunder (to Hoechst AG).
99. Ger. Offen. DE 2845905 (Apr. 24, 1980), E. Lange and M. Zur Hausen (to Chemische Werke Huels AG).
100. Jpn. Kokai Tokkyo Koho JP 02233627 (Sept. 17, 1990), S. Suzuki, H. Inagaki, and H. Ueno (to Tonen Co., Ltd.).
101. Eur. Pat. Appl. EP 373946 (June 20, 1990), S. Suzuki, H. Inagaki, and H. Ueno (to Tonen Co., Ltd.).
102. G. L. Castiglioni and co-workers, *Chem. Ind.*, (13), 510 (July 5, 1993).
103. Chin. Pat. CN 1,058,400 (Feb. 5, 1992), Y. Xiang and co-workers (to Fudan University).
104. PCT Int. Appl. WO 9116132 (Oct. 31, 1991), P. D. Taylor, W. De Thomas, and D. W. Buchanan, Jr. (to ISP Investments, Inc.).
105. Jpn. Kokai Tokkyo Koho JP 05,148,254 (June 15, 1993), T. Fuchigami and co-workers (to Tosoh Corp.).
106. Jpn. Kokai Tokkyo Koho JP 04016237 (Jan. 21, 1992), S. Nishiyama, S. Kumoi, and N. Mizui (to Tosoh Corp.).
107. Jpn. Kokai Tokkyo Koho JP 02200680 (Aug. 8, 1990), H. Wada, Y. Hara, and Y. Yuzawa (to Mitsubishi Kasei Corp.).
108. Eur. Pat. Appl. EP 339012 (Oct. 25, 1989), J. L. Dallons and co-workers (to UCB SA).
109. W. Stadig, *Chem. Process.* **55**(8), 26 (1992).
110. U.S. Pat. 4,550,185 (Oct. 29, 1985), M. A. Mabry, W. W. Prichard, and S. B. Ziemecki (to E. I. du Pont de Nemours & Co., Inc.).
111. PCT Int. Appl. WO 9202298 (Feb. 20, 1992), R. E. Ernst and J. B. Michel (to E. I. du Pont de Nemours & Co., Inc.).
112. U.S. Pat. 3,922,272 (Nov. 25, 1975), V. Lamberti (to Lever Brothers Co.).
113. V. G. Gruzdev, *J. Gen. Chem. USSR* **49**, 725 (1979).
114. S. Shelly, K. Fouhy, and S. Moore, *Chem. Eng.* **100**(12), 61 (Dec. 1993).
115. *Maleic Anhydride World Survey*, Tecnon (U.K.) Ltd., London, 1992.
116. K. T. Tchang and L. P. Barnes, Huntsman Specialty Chemicals Corp., personal communication, 1994.
117. *Chem. Mark. Rep.* **242**(11), 4 (Sept. 14, 1992).
118. *Hydrocarbon Process.* **70**(3), 164 (1991).
119. *Eur. Chem. News.* **60**(1599), 24 (Dec. 20/27, 1993).
120. M. Malow, *Environ. Prog.* **4**, 151 (1985).

121. G. Stefani and co-workers, *Chim. Ind. (Milan)* **72**(7), 604 (1990); G. Stefani and co-workers, in G. Centi and F. Trifiro, eds., *New Developments in Selective Oxidation, Studies in Surface Science and Catalysis*, Vol. 55, Elsevier, Amsterdam, the Netherlands, 1990, pp. 537–550.
122. J. Haggin, *Chem. Eng. News* **69**(44), 34 (Nov. 4, 1991).
123. G. Centi and F. Trifiro, *Chim. Ind. (Milan)* **68**(12), 74 (1986); F. Cavani and F. Trifiro, *CHEMTECH*, 18–25 (Apr. 1994).
124. P. Amorós and co-workers, *Chem. Mater.* **3**, 407 (1991).
125. G. J. Hutchings, *Catal. Today* **16**(1), 139 (1993).
126. B. K. Hodnett, *Catal. Today* **16**(1), 131 (1993).
127. I. Matsuura, *Catal. Today* **16**(1), 123 (1993).
128. U.S. Pat. 5,137,860 (Aug. 11, 1992), J. R. Ebner and W. J. Andrews (to Monsanto Co.).
129. U.S. Pat. 4,632,915 (Dec. 30, 1986), R. A. Keppel and V. M. Franchetti (to Monsanto Co.).
130. G. J. Hutchings, *Appl. Catal.* **72**(1), 1 (1991).
131. E. Bordes, *Catal. Today* **16**(1), 27 (1993).
132. Y. Zhang, R. P. A. Sneede, and J. C. Volta, *Catal. Today* **16**(1), 39 (1993).
133. J. R. Ebner and M. R. Thompson, *Catal. Today* **16**(1), 51 (1993).
134. M. López Granados, J. C. Canesa, and M. Fernández-Garcia, *J. Catal.* **141**, 671 (1993).
135. M. T. Sananes, A. Tuel, and J. C. Volta, *J. Catal.* **145**(2), 251 (1994).
136. F. Cavani, G. Centi, and F. Trifiro, *Appl. Catal.* **15**(1), 151 (1985).
137. U.S. Pat. 5,185,455 (Feb. 9, 1993), J. R. Ebner (to Monsanto Co.).
138. U.S. Pat. 4,855,459 (Aug. 8, 1989), M. J. Mummey (to Monsanto Co.).
139. J. S. Buchanan and S. Sundavesan, *Chem. Eng. Commun.* **52**(1–3), 33 (1987).
140. R. M. Contractor and A. W. Sleight, *Catal. Today* **1**(5), 587 (1987).
141. G. Emig and F. G. Martin, *Ind. Eng. Chem. Res.* **30**(6), 1110 (1991).
142. B. C. Trivedi and B. M. Culbertson, *Maleic Anhydride*, Plenum Press, New York, 1982.
143. G. Emig and F. Martin, *Catal. Today* **1**(5), 477 (1987).
144. *Fire Protection Guide on Hazardous Materials*, 6th ed., National Fire Protection Association, Boston, Mass., 1975, p. 325M-31.
145. *Eur. Chem. News* **60**(1578), 37 (July 19, 1993).
146. U.S. Pat. 4,987,239 (Jan. 22, 1991), R. Ramachandran, Y. Shukla, and D. L. McLean (to The BOC Group, Inc.).
147. U.S. Pat. 5,126,463 (June 30, 1992), R. Ramachandran, A. I. Shirley, and L.-L. Sheu (to The BOC Group, Inc.).
148. U.S. Pat. 4,691,031 (Sept. 1, 1987), G. D. Suciu and J. E. Paustian (to Lummus Co.).
149. D. O'Sullivan, *Chem. Eng. News* **70**(10), 6 (Mar. 9, 1992).
150. U.S. Pat. 4,668,802 (May 26, 1987), R. M. Contractor (to E. I. du Pont de Nemours & Co., Inc.).
151. U.S. Pat. 5,021,588 (June 4, 1991), R. M. Contractor (to E. I. du Pont de Nemours & Co., Inc.).
152. J. Dixon and J. Longfield, in P. H. Emmett, ed., *Catalysis*, Vol. 7, Reinhold Publishing Corp., New York, 1960, Chapt. 3.
153. Ger. Offen. 2,440,746 (Mar. 13, 1975), J. Ramoulle (to UCB, SA).
154. Ger. Offen. 2,530,895 (Jan. 29, 1976), J. M. Lietard and G. Matthijs (to UCB, SA).
155. *Hydrocarbon Process., Int. Ed.* **62**(11), 110 (1983).
156. U.S. Pat. 2,955,136 (Oct. 4, 1960), J. D. Sullivan, W. D. Robinson, and H. D. Cummings (to Monsanto Chemical Co.).
157. U.S. Pat. 3,380,173 (June 18, 1968), J. L. Russell and H. Olenburg (to Halcon International, Inc.).

158. U.S. Pat. 2,914,559 (Nov. 24, 1959), W. J. Stefaniac (to Allied Chemical Corp.).
159. Jpn. Kokai 73 34,121 (May 16, 1973), T. Inoue (to Nippon Steel Chemical Industry Co., Ltd.).
160. U.S. Pat. 2,704,296 (Mar. 15, 1955), E. H. Dobratz (to Monsanto Chemical Co.); U.S. Pat. 3,025,321 (Mar. 13, 1962), H. A. Lindahl and H. Hennig (to The Pure Oil Co.); U.S. Pat. 3,702,342 (Nov. 7, 1972), G. M. Kibler and C. L. Singleton (to Pfizer, Inc.).
161. "Maleic Anhydride," *Chemical Economics Handbook*, SRI International, Menlo Park, Calif., Aug. 1993, p. 672.5000A.
162. *Synthetic Organic Chemicals, U.S. Production and Sales*, U.S. Trade Commission, Washington, D.C., annual.
163. *Chem. Mark. Rep.* **242**(1), 42 (July 6, 1992).
164. *Chem. Mark. Rep.* **241**(26), 38 (June 29, 1992).
165. "D3504-91, Standard Specifications for Maleic Anhydride," *1992 Annual Book of ASTM Standards*, Sec. 6, Vol. 06.03, ASTM, Philadelphia, Pa., 1992, p. 676.
166. Ref. 165, pp. 645–646.
167. S. Siggia and N. A. Floramo, *Anal. Chem.* **25**, 797 (1953).
168. Ref. 165, pp. 589–595.
169. Ref. 165, pp. 660–662.
170. S. Siggia and J. G. Hanna, *Anal. Chem.* **23**, 1717 (1951).
171. F. E. Critchfield and J. B. Johnson, *Anal. Chem.* **28**, 430 (1956).
172. B. M. Culbertson, in J. I. Kroschwitz, ed., *Encyclopedia of Polymer Science and Engineering*, 2nd ed., Vol. 9, Wiley-Interscience, New York, 1987, pp. 225–294.
173. O. Kiser and co-workers, *J. Chromatogr.* **151**(1), 81 (1978).
174. C. Laguerie and M. Aubry, *J. Chromatogr.* **101**, 357 (1974).
175. P. J. Elving and I. Rosenthal, *Anal. Chem.* **26**, 1454 (1954).
176. I. Temesvari, E. Pungor, and G. Liptay, *Proc. Anal. Chem. Conf. 3rd* **2**, 341 (1970).
177. J. J. Wohnlich, *Bull. Soc. Chim. Biol.* **49**(7), 900 (1967); H. Thielemann, *Mikrochim. Acta*, (4), 521 (1973).
178. USSR Pat. 427,279 (May 5, 1974), V. K. Pachebula, V. F. Kashcheeva, and G. A. Kochurova.
179. A. J. W. Brook, *J. Chromatogr.* **39**(3), 328 (1969).
180. D. J. Patel, R. S. Hegde, and S. L. Bafna, *Indian J. Appl. Chem.* **33**(2), 133 (1970); J. P. Lefevre, M. Caude, and R. Rosset, *Analysis* **4**, 16 (1976).
181. R. J. Lewis, Sr., *Sax's Dangerous Properties of Industrial Materials*, 8th ed., Van Nostrand Reinhold, New York, 1992.
182. "Maleic Anhydride," *Chemical Product Synopsis*, Mannsville Chemical Products Corp., Asbury Park, N.J., Nov. 1992.
183. "Unsaturated Polyester Resins," *Chemical Economics Handbook*, SRI International, Menlo Park, Calif., July 1993, p. 580.1200A.
184. S. Stinson, *Chem. Eng. News* **70**(29), 21 (July 20, 1992).
185. PCT Int. WO 9214753 (Sept. 3, 1992), L. P. Koskan and co-workers (to Donlar Corp.).
186. U.S. Pat. 5,219,986 (June 15, 1993), T. A. Cassata (to Cygnus Corp.).
187. U.S. Pat. 5,116,513 (May 26, 1992), L. P. Koskan and K. C. Low (to Donlar Corp.).
188. U.S. Pat. 4,971,724 (Nov. 20, 1990), D. J. Kalota and D. C. Silverman (to Monsanto Co.).
189. E. Chiellini and R. Solaro, *CHEMTECH*, 29–36 (July 1993).

General References

B. C. Trivedi and B. M. Culbertson, *Maleic Anhydride*, Plenum Press, New York, 1982. Approximately 500 pages on polymers and their applications.

B. M. Culbertson, "Maleic and Fumaric Polymers," in J. I. Kroschwitz, ed., *Encyclopedia of Polymer Science and Engineering*, 2nd ed., Vol. 9, Wiley-Interscience, New York, 1987, pp. 225–294.

S. D. Cooley and J. D. Powers, "Maleic Acid and Anhydride," in *Encyclopedia of Chemical Processing and Design*, Vol. 29, Marcel Dekker, Inc., New York, 1988, pp. 35–55.
K. Lohbeck and co-workers, "Maleic and Fumaric Acids," in *Ullmann's Encyclopedia of Industrial Chemistry*, Vol. A16, 5th ed., VCH, Weinheim, Germany, 1990, pp. 53–62.
K. Weissermel and H.-J. Arpe, *Industrial Organic Chemistry*, 2nd ed., VCH, Weinheim, Germany, 1993, pp. 362–369.

TIMOTHY R. FELTHOUSE
JOSEPH C. BURNETT
SCOTT F. MITCHELL
MICHAEL J. MUMMEY
Huntsman Specialty Chemicals Corporation

MALONIC ACID AND DERIVATIVES

Malonic Acid

Physical Properties. Malonic acid, $HOOC{-}CH_2{-}COOH$ (**1**), was discovered and isolated in 1858 as a product of malic acid oxidation. The physical properties of malonic acid are listed in Table 1.

Reactions. Heating an aqueous solution of malonic acid above 70°C results in its decomposition to acetic acid and carbon dioxide. Malonic acid is a useful tool for synthesizing α-unsaturated carboxylic acids because of its ability to undergo decarboxylation and condensation with aldehydes or ketones at the methylene group. Cinnamic acids are formed from the reaction of malonic acid and benzaldehyde derivatives (1). If aliphatic aldehydes are used acrylic acids result (2). Similarly this facile decarboxylation combined with the condensation

Table 1. Properties of Malonic Acid[a]

Property	Value
mol wt	104.06
melting point, °C	135 (dec)
ionization constants	
K_1	1.42×10^{-3}
K_2	2.01×10^{-6}
appearance	white crystals
solubility	
in water at 20°C	139 g/100 mL
in pyridine at 15°C	15 g/100 g

[a]Also called propanedioic acid or methanedicarboxylic acid.

with an activated double bond yields α-substituted acetic acid derivatives. For example, 4-thiazolidine acetic acids (**2**) are readily prepared from 2,5-dihydro-1,3-thiazoles (3). A further feature of malonic acid is that it does not form an anhydride when heated with phosphorous pentoxide [*1314-56-3*] but rather carbon suboxide [*504-64-3*], [O=C=C=O], a toxic gas that reacts with water to reform malonic acid.

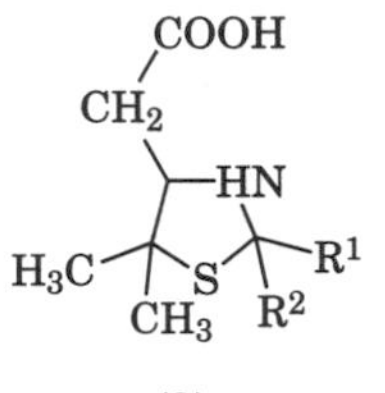

(**2**) 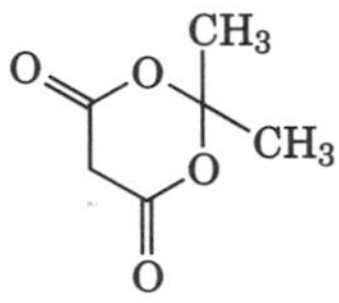 (**3**) 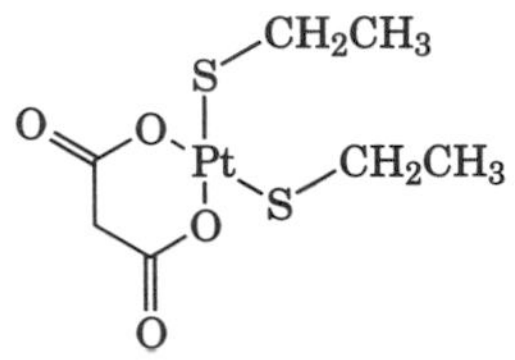 (**4**)

Reactions of the carboxylic acid groups include monoesterification, diesterification, or conversion with thiols. The synthesis of mono(*tert*-butyl) malonate [*40052-13-9*] through condensation with *tert*-butyl alcohol [*75-65-0*] (4), of di-*p*-methylbenzyl malonate through condensation with *p*-methylbenzyl alcohol [*589-18-4*] (5), or propanebis(thioic) *S*,*S'*-diesters, $CH_2(COSR)_2$, through condensation with thiols has been reported (6). Further reactions at the carboxylic acid groups lead to ring closure. Of special interest is the synthesis of 2,2-dimethyl-1,3-dioxan-4,6-dione (**3**), commonly named Meldrum's acid [*2033-24-1*], through condensation with acetone [*67-64-1*] in the presence of acetic acid [*64-19-7*] and sulfuric acid [*7664-93-9*] (7). Malonic acid also forms acidic and neutral salts as well as double and complex salts. The platinum complexes (**4**) have been investigated as antitumor agents (8).

Preparation. The industrial production of malonic acid is much less important than that of the malonates. Malonic acid is usually produced by acid saponification of malonates (9). Further methods which have been recently investigated are the ozonolysis of cyclopentadiene [*542-92-7*] (10), the air oxidation of 1,3-propanediol [*504-63-2*] (11), or the use of microorganisms for converting nitriles into acids (12).

Economic Aspects. Malonic acid is produced by Juzen and Tateyama in Japan as well as Lonza Ltd. in Switzerland and Riedel-De Haen Ltd. in Germany. It costs around \$30/kg (1993) for shipments of one to two drums.

Analytical and Test Methods. Potentiometric titration with sodium hydroxide [*1310-73-2*] is employed. Both equivalent points are measured, and the content is determined using the following equation:

$$\text{malonic acid, \%} = \frac{(E_2 - E_1) \times M \times 10.406}{W}$$

where E_1 = volume of standard NaOH at the first equivalence point, E_2 = volume of standard NaOH at the second equivalence point, M = molarity of standard NaOH solution, and W = weight of sample in grams.

Health and Safety Factors (Toxicology). No special precautions are necessary in the handling of malonic acid beyond normal safe handling measures. Due to its acidity malonic acid is classified as a mild irritant (skin irritation, rabbits). The LD_{50} value (oral, rats) for malonic acid is 2750 mg/kg. Transport classification: RID/ADR, IMDG-Code, IATA/ICAO: not restricted.

Uses. Malonic acid is used instead of the less expensive malonates for the introduction of a CH–COOH group under mild conditions by Knoevenagel condensation and subsequent decarboxylation. The synthesis of 3,4,5-trimethoxycinnamic acid, the key intermediate for the coronary vasolidator Cinepazet maleate [*50679-07-7*] (**5**) involves such a pathway (13).

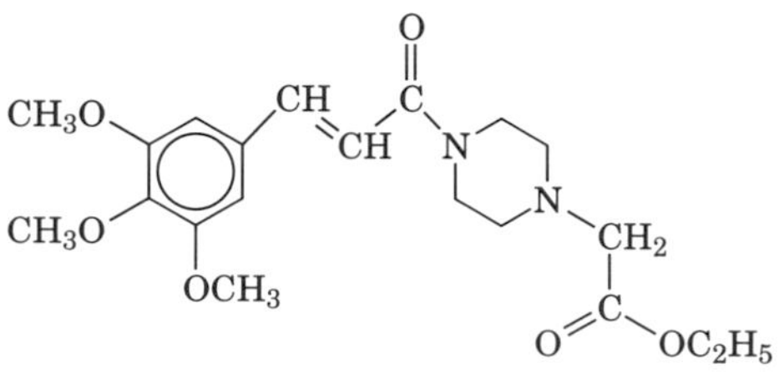

CH_3O CH_3O OCH_3 CH CH C O N N CH_2 C O OC_2H_5

(**5**)

C_5H_{11} O O

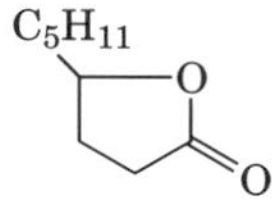

(**6**)

Knoevenagel condensation of malonic acid with heptaldehyde [*111-71-7*], followed by ring closure, gives the fragrance γ-nonanoic lactone [*104-61-0*] (**6**) (14). Beside organic synthesis, malonic acid can also be used as electrolyte additive for anodization of aluminum [*7429-90-5*] (15), or as additive in adhesive compositions (16).

Meldrum's Acid. Meldrum's acid [*2033-24-1*] (**3**) is commercially used for the production of monoesters of malonic acid and beta-keto acids (17). The chemistry of Meldrum's acid is extensively reviewed in Reference 18.

Malonates

Physical Properties. Industrially, the most important esters are dimethyl malonate [*108-59-8*] and diethyl malonate [*105-53-3*], whose physical properties are summarized in Table 2. Both are sparingly soluble in water (1 g/50 mL for the diethyl ester) and miscible in all proportions with ether and alcohol.

Reactions. The chemical properties of malonates are highlighted by the acidity of the methylene group ($pK_a \sim 13$) to such an extent that a proton can be easily detached by a strong base, usually alkoxides. Alkylation with 1-bromooctane [*111-83-1*] in the presence of sodium ethoxide [*141-52-6*] gives diethyl octylmalonate [*1472-85-1*], which can be further reduced with lithium aluminum hydride [*16853-85-3*], $LiAlH_4$, to yield 2-octyl-1,3-propanediol [*74971-70-3*] (19). The use of other starting materials which contain reactive halogen, such as 4-halo-3-methoxy-2-butenoate, gives access to 1,3-cyclopentanedione [*3859-41-4*] (20) upon subsequent cyclization and decarboxylation. The second

Table 2. Physical Properties of Dimethyl and Diethyl Malonate

Property	Dimethyl malonate	Diethyl malonate
other names	propanedioic acid dimethyl ester	propanedioic acid diethyl ester
appearance	colorless liquid	colorless liquid
mol wt	132.12	160.17
mp, °C	−62	−50
bp, °C[a]	181.4	199
d_4^{20}, g/mL	1.1544	1.0551
refractive index at 20°C	1.4140	1.4143
dipole moment, C·m[b]	7.97×10^{-30}	5.24×10^{-30}

[a]At 101 kPa = 1 atm.
[b]To convert C·m to debyes, multiply by 3×10^{29}.

hydrogen atom at the activated methylene group can be substituted analogously; thus cyclic dicarboxylic acids such as cyclopropyl-1,1-dicarboxylates (21) can be obtained starting from 1,2-dibromoethane [*106-93-4*]. The first ester function of the malonates is hydrolyzed much more easily than the second. This property can be used for synthesizing a large number of carboxylic acids by alkylation or acylation of a malonate followed by hydrolysis and decarboxylation of one ester group. This is the case for ethyl 2,4-dichloro-5-fluorobenzoylacetate (**7**) [*86483-51-4*] made through acetylation of diethyl malonate by 2,4-dichloro-5-fluorobenzoyl chloride [*86393-34-2*] (22).

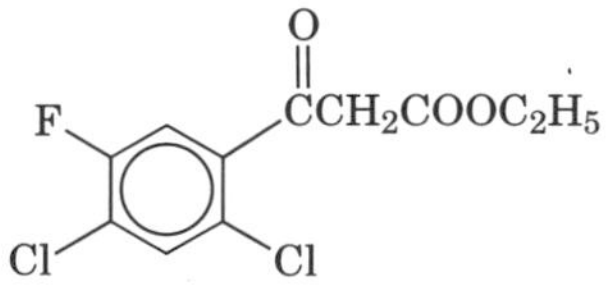

(**7**) 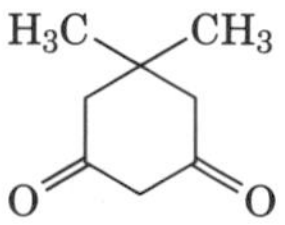 (**8**)

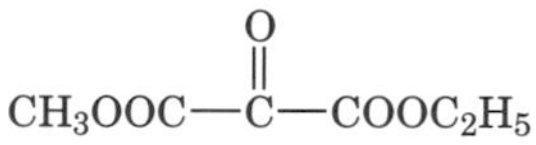

$$CH_3OOC-\overset{O}{\overset{\|}{C}}-COOC_2H_5$$

(**9**)

$$CH_3-\overset{O}{\overset{\|}{C}}-NHCH(COOR)_2$$

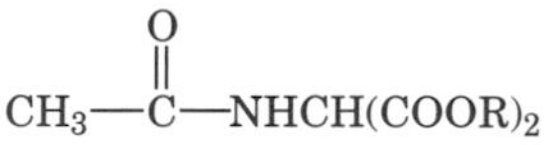

(**10**)

Further reactions on the activated methylene group involve the Knoevenagel condensation with acetones or aldehydes (23) yielding α,β-unsaturated compounds. The reaction with triethyl orthoformate [*122-51-0*] gives diethyl ethoxymethylenemalonate [*87-13-8*], $C_2H_5OCH{=}C(COOC_2H_5)_2$ (24). The Michael condensation with activated double bond containing moieties such as acrylonitrile gives the 2-(2-cyanoethyl)malonate (25). Similarly with mesityl oxide [*141-79-7*], dimedone [*126-81-8*] (**8**) is obtained upon subsequent cyclization and decarboxylation (26). The oxidation of malonates with air in the presence of a catalyst results in esters of mesoxalic acid (**9**) (27). The nitrosation with nitrous acid [*7782-77-6*] gives isonitromalonates, $HO—N{=}C(COOR)_2$, which are simultaneously acylated and hydrogenated to acetaminomalonates (**10**) (28). Besides decarboxylation and hydrogenation, the ester groups can be transesterified with *tert*-butyl alcohol [*75-65-0*] giving the mixed malonate, $H_5C_2OOCCH_2COOC(CH_3)_3$ (29). Finally, malonates can be converted into nitrogen-containing heterocycles, eg, pyrazolones such as, 2,4-pyrazolidinedione-1-phenyl [*19933-22-3*] upon reaction with phenylhydrazine [*100-63-0*] (30), or pyrimidines such as 2,4,6-trihydroxypyrimidine [*67-52-7*] upon reaction with urea [*57-13-6*] (31). Similarly 2-amino-4,6-dihydroxypyrimidine [*56-09-7*] and 4,6-dihydroxypyrimidine [*1193-24-4*] are obtained upon reaction with guanidine [*113-00-8*] (32) and formamide [*75-12-7*] (33), respectively.

Manufacture. *Hydrogen Cyanide Process.* This process, one of two used for the industrial production of malonates, is based on hydrogen cyanide [*74-90-8*] and chloroacetic acid [*79-11-8*]. The intermediate cyanoacetic acid [*372-09-8*] is esterified in the presence of a large excess of mineral acid and alcohol.

$$Cl—CH_2—COONa + NaCN \longrightarrow NC—CH_2—COONa + NaCl$$

$$NC—CH_2—COONa \xrightarrow[H_2SO_4,\ H_2O]{+ROH} ROOC—CH_2—COOR$$

A solution of sodium cyanide [*143-33-9*] (ca 25%) in water is heated to 65–70°C in a stainless steel reaction vessel. An aqueous solution of sodium chloroacetate [*3926-62-3*] is then added slowly with stirring. The temperature must not exceed 90°C. Stirring is maintained at this temperature for one hour. Particular care must be taken to ensure that the hydrogen cyanide, which is formed continuously in small amounts, is trapped and neutralized. The solution of sodium cyanoacetate [*1071-36-9*] is concentrated by evaporation under vacuum and then transferred to a glass-lined reaction vessel for hydrolysis of the cyano group and esterification. The alcohol and mineral acid (weight ratio 1:2 to 1:3) are introduced in such a manner that the temperature does not rise above 60–80°C. For each mole of ester, ca 1.2 moles of alcohol are added.

Hydrochloric acid [*7647-01-0*], which is formed as by-product from unreacted chloroacetic acid, is fed into an absorption column. After the addition of acid and alcohol is complete, the mixture is heated at reflux for 6–8 h, whereby the intermediate malonic acid ester monoamide is hydrolyzed to a dialkyl malonate. The pure ester is obtained from the mixture of crude esters by extraction with benzene [*71-43-2*], toluene [*108-88-3*], or xylene [*1330-20-7*]. The organic phase is washed with dilute sodium hydroxide [*1310-73-2*] to remove small amounts of the monoester. The diester is then separated from solvent by distillation at atmospheric pressure, and the malonic ester obtained by redistillation under vacuum as a colorless liquid with a minimum assay of 99%. The aqueous phase contains considerable amounts of mineral acid and salts and must be treated before being fed to the waste treatment plant. The process is suitable for both the dimethyl and diethyl esters. The yield based on sodium chloroacetate is 75–85%.

Various low molecular mass hydrocarbons, some of them partially chlorinated, are formed as by-products. Although a relatively simple plant is sufficient for the reaction itself, a sizeable investment is required for treatment of the wastewater and exhaust gas.

Carbon Monoxide Process. This process involves the insertion of carbon monoxide [*630-08-0*] into a chloroacetate. According to the literature (34) in the first step ethyl chloroacetate [*105-39-5*] reacts with carbon monoxide in ethanol [*64-17-5*] in the presence of dicobalt octacarbonyl [*15226-74-1*], $Co_2(CO)_8$, at typical temperature of 100°C under a pressure of 1800 kPa (18 bars) and at pH 5.7. Upon completion of the reaction the sodium chloride formed is separated along with the catalyst. The ethanol, as well as the low boiling point components, is distilled and the nonconverted ethyl chloroacetate recovered through distillation in a further column. The crude diethyl malonate obtained is further purified by redistillation. This process also applies for dimethyl malonate and diisopropyl malonate.

Other processes described in the literature for the production of malonates but which have not gained industrial importance are the reaction of ketene [*463-51-4*] with carbon monoxide in the presence of alkyl nitrite and a palladium salt as a catalyst (35) and the reaction of dichloromethane [*75-09-2*] with carbon monoxide in the presence of an alcohol, dicobalt octacarbonyl, and an imidazole (36).

Economic Aspects. Dimethyl and diethyl malonates are produced as shown in Table 3. Total capacity is estimated to be about 12,000 t/yr. Furthermore, producers are also reported in the People's Republic of China and in Romania. In bulk shipments, both malonates are available at ca $6/kg (1993).

Table 3. Malonate Production

Country	Company	Process
Europe	Hüls (Germany)	carbon monoxide process
	Lonza (Switzerland)	hydrogen cyanide process
Japan	Juzen	carbon monoxide process
	Tateyama	hydrogen cyanide process
South Korea	Korean Fertilizers	carbon monoxide process

Analytical and Test Methods. Gas chromatography is used for the quantitative analysis of malonates. Typical analysis conditions are 5% Reoplex 400 on Chromosorb G 80–100 mesh; 2 m, 0.3 cm diameter metal column; temperature for column = 120°C; detector, 150°C; and injector, 120°C. The determination of the free acid through titration and of the determination of water content according to the Karl-Fischer method are also important.

Health and Safety Factors. Dimethyl malonate and diethyl malonate do not present any specific danger of health hazard if handled with the usual precautions. Nevertheless, inhalation and skin contact should be avoided. Dimethyl malonate has a LD_{50} (oral, rats) of 4520 mg/kg and is classified as nonirritant (skin irritation, rabbits). Diethyl malonate has an LD_{50} (oral, rats) greater than 5000 mg/kg and is also classified as nonirritant (skin irritation, rabbits). Transport classification for both esters is RID/ADR: 3, IMDH-Code, IATA-ICAO: not restricted.

Uses. Dimethyl malonate and diethyl malonate have found many applications in the pharmaceutical, vitamin, agrochemical, fragrance, and dyestuff industries. Some of the many outlets in the pharmaceutical industry are the synthesis of the 5,5-disubstituted barbiturates and thiobarbituric acids such as methohexital sodium [*22151-68-4*] (**11**) (37) and thiopental sodium [*71-73-8*] (**12**)

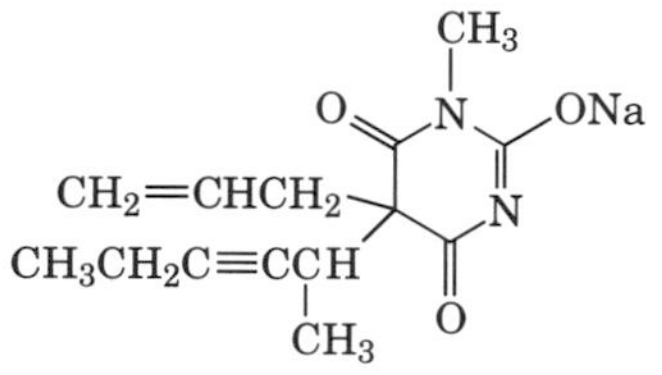

(**11**) 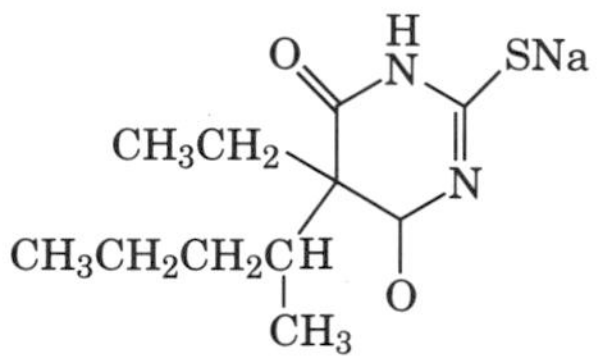 (**12**)

(38) (see PHARMACEUTICALS). Another important derivative is the vasodilatator Naftidrofuryl [*31329-57-4*] (**13**) obtained through condensation of 2-tetrahydrofurfuryl chloride [*3003-84-7*] and 1-(chloromethyl)naphthalene [*86-52-2*] (39). Other derivatives include the antiinflammatories phenylbutazone

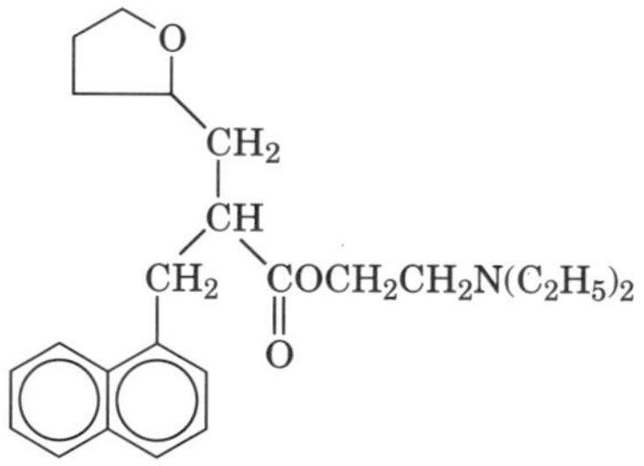

(**13**) 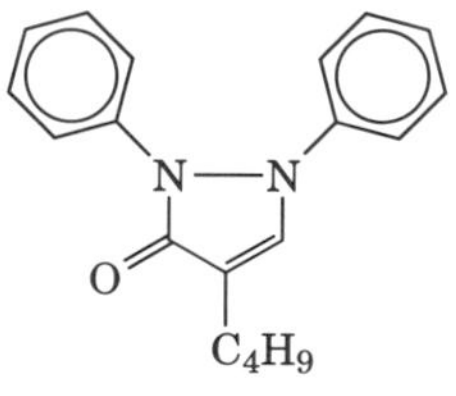 (**14**) 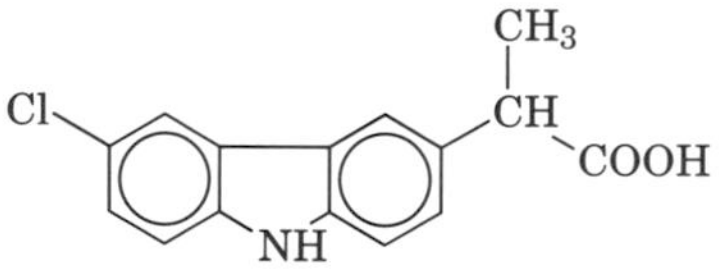 (**15**)

[*50-33-9*] (**14**), which is derived from the condensation of 2-(*n*-butyl)malonate with 1,2-diphenylhydrazine [*122-66-7*] (40), and carprofen [*53716-49-7*] (**15**) prepared from the Michael condensation of dimethyl methylmalonate [*609-02-9*] with 2-cyclohexen-1-one [*930-68-7*] (41) (see ANALGESICS, ANTIPYRETICS, AND ANTIINFLAMMATORY AGENTS). A wide range of quinolone antibacterial agents can be prepared through condensation of an amino containing moiety with diethyl 2-ethoxymethylenemalonate (see ANTIBACTERIAL AGENTS, SYNTHETIC). Since the introduction of nalidixic acid [*389-08-2*] (**16**), many new compounds have been developed such as ofloxacin [*82419-36-1*] (**17**). Several new fluoroquinolones are reported in clinical trials (42). The anticonvulsant Vigabatrin [*60643-86-9*], $CH_2{=}CHCH(NH_2)CH_2CH_2COOH$ (**18**), obtained through condensation of diethyl malonate with 1,4-dichloro-2-butene [*764-41-0*] (43), or the antiulcer Rebamipide [*90098-04-7*] (**19**), whose synthesis involves the use of 2-(acetylamino)malonate (44), are examples of new pharmaceuticals recently launched.

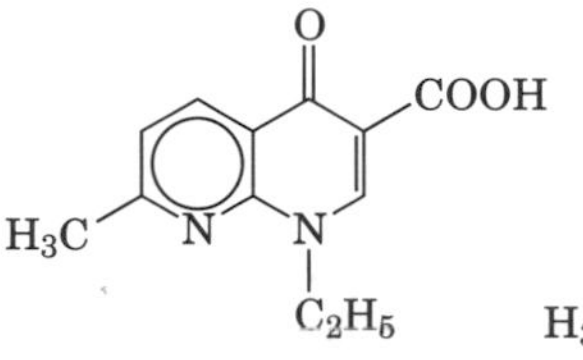

(**16**) 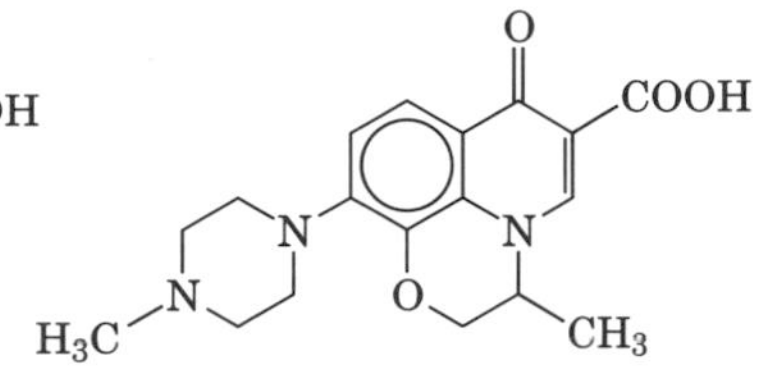 (**17**) 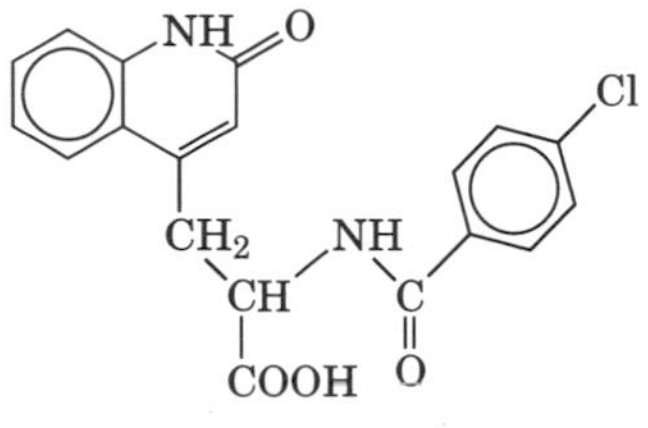 (**19**)

In the vitamins field, the synthesis of vitamin B_2 [*83-88-5*] (**20**) starting from barbituric acid [*67-52-7*] (45) is a significant outlet (see VITAMINS). Since

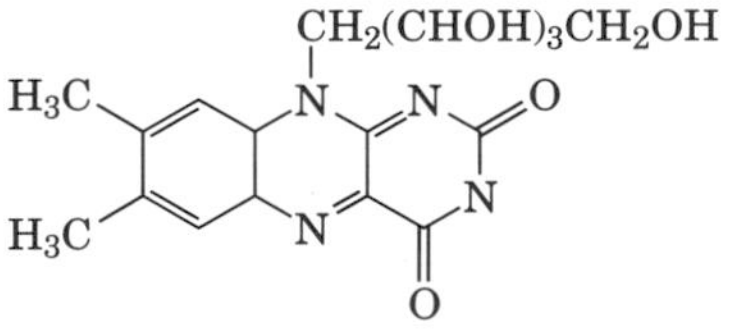

(**20**) 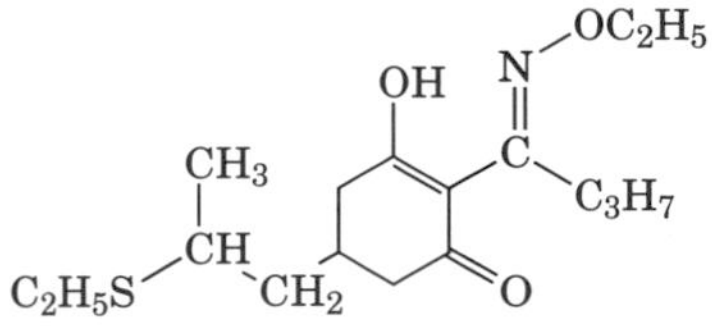 (**21**)

the 1970s the agrochemical industry has contributed to the increasing number of new applications for malonates. This is mainly due to the introduction of the new cyclohexanedione-type herbicides whose first steps of the respective synthesis resembles the synthetic method of dimedone (26), which is a Michael addition followed by ring closure and decarboxylation (see HERBICIDES). One of the first products introduced was sethoxydim [*74051-80-2*] (**21**) (46) which has been since been followed by other derivatives such as tralkoxydim [*87820-88-0*] (47). A further prominent new class has been the novel low dosage sulfonylurea herbicides deriving from 2-aminopyrimidines such as 2-amino-4-chloro-6-methoxypyrimidine [*5734-64-5*] and 2-amino-4,6-dimethoxypyrimidine [*36315-01-2*]. Whereas the former is made from the condensation of a malonate with guanidine, the latter is preferably accessible from malononitrile. Representatives of this new herbicides class are chlorimuron–ethyl [*90982-32-4*] (**22**) (48) and bensulfuron–methyl [*83055-99-6*] (**23**) (49). As of this writing more than 10 different sulfonylurea herbicides

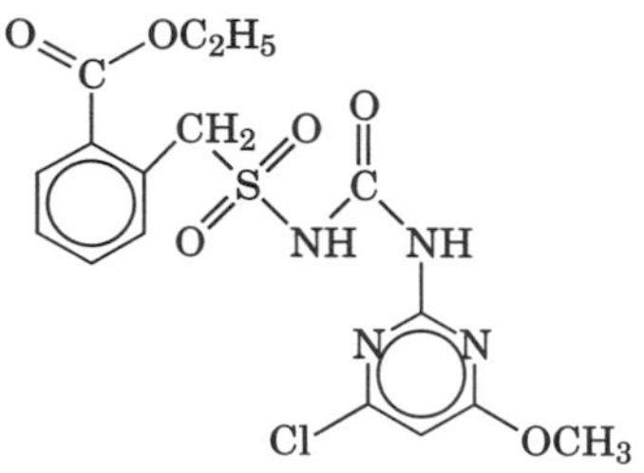

(22)

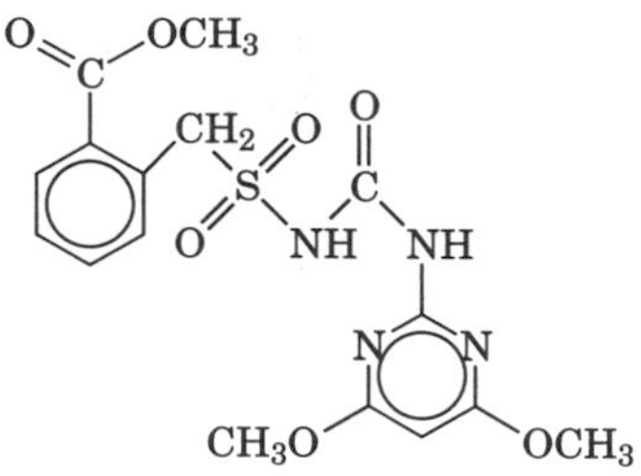

(23)

are either launched or in a late development stage. In the dyestuff industry, both the malonate derivatives barbituric acid and 2,4-dihydroquinoline [*86-95-3*] are used as coupling components for the CI Pigment Yellow 139 [*36888-99-0*] (**24**) (50) and the CI Disperse Yellow 5 [*6439-53-8*] (**25**) (51) (see DYES AND DYE INTERMEDIATES).

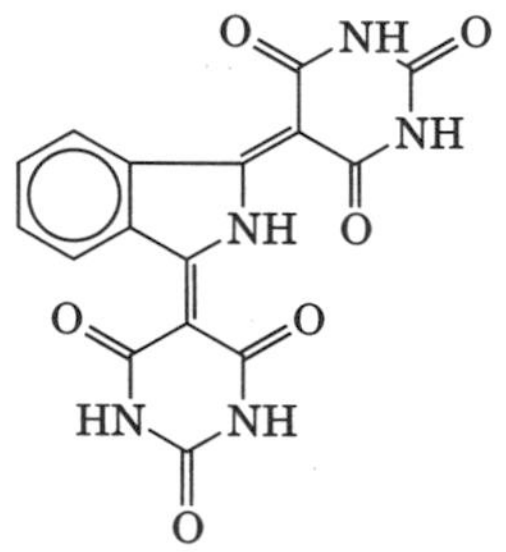

(24)

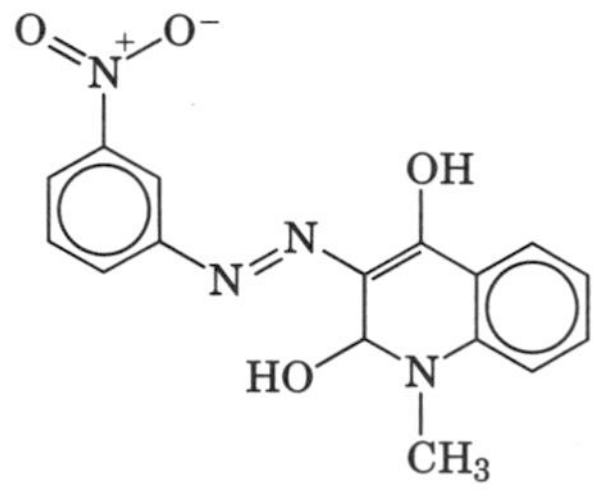

(25)

A commercially important outlet in the fragrance industry is the methyl dihydrojasmonate [*24851-98-7*] (**26**) which is made by Michael addition of a malonate to 2-pentyl-2-cyclopenten-1-one [*91791-21-8*] (52) and which is used in perfumery for blossom fragrances, particularly jasmine (see PERFUMES).

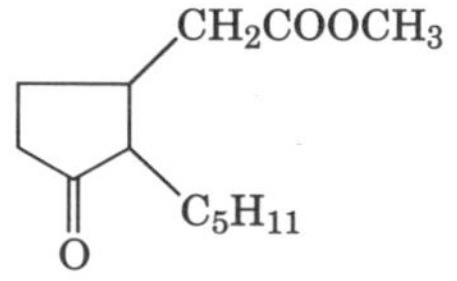

(26)

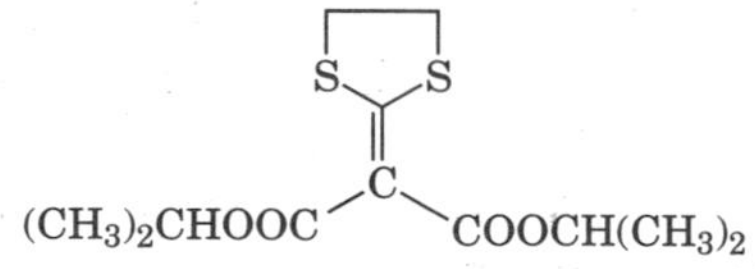

(27)

Malonates can also be used as blocking agents in the formulation of one-part urethanes. These systems, curable by moisture, are used, for example, for automotive windshield glazing (53) (see URETHANE POLYMERS).

Diisopropyl Malonate. This dialkyl malonate has gained industrial importance for the synthesis of the fungicide isoprothiolane [*50512-35-1*] (**27**) through condensation with carbon disulfide [*75-15-0*] and ethylene dichloride [*107-06-2*] (54). Diisopropyl malonate [*13195-64-7*] is produced by Mitsubishi Chemical in Japan using the carbon monoxide process.

Cyanoacetic Acid and Cyanoacetates

Physical Properties. The physical properties of cyanoacetic acid [*372-09-8*], $N{\equiv}C{-}CH_2COOH$ (**28**) are summarized in Table 4. The industrially most important esters are methyl cyanoacetate [*105-34-0*] and ethyl cyanoacetate [*105-56-6*]. Both esters are miscible with alcohol and ether and immiscible with water.

Reactions. The chemical properties of cyanoacetates are quite similar to those of the malonates. The carbonyl activity of the ester function is increased by the cyano group's tendency to withdraw electrons. Therefore, amidation with ammonia [*7664-41-7*] to cyanoacetamide [*107-91-5*] (55) or with urea to cyanoacetylurea [*448-98-2*] (56) proceeds very easily. An interesting reaction of cyanoacetic acid is the Knoevenagel condensation with aldehydes followed by decarboxylation which leads to substituted acrylonitriles (57) such as (**29**), or with ketones

O CH_2CN

(**29**)

CH_2CN

(**30**)

followed by decarboxylation with a shift of the double bond to give β,γ-unsaturated nitriles (58) such as (**30**) when cyclohexanone [*108-94-1*] is used.

Manufacture. Cyanoacetic acid and cyanoacetates are industrially produced by the same route as the malonates starting from a sodium chloroacetate solution via a sodium cyanoacetate solution. Cyanoacetic acid is obtained by acidification of the sodium cyanoacetate solution followed by organic solvent extraction and evaporation. Cyanoacetates are obtained by acidification of the sodium

Table 4. Physical Properties of Cyanoacetic Acid, Methyl Cyanoacetate, and Ethyl Cyanoacetate

Property	Cyanoacetic acid	Methyl cyanoacetate	Ethyl cyanoacetate
other names	malonic mononitrile		
appearance	colorless crystals	colorless liquid	colorless liquid
mol wt	85.06	99.09	113.12
mp, °C	66	−22	−22
bp °C/kPa[a]	108°/1.5 kPa	203°/101 kPa	206°/101 kPa
refractive index		1.418–1.419	1.417–1.418

[a]To convert kPa to mm Hg, multiply by 7.5.

cyanoacetate solution and subsequent esterification with the water formed being distilled off. Other processes reported in the literature involve the oxidation of partially oxidized propionitrile [*107-12-0*] (59). Higher esters of cyanoacetic acid are usually made through transesterification of methyl cyanoacetate in the presence of aluminium isopropoxide [*555-31-7*] as a catalyst (60).

Economic Aspects. In order to avoid the extraction and evaporation steps, most of the cyanoacetic acid derivatives are made directly from solution; therefore, only a small portion of the acid produced is traded. Cyanoacetic acid is produced by Boehringer-Ingelheim and Knoll in Germany, Juzen in Japan, as well as Hüls in the United States. When sold in tons, the price of cyanoacetic acid was ~$9/kg in 1993.

Methyl cyanoacetate and ethyl cyanoacetate are produced by Lonza in Switzerland and Hüls in the United States, as well as Juzen and Tateyama in Japan. The total production capacity is estimated to be in the range of 10,000 metric tons per year. The market price for both esters in bulk shipments was around $6/kg in 1993.

Analytical and Test Methods. Potentiometic titration is an analytical method for cyanoacetic acid. Methyl and ethyl cyanoacetates are usually analyzed by gas chromatography using the same equipment as for the malonates but with a higher column and injector temperatures, namely 150 and 200°C, respectively.

Health and Safety Factors. Handling of cyanoacetic acid and cyanoacetates do not present any specific danger or health hazard if handled with the usual precautions. Cyanoacetic acid is classified as a moderate irritant (skin irritation, rabbits) and has an LD_{50} (oral, rats) of 1500 mg/kg. Methyl and ethyl cyanoacetate are both classified as slight irritants (skin irritation, rabbits) and have an LD_{50} (oral, rats) of 3062 and 2820 mg/kg, respectively. Transport classification: cyanoacetic acid:RID/ADR: 8; IMDG-Code: 8; IATA/ICAO: 6.1. Methyl and ethyl cyanoacetate: RID/ADR: 6.1; IMDG-Code:6.1; IATA/ICAO: 6.1.

Uses. In many cases cyanoacetic acid, cyanoacetates, or cyanoacetamide can be used alternatively. The traded cyanoacetic acid is mainly intended for the synthesis of the cough remedy dextromethorphan [*125-71-3*] (**31**) (61) (see EXPECTORANTS, ANTITUSSIVES, AND RELATED AGENTS) and of the fungicide cymoxanil [*57966-95-7*] (**32**) (62) (see FUNGICIDES, AGRICULTURAL).

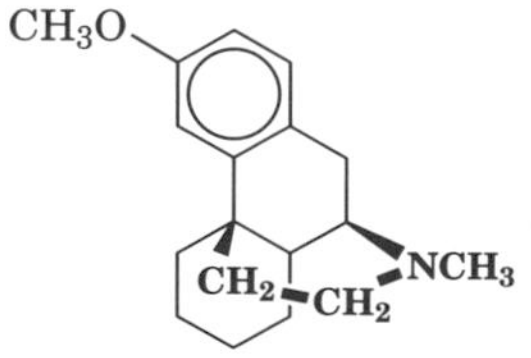

CH_3O ... CH_2 CH_2 NCH_3

(**31**)

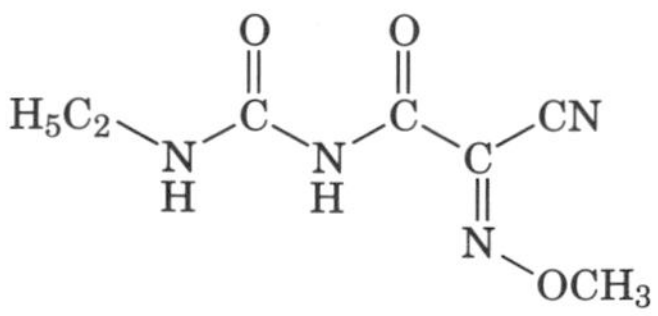

H_5C_2–NH–C(=O)–NH–C(=O)–C(=N–OCH_3)–CN

(**32**)

H_3C, O, R, N, CH_3

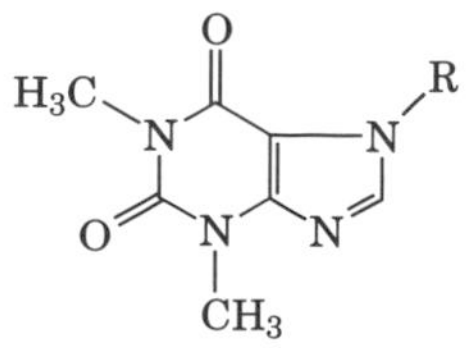

(**33**)

Otherwise cyanoacetic acid is directly converted as a solution with 1,3-dimethylurea [*96-31-1*] into 2-cyano-*N*,*N*′-dimethylcarbamoyl acetamide [*39615-*

79-7] which is further upgraded into the diuretics theophylline [*58-55-9*] (**33** where R = H) and caffeine [*58-08-2*] (**33**, where R = CH_3) (63).

The largest application of methyl and ethyl cyanoacetate is the production of the cyanoacrylate adhesives widely used within the car and electronic industries. Esters of higher alcohols such as 1-butanol [*71-36-3*] have also gained industrial importance. Basically the Knoevenagel condensation of a cyanoacetate and formaldehyde [*50-00-0*] followed by nearly spontaneous dehydration gives cyanoacrylate which undergoes an immediate polymerization to polycyanoacrylate. However, this polymerization is reversible on heating and the monomer can be formed again and subsequently purposely polymerized by traces of moisture.

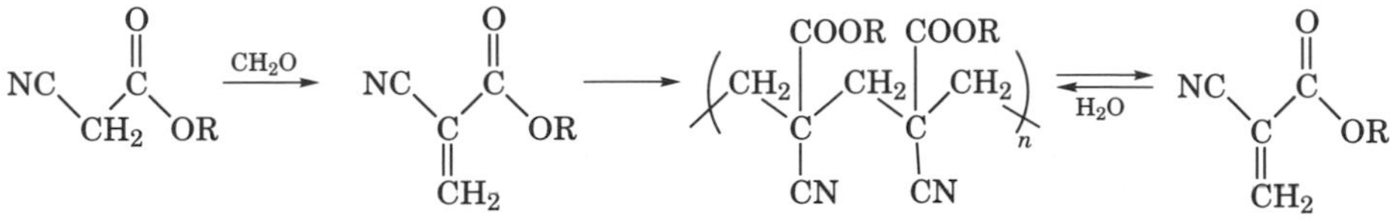

Otherwise cyanoacetates are widely used in the synthesis of pharmaceuticals such as the anticonvulsant ethosuximide [*77-67-8*] (**34**), (64) and valproic acid

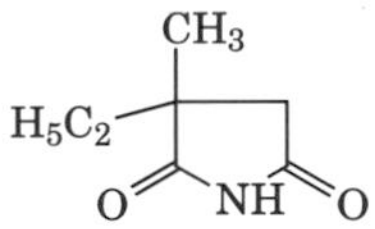

(**34**) 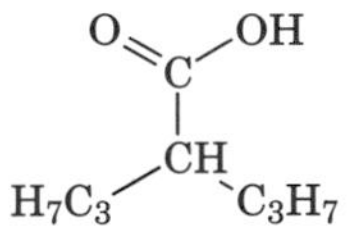 (**35**) 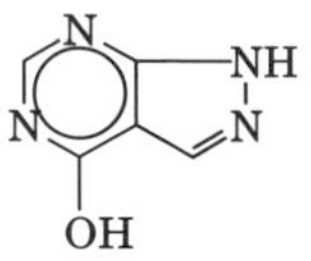 (**36**)

[*99-66-1*] (**35**) (65), the gout remedy allopurinol [*315-30-0*] (**36**) (66), or the antihypertensive amiloride [*2609-46-3*] (**37**) (67) (see PHARMACEUTICALS).

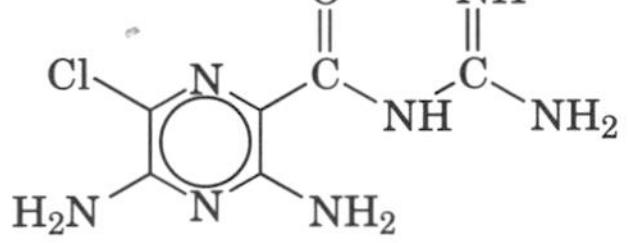

(**37**) 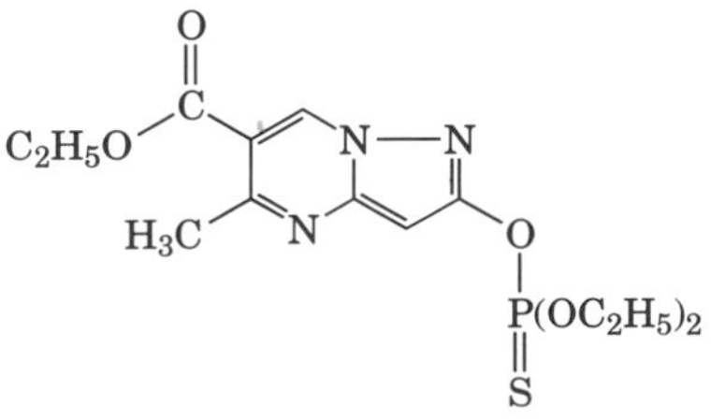 (**38**) 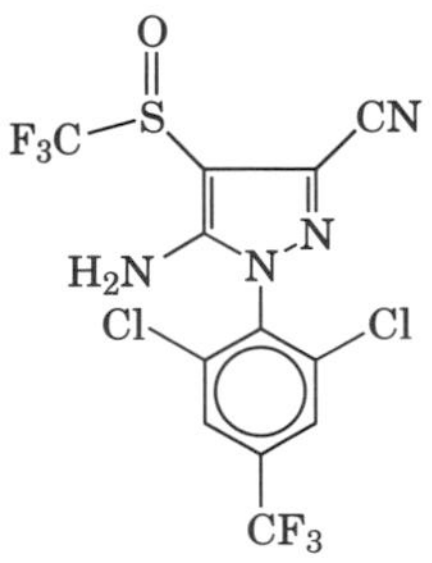 (**39**)

In the agrochemical field, outlets for cyanoacetates are the fungicides pyrazophos [*13457-18-6*] (**38**) (68) and fipronil [*120068-37-3*] (**39**) (69) (see

FUNGICIDES, AGRICULTURAL). Cyanoacetates are also used as dye intermediates, for producing Celliton Fast Yellow 7G (**40**) (70), as well as in the synthesis of uv

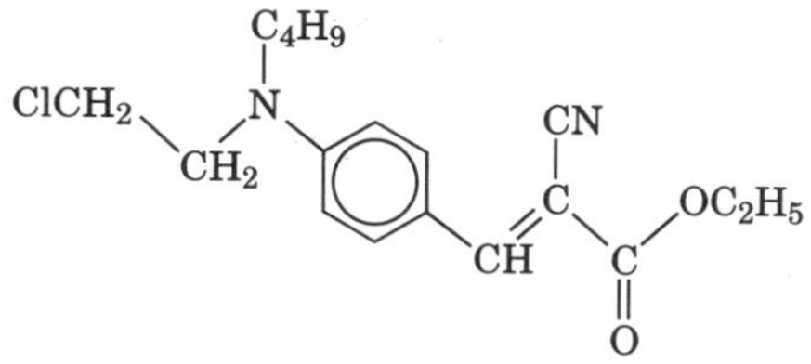

(**40**) 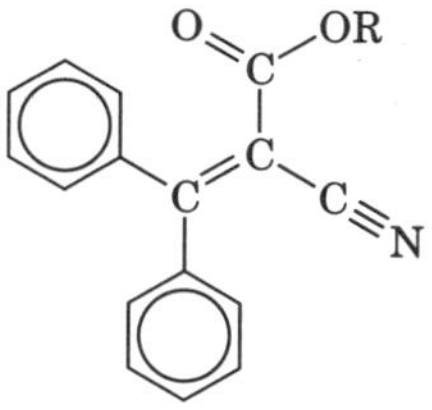 (**41**)

absorbers (71) etocrylene [*5232-99-5*] (**41**, R = ethyl) and octocrylene [*6197-30-4*] (**41**, R = 2-ethylhexyl) (see DYES AND DYE INTERMEDIATES).

Malononitrile

Physical Properties. The physical properties of malononitrile [*109-77-3*], $N{\equiv}C{-}CH_2C{\equiv}N$ (**42**) are listed in Table 5.

Reactions. The chemistry of malononitrile is reviewed in References 72 and 73. Like malonates and cyanoacetates, the chemical properties of malononitrile are determined by two reactive centers, namely the methylene group and the two cyano functions. The high acidity of the methylene group ($pK_a \sim 11$) as compared to malonates ($pK_a \sim 13$) makes the deprotonation of malononitrile quite easy even with relatively weak base. A peculiar reaction of malononitrile is the base-catalyzed dimerization leading to 2-amino-1,1,3-tricyanopropene [*868-54-2*] (**43**) (74). Reaction of malononitrile at the cyanide groups without partici-

Table 5. Physical Properties of Malononitrile

Property	Value
other names	propanedinitrile, dicyanomethane
mol wt	66.06
appearance	white crystals
mp, °C	31
bp, °C	
at 76 kPa[a]	218–219
at 1.7 kPa[a]	108–109
d_4^{35}, g/mL	1.0494
dipole moment, 25°C, C·m[b]	1.19×10^{-29}
solubility, g/mL	
in water	0.133
in ethanol	0.40
in ether	0.20

[a]To convert kPa to mm Hg, multiply by 7.5.
[b]To convert C·m to debyes, multiply by 3.0×10^{29}.

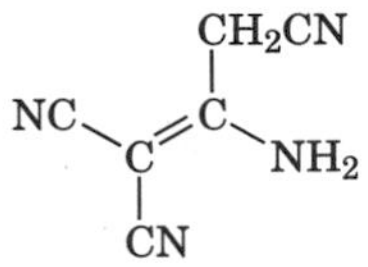

(43) 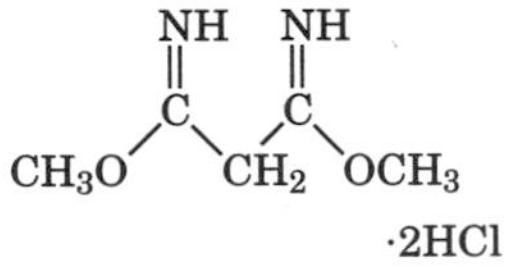 (44) 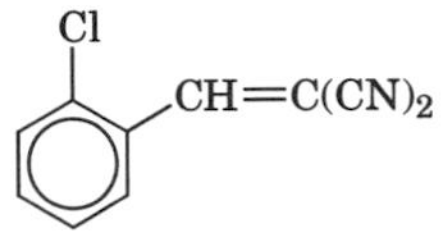 (45)

pation of the methylene group are rare. The most common one is the acid-catalyzed addition of alcohols in which one or both cyano groups are converted into the imidate group. For example (**44**) which upon ring closure gives access to 2-amino-4,6-dimethoxypyrimidine [*36315-01-2*] (75). Another one is the condensation with guanidine giving 2,4,6-triaminopyrimidine [*1004-38-2*] (76).

Otherwise, the main reactions at the methylene group are the dialkylation with alkyl halides (77), the acetylation with acetyl chloride which yields acetylmalononitrile [*1187-11-7*] (78), the Knoevenagel condensation, as well as the condensation with triethyl orthoformate, gives ethoxymethylenemalononitrile [*123-06-8*], $C_2H_5O{-}CH{=}C(CN)_2$ (79). In the Knoevenagel condensation, aliphatic aldehydes with α-hydrogen atoms usually lead to further condensation products. In the case of aromatic aldehydes this problem does not exist and dicyanostyryl compounds such as (**45**) can easily be made (80). Condensation with 1,4-cyclohexandione [*637-88-7*] gives 7,7,8,8-tetracyanoquinodimethane [*1518-16-7*] (**46**) upon further dehydrogenation (81).

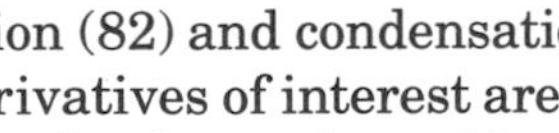

(46) 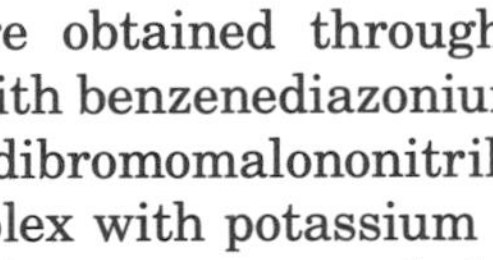 (47) 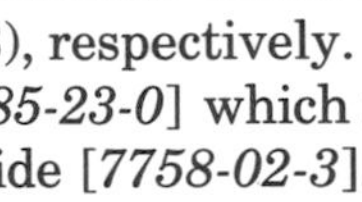 (48)

Nitrogen derivatives such as 2-aminomalononitrile [*5181-05-5*] and phenylazomalononitrile [*6017-21-6*] (**47**) are obtained through nitrosation followed by reduction (82) and condensation with benzenediazonium (83), respectively. Halogen derivatives of interest are the dibromomalononitrile [*1885-23-0*] which is isolated in the form of a stable complex with potassium bromide [*7758-02-3*]. This complex can be debrominated to give tetracyanoethylene [*670-54-2*] (**48**) (84).

Manufacture. Malononitrile can be produced batchwise by elimination of water from cyanoacetamide [*107-91-5*] with phosphorous pentachloride [*10026-13-8*] (85). It is now produced continuously starting from cyanogen chloride [*506-77-4*] and acetonitrile [*75-05-8*]. The reaction takes place at temperatures above 700°C, the reaction products are malononitrile, excess acetonitrile, hydrochloric acid, and small amounts of maleic acid [*110-16-7*], succinic acid [*110-15-6*], and fumaric acid [*110-17-8*]. The products leaving the reactor are immediately cooled to 40–80°C, and gaseous hydrogen chloride is simultaneously separated, fed into a washer, and recovered as dilute hydrochloric acid. Excess acetonitrile is removed by a combination of vacuum distillation and thin-film evaporation. The recovered acetonitrile contains very little hydrogen chloride and can be recycled without risk of corrosion.

Removal of maleic and fumaric acids from the crude malononitrile by fractional distillation is impractical because the boiling points differ only slightly. The impurities are therefore converted into high boiling compounds in a conventional reactor by means of a Diels-Alder reaction with a 1,3-diene. The volatile and nonvolatile by-products are finally removed by two vacuum distillations. The by-products are burned. The yield of malononitrile amounts to 66% based on cyanogen chloride or acetonitrile.

Other processes recently reported in the literature are the gas-phase reaction of lactonitrile [*78-97-7*] with ammonia and oxygen in the presence of molybdenum catalyst (86), or the vapor-phase reaction of dimethyl malonate with ammonia in the presence of dehydration catalyst (87).

Economic Aspects. Malononitrile of minimum 99% purity was available as a solidified melt for ca $30/kg in 1993 for ton quantities. Malononitrile is produced by Lonza Ltd. (Switzerland) using the cyanogen chloride process.

Analytical and Test Method. Gas chromatography is appropriate for the quantitiative analysis of malononitrile. Typical analysis conditions are 3% Reoplex 400 on Chromosorb G 80–100 mesh; 2 m, 2 mm diameter column; temperature for column = 60–180°C; injector, 200°C; and detector, 200°C. The solidification point is usually measured also.

Health and Safety Factors. Malononitrile is usually available as a solidified melt in plastic-lined drums. Remelting has to be done carefully because spontaneous decomposition can occur at elevated temperatures, particularly above 100°C, in the presence of impurities such as alkalies, ammonium, and zinc salts. Melting should be carried out by means of a water bath and only shortly before use. Occupational exposure to malononitrile mainly occurs by inhalation of vapors and absorption through the skin. Malononitrile has a recommended workplace exposure limit of 8 mg/m^3, an LD_{50} (oral, rats) of 13.9 mg/kg, and is classified as slight irritant (skin irritation, rabbits). Transport classification: RID/ADR: 61, IMDG-Code: 6.1, IATA/ICAO: 6.1.

Uses. Malononitrile is extensively used in the life science industry, especially for the synthesis of *N*-containing heterocycles. The synthesis of thiamine or vitamin B_1 [*59-43-8*] (**49**) from ethoxymethylenemalononitrile [*123-06-8*] through 4-amino-5-cyano-2-methylpyrimidine [*698-29-3*] (88) still represents an important outlet for malononitrile. The introduction of several new highly active

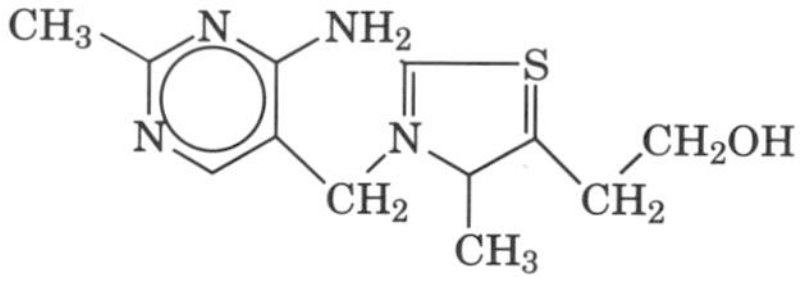

(**49**)

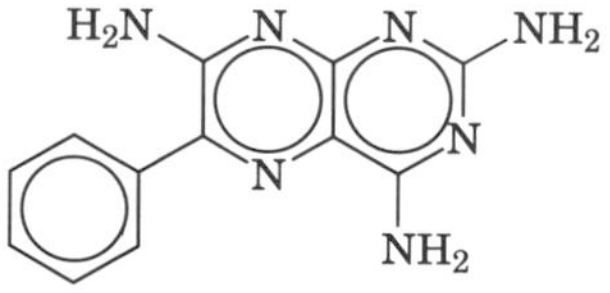

(**50**)

sulfonylurea herbicides such as (**23**) based on 2-amino-4,6-dimethoxypyrimidine has also opened an industrial outlet for malononitrile. Further life science products of industrial importance are the diuretic triamterene [*396-01-0*] (**50**) (89) (see DIURETICS) and the antineoplastic methotrexate [*59-05-2*] (**51**) (90), both deriving from 2,4,6-triaminopyrimidine, or the antihypertensive minoxidil [*38304-91-5*] (**52**) made via ethyl *N*,2-dicyanoacetamidate [*53557-77-0*], $NCCH_2C(NCN)OC_2H_5$ (91).

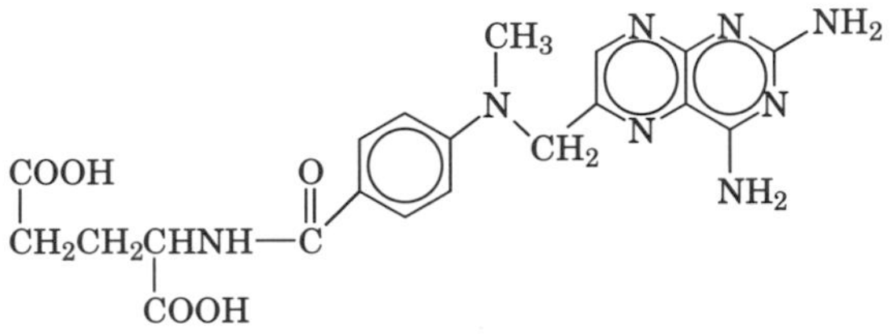

(**51**) 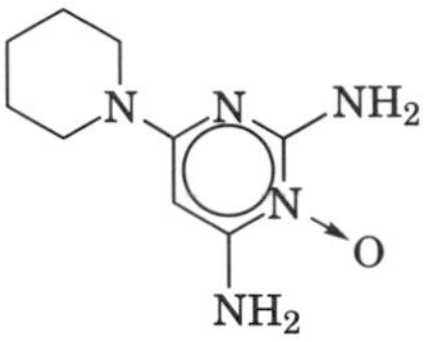 (**52**)

Another compound of interest is adenine [*73-24-5*] or 6-aminopurine (**53**) derived from phenylazomalononitrile (92). The introduction of the dicyanostyryl moiety has led to the industrialization of several methine dyes such as the CI Disperse Yellow [*6684-20-4*] (**54**) (93). The CI Disperse Blue 354 [*74239-96-6*] (**55**) also represents a new class of aminoarylneutrocyanine dyes with a brilliant blue shade (94). The dimer of malononitrile is also used for the synthesis of new dyes (95).

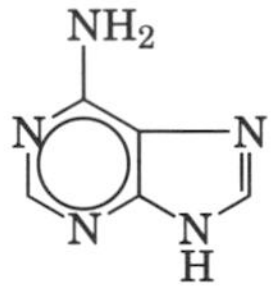

(**53**) 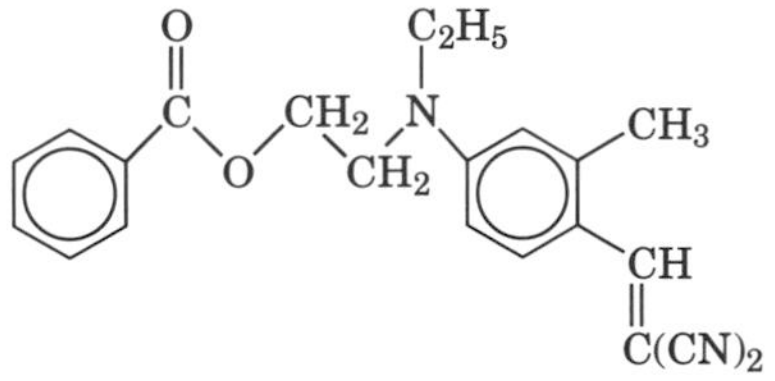 (**54**) 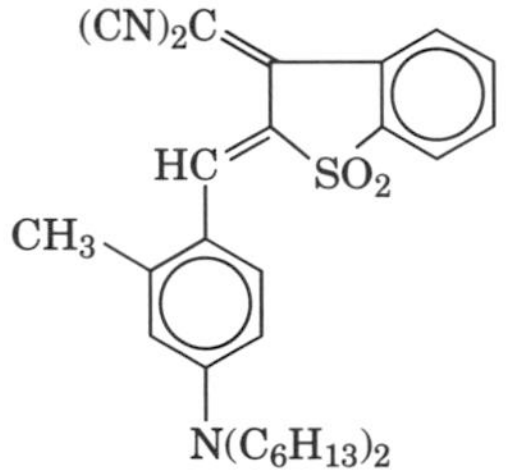 (**55**)

Other miscellaneous applications of malononitrile are the synthesis of 7,7,8,8-tetracyanoquinodimethane (**46**) which is a powerful electron acceptor in the formation of charge-transfer complexes which are of interest because of their conductivity of electricity (96), as well as of 2-chlorobenzylidene malononitrile [*2698-41-1*] (**45**) also known as CS-gas, which is a safe lachrymatory chemical used for self-defense devices (97).

BIBLIOGRAPHY

"Malonic Acid and Its Derivatives" in *ECT* 1st ed., Vol. 8, pp. 697–705, by W. Wenner, Hoffmann-La Roche, Inc.; "Malonic Acid and Derivatives" in *ECT* 2nd ed., Vol. 12, pp. 849–890, by W. Wenner, Hoffmann-LaRoche, Inc.; in *ECT* 3rd ed., Vol. 14, pp. 794–810, by D. W. Hughes, Dow Chemical U.S.A.

1. S. K. Gupta and co-workers, *J. Indian Chem. Sec.* **65**(3), 187 (1988).
2. L. Chang Kui and S. Yeung Yesp, *Org. Prep. Prod. Int.* **22**(1), 94 (1990).
3. J. Martens, J. Kintscher, and W. Arnold, *Synthesis* **6**, 497 (1991).
4. Eur. Pat. Appl. EP 481,201 (Apr. 22, 1992), T. Koga and co-workers (to Kanegafuchi Chem. Ind. Co. Ltd.).
5. Jpn. Kokai Tokkyo Koho JP 03,275,650 (Dec. 6, 1991), T. Toya, K. Ono, and S. Fujita (to Dainippon Ink and Chemicals, Inc.).
6. T. Imamoto, M. Kodira, and M. Yokoyama, *Bul. Chem. Soc. Jpn.* **55**(7), 2304 (1982).
7. U.S. Pat. 4,613,671 (Sept. 23, 1986), A. G. Relenji, D. E. Wallick, and J. P. Streit (to Dow Chemical Co.).
8. Span. Pat. ES2,004,446 (Jan. 1, 1989), R. F. Ambros and co-workers (to Ferrer Internacional SA).
9. Rom. Pat. RO86,542 (Mar. 30, 1985), S. Mager and co-workers (to Interprinderea de Madicamente "Terapia").
10. Czech. Pat. CS249,237 (Mar. 15, 1988), M. Matas and K. Fancovic.
11. Ger. Offen. DE4,107,986 (Sept. 17, 1992), A. Behr and co-workers (to Henkel KGaA).
12. Eur. Pat. Appl. EP187,680 (July 16, 1986), K. Enomoto and co-workers (to Nitto Chemical Industry Co., Ltd., Mitsubishi Rayon Co. Ltd.).
13. Brit. Pat. 1,168,108 (Oct. 22, 1969), C. Fauran and co-workers (to Delalande SA).
14. H. Pyyasalo and co-workers, *Finn. Chem. Lett.* **5**, 129 (1975).
15. D. E. Peterson, *Report*, BDX-613-2758, Bendix Corp., Kansas City, Mo., 1982.
16. U.S. Pat. 4,714,730 (Dec. 22, 1987), P. C. Briggs and D. E. Gosiewski (to Illinois Tool Works, Inc.).
17. D. H. Grayson and M. R. J. Tuite, *J. Chem. Soc., Perkin Trans. 1* **12**, 2137 (1986).
18. C. Band-Chi, *Heterocycles* **32**(3), 529 (1991).
19. Jpn. Kokai Tokkyo Koho JP62,289,574 (Dec. 16, 1987), M. Tanaka, T. Miyagawa, and H. Fuji (to Wako Pure Chemicals Industries, Ltd.).
20. Eur. Pat. Appl. EP 378,073 (July 18, 1990), R. Fuchs and J. McGarrity (to Lonza AG).
21. U.S. Pat. 4,154,952 (May 15, 1979), K. W. Schultz.
22. Ger. Offen DE3,142,854 (May 11, 1983), K. Grohe, J. H. Zeiler, and K. Metzger (to Bayer AG).
23. W. Lehnert, *Tetrahedron Lett.*, 4723 (1970).
24. Fr. Demande 2,589,466 (May 7, 1987), S. Ratton (to Rhône-Poulenc Specialités Chimiques).
25. A. A. Liebman and co-workers, *J. Heterocycl. Chem.* **11**(6), 1105 (1974).
26. Eur. Pat. Appl. EP 65,706 (Dec. 1, 1982), P. Lehky (to Lonza AG).
27. Eur. Pat. Appl. EP167, 053 (Jan. 8, 1986), R. Santi, G. Cometti, and A. Pagani (to Montedison SpA).
28. B. Arct and co-workers, *J. Environ. Sci. Health, Part. B* **B-18**(4–5), 559 (1983).
29. Jpn. Kokai Tokkyo Koho 6,239,544 (Feb. 20, 1987), N. Tachikara and co-workers (to Dentu Kagaku Kogyo KK).
30. Eur. Pat. Appl. EP 115,469 (Aug. 8, 1984), N. Yokoyama (to Ciba-Geigy AG).
31. G. Dickey, *Org. Syn. Coll. Vol. II*, 60 (1943).
32. Jpn. Kokai Tokkyo Koho JP0183,071 (Mar. 28, 1989), H. Miki (to Takeda Chemical Industries, Ltd.).

33. B. Brobanski and K. Sankiewicz, *Rocz. Chem. Ann. Soc. Chim. Polonorum* **45**, 277 (1971).
34. M. El-chahawi and U. Prange, *Chemiker Zeitung* **102**(1), 1 (1978).
35. Eur. Pat. 0006611 (Jan. 9, 1980), K. Nishimura (to Ube Industries Ltd.).
36. Jpn. Kokai Tokkyo Koho 63,170,338 (July 14, 1988), K. Watabe and H. Ono (to Mitsui Toatsu Chemicals, Inc.).
37. U.S. Pat. 2,872,448 (Feb. 3, 1959), Doran (to Lilly).
38. E. Miller and co-workers, *J. Am. Chem. Soc.* **58**, 1090 (1936).
39. Fr. Pat. 1,363,948 (June 19, 1964), E. Szarvasi and M. Bayssat (to Lyonnaise Industrielle Pharmaceutique).
40. U.S. Pat. 2,562,830 (July 31, 1951), H. Stenzl (to J. R. Geigy AG).
41. Ger. Offen. 2,337,340 (Feb. 14, 1974), L. Berger and J. A. Corraz (to Hoffmann-La Roche).
42. *Drugs Future* **17**(2), 114 (1992).
43. Belg. Pat. 873,766 (May 16, 1979), M. Gittos and G. J. Letertre (to Merrell Toraude SA).
44. Ger. Offen. 3,324,034 (Jan. 5, 1984), M. Uchida, M. Komatsu, and K. Nakagawa (to Otsuka Pharmaceutical Co., Ltd.).
45. M. Tischler, *J. Am. Chem. Soc.* **69**, 1487 (1947).
46. Jpn. Kokai Tokkyo Koho 8,175,408 (June 22, 1981), (to Nippon Soda Co., Ltd.).
47. Eur. Pat. Appl. 80,301 (June 1, 1983), R. B. Warner and co-workers (to ICI Australia Ltd.).
48. Braz. Pedido BR 8,303,322 (Feb. 7, 1984), A. D. Wolf (to E. I. du Pont de Nemours & Co., Inc.).
49. Eur. Pat. Appl. 51,466 (May 12, 1982), R. F. Saners (to E. I. du Pont de Nemours & Co., Inc.).
50. Eur. Pat. Appl. 38,548 (Oct. 28, 1981), L. R. Lerner, C. E. Shannon, and F. Santimauro (to Mobay Chemical Corp.).
51. U.S. Pat. 1,969,463 (Aug. 7, 1934), K. Holzach and G. von Rosenberg (to General Aniline Works).
52. Brit. Pat. 907,431 (Feb. 25, 1960), A. Firmenich and co-workers.
53. U.S. Pat. 3,779,794 (Dec. 18, 1973), W. G. De Santis (to Essex Chemical Corp.).
54. Ger. Pat. 2,316,921 (Oct. 25, 1973), K. Taninaka and co-workers (to Nihon Nohyaku Co., Ltd.).
55. G. Abdel and co-workers, *Heterocycles* **24**(7), 2023 (1986).
56. G. M. Karmouta and co-workers, *Eur. J. Med. Chem.* **24**(5), 547 (1989).
57. *Org. React.* **15**, 204 (1967).
58. *Org. Synth. Coll. Vol.* **4**, 234 (1963).
59. Int. Pat. Appl. 9,212,962 (Aug. 6, 1992) A. R. Bulls, J. S. Fellmann, and R. A. Periana (to Catalytica Inc.).
60. A. Said, *Chimia* **28**, 5 (1974).
61. O. Schnider and A. Grüssner, *Helv. Chim. Acta* **34**, 2211 (1951).
62. Ger. Offen. 2,312,956 (Sept. 20, 1973), S. H. Davidson (to E. I. du Pont de Nemours & Co., Inc.).
63. H. Bredereck and A. Edenhofer, *Chem. Ber.* **88**, 1306 (1955).
64. S. Sahay and G. Sircar, *J. Chem. Soc.*, 1252 (1927).
65. W. Thuan, *Bull. Soc. Chim. France*, 199 (1958).
66. Ger. Offen. 1,966,640 (Aug. 9, 1973), R. M. Cresswell and co-workers (to Wellcome Foundation Ltd.).
67. U.S. Pat. 3,313,813 (Apr. 11, 1967), J. E. Cragoe (to Merck and Co., Inc.).
68. Neth. Appl. 6,602,131 (Feb. 20, 1965), (to Farbwerke Hoechst AG).
69. Eur. Pat. Appl. EP295,117 (Dec. 14, 1988), J. G. Buntain and co-workers (to May and Baker Ltd.).

70. S. Hafenrichter, *Textil-Praxis* **16**, 273 (1961).
71. Eur. Pat. Appl. EP430,023 (June 5, 1991), S. I. Goldstein and E. F. Labrie (to BASF Corp.).
72. A. J. Fatiadi, *Synthesis* **3**, 165 (1978).
73. F. Freeman, *Synthesis* **12**, 925 (1981).
74. R. A. Carboni, D. D. Coffman, and E. G. Howard, *J. Am. Chem. Soc.* **80**, 2838 (1958).
75. U.S. Pat. 4,412,957 (Nov. 1, 1983), J. S. Gramm (to E. I. du Pont de Nemours & Co., Inc.).
76. Ger. Offen. 2,651,794 (May 18, 1978), D. Rose (to Henkel KGaA).
77. Ger. Offen. 2,329,251 (Dec. 13, 1973), H. Marketz (to Lonza Ltd.).
78. J. P. Fleury, *Bull. Soc. Chim. France* **2**, 413 (1964).
79. Ger. Offen. 2,635,841 (Aug. 10, 1976), O. Ackerman and co-workers (to Dynamit Nobel AG).
80. E. C. Rietvield, M. M. P. Hendrikx, and F. Seutter-Bulage, *Arch. Toxicol.* **59**(4), 228 (1986).
81. D. S. Acker and W. R. Hertler, *J. Am. Chem. Soc.* **84**, 3370 (1962).
82. Eur. Pat. Appl. 3,335 (Aug. 8, 1979), H. Junek and M. Mittelbach (to Lonza AG).
83. E. Sturdik and co-workers, *Collect. Czech. Chem. Commun.* **52**(11), 2819 (1987).
84. D. N. Dhar, *Chem. Rev.* **67**, 611 (1967).
85. Fr. Pat. 1,365,202 (June 26, 1964), (to lonza Ltd).
86. Jpn. Kokai Tokkyo Koho JP 5804,755 (Jan. 11, 1983), (to Asahi Chemical Industry Co., Ltd.).
87. Jpn. Kokai Tokkyo Koho JP 81,113,752 (Sept. 7, 1981), (to Ube Industries, Ltd.).
88. L. Velluz, *Substances Naturelles de Synthése* **3**, 59 (1951).
89. U.S. Pat. 3,081,230 (Mar. 12, 1963), J. Weinstock and V. D. Viebelhaus (to Smith-Kline and French Laboratories.
90. Ger. Offen. 2,741,383 (Feb. 2, 1979), E. Catalucci (to Lonza AG).
91. U.S. Pat. 4,032,559 (June 28, 1977), J. M. McCall and J. J. Ursprung (to Upjohn Co.).
92. Fr. Pat. 2,388,800 (Apr. 29, 1977), (to Merck and Co., Inc.).
93. U.S. Pat. 3,909,198 (Sept. 30, 1975), E. E. Renfrew and H. W. Pons (to American Aniline Products, Inc.).
94. Brit. Pat. Appl. 2,026,528 (Feb. 6, 1980), W. Baumann (to Sandoz Ltd.).
95. Ger. Offen. DE3,716,840 (Nov. 26, 1987), H. Matsumoto, H. Imai, and S. Tada (to Nippon Kayaku Co., Ltd.).
96. T. Nakamura and K. Kikuchi, *Kino Zairyo* **9**(11), 43 (1989).
97. *Cah. Notes Doc.* **123**, 229 (1986).

PETER POLLAK
GÉRARD ROMEDER
Lonza Ltd.

MALTS AND MALTING

Malting is essentially the same process as occurs when seeds fall to the ground or are planted, are moistened by water (qv), and germinate. During germination, rootlets (sprouts) and a nascent stem (acrospire) emerge; simultaneously, enzymes are produced or activated and the cellular structure and composition are modified, resulting in a product that can be used as a substrate for fermented beverages and as a food adjunct. The terms malt and malting can apply to any germinated grain; however, nearly all commercial malting involves barley. Because the brewing process and finished beer characteristics are a function of malt properties, malting is considered to be a part of the brewing process. Brewers' malt is designed to provide fermentable carbohydrates, assimilable nitrogen, as well as precursors for beer flavor (see BEER). Malt enzymes convert added carbohydrate, eg, corn grits or rice, into fermentable sugars.

Approximately 95% of the malt produced is used to make beer while small amounts are used as distillers' and food malts. Distillers' malt, which is used to convert starch-containing grains into fermentable sugars, is prepared almost exclusively for its enzymes, especially α-amylase (see BEVERAGE SPIRITS, DISTILLED). Food malts are sold for their flavor and/or enzyme contribution to food products.

Manufacturing and Processing

Raw Materials. Two principal types of malting-grade barley are in use, ie, six-row and two-row. Six-row barley has six kernels around the stalk, whereas the two-row variety has two kernels. Six-row kernels tend to be twisted, and the two-row grade is more symmetrical. Figure 1 is an illustration of a barley kernel and its key components. As barley is converted into malt, the acrospire for the embryo lengthens until it reaches the far end of the kernel, and rootlets also grow as if the seed is germinating into a new plant. A discussion of the structure and composition of barley can be found in Reference 1.

The main growing areas for barley are North Dakota, Montana, eastern South Dakota, and western Minnesota; six-row barley is predominant. Increasingly significant areas are California, Oregon, Washington, Idaho, and Colorado, where predominantly two-row barley is produced. Less than one-half of the barley grown in the United States is processed by the malt industry; the remainder is used as animal feed, and ca 80% of the barley used by the malting industry is the six-row variety (2–5) (see FEEDS AND FEED ADDITIVES, PETS).

Barley varieties recommended by the American Malting Barley Association, Inc. (Milwaukee, Wisconsin) are used for producing brewers' malt. Anheuser-Busch supplements these varieties with some of their own malting barley varieties, whereas Adolph Coors primarily uses their own barley variety (Moravian III). The main malting-grade varieties for two-row barley are Harrington, Moravian III, and B1202 and, for six-row barley, Robust, Excel, Morex, and Azure (6). These varieties are purchased by maltsters based on kernel appearance, germination ability, and protein content. Kernels should be plump and uniform in size and free from mold or staining. At least 95% of the kernels must germinate.

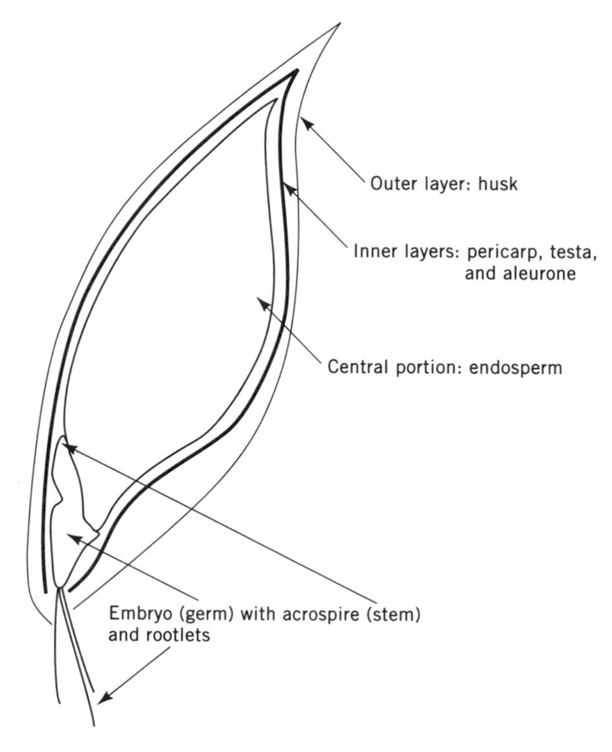

Fig. 1. Barley kernel and key components, shown in early stage of malting process.

Historically, barley has been selected for low protein content and, implicit in the low protein content, high extract.

With the exception of Canadian barleys, most of the malting barleys in the world are two-row varieties; these are characterized by larger berries, lower protein content, lower enzyme activity, and higher extract than the predominant six-row varieties used in the United States. Barley subjected to wet harvest conditions should be dried prior to storage. Too much moisture at harvest might result in unacceptable kernel staining, mycotoxin contamination, and poor storage stability.

Processing. *United States.* The malting process consists of three basic steps: steeping, germination, and kilning (Fig. 2). Prior to steeping, barley is cleaned and then sized according to kernel width. After kilning, the malt is cleaned, stored, and blended with other malt to meet customer specifications. A typical material balance for the malting process, based on raw barley solids, is as follows:

Material	Range, %	Typical percentage
barley cleanout	5.0–15.0	10
steepwater chaff and solubles	0.5–1.5	1
respiration losses	2.5–4.5	3
hulls and sprouts	2.5–6.5	4
finished malt	77.5–83.5	82

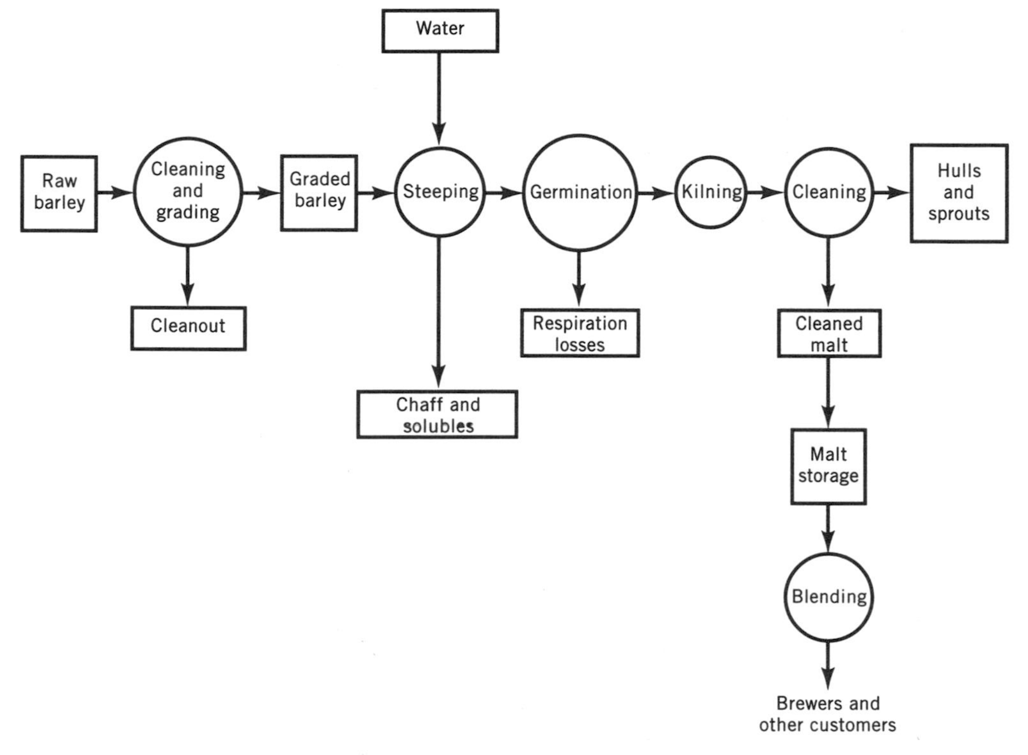

Fig. 2. Malting process.

Cleaning and Grading. Prior to malting, raw barley must be cleaned to remove tramp metal, dirt, debris, and other cereal grains, and be graded through slotted screens to produce a uniformly sized product. The grain first is passed over a magnet to remove metal and then is aspirated to remove chaff, dust, and other light materials. Next, the grain is passed over slotted screens which retain corn and large seeds, whereas barley and smaller seeds pass through and are separated on another slotted screen. Finally, the very thin barley kernels (needles) are separated from the barley by aspiration.

United States barley grades are determined according to kernel width, eg, A $>$ 2.48 mm, 2.18 mm $<$ B $<$ 2.48 mm, 1.98 mm $<$ C $<$ 2.18 mm, 1.93 mm $<$ D $<$ 1.98 mm, and throughs $<$ 1.93 mm. Grades A and B produce the highest extract and they are used for brewers' malt. Because different size kernels absorb moisture at different rates, it is desirable to process uniform kernel sizes to improve product uniformity and quality. Smaller kernels have higher protein content and are malted primarily for high enzymatic activity to meet distillers' malt specifications or to be blended with other brewers' malt. Normally throughs are not malted but are sold with the clean-out grain. About 90% of the incoming grain is malted and the remainder is sold as animal feed.

Steeping and Germination. In steeping, cleaned, graded barley from storage is immersed in water, resulting in a rise in the moisture content to 35–45% and initiation of growth. A diagram of a typical steep tank is shown in Figure 3. The size of steep tanks varies in the United States, with the smallest tanks

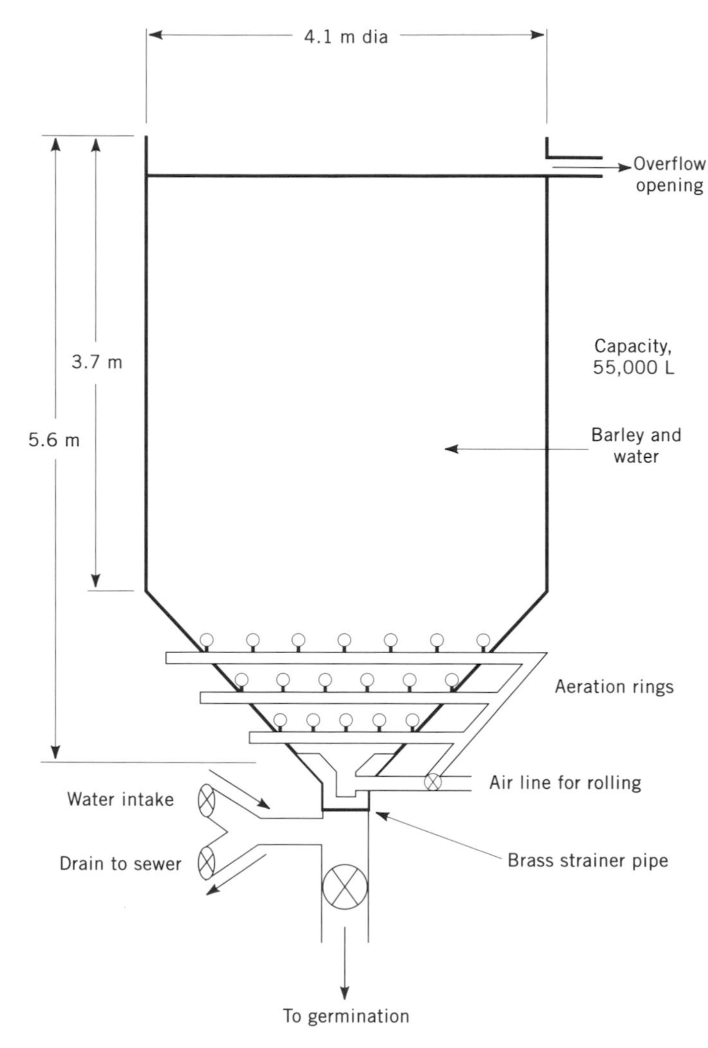

Fig. 3. Steep tank.

holding less than 6 metric tons of barley and the largest tanks holding over 40 t. The tank shown in Figure 3 holds 26 t of barley. Most tanks are fitted with a 45° conical bottom to allow the barley to flow freely from the tank. Steep tanks normally are equipped with overflow chambers which allow floating kernels, chaff, dust, and miscellaneous material to be skimmed and floated out of the tank during an immersion, ie, a water change. The bottom conical section also can be fitted with external air rings so that the barley can be supplied with oxygen via compressed air during immersions and couches (periods during which tanks are drained and aerated). Compressed air also is used to provide mixing (rolling) during immersions, either via the aeration rings or an air line located at the bottom of the tank. Where spray steeping (aerobic steeping) technology is used, steep tanks are equipped with an automatically controlled, suction-type aeration system and spray nozzles. Aerobic steeping results in substantially reduced malthouse water effluent, ie, from 4.2–17.0 to 2.3 m^3/t malt; up to a 25% decrease in effluent BOD; a reduction of up to 24 h in germination time; and an increase in malt solids recovery of as much as 1%. This technology has been engineered and installed in malthouses in the United States and abroad (7–9). A variation of aerobic steeping is activated germination malting which is used in Japan (10).

Although steeping cycles vary from maltster to maltster, they can be classified as a variation or combination of one of three processes, indicated A, B, and C. The choice of steeping procedures depends on equipment limitations, process or product specifications, and company tradition.

Process A, total immersion	Process B, multiple immersion	Process C, aerobic steeping
initial immersion with overflow	initial immersion with overflow	initial immersion with overflow
drain and refill, followed by overflow for duration of cycle	drain and couch with aeration	drain and intermittently spray and aerate for duration of cycle

In process B, the first two steps are repeated and a total of three or four immersions and couches may be used during the steep cycle.

In the 1950s maltsters began to recognize the importance of oxygen during steeping; respiration and growth could be initiated in aerated steep tanks instead of in germination compartments, thereby reducing steeping and germination time (11). Hence, maltsters began to adopt steeping processes with multiple immersions and couches. Process C represents the newest type of steep technology, developed in the mid-1970s to reduce water effluent. Typical values and ranges for key steeping parameters for all three processes are as follows:

Parameter	Range	Typical value
steeping time, h	20–50	42
water temperature, °C	2–23	15
barley steep-out moisture, %	35–45	43
barley steep-out temperature, °C	10–22	18

After steeping, barley is either wet transferred (pumped with water) or dry transferred by gravity or conveyors to germination compartments. Although still practiced by some United States maltsters, the wet-transfer process inhibits barley respiration, resulting in a substantial increase in germination time (12). A diagram of a typical germination compartment is shown in Figure 4. Whereas water is used to control temperature during steeping, preconditioned air is used to control temperature during germination. Air is saturated with water and pulled or pushed down or up through the germinating bed. During summer, spray water used to saturate germination air normally is refrigerated. Depending on ambient conditions, germination air is recycled to conserve energy. The germination fans are designed to deliver 0.1–0.16 m^3 air/t barley. The grain is turned every 8–12 h to minimize temperature differences between the top and bottom of the germination bed. A watering device is mounted on the turner

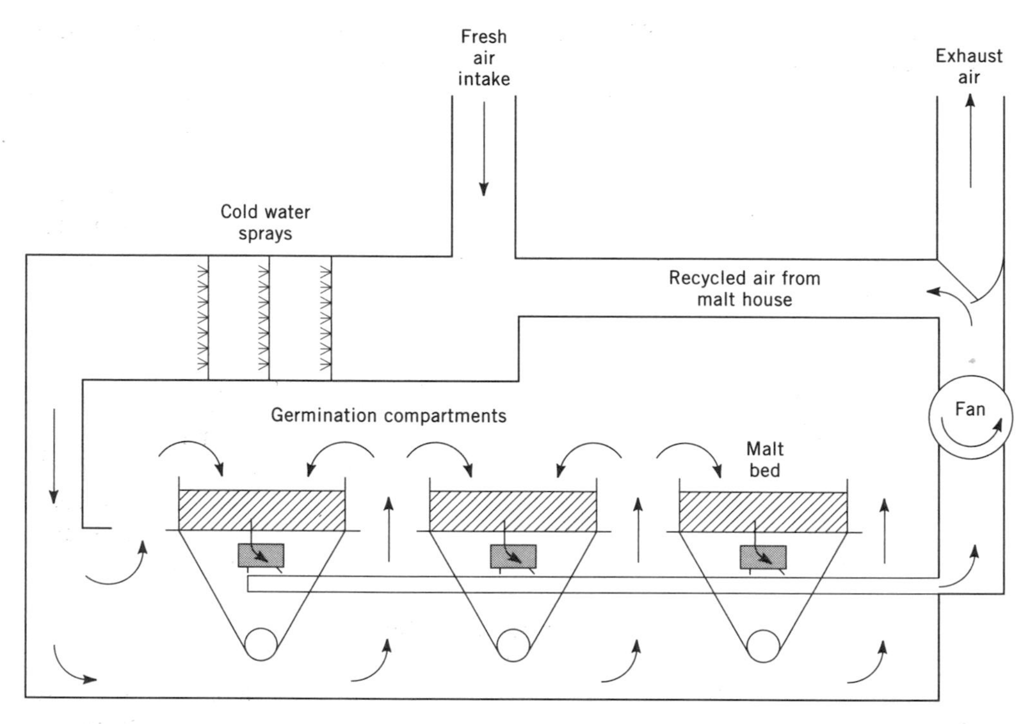

Fig. 4. Germination compartment.

and is used to increase or control green malt moisture content. Capacities of germination compartments vary between 20–200 t. Typically, germination beds are 0.75–1.0 m deep. Process parameters that are controlled during germination include the following. In general, lower temperature and moisture contents correspond to longer germination time and vice versa.

Parameter	Range	Typical value
germination time, d	3–6	4
germination temperature, °C	12–25	18
steep-out moisture, %	38–45	43
moisture after 2 d, %	44–47	46
load-to-kiln moisture, %	42–48	45

Kilning. Following germination, green malt is dried in a kiln until the moisture content is reduced to 4–6% (13). During kilning, growth and enzymatic processes are terminated and important flavor and color reactions are catalyzed by heat. Kilning is a batch process; the kilns are single-, double-, or triple-bed compartments and hot air is circulated through each compartment to remove moisture from the green malt. A typical double-deck kiln is illustrated in Figure 5. A number of U.S. maltsters also have fleximalt compartments, which are equipped for germination and kilning in the same compartment. More recently, tower malting facilities with circular kilns have been built (14).

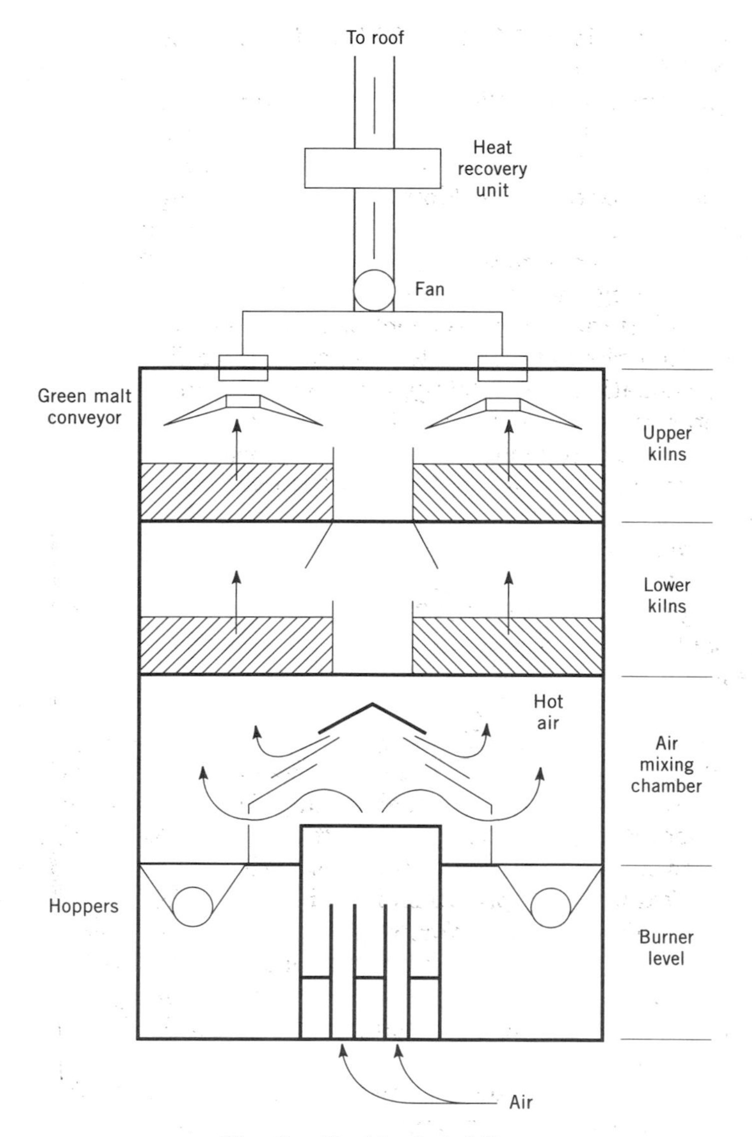

Fig. 5. Double-deck kiln.

In double-deck kilns, green malt is dried on the upper deck to 10–20% moisture with influent air temperature of 40–60°C during the first half of the cycle. The malt is then dropped to the lower deck and is dried at higher air temperature, eg, 60–85°C, to 4% moisture for brewers malt. Low temperatures, ie, 60°C, are used during the first portion of the kilning cycle while the green malt moisture is high in order to minimize color formation and thermal degradation of enzymes. Distillers' malts are dried at low temperature (60°C or less) to preserve maximum enzyme content; final moisture content is 6%.

To reduce drying time, maximum airflows are used during the first portion of the kilning cycle until the exit air is no longer saturated with moisture. Airflow then is reduced or recirculated to conserve energy. Average fuel consumption for a United States kiln is ca 5.9×10^6 kJ/t (1.4×10^6 kcal/t) malt, with a range of $2.9-10 \times 10^6$ kJ/t of malt.

Storage, Blending, and Shipping. After kilning, malt is cleaned to remove sprouts and loose hulls; it is then stored in bins according to variety and malt analytical properties. Prior to shipping, malts from several bins are blended to satisfy customer specifications. Most of the malt is shipped in railroad hopper cars; small quantities are trucked to local customers and smaller breweries. Midwestern malt is sometimes barged to the Gulf Coast for export markets.

Foreign. Foreign malting plants generally are smaller and cover a wide spectrum of configurations; eg, conventional compartments, large drums, semicontinuous (Wanderhaufen) units, towers, continuous (Domalt) units, fleximalt (combined germination and kilning), circular units, and many others (15–17). Of special interest is the French development (Nordon et Cie, Nancy, France) of a large flat-bottomed steeping unit which is provided with a reversible turner that can be used to both load and unload barley, and that is easily adapted to aerobic steeping (9).

In a unique application of gibberellic acid, barley is abraded at the distal end, opposite the germ, so that gibberellic acid is distributed more evenly and quickly throughout the kernel (18–20). In order to reduce rootlet and respiration losses by about 5%, a resteeping technique is employed in the U.K. (20–22). This process essentially consists of drowning the actively respiring barley in order to inhibit rootlet growth and adding gibberellic acid to improve modification.

New Technology. Barley Breeding. The barley breeding programs involve conventional crossbreeding techniques and have resulted in barley varieties of better yield, disease resistance, and malt quality (National Barley and Malt Laboratory, Madison, Wisconsin; American Malting Barley Association, Milwaukee, Wisconsin; Brewing and Malting Barley Research Institute, Winnipeg, Canada) (23–25). Classical breeding programs have been used for decades and improvements in agronomic and malting/brewing processing characteristics continue to be made by these techniques. Traits such as disease and pesticide resistance and enzymatic properties are early targets of genetic engineering efforts. Genetic engineering techniques are currently being explored which ultimately could revolutionize barley breeding.

A substantial breakthrough in improving malting quality through the use of mutagenic techniques is the development of experimental proanthocyanidin-free varieties which can yield beer that is colloidally stable and thus does not require stabilization in the brewery (26–30). This technique shows promise of developing other important quality characteristics, eg, reduction of the β-glucan content of barley, which causes malting modification and beer filtration problems (18).

Growth Regulators. Perhaps the most significant scientific contribution to malting technology has been the use of gibberellic acid and an increased understanding of its role in the malt modification process (18,31–33). It occurs naturally in barley at 20–150 ppb, and is obligatory for the *de novo* synthesis of α-amylase. When small amounts are added, eg, 0.01–1 ppm dry barley basis, germination time is reduced by at least 24 h or α-amylase content is increased by at least 50% (34,35). Potassium bromate can be added to inhibit gibberellic acid-induced proteolysis (36). An alternative technology to improve the efficiency of the malting process, the squeeze-malt process, is described in References 37 and 38.

Nitrosamines and Kilning. In order to avoid the formation of nitrosamines, which are attributed to the reaction of nitrogen oxides and amines that are present in barley, either indirect heating of kiln air or using low Nox burners (Maxon Corp., Muncie, Indiana) on direct-fired kilns are being applied (see *N*-NITROSAMINES). The introduction of small amounts of sulfur dioxide during the early stages of kilning also reduces formation of nitrosamines.

To recover the sensible heat content of water-saturated exit kiln air, heat exchangers are being employed to contact the exit air with incoming fresh air. Fuel savings of about 30% are being achieved.

Specialty Malts and Malt Substitutes. The amount of distillers' malt produced in the U.S. decreased substantially in the 1950s when the U.S. consumer switched to white goods (vodka and gin), which require less malt. Starting in the early 1970s, U.S. consumers have switched to lighter and lower calorie beers, which also use less malt. However, brewers' malt still comprises over 50% of the grain bill for beer production. Specialty malts, malt substitutes, and alternative brewing technology have the potential to lower brewers malt usage in favor of lower cost grain bills and processes, but no significant trends among large U.S. brewers to dramatically lower malt usage have been detected.

Economic Aspects

Malt Production and Producers. World and U.S. beer and malt production are shown in Figure 6. Because approximately 95% of malt manufactured is used to make beer, malt production follows trends in beer production. World brewers' malt and beer production in 1992 was approximately 13 million tons and 1.2 billion hectoliters and was growing at 3% per year. U.S. brewers' malt and beer production in 1992 was 2.2 million tons and 240 million hectoliters, but demand has been stagnant since 1982. Distillers and food malts account for approximately 5% of the U.S. and world malt production.

Figure 7 illustrates malt usage per hectoliter of beer. World usage is slightly higher than in the United States because U.S. brewers use higher quantities of corn and rice adjuncts than foreign brewers. Malt usage in the United States has also been declining for many years because of the popularity of lower calorie and lower alcohol beers, which require less malt to make.

Unless changes in malt composition, brewing technology, or beer products develop, it is probable that the unit usage will not decrease substantially below the current level. If too little malt is used, the low husk content in the brewers' mash results in slow lautering or mash filtration. Since malt is the only source of assimilable nitrogen, too little malt could cause variable fermentations and corresponding beer flavor problems. However, continued growth in lower calorie beer, cost pressures to substitute lower cost adjuncts for malt in the brewing process, or changes in high gravity and other brewing technology could reduce usage.

U.S. and Canadian maltsters are shown in Table 1, along with a range of estimated annual capacities. Since the 1970s, the number of malting companies in the United States has decreased due to mergers, acquisitions, and the closing of smaller malting companies and individual malthouses.

Commercial information on the U.S. malting industry can be obtained from the Beer Institute (Washington D.C.). Data on barley can be obtained from the

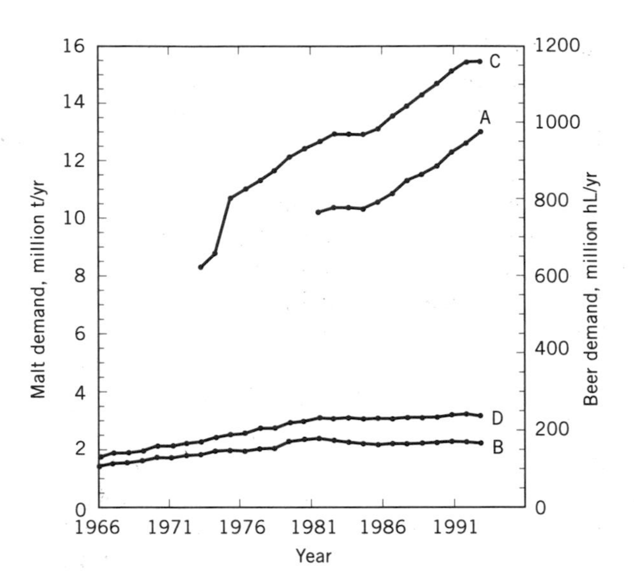

Fig. 6. Malt and beer production: A, world malt; B, U.S. malt; C, world beer; and D, U.S. beer (5,39–44 and Bio-Technical Resources files).

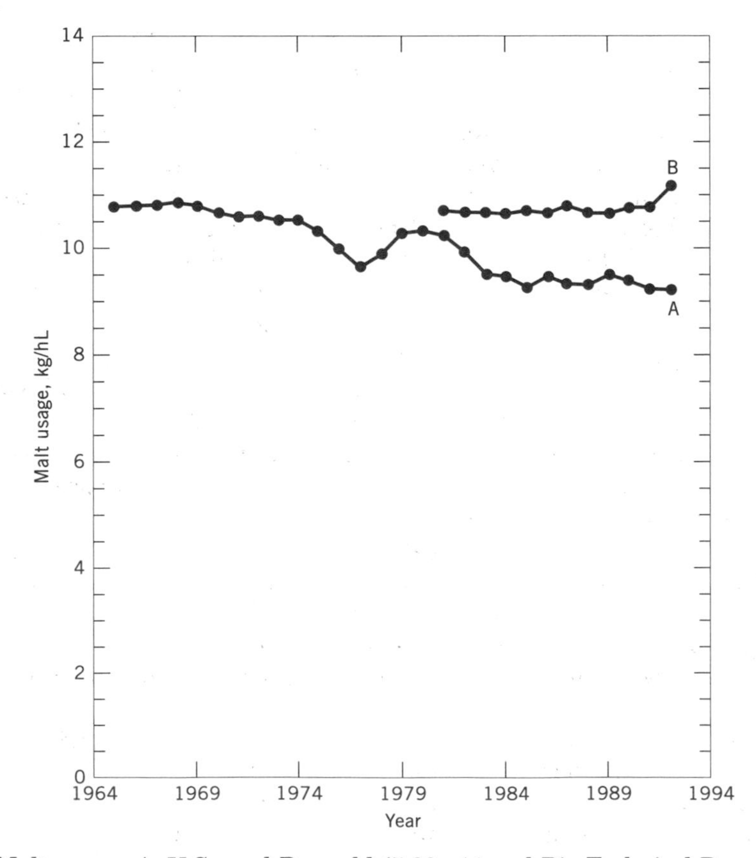

Fig. 7. Malt usage: A, U.S.; and B, world (5,39–44 and Bio-Technical Resources files).

Table 1. United States and Canadian Maltsters

Company	Plant location(s)	Estimated annual malt capacity, 10^3 t/yr
Adolph Coors Co.	Golden, Colorado	150–300
Anheuser-Busch Co., Inc.	Idaho Falls, Idaho	400–600
	Manitowoc, Wisconsin	
	Moosehead, Minnesota	
Breiss Malting Co.	Chilton, Wisconsin	<15
Canada Malting Co., Ltd.	Calgary, Alberta	400–600
	Montreal, Quebec	
	Thunder Bay, Ontario	
Dominion Malting Co.[a]	Winnipeg, Manitoba	5–150
Fleischmann Kurth Malting Co.[b]	Chicago, Illinois	300–400
	Manitowoc, Wisconsin	
	Milwaukee, Wisconsin	
	Red Wing, Minnesota	
Froedtert Malting Corp.[c]	Milwaukee, Wisconsin	300–400
	Winona, Minnesota	
Great Western Malting Co.[d]	Los Angeles, California	300–400
	Pocatello, Idaho	
	Vancouver, Washington	
Ladish Malting Co.[e]	Jefferson Junction, Wisconsin	400–600
	Spiritwood, North Dakota	
Miller Brewing Co.	Waterloo, Wisconsin	<30
Minnesota Malting Co.	Cannon Falls, Minnesota	50–150
Prairie Malt Ltd.[f]	Biggar, Saskatchewan	50–150
Rahr Malting Co.	Shakopee, Minnesota	300–400
Schreier Malting Co.	Sheboygan, Wisconsin	50–150
Stroh Brewery Co.	St. Paul, Minnesota	<30
Westcan Malting Ltd.[g]	Alix, Alberta	50–150

[a]Partly owned by ADM, Inc.
[b]Owned by ADM, Inc.
[c]Owned by Grand Malteries Modern (France).
[d]Owned by Canada Malting Co.
[e]Owned by Cargill, Inc.
[f]Majority owned by Schreier.
[g]Partly owned by Rahr.

American Malting Barley Association, Inc. (Milwaukee, Wisconsin). Additional sources are References 45 and 46. Canadian statistics are available from the Canadian Grain Commission (23) and Brewing and Malting Barley Research Institute (Winnipeg, Canada). Data on West European malt and world production are given in References 47 and 48.

Investment, Costs, and Prices for Barley and Malt. Estimated malthouse investment (1993) and costs for a new malthouse with annual capacity of 120,000 t are shown in Table 2. This malthouse is equipped with twelve 1,700-bushel steep tanks, eight 10,000-bushel germination compartments, and two double-deck kilns. The kilns are equipped with standard heat recovery units and indirect heat.

The largest cost to produce malt is the cost of barley. U.S. barley and malt prices from 1966 through 1992 are shown in Figure 8, along with the spread between malt and barley prices. The spread or margin between malt and barley

Table 2. Estimated Investment and Operating Costs for New Malting Facilities[a] 120,000 Malt t/yr

Unit	Cost	Assumptions[b]
investment (I), $	70,000,000	
operating costs, $/t malt		
capital related costs	58	includes maintenance (0.02% I), depreciation (0.05% I), insurance, property taxes, and plant supliers (0.03% I)
operating labor	10	assumes 30 plant workers, a plant manager, an engineer, a chemist, one technician, and two clerical
utilities	33	primarily consisting of kiln energy consumption at 5.9×10^6 kJ/t, electrical at 0.8 kJ/t, and wastewater at 60 hL/t
chemicals	2	
barley	133	based on choice barley at $107/t (Minneapolis), 5% cleanout barley, and 92% conversion of dry barley to malt (dry basis)
by-products credit	9	based on by 5% cleanout barley ($84/t) and 4% conversion (dry basis) of barley to malt sprouts ($75/t)
Total costs, $/t malt	*227*	

[a]Greenfield facility, with 40,000-t elevator. Utilities (electrical power and natural gas) are purchased; water comes from new wells and wastewater is treated by local municipality.
[b]To convert kJ to kcal, divide by 4.184.

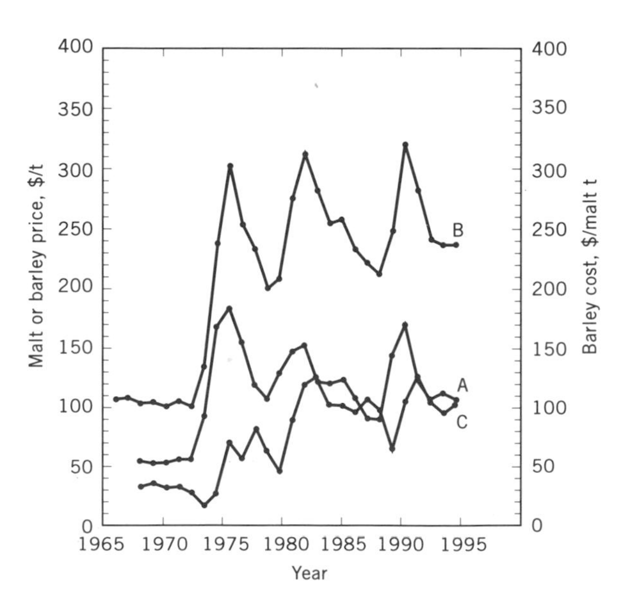

Fig. 8. Barley cost and price of malt and barley, Minneapolis basis: A, barley price; B, malt price; and C, barley cost (41 and Bio-Technical Resources files).

prices is expressed in $/t of malt and assumes 5% of barley purchased is cleanout barley and that 92% of barley solids ends up as malt. Barley and malt prices are published in *The Brewers Bulletin* (Thiensville, Wisconsin).

Malthouses being built in the United States are being designed for large batch sizes (150–200 t malt) and are highly automated to reduce labor and utility costs. However, unless the beer market begins to grow or new uses for malt are found, it is likely that few new malthouses will be built.

Specifications

Typical ranges of U.S. product specifications for brewers' malt are listed in Table 3. Moisture specifications define the drying process and imply a limit of economically undesirable material; extract is a measure of the amount of water-soluble extract in malt, which defines the amount of beer that can be made from a given amount of malt; fine–coarse difference is a measure of predicting how efficiently the extract can be obtained in the brewery; diastatic power and α-amylase are indirectly related to the fermentability of the wort; color is directly related to the color of the finished beer; wort viscosity is an indirect measurement of filterability in the brewery; soluble protein is related to yeast growth and flavor metabolites as well as foam stability of the finished beer; fine–coarse difference and the ratio of soluble to total protein is an indicator of malt modification; and assortment uniformity is necessary for predictable milling in the brewery. The analytical procedures for these parameters are given in Reference 49. Large U.S. breweries also specify all or at least part of the process conditions for manufacturing their products. Typical examples of process specifications are kiln finishing temperature and the germination time for certain malt varieties. Additional malt specifications are likely to be added in the future, primarily in response to resolving or minimizing brewing problems (50–54).

Table 3. Typical U.S. Brewers' Malt Specifications[a]

Property	Specification
moisture, wt %	4.0–5.0
fine grind extract (dry basis), wt %	73–80
coarse grind extract (dry basis), wt %	71–78
fine–coarse difference (dry basis), wt %	1.0–2.0
color, °Lovibond	1.5–2.0
diastatic power (dry basis), °Lintner	110–170
alpha amylase, 20° Dextrinizing units	35–55
viscosity, mPa·s(=cP)	1.40–1.60
total protein (dry basis), wt %	11–14
soluble protein (dry basis), wt %	4.5–6.0
ratio of soluble to total protein (dry basis), wt %	35–50
assortment, %	
kernel width >2.38 mm (6/64 in.)	80 (min)
kernal width <1.98 mm (5/64 in.)	3 (max)

[a]Ref. 49.

These rigorous specifications have created a complex blending problem, since each brewer has his own list of specifications. In order to meet such specifications, large commercial maltsters produce and store different types of malt and then carefully blend from several bins at a time.

Health and Safety Factors. Dust-control systems and good housekeeping are employed in the barley and malt elevators, steephouse, and other processing areas to eliminate or minimize the potential of dust explosions and inhalation. Although low levels of sulfur dioxide are employed (0.1–2.0 kg S/(h·t malt)), the potential of toxic sulfur dioxide concentrations resulting from process or operator error does exist. Fumigants and various cleaning agents are used routinely in the malting industry in accordance with safe operating practices.

Specialty Products and By-Products

A wide variety of special malts are produced which impart different flavor characteristics to beers. These malts are made from green (malt that has not been dried) or finished malts by roasting at elevated temperatures or by adjusting temperature profiles during kilning. A partial list of specialty malts includes standard malts, ie, standard brewers, lager, ale, Vienna, and wheat; caramelized malts, ie, Munich, caramel, and dextrine; and roasted products, ie, amber, chocolate, black, and roasted barley.

These malts vary in flavor characteristics and color among suppliers (55,56). Manufacturing protocols for specialty malts vary widely (57), with some malts being prepared in full-scale conventional kilns and others in roasters. Specialty malts for distillers have been made from rye and oats in the past, whereas wheat and sorghum (58) have been malted for wheat beers in Europe and the United States, and sorghum beer in Africa. Although almost any cereal grain can be malted, only barley has been bred to be a good quality malting grain.

Only two companies produce specialty malts in roasters or specialty kilns in North America: Breiss Malting Co. (Chilton, Wisconsin) and Extractos y Maltas (Mexico City). Other malting companies produce high dried malts in conventional kilns which are used by brewers for color or flavor purposes. Specialty malts represent less than 2% of malt sold in North America.

Malt syrups, which are extracts of conventional or specialty malts, are produced by three companies in the United States: Breiss Malting Co., Malt Products Corp., and Crompton & Knowles Corp. Malt extracts are used in a variety of food applications and by microbrewers and home brewers.

The main by-products from the malting industry are malt sprouts, cleanout material, and small-kernel barley. Malt sprouts are primarily dried malt rootlets, containing 24–26% protein, 2–3% fat, and 12–14% fiber. Since the protein is readily available, malt sprouts are used in various animal feed blends. Occasionally, malt hulls and barley chaff are blended with malt sprouts. The remainder of the cleanout material and small kernel barley is sold as feed.

BIBLIOGRAPHY

"Malts and Malting" in *ECT* 1st ed., Vol. 8, pp. 705–718, by A. D. Dickson, U.S. Dept. of Agriculture, and E. Kneen, Kurth Malting Co.; in *ECT* 2nd ed., Vol. 12, pp. 861–886, by

E. Kneen, Kurth Malting Co., and A. D. Dickson, U.S. Dept. of Agriculture; in *ECT* 3rd ed., Vol. 14, pp. 810–823, by M. R. Sfat and J. A. Doncheck, Bio-Technical Resources, Inc.

1. A. W. MacGregor and R. S. Bhatty, eds., *Barley: Chemistry and Technology*, American Association of Cereal Chemists, Inc., St. Paul, Minn., 1993.
2. W. G. Heid, Jr. and M. N. Leath, *U.S. Barley Industry, Agricultural Economic Report No. 395*, Commodity Economics Division of the Economics, Statistics, and Cooperative Service, U.S. Department of Agriculture, Washington, D.C., 1978.
3. C. A. Carter, *An Economic Analysis of a Single North American Barley Market*, Grains and Oilseeds Branch, Agriculture Canada, Winnepeg, Manitoba, Canada, 1993.
4. W. W. Wilson, *Barley Production and Marketing in the United States and Canada*, Agricultural Economics Misc. Report No. 66, North Dakota Agricultural Experiment Station, NDSU, Fargo, 1983.
5. D. Johnson and W. Wilson, *North American Barley Trade and Competition, 1993*, Dept. of Agricultural Economics, NDSU, Fargo, 1994.
6. *Quality Malting Barley Production in the Midwest*, American Malting Barley Association, Inc., Milwaukee, Wis., 1986.
7. M. R. Sfat, *Tech. Q. MBAA* **1**, 22 (1966).
8. A. Olsen, *Brygmesteren* **32**, 85 (1975).
9. J. A. Doncheck and M. R. Sfat, *Brew. Dig.* **54**, 34 (1979).
10. *Technical Report of Kirin*, No. 32, Kirin Brewery Co. Ltd., Tokyo, 1989.
11. R. V. Dahlstrom, B. J. Morton, and M. R. Sfat, *Am. Soc. Brew. Chem. Proc.*, 64 (1963).
12. D. E. Davidson and N. O. Jangaard, *J. Am. Soc. Brew. Chem.* **36**, 51 (1978).
13. M. R. Sfat, *Brew. Dig.* **40**, 50 (1965).
14. S. J. O'Donnell, C. Giarratano, and D. Burczyk, *Brew. Dig.* **60**, 13 (1985).
15. M. J. G. Minch, *Brew. Guard.* **105**, 31 (1976).
16. R. W. Haman and T. H. Hartzell, *Tech. Q. MBAA* **9**, 161 (1972).
17. G. Gibson, *Brew. Guard.* **112**, 17 (1983).
18. A. M. MacLeod, *Proceedings of the 16th Congress, European Brewery Convention, Amsterdam, the Netherlands*, 1977, p. 63.
19. G. H. Palmer, *Brew. Dig.* **49**, 40 (1974).
20. D. E. Briggs, *J. Am. Soc. Brew. Chem.* **45**, 1 (1987).
21. J. R. A. Pollock, *J. Inst. Brew.* **66**, 22 (1960).
22. A. A. Pool, *J. Inst. Brew.* **68**, 21 (1962).
23. *Canadian Barley*, Canadian Grain Commission, Grain Research Laboratory, Winnipeg, Manitoba, Canada.
24. *Proceedings of the Barley Improvement Conference*, Malting Barley Improvement Association, Milwaukee, Wis.
25. *Quality Evaluation of Barley Varieties and Selections, Quality Publications*, Malting Barley Improvement Association, Milwaukee, Wis.
26. D. von Wettstein and co-workers, *Carlsberg Res. Comm.* **42**, 341 (1977).
27. J. D. C. Figueroa, M. A. Madson, and B. L. D'Appolonia, *J. Am. Soc. Brew. Chem.* **47**, 44 (1989).
28. K. Erdal, *J. Inst. Brew.* **93**, 3 (1986).
29. D. M. Wesenberg and co-workers, *J. Am. Soc. Brew. Chem.* **47**, 82 (1989).
30. D. von Wettstein and co-workers, *Tech. Q. MBAA* **22**, 41 (1985).
31. J. G. Dickson, *Tech. Proc. Master Brew. Assoc. Am.*, 64 (1960).
32. C. M. Hollenbeck, *Cereal Sci. Today* **10**, 368, 392 (1965).
33. W. J. Pitz, *J. Am. Soc. Brew. Chem.* **48**, 33 (1990).
34. U.S. Pat. 3,116,221 (Dec. 31, 1963), M. R. Sfat and co-workers, (to Rahr Malting Co.).
35. U.S. Pat. 3,159,551 (Dec. 1, 1964), E. Sandegren and H. Beling.
36. U.S. Pat. 3,193,470 (July 6, 1965), A. Macey and K. Stowell (to Association of British Maltsters).
37. J. R. A. Pollock and A. A. Pool, in I. D. Morton, ed., *Ellis Horwood Series in Food Science and Technology: Cereals in a European Context*, VCH Publishers, New York, 1987, pp. 146–149.

38. P. C. Northam, "The Commercial Viability of Low Moisture Malting," *Twentieth International Congress*, European Brewery Convention, Helsinki, Finland, 1985.
39. U.S. Dept. of Agriculture, *Agricultural Statistics, 1990*, U.S. Government Printing Office, Washington, D.C., 1990.
40. H. M. Gauger, *Statistical Digest 1992/93*, Brussels, Belgium, 1993, pp. 40–58.
41. *The Brewers Bulletin*, Thiensville, Wis., 1966–1993.
42. World beer production figures, Beer Institute, Washington, D.C., 1993.
43. *Brew. Dig.* **67**, 16 (1992).
44. "International Report," *Brauwelt Int'l.* **IV** (1992).
45. Department of the Treasury, Internal Revenue Service, Bureau of Alcohol, Tobacco and Firearms, *Annual Report, Alcohol and Tobacco Summary*, U.S. Government Printing Office, Washington, D.C.
46. U.S. Department of Commerce, Social and Economic Statistics Administration, Bureau of the Census, *Malt, SIC 2083, Malt Beverage, SIC 2082, Distilled Liquor, SIC 2085*, U.S. Government Printing Office, Washington, D.C.
47. *Brauwelt-Brevier*, Brauwelt-Verlag, Nuremberg, Germany.
48. World malt production figures, John Barth and Son, Nuremberg, Germany.
49. American Society of Brewing Chemists, *Methods of Analysis*, 6th ed., St. Paul, Minn., 1975.
50. J. Britnell, *Tech. Q. MBAA* **23**, 115 (1986).
51. B. Axcell and co-workers, *Tech. Q. MBAA* **21**, 101 (1984).
52. G. H. Palmer, *J. Am. Soc. Brew. Chem.* **50**, 121 (1992).
53. S. Aastrup, *J. Am. Soc. Brew. Chem. 46*, 37 (1988).
54. C. E. Giarratano and D. A. Thomas, *J. Am. Soc. Brew. Chem.* **44**, 95 (1986).
55. *Brew. Techniques*, 45 (Jan.–Feb. 1994).
56. M. Griffiths, *Brew. Guard.* **121**, 29 (1992).
57. P. Blenkinsop, *Tech. Q. MBAA* **28**, 145 (1991).
58. J. P. Dufour and L. Melotte, *J. Am. Soc. Brew. Chem.* **50**, 110 (1992).

General References

A. W. MacGregor and R. S. Bhatty, eds., *Barley: Chemistry and Technology*, American Association of Cereal Chemists, Inc., St. Paul, Minn., 1993.

D. E. Briggs and co-workers, *Malting and Brewing Science*, Vol. 1, 2nd ed., Chapman and Hall, New York, 1981.

E. B. Adamic, *The Practical Brewer*, Master Brewers Association of the Americas, Madison, Wis., 1977, pp. 21–39.

A. H. Cook, *Barley and Malt*, Academic Press, Inc., New York, 1962.

J. S. Hough, D. E. Briggs, and R. Stevens, *Malting and Brewing Science*, 1st ed., Chapman and Hall, London, 1971.

K. Schuster, *Die Technologie der Malzbereitung*, Ferdinand Enke Verlag, Stuttgart, Germany, 1963.

J. Inst. Brew. **77**, 181 (1971), describes methods of analysis for barley and malt according to British Standards.

European Brewery Convention, *Analytica-EBC*, 3rd ed., Schweizer Brauerei-Rundschau, Zurich, Switzerland, 1975, describes methods of analysis for barley and malt according to European standards.

JAMES A. DONCHECK
Bio-Technical Resources, L.P.

MANGANESE AND MANGANESE ALLOYS

Manganese [*7439-96-5*], atomic number 25, atomic weight 54.94, belongs to Group 7 (VII) in the Periodic Table. Its isotopes are ^{51}Mn, ^{52}Mn, ^{54}Mn, ^{55}Mn, and ^{56}Mn, but ^{55}Mn is the only stable one. Manganese, a gray metal resembling iron, is hard and brittle and of little use alone. Its principal use in the metallic form is as an alloying element and cleansing agent for steel, cast iron, and nonferrous metals (see METAL SURFACE TREATMENTS). Manganese is essential to the steel (qv) industry where it is used mostly as a ferroalloy. After iron (qv), aluminum (see ALUMINUM AND ALUMINUM ALLOYS), and copper (qv), manganese ranks along with zinc as the next most used metal (see also MANGANESE COMPOUNDS).

The name manganese is derived from the Latin *magnes*, meaning magnet. It was once thought that manganese ores were iron ores. Manganese, recognized as an element by the Swedish chemist Scheele in 1771, was isolated in 1774 by reduction of manganese dioxide with carbon. In the 1840s, manganese was used for the first time in crucible steels. The addition of manganese as spiegeleisen (20% manganese–iron alloy) made the Bessemer process a success (see STEEL). Ferromanganese as an additive to counteract the effects of sulfur and phosphorus in steel was patented in the late 1860s. The use of manganese as an alloying agent to produce a tough wear-resistant steel was discovered in 1882.

Properties

Tables 1 and 2, respectively, list the properties of manganese and its allotropic forms. The α- and β-forms are brittle. The ductile γ-form is unstable and quickly reverses to the α-form unless it is kept at low temperature. This form when quenched shows tensile strength 500 MPa (72,500 psi), yield strength 250 MPa (34,800 psi), elongation 40%, hardness 35 Rockwell C (see HARDNESS). The

Table 1. Properties of Manganese[a]

Property	Value
melting point, °C	1244
boiling point, °C	2060
density at 20°C, g/cm^3	7.4
specific heat at 25.2°C, J/g[b]	0.48
latent heat of fusion, J/g[b]	244
linear coefficient of expansion, 0–100°C, °C^{-1}	22.8×10^{-6}
hardness, Mohs' scale	5.0
compressibility	8.4×10^{-7}
solidification shrinkage, %	1.7
latent heat of vaporization at bp, J/g[b]	4020
standard electrode potential, V	1.134
magnetic susceptibility, m^3/kg	1.21×10^{-7}

[a]Ref. 1.
[b]To convert J to cal, divide by 4.184.

Table 2. Properties of Manganese Allotropes[a]

Property	α	β	γ	δ
crystal structure	cubic	cubic	fcc	bcc
atoms per unit cube	58	20	4	2
lattice parameter, nm	0.89	0.63	0.387	0.309
transformation temp, °C	720[b]	1100[c]	1136[d]	1244[e]
latent heat of transformation, J/g[f]	36.4[b]	41.9[c]	32.8[d]	243.9[e]
density at 20°C, g/cm^3	7.43	7.29	7.18	6.3
electrical resistivity at 20°C, mW·cm	160	90	40	
thermal expansion, $°C^{-1}$	25.2×10^{-6}	43.0×10^{-6}	45.2×10^{-6}	41.6×10^{-6}
heat capacity, Cp, J/mol[f,g]				
298–980 K	$23.589 + 0.014\,T - 1.397 \times 10^5/T^2$			
980–1360 K		$32.715 + 0.0047\,T + 2.234 \times 10^5/T^2$		
1360–1410 K			$31.757 + 0.0084\,T$	
1410–1517 K				$34.221 + 0.0078\,T$

[a] Ref. 1.
[b] From α to β.
[c] From β to γ.
[d] From γ to δ.
[e] From δ to liquid.
[f] To convert J to cal, divide by 4.184.
[g] The heat capacity for liquid Mn from 1517–2000 K is $Cp = 46.024$.

γ-phase may be stabilized using small amounts of copper and nickel. Additional compilations of properties and phase diagrams are given in References 1 and 2.

Minerals and Ores

Manganese, which occurs in many minerals widely distributed in the earth's crust, constitutes about 0.1% of the earth's crust and is the twelfth most abundant element (3–6). The principal sources of commercial grades of manganese ore for the world are found in Australia, Brazil, Gabon, the Republic of South Africa, and Ukraine. The chief minerals of manganese are pyrolusite, romanechite, manganite, and hausmannite (Table 3). There is also wad, which is not a definite mineral but is a term used to describe an earthy manganese-bearing amorphous material of high moisture content.

Table 3. Common Manganese Minerals[a]

Mineral	CAS Registry Number	Composition	Mn, %
bementite	[*66733-93-5*]	$Mn_8Si_6O_{15}(OH)_{10}$	43.2
braunite		$Mn_2Mn_6SiO_{12}$	66.6
cryptomelane	[*12260-01-4*]	KMn_8O_{16}	59.8
franklinite		$(Fe,Zn,Mn)O{\cdot}(Fe,Mn)_2O_3$	10–20
hausmannite	[*1309-55-3*]	Mn_3O_4	72.0
manganite	[*52019-58-6*]	$Mn_2O_3{\cdot}H_2O$	62.5
manganoan calcite		$(Ca,Mn)CO_3$	35.4
romanechite		$BaMnMn_8O_{16}(OH)_4$	51.7
pyrolusite	[*14854-26-3*]	MnO_2	63.2
rhodochrosite	[*598-62-9*]	$MnCO_3$	47.8
rhodonite	[*14567-57-8*]	$MnSiO_3$	41.9
wad		hydrous mixture of oxides	variable

[a]Refs. 3–7.

Pyrolusite is a black, opaque mineral with a metallic luster and is frequently soft enough to soil the fingers. Most varieties contain several percent water. Pyrolusite is usually a secondary mineral formed by the oxidation of other manganese minerals. Romanechite, a newer name for what was once known as psilomelane [*12322-95-1*] (now a group name) (7), is an oxide of variable composition, usually containing several percent water. It is a hard, black amorphous material with a dull luster and commonly found in the massive form. When free of other oxide minerals, romanechite can be identified readily by its superior hardness and lack of crystallinity.

Manganite is an opaque mineral of medium hardness, ranging in color from steel gray to iron black, and having a dark, reddish brown streak. It also has a submetallic luster. Hausmannite is a black to brownish black opaque mineral, usually crystalline having a submetallic luster and a specific gravity of 4.73–4.86. It has a hardness of 5.0–5.5 (Mohs' scale) and a brownish black streak. A fresh fracture usually shows numerous bright shiny crystal faces. Rhodochrosite is the most common carbonate mineral of manganese and usually occurs with rhodonite. It is light rose in color, although other shades are not uncommon. It is a translucent mineral with a vitreous luster and a colorless streak. Cryptomelane is found in many manganese ores around the world, and may constitute as much as 50% of some ores. The K_2O content of manganese ores is usually associated with cryptomelane.

None of the natural sulfides of manganese are of any commercial importance. Some silicates have been mined. Rhodonite and braunite are of interest because these are frequently associated with the oxide and carbonate minerals. The chemical composition of some common manganese minerals are given in Table 3.

Manganese Ores. In general, only ores containing at least 35% manganese are classified as manganese ores. Ores having 10–35% Mn are known as ferruginous manganese ores, and ores containing 5–10% manganese are known as manganiferrous ores. Ores containing less than 5% manganese with the balance mostly iron are classified as iron ores.

Table 4 gives typical analyses of some of the commercial manganese ores available in the world market. Table 5 gives a breakdown of the world's total estimated manganese ore reserves that account for 98–99% of the known world reserves of economic significance. No manganese ores of commercial value are to be found in the United States.

Table 6 shows the production of manganese ores in the world by countries. In 1989, the world production represented about 89% of the rated production capacity, which in 1992 was reported to be 10.4 million metric tons of contained manganese. This percentage steadily decreased reaching 64% in 1992. Rated capacity is defined as the maximum quantity of product that can be produced in a period of time on a normally sustainable, long-term operating rate (9). This large decrease in manganese ore production is a reflection of the worldwide decline since 1979 in the volume of steel production coupled with improvements in steelmaking processes which use ferromanganese alloy additions more efficiently.

Deep-Sea Manganese Nodules. A potentially important future source of manganese is the deep-sea nodules found over wide areas of ocean bottom (see OCEAN RAW MATERIALS). At depths of 4–6 km, billions of metric tons of nodules are scattered over the ocean floor in concentrations of up to 100,000 t/km^2. Metal content varies widely. Higher grade nodules are found in the North Pacific Ocean (see Table 7). Although the prime interests in deep-sea nodules are the nickel, copper, and cobalt values, the large quantities of manganese could also be of future importance (10,11).

Several multinational private consortia were formed in the 1970s to explore and develop the mining and extraction processes for deep-sea nodules. Each of these consortia developed mining equipment deemed economically and technically feasible. However, by the mid-1980s, after collectively spending around $700 million, none of the consortia continued its development program. Additionally as of this writing none has a near-term development plan. Private sector groups are maintaining leases for the future. There are active government funded mining programs in Japan, India, and Korea (11).

Impurities. Impurities usually found in manganese ore may be classified into metal oxides, eg, iron, zinc, and copper; gangue; volatile matter such as water, carbon dioxide, and organic matter; and other nonmetallics.

A good metallurgical grade of manganese ore for smelting ferromanganese meets the specifications given in Table 8. In practice, ores are blended to provide the most economical mixture consistent with the specifications for ferromanganese. Thus individual ores can exceed these specifications.

Gangue. Oxides such as silica, alumina, lime, and magnesia are considered gangue which form slag in the smelting process. Although slag removes impurities, excessive gangue results in larger slag volumes that require more energy for the smelting process. Furthermore, because the percentage of manganese in the slag is independent of slag volume, larger slag volumes result in greater loss of manganese to the slag.

Volatiles. Manganese may contain some chemically bonded water or carbon dioxide both of which can be removed by calcining or sintering. The Mexican Molango ore (Table 4) is an example of a low grade ore that is upgraded by calcination in a rotary kiln.

Table 4. Chemical Composition of Important Manganese Ores, wt % Dry Basis[a]

Location	Typical moisture	Mn	Fe	SiO_2	Al_2O_3	CaO	MgO	BaO	P	As	K_2O	CO_2	Available oxygen[b]
Moanda, Gabon	8.8	50.2	2.8	5.0	5.9	0.05	0.06	0.25	0.101	0.003	0.81	0.03	13.9
Groote Eylandt, Australia	1.8	48.6	3.8	6.8	4.0	0.08	0.15	0.74	0.084	0.004	1.84		13.4
South Africa													
Mamatwan	1.4	38.8	4.7	4.8	0.3	13.9	4.6	0.25	0.003	0.00	0.22	13.7	4.4
Gloria	0.7	38.9	4.8	5.9	0.3	11.8	3.5	0.05	0.021		0.05	15.5	3.9
Associated 50	0.7	50.4	10.2	5.8	0.4	4.9	0.7	0.44	0.033	<0.002	0.01	1.20	6.0
Molango,[c] Mexico	1.1	38.4	7.7	15.6	3.7	9.9	9.4	0.02	0.072	<0.002	0.04	0.09	

[a]Ref. 8.
[b]Associated with Mn in excess of MnO.
[c]Calcined nodules.

Table 5. World Manganese Resources and Reserves Base, 10^6 t Manganese Content[a]

Region	Reserves	Reserves base
North America		
Mexico	4	9
South America		
Brazil	21	58
Europe		
former USSR[b]	300	450
Africa		
Gabon	52	160
Ghana	1	4
Republic of South Africa	370	4000
Total Africa	*423*	*4164*
Asia		
China	14	29
India	17	25
Total Asia	*31*	*60*
Oceania		
Australia	26	72
World Total[c]	*800*	*5300*

[a]Ref. 9.
[b]Information is inadequate to formulate reliable estimates for individual republics.
[c]Total is rounded.

Nonmetallics. During smelting, arsenic and phosphorus pass into the ferromanganese alloy. Because these elements are undesirable in steel, their content in the ore and other raw materials in the furnace charge must be kept low. Sulfur in ferromanganese is usually at very low levels because sulfur remains in the slag as manganese sulfide. Table 9 shows specifications for manganese ferroalloys.

Ore Processing. *Ore Size.* The particle size of manganese ores is an important consideration for the smelting furnace. In general, the ore size for the furnace charge is −75 mm with a limit to the amount of fines (−6 mm) allowed. Neither electric furnaces nor blast furnaces operate satisfactorily when excessive amounts of fines are in the charge.

Large amounts of fines in the ores and other raw materials promote agglomeration or caking of the charge in the furnace. This is detrimental to efficient smelting because it inhibits the downward movement of the charge and the uniform distribution of reaction gases through the burden. Furnace operators refer to this as mix bridging. In the case of the electric furnace, this can result in a hazardous situation that could lead to violent eruption of the furnace contents. Thus many of the ferromanganese-smelting operations around the world that use electric furnaces limit the amount of −6 mm ore fines in the charge to a maximum of 15%.

Ore Sintering. Most of the principal producers of ferromanganese operate sintering plants as the means of agglomeration of manganese ore fines. A mixture of manganese ore fines, fine coke, and returned sinter fines are fed to a grate

Table 6. World Production of Manganese Ore,[a] 10^3 t

Country	Range, % Mn[b]	Gross weight				
		1988	1989	1990	1991	1992
Australia	37–53	1,985	2,124	1,920	1,482	1,200
Bosnia and Herzegovina[c]	25–45					15
Brazil	30–50	1,991	1,904	2,300	2,000	1,800
Bulgaria	29–35	34	32	39	35	35
Chile	30–40	44	44	40	44	44
China	20–30	3,212	3,331	3,300	3,400	3,500
Gabon	50–53	2,254	2,592	2,423	1,620	1,556
Georgia[d]	29–30					1,200
Ghana	30–50	260	279	247	320	279
Hungary	30–33	81	84	60	30	18
India	10–54	1,333	1,334	1,393	1,401	1,400
Iran	25–35	75	81	54	48	50
Kaszakhstan	29–30					9
Mexico	27–50	444	394	365	254	407
Morocco	50–53	30	32	49	59	59
Romania	30	65	60	55	50	45
Republic of South Africa	30–48	4,023	4,884	4,402	3,146	2,464
Ukraine[d]	29–30					5,800
USSR	29–30	9,108	9,141	8,500	7,240	
Yugoslavia	25–45	40	39	51	40	
other[e]		35	33	54	43	48
Total ore		*25,013*	*26,389*	*25,252*	*21,213*	*19,929*
Total contained Mn		*8,646*	*9,273*	*8,875*	*7,273*	*6,701*

[a]Ref. 9.
[b]Estimated.
[c]Previously reported as production of the former Yugoslavia from which came all production.
[d]Previously reported as production of the former USSR; data not reported separately until 1992.
[e]Represents the combined totals of Burma, Egypt, Greece, Indonesia, Italy (from wastes), Japan, the Philippines, Thailand, Turkey, and Zambia.

(moving or stationary) where the coke is ignited and, under suction, generates enough heat to fuse together the ore particles. During the process, the MnO_2 in the manganese ore is reduced to lower oxides, eg, Mn_2O_3 or Mn_3O_4. The amount of manganese ore sinter in the electric furnace smelting charge is in the range of about 20–35% of the ore blend. Higher amounts of sinter in the ore blend for the electric furnace result in somewhat greater electrical energy consumption (12).

Since 1988, a 500,000 t/yr sinter plant has been in operation in South Africa at the Mamatwan manganese ore mine (13). The sintering process removes the 15–17% CO_2 that is contained in the Mamatwan ore. With 35% sintered Mamatwan ore replacing an equivalent amount of the unsintered ore in the furnace charge, energy consumption in the electric smelting furnace is decreased by about 16%. This lower power consumption and accompanying increase in productivity is the result of the removal of CO_2 before smelting in the furnace (see also MINERAL RECOVERY AND PROCESSING).

Table 7. Average Metal Content of Deep-Sea Nodules,[a] wt % Dry Basis

Location	Mn	Fe	Ni	Cu	Co
North Atlantic					
Blake Plateau	14.5	13.7	0.50	0.08	0.42
red clay region[b]	13.9		0.36	0.24	0.35
seamount and mid-Atlantic ridge	13.5		0.39	0.14	0.36
South Atlantic	7.2		0.14	0.09	0.05
Indian Ocean	16.3		0.54	0.20	0.26
North Pacific					
red clay regions	18.2	11.5	0.76	0.49	0.25
siliceous ooze	24.6	8.2	1.28	1.16	0.23
South Pacific					
deep-water clay region	15.1		0.51	0.23	0.34
submarine highs on seamounts and plateaus	14.6		0.41	0.13	0.78

[a]Ref. 10.
[b]1770 km east of Florida.

Table 8. Specifications of Metallurgical-Grade Manganese Ore, wt %, Dry Basis[a]

Assay	Composition	Assay	Composition
Mn[b]	48.0	As	0.18
Fe	6.0	P	0.19
Al_2O_3	7.0	Cu + Pb + Zn	0.30
$Al_2O_3 + SiO_2$	11.0		

[a]All values are maximum unless otherwise indicated.
[b]Value is minimum.

Table 9. Compositions of Manganese Ferroalloys, wt %[a]

Alloy	Mn[b]	C	Si	P	S	As
ferromanganese						
high carbon	76–80	7.5	1.2	0.35	0.05	0.30
medium carbon	80–85	1.5	1.2	0.30	0.02	0.15
low carbon	85–90	<0.10–0.50	1.2	0.20	0.02	0.10
silicomanganese	65–78	2.00	16.0–18.5	0.20	0.04	0.10
low carbon silicomanganese	63–66	0.08	28–32	0.05	0.04	0.15

[a]Values given are maximum unless otherwise noted.
[b]Range represents minimum values.

Alloy Processing

Smelting. The greatest application of manganese is in ferrous metallurgy (qv). Manganese alloys such as those listed in Table 9 are employed as refining agents as well as alloying additions. The nature of manganese raw materials and the requirements of the iron and steel industry have created an interesting and varied extractive metallurgy for producing ferromanganese alloys (see METALLURGY, EXTRACTIVE).

Process Chemistry. Manganese is combined with oxygen in its ores (see Table 3) and carbon is the most economical reducing agent for oxides. Therefore, the essential characteristics of manganese metallurgy is evident from examination of the interactions between manganese oxides and carbon (14). The highest oxide, MnO_2, decomposes to Mn_2O_3 at 507°C, and Mn_2O_3 goes to Mn_3O_4 at 1240°C. These oxides can be reduced exothermically by CO in the shaft of a furnace. This is analogous to the situation in the Fe–O–C system where Fe_2O_3 can be reduced by CO to FeO. However, the Mn and Fe systems differ in the conditions required for the final reactions:

$$FeO + C \xrightarrow{\text{air}} Fe + CO + CO_2 \quad 670°C$$

$$7\ MnO + 10\ C \xrightarrow{\text{air}} Mn_7C_3 + 7\ CO \quad 1267°C$$

Because of the relative instability of FeO, the reduction to metallic Fe occurs at a much lower temperature and appreciable CO_2 is present in the product gas. The high temperature required for the reaction of MnO and C results in the formation of essentially pure CO; the partial pressures of CO_2 and Mn are <0.1 kPa (1×10^{-3} atm). The product of this reaction is manganese carbide (7:3) [*12076-37-8*], Mn_7C_3, containing 8.56% carbon. Assuming immiscibility of the metal and carbide, Mn should be obtainable by the reaction of MnO and Mn_7C_3 at 1607°C. However, at this temperature and activity of Mn, the partial pressure of Mn vapor is approximately 10 kPa (0.1 atm) which would lead to large manganese losses.

This simplified picture must be modified to take into account the other constituents of the ore that include FeO, CaO, Al_2O_3, and SiO_2 (Table 4). In the presence of carbon at elevated temperatures, Fe is easily reduced and thus reports to the metallic phase; CaO and Al_2O_3 are unreactive with carbon under the conditions of Mn smelting and enter the slag. SiO_2 is also largely retained in the slag but under certain conditions can be reduced and contribute Si to the alloy.

Slag composition influences the amount of manganese recovered as metal (15). Basic slags, those containing CaO, increase the activity of MnO and therefore promote manganese reduction. Such slags also suppress the activity of SiO_2 thus limiting the introduction of silicon into the alloy. Although adjusting the slag composition by the addition of base can lead to higher activity of MnO, it can have a negative influence on recovery by increasing the slag volume. This latter should be kept to a minimum.

Slag volume is directly related to the SiO_2 content of the raw materials and to a lesser extent, Al_2O_3. Although the addition of base leads to higher activity of MnO, the melting point and viscosity of the slag must also be taken into consideration. The melting point of the slag increases as MnO in the slag is replaced by the CaO and MgO added to the charge mixture; therefore, complete replacement of MnO with CaO is impractical. For this reason high grade ores, ie, of high manganese content, are favored and slag chemistry is best adjusted by employing a blend of ores of different gangue compositions to attain the desired slag composition.

Increased temperature favors Mn reduction. A thermodynamic analysis of manganese smelting (16) showed that under conditions of carbon saturation on a typical smelting charge at 1400°C, the recovery of manganese as metal was low, leaving high concentrations of manganese in the slag. At 1500°C, 90% recovery of manganese as metal was predicted. However, at 1600°C, the recovery in the metal did not improve. Appreciable silicon was present in the alloy and Mn vapor appeared in the gas phase.

The silicon content of the alloy is increased by smelting manganese bearing charges containing increasing amounts of SiO_2 and by increasing smelting temperature through slag composition. Manganese vaporization is not excessive but SiO losses become significant at Si levels of greater than 25% (17). Silicon and carbon in manganese alloys saturated with carbon have an inverse relationship, as shown in Figure 1, and therefore at high silicon concentrations carbon content is low (18,19).

Because of the limitations imposed by the presence of gangue and the relatively high volatility of manganese, the initial products obtained when reducing manganese ore with carbon are a carbon saturated alloy and a slag containing some manganese. Under suitable conditions the manganese recovery can be sufficiently high to justify discarding the slag. Otherwise, the slag is smelted carbothermically using SiO_2 additions at higher temperature to produce silico-

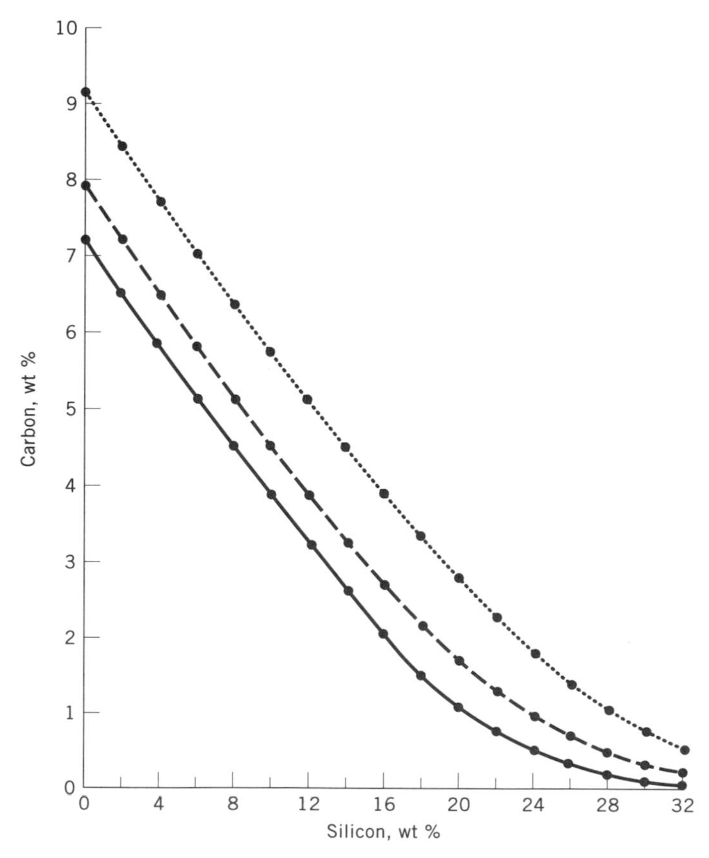

Fig. 1. Effect of silicon on carbon solubility in ferromanganese at (—) 1350°C, (---) 1500°C, and (····) 1800°C (19).

manganese and a slag of low manganese content giving a high overall recovery of manganese.

Ferromanganese low in carbon cannot be produced directly from ore and carbon because of the Mn volatility or the co-smelting of gangue constituents. Refining the primary reduction products is required. Figure 2 shows the overall scheme of reducing Mn ore to a variety of products. High carbon ferromanganese can be made by three different practices, blast furnace, discard slag electric furnace, and high manganese slag electric furnace. Medium carbon ferromanganese can be made by decarburizing high carbon ferromanganese or by the oxidation–reduction reaction of silicon in silicomanganese with manganese ore. Silicomanganese and low carbon silicomanganese are made in an electric furnace. Low carbon ferromanganese is produced by the reaction of manganese ore and low carbon silicomanganese.

The blast furnace for ferromanganese is usually located in a steelworks and is owned and operated by the steel producer. Most of the electric furnaces for manganese ferroalloy products are located in ferroalloy plants where other manganese products are made. In addition to large tonnages of high carbon ferromanganese, silicomanganese and refined (medium and low carbon) ferromanganese are also produced. Manganese-rich slags from the various ferromanganese production steps are used as all or part of the manganese charge to the silicomanganese furnace.

High Carbon Ferromanganese. Ferromanganese, also known as high carbon ferromanganese, or in the United States as standard ferromanganese, is the largest tonnage manganese alloy used in the steel industry. Of the overall average usage of manganese in steel, 75% is supplied as ferromanganese. Specifications of the various grades of ferromanganese and other manganese ferroalloys are given in Table 9. Table 10 is a listing of the countries in the world that produce ferromanganese and silicomanganese.

Ferromanganese is produced in blast furnaces and electric smelting furnaces. Economics usually determine which smelting process is chosen for ferromanganese. Both methods require about the same amount of coke for reduction to metal, but in the case of the blast furnace, the thermal energy required for the smelting process is supplied by the combustion of additional coke, which in most countries is a more expensive form of energy than electricity.

Capital requirements for a new facility generally favor the electric furnace process. However, in some countries having integrated steel industries, the availability of excess ironmaking blast furnaces, metallurgical coke, and relatively high cost of electricity, the blast furnace is an attractive choice (21). There was a decline in blast furnace production during the 1960s and 1970s. As of the start of the 1990s, however, ~30% of world ferromanganese production was made by blast furnaces. Five European countries produced ferromanganese by blast furnace representing about half of the total European production (see Table 10). More recently, Germany and the U.K. have discontinued blast furnace production of ferromanganese. For a time in the early 1980s, there was one blast furnace producer of ferromanganese in Japan. As of this writing (ca 1994) China produces about two-thirds of its ferromanganese by the blast furnace.

Blast Furnace Production. High carbon ferromanganese is produced in blast furnaces in a process similar to the production of iron (21). Ferromanganese

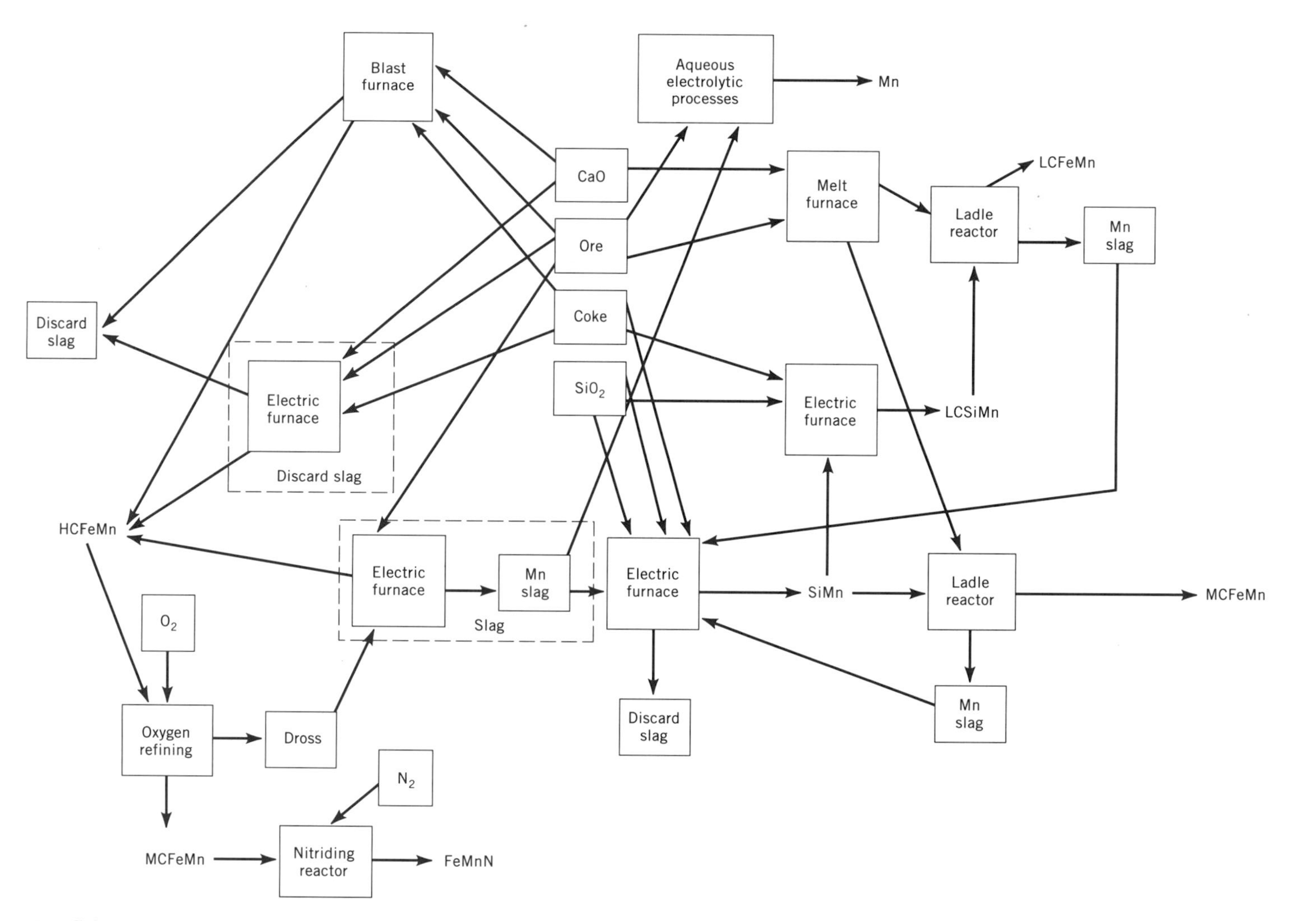

Fig. 2. Schematic diagram of manganese metallurgy where HCFeMn = high carbon ferromanganese; MCFeMn = medium carbon ferromanganese; SiMn = silicomanganese; LCFeMn = low carbon ferromanganese; LCSiMn = low carbon silicomanganese; FeMnN = nitrided ferromanganese; and Mn = electrolytic manganese.

Table 10. World Production of Ferromanganese and Silicomanganese,[a] 10^3 t gross weight

	1988			1989			1990			1991			1992[e]		
Country	FeMn[b,c]	SiMn	Total[d]	FeMn	SiMn	Total[d]	FeMn	SiMn	Total[d]	FeMn	SiMn	Total[d]	FeMn	SiMn	Total[d]
Argentina	20	12	31	24	21	46	24	22	46	24	22[e]	45	24	22	46
Australia[e]	58	44	102	67	55	122	70	65	135	55	65	120	55	65	120
Belgium[e]	95		95	95		95	90		90	90		90	90		90
Brazil	181	193	374	181	208	389	171	217	387	169	272	441	180	261	440
Bulgaria[f]	31		31	30		30									
Canada[e,f]	161		161	185		185	185		185	45		45			
Chile	7	1	8	7	<0.5	8	4	1	5	6	1[e]	7	7	<0.5	7
China[e]	309	220	529	360	240	600	370	240	610	390	250	640	400	270	670
	(220)			(240)			(240)			(250)			(250)		
Croatia[g]													35	30	65
Czechoslovakia[e,f]	95		95	100		100	102		102	90		90	90		90
France[h]	346	59	405	373	59[e]	432	357	62	418	350	30[e]	380	360	30	390
	(324)			(346)			(320)			(320[e])			(300)		
Georgia[i]													100	100	200
Germany[e,f,j]										260		260	220		220
eastern states	67		67	67		67	65		65	(220)			(190)		
western states[k]	309		309	350		350	288		288						
	(274)			(305)			(250)								
India	138	53	191	158	72	230	160	75	235	148	46	194	145	45	190
Italy	12	69[e]	81	14	47[e]	61	11	56[e]	67	14	55[e]	69	14	50	64
Japan	378	107	485	394	122	516	452	77	530	464	87	551	362	96[l]	458
Korea, North[f]	70		70	70		70	70		70	70		70	70		70
Korea, South	76		76	85		85	84		84	95		95	95		95
Mexico	165	80	246	168	99	267	186	71	257	147	67	214	150	70	220
Norway	361	233	594	221	270	491	213	223	437	173	227	400	203	213[l]	416
Peru	1		1	1		1	1		1	1		1	1		1
Philippines													5		5

Table 10. (*Continued*)

Country	1988			1989			1990			1991			1992[e]		
	FeMn[b,c]	SiMn	Total[d]	FeMn	SiMn	Total[d]	FeMn	SiMn	Total[d]	FeMn	SiMn	Total[d]	FeMn	SiMn	Total[d]
Poland[f]	95		95	92		92	76		76	62		62	55		55
	(91)			(90)			(71)			(57)			(50)		
Portugal	10	5[e]	15	13		13	12		12	12		12	10		10
Romania[e]	80	40	120	80	40	120	80	40	120	70	20	90	50	15	65
Russia[l]													200		200
													(200)		
South Africa, Republic of	447	248	695	394	258	652	404	234	638	260	235	495	270	242	512
Spain[e]	48	38	86	50	40	90	52	38	91	50	40	90	50	40	90
Taiwan	26	31	57	31	27	57	44	21	64	40	13	53	41	5	46
Ukraine[i]													250	1,000	1,250
USSR[e,m]	1,058	1,300	2,358	1,023	1,300	2,323	691	1,300	1,991	605	1,100	1,705			
	(604)			(609)			(281)			(235)					
United Kingdom	107		107	140		140	143		143	178		178	130		130
	(107)			(140)			(143)			(178)			(130)		
Venezuela		34	34		32	32		30[e]	30		30[e]	30		31	31
Yugoslavia[n]	45	47	92	34	53	86	32	61	92	20	50[e]	70			
Zimbabwe	2		2												
Total[o]	*4,798*	*2,813*	*7,611*	*4,808*	*2,944*	*7,752*	*4,437*	*2,833*	*7,271*	*3,894*	*2,609*	*6,503*	*3,661*	*2,585*	*6,246*

[a] Ref. 20.
[b] Total production process is by electric furnace unless indicated.
[c] Blast furnace production is given in parentheses.
[d] Data may not add to totals shown due to independent rounding.
[e] Value is estimated.
[f] Data for ferromanganese includes silicomanganese, if any.
[g] Formerly part of Yugoslavia; data were not reported separately until 1992.
[h] Includes silicospiegeleisen, if any.
[i] Formerly part of the USSR; data were not reported separately until 1992.
[j] Totals for Germany in 1991–1992 include data for Eastern and Western states.
[k] Data for blast furnace ferromanganese includes spiegeleisen, if any.
[l] Reported figure.
[m] Dissolved in December 1991.
[n] Dissolved in April 1992.
[o] U.S. data is proprietary and thus not included in the total.

production in a blast furnace differs from pig iron production in that larger amounts of coke are required in the former. The reduction of MnO takes place at a higher temperature than the reduction of FeO necessitating a high blast temperature which results in a high top temperature (2).

Preheating the blast and oxygen enrichment are used to reduce coke requirement. Further increase in blast temperatures is obtained by using plasma heaters, and because of the substitution of electrical energy for coke, lower production costs are obtained (22). Dolomite or limestone added to the charge raises the activity of MnO. By careful control and a method for ensuring a more uniform charge mixture in the shaft, manganese recoveries of over 90% and coke rates of 1530 kg/t are reported (23) (see FURNACES, FUEL-FIRED).

Electric Furnace Production. Electric furnaces used for smelting ferromanganese range in size up to 40 MW. Important design parameters for electric furnaces are electrode diameter and spacing, hearth diameter, crucible depth, voltage range, and KVA capacity of the transformer. For furnaces larger than 15 MW, design parameters are specific for each ferroalloy product to be produced. Because the resistivity of the burden in the electric furnace production of ferromanganese is low, low voltages between the electrodes are necessary to maintain satisfactory penetration of the electrodes in the charge. To obtain the proper power loading for the furnace, higher currents are required. Therefore, to operate within the current carrying capacities of carbon electrodes, the diameter of the electrodes for ferromanganese furnaces is larger than for other ferroalloy furnaces (see FURNACES, ELECTRIC). Large electric smelting furnaces operate at low power factors. If the factor goes below ~ 0.90, a penalty may be imposed by the supplier, therefore capacitors are installed to maintain power requirements.

Most electric furnaces have sealed covers although some are open and have a hood for fume collection. Figure 3 shows a 40 MW covered furnace for ferromanganese. The furnace gasses are drawn off and wet scrubbed by high energy Venturi scrubbers or can be dry filtered through large bag collectors if the furnace is of the open type. The sealed cover is generally preferred because it allows better fume control at lower capital cost and lower energy requirement. The crucibles are steel shells lined with refractory oxide brick and an inner lining of carbon blocks. The hearth is similar, but has a thicker carbon lining. Electric energy is supplied to the smelting reaction through three carbon electrodes, of the Søderberg, ie, self-baking, type. The electrodes for large (>30 MW) furnaces are 1.9 m in diameter. Most large ferromanganese furnaces have separate tapholes for metal and slag (see Fig. 3).

An important consideration for a successful smelting operation is the slag composition, which is largely determined by the make-up of the gangue constituents of the ores. The composition of the slag has pronounced effects on furnace resistivity, smelting temperature, recovery of manganese, and the amount of silicon in the ferromanganese. Blending of ores or the addition of fluxing reagents is often necessary to produce the desired slag composition.

High Manganese Slag Practice. Table 11 gives typical ranges of operating data for a large ferromanganese furnace using the high manganese slag practice using no flux additions in the charge, ie, a self-fluxing system. The high manganese slag practice is used by most plants where high grade manganese ores are smelted and silicomanganese is also produced. Manganese content of this slag

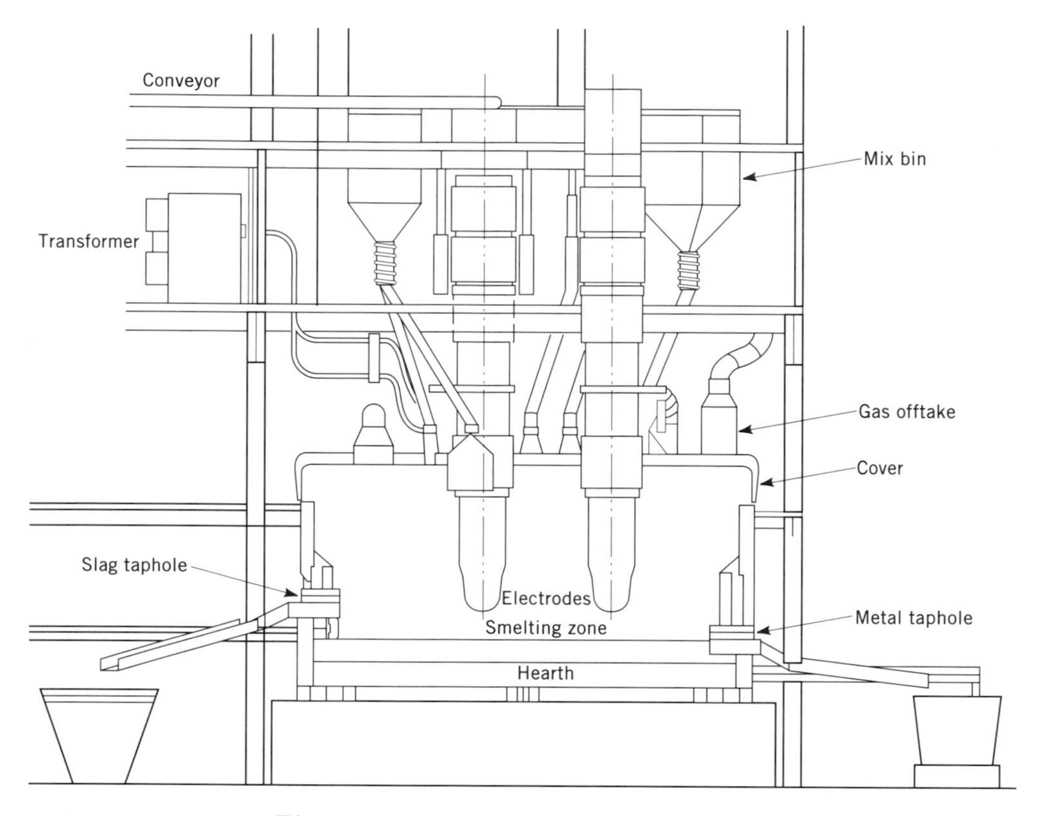

Fig. 3. Covered ferromanganese furnace.

ranges from 30–42%. Small amounts of fluxes may be used, such as dolomite or limestone, depending on the manganese concentration desired in the slag. In a silicomanganese furnace, smoother operation and higher overall manganese recovery are obtained if a high manganese slag is used as the charge.

High grade ores without flux additions (self-fluxing) produce slags containing 36–42% manganese. The ratio of slag to alloy ranges from 0.4 to 0.7, depending on the gangue level of the ore and the manganese content of the slag. Recovery of Mn in the alloy ranges from 70–80%, depending again on the gangue concentration (principally SiO_2) and the amount of flux used.

An efficient electric furnace operation for ferromanganese consumes ca 2100 kWh/t alloy. Energy needs for less efficient furnaces or when smelting lower grade ores that produce excessive slag, range from ca 2400 to 2800 kWh/t. Unlike any other electric furnace ferroalloy process, a substantial portion of the gas produced in a ferromanganese smelter is carbon dioxide. Coke and energy consumption are minimized and the CO_2 content is maximized when the CO reduction of the manganese oxides is carried to completion. The gaseous reduction is affected by smelting conditions, the skill of the furnace operator, and the charge permeability. Ore-size distribution and coke size are important for charge permeability. In this regard, excessive amounts of ore fines (<6–8 mm)

Table 11. Operating Data for a Ferromanganese Electric Furnace[a]

Parameter	Value
operating load, MW	30–35
apparent power, MV·A	58.9–61.6
power factor	0.51–0.57
electrode current, kA	130
single-phase resistance, mΩ	0.592–0.690
secondary voltage[b]	262–274
usage per ton alloy	
Mn ore, t[c]	1.969
coke, t[c]	0.417
electrode paste, kg	8.0
power, kWh	2200
ratio, slag:alloy	0.45
slag composition, wt %	
MnO	38.7
SiO_2	24.7
Al_2O_3	14.4
CaO	13.0
MgO	2.7
alloy composition, wt %	
Mn	79.0
Fe	13.0
Si	0.3
C	6.7
Mn recovery, %	
alloy	83.2
slag	13.8
fume, etc	3.0

[a]Using high manganese slag practice. Furnace has three 1.9-m Søderberg electrodes.
[b]Assuming single-phase reactance = 1.0 mΩ.
[c]Dry basis.

are undesirable and some smelters agglomerate manganese ore fines before using the smelting operation (13,24).

Discard Slag Practice. The discard slag practice is followed when the ore is of such low quality that a high degree of manganese extraction is required to achieve the alloy grade, or the ore contains base oxides, eg, CaO and MgO, which if smelted alone, leads naturally to low manganese slags. Manganese content of the slag from this practice ranges from 10 to 20% and manganese recovery in alloy ranges between 85 and 90%.

If the ores contain little CaO or MgO, the charge is made up of manganese ores, coke, and a basic flux such as limestone or dolomite. Better results are obtained when the required basic oxides (CaO or MgO) are contained in the manganese ores, eg, as in some of the South African manganese ores (see Table 4). Specific energy consumption for this practice ranges from ca 2600 to 3100 kWh/t alloy. Power consumption is higher than that for the high manganese slag practice because of the additional energy required to calcine dolomite or limestone in the charge and the greater amount of manganese extracted from the slag which results in higher CO content in the off-gas than in that of the high manganese practice.

In South Africa where most of the manganese ores contain $CaCO_3$ and $MgCO_3$, there is little choice but to smelt with the discard slag practice, resulting in higher energy consumption. The introduction of sintering of part of the carbonate ores mined in South Africa, where the CO_2 is removed in the sintering process, substantially improved the electric furnace smelting operation. Energy consumption was reduced 14.69% and production was increased by 17.37% (12).

Silicomanganese. Silicomanganese is an alloy of manganese and iron containing 12.5 to 18.5% silicon which provides the steel industry with a convenient source of the two most important alloying and deoxidizing elements (Mn,Si) consumed in the production of steel. The production process is similar to that of electric furnacing smelting of ferromanganese. It differs in the furnace charge which contains large amounts of quartz (SiO_2) and, if required to adjust slag composition, limestone or dolomite. Smelting temperature is higher and the off-gas is predominantly carbon monoxide. In a plant that produces high manganese slags from other ferromanganese products, these slags are blended with ore for smelting in the silicomanganese furnace. Power consumption ranges from ca 3.86 to 4.84 MWh/t alloy, depending on the quality of ores, slag volume, furnace efficiency and, more importantly, the silicon content of the alloy.

In plants or companies that also produce ferrosilicon products, off-grade silicon-bearing materials such as ferrosilicon fines, drosses, or ladle digouts may be available. The addition of these materials to the silicomanganese furnace charge provides a relatively inexpensive method for increasing the productivity of the silicomanganese operation because remelting such silicon materials uses less energy than smelting silicon.

A furnace designed specifically for silicomanganese has a smaller crucible, electrode diameter, and electrode spacing than required for ferromanganese at the same power loading. Provided the gas-cleaning system has the capacity, a furnace designed for ferromanganese can be operated at substantially higher power when producing silicomanganese because of higher electrical resistance. Silicomanganese contains less carbon than ferromanganese but, nevertheless, is carbon saturated. The relationship of carbon and silicon in manganese–iron–silicon alloys is an inverse one (Fig. 1).

Low Carbon Silicomanganese. A low carbon grade of silicomanganese containing 28–32% Si and $<0.06\%$ C is usually made by a two-stage process. In the first step, low iron silicomanganese containing 16 to 18% silicon and about 2% carbon is made by smelting quartz and ferromanganese slag which is depleted in iron. Subsequently, in a separate furnace process, a mixture containing the crushed low iron silicomanganese, quartz, and coal (qv) or coke is smelted in a slagless process where the quartz is reduced to silicon that displaces the carbon in the remelted silicomanganese.

This product is mainly used within the ferroalloy plant as the reducing agent in a silicothermic process to produce the low carbon grade of refined ferromanganese.

Refined Ferromanganese. Refined ferromanganese refers to alloys that are not carbon saturated and range from less than 0.10 to 1.50% maximum carbon. Medium carbon grades are used in special grades of steels where in final additions carbon control is important. The low carbon grades are used mainly in the production of certain grades of stainless steels.

In the past, all grades of refined ferromanganese were made by various modifications of multistep silicon reduction processes. Depending on the carbon content desired in the product, a manganese ore and lime mixture was allowed to react with the silicon in silicomanganese or low carbon silicomanganese in an open, electric-arc furnace. The equilibrium reaction is

$$\begin{array}{c} 2\,MnO_2 + Si \rightleftharpoons 2\,MnO + SiO_2 \\ \underline{2\,MnO + Si \rightleftharpoons 2\,Mn + SiO_2} \\ MnO_2 + Si \rightleftharpoons Mn + SiO_2 \end{array}$$

Lime is added to the reaction to increase the activity of MnO and reduce the activity of SiO_2. This allows greater extraction of manganese in equilibrium with less than 2% Si in the alloy.

In later modifications the manganese ore was reduced to MnO in an ore–lime melt by carbon reduction. The reaction of molten ore–lime with the silicon in silicomanganese was conducted in large ladles. The advantage is the use of less expensive carbon before the final reduction of MnO by silicon. Because half as much silicon is required, less SiO_2 is generated, therefore requiring less lime and producing less slag, which allows the extraction of more manganese for better recovery.

In 1971, oxygen refining of high carbon ferromanganese was introduced as a method for producing medium carbon ferromanganese (25). This manganese oxygen refining (MOR) process is similar to that used by the steel industry to produce steel in the basic oxygen furnace (BOF) (see Fig. 4). The MOR process offers substantial savings in investment costs, production costs, and energy usage but is limited to the production of medium carbon ferromanganese. Other oxygen refining processes for ferromanganese have been developed that use different methods of introducing oxygen with and without inert gases to the refining vessel.

The low carbon grade of ferromanganese must still be made by the silicon reduction method. The process is essentially the same as the silicothermic process shown in Figure 4 but uses ferromanganese–silicon, the low carbon grade of silicomanganese, as the reducing agent.

Manganese Nitride. Manganese nitride is an addition agent for steel where both manganese and nitrogen are required in certain grades of steel. Briquettes of comminuted medium carbon ferromanganese are nitrided in an annealing-type furnace in a nitrogen atmosphere. The reaction is exothermic and commences at 600°C (26). Nitrogen content of over 4% is attained. The nitrogen content is influenced by the amount of iron, carbon, and silicon in the medium carbon ferromanganese.

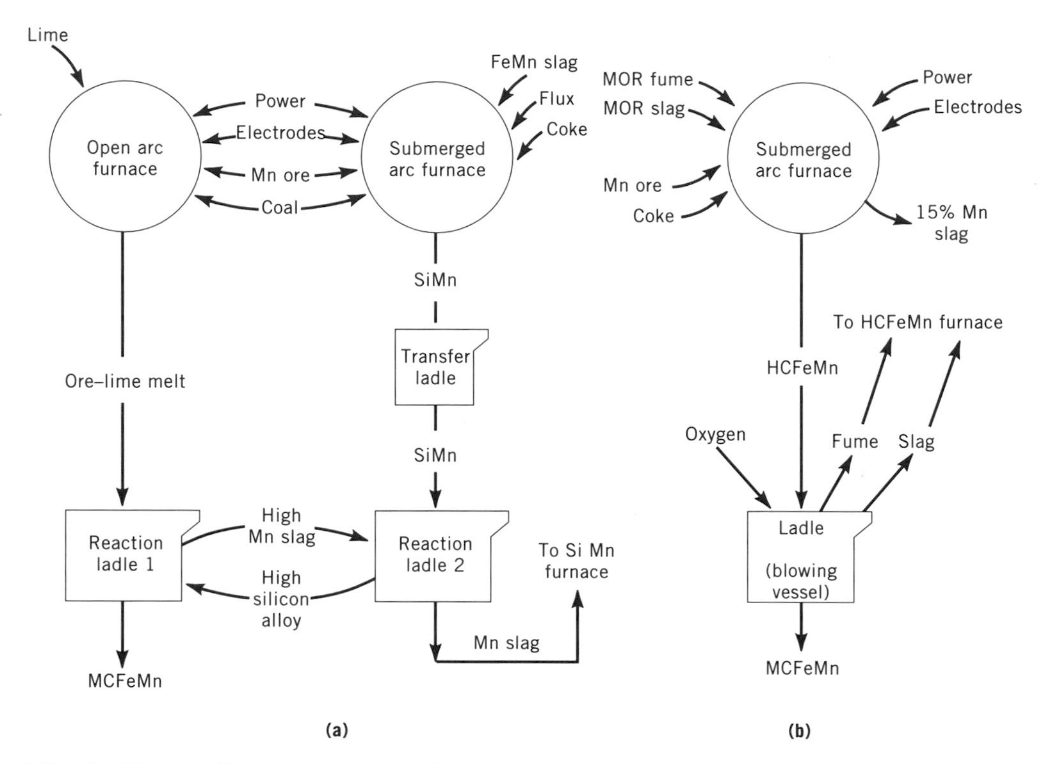

Fig. 4. Process flow sheets for (**a**) the silicothermic reduction and (**b**) the MOR process. See text.

Electrolytic Processes

Electrolysis of Aqueous Solutions. The electrolytic process for manganese metal, pioneered by the U.S. Bureau of Mines, is used in the Republic of South Africa, the United States, Japan, and beginning in 1989, Brazil, in decreasing order of production capacity. Electrolytic manganese metal is also produced in China and Georgia.

Manganese metal made by this process is 99.9% pure. It is in the form of irregular flakes (broken cathode deposits) about 3-mm thick, and because of its brittleness, has little use alone. Most of the electrolytic manganese that is used in the aluminum industry is ground to a fine size and compacted with granulated aluminum to form briquettes that typically contain 75% Mn and 25% Al.

Manganese ore is roasted to reduce the higher oxides to MnO which is acid soluble, or as practiced by Elkem Metals (Marietta, Ohio), the MnO is supplied from slag produced as a by-product from the ferromanganese smelting operation. The reduced ore or slag is then leached with sulfuric acid at pH 3 to give manganese(II) sulfate. The solution is neutralized with ammonia to pH 6–7 to precipitate iron and aluminum which are removed by filtration. After treatment with hydrogen sulfide gas, arsenic, copper, zinc, lead, cobalt, and molybdenum are removed as sulfides. Ferrous sulfide or ammonium sulfide plus air is then added to remove colloidal sulfur, colloidal metallic sulfides, and organic matter. The purified liquid is electrolyzed in a diaphragm cell. Table 12 gives the conditions of electrolysis.

Table 12. Electrolysis of Aqueous Solutions of Manganese

Condition	Value
purified feed solution, catholyte	
Mn as $MnSO_4$, g/L	30–40
$(NH_4)_2SO_4$, g/L	125–150
SO_2, g/L	0.30–0.50
anolyte	
Mn as $MnSO_4$, g/L	10–20
H_2SO_4, g/L	25–40
$(NH_4)_2SO_4$, g/L	125–150
current density, mA/cm^2	43–65
catholyte pH	6–7.2
anode composition	Pb + 1% Ag
cathode composition	Hastelloy, type 316 stainless steel, or Ti
cell voltage, V	5.1
diaphragm	acrylic[a]
current efficiency, %	60–70

[a]Usually specified as to porosity.

The impurity levels in electrolytic manganese metal are as follows:

Element	Quantity, wt %
Fe	0.0015
Cu	0.0010
As	0.0005
Co	0.0025
Ni	0.0025
Pb	0.0025
Mo	0.0010
S as sulfide	0.0170
S as sulfate	0.0140
C	0.0020
H_2	0.0150

Fused-Salt Electrolysis. Fused-salt electrolysis in many ways is similar to the Hall process for producing aluminum (see ALUMINUM AND ALUMINUM ALLOYS). The process was developed and solely used by Chemetals Corp. (26). The plant using this process discontinued production in 1985. The starting material is manganese ore reduced to the manganese(II) level in the solid state by using a rotary kiln under reducing conditions. The reduced ore is charged to the electrolytic cell which contains molten calcium fluoride and lime. Because of the ore's gangue content, additional fluorspar and lime are required as the reaction proceeds, the fluorspar to maintain the desired fused-salt composition, and the lime to neutralize the silica contained in the ore. As the volume of fused electrolyte increases, excess fused electrolyte is periodically removed.

The cell for this process is unlike the cell for the electrolysis of aluminum which is made of carbon and also acts as the cathode. The cell for the fused-salt electrolysis is made of high temperature refractory oxide material because molten manganese readily dissolves carbon. The anode, like that for aluminum, is made of carbon. Cathode contact is made by water-cooled iron bars that are buried in the wall near the hearth of the refractory oxide cell.

The cell is operated above 1300°C so as to maintain the manganese in the molten state. Molten manganese metal is periodically removed from the cell, much like tapping an ordinary ferromanganese furnace, and is cast in cast-iron pots. The metal produced by this process contains 92–98% Mn with iron as the main impurity. Manganese ore, chemically pretreated to remove iron, is used as the cell feed for the 98% Mn grade. The metal is reduced to lumps <150 mm for sale to the steel industry. Electrical energy consumption is ca 8–9 kWh/kg product. In addition to the lump form of manganese metal, refined ferromanganese products such as medium and low carbon ferromanganese also were produced by the fused salt electrolysis process at this plant.

Uses of Metallic Manganese

Manganese is essential to the production of steel. It is found in varying amounts in all steels and cast iron. About two-thirds of the manganese used in steel is as an alloying element to enhance the hardenability, strength, and other mechanical properties of steel. The remaining third of the manganese is used in steel to combine with the residual sulfur in iron to prevent hot shortness during the rolling process for steel and to improve the deoxidizing effect of aluminum and silicon in molten steel. The bulk of steel production is of the multipurpose low carbon steels containing from 0.15% to 0.8% manganese (27,28). In cast iron, manganese neutralizes the sulfur in the iron and adds strength to the final casting (29).

Austenitic Manganese Steels. The invention of a manganese-bearing steel in 1882 by Sir Robert Hadfield was the first of the austenitic group of steels. It contained 1.2% C and 12% Mn, and combined toughness and ductility with good resistance to wear. Austenitic manganese steels find use in crushing and grinding equipment and many applications that require toughness and wear resistance. The composition of the various grades of austenitic manganese steel castings all contain 11 to 14% Mn with carbon in the range of 0.7–1.35%.

Austenitic steels that are used for nonmagnetic and cryogenic applications have lower carbon content than Hadfield steels and range in composition from 15 to 29% manganese (30).

High Strength Low Alloy Steels. Steels low in carbon and containing 1 to 1.8% of manganese in combination with microalloying additions such as niobium or vanadium in amounts of around 0.10% are known as high strength low alloy steels. These steels take advantage of the effect of manganese on the austenitic transformation temperature to obtain ultrafine-grained ferrite and the increased strength obtained from the carbonitrides formed by the microalloy additions. By controlled rolling and rapid quenching, steels of high strength-to-weight ratios are obtained that are used in pipelines (qv), automobiles, and other transportation (qv) equipment, as well as structural applications (27,28).

Stainless Steels. In chromium–nickel stainless steels, 1–2% manganese improves the hot working and weldability characteristics. Since the 1950s, the 200 series of stainless steels have come into use as a partial substitute for the AISI 300 series. The 200 series consists, among others, of AISI types 201 and 202 which contain 5.5 to 7.5 and 7.5 to 10.0% manganese, respectively. The 200-series stainless steels were developed to obtain higher strength and more economical austenitic materials by using the less expensive manganese in place of nickel.

Nonferrous Uses. In the nonferrous metal industry, manganese is consumed in significant amounts for its effect on hot working and for modifying physical properties. The metals so treated include a number of aluminum and copper alloys and some nickel- and cobalt-base alloys. In the special cases of some bronzes, hardening alloys made from copper and ferromanganese are added to the molten bronze in order to add small amounts of iron that cannot be alloyed directly. Manganese has many beneficial effects when combined with copper-based alloys containing one or more of the alloying elements of aluminum, nickel, and zinc (31).

Manganese has been employed as an alloying element almost from the beginning of the aluminum industry. The amounts added are small and very seldom exceed 2% Mn and most often are in the range of 0.20 to 0.8%. In these amounts, the physical properties of aluminum are only slightly affected but the electrical properties are greatly affected, eg, resistivity increases threefold. Manganese in solution in aluminum has a strengthening effect accompanied by a small decrease in ductility (32).

Aluminum–manganese alloys are extensively used for food handling equipment, ranging from cooking utensils to frozen food trays, beverage cans, bottle caps, and any other application where good corrosion resistance and absence of food contaminants, together with strength and wear resistance higher than aluminum, are required. By far the most important use for this aluminum alloy is for beer (qv) and soft drink cans of which some 100 billion are produced annually. In the building industry the aluminum–manganese alloys are used mainly for siding and roofing for houses and buildings, sometimes painted or enameled for durability and low maintenance (31,33) (see BUILDING MATERIALS, SURVEY; CARBONATED BEVERAGES; FOOD PACKAGING).

Economic Aspects

Total world production of manganese ores in 1992 was about 6.7×10^6 metric tons manganese content, down from 9.3×10^6 in 1989. Table 13 shows the price of a kilogram of manganese in terms of constant U.S. 1987 dollars, ie, adjusted for inflation.

The price range of imported ferromanganese in the United States at the end of 1990 was $640–650/t of alloy but declined to $477–482/t alloy by the end of 1992. This corresponds to a 26% decrease in price for those two years. During the same period, the range for silicomanganese (2% C grade) went from $0.568–$0.584/kg alloy to $0.507–$0.513/kg alloy, a drop of 11.5% in price. The price of electrolytic manganese metal produced in the United States was $2.304/kg during the years 1991 and 1992 (9).

Table 13. Prices for Manganese In Metallurgical Ore[a]

	Average annual U.S. price, $/kg	
Year	Actual price	Based on constant 1987 dollars[b]
1983	0.136	0.156
1984	0.140	0.154
1985	0.141	0.149
1986	0.132	0.136
1987	0.127	0.127
1988	0.175	0.169
1989	0.276	0.254
1990	0.378	0.334
1991	0.372	0.316
1992	0.325	0.269

[a]Ref. 9.
[b]Based on Implicit Price Deflaters for GDP (1987 = 100%). Ref. 34.

Outlook. Because the world iron and steel industry consumes about 94% of the manganese mined, the market for manganese is largely dependent on the world steel market. The marked decrease in manganese ore production for the five-year period of 1988–1992 shown in Table 6 is a reflection of the worldwide decline in the volume of steel production during that same period and is illustrated in Figure 5**a**. Improvements in steelmaking processes that use ferromanganese alloy additions more efficiently along with changes in steel grades that require less manganese have also lessened manganese consumption. Figure 5**b** shows the world production of crude steel as compared to the world production of the two principal manganese addition alloys, ferromanganese and silicomanganese.

During the years 1981 to 1986, the average consumption of manganese units (as ferroalloys) for the EEC, the United States, and Japan combined, decreased from 6.5 to 5.5 kg/t of steel. For the same period in the United States, the consumption of manganese decreased from 6.2 to 4.7 kg/t of steel (33), and apparently decreased further in the years of 1990, 1991, and 1992 to 4.15, 4.11, 3.85 kg/t of steel, respectively (9). In contrast, in 1984, the steel industry of the former USSR, where 50% of steel production was still made in open-hearth furnaces, had an average consumption of manganese units of 13 kg/t steel (35).

A change in the relative consumption of silicomanganese and ferromanganese occurred during 1988–1992. World production (Table 10), ratio of FeMn:SiMn, decreased from 1.7 to ca 1.45. Silicomanganese is a convenient reagent for deoxidizing molten steel and the extent of its usage is dependent on its price vs the equivalent combination of ferrosilicon and ferromanganese.

Health and Safety Factors

Health and Environment. Manganese in trace amounts is an essential element for both plants and animals and is among the trace elements least toxic to mammals including humans. Exposure to abnormally high concentrations of

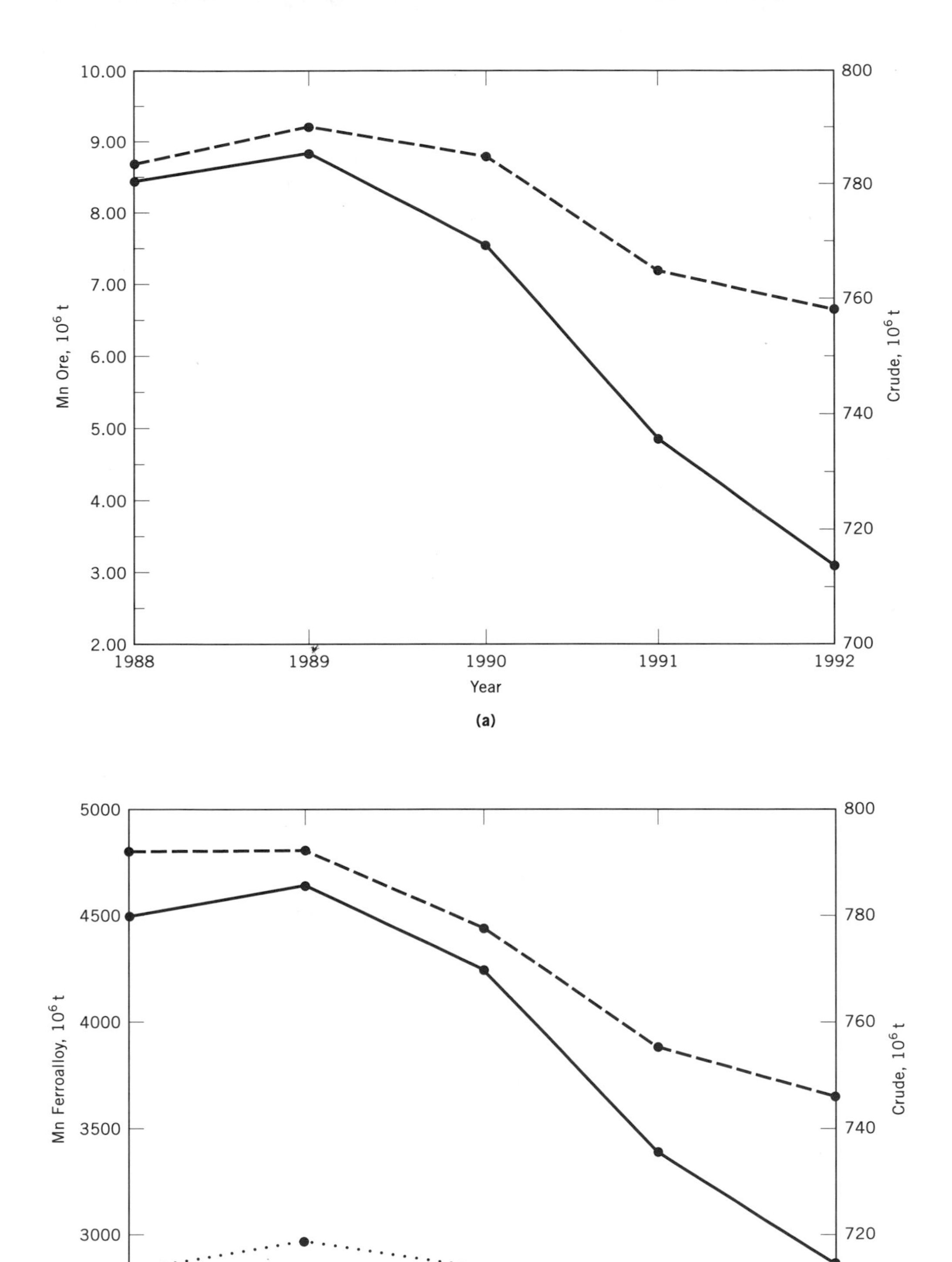

Fig. 5. World production: (**a**) (---) manganese ore and (—) crude steel; (**b**) (---) FeMn, (····) SiMn, and (—) crude steel (9,31).

manganese, particularly in the form of dust and fumes, is, however, known to have resulted in adverse effects to humans (36,37) (see MINERAL NUTRIENTS).

Two kinds of diseases owing to manganese are known in humans: manganic pneumonia and manganism. These diseases mainly concern workers occupied in manganese ore mills, smelting works, battery factories, and manganese mines. A risk also exists when welding (qv) if electrodes containing manganese are used. Short exposures to manganese dust or fume can result in manganic pneumonia, a form of lobar pneumonia that is unresponsive to antibiotic treatments.

Manganese poisoning can result in chronic manganism which is primarily a disease of the central nervous system often first manifested by disordered mentation similar to the symptoms of Parkinson's disease. Presymptomatic exposure periods have been known to range from three months to 16 years but usually cases of manganism develop after 1–3 years of exposure, although individual susceptibility plays a role in the development of the disease. Many symptoms of manganism regress or disappear quickly if the victim is removed from exposure, although disturbances in speech and gait may remain. Well-established manganism is a crippling disease with permanent disability, but is not fatal and is considered to be rare.

Airborne manganese concentrations in the United States range from 0.02 to 0.57 $\mu g/m^3$ in urban areas and 0.0017–0.047 $\mu g/m^3$ in nonurban areas. The ACGIH recommends a TLV of 5 mg/m^3.

Plant Safety. Of the many ferroalloy products produced in electric furnaces, ferromanganese has the greatest potential for furnace eruptions or the more serious furnace explosions. The severity of the explosions increases with the size of the furnace. Such incidents are infrequent, but can occur, and when they do are often disastrous. Explosions usually result in extensive damage to the furnace and surrounding area, and often severe injuries or death to personnel in the immediate area. An eruption is the sudden ejection of solids, liquids, or gases from the furnace interior. A more violent and instantaneous ejection of material, accompanied by rapid expansion of burning gas, is considered an explosion (38).

The higher oxides of manganese (MnO_2, Mn_2O_3) are strongly exothermic in the presence of carbon and CO. In a normal, well-running ferromanganese furnace, the charge travels downward at a steady rate while the CO gas formed deep in the furnace rises and reacts with the oxygen-rich ores quietly and at a steady rate. Thus the manganese ore in the charge is reduced to manganese(II) before entering the hot reaction zone of the furnace.

Most all the serious eruptions of manganese furnaces can be traced to a set of conditions that cause bridging or hang-up of the charge materials so that the normal downward movement through the furnace is disrupted or retarded. As electrical energy continues to be supplied to the furnace with little or no movement of charge owing to the bridging, a cavity forms under and around the electrodes particularly after tapping the furnace. When conditions change that allow the bridge to collapse, usually during or shortly after a tap, large quantities of unreacted mixture enter the superheated cavity and react rapidly with a sudden release of CO and the ejection of hot and molten materials.

Safe operation of ferromanganese furnaces requires careful control of raw material particle size, oxygen content of the ore blend, and charge stoichiometry (38).

Most modern furnaces are equipped with computers that log raw material usage and other operating data. Many of the larger furnaces have computer systems that are programmed to continually monitor operating data and to automatically make adjustments to obtain optimum performance. On the basis of the analysis of each tap of slag and metal, continuous analysis of furnace off-gas, electrode length and penetration, and electrical data (volts, amperes, resistance), the computer makes adjustments to the amount of carbon in the charge mixture and controls the movement of the electrode. The benefits of computer controlled electric smelting furnaces include safer and more efficient operation, as well as increased productivity and lower costs (39,40) (see PROCESS CONTROL).

BIBLIOGRAPHY

"Manganese and Manganese Alloys" in *ECT* 1st ed., Vol. 8, pp. 718–735, by J. H. Brennan, Union Carbide and Carbon Corp., and C. Longenecker, ed., *Blast Furnace and Steel Plant*; in *ECT* 2nd ed., Vol. 12, pp. 887–905, by F. E. Bacon, Union Carbide Corp; in *ECT* 3rd ed., Vol. 14, pp. 824–843, by L. R. Matricardi and J. H. Downing, Union Carbide Corp.

1. E. A. Brandes and R. F. Flint, *Manganese Phase Diagrams*, The Manganese Centre, Paris, 1980; L. B. Pankratz, *Thermodynamic Properties of Elements and Oxides*, Bull. 672, U.S. Bureau of Mines, Washington, D.C., 1982.
2. G. Volkert and co-workers, in *Metallurgie der Ferrolegierungen*, Springer, New York, 1972.
3. F. W. Fraser and C. B. Belcher, "Mineralogical Studies of the Groote Eylandt Manganese Ore Deposits," *Proceedings Australasian Institute of Minerals and Metallurgy* No. 254, June 1975.
4. I. Kostov, *Mineralogy 1*, Oliver and Boyd, London, 1968.
5. *Powder Diffraction File*, Pub. SMA-29, JCPDS, International Centre for Diffraction Data, Swarthmore, Pa., 1979.
6. C. Palache, H. Berman, and C. Frondel, *Dana's System of Mineralogy*, 7th ed., John Wiley & Sons, Inc., New York, 1955.
7. W. L. Roberts, T. J. Campbell, and G. R. Rapp, Jr., *Encyclopedia of Minerals*, 2nd ed., Van Nostrand Rheinhold, New York, 1990, p. 735.
8. Elkem Metals Co., *Ore analyses*, Marietta, Ohio, 1994.
9. T. S. Jones, *Manganese 1992, Annual Report*, U.S. Bureau of Mines, Washington, D.C., Sept. 1993.
10. *Manganese Recovery Technology*, National Materials Advisory Board, NMAB 323, 1976.
11. J. C. Wiltshire, "Seafloor Cobalt Deposits: A Major Untapped Resource," presented at *Cobalt at the Crossroads, Intertech Conferences*, Herndon, Va., June 1992.
12. P. C. Pienaar and W. F. P. Smith, "A Case Study of the Production of High Grade Manganese Sinter from Low-Grade Mamatwan Manganese Ore," *Proceedings of the 6th International Ferroalloys Congress (Infacon)*, Cape Town, South Africa, 1992, p. 149.
13. W. Gericke, "The Establishment of a 500,000 tpa Sinter Plant at Samancor's Mamatwan Manganese Ore Mine," *Proceedings of the 5th International Ferroalloys Congress (Infacon)*, New Orleans, La., Apr. 1989.
14. J. H. Downing, *Electr. Furn. Conf. Proc. AIME* **21**, 288 (1963).
15. H. Cengizler and R. H. Eric, in Ref. 12, Vol. 1, p. 167.
16. Y. E. Lee and J. H. Downing, *Canadian Met. Quart.* **19**, 315 (1981).
17. W. Ding and S. E. Olsen, *Electr. Furn. Conf. Proc.* **49**, 259 (1991).
18. J. Sandvik and J. Kr. Tuset, *The Solubility of Carbon in Ferrosilicomanganese at 1330–1630°C*, The Engineering Research Foundation, Trondheim, Norway, 1970.

19. N. F. Yakushevich, V. D. Mukovkin, and V. A. Rudenko, *Izv. Vyssh. Ucheb. Zaved., Chrn. Met.* **10**, 67–70 (1968).
20. *1992 Iron and Steel Annual Report*, in the *1992 Minerals Yearbook*, U.S. Bureau of Mines, Washington, D.C., 1992.
21. A. Kitera and co-workers, *Infacon 86, Proceedings*, Vol. 1, Tokyo, 1986.
22. D. A. MacRae, in Ref. 15, p. 28.
23. S. Suzikii and M. Masukawa, in Ref. 15, p. 149.
24. R. T. Hooper, *Electr. Furn. Conf. Proc.* **36**, 118 (1978).
25. D. S. Kozak and L. R. Matricardi, *Electr. Furn. Proc.* **38**, 123–127 (1980).
26. J. H. Downing, *Elec. Furn. Proc.* **44** (1986); J. P. Faunce and J. Y. Welsh, "The Production of Manganese Metal," presented at *105th Annual Meeting, AIME*, Las Vegas, Nev., Feb. 22–26, 1976.
27. C. P. Desforges, W. E. Duckworth, and T. F. J. H. Ryan, *Manganese in Ferrous Metallurgy*, The Manganese Center, Paris, 1976.
28. *Mn Manganese*, brochure published by the International Manganese Institute, Paris, 1990.
29. *AFS Cupola Handbook*, American Foundrymen's Society, Chicago, Ill., 1975, p. 210.
30. *Properties and Selection: Iron, Steel, and High Performance Alloys*, Vol. 1, 10th ed., ASM International, Mar. 1990.
31. G. Greetham, *Materieux et Techniques*, 105–116 (Dec. 1977).
32. L. F. Mondolfo and P. L. Dancoisne, *Materieux et Techniques*, 89–103 (Dec. 1977).
33. P. L. Dancoisne, "Past and Present Evolution of the Manganese Demand," presented at *4th ILAFA-ABM Ferroalloy Congress*, Salvador de Bahia, Brazil, Nov. 1988.
34. Department of Commerce, Bureau of Economic Analysis, Washington, D.C., 1988.
35. V. A. Strishkov and R. M. Levine, *The Manganese Industry of the USSR*, U.S. Bureau of Mines, Washington, D.C., Sept. 1987.
36. *Medical and Biologic Effects of Environmental Pollutants-Manganese*, Division of Medical Sciences, National Research Council, National Academy of Sciences, 1973.
37. *Scientific and Technical Assessment Report on Manganese*, EPA Report No. 600/6-75-002, U.S. Environmental Protection Agency, Washington, D.C., Apr. 1975.
38. J. G. Oxaal, L. R. Matricardi, and J. H. Downing, *Electr. Furn. Conf. Proc. AIME* **38** (1980).
39. C. T. Ray and A. H. Olsen, *Electr. Furn. Conf. Proc.* **44**, 217–223 (1987).
40. T. K. Leonard, *Proc. Int. Ferroalloy Congr. 5th*, 267–279 (1989).

LOUIS R. MATRICARDI
Consultant

JAMES DOWNING
Consultant

MANGANESE COMPOUNDS

Manganese is the twelfth most abundant element in the earth's crust and is the fourth most used metal following iron, aluminum, and copper (see also MANGANESE AND MANGANESE ALLOYS). Manganese use dates back to antiquity where early references relate to the use of manganese in glassmaking. The first useful manganese compound isolated was permanganate discovered by the German chemist, Glauber, in the seventeenth century. The most used manganese compound aps of the early 1990s, outside of the manganese–iron alloy, ferromanganese, used in steelmaking (see STEEL), is manganese dioxide. Use of manganese dioxide had its beginnings in 1868, when Leclanché developed the dry cell battery (see BATTERIES) which uses manganese dioxide as the primary component of the cathode mixture.

Manganese, atomic no. 25, belongs to the first transition series and is the principal member of Group 7 (VIIA). It has nine isotopes (1,2) (Table 1).

Ground-state electronic configuration is $1s^22s^22p^63s^23p^63d^54s^2$. Manganese compounds are known to exist in oxidation states ranging from -3 to $+7$ (Table 2). Both the lower and higher oxidation states are stabilized by complex formation. In its lower valence, manganese resembles its first row neighbors chromium and especially iron in the Periodic Table. Commercially the most important valances are Mn^{2+}, Mn^{4+}, or Mn^{7+}.

As the oxidation state of manganese increases, the basicity declines, eg, from MnO to Mn_2O_7. Oxyanions are more readily formed in the higher valence states. Another characteristic of higher valence-state manganese chemistry is the abundance of disproportionation reactions.

$$2\ Mn^{3+} \longrightarrow Mn^{2+} + Mn^{4+}$$

$$2\ Mn^{5+} \longrightarrow Mn^{4+} + Mn^{6+}$$

$$3\ Mn^{6+} \longrightarrow Mn^{4+} + 2\ Mn^{7+}$$

Thermodynamic data (4) for selected manganese compounds is given in Table 3; standard electrode potentials are given in Table 4. A pH–potential diagram for aqueous manganese compounds at 25°C is shown in Figure 1 (9).

Table 1. Isotopes of Manganese

Isotope	Atomic mass	Half-life, $t_{1/2}$	Decay mode[a]
^{50}Mn		0.286 s	β^+
^{51}Mn		45 min	β^+, EC
^{52}Mn		5.7 d	β^+, EC
^{53}Mn	52.9413	2×10^6 yr	EC
^{54}Mn	53.9402	303 d	EC
^{55}Mn[b]	54.9381	stable	
^{56}Mn		2.576 h	β^-
^{57}Mn		1.7 min	β^-
^{58}Mn		1.1 min	β^-

[a] β^+ = positron emission, β^- = negative beta emission, and EC = orbital electron capture.
[b] The natural abundance of ^{55}Mn is 100%.

Table 2. Representative Manganese Compounds[a]

Oxidation state	Geometry	Example
Mn^{3-}	tetrahedral	$Mn(NO)_3CO$
Mn^{2-}	square	$[Mn(phthalocyanine)]^{2-}$
Mn^{-}	trigonal bipyramid	$Mn(CO)_5^-$
Mn^0	octahedral	$Mn_2(CO)_{10}$
Mn^+	octahedral	$[Mn(CN)_6]^{5-}$
Mn^{2+}	tetrahedral	$MnCl_4^{2-}$
Mn^{3+}	octahedral	MnF_3
Mn^{4+}	octahedral	MnO_2
Mn^{5+}	tetrahedral	MnO_4^{3-}
Mn^{6+}	tetrahedral	MnO_4^{2-}
Mn^{7+}	tetrahedral	MnO_4^-

[a]Ref. 3.

There are approximately 250 known manganese minerals. The primary ores which typically have a Mn content >35%, usually occur as oxides or hydrated oxides, or to a lesser extent as silicates or carbonates. Table 5 lists the manganese-containing minerals of economic significance (10). Battery-grade manganese dioxide ores are composed predominately of nsutite, cryptomelane, and todorokite.

Worldwide reserves of manganese in ore deposits are estimated (ca 1994) at 1.8×10^9 and potential resources are 1.1×10^9 t (11). The world's supply of commercial manganese ore comes from Australia, Brazil, Gabon, and the Republic of South Africa.

Deep-sea manganese nodules represent a significant potential mineral resource. Whereas the principal constituent of these deposits is manganese, the primary interest has come from the associated metals that the nodules can also contain (see OCEAN RAW MATERIALS). For example, metals can range from 0.01–2.0% nickel, 0.01–2.0% copper, and 0.01–2.25% cobalt (12). Recovery is considered an economic potential in the northwestern equatorial Pacific, and to a lesser degree in the southern and western Pacific and Indian Oceans (13–18).

United States resources of manganese ores are estimated at over 70×10^6 metric tons. These are, however, low grade deposits having manganese contents ranging from 0.6–12% and U.S. manganese production from domestic ore stopped in 1970. The United States depends on imports for its manganese needs and maintains sizable stockpiles for emergencies.

Whereas hydrogen does not react with manganese to form a hydride, hydrogen is soluble to some extent in manganese metal. Exposure of manganese to oxygen leads to the ready formation of manganese oxides, especially at higher temperatures. Nitrogen above 740°C forms solid solutions, as well as several nitrides such as, MnN [*36678-21-4*], Mn_6N_5 [*64886-63-1*], Mn_3N_2 [*12033-03-3*], Mn_2N [*12163-53-0*], and Mn_4N [*12033-07-7*]. Manganese nitrides are used in steelmaking as nitrogen-containing intermediate alloys (19) (see NITRIDES).

Carbon reacts with molten manganese forming various carbides including $Mn_{23}C_6$ [*12266-65-8*], Mn_3C [*12121-90-3*], Mn_7C_3 [*12076-37-8*], Mn_2C_7 [*75718-05-7*], and $Mn_{15}C_4$ [*12364-85-1*]. Manganese tricarbide [*75718-05-7*] reacts with water to yield about 75% H_2, 12–15% CH_4, and 6–8% ethylene. It is an impor-

Table 3. Thermodynamic Data for Manganese Compounds at 25°C

Substance[a]	Heat of formation, ΔH_f, kJ/mol[b]	Free energy of formation, ΔG_f, kJ/mol[b]	Entropy, S, J/(mol·K)[b]
Mn^+ (g)	1002.9		
Mn^{2+} (g)	2512.1		
Mn^{2+} (aq)	− 222.6	−230.5	−71.1
$MnOH^+$ (aq)		−407.1	
$MnSO_3$ (aq)[c]	−1117.5	−987.8	37.7
Mn^{3+} (aq)		−87.9[d]	
MnO (aq)		−531.4[d]	
MnO_4^{2-} (aq)	−656.9	−504.2	58.6[d]
MnO_4^- (aq)	−543.5	−450.2	194.6
MnO	−384.9	−362.8	59.8
Mn_2O_3	−956.9	−879.1	110.5
Mn_3O_4	−1386.2	−1281.1	154.0
MnO_2	−520.5	−465.7	53.1
$Mn(OH)_2$	−698.7	619.2	96.2[d]
$Mn(OH)_3$	−924.7[d]	−795.0[d]	(100.4)
MnF_2	−795.0[d]	748.9[d]	93.1
$MnCl_2$	−483.7	−443.1	118.2
$MnBr_2$	−387.4	−372.4	(138.1)
MnI_2	−268.6	−272.0	(154.8)
$Mn(IO_3)_2$	−674.9	−522.6	255.2
MnS	−207.1	−211.3	78.2
MnS[e]	−213.4		
$MnSO_4$	−1066.5	958.6	112.1
$MnSO_4 \cdot H_2O$	−1377.4	1212.9	(154.8)
$Mn(NO_3)_2$	−698.7		
$Mn(NO_3)_2 \cdot 3H_2O$	−1485.3		
$Mn(NO_3)_2 \cdot H_2O$	−2371.9		
$Mn(NO_3)_2 \cdot 6H_2O$ (l)	−2331.7		
$MnCO_3$	−895.4	−820.0	85.8
$MnCO_3$[f]	−882.8	−811.7	113.00
K_2MnO_4	(−1179.9)		
$KMnO_4$	−839.3	−739.7	171.5
$BaMnO_4$	−1225.9	−1121.3	(154.8)
$Mn_2(CO)_{10}$	−1677.4		
$Mn_2(CO)_{10}$ (g)	−1614.6		

[a]Material is crystalline unless otherwise noted.
[b]To convert J to cal, divide by 4.184; items in parentheses are estimates.
[c]Ion pair.
[d]Approximate value.
[e]Alabandite.
[f]Solid precipitate.

tant factor in a fuel–alloy process designed to produce liquid hydrocarbons (qv) (19,20). In steel and other ferrous alloys, manganese carbides (qv) achieve the desired mechanical properties. With silicon, manganese forms a series of silicides, eg, Mn_3Si [*12163-59-6*], Mn_5Si_3 [*12033-10-2*], MnSi [*12032-85-8*], and $MnSi_{1.7}$. Manganese silicides have excellent heat-resisting properties. Manganese forms

Table 4. Standard Reduction Potentials for Selected Manganese Compounds

Reaction	Potential, E^0, V	Reference
Acid solution		
$Mn^{2+} + 2\,e^- \longrightarrow Mn^0$	−1.18	5
$Mn^{3+} + e^- \longrightarrow Mn^{2+}$	ca 1.5	6
$MnO_2\,(c) + 4\,H^+ + 2\,e^- \longrightarrow Mn^{2+} + 2\,H_2O$	1.23	7
$MnO_4^- + 8\,H^+ + 5\,e^- \longrightarrow Mn^{2+} + 4\,H_2O$	1.51	5
$MnO_4^- + 4\,H^+ + 3\,e^- \longrightarrow MnO_2 + 2\,H_2O$	1.70	5
Basic solution		
$Mn(OH)_2\,(c) + 2\,e^- \longrightarrow Mn^0 + 2\,OH^-$	−1.56	5
$MnO_2\,(c) + 2\,H_2O + 2\,e^- \longrightarrow Mn(OH)_2 + 2\,OH^-$	−0.05	5
$MnO_4^- + 4\,H_2O + 5\,e^- \longrightarrow Mn(OH)_2 + 6\,OH^-$	0.34	5
$MnO_4^- + 2\,H_2O + 3\,e^- \longrightarrow MnO_2 + 4\,OH^-$	0.59	5
$MnO_4^- + e^- \longrightarrow MnO_4^{2-}$	0.56	8
$MnO_2 + H_2O + 2\,e^- \longrightarrow Mn(OH)_2 + 2\,OH^-$	−0.05	5
$MnO_2 + H_2O + 2\,e^- \longrightarrow \gamma\text{-}MnOOH + OH^-$	0.19	5
$MnO_4^{-2} + e^- \longrightarrow MnO_4^{-3}$	0.27	5

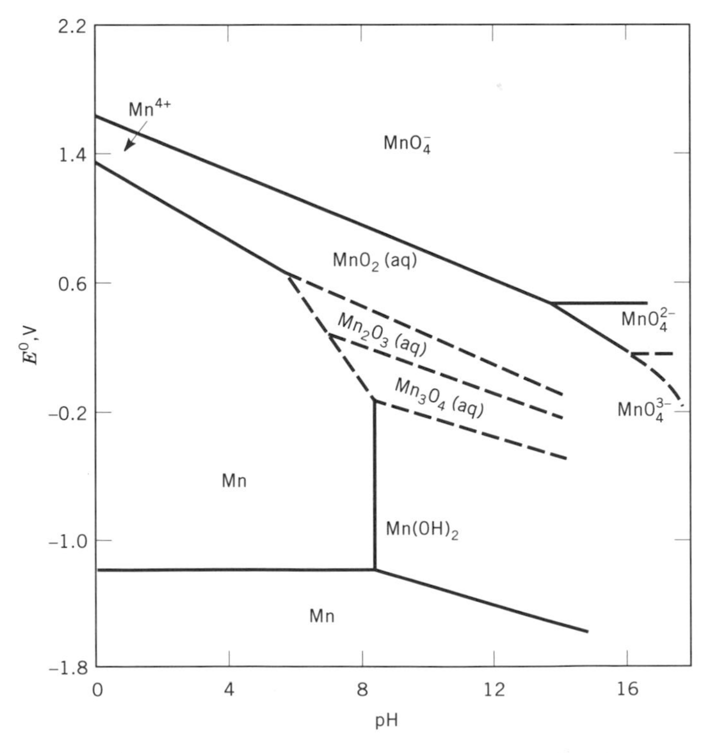

Fig. 1. pH potential diagram for manganese compounds at 25°C. Courtesy of Springer-Verlag (9).

Table 5. Manganese Minerals of Economic Significance

Mineral	CAS Registry Number	Approximate composition	Crystal	Mn%
pyrolusite	[*14854-26-3*]	β-MnO_2	tetragonal	63.2
braunite		$Mn^{2+}Mn_6^{3+}SiO_{12}$	orthorhombic	66.6
nsutite		$(Mn^{2+}, Mn^{3+}, Mn^{4+})(O, OH)_2$	hexagonal	
manganite	[*52019-58-6*]	γ-$MnOOH$	monoclinic	62
psilomelane	[*12322-95-1*]	$(Ba, Mn^{2+})_3(O, OH)_6Mn_{18}O_{16}$	monoclinic	45–60
cryptomelane	[*12260-01-4*]	$K_{1-2}Mn_8O_{16}{\cdot}xH_2O$	tetragonal	62
hausmannite	[*1309-55-3*]	Mn_3O_4	tetragonal	73
jacobsite		Fe_2MnO_4		23.8
bixbyite		$(Mn, Fe)_2O_3$		30–40
rhodonite	[*14567-57-8*]	$(Mn, Fe, Ca)SiO_3$		42
rhodochrosite	[*14476-12-1*]	$MnCO_3$		47.6
bementite	[*66733-93-5*]	$Mn_3Si_2O_5(OH)_4$		43.2
todorokite		$(Ca, Na, K)(Mg, Mn^{2+})$-$Mn_5O_{12}{\cdot}xH_2O$	monoclinic	
ramsdellite	[*12032-73-4*]	MnO_2		63

compounds only with a limited number of metals, ie, Au, Be, Zn, Al, In, Ti, Ge, Sn, As, Sb, Bi, Ni, and Pd. In the commercially important iron–manganese system, no compounds are formed.

Manganese metal reacts with many compounds (21). Although Mn is fairly stable against water at room temperature, a slow reaction accompanied by the evolution of hydrogen takes place at 100°C. Most dilute acids dissolve manganese at a fast rate. At 350–875°C, anhydrous ammonia converts Mn into nitrides. Concentrated alkalies, eg, KOH and NaOH, dissolve Mn metal at their boiling temperatures to form hydrogen and manganese(II) hydroxide.

Low Valent Manganese Compounds

A family of highly reduced metal carbonyls (qv) based on the anion $Mn(CO)_4^{3-}$ which contains manganese(−III), is obtained by the reduction of $Mn_2(CO)_{10}$ in the presence of sodium metal in hexamethylphosphoramide (22). A further example of −3 valent manganese is $Mn(NO)_3CO$ [*14951-98-5*], the volatile green crystals, mp 27°C, obtained by treating $Mn(CO)_5I$ [*14896-42-6*] with NO. The −2 oxidation state is represented by an anionic complex of Mn with phthalocyanine which can be made by reducing manganese(II) phthalocyanine [*14325-24-7*] with lithium in tetrahydrofuran (23). Manganese pentacarbonyl anion, $Mn(CO)_5^-$, which contains −1-valent Mn, is prepared by treatment of dimanganese decacarbonyl [*101070-69-1*] with sodium amalgam in tetrahydrofuran or with alcoholic potassium hydroxide.

The manganese(0) compound, $Mn_2(CO)_{10}$, has yellow crystals, mp 154–155°C, and sublimes *in vacuo* (24). The metal–metal bond is 29.3 pm and has an estimated strength of 142 kJ/mol (34 kcal/mol). This compound is obtained from the monovalent methylcyclopentadienylmanganese tricarbonyl [*12108-13-3*] (MMT), $C_9H_7Mn(CO)_3$, by reduction with sodium in diglyme in the presence of CO under pressure. Manganese carbonyl is the parent compound for a large family of manganese carbonyl compounds.

Compounds having manganese in the univalent positive state exist only as complexes. Reduction of sodium or potassium manganese(II) cyanides, eg, $K_4(Mn(CN)_6)$, in the presence of aluminum powder in alkaline solution under hydrogen, yields octahedral sodium or potassium manganese(I) cyanide, $M_5(Mn(CN)_6)$. The manganese(I) species have also been formed by electrolytic reduction of the manganese(II) cyanides. These colorless alkali metal salts are strong reducing agents. Sodium manganese(I) hexacyanide [*75535-10-3*] generates hydrogen from hot water. Monovalent Mn–cyanide complexes may also contain carbonyl or nitrosyl groups. The monovalent nitrate complex $MnONO_2(CO)_5$ [*14488-62-1*] can be obtained by the action of N_2O_4 on $Mn_2(CO)_{10}$. In the manufacture of MMT, methylcyclopentadienyl dimer is gradually added to a mixture of molten sodium metal and diethylene glycol dimethyl ether at 185–190°C to form methylcyclopentadienylsodium. After completion of this reaction, anhydrous flaked manganese chloride is added to the reaction mixture which is kept agitated at 165°C. The product formed, bis(methylcyclopentadienylmanganese), is subsequently treated with carbon monoxide at 4.3 MPa (625 psi) or 4.5 MPa (650 psi) at 193°C. The product is isolated from the reaction mixture by vacuum distillation (25). The complex MMT is a light amber liquid, mp 1.5°C, bp 233°C, having a density of 139 g/cm^3 at 20°C.

Divalent Manganese

Divalent manganese compounds are stable in acidic solutions but are readily oxidized under alkaline conditions. Most soluble forms of manganese that occur in nature are of the divalent state. Manganese(II) compounds are characteristically pink to colorless, with the exception of MnO and MnS which are green, and $Mn(OH)_2$, which is white. The physical properties of selected manganese(II) compounds are given in Table 6.

In neutral and acidic aqueous solution, the Mn^{2+} cation occurs as the pale pink hexaaqua complex $[Mn(H_2O)_6]^{2+}$. The rate of oxidation of dissolved divalent Mn cation to insoluble higher valent oxide hydrates accelerates with increasing hydroxyl ion concentration (see Fig. 1). Strong oxidizing agents such as permanganate, ozone, or hydrogen peroxide accomplish the oxidation at a lower pH than molecular oxygen, which requires a pH of 9.3–10, eg,

$$3\,Mn^{2+} + 2\,MnO_4^- + 2\,H_2O \longrightarrow 5\,MnO_2\,(s) + 4\,H^+$$

The dependence of the oxidizability of Mn^{2+} on the degree of alkalinity is an important factor in some processes for the removal of manganese from water and wastewater. Manganese(II) compounds are fairly stable, although the hydroxide and carbonate precipitated from alkaline solution tend to oxidize to MnO(OH).

In neutral solutions, the manganese salts of more common acids show very little hydrolysis, although hydrolysis may occur over a period of time. Slow precipitation of the hydroxide, which in turn oxidizes, occurs resulting in a brown precipitate being formed in the solutions. Manganese sulfate, $MnSO_4{\cdot}xH_2O$ where $x = 7, 5, 4, 2$, or 1, is the most stable common divalent manganese salt.

Table 6. Physical Properties of Manganese(II) Compounds

Compound	CAS Registry Number	Formula	Appearance	Crystal system and space group	Density,[a] g/cm^3	Mp, °C	Bp, °C	Solubility
manganese acetate tetrahydrate	[*15243-27-3*]	$Mn(C_2H_3O_2)_2 \cdot 4H_2O$	pale red crystals	monoclinic	1.589			sl sol H_2O, sol ethanol, methanol
manganese borate	[*12228-91-0*]	$MnB_4O_7 \cdot 8H_2O$	white to pale red solid					insol H_2O, ethanol, sol dil acids
manganese carbonate[b]	[*589-62-9*]	$MnCO_3$	pink solid	trigonal D_{3d}^6	3.125	dec > 200		sol prod H_2O: 8.8×10^{-11} sol in dil acids
manganese chloride	[*7773-01-5*]	$MnCl_2$	pink crystal solid	trigonal D_{3d}^5	2.977_{25}	650	1190	v sol H_2O, sol pyridine, ethanol, insol ether
manganese hydroxide[c]	[*18933-05-6*]	$Mn(OH)_2$	white to pink	hexagonal D_{3d}^3	3.26_{25}	dec 140		sol acid, sol base at higher temp
manganese nitrate hexahydrate	[*17141-63-8*]	$Mn(NO_3)_2 \cdot 6H_2O$	colorless to slightly pink crystals	rhombic D_{2h}^{16}	1.82	25.8	129.4	v sol, H_2O, sol ethanol
manganese(II) oxide[d]	[*1344-43-0*]	MnO	green	cubic Fm3m o_h^5	5.37_{23}	1945		insol H_2O
manganese sulfate	[*7785-87-7*]	$MnSO_4$	almost white crystal solid	orthorhombic D_{2h}^{17}	3.25	700	dec 850	sol 52 g/100 g H_2O, sl sol methanol, insol ether
manganese dihydrogen phosphate dihydrate	[*18718-07-5*]	$Mn(H_2PO_4)_2 \cdot 2H_2O$	almost colorless crystal solid four-sided prisms			$-H_2O$, 100		sol H_2O, insol ethanol, deliquescent

[a]Temperatures in °C of readings given as subscript.
[b]Also known as rhodochrosite [*14476-12-1*].
[c]Also known as pyrochroite [*1310-97-0*].
[d]Also known as manganosite [*1313-12-8*].

The color of the manganese ion is considered pink, although some solutions containing it are almost colorless. Many simple manganese salts can be obtained as hydrates, containing the pale pink cation $[Mn(H_2O)_6]^{2+}$. These hydrates are water soluble. Manganese(II) chlorides, nitrates, and sulfates are soluble in water, but the carbonates, hydroxides, and oxides are only sparingly soluble. A number of industrial processes require the purification of Mn(II) salt solutions, particularly those that have been prepared from gangue-bearing manganese ores. A useful technique employs the property that most other heavy-metal hydroxides, such as Fe(III) and Al, precipitate at lower pH than $Mn(OH)_2$ which precipitates at pH 8.3. Furthermore, in the presence of S^{2-} and OH^- ions, the sulfides of most other heavy metals precipitate at lower alkalinities than does manganese monosulfide [*18820-29-6*], MnS.

There are two manganese(II) sulfides, MnS and MnS_2. Manganese(II) disulfide contains a S–S bond and has a pyrite structure. When a solution of a manganous salt is treated with ammonium sulfide, a flesh-colored hydrated precipitate is formed which is comprised of MnS and $Mn(II)S_2$. This mixture very slowly changes to the more stable green-black MnS.

A number of complexes containing the manganese(II) ion are known. Acetoacetic ester, acetonedicarboxylic ester, salicylaldehyde, benzoyl acetone, and ethylenediaminetetraacetate (EDTA) are included among the number of compounds that form such complexes. All are pale pink or yellow solids insoluble in water, but soluble in organic solvents. Manganese(II) forms complex cyanides of the general type $M_4(Mn(CN)_6)$, where M is monovalent, that are similar to the ferrocyanides. These are readily soluble in water, but hydrolyze fairly rapidly. They also are readily oxidized to the trivalent complex $M_3^+(Mn^{3+}(CN)_6)$, which corresponds to the ferricyanides.

Manganese Carbonate. Manganese carbonate [*589-62-9*] occurs in nature as rhodochrosite, a reddish white to brown hexagonal–rhombohedral mineral having a specific gravity of 3.70. Pure manganese carbonate is a pink-white powder that is relatively insoluble. Rhodochrosite is the main constituent of the manganese ores used as the raw material for a portion of Japan's electrolytic manganese dioxide (EMD) production (see ELECTROCHEMICAL PROCESSING, INORGANIC).

Synthetic manganese carbonate is made from a water-soluble Mn(II) salt, usually the sulfate, by precipitation with an alkali or ammonium carbonate. The desired degree of product purity determines the quality of manganese sulfate and the form of carbonate to be used. For electronic-grade material, where the content of K_2O and Na_2O cannot exceed 0.1% each, the $MnSO_4$ is specially prepared from manganese metal, and ammonium bicarbonate is used (26) (see ELECTRONIC MATERIALS). After precipitation, the $MnCO_3$ is filtered, washed free of excess carbonate, and then, to avoid undesirable oxidation by O_2, dried carefully at a maximum temperature of 120°C.

Halides. The Mn^{2+} ion combines with each of the halogens to make compounds of the type MnX_2. The halides, except for the fluorides, are quite soluble, as are the corresponding hydrates. The chloride has hydrates of 2, 4, 6, and 8 waters and these are normally obtained by the action of hydrochloric acid on the oxides or carbonates, followed by crystallization and, if desired, dehydration. Complete dehydration is difficult to accomplish, but the anhydrous salt can be made directly by the action of HCl gas on heated manganese metal or Mn^{2+} oxide. Hydrochloric acid also reacts with higher valent manganese oxides to produce manganous chloride, but in this case chlorine is also generated. The bromide is similar to the chloride and forms the same series of hydrates. The iodide forms hydrates having 1, 2, 4, 6, or 9 waters and is very hygroscopic. These can only be dehydrated under vacuum because they lose iodine above 80°C. The fluoride is only slightly soluble. It may be prepared by action of hydrogen fluoride on metallic manganese or Mn^{2+} oxide (see FLUORINE COMPOUNDS, INORGANIC). A cyanide, $Mn(CN)_2$, is known, but this material is rapidly oxidized and has never been obtained in a pure state.

There are a number of complex chlorides of three general types: $M(MnCl_3)$, $M_2(MnCl_4)$, and $M_4(MnCl_6)$. M is monovalent in each case. Fluorine forms only $M(MnF_3)$ and the only complex bromine compound reported is $Ca(MnBr_4)\cdot 4H_2O$. There are no iodide complexes. The anhydrous salt, $MnCl_2$, forms cubic pink crystals, and three well-defined hydrates exist. Aqueous solubilities of the tetrahydrate and dihydrate are given in Table 7.

$$MnCl_2\cdot 6H_2O \xrightleftharpoons{-2°C} MnCl_2\cdot 4H_2O \xrightleftharpoons{58°C} MnCl_2\cdot 2H_2O \xrightleftharpoons{198°C} MnCl_2$$

In the presence of moist air, $MnCl_2$ vapor decomposes into hydrochloric acid and manganese oxides.

Manganese chloride can be prepared from the carbonate or oxide by dissolving it in hydrochloric acid. Heavy-metal contamination can be removed by precipitation through the addition of manganese carbonate which increases the pH. Following filtration, the solution can be concentrated and upon cooling, crystals of $MnCl_2\cdot 4H_2O$ are collected. If an anhydrous product is desired, dehydration in a rotary dryer to a final temperature of 220°C is required. Anhydrous manganese chloride can also be made by reaction of manganese metal, carbonate or oxide, and dry hydrochloric acid.

Manganese chloride is manufactured by Chemetals Corp. using a process in which manganese(II) oxide is leached with hydrochloric acid. Manganese carbonate is added after completion of initial reaction to precipitate the heavy-metal impurities. Following filtration of the impurities, the solution is concentrated and cooled and the manganese chloride is isolated. Gradual heating in a rotary dryer above 200°C gives anhydrous manganese chloride (27,28).

For top quality $MnCl_2\cdot xH_2O$ grades, the starting material is manganese metal or high purity MnO. To make anhydrous $MnCl_2$ directly, manganese metal or ferromanganese is chlorinated at 700–1000°C. Any $FeCl_3$ initially present in the product is removed by sublimation (29).

Manganese Nitrate. Manganese nitrate [*10377-66-9*] is prepared from manganese(II) oxide or carbonate using dilute nitric acid, or from MnO_2 and

Table 7. Solubility of Manganese Chloride in Water, g/100 g

	Temperature, °C						
Compound	0	20	40	50	60	80	100
$MnCl_2\cdot 4H_2O$	63.4	73.9	88.6	98.2			
$MnCl_2\cdot 2H_2O$					108.6	112.7	115.3

a mixture of nitrous and nitric acids. $Mn(NO_3)_2$ exists as the anhydrous salt [*10377-66-9*]; the monohydrate [*3228-81-9*]; trihydrate [*55802-19-2*], mp 35.5°C; tetrahydrate [*20694-39-7*]; and hexahydrate [*17141-63-8*], deliquescent needles having mp 25.8°C. Manganese nitrate is very soluble in water and decomposes at 180°C yielding manganese dioxide and oxides of nitrogen. The commercial product consists of 61 and 70% solutions.

Manganese(II) Oxide. The MnO found in nature as manganosite can be prepared by the reduction of higher valent manganese oxides by either thermal means or by the use of a reducing agent such as CO, H_2, or carbon at elevated temperatures. Manganese oxide, insoluble in water, readily dissolves in mineral acids. Manganese(II) oxide is bright green in color and is moderately stable in air. Manganese monoxide absorbs oxygen and the oxidation state of the manganese changes progressively according to the increased oxygen absorption to Mn_2O_3 (30). The activity of manganese oxide toward oxygen absorption is enhanced by low temperatures employed in its formation. Manganese(II) oxide is an important precursor of such manganese compounds as $MnSO_4$, $MnCl_2$, and EMD, as well as being an ingredient in fertilizer and feedstuff formulations (see FERTILIZERS; FEED AND FEED ADDITIVES). It is made from manganese dioxide ores through a reductive roasting process (31–33).

The Chemetals process (34) uses a stationary bed of crushed (particle size <10 mm) manganese ore which is continuously replenished from the top and into which a reducing gas, eg, CH_4 and air, is introduced from the bottom. The MnO is formed from the MnO_2 in a reaction zone immediately beneath the top layer where the temperature is controlled to 760–1040°C to avoid sintering. The manganese oxide moves downward and finally passes through an inert atmosphere cooling zone before it is ground to <74 μm. A significant portion of the MnO_2 is already thermally decomposed to Mn_2O_3 before it comes in contact with the reducing gas, thus economizing on fuel.

Other processes use rotary kilns for the reduction step. Before roasting, the finely ground MnO_2 ore is mixed with the reducing agent, such as hydrogen, carbon monoxide producer gas, or heavy (Bunker C) oil. Thus, in order to obtain better manganese yields (98–99%), the roasting temperature is kept at 800°C, even though some of the iron is also reduced to the acid-soluble FeO (35).

Manganese Hydroxide. Manganese hydroxide [*18933-05-6*] is a weakly amphoteric base having low solubility in water. $Mn(OH)_2$ crystals are reported to be almost pure white and darken on exposure to air. Manganese dihydroxide occurs in nature as the mineral pyrochroite and can also be prepared synthetically by reaction of manganese chloride and potassium hydroxide that is scrupulously free of oxygen. The entire reaction is conducted under reducing conditions (36).

Manganese Sulfate. Manganese sulfate [*7785-87-5*] is made by dissolving manganese carbonate ore (rhodochrosite) or manganese(II) oxide in sulfuric acid. The solubility of manganese sulfate in water is given in Table 8. The tetrahydrate readily converts into the monohydrate upon gentle heating; $MnSO_4 \cdot H_2O$ [*31746-59-5*] is stable up to 280°C. Manganese sulfate may also be produced by treating finely ground manganese dioxide with sulfuric acid and a reducing agent:

$$2\,MnO_2 + 2\,H_2SO_4 + C \xrightarrow{\text{air}} MnSO_4 \cdot H_2O + CO_2$$

Table 8. Aqueous Solubility of Manganese Sulfate, g/g sat'd sol'n

	Temperature, °C										
	0	20	26.1	31	40	50	60	70	80	90	100.7
$MnSO_4$	34.6	38.6	39.4	40.4	37.5	36.3	34.9	38.2	31.3	29	26.1

Manganese Ethylenebis(thiocarbamate). Maneb [*12427-38-2*], $(C_4H_6MnN_2S_4)_x$ (**1**), is a yellow powder used as a leaf and soil fungicide (see FUNGICIDES, AGRICULTURAL). Maneb is obtained by treating disodium ethylenebis-(dithiocarbamate) with an aqueous solution of maganese(II) sulfate (37).

```
 [ H    H   S            ]
 [  \   |   ||           ]
 [   C—N—C—S\            ]
 [ H/ |          Mn      ]
 [   C—N—C—S/            ]
 [ H/ |   |   ||         ]
 [    H   H   S          ]x
```

(**1**)

Trivalent Manganese

The Mn^{3+} ion is so unstable that it scarcely exists in aqueous solution. In acidic aqueous solution, manganic compounds readily disproportionate to form Mn^{2+} ions and hydrated manganese(IV) oxide, $MnO_2{\cdot}2H_2O$; in basic solution these compounds hydrolyze to hydrous manganese(III) oxide, MnO(OH). Sulfuric acid concentrations of about 400–450 g/L are required to stabilize the noncomplexed Mn^{3+} ion in aqueous solutions.

In the solid phase the most stable forms of Mn(III) are manganese sesquioxide [*1317-34-6*], Mn_2O_3, and its hydrate $Mn_2O_3{\cdot}nH_2O$, and manganese oxide, Mn_3O_4, which is thermally the most stable manganese oxide. Physical properties of manganese(III) compounds are given in Table 9.

Compounds of trivalent manganese can be made either by oxidation of corresponding manganous compounds or by reduction of the more highly oxidized compounds. The color of Mn(III) compounds in the solid state can vary from red to green. The corresponding aqueous solutions mostly have a reddish purple, almost permanganate-like appearance. Simple cationic Mn(III) compounds are olive green $MnPO_4{\cdot}H_2O$ [*14986-93-7*], dark green $Mn_2(SO_4)_3$ [*13444-72-9*], and reddish purple MnF_3.

Manganese(III) fluoride, MnF_3, a red salt, can be made by reaction of fluorine and manganese(II) iodide. The dihydrate, $MnF_3{\cdot}2H_2O$, can be made by dissolving manganese(III) oxide, Mn_2O_3, or preferably the hydrated MnO(OH), in hydrofluoric acid. Manganese(III) chloride, a brown salt giving a green solution in organic solvents, is only stable below −40°C, above which it decomposes to $MnCl_2$ and chlorine. It is formed by treatment of a suspension of MnO_2 in ether with HCl gas below −70°C and precipitating with carbon tetrachloride. The bromide and iodide, even as complexes, are not known.

The brown crystalline manganese(III) acetate dihydrate is of considerable commercial importance because it is often used as the source material for

Table 9. Physical Properties of Manganese(III) Compounds

Compound	CAS Registry Number	Formula	Appearance	Crystal system and space group	Density, g/m^3	Mp, °C	Solubility
trimanganese tetraoxide[a] α-phase[b]	[*1317-35-7*]	Mn_3O_4	black crystals having metallic sheen	tetragonal D_{4h}^{19}	4.84	1560	insol H_2O
manganese(III) acetate dihydrate	[*19513-05-4*]	$Mn(C_2H_3O_2)_3 \cdot 2H_2O$	cinnamon brown crystal solid				dec H_2O
manganese(III) acetyl-acetonate	[*14284-89-0*]	$Mn(C_5H_7O_2)_3$	brown to black crystal solid			172	insol H_2O, sol org solv
manganese(III) fluoride	[*7783-53-1*]	MnF_3	red crystals	monoclinic C_{2h}^{6}	3.54	dec (stable to 600)	dec H_2O
α-manganese(III) oxide	[*1317-34-6*]	Mn_2O_3	black to brown solid	rhombic (also cubic) D_{2h}^{15}	4.89_{25}	871–887 dec	insol H_2O
γ-manganese(III) oxide, hydrated	[*1332-64-3*]	MnO(OH)	black solid	monoclinic C_{2h}^{5}	4.2–4.4	250 dec to Mn_2O_3	insol H_2O, disproportionates in dilute acids

[a]Mixed Mn(II), Mn(III) valent compound.
[b]Also known as hausmannite [*1309-55-3*].

other trivalent manganese compounds. It can be made by oxidation of manganese(II) acetate using chlorine or potassium permanganate, or by reaction of manganese(II) nitrate and acetic anhydride.

Although the manganese(III) ion is more acidic than the manganese(II), and hence forms fewer stable simple salts, Mn(III) has a greater tendency to form stable complex compounds. Complex chlorides, chloromanganates of the type $M_2(MnCl_5)$ where M is a monovalent cation, are dark red and rather easily prepared, either by reduction of permanganate and addition of alkaline chloride, or from Mn_2O_3 and an alkaline chloride in aqueous HCl at 0°C. In water these complexes hydrolyze. Complex fluorides are of the type $M(MnF_4)$ and $M_2(MnF_5)$. These are obtained as dark red crystals by adding alkali fluorides to a solution of MnF_3.

Complex phosphates of two types, $MH_2(Mn(PO_4)_2)$ and $M_3(Mn(HPO_4)(OH)_2)H_2O$, are known. The complex with pyrophosphate $(Mn_2(H_2P_2O_7)_3)^{3-}$ is the stable product in the potentiometric determination of manganese. Manganese(III) does not coordinate with amines or nitro complexes, but it does make manganicyanides of the types $M_3(Mn(CN)_6)$ and $M_3(Mn(CN)_5(OH))$, which are similar to the ferricyanides. The K^+, Na^+, Li^+, and NH_4^+ manganicyanides have been prepared and slowly hydrolyze in water to MnO(OH).

Manganese(III) Oxides. The sesquioxide, Mn_2O_3, exists in dimorphic forms. The α-Mn_2O_3 exists in nature as the mineral bixbyite. Synthetic α-Mn_2O_3 is prepared by the thermal decomposition of the nitrate, dioxide, carbonate, oxalate, or chloride in air in the temperature range of 500–800°C. Heating above 940°C, it is reduced to Mn_3O_4. Higher temperatures or the presence of reducing agents allow the reduction to continue to MnO. The gamma form is unstable and thus does not occur in nature. γ-Mn_2O_3 may be synthesized by the dehydration of the precipitated hydrated form at 250°C in vacuum.

The mixed valent oxide Mn_3O_4 occurs in nature as the mineral hasumannite. The structure of this ferromagnetic material has been the subject of much dispute. Mn_3O_4 is the most stable of the manganese oxides, and is formed when any of the other oxides or hydroxides are heated in air above 940–1000°C. The oxidation of aqueous solutions of $Mn(OH)_2$ can also lead to the formation of Mn_3O_4.

As seen in some iron oxides, both γ-Mn_2O_3 and Mn_3O_4 have pseudo-spinel structure and tetragonal symmetry. Because of the close similarities between the structures of these two oxides the x-ray diffraction patterns are nearly identical. Both oxides can be represented by the general formula $Mn^{II}Mn_2^{III}O_4$ where the Mn(II) ions occupy tetrahedral sites and the Mn(III) ions occupy the octahedral sites of the spinel (38). Upon heating Mn_3O_4 above 1170°C, a reversible transition from the tetragonal to a cubic structure occurs. The weakly acidic manganese sequioxide forms oxo anions with alkali metal hydroxides. Thus potassium manganate(III) [*12142-17-5*], $KMnO_2$, is prepared by heating Mn_2O_3 in the presence of KOH. Needle-shaped gray crystals that are readily oxidized by O_2 to higher alkali manganates are formed (39).

Tetravalent Manganese

By far the most significant manganese(IV) compound is the dioxide [*1313-13-9*] MnO_2, found in nature as pyrolusite, a black mineral. There is also a hydrated form approximating $MnO_2 \cdot 2H_2O$ which is formed by precipitation from solutions. The dioxide is a reasonably good conductor. Its specific conductance is 0.16 Ω^{-1} at 0°C. Manganese dioxide is seldom stoichiometric and is insoluble in water. This lack of aqueous solubility is responsible for much of its stability, because the Mn(IV) ion is unstable in solution. Physical properties of manganese(IV) compounds are given in Table 10.

In acid solution, MnO_2 is an oxidizing agent, and is used as such in industry. The classic example is the oxidation of chloride in HCl, which has been a convenient means of chlorine generation, both in the laboratory and in the old Weldon process for the manufacture of chlorine.

$$MnO_2 + 4\,HCl \longrightarrow MnCl_4 + 2\,H_2O$$
$$MnCl_4 \xrightarrow{-1/2\,Cl_2} MnCl_3 \xrightarrow[\text{heat}]{-1/2\,Cl_2} MnCl_2$$

Most of the simple halides, MnX_4, are unknown except for blue manganese tetrafluoride [*15195-58-1*]. Manganese(IV) is amphoteric, appearing as the cation in salts and as an anion in compounds known as manganites, M_2MnO_3, where M is monovalent. The manganites are salts of the very weak manganous acid, H_2MnO_3, or hydrated dioxide, $MnO_2 \cdot 2H_2O$, which can be obtained by reduction of potassium permanganate in slightly alkaline solutions, then dissolved in concentrated alkalies to give the manganites. The manganites may also be prepared by fusing MnO_2 and the oxide of the desired cation.

The principal complexes of Mn^{4+} are of the type K_2MnX_6, where X may be fluoride, chloride, cyanide, or iodate, and as a group these materials are readily hydrolyzed. The hexafluoride, K_2MnF_6, a yellow hexagonal crystal, can be formed by treating either a manganate(V) or (VI) with concentrated HF. The chlorides, M_2MnCl_6, M monovalent, have dark red crystals, and are formed by reaction between permanganate and HCl gas in glacial acetic acid. A similar cyanide complex, $K_2Mn(CN)_6$, is said to be made by treatment of potassium permanganate using potassium cyanide in a saturated aqueous solution.

A triperoxymanganate(IV), $K_2H_2MnO(O_2)_2$, is said to be formed (40) when $KMnO_4$ in 30% KOH is treated with H_2O_2 at −18°C. It is a dark red-brown crystalline compound, which in water slowly evolves O_2 and precipitates MnO_2. In the dry state this material is explosive above 0°C, but under acetone at −60°C, it is stable for several days.

The oxidation of Mn(II) to MnO_2 using oxygen can be accelerated by certain microorganisms such as *Pedomicrobium manganicum* or *Hyphomicrobium manganoxydans*. This biochemical oxidation can occur in the pH range of 5.5–7.5, in contrast to the pH range of 8.5–10 normally required for the chemical oxidation of Mn(II) by O_2. Microbial action is assumed to have been involved in the genesis of some manganese ore deposits and could also contribute to the clogging of water distribution systems with manganese dioxide deposits (41).

Water purification chemistry depends on the oxidation of Mn(II) by a suitable oxidant, eg, O_2 at $pH > 9$, or at lower pH potassium permanganate or ozone,

Table 10. Physical Properties of Manganese(IV) Compounds

Compound	CAS Registry Number	Formula	Appearance	Crystal system and space group	Density, g/cm^3	Mp, °C	Solubility
pentamanganese octaoxide[a]	[*12163-64-3*]	Mn_5O_8	black solid	monoclinic C_{2h}^3	4.85_{20}	550 dec to a Mn_2O_3	insol H_2O
β-manganese(IV) oxide	[*14854-26-3*]	MnO_2	black to gray crystal solid	tetragonal D_{4h}^{14}	5.026	535	insol H_2O
potassium manganate(IV)	[*12142-27-7*]	K_2MnO_3	black microscopic crystals		3.071_{25}	1100	dec H_2O, disproportionates

[a]Mixed valent Mn(II), Mn(IV) compound.

resulting in the formation of hydrous manganese dioxide. The colloidal properties of hydrous manganese oxide have been studied (42) and exploited in the treatment of potable water. As shown in Figure 2 (43), the surface of a hydrous MnO_2 particle in a colloid solution has an outer layer of exposed OH groups capable of adsorbing charged species, such as H^+, OH^-, and metal ions. Hydrous manganese dioxide is amphoteric and can participate in surface–solution exchanges of H^+ and OH^-. The zero point of charge for hydrous manganese dioxide is in the range of pH 2.8–4.5. In the absence of other ions, the overall surface charge is largely determined by the pH of the solution. The sorptive and ion-exchange characteristics of MnO_2 solutions, as well as the parameters controlling coagulation, have been studied in some detail (44).

Manganese Oxides. Manganese(IV) dioxide rarely corresponds to the expected stoichiometric composition of MnO_2, but is more realistically represented by the formula $MnO_{1.7-2.0}$, because it invariably contains varying percentages of lower valent manganese. It also exists in a number of different crystal forms, in various states of hydration, and with a variety of contents of foreign ions.

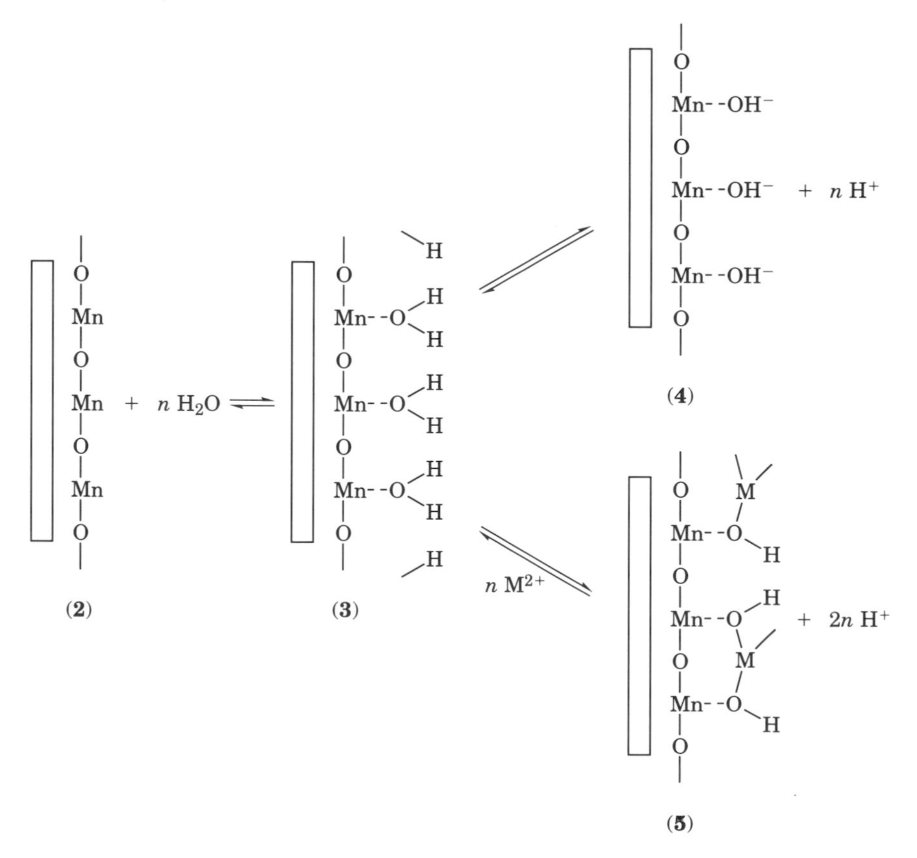

Fig. 2. Hydration of the surface structure of manganese dioxide (**2**) and subsequent reactions of hydrous manganese oxide (**3**) showing proton transfer (**4**) and ion exchange (**5**) where M^{2+} represents a divalent metal ion.

The structure of manganese dioxide can be described as being composed of MnO_6 octahedra where each manganese ion is coordinated to the six nearest oxygen atoms, each of which is located at the vertex of an octahedron. The solid structure is then composed of MnO_6 octahedra, which form chains by sharing edges and cavities or tunnels. The various chains are cross-linked by sharing corners (vertices). Thus a hierarchy can be constructed based on increased complexity of crystal structure. Examples of natural and synthetic manganese dioxides are listed in Table 11. Structures are shown in Figure 3

α-MnO_2 is isostructural to the minerals hollandite [*12008-99-0*], psilomelane [*12322-95-1*], cryptomelane [*12260-01-4*], and coronadite [*12414-82-3*]. Its 2 × 2 tunnel structure is thought to require stabilization at the centers by large cations. Typically, compounds having the general formula MnO_x, where x ranges from 1.88–1.95 are found, but x can be as low as 1.6 for the psilomelane family (47). A highly crystalline α-MnO_2 has been prepared from Li_2MnO_3 by a high (ca 90°C) temperature acid leaching process (48).

The mineral pyrolusite (**8**) is the primary example of β-MnO_2, having the general formula MnO_x where x ranges from 1.95–2.00. This mineral is characterized by a high degree of crystallinity and an almost perfect stoichiometric composition. The beta group can be visualized as being composed of single chains of edge-shared MnO_6 octahedra, cross-linked to neighboring chains, through corner sharing of oxygen atoms (Fig. 3).

The crystal structure of ramsdellite [*12032-73-4*] is similar to that of β-MnO_2 except that double chains of MnO_6 octahedra are cross-linked to adjacent double chains through the sharing of oxygen atoms located at the corners. Ramsdellite and pyrolusite are the only manganese dioxide phases where the composition approaches the stoichiometric MnO_2 formula. Heating ramsdellite to 250°C transforms it to pyrolusite.

The thermal transitions of the manganese dioxides can be illustrated by the beta phase which undergoes successive reductions and corresponding loss of oxygen as the temperature is increased (49).

$$\beta\text{-}MnO_2 \xrightarrow{600-700^\circ C} \alpha\text{-}Mn_2O_3 \xrightarrow{900-1000^\circ C} Mn_3O_4 \xrightarrow{>1250^\circ C} MnO$$

Table 11. Manganese Dioxide Crystal Phases[a]

Phase	Example mineral	Structure number	Description
α-MnO_2	hollandite group	(**6**) and (**7**)	tunnel structure of corner-shared double chains of (MnO_6) octahedra (2 × 2 channels)
β-MnO_2	pyrolusite	(**8**)	single chains of edge-shared (MnO_6) octahedra (1 × 1 channels)
γ-MnO_2	nsutite	(**9**)	regions of single and double chains of edge-shared (MnO_6) octahedra (both 2 × 1 and 1 × 1 channels)
			layers of edge-shared (MnO_6) octahedra
δ-MnO_2	ramsdellite	(**10**)	double chains of edge-shared (MnO_6) octahedra (2 × 1 channels)

[a]Structures are shown in Fig. 3; Ref. 45.

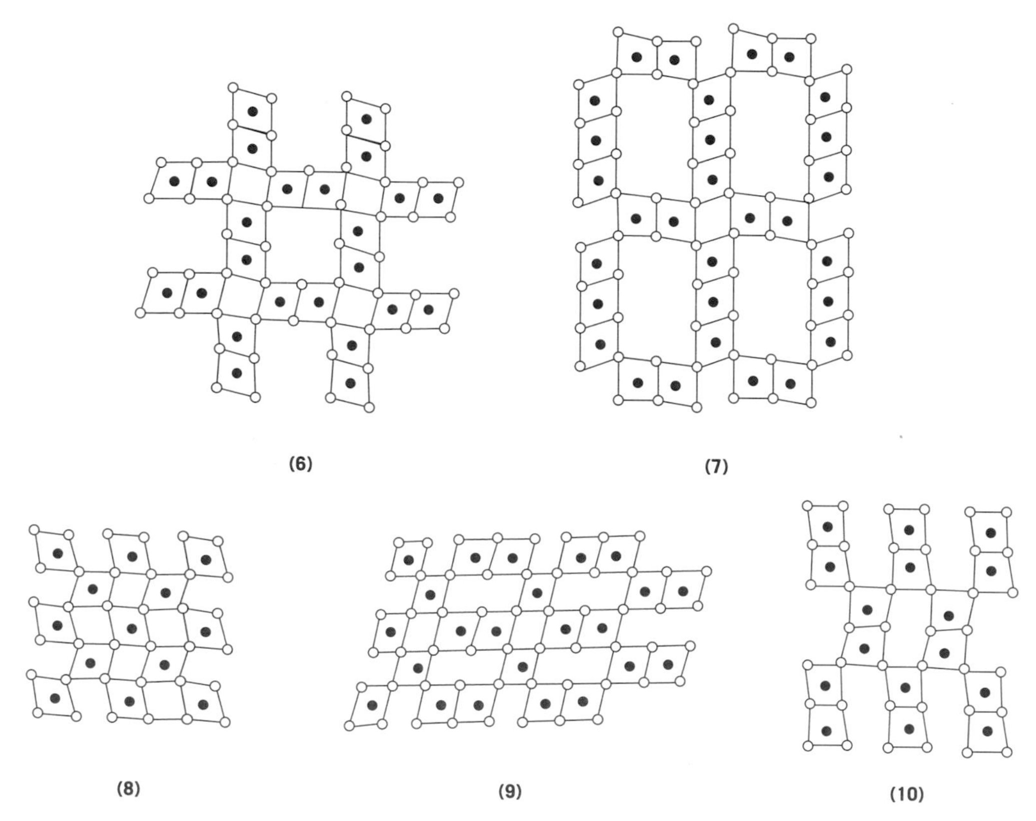

Fig. 3. Crystal structure of manganese dioxides where (•) represent Mn and (∘), O^{2-}; and the lines define the octahedra; hollandite (**6**), psilomelane (**7**), pyrolusite (**8**), nsutite (**9**), and ramsdellite (**10**) (46).

Manganese dioxides typically lose adsorbed water between 25–105°C. The complete removal of interlayer water, which results in collapse of the crystal lattice, can, however, require temperatures as high as 150–250°C. In the γ-MnO_2 phase the condensation of OH groups occurs in the temperature range of 105–500°C, resulting in the release of this most tightly bound water and collapse of the crystal lattice. The various manganese dioxide phases eventually transform into β-MnO_2, as shown in Figure 4, suggesting the relative thermodynamic stability of the beta phase (50).

The term γ-manganese dioxide is applied to a series of hydrated manganese dioxides of moderate crystallinity that are suitable for battery purposes. These occur in nature as the mineral nsutite and have optimum activity for use in dry cell batteries. The gamma group can be visualized as being composed of irregular structural combinations of β-MnO_2 (single-chain) and ramsdellite (double-chain) components.

The electrochemically active phases of manganese dioxide, ie, gamma and rho, typically contain approximately 4% by weight chemically bonded or structural water. The chemical and electrochemical reactivity of manganese dioxide has been shown to result from the presence of cation (51) vacancies in the manganese dioxide crystal lattice. The cation vacancy model suggests the general

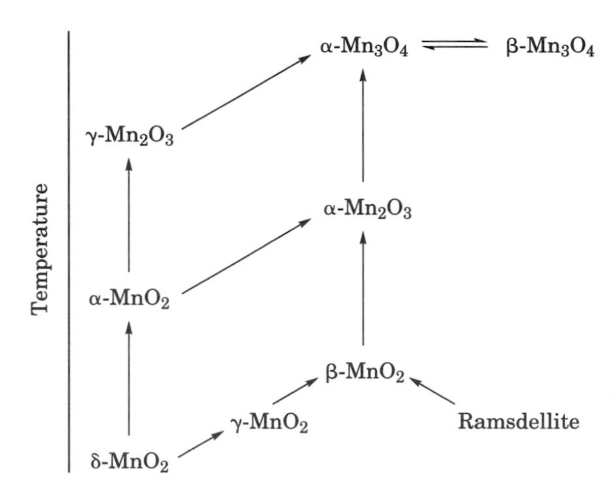

Fig. 4. Thermal transformation of manganese oxide phases.

formula for this type of manganese dioxide to be $Mn^{4+}_{(1-x-y)} \cdot Mn^{3+}_{y} \cdot O^{2-}_{(2-4x-y)} \cdot OH^{-}_{(4x+y)}$ where x is the fraction of Mn^{4+} ions missing in the crystal lattice, and y, the fraction of Mn^{4+} ions replaced by Mn^{3+} in the crystal lattice. Each vacant Mn^{4+} site is coordinated to four groups present in the lattice as OH^- ions that have replaced O^{2-} in the crystal lattice. Additional replacement of O^{2-} in the crystal lattice is caused by the charge compensation determined by the fraction of Mn^{4+} that has been replaced by Mn^{3+}. Substitution of Mn^{4+}, having radius, r (=53 pm) by Mn^{3+} (r = 64.5 pm) leads to lattice expansion. The model suggests that electrochemical reactivity is caused by manganese lattice vacancies and a corresponding water content that provides the initial concentration of protons for the transfer process.

The delta phase is generally applied to characterize a group of amorphous essentially noncrystalline hydrous manganese dioxides. Typical minerals belonging to the δ-manganese dioxide group include birnessite [*1244-32-5*], ranceite, takanelite, toddorokite, woodruffite, chalcophanite, and lithioporite. This phase is often labeled as hydrated manganese dioxide because it typically contains 5–40% of chemically bound water. These dioxides have values of x in the range of 1.7–2.0 for the general formula MnO_x and often also contain Na^+, Ca^{2+}, or Ba^{2+}.

The oxidation of divalent manganese, or the reduction of potassium permanganate, leads to the formation of δ-manganese dioxide. These manganites generally follow the composition $M_2O{\cdot}4MnO_2$, where M is an alkali metal cation. This alkali metal cation is exchangeable with other metal cations. The absorption of alkaline-earth cations by δ-manganese dioxide is shown to be similar to that reported for colloidal hydrous manganese dioxides (52). Absorption capacity increases in the series $Mg^{2+} < Ca^{2+} < Sr^{2+} < Ba^{2+}$. Absorption has been explained by an ion-exchange mechanism (Fig. 5) involving the manganese dioxide surface and hydrated divalent cations in the bulk solution. The electrostatic attraction between the solid surface and the cations is expected to increase as the radius of the hydrated cations decreases. For the alkaline-earth cations, the increase occurs as the ionic radius decreases (Table 12) (53). Transition metals

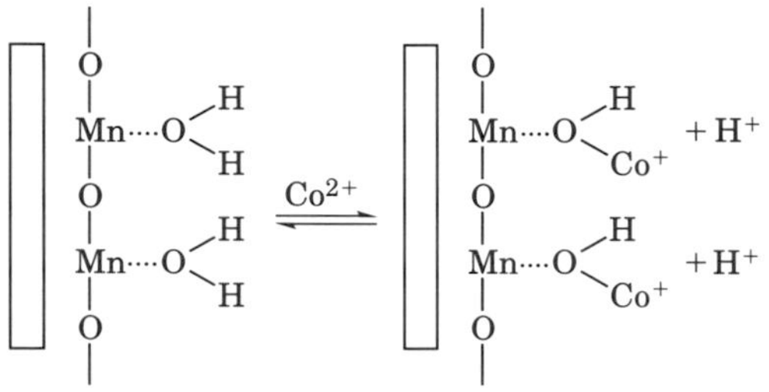

Fig. 5. Ion-exchange properties of hydrous manganese dioxide.

Table 12. Langmuir Constants for the Absorption of Alkaline-Earth Cations on Manganese Dioxide[a,b]

Ion	Ionic radius, pm	Colloidal hydrous manganese dioxide, X_m,[a] mol/mol	Hydrated radius, pm	δ-MnO_2,[c] X_m μmol/g
Mg^{2+}	65	0.100	297	58
Ca^{2+}	99	0.113	286	63
Sr^{2+}	113	0.135	277	65
Ba^{2+}	135	0.180	269	117

[a]Refs. 52 and 53.
[b]Absorption experiments conducted at pH 7.
[c]δ-MnO_2 powder exhibits a BET surface area of ca 4 m^2/g, and a zero point of charge at pH 3.33 ± 0.5.

were found to exhibit the highest degree of absorption on δ-manganese dioxides. Selectivity increased in the series Co ≥ Mn > Zn > Ni > Ba > Sr > Ca > Mg (54). The interaction of metal ions with the hydrous manganese dioxide involves the release of a proton from the surface.

Hydrous manganese oxides of composition $MnO_{1.9-1.95}$ having x-ray diffraction patterns suggesting a low degree of crystallinity characteristic of δ-MnO_2, were found to remain colloidally dispersed for several months, and to exhibit Mn(II) absorption (55) in the range of 0.2–0.3 mol/mol MnO_2 in the pH range of 5–7. This value increased to an excess of 0.5 mol Mn(II) per mol MnO_2 in the slightly alkaline range. δ-Manganese dioxides can be oxidized under alkaline conditions to potassium manganate.

The effect H^+ and OH^- ions exert on a manganese dioxide surface can be summarized by the pH at which there is zero surface charge (pH_{pzc}). This point of zero charge (pzc) for the various manganese dioxide phases increases in the series δ-MnO_2 < α-MnO_2 ≈ β-MnO_2 <γ-MnO_2. Values of pH_{pzc} are 1.5–3.0, 4.5, 4.6, and 3.3–5.0, respectively (56–58).

Synthetic active manganese dioxides, prepared by the reduction of permanganate or the pyrolysis of lower valent manganese salts, have been used as mild, selective, heterogeneous oxidation reagents. The amorphous manganese dioxides are mild enough to be used for the oxidation of α,β-unsaturated alcohols and unsaturated polyene alcohols into the corresponding α,β-unsaturated carbonyl compounds; the oxidation of benzylic alcohols into the corresponding aldehydes (qv) and ketones (qv); and the oxidative coupling of phenols (59).

Manganese dioxide, in combination with other metal oxides, forms a series of active catalysts (60) that participate in a variety of environmentally important oxidation and decomposition reactions. The manganese-based catalysts for these applications exhibit a long life and high catalytic activity. At moderately elevated temperatures, manganese dioxide catalysts (61,62) are used for the complete oxidative degradation of many organic compounds. These catalysts are particularly effective for oxygenated compounds such as alcohols, acetates, and ketones. At a contact time of ca 0.24 s, 95% hydrocarbon destruction efficiency is achieved for ethanol at 204°C; ethyl acetate, 218°C; propanol, 216°C; propyl acetate, 238°C; 2-butanone, 224°C; toluene, 216°C; and heptane, 316°C. Carbon monoxide oxidation occurs at ambient temperatures when no H_2O is present. A contact time of ca 0.36 s gives $>95\%$ destruction efficiency. Ozone decomposition, at ambient temperatures, that is $>99\%$ efficient requires a contact time of ca 0.72 s. Manganese dioxide also catalyzes the decomposition of H_2O_2 at room temperature and of alkali metal chlorates at about 270°C.

Manufacture of manganese metal or compound requires the manganese dioxide of the natural ores to be reduced to lower oxides, principally MnO or other Mn(II) salts. This reduction is usually carried out at 600–900°C by roasting finely powdered MnO_2 mixed with ground coal or heavy oil. Alternatively, the presence of gaseous reductants such as carbon monoxide or hydrogen is employed. Reduction to the Mn(II) state and conversion of the oxide to the desired salt can also be performed in the liquid phase, in a single step, by employing acids which have reducing properties. Hydrochloric acid forms $MnCl_2$ and Cl_2, sulfurous acid gives $MnSO_4$ and MnS_2O_6 [*13568-72-4*], and NO_2 gives $Mn(NO_3)_2$ [*10377-66-9*].

Oxidation of manganese dioxide to higher valence states takes place in the fusion process of MnO_2 and KOH. A tetravalent manganese salt identified as K_2MnO_3 [*12142-27-7*] (63) which disproportionates spontaneously is formed.

$$2\ K_2MnO_3 \longrightarrow KMnO_2 + K_3MnO_4$$

Both of these manganates can be further oxidized. K_2MnO_3 is black and has a density of 3.071 g/cm^3.

Synthetic Manganese Dioxides. Chemical manganese dioxide (CMD) can be prepared by various methods including the thermal decomposition of manganese salts such as $MnCO_3$ or $Mn(NO_3)_2$ under oxidizing conditions. CMD can also result from the reduction of higher valent manganese compounds, eg, those containing the MnO_4^- ion.

A commercially practiced process (64) for the recovery of high purity manganese dioxide based on thermal decomposition results in a well-defined crystalline manganese dioxide exhibiting a purity greater than 99.5%. The process involves adjusting the pH of an aqueous manganese nitrate solution using manganese oxide to a pH of 4.8–5.0. The mixture is heated to 105°C, filtered, and combined with a previously prepared manganese dioxide. The manganese nitrate decomposition occurs between 135–146°C under vigorous agitation and controlled at a rate of 0.24–0.61 kg of manganese dioxide product per liter of slurry per day.

$$Mn(NO_3)_2 \longrightarrow MnO_2 + 2\ NO_2$$

The NO_2 generated can be recycled by reaction with water:

$$2\ NO_2 + H_2O \longrightarrow HNO_2 + HNO_3$$

The acid formed then reacts with low grade manganese dioxide ore (65) forming a slurry of impure manganese nitrate to feed the process.

$$HNO_2 + HNO_3 + MnO_2 \longrightarrow Mn(NO_3)_2 + H_2O$$

In another commercial process, finely ground manganese dioxide ore is reduced to manganese(II) oxide using H_2 or CO (66). The MnO is then leached with sulfuric acid and the manganese sulfate solution neutralized (pH 4–6) and filtered to remove impurities. The addition of $(NH_4)_2CO_3$ precipitates manganese(II) carbonate which is recovered by filtration. The $MnCO_3$ is heated in air at about 450°C resulting in 80% conversion to MnO_2:

$$MnCO_3 + 1/2\ O_2 \longrightarrow MnO_2 + CO_2$$

The manganese dioxide and manganese carbonate mixture is further leached with sulfuric acid and oxidized with $NaClO_3$. Following washing and drying, the product is found to have a manganese content of approximately 60%, 90% as MnO_2. This battery-grade manganese dioxide is manufactured by Sedema (Belgium) at a scale of ca 20,000 t/yr, and sold under the trade name Faradiser M. This CMD contains approximately 90% MnO_2, has a surface area of ca 80 m^2/g, an ion-exchange capacity in the range of 1.5–1.6 meq, and a density in the range of 1.3–1.5 g/cm^3. It has been used in magnesium batteries (qv) produced exclusively for military applications, and as a replacement for EMD in Leclanché cells.

The starting material for activated native ore is a higher quality native MnO_2 ore. It is the objective of the activation process to chemically remove the top layer of the ore particle and to create a new highly porous and chemically active surface. For this purpose, the MnO_2 is first reduced to Mn_2O_3 either thermally, by heating it to 600–800°C in air, or at about 300°C in the presence of a reducing agent. The reduced mass is then treated with hot sulfuric acid and the Mn_2O_3 disproportionates to a highly active γ-MnO_2 and $MnSO_4$:

$$Mn_2O_3 + H_2SO_4 \longrightarrow \gamma\text{-}MnO_2 + MnSO_4 + H_2O$$

The newly formed γ-MnO_2 actually coats the surfaces of the particles of the solid phase; the $MnSO_4$ dissolves in the liquid phase, along with the majority of the ore impurities. The effective surface area is expanded by the etching action of the sulfuric acid. Following the acid treatment step, the slurry is filtered and the cake is carefully washed and dried at a controlled temperature.

The oxidation of *o*-toluenesulfonamide under alkaline conditions by potassium permanganate results in sulfonbenzimide (saccharin) and a hydrated manganese dioxide. The co-produced chemical manganese dioxide (CMD) can be washed and carefully dried to produce a product containing approximately 79%

MnO_2, 5–18% water, and alkali contents of 4–12% as KOH (67). The manganese dioxide powder, Permanox, has been characterized as a poorly crystalline form of δ-MnO_2. Permanox exhibits apparent densities in the range of 0.7–1.1 g/cm^3 and BET surface area at 30 m^2/g.

A thermally stable, pure todorokite has been synthesized by autoclaving a layered structured manganese oxide, initially generated from the reaction of MnO_4^- and Mn^{2+} under alkaline conditions. The synthetic manganese oxide molecular sieve (**11**) was shown to have a tunnel size, ie, diameter of 690 pm. This material was thermally stable to 500°C just as natural todorokite is (68).

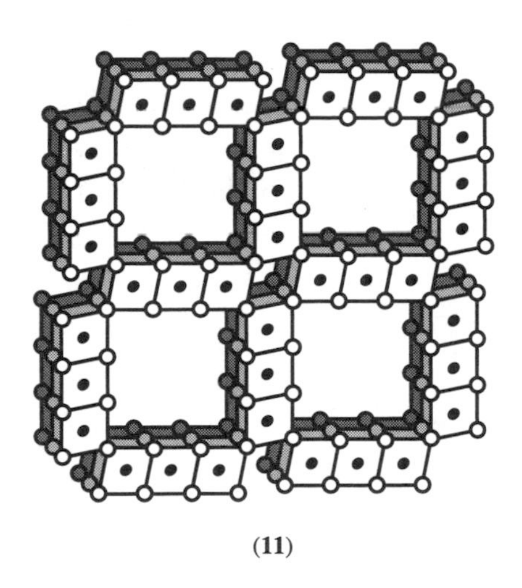

(11)

Electrolytic Manganese Dioxide. The anodic oxidation of an Mn(II) salt to manganese dioxide dates back to 1830, but the usefulness of electrolytically prepared manganese dioxide for battery purposes was not recognized until 1918 (69). Initial use of electrolytic manganese dioxide (EMD) for battery use was in Japan (70) where usage continues.

The properties of EMD are summarized in Table 13. Historically EMD was assumed to consist essentially of gamma-phase manganese dioxide, but more recent reports suggest the predominance of the epsilon phase (71). EMD is strictly a nonstoichiometric manganese dioxide containing 2–5% lower valent manganese oxides and 3–5% chemically bound water. The Ruetschi vacancy model for EMD assumes a vacancy fraction of 0.06, a Mn^{3+} fraction of 0.08, and a structure intermediate between ramsdellite and ϵ-manganese dioxide. This suggests a true EMD density of 4.55 g/cm^3, a value in agreement with experimental measurements (72).

EMD is prepared from the electrolysis of acidified manganese sulfate solution and can be summarized as follows:

Anode $$Mn^{2+} + 2\,H_2O \longrightarrow MnO_2 + 4\,H^+ + 2\,e^-$$

Cathode $$2\,H^+ + 2\,e^- \longrightarrow H_2$$

Overall $$Mn^{2+} + 2\,H_2O \longrightarrow MnO_2 + 2\,H^+ + H_2$$

Table 13. Properties of Electrolytic Manganese Dioxide

Property	Typical ranges
density, g/cm^3	4.2–4.5
bulk density, g/cm^3	1.7–2.5
particle size, μm	< 74
surface area, m^2/g	30–60
MnO_2, wt %[a]	92

[a]Minimum value.

Deposition of MnO_2 from a solution containing Mn cations on the anode is not considered the primary electrode process. Initially the Mn(III) ion is formed on the anode (73). MnO_2 formation arises from Mn(III) disproportionation:

$$Mn^{3+} + 2\,H_2O \longrightarrow MnO_2 + Mn^{2+} + 4\,H^+$$

At the anode, sulfuric acid is generated. This acid is then recycled into the leaching of $MnCO_3$ or of MnO. Although two side reactions, namely, the evolution of O_2 and the generation of Mn^{3+}, have been identified, the current efficiency on the anode is greater than 95%. The manganese dioxide formed by electrolysis is generally considered to be of the gamma variety, but at low acid concentration, some β-MnO_2 may be co-produced, and conversely, at high acid concentration, the formation of the α-MnO_2 phase is favored. The steps involved in the manufacture of EMD using either rhodochrosite or manganese dioxide ores are summarized in Figure 6.

The use of rhodochrosite ore is primarily practiced in Japan (74) and involves the use of an ore concentrate, containing about 20–25% Mn, having a particle size of $< 150\ \mu m$. Typical composition of the ore on a wt % basis is $MnCO_3$, 58–68; Mn, 30–35; SiO_2, 15–20; Fe, 2–3; Pb, 0.1–0.2; Zn, 0.2–0.3; Ni, 0.1; and Co, 0.01 (75). The starting material is treated with about 10% excess (100–150 g/L) sulfuric acid at 80–90°C according to the following reaction:

$$MnCO_3 + H_2SO_4 \longrightarrow MnSO_4 + H_2O + CO_2$$

Any divalent iron present in the reaction mixture is oxidized through the addition of finely ground MnO_2. The pH of the slurry is then adjusted to the range of 4–8 using $Ca(OH)_2$ or $CaCO_3$, which results in the precipitation of $Fe(OH)_3$ and other minor impurities such as Pb, Ni, and Co. The majority of the Ca separates as $CaSO_4$ upon cooling. The solids are removed by filtration and the filtrate is added to the cell liquor, displacing an equivalent volume of the electrolyte which is regenerated with $MnCO_3$. The composition of the electrolyte is maintained at about 80–180 g/L $MnSO_4$ and 50–100 g/L H_2SO_4 at 80–88°C. Anodes are composed of Ti, Pb, or graphite and operate at an anode current density of 0.7–1.2 A/dm^2 (see METAL ANODES).

EMD is prepared from manganese dioxide ore containing a minimum of 75% MnO_2. Initially the MnO_2 in the ore is reduced to MnO in rotary kilns,

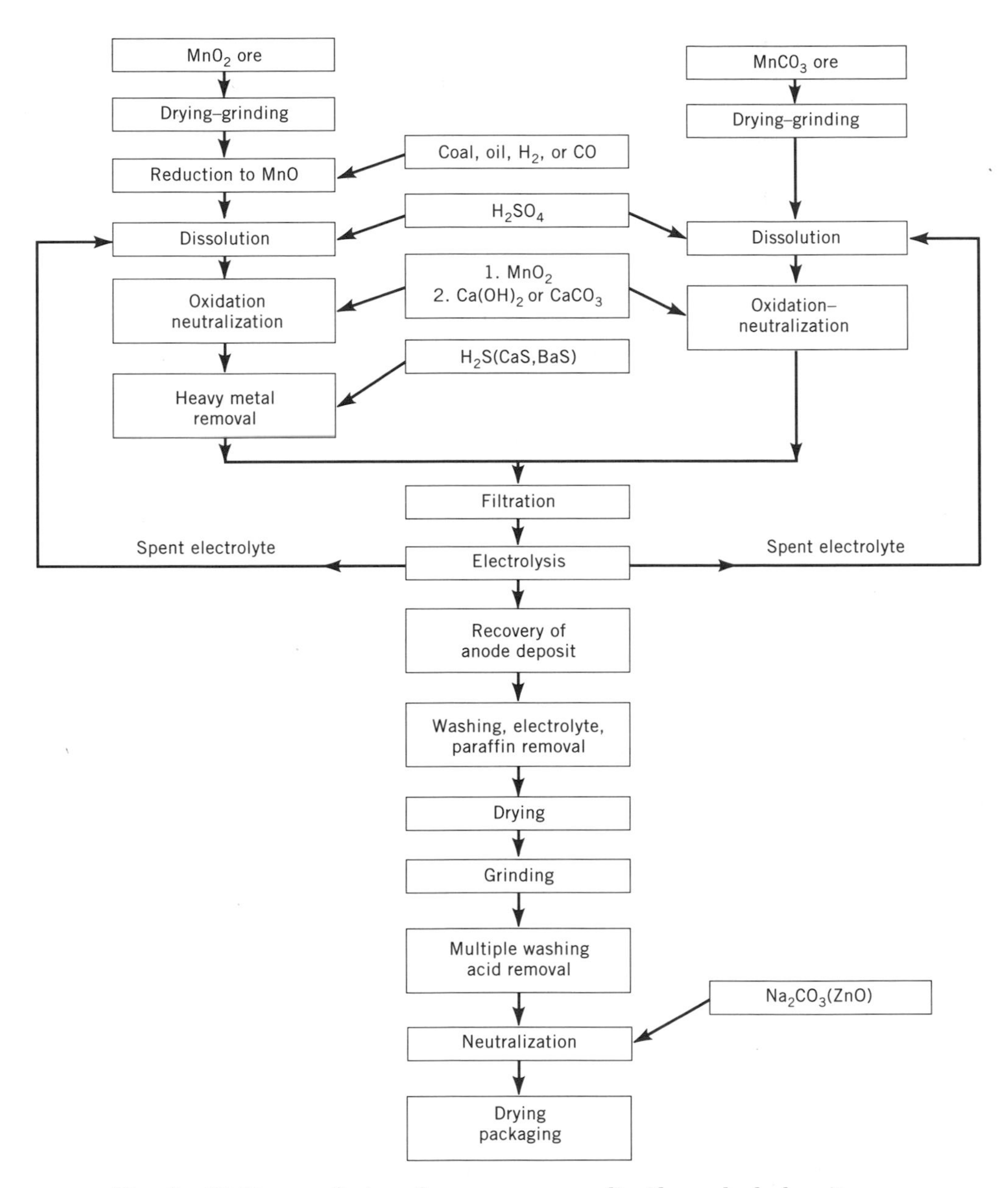

Fig. 6. EMD manufacture from manganese dioxide or rhodochrosite ore.

or with natural gas or hydrogen (76) by pile roasting (77), at about 700–900°C. Conditions can be controlled so that MnO_2 is reduced in preference to Fe_2O_3. Following the reduction, the solid mixture is allowed to cool to below 100°C in an inert or reducing atmosphere, to prevent the oxidation of the MnO.

$$MnO_2 + C \longrightarrow MnO + CO$$

The leaching step with H_2SO_4 is essentially identical to the one in use for rhodochrosite, except that owing to the generally higher potassium content in MnO_2 ores an additional purification step or steps may be required. For example, as the sulfuric acid level of the reaction mixture decreases, as a result of the leaching reaction, to approximately 0.1 *M* H_2SO_4, precipitation of potassium jarosite occurs (78).

$$K^+ + 3\ Fe^{3+} + 2\ SO_4^{2-} + 6\ H_2O \longrightarrow KFe_3(SO_4)_2(OH)_6 + 6\ H^+$$

Potassium removal is required because the presence of potassium during electrolysis reportedly promotes the formation of the α-MnO_2 phase which is nonbattery active. Neutralization is continued to a pH of approximately 4.5, which results in the precipitation of additional trace elements and, along with the ore gangue, can be removed by filtration. Final purification of the electrolyte liquor by the addition of sulfide salts results in the precipitation of all nonmanganese transition metals.

The cells are usually rectangular open steel or concrete troughs, lined with a corrosion-resistant nonconductive material such as fiber-reinforced plastic, rubber, concrete, or acid-proof brick. A design of a covered cell (79) has a hermetically sealed top with a rupture disk as a safety feature. The electrodes are flat or corrugated plates, cylindrical rods, or tubes, and the anodes are easily removed for stripping of the EMD deposits. The ratio between the effective anode and cathode areas is about 1:2, the spacing between anode and cathode ranges from 25 to 50 mm. Most cells are equipped with heating devices for the electrolyte.

The anodes can be made of graphite which tolerates high current densities without passivation, but are subject to gradual corrosive attack causing a lowering of its mechanical strength. Graphite anodes are said to be good for about 300 days of operation before they break in the EMD-removal step (80). Titanium is the dominant anode material because of its mechanical stability. This anode is better suited for the use of automated EMD-stripping systems and produces an EMD well suited for alkaline batteries. Titanium anodes have a life of several years, but are expensive. Furthermore, they tend to passivate, if current densities and H_2SO_4 concentrations are not properly controlled. Performance might be improved by application of specific surface coatings.

Cathodes are made from graphite, soft or hard lead, or copper. A production-size cell (81) may contain 44 anode units, each comprising five graphite plate electrodes 25-mm thick, 175-mm wide, and 1100-mm long.

The electrolysis is conducted at 90–95°C and an anode current density of about 50–120 A/m^2 when using lead alloy anodes and lead cathodes. Using graphite electrodes, the current density is from 70–100 A/m^2; using titanium anodes and graphite cathodes, the current density is 50–80 A/m^2 (82).

A layer of oil or paraffin wax is floated on top of the electrolyte to minimize heat and water losses. Cell voltages vary between 1.8 and 2.5 V. In one mode of operation (83) the electrolyte is circulated through the cell at a rate of about 3% of the total volume per minute with the solution usually being introduced at the bottom of the cell. The electrolyte is replenished every one to two hours by drawing off about 10–20% of the total electrolyte volume for treatment with $MnCO_3$ or MnO, followed by filtration, and is then returned to the electrolysis cycle.

In an alternative procedure (84), the electrolyte is pumped through the cells at such a rate that the outlet concentration is 50 g/L $MnSO_4$ and 67 g/L H_2SO_4. This spent electrolyte is then mixed with equal parts of make-up solution containing 150 g/L $MnSO_4$ and the mixture returned to the electrolysis step. The electrolysis is continued over a period of days and terminated when the EMD layer deposited on the anode reaches a specific thickness, usually on the order of 1–3 or 6–8 mm. Following completion of the electrolysis cycle, the entire electrode assembly is removed from the cell for removal of the deposited EMD, either manually or by an automated system (85). The product is repeatedly washed with water to extract the occluded acid (83) and dried at about 85°C in air.

The current efficiency in the electrolysis step is 90–95%, making the energy requirements on the order of 0.45 kWh/kg of EMD. Oxygen is generated at the anode as the principal by-product. This can become significant in electrolytes having high sulfuric acid concentration. At the cathode, a mole of hydrogen is generated for each mole of EMD produced.

The main disadvantage of producing EMD in the form of a solid deposit on the anode is that the process must be interrupted to remove the anodes from the cells. Stripping the MnO_2 is a labor-intensive operation. A number of continuous processes have been devised to generate the MnO_2 as a precipitate that collects at the bottom of the cell where it can be removed without interruption of the electrolysis (86–89).

Newer technology, the suspension-bath process, allows small manganese dioxide particles to adsorb on the electrode surface, thus allowing for higher current densities to be used during electrolysis (90). The resulting manganese dioxide has a macroporous structure, even though some of the micropore area (BET surface area = 10–25 m^2/g) is lost (82).

Manganese(V) Compounds

Manganese(V) appears to exist only as the oxyanion MnO_4^{3-} and is generally referred to as manganate(V); occasionally the term hypomanganate is used. The pentavalency of manganese was only recognized in the 1940s (91). Historically, the compounds were considered mixtures of Mn^{4+} and Mn^{6+}. Selected manganese(V) compounds and their physical properties are given in Table 14.

The alkali manganates(V) in strongly alkaline solution (45–50% at 0°C) are all blue. In water these manganate(V) compounds instantly disproportionate into manganate(VI) and MnO_2. Lithium manganate(V), prepared by reaction of $LiMnO_4$ and excess LiOH at 124°C, is an exception. This compound is relatively stable in 3% LiOH solution at 0°C and in absolute methyl alcohol.

The most important manganese(V) compound is K_3MnO_4, a key intermediate in the manufacture of potassium permanganate. Potassium manganate(V) is an easily crystallized salt obtained by reduction of potassium permanganate using sodium sulfite in strong sodium hydroxide solution. This was the first compound to be recognized as exclusively pentavalent.

In the early days of $KMnO_4$ manufacture, the yield was only two-thirds of the theoretical; the yield of $NaMnO_4$ never exceeded one-half theoretical. It is now known that the formation of manganate(VI) from MnO_2 passes through a

Table 14. Physical Properties of Manganese(V) Compounds

Compound	CAS Registry Number	Formula	Appearance	Crystal system and space group	Density, g/m^3	Mp, °C	Solubility
barium manganate(V)	[*12231-83-3*]	$Ba_3(MnO_4)_2$	emerald green crystals	rhombic D_{3d}^5	5.25	dec 960	insol H_2O
lithium manganate(V)	[*12201-25-1*]	Li_3MnO_4				dec > 125	sol 3% LiOH at 0°C
potassium manganate(V)	[*12142-41-5*]	K_3MnO_4	turquoise blue microscopic crystals		2.78	dec 800–1100	v sol H_2O dec, hygroscopic, sol 40% KOH at −15°C
rubidium manganate(V)	[*12438-62-9*]	Rb_3MnO_4					
sodium manganate(V)	[*12163-41-6*]	Na_3MnO_4	bluish, dark green microscopic crystals	orthorhombic D_{2h}^{16}		dec 1250	v sol H_2O, dec, hygroscopic
strontium manganate(V)–strontium hydroxide		$Sr_3(MnO_4)_2 \cdot Sr(OH)_2$					insol H_2O

manganate(V) step.

$$2\ MnO_2 + 6\ KOH + 1/2\ O_2 \longrightarrow 2\ K_3MnO_4 + 3\ H_2O$$

$$2\ K_3MnO_4 + H_2O + 1/2\ O_2 \longrightarrow 2\ K_2MnO_4 + 2\ KOH$$

$$MnO_2 + 2\ KOH + 1/2\ O_2 \longrightarrow K_2MnO_4 + H_2O$$

The reaction conditions favoring the formation of manganate(V) do not favor the oxidation of K_3MnO_4 to K_2MnO_4. This latter requires a lower temperature, lower KOH concentration, and higher H_2O concentration.

Although sodium manganate(V) can be made from NaOH and MnO_2 in an oxidizing melt, it is not possible to oxidize Na_3MnO_4 to Na_2MnO_4 using oxygen. Extracting a Na_3MnO_4 melt with water results in disproportionation to Na_2MnO_4 and manganese dioxide.

The intermediate formed in the oxidation of alkenes by permanganate ion is considered a cyclic manganate(V) ester (92). Investigations have suggested that manganate(V) intermediates play a significant role in virtually all permanganate oxidation reactions. It is therefore the further reactions of the manganate(V) intermediates, either oxidation or disproportionation, that are responsible for the dependence of the reaction rates and product composition based on basicity of the system (93). Additionally, a manganese(V) peroxo complex, considered to be a faster oxidizing agent than permanganate, is postulated as an intermediary in the photochemical decomposition of permanganate ion in solution (94).

Manganese(VI) Compounds

The hexavalent state of manganese is represented by a few alkali metal and alkaline-earth metal salts of manganic acid [*54065-28-0*], H_2MnO_4, which is known only through its sodium, potassium, rubidium, cesium, barium, and strontium salts. Properties of a few of these salts are given in Table 15. In the laboratory, manganate(IV) salts are prepared by heating an aqueous solution of alkali metal permanganate in the presence of an excess of strong alkali, excluding carbon dioxide (95).

$$4\ KMnO_4 + 4\ KOH \xrightarrow{\text{reflux}} 4\ K_2MnO_4 + 2\ H_2O + 2\ O_2$$

In a similar fashion, Na_2MnO_4 is made from $NaMnO_4$ and NaOH. Rubidium manganate(VI) [*25583-21-5*] and cesium manganate(VI) [*25583-22-6*] are prepared by mixing a solution of Na_2MnO_4 and a concentrated solution of RbOH or CsOH, respectively, whereby the less soluble Rb and Cs salts precipitate. The practically insoluble barium manganate(VI) can be made by decomposition of $KMnO_4$ in $Ba(OH)_2$ solution, or by adding $BaCl_2$ to a solution of K_2MnO_4, leading to the precipitation of $BaMnO_4$.

Potassium manganate(VI), precursor of potassium permanganate, is made commercially in either a one- or two-stage fusion reaction, or by anodic oxidation of manganese metal or ferromanganese in KOH. X-ray crystallographic studies on potassium manganate(VI) obtained from the reduction of potassium permanganate suggest the manganate(VI) forms orthorhombic crystals having the

Table 15. Physical Properties of Manganese(VI) Compounds

Compound	CAS Registry Number	Formula	Appearance	Crystal system and space group	Density, g/m^3	Mp, °C	Solubility
barium manganate(VI)	*[7787-35-1]*	$BaMnO_4$	small green to black crystals	rhombic D_{2h}^{16}	4.85	dec 1150	insol H_2O sol product 2.46×10^{-10}
potassium manganate(VI)	*[10294-64-1]*	K_2MnO_4	dark green to black crystals	orthorhombic D_{2h}^{16}	2.80_{23}	dec 190	sol H_2O dec, sol KOH
sodium manganate(VI)	*[15702-33-7]*	Na_2MnO_4	small dark green needles	rhombic		dec 300	sol H_2O dec

Pnma space group. There are four molecules in the unit cell. Cell dimensions are $a = 766.7$ pm, $b = 589.5$ pm, and $c = 1035.9$ pm. The average Mn–O bond distance is 165.9 ± 0.8 pm and the O–Mn–O bond angle is 109.5 ± 0.7°, suggesting that the structure is of a regular tetrahedron. The increase in the Mn–O bond length of approximately 3.0 pm over the permanganate ion is consistent with the molecular orbital view of the manganate ion (96), where the extra electrons are considered to occupy an antibonding orbital.

In aqueous solution, K_2MnO_4 is stable when the KOH concentration is $>1\ M\ [OH^-]$ and in a less alkaline environment, disproportionation (97) occurs according to the following:

$$3\ MnO_4^{2-} + 2\ H_2O \longrightarrow 2\ MnO_4^- + MnO_2 + 4\ OH^-$$

This disproportionation is slow under less than molar alkaline conditions, and instantaneous under neutral or acidic conditions. The equilibrium constant, K_{eq}, for the reaction at 25°C, is as follows (98):

$$K_{eq} = [MnO_4^-]^2[OH^-]^4/[MnO_4^{2-}]^3 = \text{ca } 16$$

The solubility of K_2MnO_4 in 2 M potassium hydroxide at 20°C is 225 g/L and decreases with increasing alkali concentration (99). Potassium manganate is insoluble in organic solvents. Pure potassium manganate(VI) is thermally stable to 600°C. Manganate(VI) can be converted to manganate(VII) by either disproportionation or by oxidation using hypochlorite, ozone, or anodic oxidation.

Alkali manganate(VI) salts are used as oxidants in synthetic organic reactions (100) and their reactions have been observed to be similar to permanganate, except that manganate(VI) exhibits lower reactivity. Additionally, solid $BaMnO_4$ in methylene chloride has been reported to achieve high yields for the oxidation of diols to dialdehydes (101).

Manganese(VII) Compounds

Permanganic acid [*13465-41-3*], $HMnO_4$, is conveniently prepared in the laboratory from barium permanganate and sulfuric acid, or by anodic oxidation of ferromanganese in a divided cell using H_2SO_4 as the electrolyte. The acidity of $HMnO_4$ is comparable to that of HNO_3. Aqueous solutions up to 20% are relatively stable and dilute solutions can be heated to boiling without decomposition. Solutions of permanganic acid can be distilled. This acid is assumed to undergo vapor-phase dissociation into Mn_2O_7 and H_2O. Anhydrous $HMnO_4$ and the dihydrate have been isolated (102), but the monohydrate does not appear to be stable, even at low temperatures and pressures. Anhydrous permanganic acid is hygroscopic and soluble in water, slightly soluble in perfluorodecalin, and insoluble in carbon tetrachloride and chloroform. Anhydrous $HMnO_4$ is a powerful oxidant decomposing violently at 3°C; $HMnO_4{\cdot}2H_2O$ [*24653-70-1*] decomposes at 18°C. Aqueous solutions of permanganic acid below a concentration of 3 wt % are stable over time, whereas in the concentration range of 5–15% $HMnO_4$, the decomposition rate increases with increasing initial solution concentration at room temperature (103).

The anhydride of permanganic acid, manganese heptoxide [*12057-92-0*], Mn_2O_7, is a viscous oil which looks metallic green in reflected light and red in transmitted light (Table 16). It is prepared by titrating solid potassium permanganate with concentrated sulfuric acid at 25–35°C. Manganese heptoxide is hygroscopic and forms $HMnO_4$ with water. This anhydride is hazardous. Its sensitivity to shock and temperature is comparable to that of mercury fulminate [*628-86-4*] (see MERCURY COMPOUNDS). Manganese heptoxide reacts with many oxidizable materials such as alcohol, ether, sulfur, and phosphorus with explosive force.

Manganese(VII) compounds containing the cationic permanganyl ion, MnO_3^+, are known. For example, permanganyl fluoride [*15586-97-7*], MnO_3F, is obtained as a green vapor when $KMnO_4$ reacts with anhydrous HF. The highly unstable gaseous green-violet permanganyl chloride [*15605-27-3*], MnO_3Cl, is prepared by passing dry HCl gas through a solution of $KMnO_4$ in concentrated sulfuric acid at −50°C (104).

The purple permanganate ion [*14333-13-2*], MnO_4^-, can be obtained from lower valent manganese compounds by a wide variety of reactions, eg, from manganese metal by anodic oxidation; from Mn(II) solution by oxidants such as ozone, periodate, bismuthate, and persulfate (using Ag^+ as catalyst), lead peroxide in acid, or chlorine in base; or from MnO_4^{2-} by disproportionation, or chemical or electrochemical oxidation.

The overall decomposition of solid potassium permanganate in the temperature range of 250–300°C leads to the formation of a delta-manganese dioxide (shown in brackets) and can be represented as follows:

$$10\ KMnO_4 \xrightarrow{250-300°C} 2.65\ K_2MnO_4 + [2.35\ K_2O \cdot 7.35\ MnO_{2.05}] + 6\ O_2$$

The enthalpy of decomposition has been determined to be approximately 10 kJ/mol (2.4 kcal/mol) of $KMnO_4$ (105). The decomposition has been shown to occur in two stages. In the first stage essentially all of the $KMnO_4$ decomposes into $K_3(MnO_4)_2$ and δ-MnO_2 accompanied by the release of oxygen; in the second stage the $K_3(MnO_4)_2$ decomposes forming additional δ-MnO_2 and oxygen.

Aqueous potassium permanganate solutions are not perfectly thermodynamically stable at 25°C, because MnO_2, not MnO_4^-, is the thermodynamically stable form of manganese in water. Thus permanganate tends to oxidize water with the evolution of oxygen and the deposition of manganese dioxide, which acts to further catalyze the reaction.

$$4\ MnO_4^- + 4\ H^+ \longrightarrow 4\ MnO_2 + 2\ H_2O + 3\ O_2\ (g)$$

The kinetics of the reaction are relatively slow and permanganate solutions exhibit greatest stability around a neutral pH. The decomposition rates increase below pH 3 or above pH 10. Potassium permanganate solutions are stable at elevated temperatures, up to approximately 3 *N* sodium hydroxide, above which decomposition into manganate occurs.

Table 16. Physical Properties of Manganese(VII) Compounds

Compound	CAS Registry Number	Formula	Appearance	Crystal system and space group	Density, g/m^3	Mp, °C	Solubility
potassium manganate(VI), manganate(VII) double salt	[*12362-73-1*]	$KMnO_4 \cdot K_2MnO_4$	dark, small hexagonal plates	monoclinic D_{3d}^5			sol H_2O dec
manganese heptoxide	[*12057-92-0*]	Mn_2O_7	dark red oil		2.396_{20}	5.9	v sol H_2O, hygroscopic
ammonium permanganate	[*13446-10-1*]	NH_4MnO_4	dark purple rhombic bipyramidal needles	rhombic D_{2h}^{16}	2.22_{25}	dec > 70	8 g/100 g H_2O at 15°C
barium permanganate	[*7787-36-2*]	$Ba(MnO_4)_2$	dark purple crystals	rhombic D_{2h}^{24}	3.77	dec 95–100	72.4 g/100 g H_2O at 25°C
calcium permanganate tetrahydrate	[*7789-81-3*]	$Ca(MnO_4)_2 \cdot 4H_2O$	black crystals, solutions look purple		2.49	dec 130–140	388 g/100 g H_2O at 25°C, deliquescent
cesium permanganate	[*13456-28-5*]	$CsMnO_4$	dark purple rhombic bipyramidal prisms or needles	orthorhombic at RT and atmospheric pressure	3.579	dec 320	0.23 g/100 g H_2O at 20°C
lithium permanganate	[*13452-79-7*]	$LiMnO_4 \cdot 3H_2O$	long, dark purple needles	hexagonal	2.06_{25}	dec 190	71 g/100 g H_2O at 16°C
magnesium permanganate hexahydrate	[*10377-62-5*]	$Mg(MnO_4)_2 \cdot 6H_2O$	bluish gray crystals	rhombic C_{2v}^7	2.18	dec 130	v sol H_2O, sol CH_3OH, pyridine, glac acetic acid
potassium permanganate	[*7722-64-7*]	$KMnO_4$	dark purple bipyrimidal rhombic prisms	orthorhombic D_{2h}^{16}	2.703_{20}	dec 200–300	sol H_2O, acetic acid, trifluoroacetic acid, acetic anhydride, acetone, pyriding, benzonitrile, sulfolane
rubidium permanganate	[*13465-49-1*]	$RbMnO_4$	dark purple rhombic bipyramidal prisms	orthorhombic D_{2h}^{16}	3.23_{25}	dec 250	1.1 g/100 g H_2O at 19°C
silver permanganate	[*7783-98-4*]	$AgMnO_4$	dark purple	monoclinic C_{2h}^5	4.27	110 dec	0.92 g/100 g H_2O at 20°C
sodium permanganate	[*10101-50-5*]	$NaMnO_4 \cdot 3H_2O$	dark purple crystals		1.972	36.0	v sol H_2O, deliquescent
zinc permanganate hexahydrate	[*23414-72-4*]	$Zn(MnO_4)_2 \cdot 6H_2O$	black crystals, solutions look purple	rhombic C_{2v}^7	2.45	dec 90–105	v sol H_2O, deliquescent

The density of the alkali metal permanganate salts increases with the atomic number of the Group 1 (IA) cation, whereas the corresponding aqueous solubility decreases (106). At room temperature aqueous solubility decreases from about 900 g/L for $NaMnO_4$ to 60 g/L for $KMnO_4$, and to 2.5 g/L for $CsMnO_4$. The solubility of potassium permanganate in water as a function of temperature is as follows:

Temperature, °C	Solubility of $KMnO_4$ g/L	Temperature, °C	Solubility of $KMnO_4$ g/L
0	27.81	40	125.16
10	43.93	53	182.37
20	64.95	63	225.83
30	90.55	70	286.36

The aqueous solubility of $KMnO_4$ in g/L at various solution temperatures in °C can be estimated by

$$\text{solubility}_{KMnO_4} = 30.55 + 0.796\,T + 0.0392\,T^2$$

The heat of dissolution of potassium permanganate in water varies between 41.84 kJ/mol (10.62 kcal/mol) in dilute solution to 38.5 kJ/mol (9.20 kcal/mol) as saturation is approached (107). The rate of dissolution is primarily controlled by temperature, particle size distribution of the crystalline potassium permanganate, the amount of energy supplied as agitation, and the final degree of saturation of the resulting solution.

The solubility of potassium permanganate in aqueous potassium hydroxide (108) is shown in Figure 7. Permanganates are soluble in certain nonaqueous solvents such as liquid NH_3, but not in liquid SO_2. Organic solvents such as glacial acetic acid, acetone, acetonitrile, *tert*-butyl alcohol, benzonitrile, pyridine, and trifluoroacetic acid, among others, dissolve $KMnO_4$ to some extent, but the resulting solutions are of limited stability because of the attack by the permanganate ion on the solvent.

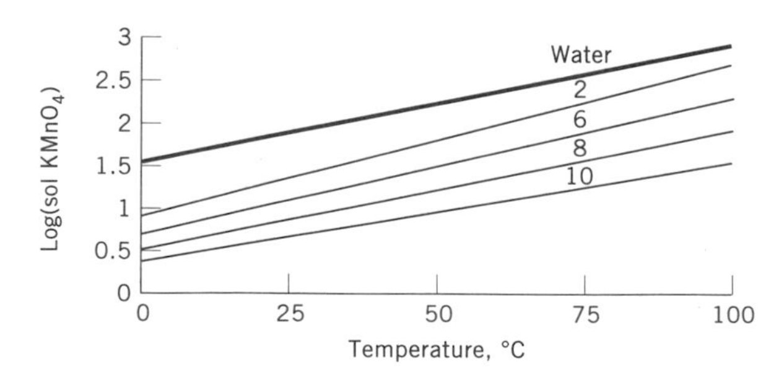

Fig. 7. Solubility of potassium permanganate, in g/L, in water and aqueous potassium hydroxide. Numbers represent KOH concentration in normality. Solubility can be approximated by the equation log(sol MnO_4, g/L) = $1.288 + 0.016\,T - 0.108\,N$ where T is temperature in °C and N is the KOH normality.

Tetraphenylphosphonium permanganate [*34209-26-2*], $(C_6H_5)_4PMnO_4$, and tetraphenylarsonium permanganate [*4312-28-1*], $(C_6H_5)_4AsMnO_4$, are nearly insoluble in water, but soluble in organic solvents, thus allowing permanganate oxidations of water-insoluble substrates to be carried out in the homogeneous phase. These salts are obtained when water-soluble forms of the quaternary onium salts, typically the chloride or the bromide, are treated with permanganates.

Crystallographic studies on potassium permanganate suggest that in the solid phase it has an orthorhombic crystal structure, space group Pnma, and four molecules per unit cell. Cell dimensions are $a = 910.5$ pm, $b = 572.0$ pm, and $c = 742.5$ pm. The average Mn–O bond distance in the permanganate ion is 162.9 ± 0.8 pm and the average O–Mn–O bond angle is $109.4 \pm 0.7°$. Thus the permanganate ion can be visualized as a manganese atom surrounded by four oxygen atoms at the corners of a regular tetrahedron (109).

Manufacture of Potassium Permanganate. Potassium permanganate may be manufactured by the one-step electrolytic conversion of ferromanganese to permanganate, or by a two-step process involving the thermal oxidation of manganese(IV) dioxide of a naturally occurring ore into potassium manganate(VI), followed by electrolytic oxidation to permanganate:

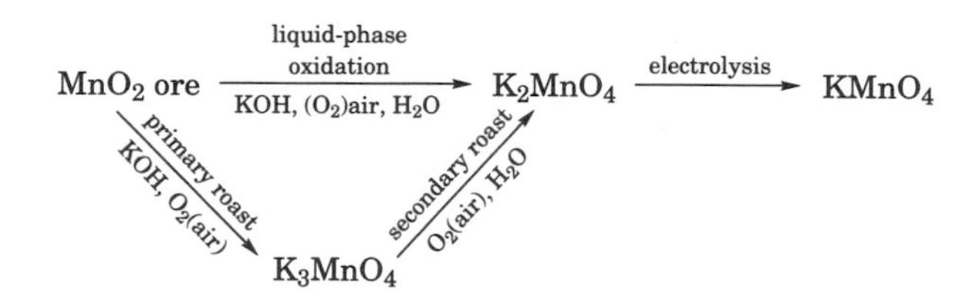

Depending on the means of conversion of manganate(V) to (VI), the process may be classified as a roasting or liquid-phase process (Fig. 8). The roasting process employs a solid reaction mixture having a molar ratio between MnO_2 and KOH in the range of 1:2 to 1:3. In contrast, the liquid-phase route operates at a higher ($\geq 1:5$) molar ratio between MnO_2 and KOH.

The roasting process, or variations of it, are most common. Liquid-phase processes are in operation, however, both in the United States and the former USSR. The former USSR is the only place where $KMnO_4$ was produced by anodic oxidation of ferromanganese. Table 17 summarizes the various $KMnO_4$ manufacturing facilities worldwide as of this writing.

Liquid-Phase Oxidation. In the early 1960s, both Carus Chemical Co. (La Salle, Illinois) and a plant in the Soviet Union started to operate modernized liquid-phase oxidation processes.

The USSR process (118–120) is discontinuous, uses turbine-agitated, low pressure reactors having a volume of 4 m^3 each, and processes 2000–2500 L/batch. Preconcentrated molten potassium hydroxide (70–80%) is added to the reactor with a quantity of 78–80% MnO_2 ore (<0.1-mm particle size) resulting in a 1:5 molar ratio of MnO_2:KOH. Air, or O_2, is introduced below the liquid level by a sparging device at such a rate that a positive pressure of 186–216 kPa (1.9–2.2 atm) is maintained. The temperature is kept at 250–320°C for the

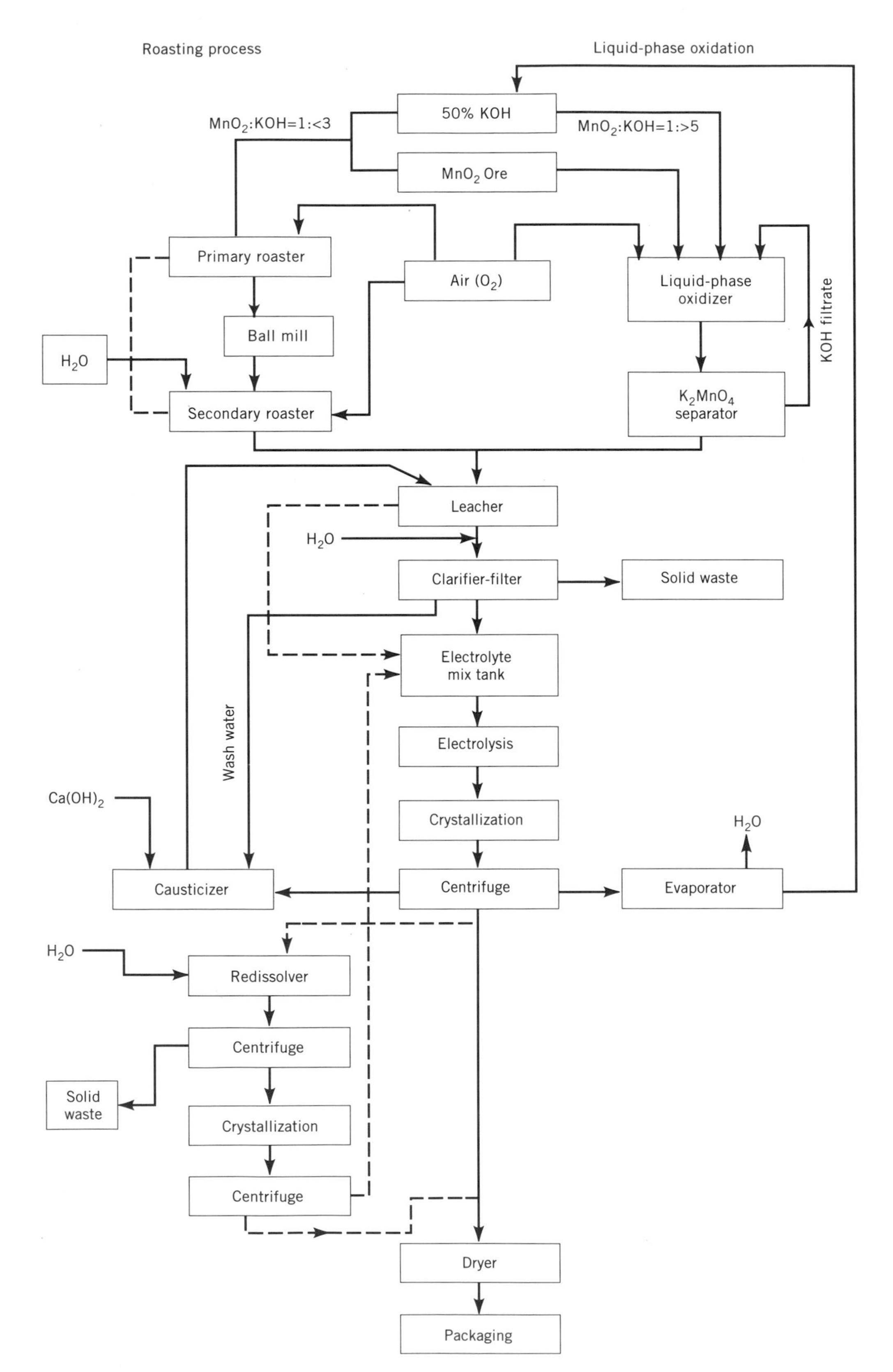

Fig. 8. Manufacture of potassium permanganate indicating both the roasting and the liquid-phase oxidation route; the latter is also known as the Carus process. The dashed line indicates an optional step.

Table 17. Permanganate Production Worldwide

Country	Approximate year installed	Manganate technology	Electrolysis/crystallization process	Estimated capacity, t/yr	Refs.
People's Republic of China	ca 10 plants, after WWII	most plants use open-hearth roasters; some have liquid-phase oxidizer systems		7,000–15,000	110
Czech Republic, Usti n.L	early 1900s	rotary hearth furnaces, requiring 40 h roasting for conversion from MnO_2 to K_3MnO_4; discontinuous operation	rectangular monopolar, divided asbestos diaphragm cells of 0.5 m^3 volume; anodes and cathodes made from iron wire mesh (filtered electrolyte); continuous operation; power requirements about 950 kWh/t $KMnO_4$ for electrolysis only	2,000	111
Spain, Trubia nea Oviedo	1972	spray chamber as primary roaster, plate reactor as secondary stage	continuous electrolysis of filtered electrolyte, continuous crystallization	2,000[a]	112
Germany, Bitterfeld	1920	two-stage rotary kilns heated internally using intermediate grinding of roast; oxidation completed within 3–4 h	cylindrical monopolar cells, 4 m^3 volume undivided; concentric Ni anodes, rod-shaped Fe cathodes; unfiltered electrolyte; batch operation; $KMnO_4$ crystallizes in cell; electrolysis energy consumption about 700 kWh/t	4,000[a]	27,113
India, Goa and Bombay	two plants built ca 1950	batch-type open-hearth roasters	cells are operated discontinuously, the $KMnO_4$ is crystallized separately in agitated tanks	1,200	114
former USSR, Saki, Crimea	1960	discontinuous low pressure liquid-phase oxidation; K_2MnO_4 separated from melt by dilution, followed by filtration	discontinuous electrolysis of unfiltered electrolyte in monopolar cells, crystallization in cells; stainless steel anodes, mild steel cathodes, cathode shields of plastic gauze; electrolysis power consumption 500 kWh/t	3,000[b]	115–117
United States, LaSalle, Ill.	1918	continuous liquid-phase oxidation (since ca 1961); K_2MnO_4 separation from liquid phase is without prior dilution	continuous electrolysis of filtered electrolyte in bipolar cells; Monel anodes, mild steel cathodes, vacuum crystallization	14,000	

[a] Periodically shuts down.
[b] Operation was shut down in 1990; restructured in 1993.

duration of the reaction, which requires approximately 4–6 h for completion. The reaction mixture, which reportedly remains fluid during the entire time, is then emptied through a siphon. Conversion of MnO_2 to K_2MnO_4 ranges from 87–94%. To isolate the potassium manganate from the melt, recycle KOH of 10–12% concentration is added until the overall caustic potash concentration is about 450–550 g/L. Upon cooling to 30–40°C, the K_2MnO_4 settles and is separated by centrifugation for use in the preparation of the electrolyte. This process is reported to work well, even with high (up to 11%) silica ores.

The Carus liquid-phase oxidation process is similar in principle; however, it is operated continuously, its oxidation reaction vessels are of a much larger scale, and the separation of the manganate intermediate from the caustic melt is accomplished without dilution by means of filtration (qv) (121–123).

Roasting Processes. The first step in roasting processes is the formation of K_3MnO_4 from MnO_2 ore. This is promoted by high temperature and high KOH and low H_2O concentration. The second step oxidizes Mn(V) to Mn(VI). A lower temperature and control of moisture in the air is used.

In a typical procedure, a slurry of 50% potassium hydroxide solution and finely ground manganese dioxide ore (MnO_2:KOH = 1:2.3–2.7) is sprayed into an atmosphere of preheated (390–420°C) air contained in a rotary kiln or a spray chamber. Formation of K_3MnO_4 occurs almost instantly, ie, within one minute or less. The rate-determining step is the removal of the water from the melt (124), but the specific reactivity of the ore also can be a factor. The material is ground in a ball mill and then transferred to another rotary kiln or a fluidized-bed reactor kept at 180–220°C. Controlled amounts of water are added in a fine spray bringing the average water content of the air to about 300 g/m^3. The retention time in the secondary reactor may be up to 3–4 h for preground roast, and probably on the order of 20 h for unground material. Conversion of the MnO_2 in the ore to K_2MnO_4 ranges from 85–90%.

The rotary kilns used in manganate production in Bitterfeld, Germany, are 5.5 m long and have a diameter of 1.75 m (125). For primary roasting, these are internally heated by hydrogen-fueled burners which minimize the problem of crust formation on the inner walls of the kiln. The use of hydrogen also reduces the amount of K_2CO_3 formed in the course of the roasting process. A total of eight kilns are available.

Newer manganate roasters are said to be significantly larger having diameters of about 2 m and lengths of 10 m. Larger dimensions permit distribution of the KOH–H_2O/MnO_2 spray over a wider area, and thus prevent undesirable agglomeration of the roast. Figure 9 shows a typical kiln arrangement.

Electrolysis. Although the oxidation of manganate(VI) to permanganate can be accomplished by chlorination, ozonation, or disproportionation (126), electrolysis is the preferred method. Reactions are as follows:

Anode $$MnO_4^{2-} \longrightarrow MnO_4^- + e^-$$

Cathode $$H_2O + e^- \longrightarrow 1/2\ H_2 + OH^-$$

Overall $$MnO_4^{2-} + H_2O \longrightarrow MnO_4^- + 1/2\ H_2 + OH^-$$

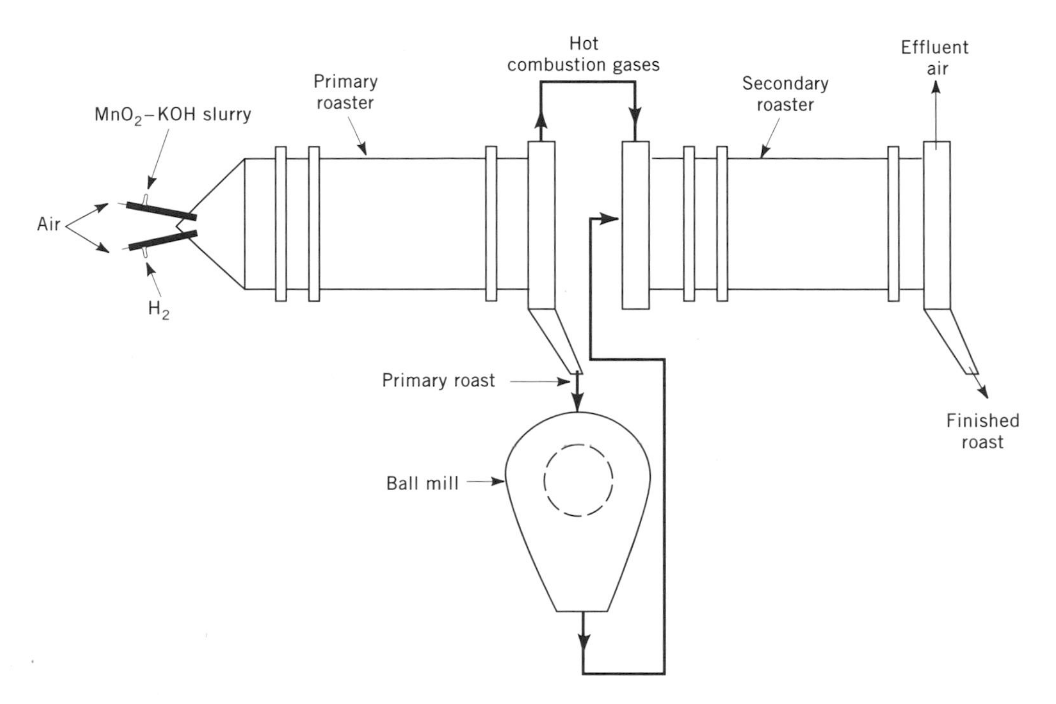

Fig. 9. Rotary kiln for the production of potassium manganate(VI).

The primary side reaction at the anode is the oxidation of hydroxyl ion to oxygen. In an undivided cell, a side reaction takes place also at the cathode, ie, the unwanted reduction of MnO_4^{2-} and MnO_4^- to lower valent manganese species.

Anode metals include nickel, Monel, stainless steel, and mild steel; cathode metal is practically always mild steel. Most cell designs do not have diaphragms and undesirable reductive effects at the cathode are minimized by making the cathode area very small in comparison to the anode area. Diaphragms, if employed, are made of asbestos (qv) sheet or plastic-impregnated fabric. Permanganate cells are mostly monopolar, ie, all anodes on the one hand, and all cathodes on the other hand, are connected in parallel. Bipolar designs are used occasionally. To prevent depletion of manganate ion at the anode, the electrolyte in the cell must be well agitated. Built-in agitators, bubbling gas through the electrolyte, or gravity or pumped electrolyte flow through the cell is used. In batch-type cells, the $KMnO_4$ is allowed to crystallize in the cell and is collected in the usually conical-shaped bottom section. Batch-type cells are frequently run using unfiltered electrolyte containing substantial concentrations of suspended solids, largely consisting of ore gangue and precipitated MnO_2. In continuously operated cells, the permanganate is crystallized outside of the cells in crystallizers specially designed for that purpose. These cells are operated only on filtered electrolyte. The operating temperature of permanganate cells is 40–60°C and cell voltage is between 2.3 and 3.0 V. For undivided cells, anode current densities range from 50–1500 A/m^2 and cathode current densities are 500–5000 A/m^2 or higher. Current efficiencies are in the range of 60–80%.

Manganate(VI) formed in the initial oxidation process must first be dissolved in a dilute solution of potassium hydroxide. The concentrations depend on the type of electrolytic cell employed. For example, the continuous Carus cell uses 120–150 g/L KOH and 50–60 g/L K_2MnO_4; the batch-operated Bitterfeld cell starts out with KOH concentrations of 150–160 g/L KOH and 200–220 g/L K_2MnO_4. These concentration parameters minimize the disproportionation of the K_2MnO_4 and control the solubility of the $KMnO_4$ formed in the course of electrolysis.

The raw potassium manganate(VI) from the secondary roaster or the liquid-phase oxidizer contains a fair amount of insoluble material such as unreacted MnO_2 and ore gangue. In most continuous processes, these insolubles are removed by sedimentation using thickeners or filtration and are disposed of as waste.

The Carus cell is of continuous flow-through bipolar design. The anode side of its rectangular bipolar electrode consists of a Monel screen attached to a steel back plate; the cathode side is mostly covered with a corrosion-resistant plastic material, except for regularly spaced exposed mild steel tips that are also connected to the back plate, and extend through the plastic. A single 4 m^3 volume cell can contain up to 60 electrodes. Cell frames and flow dividers are made of nonconductive materials. Both the electrolyte and the electric current flow are in series. The assembled cell, which contains the electrodes in the horizontal position, is mounted at an incline to allow the electrolyte and the cell gases (H_2 and O_2) to flow in the same direction. Current flow through the cell is 1.2–1.4 kA. Anode current density ranges from 85–100 A/m^2; the cathode current density is estimated to be about 13–15 kA/m^2. Current yields of up to 90% can be obtained (127–129).

The monopolar batch-type Bitterfeld cell is circular in shape having a flat bottom. The anodes consist of concentrically arranged Monel sheets; the cathodes, composed of mild steel rods, are placed in the annular space between the anodes. The closed cell has its own agitator consisting of a helix mounted on the vertical shaft and a stirring blade rotating near the bottom of the cell. The crystalline permanganate along with the sludge produced is periodically withdrawn by a bottom-discharge valve. The effective cell volume is about 4 m^3, voltage is 2.2–2.5 V, and the current flow 5000 A. Anode current density is 70 A/m^2; cathode current density is 700 A/m^2. The operating temperature is approximately 45°C, current yield is about 75%, and the energy required for electrolysis is estimated at 0.7 kWh/kg $KMnO_4$.

The electrolyte is unfiltered leach containing 200–250 g/L K_2MnO_4 and 120–150 g/L KOH. It takes about 45 h to bring the K_2MnO_4 concentration down to 20–30 g/L, at which point the electrolysis is terminated.

Crystallization can take place either within the cell container or in separate crystallizers. Using unfiltered electrolyte, the raw product which contains on the order of 90% $KMnO_4$ must still be purified by recrystallization. The crystallizers range from simple agitated tanks equipped with plate coolers or cooling jackets to continuous vacuum crystallizers. The latter, used by Carus Chemical Co., give a product that usually meets the specifications of at least technical-grade $KMnO_4$. Pharmaceutical and reagent-grade are obtained by further recrystallization.

The potassium permanganate crystals are dried at atmospheric pressure below 150°C, cooled, and packaged. Care is taken to prevent heating the product above 200°C during drying to avoid autocatalytic exothermic decomposition of the product.

In the electrolysis of K_2MnO_4, one mole of KOH is coproduced for every mole of $KMnO_4$ generated. This by-product potassium hydroxide must be recovered and utilized. For recycling, it also needs to be purified (130). Alternatively, the KOH can be converted into potassium carbonate by treatment with CO_2 in the red-lye process (131).

Electrolytic Oxidation. Electrolytic oxidation of ferromanganese or manganese metal is a one-stage process that circumvents the problem of ore impurities. Moreover, this procedure can be used with low caustic concentrations at room temperature. This process is based on the following reactions:

Anode $$2\,Mn^0 + 16\,OH^- \longrightarrow 2\,MnO_4^- + 8\,H_2O + 14\,e^-$$

Cathode $$14\,H_2O + 14\,e^- \longrightarrow 7\,H_2 + 14\,OH^-$$

Overall $$2\,Mn^0 + 2\,OH^- + 6\,H_2O \longrightarrow 2\,MnO_4^- + 7\,H_2$$

The process is most successful using pure manganese metal, but high grade ferromanganese, containing >80% Mn, is more economical. Passivation of the electrodes resulting in higher voltage drops and low current efficiencies can be a problem.

A procedure for industrial-scale (500 t/yr) production of potassium permanganate using this process was adapted for the Rustaw Nitrogen Fertilizer Plant (132). For the casting of anodes, ferromanganese having at least 80% Mn is melted with the addition of cryolite as a flux in an electric furnace. The precast is in the form of a copper coil through which cooling water is passed to keep the cell operating temperature at 20°C. Each cell is operated at 4.5 V and 6500 A. Current densities are 2300 A/m^2 at the anode and 1800 A/m^2 at the cathode. Initially, the electrolyte contains 250 g/L KOH, which gradually decreases to about 30 g/L during the course of the electrolysis. The crystalline $KMnO_4$ generated collects in the bottom of the cone, along with a sludge largely consisting of hydrous iron oxides, manganese dioxide, and SiO_2. The permanganate is separated and purified by recrystallization and the sludge, after drying, is said to have use as a catalyst, a depolarizer in dry cells, or an adsorptive agent for sulfur compounds.

The production of $KMnO_4$ by direct anodic oxidation of ferromanganese is energy intensive. Current efficiencies are on the order of 40% at best, and at the practically required voltages of 4.5 V or higher, the power consumption for the electrolysis alone is estimated to be about 15 kWh/kg $KMnO_4$. Operation at 20°C also requires a significant expense for cooling. This method is attractive under conditions where only low grade Mn ores having high percentages of alkali-soluble impurities (Si, Al) are available, or where those ores can be used to produce ferromanganese.

Oxidation Reactions. Potassium permanganate is a versatile oxidizing agent characterized by a high standard electrode potential that can be used under a wide range of reaction conditions (100,133–141). The permanganate ion can participate in a reaction in any of three distinct redox couples, depending on the nature of the reducing agent and the pH of the system. Typically permanganate oxidation reactions are conducted in an aqueous environment, or in organic cosolvents, which exhibit some degree of stability toward the oxidant. Solvents include acetone, acetic acid, acetic anhydride, *t*-butanol, ethanol, pyridine, and trifluoroacetic acid. Permanganate oxidizes hydrogen, carbon monoxide, and hydrogen peroxide under a variety of pH conditions, and the halides under acidic conditions.

Under extremely alkaline conditions, pH >12, potassium permanganate reacts involving a single-electron transfer, resulting in the formation of manganate(VI).

$$MnO_4^- + e^- \longrightarrow MnO_4^{2-}$$

In the pH range of 3.5 to 12, and in the presence of most reducing agents, permanganate reactions normally undergo a three-electron exchange resulting in the formation of hydrous manganese dioxide.

$$MnO_4^- + 2\,H_2O + 3\,e^- \longrightarrow MnO_2 + 4\,OH^-$$

Under acidic conditions, pH < 3.5, and in the presence of certain reducing agents, the permanganate ion can undergo a five-electron exchange resulting in the divalent manganese ion.

$$MnO_4^- + 8\,H^+ + 5\,e^- \longrightarrow Mn^{2+} + 4\,H_2O$$

The pH has an additional effect on permanganate reactions because the pH can cause a change in the substrate, altering the subtrate's degree of susceptibility toward oxidation. For potassium permanganate, as a general rule (138) for an anion, Z^{2-}, the degree of oxidation follows:

$$Z^{2-} > HZ^- > H_2Z > H_3Z^+$$

Additionally, the enthalpies of activation (142) for the permanganate oxidation of organic compounds is characteristically low in the range of 21–42 kJ/mol (5.1–10.0 kcal/mol). Figure 10 contains functional groups oxidized by $KMnO_4$.

One of the most well-known reactions of potassium permanganate is the oxidation of unsaturated compounds. Alkenes are readily oxidized by potassium permanganate under mild conditions which include a pH in the range of 4–8 and ambient temperature. The initial oxidation product is a diol or hydroxyketone, which with additional permanganate is further cleaved into carboxylic acids. The oxidation of alkenes (see Fig. 11) involves the formation of a cyclic manganese ester, which accounts for the observations that the hydroxyl groups are added in a cis fashion and the oxygen atoms come from the permanganate ion.

Primary and secondary alcohols are readily oxidized to aldehydes and ketones under alkaline conditions. Aldehydes, both aliphatic and aromatic, are

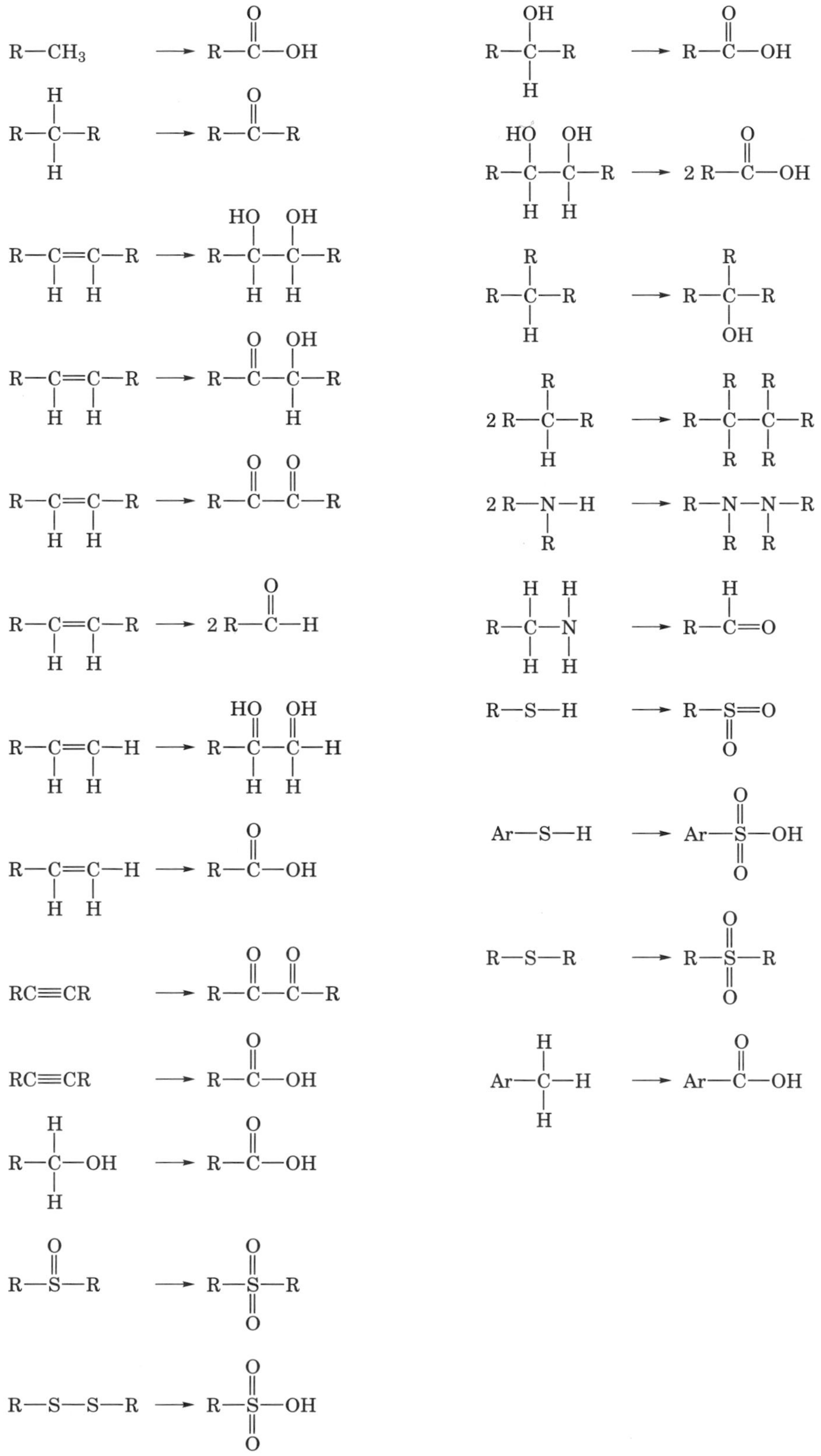

$$R{-}CH_3 \longrightarrow R{-}C({=}O){-}OH$$

$$R{-}CH_2{-}R \longrightarrow R{-}C({=}O){-}R$$

$$R{-}CH{=}CH{-}R \longrightarrow R{-}CH(OH){-}CH(OH){-}R$$

$$R{-}CH{=}CH{-}R \longrightarrow R{-}C({=}O){-}CH(OH){-}R$$

$$R{-}CH{=}CH{-}R \longrightarrow R{-}C({=}O){-}C({=}O){-}R$$

$$R{-}CH{=}CH{-}R \longrightarrow 2\,R{-}C({=}O){-}H$$

$$R{-}CH{=}CH{-}H \longrightarrow R{-}CH(OH){-}CH(OH){-}H$$

$$R{-}CH{=}CH{-}H \longrightarrow R{-}C({=}O){-}OH$$

$$RC{\equiv}CR \longrightarrow R{-}C({=}O){-}C({=}O){-}R$$

$$RC{\equiv}CR \longrightarrow R{-}C({=}O){-}OH$$

$$R{-}CH_2{-}OH \longrightarrow R{-}C({=}O){-}OH$$

$$R{-}S({=}O){-}R \longrightarrow R{-}S({=}O)_2{-}R$$

$$R{-}S{-}S{-}R \longrightarrow R{-}S({=}O)_2{-}OH$$

$$R{-}CH(OH){-}R \longrightarrow R{-}C({=}O){-}OH$$

$$R{-}CH(OH){-}CH(OH){-}R \longrightarrow 2\,R{-}C({=}O){-}OH$$

$$R{-}CR_2{-}H \longrightarrow R{-}CR_2{-}OH$$

$$2\,R{-}CR_2{-}H \longrightarrow R{-}CR_2{-}CR_2{-}R$$

$$2\,R{-}NR{-}H \longrightarrow R{-}NR{-}NR{-}R$$

$$R{-}CH_2{-}NH_2 \longrightarrow R{-}CH{=}O$$

$$R{-}S{-}H \longrightarrow R{-}S({=}O){=}O$$

$$Ar{-}S{-}H \longrightarrow Ar{-}S({=}O)_2{-}OH$$

$$R{-}S{-}R \longrightarrow R{-}S({=}O)_2{-}R$$

$$Ar{-}CH_3 \longrightarrow Ar{-}C({=}O){-}OH$$

Fig. 10. Functional groups oxidized by potassium permanganate. Ar is an aryl group.

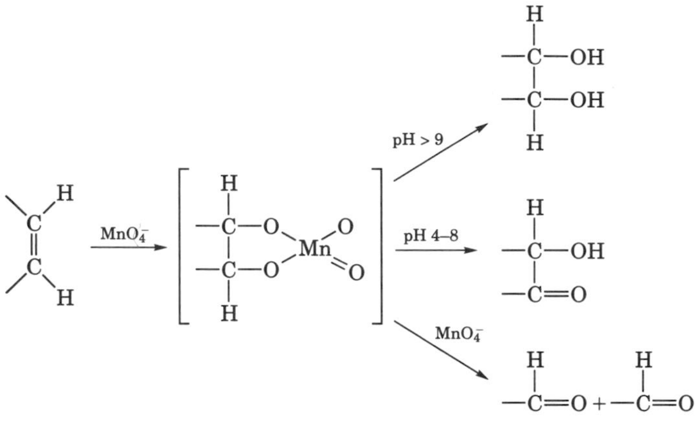

Fig. 11. Permanganate oxidation of alkenes.

converted into the corresponding carboxylic acids. Ketones are generally oxidation resistant unless sufficient alkali is present to effect enolization. The enol can be oxidatively cleaved.

Phenols as a compound class are readily oxidized by potassium permanganate and, if sufficient oxidant is added, phenol (qv) can be completely oxidized into carbon dioxide and water.

$$3\ C_6H_5OH + 28\ KMnO_4 + 5\ H_2O \xrightarrow{pH\ 8.5-9.5} 18\ CO_2 + 28\ KOH + 28\ MnO_2$$

Less than stoichiometric amounts of permanganate, added for phenol oxidation, results in ring cleavage and the formation of mesotartaric, formic, and oxalic acids as degradation products (143). Phenols having electron-withdrawing substituents such as halogen or nitro groups are more resistant to permanganate oxidation than phenol or methyl-substituted phenols.

Primary aromatic amines are readily oxidized by neutral or alkaline potassium permanganate, and complete mineralization into carbon dioxide, ammonia, and water can occur if sufficient oxidant is added (144). Aliphatic amines are also rapidly oxidized by neutral or alkaline permanganate, however multiple oxidation products are possible. Oxidation of primary, secondary, or tertiary amines containing hydrogen on carbon bonded to the amine nitrogen, are oxidized rapidly and at high yields by neutral potassium permanganate in aqueous *t*-butanol, at 60–80°C. The corresponding aldehyde or ketone results (145).

Organic sulfur compounds are readily oxidized by permanganate. Permanganate oxidizes aliphatic thiols to disulfides and aromatic thiols to sulfonic acids. Sulfides and sulfoxides are oxidized by permanganate to sulfones. Disulfides are generally resistant to oxidation by permanganate. Organometallic compounds such as tetraethyllead [*78-00-2*] (146) and tetrabutyltin (147) can be oxidized by potassium permanganate.

Under extreme pH, ie, conditions such as very high acidity, an increase in oxidation rate can result, presumably through the formation of permanganic acid, $HMnO_4$.

Permanganate salts generally exhibit a lack of solubility in the solvents which are suitable for many organic compounds. Phase-transfer reagents such as quaternary ammonium compounds (148) are used to bring the oxidant and the reactant into a common phase. Crown ethers, which complex the cation in the permanganate salt, are also used for phase transfer of permanganate ion into the organic phase (149). Complexing the permanganate's cation, or replacing it with a quaternary ammonium or phosphonium ion, allows the permanganate to be dissolved in nonpolar solvents such as benzene or methylene chloride (150). This results in a mild, but effective oxidizing agent, where olefins can be converted by $KMnO_4$ in good yields into either 1,2-diols or aldehydes.

The use of solid supports in conjunction with permanganate reactions leads to modification of the reactivity and selectivity of the oxidant. The use of an inert support, such as bentonite (see CLAYS), copper sulfate pentahydrate, molecular sieves (qv) (151), or silica, results in an oxidant that does not react with alkenes, but can be used, for example, to convert alcohols to ketones (152). A solid supported permanganate reagent, composed of copper sulfate pentahydrate and potassium permanganate (153), has been shown to readily convert secondary alcohols into ketones under mild conditions, and in contrast to traditional permanganate reactivity, the reagent does not react with double bonds (154).

Heterogeneous permanganate oxidation of alkenes employing moist alumina as the solid support and methylene chloride as the solvent also results in cleavage of carbon–carbon double bonds. Aldehydes are the products (155). In addition to mild reaction conditions, heterogeneous permanganate oxidation reactions result in easy work-up and product isolation. The spent oxidant is adsorbed onto the solid support and is easily removed through filtration. The product is recovered through evaporation of the solvent.

Solid sodium permanganate monohydrate has been shown to be a selective synthetic reagent (156). It is typically used in hexane for the heterogeneous oxidation of aldehydes, alcohols, and sulfides. Synthetic methodology based on crystal surfaces exhibited greater selectivity, higher yield, and easier work-up as compared to aqueous permanganate reactions.

Economic Factors

Manganese-containing ores, concentrates nodules, or synthetic materials are classified based on manganese content as metallurgical-grade, 38–55 wt % Mn, or chemical- or battery-grade, 44–54 wt % Mn (157). World manganese mine production capacity of manganese (Mn >35 wt %) and manganiferrous (Mn from 5–35 wt %) ores is on the order of 10×10^6 t/yr based on manganese content (158).

Total world capacity for electrolytic manganese dioxide (EMD) is estimated to be in the area of 194,500 t/yr, and annual capacity of chemical manganese dioxide (CMD) is estimated to be in the range of 40,000 t. Producers are listed in Tables 18 and 19, respectively. Capacity and process information on potassium permanganate is given in Table 17.

The United States consumption of manganese is distributed between three industries: iron and steelmaking, where 88% of the Mn is consumed; the manu-

Table 18. Electrolytic Manganese Dioxide Production[a]

Country	Producer	Location	Production capacity, 10^3 t/yr	Anode material
Japan	Toyo Soda	Hyuga	24	Ti
	Mitsui Mining and Smelting	Takehara	24	Ti
	Japan Metals and Chemicals Co.	Takaoka	18	Ti
Greece	Tekkosha Hellas	Thessoloniki	15	C
Ireland	Mitsui Denman	Cork	12	Ti
United States	Kerr McGee	Nenderson, Nev.	14.5	Ti
	Eveready Battery Co.	Marietta, Ohio	10	Ti
	Chemetals, Inc.	New Johnsonville, Tenn.	16	Ti
Spain	Cegasa	Onate	5	Pb
former Soviet Union	State	Rustavi	5	Pb
People's Republic of China	State		8	C and Ti
India	Union Carbide	Thaha	5	C
	T.K. Chemical	Trivandrum	1	C
Brazil	Union Carbide	Itapercerica	4	C
South Africa	Delta	Nelspruit	12	Pb and Ti
Australia	Australian Manganese Co., Ltd. (AMCL)	Mayfield, NSW	18	Ti
Total world			*194.5*	

[a]Ref. 159.

facture of batteries, where 7% is used; and chemical usage, which accounts for the remaining 5%. United States manganese demand is shown in Figure 12.

Purchases of manganese ore are made on the basis of user requirements, individual specifications, and availability. Most grades of manufactured manganese chemicals are subject to government and commercial specifications, many of which apply to specific uses. Prices of metallurgical manganese ore reached a high of \$3.78/t in 1990 (Fig. 13).

Analytical Chemistry

The general analytical chemistry of manganese is discussed elsewhere (162–167). A review covering more modern techniques, specifically for manganese dioxide, has also been published (168). A series of analytical techniques and procedures have been developed to study the metabolic fate of manganese (169,170).

The presence of manganese can be detected by formation of the purple MnO_4^- upon oxidation using bismuth or periodate in acidic solution. A very sensitive test is the reaction of Mn^{2+} and formaldoxime hydrochloride in aqueous alkaline solution, which also leads to the production of a purple MnO_4^- color. Modern quantitative methods for manganese determination rely heavily on in-

Table 19. Chemical Manganese Dioxide Producers[a]

Company	Location	Grade, type, or trade name	Capacity, t/yr	Applications
Sedema	Tertre, Belgium	Faradiser M	36,000	batteries
Chemetals Corp.	Baltimore, Md.	HP		ferrites, high purity lower oxides
The Boots Co., Ltd.[b]	Nottingham, U.K.	hydrated manganese dioxide	1,000	batteries
VEB Chemiekombinat[b]	Bitterfeld, Germany	Manganit		batteries
Perstorp Austria GmbH[b]	Vienna, Austria	Permanox		batteries
Shepherd Chemical Co.	Cincinnati, Ohio	808	300	curing of polymer sealants
Winthrop Laboratories[c]	New York, N.Y.	activated MnO_2		organic oxidations

[a]Ref. 159.
[b]CMD comes from saccharin manufacture.
[c]MnO_2 is priced at $30/kg in 25-kg lots or larger.

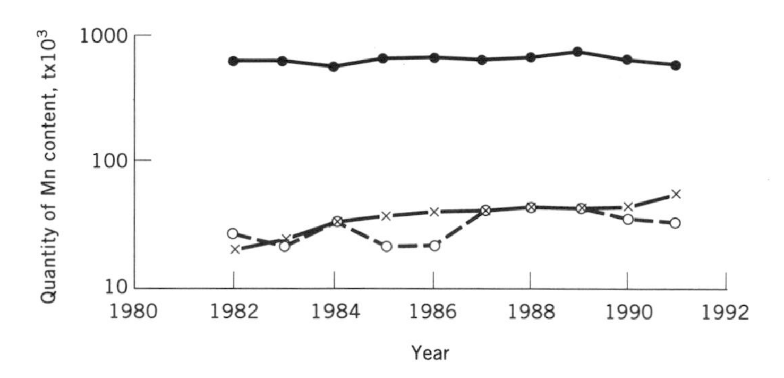

Fig. 12. United States manganese consumption where (•) represents total U.S. demand, (×) battery consumption, and (○) chemicals (160).

strumentation. For the analysis of high (>0.1% Mn) concentrations of manganese, potentiometric titration involving the titration of manganese(II) ion with permanganate ion in a neutral pyrophosphate solution is the most appropriate analytical method. This titration results in formation of a pyrophosphate complex of trivalent manganese (171).

$$4\ Mn^{2+} + MnO_4^- + 8\ H^+ + 15\ H_2P_2O_7^{2-} \longrightarrow 5\ Mn(H_2P_2O_7)_3^{3-} + 4\ H_2O$$

For low to medium (<0.01% Mn) concentration, atomic absorption/emission spectroscopy is widely used (172,173). Manganese can be determined by atomic absorption spectroscopy in an air–acetylene flame without significant interferences and a Mn concentration of 0.03 mg/L results in approximately 1% absorption at the 279.5 nm resonance line (174). The detection limit for manganese

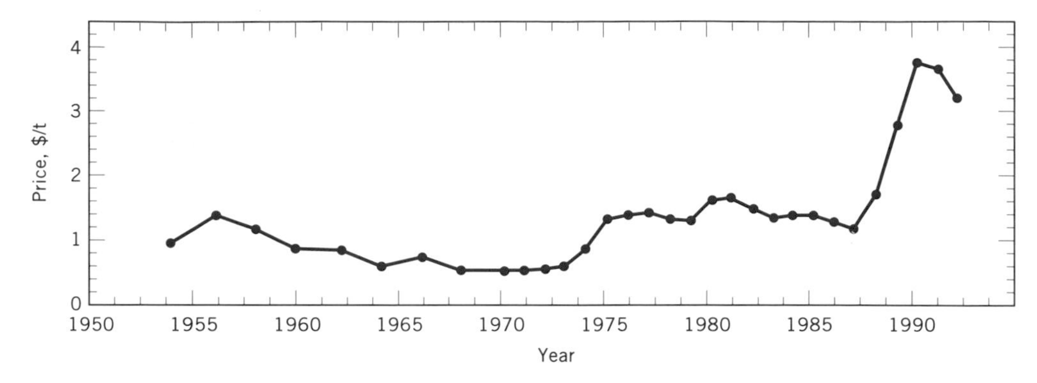

Fig. 13. Price of manganese metallurgical ore (161).

by flame atomic absorption is around 0.01 ppm, the detection limit by furnace atomic absorption is 0.002 ppm, and the detection limit via inductively coupled plasma emission spectroscopy is about 0.005 ppm. Spectrophotometry is also frequently used for the analysis of manganese using colorimetric methods. A comprehensive review has been published (175). There are a wide range of direct color-forming reactions of manganese and organic ligands.

The classical spectrophotometric method involves the oxidation of manganese to permanganate. Characteristic ultraviolet and visible spectra of the various oxymanganate ions are shown in Figure 14 (176). Convenient wavelengths for measurements of permanganate and manganate(VI) are 525 and 603 nm, respectively. Because manganese dioxide interferes, this material must first be removed by filtration or centrifugation.

Solid potassium permanganate is usually assayed volumetrically using sodium oxalate.

$$5\ C_2O_4^{2-} + 2\ MnO_4^- + 16\ H^+ \longrightarrow 10\ CO_2 + 2\ Mn^{2+} + 8\ H_2O$$

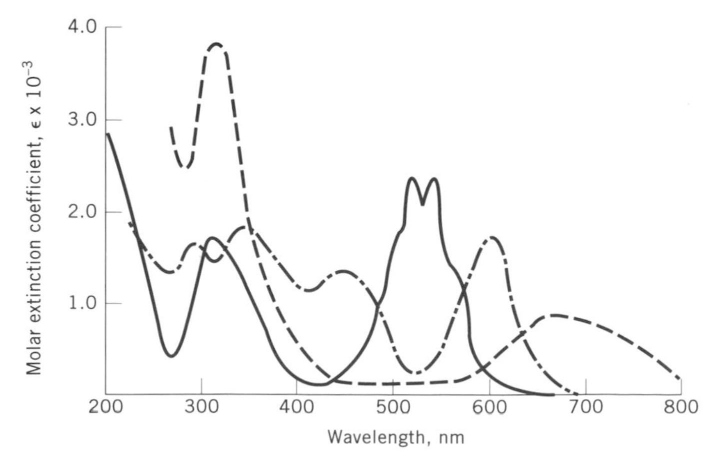

Fig. 14. Ultraviolet and visible spectrum of the oxyanions of manganese where (——) represents MnO_4^-, (—·—) MnO_4^{2-}, and (– – –) MnO_4^{3-}. Courtesy of Academic Press.

Even though the mechanism of the reaction between oxalate and permanganate is extremely complicated, titration under acidic conditions is extremely accurate. This is the recommended method for standardization of permanganate solutions.

An alternative method for the analysis of permanganate is the use of conventional iodometric methods (177) where excess potassium iodide is added to a solution of permanganate under acidic conditions. The liberated iodide is then titrated with standard thiosulfate solution using starch as an indicator.

$$2\ MnO_4^- + 10\ I^- + 16\ H^+ \longrightarrow 2\ Mn^{2+} + 5\ I_2 + 8\ H_2O$$

$$I_2 + 2\ S_2O_3^{2-} \longrightarrow 2\ I^- + S_4O_6^{2-}$$

The iodide method can also be applied to the analysis of other manganese species, but mixtures of permanganate, manganate, and MnO_2 interfere with one another in the iodometric method.

The proportion of manganese dioxide in ores and synthetic products is commonly measured by determining the samples active oxygen. This is a measurement of the amount of manganese having a valence greater than two. The weighed material is digested with a measured excess of reducing agent (potassium iodide, sodium oxalate, ferrous sulfate, arsenious oxide, or vanadyl sulfate) and the unconsumed reducing agent determined by titration (165,167). Using differential thermal analysis (DTA), it is possible to determine the presence of specific MnO_2 phases in the mixture (178).

A detailed review of the methods for determination of low manganese concentration in water and waste is available (179). A review on the speciation of Mn in fresh waters has been reported (180). Reviews for the chemical analysis of Mn in seawater, soil and plants, and air are presented in References 181, 182, and 183, respectively.

Health and Safety

Manganese appears to be an essential trace element for all living organisms (see MINERAL NUTRIENTS). Its concentration in organisms primarily depends on the species. Plants contain between 1–700 mg/kg, ocean fish between 0.3–4.6 mg/kg, and muscles of mammals 0.2–3 mg/kg (184). Manganese appears to be particularly stored and concentrated in tissues rich in mitochondria. Manganese, a cofactor for the enzyme pyruvate carboxylase (185), is also thought to act as a nonspecific activator for the enzymes succinate dehydrogenase, prolidase, arginase, alkaline phosphatase, farnesyl pyrophatase, superoxide dismutase, glycosyl transferases, and adenosine triphosphatases (186).

Manganese is considered an essential micronutrient for plants, which are thought to assimilate the element in the divalent state from soil. Deficiency of manganese can be a significant factor in reducing crop yields, which has been shown to occur in soils of high organic matter and pH values above 6.5 (187). Plant species vary widely in manganese requirements, as well as tolerance for excess manganese concentrations within their tissues. Data from a variety of species suggest the manganese deficiency occurs when plant tops contain less than about 20 ppm manganese on a dry weight basis. A concentration of up to about 500 ppm manganese is considered adequate. Manganese concentrations typically greater than about 500–1000 ppm have been reported to be toxic (188).

Studies have shown manganese to be required for good health in animals (189). Manganese deficiency in animals includes impaired growth, skeletal abnormalities (190), and altered metabolism of carbohydrates (qv) (191) and lipids. Manganese is considered an essential trace element for humans, although there is no direct evidence for this. The National Academy of Sciences recommended dietary allowance for adults is 2.5–5.0 mg/d of manganese (192). The usual daily intake of manganese for humans is in the range of 2–9 mg/d of Mn and comes primarily from food (193).

Food	Manganese content, mg/kg
nut products	18.21–46.83
grain products	0.42–40.70
legumes	2.24–7.73
fruits	0.20–10.38
vegetables	0.42–6.64
meat, fish, and eggs	0.10–3.99
beverages	0.0–2.09
milk	0.02–0.49

Groundwater concentrations of manganese are generally $< 100\ \mu g/L$, although values $\geq 1000\ \mu g/L$ are not uncommon. Surface water sources contain an average of 58 $\mu g/L$ (194) and U.S. rivers have a median value of 24 $\mu g/L$ (195,196).

Environmentally, manganese-bearing particulate matter is usually removed from air using dust collecting devices such as electrostatic precipitators, filter systems, cyclones, or wet scrubbers (see AIR POLLUTION CONTROL METHODS). In the case of liquids, soluble manganese can be removed from liquid effluents through precipitation as a hydrous oxide by adjustment of the pH to >8.3 using $Ca(OH)_2$, plus the application of an oxidizing agent such as O_2, Cl_2, ClO_2, NaClO, or $KMnO_4$. Aeration is also effective, provided the pH is raised to above 9.4. The final disposal of the resulting manganese dioxide-containing sludges depends on local conditions and regulations. These sludges are usually deposited in landfills.

Inhalation of particulate manganese compounds, such as manganese dioxide, can lead to an inflammatory response in the lungs of both humans and animals. This response is characteristic of all inhalable particulate matter, however, suggesting that the manganese compound is not specifically responsible. Inhalation of soluble manganese compounds does not cause an inflammatory response in the lungs of test rabbits (197). There are reports of increased susceptibility to lung infections reported for both humans and animals where there is a chronic exposure to manganese dusts. This is considered secondary to the irritation caused by the inhaled particulate matter. General population exposure to manganese compounds in the air in nonurban areas is about 5 ng/m^3; in urban areas, 33 ng/m^3; and in source dominated areas, 135 ng/m^3 (198). In the soil, manganese is estimated to be in the 40–900 mg/kg range. The maximum reported is 7000 mg/kg (199). The lowest observed adverse effect level (LOAEL) (200) reported for Mn by inhalation is 0.14 mg/m^3. This results in a chronic exposure by inhalation minimal risk level (MRL) of 0.3 $\mu g/m^3$. The dermal absorption of inorganic manganese compounds is not considered toxicologically significant. The primary exposure of the general population to manganese compounds is from diet.

There is conclusive evidence from human studies that inhalation exposure to high levels of manganese compounds can lead to a disabling syndrome of neurological effects termed manganism (201). Clinically, many similarities exist between the syndrome of manganism and Parkinson's disease, a progressive disease resulting from destruction of cells of the basal ganglia; however, manganism is considered a distinct and much more complex disease (202) effecting multiple systems. The onset of the illness may be expressed by slow speech without tone or inflection, dull and emotionless facial expressions, and slow and clumsy limb movements. In the later stages of the disease, walking becomes difficult and a characteristic staggering gait develops. Voluntary movements are accompanied by tremors, and a variety of behavioral changes have also been reported. The incapacitations caused by chronic manganese poisoning result in permanent disability, but the disease is not fatal (203). It has only been documented in workers exposed to high levels of manganese dust or fumes in mines or foundries, typically following several years of exposure. The lowest average concentration of manganese at which a case of chronic manganese poisoning has occurred was 30 mg/m^3, found in a manganese mill (204). The neurological damage produced by prolonged excessive exposure to manganese is mainly irreversible, but some anti-Parkinsonian drugs such as L-dopa, have been reported to reverse some of the neuromuscular symptoms (see NEUROREGULATORS) (205).

Human and animal studies indicate that inorganic manganese compounds have a very low acute toxicity by any route of exposure. The toxicity values for a given Mn compound are shown in Table 20 to depend on the species of test animal as well as the route of exposure. Manganese concentrations as high as 2000 ppm were found to be tolerated by test animals over a six-month period without any ill effects (208).

Only a small (ca 3%) fraction of ingested or inhaled manganese is absorbed, which occurs primarily by the intestines (209). Once absorbed, manganese is regulated by the liver, where it is excreted into the bile and passes back into the intestine, where some reabsorption may occur (210). Manganese is eliminated almost exclusively (>95%) by the bile in the gastrointestinal tract.

Limited laboratory studies have shown manganese to have mutagenic effects in bacteria, bacteriophage T4, and yeast (211). Tests on mice and rats have shown that neither Mn powder nor MnO_2 have any tumorigenic effects. However, manganese(II) acetylacetonate tested in the same series produced malignant neoplasms (212). No evidence of carcinogenic activity of manganese(II) sulfate monohydrate to male or female rats has been found, but male and female mice exposed to this compound had marginally increased incidence of thyroid gland follicular cell adenomas and significantly increased incidence of follicular cell hyperplasia (213). The carcinogenicity of manganese in humans is probably small (214).

Regulations and guidelines have been established in many countries for manganese and its compounds (Tables 21 and 22).

Table 20. Toxicity Data for Select Manganese Compounds[a]

Compound	CAS Registry Number	Test animal	Route of exposure	LD_{50} mg/kg
manganese(II) acetate	*[638-38-0]*	rat	oral	2940
cyclopentadienyl-manganese tricarbonyl	*[12079-65-1]*	mouse	iv	0.71
		rat	oral	22
manganese(II) chloride	*[7773-01-5]*	mouse	oral	1715
		mouse	ip	121
		mouse	intramuscular	255
		rat	oral	770
		rat	ip	700
		rat	intramuscular	700
		dog	iv	202
		rabbit	subcutaneous	180[b]
manganese(II) ethylenebis-(dithiocarbamate)	*[12427-38-2]*	rat	oral	3000
		mouse	oral	2600
		guinea pig	oral	6400[b]
manganese(II) oxide	*[1344-43-0]*	mouse	subcutaneous	1000
manganese(III) oxide	*[1317-34-6]*	mouse	subcutaneous	616
		rat	intratracheal	100[b]
manganese(IV) dioxide	*[1313-13-9]*	mouse	subcutaneous	422
		rat	intratracheal	50[b]
		rabbit	intravenous	45[b]
manganese(II) sulfate	*[7785-87-7]*	mouse	ip	332
manganese(II) sulfate tetrahydrate	*[10101-68-5]*	mouse	ip	534
manganese tricarbonyl methylcyclopenta-dienyl	*[12108-13-3]*	rat	oral	50
		rat	ip	23
		mouse	oral	230
		mouse	ip	152
calcium permanganate	*[10118-76-0]*	rabbit	iv	50[b]
potassium permanganate	*[7722-64-7]*	rat	oral	1090
		mouse	subcutaneous	500
		rabbit	oral	600[b]
		rabbit	iv	70[b]

[a]Refs. 206 and 207.
[b]Value is LD_{LO} = lethal dose low, ie, lowest dose of a substance introduced by any route other than inhalation, administered over any given period of time in a single or divided dosage that has been reported to have caused death in the test organism.

Potassium permanganate under RCRA definition meets the criteria of an ignitable waste, and if discarded is considered a hazardous waste. The reportable quantity (RQ) (220) for potassium permanganate is 45.4 kg (100 lbs) and releases into the environment greater than this value must be reported to the U.S. Coast Guard National Response Center.

Uses

The most significant nonferrous use of manganese compounds is for primary batteries, where manganese dioxide is the principal constituent of the cathode

Table 21. Occupational Standards for Manganese Compounds[a]

Country	Mn TWA, mg/m^3	Mn STEL[b], mg/m^3
United States	5[c]	
Australia	5	
Belgium	5	
Czechoslovakia	2	6
Denmark	2.5	
Finland	2.5	
Hungary	0.3	0.6
Japan	0.3	
the Netherlands	1	
Poland	0.3	
Sweden	5	2.5
United Kingdom	5	

[a]Ref. 215.
[b]STEL = short-term exposure limit.
[c]Value is a PEL:C, ie, permissible exposure limit:ceiling exposure limit.

Table 22. U.S. Federal Guidelines for Manganese Compounds[a]

Material	Mn concentration, mg/m^3	Reference
	TLV/TWA	
dust and compounds	5	
manganese tetroxide and manganese fume	1	
manganese cyclopentadienyltricarbonyl	0.1	
2-methylcyclopentadienyl manganese tricarbonyl	0.2	
	STEL	
manganese fume	3	217
chronic oral reference dose (RFD)[b]	0.1[c]	
chronic inhalation reference concentration (RFC)	0.0004	
carcinogenic classification	group D: not classifiable as to human carcinogenicity	218
secondary maximum contaminant level (MCL) water	0.05[d]	219

[a]Ref. 216.
[b]RFD = average daily intake considered adequate and safe.
[c]Units are mg/(kg·d).
[d]Units are mg/L.

mix. In the standard Leclanché cell, zinc and ammonium chloride are mixed to form the electrolyte, a mixture of carbon and MnO_2 forms the cathode, and zinc acts as the anode (221). The principal cell reaction is as follows:

$$Zn^{2+} + 2\,NH_4Cl + MnO_2 + e^- \longrightarrow Zn(NH_3)_2Cl_2 + MnOOH + H^+$$

A secondary reaction forming hetaerolite is also known:

$$Zn^{2+} + 2\,MnO_2 + 2\,e^- \longrightarrow ZnO\cdot Mn_2O_3$$

The cathode mix for a Leclanché primary battery consists of 50–60% manganese dioxide ore, 5–10% acetylene black, 10–20% ammonium chloride, and 3–12% zinc chloride. The remainder is water (see BATTERIES, PRIMARY CELLS).

Battery-active manganese dioxides include the naturally occurring manganese ores, chemical manganese dioxides, or electrolytic manganese dioxides. The highest activity for battery use has been attributed to γ-phase manganese dioxide, which exhibits poor crystallinity. Generally, the energy intensity of dry cells increases as the proportion of synthetic manganese dioxide added to the cathode mixture increases (222,223). Thus naturally occurring manganese dioxide ores have the highest use in low cost general-purpose batteries. Heavy-duty cells utilize mixtures of native ore, EMD, and zinc chloride as the electrolyte.

Primary battery system	Volume energy density, kJ/dm^3
carbon	440
heavy-duty	660
alkaline–MnO_2	470–840
Li–MnO_2	ca 1450

In the case of the alkaline manganese dioxide cell, only high quality synthetic manganese dioxide, typically an EMD (224), is used and graphite is the cathode. The anode is an amalgamated zinc. Potassium hydroxide serves as the electrolyte and the reaction can be summarized as follows:

$$Zn^0 + 2\,MnO_2 + H_2O \longrightarrow ZnO + 2\,MnOOH$$

For the alkaline manganese cell, cell voltage corresponds to the reduction (225) of the γ-MnO_2 from $MnO_{1.95}$ to $MnO_{1.5}$. Reduction in oxidation state from $MnO_{1.95}$ to $MnO_{1.75}$ corresponds to a transformation of the initial gamma phase to the δ-MnO_2 phase. Further reduction leads to the formation of MnOOH. The alkaline manganese cells are characterized by improved high load discharge characteristics, improved storage characteristics, and low temperature performance. Total consumption of battery-grade manganese dioxides, for all types of cells, is estimated to exceed 400,000 t/yr.

Manganese dioxides prepared from Li_2MnO_3 show promise for use in lithium battery applications. Capacities are reported to be in excess of 200 mA·h/g at a cut-off voltage of 2.0 V, and the electrochemical reaction is reversible. Rechargeable batteries containing a $LiMn_2O_4$ electrode and an aqueous electrolyte have been demonstrated to be a safe and cost-effective alternative to nickel–cadmium and lead–acid batteries (226). Manganese dioxide is a mild oxidizing agent. It is used for the production of various organic compounds, for the curing of polysulfide rubber sealants, and for numerous other industrial applications (Table 23).

Uses of manganese compounds (Table 24) range from a number of agricultural applications to polyester applications. The carbonate and sulfate salts act as the starting material for a variety of other manganese salts. Manganese(II) salts also act as oxidation catalysts. Manganese, in the form of oxides, is 10–30% of the composition of soft ferrite (see FERRITES).

Table 23. Uses of Manganese Dioxides

Application	Use	Reference
curing agent	oxidation of mercaptan terminals of liquid polysulfide polymers into disulfide bonds by manganese dioxide results in polysulfide rubber	227
radium removal	absorption of radium onto precipitated manganese dioxide followed by diatomaceous earth filtration removes 80–97% of Ra present in groundwaters; removal not pH sensitive in the 6.5–9.5 range and surface loading of 370–740 Bq/mg (10–20 pCi/mg) MnO_2 results in equilibrium radium activities in the treated water of 37−185 Bq/L (1–5 pCi/L)	228
brick colorant	manganese ores are used as colorants in brick manufacture; color shades from brown to red and gray to black can be obtained	229
uranium refining	manganese dioxide as oxidizing agent in the refining of uranium ore into yellowcake	
oxidation catalyst	manganese dioxide in combination with transition-metal elements serve as highly active thermally stable catalysts for volatile organic compound (VOC) oxidation; catalysts operate at temperatures significantly lower than comparable noble metal-based catalysts	230
glassmaking	as decolorizer, manganese dioxide is a common compensating agent for the yellow-green color imparted to glass by iron; as colorant, Mn(III) imparts a violet color, and Mn(II) imparts a faint yellow color to the glass; as finning agent, eliminates gas bubbles in glass melt	

The first known report on the use of potassium permanganate for taste and odor control were given (235) to the London Water Board where the chemical was used at the turn of the century to remove tastes attributed to the growth of *Tabelleria* at the West Middlesex Works (see ODOR MODIFICATION). Potassium permanganate has been readily applied to the treatment of potable water (236). It removes, through oxidation, objectionable tastes and odors present in raw water sources (237); oxidizes soluble iron and manganese into insoluble form (238−240); lessens or eliminates the need for activated carbon treatment, eg, oxidizes As(III) to As(V) which is then efficiently absorbed by alum (241); allows for the reduction in the formation of chlorinated organics (237); and inhibits infestation by the zebra mussel (242,243).

Potassium permanganate is also used for wastewater treatment. It oxidizes organic compounds that are toxic or inhibitory to biological systems (244,245). Vinyl chloride [*75-01-4*] in water can be oxidized by permanganate at a molar ratio of 3.15 parts $KMnO_4$:1 (246). Addition of potassium permanganate, prior to sludge dewatering at a pulp and paper mill wastewater treatment facility, results in the elimination of hydrogen sulfide odor and corrosivity from the dewatering belt presses and downstream areas (247).

Phenol can be oxidized and hence removed, ie, to levels $< 20\ \mu g/L$, from wastewater (248). Moreover, addition of potassium permanganate to the return

Table 24. Uses of Manganese Compounds

Compound	Use	Reference
manganese acetate	transesterification catalyst in the production of poly(ethylene terephthalate) (PETP) from dimethyl terephthalate to form polyester; catalyst for the oxidation alkanes to carboxylic acids	231
manganese oxide	source for fertilizer and animal nutrients; starting point for other Mn salts, preparation of ferrites	
manganese sesquioxide		
Mn_2O_3	production of ferrites, thermistors	
Mn_3O_4	production of soft ferrites, welding rod coatings	
manganese carbonate	production of ferrites, pigments	
manganese sulfate	micronutrient in fertilizer (ca 50% domestic consumption); trace mineral in animal feed (ca 35% domestic consumption)	
manganese soaps	manganese oleate [*23250-73-9*], manganese naphthenate [*1336-93-2*], manganese stearate [*3353-05-7*], and other manganese soaps are used as primary driers in oil or alkyd resin-based paints, printing inks, and varnishes and function as catalysts for the autooxidation of the resin; manganese octoates have been shown to act as superior driers exhibiting high gloss and extended storage performance in water-based alkyd resin systems	232
MMT	combustion improver, antiknock compound	233
manganese phosphates	phosphating primarily ferrous metals using manganese phosphate can provide a phosphate coating imparting the item corrosion protection and in other cases improving the retention of lubricant, thus improving the wear resistance of bearing surfaces	
manganese complexes from triaza-cyclonone ligands	effective catalysts for the low temperature bleaching of stains by hydrogen peroxide	234
manganese chloride	production of MMT	
Maneb, Mancozeb[a]	leaf and soil fungicide	

[a]CAS Registry Number [*8018-01-7*].

activated sludge results in reduction of odors issued from the aeration tanks of conventional activated sludge wastewater treatment plants without any change occurring to the microbiology of the system (249).

Printed circuit boards manufacture is aided by the use of $KMnO_4$. Alkaline permanganate solution is used to remove resin smeared on the interior hole wall of multilayered printed circuit boards. Additionally the hole wall is etched, resulting in a surface with excellent adhesion characteristics, for electrodeless copper (250). The alkaline permanganate etchback system containing >60 g/L $KMnO_4$ and 40–80 g/L NaOH at 70–80°C, is effective for difunctional, tetrafunctional, and polyimide resin substrates, where the level of etchback is directly proportional to the immersion time (10–20 min) (251).

Alkaline solutions consisting of approximately 160 g/L $NaMnO_4$ and 60 g/L NaOH at 75°C and contact times in the range of 15 minutes, exhibit high etch rates for printed circuit boards (252,253). The resulting manganese residues can readily be removed by acid neutralization. Addition of K^+, Rb^+, and Cs^+ as co-ions to an alkaline $NaMnO_4$ solution maintains etch rates of resin substrates comparable to solutions of higher $NaMnO_4$ concentrations (254).

Potassium permanganate has been found to be an effective (>99%) detoxicant for decontamination by several carcinogenic compounds, among which are benzo[*a*]anthracene [*56-55-3*], benzo[*a*]pyrene [*50-32-8*], 7,12-dimethylbenzanthracene [*57-97-6*], and 3-methylcholanthrene [*56-49-5*] (255). Acidic permanganate has also been used to detoxify wastewater containing the carcinogen *N,N*-dimethyl-4-amino-4′-hydroxyazobenzene [*2496-15-3*] (256), whereas alkaline permanganate detoxifies melphalan in laboratory wastes (257). High levels (>99%) of decontamination of carcinogenic laboratory wastes have been obtained using 0.5 *N* H_2SO_4. Materials oxidized included 3,3′-dichlorobenzidine [*91-94-1*], 3,3′-diaminobenzidine [*91-95-2*], 1-naphthylamine [*134-32-7*], 2-naphthylamine [*91-59-8*], 2,4-diaminotoluene [*95-80-7*], benzidine [*92-87-5*], *o*-tolidine [*119-93-7*], 4-aminobiphenyl [*92-67-1*], 4-nitrobiphenyl [*92-93-3*], and 4,4′-methylenebis-(*o*-chloroaniline) [*101-14-4*] (258,259).

Alkaline permanganate pretreatment of steel for the removal of heat scale and smut prior to acid pickling results in faster descaling and reduced metal attack (see METAL SURFACE TREATMENTS; METAL TREATMENTS). Stainless steel alloys can also be cleaned by alkaline permanganate followed by pickling in nonoxidizing acids (260).

BIBLIOGRAPHY

"Manganese Compounds" in *ECT* 2nd ed., Vol. 13, pp. 1–55, by J. O'Hay, The Harshaw Chemical Co.; in *ECT* 3rd ed., Vol. 14, pp. 844–895, by A. H. Reidies, Carus Chemical Co.

1. J. Emsley, *The Elements*, Clarendon Press, Oxford, U.K., 1989.
2. R. L. Heath, *Table of the Isotopes, Handbook of Chemistry and Physics*, 57th ed., CRC Press, Boca Raton, Fla.
3. F. A. Cotton and G. Wilkinson, *Advanced Inorganic Chemistry*, 2nd ed., John Wiley & Sons, Inc., New York, 1966.
4. T. A. Zordan and L. G. Hepler, *Chem. Rev.* **68**, 737 (1968).
5. C. C. Liang, *Encyclopedia of the Electrochemistry of the Elements*, Vol. 1, Marcel Dekker, New York, 1973, p. 349.
6. J. C. Hunter and A. Kozawa, in A. J. Bard, R. Parson, and J. Jordan, ed., *Standard Potentials in Aqueous Solution*, Marcel Dekker, New York, 1988, pp. 429–439.
7. A. K. Covington, T. Cressey, B. G. Lever, and H. R. Thirsk, *Trans. Faraday Soc.* **58**, 1975 (1962).
8. A. Carrington, *M. C. R. Symons, J. Chem. Soc.*, 3373 (1956).
9. G. L. Dettuff, Bureau of Mines Minerals Yearbook, U.S. Dept. of the Interior, Washington, D.C., 1977, p. 255; G. Charlot, *Théorie et Méthods Nouvelles d'Analyse Qualitative*, 3rd ed., Masson, Paris.

10. D. B. Wellbeloved, P. M. Craven, and J. W. Waudby, *Ullmann's Encyclopedia of Industrial Chemistry*, Vol. A16, VCH, Weinheim, Germany, 1990, p. 80.
11. G. O. Barney, ed., *GLOBAL 2000*, Report to the President, U.S. Council on Environmental Quality and U.S. Foreign Department, U.S. Government Printing Office, Washington, D.C., 1980.
12. B. W. Haynes, *U.S. Bur. Mines Bull.* **679**, 2–32 (1980).
13. G. Hubred, *Miner. Sci. Eng.* **7**, 71 (1975).
14. D. S. Cronan, *Endeavour* **2**(2), 80 (1978).
15. R. G. Burns and V. Burns, *Philos. Trans. R. Soc. London Ser. A.* **286**, 283 (1977).
16. G. P. Glasby and G. L. Hubred, *Comprehensive Bibliography of Marine Manganese Nodules*, Memoir 71, New Zealand Oceanographic Institute, Wellington, 1976, p. 55.
17. M. A. Meylan, B. K. Dugolinski, and L. Fortin, *Bibliography and Index to Literature on Manganese Nodules, Key to Geophysical Records Documentation No. 6*, National Oceanic and Atmospheric Administration, Washington, D.C., 1874–1975, p. 365, p. 60 addendum.
18. National Oceanic and Atmospheric Administration, *Description of Manganese Nodule Processing Activities for Environmental Studies*, 3 Vols., NTIS, Springfield, Va., 1977, PB-274 913, PB-274 914, PB-274 915.
19. R. D. W. Kemmit, in J. C. Bailar, H. J. Emeleus, R. Nyholm, and A. F. Trotman-Dickenson, eds., *Comprehensive Inorganic Chemistry*, Vol. 3, Pergamon Press Ltd., Oxford, U.K., 1973, p. 816.
20. U.S. Pat. 4,110,082 (Aug. 29, 1978), V. Michaels-Christopher.
21. F. H. Cotton and G. Wilkinson, *Advanced Inorganic Chemistry*, 4th ed., Wiley-Interscience, New York, 1980, pp. 736–752.
22. J. E. Ellis and R. A. Faltynek, *J. Am. Chem. Soc.* **99**, 1801 (1977).
23. R. Taube, *Chem. Zvesti* **19**, 215 (1965).
24. F. Calderazzo, R. Ercoli, G. Natta, I. Wender, and P. Pino, eds., *Organic Synthesis via Metal Carbonyls*, Vol. 1, Wiley-Interscience, New York, 1968.
25. U.S. Pat. 2,868,697 (Jan. 13, 1959), J. B. Bingeman and A. F. Limper (to Ethyl Corp.).
26. Ger. Pat. 45,071 (Jan. 5, 1966), H. J. Koch and F. Henneberger.
27. H. Marcy, in F. Matthes and G. Wehner, eds., *Anorganisch-Technische Verfahren*, VEB Deutscher Verlag für Grundstoffindustrie, Leipzig, Germany, 1964.
28. D. F. DeCraene and A. Reidies, personal communication, Chemetals Corp., Baltimore, Md., 1979.
29. Brit. Pat. 680,710 (Oct. 8, 1952), A. L. Hock and J. A. Dukes (to Magnesium Elektron Ltd.).
30. T. E. Moore, M. Ellis, and P. W. Selwood, *J. Am. Chem. Soc.* **72**, 856–866 (1950).
31. P. R. Khangaonkar and V. N. Mista, *J. Inst. Eng. (India)* **55**(Part MM), 59 (1975).
32. G. W. Kor, *Metall. Trans.* **9B**, 307 (1978).
33. S. K. Singh, T. Deb Roy, and K. P. Abraham, *Trans. Indian Inst. Met.* **27**, 87 (1974).
34. U.S. Pat. 3,375,097 (Mar. 26, 1968), J. Y. Welsh (to Manganese Chemicals Corp.).
35. W. Dautzenberg, in *Ullmanns Encyklopädie der Technischen Chemie*, 3rd ed., Vol. 12, Urban und Schwarzenberg, München-Berlin, 1960 p. 217.
36. T. E. Moore, M. Ellis, and P. W. Selwood, *J. Am. Chem. Soc.* **72**, 856–866 (1950).
37. U.S. Pat. 2,504,404 (Apr. 18, 1950), A. L. Flenner (to E. I. du Pont de Nemours & Co., Inc.).
38. J. B. Goodenough and A. L. Loeb, *Phys. Rev.* **98**, 391 (1955).
39. H. A. Lehmann and K. Teske, *Z. Anorg. Allg. Chem.* **336**, 197 (1965).
40. J. A. Conner and E. A. V. Ebsworth, *Peroxy Compounds of Transition Metals, Advances in Inorganic Chemistry and Radiochemistry*, N.7, Vol. 6, Academic Press, Inc., New York, 1961.

41. H. Gehrke, *MnO_2 Deposition in Feedlot Water Lines in South Dakota*, Completion Report No. PB-278809 for projects A-048-SDAK and A-058-SDAK, U.S. Dept. of the Interior, Office of Water Research and Technology, Washington, D.C., Jan. 1978.
42. W. Stumm and J. J. Morgan, *Aquatic Chemistry*, Wiley-Interscience, New York, 1970.
43. H. S. Posselt, F. J. Anderson, and W. J. Weber, Jr., *Environ. Sci. Technol.* **2**, 1087 (1968).
44. H. S. Posselt, A. H. Reidies, and W. J. Weber, Jr., *J. Am. Water Works Assoc.* **60**, 48 (1968).
45. R. G. Burns and V. M. Burns, *Structural Relationships Between Manganese(IV) Oxides in The Manganese Dioxide Symposium*, Vol. 1, I. C. Sample Office, 1975.
46. A. Byström and A. M. Byström, *Acta Cryst.* **3**, 146–151 (1950).
47. K. J. Euler, *J. Power Sources* **8**, 133–144 (1982).
48. M. M. Thackeray, *Prog. Batt. Batt. Mater.* **11**, 150–157 (1992).
49. R. Giovanoli, *Thermochimica Acta* **234**, 303–313 (1994).
50. M. A. Malati, *Chem. Ind.*, 446–452 (Apr. 24, 1971).
51. P. Ruetschi, *J. Electrochem. Soc., Electrochem. Sci. Tech.*, 2737–2744 (Dec. 1984).
52. H. S. Posselt, F. J. Anderson, and W. J. Weber, Jr., *ES&T* **2**(12), 1087–1093 (1968).
53. M. J. Gray and M. A. Malati, *J. Chem. Tech. Biotechnol.* **29**, 127–134 (1979).
54. J. W. Murray, *Geochim. Cosmochim. Acta* **39**, 505–519 (1975).
55. J. J. Morgan and W. Stumm, *J. Colloid Sci.* **19**, 347–359 (1964).
56. M. J. Gray, M. A. Malati, and W. M. Rophael, *J. Electroanal. Chem.* **89**, 135–140 (1978).
57. T. W. Healy, A. P. Herring, and D. W. Fuerstenau, *J. Coll. Interface Sci.* **21**, 435–444 (1966).
58. M. A. Malati, A. A. Yousef, and M. A. Arafa, *Chem. Ind.*, 459–460 (1969).
59. A. J. Fatiadi, *Organic Chem., Part 1, Synth.* (2), 65–104 (1976).
60. U.S. Pat. 4,299,735 (Nov. 10, 1981), P. G. Mein and A. H. Reidies (to Carus Corp.).
61. U.S. Pat. 5,260,248 (Nov. 9, 1993), N. Singh and K. S. Pisarczyk (to Carus Corp.).
62. U.S. Pat. 4,290,923 (Sept. 22, 1981), P. G. Mein and A. H. Reidies (to Carus Corp.).
63. H. A. Lehmann and K. Teske, *Z. Anorg. Allg. Chem.* **336**, 197 (1965).
64. U.S. Pat. 3,780,158 (Dec. 18, 1973), J. Y. Welsh (to Diamond Shamrock Corp.).
65. U.S. Pat. 3,780,159 (Dec. 18, 1973), J. Y. Welsh (to Diamond Shamrock Corp.).
66. U.S. Pat. 2,956,860 (Oct. 18, 1960), J. Y. Welsh (to Manganese Chemicals Corp.).
67. K. J. Euler and H. Mueller-Helsa, *J. Power Sources* **4**, 77 (1979).
68. Y. F. Shen and co-workers, *Science* **260**, 511–515 (Apr. 1993).
69. G. D. Van Arsdale and C. G. Maier, *Trans. Electrochem. Soc.* **33**, 109 (1918).
70. Jpn. Pat. 82344 (Mar. 27, 1929), H. Inoue and S. Haga.
71. P. M. DeWolff, J. W. Visser, R. Giovanoli, and R. Brütsch, *Chimia (Buenos Aires)* **32**, 257 (1978).
72. D. A. J. Swinkels and K. N. Hall, Prog. Batt. Batt. Mater. **11**, 16–24 (1992).
73. J. Y. Welsh, *Electrochem. Technol.* **5**, 504 (1967).
74. T. Matsumura, personal communication, Toyo Soda Manufacturing Co., Ltd., Tokyo, Japan, 1979.
75. G. Kano, *Mem. Faculty Eng. (FUKUI University)* **34**(2), (1986).
76. G. Strauss and A. G. Hoechst, personal communication, Werk Knapsack, Germany, 1978.
77. U.S. Pat. 3,375,097 (Mar. 26, 1968), J. Y. Welsh (to Manganese Chemicals Corp.).
78. C. B. Ward, A. I. Walker, and A. R. Taylor, *Prog. Batt. Batt. Mater.* **11**, 40–46 (1992).
79. Brit. Pat. 1,124,317 (Aug. 21, 1968), (to Knapsack Aktiengesellschaft).
80. T. Naguoki, in A. Kozawa and R. J. Brodd eds., *Manganese Dioxide Symposium*, Vol. 1, I. C. Sample Office c/o Union Carbide Corp., Parma Technical Center, Cleveland, Ohio, 1975.

81. Brit. Pat. 1,423,503 (Feb. 4, 1976), H. Kojima (to Mitsui Mining and Smelting Co., Ltd.).
82. E. Preisler, *Prog. Batt. Batt. Mater.* **10** (1991).
83. R. Huber, in K. V. Kordesch, ed., *Batteries*, Vol. 1, Marcel Dekker, Inc., New York, 1974, pp. 28–29.
84. W. C. Gardiner, *J. Electrochem. Soc.* **125**, 22C (1978).
85. Brit. Pat. 1,423,503 (Feb. 4, 1976), H. Kojima (to Mitsui Mining and Smelting Co., Ltd.).
86. U.S. Pat. 3,065,155 (Nov. 20, 1962), J. Y. Welsh (to Manganese Chemicals Corp.).
87. U.S. Pat. 3,702,287 (Nov. 7, 1972), H. Yamagishi and M. Tanaka (to Nippon Kokan Kabushiki Kaisha).
88. Belg. Pat. 871,683 (Feb. 15, 1979), (to Diamond Shamrock Technologies, SA).
89. E. Preisler, personal communication; E. Preisler and A. H. Reidies, *Winnacker-Küchler, Chemische Technologie*, Vol. 2, 4th ed., Carl Hanser Verlag, München, Germany, 1981.
90. E. Preisler, *J. Appl. Electrochem.* **19**(4), 540–546 (1989).
91. H. Lux, *Z. Naturforsch* **1**, 281 (1946).
92. K. B. Wiberg and K. A. Saegebarth, *J. Am. Chem. Soc.* **79**, 2822 (1957).
93. T. Chen, *The Oxidation of Alkenes, Alcohols and Sulfides by Manganates*, MS dissertation, University of Regina, Saskatchewan, Canada, 1991, p. 262.
94. D. G. Lee, C. R. Moylan, T. Hayashi, and J. I. Brauman, *J. Am. Chem. Soc.* **109**, 3003–3010 (1987).
95. R. S. Nyholm and P. R. Woolliams, in W. L. Jolly, ed., *Inorganic Synthesis*, Vol. XI, McGraw Hill, New York, 1968, pp. 56–61.
96. G. J. Palenik, *Inorg. Chem.* **6**(3), 507–511 (Mar. 1967).
97. F. R. Duke, *J. Phys. Chem.* **56**, 882 (1952).
98. H. I. Schlesinger and H. B. Siems, *J. Am. Chem. Soc.* **46**(9), 1965–1978 (1924).
99. R. Landsberg, *P. Orgel, Chem. Technol.* **13**, 665 (1961).
100. D. Arndt, *Manganese Compounds as Oxidizing Agents in Organic Chemistry*, Open Court Publishing Co., LaSalle, Ill., 1981, pp. 169–177.
101. H. Firouzabadi and E. Ghaderi, *Tetrahedron Lett.* (9), 839–840 (1978).
102. N. A. Frigerio, *J. Am. Chem. Soc.* **91**, 6200 (1969).
103. G. Lauf, personal communication, Carus Chemical Co., LaSalle, Ill., 1994.
104. R. Colton and J. H. Canterford, *Halides of The Transition Elements*, Wiley-Interscience, New York, 1969, pp. 212–213.
105. F. H. Herbstein, M. Kapon, and A. Weissman, *J. Therm. Anal.* **41**, 303–322 (1991).
106. J. P. Suttle, in M. C. Sneed and R. C. Brasted, eds., *Comprehensive Inorganic Chemistry*, Vol. 6, Van Nostrand Co., New York, 1957, p. 141.
107. *Gmelin Handbook of Inorganic Chemistry*, 8th ed., Manganese Part C 2, System Nymber 56, Springer-Verlag, New York, 1975, p. 177.
108. O. Sackur and W. Taegener, *Z. Elektrochem.* **18**, 718–724 (1912).
109. G. J. Palenik, *Inorg. Chem.* **6**(3), 503–507 (Mar. 1967).
110. A. H. Reidies, personal communication, Carus Chemical Co., LaSalle, Ill., 1994.
111. A. H. Reidies, personal communication, Spolek Pro Chemickov a Hutni Vyrobu n.p., Usti n. Labem, 1977.
112. BE 905 130 (1986), M. O. Gonzalez Garcia (to Industrial Quimica del Nalón S. A.).
113. H. Marcy in F. Matthes and G. Wehner, eds., Anorganisch-Technisache Verfahren, VEB Deutscher Verlag für Grundstoffindustrie, Leipzig, DDR, (1964).
114. A. K. Kohli and K. T. Cherian, personal communication, The Swadeshi Chemicals Private Ltd., Bombay, India, 1977.
115. V. M. Markavoa, A. P. Popova, I. M. Vulfovich, and V. M. Pisnyi, *Khim. Tekhnol. Soedin. Margantsa* (1), 78 (1975).

116. A. P. Popova, I. M. Vulfovich, and V. M. Markova, *Khim. Tekhnol. Soedin. Margantsa* (1), 66 (1975).
117. USSR Pat. 145,894 (Apr. 6, 1962), M. N. Machulkin and co-workers.
118. V. M. Markova, A. P. Popova, I. M. Vulfovich, and V. M. Pisnyi, *Khim. Tekhnol. Soedin. Margantsa* (1), 78 (1975).
119. K. Teske and H. A. Lehmann, *Chem. Technol.* **17**, 493 (1965).
120. USSR Pat. 150,828 (Oct. 25, 1962), M. N. Machulkin and co-workers.
121. U.S. Pat. 2,940,821 (June 14, 1960), M. B. Carus and A. H. Reidies (to Carus Chemical Co.).
122. U.S. Pat. 2,940,822 (June 14, 1960), M. B. Carus and A. H. Reidies (to Carus Chemical Co.).
123. U.S. Pat. 2,940,823 (June 14, 1960), M. B. Carus and A. H. Reidies (to Carus Chemical Co.).
124. W. Baronius and J. Marcy, *Chem. Technol.* **18**, 723 (1966).
125. Ref. 27, pp. 744–750.
126. D. K. Taraphdar, S. K. Ghosal, and R. N. Mukherja, *Indian J. Technol.* **12**, 260 (1974).
127. U.S. Pat. 2,843,537 (July 15, 1958), M. B. Carus (to Carus Chemical Co.).
128. U.S. Pat. 2,908,620 (Oct. 13, 1959), M. B. Carus (to Carus Chemical Co.).
129. U.S. Pat. 3,062,734 (Nov. 6, 1962), M. B. Carus (to Carus Chemical Co.).
130. U.S. Pat. 3,172,830 (Mar. 9, 1965), M. B. Carus (to Carus Chemical Co.).
131. Ref. 29, p. 753.
132. T. Kaczmarek, J. Chwalczyk, and B. Kolomyjek, *Chemik* **25**, 2, 58 (1972).
133. E. Müller, ed., *Methoden der Organischen Chemie (Houben-Weyl)*, Vol. IV/1b, Georg Thieme Verlag, Stuttgart, Germany, 1975.
134. A. J. Fatiadi, *J. Synth. Org. Chem.* (2), 85–127 (1987).
135. W. A. Waters, *Q. Rev. Chem. Soc.* **12**, 277 (1958).
136. J. W. Ladbury and C. F. Cullis, *Chem. Rev.* **58**, 403 (1958).
137. R. Stewart, *Oxidation Mechanisms*, W. A. Benjamin, Inc., Amsterdam, the Netherlands, 1964, pp. 58–76.
138. R. Stewart, in K. B. Wiberg, ed., *Oxidation in Organic Chemistry*, Part A, Academic Press, Inc., New York, 1965.
139. D. G. Lee, in R. L. Augustine, ed., *Oxidation—Techniques and Applications in Organic Synthesis*, Vol. 1, Marcel Dekker, Inc., New York, 1969, pp. 1–111.
140. D. G. Lee, *The Oxidation of Organic Compounds by Permanganate Ion and Hexavalent Chromium*, Open Court Publishing Co., La Salle, Ill., 1980.
141. L. F. Fieser and M. Fieser, *Reag. Org. Synth.* **1**, 942–952; **2**, 348; **4**, 30–31, 143; **5**, 562–563, John Wiley & Sons, Inc., New York, 1967, 1969, 1975, 1976.
142. Ref. 138, Table VII, p. 67.
143. V. N. Bobkov and Tr. VNII Vodosnabzh, *Kanaliz. Gidrotekhn. Sooruhz. i Inzh. Gidrogeol* (50), 48–50 (1975).
144. Ref. 138, pp. 60–61.
145. S. S. Rawalay and H. Shechter, *J. Org. Chem.* **32**, 3129–3131 (1967).
146. Y. Sayato, K. Nakamuro, K. Tsuji, and M. Tonomura, *Nat. Inst. Hyg. Sci., Eisei Shikenjo Hokoku (Japan)* **93**, 54–57 (1975).
147. H. Yanagibashi, *Tokyo Ika Daigaku Zasshi* **20**(1), 1–50 (1962).
148. J. Dockx, *Synthesis*, 441–456 (Aug. 1973).
149. G. W. Gokel and H. D. Durst, *ALDRICHIMICA Acta* **9**(1), (1976).
150. D. G. Lee, in W. S. Trahanowski, ed., *Oxidation in Organic Chemistry*, Academic Press, Inc., New York, 1980.
151. S. L. Regen and C. Koteel, *J. Am. Chem. Soc.* **99**, 3837–3838 (1977).
152. N. A. Noureldin and D. G. Lee, *Tetrahedron Lett.* **22**, 4889 (1981).

153. F. M. Menger and C. Lee, *J. Org. Chem.* **44**, 3446–3448 (1979).
154. A. J. Fatiadi, *J. Synth. Org. Chem.* (2), 102 (1987).
155. D. G. Lee, T. Chen, and Z. Wang, *J. Org. Chem.* **58**(10), 2918–2919 (1993).
156. F. M. Menger and C. Lee, *Tetrahedron Lett.* **22**(18), 1655–1656 (1981).
157. T. S. Jones, *Manganese 1992 Annual Report*, U.S. Department of the Interior, U.S. Bureau of Mines, Washington, D.C., Sept. 1993, p. 1.
158. Ref. 157, pp. 21–22.
159. D. F. DeCraene and K. Pisarczyk, personal communication, Chemetals Corp., Baltimore, Md., 1994.
160. Ref. 157, p. 24.
161. Ref. 157, p. 15.
162. R. Deblois, personal communication, Carus Chemical Co., La Salle, Ill., 1994.
163. N. H. Furman, ed., *Scott's Standard Methods of Chemical Analysis*, 6th ed., Vol. 1, D. Van Nostrand Co., Inc., Princeton, N.J., 1962, pp. 638–655.
164. F. D. Snell and L. S. Ettre, eds., *Encyclopedia of Industrial Chemical Analysis*, Vol. 15, Wiley-Interscience, New York, 1972, pp. 447–496.
165. A. J. Vogel, *A Text-Book of Quantitive Inorganic Analysis*, 3rd ed., Longman Group Ltd., London, U.K., 1961.
166. C. L. Wilson and D. W. Wilson, eds., *Comprehensive Analytical Chemistry*, Vol. IC, Elsevier Publishing Co., Amsterdam, the Netherlands, 1960, pp. 617–621.
167. I. M. Kolthoff, E. B. Sandell, E. S. Mechan, and S. Bruckenstein, eds., *Quantitative Chemical Analysis*, 4th ed., The Macmillan Co., London, 1969.
168. D. Glover, B. Schumm, Jr., and A. Kozawa, eds., *Handbook of Manganese Dioxides–Battery Grade*, International Battery Material Association (IBBA, Inc.), Brunswick, Ohio, 1989.
169. U. L. Schramm and F. C. Wedler, *Manganese in Metabolism and Enzyme Function*, Academic Press, Inc., Orlando, Fla., and references therein, 1986, Chapt. 18.
170. F. Baruthio and co-workers, *Clin. Chem.* **34**(2), 227–234 (1988).
171. J. J. Lingane and R. Karplus, *Ind. Eng. Chem.* **18**(3), 191–194 (1946).
172. W. Slavin, *Atomic Absorption Spectroscopy*, Interscience Publishers, a division of John Wiley & Sons, Inc., New York, 1968.
173. A. Montaser and D. W. Golightly, eds., *Inductively Coupled Plasmas in Analytical Atomic Spectrometry*, 2nd ed., VCH Publishers, Inc., New York, 1992.
174. B. Welz, *Atomic Absorption Spectrometry*, VCH, Weinheim, Germany, 1985.
175. B. Chiswell, G. Rauchle, and M. Pascoe, *Talanta* **37**(2), 237–259 (1990).
176. Ref. 138, Vol. 5, p. 14.
177. K. B. Wiberg, ed., *Oxidation in Organic Chemistry*, Vol. 5-A, Academic Press, Inc., New York, 1965, pp. 13–20.
178. H. W. Fishburn, Jr., and W. E. Dill, Jr., "A Method for the Semiquantitative Analysis and Identification of Mixed Phases of Manganese Dioxide," paper presented at the *Power Sources Conference*, Atlantic City, N.J., May 10, 1961.
179. *Standard Methods for Examination of Water and Wastewater*, 18th ed., American Public Health Association, Washington, D.C., 1992.
180. B. Chriswell and M. B. Mokhtar, *Talanta* **33**(8), 669–677 (1986).
181. E. Nakayamachal, *Anal. Sci.* **5**(2), 129–139 (1986).
182. N. C. Uren, C. J. Asher, and N. E. Longnecker, *Dev. Plant Soil Sci.* **33**, 309–328 (1988).
183. R. M. Riggin, E. J. Mezey, and W. M. Henry, *Gov. Rep. Announcement Index (U.S.)* **84**(12), 75 (1984).
184. H. J. M. Bowen, *Environmental Chemistry of the Elements*, Academic Press, Inc., New York, 1979.
185. M. C. Scrutton, M. F. Utter, and A. S. Mildvan, *J. Biol. Chem.* **241**(15), 3480–3487 (1966).

186. R. Schiele, *Manganese in Metals and their Compounds in the Environment*, VCH, Weinheim, Germany.
187. M. Alexander, *Introduction to Soil Microbiology*, John Wiley and Sons, Inc., New York, 1967.
188. National Research Council, *Manganese: Medical and Biological Effects of Environmental Pollutants*, National Academy of Sciences, Washington, D.C., 1973.
189. G. Matrone and co-workers, *Manganese Geochem. Environ.* **2**, 29–39 (1977).
190. M. O. Admur, L. C. Norris, and G. F. Heuser, *Rat. Proc. Soc. Exptl. Biol. Med.* **59**, 254–255 (1945).
191. D. L. Baly, I. Lee, and R. Doshi, *FEBS Lett.* **239**(1), 55–58 (1988).
192. *Manganese in Recommended Dietary Allowances*, 9th ed., National Academy of Sciences, Washington, D.C., 1980, pp. 154–157.
193. J. A. T. Pennington and co-workers, *J. Am. Diet Assoc.* **86**, 876–891 (1986).
194. J. F. Kopp and R. C. Kroner, *Trace Metals in Waters of The United States*, PB215680, U.S. Department of the Interior, Washington, D.C., 1967, p. 13.
195. R. A. Smith, R. B. Alexander, and M. G. Wolman, *Science* **235**, 1607–1615 (1987).
196. J. F. Kopp and R. C. Kroner, *Trace Metals in Waters of the United States*, PB215680, U.S. Department of the Interior, Washington, D.C., 1967.
197. P. Camner and co-workers, *Environ. Res.* **38**(2), 301–309 (1985).
198. *Toxicological Profile for Manganese and Compounds*, PB93-110781, Agency for Toxic Substances and Disease Registry, U.S. Public Health Service, Washington, D.C., 1992, p. 79.
199. Ref. 198, p. 80.
200. A. Iregren, *Neurotoxicol. Teratol.* **12**, 673–675 (1990).
201. Ref. 198, pp. 15–16.
202. A. Barbeau, *NeuroToxicology* **1**, 13–36 (1984).
203. H. E. Stokinger, in G. D Clayton and F. E. Clayton eds., *Patty's Industrial Hygiene and Toxicology*, Vol. 2A, John Wiley & Sons, Inc., New York, 1978, pp. 1749–1767.
204. *Scientific and Technical Assessment Report on Manganese*, PB-242-291, Environmental Protection Agency, Washington, D.C., 1975, pp. 6–7.
205. H. A. Rosenstock, D. G. Simons, and J. S. Meyer, *J. Am. Med. Assoc.* **217**, 1354–1358 (1971).
206. R. J. Lewis, *Sax's Dangerous Properties of Industrial Materials*, 8th ed., Van Nostrand Reinhold Co., New York, 1992.
207. *Registry of Toxic Effects of Chemical Substances*, National Institute of Occupational Safety and Health, Washington, D.C., Aug. 1994
208. R. J. Bull and G. F. Craun, *J. Am. Water Works Assoc.* **69**, 662 (1977).
209. I. Mena K. Horiuchi, K. Burke, and G. C. Cotzias, *Neurology* **17**, 128 (1969).
210. P. S. Papavasiliou, S. T. Miller, and G. C. Cotazias, *Am. J. Physiol.* **211**, 211–216 (1966).
211. A. Putrament, H. Baranowska, A. Ejchart, and W. Prazmo, *Methods Cell Biol.* **20**, 25 (1978).
212. A. Furst, *J. Nat. Cancer. Inst.* **60**(5), 1171 (1978).
213. *Chem. Reg. Rep.* **16**(13) (1992).
214. Ref. 198, p. 42.
215. *Permanganic Acid, Potassium Salt*, Registry of Toxic Effects of Chemical Substances, RTECS Number SD6475000, May 1993.
216. Ref. 198, pp. 97–98.
217. *American Conference of Governmental Industrial Hygienists*, 1986, 1988.
218. *Integrated Risk Information System*, U.S. Environmental Protection Agency, Washington, D.C., 1991.
219. *Code of Federal Regulations*, Title 40, Part 143.3, Washington, D.C., 1994.

220. "Superfund Ammendment Reauthorization Act," *Code of Federal Regulations*, Title 40, Section 313, Part 302, Table 302.4, 1994.
221. D. Glover, B. Schumm, Jr., and A. Kozawa, eds., *Handbook of Manganese Dioxide Battery Grades*, International Battery Material Assoc., 1989.
222. G. Blomgren, B. Schumm, Jr., and L. F. Urry, in W. Gerhartz, ed., *Ullmann's Encyclopedia of Industrial Chemistry*, Vol. A3, VCH Verlagsgesellschaft, Weinheim, Germany, 1985, pp. 352–361.
223. M. Kronenberg, in Ref. 222, p. 393.
224. S. Toon, *Indust. Miner.*, 19–35 (July 1985).
225. D. M. Holton, W. C. Maskell, and F. L. Tye, in L. Pearce, ed., *Power Sources*, Vol. 10, The Paul Press, Ltd., London, 1985, pp. 247–270.
226. L. Wu, J. R. Dahn, and D. S. Wainwright, *Science* **264**, 1115–1117 (1994).
227. J. R. Panek, in M. Morton, ed., *Polysulfide Rubbers in Rubber Technology*, 2nd ed., Van Nostrand Reinhold Co., New York.
228. R. Patel and D. Clifford, *Radium Removal from Water by Manganese Dioxide Adsorption and Diatomaceous Earth Filtration*, NTIS PB92-115260/AS, Springfield, Va., 1992.
229. B. McMichael, *Indust. Miner.*, 42 (May 1989).
230. N. Singh, "VOC Destruction at Low Temperatures Using a Novel Thermally Stable Transition-Metal Oxide-Based Catalyst," presented at the *First North American Conference on Emerging Clean Air Technologies and Business Opportunities*, Toronto, Canada, Sept. 1994.
231. U.S. Pat. 4,208,527 (June 17, 1980), G. Horlbeck, K. Burzin, and R. Feinauer (to Chem. Werke Hüls Ag).
232. J. Bieleman, *Eur. Poly. Paint Colour J.* **182**(4311), 412–416 (1992).
233. J. D. Bailie and A. H. Zeitz, Jr., in D. M. Considine, ed., *Methylcyclopentadienylmanganese Tricarbonyl Combustion Improver in Energy Technology Handbook*, McGraw-Hill Book Co., Inc., New York, 1977.
234. R. Hage and co-workers, *Nature* **369**(6482), 637–639 (1994).
235. A. Houston, *15th, 19th, 20th, 25th Annual Reports to the Metropolitan Water Board*, London, 1920–1930.
236. K. J. Ficek, in Robert L. Sanks, ed., *Water Treatment Design for the Practicing Engineer*, Ann Arbor Science Publishers, Inc., Mich., 1978, Chapt. 21, pp. 461–479.
237. K. J. Ficek and J. E. Boll, *Aqua* (7) (1980).
238. R. E. Hubel, E. W. Howe, A. Wilczak, T. A. Wolfe, and S. J. Tambini, *J. AWWA*, 43–51 (Aug. 1992).
239. W. R. Knocke, R. C. Hoehn, and R. L. Sinsabaugh, *J. AWWA*, 75–79 (Mar. 1987).
240. W. R. Knocke, J. E. Van Benschoten, M. J. Kearney, A. W. Soborski, and D. A. Reckhow, *J. AWWA*, 80–87 (June 1991).
241. G. F. Lauf and M. A. Waer, "Arsenic Removal Using Potassium Permanganate," presented at the *1993 Water Quality Technology Conference*, Miami, Fla., Nov. 1993.
242. K. J. Ficek and M. A. Waer, *The Fifth National Conference on Drinking Water*, Winnipeg, Manitoba, Canada, 1992, pp. 205–219.
243. P. L. Klerks and P. C. Fraleigh, *J. AWWA*, 92–100 (1991).
244. P. A. Vella, J. Munder, B. Patel, and B. Veronda, "Chemical Oxidation: A Tool for Toxicity Reduction," *Proceedings of the 47th Industrial Waste Conference*, Purdue University, West Lafayette, Ind., 1992.
245. P. A. Vella, J. Munder, and B. Veronda, "Chemical Oxidation: A Tool for Toxicity Reduction and Bio-Enhancement," presented at the *65th Water Environment Federation Conference and Exposition*, New Orleans, La., Sept. 1992.
246. U.S. Pat. 4,062,925 (1977), D. E. Witenhafer, C. A. Daniels, and R. F. Koebel (to B. F. Goodrich Co.).

247. J. H. Jackson, "Potassium Permanganate Solves Odor Problems at Consolided Mill," *Pulp Paper* (1984).
248. P. A. Vella, G. Deshinsky, J. E. Boll, J. Munder, and W. M. Joyce, *Res. J. WPCF* **62**(7), 907–914 (Nov./Dec. 1990).
249. R. J. Pope and R. A. Weber, *Water Environment Federation Proceedings of the 66th Annual Conference and Exposition*, 1993, pp. 447–456.
250. U.S. Pat. 4,425,380 (1984), F. J. Nuzzi and J. K. Duffy (to Kollmorgen Technologies.).
251. A. R. Del Gobbo and C. I. Courduvelis, *PC Fab.*, 54–59 (1987).
252. U.S. Pat. 4,601,783 (July 22, 1986), G. Krulik (to Morton Thiokol, Inc.).
253. N. V. Mandlich and G. A. Krulik, *Plat. Surf. Finish.*, 56–61 (Dec. 1992).
254. U.S. Pat. 4,601,784 (July 22, 1986), G. Krulik (to Morton Thiokol, Inc.).
255. M. Castegnaro, M. Coombs, M. A. Phillipson, M. C. Bourgade, and J. Michelon, *the 7th Polynuclear Aromatic Hydrocarbons, International Symposium*, Batelle Press, Columbus, Ohio, 1982, pp. 257–268.
256. J. Barek and L. Kelnar, *Microchem. J.* **33**(2), 239–242 (1986).
257. J. Barek, M. Castegnaro, C. Malaveille, I. Brouet, and J. Zima, *Microchem. J.* **36**(2), 192–197 (1987).
258. J. Barek, *Microchem. J.* **33**, 97–101 (1986).
259. J. Barek, A. Berka, and M. Muller, *Microchem. J.* **33**(1), 102–105 (1986).
260. H. S. Posselt and F. J. Anderson, Steel Wire Handbook, Vol. 2, The Wire Association, Branford, Conn., 1969, pp. 211–228.

KENNETH PISARCZYK
Carus Chemical Company

MANNITOL. See SUGAR ALCOHOLS.

MANNITOL HEXANITRATE. See EXPLOSIVES AND PROPELLANTS.

MANNOSE. See CARBOHYDRATES; SWEETENERS.

MANUFACTURED GAS. See FUELS, SYNTHETIC–GASEOUS FUELS.

MARCASITE. See IRON; IRON COMPOUNDS.

MARGARINE. See VEGETABLE OILS.

MARKET AND MARKETING RESEARCH

Market research is a long established technique used to secure data for management to use in its decision making. Market research may be short or long term. Some market analysts use the following time frames: short term, up to 18 months; intermediate term, 18 months to 5 years; long term, 5 to 10 years. In general, short-term market research is synonymous with sales analysis and is used to assist the sales manager in setting goals, measuring performance, and giving the production department operating targets.

Intermediate or long-term market research has as its objectives the quantifying of markets for a particular chemical in terms of tonnages, growth potentials, general location of markets, competitive factors, and the impact of existing or potential government regulations on the market.

Marketing research, as compared to market research, is more directly concerned with identifying existing or potential users of a product, their present sources of supply, the nature and duration of any contracts that exist between producer and buyer, competitors' strategies in product development and pricing, requirements for facilities and personnel to compete successfully, and the status of competition from producers in other countries. Also, government regulations involving production of chemicals, their transportation, and disposal of wastes and by-products have a marked influence on the profitability of most chemical process industry operations.

Market research studies usually originate in the sales or marketing groups of a company. As a general rule, the sales analysis or short-term-type study is done by in-house personnel, often on a continuous basis. Field sales personnel are often used to assist the market research group in securing data. Long-term market research studies may originate in sales or marketing groups if the company already produces the product. If a new product is involved, the study may originate in the research and development group or at the corporate planning level. Marketing research studies usually originate in the higher levels of management, eg, general manager or vice president. This is especially true if the proposed study is for a product new to the company.

Selection of an in-house group or consulting firm to do a market or marketing research study does not follow any set pattern. In some cases, an outside consulting group is retained to do an independent study of an area already researched by an in-house group. Most often the consultancy is unaware of the prior in-house study, and its report is used as a check study or an assurance to management that its report made a proper assessment.

Use of consultants to conduct chemical market and marketing research studies began in the early 1950s in the United States and grew rapidly during the mid-1960s. In the 1990s there are at least 100 well-known and capable consultants or consulting firms in the United States performing this function for individual clients or on a multiple client basis.

Western European chemical companies were slower in adopting market and marketing research tools to assist in their operations and planning. A few large companies such as ICI, Bayer, Hoechst, and Solvay were using the techniques in the late 1950s, but not until the early 1970s did European chemical companies use these studies as frequently as their U.S. counterparts.

A company decides to use a consultancy for a variety of reasons. The most frequent motivations are desire for the study to be done anonymously in order not to arouse existing producers or customers; the company is unfamiliar with the geographic area or the product line; an independent opinion is wanted without any chance of bias or preconceived ideas; the company is raising capital and wants to show the lender or underwriter an independent report; the in-house market or marketing research group is overloaded with work or the company does not have an in-house group; and an "insurance" study to compare against their own work is wanted.

Sales Analysis

It is axiomatic that sales analysis depends on detailed records of sales of a specific chemical to a specific company. Paramount to the success of such studies is the existence of data recorded on a systematic and continuous basis. It follows that these studies are done best by an in-house staff on products already produced by the company. However, on occasion, a product new to the company can be studied by the in-house group with the assistance of their field sales force. For example, a producer of polypropylene could use its people to secure data on the consumption of other thermoplastics by their customers. Such an exercise might identify opportunities for a new producer, but a more detailed marketing research study would probably be done before entry into the new product area was made.

Methodology. All internal sales and purchase records are computerized. This makes their retrieval and manipulation relatively quick and easy. In practical terms, a company can follow the trends in its markets on a monthly or quarterly basis. Changes in customer patterns in terms of quantities, grades, or payment alert management to seek causative factors and take appropriate action.

Use of Results. Sales analysis data are used in many ways by company management. The results are most useful in production planning, particularly if grade differences appear to be in the offing, and in assuring that adequate supplies are available for sales. Inventory control, raw material procurement, technical service requirements, and trends in accounts receivable are beneficiaries of good sales analyses.

Market Research

Market research in the chemical process industries differs sharply from consumer market research primarily in the so-called universe. In industry studies, the universe is quite small compared to the consumer market. For example, in some industrial markets five customers may use 80 to 90% of the total production of a given product. There are also other important differences: "industrial marketing research pays a great deal of attention to market size and potential estimation, and relatively little attention to psychological market segmentation.

Industrial market demand is derived demand. As such, it is more volatile than consumer demand. As a result, industrial marketing research shows a greater concern with business and economic conditions, raw material prices, and inventories. Industrial market demand also results from group-buying decisions to a greater extent than does consumer market demand. Organizational factors play a key role in industrial buying. Consequently, industrial buying behavior research tends to focus on different issues and to employ different research procedures than those that typify consumer research" (1).

Methodology. Practitioners of chemical market research develop individual styles and techniques. However, four elements are essential to every useful study: defining the problem, data gathering, analysis of data, and presentation of findings.

Defining the problem is often overlooked. When it is, the final report is often useless. In-house groups are sometimes given vague generalizations as the basis for a study. It is incumbent upon the market research manager to meet with the people requesting the study and with the executives approving the study to ensure that each party understands what is wanted. The better the guidance that the study users can give to the market research staff, the better the study. For example, the authorizers should state clearly what data are required, why they are needed, what will be done with the results, and what corporate decisions may be made as a result of the study. Armed with this knowledge, the market research staff can plan its program properly for in-house and external interviews, search of secondary data sources, and other steps basic to a study.

If a consultant is hired to do the study, it is equally important for the client to inform the consultant of its need in detail. Since the consultant usually does not have daily access to the client as does the in-house group, it is more important for the consultant to have an accurate statement of the client's problem and needs.

Data gathering usually consists of using both primary and secondary sources of information. Primary sources are usually the existing or potential users and/or producers of the product under study. Field work or interviews with these respondents is best done by personal meetings. Such meetings permit the interviewer to gauge the reaction of the respondent to a particular question, to encourage the respondent to check his or her records, or to contact other people in the company, and often a walk through the plant reveals useful facts, eg, number and source of tank cars on a siding.

Unlike consumer market research, chemical market research interviewers rarely use a printed questionnaire. Instead, a dialogue is sought between the respective parties interested in the same subject. There is no hard and fast rule on whether to make appointments or risk cold calls. Each practitioner must decide which course should be followed, taking into account the time and budget parameters of the study. Usually, a few key interviews in a given geographic area are scheduled and others are cold.

Telephone interviews are often substituted for personal visits or are used to supplement such visits. They do not equal personal visits but are useful if only a few questions are involved or if time and cost parameters preclude extensive personal field calls. Time and cost factors have led to a decline in the number of personal calls made by companies or consultancys in their market studies.

Telephone interviewing by its very nature implies a cold call at what may be an inconvenient time for the respondent. To overcome this limitation, some market researchers send a brief letter to the respondent outlining the subject for discussion and why the respondent will benefit from the discussion. A suggested time for the actual call is indicated. This technique has yielded good results in many studies.

Experience is the key element in successful market research interviewing. A successful practitioner knows which companies and individuals are usually receptive to inquiries, which respondents have proved accurate in the past, and which have usually been misinformed or deliberately misleading. In the latter case, the researcher should try to determine the reason for this; if successful, an important clue may be found for the final evaluation. Researchers also face the problem of weighing the opinions of different respondents in the same company with those of respondents in other companies. Judgment and experience are the only criteria that help in this difficult task.

Protocol problems may exist for market researchers in a company. Often company policy dictates that any call on a customer requires the agreement of the sales department and the presence of a sales representative at the actual meeting. Similarly, if calls are to be made on suppliers, purchasing department approval and attendance are often necessary. These rules can delay or lessen opportunities for a useful dialogue but their existence must be recognized and coped with in the field.

The number of calls to primary data sources can range from 5 to 500 or more, depending on the subject, the type of data needed, the degree of accuracy required, and the time/cost parameters of doing the study. Again, judgment and experience are important to a market researcher. It is easier to guess at the number of companies and who they are than it is to guess which people or job titles should be contacted. Often the larger the company, the greater the number of people who must be contacted in order to develop any useful data. It would be foolhardy of a market researcher studying the existing or future uses of a particular plastic in automobiles to contact one person at General Motors and conclude what GM will do. Recognition of the various centers of decision making in large individual organizations develops slowly, but the experienced practitioner has learned the hard way that casual approaches often yield sloppy and useless reports or even capital losses. An analysis of hundreds of market research studies on specific products reveals that the average number of company contacts is 50 to 60 and the number of persons contacted is 75 to 150.

Secondary sources of data are useful when they exist. Databases (qv) of published information have been assembled, and market researchers can tap them provided their company buys the service. These databases can save the market analyst many hours of work. The services also provide much of the general sociopolitical–economic background needed, such as petroleum (qv) prices, government regulations, foreign competition, etc.

Most databases secure their information from printed sources. On occasion, however, a subsequent letter to the editor of a publication by a company mentioned in the article will point out an error. Unfortunately, these corrections are not always picked up by the respective databases that entered the initial data.

It is estimated that over 6000 databases exist either as online or portable types, eg, CD-ROM. Of the total, only about 60 primarily cover chemicals (2).

There is also a growing number of specialized databases available on specific topics such as CFC Replacement, effluents and pesticides, environmental chemical data, etc. These are usually on CD-ROM or floppy disk (3).

Secondary sources also may exist within a company or consulting firm. These sources are usually unpublished reports or raw data collected at a prior time for another purpose.

Mail surveys are rarely used in chemical market research studies. However, they can be useful if the right conditions prevail. For example, if only a few questions requiring simple answers are needed from a large number (≥500) of respondents, a properly designed mail questionnaire usually generates an acceptable return. On the average, the returns are 25 to 40% of the number mailed, but some returns of 90% have occurred. Mail questionnaires can often be used as a screening device to identify possible respondents for follow-up telephone or personal visits when the market researcher is studying a totally new universe. This is especially true if the product under study is bought by many small users.

Record keeping is an essential requisite of good market research. In the chemical field, call reports or visit reports are usually written by the interviewer and become part of the report in some cases and certainly should become part of the company or consultant files for future reference. Obviously, the call report serves a valuable purpose in the analysis and writing stage. Some market researchers have also found that cross-referencing call reports over a period of time allows rapid identification of the respondents who have demonstrated the greatest ability in forecasting their company needs and/or the needs of their industry.

There are seminars and short courses on the methodology of chemical market research. Reference 4 is a useful primer, but there is no substitute for actual fieldwork experience, facing varying degrees of suspicion by the respondents, to hone the practitioner's skills.

Multiple client studies have proliferated on a world basis since the first generally recognized study of this type done in 1952 on polyethylene by the Roger Williams organization. There are directories of available multiple client studies that can direct the researcher to a source, eg, *FINDex–The Directory of Market Research Reports, Studies, and Surveys*. Multiple client studies can have budgets of $50,000 to $250,000 and can require the equivalent of several work years by the researchers.

Consultants, such as Chem Systems, CMAI, Colin Houston & Associates, Kline & Co., Arthur D. Little, Philip Townsend Associates, Inc., Stanford Research Institute (SRI), and Strategic Analysis, Inc. are a few of the many companies that prepare such reports.

Large consulting firms, usually identified as management consultants, often prepare individual client reports for chemical companies. These consultancys include Anderson Consulting, DeLoitte & Touche, McKinsey, Peat Marwick, etc.

A multiple client study may serve many useful purposes to company management. It may act as a primer for intensive study of a narrower part of the market, a guide to deciding whether to expend R&D effort in the field, or a guide to deciding whether to acquire a company already in the field. The independent study can also resolve conflicting opinions within different departments of the client company. Finally, the cost of a multiple client study is usually lower than the cost of a comparable study by a clients' in-house personnel.

Costs. There are two cost elements in doing marketing research studies: professional charges and out-of-pocket expenses. The actual cost of any study is entirely dependent upon the number of interviews and the type of interviews. In practice, a 100-person interview study will cost $30,000 to $50,000. However, if a $50 $\times$ 10^6 investment is involved, the market study is cheap insurance.

Analysis of Data. A veteran practitioner of chemical market research likened this step to the assembly of a jigsaw puzzle. There are many pieces of unequal size and importance that must be put together to make a picture understandable to everyone. Call reports, secondary data inputs, experience, and judgment are the tools used by the market researcher to analyze the data, reach conclusions, make recommendations, and write the report.

Both novice and experienced market researchers face the same problems at the outset. Two problems usually arise: certain pieces of desired information are lacking, and respondents contradict each other to a significant degree. Missing information can usually be obtained by follow-up telephone calls to the original respondents or new respondents. Of course, if highly proprietary information is being sought the chances of success are slim, and the researcher must resort to judgment based on experience or the counsel of others.

Clarification of contradictory opinions also can be obtained by follow-up calls. It has often been found that if one recontacts a respondent and bluntly says that others have contrary opinions, the respondent will answer by giving the basis for his opinion or change it.

However, the market researcher has to form an opinion based on all the data. Various methods exist for manipulating the opinions, facts, and numerical data into forecasts and conclusions. Techniques in use include statistical analysis, correlations with external factors, correlations with other products, and informed opinion.

Statistical analysis can range from relatively simple regression analysis to complex input/output and mathematical models. The advent of the computer and its accessibility in most companies has broadened the tools a researcher has to manipulate data. However, the results are only as good as the inputs. Most veteran market researchers accept the statistical tools available to them but use the results to implement their judgment rather than uncritically accepting the machine output.

Market researchers doing sales analysis usually have an excellent record of accuracy over the short term. This is a result of good data, good judgment, and the easier predictability of events a year ahead rather than three or five years ahead. In the case of a completely new product, the first year or two can be difficult and the analyst either too optimistic or too conservative.

Correlation of markets for a product with external factors is a relatively quick and easy method of analysis, useful if the markets are correlatable with factors such as population, gross national product (GNP), Federal Reserve Board (FRB), index, etc.

Correlation with markets for other products is particularly useful for a new product. For example, market growth history of an older product, eg, nylon, can be plotted on a graph to predict the probable growth for a newer product, eg, polyester fibers. Data for both products may be plotted on the same chart, though not necessarily to the same scale and with the time scale shifted to bring the respective curves in parallel.

Informed opinion is a nonmathematical technique which is often called blue-sky estimating. The two terms are not necessarily synonymous. Informed opinion may be the consensus of people inside the company, along with or without outside opinions. Such a consensus can be broadened into the Delphi technique, which uses a sequential series of questionnaires submitted to a panel of experts who have little or no contact with each other. Blue-sky answers are akin to the informed opinion except that they usually consist of subjective speculation by experts in a confined one-day meeting in which each participant knows the views of the others.

Presentation of Results. A wide variety of procedures and techniques are used to present results of market research. The technique used varies by type of study, by the source of the study within a company, and from company to company. In broadest terms, both written and oral reports cover almost every possibility.

Written reports, within a given company, are often prepared to a prescribed format. In some cases, brevity is the criterion, and management requires a single-page summary report. In other cases, a diagrammatic scheme is used in fold-out form of roughly desk-top size, and all pertinent facts are presented. Some companies have two reports prepared: a detailed presentation for reference, and an executive summary of one to eight pages abstracted from the detailed report.

Oral reports usually involve use of either slides or flip charts as visual aids. It has been found that more executives will attend an oral report presentation than will read a 100-page written report. Combinations and variations of oral and written reports are also widely used. For example, both consultants and in-house groups often present a brief oral executive summary with or without visual aids several weeks in advance of the written report.

Use of Results. Market researchers are occasionally disappointed in the use made of their reports. They cite instances where action contrary to their recommendations is taken, often with discouraging results; or where no action is taken, and another company successfully takes advantage of the opportunity. It is good practice for a market research manager to follow up a report and try to determine if management is using it in making decisions. Of course, a market research manager must recognize that management may have compelling reasons for not following the report recommendations. In some cases, the reasons can initially only be divulged to top management. For example, negotiations may be underway for an acquisition in the market under study or the company may have a temporary cash flow problem.

Market and Marketing Research Organizations

Market research in the U.S. chemical industry began to be formalized as early as 1940. In 1945, the Chemical Market Research Association (CMRA) was formed with 75 members. In mid-1965, it was renamed and became the Chemical Marketing Research Association to reflect the broadened function of its members. In 1990, the CMRA was again renamed and is now the Chemical Management and Resources Association and has about 1000 members.

The current CMRA defines its purposes to be "to promote the growth and development of marketing management, business development, business intelligence and planning in the chemical or allied process industries through industrial marketing or business/market research; to provide continuing education and foster the development of those so engaged; to contribute and make available to the public, information in the field of chemical and industrial marketing, management, and business research; to cooperate with government officials in furthering the national welfare, and to carry out such activities recognized as law for such organizations" (5).

In West Europe, a similar pattern of evolution began in the 1960s with informal meetings and was formalized in 1967 when the European Chemical Marketing Research Association (ECMRA) became a division of the European Association for Industrial Marketing (EVAF). EVAF was later renamed the Federation of European Marketing Research Association (FEMRA) and has about 500 members of which about 300 are ECMRA members.

In Asia, the Asian Chemical Marketing Research Association (ACMRA) was organized in 1987 and has about 30 members as of this writing.

The globalization of the markets for chemicals has led to an increasing number of market and marketing analysts to have a membership in all of these organizations and to attend meetings in each geographic area.

Methodology. The methodology previously outlined for market research studies is applicable to marketing research studies. However, many more elements must be considered, especially in the realm of strategy factors.

Marketing research groups in some chemical companies also conduct studies beyond their normal activities. These include assistance in the market development phases of new product introductions, searches for unfilled needs in products or services which their company may be able to meet, and searches for new uses for existing products. As in market or marketing research studies, the methods used in such search studies require contact with a broad spectrum of people in a diverse range of companies. In these studies, however, the initial work is usually qualitative in nature. The researcher strives to get as broad an exposure as possible in the hope that this yields a few recurring ideas or suggestions to be studied later if they appear to be promising opportunities. In searches for unfilled needs or new uses, it is not uncommon for the success ratio to be 1 out of 50.

New product development programs present another type of challenge to the researcher. Often the researcher has no guidelines for evaluating the new product and must formulate a unique plan for developing enough information to construct a matrix that would show the risks and rewards of the project. Reference 6 presents 10 commandments for new-product development.

A competent market research manager must be aware of the thinking and plans of top management and must impart the necessary guidance to the staff in order for them to make logical and reasonable decisions involving new-product development. Methodology for marketing research studies differs most from market research studies when the researcher evaluates the competitive forces at work and formulates a marketing strategy.

Competitors usually are readily identifiable. Their current product lines, their recent and current pricing policies, and their recent and current marketing practices are also usually discernible. However, predicting their future products and policies and their response to a new and aggressive competitor poses a significant challenge to the skills of the researcher. An exhaustive study is necessary to secure even faint clues as to the competitor's future actions and, of course, most markets surveyed have several entrenched competitors. The researcher pursues many avenues of inquiry; among them are existing or potential customers for the product, engineering companies and equipment suppliers specializing in plants for the product under study, financial analysts, and raw material suppliers. Usually, only a few clues are found and the experienced researcher has to make a judgment even with gaps in knowledge. One of the most difficult aspects of such inquiries by an in-house group is that every competitor quickly is alerted to the interest of the researcher's company. Thus, in many cases, companies retain a consulting firm for the study. Consultants often can secure a bit more information on these sensitive subjects provided the respondents know them and trust their discretion.

As a general rule, marketing research studies involving comprehensive coverage of competitive firms requires personal contacts. Telephone inquiries usually are futile unless the respondent is a friend of long-standing, the mail surveys are generally unsuitable.

Costs. Since much more personal contact work is required, the cost of marketing research studies is significantly higher than the cost of market research studies. Also, the advisability of using the most senior personnel raises the cost. It is not uncommon for in-depth marketing research studies to cost $35,000 to $100,000, depending on the complexity of the subject.

Analysis of Data. Again, the basic techniques outlined earlier for market research studies apply to marketing research studies. However, in the realm of competitive forces and the formulation of strategies, the pragmatic judgment of the experienced researcher is essential. In most cases, the researcher does not have hard data to draw on. Instead, to some degree, a series of mental images of the principal competitors has been formulated indicating their probable response to new developments or competitors in a given market.

Elements that the researcher evaluates about competitors include plants, processes, raw material costs and availability, distribution channels, product development skills, service facilities, personnel, pricing policies, eg, does the competitor lead or follow?, and practices or concessions to secure and hold large customers. All of these factors are weighed and then the researcher decides on a strategy for the company.

The classic strategies are well known: acquisition, internal development, licensing, and joint ventures.

As a general rule, acquisitions are considered for established products with above-average growth potentials. Often, entry by acquisition is more timely and profitable than internal development and subsequent plant construction. Following the latter course might take 5 to 10 years, during which time the highest return on investment (ROI) is lost.

Internal development is usually recommended if the company has a unique process or product for an evolving market or a unique, less expensive process for an existing product. Licensing is resorted to when entry is desired and no suitable acquisitions can be found but a licensor of a suitable process exists.

Joint ventures are recommended if the respective companies have complementary strengths. For example, one may have the process, the other the raw materials, a strong position in the market, or an appropriate geographic location.

Finally, the researcher must consider the kind of strategy which best fits the company taking into account management, financial, marketing, and technical resources.

Marketing Strategy Factors. Of the elements mentioned earlier as factors in determining strategy, several deserve more detailed discussion: pricing, distribution channels, applications research, technical service, and concessions to customers. It is useful to divide the products of the chemical industry into two broad groups: commodity and specialty. Commodity chemicals imply those produced in large tonnages. Although this is usually true, a more useful criterion is that a commodity chemical is used for its specific chemical structure, which can be specified, and that the material available from every producer is essentially identical.

Specialty chemicals, however, differ in that they are used for their performance properties and usually are not specified chemical entities. Products from different suppliers usually differ somewhat, and free interchangability is not always possible. Special chemical systems also exist in the market and these are formulated products that contain both commodity and specialty chemicals.

Management skills that are successful for commodity chemicals may fail for specialty chemicals and vice versa. A simplified comparison of these two types of chemicals is pertinent and illustrative.

Production Requirements. Production of commodity chemicals usually requires large dedicated plants, generally in continuous operation and often with a proprietary process. A basic raw material position is preferred and is sometimes required. Specialty chemicals, on the other hand, require small- to medium-sized batch-type plants with inherent flexibility. A basic raw material position is rarely required.

Marketing Requirements. The primary portion of commodity chemical output is often sold on long-term contracts. Selling and service costs are minimal. For specialty chemicals, service and selling costs are likely to be high.

Prices. The price of commodity chemicals is based on cost of production, capital needs for expansion, and the ratio of supply to demand. Profit margins can drop under changing conditions, and unit price tends to be low. Specialty chemical prices vary widely. They are based on the value of a product or system to the customer. Profit margins can usually be maintained, and unit price is higher than for commodity chemicals.

Research and Development. For commodity chemicals, emphasis is on the improvement of plant operation and reduction of production costs. For specialty chemicals, emphasis is on assembling a staff capable of quickly identifying and solving a customer problem under the existing plant conditions and operating procedures of the customer.

Capital Investment and Returns. Capital needs for commodity chemicals are usually very high, hundreds of millions of dollars being needed for many petrochemical plants of economic size. Return on investment (ROI) and return on sales (ROS) vary widely. Capital needs of specialty products are usually quite low, often about 50 to 60¢ per dollar of sales. ROI and ROS vary widely but are usually higher than those for commodities.

Management. Top management usually has financial or production orientation and interests in commodity chemicals. For specialty chemicals, top management often is entrepreneurial, versed in customers' needs and dedicated to solving customers' problems.

Experienced market researchers planning a strategy that involves acquisition of a specialty chemical company by a commodity chemical producer have to be aware of these differences. Many such acquisitions have failed over the years because the commodity company management failed to recognize or refused to believe these differences.

A researcher studying a specialty chemical business or a specific specialty chemical company should pay particular attention to these rudiments: innovative talents, service facilities and performance, marketing abilities, and responsiveness to customer needs. Of these, the last may be the key criterion in most cases. A company that frequently is first with the solution to a customer's problem (even if the solution is sometimes less than perfect) usually holds the customer and a dominant market share against future competition.

Finally, a researcher must assess how the differing personalities and practices of the company for which the study is being done compare to those of the specialty company. If conflict appears likely, the researcher should discourage the acquisition because it probably will fail to meet its goal.

Pricing. Chemical pricing has always been a complex subject, but rapidly escalating raw materials costs, costs of meeting government regulations, inflationary pressures, existence of competition on a world basis, and excessively high costs of capital give management more problems than ever.

In the past, commodity chemicals were generally priced on the basis of ROI. Capital cost was the most critical item, and those elements that are related to capital cost were the principal factors in the selling price (excluding raw material cost in some cases). On this basis, a satisfactory ROI resulted in acceptable values for other criteria such as ROS or sales margin. Many analysts favor ROS as a benchmark for comparison because it is up to date and simple and because it is increasingly difficult to determine a true ROI based on what profits might be on plants built under inflation and expensive capital and construction costs.

Historically, in the commodity chemical business the newest and largest plant has been the lowest cost producer. Under an inflationary economy, the newest plant is usually the highest cost producer unless it features a unique process with significantly higher product yields and/or lower production costs. Often in the past, the newest producer having the lowest cost plant bought a position in the market by lowering the price. Recently, the new producers list price is that of the marketplace.

Pricing of specialty chemicals and specialty chemical systems is, as noted, based on value to the customer. The elements of raw material and production cost enter into the producer's calculations, but extra emphasis must be given to applications research and service costs with minimal attention in most cases to capital costs.

The pricing of a new chemical that will compete against other chemicals does involve the usual cost elements that set the price. However, it has been shown that an empirical approach may be of value. One empirical approach is the exclusion chart (7,8) developed in 1979 and modified several times over the years. The chart indicates the approximate volume that existing chemicals used for the same function have achieved at their price. There are exceptions where the new chemical can command a higher price and achieve comparable volume because it possesses a unique property.

Distribution Channels. Most commodity chemicals are primarily sold by the producer to a relatively small number of very large users. However, producers of commodity chemicals also utilize distributors to reach small volume users. Distributors buy in bulk and repackage or resell in smaller amounts to a broad spectrum of users. Distributors profit by the difference between their bulk cost and their LCL (less-than-carload lots) sales plus a commission from the producer, which may be as high as 15% of the bulk price but is more often 5 to 10%.

Distributors operate on an industrial and/or geographic basis. The industrial group is especially successful in selling some types of specialties and many commodities. Geographic distributors usually handle a wider range of products from many suppliers and are more proficient in selling commodities rather than specialties.

A researcher formulating a strategy for a particular company must determine what portion of the planned output will be sold by its sales force and whether distributors or manufacturers' representatives might be beneficial. If it appears that a distributor or agent is needed, a separate evaluation of the capabilities of representative companies or individuals should be made.

Applications Research. Specialty chemical producers devote a larger share of their time and costs to applications research than do producers of most commodity chemicals. As noted earlier, the most successful specialty chemical producers have been those companies that are able to respond quickly to customer needs and problems under the conditions found in the customer's plant. This entails having, at the specialty chemical plant, equipment and procedural knowledge which closely approximate those found among customers. Tests can then be run and a solution to the problem or need may result. If successful, even in part, it can be brought to the customers and tried there. In practice, of course, each customer's plant has some variables which make a single answer or product quite unlikely. Fortunately, slight modifications by the supplier will often solve the next customer's problem.

Commodity chemical producers have varying records of performance in applications research. It is usually high on the priority list when the product is still evolving, eg, low density polyethylene in the late 1950s and early 1960s. In times of pinched profit margins, these services often have been dropped, sometimes to be reinstituted, especially if totally new uses appeared.

A researcher planning a strategy must determine if a commitment to applications research is required. If so, the cost of facilities and personnel and the time required to assemble these must be calculated and included in the overall cost of entry.

Technical Service. Technical service usually occurs at the customer level, in contrast to applications research which occurs at both producer and customer locations. Often the field sales force functions as a quasitechnical-service group backed up by specialists operating from a corporate location. In the commodities, technical service is usually found where polymers are involved. In specialties, it exists over a broader spectrum of producers and users. As a generalization, technical service more often functions to quickly solve an operating problem which arises on a given day; applications research usually involves solving problems which arise because of basic changes in the procedures or formulations used by the customer.

A researcher planning strategy must determine whether commitment to a technical service facility and personnel is required. If it is, the cost of this commitment must be determined and included in the overall cost of entering the product field.

Concessions to Customers. A researcher formulating a strategy for a client company must take into account any special situations that may exist between a seller and a buyer for a given product. These arrangements may arise for various reasons that can be entirely legal. Long ago reciprocity arrangements were the subject of federal government examination and blatant cases were banished. However, favored relationships do exist and must be searched out and evaluated.

If two or three of the principal customers are unavailable to a new supplier, the problem of selling becomes more acute. In fact, if a significant portion of the so-called merchant market is unavailable to a new producer, entry into the field could be disastrous. Special arrangements can arise because of the proximity of supplier plant to user plant, raw material availability from one firm to the other, common financial ownership to some degree, toll arrangements, etc.

Presentation of Results. As in the case of market research reports, there are multiple techniques for informing management of the results of a study. Because the marketing research study is usually more complex and more detailed, a series of reports or presentations may occur, including some or all of the following: overview oral report to top management; overview but more detailed oral report to individual departments or divisions; brief written reports for top management, highly visual in nature; brief written reports for division heads; and a complete written report for reference.

Use of Results. Since a marketing research study is often part of a total feasibility study, the results are usually evaluated by management and a decision is made as to the corporate position. It is incumbent on marketing research managers or their superiors to determine if the recommendations they made will be considered. On occasion, middle and top management personnel have ignored the marketing research report recommendations and serious problems have arisen. Marketing research personnel are not omniscient but they have to defend their conclusions and recommendations with authority in the overall corporate decision-making process.

Purchasing Research

A relatively new but growing responsibility for some market research practitioners is purchasing research, which uses most of the procedures previously discussed, but has a different objective, ie, ensuring the availability of raw materials at competitive prices for three or more years ahead. Historically, purchasing agents in large chemical companies always practiced an informal type of market research. However, the shortages and price escalations that followed the October 1973 oil embargo led to the adoption of more formalized purchasing research in many chemical companies and this practice continues.

Hardly a raw material used by the chemical industry has been unaffected by the continuing ripple effect of on-and-off inflation. Raw material prices have not risen evenly. Shifts in relative prices of competing materials continue to occur. Dealing with these shifts is a significant challenge to market research and chemical buyer personnel. It must be stressed that a total view of prices, availability, and competing demands is now required and developments must be constantly monitored.

Of course, some differences in methodology exist between purchasing research practices and conventional chemical market research. For example, the analyst must be in rather continuous contact with the operating departments of the company to keep informed of their raw material demands for five to ten years ahead. Recognition must be given to the life cycle of the company product line. Thus if the product line is primarily one of mature commodity chemicals, growth in demand is unlikely to exceed 5%/yr. However, if the product line includes a number of new products in the early stages of their life cycle, growth of 20–25%/yr is quite possible, albeit from a rather small base tonnage.

Reports in purchasing research usually differ from conventional market research reports. In many companies, a purchase profile report is prepared. It shows concisely the existing vendor capacities for the raw material, planned expansions or new producers, demands for other uses, and demand within the analyst's company. In some cases, the report includes a world supply and demand balance. A key objective of a purchase profile report is to make the buyer as well informed as the marketing manager of the seller. If this is achieved, the buyer can often secure a beneficial purchase contract.

Key contents of a purchase profile report for a specific chemical are identity, location, and capacity of primary vendors; expected additions or deletions of capacity and their timing; captive use/merchant supply status of each vendor; pricing history; pricing influences (feedstock, energy, etc); demand by use and anticipated growth; and demand in the purchaser company up to five or ten years ahead.

Company practices differ in who does purchasing research and how it is done. Several patterns are evident. The chemical buyer is responsible for preparing the purchase profiles, possibly with in-house library assistance. Market research analysts are assigned to the purchasing department and prepare some or all of the profiles needed. Outside consultants are used to prepare some of the purchase profiles or as a check on internal procedures and conclusions.

There is a difference of opinion as to whether a chemical buyer or purchasing-research analyst should be product or division oriented. Those who favor product orientation claim they achieve a broader and deeper understanding of the outlook for the chemicals they buy and this leads to sound purchasing strategy. Proponents of the division orientation claim that the product-oriented analyst has too many chemicals to follow (up to 100 specific chemicals in some companies with 10 to 15 as principal purchases). If, instead, division needs are paramount in the mind of the analyst, more profitable buys can be made. The weakness of this latter argument is that in multidivisional, multibillion dollar chemical companies, this division-oriented analyst may have as many chemicals to follow as a product-oriented counterpart.

Competitive uses exist for almost every chemical a purchasing department buys. Often the demands of other uses and users are larger and influence supply and prices. Therefore, it is necessary to monitor these uses and especially to determine whether alternative materials might come into use.

Competitive Intelligence

A few market and marketing analysts in the chemical and chemical process industries are doing competitive intelligence work. This function has its own professional organization, the Society of Competitive Intelligence Professionals, and it has about 2000 members, largely from nonchemical businesses. Competitive intelligence work requires a constant monitoring of announced and rumored developments. It is used to seek out emerging technologies that may impact on some operation of a company and affect its competitive standing.

Competitive technology intelligence (CTI) is used to monitor technologies operating globally to produce consumer and industrial products including some in the chemical process industries (9). Competitive intelligence techniques are also used in benchmarking. This involves comparing one company against other companies to determine if the company is ahead or behind in technology, marketing, R&D, customer service, rapidity of introducing new products, etc. It is also being used to develop market and marketing research information that may indicate forthcoming acquisitions, alliances, and changes in a competitor's corporate structure.

The gradual but continuing reorganization of many chemical companies began in the late 1980s. One of the groups that has been affected are staff personnel in the marketing research group. This has led to the origin of some new consultancys or the addition of the released analysts to existing consulting firms. It is uncertain as to when or if the internal marketing research groups will resume studies.

BIBLIOGRAPHY

"Marketing and Marketing Research" in *ECT* 2nd ed., Vol. 13, pp. 66–87, by C. Pacifico, Management Supplements, Inc., and R. Williams, Jr., Roger Williams Technical & Economic Services, Inc.; "Market and Marketing Research" in *ECT* 3rd ed., Vol. 14, pp. 895–910, by E. Tarnell, Roger Williams Technical & Economic Services, Inc.

1. W. E. Cox, Jr. and L. V. Dominquez, *Ind. Market. Manage.* **8**, 81 (1979).
2. *Today's Chemist at Work*, 21 (Jan. 1993).
3. *Chem. Eng.*, 145 (Feb. 1993).
4. D. D. Lee, *Industrial Marketing Research, Techniques and Practices*, Technomic Publishing Co., Westport, Conn., 1978.
5. *Bylaws*, Chemical Management and Resources Association, 1990.
6. R. G. Block, *Ind. Res. Dev.* **21**, 97 (Mar. 1979).
7. H. W. Zabel, *Chem. Eng.* **66**, 183 (Oct. 19, 1959).

8. R. Williams, *CHEMTECH*, 592 (Oct. 1973).
9. S. R. Vatcha, *CHEMTECH*, 40 (May 1993).

EDWARD TARNELL
Colin A. Houston & Associates, Inc.

MARTENSITE. See METAL TREATMENTS; STEEL.

MASS SPECTROMETRY

In its simplest form, a mass spectrometer is an instrument that measures the mass-to-charge ratios m/z of ions formed when a sample is ionized by one of a number of different ionization methods (1). If some of the sample molecules are singly ionized and reach the ion detector without fragmenting, then the m/z ratio of these ions gives a direct measurement of the molecular weight. The first instrument for positive ray analysis was built by Thompson (2) in 1913 to show the existence of isotopic forms of the stable elements. Later, mass spectrometers were used for precision measurements of ionic mass and abundances (3,4).

Ideally, a mass spectrum contains a molecular ion, corresponding to the molecular mass of the analyte, as well as structurally significant fragment ions which allow either the direct determination of structure or a comparison to libraries of spectra of known compounds. Mass spectrometry (ms) is unique in its ability to determine directly the molecular mass of a sample. Other techniques such as nuclear magnetic resonance (nmr) and infrared spectroscopy give structural information from which the molecular mass may be inferred (see INFRARED TECHNOLOGY AND RAMAN SPECTROSCOPY; MAGNETIC SPIN RESONANCE).

If the sample to be analyzed is a mixture, it is not always easy to distinguish between molecular ions and fragment ions. Fragment ions can only be used for structure determination, however, if ascribed to a particular molecular ion. Thus separation of the molecular species in a mixture before mass spectral analysis is very important and chromatographic techniques are often employed (see CHROMATOGRAPHY). Interfaces to link mass spectrometers to gas chromatographs appeared in the 1960s (5), and to liquid chromatographs in the 1970s (6). As newer chromatographic systems have been developed, interfaces for mass spectrometers have also appeared (see ANALYTICAL METHODS, HYPHENATED INSTRUMENTS).

An important alternative to chromatographic separation (7) of a mixture is the use of tandem mass spectrometry, designated mass spectrometry/mass spectrometry (ms/ms) (8). In ms/ms, a molecular ion is mass selected by a mass spectrometer for activation in a collision cell. Some of the excited ions have enough energy to fragment, and the resulting product ions are mass analyzed by a second mass spectrometer. In the analysis of a mixture, therefore, all the different molecular species can be individually selected and the corresponding characteristic mass spectra obtained without interference from the other components. This is only possible, however, for mixtures which do not contain isobaric components; therefore, there is a practical limit to the complexity of mixtures which can be analyzed by ms/ms alone. Consequently, ms/ms and chromatography are frequently used together to analyze a complex sample (9).

In the first mass spectrometers, the sample was heated to give a vapor which was then ionized by electron ionization (EI). Some compounds do not give molecular ions by EI and the development of chemical ionization (CI), an attempt to obtain molecular weight information for such compounds, followed (10). In CI, the sample is ionized by ion–molecule collisions between sample molecules and an ionized reagent gas, eg, methane. A gas-tight ion source is used to maximize the number of collisions, and the high pressure reagent gas is ionized by electron ionization. This technique produces, instead of molecular ions, protonated or deprotonated molecular ions. These ions can have very little excess energy (11) and may therefore reach the detector before fragmenting. For thermally labile materials, soft ionization techniques which do not require direct heating of the sample have been developed. These techniques also give protonated, deprotonated, or cationized ions, which may not fragment within the ion source. The resulting spectra therefore may not contain the fragment ions needed for structural studies. The ms/ms process can then be used to decompose these molecular ion species to give the product ion data needed to confirm the molecular structure.

Instrumentation

A mass spectrometer consists of four basic parts: a sample inlet system, an ion source, a means of separating ions according to the mass-to-charge ratios, ie, a mass analyzer, and an ion detection system. Additionally, modern instruments are usually supplied with a data system for instrument control, data acquisition, and data processing. Only a limited number of combinations of these four parts are compatible and thus available commercially (Table 1).

Three important parameters for mass spectrometers are mass resolution, mass range, and sensitivity. The resolution, R, required to separate two ions of mass m and $(m + \Delta m)$ is given by equation 1.

$$R = \frac{m}{\Delta m} \tag{1}$$

For example, a resolution of 500 separates m/z 501 from m/z 500, or m/z 50.1 from m/z 50. As mass resolution is increased, a more accurate measurement of ion mass is obtained, and because elements do not have integer masses, the exact mass of an ion can be used to determine its elemental formula (12). In target compound analysis, high resolution is used to separate the ions of interest from the matrix background and to confirm the ions' identity. A reference compound, eg, perfluorokerosene, which gives fragment ions that are mass-deficient, ie,

Table 1. Compatible Inlet Systems, Ion Sources, and Mass Analyzers

Inlet system[a]	Ion source/interface[b]	Mass spectrometer[c]
heatable probe, gas chromatograph	EI, CI	sector quadrupole, ion trap, ftms, TOF
hplc	TSP, particle beam (EI/CI source), continuous flow FAB, ESI	quadrupole, sector, ftms
dedicated probe or other inlet system	FAB	sector, quadrupole, ftms
	FD	sector
	MALDI	TOF, ftms, sector
	plasma desorption	TOF

[a]hplc = high performance liquid chromatograph.
[b]EI = electron ionization, CI = chemical ionization; TSP = thermospray; FAB = fast atom bombardment; FD = field desorption, MALDI = matrix assisted laser desorption.
[c]ftms = Fourier transform mass spectrometry; TOF = time of flight.

that have masses slightly less than an integer number, is used as an internal mass reference during high resolution experiments. Most high resolution data are acquired using magnetic sector instruments, but much higher resolutions are possible with Fourier transform mass spectrometry (ftms) instruments.

The term mass range refers to the range of masses of singly charged ions which can be analyzed by the mass spectrometer. Because the mass scale is actually a scale of mass-to-charge ratio, the detection of multiply charged ions makes it possible to determine the mass of a molecule outside of the mass range of the mass spectrometer. Sensitivity is generally defined in one of two ways, either as a single-to-noise ratio for a specific compound under defined analysis conditions, or as a signal strength, eg, C/μg.

Sectors. In a sector mass spectrometer, ions are formed in an ion source that is at a potential V, usually 8 to 10 kV. Upon leaving the ion source they undergo acceleration to a velocity v, where $eV = 1/2\, mv^2$, and then enter a magnetic sector which separates the beam of ions according to their corresponding momenta. The transmission of ions through such a device is described by

$$\frac{m}{z} = \frac{B^2 r^2 e}{2\,V} \tag{2}$$

where r is the radius of the magnet, B is the magnet field strength, V is the ion source voltage, m is the mass of the ion, z is the charge of the ion, and e is the charge of an electron. Varying the magnetic field sequentially focuses ions of different masses at the detector. Mass resolution is set using adjustable slits immediately after the ion source and before the detector. The smaller the slit aperture, the higher the resolution.

The addition of an electrostatic analyzer, before or after the magnetic sector, increases the maximum resolution of the mass spectrometer. These double-focusing mass spectrometers containing both a magnetic sector and an electrostatic analyzer (13) were first produced in the later 1950s. Figure 1 is a schematic diagram of a double-focusing sector mass spectrometer of Nier-Johnson (EB) geometry, where EB refers to the electric field, E, proceeding the magnetic field, B. Between the analyzers are field-free regions, eg, the first field-free region

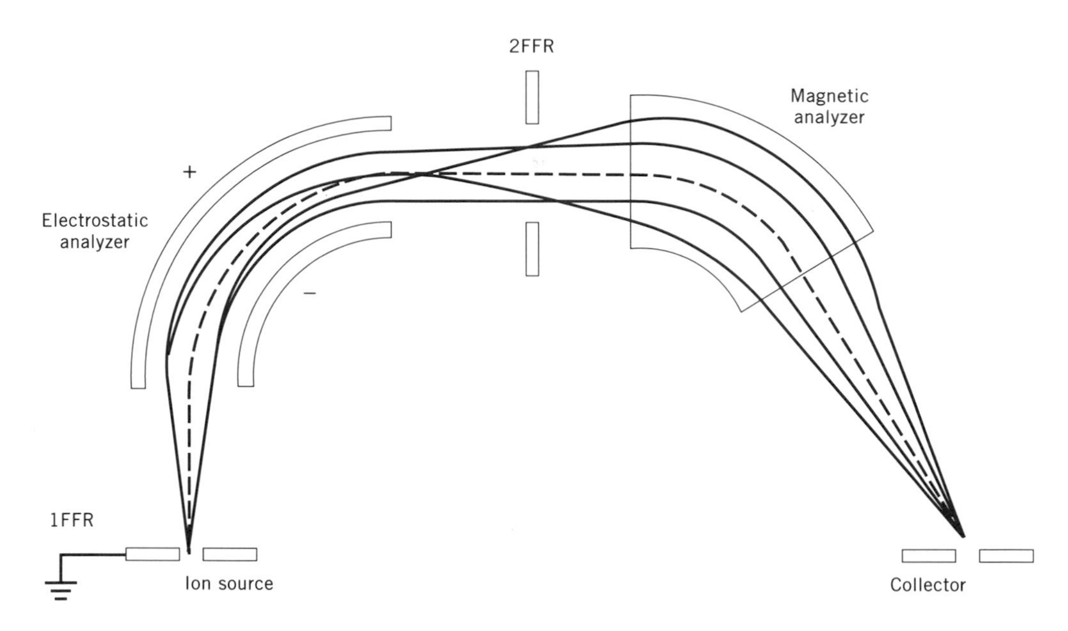

Fig. 1. Schematic diagram of a double-focusing Nier-Johnson magnetic sector mass spectrometer where (——) represents paths of ions having slightly more or less kinetic energy than the mean kinetic energy of all the ions (– – –). 1FFR and 2FFR represent first and second field-free regions, respectively (1).

(1FFR) between the ion source and electrostatic analyzer. This is where collision cells are placed for acquiring ms/ms spectra. Mass resolution for high performance sector instruments has steadily increased since their introduction, and commercial instruments offer resolutions in excess of 10^5. One application of very high resolution scans is the analysis of petroleum oil.

Environmental applications, such as dioxin analysis, have led to the development of very sensitive sector instruments which, when operating at 10^4 resolution, can quantitatively detect dioxins at femtogram levels in environmental extracts. This level of performance gives overall method sensitivities in the ppm range. Newer magnet technologies have led to instruments having mass ranges of 10,000 or more, at full sensitivity (14).

To give increased sensitivity when the analysis is not limited by chemical noise, eg, in the ms/ms mode, array detectors (15) have been developed. No collector slit is fitted, and the full width of the ion beam falls onto microchannel plates which emit electrons. The emitted electrons strike a phosphor coating on the end of a fiber-optic cable (see FIBER OPTICS). The phosphor emits photons which travel along the cable to the photodiode array. Simultaneous detection of between 4 and 40% of the mass range, gives two orders of magnitude decrease in detection limits (see PHOTODETECTORS).

Full computer control of sector instruments has made this type of instrument much easier to operate than previously. These instruments require more maintenance than lower resolution instruments, however.

Quadrupoles. The quadrupole mass filter (16,17), which became available commercially in 1967, consists of four cylindrical rods having circular or, more

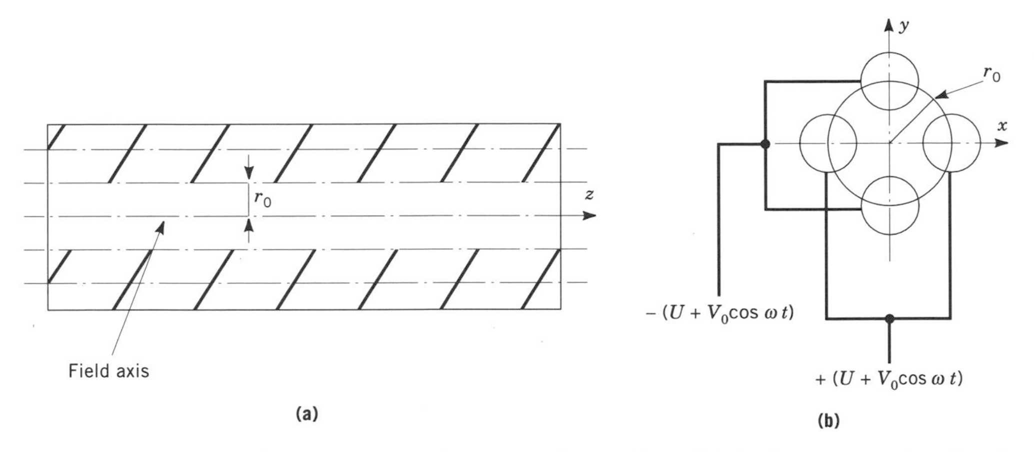

Fig. 2. Schematic diagram of a quadrupole analyzer where (**a**) is the cross-sectional and (**b**) the end-on view (1). Terms are defined in text.

recently, hyperbolic cross sections, arranged with their centers on the perimeter of a circle of radius r_0. Figure 2 is a schematic diagram of a quadrupole analyzer. The rods have alternate +/− d-c potentials, U, and a radio-frequency (r-f) component $V_0\cos\omega t$. The parameters a and q are defined as follows:

$$a = \frac{8eU}{mr_0^2\omega^2} \tag{3}$$

$$q = \frac{4eV_0}{mr_0^2\omega^2} \tag{4}$$

Only a finite range of values of a and q allow ions of mass m to travel along the central axis of the spectrometer having trajectories stable enough to reach the ion detector. The spectrometer is well suited to computer control because the range of masses transmitted is determined by scanning U and V_0 while keeping the ratio U/V_0 constant. The mass resolution is increased when U is changed so that a/q is increased. This also reduces the number of ions transmitted.

When compared to sector instruments, however, the quadrupole mass spectrometer is a low resolution device having a maximum mass resolution of only a few thousand. A big advantage of quadrupole instruments is that because the ion source is at a potential of only a few hundred volts, there are fewer problems with ion source voltage breakdown when these are interfaced to liquid chromatography systems. Initially, quadrupole systems had mass ranges of less than 1000, but high mass systems now go to molecular masses of ca 4000. The low cost, ease of use, and easy interfacing to chromatographic techniques has made the quadrupole ms the most widely used type of mass spectrometer. These can be made small enough and rugged enough to be used for direct monitoring of industrial-scale reactions (18).

Ion Traps. The mass analyzer of the ion trap mass spectrometer (19,20) is formed from three cylindrically symmetrical electrodes (Fig. 3). The two end cap electrodes are held at ground potential and contain channels through which

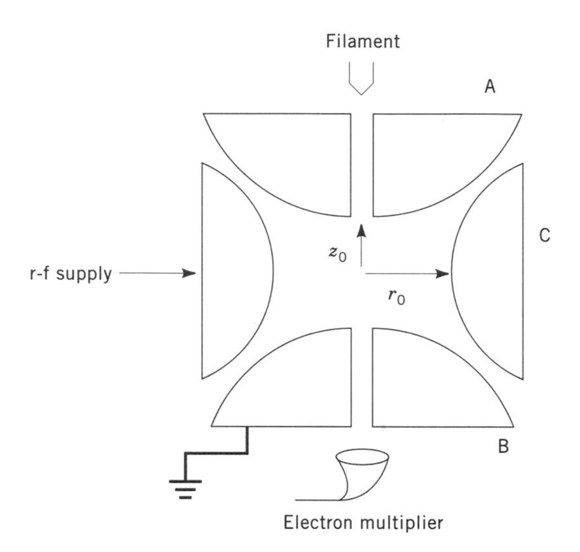

Fig. 3. Schematic diagram of an ion trap where A and B represent end cap electrodes, C the ring electrode, r_0 the internal radius of C, and z_0 the internal radius of the trap along the z-axis (1).

electrons can be introduced into the trap for sample ionization by EI. The resulting ions are expelled to the detector. The third-ring electrode is fed with an r-f voltage and sometimes a d-c voltage. Operation of the trap is similar to the quadrupole, and the parameters a_z and q_z may be calculated from equations 3 and 4, where r_0 is the internal radius of the ring electrode. Typically, the electron beam is pulsed on for a finite length of time and then switched off. The amplitude of the applied r-f voltage is then increased, expelling ions from the trap in order of increasing m/z values to be detected by an electron multiplier. The ion trap has very high sensitivity because it is an ion storage device and almost all the ions produced by the electron pulse are detected. By storing ions of a particular mass and collisionally activating them, low collision energy $(\text{ms})^n$ experiments can be performed. Sample dynamic range is limited, however, because when a large number of ions are formed in the trap, space–charge effects can occur which make the system behave in a nonideal manner. To reduce the impact of this problem, software has been developed which varies the filament on-time to give a constant number of ions in the trap.

The low cost of these instruments has encouraged modification of ion-trap mass spectrometers. Data showing high mass resolution (21), a wider mass range than the normal 600 amu (22), and the interfacing of traps to ionization techniques other than EI (23), have been produced by a number of laboratories. Commercial instruments having these features are not, however, available. A complete gas chromatography/mass spectrometry (gc/ms) ion trap, which includes the gas chromatograph and data system, can be purchased for around $80,000.

Time of Flight. In principle, time-of-flight (TOF) mass spectrometers (24) are probably the simplest (Fig. 4). Ionization of the sample in an ion source at a potential V produces ions which are accelerated to kilovolt energies on leaving

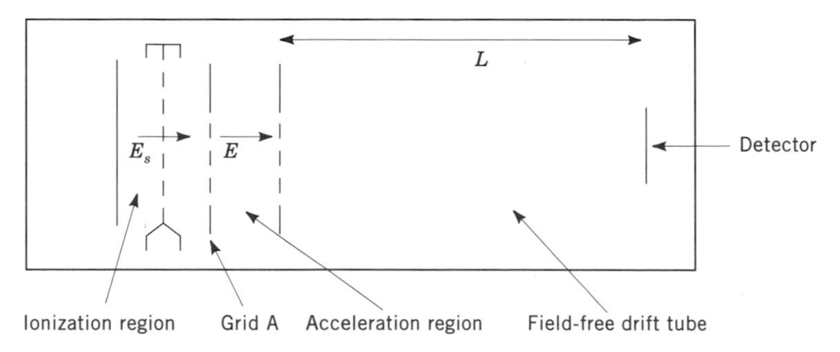

Fig. 4. Schematic diagram of a TOF spectrometer where E_s is the extraction field, E the acceleration field, and L tube length (1).

the source region, and the time, t, taken for the ions to transverse a linear flight tube of length L is proportional to the square root of the mass, m, where

$$t = \left[\frac{m}{2zeV}\right]^{1/2} L \quad (5)$$

Very high sensitivity is obtained because almost all the ions formed in the ion source are detected, and the mass range is almost limitless. TOF systems work best when pulsed ion sources are used, and the flight time of the ions is then given by

$$t = t_a - t_0 \quad (6)$$

where t_a is the time at which the ion is detected and t_0 is the time at the start of ionization. Advances in digital electronics provide very accurate timing circuits, given good time and consequently good mass resolution even from short flight times. Short flight times are important because these allow many ionization events over a short time period. Hence rapid changes in analyte concentrations can be followed. A number of groups are using TOF systems for fast gc/ms by acquiring tens of scans per second (25). A principal disadvantage of TOF instruments is the lack of mass resolution, because ions of the same mass do not all have exactly the same velocity. To improve the mass resolution, reflection TOF instruments have an ion mirror to compress the spread in velocities of ions of the same mass, but even then, unit mass resolution is only possible for ion masses up to a couple of thousand mass units.

Fourier Transform Systems. In instruments for Fourier transform ion cyclotron resonance mass spectrometry (fticr or ftms), ions are produced inside a cell within the solenoid of a superconducting magnet, as shown in Figure 5. Ions of mass m, having a charge z, inside a magnet of field strength B, move in a circular path perpendicular to the magnet field axis with a frequency of ω where when B is in Telsa and ω is in Hertz,

$$\omega = 1.537 \times 10^7 \left(\frac{zB}{m}\right) \quad (7)$$

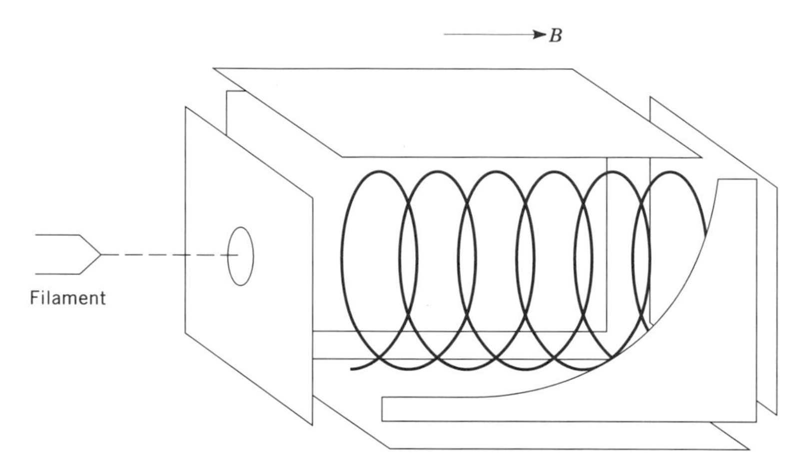

Fig. 5. Schematic diagram of an fticr cell. The direction of the magnetic field B, is shown by the arrow (1).

Applying a r-f voltage of frequency ω to the transmitter plates in the cell increases the orbital radius of the ions without changing ω, and after a few rotations the ions move together coherently and can be detected. The detection process is relatively inefficient, however, because between 10 and 100 ions are needed to produce a detectible signal. In contrast an electron multiplier can detect single ions.

To record a complete mass spectrum, a very fast frequency sweep of the voltage applied to the transmitter plates is performed. The induced image currents are detected as a time–domain signal. This signal is then converted into a frequency–domain signal by the application of a fast Fourier transform (fft). Very high ($<10^6$) resolution spectra can be obtained as long as the source pressure is low ($<10^{-6}$ Pa ($<10^{-8}$ torr)) enough. Therefore, the interfacing of high pressure techniques such as high performance liquid chromatography (hplc) to ftms has required the development of external source machines. A high pressure ion source outside the magnet is used to produce ions that are injected through a differentially pumped transfer line into the analyzer cell, which is maintained at the low pressure needed for high resolution. Because this is an ion storage device, like the ion trap, complex $(\text{ms})^n$ experiments are possible. Like ion traps, ftms instruments also suffer from space–charge effects which result in perturbation of the ion motion, and this in turn limits the system's dynamic range. These systems have benefited from the continual reduction in price of the powerful computers needed for system control and data processing.

Mass Spectrometers for Tandem Mass Spectrometry. To acquire ms/ms spectra the ion of interest is isolated from other sample ions and activated to cause it to fragment. The resulting product ions are mass analyzed to give the ms/ms spectrum. This is most easily achieved using two mass spectrometers in tandem and having a high pressure gas cell between them (26). The gas pressure is usually set to reduce the precursor ion intensity at the final detector by 50%. The resulting product ions have an abundance of only a few percent relative to the precursor ion.

The simplest example of this type of instrument is the triple quadrupole ms (27), in which Q_1 and Q_3 are used for mass analysis of the precursor and product ions, respectively, and Q_2 is a rf-only quadrupole collision cell. The maximum possible energy uptake during collisional activation, E_{CM}, is a function of the mass of the collision gas, m_g, the voltage offset between Q_1 and Q_2, E_{LAB}, and the mass of the ion m, as given by

$$E_{CM} = E_{LAB}\left[\frac{m_g}{m + m_g}\right] \tag{8}$$

For low ($E_{LAB} < 500\ eV$) energy collisions either argon or xenon is typically used as the collision gas to give the largest possible E_{CM}. A disadvantage of low energy collisions is that some structurally informative reactions, such as charge–remote reactions, are not easily accessed at low collision energies (28). Triple quadrupole instruments can easily be used to perform a wide variety of ms/ms experiments under full computer control. These are described in Table 2.

During gc/ms or liquid chromatography/mass spectrometry (lc/ms) acquisitions, it is possible to perform a mixture of the experiments described in Table 2 for different time windows, with the experimental parameters, such as the collision energy, optimized for each analyte.

All the experiments described in Table 2 can also be performed using magnetic sector instruments. Both Nier-Johnson and reverse geometry instruments are used. Collision chambers are located in the field-free regions as shown in Figure 1 for the former, and between the ion source and magnetic sector and between the magnetic sector and the electrostatic analyzer for the latter. Product ions formed in a first field-free region collision chamber of either geometry mass spectrometer can be detected by linked scan. All product ions from a given precursor ion are detected if B/E is a constant. All precursors of a given product ion are detected if B^2/E is a constant. Ions which have undergone a neutral loss of m_n mass units are detected when $B/[E^{-2} - (E_0E)^{-1}]^{-1/2}$ is a constant. Additionally, selected decomposition monitoring is possible upon appropriate B/E ratios being set for each product ion mass.

Other linked scans are used to detect product ions formed in the second field-free region of a reverse geometry mass spectrometer. The most widely used of these is the mass-analyzed ion kinetic energy scan (mikes). The precursor ion

Table 2. Scan Functions for a Triple Quadrupole Mass Spectrometer

ms/ms mode	Q_1	Q_3
all product ions from a given precusor ion	set to precursor ion mass	scan
all precursors of a given product ion	scan	set to product ion mass
constant neutral loss of m_n mass units	scan	scan so that passes masses m_n mass units less than Q_1
selected decomposition monitoring	set to precursor ion mass	set to product ion mass

is selected by adjustment of the magnet field, and product ions are detected by scanning the voltage of E. Extensive use of mikes scans has been made in studies of ion structures (29). When compared to product ions scans using the first ffr, mikes give better precursor ion resolution, but poorer product ion resolution.

The sector ms/ms machine analogous to the triple quadrupole is the four-sector instrument, which is a pair of double-focusing mass spectrometers linked by a collision chamber (Fig. 6) (30). The collision chamber is usually floated (31) with respect to earth, because this improves the collection efficiency for low mass product ions. Collisional activation occurs at E_{LAB} between 4 and 8 keV, using helium or helium–argon mixtures as the collision gas (32). The first mass spectrometer, ms1, is used to select a precursor ion, and the product ion spectrum is obtained by performing a linked scan of the magnet field strength, B, and electrostatic analyzer voltage, E, for ms2 such that B/E is a constant. The resulting spectra are structurally informative and can include fragmentation reactions that are not observed using triple quadrupole instruments.

When sequencing peptides, it is possible to distinguish between the amino acids (qv) leucine and isoleucine from the loss of a radical from the alkyl side chain, but this process is only observed at high collision energies (33). Figure 7 is the ms/ms spectrum of a small peptide, showing how the fragment ion masses confirm the amino acid sequence. Mass differences between the peaks give the masses of the amino acids eg, b_3 minus b_2 is 57 amu, a glycine residue (35).

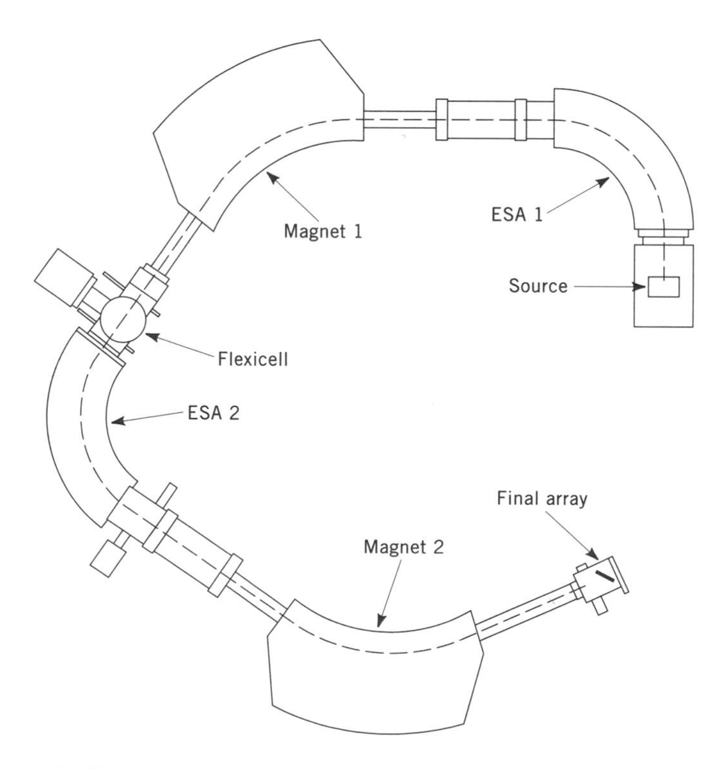

Fig. 6. Schematic diagram of a four-sector mass spectrometer of *EBEB* geometry where ESA = electrostatic analyzer (30). The flexicell is the collision chamber and the final detector is an array.

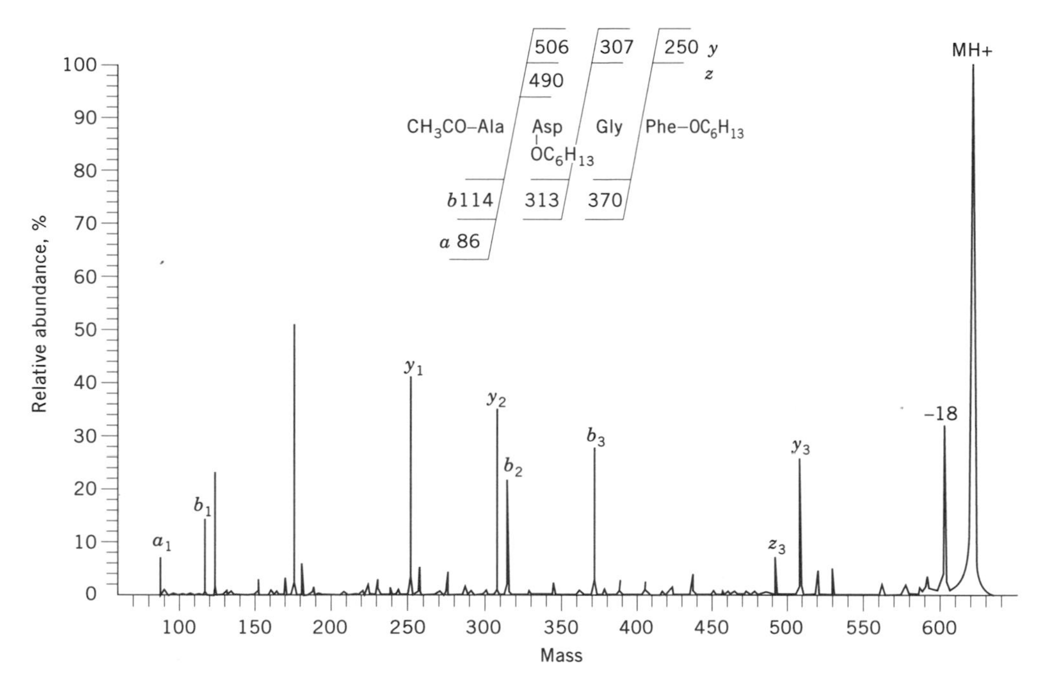

Fig. 7. High energy ms/ms product ion spectrum of a hexylated tetra-peptide (33,34).

The use of an array detector instead of an electron multiplier as the ion detector makes it possible to obtain product ion spectra for peptide sequencing from subpicomole (36) quantities of peptides. Experiments other than product ion scans can be performed but require more complex scan functions, and consequently, four-sector instruments are frequently only used for product ion scans.

There is a wide range of mass analyzers available and numerous hybrid ms/ms instruments have been produced. The most common are those having a sector ms for ms1 and a dual quadrupole system for the collision chamber and ms2, ie, *EBqQ* or *BEqQ*. High energy collision spectra are obtained as in a two-sector mass spectrometer, eg, when B/E is a constant (37), all products of a given precursor ion are detected. Low energy collisions can also be performed, by using the first rf-only quadrupole, q, for collisional activation and the second quadrupole, Q, for mass analysis. Ions are decelerated to kinetic energies between 0 and 500 eV before entering the collision quadrupole by floating both quadrupoles at a voltage close to the ion source voltage. The precursor ion can be selected at high resolution by the first two sectors, as shown by the data in Figure 8, and the analytical quadrupole gives unit–mass resolution for product ions up to around 1000 mass units.

A newer hybrid system available commercially is the magnetic sector–TOF hybrid (38). The precursor ions can be selected with better than unit–mass resolution by ms1 and the product ion ions detected at high sensitivity by the TOF ms2 (39).

Data Systems. A very important part of a mass spectrometer is the computer system used to acquire and process the mass spectral data. As of this writing, the standard computers supplied with mass spectrometers are 32-bit

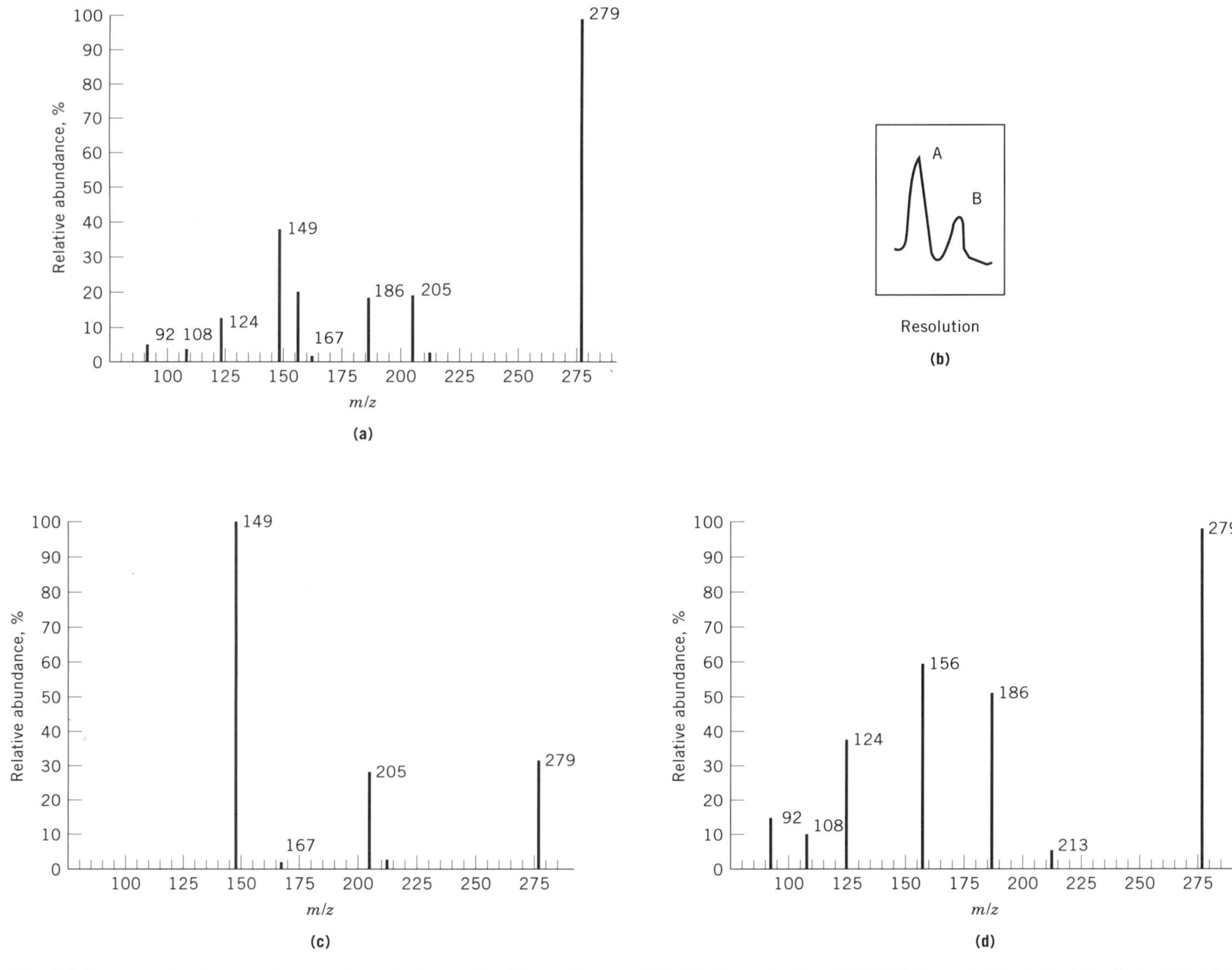

Fig. 8. (**a**) Low resolution spectrum of a mixture of sulfamethazine [*57-68-1*], mol wt = 279.0916, and dibutyl phthalate [*84-74-2*], mol wt = 279.1596; (**b**) 279 peak resolved at a resolution of 7500 on ms1 where A represents dibutyl phthalate and B, sulfamethazine; and (**c**) and (**d**) daughter ion spectra of A and B, dibutyl phtalate and sulfamethazine, respectively (1). Courtesy of Kratos Analytical.

workstations. Networking ms systems to each other and to laboratory information management systems (qv) (LIMS) is becoming more important (40). The American Society of Mass Spectrometry (ASMS) has overseen the development of NetCDF software that is a standard format for ms data. Many manufacturers have developed software to convert their data to NetCDF format so that data can be processed on other manufacturers' data systems. The use of workstations allows rapid searching of spectral libraries and databases (qv) that usually contain over 10^5 entries. True multitasking, ie, simultaneous data acquisition and data processing on a single processor, is also possible.

Commercial Mass Spectrometers

The first commercial sector mass spectrometers were produced in 1942 by CEC in the United States, and in the United Kingdom, high resolution sector instruments were produced in the 1950s by Metropolitan Vickers. Finnigan was formed in 1967 to manufacture the low resolution quadrupole mass filter, which was sold as a detector for gas chromatography. In the 1970s, many of the original manufacturers either closed or were subject to takeovers. This trend continued in the 1980s. Poor sales in the high resolution market have encouraged all the manufacturers to produce low resolution machines. Table 3 lists the types of ms

Table 3. Manufacturers of Mass Spectrometers[a,b]

Mass spectrometer[c]	Manufacturers	Unit cost, $ $\times 10^3$	World market, $ $\times 10^6$
magnetic sector	Fisons Instruments, JEOL, Finnigan MAT	300–1000	65
quadrupole	Finnigan, Sciex, Hewlett-Packard, Fisons Instruments, Shimadzu, Waters	50–600	370
ion trap	Finnigan, Varian, Teledyne	60–125	98
TOF (for MALDI)	Kratos Analytical, Fisons Instruments, Finnigan, Hewlett-Packard, Bruker, Vestec, Waters	75–250	25
ftms	Bruker, Extrel	325–750	5
inorganic ms, ie, icpms, gdms, TOF–sims	Fisons Instruments, Kratos Analytical, Perkin Elmer, Charles Evans & Associates, Finnigan, Cameca	125–300	63

[a]Courtesy of Strategic Directions International, Inc.
[b]Unit cost and market figures shown are for 1993.
[c]icpms = inductively coupled plasma mass spectrometer; gdms = glow discharge mass spectrometer; sims = secondary ion mass spectrometer.

instruments available commercially as of this writing, along with the principal manufacturers, price range of instruments, and the total market size in 1993.

Applications

Biotechnology. There has been a tremendous growth in the application of mass spectrometry in biotechnology (qv) since the 1980s. New ionization methods have steadily expanded the range of biomolecules which can be analyzed by mass spectrometry. The first of these ionization methods was plasma desorption (PD) (1976) fitted to a TOF mass spectrometer (41). The source contains californium-252 [*13981-17-4*], ^{252}Cf, which decays to give two high energy (ca 100 MeV) fission fragments and alpha particles. The fission fragments, ^{106}Tc and ^{142}Ba, are emitted in opposite directions, so one is used to start the time-of-flight measurement and the other to desorb the sample. In preparation for analysis, a few picomoles of the sample is deposited onto a metal foil coated with nitrocellulose to which peptides and proteins (qv) bond (42). The resulting mass spectra contain singly protonated molecular ions plus multiply charged ions which become more abundant as the sample loading is reduced. Prior to the introduction of fast atom bombardment (FAB) in 1980 (43), PDms was the best way of obtaining mass spectra of small peptides. It is more sensitive than FAB and has a wider mass range. For analysis by FAB, the sample is dissolved in a liquid matrix such as glycerol and then bombarded with kilovolt energy Ar or Xe atoms to give mass spectra containing mainly $(M + H)^+$ ions in positive ion mode and $(M - H)^-$ ions in negative ion mode. The use of kilovolt energy Cs^+ ions instead of Xe atoms, using liquid secondary ion mass spectrometry (lsims) (44), has allowed the desorption of proteins having molecular weights up to 20,000, as shown by the lsims spectrum of lysozyme [*9001-63-2*] in Figure 9**a**.

The desire to interface hplc systems to FABms has led to the development of continuous flow FAB (CF-FAB) (45), whereby the eluent from the hplc column is mixed with solvent that contains a few percent of the matrix, and the mixture flows along a fused silica capillary to the tip of the heated FAB probe inside the ion source. Ionization occurs as in static FAB, but gives a lower matrix signal. The direct analysis of tryptic digests separated by hplc (46), is an example of the application of this method. The biggest limitation of the CF-FAB technique is that it can only handle flow rates up to 5–10 μL/min. Thus either the column eluent must be split or microbore columns used. Cooling a standard FAB probe and carefully controlling the flux of the bombarding species produces stable beams of analyte ions for long times, which is advantageous for ms/ms analysis. FAB has also been applied to the analysis of oligosaccharides, oligonucleotides, and other polar and labile compounds such as antibiotics (qv), organometallics, and drug and steroid conjugates (47).

In 1983 an interface appeared that could handle hplc flow rates up to 1 mL/min, the thermospray interface (48). The lc eluent which contains a high concentration of a buffer, eg, 0.1 *M* ammonium acetate, passes through a resistively heated capillary tube into the ion source block. The block is directly pumped by a rotary pump to remove the solvent vapor. As the droplets get smaller, charged particles migrate to the outside of the drop. When all the solvent has evaporated, only ions remain. These ions are extracted orthogonal to the

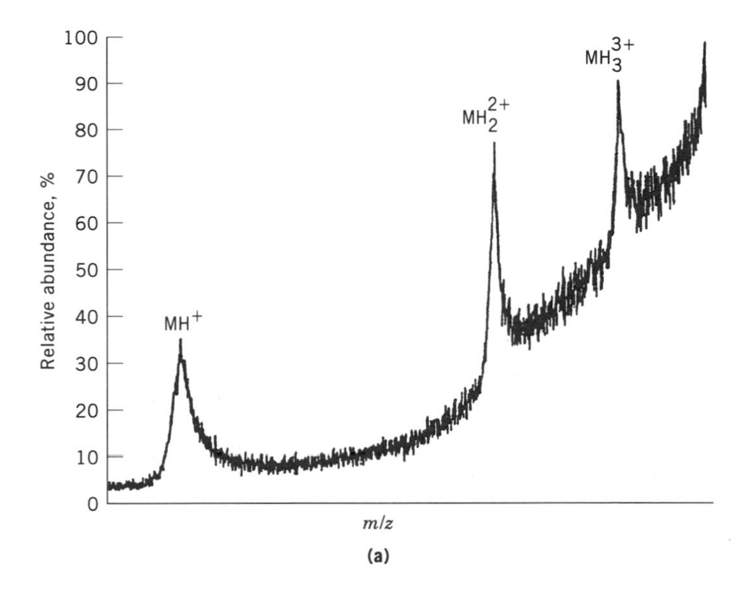

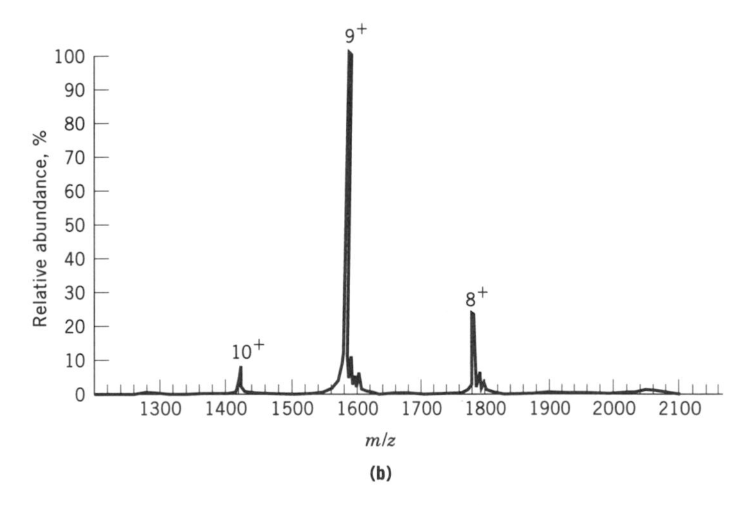

Fig. 9. Mass spectra of lysozyme having 14,305 mol wt. See text. (**a**) lsims spectrum; (**b**) ESI spectrum (1). Courtesy of Kratos Analytical.

direction of spray of the liquid. Thermospray mass spectra contain $(M + H)^+$ or $(M + NH_4)^+$ ions in positive ion mode and $(M - H)^-$ ions in negative ion mode. The need for a high buffer content is a problem, but the addition of a filament or discharge electrode to the source allows the use of mainly organic mobile phases without the buffer. Discharge assisted thermospray also reduces changes in sensitivity that occur during gradient hplc runs using pure thermospray. Figure 10 shows data from the reverse-phase gradient separation of methoxy poly(ethylene glycol) oligomer. The technique is widely used for thermally labile materials having molecular masses <1000, eg, in the pharmaceutical industry. The ther-

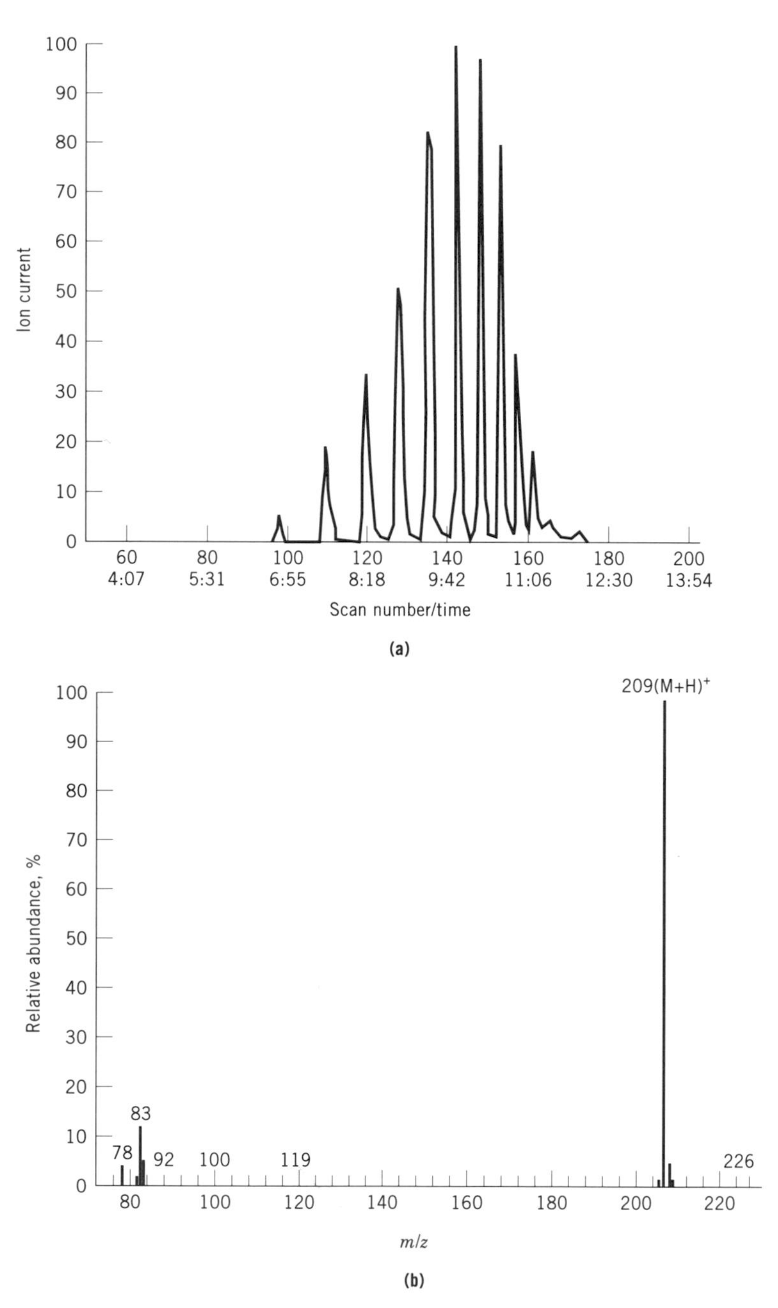

Fig. 10. (**a**) Total ion current trace for the reverse phase gradient separation of methoxypoly(ethylene glycol) oligomer, $CH_3O(CH_2CH_2O)_nH$, using thermospray ionization; (**b**) thermospray discharge spectrum of the component eluting at 8 min 14 s, $CH_3O(CH_2CH_2O)_4H$. Courtesy of Kratos Analytical.

mospray interface has also been used as a means of interfacing packed supercritical fluid chromatography (sfc) columns to a mass spectrometer (49) (see also SUPERCRITICAL FLUIDS). When the interface is operated in discharge mode, the mobile phase (often carbon dioxide) is ionized, and ionization of the sample occurs by charge–exchange reactions. The technique has advantages for the analysis of labile pharmaceuticals such as benzylpenicillin [*61-33-6*] (50).

Another big advance in the application of ms in biotechnology was the development of atmospheric pressure ionization (API) techniques. There are three variants of API sources, a heated nebulizer plus a corona discharge for ionization (APCI) (51), electrospray (ESI) (52), and ion spray (53). In the APCI interface, the lc eluent is converted into droplets by pneumatic nebulization, and then a sheath gas sweeps the droplets through a heated tube that vaporizes the solvent and analyte. The corona discharge ionizes solvent molecules, which protonate the analyte. Ions transfer into the mass spectrometer through a transfer line which is cryopumped, to keep a reasonable source pressure.

An ESI ion source produces multiply charged ions from biomolecules with high sensitivity. The analyte solution is sprayed at atmospheric pressure from a needle floated at a few kV above ground potential, and sampling cones select the center of the spray to enter the mass spectrometer via a differentially pumped transfer region. The resulting spectra contain a series of multiply charged ions each having one charge more than the next highest mass ion, and because a mass spectrometer measures the mass-to-charge ratio of an ion, ESI produces ions from large molecules which can be analyzed by such mass spectrometers as quadrupoles and ion traps. Figure 9**b**, an electrospray spectrum of lysozyme, shows that the most abundant ion in this case is the +9 charge state. Analysis conditions such as the pH of the analyte solution influence which charge state is the most abundant. For bovine albumin which has a molecular mass of 66,625, the ESI mass spectrum typically contains ions having between 20 and 45 charges, ie, m/z of 3331–1481. The mass of the analyte may be calculated using the following formula:

$$\begin{aligned} m_i &= \frac{M_R + iM_H}{i} \\ i &= \frac{m_{i+1} - M_H}{m_i - M_{i+1}} \end{aligned} \qquad (9)$$

in which m_i is the mass of the ion having i charges, m_{i+1} is the mass of the ion having $i + 1$ charges, M_R is the mass of the analyte, and M_H is the mass of a proton.

The mass of the analyte, M_R, is calculated for each charge state, and because there are many charge states, random errors are reduced and a mass measurement accuracy of $\pm 0.01\%$ is easily achieved. This technique is useful for analyzing samples which encompass a wide mass range, eg, from small drugs and their metabolites to proteins having masses >50,000. Solvent flow rates are similar to CF-FAB at around 5–10 μL/min. Direct coupling to lc, a wide mass range, and high sensitivity have made ESI a widely used tool in the pharmaceutical industry (see PHARMACEUTICALS).

When a coaxial flow of nebulizing gas is used with electrospray it is called ion spray. The main advantages of ion spray are that it can handle high flow rates, up to 1 mL/min and mobile phases containing high percentages of water give stable signals.

Prior to the introduction of ESI, ms/ms studies of peptides were generally limited to molecules mol wt <3500 (33). This limitation was a consequence of the rapid drop in precursor ion intensity from lsims ion sources with increasing mass, and the inefficiency of collisional activation. Good quality ms/ms spectra of much larger peptides have been obtained after the collisional activation of multiply charged ions from ESI, using either high (54) or low (55) collision energies. Although interpreting the spectra of unknowns is difficult, it should become easier as the rules governing the fragmentation of multiply charged ions become better understood.

The use of an external ion source fticr mass spectrometer and electrospray ionization allows use of its high resolution capabilities to determine the charge states of multiply charged ions directly (56). This is very useful in ms/ms experiments because product ions from a multiply charged precursor may have the same or a lesser charge than the precursor ion. As the mass spectrometer gives the mass-to-charge ratio of the product ions, determination of the charge allows the mass of the ion to be calculated. Subtracting this from the precursor ion mass gives the mass of the fragment lost to form the product ion.

A complementary technique to ESI is matrix-assisted laser desorption ionization (MALDI) (57), typically used with a TOF mass spectrometer. An excess of matrix, which absorbs strongly at the wavelength of light emitted by the laser, eg, 266 or 337 nm for a uv laser, is mixed with the sample. The solution is dried on a sample slide and the slide placed inside the mass spectrometer. When the sample is irradiated by the laser, protonated sample and matrix ions are produced. Figure 11 shows a MALDI spectrum of immunoglobulin. The spectrum is very simple, the highest mass ions are singly protonated sample ions, and there are also some multiply charged ions. At the low mass end of the spectrum (Fig. 11) are matrix ions (M + H, and M + H − H_2O) and metal ions, eg, Na^+. No fragment ions are seen from the analyte, which makes the technique very

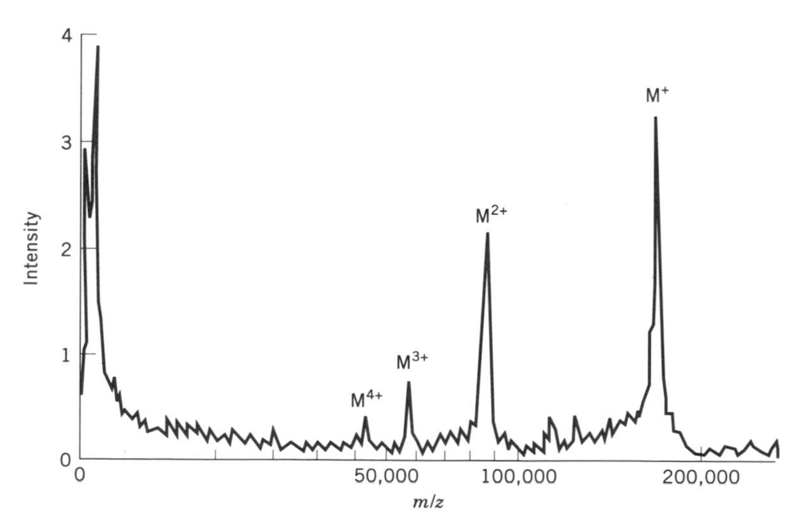

Fig. 11. MALDI spectrum of an immunoglobulin using sinapinnic acid as the matrix and laser light at 337 nm. M^+ is at 172,450 mol wt.

good for mixture analysis. The choice of matrix depends on the analyte. For complex mixtures, such as tryptic digests, sample analysis with two or more different matrices is recommended to avoid missing components from suppression effects (58). Compared with ESI, MALDI gives less accurate mass measurement ($\pm$0.1% typically, $\pm$0.01% with internal standards), as of this writing, direct interfacing to hplc is in the development stage (59), and MALDI has very low mass resolution (<2000). MALDI is however, an easier and faster technique to use, and has better high (>50,000) mass sensitivity. MALDI is also more tolerant of the salts and buffers used to keep biomolecules in solution. Its speed makes it an ideal screening technique for samples before complementary analysis by ESI. By combining MALDI with enzymatic digestion of a peptide, sequencing of subpicomole quantities is possible (60). Protocols are also being developed for the direct analysis of poly(vinylidene difluoride) (PVDF) membranes, onto which proteins separated by electrophoresis have been electroblotted (61). The MALDI technique gives an order of magnitude better mass measurement for this application than comparison of the sample position on the gel to the position of standards.

Environmental. The high sensitivity and specificity of mass spectrometry when coupled with high resolution chromatography make ms an ideal method for use in environmental analyses. A number of standard ms methods exist for target compound analysis. These methods have been written by government agencies such as the U.S. Environmental Protection Agency (EPA). For example, a method describes qualitative and quantitative analysis of municipal and industrial discharges for base/neutral and acid compounds by gc/ms is available. Pumping speeds inside mass spectrometers are high enough to allow the capillary gc column to go directly into the ion source, which gives the best sensitivity. If packed columns are used, however, an interface such as a jet separator is needed to reduce the amount of carrier gas entering the ion source (62).

Both low and high resolution ms, and ms/ms, have been combined with capillary gc analysis of environmental samples (9), and lc/ms has been used for compounds which are not amenable to gc/ms. The particle beam–lc/ms interface can be used, for example, to confirm the presence of chlorinated phenoxy acid herbicides (qv) which are only amenable to gc/ms after derivatization. This interface is derived from the monodisperse aerosol generator for interfacing chromatography– (magic–) lc/ms interface first described in 1984 (63). The lc eluent is sprayed into a heated desolvation chamber by a pneumatic nebulizer. Under the influence of a carrier gas (helium), the sample droplets enter a differentially pumped, two-stage momentum separator which removes most of the remaining solvent vapor. The well-defined sample beam then enters the ion source for ionization by electron ionization or chemical ionization. The ability to obtain library searchable EI mass spectra from the lc eluent is one of the main advantages of the particle beam interface over other lc/ms interfaces.

Quantitative mass spectrometry, also used for pharmaceutical applications, involves the use of isotopically labeled internal standards for method calibration and the calculation of percent recoveries (9). Maximum sensitivity is obtained when the mass spectrometer is set to monitor only a few ions, which are characteristic of the target compounds to be quantified, a procedure known as the selected ion monitoring mode (sim). When chlorinated species are to be detected, then two ions from the isotopic envelope can be monitored, and confirmation of the target compound can be based not only on the gc retention time and the mass, but on the ratio of the two ion abundances being close to the theoretically expected value. The spectrometer cycles through the ions in the shortest possible time. This avoids compromising the chromatographic resolution of the gc, because even after extraction the sample contains many compounds in addition to the analyte. To increase sensitivity, some methods use sample concentration techniques.

Because of the large number of samples and repetitive nature of environmental analysis, automation is very important. Autosamplers are used for sample injection with gc and lc systems, and data analysis is often handled automatically by user-defined macros in the data system. The high demand for the analysis of environmental samples has led to the establishment of contract laboratories which are supported purely by profits from the analysis. On-site monitoring of pollutants is also possible using small quadrupole ms systems fitted into mobile laboratories.

Oil Analysis. Characterization of oil samples is difficult because these are very complex. At any nominal mass, ions from different homologous series containing heteroatoms such as nitrogen, sulfur, and oxygen as well as the hydrocarbon backbone appear. Resolution of these multiplets that differ in mass by only a few hundredths or thousandths of a m/z requires very high resolution (64), which is only possible using a limited number of instruments. For example, two ions having a nominal mass of 250, but having elemental formulas XCH_2 and XN would require a mass resolution of 20,000 for separation.

To simplify oil sample spectra, data is obtained using low energy (<10 eV) electron ionization. This minimizes fragmentation and gives spectra containing mainly molecular ions (65), but there is still a significant amount of data to analyze (66). The geographic origin of an oil sample can be deduced from the types and abundances of steranes it contains (67). Sterane distributions have been determined by using ms/ms to monitor fragmentation of the sterane molecular ions which produces the characteristic ion of m/z 217 or by high resolution monitoring of m/z 217 (68).

Polymers. Mass spectrometric analysis of polymers is problematic because of the wide mass range and difficulties in ionization. One useful ionization method is field desorption (FD) (69), where the sample is applied to an emitter wire on which multipoint microneedles (dendrites) of carbon have been grown. The emitter is held at 8–12,000 V relative to a counter electrode which is approximately 2 mm from it. Ions are desorbed from the emitter by the intense electric field. The main problems are the difficulty in obtaining good emitters and the inherent fragility. Field desorption has been used, however, for polymer samples having molecular masses $>10,000$ on sector mass spectrometers (70). Because fragment ions are not seen, the molecular mass profile can be directly recorded. One promising development is the application of MALDI to polymer analysis. Like FD, MALDI does not give fragment ions. Molecular mass distributions for polystyrene samples up to PS70000 have been reported (71), albeit with low mass resolution.

Inorganic Applications. The three most widely used mass spectrometric techniques in inorganic analysis are secondary ion mass spectrometry (sims) (72), inductively coupled plasma mass spectrometry (icpms) (73), and glow discharge mass spectrometry (gdms) (74). Sims detects ions sputtered from a sample which is bombarded by a probe beam of Cs^+ ions. The secondary ions are mass analyzed using a quadrupole mass filter, a TOF mass spectrometer, or a sector mass spectrometer for high mass resolution. The instruments identify elements present in the sample, and map the ion abundances across the sample as a function of depth. The technique has found heavy usage in the semiconductor industry (see SEMICONDUCTORS). It is now also being used in biological studies such as the determination of copper in body organs (75), and in geological applications such as coal (qv) oxidation using ^{18}O and sims imaging (76).

Icpms instruments generally use quadrupole mass analyzers, although sector instruments are becoming available. Since the commercial launch of this technique in 1984, it has become very important, offering rapid, simultaneous multielement analysis and parts per billion (ppb) detection limits for some elements. Icpms also has a wide linear dynamic range, up to five or six orders of magnitude, and high precision, and is insensitive to interelement matrix effects. Sample solution at flow rates around 1 mL/min passes through a nebulizer into an argon gas plasma that is produced in a quartz torch using a 1–2.5-kW radio frequency power supply. The plasma is sampled by a series of skimmers along a differentially pumped transfer line leading into the mass spectrometer. Initially applied to geological analysis problems such as the determination of Os isotope ratios (77), the technique is also being applied to the determination of metals in environmental and biological samples such as blood and urine (78).

Gdms allows the direct determination of the elemental composition of solid samples. Originally available on sector instruments, low cost quadrupole instruments have become available and glow discharge ion sources have also appeared as accessories for organic mass spectrometers. A dc potential (−1000 V) is applied to the metallic sample surrounded by argon at a pressure near 130 Pa (1 torr). A visible glow surrounds the sample, and argon ions are accelerated into the sample, causing atoms to be sputtered from its surface. The sputtered atoms diffuse into the negative glow region, which contains energetic electrons, ions, and metastable atoms. Sample atoms are ionized either by EI (eq. 10) or by Penning ionization (eq. 11) from the impact of eg, metastable argon atoms.

$$\mathrm{M} + e^- \longrightarrow \mathrm{M}^+ + 2\,e^- \qquad (10)$$

$$\mathrm{M} + \mathrm{Ar}^* \longrightarrow \mathrm{M}^+ + \mathrm{Ar} + e^- \qquad (11)$$

A sampling orifice near the negative glow region allows ions to transfer into the mass spectrometer along the differentially pumped transfer region. The discharge may be powered by d-c, pulsed d-c, or r-f power supplies; the latter two permit the analysis of nonconducting samples. Applications of the technique include the direct analysis of metals and alloys, thin-film analysis, and the analysis of nonconducting materials as compacted samples with matrices such as graphite (79).

BIBLIOGRAPHY

"Analytical Methods" in *ECT* 3rd ed., Vol. 2, pp. 586–683, by E. Lifshin and E. A. Williams, General Electric Co.

1. J. R. Chapman, *Practical Organic Mass Spectrometry*, 2nd ed., John Wiley & Sons, Inc., Chichester, UK, 1993.
2. J. J. Thompson, *Rays of Positive Electricity*, Longmans, Green and Co., London, 1913.
3. F. W. Aston, *Phil. Mag.* **38**, 707 (1919).
4. A. J. Dempster, *Phys. Ref.* **11**, 316 (1918).
5. R. S. Gohlke, *Anal. Chem.* **34**, 1332 (1976).
6. P. Arpino and co-workers, *Biomed. Mass Spectrom.* **1**, 80 (1974).
7. E. M. H. Finlay, D. E. Games, J. R. Startin, and J. Gilbert, *Biomed. Environ. Mass Spectrom.* **13**, 633 (1986).
8. F. W. McLafferty, ed., *Tandem Mass Spectrometry*, John Wiley & Sons, Inc., New York, 1983.
9. L. Q. Huang, B. Eitzer, C. Moore, S. McGown, and K. B. Tomer, *Biolog. Mass Spectrom.* **20**, 161 (1991).
10. F. H. Field and M. S. B. Munson, *J. Amer. Chem. Soc.* **87**, 3289 (1965).
11. A. G. Harrison, *Chemical Ionization Mass Spectrometry*, CRC Press, Boca Raton, Fla., 1983.
12. M. Gross, *J. Amer. Soc. Mass Spectrom.* **5**, 57 (1994).
13. E. G. Johnson and A. O. Nier, *Phys. Chem. Rev.* **91**, 10 (1953).
14. J. S. Cottrell and R. J. Greathead, *Mass Spectrom. Rev.* **5**, 215 (1986).
15. J. S. Cottrell and S. Evans, *Anal. Chem.* **59**, 1990 (1987).
16. P. H. Dawson, ed., *Quadrupole Mass Spectrometry and Its Applications*, Elsevier, New York, 1976.
17. J. E. Campana, *Int. J. Mass Spectrom. Ion Phys.* **33**, 101 (1980).
18. N. S. Arnold, W. H. McClennen, and H. L. C. Meuzelaar, *Anal. Chem.* **63**, 289 (1991).
19. R. E. March and R. J. Hughes, *Quadrupole Storage Mass Spectrometry*, John Wiley & Sons, Inc., New York, 1989.
20. J. F. J. Todd, *Mass Spectrom. Rev.* **10**, 3 (1991).
21. J. C. Schwartz, J. E. P. Syka, and I. Jardine, *J. Amer. Soc. Mass Spectrom.* **2**, 198 (1991).
22. R. E. Kaiser, Jr., R. G. Cooks, G. C. Stafford, Jr., E. P. Syka, and P. H. Hemberger, *Int. J. Mass Spectrom. Ion Process.* **106**, 79 (1991).
23. G. J. Van Berkel, G. L. Glish, and S. A. McLuckey, *Anal. Chem.* **62**, 1284 (1990).
24. R. J. Cotter, *Anal. Chem.* **64**, 1027A (1992).
25. D. J. Erickson, C. E. Enke, J. F. Holland, and J. T. Watson, *Anal. Chem.* **62**, 179 (1990).
26. J. Bordas-Nagy and K. R. Jennings, *Int. J. Mass Spectrom. Ion Process.* **100**, 105 (1990).
27. R. A. Yost and C. G. Enke, *Anal. Chem.* **51**, 1251A (1970).
28. J. Adams, *Mass Spectrom. Rev.* **9**, 141 (1990).
29. R. G. Cooks, J. H. Beynon, R. M. Caprioli, and G. R. Lester, *Metastable Ions*, Elsevier, Amsterdam, the Netherlands, 1973.
30. F. C. Walls and co-workers, *Biological Mass Spectrometry*, Elsevier, Amsterdam, the Netherlands, 1990.
31. R. K. Boyd, *Int. J. Mass Spectrom. Ion Process.* **75**, 243 (1987).
32. J. Bordas-Nagy, D. Despeyroux, and K. R. Jennings, *J. Amer. Soc. Mass Spectrom.* **3**, 502 (1992).
33. R. S. Johnson, S. A. Martin, and K. Biemann, *Int. J. Mass Spectrom. Ion Process.* **86**, 137 (1988).

34. J. Herrmann and co-workers, *J. Biol. Chem.* **268**, 26704 (1993).
35. P. Roepstorff and J. Fohlman, *Biomed. Mass Spectrom.* **11**, 601 (1984).
36. J. A. Hill, J. E. Biller, and K. Biemann, *Int. J. Mass Spectrom. Ion Process.* **111**, 1 (1991).
37. A. P. Bruins, K. R. Jennings, and S. Evans, *Int. J. Mass Spectrom. Ion Phys.* **26**, 395 (1978).
38. E. Clayton and R. H. Bateman, *Rapid Commun. Mass Spectrom.* **6**, 719 (1992).
39. R. H. Bateman, M. R. Green, and G. Scott, *Proc. 42nd ASMS Conf.*, 1034 (1994).
40. M. J. Hayward, P. V. Robandt, J. T. Meek, and M. L. Thomson, *J. Amer. Soc. Mass Spectrom.* **4**, 742 (1993).
41. R. D. MacFarlane and D. F. Torgerson, *Science* **191**, 920 (1976).
42. P. Roepstorff, *Acc. Chem. Res.* **22**, 421 (1989).
43. M. Barber, R. S. Bordoli, R. D. Sedgewick and A. N. Tyler, *Chem. Commun.* **7**, 325 (1981).
44. W. Aberth, K. M. Straub, and A. L. Burlingame, *Anal. Chem.* **54**, 2029 (1982).
45. R. M. Caprioli, ed., *Continuous-Flow Fast Atom Bombardment Mass Spectrometry*, John Wiley & Sons, Inc., Chichester, 1990.
46. R. M. Caprioli, B. DaGue, T. Fan, and W. T. Moore, *Biochem. Biophys. Res. Commun.* **146**, 291 (1987).
47. E. De Pauw, A. Agnello, and F. Derwa, *Mass Spectrom. Rev.* **10**, 283 (1991).
48. C. R. Blackley and M. L. Vestal, *Anal. Chem.* **55**, 2280 (1983).
49. J. R. Chapman, *Rapid Commun. Mass Spectrom.* **2**, 6 (1988).
50. H. T. Kalinoski and L. O. Hargiss, *J. Chromatogr.* **474**, 69 (1989).
51. S. Pleasance, J. F. Anadeto, M. R. Bailey, and D. H. Norch, *J. Amer. Soc. Mass Spectrom.* **3**, 378 (1991).
52. J. B. Fenn, M. Mann, C. K. Meng, S. F. Wong, and C. M. Whitehouse, *Mass Spectrom. Rev.* **9**, 37 (1990).
53. A. P. Bruins, T. R. Covey, and J. D. Henion, *Anal. Chem.* **59**, 2642 (1987).
54. D. Fabris, M. Kelly, C. Murphy, Z. Wu, and C. Fenselau, *J. Amer. Soc. Mass Spectrom.* **4**, 652 (1993).
55. J. A. Loo, C. G. Edmonds, and R. D. Smith, *Anal. Chem.*, **65**, 425 (1993).
56. M. W. Senko, S. C. Beu, and F. W. McLafferty, *Anal. Chem.* **66**, 415 (1994).
57. F. Hillenkamp, M. Karas, R. C. Beavis, and B. T. Chait, *Anal. Chem.* **63**, 1193A (1991).
58. T. M. Billeci and J. T. Stubbs, *Anal. Chem.* **65**, 1709 (1993).
59. K. K. Murray and D. H. Russell, *J. Amer. Chem. Soc. Mass Spectrom.* **5**, 1 (1994).
60. B. T. Chait, R. Wang, R. C. Beavis, and S. B. H. Kent, *Science* **262**, 89 (1993).
61. M. M. Vestling and C. Fenselau, *Anal. Chem.* **66**, 471 (1994).
62. R. Ryhage, *Anal. Chem.* **36**, 759 (1964).
63. R. C. Willoughby and R. F. Browner, *Anal. Chem.* **56**, 2626 (1984).
64. C. S. Hsu, Z. Liang, and J. E. Campana, *Anal. Chem.* **66**, 850 (1994).
65. H. E. Lumpkin, *Anal. Chem.* **36**, 2399 (1964).
66. T. Aczel, D. E. Allen, J. H. Hardng, and E. A. Krupp, *Anal. Chem.* **42**, 341 (1970).
67. B. P. Tisot and D. H. Welte, *Petroleum Formation and Occurrence*, Springer-Verlag, New York, 1978, Chapt. 3.
68. G. A. Warburton and J. E. Zumberge, *Anal. Chem.* **55**, 123 (1983).
69. H. D. Beckey, *Principles of Field Ionization and Field Desorption Mass Spectrometry*, Pergamon, London, 1977.
70. K. Rollins, J. H. Scrivens, M. J. Taylor, and H. Major, *Rapid Comm. Mass Spectrom.* **4**, 355 (1990).
71. V. Buhr, A. Deppe, M. Karas, F. Hillenkamp, and V. Giessmann, *Anal. Chem.* **64**, 2800 (1992).
72. J. C. Vickerman, A. Brown, and N. M. Reed, *Secondary Ion Mass Spectrometry, Int. Ser. Mono. Chem. 17*, Oxford University Press, Oxford, UK, 1990.

73. G. Meyer, *Anal. Chem.* **59**, 1345A (1987).
74. W. W. Harrison, K. R. Hess, R. K. Marcus, and F. L. King, *Anal. Chem.* **58**, 341A (1986).
75. D. Marchai-Segault, C. Briancon, S. Halpern, P. Fragu, and G. Lauge, *Biol. Cell* **70**, 129 (1990).
76. R. R. Martin, J. A. MacPhee, *Energy Sources* **11**, 105 (1989).
77. D. C. Gregoire, *Anal. Chem.* **62**, 141 (1990).
78. P. Allain, Y. Mauras, C. Douge, L. Jaunault, T. Delaporte, and C. Beaugrand, *Analyst* **115**, 813 (1990).
79. Y. Mei and R. K. Marcus, *Anal. Chem.* **12**, 86 (1993).

COLIN MOORE
Uniroyal Chemical Company

3 5282 00422 8428